Praise for

Rising Star

"Impressive. . . . [A] deeply reported work of biography. . . . Garrow made the inspired decision to open the book on the economically ravaged South Side of Chicago in 1980—five years before Obama showed up as a novice community organizer—thus giving us a sidewalk-level view of the joblessness, environmental degradation and failing schools that formed day-to-day reality. We see right away what our hero is up against in his altruistic quest to 'create change.' . . . The depth of detail allows the reader to see familiar parts of this story with fresh eyes." —*New York Times Book Review*

"A masterwork of historical and journalistic research, Robert Caro-like in its exhaustiveness, and easily the most authoritative account of Obama's pre-presidential life we've seen or are likely ever to see. It's also a terrific read."
—David Greenberg, Politico

"Rigorous. . . . Thorough. . . . Meticulously researched. . . . Delivers insight and clarity on Obama's enigmatic personality." —*Christian Science Monitor*

"One of the most impressive presidential biographies. . . . A look at the social construction of race." —*Bloomberg* (a Must-Read of 2017)

"The authoritative biography of Barack Obama's pre-presidential years. . . . Illuminating. . . . Impressively researched. . . . Readers . . . will be richly rewarded."
—*Library Journal* (starred review)

"A convincing and exceptionally detailed portrait. . . . Political history buffs will be fascinated." —*Publishers Weekly* (starred review)

"A tour de force. . . . An epic triumph of personal and political biography."
—Paul Street, New York Journal of Books

"May endure in the American presidential canon."
—*The Globe and Mail* (Toronto)

"Gripping. . . . [A] compelling read [that] should appeal to political junkies and insiders. . . . Foundational." —Washington Independent Review of Books

"Definitive. . . . Fascinating. . . . So interesting, you'll be hard-pressed to put it down."
—Paul Lisnek, WGN Chicago

"[Contains] intriguing insight into the growing pains of a 20-something who would go on to become the leader of the free world."
—*Time*

"Immensely informative. . . . Reveals Mr. Obama in all his complexity."
—*Pittsburgh Post-Gazette*

"Garrow is a demon for research. . . . Eminently solid. . . . Consistently readable—an impressive work."
—*Kirkus Reviews*

"Very thorough, very well-researched."
—Roland Martin, *NewsOne Now*

"Any Obama fan is likely to find this in-depth portrait fascinating."
—*St. Augustine News*

"Comprehensive. . . . Contains many insights to Obama's life and character. . . . Will be the defining work on Obama's early life for years to come."
—The Federalist

RISING STAR

ALSO BY DAVID J. GARROW

Liberty and Sexuality: The Right to Privacy and the Making of Roe v. Wade

Bearing the Cross: Martin Luther King, Jr., and the Southern Christian Leadership Conference

The FBI and Martin Luther King, Jr.: From "Solo" to Memphis

Protest at Selma: Martin Luther King, Jr., and the Voting Rights Act of 1965

RISING STAR

THE MAKING OF BARACK OBAMA

David J. Garrow

WILLIAM MORROW
An Imprint of HarperCollins*Publishers*

For Darleen,
who endured

A hardcover edition of this book was published in 2017 by William Morrow, an imprint of HarperCollins Publishers.

FIRST WILLIAM MORROW PAPERBACK EDITION PUBLISHED 2018.

Library of Congress Cataloging-in-Publication Data has been applied for.

ISBN 978-0-06-264184-7

18 19 20 21 22 LSC 10 9 8 7 6 5 4 3 2 1

CONTENTS

RISING STAR

Chapter One

THE END OF THE WORLD AS THEY KNEW IT

CHICAGO'S FAR SOUTH SIDE

MARCH 1980–JULY 1985

Frank Lumpkin never forgot the first phone call that afternoon. Although he was off work that Friday with a broken foot, he had stopped by the pay office at Wisconsin Steel, where he'd labored for more than thirty years, to pick up checks totaling $8,084.57, money he was due in back vacation pay. He still had thirteen weeks of vacation coming, accumulated over five years, and he was about to take a long-planned trip to Africa. But March 28—"Black Friday," as it would be called—was about to become absolutely unforgettable.

Wisconsin Steel's hulking metal sheds stretched south and a bit eastward from the intersection of Torrence Avenue and East 106th Street in a neighborhood that most residents called Irondale, even if Chicago city maps labeled it South Deering. William Deering was a long-forgotten industrialist who had cofounded International Harvester Company in 1902; Wisconsin's oldest corporate ancestor, Brown's Mill, dated from 1875 and had been South Chicago's first steel plant. Six years later, Andrew Carnegie opened a larger mill just north from where the Calumet River flowed into Lake Michigan, and by the dawn of the twentieth century the southwestern crescent of that lakeshore, stretching from the Calumet eastward across the Indiana state line to Gary and Burns Harbor, had become the most dense concentration of steel mills in the world.

Steelmaking was dangerous and strenuous work, as Frank Lumpkin well knew, but steelworkers had significant freedoms. "There were no time clocks, and they could come out for lunch," a nearby barber recounted. "The workers figured that they could get their hair cut on company time because it grows on company time."

By 1980 the region's mills had sustained generation after generation of working-class families whose breadwinners didn't need to graduate high school to get jobs that paid three times what college graduates could earn as public school teachers.

Frank and his wife Bea had raised four children during their thirty-one years of marriage, and they had just moved to South Shore, a middle-class neighborhood a bit northwest of where Carnegie's mill—now United States Steel's huge South Works—employed almost three times the 3,450 men who worked at Wisconsin. South Shore was a comfortable area for an interracial couple—Frank was black, Bea white—and tolerant too of a couple who had spent many years as dedicated members of the Communist Party USA. Frank's fellow workers at Wisconsin—black, white, and Hispanic—didn't view him as a radical, just an outgoing man who had worked his way up through the odd series of job titles a steel mill offered: chipper, scarfer, millwright.

A little after 3:00 P.M. that Friday, Frank limped across Torrence Avenue to the Progressive Steel Workers (PSW) union hall, just north of 107th Street. The PSW was a so-called "independent" union, not part of the United Steelworkers of America or any labor federation, but it was actually no more "independent" than it was "progressive." Its first president, William Reilly, was a Wisconsin steelworker, but he left his union post to become chief labor spokesman for International Harvester, Wisconsin Steel's corporate owner. PSW's current president, forty-three-year-old Leonard "Tony" Roque, had been in office since 1973, and he had engineered an almost 50 percent increase in union dues, increased his own salary by thousands of dollars, and was overseeing a $150,000 expansion of the union hall. Roque was widely seen as nothing more than a flunky for the political king of South Chicago, 10th Ward alderman Edward R. "Fast Eddie" Vrdolyak, whose law firm received an annual retainer of $30,000 from the PSW and whose election campaigns the union also contributed to. Vrdolyak was a graduate of the University of Chicago's highly prestigious law school, but some acquaintances remembered him more for the charge of attempted murder that had been filed against him, and then dropped, during his law school years.

The phone call Frank would always remember came when he was speaking with union vice president Steve Plesha at about 3:30 P.M. The message was abrupt: Wisconsin Steel was closing, the workers were being sent home, and the gates were being locked.

"He looked at me and I looked at him because I couldn't believe it. I had just been there twenty minutes ago," Frank recalled a decade later. What's more, just the night before PSW had held a meeting where both Roque and Ronald K. Linde, board chairman of Wisconsin Steel's new owner, Envirodyne Industries, had reassured some fifteen hundred members that reports about Wisconsin possibly closing were incorrect.

But the 3:30 P.M. phone call should have been less surprising than it was. Three weeks earlier, *Crain's Chicago Business*—in fairness, a publication not often read by South Chicago steelworkers—had reported that Wisconsin faced "imminent bankruptcy" because its former corporate owner, International Harvester, which still bought some 40 percent of its steel from Wisconsin for use in Harvester's farm equipment, was increasingly crippled by an ongoing United Auto Workers strike that had begun on November 1, 1979. Ironically,

November 1 had also been the date when Wisconsin's new owner, Envirodyne Industries, secured a package of loans, guaranteed by the U.S. Department of Commerce's Economic Development Administration (EDA), to modernize its plant.

International Harvester had begun trying to sell Wisconsin in 1975, and for good reason: in 1976, Wisconsin lost $4.6 million, bringing its cumulative losses since 1970 to $77 million. By early 1977 Harvester was in serious discussions with Envirodyne, a tiny enterprise boasting just a dozen employees that focused on acquiring larger companies through stock swaps. Envirodyne had no experience in the steel industry and badly needed cash to cover an existing bank loan, but Harvester was willing to accept $50 million in notes, secured primarily by the two iron ore mines in upper Michigan, one ore ship, and coal properties in Kentucky that Wisconsin Steel also owned. More crucially, Chase Manhattan Bank was willing to provide $15 million in cash to Envirodyne. The *Chicago Tribune* labeled the purchase "a minnow trying to swallow a whale," but nonetheless the sale closed on July 31, 1977, and Wisconsin's hefty ongoing annual losses continued: $32 million on revenues of $236 million in the twelve months ending in September 1979. Thanks to EDA's loan guarantees, six insurance companies provided $75 million, and Chase Manhattan ponied up another $15 million for current operating expenses.

The *Crain's* story went on to say that "without a strong infusion of working capital," beyond the 1979 loans, "Wisconsin's collapse is unavoidable," and on March 27, the *Chicago Tribune*'s widely respected business editor, Richard Longworth, reported that the second $15 million from Chase had been expended and that the federal EDA would support additional money for Wisconsin only if International Harvester would advance Wisconsin new funds too.

Harvester, which had recently sustained losses of $225 million during the first quarter of its 1979–80 fiscal year, worried that Wisconsin's iron and coal mine assets were vulnerable to potential seizure by external creditors. Thus, earlier on March 28, Harvester foreclosed on those properties and the ore ship. But Harvester had failed to consult with Chase Manhattan before acting, and, according to Wisconsin plant manager George J. Harper, several hours later, Chase "impounded all our inventories and stopped all our shipments. At that point, we were literally dead," Harper explained. "We had to start telling the workers that we were shut down."

As one employee recounted, "We got no warning of this closing at all. I was loading boxes and the foreman came up and just told me to go home. I figured they'd run into some kind of problem with the trucks." As Harper remembered, "We just dumped them in the street with nothing to show for what they had done. . . . It was anything but honorable and anything but diplomatic. . . . It makes you feel sick inside."

Harper and the workers also didn't know that Chase had frozen Wisconsin's bank accounts. By 5:00 P.M., the foremen were instructed to tell the men on their shifts not to come back to work. Frank had returned home by that time, and his foreman telephoned him there. "Lumpkin," he said, "don't expect to come back to work. It looks bad." By early Saturday morning, the news had spread to all of the workers and their families. One woman remembers being a fourteen-year-old girl when her father worked at Wisconsin. She never forgot her mother waking her on Saturday morning to tell her what had happened: "They called the ore boat back," with the coast guard radioing it to return to South Chicago. "It was a crucial moment of rupture," she explained, when the "widespread belief in future prosperity for oneself and one's family" and the stability that flowed from that assumption was first called into question. Wisconsin's closing "would tear through a fabric that had sustained generations" and portended "the collapse of the world as I had known it in Southeast Chicago." For her dad, the mill's demise "upended the world as my father knew it."[1]

By midday on Monday, March 31, a sense of trauma, crisis, and fear had spread across the Southeast Side as people gradually realized that all of Friday's paychecks were now worthless. Both of Envirodyne's Wisconsin holding companies had filed for bankruptcy. One lawyer involved told Richard Longworth, "I pleaded with Chase to at least take care of those checks that bounced. Chase said they couldn't see any legal responsibility. I told them there's more than a legal responsibility involved here," but that was rejected.

Scores of other businesses—industrial suppliers closely tied to Wisconsin Steel and retail establishments patronized by Wisconsin workers—immediately began to suffer their own financial consequences. By Wednesday, April 2, the *Daily Calumet* was reporting that more than a thousand workers at Chicago Slag & Ballast and the Chicago West Pullman & Southern Railroad, both of which had serviced Wisconsin, had also lost their jobs. The *Daily Cal*'s editorial page predicted "a chain reaction within the community" as "unemployment will spread forth from the plant," and warned its readers to grasp "the all-too-real possibility" that Wisconsin "might never reopen." If so, "in a short time, the life and breath of the community will cease to exist, and the neighborhood will be as dead as Wisconsin Steel."

Each day the news grew worse. Before the week was out, the *Daily Cal* was reporting a total of "nearly 7,000 'ripple effect' layoffs" by employers whose businesses had been tied to Wisconsin. The federal bankruptcy court authorized the EDA to spend up to $1 million to purchase the coal that was necessary to avoid shutting down the coke ovens—which once cooled become unstable and are impossible to restart—but as Easter weekend began, former Wisconsin workers complained that there was a two-week lag in unemployment checks,

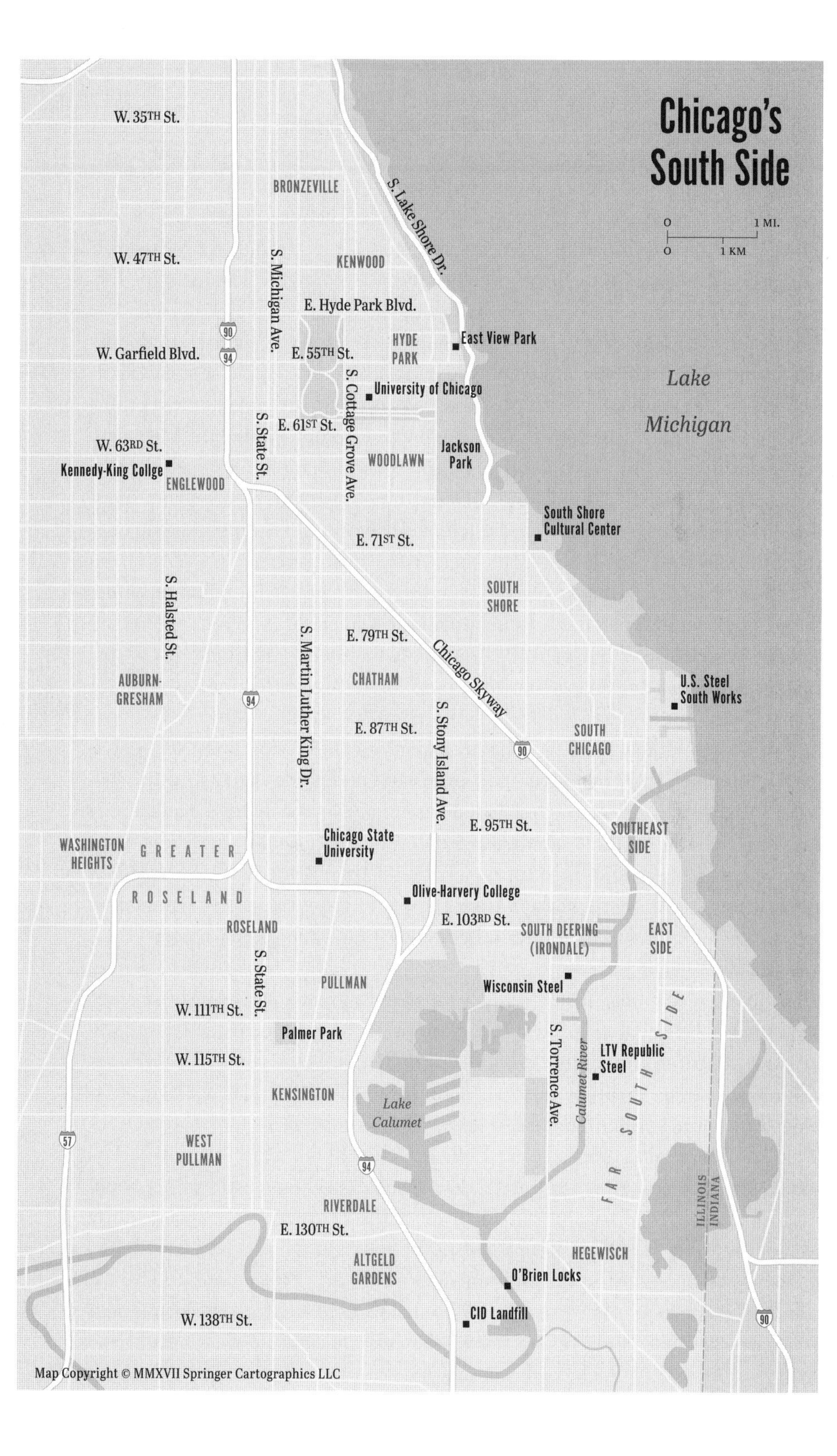

Chicago's South Side
0
1 MI.
0
1 KM
Lake Michigan
W. 35TH St.
BRONZEVILLE
S. Lake Shore Dr.
W. 47TH St.
KENWOOD
S. Michigan Ave.
E. Hyde Park Blvd.
HYDE PARK
East View Park
W. Garfield Blvd.
E. 55TH St.
S. Cottage Grove Ave.
University of Chicago
E. 61ST St.
S. State St.
W. 63RD St.
Kennedy-King Collge
ENGLEWOOD
WOODLAWN
Jackson Park
South Shore Cultural Center
E. 71ST St.
SOUTH SHORE
S. Halsted St.
E. 79TH St.
S. Martin Luther King Dr.
Chicago Skyway
CHATHAM
AUBURN-GRESHAM
U.S. Steel South Works
E. 87TH St.
S. Stony Island Ave.
SOUTH CHICAGO
E. 95TH St.
SOUTHEAST SIDE
WASHINGTON HEIGHTS
GREATER ROSELAND
Chicago State University
Olive-Harvery College
E. 103RD St.
ROSELAND
SOUTH DEERING (IRONDALE)
EAST SIDE
S. State St.
PULLMAN
Wisconsin Steel
W. 111TH St.
Palmer Park
FAR SOUTH SIDE
S. Torrence Ave.
Calumet River
LTV Republic Steel
W. 115TH St.
KENSINGTON
Lake Calumet
WEST PULLMAN
ILLINOIS
INDIANA
RIVERDALE
E. 130TH St.
ALTGELD GARDENS
HEGEWISCH
O'Brien Locks
CID Landfill
W. 138TH St.
90
94
57
Map Copyright © MMXVII Springer Cartographics LLC

that food stamp applications were being rejected if children did not have Social Security numbers, and that Wisconsin wouldn't let them into the plant to get their personal tools and work shoes. "All I feel now is hatred," one worker told John Wasik, the *Daily Cal* reporter who was chronicling the debacle. South Chicago Savings Bank announced a three-month moratorium for Wisconsin borrowers with outstanding loans, and offered new emergency loans to the former workers too. The *Daily Cal* warned that "the longer the plant sits idle, the greater are the chances it will never reopen. . . . At stake is more than dollars and cents, more than jobs and employment . . . there are people at stake."[2]

One voice that remained utterly silent even as the crisis moved into its third week was PSW president Tony Roque. But on Wednesday, April 16, about thirty men went first to the Chicago office of the federal National Labor Relations Board, and then to the Illinois State Department of Labor to complain about the PSW's utter passivity, only to be told they should file claims in bankrupty court. Their efforts made the front page of the next day's *Daily Cal,* and the story concluded by telling interested workers to call Frank Lumpkin at home. Roque responded immediately by sending letters to every member announcing a general meeting on Sunday, April 27—in the ballroom of the mammoth Chicago Hilton hotel, in the downtown "Loop," more than fifteen miles north of South Deering.

The PSW's Hilton meeting generated angry jibes—"Why have they rented the Hilton when their members can't even buy food?" one wife asked the *Tribune*'s Richard Longworth—but when testimony in the federal bankruptcy case revealed that workers' compensation coverage for the skeleton crew manning the coke ovens and blast furnace had ended on April 1, the PSW *struck* Wisconsin, pulling those workers from the plant. Only the EDA's willingness to pay the $35,000 a week in natural gas costs prevented the coke ovens from going cold.[3]

As Wisconsin's final death rattle was sounding, two progressive Chicago clergymen—Father Tom Joyce, a Claretian priest who directed the Claretians' Peace and Justice Committee, and Dick Poethig, director of the Presbyterian Church's Institute on the Church in Urban Industrial Society, decided to attend a mid-May gathering at St. Thomas More College in Covington, Kentucky, a "National Conference on Religion and Labor." One of the featured speakers was Presbyterian minister Rev. Chuck Rawlings, who talked about his recent experience as principal organizer of the Ecumenical Coalition of the Mahoning Valley (ECMV), in northeastern Ohio.

As they listened to Rawlings, Tom Joyce and Dick Poethig could tell how similar the effects of the closing of Wisconsin were to what had occurred near Youngstown, Ohio, three years earlier. They also recognized that it had been clergymen, not union leaders, business interests, or elected officials, who had led the local community's response.

On September 19, 1977, the Lykes Corporation announced the closing of Campbell Works, with a loss of more than forty-one hundred jobs. When Chuck Rawlings, who worked for the Church and Society department of the Episcopal Diocese of Ohio, heard of the closing, he called Episcopal bishop John H. Burt, who in turn phoned James W. Malone, the Roman Catholic bishop of Youngstown. An interfaith breakfast was convened, and Rawlings circulated a memorandum calling for church leaders to confront the steel crisis. In the meantime, Youngstown attorney Staughton Lynd contacted the Washington-based National Center for Economic Alternatives (NCEA), whose codirectors, Gar Alperovitz and Jeff Faux, believed the shutdown called for an infusion of investment capital from the federal government, which would require an "unusual political mobilization" featuring "a dramatic local and national moral campaign." In a *New York Times* op-ed essay, Alperovitz and Faux called for a Tennessee Valley Authority–style "development corporation" with "mixed community and employee ownership" to oversee such a federal investment.

Rawlings's band of Ohio bishops and pastors called themselves the Ecumenical Coalition of the Mahoning Valley (ECMV), and they convened a "Steel Crisis Conference," at which Alperovitz was the featured speaker. Out of that came "A Religious Response to the Mahoning Valley Steel Crisis," which was signed by more than two hundred clergy members. In this pastoral letter, the clergymen echoed Alperovitz in declaring that "this is not in any sense a purely economic problem." They were "convinced that corporations have social and moral responsibilities," and said they were "seriously exploring the possibility of community and/or worker ownership" of a reopened Campbell Works.

U.S. Steel chairman Edgar Speer condemned their efforts as "nothing short of a Communist takeover," but ECMV, taking advantage of $335,000 in federal support from the Department of Housing and Urban Development, commissioned Alperovitz to undertake a six-month study to determine if Campbell could be reopened. National newspapers like the *Times* and the *Washington Post* covered the effort, especially once Alperovitz announced a preliminary finding that about $500 million would allow Campbell to reopen with about half of its prior workforce. But White House aides to President Jimmy Carter would support only $100 million and quietly asked Harvard Business School professor Richard S. Rosenbloom to evaluate Alperovitz's analysis while postponing any decision until after the November 1978 midterm elections.

In March 1979, the White House notified the ECMV that their proposal had been rejected. Chuck Rawlings thought he and his colleagues had been "naive" to expect federal help, especially when Youngstown parishioners had remained far more silent than their pastors, but when Tom Joyce and Dick Poethig spoke with Rawlings after his presentation at the May 1980 conference about Wisconsin's demise, his advice was decisive—"Go back and organize!"—and Tom and Dick agreed to do just that.[4]

Joyce knew even before he returned to Chicago that the first person he would contact was Leo Mahon. Fifty-four years old at the time of Tom's call, Mahon had been pastor of St. Victor Roman Catholic Church in Calumet City, the first suburb just south of Chicago's southeastern city limits, since 1975. Mahon had been ordained a priest of the Chicago archdiocese in 1951. Early on, he worked with Puerto Rican parishioners and learned Spanish while also rubbing shoulders with a young community organizer named Nicholas von Hoffman and von Hoffman's well-known mentor, Saul Alinsky, the father of community organizing. Within a few years, Mahon became head of the archdiocese's Committee for the Spanish Speaking, which planned to start a mission in Panama. Archbishop Albert Cardinal Meyer, whom Leo adored, chose Mahon to lead it, and in early 1963 Leo left for Panama, where he spent the next twelve years.

The San Miguelito mission flourished under Leo's leadership, but government officials took a dim view of his pastoral defense of human rights, and pliable Catholic leaders in Panama twice put Leo on trial for heresy. After Cardinal Meyer died, in early 1965, the Vatican named St. Louis native John Patrick Cody as his successor, and Cody was far less supportive of Leo's work. When Leo returned from Panama to Chicago in 1975, Cody, perhaps out of fear of Mahon's possible radicalism, refused to take advantage of his Spanish and Latin American expertise and instead "exiled him" to Calumet City.

Leo had left San Miguelito despondent, knowing that Cody's attitude meant his long-standing expectation of becoming a bishop would come to naught, but at St. Victor Mahon found a core of energetic and committed young adult parishioners with whom he quickly bonded. Father Leo was "a breath of fresh air," Jan Poledziewski recalled, selecting female altar servers and using the *Sunday Bulletin* to advocate for the Equal Rights Amendment. "He empowered laypeople" and "everyone just adored him," Christine Gervais remembered. "He was such a charismatic person that if he asked you to do something, you just couldn't wait to help him out."[5]

Sometime in late May 1980, Tom Joyce and Dick Poethig met Leo at St. Victor and asked him to lead a clergy effort to respond to Wisconsin's closing. "It was quite obvious that the man to see was Leo," Tom later explained. "Right away, within five minutes, he says, 'Yes, we've got to do something about it.'" Dick Poethig remembered it similarly: "He had the right feeling, right off the bat." The three clergymen agreed they would invite some ecumenical colleagues on both sides of the nearby Illinois–Indiana state line to an initial meeting at St. Victor on Friday, June 6.

Come that day, sixteen clergymen and four laypeople joined the initial trio at St. Victor, and, as Joyce wrote in a memo the next day, reached "unanimous agreement that the Church or the parishes and congregations should organize in an effort to get some community say into the steel mill closings." On June 23, nine of them again assembled at St. Victor, with Tom Joyce stating that their "only model" was the clergy response in Youngstown. He went on to say that the community deserved to have "a modernized, efficient, competitive steel industry" based upon "modernization of the present plants." At a third meeting on July 7, they chose August 23 as the date to host "a workshop for key leadership people in the community, labor and church." For that session, Leo emphasized that their effort must not be seen as simply pro-union but instead be "distinctly a religious response."[6]

Later on July 7, Frank Lumpkin and twenty-four other former Wisconsin workers assembled at the union hall of United Steel Workers Local 65—which represented employees at U.S. Steel's huge but shrinking South Works—and signed a declaration that "we are tired of waiting" and that action was needed "Now." When contractors for Chase Manhattan Bank tried on July 22 to remove the existing steel inventory, which Chase had arranged to sell for $16 million, from the Wisconsin site, angry former workers blocked the plant gates; on August 5, when contractors sought to remove a crane, the ex-workers prevented that too.[7]

On Monday night, July 28, Frank Lumpkin's group of workers, now calling themselves the Save Our Jobs Committee (SOJC), were joined by Mary Gonzales, a Chicago native in her late thirties who, with her new husband, Greg Galluzzo, had begun working in the Southeast Side communities just before Wisconsin's demise. Gonzales and Galluzzo had first met eight years earlier, when Greg was a Jesuit seminarian working for Chicago's Pilsen Neighbors Community Council and Mary was married with several young children. By 1979, Mary was a single mother of three daughters and Greg was leaving the priesthood, and late that year Mary was hired by the Latino Institute as director of advocacy while Greg was working for the Illinois Public Action Council (IPA), which traced its organizational roots back to Saul Alinsky, who had died in 1972. Together they began to work in the Southeast Side's increasingly Hispanic—primarily Mexican—neighborhoods, but as of February 1980, when they married, their only office in South Chicago was their car.

Mary and Greg went person by person through South Chicago, focusing on its Catholic parishes. Their long-range goal was to have "a citywide coalition" of permanent, neighborhood-based advocacy groups. By early March, they had conducted scores of one-on-one interviews, established contact with five parishes, and had five small nascent groups of residents meeting and talking.

Mary's father had worked at Wisconsin Steel for thirty-five years, never missing a single day, before dying of brain cancer at age sixty-two. Mary had not heard any advance rumors about Wisconsin closing, but when it did, "it just reverberated through that whole neighborhood," with thousands of families losing all of their health care coverage. Doing one-on-one interviews all across South Chicago, she recalled years later, the most common refrain was "I don't have a doctor."

By June, their new organization had a name—the United Neighborhood Organization (UNO) of Southeast Chicago—and Mary had drafted a proposal to circulate to potential funders. Wisconsin's demise had created not only "tremendous unemployment," but also "psychological pressure on families." Life "has changed in a shattering fashion," and more than two dozen people out of the several hundred they had approached were now actively participating in the nascent UNO. "The staff's main function will be to train leadership," Mary's proposal said, and they hoped to publicly launch UNO as a southeast-wide organization within eighteen months. "Never since the Depression has this community been hit so hard."[8]

Leo Mahon's August 23 conference at Calumet College in Whiting, Indiana—just across the state line from southeastern Chicago's largely white East Side—was a four-hour event that attracted a good crowd and good press coverage. Leo presented a vision that was grand, or grandiose, given how little he knew about Chuck Rawlings's unsuccessful effort to save Youngstown's steel economy. A statement issued on behalf of the conveners asked how management "can morally justify divestiture" in light of "its unwillingness to invest its profits" to modernize antiquated plants. They went on to say that local parishioners must force "the industry to see its responsibility to the community rather than simply to shareholders," yet Leo confessed to a reporter, "I had one of my parishioners tell me we're two years too late."

The conveners believed the conference gave them "a mandate to organize a permanent structure," and Dick Poethig imagined they might attract $60,000 in support from the Presbyterian Church and a combined $35,000 from Cardinal Cody and the Catholic bishop of nearby Gary, Indiana. Leo suggested they name themselves the Calumet Religious Community Conference—soon changed to Calumet Community Religious Conference, or CCRC—and that they hire Roberta Lynch, who had contacted him when she heard about his efforts to mobilize the community in response to Wisconsin's closing; Roberta had two uncles who were priests in Panama, so she had long heard of Leo Mahon. Roberta brought experience from working for progressive Southeast Side Illinois state representative Miriam Balanoff. By the end of August, CCRC hired Roberta as its first staff member, at a salary of $500 a month.[9]

Two days after CCRC's conference, the PSW's attorney informed the federal bankruptcy court that PSW, Chase Manhattan, and International Harvester had reached agreement that Chase would cover 100 percent of the March 28 checks that had bounced—$1.3 million, including vacation pay—and 30 percent of an additional one week's wages—some $1.1 million—that Wisconsin workers were owed contractually. An earlier offer of just the $1.3 million had been rejected seventeen hundred to sixty-two, but the new deal elicited an angry protest by several hundred workers because this agreement also cleared the way for Chase to sell the accumulated inventory that the workers had previously blocked. The *Tribune* reported that these workers "also marched on their union headquarters," but Tony Roque "refused to meet with them."

During September, reports spread that Thomas Fleming, a businessman who had helped Envirodyne acquire Wisconsin Steel, had encouraged an African American friend, Walt Palmer, to pursue reopening the mill. Savvy journalists were highly dubious, but on October 6 Palmer, Tony Roque, and Chicago mayor Jane Byrne appeared before what the *Tribune* called "1,500 cheering steelworkers" at the downtown Auditorium Theatre to announce that Wisconsin Steel would reopen on November 1. Byrne stated that President Carter—just four weeks away from a tight reelection face-off against Ronald Reagan—had said federal support was available, and "as soon as there's agreement on the financial plan presented by Mr. Palmer, you can count on the steel mill opening." According to the *Tribune*'s Richard Longworth, Palmer was "a mesmerizing speaker," who received "a standing ovation" from the workers, many of whom "had tears of joy in their eyes as they left." But it was only a political chimera, and nothing more. By the end of October Byrne was claiming that the government would loan $10 million to rehabilitate Wisconsin's coke oven, but she admitted that Walt Palmer was "no longer in the picture."[10]

CCRC's efforts to become an active organization met with mixed success, as Roberta Lynch was having trouble organizing groups of parishioners. At the CCRC Steering Committee's monthly meeting, she said that "getting people involved at the congregation level is taking much more time than we had originally anticipated." Neither the bishop of Gary nor Cardinal Cody had offered any firm financial support, and as 1980 was ending, CCRC's clergymen worried that "a fully-staffed and functional organization" would not be in place prior to 1982, and they reduced Roberta's work to just half-time.[11]

By the end of October, Mary Gonzales and Greg Galluzzo had a small UNO office in the heart of South Chicago's commercial district. On Thanksgiving, in full alliance with Frank Lumpkin's Save Our Jobs Committee (SOJC), UNO staged its first protest action as thirty former Wisconsin workers, and their families, descended upon the "Gold Coast" block where Jane Byrne lived in a forty-third-floor condominium apartment, chanting, "The mayor is a turkey." Much of Mary's work focused on organizing parents at an overcrowded elementary school to push for construction of a new building. By the outset of 1981, she and Greg had won financial support for UNO from Tom Joyce's Claretian Social Development Fund and also from two small, progressive Chicago funders, the Wieboldt Foundation and the Woods Charitable Fund, the latter of which had just hired its first staffer, a young woman named Jean Rudd.[12]

At CCRC's first monthly meeting in early 1981, Roberta Lynch echoed something Dick Poethig had said two months earlier: "there is still not a widespread sense of crisis about the steel industry in our area." What's more, she admitted, "the vagueness of CCRC's program makes it difficult for people to see what they might accomplish by getting involved." Dick Poethig suggested that CCRC mount "a mortgage-protection campaign to prevent the unemployed in the region from losing their homes" and pursue "state legislation calling for advance notice of a plant closing" plus state funding "for retraining the unemployed."

A CCRC training session in mid-February allowed Roberta to describe why she, like Leo, rejected Saul Alinsky's confrontational approach to community organizing. She said they would not use a model where "you find a target, you look for ways to bring people quickly into confrontation with it" yet only "on a very narrow . . . basis . . . looking to win a very quick victory." The CCRC, she said, should not be "deluding people" with any easy victory "to get this or that" because that "isn't going to have meaning in terms of what the real problems are." Instead, since the church is "a tremendously vital and important force," reaching out to "clergy people in every congregation in the region" would allow CCRC to become "an organization that can go to U.S. Steel and say we represent 200 churches, 50,000 people in the Calumet region." But so far congregations' responses had been "very mixed," since "one of the big problems we have is just . . . convincing people that a problem exists." In a subsequent memo, Roberta again emphasized how CCRC needed "to identify an initial program," for "a concrete focus is essential if we are to convince people to work with us." Investing time made sense to parishioners only if they believed it was "building toward something that will have an actual impact," and she confessed, "I have certain hesitations about whether we will really be capable of carrying out sustained activity."[13]

Early in 1981 the federal bankruptcy court awarded title to the Wisconsin Steel site to the federal EDA. The EDA imagined selling the plant, perhaps for use as a "mini-mill" that would employ less than half of Wisconsin's onetime work force, but everyone realized that with Ronald Reagan's inauguration, the chances of federal action to prop up antiquated steel plants had vanished. Nonetheless, Frank Lumpkin announced that 150 former Wisconsin workers would travel to Washington, D.C., to lobby for federal action. Frank estimated that only 10 percent of the ex-employees had found new jobs, and he stressed that all benefits had now run out. When the workers visited the House gallery, six members of Congress rose to speak on their behalf, including Chicago's Harold Washington.[14]

In April, Roberta Lynch resigned to pursue a full-time job. CCRC continued to meet for the rest of 1981, but without even a part-time paid staffer, little meaningful outreach activity was taking place. In stark contrast, Mary and Greg's UNO of Southeast Chicago was receiving funding commitments from multiple sources ranging from the United Way of Metro Chicago and the Chicago Community Trust to the Wieboldt Foundation and the Roman Catholic Church's national Campaign for Human Development (CHD), a then relatively low-profile program with a social-action support mission very similar to Tom Joyce's much smaller Claretian program. Mary also contacted Jean Rudd at the Woods Fund, and by the end of 1981 UNO had scheduled a large ceremony for May 8 to publicly launch the organization. Similarly, Frank Lumpkin and his Save Our Jobs Committee, with UNO acting as their fiscal agent, successfully approached small foundations such as the Crossroads Fund for modest support to ensure SOJC's future. More significantly, thanks to progressive attorney and legendary former Chicago alderman Leon Despres, Frank secured the pro bono services of a savvy young attorney, Tom Geoghegan, so that from mid-1981 onward, SOJC would be an increasingly active participant in the legal arm-wrestling about liability for Wisconsin Steel's demise.[15]

Most important, by early 1982 Greg Galluzzo had added to UNO's staff a thirty-one-year-old organizer who quickly found his way to Calumet City to introduce himself to Leo Mahon. Jerry Kellman had grown up in the New York City suburb of New Rochelle, drifted through two years of college, first in Madison, Wisconsin, and then Portland, Oregon, and by 1971 was undergoing Alinsky-style training by the Industrial Areas Foundation (IAF) staff, the truest—and most aggressive—disciples of the late community organizing guru. That training led to organizing assignments in Chicago, suburban DuPage County, Philadelphia, and Lincoln, Nebraska, where he put together a citizens coalition made up primarily of one congregation's parishioners. By 1979, Kellman was back in Chicago and in graduate school, first at Northwestern and then at the University of Chicago. Galluzzo knew immediately that he wanted to add Kellman's faith-based organizing expertise to UNO's expanding work on the Southeast Side.

In February 1982, Leo told Tom Joyce, Dick Poethig, and his other colleagues about Kellman, and they agreed to invite him to CCRC's next meeting. The organization's bank account balance totaled $473, but UNO and the Latino Institute had Kellman's salary covered and within four weeks Jerry, Mary, and Greg sent Leo a detailed three-page memo titled "Our Suggestions for a Church-Based Organization in the Calumet Region." "We agree with you that the Calumet Region needs organizing if it is to avoid becoming an economic wasteland," they wrote, but there were two essential challenges: first, "how to organize enough strength to change the situation, rather than set people up for still another defeat," and second, "how to sustain the organizing over an extended period of time by developing the parish as a community through the organizing process."

The trio wanted to expand UNO's Catholic-parish-based organizing from Chicago's Hispanic neighborhoods southward into parishes in majority-white suburban towns like Calumet City, with Kellman doing that outreach. Once a core group of at least ten parishes was organized, the effort could expand to Protestant churches. Funding for the expansion could be sought from CHD and foundations like Woods and Wieboldt, so that by 1984–85 Kellman could add staff to do "leadership development within each parish and congregation." Then those parishes could "come together for common programs which affect the entire region. The issues start small, but grow progressively larger as the organization grows stronger and as the leaders become increasingly sophisticated." Leadership training would be ongoing, and "the professional staff is there to share what they know, not to make the leadership dependent on them."[16]

Leo took their proposal to his CCRC colleagues, telling them, "I feel that this is the kind of direction our organization must take." He half-humorously told his own parishioners that "the talk around Calumet City . . . is that the parish of St. Victor's is openly going 'Communist.'" Frank Lumpkin, the actual Communist, was continuing his work for SOJC, and the *Tribune*'s Richard Longworth published a moving profile of Frank and his colleagues, in which Frank estimated that five hundred former Wisconsin workers had left town, fifteen hundred were still unemployed, and twelve hundred or so, including his friend Daniel "Muscles" Vitas, had found some type of new job, Vitas as a school crossing guard.

UNO's May 8 founding convention was "a sight of such inspiration that few will forget it," observed Father Tom Cima, UNO's new board chairman and pastor of Our Lady Gate of Heaven Parish in Jeffery Manor—a primarily black middle-class neighborhood located between South Chicago and South Deering. UNO and SOJC collaborated in a downtown protest at which marchers chanted "We want jobs," and progressive Catholic clergy throughout Chicagoland—as most residents called the metropolitan area—were overjoyed when on July 10 Joseph Bernardin, the liberal archbishop of Cincinnati, was named archbishop of Chicago, succeeding the widely reviled John Patrick Cody, who had died on April 25.[17]

Of seemingly lesser consequence, in the summer of 1982 Mary and Greg's corps of southeastern Chicago organizers received a new recruit. The twenty-two-year-old Bob Moriarty had grown up in an Irish working-class Chicago suburb, and during his junior year at Northwestern University in Evanston, the town just north of Chicago, he had taken a community organizing seminar taught by a professor named John McKnight. A fifty-year-old Ohio native and navy veteran, McKnight had worked for the Chicago Commission on Human Relations, directed the Illinois chapter of the American Civil Liberties Union (ACLU), and headed up the Midwest office of the U.S. Commission on Civil Rights. In the latter role, McKnight had been in the room when Martin Luther King Jr. negotiated a much-criticized end to his 1966 civil rights protests that had roiled Chicago, and from that post, McKnight had moved to Northwestern.

Beginning in the mid-1970s, McKnight wrote a series of influential articles on how service economies reduce citizens to consumers and clients. Writing first in the *Christian Century* in 1975, McKnight explained that each time a social problem, or *need,* is identified, "citizens have an increased sense of deficiency and dependence." Two years later, McKnight expanded on that analysis and argued that service economies "are peopled with service producers and service consumers—professionals and clients." The former controlled the relationship, and "the client is less a person in need than a person who is needed" in order to justify the salary or income of the provider. As "the interpretation of the need necessarily becomes individualized," it disables "the capacities of citizens to perceive and deal with issues in political terms."

By 1979, McKnight had honed his analysis further. "A service economy *needs* 'deficiency,' 'human problems,' and 'needs' if it is to grow. . . . This economic need for need creates a demand for redefining conditions as deficiencies" and "the power to label people deficient and declare them in need is the basic tool of control and oppression." As government social welfare bureaucracies expand, "the professional servicers now receive more money for their help than the recipients receive in cash grants." Quite possibly, McKnight contended, "there are more people in Chicago who derive an income from serving the poor than there are poor people. . . . The welfare recipient is the raw material for the case workers, administrators, doctors, lawyers, mental health workers, drug counselors, youth workers, and police officers. Do the servicers need the recipient more than she needs them? . . . Who really needs whom?" McKnight believed that professionals willing to cast aside their own self-interest must commit themselves "to reallocation of power to the people we serve so that we no longer will need to serve."[18]

John McKnight was unquestionably the most influential social analyst in 1980s Chicago, and he brought Bob Moriarty to organizing. But Greg Galluzzo thought the twenty-two-year-old Moriarty was too young for the congregation-based organizing that UNO was moving toward under Jerry Kellman's tutelage, so by September 1982 Moriarty was going door to door in South Deering, just like Mary Gonzales had in South Chicago two years earlier. Moriarty's job was to warn residents that Waste Management Incorporated (WMI), a huge garbage conglomerate that already operated a four-hundred-acre landfill farther south, below 130th Street, had just applied for a city permit to open a new landfill on the 289-acre "Big Marsh," located just south of 110th Street and west of Torrence, only a few blocks from residents' homes and the Bright Elementary School on South Calhoun Avenue.

One day Bob knocked on the door of a home on 108th Street, hardly four blocks from the now-shuttered gates of the Wisconsin Steel plant. Moriarty introduced himself to a woman named Petra Rodriguez, who was interested in his information, but Rodriguez also had a hugely consequential recommendation for him: "You should meet my daughter." And so Bob walked around the corner to her home at 10814 South Hoxie Avenue and brought to Chicago organizing the most important recruit of the decade. The next eight years of Chicago politics would be different because he did so.

Mary Ellen Rodriguez Montes was a twenty-four-year-old stay-at-home mother of three young children. In Spanish, her name was Maria Elena, but to her family, and to the young organizers she would work with, she was simply Lena. "She was very smart, very beautiful, very tough," Bob remembered, and a "quite extraordinary person," another organizer explained. A priest who knew Lena well recalled her as "a real dynamo. She was also very attractive: great charisma and personality and very engaging."

Lena easily recalled Bob's first visit: "I remember him coming to the door." She knew about the Love Canal environmental disaster near Niagara Falls, New York, and Chicago newspapers were reporting that a company called SCA Chemical Services had asked the Illinois Environmental Protection Agency (IEPA) for permission to move toxic chemical waste from downstate Illinois to an incinerator located at 11700 South Stony Island Avenue, just southwest of where WMI wanted to locate its landfill. Lena and her husband Ray agreed to host the first meeting of Bob's recruits in their second-floor living room. Another young stay-at-home mom who attended was Alma Avalos. One year younger than Lena, she had grown up on Petra Rodriguez's block and now had two young children. Bob, Lena, and Alma then spent the next several weeks recruiting other South Deering residents to protest against the two facilities.

Before the end of October they were ready to act. They wanted a public meeting, in South Deering, with IEPA director Richard Carlson, but they got no response. Then Moriarty, along with another organizer, Phil Mullins, who had come to UNO from Pilsen, suggested taking a busload of residents, along with their children, to Governor James R. Thompson's office in downtown Chicago. Arming the children with sticky caramel apples, the group made its way to the governor's suite via an unsecured back stairway. With Lena and Alma in the lead, the group said they weren't leaving until they spoke to Carlson. Unhappy staffers got Carlson on the phone, and he promised a meeting, but by the upcoming Election Day, November 2, one still had not been scheduled. Jerry Kellman happened to know where the governor voted, so another bus trip was scheduled for what turned out to be a chilly, rainy day. After a wait of several hours, Thompson's limousine finally appeared, and the group dashed toward him brandishing picket signs. With plenty of journalists looking on, a public meeting was quickly promised.

Four weeks later, on the evening of December 6, Carlson traveled to the Trumbull Park Fieldhouse, on South Deering's northwestern flank, to speak to a crowd of more than two hundred residents. Moriarty and his recruits had prepared carefully for the session. They were concerned, however, by the presence of Foster Milhouse, a well-known precinct captain in Alderman Vrdolyak's 10th Ward political organization and a leader of the old-line South Deering Improvement Association (SDIA), a group that traced its roots back to an infamous August 1953 race riot. When the Chicago Housing Authority (CHA) accidentally assigned one black family to the housing project that adjoined Trumbull Park, SDIA's membership responded with violence. After more African American families moved in during the next four years, even greater violence erupted in July 1957. When Milhouse began heckling at the outset of the December 6 meeting, Moriarty's recruits responded quickly. "We call him Judas," Moriarty remembered, and "they just jeer him out of the hall . . . 'Judas, Judas, Judas.'" Once Milhouse was dispensed with, Carlson quickly agreed to the residents' requests, but it was Lena who emerged as the star of the evening. "I somehow kind of like blossomed in this room," she remembered. "I actually enjoyed it," and indeed "felt called to it." Also present was her husband Ray, who "really had an interest in being a lead person" and who "seemed a little bit upset about it," Lena explained, when his wife emerged as the residents' lead spokesperson. Ray "was a decent guy, but really insecure," Bob recalled, and Alma described it similarly: "jealousy."[19]

Carlson's appearance put their group on the map, and by the end of the year, they had chosen a name to distinguish themselves from the larger UNO: Irondalers Against the Chemical Threat, or IACT. In February 1983, a wary Alderman Vrdolyak met with them about WMI's proposed landfill. Lena recalled that he "met with us on the site of Waste Management's proposed dump and from where we stood, we could see our homes. He said, 'Gee, I didn't realize that it was this close to the houses.' I said, 'Does this mean you're going to oppose it?' And he said, 'Oh no, I'll reconsider and get back to you.' Well, he never did get back to us."

In the meantime, when incumbent mayor Jane Byrne, whom Vrdolyak energetically backed, finished second in the Democratic mayoral primary on February 22, Vrdolyak's political fortunes took a turn for the worse. Byrne got 33 percent, and Cook County State's Attorney Richard M. Daley—a son of the late Richard J. Daley, Chicago's powerful mayor from 1955 until his death in late 1976—placed third with 30 percent. The upset winner was African American congressman Harold Washington, who rode a tidal wave of enthusiasm among black voters to a 36 percent plurality. Washington still had to win the general election against Republican former state legislator Bernard Epton, and the racial symbolism of Chicago electing its first black mayor—or white voters uniting to stop it—cast the contest in starkly racial terms.

Washington visited the IACT activists at Bright School on March 29, and on April 12, he narrowly edged Epton, winning 51.7 percent against the Republican's 48 percent. Analysts concluded that only 12.3 percent of the city's white voters, primarily from the generally liberal lakefront wards, voted for Washington. Seventeen days later, on April 29, 1983, Chicago's first black mayor took office.[20]

Throughout the latter part of 1982 and the first five months of 1983, the outlook for Southeast Side steelworkers grew worse and worse. There was even more concern when word got out that PSW president Tony Roque had signed an agreement with Chase Manhattan in August 1980 that allowed the workers to recoup their bounced checks, but that also potentially released International Harvester from most if not all of its pension obligations to Wisconsin's former workers. Roque had not understood the legal implications of what he had signed. Some families were becoming so desperate that SOJC had initiated free food distribution twice each month and received additional funding support via UNO.

For decades, U.S. Steel's South Works, located well north of 95th Street, had been the unchallenged behemoth of the Calumet region's steel mills. Its 1973 workforce of ninety-nine hundred had shrunk to seventy-four hundred in 1979, fifty-two hundred in early 1981, and then forty-eight hundred in the spring of 1982, but in September 1982, U.S. Steel chairman David M. Roderick announced that the company would build a new rail mill at South Works thanks to concessions from both USW Local 65 and the state of Illinois. The new facility would add up to one thousand jobs, and completion was targeted for late 1983. "If we were going to be shutting down South Works, we wouldn't be building the rail mill here," Roderick assured Chicago journalists and state officials. Six months later the USW accepted an openly concessionary contract, hoping that laid-off workers would be brought back. Then, in May 1983, Roderick reversed himself and told U.S. Steel's annual meeting that South Works might indeed be closed due to the impact of environmental regulations on such an aged plant. The same week that Roderick spoke, a comprehensive survey of Southeast Side neighborhoods showed "a job loss rate of 56 percent since 1980" and "an unemployment rate of 35 percent."[21]

In late May, more than 150 IACTers and other antidumping protesters descended upon WMI's annual meeting in tony suburban Oak Brook. The protest drew significant press attention, and next the IACTers—who had revised their name to Irondalers to *Abolish* the Chemical Threat, rather than just "Against"—blockaded the entrance to WMI's large Calumet Industrial District (CID) landfill south of 130th Street, creating a backlog of scores of garbage trucks. Chicago police, unsure whether the remote location was in Chicago or instead in Calumet City, made no arrests. In the meantime, Mary Ellen Montes, who was seeking an appointment with the city's new mayor, met with his sewer commissioner on June 10, and six days later the IACTers again blocked the CID entrance. This time Chicago police had a map, and seventeen of the sixty protesters were arrested, including Lena. Her mother Petra spoke to reporters, and Moriarty and others went door to door in South Deering to raise bail money.

Everyone was released in time for a 10:00 A.M. meeting the next day with new mayor Harold Washington. At least one woman showed Washington the visible bruises she had from her arrest, and the mayor agreed to speak at an IACT meeting in South Deering. By midsummer, IACT had access to a crucial meeting place which previously had been denied it: St. Kevin Roman Catholic Church, on the east side of South Torrence, just north of the rusting Wisconsin Steel plant and by far the neighborhood's largest church. Up until early 1983, St. Kevin's pastor had been Father Bernard "Benny" Scheid, a notoriously hateful and sometimes drunken political ally of Alderman Vrdolyak. Two years earlier, when the *Chicago Sun-Times* had publicly exposed the extent of then-Cardinal Cody's financial misdeeds, Scheid wrote a letter to the paper's editor warning him to "get your affairs in order. We pray for your sudden and unprovided death every day."

Fortunately for IACT and South Deering, Scheid's successor was Father George Schopp, who for several years had worked with Greg Galluzzo and UNO as pastor of St. Francis de Sales Parish on the East Side. Schopp was inheriting a parish that included not only Lena and Alma, but, far more menacingly, Scheid's buddies and Vrdolyak precinct captains like Foster Milhouse ("we used to call him Fester Outhouse," Schopp recounted) who were "kind of a goon squad." But Schopp was already familiar with Vrdolyak's iron grip control of Southeast Side politics, and his arrival at St. Kevin dramatically altered the parish's political role, as all of Chicago would soon see.

In late August, Harold Washington announced that he was blocking WMI's attempt to open a landfill in Big Marsh as well as a proposed expansion of the nearby existing Paxton Landfill on East 120th Street. Then, on Wednesday evening, August 24, Washington came to South Deering to speak to an IACT-organized crowd of some six hundred people packed into St. Kevin's large basement hall. As the *Tribune*'s headline the next morning put it, "Washington Invades Ald. Vrdolyak's 10th Ward Turf."

Both George Schopp and Dennis Geaney, Leo's associate pastor from St. Victor, were worried about what Benny Scheid and Vrdolyak's lackeys might try to do, so Schopp asked a number of supportive priests to stay close to Scheid. As reporters scanned the crowd and a television crew set up their camera, Scheid "assured me that he would work over the crowd by telling them that Washington was an ex-convict and still a big crook," Geaney recalled a few weeks later. Once Washington arrived and the meeting got under way, Geaney happened to sit beside Petra Rodriguez, "who told me that the chairperson was her daughter, Mary Ellen Montes. This tiny woman steered the tight ship of 600 people like a seasoned sea captain. Benny and the 10th Ward Regulars never got an opening." Scheid "got up and started blustering, trying to berate Mary Ellen," Schopp recalled, but Lena was unbowed and the hecklers were silenced.

Washington was a powerful and emphatic speaker, and he took control of the crowd. "There is an over-concentration of waste facilities in this community" and the multiple dumps posed a significant danger. "I am appalled things have gone this far." Washington singled out WMI by name: "I believe this company has a horrible record of violating the public trust and endangering the public health," he said. "We'll do whatever it takes to get to the bottom of this. We're investigating this now. Apparently Waste Management has quite a bit of influence over certain key people," an obvious allusion to Vrdolyak, who had received at least $18,500 in political contributions from WMI.

As Washington concluded, the crowd rose to give him a standing ovation, but Lena, standing beside him, immediately intervened: "The meeting's not over yet, Mr. Mayor. We're not finished." She then sternly insisted that Washington give a yes or no answer to each of five specific IACT demands for city action, and as she recited them, Washington smilingly said "yes" each time. After her fifth one—"Are you committed to stopping Waste Management?" to which Washington responded, "Yes, I am"—Lena reached up, "threw her arms around him and kissed him" on the cheek, as 10:00 P.M. news viewers all across Chicagoland soon witnessed. Washington looked smitten. "Now I know why she is your leader. She's quite a politician," he told the crowd. To Lena herself, the mayor was even more complimentary: "Boy, you're a tough woman. I don't want to mess with you. I'll do anything you want me to do," and Washington gave her his private home phone number.

Observers were blown away by Lena's aplomb. "It was an impressive performance by Mary Ellen," environmental expert Bob Ginsburg remembered. "She held him there until he agreed" and "it made UNO a citywide player" operating from the veritable backyard of a powerful city council figure who had already become Washington's greatest political nemesis. "It was a big deal." Dennis Geaney felt likewise: Lena "was too astute to let him use his charisma as a substitute for hard answers." Phil Mullins, who had just succeeded Bob Moriarty as UNO's IACT organizer, was astonished. "It was an awesome meeting. . . . It was amazing. It just changed everything."[22]

George Schopp felt the backlash, "big time. . . . It sent Vrdolyak off the wall." Former alderman and Vrdolyak ally John Buchanan told the priest, "You're part of a Communist conspiracy." Different repercussions came from beyond the neighborhood. The CID landfill below 130th Street received almost two-thirds of Chicago's garbage. It and the older Paxton Landfill at 122nd Street were essential sites; after all, the city's waste had to go somewhere. The chairman of the Zoning Board of Appeals saw the Southeast Side locations as simple common sense: "from a land use point of view, that is an area that has become dedicated to this type of business." The other most relevant city official viewed WMI in economic development terms: "Their proposal is no different from a steel mill starting to expand. I'm looking at it as an industry expanding, and we need jobs." The Chicago Association of Commerce and Industry agreed, which led to a *Tribune* headline saying "Dumping Ban Called Threat to Business." Unlike with landfills, the city had no regulatory authority over the SCA incinerator at 117th Street, and in early October the Reagan administration's Environmental Protection Agency (EPA) gave what the *Tribune* called "the largest commercial toxic waste incinerator in the United States" a permit to burn PCBs inside Chicago's city limits. The chemical waste company confidently declared: "This is a state-of-the-art incinerator. It will not pose a health threat to residents."[23]

But the most pressing threat to the Southeast Side's well-being was the ongoing uncertainty of what U.S. Steel would do with South Works, where the active workforce was down to twelve hundred. Although the ongoing shrinkage of South Works' employee roster was not as sudden or dramatic as the Wisconsin catastrophe, the cumulative job loss over time was almost three times greater, and it was the result of industry-wide trends, not a series of missteps. Between the late 1970s and the mid-1980s, the domestic U.S. steel market shrank dramatically, especially because of greatly reduced demand from the U.S. auto industry and a more than 50 percent growth in the import of steel from abroad.

Throughout the fall of 1983 and into early 1984, U.S. Steel's leadership continued to threaten a possible shutdown of South Works unless environmental protections were seriously loosened and the United Steelworkers union surrendered even more far-reaching contract concessions. Two days after Christmas,

U.S. Steel announced that it would shrink the plant to just its beam mill and one electric furnace, reducing the workforce to just eight hundred. Members of Congress were joined by Archbishop—and now Cardinal—Joseph Bernardin in denouncing U.S. Steel's behavior, and *Tribune* business editor Richard Longworth wrote an angry column, declaring that U.S. Steel's behavior "violated every standard of decency and broke every obligation to the workers and the community that made it rich." The company had betrayed "every principle of economic fair play established over the last fifty years" and new federal legislation might be necessary "to protect the country from companies like U.S. Steel." Indeed, "U.S. Steel is operating so far outside the rules of normal free enterprise that it is challenging the entire American industrial system." Among those responding to the essay was St. Victor's Father Dennis Geaney, who commended Longworth and rued the damage done to "people who are being treated like obsolete machinery."[24]

While IACT and the Southeast Side steel crisis were making news throughout the summer and fall of 1983, Jerry Kellman reactivated CCRC in tandem with the Catholic parishes that stretched across Cook County's suburban townships, the area that comprised the archdiocese's Vicariate XII. Mary Gonzales and Greg Galluzzo also were expanding UNO's organizational reach into three more predominantly Hispanic Chicago neighborhoods: Back of the Yards, Little Village, and Pilsen, where Danny Solis was transforming the Pilsen Neighbors Community Council into a UNO affiliate. Before the end of the year Mary and Greg also added to UNO's staff Peter Martinez, a veteran IAF organizer. Martinez had known and clashed with Kellman a decade earlier, and his arrival not only increased tensions between Jerry and Greg, but it spurred Kellman's gradual shift from UNO to Leo Mahon's CCRC. Jerry sought support for CCRC from the archdiocese's CHD, the Woods Fund, and Tom Joyce's Claretian Social Development Fund, emphasizing that his congregational organizing would lead parishioners toward "understanding social action as part of a faith commitment."

By midsummer, Kellman had visited pastors across the vicariate and had won the support of urban vicar Father Ray Nugent. Combining the vicariate's numerical designation with the Book of Ecclesiastes' (3:1–2) well-known invocation of "time," Kellman and the pastors came up with "Time for XII" as the name for a program which would work hand in glove with CCRC to train lay leaders in each parish to listen to fellow parishioners' thoughts about the region's economic crisis.

Cardinal Bernardin gave Time for XII his enthusiastic support, and on August 29 he endorsed it at a meeting of three hundred parish leaders from across the vicariate. Kellman and Leo Mahon believed it would take until October

1984 to raise the necessary funds—in part through contributions from each church—to hire staff and begin work at the more than twenty parishes that said they would sign on. In the interim Kellman would kick off "pilot projects" at St. Victor and at Father Paul Burke's Holy Ghost Parish in neighboring South Holland.

At St. Victor Parish, Leo's—and soon Jerry's—right-hand layman was the energetic Fred Simari. Under the tutelage of Leo's first young associate pastor at St. Victor, Bill Stenzel, Fred had proceeded through the archdiocese's three-year deaconate school. At almost forty years old, Fred was six years older than Kellman, who quickly impressed him as an "incredibly hard worker" who "was great at what he did." Also involved at St. Victor were two other key parishioners, Gloria Boyda and Jan Poledziewski. Within St. Victor, "lots of laypeople got involved," Jan recalled, another of whom was Christine Gervais. "We just went into different homes and spoke to the people and then kind of brought back all of our information," Gervais remembered. We "just sat and talked," especially about what families needed. People just "refused to believe that the steel industry was going down," because for many families, the plants were the only jobs that three successive generations of breadwinners had known.

By the beginning of 1984, Kellman was expanding beyond St. Victor and Holy Ghost, and at Annunciata Parish on the East Side Kellman used a small retreat as an opportunity to explain Time for XII. Soon thereafter Jerry spoke with one young man from the parish, Ken Jania, about joining him to do further outreach. Jania, newly married and running a small, failing East Side restaurant, jumped at the chance, and by May 1984, Ken was CCRC's second paid staff member.

"My job was to connect with the parishioners," Jania recalled, "to make a presentation in front of church" and "organize and start the interview process with parishioners." On the East Side, in predominantly Polish Hegewisch, Chicago's southeasternmost neighborhood, in Calumet City and other southern suburbs, there were "hundreds of interviews that we documented." The job was harder than it sounded, for "it was very difficult for me to come from that neighborhood and to go and do those interviews, because in many cases I knew the people" going back to high school. "It was very difficult with these families" since "they'd lost everything" when a father's steel plant job disappeared. "He's got nothing," since "their skills didn't translate to anything," and that meant "absolute desperation," with prolonged unemployment signaling how "the traditional blue collar nuclear family's exploded."

As IACT's Alma Avalos explained, "I don't think the reality really sunk in until after a couple of years passed." As Christine Walley, the most poignant chronicler of the Southeast Side's disintegration, later wrote, "it sometimes felt as if our entire world was collapsing." Permanent closure of the mills, whether

Wisconsin or especially South Works, "was simply unfathomable," and for many men "the stigma of being out of work was deeply traumatic." In her household, following the Wisconsin shutdown, "my dad became increasingly depressed, eventually refusing to leave the house. . . . He would never hold a permanent job again." The Southeast Side's economic demise also "caused untold social devastation" among neighbors as another former Wisconsin worker attempted suicide and a third drank himself to death. Her father lived on, wallowing in "the deep-seated bitterness of a man who felt that life had passed him by." All across the Calumet region, it slowly dawned on people that a "world we thought would never change" had suddenly proven "far more ephemeral" than anyone had imagined possible.

Jerry Kellman's reenlivened CCRC had an expanded geographic reach thanks to Vicariate XII's archdiocesan links with neighboring Vicariate X, which encompassed all of the Chicago neighborhoods that comprised Greater Roseland. Far more crucial, however, in early 1983 Leo Mahon's protégé and former associate pastor, Bill Stenzel, was assigned to the small, struggling Holy Rosary Church at the southwest corner of 113th Street and King Drive. Stenzel had spent some previous months with Father Tom "Rock" Kaminski at neighboring St. Helena of the Cross Parish on S. Parnell Avenue at 101st Street. One of Stenzel's tasks was to merge an even weaker nearby parish, St. Salomea, a historically Polish church, into Holy Rosary, which had been traditionally Irish. Holy Rosary had some deeply committed parishioners, like Ralph Viall, a white man in his fifties, and Betty Garrett, an African American woman who had moved to Roseland in 1971, but the merger faced no opposition because there was hardly anyone either Irish or Polish or—excepting Ralph Viall and his friend Ken—indeed *white* left in Roseland.[25]

If any one neighborhood in America epitomized the experience of "white flight" in its most traumatic form, Roseland was it. The name went back to the earliest white settlers, Dutch immigrants who first arrived in 1849 to build homes and farms in the area around what would become 103rd to 111th Streets at South Michigan Avenue—the same street that fifteen miles northward becomes Chicago's "Magnificent Mile" shopping district. In 1852 the Illinois Central and Michigan Central Railroads interconnected just a little to the southeast, and the settlement that grew up there would be called Kensington. Over the next quarter century Chicago's role as major rail hub grew dramatically, and in 1880 the already-famous sleeping car magnate George Pullman chose an area just to the northeast—between what later would be 103rd and 115th Streets—to build a new manufacturing plant as well as a company town he would name after himself. By the turn of the century, Pullman's burgeoning plant employed many workers who lived in Roseland and Kensington, and in the coming decades and the World War II era, thousands of men—white men—who found well-paying jobs in the steel plants east of there, across the large geographic divide of Lake Calumet and its attendant marshes, made their homes in Roseland or the adjoining neighborhoods of West Pullman and Washington Heights, both of which, like Kensington, were often lumped into Greater Roseland.

Black people were almost nonexistent in those neighborhoods. To the north, between 91st and 97th Streets astride State Street, a small black community called Lilydale grew up in the years after 1912, and by 1937, its residents successfully protested for the construction of a neighborhood public school. At the time of the 1930 census, Kensington had 170 black residents. In 1933, when an African American woman purchased a duplex some fifteen blocks southwestward, near 120th Street and Stewart Avenue, white neighbors bombed the property. A decade later, when white real estate developer Donald O'Toole announced the construction of Princeton Park, a new neighborhood of primarily single-family homes for African Americans just west of Lilydale, eleven thousand whites petitioned unsuccessfully to block the development. The end of World War II created a serious housing shortage, and when the CHA moved the families of several black war veterans into a reconstructed barracks project on the east side of Halsted Street at 105th Street, it took more than a thousand law enforcement officers to finally end three nights of violent white protest riots.

Following World War II, Greater Roseland's racial composition changed gradually, and then incredibly abruptly. Blacks were 18 percent of the population in 1950, but the proportion increased to 23 percent in 1960, to 55 percent in 1970, and then to 97 percent by 1980. But those statistics, while dramatic, nonetheless fail to convey how stark the transformation was. In 1960, West Pullman was 100 percent white; by 1980, it was 90 percent black. Washington Heights, 12 percent black in 1960, was 75 percent so by 1970, and 98 percent by 1980. In central Roseland, the dominant church presence, reaching all the way back to the original settlers, was the four congregations of the Christian Reformed Church (CRC) and four more of the Reformed Church of America (RCA). One of the CRC churches considered reaching out to new African American residents in early 1964, but then dropped the idea in July 1968, concluding that the neighborhood was in "rapid decline" by the spring of 1969. As in other neighborhoods all across Chicago's vast South Side, the onset of the real cataclysm could be dated quite precisely: April 4, 1968, when Martin Luther King Jr. was assassinated in Memphis. "From that day on, everything changed," one resident told Louis Rosen, who wrote a powerful memoir of the transformation before becoming a successful musician. "It was rapid. It was awful," one white person recalled. "It was an exodus."

That is what happened in Roseland. In hardly twelve months in 1971–72, all four self-governing CRC churches abandoned the neighborhood and moved to the white suburbs. Of the RCA churches, one decamped in 1971, a second in 1974, and a third in 1977; the last survivor held out well into the 1980s. Of all the statistics measuring white flight, one may capture the price that the neighborhoods—and the new residents—paid more powerfully than any other: in 1960, fifty-eight M.D.s practiced in Roseland. Twenty years later, in 1980, after a population *increase* of five thousand residents, there were only eleven.

While virtually all whites fled, one Christian Reformed couple in their late thirties walked *against* the tide. Rev. Tony Van Zanten had finished seminary in the early 1960s, spent some time in Harlem and then over a decade in Paterson, New Jersey, another city experiencing serious decline. In August 1976 Tony and his wife Donna relocated to Chicago and opened Roseland Christian Ministries Center in the heart of South Michigan Avenue's once-vibrant business district. They fully realized how "the racial change in Roseland was a very radical and very swift one," maybe more stark than in any other place. "There were no social services at all," Donna remembered. "There was nothing there for the new people."[26]

Standing against the tide were the Roman Catholic parishes that for decades had stood within fifteen blocks or so of each other all across Greater Roseland. Several closings and mergers had taken place in the previous decade as the area's Catholic population shrank due to the racial turnover, but new African American members energized some parishes. Father Paul Burak was newly ordained when he arrived at St. Catherine of Genoa in West Pullman in 1972, when white flight was near its peak and the population, for the moment, was roughly 50 percent black and 50 percent white. St. Catherine's retired pastor, Father Frank Murphy, had been a forceful proponent of racial equality, but that hadn't stemmed the flight. "I experienced a lot of struggle and confusion about the parish" through the early and mid-1970s, Burak recounted. "Every weekend I would meet someone saying 'Father, this is our last weekend here.'" Burak left St. Catherine in 1978, only to return in 1981, and by then few white parishioners remained. Tom Kaminski arrived at St. Helena in 1977, and only a few elderly white people were still in the congregation.

At first glance, the massive white depopulation of these neighborhoods promised a wonderful opportunity—thousands of newly available, often well-constructed brick bungalow-style homes—for African American Chicagoans whose families had for decades been trapped within the clear racial boundaries of Chicago's South and West Side neighborhoods. But the reality of Roseland's racial transformation again made black families highly vulnerable to exploitative white real estate "professionals," this time due almost entirely to federal government policy choices. The Federal Housing Administration

(FHA), by 1968 part of the Department of Housing and Urban Development (HUD), was indisputably the villain, but until the mid-1970s, almost no one fully fathomed—or sought to expose—the consequences of government policy-making gone awry.

At first only one little-known housing policy expert, Calvin Bradford, was determined to unmask a widely ignored evil. The term "redlining" was well known, if not well understood, but in newly African American neighborhoods like Roseland, it was not lenders' refusal to make conventional home mortgage loans available to black home buyers that wreaked widespread damage, but how the FHA, starting in August 1968, made government-*insured* loans available to such purchasers—often through exploitative mortgage bankers, and even for properties of dubious quality—that ended up decimating newly black neighborhoods in which such insured loans were concentrated.

Bradford and a coauthor figured out that while seeking to "encourage inner-city lending," the FHA caused local mortgage bankers to simply maximize the number of black purchasers they could entice to buy homes. As a result, thousands of black families were issued mortgages that they were not qualified to successfully carry—especially in an urban economy where blue-collar jobs like those at the Southeast Side's steel mills were vanishing by the thousands year after year. The result was "massive numbers of foreclosed and abandoned properties," with the FHA insurance actively *encouraging* fast-buck lenders to foreclose as quickly as possible on as many properties as they could.

As Bradford explained in a subsequent essay, the federally insured loans provided "the certainty that FHA will take the property from the lender after foreclosure and pay the claim," and that led to lax underwriting and the profligate issuance of loans because "neither the mortgage companies that originated them nor the investor that purchased them—basically FNMA [popularly known as Fannie Mae, the Federal National Mortgage Association] cared about the soundness of the loans." Indeed, as Bradford later observed, "the financial incentives were so great that scores of real estate agents, lenders, and even FHA officials engaged in fraud in order to make sales to unqualified and unsuspecting minority homebuyers." The end result, as dramatically witnessed in Roseland, was "the government taking all of the losses and the communities suffering all the devastation" of foreclosed and abandoned homes.[27]

The impact of such policies and such behavior could be seen all across Roseland, both in "board-ups"—homes with plywood covering their windows—and in the rapid decline of the South Michigan Avenue shopping district. One group, the Greater Roseland Organization (GRO), founded in 1969, tried to ease the racial transition. The GRO was comprised of smaller, neighborhood-specific groups such as the Pullman Civic Organization and the Roseland Heights Community Association, and it included both older white residents, like Holy Rosary's Ralph Viall, and newly arrived African Americans, like Mary Bates and Lenora Rodgers. With funding support from both CHD and the Chicago Community Trust, GRO emerged in the early 1980s as the only audible voice speaking for the neighborhood.

The summer of 1980 saw the first stirrings of gang influence in Roseland, and in September the long-famous Gatelys Peoples Store, the largest business on South Michigan Avenue, closed. A citywide study of different neighborhoods' needs described Gatelys' closure as "psychologically . . . probably the most serious blow imaginable" to Roseland's economic well-being. Then early in 1981, in three separate incidents, three teenage students at Fenger High School were shot and killed. By this time, close to five hundred properties in Roseland and West Pullman were in foreclosure, and more than sixteen thousand people, one-quarter of Roseland's population, were receiving public aid. A study of recent job losses in the area highlighted Wisconsin's closing and stated, "Many of these workers were Roseland residents." Their prospects for new steel plant jobs were nonexistent, the report underscored: "People in these types of jobs are not merely out of work, they are out of careers."

A parallel study, focusing on men who had lost their jobs at U.S. Steel's South Works, found that 47 percent had not found new employment, but that summary statistic concealed a significant racial disparity: 67 percent of black workers were still unemployed, as compared to only 32 percent of whites. "Once laid off from their mill jobs," the study noted, "blacks in particular remain the least likely to find new jobs."

In early 1983, Lenora Rodgers mounted a renewed push to win foundation funding for GRO. She told one foundation that "community organizing and an issue-based community organization is the key to neighborhood preservation." She said GRO's greatest need was to hire organizers, since "organizers help to identify new and potential leaders" who could mobilize Roseland against the dangers engulfing it. But Rodgers's efforts were unsuccessful, and by April 1984 GRO was no longer responding to letters from potential funders.[28]

As Jerry Kellman extended CCRC's presence into Roseland early in 1984, he accepted office space at Bill Stenzel's Holy Rosary Parish in lieu of dues. IACT and UNO continued their antidumping protests, with Lena's name appearing in Chicago newspapers almost weekly. On January 30, 1984, a city council committee, spurred by Alderman Vrdolyak's desire to at least *appear* to be against dumping, approved a one-year moratorium on new landfills within the city. In mid-February Lena joined U.S. Representative Paul Simon, a Democratic candidate for the U.S. Senate seat held by Republican Chuck Percy, as he toured South Deering and its neighboring waste sites. Two days later, when the full city council approved the one-year moratorium, Vrdolyak amended the measure to exempt liquid waste handlers and transfer stations, leading Mayor Washington's backers to oppose the diluted ban.[29]

March 1984 was the fourth anniversary of Wisconsin Steel's shutdown. Mayor Washington spoke at an SOJC anniversary rally, and on March 28, Frank Lumpkin and others picketed International Harvester's downtown headquarters. Frank told one reporter he believed four hundred of the three thousand ex-Wisconsin steelworkers had died in the last four years. Later he told a U.S. congressional subcommittee that nowadays in South Chicago "the only ambition a kid can have is to steal hubcaps. There is nothing else there. There's no jobs." The *Daily Calumet* reported on the closings of more and more retail businesses; a UNO meeting on jobs drew more than five hundred neighborhood residents plus Mayor Washington's top three development and employment aides.[30]

By August 1984, Jerry Kellman was ready to publicly launch the reborn CCRC. Thanks to Leo Mahon's core parishioners from St. Victor—Fred Simari, Jan Poledziewski, Gloria Boyda, and Christine Gervais—CCRC was ready to play an active role in a retraining program for one thousand former heavy industry workers in several south suburban Cook County townships, funded with $500,000 from the U.S. Department of Labor under the 1982 Job Training Partnership Act. And thanks to Bill Stenzel's hosting of Kellman at Holy Rosary church in Roseland, Kellman was beginning to pull together a new network of virtually all-black Catholic parishes across Roseland under the distinct rubric of the Developing Communities Project (DCP), with DCP for the moment a "project" or "subgroup" of CCRC.

Kellman asked each parish for two lead representatives. From Holy Rosary came Stenzel's two most active parishioners, Ralph Viall and Betty Garrett. At St. Catherine of Genoa, Father Paul Burak suggested two people he felt had "a passion I think for social justice": Dan Lee, who had attended deaconate school, and Cathy Askew, a young white single parent with two mixed-race daughters who was teaching at St. Catherine's School. St. Catherine's senior deacon, Tommy West, was interested too, but he channeled much of his community work through another parish well north of 95th Street, St. Sabina. At St. Helena, Father Tom Kaminski volunteered himself and Eva Sturgies, an active parishioner who lived on 99th Street. From St. John de la Salle at 102nd and South Vernon Avenue, eleven blocks north of Holy Rosary, Father Joe Bennett asked Adrienne Bitoy Jackson, a young woman with an office job at Inland Steel, and Marlene Dillard, who lived in the London Towne Homes cooperative development east of Cottage Grove Avenue. Not every Roseland pastor responded with enthusiasm. At Holy Name of Mary Parish, Father Tony Vader brushed off Kellman but told his associate pastor, Father John Calicott, the only African American priest on the Far South Side, to do what he could.

The most unusual Catholic parish Kellman contacted was Our Lady of the Gardens, a church that traced its beginnings only to 1947 and which was staffed by fathers from the Society of the Divine Word. When Kellman visited Father Stanley Farier, the priest recommended two women of different circumstances: Loretta Augustine, in her early forties, who lived in a single-family home in the Golden Gate neighborhood west of the church, and Yvonne Lloyd, a fifty-five-year-old mother of eleven who lived in the Eden Green town house and apartment development just west of both Golden Gate and the sprawling public housing project from which the parish drew its name: Altgeld Gardens.[31]

Altgeld Gardens was a by-product of World War II. When Chicago faced a dire housing shortage during the war years, officials looked to the land south of 130th Street, west of the CID landfill and Beaubien Woods Forest Preserve, north of the Cal-Sag Channel (a man-made tributary dug during the 1910s) and east of St. Lawrence Avenue. This area had in earlier decades served as the sewage farm for George Pullman's eponymous town a mile northward. The Metropolitan Sanitary District's massive sewage treatment plant, opened in 1922, was located just north of 130th Street. Construction began in 1943 on a 1,463-unit "war housing development" on the 157-acre site. The first families moved in come fall 1944, and a year later, in August 1945, a formal dedication ceremony featuring local congressman William A. Rowan plus Chicago Housing Authority (CHA) chairman Robert R. Taylor, an African American, took place before a crowd of five thousand. More than seven thousand residents were already living there, the *Tribune* reported, "nearly all Negroes." An elementary school and then a high school, both named for the black scientist George Washington Carver, who had died in 1943, were soon part of the new development, but by 1951 school parents were protesting the presence of an open ditch carrying raw sewage that abutted the school grounds. In 1954 another five hundred apartments, officially called the Philip Murray Homes, were added to the Altgeld development.

For its first fifteen to twenty years, residents described Altgeld—or, more colloquially, just "the Gardens"—as "this heavenly place," "just paradise," as two different residents recalled. "It was just really a wholesome place to live," a third remembered. "There was a feeling of family throughout the entire development," said a fourth. The Gardens was its own world: almost anyone who worked had to own an automobile because public bus service to and from Altgeld was poor at best. "We felt so isolated and away from the mainstream of what was occurring in Chicago. . . . We were cut off from a lot of opportunities," one resident explained.

Loretta Freeman Augustine grew up in Lilydale, married a man from Altgeld at age nineteen, and lived in an apartment there from 1961 to 1966, when the couple moved to a single-family home in Golden Gate, just to the west. "The community had a stability" during those years. "People had nice lawns with beautiful flowers," and it was "very much a family-oriented community." The Carver High School basketball team, under Coach Larry Hawkins, won the Illinois state championship in 1963.

By the late 1960s, things had changed for the worse. Many blamed the CHA, which had evicted residents when their incomes rose above the ceiling allowable in public housing. "People were being forced out because they were over the income," one woman recalled. Dr. Alma Jones, hired as Carver Elementary School's principal in 1975, had had a similar experience some years earlier at the CHA's LeClaire Courts. "It was absolutely beautiful," but "my husband got a raise, and they put us out: excess income," she recalled. "I was devastated because I had just had twins . . . it was heartbreaking." It was also destructive. "That eliminated everybody who was upwardly mobile, because if you . . . started to progress, then they put you out, which was the worst thing that could possibly happen" because "you take out the element of folk who know how to live in a community."

One young man who grew up in Altgeld in the 1950s returned to the Far South Side twenty-five years later as a police officer. Initially "there were very few troubled families. I would say less than 5 percent. When I returned, I saw that the 5 percent was still residing in that development, but so were their children, and grandchildren, it was just a procession. The 5 percent had expanded to 85 percent." Another resident said, "It began to change in the 1970s. I don't think it really started to decline until the drugs became prevalent." A 1972 *Tribune* article marking the twenty-fifth anniversary of Our Lady of the Gardens parish described Altgeld as a place "from which long-time residents are striving to get out" and stated that drugs and crime were the Gardens' top problems. As Father Al Zimmerman commented, "With no job prospects, the temptation to turn to drugs is powerful."[32]

In April 1974 Altgeld made news in a different fashion when the entire facility was evacuated after a tank containing 500,000 gallons of silicon tetrachloride ruptured at a tank farm just ten blocks to the north. "A dense cloud of fumes half a mile wide" drifted toward the project, and almost twelve hours passed before residents were allowed to return home. More than two hundred people were hospitalized from exposure to "a heavy cloud of hydrochloric acid" that was generated when clueless workers turned firehoses on the tank, making the emissions far worse, rather than notify public officials. Chicago soon filed suit against Bulk Terminals, the tank's owner, and the *Tribune* quoted a state official as saying, "It should be a criminal offense to know of a leakage of toxic materials without reporting it immediately."

A 1982 citywide neighborhoods study found that Altgeld's needs were especially dire. "The physical isolation of this community from the rest of the city" was so great that "residents of this area are rural rather than urban poor," it noted. Tenants believe "that job training is one of their community's most important needs," but people who had never held a job needed to be taught "how to look for work." Yet "many do not own cars," and public transport was still "frightfully poor." The Gardens' one small food store shocked the outsiders. Not only did it smell "particularly bad," but rats were now regularly "visible in the food store during daylight hours." That study pointedly advised that "Altgeld Gardens needs an advocate and/or organizer to improve the coordination of the many municipal services provided to the development, and to work with the private sector to help create employment opportunities for local residents."

Altgeld residents were theoretically represented by a local advisory council, but by the early 1980s, the council had for years been dominated with an iron fist by its president, Esther Wheeler, or "Queen Esther" to many dismayed residents. "Her whole concern was nobody would take her place, or usurp her authority," Carver principal Alma Jones later explained. "She was extremely authoritarian and she owned Altgeld." But come September 1982, residents had a new opportunity to organize against the toxic waste, garbage, and sewage that surrounded them, when Hazel Johnson, a forty-seven-year-old widow and mother of seven who had first moved to the Gardens in March 1962, founded People for Community Recovery (PCR) and attracted a small band of active members.

Hazel's husband John had died of lung cancer in 1969, at age forty-one, and by the early 1980s, questions arose about the long-term health of those living at Altgeld. In late April 1984, Hazel saw a television news story about a study of cancer rates in the Far South Side. Its statistics were alarming, yet Illinois EPA director Richard Carlson brushed aside the findings. In response, Hazel contacted the IEPA, which tried to placate her with some pollution complaint forms. She responded by distributing several hundred of them throughout the Gardens over the ensuing six months.[33]

In late 1983, a new organizing effort began in the Eden Green community. Madeline Talbott and Keith Kelleher, two Association of Community Organizations for Reform Now (ACORN) organizers, had met Tom Joyce while working in Detroit, and Tom's Claretian Social Development Fund provided the initial seed money to launch ACORN in Chicago. ACORN's Eden Green organizer was Grant Williams, who had worked for ACORN in Philadelphia and St. Louis. Williams viewed Eden Green as more promising turf than Altgeld, but after a founding meeting of South Side United Neighbors, Williams expanded his work into the Gardens. He contacted Lena and her IACT colleagues about the new-dumps moratorium, but in Altgeld, the residents' biggest concern was the poor public bus service to the outside world. By March 1984 Williams had interested a reporter from Chicago's premier African American newspaper, the *Defender,* in Altgeld's transit plight, and a community meeting to oppose possible service cuts by the Chicago Transit Authority drew a good crowd.

By summer, Williams had signed up some eighty-three dues-paying members for ACORN—$16 a year—but by late August, he was moving to Detroit, and a brand-new University of Chicago graduate, Steuart Pittman, would take over in September.[34]

Before the end of summer 1984, Jerry Kellman also made his first successful forays toward enlisting some Protestant pastors to join his previously all-Catholic CCRC. His first two recruits were Rev. Bob Klonowski of Hegewisch's Lebanon Lutheran Church and Rev. Tom Knutson of First Lutheran Church in Harvey, a far-from-prosperous suburban town two miles southwest of Altgeld Gardens. Also joining CCRC was St. Anne Parish in suburban Hazel Crest, whose new pastor, Father Len Dubi, had known Kellman for more than a decade. The city parishes' DCP designees first met Kellman and their CCRC colleagues from St. Victor and the other predominantly white congregations at Tom Knutson's church in Harvey.

An early August issue of the *Daily Calumet* ran a prominent story heralding CCRC's "phoenix-like return." It quoted Kellman as saying he hoped sixty churches would join by October, and that by year's end he wanted to have "a full range of support programs for laid off workers." His top goal was to devise "a long-range plan for economic development," and within a year he hoped to have "three or four full time employees." Fred Simari and Gloria Boyda from St. Victor were appointed to a task force overseeing the new federally funded suburban job retraining program, but the program would not be able to accept applications until December or January.[35]

Throughout the fall of 1984 and early 1985, IACT, Hazel Johnson's PCR, and even ACORN appeared more visibly active than CCRC and DCP. A mid-September appearance by IEPA director Richard Carlson at St. Kevin drew an angry crowd that erupted in shouting when he insisted that the Southeast Side's bevy of waste facilities posed no threat to anyone's health. Marian Byrnes from Jeffery Manor, a widowed, recently retired schoolteacher who had founded the Committee to Protect the Prairie to avert construction on the undisturbed, 117-acre Van Vlissingen Prairie north of 103rd Street, forcefully told Carlson, "We will never believe you! You might as well go home!" That tussle was quickly overshadowed when the *Chicago Sun-Times* reported that water from at least three residential wells just south of Altgeld Gardens contained cyanide, benzene, and toluene. Most Chicago residents were no doubt surprised that anyone within the city limits had to rely upon wells for water service, but city officials had been aware of the issue for three months. Homeowners in the tiny, seven-home enclave called Maryland Manor paid city taxes but had neither paved streets nor water and sewer service. The residents were wary enough of their cloudy well water that they used it only for toilets and the like, as opposed to drinking, but the extensive press coverage was a huge embarrassment for Mayor Washington, one that would have been worse had the press known that the issue had been handed off to an intern over the summer.

In late October, just two weeks before the November general election, Lena and her colleagues successfully targeted incumbent U.S. senator Chuck Percy after he skipped a UNO candidates' forum with Democratic challenger Paul Simon. UNO followed Percy to a black radio station, WVON, and stormed the building, causing the beleaguered senator to take refuge in a women's restroom. Percy remained locked inside there for some hours, and the standoff made for memorable local television news footage. On November 6, Simon defeated Percy by fewer than ninety thousand votes out of more than 4.6 million that were cast.[36]

When the U.S. EPA denied an IACT request to review the state's finding of no health threat, Lena told the media the refusal was "quite ironic" in light of the Maryland Manor contamination. In mid-November, when state officials authorized the cleanup of an abandoned dump at 119th Street that contained 1,750 barrels of unknown chemical waste, Governor Thompson showed up wearing a protective suit, boots, and a mask to tell journalists that the site was "a monument to man's greed and disregard for the health and safety of fellow citizens." Along with Frank Lumpkin's SOJC, UNO also continued to push city officials to open a job retraining center on the Southeast Side, but environmental issues had now replaced economic ones at the top of the local agenda.[37]

ACORN's fall 1984 efforts in Altgeld Gardens underscored that shift. Once Steuart Pittman took over from Grant Williams, the small group changed its name to Altgeld Tenants United (ATU). Williams had warned Pittman that local advisory council (LAC) president Esther Wheeler was "kind of nuts," but when ATU sought to use the project's community building for a neighborhood-wide meeting, Wheeler summoned "your Leader" to meet with her executive board. ATU still drew more than one hundred residents to an October 30 meeting, but Wheeler showed up to accuse Pittman of having an intimate relationship with an elderly and devout ATU leader: "That white boy is shacking up with Maggie Davis." It was a ludicrous allegation, but Wheeler's role in Altgeld caused untold harm to the Garden's residents. As Pittman reported to ACORN's Madeline Talbott, "the grocery store"—the one whose visible population of daytime rats had astonished outsiders several years earlier—"has a plaque award for community service in it from Esther Wheeler and the LAC."

ATU reached out to both the city's sewer department and to CHA's Altgeld head manager, Walter Williams, who told the organization, "I'll resign my job before giving in to tenants' demands." The sewer department deployed workers, who told residents Altgeld's sewers were the worst they had ever seen and would take months to clean, but work was halted after one week by the CHA, which would have to foot the bill. In response, over a dozen ATU members picketed CHA headquarters in the downtown Loop on November 14 and then held a press conference.

The African American *Defender* gave them front-page coverage, and the local 9th Ward alderman, Perry Hutchinson, took an interest, telling the *Defender* that "Chicago has forgotten about south of 130th Street" and the people marooned there. But Pittman was disappointed that turnout at ATU meetings was declining. When he arranged a January 23 tour of WMI's huge CID landfill east of Altgeld, only ten people showed up. Hoping to spur greater interest, he adopted Lena and IACT's tactic from almost two years earlier, and on February 19 sixteen ATU protesters blocked garbage trucks' entry into the landfill. Pittman, the elderly Ms. Davis, and one young man were arrested. For a second blockade on March 7, only eight people participated, and the protest resulted in three more arrests. Pittman had privately given ACORN notice four months earlier that he would be leaving as of March 15, 1985, and when he departed no one immediately replaced him. At their final meeting, ATU members wondered whether they should join Hazel Johnson's PCR.[38]

In mid-January 1985, PCR received attention citywide for the first time when Hazel held a press conference to publicize the IEPA complaint forms she had circulated within Altgeld over the previous six months and to highlight that the city's one-year moratorium on new landfills would expire on February 1. One week later, Mayor Washington called a City Hall press conference, and with both Lena and Hazel standing behind him, recommended a six-month extension of the ban, which was unanimously approved by the city council. Washington also appointed a Solid Waste Management Task Force to study the city's landfill options. Lena, Hazel, and Bob Ginsburg from Citizens for a Better Environment were all named to the panel, as were 9th and 10th Ward aldermen Hutchinson and Vrdolyak and South Chicago Savings Bank president James A. Fitch; Washington administration insiders like Jacky Grimshaw and Marilyn Katz were also included to assure that the task force would not go astray.

By early spring 1985, however, rumors had gradually spread that the city administration was quietly considering an entirely different new landfill possibility, centered on 140 acres of Metropolitan Sanitary District property south of 130th Street on the east bank of the Calumet River, a location generally spoken of as the O'Brien Locks site after a nearby dam. A city Planning Department draft report had discussed the idea a year earlier, and while Mayor Washington reiterated his opposition to any dump at the 116th Street Big Marsh location when he spoke at UNO's annual convention at St. Kevin in late April, concerned residents of Hegewisch and its northern Avalon Trails neighborhood—both just east of the O'Brien property—publicly criticized Lena and UNO for not pressing Washington for a similar commitment concerning the O'Brien site.

The weekly *Hegewisch News* began to sound the alarm, with editor Violet Czachorski proclaiming that while Hegewisch residents had supported people in South Deering in opposing any Big Marsh landfill, now UNO and IACT were failing to take a similarly principled stance when a landfill was proposed for Hegewisch's backyard rather than theirs. Writing in the *News,* University of Illinois at Chicago geographer James Landing, who in 1980 had created the Lake Calumet Study Committee to help protect that body, warned that a "lack of unity among neighborhood groups . . . serves the interests of the dump companies."[39]

Harold Washington and his top aides were devoting attention to Roseland as his four-year term approached its halfway mark. In part their concern was stimulated by the Borg-Warner Foundation, whose executive director, Ellen Benjamin, had taken an interest in the neighborhood and had commissioned a "needs assessment" from a team at the University of Illinois at Chicago. Roseland had lost more than sixty-eight hundred jobs between 1977 and 1983, and loss of employment meant "many people are having trouble maintaining their houses, keeping food on the table" and avoiding foreclosure. The researchers conducted 115 interviews in Roseland, and while 33 respondents named jobs as the top problem, almost twice as many—64—described how "crime and gangs have proliferated and the feeling of insecurity has increased." Before the report was issued, Washington's top staffers were briefed on the findings. "Highest infant mortality rate in city," "highest number of foreclosed homes in the nation," South "Michigan [Ave.] business district gone," their notes recorded. With just one exception, community groups were disappointing: "Good Roseland Christian Ministries," the staff notes emphasized.

At 1:30 P.M. on Sunday, March 17, a man wearing a long dark coat and a baseball cap with the *Playboy* logo drew a gun on cashier Lavergne McDonald inside Fortenberry Liquors at 36 East 111th Street in central Roseland. She screamed, and the gunman fled. Fifteen minutes later, Roger Nelson, a seminary student who had interned at Roseland Christian Ministries, his fiancée, and his parents finished chatting with Tony and Donna Van Zanten after church services and crossed South Michigan Avenue just north of 109th Street to the lot where their car was parked. The same gunman came up to them, ordered them into the car, and instructed them to hand over their valuables. Roger's father, fifty-year-old Northwestern College of Iowa professor Ronald Nelson, was in the driver's seat, with the gunman crouched by the open driver's door. As the quartet fumbled through their belongings, Donna Van Zanten and her son Kent approached and were also ordered into the back seat. Ronald Nelson handed the man his car keys and checkbook, but the gunman angrily said, "I don't think you gave me all you have." Nelson protested, but the gunman handed back the checkbook, called Nelson a "Goddamned lying bastard," and fired one shot into the left side of Nelson's abdomen.

As Ronald Nelson lay dying at the scene, the gunman fled past two men working on a car nearby. "Brothers, you all be cool," the gunman called out. "You know them was honkies over there." One of the men was on work release for possession of a stolen car, but the gunman's appeal to race fell flat: they not only knew Roseland Christian Ministries, one of them knew Roger Nelson from his work there. After police arrived, a shaken Donna and Kent Van Zanten accompanied officers on a ninety-minute drive throughout the neighborhood while Roger and his fiancée went to a station house with detectives to look at photos of possible suspects.

Twelve days later, one of the car repairers identified a photo he believed matched the gunman; police also received an anonymous telephone tip that the man they wanted went by the nickname "Squeaky." Detectives went to 10727 South Indiana Avenue, less than four blocks from the scene of Nelson's murder, and told the older man who answered the door that they wanted to speak with Clarence Hayes. "Hey, Squeaky," he called upstairs. Hayes wasn't home, nor was he on five subsequent occasions when police stopped by, but on Sunday morning, April 14, the thirty-four-year-old three-time ex-convict and drug addict was arrested at a nearby currency exchange. That afternoon, Lavergne McDonald, Donna and Kent Van Zanten, and both of the car repairers picked Hayes out of a police lineup, as did Roger Nelson and his fiancée when they arrived in Chicago that evening.

Ronald Nelson's murder—a white victim, a black gunman, a Sunday church parking lot—drew more news coverage than anything else that had happened in Roseland in years. Eighteen months later Clarence Hayes was convicted of murder and multiple counts of armed robbery and sentenced to death; after appellate review he was sentenced to life in prison. Over a quarter century later, he was still challenging his conviction in the courts, but on the thirtieth anniversary of Nelson's murder Clarence Hayes remained safely ensconced in the maximum-security Stateville Correctional Center in Crest Hill, Illinois.[40]

March 28, 1985, was the fifth anniversary of Wisconsin Steel's sudden shutdown. Frank Lumpkin, now sixty-seven, was one of the few ex-workers whose more than thirty years at the plant meant he was collecting his full pension.

Those not so fortunate received little if anything: Felix Vasquez, age fifty-seven, was receiving $150 a month for his twenty-four years of work. Lawyer Tom Geoghegan, whose lawsuit on their behalf against International Harvester was mired in the courts, told one reporter that men like Vasquez "were cheated by a company they gave their whole lives to."

Thanks to ongoing support from the Crossroads Fund, Lumpkin's Save Our Jobs Committee (SOJC) remained active, but in South Deering, the plant was now little more than "heaps of rusted scrap." An anniversary rally drew only two hundred people, and one former worker told the *Tribune* that South Deering was now "a battered hulk of a neighborhood" strewn with "battered, empty hulks of men." Now, five years later, no one at all doubted that "Black Friday" had indeed been "the end of an era."

The former Wisconsin workers were not alone. At South Works, most of the south half of the mill had been demolished during the previous winter, and the remaining workforce was static at eight hundred. The Southeast Side's third major mill, Republic Steel, on the East Side, had a storied history—ten striking workers had been shot dead by Chicago police on Memorial Day 1937. By the mid-1970s, however, it was known to suffer from a "morale problem," and longtime United Steelworkers Local 1033 president Frank Guzzo "throws up his hands when discussing the increasing number of men who are drinking on the job." The consequences were severe: in early 1976, 46 percent of the steel shipped from Republic was "rejected because it was not up to standards," a problem Frank Lumpkin had also seen at Wisconsin.

As of early 1982 Republic had an active workforce of five thousand, but eighteen months later that number had been halved. Then, in early 1984, Republic was bought by the Ling-Temco-Vought (LTV) conglomerate, which six years earlier had acquired Youngstown Sheet & Tube in Ohio. Guzzo tried to put a bright face on the move, but workers grew increasingly unhappy with Guzzo's concessionary attitude. In April 1982, Guzzo had won reelection over a young challenger by a margin of 1,167 to 935 in a multicandidate field, but as the April 1985 election neared, a different outcome loomed.

Guzzo's top challenger both in 1982 and three years later was thirty-year-old Maury Richards, a tall, physically imposing man who was attending law school part-time and who in 1984 had mounted a credible insurgent challenge against an East Side state legislator and bar owner who was a Vrdolyak lackey. When the April 1985 ballots were tallied at Republic, it was clear that an era had ended there as well when Frank Guzzo finished fourth with just 331 votes and Maury Richards prevailed with a plurality of 538. However dim the future might be for steelmaking on Chicago's Southeast Side, the workers now had a new voice, one almost forty years younger than Frank Lumpkin.[41]

As 1985 dawned for Jerry Kellman's CCRC, Time for XII, and DCP trio, he and Ken Jania were joined by a third organizer, an old IAF colleague of Kellman's named Mike Kruglik. A 1964 graduate of Princeton University, Kruglik had spent several years as a history graduate student at Northwestern University before shifting into organizing in 1973. He spent the mid-1970s working in Chicago, but by 1979 Kruglik was in San Antonio, Texas. Then, late in the fall of 1984, the Roman Catholic Church's national Campaign for Human Development committed at least $42,000 to CCRC for 1985, and Kellman invited Kruglik back to Chicago to take the lead in building DCP. Several months later the Woods Fund, which had just designated community organizing as its "primary interest," indicated that it would provide a further $30,000 to support CCRC and DCP salaries.

When Ken Jania was offered a much better paying job and left CCRC in March, Kellman asked Adrienne Jackson, who had been conducting parishioner interviews as a volunteer, to come on board full time, and she took up outreach to new churches. Mike Kruglik focused on expanding DCP's reach across Greater Roseland; a public meeting at St. Thaddeus parish just south of 95th Street attracted both the 21st Ward alderman and Nadyne Griffin, an energetic woman in her late forties who had lived in the Lowden Homes town house project north of 95th Street for many years. She took an immediate liking to Kruglik, but other DCP members, who already found Kellman's hard-driving style to be grating, thought Kruglik was just more of the same.

In late April or early May, the tensions came to a head. "My compadres felt Mike was kind of pushy," St. Catherine deacon Dan Lee remembered. "So one night we had a little caucus, and it was just *us*. Mike wasn't there, Jerry wasn't there." The small group agreed that "we are talking about *black* issues," Dan recounted. "When we talk to Mike, it's like we can't get through. . . . We need a black person to be our mentor. We need a black person. . . . Let's talk to Jerry." Dan, Loretta Augustine, and Yvonne Lloyd went to Kellman. "Nothing against Mike, but we want somebody black over here because we are black," Dan recalled. Kellman didn't argue. "Okay, if that's what you want, that's what I'll do." From 1980 forward, the entire UNO and CCRC organizing effort had failed to employ an experienced black organizer; only parish volunteer Adrienne Jackson, just added to staff, was African American.

Kellman tried to make good on his commitment, but no plausible candidates could be found. "Jerry was busting his behind to find a black organizer," CCRC's Bob Klonowski recalled, but was "just having no luck." Reluctantly, Kellman asked the DCP members to stick with Kruglik after all, but Loretta Augustine took the lead in saying no: "He's not what we feel we need." Loretta was "a very strong-willed person," her colleagues knew, "very outspoken . . . if she didn't like something, she let you know," and her verdict on Kruglik was final.

But it was Father John Calicott, the African American associate pastor from Holy Name of Mary, who hammered the point home most forcefully. Calicott had seen the same pattern too many times before throughout the Chicago archdiocese. "I just had a problem with white folks always figuring that they knew more about what to do for us than we did," he later explained. He had had the same reaction when he first met Kellman. Jerry was "well intentioned, really wants to do the right thing, but *cannot hear,*" Calicott recalled, and when Kellman had first introduced Kruglik to the DCPers, the same dynamic reoccurred. Calicott posed several questions, asking, essentially, "Are you willing to listen to our ideas?" In essence Mike replied, "'Well, yes, but you know, this is the way we've done it before, and we know this is going to work.'" That "really left a bad taste in my mouth," Calicott recounted.

When Kellman again asked them to accept Kruglik, and Loretta said no, Calicott spoke up to second Loretta's refusal: "Let's get somebody who knows *us*!" As Loretta vividly recalled, Calicott didn't stop there. "The priest pointed his finger at Jerry, and he said, 'I don't know where you're looking, but there's got to be somebody out there who looks like us and thinks like us and understands our needs. So wherever you've been looking, you go back and look again.'" Yvonne Lloyd remembered those five words just as Loretta did: "go back and look again," but "Jerry was livid," Loretta recalled. Kellman insisted he would not jettison Kruglik, and Calicott said fine, but not for DCP. "The whole room was just absolutely quiet," Loretta remembered, but Kellman agreed that he would look again.

Mike Kruglik was not happy about what had happened. "The people said, 'We don't want you because you're not black,'" he acknowledged years later. Kellman, feeling "desperation," told CCRC clergyman Bob Klonowski he would shift gears and advertise for a "black organizer trainee" in addition to an experienced organizer. Since the late 1970s, a little-known national organization called the Community Careers Resource Center had published *Community Jobs,* a small newsprint magazine comprised mainly of want ads that came out ten times a year. *Community Jobs* did not have many individual subscribers, but many university and public libraries paid twenty dollars a year to subscribe. It was not a publication they saw any point in retaining—who could possibly want to read job ads from 1985?—and so a quarter century later only one single library would still possess the June 1985 issue containing the job ad that Jerry Kellman submitted.

Community Jobs organized its ads geographically, so on page 3, under a large "Midwest" heading and directly below an ad for "Canvass Director, North Dakota," appeared Jerry Kellman's ad with a boldface title, "**Two Minority Jobs Chicago.**"

The Calumet Community Religious Conference (CCRC) is an Alinsky organizing project in the industrial heart of Chicago. This region was once a world leader in steel production. However, in the past four years, 50,000 jobs have been lost. CCRC has pulled together 60 churches from the far Southside of Chicago and suburban Cook County to address this economic crisis. Half of CCRC's budget comes from local church dues. The project is also committed to church renewal.

APPRENTICE DIRECTOR

Duties: Help to supervise all organizing on the far Southside of Chicago, an area which is 95 percent black. Serve as consultant to local parishes; recruit and train lay leaders in listening skills, research, strategic planning, public action skills and (with local clergy) theological reflection.

Requirements: Experience with church-based or community organizing; *or* experience in leadership and church development; highly disciplined; confident; mature; reflective; able to think and act strategically; experience in black community preferred.

Salary: $20,000/year to start, negotiable for more experienced organizer. Automobile allowance; health insurance.

To apply: Send resume to Gerald Kellman, Director, CCRC, 351 E. 113th St., Chicago, IL 60628. 312/995-8182. Selected candidates will receive phone interviews. Finalists will have interview in Chicago (CCRC will cover travel expenses). Affirmative action position.

TRAINEE

Duties and Requirements: Same as for Apprentice Director but not expected to have skills in advance, must have ability to pick up skills and master them quickly.

Salary: $10,000/year to start. Similar benefits as Apprentice Director.

To Apply: Same as for Apprentice Director.

In early June 1985, the new issue of *Community Jobs* started landing on library shelves across the United States.[42]

Chapter Two

A PLACE IN THE WORLD

HONOLULU, SEATTLE, HONOLULU, JAKARTA, AND HONOLULU

AUGUST 1961–SEPTEMBER 1979

Barack Hussein Obama departed Nairobi's Embakasi Airport on the evening of August 4, 1959, bound for New York, via Rome, Paris, and London. He was twenty-five years old—not twenty-three, as he would later claim—and he was leaving behind a nineteen-year-old wife, Grace Kezia Aoko, who was three months pregnant with a second child, and a sixteen-month-old son, Roy Abon'go.

Obama's dream was to have an education beyond what was available in colonial Kenya. A possession of Great Britain since the late nineteenth century, Kenya lacked any post-secondary educational institution aside from a newly opened technical college. Three years earlier, a dynamic young Kenyan politician, Tom Mboya—who, like Obama, was a Luo, Kenya's third-largest ethnic group—had visited the United States and begun making it possible for young Kenyans to seek higher education opportunities there. Mboya was introduced to Bill Scheinman, a wealthy young businessman likewise interested in African decolonization, and thanks largely to Scheinman's personal largesse, as many as thirty-nine Kenyan students enrolled at a variety of U.S. colleges and universities during the years 1957 and 1958.

By 1958, Barack Obama and his young wife were living in Kenya's capital of Nairobi, yet the first twenty-four years of his life had been anything but easy. The second child, and first son, of Hussein Onyango Obama and Habiba Akumu, he was born near Kendu Bay in the Nyanza region of western Kenya, close to Lake Victoria. Hussein Onyango had served as a cook with the British colonial military forces, traveling widely. Hussein's third child, Hawa Auma, later recounted that "he loved all the whites, and they loved him." Another younger daughter, Zeituni Onyango, remembered Hussein as "unyielding and unapologetic. . . . My father never shed the attitude of a soldier," nor his belief in corporal punishment for wives as well as children.

When Barack Hussein was nine years old and his older sister Sarah Nyaoke about twelve, Hussein Onyango moved the family—now including a second wife, Sarah Ogwel—from Kendu Bay to the village of Kogelo, well north of Lake Victoria in the Alego area of Nyanza, where his ancestors had historic roots. But Alego was wild and rugged, and within a few months, a pregnant Habiba Akumu escaped from her husband and three children and returned to Kendu Bay. In despair, Barack and Sarah soon tried to follow her but were returned to Kogelo to live with their stepmother, Sarah Ogwel, while their father increasingly worked in Nairobi. Decades later Sarah would tell her stepgrandson that his father "could not forgive his abandonment, and acted as if Akumu didn't exist. He told everyone that I was his mother."

In Kogelo Barack attended Ng'iya Intermediate School, and in 1949, at age fifteen, he took the Kenya Africa Examination. In early 1950, he was admitted to Maseno Mission School, Kenya's oldest secondary institution. School records initially described Barack as "very keen, steady . . . reliable and outgoing," but during his senior year school administrators took a strong dislike to him and effectively expelled him. Classmates acknowledged that Obama had become "rude and arrogant" toward teachers, and the white English principal fingered him as the primary author of an anonymous letter criticizing the school's practices. Wherever the blame lay, Obama was out of school without having graduated, and a furious Hussein Onyango instructed him to move to Mombasa, Kenya's eastern port city, to earn his own living.

By some time in 1955, Barack had relocated to Nairobi, where he was a clerk typist in a law firm and also did some work for a British engineering firm. At a Christmas Day 1956 dance party back in Kendu Bay, he met sixteen-year-old Grace Kezia Aoko, and the next month, they were married and moved into Obama's Nairobi apartment. Fourteen months later, Kezia gave birth to Roy Abon'go. Soon thereafter, sometime in mid-1958, Barack met Betty Mooney, the forty-four-year-old American woman who would become his ticket to the United States.[1]

For more than a decade before arriving in Nairobi in 1957, Betty Mooney had worked closely with world-renowned literacy advocate Frank Laubach, whose "each one teach one" method had helped millions across the globe learn to read. Mooney had spent eight years in India before moving to Baltimore to oversee the training of additional literacy teachers at the Laubach-sponsored Koinonia Foundation. In Nairobi, she quickly won the active support of Tom Mboya, who introduced her to a large crowd at one of his weekly political rallies. Then, in the summer of 1958, she and Helen Roberts, another American literacy teacher, began preparing a series of elementary instructional readers in Swahili, Luo, and Kamba.

In September 1958, Mooney hired the young Barack Obama as her secretary and clerk and paid him the handsome sum of $100 monthly. Before long Obama was taking a lead role in the writing of two Luo readers Mooney's team was producing. Laubach himself visited Nairobi in November 1958; a photo published in the monthly newsletter Mooney had just launched pictured her, Laubach, and "Mr. B. O'Bama."

This was a great opportunity for Obama to perfect his own English literacy, and Mooney quickly became impressed by his abilities. "Barack is a whiz and types so fast that I have a hard time keeping ahead of him," she wrote Laubach. "I think I better bring him along and let him be your secretary in the USA." Indeed, getting to the U.S. was Obama's express goal, and by early 1959, even without a diploma from a secondary school and with only some UK correspondence courses on his record, he wrote to several dozen U.S. colleges and universities seeking undergraduate admission for fall 1959. He had read about one of them in the *Saturday Evening Post,* a weekly U.S. pictorial magazine, in Mooney's office. The University of Hawaii was described as being a "Colorful Campus of the Islands." The article praised the "multi-racial make-up" of the university's student body and emphasized that Hawaii was "one of the few spots on earth where there is little racial prejudice."

In early March, Barack Obama received notice of his acceptance from the University of Hawaii, plus a certificate to show U.S. consular officials in order to obtain a student entry visa. Classes would begin on September 21. Betty Mooney was overjoyed, and quickly wrote Frank Laubach to request his help. Barack "is extremely intelligent and his English is excellent, so I have no doubt that he will do well." Mooney wanted to pay both Obama's tuition and half of his estimated $800 annual room and board, but she wanted Kenyan officials—and apparently Barack too—to view these funds as a scholarship rather than a personal gift, and Laubach agreed to help. "I remember him very well, and agree that he is unusually smart. I have no doubt that he will do a very good job." Enclosed with his reply to Mooney was a copy of a letter addressed to the University of Hawaii, which stated that the Laubach Literacy and Mission Fund had granted Obama $400 toward his first year of studies.

Barack worked to complete the Luo primers and also advertised in Kenya's Luo language newspaper, *Ramogi,* for contributions toward his upcoming expenses in Hawaii. Gordon Hagberg, an American whose family had employed Hussein Onyango Obama while they resided in Nairobi, asked his employer, the African-American Institute (AAI), to assist with Obama's airfare, explaining that Obama "is what could be called a self-made man." In late July the U.S. consul general formally issued Barack's nonimmigrant student visa, and AAI booked and paid for his flights. Obama wrote to Frank Laubach, thanking him "for all that you have done for me to make my ways for further studies possible," including the essential $400 that actually came from Betty Mooney. Barack hoped to see Laubach during the three weeks that Betty had arranged for him to stay at Koinonia, outside Baltimore, before going to Hawaii. On

Sunday morning, August 9, 1959, Barack Hussein Obama arrived on a British Overseas Airways Corporation Comet 4 at New York's Idlewild Airport and was granted entry to the United States.[2]

Even before Obama registered for his fall semester courses on September 21, one of Honolulu's two daily newspapers, the *Star-Bulletin,* ran a photo of the twenty-five-year-old freshman in an article entitled "Young Men From Kenya, Jordan and Iran Here to Study at U.H." Obama had secured a room at the Atherton YMCA, just across University Avenue from the campus, but he told the newspaper he was already surprised by the high cost of living. He enrolled in a roster of unsurprising freshman courses—English Composition, World Civilization, Introduction to Government, Business Calculations—and as the first and only African student on campus, and perhaps the only student always wearing dark slacks and dress shirts rather than casual Hawaiian clothing, Obama was immediately a standout presence at UH.

Obama frequented a campus snack bar with lower prices than the main cafeteria, and he soon fell in with a band of friends. Neil Abercrombie was a newly arrived graduate student in sociology from Buffalo, New York; undergraduates Andy "Pake" Zane and Ed Hasegawa had grown up on Oahu—Hawaii's commercial hub—and the Big Island—Hawaii's most rural isle—respectively. Abercrombie recalled Obama as "an unforgettable presence" with a "James Earl Jones voice. It was resonant, deep, booming and rich. It carried authority. He spoke in sentences and paragraphs." Zane agreed. It was "a simply amazing voice," sometimes "mesmerizing."

But Abercrombie remembered Obama for more than just his voice. "He was always the center of attention because he had an opinion on everything and was quite willing to state it. . . . He had this tremendous smile, a pipe in his mouth, dark-rimmed glasses with bright eyes. He was incandescent." Abercrombie told journalist Sally Jacobs how Obama "talked about ambition, his ambition for independence in Africa in general, and his own personal ambition to participate in the emerging nationalism in Kenya . . . it was the central focus of his life. He was full of such energy and purpose." Obama's brimming self-confidence was usually engaging rather than off-putting. "He thought he was the smartest guy in the room, I think, and with good reason . . . everybody else thought so too," Abercrombie recalled. "I could easily call him the smartest person I've ever met."[3]

Just two weeks into the fall semester, the UH student newspaper, *Ka Leo O Hawaii,* published a story on Obama, in which he said he chose UH over other acceptances from San Francisco State College and Morgan State College in Baltimore but again referred to Honolulu's high cost of living. He spoke of his homeland's desire for independence from Britain, saying, "Kenyans are tired of exploitation." Several weeks later, *Ka Leo O Hawaii* ran a photograph on its front page of Obama talking with university president Laurence H. Snyder about UH's newly proposed trans-Pacific East-West Center. In late November, the *Honolulu Star-Bulletin* printed its second story on Obama, under the headline "Isle Inter-Racial Attitude Impresses Kenya Student." This time Obama was quoted as saying he was surprised that "no one seems to be conscious of color" in Hawaii, adding that "people are very nice around here, very friendly." He hoped to finish his degree in three years and hoped to take up some type of government work when he returned to Kenya.

Sometime in November, Betty Mooney, returning to the U.S. via Asia and the Pacific, stopped in Hawaii for several days and was "much impressed" with how well Obama was doing. So was Frank Laubach when he passed through Honolulu several weeks later. In early December Obama sought permission from U.S. immigration officials to work part-time, citing the "high cost of meals," and he was approved for up to twenty-five hours weekly. Once the 1960 spring semester began, Obama participated in a model United Nations exercise that debated race, and in early June, he submitted a strongly worded letter to the editor criticizing a *Star-Bulletin* editorial that had denounced "Terror in the Congo." "Speaking as one who has been in the Congo," he wrote, Africa needed to throw off "the yoke of colonialism" as "the time for exploitation, special prerogatives and privileges is over."

By midsummer, Obama had moved first to an apartment on Tenth Avenue east of the university, then to one on Eleventh Avenue, and finally westward to a neighborhood just north of the Punahou School. In late July 1960, he submitted a routine request to extend his student visa, noting that he was earning $5 a day as a dishwasher at the Inkblot Coffee Shop while also taking a full summer-session course load. After summer session ended, Obama earned $1.33 an hour from Dole Corporation—Oahu's principal pineapple grower—during August and September as an "ordinary summer worker."

During his time in Honolulu Obama exhibited an increasing appetite for alcohol. Drinking and talking were two of Obama's favorite pastimes, but there was also a third. As one female student later told Sally Jacobs, Obama "was always ready to engage you as a woman beyond the normal conversation, you know, to take it one step further. Today you'd call it 'coming on.'" Another woman agreed. "He was flirtatious," but "he was too close in my personal space. . . . I thought he was a little bit almost aggressive in his way of meeting and being around women." Among Obama's Luo friends in Kenya, "he-man-ship" was "no big deal," and one of his closest acquaintances later boasted that Luo men of their generation had a "habit of waylaying foreign women and literally pulling them into bed."

When fall 1960 classes began on September 26, Obama's seven courses included Russian 101. A fellow student was a virginal seventeen-year-old freshman with an incongruous first name who still lived at home with her parents. By early November 1960, however, Stanley Ann Dunham was pregnant.[4]

Stanley Ann Dunham was born on November 29, 1942, at St. Francis Hospital in Wichita, Kansas. She received her forename not from her identically named father but from her mother. Seventeen-year-old Madelyn Payne had secretly married twenty-two-year-old Stanley Armour Dunham a month before her own high school graduation in June 1940. Stanley's mother, Ruth Armour Dunham, had named her second son after the explorer Henry M. Stanley, her eldest son Ralph would later explain, and the Dunhams didn't see Stanley as "a man's name or a girl's name, it was a family name."

Ruth Dunham had committed suicide by swallowing strychnine in 1925, at age twenty-six, after learning that her husband was busy womanizing. Her sons, ages seven and eight, grew up living with their maternal grandparents in the small town of El Dorado, Kansas, and would only "very rarely" ever see their father again.

Teenaged Madelyn Dunham was also a devoted fan of the actress Bette Davis, who six months earlier, in a popular feature film titled *In This Our Life,* had played a southern belle character named Stanley Timberlake. Asked decades later why she had named her daughter Stanley, all Madelyn would say is "Oh, I don't know why I did that."

Madelyn's family had been far from pleased about her marriage to Stanley Dunham, who had failed one year of high school and whose older brother Ralph described him as "a Dennis the Menace type" given to naughty high jinks. One of Madelyn's younger brothers later said, "I think she was looking at Stanley as a way of getting out of Dodge," and the newlyweds soon set out on a road trip to the San Francisco Bay Area. By 1941 they were back in Kansas, with Stanley apparently working in an auto parts store before enlisting in the army a few months after Pearl Harbor. With her husband away and a new baby to care for, Madelyn moved in with her parents and commuted to a night shift job at a new Boeing B-29 bomber plant in Wichita. Stanley had become a sergeant by the time his unit entered France some weeks after D-Day, but in April 1945 he was reassigned back to Britain before being discharged that August, following Germany's defeat and Japan's announced surrender.[5]

Just a few weeks later, Stanley, his wife, and his daughter all arrived in Berkeley, where he began taking classes at the University of California. But academic work was not Dunham's forte. His older brother Ralph, who was working on a Ph.D. at Berkeley, remembered that Stan could not cope with the foreign language requirement. Madelyn's younger brother Charles heard from his sister that Stanley was more interested in reading murder mysteries than doing his course work, and he expected Madelyn to write his term papers for him. "What can you do when your wife won't support you in getting an education?" Stan later told Charles.

Madelyn was unhappy with their situation, and in mid-1947, Stanley, Madelyn, and four-year-old Ann drove eastward with Ralph Dunham. Following a July 4 stopover at Yellowstone National Park, Ralph dropped the young family off in Kansas, inscribing a copy of C. S. Forester's *Poo-Poo and the Dragons* for his niece: "To Stanley Ann Dunham / As a going away present from her Uncle Ralph / Summer of 1947." More than sixty-five years later that volume and Ann's other childhood books would lie well preserved in a box in Honolulu.

Stanley enrolled in several classes at Wichita State University, but within months, he had taken a sales job at the Jay Paris Furniture Store in Ponca City, Oklahoma, two hours south of Wichita. One colleague later remembered Stan as a successful, first-rate salesman, knowledgeable about both furniture and his customers. He was also remembered as "a smart guy who liked to tell you how smart he was." In Ponca City, Madelyn initially stayed home before realizing that she had to have a job. "The evening cocktail hour gets earlier every day. If I don't work, I'll turn into an alcoholic."

Ann began first grade at Ponca City's Jefferson Elementary School in September 1948, and in 1950, she transferred to another for third grade after the family moved to a different home. Then, in the spring of 1951, Stanley moved the family more than 250 miles southwest, to Vernon, Texas, when he took a new furniture store job, and Ann completed third grade there, as well as all of fourth, fifth, and sixth, before the peripatetic family again moved, this time back to El Dorado, Kansas. Stanley worked first at a Farm & Home store, then got a better job at Hellum's Furniture in Wichita, while Ann attended seventh grade in El Dorado.[6]

During the summer of 1955, the Dunhams moved yet again, this time all the way westward to Seattle, where Stanley had a job at the huge Standard-Grunbaum Furniture store. They moved into an apartment northeast of the University of Washington's campus, and Ann walked to nearby Eckstein Middle School for eighth grade. The next summer they moved to Mercer Island in Lake Washington, southeast of downtown Seattle, and Ann began ninth grade at the brand-new Mercer Island High School. They rented a nice apartment in Shorewood, and sometime in 1957 Stan changed jobs once more, working at Doces Majestic Furniture.

Throughout high school, Ann went by her given name of Stanley, or Stannie. She made a good number of friends and was taught by some outspokenly progressive teachers. One friend later recalled that Stanley showed little interest in clothes or boys; instead, she and her friends would take a long bus ride to the lively "UDub" campus neighborhood, an unusual expedition for Mercer Island teenagers. At home, tensions about money sometimes brought on loud arguments between Stan and Madelyn, who had found a job as an escrow officer at a bank in nearby Bellevue. Stanley also had a strained relationship with her father, and one high school friend said she "hated her father at the time that I knew her."

Sometime during her senior year, she and a male classmate set off on a nonromantic road trip that took them as far south as Berkeley, California, before anguished parents and law enforcement officials located them there. Stan Dunham flew down and drove them back to Seattle. Also during her senior year, Stannie saw a much-heralded foreign film, *Black Orpheus,* which was French director Marcel Camus's adaptation of the famous Greek legend, set in Rio de Janeiro, Brazil. She would still recall the movie a quarter century later, and she may have been especially struck by the film's male lead, black Brazilian actor Breno Mello.

Toward the end of Stannie's senior year, Stan heard about a job opportunity that was even farther west than Seattle—in Honolulu. Albert "Bob" Pratt, who operated Isle Wide furniture distributors, was adding a retail outlet, and he hired Stan Dunham to run it. The rental home where Pratt's family lived, at 6085 Kalanianaole Highway, had a backyard cottage, and Stan relocated there sometime before Stannie's high school graduation. On the day after commencement in June 1960, she and her mother flew to Honolulu.

Stannie had not wanted to move to Hawaii, especially given her great attraction to UDub in Seattle, but she was still five months shy of her eighteenth birthday. So, in September 1960, she enrolled as a freshman at the University of Hawaii, taking a philosophy course and perhaps others in addition to Russian 101.[7]

How Stanley Ann Dunham's relationship with Barack Obama commenced and developed remains deeply shrouded in long-unasked and now-unanswerable questions. A quarter century after she became pregnant, her son, temporarily back in Honolulu, would write to his girlfriend that "one block from where I sit, the apartment house where I was conceived still stands." By early 1961, Barack Obama Sr. was living in apartment 15 at 1704 Punahou Street, just across the street from Punahou School, and while literary license shrank three or four blocks to one, that is where Ann Dunham said her pregnancy originated in November 1960.

When the final exam for that Russian 101 course took place on January 28, 1961, Ann Dunham as well as her parents knew she was almost three months pregnant. According to later documents—no contemporary one has ever been located—on Thursday, February 2, 1961, Ann and Barack took a brief inter-island flight from Honolulu to Maui and were married in the small county seat of Wailuku, with no relatives or friends present. Obama's closest confidante, his younger sister Zeituni Onyango, recounted her older brother's version of what had occurred: "the father of Ann said that they have to marry." Stanley Dunham insisted that his pregnant daughter get married rather than give birth to a bastard. But why did they go to the time and expense of flying from Honolulu to Maui? Stanley and Madelyn likely did not want any potentially embarrassing questions arising at either Isle-Wide furniture or at the Bank of Hawaii, where Madelyn had been hired as an escrow officer. They knew that marriages on Oahu were regularly listed in both of Honolulu's daily newspapers, but ones occurring in the outer islands were not.[8]

Ann Dunham Obama did not register for spring classes at the University of Hawaii. In contrast, Obama was honored with a Phi Kappa Phi certificate for his freshman-year grade point average and then a few weeks later was named to the Dean's List because of his fall 1960 GPA. A young English professor, writing to AAI in support of Obama's request for scholarship assistance for his sophomore year, reported that "Obama has done an exemplary job of getting along with people" and called him "a genuinely enlightened twentieth-century man." Obama's friends Neil Abercrombie and Andy Zane were leading local racial equality efforts, and when a national governors' conference brought outspoken segregationist governor John Patterson of Alabama to Honolulu in June 1961, he was greeted at the airport by about two dozen picketers holding signs proclaiming "Welcome to the Land of Miscegenation." The lone black participant certainly represented the truth of that slogan, and he told a reporter that "Hawaii gives them an example where races live together," but he asked "not to be identified" other than as a UH student.[9]

But in that student's own personal context, the races actually did not live together. During her pregnancy, Ann continued to reside with her parents at 6085 Kalanianaole Highway, and Obama remained in his apartment on Punahou Street. When UH's foreign-student adviser, Mrs. Sumie McCabe, learned of Obama's new marriage some two months after it occurred, she immediately called the Honolulu office of the U.S. Immigration and Naturalization Service to tell the INS about his changed circumstances. INS agent Lyle Dahlin memorialized McCabe's call in a memo that went into Obama's file, noting that "the problem is that when he arrived in the U.S. the subject had a wife in Kenya." McCabe said Obama "is very intelligent," but he "has been running around with several girls since he first arrived here and last summer she cautioned him about his playboy ways. Subject replied that he would 'try' to stay away from the girls. Subject got his USC [U.S. citizen] wife 'Hapei' and although they were married, they do not live together, and Miss Dunham is making arrangements with the Salvation Army to give the baby away. Subject told Mrs. McCabe that in Kenya all that is necessary to be divorced is to tell the wife that she is divorced and that constitutes a legal divorce. Subject claims to have been divorced from his wife in Kenya in this method."

The INS was powerless to take any action absent a criminal conviction for bigamy, but Dahlin recommended that Obama be "closely questioned" before he was approved for another extension of his student residency visa and that "denial be considered." If Ann were to petition on his behalf, "make sure an investigation is conducted as to the bona-fide[s] of the marriage." Subsequent documents in Obama's own hand would soon demonstrate that he in no way really considered himself divorced from Kezia. He had grown up in a family and ethnic culture where multiple wives were the norm, and he was not telling the truth about that to McCabe. There are no documents or anyone's recollections to support Obama's claim that Ann Dunham intended to give birth to their child and then put it up for adoption. Obama's closest relative, his sister Zeituni, dismissed the possibility out of hand when the memo first came to light decades later: "no African especially in Kenya would think of giving his child away."

So when Dr. David A. Sinclair delivered Barack Hussein Obama II at 7:24 P.M. on Friday, August 4, 1961, at Kapiolani Maternity & Gynecological Hospital on Punahou Street, just three blocks south from where the child had been conceived, the Salvation Army was not called. Instead, Madelyn and Stan each called their siblings with the news. Madelyn's younger brother Charles recounted her description of the new baby: "He's not black like his father, he's not white. More like coffee with cream." Ralph Dunham remembered Stan calling him from the hospital and Madelyn getting on the phone too. Stan's younger sister Virginia Dunham Goeldner recalled him phoning her too and, fifty years later, expressed astonishment that some of her longtime neighbors in Maumelle, Arkansas, doubted the fact of her grandnephew's birth. "Why did Stanley call and say he was born and why were they over at the hospital? Why did he bother to call" on that Friday night?

The birth occurred exactly two years to the day, and indeed almost exactly to the hour, since Barack Hussein Obama had boarded his flight at Nairobi Embakasi. On Monday, Ann Dunham Obama signed her son's Hawaii State Department of Health birth certificate, and it was signed on Tuesday by Dr. Sinclair and the local registrar of births. Five days later, on August 13, the *Honolulu Advertiser*'s listing of "Births, Marriages, Deaths" on page B6 included in the first category "Mr. and Mrs. Barack H. Obama, 6085 Kalanianaole Hwy., son, Aug. 4." The next day's *Honolulu Star-Bulletin* carried the same listing on page 24, with copy editors at that paper spelling out "Highway" and "August" in full. The birth certificate only contained the address for "Usual Residence of Mother"; there was no request for an address following "Full Name of Father," so the newspapers presumed that the newborn's parents lived together.[10]

Less than four weeks after his son's birth, Barack Hussein Obama applied for and quickly received a routine one-year extension of his student residency visa. Lee Zeigler, newly arrived from Stanford University, had replaced Sumie McCabe as UH's foreign student adviser, and a different INS agent, William T. Wood II, not Lyle Dahlin, reviewed and approved Obama's application. Obama said he had received $1,000 in scholarship support via the African-American Institute, but again requested to work for up to twenty-five hours a week to meet the balance of his expenses. He also indicated that sometime subsequent to March 1961 he had moved from Punahou Street to 1482 Alencastre Street, well east of UH's campus. Barack listed Ann S. Dunham as his spouse, and Agent Wood's summary memo noted, "They have one child born Honolulu on 8/4/61—Barack Obama II, child living with mother (she lives with her parents & subject lives at 1482 Alencastre St.)." But Wood noted something else too: "U.S.C. spouse to go to Wash. State University next semester."

Sometime soon after Wood wrote that memo, Ann and her weeks-old son flew from Honolulu to Seattle: not so she could attend WSU, in far southeastern Washington State, but to enroll at her beloved UDub, which she had wanted to attend a year earlier. Ann and baby Barack stayed briefly with a family friend on Mercer Island before settling into an apartment at 516 Thirteenth Avenue East in Seattle's Capitol Hill neighborhood, well south of the university. According to her UDub transcript, she registered for two evening courses, Anthropology 100: Introduction to the Study of Man and Political Science 201: Modern Government. Classes began in late September.

But why did Ann Dunham Obama take her newborn and leave her husband, parents, and Honolulu for the familiar confines of Seattle? She clearly preferred UDub and its environs over UH, but she told half a dozen old high school friends, as well as a woman who also lived at 516 Thirteenth Avenue East and babysat young Barack while Ann attended classes, that she loved her husband. But the young couple never chose to live together at any time following the onset of Ann's pregnancy, and Ann relocated herself a long airplane flight away as soon as her son was old enough to travel. None of the direct participants—Ann, Obama, Madelyn, and Stan—ever offered a clear explanation that has survived in anyone's recollections a half century later.

Obama had taken to calling his son's mother Anna, not Ann, and she seems to have adopted this as well, according to both the 1961–62 Polk City Directory for Seattle, which lists "Obama Anna Mrs. studt" and her neighboring babysitter, Alaskan native Mary Toutonghi, who also remembered her as Anna. Ann did well in her fall courses, earning an A in anthropology and a B in political science; she did even better in the winter term that ran from late December 1961 through mid-March 1962, getting As in both Philosophy 120: Introduction to Logic and, interestingly, History 478: History of Southern Africa. Mary

Toutonghi babysat regularly during those months on the evenings Ann attended classes, and years later she would recall infant Barack as "very curious and very alert," "very happy and a good size." In March Ann enrolled in three regular daytime courses, obtaining Bs in Chinese Civilization and History of Modern Philosophy but changing English Political and Social History to just an audit.[11]

With Ann in Seattle, Obama launched into his senior year at UH. Only Neil Abercrombie was aware of Obama's relationship with Dunham or that he had fathered a child in Honolulu. One new graduate student, Robert Ruenitz, would later admit that "for any of us to say that we knew Obama well would be difficult. He was a private man with academic achievement his foremost goal." Another 1961 grad student, Cambodia native Naranhkiri Tith, debated nuclear arms with Obama at a widely publicized campus symposium. Obama labeled the issue not a "balance of power" but a "balance of terror" and asserted that most U.S. foreign aid took the form of weapons and other military assistance. Tith and other graduate students also partied regularly with Obama, who "loved to drink" to the point of becoming "totally drunk" at repeated parties. "He also was a womanizer," Tith recounted years later.

Even so, Obama's academic success continued apace. In mid-January he addressed the NAACP's Honolulu branch on "Changes in Africa Today," and in early February, he was featured prominently in a "Dear Friend" fund-raising appeal distributed by Bill Scheinman and Tom Mboya's African-American Students Foundation. Sent in the name of Ruth Bunche, whose diplomat husband Ralph in 1950 had been the first African American ever to win the Nobel Peace Prize, the letter briefly profiled two young men "of whom we are especially proud" out of more than five hundred African students who were then studying in the U.S. One was completing a graduate degree in engineering at Columbia University; the other was Obama, "an honor student of the University of Hawaii where he will complete a four year course in three years." The letter predicted Obama would soon qualify for the national academic honor society, Phi Beta Kappa, and in late April he was elected to membership.

With graduation only a month away, Obama was also a featured speaker at a large Mother's Day event organized by the Hawaii Peace Rally Committee to oppose nuclear weapons. The afternoon event drew hundreds to Ala Moana Beach Park. Liberal Democratic state legislators Tom Gill and Patsy Mink were joined on the speakers' platform by four clergymen and several UH professors. The crowd included powerful International Longshoremen's and Warehousemen's Union (ILWU) director Jack Hall, and conservative counterprotesters from the Young Americans for Freedom (YAF) who waved signs advocating continued U.S. nuclear weapons testing.

Speaking to the crowd, Obama denounced "foreign aid which is directed toward military conquest or the acquisition of bases." Speaking as an African, "anything which relieves military spending will help us," and if peace were to replace nuclear confrontation, "we will be able to receive your aid with an open mind and without suspicion."[12]

In early May 1962, Betty Mooney Kirk, who had married and relocated to her husband's hometown of Tulsa, Oklahoma, wrote to Tom Mboya in Nairobi to seek his help in finding someone to sponsor Obama for graduate school, "preferably at Harvard." She enclosed a copy of Barack's résumé, which she had prepared, and it stated that Obama already had applied to and been accepted for graduate study at Harvard, Yale, the University of Michigan, and the University of California at Berkeley. Harvard alone had offered financial aid, in the limited amount of $1,500, but Betty hoped Tom could find further assistance because Barack "has the opportunity and the brains." Mboya replied with congratulations, but according to Betty was "not very hopeful" about locating available funding.

Betty's colleague Helen Roberts was back in Nairobi, and, perhaps at Betty's urging, was actively assisting Kezia Obama, now the single mother of two young children—Rita Auma had been born in early 1960, six months after her father's departure for the U.S. Kezia was sometimes in Kogelo with her two children and Barack's father and stepmother Sarah, sometimes with her parents in Kendu Bay, and other times staying with her brother Wilson Odiawo in Nairobi. Roberts helped Kezia take some educational courses, and told one friend that Kezia "is very anxious to be a suitable wife for Barack when he returns." Roberts remarked, "I think Barack will notice quite a difference in her when he at last returns."

In late May 1962, Obama wrote to Mboya and apologized for not having written in a long time. He bragged about his academic achievements at UH, falsely claiming to have already earned an M.A. degree in addition to his impressive three-year B.A. and a 3.6 GPA. Reciting his Phi Beta Kappa and Phi Kappa Phi honors as well as an Omicron Delta Kappa award, he told Mboya—twice, in almost identical sentences—that these were "the highest academic honours that anyone can get in the U.S.A. for high academic attainments." What's more, he was about to leave for Harvard, "where I have been offered a fellowship for my Ph.D. I intend to take at least two years working on my Ph.D. and at most three years. Then I will be coming home." Obama closed by telling Mboya, "I have enjoyed my stay here, but I will be accelerating my coming home as much as I can. You know my wife is in Nairobi there, and I would really appreciate any help you may give her."[13]

His letter to Mboya did not mention his second wife or third child, nor did he ever say anything about them to Helen Roberts or to the hugely supportive Betty Mooney Kirk. As his eldest son would ruefully put it years later, by the end of his time in Hawaii "Barack's life was now a series of compartments."

On June 17, 1962, Obama received his B.A. degree—and not any M.A.—at UH's commencement. Three days later the *Honolulu Star-Bulletin,* in an article headlined "Kenya Student Wins Fellowship," reported how the "straight A" economics major was headed to Harvard to obtain his Ph.D. "He plans to return to Africa and work in development of underdeveloped areas and international trade at the planning and policy-making level," the story explained. "He leaves next week for a tour of mainland universities," beginning in California, prior to entering Harvard.

In Seattle, Ann's spring quarter classes had concluded, and her high school friend Barbara Cannon Rusk, who had moved to Utah after graduating, "came back to Seattle in the summer of 1962." One day, Rusk stopped by Ann's apartment on Capitol Hill. Her initial visit "was after June, and could have been as late as September. I visited her a couple of times," she recalled more than forty years later. "She wasn't in classes, and didn't have a job. I recall her being melancholy. . . . I had a sense that something wasn't right in her marriage. It was all very mysterious," as her husband was already headed to Harvard. "I didn't ask her about the relationship."

Also years later, another young woman whose Mercer Island family had known the Dunhams very well, Judy Farner Ware, would recount to Janny Scott, Ann's biographer, a distinct memory of meeting Ann and Obama in what she recalled was Port Angeles, Washington—the ferry port at the top of western Washington State's Olympic Peninsula, just across the Strait of Juan de Fuca from Victoria, British Columbia. She remembered the meeting because an openly flirtatious Obama all but hit on her. Had Obama traveled north from San Francisco to see his second wife and second son in Seattle, and then perhaps they toured the region? Ann didn't own a car or know how to drive, and neither Ann nor Obama ever mentioned such a visit to anyone in later years.

Ann and her son were still in Seattle when Obama left Honolulu for the mainland. Perhaps it should be presumed that Obama *did* set eyes on his newborn son back in August 1961 before Ann and the baby left for Seattle—though no one's surviving accounts say that *did* occur—but unless Obama made some equally unrecorded, unremembered visit to Seattle before heading eastward, he would not have seen his son for years to come. In truth, as one scholar would acutely put it, Barack Hussein Obama was only "a sperm donor in his son's life."

Almost three decades later, his eldest daughter would meet Ann Dunham and ask her what had happened between her and her father. Ann's story then was that Obama had asked her to join him at Harvard, but "she had not wanted to go. She had loved him, but she had feared having to give up too much of herself."[14]

By mid-July 1962, Obama had gotten as far east as Oklahoma, where he stopped in Tulsa to visit Betty Mooney Kirk and her husband. By no later than August

17, he was in Baltimore, at the Koinonia Foundation's campus, where he had stayed exactly three years earlier. While there, he updated his immigration papers, telling the INS his study at Harvard would be supported by $1,000 each from Frank Laubach's Literacy Fund and the Phelps Stokes Fund, in addition to his university fellowship. On his "Application to Extend Time of Temporary Stay," Obama listed himself as married, but under children entered only one name: "Roy Obama."

By September, Obama had arrived at Harvard, and Ann and her now one-year-old son had returned to Honolulu. Stan and Madelyn had moved from Kalanianaole Highway to an apartment on Alexander Street, but Ann and young Barack initially stayed at 2277 Kamehameha Avenue, close to UH. Ann sat out the fall semester, but in January 1963, she resumed taking classes as a sophomore. Sometime prior to the end of 1963, Stan and Madelyn relocated to a house at 2234 University Avenue, and Ann and her son soon moved in with her parents.

As Ann adapted to a heavier academic load, and Madelyn worked long days at her bank job, young Barack spent most of his time with his fit and youthful forty-five-year-old grandfather. Obama Sr.'s old friend Neil Abercrombie, still a graduate student at UH, saw Stan and young Barry—as his grandparents called him—around town during Barry's childhood. "His grandfather was the most wonderful guy" and it was readily apparent that "Stanley loved that little boy," Abercrombie remembered. "He took him everywhere," including to an arrival ceremony for two Gemini astronauts who had splashed down safely in the Pacific after an aborted space flight. Barack would "remember sitting on my grandfather's shoulders" at Hickam Air Force Base and "dreaming of where they had been." Abercombie recalled: "In the absence of his father, there was not a kinder, more understanding man than Stanley Dunham. He was loving and generous."

Indeed, among the dozens of photos of young Barry from his childhood, it is impossible to find one where he is not smiling broadly. Stan's boss's daughter, Cindy Pratt Holtz, remembers Stanley bringing Barry with him to the Pratt furniture warehouse. Young Obama was "so full of life, a twinkle in the eye, giggling all the time." In the fall of 1966, five-year-old Barry began kindergarten at nearby Noelani Elementary School, and Aimee Yatsushiro, one of his two teachers, remembers him similarly: "always smiling—had a perpetual smile." Obama later said, "My earliest memory is running around in a backyard gathering up mangoes that had fallen in our backyard when I was five" or perhaps four. "A lot of my early memories," he added, are "of an almost idyllic sort of early childhood in Hawaii."[15]

In the meantime, his barely twenty-one-year-old mother had found new happiness in tandem with her studies. "Lolo" Soetoro—officially Soetoro Martodihardjo, after his Javanese father's name—first arrived in Honolulu from Yogyakarta, Indonesia, in September 1962 as a twenty-seven-year-old graduate student in geography. After his first year of classes, Soetoro spent the summer of 1963 at Northwestern University in Evanston, Illinois, and at the University of Wisconsin at Madison, but that fall he returned to UH for the final year of his two-year master's program. He and Ann met each other sometime during those months. One mutual friend recalled that "he had a good sense of humor, and he loved to party." Ann would later remark how attractive Lolo was in tennis shorts. "She liked brown bums," her most outspoken friend would tell biographer Janny Scott, and by early 1964, Ann and Lolo were a public couple. Seemingly because of this new romance, on January 20, 1964, Stanley Ann Dunham Obama signed a "Libel for Divorce," as Hawaii legal process termed the form, and five days later the complaint was officially filed in Honolulu circuit court. A copy was addressed to Barack H. Obama in Cambridge, Massachusetts.

Obama had been at Harvard for almost eighteen months. He was one of thirty-five newly admitted doctoral students in the Department of Economics, and in a December 1962 letter to a friend in Hawaii, Obama confessed that "the competition here is just maddening." The heavy reading load made every week "pretty rough," and while "I find Harvard a very stimulating place at least intellectually," his focus was "my own research on the theory I am trying to build." He added, "I will stay here at least for two years to three years depending on when I am able to finish my dissertation," but after he received a C+ and two Bs in his first semester, Harvard refused to renew his fellowship to cover his second year of classes. Two senior economists nonetheless praised Obama's "intelligence, initiative, and diligence," and thanks once again to Betty Mooney Kirk and the African-American Institute, external funding allowed him to continue.

Barack first lived at 49 Irving Street before moving into a top-floor apartment at 170 Magazine Street with a Nigerian fellow, one of about eighty African students at Harvard—a vast change from his unique status in Honolulu. Obama actively mentored younger Kenyan students from around greater Boston; George Saitoti, who was eighteen years old when he knew Obama, told biographer Sally Jacobs "we looked upon him as a model. He really gave us inspiration." In the fall of 1963, Obama's brother Omar Onyango, a decade younger, arrived in Boston to attend the posh Browne & Nichols School, just west of Harvard, thanks to his older brother's social acquaintance with a young woman whose father was the school's treasurer.

That same young woman, like a number of Obama's African friends in Cambridge, also witnessed a continuation—and perhaps an intensification—of the heavy drinking and heavy-handed pursuit of women that had marked Barack's three years at UH. "He'd dance in a very suggestive way, no subtlety," that female friend recounted to Sally Jacobs. "He used suggestive, provocative language, I would say overly sexual. . . . It was kind of a God's gift to women thing." One Nigerian friend recalled telling a drunken Obama to leave a young woman alone, and an African undergraduate woman told Jacobs about consoling a fellow female undergraduate who had been an Obama girlfriend until she learned he was already married, presumably to Kezia.

In late January 1964, Rev. Dana Klotzle, who oversaw the Unitarian Universalist Association's (UUA) sponsorship of about a dozen East African students who, like Omar, were attending secondary schools around Boston, notified the local INS office of a troubling development. A young Kenyan woman who was attending school in Auburndale, Massachusetts, had suddenly flown to London on January 10 on a round-trip ticket. UUA had terminated her sponsorship and would not accept her back; an INS agent phoned the school for additional information. The dean of women said the girl had claimed she was visiting a sick sister, but there was no evidence of a sister in Britain. What's more, she had been "receiving advice from another student from Kenya, one Obama who is likely her boy friend and who is at Harvard." The Unitarians suspected she had flown to London to obtain an abortion. Obama had been phoning the school seeking her reinstatement and also had called a second school, which refused to accept her. Rev. Klotzle, the memo reported, thought Obama was "a slippery character." The Boston INS office then notified the U.S. consul in London of the girl's flight and Obama's involvement.

In Hawaii, on March 5, Judge Samuel P. King held a brief hearing on Ann's divorce petition; fifteen days later, he signed a "Decree of Divorce." Ann was "granted the care, custody and control of Barack Hussein Obama, II," with Obama Sr. having "the right of reasonable visitation." Pursuant to Ann's request, "the question of child support is specifically reserved until raised hereafter." As with Ann's initial complaint, a copy was mailed to Obama in Cambridge.

Four weeks later, Obama visited the Boston INS office to extend his student residency visa for another year. For the new application, Harvard certified that "Mr. Obama expects to be registered as a full-time student during the academic year 1964–65," but the INS agent reviewing the file noted the January contretemps and a supervisor instructed him to "hold up extension for present." The agent made several calls to Harvard, in part because Obama had left blank both the marital line and the one about employment, stating there that he could not remember where he had worked in the U.S. The agent noted: "Harvard thinks he's married to someone in Kenya and someone in Honolulu, but that possibly he belongs to a tribe where multiple marriages are O.K." Obama's doctoral qualifying exams were soon approaching, and the director of Harvard's international students office wanted to hold off on questioning Obama until those were finished.

Obama was aware of the inquiries, and he called the INS to say he now remembered working at the Institute of International Marketing in Cambridge during the summer of 1963. Harvard officials told the INS that Obama might also be married to someone in Cambridge, and in mid-May David Henry, director of Harvard's international students office, called INS agent M. F. McKeon to say he had conferred with both a graduate school dean and the chairman of Harvard's Economics Department.

"Obama has passed his general exams, which indicates that on academic grounds, he is entitled to stay around here and write his thesis," McKeon wrote in a memo memorializing the phone conversation. "However, they are going to try to cook something up to ease him out. All three will have to agree on this, however. They are planning on telling him that they will not give him any money, and that he had better return to Kenya and prepare his thesis at home." That would take several weeks, but "at this time Harvard does not plan on having Obama registered as a full-time student during the academic year 1964–1965 as stated on" Obama's application a month earlier.

On May 27, 1964, Harvard's David Henry sent Obama a life-changing letter. It began by acknowledging that Obama had completed his course work and that only his thesis remained to be completed before he could get his Ph.D. But the letter also said that neither the Department of Economics nor the graduate school had the funds to support him in Cambridge. It then said, "We have, therefore, come to the conclusion that you should terminate your stay in the United States and return to Kenya to carry on your research and the writing of your thesis." He was given until June 19—which was hardly three weeks away!—to arrange for his departure. Henry indicated that copies of the letter were going to graduate school associate dean Reginald H. Phelps, a historian of modern Germany, and Economics Department chairman John T. Dunlop, a distinguished professor who would go on to become dean of Harvard's Faculty of Arts and Sciences and then U.S. secretary of labor.

Unspoken in Henry's letter—though crystal clear in Obama's INS file—was Harvard's unwillingness to continue hosting a man whose sexual energies, whether inter-African or serially miscegenous, would not be tolerated in tony Cambridge as they had been in multihued Honolulu. Two weeks later an INS form letter instructed Obama that he had until July 8, instead of June 19, to depart the United States. On June 18, an understandably agitated Obama phoned the Boston INS office and insisted that he be given specific grounds for why his residency extension was being denied. An INS agent emphasized that the decision was final, but Obama called again the next day and asked to speak to the district director, who refused to take the call. Obama declared he lacked funds to leave the U.S., and the next day he asked a Harvard secretary to call the INS on his behalf. She too was told INS's ruling was final. At that point, Obama apparently gave up; on Monday, July 6, 1964, he departed from New York's newly renamed John F. Kennedy International Airport bound for Paris and then Nairobi, which, as of seven months earlier, was now the capital of newly independent Kenya.[16]

On the other side of the United States, Ann Dunham and Lolo Soetoro were married on Monday, March 15, 1965, on Molokai, a smaller Hawaiian isle southeast of Oahu. Neither Ann's son nor her parents attended the ceremony, which took place only three months before Lolo's current residency visa would expire. He had received his M.A. in geography in June 1964, but a month later both UH and the INS approved another one-year residency during which he could get practical experience working for local engineering and surveying firms.

INS documents indicate that Ann and Barry never moved to 3326 Oahu Avenue, where Lolo was living, but instead remained at 2234 University Avenue with Stan and Madelyn. The looming question of whether Lolo would be able to remain in the U.S. beyond June soon brought both him and Ann into extensive contacts with the INS that mirrored what her ex-spouse had experienced a year earlier.

Sometime during May or June 1965, UH's East-West Center (EWC), which had sponsored Lolo's graduate study, received a cable from the Indonesian embassy in Washington requesting Soetoro's immediate return to Jakarta. But Lolo and Ann had already taken the initiative to win an extension of his visa, and following two joint interviews at the Honolulu INS office, on June 7 Lolo's residency permit was extended until mid-June 1966. On July 2, when Lolo informed the EWC of that, he was summoned to a July 6 meeting to be reminded "that the East-West Center still retained visa sponsorship and authority" regarding his residency. Lolo said he had sought the extension because his wife was suffering from a stomach ailment that might require surgery, but later that day EWC phoned INS, which immediately summoned both Lolo and Ann to another interview on July 19. In the interim, Ann, using Dunham as her surname, applied for and received her first U.S. passport.

Officials from the EWC visited the Honolulu INS office to explain that their agreement with the Indonesian government required that "every effort will be made to return students at the completion of their grants." Thus EWC "shall appreciate any effort which you can make to insure that Mr. Soetoro will be returned to Indonesia as soon as possible."

Before the July 19 session, Lolo submitted a statement to the Honolulu INS office noting that in his homeland "anti-American feeling has reached a feverish pitch under the direction of the Indonesian communist party." This was supported by widespread U.S. press reports. Lolo asserted, "I have been advised by both family and friends in Indonesia that it would be dangerous to endeavor to return with my wife at the present time." In addition, "I would meet with much prejudice myself in seeking employment" because of his U.S. educational background, and "land belonging to my family has already been confiscated by the government as part of a communistic land reform plan," a policy that press reports again corroborated. Citing his "former compulsory association with the Indonesian army while still a student," Lolo also feared being dragooned into battlefield service in Indonesia's armed conflict with Malaysia if he returned home.

Soon after the July 19 interview, INS Honolulu recommended denial of any ongoing residency for Lolo. But almost two months later, the EWC notified Indonesia's San Francisco consulate that Lolo would return to Indonesia in June 1966—and his wife would accompany him. This was just days before Indonesia was plunged into months of bloody, widespread violence in which hundreds of thousands of the previously ascendant Communists and perceived sympathizers were slaughtered by the Indonesian army and allied militias. That turmoil commenced with an unsuccessful, Communist-backed revolt against the army leadership by a small band of junior officers on September 30, 1965.

For the next six months, the violently anti-Communist army leadership took firm control of the country and a half million or more civilians were killed. Even with knowledge of the tumult, Ann, on November 30, gave the INS an affidavit acknowledging, "I don't feel that I would undergo any exceptional hardship if my husband were to depart from the United [States] to reside abroad as the regulations require." Those rules would allow Lolo's readmission, as her husband, after two years' absence from the U.S., a preferable course to being hamstrung by EWC's deference to Indonesian authorities.

If the elimination of the anti-American Communist presence in Indonesia is what caused Lolo and Ann to change their strategy, that has gone unrecorded. Ann's affidavit did, however, say she was "living with my parents in the home which they rent" and that "my son by a former marriage lives there with us." INS's efforts to revoke Lolo's existing extension petered out, and on June 20, 1966—the last possible day—Lolo Soetoro flew out of Honolulu bound for Jakarta.[17]

After Lolo's departure, Ann took a secretarial job in UH's student government office and also began doing some temporary nighttime tutoring and paper grading. That gave her an income of about $400 per month, and she told INS officials she hoped to save enough money to join Lolo in Indonesia in summer 1967. "We figure on going and staying until my husband's time is up and then come back together." With young Barry in kindergarten at Noelani Elementary School, and Stan and Madelyn both working full-time, Ann spent $50 to $75 a month for a babysitter on weekdays from 2:30 P.M. to 5:00 P.M. In December 1966, she told INS that she expected to complete her B.A. degree in anthropology in August 1967 and would join Lolo in Indonesia that October. She was already attempting to secure employment at the U.S. embassy in Jakarta.

INS did not appear open to waiving the two-years-abroad requirement for Lolo, and in May 1967 INS agent Robert Schultz phoned Ann for an update. "She and her child will definitely go to Indonesia to join her husband if he is not permitted to return to the United States sometime in the near future, as she is no longer able to endure the separation," Schultz noted. "Her son is now in kindergarten and will commence the first grade next September and if it is necessary for her and the child to go to Indonesia, she will educate the child at home with the help of school texts from the U.S. as approved by the Board of Education in Honolulu." Unbeknownst to Ann, this description of young Barry's educational plight would set in motion a change in the INS's attitude about a waiver. Still, in late June, she applied to amend her 1965 passport, taking Soetoro rather than Dunham as her surname.

In August 1967, just as Ann was receiving her B.A. from UH, INS, layer by bureaucratic layer, gradually agreed to grant Lolo a waiver, and two months later notified the State Department of that intent. Nine months would then pass before the Honolulu INS office realized that State had never responded. In the interim, sometime in October 1967, twenty-four-year-old Ann Soetoro and six-year-old Barry Obama boarded a Japan Airlines flight from Honolulu to Tokyo. During a three-day stopover, Ann took Barry to see the giant bronze Amida Buddha in Kamakura, thirty miles southwest of Tokyo. Then they boarded another plane, headed for Jakarta via Sydney.[18]

In Honolulu, Barry had begun first grade at Noelani Elementary School, and upon arrival in Jakarta, Ann initially followed through on her promise to homeschool her son. Home was 16 Haji Ramli Street, a small, concrete house with a flat, red-tiled roof and unreliable electricity on an unpaved lane in the newly settled, far from well-to-do Menteng Dalam neighborhood. Jakarta was a sprawling metropolis, but one where bicycle cabs—*becak,* in Indonesian—and small motorbikes far outnumbered automobiles.

Outside of the privileged expatriate community, where young children attended the costly international school, "Jakarta was a very hard city to live in," said another American woman—later a close friend of Ann's—who lived there in 1967–68. One had to deal with nonflushing toilets, open sewers, a lack of potable water, unreliable medical care, unpaved streets, and spotty electricity. When Ann and Barry arrived, Lolo was indeed working for the

Indonesian army's mapping agency, though now, unlike four months earlier, he was based on the other side of Jakarta, not hundreds of miles away in far-eastern Java.

Barry would later say that "for me, as a young boy," Jakarta was "a magical place." Revisiting the city more than forty years later, he recounted how "we had a mango tree out front" and "my Indonesian friends and I used to run in the fields with water buffalo and goats" while "flying kites" and "catching dragonflies." But during the long rainy season, Jakarta was no wonderland: Barry, like others, would have to wear plastic bags over his footwear, and on one mud-sliding jaunt, he badly cut his forearm on barbed wire, a wound that required twenty stitches and left him with what he later called "an ugly scar."

In January 1968, Ann enrolled Barry, using the surname Soetoro, in a newly built Roman Catholic school three blocks from their home—"she didn't have the money to send me to the fancy international school where all the American kids went," Barry later recounted. That allowed Ann to take a paid job as assistant to the director of a U.S. embassy–sponsored program offering English language classes to interested Indonesians. Barry's school, St. Francis Assisi, as its name would be rendered in English, was avowedly Catholic: "you would start every day with a prayer," Barry later explained, but classes met for only two and a half hours on weekday mornings. His first-grade teacher there, Israella Darmawan, decades later told credulous reporters, "He wrote an essay titled, 'I Want to Become President'" during that spring of 1968, prior to his seventh birthday. She also told journalists that Barry struggled greatly to learn Indonesian; in contrast, Obama later boasted that "it had taken me less than six months to learn Indonesia's language, its customs, and its legends."

Barry's second-grade teacher, Cecilia Sugini, spoke no English, but Barry received more exposure to the Indonesian language during family visits to Lolo's relatives in Yogyakarta, in central Java. Yet even his third-grade teacher, Fermina Katarina Sinaga, later stated that eight-year-old Barry was not fluent in Indonesian. And she would also tell wide-eyed reporters that Barry, during the fall of 1969, declared in a paper, written in Indonesian, that "Someday I want to be President." One journalist, embracing Sinaga's direct quotation forty years later, would insist that Sinaga's "memory is precise and there is no reason not to trust it."

By the end of 1969, Lolo, thanks to his nephew "Sonny" Trisulo, switched to a much better job with Union Oil Company of California. Soon thereafter, he, Barry, and newly pregnant Ann moved to a far nicer home at 22 Taman Amir Hamzah Street in the better neighborhood of Matraman. Around the same time, Ann left the English teaching post, which she had come to loathe, for more rewarding work, primarily in the evenings, at a nonprofit management training school headed by a Dutch Jesuit priest.

Moving houses also meant that Barry would attend the Besuki elementary school, which traced its roots back thirty years to Indonesia's Dutch colonial government. Classes met for five hours each weekday, double what St. Francis Assisi offered. Ann's new work schedule gave her time to intensify her efforts to homeschool Barry in English using workbooks from the U.S. At Besuki, his all-Indonesian classmates found Barry—or "Berry," as they pronounced it—unique not only because of his darker complexion and chubby build but also because he was the only left-hander.

Before the spring of 1970 was out, and with a second child on the way, Ann hired an openly gay twenty-four-year-old, sometimes-cross-dressing man—Turdi by day, Evie by night—to be both cook and nanny. Neighbors thought little of it. "She was a nice person and always patient and caring in keeping young Barry," one later recalled. Turdi often accompanied Barry to and from school. Later, Turdi, at age sixty-six, told the Associated Press: "I never let him see me wearing women's clothes. But he did see me trying on his mother's lipstick sometimes. That used to really crack him up."[19]

Sometime apparently also during that spring, Barry saw something that, in his later tellings, had a vastly more powerful impact upon his young mind. A quarter century passed between the moment and Obama's first telling of it, but in his 1995 version, the memory was of paging through a pile of *Life* magazines in an American library in Jakarta and finding an article with photographs of a man of color who had paid for chemical treatments in a horribly unsuccessful attempt to make himself appear white. In Obama's 1995 account, "thousands of people like him, black men and women back in America," had "undergone the same treatment in response to advertisements that promised happiness as a white person." To him, "seeing that article was violent for me, an ambush attack," leaving his image of his own skin color "permanently altered."

In a conversation soon after writing that, Obama recounted how "after reading that story, I knew there had to be something wrong with being black." Earlier, while "growing up in Hawaii, all of the kids were kind of brown," so "I didn't stand out" and "I was too busy running around being a kid" to appreciate racial differences. At his two Jakarta schools, he experienced some normal teasing by other children, but to no obvious or remembered ill effect. "He was a plump kid with big ears and very outgoing and friendly," one of Ann's closest Jakarta friends later recalled.

Nine years later, Obama described the memory again. "I became aware of the cesspool of stereotypes when I was eight or nine. I saw a story in *Life* magazine about people who were using skin bleach to make themselves white. I was really disturbed by that. Why would somebody want to do that?" A few weeks later, Obama again recounted seeing a *Life* magazine picture of "a black guy who had bleached his skin with these skin-lightening products." That was "the first time I remember thinking about race" and worrying that having darker skin was "not a good thing."

In 2007, a reporter told Obama that no issue of *Life* magazine ever contained such an article or such photographs; this was confirmed by *Life*. "It might have been an *Ebony* or it might have been . . . who knows what it was?" a flustered Obama responded. But then *Ebony* too examined its archive of past issues and found no such story. Indeed, the other two major picture magazines of that era, *Look* and the *Saturday Evening Post,* published no such story either. Yet Obama understandably stood by his recollection: "I remember the story was very specific about a person who had gone through it and regretted it."

But *Ebony had* published a *somewhat* similar story, in its December 1968 issue, titled "I Wish I Were Black—Again." It was a profile of Juana Burke, a young African American art teacher who at age sixteen had begun to suffer from vitiligo, a disease which turned portions of her dark brown skin white as it killed off pigmentation cells. The article included photographs of her forearm and legs. Dermatologists' efforts to counteract the spread of the affliction through skin chemicals and even prolonged sunbathing failed completely, and Ms. Burke reluctantly accepted her pale new appearance.

The four-page *Ebony* spread stressed that she "retains her old sense of black pride and identifies with her people," and she continued to teach at a predominantly black school. However, becoming white had left her "very pessimistic about the future of race relations in this country." A black boyfriend had ditched her, and she was dismayed to repeatedly experience a "more courteous attitude" from white strangers than she had when she had been visibly black.

Had eight-year-old Barry actually seen that issue of *Ebony*? Who knows. But many teenagers growing up in the 1960s heard about a journalist named John Howard Griffin, a white Texan who, in the late 1950s, had undergone chemical treatments so he could pass as black and write about the experience—the obverse of Ms. Burke's deflating color change. Griffin's resulting book, *Black Like Me,* first published in 1961, was a nationwide best seller and was made into a major motion picture.

Irrespective of what magazine pictures young Barry did or did not see, the overarching question of how and why anyone would seek to alter their visible racial identity had become a staple of U.S. popular culture in the late 1960s, even if the notion of any African American becoming white was starkly out-of-date in the new era of "I'm Black and I'm Beautiful." Obama's encounter with the pictures had seemingly been a *"turning point,"* "a transformation in the life story that marks a considerable shift in self-understanding" and in "his racial identity development." The Obama of 1995, 2004, and 2007–08 certainly agreed—"Growing up, I wasn't always sure who I was"—regardless of whether at age ten, at age eighteen, or even at age twenty-seven he actually pondered the memory of those images.[20]

Sometime in the late spring of 1970 Ann Dunham, in concert with her father and no doubt her mother, decided that within a year's time, when Barry would begin fifth grade, he should continue his future schooling in Honolulu rather than Jakarta. Stan Dunham's twenty-year career as a furniture salesman had ended sometime in 1968, following changes in Bob Pratt's enterprises, and by 1969, he was one of about twenty-five agents at John S. Williamson's John Hancock Mutual Insurance agency in downtown Honolulu. Perhaps because of a decrease in income from that shift, Stan and Madelyn had left the rental home at 2234 University Avenue and relocated to unit 1206 in the Punahou Circle Apartments at 1617 South Beretania Street, just a few blocks south of Punahou School.

Ann had been aware of Punahou, and its unequaled-in-Hawaii educational reputation, since her earliest months in Honolulu. Her son was even conceived just across Punahou Street from its spacious campus. Founded in 1841 by Christian missionaries, Punahou had a student body that was still predominantly white—*haole,* in local parlance—and its alumni included many of Oahu's civic elite. Fifth grade was one of the two best opportunities—ninth was the other—for youngsters who had not started elementary school there to gain admission, as class sizes increased at the middle and then high school levels.

It is unknown when Ann first thought of sending Barry there, but Stanley had become good friends with Alec Williamson, who also worked at his father's insurance agency. Alec's dad had graduated from Punahou in 1937, and both of his sisters had gone there as well, although he had not. Punahou administered admissions tests and required personal interviews. It was "the quintessential local school," Alec's sister Susan later explained, and the Dunhams were mainlanders, but John Williamson was more than willing to recommend Stanley's bright grandson to his alma mater: "My dad wrote the letter," Alec recounted forty years later.

Sometime in the summer of 1970, eight-year-old Barry, apparently unaccompanied, flew back to Honolulu to live for some weeks with his grandparents—and, more important, to interview with Punahou's admissions office and take the necessary tests. In his own later telling, those were glorious weeks—lots of ice cream and days at the beach, a radical upgrade from daily life and school in Jakarta. Then, one late July or early August afternoon, after an appointment at Punahou, and with Barry still dressed to impress, Stan took his *hapa-haole*—half-white—grandson to meet one of his best friends, a sixty-four-year-old black man who had fathered five hapa-haole Hawaiian children of his own.[21]

During their first ten years in Honolulu, Stan and Madelyn's favorite shared pastime had become contract bridge. Madelyn's brother Charles Payne later said they played "with almost a fanaticism" and "they were really, really into it" and "worked well together." Through that hobby, they had met another bridge-playing couple: Helen Canfield Davis, a once-wealthy white woman in her early forties, and her almost-two-decades-older African American husband, Frank Marshall Davis.

By 1970, Frank Davis's publications, involvements, and activities—some self-cataloged, others invasively and meticulously collected by the Federal Bureau of Investigation from 1944 until 1963—were extensive enough to suggest that Davis had led three lives. And indeed he had: almost twenty years as a widely published, often-discussed African American poet and journalist, close to a decade as a dues-paying member of the Communist Party USA, and an entire adult life as an unbounded sexual adventurer.

Born the last day of 1905 in Arkansas City, Kansas—just sixty miles south of Wichita and the neighboring small towns where Stan and Madelyn Dunham would grow up some fifteen years later—Frank's parents divorced while he was a child. He was raised by his mother, stepfather, and grandparents; he graduated from high school, spent a year working in Wichita, and then attended Kansas State Agricultural College. Already interested in poetry and journalism, he left school in 1927 to move to Chicago and found work with a succession of black newspapers there and in nearby Gary, Indiana. In 1931 Frank moved to Atlanta for a better newspaper job, and while there, he met and married Thelma Boyd. He returned to Chicago in 1934, drawn back primarily because of an intense affair with a married white woman who encouraged him to pursue poetry more seriously. His first volume of poems, *Black Man's Verse,* appeared in mid-1935, followed by two more volumes in 1937 and 1938. By the early 1940s Davis had a reputation as an African American writer of significant power and great promise, a leading voice in what would be called the Chicago Black Renaissance.

Decades later, one scholar of mid-twentieth-century black literature would say that Davis was "among the best critical voices of his generation," but his most thorough biographer would acknowledge that "Davis's poetry did not survive the era in which it was written," in significant part because much of it was so polemically political. Another commentator observed that "even at the moments of narratorial identification with the folk, a certain distance is formally maintained." Similarly, asked years later about an oft-cited poem titled "Mojo Mike's Beer Garden," Frank readily acknowledged that his portrayal "was sort of a composite."

Starting in 1943–44, Frank also began teaching classes on the history of jazz at Chicago's Abraham Lincoln School, a Communist-allied institution aimed especially at African Americans. Frank would later complain that "only two black students" took the course in four years, but among the whites who enrolled was a twenty-one-year-old, newly married woman with a wealthy stepfather named Helen Canfield Peck. Within little more than a year, she and Frank had secured divorces and were married in May 1946.

In or around April 1943, Frank had become a dues-paying member of the Communist Party USA, according to FBI informants within the party. From mid-1946 until fall 1947, Frank wrote a weekly column for a newly founded, almost openly Communist newspaper, the *Chicago Star;* in 1948 he published *47th Street: Poems,* which scholars later said was his best book of verse.

During the summer of that year, Helen Canfield Davis, who had also joined the party, read a magazine article about life in Hawaii. Not long after that, Frank spoke about the islands with Paul Robeson, the well-known singer who shared his pro-Communist views. Robeson had visited Hawaii in March 1948 on a concert tour sponsored by the International Longshoremen's and Warehousemen's Union (ILWU) to boost the left-wing Progressive Party. Frank also heard about life in the islands from ILWU president Harry Bridges. Then that fall, Helen received an inheritance of securities worth tens of thousands of dollars from her wealthy stepfather, investment banker Gerald W. Peck. With that windfall, Frank and Helen decided to see for themselves what Hawaii was like for an interracial couple; they packed with an eye toward making this a permanent move and arrived in Honolulu on December 8, 1948.

From their hotel in Waikiki, Frank called ILWU director Jack Hall at Bridges's suggestion. The FBI had a tap on Hall's phone, and this prompted them to watch Frank as well; according to Bureau files, Frank and Helen met Hall in person on December 11. Far more important, though, Frank and Helen thought Hawaii was simply "an amazing place," and that ironically racial prejudice "was directed primarily toward male whites, known as 'haoles.'" As Frank later recounted, "Virtually from the start I had a sense of human dignity. I felt that somehow I had been suddenly freed from the chains of white oppression," and "within a week" he and Helen agreed they wanted to remain in Hawaii permanently, "although I knew it would mean giving up what prestige I had acquired back in Chicago."

By May 1949, Frank began writing an unpaid regular column for the *Honolulu Record,* a weekly paper that matched his political views. In July the FBI placed his name on the Security Index, a register of the nation's most dangerous supposed subversives, and four months on he was added to DETCOM, the political equivalent of the Bureau's "most wanted" list of top Communists marked for immediate detention in the event of a national emergency.

Frank had realized almost immediately that he would not be able to make a living as a writer in Hawaii, and in January 1950, he started Oahu Paper

Company. That same month he and Helen purchased a home in the village of Hauula, thirty miles from Honolulu in northeastern Oahu, for their quickly growing family that included daughter Lynn, who was approaching her first birthday, and son Mark, who would be born ten months later.

The FBI began constant surveillance of the Davises' mail in mid-1950, and in March 1951, a fire at Oahu Paper destroyed thousands of dollars' worth of stock. The Bureau's agents reported that Frank was fully insured, and in June 1952 an informant who had quit Hawaii's Communist Party told agents he had personally collected Frank and Helen's monthly party dues for the last two years. In early 1953 Frank became president of the small Hawaii Civil Rights Congress (HCRC), but within two years the group was "almost inactive." The FBI also noted that on Christmas Day 1955 the Communist Party's national newspaper, the *Daily Worker,* included an article by Frank on jazz.

By that time, Frank and Helen had a third child, but in April 1956, he closed Oahu Paper, filed for personal bankruptcy, and took a job as a salesman. That summer the family moved from Hauula to Kahaluu. Several months later, Eugene Dennis, general secretary of CPUSA, writing in a national newspaper, and then Frank in his weekly *Honolulu Record* column, said "there is no longer a Communist Party in Hawaii." Even so, Mississippi senator James O. Eastland, chairman of the U.S. Senate's Internal Security Subcommittee, scheduled a December hearing in Honolulu to probe Soviet activity in the balmy islands. Fearing how Davis might dress down the notoriously racist Eastland in a public hearing, the subcommittee instead subpoenaed Davis to appear at a private executive session, where he took the Fifth Amendment three times when questioned about his CPUSA ties. Just two weeks later, in another *Record* column, Frank forcefully attacked the Soviet Union for its military invasion of Hungary, calling the move "a tragic mistake from which Moscow will not soon recover."

But Honolulu FBI agents, and their informants, kept their focus on Frank. In mid-1957 he told one supposed friend that Helen had taken up with a visiting musician who was performing in Waikiki. The Bureau quickly took note of Frank's move to the Central YMCA for a month before he and Helen reconciled and the family moved to a house up in Honolulu's Kalihi Valley neighborhood. In February 1958 Helen gave birth to twin daughters, and a year later Frank started a new company, Paradise Papers.

Two years later, agents learned that Helen was working for Avon Products and "works mostly in the evenings making house calls. As a result, subject is now forced to spend most of his evenings babysitting and has little opportunity to contact his former friends outside working hours." That led Honolulu agents to request that Frank be demoted from the top-risk Security Index, but FBI headquarters refused until early 1963, when it ordered Honolulu to interview Frank about his past affiliations, and Frank met with two agents in Kapiolani

Park on August 26, 1963. Asked to confirm his CPUSA membership, Frank said the party had not existed in Hawaii for at least seven years and that it would do him no good to acknowledge his past membership. But, Frank added, he would "consort with the devil" in order to advance racial equality. With that the FBI finally closed its file on fifty-seven-year-old Frank Marshall Davis.

Frank busied himself with Paradise Papers, but it, and Helen's work, hardly provided enough money to raise a family. By June 1968, Frank's two eldest children had graduated high school, and that summer Frank earned a modest sum of money by publishing a self-proclaimed sexual autobiography, *Sex Rebel: Black—Memoirs of a Gash Gourmet,* under the pseudonym "Bob Greene." It began with an introduction, supposedly authored by "Dale Gordon, Ph.D.," which observed that the author may have "strong homosexual tendencies." "Bob Greene" then acknowledged that "under certain circumstances I am bisexual" and stated that "all incidents I have described have been taken from actual experiences" and were not fictionalized. "Bob's" dominant preference was threesomes, and he recounted the intense emotional trauma he experienced years earlier when he learned that a white Chicago couple with whom he had repeatedly enjoyed such experiences were killed in a violent highway accident.

"Bob," or Frank, championed recreational sex, arguing that "this whole concept of sex-for-reproduction-only carries with it contempt for women. It implies that women were created solely to bear children." And Frank did little to hide behind the "Bob Greene" pseudonym with close friends. Four months after the 323-page, $1.75 paperback first appeared, Frank wrote to his old Chicago friend Margaret Burroughs to let her know about the availability of "my thoroughly erotic autobiography." Since it was "what some people call pornography (I call it erotic realism)," it would not be in Chicago bookstores. "You are 'Flo,'" and "you will find out things about me sexually that you probably never suspected—but in this period of wider acceptance of sexual attitudes, I can be more frank than was possible 20 years ago." He closed by telling Burroughs, "I'm still swinging."

In June 1969, Frank moved from his family's home to a small cottage just off Kuhio Avenue in the cramped, three-square-block section of Waikiki known as the Koa Cottages or simply the Jungle. He and Helen divorced the next year, and, as his son Mark would later write, Frank "entered his golden years with glee," given what life in the Jungle offered. As Frank described it, his little studio had a tiny front porch "only two feet from the sidewalk" and "my pad is sort of a meeting area, kind of a town hall to an extent." The Jungle was "a place known for both sex and dope," and was really "a ghetto surrounded by high-rise buildings," but it was without a doubt "the most interesting place I have ever lived." Soon after moving there, Frank became known as the "Keeper of the Dolls," and he later recounted how he had written "a series of short portraits called 'Horizontal Cameos' about women who make their living on their backs."

Two of Frank's closest acquaintances from the early and mid-1970s readily and independently confirm that Stan Dunham was one of Frank's best friends during the years he lived in the Jungle. Dawna Weatherly-Williams, a twenty-two-year-old white woman with a black husband and an interracial son, was by 1970 effectively Frank's adopted daughter and called him "Daddy." She later described Stan as "a wonderful guy." She said he and Frank "had good fun together. They knew each other quite a while before I knew them—several years. They were really good buddies. They did a lot of adventures together that they were very proud of." As of 1970 Stan "came a couple of times a week to visit Daddy," and the two men particularly enjoyed crafting "a lot of limericks that were slightly off-color, and they took great fun in those" and in other discussions of sex, which Dawna would avoid.

Despite what was readily available in the neighborhood, "Frank never really did drugs, though he and Stan would smoke pot together," Dawna remembered. Stan had told Frank about his exceptionally bright interracial grandson well before August 1970. According to Dawna, "Stan had been promising to bring Barry by because we all had that in common—Frank's kids were half-white, Stan's grandson was half-black, and my son was half-black." Decades later she could still picture the afternoon when Stan brought young Barry along to first meet Frank: "Hey, Stan! Oh, is this him?" She remembers that over the next nine or ten years, Stan brought his grandson with him again and again when he went to visit Frank, and as Barry got older, Stan encouraged him to talk with Davis on his own. Obama would remember, "I was intrigued by old Frank," and years later his younger half sister, Maya Kassandra Soetoro, who was born on August 15, 1970, during her brother's visit with their grandparents in Hawaii, described Stanley telling her that Davis "was a point of connection, a bridge if you will, to the larger African American experience for my brother." Once Obama entered politics, Davis's Communist background plus his kinky exploits made him politically radioactive, and Obama would grudgingly admit only to having visited Davis maybe "ten to fifteen times."[22]

Soon after Ann Dunham Soetoro's second child was born, Madelyn Dunham, along with her grandson, flew to Jakarta to see her new granddaughter and to meet Lolo's mother and family. Within weeks nine-year-old Barry was back at Besuki school to start fourth grade. The boy who sat next to him, Widiyanto Hendro, later "said Obama sometimes struggled to make himself understood in Indonesian and at times used hand signals to communicate." The summer in Honolulu had not improved his limited grasp of the Indonesian language, and Lolo's relatives who saw Barry during his fourth-grade school year noted how much chubbier he had become during his now three-plus years in Indonesia.

For more than a year in Honolulu, Lolo Soetoro had served as Barry's off-site stepfather, often roughhousing with him and also playing chess with Stanley at the Dunhams' home. Then, in Jakarta, Barry lived with Lolo on a daily basis for just more than three years, and throughout that time the young boy was impressed with Lolo's knowledge and self-control, especially the latter. "His knowledge of the world seemed inexhaustible," particularly with "elusive things," such as "managing the emotions I felt," Obama would later write. Lolo's own temperament was "imperturbable," and Barry "never heard him talk about what he was feeling. I had never seen him really angry or sad. He seemed to inhabit a world of hard surfaces and well-defined thoughts."

Three decades later, after Obama's memoir *Dreams From My Father* was published, he would select the brief portrait of Lolo he had written when asked to give a short reading from his book. In that scene, young Barry asks his stepfather if he has ever seen someone killed, and when Lolo reluctantly says yes, Barry asks why. "Because he was weak," Lolo answers. Barry was puzzled. Strong men "take advantage of weakness in other men," Lolo responds, and asks Barry, "Which would you rather be?" Lolo declares, "Better to be strong. If you can't be strong, be clever and make peace with someone who's strong. But always better to be strong yourself. Always."

In subsequent years, Obama would believe that by 1970–71, Lolo's acclimation to his new job with Union Oil led Ann to become increasingly disillusioned with her second husband's evolution into an American-style business executive. Obama admired Lolo's "natural reserve" if not his "remoteness," and believed his mother's growing disappointment with Lolo led her to use an image of his absent father to persuade her son to pursue a life of idealism over comfort. "She paints him as this Nelson Mandela/Harry Belafonte figure, which turns out to be a wonderful thing for me in the sense that I end up having a very positive image" of my father, Obama would later recount. "I had a whole mythology about who he was," a "mythology that my mother fed me." But his memories of Lolo from 1970–71 would become dismissive. "His big thing was Johnnie Walker Black, Andy Williams records," Obama recalled. "I still remember 'Moon River.' He'd be playing it, sipping, and playing tennis at the country club. That was his whole thing. I think their expectations diverged fairly rapidly" after 1970.

Some scholars would later credit "the Javanese art of restraint, of not displaying emotions, of never raising your voice," all of which young Barry witnessed in Lolo, with deeply influencing Obama. Lolo "was as close to a father figure as Obama ever had," albeit briefly, and "the lessons Obama learned from Jakarta and Lolo," particularly not "disclosing too much about how one feels," supplied the human template for Obama's own practice and appreciation of the "benefit of managing emotions," a second commentator would conclude.

In subsequent years, when asked about the impact on him of his three-plus years in Indonesia, Obama more often cited an external perception—"I lived in a country where I saw extreme poverty at a very early age"—than any internal conclusions or emotional lessons. "It left a very strong mark on me living there because you got a real sense of just how poor folks can get," he told one questioner twenty years later. "I was educated in the potential oppressiveness of power and the inequality of wealth," he told another. "I witnessed firsthand the huge gulf between rich and poor" and "I think it had a tremendous impact on me," he explained more than once. Such an insistent theme would lead one smart journalist to assert years later that for Obama, "Indonesia was **the** formative experience."[23]

Sometime soon after his tenth birthday, in early August 1971, Barry again flew from Jakarta to Honolulu. As he had the previous summer, he would live with his grandparents, and in September he began fifth-grade classes at Punahou School, just a four-block walk up Punahou Street. Families of fifth (and sixth) graders received a "narrative conference report form three times during the school year. No letter grades are given. At the initial conference during the fall, achievement test scores, the class standing and a detailed written evaluation of progress in each subject area will be discussed." Four major subject areas—Language Arts, Social Studies, Mathematics, and Science—were supplemented by a weekly arts class, a music class, and four sessions of physical education. A year's tuition was $1,165. With two well-employed parents, plus his grandparents—Madelyn nine months earlier had been named one of Bank of Hawaii's first two women vice presidents—Obama did not receive any form of financial aid.

For Mathematics and Science, Barry was taught by twenty-five-year-old Hastings Judd Kauwela "Pal" Eldredge, who had graduated from Punahou seven years earlier and earned his undergraduate degree at Brigham Young University. For Language Arts and Social Studies, in 307 Castle Hall, Barry had his homeroom teacher, fifty-six-year-old Mrs. Mabel Hefty, a 1935 graduate of San Francisco State College who had taught at Punahou since 1947 and had spent a recent sabbatical year teaching in Kenya. At Punahou, fifth graders had homework, and after a brief period of Barry tackling it at the Dunhams' dining room table, Stan asked Alec Williamson, his insurance agency friend, to build a desk to go in Barry's small bedroom. In return, Barry offered Williamson a guitar he had lost interest in. (Williamson still had it more than forty years later.)

As an adult, Obama would praise Hefty for making him feel entirely welcome and fully at home among classmates, most of whom had been together since kindergarten or first grade. Hefty split her class into groups of four at shared desks; Barry was with Ronald Loui, Malcolm Waugh, and Mark "Hebs"

Hebing, his best friend that year. "Mrs. Hefty was a great teacher," Hebing recalled. "One of the first things we had to do" was "memorize the Gettysburg Address"—"the whole thing." In Hebing's memory forty years later, Barry was the first student to succeed.

There was one other African American student, Joella Edwards, in Barry's fifth-grade class, and she was "shocked" by the arrival of her new classmate. She would remember Barry as "soft-spoken, quiet, and reserved," but he hung back from befriending her in any way. Ronald Loui, like Joella, would recall other classmates teasing both her and Barry with common grade-school rhymes. "There were many times that I looked to Barry for a word, a sign, or signal that we were in this together," Joella later wrote, but none ever came. For the next three years too—grades six, seven, and eight—they would be the only two black students in Punahou's middle school, but no bond ever formed before Joella left Punahou come tenth grade.[24]

In late October 1971, Ann Dunham returned to Honolulu from Jakarta. It is unclear who suggested what to whom, but the timing of her trip was not happenstance because five weeks or so after her arrival in Honolulu, Barack H. Obama Sr. arrived there as well, from Nairobi.

The seven years since Obama Sr. had been forced to leave the United States in July 1964 had been eventful and often painful. Not even a week after his departure, an agitated woman from Newton, Massachusetts, Ida Baker, twice telephoned the Boston INS office to report that her twenty-seven-year-old daughter, Ruth, was so romantically infatuated with Obama she was planning to follow him to Kenya and get married. In late August, Mrs. Baker called again to say that Ruth had flown to Nairobi on August 16. An INS agent checked in with Unitarian reverend Dana Klotzle, as well as an official at Harvard, both of whom reported that Obama already had two wives, plus a child in Honolulu. Mrs. Baker acknowledged that Ruth knew of at least the wife in Kenya, but pursued Obama anyway. The agent concluded the report with: "Suggest we discourage her from further inquiries," because it was "time consuming and to no point where her daughter, an adult and apparently fully competent, is in possession of the information re Obama's marriages."

Ruth Beatrice Baker, a 1958 graduate of Simmons College, had become involved with Obama in April 1964 after meeting him at a party. "He had a flat in Cambridge with some other African students, and I was there almost every day from then on. I felt I loved him very much—he was very charming and there never was a dull moment—but he was not faithful to me, although he told me he loved me too." In June, Obama told her he had to return to Kenya, but said she "should come there, and if I liked the country we could marry. I took him at his word" and bought a one-way plane ticket despite how "devastated" her parents were. But Obama was not at the Nairobi airport to meet her, and a helpful airport employee who knew Obama took her home, made some phone calls, and Obama soon appeared. "We went off and started living together" in a home at 16 Rosslyn Close, but "right from the very start he was drinking heavily, staying out to all hours of the night" and "sometimes hitting me and often verbally insulting me," Ruth later recounted. "But I was in love and very, very insecure so somehow I hung on."

On December 24, 1964, she and Obama were formally married; by then his two oldest children, Roy and Rita, were living with him and Ruth in Nairobi. As Barack's younger sister Zeituni described the highly uncomfortable situation: "the children did not know their father, and this white mother did not speak Luo." Zeituni moved in with them to try to ease the tensions, but Obama's deepening alcoholism—Johnnie Walker Black Label was his drink of choice—and abusive behavior made for an unceasingly volatile situation.

Following his return from the U.S., Obama had a job with Shell Oil Company, but five months after Tom Mboya became Kenya's minister of economic planning and development in December 1964, Obama became a senior economist in that ministry. That involved a move to a house at 101 Hurlingham Road, and within three weeks of Obama's joining Mboya's team, the ministry issued a landmark fifty-two-page sessional paper titled "African Socialism and Its Application to Planning in Kenya." In it, President Jomo Kenyatta declared that under his KANU (Kenya African National Union) Party, Kenya "would develop on the basis of the concepts and philosophy of Democratic African Socialism" and had "rejected both Western Capitalism and Eastern Communism" as models for economic development. Kenyatta said that publication of the paper "should bring to an end all the conflicting, theoretical and academic arguments that have been going on," for political stability and confidence could not be established "if we continue with debates on theories and doubts about the aims of our society."

The paper was understood to be primarily Mboya's own handiwork, and knowledgeable commentators praised it as "a middle-of-the-road approach" aimed at tamping down strong ideological differences within KANU. When students at a left-wing institute voiced critical objections, parliament authorized an immediate takeover of the school, with Mboya seconding the motion to do so. But less than eight weeks later, the *East Africa Journal* published an eight-page critique of the paper written by Barack H. Obama.

There was no mistaking Obama's political views. "The question is how are we going to remove the disparities in our country," and "we may find it necessary to force people to do things which they would not do otherwise." In addition, "we also need to eliminate power structures that have been built through excessive accumulation so that not only a few individuals shall control a vast magnitude of resources as is the case now." Obama argued that the sessional paper was too tolerant of such "economic power concentrations" and what was "more important is to find means by which we can redistribute our economic gains to the benefit of all." Not only should government "tax the rich more" and pursue nationalization; it should do so in an explicitly racial way. "We have to give the African his place in his own country," he asserted, "and we have to give him this economic power if he is going to develop." Obama ended with a political call to arms. "Is it the African who owns this country? If he does, then why should he not control the economic means of growth in this country? . . . The government must do something about this and soon."

Obama's essay also featured some thinly veiled special pleading, observing that "we do not have many people qualified to take up managerial positions" or "who could participate intelligently in policy-making functions." What's more, "the few who are available are not utilized fully." Obama almost certainly believed he deserved a more senior job in the government. Not surprisingly, his employment at the ministry came to an end within months after his searing article was published. With that came another household move, this time to city council housing at 16A Woodley Estate.

Sometime soon after that, a drunken Obama insisted on taking the wheel of his friend Adede Abiero's new car and promptly wrecked it. Abiero died in the crash. Obama suffered only minor injuries, but his longtime friend Leo Odera Omolo later said, "Barack never really recovered from that. It had a strong impact." Even so, it did not lead to any increased self-discipline or sobriety. In November 1965 Obama contacted Harvard, seeking the university's support for a return to the U.S. so he could present his Ph.D. dissertation. But the registrar's office rebuffed his request, saying he had failed to register its title with Harvard's Economics Department. Ruth later recalled Obama telling her that his dissertation materials had disappeared following a burglary in which their television was stolen, but in any event Obama failed to pursue the matter further with Harvard, although in Kenya he would often declare himself to be Dr. Obama.

On November 28, 1965, Ruth and Obama's first child, Mark Okoth Obama, was born, but their home life remained fraught with drunken abuse. In 1966 there was increased tension in Kenya's domestic politics, beginning when left-wing Luo vice president Oginga Odinga broke from KANU and formed a new opposition party, the Kenya People's Union (KPU). That was seen as a "direct challenge to Kenyatta," and days later KANU pushed through two constitutional amendments, one mandating new parliamentary elections and another enlarging the president's national security powers to allow for detention without trial.

Kenyatta's security services turned an increasingly hostile eye toward foreigners, and particularly Americans, who were in Odinga's political orbit. The American-born wife of the first Kenyan to attain a Ph.D., Julius Gikonyo Kiano, was charged with disloyalty and expelled; some months later the focus was on a young white American woman from southern Illinois, Sandra Hansen, who had come to Nairobi as a Northwestern University undergraduate interested in African literature. While taking classes at what by then was University College Nairobi, she met a Luo student who invited her to a party at which "the center of attention," as she recounted years later, was a somewhat older Luo man, Barack Obama. Sandy found him "funny, charming," and "extremely charismatic," and they "became fast friends and spent a lot of time together" during 1966 and 1967, by which time Hansen was teaching at a boys' school. "His drinking started to be more of a problem," she recollected, but he "loved music, dancing and dressing well."

Obama was the first person Hansen turned to when Kenyan security officers told her she had seventy-two hours to leave the country or be arrested. Obama accompanied her to see some official in the security ministry, who displayed an extensive file they had collected on her. "I think, Sandy, you've got to go," Obama told her. When her day of departure arrived, Obama drove her to the airport and walked her to the boarding area. Almost fifty years later, Hansen's memories of what Mark Obama would later call "my father's warm and gracious side" are a partial counterpoint to the alcoholic rages that Ruth and his African children endured. But that side was memorialized in an indelible way too, even if for half a century only the tiniest number of people knew the story. Upon leaving Nairobi, Hansen stopped in London, where she saw her Luo boyfriend, Godfrey Kassim Owango, like Obama an economist and later chairman of Kenya's Chambers of Commerce. Back in Illinois, nine months later, Hansen gave birth to a son. She named him not for his father, but for the Kenyan man she most admired and remembered, Barack Obama.[25]

Few other people's experiences with Obama mirrored Sandy Hansen's. In September 1966, Obama had found new employment, with the Central Bank of Kenya, but he was terminated nine months later. Then Ruth, fed up with his violence, fled with one-year-old Mark to the United States. Obama flew across the Atlantic and persuaded her to return to Kenya. "He was a man I had a very strong passion for," Ruth told Sally Jacobs years later. "I loved him despite everything," but Obama's behavior hardly changed for the better. In September 1967, he secured a new job as a senior officer at the Kenya Tourist Development Corporation (KTDC), but within six weeks there were reports that he had drunkenly driven his vehicle into a milk cart one day at 4:00 A.M. By the new year, Ruth was pregnant with their second child; David Opiyo Obama was born on September 11, 1968, at Nairobi Hospital.

Sometime in late 1968 Neil Abercrombie and Andy "Pake" Zane, two of Obama's best buddies from the University of Hawaii, came through Nairobi as part of a months-long tour through Europe, the Middle East, and East Africa. "He showed us around, we stayed at his house, partied, had a good time," and met Ruth, Roy, Rita, and young Mark, Zane recalled more than forty years later, with dozens of photographs from that visit spread out before him. Abercrombie thought "he seemed very frustrated . . . that he was being underutilized" at KTDC. As Zane recalled to Sally Jacobs, "The one thing Barack wanted was to do something for his country, but he felt he could not" accomplish anything significant at KTDC. "He was angry, but it was contained." Yet Abercrombie recalled that "he was drinking constantly. It was as though the drinking was now part of his existence." But in retrospect, one other thing stood out in both friends' memories: Obama never asked about his American son or his ex-wife Ann.

At about 1:00 P.M. on Saturday, July 5, 1969, Tom Mboya was shot and killed at close range outside Chhani's Pharmacy on Nairobi's Government Road. Just moments earlier, Barack Obama had seen Mboya's car parked on a yellow line in the street and had stopped to talk and joke with his friend for four or five minutes. "You will get a ticket," he had warned.

A gunman was arrested, though it was commonly believed that Mboya's assassination was ordered by someone at or near the peak of Kenya's government. On September 8, Obama was the prosecution's final witness at the gunman's trial, testifying about Mboya's final sidewalk chat. The defendant was convicted and soon hanged, but that resolved nothing. Far more than one man had died on Government Road, for Kenya's future as a nonviolent, multiethnic, multiparty democracy died with Tom Mboya.

In June 1970, Obama was fired by the KTDC because of serial dishonesty in matters large and small. Some months later, he had another drunken car crash, and this time he suffered at least one badly injured leg that required prolonged hospitalization. Still, by the early fall of 1971, he was planning a trip to the U.S., perhaps in part because he expected that Ruth would flee from him again, this time permanently.

Rita Auma Obama, who was eleven years old by the time of her father's 1971 departure, recalled him speaking of her American brother and how Ann "would send his school reports to my father." Her older brother Roy, later Abon'go Malik, would later remember seeing "an old briefcase" that contained "the divorce letters, and Ann Dunham's letters." Even Ruth told Sally Jacobs how "very proud" Obama was of his American son. "He had a little picture of him on his tricycle with a hat on his head. And he kept that picture in every house that we lived in. He loved his son."[26]

Barack Obama Sr. arrived back in Honolulu almost ten years after he had left there with glowing credentials to earn a Harvard Ph.D. and then help guide Kenya's economic future. Now he had no doctoral degree, no job, and a visible limp. How he financed the trip remains a mystery. He planned to stay for a month, and the Dunhams had sublet an apartment downstairs from theirs where Obama could sleep.

Madelyn's younger sister Arlene Payne, who also was in Hawaii at that time along with her lifelong companion, Margery Duffey, later told Janny Scott, "I had the sense then, as I had earlier, that both Madelyn and Stanley were impressed with him in some way. They were very respectful to him" and "they liked to listen to what he had to say."

How Ann viewed Obama's visit, and whether he did suggest to his married ex-spouse that he would welcome her and their son joining him in Kenya, is unknown. Obama still referred to her as Anna, and he brought along for his ten-year-old son a trio of Kenyan trinkets: "three wooden figurines—a lion, an elephant, and an ebony man in tribal dress beating a drum." Ann, Stan, and Madelyn had prepared Barry for the visit with intensified renditions of the upbeat themes Ann had insistently sounded during Barry's earlier years. "My father was this very imposing, almost mythic figure," he recounted years later. "In my mind he was the smartest, most sophisticated person that my maternal grandparents had ever met." Then, when they first met, his father entirely lived up to his advance billing, at least in the son's subsequent retelling of it. "He was imposing and he was impressive, and he did change the space around him when he walked into a room," Barry recalled. "His capacity to establish an image for himself of being in command was in full force, and it had an impressive effect on a ten-year-old boy."

"He was an intimidating character," the son told a subsequent interviewer. "He had this big, deep, booming voice and always felt like he was right about everything." All told, it "was a very powerful moment for me," but he also confessed later that his father's visit was deeply unsettling. "If you've got this person who suddenly shows up and says, 'I'm your father, and I'm going to tell you what to do,' and you don't have any sense of who this person is, and you don't necessarily have a deep bond of trust with him, I don't think your reaction is, 'How do I get him to stay?' I think the reaction may be 'What's this guy doing here and who does he think he is?'"

One day during the first two weeks, Ann told her son that Mabel Hefty had invited his father to speak to her and Pal Eldredge's fifth-grade classes about Kenya. That news made Barry nervous, but Obama Sr. carried off the appearance in fine form, and Barry was enormously relieved. Years later, Eldredge could still picture the scene: "He seemed to be real proud, right at his side, kind of holding on to his dad's arm." Barry's classmate Dean Ando recalled it similarly: "All I remember is Barry was just so happy that day it was incredible . . . the dad and Barry had the same smile." Young Obama remembered Eldredge telling him, "You've got a pretty impressive father," and a classmate saying, "Your dad is pretty cool."

A few days after Obama's appearance at Punahou, he took his son to a Honolulu Symphony concert featuring the famous jazz pianist Dave Brubeck, who was joined by his sixteen- and nineteen-year-old sons, Daniel and Christopher, on bass and drums. It was a grand event. The Honolulu Chorale joined the symphony and the family trio to perform Brubeck's new oratorio, *The Light in the Wilderness*. Hawaii's junior U.S. senator, Daniel K. Inouye, served as narrator for the piece.

For Christmas, Obama gave Barry his first basketball. But the end of the month was fast approaching. Obama failed to look up his old Honolulu friends Neil Abercrombie and Andy "Pake" Zane, and when he was with his son, "he never pushed me to speak," Barry later recounted. "It was only during the course of that month—by the end of that month—that I think I started to open myself up to understanding who he was. But then he was gone, and I never saw him again."[27]

Right after New Year's, Ann applied for a new U.S. passport in order to "return home" to Jakarta on January 14, 1972. She listed her stay there as "indefinite," but within a month she made the first of three requests that spring 1972 for UDub to send copies of her old 1961–62 transcript to University of Hawaii's graduate school. In Honolulu, Barry immediately started putting his favorite Christmas present to good use, playing basketball with his good friend Mark Hebing, among others, sometimes at several courts on King Street only a block or so south of his grandparents' apartment building.

His math and science teacher, Pal Eldredge, would remember fifth-grade Barry as "a happy kid. He had a good sense of humor and was smiling all the time," as virtually every photo of young Obama from that time confirms. "He was a rascal too—he had a little spunk to him," Eldredge adds, but "he was always smiling" and was "a good student—he related well with everybody." Obama Sr.'s old buddy Neil Abercrombie, now at work on a Ph.D. dissertation and holding down a variety of odd jobs, would run into Stan Dunham and Stan's grandson several times that spring. "When I would see them, Stanley would offer how bright Barry was and how well he was doing in school. He had ambitions for little Barry," Abercrombie remembered. "It was obvious to everybody and certainly must have been obvious to little Barry that his grandfather not only loved him but, more importantly, liked him and liked having him around and liked him as a pal."

By September 1972, when Barry began sixth grade, Ann and now-two-year-old Maya had returned to Honolulu from Jakarta so that Ann could begin graduate study in anthropology that fall at UH, thanks to a grant from the Asia Foundation. Ann and both of her children lived in apartment #3 at 1839 Poki Street, only one short block west of Punahou. A classmate who sat beside Barry remembered a "chubby-cheeked boy" who was "articulate, bright, funny, and kind." Sixth-grade coursework added "oceanography, electricity and atomic structure" to the science class and also introduced students to "the use and abuse of drugs." In addition, one week at Camp Timberline gave the class an opportunity to try archery and horseback riding; four decades later homeroom teacher Betty Morioka still had a photograph showing a pensive Barry in an oversized gray T-shirt, a rare instance of a picture in which he was not smiling broadly. Young Obama's clearest memory was of a Jewish camp counselor who described the time he had spent in Israel.

Not long after the end of that sixth-grade year, Ann, Madelyn, Barry, and Maya set off on a long tour of the American West. They first flew to Seattle—Ann's first time back there, or anywhere else on the mainland, since her return to Hawaii eleven years earlier—and then headed south down the West Coast. From Disneyland, in Southern California, they headed east to the Grand Canyon, then to Kansas City, where Madelyn's sister Arlene was teaching at the University of Missouri. From there it was north to Chicago, then back westward to Yellowstone National Park and San Francisco before returning to Honolulu. Ann told a friend the trek was "pretty exhausting" since "we traveled by bus most of the way." Her son remembered chasing bison at Yellowstone, but also the "shrunken heads—real shrunken heads" at Chicago's Field Museum. "That was actually the highlight. That was almost as good as Disneyland."

As summer ended, Ann wrote to an old friend in Seattle to say that "I do hope to spend most of my time for the next few years in the islands, since my son Barry is doing very well in school here, and I hate to take him abroad again till he graduates, which won't be for another 6 years." In seventh grade Barry began foreign language (French) instruction, and his other classes would also now be taught by departmental specialists. Barry's homeroom was in 102 Bishop Hall with Joyce Kang; a yearbook photo of the group labeled "Mixed Races of America" declared, "Whether you're a [Sarah] Tmora, a [Pam] Ching, or an Obama, we share the same world." A girl who had pre-algebra and other seventh- and eighth-grade classes with Barry remembered him as "boisterously funny and a big, good-hearted tease" who had "a variety of friends and activities," one of which was now tennis. Throughout these years, Barry spent a good deal of time at Punahou's tennis courts, and one classmate, Kristen B. Caldwell, later wrote and spoke about one incident that remained painfully clear in her memory.

A chart of who would play whom in some tournament had just been posted by Tom Mauch, Punahou's tennis pro. Mauch, then in his early forties, had come to Punahou in 1967 from Northern California's East Bay. Barry and other students were running their fingers along the chart when Mauch told him, "Don't touch that, you'll get it dirty!" In Caldwell's memory, "he singled him out, and the implication was absolutely clear: Barry's hands weren't grubby; the message was that his darker skin would somehow soil" the diagram. "I could tell it upset Barry," she recalled, but "he said, 'What do you mean by that?' with just a perfect amount of iciness to get his point across." Mauch fumbled for a response. "Nothing—I was making a joke."

Only once, in 1995, would Obama himself expressly refer to the incident with the tennis pro. In subsequent years, aside from one unspecific allusion, Obama never mentioned the exchange to any interviewers. Contacted forty years later and asked for the very first time if he remembered Obama, Tom Mauch refused to talk about his years at Punahou.

Barry's eighth-grade year featured one semester of Government and Living in a World of Change and one of Christian Ethics instead of social studies. "Biblical faith is placed in the context of the world in which we live" while examining "the relationship between faith and the everyday experiences of life," Punahou's catalog explained. For French, Barry had his former homeroom teacher, now Joyce Kang Torrey.

In the fall, a still-chubby Barry played defensive end on the intermediate football team coached by Pal Eldredge, his fifth-grade teacher. According to Punahou's catalog, the yearlong science class stressed "human physiology and health . . . drug and sex education are part of the curriculum as the need and interest are manifested." Toward the end of the school year, on April 30, an evening open house called "Science '75" featured eighth graders' second-semester science projects. Barry's was titled "Effects of Music on Plants," though his friend Mark Bendix's "The Effect of Aerosol Spray on Plants" was probably easier to execute.[28]

During Barry's eighth-grade year, Ann finished her graduate coursework, passed her Ph.D. qualifying exams, and gave up the Poki Street apartment to return to Indonesia with four-year-old Maya. She and Lolo had informally separated in mid-1974, and Ann would later record that Lolo did not contribute to her or Maya's support after that time, though her relationship with both him and his parents remained caring and cordial. With her departure from Honolulu, Barry moved back in with his grandparents, who in 1973 had moved from their twelfth-floor apartment to unit 1008 in the same building. Barry spent the summer of 1975 in Indonesia with Ann and Maya before returning to Honolulu in August before his ninth-grade year.

Punahou spoke of its four high school years as "the Academy," and many new students entered for ninth grade, bringing each annual class to 400 to 425 students, or twenty homerooms of twenty students apiece. Barry's new homeroom teacher was Eric Kusunoki, a 1967 Punahou graduate who remembered calling the official roll the very first day and having Obama respond, "Just call me Barry." The biggest change from prior grades was the Academy's unusual six-day variable modular schedule that principal Win Healy had instituted four years earlier: days were A-B-C-D-E-F, not Monday through Friday. That arrangement left students with considerable free time between classes on some days, and Barry usually devoted as much of that time as possible to pickup basketball.

"He always had a basketball in his hands and was always looking for a pickup game," classmate Larry Tavares remembered. Barry later recalled having his worst grade ever—a D in French—that year, and his other classes ranged from speech to boys' chorus to one on Europe. Classmate Whitey Kahoohanohano recounts that "Barry was happy-go-lucky. A prankster. A tease. He liked to have fun. I remember him giggling a lot. He was real pleasant" and "smart." Another, Sharon Yanagi, indicates that Barry's basic persona had not changed at all from previous years: "he was always smiling."

During his ninth-grade year, Barry began a serious friendship with two older African American students, senior Tony Peterson and junior Rik Smith. Tony was only in his second year at Punahou, but as one younger student stressed, "people looked up to Tony. He was a real smart guy." One day a week, Tony, Rik, and Barry would meet up on the steps of Cooke Hall, right outside the attendance office. Tony later said that much of their interaction involved "standing around trying to impress each other with how smart we are."

Although biracial, Rik already firmly identified as black and felt that racism most definitely existed in Hawaii. "Punahou was an amazing school," he said years later, "but it could be a lonely place." In his mind, "those of us who were black did feel isolated." Tony did not entirely share Rik's attitude. "For black people, there was not a lot of discrimination against us." The three of them "talked about race but not, I thought, out of a deep sense of pain," he explained.

One spring morning, to help with an English assignment, Tony recorded some of the trio's conversation. Rik asked "What is time?" and fourteen-year-old Barry responded that "time is just a collection of human experiences combined so that they make a long, flowing stream of thought." At the end of that school year, Barry wrote in Tony's 1976 *Oahuan* yearbook: "Tony, man, I am sure glad I got to know you before you left. All those Ethnic Corner trips to the snack bar and playing ball made the year a lot more enjoyable, even though the snack bar trips cost me a fortune." Playing off of some prior conversation, Barry also told Tony to "get that law degree. Some day when I am a pro basketballer, and I want to sue my team for more money, I'll call on you."[29]

Ann had intended for Barry to once again come to Indonesia for the summer. She and Maya had been living with Lolo's mother in Jogyakarta rather than the capital so she could pursue her doctoral research. "What an enjoyable city it is, especially as compared with Jakarta!" she wrote her University of Hawaii dissertation adviser, Alice Dewey. But in May, she had changed their plans, and in mid-June she and Maya flew to Honolulu, staying at Dewey's home while Barry continued to live with his grandparents. Stanley was still working at the insurance agency, but his two best friends there, Alec Williamson and Rolf Nordahl, could tell how unfulfilling and oftentimes unpleasant he found the work. "During the day, there wasn't a whole lot of business" with potential customers not at home, Nordahl recalled, and he and Stan would chat and often at lunchtime go make sandwiches at the Dunhams' apartment. More than once, Rolf heard Stan mention the Spencer Tracy–Katharine Hepburn film *Guess Who's Coming to Dinner*. Released in December 1967, it starred the black Bahamian American actor Sidney Poitier as Dr. John Wade Prentice of Hawaii, whose white fiancée brings him home to meet her parents. "Well, I lived it," Stan would explain.

On evenings when the two men were finished with customer calls, they often went to Bob's Soul Food Place or the Family Inn bar on Honolulu's Smith Street, in the city's well-known red-light district. "Stanley did not have a great deal of success" selling life insurance, mainly because of his "call reluctance," Nordahl explained. "There's nothing worse than calling somebody and wanting to talk to them about life insurance . . . it's the last thing anybody wants to talk about." But Stanley was committed to sticking with the job and wanted to "come up to snuff with Madelyn . . . I know that bothered him." To Nordahl, "he spoke very fondly of her" and gave no sign that his job difficulties altered his personality. "He always had a joke" and seemed like "a very, very happy man—always a big smile. I wouldn't say that I saw any unhappiness at all."

Stanley also "wanted to learn more about black people," Rolf knew, and that influenced his and his grandson's ongoing visits with Frank Marshall Davis. Barry later described Frank's "big dewlapped face and an ill-kempt gray Afro that made him look like an old, shaggy-maned lion. He would read us his poetry whenever we stopped by his house, sharing whiskey with Gramps out of an emptied jelly jar." Stan's close relationship with Frank also generated his own interest in writing poetry, something he regularly talked about with Alec Williamson.

"He loved science fiction," Williamson recalled, and "we talked a lot of politics." Stan "did not like Nixon," would "argue the liberal side," and often brought his grandson by the office during his late middle school years. Barry "was a good kid . . . well-educated . . . I liked him." Stan was indeed "something of a poet," and more than thirty-five years later Williamson still had copies of, and indeed could recite, two deeply poignant ones:

Life

Oh, where have they gone
Those days of our youth
With those wonderful dreams
Of worlds to be won
When life was a search
For the ultimate truth
Full of adventure
And, Oh, so much fun
Win all our battles
We just couldn't fail
For then right was right
It just had to prevail
Then came life's middle years
Impending old age with all of its fears
The many missed chances
The oft shed tears
Till hope at last dwindles
And disappears
Then comes rebirth
For 'tis Nature's way
The circle's full round
Life's dawned a new day
Erase life's slate clean
But sell not your shares
For hope still survives
In our children, and in theirs
And if not, SO WHAT!

—STANLEY A. DUNHAM

The second, brief, untitled one spoke to home:

Man can span the oceans of space
Split the atom. Win the race

But all is for naught when against his wishes
He has to help with the dinner dishes.

—STANLEY A. DUNHAM

Williamson and Nordahl agreed that "Stan was a great guy," and "we had a lot of good times together." Pal Eldredge at Punahou had exactly the same impression. Both Stan and Madelyn came to "most of the activities" and "any kind of performances we had." Stan "was a fun guy" and "they were always here with" Barry, Eldredge remembered. "It was always good to be around him because he was always joking with people." Ann's mentor Alice Dewey felt similarly: Stanley was a "very charming and fun person, and very affectionate."

Madelyn, particularly at work, was far less outgoing and seemingly far less happy, though her professional success far eclipsed her husband's. One young management trainee from the 1970s, who later became Bank of Hawaii's vice chairman, bluntly acknowledged, "I was afraid of her. She definitely intimidated me. If you were new and still learning, she was like a drill sergeant." Another young man remembered similarly: "We were afraid of her because she was so gruff." Two women had comparable experiences. To Naomi Komenaka, Madelyn was "demanding, sharp and feisty"; to Myrtle Choan, one of Madelyn's direct deputies, "she was a tough lady. Tough, tough lady . . . I was so afraid of her. I called her Mrs. Dunham, never by her first name."

At home, though, Madelyn was as devoted to Barry as Stan was; Barry called her "Tut," with a long "oo" sound, after a common Hawaiian term for grandmother, *tutu*. Her brother Charles recalled her telling him well before Barry's high school years that he was a genius; on one of the few occasions she ever spoke publicly about her grandson, she remembered him as "just a basketball-happy little boy. . . . I think his ambition when he was young was to be a pro basketball player, but he didn't grow tall enough."

In their modest apartment, Barry's tiny bedroom was hardly six feet by eight feet, according to Stan's brother Ralph, who visited them in Hawaii at that time. "It was about the size of a jail cell." Stan took Ralph along to his regular chess and checkers club, and also took him to meet Frank Marshall Davis. Along with Stan and young Barry, "we had a terrific time," Ralph recounted years later. Even close family members did not understand why Stan and Madelyn remained in the small apartment at 1617 South Beretania. Madelyn's brother Charles believed Stan thought it was perfectly fine, but that his sister did not like it. It was a "pretty depressing" building and "not where a bank vice president would live."[30]

For Barry's tenth-grade year, Literature & Writing used the traditional *Warriner's English Grammar and Composition*. A full year of science was required, as was at least one semester of a course on Asia. Tony Peterson had graduated, and seemingly taking his place for Barry was Keith Kakugawa, now a senior, with whom Obama had been acquainted since Keith first arrived at Punahou three years earlier. With a half-Japanese, half–native Hawaiian father who worked as an exterminator, and a half-black and half–Native American mother, Kakugawa personified the islands' rich mix of ethnicities. Bright, an excellent athlete, particularly in track, and blessed with an excellent memory, Kakugawa lived well west of Honolulu in working-class Pearl City, just north of the famous Pearl Harbor navy base.

Obama later wrote that Keith possessed "a warmth and brash humor" that led to "an easy friendship," but around Punahou, opinions on Kakugawa varied widely. Tony Peterson thought "his social skills weren't the best," and Pal Eldredge felt "he had a chip on his shoulder." Keith's longtime friend David Craven later commented that "Barry was the nice guy he hung around with," but Keith's best friend at Punahou, Marc Haine, remembered him as "a popular athlete, a popular figure."

To Kakugawa, fifteen-year-old Barry was "very, very quiet" and "very, very shy . . . I wouldn't say introverted, but he was just a very shy, cautious kid." Keith took a great liking to Stan Dunham. "Gramps was so great to all of us," he recounted years later. "He was everyone's grandfather."

Tenth grade was also the first time Barry played on an actual basketball *team*. Punahou's junior varsity one was coached by 1961 Punahou graduate Norbie Mendez and played five preseason and fourteen regular season games between December 1976 and February 1977. Barry's friends Mark Bendix, Greg Orme, Tom Topolinski, Joe Hansen, and Mark Heflin were also on the team. Obama never cracked the starting lineup, but he ended up as the third leading scorer as the team won nine of its fourteen official games.

Yearbook photos that year show Barry with a bushy Afro and sometimes more than a little extra weight. A classroom picture captures a decidedly chubby Obama, whereas the ones for JV basketball and concert choir, perhaps taken later, show a visibly more mature fifteen-year-old. Sometime midyear Barry also spent significant time reading *The Autobiography of Malcolm X*. Keith Kakugawa remembers Obama pointing out the book in Punahou's library, and Mark Hebing recalls Barry recommending it to him. Barry would later acknowledge absorbing the book that year, saying that Malcolm's "repeated acts of self-creation spoke to me . . . forged through sheer force of will."

One Saturday evening sometime in the late spring of 1977, Kakugawa, Mike Ramos, a junior athlete on Punahou's varsity basketball team, Obama, and fellow sophomore Greg Orme headed to a party at Schofield Barracks, a large U.S. Army base almost twenty-five miles northwest of Punahou. Most of Oahu's African Americans were from military families, and the people at this party were predominantly black men and women several years older than the teenage Punahou quartet. Barry didn't yet drink, but the group stopped to buy a case of Heineken on the way. Once there, the Punahou youngsters were not welcomed with open arms by everyone present.

"The place was packed" and "it was dark," Ramos remembered. He was more than a year older than Obama, and he was from a Filipino family of modest means; he had first met Barry a year earlier at a party where they discovered they were both fans of jazz saxophonist Grover Washington Jr. Orme was white, and he had known Obama since their seventh-grade year. Barry had passed along his interest in jazz to several friends that year. But at the Schofield party "everyone else, except the four of us, was dressed for a night at Studio 54, and we were dressed for a luau at the beach," Kakugawa later said.

Especially if they were haoles, Punahou students were looked upon "as the snobs, the rich kids" in many circles on Oahu. "Everyone on the island treated you differently once they knew you were from Punahou," Keith said. Ramos remembers that Kakugawa, who was known to some of the people at the party, took offense whenever a verbal putdown of Punahou was uttered, and Ramos also remembered Keith "doing a bunch of trash talking" that night in response. Kakugawa admitted "we got in an argument because we were from Punahou," and the four of them were headed for the door in less than an hour.

Ramos was confused. "I was having a pretty good time"—"Why are we leaving?" For Orme, it was the first time he had been one of the few white people in a mostly black setting, and during the car ride back, he mentioned that to Barry. "One of us said that being the different guys in the room had awakened a little bit of empathy to what he must feel all the time at school," Orme later recalled. Ramos agreed. "For the haole guys in our group, it was a kind of eye-opening experience for them." But for whatever reason, Obama was upset by Orme's comment—"he clearly didn't appreciate that," Greg remembered. Kakugawa thought Barry was bothered that one or more girls at the party had refused to dance with him, but Barry had been the youngest person there. Years later, he would describe the evening as a racial coming-of-age moment for him, but Ramos and especially Orme, who would become Obama's closest friend during their two remaining years at Punahou, never heard or saw anything of the sort. Barry "would bring up worldly topics far beyond his years. But we never talked race."

In late April or early May 1977, not long after the party, Ann and Maya returned to Indonesia so that Ann could resume her dissertation fieldwork. Keith Kakugawa starkly remembers the day they departed. Ann told her son she was headed "home," and "Barry was disgusted" after Ann and Maya were dropped off at the airport. "You know what, man? I'm really tired of this,'" Obama complained. Kakugawa told his friend that Ann was just doing her job, but Barry almost spat out his response: "Well, then, let her stay there and do it." Keith's buddy Jack McAdoo said he remembers that day too and recalls that "there was a lot of pain there" for Obama. Kakugawa knew that Barry "was going through a tough time" that spring and was experiencing a lot of "inner turmoil," but "it wasn't a race thing . . . Barry's biggest struggles then were missing his parents. His biggest struggles were his feelings of abandonment. The idea that his biggest struggle was race is bullshit." The crux of what his friend was wrestling with was "the hurt he felt about being abandoned by his mother" on top of his long-absent father.

In later years, Obama would almost always suppress his past feelings about his by-then-deceased mother, but occasionally a highly revealing comment could slip out. "When I was a kid, I don't remember having, I think, one birthday party the whole time I was growing up," and he admitted, "I spent a childhood adrift." But most of Barry's classmates that spring were not aware of what Orme, Ramos, and especially Kakugawa could sense. "I was probably the only one who didn't always see him smiling," Keith recounted. To Kelli Furushima, an attractive Asian classmate whom Obama sought out at the once-per-class-cycle chapel sessions, Barry seemed "a happy guy, comfortable in his skin." She enjoyed his "casually flirting" with her; "he was very friendly, very warm and had a great sense of humor." When the school year was ending and everyone was signing each other's 1977 yearbooks, Obama's note on Kelli's copy likewise reflected no angst: "Our relationship is still young so I am looking forward to picking it up where it left off next year. Your [sic] a small but dynamic person. Have a beautiful summer and see you next year. Love, Barry."[31]

Obama would not turn sixteen years old until August 4, 1977, so getting a summer job was a challenge, though years later he would say he had worked bagging groceries. But that birthday brought with it a driver's license, and he began driving Stan's reddish-brown Ford Granada, a car he would look back on with no fondness. One day that summer Keith Kakugawa and Marc Haine took Barry out paddling—Hawaiian for canoeing—and after they were back on dry land, beer was at hand. "I distinctly remember cajoling Barry into getting drunk with us." He said, "I don't drink," but Keith corrected him: "You're gonna drink." Kakugawa boasted, "I was the one responsible for making Barry take his first drink, but Marc Haine was the one that handed it to him."

When Barry's junior year began, a full-year course in American history was mandatory. Barry had the regular class, not the advanced placement version, taught by his classmate Kent Torrey's father Bob, who knew Barry pretty well but remembers him as "a totally average" student. Another full year of English was required, but in addition to American Literature, the students chose from Punahou's almost collegelike breadth of electives. With the standardized college-entry SAT exam scheduled for November, the fall kicked off with eight weeks or so of Saturday-morning preparation classes; instructor Bill Messer recalled Obama as "affable and pleasant" but "oddly quiet in class." Barry later claimed that "art history was one of my favorite subjects in high school," but he also took drama. He and four classmates produced a short film they titled *The Narc Squad,* a parody of *The Mod Squad*. Linne Nickelsen, who had "long, straight blond hair and a closet full of miniskirts," played the Peggy Lipton character, and Barry imitated the African American actor Clarence Williams III. Plenty of surfing footage was included, and Barry added a dashiki to his bushy Afro. Nickelsen later took credit for luring "Barry out of his dashiki for the pool party scene. . . . I must admit to being disappointed at not having received even passing credit for instigating that disrobing." No screenings of the film have been reported for decades.

Basketball season began in December. Barry and his closest buddies had been playing ball whenever and wherever possible, including evenings and weekends, sometimes going up against adult men and occasionally heading up to UH's Manoa campus to play there. Outside of basketball season, Punahou's mandatory after-school phys ed class was another venue for practicing "hack league" skills. Classmates and teachers all unanimously agreed that basketball was Barry's real passion during all of his high school years, but that December Obama was relegated to the number-two, A-level varsity squad rather than making the cut for the top AA team. Greg Orme and Mark Bendix were also on the A team, but second-class status meant their practices were held every morning from 6:30 A.M. to 8:00 A.M. Once games got under way, the A ballers stumbled to an unmemorable record of seven wins and ten losses. A story in the student newspaper *Ka Punahou* asserted that the players were "having a good time" nonetheless. Coach Jim Iams insisted the season was "successful," but decades later, Iams had no recollection of Obama being on his team.

By the fall of his junior year, Barry was a more memorable member of another Punahou assemblage, known as the Choom Gang. It's not clear when Obama first started to smoke cigarettes, but his friend Mark Hebing can picture Barry coming out of his grandparents' apartment building on their way to go bodysurfing at Sandy Beach, east of Honolulu, carrying only a towel and a carton of cigarettes. Stan Dunham smoked three packs a day—"always Philip Morris," his brother Ralph recalled—and Madelyn smoked even more. But the Choom Gang didn't choom tobacco, they choomed pakalolo, the Hawaiian word for marijuana.

Barry's friend Mark Bendix was seen as "the ringleader" of the group. Tom "Topo" Topolinski, who was half-Polish, half-Chinese, explained that "everything centered around him. He always had the idea first, or he had a stronger opinion, or he wanted to do something rowdier. That was Bendix." The other core members—Russ Cunningham, Joe Hansen, Kenji Salz, Mark "Hebs" Hebing, Greg Orme, Mike Ramos, and eventually Wayne Weightman and Rob Rask—enjoyed drinking beer, playing basketball, bodysurfing when the waves were up, and getting high whenever they had enough money. When they did, Bendix, Topo, and/or Barry would head over to Puck's Alley on the east side of University Avenue where Ray Boyer, their go-to drug dealer, worked at Mama Mia's pizzeria.

Boyer was haole, but just as visibly, he was gay. "Let's just say if he was closeted, he wasn't fooling anybody," Hebs said later. The Choom Gang called him "Gay Ray." He was twenty-nine years old, and he lived in an abandoned bus inside a deserted warehouse in Kakaako, a then-desolate neighborhood west of Waikiki. Topo remembered that the scene there "was very scary. . . . No one in their right mind would live there." Ray also had another main interest: porn. "I think he was looking to convert some people," Topo said years later. "He would bring them back to his bus and stone 'em, with porn movies on—they were heterosexual porn movies, but it was still really creepy." But Ray always had good-quality pakalolo on hand, so the connection was important. "Ray freaked me out. I was afraid of the guy," Topolinski said. "But he did befriend us, and he was our connection . . . there were times where he would take us to a drive-in movie. . . . He partied with us, but there was something about him that never made me feel comfortable."

"We were potheads. We loved our beer," Topo said, but the Choom Gang was not entirely about getting drunk or high. "We loved basketball so much that we couldn't get enough of it," and that was especially true of Barry. "For us, it was just all fun and games and basketball and hanging out, listening to music and going to the beach," Topo emphasized. Bendix's mother taught at Punahou, and they would quietly borrow her car on days when they had a break in their Academy schedules or when they simply decided to cut class. "We could have been easily terminated for what we did," Topo said. "We did it all the time" since the lure of the beach oftentimes was too great. For most Choom Gangers, parental supervision was lax at best; when Topo's parents discovered sand in his shorts one weekday and angrily confronted him about how their tuition payments were being wasted, his lesson was obvious: "I just learned to rinse my shorts out better."

Even for Topo, who had two parents at home, "the Choom Gang became more of a family to me than my own family." For Barry, whose grandmother left for the bank each weekday morning at 6:30 and whose grandfather often was trying to sell life insurance in the evenings, his buddies and especially their devotion to basketball became the centerpiece of his daily life. His friend Bobby Titcomb, a year younger and not a Choom Ganger but whom one upperclassman called "a bit of a badass," has vivid memories of "Obama dribbling his ball, running down the sidewalk on Punahou Street to his apartment, passing the ball between his legs. . . . He was into it." Topo saw the exact same thing, with Barry "dribbling his basketball to class every day. He was married to that thing."

Mike Ramos's younger brother Greg, who was a year behind Barry, and Greg's best friend Keith Peterson shared the gang's love of basketball and thought Obama was a visibly much happier teenager than most of his friends, including Mike, Greg Orme, Mark Bendix, and Joe Hansen. Keith was Tony Peterson's younger brother, and by 1977–78 Keith and Barry were the only two black males in the Academy's sixteen-hundred-plus student body. To Keith, Mike Ramos was "brooding, unpleasant . . . just mean." From both Mike and Orme, Greg Ramos and Keith "got the full big brother to little brother treatment." Orme was usually just "a jerk. Greg would challenge us to basketball games just for the pleasure of beating us to death."

Barry and Mike "were very close," but Obama took part in none of the taunting the younger boys suffered from the older ones. Instead Barry manifested "a level of kindness and genuine caring" which "was pretty unusual, in particular with that group." Hebs's entire family felt the same way about how Obama treated Hebs's younger brother Brad. Both Keith and Greg Ramos also felt that Topo and Bobby Titcomb were each "a great guy," but they thought Obama "was always a happy guy." Indeed, as Keith Peterson puts it, Barry "stands out in my mind as being the happiest of that group."

The Choom Gang had several regular off-campus hangouts—no one dared to choom at school. Each year's catalog emphasized that Punahou "will not condone" drugs or alcohol and expressly prohibited even tobacco smoking "on or in the vicinity of" campus. Everyone understood that they would get expelled from school if they were caught. "We always gravitated to areas that were secluded," Topo explained, and one of their favorite spots was a glade named the Makiki Pumping Station, near the Round Top Drive loop road that circles Mount Tantalus, just a bit northwest of Punahou. "It was a very tucked away, beautiful place," Topo recalled. "It was kind of like our safe haven."

One evening the group headed up there in Mark Bendix's Volkswagen van and Russ Cunningham's Toyota. "We pulled over at the beginning of the hill and we puffed away," Topo remembered. Then Mark and Russ decided their two vehicles should race. Barry and Kenji Salz were with Cunningham, Topo and Joe Hansen with Bendix. "'On your mark, get set, go,' so we took off, and we pulled ahead, and we made a turn, and then nothing happened. We're up there, and we parked, rolled another fatty, and another one," Topolinski said. "We're not that much faster than them. Where the hell did they go?" So "we stayed up there for about twenty minutes and then we decided to go back down just to see what's going on, and we're about halfway down, and we see Barry running up the road, halfway in hysterics. 'What the hell's going on? Barry, where is everybody?'" Obama's answer was startling: "'Kooks rolled the car.' 'What do you mean he rolled the car?' 'It's upside down in the middle of the road.' 'What?' And he's laughing. 'Okay, well we need to go down there.' So Barry got in the van, and we drove to the accident site and sure enough his little Toyota was on its roof in the middle of the road."

Cunningham had a bloody nose, but Kenji, like Barry, was fine. "We didn't want to get in any more trouble so the rest of us left Russell there by himself," Topo recalled, "and we piled in the van to go buy more beer." After downing some, they decided, "Let's go back up and see what's up." At the accident scene, "there's fire trucks, flares," even an ambulance. "We wanted no part of that," so Bendix made a quick U-turn, and the Choom Gang headed for home.[32]

Given the Choom Gang's intake of both beer and pakololo, it was fortunate that nothing worse than a bloody nose and a totaled Toyota resulted from their many outings. In the middle of the 1977–78 school year, Ann Dunham and Maya returned to Honolulu, once again living at Alice Dewey's home, so that Ann could take her doctoral candidacy exams in May. Lolo had been stricken with a serious liver disease, and Ann had taken the lead in forcing Union Oil to send him to Los Angeles for treatment before he then joined Ann and Maya at Dewey's home to recuperate. It is unclear how much Ann saw of her son those months while he continued to live in his grandparents' tiny apartment. Alice Dewey remembers Barry coming by one Sunday to take his mother, sister, and stepfather out for lunch. In late May, the three of them left to return to Indonesia, but no one recalls any intense bitterness like what Keith Kakugawa had witnessed when Ann left a year earlier.

As school ended, Barry wrote another flirtatious message in Kelli Furushima's 1978 yearbook. Tom Topolinski wistfully recalled that Kelli was so popular "there was a line for her" and that Barry "wasn't very forward" with girls. But to Kelli there had been no change in Barry's demeanor. "He was very funny. He was really warm, friendly," she recalled. "He just seemed happy all the time, smiling all the time."

Over the summer, Barry worked at a newly opened Baskin-Robbins at 1618 South King Street, less than two blocks from his grandparents' apartment. Owners Clyde and Teri Higa remember him clearly as "a very good-natured young man, quick with a smile." He worked alongside Punahou classmate Kent Torrey, whose dad had just taught Obama's junior-year U.S. history course; rising junior Annette Yee worked there as well. Clyde Higa said Barry was "the tallest employee we ever had" and thus "seemed to have great difficulty bending over and reaching into the ice cream cabinets to scoop the ice cream." Obama also remembered it was "tough work" behind the counter, and he did not like the mandatory uniform and the accompanying paper cap. He did have an easier time than tiny Annette Yee, who never forgot falling "head first into the near tub" and how Barry "hauled me out." Teri Higa remembered seeing Barry "gazing out the front store window at times" as if "wishing he was at the beach instead of working." One customer who recognized Barry behind the counter was Frank Marshall Davis's dear friend Dawna Weatherly-Williams. "He was a wonderful kid" and even behind the counter, he always "had this beautiful grin on his face."

By the beginning of Barry's senior year, Punahou's tuition and fees stood at $2,050. For a senior, one semester of economics was required; Barry and Mark "Hebs" Hebing both had it with instructor Stuart Gross. But Barry also enrolled in Punahou's most demanding senior year elective, Law and Society, which was taught by Honolulu attorney Ian Mattoch at 7:15 A.M., three mornings a week. Punahou's catalog said the course would "enable students to conceptualize the legal framework of his society, to analyze the terms of the social contract between the individual and the society. Emphasis on the study of the various rights afforded the individual in the Bill of Rights and Constitution and consideration of the bases and characteristics of the executive, judicial, and legislative processes." Mattoch, a member of Punahou's class of 1961, had graduated from Occidental College in Los Angeles in 1965 and Northwestern University's law school in 1968. He had begun offering the course in 1970, and as its syllabus readily revealed, the content was "about a sophomore in college level because Punahou students are eminently capable of doing work at that level," Mattoch explained years later. A 1976 article in Punahou's student newspaper had noted that there were also "optional 'law labs' held on Saturday mornings at Mr. Mattoch's law office" in addition to field trips to courtroom trials. Mattoch said his goal was for students "to understand that law is a product of men and institutions."

Punahou alumni who went on to successful careers in the law testified that Mattoch's class had been directly helpful to them during law school. Mattoch began by asking "What is law?" with readings ranging from Roscoe Pound to Sigmund Freud. Then he moved to "the organizational basis of the legal system," with students expected to master the first hundred-plus pages of a text by a well-known University of Wisconsin law professor. A midterm exam might have as many as seven questions; optional "extra reading reports" on books such as Benjamin Cardozo's *The Nature of the Judicial Process* could earn students additional credit. "How a bill becomes a law" and "basic techniques of legal research" were followed by a study of notable Supreme Court constitutional decisions ranging from *Griswold v. Connecticut*'s 1965 recognition of a right to privacy to criminal procedure rulings. Students then submitted reports comparing the Warren Court of the 1960s to the Burger Court of the 1970s. The final exam lasted ninety minutes and featured more than two dozen questions.

Mattoch remembered Obama as "relatively shy and nonassertive" in class, but he showed up early each morning and did first-rate work. His Punahou transcript indicates an A- in Law and Society, perhaps the single best grade he earned during his high school years. Barry also took a creative writing class that included poetry that fall of his senior year, and in mid-December, the student newspaper published a poem he had written entitled "The Old Man":

I saw an old, forgotten man
On an old, forgotten road
Staggering and numb under the glare of the
Spotlight. His eyes, so dull and grey,
Slide from right, to left, to right
Looking for his life, misplaced in a
Shallow, muddy gutter long ago
I am found instead.
Seeking a hiding place, the night seals us together.
A transient spark lights his face, and in my honor,
He pulls out forgotten dignity from under his flaking coat,
And walks a straight line along the crooked world.

While it is hard to imagine seventeen-year-old Barry being inspired to take up poetry by the example of grandfather Stan, his regular visits to see Frank Marshall Davis, a well-published poet of considerable repute, and by 1978 a man of seventy-two, are a more plausible inspiration, even if Davis's personal history would prevent his role from ever being fully acknowledged.[33]

In December 1978, Barry finally played his way onto the twelve-man roster of Punahou's top AA varsity basketball team. The coach was thirty-two-year-old Chris McLachlin, a 1964 Punahou graduate with a master's degree from Stanford who was a devout student of the highly structured style of play that had been successfully pioneered by UNC Chapel Hill's Dean Smith and especially UCLA's John Wooden, with whom McLachlin had personally conferred several years earlier. McLachlin's 1975 team had won the Hawaii state championship in his first year as head coach, and Punahou's student newspaper commended the "high quality, hustle-oriented basketball which has become a McLachlin and Punahou trademark." The 1979 team featured seven returning seniors, including starting guards Larry Tavares and Darryl Gabriel and forward Boy Eldredge (Pal's nephew), plus star junior forward John Kamana and six-foot-five-inch sophomore Dan Hale at center. Tom Topolinski was often the first man off the bench. The four new members also included Barry's best friend Greg Orme and junior guard Alan Lum.

Mike Ramos had left for college on the mainland, but his younger brother Greg was an AA team manager. Mike's departure led to both Greg and his best friend Keith Peterson spending more time with Barry, and Greg immediately saw that just because Barry was taking a class like Ian Mattoch's, it did not mean the Choom Gang had gone on hiatus or lost its connection to Gay Ray. "When Mike went to college, the first thing Barack did was take me up to the mountains to try to get me stoned, because Mike protected me," Greg confessed years later. But from December through mid-March, basketball dominated their lives on a daily basis.

A Punahou student newspaper profile of "Coach Mac" said that McLachlin possessed a "great ability to get along with students" and quoted him as explaining that "my obligation is not only to teach the skills and strategy, but also to build character and to develop a sense of team and individual pride." Practices were six days a week, two hours each time, but, unlike the A team's, they took place in the late afternoon rather than at dawn. Games ran just thirty-two minutes: four eight-minute quarters. The veteran players understood and respected McLachlin. He was "a very tough coach who knew a lot about the game and made sure that we knew about it as well," Topo explained. McLachlin told the student newspaper: "I'd like to think of myself as a teacher first and a coach second. A ball team is like class after school. I try to teach things like punctuality, industriousness and honesty," as well as relaxation techniques, "since relaxation is very important in situations when the pressure's on, like at the free throw line."

As Topo put it, Punahou's 1979 team was "just loaded with talent," and while McLachlin acknowledged that Punahou's reserves "could have started for any other team in the state," he also "did not have an everybody plays approach," Dan Hale remembered. Like so many others, McLachlin had seen Barry dribbling his basketball and shooting baskets whenever possible, and he respected Obama's "real love and passion for the game." Ironically, though, Obama's many hours of pickup game experience on local courts worked against him with McLachlin. As Alan Lum described it, Barry was "a very creative player," but "his game didn't really fit our system. . . . We ran a structured offense. We were very disciplined." Obama would later assert, "I had an overtly black game," but that misstated the core of basketball's deep appeal to him, which he expressed far better when he compared his favorite sport to his favorite music, jazz. "There's an aspect of improvisation within a discipline that I find very, very powerful."

Team play got under way with an invitational tournament victory on Maui followed by five straight regular-season wins on Oahu before a one-point loss to

University High School. Two victories preceded a defeat by rival Iolani School, then two more wins were followed by a second one-point loss to University High.

Throughout that schedule of games, Barry Obama got little playing time; some days only seven of McLachlin's twelve players saw game action. At one point during those weeks, Obama, along with Alan Lum and Darin Maurer, made an appointment with McLachlin to request that they receive more playing time. McLachlin remembers the meeting as "nonconfrontational and respectful." Barry "basically represented the group" and "spoke for them. . . . It was 'Coach, what can we do to garner more playing time?'" Years later, though, Obama recounted a far angrier scene. "I got into a fight with the guy, and he benched me for three or four games. Just wouldn't play me. And I was furious." Lum had been surprised at how direct Barry was with Coach Mac, and McLachlin acknowledged that Barry was clearly "disgruntled." But Obama was convinced the coach was treating him unfairly. "The truth was, on the playground, I could beat a lot of the guys who were starters," he later claimed. To team manager Greg Ramos, that was just a "total rationalization. . . . My perception at the time was that people were where they should have been, and Barack always thought he should have played more than he did." Even with that tension, Barry's teammates all remember his usual sunny self. "Very happy, very outgoing," "a very, very pleasant person to be around," and "always" with "that smile on his face."

Punahou was in second place in the AA standings prior to an early March league playoff in which the team defeated University High by one point in overtime before a crowd of more than twenty-two hundred. The student newspaper reported that Obama "gave the team a lift as a second half sub, scoring six points on offense and hustling on defense," but the box score in the *Honolulu Advertiser* had Barry missing three free throws. That win gave Punahou top seed in the upcoming three-round Hawaii state tournament. A blowout 77–29 win in their first game included three points by Obama; he played briefly in the second and did not score, but the game still ended in victory for Punahou. The ultimate championship game, against Moanalua on Saturday evening, March 10, 1979, was preceded by a midday team meal at Dan Hale's home. McLachlin's wife Beth made "super burgers" that supposedly aided players' ability to jump.

At Blaisdell Arena in downtown Honolulu, an astonishing crowd of more than sixty-four hundred awaited the contest. "The players majestically strode into the arena as their little admirers flocked around them as if they were blue-clad gods," Punahou's student newspaper claimed in its next issue. "They willingly signed autographs and received handshakes from parents and well-wishers," including of course Stan Dunham. Punahou jumped out to an early lead of 18–4, then ran the score to 32–7. As Alan Lum remembered, "the game was over in the first half" and Coach Mac began substituting liberally. Barry Obama sank one field goal, missed his sole free throw, and finished with two points, but was ecstatic at the 60–28 victory.

"These are the best bunch of guys. We made so many sacrifices to get here," he told Punahou's student reporter. McLachlin was equally happy, telling the *Advertiser* that "we played as near-perfect a game as ever," including how "the subs came in and played as great a team defense as the regulars." Within the world of Punahou, winning the state AA championship was just "huge," as Topo remembered. On the bus ride back to the school, Coach Mac told his team that they had just played "as good a game as I've ever seen a high school basketball team play. You played a perfect game, and that included everyone who stepped on that court. This is the finest effort by twelve young men that I have ever seen."

To Barry Obama, the team's championship was such an enormous achievement he wrote a tribute to their season for Punahou's 1979 yearbook, a brief essay that somehow remained utterly undiscovered even a decade after dozens of journalists had traipsed their ways through all manner of various real and imagined details of Obama's early life. Titled "Winner," the piece is in no way remarkable, but it most certainly captures how central that team experience was for seventeen-year-old Barry:

> *A lot of words are thrown around in basketball: unity, character, determination, and sportsmanship. Well, this is a team that lived up to these clichés, both on and off the court. When the season started, we all felt the electricity of something special; through sacrifice, trust, hard work, and a lot of help from coaches and managers, a group of diverse individuals joined together to truly become a team. At times we've had problems playing together, but we've never had any difficulty getting along; I have never seen a closer bunch of guys. Each player carried his weight and supported the others when they were down; if Gabe wasn't hot, then John would do it; if Danny's dunks didn't beat you, E's defense would. Some people think that it's the win/loss record that is important; others think that it's how you play the game that is important; no matter how you think of it, though, this team was a winner in every sense of the word.*

Obama's memory of his role in the team's season would grow rosy with age. "My senior year, when we won the state championship, there were a couple games where I think I was a difference maker." He said his grandfather would recount how impressive a broadcaster had made Barry's one successful jump shot in that final game sound, but he also acknowledged how those four months had taught him "a lot about discipline, about handling disappointment, being team oriented, and realizing not everything is about you."[34]

The overall picture Obama would usually paint of his final year at Punahou bore no resemblance to that long-undiscovered essay in his senior yearbook or to the A- he earned in Ian Mattoch's exceptionally demanding early-morning Law and Society class. "I'm playing basketball, I'm getting high, and I'm not taking my work seriously at all," he recalled on one occasion. "We'd have basketball practice get over about six, maybe six-thirty, and we'd go get a six-pack . . . and go out to the park and just screw around. . . . Then you'd be waking up in the morning and you hadn't done the reading." In some tellings, Obama admitted to taking his schoolwork more seriously his senior year—"Man, I should try to go to college, so let me focus a bit more"—but then his senior year was the only one with afternoon basketball practice every day. In another version, he has his mother, ostensibly back in Honolulu early in his senior year, upbraiding him about his grades and his disinterest in applying to colleges, calling him a loafer and voicing her disappointment in him. Presented with that account, one interviewer responded that it made Barry sound like a hood, a hoodlum, to which Obama responded, "That's basically it. In fact I think my mother referred to me as such at one point." On another occasion, Obama went even further, claiming, "I think I was a thug for a big part of my growing up."

The notion that *anyone* at Punahou or among his friends and other acquaintances ever thought of Barry Obama as a "thug" or hoodlum could not be further from the way he is remembered. Scores of them *did* see him as "a pretty good jock," as Obama also called himself; tiny Annette Yee from their summer at Baskin-Robbins thought of him as "a basketball jock"; their mutual friend and ice cream coworker Kent Torrey likewise recalled Barry as "one of the biggest jocks on campus." But his intense love of basketball, as his yearbook tribute vividly captured, carried no negativity, and anyone who experienced teenage bullying from one or more of Barry's closest friends without exception recalls Obama as never manifesting any bad attitudes.

By the onset of Barry's senior year, Greg Orme was unquestionably his closest friend. Many days after basketball practice, they did *not* "get a six-pack . . . and go out to the park and just screw around," but instead walked down to 1617 South Beretania. "It was Tut and Gramps in this small apartment and us two six-footers," Orme recalled. "We'd raid the refrigerator and then go to his room. He'd put on his earphones. He liked to listen to Stevie Wonder and jazz, like Grover Washington. So he'd have the earphones on and read his books." Tom Topo came over "once or twice a month." He remembered that visitors could barely "set foot in his room—he lived off the floor—everything was on the floor. . . . He always had a Stevie Wonder record on the phonograph," like

Songs in the Key of Life, and sometimes "like a week-old pizza under his bed . . . he was a messy person."

But the Choom Gang was no less active than the year before. Indeed, come January 1979, Punahou's student newspaper, in a humorous survey of different student "species," profiled one it called "Cravius Cannabis." Recognizable from "their bloodshot eyes . . . Cannabi migrate in vans to 'country,' where they indulge in enlightening tribal rituals. They have developed specialized language to deal with their common interests: agriculture and commerce." Years later, on the one occasion when anyone was able to ask Madelyn Dunham about Barry's high school drug use, she admitted, "I had a few hints, and I think I talked to him a little about it. But it didn't seem overwhelming or prolonged." Obama would later make light of chooming, readily admitting, "I smoked pot as a kid, and I view it as a bad habit and a vice, not very different from the cigarettes I smoked" all through those years.

Once Obama seemingly copped to something more than just pakalolo during his Punahou years, referencing "maybe a little blow when you could afford it" while relating how a food service worker supposedly had offered Barry "smack" (heroin) as well. The Choomers' relationship with Gay Ray went well beyond just "commerce," and some of them spent significant social time with him, as Topo readily acknowledged. "The guy was a maniac on the road. He tailgated everybody. I was afraid for my life every time I rode with him." Yet cocaine had zero presence among the Choom Gang, and multiple friends firmly say they never set eyes on it up through 1979. But at least one Choomer knew for sure that "Barry started to experiment with cocaine." While Topo recalled that "the police were really, really, really relaxed" regarding pakololo, that was not the case with harder drugs, and in Obama's own later telling, his mother asks him about his friend "Pablo," who "was just arrested for drug possession" involving cocaine.

By Barry's senior year his second-closest friend, after Orme, was Bobby Titcomb, who was one year behind before leaving Punahou without graduating. Titcomb came from a distinguished local family. An ancestor had "married into Hawaiian royalty," according to the *Honolulu Advertiser,* and his father was a longtime local judge who had won two Bronze Stars during World War II and lost a congressional race against future U.S. senator Daniel Inouye. Perhaps most notably, the elder Titcomb also frequently appeared on the locally filmed CBS television show *Hawaii Five-0,* which had begun airing nationally in 1968. Bobby Titcomb had first met Barry back in the fifth grade, and the two young men would hike Rocky Hill, just above Punahou's campus, before extending their reach to Peacock Flats, on the west side of Oahu. Obama's private, extremely close friendship with Bobby would remain a constant in his life for many years after 1978–79.[35]

In his most starkly dramatized account of his supposedly angry and antisocial high school years, Obama wrote about "the intimation of danger that would come upon me whenever I split another boy's lip or raced down a highway with gin clouding my head." Not one of Barry's friends and acquaintances can recall any angry exchanges, never mind any busted lips, and no one remembers him having a taste for gin. In that same passage, Obama also recounted "the swagger that carried me into a classroom drunk or high, knowing that my teachers will smell beer or reefer on my breath, just daring them to say something."

Paula Miyashiro, Punahou's 1979 class dean since their freshman year, never knew Barry to have any disciplinary issues and remembers "his infectious, genuine smile" and his "happy," indeed "jovial," demeanor. She met with him at least annually to discuss his upcoming year's classes, and then more frequently as college applications approached. "I remember talking to him about the particular schools he was interested in." But above all, Miyashiro—now Paula Kurashige—stresses that the one person who saw the most of Barry during high school at Punahou was his homeroom teacher, Eric Kusunoki: "for four years, he saw Barry every single day."

Even during his junior year's early-morning basketball practices, or his fall senior year's three-times-a-cycle early-bird Law and Society classes, Barry was in Kusunoki's homeroom every morning. Kusunoki was no innocent. "To say there was a lot of drugs going on back then is a fair statement, maybe an understatement," he frankly acknowledges. But "it wasn't like guys were smoking dope on campus and coming to school high," he added. "If they did, it would have been pretty obvious." Barry without fail would greet "Mr. Kus" with "Good morning," always "very positive, very pleasant," and sporting a "big smile." Barry was "very personable, very respectful," a "bright presence in the classroom," and "never got in trouble." Classmates all continued to have an identical view. To senior-year homeroom colleague Bart Burford, "Barry was one of the more buoyant personalities on campus"; to longtime friend Kelli Furushima, Barry still "just seemed happy all the time. Smiling all the time."

Decades later, as scores of journalists plumbed the question of Obama's racial consciousness during his high school years, no more than two would take the trouble to even telephone, never mind visit, the only other black male student in the Academy during both of Obama's last two years, his friend Keith Peterson. And not a single one would even call the lone black female student, Kim Jones—like Keith, one year behind Barry—who was in the Academy those two years.

Keith Peterson puts it simply. "There was no blackness at Punahou," only three black students among more than sixteen hundred total there. Across the full range of Barry's friends and acquaintances, Keith's perception wins unanimous concurrence. To Bobby Titcomb, Obama "was just another color in the rainbow." Mark Hebing explains, "We didn't think about his blackness." Mike Ramos's younger sister Connie, who was in the class of '79, recalls, "I never once thought of Barry as 'black.'" Barry's friend John Kolivas, who was half Korean and half Greek, says "we didn't think of each other in terms of race."

To Keith Peterson, Barry's "parental piece was a total and complete mystery," and one Barry never spoke of. "I knew nothing about his father" and "I can't say that I knew he had a mother at all." Barry "lived with this older white couple" and "both of his parents so to speak were white." Indeed "I thought that he was actually adopted by a white family," that this "older couple had adopted him." And, in reality, Keith's impression was not at all wrong. As Madelyn's younger brother Charles put it, Barack Obama "was raised in a white family." Bobby Titcomb knew that Stan and Madelyn were Barry's grandparents, but he understood that actually they were more than that. "You call them grandparents, but they were his parents growing up."

Yet no one speaks more powerfully, more movingly, about what it was like to be black at Punahou in the late 1970s than Kim Jones Nelson. "It is different being black growing up in Hawaii" than anywhere on the U.S. mainland, she explains. Even more significantly, "It's an enormous privilege to be a black person in America and grow up in Hawaii. There is no other place in the entire U.S. that provides you with that experience where the color of your skin, the darker your skin is, is not a bad thing." Just as Frank Marshall Davis had realized thirty years earlier, "the whole race dynamic is turned on its head in Hawaii," for "in 1970s Honolulu," as one veteran Asian American journalist would recount, "white people were routinely the target of discrimination," not African Americans.

Kim Jones had arrived at Punahou in 1976 for ninth grade, and it was "such a multicultural world" that "I never thought about" being the only black female among sixteen hundred students. Hawaii has "just a different worldview," and "Punahou's a reflection of the island culture, which is an extremely inclusive culture" of so many countless mixed ethnicities that "I would have had no idea what somebody was." Across four years at Punahou, Kim had no experience of discrimination. "Not ever. Never, ever, ever. Certainly nothing associated with race." What instead stood out was "the quality of the teaching" and the richness of the curriculum: the Novel and Film, art history, creative writing. "I loved Punahou." She barely knew Barry. "He was part of that jock group," and "I can't recall any personal interactions with him."[36]

On at least one occasion twenty years later, Obama acknowledged "the carefree childhood I experienced in Hawaii" and "how truly lucky I was to have been raised" there. He would give thanks as well for "the wonderful education I received at Punahou" and how "Punahou gave me a great foundation," especially in terms of "values and ethics," whose "long-term impact on the trajectory of my life" he would appreciate only a decade later. As a prominent African American, Chicago-based theologian who worshiped in the same church later emphasized, above all else, including color, complexion, and race, first and foremost Barack "Obama is Hawaiian."[37]

Eight or ten weeks in advance of Punahou's June 2, 1979, graduation ceremony, seniors had to submit whatever they wanted published on the one-quarter page they would each have in the 1979 *Oahuan*. Mark Bendix's would contain sketches of his VW bus, the "choom van," and the Koolaus mountain range above Pearl City, plus a "thanks for everything" to classmates whose initials readily translate into their full names: Barry Obama, Kenji Salz, Joe Hansen, Greg Orme, Russ Cunningham, Tom Topolinski, Wayne Weightman, Mark Hebing, and others. Russ Cunningham's page featured a trio of small photos of Barry, Kenji, and Greg, and "Special Thanks to Friends, Family, Choom Gang." Both Hebs's and Topo's were free of any such allusions, but Orme's read, "Many thanks to all my friends. Especially the Choom Gang," with initials for Bendix, Barry, Hansen—who had left school—Kenji, and Cunningham. Kenji's featured a photo that included Barry and captioned "Ooooochoom Gangooooooo"; his acknowledgments included Bendix, Orme, Barry, Hansen, Cunningham, Hebs, and Weightman. Wayne's featured the slogan "Fellow students: it's time to choom!" and a reference to "Pumping station blues."

Barry Obama's quarter-page was by far the most striking of all.

Its upper-right corner featured a handsome photo of Barry in a jacket and wide-collared shirt that could have been borrowed from the 1977 dance film *Saturday Night Fever*. At upper left was a picture of a happy, smiling Obama on a basketball court, captioned "we go play hoop." At the bottom was a photograph labeled "still life" that included a beer bottle, a record turntable, a telephone, and rolling papers. In the middle was Barry's chosen message: "Thanks Tut, Gramps, Choom Gang, and Ray for all the good times."

Decades later, that sentence would receive far less public attention and discussion than it should have. Barry, alone of all the Choom Gang, had singled out their weird, gay, porn-showing drug dealer by name and thanked him "for all the good times." As Tom Topo most frankly acknowledged, the Choom Gangers *had* spent plenty of time with Gay Ray over the previous two years, but a public—and permanent—thank-you to their drug connection was something that all the others, even Mark Bendix, did not go so far as to put into print.

Although decades would pass before Obama would learn of his fate, on New Year's Day 1986, a sleeping thirty-seven-year-old Ray Boyer was bludgeoned to death with a hammer by an angry twenty-year-old male prostitute in an apartment less than two blocks south of 1617 South Beretania Street.[38]

Several days in advance of the graduation ceremony, Ann Dunham returned to Honolulu from Indonesia for the first time in a year. In late 1978, she had completed her fieldwork for her dissertation but, low on funds, had taken a well-paying job with a USAID contractor, Development Alternatives Inc. Based in the Central Java city of Semarang, the job came with a house, servants, and a driver. She and Lolo were on the verge of formally divorcing, and, staying once again at Alice Dewey's home, she soon would adopt Dewey's suggestion that she keep Soetoro as her surname rather than revert to Dunham. But she also made a change from Lolo's colonial Dutch spelling to the Indonesian "Sutoro."

Preparations for the Saturday-night commencement required extensive choral rehearsals on the part of the entire graduating class. Punahou's senior prom took place the night before, Friday, June 1. Greg Orme and his steady girlfriend, red-headed Kelli McCormack, hosted Barry and his date, Megan Hughes, a student at La Pietra School for Girls near Diamond Head, for champagne at her family's home before the two couples headed to the dance and then an afterparty. Decades later Kelli would describe Barry and Greg as "like brothers" and described Barry as "very intelligent and witty." Megan's presence that evening was the first time Kelli had seen Barry with a date, but in Kelli's memory, Megan was "gorgeous. . . . She had the face of an angel and the body of a goddess." But Barry's relationship with her was short-lived. In 1983 Megan would have a brief appearance in one episode of the television series *Magnum, P.I.,* which was filmed on Oahu. A decade later Hughes would appear as Terence Stamp's girlfriend in a movie called *The Real McCoy,* starring Kim Basinger. Two years later Megan had her own starring role in an R-rated "erotic adventure" film titled *Smooth Operator,* but her topless appearance failed to make the movie a popular or commercial success.

Yet in 1979, with Orme already scheduled to be away from Hawaii that summer, Barry betrayed more than a hint of desire for his best friend's girl in the message he wrote in Kelli's yearbook. "It has been so nice getting to know you this year. You are extremely sweet and foxy. I don't know why Greg would want to spend any time with me at all! You really deserve better than clowns like us; you even laugh at my jokes! I hope we can keep in touch this summer, even though Greg will be away." Inscribing his grandparents' phone number, Barry encouraged Kelli to "Call me up and I'll buy you lunch . . . good luck in everything you do, and stay happy. Your friend, Love, Barry Obama." McCormack soon broke up with Orme, and she did like Barry. "He and I really clicked. We had great vibes between us," she recounted years later. But she never called him that summer.[39]

On Saturday evening, June 2, Punahou's 412 graduating seniors, all dressed in matching blazers for the men and long dresses for the women, filed into Honolulu's Blaisdell Arena, where three months earlier Barry's AA basketball team had won their championship. A prayer opened the ceremony, followed by the entire class singing a school song. Three seniors—Byron Leong, Annabelle Okada, and class president Dennis Bader—had major speaking roles, interspersed among four more choral selections. Bader's impressive remarks, in which he told his classmates to follow "your pilot light," drew prolonged applause from the families and friends seated on the arena's main floor. Class dean Paula Miyashiro welcomed the graduates, and Academy principal Win Healy invoked his personal tradition of choosing one adjective to describe each year's class. Commending the 1979 graduates for making 1978–79 "the smoothest and best year of the 1970s" at Punahou, he said the best word to describe them was "harmonic." President Rod McPhee commended Paula Miyashiro on her "great job" with the class, and then presented each of the graduates with their diploma. As the ceremony was ending, Barry ran into his former Baskin-Robbins coworker Kent Torrey. "Kent, I've got to tell you, your dad was one major S.O.B. of a teacher, but at least I learned something from him" in junior-year U.S. history. "What a cool, backhanded compliment from one of the bigger jocks on campus," Kent thought.

Several of Barry's friends remember a cohosted graduation party at Kenji Salz's family's home with Stan Dunham serving as greeter. "Gramps" was "a great guy" who would always "make sure everybody's being included," Greg Ramos remembered. None of Barry's friends have any clear recollections of Ann Dunham from that weekend. Some, like Mike Ramos and Dan Hale, believe they met her then or at some other time, but as Mike put it, "she lived in Indonesia" and "was just in and out" when visiting Honolulu. Mike's brother Greg knows he never met her. "She was not a part of his life" during those final years at Punahou. "His grandparents raised him."

Ann remained in Honolulu for five weeks before returning to Indonesia. Early that summer, Barry hoped to get a job at a pizza parlor—not Mama Mia's with Gay Ray—and Mark Hebing gave him a ride to the interview. But Barry quickly came back out. The place served beer, and Barry was still two months shy of being eighteen—too young to serve beer if not to drink it. Barry sent eighteen-year-old Hebs in, and he was hired. Barry later recalled making $4 an hour painting instead and also working as a waiter at an assisted-living facility.

By the time of his Punahou graduation, Barry knew that in the fall he would be attending Occidental College in Los Angeles—more precisely, in a far northeastern neighborhood called Eagle Rock, close to the small city of Pasadena. Obama later once half-claimed he chose "Oxy" because he had met some girl on vacation in Honolulu who was from Brentwood—far on the opposite side of sprawling Los Angeles—but the choice may also have been influenced by the hope that he was good enough to play college basketball. Punahou teammate Dan Hale remembers that Barry "*really* wanted to play college basketball" and as of spring 1979, he believed he "had an opportunity to play there" on an NCAA Division III team.

Occidental recruiter Kraig King, a 1977 Oxy graduate who had combined a stellar academic record with four years of standout play as a starter on Oxy's varsity basketball team, had visited Punahou back in mid-November 1978, at the same time that Barry was doing so well in Occidental graduate Ian Mattoch's Law and Society class. Obama later publicly thanked Paula Miyashiro Kurashige as "my dean who got me into college," and Greg Ramos has a clear memory of Barry being disappointed at how his college applications had turned out. Oxy "was clearly a second choice for him," especially with another basketball teammate, Darin Maurer, headed to Stanford. Years later Obama said that Swarthmore College in Pennsylvania also rejected him. Oxy required two letters of recommendation; might one from an alumnus who could testify that Barry's A- in Law and Society was a better predictor of his academic potential than the rest of his Punahou transcript have been decisive? If so, no copy survives.

One afternoon in early September 1979, just a few days before Obama was leaving for Occidental, he paid another visit to grandfatherly Frank Marshall Davis. Frank asked him what he expected to get out of college, and Barry, at least as he later recounted their conversation, replied that he didn't know.

"That's the problem, isn't it? You *don't know*. . . . All you know is that college is the next thing you're supposed to do." But Frank had a warning. "Understand something, boy. You're not going to college to get educated. You're going there to get *trained*. They'll train you to want what you don't need. They'll train you to manipulate words so they don't mean anything anymore. They'll train you to forget what you already know. They'll train you so good, you'll start believing what they tell you about equal opportunity and the American way and all that shit. They'll give you a corner office and invite you to fancy dinners, and tell you you're a credit to your race. Until you want to actually start running things, and then they'll yank on your chain and let you know that you may well be a well-trained, well-paid nigger, but you're a nigger just the same."

Barry was confused. Was Frank saying he shouldn't be going to college? Frank sighed. "No. I didn't say that. You've got to go. I'm just telling you to keep your eyes open. Stay awake." With those words of paternal advice, the only African American adult eighteen-year-old Barry Obama had ever known bid him farewell for the West Coast mainland.[40]

Chapter Three

SEARCHING FOR HOME

EAGLE ROCK, MANHATTAN, BROOKLYN, AND HERMITAGE, PA

SEPTEMBER 1979–JULY 1985

Eighteen-year-old Barry Obama arrived on Occidental College's campus in Los Angeles's far northeastern Eagle Rock neighborhood on Sunday, September 16, 1979. Upon arrival, he learned that his dormitory assignment was Haines Hall Annex, room A104, a small, three-man "triple." Oxy, as everyone called it, had expected 425 entering freshmen but, on the day before Obama arrived, the number hit 434 before growing to 458. Of them, 243 were men and 215 were women. The students noted the unexpectedly tight quarters, but college officials were overjoyed, because for several years Oxy had been having a hard time both attracting and retaining academically qualified undergraduates.

Eighteen months earlier, college president Richard C. Gilman had told the faculty that the freshman class target was being reduced from 450 to 425 because for the last three or four years "admission had been offered to every qualified applicant," Oxy's student newspaper reported. Gilman confessed that some who were admitted "may not have been fully qualified." Out of 1,124 applicants for the class of 1980, only 179 had been refused admission. To raise Oxy's standards, a new dean of admissions and new staff were hired in mid-1977, but as of 1978 only 54.2 percent of students admitted to the previous four graduating classes had graduated in the normal four years. The class of 1980 was distinguished by the number of dropouts and students transferring to larger institutions. Oxy's student newspaper interviewed sophomores about their plans, and in a front-page story reported that "it seems that at least half of them are not planning to return next year." The "primary complaint is that the college is too small and limited" academically; other issues were "the limited social atmosphere, the immaturity of the student body and the lack of privacy on campus."

Privacy didn't get any easier with the advent of three-person triples, and Haines Annex and a second dorm each had three of these freshman rooms interspersed on hallways otherwise housing upperclassmen. Barry's two roommates had arrived before him: Paul Carpenter had grown up in nearby Diamond Bar, California, and graduated from Ganesha High School in Pomona. Imad Husain was originally from Karachi, Pakistan, and he and his family now lived in Dubai; he had graduated from the Bedford School in England. Imad was one of many international students at Oxy; in contrast, this freshman class of 1983 included only twenty-plus African American students, mostly from heavily black neighborhoods in nearby South Central Los Angeles. Oxy's tuition for the 1979–80 year was $4,752, with room and board adding another $2,100, for a total of just under $7,000.

Barry's mother Ann Sutoro was earning a respectable salary from DAI in Indonesia—and, as the IRS would charge six years later, was failing to pay her U.S. taxes on it—and Madelyn Dunham still worked as a vice president at Bank of Hawaii. Decades later, an article in an Occidental publication would print Obama's statement that he had received "a full scholarship" that he recalled totaling $7,700, and journalists would repeat that pronouncement as an unquestioned fact. But Occidental awarded financial aid *only* on the basis of financial need and, like Punahou, made work-study employment a part of any recipient's financial aid package. There is no evidence in Oxy's surviving records that support Obama's statement about financial aid, and none of his Oxy classmates remember him working any on-campus job.

Classes began on September 20. Occidental operated on a quarter, or more accurately, trimester system—fall, winter, and spring. Most freshmen followed Oxy's Core Program. A freshman seminar covered the basics of how to use the library and write a paper, easy indeed for anyone from Punahou. Distribution requirements mandated a sampling of American, European, and "World" culture courses across freshman and sophomore years, plus a foreign language—Spanish in Obama's case, after his unfortunate early encounter with French at Punahou—but within that framework students had a great deal of choice. That fall Obama selected Political Science 90, American Political Ideas and Institutions, a lecture class of about 120 students taught in two five-week segments. The first, covering American political thought from Madison and Jefferson through Lincoln, the Progressives, the New Deal, and the mid-twentieth-century debate over pluralism versus elitism, was handled by Roger Boesche, a young assistant professor who had received his Ph.D. from Stanford in 1976. The second, covering the structure and powers of the federal legislative, executive, and judicial branches, was taught by Richard F. Reath, a soon-to-retire senior professor. Boesche in particular impressed Obama. One of his other enduring memories from freshman year was reading Toni Morrison's *Song of Solomon,* published just two years earlier, in an introductory literature course.

One hope Barry brought with him to Oxy was quickly diminished, namely any future as a collegiate basketball player, even at a small Division III school. Midday pickup games at Rush Gymnasium well in advance of preseason team practice—"noon ball" in the school's parlance—quickly demonstrated that Oxy had plenty of players with talent well beyond what Obama had seen in Hawaiian high school AA games. Barry was still not a good outside shooter, and he was so left-handed he always drove leftward and could not cross over. His love of basketball would remain, but his final official team game had taken place in Honolulu six months earlier.

Instead Obama's freshman year revolved around Haines Annex, with its mix of students crammed into tiny rooms on a narrow hallway with one alcove that offered an old couch of uncertain color. In addition to Barry, Paul, and Imad, a second freshman triple included Phil Boerner, a graduate of Walt Whitman High School in Bethesda, Maryland, whose father was a foreign service officer stationed at the U.S. embassy in London. Another triple just around the corner housed Paul Anderson from Minneapolis, a track-and-field athlete. A sophomore triple right across the hall had two Southern Californians, Ken Sulzer and John Boyer. A second next door included Sim Heninger, a North Carolina native who had grown up in Bremerton, Washington; a third had Adam Sherman, from Rockville, Maryland. Tight quarters made for open doors and quick, close friendships. One night early on Barry, Paul Carpenter, Phil Boerner, John Boyer, and others drove to Hollywood to see the movie *Apocalypse Now,* which had opened just a few weeks earlier. There was a long line; right in front of the Oxy crew was the well-known musician Tom Waits, who Phil remembered was "quite wasted."

Getting wasted happened at Oxy too, and perhaps more in Haines Annex than in any other dorm. Loud music helped set the tone, and as Ken Sulzer drily recalled, "if there was an alcohol restriction in the dorms, I wasn't aware of it." But drinking wasn't the half of it. "Choom" and "pakalolo" weren't part of mainland vocabulary, but partaking was even more common in Haines Annex that 1979–80 school year than Barry's trips up to Pumping Station had been a year earlier. Adam Sherman, who was an enthusiastic participant in what he later would acknowledge was a "very wild year," wrote a short story describing the group of regulars who gathered at least four or five nights a week in the hallway alcove that Sim Heninger termed "a male sanctum." The "threadbare couch" sat on a "cigarette-scarred" "aquamarine carpet which is littered with broken, stale potato chips" and "a few mangled and crushed beer cans." Drawing from a ceramic "crimson bong," "the dope" is passed from Paul Carpenter, whose blue eyes are "glazed over in pink, dilated inebriation," to Imad and then to Sim Heninger. The early-morning scene ends with Paul waking Adam from a sound sleep on the hallway floor. Other nights proceeded more energetically, with Phil Boerner ruefully recalling how the regulars would "repeatedly break the fire extinguisher glass during late-night wrestling matches."

Barry Obama was a nightly participant in the hallway gatherings. John Boyer, who kept an irregular journal over the course of the year, recorded how Adam was upset after one holiday break when Obama failed to bring something back for the group from the lush environs of Oahu. Carpenter remembered Obama as someone who "listened carefully" during hallway discussions; Michael Schwartz, a good friend of Carpenter's and Anderson's, remembered Barry as "reserved" and can picture him drinking beer out of a paper cup. Samuel Yaw "Kofi" Manu, a Ghanaian student who met Obama in the introductory political science class, recalls how "extremely friendly" Obama was; sophomore Mark Parsons, a fellow heavy smoker, remembers Barry telling him, "I smoke like this because I want to keep my weight down." John Boyer still has an image of Barry and Adam having long, late-night conversations on the decrepit couch. Obama was "personable" and "quick to laugh," with "a great sense of humor." But Boyer notes that Barry was "always vague" about his family, and even during those late-night discussions, with beer and marijuana relaxing most everyone's demeanors, "there was always kind of a wall" on Barry's part. It was "not really aloofness," Boyer explained, but something self-protective; Obama was more an observer than a spontaneous participant. "'Remove' is a good word," Boyer concluded.

One Friday night in mid-October Barry, Sim, and a sophomore woman were sitting in Haines Hall proper, all under the influence of mood enhancers. In Barry's case those included psychedelic mushrooms, and as a result, Obama "just came unglued. He was a mess." Sim believed that Barry had been adopted and raised by an older white couple whose photo he once displayed, but this night, as Obama babbled about identity and nudity and not wanting to experience rejection, it seemed as if he "was pretty troubled" and was experiencing a "big crisis." At bottom Barry seemed "uncomfortable and frightened," as well as "hysterical and angry," Heninger remembered. "There was no barrier between us in this moment," but for Sim "it was difficult and uncomfortable" in the extreme. Eventually Barry "scraped himself together." Given everything that had been consumed, Sim later mused that neither Obama nor the young woman probably remembered the experience at all, even though in Heninger's eyes it was "a big deal. We called it a day and it blew over," and "I never talked to him about it" again.[1]

For Thanksgiving 1979, just like on weekends "if there was wash to be done, or refrigerators to be raided," Barry joined Paul Carpenter at Paul's family home about thirty miles away. Mike Ramos, in college in Washington State, remembers some holiday in late 1979 when he picked up Greg Orme in Oregon and then rendezvoused with Barry and some others at Mike's younger sister's apartment in Berkeley, just east of San Francisco. Oxy had an almost four-week break after fall exams ended on December 6 and before winter term classes began on January 3, 1980, so Barry probably returned to Honolulu for a good chunk of that time, when Oxy's dorms were closed.

When winter term commenced, Obama took the second course in the political science introductory sequence, Comparative Politics, which that year was taught by the campus's most easily recognized and outspoken young faculty member, openly gay assistant professor Lawrence Goldyn. A 1973 graduate of Reed College in Oregon who had earned his Ph.D. from Stanford just months earlier, Goldyn was an unmistakable figure on Oxy's campus. To say that Goldyn was out "would be an understatement," political science major Ken Sulzer recalled. Goldyn was "funny, engaging," and wore "these really tight bright yellow pants and open-toed sandals." Gay liberation was not part of the Comparative Politics course, but Goldyn drew "a good-sized crowd" one evening during that term when he spoke on gay activism, and a column he wrote for the student newspaper ended by declaring that "the point of liberation, sexual or otherwise, is to rewrite the rules."

Goldyn made a huge impact on Barry Obama. Almost a quarter century later, asked about his understanding of gay issues, Obama enthusiastically said, "my favorite professor my first year in college was one of the first openly gay people that I knew . . . He was a terrific guy" with whom Obama developed a "friendship" beyond the classroom. Four years later, in a similar interview, Obama again brought up Goldyn. "He was the first . . . openly gay person of authority that I had come in contact with. And he was just a terrific guy," displaying "comfort in his own skin," and the "strong friendship" that "we developed helped to educate me" about gayness.

Goldyn years later would remember that Obama "was not fearful of being associated with me" in terms of "talking socially" and "learning from me" after as well as in class. Three years later, Obama wrote that he regarded homosexuality as "an attempt to remove oneself from the present, a refusal perhaps to perpetuate the endless farce of earthly life. You see, I make love to men daily, but in the imagination. My mind is androgynous to a great extent and I hope to make it more so." But there is no doubting that Goldyn gave eighteen-year-old Barry a vastly more positive and uplifting image of gay identity and self-confidence than he had known in Honolulu.[2]

Gayness was not one of the subjects discussed every night in Haines Annex's grungy alcove in early 1980. But the residents did talk about the Soviet Union's recent invasion of Afghanistan. Then, on January 23, President Jimmy Carter in his State of the Union speech announced that he would ask Congress to register young men in preparation for possibly reinstituting a military draft to augment the U.S.'s all-volunteer forces. That news gave Oxy's small band of politically conscious students a new issue to use to regenerate significant student activism.

Two years earlier, a trio of Oxy students—Andy Roth, Gary Chapman, and Doyle Van Fossen—had responded to a challenge posed by the well-known political activist Ralph Nader during an early 1978 campus speech. Occidental,

Nader noted, had some $3 million of its endowment invested in more than a dozen corporations such as IBM, Ford, General Motors, and Bank of America that did business in South Africa, which was known for its harshly racist system of apartheid. In reaction, the undergraduate Democratic Socialist Fellowship formed a Student Coalition Against Apartheid (SCAA) to demand that Oxy divest its stock holdings in companies that continued to operate there.

SCAA quickly gathered more than eight hundred student signatures on a petition calling for divestment, but in early April 1978, Oxy president Gilman rebuffed the students' request. A week later a protest rally of more than three hundred, including Oxy's only African American faculty member, Mary Jane Hewitt, greeted a board of trustees' meeting that affirmed Gilman's refusal. Several weeks later Hewitt resigned from Oxy after she was denied promotion to a higher rank, and the trio of student leaders submitted an angry letter to Oxy's weekly newspaper saying that in light of those two outcomes "we are forced to conclude that a racial bias permeates this institution."

By the 1978–79 academic year, the trustees' finance committee chairman, Harry Colmery, debated Gary Chapman, head of the newly renamed Democratic Socialist Alliance (DSA) at a campus forum, but then the board announced it had ceded investment decisions to a mutual fund, thus ostensibly rendering the entire issue moot. Oxy's faculty responded in May 1979 by adopting a resolution condemning the board's action, but in June the trustees again reaffirmed their refusal to divest.

In early 1980, the *Los Angeles Times* ran two stories about Oxy and its students that highlighted how significant increases in tuition and room and board fees would raise an undergraduate's annual tab to $8,200 the next fall. Oxy's student body was called "introspective" by one senior, and the reporter stated that "student life today" in Eagle Rock "seems placid, serene, contemplative." Given that portrait, a turnout of more than five hundred students at an afternoon protest rally just a week after Carter's nationally televised speech was a dramatic triumph for Oxy's DSA. But a second meeting drew only 150 students, and a teach-in two weeks later attracted just sixty. As winter term ended in mid-March, a student newspaper headline signaled the short-lived movement's demise: "Anti-Draft Activism Fades with Finals."[3]

Oxy's small black student population, about seventy in 1979–80, represented a marked decline from more than 120 just three years earlier. Academic attrition was high, the student paper reported, and after Mary Jane Hewitt's resignation, two brand-new assistant professors, one in French, the other in American Studies, represented Oxy's entire black faculty. A young black graduate of Vassar College was a newly hired assistant dean, but by spring she had submitted her resignation before a student petition effort led Oxy to successfully request that she withdraw it. Two black male sophomores, Earl Chew and Neil Moody, petitioned to establish a chapter of the Kappa Alpha Psi fraternity, citing "a serious social and cultural problem on campus" for minority students. Chew, a St. Louis native, had graduated from tony Phillips Exeter Academy in New Hampshire and weeks earlier had taken the lead in creating an Oxy lacrosse team. But blacks at Oxy, Chew and Moody said, suffered from "a lack of cohesiveness, a generally present personal sense of being members of an ethnic minority group which cannot engage in collective achievement." Indeed, when Oxy's yearbook, *La Encina,* scheduled its 1979–80 photo of Ujima, the African American undergraduate group, only fourteen students showed up to appear in the picture. Barry Obama was not one of them.

Haines Annex's short hallway was home to three other black male undergraduates besides Barry, sophomores Neil Moody and Ricky Johnson and freshman Willard Hankins Jr., but Obama did not develop relationships with any of them like he did with the crew of late-night alcove regulars. Most Oxy black students, particularly those from greater Los Angeles, stuck pretty much together. "There is a certain amount of minority segregation in the dining hall and in the quad," the student paper observed. Black students who did not follow that pattern stood out.

Judith Pinn Carlisle's African American mother had graduated from Howard University, her white father from the U.S. Military Academy at West Point. She and her two siblings all attended junior high schools in Greenwich, Connecticut, but a family financial setback during Judith's high school years had her living in South Central Los Angeles while she attended Oxy. Shy and quiet, "I kept to myself," she said, and as a result, "I was challenged by many people on that campus as to my black legitimacy," notwithstanding how "I'm living at Crenshaw and Adams" in the heart of black L.A. Earl Chew was a particular antagonist, treating Judith as if she was "a sellout," she recalled. Chew "did not like me" and the hostility "was very disturbing."

Sophomore Eric Moore's mother had also graduated from Howard, and Eric grew up in mostly white, upper-middle-class Boulder, Colorado. "There weren't that many black students on campus and there weren't that many that went outside of the black clique," he recounted. Eric had a diverse set of friends, including a junior from Karachi, Pakistan, Hasan Chandoo, who had grown up largely in Singapore and transferred to Oxy after a freshman year at Windham College in Vermont. Also from Karachi was sophomore Wahid Hamid, who roomed with French-born sophomore Laurent Delanney. Both Hamid and Chandoo had long known Imad Husain, Barry Obama's roommate. By the spring of 1980 Chandoo was living off-campus with Vinai Thummalapally, an Indian graduate student who along with Hasan's cousin Ahmed Chandoo was attending California State University and whose girlfriend, Barbara Nichols-Roy, who had also grown up in India, was an Oxy junior. As Eric Moore later said, they had "our own little UN there."

Eric remembers Obama as always having "that big beaming smile," and says he was "always in a Hawaiian shirt and some OP shorts and flip-flops." Indeed, he says, Barry seemed "more Hawaiian and Asian and international in his acculturation than certainly he was African American." Obama "hadn't had an urban African American experience at all," and at Oxy "many of the local Los Angeles African Americans were not as receptive to the cultural diversity" on campus as Eric was. Barry was "a little isolated from that group," and just as Judith experienced, "there was some pushback from certain individuals."

Earl Chew was the most widely visible African American student on campus, and while some found him hostile, Hasan Chandoo considered him a "really wonderful friend." African American freshman Kim Kimbrew, later Amiekoleh Usafi, remembers Earl as "a bright and shining person" who was "just completely committed" to black advancement. Like Eric, she viewed Barry as "a really relaxed boy from Hawaii who wore flip-flops and shorts." Obama once asked her to "come over here and talk to me," Kim recalled. "I don't know if he'd ever really been around black women at that point" and in terms of pursuing women, "he seemed to keep himself away from all of that."[4]

Spring term began the last week of March and lasted until early June. Barry, along with Paul Carpenter, was in a third core political science course, this one on international relations and cotaught by professors Larry Caldwell and Carlos Alan Egan. Junior Susan Keselenko found Egan "a very romantic figure," but the course itself was "really tedious." A significant portion of it involved pairs of nine-student teams contending with each other in a multistage group paper exercise that Keselenko would remember as "very kind of mechanical." Susan and a fellow junior, Caroline Boss, ended up in "Group Y" along with Barry; Paul Carpenter was in the opposing "Group A." Boss, a political science major and active DSA member who as a freshman had run on the progressive slate for Oxy's student government offices, served as the group's informal leader. In mid-May Caroline and Susan orally presented Group Y's six-page paper, "The MX Missile: Bigger Is Not Better."

In the January State of the Union speech that had generated Oxy's draft registration protests, President Carter also had proposed spending as much as $70 billion to build two hundred mobile, ten-warhead-apiece MX missiles that would be deployed all across the U.S. Southwest. Attacking Carter's proposal as "an unnecessary, economically and environmentally devastating venture," Group Y said that if implemented, the MX project "will destabilize the international balance, accelerate the arms race, and increase the likelihood of nuclear war"—the same themes that one group member's father had publicly articulated exactly eighteen years earlier!

Whichever instructor gave it a C was not impressed. The paper had "a certain superficial fluency or glibness," he wrote, but "it demonstrates a very great disregard for careful thought, little concept of how one analyzes an issue, and fails to make a persuasive argument." Ouch. Carpenter's Group A was hardly kinder in their critique, asserting that "Group Y's paper as a whole lacked original analysis" and that "vital contradictions . . . undermined their thesis considerably." Y then penned a rebuttal as well as their critique of Group A's own paper, which addressed the 1978 Camp David Accords. The critique received an A even though it contained multiple obvious spelling errors, including "Palestenians," and creation of the verb "abilitated." Obama appears to have orally presented Group Y's critique, for in the margin alongside their paper's statement that "A settlement amenable to the oil producing Arab states does not insure an improved position for the U.S. in regard to oil," one of his fellow students penned "Barry > abandon Israel will not protect U.S. oil access."

Outside of class, regular activities from earlier in the academic year continued apace. Humorous event listings in the somewhat tardy April Fools' issue of the student newspaper included one announcing that "Haines Annex will host a religious revival this Wednesday at 8:00. Participants will be asked to let their hair down for one night in an effort to communicate with extra-terrestrial Gods utilizing the means of herbal stimuli." Just as at Punahou a year earlier, there was hardly anything secretive about some students' recreational preferences.

One Saturday Barry, Eric Moore, and seniors Mark Anderson and Romeo Garcia went to a music festival in nearby Pasadena Central Park. Eric remembered that "we were culture and music hounds," but with Oxy being an "island in the barrio" of surrounding Eagle Rock, Obama would join him on drives to South Central Los Angeles to get their hair cut. Sometimes the police pulled over Eric and Barry. "It was par for the course," Moore explained years later.

The April 4, 1979, execution of former Pakistani president Zulfikar Ali Bhutto, who had been deposed two years earlier in a military coup led by General Muhammad Zia-ul-Haq, had greatly intensified Pakistan's political turmoil and also caught the attention of several Oxy students. Hasan Chandoo had grown up in a politically aware family, and his mother was a distant relation of Pakistan's revered founder, Muhammad Ali Jinnah. Chandoo's girlfriend Margot Mifflin, a sophomore who had started an Oxy field hockey team that Hasan volunteered to coach because he found its star player so attractive, knew best how "passionate" Chandoo's hatred was for the military dictatorship. "I think I recall Hasan spray painting 'Death to Zia' somewhere on campus," she later recounted. Pakistan was a regular topic of discussion in the Haines Annex alcove and even more so in the Freeman student union snack bar that everyone called the Cooler. Open during the day and then again from 8:00 P.M. to 11:30 P.M., the Cooler was the favorite hangout for Oxy's most politically conscious students, like Caroline Boss, as well as for self-identified literati like Chuck Jensvold, a junior transfer from a community college who was five years older than his classmates.

By spring term 1980, Barry was an evening regular there too. "Obama always seemed to be in there," smoking and drinking coffee, "just jousting back and forth with whoever would come," Eric Moore remembered. One day when Barry walked into the Cooler, Caroline Boss from his political science class introduced him to Susan Keselenko's roommate, junior Lisa Jack, an aspiring portrait photographer. Lisa already had been told that Barry was this "hot" guy, and seeing that he indeed was "really cute," she asked if she could take a roll of photos of him. Barry readily agreed, and a few days later he walked over to Lisa and Susan's nearby apartment. Wearing jeans, a dress shirt, and a leather bomber jacket with a fur collar, Barry also wore a ring on his left index finger, a digital watch on his left wrist, and a bushy Afro that was in need of a drive to South Central. Jack's first fifteen photos captured Obama smiling and smoking while sitting on a simple couch. Then Obama doffed the jacket, rolled up his shirtsleeves, and put on a colorful Panama hat he had brought along. Jack shot eighteen pictures of Obama wearing the hat, then a final three of him bareheaded. Throughout them all, Obama looked without question happy, carefree, and very young for eighteen years of age.

Over a quarter century later, when Jack discovered her old negatives and sold publication rights for some of them to *Time* magazine, former Oxy classmates who had clear memories of Obama's daily appearance during those years said the guy in the photos bore little resemblance to how they remembered him. "That's not how he looked or dressed," Eric Moore commented. John Boyer was even more succinct: "That's not him."[5]

After spring term exams, Barry spent some of the summer living with Vinai Thummalapally in the apartment Vinai and Hasan Chandoo had shared. Barry returned to Hawaii for at least part of the summer, and on July 29 he registered for the reinstituted military draft at a Honolulu post office. His almost ten-year-old sister Maya had landed in Hawaii twelve days earlier; their mother, Ann, apparently had arrived some weeks previously, because on June 15, a local attorney had filed her signed divorce complaint against Lolo in the same court where sixteen years earlier Ann had divorced Barry's biological father. Ann's filing said that "the marriage is irretrievably broken"; in a supporting document she stated that "husband has not contributed to support of wife and children since 1974," was "living with another woman" and "wishes to remarry." Ann reported that she was living "in 4-bedroom house provided by" DAI, her employer, and that she had "2 full-time live-in domestics." The decree she and Lolo signed stated that Lolo "shall not be required to provide for the support, maintenance, and education" of Maya.

Ann and Maya were again staying with Alice Dewey, but Barry was back with his grandparents in their apartment near Punahou. As Obama later told it, one morning Stan and Madelyn argued over her wanting him to drive her to the Bank of Hawaii instead of her continuing her years-long pattern of taking a bus. Madelyn said that on the previous morning an aggressive panhandler had continued to confront her even after she gave him a dollar. Barry offered to drive her downtown, but Stanley objected. He said Madelyn had experienced this before and had been able to shrug it off, but now her fear was greater simply because this panhandler was black. That angered Stan, who refused to take her.

In Obama's later telling, Stan's use of the word "black" was "like a fist in my stomach, and I wobbled to regain my composure." Stan apologized for telling Barry, and said he would drive Madelyn downtown. Then they left. Stanley's obvious comfort with people of color, as well as his liberal political leanings, may not have been fully shared by now fifty-seven-year-old Madelyn, and in Obama's recounting years later he added that never had either grandparent "given me reason to doubt their love." Yet he was struck by the realization that men "who might easily have been my brothers" could spur Madelyn's "rawest fears," at least when they aggressively approached her at close quarters.

Obama says he went that evening to see Frank Marshall Davis, who was now approaching his seventy-fifth birthday. Frank's poetry from the years before his 1948 move to Hawaii was now being rediscovered and studied by a younger generation of African American literature scholars, several of whom had interviewed Frank about his long and fascinating life. Barry recounted his grandparents' argument, and Frank asked if Barry knew that he and Barry's grandparents had grown up hardly fifty miles apart in south central Kansas at a time when young black men were expected to step off the sidewalk if a white pedestrian approached. Barry hadn't. Frank remembered Stan telling him that when Ann was young, he and Madelyn had hired a young black woman as a babysitter and that she had become "a regular part of the family." Frank scoffed at that patronizing, but told Barry that Stanley was a good man even if he could never understand what it felt like to be black and how those feelings could affect black people.

"What I'm trying to tell you is, your grandma's right to be scared. She's at least as right as Stanley is. She understands that black people have a reason to hate. That's just how it is. For your sake, I wish it was otherwise. But it's not. So you might as well get used to it." In Obama's telling, Frank then fell asleep in his chair, and Barry left. Walking to the car, "the earth shook under my feet, ready to crack open at any moment. I stopped, trying to steady myself, and knew for the first time that I was utterly alone."

That night was apparently the last time Obama saw Frank Marshall Davis. But no matter how overdramatized Obama's later account may have been, his previous nine months at Oxy had exposed him for the very first time to mainland African Americans who had a racial consciousness that a Hawaiian who had hardly ever experienced even minor racial mistreatment could not grasp any more than his sixty-two-year-old white grandfather could understand what four decades of being a black man in mainland America had taught his friend Frank. And black Oxy students from South Central L.A. or St. Louis had more trouble feeling at ease in a 90 percent white institution than someone from Punahou could. Barry Obama's Oxy classmates were not being racially obtuse when they saw their happy, relaxed, and reserved friend as a multiethnic Hawaiian rather than a black American.[6]

Oxy's fall term classes began at the end of September 1980. By then, Barry had accepted Hasan Chandoo's invitation to share a two-bedroom ground-floor apartment in a small two-story, multiunit building at 253 East Glenarm Street in South Pasadena, a fifteen-minute drive from Oxy. Before the dorms opened, Hasan's younger friend Asad Jumabhoy, an Indian-origin Muslim also from Singapore who was an entering freshman, crashed on their living room couch. Hasan had a yellow Fiat 128S, and Barry soon acquired a beat-up red Fiat coupe. Vinai Thummalapally and an Indian roommate lived upstairs, and Barry sometimes gave Vinai's girlfriend Barbara a lift to or from Oxy.

Living off campus, Barry spent more time hanging out in the Cooler between and after classes. Cooler regular Caroline Boss cochaired Oxy's Democratic Socialist Alliance, in which Hasan was active, and Hasan was also still coaching Margot Mifflin's field hockey team. As Margot and Hasan got more involved, Margot and her roommate, Dina Silva, spent increasing time at Hasan and Barry's apartment. "They had great social gatherings, parties, dinners," Dina recalled, and Imad Husain and Paul Carpenter, still living in Haines Annex, plus Paul's girlfriend Beth Kahn, were among the regulars. "They used to throw a great party there," Paul agreed. "Food and dancing and a great mix of folks," including Bill Snider and Sim Heninger from the old Haines Annex crowd plus Wahid Hamid, Eric Moore, and Laurent Delanney. Barry and Hasan went on outings with Wahid or Vinai and Barbara to places like Venice, where Margot took a photo of Barry, Wahid, Hasan, and Hasan's cousin Ahmed all wearing roller skates.

Whether in the Cooler or at Glenarm, Hasan's passionate interest in politics dominated many discussions. Hasan was "very outspoken about his political views, very aggressive, opinionated, extroverted," Margot remembers, and identified himself as a Marxist—at least "to the extent that any of us knew what we were talking about," as Susan Keselenko sheepishly puts it. Asad Jumabhoy concurs that "Hasan was very radical at the time" and "had very strong views and he could support his argument very well." To Chris Welton, who returned to Oxy that fall after a year abroad and soon became a close friend after meeting Hasan in one of Roger Boesche's classes, what everyone in Hasan's circle shared was "an outlook" that contemplated the wider world beyond "the borders of the United States." The crux of their orientation was "international, period," or what Caroline Boss called a "more globalized perspective" than undergraduates who had experienced only the mainland U.S. could envision.

Irrespective of the venue, Hasan was "a force to be reckoned with," Paul Carpenter recalls; Sim Heninger terms him "just a domineering personality." Compared to Hasan, who "cursed like a sailor" while smoking incessantly, everyone saw Barry as quiet, measured, and reserved. Chris Welton remembers him as "a keen observer," Caroline Boss would call him "mainly an observer." Dina Silva thought of Barry as "quiet," "thoughtful," and "contemplative" during conversations. Obama "was listening and absorbing everything much more than being demonstrative," Paul Anderson recalls. "He would watch people—that is what I remember," artist friend and junior Shelley Marks recollects. "I specifically remember him being quiet and watching and observing."

During the 1980–81 academic year, Barry and Hasan became the closest of friends. Hasan's girlfriend Margot describes it as "an affectionate relationship," one that "wasn't hampered by masculinity issues. They were open with each other, affectionate with each other," for Hasan was "an open, intimate, direct person." Often the two of them would sit and study in their kitchen; Chandoo can picture Obama reading Martin Luther King Jr.'s "Letter from a Birmingham Jail" at that kitchen table. Some nights Barry studied in a glass-enclosed area in the library's basement that everyone called the Fishbowl. Other nights Margot Mifflin and Dina Silva would join Hasan and Barry to study on East Glenarm. Marijuana was a regular though perhaps not nightly relaxant for Hasan and Barry. "I got stoned with him many times," Margot acknowledged when asked about that 1980-81 year. And, on a less regular basis, "we did occasionally snort cocaine" at Glenarm as well, although it was "not a routine part" of their lives at that time. Sim Heninger can remember nights of "uproar and hilarity" at Barry and Hasan's apartment, but also at least one scene that was "a little scary for me." Bill Snider, reflecting back on both Obama's freshman and sophomore years, deftly remarks that "his memory may be a little *hazy*" both from those nights at Haines Annex and from the subsequent regular parties on Glenarm.[7]

In mid-October 1980 Roger Boesche and faculty colleague Eric Newhall failed badly in an effort to persuade Oxy's faculty to adopt a resolution demanding that the college divest itself of stock holdings in companies still doing business in South Africa. Oxy's student newspaper immediately noted that two years earlier student activism had forced the issue to the top of Oxy's agenda. Just days later, Caroline Boss announced that she and friends were reviving the Student Coalition Against Apartheid (SCAA). Oxy president Richard Gilman dismissed divestment as "an altogether too simplistic solution" while nonetheless acknowledging "the racist conditions in South Africa," but a young sociology professor, Dario Longhi, who had studied in Zambia, took the lead in organizing a series of expert visiting speakers on South Africa for late in the fall term. Earl Chew took an active role while also complaining that Oxy lacked a "multicultural curriculum and social life" and needed far greater diversity.

Sometime late in the fall term, Eric Moore and Hasan Chandoo had conversations with Barry Obama about his name. Eric had spent part of the previous summer in Kenya as part of Crossroads Africa, a student educational program that dated from 1958 and in which Oxy was an active collegiate participant. "What kind of name is Barry Obama—for a brother?" Eric asked him one day. "Actually, my name's Barack Obama," came the answer. "I go by Barry so that I don't have to explain my name all the time." Moore was struck by Barack. "That's a very strong name," he told Obama, who then raised the issue with Hasan one day while they walked across campus. Chandoo agreed, and liked Barry's middle name too. While most close friends like Paul Carpenter and Wahid Hamid had called him "Obama" instead of Barry and stuck with that usage, from that day forward Eric, joined only by Bill Snider, began addressing him as Barack while Hasan, being Hasan, would sometimes say "Barack Hussein," as Asad Jumabhoy clearly remembers even thirty years later. Margot Mifflin too "can remember Hasan saying 'He goes by Barack now,' and I said, 'Well, what is Barack?' and he said 'That's his name.'" At the time, she recalls, "it was jarring."

Over a quarter century later, Obama would say that he saw the change from Barry to Barack as "an assertion that I was coming of age, an assertion of being comfortable with the fact that I was different and that I didn't need to try to fit in in a certain way." With his Oxy friends "he would never correct you" if he was addressed as Barry, Asad explains, but when Obama returned to Honolulu for Christmas 1980, he told his mother and his sister that from now on he would no longer use his childhood nickname and instead would identify himself as Barack Obama. But to his family, just as with Hasan, Eric, and Bill, the name change signified no break in who they thought he was. As Snider explained, "I did not think of Barack as black. I did think of him as the Hawaiian surfer guy."[8]

Long breaks between academic terms gave Barack and his best friends plenty of opportunities to travel. One week Barack and Wahid Hamid headed down to Mexico, then northward to Oregon, in Obama's red Fiat. Two days before the end of fall term exams, Hasan and Barack showed up in the Oxy library with a surprise birthday cake for Caroline Boss, who was hard at work on her senior thesis and whom they spoke with almost every day in the Cooler. Caroline invited the duo to stop by her family home in Portola Valley, near Stanford, over the holidays when Hasan and Barack would be on the road in Hasan's yellow Fiat. Either before or after a New Year's Eve party in San Francisco at which Hasan introduced Barack to another Pakistani friend, Sohale Siddiqi, Hasan and Barack arrived at midday at Boss's home.

Her boyfriend John Drew, a 1979 magna cum laude Oxy political science graduate, was also there; he was in his second year of graduate school at Cornell University. Boss had spent the summer of 1980 in Ithaca taking summer classes, and Drew knew Caroline as "a fun, scintillating, hyper-extroverted," and "intellectually vibrant" young woman who, despite her adoptive parents' significant wealth, worked cleaning an Oxy professor's home. That winter day in San Mateo County, the four young people headed out to lunch with Caroline's parents; Drew recalled much of their conversation focusing on Latin America and particularly El Salvador. Back at the Bosses' home, as Drew remembered it, he and Obama got into a "high-intensity" argument about the relevance of Marxist analysis to contemporary politics. Drew's "most vivid memory" was how strongly Obama "argued a rather simple-minded version of Marxist theory" and that "he was passionate about his point of view."

Drew recalled Obama citing the work of the late French Caribbean decolonization scholar Frantz Fanon. At Oxy Drew had been active in the democratic socialist student group, but a course that past fall with Cornell's Peter Katzenstein had significantly altered Drew's views. "I made a strong argument that his Marxist ideas were not in line with contemporary reality—particularly the practical experience of Western Europe," Drew would recount years later. In Drew's memory, "Caroline was a little shocked that her old boyfriend was suddenly this reactionary conservative," but Obama shifted to downplay their degree of disagreement, conceding that there was validity to some of Drew's points. Drew briefly saw Barack three more times during the remaining six months of his relationship with Boss before finishing his Ph.D., teaching at Williams College, and evolving into an ardent Tea Party conservative.[9]

Oxy's 1981 winter classes began on January 6. For the first two weeks of January, Hasan's friend Sohale, who now lived in New York City, joined them in the Glenarm apartment as they hosted almost nightly parties. For Barack, though, this new term would be by far the most academically and politically engaging ten weeks of his collegiate career. One course he chose, Introduction to Literary Analysis, was taught by English professor Anne Howells, who had been at Oxy almost fifteen years. Utilizing the popular *Norton Introduction to*

Literature, Howells had her fifteen students devote the first five weeks of the term to a "close reading of poetry—old-fashioned textual analysis" and then five weeks to reading short stories. There was "not very much reading, but a lot of writing"—five papers in the course of ten weeks. Barack "spoke well in class," Howells would recall, and submitted well-written papers, but "he wasn't a really committed student" and was late with assignments more than once.

Obama also enrolled in English 110, Creative Writing, which met Tuesday and Friday mornings 10:00 A.M. until 12:00 P.M., with David James, an Englishman who had graduated from Cambridge in 1967 and earned his Ph.D. at the University of Pennsylvania in 1971. After teaching for nine years at the University of California at Riverside, James had arrived at Oxy just four months earlier. He required interested students to submit a writing sample prior to registration, an indication of the seriousness he brought to his teaching. At least five of the dozen or so students in the small class were earnest aspiring writers: Jeff Wettleson, Mark Dery, Hasan's girlfriend Margot Mifflin, and Bill Snider and Chuck Jensvold, the older transfer student, both of whom Barack had known since his freshman year.

David James was "a marvelous character," Dery recalled, "a classic British Marxist film theory jock" who was "a very penetrating analyst of poetry." James handed out copies of contemporary poems he believed students would find stimulating, including ones by Sylvia Plath, W. S. Merwin, and Charles Bukowski, but viewed the course as "essentially a workshop to facilitate the students' own compositions." Jensvold's presence was especially generative, for writing "seemed to define his whole being," James remembered. In particular, Jensvold had an acute "eye for concrete detail" and knew that "amassing an inventory of details" was invaluable to a creative writer. Dery also appreciated Jensvold's presence in what became "a very invigorating class." Chuck was "an exemplar of the serious writer," and as an older student he was "almost our mentor." Dery and another classmate each referred to Jensvold as "hard-boiled," and that classmate warmly remembered Chuck as "the Bogart of Occidental."

Dery also recalled James as "a strict disciplinarian" who "didn't suffer fools gladly" and had "zero tolerance for undergraduate lackadaisicalism." A growing problem as the term progressed was students "straggling into class late." One morning an angry James announced, "I am going to lock the door." Soon a figure appeared outside the frosted glass door, unsuccessfully trying the handle. James did not react, nor to an ensuing tap or two on the classroom window. Finally Dery took the initiative to open the door, and in strolled Barack Obama. To Dery, Barack was "some species of *GQ* Marxist," but an "almost painfully diffident" one whose "caginess," even in the Cooler, made it hard to know "what he truly thought."

A third course that winter was Political Science 133 III, the final trimester of Roger Boesche's upper-level survey of political thought, this one covering from Nietzsche through Weber to Foucault. The fifteen or so students included Hasan as well as Barack's former Haines Annex neighbor Ken Sulzer. Although memories three decades later would be hazy on the exact details, the results of two different assignments were notable in disparate ways. Sulzer and his friend John Boyer recalled seeing Obama heading into the library late one evening before an exam or paper was due. A day or so later, pleased with his own A-, Sulzer asked Barack what he had gotten, but Obama demurred. Sulzer grabbed Barack's blue book and was astonished that Obama had gotten a higher A than he had.

Some weeks later, though, Boesche returned a paper on which he had given Obama a B. As both Barack and Boesche later recounted, within days, Obama encountered his young professor in the Cooler and asked, "Why did I get a B on this?" Boesche viewed Obama as "a student who gives incredibly good answers in class" but failed to consistently live up to his ability level on written assignments that required sustained preparation. In the Cooler, Boesche told Obama that he was smart, but "You didn't apply yourself," that he "wasn't working hard enough." Barack responded, in essence, "I'm working as hard as I can." Boesche knew better than to believe that, but Obama would remain irritated about that B even a quarter century later, especially because Hasan Chandoo, not known for his academic diligence, received a higher grade for the term than did Barack: "I knew that even though I hadn't studied that I knew this stuff much better than my classmates." Obama believed Boesche was "grading me on a different curve, and I was pissed."

Obama would allude to that experience a half-dozen times in later years, often not expressly mentioning Boesche but recounting how "I had some wonderful professors . . . who started giving me a hard time. . . . 'Why don't you try to apply yourself a little bit?' And that made a big difference" in later years as Obama gradually came to appreciate that he was much smarter and far more analytically gifted than anyone who knew him at either Punahou or Occidental fully realized in those times and places.[10]

Early 1981 was just as significant politically for Obama as it was academically. A listless rally on the first day of classes resulted in an Oxy newspaper headline reporting "Students Lack Interest on Draft Issue." Six days later, an evening appearance by Dick Gregory, the comedian and political activist, that Hasan played a lead role in engineering, attracted a huge crowd of 550. Greeted by the "thunderous applause," Gregory then spoke for more than two hours, offering up a pastiche of loony assertions about election tampering and CIA and FBI involvement in the killings of both Kennedy brothers and Martin Luther King Jr., all somewhat leavened by countless humorous asides.

Two days later, a far more serious student forum explored all manner of prejudices at Oxy itself, with African American junior Earl Chew confessing, "Coming here was hard for me. A lot of things that I knew as a black student, that I knew as a black, period, weren't accepted on this campus." Three days later, in response to a flyer distributed on campus, Hasan and Barack drove to Beverly Hills to join 350 others in a silent candlelight vigil protesting the opening of a new South African consulate on Wilshire Boulevard. Relocation of the office there, from San Francisco, had attracted hundreds of protesters three months earlier when the consulate first opened. The vigil was cosponsored by the Gathering, a two-year-old South Central clergy coalition led by Dr. King's closest L.A. friend, Rev. Thomas Kilgore, the L.A. chapter of King's old Southern Christian Leadership Conference, and the antinuclear Alliance for Survival.

Four days after that event, Thamsanqa "Tim" Ngubeni, a thirty-one-year-old South African member of the African National Congress living in exile in Los Angeles, addressed a lunchtime crowd of students on Oxy's outdoor Quad. Born on the outskirts of Johannesburg in 1949, Ngubeni had joined the South African Students' Organization in his early twenties and moved to Cape Town. A friend of prominent student leader Steve Biko, Ngubeni was arrested by South African authorities and imprisoned for several months before leaving South Africa and eventually making his way to L.A. in 1974, three years before Biko was killed while in South African police custody. A soccer scholarship enabled Ngubeni to attend UCLA, where he helped initiate the Afrikan Education Project and led demonstrations against the Bank of America's involvement in South Africa.

In his January 21 speech at Oxy, Ngubeni told the students that South African apartheid "causes human beings to be considered as second-class citizens within their own country, the place where they were born and raised." ANC's goal was for black South Africans to be "recognized as human beings" in a country that had "more prisons than schools." Ngubeni defended the ANC's own use of violence against the South African regime, emphasizing that "we've been negotiating with them all these years while they were shooting us down in the streets." He challenged Oxy's students to reconsider which banks they patronized given how Bank of America and Security Pacific, like IBM and General Motors, continued to do business in South Africa.

Ngubeni often preached that "if you live for yourself, you live in vain. If you live for others, you live forever," and his remarks that day certainly made an indelible impression on at least one of his young listeners. Barack Obama would tell two student interviewers a quarter century later that his meeting an ANC representative was the first time he thought about his "responsibilities to help shape the larger world." Hasan's intense politicization and their drive to Beverly Hills for the candlelight protest against South African apartheid were, like Roger Boesche's professorial reprimand, the beginnings of an evolution that would flower more fully in the years ahead.[11]

As winter term approached its midpoint, political events took place almost nightly. On Sunday evening, February 8, Hasan and Barack both attended a dinner held by Ujima as part of Black Awareness Month at Oxy. The next night Lawrence Goldyn delivered a scintillating talk, deftly titled "Why Homosexuals Are Revolting." On Tuesday evening, Phyllis Schlafly, the conservative Equal Rights Amendment opponent, spoke at Oxy and was met with heckling from a trio of young men: Hasan Chandoo, Chris Welton, and Barack Obama. But the activist students' primary focus was on the Student Coalition Against Apartheid's upcoming divestment rally on February 18, scheduled to coincide with the next meeting of Oxy's board of trustees. Oxy's paper urged all students to attend since it "has the potential to be the most effective display of student initiative in recent years."[12]

Three students took the lead in organizing the rally: Caroline Boss, Hasan Chandoo, and Chris Welton. Caroline and Hasan decided the roster of speakers, with Hasan recruiting Tim Ngubeni to return as their keynote speaker, while Chris and Hasan handled the logistics for the noontime gathering just outside Oxy's administration building, Coons Hall. No one can remember who first had the idea of opening the rally with a "bit of street theater," in which two supposed South African policemen dragoon a young black speaker who wants to quiet the crowd, but Hasan and Caroline were two of Barack Obama's closest friends. Obama later wrote that he prepared for what he expected would be two minutes of remarks prior to being dragged off.

Margot Mifflin and Chuck Jensvold videotaped the rally for a class project. A large banner calling for "Affirmative Action & Divestment NOW" hung in the background. Two folksingers played "The Harder They Come" and the crowd sang along before Barack, wearing a red T-shirt and white jeans, stepped up to the microphone. It was too low, forcing him to hunch over it. Barack asked "How are you doing this fine day?" before declaring that "We call this rally today to bring attention to Occidental's investment in South Africa and Occidental's lack of investment in multicultural education." The crowd cheered and clapped. Barack, with his right hand in his front pants pocket, nodded and resumed speaking. "At the front and center of higher learning, we find it appalling that Occidental has not addressed these pressing problems." The crowd cheered again, and Barack continued, "There is no—" before Chris Welton and another white student suddenly grabbed him from behind and wrestled him offstage, much sooner than Barack had expected.

"I really wanted to stay up there," Barack later wrote, "to hear my voice bouncing off the crowd and returning back to me in applause. I had so much left to say." In his own fictional retelling, he spoke much longer than actually was the case. "There's a struggle going on," he imagined having said. "I say, there's a struggle going on. It's happening an ocean away. But it's a struggle that touches each and every one of us. Whether we know it or not. Whether we want it or not. A struggle that demands we choose sides." In his recounting, Barack had gone on for another seven or more sentences, drawing cheers from the crowd and imagining that a "connection had been made." But as Margot Mifflin later wrote after watching the videotape she had long retained, Barack's version was "factually inaccurate" and "Obama's speech was not long enough to be galvanic, or really even to be called a speech." A detailed account of the rally in the next issue of Oxy's student paper did not mention the opening skit at all. "Led by chants of 'money out, freedom in,' and 'people united will never be defeated,'" the story said the ninety-minute rally attracted a crowd of more than three hundred, plus several local TV news crews.

As a few Oxy trustees and even President Gilman watched, Caroline Boss introduced Tim Ngubeni, who spoke briefly before giving way to senior Sarah-Etta Harris, a Cleveland native, Philips Exeter graduate, and Ujima leader with a glowing résumé that included a semester's study in Madrid and a summer fellowship in Washington, D.C. A photograph taken by sophomore Tom Grauman during Ngubeni's remarks captured a tall, regal Harris standing well apart from Caroline, Hasan, Obama, and other friends, including Wahid Hamid and Laurent Delanney.

Harris warned the crowd that by continuing to invest in companies that did business in South Africa, Oxy's trustees "are telling us that they support oppression over there and also over here." She then spoke directly about Oxy. "I can count the number of black faculty on two fingers, and black student enrollment has been going down steadily." Harris drew cheers from the audience when she said she found it "hard to believe that the trustees cannot redirect their investments to correct some of the problems we have right here on campus."

Many viewed freshman Becky Rivera as the day's star speaker. "We're upset," she announced, and they were "demanding answers and demanding action." Margot Mifflin remembers several African American women jumping to their feet during Rivera's speech and exclaiming, "Say it!" Rivera closed by declaring that "students are responsible for revolutions. Students have power. It starts on the campuses."

The rally's final speaker was Ujima president Earl Chew. Caroline Boss remembers him as "so angry and on fire" but also "a kind person." Chew was a complicated figure because he "had this prep school background, but at the same time he was very street." After almost three years at Oxy, he was "very disillusioned" with a college that was "so painfully white." Chew denounced Oxy's idea of a liberal arts education as "a farce," and excoriated the college for "taking our tuition and investing it in the oppression of our ancestral people." Divestment "may not change the apartheid regime, but it's letting our brothers and sisters in South Africa know that we . . . know better than to oppress other humans for economic gain." As the rally broke up, Rivera sought out Obama to congratulate him on his role. "He really had been on the fringes politically up until that point," Rivera recalled. She told him, "I wish you would get more involved," but rather than thanking Rivera, Obama was simply "noncommittal."[13]

That evening Hasan and Barack hosted a party to celebrate everyone's efforts. Caroline Boss remembers her exchanges with Obama that evening and, like Rebecca Rivera earlier, she was annoyed that he was openly moping rather than savoring his role. "I was really annoyed with him" when he started "yapping about how 'I didn't do a good job, and I could have said it better.'" As Boss recalled, "we all sort of went 'Shut up! It was great. It was fine. You did what you were supposed to do. Move on.'" She was irritated that Barack viewed their group effort only in terms of himself. "The rally wasn't about you developing your technique," she spat out. "It was about South Africa, not you."

Obama would recall a similar conversation with Sarah-Etta Harris, who had quietly befriended him a year earlier. Boss's annoyance was grounded in the many discussions she had had with him in the Cooler, including ones about Obama adopting Barack in place of Barry. "A lot of the year's conversations when they weren't about politics was about identity, me talking about being adopted and him talking about sort of this weird experience of who's his father and where's his mother." Caroline's adoptive parents came from Switzerland, and were now wealthy, but Caroline's maternal grandparents had been peasants who worked as a janitor and maid in an Interlaken bank. "I was very forthright about my own feelings about my adoptive state," Boss remembers, and Barack was "absolutely" clear that he was wrestling with his own feelings about having been abandoned by both his birth parents. "That was something where he and I had a kind of a common understanding," she explains, "of what it means to try to figure all that out." Regarding his mother, Barack "was just very conflicted that she was absent so much," yet given her resolute independence "he admired her enormously and of course found her irritating."

Obama and Boss also discussed "class and race," with Boss citing her grandmother's story to argue the primacy of the former over the latter. Her grandmother had "a royal name," Regina, despite her humble life circumstances, and Boss made Regina "a prominent part of some very intense conversations concerning the relationship between class and race." She said Barack talked "about a lot of things that he's seeing and feeling" with regard to race in the

U.S. Boss said she would reply, "Yeah, but my grandmother in Switzerland—you have to see this internationally—my grandmother's scrubbing those floors, and my mother and her brothers aren't allowed to go into most of the town. The police will come and get them because they're in the tourists' place because they're just these little local brats as far as the town council was concerned, so class is huge."

Fifteen years later, Obama combined these reprimands into an account of how a woman named "Regina" upbraided him after he responds to her kudos about his skit with cynical sarcasm, saying that it was "a nice, cheap thrill" and nothing more. Regina replies that he had sounded sincere, and when Obama calls her naive, Regina counters that "If anybody's naive, it's you" and tells him his real problem: "You always think everything's about you. . . . The rally is about you. The speech is about you. The hurt is always your hurt. Well, let me tell you something . . . It's not just about you. It's never just about you. It's about people who need your help."

In Obama's telling, that night at the party another friend approaches and recalls the awful messes they had left in the Haines Annex hallway a year earlier for the poor Mexican cleaning ladies, to which Barack manages a weak smile. "Regina" angrily asks Obama why he thinks that's funny. "That could have been my grandmother," she said. "She had to clean up behind people for most of her life. I bet the people she worked for thought it was funny too." In truth, it was not Harris but another African American woman, young American studies professor Arthé Anthony, who had objected when someone mentioned the dorm messes during a conversation in Barack and Hasan's kitchen.

In future years, Obama would embrace the mantra of "It's not about you" as a core life lesson and as a powerful antidote to what he described as "the constant, crippling fear that I didn't belong somehow." He would invoke the story of a cleaning lady "having to clean up after our mess" on subsequent occasions both obscure and prominent, and would cite a friend's rebuke about her grandmother having cleaned up other people's messes. With one interviewer, Obama would fuzz the details, saying, "I remember having a conversation with somebody and them saying to me that, you know, 'It's not about you, it's about what you can do for other people.' And something clicked in my head, and I got real serious after that."

Obama would recite the moral of the story—"It's not about you. Not everything's about you"—without naming Harris, Boss, or Anthony. In one rendition, Obama said the rebuke had come from a female professor, but in his written account of that night, he gave Regina the biography of the tall, regal Sarah-Etta Harris. His physical description of Harris was distorted—only her "tinted, oversized glasses" match up to the Harris of 1981—and he says she was brought up in Chicago, not Cleveland, but the other history he attributes to "Regina" is drawn from Harris's own undergraduate achievements. Most of Obama's Oxy friends have difficulty remembering Harris, but "Regina" leaves Oxy "on her way to Andalusia to study Spanish Gypsies." A 1981 Occidental promotional prospectus highlighted Harris's receipt of a Watson Fellowship and said she "will travel to Hungary, France, Italy and Spain to study the socio-economic problems of sedentary gypsies of those countries." Obama's account accurately details the lifelong impact that Harris's, Boss's, and Anthony's rebukes of his self-centeredness had on him, but the "Regina" story merges no fewer than three conversations into one.[14]

The same issue of the Oxy newspaper that covered the rally on its front page also featured "Rising Above Oxy Through Columbia." Junior Karla Olson wrote that a year earlier she had been "stuck in a rut" at Oxy. Deciding that she needed to try another institution, "I tried to pick a college the polar opposite of Oxy" and "within a month I had applied and been accepted as a visiting student to Columbia University in New York City." Upon arriving there, she had been presented with "Columbia's catalogue of 2,000 or more available classes," a stark contrast to tiny Occidental. Olson knew Columbia was "an Ivy League school," but "I soon realized that I wouldn't have to work nearly as hard as I do at Oxy." Olson lived in Greenwich Village, where she had easy access to "museums, art galleries, Broadway, Fifth Avenue, Central Park, clubs, bars, restaurants, and a myriad of other diversions." It seemed as if "95 percent of Columbia students live off-campus, and most go straight home from class . . . making it really hard to meet people," but "my overall experience at Columbia was fantastic. . . . I learned a lot from my classes and benefited even more from the opportunities New York offered in my spare time."

At Occidental, many if not most students thought about transferring to larger institutions. Eric Moore, who tried unsuccessfully to transfer to Stanford, believed "most people were trying to transfer from Oxy," and Phil Boerner, Obama's across-the-hall friend from Haines Annex, "wanted to attend a larger university" and one that was less "like Peyton Place" than Oxy, where "everybody knew who was dating who."

Nineteen-year-old Barack Obama had never even passed through New York City, and he knew no one there aside from Hasan's friend Sohale Siddiqi, but one day he asked Anne Howells, his literature professor, if she would write a letter of recommendation for him to Columbia University. "He wanted a bigger school and the experience of Manhattan," Howells recalled years later. "I thought it was a good move for him." Oxy assistant dean Romelle Rowe remembers Barack discussing a transfer to Columbia and saying he wanted "a bigger environment." Years later Obama would say he transferred "more for what the city had to offer than for" Columbia, that "the idea of being in New York was very appealing." Crucial too was how his closest friends were soon leaving Occidental. Hasan and Caroline were graduating in June and both were headed for London, Hasan to join his family's shipping business and Caroline to study at the London School of Economics. Wahid Hamid, in a dual degree program, was about to shift to Cal Tech.

Caroline Boss, with whom Barack spoke most days, agrees that "he did have a lot of us who were graduating," but "he definitely felt the need to transfer," and not just because his closest friends were leaving. The biggest factor was Obama's emerging belief that he ought to make more of himself than he felt able to do at Oxy. When Caroline and Barack discussed their families, he always said "that his father was a chief" of some sort among Kenya's Luo people and that that lineage represented something he should be proud of, "something to live up to." A quarter century later, Obama would reflect that if "you don't have a sense of connection to ancestors . . . you start feeling adrift and . . . you start maybe devaluing yourself and internalizing" self-doubts. Pride in one's roots can offer "a more powerful sense of direction going forward." Obama's conversations with Boss gave him his first-ever opportunity, far more so than with any friend from Punahou or even the voluble Hasan, to verbalize his developing thoughts about the kind of person he thought he should try to become.

Boss remembers "he basically said, 'I've got to bail. I've got to get myself to where it's cold. I have to be in the library'" and someplace where he would have "'access to a black cultural experience that I don't actually know'" and that would "'make me an American and not just this cosmopolitan guy.'" As Boss remembers, they "talked about why it was important to be Barack. We talked about why it was important to be a man, and why that meant leaving Oxy because he was just going to hang around being a stupid little boy, smoking cigarettes and . . . never getting the A that he knows he's perfectly capable of because you kind of slide."

Obama also realized that the beer drinking, pot smoking, and cocaine snorting that Oxy, like Punahou, offered him, and that had cemented his reputation as "a hard-core party animal" to some friends, was incompatible with any self-transformation into a more serious student and person. Sim Heninger and Bill Snider believed that Obama's decision to apply to Columbia sprang from a desire for greater self-discipline, and over a quarter century later Obama would remark, "I think part of the attraction of transferring was it's hard to remake yourself around people who have known you for a long time." He knew he was at a "dead end" at Oxy and needed a fresh start, that "I need to connect with something bigger than myself." So when Barack mailed his transfer application sometime just before Oxy's spring break began on March 20, at bottom he was making "a conscious decision: I want to grow up."[15]

For Oxy's spring term, Barack, along with scores of other students, enrolled in Lawrence Goldyn's PS 115, Sexual Politics, which met Tuesday and Thursday afternoons. His old Haines Annex friend Paul Anderson took it too and recalls "many times having to stand in the back because there weren't any chairs left."

A new regular for the daily conversations in the Cooler was sophomore Alex McNear, who had arrived at Oxy the previous fall as a transfer from Hunter College in New York City, where she lived. By the end of winter term, she and fellow sophomore Tom Grauman had decided to start a literary magazine at Oxy. They announced the launch of *Feast* in the Oxy newspaper and invited submissions of short stories and poems for spring term.

Obama had composed two poems in David James's winter term creative writing seminar, and he had presented each in class sometime late in the term. One, a twelve-line composition entitled "Underground," may be no more comprehensible now than it was in 1981:

Under water grottos, caverns
Filled with apes
That eat figs.
Stepping on the figs
That the apes
Eat, they crunch.
The apes howl, bare
Their fangs, dance,
Tumble in the
Rushing water,
Musty, wet pelts
Glistening in the blue.

The second, titled "Pop," made enough of an impression on his listeners in March 1981 that at least two of them still recalled that morning three decades later.

Sitting in his seat, a seat broad and broken
In, sprinkled with ashes,
Pop switches channels, takes another
Shot of Seagrams, neat, and asks
What to do with me, a green young man
Who fails to consider the
Flim and flam of the world, since
Things have been easy for me;
I stare hard at his face, a stare
That deflects off his brow;

I'm sure he's unaware of his
Dark, watery eyes, that
Glance in different directions,
And his slow, unwelcome twitches,
Fail to pass.
I listen, nod,
Listen, open, till I cling to his pale,
Beige T-shirt, yelling,
Yelling in his ears, that hang
With heavy lobes, but he's still telling
His joke, so I ask why
He's so unhappy, to which he replies . . .
But I don't care anymore, 'cause
He took too damn long, and from
Under my seat, I pull out the
Mirror I've been saving; I'm laughing,
Laughing loud, the blood rushing from his face
To mine, as he grows small,
A spot in my brain, something
That may be squeezed out, like a
Watermelon seed between
Two fingers.
Pop takes another shot, neat,
Points out the same amber
Stain on his shorts that I've got on mine, and
Makes me smell his smell, coming
From me; he switches channels, recites an old poem
He wrote before his mother died,
Stands, shouts, and asks
For a hug, as I shrink, my
Arms barely reaching around
His thick, oily neck, and his broad back; 'cause
I see my face, framed within
Pop's black-framed glasses
And know he's laughing too.

David James can remember Obama "reading that or an early version of that," and particularly the "amber stain" reference to urine, because of "how powerful an image it was." But James thought the poem seemed "dispassionate" because it was "neither sentimental nor cruel." Margot Mifflin, no doubt with "Underground" in mind, noted that Obama's "previous poems had been more abstract and fanciful," but "Pop" made a stronger impression because of its "honest ambivalence and because it was so unabashedly personal, especially coming from someone who tended to be reserved." Mifflin was also impressed that "it wasn't sentimental. It had an edge of darkness to it, and that made it genuine."

Alex McNear and her colleagues accepted both of Barack's poems for *Feast*'s inaugural issue. When the fifty-page magazine arrived on campus in May, a review in the student newspaper described it as a "most outstanding collegiate example of writing talent" and said copies "should be sent to other colleges." McNear and her contributors appreciated that praise, and over a quarter century later, *Feast* would be discovered by a new generation of readers who sought to understand Obama's poems. Given both the title and the reference to "black-framed glasses," most commentators presumed that Obama had written about his grandfather, Stan Dunham, not Frank Marshall Davis. But hostile critics focused on how the subject "recites an old poem he wrote before his mother died" and noted that Stan's mother had killed herself when he was eight years old, yet Barack would forcefully reject the Davis hypothesis. "This is about my grandfather."[16]

Alex McNear was one of two Occidental women whom male students immediately remembered three decades later. The list of men who actively sought her attention included 1980 graduate Andy Roth, who was in Eagle Rock through February 1981; Phil Boerner, who invited Alex to brunch multiple times; and *Feast* cofounder Tom Grauman, who found her "a magnetic force" before shifting his gaze to Caroline Boss. But interest in Alex ranged far wider. One upperclassman imagined she was "the most beautiful lesbian I ever knew," and Susan Keselenko recalled that "everybody had a crush on Alex."

McNear was unaware of the full degree of interest in her, but she was curious about one fellow sophomore she got to know in the Cooler during spring 1981. To Alex, Barack was "intriguing and interesting and smart and attractive," and they spoke regularly as the school year wound down. Some friends believed the interest was mutual. "I thought he was pursuing her," recalled Margot Mifflin, Occidental's other unforgettable woman. She and the dashing Hasan Chandoo had been a steady couple the entire year, but that did not lessen Margot's own "magnetism," Tom Grauman noted. "It's hard to forget Margot," said Paul Anderson. "She had a way about her that was captivating." Chandoo too was a compelling presence. When one classmate compared Hasan to the singer Freddie Mercury, Margot coolly suggested the handsome actor Omar Sharif instead. But perhaps Tom Grauman's most striking photograph from that spring captures an enchanting Margot addressing a slightly blurred Barack. In contrast, a survey publicized in the student newspaper reported that 32 percent of Oxy women and 17 percent of men admitted that they had never had sex.[17]

Oxy's black students remained politically marginalized despite the almost nonstop efforts of Earl Chew, who by April was trying to get support for a Black Theme House dormitory. Only nine African American students turned out for Ujima's annual photo, and when a five-page Black Student Directory was distributed during winter term, one perplexed undergraduate sent a letter to the student paper that asked, "Why does the Oxy black community need their own directory; don't they know who they are?" Whoever compiled it did not know Obama well, because it spelled his given name as "Barrack."

Oxy president Gilman announced there would be no further discussion of divestment, yet the controversy continued to percolate even years later. In May Earl Chew, along with Becky Rivera, ran successfully for top student government positions, but Chew's close girlfriend from that year later said she had no memory of Barack working with them at all.[18]

Instead Barack and Hasan channeled their energies into a newly created Oxy chapter of the Committee in Solidarity with the People of El Salvador. CISPES had been founded six months earlier to oppose U.S. aid to El Salvador's military government, whose violent death squads had assassinated Roman Catholic archbishop Oscar Romero while he was offering mass in March 1980. CISPES supported El Salvador's left-wing opposition, the Farabundo Martí National Liberation Front (FMLN), whose ties to the Salvadoran Communist Party had led the FBI to investigate CISPES even before Barack and Hasan helped start Oxy's chapter in early March 1981.

One of Barack's favorite professors, Carlos Alan Egan, had prompted the student interest in El Salvador, and on Saturday, April 18, Barack, Hasan, and Paul Anderson drove to MacArthur Park in L.A.'s Westlake neighborhood for a CISPES event led by actors Ed Asner and Mike Farrell. The short march and ensuing rally attracted a crowd of three thousand as well as counterprotesters from Rev. Sun Myung Moon's Unification Church. "We were walking around and absorbing it all," Anderson remembers, and particular enjoyed "exchanging barbs" and "verbally going back and forth" with the hostile Moonies. Scores of banners and signs made for a colorful event. "Long Live the Revolutionary Democratic Front," an FMLN ally, read one; "U.S. Out of El Salvador" demanded another.

With Egan's assistance, Barack, Paul, and Hasan helped organize a mid-May symposium on El Salvador that drew seven prominent speakers to Oxy, including Sister Patricia Krommer, one of the most prominent U.S. supporters of the Salvadoran left. Several academics, a U.S. State Department officer, and a refugee doctor whom Paul would remember as "actually an El Salvadoran revolutionary" were also on the panel. The event drew a crowd of more than three hundred—a larger number than attended the February divestment rally—and the Oxy paper devoted a front-page photo and a lengthy story to the symposium. Paul recalls several right-wing hecklers showing up, and the student organizers "had to have them extracted." Multiple off-campus CISPES events took place during late May and early June, with flyers for all of them distributed on the Oxy campus by student CISPES supporters.[19]

Hasan also organized a Sunday forum featuring former Pakistani Supreme Court justice Ghulam Safdar Shah, who had opposed the military dictatorship's execution of former president Bhutto and been forced from office in late 1980, plus prominent politician Afzal Bangash, who had founded Pakistan's most militant Marxist party in 1968 but was living in exile following Zia-ul-Haq's 1979 military coup. This event too was a success, attracting an audience of more than two hundred, and Barack and Hasan's friend Chris Welton wrote a front-page account of it for Oxy's newspaper.

That same week, Oxy students were angered when they learned that the Political Science faculty had refused to reappoint Lawrence Goldyn, the department's most popular professor. Goldyn denounced the senior faculty's conduct as "completely unethical and certainly unprofessional," but he was also aware of the effect he had had on countless Oxy students. Speaking with Susan Keselenko for a long, two-page profile in the year's final issue of the newspaper, Goldyn said that "frankly I think I've probably had as much or more impact on straight people at this school than I have on gay people."

Goldyn believed he had had "a rather profound effect on a lot of people, but there's no way of measuring that or knowing that," at least not for years to come. Many students have "studied with me, gotten close to me, and really been supportive of and loyal to me. I can't tell you how much respect I have for people like this." Notwithstanding that realization, Goldyn would abandon political science after this and enter medical school, earning his M.D. from Tufts University in 1988 and becoming an internal medicine/HIV specialist in Northern California. Thirty-three years after his dismissal from Oxy, one former student would thank and laud him in one of Washington's most august settings.[20]

Obama's belief in his potential as a serious writer suffered a painful blow when Tom Grauman told him a short story about driving a car had been voted down for publication in *Feast*'s second issue on the grounds that it needed more work. Obama was livid. Caroline Boss appreciated how Barack took pride in "his writing and he was trying to hone that as an actual craft," especially after taking David James's seminar. But to be told by Grauman that his story had been rejected was more than he could bear. "We had a fairly frank argument about it," indeed "a shouting match," Grauman recounted years later. "I just said it wasn't done," and "I didn't want to finish it for him." Obama "was peeved" and "you could see that his feelings were hurt," in part because *Feast* "was a social thing as well as a literary thing," and "he felt he deserved to be a part of this group."

But Obama had more important things to consider. Sometime before Oxy's spring term exams ended on June 9, he and Phil Boerner learned they had each been admitted to Columbia University. Boerner's Oxy GPA was a less than robust 3.25, or a high B, but Columbia College—the men's undergraduate portion of the larger university that also included all-female Barnard College—had received only 450 transfer applications that spring, and accepted sixty-seven. Columbia admissions officials were unhappy about the quality of those applicants as well as the quantity. Dean Arnold Collery believed they had not attracted stronger applicants because "we're not housing them," instead leaving them to fend for themselves in Manhattan's rental apartment market. Assistant Dean of Admissions Robert Boatti also cited the limited financial aid as well as Columbia's policy of requiring all transfers to take core courses that other students had completed as freshmen and sophomores. As a result, a majority of transfer students came from the New York area, many from community colleges. Those who had been accepted had GPAs of about 3.0 and combined SAT scores of 1,100, Boatti said, while entering freshmen averaged well over 1,200.

Boerner and Obama were excited by their acceptances. Eric Moore heard the news and tried unsuccessfully to persuade Barack to stay at Oxy. Sim Heninger, who had seen so much of "Barry" in Haines Annex almost two years earlier, spoke with him in the Cooler one day at the end of spring term and was struck by how "polished and funny" Barack now was. He had changed "very swiftly," Sim thought, from the emotionally fragile youngster who had almost come apart in front of Heninger in October 1979. "He was a different person by the time he left," Heninger said. Kent Goss, who had played basketball with Barry during fall 1979, saw the same thing. "He was different. He was more serious . . . there was a shift somewhere in those two years."

Alex McNear sensed something similar. "I think he had like a broader vision of something. . . . He thought there was going to be a different kind of opportunity" at Columbia. Caroline Boss would remember Obama saying "very directly . . . 'I don't know what it is, but I feel I have a destiny, and I feel I have a purpose. I feel I have something that is being asked of me to do; I just don't know what it is, and I need to . . . go someplace that challenges me, that forces me to focus and that gives me a sense of direction and personal identity, some kind of avenue to ask myself good questions,'" so that "'I will be prepared . . . I need to be prepared so that if the moment comes, I'm ready for it.'"[21]

When spring term exams ended on June 9, Hasan and Barack hosted a graduation party for all their friends and friends of friends before everyone left for the summer. Margot and her roommate Dina Silva, plus Dina's boyfriend Chuck Jensvold, were there. Caroline Boss brought her soon-to-be-ex-boyfriend John Drew, who was spending the summer in D.C. working at Gar Alperovitz's National Center for Economic Alternatives, just two years after its failure to win a reopening of Youngstown's Campbell Works. The next Saturday, Barack and Imad Husain went to Paul Carpenter's family home for his younger brother's birthday party, and Barack told friends he and Hasan were leaving L.A. the next Thursday, June 18.

After stopping in Honolulu to see Stan and Madelyn, they continued westward, with Barack heading to Jakarta to spend a good part of the summer with his mother and sister Maya before meeting up with both Hasan and Wahid Hamid in Karachi, Pakistan. From Pakistan, Obama would fly to France, and then to London, before arriving in New York City for Columbia's fall semester. Carpenter told Phil Boerner that "knowing O'Bama, he will reach the shores of the 'empire state' at the last possible moment if not later."

In Honolulu, Hasan was struck by how "very small" Barack's grandparents' apartment was. Madelyn Dunham was pleased to see her grandson and later told an interviewer "he seemed to have gotten some purpose in life during those two years at Occidental." Barack ran into Keith Kakugawa and his two young sons. Keith remembered that Barack "told me he was leaving Oxy and going to Columbia." From there, Barack headed to Indonesia.

Six months earlier, Ann had left USAID contractor Development Alternatives for a two-year stint as a program officer in the Ford Foundation's regional office in Jakarta. That job came with a comfortable Ford-owned home at Jalan Daksa I/14 in the lush South Jakarta neighborhood of Kebayoran Baru, a considerable step upward from the two houses Barry had lived in with her when she was married to Lolo. The Ford salary gave Ann enough to send ten-year-old Maya to the Jakarta International School, and Ford appears to have paid for Barack's summer 1981 plane fares after Ann requested that her educational travel for dependent children benefits be used for that purpose.

Ann knew her son needed a place to stay when he arrived in Manhattan in late August, and she heard from a fellow expatriate about an apartment for sublet at 142 West 109th Street, just a few blocks south of Columbia's campus. Barack spent most of July living with his mother and sister, and he later wrote that he "spent the summer brooding over a misspent youth." Then he flew to Karachi, where Hasan and Wahid met him at the airport. He stayed for a few days at the Chandoo family's Karachi home, then for a week or more at Hamid's before the trio and several relatives set out on a road trip that took them northeastward to Sindh province's second-largest city, Hyderabad, and then well north to Larkana in interior Sindh. They visited a high school friend of Wahid's whose famous family, the Talpurs, had played a major role in Sindh's earlier history and whose feudal lands were still worked by peasants. Barack met one such older man of African ancestry, and traditional hospitality also led to a partridge hunting outing for the visitors.

Obama then flew westward from Karachi. How much time he spent in either France or London is unrecorded, and in subsequent years, he did not write about his three weeks in Pakistan nor did he often mention the visit when recounting his life experiences. But no later than August 24, Barack Obama arrived in New York City for the first time, having just turned twenty years old—five years younger than his father had been when he first arrived there twenty-two Augusts earlier.[22]

Obama later wrote that he "spent my first night in Manhattan curled up in an alleyway" off Amsterdam Avenue after no one was present at 142 West 109th Street's apartment 3E when he arrived a little after 10:00 P.M. He did not have enough money for a hotel room, and while he did have Hasan's friend Sohale Siddiqi's phone number, Sohale worked nights at a restaurant. After waiting futilely on the building's front stoop for two hours, he "crawled through a fence" to an alleyway. "I found a dry spot, propped my luggage beneath me, and fell asleep. . . . In the morning, I woke up to find a white hen pecking at the garbage near my feet."

In the morning, Barack called Sohale, who told him to take a cab to his building on the Upper East Side. Sohale remembered Obama arriving "totally disheveled," and after breakfast at a nearby coffee shop, Obama crashed on Siddiqi's couch. Within a day or two, Barack gained access to the two-bedroom apartment on 109th, and he immediately invited Phil Boerner, who was still searching for housing, to share it with him: $180 a month per person. Phil arrived on Saturday, August 29, two days before Columbia's orientation, and his top priority was retrieving a bed from family friends. Phil and Emmett Bassett, a sixty-year-old African American medical school professor, got the bed to 109th Street on top of Bassett's station wagon. Phil remembers "Barack and I huffing and puffing up the stairs to get half the bed" up to their third-floor apartment; in contrast, "Emmett just grabbed the mattress with one arm and hauled it up no problem." The next weekend, Phil and Barack spent Labor Day weekend at the Bassetts' country home in the Catskills. Phil said later that Emmett "was the most impressive person I've ever met," but years later Bassett could not recall Obama at all.

That weekend was a welcome respite from both the sorry state of apartment 3E and the strictures Columbia imposed upon junior transfers. On 109th Street, the downstairs buzzer did not work, forcing visitors to shout their arrival from the sidewalk, as Phil had discovered when he first arrived. The apartment door featured at least five locks, perhaps not a bad idea, as the unit next door was vacant and burned out. The four-room apartment was a railway flat: the front door opened into the kitchen, which led first to Barack's bedroom, then Phil's, and lastly their living room. With no interior hallway, there was no privacy whatsoever. The bathroom had a tub, but no shower, and hot water was a rare treat. Hot showers could be taken at Columbia's gym—phys ed was another requirement—and once the weather turned cold, keeping coats on indoors and taking refuge in sleeping bags could partially compensate for the usually stone cold radiators.

Barack had learned to cook one or two chicken dishes from Oxy's Pakistanis, but in stark contrast to daily life back on Glenarm, Phil and Barack hosted parties only when someone from Oxy like Paul Carpenter and his girlfriend Beth Kahn visited them. Sometime in September, Earl Chew and a friend came to stay for several nights, but "spent most of their time at a porn theater," Phil remembered. Chew would not complete his senior year at Oxy and next surfaced eleven years later when he was arrested on a charge of attempted murder outside a Santa Monica reggae club. Jailed for six months before being convicted, Chew pled guilty to misdemeanor assault after his initial conviction was overturned. He was last known to have disappeared from an L.A. halfway house in 2007, three decades after his graduation from Philips Exeter and his arrival at Oxy.[23]

Registration and the start of classes was not any more welcoming for Barack and Phil than their living conditions on 109th Street. Columbia required them to take a full year of both Humanities and Contemporary Civilization plus a semester apiece of art and music. Two semesters of physical education were reduced to one if you passed a swimming test. A foreign language was mandatory; Barack enrolled in an intermediate Spanish class. In all of those courses his classmates were almost entirely freshmen and sophomores. Additional requirements included two natural science courses plus whatever was necessary for a departmental major.

Columbia would accept up to sixty transfer credits from Oxy toward the 124 needed to graduate, and Phil and Barack each met with assistant dean Frank Ayala to determine which of their Oxy courses satisfied Columbia's many requirements. Tuition was $3,350 per semester, and with other fees, the cost of a full year was $8,620, independent of food, lodging, and books. Obama may have received some financial aid in the form of federal Pell Grant assistance, and he may have taken out a modest amount of student loans.

A series of headlines in the *Columbia Spectator,* the excellent student newspaper, told the story of life for the Columbia undergraduate: "Alienation Is Common for Minority Students," "Students Label CU Life Depressing," "Striking Tenants Demand Front-Door Locks." The second of those stories described "the crime and poverty surrounding the school and the immensity of the university bureaucracy." One classmate later said, "Columbia was a very isolating place," and another rued "a culture at the college and in the city that wasn't exactly nurturing." Several years later Obama wrote to Boerner, "I am still amazed when I think of what we put up with there" on 109th Street. Living in Manhattan and going to Columbia was nothing at all like that glowing account the Oxy student newspaper had offered up eight months earlier.[24]

Luckily there were interesting events to attend. One flyer advertised "A Forum on South Africa—Including the Film *The Rising Tide*," with speakers such as David Ndaba, the nom de guerre for Dr. Sam Gulabe, the African National Congress's representative to the United Nations. Boerner said he and Barack attended that forum, but not a mid-November speech by black Georgia state legislator Julian Bond. The two apartment mates often ate breakfast at Tom's Restaurant on Broadway at 112th Street, and they sometimes had dinner at nearby Empire Szechuan. Drugs were no part of the depressing scene at 142 W. 109th, but they would go drink beer at the venerable West End bar on Broadway or with Phil's cousin Peregrine "Pern" Beckman, a Columbia sophomore, at his nearby apartment. To Pern, Phil's friend seemed "diffident," a "shy kid who spoke when spoken to" while "nursing a beer."

Looming over their daily lives was how unlivable their dire apartment was, especially as winter closed in. They could let their sublease expire on December 7 and simply remain until fall exams concluded on December 23; given New York's arcane housing code, they could receive no punishment for that brief time. They began searching without success for a place to live in January. Phil had family friends in Brooklyn Heights where he could stay, but they still had no solution for both of them when Phil left to spend Christmas with his parents at their home in London.

Columbia's winter break extended from Christmas Eve until spring registration on January 20. Obama spent most of that time in Los Angeles, seeing Wahid Hamid and Paul Carpenter and encountering old friends like Alex McNear when he visited Occidental, whose winter term started two weeks before Columbia's spring semester. Obama returned to New York on January 15 and slept on the floor of a friend named Ron on the Upper East Side while he again searched for an apartment. At registration, Barack ran into Pern, who gave him Phil's number in Brooklyn, and when Barack called two nights later, he told Phil he had found only a $250-per-month one-person studio just south of 106th Street that he could soon move into as a sublet.

On Sunday afternoon, January 24, the day before Columbia's spring classes began, Barack and Ron went to visit Phil where he was staying, at 11 Cranberry Street in Brooklyn Heights. They drank beer and ate bagels while watching the San Francisco 49ers defeat the Cincinnati Bengals 26–21 in Super Bowl XVI. On Monday Barack was again saddled with Columbia's core courses and another semester of Spanish. That Friday night, Phil met up with Barack at Columbia; the two then headed downtown on the subway to rendezvous with

Ron and his girlfriend at a Lower East Side bar before having dinner at the well-known Odessa restaurant on Avenue A, just across from Tompkins Square Park. Then Barack and Phil headed back to Morningside Heights, and Boerner spent the night on Obama's floor rather than take the subway back to Brooklyn after midnight. The next Friday Barack and Phil ate an early Chinese dinner before taking the subway to Phil's place and polishing off a bottle of wine while watching the New Jersey Nets play the Philadelphia 76ers. Barack still occasionally played pickup basketball in Columbia's Dodge gym; a female graduate student years later remembered him playing pickup soccer on the lawn in front of Columbia's imposing Butler Library.[25]

But Barack's life during those early months of 1982 was radically different from his daily routine one year earlier. At Oxy, living with Hasan was an almost nonstop party with a band of close friends. Rallies, protests, and political events occurred almost weekly, and the days were filled with energetic debates and conversations in the Cooler. Now, in dreary Morningside Heights, Obama faced a daily schedule of core classes and perhaps a once-a-week meet-up with Phil to have dinner, drink beer, or watch a game. A quarter century later, Obama remembered that time as just "an intense period of study. . . . I spent a lot of time in the library. I didn't socialize that much. I was like a monk." He started keeping a journal, less a diary than a descriptive collection of city scenes and characters that caught his eye. In retrospect, he believed it was an "extremely important" period "when I grew as much as I have ever grown intellectually. But it was a very internal growth," one that left him "painfully alone and really not focused on anything, except maybe thinking a lot." He had been "comfortable in my solitude, the safest place I knew."

"It was a pretty grim and humorless time that I went through," Obama remembered, an "ascetic" and "hermetic existence . . . I literally went to class, came home, read books, took long walks," and wrote in his journal. "It's hard to say what exactly prompted" such a stark change from his attitude toward life both at Oxy and back in Hawaii, Obama told the journalist David Remnick. Yet he gained "a seriousness of purpose that I had lacked before."

Spring exams ended by the middle of May, and sometime soon after that, Obama lost both his studio apartment and his security deposit when the actual leaseholder informed him that his sublet was invalid. Thus sometime in early or mid-summer 1982, Barack moved in with Sohale Siddiqi in a two-bedroom apartment at 339 East 94th Street, just west of First Avenue. Siddiqi managed to score the lease on the $450-a-month sixth-floor walk-up by exaggerating his own income, and though Barack was now living on Manhattan's Upper East Side, their immediate neighborhood was anything but fashionable. Sohale recalled that it was "a scary street," with a corner gas station "patrolled by this Doberman Pinscher with a beer bottle in his mouth." Their own building was "a hovel . . . the hallways were dingy. Everything was beat up and gray and dimly lit. The front door didn't lock completely." Up in 6A, "you would enter through a kitchen, which would lead into a living room on one side and a bathroom on the other." The wooden floors "were all warped . . . with big gaps between the planks." One bedroom "was really a closet with a window." A friendly footrace was used to determine who got the decent bedroom, and Obama, now a regular runner, won. As on 109th Street, "there was never hot water when you wanted it," but in contrast the heat was always on, "so we used to have our windows wide open, just to cool down." Outside Barack's bedroom window was a fire escape, which Sohale said served as "our balcony." All in all, it was just "a horrid place."

Siddiqi also witnessed a "transformation" from the "fun-loving . . . easygoing" Obama he had met eighteen months earlier in South Pasadena to someone who was now "very serious and less lighthearted." Yet Barack had a stereo and "a huge record collection. Bob Marley was big, Stevie Wonder . . . also plenty of Bob Dylan, Van Morrison, Talking Heads, and this group who I had never heard of before . . . he had at least twelve of their albums," the Ohio Players. But once Barack arrived on East 94th Street, "I don't think he bought any more albums . . . or even played the stereo much." He did have "to remind me a few times to call him Barack" rather than Barry.

Sometime soon after Barack moved in with Sohale, his mother Ann and almost twelve-year-old sister Maya arrived in New York from Jakarta, where Ann still worked for the Ford Foundation. They too were struck by the change in the young man they called "Bar." Maya later recalled that "he seemed more serious. He seemed more pensive. He was reading a great deal." She believed "he had started taking himself very seriously" and often appeared to have "wrapped himself in his own solitude." Obama later wrote that he took a summer job "clearing a construction site on the Upper West Side," and he also described, in a patronizing manner, his mother's insistence that they see *Black Orpheus,* a film she had loved as a high school senior twenty years earlier that was playing at a revival theater. Obama told a subsequent interviewer that during that visit "my mother used to tease me and call me Gandhi" because of his newly ascetic life, but Barack did not deny that he had become "deadly serious during those late college years . . . People would invite me to parties, and I'd say, 'What are you talking about? We've got a revolution that has to take place.'"

Andy Roth, who had lived in Oakland after leaving Eagle Rock, moved to Manhattan in the late spring of 1982 and got an apartment on East 95th Street, hardly two blocks from Barack and Sohale. Siddiqi was nowhere near as political as Chandoo, nor as intellectually curious as the new Obama, but when Andy went for dinner at their apartment with several other Pakistanis, he remembered seeing "a portrait of Bhutto on the wall."

The most significant new acquaintance that year was Mir Mahboob Mahmood, known to friends like Sohale, Wahid, and Hasan as "Beenu." A 1981 graduate of Princeton University, Beenu was working as a paralegal in Manhattan while preparing to attend law school. At Princeton, Beenu had written a one-hundred-plus-page senior thesis on Mohandas Gandhi, and he was hugely influenced by the teaching of political theorist Sheldon Wolin. As Beenu remembered it, he and Barack began an on-and-off program of reading and discussing a half a dozen or so significant books, interrupted partly by Beenu's one year of graduate study in political science at Johns Hopkins University. Beenu reconstructed their reading list as beginning with John Locke's *Two Treatises of Government* (1690), then Albert O. Hirschman's *The Passions and the Interests: Political Arguments for Capitalism Before Its Triumph* (1977), followed by Barrington Moore Jr.'s famous *The Social Origins of Dictatorship and Democracy: Lord and Peasant in the Making of the Modern World* (1966) and E. H. Carr's *The Twenty Years' Crisis, 1919–1939: An Introduction to the Study of International Relations* (1939).[26]

At Oxy in 1980–81, Andy Roth had been one of the many men interested in Alex McNear. When Alex returned to her mother's apartment at 21 East 90th Street in June 1982 to spend the summer taking a theater course at NYU and interning at a publishing company, they got in touch, and whether from Andy or another Oxy friend, Alex got Barack's phone number and called him. Sometime in midsummer, they had dinner at an Italian restaurant on Lexington Avenue, and although that evening remained chaste, according to Alex, they "started to see a lot of each other" that summer, becoming "really very close."

Alex's parents had divorced when she was four years old. Her father had remained in Chicago, and her mother, originally from Wisconsin, got a graduate degree from UW–Madison before moving to New York. Her comfortable apartment, with its library of "thousands of books" in a tall building just off Madison Avenue, was a world away from Barack and Sohale's dismal quarters only eight blocks to the northeast. Alex remembers that apartment as "sparsely decorated" with "very little furniture" and recalls "opening the refrigerator and seeing like virtually nothing in there."

Understandably Alex and Barack "didn't spend a lot of time there," and apart from one visit with Phil and a party that also included Andy, they socialized only with each other. After going out to dinner, a museum, or a Broadway play like Athol Fugard's *"Master Harold" . . . and the Boys,* about South African apartheid, "he'd come over, and we'd stay at my apartment." Barack "did not seem like someone who was at all experienced" or indeed "terribly driven," but his "shyness" and "lack of experience" matched the novelty of their intimacy for her too.[27]

Barack's senior-year classes at Columbia began on September 8. One entering freshman remembered Obama as a fellow student in his yearlong Contemporary Civilization core course, and a junior political science major recalled chatting with Barack a number of times in a hallway in the School of International and Public Affairs (SIPA) building before one or another political science class. Barack, Phil, and Phil's cousin Pern all enrolled in C3207, Modern Fiction, taught by Edward Said, a Palestinian American literary scholar whom Columbia's student newspaper months earlier had called a "bitter" critic of Israel at "a largely Jewish university." Said would "come in and ramble on for an hour about who knows what," Pern remembered. The reading list for the twice-a-week, one-hundred-student class included Joseph Conrad, Franz Kafka, and William Faulkner. Pern thought Said could be "brilliant," and "when he was focused, he was incredible." But preparing for class was clearly not one of Said's priorities, and Phil and Barack did not share Pern's opinion. "We did not think highly of Said," Phil recounted. "We thought his class was pretty worthless."

Having settled on political science as his major, Barack had to take a two-semester senior colloquium and seminar in one of the department's four subfields. Barack's course credits fit best with international relations, and he and seven fellow seniors ended up in W3811x and y, taught by Michael L. Baron. A young instructor who had completed his Columbia Ph.D. dissertation on U.S. policy toward China after World War II two years earlier, Baron was returning to Columbia after a year of teaching in Beijing. Baron's dissertation had argued that President Harry S. Truman "was an activist in foreign policy, basing decisions primarily on personal proclivities," and this course focused on U.S. foreign policy decision-making rather than a particular topical area.

The assigned readings during the fall semester analyzed how past important decisions had been made: the 1961 Bay of Pigs invasion of Cuba, the fall 1962 Cuban missile crisis, and the escalation of the Vietnam War. Readings included works by Joseph Nye and Ernest May, as well as Irving Janis's famous 1972 *Victims of Groupthink: A Psychological Study of Foreign-Policy Decisions and Fiascoes*. Baron recalled that "we definitely focused on groupthink." Throughout the fall, their focus was historical and practical, not theoretical: "What do you need to do to make a good decision? Which presidents were making good decisions," first and foremost by "taking advice from people outside the inner circle?"

Baron as well as two other students in the small class recall Obama as a standout performer. "He was a very bright student," Baron said years later, "clearly one of the top one or two students in the class." One classmate remembered Barack as "a very, very active participant" in class discussions, and another who "didn't think the seminar was that great" nonetheless was "impressed with Barack. . . . There was a maturity about what he said and how he said it" that surpassed the comments offered by most if not all of his fellow seniors.

One way of satisfying half of Columbia's science requirement was Elementary Physics, C1001x, taught by Gerald Feinberg, a professor in his late forties who had taught at Columbia for more than twenty years. The catalog described the course as "an introduction to physics for students with no previous background in physics" and stated that "very little mathematics is used." Indeed, Feinberg's popular 1977 book, *What Is the World Made Of? Atoms, Leptons, Quarks, and Other Tantalizing Particles,* which Feinberg assigned that fall, declared, "I am convinced that a substantial comprehension of modern physics can be obtained without advanced mathematical training." That book, like Feinberg's syllabus, focused on "the study of atoms and of their subatomic constituents." It explained quantum theory, the special theory of relativity, and especially particle physics, "the main feature of the physics of the last twenty-five years." The course was essentially "a history of physics," Feinberg's son Jeremy recalled from personal experience, and his dad "called it physics for poets."[28]

In mid-September, Alex McNear left Manhattan to return to Oxy for the fall quarter, and on September 26, Barack sent her a long letter reporting on his first two weeks of classes after receiving a note from her. "I sit in the campus cafe drinking V-8 juice and listening to a badly scratched opera being broadcast. I am taking a break from studying a theoretical analysis of strategic deterrence in the international arena, muddling through concepts like first strike, mutual assured destruction, nuclear payload, and other such elaborated madness. So forgive the dryness and confusion that have undoubtedly rubbed off."

Describing himself as trying to stretch "across as many disciplines as possible," Obama sounded as if he had sat in on a number of courses while considering which ones to take. "My favorite so far is a physics course for non-mathematicians that I'm taking to fulfill the science requirement. We study electrons, neutrons, quarks, electromagnetic fields and other tantalizing phenomenon under the auspice of a Professor Fienberg [sic]. He embodies every stereotype of a science professor from the bow-tie to the sparse, balding hair combed back and monotonous Midwestern twang. Behind the thick glasses that pinch his nose, one can sense the passion he brings to the topic and the quiet, unobtrusive cockiness you find in scientists, certain that no one knows any more than they do."

Obama's descriptive sketch of Feinberg confirms the impression that Alex and Phil each had of Barack in 1982: that he wanted very much to become a writer. Presented with that depiction of his father years later, Jeremy Feinberg was impressed: "the physical description . . . is spot-on" and even Obama's characterization of his voice was accurate. In Barack's letter, he then cited what he called "the frustrations of studying men and their frequently dingy institutions" while adding that "the fact that of course the knowledge I absorb in the class facilitates nuclear war prevents a real clean break."

Then he shifted gears. "A steady flow of visitors camped in our living room for two weeks or so, and I had a chance to catch up with friends and play host." Hasan and his cousin Ahmed were among them, and Barack told Alex that "Hasan will be marrying" soon and taking over the family business. With "old friends receding into the structures pre-conscribed for them," relationships became "more reserved" and "the idealistic chatter of college is diplomatically ignored."

Barack had also heard from Greg Orme in Oregon, who had a new car, a television, and a hot tub. "I must admit large dollops of envy for both groups, my American friends consuming their life in the comfortable mainstream, foreign friends in the international business world. Caught without a class, a structure, or tradition to support me, in a sense the choice to take a different path is made for me. The only way to assuage my feelings of isolation are to absorb all the traditions, classes, make them mine, me theirs. Taken separately, they're unacceptable and untenable."

Obama's statement that he felt isolated was unsurprising given his life over the previous twelve months, nor were his remarks about Hasan and Greg. He did not respond to what Alex had written to him about herself, "since I spent so much of my mental energy with you and now need to refuel" and return to Butler Library. "I trust you know that I miss you, that my concern for you is as wide as the air, my confidence in you as deep as the sea, my love rich and plentiful. Please comfort me with another letter when you get a chance. My regards to everyone. Love, Barack."

A postscript said he was enclosing a *New York Times* book review from two weeks earlier of *Becoming a Heroine* by Rachel Brownstein, which he thought would interest her, as well as an excerpt from W. B. Yeats's 1928 poem *The Tower* about "a woman won or lost." He asked Alex, "Who is the 'woman' for you?"[29]

Back in Butler Library, Barack studied for Gerald Feinberg's first physics exam on October 12. Students had to answer two out of three questions: "(1) Discuss the present atomic theory of the structure of ordinary matter . . . (2) Describe the photon theory of light . . . (3) Discuss how some of the estimates of the number of various subatomic particles . . . are obtained. Your discussion need not be precise on numbers. . . ."

In early November, Columbia's Coalition for a Free South Africa hosted the prominent white anti-apartheid activist Donald Woods, and Obama may have been among the large crowd. Two weeks later Columbia's student newspaper published another front-page article reporting how "this is not an easy place to go through as a minority student." One black senior complained that white students presume the admissions standards for black undergraduates were lower and that "we can't do the work"; another commented that Columbia's core curriculum ignored African Americans and "doesn't prepare you for the real world."

Alex McNear wrote to Barack in mid-November, and he quickly replied. "You speak with force, Alex, calm and confident, and I'm frankly amazed, not by the brimming talent, not by the thoughts in themselves but by the sureness of the words." Admitting that "mixed in with those feelings are bits and pieces of envy, uncertainty, some intimidation," he confessed that "the prejudiced, frightened male makes its atavistic appearance." Writing cryptically about what he termed "the habits and grooves of separate existence," he declared that "we will talk long and deep, Alex, and see what we can make of all this."

Rereading his letter a quarter century later, McNear noted Obama's "formalism," and wondered "were mine very formal" as well? "I just wonder where the tone came from." Barack also wrote, "I think you may have caught me being a fool in my last letter," and he referred to a conversation they had had outside the Metropolitan Museum of Art. "When we spoke in front of the Met, I insisted that I made choices, that I wasn't kind out of necessity, because it can certainly be argued that I'm compelled by my past to be that way, that no choice is involved . . . and my insistence arises from my fear of emasculation, that if I can't be cruel any longer, then I must not be a man." Citing "the past of my ancestors," Obama wrote, "I see that in a real sense my gravestone is already planted, the feeble eulogic etchings overgrown with moss, blurred and forgotten."

Barack went on that "this cordoning off of individuals into compartments is something I fight every day," and said that Virginia Woolf's 1925 novel *Mrs. Dalloway* expressed his point better than he could. After some musings about birth and death, Barack asserted that "the betrayal lies in separation," that "we feel betrayed by this act of separation," and that "because the initial act of separation has traditionally been from the mother, men's retaliation is indeed towards women."

Alex understood that "he's clearly enjoying writing: writing it out, toying it out, figuring it out." Barack continued that "some choose to escape the pain by limiting their interaction with the world. They abstract themselves . . . which is perhaps the most tolerable option." He quoted Friedrich Nietzsche's statement that "in all desire to know there is a drop of cruelty" before acknowledging that "I'm running out of steam and the thoughts are becoming blurred." Yet he had two points for subsequent discussion: one, "I see that I have been made a man, and physically, in life, I choose to accept that contingency." Second, "there is a reason why western man has been able to subjugate women and the dark races,

Alex; the ideology they present is backed by a very real power. We will speak of this too."

Rereading that passage, McNear observed how "very professorial" Obama sounded. Barack then added, "I've never tried to put down so comprehensively my views. They come out muddled and incomplete, and I wish I could explain more fully." To that, Alex would wonder "maybe . . . that was as close as he could get to intimacy."

Finally Obama concluded by returning to the real world a month hence, after Columbia's exams ended. "I arrive in L.A. on December 23rd, and expect to be at Wahid's apartment that evening. I'll call you upon arrival . . . see you then. Love, Barack."[30]

Three or four days after Barack mailed that letter to Alex, Sohale answered their phone on East 94th Street and an unfamiliar, foreign-sounding woman asked for Barack. It was Kezia Obama's sister Silpa Obonyo, known throughout the family as Aunt Jane, a Nairobi telephone operator able to make international calls. She had just phoned Auma Obama, now a twenty-two-year-old student in Germany and no longer using "Rita," and her reason for calling Barack was the same as for dialing Heidelberg: Barack Obama Sr. was dead at age forty-eight.

He had died in the early-morning hours of Wednesday, November 24, when the vehicle he was driving had gone off Elgon Road and struck a large tree stump. Drunken driving had indeed finally killed him. "The body had to be wedged out of the car," the *Nairobi Times* reported. The crash site was in Upper Hill, the Nairobi neighborhood where Obama had been living with twenty-two-year-old Jael Atieno, who six months earlier had given birth to George Hussein Obama.

Barack later wrote, "I felt no pain, only the vague sense of an opportunity lost," and he called his mother in Jakarta, then his uncle Omar Onyango in Massachusetts. A year earlier Barack and his father had corresponded about his visiting Kenya to meet his relatives there once he graduated from Columbia, but now that idea was on hold.

The Nairobi paper said Obama Sr. "leaves four wives and several children," and the eleven years that had passed since he last saw his second-eldest son in Honolulu in 1971 had been no happier than the seven that had preceded it. Soon after Obama returned to Nairobi in early 1972, Ruth divorced him, but she did not take her two sons and actually leave him until later that year after Obama put a knife to her neck and struck the youngest boy, David Opiyo. Unemployed and still drinking heavily, Obama was able to secure a job offer from the World Bank for a post in the Ivory Coast, but when his sister Zeituni took him to the Nairobi airport, government officials turned him away and canceled his passport. "He left the airport crying, angry, and frustrated," Zeituni recalled. Some months later, in mid-1973, yet another drunken car crash left him with two badly broken legs and a shattered kneecap. Obama remained hospitalized for six months, sometimes "in real pain," Zeituni remembered. His teenage daughter Auma was repeatedly "sent home from school" due to unpaid fees and bounced checks. She and her older brother Roy, alone without a mother or stepmother after Ruth left with Mark and David, relied upon Zeituni for their survival. She "regularly brought us something to eat," Auma remembered.

Not long after Obama left the hospital, his old Honolulu friend Andy "Pake" Zane and his partner Jane visited Nairobi and were astonished at Obama's condition. "He was a broken spirit," just "a shell" of the man Zane and Neil Abercrombie had visited in 1968. "Why do you have a limp?" Zane asked. Obama replied, "They tried to kill me," asserting that the conspirators behind Tom Mboya's 1969 assassination tried to rub out a possible witness. "He was very depressed," Zane remembered, and "he got very drunk and very angry" each night they saw him. By then he had been evicted from the council house at 16A Woodley Estate for nonpayment of rent and had a roof over his head only thanks to his friend Sebastian Peter Okoda, who allowed Obama to join him in his flat at Dolphin Court for more than a year.

On November 25, 1975, Obama's father, Hussein Onyango Obama, died at age eighty. Barack Sr. attended the funeral accompanied by his then-girlfriend, Akinyi Nyaugenya. Sometime soon after, Obama's old friend Mwai Kibaki, now Kenya's minister of finance, hired him into a post there. His drinking was as heavy as ever, and colleagues often saw him headed to a bar before noontime. "I can't stand it anymore. Let's get a drink." Once Obama received a large cash travel advance a day before his scheduled departure on a business trip, but before the night was out, he had spent the entire sum treating colleagues, including Okoda, to rounds of drinks at a fancy Nairobi bar.

One Kenyan academic who had first met Barack Sr. in the U.S. found his deterioration sad. "Before, he was everyone's role model. With that big beautiful voice, we all wanted to be like him. Later, everybody was asking what happened." A younger female government colleague knew that Obama was "disillusioned and discouraged and depressed" but mentioned to him her desire to get a Ph.D. She was astonished at his response: "It's useless doing a Ph.D. What do you want a Ph.D. for? It's just academic." Obama "was very, very strong about that: 'Don't do a Ph.D.'"

By mid-1980, the forty-six-year-old Obama was living with Jael Atieno, a friend of his younger sister Marsat who was the same age as his daughter Auma. Obama had visited Auma in Europe two years before his death, but "I didn't want to see him," she recounted. He "appeared broken" and "seemed defeated." Auma felt "betrayed by my father, blaming him for not holding the family together." When Aunt Jane telephoned her in November 1982, "I scarcely felt anything for him."

Ruth Baker Obama Ndesandjo had remarried to a stable and reliable man, Simeon Ndesandjo, and in April 1980, she had opened a preschool, the Madari Kindergarten. She learned of Obama's death from that November 30 newspaper story. "I was not surprised, because he had been heading that way for a long time," Ruth said. She told her son Mark more than once that his father had been "a brilliant man, but a social failure."

Zeituni Onyango, the person to whom Barack Obama Sr. was closest and who loved him most profoundly, understood more deeply than anyone the immense tragedy and lost promise of his life. Despite Barack's remarkable intelligence, he had ended up living a "miserable life" that was "ruined by alcohol." Yet even in his last days, one thing had never changed since his unwilling 1964 return to Nairobi without the Ph.D. from Harvard that had been his dream. Even though almost none of his friends ever heard him mention a son in the U.S., "Barack's picture was always next to his bed," according to Zeituni.[31]

In later years, the younger Barack's comments about his father's death would vary considerably. Sometimes he incorrectly said he began using the name Barack instead of Barry at that time, rather than almost two years earlier. On other occasions, he mused that he had been motivated by his father's death. "I think it's at that point where I got disciplined, and I got serious," but the few people who knew Barack well during his first fifteen months in Manhattan—Phil, Sohale, and Alex—had all witnessed that transformation take hold many months earlier.

Obama also later wrote that "my fierce ambitions might have been fueled by my father . . . by my resentments and anger toward him," but in 1982 those who knew him well did not see evidence of any fierce ambitions. Obama's belief that in some ways he had raised himself would not be questioned by anyone from Punahou or Occidental who had noticed the absence of his birth parents in his life, but his most acute and accurate comments about the father he saw for only a few weeks when he was ten years old acknowledged how "I didn't know him well enough to be angry at him as a father. Mostly I feel a certain sadness for him, and the way that his life ended up unfulfilled, despite his enormous talents." In time he realized that "I was probably lucky not to have been living in his house as I was growing up."[32]

After Aunt Jane's telephone call, Barack said little about his father's death. "I had no clue," Sohale Siddiqi recalls. "Not a word from" Barack referenced it. "I never heard about his father from him. I would hear about" Stanley Dunham in particular. "He brought up his grandparents plenty," and Sohale remembers an off-color note Stan had sent his grandson along with a jogger's wristband, suggesting there were multiple appendages upon which he might wear it.

Sometime in early December, Keith Kakugawa was in New York for several days and managed to meet up with Barack for lunch near Columbia and then one night for dinner. Obama told Keith that "his dad had just died in a car wreck." When Barack arrived in Los Angeles just before Christmas and began seeing Alex McNear almost daily, Alex too remembers him telling her about his father's death, but, she says, "it was not an emotional telling." In a letter to Phil Boerner, Obama described his three weeks of winter vacation in Los Angeles. Other than when he was with Alex, Barack stayed with Wahid. He told Phil that his days there were "standard fare—good relaxation with the Paki crowd, dinners with Alex, lunches with Jensvold, tennis with Imad, and pipe tokes with Carpenter. They all seem to be doing well enough and have themselves set up in big cushy apartments. The whole process was like a spiral back in time; nothing had changed except my perceptions, it seemed." Barack's close friend Mike Ramos had moved from Honolulu to Orange County six months earlier, and he remembers driving up to Pasadena and eating curry with Barack and the Pakistani trio of Wahid, Imad, and Asad.

Other than this unrevealing reference in his letter to Phil, the extent of Barack and Alex's relationship was kept entirely private, with all of their time together involving just the two of them and not any of Barack's other friends. "No one really knew that we were having a relationship," Alex later explained. Barack "seemed to be incapable of bringing a relationship into the rest of his world," and their quiet dinners often featured long conversations about Alex's immersion in French literary theory and especially its focus on the concept of difference. "I'm really much more interested in how people are similar," Alex remembers Barack saying in response to one such discussion. Alex recalls writing in her diary at that time, "I realized how much I loved him, and that he was like my closest friend," but she also expressed doubts that their relationship could blossom. Barack seemed "very controlling" and "so self-conscious." On January 17, they had a final dinner before Barack flew back to New York City for his final semester as an undergraduate.[33]

Barack was still taking Spanish, and the spring semester of his yearlong political science senior seminar would have fewer class meetings in lieu of students writing a paper due at the end of the term after multiple one-on-one consultations with Michael Baron. But in addition, Barack was able to choose three upper-level elective classes. One was a seminar taught by a young English professor whom Pern Beckman had recommended as a "really cool guy." Lennard "Len" Davis's first book, *Factual Fictions: The Origins of the English Novel,* a revision of his 1976 Columbia Ph.D. dissertation under Edward Said, was just being published by Columbia University Press. The Columbia catalog said

Davis's seminar Ideology and the Novel would examine "the nature of ideology in Marxist and sociological thought." Theoretical readings included works by Raymond Williams and Louis Althusser as well as Karl Marx's *Economic and Philosophical Manuscripts of 1844*. Novels to be read started with Daniel Defoe's *Robinson Crusoe* (1719) and moved on to Elizabeth Gaskell's *North and South* (1855) and Charles Dickens's *Hard Times* (1854). Only about twelve students enrolled in the seminar, and years later Davis readily volunteered that "it was definitely a Marxist course." His purpose, Davis explained, was to make students realize that while any individual novelist "felt free to improvise and create," nonetheless the surrounding "culture and its ideology would ultimately determine the novelist's innovations."

Davis devoted a chapter of his new book to analyzing *Robinson Crusoe,* and in teaching Defoe's novel, he asked students what "overt ideological statements" they could detect in *Robinson Crusoe.* An erudite and well-spoken teacher, Davis eventually answered his own question by saying "it's a philosophy of stasis . . . you reconcile yourself to the world as it is . . . in a giant way it is against change . . . it's essentially an ideology that reconciles you to middle-class life." That was one illustration of the overarching argument Davis offered in his book: in those eighteenth- and nineteenth-century English novels, "the novel's fictionality is a ploy to mask the genuine ideological, reportorial, commentative function of the novel." At bottom "the inherent confusion in any factual fiction" would give rise to "the later nineteenth-century assumption that literature is a more penetrating depiction of life than life." Four years later, in his second book, *Resisting Novels: Ideology and Fiction,* Davis singled out the "particularly wonderful" 1983 seminar on Ideology and the Novel in helping him advance and distill his analysis.[34]

A second course Obama took that spring was International Monetary Theory and Policy, taught by Maurice Obstfeld, a young associate professor of economics who had received his Ph.D. at MIT four years earlier. Barack had already taken Intermediate Macroeconomics, and Obstfeld's class focused on "the evolution of the world monetary system since 1945" and "monetary problems in international trade." The third upper-level elective he chose was Sociology W3229y, State Socialist Societies, taught by Andrew G. Walder, a brand-new assistant professor in his first year of teaching who had earned his Ph.D. at the University of Michigan in 1981. Walder's syllabus explained that the course would be "an analysis of state socialism as a system. The primary focus will be on the central features of these societies that distinguish them from others," i.e., "the core features of state socialism," and "we will spend most of the term comparing China and the Soviet Union."

Walder recommended that students purchase seven books from which most of the required reading would be drawn, and recommended half a dozen others.

The semester began with "The Origins of Russian and Chinese Communism," and students read Alex Nove's *Stalinism and After,* a survey of Soviet political history that focused on how deadly Soviet rule had proven for the regime's many victims, including the early Bolsheviks. The second major topic was "The Communist Party as an Organization," and over several weeks students read Elizabeth Pond's *From the Yaroslavsky Station: Russia Perceived,* a journalistic work structured around the author's rich and lengthy Trans-Siberian Railway journey.

Prior to Walder's seven-to-ten-page typed, take-home midterm exam, the class covered "The Communist Party as a Status Group" and "Organized Surveillance and Repression." Students read Hedrick Smith's well-known *The Russians* as well as Jonathan Unger's new *Education Under Mao.* Fox Butterfield's *China: Alive in the Bitter Sea,* Miklos Haraszti's *A Worker in a Workers' State,* and Roy Medvedev's *Let History Judge* were three other titles on the syllabus. The second half of the semester covered topics such as "Bureaucracy, Office-Holding, and Corruption" as well as "Social Stratification and Inequality." Everyone read David Lane's *The End of Social Inequality? Class, Status and Power Under State Socialism,* which argues that "inequality is a characteristic of state-socialist society as it is of the capitalist."

The final assigned reading prior to Walder's ten-to-twelve-page typed, take-home final exam was Milovan Djilas's 1957 classic *The New Class.* Smuggled out of Yugoslavia and published in the U.S. while the author languished in a political prison, the book's publication was "an immediate sensation," according to the *New York Times.* Djilas in the late 1940s had served as Yugoslavian ruler Josip Broz Tito's personal intermediary to Joseph Stalin, but he was expelled from Yugoslavia's Communist Party after expressing heretical thoughts. His best-selling book gave voice to those insights. Once a Communist Party "has consolidated its power, party membership means that one belongs to a privileged class. And at the core of the party are the all-powerful exploiters and masters." Experiencing that in Yugoslavia had transformed Djilas into a democratic socialist. "To the extent that one class, party, or leader stifles criticism completely, or holds absolute power, it or he inevitably falls into an unrealistic, egotistical, and pretentious judgment of reality."

The New Class was a powerful way to end a semester, or indeed to complete one's undergraduate education. As one student said with some understatement years later, "Walder was anything but a Marxist," and Walder himself drily observed that what his syllabus presented "was not a flattering portrayal of political and social life in these now-thankfully defunct systems." Only an extreme control freak would withhold his academic transcripts from public view simply so as to avoid any public discussion of what possible ideological influence either Walder's impressive reading list or Len Davis's teaching about the political uses of fiction might have had upon an intellectually hungry twenty-one-year-old mind.[35]

Several weeks into that spring semester, Obama wrote another lengthy letter to Alex McNear in Eagle Rock. "I run every other day up at the small indoor track" at Columbia, a bit of news that Barack then spun into a lengthy descriptive portrait of the act of running. "After getting clean, I go to the Greek coffee shop . . . and have the best bran muffin in New York City—dark and fibrous . . . and coffee and a glass of water. I light a cigarette, make some talk with the Ethiopian cashier with big murky eyes and the sly smile and a small tattoo on her right hand in foreign code."

Rereading this decades later, McNear wondered whether her letters to Barack "were just as pretentious . . . equally as convolutedly long and laborious." Barack wrote, "I enjoyed your letter. I like the way you use words," and then proceeded to a long disquisition on the concept of resistance against a "bankrupt" and "distorted system. . . . But people are busy keeping mouths fed and surroundings intact, and it is left to the obsessed ones like us to make the alternatives more tangible . . . so that resistance and destruction arrive in the form of creation." He then paused to say "excuse the sermons," but "I also have thought about us and conclude that I like what we have. . . . Perhaps what I've been after is a correspondence, a union, yes—but never exclusive, cocoon-like, ingrown. Rather something outwards, a point of extension."

With graduation approaching, he had more mundane issues to discuss. "I've been sending out letters to development and social services agencies, as well as a few publications, so I should find out in the next few months what I have to work with next year. Classes are the average fare. A class called Novel and Ideology has an interesting reading list covering several of the things we spoke of" when Barack was in Los Angeles. "Of what I've read so far I recommend *Marxism and Literature* by Raymond Williams. A lot of it is simplification, but it generally has a pretty good aim at some Marxist applications of cultural study. Anyway, it might be a good point of departure for further haranguing between us."

Obama then ends the letter, but the next day he added a long postscript that began by mentioning homeless panhandlers he often saw. "I play with words and work pretty patterns in my head, but the hole is dark and deep below, immeasurably deep. Know that it's always there, rats nibbling at the foundations, and it can set me to tremble." Then he thought to respond to something Alex had said about T. S. Eliot's 1922 poem *The Waste Land* and told her to read Eliot's 1919 essay "Tradition and the Individual Talent" as well as his *Four Quartets*. "Remember how I said there's a certain brand of conservatism I respect more than bourgeois liberalism." Barack finally concluded by saying, "I can't mobilize my thoughts right now, so . . . I leave you to piece together this jumble."[36]

The label "jumble" could also apply to an article Barack submitted to *Sundial,* a weekly Columbia student newsmagazine. Just prior to its publication, he wrote to Phil Boerner in Arkansas and mentioned he had "been sending out some letters of inquiry to some social service organizations and will also be making up a resume (no comment) soon. I've also written an article for the *Sundial* purely for calculated reasons of beefing up the thing. No keeping your hands clean, eh."

The article, titled "Breaking the War Mentality," began by asserting that "The more sensitive among us struggle to extrapolate experiences of war from our everyday experience, discussing the latest mortality statistics from Guatemala, sensitizing ourselves to our parents' wartime memories, or incorporating into our frameworks of reality as depicted by a Mailer or a Coppola. But the taste of war—the sounds and chill, the dead bodies, are remote and far removed. We know that wars have occurred, will occur, are occurring, but bringing such experiences down into our hearts, and taking continual, tangible steps to prevent war, becomes a difficult task."

Following that introduction, Obama cited what he called "the growing threat of war" while profiling the first of two campus antinuclear groups, Arms Race Alternatives. "Generally, the narrow focus of the Freeze movement as well as academic discussions of first versus second strike capabilities, suit the military-industrial interests, as they continue adding to their billion dollar erector sets," he opined. "One is forced to wonder whether disarmament or arms control issues, severed from economic and political issues, might be another instance of focusing on the symptoms of a problem instead of the disease itself."

Obama also referenced a recently adopted federal law, set to take effect on July 1, that required every male recipient of federal student aid to demonstrate that he had registered for the draft. Some voices were calling for noncompliance, and Obama observed that "an estimated half-million non-registrants can definitely be a powerful signal" that could herald a "future mobilization against the relentless, often silent spread of militarism in the country." He then described the second campus group, Students Against Militarism. Declaring that "perhaps the essential goodness of humanity is an arguable proposition," Barack contended that "the most pervasive malady of the collegiate system specifically, and the American experience generally, is that elaborate patterns of knowledge and theory have been disembodied from individual choices and government policy." He ended by commending both groups, saying that by trying to "enhance the possibility of a decent world, they may help deprive us of a spectacular experience—that of war."

New Yorker editor David Remnick would later characterize Obama's article as "muddled," but when it returned to public view twenty-six years after it was first published, it generated astonishingly little discussion of what it said about the political views Obama had held on the cusp of his graduation from college. In his letter to Phil, Barack belittled how "school is just making the same motions, long stretches of numbness punctuated with the occasional insight." Referencing their disappointing fall semester course, he complained that "Said still didn't have the grades out for his class until a month into the term, and he cancelled his second term class, so we should feel justified in labelling him a flake."

As he had in his earlier letter to Alex, Barack singled out Davis's Ideology and the Novel as an "interesting" course, one "where I make cutting remarks to bourgeois English majors and can get away with it." But overall, "nothing significant, Philip. Life rolls on, and I feel a growing competence and maturity." After insulting what he called Phil's "sojourn to Buttfuck, Arkansas," Barack closed by saying, "Will get back to you when I know my location for next year."[37]

In Boerner's absence, Barack had been spending time with old Oxy friend Andy Roth, who was working at the John Wiley publishing house. On Friday, April 1, he and Andy attended the first day of the inaugural Socialist Scholars Conference, held at the famous Cooper Union. Roth remembered them both attending an "interesting" talk by sociologist Bogdan Denitch, one of the most committed leaders of Democratic Socialists of America. But rather than attend the conference's second day, Barack wrote another long letter to Alex McNear. "There are moments of uncertainty in everything that I believe; it's that very uncertainty that keeps my head alive," Barack explained. "In pursuit of such hopes, I attended a socialist scholars conference yesterday with Andy. A generally collegiate affair with a lot of vague discourse and bombast. Still, I was pleasantly surprised at the large turnout, and the flashes of insight and seriousness amongst the participants. As I told Andy, one gets the feeling that the stage is being set, that conduits of word and spirit are being layed across diverse minds—feminists, black nationalists, romantics. What remains to be had is a script, a crystallization of events. Until that time a pervasive mood of unreality hangs over such events, a mood that you can see the people fighting against in their eyes, their tone."

A decade later Obama wrote passingly but imprecisely about "the socialist conferences I sometimes attended at Cooper Union." His erroneous use of the plural gave future critics fodder to imagine that the "impact of these conferences on Obama was immense" and that listening to Denitch and similar speakers had "turned out to be Obama's life-defining experience," a notion that his letter the very next day to Alex utterly rebuts.[38]

At the outset of that letter, Barack wrote of "churning out assignments" on "state communism" and "the international monetary system" before pausing to reflect on "the wilderness we call life" while smoking and drinking scotch. He reported that he had given his papers to "an elderly woman with a hair-lip and hoarse voice" for typing. This lady was Miss Diane Dee, who was a famous figure around Columbia for decades. A later *New York Times* profile said "her advertising flyers" are "indigenous to campus walls," but that her "unkempt hair" and "tired face" made her seem "half-crazed." One 1983 Columbia graduate believed "she was crazy," and Gerald Feinberg's son Jeremy recalled her as "a colorful character—someone I'd expect to appear in a conspiracy theory movie or a Michael Moore film, or both."

Barack told Alex that Miss Dee was someone "upon whom my graduation depends," and he was starting to panic because she "has missed two deadlines so far and now seems to have disappeared . . . no one has answered the phone at her apartment for four days." He worried that "she's a mad woman who lures unsuspecting undergraduate papers into her home and then burns them, or uses them to line the bottom of the goldfinch cage." Following his description of the Socialist Scholars Conference, Barack told Alex that a "sense of unreality describes my position of late. I feel sunk in that long corridor between old values, actions, modes of thought, and those that I seek, that I work towards . . . this ambivalence is acted out in my non-decision as yet about next year" following graduation.[39]

Three days after Obama wrote that letter, Columbia's Coalition for a Free South Africa, in tandem with Students for a Democratic Campus, held a divestment rally outside Low Library, Columbia's administration building. Student leaders Danny Armstrong and Barbara Ransby had been pressing the issue for months, and the scheduling of a university board of trustees meeting on the fifteenth anniversary of Martin Luther King Jr.'s assassination offered an occasion for an afternoon protest. Barack convinced his largely apolitical apartment mate Sohale Siddiqi to attend the rally with him. But Siddiqi thought that compared to the black New Yorkers Siddiqi knew from the restaurant where he worked, "Barack didn't seem like one of them." Obama "was soft-spoken and gentle" and used "clean language." Indeed, "I didn't consider him American," never mind African American. Siddiqi believed Barack "seemed like an international individual."

Siddiqi recalls having witnessed an intensifying change in Barack over the preceding eight months. "He had kind of gone into a bit of a shell and wasn't as talkative or outgoing as in his earlier days," Sohale said. When Barack did speak, "he'd give me lectures" about "the plight of the poor" and downtrodden. "He seemed very troubled by it," and "I would ask him why he was so serious." Sohale's interest in drinking, picking up women, and enjoying cocaine held little appeal for the newly abstemious Barack. "He took himself too seriously," Sohale felt, and "I would find him ponderous and dull and lecturing."

The April 4 rally drew a disappointing crowd of about a hundred, but afterward a core group of about fifteen student coalition members kept up a 9:00 A.M. to 5:00 P.M. weekday vigil to highlight Columbia's refusal to divest. Barack was not among them, and few of Columbia and Barnard's black undergraduates from 1981 to 1983 have any recollection of Barack Obama. Coalition leader Danny Armstrong said, "I recall seeing Obama on campus" but never interacted with him. Barbara Ransby, a graduate student who managed the coalition's contact list and chaired most of its meetings, had "utterly no recollection of Obama." Verna Bigger Myers, president of the Black Students' Organization (BSO) in 1981–82, remembered that on a campus with so few minority undergraduates, "all the black people see the other black people" even if they did not really know them. Myers's close friend Janis Hardiman, who a decade later would be Obama's sister-in-law, remembered Barack as "sort of a phantom who just kind of walked into" BSO meetings but "did not have an active role" or participate in any the group's activities.

Wayne Weddington, a junior in 1982–83, remembers seeing Obama at BSO meetings, and Darwin Malloy, a year ahead of Obama, recalled meeting him in the cafeteria in John Jay Hall but agrees that "most people would only remember him as a familiar face." Malloy believed his friend Gerrard Bushell "probably had more interaction with him than anyone," but Bushell said he "would see him periodically" and "remember him by face" but no more.

Obama would dramatically exaggerate his involvement in Columbia's divestment activism on several occasions, telling one interviewer that "I was a leader on these issues both at Occidental and at Columbia." Talking about his two years there to a second questioner, Obama asserted that "while I was on campus, I was very active in a number of student movements" and particularly "I was very active in the divestment movement on campus." Several years later, on his first visit to South Africa, Obama declared that "I became deeply involved with the divestment movement" and "I remember meeting with a group of ANC leaders" or at least "ANC members one day in New York City." There are no contemporary records or other participants' memories that attest to any such encounter.

Columbia's African American students were also acutely aware that the Faculty of Arts and Sciences included only four black professors. By far the most visible was the handsome, bow-tie-wearing Charles V. Hamilton, best known as the coauthor of Stokely Carmichael's 1967 book, *Black Power: The Politics of Liberation in America*. Hamilton had arrived at Columbia in 1969, was named to a chaired professorship two years later, and in 1982 was the recipient of the university's award for excellence in undergraduate teaching. "Everyone knew who Hamilton was," one 1983 political science major recalled. In addition, as one younger colleague said, "Hamilton was always approachable. The hallway outside his office at the southwest end of the SIPA building was often filled with students."

But, thirty years later, when asked for the first time whether one particular 1983 African American poli sci major had ever taken one of his courses or sought his counsel, Hamilton said, "I didn't know him at all." Black history professor Hollis R. Lynch also has no recollection of Obama, nor does the entire roster of senior political science faculty. Even within Obama's particular area of concentration, international relations, neither Warner Schilling, Roger Hilsman, Zbigniew Brzezinski, nor John G. Ruggie remembers him. "Never laid eyes on him," Ruggie said. Indeed, apart from young Michael Baron, "nobody really knew him," one thirty-year veteran of the department reported. Obama would earn an A on the senior paper he wrote for Baron. It analyzed the decision-making during the arms-reductions negotiations between the United States and the Soviet Union. But Baron would discard the paper years before the world beyond Morningside Heights would yearn to read it.[40]

One week before the end of classes, Obama wrote another long letter to Alex McNear, who had taken leave from Oxy for spring term and was working in Pasadena. Citing "the method of negation" and "comparing what is to what might be," Barack for a second time referred to something he had said in an earlier letter. Then he referenced their time in L.A. four months earlier: "a young black man" and "a young white woman" "that night in Wahid's apartment in a timeless reddened room." He also said his job prospect letters had not produced any firm leads, and that "I feel like forgetting the whole enterprise and taking you with me to Bali or Hawaii to live." McNear years later recalled no actual invitation, and then Barack's letter descended into vague declarations. "I am often cruel, and my mind will flash on the screen scenes of violence or petty malevolence or betrayal on my part." McNear had no idea what he was referencing, and then Barack asked, "am I a blathering chump to you right now, or do you glean some sense from this mess?"

Barack continued on similarly, invoking "the necessary illusion that my struggles are the struggles of the first man, the river is the original river . . . What else? I enjoy my body, even when it frightens or disgusts me . . . I recall you saying that you still believe the mind is stronger than the body. A dangerous distinction, Alex, a vestige of western thought." Rereading that passage years later, McNear felt it was "condescending." Finally leaving his "river" for firmer ground, Barack wrote that it "looks like I will take a two month vacation to Indonesia and Hawaii next month. Will be stopping in L.A. either on the way over or back, if I come back. Will get in touch before I leave . . . Love, Barack."

Photos from sometime around Columbia's May 17 commencement show that Stan and Madelyn Dunham made the long trip from Honolulu to New York City to see their grandson, although Barack years later explained that "I actually didn't go to my own graduation ceremony" since "my parents couldn't come." Soon after, Barack flew from New York to Los Angeles, where he stayed for several days with Alex McNear in her apartment. "We had this kind of picnic lunch on the floor of my living room," Alex recalled. Even with Barack's invocation of "black man" and "white woman" in his letter, blackness and racial identity "really was not something that he talked about a lot," she remembered. "It barely ever came up." Reflecting on that visit years later, "I felt he was less engaged" than he had been four months earlier. She thought there was "an enigmatic quality" to Obama, and "most of this relationship really revolved around these letters," irrespective of their clarity.

Obama then flew to Singapore, where he spent five days with Hasan Chandoo and his family, a visit that coincided with Asad Jumabhoy's appearance in the championship polo match of the Southeast Asian Games. In a letter to Alex, Barack wrote that Hasan "seemed fine, if more subdued than you remember him." Barack found Singapore "an incongruous place . . . slick and modern and ordered, one vast supermarket surrounded by ocean and forest and the poverty of ages. Mostly peopled with businessmen from the states, Japan, Hong Kong as well as various family elites of Southeast Asia." Hasan, Asad, and Barack went to a discotheque or two, but Barack wrote Alex that he remained largely silent when Hasan talked about the choices he was facing, "primarily to leave a space for our friendship should he move into the business world permanently."

From Singapore, Barack flew to Indonesia and stayed at his mother's comfortable Ford Foundation home in South Jakarta. Anthropologist friends of Ann's often stayed there too, and among the guests that summer was a Rutgers University graduate student, Tim Jessup. Ann's work at Ford had kept her busy throughout 1982 and into 1983. She had written a brief paper entitled "Civil Rights of Working Indonesian Women" and delivered a lecture entitled "The Effects of Industrialization on Women Workers in Indonesia" to the Indonesian Society of Development. In mid-May, she had spent four days in Kenya, of all places, less than six months after her first husband's death, but Ann never spoke about this trip to friends.

When she wrote a long priorities analysis for Ford's continued work in Indonesia, Ann recommended that they "focus on a relatively new program area, women and employment," and particularly "the role of poor women as workers and income-earners." The ideal grantee would be "a non-governmental social action group founded by women for women," but she rued "the general weakness and lack of leadership within the women's movement in Indonesia." As Alice Dewey and a colleague wrote years later about Ann, "Java"—Indonesia's principal island—"was as much her home"—if not far more—"as Honolulu."

In his letter to Alex, Barack wrote that "my mother and sister are doing well," but with Ann "the struggling seems out of her, and the colonial residue of her life style—the servants, the shopping at the American supermarket, the office politics of the international agencies—throw up continual contradictions to the professed aims of her work." Barack was writing not from Jakarta but "from a screened porch somewhere on the northwest tip of Java." He confessed that "I can't speak the language" and that Indonesians treated him "with a mixture of puzzlement, deference and scorn because I'm American, my money and my plane ticket back to the U.S. overriding my blackness," as Obama was now old enough to perceive how many Indonesians loathed his skin color. But he closed by saying, "I feel good, engaged, the mystery of reality, the reality of mystery filling me up."

From that same porch, sitting "in my sarong, sipping strong coffee and drawing on a clove cigarette, watching the heavy dusk close over the paddy terraces of Java," Obama also wrote a postcard to Phil Boerner. "Very kick back, so far away from the madness. I'm halfway through my vacation, but still feel the tug of that tense existence. . . . Right now, my plans are uncertain; most probably I will go back" to New York City "after a month or two in Hawaii." In early July, he and apparently also his almost thirteen-year-old sister Maya flew from Jakarta to Honolulu. Only then did he actually mail his letter to Alex, writing on the back of the envelope that "with the postal system in Indonesia we'd be dead and gone before it arrived."[41]

Barack did not write to Alex again until September 1, and by then, his tone was dramatically cooler, almost palpably angry. Reacting apparently to something Alex had written to him, Barack wrote that "you are correct when you say that initially you were to me nothing more than a lovely wraith I had shaped to fit my needs, and I fought against this and you taught me yourself and I feel I showed progress eagerly, like a repentant student coming home with high marks. All this I have admitted to you in my letters—go back and reread some." Indeed rereading them three decades later, McNear reacted immediately to Obama's self-characterization: "student? . . . They all seem more like the professor," and that he was writing lectures. She said they are "not romance letters at all."

Then Barack's tone turned almost hostile, or ugly. "When I see you, the palpitations of the heart don't boil to the surface," he told Alex. "I care for you as yourself, nothing less but also nothing more. Does this anger you?" Again, "Does this anger you? When I sit down to write I no longer feel the need to bleed for brilliance on the page." Years later, McNear does not know why Obama's attitude had changed so starkly during these two summer months in Honolulu. "Here was someone who I felt that I cared a lot about. I loved him, I enjoyed seeing him," but his interest in her had somehow waned. "I trust the strength of our relationship enough that I can show myself with rollers in my hair," Barack wrote before again asking, "Does this anger you, Alex? It shouldn't. Friends feel weary sometimes. When I said that we will ever want what we can't have, I missed nothing." He referenced "the bitterness that plagues my grandparents," a comment McNear later said contradicted everything else she recalled him saying about them.

"I seek something in myself using the clues of this wind, that boy, my mother, your pain, perhaps the world," Barack wrote, returning to his prior form. "Yes, this requires a monumental arrogance, and of late I feel it whittling away. If my arrogance (which has always been confessed—run back the tapes) angers you, then my last letter, our last meeting, should douse it. It is precisely the arrogance, the sense of destiny, that has been absent. I no longer feel compelled to try to shackle you in my abstruse dreams." Addressing something Alex apparently had written, "When you doubt my honesty, you give me more credit in the past than I deserve. As though I calculated to deceive you in some way, represented myself as something I didn't believe I was. You judge me badly; I think I have been as upfront about my doubts and demons as I knew how to be."

Barack's defensiveness was more than evident. "And when you question my sincerity (a word much abused, like democracy and justice) . . . you haven't been listening very well when I spoke of myself." He closed the letter by saying, "my plans are still uncertain right now" and that he had been "typing up letters to perspective [sic] employers for the last two hours with maniacal tidiness." But "unless a job of some interest pops up soon, I'll be flying back to New York at the end of this month" and will stop in Los Angeles. Almost three months would pass before Barack wrote her again.

Sometime in the first half of October, Obama left Honolulu for New York. If he stopped in Los Angeles, he did not contact Alex. Arriving in New York, he stayed for one week with Wahid Hamid, who now had a job with Siemens, in a second-floor apartment in suburban Long Island. He then returned to the familiar confines of 339 East 94th Street #6A, where he crashed on the living room couch of Sohale and his new roommate. "I felt a slight longing to move back into the known quantity," Barack wrote Alex a month later, but "the howling and drinking and haze of my short stay smothered any productive impulses I may have had, so that moving to a new situation came easily."

Obama next found a room in a three-bedroom apartment at 622 West 114th Street #43. He told Alex that out on Long Island he had spent his days "in seclusion sorting through various letters and timetables, spying on the damp, A-frame life of suburbia, wandering along the shoulders of roads" that lacked sidewalks. Wahid was "absorbed with his work" but retains "his integrity and curiosity for the strangeness of life, and I left his apartment certain that our friendship can straddle the divide of our different choices." His time at Sohale's had presented "cash flow problems" that had hindered getting his "job hunt in motion" as one week he was unable to pay for postage to mail out résumés, and "the next I have to bounce a check to rent a typewriter." To remedy that shortcoming,

> *I took a one week stint supervising a project to transfer the files of the Manhattan Fire Department into a new facility, a fascinating experience affording me a taste of the grinding toil of low-rung white collar jobs, as well as the ambivalent relationships established between employers, employees, and personnel agencies with their shifting mix of loyalty, manipulation, abuses of power, rebellion and concession. The workers, an odd assortment of lower income kids, elderly women and unemployed liberal arts majors, struck me as some of the best people I've met; as the 12 hour shifts (uh-huh) wore on I watched much subtle straightforward contact being made and greater political perception than I had expected (although rarely framed in political terms). I felt a greater affinity to the blacks and Latinos there (who predictably comprised about three-fourths of the workforce there) than I had felt in a long time, and it strengthened me in some important way. My role as supervisor clouded the relationship between myself and them, however, since I felt that the company was using it as leverage to extract cooperation from the people for sometimes unreasonable demands. I tapped a feeling of community that comes in people acting in concert on a certain process; yet I felt frustrated that the project was imposed from above, structured according to bids and contracts and the bookkeeper, without lasting benefit beyond the pint-sized paychecks for the workers involved.*

At 622 West 114th, "I occupy a room in the apartment of a woman in her late twenties," Dawn Reilly, who was "a dance instructor and taxi cab driver. The arrangement is fine for now. The place is large and warm, we see little of each other and when we do we maintain a pretty good patter. I suspect I may move if and when the opportunity for a lease arises, though, simply because the rent is a bit steep, and I feel obliged to keep the kitchen clean and air out the living room when I smoke." Since salaries in "community organizations are too low to survive on right now . . . I hope to work in some more conventional capacity for a year, allowing me to store up enough nuts to pursue those interests the next."

He closed by remarking that "I feel lonely yet surefooted, and hope all goes well for you . . . Get in touch when you get a chance, or impulse. Love, Barack." A month earlier, Alex had begun seeing a new young man, and she believes she did not reply. Six months would pass before she next contacted Obama.[42]

Within days of mailing that letter, Barack saw a posting at the career office of Columbia's School of International and Public Affairs for a job at Business International Corporation, an international finance information and research firm founded in 1956 that published numerous analytical and data service periodicals from its offices on the seventh floor of One Dag Hammarskjold Plaza, on the west side of Second Avenue between 47th and 48th Streets in midtown Manhattan. The job had been posted by BI's Cathy Lazere, a 1974 graduate of Yale who had earned an M.B.A. at New York University before joining BI's Global Finance Division eight months earlier, in February 1983. Lazere was responsible for a bimonthly newsletter entitled *Financing Foreign Operations—Interest Rate & Foreign Exchange Rate Updater,* a four- or five-page publication that cost $900 to subscribe to annually but helped BI pull in a tidy profit as one of a bundle of services that major corporations purchased through individual client programs. In November Lazere was being promoted by division vice president Lou Celi to oversee both *FFO* and its sister reference publication, *Investing, Licensing & Trading (ILT),* which was edited by Beth Noymer, a 1983 graduate of Franklin & Marshall College who had joined BI just four months earlier.

Lazere had a roommate at Yale who was from Hawaii and had graduated from Punahou, so when a résumé arrived listing that as well as Columbia, Lazere invited him for an interview. Obama impressed her as "articulate and bright," and knowing he had attended Punahou, "I assumed he came from a privileged background." Cathy introduced Barack to Beth Noymer and telephoned Lou Celi to get his approval before offering Barack the position. Obama's salary would be about $18,500—a respectable sum for a newly minted B.A. in 1983—and he would be expected to write for the finance unit's flagship newsletter, *Business International Money Report (BIMR),* as well as to do the research and copyediting necessary to churn out each issue of *FFO.* A few years later Barack would tell a questioner that he took the job at BI because "I wanted to know how money worked."

More than three dozen correspondents all around the world submitted the data and material that Lou Celi's unit sliced and diced to produce their multiple publications. *FFO*'s content, as its title made clear, was both arcane and impenetrable. Celi's top deputy and sidekick, Barry Rutizer, had started out at BI doing *FFO,* and he said, "I couldn't even read it when I was editing it." Cathy Lazere agreed. "I was certainly bored when I was editing that stuff." At the time of Obama's arrival, issues of *FFO* consisted of lengthy country-by-country lists of exchange rates accompanied by brief comments and a summary table of "Foreign Exchange Rates of Major Currencies." Barack had his own office, but the composition and production of BI's publications took place on a central word processing system that relied upon Wang terminals scattered around the office rather than individually assigned. As a result "we sort of duked it out over Wang time," Beth Noymer said, with almost everyone regularly moving around the roughly sixty-person office. "The Wangs were a big part of our lives," Celi's assistant Lisa Shachtman Hennessey recalled, and "every ten minutes" someone seemed to call out from the bullpen area that "the Wangs are down." Smoking was more than allowed—"there were ashtrays everywhere," Lisa remembered—and Obama regularly smoked Marlboros while editing manuscript copy by hand. "There was almost no way to get all your work done between nine and five," Beth explained, especially on days when the final content had to be sent down to the print shop that BI veteran Peggy Mendelow oversaw in the building's basement.

Much of BI's information gathering required telephoning various midlevel officials at corporations and banks. When calls were returned, BI's switchboard operator announced the call over an office-wide paging system if someone was not at their desk. Brenda Vinson, an African American woman in her late thirties, worked in the library and often covered the switchboard. Obama was the first black college graduate to work at BI, and his unfamiliar first name was a challenge to pronounce. Vinson remembers Barack as "very personable" toward her and her cousin, who were BI's only other black employees; a Puerto Rican father and son staffed BI's mailroom. There was a good bit of socializing among the young professionals who worked at BI. A nearby Irish pub was one regular destination, and Beth Noymer later described BI's office culture as "a hotbed of young singles."[43]

Obama did not socialize with his BI workmates, but sometime prior to New Year's Eve, his friend Andy Roth invited him to a party that Andy's brother Jon was hosting in their sixth-floor apartment at 240 East 13th Street. Jon worked at Chanticleer Press, a publisher that helped produce National Audubon Society guides, and other invitees included Genevieve Cook, a twenty-five-year-old Swarthmore College graduate who had worked at Chanticleer before beginning coursework toward a master's degree in early childhood education at Bank Street College of Education. She was born in 1958 to parents who were Australian: Helen Ibbitson, the daughter of a Melbourne banker, and Michael J. Cook, a conservative diplomat who would go on to head up Australia's top intelligence agency before serving for four years as ambassador to the United States. Her parents had divorced when Genevieve was ten years old, and Helen then married Philip C. Jessup Jr., an American lawyer and executive whose International Nickel Company post had him and Helen living in Jakarta during the 1970s. Genevieve attended multiple boarding schools in the U.S. before graduating from the Emma Willard School near Albany, New York. While she was at Swarthmore, her mother and stepfather Phil had moved from Jakarta to New York, and by late 1983 Genevieve was temporarily living in their spacious apartment on Park Avenue just below 90th Street after breaking up with a Swarthmore boyfriend with whom she had lived in Manhattan's East Village while student teaching that year at the Brooklyn Friends School.

Her four years at Swarthmore were the first time Genevieve attended the same school for more than two years, and it was her first time in one country for more than three straight. "At Swarthmore, I was very drawn to . . . the drug oriented counterculture" and "its ritualized pot smoking," she wrote in her impressive 1981 senior anthropology thesis, "Dancing in Doorways." For the thesis, she interviewed fifteen fellow students who were also the children of expatriates, "people who spent their lives from the time they were born moving around from country to country, who are not members of any one culture, who come from nowhere in particular, and who do not really belong anywhere. You will always know them when you meet them."

At the Roth brothers' party, Genevieve did know one when she met one. She and Obama struck up a conversation that lasted several hours after they discovered their mutual ties to Indonesia and expatriate similarities. "I remember being very engaged, and just talking nonstop," she later wrote. "We both had this feeling of how bizarre and exciting it was that we'd both grown up in Indonesia, and we felt we very much had a worldview in common." Barack was in no way aggressive. "If anything, he struck me as diffident . . . although also at ease with himself" and "clearly interested in pursuing this conversation with me." She found him "just really interesting, intellectually," and she later reflected that "the thing that connected us is that we both came from nowhere—we really didn't belong." Before the night was out, he handed her a small scrap of paper—"Barack Obama 866-8172 622 W. 114th #43"—that she still retains thirty years later. After a phone call the next week, she agreed to meet him at his apartment for dinner, where Barack cooked for the two of them. "Then we went and talked in his bedroom. And then I spent the night. It all felt very inevitable," she wrote in a private memoir.

That evening stood in sharp contrast to Genevieve's rejection of a dinner host seven months earlier. That spring she had taken a course at Bank Street that involved having the students share recipes with their classmates. Three decades later Genevieve still had the "Floating Island Pudding" she shared as well as "Zayd's Catsup," a contribution from a thirty-seven-year-old classmate. At the end of the semester, that classmate invited her to dinner at his apartment at 520 West 123rd Street #5W. Five-year-old Zayd and his two-year-old brother Malik were asleep, as was Chesa, another almost two-year-old member of the household, whose mother and father were both in prison.

Genevieve was initially surprised that the only food her host had for dinner was grapes, but it quickly became clear what he wanted for dessert. He explained that he was in an open relationship; he and his partner had been leading figures a decade earlier in a group whose slogans included "Smash Monogamy!" His partner was not coming home that night; indeed she was residing involuntarily at the Metropolitan Correctional Center in lower Manhattan, where she would remain for another six months. "He gave getting me into bed quite a good go," Genevieve recalled, but with a thirteen-year age difference between them, he "seemed awfully old to me!" Her host "was quite miffed that I was not impressed by his 'status'" as a notorious former radical, albeit one whose FBI "Wanted" poster made him appear to have just fallen out of bed rather than striving to get into one. He "backed off when I wasn't interested," and Genevieve's rebuff may have had a greater impact than she realized.

Four months later a federal judge gave her host's partner a weekend furlough so the two former anti-monogamy advocates could marry, and two months after that, the judge allowed her to return to 520 West 123rd Street on a Christmas furlough. Four days after New Year's, the judge granted a motion to vacate the contempt citation that had kept Bernardine Dohrn jailed since May 19, and she was free to remain with her two sons and now husband, Bill Ayers. As Genevieve would pluperfectly capture the essence of the story, sometimes indeed the "truth is so much stranger than fiction!"[44]

On Monday, January 9, Genevieve spent a second night with Barack on 114th Street, and the next day wrote in her journal, "I have not experienced the kind of intellectual stimulation Barack offers me since I left college." She expressed similar feelings in a letter she wrote to him but did not mail, a letter she still had three decades later. "You are the first person I've met since being in college who has in some way engaged me in a process of self-intellectual questioning. It is a shock to recognize that my engagement with Bank St., education, friends I've made through Bank St. & teaching & the kind of process I've touted, of teaching forcing you to be self-evaluative, has all been 'professional.'"

Over the next four weeks, Genevieve continued to record in her journal her reactions to Barack. Intercourse was pleasant, and in bed "he neither came off as experienced or inexperienced," she later recalled. "Sexually he really wasn't very imaginative, but he was comfortable. He was no kind of shrinking 'Can't handle it. This is invasive' or 'I'm timid' in any way; he was quite earthy." In one late January entry, she wondered "how is he so old already, at the age of 22?" and she wrote two poems for Barack, one teasing him by way of implicit comparison to the mythical Orion. The second, alphabetical in form, progressed from "B. That's for you" to "F's for all the fucking that we do" to "L I love you . . . O is too."

Obama spoke excitedly about Genevieve during one phone conversation with Hasan Chandoo in London, yet in a "rather mumbled" one with his mother Ann in Jakarta he made no mention of Genevieve but talked about BI. "Barry," as Ann called him in a letter to Alice Dewey, "is working in New York this year, saving his pennies so he can travel next year . . . he works for a consulting organization that writes reports on request about social, political and economic conditions in third world countries. He calls it 'working for the enemy' because some of the reports are written for commercial firms that want to invest in those countries. He seems to be learning a lot about the realities of international finance and politics, however, and I think that information will stand him in good stead in the future."

With Genevieve, Barack spoke of BI "only to be disparaging" and only "very rarely" did he speak about his actual five-days-a-week work tasks. "His entire attitude was bearing a necessary burden" and one he was deeply uncomfortable with. "Even just putting on the clothes to go to work in that environment was a political divide—it divided him from how he really saw himself," she later explained. "He definitely wore it like a penance." On weekday evenings, Barack's posture was "They're the enemy, and I've just spent all day at work" but also that "I'm above being emotionally affected by my job."

His coworkers at BI quickly picked up on Barack's emotional distance. Cathy Lazere, his immediate supervisor, noted his "aloofness." "He came across as someone not interested in other people," she said. Beth Noymer, his closest colleague, said he was "quiet and kind of kept to himself." To vice president Lou Celi, Barack came across as "shy and withdrawn" and "always seemed aloof." Lou's deputy Barry Rutizer thought Obama "was kind of self-involved" and "somewhat withdrawn." Lou's assistant Lisa Shachtman remembered Barack as someone who was "sitting back and observing" others rather than interacting with them. *BIMR* editor Dan Armstrong recalled that Obama "really just kind of kept to himself" and "never joined us" when everyone went out after work. Dan thought "diffident" was the perfect adjective, and in Peggy Mendelow's memories Obama "had an obvious 'Do Not Disturb' sign on him." Years later one colleague would describe Barack as "the whitest black guy I've ever met."[45]

Barack's life with Genevieve on 114th Street, where she came on Thursdays after a class at Bank Street as well as on weekends, was a separate world from his workday week. In mid-February, Genevieve recorded that Barack the night before had "talked of drawing a circle around the tender in him—protecting the ability to feel innocence and spring born—I think he also fights against showing it to others, to me." The next day, Presidents' Day, she recorded her worries about her "unwillingness to believe that he really does like the time spent with me, that he likes me." Alluding to BI, it "makes me cross that he's sitting there in that office while all the others take the day off—too soon to scorn the blatant taking advantage of the least powerful on the totem pole—but I hope he will soon find a way to make it clear he sees limits to what can be thrust on him—the loose ends, the overtime." Her thoughts returned to their weekend together. "It means so much more than lust, after all, all this fucking we do." Four days later, after Thursday night together, she wrote of "making love with Barack, so warm and flowing and soft but deep—relaxed and loving—opening up more."

But Genevieve had her own self-doubts. "It's all too interior, always in his bedroom without clothes on [or] reading papers in the living room. I'm willing to sit and wait too much." Barack did go out to Long Island one weekend to see Wahid, who had now married his longtime girlfriend Ferial "Filly" Adamjee, yet with Genevieve their time together was interior indeed. She worried "that what men reach for and stay with is the availability to them of a warm time in bed, while they put up with or dismiss the 'noise' women 'indulge' themselves in." While "the sexual warmth is definitely there . . . the rest of it has sharp edges, and I'm finding it all unsettling and finding myself wanting to withdraw from it all. I have to admit that I am feeling anger at him for some reason, multi-stranded reasons. His warmth can be deceptive, tho he speaks sweet words and can be open and trusting, there is also that coolness . . . Both of us wary . . . and that is tiring."

The next weekend found a return to warmth, with Genevieve pleased by Barack's compliments, "his saying 'You're sweet to me,' that I'm kind, as if he's not accustomed to that, has not had much of that." One evening at dinner, Genevieve recorded, she had an image "of being with Barack, 20 years hence, as he falters through politics and the external/internal struggle lived out." One week later, with Genevieve experiencing a bout of tearful self-pity, "Barack said he used to cry a lot when he was 15—feeling sorry for himself." Fifteen would have been his tenth-grade year at Punahou, when Barack was closest to Keith Kakugawa, but Genevieve was happy that he "won't let important things go unspoken. 'Speak to me.'"

Yet more often Genevieve felt a self-distancing on Barack's part, "a sense of you biding your time and drawing others' cards out of their hands for careful inspection—without giving too much of your own away—played with a good poker face. And as you say, it's not a question of intent on your part—or deliberate withholding—you feel accessible, and you are, in disarming ways. But I feel that you carefully filter everything in your mind and heart . . . there's something also there of smoothed veneer, of guardedness . . . I'm still left with this feeling of . . . a bit of a wall—the veil."

In early March, Genevieve moved from her parents' Park Avenue apartment to a top-floor, two-bedroom apartment in a Park Slope brownstone at 640 2nd Street that was owned by relatives of a Brooklyn Friends School secretary. One mid-March morning, she telephoned West 114th Street and for a moment mistook the voice of the third roommate, Michael Isbell, for Barack. Michael's firm "I'm good" in response to Genevieve's "How are you?" made her realize "that Barack often doesn't feel firmly good."

Barack and Genevieve saw even less of Michael than they did of Dawn Reilly, whom they mostly interacted with on Sunday mornings. Michael worked in advertising, had a full-time girlfriend, and remembered Barack as a quiet, studious smoker. Michael did not get along with Dawn as well as Barack did, and Genevieve believed Dawn "had this kind of motherly attitude towards" Barack, even though she was just five years older. Genevieve remembers Dawn as "a vivacious character . . . very fond of Barack" and "she thought we were a very cute couple. She saw enough of us that she was very aware when things were good" and also when "things got a little bit strained." Genevieve recalled that in mid-March, Dawn told her, "'I feel more tension between you two'" and said she believed Genevieve was good for Barack, whom she thought was confused about what he wanted. To herself that day Genevieve wrote, "I'm a little worried about Barack. He seems young and defenseless these days."[46]

Barack intrigued Genevieve greatly, but there is "so much going on beneath the surface, out of reach. Guarded, controlled," she wrote to herself. Genevieve took the initiative to buy them tickets to a one-woman performance of three short plays by Samuel Beckett at his namesake theater on West 42nd Street. In the last of them, *Rockaby,* actress Billie Whitelaw onstage uttered just one word, "more." In between her four increasingly fearful incantations of that one syllable, the audience heard "the tortured final thrashings of a consciousness, as recorded by the actress on tape." At its close, the audience experienced "relief" as "death becomes . . . a happy ending," *New York Times* critic Frank Rich wrote, saying it was "riveting theater . . . that no theatergoer will soon forget." In her journal, Genevieve wrote that "Billie Whitelaw was superb."

At the end of March, Hasan Chandoo arrived in New York to prepare for his move from London to Brooklyn in early summer. The 1984 Democratic presidential race was in full swing, and on Wednesday evening, March 28, top contenders Walter Mondale and Gary Hart were joined by third-place contestant Jesse Jackson for an intense televised debate from Columbia's Low Library, moderated by CBS's Dan Rather. By that time Obama was actively interested in Jackson's campaign, and the next Saturday, March 31, he persuaded Hasan, Beenu Mahmood, now in his first year at Columbia Law School, and even Sohale, to join him in attending a Jackson campaign rally on 125th Street in central Harlem. Three days later, Jackson finished a strong third in the New York primary with over 25 percent of the vote, including presumably Barack's. Both Hasan and Beenu also remember that spring and summer that Barack often carried with him a well-worn copy of Ralph Ellison's famous 1952 novel *Invisible Man,* which he was reading and rereading. Beenu believed that "*Invisible Man* became a prism for his self-reflection," and in retrospect Beenu thought that over time "Ellison assisted Barack in reaching a fork in his life."

But that fork was more than a year away. Later that night, Barack and Genevieve joined Hasan and Sohale at the latter's East 94th Street apartment, where Barack had lived a year earlier. "Long time friends, easy with each other but also challenging," Genevieve wrote in her journal the next day. Everyone "did several lines of cocaine, which added an edge to it all." For Barack, the almost three years that had passed since he lived with Hasan in South Pasadena in 1980–81 had been almost entirely drug-free, certainly when compared to that year at Oxy and indeed the three previous as well. His twelve months living with Sohale in 1982–83 had seen plenty of "partying" by Siddiqi, but only with Hasan's incipient move to New York would Barack feel compelled, on account of their friendship, to reengage in something that without Hasan he felt no need to seek out. He was seriously involved with a woman who smoked pot daily, but except when they were at parties with Hasan, Sohale, and Imad, Barack and Genevieve did not partake of such pursuits.[47]

Before the end of that weekend, Barack told Genevieve that "I really care very much about you" and that "No matter how things turn out between us, I always will." She wrote in her journal that he talked to her as well about his "tendency to be always the observer, how to effect change, wanting to get past his antipathy to working at BI." A week later "Barack talked of his adolescent image of the perfect, ideal woman—searching for her at the expense of hooking up with available girls." Presently she imagined him "opting for dirtying his hands in the contradictions and overwhelming complexities this city offers" and resolved that "I must enjoy Barack while I can." Recounting a scene Barack had described to her, "The image of Barack shaking his grandpa by the shoulders and asking 'Why are you so damn unhappy?' really struck me."

In early April, Barack received a call from Alex McNear, who was still living in Eagle Rock, and soon after, he sent her a long letter, one that portrayed his role at BI somewhat differently from how his coworkers and Genevieve did. "I've emerged as one of the 'promising young men' of Business International, with everyone slapping my back and praising my work. There is the possibility that they offer the job of Managing Editor for one of the publications, which would involve a hefty raise, but an extended stay," Barack asserted. "The style and substance of what I write" was such that "I can churn out the crap without much effort" yet "the finished product confronts me as an alien being, not threatening, but a part of another system, another sensibility."

Barack's description of some of his interactions with colleagues beggared belief. "Without effort, I find I can perform with flawless grace, patching up their insecurities, smoothing over ruffles among the co-workers." Yet he described his own attitude with considerable accuracy. "All of them, including my superiors, sense some sort of tethered fury, or something set aside, below the calm surface . . . so that I remain somewhat alien to them." Indeed "the implacable manner is not an act, nor is the anger underneath," but Barack acknowledged that his colleagues "are good people, warm and intelligent." He told Alex, "I've cultivated strong bonds with the black women and their children in the company, who work as librarians, receptionists," and reported that the only other black men "one sees are teenage messengers."

Barack admitted "the resistance I wage does wear me down—because of the position, the best I can hope for is a draw, since I have no vehicle or forum to try to change things. For this reason, I can't stay very much longer than a year. Thankfully, I don't yet feel like the job has dulled my senses or done irreparable damage to my values, although it has stalled their growth." But, "like other malcontents, I have my other life as opposed to my working life . . . weeknights I spend a few hours writing, a few hours eating, and take occasional walks along the river. I recently finished the first fiction piece I've attempted in over a year, and I got some good feelings doing it, even though it's not top quality. I still have a certain ambivalence towards writing/art as a vocation." As for "my political reading/spectating—my ideas aren't as crystalized as they were while in school, but they have an immediacy and weight that may be more useful if and when I'm less observer and more participant. On weekends I see Sohale, Wahid et al. fairly frequently and let myself slip back into old comfortable activities like bullshitting and watching basketball," though to Alex, Barack made no mention of his reintroduction to cocaine. He confessed that "I've also become quite close to an Australian woman who teaches in a Brooklyn grade school. She doesn't put up with a lot of my guff, and has a good sense of humor without any cynicism, which is a good tonic for my occasional attitude problems." Obama ended the letter by saying "look forward to seeing you in the summer if you choose to come back East. Love, Barack."[48]

The divergence between how Obama described his interactions with his BI colleagues and how they viewed him was great indeed. Eugene Chang, one of the two editors of the finance unit's lead newsletter, *Business International Money Report,* made an effort to get to know Obama, inviting him to lunch at a Korean restaurant and mentioning how he jogged. Barack's responses were chilly and abrupt: "I don't jog, I run." Susan Arterian, a decade older than Obama, thought "there was a certain hauteur about him and a somewhat cultivated aura of mystery." To her, "BI was a friendly place" with lots of "wonderfully quirky characters," and Eugene's *BIMR* coeditor, Dan Armstrong, saw Barack as "reserved and distant towards all of his coworkers," notwithstanding how BI was "not a corporate place in any way."

Bill Millar, a 1983 graduate of CUNY's Baruch College, found Obama "arrogant and condescending," someone who "treated me like something less than an equal" even though Millar was a higher-ranking assistant editor. Millar once argued with Barack about corporations that did business in South Africa, and another colleague, Tom Ehrbar, recalled Barack quarreling about the CIA with another coworker who did not remember the exchange. As Peggy Mendelow described Barack, he "kept very much to himself" and "didn't seem to want to be there."[49]

Barack was far more interested in old friends than in making new ones. Genevieve described "Barack's face opening up in a broad grin after talking with Bobby [Titcomb] on the phone in Honolulu." She also described their sexual interactions positively: "really communicating instead of merely getting off." At the end of April, Genevieve wrote, "I'm falling in love with Barack. . . . Spent Sunday with Barack in the park." They saw a boy in a sandbox "with his Superman cape on, and I launched into some kind of spiel about kids and imagination and fantasy, and he launched into this thing about superheroes and was revealing about some relationship he had to superheroes, and I thought, 'Oh my God, that's fascinating, I've never heard him come out with that before,' and I pounced on it and wanted to really like push an exploration of it" but Barack gently rebuffed her.

At the end of April, old Oxy friend Sim Heninger came through New York and stayed with Phil Boerner, who was just about to finish his degree at Columbia. One evening, Sim, Phil, and Phil's girlfriend Karen had dinner with Barack and Genevieve. Sim in particular was struck by the seriousness of their romantic attachment. Early in May, however, Genevieve detected a "deliberate distancing" on Barack's part and wrote, "I think I am probably being rejected more for what I represent in Barack's mind than for who I am." She imagined that Barack would be more comfortable with a black woman, and she wrote in her journal, "I think I've known all along that he plots this into his life as something temporary—not open-ended as he had said." She wondered if they were just "using each other," yet understanding what was going on was difficult, because "he is so wary, wary. Has visions of his life, but in a hiatus as to their implementation—wants to fly, and hasn't yet started to take off."

Within a week their relationship had righted itself, although Genevieve was feeling "depressed about teaching" as the school year was ending. "It so delights me that from time to time, Barack will talk about the more private, inner aspects of what he sees and feels of our relationship." In late May, Barack told her one night "of having pushed his mother away over the past 2 years in an effort to extract himself from the role of supporting man in her life—she feels rejected and has withdrawn somewhat." By then he knew his mother and sister were moving back to Honolulu in mid-August. Ann had learned in February that her Ford Foundation post would expire in six months. She had resolved to make the best of that by returning to her long unfinished Ph.D. at the University of Hawaii, a move that would allow soon-to-be fourteen-year-old Maya to begin ninth grade at Punahou. Ann wrote the chairman of UH's Anthropology Department to say that "the major reason" for her long absence from the program had been "the need to work to put my son through college," and with his graduation, "I'm now free to complete my own studies."[50]

Once classes ended at Brooklyn Friends School, Genevieve left New York to spend a week at her stepfather's family's estate in Norfolk, Connecticut. She dreaded the next school year, when she would be teaching first grade at PS 133 in Park Slope. "I'm feeling really bad about myself in general," she wrote in her journal, and by phone Barack sought to reassure her. By early June, Hasan Chandoo had an apartment at the Eagle Warehouse building on Old Fulton Street underneath the Brooklyn Bridge, and both there and at Sohale's apartment on East 94th Street, Barack and Genevieve joined some assortment of the Pakistani friends almost every weekend. If Wahid came into the city, or if Beenu and his girlfriend Chinan were present, drinks and dinner would be the centerpiece of an evening. But with Hasan, Sohale, and Imad, pot and cocaine were usually involved, though Barack's ambivalence about those activities was crystal clear to Genevieve and obvious to Hasan too.

Sometimes Barack would beg off, but most times he asked Genevieve to come along—"We'll go together," he would say—knowing that one or both of them would try to leave before the evening got too late or the activities got too "out of control and manic," as Genevieve described it. For all her pot smoking, Genevieve did not care for cocaine, yet Barack "didn't like it when I said 'Well I'm going to leave now' or 'I don't feel like coming'" because "that made it harder for him to ignore the fact that he didn't really want to go either, that he would have rather stayed home and read."

But Barack's bond with Hasan was stronger than his self-discipline, and Genevieve thought "it seemed important to Barack that I bolster him in his desire to maintain allegiance to the guys." To her, Barack's indulgence "was definitely out of loyalty and an inability to kind of give the flick to people who had been so incredibly loyal and embracing" of "this lost boy, who had no group, who had no community, and they knew him from before," from Oxy, "and embraced him warmly." Hasan was "absolutely" the driving force, not Sohale or Imad, and while that trio was "doing lots of cocaine," Barack "did not do as much as they did." Indeed, Barack did "a lot less of everything, like for every five lines that somebody did, he would have done half, and for every scotch that Hasan poured, he would have had one out of every ten compared to what Hasan was drinking."

In a more understated voice, Hasan agreed with Genevieve. "We dabbled in drugs," but with Barack "there wasn't anything excessive by him, by my standards." As of that 1984 summer, Obama was "much more serious" than the college sophomore Hasan had lived with three years earlier, and at times Barack "would tell me to go easy on my drinking or my smoking pot, and I'm saying 'What a change!'" Genevieve recognized a tension between Barack's loyalty to his Pakistani friends and his emerging realization that "somehow splitting himself off from people is necessary to his feeling of following some chosen route which basically remains undefined." She continued to worry about "veils and lids and control," but Genevieve enjoyed being "cosseted in Barack's apartment" on weekend nights before returning to her Brooklyn apartment and a new roommate whose presence she found irritating. Genevieve found her intimate time with Barack special and uplifting, but she was sometimes troubled by his behavior toward the Pakistanis, writing one night that "the abruptness and apparent lack of warmth with which Barack left them was jarring."[51]

A few days later Hasan and Barack had dinner at Genevieve's apartment, and she remained fascinated with Barack's deeper, preoccupying thoughts. He talked about Ernest Hemingway "and the integrity of grasping for those times, those visions that are ones of true magnificence and profundity," but "when Barack speaks of missing the signs of some central, centered connection with the powerful maelstroms of deep feeling, grand scopes, I have responded with comments such as 'Maybe you need not to look for them at such dizzying heights, but on other levels.'"

In mid-July, Genevieve took offense at "all the artifice in his manner," but a large Saturday-night dinner at Hasan's that included his cousin Ahmed, Beenu and his sister Tahir, and Wahid and his wife Filly left Genevieve impressed with Filly's intelligence and independence. Yet Genevieve's persistent self-doubts continued to trouble her feelings about Barack. "How long will it take him to see that I am silly and insecure and inarticulate in a way he will find repulsive rather than acceptable?" she wrote in her journal.

A trio of cheap photo booth pictures the couple took of themselves that summer shows Genevieve looking exceptionally energized, striking, and happy, and a somewhat full-faced Barack looking pleased and happy as well. On the first Sunday in August, Genevieve challenged him to a footrace in Prospect Park near her apartment. Barack greeted the challenge with gently mocking bemusement, but then, to his utter amazement and chagrin, Genevieve won, demonstrating that he had seriously underestimated her. "Barack couldn't really believe it and continued to feel a bit unsettled by it all weekend" as they showered and then went to see a new film, James Ivory's *The Bostonians,* starring Vanessa Redgrave and Christopher Reeve. "Being beaten by a woman," especially when Barack prided himself on his almost daily running, "*really* unsettled him," Genevieve recalled.

Genevieve believed that Barack's running was motivated by unpleasant memories of having been a chubby boy in his pre-basketball years. "There was still quite a bit of 'I was a fat boy' feeling lurking underneath his resolve to be so disciplined with the running. He was very trim, except for a bit of pudgy tummy," which "he couldn't get rid of" and "was quite self-conscious about," she remembered. "That's why he ran," to "get rid of that last little bit of being a pudgy boy."

That did not hinder what she described as "passionate sex," and after a week apart when Genevieve went to London, she returned to find Barack troubled after having been told by his African sister Auma, who hoped to visit New York in November, about a rumor that their father may have been murdered rather than killed by his own drunken driving. Barack and Auma had become irregular correspondents following Obama Sr.'s November 1982 death, and either just before or just after this latest word from Auma, Barack had a memorable dream about his father that he shared with Genevieve, who had also "grown up without my dad." But now he also told her that a month earlier he had cried when he saw television news coverage of a mass murder that claimed the lives of twenty-one people at a fast-food restaurant near San Diego. "Interesting that he was connecting the two," Genevieve wrote in her journal, "when in fact the tears he cries are, I'm sure, buried tears over his dad, and the loss over all the years without him. He was very subdued" for the balance of that weekend.

To Genevieve, who was "constantly looking for an explanation for this wary guardedness" she so often felt from Barack, the answer lay in how "he was not in touch with how deeply wounded he was by his mother's and his father's relationships with him." In her mind, Barack's "woundedness" and "abandoned child persona" meant "the amount of suppression of negative emotion is just heroic" and explained why "there was a 'no go' zone very, very quickly" whenever talk about deep personal feelings threatened to undermine all of that successful suppression.[52]

In late August, Alex McNear called Barack to say she would be arriving in New York on August 23. The two of them had dinner that night, although years later Alex would have no memory at all of that evening. Genevieve was not looking forward to the start of her school year at PS 133, but in her journal she again wrote, "I love him very much." Obama met up with Mike Ramos for a beer one night when Mike came to New York for the first of two training events for his job at a large accounting firm. Barack talked about quitting his job at BI so he could do something more rewarding, and Mike, impressed with his friend's courage, ended up crashing at 114th Street rather than making it back to his hotel. Either during that visit or when Mike returned to New York just before Halloween, they had dinner one night with Genevieve, whose unusual name Mike would remember years later.

Early in the fall, Phil Boerner, Barack, and another old Oxy friend, Paul Herrmannsfeldt, who was working at a publishing house, started a book discussion group at Paul's seventh-floor apartment in Soho. Their first selection was Samuel Beckett's 1938 novel *Murphy,* and Phil's girlfriend Karen plus several friends of friends attended two or three subsequent meetings, but the group petered out within two months. Barack also attended a reading by several writers at the West End bar on Broadway just south of 114th Street that his apartment mate Michael organized just prior to moving out, but his attempt to interest Michael in his own work failed. As Phil later said, they all found Barack "an interesting yet unremarkable person," a young man whom some saw as "a bit smug" but whom no one imagined would ever be seen as an exceptional individual.[53]

At BI Barack's colleagues felt similarly. In early fall, Lou Celi and Cathy Lazere launched a new series of "Financial Action Reports" that required updating BI's data on companies' cash management strategies in particular foreign countries, with new information gleaned from interviews with corporate treasurers. The first two countries were Mexico and Brazil, and Obama and the slightly more senior Michael Williams were given a task that Williams remembered as "my least favorite project" at BI. About twenty treasurers had to be contacted either by phone or in person in New York, and the thirty-minute interviews had to be transcribed. Williams and Obama each took half, and though Williams recalled transcribing his own tapes, Lou's assistant Lisa got newly arrived editorial assistant Jeanne Reynolds to transcribe at least one of Obama's more difficult ones. Williams remembered Barack as someone who "kept to himself," spoke only when necessary, and never seemed "fully engaged." That was atypical indeed at "a very friendly place" with "a pretty hip crowd" that offered great opportunities for advancement "if you wanted them."

Jeanne Reynolds recalled Barack as "quiet, reserved, polite," and Barack's copy editor on the Mexico and Brazil reports, newly arrived Maria Stathis, would likewise remember him as "very quiet." Another new arrival, Gary Seidman, remembered Barack teaching him to use the Telex machine that was cheaper than the telephone for international communication. Barack seemed "aloof," a stark contrast to his "vibrant" coworker Beth Noymer. When Obama gave Cathy Lazere formal notice one day in November that he was quitting effective early December, Cathy mused that "it must have been a little lonely for him to work at a place for a year and not be fully engaged in the world around him."

A few days earlier, his sister Auma called from Nairobi to say she was canceling her New York trip because their younger brother David Opiyo had just been killed in a motorcycle crash at age sixteen. That news may have strengthened Barack's resolve to leave a job he found so foreign to his political views, and although he told Cathy "he wanted to be a community organizer because he didn't find business that meaningful," he also was leaving BI without a new job in hand. Cathy, Gary Seidman, and the young man Cathy interviewed and then hired as Barack's replacement, Brent Feigenbaum, all had the impression that Barack was considering law school in addition to community organizing. In Barack's exit interview, Lou Celi told him, as he told everyone leaving BI, that he was making a big career mistake, and when Barack told editor Dan Armstrong he did not yet have a new job, Dan asked, "Are you crazy?" He also told Barack he at least should "get another job before you quit." In Armstrong's memory, Barack simply shrugged. Feigenbaum spent one day working alongside Barack and recalled him as "remote . . . not a terribly warm person." Beth Noymer's monthly calendar for December 1984 would show "Barack lunch" on Friday the fourteenth, but neither she nor Cathy nor anyone else had any memories of a farewell meal.

Asked two decades later what he recalled from his time at BI, Barack answered "the coldness of capitalism." He told an earlier questioner, "I did that for one year to the day," a clear indicator of his desire to leave that world for something he found more fulfilling. But giving up his BI paycheck meant leaving the apartment on West 114th Street, and on the weekend of December 1 and 2, Barack temporarily moved in with Genevieve on the top floor of 640 2nd Street in Park Slope.

Earlier in the fall, they had taken the bus to her family's estate in Norfolk, Connecticut, where they slept in an open-air cottage and joined Genevieve's mother and stepfather for one meal. Barack later recounted paddling a canoe on a nearby pond, and a photo shows a happy and relaxed young couple outdoors in the morning sun. Their first week together in Genevieve's cramped quarters produced minor irritations, but a nice weekend then included seeing the Eddie Murphy film *Beverly Hills Cop* in downtown Brooklyn. Genevieve was the only white person in the audience, but she says she and Barack never experienced any hostility or rudeness toward them as an interracial couple.

In the days just before Barack left to spend the holidays in Honolulu, their feelings of being in each other's way multiplied, with Barack saying, "I know it's irritating to have me here," and telling Genevieve that she was being "impatient and domineering." But they exchanged Christmas gifts, with Barack embarrassed when Genevieve bought an expensive white Aran cable-knit wool sweater for him at Saks Fifth Avenue. When he asked her what she wanted, Genevieve suggested lingerie, which she says "threw him into an absolute tailspin" before he returned with something that Genevieve privately thought was "incredibly tame."[54]

A week before Christmas, Obama flew to Honolulu, and he spent much of his time in transit reading a book by Studs Terkel, most likely his newly published *The Good War: An Oral History of World War II*. On New Year's Day, he wrote to Genevieve that "my trip has progressed without any notable events" but that "I was foolish to think that I'd have the time or energy to work on my writing" in Hawaii because "reacquainting myself with the family has proven to be a fulltime job . . . they all have used me as a sounding board for all sorts of conflicts and emotions that have previously stayed below the surface. . . . I've been the catalyst for tears, confessions, ruminations, and accusations."

Genevieve had never met any of Barack's family, but he offered her sketches of them all. "My mother is as I last saw her, gregarious and sensitive, although she's undergoing some difficult changes after uprooting herself from Indonesia" to live in a visibly humble cinder-block apartment building at 1512 Spreckels Street, where she and fourteen-year-old Maya shared the two-bedroom unit 402, less than a block from Maya's ninth-grade classrooms at Punahou.

"My grandfather," Barack went on, now age sixty-six and retired, "appears immutable. He looks more robust than ever, even while eating donuts, smoking a cigarette, and drinking whiskey simultaneously. My grandmother is doing less well—she continues to drink herself into oblivion," indeed "incoherence," when not working. "Her unhappiness saddens me deeply. . . . It may be my helplessness in the face of her problem that angers me more than the problem itself. But she retires next year, and if the fortitude she's channeled exclusively into her work can't be transferred into the remainder of her life, not much life will remain." With Maya, "I watch with joy her development into a fine person," but being back in the all too familiar tenth-floor apartment at 1617 South Beretania meant that "ghosts of myself and others in my past lurk around every corner."

Obama wrote that his relatives all "think I'm too somber. . . . My mother explains that I was normal until 14, from there went directly to 35." Stan joked that Barack is "as mean to himself as he is to everyone else. The only difference is he likes it." But Obama was clearly discomforted by this return to his childhood surroundings. "I have trouble fitting into these Island Ways," he told Genevieve. Within blocks were "the apartment house where I was conceived . . . the hospital where I was born . . . and the school where I spent a third of my life." But nonetheless "I'm displaced here, it's not where I belong—sometimes I think my only home is on the road towards expectations, leaving what's known, complete, behind. I no longer find that condition romantic—at times I resent it deeply—but I accept it."

That sentence was as self-revealing as any Obama had ever written, but the contrast between "the incredible isolation of people here" in far-off Hawaii and "the nervous energy or self-consciousness you find in New York" was discombobulating. "My contradictory feeling for Hawaii reflects itself in my relationship with Bobby, who embodies the beauty and limitations of the place. He's making a comfortable living running a concession at a local high-school, and supplements his income with cash from a few big cocaine deals he was involved with last year"—an aspect of a best friend's life that was unobjectionable only if you viewed cocaine use itself as unremarkable.

Obama observed that Titcomb "jokes about his appetite for food and women, exhibiting a charm and flair in everything he does, but a sustained commitment or depth in nothing." But Barack added, "He loves me and I love him, but he senses different priorities in me now," though "I admire and envy his easy manner and fluid grace." One day the two old friends went scuba diving two miles off Oahu, fifty feet down. On another, Ann's mentor Alice Dewey "argued in husky tones with me and a few other of my mother's friends over politics, sexual relations, art and the economy." Only "after five hours and four cups of coffee" did Barack drive her home.

Alluding most likely to how they had met exactly one year earlier, Barack asked Genevieve, "How did you spend New Year's Eve? Mine was not as eventful as the last one." He told her he was flying back to New York on January 22 and would likely stay with Hasan and his cousin Ahmed at the Eagle Warehouse apartment "until I find a place. I confess to a fear of failing to find a useful gig for myself upon my return, but have no thoughts of doing anything else. I expect the transition may be tough on me," as his prior attempts to obtain a politically satisfying job had failed, "but I expect you to have some patience with my foolishness and kick me when I get out of line. I miss you very much, and hope your enthusiasm for school stays high." He signed off "Love, Barack."

The same day Barack wrote that letter to Genevieve, his mother Ann privately recorded her own plans for the New Year. Many of her jottings concerned the multiple debts she owed her parents, including $1,764 per semester for Maya's Punahou tuition, and $175 for an airplane ticket for "Barry." The "$4,846 withdrawn from account by Toot" was later updated with "$3,940 repaid 2/6/85." A long numerical "People List" began with Maya as #1, Ann's Indonesian lover Adi Sasono #2, "Bar" #3, her parents #s 4 and 5, and included former brother-in-law Omar Obama as #175. The "Long Range Goals" she listed on New Year's Day began "1. Finish Ph.D. 2. 60K 3. In shape 4. Remarry 5. Another culture 6. House + land 7. Pay off debts (taxes) 8. Memoirs of Indon. 9. Spir. develop (ilmu batin) 10. Raise Maya well 11. Continuing constructive dialogue w/ Barry."[55]

Once Obama returned on January 22, Genevieve was disappointed that having him back in her daily life was "so disruptive, instead of a sweet remeeting." Given how challenging teaching first grade at PS 133 was, "I actually find his interruption of my focus on school as damaging, disconcerting," but "he's really into travelling his path with concentrated determination as well. It is still true that I want to live alone." Obama later wrote of refusing the offer of a well-paid job from an impressive black man who headed a New York City civil rights group and had recently dined at the White House with "Jack," the secretary of housing and urban development. Arthur H. Barnes headed up the New York Urban Coalition, but African American New Yorker Samuel R. Pierce was HUD secretary; only four years later did Jack Kemp succeed him.

Instead Obama focused on a job ad from the New York Public Interest Research Group (NYPIRG), for its Project Coordinator post at the City College of New York (CCNY) in West Harlem. NYPIRG, founded in 1973, was based on a campus chapter model first outlined in *Action for a Change. A Student's*

Manual for Public Interest Organizing, a 1971 book written by famous consumer advocate Ralph Nader and three coauthors, one of whom, Donald K. Ross, became NYPIRG's initial executive director. By 1985 NYPIRG had chapters at most campuses of New York State's two public college systems, the predominantly white State University of New York (SUNY) and the largely minority City University of New York (CUNY). Campus referenda that authorized a $2-per-student-per-semester fee provided NYPIRG's financial base.

The project coordinator post paid only about $9,200, half of what Barack had been making at BI, and the CCNY job was open at midyear because of the departure of a young woman whose fall tenure had been unsuccessful. Yet the CCNY chapter boasted one of NYPIRG's most experienced student leaders, Buffalo native Diana Mitsu Klos, who had moved to New York City eighteen months earlier upon being elected NYPIRG's student board chair for the 1983–84 academic year. NYPIRG's campus projects statewide were overseen by Chris Meyer, and 1983 Yale graduate Eileen Hershenov supervised CCNY and other Manhattan chapters from NYPIRG's tumbledown office at 9 Murray Street in lower Manhattan. On some day in late January or early February, Obama appeared there for a job interview with Hershenov and Meyer, who were "enormously impressed" with him, particularly since NYPIRG was "desperate to diversify" its predominantly white staff, especially at such a heavily minority campus as CCNY.

Obama's hiring was all but immediate, as CCNY's spring semester classes began on Monday, February 4. Eileen accompanied him up to City and introduced him to Diana Klos and seven or eight other core chapter members, including Alison Kelley, who thought Obama was "very poised, very together" right from day one. The NYPIRG chapter had a small office with desks and a telephone in a homely metal trailer known as the Math Hut that sat between CCNY's iconic Shepard Hall and the college's low-rise administration building south of 140th Street on the east side of Convent Avenue, just across from the North Academic Center (NAC), a hulking modern gray-brick behemoth that housed City's humanities and social science departments.

The key to NYPIRG's student recruitment efforts was "class raps," where the project coordinator would ask faculty members to give up five minutes of class time so that students could hear a brief pitch about NYPIRG. As of February 1985, NYPIRG's top statewide issue was its Toxic Victims Access to Justice Campaign, which sought passage of state legislation that would allow women and their offspring who had been harmed by the synthetic estrogen DES to file civil damage suits. New York was one of only seven states where such a right to sue did not exist, and generating citizen pressure on state legislators was a major focus of Hershenov and her colleagues' work.

Barack and Diana were present in NYPIRG's trailer office every day, and throughout February they concentrated on doing as many "class raps" as possible in advance of a late February "general interest meeting" intended to attract several hundred students. In some departments, like African American Studies, where senior professors like Leonard Jeffries Jr. and Eugenia "Sister" Bain were notorious for showing up late, if at all, for many scheduled classes, getting "class rap" time was easy. In others, like Political Science, Barack's efforts met with mixed success. Frances Fox Piven, who taught Politics and the Welfare State, and Ned Schneier, whose Congress and the Legislative Process also met on Tuesday and Thursday mornings, both said yes. A young associate professor teaching civil rights and civil liberties would have been less amenable, replying that it was inappropriate to sacrifice classroom time for nonacademic matters.

Like all CUNY colleges, CCNY was entirely a commuter campus, and one with "nowhere to hang out anywhere near there," as Alison Kelley recalled, so generating student participation in such a setting was a considerable challenge. Eileen Hershenov pitched in at least one day a week, and every Friday Obama joined project coordinators from NYPIRG's ten other southern New York schools for an afternoon staff meeting at 9 Murray Street. NYPIRG had an active relationship with Saul Alinsky's inheritors at the Industrial Areas Foundation, and one Friday Michael Gecan, one of IAF's four top organizers, spoke to the group and also spoke individually with Obama.

Eileen Hershenov was Barack's closest staff colleague, and they had several conversations about different models of organizing. She had read Clayborne Carson's 1981 history of the Student Nonviolent Coordinating Committee (SNCC), *In Struggle,* and could remember even two decades later at least one conversation with Obama about SNCC's Mississippi grassroots organizing, which Bob Moses came to personify. "It was a very abstract intellectual conversation that I had with him," she recalled, one that contrasted SNCC's approach versus charismatic leaders, but all toward the end of "how do you empower people?"[56]

Even with his full-time weekday job up at CCNY, Obama spent all of February and early March splitting his time between Hasan's apartment below the Brooklyn Bridge and Genevieve's top-floor apartment in Park Slope, quite a commute on New York's far from reliable subways. "Who is this boy/man/person, Barack Obama?" Genevieve wrote in her journal in early February. "We communicate, we make love, we talk, we laugh. I insulted him the other night—a retaliatory 'fuck you'" for his complaining about Genevieve "always wimping out on dinners with the gang" or saying "You stay, I'm leaving" as a night drew on. "Both of us feeling dissatisfied, wanting something more—but he from himself, and me from the pair of us. . . . I don't really know or understand how he feels, privately, about me, us," given his "veiled withholding."

The next day she wrote that "since I've known him, he has not yet developed a concrete sense of direction," and in retrospect Genevieve remembered that Barack "came back from Hawaii definitely exuding impatience and frustration and dissatisfaction with the life he was leading." She was increasingly stressed by her teaching and by the shared space at 640 2nd Street. In mid-March her unhappiness led Barack to remark, "You like to make trouble." When that led to tears, Genevieve wrote that her own emotional insecurity "all relates back to my father, and his 'abandonment' of me and wanting desperately to have some-one love me like a father." But she also believed that "all of this insecurity" is "a product of the conversation" she and Barack had had "about living together," coupled with all of the "distance on his part." Before the end of March, Gene-vieve found a better apartment at 481 Warren Street #4A in Brooklyn's Boerum Hill neighborhood.

Obama was still in touch with Oxy friends Phil Boerner and Andy Roth, and by early 1985 Andy was living with musician friend Keith Patchel in an informal sublet at 350 West 48th Street in Manhattan's rough Hell's Kitchen neighborhood. Patchel had left for Sweden at the first of the year to work on a record with his friend Richard Lloyd, a member of the underground band Tele-vision, and in March Andy was headed to Managua, Nicaragua, for more than two months. Apartment 4E had been home to the grandparents of neighbor Nick Martakis, a friend of Television's manager, and Keith and Andy simply paid Nick $200 or $250 a month in cash—"there was no lease" and "every-thing was on the down-low," Keith explained. "Keith and I were expected to keep a low profile," Andy remembered, and even incoming mail had to be ad-dressed "c/o Martakis." The building was "decrepit," a step down even from Sohale's apartment on East 94th Street, but it was relatively spacious. It was also a shorter subway trip to CCNY in West Harlem, and on Sunday, March 31, Barack moved in there, while continuing to spend each weekend at Genevieve's Warren Street apartment in Brooklyn.

Two visitors whom Barack had met two years earlier in Jakarta—his mother Ann's anthropologist friends Pete Vayda and Tim Jessup—came for dinner one weekend. Tim was Genevieve's older stepbrother—the son of her stepfather, Phil Jessup. Barack did not say much to Genevieve about his NYPIRG work, and he never introduced her to any staff or students: "that compartmentaliz-ing thing," she later remarked. Genevieve continued to wrestle with her own issues, and in mid-April wrote in her journal, "I am making a commitment here and now to stop smoking pot. I must. Because I am debilitating myself."

Genevieve's ongoing anxieties about teaching were always close to the sur-face. One mid-April weekend Genevieve told Barack that the older teachers at PS 133 had said, "Just stick in there. Nobody has a good first year, and the pen-sion's really good." That pension reference so offended Barack he almost yelled in response. "It just *really* set him off. I had never seen him so upset," Genevieve recounted. "He was almost thumping the table he was so upset—the idea that you would sell out for security" made him "so angry." The next weekend Genevieve and Barack walked from Boerum Hill into Brooklyn Heights, and by Sunday afternoon, he was "acting a tad hostile. When he talks of enjoying being alone, I wonder that he so regularly attends this weekend pattern of ours," she recorded.[57]

On Wednesday, May 1, the CCNY chapter of NYPIRG held a demonstration at Broadway and 137th Street to protest the abysmal condition of that IRT #1 line station. This was part of NYPIRG's Straphangers Campaign for better subway service across the city. The next day a noontime rally outside NAC drew attention to how the NYPIRG chapter, in tandem with CCNY's student government, had gathered more than a thousand handwritten letters to members of Congress opposing the Reagan administration's proposed budget cuts for Pell Grants and guaranteed student loans in the pending Higher Education Reauthorization Act.

That weekend Barack and Genevieve went to Hasan's apartment on Saturday night and got "high on coke," she recorded in her journal. On Sunday Barack sat around her Warren Street apartment reading the *New York Times* and watching basketball. As he left that evening, Barack "said he felt strained," and then told Genevieve, "This apartment is alien to me," a comment she found hurtful. "Barack is discovering the ennui of life being uneventful and unfulfilling and not knowing where to look for the source of it," she wrote. In retrospect she thought he was "depressed" and "not really talking that much." She sensed "a great deal of disappointment and more emotional involvement" with his NYPIRG job than at BI. Barack's inability to distance himself from his job as he had at BI made him "significantly more troubled by what was going on at work." Genevieve thought that "as long as he was still at BI and paying his dues," Barack had believed that a community organizing job would "be the fulfillment of his dream" and gave him hope for the future after his self-imposed 365-day sentence was up. But "the disappointment with how his actual experience" at NYPIRG had played out was leaving him "very broody."

Yet the CCNY students and NYPIRG staffers who worked most closely with Barack could not have been happier with him. On Tuesday, May 7, NYPIRG's annual Lobby Day in Albany drew almost two hundred students from around the state, and the senate leader who previously had blocked passage of NYPIRG's Toxic Victims Access to Justice bill told reporters, "I am convinced we will have an agreed-upon bill before the session ends." The next afternoon CCNY's NYPIRG chapter hosted a well-publicized community forum on federal budget cuts featuring Frances Fox Piven. CCNY's classes ended one week later, on May 15, and the NYPIRG chapter held an end-of-semester pizza party in their Math Hut office.

Alison Kelley had interacted with Barack as much as anyone over the preceding three months, and Obama had encouraged her to overcome her shyness and learn new skills, accept an invitation to join a singing group, and run for a seat on NYPIRG's state board. "He was relentless," and he had "a huge impact," even if he did not realize it. "He changed the vector of my life in *so* many ways," Alison later recalled. Barack was also "constantly bringing stuff in for us to read," including publications on South Africa. In return, "Every day I'd hound him because he smoked," and Barack's usual reply was "We all have flaws." Nonetheless, "every single female had a crush on him" and "everyone thought he was cute," but "he was just always warm and friendly . . . in a very professional way."

At the end of the semester, "we begged him to stay. We loved him," because "we all felt that he was the perfect person for our chapter, to meet our needs," Alison recounted. But Barack was explicit that he would not return to CCNY for the fall semester nor stay with NYPIRG over the summer. "He talked about being frustrated, that he wasn't moving fast enough," Alison remembered. "It was so clear that he wasn't sure what he wanted to do, and he didn't feel that he fit in anywhere." Barack "was just deeply searching for his niche in the world . . . searching in terms of his own psyche." He "seemed unsure of where he belonged" and "didn't know where he was going."

Eileen Hershenov "desperately wanted him to stay," and her boss Chris Meyer remembered "having a conversation with Barack" after "Eileen put me up to trying to convince him to stay." To Eileen it was clear that Barack "was interested in organizing but not this kind of organizing." NYPIRG executive director Tom Wathen concluded that Barack viewed NYPIRG's issues as "too vanilla," and "we were disappointed, but not surprised" when he resigned at the end of CCNY's spring semester.

NYPIRG's top priority, the Toxic Victims Access to Justice bill, would be signed into law a year later by New York governor Mario Cuomo, but in Obama's own subsequent references to his time at NYPIRG, he never once mentioned that signal achievement. The first time he described his experience, nine years later, he stated that "when I first got involved in organizing, in Harlem, I was out there because of liberal guilt, and I got disabused of that real quickly." NYPIRG "sent me out into the middle of Harlem to try to get people involved in environmental and recycling issues," but "the folks in Harlem weren't all that interested . . . and I would guilt them all the time." That description bore no relationship whatsoever to Obama's actual work on CCNY's campus, nor would the single phrase he devoted to his NYPIRG work in his subsequent book—"trying to convince the minority students at City College about the importance of recycling"—be much better.

In a trio of cable news show interviews in subsequent years, Obama would say that he was "an organizer in Harlem" for "about a year," that for "six months" he "recruited students out of the City College of New York . . . to work in the community," and that he left NYPIRG because "the organization ran out of money." None of these statements were accurate.[58]

At much the same time that his work at CCNY ended, in mid-May 1985, Barack's relationship with Genevieve deteriorated seriously. Genevieve wrote a brief poem entitled "Where's the Beef?" invoking a famous 1984 political aphorism. That same day she told her closest friend at PS 133, "I just wanted to chop his dick off," and the next day, she recorded in her journal, "I called him a prick." Years later she had no idea what had made her so "aggressively angry at him" or why her feelings had been "so vicious." That weekend, however, they once again partied with the Pakistanis, but less than two weeks later, Genevieve wrote in her journal about "Barack leaving my life—at least as far as lovers go. In the same way that the relationship was founded on calculated boundaries and carefully, rationally considered developments, it seems to be ending along coolly considered lines."

Once again, Genevieve's own issues were front and center, as she acknowledged, after rereading her journal entries, "how consistently I mention having been drunk and how many times I've said I was giving up pot." Yet "from the beginning what I have been most concerned with has been my sense of Barack's withholding the kind of emotional involvement I was seeking. I guess I hoped time would change things, and he'd let go and 'fall in love' with me. Now, at this point, I'm left wondering if Barack's reserve, etc., is not just the time in his life, but, after all, emotional scarring that will make it difficult for him to get involved even after he's sorted his life through with age and experience. Hard to say."

But within just a few days, on Genevieve's birthday, June 7, Barack gave her a huge philodendron. Even so, a few days later, Genevieve composed another deeply critical poem about him:

You masquerade, you pompous jive, you act,
but clothes don't make a man,
and I know you just coverin' a whole lot of pain and confusion.
You think you got it taken care of,
but I'm tellin' you bro, you don't.
You masquerade, you pompous jive, you act.

A week later, Genevieve wrote a letter to a friend that she never sent. Only one week of school remained, and "I don't know how I've managed to survive this year: it has been horrendous." She was continuing to debate whether to teach again in 1985–86, but she was happy at 481 Warren Street, where "I'm making a home for myself. I had wanted to live with Barack—we were lovers for a year—but the relationship as far as steadiness goes has dropped away. The fact that he's 23 put us in very different places, and I was demanding more than he could give. So I just quietly stopped asking, and it all fell away. We are friends and feel close, and suddenly I find that it's not me who's the confused, needy person in a relationship."[59]

By the second week of June 1985, Barack Obama was unemployed again, and he had broken up with the woman with whom he had had the closest and most intimate relationship of his entire life. He was also living alone in an unfurnished apartment that Genevieve described as "creepy and very dark" under "very sketchy" circumstances that left Barack visibly nervous the few times Genevieve ever visited him there.

But his NYPIRG experience at CCNY had not extinguished Barack's desire to pursue some different kind of organizing, and with time on his hands, he spent a good amount of it at the Mid-Manhattan Public Library, on the east side of Fifth Avenue just south of 40th Street. "I started casting a wide net to see if there were jobs available doing grass roots organizing all across the country," he later recounted. As Obama remembered it, he felt "a hunger for some sort of meaning in my life. I wanted to be part of something larger," something "larger than myself," indeed something less "'vanilla'" than NYPIRG. One resource he carefully perused was the June issue of *Community Jobs*. "I wrote to every organization" that advertised, and one résumé and cover letter he put in the mail some time in mid-June was addressed to Gerald Kellman, Director, Calumet Community Religious Conference, 351 E. 113th St., Chicago, IL 60628.

Jerry Kellman remembered receiving Obama's résumé, seeing his surname and Hawaiian background, and asking his Japanese American wife April whether "Obama" might be Japanese. "Sure, it could be," she replied. Within a day or so Jerry telephoned Barack in New York, and early in that conversation, it was clear that Obama was African American—just what Jerry's ad, and his DCP leadership, hoped to find. Jerry's father lived on the Upper West Side, and Jerry was already scheduled to visit him two weeks later. He told Barack they should talk in person in Manhattan.

Barack was ecstatic and nervous about his upcoming meeting with Kellman. "He was very much keyed up about it," Genevieve remembered, "with a very strong sense of wanting to impress and be found suitable . . . but also a great deal of angst about how the future of his entire life hinged on this meeting." There was the deep attraction of a real community organizing job, but also, visible yet unspoken, was a strong desire to break free from the weekly pattern of "partying" with Hasan, Sohale, and Imad. "He felt trapped by the Pakistanis and their expectations that he would continue to party with them," Genevieve realized. Barack "had zero drive to substance use from within himself. It was just an 'If I don't, they'll think I'm stuck up'" fear on his part. "He was only doing it so as not to rub it in their face that they were still doing the same-old same-old and he wasn't interested." That problem did not present itself when the group gathered at Beenu Mahmood's apartment on Riverside Drive, or when Wahid attended, but otherwise the weekend cocaine parties extended right up through the spring and summer of 1985—"nonstop—without a doubt—continuous," Genevieve replied when asked about that time. "It's uppity to decline because you're being superior, and he just didn't want to." She believed Barack "was very uncomfortable with what he felt was incredibly deep division between where they were going and how they chose to conduct themselves. . . . If he had been capable of hurting their feelings and being disloyal, he would have stopped engaging" in the weekend gatherings, but he was not, so the prospect of *having* to leave New York for an organizing job elsewhere held out a promise of freeing him from the bonds of a friendship he could not bring himself to break but very much wanted to sunder. "He mostly wanted to get away from the Pakistanis," Genevieve believed, for "the restrictions the Pakistanis put on his sense of expanded identity and hopes for the future" had become too much.

Hasan Chandoo sensed much the same thing, especially come that spring and early summer of 1985. Barack "had a tough life in New York. No money, hard work," and in private he had begun to lecture Hasan about how his profligate use of pot and cocaine was part and parcel of a criminal drug economy that was doing untold damage to black young men and black neighborhoods. "He's telling me as a friend, 'Stay away from this shit'" and become "more disciplined," Hasan recalled. "I listened to him" and "I was taken by his maturity," but "I didn't stop completely."

Just before the July 4 holiday, Keith Patchel returned to 350 West 48th Street from Stockholm. Keith was eight months into the sort of lifestyle change Barack was advocating to Hasan, and as a result "I wasn't there very much" since "I was going to a lot of meetings." Apartment 4E was "such a casual, come and go kind of place" that "there were endless occasions when I wouldn't see my roommates for days at a time, especially" given the apartment's layout. "I do remember somebody being there" when Keith returned, "somebody in a Hawaiian shirt," but "we could both be living there and not see each other for days at a time." Only after being told who that had been did Keith come to the realization that "I do have a recollection of meeting Barack."

Sometime soon after the July 4 holiday, Jerry Kellman arrived in Manhattan and met Obama at a coffee shop for a good two hours. "The whole purpose of this interview for me was to ascertain his motivation," Kellman recalled, "because it seemed like he must have had so many opportunities" given Barack's résumé. "He was good-looking and articulate and obviously very, very bright." When Jerry asked him what he knew about Chicago, Barack's immediate reply—"Hog butcher for the world," the opening line of Carl Sandburg's famous 1914 poem "Chicago"—demonstrated just how much literature Barack had absorbed during his college years.

Kellman knew from his years in organizing that "people who were as young as" Barack—not yet twenty-four—sometimes "burn out very quickly" when they "for the first time in their life, encounter significant failure" as novice organizers. For that reason, Kellman wanted to understand why a Columbia graduate who had earned almost $20,000 at BI was applying for this trainee position and its advertised salary of $10,000. Kellman explained CCRC's hope of staunching the Calumet region's loss of steel industry jobs, and how church congregations were CCRC's organizational base. He also made clear that for DCP's Chicago parishes, he inescapably needed a black organizer. But Jerry had to know why this animated Barack.

"What he told me was that he wanted to make positive changes around economic equality," Kellman recalled. Barack "was clearly an idealist" and "was very hungry to learn." Indeed Obama "challenged me on whether we could teach him" more than he had experienced at NYPIRG. "'How are you going to train me?' and 'What am I going to learn?'" were questions Jerry could answer convincingly given his organizing experience stretching back to IAF. Barack wondered too "how would he survive financially" on $10,000, and Kellman suggested Hyde Park as a place to live while making it clear that if Barack rose from his initial trainee status, his salary would go up as well. Jerry then offered Obama a job on the spot, which he accepted, and upon discovering that Barack did not own a car, which would be essential on the Far South Side, Kellman offered him an additional $1,000 to buy a car. A check would be in the mail as soon as Jerry returned to Chicago, and Barack agreed to make the drive to Chicago as soon as he acquired a car that would get him there. Beenu Mahmood was in Chicago, living in Hyde Park while working as a summer associate at the law firm of Sidley & Austin prior to his final year of law school, so Obama knew he had an initial place to stay.

Either that day or the next, Barack went to Brooklyn to see Genevieve at her Warren Street apartment. "He was very, very sure that Chicago would offer him the organizing experience that New York" had not, she recalled, but what she remembered most clearly was the question he posed to her: "Do you want to move to Chicago with me?" Given how their relationship had "devolved" over the previous two months, Genevieve was surprised, but "I was so sick of the withheldness that I never even hesitated with 'No.'" Years later she would debate with herself whether Barack asked only because he was certain she would decline, or whether he still felt a deep attachment to her.

Only in retrospect would she come to believe that "all the tension he was feeling that May and June had very little to do with me" as opposed to Barack's need to find a purpose in his life. Now he was headed to a predominantly African American working environment, even though, in her view, "he had zero experience of black culture." Genevieve had long recognized and teased him about "the grandness of his vision," even though "it wasn't at all an articulated vision." But now Barack was undertaking the most consequential decision of his still young life, leaving a city to which he repeatedly had returned following his college graduation two years earlier for a new metropolis he had glimpsed only as a young child for whom the Field Museum's shrunken heads were Chicago's most memorable attraction.

By July 16 Barack was looking to buy an affordable used car, and within a week, he acquired an "old, beat up" blue Honda Civic for the grand sum of $800, $200 less than what Kellman had sent him. Genevieve recorded in her journal "the thought of being alone again and somehow defenseless once Barack's gone." He had a good-bye lunch with Andy Roth on the Upper East Side, and early on Friday afternoon, July 26, 1985, Barack pulled away from 350 West 48th Street with all of his worldly possessions in his "raggedy" Honda.

"My radio/cassette sprang to life with a slight touch of the antenna, just as I was about to enter the West Side Highway," he wrote some days later. "I tuned into the jazz station and drove over the George Washington Bridge, straining my neck to catch a last glimpse of the Manhattan skyline. It was overcast." Once across the soaring span, he bore right when the expressway forked, with the New Jersey Turnpike bending south and Interstate 80 heading west as the urban clutter of Hackensack and Paterson gave way to the rural countryside of northwest New Jersey before the road descended to the Delaware Water Gap and the historic river that marked the Pennsylvania state line. It was a road he had never traveled, and as Pennsylvania passed by with nary a single city to be seen, all Barack would remember of the transit was "hazy green."

In Boerum Hill, Genevieve mourned the departure of a man she would never see again, "sitting in a chair weeping about the fact that he had left." She wrote in her journal, "So. Alone again . . . Barack's leaving—now being goneness." A chapter in both of their lives had closed, and for Barack a brand-new one was about to open.[60]

"Around 9:00PM, too tired to drive further," Barack turned off of I-80 at the last exit before the Ohio state line. Leaving the interstate, a local highway afforded an easy right turn onto South Hermitage Road. A Holiday Inn was brightly visible, but so was a sign advertising a budget motel a few hundred yards farther north, on the west side of the road across from the Tam O'Shanter Golf Course. Less than eight weeks earlier a tornado, rare in the Shenango

Valley, had leveled many surrounding trees, but the funnel cloud had inflicted only incidental damage on the Fairway Inn.

"I rang the bell at a small, ill-lit lobby, and out came a tall, gangly man with a checkered shirt, plaid jacket and golf hat. He looked like an overgrown leprechaun. . . . He pulled out a slip of paper and ran off the nightly rates in rapid fire. I told him I would take the cheapest room and gave him my driver's license," which featured Barack's full name. The owner "was struck by my name"—"Hussein . . .isn't that some bad guy there in the Middle East?"—"and asked me what I did for a living. I explained my new job, and he went into a ten-minute monologue."

Ten-minute monologues were not unusual for Bob Elia, but the one he delivered that evening to Barack Hussein Obama would replay itself again and again in Obama's mind in the years to come. Barack recounted Bob's monologue in a letter to Genevieve two weeks later, and he would allude to it three years later in a magazine essay. He would also transmogrify Elia into a fictional black security guard in his first book, and he would recount Elia's monologue virtually word for word almost half a dozen times to diverse audiences more than twenty years later.

Many individuals who come to believe that their lives stand for more—sometimes much more—than the sum of their own personal experiences retrospectively identify one signal event, one single conversation, as representing the moment when they first knew that they could contribute to the world something more eternal than their own individual fate. Such experiences occur in places sacred, historic, and profane: the kitchen of a parsonage at 309 South Jackson Street in a southern capital city on a January night, the front yard of Coffin Point Plantation on St. Helena Island on a balmy New Year's Eve, or the lobby of a budget motel at 2810 South Hermitage Road in Hermitage, Pennsylvania, on a warm July evening.

A minister of the gospel might understandably believe he was communicating with a higher being—"He promised never to leave me, never to leave me alone. No never alone. No never alone." Someone less religious might hear a voice from history's recent past, offering fortitude against the national security state. But when Barack Obama's foundational experience occurred, the voice he was hearing was indisputably that of fifty-two-year-old, six-foot-two-inch Bob Elia.

Bob grew up in nearby Farrell, Pennsylvania, where most of the racially diverse population drew its paychecks at Sharon Steel's Roemer Works. Bob had a number of black friends across the early decades of his life. By 1985 the local *Sharon Herald* had for several years been publishing the syndicated columns of conservative black economists Thomas Sowell and Walter E. Williams, and Bob had taken a hankering to both men's writings, regularly clipping and saving

Williams's essays. Bob often cited their analyses when telling acquaintances how they might better themselves, and operating a motel gave Bob a never-ending population of new guests to whom he could offer his thoughts on how they could improve their lives.

Barack's brief response that he was headed to Chicago to become a community organizer was fuel for Bob Elia's fire. "Look here, Mr. Hussein, I'm going to give you the best advice anyone's ever given you. . . . Drop this public service crap and pursue something that's going to get you some money and status . . . and then maybe you'll have the power to do something for your people. I'm telling you now because I see potential in you . . . you got a nice voice, you can be one of them T.V. announcers, or one of those high-priced salesmen. Peddling bullshit, but look here, bullshit's the American way." Bob cited Williams or Sowell in telling Barack that political and social influence follow from economic strength, period. "Black people need more like this fellah, not Jesse with rhymes and jive. . . . Most folks at the bottom can't be helped by you" and "most of 'em don't want your help."

Finally Barack was able to interject a more pressing question: Where could he get dinner? Bob recommended the West Middlesex Diner, back down South Hermitage Road just south of I-80. Wanting some fresh air and time to reflect, "I took the man's words on a long walk along the highway to a small all-night diner and had supper," Barack wrote Genevieve. After eating, "the walk back was cool and silent, the stars cluttering the sky as they hadn't in five years. Back in my room, *The Year of Living Dangerously*"—a 1982 film set in a city that Barack himself knew, Jakarta on the eve of the mid-1960s mass killings—"was playing" on the black-and-white television. If Bob Elia's fervent plea to reconsider his new job had given him something to think about, being reminded of his years in a truly foreign land while in the middle of *this* drive offered even more.

Elia's comments would echo in Barack's memory for decades to come. Three years later, some of Bob's advice would be attributed to a black female school aide. "Listen, Obama. You're a bright young man. . . . I just cannot understand why a bright young man like you would . . . become a community organizer . . . 'cause the pay is low, the hours is long, and don't nobody appreciate you." In its next rendition, Bob's remarks would be made by "Ike," a fictional black security guard: "Forget about this organizing business and do something that's gonna make you some money. . . . I'm telling you this 'cause I can see potential in you. Young man like you, got a nice voice—hell, you could be one a them announcers on TV. Or sales . . . making some real money there. That's what we need, see. Not more folks running around here, all rhymes and jive. You can't help folks that ain't gonna make it nohow, and they won't appreciate you trying."

More than twenty years later, Barack would accurately recount how "I stopped for the night at a small town in Pennsylvania whose name I can't remember any more, and I found a motel that looked cheap and clean, and I pulled into the driveway, and I went to the counter where there was this old guy doing crossword puzzles" who "asked me where I was headed, and I explained to him I was going to Chicago because I was going to be a community organizer, and he asked me what was that." Then came Bob's monologue: "You look like a nice clean-cut young man, you've got a nice voice. So let me give you a piece of advice: forget this community organizing business. You can't change the world, and people will not appreciate you trying. What you should do is go into television broadcasting. I'm telling you, you can make a name for yourself there."

"Objectively speaking, he made some sense," Barack would reflect in that retelling. Two weeks later, he repeated the centerpiece of Bob's monologue word for word. Thirteen months later Obama recited Bob's message once more, as he did again five months after that. In May 2008, Obama recounted his memory of that night yet again: "You've got a nice voice, so you should think about going into television broadcasting. I'm telling you, you have a future there."

In contrast to Barack's indelible memory of their conversation, twenty-nine years later Bob Elia had no recollection of young Obama. When asked, though, he almost immediately told a caller, "You've got a nice voice" before launching into a ten-minute monologue on the life-extending powers of a trio of multisyllabic nutritional supplements.

But in 1985, on the next summer morning, Barack checked out of the motel without reencountering Bob, and after heading down South Hermitage Road and bearing right onto I-80, the Ohio state line was just a few miles ahead. I-80 became the Ohio Turnpike and then the Indiana Toll Road once it crossed another state line. Chicago lay six hours ahead, but as Barack Obama drove west, he was headed toward a place he *really* had never been, indeed toward a place he really had never *known*: he was heading west toward *home*.[61]

Chapter Four

TRANSFORMATION AND IDENTITY

ROSELAND, HYDE PARK, AND KENYA

AUGUST 1985–AUGUST 1988

West of South Bend, the Indiana Toll Road slides southward as the shoreline of Lake Michigan draws near. The Indiana Dunes give way to Burns Harbor and its huge steel mill, which marks the eastern edge of the Calumet region's industrial lakeshore. Gary and East Chicago offer a gritty industrial visage before the highway turns sharply north as the Illinois state line approaches. There the interstate becomes the Chicago Skyway, with the East Side, the Calumet River, and then South Chicago flashing by underneath the elevated roadway.

On Saturday afternoon, July 27, Barack Hussein Obama took the next exit heading for Hyde Park, turning northward on the broad boulevard of Stony Island Avenue. At 67th Street, Jackson Park appeared on the east side of the road, offering sunlit greenery all the way to 56th Street. Beenu Mahmood's summer apartment at 5500 South Shore Drive was just a few blocks away.

Obama stopped at a pay phone but discovered he had miswritten Beenu's number, and he called Sohale to get it right. Then Beenu met Barack in front of the tall luxury building, whose tenants had access to a heated swimming pool plus an on-site deli—"not exactly the setting I had envisioned for launching my career as selfless organizer of the people," Barack wrote Genevieve a few days later. "The discordance only increased when we went to a fancy outdoor café downtown to feast on barbecued ribs."

Beenu's fiancée, Samia Ahad, was in Chicago too, and after a restful Sunday Barack drove south to Roseland on Monday morning, while Beenu headed to Sidley & Austin's downtown office. At Holy Rosary's rectory, on 113th Street across from the sprawling Palmer Park, Barack met his Calumet Community Religious Conference and Developing Communities Project coworkers. Mike Kruglik, he wrote Genevieve, "reminds me of the grumpy dwarf in Snow White" with "a thick beard and mustache. He speaks with the blunt, succinct clip of working class Chicago." That first day "he barely acknowledged my presence" but as the week went on it became clear that Mike is "both competent and warm." Adrienne Jackson was "prim," "helpful and committed," with "polished administrative skills," and Obama quickly determined that she, like himself, had been "hired as much to give the staff a racial balance as she was for her abilities." Of Jerry Kellman, Barack told Genevieve, "In his rumpled, messy way, he exhibits a real passion for justice and the concept of grassroots organizing. He speaks softly and is chronically late, but is real sharp in his analysis of power and politics, and is also disarmingly blunt and at times manipulative. A complicated man . . . but someone from whom I expect I can learn a few things."

Obama also wrote that he "made full use of the amenities" that Beenu's building offered "without guilt." Samia was on her way to becoming a professionally acclaimed chef, and one evening she cooked a Pakistani dinner; Beenu's friend Asif Agha joined them, even though the apartment had no dining table or chairs. Asif, like Beenu, had graduated from the famous Karachi Grammar School before receiving his undergraduate degree from Princeton University. He was the same age as Barack, and he had arrived in Hyde Park two years earlier to begin graduate study in languages reaching from Greek to Tibetan, under the auspices of the University of Chicago's interdisciplinary Committee on Social Thought. With Beenu about to return to Manhattan for his third year at Columbia Law School, Asif was another smart and outgoing member of the Pakistani diaspora that had provided Barack's closest male friends for the past six years.

Tuesday morning Barack discovered that he had left his car lights on all night, and he needed a jump start so he could meet his day's schedule. Wednesday morning the car again failed to start, and Beenu, Samia, and Asif all helped push it to get it going. A deeply embarrassed Barack confessed to Genevieve that "I appeared to have left my brains back in N.Y." because "similar lapses have repeated themselves." By the next weekend, he had signed a lease for a small $300-a-month studio apartment, #22-I, at 1440 East 52nd Street, in the heart of Hyde Park. He also told Genevieve about the "pang of envy and resentment" he felt toward Beenu's "prestigious, well-paying and basically straightforward work as a corporate lawyer" and how it contrasted with his far more precarious existence.

Down in Roseland, Kellman's first goal was to teach Obama community organizing's defining centerpiece, the one-on-one interview, or what IAF traditionalists called "the relational meeting." All Alinsky-style organizing recognized the cardinal principle that first "an organizer has to . . . listen—a lot." According to Industrial Areas Foundation veteran and United Neighborhood Organization adviser Peter Martinez, the ability to listen is *the* "critical skill," for it enables an organizer "to synergize all of the things that they're experiencing so that they can incorporate that into their thinking in a way that when they talk with people, people can hear themselves coming back within the structure of what it is you're suggesting might be done." This, Martinez said, would keep people from feeling like the organizer is putting something "on top of them."

Kellman knew the first month was "very crucial" with any new recruit, and particularly with someone who "had never encountered blue-collar and lower-class African Americans." During Obama's first few days, Jerry took him along so that Barack could watch a veteran organizer ask people to talk about their lives and to say what they thought were the community's problems, listening especially for how that person's own self-interest could motivate them to take an active part in DCP. Obama "struggled with this in the beginning," Kellman recalled, as his connections with people "could be superficial," and "I would challenge Barack to go deeper, to connect with their strongest longings." But Jerry was too busy to do this full-time, and so "very quickly he was out on his own, just talking to people, day after day," with the expectation that each week Barack could conduct between twenty and thirty such one-on-ones with pastors and parishioners from DCP's Roman Catholic churches. Among the first pastors Obama called upon were Father Joe Bennett at St. John de la Salle, the church that Adrienne Jackson attended, Father Tom Kaminski at St. Helena of the Cross, and Father John Calicott at Holy Name of Mary—who had been so responsible for Jerry's ad in *Community Jobs*. Bennett remembered Barack asking him "all kinds of questions," and when he left Bennett thought "what a sharp, brilliant young man."

Kellman also took Obama on a driving tour of the neighborhoods DCP and CCRC serviced. More than two decades later, Barack could spontaneously describe what he saw when Jerry drove east on 103rd Street past Trumbull Park before turning south on Torrence Avenue. "I can still remember the first time I saw a shuttered steel mill. It was late in the afternoon, and I took a drive with another organizer over to the old Wisconsin Steel plant on the southeast side of Chicago. . . . As we drove up . . . I saw a plant that was empty and rusty. And behind a chain-link fence, I saw weeds sprouting up through the concrete, and an old mangy cat running around. And I thought about all the good jobs it used to provide." As Kellman had told him, "when a plant shuts down, it's not just the workers who pay a price, it's the whole community."

The people of South Deering had been living for more than five years with what Obama saw that afternoon. "The mill is just like a ghost hanging over the whole community like a cloud," one St. Kevin parishioner explained, indeed "the ghost of the Industrial Revolution," another resident realized. "It just sat there and rotted before everyone's eyes," Father George Schopp explained, and a beautifully written article in the August issue of *Chicago Magazine*—one that would have made a memorable impression on any aspiring young writer who read it—described South Deering in the summer of 1985 as "the essence of the Rust Belt . . . along Torrence near Wisconsin Steel, the stores are empty; only a few taverns remain."

Kellman also took Obama southward from Roseland, to show him the brick expanse of Altgeld Gardens, which was so distant from all the rest of Chicago apart from the huge Calumet Industrial Development (CID) landfill just to the east where all of the city's garbage was dumped, and the Metropolitan Sanitary District's 127 acres of "drying beds" for sewerage sludge just north across 130th Street. Neither the dump nor the sewer plant ever drew much attention, yet just to the northeast, near the older Paxton Landfill, as *Chicago Magazine*'s Jerry Sullivan wrote, "the greatest concentration of rare birds in Illinois" was spending the summer "squeezed between a garbage dump and a shit farm."

Obscure scientific journals with names like *Chemosphere* occasionally published studies that detailed how the presence of airborne PCBs was "significantly higher" around "the Gardens" than anywhere else in Chicago, but just a week after Obama's introductory tour of the area, an *underground fire* at an abandoned landfill abutting Paxton—a weird and unprecedented event—drew camera crews, reporters, and city officials to the Far South Side's toxic wasteland. The chemical conflagration took more than twelve days to finally burn itself out, and region-wide press coverage featured UNO's Mary Ellen Montes criticizing the U.S. Environmental Protection Agency for its lack of interest. City officials suggested *blowing up* the remaining metal drums with unknown contents, and UNO filed suit in federal court against the EPA in hopes of jarring federal officials into action.[1]

In Roseland, more than a mile northwest of the underground fire, the only newsworthy event in the eyes of Chicago newspapers during Barack's first weeks was the arrest of an initial suspect in a cold-blooded killing that was all too reminiscent of Ronald Nelson's murder just five months earlier. One Saturday evening, shortly before Barack's arrival, forty-nine-year-old factory foreman Enos Conard and his twenty-three-year-old son were manning an ice cream truck on 105th Street, hardly five blocks from Father Tom Kaminski's St. Helena Church. Two men in their early thirties approached and ordered ice cream before one suddenly drew a gun. As Conard reached for his own weapon, the lead assailant fired a single fatal shot into Conard's chest. Conard left behind a widow and four children, and a fellow vendor complained to reporters that "so many people have guns."

A mortician from nearby Cedar Park Cemetery had drawn press attention by publicizing his offer of free funerals and burials to victims of Far South Side gun violence, and Conard's family became the seventh to accept. In late August, police arrested L. C. Riley of Roseland as one of the assailants, and two years later both Riley and triggerman Willie Dixon, also from Roseland, were convicted of Conard's murder. Dixon was sentenced to life in prison, and three decades later, Dixon remained in the same cellblock at Stateville Correctional Center as Nelson's killer, Clarence Hayes.

Also in Roseland, although invisible to Kellman and Obama, ACORN's

Madeline Talbott had hired a new organizer, Ted Aranda, who had worked previously for a year under Greg Galluzzo at UNO, to revitalize ACORN's Far South Side presence after months of inactivity following Steuart Pittman's departure in mid-March 1985. Unlike CCRC's and DCP's church-based organizing, ACORN went block by block knocking on doors to get residents together to tackle community problems. Aranda explained years later that "most of the people that I got involved in the organization were always women." The goal was to attract enough recruits to hold a community meeting, and Aranda's Central American heritage was a significant asset, because his dark complexion led most Roseland residents to assume he was African American. "That's not my identity" once "you look beyond my skin color," Ted said, but he also had an advantage in Latino neighborhoods, where "they took me for a Hispanic."

Aranda learned that Roseland residents were angry that the city had not provided them with garbage cans, and Pullman people were upset about a disco patronized by gang members. By early September, the Roseland group, COAR—Community Organized for Action & Reform—had drawn enough interest that Talbott convened a meeting at King Drive and 113th Street—i.e., Holy Rosary. In late October, the Pullman group, ERPCCO—East Roseland Pullman Concerned Citizens' Organization—succeeded in closing the disco. But by the winter, Ted Aranda "became disenchanted with community organizing as a viable model for radical change," and he resigned. By the standards of Alinsky-style organizing, Aranda's months in Roseland had been successful, but as he explained years later, he "was more convinced than ever by the end of my short organizing stint that the political system itself was the problem." And even though COAR and DCP were working in the very same neighborhood, "Barack Obama I never met at all."[2]

DCP's board met on the second Tuesday evening of each month, and at the August meeting in St. Helena's basement, Kellman introduced Obama so the members could formally ratify his hiring. Virtually everyone was taken aback by how youthful he seemed. "My first thought was 'Gee, he is really young,'" Loretta Augustine recalled years later, and she whispered that to Yvonne Lloyd sitting next to her. Yvonne's first reaction was just like Loretta's: "We had children older than he was." The always outspoken Dan Lee said aloud what they all were thinking: "Whoa, this is a baby right here." Obama smiled and acknowledged that he looked young, but once he spoke to them about himself and responded to their questions, he quickly won them over.

"He was very candid in his answers—straightforward," Loretta remembered. "The impressive part was that he seemed to really understand what we were saying to him," which she considered a marked change from both Mike and Jerry. "When we talked about certain things that he didn't know about, he didn't lie. He basically said, 'You know what, I'm not really familiar with that. However, these are things that we can learn together.'" In short order, "we knew he was the right person for us," Loretta recalled, and though "his honesty has a lot to do with it," so did Barack's appearance. "His color did make a difference to us, because it's important for us and our children and everybody else to understand that people who look like us can do the job."

The day after that meeting, Barack wrote the letter to Genevieve that described his trip from New York—and his unforgettable conversation with Bob Elia at the Fairway Inn—as well as his first weeks in Chicago. Genevieve had called Barack several days earlier, and he began by apologizing for "my phone manner. You know I dislike the telephone. . . . Combined with the lingering pain of separation, I'm sure I sounded guarded and stand-offish. I'm better with letters . . . (yes, more control)." Barack said that Jerry "has thrown me into" several neighborhoods, including Altgeld Gardens and Roseland, "without much . . . guidance" at all. "There are some established leaders with whom I can work, but I must say that for now, I'm pretty confused and feel my inexperience acutely." He realized that having "a trustworthy face" worked to his advantage, as did "the dearth of educated young men in the area who haven't gone into the corporate world." Barack was pleased with the job, but questions remained. "For all the kindness and helpfulness the communities have offered me so far, I can see the thoughts running through their heads—'another young do-gooder.' I know it runs through mine." So "doubts of my effectiveness in such a setting remain, but at least I feel like I'm in one of the best settings to really test my values that I could hope to find right now."

Overall, "the work offers neither more nor less than I had anticipated," which he found reassuring. He characterized Hyde Park as "a poor man's Greenwich Village," but he was pleased with his apartment and "the cheap prices in restaurants" though not "the disappointing newspapers." But another contrast from New York was more striking. "Blacks seem more plentiful, and more importantly, seem to exude a sense of ownership, of comfortable dignity about who they are and where they live than do blacks" in New York. Chicago offered "a much more visible well-to-do and middle class black population who still live in a cohering black community," and "black culture here is more closely rooted to the South; the neighborhoods have a down home feel. . . . Even the poorest black neighborhoods seem to have a stronger social fabric on which to rely than in NY" and "as a result, the young bucks, though no less surly and pained than their NY counterparts, appear to feel less need to constantly assert themselves against the respectable, and in particular, the white, world." Obama wondered whether "these strands of self-confidence" were due to Mayor Harold Washington, whose "grizzled, handsome face shines out from many store front windows in the areas I work."

Barack wrote that he already had swum in Lake Michigan, but confessed "the almost daily thump in my chest, pain and longing when I think of Manhattan, and the Pakistanis, and when I think of you." He wrote out his address and phone number and told Genevieve, "I expect you to make use of this information frequently." He enclosed a $130 check for money he owed her, and closed by telling her about his African sister who had canceled her trip to New York ten months earlier at the last minute: "Auma did get in touch with me and will be coming through Chicago in two weeks. Very excited."[3]

By late summer 1985, Auma Obama was still in university at Heidelberg, but her closest German friend was now studying at Southern Illinois University in Carbondale, a small town southeast of St. Louis, Missouri, which was far from Chicago. Auma traveled to Carbondale for two weeks, calling Barack once to update him on her plans, and then took a long train trip to Chicago, where he met her at the station and then cooked a South Asian dinner for them in his small apartment, where Auma would bunk on the living room couch. Obama was eager to have his sister tell him about their late father, and for the next ten days—interrupted only by his work—the siblings spoke for long hours about Barack Hussein Obama Sr. One day Auma went with him to work at Holy Rosary, where she met Jerry and several parish volunteers. Back in Hyde Park, one evening Auma went beyond her somewhat-edited comments about Obama Sr. and told her brother that he had been fortunate not to have grown up in his father's household, particularly after Obama Sr. married Ruth. Auma showed family photos to Barack, but she also spoke about Obama Sr.'s drinking problem and the suffering his older children endured as a result of his financial irresponsibility, Roy Abon'go even more than her. Auma also mentioned "the old man's" auto accidents and job-loss experiences, and told Barack, "I think he was basically a very lonely man." Barack generally said little in response, but he took time to show Auma Chicago's downtown sights and museums before the Carbondale friend and her boyfriend arrived in Chicago to take Auma with them to Wisconsin. Before she left, Auma urged Barack to visit her once she returned to Kenya.

Auma later remarked that the visit was "a very intense ten days together" and that "I was very conscious of trying to give him a full picture of who his father was." In the immediate aftermath of her visit, Barack said little about the new and sad portrait Auma had painted of Obama Sr. to his coworkers or to his only regular outside-work acquaintance, Asif Agha. Barack and Asif had drinks and dinner almost every Thursday night at a restaurant on 55th Street in Hyde Park. "We hit it off . . . and we saw each other extremely regularly," Agha recalled years later. "He didn't know anyone" beyond DCP, and it was obvious that "the work was stressful, and he was discovering himself." Barack did not talk much about DCP to Asif. "The only person he ever told me much about" was fellow Princeton graduate Mike Kruglik, whom Obama clearly liked. "He came up frequently." Mostly the two twenty-four-year-olds talked about writing. "I used to write poetry, and he used to write short stories," and each Thursday "we would share whatever we had been writing." According to Asif, Barack "was very serious about writing" and regularly turned out short sketches of six to ten double-spaced typed pages, but there was no real suggestion that he would pursue writing as a career. These dinners were sometimes leavened with shots of tequila, giving Obama at least one regular outlet from the stresses and strains of being a real organizer.

A decade later, Obama offered a sketch of Auma's visit that had her arriving at Chicago's O'Hare Airport, not Union Station, but that did describe her telling him about their father's tragic latter years. "Where once I'd felt the need to live up to his expectations, I now felt as if I had to make up for all of his mistakes." Another decade later, during the first six months of his emergence as a nationally known figure, Obama several times opened up about his recollections of Auma's visit. "Every man is either trying to live up to his father's expectations or making up for his mistakes," he told one questioner. "In some ways, I still chase after his ghost a little bit."

In a long interview with radio journalist Dave Davies, Obama spoke more extensively about his father than at any other time in his life, stating that during Auma's visit, he learned that his father had had "a very troubled life." He understood that some of Obama Sr.'s employment problems had occurred "in part because he was somebody who was willing to speak out against corruption and nepotism" within the Kenyan government, but the portrait his mother had so insistently painted "of this very strong, powerful, imposing figure was suddenly balanced by this picture of a very tragic figure who had never been able to really pull all the pieces of his life together." What Auma had told him was "a very disquieting revelation" that really "shook me up" and "forced me to grow up a little bit." While "in some ways it was liberating" relative to the implicit expectations Ann's glowing comments plus Obama Sr.'s own self-presentation to his ten-year-old son back in December 1971 had created, "it also made me question myself in all sorts of ways" because "you worry that there are elements of their character that have seeped into you . . . and you've got to figure out how you're going to cope with those things," particularly how Obama Sr. had behaved toward women and the offspring he sired.

Asked on camera by Oprah Winfrey about his father, Obama said, "he ended up having an alcoholism problem and ended up leading a fairly tragic life." When a friend asked Barack to quickly compose some uplifting advice for young black men, Obama e-mailed that "none of us have control of the circumstances into which we are born" and that some will have to "confront the failings of our own parents." But "your life is what you make it." A few years later, Barack admitted that "part of my life has been a deliberate attempt to not repeat mistakes of my father," whom he acknowledged "was an alcoholic" and "a womanizer." Obama acknowledged that Auma had revealed how their father had "treated his family shabbily" and had lived "a very tragic life." Even though Barack did not speak about this disquieting news to Asif, Mike, or Jerry, the long-term impact of Auma's truth telling would be profound. "This was someone who made an awful lot of mistakes in his life, but at least I understand why."[4]

In early September 1985, Chicago's public school teachers went on a citywide strike. It was the third straight fall, and the eighth in eighteen years, that school days were lost to a labor dispute. The Chicago Teachers Union was demanding a 9 percent pay raise and the city had offered 3.5 percent; only 14 percent of union members had actually participated in the strike vote. Independent observers, such as education researcher Fred Hess, told reporters that both sides were being unreasonable, and a *Chicago Sun-Times* editorial described the union's behavior as "unconscionable." Quick intervention by Illinois governor James R. Thompson led to a 6 percent settlement and the loss of only two school days.

That fall, Jerry Kellman was still savoring the triumph he had experienced in early July when the Illinois legislature appropriated $500,000 to fund a computerized CCRC jobs bank that would assess unemployed workers' skills and market their résumés to potential employers. The big pot of money had been obtained by Calumet City state representative Frank Giglio, a close friend of Fred Simari, the St. Victor parishioner who had been volunteering virtually full-time for Kellman, as well as Hazel Crest state senator Richard Kelly.

The half-million dollars would allow Governors State University (GSU) to hire twenty job-skill-assessment interviewers for ten months to create résumés for unemployed individuals. In news articles about this, Kellman said the program's success was dependent upon "hundreds" of volunteers stepping forward and pressing employers to hire those workers. Once the funding was confirmed, Jerry made plans to shift Adrienne Jackson to help oversee the new program and began aiming for a massive public rally to kick off the enterprise. Before the end of August, he hired Sister Mary Bernstein, a forty-year-old Catholic nun and experienced organizer, and assigned her to St. Victor to handle CCRC's Catholic parishes in the south suburbs. In tandem with Mike Kruglik and DCP, the immediate goal was to mobilize as large a crowd as possible for the kickoff rally Kellman scheduled for Monday evening, September 30, a day before the program office at GSU would open officially.

Kellman arranged for the two most powerful individuals in Illinois—Governor "Big Jim" Thompson and Archbishop Joseph Cardinal Bernardin—to speak at the event. Choirs from Joe Bennett's St. John de la Salle and John Calicott's Holy Name of Mary would perform, and the invaluable Fred Simari would preside as master of ceremonies. Also featured on the podium would be Lutheran bishop Paul E. Erickson, Methodist bishop Jesse DeWitt, Presbyterian executive Gary Skinner, and the towering young Maury Richards from United Steelworkers Local 1033, whom advance press reports described as "president of the state's largest"—they might have added "remaining"—steel workers local.

Five days before the rally, the food processor Libby, McNeill & Libby announced that within the next year, it would close its Far South Side Chicago plant on 119th Street; that meant a loss of 450 good jobs. The Sunday before the rally, Leo Mahon praised his St. Victor parishioners like Fred Simari and Gloria Boyda for the time they gave to CCRC's employment efforts and reminded his congregation that scripture teaches that "the desire for money is the root of all evil."

On Monday evening, CCRC vice president Rev. Thomas Knutson hosted a pre-rally dinner for the almost two hundred program participants at his First Lutheran Church of Harvey before the 8:00 P.M. rally kicked off at nearby Thornton High School. A racially diverse crowd of more than a thousand, including a watchful Barack Obama and dozens of people from DCP's Chicago parishes, filled the gymnasium as Fred introduced the speakers, including Loretta Augustine on behalf of DCP. After Maury Richards told the audience, "we've lost forty thousand jobs in the past few years," the governor came forward and began by saying, "My name is Jim Thompson. My job is jobs." He went on to declare that "jobs are more important than mental health or law enforcement, because unless people are working and paying taxes, there won't be any resources to pay for those services."

But the evening's real star was Cardinal Joe Bernardin, who denounced racism and called for "cultural and ethnic unity in the Calumet region." He noted how unemployment "cuts across racial and ethnic lines," and he promised that "the church is here to help you" while stressing that "the real leadership must come from the laity." Sounding at times like Leo Mahon, Bernardin declared that "every person has a right to a decent home" and vowed that "the cycle of poverty can be broken and community decline can be turned around." The archbishop pledged further church support for CCRC, and the rally ended with a white female parishioner from Hazel Crest asking the crowd: "Do you want to be part of a community that controls its future?" The audience responded with lengthy applause.

For Obama, the rally and the bus ride back to Holy Rosary provided an opportunity to make some new acquaintances, such as Cathy Askew, who had sat quietly through their introductory meeting at St. Helena. He was also able to talk more with the dynamic Dan Lee, DCP's board president, and with Dan's fellow deacon at St. Catherine, the vigorous Tommy West. For Jerry, Fred, Gloria, and most of all Leo Mahon, the rally was a wonderful culmination of their efforts that reached back over five years. Harvey Lutheran pastor Tom Knutson described the rally as "a tremendous experience for the local community."

Now CCRC's challenge was to get the new "Regional Employment Network" (REN) up and running. GSU planned to have some skills assessors ready to begin interviewing unemployed individuals by early November, but in early October news broke that an Allis-Chalmers engine plant and an Atlantic-Richfield facility would soon be closing, costing up to nine hundred more good jobs.

Kellman privately had been told a few days before Bernardin's appearance that the national Catholic Campaign for Human Development (CHD) would be awarding CCRC an additional $40,000 to support the REN program, with an event on Saturday, October 26, marking the public announcement. Obama joined Kellman at the ceremony, and a story in Monday's *Chicago Tribune* marked his first appearance in the Chicago press: "Barack Obama, who works with the Calumet Community Religious Conference, said its grant will be used to assess skills of unemployed workers and to aid them in finding jobs." The first actual assessment sessions kicked off at St. Victor in Calumet City on November 14 and 15, attracting eighty-six applicants ranging in age from nineteen to sixty-seven years old. Six skills assessors prepared a fourteen-page information sheet on each applicant, and Adrienne Jackson wishfully told a local reporter, "There are hundreds of employers out there who need people." She predicted that REN would interview more than thirteen thousand job seekers during the next eight months.[5]

Looming most dauntingly was the future of LTV's East Side Republic Steel plant, where the thirty-three-hundred-person workforce included twenty-four hundred members of Maury Richards's United Steelworkers Local 1033. Since midsummer, LTV executives had been demanding tax abatements from Governor Thompson and Mayor Harold Washington; the city had responded with proposed investment incentives, as the East Side plant, just like Wisconsin and South Works before it, desperately needed significant modernization if there was any chance of long-term survival. LTV had more than $2 billion in debt, had lost more than $64 million in 1984, and its losses for 1985 could be triple that figure. Richards told reporters the only way to save aging U.S steel plants was a commitment from the federal government.

Frank Lumpkin told a congressional subcommittee that the underlying issue was more fundamental: "jobs or income is the basic human right, the right to survive." Throughout the late summer of 1985, Frank had been writing to Chicago's daily papers, saying that unemployed workers "are fed up with programs for training and retraining when jobs don't exist and with job search programs that only provide employment for those who run them." His bottom-line demand was clear: "the federal government must take over and run these mills—nationalize them—for the good of our country and our community."

In the *Chicago Tribune,* Mary Schmich profiled former Local 65 president Don Stazak, who now worked as toll collector on a nearby interstate rather than for U.S. Steel. "I thought of the company as a father," he told Schmich. In early November, Maury Richards and his 1033 colleagues decided that their situation was so dire that previously full-time officers like the local's president would return to work in the mill rather than draw union paychecks.[6]

In late September, Obama got news that was almost as disquieting as Auma's revelations about their father: Genevieve wrote to confess that she had become sexually involved with Sohale Siddiqi. Soon after Barack's late July departure for Chicago, Genevieve had flown to San Francisco to visit a friend before returning to New York on August 14. That evening she and Sohale "went to a Bonnie Raitt concert together and did ecstasy, that's what did it," she later recounted. Her own struggles with alcohol had not improved in the wake of Barack leaving and with the beginning of another school year at PS 133, yet Barack thanked her for "your sweet letter" when he wrote back to her. "The news of Sohale and you did hurt . . . in part because I was the last to know—the Pakis were sounding awfully stiff the last time I spoke to them. But mainly the hurt was a final tremor of all the mixed-up pain I had been feeling before we parted—watching something I valued more than you may know pass from what is, what might still be, to what was."

But Barack's first two months in Chicago leavened his heartache. "It seems that we have both ended up where we need to be at this stage in our lives," so while "the pain of your absence is real, and won't lessen without more time, I feel no regrets about the way things have turned out." He ended by saying he hoped to get back to New York sometime in the months ahead. "All my love—Barack." Reflecting back years later on what had transpired, Genevieve mused that Barack was probably "very disappointed with me," for given Siddiqi's dismissive attitude toward life, Barack no doubt "thought Sohale was an empty shell for a man."

In Chicago, Barack's work environs offered him better opportunities for self-reflection than his once-a-week reimmersion in the easy camaraderie of the Pakistani diaspora when he met up with Asif Agha. Jerry Kellman's invaluable sidekick Fred Simari saw Barack at Holy Rosary almost every weekday that fall. Simari recalled Obama as "quiet, laid back," "extremely bright," and as someone who "seemed like he really studied everything." Father George Schopp had the same impression: Barack was in "learning mode," just "watching and reflecting." In addition to his daily work discussions with Jerry, Mike, and Mary Bernstein, Barack also interacted with Holy Rosary secretary Bonnie Nitsche and the parish's most committed volunteer, Betty Garrett. "We took him as our son from jump street," Betty said of herself and Bonnie. The two regularly pestered Barack and Holy Rosary's forty-one-year-old pastor, Bill Stenzel, about their cigarette smoking, which was allowed indoors only in the rectory's kitchen. This addiction brought Barack and Bill together more than would otherwise have been the case, but that fall Obama also visited every week with St. Helena's forty-five-year-old Father Tom Kaminski. Like Leo Mahon and George Schopp, Bill and Tom were both progressive and challenging priests, men whose religious faith accorded far more closely with Joe Bernardin's Catholicism than with the hierarchical, top-down archdiocese that John Patrick Cody had ruled. Bonnie Nitsche and her husband Wally thought that Bill's strong but gentle spirit made Holy Rosary's small multiethnic congregation into "a microcosm" of what a community would be if you "got rid of prejudices." The rectory "was like one big office," Bill remembered, with the organizers on one side of the first floor, and Bill and Bonnie on the other.

From the beginning, Bonnie thought Barack "was more together, more poised" than his older coworkers, and Bill recalled that Barack became "very curious" about religious faith while suddenly being surrounded by so many committed Catholics. "He had a curiosity about what's this phenomenon" and a "very respectful" attitude toward faith. Obama asked if he could attend Sunday mass, and Bill can recall him sitting with the congregation. Jerry Kellman was about to convert to Catholicism, and he understood how "the churches we dealt with were extended families," ones that exposed Obama to "a broad sense of religion." For Barack "his sense of church and his sense of God became very much a community experience," and "it was a very formative period" for him, Jerry explained. Obama often drank coffee with Tom Kaminski, and they talked "about all sorts of things," including family, but Fred Simari believes that Obama's time in the kitchen at Holy Rosary had the most impact. "Bill Stenzel spent a tremendous amount of time with Barack," and "some of that spiritual-type formation" that Bill exuded "wore off on Barack, there's no doubt."

That fall, Barack continued his one-on-one conversations with pastors like Bob Klonowski of Lebanon Lutheran in Hegewisch and Catholic parishioners like Loretta Augustine at her home west of Altgeld Gardens. "It was surprising how receptive people were to talking with him," Loretta remembered. Tom Kaminski noted "what a terrific listener he was" and watched as Barack's acceptance spread. At the three-month mark, Obama's $10,000 trainee salary was doubled to $20,000, the apprentice director salary that Kellman had advertised five months earlier.[7]

Meanwhile the warfare between Harold Washington and the city council majority opposed to him, led by South Chicago's 10th Ward Alderman Ed Vrdolyak, was constantly in the headlines. Washington had accepted UNO's invitation to speak at its annual fund-raising banquet on October 30, where the mayor would present a thank-you award to the archdiocesan Campaign for Human Development (CHD). Attendees were greeted outside by picketers led by South Deering Improvement Association president Foster Milhouse, who told reporters, "We want UNO out of our neighborhood, and we want Father Schopp and UNO out of St. Kevin's." The far right's complaints continued with a letter to the editor of the *Daily Calumet* denouncing CCRC and UNO and calling for concerned citizens to "rid their communities of these revolutionaries." An anti-UNO rally at the Calumet City American Legion hall featured Foster Milhouse and attracted a crowd of about a hundred, and another letter to the *Daily Cal* thanking the paper for its coverage warned of the philosophy of the "anti-God idolizer of Lucifer" Saul Alinsky.

With UNO adding affiliates in other Hispanic neighborhoods, Mary Ellen Montes chaired an evening meeting that drew a crowd of two thousand. Mayor Washington, Governor Thompson, and powerful Illinois House speaker Michael J. Madigan all joined Lena on stage, but afterward she denounced Thompson's refusal to commit $6 million for a new West Side technical institute. UNO and other Southeast Side groups continued to fight against any expansion of the area's overflowing landfills, but with city officials all too aware of Chicago's looming garbage crisis, Washington's aides maintained an ominous silence on the issue. City officials had finally acknowledged that the well water samples from the isolated Maryland Manor neighborhood south of Altgeld Gardens "definitely contain cyanide," but the projected cost of $460,000 to extend water and sewer lines to those taxpayers' homes postponed any remedial action, even though the *Tribune* and the *Chicago Defender* ran prominent news stories about the problem. More than a year would pass before the work was carried out.[8]

In mid-November Obama was finally able to write a long letter to Phil Boerner. "My humblest apologies for the lack of communication these past months. Work has taken up much of my time," but now "things have begun to settle into coherence of late." Barack described Chicago for Phil, calling it "a handsome town" with "wide streets, lush parks," and "Lake Michigan forming its whole east side." Although "it's a big city with big city problems, the scale and impact of the place is nothing like NY, mainly because of its dispersion, lack of congestion." Chicagoans are "not as uptight, neurotic, as Manhattanites," and "you still see country in a lot of folks' ways," but "to a much greater degree than NY, the various tribes remain discrete. . . . Of course, the most pertinent division here is that between the black tribe and the white tribe. The friction doesn't appear to be any greater than in NY, but it's more manifest since there's a black mayor in power and a white City Council. And the races are spatially very separate; where I work, in the South Side, you go ten miles in any direction and will not see a single white face" excepting in Hyde Park, a considerable exaggeration on Obama's part. "But generally the dictum holds fast—separate and unequal."

Obama said his work took him to differing neighborhoods, with residents' concerns ranging from sanitation complaints to job-training programs. "In either situation, I walk into a room and make promises I hope they can help me keep. They generally trust me, despite the fact that they've seen earnest young men pass through here before, expecting to change the world and eventually succumbing to the lure of a corporate office. And in a short time, I've learned to care for them very much and want to do everything I can for them. It's tough though. Lots of driving, lots of hours on the phone trying to break through lethargy, lots of dull meetings. Lots of frustration when you see a 43% drop out rate in the public schools and don't know where to begin denting that figure. But about 5% of the time, you see something happen—a shy housewife standing up to a bumbling official, or the sudden sound of hope in the voice of a grizzled old man that gives a hint of the possibilities, of people taking hold of their lives, working together to bring about a small justice. And it's that possibility that keeps you going through all the trenchwork."

But Obama's most vivid image was the one Kellman had shown him three months earlier in South Deering: "closed down mills lie blanched and still as dinosaur fossils. We've been talking to some key unions about the possibility of working with them to keep the last major mill open, but it's owned by LTV," which "wants to close as soon as possible to garner the tax loss," Barack told Phil. He ended the letter by saying his apartment was "a comfortable studio near the lake" and that "since I often work at night, I usually reserve the morning to myself for running, reading and writing." He enclosed a draft of a short story dealing with the black church and asked Phil to mark it up and return it. "I live in mortal fear of Chicago winters," and "I miss NY and the people in it . . . Love, Barack."

The letter documented several significant turning points in Barack's new life. Most important of all was the emotional attachment Barack had already developed toward the people on whose behalf he was working: I "care for them very much and want to do everything I can for them." The second was that now, more than three months in, Obama was much more comfortable with his weekday work. As his close friend Asif Agha remembered, "in the beginning, he was enormously frustrated because the whole scene was completely chaotic," but in "struggling with those frustrations," Asif witnessed Barack "coming together around them as a purposeful person." Kellman was focused on the jobs bank and his initial contacts with Maury Richards's Local 1033 at LTV Republic Steel, but Obama's workday involved interactions with DCP's core participants—Loretta Augustine and Yvonne Lloyd from Golden Gate and Eden Green, down near Altgeld Gardens, Dan Lee and Cathy Askew from St. Catherine of Genoa in struggling West Pullman, Betty Garrett and Tom Kaminski from central Roseland, and Marlene Dillard from the solidly working-class London Towne Homes well to the east—which presented daily challenges that were mundane as well as relentlessly unceasing.

Yet as close as Obama became with Loretta and Dan, as well as to white priests like Bill Stenzel and Tom Kaminski, he had no closer a bond with anyone than Cathy Askew, the white single mother of two "hapa" daughters—half black, half white. Askew was a small-town Indiana native whose family had shunned her after she married an African American who fathered her two daughters, then disappeared entirely from their lives. Cathy taught school at St. Catherine's, and one of her daughters, Stephanie, had a congenital heart condition that left her less than fully robust. Early on Obama told Cathy about his own father and mother. As a result, "I don't think of him as really like African American. He's African, and he's American," she said. As he later wrote, Barack also recognized "the easy parallels between my own mother and Cathy, and between myself and Cathy's daughters, such sweet and pretty girls whose lives were so much more difficult than mine had ever been."

Having lived with a black man, Cathy saw Barack as something else: "he didn't look African American to me. I was really glad to find out he was a hybrid, because my kids are hybrids." From the beginning, she thought Barack "was gawky . . . very quiet . . . very thoughtful," she remembered. In those early months, he "was at a loss for a long time, I think, over where to focus. . . . I saw him doing a lot of listening to people and trying to pull things out of people. There was no focus. Everybody seemed like they wanted something different and nobody had anything in common."

Early that winter, Obama organized a meeting at St. Catherine for area residents to voice their unhappiness about police responsiveness to the district commander. Hardly a dozen people showed up, and the commander was a no-show too. Years later, Obama would repeatedly recall that evening. Loretta had been there, as well as Cathy. "He got frustrated. I think he almost quit once," Cathy said. Obama would recount even his core members saying they were tired and ready to give up, and "I was pretty depressed." As he remembered, across from the church, in an empty lot, two boys were tossing stones at an abandoned building. "Those boys reminded me of me. . . . What's going to happen to those boys if we quit?"[9]

At the beginning of December, Chicago's steel industry was back in the news. Eighteen months earlier, Mayor Washington had appointed a Task Force on Steel to study the industry's future, and its report demonstrated that the entire effort had been a waste of time and energy. Early on Frank Lumpkin and his Save Our Jobs Committee colleagues had hoped the task force would show that their ideas about government investment could revive one or more Southeast Side mills. Washington sympathized with these hopes, but the task force brushed aside Lumpkin and his sympathizers. Its primary academic consultant, Ann R. Markusen, asserted that "Reaganomics and industrial leadership (or lack of it) deserve the major blame" for the plant closings and added that Chicago was "a city in deep, long-term trouble." She also acknowledged that "national and international forces beyond the grasp of local governments are important determinants of steel job loss." Task force member David Ranney stressed that in retrospect the "idea that the steel industry in Chicago might recover appeared absurd" and that what the Far South Side experienced during the 1980s "highlights the limits of local electoral politics."[10]

During November and December, Maury Richards from USW Local 1033 and Jerry Kellman from CCRC met four times with Chicago's economic development commissioner, Rob Mier, one of Washington's most influential aides, to determine how they should react to the loss of a thousand or more jobs if there were even a partial shutdown of LTV's East Side mill. Mier knew that this would be "a severe blow" to the city because a majority of the steelworkers lived in Chicago, and the mill paid $18 million a year in city and state taxes while contributing $300 million to the city's economy. The union, CCRC, and Mier agreed that everything possible had to be done to convince LTV to invest $250 million to rebuild a blast furnace and acquire a continuous caster, or alternatively to sell the aging plant to a company that would. "Most of the discussants seem to realize that neither of these outcomes may be possible," Mier told the mayor. Noting that LTV held defense contracts worth $1.3 billion, Mier reported that "both CCRC and Local 1033 are committed to coordinating a corporate campaign against LTV if LTV refuses" to invest or sell.

Obama, Loretta Augustine, and Marlene Dillard went with Jerry to one or more of his meetings with 1033's officers and city officials, though no one except Jerry spoke on CCRC's behalf. On December 19 Kellman and Richards sent a formal letter to the mayor, copied to Mier, asking Washington to "provide the leadership," in coordination with local congressmen, that would pull together a package of city, state, and federal financial incentives for LTV. The Christmas holidays stalled these actions, but in the press, LTV called its Chicago mill "more vulnerable" than its remaining ones in Ohio and cited its massive $2.6 billion debt burden, intimating that bankruptcy was far likelier than any investment of capital in the East Side plant.[11]

Over the holidays, Barack took off almost two weeks, flying first to Washington, D.C., to meet his—and Auma's—older brother Roy Abon'go, who had married an African American Peace Corps volunteer named Mary. Before leaving

Chicago, Barack told Tom Kaminski how apprehensive he was about seeing Roy, and the visit got off to a bad start when Roy failed to meet him at the airport. When Barack telephoned, Roy said a marital argument meant that Barack should find a hotel room rather than stay with Roy. The two brothers did have a long dinner that night, plus breakfast in the morning, before Barack headed to New York, where he would rendezvous with his mother and sister and where Beenu Mahmood and Hasan Chandoo were happy to offer free lodging and renew their close acquaintances.

Maya was now a tenth grader at Punahou, and Ann was still living in Honolulu, trying to finish her Ph.D. dissertation. Two months earlier, the Internal Revenue Service had levied a $17,600 assessment against her for unpaid taxes on her 1979 and 1980 income from USAID contractor DAI, but Ann would leave the levy unpaid for years to come. Barack ended up spending more time with Hasan, Beenu, and Wahid Hamid than with his mother and sister, and though he did not see Genevieve in person, a letter he wrote to her soon after New Year's recounted an emotional phone conversation they had had, albeit one she would be unable to recall in any detail years later.

> Hard guy that I am, I've managed to stay embittered and sullen towards you for a whole week and a half. But that's about it. I won't try to analyze whether what I did was correct or incorrect, right or wrong, for you or for me. I do know that I had to vent my feelings fully; otherwise I would have choked off something important inside me, permanently. Had to get my head and heart in better communication with each other, in better balance. The consensus seems to be that the whole episode was good for me. My mother and Maya enjoyed comforting me for a change. Asif, my linguist friend in Chicago, says I need the humility.

Whatever had transpired, Barack wrote, "reminded me of the rare, fleeting nature of things. My own dispensability," and "perhaps I'm more apt to believe now something you seem to have understood better than I—when happiness presents itself . . . grab it with both hands." But "I still feel some frustration at the fact that you seemed to have wrapped me up in a neat package in our conversations. Stiff, routinized, controlled. The man in the grey flannel suit. A stock figure. It felt like you had forgotten who I was."

Friends, Barack wrote, "recognize who you are . . . even when you're acting out of character," as Barack apparently had. "I hope I haven't lost that with you. I hope I remain as complicated and confusing and various and surprising in your mind as you are in mine." He closed by saying that "phone calls will still be tough on me for the time being, but cards or letters are welcome." He hoped to get back to New York in the summer to see the child whom Wahid and his wife Filly were expecting, and "hopefully we can spend some time more productively than this last time out. Some fun, maybe. Laughter. Ambivalently yours. But w/ unconditional love—Barack."

More than six months since they had parted, the depth of Barack's emotional tie to Genevieve remained powerful indeed.[12]

While Obama was away, two major developments upended Chicago politics. The *Chicago Sun-Times* gave Harold Washington a stinker of a Christmas morning gift by revealing that an undercover FBI informant, working at the behest of the local U.S. attorney, had made cash payoffs to several city officials and aldermen. As the story played out over the coming weeks, Michael Burnett, aka Michael Raymond, had been introduced to his targets by a "friendly, easy-going" young lobbyist, Raymond Akers, whose car sported a personalized license tag: LNDFLL. Akers was the city council lobbyist for Waste Management Incorporated (WMI), which had 1985 revenues totaling $1.63 billion.

Four months earlier two administration appointees had accepted as much as $10,000 in cash from Burnett, and on December 20, FBI agents had confronted 9th Ward alderman Perry Hutchinson at his Roseland home. On two occasions in early October, Hutchinson had accepted a total of $17,200 from Akers in a lakefront apartment near Chicago's Navy Pier. Unbeknownst to Hutchinson, FBI agents in the apartment next door filmed the encounters with a camera inserted through the common wall. A week later Hutchinson accepted another $5,000 from Akers in Roseland. All told, Hutchinson had received $28,500, and he told journalists, "I figured as long as the guy was dumb enough to give me all of that money, I'd be smart enough to take it." Hutchinson claimed he used $8,500 to hire an additional staffer and had distributed the remaining $20,000 to schools and community groups in Roseland. Reporters were unable to identify any recipients.

Soon it was revealed that a second black alderman and mayoral supporter, Clifford Kelley, who had led a city council effort to discredit a top WMI competitor, had accepted cash bribes too. Then news broke that city corporation counsel James Montgomery allegedly had been aware of at least one of these payoffs months before Washington first learned of the bribes on Christmas morning. The *Chicago Tribune* described this as "a widening scandal that some believe could cost [Washington] re-election" a year later. Montgomery quickly resigned, but several weeks later a *Tribune* story headlined "Lobbyist Paid for City Aide's Vacation" showed that a year earlier, Akers had given a travel agent $4,200 in cash to cover a weeklong trip to Acapulco for Montgomery and his family. No charges ensued.

Independent white voters who loathed Chicago's long history of public corruption had been essential to Washington's 1983 triumph, and they would be needed for his reelection in 1987. Ironically, as the controversy built, mayoral opponents like Ed Vrdolyak remained largely silent. One opposing alderman explained to the *Tribune*: "If we do nothing, the mayor might ultimately bury himself."

Despite Washington's huge popularity among African Americans, his first three years in office had been anything but successful. Some months earlier, *Chicago Magazine*—not a bastion of Vrdolyak supporters—had published a thoroughly negative report on Washington's record to date. He "has been miserably inept at communicating his ideas to the city" and "his administration is plagued by excessive disorganization," *Chicago* reported. Washington had "gained 30 pounds" and looked "physically run-down." One black activist who had championed his election back in 1983, Lu Palmer, complained that Washington was relying upon "apolitical technocrats" who were "barricaded on the fifth floor of City Hall. The people aren't part of it." The magazine also said Washington "may be the least powerful Chicago mayor in recent history" and singled out the Chicago Housing Authority (CHA) as a "full-fledged disaster." Politically, Washington's agenda was "to a great extent, stalled," primarily because Vrdolyak's city council majority kept the mayor largely "on the defensive."

When the bribery scandal broke, Washington's administration was already besieged. Yet a federal appeals court ruling in August 1984 that black and Hispanic voters were so underrepresented by the city's existing ward map that the Voting Rights Act was being violated seemed to provide an opening for Washington, because new elections could overturn Vrdolyak's 29–21 council majority. In June 1985, the U.S. Supreme Court refused to review the appellate finding, and the case was sent to a district judge who instructed the opposing lawyers to redraw seven wards, all of which were represented by Washington opponents. On December 30, the court ordered new elections in those wards to be held on March 18. In the interim, Washington announced that Judson H. Miner, the forty-four-year-old civil rights lawyer who had litigated the redistricting challenge, would be his new corporation counsel.[13]

But on the Far South Side, the fate of LTV Republic's East Side steel mill was still in question. In early January, a front-page *Wall Street Journal* story described the company's prospects as "dim at best," and a week later *Crain's Chicago Business* published a long report that the *Daily Calumet* said "sent shock waves through the Southeast Side." Jerry Kellman told the local paper that LTV should put the mill up for sale, but Maury Richards said that the 1033 union understood that the East Side facility was "losing between $5 million and $8 million every month."

In mid-January Obama attended a three-day training event in Milwaukee for minority organizers sponsored by the Campaign for Human Development and the Industrial Areas Foundation. Soon after he returned, Chicago headlines confirmed Jerry and Maury's fears: "LTV Announces 775 Layoffs" plus the closing of the East Side mill's most modern blast furnace. Kellman told reporters that Illinois officials had "written off the steel industry" and GSU's Regional Employment Network immediately scheduled four days of skills-assessment interviews at Local 1033's office in an effort to find new jobs for laid-off workers. Yet everything Frank Lumpkin and his colleagues had experienced in the years since Wisconsin Steel's sudden closure told them how scarce jobs were on the Far South Side and in the south suburbs.

Loretta Augustine and Yvonne Lloyd encouraged Barack to focus on the sprawling Altgeld Gardens public housing project, just east of where they lived. They introduced him to their pastor at Our Lady of the Gardens, Father Dominic Carmon, and Obama also met with parents whose children attended Our Lady's small Catholic grade school. He was introduced to Dr. Alma Jones, the feisty principal of Carver Primary School and its adjoining Wheatley Child-Parent Center, virtually all of whose students came from within Altgeld. Jones was immediately impressed with Barack. "Talking to him, he was so much older than he was. It was like talking to your peers rather than somebody the age of your children." In a community where few people could imagine meaningful change, Jones stood out as an important voice of encouragement for a young organizer venturing into unfamiliar territory.

Despite Altgeld residents' letters to Mayor Washington complaining about "heavy drug traffic" seven days a week, no city residents were more completely ignored and forgotten than the tenants of Altgeld Gardens. Dr. Gloria Jackson Bacon, who almost single-handedly provided medical care to Altgeld residents for decades, explained that by the 1980s "many of them did not venture outside. Many of them lived almost like insular lives inside of Altgeld." Bacon recalled others speaking pejoratively of "'those people out there, those people out there,' and I'd say, 'These are *your* people.'" Loretta Augustine remembered once taking some Altgeld schoolchildren to the zoo and realizing that "one or two of the kids had never been downtown before." Loretta referred to Lake Michigan, and "the kid responded, 'Chicago has a lake?'" As Alma Jones told one reporter who visited her school, "Altgeld Gardens is an isolated, enclosed island. We have no stores, no jobs and one traffic signal."[14]

Barack continued his weekly get-togethers with Asif Agha, but his social life was so meager that Kellman and Mary Bernstein discussed ways to help him meet more people his own age. "He felt to me like a nephew," Mary remembered, with Barack calling her "Sistah," and she calling him simply "Obama." In her eyes, Barack "was always serious," indeed "driven," but above all "he was very solitary." Bernstein recalled that Barack once asked her, "How am I going to get a date?" working in Roseland and spending his evenings at meetings or visiting DCP parishioners. "You don't want to date any of the women I know,"

Mary humorously replied. "They're all old, and they're all nuns."

Loretta Augustine, Yvonne Lloyd, and Nadyne Griffin were all looking out for the young man's welfare. "I felt very protective, very motherly towards him," Loretta later told journalist Sasha Abramsky. "We were worried that he wasn't eating enough," Yvonne Lloyd recalled. "We were always trying to make him eat more." Loretta could see that Barack was "very focused" and "very serious," and more than once she suggested he lighten up. "You shouldn't be so somber and uptight and serious all the time." Obama later said he was indeed "very serious about the work that I was trying to do." Marlene Dillard's strong interest in jobs had her spending as much time with Barack as anyone, and though she found him "very dynamic" and "very sincere," his maturity meant she "never looked at him as a son." But Nadyne Griffin felt just like Loretta did: "He was just like a son to me," and Tom Kaminski found the church ladies' solicitude heartwarming: "Everyone wanted to be his mother, everyone thought she was his mother," and "I felt like an uncle."

Barack stayed in touch through regular long-distance phone calls with old friends like Hasan Chandoo, Wahid Hamid, and Andy Roth, and in late February, he sent Andy a long letter that was similar to the one he had written Phil Boerner three months earlier:

> As I may have told you on the phone, when you're alone in a new city, the fullness or emptiness of the mailbox can set the tone for the entire day.
>
> Work continues to kick my ass. A lot of responsibility has been dumped on me: I'm to organize an area of about 70,000–100,000 folks and bring the local churches and unions into the action. I confront the standard stuff: the turpitude of established leaders (i.e. aldermen, preachers); the lassitude of the masses; the "we've seen middle class folks come in here before and make promises and ain't nothin' happened" attitude, which is true; my own inhibitions about playing for power and manipulating folks, even when it's for what I perceive to be their own good. At least once a day I think about what I'm doing out here, and think about the pleasures of the upwardly mobile (though still liberal Democratic) lifestyle, and consider chucking all this. Fortunately, one of two things invariably snap me out of my brooding: 1) I see such squalor or degradation or corruption going on that I get damned angry and pour the energy into work; or 2) I see a sign of progress—one of my leaders, a shy housewife, dressing down some evasive bureaucrat, or a young man who's unemployed volunteering to help distribute some flyers—and the small spark will keep me rolling for a whole day or two.
>
> Who would have ever believed that I'd be the sucker who'd believed

all that crap we talked about in the Oxy cooler and keep on believing despite all the evidence to the contrary. Speaking of contrary, I'm in such a state for lack of female companionship, but that will require a whole separate exegesis.

A few weeks later, Barack sent a postcard to his brother Roy and his wife Mary, and a longer message to Phil Boerner. He thanked Phil for his encouraging comments about the short story Barack had sent him, but he emphasized what a "discouraging time" he was having:

Unfortunately, I haven't had much time for writing (stories or letters) lately, what with this work continuing to kick my ass. Experienced some serious discouragement these past three weeks, mainly because of the incredible amount of time to get even the smallest concrete gain. Still, I'm putting my head down and plan to work through my frustrations for at least another year. By that time I should have a fairly good perspective on both the possibilities and limitations of the work.[15]

No one in mid-1980s Chicagoland had anywhere near the degree of success Jerry Kellman did in winning major grants from the Campaign for Human Development, the Woods Fund, the Joyce Foundation, and Tom Joyce's small but always-pioneering Claretian Social Development Fund. Grant makers regularly visited the organizations they supported, and by early 1986, Jerry had been introducing Barack as a new mainstay in DCP's Far South Side organizing work. Archdiocesan CHD staffers Ken Brucks and Mary Yu met Barack through Kellman, but the two most influential funders Barack got to know that winter were Woods Fund director Jean Rudd and program officer Ken Rolling. Jean had become Woods's first staffer five years earlier. Ken, like Greg Galluzzo, was a former Catholic priest who had spent more than half a dozen years in organizing before joining Jean at Woods in 1985. Woods's commitment to organizing was reaching full flower just as Jerry and then Barack arrived on the scene. As Jean deeply believed, "community organizing is intended to be transformational for 'ordinary' people. Through its training and actions, people recognize their worthiness, their legitimacy, their place in a democracy, their power, their voice." That was the work, and the teaching, that she and Ken wanted to support and champion.

Decades later Jean remembered when Jerry first brought Barack to meet them. "In that first meeting, I was very, very impressed. . . . He was very, very reflective, very candid . . . very winningly . . . humble about what he had to catch up on" about organizing and about Chicago. "I believe I said to my husband, 'I've really met the most amazing person today.'" But most of Woods's actual contact with DCP, CCRC, and other grantees like Madeline Talbott's ACORN was handled by Ken Rolling, who was even more impressed upon first meeting Barack. "I've just met the first African American president," he told his wife Rochelle Davis that evening. Ken said much the same thing to CHD staffer Sharon Jacobson, who a quarter century later remembered it just as Ken and Rochelle had: "I want you to watch this guy, Sharon. He's going to be president of the United States one day."

Jacobson played a leading role in CHD's grant making in Chicago, and Renee Brereton, based in Washington, was the crucial staffer for allocating national funds. She had directed $42,000 to CCRC in 1984, $40,000 in 1985, and by early 1986, Brereton and Jacobson were overseeing that year's grant to DCP. She gave DCP's application 90 out of a possible 100 points, citing as the only shortcoming the confusing organizational overlap between CCRC and DCP. Brereton believed that "DCP is expanding its power base through coalition work with unions, and public housing projects," and she was impressed with the leadership training Kellman had done with parishioners from DCP's Catholic churches. "The staff is strong with a commitment to hiring minority staff," and she recommended at least another $30,000.

Sharon Jacobson oversaw the local Chicago committee that ratified Brereton's recommendations, and she noted DCP's intent to develop an employment training and placement program for residents of Altgeld Gardens as well as its desire to improve Far South Side public schools. She wrote that "the large geographical area" DCP sought to cover "is too broad" and threatened to dissipate DCP's efforts rather than focus them, but DCP was Chicagoland's "strongest organizing project. In the past year, we have witnessed thorough leadership training, successful multi-issue campaigns, and widespread grassroots community support," and $33,000 was committed to DCP.

Jacobson also wrote to Brereton that Obama's attendance at the CHD-IAF minority organizer training in Milwaukee had proven notable; out of twenty attendees, he and one other "had demonstrated the most potential," leading Jacobson and Mary Yu to recommend that Barack be invited to attend IAF's premier training event, a ten-day course that took place each July at Mount St. Mary's College in the Santa Monica Mountains above Los Angeles. CHD would pay Barack's $500 tuition, $400 room and board, and also cover his travel expenses.[16]

On March 18, special elections were held in the seven redrawn city council wards. Washington's backers captured two seats from Vrdolyak's 29–21 majority, but two other pro-Washington candidates fell short of the necessary 50 percent plus 1 and were forced into runoff contests to be held on April 29. Victory was assured for the Washington supporter in the black-majority 15th Ward, but in the 26th Ward, two Puerto Rican candidates, one of whom was sponsored by powerful Vrdolyak ally Richard Mell, faced off amid a cascade of election-misconduct allegations. The *Chicago Tribune* labeled the 26th Ward contest "the most closely watched election in Chicago history," with both Washington and Vrdolyak campaigning there two days before the rematch, and Washington's young election lawyer, Tom Johnson, a 1975 graduate of Harvard Law School, keeping a close eye on the proceedings. When Washington's ally, Luis Gutierrez, prevailed by a surprisingly comfortable margin of more than 850 votes, the mayor attained a 25–25 city council split, putting him in position to cast a decisive tie-breaking vote—at least until the next regularly scheduled city elections just one year later.

While most of Chicago focused on the Washington vs. Vrdolyak contest, another vote—by United Steelworkers members on a dramatically concessionary new contract with LTV—was building to its own climax on April 4. One week earlier, on March 28—the sixth anniversary of Wisconsin Steel's closure—Harold Washington met with Frank Lumpkin's Save Our Jobs Committee. At LTV Republic's East Side mill, Maury Richards campaigned against the proposed 9 percent *reduction* in workers' hourly wage rates and benefits. But even though 1033's members voted against the new contract 1,254 to 750, well over 60 percent of LTV's nineteen thousand steelworkers in other states approved it.[17]

As winter turned to spring, DCP began to focus on the forlorn state of the Far South Side's public parks, a visible example of basic city services being denied to black and Hispanic neighborhoods but not white ones. The city's parks were overseen by a quasi-independent entity, the Chicago Park District (CPD), which remained a notorious nesting ground for white Democratic ward organization loyalists, even after Washington's three years in office. Two energetic DCP members—Nadyne Griffin, who had a special interest in Robichaux Park, up at 95th Street, and Eva Sturgies, who lived across the street from Smith Park, at 99th and Princeton—had brought this issue to Obama's attention. DCP began distributing leaflets in the solidly middle-class blocks around Smith Park, encouraging the community to attend a meeting to address the problem. Aletha Strong Gibson, a college graduate homemaker in her early thirties who lived one block south on Princeton, knew that Smith Park "really wasn't very safe or conducive for young children" like her six- and four-year-olds. Gibson went to the meeting and spoke up. Afterward Barack "said he'd like to come meet with me about doing some more work on the parks issue," and following that one-on-one Aletha became a key recruit.

One day early that spring, when Barack was visiting the handsome old Monadnock Building in the downtown Loop, which housed many small progressive organizations, he stopped into the offices of the ten-year-old Friends of the Parks (FOP) and introduced himself to John Owens, a twenty-nine-year-old army veteran who had become FOP's community planning director a year earlier after finishing a degree in urban geography at Chicago State University. Owens was immediately impressed with Obama. "This guy sounds like he's president of the country already," Owens recounted just four years later. "He had an air of authority and a presence that made you want to listen." Barack talked about the discriminatory treatment accorded Far South Side parks, and Owens explained some of what he knew about "the ins and outs of the Chicago Park District." Barack had "all kinds of personality," and "we sort of clicked," Owens explained.

Barack invited John down to Roseland, and they worked together to start compiling a list of parks the CPD was ignoring: Abbott Park, east of the Dan Ryan Expressway; West Pullman Park, on Princeton Avenue; Carver Park, down in Altgeld Gardens; and the huge Palmer Park, just north of Holy Rosary. One afternoon in Palmer Park, gunshots sounded nearby, and they both ducked behind parked cars. Owen recalled Obama saying, "'You hear that? Whoa!'" and remembers thinking, "'Well, he hasn't been around here very long.'"

John and Barack hit it off. Some evenings they went to music clubs together. "I could see he was somebody that I could learn a lot from," said Owens, and Obama also could learn from Owens, a native of the South Side's middle-class Chatham neighborhood, about his life experiences as a black man who had grown up in Chicago. Johnnie—as he was often called—quickly became Barack's first truly close black male friend, at least since the cosmopolitan Eric Moore at Oxy. Before the end of April, Barack asked Johnnie to attend the upcoming July IAF training in Los Angeles, and Owens readily agreed.[18]

In early May, Jerry Kellman's CCRC got another major grant: $30,000 from the Joyce Foundation to support Mike Kruglik's organizing work in Chicago's south suburbs. But Kellman also used his connections within Chicago's Catholic archdiocese to re-create, in somewhat different form, Tom Joyce and Leo Mahon's original vision of CCRC as an organization straddling the Illinois–Indiana state line to encompass the entire Calumet industrial region. Jerry had been thinking about restarting work in Indiana even when Barack first arrived, but underlying that was a fundamental truth that Leo and Tom had experienced and that Greg Galluzzo best articulated in explaining that "what was joined together industrially and geographically is not together politically."

Chicagoans, especially black Chicagoans, who were deeply proud of their first black mayor, identified with their city. Residents of the south suburbs, many of whom had fled Chicago in earlier years, focused on their own townships and the larger overlay of Cook County, not the city's government. Indiana residents paid little attention to Chicago politics and even less to Illinois. These regional identities were more significant than Leo and Tom's belief in a "Calumet community." Also the diocese of Gary, in northwest Indiana, was organizationally separate from the Chicago archdiocese, with independent financial access to local and national CHD resources.

By early May, Kellman had successfully attained funding from Father Tom Joyce's Claretian Social Development Fund for what he called the "Northwest Indiana Organizing Project," and thanks to Bishop Wilton Gregory, one of Cardinal Bernardin's deputies, Bernardin requested that Norbert F. Gaughan, who had become bishop of Gary in October 1984, meet with Gregory and Kellman so that CCRC's work could be reborn east of the Illinois state line. Gaughan agreed to commit his own diocesan funds to the effort, and sent a letter to the pastors of all his parishes, encouraging them to work with Kellman, who began establishing ties with existing groups such as East Chicago's United Citizens Organization. But neither UCO nor the larger Calumet Project for Industrial Jobs, with a predominant union focus, used the church-based organizing model Kellman had deployed so successfully in creating DCP. Kellman's and Kruglik's Illinois efforts outside Chicago would be given a new, regionally distinct identity as the South Suburban Action Conference (SSAC), with Jerry as its executive director until his shift to Gary was complete. Everyone agreed that all three pieces—Barack's, Mike's, and Jerry's—would flourish better on their own than under the old "Calumet Community" rubric.

By late spring 1986, it was clear that Frank Lumpkin's cynical complaint that CCRC's Regional Employment Network only created employment opportunities for its own employees was true. One headline announced that "Regional Employment Network Reports Initial Success," but, as the article made clear, the organization's definition of "success" was its data bank of 1,350 job seekers, not actual job offers. Governors State University appealed to state officials for an additional $375,000 to extend the program past the summer of 1986, but the request received no support. Obama would later remember REN as "a bust" that failed to find work for even one applicant. The savvy Fred Simari recalled how "there was an elaborate system to assess their needs," but it "was all smoke and mirrors, the whole thing."[19]

By mid-April, Barack had been working for several months to broaden DCP's outreach with Altgeld Gardens residents, but with only modest success. Then one lady, Callie Smith, handed him an ad she had seen in April 14's *Chicago Sun-Times*. "Specification No. 8632" sought bids for the "Removal of Ceiling and Pipe Insulation Containing Asbestos at the Management Office Building of Altgeld Gardens, 940 E. 132nd Street." The bids were due April 30, and specifics could be obtained at Chicago Housing Authority (CHA) headquarters in the Loop. Potentially cancer-causing asbestos had been discovered in December 1985, but Altgeld was not the only CHA property with such a problem: asbestos had just been uncovered in two apartments at the Ida B. Wells

Extension Homes in the Bronzeville neighborhood. Linda Randle, an organizer at the nearby Centers for New Horizons and who lived in another part of Wells, told her friend Martha Allen, who wrote for the Community Renewal Society's (CRS) monthly *Chicago Reporter,* about the asbestos. Allen arranged for laboratory testing of a sample from Wells, and the results were shocking: "to find that much amosite [a type of asbestos] there is astounding," one scientist stated.

Obama met Randle at a CRS-hosted meeting of organizers, and when Linda mentioned the discovery at Wells, Barack pulled out the *Sun-Times* ad and said, "the same thing is happening in Altgeld." Linda and Barack agreed to be in touch, and back in Altgeld, Callie Smith called CHA manager Walter Williams to ask if the CHA had determined whether or not asbestos was present throughout the hundreds of homes as well as in the management office. In his own later telling, Obama accompanied Callie to a meeting where Williams said the CHA had checked and none had been found. Smith and Obama understandably doubted that assertion and sought documentation to back up Williams's claim.

On May 9, several residents met with Gaylene Domer, executive assistant to CHA executive director Zirl Smith, to request immediate, independent testing of Altgeld residential buildings and public release of the results. They also asked that Smith appear at an Altgeld community meeting to respond to residents' concerns. After a week with no response, Callie Smith, Loretta Augustine, and two other members of the Altgeld Developing Communities Project sent a Western Union Mailgram to Smith, with a copy to Mayor Washington. Citing the May 9 meeting, followed by CHA's silence, their message repeated the two requests and asked for a written response within five days. On May 20 they wrote to Washington on CCRC letterhead and asked his staff to intervene.

Coincidentally or not, that same day CHA contacted a testing firm, and within twenty-four hours two vacant apartments and two boiler rooms were surveyed, with asbestos readily apparent in three of the four locations. Although most of the asbestos pipe insulation was in good condition, the inspectors warned that in residences it "is highly subject to damage" and should be removed whenever apartments become vacant.

That same day a Developing Communites Project press release noted the Altgeld complaints and said residents would visit CHA's downtown headquarters the next morning. The *Chicago Defender* quoted liberally from Callie Smith's statements in the release: "Basically we feel like we've been lied to and given the run-around," she said. "We think it's typical of the arrogance of the CHA to remove hazardous materials from its own offices without even checking to see if residents have the same problems." WBBM Newsradio 780 began covering the story from daybreak onward.

Obama had booked a yellow school bus to take his community members to CHA headquarters downtown, and he had multiple copies of an outline of the residents' demands. But only a modest number of people, including Callie Smith and Hazel Johnson of PCR, plus several children, showed up for the trip. When they arrived, they were brusquely told that Zirl Smith was unavailable, but the presence of one or more TV crews motivated officials to promise that testing would move ahead and that Smith would attend a community meeting in Altgeld on June 9.

Obama, in his own account nine years later, gave the CHA visit an oddly outsized importance, writing that "I changed as a result of that bus trip, in a fundamental way," since it had suggested "what might be possible and therefore spurs you on. That bus ride kept me going, I think. Maybe it still does." He also wrote that only eight people, rather than "about 20 Altgeld residents" as reported in the press, made up his group.

On the next night's 10:00 P.M. WBBM Channel 2 newscast, reporter Walter Jacobson recounted his inability to get anyone from CHA to respond to residents' complaints about what "literally may be a question of life or death." Instead, "the public affairs director of the CHA is getting her lunch while the people who live in the CHA continue getting poisoned by asbestos." It was powerful television.

A CHA press release the next day said the issue was "resolved" and that test results demonstrated "no asbestos exposure danger." Zirl Smith appeared on WBBM's 10:00 P.M. newscast and insisted that "residents know we're here to serve them" and that "we are good managers: we feel a responsibility to our residents." Three days later, WBBM revealed on its 6:00 P.M. show that it had paid for testing at both Altgeld and Wells and had gotten dire results. "This is definitely a threat to human health, a threat to the health of the people who live there," a medical expert told viewers. "It's a situation that should be corrected as soon as possible."

Within an hour, CHA ordered emergency inspections, and on WBBM's 10:00 P.M. news, 2nd Ward alderman Bobby Rush called CHA's behavior "criminal." When the CHA's inspectors finally began work at Altgeld on June 4, they found "many samples of exposed asbestos," Walter Jacobson told WBBM viewers. While that was taking place, Obama and three Altgeld residents—Hazel Johnson, Evangeline "Vangie" Irving, and Cleonia Graham—were at City Hall trying to invite Washington to attend the Altgeld community meeting. With TV cameras rolling, the four were sent to an impromptu meeting in a tiny room with Washington's city council floor leader, 4th Ward alderman Tim Evans. "We are sincere about having this taken care of," Vangie Irving explained. Hazel Johnson added, as journalists looked on: "We're asking the mayor to save our children and ourselves. We don't have faith in promises from CHA management. With the mayor's support, I think we'll get some action."

Evans promised to take their concerns to the mayor, and Obama was mentioned in the next day's *Sun-Times* and *Defender*. The former identified him as a "community organizer," but Evans would remember having the impression Barack was "related to someone who actually resides in CHA because that was the way he was relating to these people, like a member of their family. It was clear they had talked before that meeting about who would deal with what aspect of the plight" and when one participant got nervous, "this young man got up and sat next to the person who was supposed to speak. . . . 'Let Alderman Evans know what concerns you have.' He was not there to impose himself on them, he was there to facilitate their discussions with the so-called powers that be."[20]

The next day Zirl Smith made a bad situation worse by admitting to a city council committee that CHA had discovered the asbestos six months earlier and then blaming residents "for disturbing the asbestos-covered pipes, creating a crisis themselves." Bobby Rush responded that it was "very irresponsible" to allege that tenants were purposely "exposing themselves and their children to possible cancer," and after the session, Smith hid in a men's room to avoid WBBM reporter Jim Avila and his camera crew.

On Monday night, Avila as well as crews from two other TV stations were waiting at Our Lady of the Gardens school gymnasium for Smith's 6:30 P.M. appearance. Callie Smith and others had gone door to door in Altgeld, encouraging residents to turn out, and approximately seven hundred people jammed into the "hot, steamy" gym. But by 7:30 P.M. there was no sign of Smith, and the crowd was "boiling over" when he finally arrived, seventy-five minutes late. Once he was at the podium, moderator Callie Smith asked him, "Do you have a plan to remove the asbestos that's in our homes?" and handed him the microphone. In the *Tribune*'s description of the scene, "Smith shrugged and said, 'I don't know. We have not completed all the tests on the apartments. As soon as this is complete, we will start the abatement process.'"

That response infuriated the crowd, and shouts of "No" drowned him out. Obama had told Callie and her colleague Vangie Irving, "Try not to let the director hog the microphone," but Smith resumed his answer—"We've started in Altgeld and we're going from apartment to apartment to determine the severity of the asbestos." Callie, standing to Smith's left, reached her right hand to take the microphone from Smith, who parried her with his left hand—"We will install an abatement plan that you'll be—excuse me, excuse me" and sought to retain the mike as the crowd began chanting, "Take it away, right away."

Then Vangie Irving reached in, grasped it, and handed it to Callie. As she did so, Smith stood up and slowly walked out. The uproar grew, an older man in the crowd collapsed, and the meeting was over. Outside, Smith had his driver call for an ambulance as some residents chanted, "No more rent." Smith told reporters, "I'm perfectly willing to meet with them, but I can't under these circumstances." Tuesday's *Sun-Times* reported Smith saying "that 'people who do not live in Altgeld' were behind the meeting."

As Smith's black Ford LTD pulled away, Obama called his volunteers together and told Callie that he had messed up and should have coached her on how to deal with Smith and the microphone. Dan Lee told Barack not to blame himself, that they had prepared for every possibility except Smith physically commandeering the mike. Adrienne Jackson believed the crowd had felt empowered by what had happened, that "they had stood up to somebody," and she viewed the evening "as a success." The 10:00 P.M. newscasts offered a different verdict on the meeting. WBBM's coverage said the residents had kept Smith from speaking; WMAQ's Carol Marin said the moderators had not allowed Smith to use the microphone, and so "Smith walked out."

Barack was more disconsolate than he should have been, and when he got back to his Hyde Park apartment, he called Johnnie Owens, who had turned down Barack's invitation to attend. "He was certainly very clearly upset over what had happened," Owens recalled. "He sounded angry at himself . . . and felt there was more that he could have done to prepare the leadership and sort of anticipate the problems that occurred." To everyone else, the Altgeld meeting was a blip in the daily news cycle, and coverage of CHA's testing and abatement work—the asbestos was far worse at Ida B. Wells than at Altgeld—continued throughout the summer as CHA sought federal funding to help cover the costs.[21]

In the midst of the asbestos campaign, Obama bought his airplane ticket to Los Angeles for the IAF training in mid-July, with the Chicago archdiocese's CHD staff reimbursing its $196 cost. He also sent a late May postcard to Phil Boerner, writing, "Work continues to be rough, but I'm learning at a steady clip and am starting to see some results. We were on the tube and made the papers this week" and "I took a busload of public housing tenants downtown to protest living conditions. Still following up on this." A few days later, he sent a lengthier greeting card to Genevieve, whose birthday was on June 7:

> The pace of my life has quickened these past few months; feel like I've broken through the lengthy "Buddhist" phase—acting more forcefully, letting myself make mistakes. I've stopped eating peanuts, I'm working like a bitch, still writing when I have the time.
>
> Made some good new friends; still miss my old ones. And trying to develop a new kind of discipline in myself—not the stiff martial discipline I'd let myself get locked into, but more the discipline to decide

> what feels right, to dig deep, take risks and make sure that I'm enjoying myself. A good time, a hungry time, and I give much credit to you for it (your pokes and prods had a subtle but sure effect).

Barack closed by asking, "Still going to stop teaching next year?" and offering "Regards to Sohale, the family, etc. Love—Barack."

A week or so after the Altgeld meeting, Barack went to dinner with Asif Agha, his girlfriend Tammy Hamlish, an anthropology graduate student, and a mutual friend and classmate whom they wanted to introduce to Barack. Almost a year had passed since twenty-four-year-old Barack had last enjoyed any "female companionship," as he put it in a note three months earlier to Andy Roth. But Asif and Tammy's initiative would have momentous consequences, almost as great as Barack's decision a year earlier to leave New York and find an entirely new life in Chicago.[22]

Sheila Miyoshi Jager was two years younger than Barack and a 1984 graduate of Bennington College in Vermont. She had attended two years of middle school in France and had spent 1984–85 in Paris, writing a thesis in French on Claude Lévi-Strauss toward an M.A. from Middlebury College. Lévi-Strauss suggested she pursue anthropology and recommended she study Korea, a country he had just visited, and he recommended her highly to his friend Marshall Sahlins, a well-known anthropologist at the University of Chicago. In September 1985 Sheila arrived there to begin the doctoral program in anthropology.

Sheila grew up in Northern California, where her father, Bernd, a polymath psychologist whose Ph.D. dissertation at Duquesne University had dealt with Freudian psychoanalysis, worked as a clinical psychologist at two state hospitals and taught psychology at California State University's Sonoma County campus. Born in Groningen, the Netherlands, in 1931, Bernd was not yet nine when the Nazis occupied Holland in May 1940. Three years later he watched "hundreds of our Jewish neighbors being herded like cattle to the train station" in the middle of the night, and he would always "vividly remember the German soldiers as they loudly goose-stepped" through town. Bernd's father, Hendrik, played a major role in an underground network that sheltered dozens of Jewish children from the Nazis. One young girl, Greetje de Haas, lived with Hendrik, his wife Geesje, and their two sons for three years. The Jagers' courageous involvement was posthumously recognized when Israel's Yad Vashem honored them with inclusion on the Wall of Honor in the Garden of the Righteous in Jerusalem.

After World War II, Bernd studied at the Royal Institute for Tropical Agriculture outside Amsterdam before spending two years working under Albert Schweitzer at Lambaréné, in what was then French Equatorial Africa. He fell in love with an African woman whom he sought to marry, but that came to naught. From there Bernd came to the U.S., and during a year's study at Berea College in Kentucky, he met and married Shinko Sakata, a Japanese woman six years his junior. Together they moved to Groningen, where Bernd earned an M.A. before they returned to the U.S., eventually settling into a handsomely situated three-bedroom home in Santa Rosa, ten miles north of Cal State's Rohnert Park campus. Five years after Sheila's birth, a younger brother named David joined the family, and in time Shinko Jager became a well-known Sonoma County potter. Bernd's closest friend, Michael Dees, who also worked at one of the hospitals, witnessed all of Sheila's childhood. "I've known her since she was one," he recounted years later. Bernd and Shinko "had strict parameters with her," he recalled, and "did a very good job raising Sheila." Mike would remain close to her even half a century later: "She's like a daughter to me."

Barack quickly became deeply taken with the bright, beautiful, and intense half-Dutch, half-Japanese Sheila. She wore her dark hair in a short pageboy-style cut, and as Barack would later write, she had "specks of green in her eyes." Three or four times in June and early July, they went on double dates with Asif and Tammy. Hanging out with a trio of anthropologists was no problem for the son of Ann Dunham, Asif recalled, because Barack "knew the idiom, he knew the concerns, and he was right at home." Asif encouraged Barack's interest in Sheila, joshingly telling him from recent experience that asking "Can I kiss you now?" was a surefire way to pose the question. As Asif remembered it, Barack soon let him know it had worked. Barack and Sheila—"two mixed-race kids," she would later say—were together almost every day in the two or three weeks before he left for Los Angeles on July 8.

In the aftermath of the asbestos campaign, Barack finally had time to turn his attention to two other DCP efforts. Prior to asbestos, Barack had thought that DCP's best chance to organize within Altgeld was to focus on residents' lack of employment opportunities and on how the city's primary jobs-referral agency, the Mayor's Office of Employment and Training (MOET), offered no services on the Far South Side below 95th Street. DCP had made contact with MOET officials, and finally a visit to Our Lady of the Gardens in Altgeld by MOET director Maria B. Cerda was scheduled for August.

By early July, the DCP women interested in tackling the discriminatory behavior of the Chicago Park District (CPD) were also ready to move. Harold Washington, after his late April city council victory, was now able to win approval of two new CPD board members. Washington's appointees, architect Walter Netsch and African American arts figure Margaret Burroughs—Frank Marshall Davis's old friend "Flo" in *Sex Rebel: Black*—would give him a board majority that could sideline CPD superintendent Ed Kelly and transfer his authority to Washington aide Jesse Madison. This would effectively undermine the core of Chicago's traditional Democratic machine, for journalists believed that a thousand of the CPD's thirty-four hundred full-time employees were precinct captains and political operatives, particularly from Kelly's own 47th Ward. When, on June 16, the new board majority did just that, Kelly filed suit challenging the action's legitimacy. In the meantime, on July 2, DCP gave Netsch and Burroughs a tour of disheveled Far South Side parks before they spoke at a DCP community meeting at St. Helena of the Cross. The two new members told DCP to bring a list of needed repairs that CPD staffers had failed to make to the next CPD board meeting on July 10.

The next day a judge upheld the board's removal of Ed Kelly, who lashed out angrily at the mayor. "Washington's ruined this city," Kelly told reporters. "He's going to be gone. We've got to get him out." On July 10, after Obama was already in Los Angeles, St. Helena parishioner Eva Sturgies took charge of DCP's presentation. She had carefully prepared a list of forty-one requests regarding eight different parks that DCP had given to CPD employees two months earlier—and only ten had been completed. At the meeting, Sturgies summarized the disappointing inaction: "The summer is already half over, and we have only seen work done on the most cosmetic aspects" of DCP's list. She indicated that DCP knew, thanks to Freedom of Information Act requests it had submitted, that CPD staffers had not even prepared work orders for the other items. Sturgies said that Far South Side citizens want "the same kind of service that other neighborhoods in the city get."

Four days later, a court ruling reaffirmed the authority of Harold Washington's new appointees, and the next week Ed Kelly surrendered his post, promising, "I don't get angry. I get even." Washington quickly named George Galland, corporation counsel Judson Miner's former law partner, as the Park District's new attorney. Galland announced that thousands of supposed jobs on the CPD's payroll were "blatantly illegal," and vowed to clean house. "It has been one of the most valuable havens for employment by the Democratic organization," Galland tartly commented.

Before Obama left for Los Angeles, he and Jerry Kellman attended a ceremony where Joseph Cardinal Bernardin publicly presented the CHD grants, including the latest $33,000 to DCP for organizing in Altgeld Gardens and for Barack's prospective plan "to improve high school and college opportunities for minority students," a CHD press release noted. Then, following the holiday weekend, Barack left for Los Angeles on Tuesday, July 8.[23]

In the fourteen years following Saul Alinsky's 1972 death, IAF had abandoned its Chicago roots, but attending an IAF training was still a rite of passage even for experienced organizers. Three years earlier Greg Galluzzo, Mary Gonzales, and their younger UNO colleague Danny Solis had taken it thanks to Peter

Martinez, and in 1985 UNO's Phil Mullins attended as well. Mount St. Mary's College was IAF's long-standing summer location, with trainees arriving on a Tuesday afternoon. Wednesday was devoted to two basic staples of IAF teaching: "World As Is/World As Should Be 'Power' Session" and "Power and Self-Interest." The 125 or so attendees were divided into groups of about 25, with the IAF's five senior "cabinet members"—Ed Chambers, Ernie Cortes, Mike Gecan, Arnie Graf, and Larry McNeil—rotating among them.

Obama was already familiar with Alinsky's major themes and principles, thanks to Kellman, Kruglik, and the three-day Milwaukee event six months earlier. Ed Chambers, Alinsky's lead inheritor, had articulated them in a small 1978 volume titled *Organizing for Family and Congregation*. Alinsky's best-known principle was that "power tends to come in two forms: organized people and organized money." But Alinsky had never fully grasped a second point that was now emphasized by Kellman, Galluzzo, and Chambers: "one of the largest reservoirs of untapped power is the institution of the parish and congregation," because "they have the people, the values, and the money."

Mike Gecan acknowledged that Alinsky had failed to "create organizations that endured," and by 1986 IAF was striving for permanency through growth: "as the number of local churches in the organization increases, the organization becomes increasingly self-sufficient" thanks to each congregation's financial support. Obama already knew that the number of organizers a group could hire was dependent on its funding. Barack understood how organizers had to prioritize "the finding and developing of a strong collective leadership," and thanks to the many one-on-ones Kellman had had him conduct, he appreciated how "the single most important element in the interview is the interviewer's capacity to *listen*."

As Jerry and Mike had made clear, it was crucial to "select and engage in battles that can be won," instead of larger problems—like the closure of Wisconsin Steel—that were obviously insoluble. Resolving small issues would "season people in victory." Failures sapped morale and enthusiasm, whereas with small victories "a sense of competence and confidence grows."

For individuals already working as organizers, the ten-day IAF training was "designed to force reflection on what you have done, what you didn't do, and ways and means of approaching reality and institutions based on a clear understanding of the tension between self-interest and self-sacrifice. The first week is spent in getting people to look at who they are, what their visions are, are they willing to reorganize themselves and their visions." The first day's sessions would "force people to reflect and open up to the world as it is against the world as it should be," and participants read a portion of Viktor Frankl's famous 1946 book *Man's Search for Meaning*. From there, the training explored "the difference between issues and problems" before sessions on "how to spot and develop leaders" and "how to do a power analysis of a particular community." Next were the more confrontational elements of the Alinsky model: choosing "enemies" and "creating crises."

The fourth day was devoted to individual meetings between attendees and the IAF trainers. Mike Gecan, who remembered Obama from their brief meeting eighteen months earlier at NYPIRG, was struck by how Barack now seemed "very intellectual, very abstractly intense. He'd be very intense about an idea," such as identity, but "very detached about the people." Instead of personalizing a subject, "he'd be lost in the idea," exhibiting "very little connection" and seemingly "abstracted from relationships and others," Gecan recalled.

Early in the training, before a session with Arnie Graf, Obama learned from either Gecan or Ernie Cortes that Graf was married to an African American woman and had several interracial children. Graf remembers that Barack sought him out, and from Friday onward, the two had "a series of conversations . . . that were not related to the training. He had lots of questions . . . he's very curious about interracial relationships and children and how you raise a child." Graf remembers Obama wanting to know how Arnie felt "as a white person, raising interracial children and being in a solidified marriage." Barack did not say much about his own childhood, but he wanted to hear about Arnie's experience: "How do my children see themselves, how do they identify themselves and how do I feel about that and how does my wife feel about it?"

Arnie and his wife Martha had met in graduate school, and by 1986 they had been married for thirteen years and had three children: two boys and a girl, ages ten, seven, and three. Barack wondered how they raised them "to understand who they are?" Arnie replied that given his children's ages, "I'm not sure how they see themselves quite yet," but that they knew their racially distinct grandparents well. Barack explained, "People look at me and approach me as an African American, but I don't have that history" and "I have to develop a way of understanding because that isn't my experience." Individuals who did not know Obama viewed him as black, but until he arrived in Chicago, the only African Americans he had really known were Frank Marshall Davis, Keith Kakugawa, and Eric Moore.

Obama and Graf's initial conversation about race and identity ran close to two hours. Barack's friend Johnnie Owens was at the training too, and on Saturday afternoon, they went to Will Rogers State Beach to try out the Pacific Ocean. Johnnie was already impressed by Barack's insistence on some form of exercise "every single day," and one evening, Obama reproached Owens for eating dessert. Once in the water, Johnnie was astounded by Barack's self-confidence as a swimmer: "he goes way out there," Owens remembered. "He's used to being in that ocean," as all his Punahou friends could attest. Owens also went along when Obama asked for a tour of some of South Central L.A.'s most gang-infested neighborhoods. "It was a real experience. I was freaked out," Owens recalls.

In the second week, Obama had another long conversation with Arnie Graf, again focused on "family and race." But Barack also asked Arnie to talk about his experiences helping build a chapter of CORE, the Congress of Racial Equality, as a college student in Buffalo in the early 1960s. The IAF wanted to add minority organizers to its staff, and Arnie suggested that Barack come work with him in Baltimore. Barack said he had moved around a lot in life and wanted to stay in Chicago: "I have to have some place where I want to be that feels like home." Graf asked whether Barack envisioned a career as an organizer, and Obama said no. "I'd like to organize for another couple of years, because I think I need to get that under my belt. I need to understand on the ground how to relate." Then, according to Graf, Obama said he thought he would go to a top law school and become a civil rights lawyer and perhaps a judge, a career his grandmother and mother had repeatedly mentioned to him.

Monday was devoted to analyzing how an Alinsky organization chose an "enemy" and how it could use a confrontation in a way that leads to a relationship with someone who had been ignoring you. "It's a relational tool, not a tactic," Mike Gecan explained. "The purpose of polarizing is to get into a relationship and then depolarize it." Tuesday focused on values and congregations, Wednesday on IAF as an organization. Before the training ended at midday Thursday, Obama spoke again with Arnie Graf and said he had most enjoyed the theoretical basis that underlay the world-as-it-is-versus-the-world-as-it-should-be dialectic, but that he was uncomfortable with how IAF conceptualized enemies and confrontation. Both Graf and Gecan wanted to recruit Barack to IAF, even though they worried that he seemed to grasp everything "more in the intellect than in the gut," as Graf put it. "There's something missing here," Graf thought, because Barack "always seemed one step removed from himself."[24]

Thursday afternoon Obama flew back to Chicago. It had been an edifying ten days, an experience that underscored how "the key to Alinskyism is a kind of pragmatic rationality" and that an organizer "must be pragmatic and nonideological." In Chicago, Barack was met by a *Tribune* front page that announced: "LTV Files for Bankruptcy." Financial analysts said this "virtually assures" the closure or sale of the mill, and on Saturday, the news turned worse when LTV terminated the medical benefits and life insurance of its more than sixty thousand retired workers nationwide, an action that Local 1033 president Maury Richards denounced as "outrageous and inhuman." On Tuesday, when reports circulated that U.S. Steel would soon lay off up to two-thirds of the 757 men still employed at South Works, the USW threatened to strike. In quick succession, the USW then struck LTV's profitable Indiana Harbor mill in East

Chicago, but not the East Side plant, where several retired *managers* who also had lost their health benefits joined Richards and hundreds of 1033 pickets while other colleagues kept the mill running. When a federal bankruptcy judge ordered LTV to restore the retirees' benefits, the USW terminated its strike, but then two days later struck U.S. Steel, and South Works shut down. The next week LTV announced that it would lay off 1,650 of the 2,300 remaining workers at the East Side mill before the end of the year.

This meant the end for Chicago's last integrated steel plant. Richards told reporters that many 1033 members "feel helpless and without hope," a familiar refrain to everyone who had witnessed Wisconsin's closure six years earlier. Jerry Kellman said LTV's East Side mill had no future "unless the governor takes action," but Thompson gave no sign of doing so and Kellman's outreach to Local 1033 lessened. "It didn't lead to any lasting working relationship there," Richards remembered.

The ripple effects were everywhere. A small-business owner in south suburban Dolton who had lost $1,000 when Wisconsin closed in 1980 told the *Daily Cal* that LTV owed him $8,000 he was unlikely to recoup. At a 1033 meeting, with members anxious that the local would lose its union hall for nonpayment of rent, the official minutes recorded an incident in which one agitated officer "threatened M[aury] Richards with physical harm." A *Chicago Tribune* feature story, referring to what had happened at Wisconsin Steel, South Works, and now LTV, described "chilling levels of alcoholism, emotional stress, and physical illness" among the unemployed and their families. There was no denying that over the past six years "the deterioration of the Southeast Side has been catastrophic." The *Daily Cal*'s superb steel reporter, Larry Galica, toured the largely silent South Works and pronounced it "a modern ghost town," and Richards blamed the steel companies' "failure to reinvest in their facilities to modernize them" as the reason why tens of thousands of people had suffered so traumatically since March 28, 1980.[25]

Amid this latest steel crisis, Harold Washington formally kicked off his reelection campaign before a crowd of four thousand supporters at a Loop hotel. Weeks earlier, in late April, Washington had reiterated his opposition to any new landfills, even though he knew there was no solution to Chicago's garbage crisis other than finding additional landfill capacity somewhere. In late May he quietly approved what the city insisted was a "reconfiguration," and not "an expansion," of the huge CID landfill just east of Altgeld Gardens. Environmental purists like Hazel Johnson were understandably not happy, but a heavy majority of his administration's Task Force on Solid Waste Management endorsed his action, including IACT's Mary Ellen Montes.

An even more imminent threat to what remained of the Southeast Side was the disappearance of gainful employment. Washington's aides reported, "Stores and local businesses are closing down because the only purchases are for bare necessities. The area is becoming blighted and people's attitudes are of hopelessness." Additionally, "there is a great need for more police coverage," since "there are not enough police officers on the street patrolling the neighborhoods."[26]

In August, Maria Cerda, the director of the Mayor's Office of Employment and Training (MOET), finally appeared at a community meeting in Altgeld Gardens to respond to DCP's request that MOET open an office within reasonable travel distance of the Gardens. Obama had prepared Loretta Augustine to chair the meeting, but almost immediately, Cerda became "very aggressive and domineering," according to Loretta. "I was supposed to introduce the issue, and she tried to take over," and became openly patronizing, asking Loretta, "Do you even know what we do?" Then, from the back of the room, came Barack's voice: "Let Loretta speak! We want to hear what Loretta has to say!"

Obama was determined to avoid another breakdown like the Zirl Smith session, so he put aside his own rule about remaining quietly in the background, and this time intervened forcefully if anonymously. Loretta remembered that "people kind of picked it up," chanting "Let Loretta speak!" In the end, Cerda agreed that MOET would open an office on South Michigan Avenue in central Roseland, a ten-minute bus ride from the Gardens, before the end of November.

By midsummer, Greg Galluzzo and Mary Gonzales were expanding UNO's reach by linking up with and reactivating the Gamaliel Foundation, founded in the 1960s but long dormant. Greg had been seeking funding for this new vision since early 1986, and he believed that Gamaliel could serve as a training institute that would "generate a flow of leadership for the city's future." Mary and Greg saw community organizing as "the way people can move from a sense of helplessness and isolation to active participation in the decisions affecting their lives." Greg felt that Chicago's vibrant "movement" activity in earlier years had "distracted people" from pursuing long-lasting change. "Compared to the work of real community organizations, movement activity is much less grounded in communities," he wrote. By September Greg had raised almost $100,000 for Gamaliel, with the two largest grants—$40,000 and $30,000—coming from the Joyce Foundation and the Woods Fund.

Within UNO, Danny Solis and Phil Mullins wanted to train the parents of Chicago public school students: "Improving the quality of education in Chicago is the city's greatest challenge and clearest need." Citing the research of Fred Hess on Chicago dropout rates, Gamaliel noted how "students are simply not prepared to handle high school classes." By 1986, that represented a bigger problem than ever before. "Ten years ago in S.E. Chicago, a student at Bowen

High School could drop out of school and get a job in the steel mills making more money than their teachers." That world was gone, and "the major obstacle" to improving outcomes for current students "is the ineptness and mismanagement at the Board of Education level."

Like IAF, Gamaliel would do trainings, and in early August, a one-week course for sixty community group members took place at Techny Towers, a suburban retreat center owned by the Society of the Divine Word. Obama, having the full IAF training to his credit, delivered a presentation there one day, and among those in attendance was Mary Ellen Montes. "Everyone was awed by Barack," Lena remembered. "I was getting divorced," and as a single mother with three children, she now had a full-time job at Fiesta Educativa, an advocacy group for disabled Hispanic students. She was immediately taken with the young man three years her junior. "Barack was extremely charismatic," and she wanted to see more of him. "We talked quite a bit after I met him at Techny Towers," and with Sheila Jager in California visiting her family, Barack and Lena spent a number of late-summer evenings together in Hyde Park.

"We went out to eat a few times" and "we just enjoyed talking to each other," Lena recounted. "He was a lot of fun to talk to and we really enjoyed each other's company." Obama would remember some intense making-out, while Lena explained, "I'm a passionate person." What she termed "the relationship/friendship that we had" became a close one as Barack became part of the UNO–Gamaliel network. Greg Galluzzo believed that "the continuing development of community organizing as a profession is mandatory," and with Gamaliel often bringing its organizers together, Barack and Lena were now professional colleagues as well as intimate friends.[27]

Obama had been in Chicago for more than a year, and it had been a more challenging and instructive period than any other of his life. He had learned a great deal about others and himself. He had bonded with Mike Kruglik, Bill Stenzel, and Tom Kaminski, as well as with DCP leaders, such as Loretta Augustine, Dan Lee, and Cathy Askew. He had also made two good friends around his age: Asif Agha and Johnnie Owens. But more changes were coming.

Jerry was shifting his focus to Gary, after being bruised by the failure of his Regional Employment Network, which could not claim to have placed even one unemployed worker in a meaningful new job. Given what was now transpiring at LTV's East Side mill and at South Works, jobless workers on Chicago's Far South Side faced even dimmer prospects for employment in the immediate future. Obama later wrote that Kellman had told him that CCRC's region-wide aspirations, like DCP's highly disparate neighborhood composition, were fundamentally ungainly and that he "should have known better." Jerry asked Barack to come with him to Gary, but Barack said no: "I can't just leave, Jerry.

I just got here." Kellman warned him, "Stay here and you're bound to fail. You'll give up organizing before you give it a real shot," but Barack stood firm. He saw a fundamental human difference between them. Barack in twelve months had established meaningful emotional bonds with half a dozen or more colleagues, while Jerry, as Obama later wrote, had "made no particular attachments to people or place during his three years in the area."

Obama also made a personal decision that was unlike anything he ever suggested to Genevieve: he asked Sheila Jager to move in with him. "It all seemed to happen so fast," she later explained. Earlier in the summer, Barack had renewed his one-year lease on his studio, but it was too small for two people. Johnnie Owens was living with his parents and looking for a place, so Barack suggested he sublet the studio. Sheila had no support apart from her graduate fellowship, but Barack offered to cover the $450-a-month cost of apartment 1-N at 5429 South Harper Avenue, a quiet, tree-lined block of three-story brick buildings.

Sheila thought it was "really spacious," with a living room, an open study, "a good-sized bedroom," and a large, eat-in kitchen. "I thought it was a bit plush for a struggling couple," but with Barack's $20,000 salary and $100 a month from Sheila toward food and shopping, by early October the young couple had set up house. Barack continued to see both Asif and Johnnie regularly, with Barack taking Johnnie to see a fall exhibit featuring the work of French photojournalist Henri Cartier-Bresson at the Art Institute of Chicago.

Soon after moving in with Sheila, Barack tardily responded to a letter and short story Phil Boerner had sent him. Barack confessed to feeling older, or at least overextended. "Where once I could party, read and write, with a whole day's worth of activity to spare, I now feel as if I have barely enough time to read the newspaper." He counseled Phil about writing fiction, recommending that he focus on "the key moment(s) in the story, and build tension leading to those key moments." He also suggested that Phil "write outside your own experience," because "I find that this works the fictive imagination harder." Barack spoke of thinking about and missing Manhattan, "but I doubt I'll be going back."

Then he introduced Sheila. "The biggest news on my end is that I have a new girlfriend, with whom I now share an apartment. She's half-Japanese, half-Dutch, and is a Ph.D. candidate" at the University of Chicago. "Very sweet lady, as busy as I am, and so temperamentally well-suited. Not that there are no strains; I'm not really accustomed to having another person underfoot all the time, and there are moments when I miss the solitude of a bachelor's life. On the other hand, winter's fast approaching, and it is nice to have someone to come home to after a late night's work. Compromises, compromises."

Obama's pose that fear of cold temperatures underlay his desire that Sheila and he live together downplayed his own decision and initiative. Then he updated Phil on how DCP was now becoming a freestanding organization, "which gives me no one to directly answer to and control over my own schedule. But the downside is that I shoulder the responsibility of making something work that may not be able to work. The scope of the problems here—25 percent unemployment; 40 percent high school drop out rate; infant mortality on a par with Haiti—is daunting; and I often feel impotent to initiate anything with major impact. Nevertheless, I plan to plug away at it at least until the end of 1987. After that, I'll have to make a judgment as to whether I've got the patience and determination necessary for this line of work."[28]

The breakup of CCRC, with DCP and the South Suburban Action Conference (SSAC) taking its place, coincided with Leo Mahon leaving St. Victor. He had announced seven months earlier his September departure, and he had felt his energy flagging before that. In one of his last pastoral letters, Leo addressed the LTV and South Works news in words that echoed across the previous six years. "Catholic social teaching insists that the workers have first right to the fruits of their labor," in line with "the Christian principle that workers, not money, come first." Corporations should put "human concerns ahead of profit and dividends," but LTV and U.S. Steel had made "clear that power and profit are both more important than people and jobs." Leo's advice was the same as in 1980: "let us translate our concern and our outrage into protest and into political action."[29]

Following Kellman's shift to Gary, Greg Galluzzo took on a formal mentoring role to Obama by becoming DCP's "consultant." He introduced Barack to a young tax attorney, Mary-Ann Wilson, who at no charge—"pro bono"—would handle the state and federal paperwork necessary to establish DCP, like UNO before it, as an independent, nonprofit corporation.

Illinois, like the Internal Revenue Service, required a bevy of forms and submissions that Barack and DCP members would spend a good portion of that fall completing and signing. A set of bylaws was copied and updated from an earlier version Kellman had drafted in 1984. Illinois articles of incorporation required three directors—Dan Lee, Loretta Augustine, and Tom Kaminski were named—as well as a registered agent. Barack, as "project director," left blank the space asking for his middle name. A separate, full-fledged board of directors listed Dan as president, Loretta as vice president, and Marlene Dillard as treasurer, along with everyone who was active in DCP: Cathy Askew, Yvonne Lloyd, Nadyne Griffin, Eva Sturgies, and Aletha Gibson, plus several women who had signed in at a meeting or two but not reappeared.

Simultaneous to his work with Mary-Ann Wilson, Barack was conducting one-evening-a-week training for DCP members at Holy Rosary, while also drafting his own initial grant applications. The trainings, informed by what he had learned during his ten days with IAF, involved DCP veterans such as Dan

Lee and Betty Garrett plus relative newcomers such as Aletha Strong Gibson. They also attracted new faces such as Loretta's neighbor Margaret Bagby and Aletha's close friend Ann West, a white Australian whose husband was black and president of the PTA at Turner-Drew Elementary School. Another important PTA figure, Isabella Waller, president of the regional Southwest Council, brought along her best friend, Deloris Burnam. Ernest Powell Jr., the politically savvy president of the Euclid Park neighborhood association and someone Barack had recruited over the summer, came as well. Close to twenty people attended the weekly sessions, at which Barack always took off his watch and put it on the table so he could see it during the training.

For Obama, the grant applications were a major concern. The first would go to Ken Rolling and Jean Rudd at the Woods Fund, which he hoped would renew the $30,000 given to CCRC for DCP in 1986. The ten-page, single-spaced document offered a retrospective account of the past fourteen months, and made clear, as he had stressed previously, that the Far South Side's "two most pressing problems" were a "lack of jobs, and lack of educational opportunity." Obama had one especially audacious goal for 1987: a "Career Education Network to serve the entire Far South Side area—a comprehensive and coordinated system of career guidance and counseling for high school age youth" with "a centralized counseling office" augmenting in-school counselors so as to reduce the dropout rate and channel more black high school graduates into higher education. "Youth in the area," he warned, "are slipping behind their parents in educational achievement."

Obama already knew that to build DCP two types of expansion were necessary: an outreach to congregations beyond the Roman Catholic base that Kellman had developed and the recruitment of more block club and PTA members like Eva Sturgies, Aletha Strong Gibson, Ann West, and Ernie Powell. DCP's neighborhoods, from Altgeld northward to Roseland, West Pullman, and even solidly middle-class Washington Heights, were ill served by their elected officials, who "have generally poured their efforts into conventional political campaigns, with almost no concrete results for area residents." Obama spoke of the need to "dramatize problems through the media and through direct action," but DCP's most pressing need was funds to hire another African American organizer. He realized a $10,000 entry-level salary would not attract experienced candidates, so DCP's proposed 1987 budget called for $20,000 for that position and an increase in Barack's salary to $25,000.

Ken and Jean at the Woods Fund told Barack to introduce himself to Anne Hallett, the new director of the small Wieboldt Foundation, whose interests closely paralleled those of Woods, and also to consider applying to the larger MacArthur Foundation. Obama visited and spoke with Hallett in November, and soon thereafter a front-page story in the *Chicago Defender* highlighted

weaknesses," one that emphasized "deficiency rather than capacity." Service professionals' control over the lives of the poor magnified rather than alleviated their poverty, for "if you are nothing but a client, you have the most degraded status our society will provide."

Health services were the most dire problem, and publicly funded "medical insurance systematically misdirects public wealth from income to the poor to income to medical professionals." Yet change would be difficult because "many institutional leaders" had come to see communities as nothing more than "collections of parochial, inexpert, uninformed, and biased people" who understood their own needs far less well than did service professionals. "Institutionalized systems grow at the expense of communities," and America's "essential problem is weak communities."

John McKnight's influence on people exposed to his social vision was profound even if unobtrusive. Ellen Schumer, a fellow Gamaliel board member and UNO veteran, said that within the world of Chicago organizing, McKnight "really challenged the old model." His impact was also felt throughout Chicago's progressive foundations, with Wieboldt acknowledging that it was "fortunate to have John McKnight join us" at the board's annual retreat "to talk about the substantive changes in Chicago neighborhoods today that require new strategies for organizers."[32]

If McKnight was Chicago's most significant social critic, the front pages of the December 3, 1986, editions of all three Chicago daily newspapers announced another: G. Alfred "Fred" Hess, a little-known policy analyst, education researcher, and former Methodist minister. After completing a Ph.D. in educational anthropology at Northwestern University in 1980, with a dissertation on a village development project in western India, Hess had joined the foundation-funded Chicago Panel on Public School Finances, and in 1983 he succeeded Anne Hallett as executive director.

Hess had made the front pages of Chicago's daily papers nineteen months earlier thanks to a lengthy study entitled "Dropouts from the Chicago Public Schools: An Analysis of the Classes of 1982–1983–1984." The *Chicago Sun-Times* had summarized Hess's findings in a boldface banner headline: "School Dropout Rate Nearly 50 Percent!" His discovery that 43 percent of students who had entered Chicago high schools in September 1978 dropped out prior to graduation was a percentage that Obama had accurately cited a year earlier when he had written to Phil Boerner about the worst ills plaguing the neighborhoods where he worked. Hess's data readily showed that high schools with predominantly minority populations had a dropout rate as high as 63 percent.

One *Tribune* story highlighted how at one Roseland high school, remaining in school did not mean students were studying. "We were sitting at a lunch table in the cafeteria rolling joints one morning," a seventeen-year-old girl told reporter Jean Latz Griffin. "There was a security guard right next to us, but he didn't say anything. People smoke it anywhere. Some teachers say 'Put it out,' but no one really does anything. You can't snort coke inside school though. That would be too obvious." Out of Corliss High School's eighteen hundred students, only forty-eight were taking physics, only seventeen had qualified for the school's first-ever calculus class, and almost 50 percent were enrolled in remedial English. A report similar to Hess's, from a parallel research enterprise, Donald Moore's Designs for Change, revealed that only three hundred out of the sixty-seven hundred students who had entered Chicago's eighteen most disadvantaged high schools in 1980 had been able to read at a twelfth-grade level if they graduated in 1984.

In a prominent *Tribune* feature four weeks before Obama arrived in Chicago, Hess had warned that Chicago Public Schools (CPS) were damaging the city's future. "We are in danger of creating a permanent underclass that is uneducated and unable to advance. It means a set of neighborhoods in which the majority of people are constantly unemployed and a strain on the social system in welfare and the high cost of crime. If we don't take strong action now, we'll pay for it later." Don Moore emphasized how job losses were magnifying the consequences of CPS's failures. "Parents are feeling a real desperation now because they are seeing unskilled jobs disappearing . . . 20 years ago the manufacturing industries in Chicago didn't depend on employees being terribly literate. The economy has changed." Former Illinois state education superintendent Michael Bakalis wondered to the *Tribune* why "the general community of Chicago, particularly the business community, has not been absolutely outraged by the performance of the Chicago Public Schools." In July 1986, Hess told *Crain's Chicago Business* that the way to reform public schools was to develop "real power for local citizens to control their local schools. Let local citizens hold local school officials accountable for the effectiveness of their schools. This would be real and significant decentralization of power that could make a difference."

Illinois state law mandated that high school students receive at least three hundred minutes of daily instruction. Beginning in spring 1986, Hess's Chicago Panel discovered that many Chicago high schools had fictional "study hall" classes in their students' otherwise skimpy schedules that created a false paper trail to meet that requirement. The resulting report, "Where's Room 185?," released on December 2, created an immediate uproar. The *Tribune* editorialized that CPS "administrators are deliberately and illegally cheating students of part of the education to which they are entitled." It also showed that in an overwhelming majority of actual classes, teachers actively *taught* for less than half of the class period. Noting how recently promoted CPS superintendent Manford Byrd's response was to "hunker down and criticize the design of the study," the *Tribune* declared that it was "inexcusable that it took an outside research panel" to uncover CPS's "fundamental failure."

The *Chicago Defender* ran a prominent page-three story, "Dropout Rate Irks Parents," publicizing DCP's desire to meet with board of education president George Munoz to discuss the problem. "We have a lot of dropouts," DCP project director Barack Obama told the *Defender*. "We acknowledge that the dropout question is complicated, and there are no quick fixes to keep kids in school. We are urging Munoz to study the counseling system, recommend changes, and expand the numbers. This will have a direct effect on kids in the schools now." Obama also said that students' preparation prior to their high school years should be examined, "because there are a lot of possibilities out there."[33]

In addition, two public controversies highlighted Altgeld Gardens residents' problems with both the CPS and the CHA. On December 4, one day after "Where's Room 185?" debuted in Chicago newspapers, the *Defender*'s Chinta Strausberg reported on six teachers at the Wheatley Child-Parent Center who had transferred to other schools because they feared that Wheatley was permeated with asbestos. Superintendent Byrd insisted that tests had shown the *air* at Wheatley was "safe," but Wheatley parents threatened to boycott the school and demanded an immediate inspection. On December 15 just 37 of Wheatley's 377 young children attended school, and for the next two days that number dropped to 17 and then 11. One parent told the *Tribune,* "We can't help but notice that the kids go to school all week and come home with rashes and wheezes. When they are home for the weekend, it all clears up."

The boycott continued until the Christmas–New Year's break, but being "home" at Altgeld was no picnic either, as a *Tribune* series detailing what it called the CHA's "national reputation for mismanagement" documented. "If I had a job, I wouldn't be here. This is not a good place to raise kids," one young Altgeld mother told the newspaper while complaining about the prevalence of youth gangs. Two days later the *Tribune* reported that Zirl Smith was resigning as CHA executive director, and the next day's paper emphasized how "Chicago has used the CHA as a way to isolate blacks."

When news broke that administrative ineptitude had cost CHA $7 million in federal aid, CHA's board chairman resigned as well. Fifteen months would pass before asbestos removal finally began in some 575 Altgeld homes, almost two years after Smith's tumultuous visit to the Gardens. Harold Washington's press secretary, Al Miller, later recounted the mayor remarking that "he didn't believe there *was* a solution" to the CHA's profound problems.[34]

Shortly before Christmas, Barack Obama and Sheila Jager flew to San Francisco to spend a holiday week at her parents' home in Santa Rosa. Although they had been living together for hardly four months, their relationship had quickly become one of deep commitment—indeed, so deep that for several weeks they had been discussing getting married during the trip to California.

Asif Agha had watched their relationship grow. Over the previous sixteen months, Asif had seen "Barry"—as he alone called Obama—acclimate to Chicago. "We were kind of an anchor point for each other," and "Barry" spoke frankly to Asif about his acculturation. "I am the kind of well-spoken black man that white organization leaders love to give money to," Asif remembered Obama remarking. Asif saw Obama with the eyes, and ears, of a linguistic anthropologist. "In terms of his performed demeanor, diction, speech style, he was white, not black," Asif observed. Obama was open enough with Asif for him to know that Barack's significant girlfriends prior to Sheila had been white, and Asif appreciated the underlying duality of Obama's Chicago experience. His weekday work in Altgeld and Greater Roseland immersed him in African American life in a way that no prior experiences ever had, but in Hyde Park, his home life with Sheila and their occasional socializing with other anthropology graduate students was entirely multiethnic and international, just like his Punahou and Pakistani diaspora life had been in Honolulu, Eagle Rock, and New York.

Asif Agha. Eunhee Kim Yi. Arjun Guneratne. Their names alone, just like Sheila Miyoshi Jager, highlighted their international and ethnic diversity. Tania Forte was Egyptian, Jewish, and had grown up in France. Chin See Ming was born and raised in Malaysia before graduating with honors from Rice University in Houston. It was a "very, very cosmopolitan" group, Ming recalled, and when Sheila one day introduced Ming to Barack, I "just assumed he was a graduate student."

For Sheila and her classmates, the first two years of the graduate anthropology program were "like boot camp," Ming explained. Everyone had to take a double-credit introductory course called Sociocultural Systems, taught by Marshall Sahlins, a prominent anthropologist but "not a warm and cuddly person" and indeed "a very, very scary man" to some. Sheila coped far better with Sahlins than most of her classmates, and in her dissertation she wrote that "my greatest intellectual debt" was to Sahlins. "There was a very strong esprit de corps among the grad students" and "people worked very hard," Ming recalled. "You were never off," and everyone knew that student attrition would reach 50 percent.

Asif knew Sheila as "a very wonderful, wonderful person," someone who was "passionate" about her work as well as her relationship with Barack. One evening the three of them accompanied Asif's girlfriend Tammy Hamlish to a talk that her aunt Florence Hamlish Levinsohn, an outspoken local writer, was giving. Three years earlier Levinsohn had published a "patchy, parochial, frankly admiring" biography of her university classmate Harold Washington just after his election as mayor. Tammy had wanted Asif to meet Aunt Florence, but the evening quickly devolved into an unmitigated disaster. Asif remembered that he "started giggling at what the lady was saying, and Barry and I made eye contact, and that was fatal, because then for the next ten minutes we kept uncontrollably giggling and couldn't control it and almost falling off our chairs because what the lady was saying was just absurd. And neither of us could control it, and because we were sitting next to each other and kept making eye contact, we're triggering each other over and over and over. Tammy meanwhile is turning red," and Levinsohn took note of their behavior too. She "was most upset" and "Tammy was mortified," Asif recalled.

Apart from that embarrassing scene, Barack and Sheila were familiar faces at anthropology graduate students' occasional parties. Sheila Quinlan, a Reed College graduate who was a year ahead of Sheila, remembered how "everyone thought they were a very sincere couple." Barack was "quiet," "friendly," and "a sweet boy." Indeed, as Chin See Ming put it, Barack "fit into the scene," just as he likewise had learned to do in Roseland and even down in the Gardens. Jerry Kellman watched as Obama comfortably embraced his dual lives. "He found a way to be part of the black community and live beyond the black community," Kellman explained. "He discovered he could live in both worlds."[35]

But in December 1986, and for almost two years thereafter, the looming and overarching question was whether Barack Obama could live in both worlds with Sheila Miyoshi Jager as his wife. Five months earlier, before asking Sheila to live with him, he had inquisitively questioned Arnie Graf about the long-term dynamics of interracial marriage and raising half-black, half-white children. And although most passersby and even most anthropology students would not see them in such a way, Sheila knew that "Barack is as much white as I am." With her half-Japanese ancestry paralleling his half-Kenyan, she and Barack were *equally white*—one half apiece.

"Marriage was THE vital issue between us and we talked about it all the time," Sheila explained more than two decades later. Barack "kept work matters and his private life separate," so their marriage conversations, while known to Asif, were not something Kellman, Kruglik, Tom Kaminski, or Cathy Askew ever heard about from someone so "very private" as Obama. In their time alone together, Sheila saw someone whom no one in DCP ever did, someone with a "deep seated need to be loved and admired." In their evenings at the spacious apartment on South Harper, Barack read literature, not history, while Sheila had more than enough course readings to occupy her time. And, of course, there was another dimension as well. Barack "is a very sexual/sensual person, and sex was a big part of our relationship," Sheila later acknowledged.

Everything had come together fast. In "the winter of '86, when we visited my parents, he asked me to marry him," Sheila recounted. Mike Dees, her father's closest friend and Sheila's virtual uncle, had been told by Bernd in advance that Barack was a prospective son-in-law. "He and Sheila . . . were going to get married," Dees explained. "They were coming out, they wanted to get married, and so they called me to come up and look the guy over and see what I think."

One day right after Christmas Mike drove up to Santa Rosa, and "I ended up with Barack for an afternoon," he recalled. "We just visited." Barack was clearly a "very bright guy," and, complexion aside, came across like "a white, middle-class kid." Then, after dinner, Bernd and Mike had "a big political thing with Barack" while Sheila and her mother Shinko were occupied elsewhere. The two older men and Barack found themselves on "completely opposite sides of the fence," Mike explained, because "we're both conservative Republicans." Barack "kind of thought he was going to lecture to us" about politics and ideology, "and we kind of shot him down." It got "really heated" and "went on for quite a while." Barack seemed "very taken aback by it," Mike recalled. "Barack kind of thought he was going to sit down and get anointed. He's very self-centered, and he ended up getting beat up."

Although Barack "didn't do very well," to Mike "it was just a political argument. I think to the father and Barack it was more than that," indeed, much more. "For Barack it was a big deal, for Bernd it was a big deal," because Barack "was going to be the chosen one that night, and it didn't work out" that way. It was readily apparent that the older man was sitting in judgment on the younger, perhaps not so differently from the time twenty-five years earlier when Stan Dunham had first met Barack Obama Sr. But here the verdict was negative, not positive. "I don't know whether his color entered into the picture or not," Dees said about Bernd's attitude toward Barack.

By the end of the evening, and again the next morning, Bernd made his view of Barack and his marriage proposal crystal clear. "Bernd was against it," because he felt that Barack was unworthy of his daughter's hand. Sheila remembers that "my father was concerned over his 'lack of prospects' and wondered whether, as a community organizer, he could even support me, something that deeply offended Barack. My mom liked Barack a lot, but simply said I was too young" to get married. "I went along with their judgment, basically saying 'Not yet,'" she recalled. The unsettling holiday visit concluded, and "they ended up going back without getting married," Mike Dees affirmed.

Barack and Sheila would revisit that decision again and again over the twenty months that lay ahead.[36]

During the first week of January 1987, Gamaliel held a three-day retreat for its seventeen organizers, plus Ken Rolling from Woods, at a Holiday Inn in south suburban Matteson. More important, Obama was now meeting for at least an hour a week, one on one, with Greg Galluzzo to talk about his Developing Communities Project work. Greg's monthly calendar recorded the regularity of their discussions: Thursday morning, January 15; Wednesday lunchtime, the 21st; Friday morning, the 30th. On Wednesday, February 4, they spoke for two and a half hours; on Monday, February 9, for two more. Wednesday, the 11th, Greg came to Barack's apartment; Tuesday, the 17th, they met for another ninety minutes—ten hours of conversation in just thirty days.

As anyone who knew Galluzzo would testify, he was so intense that no discussion with him ever devolved into idle chitchat. Deeply committed to his belief that "the essence of organizing is the transformation of the person" into someone who was "serious about being a power person in the public arena," Greg's self-confidence and challenging interpersonal style was robust indeed. True to the Alinsky tradition, Greg's insistence that people unapologetically seek power for themselves was completely nonideological and entirely pragmatic: "Find out what people want," he said, "and then we'll decide what we're going to do about it." He also agreed with the Alinsky tradition of avoiding insoluble "problems" while constantly searching out winnable "issues." Greg knew it was essential to "learn politics as the art of the possible." He believed that, properly taught, "community organizers are pragmatists" who firmly appreciate that "in 99 percent of the cases the end does justify the means." In addition, an Alinsky organizer assumed a dual public and private life. As one veteran put it, "you compartmentalize life and then it allows your consciousness to go to the beach."[37]

On the Far South Side, parents in Altgeld Gardens, aided by Barack, maintained their boycott of the Wheatley Child-Parent Center until January 20, when the CPS provided documentation that no asbestos had been found in the school. The boycott energized Altgeld residents' interest in Hazel Johnson's People for Community Recovery, and Mayor Washington spoke at a PCR meeting that drew several hundred people. The *Tribune* reported a new U.S. Environmental Protection Agency study that had found that "more than 20,000 tons of 38 toxic chemicals are emitted into the air each year in the Lake Calumet region on the Southeast Side." The *Daily Calumet* reported that Waste Management's 1986 quarterly earnings were up an astonishing 349 percent, and a WMI representative, Mary Ryan, had quietly begun speaking with interested parties about WMI's desire to add the O'Brien Locks site as an adjoining landfill to its huge CID dump just east of Altgeld. Ryan spoke first with Bruce Orenstein, UNO's new Southeast Chicago organizer, and Mary Ellen Montes. "Waste Management would be very willing to bring benefits to the community, gifts to the community, for your acquiesence, your agreement to allow us to dump there," Orenstein remembers Ryan explaining. He and Lena were intrigued by Ryan's offer but wanted to ponder what dollar amount would be appropriate and how such a sum should be managed and invested. Ryan also called on George Schopp at St. Kevin, Dominic Carmon at Our Lady of the

Gardens, Hazel Johnson of PCR, civic activists ranging from Marian Byrnes to Foster Milhouse, and Obama on behalf of DCP. In mid-February Lena, Bruce, Hazel, and Barack met to discuss how to respond to WMI's bold initiative.[38]

But Obama's top priority in early 1987 was recruiting supporters for his Career Education Network. In addition to Dan Lee, Aletha Strong Gibson, and Isabella Waller, the three DCP members most interested in the proposal, he also discussed his idea with four people he already knew: Dr. Alma Jones, the indomitable principal of Carver Primary School in Altgeld; John McKnight, whom he was getting to know well through Gamaliel; Anne Hallett of the Wieboldt Foundation, with whom he had spoken in November and who had given DCP a $7,500 grant; and John Ayers, a close friend of Ken Rolling's who had observed DCP's "Let Loretta speak!" meeting with Maria Cerda and worked for the prestigious Commercial Club. McKnight and/or Hallett suggested that Barack approach Fred Hess, and Hess recommended that Barack introduce himself to Gwendolyn LaRoche, the education director of the Chicago Urban League. LaRoche scheduled a thirty-minute appointment, but her conversation with the "very polite young man" lasted "about two hours." LaRoche explained that much of the federal and state funding that was supposed to go to schools with high concentrations of poor students was instead being spent on administrative jobs at CPS headquarters. "Barack was very much interested in" how "our kids were being cheated" by CPS, Gwen LaRoche Rogers recalled years later. The Urban League was already sponsoring after-school tutoring programs in churches, an approach that would fit perfectly with DCP's congregational base.

Obama reached out as well to Homer D. Franklin at Olive-Harvey, a fifteen-year-old community college just north of 103rd Street that was named for two African American Medal of Honor recipients, and to George Ayers at Chicago State University, on the south side of 95th Street. Olive-Harvey would be crucial to Barack's hope of establishing a job-training program focused upon public aid recipients and especially Altgeld Gardens residents. DCP also won board of education president George Munoz's agreement that CPS counseling specialists would cooperate with DCP's effort.

Obama would later say that his Far South Side work was inspired by what he knew about the history of the black freedom struggle during the 1950s and 1960s, and in late January 1987, Chicago's public television station, WTTW, joined in the nationwide PBS broadcast of the six-part landmark documentary series *Eyes on the Prize* on Wednesday evenings at 9:00 P.M. These first six episodes covered only the years from 1954 to 1965—a second set of eight programs three years later would include an episode focusing on Chicago's own 1965–1967 Freedom Movement—but the *Defender* and the *Tribune* publicized and praised the early 1987 *Eyes* telecasts. The *Defender* also accorded a full-page headline to a review of a new biography of Martin Luther King Jr., calling the book "a magnificent and unvarnished study" based upon "exhaustive research." Written by one of *Eyes'* three senior advisers, the book devoted the better part of two long chapters to King's role in the Chicago Freedom Movement. Those chapters highlighted how one particular Chicago civil rights activist, thirty-two-year-old schoolteacher Al Raby, had convinced King to come to Chicago and had been at his side throughout every day of the 1966 campaign.

In subsequent years, Raby played a significant role in the 1970 revision of Illinois's state constitution before running unsuccessfully for 5th Ward alderman in Hyde Park. Eight years later, Harold Washington's top political adviser, Jacky Grimshaw, persuaded Washington to name Raby his campaign manager for his successful 1983 primary campaign against incumbent mayor Jane Byrne. But as Raby's longtime closest friend, Steve Perkins, later explained, "Al and Harold just didn't have good chemistry" and indeed "rarely communicated." Once Washington triumphed, Raby was "moved aside" and "totally marginalized." Raby ran for Washington's now-vacant congressional seat, but the new mayor backed labor veteran Charles Hayes, and Raby's loss seemingly left him with "no future in Chicago." He accepted the number two post at Project VOTE!, a trailblazing voter registration organization, and moved to Washington, D.C.

Eighteen months later, in early 1985, Raby returned to Chicago after Jacky Grimshaw convinced Harold Washington to name him as the new head of the city's Commission on Human Relations. But Al Raby was no one's bureaucrat. Instead, as Steve Perkins would emphasize, notwithstanding how Al was "an intensely private person," he was also "a continual talent scout, always looking for new activists with special gifts." Judy Stevens, Raby's deputy at the commission, also watched as "he mentored people." By early 1987 Al was living in Hyde Park with Patty Novick, who two decades earlier had introduced Al and Steve.

One morning Obama joined Patty and Al for breakfast, and soon after, Al called Steve: "he wanted me to come have breakfast with Barack, and so we had breakfast at Mellow Yellow," a well-known Hyde Park eatery. "Barack was fabulous. He was smart, he was articulate—it was a joy to meet him," Steve remembered. Patty recalled how "I want you to talk to" was one of Al's favorite phrases, and one day in early 1987 Raby took Barack to City Hall and introduced him to Jacky Grimshaw, who was serving as Washington's director of intergovernmental affairs. According to Jane Ramsey, the mayor's director of community relations, "Al was all over the place," and Grimshaw can picture Barack that day: "he was a kid," although a "very serious" one. But Raby was far more influenced by his experiences with King than by city politics. "Al was determined to create a base for a long-term movement," Steve Perkins emphasizes.[39]

In early 1987, Harold Washington's aides were focused on his reelection, and Jane Byrne was mounting a stiff challenge in the upcoming Democratic primary. The mayor's employment aide Maria Cerda had made a point of renting a portion of Local 1033's East Side union hall as a new jobs-referral site. Maury Richards said that income "was just totally instrumental in keeping us afloat" and "at a critical time." Soon Richards was put on the city's payroll part-time, allowing him to hold the local union together while also staffing the Dislocated Worker Program. Washington needed all the votes he could get in Southeast Side precincts, and Maury was deeply thankful to Washington, whom he described as "a great guy." Eight days before the election, steel was poured at South Works for the first time in seven months, another small victory for South Side steelworkers.

Just before midnight on February 25, Byrne conceded defeat after mounting a "powerful challenge" to Washington, whose 53.5 to 46.3 percent victory "was much closer than the mayor's strategists had been counting on," the *Tribune* reported the next day. But the headline "Mayor's Tight Win Just Half the Battle" pointed to the April 7 general election, when Washington would face two white Democrats running as independents: Cook County assessor Tom Hynes, a relative "Mr. Clean" by Chicago standards, and 10th Ward alderman Ed Vrdolyak, the mayor's worst nemesis. Fewer white voters had turned out to support Byrne than had voted against Washington in 1983, though the mayor's own white support had barely risen from four years earlier. A larger, more heavily white electorate was possible on April 7, but Washington was "a heavy favorite" as long as both Hynes and Vrdolyak remained in the race.[40]

On the Far South Side, Obama was trying to generate more grassroots support for his Career Education Network idea. Five high schools drew students from major portions of DCP's neighborhoods: Harlan at 96th Street and South Michigan Avenue, Corliss on East 103rd Street, Julian at 103rd and South Elizabeth in Washington Heights, Fenger on South Wallace Street at 112th Street, and Carver down in Altgeld Gardens. Barack and a number of DCP's most active members, including Dan Lee and Adrienne Jackson, began staging modest street corner demonstrations near the troubled high schools to draw attention to the schools' horrible dropout rates, especially among young black men. One morning that spring, they held a rally on busy 111th Street a block north of Fenger, where one year earlier a *Defender* photo of the school's top academic achievers had pictured one dozen African American *women*. Among the passersby was Illinois state senator Emil Jones Jr., whose office was less than two blocks away. "I stopped to see what they were out there for," Jones later explained. He knew Adrienne Jackson, and she introduced Jones to Obama. "Barack was part of the group," Jones recalled. "I met him on the corner." But

Jones was not happy to see protesters just down from the 34th Ward headquarters. "You have a lot to learn," he told Adrienne and Barack. "You'll get more flies with honey than you will the way you all are doing this."

Jones viewed them as an "in-your-face type of group," but invited them to his office, where they laid out their dropout prevention goals, and particularly Obama's hope of winning state funding for his Career Education Network. Jones thought Obama was "very bright and intelligent and very sincere" but also "very aggressive and somewhat pushy." More seriously, Jones thought Obama "was naive as related to the political situation." He did not know that in most any Chicago ward organization, real power was with the ward committeeman, who often doubled as alderman, and not with state legislators, even one with the grand title of senator. In addition, the Illinois legislature was made up of two very separate chambers.

"We can work together," Jones told Obama, but "you haven't got a deal on the House side" until a supportive state representative was recruited. But with Jones's backing, Barack now had a significant state political figure, in addition to Al Raby's City Hall connections.[41]

But still Obama's biggest challenge was expanding DCP's base beyond Roman Catholic parishes like Holy Rosary and St. Catherine and PTA groups from middle-class Washington Heights. His first significant recruit was Rev. Rick Williams, the Panamanian-born pastor of Pullman Christian Reformed Church (PCRC) on East 103rd Street. PCRC had been founded in 1972 as a "mission" church when Roseland's four long-standing Christian Reformed churches left the neighborhood in the wake of its rapid racial turnover. Williams arrived at PCRC in 1981, and by early 1987 PCRC possessed the most racially integrated congregation on the Far South Side.

One day Obama and Adrienne Jackson called on Williams, who was immediately impressed by Barack's "humility" and "his ease with people." Williams also saw that Obama's focus on growing DCP was rooted in an IAF-style worldview: "they wanted to work with churches because churches have values and churches have people and churches have money." But Williams also knew that building an ecumenical base for DCP would be difficult because "these churches are of different persuasions, denominations, ways of thinking. . . . Creating community out of these churches" would be "a very complicated thing," and even more difficult for some pastors because Barack himself was "not a church-going person." But because Obama was "a principled person," Williams readily signed on, telling Barack, "You are wise beyond your years," when he and Adrienne departed.

Just a block west of Holy Rosary was Reformation Lutheran Church. One young woman from that congregation, Kimetha Webster, had been active in DCP for months, and sometime that spring, she took Barack and Bill Stenzel there and introduced them to the church's new young pastor, Tyrone Partee, as well as her father, John Webster, a congregation mainstay and the church's caretaker. If Obama's Career Education Network became a reality, its after-school counseling and tutoring efforts would require more space than Holy Rosary alone could offer. Obama explained DCP's aspirations before asking, "Pastor, do you think it's possible that we could do some things here at the church?"

Partee was, like Barack, just twenty-five years old, and he came from a politically active family. His uncle Cecil Partee, the longtime committeeman of the 20th Ward, had served for two decades in the Illinois state legislature, including one term as president of the state Senate, a landmark achievement for an African American in the thoroughly white Illinois state capitol. Cecil Partee also was a crucial supporter of Harold Washington and now served as city treasurer. Tyrone immediately offered Barack Reformation's support and space in its Fellowship Hall. "I believed in what he was doing for our community," Partee said. But getting to know John Webster was even more valuable because he offered to show Barack around Roseland. "Everybody knew Mr. Webster," Partee recalled. "He knew the good and the bad on everything."

A third pastor Barack called upon was Alonzo C. Pruitt, a former Chicago Urban League community organizer and now the young vicar of St. George and St. Matthias Episcopal Church on 111th Street. St. George was known for its weekday program that each morning fed about forty hungry people, some of whom lived at the nearby Roseland YMCA and others in the neighborhood's abandoned buildings. Pruitt was also impressed with Obama and agreed to lend his name to DCP's efforts.[42]

Among Roseland's many churches, the faith most widely represented was not Catholic, Lutheran, Christian Reformed, or Episcopal; it was Baptist. Baptist churches were freestanding and independent, not tied to any denominational hierarchy or bishop, and their pastors could be as iconoclastic as they chose to be. By early 1987, central Roseland's most immediately pressing problem, as Pruitt's feed-the-hungry program highlighted, was the increase in the number of homeless people. That problem had its roots in the foreclosed loans and boarded-up homes that had increased dramatically in the past seven years due to the loss of tens of thousands of jobs, in the steel mills and also at previously vibrant manufacturing firms, from Dutch Boy and Sherwin-Williams paints to Carl Buddig meats and the Libby, McNeill & Libby food cannery.

Late in 1986 the *Daily Calumet*'s superb steel reporter, Larry Galica, in an article about the human costs of unemployment, quoted Alonzo Grant, a black Roseland homeowner with a wife and three children who had lost his job at South Works and not found a new one. "I have no income whatsoever. I can't receive public aid. I'm three months behind in my house mortgage payments, I'm two months behind in my car payments, and I'm behind in my utility bills."

Starting in late 1985, Neighborhood Housing Services (NHS), a ten-year-old foundation-supported organization whose mission was to help homeowners in declining, heavily minority neighborhoods, began planning a Roseland program at the request of Ellen Benjamin, executive director of the Borg-Warner Foundation. Benjamin had been interested in Roseland for several years, and within six months, the Borg-Warner Foundation committed $450,000 to NHS Roseland. Chicago's Department of Housing soon matched that with $500,000 in city funds, and the state of Illinois contributed $200,000. By late 1986, NHS had named a neighborhood director and had appointed a local board that included Salim Al Nurridin, a Roseland civic activist and native of Altgeld Gardens who had converted to Islam years earlier.[43]

Early in 1987, with Alonzo Pruitt of St. George in the lead, six Roseland churches announced they were offering overnight shelter to any needy person on evenings when the temperature fell well below freezing. Also participating was Mission of Faith Baptist Church, whose pastor, Rev. Eugene Gibson, was president of the Roseland Clergy Association (RCA), and Fernwood United Methodist Church, whose pastor, Rev. Al Sampson, was a forty-eight-year-old veteran of Martin Luther King Jr.'s Southern Christian Leadership Conference whom King himself had ordained as a minister. Those three clergymen were taking the lead in protesting the lack of black professionals at a heavily patronized bank in Beverly, a largely white neighborhood immediately west of Washington Heights. They were also demanding youth employment opportunities at a large shopping plaza west of Beverly in Evergreen Park.

Pruitt, Gibson, and Sampson's efforts received prominent coverage in the *Defender,* and sometime in early 1987 Obama got Gibson on the phone and won an invitation to the RCA's next regular meeting. As Obama later told it, he made a brief presentation to the ten or so clergymen before someone else arrived late to the meeting. "A tall, pecan-colored man" with straightened hair "swept back in a pompadour," wearing "a blue, double-breasted suit and a large gold cross across his scarlet tie" asked Barack whom he represented.

When Barack said DCP, the minister said that reminded him of a white man who had called on him many months before. "Funny-looking guy. Jewish name. You connected to the Catholics?" When Barack said yes, this person whom Obama called "Charles Smalls," responded that "the last thing we need is to join up with a bunch of white money and Catholic churches and Jewish organizers to solve our problems . . . the archdiocese in this city is run by stone-cold racists. Always has been. White folks come in here thinking they know what's best for us. . . . It's all a political thing." Smalls knew Obama meant well, but Barack wrote that he felt he was "roasting like a pig on a spit."

Years later, a journalist named Al Sampson as the intolerant preacher, and Obama later confirmed that identification, explaining that he had just changed the appearance of the short, stout, and dark-skinned Sampson. Asked for the first time about the allegation, Sampson said he did not recall ever meeting Barack Obama in the late 1980s, but in a 2002 video interview Sampson had expressed his admiration for the notoriously bigoted Louis Farrakhan.

More than a quarter century later, Alonzo Pruitt still had a "vibrant memory" of that RCA meeting, with Barack wearing "an open-necked pale yellow shirt" and light brown dress shoes. Pruitt could picture Barack "carefully listening" and "responding with courtesy and restraint even when" others "did not practice courtesy and restraint. I was impressed that he was not defensive or hostile even when a reasonable person might choose the latter. At first I thought he was aloof, but as the meeting went on I realized that his getting angry would simply create a new issue with which to deal, and he was focused on what he perceived to be the heart of the matter."[44]

Obama received a dramatically warmer welcome when he visited Trinity United Church of Christ (TUCC) on 95th Street. Trinity was well known to every minister on the South Side because its pastor, Jeremiah A. Wright Jr., had grown his congregation from just eighty-seven members when he started there in 1972 to more than four thousand by the day Obama first visited. Almost two years earlier, Adrienne Jackson had tried unsuccessfully to interest Wright in DCP, and in Obama's own later account, an aged "Reverend Philips" with a dying church first recommended he visit Wright. Yet among Chicago's black preachers, an undocumented consensus would emerge that it was Rev. Lacey K. Curry, the dynamic pastor of Emmanuel Baptist, a vibrant church in the Auburn Gresham community north of DCP's self-defined 95th Street boundary, who had told Barack to go see Wright.

But Wright would attest to much of Obama's account of his first visit to Trinity, where Wright's attractive secretary, Donita Powell Anderson, was at least as interested in the young gentleman caller as was her pastor. "She was smitten," Wright smilingly remembered. In Barack's telling, Wright's first words to him were a humorous greeting: "Let's see if Donita here will let me have a minute of your time."

As of March 1987, forty-five-year-old Jerry Wright had already lived an eventful life. Raised by two well-educated parents in the Germantown neighborhood of northwest Philadelphia, Wright knew the black church from his earliest years because his father, Jeremiah Sr., was pastor of Grace Baptist Church. Years later, in a long interview, Wright would confess to misbehavior during his high school years—including an arrest for car theft—that was more serious than any of Obama's indulgences while at Punahou. Jerry followed his father's and mother's footsteps and began college at Virginia Union University in Richmond before dropping out and enlisting in the marines. After two years, he changed uniforms and became a navy medical corpsman, ending up at Lyndon B. Johnson's Bethesda bedside when the president underwent surgery in late 1966.

Upon leaving the service, Wright enrolled at Howard University to complete his undergraduate degree and also earn a master's. Reconnecting to religious faith, Wright entered the University of Chicago Divinity School before becoming an assistant pastor at Beth Eden Baptist Church, in the Morgan Park neighborhood west of central Roseland. By late 1971, that affiliation had ended and Wright was searching for new employment when an older friend and mentor, Rev. Kenneth B. Smith, mentioned that the small congregation of Trinity UCC, where Smith had been the founding pastor in 1961, was searching for a new minister. Wright was interviewed by Vallmer Jordan, one of TUCC's most dedicated members, and on March 1, 1972, Wright became Trinity's pastor.

Wright inherited a small congregation and an annual budget of just $39,000, but the church had something almost equally valuable: a newly coined church slogan that declared Trinity as "unashamedly black and unapologetically Christian." When the United Church of Christ created Trinity, it was aiming for a "high potential church" that would attract "the right kind of black people," according to longtime Trinity member and staffer Julia M. Speller in her University of Chicago Ph.D. dissertation. "The class discrimination exhibited by the denomination" was stark, and soon after arriving at Trinity, Wright complained publicly that his new congregation had become "a citadel for the ultra-middle-class Negro." He later quoted one founding member as confessing that "we could out-white white people," and he also wrote that a "'white church in a black face' is exactly what we had become!"

Just two blocks east of Trinity were the Lowden Homes, where DCP's Nadyne Griffin lived, and Wright later remembered that when he arrived at Trinity, "we first had to stop looking at the neighbors around the church as 'those people.'" Within eight months, he had introduced a new youth choir, and not long after that, he told the senior choir to expand its repertoire to embrace gospel music. Those innovations caused almost two dozen of Trinity's existing members to leave the church, and Jerry later wrote that "eighteen months into my pastorate . . . I felt as if I were a failure. It seemed to me as if everyone was leaving our church."

But these changes brought in new members, and by 1977 Trinity's congregation had grown to four hundred. In late 1978, the church moved into a new building with a seven-hundred-seat sanctuary, and in 1980, with Wright's powerful sermons now being broadcast on the radio, Trinity's membership began a rapid climb, reaching sixteen hundred by early 1981. The congregation included a number of prominent black Chicagoans, such as well-known Illinois appellate judge R. Eugene Pincham and Manford Byrd, like Val Jordan a charter member since 1961. In early 1981, when Byrd was passed over for promotion from deputy to superintendent of the Chicago Public Schools in favor of a black woman from California, Trinitarians were among the many black Chicagoans who vocally protested the denial of what Trinity called Byrd's "earned ascension" in favor of an outsider. In response, Val Jordan and several others drafted a wide-ranging statement of values, modeled in part on the Ten Commandments, as a way of honoring Byrd at an August 9, 1981, ceremony. Trinity's twelve-point "Black Value System" was notable for its powerful "disavowal of the pursuit of middleclassness" and an attendant warning against thinking "in terms of 'we' and 'they'"—i.e., "those people"—"instead of 'US'!"

By fall of 1982, Trinity had reached twenty-eight hundred members and its annual budget was now $700,000. In response, Wright and a trio of academically oriented members—Sokoni Karanja, Ayana Johnson-Karanja, and Iva E. Carruthers—drafted an almost two-hundred-page "compendium text for church-wide study." Wright wrote an eleven-page statement of Trinity's mission, beginning with a forceful call for "a conscious cutting across class and caste lines and so-called economic levels" and "utterly abandoning or rejecting the notion of the 'middle class' as the proper vineyard into which God has called us to labor."

Wright also called out the usually unspoken dangers that "Black self-hatred" posed in African American communities, and he later recalled with some embarrassment how he had been entirely ignorant of the harm that youth gangs were doing in neighborhoods like Roseland until his eldest daughter Janet and her boyfriend were robbed at gunpoint in 1982 on Halsted Avenue, less than ten blocks from the Wrights' home, by several Gangster Disciples. But perhaps equally daunting was how his daughter got her property returned, along with an apology, in just three hours after complaining to a next-door neighbor who knew who to call.

In that 1982 essay, Wright emphasized that Trinitarians "start from the cultural strengths already in existence within the Black tradition," a view in keeping with John McKnight's social capital emphasis. Throughout the decade, Trinity's outreach ministries would grow along with the church, with a food co-op and a credit union being joined by a housing ministry that addressed the problem of foreclosed, boarded-up homes plus a high school counseling project and Saturday youth programs. "Educating constituents as to all the nuances and subtleties of the racist political system operative in Chicago," Wright wrote, "is a very definite part of our ministry at Trinity."

In 1983, Wright took a lead role, along with eight other black churchmen including Al Sampson, in fervently endorsing Harold Washington's mayoral campaign. Borrowing Trinity's own "unashamedly black and unapologetically Christian" slogan, the statement was supported by more than 250 members of the clergy. By 1986 Trinity had more than four thousand members, twenty-eight of whom were preparing for the ministry, and Wright was preaching at two separate Sunday services to cope with the growth. One charter member cited Wright's "ability to call all his parishioners by their names, even as the church membership grew into the thousands," as one more of his impressive gifts. Julia Speller wrote that by 1986 "a definite mission-consciousness began to emerge at Trinity," and Jerry was pursuing a deepening interest in black Americans' African cultural roots. Wright had been profoundly influenced by the pioneering black liberation theologian James H. Cone's landmark 1969 book *Black Theology and Black Power,* although he strongly faulted Cone for calling African Americans "a people who were completely stripped of their African heritage." Trinity, Wright wrote, "affirms our Africanness," including "the premise that Christianity did not start in Europe. It started in Africa," and "we affirm our African roots and use Africa as a starting point for understanding ourselves, understanding God, and understanding the world." Indeed, "we understand Africa as the place where civilization began."

By the time of Obama's visit to Trinity in March 1987, word about Jeremiah Wright's church had spread well beyond Chicago. A PBS *Frontline* television crew and well-known black journalist Roger Wilkins had just spent days at Trinity preparing an hour-long documentary on the church that would be nationally broadcast ten weeks later. "The rooms of Trinity are crammed full of its members all day, every day," Wilkins told viewers while describing the church's outreach ministries and Bible-study classes. "Trinity is one of the fastest growing and strongest black churches in America."

Responding to Wilkins's questions, Wright spoke colloquially and bluntly. For black teenagers in Roseland, Wright said, "You can't be what you ain't seen. . . . So many of our young boys haven't seen nothing but the gangs and the pimps and the brothers on the corner," and in their daily lives "they never have their horizons lifted." But Wright also emphasized black Americans' lack of self-esteem. "If I can somehow be white: a lot of black people have that feeling. If I can somehow be accepted. And Africa is a bad thing. I'm not African. I'm not African. I'm part Indian. I'm part Chinese. I'm part anything."

That part of Wright's worldview would resonate deeply with Barack, but his perspective on the breadth and depth of American racism matched that of Martin Luther King Jr. "How do we attack a system, get at systemic evil and realize that it's not the individuals, it's the system," he told Wilkins. "You hate the sin and not the sinner." Wilkins closed the telecast with a prophetic description of Trinity's importance. "This church will be measured by how much of its power will reach beyond its own doors, and by how much its members will reach back, back to those left behind." The day of the broadcast, the *Sun-Times* told Chicagoans not to miss "a compelling and moving portrait of one Chicago clergyman who has made a difference." Jeremiah Wright "sets a standard of excellence that should inspire clergy of all faiths."

"The first time I walked into Trinity, I felt at home," Barack later told Wright's daughter Janet. Furthermore, Obama recalled, "there was an explicitly political aspect to the mission and message of Trinity at that time that I found appealing." In their first 1987 conversation, Barack tried to sell Wright on DCP's program. "He came with this Saul Alinsky community organizing vision," Wright recounted years later. "He was interested in organizing churches," yet Barack's depth of knowledge about the black church was woeful indeed. "He didn't know who J. H. Jackson was," Wright remembered, naming the conservative, dictatorial president of the National Baptist Convention who pastored Chicago's Olivet Baptist Church and was infamous for changing Olivet's address from 3101 South Parkway to 405 East 31st Street when Parkway was renamed Martin Luther King Drive.

Wright remembers that Obama "had this wild-eyed idealistic exciting plan" of "organizing pastors and churches" all across Roseland in support of his Career Education Network. "I looked at him and I said, 'Do you know what Joseph's brother said when they saw him coming across the field?" Obama, utterly unfamiliar with the Bible, said no. "They said 'Behold the dreamer.' You're dreaming. This is not going to happen," Wright told him. "You're in a minefield you have no concept about whatsoever in terms of trying to get us all to work together, even on something as important as the educational issues in the Roseland community," Wright explained, citing the twin evils of denominational divides and local Chicago politics.

Given Wright's busy schedule, that first conversation ended after an hour. But Obama soon returned, talking first with Donita before sitting down with Wright, who remembers he had "questions about this unknown entity, the black church, and its theology. . . . I had studied Islam in West Africa, and he wanted to know about that." In addition, "we talked about the difference between theological investigation, rabbinic study, and personal faith, personal beliefs, and how I separated those two," Wright recalled. "Our visits became more of that nature and that level than the community organizing piece, because I said, 'That ain't going to happen. If you *mention* my name, I can tell you preachers who are not coming in the Roseland area.'" That surprised Obama, and Wright also spoke about the black church's "rabid anti-Catholic" sentiment. "We would spend time talking about religious stuff like that to help him understand that brick wall he's running up against in terms of organizing churches." So "most of the time . . . we talked about how insane" religious antipathies could be, "more so than community organizing."

Barack continued to visit Wright in the months ahead, and their conversations gave Obama a greater understanding of why almost all of the people with whom he was working held their religious faith as a source of strength that could give them courage. It not only "bolstered them against heartache and disappointment" but could be "an active, palpable agent in the world," undergirding their involvement by offering "a source of hope." Witnessing that, Barack remembered, "moved me deeply" and "made me recognize that many of the impulses that . . . were propelling me forward were the same impulses that express themselves through the church."[45]

Obama was even more warmly welcomed by Father Michael Pfleger at St. Sabina Roman Catholic Church in the Gresham neighborhood, well above DCP's northern boundary. The thirty-eight-year-old Pfleger had been a seminary classmate of Holy Rosary's Bill Stenzel, had first met Jerry Wright five years earlier in an Ashland Avenue barber shop, and was well acquainted with Deacon Tommy West, the energetic DCP member from St. Catherine's who spent more time at St. Sabina than at his home church. Pfleger had grown up barely a mile west of St. Sabina on Chicago's Southwest Side, and in 1966, at age seventeen, he had watched as an angry white mob attacked an open-housing march being led by Martin Luther King Jr. in Marquette Park, just a few blocks north. By 1987 Pfleger had been at St. Sabina for twelve years, and although his congregation was nowhere near the size of Trinity's, no church in Chicago, and certainly not one with a white priest, offered as vibrant a Sunday service as Mike Pfleger did.

Years later, Pfleger recalled that Obama "came in and introduced himself and what he was doing." He spoke about how churches "were the most powerful tool in the community for social justice and for equality" and how they should be actively pursuing those goals, not watching from the sidelines. "I was amazed by his brilliance," Pfleger recalled; he was struck as well by "his aggressiveness." Pfleger asked Barack "what was his church," because "people that want to work with churches ought to be in a church." Barack replied that "he was still looking, had been visiting some places, Trinity being one of them." Pfleger had expected a twenty-minute conversation, "and it went much longer." He offered Obama his full support, and after Barack left, Pfleger could remember "walking out of this room saying, 'That's somebody to be watched. He's going places.'"[46]

For Obama, these early months of 1987 were intense as he expanded his horizons and added to his growing set of influential acquaintances. On March 2, in faraway Jakarta, Lolo Soetoro died of liver disease at age fifty-two. If Ann called Barack with the news—"they did not talk often," Sheila recalled—he did not mention it to her or anyone else. He also "never talked that much about his dad" or his death to Sheila, and as best she could tell, "Barack's father played virtually no emotional role in Barack's life." He continued his weekly conversations with Greg Galluzzo—an hour on March 4, ninety minutes each on March 13 and 20, another hour on March 24—and he also introduced his good friend Johnnie Owens to Galluzzo.

Barack and Johnnie had begun discussing whether Johnnie would leave Friends of the Parks and join Barack at DCP, but Owens needed a salary much like Barack's $20,000, and that meant Barack would have to add the MacArthur Foundation as a funder in addition to CHD, Woods, and Wieboldt.

By mid-March, Barack's most pressing concern was on the jobs front, and on Monday, March 23—just two weeks before Election Day—Mayor Washington was coming to Roseland to open the much-delayed new Far South Side jobs center that his employment deputy Maria Cerda had agreed to establish more than six months earlier. In the run-up to that ceremony, Barack dealt extensively with Salim Al Nurridin, a politically sophisticated Roseland figure whose Roseland Community Development Corporation (RCDC) was relatively low profile but whose long-standing acquaintance with one of Barack's new mentors allowed for an easy introduction to this young organizer who was "under the tutelage of Al Raby."

As a native of Altgeld, Salim knew Hazel Johnson, and he had significantly helped 9th Ward alderman Perry Hutchinson, now well known for his star role in the FBI's sting operation, win the seat he was now in danger of losing on April 7. Salim had become a Muslim under the influence of Roseland's least-known figure of quiet political significance, Sheikh Muhammad Umar Faruqi, who oversaw a mosque on South Michigan Avenue, but Salim was not a Nation of Islam "Black Muslim." The new jobs center would be located in a building that Faruqi and Roseland's low-key Muslim community had acquired. Salim worked easily with Barack, whom he saw as "a very energetic and purposeful young man, with a passion to do things effectively."

On that Monday morning, Washington and his two-man security detail arrived at the RCDC office at 33 East 111th Place. The mayor had been told he would be greeted by Loretta Augustine on behalf of DCP as well as Salim, and that Maria Cerda, Perry Hutchinson, DCP's Dan Lee, and "Barac" Obama would be there as well. A city photographer snapped away as the hefty Washington, holding his own notes and with his trench coat thrown over his left arm, shook hands with local well-wishers as a beaming Loretta stood to his right clad in a handsome white coat. The mayor and Cerda listened carefully as Loretta thanked him for coming. One photograph captured Sheikh Faruqi a few paces behind Washington; three different photographs include a tall young man with a slightly bushy Afro standing in the rear of the small room, listening intensely to Washington and Loretta.

Obama would later quote the mayor as saying to Loretta: "I've heard excellent things about your work." Then the entire group walked outside and turned south on South Michigan Avenue. With traffic blocked off and the sun in their faces, Washington and Loretta led the procession a little more than a block to the new office at 11220 South Michigan. Sheikh Faruqi trailed slightly to Loretta's left; Cerda, 34th Ward aldermanic candidate Lemuel Austin, and state senator Emil Jones Jr. trailed to Washington's right. At the front door of the new office, Washington, Cerda, Loretta, and a camera-hogging Perry Hutchinson posed with a white ribbon and a pair of scissors. The mayor spoke to the crowd, and then the ribbon was cut. The mayor climbed into his car for a short drive to 200 East 115th Street, where he broke ground for a future Roseland health center. Barack had emphasized repeatedly to Loretta that she should press Washington to attend a DCP rally for their Career Education Network program, but Loretta had not gotten a commitment.

In his own, overly dramatized retelling of the morning, Barack cursed in anger at her failing and stomped off while Dan Lee tried to calm him down. Loretta remembered no such scene, saying she had "never seen him angry" even when he must have been. "I've seen him drop his head," but, beyond that, "he never showed it." Tommy West agreed. "You could never see him angry." Later that day, Barack had an initial appointment with a Hyde Park physician, Dr. David L. Scheiner, who would remember Obama exhibiting no emotional turmoil during his office visits.[47]

On March 28, four hundred former Wisconsin Steel workers attended a seventh-anniversary rally in South Chicago, where their pro bono lawyer, Tom Geoghegan, told them he hoped their lawsuit against International Harvester—which had just renamed itself Navistar—would soon go to trial. A day earlier Frank Lumpkin and others had picketed outside Navistar's annual meeting at the Art Institute of Chicago. Envirodyne Industries, to which Harvester had sold Wisconsin before its sudden closing, was also suing Navistar, "alleging fraud and racketeering," the *Tribune* noted. The U.S. Economic Development Administration had recouped a tiny portion of its $55 million loan to Envirodyne by selling the mill as scrap to Cuyahoga Wrecking for $3 million, but Cuyahoga went bankrupt before clearing the site, leaving the rusting shell of one mill as a haunting symbol of South Deering's past. "Frank Lumpkin deserves a spot in the organizers' hall of fame," the *Tribune* rightly observed.

On April 2, Obama joined Mary Ellen Montes and Bruce Orenstein for a joint UNO–DCP press conference in response to Mary Ryan's private approaches on behalf of Waste Management Inc. Barack, Lena, and Bruce had decided that playing hard to get—indeed, very hard to get—would maximize the price WMI had already indicated it was willing to pay to expand its Southeast Side landfill capacity. UNO and DCP publicly embraced a no-exceptions moratorium on any new or expanded landfills, while calling for WMI to "commit to a long-term reinvestment program" for the "economic development of neighborhoods around its landfills," the *Sun-Times* reported. UNO and DCP were sending a clear message they *were* willing to make a deal, but WMI had to be generous in purchasing their assent.

The next day, Barack again met for an hour with Greg Galluzzo—in April, as in March, they would have five hours of conversation spread over four weekly meetings. Most of Chicago was consumed by the mayoral contest that would climax on April 7. Ed Vrdolyak was running a surprisingly populist, multiethnic campaign, while Tom Hynes seemed focused on trying to take down Vrdolyak rather than targeting Washington. The *Tribune* heartily endorsed the mayor, saying that "Chicago is in better shape today than it was when Mr. Washington took office" in 1983. Uppermost among Washington's "unfinished business," the paper pointedly added, was "helping depressed neighborhoods get better housing and more jobs."

Two days before the election, Tom Hynes dropped out, in a strategic attempt to unite white voters behind Vrdolyak. On Election Day, Washington swept through Altgeld Gardens in a voter-turnout effort, and he ultimately triumphed with almost 54 percent of the vote. One poll showed him winning 15 percent of white voters and 97 percent of blacks. In Washington's best precinct, Marlene Dillard's London Towne Homes, a young 8th Ward precinct captain named Donne Trotter was given credit for the mayor winning 795 out of the 798 votes cast. Citywide, Ed Vrdolyak received 42 percent, which the *Tribune* noted was "better than anyone had predicted." There also was no question that "the biggest loser" was Tom Hynes, whose sullying of the family name among black Chicagoans would redound in another election two decades later. Another loser was Alderman Perry Hutchinson, who was narrowly edged out by his predecessor, Robert Shaw, in what one observer termed "a choice between two snakes." An indictment followed just six days later, and Hutchinson was soon on his way to federal prison, where he died at age forty-eight.[48]

Obama had attained such a glowing reputation among the CHD staffers at the Chicago archdiocese that Cynthia Norris, the thirty-year-old director of the Office of Black Catholic Ministries, requested that he conduct a training session for the eighteen delegates Chicago was sending to the National Black Catholic Congress in Washington in late May. Norris wanted the delegates to be well prepared to represent Chicago at the huge assembly, the first such gathering since 1894, and Obama trained them in the basement of Holy Name Cathedral, the famous diocesan seat just west of downtown's Magnificent Mile.

At the end of April, Gamaliel hosted a second weeklong training session at Techny Towers in suburban Northbrook. DCP's Margaret Bagby was among the forty or so community members who attended, along with Lena, Mike Kruglik, and CHD's Renee Brereton, plus younger organizers such as Linda Randle. Augustana College senior David Kindler, a young trainee who had already gotten a taste of organizing work in the Quad Cities area where Rock Island, Illinois, and Davenport, Iowa, face each other across the Mississippi River, would remember Mary Gonzales as the star performer among an otherwise all-male and largely macho cast of trainers: Greg Galluzzo, Peter Martinez, and Phil Mullins. "Hard-assed" and "maternal," Mary was just "phenomenally good." Barack took charge of at least two sessions, and Kindler would recall him as someone who "likes everybody to love him."

Galluzzo wanted to nurture and develop new, full-time organizers, and he was regularly petitioning every possible foundation to contribute to the first-year salaries of beginning organizers, just as Kellman had done with CHD and Woods when he hired Barack. Galluzzo knew that organizers must develop "sensitivity, patience and inventiveness" and understand that "he or she is there as a facilitator" who has to motivate community organizations composed entirely of volunteers. "Since every church is an important potential organizing base," Greg said, "an organizer needs to know something of the theological and institutional characteristics of the churches in the community."[49]

Barack's top priority was still his Career Education Network, and his goal was to win Washington's support for the program. Thanks to Al Raby's introductions at City Hall, Barack had already spoken with Luz Martinez, a relatively junior aide, about a mayoral endorsement, and in early May a seven-page document entitled "Proposal for Career Education and Intervention Services in the Far South Side of Chicago" was sent to Washington with a cover letter bearing the names of DCP president Dan Lee and now "Executive Director" Barack H. Obama. The cover letter said DCP wanted "to identify concrete ways that we can positively impact our schools" and emphasized that they "are not seeking any City funding for our program," although they did want the mayor's "whole-hearted support and endorsement of our program" and requested that he meet with DCP leaders sometime in the next month. They also asked that Washington "keynote a large meeting of parents and church leadership," which DCP hoped to convene in mid-June.

In Barack's own letter to Luz Martinez, he volunteered that Al Raby might have already mentioned DCP's request to her or to her immediate boss, Kari Moe. The proposal said the number of blacks graduating from college in Illinois had declined since 1975, and that the dropout rate at the five Far South Side high schools was more than 40 percent. The scale of what Obama and DCP envisioned was grandiose, with "two central offices" coordinating the work of staff representatives at each high school plus supplementary personnel in various churches and social service agencies. The document said the program would give "individualized attention to at-risk students" and offer "incentives for student performance." It would be "administered by the Developing Communities Project," would have thirteen full-time employees as well as twenty part-time tutors, and required an annual budget of $531,000. "State funds would be used to fund this first year of the program," with corporate and foundation support increasing the projected budget to $600,000 and then $775,000 in the two subsequent years. Obama's plan might have seemed familiar to anyone who recalled Jerry Kellman's Regional Employment Network and its initial $500,000 in state funding, but in this case underperforming high school students were taking the place of unemployed steelworkers.

Barack believed a key ingredient was his "Proposed Advisory Board," a list of fifteen people who "have been invited or have already accepted" a request to participate. His list was headed by Albert Raby and state senator Emil Jones, but it also included Carver-Wheatley principal Dr. Alma Jones, Chicago State president Dr. George Ayers, and Olive-Harvey president Homer Franklin. They were followed by Rev. Jeremiah Wright, Dr. Gwendolyn LaRoche of the Chicago Urban League—whose name was misspelled—and Father Michael Pfleger. Also on the list were Ann Hallett of the Wieboldt Foundation, education researcher Dr. Fred Hess, Northwestern University's Dr. John McKnight, and John Ayers from the Commercial Club. The list concluded with three of DCP's most committed members: Dan Lee, Aletha Gibson, and Isabella Waller.

Obama's proposal did not go over well at City Hall. Three of the mayor's aides marked up the document, highlighting its astonishing scale, eye-popping budget, and the preponderance of professionals on the proposed board. One staffer wrote that it needed "more parents/local community residents, student(s), employer(s)," but even a *Sun-Times* story headlined "'85 Dropout Rate Topped 50% at 29 City High Schools" failed to elevate DCP's request among staff priorities. Fred Hess emphasized in the *Tribune* how the utmost priority should be "to make the schools more accountable at the local level," and by May powerful Illinois House speaker Michael J. Madigan, along with Danny Solis and Mary Gonzales of UNO, had embraced a Hess-drafted school autonomy pilot program, House Bill 935.

That plan would allow up to forty-six schools to operate independently of the CPS's hierarchical bureaucracy, and when Washington appeared at UNO's twenty-eight-hundred-person annual convention at the Chicago Hilton on May 21, he was pressed to support the bill. Washington told the crowd that UNO had "hit the nail on the head" in demanding more local autonomy, which Hess and others interpreted as an endorsement. The bill passed the Illinois House the next day, but Washington's top aides quickly signaled that the mayor was actually opposed to such a "drastic" decentralization of CPS. Rival researcher Don Moore at Designs for Change opposed it too, and when Education Committee chairman Arthur Berman killed the measure in the state Senate, UNO acquiesced. Hess was furious, arguing that far-reaching educational changes during "the early years are the most crucial" if there was to be any hope of reducing sky-high dropout rates during high school.

Barack still sought a response from the mayor's office to his plan, and he contacted Joe Washington, a young staffer who was a Roseland native, but made no headway. Disappointed at City Hall's lack of interest, Barack wrote another letter to the mayor, this one featuring the names of fifteen additional signatories in addition to DCP president Dan Lee. Three were DCP members—Aletha Gibson, Isabella Waller, and Ellis Jordan, a fellow PTA leader—and twelve were Roseland clergymen: Bill Stenzel, Rick Williams, Tony Van Zanten, Paul Burak, Tom Kaminski (whose surname was misspelled), Eddie Knox (a new DCP recruit who was the recently arrived pastor of Pullman Presbyterian Church), Joe Bennett, Alonzo Pruitt, Tyrone Partee, and three more.

Obama's inclusion of these new names suggested that a demonstration of DCP's interdenominational support would impress either Joe Washington or the mayor. As pastors of "representative religious institutions of the Far South Side," the signers warned that "high school age youth have been hit hard by the problems of the Chicago school system. In our area, we have seen too many youth drop out, join gangs, and turn to drugs and teen pregnancy instead of staying in school and going on to stable and successful careers." The letter again requested a brief meeting with the mayor to discuss what was now called "a pilot Career Education and Intervention Network." Noting that it would complement Washington's nascent Mayor's Education Summit, it said, "we see the urgent need for this program. We also see the need for your leadership and support in getting it started."

But invoking the twelve Roseland pastors was not any more successful for Barack. One Washington aide jotted on the letter: "Mr. Obama is paid staff person. From Roseland upset w/ Joe." A note to Kari Moe's secretary instructed, "Do not schedule meeting," and two weeks later the office file on DCP was marked "Close." Months would pass before Barack was able to meet with one of Washington's top aides.[50]

By late May, Barack and DCP's board decided to concentrate on the education project and pull back from any further employment focus. The new MOET office had been a signal achievement, but DCP's visits to major local employers—Libby, McNeill & Libby, Carl Buddig, and Sherwin-Williams—to request that they hire local residents had only uncovered news that all three were soon closing their Far South Side plants. No one in DCP was more focused on jobs than Marlene Dillard, but this shift to education opened up tensions within the organization that began with Jerry Kellman's initial decision to have DCP cover such a wide group of different neighborhoods. The southern trio of Altgeld Gardens, Eden Green, and Golden Gate were geographically separate from Roseland and West Pullman, and the westernmost and easternmost neighborhoods, Washington Heights and London Towne Homes, were not eager to be associated even with Roseland and especially not with the Gardens.

These divisions were personified by the differing perspectives of Loretta Augustine and her two close friends, Yvonne Lloyd and Margaret Bagby, each of whom lived just west of Altgeld, and the two different women who represented St. John de la Salle parish on DCP's board, Marlene Dillard and Adrienne Jackson. "Certain issues I was not interested in," Dillard explained years later. "I couldn't center myself around individuals who were in Altgeld Gardens." Residents of London Towne were "not on the poverty line," and although they worried about job loss, they did not require the most basic job training skills that most Altgeld residents needed. In addition, "my son went to a private school," so Roseland's failing public high schools likewise were not a high priority. "I don't feel that London Towne and Roseland can be linked together," for "we have different values and different interests."

Yvonne Lloyd, who lived near Altgeld in Eden Green, agreed with Dillard's explanation. The areas "had different problems" and indeed were "totally different" because solid residential areas like London Towne had "facilities that Altgeld didn't." She, like Margaret and especially Loretta, believed Altgeld's scale of deprivation meant it should be DCP's top concern, because "those were the people we were really, really concerned about" the most. Betty Garrett, the gentle mainstay of Bill Stenzel's congregation at Holy Rosary, watched as the divide deepened between Loretta and Marlene. "They fought constantly," she recalled, mostly over Dillard's emphasis on jobs. "Loretta wanted it to be more widespread." Marlene understood that Loretta "was more interested in poverty issues" than she was, and over time her attitude became "let Loretta and them take care of Altgeld Gardens."

Barack was very close to both Loretta and Marlene, often talking with Marlene's mother and helping Marlene when she ran for election to London Towne's board of directors. Yet by May 1987, there was no getting around the power struggle within DCP, and how Loretta's viewpoint was more widely shared than Marlene's. "Barack was the person who held it together" as long as it did hold, Marlene recalled, but after the May meeting, she shifted her attention to DCP's nascent landfill alliance with UNO's Southeast Chicago chapter.

If DCP hoped to make Barack's Career Education Network even a modest-sized reality, it needed a second full-time organizer, such as Johnnie Owens, and the money to pay his salary. By late spring 1987, Barack had submitted his grant proposal to the MacArthur Foundation, where it went to Aurie Pennick, an African American and South Side native. MacArthur had little experience with community organizing, but soon after Pennick's arrival in 1984, she had initiated a program called the Fund for Neighborhood Initiatives, which would direct about $700,000 annually toward "revitalizing some of Chicago's poorest communities." The small world of Chicago philanthropy was highly interactive, and Pennick had heard Ken Rolling speak glowingly about DCP and was aware that it was being funded by Woods, Wieboldt, and CHD. Pennick lived in West Pullman and her daughter attended Reformation Lutheran's small school on 113th Street, so she knew of DCP's connection there too. But when she read Barack's proposal, she was "underwhelmed" by it. Pennick was deeply averse to the "top-down" type of projects that often won CHD support, and instead favored indigenous activists such as Hazel Johnson from Altgeld. She met with Barack and a trio of DCP members—Loretta, Marlene, and Yvonne—but came away with mixed reactions.

In every such meeting, as with city council leader Tim Evans almost a year earlier, Barack insisted that his community members take the lead while he remained almost silently in the background. "He would never speak. He always put us out front," Cathy Askew explained in recalling a time when she and Marlene accompanied Barack to a meeting with Jean Rudd at Woods. All of the DCP women remember Barack picking them up in his small blue car; wintertime appointments downtown were more memorable than summer ones because Barack's car had a hole in the floor and little if any heat. Yvonne Lloyd remembered the preparations for the MacArthur meeting, with Barack insisting that she, Loretta, and Marlene have the speaking parts and not him. "'You have to be the ones to actually do it because this is your community, not mine,'" Lloyd recalled him saying. "'You can tell your story better than I can.'"

Aurie Pennick found Loretta Augustine "very articulate, very smart" at DCP's meeting with MacArthur. "I was impressed with her. Barack was a little skinny guy in the back, said very little." Yet Pennick's South Side roots left her uncomfortable with how DCP "was very much noninclusive of lower-income folk," such as the actual residents of Altgeld. She also detected a "kind of classist thinking" in some of the DCP members' statements. Pennick told them she had not heard about them being active in West Pullman and asked Barack if DCP had held community meetings there. He assured her that DCP did have a presence there, but when the meeting ended, Pennick "wasn't sure whether MacArthur would make a grant." In subsequent days, when Pennick asked her immediate neighbors if they were familiar with DCP, no one was, and she decided that DCP was "too new and lightweight" to merit MacArthur funding.

To bolster DCP's dropout prevention focus, Barack wanted to generate parental interest in his CEN idea before school ended in early June. He and his two most energetic education volunteers, Aletha Strong Gibson and Ann West, called on the principals of all five Roseland area high schools and asked if they could hold "parent assemblies" in Roseland, Altgeld, Washington Heights, and West Pullman. Barack "was very professional . . . very articulate," Ann West recalled. "He was driven, and he was committed. . . . It didn't appear to be just a job." Aletha felt similarly, describing Barack as "heartfelt" and "committed to the people" as well as "very charismatic." He spent many hours with Aletha and Ann, but even though he knew Aletha had spent her junior year of college in Kenya, and that Ann was a white Australian woman married to an African American, Barack never said anything about his father or about Genevieve. "He was so private," Ann remembered, and they knew nothing of his personal life. "He didn't mix the two."

DCP's members collected a repertoire of Obama's stock expressions, which became something of a running joke. Margaret Bagby remembered, "Whenever he tells you, 'I don't think,' he's telling you that he knows what he wants. And you really need to look out when he says, 'My sense is that.'" Ann West recalled, "He would say to us, 'This is what we need to do,'" and if he were asked a question and he didn't know the answer, he would reply with one or both of these phrases: "Let me look into it" and "I'll research it," Yvonne remembered. They could all see, as Yvonne explained, how "precise and thorough" Barack was in making plans.

One Saturday morning before DCP's "parent assembly" in West Pullman, Aurie Pennick was doing yard work in front of her home, when "all of a sudden I hear 'Miss Pennick'" from someone who recognized her from behind. It was Barack, passing out flyers for DCP's upcoming meeting. "This is a smart man. He probably figured out where I live," Pennick immediately realized. DCP's leafleting was extensive, and Pennick remembers going to the meeting "and it's *packed*. . . . They really had thought it out," and Barack's careful strategizing paid off just as he had hoped. "I of course made the grant," Pennick explained, and in September DCP received $20,000 from MacArthur for general operating support—exactly what was needed to pay Johnnie Owens's salary.

Owens came from a working-class black family, and for him wearing a shirt and tie to Friends of the Parks' downtown office was far more inviting than working out of DCP's windowless office on the ground floor of Holy Rosary's small rectory in Roseland. But Obama was determined to hire him, and Owens recalls Barack challenging him by saying things like, "'If you're really interested in changing neighborhoods and building power, you can't do it from downtown.'" To sweeten the deal, Barack gave Johnnie money from DCP to buy a car, and yet was royally pissed when Owens got a brand-new Nissan Sentra, which was far swankier than the rapidly aging Honda Civic Obama was still driving. "We always had a little tension about that," Owens remembered, but Barack was exceptionally happy to have Johnnie start in July.

DCP's work in West Pullman had attracted some new members, including Loretta's friend Rosa Thomas and a young housewife, Carolyn Wortham, but Barack's grand plans for a half-million-dollar-a-year CEN depended on support from Emil Jones and the Illinois state legislature, which would be struggling with the state budget through June. Barack organized a lobbying trip to Springfield and took some of DCP's most devoted members—Dan Lee, Cathy Askew, and Ernie Powell, Loretta and her friend Rosa Thomas, several other ladies, Ellis Jordan, as well as Loretta's young daughter and both of Cathy's. Emil Jones was a gracious host, posing with the whole group for a photo in his office. Dan Lee's dark jacket and white pocket square matched his mod eyeglass frames all too well, Loretta looked lovely in a stylish white dress, and Ernie Powell personified strong workingman dignity with a well-knotted tie below one of Illinois's more impressive mustaches. Barack wore a blue blazer, a white shirt, and no tie, but he closed his eyes when the camera clicked. Barack's dream of obtaining a $500,000 state appropriation remained just that, although Jones arranged for the Illinois State Board of Education to give DCP a $25,000 planning grant that gave Barack enough to get a semblance of CEN started in early 1988.[51]

Back in Chicago, Obama continued his almost weekly discussions with Greg Galluzzo, who told numerous organizing colleagues that Barack was "really special." But even though Greg spent more time with him than any other person in Barack's workday world, he knew almost nothing of Barack's home life, and he met Sheila Jager only once in passing.

The young couple's first nine months of living together had melded two intensely busy lives into an increasingly cloistered relationship where Sheila saw almost no one from Barack's day job, and Asif was their only regular contact in Hyde Park's graduate student community. Barack's heavy smoking was a regular topic of comment within DCP, and Reformation Lutheran pastor Tyrone Partee nicknamed him "Smokestack." Sheila said that at home, "he actually introduced me to smoking, so we smoked like chimneys together." She wanted a cat, and after Barack relented, "Max" joined their household and became a less-than-fully-welcome presence in Obama's life. "He drove Barack crazy because the cat would always pee" in their one large houseplant.

Sheila recalls the early months of 1987 as a time when she witnessed a profound self-transformation in Barack. "He was actually quite ordinary when I met him, although I always felt there was something quite special about him even during our earliest months, but he became someone quite extraordinary . . . and so very ambitious, and this happened over the course of a few months. I remember very clearly when this transformation happened, and I remember very specifically that by 1987, about a year into our relationship, he already had his sights on becoming president."

This change in Barack encompassed two interwoven themes: a belief that he had a "calling," coupled with a heightened awareness that to pursue it he had to fully identify as African American. The "'calling' had more to do with developing a sense of purpose in the world," Sheila later explained, and even two years earlier, Genevieve Cook had sensed an incipient presence of the same thing. She remembered thinking that "all along he had some notion of testing his own mettle and potential for greatness, and that it was as much about that personal journey as it was finding the best way to effect the maximum positive social change. Those two aspirations, the personal and the heroic," were "melded from very early on." Yet by early summer 1987, Barack's understanding of his "calling" was as "something he felt he really had no control over; it was his destiny," Sheila explained. "He always said this was destiny."

By then, Barack had gotten to know Al Raby and John McKnight, whose political roots lay in the civil rights struggles of the 1960s, and he had a relationship with Jeremiah Wright, whose theology sprang from that same soil. Barack had also developed an acquaintance with Emil Jones, a savvy politician, and he had witnessed at close range the charm and aura Harold Washington possessed even in a nondescript storefront. But as Sheila experienced and reflected upon what occurred, she realized the stimulus for Barack's transformation lay not in one or another precinct of black Chicago, but in the disparaging evaluation he had received from her father back at Christmas. "After that visit, and over the course of spring '87, he changed—brooding, quiet, distant—and it was only then, as I recall, that he began to talk about going into politics and race became a big issue between us." Once "we got back from California," Barack "became very introspective and quiet," Sheila recalled. "I remember very specifically that it was then he began to talk about entering politics and his presidential ambitions and conflicts about our worlds being too far apart."

Meanwhile, "the marriage discussion dragged on and on," but it was affected by what Sheila describes as Barack's "torment over this central issue of his life," the question of his own "race and identity." The "resolution of his 'black' identity was directly linked to his decision to pursue a political career," and to the crystallization of the "drive and desire to become the most powerful person in the world."

Eight years later, Obama would say that through organizing "I think I really grew into myself in terms of my identity," and that his community work "represented the best of my legacy as an African American." It had allowed him to feel that his "own life would be vindicated in some fashion," and his immersion in black Chicago gave him "a sense of self-understanding and empowerment and connection." Obama's daily experiences on the Far South Side had reshaped him. "I came home in Chicago. I began to see my identity and my individual struggles were one with the struggles that folks face in Chicago. My identity problems began to mesh once I started working on behalf of something larger than myself." He also explained that organizing had "rooted me in a specific community of African Americans whose values and stories I soaked up and found an affinity with." And most specifically, "by the second year," he told one interviewer, "I just really felt deeply connected to those people that I was working with."

Sheila was convinced that "something fundamentally changed" inside Barack during the first half of 1987 that had transformed him into a "powerfully ambitious person" right before her eyes. "We lived so cut off from everyone else" that no one else was privy to her perspective, and Barack's ability to "compartmentalize his work and home life, to the extent that the two worlds were never brought together physically" or in any social setting, meant that their increasingly stressed and intense relationship existed as "an island unto ourselves."

In later years, Obama once said that his experiences in Chicago had "converged to give me a sense of strength." At an expressly religious event, he cited Roseland as where "I first heard God's spirit beckon me. It was there that I felt called to a higher purpose." Sheila caviled at that, saying she "would not call him religious. Perhaps spiritual is a better description" of the man she lived with. "Barack was definitely not religious in the conventional sense. He talked about God in the abstract, but it was mostly in terms of his destiny and/or some spiritual force."

Early in the summer, Barack's older brother Roy visited Chicago and met Sheila briefly, but Barack went alone to the Chicago home of his maternal uncle Charles Payne for his nephew Richard's high school graduation. His sister Maya had just completed her junior year at Punahou, and she wanted to visit a number of mainland colleges before submitting her applications. Ann Dunham was attending the Southeast Asian Summer Studies Institute being held at Northern Illinois University, west of Chicago in DeKalb, so she and Maya arrived in Chicago before Sheila left to see her parents in California and make a brief initial research trip to South Korea.

This was the first time Barack and Ann had seen each other in eighteen months, and Ann had gained a tremendous amount of weight and now seemed "very matronly." She had not made much headway on her Ph.D. dissertation, in part because she had spent half of 1986 in the Punjab, working as a consultant for the Agricultural Development Bank of Pakistan. But her analysis was coming together, and one of her closest academic colleagues described her conclusions in words that echoed what her son had learned from John McKnight. "Anti-poverty programs . . . only reinforce the power of elites" and "it is resources and not motivation that poor villagers lack," in Indonesia as elsewhere. Once Ann's summer institute was complete, she would return to Pakistan for three more months of work. Barack's close friend Asif Agha recalled playing volleyball out at Indiana Dunes during Maya's visit, and Jerry Kellman remembers Barack bringing Maya along to a barbecue at Jerry's home. During this trip, Maya wanted to see the campuses of the University of Michigan and the University of Wisconsin, where Asif spent much of the summer studying Tibetan.

Early in the summer Barack decided that he and Sheila should acquire a new Macintosh SE computer, which had debuted just three months earlier. "It was the latest model and very fancy," Sheila remembered, and as with their rent, Barack happily footed the bill. "Barack said we both needed it," and they each used it a lot, but when it first arrived, Barack had no idea how to operate the mouse. A call to Asif resulted in the dispatch of Asif's other best friend, Doug Glick, a fellow linguistic anthropology graduate student who quickly showed Barack that you do not hold the mouse up in the air.

By the end of June, with Sheila away from Chicago and Asif up in Madison, Barack and Doug on several weekends made the three-hour drive up to where Asif was housesitting in some Wisconsin professor's lakefront home. Obama years later would publicly joke that he had had "some fun times, which I can't discuss in detail," on those visits and "some good memories," but Glick clearly remembered the long drives in Barack's noisy Honda. Barack was "just a regular guy," an "incredibly friendly guy" with "a great sense of humor." During the road trips, Barack talked "about wanting to write the great American novel. . . . I spent an awful lot of time in the car with him. These are long drives, he can talk," and "he doesn't shut up." At least once Obama mentioned an interest in law school, but he rarely talked about his DCP work. "I never heard him talking about community work and public service as the driving force of who he was." Sometimes "I'm making fun of him," asking, "'If I whisper "shut up," will you hear it with those ears?'" But Barack clearly had "tremendous intelligence, tremendous charisma," and indeed "a certain kind of aura to him," Glick thought. Doug, like Asif, felt that "Barack is not that black," but it also seemed as if "he was ideologically loaded a little."

During one drive, Glick recounted, "we had a god moment. The strangest thing that has ever happened to me in my life happened with him." Every trip included a pit stop to get gas and pick up something to drink, and on one occasion Barack was "sitting in the car in the driver's seat" with the window open as Doug returned carrying snacks and bottles. "I tripped. The Snapple goes flying. . . . We both watch as it hits the ground, breaks like an egg, goes up through the air, goes through the window and both of the things land on his legs face up." Yet somehow Obama was completely dry. "We're never going to forget that," Barack said to Doug. "Religions start at moments like that."[52]

In Chicago, the battling intensified over the Southeast Side landfills and their toxic impact on nearby residents. The *Sun-Times* published a six-part, front-page series of stories titled "Our Toxic Trap" which focused on CID, the Metropolitan Sanitary District's (MSD) "shit farm" just north of Altgeld Gardens, and older, more mysterious dumps like the Paxton Landfill. Hazel Johnson was quoted on the "nauseating stench" of sewage sludge permeating Altgeld and said, "it smells just like dead bodies." In response, the state legislature created a special joint committee to investigate the problems, and the MSD's board pledged its own study after an angry public meeting during which Johnson called one African American MSD commissioner an "Uncle Tom." After a large illegal dump was discovered in a remote corner of Auburn Gresham, four city sanitation workers who were excavating the waste for transfer were "overcome by noxious garbage fumes" and hospitalized.

On June 30, WMI's Mary Ryan proposed to Chicago's city council that if the city would set aside its existing moratorium on landfill growth, WMI would move forward with an "economic and community development assistance program" that could be a huge "'catalyst' toward revitalizing" the entire Southeast Side. But Ryan's proposal only intensified the fury of local activists like Marian Byrnes, Vi Czachorski, and Hazel Johnson over a possible deal between the city and WMI to expand landfill capacity. "Perhaps Washington is Vrdolyak in disguise on dump issues," James Landing, the chairman of the Lake Calumet Study Committee (LCSC), told his fellow allies.

With a new and energetic Chicago chapter of the international environmental group Greenpeace eagerly joining in, Southeast Side activists prepared for a July 29 blockade of all dumping at CID. A large rally drew media coverage, and a dozen or more Greenpeace members and local activists would chain themselves together to CID's entrance gate to block waste trucks from entering the landfill. By the morning of the twenty-ninth, DCP's Dan Lee, Cathy Askew, Margaret Bagby, Loretta Augustine, Betty Garrett, and Obama joined the protesters. Cathy recalled years later, "He led that. He led that in the background. He had to be there to bail them out."

The day was "beastly hot," one young Greenpeace member remembered, and "wearing a media-friendly buttoned-down shirt" became a sweaty mistake. But the blockade was a grand success. The *Daily Calumet* reported that a crowd of 150 people gathered and said it was "the largest environmental protest in years." As many as a hundred waste trucks were backed up on the nearby expressway and unable to enter CID, as protesters chanted, "Take it back!" They blocked the entry gate from 10:00 A.M. until midafternoon. One photo caption said: "Wearing a gas mask, Deacon Daniel Lee of the Developing Communities Project . . . makes a point about odors and toxic wastes."

Some reporters lost interest as the day dragged on, and the next morning's *Sun-Times* erroneously reported that "there were no arrests." Leonard Lamkin, an East Side activist who joined the chain-in, said that "when the media went away, that's when they made the arrests." Hazel Johnson and Marian Byrnes were among the sixteen participants taken into custody, and the women remained jailed for six hours even though the men were released in less than two. Scott Sederstrom, the young, overdressed Greenpeacer, thought "it was almost an act of mercy by the Chicago police to take us into their air-conditioned precinct house for booking. . . . The Cubs game was on TV in the station" and comments about baseball leavened the fingerprinting process. "As a further act of generosity, they let me stay in the air-conditioned area watching a little more of the game" instead of moving Sederstrom to a holding cell. Given the Cubs' all-too-typical performance, though—they were trailing 10–0 by the seventh-inning stretch—interest in the game understandably waned. Weeks later the charges were dropped against defendants who agreed not to enter any WMI properties for one year.[53]

Before the summer ended, Hazel Johnson and Marian Byrnes staged three more protests near the CID entrance, taking care not to get arrested. They also testified before the special joint legislative committee, chaired by Emil Jones. Assisting Hazel was a thirty-five-year-old black man who had just returned to Chicago after working toward a master's degree at Harvard University's Kennedy School of Government and who had earlier won a Rhodes Scholarship to attend Oxford University in England. Like Jones, Melvin J. "Mel" Reynolds was eyeing a challenge against incumbent U.S. congressman Gus Savage in the spring 1988 Democratic primary, and both men—like Savage—were eager to raise their profiles among district residents angry over authorities' inability to take meaningful action against the Southeast Side's multiple toxic threats.

Several months earlier, Harold Washington had elevated Howard Stanback to an influential post as his assistant in charge of the city's infrastructure. An African American economist, Stanback had taught at New York's New School for Social Research until he came to Chicago as Maria Cerda's deputy at MOET. His new appointment made him the mayor's primary adviser on Chicago's landfill crisis, and in late summer 1987, Stanback gave Washington a memo that laid out the city's options. Chicago had done "very little towards developing and implementing alternatives to dumping," making the city almost completely dependent on landfill availability. The only solution was "the O'Brien Locks property currently owned by the Metropolitan Sanitary District," but using that property would mean lifting the moratorium and incurring a huge uproar from Southeast Chicago.

Stanback gave the mayor two options for how to proceed. One would be to convey the land to WMI, to whom "the site is probably worth $1 billion," and in order "to neutralize opposition to lifting the moratorium," WMI would contribute sufficient funds to the surrounding neighborhoods, just as Mary Ryan's outreach efforts envisioned. Stanback believed that this could succeed, even with WMI's "negative image," and that this was superior to the second option, which would involve the city itself operating a landfill on the O'Brien Locks property. Stanback believed political opposition would be higher to this scenario because it would not include WMI's contributing to community revitalization projects. "Operating a landfill is not a business the City should enter," Stanback recommended.

One Saturday, Stanback drove to South Chicago to meet Bruce Orenstein at UNO's East 91st Street office. Also there that morning was DCP's Barack Obama, whom Bruce had asked to join them. Stanback described the city's thinking regarding the landfill and WMI's proposal, but he also explained that Washington wanted to be sure that WMI's big gift would not be controlled by the Southeast Side's traditional power brokers, particularly South Chicago Savings Bank president James A. Fitch, a longtime backer of mayoral rival Ed Vrdolyak and the dominant figure in the four-year-old Southeast Chicago Development Commission (SEDCOM). If a deal could be cut with WMI, the mayor wanted his allies—such as UNO, with whom Washington had worked in close alliance for four years—to take charge of the windfall.

"Barack in particular, his eyes got so bright," Stanback remembered. "He said, 'This can be one of the biggest community development coups of all time.' I said, 'You're right,'" but UNO at present had no development capacity. "We agreed that nothing was going to happen anytime soon," Stanback recalled, but "we agreed in principle" that UNO, DCP, and the city would closely coordinate as discussions moved forward. When Fitch then wrote to another Washington aide, budget director Sharon Gist Gilliam, to initiate a discussion of lifting the moratorium to allow for WMI's use of the O'Brien parcel, Gilliam waited twelve days before sending Fitch a cold, rude reply stating that she had given his letter to Stanback.[54]

For Obama, the late summer of 1987 was a busy and intense time. Throughout July, his hourly consultations with Greg Galluzzo were more than weekly, but from August forward the two men met only twice monthly, as Greg began having ninety-minute or longer sessions with Johnnie Owens almost weekly. One weekend, Barack met Ann and Maya in New York, where Maya was looking at Barnard College and Ann was visiting friends before returning to Pakistan via London. A rooftop photograph shows Ann and Barack with several of her anthropologist friends, including Tim Jessup, who had first met Barack in Jakarta four years earlier and had seen him again in Brooklyn in 1985 with Genevieve. Barack stayed with Hasan and Raazia Chandoo in Brooklyn Heights, playing basketball nearby with Hasan and walking across the Brooklyn Bridge into Manhattan.

Hasan recalled that "by that time he knows for sure he wants to be a political person," and Beenu Mahmood, then a lawyer in Sidley & Austin's New York office, remembered the visit similarly. "By that time he was very clear that he was going into politics" and "it was very clear that law would be the vehicle for getting into politics for him." Several nights Raazia cooked dinner, but one thing had most definitely changed: by 1987 Barack never again "partied" as he had so many times in 1984 and 1985. Hasan recalled Barack mentioning how brutally cold Chicago winters were and also remembers him describing the time he and Johnnie had to duck behind a car when they heard gunfire nearby in Palmer Park. Raazia, five years younger, found "Barack a little bit arrogant"—just "intellectually arrogant," Hasan interjected—"so I didn't want much to do with him."

Obama was back in Chicago by the end of the second week in August, and he may or may not have seen a prominent headline in the *Defender* that would have reminded him of an influential relationship from earlier in his life: "Frank Davis Dead at 81." During Sheila's midsummer visit home, she told her mother much of what Barack had said to her in recent months, and Shinko Jager in turn recounted Sheila's comments to Mike Dees, the family's closest friend, who had met Barack months earlier during the Christmas holiday. Barack's marriage proposal still loomed, and "if Sheila went with Barack, she would have to follow his lead. He wanted to be president." Shinko remained opposed to the marriage, "but she never gave a reason," Mike recounted. "I was against it because I thought they were two ambitious people, and I knew they wanted their own separate careers, and he was talking about being president, which I thought was a little strange" for a twenty-five-year-old community organizer. But there was also something more, something Barack had begun to articulate to Sheila. "There was a problem there," Mike recalled. "He was concerned if he was going to take the steps to the presidency with a white wife."

One August Friday, Sheila joined Barack for the trip to Asif's summer house in Madison. Sheila was "very quiet" and slept in the back of the car most of the drive, but an unusual tension was present. By Saturday morning, the problem broke into the open, and Barack and Sheila kept pretty much to themselves upstairs. But according to someone there that weekend, "it's the summer . . . these houses are old. You'd die if you closed windows. Everything is open." From morning onward "they went back and forth, having sex, screaming yelling, having sex, screaming yelling." It continued all day. "That whole afternoon they went back and forth between having sex and fighting."

Others remember "moving around to the other side of the porch just to be able to talk." It "was a long weekend" and "an incredibly unpleasant one," one person recalled. "It was so stressed and tense." Barack tried "to be more social about it," and "they came down a few times to grab a beer, to eat," but "then they went back up to scream or fight." Sheila was "a very sweet person . . . very mild-mannered," and "certainly exotic" in her looks, but "shy and withdrawn" that "extremely emotional" weekend.

"They called truces here and there, but it kept popping back up" that Saturday afternoon, as "she screamed and they fought." Sheila's voice came through loud and clear: "That's wrong! That's wrong! That's not a reason," she was heard saying. As the others talked quietly, the explanation of what they were hearing was shared: Barack's political destiny meant that he and Sheila could not have a long-term future together, no matter how deeply they loved each other. But she refused to accept his rationale: "the fact that it was her race." It was clear—audibly clear—that "she was unbelievably in love with him," that "the sex for her was the way to bring it back." Barack "was very drawn to her, they were very close," yet he felt trapped between the woman he loved and the destiny he knew was his. According to one friend, Barack "wasn't black enough to pull that off and to rise up" with a white wife.

Sunday afternoon Barack drove them south to Chicago, with Sheila again napping in the back seat. "Thank god it's a crappy car that made a lot of noise!" A quarter century later, Sheila had almost no recollection of going with Barack to Madison, but she unhesitatingly characterized their relationship as "a very tumultuous love affair." No matter how others saw Obama, "the Barack I knew was not emotionally detached, in control, and cool," she stressed. Instead, Barack was "a very passionate, sentimental" and "deeply emotional person," indeed "the most overwhelmingly passionate and caring person I ever knew."

In Barack's workday world, Cathy Askew, the white single mother of two half-black daughters, witnessed most starkly Barack's newly articulated racial identity. With a new school year about to begin, Cathy told Barack about what she considered a perplexing and offensive racial conundrum: although the Chicago Board of Health said her daughters were black, their school in West Pullman wanted to count them as white. Cathy rejected this binary view of racial identity—"50-50 is a good term"—and expected biracial Barack to agree. She was astounded when he rejected any middle ground, especially since he had spent most of his childhood in a predominantly "hapa" world. But he did. "He said, 'Well, there comes a time when you have to pick a side, you have to choose a side,'" Cathy remembered him saying. And Barack repeated it: "You have to choose."

More than a decade later, Barack again gave voice to the sentiment he had expressed to Cathy. "For persons of mixed race to spend a lot of time insisting on their mixed race status touches on a fear" that a darker complexion is innately inferior to a lighter one, and too many nonwhite people are "color-struck. . . . There's a history among African Americans . . . that somehow if you're whitened a little bit that somehow makes you better, and that's always been a distasteful notion to me," perhaps ever since seeing that magazine story one day in Jakarta. "To me, defining myself as African American already acknowledges my hybrid status," and "I don't have to go around advertising that I'm of mixed race to acknowledge those aspects of myself that are European . . . they're already self-apparent, and they're in the definition of me being a black American." Any other mind-set intimated racism. "I'm suspicious of . . . attitudes that would deny our blackness."[55]

Fred Hess knew well ahead of time that September 1987 would witness a train wreck of historic proportions for Chicago Public Schools. The board of education needed to negotiate new contracts with multiple unions, most important the Chicago Teachers Union (CTU), and in midsummer, Hess had told the board that it could afford to offer a pay increase of 3.5 percent to teachers. Instead, the board announced a three-day reduction in the upcoming school year, thereby *cutting* teachers' pay by 1.7 percent. The CTU demanded a 10 percent salary increase, and Superintendent Manford Byrd asserted that CPS could afford no raise at all. Hess labeled that claim "a bunch of bull," and on September 4, four days before schools were to open, more than 90 percent of CTU members voted to strike.

The strike began on September 8, with Hess warning the *Sun-Times* that "it's going to be a long strike because it looks like it has turned into a question of principle," especially for Byrd. In a subsequent op-ed in the *Tribune,* Hess denounced "an administration that insists on adding central office administrators while cutting other employees' salaries." More than 430,000 students were out of school, and on September 11 a large group of parents and students from sixteen community groups picketed CPS headquarters, with the group's spokesman, Sokoni Karanja, telling the news media that "our children are victims of a school system that is failing to educate them."

Karanja was a forty-seven-year-old African American native of Topeka, Kansas, who had earned a Ph.D. from Brandeis University in 1971 before moving to Chicago's historic Bronzeville neighborhood, joining Trinity UCC, and founding a family-aid organization he christened the Centers for New Horizons. Nine months earlier, Karanja had become the Woods Fund's first black board member, but this September 11 was significant because it illuminated the deepest social and political chasm in black Chicago: African American families with children in heavily minority public schools were fighting against a system made up of 41.6 percent black administrators and 47 percent black teachers.

Those numbers explained why Harold Washington had been so ambivalent four months earlier when he seemingly embraced school reform. The city's public schools provided middle-class salaries and middle-class status to thousands of black Chicago families, while at the same time they were shortchanging tens of thousands of black students. The human toll was staggering: more than 40 percent of African American men in Chicago in their early twenties and 35 percent of young African American women were unemployed, many of them lacking basic job skills because the CPS had not offered them real schooling.

"Our education system is in shambles," Rob Mier, Washington's economic development commissioner, publicly acknowledged. "We're producing a legion of functional illiterates." Some critics described "a vicious circle of incompetence," with CPS hiring teachers who had graduated from weak nearby state universities, which enrolled mainly ill-prepared graduates of Chicago high schools. As Fred Hess kept emphasizing, the abysmal state of public education in Chicago was the result not of a lack of resources, but instead of their dramatic misallocation: during the 1980s, "the number of central office administrators rose by 29 percent while the number of staff members in the schools rose by 2 percent." By 1987, CPS's central bureaucracy had reached an astonishing thirteen thousand employees.

On Thursday, September 17, two hundred angry parents picketed outside Harold Washington's Hyde Park apartment building while others targeted the homes of Governor Jim Thompson, new board of education president Frank Gardner, and CTU president Jacqueline Vaughn. The next morning more than a thousand parents and children picketed outside CPS headquarters, but as the strike moved into its third week, neither the board nor the union were showing any signs of compromise. Superintendent Manford Byrd, furious at Fred Hess's disparagement of CPS's central administration, wrote his own op-ed for the *Tribune*, claiming that the system's fundamental problem was "the extraordinary special needs of most Chicago public school students."

That remarkable assertion—a school superintendent labeling the majority of his system's children as "special needs" students—laid bare the deep class divide within black Chicago between middle-class professionals and "those people." As Hess later explained, again echoing John McKnight's analysis, Byrd's explicit articulation of a "deficit model of 'at-risk'" children revealed the attitudes of administrators "whose jobs depend on the existence of a pool of 'at-risk' students as clients." Unable to see that students might bring strengths to school, CPS bureaucrats could not imagine that Chicago's schools, "rather than the students themselves, might be to blame for students' lack of success," Hess explained. Only by reallocating power from CPS headquarters to local schools so as to "reemphasize local communities rather than large, hierarchical bureaucracies" could meaningful educational improvement be attained.

While most parents and community groups were angry with the CTU as well as the board, UNO simply backed the teachers' demands. As the strike entered its third week, the parents' groups, with support from Anne Hallett at Wieboldt and input from Al Raby, came together as the People's Coalition for Educational Reform (PCER). On Thursday, October 1, in what the *Tribune* called a "dramatic confrontation," a predominantly black group of PCER representatives, many of whom were close allies of Harold Washington, told both the board and the CTU that a 3 percent raise must be agreed upon to end the strike. First the board, overruling Byrd, and then a reluctant CTU, bowed to the coalition's demand, and on Saturday morning it was announced that schools would reopen on Monday. Nineteen days of classroom time had been lost in "the longest public employee strike in Illinois history."

On Sunday, Harold Washington summoned Casey Banas, the *Tribune*'s education reporter, to his Hyde Park home. The mayor wanted everyone to know he understood the significance of the parents' protests. "Never have I seen such tremendous anger and never have I seen a stronger commitment on the part of people" to make Chicago a better city. But Washington also did not want to alienate the black educators Manford Byrd represented. "There's no discipline in many homes. There's no stimulus in many homes. And even though the educational structure is a poor substitute to supply what the family is not supplying, it must be done," the mayor declared. He said he would personally lead a new, far more inclusive iteration of his Education Summit to reform Chicago's public schools. It would begin the next Sunday at the University of Illinois's Chicago Circle campus (UICC). "We're going to have a massive forum," Washington promised.[56]

Obama and DCP had remained entirely on the sidelines throughout the protests against the shutdown. On the Monday of the strike's final week, and after three months of trying, Barack finally had an appointment with Harold Washington's top policy adviser, Hal Baron, to pitch his Career Education Network plan to the mayor. It was entirely thanks to John McKnight that Baron consented to see Obama, despite education aide and Roseland native Joe Washington's negative attitude. "I did it purely as a favor to John McKnight," Baron recalled. "John was just so high on him. . . . I think John wanted me to get him a meeting with the mayor, but he pestered me and I finally did a meeting with him myself." Joe Washington joined Baron in Hal's City Hall office, but Barack's presentation was anything but persuasive. Baron remembers it as "cocky standard Alinsky bullshit" and "very glib." He also recalls Obama saying, "'We've got Roseland organized.'" After Barack left empty-handed, Joe Washington told Baron what he thought: "That guy doesn't know shit about Roseland."

Far more fruitful for Obama was his relationship with John McKnight. Barack's reactions to now two years of immersion in the macho, conflict-seeking world of Alinskyite organizing—including his first year with Jerry Kellman, dozens of hours with Greg Galluzzo, and his warm, almost brotherly relationship with fifteen-years-older Mike Kruglik—had drawn him toward McKnight's alternative vision of human communities. At least twice Barack drove up to McKnight's second home in Spring Green, Wisconsin, west of

Madison, for weekend-long discussions. He was the only young organizer other than Bob Moriarty with whom McKnight had developed a truly personal relationship, and one weekend Barack brought Sheila along with him. Jerry, Greg, Johnnie, Mary Bernstein, Bruce Orenstein, and Linda Randle would all remember meeting Sheila a time or two, usually by chance in Hyde Park or at her and Barack's apartment on South Harper. "She's a cutie," Bruce remembered, but virtually never did Barack seek for anyone in his workday world to meet or know the woman with whom he lived. Mike Kruglik never laid eyes on her.

But one Friday Barack drove north with Sheila, and McKnight remembered Barack calling him from the Penguin, a working-class bar in nearby Sauk City, to say that his shabby Honda had broken down. John drove in to pick them up, and Barack introduced him to Sheila, whom McKnight thought was "absolutely stunning." The regulars at the Penguin "must have been pretty surprised when that couple walked in," McKnight suggested.

That weekend, like others, was spent mostly "talking about ideas," McKnight recounted. Barack had a "set of things he was concerned about that he didn't think he could talk with Kellman or Greg about that was outside the true faith." McKnight, as a longtime Gamaliel board member, knew how "absolutely rigid" Greg was about the Alinsky model's view that the way to make people "feel powerful was their anger," instead of feeling "powerful because of their contributions." Barack was deeply averse to anger and confrontation, and therein lay his difficulty with the attitude that Alinsky organizing sought to inculcate in its young initiates. McKnight remembered these discussions were "mostly my . . . responding to him about questions he had," with McKnight talking about how his asset orientation differed from the "really true faith people" like Greg. McKnight had just written a powerful new paper, "The Future of Low-Income Neighborhoods and the People Who Reside There," addressing what he termed "client neighborhoods." Anyone who had spent time in Altgeld Gardens would have appreciated McKnight's analysis. Such areas are "places of residence for people who are not a part of the productive process" and "have no hopeful future." What was needed was "a new vision of neighborhood that focuses every available resource upon production." In a sentence that reached directly back to the quintessential lesson of the 1960s' southern freedom movement, McKnight insisted that for a group like DCP, "it is the identification of leadership capacities in every citizen that is the basis for effective community organization."

As McKnight and Obama continued to talk throughout 1987, what became "*very* clear to me was that I was talking to a young man who had not bought the true faith," McKnight remembered, one who had come to realize that Alinskyism "is ultimately parochial" and offered no prospects whatsoever for attaining large-scale social change. As the best historian of Chicago racism, Beryl Satter, would incisively note, Alinskyism's "insistence on fighting only for winnable ends guaranteed that" community organizing "would never truly confront the powerful forces devastating racially changing and black neighborhoods."

McKnight's long-term impact on Obama would be profound, irrespective of how much Barack later remembered of their conversations or how few commentators were knowledgeable enough to hear the readily detectable echos. "If you want to see an intellectual influence on Obama's thinking, it's John McKnight," citizenship scholar Harry Boyte told one Washington audience two decades later. "A lot of things started in part through John McKnight," observed Harvey Lyon, a Gamaliel board colleague whose political roots also reached back to the 1960s. Stanley Hallett, an influential urban development pioneer, a onetime theology school classmate of Martin Luther King Jr., and the husband of Wieboldt's Anne Hallett, was also deeply influenced by McKnight. For Hallett, directing public funds to service poor people's professionally identified needs is "money spent to maintain people in a condition of dependency," Hallett told one interviewer. Progressive public officials needed to stop "looking at people in terms of their problems instead of their capabilities."[57]

The People's Coalition for Educational Reform issued its demands in advance of the mayor's Sunday, October 11, summit. Calling for local school-based management, greater parental involvement in schools, a requirement that 80 percent of students perform at or above grade-level standards, and deep cuts in CPS's huge central bureaucracy, PCER made clear its student-centered focus: "We do not want cuts to affect anyone providing direct service to students." On Sunday, almost a thousand people packed UICC's Pavilion as the mayor slowly made his way through the crowd. The *Tribune* described the almost four-hour program as having a "town-meeting atmosphere," and called it "the most remarkable gathering to focus on the Chicago public schools in at least 25 years." Washington proclaimed that "a thorough and complete overhaul of the system is necessary," promised to appoint a fifty-member parent/community advisory council within two weeks, and pledged to present a unified reform plan within four months.

Washington soon convened the first meeting of his Parent Community Council, promising it would begin holding neighborhood forums before the end of November. The city's most influential biracial coalition of business leaders, Chicago United, also shared its own school reform plan with the *Tribune*. Angry at Manford Byrd for his arrogance and disinterest in meaningful change, Chicago United had adopted the belief of Fred Hess and Don Moore that the best path toward improved student performance was through parental and community control of local schools. Patrick J. Keleher Jr., Chicago United's public policy director, said, "we think the leadership could be found" and that organizations such as UNO could recruit and train interested parents.[58]

At the same time, Barack's DCP work remained often frustrating. Some Saturday mornings he met with the Congregation Involvement Committee at Rev. Rick Williams's biracial Pullman Christian Reformed Church, while he and Johnnie Owens also met with two Chicago State University (CSU) professors who had just started a Neighborhood Assistance Center. The only result of that was Barack being invited to join a CSU-sponsored panel at a mid-October conference on "Developing Illinois' Economy." One of his fellow panelists, CSU's Mark Bouman, addressed "the spectacular fall of big steel," and another speaker was *Chicago Tribune* business editor Richard Longworth, who had written so forcefully about Frank Lumpkin's efforts on behalf of Wisconsin Steel's former workers. Whatever Barack contributed to the session was unmemorable, for Longworth, looking two decades later at a photograph of himself and Obama sitting side by side, had no recollection of the event whatsoever.

In Roseland, significant economic development news came from the low-profile Chicago Roseland Coalition for Community Control (CRCCC), a twelve-year-old organization that had successfully followed through on the interdenominational protests against Beverly Bank that Alonzo Pruitt and Al Sampson had mounted seven months earlier. When the bank announced plans to open a new branch in suburban Oak Lawn, the little-known Community Reinvestment Act (CRA) of 1977 allowed CRCCC to petition the Federal Deposit Insurance Commission (FDIC) to block that expansion. Three years earlier Beverly Bank had stopped making home mortgage loans, leaving it vulnerable to FDIC enforcement of the CRA's community service requirements that proscribed disinvestment in older neighborhoods. Advising and guiding CRCCC's strategy was the similarly low-profile Woodstock Institute, a nonprofit fair-lending organization created in 1973 by five founders, three of whom were John McKnight, Al Raby, and Stan Hallett. Woodstock vice president Josh Hoyt, an organizer who previously had worked under Greg Galluzzo and Mary Gonzales at UNO's Back of the Yards and Pilsen affiliates, brought the Beverly situation to the attention of U.S. Senate Banking Committee chairman William Proxmire's staff, and Proxmire wrote to FDIC chairman William Seidman. Soon Beverly Bank was in negotiations with CRCCC president Willie Lomax, and the national *American Banker* publicized the tussle. By mid-September Beverly had agreed to commit $20 million worth of loans to low-income Far South Side neighborhoods over the next four years and to open central Roseland's first ATM. "We made use of some tools of the law to get the bank here," the courageous Lomax told the weekly *Chicago Reader*'s excellent political reporter Ben Joravsky. "We had to play hardball."[59]

The "tools of the law" were increasingly on Barack's mind by October 1987. Going to law school had been a possibility for years, ever since his graduation from Columbia. His grandmother more than once had spoken of a career in law and then a judgeship, and Barack had never seen his work as a community organizer as something long term. He had mentioned going to law school in passing to several people in recent years—IAF's Arnie Graf being one—and early in 1987 Barack had heard a National Public Radio interview with Supreme Court justice William J. Brennan Jr. Brennan told NPR's Nina Totenberg that the guarantees in the Constitution's Bill of Rights are "there to protect all of us" and to "protect the minority from being overwhelmed by the majority." Barack would later recall what he termed "the wisdom and conviction" of Brennan's words.

Barack's embrace of his "destiny," as he described it to Sheila, had quickened his thinking for the last six months. Kellman's move to Gary had left DCP entirely in Barack's hands until he had been able to hire Johnnie Owens, but once Johnnie was on board, Barack's references to law school—to Sheila, to his close friend Asif Agha, to Asif's friend Doug Glick on their long drives up to Madison and back—became far more frequent. Early that fall Bobby Titcomb, Barack's closest Hawaii friend, passed through Chicago, and he remembers Barack talking then about wanting to get a law degree. Asif left Chicago in early November for six months in Nepal, and one October evening, not long before he left, he went to Barack and Sheila's apartment on South Harper. "We used to cook dinner for each other a lot," said Asif, and that evening Barack was out on one of his several-days-a-week runs along Hyde Park's lakeshore. "I was sitting with Sheila in the kitchen, and he walked in all sweaty," wearing shorts, and "I remember him saying something about his law school essay, and that he had been mulling it over for many days, maybe weeks." Mary Bernstein remembered Barack mentioning it to her as well. He had cited Harvard in particular to Asif—"his dad went to Harvard," Asif remembered, "so he had some interest in Harvard" and "that was his first choice." Asif had not heard what Barack had told Sheila about a "destiny," but after two years of weekly conversations, he knew Barack "was an ambitious person."

Several weeks earlier, Jerry Kellman told Obama about a conference titled "The Disadvantaged Among the Disadvantaged: Responsibility of the Black Churches to the Underclass." It would be held in late October at Harvard Divinity School in Cambridge, Massachusetts, and Barack agreed they should go. The weekend of "study, reflection, and worship," as the divinity school's dean termed it, for the two hundred attendees began on Friday evening with a sermon by Samuel D. Proctor, pastor of New York's Abyssinian Baptist Church and one of the most famous living black preachers. Proctor's sermon "chilled us," one listener recounted, with its description of "the ordinary terrors of normal life in Harlem."

Saturday morning featured a powerful and provocative lecture by Philadelphia congressman and clergyman William H. Gray. "Will we have a two-tiered society of the haves and the have-nots . . . or will we have a society . . . that allows people to move from the underclass on up?" Gray asked. For black Americans, "education is an absolute essential," as the record was clear: "no education, no advancement," and "blacks with no education have very little hope" of bettering their lives. Gray did not shy from naming other ills. "The leading cause of death among young black males is black-on-black crime. That's us. Superfly selling drugs and coke in our community is not someone else. It's us. Those who are mugging us, raping us, and robbing us are not coming from somewhere outside. It's us. The teenage pregnancy problem is not the white man's problem, it's our problem. . . . You cannot ever escape poverty with children having children."

Two sets of four concurrent workshops bracketed a lunchtime lecture by sociologist William Julius Wilson. After the workshops ended, Jerry and Barack took a walk around the Harvard campus. "We didn't spend a lot of time with other people at the conference," Kellman recalled, "just with each other." During their stroll, Barack told him that he was applying to law schools. Kellman remembers Barack saying, "'I owe it to you to let you know as soon as I can,'" and Kellman recalls being "very surprised." Kellman also remembers Barack saying he did not think that community organizing was an effective means for attaining large-scale change or a practical method for influencing elected officials who potentially could. Years later Obama described his thoughts similarly, saying that "many of the problems the communities were facing were not really local," and that he needed a better understanding of how America's economy worked "and how the legal structure shaped that economy." He also "wanted to find out how the private sector works, how it thinks." By "working at such a local level" he had learned that problems like joblessness and bad public schools were "citywide issues, statewide issues, national issues" and he wanted to "potentially have more power to shape the decisions that were affecting those issues." In a subsequent interview he would specifically cite decisions that "were being made downtown in City Hall or . . . at the state level." He had been able to get "outside of myself by becoming a community organizer," but the experience also had taught him that "community organizing was too localized and too small" to offer significant promise to those it sought to empower.

Saturday night's dinner speaker was Children's Defense Fund president Marian Wright Edelman, who one listener said presented a "terrifying analysis" and emphasized the need for "an ethic of achievement and self-esteem in poor and middle-class black children." Sunday morning the symposium concluded with a sermon by Harvard professor and pastor Peter J. Gomes, who worried that after two days of "some pretty depressing statistics, some very grim predictions, some very sobering analyses" the discussion had become "so intimidating, so daunting as to lead to paralysis rather than action. With more conferences like this, one will be terrified of thinking about any form of response." But Gomes believed the weekend's clear message was that "money, programs, and advocacy alone will not solve the problems of the underclass." Instead, a consensus had identified "despair as the root and fundamental disease: despair, the loss of hope, the loss of any sense of purpose, or worth, or direction, or place," an analysis all too true to someone familiar not only with Altgeld Gardens but also Chicago's Far South Side high schools.[60]

Back in Chicago on Monday, Barack wrote a congratulatory postcard to Phil Boerner, who had announced his upcoming marriage to his longtime girlfriend Karen McCraw. "Life in Chicago is pretty good," Barack wrote. "I remain the director of a community organization here, and I'm now considering going to law school. Makes for a busy life: not as much time to read and write as there used to be. Seeing a good woman, a doctoral student at Univ. of Chicago."

The next weekend was the beginning of Gamaliel's third weeklong training at suburban Techny Towers, with about sixty trainees joining the network's usual roster of trainers: Greg, Mary, Peter Martinez, Mike Kruglik, Phil Mullins, Danny Solis, and Barack. David Kindler now worked under Mike at SSAC in Cook County's south suburbs, and Kindler brought his Quad Cities friend Kevin Jokisch to the early November training. "Who you brought to training was a reflection of your ability as an organizer," Kindler explained, and Jokisch already had hands-on experience.

"Barack was always a great trainer," Danny Solis remembered. "He had a presence." After sitting in on two of Obama's sessions, Jokisch agreed. "Barack was totally in control without appearing to be in control. . . . His sessions flowed, were all lively, and engaged participants. He drew people out, had them tell their stories in the context of the session he was leading. His style was different than many of the other trainers. Most of the trainers were aggressive, in-your-face types." In contrast, "Barack and Mary Gonzales had a very similar way of moving, speaking, drawing people out, utilizing humor, and probably most importantly being trusted by those in the room."

Most nights some of the trainers adjourned to an Italian restaurant and bar just down the road. "Obama showed up twice to have beers with us. The after-hours sessions were very lively," Jokisch recalled. "Most of the debates were around politics and politicians," with Barack jumping in. Phil Mullins remembers how Barack "saw the limitations of just pure community organization" and asked, "How do you get at these bigger issues?" Given Gamaliel's IAF worldview, that meant Barack "was kind of out there on his own," and he was "constantly asking himself what he actually thinks about something," another colleague recalled. Phil, like Kindler, also knew that among all the organizers, Barack's real relationship was with Kruglik. "That's the tighter person-to-person relationship" for Obama, more so than Jerry or even Greg, Mullins recounted. It was "more personal," because as everyone could see, "Kruglik's a warmer personality." Given his Princeton undergraduate education, plus his graduate school history background, Kruglik's intellectual depth and acumen were unique among the Gamaliel network. He also possessed an uncommonly superb memory.

By November 1987, Mike and Barack had known each other well for more than two years, and with Mike Barack could be spontaneous and frank to a degree he rarely was with other organizing colleagues. Even a quarter century later, Mike remembered some of what Barack said to him that week. Barack's time in Roseland had placed him "in the armpit of the region, as far away from the center of power as you can get," and he saw a prestigious law degree as the first step on the road to true power. Mike disagreed, telling Barack not to leave organizing and instead to commit himself to building a citywide network of organizations broader than UNO's set of community groups, a network so sweeping that it would represent the fulfillment of what King and Al Raby had hoped the 1966 Chicago Freedom Movement would become.

Barack demurred. A top law school would give him entrée to the corridors of power. "I can learn from these people what they know about power," and a legal education would allow him to understand "financial strategies and banks and how money flows and how power flows. Then I can come back to Chicago and use that knowledge to build power for ordinary people." But Barack was imagining more than just building a powerful network of community groups, Mike recalled. "He said to me, 'I'm going to become mayor of Chicago. I'm thinking I should run for mayor of Chicago.'"

Barack believed that Chicago's mayor was the most powerful of any U.S. city's, one who with widespread grassroots support could begin the rebirth of neighborhoods like Roseland. "'At the end of the day, the question is, How do we lift people out of poverty? How do we change the lives of poor people, in the most profound manner?' That's what he was interested in," Kruglik recounted.

And by November 1987, Barack had a specific political template in mind. "Harold was the inspiration," Mike recalls. "Obama saw himself in a very specific way as following in the footsteps of Harold Washington . . . following the path to power of Harold Washington." Ever since the mayor's appearance at the opening of the Roseland jobs center, Barack's attempts to win direct access to Washington had foundered. "It's almost like he was saying to himself . . . 'I'm limited by the power of Harold Washington the mayor. Therefore the answer is, I'll be Harold Washington.' That's what happened," Kruglick explains. "The Harold path was to become a lawyer, become a state legislator, become a congressman, then become mayor. That's the Harold path."

Barack was "fascinated with Washington," Mike believed, and "replicating

Washington step by step" was his game plan. "That was in his mind. He talked about that." Barack "was constantly thinking about his path to significance and power," and "Harold Washington inspired him to think about becoming a politician."[61]

The next Thursday was the launch of what Barack and DCP were now calling the Career Education Network's "Partnership for Educational Progress," a label borrowed at least in part from Chicago United's blueprint for improving the employment skills of public high school graduates. Ever since DCP's May decision to make education and particularly high school anti-dropout efforts a priority, Barack, Johnnie, and top DCP education volunteers like Aletha Strong Gibson, Ann West, and Carolyn Wortham had been in contact with Far South Side high school principals and guidance counselors and with officials at both Olive-Harvey College and Chicago State University. Thanks to Al Raby, who had just left his city human relations post so he could work for school reform at a newly founded consultancy called the Haymarket Group, Barack had recently met reform proponents Patrick Keleher, of Chicago United, and Sokoni Karanja.

In a fall proposal Obama would submit to multiple funders, he wrote that "the condition of the secondary school system called for a wider and more intensive campaign than we originally envisioned." But beyond the initial $25,000 that Emil Jones had obtained from the state board of education, CEN's only other support would be an additional $25,000 that the Woods Fund would soon officially announce.

Ever since early summer, Obama had met ambivalence among DCP's congregations about forcefully targeting the sorry state of Chicago's public schools. He later acknowledged that "every one of our churches was filled with teachers, principals, and district superintendents," or, as UNO's Phil Mullins more pungently yet properly put it, "if you removed every education bureaucrat from Reverend Wright's church, it would go under." Barack spent much of the last months of 1987 trying to expand DCP's base by approaching the pastors of largely Protestant, and mainly Baptist, Greater Roseland churches. He realized that these congregations "have had no direct involvement in the issues surrounding the public school system," and DCP wanted to enlighten them about "the need for broader reform in the school system." Jerry Kellman knew that Mary Bernstein's father, a senior Teamsters official, was a close colleague of Robert Healey, a former Chicago Teachers Union president who was now head of both the Chicago Federation of Labor and the Illinois Federation of Teachers. Mary arranged for Obama to meet with Healey, but Healey had no interest in aiding a movement that would empower parents.

Roseland's five high schools were in sorry shape. One study revealed, "High school students in Roseland are testing more than ten percent below the city-wide average," and that average was more than 30 percent below grade-level norms. The dropout rates at the two weakest schools, Harlan and Corliss, were rapidly increasing. Corliss's principal published an essay in the *Journal of Negro Education* describing her efforts to combat "gang activity and vandalism," "low teacher morale," "disrespectful attitude and behavior of students," and "student apathy and high failure rate" at her school. Julian, named after the pioneering black chemist Percy Julian, was considered to be the best of the five, but the principal there, Edward H. Oliver, still had a serious problem with ganglike female "social clubs."

A crowd of three hundred showed up for CEN's November 17 kickoff rally at Tyrone Partee's Reformation Lutheran Church. The *Defender* covered the event, where DCP announced it would begin offering tutorial and counseling services at both Holy Rosary and Reformation in early 1988 to students referred by Carver, Fenger, and Julian high schools. Olive-Harvey College president Homer Franklin as well as Chicago State University vice president Johnny Hill pledged their institutions' assistance, and Chicago United policy director Patrick Keleher said that his organization would arrange for employment internships.[62]

Barack was also keeping up with the city's landfill crisis. When the U.S. Environmental Protection Agency issued a report attributing the South Side's poor air quality to highway traffic and wood-burning stoves without mentioning landfills, the *Tribune* said that UNO's Mary Ellen Montes "laughed when told of the EPA's findings. 'That's crazy. Wood-burning stoves? Are there any left?'" Then the *Sun-Times* revealed that the Paxton Landfill had been operating without the necessary permits since 1983. The newspapers had a field day at the Washington administration's expense, but mayoral aide Howard Stanback remained focused on the O'Brien Locks issue.

On November 16 South Chicago Savings Bank president Jim Fitch convened an initial meeting of all interested parties, ranging from South Deering's Foster Milhouse to Bruce Orenstein and Mary Ellen Montes from UNO and hard-core landfill opponents Marian Byrnes and Hazel Johnson, who did not like anything they heard. Four days later Lake Calumet environmentalist James Landing distributed a letter warning that the Washington administration "is making prodigious attempts" to win over opponents of a new O'Brien Locks landfill.[63]

On Wednesday morning November 25, the day before Thanksgiving, a *Chicago Tribune* headline announced "948 School Jobs Axed for Teachers' Raises." In order to meet the pay raises in CPS's new union contracts, 167 elementary school teachers had been terminated. Then, at 11:01 that morning, Harold

Washington collapsed with a heart attack in his City Hall office. The sixty-five-year-old mayor was seriously overweight, and attempts to revive him were unsuccessful. An official announcement was delayed for more than two hours, but word that Washington had died spread rapidly throughout the city, with tearful crowds gathering outside City Hall.

"Mayor's Death Stuns City" read one headline the next day. Black Chicago's loss was especially painful and heartfelt. Seven months earlier, when Washington was reelected, the *Tribune* editorialized that "he has been a symbol more than a leader," but he was also the greatest "symbol of black empowerment" the city had ever seen. Only with his April 1986 erasure of Ed Vrdolyak's city council majority had Washington truly become Chicago's mayor, and as one *Tribune* story poignantly declared, "Washington's legacy is not what he did, but what he was on the verge of doing."

The next morning the *Tribune* lauded Washington as "a symbol of success and dreams realized for people who felt they had little reason to dream, let alone achieve," while again noting that "his tangible record of accomplishments is a short list." Economic development commissioner Rob Mier would write that "many of his goals and plans remained unfulfilled or barely started." Mier also recognized that Chicago's loss was greater because Washington had "died at the peak of his power."

Tribune reporter John Kass highlighted how Washington had been "an incredibly charismatic leader," but one of Washington's most fervent early backers identified the mayor's greatest mistake. "He took the power to himself, almost like Mayor Daley" in earlier decades, "and the political maturity of black politics stopped while he increased his power." White 49th Ward reform alderman David Orr, who became interim mayor upon Washington's death, had articulated the underlying problem months earlier: "There's a large group of black aldermen . . . who don't support reform but who have to vote with the mayor because he's so popular in their wards." By tolerating rather than purging those black aldermen who professed to support him while nonetheless remaining fully loyal to the Democratic party machine, Washington had advanced "his own political self-interest at the expense of institutionalizing his reform movement," wrote historian Bill Grimshaw, the husband of Washington's top political aide, Jacky Grimshaw.

The enormity of Washington's failure became clear within the first hours after his death, as his city council majority sundered into two angrily hostile camps. Washington's true supporters rallied behind the mayor's council leader, 4th Ward alderman Tim Evans, with whom Barack and DCP's Altgeld asbestos protesters had met eighteen months earlier. Washington's opponents eagerly reached out to the black machine aldermen, who now controlled the balance of power in a political world where they no longer had to bow to a singularly charismatic leader. Washington's political base "unraveled immediately after he was pronounced dead," John Kass wrote, and Rob Mier also rued "the immediate collapse of his political coalition."

Over the Thanksgiving holiday weekend, the two factions warred publicly as Washington's body lay in state for a fifty-six-hour around-the-clock wake in the lobby of City Hall. Monday night at the UICC Pavilion where Washington had hosted his Education Summit just seven weeks earlier, his official memorial service turned into a political rally for Evans. Yet the only votes that would count were those of the fifty city council members, and by Tuesday morning, there was little question that 6th Ward black machine alderman Eugene Sawyer would become Chicago's next mayor thanks to Washington's hard-core opponents plus at least five black aldermen who would support Sawyer over Evans.

As Tuesday night's council meeting convened, a crowd of thousands stood outside City Hall, chanting "No deals" and "We want Evans." Clownish behavior marred the council's proceedings as protesters mocked black Sawyer allies like 9th Ward alderman Robert Shaw, and Sawyer was not formally elected as Harold Washington's successor and sworn into office until 4:00 A.M. Wednesday.

In skin color Gene Sawyer was just as black as Harold Washington, but as the angry crowd well knew, Chicago now had a completely different mayor than the one it had just buried. When Sawyer was asked by a historian some months later about Washington nemesis Ed Vrdolyak, his answer highlighted the chasm: "He's a fun guy!"[64]

"I loved Harold Washington," Barack blurted out years later when asked what he had thought of the mayor. He once wrongly but perhaps wishfully stated that in 1985 "I came because of Harold Washington," and at another time, he mused that "part of the reason, I think, I had been attracted to Chicago was reading about Harold Washington." There was no doubt that Washington, or more precisely Washington's treatment at the hands of the Vrdolyak majority during Barack's first nine months in Chicago, contributed in some degree to Barack's own embrace of a resolutely black racial identity. "Every single day it was about race. I mean every day it was black folks and white folks going at each other. Every day, in the newspapers, on TV, in meetings. You couldn't get away from it," Obama later recounted. "It was impossible for Harold to do anything."

Upon his arrival in Chicago, and throughout all of his Oxy and Columbia years prior to Jesse Jackson's 1984 presidential run, Barack had been "skeptical of electoral politics as a strategy for social change," he later acknowledged. "I was pretty skeptical about politics. I always thought that the compromises involved in politics probably didn't suit me." Jerry Kellman, Greg Galluzzo, and the Alinsky tradition of organizing certainly did not teach respect or admiration for elected officials. But watching Washington week after week, even if he had never been physically closer than in that nondescript Roseland storefront eight months earlier, had fundamentally changed Barack's mind. "You just had this sense that his ability to move people and set an agenda was always going to be superior to anything I could organize at a local level," Obama explained in 2011.

In his own telling years later, Obama was in that angry, chanting crowd outside the city council chambers that Tuesday night, witnessing what he called Washington's "second death." Yet even at the time he wrote that, Barack understood Washington's fundamental error, just as Bill Grimshaw had explained it. "Washington was the best of the classic politicians," Obama told an interviewer. "But he, like all politicians, was primarily interested in maintaining his power and working the levers of power. He was a classic charismatic leader, and when he died, all of that dissipated. This potentially powerful collective spirit that went into supporting him was never translated into clear principles, or into an articulable agenda for community change," Obama rightly stated in words that would echo painfully two decades later. "All that power dissipated."

Yet in those last weeks of 1987, Washington's death strengthened Barack's belief that now was the time to leave DCP for law school. He had "a sense that the city was going to be going through a transition, that the kinds of organizing work that I was doing wasn't going to be the focal point of people's attention because there were all these transitions and struggles and tumult that was going on in the African American community," he recalled in 2001. "So I decided it was a good time for me to pull back" and attend law school.

Barack's long-pondered personal essay was finished, but completing his application required soliciting letters of recommendation as well. Al Raby was a recognizable name to anyone who knew the history of the 1960s, and Michael Baron had given him an A for his senior year paper at Columbia, the most serious piece of coursework Barack had ever tackled. Now working at SONY, Baron readily agreed to write a letter. But Baron's knowledge of him was now more than four years dated, so Barack also went to see John McKnight, asking him to keep their conversation confidential, especially from Greg and Mary. "Would you write me a letter of reference? You're the only professor I know." McKnight immediately said yes, but asked Barack what his plans were. "I want to go into public life. I think I can see what can be done at the neighborhood level, but it's not enough change for me. I want to see what would happen in public life" and "I think I have to go to law school to do that." McKnight questioned whether Barack understood how fundamentally different life as an elected official would be from that of an organizer. While the latter was quintessentially an advocate, "my experience is that legislators are compromisers,"

McKnight observed, people who synthesize conflicting interests. "You want to go into a world of compromise?" he asked. Barack responded affirmatively, saying, "That's why I want to go into public life" and to pursue a role quite opposite that of a confrontational Alinsky organizer. "It's clear to him he's making a decision that that's not the way he's going," McKnight remembered. "He left for a different mode of seeking change."[65]

With Gene Sawyer uncomfortably ensconced in City Hall, the Parent Community Council's ten public forums were surrounded by uncertainty. At the first one, Chicago United's Patrick Keleher reiterated the business community's demand for dramatic reforms, and at the third Sawyer pledged "my commitment to the Washington reform agenda." In response, Manford Byrd protested that CPS's "many needy students" meant that any improvement in schools' performance would require "major additional funding" for more teachers, counselors, and, of course, "other professionals." DCP's Aletha Strong Gibson told one reporter that the reform movement would be undeterred by the mayor's loss. "Harold Washington did not move the community. The community moved Harold Washington," she declared. "It is incumbent upon us to keep our voices raised. We have to take back ownership of the schools."

Sawyer also inherited Chicago's landfill problem, with Howard Stanback seeking to quickly explain the city's O'Brien Locks strategy to the new mayor. Environmental activist James Landing realized that Washington's death would not alter the situation, and even he wondered whether the opponents should give up and join Jim Fitch's effort to agree on what the Southeast Side neighborhoods should demand from WMI. Senator Emil Jones's special legislative committee concluded its investigation by ruing "the lack of one centralized authority to address all environmental problems" in the area, but also bluntly acknowledging that continuing Chicago's ban on landfill expansion "is irresponsible unless the city is able to implement a successful citywide waste-disposal plan" through massive recycling.

At DCP, Barack and Johnnie were preparing the CEN tutoring program to start in early 1988, in both Roseland and Altgeld, undeterred by the December murder of an eighteen-year-old Gardens youth by five fellow teenagers, all members of the infamous Vice Lords gang. Weeks earlier Barack had asked Sheila to go with him to Honolulu for the Christmas holidays so that his relatives could meet her, the first time Barack had ever introduced a girlfriend to his family.

Madelyn had just turned sixty-five, and a year earlier had retired from the Bank of Hawaii. Stanley, almost seventy, was now retired too. Ann Dunham, still struggling to complete her dissertation, told her mentor Alice Dewey how much she was looking forward to meeting Sheila, and Sheila would always remember how exceptionally warm Ann was to her throughout the couple's visit to Honolulu. Ann was fascinated that Sheila had written her master's thesis on Claude Lévi-Strauss, and Sheila remembers the two of them "talking about that thesis, my time in Paris, and my work at Chicago." Ann was "genuinely very interested and warm and inquisitive. She was extremely generous with us and treated Barack with reverence. She really admired him and thought the world of him."

Even though Ann and Sheila liked each other very much, Sheila felt that Ann was not in favor of her and Barack getting married. She wondered if Ann "sensed what we already knew, that we were too isolated and would sophisticate each other." Sheila was especially struck by how everyone in Hawaii—Gramps, Toot, Maya—called Barack either "Bar" or "Barry." In Chicago, only Asif used "Barry." When Sheila called him "Barry for fun one day, just because everyone else was calling him that name," Obama's reaction was unforgettable. "He got so angry at me. Irrationally furious, I'd say. He told me that under no circumstances was I ever to use that name with him." Perhaps Obama had some deep boy-versus-man association to the two names, but Sheila understood that he "was very sensitive about this aspect of his life and wanted me walled off from it—like a lot of other things in his life."[66]

Back in Chicago, the mayor's Parent Community Council moved toward recommending that each Chicago public school be controlled by a locally elected governing board, but without Harold Washington present to embrace its conclusions, school reformers started arguing over which of several just slightly different proposals should be introduced in the state legislature. DCP participated tangentially, following UNO's lead, with Danny Solis in charge and Johnnie Owens rather than Barack following developments most closely.

In late January Barack wrote to Phil and Karen Boerner to apologize for missing their wedding, and updated them on his plans: "I've decided to go back to law school this fall—probably Harvard." At the end of the month, Barack also wrote to Anne Hallett at Wieboldt to submit DCP's 1988 grant application, but he gave no indication that he might be leaving DCP anytime soon. He wrote that "DCP has come a long way in the past year," most notably in hiring Owens, "someone with both the talent and background to become a lead organizer in his own right." Because of DCP's "success with the education issue . . . we have the potential in the coming year to become a truly powerful advocate for change not only in the area, but citywide," Barack boastfully asserted. "Whether we fulfill that potential will depend on two things: how well we parley the Career Education Network into a vehicle for organizing parents and community, and whether the relationships we have established with the major Black churches in the South Side translate into their making a full commitment" to DCP by contributing financially so that the organization could begin to wean itself from outside funders. "If we succeed, I envision us having 30 new churches involved by the end of 1988."

That was optimistic, because Obama's ongoing efforts to connect with dozens of black pastors had garnered polite conversations but few enlistments. If CEN was to grow beyond a small pilot program in which DCP housewives tutored dozens of high school students in a trio of church halls, funding was necessary for "significant expansion of the program by the State Legislature." Barack also still hoped that Olive-Harvey could reallocate resources "to create a comprehensive job training program with specific emphasis on Public Aid recipients and with the outreach and satellite facilities necessary to target the Altgeld Gardens population."

DCP would soon hire a CEN project coordinator, thanks to Emil Jones's state money and the Woods Fund, and hoped to approach major corporations through Chicago United. Barack's success at fund-raising had let him raise his own salary to $27,250 and Johnnie's to $24,000. Ideally Olive-Harvey would foot the bill to house CEN, but expanding the program for the 1988–89 school year depended on state board of education officials and state legislators.

Obama believed that "parents and churches" were "the most crucial ingredients" for "a long-term process of educational reform." Gamaliel and Don Moore's Designs for Change, now a top player in the citywide school reform movement, could be asked to provide parental training. The city's community colleges were responsible for vocational and general educational development (GED) training, but their actual track record was even worse than that of Chicago Public Schools. "Only 8 percent of the 19,200 persons enrolled in GED preparation classes in 1980 actually received certificates," Barack had discovered, and "only 2.5 percent of those enrolled in City College basic education programs end up pursuing additional vocational or higher education." Instead, just as at CPS, "funds go into central administrative tasks rather than student instruction."[67]

Barack's church-recruitment efforts continued throughout the winter of 1987–88. One successful visit was to a small church on West 113th Street just across from Fenger High School. Thirty-two-year-old Rev. Alvin Love had arrived at Lilydale First Baptist Church four years earlier, inheriting an "elderly congregation" that was "comfortable sitting and doing nothing." Love wanted to "get this congregation engaged in their community," and he was happy when Obama "just walked up to the door and rang the bell" and asked if Love would tell him what he thought Roseland needed. Love, like other pastors before him, asked Barack which church he belonged to. Love warned Barack that his standard response—"I'm working on it"—wasn't going to be acceptable to all black clergymen. Barack invited Love to a box-lunch gathering of other interested pastors, and he was slowly beginning to expand DCP's ties to freestanding Protestant congregations.

The clergymen with whom Barack was having the most contact, however, were not involved at all in DCP. One was Jeremiah Wright at Trinity United Church of Christ, whom Barack continued to visit on a regular basis. Barack also began speaking with Sokoni Karanja, a longtime Trinity member whom he had met through Al Raby when Karanja emerged as one of the most outspoken African American voices calling for school reform. Barack "was trying to think about ministers and how to organize ministers across the whole city—African American ministers—because he felt like the power that was needed for the community to get its just due was through that," Sokoni remembered. "He was trying to think about churches and how to organize the churches." Barack also told Sokoni about his plan to attend law school. "One of the things I notice is that a lot of these politicians get in trouble because they don't know the law," Barack commented. Becoming a politician was indeed his goal. "He was talking about becoming mayor of Chicago," Sokoni—just like Mike Kruglik—remembered. "We had a lot of conversations about it," and "that's what he emphasized to me. It sounded like a good idea."

The other church Barack visited frequently was St. Sabina. DCP members like Cathy Askew, Nadyne Griffin, and Rosa Thomas knew Barack "was a big admirer of Father Pfleger," but Cathy's fellow St. Catherine's parishioner, Deacon Tommy West, saw the most of Barack at St. Sabina. One of St. Sabina's wintertime ministries involved a tangible outreach to the homeless, many of whom lived in the relative warmth of Lower Wacker Drive, the underground level of a downtown Chicago roadway. "We brought clothes and food down there for them," West explained. "My wife made corn bread" and "minestrone soup." One evening Barack joined Tommy and Mike Pfleger for the trip downtown. "He was with us underground, feeding the homeless," West recounted. "Lower Wacker—I remember him going to that with us," Mike agreed. Tommy also remembered Barack saying he was about to go back to school. He "told me when he went down to Wacker Drive with us and we had the soup and the coats." Tommy had asked why, and he recalls Barack saying, "I just get tired of getting cut off at the pass" by government officials, and "I need some other stuff. I've got to go. This is something I have to do."

But one day in those early months of 1988, Barack went with Mike Pfleger and Tommy West on a different sort of outing. Years later Barack would say that "the single most important thing in terms of establishing" safe neighborhoods "is having a community of parents, men, church leaders who are committed to being present and getting into the community and making sure that the gangbangers aren't taking over, making sure that there's zero tolerance for drug dealing." In 1988 Pfleger shared that sentiment as strongly as anyone could, and he had adopted an interdiction technique that was nonviolent direct action at its most aggressive: walking into a known drug seller's home in clerical garb, asking if he could use the bathroom, then grabbing whatever narcotics he could and heading for the toilet. Only "a very small group" of men accompanied Mike on such drug raids, and Tommy West—wearing a deacon's collar—was a regular participant.

Pfleger cannot remember Barack going with him on such a sortie, although "he very well may have." But Tommy West recalls Barack's involvement with great clarity. "One time he went with us to this drug dealer's house, and Father Mike grabbed up the guy's drugs and ran in the washroom—first he asked the guy, could he use the washroom," and once in there Mike "dropped it and flushed it, and the guy pulls out his pistol. I said, 'Look, put that up. He's a priest—you don't want to hurt him. Use it on me.' He said, 'No, he's the one who did this.' I said, 'Don't touch him. That's God's man. You don't want God to be down on you the rest of your life. Leave him be and try to find you somewhere else.' He said, 'I think I will, because I'm ready to kill him.' So Barack said, 'Woo.'" Mike Pfleger looked calmly at his colleagues. "I'm doing God's work. God is taking care of me."

"He was around for that one," Tommy affirmed. "That was the first time I remember him going inside of the drug dealer's home with us to see what Father Mike did." Then, with the dealer still holding his gun, Tommy again asked the man to leave: "'then we can get out of here.'" The dealer did take his gun and leave, and as the raiding party exited as well, "everybody was kind of shook up," Tommy West recalls with considerable understatement.

Barack had developed significant relationships with Mike Pfleger and Jeremiah Wright, just as he had two years earlier with Bill Stenzel and Tom Kaminski. But at home on South Harper, Sheila Jager never heard a single word about any of them, or about institution-based religious faith. "I don't think Barack ever attended church once while we were together," and he "certainly was not religious in the conventional sense," Sheila reiterated. Barack "never suggested going to church together," and "I had no awareness of Jeremiah. He certainly never mentioned him to me."

Barack's closest Hyde Park male friend, Asif Agha, remembers their conversations similarly. "I always assumed he was an atheist like me." Barack "had no interest in religion" and knew "nothing about Islam." Asif's friend Doug Glick, Barack's companion on all those long summer drives to and from Madison, bluntly concurred. "I do not believe he believes in Jesus Christ." Asif states the trio's consensus succinctly: Barack "did not have a religious bone in his body."

In Barack's daily life with Sheila, "he didn't tell me a whole lot about" who he was dealing with in his work, Sheila explains, "although he did bring a lot of their ideas home." Barack "never compartmentalized ideas, which we discussed freely." At home "we spent a great deal of our time discussing/talking about all sorts of things, which was one of the things I liked so much about him." And, just like at DCP, Barack "was also a very good listener." Yet "we lived a very isolated existence," and even their immediate neighbors at 5429 South Harper barely remember either Barack or Sheila.

One longtime older resident had no recollection at all, and custodian Joe Vukojevic, who lived directly across the hall from Barack and Sheila, only remembered being introduced to Barack's mother Ann during her summer 1987 visit. Barack and Sheila's immediate upstairs neighbors, John Morillo and Andrea Atkins, thought Joe the janitor was "a very nice guy" but "practically never" saw Sheila or Barack. They would remember the unusual name on the mailbox, and saying hi to Sheila, but Barack was "just another guy doing laundry" in the basement laundry room.[68]

On January 31, the *Daily Calumet* reported that South Chicago Savings Bank president Jim Fitch was privately brokering talks to allow Waste Management to use the O'Brien Locks site as a new landfill in exchange for a $20 million community trust fund. Fitch then called for another gathering of neighborhood representatives at 7:30 P.M. Monday, February 8, at his bank to discuss "the structure of the community trust." Since the *Daily Cal* story "events have been evolving rapidly," Fitch explained, and "conversations with Waste Management have continued."

UNO's Bruce Orenstein and Mary Ellen Montes had previously attended Fitch's conclaves, but now they grew concerned. If UNO and the city's Howard Stanback hoped to have the community fund managed by mayoral allies, they could not allow Fitch to continue driving the conversation with WMI. Bruce and Lena called on Fitch at his bank in advance of the meeting and told him he lacked the authority to have this gathering because he had not been involved in the fight to stop the landfills. Bruce recounts telling him, "We think you're undermining our agenda, you're undermining the agenda of the community, the wishes of the community." Bruce and Lena "really asked him to stop," but Fitch was no rookie at power politics. He recognized their raw grab for control and brusquely dismissed them.

Orenstein next called Barack to ask for DCP's support in a confrontational move straight out of the Alinsky-style community organizing playbook. Local print and broadcast journalists were notified, a press release was prepared, and late Monday afternoon prior to the scheduled meeting, Barack, Loretta Augustine, and other DCP members drove to UNO's office on East 91st Street, less than two blocks from Fitch's South Chicago Savings Bank. "We met there, we practiced," Bruce remembered. "Barack and I and Mary Ellen and Loretta have devised this thing," and shortly after 7:30 P.M., Mary Ellen led a column of more than one hundred participants, including some children, out the door and down the street toward the bank's second-floor conference room. The goal was "to let Fitch know in *no* uncertain terms he does not represent the will of the community on this issue," Bruce recounted. "We just wanted to tell him that in front of everybody else who was around the table," including UNO and DCP's old friend Bob Klonowski, pastor of Hegewisch's Lebanon Lutheran Church. Bruce remembered the group very quietly "walking up the stairs and being actually quite nervous," for "it was quite a confrontational approach to things."

But with Mary Ellen in the lead, the "ambush"—as the *Daily Cal*'s front-page headline the next day called it—worked to perfection. "Mary Ellen led the charge and we walked in and the camera lights went on," Bruce recalled. "Barack and Loretta and the TV cameras and the *Chicago Tribune*" reporter Casey Bukro all followed Lena into the conference room. "Everybody looks up" as Mary Ellen approached Fitch. "Tonight we publicly take the position 'No deals Jim Fitch, no deals Waste Management,'" Lena proclaimed. "We will fight you every step of the way." As a confrontational leader, "she was fantastic," Bruce knew. "Then we turned on our heels and walked out." It was a "very memorable action," even "a hoot of an action," and from Bruce's perspective, Barack had shown no discomfort with these tactics: "I certainly think he enjoyed it."

UNO and DCP's press release claimed that "We are diametrically opposed to any 'buyoff' deal . . . with Waste Management," while acknowledging that "communities are due substantial reinvestment." A simultaneous one by Howard Stanback on behalf of the mayor's office was deceptively titled "City Reaffirms Landfill Moratorium" and asserted that "Waste Management is misleading the public if they suggest that they are in a position to acquire O'Brien Locks for landfilling."

As Bruce Orenstein told the *Daily Calumet,* "If Waste Management gets ahold of that property, out the window goes any protection for the community." The real issue was not the fate of the O'Brien Locks acreage, whose future looked preordained, but who on the Southeast Side would control whatever community trust fund would receive the $20 or $25 million that Waste Management was clearly willing to pay. Only *Hegewisch News* editor and community activist Vi Czachorski gave readers a clear understanding of what was happening. "Montes wants a landfill at the O'Brien Locks," directly across the Calumet River from Hegewisch, and in exchange, she gets "a trust that UNO would operate" rather than Jim Fitch and other traditional Vrdolyak loyalists. But UNO's Alinskyite "ambush" not only infuriated Fitch's nascent coalition, it also angered the trio of hardcore landfill opponents—Jim Landing, Marian Byrnes, and Hazel Johnson—who had been reconsidering their own position. Within a week, they were picketing outside Chicago's City Hall, clear evidence that Bruce and Barack's strategy of blowing up Fitch's negotiations had torpedoed any prospect of achieving a community-wide consensus.

For Barack, all this tussling created a fundamental personal tension. He had made clear to John McKnight that he rejected the confrontational politics of the Alinsky tradition, even though he had just helped lead an action that was so pugnacious it had made even the hard-bitten Orenstein nervous. It unquestionably was the "most confrontive meeting he had ever been involved in," Bruce acknowledged, and "left to his own devices, I don't think he would have designed an action like that." True enough. What then accounted for so stark a contradiction?[69]

A weekend or two after the South Chicago ambush, Barack took Sheila to see a movie that had debuted the Friday before the Fitch action: *The Unbearable Lightness of Being,* an adaptation of Czech writer Milan Kundera's 1984 novel. Set in Czechoslovakia in the years leading up to the Prague Spring of 1968, the film featured three leading characters: Tomas, a young doctor, his partner Tereza, and his additional lover Sabina. The movie was not necessarily loyal to the spirit of Kundera's book, and, at almost three hours' length, it had not been praised in prominent reviews in the *Chicago Tribune* and the *New York Times*. The *Tribune*'s critic wrote that "the film has no vision and no life," and warned that it "is likely to be incomprehensible to anyone who hasn't read the novel."

Vincent Canby's *Times* review was more revealing. Tomas, played by Daniel Day-Lewis, lived a compartmentalized life, with "one part of his mind" analyzing something while another "part that's outside it criticizes" the first. Tereza, played by Juliette Binoche, "falls profoundly in love with" Tomas "without knowing anything about him." In turn, "Tomas is drawn against his will into commitment to Tereza," yet with Sabina, played by Lena Olin, he indulges in "a passion for . . . sex that excludes serious emotional commitment . . . while always remaining a little detached." Tomas "remains committed to Tereza, though still unfaithful." The film conveyed "an accumulating heaviness" accentuated by its "immense length."

Almost a quarter century later, Sheila Jager described the film as an indelible memory, explaining that it could offer "some insight into our relationship. . . . Although Barack did not fool around (not that I know of), I remember being powerfully moved by that film when we saw it together because Tomas and Tereza's relationship seemed to so uncannily mirror the dynamics of our own—Tomas's 'neurosis' like Barack's 'calling.'" She believed that perhaps was "why I reacted so hard when I saw that film. Because it was mirroring reality in an eerie sort of way, and I somehow understood what was happening even if I was unaware of what was going on. I remember feeling so trapped and suffocated back then, just like poor Tereza and her cheating husband. I'll never forget that feeling of desperation, and wondering what I was going to do. I remember him telling me how he wished he could take me to the countryside and live with me," just as Tomas does with Tereza, "but he couldn't do that, no matter how much he loved me," because his destiny inescapably must trump love. "I always knew that I couldn't marry him," yet in those early months of 1988, Sheila never doubted Barack, in part because something happened between them, something Barack subsequently never spoke about.

Barack was also close with the almost thirty-year-old Mary Ellen Montes—Lena—and he told her too about the vision of his future that otherwise he had only shared with Sheila. "He wanted to be the president," Lena explained. "He used to say that his goal was to be the president of the United States." Their ambitions were mutual. "By the time I met Barack, I was thinking about politics as well, with aspirations of being the mayor." Lena told him, "I could see myself being the mayor of the city of Chicago. That's where I'd want to end it, and his thing was oh no, he wanted to go on to be the president." While Barack had told Mike Kruglik and Sokoni Karanja that being mayor of Chicago was his ultimate goal, Lena firmly declared, "That's not what he's telling me."

Lena understood their similar trajectories. "You start to feel and realize your potential, so as you're growing in this arena, why wouldn't you think about those things? . . . That's why I thought about the mayor," and for Barack "it's because of what he realized as he's growing in the public arena and realizing his potential." She knew that getting a law degree was "absolutely" his first step toward electoral politics. Across those months, "our conversations—they were real. They were genuine, sincere conversations about ourselves and what we wanted to do, what we were doing, what we were thinking of." Lena knew Barack lived with someone. "I hear of her," Lena explained. "Asian woman." No, "I never met her. . . . I remember him saying she was Asian." What did Barack tell Lena about that relationship? "He gave the impression that they lived together more because of convenience—they both needed a place to stay."

Barack's diminution of his life with Sheila to Lena was reminiscent of how he had characterized it to Phil Boerner eighteen months earlier: "winter's fast approaching, and it is nice to have someone to come home to," given his "mortal fear of Chicago winters." After years of distancing himself from his mother, Barack's identification as *African* American—not international, not hapa, not biracial—was now complete. This transformation had been immensely aided by his exposure to and ease with strong *black* women like Loretta Augustine, Marlene Dillard, Aletha Strong Gibson, and Yvonne Lloyd, but this success came at a high price, one visible only in the light of the distance, the unknowable distance, that was always impenetrably there. That distance, that *lightness,* would extend well beyond 1988.[70]

Two days after the action against Jim Fitch, a remarkable, substantive victory was announced by attorney Tom Geoghegan: Navistar, the renamed International Harvester, would pay $14.8 million to Frank Lumpkin and twenty-seven hundred other surviving former Wisconsin steelworkers. The largest individual payment would be $17,200, though Frank, with a better-protected pension, would receive only $4,000.

No one in Chicago doubted that Frank deserved the most credit for this achievement, and the ex-workers approved the settlement in an overwhelming vote of 583 to 75. But Frank was never someone to pat himself on the back. "It is a victory of sorts," he told the *Daily Cal*. "It was the best we could get, and that's the way everyone who voted for it felt. But we appreciate the feeling that the little guy has won and that giants can fall." *Tribune* business editor Richard Longworth wrote a wonderful tribute to Frank, describing him as "an amazing man . . . who is probably as close to a saint as Chicago has these days." Tom Geoghegan is "the only other real hero," a commendation underscored when James B. Moran, the federal judge handling the Wisconsin litigation, publicly praised Tom's "dedication," "professionalism," and "modesty in seeking fees." Frank also represented the last of a dying breed: at South Works hardly seven hundred men were still working, and Republic LTV was down to 640. Maury Richards would soon be reelected as Local 1033's president, but even as dedicated a steelworker as Maury was beginning to wonder what his next career would be.[71]

By mid-February, UNO, hoping to take the lead in Chicago's fractured school reform movement, distributed a twenty-nine-page proposal to compete with a much more detailed plan championed by Don Moore's Designs for Change, Sokoni Karanja, and Al Raby from Haymarket. DCP was listed as an organizational backer of UNO's plan, but no DCP member was among the twelve names credited with preparing the document. Danny Solis, Peter Martinez, and Lourdes Monteagudo, an elementary school principal now working closely with UNO, were among them, as was an education professor at the University of Illinois at Chicago, Bill Ayers, who had arrived there six months earlier and met both Danny and Anne Hallett the previous fall. Acknowledging that "good schooling is an expensive proposition," the UNO proposal called for the hiring of an astonishing 14,563 additional educators for Chicago's elementary schools, at a cost of $442 million. UNO envisioned an annual CPS budget increase of $584 million, and called for a $481 million increase in state funding to support it.

On February 18, the same day that the *Sun-Times* gave the proposal prominent coverage, UNO and DCP brought busloads of members to a school reform hearing at board of education headquarters, but the overflow crowd intimidated officials and the meeting was adjourned. Soon a third major plan, this one backed by Fred Hess's Chicago Panel, Gwendolyn LaRoche from the Chicago Urban League, and Patrick Keleher from Chicago United, joined the confusing fray. DCP concentrated on getting its Career Education Network off the ground, with Obama and Owens hiring an African American woman in her early thirties, Cassandra Lowe, who had been working as a college recruiter for nearby St. Xavier University, to oversee it. By early March, afternoon counseling sessions for fifty or so high school students were finally under way at Reformation Lutheran and at Our Lady of the Gardens. Asked about DCP's 1987–88 change from an employment emphasis to its new concentration on secondary schooling, Owens explained that "the focus shifted to the more fundamental question of preparing people for jobs in a changing society."[72]

By the end of February, Barack had to decide both about law school and about making his long-mulled trip to Kenya before the fall 1988 academic year began. His sister Auma had returned home from Heidelberg and would eagerly host a midsummer visit.

Barack later would write that he applied to Harvard, Yale, and Stanford, but he also applied elsewhere, including to Northwestern University's law school, right in downtown Chicago. Acceptance letters had arrived from both Harvard and Northwestern, but with one huge difference: Harvard's financial aid package would require him to take out loans of well over $10,000 a year, while Northwestern's offer, the Ronald E. Kennedy Scholarship, would allow him to attend a top-twenty law school in Chicago for free. Debating his choice of school, Barack asked Jean Rudd and Ken Rolling at the Woods Fund about attorneys from whom he could seek advice. Jean's husband Lionel Bolin was a descendent of a famous African American family, a 1948 graduate and now a trustee of prestigious Williams College, and a successful broadcast executive who, after serving in the U.S. military, had graduated from low-cost New York Law School. Woods Fund board member George Kelm, a low-key civic activist, had been managing partner of a prominent Chicago law firm, Hopkins and Sutter, before becoming president of the Woods family's Sahara Enterprises investment firm.

Barack "was trying to make a strategic choice about which school," Jean Rudd recalled, and Jean and Ken remember Barack telling them about Northwestern's full-scholarship offer. George Kelm was a Northwestern Law School alumnus and a past president of its alumni association, and he strongly advised Barack against attending Harvard. Northwestern was so interested in persuading Barack to accept its Kennedy Scholarship, named after an African American faculty member who had died four years earlier at the age of forty-two, that the admissions office asked the law school's dean, Robert W. Bennett, to speak with Obama. "The admissions people came to me and they said, 'We've got a fantastic prospect for this scholarship'" and "'we want you to try to talk him into taking it,'" Bennett recounted. "Barack was brought to my office" and "I tried to talk him into taking this Ronald Kennedy Scholarship." Bennett was a 1965 cum laude graduate of Harvard Law School, and Barack was "the only applicant that the admissions people ever" asked him to help recruit during a full decade as dean.

Neither Kelm and Bennett's efforts nor the full three-year scholarship were sufficient to outweigh Barack's belief in his destiny. Harold Washington had graduated from Northwestern's law school, and only once had a Harvard law graduate become president—Rutherford B. Hayes, in 1877. Northwestern law alumni had been major party presidential nominees five times, but William Jennings Bryan was a three-time loser and Adlai Stevenson had lost twice. It would be a costly decision for Barack—a cumulative difference of more than $40,000—but his choice was evidence of how deeply he believed what he so far had shared only with Sheila and Lena.

The only person in Barack's workday world, other than Lena, to whom he spoke about leaving was Johnnie Owens, whom he had recruited to DCP with at least half an eye toward this decision. Johnnie remembered the moment clearly. "I didn't have a clue until one day he asked me, 'Are you ready to lead?' I'm like 'What do you mean? What are you talking about?' 'I've been accepted at Harvard Law School,'" and he would be leaving DCP to attend its three-year J.D. program. "And I'm like 'What?'" Owens remembered, for there had been no prior indications that Barack was contemplating such a future. "Nothing. Absolutely nothing: about applying, that he was interested, anything like that. And so he began explaining to me how he'd been struggling with the thought of maybe going into the ministry versus law school." Neither Sheila nor Lena ever heard him talk about the ministry, but as Johnnie remembers it, Barack "said he had ideas and thoughts about going into the ministry and that he had actually talked to Reverend Wright about some of this."

Barack asked Johnnie to succeed him as DCP's executive director, promising not only to work with the members on the transition, but also to introduce Owens to the trio of women who were DCP's most important funders: Jean Rudd at Woods, Aurie Pennick at MacArthur, and Anne Hallett at Wieboldt. Owens agreed, but several weeks passed before Barack was ready to tell DCP's volunteer leaders about his upcoming move.[73]

On March 5, ten days before Democratic ward-level and congressional primary elections across Chicagoland, the *Chicago Tribune* reported that Waste Management had fired two managers at its SCA chemical waste incinerator at 11700 South Stony Island Avenue for repeatedly disconnecting air-monitoring devices designed to measure the facility's destruction of highly toxic PCBs. WMI insisted that the misconduct "did not threaten health or safety," but Marian Byrnes, Hazel Johnson, and congressional candidate Mel Reynolds picketed the plant, demanding it be closed. Metropolitan Sanitary District officials pulled back from a plan to dump eighty thousand cubic yards of sewage sludge in a wetlands property five blocks south of SCA.

Howard Stanback, Bruce Orenstein, and Barack were working on plans to have Mayor Sawyer attend a postelection March 17 rally at St. Kevin to announce publicly the city's strategic alliance with UNO and DCP regarding landfills. On Election Day, four African American ward committeemen who were allied with Sawyer were defeated, an unsurprising verdict on the process that had made Sawyer Harold Washington's successor. Two successful challengers were forty-eight-year-old educator Alice Palmer in the 7th Ward, who defeated organization loyalist William Beavers in a virtual landslide, and young West Side activist Rickey Hendon in the 27th Ward. Another winner, in a South Side state representative contest, was 8th Ward precinct captain Donne Trotter, who a year earlier had turned out such an impressive victory margin for Harold Washington at London Towne Homes. One of the few challenges to a Sawyer loyalist that failed was Salim Al Nurridin's 9th Ward committeeman contest against Bill Shaw, whose twin brother Bob, the 9th Ward alderman, bizarrely alleged that Salim operated a harem full of welfare recipients. Only slightly more uplifting had been Emil Jones and Mel Reynolds's unsuccessful challenges to incumbent 2nd District congressman Gus Savage.

When the *Tribune* reported that the thirty-six-year-old Reynolds had voted only twice since he turned twenty-one, Reynolds claimed that plotters had altered his voting records. *Tribune* political reporter R. Bruce Dold commended Reynolds for running "a surprisingly effective first-time campaign" and praised him as "a walking role model for black achievement." But when the votes were tallied, Reynolds received only 14 percent, Jones 24 percent, and Savage won renomination with just 53 percent.[74]

Stanback, Orenstein, and Obama carefully scripted the St. Kevin evening rally where Gene Sawyer would eagerly agree to UNO and DCP's demand for a new mayoral task force to study the city's landfill options. Unlike Jim Fitch's committee, this new group would be heavily stacked with UNO and DCP loyalists. Howard, Bruce, and Barack jointly drafted Sawyer's remarks, and then both organizers, along with Lena and Loretta, met with Sawyer, Stanback, and other mayoral aides at City Hall. A young assistant to Stanback, Judy Byrd, remembered being struck by Barack, who "spoke with such command and such clarity." This was in stark contrast with Bruce's impression of the new mayor. "I remember in that meeting talking to Sawyer," Orenstein said, "and feeling like no one's there, no one's home." Yet the central trio worked exceedingly well together. Bruce found Barack "very collaborative and very easy to work with," and was repeatedly impressed by how Barack made sure that his top leader was never ignored or left out: "he was looking out for Loretta." In addition, "Stanback's a full partner. I remember Stanback saying at the time that he's never had a more collaborative relationship with a community organization, and he really appreciated it."

At Stanback's insistent urging, Orenstein also tried to convince some hardcore landfill opponents like Marian Byrnes to take part in the new process, but Byrnes realized that this was all leading to two predetermined ends: a new landfill at O'Brien Locks that the city desperately needed, and a $20 to $25 million community trust funded by Waste Management that would be controlled by UNO and DCP, not Jim Fitch and the wider community.

Angry but determined, Marian, Hazel Johnson, and others picketed St. Kevin that evening, distributing a no-more-landfills flyer that invoked the title and featured song from the *Eyes on the Prize* civil rights documentary that had aired a year earlier. UNO members tried to obstruct the leafleting, and when *Hegewisch News* editor Vi Czachorski, a UNO opponent, tried to enter the basement, UNO's Phil Mullins physically blocked her. "As I descended St. Kevin's stairs, Mullins put his arms across the narrow stairway and said 'You can't attend this meeting.' I tried to continue, crowds pushed. Mullins said 'I'm getting the police. I'll charge assault!'" Czachorski wrote in the next issue of her weekly newspaper. "I left."

UNO and DCP's own dueling flyer demanded that Sawyer name a new task force "made up entirely of residents who live in communities affected by landfills." Only such a group can "take this issue out of the backrooms and into the light of day." DCP also distributed a statement in Loretta Augustine's name denouncing "backroom deals that ram landfills down the communities' throats and send the enormous profits from such dumping into corporate and city coffers."

As a crowd of more than six hundred filled St. Kevin's basement, Orenstein paced nervously while Barack was "relaxed and cool." Sawyer carried with him a briefing memo summarizing the remarks that Loretta and Mary Ellen would make as well as his own speech, typed out in large, bold capital letters. A seven-piece mariachi band provided entertainment as multiple TV camera crews set up their equipment. DCP president Dan Lee and St. Kevin pastor George Schopp joined Loretta, Lena, and the mayor on the stage.

DCP's Loretta Augustine opened the meeting. "We, as residents, have had no control over what has happened in our community. We are tired of being victims. We are taking control of our own community." Then Lena spoke, followed by Sawyer. "Waste disposal and landfill decisions will no longer be made in the back room, at a table full of politically connected financial opportunists," the mayor read, his text sounding far more like Bruce Orenstein than Gene Sawyer. "Whatever happens here will be because you decide."

With Lena and Loretta flanking the mayor, Lena then took charge of the traditional IAF-style colloquy with Sawyer, just as she had with Harold Washington almost five years earlier on that same stage. UNO and DCP had encouraged their supporters to be boisterous, and one reporter called the crowd "raucous." Lena enjoyed her role to the hilt, and she began reciting the formal demand that the mayor appoint a new task force within ten days. She warned Sawyer to "be careful how you respond because this is an angry group of people tonight." The mayor stuck to his script and pledged full acceptance of UNO and DCP's demands. At that point, Lena turned to the cheering crowd and declared, "I'm going to take it for granted that we will have all the power we want!" As one veteran organizer later remarked, five years as a quintessential Alinsky leader had made Mary Ellen Montes into "one of the most macho women I had ever met."

As the gathering concluded, Bruce and Barack were ecstatic about the meeting. But UNO and DCP's Alinsky-style power grabs—first blowing up the Fitch talks, then bringing a sad sack mayor to heel before an excited crowd—had fractured the Southeast Side community. Bruce, Lena, and Barack had succeeded in infuriating and alienating the local business leadership and the true environmentalists, two groups that just weeks earlier had been prepared to join forces in a true community consensus. Ed Vrdolyak quickly put Sawyer on notice, objecting to the city allowing UNO and DCP to control negotiations with Waste Management: "For certain community organizations who without question do not truly represent the vast majority of homeowners, residents, and taxpayers to submit their community buyout (sellout) wish list is totally and completely wrong."

But Vrdolyak's public protest bore no political fruit, and a week later, Sawyer and Stanback announced a new sixteen-member Task Force on Landfill Options: Mary Ellen Montes led a group of five UNO supporters, including Father George Schopp; five other appointees were DCP members: Loretta Augustine, Dan Lee, Marlene Dillard, Margaret Bagby, and Father Dominic Carmon. Bob Klonowski was another ally, and no more than three appointees, including Marian Byrnes and Hazel Johnson, were likely dissenters. It was hard to imagine a more politically unrepresentative group.[75]

In late March Barack announced his upcoming departure. He went to see Loretta first. "He told me he was leaving and he needed to go back to school." Most DCP members learned the news at a meeting where Barack spoke of a smooth transition to Johnnie Owens as his successor. Dan Lee recalls that "I wanted to cry" and "we all got teary-eyed. . . . He was like a brother." Tommy West called out, "No, you can't go," but they all realized that Barack's potential reached well beyond Roseland. "We hated to see him go," Yvonne Lloyd remembered. "It was very sad," but they all appreciated, as Betty Garrett explained, that "if he could better himself, then we wanted him to go." Barack remembered overhearing Yvonne remark how different he seemed now than he did on that August day two and a half years earlier when Jerry Kellman had first introduced him. "He was just a boy. I swear, you look at him now, you'd think he was a different person." Of course, in many ways indeed he was.

Cathy Askew was the most emotional about Barack's announcement. "I was really upset," she recalled. "I thought we were friends." Barack remembered Cathy expressing her disappointment and saying. "What is it with you men? Why is it you're always in a hurry? Why is it that what you have isn't good enough?" Yet they all understood how frustrating the past year had been for Barack. "For the leader or organizer who feels expected to bring some change and improvements to the community, the day-to-day litany of roadblocks and resistances makes it hard," one close observer of Chicago organizing wrote that spring.

Marlene Dillard had watched Barack experience repeated setbacks while always trying to hide his disappointment from DCP's members. The outreach to Local 1033 at Republic LTV had gone for naught, the efforts in Altgeld Gardens had led to little, and only now was a tiny version of CEN getting under way. Again and again, "I always felt that it was a disappointment to him." Whenever she and Barack visited a funder like Woods, "he was trying to project how great we were doing." Then, "when we were leaving," he would turn to her and apologize for his braggadocio: "Well, we're trying." Overall, "I think it weighed very heavy on him. . . . He was leading people, and he was getting nowhere." Indeed, Marlene came to believe "that he felt 'If I could just become the mayor of Chicago, I would be able to do this.'"

Ernie Powell saw the same thing. "I think Barack got a little frustrated with that, and he felt like he had to get into the seat of power." The DCP pastors who interacted regularly with Barack understood likewise. With CEN operating out of Reformation Lutheran, Tyrone Partee saw Barack almost daily and remembered him saying, "I'm going to law school." Barack knew Tyrone was from a political family, and to him, Barack was "clear that he wanted to go into politics. 'I believe that's what I'm called to do.'" Alvin Love was caught off guard by Barack's announcement, but he realized Barack was "frustrated with the speed of change" and had concluded that "there might be a better way to do" things.

Barack went to see both Rev. Eddie Knox at Pullman Presbyterian Church and Rev. Rick Williams at Pullman Christian Reformed Church in person. With Knox, Barack presented Harvard as an opportunity he was pondering, and Knox smilingly replied, "There isn't much to think about." Harvard was such "a golden opportunity" and Knox believed "You're going to go far." Rick Williams reacted similarly. "I'm happy for you," Rick remembered saying, "but I'm also sad, because this kind of work needs people for the long haul, people like yourself." Rick understood Barack's hope of building a truly large, multi-congregational alliance to pursue educational reform and employment opportunity all across Chicago, but, just as Jeremiah Wright had sought to explain a year earlier, bringing people and churches together behind such an agenda was far more complicated than Barack could imagine. Rick told Barack that Harvard was "a wise decision" and wished him well. "You are going to do more for more people getting a law degree from Harvard than you would do here." Barack had "a passion for making life better for lots of people," Rick remembered, and to do that, "you've got to have power."

Barack also visited Emil Jones at his office on 111th Street. No elected official had done more for Barack and DCP, and Jones said he was sorry to see him go. To Jones as to others, Barack emphasized that there was no question he would return to Chicago after law school. He called Renee Brereton and other CHD staffers to tell them, plus organizing colleagues like Linda Randle. Barack apologized to Howard Stanback for pulling up stakes during their landfill effort. Stanback was surprised by Barack's choice. "'Why are you going to Harvard?' He said, 'Because I need to.' I said, 'Are you coming back?'" to which Obama said, "I'm absolutely coming back."[76]

Barack, Loretta, and Yvonne Lloyd all attended Lena's thirtieth birthday party, but by early April, there was not much to celebrate regarding UNO and DCP's position in the Southeast Side's landfill war. Anger at Sawyer's new task force was white hot, especially in Hegewisch, just across the Calumet River from the O'Brien Locks site. *Hegewisch News* editor Vi Czachorski asked Howard Stanback, "Why should UNO decide if there will be a landfill in our backyard?" and Marian Byrnes, Hazel Johnson, and three allies called the task force unrepresentative and called on Mayor Sawyer to disband it. That group held its first public hearing at St. Kevin on April 7, and this time UNO's critics made it into the basement meeting room, mocking cochairwomen Lena Montes and Loretta Augustine with chants of "No deals," the same slogan Lena had used during the Fitch ambush two months earlier.

The *Daily Cal* reported that Loretta defended the panel's "makeup and goals" as a "positive development for the community and said the mayor had pledged to abide by the task force's findings," with its report due at the end of May. "The panel approved reopening discussions with Waste Management," with Lena declaring, "What is different about it is that it will be talked about in open hearings, not behind closed doors." But when old foe Foster Milhouse rose to speak, "Montes quickly closed" the meeting.

The *Daily Calumet* editorialized against any reopening of negotiations with WMI, and UNO's opponents advocated for a popular referendum vote against landfill expansion on the upcoming fall general election ballot in Southeast Side wards. One week later, the task force convened its second hearing at Our Lady of the Gardens gymnasium in Altgeld Gardens, where the tumultous

Zirl Smith meeting had occurred two years earlier. Father Dominic Carmon, a task force member and Our Lady's pastor, remembered that Barack "was there listening," as at the previous St. Kevin session too. Howard Stanback, Jim Fitch, and WMI's Mary Ryan all spoke to the panel as a crowd of 125 chanted "No more dumps." For the first time, Stanback publicly acknowledged that the city *did* want the O'Brien Locks site to become a new landfill. Mary Ryan said WMI would immediately place $2 million into a community trust fund, with similar sums to be added every year, and it would give up title to two other parcels of land, including the marshland just below South Deering whose vulnerability had led Harold Washington to impose the initial landfill moratorium. That was followed by numerous single community members who spoke fervently against the deal. In the days that followed, the *Daily Cal* kept up a regular drumbeat against the task force. "Why study something no one in the community wants or asked for?" political columnist Phil Kadner queried.

As the UNO–DCP landfill gambit drew more and more flak, the task force's third hearing, scheduled to take place at Bob Klonowski's Hegewisch church, was postponed and moved to Mann Park's field house. When it finally convened, Klonowski welcomed the crowd in a calm tone, but, according to Vi Czachorksi, when a city representative again explained why O'Brien Locks was the best option available, "500 angry, frustrated people shouted down the city proposal of a new landfill." The meeting "ended abruptly when Mary Ellen Montes lost control, stating 'This has turned into a war.'" Declaring that "It doesn't appear we can conduct this in a civilized manner," she dissolved the hearing, leaving the entire UNO–DCP–Stanback strategy in tatters. The mayor's office named several additional task force members and postponed its reporting date until later in the summer, but the entire venture was now dead.

Years later, Bruce Orenstein acknowledged that his and Barack's game plan had gone entirely awry. When "we make ourselves the center of authority . . . we make ourselves the target" for large numbers of Southeast Side residents who for years had been opposed to the city using their neighborhoods as a dumping ground. Orenstein mused that if Harold Washington had not died, perhaps the O'Brien Locks deal with Waste Management could indeed have netted the community a $25 million trust fund, just as he, Obama, and Stanback had envisioned. But Gene Sawyer had no public stature as mayor. "With a very strong mayor" like Washington, a successful outcome was highly plausible, but "now we have a very weak mayor."[77]

During April's landfill warfare, Gamaliel held its fourth weeklong training at Techny Towers, and Barack drove out there for several days' sessions. By then, everyone knew he was leaving. Mary Gonzales was "pretty upset" when she heard, and David Kindler recalled thinking: "there goes one of our best and brightest." Kindler's friend Kevin Jokisch remembered telling Greg Galluzzo that Gamaliel would certainly miss Barack, with Greg responding, "We held on to him about as long as we were going to. Barack will probably end up being a United States senator."

One evening that week, Barack drank beer with Mike Kruglik and talked about what he wanted to do after law school. "Obama is talking about his vision for a very powerful, sweeping organization across black Chicago, of fifty to 150 congregations, that would be highly disciplined, highly focused, professionally organized," Mike remembered. "This vision . . . had such a claim on his mind" and Barack was explicit about "coming back to Chicago after Harvard and re-engaging in community organizing in a more powerful way."

Barack was thinking ahead in part because Ken Rolling and Jean Rudd at Woods had decided to fund and commission a series of articles about community organizing in *Illinois Issues,* the state's premier public policy journal. Barack accepted an invitation to write one, but a quick deadline loomed. He had decided that he would travel to Nairobi to see Auma and meet the other members of his Kenyan family, but before that he wanted to spend at least three weeks on his own touring the big cities of Europe. Thus he needed to write his article before he left Chicago in late May. His essay would document how his thinking had evolved. Recent African American activism, Barack wrote, had featured "three major strands": political empowerment, as personified by Harold Washington; economic development, of which black Chicago had seen very little; and community organizing. Barack argued that neither of the first two "offers lasting hope of real change for the inner city unless undergirded by a systematic approach to community organization." Electing a black mayor like Washington was "not enough to bring jobs to inner-city neighborhoods or cut a 50 percent drop-out rate in the schools," though such a victory did have "an important symbolic effect."

At the community level, "a viable organization can only be achieved if a broadly based indigenous leadership—and not one or two charismatic leaders—can knit together the diverse interests of their local institutions." Barack claimed that DCP and similar groups had attained "impressive results," ranging from school accountability and job training programs to renovated housing and refurbished parks. Those assertions echoed what Marlene Dillard had heard Barack boast about during their visits to DCP's funders, yet when he wrote that "crime and drug problems have been curtailed," he was making his wishfulness give way to fantasy. It was true that "a sophisticated pool of local civic leadership has been developed" thanks to DCP's recruitment and training efforts, but he admitted that organizing in African American neighborhoods "faces enormous problems." One was "the not entirely undeserved skepticism organizers face in many communities," as he had experienced; a second was the "exodus from the inner city of financial resources, institutions, role models and jobs," as he had seen all too well in Roseland and especially in Altgeld. Third, far too many groups emphasized what John McKnight called "consumer advocacy," and demanded increased services rather than "harnessing the internal productive capacities . . . that already exist in communities." Lastly, Barack declared that "low salaries, the lack of quality training and ill-defined possibilities for advancement discourage the most talented young blacks from viewing organizing as a legitimate career option."

Barack also argued that "the leadership vacuum and disillusionment following the death of Harold Washington" highlighted the need for a new political strategy. "Nowhere is the promise of organizing more apparent than in the traditional black churches," *if* those institutions would "educate and empower entire congregations and not just serve as a platform for a few prophetic leaders. Should a mere 50 prominent black churches, out of the thousands that exist in cities like Chicago, decide to collaborate with a trained organizing staff, enormous positive changes could be wrought in the education, housing, employment and spirit of inner-city black communities, changes that would send powerful ripples throughout the city."

Barack ended his essay on a revealingly poetic note, writing that "organizing teaches as nothing else does the beauty and strength of everyday people." When the entire series of *Illinois Issues* articles was subsequently republished in book form, one reviewer quoted that sentence as the single most powerful statement in the entire volume. But another of Barack's sentences about organizing was the most revealing of all, for when he wrote that through their work "organizers can shape a sense of community not only for others, but for themselves," he was publicly acknowledging the self-transformation he had experienced in the homes and churches of Greater Roseland.

As early as his second year at Oxy, Barack had felt "a longing for a place," for "a community . . . where I could put down stakes." The idea of *home,* of finding a real *home,* "was something so powerful and compelling for me" because growing up he had been a youngster who "never entirely felt like he was rooted. That was part of my upbringing, to be traveling and always . . . wanting a place," "a community that was mine." His "history of being uprooted" allowed Barack to develop in less than two years what Sheila knew was "his deep emotional attachment to" Chicago, one that was almost entirely a product of Greater Roseland, not Hyde Park.

"When he worked with these folks, he saw what he never saw in his life," Fred Simari explained. "He grew tremendously through this," through what he acknowledged was "the transformative experience" of his life, through what Fred saw was "him getting molded." Greg Galluzzo saw it too and said that Barack "really doesn't understand what it means to be African American until he arrives in Chicago." But, working with the people of the Far South Side,

Barack "recognizes in them their greatness and then affirms something inside of himself." Through "the richest experience" of his life, through discovering and experiencing black Americans for the first time, Barack "fell in love with the people, and then he fell in love with himself."

Years later, Barack admitted that "the victories that we achieved were extraordinarily modest: getting a job-training site set up or getting an after-school program for young people put in place." And he also knew that "the work that I did in those communities changed me much more than I changed the communities." Ted Aranda, who had worked for Greg and in Roseland before Barack and whose Central American heritage made it possible for him to be accepted as black or Latino, came to the same conclusion as Barack. "I'm not sure that community organizing really did that much for Chicago," he reflected. "I don't know that we had any really tremendous long-term effect." But Greg, looking back on a lifetime of organizing, understood the great fundamental truth of Barack's realization: "it's the people you encounter who are the victories." For Ted, the disappointments and frustration of organizing radicalized him. A quarter century later, deeply devoted to Occupy, Ted was driving a cab. Greg understood as deeply as anyone that "the great victory of the whole thing is Barack himself."[78]

In mid-April, the Spertus museum, part of a historic Jewish cultural center on South Michigan Avenue in downtown Chicago, opened a seven-week exhibit depicting the 1961 trial of Adolf Eichmann, the Nazi henchman who had played such a central role in the anti-Semitic effort to exterminate European Jews. The centerpiece of the exhibition was a continuous film of the trial, supplemented by large photographs and illustrations of newspaper stories plus Jewish artifacts documenting the culture that the Nazi Holocaust had sought to destroy.

The *Tribune* publicized the opening, and then, less than three weeks later, a front-page *Tribune* story revealed that anti-Semitism was alive and well even in Chicago's City Hall: "Sawyer Aide's Ethnic Slurs Stir Uproar," read the headline of a story about mayoral assistant Steve Cokely, who had recently delivered four "long and frequently disjointed" lectures under the auspices of Louis Farrakhan's Nation of Islam (NOI). Tapes of them were on sale at an NOI bookstore, and while anti-Semitism lay at the center of Cokely's often incoherent ramblings about a "secret society," he also called both Jesse Jackson and the late Harold Washington "nigger." Even worse, it was revealed that Mayor Sawyer's office had known about the recordings for more than four months, and three weeks earlier representatives of the Anti-Defamation League had met with Sawyer about the lectures. But Cokely was still on the mayoral payroll.

Well-known Catholic monsignor Jack Egan labeled Cokely's retention a "travesty," but a number of prominent black aldermen defended Cokely. Danny Davis, a supposed reformer from the 29th Ward, called Cokely "a very bright, talented researcher with an excellent command of the English language." The 9th Ward's Robert Shaw, citing voters he knew, said, "I don't think it would be politically wise for the mayor to get rid of Mr. Cokely." But the *Tribune* published a blistering editorial, denouncing Cokely as "a hate-spewing demagogue" and "a fanatic anti-Semite" and also lambasting Davis. After five days of feckless indecision, Gene Sawyer finally fired Cokely, but the damage to Chicago, never mind to Sawyer's indelibly stained reputation, was already done. That evening, at a large West Side rally, Roseland's Rev. Al Sampson introduced Cokely to a cheering crowd as "our warrior" and declared that "this is a case of Jewish organizations trying to stop one black man from having the right to speak."

In the middle of this, Barack took Sheila to see the Eichmann exhibit. Both of them would long remember what ensued. In Obama's later account, in the one single public reference he would ever make to his 1980s girlfriends, he created a character who was a conflation of Alex, Genevieve, and mostly Sheila who goes with him to "a new play by a black playwright." Several weeks earlier Barack had taken Sheila to see a Chicago amateur production of August Wilson's powerful 1985 play "Ma Rainey's Black Bottom," but Sheila would remember the aftermath of the Eichmann exhibit more vividly than Wilson's play. As they left, she asked Barack not about Eichmann, but about Steve Cokely and why so many prominent black Chicagoans were defending him rather than denouncing his moronic anti-Semitism. In Obama's version, his white girlfriend asked about black anger, and he replies: "I said it was a matter of remembering—nobody asks why Jews remember the Holocaust, I think I said—and she said that's different, and I said it wasn't, and she said that anger was just a dead end. We had a big fight, right in front of the theater," and "When we got back to the car, she started crying. She couldn't be black, she said. She would if she could, but she couldn't. She could only be herself, and wasn't that enough."

Obama would admit that "whenever I think back" to that argument, "it somehow makes me ashamed." Sheila and Barack did argue angrily that early May night on South Michigan Avenue, but it was because "I challenged him on . . . the question of black racism," and his response was so disappointing that their argument became "pretty heated." As Sheila recalled it, "I blamed him for not having the courage to confront the racial divide between us," but in retrospect, she concluded that the chasm between them was not racial at all. Instead it lay in the profound tension between Barack's insistence on "realism," on pragmatism, and what she believed was simply a lack of courage on his part. "Courage was a big issue between us," and their arguments over her belief that he lacked it were "very, very painful."

In early May, Sheila and Barack's mutual friend Asif Agha returned to Hyde Park after six months in Nepal. He remembers thinking at the time that "they had a good relationship. They were really tight, really solid," but he also noted that the tensions between them were even greater than they had been during that tumultuous weekend in Madison nine months earlier. Asif thought Sheila had a deeper commitment to their lives together than did Barack, and now, listening to Barack talk about his goals, Asif understood that his friend "wanted to have a less complex public footprint" as a future candidate for public office, particularly in the black community. Asif recalls Barack saying, "The lines are very clearly drawn. . . . If I am going out with a white woman, I have no standing here."

Asif realized just how profound the tension had become for Barack between the personal and the political. "If he was going to enter public life, either he was going to do it as an African American, or he wasn't going to do it." When asked if Barack had said he could not marry someone white, Asif assented. "He said that, exactly. That's what he told me."[79]

Even as his time in Roseland was ending, Obama still had to keep up with DCP's school reform alliance with UNO. Johnnie, Aletha Strong Gibson, and Ann West were more involved than he was, but DCP continued to follow UNO's Danny Solis and Lourdes Monteagudo. By late April, what was left of Harold Washington's official Education Summit had failed to endorse reform legislation that was muscular enough to satisfy top reformers like Don Moore, Fred Hess, and Pat Keleher of Chicago United. So UNO and DCP were now formally backing Chicago United's proposal, which was introduced in the state legislature by Senate Education Committee chairman Arthur L. Berman as S.B. 1837.

When Berman's committee held a daylong hearing on four competing bills on Tuesday, April 26, Barack and a small group of DCP members including Loretta Augustine, Rosa Thomas, Aletha Strong Gibson, and Ann West traveled to Springfield to lobby legislators and to hear Lourdes Monteagudo testify on behalf of UNO and DCP in support of S.B. 1837. Writing in the *Tribune,* CPS superintendent Manford Byrd once again energized reform advocates by decrying their attacks on "some monolithic, intractable bureaucracy which in fact does not exist" and claiming that "the school system is broadly understaffed." First the *Sun-Times* and then the *Tribune* began publishing multipart exposés on CPS's failings. The *Trib* series debuted with a long feature on one elementary school, "a hollow educational warehouse" that is "rich in remedial programs that draw attention to a child's failures." An accompanying editorial warned that "Chicago's public school system is failing its children and jeopardizing the city's future." The next day, in a culmination of negotiations that Don Moore's Designs for Change colleague Renee Montoya had been conducting with UNO's Danny Solis, DFC's reform bill, H.B. 3707, sponsored by African American Chicago representative Carol Moseley Braun, was strengthened with the addition of provisions from Chicago United's S.B. 1837. UNO and DCP joined in publicly shifting their support to the Braun bill, whose cosponsor was progressive Chicago Puerto Rican state senator Miguel del Valle, and reform energies increasingly coalesced behind the Braun–del Valle measure. In one *Trib* story, powerful 14th Ward alderman Edward Burke declared that "nobody in his right mind would send kids to public school." In another, Manford Byrd called himself "probably the most gifted urban administrator in this country" while once again dismissing CPS's obligations to its students: "When you're all done, the learner must learn for himself."[80]

In the final ten days before Barack's departure from Chicago for his two-month trip to Europe and then Kenya, things came completely apart at 5429 South Harper Avenue. Ever since Barack had transferred to Columbia almost seven years earlier, he had kept a journal, using it to record vignettes that might find their way into a future book and also sometimes for creating drafts of short stories and even letters to friends. Sheila knew of Barack's practice, and sometime after their heated argument outside the Spertus museum, she decided to take a look at it. Lena Montes heard from Barack what ensued. "She reacted to this journal that he kept under his bed or mattress," Lena recalled. "I remember when he says that she found some journal, and he talks about somebody in this journal and that she's upset" after she read it. Barack did not tell Lena whether she was that someone. "Was it a straw that broke the camel's back? I'm not sure. I just remember him saying that she . . . was leaving because of this journal."

Just as Barack was about to leave Chicago, Sheila moved out of their apartment and moved in with her younger friend Simrit "Sima" Dhesi, who had just completed her undergraduate degree at the University of Chicago, and her sister in an apartment four blocks away at 5324 South Kimbark Avenue. Sheila later said that May 1988 "was kind of a blur for me." Barack mentioned what was happening not only to Lena, but also to Loretta Augustine and even to his archdiocesan friend and Hyde Park neighbor Cynthia Norris. Cynthia understood that Sheila "was upset," and that the tensions between her and Barack were "because of her race. . . . Yes, I do remember that." Norris knew that Barack "had a lot of respect for" Sheila, and from what she knew, "I thought he handled things very, very well." Loretta remembered it similarly. "He talked to me about her," she recalled. "We had some really open and candid conversations" about the turmoil. "He obviously cared for her," and was disturbed by what was happening. "I remember telling him, 'If it's really real, what you all have, you'll come back'" after his trip and revive the relationship, "'and if it's not, you'll go forward.'"

A number of small going-away parties occurred during Barack's last week before he departed. Reformation Lutheran caretaker John Webster remembered one there, which was also where the small CEN program was now centered; Margaret Bagby recalled another one at St. Catherine's with catered food. One evening everyone from DCP was invited to a quiet party at a small restaurant in suburban Blue Island. Greg Galluzzo, who had spent so many hours with Barack over the previous eighteen months, "bought him a briefcase and had it engraved" with just "Barack" as a useful going-away present for a law student.

On another night, Barack and Bruce Orenstein went out to drink beer, and Barack asked Bruce what he would be doing ten years from now. "I'm going to be making social change videos," Bruce answered. Bruce in turn asked Barack the same question. He said he intended to write a book about his upcoming trip to Kenya, and then "I'm going to be mayor of Chicago." Bruce was taken aback. "That was the first I heard of it," and "I thought it was a lot of moxie to say that he was going to be mayor."

A few days later, with Sheila having moved from their apartment, Barack flew east. Almost that same day, a dinner announced several weeks earlier was taking place to honor Frank Lumpkin for the eight years he had devoted to winning recompense for the Wisconsin steelworkers who had been thrown out onto the streets of South Deering back in March 1980.

Frank's loyalties had not changed. The banquet's proceeds would "benefit the *People's Daily World,*" the newspaper of the Communist Party USA, but that did not deter a trio of notable figures from signing on as public patrons. State Senator Miguel del Valle, sponsor of the pending school reform bill, was one, 22nd Ward reform alderman Jesus "Chuy" Garcia was a second, and Monsignor Leo T. Mahon was a third. "Best wishes to a man who fights for justice," read Leo's greeting in the banquet program. Maybe Foster Milhouse and the other right-wing zealots had been right all along, that social justice Catholicism and grassroots communism were indeed one and the same.

Roberta Lynch, CCRC's first staff organizer, and *Tribune* business editor Dick Longworth both sent their apologies for being out of town, but a crowd of more than four hundred attended, including U.S. congressman Charles Hayes, a veteran of both labor struggles and the 1966 Chicago Freedom Movement. The *Daily Calumet* gave the event glowing coverage—"Lumpkin Honored at Dinner"—and three days later editorialized in his honor, simply and accurately calling him "a hero." After eight years of organizing, Frank Lumpkin had prevailed, and even triumphed.

Just short of three, Barack Obama was headed toward Harvard Law School with the intent of becoming not just mayor of Chicago but eventually president of the United States.[81]

Barack had scheduled a full month to see the great cities of Europe all by himself: Paris, Madrid, Barcelona, and Rome, then to London and from there to Nairobi. He later wrote that he anticipated "a whimsical detour, an opportunity to visit places I had never seen before," but his memory would be that "I'd made a mistake" in allocating so many days for his European grand tour, because "it just wasn't mine." After almost three years of interacting with a dozen or more people almost every single day, now he was entirely alone in countries where he knew virtually nothing of either the language or the culture.

Before the end of May he was busily dispatching postcards to Sheila, Lena, Cathy Askew, and Cynthia Norris from Paris, "some quite humorous," Sheila remembered. To Cathy, he wrote about how beautiful the buildings were, to Cynthia he described the city's astonishing appeal: "I wander around Paris, the most beautiful, alluring, maddening city I've ever seen; one is tempted to chuck the whole organizing/political business and be a painter on the banks of the Seine. You'll be amused to know that since I don't know a word of French, I'm left speechless most of the time. Wish you a fruitful summer. Love, Barack."

Traveling largely by bus and train, Barack was also reading a journalistic account of modern Africa in preparation for his visit to Kenya. He later told about meeting and trying to converse with a Senegalese traveler on the way from Madrid to Barcelona, but in subsequent years, Barack never referred to any experiences or memories from his four weeks on the continent. By the last weekend in June, he was in London, where Hasan and Raazia Chandoo had moved six months earlier from Brooklyn. The three of them had lunch in a brasserie before seeing Wim Wenders's new film, *Wings of Desire,* at a cinema in Notting Hill.

From London Heathrow, Barack flew to Nairobi. His sister Auma and his Aunt Zeituni, the family member who had been the closest to his late father, met him at the airport, but Barack's suitcase did not arrive until several days later. In preparation for his visit, Barack also had read a brand-new book on Dedan Kimathi, a Kenyan anticolonial warrior of the 1950s whom the British had executed in 1957, but instead of discussing Kenyan history while he stayed at Auma's apartment, sleeping on her couch, they talked mostly about their extended clan of relatives. "There was never a moment of silence or embarrassed awkwardness" between them, Auma recalled. But with Barack wanting to meet as many family members as possible, "It wasn't all nice. Sometimes he wanted to see a relative I didn't really get along with, and he'd be like 'It's my right, and I need to see them, and I'm not going alone, and you're coming with me,'" she explained.

Auma always assented, although her thoroughly unreliable Volkswagen Beetle was their primary mode of transport around sprawling Nairobi. One drive took them to see Auma's mother Kezia and her sister Silpa Jane, who six years earlier had been the telephone caller who told Barack of his father's death. On another, Zeituni took Barack to meet her older sister Sarah, who was living in a scruffy slum. But the most difficult visit was to Ruth Baker Ndesandjo, whose son Mark—Barack's younger brother—was home for the summer and about to begin graduate school at Stanford after having just graduated from Brown. Barack later imagined that Ruth had invited him and Auma to come for lunch, but both Ruth and Mark convincingly remember Barack and Auma turning up with no forewarning. As Mark recounts, a "very awkward, cold" encounter ensued, as Ruth found the unexpected arrival of her ex-husband's namesake "pretty traumatic." Years later she recalled, "I closed up. I had nothing to say," for "I didn't have the capacity to talk with him or exchange with him because he was a reflection of a man I hated. So I didn't want anything to do with him."

Barack wanted to see more of Mark, and they arranged to have lunch a few days later. Mark remembered thinking that Barack had a "cold" demeanor, "absolutely no sense of humor," and "wanted to shut out any emotional involvement" with his likewise half-Luo, half-white American brother, especially when Mark bluntly stated that "our father was a drunk and he beat women." Barack later admitted that meeting Ruth and Mark, and grasping at least in part how abusively his father had treated them, affected him deeply. "The recognition of how wrong it had all turned out, the harsh evidence of life as it had really been lived, made me so sad," far more so than he had been three years earlier when he had learned during Auma's visit to Chicago just how deeply tragic a life Barack Obama Sr. had led.

During Barack's second week in Kenya, Auma took him on a wild-game safari, and then the two of them plus Zeituni and Kezia took a train northwest to Kisumu and then a jitney bus to the family homestead at Nyang'oma Kogelo. There Barack met Sarah, the stepmother who had raised his father. Barack asked her if anything of his father's still survived. "She opened a trunk and took out a stack of letters, which she handed to me. There were more than thirty of them, all of them written by my father," carbon copies "all addressed to colleges and universities all across America" from when Obama Sr. was seeking admission to the University of Hawaii and other colleges. Holding them reminded Barack of the letters he had written a few years earlier, "trying to find a job that would give purpose to my life."

That moment, even more than standing at his father's unmarked grave in the side yard of Sarah's small, tin-roofed brick home, marked the most powerful and direct paternal link Barack had ever experienced. The connection was underscored when relatives remarked that Barack's voice sounded "exactly how his father spoke." Barack's visit to Nyang'oma Kogelo, like his earlier one to Aunt Sarah in that Nairobi slum, made Barack realize that his Kenyan relatives' lives sometimes paralleled those of people whom he knew in Altgeld Gardens. Pondering John McKnight's analysis, in Kogelo Barack wondered whether "the idea of poverty had been imported to this place, a new standard of need and want."

Barack later wrote that while in Kenya, "for the first time in my life, I found myself thinking deeply about money," perhaps foreshadowing the debt load he was about to take on to follow his father's footsteps to Harvard rather than attend Northwestern cost free. He told Auma that his decision to attend law school was a response to his experiences in Chicago, because as a community organizer he could not "ultimately bring about significant change." Back in Nairobi, everyone went to a photography studio so that a family portrait could be taken, and Barack and Auma made a brief visit to an elementary school so that he could meet his youngest sibling, George, the son of Jael Atieno.

One night Auma took Barack to meet one of her former teachers, a woman who had also known their father. Barack recounted her presciently telling him that being a historian "requires a temperament for mischief" and instructing him that when confronted with roseate fictions, "truth is usually the best corrective." For his final weekend in Kenya, Auma and Barack took a train eastward to the country's second largest city, coastal Mombasa. Overall, "it was a magical trip," Barack remembered, an immersion into the life of the father who had abandoned him, an immersion that "made me I think much more forgiving of him" than he had been before experiencing Luo life and culture for the very first time.[82]

Before the end of July, Barack returned to Chicago, where he was all alone in the 5429 South Harper apartment until the lease expired in early August. When Asif's friend Doug Glick stopped by, he was struck by how different the place looked without Sheila. But Barack and Sheila's two months apart had not ended the relationship between them. Johnnie Owens had run into Sheila in Hyde Park during the summer and remembers how "upset," even "brokenhearted" she seemed about what had happened just before Barack's trip. But Barack's more-than-weekly letters from Europe and Kenya had rebuilt much of their bond. Even so, in little more than two weeks he had to head to Massachusetts before the start of Harvard's fall semester. He needed to purchase a better car than the blue Honda Civic that he had driven from Manhattan just over three years earlier, and he replaced it with an off-yellow 1984 Toyota Tercel hatchback he bought for $500 from a suburban police officer.

Barack also wanted to be sure that Johnnie's transition was going as smoothly as possible, not just at DCP, where everyone already knew Johnnie, but also with DCP's downtown supporters. "He made sure that the leaders were comfortable," Johnnie remembered, but "the main thing he did was transfer those financial funder relations" with Jean Rudd at Woods, Anne Hallett at Wieboldt, and Aurie Pennick at MacArthur. Just a few weeks later, MacArthur would award DCP its second annual $20,000 grant. Barack "left the organization in a very good position," Johnnie explained, and the effort Barack put into the transition further showed that he intended to return to Chicago after law school.

Barack's greatest disappointment was the lack of state funds to expand DCP's after-school tutoring program, but with state government consumed by the struggle over Chicago school reform, the legislature had remained in session beyond its normal end-of-June adjournment to pass a compromise bill. At the end of May, reform forces had reunified themselves into a new coalition called the Alliance for Better Chicago Schools, or ABCs, with UNO playing a lead role and DCP a minor one. A massive June 6 rally had called upon reform supporters to make their case in person to state legislators in Springfield, and the slogan "Don't come home without it!" became reformers' new rallying cry.

The Chicago City Council approved a vote of no confidence in Manford Byrd by 39 to 4, soon followed by the resignation of a board of education member who now felt similarly. "I assumed education was the first priority of the whole system, and it is not," retired business executive William Farrow announced. Down in Springfield, House speaker Michael J. Madigan brought the interested parties together for marathon negotiation and drafting sessions in his office. Danny Solis and Al Raby both took part, but the most influential participant was one of Madigan's deputies, Chicago state representative John Cullerton. By the end of June Al Raby was proclaiming that "real reform of Chicago schools is within our grasp," and on Saturday, July 2, both houses of the state legislature passed a compromise bill. Failure to adopt the measure by the end of June would delay its effective date for a year, but Chicago United's Patrick Keleher said the bill represented "as much if not more than any of us had hoped for."

Barack heard far less encouraging news about his other top concern, the Southeast Side landfill tussle. Mayor Sawyer's new UNO-and-DCP-dominated task force had held additional public hearings during June, and in late July met privately with both city and Waste Management representatives. By the end of the summer, its report to Sawyer was complete, though several weeks would pass before cochairs Loretta Augustine and Mary Ellen Montes joined the mayor at a City Hall press conference. "The city of Chicago is facing a crisis," Sawyer announced. "We're going to have to bite the bullet and do some things that we would prefer not to do," namely allow the O'Brien Locks site to become a landfill. That outcome had always looked inevitable, but Bruce, Barack, and Mary Ellen's strategic success in blowing up the Fitch negotiations meant that any landfill now would be controlled not by Waste Management but by the Metropolitan Sanitary District.

Alinsky-style warfare had not only destroyed a likely Southeast Side consensus to accept a quid-pro-quo deal with Waste Management, it had deprived those neighborhoods of the multimillion-dollar bounty WMI had been willing to pay. It was a debacle all around. A decade later, Jim Fitch would be sent to federal prison for eighteen months and fined $1 million for looting bank funds throughout the 1980s in order to contribute to Southeast Side politicians. Another decade further on, with UNO having abandoned South Chicago and transformed itself into something that bore no resemblance to the organization that Mary Gonzales, Greg Galluzzo, and Mary Ellen Montes had originally built, UNO would endorse a Waste Management effort to expand Southeast Side landfills.

In Barack's final days before he left for Harvard, the DCP members to whom he had become closest held a small barbecue for him at Loretta Augustine's home. He assured them, "If you have problems, you can contact me and I'll do what I can." He also had a trio or more of presents for them, wooden figurines he said he had brought back from Kenya. Dan Lee remembered that Barack gave him "a statue of a warrior with a chipped beard." Cathy Askew recalled admiring a giraffe, but "I think he gave me the zebra because of the mixed black and white stripes," an acknowledgment of their disagreement about biracial identity.

Mary Ellen Montes did not attend that party, but one evening Barack took her out to dinner at a downtown restaurant. "We had our own kind of little going-away party, Barack and I, and it was just Barack and I," she remembered. Barack promised to write to her, and he would, but that night, or the next morning, was the last time Lena and Barack would see each other in person.

With the lease on the South Harper apartment expiring several days before Barack planned to leave, he joined Sheila in her apartment on South Kimbark. She was preparing to leave Chicago soon too to begin her dissertation fieldwork in South Korea thanks to a Fulbright fellowship, and when she joined Barack for a farewell visit to Jerry Kellman's home, they had a question for Jerry and his wife. "They come to dinner at our house," Jerry remembered, "and they say 'Could you please keep this cat?'" The Kellmans willingly agreed to give Max a new home. "Barack was not sad to give Max away," Sheila explained, but for Max the transition was all to the good, and he would enjoy eight years of love with the Kellmans.

Jerry knew that Barack and Sheila were not breaking up, just headed in different geographical directions. He thought "they had a great, healthy relationship," although one that was now constrained by Barack's career plans. "By the time Barack left to go to law school, he had made the decision that he would go into public life," Jerry realized. Indeed, "my sense is that Barack's dream was to come back and possibly become mayor of Chicago."

But Barack was still trapped between his belief in his own destiny and his deep emotional tie to Sheila. One day that final week "he said that he'd come to a decision and asked me to go to Harvard with him and get married, mostly, I think, out of a sense of desperation over our eventual parting and not in any real faith in our future," Sheila recalled. Her memory was reminiscent of Genevieve's from three summers earlier, when Barack had asked her to come to Chicago with him, and Genevieve had thought Barack was asking her only because he was certain her answer would be no.

Now, once again, the answer was no, and Sheila was upset by Barack's presumption that she should postpone, if not abandon, her dissertation research in order to accompany him to Harvard. A "very angry exchange" followed, with Sheila feeling that Barack believed that his career interests should trump hers.

Barack started eastward within a day or two, knowing he could stay temporarily with his uncle Omar, who had remained in greater Boston ever since first arriving there thanks to his older brother a quarter century earlier and whose phone number and address Aunt Zeituni had given Barack while he was in Nairobi. But before heading back down Stony Island Avenue to the Skyway and then the Indiana Toll Road, Barack needed to have one other conversation.

"I was a little troubled about the notion of going off to Harvard. I thought that maybe I was betraying my ideals and not living up to my values. I was feeling guilty," he told a college audience just six years later. Barack called Donita at Trinity and made an appointment to see Jeremiah Wright to seek his counsel about those doubts. In a way, it was just like the conversation he had had nine Augusts earlier, in Honolulu, when eighteen-year-old Barry had gone to visit Frank Marshall Davis before leaving for Occidental and life on the mainland.

Years later, before their relationship was torn apart, Wright would say that Barack was "like a son to me." One of the most knowledgeable and savvy women in black Chicago would make the same point: "Jeremiah Wright was the black male father figure for Barack," she emphasized. "Don't underestimate the influence that Jeremiah had on Barack." Wright would not specifically remember their conversation that August day, but Barack always would. In a way, it was a three years' bookend to the admonishing monologue about being a do-gooder that Bob Elia have given him that night in the motel lobby on South Hermitage Road.

That exchange would stay with Barack always, as would this one, but the substance of Wright's message was identical to the warning that old Frank had voiced: "You're not going to college to get educated. You're going there to get *trained,*" trained "to manipulate words so they don't mean anything anymore," trained "to forget what you already know." Wright's message was just five words, ones that would ring in Barack's ears for the entire two-day drive eastward: "Don't let Harvard change you!"[83]

Chapter Five

EMERGENCE AND ACHIEVEMENT

HARVARD LAW SCHOOL

SEPTEMBER 1988–MAY 1991

Omar Onyango Obama, the younger brother whom Barack Obama Sr. had helped come to the United States in 1963 to attend high school, was by August 1988 an unmarried forty-four-year-old store clerk living in central Cambridge. Omar's apartment at 48 Bishop Allen Drive was his twenty-seven-year-old nephew's first destination when Barack exited the Massachusetts Turnpike upon arriving from Chicago. Omar had never completed high school, but his mundane work life had not kept him from developing keen political interests. "I am a Pan-Africanist," he wrote in a letter to *Ebony* magazine, one who shared "the dreams of brothers Marcus Garvey, W. E. B. Du Bois and Malcolm X." Barack had never met his uncle Omar, but Omar happily hosted him while Barack scoured the *Boston Globe*'s classified ads looking for a place of his own.

John "Jay" Holmes owned the handsome, almost century-old Queen Anne–style Langmaid Terrace apartment building at 359–365 Broadway in Somerville's Winter Hill neighborhood. Harvard Law School was a twelve-minute, two-and-half-mile drive away, but for $700 a month a one-bedroom basement-level "garden" apartment there offered more private and spacious quarters, including a nice exposed-brick living room, than other rentals closer in. By the end of August, Barack was happily settled at 365 Broadway #B1, and on September 1, Harvard Law School (HLS)'s two-day registration and orientation for first-year students—"1Ls"—got under way on its multibuilding campus on the east side of Massachusetts Avenue, a short distance north of Harvard Square.

The entering class of 1991 numbered 548 students, selected from among seventy-one hundred applicants. Forty percent of the new class were women, and 22 percent were nonwhite, including fifty-seven African American students, of whom almost two-thirds were women. That was impressive diversity: the 2L and 3L classes had started with sixty and seventy-three African Americans, respectively. While the average age of the new 1Ls was twenty-three, with an overwhelming majority coming directly from their undergraduate studies, 5 percent of the class was over the age of thirty, including "many students who experienced the 'real world' before coming to law school," according to the weekly *Harvard Law Record*. Word among the faculty was that "preferences for applicants who had taken time off, engaged in public works, or participated in other significant outside activities or experiences" had played a significant role in admissions decisions.

The school had an illustrious reputation but a deeply troubled internal culture. For three years, the faculty had been embroiled in toxic ideological warfare that had seen four junior faculty members denied permanent appointments, the first such tenure denials in seventeen years. That quartet was seen as "crits," or proponents of "critical legal studies" (CLS), a left-wing school of thought that viewed legal rules and doctrines as inherently conservative rather than politically neutral. While several prominent crits, including CLS's most erudite proponent, Roberto Mangabeira Unger, a Brazilian legal philosopher, were senior members of the Harvard law faculty, other full professors, including Robert C. Clark and David Rosenberg, were perceived as conservative "law and economics" devotees who were generally hostile to the work of crits.

James Vorenberg, dean of the law school since 1981, had announced his upcoming departure from that post in April 1988, just as tensions over a second issue of faculty composition—racial and gender diversity—were reaching a new peak as well. Seven years earlier, the roughly sixty-member Harvard law faculty had included just one tenured woman and a single tenured black male, who passed away in 1983. By 1988 there were five senior women, all white, and two tenured black men—Derrick Bell and Christopher Edley—with one more woman and three additional black men—Charles Ogletree, Randall Kennedy, and David Wilkins—all on the cusp of consideration for permanent appointments. In May 1988, the Black Law Students Association (BLSA) occupied Vorenberg's office in a "study vigil" to protest the school's failure to appoint additional minority faculty, with Professor Bell and legendary civil rights organizer Robert Moses addressing a student rally the next day.

Even with their healthy representation in each entering class, Harvard's black law students felt a constant need to prove they were just as qualified as their white classmates to succeed at HLS. "Everything here says you can't do it," graduating BLSA president Verna Williams remarked in a fifty-page 1988 BLSA publication. "BLSA fights that negative attitude, and that builds you up." The 1988 class had found BLSA in "disarray" upon their 1985 arrival and had worked to show that black students could attain leadership positions in multiple student organizations. By their 3L year, they had succeeded, with 1987–88 marking "the first time in HLS history that black law students, in significant numbers, had not only gotten involved in campus organizations outside of BLSA, but had done so with such success and capability that their peers . . . asked them to lead the organizations."

But law students of all races were deeply unhappy with the institution's culture. In the late 1980s "attending Harvard Law School was a miserable experience for the majority of its students," one highly supportive alumnus later acknowledged, and a 1988 forum soliciting student input regarding the selection of Vorenberg's successor instead focused on "the alienation students have felt from the law faculty," the *Harvard Crimson* reported. "I don't always feel that professors are here who can teach," one 3L woman told the paper. "Law school seems to exist primarily for the professors." Another 3L wrote in a subsequent memoir, "I didn't know any of my law school professors," and described how "the one thing that actually drew the student body together was a widespread disenchantment with our teachers." He also recounted that although 70 percent of his classmates had arrived at Harvard expressing an interest in practicing public interest law, only six out of 474 actually accepted legal-services jobs after graduation.

Four years earlier a young sociologist, Robert Granfield, had begun questioning scores of Harvard law students to analyze the "complex ideological process that systematically channels students away from socially oriented work." He concluded that "students come to believe that effective social progress occurs primarily through the use of elite positions and resources." Graduates "feel that they have emerged with their altruism intact, while having actually been co-opted" into joining large corporate law firms as a result of their acculturation at what he called "a tremendously powerful co-optive institution." The bottom line was that "employment in organizations designed principally to serve the corporate rich came to be seen as highly compatible with public service ideals."

BLSA's long, mid-1988 report included a four-page essay entitled "Minority and Women Law Professors: A Comparison of Teaching Styles," written by a graduating 3L who had accepted a job at the Chicago office of Sidley & Austin, one of the country's most prominent corporate law firms. Michelle Robinson had taken Criminal Law with Charles Ogletree and Family Law with Martha Minow, a young professor who had been awarded tenure in 1986 and whose father was one of Sidley & Austin's best-known partners. Robinson briefly interviewed both Ogletree and Minow, as well as David Wilkins, because she saw each of them as highly atypical faculty members: "each one of these professors spends an enormous amount of time with students—particularly minority and women students," she wrote. Robinson clearly admired Minow, writing that "sitting in her Family Law class is like sitting in the studio audience during a taping of the *Phil Donahue Show*." Not all compliments are actually flattering, but it was David Wilkins, in just his second year of teaching at Harvard, who supplied the real gist of Robinson's essay. "The problem here is that only about ten percent of the professors care enough about students to spend time with them," he bluntly told her. "Consequently, this ten percent gets a disproportionate share of students to counsel." He pointed out further that they bore that heavy ancillary burden while also striving to write the law review articles necessary for promotion to tenured full professor.

Like many of her fellow students, Robinson lamented the law faculty's relative lack of "people who possess the enthusiasm, sensitivity and ingenuity necessary to bring excitement back into the classroom." She also decried the school's lack of interest in hiring a more diverse faculty and said it "merely reinforces racist and sexist stereotypes," attitudes that also were manifest in "the rude statements and questions posed by students"—white students—"to Professors Wilkins and Ogletree." But as Robinson's acceptance of Sidley & Austin's highly remunerative job offer underscored, another defining element of the law school's culture was that the high cost of a Harvard education provided a powerful incentive for graduates to take the best-paying jobs available.

Several hundred graduating students had completed a Robert Granfield questionnaire that probed that subject, and a heavy majority acknowledged that "the necessity of repaying educational loans" had been a decisive factor in their own job decisions. With most graduates leaving Harvard saddled with tens of thousands of dollars in debt, law students "are co-opted into futures of providing service for the corporate elite," Granfield concluded. Verna Myers had arrived at the law school straight from her 1981–82 senior year as president of Columbia and Barnard's Black Student Organization. "I went to Harvard expecting to change the world," but in 1985, "I left Harvard thinking 'How can I pay my loans?'"[1]

The first day of Harvard's fall 1988 registration for 1Ls featured a 1:00 P.M. financial aid session. Tuition for the year was $12,300, with total costs for a student living off-campus estimated at $22,450. That day or the next, the new students also received their course schedules for the fall semester. The entering class was divided into four numerically identified sections—I, II, III, and IV—of about 140 students apiece. Each section had five core first-year courses: Civil Procedure, Contracts, Torts, Criminal Law, and Property. Within each of those four sections, each student was also assigned to a thirteen-member "O Group"—O as in Orientation—which would then become their ungraded, fall semester one-afternoon-a-week Legal Methods class, taught by an upper-class student. Each O Group had a faculty mentor, and for Section III, to which Obama was assigned, one of the five substantive courses, Torts, would be divided into a trio of smaller, forty-five student classes. For Section III, Civil Procedure and Contracts would last the entire year, three hours a week. Torts and Criminal Law would meet five hours per week in the fall semester. In the spring, when 1Ls could choose a single elective from among half a dozen choices, Section III would have Property five hours per week, and the Legal

Methods class would morph into a prolonged "moot court" exercise in which two-person teams briefed and argued faux court cases. One odd feature of Harvard's first-year schedule was that final exams for the fall classes took place only in mid-January, well after the Christmas and New Year's break.

The first classmates the new students met were their fellow members of the O Group, who introduced themselves alphabetically. Richard Cloobeck was a twenty-two-year-old Dartmouth graduate from Los Angeles. African American Eric Collins, from North Carolina, had just graduated from Princeton. Twenty-three-year-old UCLA graduate Diana Derycz had just finished a master's degree at Stanford. Then came a trio of visibly older men. Rob Fisher had grown up on a farm in Charles County in southern Maryland before graduating from Duke in 1976 and completing a Ph.D. in economics there six years later. From 1982 until a few months earlier he had taught economics at the College of the Holy Cross in western Massachusetts. Mark Kozlowski, a 1980 graduate of Sarah Lawrence College, had spent the past eight years studying political theory at Columbia University and was about to submit his Ph.D. dissertation on the constitutional thought of American framer and president James Madison. When Barack Obama introduced himself, "the way he spoke was very, very memorable," Richard Cloobeck recounted. "He was talking about black folks and white folks" and "what really, really struck me is that I got the impression he was neither." Mark Kozlowski remembered Barack mentioning both Hawaii and Indonesia when he spoke. "I was immediately very impressed with Barack" and "I was immediately taken with the voice." The trio of Fisher, Kozlowski, and Obama made Cloobeck realize "there was a big distinction between the older people and the younger people," and Diana Derycz remembered similarly, thinking "how brainy people were" in this new northeastern environment.

After Obama was a twenty-two-year-old Tufts graduate. "I'm Jennifer Radding. I have no idea how I got in," she said. She remembers that Barack was among those who laughed, and "he knew I was a bit of a jokester," Jennifer explained. Jeff Richman had just graduated from the University of Michigan, and Sarah Leah Whitson was a brand-new graduate of UC Berkeley with family roots in the Middle East. The high-powered group featured strikingly diverse experiences. Diana Derycz had spent five years in Mexico as a child and was working part-time as a professional model. Fisher had studied in Australia for three years during and after graduate school, Barack had lived almost three years in Indonesia, and Whitson had spent summers with relatives in Syria, Jordan, and Lebanon.

O group instructor Scott Becker, a twenty-five-year-old 3L who was a graduate of the University of Illinois, found himself supervising a group of 1Ls some of whom—Rob, Mark, and Barack in particular—were older and more worldly-wise than he. The first day's mandate was to conduct some sort of attorney-witness Q&A exercise, and when it was over, Barack immediately introduced himself to the obviously brilliant Rob Fisher, who was seven years his senior. "We instantly became friends," Fisher recalled. Rob's intellectual enthusiasm and energy belied his thirty-four years. In 1986, New York University Press had published his book *The Logic of Economic Discovery: Neoclassical Economics and the Marginal Revolution.* The book's title purposely mirrored Karl Popper's famous *The Logic of Scientific Discovery,* and the major figure in Fisher's work was the late Hungarian philosopher of science Imre Lakatos, whose scholarly work in Popper's empiricist tradition was vastly better known than his role as a hard-core Stalinist in Hungary's post–World War II Communist Party. Fisher's focus was what he termed "A Lakatosian Approach to Economics," and his volume had garnered enthusiastically positive reviews in scholarly journals. "The book effectively demonstrates the applicability of the philosophy of science to the history of economics," one reviewer wrote, adding that "the argument is crystal clear and largely persuasive." A second observed that "the reason neoclassical economists like Lakatos so much is that his philosophy is intended fundamentally to explain the evolution of physics," since "neoclassical economics wants desperately to be a social physics."

Fisher remembered from the first day that Barack "recognized he was in a playground for ideas," one where he could exercise "his love for intellectual argument." It was a passion that Rob shared, and from that day forward the somewhat unlikely duo would become an inseparable pair whom scores of classmates recall as the closest of friends. Barack remembered his initial experience of Harvard Law School similarly. "I was excited about it. Here was an opportunity for me to read and reflect and study for as much as I wanted. That was my job," and a much easier one than what he had confronted at the Developing Communities Project. "When I went back to law school, the idea of reading a book didn't seem particularly tough to me."

The four fall 1L courses gave Barack some large books to read, and he needed to read them with painstaking care. In Contracts, Professor Ian Macneil, a 1955 Harvard Law School graduate who was visiting for the year from Northwestern University, had assigned his own *Contracts: Exchange Transactions and Relations,* 2nd ed., which was more than thirteen hundred pages long. In Civil Procedure, the text for the year, Richard H. Field et al.'s *Materials for a Basic Course in Civil Procedure,* 5th ed., came in slightly shorter, at 1,275 pages. Section III's fall semester of Civ Pro was taught by Professor David L. Shapiro, a 1957 summa cum laude Harvard Law graduate and former Supreme Court clerk, but in December Shapiro would become deputy solicitor general of the United States, and the course would be taken over for the spring semester by Stephen N. Subrin, a visiting professor from Northeastern University Law School.

In Criminal Law, Professor Richard D. Parker, a 1970 Harvard Law graduate and another former Supreme Court clerk, assigned Paul H. Robinson's brand-new *Fundamentals of Criminal Law,* a mere 1,090 pages. And in Torts, Professor David Rosenberg, a 1967 graduate of NYU's law school who had taught at Harvard since 1979, would use Page Keeton et al.'s *Cases and Materials on Tort and Accident Law,* which at 1,360 pages was the bulkiest of the four.[2]

After the long Labor Day weekend, 1L classes commenced on Tuesday morning, September 6. For Section III, their first class session was Contracts with Ian Macneil, whom the members of Barack's O Group had already met for a brown-bag lunch in his role as their faculty mentor. Since Macneil was almost as new to Harvard as they were, the assignment had little practical value, but the Scottish Macneil—in private life he was the eminent forty-sixth chief of Clan Macneil—seemed like a nice man.

But in the classroom, Macneil's use of the traditional Socratic method brought out an utterly different personal demeanor. Most members of Section III would remember their first class at Harvard Law School for the rest of their lives. They had been informed five days earlier, in the semester's initial number of the weekly, mimeographed *HLS Adviser* newsletter, by both Macneil and David Shapiro, what to read in advance of Tuesday's classes. Macneil's reading included the preface to his casebook, in which he repeatedly used the Latin abbreviation "e.g." and also stated his somewhat iconoclastic definition of the course's core concept: "contract encompasses all human activities in which economic exchange is a significant factor—marriage as much as sales of goods." That highly inclusive definition distinguished his text from others. "Thus the range of activities treated goes beyond those traditionally associated with the doctrinally structured first year contracts course." Organized with an eye toward "functional patterns, rather than doctrinally," the book "focuses on the negotiational and remedial processes that continuing exchange relationships generate."

Ian Macneil began Tuesday's 9:00 A.M. class by telling Section III's 140 students that no matter where they had gone to college, no undergraduate education had taught them to read as carefully as they would have to in order to succeed at Harvard Law School. "For example, I bet you don't know what the difference between i.e. and e.g. is," he remarked while glancing at his seating chart and randomly calling upon Alicia Rubin, a 1987 graduate of Harvard College.

Rubin knew enough Latin to nail the answer cold: "i.e. stands for id est, that is, and e.g. is exempli gratia, for example," she replied. "That really ticked him off," she recalled, because Macneil had not anticipated an immediate correct response. "Well, Miss Rubin," he said, "let's see if you've done the reading for today. Can you define a contract?" Rubin "had not even bought my books yet," and "so I fumbled around" trying to answer. Seated next to her was Michelle Jacobs, a 1988 Harvard College graduate who quickly realized that John Houseman's fictional contracts professor in the 1973 film *The Paper Chase* was, at least in Ian Macneil's classroom, no Hollywood exaggeration. "Oh my god! It really does happen!" Jacobs remembered thinking.

Jacobs knew that Macneil "had this bizarre definition of contract that was in our reading," and "I had written it down, and I kept trying to give it to her," but Rubin was focused only on Macneil. "Then he said something to the effect of 'Don't embarrass yourself further. Please leave the room, do the reading you were supposed to do, and then come back and see if you can take a better stab at it.'" David Attisani, a 1987 graduate of Williams College, watched the exchange and thought it "idiotic." So did Eric Collins, the African American Princeton graduate who was in Obama's O Group. He recalls thinking that "if this is how law school is, I've made a mistake." Rubin said she rose, "borrowed somebody else's textbook," and headed for the door. Richard Cloobeck also remembered the scene. "It was a very, very powerful moment, and he was such an asshole."

In the hallway, Rubin read through the preface of Macneil's casebook. "It seemed like an eternity, but it probably was only ten or fifteen minutes." She went "back into the room," and "then he calls on me again." Rubin tried "to stumble through some kind of response," but Macneil "picked me apart again." She felt "absolutely mortified and miserable," but finally the class ended. "It was just a horrible first day of class," Rubin recalled, but a few minutes later, in a nearby hallway, an older student she did not know came up to her. "He put his hand on my shoulder and was like 'You know, you did a really nice job.'" He introduced himself as Barack Obama. "He was giving me this little pep talk," and Ali Rubin was hugely appreciative. Barack's comments were "really nice."

David Shapiro's first session of Civ Pro featured a discussion of *Goldberg v. Kelly,* a 1970 U.S. Supreme Court decision protecting the due process rights of welfare recipients. Richard Parker's Crim Law and David Rosenberg's Torts also started the semester without incident, but each morning Section III's Contracts class met with Macneil, the room was riven with tension as the students waited to see whom he would call upon next. "If someone's terrified, they're not learning very well," recounts Amy Christian, a magna cum laude 1988 graduate of Georgetown University who later became a law professor herself. She "couldn't stand" Macneil's classroom style, and most of her classmates soon felt likewise.

Making things worse, Macneil demanded that students initial an attendance roster each morning to demonstrate who was present. American Bar Association accreditation standards for law schools mandated that students actually *attend* most of their classes, but in 1988 the majority of law professors, at Harvard and elsewhere, paid little attention to this requirement. Macneil's policy, followed by angry warnings when it was discovered that some people were signing in for absent classmates, roiled Section III even further. "People felt it was juvenile," Lisa Hay, a 1985 summa cum laude Yale graduate, recalled. The practice was "really juvenile for a professional school," agreed Martin Siegel, a 1988 highest honors graduate of the University of Texas.

"Relations between Macneil and the class broke down very severely very quickly," Mark Kozlowski explained. In class Macneil came across as "a very angry person," remembered Morris Ratner, who had just graduated with distinction from Stanford. The "toxic atmosphere" made the course "a disaster," the exact same word that David Troutt, a biracial African American 1986 graduate of Stanford, used when recalling the day Macneil "cursed at me." Aside from Macneil's attendance policy, there was also the burden of "how demanding he was in terms of the volume of material that you had to cover and also because of how ruthless and unforgiving he was of people who weren't prepared," explained Paolo Di Rosa, a 1987 magna cum laude graduate of Harvard College. "He really humiliated people who weren't prepared." Michelle Jacobs came to believe that Macneil "liked humiliating people." Jennifer Radding remembered the day Macneil called on her about secured transactions. "I'll never forget it—obviously I'm still traumatized all these years later," she joked equivocally. Greg Sater, like Ratner a 1988 Stanford graduate with distinction, felt Macneil was "really verbally abusive to people," and Sarah Leah Whitson found him "so angry and critical and disparaging of students." Di Rosa used the same adjective as Sater—"abusive"—to characterize Macneil, who was "just a monster," and Di Rosa agreed with Amy Christian that the result was "more fear than respect."

Rob Fisher was the only student who had spent six years at the front of a college classroom, but he fully shared his classmates' view that Macneil was "a *horrible* professor" and Contracts "was the most *painful* course." Fisher and Mark Kozlowski, also older, both quickly grasped that Macneil's idiosyncratic approach was profoundly simplistic. "He had one idea," Fisher recalled, "that contracts are relationships." Kozlowski was even more dismissive, viewing Macneil's focus as "a pointless effort to apply absolutely standard social science rhetoric to contracts." Even worse, "he spent a lot of time on his theory and not a great deal of time on learning contracts," which quickly led many in Section III to believe they were being shortchanged relative to friends in other sections. "There was a sense that Section III was not learning contracts at the same level as the other students and learning the doctrine," recalled Tim Driscoll, a 1988 summa cum laude graduate of Hofstra University. "People just really thought that they were not learning contracts law, and they were getting cheated," agreed Gina Torielli, an older student who had graduated from Michigan State with high honors.

Jackie Fuchs, a 1988 summa cum laude graduate of UCLA who before college played bass guitar alongside Joan Jett in the Runaways, agreed that Macneil was "very hard to get along with," but she felt he simply "did not know how to relate to people in their twenties." Kozlowski was less charitable. Not only was Macneil "extremely antagonistic in class," he repeatedly "would simply insult people to their face," and not just young women like Ali Rubin. Mark remembered a conservative white male politely saying he did not understand a question, yet Macneil responded, "It's a point that would be obvious to any intelligent person. It's not obvious to you." Shannon Schmoyer, who had graduated summa cum laude and first in her 1987 class at the University of Oklahoma, summed up Section III's view of Ian Macneil: "he was an embarrassment to Harvard Law School."[3]

Not all of Barack's time and energy during the fall semester's first weeks were devoted to his casebooks and class time. The law school's Hemenway gym was indisputably a dump—a word that a dozen alumni used for the facility—but its convenient location made finding pickup basketball games easy, as long as yoga classes or volleyball matches were not taking place. Only a few hours after reassuring Ali Rubin how well she had coped with Macneil's onslaught, Barack late that afternoon went to Hemenway and found a group that included twenty-two-year-old African American 1L Frank Harper, a magna cum laude graduate of Brown University, and four African American 2Ls: Kevin Little, Leon Bechet, Frank Cooper, and Kenny Smith. African American 1L David Hill, a 1986 high honors graduate of Wesleyan University, soon joined the regular late-afternoon mix.

When a student intramurals league got started shortly into the semester, a half-dozen of the black Hemenway regulars, including Barack, came together briefly on a team they called BOAM: Brothers on a Mission. "We weren't very good," remembered Kevin Little, perhaps HLS's top basketball devotee. All during that 1L year, Barack spent a good deal of time on the basketball court, but "he didn't want to hang out afterwards," Leon Bechet recalled. "It seemed like he always had a schedule . . . like he had a mission and he had a purpose."

With the 1991, 1990, and 1989 classes together, Harvard had more than 150 black law students on campus that fall, and the BLSA chapter seemed to have as much energy as their 1988 graduating class. BLSA scheduled its own orientation event for 1Ls soon after their arrival on campus. David Hill remembered how "Barack is holding court" and other new students "thought he was not a 1L" as they listened to him. "People were very surprised that this guy was a 1L simply because of how he was handling himself." An evening or two earlier, standing in the checkout line at the Star Market in Porter Square, just up from the law school, Barack had introduced himself to an African American woman he recognized from the first day's financial aid orientation session. "I think we're both at Harvard," he said to Cassandra "Sandy" Butts, a 1987 graduate of the University of North Carolina at Chapel Hill. Butts was in Section I, but "we certainly bonded over some of our challenges in the financial aid office," and she, like Rob Fisher, became a close friend of Barack's in their first days at Harvard.

Butts was with Obama and David Hill at that BLSA gathering when 3L Sheryll Cashin approached. Cashin too recalled other students listening as Barack spoke. "Within a couple of minutes, all of the people around were just hanging on his every word." Barack was "dressed kind of shabbily," and "I remember him talking about community organizing." Also listening was one of the youngest 1Ls, Christine Lee, a 1988 Oberlin College graduate who was still a few days shy of her twenty-first birthday. Lee had spent her early childhood in Paris, but when she was ten, her white mother left her alcoholic African American father and took her to live in Boerum Hill, Brooklyn, where she grew up as the black daughter of a white single mother. From Christine's youthful perspective, Barack sounded "kind of full of himself," because "he acted like he'd come from a whole life of work in the trenches rather than just a little stint after college." Barack "seemed pretentious," and "I immediately did not like him."

Barack remembered listening to and first meeting Derrick Bell at that BLSA orientation, but he was not immediately drawn to any of the five African American men on the law faculty. Even though Charles Ogletree was an untenured visiting professor, the 1978 Harvard Law graduate was an enthusiastic cheerleader for the school's black law students, and by mid-September, Ogletree had begun convening the first of four fall semester "Saturday School" sessions during which he hoped to reassure black 1Ls that they should not allow self-doubts to prevent them from succeeding at Harvard Law School. Derrick Bell and David Wilkins joined Ogletree, as did Charles Nesson, a white 1963 summa cum laude Harvard Law graduate who had worked in the U.S. Justice Department's Civil Rights Division before joining the Harvard faculty. Ogletree was at pains to stress that the Saturday-morning sessions were "not a remedial course," and BLSA leaders, like Ogletree, used the gatherings to encourage black 1Ls to become active in and seek out leadership roles in a variety of student organizations. About forty students attended the initial meeting, and Obama's black classmates remember that he rarely if ever attended "Saturday School."[4]

By mid-September, something else occurred in Barack's life that he did not mention to any of his new law school friends. Sheila Jager was staying with him at his Somerville apartment. Her Fulbright fellowship was for her to spend twelve months doing her dissertation fieldwork at Seoul National University in South Korea. After Barack left for Harvard, Sheila had intended to leave Chicago in mid-September and go to Seoul. But Barack wrote to her soon after his arrival in Cambridge, and after some phone calls, Sheila agreed to fly to Boston before going to South Korea.

"Why did I end up with him for a month in his Somerville apartment before leaving for Korea?" Sheila asked herself after recounting how she had refused to accompany Barack to Harvard. "I never said that I wouldn't visit him," she remembered, and she also recalled how her parents, who were then in Japan, were "really angry with me" when they learned she was staying with Barack and would not arrive in Seoul until after they had left the Far East.

Barack and Sheila's weeks together in the private basement apartment became a replay of their final months in Chicago before Barack's trip to Europe and Kenya. "I felt smothered by Barack, by his neediness to be the center of my world, by his sheer overpowering presence, and by the isolation I felt because we were always alone," Sheila recalled. Barack recounted what he most liked about his new life as a law student. "I do remember Barack telling me about Rob, although we never met. He described him, as I recall, as 'a bear of a man' and talked about him with great warmth, as a kind of older brother. . . . He said the two of them were the smartest people in their class at Harvard and ran rings around everybody else." Many of Rob and Barack's Section III colleagues were coming to share that characterization as the fall semester progressed, but after a month in Barack's apartment, Sheila was indeed ready to leave for Korea.

"He was like a huge flame that sucked up all the oxygen, and toward the end of our relationship, I felt breathless, exhausted really. I remember getting off the plane in Seoul and feeling like I could breathe again." Deep down, Sheila's feelings for Barack remained the same as in Chicago. "I knew at that point that the relationship could not work in the long term even though I loved him very deeply. He, of course, realized this as well. When I left for Korea, I felt as if I had abandoned him, although the split was completely mutual."

But even when Barack drove Sheila to Logan Airport, their relationship had not seen its true end.[5]

As September turned to October for the 1L students in Section III, their distaste for Ian Macneil gave way to gallows humor as they realized they could do nothing to alter their fate. The section took on the mordant nickname of "the Gulag." Lisa Hay, who had worked for Democratic presidential nominee Michael Dukakis's campaign before beginning her 1L year, began anonymously producing a weekly mimeographed newsletter chronicling life in Section III entitled *Three Speech*. It recorded odd or humorous statements from their reading and from classroom comments. Macneil, of course, was a prime target. One headline was: "You Make the Call: Fact? or Fiction?," followed by a quotation from Macneil's *Contracts* casebook: "Habit, custom and education develop a sense of obligation to preserve one's credit rating not altogether different from that Victorian maidens felt respecting their virginity."

Hay quoted Macneil's maladroit attempts at classroom humor that often backfired. For example: "The sexual relation is a good one for you to think about—in the context of this course, I mean." Or when Macneil asked, "Give me an example of a voluntary exchange," Richard Cloobeck, one of Section III's best provocateurs, volunteered, "Having sex with your girlfriend," to which Macneil replied, "Give me an easier one." The next week, when the concept of inadequacy arose, Macneil bluntly asked one student, "Does your girlfriend ever say you are inadequate?" and a few days later, he insultingly told another, "Maybe you don't have the skills to be a law student."

When Hay crafted a top-ten list of reasons for not doing Macneil's assigned readings, entries included "slept through class," "was laughing too hard at footnotes to read the text," and a joke about Macneil's almost daily references to the *Encyclopedia of Philosophy.* Most pointed of all was "Proportionality: Contracts homework out of proportion with length and value of class." In October, Hay reported a joking exchange between two classmates: "'If I learned I only had a week to live, I'd want to spend all my time with Macneil.' 'Why's that?' 'Because he makes every minute seem like a year!'"

With the November presidential election, pitting Massachusetts governor Dukakis against the Republican nominee, Vice President George Bush, drawing near, Barack registered to vote on October 4. Starting the next day, he also began racking up a remarkable run of thirteen parking tickets over the next four weeks on his rusty yellow Toyota, with most of them happening on Mass Avenue just west of the law school. On campus, student politics were roiled when the conservative Federalist Society chapter objected to the student government's reserving two committee seats for the Coalition for Diversity. Jesse Jackson addressed a pro-diversity rally of twelve hundred people, and Derrick Bell renewed his calls for greater minority representation on the law faculty.

David Shapiro's Civ Pro course was Section III's most vanilla class, but most students thought Shapiro was an excellent, witty teacher. Classmates and Shapiro recall Obama speaking only irregularly in Civ Pro, but they do remember him as one of the most talkative voices in Richard Parker's Crim Law classroom. Parker had "a class discussion" style that stood in contrast to Macneil's classic Socratic method, recalls Kenneth Mack, one of Section III's other African American men. Mack thought Barack was a standout presence from the first week of class. "He was a striking figure," Mack said. "He spoke very well, and very eloquently," and seemed "older and wiser than the three years that separated our birth dates. . . . It seemed like I was twenty-four, and Barack was thirty-four," like Rob Fisher. "He and Rob seemed like they were the same age." Shannon Schmoyer agreed: Barack "just seemed so much more mature" than most other 1Ls.

Parker was an entertaining teacher, and having Peter Larrowe, a former police officer, in the class sometimes added a bracing dose of reality to discussions. Several students recalled Parker saying, "If you're going to a protest, always take your toothbrush," and Lisa Hay's *Three Speech* recorded one exchange after Larrowe mentioned his background to Parker: "'Does everyone know that you were a police officer?' 'They do now. . . . No more undercover work,'" Larrowe joked. Fisher also vividly remembered Parker querying students about their own experiences with police officers and one attractive woman describing how she was patted down outside a bar. "How did that make you feel?" Parker asked. "Oh, it was kind of nice," she replied as the entire class erupted in laughter.

Sarah Leah Whitson remembered Obama once making some real-world rather than doctrinal point in a colloquy with Parker, and Richard Cloobeck recalled a similar scene where Barack "spoke from the black perspective in Crim class because something had happened to him where he'd experienced racial discrimination in profiling, and it was very personal." Sherry Colb, valedictorian of Columbia's 1988 graduating class and Section III's most loquacious female voice, remembered that when the subject of acquaintance rape came up in class, Barack expressed displeasure with what others had said. "'I don't even understand why we're debating this. Why is silence enough? Why aren't people looking for "yes"?'" Sherry recalled Barack asking. "I think the women in the class really appreciated that because there were other males in the class who took a more reactionary position."

Sima Sarrafan, an Iranian American honors graduate of Vassar College, realized that Obama "had a more pragmatic view of the law" than most classmates or professors. Roger Boord, a 1988 magna cum laude graduate of the University of Virginia, remembered Barack as "this voice of authority . . . his voice was like Walter Cronkite." But especially in Parker's class, Barack, David Troutt, and Sherry Colb spoke up so regularly that it generated irritation and derision from some fellow students. Most Section III students spoke only when asked to. "The last thing I ever wanted would be to be called upon," Greg Sater explained. "Most of my friends were the same way, and we would never in a million years ever raise our hand."

In stark contrast, Section III's most self-confident voices, or "gunners," in longtime law student parlance, raised their hands almost every day in participatory classes like Richard Parker's. Dozens of Obama's classmates remember him consistently waiting until a discussion's latter part before he chimed in, with comments that he thought synthesized what others had said. "He never really took a very strong, argumentative position," Ali Rubin recalled. Dozens laughingly recalled his insistent usage of the word "folks" as well as his regular introductory refrain of "It's my sense" or "My sense is," phrases that DCP members remembered hearing regularly during his time in Chicago. Barack "particularly loved to engage with Professor Parker," Haverford College graduate Lisa Paget recalled. She has a "vivid" memory of Barack remarking, "Professor Parker, I think what the folks here are trying to say is" so as "to synthesize what other people were saying." Barack "clearly liked to speak," but "sometimes people got frustrated because they didn't feel like they needed him" to speak for them. "'Say what your own thinking is, don't tell him what we're thinking!'"

Decades later, particularly for classmates who had become jurists or prominent attorneys, recollections of just how intensely irritating Obama's classroom performance had been were burnished with good humor. But even though he was always prepared, always articulate, and always on target, many fellow students tired of Obama's need to orate. Barack "spoke in complete paragraphs," Jennifer Radding recalled, but "he often got hissed by us because sometimes we would all make comments" and then "he would raise his hand and say 'I think what my colleagues are trying to say if I might sum up,' and we'd be like 'We can speak for ourselves—shut the fuck up!'" Radding thought Obama was "a formal person, reserved" but "always friendly." Yet "I'm not sure he related to women as well on a colleague basis" as he did with older male friends like Rob, Mark Kozlowski, and Dan Rabinovitz, a former community organizer interested in politics. Jennifer remembered Barack asking Rob and Dan substantive questions, and "then he'd ask me did I party over the weekend." One day "I called him on it, and I just said, 'You've got to be kidding me. You're someone who's so liberal, and so women's rights, and you talk to women like they're not on the same level.' It horrified him to hear that," and "I wasn't the only one who felt this way."

The joking about Obama's classroom performance intensified as the fall semester progressed. One classmate, Jerry Sorkin, christened him "The Great Obama" because "he had kind of a superior attitude," Sorkin's friend David Attisani remembered. "Barack would start a lot of his speeches with the words 'My sense is,' and Jerry would walk around kind of stroking his chin saying 'My sense is.'" Gina Torielli recalled that when Obama or especially Sherry Colb raised their hands to speak, more than a few of the younger men would "take out their watches to start timing how long" they talked. In time it became a competitive game, one played at many law schools over multiple generations, and often called "turkey bingo," in which irritated classmates wager a few dollars on how long different gunners would exchange comments with the professor. Section III named its contest "The Obamanometer," Greg Sater recalled, for it measured "how long he could talk." But Sater explained how there "was a great feeling of relief to all of us whenever he would raise his hand because that would take time off the clock and would lower the chances of us being called upon."

No one questioned the value of what Barack, Sherry, or David Troutt had to say, but much of Section III got tired of hearing the same voices day after day. With Barack, Greg said, "we were envious of him in many ways because of his intellect," self-confidence, and poise, but that did not stop the Obamanometer. "We'd kind of look at each other and tap our watch," he recalled. "You might raise five fingers," predicting that long a disquisition, "and then your buddy might raise seven." Jackie Fuchs remembered the label a little differently, explaining that students would "judge how pretentious someone's remarks are in class by how high they rank on the Obamanometer."[6]

Criminal Law was the course in which Barack "pontificated"—as Jackie called it—the most, but the class and professor that Barack and Rob found the most intellectually stimulating was Torts, with David Rosenberg. With only one-third of Section III's students in that subsection, it was more intimate than Contracts, Civ Pro, or Crim, and the forty-five students responded enthusiastically to Rosenberg's high-energy, in-your-face style and tough-minded libertarian economics. Everyone remembered Rosenberg's invitation to a 6:30 A.M. law library tour, when he would discuss the practicalities of thorough legal research. Amy Christian, Richard Cloobeck, Diana Derycz, and Barack were among the half dozen or so students who showed up, and *Three Speech* recounted Rosenberg insisting during it, "I am not a masochist."

David Attisani "loved" Rosenberg's class. "He was my most useful professor by a wide margin," and Rosenberg also held a volleyball clinic and took a large group of students out to dinner. Richard Cloobeck agreed that Rosenberg was "the most practical and realistic and effective" of Section III's professors. He also was often the most entertaining. Mark Kozlowski remembered Rosenberg asserting that Harvard-trained lawyers taking legal-services jobs was the equivalent of MIT engineering graduates becoming appliance repairmen. *Three Speech* captured an exchange in which one student asked Rosenberg if a passage in one case opinion was dicta, or beside the point; Rosenberg responded, "No! That's crap!"

Kozlowski and Cloobeck remembered how captivated Fisher and Obama were with Rosenberg, and Ali Rubin thought that "from the beginning Rosenberg treated Rob and Barack differently" than other students. Cloobeck believed Obama spoke just as much in Torts as he did in Crim, demonstrating to all how "incredibly, intensely smart and thoughtful" he was, even relative to classmates who had graduated at the top of their classes from the nation's best colleges and universities. Obama was "intellectually curious" and "sincere in his academic passion," Cloobeck thought, but he also seemed "extremely arrogant, very conceited." Yet he admitted that Barack and Rob spoke "at a level that was just beyond my comprehension."

David Rosenberg remembered Barack as "one of the most serious students I've ever encountered." Rosenberg's approach to Torts involved "applying social and natural science to social problems" in a heavily economic, functionalist manner. He recalls that "Obama and Fisher were determined to figure out what was going on, absolutely determined." The two came by his office "almost twice or three times a week, not to talk about the course in ways that would translate into a better grade, but to talk about the actual problems that I was raising and the approach." Rob and Barack were always together—"you couldn't separate them," Rosenberg explains—but "they weren't a duo in their mindsets," and "it didn't seem like one was dependent on the other. They were quite independent," and always raising "social policy questions" when they came by. Rosenberg remembered that in class, Obama "asked good questions, he fought through hard problems. I thought he was going to be a top academic."

Rob remembered that "Barack was very active" in Torts and "loved that class." Rosenberg "met argument with argument, and valued creativity," and "Barack and I just had a great time in that class. We were constantly arguing and talking and enjoying it and going to visit Rosenberg," and Torts overall was "an absolute blast." Rosenberg "had this intense intellectual passion" and a "creative style of lawyering that greatly appealed to us both." Rob believes that "Rosenberg had a vastly bigger influence on Barack's and my thinking about law" than their other 1L instructors combined.

Four years earlier, Rosenberg had published a major article in the *Harvard Law Review* calling for a restructuring of tort litigation in mass exposure cases like asbestos through the courts' use of "aggregative procedural and remedial techniques," an argument he revisited in a shorter 1986 *HLR* essay. In fall 1988, Rosenberg was writing an additional commentary, calling for "collective processing" and comprehensive settlements in mass tort cases rather than "legislative insurance schemes." He was already so impressed by Rob and Barack that he sought their input on his draft manuscript. When it was published in May 1989, the first page included thanks to Rob and Barack for their "substantive criticism and editorial advice," a remarkable acknowledgment for a Harvard law professor to bestow upon two 1L students.[7]

Amy Christian remembered a day in Torts when "one of Barack's two front teeth was broken diagonally so that like half the tooth was gone," a casualty from the previous afternoon's basketball game, he told her. "A couple of classes later he came in, and his tooth looked totally normal." Kenny Smith, a 2L, remembered Barack as "a good player," but two recollections of Obama from among the wider population of classmates who went up against him either in intramural matchups or the almost daily pickup games were his penchant for "trash talk" insults to opposing players and his tendency to call "baby" fouls in self-refereed games. Barack "was cocky as a basketball player, he was not as a regular person," 1L Brad Wiegmann thought. Martin Siegel recalled "being in a game with him where he called a foul" and "he just headed to the other end of the court." Oftentimes, "being a law school, if people called fouls there was a tendency to have that devolve into a crazy argument." Not so with Obama. In "a sign of his status," everyone "followed him without objection. It was as if he had said it, and therefore it was so."

Greg Sater had formed a volleyball league, and their court time was just after basketball, but the basketball players would refuse to give up the floor. "They would always give me shit," Sater recalled, and then Barack would "sweet-talk me into giving them an extra few minutes, always." He was "very disarming" and "very good at defusing situations and being a peacemaker." Rob Fisher had similar memories from pickup basketball games, but they spent far more time studying than on the basketball court.

The press of coursework led Barack to give up on the irregular journal he had kept ever since his first year in New York, and he and Rob often studied at his Somerville apartment rather than in Harvard's law library. Rob remembers that Barack was very proud of "three Filipino gun cases," big long wooden boxes he had bought to serve as bookshelves, but otherwise the "apartment was very sparse," with "a little TV," some "spare furniture," and overall an "ascetic" feeling. "He and I would sit around his apartment," Rob recalled, "and just bang ideas back and forth for hours."

Rob appreciated that Barack was "very much a synthetic thinker," just as other classmates had sensed from his many summary expositions, and Rob realized as well that virtually everyone around them "recognized from the start" that Barack was "exceptional." Rob understood that Barack's time in Chicago had been "an extraordinary experience for him," and although Rob has no memory of any such conversation, one of his closest friends vividly remembers a phone call early in Rob's time at Harvard in which Rob said he had just met the first African American president of the United States.

"Barack and I both looked at law school as an intellectual playground, a place to develop ideas, to have fun with ideas," Rob explained. "There is no question, when he was in law school, his path in his mind was to be a politician. That is where he was going, and that was crystal clear from day one of law school," along with definitely returning to Chicago upon graduation. During their conversations, Rob learned that "Barack recognized he had exceptional talents, and that that was a gift from God, that was something special." Barack "felt a great moral obligation to use" his "special gifts" to "help people . . . and that was very palpable, very real, and very deep." Just as Sheila Jager had heard Barack speak about his destiny, Rob too understood that Barack had a sense of himself that was rooted in his experiences in Roseland. "The way he described it to me," Rob recalled, during "long deep conversations . . . he pictured an elderly African American woman sitting on a porch and just saying to him, 'Barack, you've got special gifts. You need to use them for people.' That was a very deep-seated belief. There's no question about that," even in the fall of 1988. Rob also knew that Barack's identification as black "was a choice," and that "making choices about identity did limit personal choices, and that pained him."

Barack "spent so much time together" with Cassandra "Sandy" Butts that a mutual friend explained how "everybody thought they were dating, even though I don't think that was ever the case." As Jackie Fuchs forthrightly put it, voicing a perception widely shared among female classmates, Barack "gave off zero sexuality. . . . He came off as completely asexual." His relationship with Sandy, who was four years younger and whose parents had divorced before she was a year old, had an older brother–younger sister closeness.

"When I first met Barack, I thought he was this black guy from the Midwest, and he did not volunteer his background, other than coming from Chicago," Cassandra explained. "I didn't know that Barack's mother was white until a couple of weeks into knowing him. It wasn't something that he volunteered." Sandy was interested in Africa and had visited Zimbabwe and Botswana as an undergraduate, and only after Barack mentioned his trip to Kenya did his family story emerge: "My mother is white." Like Rob, Sandy also soon realized that Barack's years in Chicago had been "the most formative" of his life, and "he talked about how powerful the position of mayor of Chicago was," just as he had told Mike Kruglik and Bruce Orenstein months earlier. Barack "certainly saw in Harold Washington a model," one who was "incredibly influential," and "his ambition was to eventually run for mayor." In Cassandra's memory there was no question Barack "wanted to be mayor of Chicago, and that was all he talked about as far as holding office. . . . He only talked about being mayor, because he felt that is really where you have an impact. That's where you could really make a difference in the lives of those people he had spent those years organizing." She remembered too that "Barack used to say that one of his favorite sayings of the civil rights movement was 'If you cannot bear the cross, you can't wear the crown,'" a rough approximation of Martin Luther King's January 17, 1963, statement—"The cross we bear precedes the crown we wear"—that was featured in the epigraph of the King biography that had been published midway through Barack's time in Chicago.

Cassandra believed Barack's intellectual seriousness made him "a bit of a geek in law school," but his Chicago experiences gave him a real-world grounding most of his younger classmates lacked. She remembered a 1L discussion among black students about whether the preferred label should be "black" or

"African American." "Barack listens to all of this, and near the end . . . he basically makes the point that it was kind of this false choice, that whether we're called 'African Americans' or 'black,' it really doesn't matter, what matters is what we're doing to help the people who are in communities that don't have the luxury of having this debate." Cassandra thought of Barack as "African and American," with "a sense of direction and focus" that distinguished him. Compared to his younger classmates, Barack seemed "incredibly mature," with a "very calm" demeanor, and Cassandra thought "Barack was as fully formed as a person could be at that point in his life."

An informal study group emerged that included David Troutt, Gina Torielli, and Barack. They met sometimes at the BLSA office or at Torielli's apartment. On one occasion, when Barack was outside smoking, Gina worried that "my landlord was going to call the police" after seeing a black man on the front porch. On another, when everyone was saying what their dream job was, Barack "said he wanted to be governor of Illinois." Still, most of Barack's study time was spent with Rob, and when the law school's 1989 *Yearbook* appeared some months later, an early page featured an uncaptioned photo of Barack sprawled on a couch listening as a smiling Rob spoke while holding a loose-leaf notebook. "When I think of those two that year," Gina Torielli explained, that picture "is how I remember often finding them." The photo captured the ring on Barack's left index finger that he had worn ever since his first year at Oxy. Rob remembered precisely when the picture was taken. "I was explaining macroeconomics to Barack, and . . . we also had an extended discussion about the national debt in that session and how and whether it mattered." It was no accident that Cassandra thought Barack something of a geek, or that virtually everyone in Section III looked up to Rob and Barack as the smartest minds in their midst.

Barack was more forthcoming with Rob than with anyone else at Harvard, indeed with anyone other than Sheila and Lena. Even though Barack "never mentioned mayoral politics" and never said anything "that would suggest to me that he had his sights on mayoral politics," it was clear "he was just extremely politically ambitious" and "wanted to go as far as he could. There was no doubt in my mind he was thinking presidency" and "he shared that with me at the time."

Obama's classmates could see that too. We "knew he'd be in politics. That was obvious right from the start," Sherry Colb explained, and Lisa Paget agreed that Barack "was clearly going to be a politician." David Attisani thought that Harvard's classrooms were "something of a rehearsal for him for public life," that "he was getting himself ready" and seemed to be "self-consciously grooming himself for . . . some kind of public life." In retrospect, Jackie Fuchs thought Barack "had already decided that he was a future president," and wondered if his self-transformation mirrored that of her past bandmate, Joan Jett. "I'm sure

Barack as a child was perfectly ordinary, just like Joan was. Until the moment he decided that he was a star." Fuchs was not enamored of the 1988 Barack—"in law school the only thing I would have voted for Obama to do would have been to shut up"—but among the 1Ls who socialized together, David Troutt believed there was "none more careful, more guarded about his personal life than Barack." David Attisani likewise viewed Barack as "a very private guy" who was "quite cautious about where he appeared socially."

One Thursday evening, several young members of Section III, including Scott Sherman, a 1988 highest honors graduate of the University of Texas at Austin, decided to head down to Harvard Square to drink. Seeing Obama studying nearby, Sherman invited him to join the group, but Barack demurred. "No, you young guys go on down and have fun. I have work to do here," and Barack's comment left Sherman "feeling like a sophomoric frat boy. He was serious" and "he was not wasting his time at the law school," Sherman recalled. "He was there for a reason," and there was no gainsaying the five-year age gap between Barack and most of his fellow students.

Years later, classmates pictured how Barack appeared back then. "He always wore the same ugly leather jacket," Jennifer Radding said, and everyone remembered seeing him smoking outside Harkness Commons—"the Hark"—or, during the winter, in a basement smoking room that was one of the few authorized locations after a Cambridge antismoking ordinance had taken effect eighteen months earlier. "He really did smoke a lot," Mark Kozlowski recalled, and sometimes in the basement, Barack talked with Kenyan LL.M. student Maina Kiai, who also had arrived that fall. Kiai remembered him "always asking questions about Kenya and Africa," and "we talked a great deal about poverty in the USA." Diana Derycz also recalled Barack as a "big smoker" who was "outside smoking" before classes even in winter. Sarah Leah Whitson believed that by 1988 at Harvard Law smoking was a symbolically transgressive act that set one apart from the student mainstream. Rob Fisher thought that Barack "enjoyed it," and Lisa Paget remembered that when Barack was "walking on campus in his leather jacket with his typical cigarette in his hand, he had swagger."

Barack was also among several dozen 1Ls who signed up to do scut work—"sub-citing," as in substantive citation checking—for the *Harvard Civil Rights–Civil Liberties Law Review,* one of the law school's student-run journals that welcomed 1L participation. At the end of October, Barack's sister Maya, who was a freshman at Barnard College in Manhattan, came up for a weekend, including a Halloween-evening dinner party at Barack's apartment for which ten or so people were encouraged to come in costume. Rob and Barack's 1L friend Dan Rabinovitz and his buddy Thom Thacker came wearing whale outfits inspired by the freeing that week of two creatures that had been trapped in the ice off Point Barrow, Alaska.

"I remember from early on thinking that Barack was the single most impressive individual I'd ever met," Rabinovitz recalled, and that his Somerville apartment "seemed incredibly hip." Barack had "*made* it a great place to live," and his taste for Miles Davis and similar musicians was evident from "the wonderful jazz playing in the background." Dan had also been an organizer, and he agreed with Barack "that to really deal with the fundamental issues in people's lives, you needed to engage the political system." Dan too remembers how Barack "made clear absolutely it was his intention to go back to Chicago and to be involved politically." Thom Thacker vividly recalls that Halloween evening, perhaps because Barack "possessed a magnetic charm" or more likely because Maya was now "drop-dead gorgeous." But either that evening or a few days later, Thom would recount how "I remember Dan remarking to me that he thought Barack would be president of the United States one day."

On November 8, 1988, George Bush handily defeated Michael Dukakis to keep the presidency in Republican hands. "Dukakis's loss was a major loss, and we were feeling it," David Troutt recalled, but Mark Kozlowski remembers David Rosenberg beginning Wednesday's Torts class with a caustic quip about the Supreme Court aspirations of one of his favorite colleagues: "Larry Tribe can unpack!" Rosenberg's respect and affection for Rob and Barack was now expressing itself in a different way, because when their favorite NBA team, the Chicago Bulls—"We were both big Michael Jordan fans," Rob says—was in Boston that night to play the Celtics, Rosenberg gave them "like second row" tickets so they could watch an "awesome" 110–104 Bulls victory in which Jordan scored fifty-two points.[8]

By early November, Rob, Barack, and Mark Kozlowski had agreed to work together on the upcoming spring semester Ames Moot Court exercise, and they also enlisted Lisa Hertzer, another member of their small Legal Methods class and a 1988 Phi Beta Kappa graduate of Stanford. The Legal Methods course had proven more daunting for 3L instructor Scott Becker than for most of the 1Ls. "Barack was so far ahead of the curve intellectually," Becker remembered, that often sessions featured the "hyper-ambitious" Obama explaining, "I think this is what Scott means by that," although never in a way that embarrassed Becker.

Barack and Rob were also beginning to think about summer jobs for 1989, and Barack wanted to return to Chicago. Barack's old friend Beenu Mahmood, who had welcomed him there in 1985 while he was a summer associate at Sidley & Austin, was now in his third year as a lawyer in Sidley's New York office, and "I suggested that he seriously look at Sidley," Beenu remembers. Barack had kept in regular touch with Beenu, who believed by 1988 that "Barack was the most deliberate person I ever met in terms of constructing his own identity."

Sidley actively sought out top 1Ls, and Beenu recalls speaking with Sidley managing partner Thomas A. Cole about Barack. Well before Christmas, his résumé arrived at Sidley's Chicago office, where 1972 Harvard Law School graduate John G. Levi oversaw the firm's recruiting at his alma mater and 1976 Northwestern Law School graduate Geraldine Alexis headed up Sidley's minority associate recruiting effort.

On the Friday after Thanksgiving, Barack learned the surprising news in that day's *New York Times:* "Albert Raby, Civil Rights Leader in Chicago with King, Dies at 55." A memorial service at the University of Chicago's Rockefeller Chapel attracted more than a thousand mourners, and Teresa Sarmina, one of Raby's former spouses, spoke of how he "would get excited about a person and the potential he could see in that person."

In Cambridge, the last two weeks of fall semester classes featured a December 7 meeting with Ian Macneil that about 60 percent of Section III attended. "A wide range of complaints were voiced," Macneil recalled, and tensions were raised further because the final exam in the yearlong course would not take place until late May. Only in their final week of classes in mid-December did Section III learn that its other exams would take place on the afternoons of Monday January 9, Wednesday the 11th, and Friday the 13th. With Civ Pro teacher David Shapiro leaving for Washington, students' marks on the two-and-a-half-hour open-book midyear test would constitute half of their eventual grade on Harvard's somewhat odd A+, A, A-, B+ B, B-, C-, and D eight-point scale. Shapiro posed only two essay questions of equal weight, one involving securities fraud and the other concerning an interlocutory (interim) appeal involving hundreds of lawsuits stemming from a hotel fire. No one could have found it easy. Two days later, Richard Parker's three-and-a-half-hour open-book Crim Law exam posed three questions weighted at 50, 25, and 25 percent. The first posed a hilariously complicated fictional scenario involving a racially profiled terror suspect carrying cocaine who bumps into a knife-wielding drunk who then stabs a passerby. Students were instructed to "respond *specifically* to all" of six analytical issues. The shorter second and third questions were visibly easier, with the former offering as one of two options an essay on the Burger and Rehnquist Courts' rulings on searches and interrogations. The third requested a response to a quotation asserting that changes in substantive criminal law doctrines offered a better chance of combating racism than did procedural reforms. On Friday the 13th, Section III's exam week ended with a three-hour, open-book, two-question Torts exam from David Rosenberg. "Careful organization will be highly valued, as will conciseness and clarity of presentation," the exam advised. One question dealt with a gang fight prompted by a movie about gang warfare, the second probed a manufacturer's liability after a polio vaccination of a child infected the youngster's father.[9]

With his first three exams complete, Obama flew to Chicago for ten days before spring semester classes began. Staying with Jerry Kellman's family, Barack immediately learned that Al Raby's death was not the only sadness that had befallen his Chicago friends. When Barack left Chicago in mid-August, Mike Kruglik replaced Greg Galluzzo as DCP's consultant-adviser, and in Mike's first heart-to-heart conversation with John Owens, Johnnie confessed that he was losing a struggle with cocaine addiction. "I was shocked," Kruglik recalled, but he immediately arranged for Owens to enter a thirty-day residential treatment program on Chicago's North Side and drove him there to help him check in.

A month later, Johnnie was back at DCP, but the group's core members felt he had not been prepared to fully shoulder the weight of being executive director. "John was good, but he was not Barack," Betty Garrett explained. "There was no one in my eyesight that would have been able to fill Barack's shoes." Aletha Strong Gibson remembered that they all realized that Owens "wasn't quite ready" to assume such "a high-pressure position."

Jean Rudd and Ken Rolling at Woods felt similarly. "When Barack left, Johnnie was really very feeling kind of abandoned," Jean recalled. "He wasn't quite ready to be in charge yet" and "just wasn't confident at that point." Ken agreed that "John became director before his time." When Barack learned what had happened, he discussed the situation with Jean, Ken, and Jerry, who remembered that he was "deeply concerned about it" because "he's feeling responsible." Kellman also thought that if "DCP blows up in a scandal," it could hurt Barack's reputation. Owens recalled that Barack "suggested that I change the name of DCP" because he "figured if I didn't do a good job with it or something went wrong, it wouldn't come back to haunt him." Johnnie saw Barack as deeply strategic about his own future, but Barack refused to acknowledge that. "I was pressing him one time, and he got angry. 'No! I said no!'" But Barack's exceptionally rare outburst did not alter Owens's firm belief.

During Obama's first semester at Harvard, the Illinois legislature had finally approved a comprehensive Chicago school reform bill, which Chicago United's Patrick Keleher praised as "a fantastic bill," one that called for every Chicago public school to be governed by an eleven-member Local School Council composed of six parents, two area residents, two teachers, and the principal. UNO and Gamaliel had played a significant role in the victory, but the hard work of implementing it still lay ahead.

Barack had corresponded regularly with Mary Ellen Montes throughout the fall. "The letters came for a while," Lena remembered, but then there was "a determining letter that was sort of like we weren't going to write to each other anymore, and so we didn't." By January 1989, Lena was involved with someone else, and she and Barack were never again in touch. In Lena's and Sheila's absence, and with his friendship with Johnnie now seriously strained,

Barack's two strongest Chicago relationships were with Kellman and Kruglik, and during his ten days back in Chicago, he readily helped both of them with their ongoing organizing work.

Jerry was now fully occupied in Gary, Indiana, and thanks to both Woods and the Diocese of Gary, he had just publicly launched Lake Interfaith Families Together (LIFT), named after the county that encompassed much of northwest Indiana. In Chicago's south suburbs, Mike was rapidly growing the South Suburban Action Conference (SSAC), and in January 1989, he hired a new young African American organizer, Thomas Rush, a 1988 graduate of Haverford College.

Rush recalled that even during his first long conversation with Kruglik, Mike mentioned Barack, with the implication being that "this guy was special within organizing." Mike asked Barack to call Thomas, and the next morning they met for forty-five minutes over coffee. Knowing that Barack was at Harvard, Rush expected someone arrogant, but instead he thought Obama was calm and self-assured, with an "even temperament." Rush remembered that when Barack mentioned Jeremiah Wright, it was "almost like his mind left for a minute" as Barack looked away. When Thomas asked how attending Harvard Law School would connect to further organizing work, Barack said, "I don't know that I'll be directly involved, but this will always be a process that I support, whatever I do."

A few days later, during a LIFT training in Gary, which Kruglik and Rush attended, Obama, along with Kellman, led the day's sessions for a group of some forty people. At the end of the day, Rush rode back into Chicago with Barack. As Thomas recalled it, Barack mentioned "that he would like to find a good relationship," ideally "a woman with the body of Whitney Houston and the mind of Toni Morrison."

Before Barack returned to Cambridge, he had his summer job interview at Sidley & Austin's Chicago office. "We brought in all of the kids who had a Chicago connection," John Levi recalled, for twenty-minute conversations with two or more Sidley lawyers. African American applicants generally were seen by a trio of Levi, Geraldine Alexis, and Alexis's chief lieutenant, Michelle Robinson, the 1988 Harvard Law graduate who had played such an active role in BLSA before joining Sidley eight months earlier.

"Michelle I distinctly remember saying, 'I cannot see him. Will you please make sure you can see him?'" Levi remembered. He did, and "I was wowed" by Obama. "I thought he was phenomenal. He was one of the best interviews I've ever had, really." He recalls that Obama demonstrated "poise and sparkle" and clearly was "a compelling person." Geri Alexis had a similar reaction. "I remember very vividly" speaking with Barack. "He impressed me so much that I called down to the recruiting office, and I said, 'We really need to give this guy an offer before he leaves the building,' and they said, 'Well, we don't do that for 1Ls,' and I said, 'You're going to do it for this one,'" and they did.

John Levi concurred. "I called Michelle later in the day and said, 'Boy, did you miss a good one,'" and Robinson replied, "'That's what everybody is telling me.'" As Levi recalled, Barack "accepted quickly too."[10]

Spring classes began on January 25, and Barack, Rob Fisher, Mark Kozlowski, and Cassandra Butts all chose for their elective 18th and 19th Century American Legal History, taught by assistant professor William "Terry" Fisher, a 1982 Harvard Law graduate who had clerked for Justice Thurgood Marshall. The three-hour-a-week lecture class covered "the formative era of American law," with emphasis on the changes between the Revolution and the Civil War, especially regarding slavery. Famous cases like *Marbury v. Madison* and *McCulloch v. Maryland* were supplemented by doctrinal-specific readings addressing contracts, torts, property law, criminal law, and the status of women. The semester's final three weeks were devoted to slavery, with recent articles by leading scholars like Paul Finkelman and Robert Cottroll playing a central role.

Section III's Civ Pro also met for three hours a week, with Northeastern's Stephen Subrin replacing David Shapiro. Some students found Subrin likable, while others felt he was a letdown compared to Shapiro. Contracts with Ian Macneil continued for three hours a week while *Three Speech* reported an unsupported rumor that "the Scowling Scot" might remain at Harvard more than one year. Lisa Hay occasionally included a crossword puzzle, and one had the clue, "He pauses before speaking." The correct answer was "Obama."

Rob Fisher had stopped attending Contracts—"It was so horrible that I basically skipped all of it"—but Barack remained a regular if sometimes tardy presence. *Three Speech*'s list of "highlights" included "February 21: Obama knocks on contracts door, 9:21 A.M.," more than twenty minutes late. Ken Mack remembered that Barack was one of Macneil's "favorite students," and Macneil recalled that he had "such a commanding presence. . . . I was always a little too impatient in class, so if students went off the track, I would interrupt before I should. When I did that with Barack, he said 'Let me finish.' He wasn't rude, just firm.'"

Spring's most weighty course was five hours a week of Property Law, taught by Mary Ann Glendon, a 1961 graduate of the University of Chicago Law School who had joined the Harvard faculty in 1987. She assigned the class A. James Casner et al.'s 1,315-page *Cases and Text on Property,* 3rd ed. On the first day, Glendon called on Barack, mispronouncing his surname by making it rhyme with Alabama. Barack corrected her. Paolo Di Rosa thought that "was kind of a nervy thing to do," but Barack "had the confidence to do that without being rude about it."

Three Speech regularly captured how Glendon's excellent sense of humor made for a relaxed classroom atmosphere. "I assume that a lot of you are in some relation to Harvard," Glendon suggested. "What's that relation? What? No, I don't mean 'serfdom.'" On another occasion, a student asked, "How long do you have to, you know, ah, live together, for these common-law marriages?" and Glendon drily responded, "Why do you ask?" Glendon was a widely popular teacher, and Rob Fisher remembered her as "an excellent professor" whom he and Barack visited for some "very open-ended, interesting intellectual discussions." Fellow students recall both Barack and Rob as regular classroom participants. Ken Mack thought Glendon was "very interested in what Obama had to say in class" and "liked him a lot." Edward Felsenthal, a 1988 magna cum laude graduate of Princeton, would "vividly remember" Barack as someone who "talked all the time" in Property. There were some "heated battles between Barack and Mary Ann Glendon," Felsenthal recalled, because Obama "objected to some pretty core tenets of the common law of property." So "they went at it," and "nobody else sparred like he sparred with her." Rob remembered a time when Glendon asked why a court had set aside a condominium bylaw barring children as residents. Barack spoke up, saying, "Folks gotta live someplace," and "everyone laughs," but the crux of his response—the reasonableness standard—was indeed key. That "tells you a lot about how he was thinking about" legal questions while at Harvard, Rob explains. He and Barack "loved" Glendon's "great" class, even though twenty years later Glendon would refuse to appear on the same platform with her former student because of her intense opposition to abortion.[11]

In early February, the *Harvard Law Record* and the *Harvard Crimson* gave front-page coverage to news that the student-run *Harvard Law Review* had elected an Asian American 2L as its new president and an African American woman as one of its two supervising editors. Crystal Nix, a 1985 Princeton graduate who had been a *New York Times* reporter prior to law school, was the first black person ever elevated to one of the *Review*'s top masthead positions.

Far more controversial news landed a week later when Harvard president Derek Bok, a 1954 graduate of the law school who had served as its dean for three years before being elevated to the presidency in 1971, unexpectedly chose forty-four-year-old Robert C. Clark as the new law dean. Clark was "generally well-liked by students," the *Crimson* reported, because he was an excellent classroom teacher, but some of his more liberal colleagues complained about the selection of the conservative law and economics proponent. In the *Boston Globe* and the *New York Times,* Gerald Frug called Clark "a terrible choice" and Morton Horwitz denounced the selection as "a disaster for the law school," asserting that Clark had opposed the appointments of women and minority professors. Clark rebutted Horwitz's claims as "untrue" and "terribly unfair," while three more professors attacked Clark. In contrast, prominent liberal constitutional scholar Laurence Tribe spoke positively about Clark, and a *Wall Street Journal* editorial praised the selection. Several weeks later the controversy seemed to subside when the faculty unanimously promoted Randall Kennedy and Kathleen Sullivan to full professor, the third black male and the sixth white woman holding tenured appointments.[12]

At 12:30 P.M. on Thursday, February 23, the 1Ls' exam grades were finally distributed. Barack and Rob had mixed reactions. In David Rosenberg's Torts, Rob earned a straight A, and Barack something similar, but in Richard Parker's Crim Law, Rob received only a B+, and as he remembered, "Barack and I both didn't like the grade we got." Many classmates would have been overjoyed to receive a B+, as they confronted lower grades than they had ever before received. For some, the "effect was devastating," but not so for Barack and Rob, who plunged into the Ames Moot Court assignment that would culminate with a faux oral argument on Thursday evening, March 23, the night before spring break began.

The ungraded Ames exercise had four written assignments: an initial "issues analysis" of the faux case, an outline of the brief to the three-judge faux court, a draft of the brief, as well as the final brief. Throughout the six weeks, students had four conferences with Scott Becker, the twenty-five-year-old 3L from Illinois who had led their fall Legal Methods class and was now the teams' adviser. Barack and Rob took opposing sides, with Mark Kozlowski as Barack's partner and Lisa Hertzer as Rob's. Their case involved two issues related to inside information and the stock market: the first was whether a clerical employee had violated the Securities Exchange Act of 1934 in buying stock based on a research report she had proofread, and the second was whether sharing that information with a friend with whom she then twice purchased stock represented a conspiracy punishable under the RICO Act of 1970. Each student team was given a twenty-two-page faux indictment of the two defendants, "Janine Egan" and "Jennifer Cleary," that gave the facts that had led to their convictions, which were now on appeal.

Early on, Barack and Rob took the exercise "extraordinarily seriously," Mark thought, and exhibited great determination to win the competition. But once it sank in that it was ungraded, their emerging desire to graduate magna cum laude—5.80 or better, with 6 representing A minus—took precedence. Rob knew that for Barack "it was very important to him to get magna cum laude . . . to demonstrate that things"—i.e., a Harvard Law School diploma—"weren't given to him" as a result of how affirmative action may have helped him win admission. Mark Kozlowski realized that Barack and Rob "both decided they were going to make magna cum laude," and that made them "less serious about" the Ames exercise as it proceeded.

In advance of the two pairs' oral arguments before Professor Hal Scott and two other faux judges, Mark and Barack completed their thirteen-page brief, with Barack writing the insider trading argument and Mark handling the RICO question. Barack contended that Egan's work "inevitably" and "necessarily" would have made her aware that the materials she had proofread were nonpublic information, which she then misappropriated in violation of the 1934 statute. Barack flubbed badly in referencing the U.S. "Court of Appeals of the Second District," when he should have written "for the Second Circuit," and a careful eye would have caught misspellings and bad grammatical errors, such as "recieve," "the harm done by insider trading are diffused," and "employes like Egan regarding the in no way mitigates." It was visibly sloppy work, especially compared to what Mark offered in the RICO section. "The Egan-Cleary enterprise victimized not merely" Egan's employer, they argued in conclusion, "but the integrity of the securities market as a whole." Oral argument turned into "a bit of a fiasco," Kozlowski remembered, when he "got into a fight" with Professor Scott. "Barack was somewhat angry with me afterward," Mark recalled, but "by that point" both Barack and Rob were happy to leave moot court behind them.[13]

That same day, Ian Macneil received a letter from the Women's Law Association, complaining that Section III's Contracts reading earlier that week had contained sexist material. In dealing with a convoluted contracts problem known as "the battle of the forms," Macneil's casebook invoked the phrase "jockeying for position" and then quoted a couplet from Byron's "Don Juan": "A little still she strove, and much repented, / And whispering, 'I will ne'er consent,'—consented." Given Section III's history with Macneil, some classmates knew as soon as they read those lines that controversy would follow. Bonnie Savage, the WLA chair, wrote, "Repeated instances of sexism in both your contracts textbook and your classroom discussions have been brought to the attention of the Women's Law Association." She said the Byron quote reflected "sexist attitudes" and "has no place in a contracts textbook." Indeed, "by using sexist language, you encourage sexist thought and, in essence, promote hostility against women."

Macneil later acknowledged, "I knew the class considered me a first-class bastard," but the WLA aspersions were "a bolt out of the blue." When classes resumed after spring break on Monday, April 3, every member of Section III and every member of the law school faculty received an eight-page, single-spaced letter of rebuttal that Macneil addressed to the WLA. "Throughout the year I have had a great many complaints about the course from students of both sexes," but only two had anything to do with gender, and at the December 7 grievance session, "not a word was said about any alleged sexism." Macneil declared that "this whole affair . . . reeks of McCarthyism" and said its roots lay in his efforts "to insist that students act like professionals in the classroom respecting participation, preparation, attendance, and promptness." Two days later, Macneil reiterated his mandatory sign-in policy, saying he had referred the names of three regular absentees to law school administrators. A number of students spoke up in support of Macneil, and others asked about the upcoming, much-feared final exam.

A week later, the *Boston Globe* published a lengthy story, highlighting "Macneil's tough classroom manner" plus "his volatile temper and argumentative style." The *Globe* quoted Bonnie Savage as saying the WLA feared Harvard might give Macneil a permanent appointment. More significant, Jackie Fuchs told the paper that Macneil "goes out of his way to avoid being sexist," and that she, like a good many other Section III women, felt that WLA's "letter was really out of line." The *Harvard Law Record*'s own extensive coverage included multiple students noting that they had been given no notice that the law school's long-dormant attendance policy might suddenly be enforced.

Within Section III, the news coverage generated something of a pro-Macneil backlash. Brad Wiegmann remembered feeling badly for Macneil because "people were treating it as if it was the civil rights movement over whether you had to attend your Contracts class." Lisa Hay believed Macneil "got a raw deal" from Section III, and David Troutt agreed he was "a decent guy" whose relational view of contracts "actually was probably a good theory." Among students who later became contracts lawyers, some, like David Attisani and Shannon Schmoyer, said that Macneil's course had been of no professional value, but an equal if not greater number strongly disagreed. Steven Heinen "really appreciated his very practical approach to contracts," and said Macneil's teaching has "served me well" in later years. David Smail, who would become general counsel of a prominent international hotel corporation, remembers Macneil as "a pompous asshole," but "as much as I despised the man," Macneil's "relationship approach to contracts is one that I really fervently believe in, and I preach it every day in my business." As a general counsel, "you're living with the contract rather than just drafting it," and Macneil's perspective "is a very powerful way of looking at contracts."

Soon after the news stories, law school administrators convened a small meeting with Macneil and several students. Ali Rubin remembered that Barack had "distanced himself from" the complaints, and Mark Kozlowski knew that Barack felt "there was nothing we could do about it." Rob remembered that "Barack and I both felt that the revolution was a little over the top," and by April, Obama was playing "a mediating role" and was "calming people down."

Lauren Ezrol, one of Section III's official representatives on the Law School Council, recalled being summoned to the meeting. "It was Barack, me, and then everyone else in the room was an administrative type, like a dean of students, and Macneil was there." They were "supposed to try to work out the issues," but rather than some dean taking the lead, instead Barack "was the one who with great confidence talked to Macneil" and "worked out whatever resolution" was agreed to. Ezrol thought Obama was "unbelievably impressive," and his interaction with Macneil was entirely amicable. "I was just blown away," Ezrol remembered, "to see someone operate with such ease and confidence and maturity. It was remarkable." Years later, just prior to his death, at age eighty, in 2010, Macneil recalled that Obama was "a calm person in a class that was not altogether characterized by people being calm," adding that it was "obvious that his class respected him."[14]

Barack and Rob continued to visit David Rosenberg even after their fall Torts class ended. Rosenberg believed they should expose themselves to "the best minds on the faculty," and as the time for choosing 2L fall courses drew near, Rosenberg recommended well-known constitutional law teacher Laurence H. Tribe. For most 1Ls, it would take a good deal of gumption to approach one of Harvard's most prominent professors, Rosenberg knew, but on Friday, March 31, Obama did so, and Tribe was immediately taken by this heretofore unknown student. As Tribe remembered it, their first conversation lasted for more than an hour, and before it ended, Tribe asked Obama to become one of his many research assistants, which Tribe had never asked of a 1L who had yet to take one of his courses. Tribe wrote his name—"Barack Obama 1L!"—on that day's desk calendar page, and soon Barack drew Rob Fisher into this relationship with Tribe as well.

When word spread among their close aquaintances that both Obama and Fisher were now working for Tribe, Ken Mack can remember thinking how astonishing that was. Tribe was so impressed by Barack that he mentioned him to his colleague Martha Minow, a 1979 Yale Law School graduate who had clerked for Justice Thurgood Marshall before joining the Harvard faculty in 1981. Minow was always interested in students from Chicago, because she had grown up in the city's northern suburbs and her father, Newton Minow, who had served as chairman of the Federal Communications Commission during the Kennedy administration, was now a Sidley & Austin senior partner. Martha told her father about Tribe's impression of Obama, and he picked up the phone and called John Levi to ask if Sidley's Harvard recruiters knew about this 1L who had wowed Laurence Tribe. Levi had a ready answer: "We've already hired him. . . . He's coming here for the summer."

Before Barack began studying in earnest for late May's final exams, he faced a trio of choices. Wednesday night, April 12, was the informational meeting for 1Ls interested in the weeklong writing competition that would take place in early June to select thirty-eight members of the Class of 1991 for the prestigious *Harvard Law Review*. The next evening was the initial editors' meeting for the upcoming year's *Harvard Civil Rights–Civil Liberties Law Review*, which

Barack had worked on as one of thirty 1Ls during the preceding months. One 2L editor, Sung-Hee Suh, remembered Barack "talking very calmly" during a sometimes heated discussion of hate speech, "articulating both sides" of the debate as students of color and free speech absolutists disagreed vehemently on the issue. She recalled thinking he was very "calm, cool, and collected." Outgoing managing editor Wendy Pollack noted that Barack did not participate in the journal's election for the upcoming year's board, a clear sign that he intended to try for the premier *Law Review* instead. Barack had asked David Rosenberg whether he should do so, and Rosenberg said, "Yes, of course."

April 13 was also preregistration for fall classes, but the 1Ls would not learn until May 8 whether or not they had gotten into popular, oversubscribed courses, such as Laurence Tribe's section of Constitutional Law. Cassandra Butts remembered that "Barack did not get in," but some "special pleading" with Tribe quickly succeeded.

Monday, May 15, was the last day of spring classes, and Lauren Ezrol and Michelle Jacobs both recall Obama telling their Contracts class that anyone who would like a copy of the outline he had prepared from both semesters' worth of notes was welcome to it. Jacobs remembered that the document was easily 100, if not 150, typed pages and "had little jokes in it. It was great." Barack had many takers, because almost everyone agreed with Greg Sater's assessment that Barack "was the smartest in the class." Lauren Ezrol agreed: "he just seemed smart and impressive and so sure of himself and well-spoken and older."

Barack had his Legal History take-home exam on Monday, May 22, followed by Section III's trio of tests: Property on Wednesday afternoon, Contracts on Friday morning, and Civ Pro on Tuesday, May 30. Property and Civ Pro would cover just the spring, but Macneil's Contracts exam would determine everyone's entire grade for the full year, six credits, and "people were really worried," Brad Wiegmann remembered. David Attisani asserted that Section III suffered "mass hysteria" in advance of the Macneil final.

Mary Ann Glendon's Property exam was an unpleasant surprise to many members of Section III, with Mark Kozlowski remembering that some of the questions "had nothing to do with the stuff we had covered in class." Steven Heinen agreed, recalling that students first encountered the word "easement" only on the test. Half of the test, involving six specific questions, concerned Native Americans on "Vintucket" island who were being pressured to sell their seaside property to an aggressive celebrity. An estranged, unmarried couple's frozen embryos, a building lease, and a "takings" issue constituted the balance of Glendon's exam.

Friday morning, May 26, brought literal shrieks of horror when angry students confronted Ian Macneil's bizarrely demanding Contracts exam, which instructed them to write their answers "ONLY INSIDE THE BOX AFTER EACH QUESTION" on the twenty-five-page test. The first, three-part, twenty-minute question asked whether a U.S.–Soviet nuclear missile treaty did or did not constitute a contract, depending on how that concept was defined. An answer of up to fourteen lines was permitted. Ten subsequent questions, ranging from ten to twenty-five minutes apiece, involved, among other topics, a nephew cut out of a will on account of his race and the sale of fourteen hundred convertible sofas. Student complaints raged. Following the Memorial Day weekend, Section III's "very intense year," as Paolo Di Rosa termed it, ended with a straightforward two-question open-book Civ Pro exam from Stephen Subrin.[15]

Two hundred eighty-five 1Ls—179 men and 106 women—had decided to enter the *Harvard Law Review*'s writing competition, and only one day separated the end of final exams and the distribution of the *HLR* materials. The weeklong assignment consisted of two parts. First was a *Bluebook* test—*The Bluebook: A Uniform System of Citation,* then in its fourteenth edition, dated back to 1926, and was the most widely used style manual in legal academia. First-year students had been introduced to the volume's arcane instructions, especially for abbreviations, the previous fall in Legal Methods, but this *HLR* competition was their first real opportunity to apply its editing rules.

The second and far more demanding part entailed writing a lengthy essay—in law review parlance a "note"—weighing the merits and demerits of the U.S. Supreme Court's recent decision in *DeShaney v. Winnebago County.* A 6–3 majority had rebuffed a claim that a child who suffered severe brain damage at his father's hands could sue the local child protection agency, alleging a deprivation of his Fourteenth Amendment right to "liberty." The Fourteenth Amendment covers only *state* action, and the majority held that state officials do not have a constitutional obligation to protect children from their parents. Justice Harry A. Blackmun filed an emotional dissent.

Barack had a full week to work on the two-part assignment before it had to be postmarked and mailed to the *Review*'s offices. But on the day it was due, his car would not start, and time was short. He called Rob Fisher, who was just heading out the door, but Rob immediately drove to Somerville and took Barack to the closest post office, where there was a long line. But Rob remembered Barack successfully sweet-talking his way forward, explaining to people that his envelope had to be postmarked by noon.

On the cover sheet that accompanied the submission, the applicant was asked for their summer contact information. They also had the option of indicating their gender and race, and they could provide a brief personal statement if they had experienced any major hardships in life. The writing competition dated back to 1969, but only in 1981 did the *Review* introduce affirmative action into the process to increase diversity. Only a trio of rising 2L editors who oversaw the selection process would know whether and to what degree those indicators were considered when all entrants' scores were collated.

At the *Review*'s offices on the second and third floors of Gannett House, a wooden Greek Revival structure that dated to 1838, newly hired twenty-four-year-old editorial assistant Susan Higgins collected the 285 entries, assigned each a number, removed their cover sheets, and began sending batches of the submissions to the three dozen 2L editors, now scattered at their summer jobs all across the country. They would grade the numbered papers and send them back to Cambridge. Some of those 2Ls wondered whether the choice of *DeShaney* inadvertently gave 1Ls of one particular political stripe an advantage on their essays. "There were more interesting arguments to be made on the conservative side," thought Dan Bromberg, a 1986 summa cum laude graduate of Yale, "and so more points to be won" there than for 1Ls who sided with the dissenters. The grading was a multiweek process, at the end of which the 2L officers, including President Peter Yu, the *Review*'s first Asian American leader, collated the results and factored in the contestants' 1L grades that Susan Higgins had gotten from the registrar's office.

Half of the thirty-eight slots would go to the top writing competition scorers; the other half would be determined by an equation based 30 percent upon those scores and 70 percent upon their 1L grades. If a contestant had indicated their minority status, that fact, but not their names, was flagged on their submission. Gordon Whitman, incoming cochair of the *Review*'s Articles Office and a member of the selection trio, recalled that race was "a little bit of a featherweight on the process," but it did not count "a whole lot." Only when the numerical selections had been made did Susan Higgins match the chosen numbers with the individuals' names: "I was the keeper of the names with the numbers," she explained.

In midsummer, the successful contestants—twenty-two men and sixteen women—received a phone call, and then a confirming letter, telling them they had been selected. But there was a tangible cost: they would have to leave their highly remunerative summer jobs three weeks early to return to Cambridge before the beginning of fall classes, to start work on the *Review*'s first issue of the new academic year.[16]

In preparation for the summer, Barack had "sublet the cheapest apartment I could find," hardly a block north from where he and Sheila had lived in Hyde Park. He also had "purchased the first three suits ever to appear in my closet and a new pair of shoes that turned out to be half a size too small and would absolutely cripple me for the next nine weeks." Working at a corporate law firm like Sidley reintensified his fears from a year earlier that going to Harvard "represented the abandonment of my youthful ideals." But choosing Harvard over Northwestern's full-cost scholarship meant that "with student loans rapidly mounting, I was in no position to turn down the three months of salary Sidley was offering."

Obama arrived late for his first day at Sidley's offices at 10 South Dearborn. It was a rainy June morning. Some days earlier, he had spoken by phone with Michelle Robinson, whom Geraldine Alexis and senior associate Linzey Jones had assigned as Barack's summer adviser because of their mutual Harvard ties. Obama remembered that "she was very corporate and very proper on the phone, trying to explain to me how the summer program at Sidley and Austin was going to go."

Barack was shown to her office that first day, and Michelle recalled in 2004 that "he was actually cute and a lot more articulate and impressive than I expected. My first job was to take him to lunch, and we ended up talking for what seemed like hours." She had expected a biracial, Hawaiian-raised Harvard Law student to be "nerdy, strange, off-putting" and even "weird," but instead Barack was "confident, at ease with himself . . . easy to talk to and had a good sense of humor," she recounted. "I was pleasantly surprised by who he turned out to be." She did however recall that "he had this bad sport jacket and a cigarette dangling from his mouth."

Rob Fisher was also working at Sidley's Chicago office for the summer, and he remembered Barack coming by his office soon after they arrived. "He came in one day and said, 'My mentor is really hot.'" As Michelle's friend and Sidley colleague Kelly Jo MacArthur recalled, Barack wasted little time in making his interest clear. "He would try to charm her, flirt with her, and she would act very professional. He was undeniably charming and interesting and attractive," but Michelle rebuffed Barack's repeated suggestions that they do something together. "She was being so professional, so serious," MacArthur remembered, and she also knew that Michelle was someone with "conservative morals."

When Barack pressed, Michelle was characteristically direct and told him that as his adviser, it would look bad if they began going out together. Instead, she tried to set him up with several of her girlfriends, just as she did with Tom Reed, another African American 1L summer associate and Chicago native who had been one year behind her as an undergraduate at Princeton. But Barack persisted. Finally, toward the end of June, Michelle reluctantly agreed. "OK, we will go on this one date, but we won't call it a date. I will spend the day with you."

On Friday, June 30, they left Sidley's offices at about noon and walked the few blocks to the Art Institute of Chicago on South Michigan Avenue to have lunch. "He was talking Picasso," Michelle remembered. "He impressed me with his knowledge of art." Then they "walked up Michigan Avenue. It was a really beautiful summer day, and we talked, and we talked."

Barack had a plan. Opening that evening was a movie that already had been the subject of three different articles in the *Chicago Tribune:* young African American director Spike Lee's *Do the Right Thing.* It was an ingenious idea, but Michelle's concern about appearances turned all too real when they saw Sidley's Newton Minow and his wife Jo at the theater. "I think they were a little embarrassed" at being seen together, Minow recalled. After the film, Michelle remembered, "we had a deep conversation about that" and then ended the day "having drinks at the top of the John Hancock Building," which "gives you a beautiful view of the city."

"I liked him a lot. He was cute, and he was funny, and he was charming," Michelle remembered thinking, but after their encounter with the Minows she was all the more determined not to become a subject of office gossip. But when Barack invited her to accompany him to a training he had agreed to do for DCP at one of the Roseland churches, she agreed to go along.

The small group was "mostly single parent mothers," but she recalled that Barack's "eloquent" presentation "about the world as it is and the world as it should be" was one she would never forget. "To see him transform himself from the guy who was a summer associate in a law firm with a suit and then to come into this church basement with folks who were like me, who grew up like me," Michelle recounted, "and to be able to take off that suit and tie and become a whole 'nother person . . . someone who can make that transition and do it comfortably and feel comfortable in his own skin and to touch people's hearts in the way that he did, because people connected with his message," was remarkably impressive. "I knew then and there there's obviously something different about this guy," something "special," and "it touched me . . . he made me think in ways that I hadn't before," Michelle explained. "What I saw in him on that day was authenticity and truth and principle. That's who I fell in love with" in that church, and "that's why I fell in love with him."

Every summer Linzey Jones hosted a picnic at his home in south suburban Park Forest for all of Sidley's minority attorneys and summer associates. Events like this were standard fare because summer programs at big law firms were aimed at enticing the students into eventually accepting a permanent job offer. Evie Shockley, an African American 1L from the University of Michigan Law School who shared an office with Barack for part of that summer, recalled attending a Cubs game, seeing *Phantom of the Opera,* and other theater outings. "It was easy to feel like you weren't working," she explained.

That weekend day, Linzey Jones remembered Barack and Michelle being "very friendly with each other," but Barack joined in when many of the men went to a nearby junior high school to play basketball for an hour. Michelle still lived with her parents in their South Shore home, a few miles below Hyde Park, and when she drove Barack back to his apartment, he offered to buy her an ice cream cone at the Baskin-Robbins on the north side of East 53rd Street. Michelle accepted, and ordered chocolate. Sitting outside, Barack told her about working at Baskin-Robbins in Honolulu "and how difficult it was to look cool when you had the apron and the little brown cap on." Then, in a direct reprise of a question he had posed three summers earlier, also in Hyde Park, he asked Michelle "if I could kiss her. It tasted of chocolate."

"We spent the rest of the summer together," Barack later wrote, but a mid-July phone call informed him that a letter inviting him to join the *Law Review* was in the mail. That good news meant he had to be back in Cambridge by August 16 to work on the *Review*'s first issue. He took several days to ponder his choice. "We had a conversation about whether or not he was going to do *Law Review,*" fellow summer associate Tom Reed recalled. "'I'm not sure if I'm going to do it,'" Tom remembered Barack saying. "He was clearly on the fence," and "there was a moment where he was considering whether that was appropriate for his path." But finally he told *HLR* as well as Sidley that he was accepting the offer.

Barack and Michelle kept a very low profile at the law firm, and neither Tom Reed nor Evie Shockley had any idea they were dating. Michelle told only Kelly Jo MacArthur. "When she met Barack, things happened pretty quickly," Kelly Jo remembered. Michelle recalled years later during a joint interview her memories of "the apartment you were in when we first started dating," the sublet near Baskin-Robbins. "That was a dump." But bumping into people they knew seemed inevitable. Jean Rudd of the Woods Fund recounted, "I have a very vivid memory of having lunch on Dearborn Street at an outdoor café there, and Barack and this tall, beautiful woman walk by. And he stopped and introduced us and said that 'This is my boss.' . . . We chatted a little while," and when they left "I remember saying, 'What a couple.'"[17]

One late July evening, Michelle invited Barack home for dinner to meet her parents and brother. Craig Robinson, at twenty-seven years old, was two years older than his sister and also had attended Princeton University. As a senior he was Ivy League basketball's 1983 Player of the Year, and after graduation he had played professional basketball in Europe for several years before returning home. Craig met Janis Hardiman, a 1982 Barnard College graduate, soon after she moved to Chicago in 1983, and by 1987 they were engaged and living together in Hyde Park while Craig took classes toward an M.B.A. degree at the University of Chicago. Janis and Craig married in August 1988, soon after Michelle's graduation from Harvard Law School. Ever since Michelle's senior year of high school, Craig had known that his sister was quick to dispose of boyfriends, so he made a point of being at the Robinson family home at 7436 South Euclid Avenue to meet this newest suitor.

Craig and Michelle's parents, Fraser and Marian Robinson, were, like their children, lifelong residents of Chicago's South Side. Both high school graduates, they had married in October 1960, but Fraser's hope of finishing college was dashed by insufficient funds. In January 1964, just a few days before Michelle's birth, Fraser was hired by the city water department, and Marian became a stay-at-home mom. In 1965 the young family moved from the Parkway Gardens Homes in Woodlawn to the cramped top floor of Marian's aunt's home on Euclid Avenue, in solidly middle-class South Shore. Craig and Michelle attended nearby Bryn Mawr Elementary School, where Craig skipped third grade and Michelle skipped second. Separate small bedrooms and a common study area gave them their own modest spaces at home. At work, tending steam boilers, Fraser won two promotions along with salary increases, but an increasingly dark cloud hung silently over the happy young family: at age thirty, Fraser had been diagnosed with multiple sclerosis, though, as Craig later wrote, "we never had an in-depth discussion at home about the frightening course that MS was known to take." In time, Fraser needed to use a cane and then crutches to help him walk, but he stuck with his job. "We saw him struggle to get up and go to work," Michelle recalled. "He didn't complain—ever. He put his energy into us."

When Michelle reached ninth grade and was admitted to Whitney Young High School, west of downtown, she spent hours a day riding city buses to and from school. Craig won admission to Princeton in 1979, and his father insisted the family would make the necessary financial sacrifices for Craig to go there rather than accept a full scholarship from some less prestigious institution. Michelle grew up thinking she was smarter than her brother, and although Craig had a difficult freshman year, Michelle resolved that if he could attend Princeton, so could she. Her mother knew that test taking was not her forte, and a high school counselor discouraged her interest in Princeton, but Michelle applied anyway and was admitted. The difference between the South Shore world from which Michelle came and the privileged backgrounds of Princeton's overwhelmingly white and often wealthy student body was profoundly stark.

"The first time when I set foot in Princeton, when I first got in, I thought 'There's no way I can compete with these kids . . . I got in but I'm not supposed to be here,'" Michelle recalled. "I remember being shocked by college students who drove BMWs. I didn't even know parents who drove BMWs." In addition, black undergraduates realized that Princeton's racial climate, even in 1981, left much to be desired. Angela Kennedy, one of Michelle's closest friends, with whom she spent one summer working as counselors at a girls' camp in New York's Catskill Mountains, recalled that "It was a very sexist, segregated place. Things reminded you every single second that you're black, you're black, you're black."

Michelle thrived in Princeton's classrooms, and by the beginning of her senior year, she was applying to Harvard Law School. Yet in a reprise of high school, her faculty adviser on her senior thesis downplayed her chances. After initially being wait-listed, in late spring of 1985 she was accepted to Harvard.

Michelle's college thesis, "Princeton-Educated Blacks and the Black Community," was a powerfully self-revealing document. "My experiences at Princeton have made me far more aware of my 'Blackness' than ever before," Michelle wrote. "I have found that at Princeton . . . I sometimes feel like a visitor on campus; as if I really don't belong." Growing up in South Shore, neither of her parents had been especially outspoken about race, but Marian Robinson's father Purnell "Southside" Shields, who died in 1983, "was a very angry man," Michelle's mother explained. "I had a father who could be very angry about race," and Marian had given Craig the middle name Malcolm after the early 1960s' angriest racial firebrand. Marian was likewise wary of interracial relationships. "I worry about races mixing because of the difficulty," she confessed years later. "It's just very hard."

But Princeton made Michelle understand that "I'm as black as it gets." In her thesis, she observed that "with Whites at Princeton, it often seems as if, to them, I will always be Black first and a student second." Looking ahead, that left her fearful. "The path I have chosen to follow by attending Princeton will likely lead to my further integration and/or assimilation in a White cultural and social structure that will only allow me to remain on the periphery of society; never becoming a full participant." She confessed that "my goals after Princeton are not as clear as before" and she rued how "the University does not often meet the social and academic needs of its Black population." In addition, "unfortunately there are very few adequate support groups which provide some form of guidance and counsel for Black students having difficulty making the transition from their home environments to Princeton's environment," as both Michelle and her brother had. She now knew that Princeton was "infamous for being racially the most conservative of the Ivy League Universities." And her exposure to some fellow students had taught her something else, something prescient indeed: "a Black individual may be unable to understand or appreciate the Black culture because that individual was not raised in that culture, yet still be able to identify as being a Black person."

Harvard Law School did not offer a much different experience. Czerny Brasuell, Michelle's one black female Princeton mentor, recounted Michelle telling her by telephone from Cambridge that "If I could do this over, I'm not sure that I would." A female classmate told Michelle's biographer Liza Mundy that Harvard "was not a friendly, happy atmosphere." But once again Michelle persevered. After her 1L year, she returned to Chicago as a summer associate at the law firm of Chadwell & Kayser, working for female partner Jan Anne Dubin and staying with her parents in the home that her great-aunt had deeded to the Robinsons several years earlier, prior to her own death.

Back at Harvard for her 2L year, Michelle volunteered significant time at the Legal Aid Bureau, located one floor—and many status rungs—below the *Harvard Law Review* on Gannett House's ground level. After her 2L year, she was a summer associate at Sidley & Austin's Chicago office. When Sidley offered her a position once she graduated, Michelle readily accepted. At Harvard, "the plan was you go into a corporate firm. So that's what I did. And there I was. All of a sudden, I was on this path."

At graduation, her parents paid for a teasingly congratulatory message in the 1988 Harvard Law School *Yearbook:* "We knew you would do this fifteen years ago when we could never make you shut up." That summer, Sidley paid for the bar review class she took alongside a friend of her brother's, Alan King, but only on May 12, 1989—after taking the Illinois exam a second time—did Michelle become a member of the Illinois Bar. Working in Sidley's intellectual property group, Michelle yearned for meaningful assignments. Given her Harvard loans, her Sidley salary was attractive, but she had not really intended to be a corporate lawyer. "I hadn't really thought about how I got there," she recalled. "It was just sort of what you did."

Craig Robinson recalled the late July evening when Michelle introduced Barack to her family. "My sister brought him over to my mom and dad's house. We all met him, had dinner. They left to go to the movie, and my mom and dad and I were talking: 'Oh, what a nice guy. This is going to be great. Wonder how long he will last?'" Craig thought Barack was "smart, easygoing, good sense of humor," but given Michelle's proclivity for discarding boyfriends, Craig remembered thinking, "Too bad he won't be around for long." Marian Robinson was also impressed because "He didn't talk about himself," but instead drew out the Robinsons about their own lives and interests. "I didn't know his mother was white for a long time," Marian recalled. "It didn't come up."

Barack's taste in movies ran to the realistic, and opening that weekend was *Leola,* the story of a bright seventeen-year-old African American Chicago girl whose desire to attend college was endangered when she became pregnant. Filmmaker Ruby Oliver was a fifty-year-old former day care operator, and seven weeks after they saw it, Barack talked about the ninety-five-minute movie—later retitled *Love Your Mama*—while addressing the real-life challenges confronted by black youths. Michelle and Barack continued to see each other almost every day, and when they went out, Michelle usually paid. "He had no money; he was really broke," she remembered, plus "his wardrobe was kind of cruddy." Barack's Occidental roommate Paul Carpenter was visiting Chicago that August, and he heard about Michelle when the two old friends and Paul's wife Beth had dinner one evening. Another night, Craig and his wife Janis dropped off Michelle at Barack's sublet in Hyde Park, and Janis and Barack recognized each other from their time at Columbia. "He came out of his apartment to get Michelle, then he and I both said, 'Oh my gosh, I remember you,'" Janis recounted.

Before Barack's return to Cambridge, Michelle told Craig, "I really like this guy" and made a request. She had heard her father and Craig say that "you can tell a lot about a personality on the court," something Craig had learned from Pete Carril, his college coach at Princeton. Michelle knew that Craig played basketball regularly at courts around Hyde Park, and he remembers her asking: "I want you to take him to play, to see what type of guy he is when he's not around me." Craig agreed to take on this task, but he recalled, "I was nervous because I had already met Barack a few times and liked him a lot."

Craig quickly scheduled a meet-up, and they played "a hard five-on-five" for more than an hour. Craig's nervousness quickly fell away because he could see that Barack was "very team oriented, very unselfish," and "was aggressive without being a jerk." Craig was happy he could "report back to my sister that this guy is first rate," and Michelle was pleased. "It was good to hear directly from my brother that he was solid, and he was real, and he was confident, confident but not arrogant, and a team player." Craig saw only one huge flaw in Barack's skill set, but it was not relevant to Michelle's question. "Barack is a left-handed player who can only go to his left."

Before Obama headed back to Cambridge in mid-August, he knew that this new, two-month-old relationship with Michelle Robinson was perhaps on a par with his now-truncated, three-year-old involvement with Sheila. For Barack, the differences were huge. Sheila was also the biracial offspring of international parents; she had lived in Paris, spoke French and now Korean, and was comfortable around the globe—just like Australian-born, Indonesian-reared diplomat's daughter Genevieve Cook before her. Michelle Robinson was a graduate of Princeton and Harvard Law School, but she was a 100 percent product of Chicago's African American South Side, just like so many of the women and men who had revolutionized Barack's understanding of himself during his transformative years in Roseland.

Barack's prior relationships had been with women who, like himself through 1985, were citizens of the world as much as they were of any particular country or city. Before Princeton, Michelle Robinson had spent one week each summer with her family at Dukes Happy Holiday Resort, an African American forest lodge in White Cloud, Michigan, forty miles north of Grand Rapids. But if Barack truly believed that his destiny entailed what he thought, he knew full well the value of having roots in one place and having that place be essential to your journey. And who more than Michelle Robinson and her family could personify the strong, deep roots of black Chicago?

Although Michelle would not know that Barack had shared his deeply private sense of destiny first with Sheila and then with Lena, before he left for Cambridge, Barack told Michelle about his belief about his future role. "He sincerely felt, from day one that I've known him, that he has an obligation," Michelle explained, "because he has the talent, he has the passion and he has been blessed."[18]

The *Harvard Law Review,* founded in 1887, was in 1989 the oldest and most prestigious legal publication in the United States. Edited entirely by students—beginning with thirty-eight from the rising 2L class, supplemented each successive summer by several top-GPA 3L "grade-ons" for an annual total of about forty—the *Review* published eight hefty issues a year—November through June—with the law students contributing twenty to forty or more hours of work weekly, aided by a trio of female office staffers and a quintet of part-time undergraduate work-study students. Beyond the law students, there was an oversight board of two professors, the dean, and an alumnus, but they played only a nominal role. In addition, playing an obtrusive role in the *Review*'s life was eighty-five-year-old eminence grise Erwin N. Griswold, the law school's dean from 1946 to 1967 (and himself the *Review*'s top officer—president—in 1927–28), who critiqued every issue and was available to hector the student editors.

The mid-August return to Cambridge served two long traditions. One was to initiate the new 2L editors into the sometimes-complex internal workings of the *Review.* The masthead—the president, treasurer, managing editor, supervising editors (SEs), and executive editors (EEs)—oversaw the work of five "offices": Articles, which reviewed scores of long manuscripts submitted by law professors nationwide and chose a dozen or so per year for publication; Notes, which selected and edited substantive analyses written by the *HLR* editors themselves; Book Reviews and Commentaries, which assigned and handled shorter pieces; "Devo," or Developments in the Law, which prepared a major team-written study of some cutting-edge topic for publication in each year's May issue; and Supreme Court, which oversaw the annual November issue and its several dozen student-written synopses of significant cases decided during the prior term of the U.S. Supreme Court. The November issue also contained the *Review*'s two top-status faculty contributions: the foreword, written every year since 1951 by an emerging star chosen by the editors, and a major case comment authored by an eminent academic, a feature added in 1985.

In *HLR*'s very elaborate editing system, overseen by the managing editor, student-written work moved from the offices to the SEs and then the EEs; faculty pieces went directly from the offices to the EEs. Everything also went through a "P-read," in which the *Review*'s president recommended editorial changes. Each fall and winter "the 2Ls are the labor, the 3Ls are the management," 2L editor Brad Berenson recounted, until a new masthead for the upcoming year was chosen from among the 2Ls early in February.

The second reason for the pre-semester start was that the November issue had to go to the printer by mid-October. The 2Ls needed an intensive refresher course in *Bluebook* legal citation style, followed by an introduction to two other common tasks: sub-citing, in which the accuracy of every quotation and footnoted reference in each piece was confirmed by checking the original source, and roto-pool, in which every faculty-submitted manuscript was read and evaluated by several editors before full consideration by the Articles Office.

Most 2Ls spent their first *HLR* semester in "the pool," where almost every weekday morning a pink slip of paper from managing editor Scott Collins would appear in their pigeonhole mailbox in the editors' lounge on the second floor of Gannett House, telling them what their work assignment was. Editors were enticed there each morning by a spread of free muffins and bagels. "My chocolate chip muffin was the mainstay of my morning," 3L editor Diane Ring recalled. Thanks to the hefty income the *Review* received from sales of *The Bluebook,* free pastries seemed like "a very interesting strategy to make sure you got all those second-years on the doorstep every morning getting their assignment, doing the work," Ring explained. Patrick O'Brien, also a 3L, remembered that "a lot of my law review involvement had to do with free bagels and cream cheese. It would get me there every morning for a free breakfast." There also were free evening snacks for those who worked late, and as a result, Berenson recalled, Gannett House became "a gathering place," "almost like a fraternity house for the editors." Marisa Chun, a 2L, explained that the editors' lounge and its television served as "our living room." With everyone's classes in nearby buildings, popping in and out was a constant feature of *HLR* life. For some editors, Gannett House became the center of their daily lives, while for others, especially those who were already married, the *Review* was more like a demanding part-time job.

The 2Ls had three ways out of the "hideous experience" of doing pool work: join the five-person "Devo" team, whose work would satisfy the law school's written work requirement; join a multiperson group assigned to edit an especially difficult article; or write a note of one's own, an option often postponed until the 3L year and almost always done for independent study credit under faculty supervision. Gordon Whitman, the 3L Articles Office cochair, had worked for a year as a community organizer in Philadelphia before starting law school and had successfully pushed for the acceptance of a manuscript that argued that the real-world theology of Martin Luther King Jr. offered a superior perspective for examining the contentions made by "critical legal studies" scholars.

The article's author, Anthony E. Cook, was an African American associate professor of law at the University of Florida who had graduated magna cum laude from Princeton before getting his law degree at Yale. Cook's dense and complicated analysis looked especially daunting, and Whitman recruited four new 2L editors to work on it: Christine Lee, the young Oberlin graduate who was just about to turn twenty-two, the now twenty-eight-year-old Barack Obama, whom Lee had disliked from their first introduction a year earlier, and two other visibly sharp 2Ls, Susan Freiwald, a 1987 magna cum laude graduate of Harvard College, and John Parry, a 1986 summa cum laude graduate of Princeton. The lengthy manuscript would require weeks of work—it was not scheduled for publication until the March issue—and these four knew by early September what their fall work for the *Review* would entail.[19]

Fall semester classes began on September 6. Barack and Rob had carefully debated their choices. In Laurence Tribe's huge and oversubscribed Constitutional Law section, which met five hours per week, they were joined by a number of their 1L Section III classmates. The assigned casebook, William B. Lockhart et al.'s *Constitutional Law: Cases and Materials,* 6th ed., was the best available. Word among students was that African American professor Christopher Edley, who was back after serving as issues director for Michael Dukakis's presidential campaign, was "refreshingly good." His class, Administrative Law, might sound dry and arcane, but Edley taught the sixty-five-student, four-hour-a-week class as an entirely practical "this is how the public policy process works" course, and he supplemented the main text, Walter Gellhorn et al.'s *Administrative Law,* 8th ed., with various other materials.

Rob and Barack had "an extended discussion" about taking Corporations, as most 2Ls did, weighing the upside value of "understanding the world" versus how their grades in the four-credit class might harm their goal of graduating magna cum laude. But they began it and kept it, finding that Professor Reinier Kraakman, a Yale law graduate with a Harvard sociology Ph.D., "had an interesting mind and approach." Kraakman focused on "the control of managers in publicly held corporations" and emphasized "the functional analysis of legal rules as one set of constraints on corporate actors." Rob and Barack found it "a good class, really cool," and "really enjoyed it."

Edley's Ad Law was "almost as exciting as the Torts class" a year earlier with David Rosenberg. "Barack and I loved the Ad Law class" and "had a tremendous amount of fun" in it, Rob remembers, for Edley was "inspiring" and had "this very functional, argumentative intellectual approach." The course "stimulated a lot of deep discussions," plus a number of office-hours conversations with Edley, whom Rob thought was "truly a great teacher."

Mark Kozlowski recalled that Tribe's Constitutional Law had "a really lively atmosphere." As in all sizable classes, seats were preassigned, and Scott

Scheper, a 1988 summa cum laude graduate of Case Western Reserve University, found himself in a second-row aisle seat, at the front of the bowl-style classroom, with Obama just to his right. Like Scheper, Laura Jehl, a 1986 highest honors graduate of the University of California at Berkeley, had been in a different 1L section than Obama. But she already knew Tribe from her work on U.S. senator Edward M. Kennedy's Judiciary Committee staff, and she was also already one of Tribe's research assistants. Jehl took note of Obama from the outset of Tribe's class. He "spoke up and said something eminently reasonable, eminently thoughtful," and with an "absolutely amazing voice."

Section III survivor Jennifer Radding remembered that Barack "was exceptional in Tribe's class" and that "Tribe was like in love with him in a very intellectual way." Sarah Leah Whitson witnessed it too. "Barack seemed to be operating at another level. . . . His rapport back and forth with Tribe felt more like a dialogue among equals." Kevin Downey, a 1988 magna cum laude graduate of Dartmouth, noted how Obama spoke in "narrative-based" style that often included references to his own experiences, while Rob Fisher, who also stood out, made "more analytic comments." Downey thought they "were leagues beyond the rest of us."

Seated next to Obama, Scott Scheper had as close a view as anyone. Tribe was "a whole lot more theoretical" than he had anticipated, more interested in "What's right? What should be?" than in "How is? What is?" Obama was "facile and adept," and "talked more than any other single individual" in the class. Scheper recalled that "Tribe spent a whole lot of time not six feet from me in what almost became personal dialogue between him and Barack. . . . He would leave the lectern and come over . . . to our side of the class and be right in front of the front row and then Barack would be talking to him." The scene made Scheper "sort of self-conscious that I had to maintain my posture because the whole class was looking right at me because that's where the focus of the dialogue was."

Several weeks in, Scheper's demanding Trial Advocacy Workshop kept him away from Con Law for several classes. When he returned, Rob Fisher was in his spot, and Scheper realized that Obama "gave away my seat because I didn't come to class." Barack immediately apologized: "I thought you dropped the class." Then the regular pattern resumed. Obama "was always engaged in these esoteric discussions with Professor Tribe. . . . They spent a lot of time talking about what the law should be." All told, Rob explained, with Tribe plus Edley and Kraakman, fall 1989 "was a pretty fun semester."[20]

That fall was Robert Clark's first as dean of the law school. The school had been in the news over the summer because of the arrest and suspension of an African American 3L accused of raping a Harvard undergraduate. But Clark played right into the hands of his detractors when he terminated the school's public interest career counselor before the semester began. Clark called the move "a reorientation of resources" away from something that served only "symbolic, guilt-alleviating purposes," but progressive students reacted immediately. A protest rally attracted a crowd of three hundred, with Barack's close friend Cassandra Butts telling the *Harvard Law Record*, "I came to law school in particular because I was very much interested in helping people who don't have access to the law and who see the law as being more of a hindrance than a help to them."

Students viewed Clark's move as a tangible, public rebuke of those motives, and quickly created the Emergency Coalition for Public Interest Placement. National legal publications and the *Boston Globe* all covered the controversy, but the *Harvard Crimson* highlighted how tiny a percentage of Harvard Law students actually took public interest jobs once they graduated. Butts told the *Crimson* that many students arrived with such an interest, "but with the emphasis here on corporate law, they don't always leave with that attitude." Given the "insurmountable number of loans they need to pay off," students may "choose to go into corporate work, but they will be more sensitive to the need for pro bono lawyers." As the fall semester progressed, eight hundred law students signed letters protesting Clark's move, then dumped them outside the dean's office during a one-hundred-person rally that the *Record* said had "an emotional, near confrontational tone." As Christopher Edley ruefully recalled, Clark "was screwing up massively."[21]

In late September, Barack flew back to Chicago to take part in a Friday-afternoon roundtable discussion on community organizing. Funded by Ken Rolling and Jean Rudd at Woods, the event built on the commissioned essays Obama and others had written for *Illinois Issues* a year earlier. Sokoni Karanja, Wieboldt's Anne Hallett, and several local academics were part of the group, and, as he had in other settings, Barack refrained from talking until the discussion was well under way. When he did speak, Barack highlighted what he called "the educative function of organizing," for "at some point you have to link up . . . with the larger trends, larger movements in the city or the country. I think we are not very good at that." He suggested that "I am not sure we talk enough in organizing" about organizing's internal culture, and "we don't understand what the relationship between organizing and politics should be. . . . I would like to think that ideally you would focus on the local but educate for the broader arena, and that you are creating a base for political or national issues." Barack expressed disappointment that organizing had a "suspicion of politics," for "politics is a major arena of power" and "to marginalize yourself from that process is a damaging thing, and one that needs to be rethought." His critique was fundamental, and strong. "Organizing right now doesn't have a long-term vision."

In the 1960s, "a lot was lost during the civil rights movement because there was not enough effective organizing consolidating those gains," but now organizers were ignoring the potential of working with movement-style efforts, and "that long-term vision needs to be developed." Barack returned to organizing's educative mission. "How do you educate people enough so that they can be forcing their politicians to articulate their broader views and wider horizons?" he asked. "People expect politicians to express some long-term interests of theirs and not just appeal to the lowest common denominator."

Barack sat back before again weighing in. "There is this big slippery slope of folks and communities that are sinking," he reminded the group. How can organizing help them? "How do you link up some of the most important lessons about organizing . . . with some powerful messages that came out of the civil rights movement or what Jesse Jackson has done or what's been done by other charismatic leaders? A whole sense of hope is generated out of what they do. Jesse Jackson can go into these communities and get these people excited and inspired. The organizational framework to consolidate that is missing," especially given the lack of minority organizers. "The best organizers in the black community right now are the crack dealers. They are fantastic. There's tremendous entrepreneurship and skill," all being used to distribute illegal drugs. To help black neighborhoods, "organizing in these communities . . . can't just be instrumental . . . it has to be recreating and recasting how these communities think about themselves."

After a pause, Barack turned to one of his chief takeaways from his time in Roseland. Harold Washington "was an essentially charismatic leader," although "his election was an expression of a lot of organizing that had been taking place over a long time." All indications were that "to a large extent" Washington "wanted to give back to that process. He wanted to give those groups recognition and empower them in some sense," as he had done so visibly with Mary Ellen Montes and UNO, but "real empowerment was not done." An African American historian on the panel objected to Barack criticizing Harold Washington. Sokoni Karanja agreed with the angry historian, but Anne Hallett sided with Barack, who pursued his point. "When you have a charismatic leader, whether it's Jesse Jackson or Harold Washington . . . there has to be some sort of interaction" that moves all that energy back "into the community to build up more organizing . . . more of that needs to be done." Then the conversation shifted, but Obama made one final point: "Organizing can also be a bridge between the private and the public, between politics and people's everyday lives."

Barack's comments revealed how profoundly he disagreed with the worldview of IAF and Greg Galluzzo, and how convinced he was that social change energies should be focused on the political arena. In later years hardly anyone would appreciate the significance of what Obama said that day. Ben Joravsky, a fellow participant who was already on his way to becoming one of Chicago's most perceptive political journalists, later recalled Obama's "veneer of cool" but dismissed his comments as those of "a windy sociology professor with nothing particularly insightful to say." Only journalist John B. Judis, examining this moment many years later, would highlight how Obama had voiced "a litany of criticisms of Alinsky-style organizing" and note that he had "rejected the guiding principles of community organizing: the elevation of self-interest over moral vision; the disdain for charismatic leaders and their movements; and the suspicion of politics itself." But, Judis wrote, Obama "did so in a way that seemed to elude the other participants," who objected only to Barack's remark that Harold Washington had not left behind any *tangible* political legacy.

Judis mused that even Obama "seemed initially oblivious to the harsh implications of his own words," but Washington's fundamental failure should have been obvious to everyone in the room, especially because the mayor's political base had so quickly fallen apart after his death, leaving Chicago with a white, Democratic machine mayor with an all-too-familiar surname. Six months earlier, Cook County state's attorney Richard M. Daley had defeated Gene Sawyer in the Democratic primary by 55 to 44 percent, and five weeks later, Daley was elected mayor, besting Alderman Tim Evans, running as an independent, by 55 to 41 percent. Ed Vrdolyak, now a Republican, garnered 3 percent and soon added a sideline as a radio talk show host to his lucrative South Chicago law practice.[22]

The Chicago trip was also a chance for Barack to spend a weekend with Michelle Robinson, and either then or soon after, he asked her to accompany him to Honolulu over the winter holidays. Back in Cambridge, Barack kept his distance from the burgeoning student protest campaign, despite his friendship with the outspoken Cassandra Butts, who, second only to Rob Fisher, was his best friend in Cambridge. Laura Jehl, who was working with Cassandra on a manuscript for the *Harvard Civil Rights–Civil Liberties Law Review,* knew that Barack and Rob were "inseparable," and she also saw how Barack and Cassandra "were around together a lot but they didn't appear to be together," as she put it. "It did not seem to be romantic" and "it did not appear to be sexual." Another female friend concurred: "the vibe they gave off was fraternal." Laura thought that as attractive as Barack was, "there was also absolutely no body language of him that I was aware of towards anybody," and other women all agreed: "I didn't see any sexual energy from him" said one, and "never any sense" at all, recalls another.

On evenings when Barack worked on the Anthony Cook manuscript at Gannett House, he and fellow 2L African American editor Ken Mack often walked to a sandwich shop in Harvard Square for dinner. On several nights, Gordon Whitman gave Barack a lift home and spoke about how he was volunteering at the Massachusetts Affordable Housing Alliance, a Boston group headed by veteran community organizer Lew Finfer. After Barack mentioned his Chicago experience, Whitman told Finfer he should meet him. Finfer called Obama, and they met up one day at a Harvard Square coffee shop. Finfer found Barack "cool" and "dispassionate," but hoped to interest him in a return to organizing after law school. Barack politely said no. "I have a plan to return to Chicago and go into politics."

One mid-October night, 3L executive editor Tom Krause, a U.S. Navy veteran who was overseeing the group edit of the Anthony Cook article, hosted a party following a lecture by Alex Kozinski, the well-known federal appellate judge for whom Krause would be clerking after graduation. Krause invited a number of editors of all political persuasions, and Barack attended. *Review* editors were prized candidates for clerkships with top federal judges like Kozinski, and an astonishing 102 members of the law school's 1989 graduating class had won clerkships. Each fall 2Ls began eyeing and discussing which jurists they would apply to in the spring, and Ken Mack was astounded when Barack told him one evening that he was so focused on returning to Chicago after graduation that he would forgo applying for clerkships. When this news spread among African American students, there was open disbelief that such a top performer would pass up so prestigious an accolade. Kenny Smith was surprised and impressed, but others sensed an attitude of group disappointment. As Frank Harper put it, there were "these steps you're supposed to take" and "people thought that he was making a catastrophic error by not clerking."

Obama was a semiregular presence at BLSA meetings and parties. Cochairing a BLSA committee made him a formal member of BLSA's executive board, but the group's style was decidedly informal, with its annual spring conference the one major event requiring time and attendance by its members. After some BLSA gatherings Barack, Ken, and basketball buddies Frank Harper and David Hill would go to a pizza parlor on Mass Ave a bit north of the law school. Often joining them were two new 1Ls. Karla Martin was an African American 1987 Harvard College graduate; Peter Cicchino was white, gay, a year older than Barack, and had spent six years as a lay member of the Jesuits. Cicchino would become a defining member of the school's public interest community and a landmark figure in the public emergence of gay people at Harvard Law. Karla remembers Barack saying that he "wanted to be a change agent" notwithstanding his absence from student protest ranks. "It was clear he had ambitions," but "how that was going to play out was not clear."

Barack continued to spend more time on the basketball court than he did just hanging out. The new 1L class brought some new faces plus a familiar one into Hemenway gym's late-afternoon mix. Nathan Diament, an honors graduate of Yeshiva University, was short and fast. Greg Dingens had played defensive tackle at Notre Dame and three times won Academic All-American honors before graduating magna cum laude in 1986. Tom Wathen had been NYPIRG's executive director during Barack's four-month stint at City College four years earlier. Wathen recalled that they both did "a double-take" when they first saw each other. Compared to early 1985, when Barack was twenty-three years old, he of course seemed "more sophisticated" now at twenty-eight. "I was very impressed with him," Wathen remembered.

One day early that fall, Frank Harper, cochair of BLSA's community outreach committee, read a letter sent to the BLSA by Ronald A. X. Stokes, an inmate at Massachusetts's maximum-security state prison at Walpole, about thirty miles south of Cambridge. According to Harper, Stokes's message was a challenge: "you black students at Harvard should be ashamed of yourselves. There's a huge African American prison population; we never see hide nor hair of you." Harper called him at the prison and then "he calls me collect." Stokes sounded "very sincere" and mentioned that "we play a lot of basketball at Walpole."

That gave Harper an idea: "Let's have a basketball game." Harper called the warden, who thought, "This is very odd," but agreed to allow it if Harper could recruit five Harvard players. "Then I approach Barack. 'This isn't going to be an easy task,'" but they recruited black classmate Andre Nobles, a white Hemenway player, and a black poli sci grad student. A date was set, a van was rented, and when they arrived at Walpole, the warden gave them a stern briefing. "You're entering general population" and guards "are not going to be down there with you," only in sentry towers, Harper remembered. "If something happens, go to the corners, because we won't fire shots in the corners."

For the inmates, the game was major entertainment. "The entire prison surrounded the court to watch" as "the Walpole All-Stars" hosted five nervous Harvard gym rats. "We had a good team," Harper recalled, and "it was definitely a competitive game," at least until halftime. The Harvard players later joked that Obama played well until he asked the inmate who was guarding him what he was in for. "The brother said double murder, and Barack didn't take another shot," Harper remembered. "They won the game," but the inmates were grateful for the students' visit. Stokes told Harper that "he was getting out" within a few months, and "I stayed in contact with him after the game."[23]

By November, Rob and Barack's relationship with Larry Tribe had far outstripped the normal research assistants' role, especially for first semester 2Ls. The lead piece in the *Harvard Law Review*'s November issue was an essay by Tribe, rather than the annual foreword, and on page one, Tribe's first footnote stated that "I am grateful to Rob Fisher," top 3L Michael Dorf, two postgraduates, "and Barack Obama for their analytic and research assistance." It was a remarkable commendation, and Tribe had already invited Fisher and Obama to enroll in a spring seminar, limited to fifteen students, that would further consider the ideas expressed in his thirty-nine-page *Review* essay, "The Curvature of Constitutional Space: What Lawyers Can Learn from Modern Physics."

Tribe's description of the seminar invoked two contrasting conceptions of the U.S. Constitution: first that the original 1787 document was "Newtonian" in its promulgation of three branches, featuring "carefully calibrated forces and counter-forces," and second, that the twentieth-century idea of an evolving "living Constitution" was Darwinian. Tribe proposed exploring a third conception, one modeled upon the work of physicists Alfred Einstein and Werner Heisenberg, "focusing on how observers alter the nature of what is observed" and considering "the concrete geometry of the space-time continuum."

As obscure as that might sound, Rob and Barack were hooked after listening to Tribe at a late-October "organizational meeting" that began to sketch out what the selected participants would tackle. Rob wrote to Tribe that he had a "particular interest in . . . [t]he nature of the enterprise itself, that is, Law: This is the main focus of my thinking right now. Barack and I have been working on a meta-theory of the law—let's call it post-modern epistemology. (Though Barack hasn't seen this memo, so don't blame him for anything in it.) I will be looking at the Constitution as both a test for and an inspiration to that meta-theory . . . the role I see it potentially playing in the seminar is as a gadfly to your physics metaphor" but "not necessarily inconsistent" with it. It was no wonder that classmates marveled at Barack's "esoteric discussions" with Tribe and felt that he and Rob "were leagues beyond the rest of us."

At the *Law Review,* editing work continued on the dense Anthony Cook article, which would not be published until March. The November issue that featured Tribe's essay also contained the traditional foreword, authored by Erwin Chemerinsky of the University of Southern California, and case comment, by Frances Olsen of the University of California at Los Angeles. One 3L editor had been hugely impressed by Chemerinsky, a mesmerizingly intense speaker, and successfully lobbied for his selection for that prized role. Olsen was a well-known feminist scholar who had published a major article in the *Review* six years earlier. Her piece had undergone a very difficult edit, including the editors' insistence that the *Review* would not publish the word "bitch." As 2L editor Susan Freiwald, who witnessed one exchange, said, Olsen's experience at the "P-read" stage exemplified how *Review* president Peter Yu "thought he was smarter than everyone else, including professors." Yet as 3L Articles Office co-chair Andy Schapiro stressed, the perception that Yu was indeed "the smartest" had been the decisive factor in his election as president nine months earlier. Editor Pauline Wan, a 3L, agreed. "People wanted to elect the smartest person in the room," and "everyone felt Peter was the smartest person in the room."

But Yu's presidency was getting decidedly mixed reviews. During the summer, he had overseen the installation of a new computer system for the *Review*, but as the fall commenced, tensions grew. Chad Oldfather, who worked up to twenty hours a week at the *Review* as a work-study undergrad, remembered Yu as "not a warm and fuzzy guy," and 3L editor Barbara Schneider realized that he was "not a people person." Pauline Wan found him "remote," and Brad Berenson, one of the most involved 2L editors, felt Yu was "a slightly aloof figure." Kevin Downey, an active 2L, thought Peter was "not at all approachable" and "not a good leader." Supreme Court Office cochair Frank Cooper, one of four African American 3Ls, viewed Peter as "a quiet intellectual who was focused on the academic rigor" of each issue, but "he was not even a quiet leader."

Yu had a particularly strained relationship with Articles Office cochairs Gordon Whitman and Andy Schapiro, both of whom were lefty-liberals, in part over their selection of articles like the one by Anthony Cook. Susan Freiwald remembered Yu editing one piece and remarking that "every good idea in here comes from me." But Freiwald thought Yu's interactions with Frances Olsen were inexcusable, that he was "intellectually beating up" on her. During one loud, angry phone exchange, Freiwald remembers Yu "just screaming at her and her screaming back," and it left Freiwald thinking that "Peter was an asshole."

As the editing of the Cook manuscript proceeded, more and more of the work fell to Christine Lee. Articles Office member David Goldberg, a 2L, realized the piece "was way too long" and "a little jargony," and fellow 2L editor John Parry saw that Christine "worked very hard" and "reorganized it, and I think made it coherent." Christine willingly put in lots of time, but as the semester progressed, she concluded that Barack was more interested in playing basketball than doing his share of the necessary sub-citing and other work on the Cook manuscript. "We were a team," but "he would play basketball religiously, including when there was a sub-cite due," Christine remembered. "He definitely did the bare minimum," and "it was just building my resentment" as "other people were covering for the drudge work he wasn't doing."

Executive editor Tom Krause noted that at the outset, "Barack was like the primary editor," but "somehow it kind of was taken over by Christine, who ended up doing all the work." Krause was not a fan of the article, but in his eyes, Christine, "to the detriment of her own grades and class work, was doing work he"—Obama—"could have been doing." As the piece moved forward, Krause stepped up his own involvement, and he and Christine—polar political opposites—became personally close "doing the things that Barack had not done," Christine recalled. In retrospect, even Gordon Whitman, Cook's main proponent, realized the article was "pretty impenetrable." Krause thought "it would have been worse if Christine and I hadn't worked on it," but Whitman's verdict was appropriately biting: "In hindsight, what the hell was that all about?"

The real meat of each November issue was the individual case comments, which were anonymously authored by 3L editors. The 1989 issue surveyed twenty-five Supreme Court decisions, which brought the issue to a robust 404 pages. This earned Peter Yu a stern rebuke from Erwin Griswold: "I would like to suggest that one of the major tasks of the President is to edit out vast quantities of unnecessary words, and to keep the overall size of each issue, and of the volume, under control." Yu responded that he shared that concern, but he cited "the extraordinary number of leading cases from this past Term" and promised that the entire volume would not be any larger than in years past.

Griswold replied that he was "not wildly enthusiastic about" the Tribe essay; "essentially the same arguments can be made without any . . . farfetched allusions to concepts developed in the field of physics." In contrast, Griswold found Chemerinsky's foreword "both interesting and powerful" and thought Olsen's comment was "imaginative and stimulating." He thought some of the case comments were "very good," but "surely much too wordy." Griswold was happy that the December issue, at 196 pages, would be less than half the size, but again he said the sole article, a 105-page piece on statutory interpretation, was "awfully wordy." A thirty-one-page review of a book by philosopher Richard Rorty was "much too long" and "quite indigestible," a "depressing" example of something "written by a professor for the professors" rather than for "active, practicing lawyers."

Griswold was happy with the two six-page "Recent Cases" summaries, one of which, by Barack's basketball buddy Tom Perrelli, a 1988 magna cum laude graduate of Brown, exemplified how 2L editors were expected to do at least one brief piece of individual writing in addition to their pool or team assignments. Barack may not have been doing his share of the work on the Cook edit, but by late October, he was well under way writing a comment on a late 1988 decision by the Illinois Supreme Court, which held that a young child injured in an auto accident while she was still a fetus cannot sue her mother for monetary damages. Laurence Tribe was completing work on a book surveying the ongoing U.S. political debate over abortion, and Barack was one of more than a dozen of his research assistants whom Tribe had asked to read and summarize relevant materials.

Barack mentioned the case comment he was writing to his mother at the same time he told her that, unlike the previous year's Christmas holiday, he was coming to Honolulu and Michelle Robinson was coming with him. Maya had not returned to Barnard for her sophomore year and had joined her mother in Jakarta before returning to Honolulu, where she was waitressing at a restaurant. Barack had not seen his mother, who was now forty-six years old, since Christmas 1987, when he had taken Sheila Jager to Hawaii to meet his family. Just after he started at Harvard, Ann had moved back to Jakarta to take a job at the People's Bank of Indonesia—Bank Rakyat Indonesia, or BRI—the country's oldest financial institution and one specializing in microfinance, small loans to artisans and retailers. Ann had lost interest in completing her long-delayed dissertation, telling her old friend Julia Suryakasuma that "the creative part was over long ago, and it's just a matter of finishing the damn thing."

But her new work put her in direct contact with the people whose artisanal pursuits she had devoted her fieldwork to studying. Ann also had begun a fulfilling relationship with Made Suarjana, a married journalist eighteen years her junior, and in November, she wrote to Alice Dewey, her longtime mentor, to tell her that she would be returning to Honolulu for the holiday. She told Dewey that Maya's time in Indonesia "seems to have done her a lot of good, unstressed her and renewed her self-confidence" after a rocky first year at Barnard. "Barry is also coming at Christmas with a new girlfriend in tow. He is still enjoying law school and writing pro-choice opinions of the abortion issue for the *Law Review*."

No matter what Barack told his mother about his case comment, which went to press in early December, its pro-choice stance was extremely measured. The Illinois court had defended its holding on the grounds that otherwise the mother and her fetus would be "legal adversaries from the moment of conception until birth." Obama commended the court for its "thoughtful approach" and wrote that the case "highlights the unsuitability of fetal-maternal tort suits as vehicles for promoting fetal health." He said the suit "also indicates the dangers such causes of action present to women's autonomy, and the need for a constitutional framework to constrain future attempts to expand 'fetal rights,'" because "fetal-maternal tort suits affect even more fundamental interests of bodily integrity and privacy" on the part of women. He wrote that the state may have "a more compelling interest in ensuring that fetuses carried to term do not suffer from debilitating injuries than it does in ensuring that any particular fetus is born," but again stressed the primacy of "women's interests in autonomy and privacy." Obama concluded the piece by recommending that "expanded access to prenatal education and health care facilities will far more likely serve the very real state interest in preventing increasing numbers of children from being born into lives of pain and despair."[24]

Before fall classes ended on December 8, two Chicagoans visited the law school. Elvin Charity, a 1979 Harvard Law graduate who had worked for Harold Washington, was now one of two black partners at the Chicago law firm of Hopkins & Sutter. Charity was on his firm's recruiting committee "to help to promote the hiring of African-American lawyers," and Harvard was an annual target. Second-year students traditionally split their summers between a pair of firms, and although Barack would return to Sidley & Austin for part of the upcoming summer, Hopkins & Sutter was another attractive Chicago firm. To Charity, "Barack was particularly striking" in part because he showed up for his interview "pretty casually dressed," rather than in a suit and tie. "He just seemed so nonchalant about the whole thing," with "a calmness and a seeming maturity beyond his years." Charity knew the significance of Barack's *Law Review* membership, but Obama "did not seem to be full of himself" and an offer was quickly extended and accepted.

The second Chicago visitor was Michelle Robinson, who along with an Asian American colleague was there on behalf of Sidley to speak to minority 1Ls about job search techniques at an evening session sponsored by all three ethnic—black, Latino, and Asian—student groups. Several '91 and '92 BLSA women who had heard Barack talk about Michelle realized this would be a perfect opportunity to "go see who this person is," and "we were very impressed with her," Jan-Michele Lemon remembered. "She was very engaging and warm and friendly."

Michelle's Wednesday visit came just two days before the end of classes, and the next Monday Barack faced the first of three fall semester final exams. Upper-class exams took place in December, not mid-January, but Barack had Edley's Ad Law and Tribe's Con Law take-homes back-to-back on the first two days of exam period. Edley's emphasized that "it is important to demonstrate that you have synthesized the assigned readings and course materials," and students had to answer six out of nine questions, with responses limited to 150 to 500 words apiece.

Tribe's open-book final posed just one essay question, with answers limited to 1,750 words. The hypothetical situation involved an abortion clinic employee who had been fired for telling a woman the sex of her fetus. The woman then aborted the fetus, citing avowedly religious reasons, because it was female. The fired employee had been denied unemployment benefits, and Tribe's hypothetical played off a recently argued U.S. Supreme Court case, *Employment Division v. Smith,* which also involved a state's refusal of unemployment benefits to a worker fired because of religion-based conduct. Tribe asked students to write their best decision of the case. Six days later, on December 18, Barack and Rob had their last fall exam, for Kraakman's Corporations class. Then Barack was free to head to Honolulu, though upper-class students' single winter session course would begin promptly on Tuesday, January 2.[25]

Barack's introduction of Michelle Robinson to Ann, Toot, Gramps, and Maya signified, just like Sheila's visit two Christmases earlier, how serious this six-month-old relationship was. After Michelle's early December visit to Harvard, "it was clear that they were 'a couple,'" her close Sidley friend Kelly Jo MacArthur recalled, and "the discussions turned fairly soon to whether they were going to get married or not." That clearly came up during their time together in Honolulu, because soon after they flew back to the mainland, Ann Dunham sent her close Indonesian friend Julia Suryakusuma a description of

Michelle. "She is intelligent, very tall (6'1"), not beautiful but quite attractive. She did her BA at Princeton and her law degree at Harvard. But she has spent most of her life in Chicago," so she was "a little provincial and not as international as Barry. She is nice, though, and if he goes ahead and marries her after he finishes law school, I will have no objections."

Almost the first day he returned to Cambridge, Barack wrote to Sheila Jager, who two months earlier had returned to the U.S. from South Korea. Barack had written to Sheila regularly throughout her time abroad, once sending her a short story he had written and another time "an annotated copy of Tocqueville's *Democracy in America*." By the time she returned, Sheila had told Barack she had applied for and won a Harvard teaching fellowship in Asian Studies with the eminent scholar Ezra Vogel. Before the end of January, Sheila would be moving to Cambridge and renting an apartment at 5 Crawford Street. She recalled that "the job at Harvard had nothing to do with Barack being there, per se, although I can't rule out that I unconsciously applied to Harvard to be near him." But she also explained, "Then again, I wasn't going to turn down a job there simply because he was there," and during their time apart, she had begun a relationship with someone else.

In his January 2 letter, Barack told Sheila about Michelle and her Chicago roots at some length, but as he had in his 1989 letters, he also wrote about "his turmoil about our relationship" and his thoughts about "who he was and what he wanted to become." Sheila said that Barack described "his thoughts about his love for me, marriage, his 'destiny,' his race/identity, and how all this affected his relationship" with Michelle. But once Sheila arrived in Cambridge in early 1990, "we continued to see each other occasionally," despite the deepening of Barack's relationship with Michelle. Years later Barack would write that 1989–90 "was a difficult, transitional period in my life." Indeed it was, for it was a period characterized throughout by two powerful, overlapping relationships.[26]

For the 2 and 3Ls' three-week January winter term, Barack and Rob, along with Section III friends Gina Torielli and Jennifer Radding, took Evidence for three hours each weekday morning with Professor Charles R. Nesson, one of the law school's best-known faculty members. *Problems, Cases, and Materials on Evidence,* by Nesson and Eric Green, was the text for a course that introduced "what the trial system is about." Nesson's open-book exam took place on January 26, by which time Barack and Rob were already into their work for Tribe's upcoming seminar. They also jointly selected four additional courses for the spring semester that would begin January 31: Taxation, which was strongly recommended for all 2Ls, with Professor Alvin Warren; Local Government Law, with Professor Gerald Frug; Jurisprudence, with the world-renowned Roberto

Mangabeira Unger; and Law and Political Economy, with Chris Edley, whom they had enjoyed so much in the fall.

By early January, the *Law Review*'s 2L editors were focusing on the upcoming election of a new president, which would take place on the first Sunday in February, and then the subsequent selection of a new masthead and office cochairs. In the eyes of their colleagues, some 2L editors, like Julius Genachowski, a 1985 magna cum laude graduate of Columbia, had long seemed like all-but-certain presidential candidates. Others who had logged long hours of scut work at Gannett House, like David Goldberg and Christine Lee, were presumed to be natural contenders. But 2Ls who thought Peter Yu's style was far from ideal pondered who would be the *Review*'s best leader, irrespective of who might be the smartest editor. In early January, Susan Freiwald, who had been so deeply offended by Yu's behavior toward Fran Olsen, asked Barack to have lunch with her at Au Bon Pain in Harvard Square. "I told him that I thought he ought to run for president" and offered to encourage others to support him if he did. Barack replied that "he was going to think about it," and he raised the question with David Rosenberg. "You're an idiot. It's a waste of time," Rosenberg remembered telling him, explaining that being president would preclude him from writing a note, which anyone headed toward academia should do. Neither of Barack's two closest non-*Review* friends, Rob and Cassandra, thought he had any interest in running for the position, nor did Michelle Robinson and her family.

In the previous year's election, all four African American editors—Crystal Nix, Frank Cooper, Jennifer Borum, and Ira Daves—spurred by earlier BLSA achievements, had tried to win top positions on the *Review*. Nix's selection as a supervising editor, a top-five post, and Cooper's role cochairing the Supreme Court office, left them strongly committed to see that success continue. Robert Granfield, the sociologist who had perceptively interviewed so many Harvard law students, was aware of a painful downside of the *Review*'s publicly acknowledged affirmative action selection policy. A prior 3L African American editor had told Granfield that the policy created "a real stigma" that left black editors' status open to question. "Before law school, I achieved on my own abilities. On *Law Review,* I don't feel I get respect. I find myself working very hard and getting no respect." Ken Mack agreed. "Being on the *Law Review* was the most race-conscious experience of my life," he recounted. "Many of the white editors were, consciously or unconsciously, distrustful of the intellectual capacities of African American editors or authors."

Attaining leadership posts on the *Review* disproved any such derogatory attitudes, but Nix's SE position was the highest post a black editor had yet won. No African American had ever been president or treasurer, and the informal rule was that anyone aspiring to a masthead position needed to compete for the presidency as a way of demonstrating the seriousness of their interest.

Of the five African American 2Ls, Princeton graduate Monica Harris had spent little time in Gannett House, and young Christine Lee's outspokenness counterbalanced her demonstrated work ethic. Quiet Ken Mack, whose magna cum laude undergraduate degree from Drexel University was in electrical engineering, seemed an unlikely fit, but both Obama and Rebecca Haile, an Ethiopian-born 1988 Williams College graduate whose family had been forced to flee their homeland when she was ten years old, were well liked among their 2L colleagues. Nix, Cooper, Borum, and Daves pressed Lee, Mack, Obama, and Haile to try for masthead posts, but Barack continued to exhibit disinterest until the eve of the deadline when presidential contenders needed to file a one-page statement of candidacy.

That night the black *HLR* members and several African American 2L veterans of Section III gathered for dinner at Barack's apartment. Nix, Cooper, Borum, and Daves came with an agenda, even if others present were unaware of it: to convince Obama to join Mack, Lee, and Haile in running for the presidency. Crystal Nix, a former *New York Times* reporter, who had known Michelle Robinson at Princeton, was the leader. She "was extremely active in encouraging him to do it," Frank Cooper recalled. Jennifer Borum remembered "Barack sitting on the floor" as the conversation proceeded. Mack listened closely as the 3Ls stressed how there had never been an African American president of the *Review,* but they believed Barack could win.

John "Vince" Eagan, an older, generally quiet Section III veteran, was also there that night. Eagan knew that Barack was "one of the smarter guys around," but he was surprised by some of the books on Barack's shelves because "he didn't really come off as a leftist in any sense." As the 3Ls talked about Obama's chances, "I just said 'I think you should kick that door in.' That's all I remember saying. It was just like a throwaway comment." Jennifer Borum recalled that before the evening was over, Barack had changed his mind. "I remember him saying, 'I think I'm going to throw my hat into the ring.'" Of the black 2Ls, "he was the last one" to agree that all four should run. A few weeks later, Barack said that Vince Eagan's comment had been decisive. "I said I was not planning to run, and he said 'Yes you are, because that is a door that needs to be kicked down, and you can take it down.'" Shortly thereafter, Barack again recalled Eagan's comment: "'There's a door to kick down,' the friend argued, 'and you're in a position to kick it down.'"

An astonishing nineteen 2Ls—half of the 1991 class of editors—filed to run for president, and one evening at the end of January, a candidates' forum took place in a tiered classroom. The contestants each made a personal statement, and then they responded, one by one, to questions from other 2Ls and interested 3Ls. One asked what leadership skills they had, another inquired which Supreme Court justice they would most like to clerk for. Sarah Eaton had started law school with the now-3L class and made law review before taking a year off, so her 2L classmates that fall were all new faces. She, along with Section III veteran Lisa Hay and Devo office 2L Julie Cohen, had agreed to run after noting the absence of interested women. Eaton barely knew Obama, but that evening "he was amazingly impressive," she recalled. "He just blew everyone away." Likewise, prior to the forum, Barack "was somebody that frankly I hadn't really noticed," 3L Jane Catler explained. "But on that day, wow," she remembered. "He just leapt out of the pack. He was charismatic."

Then a question was posed about whether affirmative action should remain a part of the *Review*'s selection process. Jim Chen, the most outspoken of the 2L conservatives and a summa cum laude graduate of Emory University whose family had immigrated to the U.S. from Taiwan almost empty-handed, forcefully advocated abolishing it. As some remember, Barack clearly supported affirmative action, but he also said he understood and appreciated the principled arguments against it and suggested that the policy be revisited in full after the new officers were selected. Ken Mack remembered that "that really made an impression on a lot of people," and most especially "a huge impression on the conservatives" who comprised one-fifth or more of the 2L editors. Fellow candidate Julius Genachowski recalled that Barack "was really unique in being able to have other people believe that he understood their point of view." Brad Berenson, a 1986 summa cum laude Yale graduate who was the informal leader of the 2L conservatives but not a presidential contender, had a further perspective. "Barack told at least me, and maybe others, that he had declined to identify his race on his *Law Review* application," in what Berenson viewed "as a way of signaling, 'Hey, I get why you might be opposed to affirmative action and why you might not even think it's in the best interests of its supposed beneficiaries.'"

Prior to the election on Sunday, February 4, Barack did very little presidential campaigning: only 2L editor Kevin Downey recalled Barack directly asking for his support. "I was a little surprised," because during the fall Kevin, like others, had sensed no interest in it from the "very mature" and "somewhat remote" Barack. "I can clearly remember him saying to me that he thought he'd like to be the president of the *Law Review* but if he didn't get that, he wasn't going to do anything else." Instead of politicking, just three days before election, Barack and Rob gave Laurence Tribe a powerful critique of Tribe's latest manuscript using physics for constitutional analysis. "We basically completely trashed his ideas," Rob recalled. Tribe's response indicated what a remarkably unique relationship had developed between the eminent professor and his two older but nonetheless precocious second-year students. "Oh my gosh," Tribe wrote to them. "Your memo pretty much sweeps me off my feet. The trick will be to land on some place better than my back. I will count on the two of you for help. Would you consider a coauthored essay, perhaps with our names listed alphabetically? Who knows where it should appear but it deserves to be written." Tribe signed his name simply "Larry."

The presidential election proceedings commenced at 8:30 A.M. on February 4, in the Ropes-Gray Room in Pound Hall. In keeping with another *Law Review* tradition, the nineteen candidates were sequestered in the adjoining kitchen, where they prepared successive meals for the fifty-plus voters. Kevin Downey remembered his home phone ringing and managing editor Scott Collins telling him, "We really need your participation." Another 2L, Michael Weinberger, who was already there, recalled his wife contacting him to say that their car had been stolen, but he stayed for the election.

Many 2Ls were surprised to see small boxes containing each candidate's pool and editing work there in case anyone wanted to examine it. The 3Ls who had worked with each candidate offered introductory reports about them before the floor was opened for comments and questions. With nineteen contenders to consider, the entire morning was consumed by those discussions, during which informal lobbying took place, especially by 3Ls who favored a particular candidate. Who would win the presidency was not the only consideration, because the discussions of the candidates that day would redound a week later when the rest of the masthead was selected and office cochairs were elected.

From the semiautonomous Devo office, 3L cochairs Debbie Brake and Audrey Wang sought to promote their 2Ls, particularly Julie Cohen and also Rebecca Haile and Jennifer Collins. Articles Office cochairs Andy Schapiro and Gordon Whitman were energetic proponents of David Goldberg, who was a top-notch contender. Frank Cooper recalls that he, Crystal Nix, and Jennifer Borum were "hyper-focused on Barack. . . . We all encouraged other African American editors to run because we felt like we needed that representation" on the masthead, "but the focus and the pressure was on getting Barack to win." One by one, Crystal and Frank wooed fellow 3Ls to support Barack. "We felt like we had enough of a coalition that we had a chance to win. No one knew if we could actually pull it off, but we had a chance to win," Cooper explains. "People were conscious of the fact that he would become the first African American president of the *Harvard Law Review*. No one really said it" and "for the most part it wasn't talked about," he adds. "We were also very conscious that we needed a group of people who were not African American on the *Law Review* to be vocal about it" and "we all kind of worked several people to see who would be open to it. We knew there were some people who were clearly in line with that right away" and who became "part of our coalition," particularly 3Ls Radhika Rao, Dan Bromberg, and Micki Chen.

The descriptions and ensuing discussions of all nineteen candidates lasted until noon. Few participants recalled the initial voting process in precisely the same way, but with so many candidates, the first round of paper ballots called for voters to select five candidates. As a time-consuming tally took place, the atmosphere was decidedly relaxed, with some editors playing Scrabble, others reading the Sunday *New York Times,* and a trio of women calmly knitting. Once the 3L officers had cumulated the votes, there was a clear and substantial gap between the top numbers and the lower half. When no dissent was registered, ten of the nineteen were told that their cooking duties were over, and they were now voters rather than contenders as the second round of discussions commenced.

Three of the four African American candidates—Barack, Rebecca Haile, and Christine Lee—made the cut. So did two other women, Lisa Hay and conservative favorite Amy Folsom Kett, a 1984 Oberlin graduate who had gotten a Harvard music M.A. before law school. Until then, "I was pretty apolitical" and "pretty middle of the road," Kett said, but the law school's intense conflicts left her feeling like a conservative. The ten 2Ls who had been eliminated uneasily joined the larger group, and their presence swung the group away from the almost two-to-one imbalance in favor of the 3L class that had characterized the first round. Even with the slate of candidates cut in half, voters could still cast their ballots for multiple contenders, and the afternoon proceeded slowly as the nine remaining contenders were discussed and debated.

Shortly before dinnertime, ballots were again distributed, and after a lengthy tally, the individual totals showed a clear gap separating a top six from a lower trio. Tom Perrelli, Christine Lee, and Mike Froman, a 1985 Princeton graduate who had earned an Oxford doctorate before law school, left the kitchen and became voters. Of the six remaining semifinalists, three were women: Amy Kett, Lisa Hay, and Rebecca Haile, and two, Haile and Barack, were African Americans. With the field so reduced, the tenor of some comments shifted. Michael Cohen, a 2L, remembered that "at first people would speak for folks," but as the day progressed "people would start speaking against someone." Without naming anyone specifically, Mike Froman said the president should possess the best analytical skills and intellectual ability, rather than being selected because of "his charm and his poise." Listeners thought Froman was speaking on behalf of David Goldberg, but they understood whom Froman was putting down. In contrast, Christine Lee, who had the most negative perspective on Obama's editing contributions, put those feelings aside in the interest of African American advancement. Christine became "a very big and vocal backer of Barack," 3L executive editor Tom Krause recalled.

Amy Kett's image as a conservative and presumed opponent of affirmative action attracted criticism, and after the third round of balloting, she and Lisa Hay exited the competition. This left Obama, Rebecca Haile, David Goldberg, and Jean Manas, a 1987 summa cum laude Princeton graduate and joint degree student who had begun law school with the 3L class. Haile was pleasant and mild-mannered, and generated no intense feelings, but Manas, who had grown up overseas, was seen as an outspoken leftist and was much better known among 3Ls than by his 2L classmates. The perception that Manas was the 3Ls' favorite, coupled with the conservatives' belief that "he was just basically the Antichrist," as Brad Berenson recalled, was a decisive combination. With everyone still able to cast multiple votes, and with a good many of David Goldberg's most enthusiastic supporters also positively inclined toward Obama, first Haile and then Manas were eliminated.

After 9:00 P.M. the field was reduced to David Goldberg and Obama, and for the first time all day, the now seventy-plus voters each had a single vote. Frank Amanat, a 2L, remembered that "the feeling of drama in the room increased, in a way that I think was palpable to most of the people in the room." Frank Cooper recalled that when it got to the final two "it was protracted. There were many, many discussions. . . . Everyone in the room, at some point, weighed in and spoke." For many of them, the issue was *what* the president should be, above and beyond *who*. In the 1989 election, tradition suggested that the president should be the best possible *editor,* someone who could earn the respect of the professorial authors whose work he would "P-read," as well as *HLR* colleagues. John Parry, a 2L, knew that David Goldberg was "definitely seen as one of the intellectuals on the *Law Review,*" as well as being "well liked." David Blank, the most outspokenly conservative 3L, championed the traditional view. "Blank was a strong proponent for Goldberg," Frank Cooper remembered, and "it got heated" regarding Obama "because some people felt like he did not represent the traditional view of *Harvard Law Review:* academic rigor . . . that he was almost too worldly." Some argued that a president with "a principled approach to the law" was more important than someone with "experience outside of academia." But Barack's performance in Section III, and in Tribe's Con Law, left no doubt at all that he was "strong academically" as well as older and more mature.

Peter Yu's presidency had suggested that intellect, editing prowess, and *Bluebook* skills were no sure recipe for success. "I think his presidency and his style paved the way for Barack," Frank Cooper explained. Diane Ring, a 3L, saw it similarly. "They're looking for something different next time around," she thought, "someone with more people skills." Jackie Scott, a 2L, concurred. "You needed someone who wasn't just that academic person, who could bring the *Review* together. That was an overriding concern," and one the conservatives shared. "David was regarded as a very political person, and Barack was not," conservative quarterback Brad Berenson explained. Liberal 2L Scott Siff agreed that "Barack was perceived as less left than David," but it was not only the conservatives who shifted decisively to Barack as the clock neared midnight.

"A lot of the conversations also happened in separate groups," Frank Cooper explained, and as the evening progressed, both Cooper and Crystal Nix quietly mustered their forces. As an SE, Crystal had been a "kind of friend-mentor" toward nonpolitical 2L Anne Toker, and Toker remembered Nix "coming to talk to me and kind of making the pitch for Barack." Crystal likewise lobbied 3L Mark Martins. In terms of leadership experience, no editor could trump Martins. Valedictorian of his West Point graduating class, Martins had gone to Oxford as a Rhodes Scholar before serving two years as a platoon leader in the 82nd Airborne Division. When Martins had drawn the only pool assignment more loathed than a sub-cite—cleaning Gannett House—classmates laughed when he brought his own vacuum cleaner. But Martins too agreed with Nix's argument. "We're not merely trying to pick the best editor, we're trying to pick the best leader was how I saw it," and Nix "mentioned the importance of that to me . . . before I voted."

Along with Radhika Rao and Micki Chen, Frank Cooper was able to enlist both Dan Bromberg, his former roommate, as well as his present one, Jean Manas, following Manas's elimination. "Toward the end, you could see it start to swing," Cooper remembered. "People like Micki and Radhika and Dan, they were absolutely critical in swinging it because you really needed someone who was not African American to create momentum behind the fact that this person could be the right president of the *Law Review* for everybody on the *Law Review*." Tom Perrelli saw it similarly. "What often moved that room was an unexpected person who hadn't said anything stepping up and saying 'You know, I really like'" someone of a different political stripe. Daniel Slifkin was audibly British and had arrived at Harvard with advanced standing after earning two First Class Honors undergraduate degrees at Oxford. Hours earlier Slifkin had summarized Tom Perrelli's work, but during the fall, Slifkin had observed that Obama "was a very exceptional person from the get-go: more mature, more poised, more articulate," as well as "very smart." In his distinctive voice, Slifkin now stated the case and declared, "we should make Barack Obama the president of the *Law Review*."

Earlier Amy Kett had been tarred as an opponent of affirmative action, but now she spoke up to say, "I think the obvious choice is Barack." Not knowing the context, Amy did not understand why Brad Berenson and fellow conservative Adam Charnes were nodding vigorously in approval as she softly declared, "I would love to be represented by Barack." The African American editors knew the symbolic importance of the approaching vote, but so did editors of every ethnicity. Jorge Ramírez, a 3L, had spent seven years in Cambridge, first as an undergraduate, and among the 3L editors "there was a core group that saw this as an historic opportunity." Tracy Higgins, a 3L who had known Crystal Nix as a fellow Princeton undergraduate, fully appreciated that "it would be a kind of historic moment for the *Law Review*." Pauline Wan agreed. "I think people wanted to elect a black president," for "that would be a coup for our year." To her "the personality of the man" loomed far larger than race, because Barack's "personality is capable of uniting a group of very disparate individuals who all think they're smarter than everybody else in the room. To be able to do that is a huge talent." Barbara Schneider, a 3L, agreed that "it was his personality" rather than race or politics that proved decisive, as did 3L Patrick O'Brien. "He had a gravitas and wisdom and maturity about him," and without question Barack "was a race-transcending guy."

Frank Cooper knew that "you had some people who were just torn" between the pair of finalists, but in the end, even David Goldberg's two most ardent backers voted for Barack. "We bailed on David because we wanted to elect the first black president of the *Harvard Law Review*. I don't think it's that complex a story," remembered Articles Office cochair Gordon Whitman. "It was a sense that I could be part of history. We knew we were breaking a big barrier, and that we'd be famous for it." Throughout the day Whitman had been snapping photos to capture the occasion, but "I've always felt bad for David." Andy Schapiro liked both finalists, and "I don't recall having to choose between Goldy and Barack until late," he recalled, using David's nickname. "Goldy was my best friend," but Schapiro too voted for Obama.

Dinner had ended hours earlier, and with the day's cooking complete, the finalists had retreated to the editor's lounge on the second floor of Gannett House to await the result. David Goldberg remembered that a bottle of vodka was there. "We drank shots of vodka," with Barack remarking, "Whatever doesn't kill you makes you stronger." They were alone except for undergraduate assistant Chad Oldfather, who would remember Peter Yu finally coming up the stairs to ask the two finalists to come with him to Pound Hall. It was now after midnight, and Oldfather left without knowing who had won.

In the Ropes-Gray Room, the weary electors knew the runner-up would enter the room first, and everyone applauded when David Goldberg entered. Goldberg had been "quite sure" for hours what the outcome would be, and so his weary smile—captured in one of Gordon Whitman's photos—was entirely heartfelt. Then Goldberg stood aside and Obama entered the room to even greater applause.

"I remember the look on his face," Jennifer Borum recalled. "It was like he was in shock." Ken Mack remembered that he was by the door, "and so I just sort of stepped forward, and Barack stepped forward, and we just spontaneously hugged. We hugged for a very long time." Mack knew when Goldberg had appeared first that Obama's victory would be an "immensely powerful symbolic breakthrough" for African Americans at the country's most famous university, but only during Ken's embrace did Barack grasp its meaning. "It was a hard hug and it lasted for a while," he recounted a few days later. "At that point I realized this was not just an individual thing" and "not about me," or at least "more than just about me. It was about us," about an African American for the first time attaining the pinnacle post at the world's top law school. As they hugged, "tears rolled down both of our faces," Mack remembered, and Gordon Whitman caught the moment in a dimly hued photo as Barack's two hands grasped the back of Ken's white shirt.

It had been "an exhausting, draining day," 3L Barbara Eyman explained, sixteen hours all told, and "it wasn't really until after the vote was over that we collectively felt the historic nature of the vote." Mack understood how "people really got emotional at that moment," with Christine Lee and the other African American editors erupting in joy. This jubilation was puzzling to any 2Ls who had not been aware that Obama's election would represent a pathbreaking step. Michael Weinberger "had not realized that this would have any kind of a racial meaning. It's just a law review." Only when he saw the African American editors "crying and running and hugging" did Weinberger recognize the moment's symbolic importance to them. "That just had not been part of my consciousness" when voting for Obama. "I was completely clueless."[27]

Derrick Bell awoke to his ringing telephone. The clock showed 12:50 A.M. It was Crystal Nix, calling to tell the law school's most outspoken black professor about their victory. "All of us knew that if we pulled it off, it would be big news," Frank Cooper realized. "It was such a moment of high for every African American on that campus," David Hill remembered. Frank Harper explained that everyone felt "a sense of pride and achievement," that a black student could attain such an eminent position. By the time that Obama finally strolled into Gannett House, well after 12:00 P.M. on Monday afternoon, editorial assistant Susan Higgins had been swamped for hours with press calls. She had arrived that morning wondering "Who's my new boss?" and was surprised it was Obama because "he was definitely not in the mold of everybody else" who wanted a masthead position as a springboard to appellate and then Supreme Court clerkships. Many editors were bemused as well as elated by the press calls, and joshing began around a cast list for "*The Barack Obama Story,* a Made-for-TV Movie, Starring Blair Underwood as Barack Obama."

One of Monday's callers was 1989 *Law Review* member Sheryll Cashin, who was clerking for Judge Abner Mikva, a former Chicago congressman who now sat on the prestigious U.S. Court of Appeals for the District of Columbia Circuit. Cashin and Crystal Nix were close friends, and though Nix had not awoken her in the middle of the night, she had called early that morning with the historic news. Cashin in turn told Mikva, who in his ten years on the court had become known as a "feeder" judge, someone whose law clerks often went on to work for Supreme Court justices such as William J. Brennan and Thurgood Marshall. Judicial guidelines prevented formal clerkship offers from being extended until later in the spring, but Mikva was immediately interested in hiring the *Review*'s first African American president. Cashin knew Obama from the previous year, and the message Mikva asked her to convey was clear. "I was charged with letting him know that if he wanted a clerkship with him, he had it," she remembered.

Sometime Monday Cashin reached Obama. "I'm calling you because Judge Mikva's very interested in you" and "very much would like you to apply" to clerk for him come 1991. Barack's response was immediate: "I'm flattered, but no thanks—I'm going back to Chicago." Cashin paused. "I was shocked—this was unheard of," for especially among *Review* officers, "nobody didn't clerk." Cashin tried to persuade Obama to at least apply, but he was adamant. "Tell the man thank you, but I'm going back to Chicago," Barack reiterated. "I was just floored," Sheryll recalled, especially because a clerkship with Mikva virtually guaranteed a subsequent Supreme Court clerkship, and large law firms were known to pay newly minted high court clerks a $35,000 bonus. Cashin remembered that Mikva reacted with "amused surprise" when she recounted the astonishing conversation.

In Chicago, Michelle Robinson reacted similarly when Barack called her. "You're not going to clerk for them? You're kidding me!" and "He's like 'No, that's not why I went to law school. If you're going to make change, you're not going to do it as a Supreme Court clerk.'" When Mikva visited Harvard a few weeks later and met Obama, "I teased him about not interviewing with me." Barack cited his desire to return to Chicago and made it "absolutely" explicit he intended to run for office. Obama mentioned Mikva's offer to Laurence Tribe, seeking reassurance, and Tribe agreed, but Cashin and Michelle Robinson were far from alone in feeling baffled by his lack of interest in attaining a Supreme Court clerkship.

Obama spent the balance of Monday giving interviews, and by mid-evening the Associated Press had sent out a story headlined "First Black President of Harvard Law Review Elected" all across the country. Obama told the wire service, "I wouldn't want people to see my election as a symbol that there aren't problems out there with the situation of African-Americans in society," with Roseland high schoolers clearly in mind. "From experience, I know that for every one of me there are a hundred, or thousand, black and minority students who are just as smart and just as talented and never get the opportunity." He also stressed that his selection "sends a signal out that blacks can excel in competitive situations like scholarship." Again alluding to his experience in Chicago, Obama added, "I want to get people involved in having a say in how their lives are run. More and more of that needs to be done." The AP stated that Obama "has not ruled out a future in politics."

Tuesday morning Obama's election made news in the *New York Times*, the *Boston Globe*, and the *Harvard Crimson*, as well as in papers that ran the AP story. The *Times'* story showed that Obama was a disciplined interviewee, for his comments echoed what he told the AP. Reporter Fox Butterfield wrote that "Mr. Obama said he planned to spend two or three years in private law practice and then return to Chicago to reenter community work, either in politics or in local organizing." His quotes in the *Globe* also tracked his statements to the AP, but he told the *Crimson* that his election marked "a significant change from the Harvard Law School of the past," and the paper reported that Obama "plans to continue working in the public sector when he graduates from law school, but is unsure what form that work would take. He said he will consider anything from running for elected office to setting up community service programs."

One day late, the *Chicago Tribune* highlighted the hometown tie. "Over the long run, the way to improve the conditions in the cities and schools—to fight crime and drugs—is to work on the local level," Obama stated. In an ironic segue, the *Trib* then quoted Johnnie Owens saying that while at DCP, Obama "honestly evaluated his performance and made up his mind to do better." The *Trib* said Obama will "spend a couple of years practicing law after graduation next year and then it's most likely back to community organizing, maybe politics." But Obama emphasized that "I'll definitely be coming back to Chicago," which he called "a great town" and "an ideal laboratory."

In far-off Honolulu, the *Star-Bulletin* headlined "Ex-Islander Gets Prestigious Harvard Post," but a conversation a reporter at the newspaper had with Madelyn Dunham, now sixty-seven years old, produced a potpourri of misinformation. Obama Sr. had never been "Kenya's finance minister," Barack's first job in New York was not as "a social worker," and in Chicago he had not formed "a consulting firm to advise people who wanted to set up small businesses."

Two student journalists at the law school's own *Harvard Law Record* wrote a lengthy profile, speaking with Rob Fisher, Christine Lee, Cassandra Butts, and Laurence Tribe in addition to Obama, who again repeated almost verbatim what he had said to the AP and then the *Times*. But for the law school audience, he made other points too. "My election is a positive sign in that it shows people are ready to put in leadership positions black folks who have strong concerns about black issues," he asserted. "What happens at Harvard really gets magnified," he knew, "but there is so much more left to do in terms of hiring more minority faculty, in terms of dealing with the disaffection blacks feel in the university, and the need for more diverse career opportunities."

Obama also thanked BLSA and a trio of black professors—Derrick Bell,

Christopher Edley, and Charles Ogletree—as "ground breakers" who had paved the way for him to walk "through doors other folks broke down," the same metaphor Vince Eagan had used two weeks earlier. Obama said BLSA and public interest advocates had helped create "an atmosphere that allows a person of my interests and perspectives to be in the mainstream. It means white conservatives can trust me," and he called for the law school to "start thinking more about its relationship to the larger society and about the kind of commitment the school should make to assure kids like me get in these positions again." Implicitly citing Chicago, he said that "those interested in public policy have to think about how the private sector can be harnessed to promote urban development."

Laurence Tribe told the *Record* about Obama's work on both his physics article and his forthcoming book *Abortion: The Clash of Absolutes* and expressed his hope that Obama would be "a future legal scholar." The president of BLSA, 3L Tynia Richard, said Obama's victory was "a momentous event," and Chris Edley termed it a "milestone." Christine Lee seemed to be attempting some misdirection by asserting that "race was not a factor" in the election and that "It was not strategized politically." But 2L editor Frank Amanat accurately said that "the biggest thing he brings to the *Review* is his maturity. He's had a lot of exposure to the outside world." Most significant, the *Record* spoke with Rob Fisher, and introduced Rob's comments about Barack by writing that "those who know him say his charisma and self-confidence are most at home in the political arena." Rob explained, "I don't think Barack sees this as a steppingstone to the academic aspects of the law. But whatever he does, he is extraordinarily committed to making a contribution to the resolution of social problems in this country. There is no possibility that this will send him in a different direction."

As the week progressed, multiple journalists prepared Obama profiles and even *Time* magazine reported on his election. Obama told the *Boston Globe* he would return to Chicago after graduation because "I have a certain mission to make sure that the gifts I've received are plowed back into the community." The *Globe* called Johnnie Owens at DCP, who recounted thinking four years earlier that "This guy sounds like he's president of the country already" when Barack first spoke to him. The *Globe* also contacted Maya, who had returned to Barnard for her third and final semester there. "He has been like this as long as I can remember," she said about his seriousness. "I wish I could shed light on exactly what has made him so old at such a young age."

In retrospect, Barack believed his election to the *HLR* presidency was as much the result of his performance in Section III as anything else. "I had established a presence in the classroom and in other activities during my first year," he recounted years later, incorrectly citing "actively campaigning on issues of diversity in faculty hiring" as among his 1L efforts. He also badly misremembered his less than six months on the *Review* preceding the election, imagining that "by the time I was elected . . . the peers who voted for me had worked with me in close quarters for over a year."

One later commentator correctly noted that "Obama's fame began precisely with this achievement." That was well understood by his Harvard contemporaries, and the slew of media requests made that clear to everyone at the *Review*. A *Philadelphia Inquirer* profile had Barack claiming, "I know Philly, and I like it," but he emphasized his resolve to return to Chicago and neighborhoods like Roseland. "There is a lot of talent there, and a lot of energy," but "also a lot of sadness, a lot of young people whose dreams get crushed." He cited the importance of major corporations "building plants in these areas," but said, "I don't think I'll return to the Developing Communities Project." Instead "I will try to get connected with some city-wide agency. I would also like to get involved in the political scene in Chicago." As an activist, or as a candidate? "I don't know. Either one, I guess. It all depends on where I can do the most good," just as Rob had said.[28]

Before the week was out, Barack realized he had acquired a new, almost full-time job on top of a demanding course load. The first two *Review* tasks, scheduled for the next Sunday, involved only the 2L editors: the election of a new masthead and office chairs, and the completion of the transition to 2L management by re-adopting or altering a number of governing policies inherited from the now-sidelined 3Ls. Early on, following traditional form, Barack asked runner-up David Goldberg if he wanted to be treasurer, the ostensible number-two post. Goldberg quickly said no, as he knew from experience that he wanted to be in the Articles Office, which chose the *Review*'s premier content. Semifinalist Lisa Hay did want to be treasurer. The most crucial job was managing editor, because the ME oversaw the distribution of everyone's work and made certain that issues went to press in time for delivery by the tenth day of the issue's month. Meeting that date was Erwin Griswold's top fixation. Exiting ME Scott Collins believed strongly that the 2L with the best skills to take on that demanding role was semifinalist Tom Perrelli, who was willing to accept.

Office cochairs were chosen by a simple vote of all 2Ls, with some positions contested and some not. The process was more complicated for the masthead's supervising editors and executive editors. For SEs, as for treasurer and ME, 2L votes would be tallied, but the president could choose from among the top finishers. EE contenders had taken a *Bluebook* test, graded by the departing 3L EEs, but again the president was not bound by the rank order of the results. Executive editor was an impressive title to have on a résumé, but the actual work was essentially copyediting, with *Bluebook* proficiency being an EE's stock-in-trade. Indeed, a new, updated fifteenth edition of *The Bluebook* needed to be produced. SEs, meanwhile, edited their fellow students' work and worked more closely with their classmates than anyone on the *Review*.

On Sunday morning before the masthead and office elections began, a number of *Review* editors watched television as major news broke in South Africa. Barack, knowing that a difficult day lay ahead, began his introductory remarks with "Let's try and keep this in perspective. Nelson Mandela was freed from prison, and what we're doing is not that important." Tallies to name Lisa Hay as treasurer and Tom Perrelli as ME proceeded without difficulty. The SE contenders presented Barack with a list that included presidential third-place finisher Jean Manas plus four hardworking women: Julie Cohen, Anne Toker, Rebecca Haile, and Christine Lee. Barack had asked Julie to take the EE test, but she had demurred, since "I really didn't want to do it." Yet the EE test results, as Barack had feared, presented a political problem because the top three finishers, Jim Chen, Adam Charnes, and Amy Kett, were all conservatives.

The office cochair results created no controversy: David Goldberg and Jackie Scott for Articles, conservative Brad Berenson and liberal John Parry for Supreme Court, Julius Genachowski and Jenny Collins for Notes, liberal Susan Freiwald and conservative Mike Guzman for Book Reviews and Commentaries, and Mike Cohen and Marisa Chun for Devo. Chun was also fourth on the EE list, and Barack had to decide if he would follow rank-order results and have three prominent conservatives in the coveted EE positions, as well as decide whom to choose from among the SE contenders. Manas's third-place presidential finish created a certain presumption, but Anne Toker, unlike Christine Lee and Julie Cohen, had not run for president, as masthead aspirants had traditionally done. Christine, well known for her hard work as well as for her outspoken racial views, had also actively supported Obama's candidacy. "I really, really thought I had it in the bag" for SE, Christine remembered, thus becoming a black female member of the masthead, like Crystal Nix a year earlier.

Before Barack would announce those choices, the 2L editors had to affirm or amend multiple policies. Some were trivial, but still generated debate, such as what time the lavish spread of free bagels and muffins should arrive. But the most explosive issue was the *Review*'s use of affirmative action in selecting new editors. Conservative leader Brad Berenson described it as "the deepest, darkest secret of the *Review*: how affirmative action was done, to what extent," and each year only the president and two editors actually knew the truth. Amy Kett asked if the existing policy included women—it did not, because women in previous classes had opposed their inclusion—and 2L Dave Nahmias proposed expanding the categories to include "other people who've been disadvantaged," whether natives of Appalachia or Vietnamese refugees. Nahmias was perceived as a conservative, so the self-identified progressives opposed his idea.

When Jim Chen moved to end the policy completely, conservative and liberal editors alike found Chen's deeply personal attack on affirmative action "eloquent," "impassioned," and "quite memorable." Some recalled Chen citing common ethnic epithets and warning that "This is what affirmative action does—it calls people something." Progressive Susan Freiwald remembered feeling torn as Chen pleaded that "I don't want to be stigmatized, and you don't know how many people are stigmatizing me." But even hard-core conservatives like Adam Charnes realized that when someone was "saying we should abolish affirmative action, other people can construe that as an accusation that they shouldn't be sitting in the room."

Tom Perrelli recalled that Barack "was leading the discussion, but he wasn't trying to impose his own perspective." Then Shaun Martin, an impish, unpredictable character whom classmates viewed as a highly eccentric genius, quasi-seriously followed up on Chen's motion to abolish affirmative action by moving to abolish the *Review* itself. Even that managed to engender debate, but in the end, the 2Ls voted to make no changes in the president's discretionary power to use affirmative action when the next class of editors was selected.

Following the meeting, Barack mulled the difficult choices regarding the EE and SE positions. Tom Krause and the other 3L EEs told him that they believed the president should abide by the test results and not displace Chen or any of the other conservative top trio. "It was a tough decision for Barack to ratify," Krause recalled, yet the EEs, even more than the SEs, "had virtually no substantive input" into the *Review*'s content, Amy Kett realized, only final stylistic authority over everything. The conservatives were elated when Obama ratified the exam outcome. "There was a *strong* push for Barack to use his power to go down the list," Brad Berenson explained, and "the fact that he didn't spoke volumes to the conservatives on the *Review,* earned him a tremendous amount of goodwill and respect, because it was seen as a loud and clear indication that no matter what else happened, the best interests of the *Law Review* were going to be the guiding principle."

Traditionally there had been only two SEs, but Barack made two unexpected moves: first he named three SEs, and then he appointed an additional masthead member to oversee the *Bluebook* revision. But his choices—and whom he passed over—created an immediate firestorm. To the three SE positions he named Jean Manas, Julie Cohen, and Anne Toker, and as *Bluebook* editor, he named Ken Mack, his best friend among the 2L editors and fellow African American. When Obama provided the names to the *Harvard Law Record,* his appeared first, followed by Lisa Hay and Tom Perrelli. Then Mack's name was fourth, above the SEs and EEs. The ten-person masthead thus included two black men, four white women, and Asian American Jim Chen. There were no African American women, even though Rebecca Haile and Christine Lee had been among the top nine contenders for the presidency. Both women were shattered, and "Christine felt extremely betrayed," Tom Krause knew.

Barack's siding with the conservatives on one major decision, and slighting black women on a second, while elevating Ken Mack, left even Barack embarrassed. African American 3L Jennifer Borum remembered Barack telling her "'I guess you're mad at me too,'" and she told him he had violated the *Review*'s own traditions. Mack knew that there was "a lot of recrimination and criticism," and Lee's anger was visceral and verbal. She believed that "the quality of my work" merited an SE post, and "I felt betrayed, because I had worked so hard." Looking back, she realized that she "saw everything in terms of race," "didn't play well with others," and thus "had not earned it as far as the quality of my diplomacy with the other students." Her relationships with some colleagues would have required channeling "certain people's work to me and other people's work" to other SEs, and that alone was "a valid reason to not give me the position." Tom Krause, with whom Christine had become involved during their work on the Anthony Cook article, believed that her contemporaneous fury at Obama "was semi-justified" because he had failed badly at "rewarding loyalty." But Krause also thought that given the politics of the *Review,* Barack's treatment of Lee "as a pragmatic matter makes perfect sense."[29]

The week after Obama's election, Peter Yu wrote Erwin Griswold with the news. In his previous letter, Yu had described the *Review* as "a scholarly (rather than professional) publication," while also telling Griswold that the *Review*'s royalty income from burgeoning full-text data services like Lexis was now more than $100,000 a year. February's lead article had created a "huge fight" at the *Review* because of author Katharine Bartlett's insistence on using first as well as last names of authors whom she cited, in violation of a *Bluebook* rule that barred the former. The Articles Office had sided with Bartlett, but the EEs vetoed that, and in protest Bartlett's article featured a memorable initial footnote denouncing the ruling. "I had wanted to humanize and particularize the authors whose ideas I used in this Article by giving them first as well as last names. Unfortunately, the editors . . . insisted upon adhering to the 'time-honored' *Bluebook* convention. . . . In these rules, I see hierarchy, rigidity, and depersonalization, of the not altogether neutral variety. First names have been one dignified way in which women could distinguish themselves from their fathers and husbands. I apologize to the authors whose identities have been obscured in the apparently higher goals of *Bluebook* orthodoxy."

Gordon Whitman, who had sided with Bartlett, recalled the *Review*'s "obsession about form over substance" and how often the editors' behavior toward authors was "like the kindergartners running the preschool." A similar controversy had occurred with a book review in the March issue, which was just going to press, the last number for which the departing 3Ls bore "complete responsibility." Floyd Abrams, the country's most distinguished First Amendment litigator, had been asked to review *A Worthy Tradition: Freedom of Speech in America,* a posthumously published book written by the well-known scholar Harry Kalven Jr. Abrams had been astonished when the *Review*'s EEs told him *The Bluebook* prohibited the capitalization of "first amendment." Like Bartlett, Abrams voiced his unhappiness in a first-page footnote, "objecting to the transformation in law reviews of what I understand to be the First Amendment to the 'first amendment.'" Abrams cited another prominent scholar's call that "Those who believe in freedom of speech should begin by rejecting the tyranny of the *Uniform System of Citation*." Or, "to put it another way, since Kalven wrote about the First Amendment, why can't I?"

Along with Bartlett, the February issue had also featured a twenty-five-page commentary by Kenneth Lasson that mocked the oversized role that law reviews had in American legal education. Lasson cited a number of recent articles in major law reviews which had hilariously impenetrable titles, such as "Epistemological Foundations and Meta-Hermeneutic Methods: The Search for a Theoretical Justification of the Coercive Force of Legal Interpretation." Editor Brad Wiegmann, a 2L, remembered Lasson's essay as "fabulous." In a similar voice, the *Review*'s own March issue led off with Anthony Cook's "Beyond Critical Legal Studies: The Reconstructive Theology of Dr. Martin Luther King, Jr." An unhappy Erwin Griswold wrote Yu, "I wonder how many of your readers will actually ever read the leading article," but he did think Floyd Abrams's review was "very good."[30]

The Monday when the media descended on Barack was also the start of the first full week of spring semester classes. Barack and Rob had signed up for a significantly heavier course load than they had had in the fall. They were familiar with the subject matters in Larry Tribe's seminar and Chris Edley's small class, which featured once-a-week sessions with a guest speaker, ranging from liberal Massachusetts congressman Barney Frank to conservative law professor Charles Fried, who was just back after serving as President Reagan's solicitor general. But their three large courses—Taxation, Local Government Law, and Jurisprudence—were going to be demanding, especially for someone who had just taken on a burdensome new management position. Tax—i.e., federal income tax law—met four hours per week and was straightforward if highly technical. Alvin Warren was "a great Socratic professor," Obama's fellow Section III veteran David Smail believed, and Rob Fisher agreed, remembering Warren as "really excellent."

Gerry Frug's Local Government Law class used his 1,005-page casebook, which offered a wide array of multidisciplinary readings. Frug's unifying theme was "The City and Democratic Theory," and it examined the early origins of cities as legal entities. Frug's underlying emphasis was on "the relative desirability of centralization and decentralization" in democratic government, and he made no secret of his strong belief that "decentralization, participatory democracy, and civic transformation" go hand in hand. Barack was once again the most talkative student in the class. *Law Review* colleague John Parry recalled some classes as being "almost these conversations" between Frug and Barack, and Frug remembered how he and Obama "had a lot of one-on-one discussions," often stemming from Barack's organizing experience, which was "a perspective that I hadn't otherwise thought about." Their exchanges played a "really prominent role" in class, and indeed were "the major thing that happened in the course."

Barack and Rob also took Jurisprudence with Roberto Mangabeira Unger, a highly abstract thinker who rarely drew explicit links between his classroom lectures and the assigned readings. Unger viewed law as "the detailed expression of an institutionalized form of social life," and his course focused on competing theories of law. In the introduction to his assigned readings, he warned that "These materials have an oblique relation to the argument of the course. My plan will not track closely the sequence of the readings, nor shall I discuss them in detail. Study of them is nonetheless important." Radhika Rao, a 3L on the *Law Review,* found Jurisprudence to be "a weird class, because Unger would just speak in these incredibly long sentences and you didn't understand what he was saying. He was impenetrable." *Law Review* 2L Susan Freiwald found it challenging too, but she could see how Rob and Barack were "just so into it." Everyone thought Rob was "a great guy," but everyone was bemused by the intellectual intensity that he and Barack brought to so many of their classes. Rob thought Unger "was the most brilliant (and deepest) lecturer at HLS by far," even if many students often could not fathom what was being discussed.

On February 15, Rob and Barack presented their critique of Larry Tribe's constitutional paradigm at the third weekly meeting of Tribe's fifteen-student seminar. A bevy of *Law Review* 3Ls were in it too—Crystal Nix, Andy Schapiro, and Jack Chorowsky—plus Tribe's top 3L protégé, Michael Dorf. As with Unger, Tribe's post-Newtonian analysis was not always clear. Peter Dolotta, a 3L, remembers "Tribe throwing stuff up against the wall and then seeing if it stuck." Often it did not. "We were generally pretty skeptical," Peter explains, and Rob Fisher concurs. "My sense of what he was talking about was that it wasn't real coherent," though it was still "a fun seminar." Jack Chorowsky agrees. "It was a funky class. Frankly I'm not sure that anybody knew exactly what it was about, but it was fun to sit around and talk Con Law with Larry Tribe. That was the point."

Rob took the lead in making his and Barack's joint presentation. The headings on his outline captured its flavor: "Heuristic Analysis and the Constitution" . . . "The Constitution as Framework for Social Discovery" . . . "Heuristics, the Enlightenment and the Constitution Versus Positivism and the Administrative State." When Rob finished, Barack spoke up, asking the seminar "My most important question to you: What's interesting here and what's boring and why?" Rob jotted notes as Barack continued. "How can a govt. of *limited powers* possibly succeed in modern vision?" The class discussed whether "In the law, do the arguments drive the words or do the words drive the arguments?" The conversation was not entirely disconnected from constitutional analysis, for as Rob noted on the back of their outline, Justice Oliver Wendell Holmes Jr. had made a similar point in 1918: "A word is not a crystal, transparent and unchanged, it is the skin of a living thought and may vary greatly in color and content according to the circumstances and the time in which it is used."

Recalling that seminar, Michael Dorf explained how Rob and Barack "seemed sort of like a pair" and "I thought of them almost interchangeably." One participant believed that "Barack wasn't that impressed with Tribe," that "he's not really a deep thinker" like Unger or even David Rosenberg. Yet when Tribe and Dorf some months later published the much-refined results of Tribe's project as a slender book titled *On Reading the Constitution,* they twice expressly thanked the duo: "Robert Fisher and Barack Obama have influenced our thinking on virtually every subject discussed in these pages," the coauthors wrote in their acknowledgments, and a footnote stated that "We are grateful to Robert Fisher and Barack Obama for the metaphor of constitutional interpretation as conversation."[31]

Interest in Barack by journalists continued, and Crystal Nix encouraged Tammerlin Drummond, a black female reporter friend who had just joined the *Los Angeles Times,* to come interview him. Drummond flew to Boston and spent two days with Barack at the *Review.* He told her about the tens of thousands of dollars of student loan debt he now had thanks to Harvard, but cheerfully said, "One of the luxuries of going to Harvard Law School is it means you can take risks in your life. You can try to do things to improve society and still land on your feet. That's what a Harvard education should buy—enough confidence and security to pursue your dreams and give something back." Drummond wrote that Obama envisioned spending "two years at a corporate law firm, then look for community work. Down the road, he plans to run for public office."

Drummond also spoke with Christine Lee and reported that Obama "has come under the most criticism from fellow black students for being too conciliatory toward conservatives and not choosing more blacks to other top positions on the law review." Christine complained that Obama is "willing to talk to" the conservatives and "has a grasp of where they are coming from," neither of which was true for her. "His election was significant at the time, but now it's meaningless because he's becoming just like all the others." Radhika Rao told

Drummond that Obama is "very, very diplomatic" and "has a lot of experience in handling people, which stands him in good stead" at the *Review*. By far the most difficult challenge was Christine, who was "taking passive aggressive swipes" at him every chance she could get. "The tension between us was just so high," she remembered. "I kind of felt like he was full of shit, and I could see exactly the ways in which he was full of shit."

Lee recalled that one day Obama called her into his small top-floor office, and that "as soon as I got in there I started to cry." Barack told her, "We have to fix this problem between us" so that they could work together on the *Review*. He made some reference to both of them being half-white, but Christine's anger was so intense, she had no interest in ameliorating their situation. "I don't care what our problem is. We're not likely to resolve it, nor am I interested in resolving it." Deeply depressed at how *Review* politics had denied her the masthead post her work had merited, Christine began seeing a campus counselor. "Half of our meetings were about Barack," she recalled.

Before the end of February, Barack had to do his first "P-reads" on the contents of the forthcoming April issue: one article, five student notes, including one coauthored by Crystal Nix, Jean Manas, and a third editor, another one by Jim Chen, two book notes, including one by Susan Freiwald, and two recent cases similar to his own January piece. "He just did a great job," Freiwald remembered, and most authors agreed with her about Barack's "light touch" as an editor. Even Erwin Griswold was happy, praising the April issue as "extraordinarily fine" and stressing that "There is nothing in this issue which smacks of professors talking to themselves."

When a young black journalist from *American Lawyer* magazine came to interview him, Obama readily voiced a Griswold-like critique of the journal he now headed. "Generally, law review articles are written to confuse rather than illuminate," he said. "While some of that is due to the complexity of the law, some of it is just plain pretense. The average practitioner isn't interested in wading through pages and pages of drivel." Barack also told her he aspired to influence national issues, such as "whether money goes to the Stealth bomber or to needy schools."

Obama was never more loquacious than when Allison Pugh, an Associated Press reporter, interviewed him in the "Hark," the law school's student center. Perhaps remembering his 1983 stay at Wahid Hamid's Long Island apartment, he declared that he was "not interested in the suburbs. The suburbs bore me. And I'm not interested in isolating myself." Instead, "I feel good when I'm engaged in what I think are the core issues of the society, and the core issues to me are what's happening to poor folks in this society." Referencing both Indonesia and Kenya, Barack noted that "I lived in a country where I saw extreme poverty at a very early age" and "my grandmother still lives in a mud-walled house with no running water or electricity." He continued to stress that his election was not evidence of broader black progress and that people should not "point to a Barack Obama any more than you point to a Bill Cosby or a Michael Jordan and say 'Well, things are hunky dory.'" Politically, "it's critical at this stage for people who want to see genuine change to focus locally," and in inner cities "it is crucial that we figure out how to rebuild the core of leadership and institutions in these communities." Stating that "I'm interested in organizations, not movements, because movements dissipate and organizations don't," Barack believed that the U.S. needed "some new renewed sense of purpose and direction. . . . Hopefully, more and more people will begin to feel that their story is somehow part of this larger story of how we're going to reshape America in a way that is less mean-spirited and more generous." Pugh observed that Obama "radiates an oddly self-conscious sense of destiny," and her story ended with him announcing that "I really hope to be part of a transformation of this country." To another interviewer, Obama added that "if I go into politics it should grow out of work I've done on the local level, not because I'm some media creation."[32]

Before his *Review* victory, Barack had agreed to join two 3Ls in organizing BLSA's spring conference, the group's big annual weekend event. "The New Decade: The Mission of Black Professionals in the 1990s" was this year's theme. Two days of panel discussions and social events drew scores of African American alumni back to Cambridge and kicked off on Friday, March 9, with an employer forum featuring representatives from thirty-five major law firms. Evening receptions and speeches took place at Cambridge's fanciest venue, the Charles Hotel. Obama's newfound fame was threatening to overload his already jam-packed schedule. Thursday evening he moderated a debate about affirmative action at Harvard's JFK School of Government featuring the eminent sociologist Nathan Glazer versus the university's relevant policy coordinator. On Saturday afternoon at the BLSA conference, Barack moderated a panel discussion on economic development, and his much-heralded election led 3L BLSA president Tynia Richard to ask him to introduce that evening's dinner speaker, NAACP Legal Defense Fund attorney Elaine Jones.

Barack's role was intended to be modest, but even two decades later virtually everyone could recall and quote from his remarks. The *Harvard Law Record* said Obama "told the audience that despite the achievements blacks as individuals may realize, one must never forget where one came from or lose sight of the community goals. Obama stated that only when we acknowledge that one is part of a communal enterprise, can one hope to effect lasting change." That summary omitted the clarion call phrase almost all listeners remember Obama invoking multiple times, as the words of Jeremiah Wright echoed through the luxury hotel's ballroom: "Don't let Harvard change you!"

Tynia Richard, who did not know Barack well, had expected him to sound like "a nerd," but instead found his speech "stunning." Another 3L, Alan Jenkins, remembered Obama warning that Harvard could change you in ways both bad and good. "It was a very, very striking moment," he thought, "not just the eloquence but the self-possession." To Jenkins, Barack's remarks were "the best speech I had ever heard in person at that time." Nicole Lamb, a 2L, found it "impressive and articulate," and could see that Barack "spoke without notes." His "Don't let Harvard change you!" refrain just "mesmerized the audience." Ken Mack recalled that Barack talked about being accepted by Harvard while working in Chicago, and that some older black man had given him that advice: "Don't let Harvard change you!" He "kept returning to the theme," and "the enthusiasm in the room keeps growing as he keeps coming back to that refrain." It was an "incredibly compelling speech," and "when he was done, people just broke out into wild applause." The law school's black faculty were there too. David Wilkins remembered Barack's remarks as "just mesmerizing" and Randall Kennedy recalled the "standing ovation" when it concluded.[33]

Review work and the bevy of requests flowing from his election left Barack increasingly stretched and strained. Throughout his first three semesters at Harvard his course attendance had been regular if not perfect, but now, especially with Tax, which met Monday, Tuesday, and Wednesday mornings at 8:15 A.M., Obama began missing a heavy majority of classes. The presidency was "a grinding job," Mack could see, and by the third week of March, Barack was so exhausted that he jokingly remarked that if *Review* editors wanted to revolt, he was ready to be deposed.

Another day Ken remembered Barack saying, "I've got too much energy this morning. I actually got eight hours' sleep last night." Barack vented to Rob about *Review* problems, and Rob remembered repeatedly hearing about "a particular younger African American woman who *really* thought he was evil." Speaking requests included one from black graduate students at the Harvard Business School, and a listener from MIT was wowed. Barack's "ability to articulate . . . the issues" was "just head and shoulders above anyone else," Bernard Loyd thought. All the press coverage brought letters from both old friends, like Andy Roth, who was still in New York, and complete strangers. U.S. senator Paul Simon of Illinois wrote after seeing the *Philadelphia Inquirer* profile of Obama. "I was very interested to learn of your work with the Developing Communities Project in Chicago," Simon remarked, and "I was especially heartened to learn of your plans for a career in public service." Phone calls included a congratulatory one from organizing buddy Bruce Orenstein in Chicago plus one from a New York literary agent, Jane Dystel, who asked if Obama would consider writing a book. His response was "cool and reserved," Dystel remembered, but Barack agreed to speak with her when he was next in Manhattan to see his sister Maya.

All told, the six weeks following his *HLR* election were exhilarating, as his BLSA speech so powerfully reflected, but it was also debilitating. "I like to read novels, listen to Miles Davis," he told one interviewer. "I don't get to do that anymore." *Law Review* colleagues would recall hearing "Miles Davis on in the background" when they called Barack at his Somerville apartment, and Cassandra Butts remembered Barack playing a Wynton Marsalis album—the 1989 release *The Majesty of the Blues*—that also featured Rev. Jeremiah Wright, narrating a sixteen-minute track entitled "Premature Autopsies." "That's my pastor," Barack told Cassandra.[34]

The law school's spring break the last week of March gave Barack a much-needed respite and an opportunity to see Michelle Robinson in Chicago. Six months had passed since his last visit. The previous October, the first elections were held for the new Local School Councils (LSC) created by the 1988 school reform law. The *New York Times* quoted Obama's old Roseland colleague Salim Al Nurridin as insisting that those "are going to be more important to us than the election of Harold Washington," and UNO's Danny Solis proclaimed them "the largest experiment in grassroots democracy the country has ever seen."

More than 313,000 votes were cast in the school-by-school contests, and in Altgeld Gardens a trio of women who had worked with Barack were elected to Carver Primary School's council. Proponents were ecstatic, with UIC professor William Ayers, increasingly the city's leading voice for school reform, celebrating what he called "the most far-reaching change in governance ever envisioned in a modern big-city school system." In a *New York Times* op-ed, Ayers said he believed that "successful changes in schools tend to come from the bottom, not the top." Business leaders whose support for the reform law had been so crucial established a new organization, Leadership for Quality Education, to support its implementation, but Patrick Keleher, so central to the 1987–88 efforts, feared that more than a revolution in local school governance would be needed to solve "the public education disaster in this city."

One casualty of the LSCs' emergence was UNO, which had become a powerful network of Hispanic neighborhood groups. Greg Galluzzo and especially Mary Gonzales believed that church-based organizations could not in the long run emphasize public education issues, while Danny Solis and Phil Mullins emphatically disagreed. Personality clashes magnified an angry and ugly power struggle, one that ended with UNO and Gamaliel formally divorcing. Barack followed this from Cambridge, grilling Bruce Orenstein for details but firmly rebuffing Phil's effort to draw him to their side. DCP was troubled too. Johnnie Owens had passed through Boston for an evening and spent the night at Barack's apartment, where the two old friends talked into the early-morning hours. But Johnnie was also struggling. A state-funded evaluation team had been hired to work with DCP, but its leader, UIC professor James Kelly, found the assignment difficult to perform. "Six months," he wrote, were "devoted to establishing contact with" Owens. "That was no small or easy task. . . . Meetings were scheduled. John did not appear," a pattern that recurred at least four times.

Barack told Rob Fisher "that things weren't going that well for the organization" and "he was just trying to make sure things didn't fall apart," but there was little Barack could do from faraway Cambridge. Things were going vastly better for Mike Kruglik's SSAC in Cook County's south suburbs and for Jerry Kellman's growing LIFT in northwest Indiana. Joseph Cardinal Bernardin helped SSAC launch a multimillion-dollar effort to rehabilitate abandoned homes, and LIFT received over $1 million in grants to begin a similar program in Gary.[35]

Chicago politics, increasingly dominated by newly elected mayor Richard M. Daley, also included another challenge to South Side African American congressman Gus Savage by the tenacious Mel Reynolds. Nine months before the primary, Reynolds won the support of Al Johnson, the African American Cadillac dealer who had been Harold Washington's top fund-raiser. The *Tribune* described Johnson as "a major catch," but then Reynolds was accused of offering $6,000 to a twenty-year-old college student to have sex. The young woman had ties to another possible Savage challenger, and the charges were soon dismissed.

Seeking to highlight Savage's well-known erratic reputation, Reynolds called for all candidates to submit to multiple drug tests and then alleged that his campaign office had suffered a break-in. With national journalists interested in the race, *The New Republic* reported that Savage "is widely considered a buffoon" on Capitol Hill.

Amid claims that the Daley-controlled Democratic organization was quietly backing Reynolds, the former Rhodes Scholar told reporters, "I'm not an establishment guy. I have establishment credentials, but I'm a poor kid from a poor background." Al Johnson told the *New York Times* that "Mel is a sterling example of everything we ask our young people to be." In a lengthy *Chicago Reader* profile of Reynolds, Johnson praised Reynolds for returning "to the community to make a contribution" after attending Oxford and Harvard. A black pastor backing Reynolds praised him for having "not only a command of the needs of the district but also a broader perspective about the issues in the state and the nation." The *Reader* noted Savage's contention that Reynolds "is the candidate of the whites" and quoted a union leader decrying another anti-Reynolds theme: "The terrible thing in this campaign is that some people have tried to make a negative issue of Mel's education. . . . Some people are saying that he's an 'Oxfordian' and make it a negative." But *Reader* correspondent Florence Hamlish Levinsohn was ambivalent about Reynolds, writing that upstart challengers often "are egotists convinced of their ability to outshine the incumbent." She quoted an unnamed activist recounting how Reynolds had reacted when asked to run for alderman. "I'm not starting down at that level. I'm going to be President." Levinsohn concluded that "there is in Reynolds's seriousness, puritanism, and piety an ambition so overarching that it is a little bit frightening." She also saw him as "a highly educated man who is immensely proud of the moderation that his education has taught him."

On election day, Savage won a narrow 51 to 43 percent victory. That night he thanked Nation of Islam leader Louis Farrakhan for his support and lambasted the "white racist press." Ninth Ward alderman Robert Shaw asserted that Reynolds "is identified with those people from Boston and with the elitists of this town" and warned that "black people will not permit the white establishment or the newspapers or electronic media to pick a congressman for them."

A *Tribune* recap stated that Savage's efforts to portray Reynolds "as an outsider bankrolled by sources unfriendly to blacks" came "perilously close to anti-Semitism." Reynolds said of Savage that "everyone who runs against him is not black enough," and a sidebar controversy ensued as Chicago's white progressives questioned why liberal black columnist Vernon Jarrett had insistently attacked Reynolds. Jarrett called Reynolds a "mysterious phony," who was "very ambitious to go to the top quickly." In Cambridge, *Review* 3L Andy Schapiro, who came from the district's southernmost reaches, had mentioned to Obama how excited he was to see Reynolds taking on Savage. Barack shook his head dismissively while replying, "I know Mel."[36]

Mid-April was an important time in the *Law Review*'s annual calendar. All 1Ls interested in the end-of-semester writing competition were invited one evening to an "ice cream social and general information meeting," where the *Review*'s selection process was explained. Among those attending was Lee Hwang, a 1989 magna cum laude graduate of Dartmouth, who remembered the "captivating presence" of the *Review*'s "already famous" president. In subsequent weeks, there were three follow-up receptions—for women, African Americans, and political conservatives—hosted by respective editors.

Two nights after the initial 1L session, the *Review*'s 103rd annual banquet took place at the Harvard Club in downtown Boston. Each year featured an eminent keynote speaker, and the 3Ls had secured famous Harvard economist John Kenneth Galbraith. But the real highlight was the distribution of the *Harvard Law Revue,* or "review-ee," an acerbic and often stiletto-sharp compendium of jokes and insults presented in the form of a shortened issue of the *Review*. Editors and authors were targeted, and everyone found the twenty-to-thirty-page printed booklet on their chairs when the evening commenced.

The 1990 edition, written in significant part by Articles Office cochair Andy

Schapiro, skewered disappointing or argumentative authors while humorously commending or embarrassing almost all student editors. A faux contents page had "Vermin Chemerinsky" "Wrapping Up a Five-Minute Foreword," Fran Olsen required "Editing Miss Crazy," and "Anthony Kook, 'A Backhand Slice at Critical Legal Studies: Billie Jean King's Philosophy of Love'" spoke for itself. Editors believed to have done good work on difficult pieces—Dan Bromberg, Jorge Ramírez, and Christine Lee for "COOKing Her Goose"—were commended, while others were teased: Crystal Nix for working all night, Brad Berenson for being "Right of Rehnquist," David Goldberg for his addiction to Diet Pepsi.

Local celebrities were also targeted. "Laurence DiaTribe & Mike Dork, 'Constitutionally Spaced Out: What Law Professors Can Steal from High School Science Textbooks'" teased Tribe's November comment for being "Lost in Constitutional Space." Obama received his share: "Barack Obama, The Semicolon Goes Where?, 666 Bluebooking's Not Important for Celebrities L. J." preceded a footnote invoking freedom of speech: "Even incredibly annoying speech and speech habits are protected by this guarantee. See, e.g., Obama, Why I Have a Constitutional Right to Say the Word 'Folks' in Every Sentence."

Three pages labeled "Obamania" were devoted to Barack, with his night-before change of heart highlighted by "see Obama, I'm Not Going to Run for President, 103 About Face L. Rev." Some jokes were well targeted: "As you folks all know, I am extraordinarily mature" was followed first by a reference to Chicago—"There I discovered I was black, and I have remained so ever since"—and then an election jab: "some folks may think I'm crazy for appointing a bunch of conservatives to the masthead, and others may think I've been co-opted." His rationale appeared too: "Obama, But They Did the Best on the Test, in Democracy and Distrust (3d irasc. eds. 1990)."

Peter Yu was indicted for using "unintelligible jargon" and calling people "jerkface," while Barack was teased with "Obama, Hey, How's It Going, Folks?" with "folks" meaning "you still don't have to learn their names." He was complimented for "a series of articulate and startlingly mature interviews," while Christine Lee was zinged with "see Lee, Bearing My Fangs, L.A. Times." University of Chicago Law School professor Michael W. McConnell, author of the forthcoming May issue's lead article on the "Free Exercise of Religion," was criticized for "Freely Exercising His Vacation," and other authors were mocked for either insistently rejecting or passively accepting all editorial changes. The *Revue* concluded with a back page "Doers' Profile" of Barack, modeled on a well-known whiskey ad. Barack's age was given as "extraordinarily mature" and his latest accomplishment was "Deflecting Persistent Questioning About Ring on Left Hand" that he had worn for years. "Quote: 'Engage, empower, smoke Marlboro.' Profile: Shy, awkward, insecure. Not interested in politics."

At the banquet, Peter Yu and Barack offered remarks before turning the podium over to Professor Galbraith. Barack's comments echoed his speech at the BLSA event six weeks earlier. He again said his election was not simply about him, and he mentioned sitting on Gannett House's porch with a friend who was the son of a white southern sharecropper—Rob Fisher. Jorge Ramírez, a 3L, was astonished. "My jaw dropped on the ground—the guy spoke extemporaneously, absolutely no notes." Treasurer Lisa Hay felt it was "an incredible, moving speech," "just the most moving speech I'd heard from any speaker at that age." Lisa's boyfriend and husband-to-be Scott Smith was also there, and Lisa remembered that "after Barack's done, he kicks me under the table and says in a loud whisper, 'It's a good thing you didn't win!'" Lisa watched as the black waitstaff were "slowly picking things up so that they could listen" to Obama, and how they joined in the applause at the end, with one waiter going forward to shake his hand.[37]

While Barack was juggling his *Review* job and his course load, some students refocused on the fall's political battles over faculty diversity and Dean Clark's disregard for public interest lawyering. In February, when the school announced that a young African American male would join the faculty in 1991 after completing a clerkship with Justice Thurgood Marshall, Obama told the *Crimson* it was an "encouraging sign," but BLSA president Tynia Richard pointed out that there were still "no black female professors." The public interest controversy was equally intense, with Clark facing what the *Crimson* called "heated questioning" at an early-March forum with several hundred law students.

In early April, eighty protesters twice occupied the dean's office in overnight sit-ins, and as student energies swelled, Derrick Bell said he was stepping down from active teaching until a black woman was named to the school's permanent faculty. A front-page *New York Times* story reported Bell's move on the same day that student protest leaders scheduled a noontime rally outside the Hark, where Bell would publicly announce his departure. Some participants noted "how not involved" the law school's best-known minority student was, and Mark Kozlowski perceived "a certain degree of bitterness amongst black students who said Barack was not sufficiently black." Ken Mack recalled that there was "huge criticism" of Barack for his absence from protests, but Ken understood Barack's concern that protests over faculty diversity could "actually end up harming the people we're trying to help" because professors might begrudge the calls for minority hiring.

Only in retrospect did protest leaders like 2L Linda Singer realize that "we were asking the law school to change something that they would never concede in the face of student demands." Kozlowski believed Barack's nonparticipation "quite frankly took not a small amount of courage at the time." But with Bell's announcement drawing national news coverage, protest leaders asked Cassandra Butts to help recruit Obama. Linda Singer and her colleagues wanted both "the power of his speaking" and "the symbolism of his role." Barack offered "our little movement a star quality and a credibility that we didn't have," and BLSA president Tynia Richard agreed that they needed "the highest-profile guy to participate." The diversity coalition included the law school's nascent gay students group, and gay 2L Morris Ratner agreed with Singer about "the star power" Barack could bring. "He became more than an individual, he became a symbol, the minute he became president of the *Harvard Law Review*."

Butts was successful, and as hundreds of law students gathered outside the Hark a little before noon on April 24, Paolo Di Rosa, who knew Obama from Section III, found himself standing next to Barack. In earlier times Obama and Di Rosa, whose dad had been tortured by Paraguayan dictator Alfredo Stroessner's regime before going into exile, had "compared notes about our respective fathers," but on this day, when Di Rosa turned and said he hadn't seen Obama in the gym lately, Barack "didn't really even respond." That was unlike him, but Di Rosa realized "he was just in a zone," and only when the proceedings got under way did Di Rosa see that Obama was going to speak and had been "preparing the speech in his head."

A variety of protest signs dotted the crowd of seven hundred: "Where Are Our Tenured Black Women Professors?" "Is the United States 95% White and 91% Male? HLS Faculty Is." "Come Out of the Ivory Tower." "Harvard Law School on Strike for Diversity." Derrick Bell arrived with BLSA president Tynia Richard and was greeted with light applause. As TV cameras recorded the scene, Obama, wearing a light blue dress shirt and tan pants, stepped forward to introduce Bell. He recalled the scene almost two years earlier, when "the black law students had organized an orientation for the first year students, and one of the persons who spoke at that orientation was Professor Bell, and I remember him sauntering up to the front and not giving us a lecture but engaging us in a conversation, and speaking the truth." Barack went on to say he had carried Bell's words "with me ever since. Now how did this one man do all this? How's he accomplished all this? He hasn't done it simply by his good looks and easy charm, although he has both in ample measure." Richard and others laughed at Obama's poetic praise. "He hasn't done it simply because of the excellence of his scholarship, although his scholarship has opened up new vistas and new horizons, and changed the standards of what legal writing is about." Applause began. "Open up your hearts and your minds to the words of Professor Derrick Bell."

Bell stepped forward and the two men hugged. Years later, Bell recalled, "I was surprised he was there, I was surprised that he spoke," because Obama had had almost no contact with Bell during his two years at Harvard. Bell began his own prepared remarks with a self-deprecating compliment. "It may be significant that the student stands here and delivers a mighty address without notes, while the teacher"—Bell paused as laughter built. Listeners agreed with Bell's description. "It was a spectacular speech," Paolo Di Rosa recalled, and made with "no notes whatsoever." Like Bell, *Law Review* 3L Micki Chen was "so struck that the student was so much more eloquent" than a noted professor speaking from a text. "That day is really seared in my mind."

Bell said he was "removing myself from the payroll as a sacrificial, financial fast" since "I cannot continue to urge students to take risks for what they believe if I do not practice my own precepts." He noted that the law school had 107 black women students, yet not one African American female on the permanent faculty. Bell also declared that "the ends of diversity are not served by persons who look black and think white," a remark that offended many listeners. Nathan Diament, Obama's occasional basketball teammate, remembered "that was the moment when I said to myself, 'This guy just lost me.'" Most of the crowd believed the insult was aimed at younger black law professor Randall Kennedy, who a year earlier had published a powerful, well-received critique of race-centered legal arguments including Bell's recent work. Later that day a "visibly upset" Kennedy confronted Bell, who denied intending a personal insult. In the *Crimson,* a 3L said that Bell's comment "represents intolerance in its purest form. It establishes an orthodoxy of thought, and Blacks must embrace it or else face excommunication for their 'white' views." As the rally wound down, a letter from Dean Clark was read expressing respect for Bell but saying his departure was not "an appropriate or effective way to further the goal of increasing the number of minorities and women on the faculty." The *Boston Globe* reported that "students hissed as the remarks were read," and after the rally 150 students marched through the law school's campus "yelling 'Wake up, wake up, wake up, Dean Clark, wake up.'"

The next morning Bell and Tynia Richard appeared on CBS TV's national news show, and Richard spoke of her experience attending Harvard as a black woman. "I had six professors in my first year, all of whom were white and all of whom were male." As the year progressed, "none of the issues of race and gender or sexual orientation were being addressed in the classroom." She poignantly added that "I take out $20,000 a year in loans. They make five times" that amount or more.

But Obama's tender introduction of Bell marked the end of his involvement in the law school's student protests. Two weeks later, when Jesse Jackson visited the law school to speak at a far smaller rally and meet briefly with a reluctant Clark, Barack was not visible in any of the television footage. As Rob Fisher knew, Barack had "just zero interest" in campus disputes. "He was interested in real politics with real people."[38]

There was no let up in intensity at the *Law Review* as the spring semester neared its end on May 7. The June issue, the last of the academic year, was still in process, but a trio of decisions had to be made regarding content for the 1990–91 year. Each November's Supreme Court issue featured both the prestigious foreword plus a faculty case comment, with the Articles Office taking the lead in selecting the former and the Supreme Court Office the latter. In addition, several highly problematic article submissions by untenured Harvard younger faculty were being worked on, with 2L Michael Weinberger having spent much of the spring dealing with a manuscript from African American professor David Wilkins. Some months earlier the *Review* had lost out on a submission "we loved" from Robin West, a 1979 graduate of the University of Maryland Law School who had joined the faculty there in 1986 after several years at the Cleveland-Marshall College of Law. *The University of Chicago Law Review* had beaten Harvard to the punch, and West's highly regarded "Jurisprudence and Gender" had been their lead article, a signal honor, particularly for someone whose academic pedigree came from midrank public institutions rather than Harvard, Yale, or Chicago. New Articles cochair David Goldberg had rued that loss, and was now pressing for West to author the foreword for the upcoming November issue. Only once before had a woman, Harvard's own Martha Minow, authored a foreword, and with some trepidation the 2L editors ratified Goldberg's choice.

West received an unexpected phone call from Obama extending the *Review*'s offer, but West's "first reaction was 'Nope, can't do it,'" because she had just had her first child. Barack was nonplussed. "People don't really turn this down," he told West, and gradually he convinced her to accept, telling her she could write anything she wanted. *Review* tradition meant that Obama, as president, would be her primary editor. West was intrigued by Václav Havel, the new president of Czechoslovakia. "I wanted to write about authenticity and personal responsibility for authenticity," and Obama was entirely encouraging. West would have to submit her complete manuscript no later than August, but the upcoming volume's most prominent commitment was now nailed down.

Far more difficult were manuscripts that had been submitted by two of the law school's most popular young teachers, David Charny as well as Wilkins. David Goldberg's 1L section had had both professors, and Charny in particular had impressed Goldberg as a "brilliant, brilliant guy" and "just an amazing teacher." Neither manuscript was anywhere near as impressive as their authors, yet *Review* editors knew that refusal to publish either young scholar's article could prove fatal to their hopes for promotion and tenure. "The articles process was supposed to be a blind process," 3L Articles cochair Gordon Whitman had told Goldberg and other 2Ls, but Whitman fully appreciated how "the tension on *Law Review* is always what comes through the back door and what comes through the front door." Goldberg and the 2Ls thus inherited "a significant issue" concerning the two "really problematic" Wilkins and Charny manuscripts. "The question was, 'Is there a special rule for Harvard faculty?,'" i.e., a lower bar than external submissions faced. Goldberg confronted the issue straightforwardly, and "that's one of the few things I ever really talked about with Barack in a formal" way. "Based on what they were sending over to us, if they'd sent that out it wouldn't have been accepted anywhere remotely comparable and it would have been really bad for them" both, Goldberg explained. "I felt we had to temper our principles for a little humanity," especially because "both of them were people who I liked and respected."

Obama did not object to Goldberg's stance, but 2L Scott Siff took the lead in arguing that a double standard favoring Harvard professors over everyone else was untenable. Wilkins's manuscript was the more problematic of the two, for "If this had come in over the transom, it wouldn't be one that would meet our normal standards," Siff recalled, and indeed "it wasn't one that was even 'Well, it's kind of close to our normal standards.'" In addition, "it was a particularly charged thing because of his being African American," and everyone on the *Review* agreed on the "need for more diversity on the faculty." Supreme Court Office cochair John Parry remembered that "you could hear shouting through the door" during "impassioned" Articles Office meetings, but over the course of the spring, a consensus emerged that for Harvard junior faculty submissions, the question would be "Does it embarrass the *Law Review*?" That was "the standard we agreed on for faculty tenure pieces and nothing else," Goldberg explained. "Don't embarrass the *Review*."

One week before spring classes ended, a controversy far more vitriolic exploded when a number of left-wing editors objected to the Supreme Court Office's decision to invite former solicitor general Charles Fried to write the November issue's faculty case comment. Office cochair Brad Berenson had championed Fried's selection, and both well-liked liberal cochair John Parry plus other editors in the office had willingly agreed. But Jean Manas, Christine Lee, and Christine's closest friend on the *Review*, Mike Froman, with support from David Goldberg, angrily protested giving this honored role to someone who had been the Reagan administration's top Supreme Court advocate. Parry was summoned to Obama's small office on the top floor of Gannett House, where he was confronted by Froman and Manas. "They were mad," and "I was definitely blamed: 'How could you let this happen?'" Parry remembered. "Mike in particular seemed really mad," and "there was pressure brought to bear." Barack's role was "relatively passive," but "I certainly felt like I was on the spot."

Someone proposed that liberal Harvard professor Kathleen Sullivan be paired with Fried in a duo of comments, and Parry agreed to put that before the Supreme Court Office. That pairing would then be submitted to all the 2L editors for approval, but Berenson was furious at what he thought was an abuse of process. "There was no real precedent for the rest of the body second-guessing the major decisions that the offices made," but even adding Sullivan was not enough to satisfy critics' complaints. With thirty-seven voters, excluding Obama, eligible to cast paper ballots, an inconclusive tally resulted when six 2Ls did not participate. "The result was 18–13 against the proposal," but due to "a presumption in favor of approving the proposal of the Supreme Court Office," any nonvoters were "counted as yes votes," resulting in a 19–18 outcome, a summary memo reported.

John Parry strongly recommended reopening the matter and inviting all interested 2L editors to offer suggestions. "We have to avoid blowing this into a controversy that will split both the office and the *Review* beyond working ability," he told Berenson. Parry believed the balloting reflected "the uninspiring combination of Fried and Sullivan, and the question of voice inclusion," and presciently observed, "I think that a different pairing of Fried with a person of color would satisfy the vast majority of 2Ls" and perhaps everyone.

The infuriated Berenson was also worried because the controversy was delaying the issuance of invitations to Fried and whomever else. Deeply frustrated, Berenson told Parry "to do whatever Jean & Christine want. The Teenage Mutant Ninja Turtles could write for all I care." On May 4, a resolution was reached to pair Charles Fried with Patricia Williams, an African American associate professor at the University of Wisconsin. Parry thought that Barack had remained "a bit detached" from the entire controversy, and both Fried and Williams accepted the invitations.

Like Parry, the graduating 3Ls likewise perceived a relative detachment by Obama during his first three months. Unlike some editors, he did not spend the better part of each weekday at Gannett House, and more people remember him out in front, smoking, because it was not allowed inside. "I thought he was incredibly cool," Radhika Rao recalled. "He had a charisma to him," and she remembered chatting with him and Rob Fisher on Gannett House's porch. "He was often with Rob," and he would use his small top-floor office mostly in the evenings. "I remember Barack being a late-night person," Barbara Schneider recalled. "His feet would be up on the desk and he'd be smoking," irrespective of Cambridge's city ordinance. Debbie Brake thought Barack seemed "super-cool, super-collected. He had the aura of leadership about him." Brian Bertha believed Barack had "an unusual level of gravitas" and seemed "very skilled at forging a consensus." The day-to-day running of the *Law Review* was mainly in managing editor Tom Perrelli's hands, and he was "a nice person and a hard worker," Christine Lee recalled. "No one else had that combination."

Erwin Griswold responded to Obama's letter about the May issue by commending the Michael McConnell article as "very good" but, at 111 pages, "much too long." Griswold praised the annual, five-student "Devo" on "Medical Technology and the Law" as "one of the best Developments in my memory," but when Griswold saw the June issue, his normal dyspepsia returned. He liked neither the sole article nor an underwhelming five-contributor "Colloquy" responding to Randall Kennedy's article a year earlier. "Sometimes I wonder how much of it is actually read and useful," Griswold wrote Obama, "and whether they will be remembered even five years from now." Indeed, the most striking item in the issue was an eight-page critique by Radhika Rao of former federal appellate judge Robert H. Bork's jeremiad *The Tempting of America,* which savaged the U.S. Senate's 1987 rejection of his nomination to the U.S. Supreme Court. Obama's "P-read" on Rao's essay had been "cursory," and rightly so, because a quarter century later it would remain a powerful and insightful appraisal. Noting the book's "sanctimonious anger," Rao chastised Bork for his "refusal to confront the political choices implicit in originalism" and emphasized how "Bork's doctrine of selective stare decisis . . . marks the same kind of departure from objectivism that he decries in other contexts." Her final line—"In a world where objectivity and neutrality are dead, the faith that Bork professes turns out to be hollow"—was followed by a footnote that represented a coda to the year's entire eight issues: "Having begun with an essay heralding the rebirth of constitutional science"—Larry Tribe's physics essay—"this volume of the *Harvard Law Review* ends, perhaps too appropriately, with a Book Note reaffirming the death of constitutional theology." Erwin Griswold, no raving liberal, wrote to Obama that Rao's piece was "fair and persuasive," and two decades later Ken Mack remembered it as "the best thing in the volume."[39]

Before heading to Chicago for the summer, Barack had to confront his end-of-semester exams. Taxation was a looming problem. Dan Rabinovitz remembered that "Barack was so busy with *Law Review* he maybe went to one or two classes." Dan knew Barack "was incredibly thoughtful about how he spent his time," but this seemed risky, especially for someone who, like Rob Fisher, wanted to graduate magna cum laude. Rabinovitz had attended every class and "took copious notes . . . so at the end of the semester, I made a copy of my notes, and I gave them to Barack." Alvin Warren's three-hour in-class, open-book exam posed four questions, and Rabinovitz, sitting near Obama, had written perhaps one-third of an exam booklet when he saw that Obama, even with his tiny handwriting, had already filled almost an entire one. Dan appreciated that Barack was "extraordinarily intellectually gifted," but only after grades were issued did that fully sink in. "He got an A in the class," Rabinovitz learned. "I did not get an A in the class."

Gerry Frug's all-day Local Government Law take-home posed just one question, with answers limited to two thousand words. Frug's question was utterly remarkable because it cited classroom comments made by one particular student: Barack Obama. Frug began: "In an article of mine entitled 'The City as a Legal Concept,' I argued as follows," with six paragraphs reprinted. "In a recent article, Professor Richard Briffault has provided a very different picture" regarding suburbs, and Frug provided four paragraphs. "Write an essay critically discussing these contrasting views of local government law," and analyze "a specific issue that will illuminate the extent of city and suburban power in America." Students could choose most anything, "or you could discuss the following issue, one that was raised in a class discussion in March. In March, a member of the class"—Barack—"argued that there were two fundamentally different strategies that African-American communities now living in large American cities could adopt to better their social and economic conditions. On the one hand, he argued, members of the community could stay in the city, organize together to gain political control of the city government and, once this was achieved, mobilize city power and resources to begin to combat problems such as unemployment, crime, homelessness, and drug addiction. On the other hand, he said, community members could abandon the inner city, integrate suburbs and other prosperous areas, and thereby destroy the kind of inner-city environment that, some contend, fosters the kind of problems mentioned above. Given the arguments advanced by Professor Briffault and me, what impact does local government law have on these alternative strategies?"

Frug's question told the students to discuss as many specific doctrines of local government law as possible, but anyone with experience in higher education can appreciate how astonishing it was for a student taking an exam to see his own classroom comments comprising a significant portion of the test. Frug recalled two decades later, "Never before and never again have I ever referred to a student in an exam. It was a real statement . . . of his impact on the course, that I did this." Rob Fisher had prepared carefully for the exam, and followed his and Barack's "rule that you find the three intellectual tricks the professor likes" and include them in your answer. "I also wrote the word 'folks' a lot, and I got a very high grade," Rob explained. He got an A, as did Barack.

Before driving west to Chicago, Barack stopped in Manhattan to meet Jane Dystel, the literary agent who had telephoned him three months earlier. "We talked. He was attractive, charismatic, passionate about public service. I told him there was story interest" on the part of publishers, "but, despite liking the idea, he said, 'I can't even begin to write this until I graduate law school,'" Dystel later recounted. Barack's recollection was that Dystel "really didn't know anything about me" apart from his presidency of the *Review*. She hoped he could write "a memoir" that would be "a feel-good story" about his success at Harvard, but Barack downplayed the significance of his law school experience. "My life's a little more complicated than maybe what you would anticipate," he told her, but he agreed to draft a descriptive proposal over the summer. It would address "the theme that I'm interested in, which really doesn't have too much to do with Harvard" and instead involved family and race. Dystel came away from their talk thinking that Barack "suggested three parts: childhood, law school, and the search for his father," but Barack had little interest in writing about law school.[40]

The view from the Chicago Skyway in early June 1990 had changed little from five years earlier, but now Barack knew the landscape that whizzed past him. What had changed this time was that his destination was not Hyde Park, but Michelle Robinson's family home at 7436 South Euclid Avenue in South Shore, where he would spend the summer with her and her parents. Fraser Robinson was still two months shy of his fifty-fifth birthday, but the ravages of multiple sclerosis left him dependent on two arm-fitted crutches at home and at work. Barack had envied the Robinson family's stability and Chicago roots since first meeting them a year earlier, and he envied too that Michelle's father was "a great family man," in such huge contrast to his own father.

With Michelle at Sidley, where she was now overseeing the summer associate program, and Barack's summer job at Hopkins & Sutter less than a block away, each weekday the young couple headed downtown to the Loop. Just as he had the previous summer, Barack also regularly drove down to 95th Street to see Jeremiah Wright. "When I went back during the summers, I'd go back and visit Trinity. The Rev and I, my pastor, became very close," Barack later explained. Another day in mid-June, he drove out to suburban Oak Brook to speak as one of "McDonald's Black History Makers of Tomorrow" at the food chain's Hamburger University, along with twenty-five-year-old Jesse Jackson Jr., who was about to begin law school at the University of Illinois. Just a few days later, Michelle was devastated when her closest friend from Princeton, Suzanne "Suzy" Alele, died of cancer in suburban Washington at the age of twenty-six.

Like Sidley, Hopkins & Sutter was "very highly regarded," Sidley's Geraldine Alexis explained, and along with Sidley, it was one of Chicago's top two large firms when it came to minority hiring. Hopkins had two African American partners, both 1979 Harvard Law School graduates: Elvin Charity, who had interviewed Barack six months earlier, and Albert Maule, who would die of cancer just five years later at age forty. Obama's *Review* presidency meant he was now "a big deal" and indeed "a celebrity" at both Hopkins and Sidley, where he would return for a brief one- or two-week stint later that summer.

Sidley's top partners were still exceptionally interested in luring him to the firm once he finished Harvard, and they strongly advised him to pursue a Supreme Court clerkship. But to both Newton Minow and Gerry Alexis, just as to Sheryll Cashin previously, Obama said, "I just don't have the time to do that."

Alexis told him he was "crazy" not to clerk, and "we went back and forth on that" before Alexis gave up. "I wanted him to take the right career steps," but in their conversations that summer Barack "made it clear he wanted to go into politics" and may have mentioned "becoming governor of Illinois." Someone teased him that he was "not going to become a governor in Illinois unless you get an apostrophe after the O," for an Irish O'Bama, but Alexis expressed dismay that someone with such illustrious career possibilities in law would instead choose politics. "He just looked me straight in the eye, and he said, 'Well, Gerry, somebody has to do it.' . . . I was very impressed by that."

Early that summer the progressive *Chicago Reporter* conducted a survey of minority employment at the city's private law firms. Sidley and Hopkins & Sutter fared the best, and the *Reporter* interviewed Obama. He stressed the particular difficulties black students faced in a profession where many young lawyers were unhappy. "Because they don't have the support networks or because their abilities may be questioned due to racism," African Americans often feel "under the gun," but he also said, "It's a great time to be a young black law school graduate—if you're from Harvard and in the top quarter of your class. But the point is that there are a lot of talented young minorities who may not have been able to go to the top schools . . . due to financial constraints." Firms needed to look beyond "the most prestigious schools" and realize that students of color with "the intelligence and the energy to do terrific work" could be found at state law schools too. The *Chicago Tribune* reported on the survey results, and Obama made the same point there. "I think there are a lot of excellent minority law students who have families that can't incur a $25,000 debt for a law school education. So they don't go to nationally ranked schools."

At least once that summer, Barack did a training for Jerry Kellman in Gary, and Kellman met Michelle for the first time when he stopped by the Robinsons' home in South Shore. Some time that summer, Obama also had lunch with David Brint, a young housing developer who had called him at the *Law Review* after reading one of the profiles in which Obama said he wanted to pursue community development work in Chicago. Brint was one of three officers at Rezmar Inc., a year-old low-income housing development company, and at a second meeting, Brint introduced Barack to his two colleagues, Dan Mahru and Antoin "Tony" Rezko.

Rezko had an interesting life story. Born in Aleppo, Syria, in 1955, he had arrived in Chicago at age nineteen to study at the Illinois Institute of Technology. Even before completing a business management degree, Rezko began working for Crucial Concessions, a food service company owned by J. Herbert Muhammad, whose late father Elijah had been head of the Nation of Islam and who was well known as boxing champion Muhammad Ali's manager. In 1985 Crucial made Rezko general manager of the newly reopened South Shore

Cultural Center (SSCC), a former country club that the Chicago Park District (CPD) had acquired a decade earlier.

A profile in the *Hyde Park Herald* described Rezko as "a gentle-mannered, soft-spoken gentleman" who "always is on hand to supply the menu and beverage lists" and who stressed that 60 percent of SSCC's employees and two-thirds of its patrons were black. The *Herald* soon questioned why a company with "no catering experience and no facilities to prepare or transport food" had won the food service contract at SSCC, and the CPD cited Herbert Muhammad's "good name in the black community" as reason enough. Crucial and Rezko were actually subcontracting out all of the SSCC's food service, and from 1987 through 1989, the *Herald* ran a trio of stories citing "frequent complaints about the service and quality of the food" as well as high prices. At SSCC, Rezko met Daniel Mahru, an attorney whose Automatic Ice company supplied both cubes and machines, and in 1989 they joined together to create Rezmar. Mahru saw Rezko as "a humble, hard-working immigrant success story," as did Brint, and while Rezko's connections would help Rezmar, nothing came of their summer 1990 conversation with Obama.

The Harvard 1Ls who wanted to join the *Law Review* had spent the first week of June completing the five-day writing competition. Treasurer Lisa Hay had organized the exercise, and its substantive portion asked contestants to analyze *Austin v. Michigan Chamber of Commerce,* in which the Supreme Court had upheld the constitutionality of a state statute prohibiting the use of company funds in political campaigns. "Corporate wealth can unfairly influence elections," Justice Thurgood Marshall had written on behalf of a six-justice majority including conservative chief justice William H. Rehnquist. At Gannett House, Susan Higgins had mailed out the 286 entries to 3L editors scattered around the country. As the 2Ls' scores came back, Higgins also obtained the GPAs that would be used to determine which additional members of the 1991 class would be invited to join the *Review* as grade-ons.

By mid-July, Obama was telephoning his high-achieving classmates. Several accepted right away, but almost everyone had to weigh the loss of three weeks' worth of summer pay. "I debated not doing it," Darin McAtee recalled before accepting. But Leonard Feldman, whom Barack knew from 1L Section III together, politely refused the costly honor.

But the writing competition results for the 1992 class produced a troubling disparity: women as a group had scored significantly below men. While the thirty-six new 2L editors would include five African Americans, just as in the 1991 class, the gender breakdown showed twenty-seven men and only nine women. There was nothing that Lisa and Barack could do, and on July 23 three dozen invitation letters from Obama went out in the mail.

Almost every recipient had already received a phone call from Barack or a

3L editor they knew. Obama's letter warned that the *Review* required "generally 30 to 40 hours a week" and would be "a big part of your life over the next two years." He acknowledged the "financial sacrifices" imposed by "coming back early" but claimed that "*Review* work is more creative, and often more challenging, than second and third year classes." He optimistically cited "the camaraderie that you develop with your classmates" and advertised Gannett House as "a friendly on-campus 'base' where you can relax, read newspapers, get breakfast." Work on the upcoming November Supreme Court issue, which the Supreme Court Office editors had already outlined, would be "particularly interesting" given the Court's 1989–90 decisions. New *Review* staffers "must arrive in Cambridge" by midday Wednesday, August 16, and should call Barack at either Sidley or the Robinsons' home number immediately if they had not already conveyed their acceptance.

Before returning to Cambridge, Barack and Michelle went to Washington, D.C., to attend his brother Roy Abon'go's early-August wedding to his second wife, Sheree Aleta Wood. Roy's split with his first African American wife, Mary K. Cole, had occurred several years ago, but only in mid-July had he finally filed for divorce, and a judgment would not actually be entered until well over a year after his second wedding. But this wedding was the greatest Obama family gathering ever. Kezia flew from Nairobi to Washington, and Ann Dunham arrived from Honolulu.

Ann had been hurt and dismayed when the profiles of Barack six months earlier had omitted her almost entirely. Ann's relationship with Made Suarjana remained a bright spot in her life, but she still struggled with her UH dissertation. This Washington gathering was the first time Barack Obama Sr.'s first two wives had ever met, and Barack's sister Auma, who attended as well, was overjoyed at how the two women immediately bonded. "They were sitting close together, holding each other's hands, reliving the wonderful times they had spent with our father and assuring each other what a great man he had been," Auma recounted. "Two women who came from completely different worlds clearly still loved him even after so many years." Barack stood up as Roy's best man at the ceremony, and then much of the extended family followed Ann and Maya to New York City for a celebration of Maya's twentieth birthday.[41]

Back in Cambridge, Barack presided as the 3L editors welcomed the new 2L members to the *Review*. Work on the November Supreme Court issue was well under way, with most 3Ls assigned to author a case note on one or another of the 1989–90 rulings, but the May imbroglio over Charles Fried was far from forgotten. When Mike Froman and Christine Lee, two of the most vociferous complainants about Fried, arrived in Cambridge several days late, Supreme Court Office cochair Brad Berenson formally proposed that the *Review* remove or otherwise punish them. "Everyone else had had to leave their jobs early and come back and start working," Berenson remembered, and only in highly unusual circumstances were exceptions allowed.

Jean Manas launched an angry rebuttal, and the newly arrived 2Ls watched as the 3Ls savaged and insulted each other. Carol Platt, the sole conservative among the nine 2L women, remembered "thinking 'Okay, this is about them coming back late, but this is also about so much more . . . this is not really about this, it's about things we don't even know about.'" Lori-Christina Webb, a 1986 Bryn Mawr graduate who had taught for three years in the South Bronx before entering Harvard and becoming one of the five black 2Ls on the *Review,* immediately realized "there's all these tensions." Despite Obama's sunny letter, she could see that the *Review* "was not a very welcoming environment," and this leadoff debate set the tone. "The coming back late is not really about coming back late, but we don't know what the subtext is because we're brand new, but there's always a subtext, and in the end the subtext is always political, it's always us against them." The tenor demonstrated that "there was a license to be disrespectful of one another . . . that was very shocking to me."

Anger about the duo's tardiness was not limited to the *Review*'s conservatives. Liberal Michael Cohen told Obama it was unfair, but Barack responded coolly, "Some people do the minimum, and others do more." Cohen walked away thinking that Obama was "fifty years older than me in terms of his maturity and equanimity." The meeting of all the editors to debate Berenson's motion got quite heated, although Obama made no attempt to shut down the debate. Daniel Slifkin, a 3L, found it all "ridiculous," especially when someone suggested fining Froman and Lee. "It all struck me as very, very silly," as no one actually had any fining power, and Slifkin wondered why Obama let it go on well into the evening.

"We had no clue what was going on," 2L Sean Lev recalled, and fellow new editor Charlie Robb quickly realized "how weirdly political and angry" the unpleasant exchanges were. "The meeting went on *forever,* and I had no idea where all this angry energy was coming from." Robb was five years older than his classmates, and his friend Trent Norris, another new 2L, was also older, having taken three years off before law school. Norris recalled that Obama presided while "never disclosing his viewpoint on this—not a hint of it." Robb also felt Obama manifested "a very cool demeanor," even when the argument turned to a dispute about how the vote should be taken.

Bruce Spiva, an African American 2L, remembered that "there was this huge question of whether people should vote with their heads up or whether they should put their heads down." A vote on that penultimate question favored semisecret heads down, and eventually Berenson's final motion, to censure Froman and Lee rather than drop them, was defeated. Norris remembered "walking out of that meeting and thinking 'God, what a mess. What have I gotten myself into?'" New 2L Howard Ullman recalled "a palpable sense of divisiveness and polarization," with Gannett House often so tense that "various folks would not talk to other folks." Michael Cohen remembered "Christine getting a lot of friction when she came back and feeling really unwelcome." Lee could not believe Berenson had been serious, and "I was mad at Barack for even allowing it—he could have just squashed the whole thing," or so she thought. "After that, Froman and I just have nothing but contempt for Barack," she recalled. "We hated Barack together."[42]

Obama's cool disinterest in his colleagues' angry passions contrasted with his warmer one-on-one demeanor, even toward the *Review*'s undergraduate work-study students. Senior Paul Massari had worked at Gannett House ten hours per week, including summers, ever since his freshman year, but he was "just really struggling" that August because he had no idea what to do post-graduation. Sitting on Gannett House's front steps one lunchtime looking miserable, Massari was joined by Obama, who "sat down and lit up a cigarette and just kind of talked to me and asked me what was going on." Massari said he was embarrassed by feeling unready for graduate school, and Obama talked about how "he had taken a bunch of time off" and had done fine. "He was very kind to me" and having his validation "really helped me a lot." Another undergraduate worker, Lois Leveen, also remembered Barack as "incredibly personable and engaging," with an "ability to connect with people" and a "responsiveness that felt genuine," even if he always did address any group of editors as "folks."

In addition to the controversy over Froman and Lee arriving late, Obama had two health-related issues to confront. New 2L editor Jeff Hoberman had spent the summer in Caracas, Venezuela, and within two weeks of returning to Gannett House, Hoberman was diagnosed with hepatitis and sent home for a month. In an excess of caution, Obama decided that everyone at the *Review* had best receive painful gamma globulin injections at University Health Services. When Hoberman returned, he was repeatedly greeted with: "We got shot in the butt because of you."

More problematic was how Patricia Williams had contracted Lyme disease while spending the summer on Martha's Vineyard, leaving her unable to meet her deadline for the case comment that was to be paired with Charles Fried's in the November issue. This reignited the warfare from months earlier, with Brad Berenson's conservatives arguing that "she didn't meet the deadline: she's out," while Williams's proponents wanted to give her more time and include a note in the November issue that her response to Fried would appear in the next issue. That led to another "very acrimonious debate" about *where* in the November issue the reference to Williams would appear. Different participants came away with different feelings about Barack's role in the process. Brad Berenson saw him as "a pretty honest broker, someone who genuinely understood and empathized with the views and perspectives of both sides." He "gave everybody a fair hearing" and "tried to do something sensible, in the best interests of the publication" while "maintaining harmony among a very fractious group of editors." Yet most of Berenson's conservative allies viewed Obama "as essentially two-faced, as telling everybody what they wanted to hear, making everybody believe that he agreed with them" while in the end always coming down on the left.

Christine Lee was not the only liberal 3L who regretted supporting Barack for president. Masthead meetings took place at 8:00 P.M., and they could go on for hours, often featuring the same sort of bickering that had occurred over Williams. Liberal Susan Freiwald remembered being called by SE Jean Manas before one meeting and being warned that Obama was not going to be on their side. "You've got to watch out! Barack's not going to be totally on our side!" Manas "was just up in arms. 'This is terrible! I thought we won!'" Freiwald was more irritated by Barack's infinite tolerance for his colleagues' long-windedness. "Barack, I need to go home before midnight. It is not safe for me to walk home in the middle of the night. You need to limit people. . . . If this isn't over at 11:00 P.M., I'm leaving no matter what. I'm not sacrificing my safety for these blowhards."

Not having Williams's comment was not the only problem with the November issue. Robin West, writing on a tight deadline while coping with her first infant, submitted her manuscript on time, although even she realized it was not as good as her previous article. At Gannett House, the 2Ls sub-citing her piece were dismayed. Bruce Spiva remembered one evening "Barack coming in to check on the new people . . . 'How's it going, folks?'" and Bruce vented about West's manuscript. "This is in awful shape. We ought to tell these professors that they shouldn't turn in stuff like this. This is just unacceptable. We're having trouble finding what she's talking about, and it's poorly written.'" Just as with Michael Cohen's earlier complaint about the late arrivals, Barack downplayed the problem and told Bruce, "You just kind of have to get used to this" while "giving us a pep talk" about how they should not take the *Review* too seriously. "Don't worry about it, Bruce. Nobody's going to read it."

West found Barack thoughtful and polite. "Every time he called," she remembered, "he asked how the baby was doing" and was Robin getting enough sleep. Barack was "very sensitive," a "quite unusual" trait in law review editors. Barack was "very gracious about the edits" and "knew this was burdensome" for her. "He did not have sort of typical left-progressive views on identity politics," West realized, and while "he was always very formal on the phone," eventually "he started calling me Robin."

Before classes began, Obama spoke at a BLSA orientation event for African American 1Ls. Artur Davis, a magna cum laude graduate of Harvard College, remembered being struck by Barack's "extreme sophistication" and thought "he was either going to be a Supreme Court justice or a political figure." Tom Perrelli, who as managing editor was Barack's closest *Review* coworker, believed Barack was "the most fully formed person I met" during three years of law school. "He seemed to be a pretty much fully formed grown-up," SE Julie Cohen agreed, "and you would not have said that about most people" on the *Review*. 3L Devo cochair Marisa Chun likewise thought Barack "was a fully formed adult by the time he was on the *Law Review*," someone whom others looked up to as "wiser, more mature, more worldly, because he was." Sean Lev thought Barack was "extraordinarily impressive," and fellow 2L editor Jonathan Putnam agreed that Barack was "incredibly impressive," indeed "the most impressive person I had ever met in my life." Conservative 2L Carol Platt believed Barack "did a good job at projecting this image of being the adult in the room," and manifested "an absolutely preternatural self-confidence," but he "was just never that invested in" the *Review* itself. "You just got a sense that he didn't give a damn about too much."[43]

For fall classes, Rob and Barack, like most of the 3L editors, enrolled in Charles Fried's Federal Courts course. Fried used the traditional Harvard text, Henry M. Hart Jr. and Herbert Wechsler's *The Federal Courts and the Federal System*, 3rd ed., and the course examined federal jurisdiction, habeas corpus, and civil rights actions that could be brought pursuant to 42 U.S.C. §1983, which dated from the Reconstruction-era Enforcement Act of 1871. Obama's basketball buddy David Hill remembered that Barack "enjoyed sparring with Fried, and Fried enjoyed sparring with him." The students thought Fried was an excellent professor, and once it became clear that despite his service as Reagan's solicitor general, he was no "Republican apparatchik," the festering controversy about his *Review* comment quickly faded. The class met for two hours in the morning, twice a week, and Fried took a short break halfway through. Gannett House was close enough that *Review* editors could dash there and return with a bagel, donut, or muffin. The muffins were especially popular, and, as 3L editor Lourdes Lopez-Isa remembered, were "huge, like the size of your head."

ME Tom Perrelli recalled that there was no question that "muffin flaunting" took place. A year earlier, when it had become known who had made *Law Review* and who had not, those results, even more than 1L grades, painfully challenged Harvard students' "sense of shared eminence," as Robert Granfield explained. There was no denying that *Review* membership elevated editors above their classmates, and access to these luscious goodies distinguished the super-elite from the also-rans. It was as if "the rest of us didn't count," recounted Section III veteran Michelle Jacobs. "The rest of us sat there" while the *Review* members returned with their pastries and munched away as class resumed. But one morning Fried asked, "Why do certain people have these muffin thingees? Where are all these lovely muffins coming from?" And, most memorably, "Can I have a lovely muffin? Who can bring me one?" By calling out "the chosen few who got the bagels and the muffins," Christine Lee recalled, "he put it out in the open."

Editor Dave Nahmias, a 3L, was in Fed Courts and wondered "why we were spending our money to buy breakfast for the *Law Review*." Nahmias thought the law school's "poisonous and elitist" atmosphere was not helped by the editors' conspicuous consumption. "The *Law Review* people are not particularly popular with a lot of the other law students," and their "lording of this benefit" over their classmates rubbed many people the wrong way. "There had been grumbling before," 3L editor Chris Sipes remembered, "but it really reached a pitch after Fried commented on it." The "bagel war erupted," and "this was a riven issue," Brad Berenson drily observed. A full-body meeting was called, with Obama presiding. Nahmias moved to eliminate the morning spread, though not the evening snacks, because at night editors were actually working on the *Review*. Only a few people sided with Nahmias, but at one point Barack unintentionally brought the discussion to a halt by using a word—"precatory"—that no one in the room understood. "C'mon, folks," Barack responded, explaining its usage. "He always said folks," 2L Sean Lev recounted. ME Tom Perrelli spoke up to encourage editors not to consume the fruits of their privilege in front of less fortunate classmates, and Nahmias's motion was overwhelmingly defeated. "People liked the perk, and nobody cared if everybody else in the law school thought they were a bunch of elitist jerks," Nahmias explained. But at least in Fed Courts, Perrelli's advice was widely ignored. "Someone always had to bring donuts to Professor Fried," Michelle Jacobs remembered. "He ended up with donuts every day."

Barack and Rob debated taking Corporate Taxation before deciding against it. Rob's economics background drew him to Antitrust, which Barack passed on so he could take a Kennedy School of Government course on persuasion and politics taught by the well-known political scientist Gary Orren. Barack and Rob also enrolled in a once-a-week seminar, Civil Society, that Mary Ann Glendon, whom they had had for Property, was offering in conjunction with Martha Minow and a third professor, Todd Rakoff. Much of the semester would be spent listening to guest speakers, but they could use the seminar as a basis for an additional two-credit independent writing project that would give them the opportunity to pursue an idea they had been discussing for more than a year: to coauthor an intensely analytical book examining several present-day policy problems and critiquing overly ideological approaches to them.

The Glendon-Minow-Rakoff seminar began with four weeks of readings. "They tried to teach us very quickly sociology," 3L Eric Posner explained. They began with some of Thomas Hobbes's *Leviathan* plus an excerpt from British philosopher Dorothy Emmet's 1966 book *Rules, Roles and Relations*. Further readings from Émile Durkheim, Karl Marx, and Max Weber were followed by classes with Catholic labor activist Monsignor George Higgins; feminist ethicist Carol Gilligan, well known for her 1982 book *In a Different Voice*; social critic Alan Wolfe discussing Richard Rorty; and Anthony Cook speaking about liberation theology. Each week students prepared two-page papers reacting to the assigned readings, but Posner recalled that the course "didn't work" and the three professors "understood it didn't work."[44]

The sudden retirement of Supreme Court justice William J. Brennan added additional work for the *Review*'s November issue. Traditionally, after every justice's departure from the court, the *Review* published a series of brief appreciative tributes, and 2L Jennifer Collins took the lead in quickly commissioning a half-dozen such pieces. Justice Thurgood Marshall and Judge Abner Mikva led off the feature, which occupied the issue's first thirty-nine pages. Two contents pages then detailed the Supreme Court material: West's foreword, "Taking Freedom Seriously"; Fried's comment, "*Metro Broadcasting, Inc. v. FCC*: Two Concepts of Equality"; and twenty-seven separate case notes of about ten pages apiece, each written by a different editor. At the bottom, in the smallest type possible, was the note "An additional Comment on *Metro Broadcasting, Inc. v. FCC*, by Professor Patricia Williams, will appear in the December 1990 issue."

A huge wall chart in the office shared by ME Tom Perrelli and the three executive editors tracked the editing process for the November pieces, and Barack actively shepherded many of the case comments. Kevin Downey was tackling *Cruzan v. Director*, a "right to die" case that had attracted great attention, and Downey's long final footnote included a quotation from Leo Tolstoy's *The Death of Ivan Ilyich*: "In fact, the most profound moments of Ivan's life occur when others would prefer that he be relieved of his suffering." Downey remembered that "Obama really wanted me to take it out," and they wrestled over it for some time before Downey prevailed. Obama also took an active role in 3L Frank Amanat's note on *Rutan v. Republican Party*, an Illinois case in which the Supreme Court had ruled forcefully against partisan political patronage in public sector hiring. Amanat believed it was "a great decision," but a fellow 3L editor "had a real problem" with Amanat's analysis and "I took umbrage." Their disagreement became "an intractable problem" and "I took it to Barack," who "asked me to work it out and asked Tom to mediate." Perrelli persuaded Amanat to make some changes and got the other editor to back off. "That's an example of Barack's management style" as *Review* president, Amanat explained. "He always took a very problem-solving attitude."[45]

Almost weekly that fall there were student protests over faculty hiring as well as charges that a large law firm that interviewed annually on campus had engaged in discriminatory behavior at another school. Cassandra Butts was again in the lead, but most students did not pay the protests much attention. "It's gotten to the point where they're protesting everything," 3L Chris Lu told the *Crimson*. Then the Women's Law Association, having learned that only nine of the *Review*'s thirty-six new 2L editors were women, called for the *Review* to expand its affirmative action policy. Obama told the *Harvard Law Record* that editors believed the imbalance was simply "an aberration," and on the issue of expanding affirmative action, "I don't have a clear sense of whether it's needed or not." Historically everything at Harvard was "somewhat discriminatory," Barack said, but he encouraged more women to enter the writing competition, regardless of "this year's blip." The WLA pushed back, complaining that it was "even more deeply disturbing if the people running *Law Review* don't care enough to correct it."

Temperatures rose exponentially when Jim Chen, the most combustible of the *Review*'s 3L conservatives, wrote to the *Record* to lambaste the WLA's position. "The mere presence of affirmative action cheapens membership for every editor who falls within the existing groups," and advocating expansion simply "affirms the invidious belief that women and other 'beneficiaries' of affirmative action can't compete on a level field." Citing an experience when an interviewer asked him whether the *Review*'s existing policy included Asians, Chen wrote that "I felt the stigma" of that question and called for everyone to "eliminate affirmative action root and branch."

WLA leaders declared they were "completely livid" at how the *Record*'s initial story had simplified their views while also taking offense at Chen's remarks. Another full-body meeting was convened at the *Review* to discuss the situation, with the 3L editors' positions unchanged from nine months earlier. Affirmative action opponents were most angered by the *Review*'s policy of operating in complete secrecy, with the president and two other editors exercising what Brad Berenson called "totally unchecked discretionary power."

At the *Review*, 3Ls and 2Ls recognized a consistent pattern in Barack's style as presiding officer, whether the argument was about editors' tardiness, muffins, or affirmative action. In Berenson's eyes, "Barack tended to treat those disputes with a certain air of detachment and amusement. The feeling was almost 'Come on, kids, can't we just behave here?'" Obama always seemed "a little bit separate and apart from the fray, observing events from a slight emotional and intellectual distance," Berenson believed. Even Jim Chen felt that Obama "was very savvy, a good listener," someone who "always left the impression that he heard you out." Michael Cohen thought that Barack "did have a real talent for conciliating these disputes," and Marisa Chun would recall "less about his own opinions than the fact that he was masterfully able to make sure everyone else spoke their piece." African American 2L Nancy McCullough said, "I can't think of a particular issue that I can remember him feeling strongly about" and reflected that she "actually would have been happier for him to say sometimes, 'This is how we're doing this, and shut up!'" rather than letting everyone argue for hours. Lourdes Lopez-Isa agreed that with Barack, "You would never know what he thought about things." Ken Mack recalled, "I never saw Barack lose his cool, get angry, have a fit of temper, raise his voice. . . . He was a very cool character, a very cool customer in all senses of that word."

On affirmative action, Obama wrote a letter to the *Record,* explaining that the writing competition entries were graded by at least three editors. The *Review*'s policy simply said that the officers "*may* take race or physical handicap into account . . . *if*" they believe that will "enhance the representativeness of the incoming class," should the raw scores make that desirable. In response to Jim Chen, Obama described himself "as someone who has undoubtedly benefited from affirmative action programs during my academic career, and as someone who may have benefited from the *Law Review*'s affirmative action policy when I was selected to join the *Review* last year." However, 2L Jonathan Putnam recalled that among fellow editors "it was a fact in wide circulation that Barack himself had not checked off the box on the application." But Barack wrote, "I have not personally felt stigmatized either within the broader law school community or as a staff member of the *Review.* Indeed, my election as President of the *Review* would seem to indicate that at least among *Review* staff . . . affirmative action in no way tarnishes the accomplishments of those who are members of historically underrepresented groups."[46]

Barack and Rob Fisher were still playing basketball regularly, and 3L editor Julius Genachowski recalled one day when Barack "showed up to *Law Review* with a broken nose," wearing "a pretty serious white bandage thing." Rob remembered another time when some "really obnoxious guy . . . was right in my face yelling at me" and then "Barack was immediately in the middle, stopping it, calming everybody down." On election night 1990, Barack invited a number of friends—Rob, Julius, and the outspokenly conservative Brad Berenson—to his Somerville apartment to watch the returns. That quartet often played poker together, with Rob explaining that Brad "was definitely a friend," notwithstanding their differing political views.

Berenson remembered how "Barack was rooting very hard for Harvey Gantt" in a hotly contested North Carolina race, and photos from that night show Barack wearing a "Harvey Gantt for U.S. Senate T-shirt." Gantt was a highly promising African American Democrat with a nationally touted future who was challenging incumbent Republican Jesse Helms, whose atrocious record on race included die-hard opposition to a federal holiday celebrating Martin Luther King Jr.'s birthday. But the evening ended badly, with Helms eking out a narrow 52 to 47 percent victory after investing heavily in a television ad that made an explicitly racist appeal to working-class whites: "You needed that job, and you were the best qualified. But they had to give it to a minority because of a racial quota."[47]

Barack was so busy that fall he failed to send Erwin Griswold the traditional letter introducing the November issue. Even so, Griswold found it "very fine" and said it was "a marvelous performance." He especially enjoyed the tributes to Justice Brennan and thought Robin West's foreword was "interesting, though sometimes a little ethereal." Griswold took a much dimmer view of the comment, writing that "Fried's view is not only too rigid and mechanical, but represents such a narrow vision that it is quite unrealistic." Even before Griswold weighed in on November, the December issue went to press, notwithstanding a late-night computer crash that led Barack to call 3L techie Frank Amanat to ask him to return to Gannett House. At "2:00 A.M., I finally get the system up and running," Amanat remembered, and a trio of exceptionally difficult pieces—David Charny and David Wilkins's articles, plus Patricia Williams's comment—were finally off to the printers. Multiple *Law Review* colleagues recall Barack investing more time and energy in the Wilkins article than in any other piece to date, but Wilkins remembered interacting with Jim Chen, one of the EEs, not Barack. Obama took some interest in Williams's comment, which Supreme Court Office cochair John Parry had to protect from critical comments inserted by Patrick Philbin, one of the 2Ls working on the edit. "Barack had to preside over all this bickering," Parry recalled, and conservative EE Amy Kett was impressed by Barack's P-read on the Williams piece. "He seemed to be very hardworking" and was also "a very good manager of people." The December issue also included an eight-page case note by 2L Jonathan Putnam, who remembered that Barack spent "a whole weekend day, with him painstakingly, and painfully for me, going through line by line what I had written." Putnam thought this was "a lot of his energy to expend on what in the scheme of things was a tiny part of the editorial product."

Somewhat tardily, Obama sent several copies of the November issue to Justice Brennan, along with an unusual cover letter that described the impact Brennan's 1987 NPR interview had had on him. "I recall harboring considerable doubts about leaving my grass-roots work to become a lawyer—I felt concerned that too often the law served the interests of the powerful, and not the powerless. In the midst of my internal debate, I was fortunate enough to hear your interview," which "helped to convince me that legal practice could in fact be a worthwhile pursuit." More recently, he wrote, as a law student he had read a "countless" number of Brennan opinions, and he had been impressed by the justice's "unwavering commitment to 'the little guy,' the underdog, and the less fortunate. Your blend of 'reason and passion'"—a phrase Brennan had championed in a well-known 1987 lecture—"has been a source of inspiration in the development of my own legal ideas." Citing his status as "an African-American who a generation ago might not have even attended Harvard Law School, much less served as the President of its *Law Review*," Barack acknowledged "the debt that many of us owe to you in helping America to live up to its ideals of opportunity for all its citizens." Brennan replied that he was "most grateful to you for the kind comments," and Barack not only kept Brennan's letter to him but had it framed.

Obama was also involved in contract negotiations for the new edition of *The Bluebook* with the law journals at Columbia, Yale, and the University of Pennsylvania, which cosponsored publication of *Law Review*'s widely used cash cow. Harvard had a 40 percent interest in the enterprise and was performing an overwhelming majority of the work being done on the new edition. Ken Mack and Frank Amanat were leading the *Review*'s effort, but the *Yale Law Journal* had exercised its right to demand a new contract. The *Law Review* was a sizable business, with annual subscriptions bringing in approximately $300,000 a year. That alone produced a modest operating profit, but *The Bluebook* generated more than $500,000 annually, with a profit of close to $200,000. That paid for much of the salaries of office manager Dodie Hajra and circulation director Mary DiFelice, as well as those warm bagels and larger-than-life muffins. An arm's-length quartet of external trustees—the law school's dean plus two professors and an alumni "graduate treasurer"—formally oversaw the Harvard Law Review Association and received annual audit reports. The president traditionally delivered copies of each issue to the trustees, including Richard Parker, Barack's 1L Crim Law teacher, who recalled no problems whatsoever arising during Obama's presidency.[48]

In mid-November Barack and treasurer Lisa Hay went to New York City for a contentious renegotiation session that resulted in the other three journals using their 60 percent majority interest to insist upon a 7.5 percent reduction in the *Review*'s share of *The Bluebook*'s annual net profit. In Cambridge, some conservative editors groused that Obama had been too pliant a negotiator, but Barack had something more personally significant on his mind during his trip to Manhattan. A month earlier, Jane Dystel, the literary agent whom he had met with in May, had agreed that the book proposal they had been crafting since midsummer was now ready to submit to publishing companies. Dystel, no shrinking violet, had been impressed by her exchanges with Barack. "I liked his authoritative voice," she recalled years later. "He was so focused. Mature, really together, already sure of himself. Not your typical kid. No question whatsoever he was going someplace."

The proposal, entitled "Journeys in Black and White," self-confidently compared the story Obama would tell to no fewer than eight commercially successful books, most of which had been published in the early and mid-1980s. "The texture and spirit of the writing will derive from the tradition of the autobiographical narrative, typified by such works as Maya Angelou's *I Know Why the Caged Bird Sings,* Maxine Hong Kingston's *The Woman Warrior,* John Edgar Wideman's *Brothers and Keepers,* Wole Soyinka's *Ake,* Mark Mathabane's *Kaffir Boy,* and Russell Baker's *Growing Up,* as well as such travelogues as William Least Heat Moon's *Blue Highways* and V. S. Naipaul's *Finding the Center.* Such works take on the narrative force of fiction, and invite the reader to share in the hopes, dreams, disappointments and triumphs of individual characters, thereby soliciting a sense of empathy and universality that is absent in too many works on race in America."

Among the editors to whom Dystel submitted the proposal was Elaine Pfefferblit at Simon & Schuster's Poseidon imprint, an editor Dystel often approached. Two months earlier, Pfefferblit had hired a new young assistant, Laura Demanski, and she was tasked with reading the proposal and a brief sample chapter and recommending whether it had merit. Demanski's October 25 report to Pfefferblit could not have been more enthusiastic: "This proposal is very impressive. Barack Obama is uniquely qualified to write what promises to be a big book about race issues, particularly black-white relations in America today. . . . His observations and reflections come from experiences both here and in Africa," as the book would recount Barack's 1988 trip to Kenya. "Obama has an unusually broad perspective," and "even better, he analyzes his experiences and presents his ideas unusually clear-headedly."

Demanski wrote that Barack "seems to work with feeling *and* intelligence" and said his "plans for the book are concrete and well-organized . . . the sample chapter demonstrates an impressive skill at weaving detailed personal narrative with reflective and analytic passages, at honing in on specific characters or incidents and then panning out gracefully to place these within the big picture." Demanski believed Obama was "clear about what he wants the book to avoid being—confined to the anecdotal and shying away from discussion of wider events, on the one hand, but also sacrificing intimacy and inadequately capturing the 'private demons and sources of inspiration that lie at the heart of the race issue.'"

Demanski's enthusiasm was soon shared by Pfefferblit, as well as by Poseidon's publisher, Ann Patty, a well-regarded book-business figure whose list of titles the *New York Times* praised as "funky, unpredictable, low on marquee authors, rich with new talent." By the time Barack stopped by Poseidon's offices in mid-November, both Patty and Pfefferblit very much wanted to sign up "Journeys in Black and White." In person, Barack was "very bright and very personable and very charming," and Demanski remembered that "we were all really impressed." In retrospect, Ann believed "there must have been other people bidding on it," and she remembers authorizing Pfefferblit to offer $125,000 to acquire the rights. "We did take risks like that," Patty explained, and that sum was much higher than what a twenty-nine-year-old first-time nonfiction author normally could hope to receive. With the initial payout of $40,000, Barack could take a full year after graduation to write the book. The balance, half payable on submission of a complete manuscript and the remainder upon publication, would more than cover the tens of thousands of dollars in student loans Obama had incurred while attending Harvard. Less than two weeks later, on November 28, Poseidon issued the contract. Barack's due date was June 15, 1992.

Back in Cambridge, Barack was visibly ecstatic about this news. Michael Weinberger, who had done yeoman work for months on the David Wilkins article, invited Barack home to have dinner one evening with him and his wife, to whom he had said, "this guy's going to be famous" as "an important politician." Dinner was nothing fancy—spaghetti with three sauces—but the normally circumspect Barack could barely stop burbling. "He had just gotten a book contract," Weinberger recalled, and "a lot of the talk had to do with the agent." All in all, Barack "was pretty excited about it."[49]

Before classes ended, the editors had to get the unusually thin January issue—only 150 pages—off to the printer. The sole article dealt with antitrust law; a twenty-page student note by Christine Lee, which Erwin Griswold found "stimulating," addressed employment discrimination against black men; and a review considered Chris Edley's new Yale University Press book, *Administrative Law: Rethinking Judicial Control of Bureaucracy*. With the semester about to end, Barack and Rob had to finish a detailed outline of the lengthy paper that was due to Martha Minow that spring and which they hoped could become a coauthored book.

"Transformative Politics," or "Promises of Democracy: Hopeful Critiques of American Ideology" reflected remarkable aspirations and envisioned three major sections. The first would explicate a trio of topics: "How Has the American Left Failed?," "How Has the American Right Failed?," and "The Shared Assumptions of the Left and the Right." The center of the book would be three analytical chapters. "Plant Closings—The Viability of the Regulated Market" would be followed by "Race—The Limits of Rights Rhetoric" and then "Public Education—Balancing Local Control With Quality Control." Finally, their conclusion would discuss "The Importance of Democratic Dialogue."

Barack and Rob's outline was remarkably substantive and provides significant insight into how more than two years' worth of almost daily debates had shaped their thinking about a great many social and political questions. The first chapter, on the failures of the American Left, cited a "Lost Faith in Democratic Discourse" and targeted "Rudderless Pragmatism—a refusal to articulate principles that can guide debate and create lasting coalitions." Echoing John McKnight, it also indicted an "Expertist Ethos," an excessive "belief that the judiciary is the principal arena for social change," and an "insistence on bureaucratic centralization." There was also "A Failure to Understand and Use the Market" as "a tool for decentralization of power" and as "a potential promoter of equality."

The failures of the American Right began with its "lack of faith in collective action," its belief that "we are all self-contained units," and a "lack of faith in the possibility of progress for the masses." An "institutional fetishism" led to a "failure to see the variety of ways that 'civil' society in fact acts to structure oppression, hegemony of the dominant culture, etc. (the classic example being the assumptions underpinning *Plessy*)." Third was a "Failure to Recognize the Market's Destructive Power," most especially "persistent underinvestment in . . . human capital."

"The Shared Assumptions of the Left and the Right" included a "rigid dichotomy" between individuals and society. While the Right embraced the "self-interested individual," the Left "insists that groups once formed (along lines of class, gender, race, etc.) then serve as the static, self-interested actors" of pluralist theory. While "the Right and the Left become captive to their categories," the "proposition of this paper" would be "that even as we abandon the notion of the public/private distinction," a belief that society "must act to promote social change" must be embraced at the same time that "the private sphere . . . is understood as a precious social construct through which we decentralize power." Underlying everything was the belief that "the vital mechanism by which such a vision operates is deliberation—principled democratic debate, about things, and not just words."

The first of their topical chapters, on plant closings, would confront how the "complete mobility of capital" could, as it had in Southeast Chicago, lead to the "destruction of community and family" while also emphasizing how "corporate managerialism cannot take into account these externalities." Long-term solutions would require "worker participation and ownership" and "restructuring the corporate form" due to "the costs of defining corporations as persons under constitutional analysis."

In the second chapter, on race, they would address "the limits of the assimilationist vision" that had reigned during "the Golden Days of Civil Rights." It would acknowledge "the incompatibility of group redress with received constitutional understandings" and how "the language of rights" is "too blunt a tool to capture the complexity of social relations" because "rights don't require compromise or mutual understanding." It would also argue that "the equation of racial liberation with government centralization, bureaucracy, judicial intervention" means ignoring "the fundamental strengths of the black community—self-reliance, religious cohesion, community support, family." This chapter would propose "Terms for the New Debate" that would shift "from rights to responsibilities" and also involve "a reexamination of integration." A "potential restructuring of political dialogue around leveraging resources into black communities because they are poor" would not ignore how "industrial development that means jobs for blacks may also mean environmental problems that hit blacks hardest."

The final topical chapter, on public education, would confront how "the legacy of racism and skewed resource distribution promotes [the] idea that only a single, unified school system run by bureaucratic experts and controlled from the top can ensure equality of opportunity." The entire outline was informed by Barack's experiences in Chicago, and in the education context, Barack and Rob rued the exit of the "wealthiest parents" and "the most informed, participatory parents and innovative teachers" from declining public school systems. A further threat was vouchers, because "having the best parents exit . . . further erodes [a] basic commitment to public education." The way forward would entail "community ownership of schools" and "a recognition that schools are inseparable from communities." It would also require "a consensus built around equalization of resources, not equalization of teaching."

Barack and Rob seemed to have their conclusions already well in hand. "The rigid choice between individual and state has spawned a variety of false dichotomies that continue to plague the existing political debate," with neither side "capturing the reality of the problems facing today's society." In addition, "the importance of democratic dialogue" would be "the critical medium through which this transformation will take place. The quality of our deliberation determines the extent to which we can simultaneously maintain our faith in individual development, idiosyncrasy, and our faith in our capacity to act collectively." In a concluding note to Martha Minow, Barack and Rob added that "we are still in the process of making some choices about how to boil down the structure of the book idea into the paper. We suspect that . . . we may not be able to do everything outlined here. For example, we may deal with only two examples, rather than three."[50]

On November 20, three deeply committed 2Ls took the lead in organizing the most audacious student protest yet when, in the name of the HLS Coalition for Civil Rights, they filed a lawsuit in state court alleging Harvard Law School had personally harmed them by not hiring a more diverse faculty. Almost simultaneously, Harvard-Radcliffe Alumni/ae Against Apartheid (HRAAA), a five-year-old group that had already elected four of its candidates, including world-famous South African archbishop Desmond Tutu, to six-year terms on the university's Board of Overseers, tried to recruit three more well-known names to run in the upcoming 1991 alumni election. When several "famous people" declined, the group asked Obama to be a candidate, and after thinking about it for forty-eight hours, he agreed. Joining him on the HRAAA's slate would be well-known free speech advocate Nadine Strossen and prominent education critic Jonathan Kozol.

On Friday morning, December 7, Barack's 3L basketball buddy Frank Harper was on the Boston subway heading to Harvard Square when he saw a front-page *Boston Globe* headline: "Recent Parolee Is Charged in Assault-Weapon Killing." The story reported that three people had been shot, one fatally, in Boston's Dorchester neighborhood by a forty-year-old gunman wielding an AK-47 assault rifle. Just twenty-two days earlier, the killer had been released on parole after serving almost the entire previous fourteen years in Massachusetts state prisons, including Walpole. The slayer was Ronald A. X. Stokes, organizer of the Walpole All-Stars, against whom Frank and Barack had competed just a year earlier. Stokes's early 1990 assault on a prison guard—far from his first—should have blocked his release, but this time multiple convictions for the Dorchester shooting would keep him in prison until his death in 2010.

Barack faced only one in-class final exam, for Fried's Fed Courts course, before he flew to Honolulu, where Michelle Robinson again joined him for Christmas. But Barack had to return to Cambridge by New Year's Day because the law school's three-week winter term classes began on January 2. Most 3Ls met their "professional responsibility" requirement by taking a legal ethics course titled Legal Profession, and Barack, Rob, and Gina Torielli, their friend from Section III, enrolled in a winter term section of that course taught by Jeffrey Kobrick, a veteran Boston legal-services attorney who was now a full-time "visiting professor from practice." It met every weekday morning for three hours, and Kobrick used Andrew Kaufman's *Problems in Professional Responsibility*, 3rd ed., as his core text. Legal Profession addressed real-world "ethical and moral dilemmas" encountered by attorneys.

One centerpiece of Kobrick's class was an ABC News documentary, *The Shooting of Big Man: Anatomy of a Criminal Case*, first broadcast in 1979 and produced in part by a 1972 Harvard Law grad who had worked as a public defender in Seattle. The film had created a national controversy among public defenders because it showed attorney David Allen coaching his client Jack Jones on what he should say when he was testifying in his trial for the shooting of Raymond "Big Man" Collins in a Seattle flophouse. Jones was acquitted, but some fellow public defenders accused Allen of "unethical conduct." The film ended with Harvard professor Charles Nesson—whom Obama had had for Evidence a year earlier—saying Allen had not been "coaching" Jones, but simply helping his client testify "as effectively as he possibly can."

On the first day of class, Kobrick recognized Barack from earlier press coverage, and he was pleasantly surprised when the *Review*'s president attended every single day. "He would not speak in every class, but when he spoke he had something to say," Kobrick recalled, and "everybody would listen." The day that

Kobrick showed an excerpt from *Big Man,* Obama "spoke last. He spoke after everyone else," and "to a person, nobody defended the Seattle public defender" before Obama did. "They all thought he was coaching the witness too much, and then Barack spoke, very forcefully, in disagreement," and at some length. "'I don't agree,'" Kobrick recounted Barack saying. "'First of all, they got his story out of him initially without giving him any leading questions at all. They just asked him what happened.'"

Barack "was extremely specific," Kobrick recalled, arguing that Jones "did not speak the language of an all-white jury." Without Allen's firm instructions about how to describe what had happened, "the guy would have been toast," Obama stated. On the last Friday in January, Kobrick's two-part, eight-hour take-home exam limited each answer to sixteen hundred words. Kobrick's second question, concerning a welfare recipient living in public housing who faced a criminal charge that threatened both those benefits, must have sounded familiar to anyone who had spent time in Altgeld Gardens.[51]

Obama's term as president of the *Law Review* was almost over as Kobrick's class ended. By early January, the *Review*'s February issue had gone to the printer, but its one article, "Fair Driving: Gender and Race Discrimination in Retail Car Negotiations," by Northwestern University law professor Ian Ayres, was already the subject of *New York Times* and *Chicago Tribune* stories plus a nationwide Associated Press one. Within the *Review,* Ayres's work had been a topic of dispute because even its most enthusiastic Articles Office proponent, Scott Siff, acknowledged "there wasn't so much law in it" and "the methodology wasn't that good."

Siff had argued successfully that Ayres's "very interesting, provocative analysis" met his "*New York Times* test for the articles," and there was no question that the article was the *Review*'s "highest-profile piece" during Barack's presidency. The *Times* headline—"White Men Get Better Deals on Cars, Study Finds"—attracted attention, but Ayres's broader argument, that federal civil rights laws still did not adequately protect people of color and women from discriminatory treatment, was actually based upon just 165 faux sales negotiations at Chicagoland automobile dealerships. Obama told Erwin Griswold that Ayres's article "is a nice change of pace from our more typical theoretical fare," but Griswold was decidedly underwhelmed, calling it "not . . . exactly a blockbuster." Griswold did praise the *Review*'s overall content, telling Obama that "your volume has been an excellent one."

The *Times* also reported that an enterprising 3L had produced a 1991 "Black Men of Harvard Law School" calendar, archly noting that "so intense was the competition" for inclusion "that even Barack Obama," president of the *Review,* "did not make the cut," although his friends Ken Mack and David Hill did.

Some classmates wondered whether Barack's virtual absence from student diversity protests explained his omission.

Inside the offices of the *Review,* Obama's final weeks were taken up by multiple controversies and a heavy editing load. One angry debate concerned who among the 3L editors would get the slots for individual notes in the year's four remaining issues. Obama had named Anne Toker to the masthead as a supervising editor without Anne's knowing the unwritten rule that SEs, as editors of their fellow students' work, were not supposed to publish notes, and a nasty editors' meeting took place before Toker's inclusion in the April issue was affirmed. Susan Freiwald remembered that as usual, "Barack was in the background," not taking any clear position, and he took the same stance at a subsequent full-body meeting that debated whether the word "black" would or would not be capitalized in an upcoming article.

Dorothy Roberts's manuscript on pregnant women who were drug abusers had originally used a capital *B,* but by January, tensions between the EEs and many colleagues were pervasive. Conservative EE Adam Charnes explained, "We viewed ourselves as the guardians of the traditions of the *Review,*" and no one took that mission of rigid adherence to *The Bluebook*'s rules more seriously than Jim Chen. "He loved *The Bluebook* the way evangelicals love the Bible," Brad Berenson explained. But Chen often altered passages that authors believed had been settled earlier in the editing process, "and some authors would go ballistic." In the case of "black," Lori-Christina Webb and Roberts's other editors had sent the article forward with an uppercase *B,* and then the EEs made the *b* lowercase. A lengthy *Review*-wide debate ensued. "The arguments were so abstract and passionate," Webb remembered, yet by this time in the year, a good many 3Ls and 2Ls were distancing themselves from the *Review*'s endless controversies. To 3L Mike Guzman, "the combination of it being presumptuous and pointless" made "all the fighting" seem like a waste of everyone's time. Guzman viewed Barack as "a good conciliator," and during the most bitter arguments, "I don't remember him ever saying a word." In the argument over "black," Obama once again commented, "Just remember, folks. Nobody reads it."

Three years later, Barack cited the B-in-Black controversy as a notable law school memory. "One of the most frustrating things about student life was I guess what's called political correctness," he explained. "Young people tend to jump with both feet on a whole lot of symbolic issues," and "when I was the manager of the *Law Review* at Harvard, I had a young black woman come in to me and complain vehemently about the fact that the word 'black' was not capitalized in an article." A white editor said that since white was not capitalized, neither should black. The dispute was "a matter of symbols, not substance," Barack explained, and thus had not interested him.

As Barack's year as president came to an end, conservatives, leftists, and everyone agreed that managing editor Tom Perrelli had been the *Review*'s MVP. Perrelli was "*the* figure for us," 2L Sean Lev remembered, and fellow 2L Jon Molot recalled that "everybody loved Tom." EE Adam Charnes praised him as "the nicest guy in the world" as well as the person most responsible for keeping the *Review* on schedule. Perrelli remembered sitting in Barack's "little cave of an office" going over edits, but most other editors knew that "if you wanted to go talk to Barack or hang around with Barack, you had to go out on the front porch, because he liked to smoke," Mike Guzman recalled. "He was almost always on the stoop, smoking," 2L Charlie Robb agreed. Lori-Christina Webb, one of the *Review*'s few other smokers, joined Barack "on the steps of Gannett House, smoking Marlboros. . . . That is, in my memory, where he's fixed, not in the midst of these battles."

Bruce Spiva recalled that Barack "tried to step back from a lot of the day-to-day rancor," and 2L Howard Ullman felt Gannett House "was not a terribly pleasant place to be because it was so fractious." Many editors used the exact same phrase—"above the fray"—to characterize Barack's reaction to the *Review*'s battles. "He did not seem ever stressed," Anne Toker remembered, and Adam Charnes thought Barack's strategy was to "let people argue a lot and see if they work it out," although if he had been more assertive, "it probably would have saved hours and hours" of time on multiple occasions. One EE recalled having a spirited disagreement with another editor. "We each independently went to Barack, and he said to me, 'I agree with you,' and he told the other guy, 'I agree with you.'" Then the two continued their dispute and were "outside in front of Gannett House arguing when Barack came out. . . . We called him over and said, 'You said.'" An embarrassed Obama realized he had been caught. As 2L Jeff Hoberman explained about another big argument, "everybody kind of thought Barack was on their side."

Grade-on Chris Sipe, a 3L, believed "you really saw him gaining in confidence as the year went by," that Barack "was much more confident at the end of the year than at the beginning." By January, most 3Ls believed the 2L editors had made the *Review* even more contentious. Anne Toker felt that "the first year we were all on *Review* had a nicer feel to it for me than the second year." Adam Charnes agreed: "There was more of a hard-edged tone on both sides than there was the previous year" and "it got pretty nasty at times." Most editors thought this was due to the different temperaments of the 3L and 2L conservatives. Liberal Scott Siff thought the 3L conservatives were "a fairly impressive bunch," while the 2L conservatives were "much more yappy." With Brad Berenson being "the George Will of the group," as Charlie Robb put it, the "more polished" 3Ls maintained the respect of those who opposed them. Liberal 3L Michael Cohen remembered "a lot of unpleasantness with the 2L class," because the 2L conservatives were "an abrasive bunch" who could be "really loud and kind of mean-spirited," particularly "in the way they talked to people."

The overall assessment of Barack's eight-issue year was that he did "a very able job as president" and put out "a very good volume of the *Review*," as even Erwin Griswold agreed. Almost everything in the editorial pipeline for the March, April, and May issues came to Barack in the early weeks of 1991, starting with the lead article for March, an analysis of the constitutional implications of government funding of both religious schools and abortions, written by repeat author Michael W. McConnell. Articles Office cochair David Goldberg had championed the McConnell manuscript, notwithstanding some liberal discontent, and the primary editor for it, conservative 3L grade-on Darin McAtee, thought it was a "*fan*tastic" piece of work. The Articles Office believed it could be reduced in length, but editor Kevin Downey quickly realized that "McConnell was not going to accept the article being reduced in size."

McAtee knew Obama had batted away political objections to McConnell's argument, and Downey was aware that Barack was investing more time in McConnell than in any other author except David Wilkins. Downey believed Barack "had a very vigorous back and forth" with McConnell that "was somewhat exasperating," but McConnell came away impressed with what "an usually good editor" Obama was. "We had the opportunity of chatting quite a bit," McConnell recalled, and Barack "helped me to make it a better article from the point of view of what I wanted it to be. He had some very intelligent organizational suggestions and was just very impressive."

Student authors found that Obama's editing touch remained light. Marisa Chun, whose note was running in the March issue, thought Barack was "a really good editor" who "did not overedit." Scott Siff remembered that Obama "did a lot of work" and made a lot of written comments on his note on international law for April. When they met to discuss it, Barack warned Scott that "you've gone so far left you're not going to be credible." Instead, "the better way to do it is to take a much more balanced approach, present the ideas on both sides . . . and let the readers get there" on their own. "I thought it was a brilliant perspective," Scott recalled. The May issue contained the five-part Developments in the Law package on international environmental law, and Obama met with all five 2L authors. Trent Norris's contribution covered international organizations, and Barack "didn't seem particularly interested in the topic," Norris recalled. "I expected more searching comments from him than I got." His manuscript "was loaded with acronyms," and Barack asked, "Is there a way to make this look a little more like English?"[52]

The law school had a two-day break before spring classes began on January 30, and Barack flew to Atlanta to take part in the taping of a two-hour African

American *Summit for the '90s,* which Turner Broadcasting would telecast in late February. Other participants on the ten-person panel included NAACP executive director Benjamin Hooks, former congresswoman Shirley Chisholm, and Southern Christian Leadership Conference president Rev. Joseph E. Lowery. Obama waited until the entire first hour had passed before speaking up.

"Whenever we blame society for everything, or blame white racism for everything, then inevitably we're giving away our own power to some extent, the possibility that we can take responsibility and take action. At the same time, we can't let the federal government off the hook or the judiciary system off the hook. I mean the fact of the matter is we can take individual responsibility but still mobilize as communities to make the kinds of structural changes that are going to be necessary."

Eight minutes later, Barack stated that "someone who is using drugs is using drugs because they don't have hope. They don't have opportunities, they don't have jobs, education, some sense of meaning in terms of what they're going to be doing with their lives. It's not enough . . . to tell these people to say 'No' to drugs, and what you need to be able to do is to tell them what to say 'Yes' to, what to affirm." A moment later, Barack said, "it's going to be impossible, I think, over the long term to deal with some of these issues until we do tie them to issues of employment, issues of education." He worried that the Supreme Court might mandate "formally color blind" government policies, thereby ending affirmative action and "sending a signal . . . that enrollment of blacks in universities, colleges is less important, less of a priority for America."

More than fifteen minutes passed before Barack responded to another panelist's mention of self-esteem. "Buying in black stores is important, shopping in black areas is important, at some point we all have to make a commitment to live in these black areas. I think the fact of the matter is, and it's already been mentioned, that middle-class folks have a tendency to move out and . . . that means we take the money out, means we shop in suburban malls, etc., etc., and young people my age, I know, have a great deal of difficulty committing themselves to moving back into these communities and dealing with a whole range of these issues, whether it be health, education, economic empowerment, etc. So one thing I think we do need to think about is geographically how are we living and are we making a commitment to these inner cities. That's a hard thing to do. It's a hard thing for us to experiment with sending our kids to public schools as opposed to taking them out and putting them in private schools, but at some point when we talk about leadership, that's something that we are going to have to think about. That's something that we are going to have to do."

Barack quickly continued. "Right now what I hear around this table is the need for some sort of comprehensive political strategy or economic strategy that does get beyond the either/or approach. Two specifics that I keep on hearing is we still tend to be caught up in a division between either it's-all-government's-fault or it's-our-fault. Either blame the victim or blame white racism. It's clear that in our discussion both of those things are going to have to operate if we're going to come up with a strategy."

Barack paused only briefly before making his final contribution to the broadcast. "The second thing is, this whole issue of integration versus segregation to me seems to be a nonstarter. It's clear that we can develop productive communities on our own. It is also clear to me that we've got a complex, interdependent economy, a complex, interdependent society that on the one hand we want to make sure that young black males have a lot of pride and see role models in the schools, at the same time they're going to have to learn how to read and write and do the same mathematics that a Japanese child does or a white child does if they want to succeed economically. So if we can start getting beyond some of these divisions and look at the possibilities of crafting pragmatic, practical strategies that are focusing on what's going to make it work and less about whether it fits into one ideological mold or another, I think that would be the most important thing on all these issues."

Barack's *Summit* comments, especially his final remarks, were a clear distillation of everything he and Rob Fisher had been debating since the beginning of their 1L year. "Crafting pragmatic, practical strategies" while ignoring ideological molds was the essence of their sweeping book outline, with echoes of John McKnight—"productive communities"—melding with David Rosenberg's economic perspective and lessons Barack had learned in Roseland's troubled high schools. While in Atlanta, Obama also taped a sixty-second Black History Month tribute to pioneering civil rights lawyer Charles Hamilton Houston. After cataloging Houston's achievements, he ended the brief TV spot by saying, "I'm Barack Obama, remembering Charles Hamilton Houston and celebrating a great moment in our history."

Back in Cambridge for the beginning of spring classes, Barack, Rob, Cassandra Butts, and Mark Kozlowski all attended the first session of Randall Kennedy's Race, Racism, and American Law. "Barack entered the class with skepticism about whether it was a good idea," Cassandra explained, because Kennedy's relentless independent-mindedness had made him something of a campus lightning rod, as Derrick Bell's public denigration of him a year earlier had so egregiously highlighted. Barack feared that the course "was going to be kind of a slugfest, and it took him one class to have that confirmed," Cassandra recalled. "You're just not going to get anything out of this," Barack told her. "They're not going to give Randy an opportunity to really explore the issues and have an interesting nuanced discussion." Rob and Mark stayed in the course, but Cassandra soon followed Barack in dropping it.

On Sunday, February 3, Obama's successor would be elected and his term as president would end. In preparation, the 2L editors discussed adopting a new set of bylaws to govern how decisions were made and "reduce the level of tension at body meetings that results from a free-flow, ad hoc form of procedure," as everyone had experienced. Just like a year earlier, a candidates' forum preceded the daylong election, and when Bruce Spiva asked Barack for pointers about running, his reply was "not to take yourself too seriously. 'Don't worry about it. Have a beer before you go to sleep.'" Carol Platt, one of several conservatives running, remembered that at the forum "the big debate question was, 'What would you do about affirmative action?'" Almost as many editors stood for president as had a year earlier, and with Obama presiding and Tom Perrelli tallying the paper ballots, Sunday's first round of voting reduced the number of contenders to eight. A second round halved that number as the conservatives sought to muster their votes behind the strongest candidates. One possibility, Patrick Philbin, was acerbically criticized by liberals, and in the end, the final choice came down to Bruce Spiva versus David Ellen, whom fellow 2Ls viewed as "very scholarly, very soft spoken, very intellectual." The question of whether the *Review*'s president should be a leader or an editor "was definitely part of the debate," and when Ellen triumphed, more than a few 2Ls believed that "people are reacting to what was lacking in Barack."

Following tradition, Ellen and Spiva waited at Gannett House before Obama and David Goldberg came to get them. Then Goldberg escorted Spiva back to the full body, followed by Obama with David Ellen. Like Goldberg had, Spiva preferred to take on Articles, but when Ellen said he needed diversity at the top of the masthead, Spiva reluctantly agreed to become Treasurer. Carol Platt came in first for managing editor and rebuffed Ellen's request that she instead become an EE. On Monday, Obama had lunch with Ellen and told him not to take the *Review* too seriously and to keep things in perspective. On Tuesday, when the formal transition took place, exiting ME Tom Perrelli told his fellow 3Ls they needed to continue working on the *Review,* and before week's end, Obama wrote to Erwin Griswold and introduced Ellen as "a fine young man with an extraordinary combination of intellectual depth and attention to detail." Barack also thanked Griswold for his letters, which he said had "made a difficult job easier."[53]

With this time-consuming and often draining role behind him, Barack sat down for a long interview with the *Harvard Law Record*. "It's been a good year. I'm glad I did it," Obama said, while acknowledging that "at times it has been a bumpy road." Regarding his own election, "I think that I was part of a trend. I don't think that I was, necessarily, so exceptional. I happened to be there when the *Law Review* was already going through some changes," as with Crystal Nix and Peter Yu's roles on the previous masthead. "The elections this year prove that the trend continues. I'm heartened by how well minorities did in this year's elections, and how well women did," with Spiva, Platt, and a black female EE, especially given the small number of 2L women. "The elections show that women and minorities are afforded a great deal of respect, that their voices are listened to, that people recognize that they can do a good job, and that they are deserving of leadership positions." Barack was echoing BLSA's efforts three years earlier, and also his rebuttal of Jim Chen's stigma alarum the previous fall, but he pursued his point further. "Our selection process works," and "I don't think there is anybody at the *Review*," irrespective of their views on affirmative action, "who would question the fact that the editors this year were top-notch." Privately, some did, but Barack stressed that "I tend to be very much a pragmatist."

Asked if being president had changed him, Barack replied, "Certainly I have changed. It has been a terrific learning experience for me. The sheer volume of work . . . that you have to do forces you to be a better writer" and "forces you to learn actively rather than passively." Regarding all the press interest, "the publicity has been instructive in that it forced me to articulate publicly things that I had been thinking about for a long time. . . . I like to think that I haven't become more cautious in my opinions." But Barack again downplayed the significance of it all, cautioning that "we are students and thus our ambitions for something like the *Review*, which has a long tradition, tend to be very modest." He hoped his year had witnessed "a shift in tone" toward "making the *Review* a little less hierarchical, pulling in a larger group of editors into the decision-making process," albeit through many hours of angry arguments. He also hoped editors were "taking ourselves a little less seriously," and volunteered that "I am skeptical of the idea that it is somehow superior to other activities . . . or other student organizations." He noted that "the *Review* has a very short institutional memory" and rued "the tremendous time commitments" it required. "I would have liked the luxury of being more strategic about my tenure" and "able to implement some management changes."

Lastly, Barack was asked, "where do you see yourself" in five, ten, or twenty years? "Of course, I can't project twenty years out," but "I will be spending next year writing a book on issues of race, politics, and race relations in this country. It will be a series of reflections on where we are in terms of race relations . . . drawing on my experiences as an organizer in Chicago, the experiences of my family in Africa, as well as some of the work that I've done here at the law school. After that, I will work for three or four years as a lawyer. Eventually, I will return to the public sector, either in government or as an organizer. I'd like to address ways to redevelop inner cities, and how to get corporations to locate in low income areas."

The *Record* also published an oddly misdirected critique of the *Review*, calling it "a profoundly bland, unimaginative, and conformist magazine" that made it "one of the least stimulating experiences available to the contemporary magazine browser." An irritated Erwin Griswold wrote to Barack noting that the *Record* had made two glaring errors in its critique, but he praised Michael McConnell's article on which Obama had worked as "an interesting and rather innovative analysis." A few weeks later, Griswold likewise expressed great enthusiasm for the year's "Devo," saying he was "much impressed by its breadth and depth."

The major challenge at the *Review* continued to be the revision of *The Bluebook*. By the time of the presidential transition, Ken Mack and Frank Amanat had made only modest headway, and Mack and his masthead successor, Trent Norris, alerted everyone to the extent of the forthcoming changes. The anger of the departing EEs over the planned modernizations—including authors' first names and Floyd Abrams's First Amendment—was so intense that Jim Chen began preparing a rebuttal, which he would publish in *The University of Chicago Law Review,* the *Review*'s archrival. Norris described the previous officers' "failure to recognize the enormous responsibility we bear as publishers of *The Bluebook,*" especially considering that "the revenues from *The Bluebook* support everything else we do." Chen's resentful critique declared that "interest group appeasement dominated the revision," but reviews in other top law journals particularly praised the addition of authors' first names. "In this age of multiple Smiths, Joneses, and Dworkins, first names help," especially when "there are more than forty law professors named Smith."

Perrelli had stressed that 3L editors had to continue working on the *Review*, but one morning new ME Carol Platt discovered that 3L Monica Harris had taped over her mailbox and appended a note asking Platt to please speak to her before giving her any further work. Both women would remember being "very friendly with each other," as Harris put it, ever since being fellow undergraduates at Princeton, but now Platt "tore the tape off from her box" and left Harris a note, giving her a proofread, the easiest possible assignment. Fellow editors knew Harris as "a very big personality" but also as someone who was rarely seen in Gannett House, notwithstanding how she had a note scheduled for publication in April. That Harris was African American exacerbated some editors' unhappiness, since "she was kind of the poster child for the anti–affirmative action feeling that existed among" conservative editors, one 3L recalled.

Platt remembered that "I was nervous" after removing Harris's tape-over, but *Review* work was "a zero sum game," and if some editors did not contribute, the ME would be left with more assignments to distribute among everyone else. Before long, Platt saw Harris and 2L editor Nancy McCullough at the *Review*'s copy machine, not far from the ME and EEs' office. "'Why are you giving Monica work?'" Platt recalled McCullough asking. "Monica told you she didn't want any more assignments." Platt replied that if so, Harris should stop using the *Review*'s copy machine, stop eating free bagels, leave Gannett House, and remove the *Review* from her résumé. At that, Harris followed Platt into the ME's office and "Monica kind of nudged me," Platt recalled, telling her, "'I don't appreciate you speaking to me like that in public.'" Platt replied that "I wasn't embarrassed by anything I was saying," and then Harris "nudged me again and she said, 'Let's take this outside,'" Platt remembered.

While Harris was robust, Carol stood five foot one and barely topped a hundred pounds, but "we go down the stairs, we go outside in front" of Gannett House, and "Monica starts doing the dance of rage . . . and she's got her finger in my face. 'You had better take that fucking finger out of my face or I am going to rip it off,'" Platt remembered saying. Harris recalled the confrontation similarly. "I just remember her getting up in my face . . . almost like she was trying to get me to get physical with her. 'Go ahead. You know you want to hit me,'" Harris remembered Platt saying. From inside Gannett House, both at the first-floor Legal Aid Bureau as well as upstairs, the cry went out: "Oh my god! It's a catfight at the *Law Review*!" At the Bureau someone picked up a phone, and as the confrontation continued, "all of a sudden the Harvard police show up," Platt remembered. "They handled it very nicely. They said, 'Is there a problem here, ladies?' Monica was still doing the dance of rage, and I said, 'Well, yes, officers, actually there is. I am trying to deal with a recalcitrant pool worker.'" That of course meant nothing to the police. Looking back on this years later, Platt wondered, "What planet was I on? That's what I said."

Then two or three editors arrived, and Harris walked away. Platt told David Ellen, "If people aren't going to do their work, I can't do mine," but the fracas was over. Then a few days later, Obama approached Platt. "What do you say we go out back and chat?" Barack "had never shown any particular interest in chatting with me" before, and Carol thought, "This is going to be interesting," as they went to Gannett House's back steps, which faced away from the law school. "How are things going?" Barack asked. Carol said fine. "You catch more flies with honey than with vinegar," Barack remarked, and Carol replied, "Yes, but how do you get people to do their work?" Barack's response sounded a familiar theme: "Why do you care? Why sweat it?" he told her. "It just struck me as so cynical," and Barack's demeanor seemed infinitely distant. "I've never met anyone who was more impenetrable," Platt explained. Obama was "one of the most opaque people I've ever met."[54]

One morning in early March, Michelle Robinson phoned Barack from Chicago. Her father Fraser, who recently had had kidney surgery, had collapsed on his way to work and was dead at age fifty-five. Barack quickly flew to Chicago, missing BLSA's annual spring conference, during which Derrick Bell stunned the crowd by announcing he had accepted an appointment at New York University's law school. Once Barack returned to Cambridge, he wrote Bell a fascinatingly odd thank-you note, voicing disappointment at Bell's upcoming departure from Harvard. "Although my general feeling is that the loss will be Harvard's, and not yours, I do worry that nobody will be able to fill the role of 'Harvard's conscience' that you've served these years and that the educational experience of *all* students will be impoverished by your absence." Then Barack's note became even more fascinating:

> I may have mentioned this to you before, but it was your presence here that in large part brought me to Harvard in the first place; in reading your treatise the year before I enrolled, I was inspired with the belief that at Harvard, I would meet people like yourself who had a commitment to the struggle for equality and the experience and erudition to translate broad goals into concrete practice. And for all the good fortune I've experienced at Harvard, my single regret is that I did not have the opportunity to take a class from you or work with you more closely. Nevertheless, I doubt that I could have maintained my moral compass over the past three years had I not had you there, speaking out and challenging the conventions that we all have a tendency to take for granted.
>
> I've tended to take a backseat on some of the student activities surrounding diversity this year, in part because of my position at the *Review,* in part because of a desire to let the excellent student leadership that has emerged take the lead. Still, I will be working to support any efforts that the Coalition or BLSA plan for the coming months to both bring you back here and bring in more of the folks that need to be here. I assume that the student leaders are in contact with you to find out your feelings on the matter at this stage, but I hope that in the next two months you might find the time to have lunch with me and share some of your reflections on what has happened and what needs to happen in the future.

Years later, Bell would not remember a lunch date taking place, and he did not recall any contact with Barack, except for the younger man's warm introductory remarks outside the Hark eleven months earlier. A few weeks earlier, a local Superior Court judge had dismissed the faculty diversity lawsuit filed by the Coalition for Civil Rights on the grounds that the law students lacked legal standing to sue, but Barack's more than two years of distance from the protest efforts was of a piece with his up-till-now lack of interest in developing any relationship with Bell or Randall Kennedy, in the way that he had done with David Rosenberg, Larry Tribe, Chris Edley, and now, in his final semester before graduation, Roberto Mangabeira Unger.[55]

Roberto Unger's intellectual breadth and depth was truly astonishing, even at Harvard. In 1987 alone, Cambridge University Press had published a trio of books written by Unger, and their titles—*Social Theory: Its Situation and Its Task, False Necessity: Anti-Necessitarian Social Theory in the Service of Radical Democracy,* and *Plasticity into Power: Comparative Historical Studies in the Institutional Conditions of Economic and Military Success*—demonstrate Unger's range as a critical social theorist. Barack and Rob had taken Jurisprudence with Unger a year earlier, and in spring 1991, they eagerly enrolled in his small and expressly political Reinventing Democracy. Unger said the course would have three themes: "the remaking of certain central institutions," both governmental and economic; the relationship of that "program of social invention to theories of social change"; and "the significance of this program for the conception and allocation of legal rights." Unger said his emphasis would be "programmatic rather than analytic, critical, or explanatory," and that his overarching purpose was "to explore the long-term options open to progressive liberals and democratic socialists in the North-Atlantic democracies."

Rob remembered that "Barack and I loved" Reinventing Democracy as "a terrific class" taught by "a powerful thinker" who was "intellectually far deeper" than any other professor. Unger "would take you step by step through an argument about the nature of law, the nature of society," speaking in "these beautifully formed sentences that were almost Euclidian in their clarity." Steve Ganis, a 2L, felt similarly. Reinventing Democracy was "a phenomenal course" taught by an "incredibly erudite" and "very powerful teacher" who was capable of "speaking in paragraphs." Punahou graduate Ian Haney López, a 3L, found Unger "very impressive to listen to," especially since Unger spoke entirely without notes throughout the two-hour, twice-a-week classes. Rob thought Unger was "profound" and "just absolutely brilliant," although his "hyperanalytical" and "heavy post-Marxist style" made many students nervous and uncomfortable.

Halfway into the semester, two students spoke up and complained that while Unger preached about deconstructing hierarchies, his actual classroom practice placed students in a totally hierarchical relationship to himself. As Steve Ganis remembered, Unger said they made a "very good point," and he offered to let students teach several class sessions if they were willing to do the work. Then "Obama spoke and said, 'Well, I think I speak for a silent majority of us here who were okay with the way that the class was going. It's unconventional, it's not what we're used to,'" because Unger lectured at great length rather than questioning students Socratically. "I think we'd like to keep going as this was.

We thought that we knew what we were signing up for." Ganis recalled too that Obama was among a handful of students who asked the most questions, and at times "he drew on his experience from being an organizer," because in Unger's class Barack "was the only guy who really had significant experience with the disempowered."

Unger remembered that his intent with Reinventing Democracy was "to develop an argument as forcefully as I can and then to engage the students in confrontation about the argument" in the manner of a "zealot addressing the skeptics." Unger recalled that another talkative student that semester, "who always chimed in with Obama," was George Papandreou, a future prime minister of Greece. Obama "was very engaged," "appreciative of ideas but cautious about them," and "conventionally smart." Steve Ganis believed that "Obama was very impressed by Unger," and Rob recalled that he and Barack chatted "with Unger for long periods of time." Unger remembered Barack as someone who combined "a cheerful, impersonal friendliness with an inner distance" and whose "guarded personality" made him seem "somewhat enigmatic and distant."

Reflecting on Barack's time at Harvard, Gerry Frug emphasized how "the fact that he took two courses from Roberto is a statement about what he was like as a student. That's just a statement of where your intellect is." Beyond Reinventing Democracy, Barack and Rob's focus that spring was on completing the lengthy paper they owed Martha Minow. In the end, they gave Minow only two of the planned topical chapters in their grand book outline. "Plant Closings: Creative Destruction and the Viability of the Regulated Market" was a hefty 103 pages, and although the well-written paper was—and is—substantively dense and demanding, it unquestionably represented an analytical and political capstone of Barack and Rob's three years of intellectually intense policy debates.

They might not have matched Roberto Unger's analytical depth, but the breadth of their sources and citations was impressive. The first paragraph included references to Paul Kennedy's *The Rise and Fall of the Great Powers* (1987), Allan Bloom's *The Closing of the American Mind* (1987), Samuel Bowles et al.'s *Beyond the Waste Land: A Democratic Alternative to Economic Decline* (1983), and Barbara Ehrenreich's *The Worst Years of Our Lives* (1990). Rob's background as an economist was evident in their third paragraph's mention of "a Kondratiev business cycle," but their policy argument was straightforward:

> *The Reagan administration added more to the national debt than all previous administrations combined. The gullibility, or at least the wishful thinking, of the American people reached unprecedented heights as they embraced Ronald Reagan's contradictory promises of more military spending, lower taxes, and a balanced budget. The eighties recovery benefited the*

"few" far more than the "many." Distribution of income and wealth became dramatically more skewed.

They went on to cite "the virtual elimination of the progressivity of federal income taxes" as another factor in "the unsettling mixture of economic successes and failures of the eighties." As they noted in their outline, "the issue of plant closings highlights the tension that arises within our current institutional framework between the desirability of community stability and the need for capital and labor mobility." Promising a "broader discussion of the evolutionary nature of capitalism," the paper examined plant closings from neoclassical, Keynesian, and Austrian school economic perspectives. "True experimentalism," they wrote, "demands . . . that we experiment with the legal institutions that shape the economy," and Rob later explained that "experimentalism" connoted the "blend of Schumpeter and Popperian epistemology with a progressive slant" that they deployed in their analysis. "The quest is to develop guidelines on how politically progressive movements can use the market mechanism to promote social goals," they wrote.

Some of their analysis rebutted everything Frank Lumpkin and Jerry Kellman had once championed. "Those who myopically focus upon the destruction entailed by the closing of a plant avert their gaze from the creative half of the story. Plant closings occur because undesired products and outmoded techniques are driven from the market by innovative competitors." They decried "the impoverished, feudal world we would inhabit if we lacked resource mobility" yet acknowledged that "the unfettered market may result in an 'over-supply' of plant closings."

A clear stylistic shift from one coauthor to the other appeared at the top of page 30. "Capitalism cannot function unless peopled by folks who are honest and fair in their dealings, avoid overreaching, lend a helping hand when it's reasonable to do so, and follow the rules," they—or Barack in particular—wishfully imagined. After twice invoking "moral integument," a little-used phrase first coined in an 1892 magazine essay, they stated that "our moral well-being demands some sense of connection to community." They admitted that "the market does not provide a forum in which an appropriate value might be attributed to community stability" and recognized that "what starts as a snowball of unemployment ends as an avalanche of economic chaos and social despair" and can "cause a permanent capital and labor flight from a region."

Twice quoting Roberto Unger regarding the "non-innovative, risk-averse managerial class," they emphasized that an "overemphasis on the short-run bottom line is probably America's most serious systematic business problem," because in corporations "flatterers and artificers rise to the top. . . . As bold experimentation fades from American boards, others who combine daring investment philosophies with long-term agendas will surpass the United States in technological and organizational leadership." Their use of a quotation from law school dean Robert Clark seemed to belie any loyalty to the protest movement that had demonized him. Yet "there is no reason to believe that the legal status quo is the best in terms of either efficiency or fairness, and it behooves us as a society to continue the effort at improvement and innovation in the legal realm as well as in the realm of the purely economic."

Compared with mass production, "craft technologies have the positive externality of maintaining community." A reference to Youngstown led them to observe that "systemic governmental subsidies to large corporations tend to undermine community itself." Criticizing Joseph Schumpeter's celebration of entrepreneurs, "we reject this 'great man' theory of economic development." Contending that "a dual experimentalism of both economic and legal institutions is necessary," they moved to their twenty-page conclusion and endorsed legislation requiring advance notice of plant closings. Yet "the ultimate answer to plant closings . . . lies in making workers more flexible, and their human capital more transferable," and thus "policy makers should seek ways to increase the capacity of workers to move from one employment to another." Championing their theme that "experimentalism is the key to a healthy economy," they called for "a proliferation of legal forms in which to organize production." That would include ones that "allow for worker ownership and control," and in situations like Youngstown, "where there was massive support for an experiment, the government should be ready to step in and provide an infrastructure as well as ensure fair dealing by the corporation that is departing."

Barack and Rob's lengthy chapter certainly expressed progressive if not redistributive values. "While Yuppies can afford the expensive frivolities provided by The Sharper Image, others receive insufficient nutrition to allow their minds to develop properly." They proclaimed that "the political left should embrace markets as a weapon to wield in the best interests of the people," and Barack revisited his summer 1988 trip to Europe in an unusual riff on supply and demand. "Suppose that on an impulse, I fly to Paris. Having arrived in Paris, I find comfortable accommodation near the Eiffel Tower and have a fine dinner on the Rue Monge. Almost magically, these pleasures await my arrival even prior to my own knowledge that I will pursue them." Reiterating how markets "must be more consciously employed to serve the needs of the people," they warned that "if the American economic debate continues along its current lines with the right celebrating the market and the left resisting, the left's political marginalization will continue."

One unusual passage seemed unlikely to come from Rob Fisher. "The first step to radical consciousness is to realize that the world could be different and that we have the power to make it so. There is a great deal of wisdom in this idea, but also great danger. The wisdom is that it carries us beyond our own individual experience of isolation," something Barack had long known well. "I envision a jungle-gym made of ropes on which I once played" was another clear personal throwback, but after a citation to Gerry Frug, the paper ended by again underscoring how the Left must fight on the field that was inevitable. "The market is perfectly consistent with greater participation by workers in management, worker ownership, progressive income taxation, state subsidized child care, national health insurance, high inheritance taxes, etc. . . . It is even consistent with governmental ownership of banks." Their orientation was political, not legal. "The battle is over what kind of market we will have, not whether we will have one. If the Left"—now for the first time capitalized—"does not come to this realization they will be relegated to occasional footnotes in future histories—and properly so."[56]

Sometime in April, Barack and Rob completed their second, even lengthier book chapter. "Race and Rights Rhetoric" was 144 highly polished pages, seemingly all written in one consistent voice, and not that of a trained economist. "This chapter evaluates the utility of rights rhetoric . . . as a vehicle for black liberation," because the authors believed that such a focus "has impeded, rather than facilitated" the achievement of "black empowerment." They observed that "it has become increasingly apparent that the strategies rooted in the Sixties have not led blacks to the promised land of genuine political, economic and social equality," because once that decade was in the past, "political mobilization . . . ground to a halt as blacks became increasingly reliant on lawyers and professional civil rights leaders and organizations with only minimal institutional presence in local communities."

Barack and Rob acknowledged how "racism against African Americans . . . continues to exist throughout American society," which is "an admittedly racist culture." Indeed, "race relations appear to have made a turn for the worse in recent years, with the growth of intolerance evident on both big city streets and ivy league campuses." Their footnotes reflected a familiarity with well-known histories such as *The Origins of the Civil Rights Movement, Bearing the Cross,* and *The FBI and Martin Luther King, Jr.,* and they wondered "whether liberty in America must be fundamentally redefined." For example, "might we redefine 'liberty' under the due process clause to require government expenditures on enhancing the education black children receive in inner city schools," because there was a "need to centralize public school financing to achieve redistribution on a state wide or national level." Although "private property arrangements and resulting inequities in wealth and power do not devolve from divine providence," it was inescapably true that Americans have "a continuing normative commitment to the ideals of individual freedom and mobility, values that extend far beyond the issue of race in the American mind. The depth of this commitment may be summarily dismissed as the unfounded optimism of the average American—I may not be Donald Trump now, but just you wait; if I don't make it, my children will."

But "those of us on the political left" often forget "the degree to which coalition and consensus-building among the American electorate has necessarily preceded any major federal program to reform or restructure America's economic and political landscape." Therefore African Americans needed to understand how "the indiscriminate use of rights rhetoric in conventional political battles only adds fuel to the suspicion of the average white that all claims of right are nothing more than hypocritical attempts on the part of blacks to disguise their pluralist self-interest in the language of prophesy." Thus "rights rhetoric will be effective to the extent that it conforms to the aspirations of a color-blind society," and African Americans should grasp the pragmatic need for "a shift away from rights rhetoric and towards the language of opportunity." Underscoring the crux of their argument, Barack and Rob wrote that "*the American commitment to equality of opportunity, mythical though it may be, offers the most powerful analytic framework for moving beyond the current stalemate.* Pushing the logic of equality of opportunity as far as it can go" would best boost black advancement. "*Precisely because America is a racist society* . . . we cannot realistically expect white America to make special concessions towards blacks over the long haul," they emphasized, while also acknowledging that "the greatest testimony to the force of racist ideology in American culture is that it infects not only the mind of whites, but the mind of blacks as well." In the chapter's concluding pages, Barack and Rob declared that "the energy for change in race relations in America will come from a bolder political vision . . . rather than a bolder legal theory." That vision would require "framing the political debate in terms of opportunity," for "the language of opportunity is better suited for the immediate tasks at hand, precisely because it better conforms to the underlying logic of American liberalism."

The two friends had written a pragmatic call that realism should trump idealism, and Rob recalled how he and Barack had each "intensely edited all parts of the documents," for it "was very much a product of both of our minds." Reflecting on what was the most intellectually intense friendship Barack would ever have, Rob saw the two chapters as the culmination of their three years of almost daily interaction. "This belief in the power of rational discourse to find better answers, and the enjoyment we took in that process, was certainly a major foundation to our friendship."

Martha Minow was very impressed. "This is going to be an important book/work," she wrote Rob and Barack. While she found their argument "powerful and valuable," she did have some smart suggestions on how they could improve their first chapter. She thought its organization was "perplexing," as it needed "some more pronounced structure." In an unusual thank-you from a law professor, Minow volunteered that "I enjoyed the chance to talk with you and work with you both." She added that "as I told Rob, I have less to say now about the race section; perhaps we can get together for a meal or drink to talk about the ideas. I think parts of it are just beautiful."

Minow remembered once going to Barack's basement apartment for some small group discussion and informal dinner, and she knew that Barack "wanted to go into politics." For Barack, there was a direct connection between his experiences in Roseland and the arguments he and Rob articulated in their manuscript, where they addressed "the same line of questioning that I had been engaging in as a community organizer: 'How do we bring about a more just society? What are the institutional arrangements that would give people opportunity?'"

The other piece of writing Barack was focusing on that spring was "Journeys in Black and White." *Review* classmate Scott Siff, who had tried writing a novel, remembered Barack giving him a fairly extensive manuscript to read. It was certainly not either of Barack and Rob's two prospective book chapters, because, as Siff recalled, "it was this kind of memoir-ish thing," although some parts "were more sociological commentary." Yet "the social commentary was a little light" and "it didn't hold together," for "these long discursive . . . observations about society" did not work nearly as well as the more appealing memoir portions, but those "were much more limited."

In order to expand those, Barack could draw from his journals from years past, and sometime that spring he asked Sheila Jager if she had kept the letters and postcards he had sent to her during his trip to Kenya almost three years earlier. Barack and Sheila had continued to see each other irregularly throughout the 1990–91 academic year, notwithstanding the deepening of Barack's relationship with Michelle Robinson. "I always felt bad about it," Sheila confessed more than two decades later.

That spring, Sheila met a thirty-three-year-old Korean American graduate student and U.S. Army officer, Jiyul Kim. A Boston native and 1981 graduate of the University of Pennsylvania, Kim had spent five years as an intelligence officer before becoming an East Asia specialist who hoped to complete a Harvard Ph.D. Sheila's relationship with Jiyul quickly began to blossom just as Barack's final weeks of law school approached. "As much as I loved him, I was relieved when our paths finally parted," Sheila explained, because "we went through many painful things together," and "I really traveled to some dark places during that period of my life."

When they saw each other for the final time late that spring, they parted "on the best of terms," she recalled. She remembered Barack as "a profoundly lonely person at heart," someone who "was and is truly unreachable," but she always believed "that with me he was able to let go of all those masks of his."[57]

The last week of March was the law school's spring recess and gave Barack a chance to return to Chicago. The publishing contract for "Journeys in Black and White" gave him the funds necessary to postpone deciding whether or not to accept the job offer from Sidley that he had in hand. Even so, Sidley was happy to accommodate him that coming summer as he studied for Illinois's moderately difficult bar exam. Kelly Jo MacArthur, Michelle Robinson's close friend at Sidley, had "felt like they were very committed that year," but as Barack's graduation approached, "Michelle wanted to know what was going to happen" above and beyond Barack once again living with her and her mother Marian, at the Robinson family home in South Shore. "It all kind of came together at the end of his third year," MacArthur remembered; "that's about when things kind of locked in," even though nothing expressly definitive had yet been said regarding marriage.

Two phone calls Barack had received from Chicago that spring also factored into his thinking. One had come as a direct result of how impressively he had performed his P-read edit on Michael McConnell's *Law Review* article. McConnell had mentioned what an "impeccable editor" the *Review*'s president was to his University of Chicago Law School colleague Douglas Baird, who had already heard Obama's name. Baird's service as chairman of the university's Library Board made him well acquainted with the senior librarian who for a quarter century had been overseeing the library's increasing use of computer automation, Charles Payne—Madelyn Dunham's younger brother. Some months earlier, law school librarian Judith Wright had asked Baird if he knew that Payne's grandnephew was the *Harvard Law Review*'s first African American president, which Baird did not and at first believed was impossible.

Baird also chaired the law school's Appointments Committee, and Chicago was actively interested in diversifying its virtually all-white faculty. So, with the consent of law school dean Geoffrey Stone, "I just made a cold call to the *Harvard Law Review* and asked to speak with Barack Obama," Baird remembered. Baird asked Obama, "Do you have an interest in law teaching?" and Barack said no, that he had a contract to write a book on race-related issues such as voting rights and that he would be returning to Chicago to work on it. Baird then invited Obama to write his book at the U of C's law school, explaining that he could be a Law & Government Fellow, just as there previously had been Law & Economics Fellows. The appointment would be unsalaried, but it would give Barack a small office and word processor as well as access to the university's libraries. After a quick series of follow-up phone calls, Barack was assured of a quiet place to write just a fifteen-minute drive from the Robinson family's home.

The second call came from Judson Miner, who four years earlier had been Chicago's corporation counsel, Mayor Harold Washington's top lawyer. Miner had read one or another story about Obama, perhaps the *Chicago Reporter* one from some months earlier noting Obama's intent to practice law in Chicago, and Miner also called Obama at the *Law Review* to try to interest him in the small civil rights litigation firm Miner and Allison Davis had founded twenty years earlier, Davis, Miner, Barnhill & Galland. Obama mentioned his book contract but agreed to meet Miner for lunch once he arrived in Chicago, when Miner took him to the Thai Star Cafe on North State Street, just around the corner from the handsome West Erie Street town house that housed the firm's offices.

Rather than "talking about his coming to work here," Miner and Obama instead had "quite a comprehensive discussion that focused primarily on how gratifying did I find being a civil rights lawyer," Miner recalled. They agreed to keep in touch, and immediately after the two-hour conversation, Miner called his wife Linda and told her, "I just had lunch with the most impressive person I've ever met."[58]

Back at Harvard, the final five weeks of spring classes began with an array of protests mounted by the Coalition for Civil Rights, the student group that had unsuccessfully sued Harvard to try to advance faculty diversity. An April 4 student "strike" drew several hundred supporters, but a new tactic—quietly invading classes—angered and alienated many students. One band of poster-carrying CCR members silently encircled Kathleen Sullivan's 1L Crim Law classroom for twenty minutes and then returned to repeat the stunt for the class's final ten minutes. That evening BLSA mounted an overnight sit-in in Dean Clark's office, but called off the occupation when word arrived that Gerry Frug's wife Mary Joe, herself a law professor at the New England School of Law, had been brutally stabbed to death on a Cambridge sidewalk at 8:45 P.M. the previous evening. Frug's killer was never caught, and the following Wednesday CCR resumed its protests by physically blocking all access to the dean's office. CCR leader Keith Boykin called Clark "undemocratic, virtually authoritarian," and the dean responded to the blockade with a letter threatening administrative punishment for such actions. David Troutt termed that warning "Gestapo-like," while Clark told the *Record* that "certain behaviors can't be accepted" and confessed, "I see myself as being excessively nice." The *Record*'s final issue of the year closed with one letter from Law School Republicans calling for Clark to discipline the protesters and another from a pair of BLSA members declaring that the *Record* was "beginning to resemble a fascist version of the *National Enquirer*."

Amid such cheery exchanges, the *Law Review*'s editors assembled at Boston's Harvard Club for their annual banquet. Obama spoke briefly before David Ellen introduced Wayne Budd, the African American U.S. attorney for Massachusetts. Barack advised everyone to spend less time debating whether to capitalize the *B* in "black" and more on real-world issues. The *Revue* made fun of Barack's book advance, his use of "folks," and his work on Larry Tribe's abortion book. Charles Fried's request for a muffin was revisited, and problematic authors like Ian Ayres and Patricia Williams suffered send-ups, as did 3L editor Monica Harris. Ayres's treatment reflected a clear case of buyers' remorse: "Ian Errs, Taking the *Revue* for a Ride: Recriminations from Half-Baked Empirical Research," and the Charny and Wilkins "Don't Embarrass the *Revue*" standard was lampooned as well.

But Barack was a target throughout the longer-than-usual thirty-two-page booklet. "Obama, Supervised Reading & Research Is Hard Work, 1 Class a Week, Tops" reflected how undemanding his 3L course schedule had been, and quickly was followed by "Look on the bright side. At least you don't have to worry about having to listen to me talk in any of your classes." Barack's avoidance of clear positions on tough issues was twice mocked—"Chief Justice Obama neither concurred, nor dissented, and gave no reasoning," and "Barack quickly defused the situation by agreeing wholeheartedly with each of the competing people in turn"—as was his press attention: "International celebrity and bon vivant . . . mediocre bluebooker." His and Rob's plans were noted: "Await R. Fisher & B. Obama, Reinventing Democracy (forthcoming)," as was his relative absence from Gannett House: "Compare Ellen, Burning the Midnight Oil . . . with Obama, Is Gannett House Really Open All Night?, What a Waste of a Perfectly Good Office . . . ('I work at home')." But there was also clear praise: "Like me, you probably think that Barack has done a pretty decent job."[59]

As the semester was ending, Barack looked back on his three years at Harvard in a brief piece for the law school's upcoming *Yearbook*. Everyone should "question our assumptions, listen to other viewpoints, and articulate our values in a spirit of mutual respect and tolerance," something that had been sorely lacking throughout that spring. He also sensed "a continuing struggle on the part of many of us to infuse our role as lawyers with a larger sense of meaning," because "we all have a responsibility to use our legal training in ways that make this country work better." He worried about "an increasing anxiety among students about prospects for employment and economic security," but hoped that "idealism—a hard-headed, unromantic idealism that does not expect change overnight" would prevail. In another, expressly racial context, Barack remarked that "people like myself are learning a certain language of mainstream society, of power and decision making. We have an obligation to go back to the black community, to listen and learn and help give our people a voice."

Fully two-thirds of Harvard Law School's 1991 graduating class were heading into law firms, one-quarter had accepted judicial clerkships, and only 2.9 percent were taking public interest or legal services jobs. Section III classmate Jonathan King, an accomplished jazz pianist who was heading to a top Boston firm, remembered Barack telling him, "Don't give up jazz," and adding, "Do not let them squelch the poetry out of you." Barack, like Rob, owed Roberto Unger a long paper for Reinventing Democracy, but with their coauthored work for Martha Minow complete, they would both graduate magna cum laude, just as they had long intended. On the official roster, his full name would appear as it never had before: "Barack Hussein Obama, 2d."

Looking back on his standout academic success at Harvard, Barack credited his five-year age advantage over most of his classmates in addition to the seasoning and experience he had gleaned in Roseland. "As an older student," he recalled a decade later, "I knew why I was there and what I wanted to get out of it." Plus, in many classes, "I had great enthusiasm for the subject matter," and "when you're interested in something, you end up doing well on it." But part of his success also stemmed from his having "learned as an organizer to be able to articulate a position and express myself in clear ways that served me well as a law student." At the Developing Communities Project, "my whole job was persuading people to do things differently, but not being able to pay them," which meant "I had to be pretty persuasive, and I think it taught me to be able to focus in on those issues that were important to people and be able to describe them in ways that people found compelling, and I think that probably had something to do with my success at law school."

Barack's decision to compete for membership on the *Law Review* was motivated in significant part by a desire to squelch any perception that his Harvard J.D. was proof of affirmative action at work, and Rob remembers the morning when he had stopped to answer his phone on his way out the door and was able to get Barack to the post office just in the nick of time. Otherwise, "he never would have been president of the *Harvard Law Review*." Barack's experiences on the *Review* had helped him become "enormously skilled at finding the center of the room," one colleague believed, but neither Ken Mack nor Rob Fisher believed that either the *Review*, or indeed Harvard Law School, had significantly changed someone for whom they were his closest friends. "I actually think the *Law Review* was very peripheral" in Barack's experience at Harvard, Rob said. "He focused on it as a management problem," and "I don't think it was near his heart in the least. I don't think the various little controversies shaped him very much" at all. Ken, who, like Rob, had been friends with Barack since the beginning of their 1L year, agreed. "I don't think he changed much in law school. I don't think he reoriented himself and redirected himself in any way."

Commencement was a full two weeks after the law school's final exam period, and neither Barack nor Rob saw any reason for remaining in Cambridge to attend. All that Barack would miss was the news that he had lost his election bid to Harvard's Board of Overseers. He gave notice to his landlord, who was disappointed to see a model tenant leave; had a mandatory exit interview with the financial aid office about "the $60,000 worth of loans" he had taken out over three years; and advertised a major yard sale. "I remember Barack selling everything in his apartment," his bête noire Christine Lee recalled. "I bought his TV at graduation. It's still the TV I watch in my bedroom," incongruous as that seems. Over two decades later, "even the remote works."

One morning in late May, Barack packed up his mustard yellow Toyota Tercel and headed west on the Mass Pike. He left Harvard "with a degree and a lifetime of debt," he would say years later, but he also left with something else, just as Jeremiah Wright had told him almost three years earlier: "Don't let Harvard change you!" "I went into Harvard with a certain set of values," Barack recounted. "I promised myself that I would leave Harvard with those same values. And I did." Chicago lay almost a thousand miles to the west, but this time Barack knew exactly where he was headed: 7436 South Euclid Avenue, Michelle Robinson's family home.

Six summers earlier, on that July night at the Fairway Inn, Bob Elia's monologue had left a homeless twenty-three-year-old feeling existential self-doubt he would never forget. His years in Roseland, in Altgeld, and in Hyde Park had staunched those fears and replaced them with a deep and abiding sense of destiny. But now Sheila Jager, just like Harvard Law School, was entirely in the rearview mirror. This time, as the Ohio Turnpike gave way to the Indiana Toll Road and finally the Chicago Skyway, Barack Hussein Obama knew exactly where he was headed, to the place he knew was essential for the destiny that awaited him: he was heading west, *west toward home.*[60]

Chapter Six

BUILDING A FUTURE

CHICAGO

JUNE 1991–AUGUST 1995

The Chicago Barack Obama returned to in late May 1991 was a significantly different city than the one he had left three summers earlier. South Shore, where he would now be living with Michelle Robinson and her mother Marian, was neither the leafy, university-dominated enclave of Hyde Park he knew so well from 1985 to 1988, nor struggling Roseland, where violent crime was increasing. "Upon my return," Barack remembered, "I would find signs of decay accelerated throughout the South Side," but solidly middle-class South Shore was generally tranquil.

Far more momentous—and depressing—were the changes that had overtaken Chicago politics. Less than two months earlier, Mayor Richard M. Daley had been reelected to a full, four-year term of office after crushing first West Side African American Cook County commissioner and former 29th Ward alderman Danny Davis 63 to 31 percent in the Democratic primary and then black former appellate judge R. Eugene Pincham 68 to 24 percent in the general election. In the wake of those two landslide victories over well-known opponents, one astute political observer wrote that "Richie" Daley had "totally captured political control of Chicago." Harold Washington's political legacy had been almost completely eviscerated.

In addition, 4th Ward alderman Tim Evans, who four years earlier had been anointed as Washington's rightful successor by Chicago progressives, lost his seat by 109 votes to three-time challenger Toni Preckwinkle, a liberal teacher whom Evans charged was backed by gentrification-oriented developers. That race had been marred by a bizarre backfire when neighboring 5th Ward Democratic committeeman Alan Dobry, a Preckwinkle supporter, had been caught posting racist placards that he said the Evans campaign was distributing so as to turn voters against Preckwinkle, a black woman whose husband was white.

That attempt to embarrass Evans's campaign set off a citywide controversy, with a pair of *Chicago Sun-Times* columnists, African American Vernon Jarrett and white Steve Neal, mounting a sustained but unsuccessful crusade to force Dobry's resignation from his influential ward committeeman's post. Dobry, a Hyde Park progressive independent whose role in local politics reached back more than two decades, stood his ground based upon his long record of sustained opposition to the Democratic machine. Two years earlier, when local state representative Carol Moseley Braun won election to a countywide post, Dobry had refused to concur in powerful 8th Ward committeeman John Stroger's insistence that his organizational protégé Donne Trotter be appointed her successor. With the 5th Ward comprising 40 percent of the district, Dobry held the largest hand, but progressive 7th Ward committeeman Alice Palmer, who three years earlier had defeated machine loyalist William Beavers, nonetheless sided with Stroger. Their combined 60 percent share put Trotter in the state legislature, with Dobry casting his 40 percent for famed steelworkers leader Frank Lumpkin.

Less than two weeks after Preckwinkle's victory over Evans was confirmed, Hyde Park's sixty-seven-year-old African American state senator, Richard H. Newhouse Jr., who had held his seat for almost a quarter century, announced his departure from the legislature amid press reports that he was suffering from Alzheimer's disease. Newhouse had been a courageous pioneer in independent politics, defeating a machine loyalist at the height of the Chicago Freedom Movement's civil rights protests in mid-1966.

Newhouse played a decisive role in making his colleague Cecil Partee the first African American president of the Illinois Senate, and in 1975 Newhouse had been the first black candidate for mayor of Chicago, winning 8 percent of the vote in a protest campaign against longtime incumbent Richard J. Daley. When his friend Harold Washington won the mayor's seat eight years later, Newhouse stood beside him on the victory podium, holding their hands high, but throughout all those years, the unspoken question always remained "Why, after all this time, is Newhouse still in the first political office he ever sought?"

The answer was simple, and reflected one of the core truths of African American politics: "Newhouse has a white wife," Kathie, whom he had first met in 1954. Across the South Side, the word on the street was that Newhouse "talks black but he sleeps white," and one black nationalist explained that "the perception now is that an African American married to a white is unable to give 100 percent to the cause." Carol Moseley Braun, who for almost a decade had had a white husband, frankly confessed that "an interracial marriage really restricts your political options. The blind reaction of some people is just horrible." The *Chicago Tribune* celebrated Newhouse's "impressive record" and mourned how his retirement "marked the passing of an era for Chicago's black progressive political movement."

As with Moseley Braun's smaller former House seat, the committeemen whose wards made up Newhouse's Senate district would select his successor in a weighted vote. The 8th Ward's John Stroger aimed to elevate Trotter, but organization loyalists controlled only 47 percent of the district. Alan Dobry held

30 percent, and together Tim Evans, still 4th Ward committeeman, and the 7th Ward's Palmer had 23 percent.

Dobry and Evans had long respected Palmer, who had moved to Chicago at age thirty after completing her undergraduate degree at Indiana University and teaching high school English for three years in Indianapolis. With a father who was an MIT-trained architectural engineer, a mother who also was a teacher, and a grandfather who had been a prominent African American physician, Alice Roberts Robinson Palmer's interest in education had strong family roots. In Chicago she worked at several local colleges while finishing a master's degree at Roosevelt University. She also remarried, taking the surname of her second husband, Edward "Buzz" Palmer, a former Chicago police officer well known in the black community as one of the cofounders of the pioneering Afro-American Patrolmen's League.

In 1977 Alice became associate dean for African American student affairs at Northwestern University, and two years later she completed a Ph.D. in educational administration at Northwestern. Her dissertation, "Concepts and Trends in Work-Experience Education in the Soviet Union and the United States," was a sophisticated and erudite comparative analysis informed by Palmer's own research in the USSR as well as her readings of both Karl Marx and John Dewey. During the 1980s Alice and Buzz's travels included a visit to Grenada during the short reign of Marxist revolutionary Maurice Bishop plus multiple visits to Moscow, Prague, and other Soviet-bloc capitals. They founded a Chicago entity called the Black Press Institute, and Alice's membership in the U.S. Peace Council plus her active role in the International Organization of Journalists indisputably reflected pro-Soviet sympathies. In June 1986 the *People's Daily World,* the newspaper of the Communist Party USA, reported that she was "the only Black U.S. journalist to attend the 27th Congress of the Communist Party of the Soviet Union" three months earlier.

By the time of her election as 7th Ward committeeman in 1988, Palmer had left Northwestern for the Metropolitan YMCA, and by 1991 she was heading up the Chicago affiliate of Cities in Schools, a nationwide dropout-prevention program. Her international travels and affiliations attracted no attention or concern in Chicago, and she later said she had run for the 7th Ward post reluctantly, only because her well-known, outgoing husband had declined Harold Washington supporters' requests to do so. "I was the quiet one," she explained. "If the well-meaning citizens of Chicago don't go out and get their hands dirty in political action, then dirty politics will never be cleaned from the streets," she told her friend Vernon Jarrett, the *Chicago Sun-Times* columnist.

Alan Dobry and Tim Evans fully agreed that longtime independent Dick Newhouse's Hyde Park Senate seat could not be filled by someone allied with Stroger's black Democratic machine. They also concurred that Alice Palmer was the best available individual, and together they called her to make that case. An "initially reluctant" Palmer accepted, and the trio brushed aside Stroger's argument that they elevate Donne Trotter and name Palmer to his House seat. On Thursday evening, June 6, the committeemen assembled at Stroger's 8th Ward office. After brief remarks by both Trotter and Palmer, they voted 53 to 47 percent to make Palmer a new member of the Illinois state Senate before making the formal tally unanimous. Five days later, with the legislature's annual spring session running overtime into early summer, Palmer headed to Springfield and was immediately sworn into office. In the *Sun-Times,* Vernon Jarrett praised the new senator as "a role model for other individuals of her race who have attained high degrees" and declared that "there are fewer than a dozen individuals in the Illinois Legislature or the City Council who match her combination of academic attainment and commitment." The weekly *Hyde Park Herald* commended Palmer's selection, praising her independence as well as "her dedication to education." A week later another *Sun-Times* columnist reported that Palmer was already "getting high marks from legislative colleagues for her baptism of fire in Springfield. 'She's jumped right in and is having no difficulty staying in the fast lane,'" one unnamed observer stated. Before the month was out a clearly happy Palmer told the *Herald* that she "plans to make a career of the General Assembly."[1]

Like Hyde Park, Barack's new home base in South Shore sat squarely within the Newhouse-Palmer Senate district. But to whatever degree Barack noted Alice Palmer's emergence as a significant figure in South Side politics, his undisputed focus throughout all of June and July was on the Illinois bar exam, which would be administered on the last two days of July at Northwestern University's law school. Sidley & Austin had footed the tab for the intensive bar review class that Barack, like almost all other applicants, took in the run-up to the highly demanding, two-day test, and on weekdays Barack often accompanied Michelle downtown, studying at Sidley and attending review sessions. Sidley still hoped Barack would join the firm, but at one large summer dinner another guest mistook Barack for a waiter and asked, "Can I have more tea?"

On Barack's first Saturday back in Chicago, Michelle took him to the wedding of one of her best friends' younger brother at the family's Jackson Park Highlands home. To Barack, Jesse Jackson was the former presidential candidate for whom he had cheered back in 1984 in New York; to Michelle he was the father of her friend Santita, and someone she had met many times during her high school years. Twenty-six-year-old Jesse Jackson Jr. was marrying fellow University of Illinois law student Sandi Stevens, and a Jackson family videotape of the happy event captured a glimpse of Barack, whom Santita introduced to her brother.

Marriage was a subject that was also on Michelle Robinson's mind, but so was the fact that she was increasingly conscious of how unfulfilling she found her highly remunerative work at Sidley. Her close colleague Kelly Jo MacArthur remembered how the view from Michelle's office on the forty-something floor looked southward, and that unbroken perspective—"That's where I really come from"—helped fuel Michelle's "internal dissonance" about her life. "There was always some sense, from the first time that I met Michelle, that 'I'm not sure that I really belong in this corporate law firm,'" Kelly recounted, and notwithstanding how much relatively interesting work partners like Newton Minow, Charles Lomax, and Quincy White gave Michelle, including pro bono assignments, her dissatisfaction grew. "I couldn't give her something that would meet her sense of ambition to change the world," White recalled. Even before that summer, Michelle had quietly begun searching for a new job. One person she sought counsel from was Gwendolyn LaRoche Rogers, the effective stepmother of her brother Craig's best friend, fellow Princeton graduate John Rogers Jr., who, like Craig, was pursuing a career in finance. Eight years earlier Rogers had founded his own investment firm, Ariel Capital Management, with start-up funds from his mother and a Chicago couple who were lifelong family friends, James and Barbara Bowman.

Four years earlier Barack had approached Gwen LaRoche, the Chicago Urban League's education director, for help with his Career Education Network, and now Michelle Robinson sought advice too. "Michelle came to me and said that she was not happy at Sidley & Austin. She did not want to be working in a law firm. She wanted to do something for the people," Gwen LaRoche Rogers explained. Michelle remembered it similarly: "I wasn't happy" and wanted work with "a community-based feel" that would "benefit others." She asked Gwen if the Urban League could use her, but Gwen, knowing what Sidley paid, told Michelle that she would cry every night over what the League could afford. So Michelle continued her outreach, realizing that when she asked herself, "Can I get pumped up every day about coming to practice corporate law? The answer was simply 'no.'"

Barack also knew that even after devoting the next year to writing "Journeys in Black and White," practicing law at Sidley was not what he wanted to do either, no matter how well it paid. Michelle also was repeatedly raising the question of would they indeed commit to getting married, but throughout June and July, Barack kept putting it off. Then, with Barack's bar exam complete on Wednesday, July 31, the couple scheduled a celebratory dinner, and Barack chose the stylish, three-star Gordon at 500 North Clark Street, which the *Chicago Tribune* had praised as one of the city's best restaurants just weeks earlier and was best known for its flourless chocolate cake. Early in the meal Michelle mentioned marriage, and Barack once again parried. Then, after they ordered dessert, what appeared in front of Michelle was not chocolate cake but a small box. As she opened it and discovered an engagement ring, Barack asked, "Will you marry me?" and Michelle immediately replied, "Yes, yes," before being rendered otherwise speechless. "I was completely shocked," she later explained, because Barack's long-standing equivocation left her truly surprised by his proposal. Whatever Michelle had ordered for dessert was ignored. "I don't think I even ate it. I was so shocked and sort of a little embarrassed because he did sort of shut me up" after weeks of persistent questioning.

In the immediate wake of Barack and Michelle's engagement, events moved quickly. Almost simultaneous with their dinner, a copy of Michelle's résumé made its way to assistant Chicago corporation counsel Susan Sher. Just weeks earlier Sher's fellow assistant corporation counsel Valerie Jarrett, the thirty-five-year-old former daughter-in-law of columnist Vernon Jarrett and the daughter of James and Barbara Bowman, had been named Mayor Richard Daley's new deputy chief of staff. Sher, highly impressed with Robinson's Princeton and Harvard Law School credentials, passed Michelle's résumé to Jarrett, who responded similarly and quickly scheduled an interview. In person, Robinson was just as impressive as on paper, and as their initial ninety-minute conversation ended, Jarrett spontaneously offered Michelle a job as assistant to the mayor without clearing it with either Daley or her boss, chief of staff David Mosena. Michelle bonded almost immediately with Jarrett—"She understood how I felt. It was difficult to find people who understood my desire to leave a high-paying corporate job"—but replied that she wanted to think it over. A few days later Michelle called Jarrett and asked, "Would you be willing to have dinner with my fiancé so that the three of us could talk about it?" Jarrett agreed, and an evening or two later, they met up at a seafood restaurant in the downtown Loop.

Valerie Bowman Jarrett's background was almost as exotic as Barack's. Five years his senior, Jarrett had been born in Shiraz, Iran, while her father, a pathologist who specialized in blood diseases and abhorred the racial discrimination he had experienced in U.S. medical institutions, was working there for six years. Jim Bowman had met and married Chicago native Barbara Taylor, whose father Robert was the first African American chairman of the Chicago Housing Authority, during a Chicago residency. From Iran, the small family moved to London for one year before returning to Chicago, where Bowman soon joined the University of Chicago's medical faculty.

Valerie grew up in Hyde Park, completed her secondary education at the Northfield Mount Hermon School in Massachusetts, and graduated from Stanford University in 1978 before entering law school at the University of Michigan and receiving her J.D. in 1981. After six years in private practice with two Chicago law firms, Judson "Judd" Miner, Harold Washington's corporation counsel, hired Jarrett, and she remained in that office as first Eugene Sawyer and then Richard Daley became mayor. Divorced in 1988 after a five-year marriage to a childhood friend, Vernon Jarrett's son William, by 1991 Jarrett was a single parent raising a six-year-old daughter and experiencing what the *Tribune* called a "meteoric rise" in Mayor Daley's inner circle.

At dinner that night, Barack and Michelle questioned Jarrett about the mayor's office. "You know, I've never been interviewed by someone's fiancé before," Jarrett joked. "I'm the one who's supposed to be doing the interviewing." But much of their conversation was devoted to Barack and Valerie's unusual early lives. "That night we talked about his childhood compared to my childhood," Jarrett recalled. "We were comparing Indonesia to Iran." Valerie also quickly realized that the fiancé was just as impressive as Michelle. "I remember that night, sitting across the table from him at dinner, and I thought to myself, 'This is one extraordinary young man.'"

By mid-August, Barack as well as Michelle had a decision to make. Barack had had several more lunches with Judd Miner, and also spoke with Allison Davis, the other cofounder of Miner's twenty-year-old, now twelve-attorney law firm. Unlike Sidley, whose roster of corporate clients meant the firm was often opposing employment discrimination claims, Davis, Miner, Barnhill & Galland (DMBG) combined civil and voting rights litigation with Davis's strong interest in housing development. With DMBG as a definite employment option once *Journeys* was complete, Barack felt entirely comfortable in telling Gerry Alexis, John Levi, and Newton Minow that he would not be accepting Sidley's job offer. "Gerry, I want to tell you that I've decided that I'm not going to come to Sidley after all," she remembered Barack telling her. Rob Fisher and his girlfriend Lisa recalled hearing about Michelle mulling whether to leave Sidley for the mayor's office, with Michelle's only hesitation being the "50 percent cut in pay." "I was going to walk away from a pretty huge salary," Michelle later recounted. But within a day or two of Barack's apology to Gerry Alexis, "Michelle came and told me that she was leaving and she was going to work in the mayor's office," Gerry remembered. John Levi recalled, "I thought I was going to have a coronary" upon hearing the double-barreled bad news. Newt Minow learned about the pair of departures when Barack came to his office to explain that he would not be joining Sidley because "'I think I'm going to go into politics.' I said, 'Well, that's good,'" Minow recounted. "'We'll try to help you.'" Knowing the other news he had to impart, Barack remarked half-jokingly that "I don't think you're going to want to help me when I tell you the rest of the story" and suggested they both sit down. "'You know Michelle?' I said, 'Of course I know Michelle.' He said, 'Well, I'm taking Michelle with me.'" Minow sputtered in dismay before Barack interjected, "'Hold it—we're going to get married.' I said, 'That's different,'" a relieved Minow remembered. Back at Davis Miner's town house office, Barack told Judd and Allison that he indeed would join the firm once his book was complete. "There's only one condition," Davis remembered Barack adding. "You've got to reimburse Sidley for the bar review fee."[2]

With the bar exam over, Barack could turn his full attention to "Journeys in Black and White." He took the time to register to vote at the Robinsons' home address, and one weekend he and Michelle drove out to Beverly Shores, near the Indiana Dunes on the shore of Lake Michigan, to see Jean Rudd, whom Michelle had met briefly two summers earlier, and her husband Lionel Bolin, who was black. Barack also had lunch one day with Douglas Baird, who had taken the lead in arranging his new appointment as a Fellow in Law and Government at the University of Chicago Law School (UCLS). Several other faculty members joined them, and Baird also took Barack to meet law school dean Geof Stone, who found him "very impressive." With only one single black male instructional staff member—clinical professor Randolph Stone—the law school was "looking to find diversity," Geof Stone explained, and "our goal was *maybe* this is somebody we should be thinking about for a faculty appointment" once Barack finished his book. Stone's secretary Charlotte Maffia shared her boss's reaction, remarking to Stone that "he's going to be governor of Illinois some day!"

By the beginning of September Barack had a small office, room 603, on the top floor of the law school's cubical, glass-sheathed main building, designed by famed architect Eero Saarinen in the late 1950s and significantly expanded in 1987. His appointment carried no salary or benefits, but the school's small cadre of black law students—no more than about a dozen in each of the three current classes—were pleased to hear about Barack's arrival, even if the law school's annual directory, *The Glass Menagerie,* erroneously stated that at Harvard, he had been president of BLSA, not the *Law Review.* Barack's biographical sketch accurately summarized his work at DCP and said he "will spend this year writing a book on issues of race and politics." Some weeks later, when Barack remarked to Douglas Baird that his book manuscript would be in part autobiographical, Baird was greatly perplexed by exactly what Barack was up to.[3]

Either just before or soon after Michelle began work in mid-September as a $60,000-per-year assistant to the mayor, Barack had a journey for them to take together. He wanted to introduce his new fiancé to his Kenyan family, and Barack arranged with his sister Auma, who was still living in Germany, to meet him and Michelle in Nairobi. From there, they traveled to the family homestead in Kogelo, where brother Roy, who had adopted Abon'go Malik as his given names instead of Roy Abon'go, was also visiting from the U.S. Barack and Michelle spent almost a week living at stepgrandmother Sarah's humble home, which had no electricity or running water, yet Michelle adopted easily to traditional Luo customs such as eating with one's hands, and Barack later joked that her ability to learn basic Luo words quickly surpassed his. Michelle was taken aback, however, when someone asked her, "'Which one of your parents is white?' which shocked her," Barack later recounted. "'Why would you think that one of my parents is white?'" an astonished Michelle replied. Her surprise at being asked that by a true African illuminated "how American African Americans are," Barack realized.

Back in Nairobi, Aunt Zeituni loaned the trio the decrepit Volkswagen that Auma had driven three years earlier during Barack's first visit to Kenya. One day a highway breakdown left them on the roadside until a pair of mechanics made repairs, with Barack jotting down observations in the notebook he carried throughout the trip. Michelle had to head back to the U.S. for work well before Barack did, and after two weeks in Nairobi Barack and Auma returned to Kogelo for another visit with granny Sarah as Barack recorded additional family details to use in "Journeys in Black and White."[4]

Upon his return to Chicago, Barack learned that he had passed the Illinois bar exam. Much like Auma's Beetle, Barack was still driving the rusty off-yellow Toyota Tercel that had carried him to and from Harvard, and which now, just like his previous car during his time at DCP, sported a floorboard hole large enough so that passengers could watch the ground go by. Early on November 1, Barack headed north to Racine, Wisconsin, for a three-day conference to which Jacky Grimshaw, whom the late Al Raby had introduced to him in 1987, had arranged an invitation. The National Center for Careers in Public Life was the brainchild of two young women, Vanessa Kirsch and Katrina Browne, who had met in Washington, D.C., and wanted "to create a national mechanism to recruit, place and train the next generation of issue leaders." Browne, a 1990 graduate of Princeton University who was working for a newly founded, alumni-sponsored public interest endeavor, Princeton Project '55, won support from the Johnson Foundation for her and Kirsch's plan, and their kickoff gathering was taking place at Johnson's lush Wingspread Conference Center.

Kirsch and Browne's starting point was "the belief that a project *for* young people should be designed *by* young people," but many of the four dozen attendees were, like Barack and Jacky, now past their thirtieth birthdays. His listing was simple indeed—"Barack Obama, Writer, 7436 South Euclid," Chicago—and years later no other attendees would recall Barack speaking up during the Friday-Saturday-Sunday set of meetings. At the outset, Browne and Kirsch had a catchier name in mind for their new organization—"Public Allies," playing off the well-known music group Public Enemy—and also had a seventeen-member board lined up, including Barack. The two women envisioned a regional structure for the group, with an initial arm in D.C. that would recruit, train, and place in public interest organizations several dozen young "allies" who would be mentored for a year with the explicit expectation that they would pursue careers in public service. Chicago, given its set of enthusiastic local backers, including John McKnight and his Northwestern University colleague Jody Kretzmann as well as Jacky Grimshaw, was envisioned as Public Allies' second start-up a year or so hence.[5]

In addition to his Kenyan family, Barack had one particular piece of his Chicago life he wanted to introduce Michelle to: Jeremiah Wright's Trinity United Church of Christ. The Robinson family had never been regular church attendees during Michelle's childhood, and in black Chicago "church weddings are normally at the wife's church," Jerry Wright explained. With Michelle and Barack now thinking ahead toward their own wedding sometime in 1992, joining a church in order to have a church wedding was very much on their agenda. Trinity's growth had continued apace during Barack's three years at Harvard, and another capital campaign to build yet a larger new sanctuary was well under way.

Trinity required aspiring members to attend new members' classes "to teach you, 'What have I joined?'" Jerry explained, and he strongly encouraged them to enroll in one of Trinity's weekly, semester-long Bible classes. Aspirants also had to choose which of Trinity's many service ministries they would invest time in, such as the legal ministry, which dated to 1979: "Where do you want to work in this church?" Tithing was aspirational indeed, since at Trinity that meant not just 10 percent of a member's income "but 10 percent of your time," yet Wright understood full well that not everyone who joined Trinity did so in order to personally contribute to the church. "Some of the members of Ebenezer," Jerry explained, "belong to Ebenezer church because it's Ebenezer church in Atlanta," the home church of Martin Luther King Jr. and his family. "Same thing at Riverside, same thing at Abyssinian," upper Manhattan's two best-known churches, and similarly for Trinity on the South Side of Chicago.

A good many Trinity members, like Barack's friend Sokoni Karanja, were deeply devoted to the betterment of black Chicago, but just as many "were there when it's not golfing season," Jerry realized. New members were admitted at the evening service on the first Sunday of a month once they had completed new members' classes, and their chosen ministry would be announced as well, with contact information passed along to those leaders the next morning.

Barack later explained that for him, joining Trinity "wasn't an epiphany. . . . It was an emotional and spiritual progression, as well as an intellectual one. And it didn't happen overnight." Barack also would say, "What I love about Trinity is that everyone is involved in something—it's not just the pastor doing everything." Wright's preaching appealed to university professors as well as to activists like Karanja and public sector bureaucrats like former Chicago school superintendent Manford Byrd. Young University of Chicago theologian Dwight

Hopkins, who had completed his Ph.D. at New York's Union Theological Seminary under the guidance of James H. Cone, the founder of black liberation theology, viewed Wright as a "humble and unassuming" pastor with "a remarkable grasp of the Bible." One perhaps jaded local black journalist believed Wright "provided kind of a vicarious militance for Chicago's black elites" such that "they could get a dose of militance on Sunday and go back home and feel pretty good about doing their part for the black movement." Yet Jim Cone had a far less cynical view: "I would regard Jeremiah Wright's church as the really contemporary embodiments of all the things I've tried to say."

Barack would state on multiple occasions over many years that he was first drawn to Wright's ministry by one particular sermon he heard Wright preach early in 1988, called "The Audacity of Hope." Yet Wright did not preach that sermon at Trinity in 1988; he had preached it there exactly three years earlier, on February 17, 1985, before Barack arrived in Chicago, and he would next preach it at Houston's Wheeler Avenue Baptist Church in January 1991. A tape of the 1985 rendition was available for purchase at Trinity, and the 1991 Texas version soon appeared in a collected edition of Wright's sermons, so Barack knew the sermon even though he had not heard Wright deliver it in person.

Years later critics who knew little about African American Christianity would claim that Trinity was "arguably the most radical black church in the country" when Barack and Michelle formally joined on Sunday evening, February 2, 1992, but anyone who knew Trinity and its congregation would all but laugh out loud at that assertion. Newly elected 4th Ward alderman Toni Preckwinkle later explained that Trinity "was a good church for an aspiring, upwardly mobile politician to join," and many people who knew Barack and viewed him as "a very pragmatic person" felt likewise. Barack later declared that he was "offended by the suggestion" that careerist considerations had played a significant role in his choice of Trinity, but among the people who had spent the most time with Barack in Chicago, that belief was widespread if not universal.

Barack was "an immensely pragmatic person," Greg Galluzzo realized, and his fundamental life choices in that vein dated back to when he had first told Sheila that his destiny would not allow them to marry no matter how deeply they loved each other. Galluzzo asked, rhetorically: "Did he leave her for pragmatic reasons? Did he go to Jeremiah's church for pragmatic reasons? Did he go to Harvard for pragmatic reasons?" He stopped before uttering the next, obvious question, remarking instead that Michelle was "the ideal person" with whom he could "start a base in the black community." Barack readily admitted that "settling in Chicago and marrying Michelle was a conscious decision to root myself." Many black Chicagoans agreed with Galluzzo, but Barack, confronted with such sentiments years later, would dismiss them as "pretty cynical" as well as "pretty offensive to me."

Marrying Michelle "was the single most important step that he was to make in his journey to define himself and reconcile his search for racial identity," one local African American scholar rightly thought, and he was far from alone in his conclusion. "Michelle is a representation of Barack Obama's choice to aggressively move toward blackness," a black female professor agreed. "He could have actively made a choice to move toward whiteness or something else." Indeed, a month after Barack and Michelle joined Trinity, they scheduled their wedding there for Saturday, October 3, and on March 9, Barack wrote to Sheila for the first time since they had last seen each other in Cambridge to tell her that he and Michelle were engaged and would marry in October.

Trinity members who remembered Barack from DCP, like Carolyn Wortham and Deloris Burnam, plus Patty Novick, the late Al Raby's girlfriend, recalled often seeing Barack and Michelle at the 11:00 A.M. Sunday service around the time that they first joined. Barack remembered similarly, but he and Michelle were not there the Sunday after they joined, because on the previous day Stanley Dunham had died at age seventy-three after a prolonged bout with prostate cancer. Barack and Michelle arrived in Honolulu before Stan expired, yet for Barack his passing represented the loss of the most sustained male presence he had known in his life, one far more akin to a real father than a grandfather. Indeed, as Barack's sister Maya said of Stan and Madelyn, "they raised him." For Madelyn, the loss was profound, for despite perceptions of her disappointment and at times unhappiness with Stan, her brother Charles had no doubt that they were not only "extremely attached" to each other but "dependent on each other." For Ann too, it was a grievous loss, for "clearly her stronger emotional bond was with her father" rather than her mother, one close friend explained.

Ann's oral defense of her now finally completed Ph.D. dissertation was just two weeks away, and for that she owed tremendous gratitude to her endlessly supportive mentor Alice Dewey, who had loaned her money as well as convinced Ann to narrow and trim the huge ethnographic study. The Internal Revenue Service was still seeking more than $17,000 from Ann thanks to her unpaid income taxes from twelve years earlier, but Ann was ecstatic to finally complete her long-sought doctorate. She dedicated the dissertation to Madelyn, Alice Dewey, "and to Barack and Maya, who seldom complained when their mother was in the field." But among Ann's closest friends, there was no mistaking how "she felt a little bit wistful or sad that Barack had essentially moved to Chicago and chosen to take on a really strongly identified black identity," because that "had not really been a part of who he was when he was growing up." Ann viewed it as "a professional choice" on Barack's part, and although "it would be too strong to say that she felt rejection," she certainly felt "that he was distancing himself from her."[6]

In Chicago, the 1992 election year was well under way, with two serious primary challengers to incumbent Democratic U.S. senator Alan "Al the Pal" Dixon. Mel Reynolds was again challenging U.S. representative Gus Savage, but this time Reynolds's odds of success looked much improved thanks to a redrawn, more suburban district following the 1990 Census. The South Side's other black U.S. representative, Charlie Hayes, was facing a difficult reelection battle against alderman and former Black Panthers leader Bobby Rush. Senator Dixon's first challenger was Cook County Recorder of Deeds Carol Moseley Braun, who had been outraged when Dixon in October 1991 voted in favor of U.S. Supreme Court nominee Clarence Thomas, who won confirmation by only a 52–48 margin. Moseley Braun's effort was troubled by the behavior of her boyfriend and campaign manager, Kgosie Matthews, who had been accused of sexual harassment. While polls a week before the primary showed Dixon with a 20-point lead, Braun's "engaging warmth" during a final televised debate won statewide praise. A final poll showed Dixon's lead down to 12 points, while in the congressional primaries both Rush's challenge against Hayes and Reynolds's against Savage looked likely to prevail. The *Chicago Tribune* praised Reynolds as "bright, capable and promising," and on the South Side Donne Trotter was poised to join Alice Palmer in the state Senate thanks to the new redistricting. The *Hyde Park Herald* endorsed Rush over Hayes and praised Palmer as an "extraordinary woman" who while "basically a novice . . . has done a fine job" during her first nine months in office.

Several weeks before the March 17 primary, Barack received a call from Sanford "Sandy" Newman, a Washington attorney who a decade earlier had founded a nationwide voter registration project called Project VOTE! The organization concentrated its efforts on major states in presidential election years; in 1984 Al Raby, then in exile from Harold Washington's Chicago, had served as Newman's codirector before becoming the group's board chairman. Project VOTE! was overtly nonpartisan while focusing on the registration and turnout of minority citizens who were likely to become Democratic voters. In 1992 President George Bush was running for a second term, and in 1988 Bush had won Illinois's twenty-four electoral votes after carrying the state by just ninety-five thousand votes out of over 4.5 million that were cast. Illinois was at the top of Newman's list, and early in 1992 he asked Raby's old friend Jacky Grimshaw who might best direct a Chicagoland effort that summer and fall. Grimshaw's recommendation was the same she had made to the young Public Allies women, and Sandy claimed to have written the surname as "O'Bama" while insisting "I didn't want to rule him out just because he was Irish." When Sandy called, however, Barack immediately cited his book contract and its June 15 due date. He asked Newman if he could do it on a part-time basis, and Sandy said no, it would be more like sixty hours a week once the registration drive got under way. Barack declined, but by March 17 Newman had not hired anyone else.

That evening, as returns came in, Mel Reynolds easily defeated Gus Savage by a margin of 63 to 37 percent, and Bobby Rush similarly ousted Charlie Hayes by a much narrower 42 to 39 percent tally. But the big news was in the Senate primary, where Carol Moseley Braun won 38 percent of the vote in the three-candidate field and defeated Senator Dixon by more than fifty thousand votes. The *Washington Post*, the *Chicago Sun-Times*, and the *Chicago Tribune* all used the exact same phrase to describe Moseley Braun's victory: a "stunning upset." The *Post* further characterized it as a "political earthquake," and the *Tribune* called it "one of the most stupendous upsets in Illinois political history." Not only had a two-term U.S. senator been bested by a county officeholder, but a previously little-known African American woman was now on the verge of becoming the first black female senator in U.S. history.

As the historical significance of Moseley Braun's statewide win sank in, Sandy Newman again called Barack, and this time the answer was yes. Following Moseley Braun's victory, "I realized that this presented an opportunity in Illinois to enfranchise and engage a lot of African-American voters that previously had not been involved," Barack later explained, "and the work needed to be done." Sandy flew to Chicago to meet Barack before formally hiring him as Project VOTE!'s Illinois state director, and Sandy remembered that it was "as much him interviewing me as me interviewing him." Barack alerted Judd Miner and Allison Davis that this temporary job that would run through October meant that he could not join Davis Miner until late in the year. "Journeys in Black and White" would largely be put on hold too, yet rare indeed was it for a publishing house to hold an author strictly to a contract's due date.

At the suggestion of Bettylu Saltzman, a well-heeled Chicago Democratic activist who had formerly been U.S. senator Paul Simon's local chief of staff, Sandy already had lined up John R. Schmidt, a progressive, mainstream attorney who had served as Mayor Richie Daley's first chief of staff, to lead Project VOTE!'s Illinois fund-raising effort. Schmidt in turn had recruited John Rogers Jr., the young black investment executive, as his cochair. Once Barack was on board, Sandy quickly had him meet with Schmidt and Rogers, with whom he immediately meshed very well. "We wanted the state directors to be able to focus on organizing," Sandy said, so raising money would not be Barack's responsibility. In a savvy move to attract independent black political support in addition to Daley backers like Schmidt and Rogers, Newman recruited former mayoral challenger Joe Gardner to be Illinois Project VOTE!'s official chairman.

A raft of progressive grassroots groups would be ready to work with Project VOTE! if funding was available, and Barack quickly had to begin building a citywide organization to fully coordinate an enterprise that would be vastly larger than either DCP or the *Harvard Law Review*. Barack's own Chicago experience was limited to just some parts of the huge South Side, and equally important in black Chicago was the similarly spacious but economically weaker West Side. Far west 29th Ward alderman Sam Burrell's political organization was the West Side's best voter-outreach operation, and one day in early April Barack headed there to meet Burrell and his top staffer, Carol Harwell. Upon arrival, Barack apologized for being late. "I never realized Chicago was this big. I thought I was never going to stop driving."

Carol remembered Barack carrying two lightweight bags, one with Project VOTE! materials while saying of the second one that "this is my book," which he tinkered on when he could. Harwell was eager to pitch in, and quickly introduced Barack to another grassroots activist, Bruce Dixon. Given the size of Chicago, dividing the city up into sectors was mandatory. Carol would take the West Side, Bruce the North Side, and Brian Banks, a 1978 Harvard College graduate who had worked for more than a decade in the computer industry, would handle the South Side. With that trio of key staffers on board and office space on the fifth floor of 332 South Michigan Avenue provided by the Community Renewal Society (CRS), "we sat down and tried to hash out how we were actually doing to do this project," Carol recalled. "Barack was big on community organizations," and wanted to do as much registration work as possible through them, rather than get involved with committeemen's ward organizations, which always paid a cash bounty to workers based upon the number of people they signed up. When Carol explained the traditional Chicago way, "Barack was appalled" and quickly said, "We're not doing that."[7]

Before the end of April, a letter cosigned by CRS executive director Yvonne Delk, Joe Gardner, and Barack went out to several dozen grassroots activists inviting them to an 8:30 A.M. meeting on Tuesday, May 5. Attendees would create a Cook County–wide steering committee for the Project VOTE! Chicago Coalition, whose goal would be to "register at least 130,000 new voters, primarily targeting low income and minority persons." Some two dozen invitees attended, including some Barack already knew: Johnnie Owens from DCP in Roseland, Father Mike Pfleger from St. Sabina in Auburn-Gresham, Sokoni Karanja from the Centers for New Horizons in Bronzeville, and Alderman Burrell from the Far West Side. Others represented established community groups like the Kenwood-Oakland Community Organization, the Greater Grand Crossing Organizing Committee, the Woodlawn Organization, and Jesse Jackson's Operation PUSH. Timuel "Tim" Black, a well-known black political sage and historian, had met Barack briefly during his Hyde Park years; Rev. Princeton McKinney of Christ Temple Community Baptist Church in Markham was a mainstay figure in Mike Kruglik's flourishing South Suburban Action Coalition. The husband and wife team of Keith Kelleher and Madeline Talbott had launched ACORN in Chicago a decade earlier before Keith had begun organizing home health care workers into SEIU Local 880.

The May 5 "founding meeting" began with a CRS welcome, a brief statement of purpose from Joe Gardner, and then twenty minutes of remarks by Barack detailing his "Proposed Plan." The "Target Goal" would be "150,000 new minority and low-income registrants statewide," with 130,000 of them in Cook County. Project VOTE! would reach out to church congregations, who would be asked to register 100 percent of their members, to Chicago Housing Authority tenants' councils, as some fifty thousand CHA residents were unregistered, and to forty public high schools before the school year ended. Activities would get under way in late May, with a goal of twenty thousand new registrants for June, thirty thousand during July, and forty thousand a month for both August and September. Community groups needed to train their volunteers as deputy registrars who were able to enroll new voters, and groups could apply to Project VOTE! for four-figure grants. Everyone's expenses could be reimbursed, but no group could pay workers on a per-signature basis.

Before the steering committee's second meeting on May 19, Barack obtained Jeremiah Wright's agreement that Trinity members would actively participate. Barack also sought advice from retired state senator Dick Newhouse, who offered to buy him a meal, and Operation PUSH activist Jamillah Muhammad was recruited to help Brian Banks with Project VOTE!'s South Side work. In advance of a third meeting on June 3 where the grant-making process would be detailed, Barack drafted the group's first press release, targeted for distribution at an upcoming press conference in Daley Plaza, Chicago's premier outdoor space in the center of the downtown Loop.

In his statement, Barack addressed the Southern California urban disorders that had followed the April 29 acquittals of four police officers who had been videotaped assaulting black motorist Rodney King. "The Los Angeles riots reflect a deep distrust and disaffection with the existing power pattern in our society," Barack wrote. "We must make political leadership in this country more responsive to the pressing issues," and in order to do so "voter registration is a necessary, constructive first step." In particular, "it's critical that people on the bottom of the economic ladder participate in the process."

That June press conference was not covered by any of Chicago's newspapers, though Barack did win the enthusiastic cooperation of Marv Dyson, the general manager of WGCI, the city's top black radio station, who "agreed to publicize our efforts and recruit the other black radio stations" to do likewise. That was all to the good because throughout the summer the *Chicago Defender*, the city's premier black newspaper, made no mention of Project VOTE! Only in early August did two *Sun-Times* columnists mention the project, with Vernon Jarrett highlighting how Edward Gardner, owner of Soft Sheen Products, a large personal-care-products firm who had a long-standing interest in voter registration, had become a major contributor to Project VOTE!

Throughout June and July, new registration numbers ran about one-third below Barack's initial, overly optimistic monthly targets. Some of the most successful allied groups, like Keith Kelleher's Local 880 and Madeline Talbott's ACORN, were buttressed with $500 apiece each week from Project VOTE! DCP, now headquartered on 95th Street, helped in Roseland, multiple members of both Trinity and St. Sabina churches worked as deputy registrars, and Operation PUSH volunteers did yeoman work. In the notorious Robert Taylor Homes, named for Michelle's boss Valerie Jarrett's grandfather, super-energetic Rita Whitfield knocked on every door in the sixteen-story towers. On the West Side, squeaky-clean 28th Ward alderman Ed Smith's organization was the only Democratic Party arm Project VOTE! funded, and on the North Side, seventy-two-year-old retired machinist Lou Pardo, Chicago's best-known voter registration volunteer, pitched in as usual. Pardo's mentor, progressive state senator Miguel del Valle, recalled first hearing Barack's name from Lou. "I just met this young man who is *so* impressive," Pardo told him. "He's gonna be *big* someday."

Only once a week did Barack's top staffers—Carol Harwell, Bruce Dixon, and Brian Banks—sit down with him in the office for a formal staff meeting. Otherwise everyone was out and about, shuttling volunteers to retail stores and other locations with large numbers of passersby. Uppermost in Barack's mind were the weekly and monthly numerical targets he and Sandy Newman had agreed upon. "It was very competitive in the office, West Side versus North Side versus South Side," Carol recalled, declaring that "the West Side won every week." Barack was "managing to the number," Brian explained, always emphasizing that "We've got these numbers to make." Barack also continued to work on his *Journeys* manuscript when he could, and Carol remembered that "he wrote the first couple of chapters in longhand" and then asked her if she could type up the initial part. Carol agreed, even though Barack's small, left-handed penmanship was difficult to read. Carol also offered punctuation and word-change suggestions, which Barack had no interest in hearing. "I didn't ask you to read it, I asked you to type it," she remembered him barking. "Gimme my book. I'll type my own book."

With Project VOTE! taking up most of his time, Barack's presence at the University of Chicago Law School became very infrequent, but by midsummer one colleague he had first met months earlier had a request for him. Jim Holzhauer was a former Supreme Court clerk who had taught full-time at the law school for several years before moving to Mayer Brown, a top-ranked law firm, while continuing to teach several courses, including Current Issues in Racism and the Law. Chicago operated on a trimester schedule—Fall, Winter, and Spring quarters—and although Holzhauer had taught Racism six months earlier, in Winter 1992, a lengthy out-of-town trial was already on his calendar for early 1993. So "I asked Barack to take over the course," and Barack agreed to do so for Spring 1993—well after he would have completed *Journeys* and begun practicing law at Davis Miner. With that would come a change in title from Fellow to Lecturer in Law, as UCLS politely called its adjunct instructors. Barack had to submit a course description before the 1992–93 academic year got under way, and he took Holzhauer's catalog copy—"How have past and present legal approaches to racism fared?"—and expanded it significantly.

> *Has the continued emphasis on statutory solutions to racism impeded the development of potentially richer political, economic, and cultural approaches, and if so, can minorities afford to shift their emphasis given the continued prevalence of racism in society? Can, and should, the existing concepts of American jurisprudence provide racial minorities more than formal equality through the courts?*

Holzhauer had said that "students will prepare papers" rather than take an exam, but Barack specified that the papers should "discuss the comparative merits of litigation, legislation and market solutions to problems of institutional racism." Of course neither Rob Fisher nor David Rosenberg were mentioned, but the impact and legacy all of those Cambridge discussions were writ large in the first course description Barack had ever crafted.

At the same time, UCLS's annual *Glass Menagerie* corrected Barack's biographical sketch to note that he had been president of the *Law Review*. Barack, in thinking ahead to the fall, added that he was now "an associate at the firm of Davis, Miner, Barnhill & Galland" and had "recently completed a series of essays on race and politics that will be published this fall by Simon & Schuster."

By early August, Project VOTE! was in strong financial shape, largely thanks to John Schmidt's fund-raising efforts, but the project badly needed to raise its visibility throughout black Chicago, and especially among younger nonvoters, if it was to catch up to its numerical targets. Saturday, August 8, would witness the annual Bud Billiken Day Parade on the Near South Side, a six-decade summer tradition and "the biggest voter registration opportunity of the season" for Project VOTE! Ed Gardner of Soft Sheen had drawn his daughter Terri, who handled their company's marketing and media efforts, into the campaign, and in tandem with Bruce Dixon, Terri designed a handsome black and yellow poster plus a button featuring a memorable slogan: "Register to Vote—It's a POWER Thing!," implicitly playing off of the then-current saying "It's a black thing—you wouldn't understand."

The Billiken Day outing was a great success, with more than two thousand new registrants enrolled, and in its wake, the *Sun-Times*' Vernon Jarrett devoted an entire column to Barack's efforts. "There's a lot of talk about 'black power' among the young but so little action," Barack complained, highlighting how lack of interest among younger nonregistrants was why Project VOTE! was enrolling only seven thousand people a week rather than the hoped-for ten thousand. "We see hundreds of young blacks talking 'black power' and wearing Malcolm X T-shirts, but they don't bother to register and vote. We remind them that Malcolm once made a speech titled 'The Ballot or the Bullet,' and that today we've got enough bullets in the street but not enough ballots."

The key to Project VOTE!'s entire effort was the scores of unheralded volunteer deputy registrars, individuals like Rita Whitfield on the South Side and Lou Pardo's multiethnic band of North Side regulars, five of whom would tally more than one thousand new registrants apiece before the rolls closed on October 5. On the last Wednesday in August, Project VOTE! hosted a "Deputy Registrar Appreciation Affair" with free food at a well-known center-city jazz venue, and by late summer the effort was getting significant boosts from labor unions like the United Auto Workers and District Council 31 of the American Federation of State, County and Municipal Employees (AFSCME), which pledged $50,000 to the drive. By Labor Day weekend the confirmed Chicagoland registration tally was up to sixty-three thousand new registrants, and ten days later, with the *Defender* finally according Project VOTE! some much-needed coverage, the total was up to seventy-one thousand.

"There were some real tough times in terms of whether we were making those numbers or not," Brian Banks recalled. Carol Harwell's widespread contacts meant "she certainly was the political brains of it," Brian believed, but Barack was the one answerable to Sandy Newman, and as they headed into their final four weeks, Barack would "smoke packs of cigarettes" each day. By mid-September they had their weekly tally up to about nine thousand, and an overall total of eighty-one thousand. Project VOTE!'s staff and volunteers regularly interacted with both Democratic presidential nominee Bill Clinton's Illinois campaign, and even more so with Carol Moseley Braun's Senate race workers, but everyone was fully schooled on how Moseley Braun's workers had to be kept at arm's length. Moseley Braun herself understood "there was every effort . . . to keep a firewall between Project VOTE! and the actual campaign," for "we didn't want to get in trouble." Yet ACORN's Madeline Talbott knew that "We were working to elect Carol . . . but that was never stated and we knew you couldn't." In later years, Talbott's husband Keith Kelleher would write of how they and Project VOTE! had mounted "a large-scale voter registration program for U.S. Senator Carol Moseley Braun," and on multiple occasions Barack too would later bluntly state that the purpose of Project VOTE! was "to get Bill Clinton and Carol Moseley Braun elected."

On Sunday, September 20, Clinton and Moseley Braun appeared together at a late-night "voter registration rally" at the New Regal Theater on 79th Street. Project VOTE! chairman Joe Gardner proudly told the crowd that he knew of "85,000 reasons" why they would help elect Moseley Braun and Clinton. Gardner vowed to add more in the final two weeks, and Clinton then thanked the "packed house of approximately 1,000 supporters for registering more than 115,000 new voters," Monday's *Defender* reported. That 115,000 number represented Chicago's total number of new registrants since mid-March, with eighty-five thousand of them submitted by Project VOTE!'s network. Six days later the *Tribune* stated that more than half of new registrants lived in predominantly black wards, and on Saturday, October 3, with forty-eight hours to go, the *Sun-Times* reported that "radio spots using the 'Power Thing' slogan have been blitzing stations serving predominantly black audiences."[8]

October 3 had been set as the date of Barack and Michelle's wedding weeks before he had agreed to direct Project VOTE! With Monday the 5th as the final registration day for November's election, Barack had little time to devote to wedding preparations. He had called old friends like Mike Ramos and encouraged them to attend, and Michelle took the initiative to send Barack's sister Auma a black bridesmaid's dress for her role in the ceremony. Both a limousine and a tuxedo were rented, but at Project VOTE!'s office, Carol Harwell had to keep reminding Barack of other necessary tasks. In particular, "he kept forgetting to go buy his shoes," Carol remembered. "'Barack, you gotta go get shoes,'" and also "'You need to get this hair cut,'" she reminded him. "Barack didn't buy his shoes until Friday night for the wedding on Saturday."

By Friday a large raft of friends from every chapter of Barack's life were gathering in Chicago. From Punahou came Mike, Greg Orme, and Bobby Titcomb; from Oxy came Hasan Chandoo, Wahid Hamid, Vinai Thummalapally, and Laurent Delanney, plus Beenu Mahmood of New York. From the world of organizing came Jerry Kellman, Greg Galluzzo, Mike Kruglik, David Kindler, Sokoni Karanja, Ken Rolling, and Jean Rudd; from DCP came Johnnie Owens, Loretta Augustine, Yvonne Lloyd, and Margaret Bagby. Rob Fisher, Mark Kozlowski, Dan Rabinovitz, Ken Mack, Tom Perrelli, and Jan-Michele Lemon represented Barack's Harvard years, plus a number of Michelle's women friends from the law school's 1988 class. Mutual friends Gerry Alexis, Tom Reed, and Kelly Jo MacArthur came from Sidley, plus soon-to-be colleagues Allison Davis, Paul Strauss, and Laura Tilly from Davis Miner—"he must have invited the whole firm," Laura said. From Project VOTE! came John Schmidt, Carol Harwell, and Brian Banks, and from among Michelle's mayoral office colleagues there were Valerie Jarrett, Yvonne Davila, and Kevin Thompson. In addition to Auma, Abon'go came from D.C. and Ann, Madelyn, and Maya from Hawaii, where Maya had now resumed her undergraduate education at the U of H. Madelyn's brother Charles Payne and his wife Melanie attended, as did Marian and Craig Robinson, Craig's wife Janis, Michelle's close high school friend Santita Jackson and her brother Jesse Jr., and Michelle's Princeton friend Angela Kennedy.

On Friday night, Kelly Jo MacArthur took Maya, Auma, and the Punahou guys out to a jazz club in her red BMW; Hasan Chandoo played cards with some of the Oxy crowd and drank so much that the next day "at the ceremony I was so fucking hungover." Auma stayed with Barack and Michelle at the Robinsons' family home, and at midday on Saturday, family members gathered there prior to the late-afternoon ceremony at Trinity United Church of Christ on 95th Street.

Barack remembered noticing how uncomfortable Madelyn seemed, with her daughter Ann being the only other entirely white person present at the Robinsons' house, until she realized how familiar all of Marian Robinson's food dishes were. "There was an immediate connection there," Barack wrote, because Marian's "stoicism" toward life "reminds me very much of my grandmother." At Trinity, Jeremiah Wright was joined in his office by Barack and his two best men, Abon'go and Johnnie Owens, a duo that many guests viewed with surprise. Barack had never grown close to his older brother, and his relationship with Johnnie was by now "pretty strained" given Johnnie's ongoing personal struggles and the problems DCP was having as a result. Multiple guests recalled Reverend Wright speaking at considerable length, and his metaphor comparing the partners in a marriage to the pillars of Trinity's sanctuary veered toward the phallic when Jeremiah described Barack as a fine pillar of African American manhood.

With the ceremony complete, a limousine drove Barack and Michelle to the South Shore Cultural Center, just north of 71st Street, where a large and eclectic crowd gathered to celebrate the Obamas' marriage. Barack spoke briefly and movingly about how sad he was that the man who had raised him, Stanley Dunham, was not there with them. Carol Harwell remembered how good the champagne was, and how after multiple rounds of toasts to the new couple, she had offered one telling Michelle she should enjoy their two-week honeymoon in Northern California because she had married someone who was going to go far and whose name one day would be in lights. Michelle seemed less than thrilled by Carol's prediction, but to Laura Tilly the crowd at the reception seemed "like the who's who of African American Chicago and liberal Chicago." Craig Robinson's wife Janis remembered it as "one of the best weddings I've ever been to," yet to Laura it also "was kind of like a coronation: here's a man who was already designated by many people as somebody of huge potential."

Barack and Michelle had chosen Stevie Wonder's "You and I" as their wedding song, and as everyone milled about and sat down to eat, old friends who were strangers to each other mixed and mingled. Gerry Alexis was struck by "what a rainbow of people" were there and was especially happy to meet Ann Dunham. Barack and Michelle, Gerry said, "introduced me as the person that brought them together" three years earlier at Sidley. To Barack's Punahou friends the reception "seemed kind of formal," but Mike Ramos recalled that "we had a blast." Loretta, Yvonne, and Margaret were sitting together at Table 4—Yvonne saved her place setting—and Barack "came over to us and asked if there was anything we needed," Loretta recalled. "I just said 'Barack, all I want is a ticket to your inaugural presidential ball,'" and he laughingly replied, "You've got it."[9]

With Monday the final day for preelection registration, Project VOTE! deployed hundreds of deputy registrars at fast-food restaurants across Chicago and staged a midday event in the Loop to attract holdouts. Only then did Barack and Michelle fly to San Francisco to begin their honeymoon in the Napa Valley and along the California coast. In Barack's absence, Carol Harwell took charge of sending the final registration numbers to Sandy Newman in Washington, and the verified total—111,000—was so good that Newman had to be convinced that deputy registrars indeed had not been paid per-signature bounties. With Illinois fund-raising having reached $200,000, enough money remained for Carol to plan large-scale get-out-the-vote phone banking for the final days before the fall election. Each new registrant had supplied a phone number, and working largely from Soft Sheen's offices, Project VOTE! callers used a nonpartisan script while firmly asking, "You do plan to vote on Election Day, don't you?"

Some new registrants received multiple calls, and with Project VOTE!'s Illinois success matched by similar numbers in Pennsylvania and Michigan, top national political reporters like David Shribman and James Perry at the *Wall Street Journal* wrote front-page stories about how "surging registration" in big cities in battleground states would mean tens of thousands of new votes for Democrat Bill Clinton. In Project VOTE!'s own publicity materials, that 111,000 documented tally was supplemented by Barack and Joe Gardner's "estimate" that news of their effort had "produced an extra 40,000 registrations" through other channels, thus allowing Illinois Project VOTE! to claim that it had enrolled 150,000 new voters.

Everyone who had interacted with Barack during those six months came away impressed. John Schmidt, who soon took the Justice Department's third-ranked post in incoming president Bill Clinton's new administration, felt that Barack had a "remarkable presence" and "was extraordinary." AFSCME political director Bill Perkins found Barack "very impressive" and was struck by how utterly legitimate Illinois Project VOTE!'s registration numbers were. Twenty-ninth Ward alderman Sam Burrell told a local journalist that "the sky's the limit for Barack," and new 4th Ward alderman Toni Preckwinkle took note of him too. *Chicago Reporter* editor Laura Washington, whose office was also at CRS, thought Barack was a "young impressive guy." Father Mike Pfleger, whose St. Sabina Youth Organization had averaged 125 new registrants a week for Project VOTE!, was just as dazzled as he had been five years earlier.

Phil Mullins, still at UNO, realized Barack was "more detached" than he had been five years earlier, that "he had internalized something." AFSCME's Roberta Lynch felt "he was a different person" than the young organizer she had first met in 1986. At DCP new organizer Michael Evans found Barack "a regular guy" who took the time to "shoot hoops with us on a Friday night," and Barack still regularly got to the UC's gym as well. A black female U of C graduate student who was contemplating law school took note of an African American man whose "Real Men Marry Lawyers" sweatshirt indicated a connection to Harvard Law School. When she asked, he confirmed that his wife had gone there but said nothing about himself, and he told the younger woman that she too could succeed at Harvard. For Erika George it was a life-altering conversation because she followed the stranger's advice and became a member of Harvard Law's 1996 class. Only later would she realize who her self-effacing adviser was.

Bettylu Saltzman, who had first pointed Sandy Newman toward John Schmidt, remembered her first conversation with Barack that year. "I immediately thought, he's going to be president some day." Bettylu said exactly that first to her husband and then to several close friends, including Mayor Daley's media consultant, David Axelrod, who like Saltzman had worked for U.S. senator Paul Simon. "I've just met the most remarkable young guy, and I think you ought to meet him," Bettylu told Axelrod the next morning on the phone. "I know this is going to sound silly, but I think he could be the first black president." Axelrod thought, "Well, that's a little grandiose," but he agreed to have lunch with Barack and remembered the thirty-one-year-old's "earnest self-assurance."

Other people who met Barack during Project VOTE!, including Kennedy-King community college math instructor Ron Davis, a black 1974 UC graduate who had played a significant role in Harold Washington's 1987 reelection campaign, carried none of the national cachet of a major Democratic donor like Saltzman. But as grassroots organizers like DCP's Michael Evans realized, Project VOTE!'s noteworthy success gave Barack "some significant political notoriety" among a wide range of Chicago political players. A journalist for *Chicago Magazine* asked Barack whether he would run for office sometime in the future. "Who knows?" he replied. "But probably not immediately," though he hoped to have *Journeys* finished by January. "Was that a sufficiently politic 'maybe'? My sincere answer is, I'll run if I feel I can accomplish more that way than agitating from the outside. I don't know if that's true right now. Let's wait and see what happens in 1993. If the politicians in place now at city and state levels respond to African-American voters' needs, we'll gladly work with and support them. If they don't, we'll work to replace them. That's the message I want Project VOTE! to have sent."

A few years later, Barack acknowledged that Project VOTE! was "really how I started to make a lot of connections with . . . political operatives and elected officials around the city" and how it had "immersed me . . . in the mechanics of the political process" for the first time. "It was a useful education for me in learning how politics worked." Earlier, "when I was organizing, I was always pretty suspicious of politicians and politics," but by 1992 Barack had come "to appreciate the need for a more effective political movement within the African American community, the need to mobilize around an agenda and not just individuals," just as he had said three years earlier at that roundtable discussion of organizing. And that *Chicago Magazine* reporter had no doubts as to what Barack's future would entail. Obama was "the political star the Mayor should perhaps be watching for" in a reprise of Harold Washington's multiethnic reform triumph a decade earlier.[10]

Months earlier, Mayor Daley had promoted Valerie Jarrett from deputy chief of staff to commissioner of the newly conjoined departments of Planning and Economic Development. Shortly before the Obamas' wedding, Jarrett had offered Michelle the opportunity to join her as economic development coordinator, and Michelle readily accepted. She would be "the new point person responsible for monitoring the city's major business expansion and retention effort," the *Sun-Times* reported during the Obamas' Northern California honeymoon. When they returned, Michelle took up her new post, but for Barack their return home immediately presented him with truly disastrous news: on October 20, Simon & Schuster had canceled his book contract. Rather than the pleasant prospect of receiving the additional $85,000 of his advance once his manuscript was complete and then published, Barack now *owed* Simon & Schuster's Poseidon Books the initial $40,000 it had paid him two years earlier when he had signed the contract. Agent Jane Dystel reassured him that she almost certainly could resell the proposal, although not likely for the grand $125,000 figure Poseidon had offered in 1990. Barack now faced the grim fact that he needed to finish "Journeys in Black and White" if he hoped to recoup from a horrific financial disaster.

Prior to the end of Project VOTE!'s registration drive, Barack had sent a good chunk of manuscript to Poseidon's Elaine Pfefferblit, conscious that he ought to turn in as much as he could before too many months elapsed after his June 15 deadline. He also had made firm plans to turn his attention to the manuscript full-time once he and Michelle returned from their honeymoon. "He wanted to get away and write," Jean Rudd remembered, and Jean and her husband Lionel Bolin had a friendly neighbor out in Beverly Shores who went south each winter. "I told Barack that he could use her house," but Project VOTE! fund-raising chair John Schmidt and his brother shared an even more distant vacation house on Lake Geneva, in southern Wisconsin. "Barack had decided that he wanted an isolated place where he could go by himself and finish the book," Schmidt recounted. "He had the idea that he would get married, he'd come back from his honeymoon, and then he would go by himself up to Lake Geneva." Indeed, "my brother and I drove him up to Lake Geneva prior to the wedding to show him the house, and we gave him the key."

Barack's plan was to head up there for as long it took him to finish, and then he would begin work at Davis Miner. But the dire news from New York threw him for a loop. Most but not all of the fault for this turn of events was Barack's. What he tardily had submitted to Elaine Pfefferblit represented only about half of the promised manuscript, and the combination of personal storytelling and civil rights policy discussions left it bloated in some parts and dull in others. Both Elaine and her assistant Laura Demanski read it, and agreed that "it wasn't a publishable manuscript," Ann Patty was told. "It sort of read like it was phoned in."

Normally an incomplete manuscript that is three or four months late would not prove fatal, but Barack's failure to execute what his earlier proposal had promised came at a time when Poseidon's parent, Simon & Schuster, was actively seeking to reduce the number of titles published each year by a good 15 percent. A few months earlier a *New York Times* profile of Patty reported that "her bosses have reassured her that Poseidon is safe," yet not long after Barack's contract was canceled, Elaine Pfefferblit's job was eliminated, and a month later Patty left as well. Poseidon was no more.

Barack now had to rethink just what *Journeys* would be, and he turned to Rob Fisher for critical input. The two friends had continued to talk occasionally about the long-discussed policy book they had begun coauthoring back in Cambridge, and months earlier Rob had read the chapters of *Journeys* that Barack planned to send to Poseidon. Rob felt strongly that the personal parts read far better than the policy ones. "The best story here is about you. Make it about you," Rob argued. "Those are the best chapters, those are the chapters that sing." Rob realized that the book had become "a real struggle" for Barack, and "we talked about it endlessly, and I gave him a tremendous amount of editorial notes," both "detailed line-by-line edits and big-picture edits."

Rob's law firm colleague Tom Reed, who had known Barack at Sidley, and Rob's girlfriend Lisa both knew that Rob was "looking at drafts," and Lisa heard enough about the process to realize that "it was a bunch of different books, and the struggle was to figure out which book it really was." Rob had no doubt what the right answer was—"I edited it, and then gave it to him"—and in the wake of the cancellation, Barack acknowledged that he faced a forebodingly difficult challenge he had to overcome before he could begin work at Davis Miner.

Barack told an apologetic Sandy Newman what had happened while they were wrapping up Project VOTE!, but when he headed north to John Schmidt's home on Lake Geneva, he quickly concluded that the setting actually would not be conducive. "He drove up there by himself, and he lasted one night," Schmidt remembered. Just like Rob, Michelle's boss Valerie Jarrett knew "how much he was struggling" with what the book should be. "I just can't get it down on paper. I'd much rather hang out with Michelle than focus on this," Barack confessed to Jarrett. "He wanted to be monkish—he wanted to be *away,*" Jean Rudd recalled, and "he wanted to get away from Michelle too." From both his years in Indonesia and his friends in Hawaii, one perfect place beckoned, a place where his mother Ann had said she would be headed soon after attending the Chicago wedding: Bali.

Ann had been through there several times in the previous two years, as one of her closest friends, Bron Solyom, and her husband Garrett had been living there on Sanur Beach, next door to the Tandjung Sari Hotel, since 1991. Barack's old choom gang friend Kenji Salz was there too, in Denpasar, and "Bobby Titcomb went to Bali all the time," another Punahou friend explained, and shipped back furniture and whatever else to Honolulu. A few years later, Barack immediately said "Bali" when asked his favorite vacation place, and one late 1992 day Ann visited Bron and Garrett Solyom and told them that Barack "wanted to find a spot where he can just sit down and finish this book he's writing." So "we went up and down Sanur," Bron recounted, between the Tandjung Sari to the south and the Bali Beach Hotel to the north, "and we looked at a few places and Ann found a place which she felt would be fine, not very far away from us, and so she rented it—I think it was to be for, say, two months."

Barack mentioned his decision to several friends before informing his new wife. "What do you think Michelle's going to say when I tell her I've got to go?" Valerie Jarrett recalled Barack asking her. Allison Davis had a similar memory: "I can't finish this thing here. There are too many interruptions. I'm just going to take off so I can finish this." Allison asked where he was going, and when Barack answered, he enthusiastically congratulated him—"My man!"—before asking, "What's Michelle going to do?" Allison also wondered what Michelle would think about her husband choosing a locale where there would be "all these Balinese women floating around." Barack responded, "She's going to be angry. She's going to be really pissed," but when Barack finally told her what his mother had arranged, Michelle's reaction was surprise—"Didn't we just get married?"—but not anger. The book had become "the bane of their existence," Michelle's friends knew, and "he needed to go and get it done so that we could move on with our lives," Michelle later explained.

Their closest family members agreed. "He felt it important to go away and just immerse himself in the writing," Janis Robinson recalled. "He really could not do it here," and "he went away so that he could really just focus on the book." Years later Barack remarked, perhaps with Bali in mind, that "sometimes, particularly when we were early in our marriage, I wasn't always thinking about the fact that my free-spirited ways might be having an impact on the person I'm with."

So rather than spend all of December, January, and February braving winter in Chicago, or farther north on Lake Geneva, Barack instead flew westward across the Pacific. "Barack came, and we made some arrangements for him to bring his laundry—we had one household staff member who took care of everything," Bron Solyom explained. "He'd bring the laundry," and "so we'd see him on and off" when Barack stopped "for a drink and a chat." But now Barack's discipline was such that "we could never persuade him to sit and have a meal," Bron recalled. "He was just totally thinking about what he was doing, he was totally focused, and I think that's really all he did, and he was done with it faster than he thought."

Rather than the full two months, he was done within "five or six weeks," she and Garrett believed. "He had a pile of yellow notepads the one time we stopped by his house with Ann," who was in Bali before moving to New York in late January 1993 to begin a new job at Women's World Banking. Carrying his notepads, Barack also flew back to the U.S., and he stopped for several days in San Francisco before returning to Chicago. Barack had not seen his younger brother Mark in almost five years, not since they had first met in Nairobi in 1988, but Mark now lived in San Francisco, and he and Barack met for lunch one day. "He seemed humbler, less arrogant this time than he was back in Kenya," Mark thought, "a different, warmer" person than he had been then.

Mark later realized that Barack's sudden interest had no doubt been stimulated by Barack's pondering over how to address the full story of his far-flung Kenyan family in the final part of his book manuscript. But now Barack was on the road to completion, and when late that winter he arrived back in Chicago, Michelle and his in-laws were happy to learn that the odd venture had proven entirely successful. "Being away was very beneficial . . . it definitely helped," Janis Robinson remembered.[11]

Back in Chicago, Barack faced a trio of responsibilities: he could finally begin full-time work at Davis, Miner, Barnhill & Galland, he would continue to work on his book manuscript so that he would have something far better than what he had sent Poseidon for Jane Dystel to show to possible new publishers, and at the end of March, he would start teaching at the University of Chicago Law School (UCLS). As Davis Miner's newest and most junior member, Barack inherited a narrow office on the second floor of the 14 West Erie Street town house. With Allison Davis focused on housing development work, third-rank partner Chuck Barnhill based in Madison, Wisconsin, and George Galland specializing in health law issues, Judd Miner and fellow associate Jeff Cummings, the firm's other African American attorney, would be the colleagues with whom Barack would work the most.

Judd was litigating two major city council redistricting challenges, *Barnett v. City of Chicago* and a pair of cases in St. Louis. All three had highly complicated procedural postures stemming from the involvement of multiple attorneys and the filing of successive complaints, and both cities had petitioned the respective federal courts to dismiss the cases. As of March 1993, Judd was trying to stave off initial defeats in both cities, and Barack immediately was thrown into the thick of two difficult, complex challenges. Virtually all of his time was devoted to researching and drafting arguments for Judd's various motions and responses, and Judd later said that "Barack's work was enormously thoughtful and always well done."

With another associate, Mark Kende, leaving in April to become a law professor, Barack also inherited a federal district court case in which the firm was representing Ahmad Baravati, a stockbroker who had won a National Association of Securities Dealers arbitration award of $180,000 against the firm that previously had employed him and now was contesting that ruling. Additional filings were necessary because the firm, Josephthal, Lyon & Ross, was challenging a standard industry practice.

Barack also continued to devote as much time as he could to trimming and polishing "Journeys in Black and White." Allison Davis, whose spacious office adjoined Barack's narrow quarters, often saw him with his feet up on the desk and a keyboard in his lap, typing away on some part of the manuscript. George Galland, the most businesslike of the Chicago partners, sometimes expressed concern about Barack's priorities, but Barack's easy manner—"very self-assured, very directed, very bright, good sense of humor," Davis said—made for an easy transition to his primary new job. In addition, everyone at the firm seemed to realize that practicing law would occupy Barack for only so long. As young partner Laura Tilly put it, "from the first day he walked in, it was clear that this was not the end of his career." Everyone knew "this guy is not going to be here drafting loan documents in five years."[12]

UCLS's unusual eight-week quarters meant that Barack's Current Issues in Racism and the Law would meet once a week for two hours from late March to late May. Barack had been an occasional presence at the law school throughout the fall and winter, and Douglas Baird, the colleague who knew him best, recommended Barack to Jesse Ruiz, a first-year student who had grown up in Roseland and whose father had worked at Wisconsin Steel. Ruiz introduced himself to Barack, who mentioned his own Roseland background, and Ruiz realized that three years earlier he had read the *Tribune* story about Barack's *Law Review* presidency while he was considering law school. As word of the new instructor spread among UCLS's minority students, "we were excited about having a black professor," 2L Gabriel Gore recalled. That was unsurprising given a setting in which there were so "very few" black faces that "law students of color were questioned about whether they were in fact law students," 2L Jeanne Gills explained. A half-dozen African Americans were among the twelve or so law students who were joined by several social science graduate students who signed up for Barack's class.

Barack structured the course in two parts. For the first four weeks, he assigned a heavy load—some 550 pages total—of selected readings, what he called "a basic primer" surveying the role of race in American law from the colonial period and slavery to Reconstruction and then the twentieth century. Barack had retained a copy of Randall Kennedy's Harvard Law School syllabus for the Race, Racism, and American Law course he had briefly attended, and essays by well-known historical figures—Booker T. Washington, W. E. B. Du Bois, and Marcus Garvey—were followed by *Brown v. Board of Education* (1954), Martin Luther King Jr.'s "Letter from a Birmingham Jail" (1963), and a trio of well-known Supreme Court decisions from the 1970s: *San Antonio v. Rodriguez* (1973), *Washington v. Davis* (1976), and *Village of Arlington Heights v. Metropolitan Housing Development Corporation*. The readings ended with brief excerpts from contemporary race commentators like Derrick Bell and Shelby Steele, and most of the packet was made up of selections that Barack had previously read and annotated. A cover note said, "You'll see that much of the material has been marked up," but "my wife tells me that she wouldn't have minded getting the professor's notations on her reading material when she was in law school."

The final four weeks featured topical presentations by three- or four-student groups on a current issue of their choice. Barack's syllabus offered a dozen or so possibilities, such as racial reparations. "Do such proposals have any realistic chance of working their way through the political system?" he asked. University hate speech regulations was another. "Are such codes a reasonable measure to protect minorities from harassment, or is the cure worse than the disease?" A third was affirmative action and "the meaning of merit. . . . Do minorities gain or lose when fixed notions of merit give way to more flexible standards for allocating goods and privileges?" The presentations should "draw out the full spectrum of views on the issue you're dealing with" and do so with "rigor and specificity." Each student also would submit a paper of at least twelve pages, and Barack requested "a focused, tightly-crafted argument, and analytic rigor in working through the legal or policy problems raised by your topic." He also expected "a thorough examination of the diversity of opinion that exists on the issue or theme" and "a willingness . . . to take a stand and offer concrete proposals or approaches." The presentation and the paper each would count for 40 percent of a student's grade, with a fifteen-minute quiz on the last day of class constituting the other 20 percent. It would require identifying five or six quotations "whose sources will be obvious to anyone who has done the reading" and absorbed the presentations.

Julie Fernandes, a 2L, remembered the class as "a professor-led conversation that was very inclusive and very exploratory, very intellectual in a really good way." The readings were instructive for students who "weren't aware of the foundational nature of the race question" in American law. Brett Hart, another 2L, found it "much more conversational" than other courses and 2L Jeanne Gills remembered "very constructive conversations," particularly about race. Gabe Gore remembered that the social science grad students "would make comments that would make all of us law students kind of laugh," but Barack explored them and seemed to "find them relevant and interesting." Julie Fernandes thought it was "a great class" in part because Barack "didn't allow me to rest on my assumptions about the world," and 2L Julia Bronson said Barack was "very easygoing" with a "confident demeanor" yet "friendly and pleasant." Elaine Horn, another 2L, remembered that her paper argued against adding a new category for multiracial individuals to the U.S. Census, and only two years later did she learn that Barack was biracial.[13]

By April 1993 Michelle Obama had worked for the city of Chicago for eighteen months, but her shift the previous fall to Valerie Jarrett's Planning and Development department had failed to satisfy her need for truly fulfilling work. Leaving Sidley for the public sector "still wasn't enough, because city government is like a corporation in many ways," Michelle later explained. Like Barack, "we both wanted to effect the community on a larger scale than either of us could do individually," and "we wanted to do it outside of big corporations."

Barack's service on the national board of the nascent Public Allies network had entailed little in the sixteen months since the initial late 1991 conference in Wisconsin, but by early 1993, in part thanks to grants from the Woods Fund and the Joyce and MacArthur Foundations, Public Allies was about to supplement its first chapter in Washington, D.C., with a second one in Chicago that would debut in September. Field director Jason Scott, Public Allies' third national staffer after cofounders Katrina Browne and Vanessa Kirsch, "started going out to Chicago" by early fall 1992, relying primarily on John McKnight and his Northwestern colleague Jody Kretzmann, plus Jacky Grimshaw, to tell him whom to see.

Barack was high on the list, and Jason met him for coffee in a dreary basement cafeteria near Project VOTE!'s South Michigan Avenue office. Barack was "pretty impressive," but Public Allies' band of mid-twenties enthusiasts was having a difficult time selling their training and mentorship model to leaders whose organizations would host the young recruits—"allies"—who were eager to begin public service careers.

In February, Public Allies sublet a small office at 332 South Michigan—the same building from which Barack had run Project VOTE!—and lined up Tim Webb, a University of Chicago graduate student, to head the incipient Chicago project. Scores of eighteen-to-twenty-somethings eager to become "allies" were easily recruited, in part because the several dozen who eventually would be chosen would make $14,000 or more during their year of service, with the hope being that the host organizations would so love the highly motivated, well-mentored young people that they would add them to their permanent staff after the initial sponsored year ended. A kickoff event called "Tomorrow's Leaders Today" at which those young people would be introduced to potential organizational sponsors was scheduled for April 20. On the last day of March, the *Chicago Tribune* ran a feature story heralding the new effort. April's *Chicago Magazine* ran a brief profile of Webb, but behind the scenes, Barack told Jason Scott and Vanessa Kirsch that they should try to recruit an older, far-better-credentialed Chicago director than the twenty-four-year-old Webb, namely Barack's wife Michelle. So "Vanessa and I went to see Michelle" with that idea, and "she more or less threw us out of her office," Scott recalled. "I have this great job. Why would I leave? I can't believe Barack put you up to this."

At home on South Euclid, however, the newlyweds renewed their conversation about wanting to "affect the community on a larger scale." Barack now had a respectable monthly paycheck from Davis Miner, plus extra income from his UCLS teaching. Michelle was earning almost $67,000 at her city job; would Public Allies' foundation support allow her to avoid too great a reduction if she followed through on Barack's gambit? Jason Scott was the second if not the first to learn what she had decided. Michelle "literally called back two weeks later and took the job," which was doubly surprising because "we hadn't actually offered her the job," Jason recalled. "All I remember is like 'I'll take the job!'" The top trio's astonishment turned quickly to joy because "we were just so honored and excited to be able to hire her," Katrina Browne explained.

Things came together quickly thanks to support from Echoing Green, a nonprofit sponsor of social entrepreneurs that already was assisting Public Allies. Echoing Green agreed to cover Michelle's student loan payments, and the Chicago participants celebrated because "getting her was quite a coup." The five-year age gap between the twenty-nine-year-old Michelle and the recent college graduates seemed as large as Barack's age difference had at Harvard, but with Michelle on board, the long-scheduled April 20 kickoff event took on new significance and quickly acquired an all-star roster of speakers and attendees. Barack served as master of ceremonies, and keynote remarks were offered by Father Mike Pfleger and young Jesse Jackson Jr. Mayor Richard Daley and Illinois governor Jim Edgar were in attendance, and Barack introduced Daley's commissioner of planning and development, Valerie Jarrett, who then introduced Public Allies' new Chicago director. Only when Michelle Obama stepped forward did young participants like Bethann Hester discover who the new boss would be. Everyone was ecstatic to have her, and as national finance director Julian Posada explained, for all of Public Allies, "the bar when she walked in was raised, from day one."[14]

By late spring, thanks to both his own improved discipline and agent Jane Dystel's efforts, Barack had a brand-new lifeline for "Journeys in Black and White." The manuscript was still not complete, but all the work Barack had done in Bali and upon his return gave Dystel renewed confidence that a new publishing house would find the manuscript highly appealing. She telephoned Henry Ferris, a young editor at Times Books who earlier had worked at Simon & Schuster and then Houghton Mifflin. Dystel explained what had happened at Poseidon and raved about Barack's background and his now highly personal story. Ferris agreed to read it, and when he did, he was struck by Barack's "incredible gift for language and storytelling." The manuscript was still "quite raw" and much too long, even though Barack still had not completed the final section recounting his 1988 journey to meet his Kenyan relatives.

Thanks in significant part to Rob Fisher's insistent advocacy, Barack had "radically restructured" the book into a far more personal narrative than he initially had envisioned. "Originally it was going to be very much an academic book, focusing on civil rights law and civil rights policies," Barack explained a few months later. But "as I wrote it, it ended up becoming much more of a personal reflection on what it means for me to be an African American, and what my relationship is to this community as well as to Africa." Several years later Barack confessed that he had "sometimes thought 'Was it worth it for me to write the book?'" but Rob's encouragement, plus the $40,000 he owed Simon & Schuster, had helped force him to make the successful expedition to Bali to carry out the restructuring Rob had advocated. "In retrospect I can't imagine anything harder actually than writing a book," Barack later explained, "because it's very hard to feel confident consistently that what you have to say about your life is interesting in any way whatsoever."

But Henry Ferris's reaction made that self-doubt moot, and when Henry shared the manuscript with his immediate boss, Times Books publisher Peter Osnos, his response was similar. "I liked it, Henry liked it," and they told Dystel they would offer a $40,000 advance, but they wanted to meet Barack in person before issuing a contract. Sometime late that spring, Barack, and apparently Michelle, flew to New York so that Barack could introduce himself to Ferris and Osnos at Times Books' offices on East 50th Street. Their conversation was "spectacularly routine," Osnos later recounted, and "led fairly rapidly to the acquisition of the book, whereupon it moved into Henry's hands" to edit. By early summer Ferris was immersed in recommending condensation after condensation to Barack so as to shrink and tighten the oversized manuscript, and a formal contract was executed. Times Books would pay that $40,000 not to Barack, but directly to Simon & Schuster, thereby removing a huge load from the finances of a young couple who still faced substantial Harvard Law School loan debts. There had been no reason for Michelle to accompany Barack to Times Books' offices, but a most unlikely witness clearly remembered seeing them together on that visit.

"It was a sunny day," somewhere "in the 50s" on Manhattan's East Side. "I saw him walking down the street, Madison Avenue, and he was with a young African American woman . . . and they had shopping bags, and I was walking towards them," Alex McNear recalled, "and all of a sudden they crossed the street." Presumably Alex was a beautiful blond part of his past that Barack did not want to revisit, at least not with Michelle in tow, and Alex always remembered the sudden avoidance. "I can almost picture it."[15]

With their modest finances now in relatively good order, thanks primarily to Barack's $70,000 salary at Davis Miner, the young couple began to look for a home of their own as they approached the two-year anniversary of their co-occupancy of Marian Robinson's modest house in South Shore. They focused on Hyde Park, and by midsummer, they were ready to close on a spacious, 2,200-square-foot, four-bedroom first-floor condominium in handsome East View Park. Built in the 1920s as a rental development and converted to condominiums in 1976, East View's eleven three-story brick and stone buildings each had six flats of side-by-side floor-through units that faced a handsome private park with convenient, resident-only roadway parking. From the entrance of the Obamas' unit, a living room with a beautiful green-tiled fireplace plus an adjoining front sun parlor preceded three private bedrooms and a pair of bathrooms. A spacious dining room led into the kitchen, off of which were a small maid's room, a half bath with laundry facilities, and a small outdoor rear porch.

The price was $277,500, and with a 40 percent down payment—$111,000—the Obamas obtained a good mortgage rate. Barack later said that his grandmother Madelyn "helped a little bit on the down payment," which was likely quite an understatement: neither Barack nor Michelle had any significant savings, and a retired, now-widowed bank vice president who for the last quarter century had lived in a modest apartment may well have provided the *entire* $111,000. Barack later acknowledged that "our combined monthly student loan" payments were "more than our mortgage," and although Michelle had a nice Saab, which Barack borrowed whenever he could, Barack was still driving the rusty, off-yellow Toyota Tercel that had carried him to Harvard in 1988 and offered passengers a clear view of the pavement below.

Before moving in, Barack and Michelle wanted to modernize some features, particularly the kitchen, and Barack's uncle Charles Payne, whose own Hyde Park home remodeling they much admired, gave them the names of his contractors. Some months later, when a prospective purchaser of another East View unit asked his realtor about the prospects of updating the apartment, the Obamas "graciously and politely" gave the agent and her client a tour of their home. "Their apartment was shown to us as an example of how you can take a dump and make a gorgeous pearl out of it. It was breathtakingly beautiful," he recalled. "Most striking" of all was the Obamas' kitchen, which was "gorgeously renovated."

The move to Hyde Park, where all manner of retail stores, casual restaurants, lakefront parkland, basketball courts, and even the law school were within comfortable walking distance, represented a significant quality-of-life upgrade from South Shore, whose principal shopping street, East 71st, was littered with countless tattered storefronts. The barbershop Barack had patronized since he first arrived in Hyde Park was just a few blocks away, and a wide variety of local men, from grad students to old Harvard acquaintances to UC security guards, remember Barack as a regular basketball player at the UC's gym, at a lakefront court a few blocks south, and at a small church across from Jesse Jackson Sr.'s Operation PUSH headquarters.

Soon after moving to East View, Barack also became a customer at the local movie rental store, Video Good Guys on East 55th Street. Owner Garland Cox noted that the new patron had the same first name as a friendly regular, to whom he mentioned the coincidence later that day. "We just had another Barack sign up today," Cox told Barack Echols, whose mother, Sandra Hansen, twenty-six years earlier had named her son after the impressive friend who had helped her flee Nairobi and return home to Illinois. Echols had spent four years in the U.S. Army before resuming his undergraduate studies at the University of Chicago, and he had read about the younger Obama when Barack was elected *Law Review* president.

"Is it Barack Obama?" Echols asked Cox, who confirmed his guess. "I'm actually named after his father," Echols explained, mentioning his mother's time in Kenya. Echols lived right across the street and told Cox to call him the next time Obama came in. A week or so later, Cox telephoned, and Echols ran across to introduce himself. "My name is Barack too," he told an astonished Obama, and pulled out his driver's license as proof. "I heard all these stories about 'big Barack' growing up," he told Obama. "Wow, that's really interesting. What an incredible story," Obama replied. "'We should talk again,' which of course we never did," Echols later recounted, for in truth Barack was no more interested in revisiting his father's years in Kenya than he was in running into Alex McNear.

The move to East View put Barack and Michelle just a block from where her brother Craig and his wife Janis lived. Given their proximity, Janis recalled that Michelle "would come over all the time," and Janis and Craig regularly visited Barack and Michelle's "huge" apartment. Craig remembered that at one family gathering, Barack said that some day he might run for public office, and when Craig mentioned alderman, Barack demurred. "He said no—at some point he'd like to run for the U.S. Senate. And then he said, 'Possibly even run for president at some point.'" Craig was puzzled. "President? President of what?" he asked. Barack made his intent clear, but Craig balked. "Don't say that too loud. Someone might hear you and think you were nuts."

Craig, as Michelle's only sibling, and Maya Soetoro, as her brother's closest relative, each believed that Barack and Michelle were very good for each other. Maya had just married a fellow University of Hawaii undergraduate, Gary Forth, a New Zealand–born rugby player four years her senior, and Barack later starkly regretted something he said to her about it. "The worst piece of advice that I have given to anybody was to my younger sister," he replied to a questioner. "I once told her that she should—well no, actually, I probably shouldn't tell that story . . . Let me pull back on that . . . I think it's too personal." By the next spring, Gary would file for an uncontested divorce, and the short-lived union formally dissolved some months later.

Maya viewed Michelle as "fiercely pragmatic," and thus "much more like our grandmother" than their mother Ann. Craig believed it was "readily apparent" that Barack "could stand up to" Michelle's outspoken demeanor, but he thought the real key to Barack and Michelle's relationship was that Barack "knew how to manage my sister's personality," and Michelle's startling move to Public Allies clearly reflected the impact of her husband's counsel.[16]

Michelle now faced the challenge of running a start-up organization whose first class of twenty-seven Public Allies would be chosen and introduced to their organizational sponsors by early September. On many Sundays Barack and Michelle attended the 11:00 A.M. service at Trinity United Church of Christ, but neither of them became active in any of Trinity's extensive range of volunteer ministries. An *Ebony* magazine survey of "The 15 Greatest Black Preachers" ranked fifty-two-year-old Jeremiah Wright as number two in the nation, and Trinity's membership now approached eight thousand.

Barack also disdained any serious involvement in a new political venture that organizing veterans Keith Kelleher of SEIU Local 880 and Madeline Talbott of ACORN were trying to launch in Chicago. The "New Party" was the brainchild of two progressive electoral strategists, Danny Cantor and Joel Rogers, the latter of whom was married to Sarah Siskind, a Madison-based partner in the Davis Miner law firm. The founders initially envisioned the party as a left-wing "fusion" presence in New York State, where candidates could run for office on multiple parties' ballot lines. They hoped to build an entity whose independent presence would encourage Democratic politicians to move leftward so as to secure supplementary support. In Illinois they would also launch a sibling group, "Progressive Chicago," as a "support organization to . . . bring people into the New Party."

Once again, Jacky Grimshaw told Kelleher to be sure to contact Barack, and at a late-July meeting, Keith explained both vehicles. They wanted to recruit signers for an initial introductory letter that would announce the creation of Progressive Chicago, and Kelleher noted that Barack is "more than happy to be involved" but was busy with Davis Miner's work on the Chicago City Council redistricting case. Barack thought Chicago politicians would see value in the fusion concept—"some of those guys would love to get support outside [their] normal constituency"—but "will be cautious if it offends regular Democrats" and you "don't want to have to force people into [the] New Party." Barack recommended to Keith a number of people whom he had met during Project VOTE!—Sam Burrell and several other black aldermen, Carol Harwell, South Side political activist Ron Davis, and South Side state senator Alice Palmer. Barack wanted to see the initial Progressive Chicago letter his name would go on, and he said he "could make a mtg. but can't put too much into it."

Some weeks later the letter, featuring former Project VOTE! chairman Joe Gardner's name atop a modest list of signers that also included Talbott, Kelleher, Harwell, and Davis, explained that the new group would "unite progressive activists and organizations for progressive, grassroots electoral activity in local elections" in "a renewal of the old Harold Washington coalition." But national cofounder Joel Rogers, writing in the socialist weekly *In These Times,* emphasized far grander plans. Since "both major parties are dominated by big business interests" and offered only "permanent candidates" who are "literally addicted to the private money that regularly returns them to office," the New Party hoped "one day to be the new majority party of the United States." Repeated efforts by Kelleher, Talbott, and Rogers to interest Barack in the New Party fell on entirely deaf ears.[17]

Before summer's end, Barack agreed to again teach Current Issues in Racism and the Law in the spring of 1994, and he updated his UCLS biographical sketch to indicate that his forthcoming book would be published by Random House—Times Books' parent company—rather than Poseidon's Simon & Schuster. Barack turned down an offer from Northwestern University law dean Robert Bennett, who five years earlier had been unable to convince him to accept that full-tuition scholarship, to pursue a full-time faculty post there, although Barack did "jokingly tell me that he had made a bad mistake" in not accepting the previous offer. On the first Saturday in September, Barack and Michelle drove out to Oak Park for his organizing colleague David Kindler's wedding, and the organizers all ended up in the balcony of the Frank Lloyd Wright–designed Unity Temple.

"Very shortly after the wedding began, two doves began to procreate on the ledge of the rafters," Kevin Jokisch recalled. "The doves were very active, feathers were flying," and they were "very visible and close to us." Barack "was one of the first to notice and point out the doves," and stage-whispered that "this marriage is *truly* blessed." Those around him tried to suppress their laughter as Michelle shooshed her husband and "told all of us to be quiet" and watch the proceedings, not the copulating birds.

Several weeks later *Crain's Chicago Business* published a prototypical "40 Under Forty" feature, and cited Barack for having "galvanized Chicago's political community as no seasoned politician had before" with Project VOTE!'s 1992 success. In a comment starkly illuminating his view of law practice, Barack told *Crain's*, "it is an accomplishment to make a difference," while "it's no accomplishment to be a partner in a law firm." When another interviewer asked about Barack's political plans, his answer was similarly revealing:

> *My general view about politics and running for office is that if you end up being fortunate enough to have the opportunity to serve, it's because you've got a track record of service in the community, and I think right now I'm still building up that track record, and if it, a point comes where I think that I can do more good in a political office than I can doing the things I'm doing now, then I might think about it, but that time is certainly in the future.*

Late that year, Barack accepted invitations to join two boards. The Center for Neighborhood Technology (CNT) had an eclectic interest in grassroots development initiatives and was now home base for both Jacky Grimshaw and the late Al Raby's closest friend Steve Perkins. Barack attended a few meetings, but never connected substantively with CNT's work. Barack's good friend Jean Rudd initiated a similar invite from the Woods Fund, which had just added a third professional staff member, Kaye Wilson, to direct a new grant program focusing on welfare-to-work policies that impacted poor families. "You don't decide to go on a foundation board. You get asked," Jean emphasized, noting how "it was quite an honor for someone of that age to be invited on."

On one snowy Friday that winter, Barack did a traditional Alinsky-style power analysis training for Michelle's first-year group of twenty-seven Public Allies. "He talked about what power meant," Michelle's assistant Julie Alfred remembered, "money or numbers." Michelle had spent countless hours obtaining organizational placements for her trainees, plus moving the group to better quarters and cadging secondhand furniture from McKinsey & Co.'s Chicago office thanks to old Harvard acquaintance Bernard Loyd, for whose recent home purchase Barack had done the legal work. Along with her top aide Julian Posada, Michelle also worked to assemble a local advisory board of interested civic figures. Foundation director Sunny Fischer was an early recruit, as were two executives from the investment firm William Blair & Co., which provided financial assistance as well. Posada also approached attorney Jan Anne Dubin, who was surprised when Michelle Obama turned out to be the former summer associate who had worked for her eight years earlier at Chadwell & Kayser as Michelle Robinson.

With her young staffers and trainees, Michelle's "demeanor was always tough love," Posada explained. She was "a great boss" but also "pretty tough," Julie Alfred agreed. Bethann Hester, a 1993–94 Ally who then went on staff, recalled how at 10:00 A.M. they often saw Barack reading the sports section at a Starbucks equidistant from both Public Allies' new office and Davis Miner. When writing legal briefs, Barack explained, "you have to relax your brain." Michelle regularly had her staffers over to East View, where Barack was working on his book, and Julie Alfred remembered being "really taken aback that he had a pack of Marlboro Reds there." Barack's heavy smoking—always heavier than he was willing to admit—was the subject of recurring objections from Michelle. "They were constantly making deals" about it, Bethann recalled. "He could smoke until he finished" the book, and "he had to go outside," on the wooden porch "in the back of the house."

Michelle's vices were limited primarily to television series like MTV's *The Real World,* which she told Public Allies colleagues counted as professional development work. When Barack's old friend Phil Boerner wrote in early March 1994 to announce the birth of his and his wife Karen's first child, Barack replied with an informative note that confirmed the Allies' impressions. "I must say that I'm thinking about a family of my own. In fact, I'm more than half-way there, since I got married last year," he told Phil while enclosing a picture of Michelle. Barack mentioned his law practice and his course at UCLS, and said, "I'm finishing up a book that will hopefully be published next year." Their old freshman roommate Imad Husein "seems to be doing fine in Pakistan," and Barack had lost touch with old friend Paul Carpenter. "I still have the nicotine habit, but Michelle has made me promise to quit as soon as the book is done. Otherwise, no babies!"[18]

At UCLS, Barack's second iteration of Current Issues in Racism and the Law began on Tuesday afternoon, March 29. Given his heavy workload at Davis Miner, Barack otherwise spent little time in his law school office, now room 403, but his colleague David Strauss recalled Barack speaking about race at an early 1994 faculty workshop where someone asked him about journalist Ellis Cose's new book *The Rage of a Privileged Class,* which addressed the anger felt by many black professionals over recurring experiences with discrimination. Barack acknowledged that Cose "has got a point," although such books are "all a little whiny." Strauss was struck by how "undogmatic" Barack was about race. "This guy's playing it straight—he's not striking a pose."

Barack's students from his first time teaching and now the second had a similar impression of him. One 2L who was among the fourteen students who enrolled asked a female classmate who had the course a year earlier what she thought of it. Instead she described Barack. "He's so beautiful. He looks like *food*. It looks like someone *made* him." Once she "got beyond how good-looking he was, she also described what a terrific teacher he was," Adam Gross recalled.

Barack's syllabus and readings were almost entirely unchanged from a year earlier—only a brief recent essay by Cornel West was added at the end. Gross found it to be a "really interesting and provocative reading list," and Barack "would frequently show up wearing a suit," coming straight from Davis Miner to their basement-level seminar room carrying a "battered satchel." Susan Epstein, a 2L, realized he was "very bright" and an "excellent teacher," and Adam Gross thought he was "this luminous guy" whom classmates viewed as "sort of a rock star." One day that image took a serious hit when "his bag of stuff got sort of knocked over and a pack of cigarettes came out and people were like 'Ugh!' shattered," Adam remembered.

Gross recalled how Barack's "tie would always be off by the time the class started" and "he was really insistent, in every class and on every subject, that we clearly articulate all of the arguments on all sides." Susan Epstein was struck by how it was "a very diverse class in an otherwise largely white school," a class with "a real wide range of political perspectives." Even with each session featuring what Adam remembered as "really hard discussions about incredibly complicated issues," everyone was always "incredibly respectful" and Barack did "an amazing job" of "making certain we teased out all sides," that "every angle and every nuance" was covered.

In sharp contrast to many Chicago professors, particularly the fiery and intense Richard Epstein, whose Socratic questioning never left any doubt as to what the right answer was, Barack's approach was "incredibly refreshing." Susan Epstein realized that Barack "never really said what his view was," and 2L Rob Mahnke believed Barack "was very guarded about what he was thinking," as if "he was masking his own views." During the second half of the quarter, when the students' group presentations took place, discussions were "very student-directed" and Barack "didn't talk that much," Adam remembered.

When Barack did speak up, he talked about "what policies could be enacted that would make a true difference," Susan recalled. Barack "spent a fair amount of time talking about political rhetoric," asking that "even if this might be a good policy, how would you get it enacted—what pragmatic steps would you need to take in the framing of it?" Yet despite eight weeks of persistent questioning by Barack, "I couldn't tell you at the end of it what his views were." That did not count against Barack when course evaluation forms were distributed at the end. On a ten-point scale, the students' overall evaluation of Barack's teaching performance was 9.1.[19]

At Davis Miner, the December 1993 hiring of another young associate, John Belcaster, a 1992 Yale Law School graduate, meant that Barack no longer was the firm's most junior member. Barack also pressed his best friend Rob Fisher, who had been doing so much yeoman work on Barack's book manuscript, to consider moving to Chicago in order to join the firm. Rob visited and formally interviewed, meeting and impressing everyone and immediately receiving an offer. Rob also was impressed with Barack and Michelle's "beautiful" and "spacious" East View Park apartment, where old friend Ken Mack also stayed during a spring 1994 visit. But Rob finally decided that moving to Chicago would take him too far away from his southern Maryland family.

On Thursday, April 7, the U.S. Seventh Circuit Court of Appeals heard oral argument in *Baravati v. Josephthal, Lyon & Ross,* the arbitration award case Barack had inherited a year earlier. It was Barack's first significant courtroom appearance, and his seventeen minutes of remarks and answers to a three-judge panel headed by prominent chief judge Richard A. Posner demonstrated that Barack was well prepared and fully conversant with the issues presented. He spoke clearly and responsively when Posner and his colleagues asked questions, and at the end Posner commended Barack and his opposing counsel for presenting "very good arguments." Barack's client, Ahmad Baravati, thought he was "a very smart, innovative, skilled, relentless advocate," and less than three months later, the panel unanimously affirmed the district court's award of $180,000 in both actual and punitive damages to Baravati.

A few weeks later, Barack joined Judd Miner as Judd sought to revive the firm's case against the 1991 redistricting of Chicago's city council before another three-judge panel also headed by Posner. But Barack spent almost all of his work hours at Davis Miner, not in Chicago courtrooms. That *Barnett* case was a recurring constant in Barack's workday life, and Judd remembered one occasion when Barack was "really upset" and "unhappy" that one brief was not as good as he had hoped. Barack also "was spending as much, perhaps more" of his time "on the other side of our practice," in the transactional work handled by Allison Davis, Bill Miceli, and Laura Tilly. Davis represented Bishop Arthur M. Brazier's Woodlawn Preservation and Investment Corp., a major development presence in the struggling Woodlawn neighborhood just south of Hyde Park. Davis also was interested in development prospects in the deeply troubled Near North Cabrini-Green area. On one occasion, Davis remembered, "I asked Barack if he wanted to represent" one or another development entity there, "and he said 'No, I don't want to see poor people thrown out.'"

Barack felt no such qualms about Section 8 bond transactions for the Chicago Metropolitan Housing Development Corporation, loans for the nonprofit Local Initiative Support Corporation (LISC), or work for the Near North Health Service Corporation. Laura Tilly remembered that Barack "carried himself with such confidence and authority" and how he "was just terrific with clients. They just loved him." Some days Barack, Jeff Cummings, Paul Strauss, and John Belcaster ate lunch in the firm's basement conference room, with Barack reading the newspaper. On other days they went around the corner to Thai Star or across the street to PK's Place, a small, family-owned Greek diner.

The four men talked about sports more often than legal issues or politics, but a regular presence in their lives were the residents of the adjoining single-room-occupancy Dana Hotel, at 666 North State Street. One fellow named Tony regularly drank in Davis Miner's small parking lot and enjoyed greeting Barack, but one day after lunch "right in the middle of State Street is this white-haired older fellow who we see every day," John Belcaster recounted, "wearing nothing but a shirt." "We all know him, he's a recognizable face," but "my initial reaction nonetheless is to move away from him," and ditto for Jeff and Paul. Yet Barack "takes this older white fellow by the elbow and walks him to the curb. He's the only one among us to literally extend the hand of human kindness to get this guy out of harm's way."

Mental impairment rather than drink was at issue, and as the men walked back to the law firm, Paul Strauss voiced a rhetorical question. "You know what's really sad about what we've just seen? What's sad is that that guy at one point was a youngster like our young kids who had parents who loved him, and now look what's happened to him." Barack disagreed. "No, Paul. You know what's really sad about this? That guy quite likely never had anybody at home who loved him. That's what's really sad."

Barack's childhood still offered him perspectives different from those of the offspring of nuclear families, but Davis Miner was a group of attorneys whose views of the legal profession were highly atypical, even among self-identified civil rights lawyers. "All of the folks at that firm had a certain ambivalence about the practice of law," John Belcaster explained, and indeed "a certain skepticism about . . . litigation in particular."

Barack's fundamental attitude, as John remembered it, was wholly in keeping with everything Barack and Rob Fisher had written back at Harvard. "The pathway to civil rights and justice in the past generation has been the courthouse. That in many ways has been where the battles have been won and fought," John recalled Barack saying. But "going forward, I don't see the battles being won and fought in the courthouse from this point forward." Instead, "from my vantage, the problems of our times are going to be solved not in the courthouses of America but in the statehouses of America. The problems of our time are going to be addressed by folks organizing, folks petitioning, folks legislating through problems."

"That stuck with me," Belcaster recounted, because "it's somewhat uncanny to hear that in a law firm context." Yet Judd Miner, who had been patiently litigating difficult and lengthy race discrimination and voting rights cases for several decades, was anything but romantic about lawyers' lives. "Practicing law is a pretty shitty job," he later explained. "You spend an enormous amount of time on keeping your head above water. There are just a million skirmishes that are unrelated to the end result. They're really tedious, it's irritating. It's not what anybody wants to do. I think it irritated Barack a lot," because "there was just a lot of this skirmishing that he didn't like. He didn't like just the game of lawyering," especially since "there's a certain combativeness that comes along with practicing law, and that's not Barack's instinct." As Barack similarly told Belcaster, "I joined this firm not to become some scorched-earth litigator," and Barack later bluntly declared that being a trial lawyer "was never attractive to me."[20]

In the late spring of 1994, Barack took off six weeks or so from Davis Miner to finish work on his book. Thanks in large part to the copies of his summer 1988 letters to Sheila Jager, the final portion of the manuscript, recounting his trip to Kenya and the stories of his African family, was now complete. Thanks to Henry Ferris's cutting and Rob Fisher's continued input, the earlier sections were no longer bloated, and the normally self-effacing Rob admitted, "I was deeply involved with helping him sort of shape it" and "had a big influence on" the final manuscript.

One particular issue Barack and Rob debated was how forthcoming the book should be about some of Barack's experiences before he first moved to Chicago. Rob knew that Barack "already had political ambitions" and that "the book was written with that in mind, no question about that," so "we certainly struggled with the question of whether to include the fact that he had used illegal drugs." Rob did not know the full extent of that, particularly from Barack's post-Columbia years in New York, but "I was on the side of being honest about it," and "there was no doubt that we wrestled with that, about what to put in and how much to put in." Rob understood that Barack also had changed people's names and melded some individuals into composite characters, but the manuscript now read more like a book than a series of journal entries from years past.

Michelle was ecstatic that the end of what was now a four-year-long struggle was finally in sight. "That book was very cathartic for him, and it was a hard book to write. It was very hard for him to get all the pieces and make sense of them." Barack flew to New York to deliver his final revised typescript to Henry Ferris in a quick handoff in the elevator lobby outside Times Books' offices. A few weeks later, after signing off on the final line edits, Henry called Barack to explain that in July he would be leaving Times Books, as well as getting married, and that another editor, Ruth Fecych, would shepherd the manuscript toward publication. Ruth read it herself, and remembered "suggesting cutting paragraphs he wanted to keep," but by summer's end the manuscript was in final copyediting and the end of Barack's most difficult challenge in life was finally in sight.[21]

Barack was happy to tell colleagues like Keith Kelleher, who was striving to emulate Project VOTE!'s 1992 success with a 1994 voter drive, that his book was done. For the first time in four years, Barack could now consider other opportunities and involvements without having that burden looming over him. In his new role on the Woods Fund's five-member board, Jean Rudd and Ken Rolling asked Barack to chair a special panel that would review Woods's past decade of grants in support of organizing so as to inform how Woods could best aid "the Future of Organizing." Barack was well acquainted with most of the other participants, who included Mary Gonzales, Jacky Grimshaw, Madeline Talbott, and Sokoni Karanja. That year alone Woods would allocate $105,000 to Madeline's Chicago ACORN and $72,000 to Sokoni's Centers for New Horizons, for "a new African-American leadership development institute," plus another $20,000 to Public Allies Chicago.

The panel assembled for the first of four long discussions on Wednesday, June 29, with Ken Rolling and Barack describing the task ahead. Barack then posed three fundamental questions: "what is the definition of 'community organizing'?," "what is organizing effectiveness?," and how effective is organizing in fostering leadership development? Barack "raised the concern that no one can name a leader in many community organizations, but everyone can name the lead organizer." He pointedly asked, "have we built a cadre of community leaders, as the civil rights movement did?" Barack also questioned why local organizations were not more involved in policy arenas, and asserted that community organizing "has bought into three myths: 1) that self-interest needs to be small, 'rinky-dink stuff' . . . ; 2) that results need to be short-term to sustain interest by members and funders . . . a 'quarterly report' mentality; and 3) that leadership development 'slows us down.'"

Ken Rolling acerbically noted that "the brains of some organizers are not as big as their egos," and Barack agreed with Sokoni and Jacky that organizing tried too hard to fit leaders "into a mold rather than to tap their diverse talents." When Barack asked, "do we have to manufacture conflict?" Sokoni said that he saw "seizing schools as the next real chapter in school reform."

In advance of the second meeting, the panel's secretary, Sandy O'Donnell, sent Ken and Barack some suggested topics, including whether organizing is "becoming strong enough to sustain quality career organizers? (or are they inevitably lost to downtown law firms & private foundations—just joking, you guys)." Barack sent panel members some agenda suggestions, including whether "organizing lacks the tactics to be effective in the policy arena," and Barack focused their second discussion on "community organizing and public policy making." Sandy's notes recorded how "Barack challenged the group to think about a major problem with policy implications—overcrowding in the County Jail—and ask itself whether organizing would say 'there's no constituency (for change), so I won't organize around it.'"

Responses were inchoate, so "Barack pushed further—citing health-care and welfare reform as two more examples of significant policy issues on which the voice of organizing has been silent. Why, he asked, especially when we know—having observed business roundtables & lobbying—how to be effective in the policy arena?" Again discussion faltered, so Barack asked "are there constraints in this (organizing) model that preclude longer term vision development?" and "whether lack of ambition is a problem." Barack then expressly cited how in his own organizing experience, "despite his desire to get involved in the larger arena, 'something was keeping me narrow'" and too locally oriented.

Barack began the group's third session by asking: "is it the role of organizing to change the culture in a community?" Sokoni Karanja and veteran organizer Josh Hoyt said yes, and Barack noted how "this kind of goal—more specifically, having regular theological reflections on their work—was incorporated by community leaders early in his organizing career, but that it slipped off the agenda." Barack acknowledged that "time constraints with leaders are intense," but added that organizing has not "thus far been good at" framing "materialistic issues . . . in ways that build 'internal capital—people having a sense of their importance' in the civic sphere. Material issues can be used to overcome a 'poverty of spirit.'" Barack asked whether issues should be chosen so as to optimize "the building of community spirit," and Madeline Talbott observed that lots of attention was paid to "leadership development, but we don't focus on membership development."

The fourth and final discussion began with Madeline describing how ACORN recruited organizers from within their own neighborhoods, but Barack questioned whether such recruits "have staying power" and "strategy skills." Madeline said "we've professionalized organizing to the point where we can't make it a mass-based movement," but Northwestern's Jody Kretzmann wondered whether Woods and similar foundations could fund entry-level organizing jobs for "graduates of Public Allies." Sokoni said he had "found supervising Public Allies enormously labor-intensive," and Madeline stated that the fundamental problem was how difficult organizing was, not what it paid. The normally quiet Jean Rudd asked whether Woods should fund a "formal recruitment plan" for organizers, and Barack joined Madeline in demurring, saying "good recruiters must be good organizers. But good organizers think they have to be organizing all the time."

Madeline noted how policy success depended upon talented specialists, like Don Moore in the field of school reform, and Barack suggested having organizers from all around Chicago meet monthly over dinner to discuss specific issues. Reflecting that a powerful five-part *Chicago Sun-Times* series had painfully detailed how Roseland, with 21 percent male unemployment, was "losing the war against gang intimidation and recruitment" as violent crime there "skyrocketed," "Barack, Sokoni, and Madeline all thought that success in creating alternative structures to the gangs is highly dependent on moving the jobs issue."

When Sandy O'Donnell wrote the panel's report, one strongly worded "finding" encapsulated Barack's sustained critique: "organizing has *not* been effective in many of the more fundamental scourges facing our low income communities" as economic change "eroded the jobs and wage bases" and "drugs, gangs, guns, crime," etc. increased. "The prevailing organizing model reinforces localized, short term thinking," and Woods resolved that organizing had to "move beyond individual neighborhood successes" and develop "links with broader coalitions." Barack's analysis had carried the day on all points, and the discussions repeatedly reflected strong continuities between the arguments he had articulated while at Harvard and his current political outlook.[22]

In the wider world of Chicago electoral politics, the spring Democratic primary had seen U.S. representative Mel Reynolds defeat state senator Bill Shaw and 17th Ward alderman Allan Streeter with a respectable but not overwhelming 56 percent of the vote. African American state attorney general Roland Burris lost the Democratic gubernatorial nomination to sixty-eight-year-old state comptroller Dawn Clark Netsch, who was now a decided underdog in the general election matchup against incumbent Republican governor Jim Edgar. Exiting Cook County Board president Richard Phelan had finished third in that gubernatorial primary, and in a major upset, African American 8th Ward committeeman and longtime county board member John Stroger won the nomination for the county board presidency, defeating two well-known female contenders.

A *Chicago Sun-Times* headline reported "Last-Minute Cash Netted Stroger Win" over a story detailing how two principal contributors were responsible for Stroger being able to mount a huge TV advertising blitz. Automobile dealer Al Johnson had been a crucial backer of the late Mayor Washington before becoming Mel Reynolds's finance chairman, and now John Stroger's. Johnson had given Stroger more than $50,000 and loaned his campaign another $5,000, but Stroger's largest contributor by far was Syrian-born housing developer Antoin "Tony" Rezko, who had contributed more than $69,000 and loaned Stroger an additional $98,500. Rezko also had emerged as the top contributor to 4th Ward alderman Toni Preckwinkle, who had received over $17,000 from a quintet of Rezko's entities.

City housing officials told the *Hyde Park Herald* that Rezko's Rezmar Corp., which had received more than $12.8 million in city funds to rehab hundreds of apartments, was a "top-notch" developer managing "some of the best-maintained buildings in the city." Preckwinkle said she was "grateful for the work he's done" in her Kenwood neighborhood, and later explained that affordable housing development "in my mind is God's work." That view was shared by Barack's colleagues at Davis Miner, who represented several nonprofit organizations that were partnering with Rezmar, which, as Laura Tilly remembered, "was a well-respected developer of affordable housing."

Looming over everything was the question of whether any truly credible candidate would step forward to challenge Mayor Richard Daley's upcoming spring 1995 reelection bid, and both Judd Miner and former Project VOTE! chairman Joe Gardner had prominent roles in a *Chicago Tribune* story headlined "Anti-Daley Forces Start to Beat Campaign Drums." State senator Alice Palmer, the *Tribune* said, has been "rumored for months to be a possible contender," just as the *Sun-Times* earlier had asserted that Palmer "could have sent Reynolds into early retirement" had she, rather than Shaw and Streeter, challenged the congressman in the Democratic primary.

Barack had met Preckwinkle and Palmer during Project VOTE!, and he had first met Tony Rezko four years earlier, in the summer of 1990, when David Brint had introduced Barack to his business partners. By summer 1994, Rezko was talking up Obama to political friends like Fred Lebed, who had managed Attorney General Roland Burris's gubernatorial campaign before shifting into a similar role for John Stroger's general election race. At one 7:00 A.M. Stroger finance committee meeting, Rezko turned to Lebed and said, "I know someone you should meet. He reminds me a lot of you—a guy named Barack Obama. I want you to meet him—he's got political aspirations." Fred's reaction was what "a really goofy name," but Tony "sets up a meeting," and Lebed joined Obama at PK's Place, the Greek family diner just across from Davis Miner, at 8:30 A.M. on Tuesday, August 9. "Within five minutes, I'm blown away," Fred remembered, "just absolutely blown away."

Barack was interested in learning more about Chicagoland Democratic politics. Law firm colleague Paul Strauss remembered Barack talking about his interest in the congressional seat that Bobby Rush was challenging Charlie Hayes for back in 1992 when they ran into each other in Hyde Park, even before Barack joined Davis Miner, and George Galland thought Barack "was reasonably candid from the beginning that he had political ambitions." Project VOTE! fund-raiser John Schmidt was "pretty sure that in 1994 Barack talked to me about possibly running for the Cook County Board," which "I thought was a terrible idea" because "it wasn't a place that allowed any scope for talent." With a major donor like Tony Rezko talking up Barack to influential Democratic players like Fred Lebed, the purpose of their breakfast meeting "was to get to know each other," and step by step Barack began exploring the possibilities that Chicago politics offered.[23]

Just two days later South Side electoral politics were upended when Congressman Mel Reynolds held a "raucous press conference" at O'Hare Airport to denounce the Republican Cook County state's attorney for investigating allegations by a young woman that Reynolds had had sex with her two years earlier when she was sixteen years old. Reynolds sought to discredit her as a "homosexual lesbian" and an "emotionally disturbed nut case," and distributed copies of a sworn affidavit in which she recanted her story. Reynolds blasted the "racist justice system" for surveilling his telephone calls and told reporters, "this illustrates how . . . if you are an African American male, you are always vulnerable."

Two days later the *Tribune* reported that eighteen-year-old Beverly Heard was not Reynolds's only problem. Federal investigators had discovered that Reynolds had four campaign bank accounts, including one containing as much as $85,000, that he had failed to report to the Federal Election Commission. "Chicago's top business families, from the Pritzkers to the Crowns," the *Tribune* noted, have "poured money into Reynolds' coffers," and so long as the charges proved unfounded, the former Rhodes Scholar "has the potential to be an important figure in Washington," the *Tribune* editorialized.

Then, eight days after Reynolds's preemptive press conference, a Cook County grand jury indicted the congressman on twenty felony counts, ranging from sexual assault to obstruction of justice, involving his relationship with Beverly Heard. *Tribune* reporters expressed surprise at the "sweeping nature" of the charges, saying that "prosecutors have worried privately about the strength of the case and the veracity of" Heard. National coverage noted how Reynolds was "a rising star" who had been "the first black person from Illinois to win a Rhodes scholarship." *Sun-Times* political columnist Steve Neal, who months earlier had praised state senator Alice Palmer as "among the brightest and most capable members of the Illinois General Assembly," soon reported that Palmer was "emerging as the leading candidate" to replace or challenge Reynolds, who would be unopposed on the November general election ballot.

Neal again commended Palmer as "one of the more intelligent, articulate, and productive members" of the state legislature, and the *Tribune*'s John Kass highlighted how the prospect of Reynolds's congressional seat might dissuade Palmer from any further mayoral thoughts, even though she stood alone among black legislators in "never having sought an accommodation with Daley."[24]

Four days after Reynolds's indictment, the federal appeals court handed Judd Miner and his colleagues a major victory by unanimously reinstating the lawsuit challenging Chicago's existing city council districts. That would mean more voting rights work for Barack, but Barack also had an unusual paid speaking appearance coming up in Lincoln, Nebraska, as part of Nebraska Wesleyan's "University Forum." Barack's topic was "Community Revitalization," and he was identified as the author of the forthcoming book *Mixing Blood: Stories of Inheritance.* Barack began informally, acknowledging the importance of football in Nebraska, remarking, "I often don't trust Bill Clinton talking about values," and confessing that his wife enjoyed watching old TV reruns of *The Dick Van Dyke Show* and *Ozzie and Harriet.* But the subject Barack wanted to focus on was values. "It's easy to believe in civil rights and integration and busing if you send your kids to a private school," but "it's harder to believe in these values when you actually test them out." Barack's fundamental point was a challenging one: "it's not easy to live up to your ideals and your values. It requires sacrifice." He argued that point strongly:

> *If you don't live out those values, they don't mean much. And I think part of the cynicism we have about politics right now is a politics of symbolism that talks about values but does not live them out. And we all know that, and that's why people don't listen to politicians, and we don't get involved, because we know that we say one thing, and we do another thing, both personally and in the society. And so I would challenge you, first to think about what your values are, and think about whether you're living them out . . . whether you're contributing to the promotion of these values, or whether you're just giving lip service to them and not living them out.*[25]

On October 7, Barack tardily updated his voter registration to reflect his move to East View Park from South Shore over a year earlier. A week or so later, ACORN formally retained him as its attorney in anticipation that Illinois Republican governor Jim Edgar's administration would fail to implement the new National Voter Registration Act, commonly called the motor voter law, which required states to offer the opportunity to register to vote to all citizens applying for or renewing a driver's license or public assistance and which would take effect on January 1.

An editor at National Public Radio's flagship program *All Things Considered* invited Barack to prepare a broadcast review of social scientists Richard J. Herrnstein and Charles Murray's brand-new book *The Bell Curve: Intelligence and Class Structure in American Life,* which was generating huge controversy over its discussion of the relationship between race and human intelligence. Barack was happy to sign on, but when he submitted his review, an NPR editor objected to his statement that race was not a biological construct, only a social one. She "wouldn't let him make the point," and "he was very offended," Rob Fisher recalled. Barack vented his dismay about "Some Things Considered" to Rob, but NPR did allow Barack, who was introduced as a "civil rights lawyer," to state that *The Bell Curve* featured "good old-fashioned racism" that was "artfully packaged." But Barack also said that the authors were

> *right about the growing distance between the races. The violence and despair of the inner city are real. So's the problem of street crime. The longer we allow these problems to fester, the easier it becomes for white America to see all blacks as menacing and for black America to see all whites as racists. To close that gap . . . we're going to have to take concrete and deliberate action. For blacks, that means taking greater responsibility for the state of our own communities. Too many of us use white racism as an excuse for self-defeating behavior. Too many of our young people think education is a white thing, and that the values of hard work and discipline and self-respect are somehow outdated.*

But at the same time, Barack went on, Americans must acknowledge "that we've never even come close to providing equal opportunity to the majority of black children," such as having "well-funded and innovative public schools for all children" and jobs "at a living wage for everyone . . . jobs that can return some structure and dignity to people's lives." Such "ladders of opportunity," Barack concluded, would be costly in the short term, but a wise investment in the long run, and a national failure to pursue them would reflect not "an intellectual deficit" but "a moral deficit."[26]

With Mel Reynolds awaiting his upcoming criminal trial sometime in mid-1995, Alice Palmer on November 21 announced her all-but-official candidacy for Reynolds's 2nd District congressional seat. Taking the high road, Palmer made no mention of Reynolds's legal problems and instead said she was "very disappointed" with his voting record, particularly his support for the North American Free Trade Agreement. *Sun-Times* political columnist Steve Neal proclaimed that Palmer had "preempted the field" of possible Reynolds challengers with her announcement. He also reported that Jesse Jackson Sr. had been "gently warned by South Side activists that he shouldn't promote his son" Jesse Jr., who recently had bought a home in the district after finishing law school, "as an alternative to Palmer." Once again praising Palmer as "probably the most intelligent member of the Illinois General Assembly," Neal declared that Jackson Jr. "would be viewed as an interloper by the South Side black community" if he ran but that if "Jackson joins forces with Palmer," he "could win Palmer's seat in the Illinois Senate." Palmer's announcement sounded like a warning shot aimed at the twenty-nine-year-old Jackson: "I am not a newcomer to hard work . . . I am not a newcomer to coalition building . . . I am not a newcomer to independent politics . . . I am not a newcomer to the 2nd District."

Ten days after Palmer's announcement, Barack had a long conversation with his friend Ellen Schumer, a former UNO organizer with whom he had been speaking regularly for months as she began building a new organization of her own, Community Organizing and Family Issues (COFI). Schumer previously had worked for former congressman (and now judge) Abner Mikva and U.S. senator Paul Simon, so Ellen and Barack often talked about politics as well as COFI. Ellen's notes from that conversation reflected that they "Spoke re: Election—B.O. interest in 'organizing' or politics. How to build or renew a movement—turn around Dem. party? B.O. not interested in Spfld [Springfield] or in running against Alice Palmer."

Barack instead talked more about his discussions with Sokoni Karanja about a prospective leadership training and development project that the Bronzeville-based Karanja was calling the "Hope Center," in honor of early-twentieth-century African American social reformer Lugenia Burns Hope. Eighteen months earlier Sokoni had been awarded a $320,000 MacArthur Foundation "genius grant" and the Woods Fund already had committed a further $72,000 toward his effort. Karanja wanted the Hope Center to rebuild the strength of the once-vibrant Bronzeville, the social center of black Chicago during the age of segregation. "This community was literally vacuumed by the integration movement," Sokoni had told a *Washington Post* reporter eighteen months earlier. "All of these people—the doctors, the lawyers, the business owners—moved out of here, and all they left were people."

Barack was still doing a Friday training and offering other supportive input as Michelle, who that summer had shifted her law license to inactive status, shepherded her second-year class of forty Public Allies through their organizational placements. She also was busy building up an advisory board to help expand the program's reach, one that included personal friends and former city colleagues Valerie Jarrett, Yvonne Davila, and Cindy Moelis, as well as early supporters such as Jacky Grimshaw and Sunny Fischer, young banker Ian Larkin, and youthful Morehouse College graduate Craig Huffman.

Barack's range of acquaintances was growing too, and one of the most significant new ones was Deborah Leff, a 1977 University of Chicago Law School graduate who had become an award-winning producer at ABC News before being recruited as president of the Joyce Foundation in 1992. Thanks to a family timber fortune, Joyce for a quarter century had been a significant grant maker in the Great Lakes region that was its predominant focus. Several partners from Chicago's Lord, Bissell & Brook law firm, including Harvard Law School graduate John T. "Jack" Anderson, and two Kennedy administration alumni, Richard K. Donahue and Charles U. Daly, Joyce's president from 1978 to 1986, formed the core of the foundation's board. Daly's friend Lewis Butler, a former corporate lawyer greatly interested in the environment and health policy, and African American businessman Carlton Guthrie, a member of Trinity Church who, like Jack Anderson, had played high school basketball in Gary, Indiana, were also active mainstays, as was Paula Wolff, the president of south suburban Governors State University, who had a University of Chicago Ph.D. and strong ties to a trio of Republican Illinois governors.

Butler and Donahue shared Anderson and Guthrie's passion for basketball, and Debbie Leff's strong interest in adding diversity to Joyce's board led her to University of Wisconsin labor law professor Carin Clauss and to Barack. Board members like Lew Butler thought that Debbie was an "extraordinary woman," and when Leff recommended Barack to Jack Anderson as a desirable new board member, Anderson was duly impressed by his *Law Review* presidency and by Barack at an introductory breakfast. Barack's election to Joyce's board was announced just after Thanksgiving 1994.

Equally significant was Debbie Leff's recommendation of Barack to a pair of fellow presidents of prominent Chicago foundations, Adele Smith Simmons at MacArthur and Patricia A. Graham at Spencer. One year earlier publishing billionaire Walter Annenberg had pledged $500 million from his personal fortune toward improving America's public schools. Brown University and its president, Vartan Gregorian, would oversee how most of the money would be spent over a five-year period. Just months earlier, Wieboldt Foundation executive director Anne Hallett had left that post to create a new national network of urban school reformers, and University of Illinois at Chicago education professor Bill Ayers, one of the city's most energetic reform advocates, immediately queried Hallett and new Joyce Foundation education program officer Warren Chapman about Annenberg's potential for Chicago.

Over a meal at a Thai restaurant the trio discussed how local reformers could pursue some of the Annenberg funds. By May the Chicagoans had met with officials from Brown, and by late summer they submitted a draft proposal written by Hallett. The response from Brown's Robert McCarthy was encouraging, and in early November, they submitted a final proposal to Gregorian requesting $49.2 million. The sum was not outlandish—proposals from New York, Philadelphia, and Los Angeles were requesting $50 to $53 million—and Annenberg would structure them all as challenge grants, with local recipients expected to raise two times as much in local matching funds. With that requirement looming and formal approval pending, Hallett and Ayers reached out to Patricia Graham at Spencer, a former dean of Harvard's Graduate School of Education now at an education-focused foundation.

A Chicago entity would be needed to administer the Annenberg grant and raise the matching funds, and Graham in turn asked her fellow foundation presidents Adele Simmons and Debbie Leff to join her. The three women met for breakfast in mid-December, and then held several more meetings with Hallett and Ayers in early January, by which time the Annenberg commitment was firm and scheduled for public announcement two weeks hence. A local board for the new organization would have to be chosen, and Debbie strongly recommended Barack, whose name Pat recognized when Debbie cited his *Harvard Law Review* presidency. For board chair, "we did not want someone whose career was education," Adele explained, yet they had to be "smart enough to understand what was going on and what the issues were." Anne Hallett knew Barack well from when Wieboldt had funded DCP, "and all of us thought he was very impressive." Pat Graham agreed to ask Barack, and invited him to meet her for a mid-February dinner at a well-known Italian restaurant, Avanzare.

During the first part of the evening, Barack spoke about his book, as prepublication galleys had arrived just days earlier from Times Books. He recounted the Poseidon cancellation and "talked about the perils of publishing this book," Pat remembered. Barack also "had just learned that his mother had very serious ovarian cancer, and he was very concerned that the book was more about his father and should have included more about his mother."

Six months earlier Ann Dunham had returned to Jakarta and a new job with her old employer Development Alternatives Inc. (DAI) after becoming deeply dissatisfied with her Women's World Banking post in New York. In late January she had found herself in such serious pain that she had flown to Honolulu. Ann had felt increasingly unwell since her return to Indonesia, and a diagnosis of appendicitis and surgery to remove her appendix had not alleviated her discomfort. Only in Hawaii did an oncologist discover she had stage 3 uterine and ovarian cancer, and on February 14, she would undergo a total hysterectomy and begin recuperation under her mother's care in the same small bedroom that for so many years had been her son's. Maya had just returned to New York to start a master's degree at NYU, and Ann would begin monthly chemotherapy treatment in an attempt to prevent her cancer from spreading further.

At dinner that night with Barack, Pat Graham "was so impressed with how articulate he was" and they talked long into the evening. "I've never closed a restaurant before," Pat recalled. Toward the end, she finally popped the question, explaining how she, Debbie, and Adele all wanted him to chair the new Chicago Annenberg Challenge board. "I will if you'll be the vice chair," Barack responded. "Done," Pat replied, and at another breakfast meeting, Pat, Debbie, and Anne introduced Barack and Bill Ayers to each other, the first time the two men had met. The fifty-year-old Ayers was well known in Chicago for the dedication to school reform he had demonstrated since he joined UIC's education faculty in 1987. His father, Thomas G. Ayers, who was then eighty years old, had for years been one of the city's most respected business titans as the longtime head of Commonwealth Edison, and Bill's earlier life as a student radical had been chronicled in a 1990 *Chicago Reader* profile, while Barack was at Harvard, and in a 1993 *Chicago Tribune* story.

The Annenberg organizers would have to recruit "a board of Chicago worthies," as Pat put it, a much "broader-based group" than Anne and Bill's band of reformers. That group was already busily outlining how Chicago public schools would have to band together in small groups, and find a local organizational partner, before applying for Annenberg funds to support innovative new initiatives. That informal group, calling themselves the Chicago School Reform Collaborative, would design the entire Chicago Annenberg program prior to an actual executive director being selected.

In a coauthored *Chicago Tribune* essay, Ayers, Hallett, and Chapman highlighted "the unworkable size of large schools and classrooms" as a crucial problem. In a massive school system like Chicago's, the Annenberg millions "will not produce miracles," but "the reinvented schools we envision will be places that provide a personalized, more intense and flexible learning experience for students." By mid-March a board that included retiring University of Illinois president Stanley Ikenberry, whom Bill Ayers had strongly pushed, and former Northwestern University president Arnold Weber was in place, and at its first meeting, Pat Graham's motion to elect Barack as their chair was unanimously adopted.[27]

Beyond his new Joyce and Chicago Annenberg Challenge roles, Barack was immersed in four cases in his workday life at Davis Miner. He also needed to round up prepublication blurbs from well-known individuals for use on his book's dust jacket and in whatever print advertising Times Books placed. Davis Miner senior partner Allison Davis took a few copies of the 564-page bound typescript and contacted his friends Vernon Jordan and Charlayne Hunter-Gault. Barack asked the well-known author Alex Kotlowitz, whom Davis Miner represented, to read one, and Barack mailed another copy to Derrick Bell along with an apologetic cover letter. "It's been a long time since we last spoke," but Barack had read *Confronting Authority: Reflections of an Ardent Protester,* which had appeared four months earlier. "I enjoyed your new book very much—inspiring, in fact." Barack mentioned his Davis Miner work and his law school course, plus this "book that I've been writing, on again, off again, for the past two years. Originally, the book (called *Dreams of My Father*) was going to be a series of essays on issues of race and class, but as it has evolved, it's become a memoir of my family and my experiences as an organizer." Now it was "finally finished," and Barack hoped Bell could read it and pen a brief endorsement. Six weeks later Bell did just that, and Barack again wrote him to say thank you "for the wonderful blurb. It's both eloquent and generous."

At Davis Miner, Barack in early January had filed suit in federal court against Illinois governor Jim Edgar on behalf of ACORN, over the state's refusal to obey the new National Voter Registration Act. Three parallel actions—by the League of Women Voters, a trio of Hispanic groups, and the U.S. Justice Department—were also filed, and the district judge assigned the cases, Milton Shadur, wasted no time in dismissing Illinois's argument as a "sleight of hand" and ruling that the "motor voter" mandate "plainly passes constitutional muster." The state appealed, and the Seventh Circuit scheduled oral argument within weeks.

One party plaintiff remembered that "Barack was tremendously lazy on that case: half the time didn't show up when he was supposed to, was never prepared on the briefs," and that caused "a lot of frustration amongst the other attorneys." At the appellate hearing, three plaintiff attorneys spoke—those representing the League, the Hispanic organizations, and the Justice Department, with ACORN relegated to fourth-place status. The appeals court quickly affirmed Judge Shadur's ruling, and a month later at their annual Independents' Day Dinner, Illinois's top group of non-organization Democrats, the Independent Voters of Illinois–Independent Precinct Organization (IVI-IPO), gave their Legal Eagle Awards to four attorneys who had worked on the cases, including Barack. Senator Paul Simon and U.S. judge Abner Mikva were the guest speakers, and the plaintiff underwhelmed by Barack's previous contributions remembered the scene: "he's there for the press conferences!"

Judd Miner failed to win reinstatement of the first of the two St. Louis aldermanic redistricting challenges that Barack was helping him with. However, an employment discrimination case that Judd, Barack, and Jeff Cummings filed in federal court in late January against Comdisco, a highly profitable computer leasing company, on behalf of a black former employee was hugely successful within less than nine months. William Donnell had worked for more than a decade as the only black individual on Comdisco's 150-person sales force. The company was well known among business journalists for its "locker-room atmosphere," and Comdisco strenuously fought discovery requests to compel deposition answers from one executive who Donnell said knew how he had been racially targeted.

Forbes magazine reported that Donnell had identified Comdisco CEO Jack Slevin, whom one colleague called "an equal opportunity insulter," as a primary offender, and *Forbes* warned that "a trial could be embarrassing." Soon after, Judge Suzanne Conlon ordered that the deposition answers be given, because they may "provide evidence of Comdisco's racial animus and hostile work environment." Within a few weeks of that order, Comdisco agreed to settle the case under seal. Jeff Cummings recalled that the depositions documented "a very strong case of discrimination," and Barack later publicly recounted how Slevin in Donnell's presence had told a customer, "I don't have to worry about the EEOC, I've got my nigger in the window."

Judd Miner said the damages paid by Comdisco represented "a very successful settlement," and affirmed that Barack had carried "most of the load" on the case, "from the research and drafting of the initial complaint, through all phases of discovery." Barack ended up working almost six hundred hours on *Donnell v. Comdisco Inc.*, and at Barack's billing rate of $165 an hour—an unremarkable number, as Judd's was $285—Comdisco paid almost $100,000 to Davis Miner for the time Barack spent suing them on behalf of William Donnell.

Judd, Jeff, and Barack also joined Chicago attorney Fay Clayton, who previously had worked at Davis Miner, in litigating a class action "redlining" suit against Citibank on behalf of almost eight hundred African American mortgage loan applicants. Federal district judge Ruben Castillo was unimpressed with Citibank's defense claims, writing in one order that "once again, Citibank misses the point" and ruling in another that "Citibank's contentions are meritless." He soon appointed a retired federal judge as a special master to handle the heavy burden of loan application evidence, and eventually this case too settled, with Citibank paying far more in plaintiffs' legal fees—$950,000—than in money damages—a total of $360,000—to the black loan applicants. Judge Castillo nonetheless termed the settlement "a significant legal victory," called the fees "eminently reasonable," and commended Miner, Clayton, and the other plaintiffs' attorneys for their "professionalism and sacrifice in this case."

Barack invested 138 hours of work in *Buycks-Roberson v. Citibank*, a modest $23,000 of Citibank's settlement tab, but by far the most memorable moment of that case or any other occurred during a deposition in which Barack and Jeff Cummings were participating by phone from Davis Miner's basement conference room. In addition to the SRO hotel and its unpredictable residents, Davis Miner's immediate neighbors also included a restaurant that "didn't tend to their garbage so well," Jeff explained, and the rodents who enjoyed that opportunity also burrowed into Davis Miner's basement, making rat traps a necessity in the modest conference room. Barack was holding the phone when one rat emerged into the light of day and "runs and jumps" onto Barack's pants leg, briefly grabbing hold. Barack "calmly shook the rat off his leg" and "kept right on talking" as Jeff watched in amazement. No one on the call knew what had occurred, and there was no question Barack "always kept his cool," Jeff commented.

Judd Miner was twenty years Barack's senior, and although no one called their relationship paternal, there also was no question that Barack enjoyed favored son status in Judd's eyes right from the moment he arrived at Davis Miner. Judd and the firm's longtime data analyst, Whitman Soule, had for years played golf on Thursday afternoons, and Judd soon invited Barack to join them. "It was kind of a cheeky thing that he was playing golf with us," Whit explained, given how junior Barack was, but Whit had heard Judd say that "this guy is special" even before first meeting him. Barack "wasn't that good," but "he knew how to play golf. He learned how to play golf from his grandfather."

Madison-based partner Chuck Barnhill viewed Barack as "a mediocre golfer" but "a great trash talker," and Barack's proclivity for improving the lie of his balls would become what Whit termed "a subject of derision." Barack also "always seemed to want a little action" on their games, small bets of a quarter on one or another shot. Whit worked with Barack on *Donnell v. Comdisco,* analyzing the company's personnel records and sitting in when Barack deposed Comdisco officials.

It was immediately obvious that Barack did not enjoy litigation. "Barack had to go through one of these depositions, and you could just see this was so painful" for him, even though he was just posing the questions. Whit also served as the firm's IT guru, visiting East View to set up Barack's home computer, and he was repeatedly struck by how "preternaturally calm" Barack was, sometimes in a seemingly "lonely" way. "He always seemed alone, and he always seemed to be focused on the very long term."

After every golf game, Barack would "smoke a cigarette or two," which Whit found surprising. "I always thought it odd that somebody with this level of control was smoking cigarettes," especially because Barack "was very aware that it was becoming unacceptable" to do so, as the reactions of Michelle's young Public Allies colleagues and his UC law students also reflected.[28]

In late March 1995, Barack began teaching Current Issues in Racism and the Law for the third time. His now two years of experience as a practicing lawyer informed his classroom style, with students finding him "an engaging professor" who was "down to earth and practical." His syllabus remained unchanged, with "a considerable amount of reading" across the first four weeks and group presentations during the last half of the eight-week spring quarter. Some students were excited to encounter "a real practitioner" who was "a practical and straightforward teacher who encouraged us to think about real world problems and scenarios," including "some that perhaps could not be addressed by laws."

Marni Willenson's group tackled inequitable school funding, and even though Barack challenged students to consider all perspectives, "you would not have walked out knowing what his position" was on that or any other subject.

"He was very respectful of our views, listened to all of our ideas," but did not "share many of his own personal views on the topics we discussed," Susannah Baruch agreed. Students wanted to hear "what this smart young lawyer from a civil rights boutique law firm had to say," and many students found the discussions compelling. Tiffanie Cason, one of the few black women students at the law school, had been upset when her young Constitutional Law III professor, Elena Kagan, had asked Cason to address the question of whether affirmative action personally harmed its supposed beneficiaries. Dismayed by that classroom experience, "I wrote a paper about it in the Obama class," Tiffanie recalled, "how do we talk about race." Marni Willenson remembered it as "a great class," and course evaluation responses bore that out: on a 10 point scale, the twenty-one respondents gave Barack's teaching a 9.38, up from 9.1 a year earlier, and their overall mark for the course was a resounding 9.71.[29]

In early April, the two trade periodicals that published advance reviews of forthcoming books both praised Barack's. Since his letter to Derrick Bell two months earlier, however, one small but significant change had been made: the book was now titled *Dreams From My Father*. The initiative for this change had come, as had so many things, from Rob Fisher, who recalled "Barack going through a lot of different names for the book. He and I chatted about it quite a bit on the phone," and in the end "he and I came up with the title" during "quite a long conversation" after Rob had seen the bound galley. Rob suggested "shifting from *Dreams of My Father*—which seems kind of trite to me—to *Dreams From My Father,* which better suited the story of him aspiring to construct his dad, and those dreams, from what he had."

Otherwise, the final run-up to actual publication was uneventful. Ruth Fecych, who had inherited Barack from Henry Ferris yet never met him, remembered Barack as "accessible" and "articulate" on the phone. Jacket designer Robbin Schiff recalled Barack's enthusiastic reaction to her idea of using a pair of old family photos on the book's cover, one picturing young newlyweds Stan and Madelyn Dunham during World War II and the other showing a young Barack Sr. in the lap of his mother, Habiba Akumu. Times Books' own catalog called *Dreams* a "compelling autobiography" of "startling emotional and intellectual clarity" and described Barack as "a thrilling new voice." Production editor Benjamin Dreyer recalled that the lengthy manuscript "went through copyediting and proofreading virtually without incident." *Publishers Weekly*, the industry bible, called *Dreams* "a resonant book," while noting that "his mother is virtually absent." *Kirkus Reviews* was even more enthusiastic, saying the "honest, often poetic memoir" was an "affecting study of self-definition" that "records his interior struggle with precision and clarity."

The publication date for *Dreams* was still several months away, and in early April Chicago mayor Richard Daley was reelected to another four-year term.

Daley's increasing dominance of city politics easily smothered two conflicting black efforts to mount credible challenges against his new machine. Former Project VOTE! chairman Joe Gardner had launched a Democratic primary run, but exiting attorney general Roland Burris announced he would oppose Daley in the subsequent April 4 general election. The *Tribune*'s front-page story on Daley's formal announcement that he was seeking reelection stated that "violent crime remains at epidemic proportions" across Chicago. Yet proof that neither challenger would gain any meaningful traction against the mayor came when U.S. senator Carol Moseley Braun gave Daley's candidacy "a ringing endorsement."

Three years earlier, Joe Gardner had backed Moseley Braun in her primary challenge against incumbent Al Dixon, and civil rights lawyers led by Judd Miner were still in court challenging the Daley regime's city council redistricting, but Moseley Braun's embrace of Daley extended African American support for the mayor beyond his small band of early supporters like Valerie Jarrett and financier John Rogers Jr. Jesse Jackson Sr. campaigned for Gardner, but with Daley's war chest at $4 million and Gardner's at $171,000, mayoral media consultant David Axelrod had an easy campaign to oversee.

On primary Election Day, Daley swamped Gardner 66 to 33 percent amid "one of the lowest turnouts ever among Chicago voters" while winning about 25 percent of black voters. A month later Daley defeated the better-known Roland Burris 60 to 36 percent, and political historian William Grimshaw—Jacky's husband—told the *New York Times* that Daley was indeed "rebuilding the machine" that a generation earlier had made his father mayor-for-life. As another knowledgeable observer explained, Daley's emboldened reign marked "a lessening of democracy and the centralizing of power in a new machine."[30]

Receiving almost as much coverage as the mayoral contest was the ongoing criminal prosecution of U.S. representative Mel Reynolds. In mid-January young accuser Beverly Heard again recanted her allegation that Reynolds had sex with her when she was just sixteen years old, and a *Tribune* editorial asked if Reynolds was "the victim of an overzealous, irresponsible prosecution." The morning after Daley's reelection, though, a *Tribune* headline announced, "Reynolds Accused of Sex with 2nd Teen," and the paper spoke of his political career in the past tense, saying he "had been a rising star in the Democratic party." As reports swirled that Reynolds may have paid Heard for her recantation, prosecutors filed additional charges and the congressman's trial was scheduled for July. News reports predicted that tape recordings of 1994 phone conversations between Heard and Reynolds would be a key part of the case.

Waiting quietly in the wings was state senator Alice Palmer, who would challenge Reynolds in the early 1996 Democratic primary if the congressman survived his legal shoals. Should Reynolds instead be convicted and forced to leave office, a special election could take place sooner. Palmer formed a federal campaign committee in late January but did little active fund-raising. Progressive women like Jean Rudd, Anne Hallett, Adele Simmons, Aurie Pennick, and Bettylu Saltzman, plus education researcher Fred Hess and local state representative Barbara Flynn Currie, all made modest, three-figure contributions. Less than a dozen people contributed as much as $1,000, including Palmer's campaign chairman, Hal Baron, and his wife, who knew Palmer and her husband Buzz from the Harold Washington era. At least half of Palmer's remaining four-figure contributions came from employees and relatives of developer Tony Rezko.

Rezko was actively promoting Barack's future prospects as well as boosting Alice Palmer's. Tony was an enthusiastic supporter of St. Jude Children's Research Hospital, the prominent institution founded by entertainer Danny Thomas, and each May a Chicago fund-raising dinner took place in the Drake Hotel's Gold Coast Ballroom overlooking Lake Michigan. Young Andy Foster was a top aide to Republican governor Jim Edgar, and several times during Edgar's 1994 reelection race, Rezko had called to offer support. Andy thought Tony "seems like a very nice guy," with a "very innocent" interest in politics, and a few months later, Foster happily accepted when Rezko said he had a full table of seats at the St. Jude dinner. Tony's wife Rita was there, as was Cook County Board president John Stroger, his top aide, Orlando Jones, and the wife of a young state representative who was about to run for Congress, Patti Blagojevich. Tony also introduced Andy to Barack and Michelle Obama, and Andy recalled that "Obama's embarrassed" as Tony told Andy what an important person Barack would become. Andy believed Tony had clearly taken "a shine to Barack," who was "very impressive," and "it was a delightful evening" even though "Patti was a little annoying."

Barack had first been introduced to Alice Palmer by his Project VOTE! deputy Brian Banks, and as Palmer started pulling together an informal steering committee in the late spring of 1995, Barack began taking part in it. If Palmer challenged Reynolds in next year's Democratic primary, she could not simultaneously run for reelection to the Senate, although in a special election, she could run for Congress while keeping her seat. In a pair of April and May conversations with his friend Ellen Schumer, Barack talked about his new Annenberg role and Alice Palmer as well as Ellen's COFI launch.

A May 1 lunch was rescheduled for May 11 because Barack had to prepare for a deposition, but Schumer's notes from that Thursday lunch document Barack's focus. Ellen jotted three large headings: "(1) COFI, (2) Annenberg, (3) Alice Palmer," but only blank space was under the first two. Under the third, she wrote: "Cong. seat—Senate seat??—What exactly are the boundaries—includes all—Support from A.P.—Toni Preckwinkle?," the 4th Ward alderman, and "Arenda Troutman," the 20th Ward alderman whose district encompassed Woodlawn. "Senate seat" and "Support from A.P." were more than suggestive, and then just four days later, law school debts and mortgage payments notwithstanding, Barack made his first political contribution, writing a check for $500 to Friends of Alice Palmer. A second for an additional $500 followed four weeks later.

As he thought about seeking Palmer's state Senate seat, Barack drove down to Springfield in Michelle's hand-me-down Saab, which had supplanted the rusty old Toyota as his regular car. For Senate Democrats, the 1995 spring session had proven exceptionally unpleasant, with Republican Senate president James "Pate" Philip and his caucus, who held a 33–26 majority, freezing the minority party out of any role in state budget decisions. Angry floor statements by normally mild-mannered progressive Democrats like Miguel del Valle, Jesus "Chuy" Garcia, and Palmer all testified to what a strained atmosphere characterized the Senate under Philip's resolute rule.

Barack later recalled, "I had gone down there to visit before I made the decision to run and had spent a couple of days down there," and a wide range of people remembered meeting Barack in Alice Palmer's company, either during the final days of that spring session that ended unusually early on May 26 or in the weeks immediately following. Down in Springfield, young representative Jeff Schoenberg, who had met Barack three years earlier during Project VOTE!, ran into him in the legislative office building that adjoined the handsome State Capitol, and recalls that Barack mentioned Palmer and explained that "he was working with her." African American lobbyist John Hooker, a onetime student of the late Al Raby, later Commonwealth Edison's liaison to Mayor Harold Washington, and by 1995 ComEd's director of governmental affairs, first met Barack when he "worked closely with" Palmer. Indeed, in Hooker's eyes, Barack was "a real close confidant and pal of Alice Palmer," and Hooker's ComEd colleague Frank Clark felt that Palmer "saw him more as a protégé." Likewise, African American AFSCME lobbyist Ray Harris knew that "she and Barack were close" and that Palmer indeed "mentored him to an extent."

As the summer began, Mel Reynolds's legal prospects looked increasingly dim. A long *Tribune* retrospective on Reynolds's career said "he was eloquent and cared deeply about the underprivileged and dispossessed," and the head of the Rhodes Scholarship program at the University of Oxford recalled Reynolds as "a very ambitious man who was determined to be a politician." A former pollster said that Reynolds "felt like he had to be in Congress. He almost felt he was chosen." The list of unpaid debts that Reynolds had accumulated, however, now approximated $145,000, and another former Rhodes Scholar who knew him well explained that "you were never quite sure who the real Mel was. You could never put your finger on whether he was a real operator or he was a genuine nice guy."

For Barack, the prospect of succeeding Palmer if and when she took Reynolds's place in Congress was one he mulled all throughout June. He mentioned the possibility to Allison Davis, who responded with astonishment. "You must be crazy. Have you ever been to Springfield?" In contrast, Judd Miner was supportive because he knew how Barack "was struggling with what he could do to have the greatest impact on things he cared about."

Given Judd's own deep ambivalence about litigation, he was also sympathetic to Barack's view, as Barack later put it, that "the reason I got into politics was simply because I saw the law as being inadequate to the task" of achieving social change. "The idea of participating in the legislative process," Barack recalled a few years later, "became more interesting the more I saw the limits of social change through the courts." That was "something that I had been aware of when I had been studying law and legal history, but became much more apparent to me by the time that I was actually a practicing lawyer." Issues like "providing adequate funding for education or creating job opportunities . . . weren't really amenable to change through the courts." Succeeding Palmer would be a "fairly attractive starting point," Judd appreciated, because "it was the Senate as opposed to the House," and "it wasn't being alderman."

Allison Davis remembered, "I saw Alice coming and going all the time" to Barack's small office right next to his, and Steve Derks, a young health lobbyist for Lutheran General Health System, which Davis Miner represented, was asked to talk with Barack about the Senate. Derks recalled sitting in Barack's "crummy little office" and saying that the number one thing he had learned in Springfield was "the enormous influence of money in electoral politics." Steve suggested to Barack that "he might do better to stay in the private sector, focus on the things he cares about" and "create a difference that way." Barack "ultimately asked if I would help him if he decided to go forward," and Steve reluctantly agreed. But he still "tried to tell him not to do it," as did Judd's old friend Marilyn Katz, who had also worked for Harold Washington. "Don't run for the Senate," Marilyn told Barack, arguing that state senators were virtually invisible in Chicago. In rebuttal, Barack raised "the Harold Washington example," describing how Washington had moved from the legislature to Congress and then to the mayor's office.

Barack also discussed the idea with Jean Rudd and her husband Lionel Bolin, asking the politically experienced Lionel to be his campaign treasurer if he went ahead. Lionel agreed, and they invited Barack and Michelle over one night to meet their friend Larry Suffredin, an attorney and lobbyist. Suffredin was "so impressed" by Barack that "I thought this is a guy who could be the next mayor of Chicago," and Larry invited Barack to lunch with his African American law partner Paul Williams, who had served three terms as a state representative after working for Harold Washington in Springfield. At lunch,

"I tried to talk Barack out of running for the state Senate," because "I didn't see a mayor really coming from the general assembly," especially given the Republican control of the state Senate. Williams was even more wary of Barack's belief that he could inherit Palmer's seat when she ran for Congress, warning him that if Palmer did not defeat Reynolds, she would want to retain her Senate seat. But neither man's comments dissuaded Barack. "He was looking at should he run," Larry recalled, and "he made it clear that he wanted to."

Judd Miner had introduced Barack to his friend Matt Piers, a 1974 UC Law School graduate who had been Judd's deputy corporation counsel in Harold Washington's administration. Matt and his wife Maria Torres agreed to invite Barack and Michelle for dinner along with Chuy Garcia and his wife Evelyn so that Barack could meet another progressive state senator. Matt knew Barack was "much more interested in a political career than a legal career," and that night at dinner, after "everybody had had a lot of very good red wine to drink," the conversation turned to how "politics are really sleazy here." That theme was music to Michelle's ears, because when Barack had first broached the possibility of succeeding Palmer, Michelle's reaction had been highly negative.

"I married you because you're cute, and you're smart, but this is the dumbest thing you could have ever asked," Michelle later said she told Barack. "We would always have discussions about how do you create change," debates that had stimulated her move from her city job to Public Allies, but "politics didn't come into the discussion until the seat opened up," Michelle later said. "I wasn't a proponent of politics as a way you could make change," and she believed "that politics is for dirty, nasty people who aren't really trying to do much in the world." So Michelle's response had been "No, don't do it," an attitude also informed by her strong desire to have children and a strong belief that her offspring should enjoy the same sort of upbringing she and her brother Craig had had.

"They were a very, very close-knit family," her sister-in-law Janis explained, "and Michelle really liked the idea that her parents were present for whatever she was involved in. So she had the same expectation for her husband, so she was conflicted because she knew that Barack was going to have some kind of job that would take him away" if he ran for a state office. "So she had a little bit of hesitation," and that night at dinner, with Matt in particular kidding Barack about wanting to be a politician, Michelle enthusiastically chimed in.

"Barack was looking a little somber," and when Matt went to the bathroom, Barack waylaid him before Matt could return to the dining room. He "grabbed me by my shirt and he pushed me into" another room "and he said 'Listen, god damn it, Michelle doesn't want me to get involved in politics, and I've already made the decision that I have, and now if I can't count on my friends to help me with this, I'm really going to get nowhere, so I'd appreciate it if you would be quiet and stop everybody from talking about it.'" It was clear that "the wine was kind of talking," yet Matt recalled, "I was quite shocked. . . . There was both an assertiveness and a familiarity" in Barack's manner, and while he was "not angry," there was no doubt that "he was real serious." Michelle's comments that night "made it clear she wanted him to get on a tenure track at UC Law School" and that "she was not at all interested in being a politician's wife."

Years later Michelle remembered, "I thought Barack would be a partner at a law firm or maybe teach or work in the community," but by late June 1995, there was no question that Barack had his eyes firmly set on succeeding Alice Palmer in the Illinois state Senate if she gave up her seat to run for Congress. Barack later said that Michelle finally relented, telling him, "Why don't you give it a chance?"

Just as Barack had told Ellen Schumer, thanks to Tony Rezko he won a warm introduction to 4th Ward alderman Toni Preckwinkle, who remembered Barack from Project VOTE! Preckwinkle told him that Palmer's public endorsement was critical, and that he must also introduce himself to newly elected 5th Ward alderman Barbara Holt, whose district included most of Hyde Park and who also had succeeded Alan Dobry as 5th Ward committeeman. Holt remembered Alice Palmer telephoning, and "she asked me if I would meet with him." Holt agreed, Barack visited her ward office, and "I decided that I would support him." Similar outreach took place with Hyde Park state representative Barbara Flynn Currie, who knew Michelle from Public Allies, and Currie too agreed to back Barack because of Palmer's endorsement.[31]

As Mel Reynolds's criminal trial approached, political jockeying intensified. The *Tribune* reported that Gha-Is F. Askia, a Muslim employee of the state attorney general's Chicago office who six years earlier had won election to a South Shore elementary school council, would run for Palmer's Senate seat. In Hyde Park, rumors circulated that Janet Oliver-Hill, who that spring had lost the 5th Ward aldermanic race to Holt, might be interested, but she declined. Then, on June 25, a *Sun-Times* gossip columnist broke the news that Alice Palmer would formally announce her candidacy on June 27. At a Loop hotel, Palmer declared, "I am a hands-on, roll-up-your-sleeves legislator," and she avoided bashing her ostensible opponent. "Pray for Mel Reynolds and vote for me!"

The *Tribune*'s story stated that Palmer "will be giving up her legislative seat to run for Congress," as would be necessary if she ran in the 1996 primary, and the weekly *Hyde Park Herald* reported that "talk of who might replace Palmer, assuming she wins the race, has already begun. One front-runner might be Palmer supporter Barack Obama, an attorney with a background in community organization and voter registration efforts. Obama, who has lived 'in and out' of Hyde Park for 10 years, is currently serving as chairman of the board of directors of the Chicago Annenberg Challenge. Obama said that even though the election was a year away, 'I am seriously exploring that campaign.'"

Less than forty-eight hours later, a *Tribune* gossip column predicted that "Barack Obama will announce he's running for the state Senate seat occupied by Alice Palmer . . . Obama, who has worked with Palmer, is an attorney at Davis, Miner, Barnhill & Galland—and newly published author of *Dreams From My Father*," although publication was still a few weeks away. The report was correct, and six days later, Barack received a check for $5,000 as a loan to cover campaign start-up expenses from prominent, just-retired seventy-five-year-old Cadillac dealer Al Johnson, who previously had been county board president John Stroger's top contributor, Mel Reynolds's finance chairman in his successful 1992 race, and one of the late Harold Washington's closest supporters.

Johnson, whom friends would call "a dapper gentleman" and "the life of the party," graduated from Lincoln University and earned a master's degree in hospital administration from the University of Chicago. Johnson had begun working part-time as an automobile salesman in the early 1950s, when no black person, whether salesman or customer, was allowed on a dealership floor. In 1967, as the black freedom struggle crested, General Motors made Johnson its first African American franchise owner, and four years later, he switched from selling Oldsmobiles to offering Cadillacs. Al's son Don remembered his dad as a man of "expensive tastes" who lived in a lakefront condominium in Chicago's Gold Coast but "never wanted the limelight" despite his sustained commitment to black political advancement.

Progressive former congressman Abner Mikva explained how very few political donors like Johnson "were in it for the right reasons. He was one of them." Don Johnson also remembered his dad remarking about the young attorney whom Tony Rezko had introduced him to some months earlier. "There's something special about this guy," Al told him. "This guy is going to go places."

In mid-July, Barack attended a Chicago gathering of the nascent New Party, the progressive "fusion" venture Keith Kelleher and Madeline Talbott hoped to launch. The fifty attendees heard "appeals for NP support from four potential political candidates," including a newly elected alderman who was there to express thanks. The other three included "Barack Obama, chief of staff for State Sen. Alice Palmer. Obama is running for Palmer's vacant seat," the meeting notes recorded, and Senator Miguel del Valle's top aide, Willie Delgado, who also was there, remembered first encountering Barack as an "assistant" to Alice Palmer.[32]

In close tandem with Pat Graham, Barack also was pitching in on the actual launch of the Chicago Annenberg Challenge (CAC). He suggested to Pat that perhaps he should resign as board chairman if he ran for Palmer's seat, and Pat immediately said the state Senate was not significant enough to merit that. Barack and Pat also experienced disagreement, and attendant delay, over whom to name as executive director of the CAC. At the end of May, the board officially asked Spencer, Joyce, MacArthur, and the Chicago Community Trust to pony up equal fourths of an initial $90,000 in operating funds because the first Annenberg money would not arrive until $500,000 in local matching funds was raised.

Those grants were quickly made, and an executive search consultant was retained. In the interim, "since there is no actual director," board minutes noted, "Anne Hallett and Bill Ayers had worked diligently" to write the request for proposals that CAC made public in mid-June, with applications due by August 1. For an executive director, "what was most important to Barack, given the politics of the city of Chicago, was that this person be clean and honest," Pat recalled, and Barack's desire was to hire Ken Rolling, the Woods program officer whom he had known well for almost a decade. Pat wanted "someone who knew something about education," and was uncomfortable about how clubby it seemed. "I thought that was a little too much old boy network for my taste," and so "it took us a while to agree on Ken," whom Pat quickly realized was "absolutely sterling."

Ken would not start until September 1, and in the meantime the governance of Chicago's public schools suddenly shifted in a truly astonishing way. The Republican triumph in Illinois's November 1994 general election had included Republican capture of the state House of Representatives, which gave Republicans full control of both houses plus the governorship for the first time in a quarter century. Within days, both Republicans and business-oriented Chicago reformers like John Ayers, Bill's younger brother, began discussing how a possible governance shake-up could kick-start the energies for reform that had largely abated in the five years since the seemingly landmark 1988 school legislation shifting authority to local school councils had taken effect in 1989. In late April 1995, Governor Jim Edgar debuted a bill that would hand direct control of Chicago's schools to the city's mayor, and a month later Republican lawmakers, joined by a handful of progressive white Democrats, passed the measure on what the *Tribune* called "a day of historic change for a system long mired in status-quo politics."

Edgar quickly signed the bill into law, which charged Mayor Richard M. Daley with naming a new, five-member board of education and a new chief executive officer. As everyone waited for Daley to act, the *Tribune* publicized CAC's request for innovation proposals from groups of Chicago schools. Barack emphasized, "If we're really going to change things in this city, it's going to start at the grass-roots level and with our children." Six days later Daley named his chief of staff, Gery Chico, as president of the new board and highly respected city budget director Paul Vallas as CPS's new CEO, supplanting the incumbent superintendent.

The new law would allow the hard-charging Vallas to terminate administrators and non-educational employees with just two weeks' notice, and Vallas promised to eliminate the system's yawning budget gap without any impact on classroom education. One school principal, a twenty-five-year veteran of the Chicago schools, told the *Tribune* that "this is the best news I've heard in all the years I've been in the system." Less than three weeks later, Vallas announced that the budget deficit had indeed been eliminated, with more than $161 million saved by the termination of seventeen hundred non-instructional staffers. The *Tribune*'s African American education reporter Dion Haynes called the speed and scale of Vallas's changes "miraculous" and remarked how "radical" it was for school leaders to be "putting students—not bureaucrats—first." Editorially the *Tribune* praised Vallas for "a dazzling display" of "magic," and highlighted how he also had set aside $206 million to fund "ambitious new educational initiatives." That was exactly CAC's mission too, and by the August 1 deadline, more than 170 proposals were submitted, a number that Hallett, Ayers, and their colleagues found astonishing. The Collaborative volunteers quickly winnowed that huge batch almost in half, and a mid-August board meeting ratified invitations asking the ninety strongest applicants to submit full proposals by October 13.[33]

On Saturday, July 22, Barack and Sokoni Karanja made opening remarks at the first small gathering aimed at finally launching their long-discussed Lugenia Burns Hope Center. For more than a year, they had been discussing "the need for an organizing institute on the South Side," conversations that in part grew out of their joint service on the Woods Fund panel that had critiqued organizing's shortcomings. Sandy O'Donnell, who had staffed the Woods panel, played the same role in the Hope Center planning, and she noted that Barack cited the "political disaffection of the African American community in the wake of Harold Washington's death" and "the lack of an active voice among African Americans in various major policy debates—school reform, the Chicago Housing Authority, and the criminal justice system" as primary reasons for why the Hope Center was needed.

Barack explained that the center's mission was to start "an organic process to duplicate leadership—to help people become actors rather than consumers," language anyone familiar with John McKnight's work would recognize. People manifested "lots of hunger for information, but outreach [was] sorely needed—especially to young black men." Now the key would be to nail down the curriculum for the twelve-week training sessions they planned to begin at Sokoni's Centers for New Horizons in Bronzeville in the months ahead. Barack "agreed that the African perspective has to be built into the idea of 'centeredness,'" but the overall goal was that neighborhood residents who completed the training would "become organizers or even community leaders."

On Monday, July 24, opening arguments took place in the Cook County

Circuit Court trial of U.S. representative Mel Reynolds. Lead prosecutor Andrea Zopp, a 1981 graduate of Harvard Law School, was a black woman whose presence negated any claim that Reynolds was being persecuted for his sexual exploits by white racist prosecutors. Three days of jury selection, for a case that would take four weeks to try, yielded a jury of six men and six women, six black and six white. The *Tribune* observed that "regardless of what happens" when the jury reached a verdict, "Reynolds' career is probably over." The *Tribune*'s news coverage was so opinionated that it labeled Reynolds "an arrogant manipulator" while explaining that "a special election will be in order if he is convicted." That was quickly followed by a statement that the only declared candidate was "the estimable State Sen. Alice Palmer."

Jurors heard that Reynolds's sexual relationship with then-sixteen-year-old Beverly Heard had begun in June 1992, and that one week after first meeting her, he took her to dinner at the tony East Bank Club, the favorite exercise and watering hole for upper-class black Chicagoans. Two years then passed before Heard told police about her underage involvement with the now-congressman. Reynolds did not take the stand until mid-August, but a *Tribune* reporter mocked a defense attorney's assertion that Reynolds simply "has the most active libido in Cook County." Thomas Hardy told readers that "Mel Reynolds makes me sick" and "doesn't belong in public office."

As Reynolds's legal prospects darkened, Alice Palmer appeared to have competition for Reynolds's congressional seat. *Sun-Times* political columnist Steve Neal publicly named two potential candidates: state Senate minority leader Emil Jones Jr., with whom Barack had worked back in 1987–88 and whose 34th Ward Democratic organization, in tandem with other wards, guaranteed him black machine supporters, and thirty-year-old Jesse Jackson Jr., who had attended Barack and Michelle's wedding, just as they had attended his because of Michelle's close friendship with his sister Santita. Neal, a huge fan of Palmer's, warned that "her prospects are uncertain if Jones or Jackson joins the race," and although the *Tribune* reported that Palmer already had raised $51,000, reporters presumed that Jackson, as the well-spoken namesake of America's best-known black political figure, could be a strong contender.

Behind the scenes, however, both Palmer and Jackson were facing significant hurdles. Barack's former Project VOTE! deputy Brian Banks had been recruited as Palmer's campaign manager, but even before the end of July, campaign chairman Hal Baron had pushed Banks out as internal disagreements roiled Palmer's small staff. Media consultant Kitty Kurth had recruited a D.C.-based friend, Darrel Thompson, as the campaign's finance director, and Thompson now took charge of Palmer's nascent campaign.

"You've got to meet this guy named Barack Obama," Alice Palmer told Darrel, who visited Barack at Davis Miner and found him to be "a really sharp guy." Kurth first met Barack at a campaign discussion in Palmer's kitchen, and Thompson remembered Barack being "involved but not overwhelmingly involved" as Palmer's campaign took shape. But despite being both a state senator and 7th Ward committeeman, Palmer had no political organization to speak of beyond her husband Buzz and son David.

Jackson and his closest friend, Marty King, plus Jesse's wife Sandi, had been discussing a potential congressional race ever since the young couple purposely had bought a home in the 2nd District. But Jesse Jr. faced strong opposition that no one except his wife and Marty knew about, namely from his father Jesse Sr. The elder Jackson had been serving for more than four years as a "shadow" U.S. senator for the District of Columbia, a symbolic, unpaid role as an elected lobbyist for D.C. statehood, and the father refused to embrace the prospect that his son could become a real, voting member of Congress.

Jesse Jr., Alice Palmer, and Palmer's campaign chairman Hal Baron all remember the summons the two congressional aspirants received from Jesse Sr. "As far as I could see, it was all green, and then we get this call to go to Jesse's house," Palmer recounted. "Junior is relegated to sit in a corner." Baron knew what Sr.'s intent was, but Jr. was caught by surprise. "I remember being to my father's house for a meeting," Jr. explained. "He said, 'Jesse, come to my house. I want to talk with you.' 'All right, Reverend, I'll be there in a few minutes.'"

The father-son relationship had become seriously strained more than a year earlier by an intense disagreement about how best to expand Sr.'s Rainbow Coalition. But with Jackson Sr. living primarily in Washington rather than in the family's Chicago home, tensions had abated. So "I get to my father's house and sitting in his living room is Alice Palmer," Jesse Jr. recalled. "I said, 'Hey, Dad, what's going on?' and he said, 'This is what I'm thinking. I'm thinking that we should support Alice Palmer for Congress, and I want her to consider supporting you for the state Senate seat.' And I said, 'I appreciate that, Dad, but I've not had this conversation with you before.'" Yet Junior immediately understood his father's message: "He's saying to himself that Congress is too much for his boy."

But Alice Palmer was not any more interested in Jackson Sr.'s idea than was his son. "Reverend Jackson, I am sorry, but I am not interested in your deal. I have committed my seat to Barack Obama should I be elected to Congress, so there is no chance I can support your idea," Junior recalled Palmer saying. "I looked at my father and said, 'See?'" Jesse Jr. added. "'Reverend, that's not going to happen. I'm not interested in that,'" and his father's attempt to sandbag a congressional race crystallized Jesse Jr. and Marty King's resolve to go ahead and run.[34]

On July 31, Barack received his first campaign contributions, two $1,000 checks from a pair of Tony Rezko's companies. One week later, Friends of Barack Obama was formally created, with Jean Rudd's husband Lionel Bolin as treasurer, and the checks deposited. David Brint, who had first introduced Tony to Barack five summers earlier, was no longer working with Rezko, yet "I liked him. Everybody liked him." Tony "was soft-spoken" and "I thought he was an okay guy." Cook County Board president John Stroger, Rezko's most important political patron, viewed him similarly. "Tony's the type of guy who is always helping host fund-raisers and getting his friends to get you money. But he's real upfront with you, and I've never known him to ask for favors, which is unusual in this business," Stroger explained.

With Barack's odds of an actual candidacy increasing, he called Carol Harwell, his former Project VOTE! deputy. "I was kind of shocked when he called," Carol remembered, both by Barack's news that Alice Palmer "wants me to take her place" and by his desire to do so. Barack also wanted her to be his campaign manager, and Carol agreed to come over to East View to discuss the district over a pasta salad that Michelle prepared. "She still wasn't quite on board," Carol recalled, but Barack's desire to forge ahead was crystal clear, and he also asked his sister-in-law Janis Robinson to work on his campaign. Both women agreed.[35]

But Barack had to devote his next eight days to something entirely different, because finished copies of *Dreams From My Father* were now in stores, and he was going on a book tour that would take him to New York, Washington, San Francisco, and Los Angeles. Times Books had printed seventy-five hundred hardback copies of the 403-page volume, priced at $23.00. Barack's debut appearance was about ten blocks from home, a Friday-evening book signing at Hyde Park's 57th Street Books. About thirty people showed up, including two of Barack's favorite former DCP leaders, Loretta Augustine and Yvonne Lloyd. He inscribed, signed, and dated their copies with an identical phrase—"To Loretta" and "To Yvonne, one of my favorite people in the whole, wide world!" Barack also called Dan Lee to say that he had a copy for him. Barack explained that in the book Dan's character had been given the pseudonym "Wilbur Milton," a play on Dan's middle name, Willard, and Dan laughingly objected: "Oooh, that is a crappy name!" Barack made up for it with a warm inscription: "To Dan, You have been an inspiration to me all these years. I know our paths will meet in the future for peace and justice."

Dreams From My Father featured pot-smoking deacon "Wilbur Milton," but when Dan read the book, he had no complaints. "That's not a fictitious person. That is me." The book was organized into three major sections of roughly equal length. "Origins" started with the 1982 news of Barack's father's death and then traced his own life up through his undergraduate years at Columbia. "Chicago" commenced with a brief account of his postcollege time in New York before moving to a story-filled account of his work at the Developing Communities Project. "Kenya" was an extremely detailed recounting of his summer 1988 visit there.

Times Books' jacket copy introduced the book as a "lyrical, unsentimental, and compelling memoir," one that was "filled with emotional insight and intellectual clarity." The two-sentence author bio highlighted how he "served as the president of the *Harvard Law Review*," and the back cover featured blurbs by a trio of well-known African Americans plus author Alex Kotlowitz. Marian Wright Edelman termed it "an exquisite, sensitive study of this wonderful young author's journey into adulthood" that was "perceptive and wise." Derrick Bell called it "a beautifully written chronicle of a gifted young man," a "soaring book" that taught how "survival demands resilience in the face of frustrated expectations." Charlayne Hunter-Gault said it was "one of the most powerful books of self-discovery I've ever read," "beautifully written, skillfully layered, and paced like a good novel." Kotlowitz described it as "a book worth savoring" and "incisive."

Dreams commenced with a biblical epigraph that even in later years attracted no comment at all: "For we are strangers before thee, and sojourners, as were all our fathers . . . ," from 1 Chronicles 29:15. In *Dreams'* introduction, Barack noted how "Harvard Law School's peculiar place in the American mythology" and his election as the first black president of the *Review* had led to an opportunity to write a book about "the current state of race relations." He recounted how he initially had "strongly resisted" addressing instead his own life story, in part because "I learned long ago to distrust my childhood and the stories that shaped it." He confessed to having weathered "the periodic impulse to abandon the entire project" and warned of the inevitability of "selective lapses of memory." He explained that "although much of this book is based on contemporaneous journals or the oral histories of my family, the dialogue is necessarily an approximation of what was actually said" and stated that "for the sake of compression, some of the characters that appear are composites of people I've known, and some events appear out of precise chronology. With the exception of my family and a handful of public figures, the names of most characters have been changed for the sake of their privacy." He thanked Jane Dystel "for her faith and tenacity," Henry Ferris, Ruth Fecych, and "especially Robert Fisher" for their help with the manuscript, and Michelle and his family members for their support.

In "Origins," Barack described how his own birth stemmed from "my mother's predicament" and admitted that how his parents came to marry "remains an enduring puzzle to me," with even the supposed facts "a bit murky." He offered a decidedly uncharitable portrait of his grandfather, criticizing Stanley's "often-violent temper" and "his crude, ham-fisted manners." Ralph Dunham, Stan's far from uncritical older brother, later spoke with dismay about *Dreams*, saying Barack "was a bit unfair with Stanley," and a later critic lambasted Stan's portrayal as "unmistakably ironic and patronizing." Lolo Soetoro fared significantly better, with Barack citing his "imperturbable" temperament and writing that "I had never heard him talk about what he was feeling. I had never seen him really angry or sad." Of his own father during his childhood, Barack wrote that "my father remained a myth to me," and even after Obama Sr.'s 1971 visit he "remains opaque." Once Barack's narrative reached his high school years, an angry character modeled on an exaggerated version of Keith Kakugawa was given the pseudonym "Ray," no doubt by an author fully aware that his senior yearbook entry had thanked "Ray"—Ray Boyer—for "all the good times." Smoking "pot" was expressly acknowledged, as was "a little blow when you could afford it."

Frank Marshall Davis figured in the book as "a poet named Frank," although with no full identification, and one ostensible lesson of black male adolescence was that "you didn't let anyone . . . see emotions . . . you didn't want them to see." His mother Ann appeared episodically, including once to complain that "one of your friends was just arrested for drug possession" and to demand "the details of Pablo's arrest," the character based on Bobby Titcomb. Barack wrote that Malcolm X's autobiography affected him because Malcolm's "repeated acts of self-creation spoke to me" given how they were "forged through sheer force of will."

Dreams' account of Barack's time at Oxy featured his reprimand by "Regina," and presenting "Regina" as a black Chicagoan allowed her to stimulate "a longing for place, and a fixed and definite history" that Barack indeed "envied" and finally attained once he moved to Chicago. "Strange how a single conversation can change you. Or maybe it only seemed that way in retrospect."

Scholars would write that "identity may be as determined by events we believe happened to us as ones that did," an insight that could also apply to *Dreams'* account of the traumatic impact the magazine photos of a black person who wanted to be white had had on him at age nine. In *Momentous Events, Vivid Memories,* developmental psychologist David Pillemer writes that there can be no doubting the importance of a "person's *beliefs* about what happened," for "psychic reality is as important as historical truth" given how "memory is an active, reconstructive process" and not a fount of empirical facts. "Memories of personal life episodes are generally true to the original experiences, although specific details may be omitted or misremembered."

Dreams' presentation of Barack's unforgettable 1985 conversation with Bob Elia in the lobby of the Fairway Inn that summer 1985 night on the road to Chicago was transmogrified into an exchange with "Ike, the gruff black security guard in the lobby" of BI's office building at One Dag Hammarskjold Plaza. No one else at BI remembered any security guard, but that was subsidiary to how impactful Bob's challenge was on the young man on the road to a new life in a place where he had really never been. "People frequently trace the beginning of a life path, or the birth of a set of enduring beliefs or attitudes, to a single momentous event. Memories of originating events need not carry an explicit, rule-structured directive," Pillemer writes, because "they inspire rather than prescribe. The memories are a source of motivation and reorientation in the pursuit of life goals."

Far more so than "Regina," Barack's conversation with Bob Elia was one he would recount not only in *Dreams* but for years afterward. "Memories of originating events convey a sense of enduring influence and even causality," Pillemer explains. "The initiating episode is seen as the original motivating force behind a momentous decision. But it is possible," as was true with Barack, "that the memory simply represents the first conscious acknowledgment of a gradually shifting life path. Yet another possibility is that originating events are identified and selected only retrospectively" as "a coherent and orderly life story is constructed in which past events are spuriously linked to known outcomes."

But at a minimum, as with Barack's experience in Hermitage, Pennsylvania, "an originating event is perceived as creating, or at least contributing to, a new life course." Such a turning point "appears to alter or redirect the ongoing flow of the life," and autobiographies like *Dreams* "frequently draw implicit or explicit causal connections between specific early episodes"—as with Bob Elia—"and the subsequent direction of the life course." Northwestern University psychologist Dan P. McAdams stresses that someone's life story "is more like a personal myth than an objective biography, even though the subject believes the story to be true." As with Barack, "life stories provide modern men and women with *narrative identities,*" and although childhood experiences "provide material for the life story," the "story itself does not begin to take shape until society demands that a person begin to formulate a meaningful and coherent life," as Barack experienced during his young adulthood in Roseland. In *Dreams,* Barack created a life story that was a carefully crafted public version of a "developing person's own internalized and evolving narrative of the self."

In *Dreams,* Ike the security guard was just as fictional as the story of how at BI Barack had "my own secretary," and how at NYPIRG his job had been "trying to convince the minority students at City College about the importance of recycling." Yet when *Dreams'* Part Two brought Barack to Roseland, its accuracy quotient soared, if not its chronology, which made a jumble of his actual work there, but with characters directly and mostly accurately modeled not just on Dan Lee but on Cathy Askew, Loretta Augustine, Yvonne Lloyd, Margaret Bagby, Eva Sturgies, Maury Richards, Salim Al Nurridin, and Alma Jones. Jerry Kellman, Mike Kruglik, and Greg Galluzzo would be melded into a single "Marty Kaufman" who was in large part Jerry, and Barack apologized to Mike for his absence from *Dreams,* explaining that editors had eliminated some characters. Similarly, although his sister Auma's visit to Chicago was accurately rendered except for her mundane arrival by train from southern Illinois, and not at O'Hare Airport, Alex McNear, Genevieve Cook, and Sheila Jager were condensed into a single woman whose appearance in the book was fleeting indeed, and Mary Ellen Montes, a married Mexican American mother of three, did not appear at all.

Learning from Auma about how deeply tragic Obama Sr.'s later life was had transformed Barack's relationship with his deceased father. "Where once I'd felt the need to live up to his expectations, I now felt as if I had to make up for all his mistakes," he wrote in *Dreams*. The composite girlfriend "was a woman in New York that I loved. She was white. She had dark hair, and specks of green in her eyes. Her voice sounded like a wind chime. We saw each other for almost a year. On the weekends, mostly. Sometimes in her apartment, sometimes in mine," living in their "own private world. Just two people, hidden and warm." This was largely Genevieve, except the "specks of green" were in Sheila's eyes, not Genevieve's. "One weekend she invited me to her family's country house," and "I realized that our two worlds" were seemingly distant. "I knew that if we stayed together, I'd eventually live in hers," and so "I pushed her away. We started to fight. We started thinking about the future, and it pressed in on our warm little world. One night I took her to see a new play by a black playwright," and after it "we had a big fight, right in front of the theater. When we got back to the car, she started crying. She couldn't be black, she said. She would if she could, but she couldn't. She could only be herself, and wasn't that enough."

Barack and Genevieve had argued, but the "big fight" on the street had been with Sheila in Chicago, not in New York with Genevieve. The summary treatment of the unnamed New York girlfriend was unremarkable, as *Dreams* both overstated—as with BI—and mentioned not at all—e.g., the almost weekly "partying" with Hasan—major aspects of Barack's postcollegiate years there. Yet apart from one passing reference to "someone staying over" one night, *Dreams* omitted entirely the existence of *any* girlfriend in Chicago, never mind one Barack had lived with for almost two full years.

"I am completely missing from" *Dreams,* Sheila observed, and "I never understood why he wrote it that way." *Dreams* made quite clear how Barack had solidified his identity during those years in Chicago, yet she highlighted how "this transformation happened over the course of a very tumultuous love affair which, intriguingly, and paradoxically, Barack does not even mention in his memoir." Sheila mused that perhaps she had just been edited out. "I wonder if the unedited *Dreams* is as inaccurate as the published version."

"Johnnie"—Owens—and Jeremiah Wright appeared as true-to-life characters under their own names, yet Wright did not preach his "Audacity of Hope" sermon at Trinity any time in early 1988. Part Three, about Barack's summer 1988 trip to Kenya, was the most detailed part of the book by far, as Sheila appreciated better than anyone when she read *Dreams*. "There are whole passages from the book that are essentially copies of his letters to me," the ones Barack wrote during his trip and then sought copies of after he signed his first book contract. "I always found it ironic that he was using his love letters to me to write his book and then completely omitted me from the entire account."

Dreams inaccurately had Ruth Baker Ndesandjo inviting Barack and Auma to her home, instead of them appearing unannounced, but the book did not shy from confronting how troubled were the relationships Obama Sr.'s life and legacy impacted. On the last page of Part Three, Barack noted "the sense of abandonment I'd felt as a boy," but in the preceding several pages he reenvisioned both his grandfather's and his father's lives through the lens of his own. Onyango "will have to reinvent himself" and "through force of will, he will create a life." Likewise, Barack Sr. "too, will have to invent himself" and when accepted by the University of Hawaii "he must have known . . . that he had been chosen" thanks to "the blessings of God." Pierre-Marie Loizeau, an insightful literary scholar, would note that "this repetition is neither fortuitous nor arbitrary," that "in fact, Obama might just be talking about himself" and that *Dreams* was "an exercise of self-invention" whereby "blackness . . . is something to be achieved and continuously reinvented."

Dreams stated that Barack Sr.'s inability to acquire "a faith in other people" had been his fatal flaw, that "for lack of faith you clung to both too much and too little of your past," that "for all your gifts . . . you could never forge yourself into a whole man" because of his failure to exhibit "loyalty" and "a strong, true love." That was Barack's closure with the "paternal ghost" who had "haunted his son's life from childhood" right through to the final completion of his book. Likewise, perceptive philosopher Mitchell Aboulafia would realize that the insistent advice that a Kenyan teacher who had known Barack's father gives Barack in *Dreams*—"we must choose," just as Barack had insisted to DCP's Cathy Askew—was "actually Obama's" own stance, and that Obama Sr.'s "failures could in large measure be attributed to his uncompromising commitment to principles, which undermined his ability to compromise and make choices."

Only in *Dreams*' last five pages did Barack write about life after August 1988. He downplayed his enjoyment of law school, calling it "disappointing at times, a matter of applying narrow rules and arcane procedure . . . a sort of glorified accounting." His most recent years had been "a relatively quiet period, less a time of discovery than of consolidation," and the book concluded with a two-page account of Barack and Michelle's 1992 wedding. Much of it was devoted to praising Abon'go's demeanor at the reception, and the narrative ended with Abon'go offering a toast, "To those who are not here with us."[36]

Times Books had bought a third-of-a-page ad for *Dreams* in the Sunday *New York Times Book Review*, featuring the glowing endorsements by Charlayne

Hunter-Gault, Derrick Bell, and Marian Wright Edelman plus *Kirkus Review*'s advance praise and a small, grainy photo of Barack. Ironically, the ad ran alongside a review of a book about John Dewey, father of Ann Dunham's mentor Alice. Far more important, four pages earlier there also appeared a top-half-of-the-page review of *Dreams* entitled "A Promise of Redemption." Written by a young white novelist, Paul Watkins, the review characterized *Dreams* as a "provocative" book that "persuasively describes" growing up amid "a bewildering combination of races, relatives and homelands." Barack experienced "the pain of never feeling completely a part of one people or one place," but when he moves to Chicago "he quickly becomes the pawn of professional organizers, intent on profiteering from money gouged out of the city budget." Watkins obviously had not given *Dreams* a careful reading, and although he claimed that Barack's "story bogs down in discussions of racial exploitation," Watkins acknowledged that many scenes "are finely written." He observed how "Obama is no more black than he is white," yet *Dreams* offers "no emotional investigation into his other half, his 'white side,'" only his African American one. Watkins wondered if he was right in thinking that Barack's underlying argument was "that people of mixed backgrounds must choose only one culture in which to make a spiritual home" and "that it is not possible to be both black and white."

Other early reviews called *Dreams* a "poignant, probing memoir," a "lively autobiographical conversation," and "a moving and articulate exploration of what it means to be a black American" by an "eloquent and thoughtful" writer. The adjective "compelling" appeared in multiple reviews, and a day after the *New York Times* review ran, there was another in the *Los Angeles Times* that praised *Dreams* for its "sharp eye" and "generous heart." A *Boston Globe* column announced the book's publication and reminded readers that the author "was a local celebrity of sorts five years ago" as president of the *Review*.

Barack's modest book tour took him first to Manhattan and a Tuesday-night reading at a Barnes & Noble store on Astor Place. At least two familiar and friendly faces were present: Ann Dunham's old friend Pete Vayda, whom Barack had met years earlier, plus his dear friend Hasan Chandoo. Barack inscribed and signed a copy "To my brother Hasan who helped me through some tough times with love and loyalty." Barack's next stop was Washington, where he was interviewed about how *Dreams* had become so personal a narrative. "What I realized . . . was that the starting point for any insights I might have really had to do with the story of my own family," yet "I have not been completely comfortable with calling it an autobiography." Barack later reiterated that point, remarking that *Dreams* "wasn't really an autobiography," yet he told journalist Bill Thompson that "probably the most difficult part of writing it" was recounting Stan and Madelyn's argument about her fearful reaction to the aggressive black panhandler.

From Washington, Barack flew to San Francisco and a bookstore appearance way out near Golden Gate Park. Then he headed to Los Angeles and a pair of television interview tapings prior to another evening bookstore event. On *Connie Martinson Talks Books,* Barack claimed that "it was extraordinarily difficult to grow up in both Hawaii and then later in Indonesia where I was often the only person of African extraction there, and I did not have any male role models." He acknowledged that his father "dies a very bitter and lonely man," and when Martinson asked about Jeremiah Wright, Barack called him "a wonderful man" who "represents the best of what the black church has to offer."

In a second, far longer interview with Marc Strassman for *Book Channel,* Barack spoke of "the importance of making choices" when confronted with family traditions, such as "the oppressive sexism which exists in Africa," choices "about what traditions we want to cling to" and "which ones we're willing to discard." He mentioned his UC Law School class, although he mispronounced W. E. B. Du Bois's surname, wrongly using a French enunciation, and spoke about the importance of "building up human capital." Social progress would require "a grassroots mobilization . . . at the level of the street," and black America had to "break loose of this either-or mentality" in arguing about what was responsible for black inequality. "Inner-city black males who are running around getting young ladies pregnant and then leaving them" was indeed a problem, for "values do matter" and "moral purpose does matter."

Barack spoke expansively with Strassman, saying that in "my journey to understand who I was . . . it wasn't until I moved to Chicago and became a community organizer that I think I really grew into myself in terms of my identity." He admitted that "the accomplishments of my organizing were modest," but "I connected in a very direct way with the African American community in Chicago." That allowed him to "walk away with a sense of self-understanding and empowerment and connection." But Barack had come to realize that "the kind of community organizing that I did" in Roseland "did not take into account the major structural barriers to change that exist in this society," including "the international economy" and "the power of the media to shape reality."

He believed that "American history moves in waves and cycles" and that "notions of common ground are the glue that hold our society together." Strassman pressed him with a foreshadowing question: "You're willing to stake your political career on there being a common ground?" Barack replied, "that's the core of my faith." Asked about being an author, Barack said his book tour had been "exhausting," and that "I had a more romantic notion of the publishing and book tour process." He quickly added that "the reception has been terrific," but that evening at Eso Won Books, an African American shop in Leimert Park, only nine people, including one of the owners, attended Barack's reading.[37]

On Saturday morning Barack flew back to Chicago, and that evening Valerie

Jarrett was hosting a book party for him in the backyard of her parents' spacious Hyde Park home. Michelle and Valerie's good friend Susan Sher remembered that no more than ten people attended, though Valerie insisted it was closer to forty. Gwen LaRoche, Judd Miner, and 5th Ward alderman Barbara Holt were among those in attendance, and Barack soon did another signing at a bookstore at the Ford City Mall in Chicago's West Lawn neighborhood owned by the family of Michelle's Public Allies board member Craig Huffman.

Barack continued to try to win attention for the book, and Woods Fund program officer Kaye Wilson recommended he reach out to well-known Chicago editor Hermene Hartman to request a review in her *N'Digo* magazine. That came to naught, but Barack did score an hour-long morning "drive time" talk show appearance on WJJD-AM with an unlikely pair of interviewers: African American radio professional Ty Wansley and his "lightning rod" sidekick, former 10th Ward alderman and mayoral candidate Ed Vrdolyak. Barack already knew the show's producer, Barack Echols, his father's namesake whom he had met two years earlier at the Hyde Park video store. Wansley asked most of the questions, and one participant thought that the usually fiery Vrdolyak was "a little bit intimidated" by as erudite a guest as Barack. The weekly *Hyde Park Herald* also interviewed the neighborhood's newest author. Barack said that visiting Kenya to learn more about his father was "a healing process," and that "for all of his absences and faults, my father provided a positive image to live up to or disappoint. Have I made my peace with him? Yes, I think I have."

Published reviews continued to be laudatory. A Philadelphia tabloid praised the book's "mind-tingling prose" and the *Boston Globe* termed *Dreams* "a life-affirming, inspiring book." Yet another paper called it "compelling," and a prominent black monthly said *Dreams* was "wonderfully written in a bright, louvered style that indulges in a little too much detachment." A southern daily paper said Barack "shows wisdom well beyond his years," and the NAACP's monthly magazine, *The Crisis,* lauded *Dreams* as "eloquent and thoughtful." An accompanying interview quoted Barack as saying that "while I'm not advocating segregation, in segregated black communities, historically, there was a lot to be grateful for. There were black doctors, black lawyers. They didn't leave the black community when they became successful, like so many successful blacks do today."

While *Dreams From My Father* garnered heaps of praise from reviewers unacquainted with Barack, most people who had known him well during his life, particularly those from his time at Punahou, were astonished or disbelieving when they read *Dreams'* depiction of how racial anger and disaffection had dominated his school years. Both Greg Orme and Mike Ramos, Barack's two closest teenage companions, felt many passages, especially the account of their own supposedly acute racial discomfort at the mostly black Schofield Barracks party, were at best overstated exaggerations. "I never knew, until reading the book later, how much that night had upset him," Greg explained, because Barack "never verbalized any of that" racial anger which *Dreams* so forcefully dramatized. Bobby Titcomb agreed. "I didn't know that Barack was having those internal struggles until I read his book." Barack's Punahou basketball teammates agreed. "We had no clue," stated Alan Lum, and Tom Topolinski said, "I never heard or felt or sensed any kind of identity crisis." Kelli Furushima, the cute classmate with whom Barack flirted and whose yearbooks he warmly inscribed, said he "seemed to be a happy guy" and "gave no indication of feeling uncomfortable." Reading *Dreams,* Kelli found it "hard to imagine that he felt that way."

Former teammate Dan Hale explained that "the book was not, shall we say, highly thought of" among Punahou staff members because its portrayal of pot-smoking, beer-guzzling students "did not paint a very flattering picture of the school." More significant, teachers like Pal Eldredge, who knew "Barry" year upon year, and Eric Kusunoki, who saw him every morning throughout high school, agreed with Mike, Greg, Bobby, Kelli, and Barack's other fellow students about the book. Eldredge thought *Dreams* was "shocking" because "I had no clue he felt like that." Kusunoki reacted similarly. Bitter angst "never came out in his behavior. He never verbalized it, he never acted it out," and "I was totally unaware" of the disaffected attitude chronicled in *Dreams.*

When Barack's former coworkers at Business International Corporation first discovered *Dreams* a decade after its publication, they were astonished by how Barack described the company. Jeanne Reynolds Schmidt "immediately thought this was not the same place I worked! It was not a high-level consulting firm," and "it was hardly an upscale environment. And I laughed when I read in the book that he had his own secretary." Susan Arterian Chang agreed. "The BI that Barack describes in his memoirs is unrecognizable to me," and Cathy Lazere, Barack's immediate supervisor, called *Dreams*' portrayal of BI "fabricated."

Reactions of complete astonishment reached all the way to Barack's closest friends during law school. Cassandra Butts had seen an initial draft of the Kenya chapters, but when she read *Dreams*' account of Barack's pre-Chicago life, she realized that "that wasn't the person that we knew" at Harvard. "He was so far removed from that person and from that search when we knew him in law school, that was shocking, actually" to read. In subsequent years, few observers ever focused on just how glaring the contradiction was between everyone's memories of Barack, particularly from his Punahou years, and how *Dreams* portrayed a deeply troubled life. "The serene man his friends describe could not be more different from the person Obama himself describes in his memoir," Larissa MacFarquhar noted. Barack would stress that *Dreams* "was not journalism," and that "the main thing I was trying to do was just write a good enough book that I wasn't embarrassed." Yet MacFarquhar rightly observed that "the contrast between the Obama of the book and the Obama visible to the world is nonetheless so extreme as to be striking." In rebuttal, Barack insisted to her that "that angry character lasts from the time I was fifteen to the time I was twenty-one or so," but another journalist rightly wondered "how to interpret the fact that this turmoil was not evident even to his closest friends" at either Punahou or Oxy.

One careful reader concluded that "there's a very oddly detached quality to the book, almost as if he's describing somebody else." But only Jonathan Raban, an experienced world traveler as well as a distinguished novelist, would accurately take *Dreams'* full measure. The book "is less memoir than novel," he realized, an insight that the historian David Greenberg later echoed in calling *Dreams* "semi-fictional." In truth, as Barack's actual life story from the 1960s to the 1990s would subsequently reveal, *Dreams From My Father* was neither an autobiography nor a memoir. A prescient reader like Raban would note how many characters are "composites with fictional names," how *Dreams'* "total-recall dialogue is as much imagined as remembered," and how "its time sequences are intricately shuffled" while reflecting upon how novelistic *Dreams* actually was.

Less than a decade later one journalist rightly emphasized that Barack "was already weighing a political career when he wrote the book." Keith Kakugawa, just like Mike and Greg, wondered why *Dreams* so dramatically magnified their teenage years' racial tensions, but the explanation for their and others' puzzlement about *Dreams'* depiction of the Punahou years was transparently obvious *if* one realized how Barack's embrace of his own blackness during his initial three years in Chicago, and then in Kenya, had retroactively led him to reshape his entire self-presentation of the first twenty-four years of his life.

As the multiracial author Gary Kamiya perceptively put it, in *Dreams* Barack "made himself black." And, as Greg Galluzzo realized in comparing the Barack of *Dreams* to the young man with whom he had spent scores of hours, "his book could be called *A Journey to Blackness*." Barack enthusiastically told one later questioner that "I love to write," that "I love fiction, I love to read fiction, but I'm not sure I have enough talent to write fiction." Yet for once in his life, if only for one sole time, Barack Obama sold himself short. *Dreams From My Father* was not a memoir or an autobiography; it was instead, in multitudinous ways, without any question *a work of historical fiction*. True to that genre, it featured many true-to-life figures and a bevy of accurately described events that indeed had occurred, but it employed the techniques and literary license of a novel, and its most important composite character was the narrator himself.[38]

Back in Chicago following his book tour, Barack had lunch with Jesse Ruiz, his former law student who had dutifully bought a copy of *Dreams* for Barack to sign. "Hey, you might be famous one day," Jesse teased him, and Barack told him that soon he would be running for Palmer's state Senate seat. "Guys like you can help me become mayor one day," but Jesse responded dubiously: "Don't be hanging your hat on mayor."

The week of August 14 commenced with Mel Reynolds testifying in his own defense, insisting that the prosecution's tape recordings of his 1994 phone calls with Beverly Heard proved only that he had engaged in phone sex with her, not that they had had a physical relationship when she was underage. In his fourth day on the stand, however, Reynolds "unraveled" during a prosecutor's questioning, and "jurors shook their heads" during the congressman's "rambling tirades," the *Tribune* reported. Equally significant, by the middle of that week word began to spread that Jesse Jackson Jr. was definitely going to join Alice Palmer in running for Reynolds's seat, which a conviction might leave vacant well before the 1996 Democratic primary.

Monday, August 21, featured closing arguments in Reynolds's trial, with lead prosecutor Andrea Zorn mocking him as "Mr. Oxford law degree." Later that day the case went to the jury, and the next evening the jurors returned guilty verdicts on every count. Reynolds faced a minimum of four years in prison, and while he could appeal, U.S. senator Carol Moseley Braun joined the *Tribune* in calling for Reynolds's immediate resignation from Congress.

Attention quickly turned to the race to succeed him. State representative Monique Davis, a member of Trinity United Church, immediately declared her candidacy, and state senators Emil Jones Jr. and Bill Shaw, as well as 17th Ward alderman Allan Streeter, all mulled entering the race too. That Sunday's *Chicago Sun-Times* reported that "most strategists consider Palmer the early front-runner" but said Jones "would be the candidate to beat" if he entered the race. That afternoon Alice Palmer's campaign held a backyard barbecue in the Beverly neighborhood, and among the attendees was Republican state senator Steve Rauschenberger of Elgin, who already had given Palmer a $100 contribution two weeks earlier.

The thirty-eight-year-old Rauschenberger was one of the Illinois Republican Party's brightest stars, a top member of the so-called Fab Five of young senators who had joined the chamber after the 1992 election. Earlier that year GOP Senate president James "Pate" Philip had named Rauschenberger chair of the powerful Appropriations Committee, notwithstanding his lack of seniority, and *Crain's Chicago Business* said that Rauschenberger and other young Republicans, like thirty-four-year-old Palatine banker Peter Fitzgerald, "have transformed the Illinois Senate." Unquestionably one of the sharpest members of the body, Rauschenberger respected Alice Palmer as a "very bright" and "eloquent" colleague, someone whose congressional aspirations in a heavily Democratic district he had no hesitancy in supporting with another $100 check.

About 150 people attended the event, including Barack, and during the event Palmer spoke to the crowd and explained that Barack was her desired successor in the state Senate. Steve thought "that was the first time she publicly introduced Barack" and voiced her blessing. Progressive African American lobbyist William McNary remembered it too, as did Michael Lieteau, another experienced African American lobbyist who represented AT&T. Palmer's prospects seemed bright, for as former Harold Washington aide Tim Wright explained, "we all thought she was going to be congresswoman until Jesse" Jackson Jr. entered the race.

The next morning the *Chicago Defender* broke the news that Jackson would formally announce his candidacy on September 9 at a noontime rally at Salem Baptist Church. Other observers believed Emil Jones, not Jackson, might offer the stiffest competition. *Sun-Times* columnist Steve Neal, an outspoken fan, asserted that Palmer "is in the strongest position for a special election" following Reynolds's expected resignation, but other prognosticators noted that Palmer's 13th Senate District constituted only 7 percent of the 2nd Congressional District, far less than city wards like the 9th and 34th, whose Democratic organizations would support Jones.

A few days later Neal reported that Jones was definitely running and that former mayoral contender Joe Gardner would back him, which was ironic in light of reports that Mayor Daley, worried about Palmer as a potential citywide challenger should she win election to Congress, also supported Jones. Mel Reynolds appeared on *Larry King Live* to announce that he indeed would resign from Congress, effective October 1, and on September 11 Governor Jim Edgar set the primary election for Tuesday, November 28, with the pro forma general election to follow two weeks later.[39]

Barack and Carol Harwell had been busy setting up his campaign. Carol had wanted to rent an office on East 71st Street in South Shore that was already wired for multiple phone lines, but Michelle Obama insisted on a better-looking but unequipped space a few blocks west at 2152 East 71st. Barack rented computers and printers, and the Obamas provided a fancy coffee machine they had received as a wedding present to the humbly furnished storefront. The campaign also managed to secure a quite numerically appropriate phone number: 312-363-1996. Carol asked a friend to be office manager and recruited forty-three-year-old Ron Davis, the Kennedy King college math instructor whom she and Barack both knew from Project VOTE!, to be the campaign's field director.

Once Reynolds announced his resignation, guaranteeing a special election, Barack asked Palmer whether this scenario—where she now could run for

Congress without having to surrender her Senate seat—meant that he should hold off on publicly announcing his own candidacy. "If you want to hedge your bets and wait and see if you win, then I'm comfortable with that, and I can sort of keep my campaign in a holding pattern until we see what happens," Barack said he told her. "No, Barack," Palmer responded. "I'm going to win this congressional race, and so you have my blessing and my go-ahead."

Word of Palmer's commitment appeared in the *Hyde Park Herald,* which reported that "according to sources close to her campaign," Palmer "has no plans to try and recapture her Senate seat if her bid is unsuccessful." Years later, Palmer readily confirmed those reports: "I certainly did say that I wasn't going to run. There's no question about that." The *Herald* identified Barack as a "neophyte" politician who replied "warily" to questions and hoped "grass roots folks" would back him. "Obama has the support of Palmer" and also was endorsed by state representative Barbara Flynn Currie, the *Herald* stated. Carol Harwell scheduled a formal announcement for September 19 at Hyde Park's Ramada Inn, known to many as the place where Harold Washington had launched his successful mayoral candidacy, and Alice Palmer readily agreed to introduce Barack and offer a ringing endorsement.

Then daunting news arrived from Hawaii: chemotherapy had not stopped the spread of Ann Dunham's cancer, and Ann and Madelyn were about to fly to New York so that Ann could be examined by an oncologist at the Memorial Sloan Kettering Cancer Center. Maya, in grad school at NYU, met them at the airport, where Ann required a wheelchair. Over the summer Ann had kept Barack apprised about her disability insurance difficulties with Cigna, which was trying to deny her benefit claim, and within a day or two of Ann's arrival in New York, Barack and Michelle flew in from Chicago.

Ann had told at least one of her friends about her unhappiness over how her son had portrayed her in *Dreams,* particularly his rendition of the time in 1982 when he and Maya had accompanied her to the screening of *Black Orpheus.* But in New York, Ann made no mention of that, and one day Barack and Michelle joined her for lunch with her old friend Pete Vayda and his wife.

Barack also spent an hour speaking with broadcaster Bill Moyers, who at Debbie Leff's suggestion wanted to ask Barack to join the board of the Florence and John Schumann Foundation, of which Moyers was president. Then, on September 15, the oncologist informed Ann that her cancer was now stage 4 and her odds were poor, but he recommended a change to a different chemotherapy regime when she returned to Honolulu.[40]

Back in Chicago, Barack and Michelle were startled to learn that their good friend Valerie Jarrett had been bounced as Chicago's commissioner of planning and economic development. Jarrett would become executive vice president of a real estate management firm, the Habitat Co., and also take on a part-time role as chairman of the Chicago Transit Authority.

On the electoral front, state senator Emil Jones Jr. announced he soon would formally enter the congressional race. Alice Palmer supporters proclaimed that Jones was more valuable as Senate Democratic leader than he would be in Congress, and that Jesse Jackson Jr. was too inexperienced for such a post. As Barack's own Tuesday-evening announcement neared, Carol Harwell recruited former alderman Cliff Kelley, now a prominent radio host, to MC the event, and Tuesday's *Chicago Defender* publicized it with an unfortunately worded page 3 headline: "Harvard Lawyer Eyes Palmer Seat."

As the 6:00 P.M. affair got under way, "a standing-room-only audience of 200 supporters" packed the modest-sized room. Aldermen Toni Preckwinkle and Barbara Holt were present, as were Barack's old organizing colleague David Kindler and Robinson family friend John Rogers Jr. Alice Palmer had asked her most committed supporters to attend, and for longtime Hyde Park political activists Alan and Lois Dobry, it was their first opportunity to meet Barack. Kathy Stell, a young UC dean who previously had worked at the law school, had mentioned Barack and the kickoff to Will Burns, a twenty-two-year-old UC graduate who was interested in politics. When Burns arrived, field director Ron Davis was staffing a sign-in table replete with volunteer forms, and as Will filled one out, Ron asked him to specify "Field" rather than "Policy" as his preference.

Alice Palmer was the leadoff speaker, and she began by reminding the crowd that "In this room, Harold Washington announced for mayor. It looks different, but the spirit is still in the room." Then she offered her protégé a ringing endorsement. "Barack Obama carries on the tradition of independence in this district, a tradition that continued with me and most recently with Senator Newhouse. His candidacy is a passing of the torch, because he's the person that people have embraced and have lifted up as the person they want to represent this district."

Then Barack stepped to the podium and almost apologetically announced his candidacy. "Politicians are not held to highest esteem these days. They fall somewhere lower than lawyers," perhaps an implicit reference to Mel Reynolds's tattered legacy. But Barack promised to be better. "I want to inspire a renewal of morality in politics. I will work as hard as I can, as long as I can, on your behalf." John Rogers had never heard Barack speak, and he was impressed by his "total confidence" and his ability to "capture the audience the way he did."

Neither the *Tribune,* the *Sun-Times,* nor the *Defender* covered Barack's declaration of candidacy, and only a modest page-3 story in the weekly *Hyde Park*

Herald chronicled the event: "Hyde Parker Announces Run for State Senate Seat." The next day, the newly declared thirty-four-year-old politician headed to the airport to fly to Boston for a Wednesday-night book reading, but on the way he stopped at the Chicago Board of Elections to perform his first act as a candidate for office. Four years earlier, when he registered to vote after returning from law school, Barack had signed his name "Barack H. Obama." On September 20, 1995, he amended that registration to read simply "Barack Obama." For a politician, having "Hussein" as your middle name clearly merited a calculated effort to minimize the odds of anyone potentially asking a worrisome question: "What's the 'H' stand for?"[41]

Chapter Seven

INTO THE ARENA

CHICAGO AND SPRINGFIELD

SEPTEMBER 1995–SEPTEMBER 1999

The newly declared candidate's trip to Boston featured two attractive speaking venues: a Wednesday-night reading at the Cambridge Public Library and a Thursday-night appearance at the nearby Concord Festival of Authors. Barack realized when he arrived that a far more prominent African American author was also in town promoting his brand-new autobiography. Former general and potential presidential candidate Colin Powell, who was signing his *My American Journey* at a downtown bookstore, attracted a midday crowd of two thousand.

That evening Barack drew several dozen interested listeners to a modest-sized room at the Cambridge library, where he told them that *Dreams From My Father* was "not so much a memoir, I think, as sort of a journey of discovery for me." Barack read *Dreams*' exaggerated account of Mike Ramos and Greg Orme's supposedly intense racial discomfort at the mostly black Schofield Barracks party. "This sense of betrayal" that he felt from their uneasiness, Barack explained, mirrored his reaction to Stan and Madelyn's argument over the aggressive black panhandler: "I'm feeling that same sense of betrayal from my family."

Barack also read *Dreams*' account of his subsequent conversation with his sole African American confidant, whom he named in full—"Frank Marshall Davis"—and described as "a close friend" of his grandfather, whom he expressly characterized as "poor white trash." One questioner remarked that earlier that day he had heard Cornel West speak, and that West sounded very pessimistic compared to Obama.

"Cornel West has to go back to his Bible," Barack responded. "You've got to have faith," and the audience laughed as Barack quickly added, "I say that facetiously." Returning to *Dreams,* he explained that "as I was writing, I really wanted this to be a work of the imagination," and "I wanted it to read like a story and not a memoir."

In Chicago, political eyes were fixed on the contest to replace Mel Reynolds, although Barack was momentarily distracted a day after returning when he received a $75 traffic ticket for failure to stop. *Sun-Times* political columnist

Steve Neal wrote that state Senate Democratic minority leader Emil Jones Jr. "has a fair chance to win" the November 28 primary, in part because state representative Monique Davis's long-shot candidacy might cut into Alice Palmer's presumed strength among women voters, thereby helping Jones. When Jones finally announced his candidacy on September 30, the *Tribune* reported that "he is widely viewed as the person to beat" and is "heavily favored to win," especially because his Senate district comprised one-quarter of the congressional one.

Three days later, everyone's attention turned to California when a Los Angeles jury astonishingly acquitted former football star O. J. Simpson of murdering his former wife Nicole Brown Simpson and her friend Ronald Goldman, with many African Americans celebrating that verdict. Barack was crestfallen, later remarking that it "was pretty clear that O.J. was guilty, and I was ashamed for my own community to respond in that way."

Alice Palmer's congressional campaign bragged that it had raised more than $100,000, with major unions like the Service Employees (SEIU) contributing as much as $5,000, but Palmer booster Steve Neal was dismayed by the candidate's poor performance during a televised debate, writing that Palmer "talked in slogans and was unresponsive to questions." Far worse news for Palmer came from polling that all three major campaigns—hers, Jones's, and young upstart Jesse Jackson Jr.'s—were conducting. Neal, privy to the results, reported that "Palmer has dropped to a distant third" as "voters seem to know a lot more about Jackson and Jones." Palmer had been in the race longer than her opponents, but, Neal wrote, "after nearly a year of campaigning, she has failed to make much headway. . . . Jackson is known by 85 percent of probable Democratic voters in the 2nd District, and Jones is known by 67 percent," while "Palmer's poor performance is largely due to her lack of name identification." Jones's own polling showed Jackson with 24 percent, Jones with 20, and 29 percent of voters undecided. Still, Palmer told reporters she was not worried, citing a thirteen-thousand-piece mailing, ads on three radio stations, and the endorsement of the independent, anti-machine Independent Voters of Illinois–Independent Precinct Organization (IVI-IPO) coalition. A *Sun-Times* op-ed praised her "serious demeanor" and said she "clearly is the superior choice."

For Barack, Michelle's longtime friendship with Jesse Jackson Jr.'s sister Santita, plus their attendance at each other's weddings, added tension to a relationship that was now outweighed by Barack's and Alice Palmer's public support of each other's campaigns. Barack was "in a really difficult dilemma," Jackson explained years later, and one of Michelle's Public Allies alumni, Malik Nevels, who worked on Jackson's campaign, was present when Barack told a sympathetic Jackson that he could not endorse him. Indeed, when Jackson's best friend, Marty King, told one young woman that Jesse's backers could not support someone who was not supporting them, Jr. was so displeased by Marty's tone that he called the Obamas and had Marty apologize in person later that same day at a Hyde Park restaurant they all liked, Dixie Kitchen & Bait Shop. Jackson later explained that "we could think short term, or we could think long term, because we'll be in Chicago together for a long time."

As Alice Palmer's campaign headed into its final month, friends and supporters more politically savvy than Palmer told her to reconsider the race. African American lobbyist Mike Lieteau, who had attended the late August barbecue at which Palmer had introduced Barack as her successor, told her that Jackson's high-profile candidacy meant she should stand down. Research consultant Don Wiener and longtime progressive activist Don Rose both advised likewise. Fifth Ward activists Alan and Lois Dobry knew that Alice had "never built up a political base in the 7th Ward," and African American AFSCME lobbyist Ray Harris believed that "Alice was really naive when it came to politics." Her state Senate colleague Lou Viverito admired Palmer as "a very dignified lady," but he spoke for many when he said she "was not the most astute political animal." Campaign chairman Hal Baron realized that young, new-to-Chicago campaign manager Darrel Thompson had relatively little to work with given the skimpiness of Palmer's own political base, plus Alice herself was "not good at organization." Her most influential adviser was her husband Buzz, and despite a good number of upper-middle-class supporters both black and white, the actual Palmer campaign was not well organized. As much as Baron admired Alice Palmer, there was no denying that Buzz was just a classic Chicago street operator, Baron explained.

Some observers and Palmer backers were perplexed by her promise of her Senate seat to Barack when she had no need to surrender it. African American lobbyist and former state representative Paul Williams warned Barack that "there's going to be a moment when she's going to realize what she's done," and that "it made absolutely no sense" to give up her state post if Jones or Jackson won the congressional primary. Cook County Board member Danny Davis, the African American favorite for a West Side congressional seat from which the incumbent was retiring, bluntly said: "Who's this Barack Obama? What plane did he come off of?"

On the South Side, the *Hyde Park Herald* reported that Gha-Is Askia, the African American Muslim who had announced his candidacy for Palmer's state Senate seat four months earlier, was endorsed by boxer Muhammad Ali as well as by Emil Jones and black state representative Connie Howard. In addition, Marc Ewell, a thirty-year-old Howard University graduate who was the son of former state representative Raymond Ewell, a machine loyalist and sixteen-year legislative veteran whose surname was well known on the South Side, was also entering the race. In Barack's modest campaign office on East 71st

Street in South Shore, sister-in-law Janis Robinson, now mothering a three-year-old son, joined campaign manager Carol Harwell part-time to schedule more than a dozen small receptions in supporters' homes. Barack also traveled to Washington, D.C., to attend the October 16 Million Man March that Nation of Islam leader Louis Farrakhan had taken the lead in organizing. Some of the lesser-known speakers, such as Rev. Al Sampson and former congressman Gus Savage, were figures Barack was familiar with from Chicago. The historic gathering attracted widespread coverage even though only about four hundred thousand actually attended.[1]

Barack's campaign account, Friends of Barack Obama, was in decent shape thanks to $2,000 apiece from enthusiastic backers Al Johnson and Tony Rezko, as well as $1,000 donations from other developers like William Moorehead and Leon Finney and from personal friends such as Judd Miner and Bernard Loyd, plus smaller contributions from a bevy of relatives and colleagues including maternal uncle Charles Payne, Father Mike Pfleger, and the Hope Center's Sokoni Karanja. Michelle's Public Allies board member Sunny Fischer hosted a fund-raiser up in Evanston that drew more than twenty people, Barack's former law students Jesse Ruiz, Jeanne Gills, and Brett Hart hosted one at Ruiz's girlfriend's apartment that attracted about ten, and Davis Miner colleagues Jeff Cummings and John Belcaster, with Harvard basketball buddy Greg Dingens, cohosted a small late-October one at mutual friend Steve Derks's River City apartment building, where the only guest Barack did not already know was Jeff's friend Michael Parham.

Alice Palmer actively aided Barack's efforts. Davis Miner's Allison Davis remembered "her coming up to the office and asking if we would support Barack, give him money," and former 4th Ward alderman Tim Evans likewise recalled that "at the time, one of his biggest boosters was Alice Palmer." Palmer also called her fellow education advocate Bill Ayers to ask that he and his wife Bernardine Dohrn host a coffee where she could introduce Barack to Hyde Parkers who did not yet know him. Ayers had met Barack multiple times thanks to his lead role in staffing the Chicago School Reform Collaborative, whose evaluation committee played a far more consequential role in deciding which proposals the Chicago Annenberg Challenge would fund than did the CAC board of "Chicago worthies" that Barack chaired. Bill thought Barack was "enormously intelligent, bright, attractive," and so "of course" they readily agreed.

Dohrn, a 1967 University of Chicago Law School graduate who had worked at Sidley & Austin from 1984 to 1988, had met Michelle there and from 1993 forward had supervised one of Michelle's young Public Allies at Northwestern University Law School's Children and Family Justice Center, which she directed. Bill's infamous role in the Weather Underground two decades earlier had not hindered his professorial status at the University of Illinois at Chicago nor his activism in Chicago school reform circles, but Bernardine's criminal record had prevented her admission to the Illinois State Bar. A lengthy 1993 profile in *Chicago Magazine* had described them as "a lively, thoughtful, loving couple" who host "high-profile dinner parties" and stated that all across the city there was "a widespread willingness to disregard" their radical pasts.

Bill and Bernardine invited twenty to thirty people to their town house on East 50th Street, which they had purchased a year earlier. "It was Alice's initiative to have the event," Bill explained, and Bernardine recalled that "the room was packed." Bernardine welcomed everyone and introduced Alice Palmer, who, as Bill remembered, "gave a little talk" saying that Barack's "a great guy. He's wonderful. Support him." Dr. Quentin Young, a progressive physician whose partner Dr. David Scheiner had been Barack's doctor since 1987, recalled Palmer saying that Barack was "a very promising young man who will serve the district well." Young thought Obama was "extremely impressive," but a young black woman whose husband had joined UC's English faculty four years earlier later wrote that Barack's "bright eyes and easy smile" struck her "as contrived and calculated."

This gathering led to several other similar Hyde Park events. Local activists Sam and Martha Ackerman hosted one in their Hyde Park condo. Martha was exceptionally impressed with Barack, telling her husband that she could imagine him as a future president of the United States. Rabbi Arnold Jacob Wolf, a civil rights supporter who had served as Yale University's Jewish chaplain before returning to Chicago in 1980 to lead KAM Isaiah Israel Temple, the city's oldest synagogue, hosted another one and was almost as impressed as Martha Ackerman. "I said to him, Mr. Obama, someday you will be vice president of the United States," and he remembered that Barack responded, "'Why *vice* president?'"

But running for a seat in the Illinois state legislature was a decidedly low-profile and mundane task. One crucial requirement was getting the signatures of at least 757 Democratic voters registered at addresses within the 13th Senate District in order to appear on the upcoming March 19 Democratic primary ballot. Chicago's Democratic machine was famous for disqualifying independent challengers whose petition circulators did not follow all the arcane rules while gathering these signatures at front doors and on street corners. But Ron Davis, whom Carol Harwell had recruited as field director and whom Barack had known since 1992's Project VOTE!, was a master of the intricacies of the nominating petition process.

Before the end of October, Barack reached out to the city's top progressive election lawyer, Tom Johnson, whom he also had first met during Project VOTE! Judd Miner and George Galland at Davis Miner knew Johnson well, because in the 1980s, Johnson had represented first Harold Washington and then Tim Evans during the political warfare of those years. Between Davis and Johnson, Barack's petition circulators were in golden hands, and by the end of October, Barack, plus young U of C graduate Will Burns, former law students like Brett Hart and Jesse Ruiz, and young U of C dean Kathryn Stell were going door to door collecting the best possible signatures, those of people who clearly lived at the addresses the petition circulators were visiting.

Also tagging along was journalist Hank DeZutter, whom the weekly *Chicago Reader* alternative newspaper had commissioned to write a long profile of Barack, as well as photographer Mark PoKempner. DeZutter sat in one day when Barack did a Hope Center training session for eight Bronzeville women, and PoKempner snapped photos of Barack, who wore a winter coat and scarf, holding a clipboard as a similarly attired middle-aged black woman signed his petition sheet. One weekday morning, PoKempner also photographed a pensive but relaxed Barack talking on the phone in his office at Davis Miner, a framed photograph of Harold Washington and a 1992 Project VOTE! "It's a POWER Thing" poster on the wall above his computer station.

Sometime early in November, ten or so of Alice Palmer's most devoted supporters, worried about her congressional prospects, summoned Barack to the Bronzeville home of state representative Lovana "Lou" Jones, Palmer's closest legislative colleague. Jones was "the senior person there," Timuel "Tim" Black, a highly respected teacher and historian, recalled, and "Lou was very anxious to have Alice back" in Springfield if indeed she lost. "We asked him if he would step aside" if that came to pass, and "we told him, 'Whatever else you run for, we will be with you.'" Black said that Jones laid out Palmer's superior merits. "Our rationale was she's got a reputation, she's got a record already, she's well known, you are not. We want her to stay there, and we will support you in anything else."

Barack listened patiently and said he would consider their points, but he would not withdraw. "I can't do that," Barack remembered replying. "I told them we were already far into the race," that his campaign had "spent a good deal of money," that "I've been having these conversations with her for quite some time now," and that Alice had never given any "indication she was going back on her commitment."

Black recalled Barack's response similarly: "He said he had already begun to organize, and he would not be able to do that." Barack "was matter-of-fact. He was not hostile," and "it wasn't argumentative. He was very firm." No extensive discussion ensued. "It was a very brief meeting," Black explained. However, Carol Harwell recalled Barack telling her "they talked to me like I was a kid." Carol realized that by trying "to bully Barack," the Jones group's effort had backfired and "had they gone at it a little differently," perhaps Barack might have agreed to stand aside. Instead their request strengthened his resolve to stand firm.[2]

On Tuesday evening, November 7, Maya called from Honolulu. A few days earlier, she had telephoned their mother, who was in a Honolulu hospital, from New York, where Maya was now in graduate school. When she realized just how weak Ann sounded, Maya flew immediately to Honolulu. Barack had spoken with his mother several days earlier and had also realized that her time was relatively short. Sidley's Gerry Alexis remembered Barack "calling me very anguished about his mother and that she had cancer and that she was dying," but Maya's call indicated that Ann's time frame was now a matter of hours, not days. Barack booked a flight out of Chicago for the next day, but even before he could leave for the airport, word came that Ann had died hours earlier at 11:00 P.M., Honolulu time, with Maya at her bedside.

That morning Barack left a message for his friend Ellen Schumer, telling her what had happened and saying he would be away from Chicago for the better part of two weeks. Years later, asked what had been the worst experience and biggest mistake in his life, Barack cited "the death of my mother" in answer to both questions. "The biggest mistake I made was not being at my mother's bedside when she died," because she died sooner than Barack expected. "I didn't get there in time." Ann's body was cremated, and her children, her mother, and her Honolulu friends held a memorial service in the Japanese garden at the U of H's East-West Center before taking her ashes to the rocky Halona lookout point near Hanauma Bay and Sandy Beach on Oahu's southeast shore. There Barack cast Ann's remains into the water after a huge wave washed over him and Maya.

Barack stayed in Honolulu for more than two weeks, helping Maya and Madelyn with Ann's difficult affairs. The federal lien of more than $17,000 in income taxes from 1979 and 1980 remained unpaid. Barack also felt guilty that *Dreams From My Father* had largely ignored her role in his life while portraying her as being clueless about her son's experience, and now identity, as a black man in America. Barack later wrote that she had died "with a brutal swiftness," and he recounted that she did not "comment on my characterizations of her" when she had seen drafts of *Dreams* but had been "quick to explain or defend the less flattering aspects of my father's character."

Barack would insist that "what is best in me I owe to her," and later expanded that posthumous sentiment to say, "Everything that I am I owe to her. She was the kindest, most generous person I ever met," and "when I am confronted with difficult choices, I have to ask myself, what would she expect of me?" He had never spoken so glowingly during Ann's lifetime of her impact on his life, but in the years following her death at age fifty-two, his memories of her became far warmer than they had ever been when she was alive.[3]

On November 21, Barack left another message for Ellen Schumer, telling her he would be back in Chicago by the end of Thanksgiving weekend. While he was away, the congressional race between Emil Jones Jr., young Jesse Jackson Jr., and Alice Palmer intensified as Election Day neared. Some anonymous "opposition research," almost certainly from Jones's camp, revealed that Jackson's salary at his father's Rainbow Coalition had come from the Hotel & Restaurant Employees International, or, as a *Tribune* headline put it, "Mob-Linked Union Paid Jackson." But Jackson was demonstrating that he was the most adept of the contenders. When he and Jones were on the radio together, and Jones cited his impressive state legislative record, Jackson immediately said, "That's why we need you to stay in Springfield." After another joint taping turned "raucous," Jackson's campaign quickly created a television ad, with Jackson saying "they're experienced politicians, but this is how they behave." Then Jones was heard remarking, "Jesus Christ," while Palmer uttered the phrase "sour grapes." Then Jackson says: "Instead of talking about the issues, they're talking about each other. Let's put an end to politics as usual. The people of the 2nd Congressional District deserve better."

The *Tribune* "warmly endorsed" Palmer, praising her "political independence and wisdom." The next day a *Tribune* profile cited her reputation "as one of the more intelligent, productive and respected members of the General Assembly," although it also noted that she could seem "dull and deliberate." In an audience vote after a candidates' forum at Palmer's church, Jackson outpolled her 43 to 26. A *Tribune* poll two weeks before the November 28 election confirmed that Jackson's name recognition had helped him jump out to a healthy 33 to 20 percent lead over Jones, with Palmer trailing badly at 8 percent. Ninety-seven percent of voters knew Jackson, but Jones and Palmer were known by only 69 and 61 percent, respectively.

Jones and Palmer appeared financially competitive with Jackson, and all three campaigns had raised more than $200,000. The *Sun-Times* endorsed Jones over Palmer, calling him "the most experienced and capable lawmaker" in the race. On the Sunday before Election Day, all three candidates visited dozens of churches, with Jackson attracting the most attention by invoking the language of Martin Luther King Jr. "I dream that one day the South Side and the south suburbs will look like the North Side and the north suburbs." On election morning, the *Sun-Times* reported that "contributions to Palmer all but dried up during the past three weeks" after polls had shown her trailing. *Sun-Times* political columnist Steve Neal wrote that "a close finish is anticipated" between Jones and Jackson, while Palmer and long-shot contender Monique Davis "have been left in the dust." Neal, who had been a booster of Palmer's, now underscored how Jesse Jackson Jr. had outshone his older, more experienced rivals, and he called Jackson "the most impressive newcomer on the local political scene."[4]

Kitty Kurth, Palmer's media consultant, later recalled how on election morning there were still more than a thousand Palmer yard signs sitting in the campaign office. Fifth Ward activist Alan Dobry realized that "Alice didn't have enough people to cover the precincts" all across the district. Even worse, at a South Shore precinct in Palmer's own Senate district that Dobry was manning, "people there didn't know who Alice was. She had built no base there," while "everybody knew Jesse." A good many Democratic state Senate staffers had been working on Emil Jones's congressional campaign during vacation time and available weekends, and on Election Day, they realized that Jackson had "an amazing organization" with "such a coalition of people" that it seemed like "they were everywhere," young John Charles remembered. That morning Barack and his campaign manager, Carol Harwell, aiming to secure more good signatures on his own nominating petitions, "traveled around, and we actually went to Alice's home precinct," Carol recalled. Voter turnout there was very low, and they realized "This is not good" for Alice's already dim prospects. Indeed, every precinct they visited had handfuls of older women all wearing Jackson buttons.

Jackson's campaign had used a software program to identify every registered Democrat who had voted in both the 1994 and 1992 elections. Those fifty-eight thousand people, from forty-four thousand households, were concentrated in seventy-eight particular precincts, which Jackson's campaign then targeted. In the final weeks, Jackson's father also began actively supporting his campaign. One person privy to the family's internal dynamics explained, "Jesse Jackson Jr. going to *Congress* is a problem for another Jesse Jackson, who doesn't support us until *late* in the campaign when the momentum is so overwhelmingly on our side that he supports us."

On Election Day afternoon, Barack chaired a meeting of the Chicago Annenberg Challenge (CAC)'s board of directors. The first actual grants to Chicago schools and their organizational partners were to be announced in three weeks, but board members plus executive director Ken Rolling were "all concerned about how things were proceeding," the minutes noted. One concern was that "the Challenge may not be seeing new or different attempts at school reform." Pat Graham emphasized how CAC's top priority was "direct services to students," and the board was "'underwhelmed' by the current proposals." Although board members saw only summaries, and not the full applications, "they felt the proposals did not reflect much imagination or 'breaking the mold' in school reform efforts."

That evening Barack, Carol Harwell, Ron Davis, and campaign volunteers like Jesse Ruiz monitored the returns in their campaign office before Barack went to Alice Palmer's election-night gathering. The results were even more dramatic than preelection polls had suggested: Jesse Jackson Jr. triumphed with more than 48 percent of the vote, Emil Jones was second with a respectable 38 percent, and Palmer trailed badly, winning only 10 percent. Monique

Davis received just 2 percent. Jones and Jackson had run neck and neck in the south suburban part of the district, but within Chicago, Jackson carried ten of eleven wards, and even in Jones's home ward, he prevailed by only 4,973 to 4,911. Jones accepted his opponent's argument that voters "didn't want to lose me in Springfield. That's what I think cost me the election."

Wednesday morning's *Tribune* reported that a "disappointed" Alice Palmer "told a small gathering at a Harvey hotel that she wouldn't seek reelection to the state Senate," and Barack was present to hear that. He told Hal Baron, Palmer's campaign chairman, that he definitely would remain in the race, but minutes later, Palmer's husband Buzz asked Baron for help in getting Alice to change her mind. Baron declined, but Buzz began recruiting other Palmer supporters who agreed that she should reclaim her seat: political science professors Robert Starks and Adolph Reed plus Alice's legislative friend Lou Jones, an outspoken woman who forcefully upbraided her, "Are you a damn fool? Are you crazy?" Palmer continued to refuse the entreaties, but radio host and former alderman Cliff Kelley, who had presided at Barack's announcement two months earlier, called Barack and asked him to step aside. Barack's two top aides, Carol Harwell and Ron Davis, plus Michelle Obama, all argued that he should reject the unfriendly invitation. As Carol remembered, "I really saw turmoil in his face," and Barack said, "we may think about not doing it." Old Southeast Side organizing friend Rev. Bob Klonowski recalled having breakfast with Barack one morning as he debated, "what do I do about this?" But Carol and Ron kept reminding him how much money the campaign had raised and spent, and that Palmer had repeatedly and publicly said she would surrender her state Senate seat even if she lost the congressional race. Michelle Obama, no matter how ambivalent she was about her husband's prospective political career, firmly agreed with Carol's argument: "She gave her word. . . . You don't let people just walk over you like that."

Many people in Chicago independent politics viewed stepping back in deference to someone senior as an almost automatic response. One progressive legislator said Barack's refusal to do so was "a mortal sin," and that, as Lou Jones had said a few weeks earlier, "your time will come." Former legislator Paul Williams explained how the black nationalists who were close to Palmer shared that view, for "under the nation concept, Barack maybe should have stepped back because for the sake of the nation: Alice had the seniority, Alice had the ability to do things." But savvy black onlookers believed that a decisive factor in why Barack did not do so was that he had not grown up black in a city like Chicago. "I think his attitude was different than a black man raised in black America," Tim Black reflected. "We grew up in black communities and black neighborhoods," so "we came up with self-imposed limits," Paul Williams explained. "Barack didn't have all those mental limitations that we had." Carol

Harwell, a quintessential strong black woman, put it even more bluntly: "Barack doesn't know what it's like to be a black man," she recalled. "He grew up in a white world," not in "a neighborhood where he feared the police officers," she recounted. "He's an African and an American, not an African American," because "he doesn't understand the hurt that black men feel."

Political scientist Adolph Reed, who had attended the earlier Lou Jones meeting, remembered that Barack "struck me then as a vacuous opportunist, a good performer" mouthing "empty rhetoric." But more politically experienced Palmer backers, such as Lois and Allan Dobry, like Hal Baron, did not support her reentry into the Senate race and pulled back. Yet a small band of obstinate supporters—Professors Reed and Robert Starks, along with Lou Jones—continued to press Palmer to reclaim her seat. On Monday, December 4, they went public, and a story in the next morning's *Chicago Defender* was headlined "Draft Palmer Campaign Launched." Jones told the paper that "if she doesn't run for the Senate, the organization gets the seat. I'm asking Obama to release Palmer from her commitment. We need her in Springfield."

Robert Starks said, "If she doesn't run, that seat will go to a Daley supporter," which was possible because many in Hyde Park had seen an "Elect Marc EWELL" vote-mobile advertising the candidacy of someone whose surname clearly marked him as an organization loyalist. "We have asked her to reconsider not running," Starks explained, "because we don't think Obama can win. . . . He hasn't been in town long enough" and "nobody knows who he is." Anyone with a memory could recall how nine years earlier two little-known Democratic statewide primary candidates with unusual surnames, George Sangmeister and Aurelia Pucinski, had lost to Lyndon LaRouche supporters with Anglo-Saxon, plain vanilla names: Janice Hart and Mark Fairchild. Marc Ewell's familiar surname might serve him very well against Barack Obama.

Barack told reporter Chinta Strausberg he was meeting with Alice Palmer Tuesday morning, and he planned to file more than three thousand petition signatures in Springfield the following Monday. The next day Barack told his friend Ellen Schumer that he had met with Palmer, and years later Barack said this ending of his relationship with Palmer "left a little bit of a bitter taste in my mouth." He learned that "if you're going to be involved in this process that you end up having to play hardball and battle it out."

Palmer has no memory of that meeting, but three days later, she agreed to enter the race to retain her seat. Her closest Springfield friend, longtime Senate Democratic policy staffer Nia Odeoti-Hassan, said, "She did not want to renege," but Palmer bowed to her supporters' insistence. Asked if "you're never taking the initiative individually yourself?" Palmer replied "absolutely not." After the congressional loss, "I wasn't particularly engaged in this." Palmer's district aide, Constance Goosby, who supported the draft effort, attempted to justify Palmer's move by saying that Barack "was supposed to help Alice with the congressional race by recruiting workers," and "he didn't fulfill his obligation at all."[5]

The same day Palmer affirmed her desire to keep her seat, the weekly *Chicago Reader* published an eight-page profile of Barack, accompanied by a long excerpt from *Dreams From My Father.* Hank DeZutter's article offered a richly detailed portrayal of the first-time candidate's political views, and Barack's frank and revealing comments resonated with multiple echoes from his organizing days and from his years at Harvard. He also made clear how disappointed he was with Chicago electoral politics. "I am surprised at how many elected officials—even the good ones—spend so much time talking about the mechanics of politics and not matters of substance. They have . . . this overriding interest in retaining their seats or in moving their careers forward, and the business and game of politics, the political horse race, is all they talk about. Even those who are on the same page as me on the issues never seem to want to talk about them. Politics is regarded as little more than a career," and Barack wished officials instead would focus on "ways to use the political process to create jobs for our communities."

DeZutter had read *Dreams* in its entirety, and he commented that the book "reads more like a novel than a memoir." DeZutter also located Johnnie Owens, who was no longer at the Developing Communities Project, but who praised Barack effusively. "He's not about calling attention to himself. He's concerned with the work. It's as if it's his mission in life, his calling, to work for social justice. . . . I'm one of the most cynical people you want to see," Owens volunteered, "but I see nothing but integrity in this guy."

DeZutter also spoke with Jean Rudd at the Woods Fund and ACORN's Madeline Talbott. Jean described Barack as one of Woods's "most hard-nosed board members in wanting to see results. He wants to see our grants make change happen, not just pay salaries." Madeline said that "Barack has proven himself among our members." DeZutter recounted Barack conducting a Hope Center training: "We talk 'they, they, they' but don't take the time to break it down. We don't analyze. Our thinking is sloppy. And to the degree that it is, we're not going to be able to have the impact we could have. We can't afford to go out there blind, hollering and acting the fool, and get to the table and don't know who it is we're talking to—or what we're going to ask them—whether it's someone with real power or just a third-string flak catcher."

One on one, Barack addressed what needed to change in black Chicago, echoing what he had told Mike Kruglik years earlier. "All these churches and all these pastors are going it alone. And what do we have? These magnificent palatial churches in the midst of the ruins of some of the most run-down neighborhoods we'll ever see. All pastors go on thinking about how they are going to 'build my church,' without joining with others to try to influence the factors or forces that are destroying the neighborhoods. They start food pantries and community-service programs, but until they come together to build something more than an effective church, all the community-service programs, all the food pantries they start will barely take care of even a fraction of the community's problems."

Barack voiced a larger vision. "In America, we have this strong bias toward individual action . . . but individual actions, individual dreams, are not sufficient. We must unite in collective action, build collective institutions and organizations." Indeed, "what we need in America, especially in the African-American community, is a moral agenda that is tied to a concrete agenda for building and rebuilding our communities. . . . We must invest our energy and resources in a massive rebuilding effort and invent new mechanisms to strengthen and hasten this community-building effort." Wasted potential was a recurring theme. "In every church on Sunday in the African-American community, we have this moral fervor; we have energy to burn. But as soon as church lets out, the energy dissipates. We must find ways to channel all this energy into community building. The biggest failure of the civil rights movement was in failing to translate this energy, this moral fervor, into creating lasting institutions and organizational structures."

Barack also addressed Harold Washington's legacy, calling him "the best of the classic politicians . . . but he, like all politicians, was primarily interested in maintaining his power and working the levers of power. He was a classic charismatic leader, and when he died all of that dissipated. This potentially powerful collective spirit that went into supporting him was never translated into clear principles, or into an articulable agenda for community change. The only principle that came through was 'getting our fair share.' . . . When Harold died, everyone claimed the mantle of his vision and went off in different directions. All that power dissipated. Now an agenda for getting our fair share is vital. But to work, it can't see voters or communities as consumers, as mere recipients or beneficiaries of this change," a view that echoed John McKnight. People must be "producers of this change. The thrust of our organizing must be on how to make them productive, how to make them employable, how to build our human capital, how to create businesses, institutions, banks, safe public spaces—the whole agenda of creating productive communities," just as McKnight had long argued.

Barack riffed onward. "What if a politician were to see his job as that of an organizer, as part teacher and part advocate, one who does not sell voters short but who educates them about the real choices before them? As an elected public official, for instance, I could bring church and community leaders together easier than I could as a community organizer or lawyer. We could come together to form concrete economic development strategies . . . and create bridges and bonds within all sectors of the community. We must form grassroots structures that would hold me and other elected officials accountable for their actions."

Alluding to the pressure to defer to Palmer, Barack said, "I am also finding people equivocating on their support. I'm talking about progressive politicians who are on the same page with me on the issues but who warn me I may be too independent." In his own campaign, "I want to do this as much as I can from the grassroots level," relying on small contributors rather than large ones. "But to organize this district I must get known," and "before I'm known . . . I'm going to have to rely on some contributions from wealthy people." Apart from the early $2,000 each from Al Johnson and Tony Rezko, the sole contribution of more than $1,000 to Barack's campaign had come in the name of Janet Gilboy, wife of Project VOTE! fund-raiser John Schmidt, now a top U.S. Department of Justice official. But "once elected, once I'm known, I won't need that kind of money, just as Harold Washington . . . did not need to raise and spend money to get the black vote," Barack said.

He concluded by critiquing the Million Man March. "What was lacking among march organizers was a positive agenda, a coherent agenda for change. Without this agenda a lot of this energy is going to dissipate" rather than be directed toward the top problem he had cited at the outset to DeZutter: "our unemployment catastrophe."

In a shorter but no less revealing interview with a young white contributor to the weekly *Hyde Park Citizen,* Barack spoke more as the author of *Dreams.* "If I'm trying to pick up a cab in New York City, I can't hold up a sign saying I'm multicultural," Barack volunteered, but likewise "it's damaging when blacks refuse to speak proper English or read correctly in an attempt to 'out black' each other. Some of our youth view education as white, and that's damaging. We all need to get to the point where we can be diverse . . . we can listen to Mozart as well as Miles Davis, we can play chess as well as basketball." Barack was especially disappointed with "youth who have opportunity yet try to adopt this . . . urban street gang mentality," youth "trying to prove their blackness" who "reject the values of self-discipline, respect, and love which has been the glue that has kept the African-American community together."

Yet ultimately, "it's about power. My travels made me sensitive to the plight of those without power and the issues of class and inequalities as it relates to wealth and power. Anytime you have been overseas in these so-called third world countries, one thing you see is the vast disparity of wealth [between] those who are part of [the] power structure and those outside of it." But in any "environment of scarcity, where the cost of living is rising, folks begin to get angry and bitter and look for scapegoats. Historically, instead of looking at the top five percent of the country that controls all the wealth, we turn towards each other, and the Republicans have added to the fire."[6]

On Monday, December 11, one week before the final deadline, Barack filed more than four times the necessary 757 petition signatures in Springfield. Ron Davis "was a stickler for how to do petitions," campaign volunteer Will Burns remembered, and two weeks earlier, attorney Tom Johnson had reminded Davis and Carol Harwell that "the Chicago Board of Elections did its biggest purge ever this year. As a result, lots of people who think they are registered are, in fact, not registered." Tom, Ron, and Barack reviewed and blessed all of their petition pages, and only one other candidate, virtually unknown 17th Ward resident Ulmer D. Lynch, also filed early. The next day Jesse Jackson Jr. swamped a Republican opponent in the special general election, winning 76 percent of the vote, and less than forty-eight hours later he was sworn in as Mel Reynolds's congressional successor. The night before that ceremony, campaign consultant Delmarie Cobb joined Jackson and his wife Sandi for a quick meal at a D.C. café that doubled as a bookstore, and Jesse remarked that he needed to find a copy of Barack Obama's *Dreams From My Father*.

Barack's campaign scheduled a Saturday-afternoon Christmas party for December 16 at its headquarters, and more than two dozen attendees wrote out contribution checks for $150 or more. Friends like Elvin Charity and Bernard Loyd were joined by newer acquaintances like Robert Blackwell Jr., another young black professional, Dr. Quentin Young, and even *Chicago Reader* author Hank DeZutter, who had contributed $300 two weeks earlier. Then, as Carol Harwell remembered, "somebody came in and said, 'Alice Palmer's on the corner circulating petitions.'"

Harwell immediately went to have a look. "They weren't asking anything," she said of the people who were soliciting. "I didn't live in the district," but "I signed the sheet, gave them a bogus address. They never said anything, they never looked at it." With just a week before the submission deadline, the Palmer loyalists had only begun to collect the signatures necessary for Palmer to appear on the March 19 primary ballot. Palmer did not ask her Senate leader, fellow congressional also-ran Emil Jones, or Jones's chief of staff, Mike Hoffmann, for help with her petitions. Instead, Palmer's two secretaries, Beverly Criglar in Springfield and Constance Goosby in Chicago, informally asked several Senate Democratic staffers who had worked on Jones's congressional campaign to help out, but the actual on-the-street circulating fell to Goosby and a handful of neighborhood youngsters. Asked years later who had circulated her petitions, Palmer said, "I don't know," adding that whatever transpired was "pretty much removed from me."

On Monday morning, December 18, Palmer formally announced her entry into the race. Emil Jones and fellow state senators Donne Trotter and Art Berman all endorsed her, as did Lou Jones, Robert Starks, and 5th Ward alderman Barbara Holt. From there Palmer left directly for Springfield, where she submitted petition sheets containing 1,580 signatures. Barack told reporters that "my attitude is one of disappointment on a personal basis," since "I am disappointed that she's decided to go back on her word to me." He said Palmer's decision was "indicative of a political culture where self-preservation comes in rather than service," even by someone with "her reputation for integrity." Previously announced candidates Gha-Is Askia and Marc Ewell also filed their petitions, and Barack said, "I intend to run a positive campaign and win in the March 19 primary." At lunch on December 20, Ellen Schumer told Barack that "this is going to burn you," since in Chicago politics "You've got to pay your dues. I told him to back off and let Alice stay in office." But Barack's resolve, and what Ellen saw as his ambition, were fully crystallized, irrespective of what his prospects might be. Palmer booster Robert Starks told the *Defender* that "he's in and so is Alice and she will win." Even Emil Jones, citing Palmer's "courageous representation," stated that "I'm very pleased that she has decided to run again."

Palmer's decision to enter the race angered a number of people who admired both her and Barack. St. Sabina's Father Mike Pfleger made a "very painful" phone call to Alice to say, "You can't do this. . . . I think you're making a mistake. . . . You stepped out, and understand I'm going to stay with Barack." But Palmer faced a more serious problem, because Carol Harwell and Ron Davis knew how carelessly her petition signatures had been gathered. On the same day Palmer submitted her sheets, Davis went to Springfield to do what experienced political operatives always did: obtain photocopies of her petitions, as well as those submitted by Ewell, Askia, and Lynch, so that the Obama workers could review them to see if there were as many invalid signers as Harwell expected.

Harwell was waiting at the Chicago train station when Davis returned with the copies, and after a quick stop at Harold's Chicken, a famous Chicago chain, Carol and Ron began reviewing Palmer's filings. It did not take long to confirm that her sheets were "just pure garbage." Volunteers Will Burns, Kathy Stell, and Lois and Alan Dobry pitched in too, and it soon became clear that Ewell's and Lynch's petitions were almost as error strewn as Palmer's, leaving all three candidates, and perhaps Askia too, vulnerable to a formal challenge before the Cook County Electoral Board.

Barack was initially nonplussed when Ron and Carol told him that Palmer, and perhaps the other candidates, might be knocked off the March primary ballot. "I don't think he thought it was, you know, sporting," Will Burns remembered, although the veteran Davis "was more than happy to knock someone off the ballot." Once Barack saw how bad Palmer's petitions were, his discomfort gave way to acceptance. Even apparently good signatures were from voters whose residences were outside the boundaries of the 13th Senate District, and as the evidence of faulty signatories grew, Tom Johnson on Tuesday, December 26, filed the notarized declaration and objector's petitions that Ron Davis had executed four days earlier, asserting that all of Obama's potential opponents had fewer than the 757 valid signatures required to qualify for the primary ballot.

Tom Johnson remembered Davis (who died at age fifty-six in 2008) as a crucial figure in Barack's early political education. Ron "really knew election work" and was also "a link to a lot of the South Side I think that Barack did not know." Davis was "really his adviser" and "was very close to him—not just really petitions." Worries about Palmer's signatures spread when two of Emil Jones's Springfield Senate staffers learned of Johnson's filing and looked at copies of Palmer's sheets. "When you saw her submission, you knew she was in trouble," deputy chief of staff Dave Gross recounted.

After a break for the holidays, on Tuesday, January 2, the Cook County Electoral Board hearing on the Obama campaign's four petition challenges began with Alan and Lois Dobry, Kathy Stell, and longtime Hyde Park independent Saul Mendelson sometimes spelling Davis during the tedious, signature-by-signature review before hearing officer Lewis Powell that ran for five full days. West Side state senator Rickey Hendon happened to glance in on the proceedings, and he recognized the Dobrys as Palmer supporters who were now helping Barack in trying to have her thrown off the ballot.[7]

Also on January 2, the South Side's longest-running social tragedy came to a bittersweet end in a South Chicago meeting hall. Three weeks earlier, the last lawsuit over the shutdown of Wisconsin Steel more than fifteen years earlier had finally settled. In 1988, attorney Tom Geoghegan had won $14.8 million for the former Wisconsin steelworkers from Navistar, the successor corporation to International Harvester. In 1991, in *Frank Lumpkin v. Envirodyne Industries,* Geoghegan, along with fellow attorneys Tom Johnson and Leslie Jones, Johnson's wife, won a ruling from the U.S. Seventh Circuit Court of Appeals holding that the tiny firm that had bought Wisconsin Steel from Harvester was also liable for former workers' unpaid pensions.

Through it all, Frank Lumpkin's Save Our Jobs Committee (SOJC) had held annual gatherings to mark the anniversary of the 1980 closing that initiated the Southeast Side's decline. Richard Longworth of the *Chicago Tribune* also remained dedicated to the workers' struggle, writing in March 1995 about "the bitter nostalgia that still rules their lives" and then writing a long profile of the seventy-eight-year-old Frank Lumpkin. "There are people—other steelworkers, lawyers and politicians who have worked with him, reporters who have written about him—who consider Frank Lumpkin the best man in Chicago," Longworth wrote. Months later, in early December 1995, Longworth finally was able to announce the last chapter in the long-running saga: "Wisconsin Steel Deal May End Ex-Workers' Agony."

Envirodyne would sign over nine hundred thousand shares of its own stock, valued at about $3.37 million, to the former workers. Envirodyne now was worth far less than when founder Ronald K. Linde had sold his stake six years earlier, walking away with $75 million, but the roughly $18.5 million in total recompense from Navistar and now Envirodyne was a victory for the aging former workers. "Altogether, from these two, they may have gotten back more than half of what they lost," Geoghegan told Longworth. Individual workers would receive $3,000 to $6,000 apiece from the new settlement, but there were no complaints whatsoever at SOJC's celebratory rally on the night of January 2 in South Chicago. "We sure got a whole lot more than anybody thought we were going to get in 1980," Geoghegan remarked.[8]

Ron Davis and Carol Harwell gave Barack a nightly update on each day's developments at the electoral board. Volunteer Saul Mendelson recalled that signatures were struck if they were obvious forgeries, if names were printed rather than signed, if the individual had listed an address outside the 13th Senate District, or if the individual was not registered at the specified address. A routine purge by the Chicago Board of Election Commissioners in 1995 had dropped 15,871 names from the 13th District's rolls, which meant that many people who had been registered but had long failed to vote were no longer on the rolls. As the days passed, it became increasingly clear that Alice Palmer and Ulmer Lynch would fall well below the 757 minimum, and it was close for Marc Ewell and Gha-Is Askia. As the chances of striking all four from the ballot grew, Barack worried out loud to Ron and Carol about how "shotgunning" all of his opponents might look.

At Saturday's IVI-IPO endorsement session, Northwestern political science professor Adolph Reed, who had just moved to South Shore, plus 5th Ward alderman Barbara Holt, likewise a relatively new member of the organization, argued that the group should endorse Alice Palmer rather than Barack. Saul Mendelson, who had worked on the petitions challenge and knew that many of Palmer's signatures were faulty, expressed high regard for Palmer yet utter disdain for those who—like Reed—had dragooned her into the race but then left her victim to amateurish signature gathering. Afterward Reed wrote an op-ed piece for New York's *Village Voice* in which he excoriated an unnamed candidate whose "vacuous-to-repressive neoliberal politics" included "his fundamentally bootstrap line" about the needs of black America. The allusion to Booker T. Washington was an academic's way of calling Barack an Uncle Tom, one whose base is "mainly in the liberal foundation and development worlds," and Reed rued how "I suspect that his ilk is the wave of the future in U.S. black politics."[9]

On Sunday Barack completed an issues questionnaire from a group whose endorsement he sought. "Impact" was Illinois's Gay and Lesbian Political Organization, and Barack hand-printed his responses. He said he supported domestic partnership legislation, and he was for outlawing discrimination based on sexual orientation. "I oppose restrictions on a woman's right to choose, including various notification statutes, and support public funding of abortions." He opposed any restrictions upon gay people as foster or adoptive parents and "I plan to set up a gay/lesbian task force in the district to identify and promote issues." He also wrote that a state "should not interfere with same-gender couples who chose to marry" and "I would support such a resolution" allowing them to do so. In a subsequent letter to a gay newspaper, Barack wrote, "I favor legalizing same-sex marriages, and would fight efforts to prohibit such marriages." He also reiterated, "I support a woman's right to choose an abortion, favor Medicaid funding of abortions for poor women, and oppose parental notification laws."[10]

On Monday afternoon, Tom Johnson jotted notes as the Cook County Electoral Board hearings concluded. Of Alice Palmer's 1,580 signatures, objections to 1,031 had been sustained, leaving her with 549 valid ones, and a recheck of the so-called "master file" had added back only 10 more, giving her a grand total of 559, 198 less than the 757 necessary to make the ballot. Ulmer Lynch had also fallen woefully short, while Marc Ewell was 86 shy of 757 and Gha-Is Askia 69. Wednesday's *Hyde Park Herald* was first to report that the Senate race "may become no contest at all" because Obama's opponents "may not have the required amount of valid nominating petition signatures." The next morning a *Sun-Times* gossip columnist picked up the news: "Watch for state Sen. Alice Palmer to be knocked off the ballot in her re-election bid" as "she is going to be a couple of hundred short. Unbelievable."

Later that day Barack met with another group whose backing he sought, Madeline Talbott and Keith Kelleher's ACORN-based Chicago chapter of the nascent "New Party" progressive fusion group. The meeting's minutes recorded that Barack made a statement and answered questions. He "signed the New Party 'Candidate Contract'" and requested their endorsement. He "also joined the New Party." One attendee later remembered that Barack "was blunt about his desire to move the Democratic Party off the cautious center where Bill Clinton had it wedged," but another member recalled that he "seemed to measure every answer to questions put to him several times" before finally responding. Years later critics tried to make Barack's "membership" in the New Party a matter of controversy, but neither Keith Kelleher, who tried multiple times to interest Barack, nor national cofounder Joel Rogers, whose wife Sarah Siskind was a Madison-based partner in the Davis Miner law firm, ever had succeeded in winning Barack's cooperation. "I think Barack always was very strategic," Keith's wife Madeline Talbott explained, and "it was obvious that he was not going to participate in the New Party."[11]

On January 17 the *Hyde Park Herald* reported that Alice Palmer would not be on the primary ballot unless her supporters produced at least 198 "signed and notarized affidavits from the registered voters who signed her nominating petitions." Realizing how impossible that was, Palmer announced prior to the electoral board's final review that she was withdrawing. Senate colleagues like Howard Carroll were "shocked" that such a "wonderful person" was now a lame duck, and Barack told the *Defender,* "I got involved in this race based on Alice's original endorsement, and I continue to respect her" and the "significant contributions" she had made as a senator. Palmer's former consultant Don Wiener noted that "Alice could have won if someone competent had handled her petitions," and Robert Starks later confessed that "we were in too much of a rush and that leads to bad signatures." As Palmer's Senate colleague Rickey Hendon stressed, "no one thinks Barack could have ever beaten Alice Palmer" had she made it onto the March 19 primary ballot.

Like Palmer, Gha-Is Askia stood down, admitting he had paid his petition circulators $5 per sheet, and that "they round-tabled me," signing most of the names themselves. By January 20, the electoral board's review confirmed that Marc Ewell was short by eighty-six signatures, and on January 24, Ulmer Lynch filed a somewhat illiterate *pro se* lawsuit in federal district court against the state and city boards of election, as well as Tom Johnson and "Barak," alleging that "Johnson, and Mr. Obama, used there political connections to influence some Board of Election Commissioner officers and employees, to tamper with authority to remove voter's files out of board binders." The board soon filed a motion to dismiss, which led Tom to tell Barack that they could "sit back and see what happens."

Marc Ewell, represented by his father Ray, tried to argue that registrants whose names appeared on the prepurge, February 1995 list of voters should count as valid signatories, but on February 9 the board formally held that Ewell had submitted only 671 valid signatures. Four days later, Ray Ewell filed suit in federal court against the city election board, challenging its purge of previously registered voters and seeking an injunction to prohibit Marc's name from being taken off the ballot. Ewell's filing was a more professional effort than Lynch's, but it referenced "Barach Obamo," and when the case was assigned to U.S. district judge Ruben Castillo, Ewell's filings repeatedly cited "Judge Castello."[12]

Obama's successful challenge against Palmer received public blowback in both the *Hyde Park Herald* and then the *Chicago Sun-Times*. An angry Adolph Reed denounced the challenge as "vicious and underhanded" and lambasted the IVI-IPO for failing to support "open access to the ballot." Lois Dobry said that "the sad thing is that we've lost a person of this caliber" from elected office, and prominent black Chicago political commentator Salim Muwakkil authored a *Sun-Times* op-ed headlined "Candidate Not What He Seems, Foes Insist." Muwakkil wrote that Robert Starks "says Obama is the tool of forces outside the black community," and quoted Alice Palmer as saying of Obama that "I've since discovered that he's not as progressive as I first thought."

Palmer refused to say whether she would support Barack if he became the Democratic nominee, but she and Senate colleague Donne Trotter introduced SB 1554 to require that boards of elections publish the names of voters whom they purged from registration rolls. The bill went nowhere, but the animosity Palmer's nationalist supporters felt toward Barack would endure for years. Barack's Project VOTE! mentor John Schmidt remembered Barack mentioning how "the black nationalists were really going after him," but that "he found it kind of liberating because he realized that he didn't need them."

The Chicago Annenberg Challenge board chairmanship continued to take a lot of Barack's time, but whenever someone, like his friend Ellen Schumer, asked how applications should be crafted, he emphasized that proposals should focus solely on improving "student academic achievement in the classroom." In late February Barack joined a Sunday-evening "town meeting" on "Economic Insecurity: Employment and Survival in Urban America" sponsored by the University of Chicago's student chapter of DSA, the Democratic Socialists of America, whose parent Chicago-wide chapter had endorsed Barack's state Senate candidacy.

Barack, or "Bracco Obama," as the university's student newspaper spelled his name, reminded the crowd of more than three hundred how "jobs, workers, and unions were at the front and center" of the civil rights movement's famous 1963 March on Washington. Since then, "this emphasis on the centrality of jobs" had been lost. Barack stated that "resources, training and education . . . are the bare minimums of the American Dream . . . and right now these are not being provided." Referencing the federal welfare reform proposal President Bill Clinton was backing, Barack criticized "welfare elimination and reduction under the guise of reform." He argued that "it is not enough to maintain the status quo . . . it is not tolerable" and declared that "how jobs are allocated among respective geographic and ethnic communities" was a pressing issue.[13]

At a February 21 hearing, Judge Ruben Castillo expressed serious doubt as to whether Raymond Ewell's challenge of the electoral board's decision that his son was short eighty-six signatures should have been filed in federal as opposed to state court. James M. Scanlon, representing the board, and Tom Johnson, for Obama, highlighted how Illinois state law expressly gave rejected candidates ten days to challenge such a decision in Cook County Circuit Court. Ewell belatedly filed suit there on February 23, albeit in an incorrect division, and failed to notify Scanlon and Johnson of the filing, but on March 4, Judge Castillo granted Scanlon and Johnson's motions for summary judgment and dismissed the Ewells' lawsuit.

Tom Johnson immediately faxed Castillo's ruling to Barack with a handwritten cover note declaring "Victory!" One week later, the federal judge handling Ulmer Lynch's case dismissed his claims, and the elections board moved to dismiss the Ewells' tardy state court suit as untimely. One South Side biweekly newspaper noted that a race that had looked like "a major task" for Barack now had him "running unopposed for his first elected job of many public service stops." Only a Saturday traffic ticket for an illegal U-turn, resulting in a $95 fine and four hours of mandatory attendance at traffic school, marred Barack's run-up to Election Day.

Barack deployed campaign volunteers to help get out the vote, telling the *Defender* that "It's important to let people know who I am." His Springfield agenda "will focus on public school financing, ensuring fairness if welfare and Medicaid are bloc-granted to a state level" by President Clinton's reform legislation, and "job creation and economic development."

Six months to the day after Barack first declared his candidacy, the election night tally showed him receiving 16,279 votes—100 percent of those cast. Barack owed his unanimous triumph both to the professionalism of his three top advisers—Carol Harwell, Ron Davis, and Tom Johnson, who never even billed him for all his work—and also to the sloppiness of Palmer's and Ewell's signature gatherers.[14]

With only pro-forma Republican opposition anticipated for November's general election, Barack now had almost ten months as a senator-elect prior to his actual swearing-in in January 1997. He shuttered his campaign office and turned his attention to his U of C teaching, as the fourth iteration of Current Issues in Racism and the Law got under way on March 25 with about two dozen students enrolled. Mary Ellen Callahan, a 2L, found it a "provocative class," with Barack ready to challenge "traditionally liberal assumptions." She found Barack "quite approachable," as did 2L Hisham Amin, who thought he was "very open, very genuine, very thoughtful." The syllabus remained the same as before, and when the quarter concluded on May 21, the students' evaluations gave the class a rating of 9.5 on UCLS's 10-point scale.

As soon as classes ended, Barack, Michelle, and Maya flew to Europe. He and Maya had talked about such a trip in the wake of their mother's death, and Bob Bennett, who had stepped down as Northwestern Law School's dean and was on leave for the year, invited the Obamas to spend a week with him and his wife Harriet Trop in the Italian countryside near Grassina, southeast of Florence. Starting in Paris, they took the train to Nice, staying with Barack's old Occidental friend Laurent Delanney at his home in Roquebrune-Cap-Martin, near Monaco, before continuing eastward to Florence. "He was a great big brother," Maya recalled, and the trip was a good tonic in the aftermath of Ann's painful demise.

Knowing that his state Senate role would require much of his time, Barack moved to recalibrate his two other main professional commitments. Judd Miner and everyone at Davis Miner had supported Barack's Senate race, envisioning that he could move to a flexible, part-time "Of Counsel" role. Barack had continued working on Judd's big municipal redistricting lawsuits concerning Chicago and St. Louis's city councils, and he also remained a secondary participant in the ongoing "motor voter" registration case that progressive groups including ACORN were successfully concluding against the recalcitrant Illinois governor, Jim Edgar. Only after another court order in May 1996 did state officials begin exhibiting "the kind of compliance with their duties that should have been forthcoming from the outset," wrote U.S. district judge Milton Shadur. John Belcaster, George Galland, and Barack were also assisting a Peoria attorney with a sexual harassment lawsuit against a Mitsubishi auto plant in Normal, Illinois, and in April 1996 the Equal Employment Opportunity Commission joined the litigation with a class action suit of its own against Mitsubishi.

The University of Chicago Law School had always wanted more from Barack than his one-course-per-year adjunct lecturer role, and in the spring of 1996, Barack approached Douglas Baird, who had succeeded Geoffrey Stone as dean after Stone became university provost, and asked if they could have dinner. They met downtown at the Park Avenue Café, and Baird remembered that Barack wore an Armani tie and made a surprising proposal. Once in office, he would no longer draw his full Davis Miner salary, and the modest pay accorded a state senator would not make up the difference. Barack wanted to combine his new post with a larger law school teaching load, because "Michelle insisted on making the numbers work," and "the whole point of the dinner is he has to figure out how to make the numbers work in order to become a state senator," Baird recalled. "I'm frankly a bit skeptical," and "I spend a lot of time during that dinner saying 'okay, wait a second, what's your schedule in Springfield going to be like? Are you going to have to cancel too many classes?'"

The Illinois Senate's spring session began in January each year and expanded some during February and March before intensifying significantly in April and May. But until late each spring, the Senate usually met only Tuesday through Thursday, and each fall the Senate convened only for a brief six-day "veto session" to reconsider bills the governor had rejected. Given the law school's schedule of fall, winter, and spring quarters, fall and winter classes, especially if timed to avoid midweek sessions, could comport quite nicely with a state senator's schedule.

Satisfied that the timing might work, Baird raised two other issues. One was curricular. Racism and the Law, just like a small Voting Rights course Judd Miner was teaching, was a worthwhile elective, but in Baird's eyes Barack would need to teach something central to the law school's curriculum, not ancillary, in order to merit an expanded appointment. Unique among law schools, Chicago divided Constitutional Law into four distinct courses, one of which, Con Law III, covered the Fourteenth Amendment doctrinal issues of equal protection and substantive due process. For Baird, Barack's agreeing to teach an annual section of Con Law III "was absolutely essential . . . because that's a real course" where "he's really holding down part of the curriculum."

There was also the question of title, because as Geof Stone put it, "titles matter." During dinner, Baird told Barack, "We can't make you an assistant professor," the normal title for a beginning appointment, "unless you do scholarship," and Barack immediately responded, "That's not me." Baird would remember that Barack "was crystal clear" that any "idea that he was going to be a law scholar was just not him." The law school did have one special title, senior lecturer, for instructors with significant teaching roles who were not full-time law faculty, but that august rank was held only by three members of the U.S. Court of Appeals for the Seventh Circuit—Richard A. Posner, Frank H. Easterbrook, and Diane P. Wood, all of whom had been on the faculty before becoming federal judges—and the distinguished legal historian Dennis J. Hutchinson, who held a joint appointment. Baird knew it would be "very unusual" to give Barack that title, along with a decent salary and faculty benefits not accorded mere lecturers, and "I talked about this extensively with Geof Stone," who as provost "was completely supportive."

Baird and Barack tussled over how long the initial appointment would be, and Baird "insisted on it being two years rather than four years because I was worried that he would simply be a touch-and-go lecturer and not really be involved." But by late May 1996, the terms of Barack's expanded new appointment as a senior lecturer were set, and that fall, he would teach Con Law III and Racism and the Law; in winter 1997, he would take on Voting Rights as his third regular course.[15]

While Barack was negotiating a new relationship with the university, so too was Michelle. In her three years as executive director, she had built Public Allies Chicago into a highly regarded internship and training program for young people interested in public service, and the Chicago chapter was "the most mature and advanced" Public Allies site of all. Chicago board chairman Ian Larkin knew the chapter was "this huge bright star" within the Public Allies network, and as a result Michelle had become the de facto leader of the chapter directors who pushed back against the excessive "top-down" behavior of D.C.-based co-founder Vanessa Kirsch. Michelle "was the leader of the cabal," North Carolina executive director Jason Scott explained, and although Michelle "knew that she was basically unfireable," the conflict between the five local chapters and the national office was "really stressful, and it took a lot out of you."

In Chicago, Michelle had created a warm, yet challenging esprit de corps among her mid-twenties staffers, inviting them to East View for potluck dinners that sometimes involved Barack cooking the chicken dish he had learned from his Pakistani friends years earlier. Michelle was "paradoxically direct and warm," Milwaukee director Paul Schmitz recalled, while Barack, who each year still did a training for Allies, was just a "somewhat aloof guy married to a colleague." Larkin had a similar view, thinking Michelle's "star was about a thousand times brighter than Barack's." In early 1996, Kirsch was replaced by Chuck Supple, who made "a pilgrimage to Chicago" to meet the best-regarded chapter director. He thought Michelle was "this wonderful mix of poise and professionalism and competency with an activist's passion and desire to bring about change."

For Michelle's third year, 1995–96, she had forty-eight young Allies, and it was by far her most draining, primarily because of the sudden illness and all-too-fast death of Michelle's beloved young assistant. Twenty-four-year-old Richard Blount was a highly active but largely closeted gay man who was "Michelle's right hand" for much of her time at Public Allies. Richard's illness became apparent in November 1995, right after Ann Dunham's death, and Public Allies board member Jan Anne Dubin remembered that she and Michelle were "both sobbing on the phone" over Thanksgiving as they discussed Richard's condition. Another of Michelle's young protégés, Craig Huffman, recalled that time as "enormously stressful for her," with financial worries on the home front also pressing in. Richard died on February 8, and Dubin described it as "like losing a child," while emphasizing that Richard was equally close to Michelle. Richard's funeral service was held at Trinity United Church of Christ, with Barack there and Michelle as one of the speakers.

Barack and Michelle's attendance at Trinity's 11:00 A.M. Sunday service had become less frequent, and neither had become active in any of Trinity's vibrant service ministries. Michelle's thoughts increasingly turned toward wanting to have her first child, and she believed that once she had a child, she could work only part-time, something that would be impossible as executive director of Public Allies Chicago. She hoped to find a position that offered financial security and was closer to home, and that spring the University of Chicago advertised for a new associate dean charged with maximizing student volunteer opportunities. Michelle applied, despite strong childhood memories of the university's total disconnect from the African American neighborhoods just to its south and west. "I grew up five minutes from the university and never once went on campus," Michelle recalled. "All the buildings have their backs to the community. The university didn't think kids like me existed, and I certainly didn't want anything to do with that place." For four years during Michelle's childhood, her mother Marian worked as a secretary at the U of C, but Michelle had continued to think of it as "a very foreign place" and one she "never knew anything about."

Ian Larkin "was surprised" when Michelle told him she was leaving Public Allies, but she did give three full months notice before her September 1 start date at the U of C. The university's press release quoted vice president Arthur Sussman as saying that "Michelle understands the Hyde Park community and the world of public service," and Public Allies gave her a friendly send-off with a farewell party at board member Sunny Fischer's Evanston home.

In late May, when Mariana Cook, a New York photographer working on a book about couples in America, visited Michelle and Barack at East View, the informally posed pictures as well as Michelle's interview comments portrayed a woman focused on defending her privacy from the uncertainties of the external world. Michelle's Public Allies colleague Craig Huffman had heard her say "very directly" that politics was too low a calling for her husband, and she voiced those feelings very strongly to Cook. "There is a strong possibility that Barack will pursue a political career," but "there is a little tension with that," because "I'm very wary of politics" and "I don't trust the people" in it. She believed "he's too much of a good guy for the kind of brutality" politics entails, and she worried that "when you are involved in politics, your life is an open book, and people can come in who don't necessarily have good intent. I'm pretty private," and "in politics you've got to open yourself to a lot of different people."

Michelle told Cook, "I want to have kids and travel," and she expressed a greater openness apart from politics: "it'll be interesting to see what life has to offer. In many ways, we are here for the ride, just sort of seeing what opportunities open themselves up." She was happy she was not at a law firm, and she was pleased that "Barack has helped me kind of loosen up and feel comfortable with taking risks and not doing things the traditional way" thanks to "how he grew up." Barack told Cook that "all my life, I have been stitching together a family, through stories or memories or friends," and "imagining what it would be like to have a stable, solid, secure family life" like the one Michelle had enjoyed during her upbringing.[16]

Once Barack's campaign obligations lessened after his primary win, he and Michelle could enjoy more leisurely evenings than they had during the months of his uncertain candidacy. Back in the early fall, Bill Ayers and Bernardine

Dohrn had heeded Alice Palmer's request to introduce Barack to their friends, and after that gathering, Barack and Michelle began to see a great deal more of not only Bill and Bernardine but also their three closest friends, Rashid and Mona Khalidi and Carole Travis.

Just before Bill and Bernardine moved to Chicago from Manhattan in the summer of 1987, mutual friends had told them about the Khalidis, another New York couple moving to Hyde Park who were also raising three children. Rashid, a well-known historian of Palestine, had been named a professor in the U of C's Department of Near Eastern Languages and Civilizations. "I proposed, the second or third week we were out here, that we have dinner once a week at each other's houses," Bernardine recalled, as the Khalidis lived just five blocks from Bill and Bernardine. The arrangement worked so well, thanks in part to Mona and Bill each being excellent cooks, that "we had dinner together ultimately almost every night," three evenings per week at each home.

By 1995 their regular dinners also included their friend Carole Travis, a former United Auto Workers local union president and law school graduate who was political director of the Service Employees International Union's Illinois state council. Carole had a great apartment on the top floor of 5300 South Shore Drive—the building where Harold Washington had lived—and she often hosted parties there. Carole and Bernardine had known each other as fellow political activists since the late 1960s, and while Carole appreciated that Bill and Bernardine had been "totally fucking crazy" during their years in the violent Weather Underground, in subsequent years Bill, like Bernardine, had "reinvented himself" as a well-respected educator.

Bill and Bernardine had quickly made many good friends in Hyde Park, including Michelle and Barack's friend Valerie Jarrett's parents the Bowmans. "Barbara and Jimmy were close friends of ours right away," Bernardine remembered. Bill still called himself "an unreconstructed lefty," and their prior affiliations were no secret, as a mid-April *Chicago Tribune* story on a joint speaking appearance at Lake Forest College highlighted. But their calls for renewed student activism—"What will your grandchildren remember you for?" Bernardine asked the undergraduates—were well within anyone's liberal mainstream.

Carole, Mona, and Rashid first met Barack and Michelle at Bill and Bernardine's dinner table, which most evenings also included the two couples' six children. Carole remembered one early dinner conversation at the height of the Alice Palmer contretemps. "Michelle didn't like him doing politics, but Michelle, in that instance, was 'I was there. She said, "You've got my endorsement," and we've begun working on it. . . . We're not giving up.' . . . Michelle was adamant on that." By the spring of 1996 Barack and Michelle were a regular presence at the two couples' "very informal" dinners. "I would invite them often," Mona recounted. "We used to do a lot of dinners together," and "they came to our house often."

Bernardine recalled an evening when Michelle was mulling her decision to leave Public Allies. "She wanted to be closer to home" and the shift "was definitely tied to having children" so that Michelle could have "more flexible hours." Carole remembered that "Mona and Rashid always had a big table, and there were tons of people who came through that table," which became an "intellectual center" of the Hyde Park community. "Bill and Bernardine's table was a political table," Carole added, but when Barack joined in "he was never ideological."

Scott Turow, the well-known writer and attorney, first met Barack and Michelle at a summer 1996 dinner party at Bill and Bernardine's just after Michelle had accepted her U of C job. He was seated next to Michelle, and "she just blew me away," offering a memorable account of how her grandmother, who died in 1988, had on her deathbed warned Michelle, "Don't you start the revolution with my great-grandchildren. I want them to go to Princeton too!" Scott thought Michelle was "an amazing person," and when dinner was over, she introduced Scott to Barack, who of course was familiar with Scott's famous first book. "He began talking to me about *One L*," and mentioned his own book. They spoke for twenty minutes about writing, and Barack seemed so "deeply interested in literature" that Scott was uncertain "whether his career was going to be in politics or in literature." Rashid Khalidi "came in after dinner," and "he and Barack clearly had a prior acquaintance, because they were kidding each other about something," Scott recalled.

Carole Travis's parties at her apartment were larger affairs, and "Barack was at a number of them," although not Michelle. Tom Geoghegan, fresh from his successful resolution of the Wisconsin steelworkers' fifteen-year legal struggle, recalled one crowded gathering at Carole's where Jesse Jackson Sr. was among the guests. Tom spent much of the evening talking with Barack, whom he had met previously and who was familiar with Tom's own reflective, 1991 memoir about his work as a labor lawyer, *Which Side Are You On?* Tom remembered how "he was such a pleasure to talk to."[17]

Barack's role as chairman of the Chicago Annenberg Challenge board continued to be a recurring drain on his time. He remained convinced that "the spirit of the Challenge is to fund more than the status quo," and he agreed with executive director Ken Rolling's view that the proposals funded so far were neither creative nor inventive. In mid-May, Brown University president Vartan Gregorian, the Annenberg effort's top overseer, visited Chicago to have dinner with Barack, Pat Graham, and two other CAC board members to strengthen the link between the Chicago venture and the national program. For nine months the Chicago board had insisted that Rolling alone, aided hugely by the all-volunteer Chicago School Reform Collaborative, could administer the project, but at dinner Gregorian won the Chicago board leaders' agreement to hire several support staffers. Gregorian, like Rolling, hoped for bolder proposals, and he pressed the Chicago board members to raise additional matching funds.

On the Joyce Foundation board, Barack's first eighteen months left colleagues like Wisconsin law professor Carin Clauss quite aware of his strong interest in education policy. "His confidence, his sense of self" was readily apparent during the thrice-a-year all-day meetings, although another member explained that "the Joyce Board has a certain tenor and Barack came in without regard to the tenor. . . . Initially some people were taken aback" because Barack is "a pontificator." By 1996 Barack had become "an excellent board member" with "a lot of outstanding insights," yet a colleague like Clauss, who did not share the men's great interest in basketball, was struck that "with the guys" Barack "was always talking basketball," not policy issues.

Two wedding trips bracketed Barack and Michelle's summer. In late May they traveled to southern Maryland for Rob Fisher's marriage to his longtime girlfriend Lisa. In August, they flew to London for Barack's sister Auma's marriage to Ian Manners, a white British businessman. Madelyn Dunham came all the way from Honolulu to attend, so Barack was able to see much of his far-flung family. One evening prior to the wedding, Ian's brother-in-law Brian took Barack and a few others on a pub crawl in Wokingham, a Berkshire market town west of London. Barack was unimpressed with English bitter, and bailed entirely when a female stripper showed up to entertain the all-male group.

Back in Chicago, Barack's status as a senator-to-be was gradually becoming a part of his life. He checked with Emil Jones's Springfield staff about the legislature's 1997 schedule before setting his upcoming law school class times. Impressed with Cynthia Miller, a young black woman who was temping at the law school for the summer while finishing a graduate degree, Barack told her he was looking for a part-time district director to staff the empty 71st Street campaign office that in the fall would be his legislative office. As Alice Palmer and Dick Newhouse had done before him, Barack thought it was important to be in South Shore rather than Hyde Park, and Cynthia agreed to sign on as of October.

Senate Democrats invited Barack to join them on a Chicago boat cruise one evening, and powerful downstate senator Vince Demuzio warned his colleague Pat Welch that the new member was "a bomb thrower." Welch, expecting someone in a dashiki, was pleasantly relieved upon meeting Barack. Over a midsummer lunch, Carole Travis introduced Barack to Michael McGann, SEIU's Springfield lobbyist, and McGann recalled thinking that Barack "was already oozing ambition all over the place." A group of four other lobbyists asked to see the incoming new senator at Davis Miner. Former Democratic Senate staffer Jeff Stauter explained that they did not know anything about him before that meeting, except that Barack had vanquished Alice Palmer in "a pretty bold and cold move." After introducing themselves to Barack in the firm's basement conference room, "we spent quite a bit of time there," Stauter recalled, and "right away there was a recognition among all of us that this guy was going somewhere."

Earlier in the year, Carol Harwell had told Barack that as an Illinois legislator "you can't drive a foreign car," as Barack was still using Michelle's hand-me-down Saab. "He was pissy about that," Carol remembered, but Barack acquired a black Jeep Cherokee, while remaining reluctant to get rid of the Saab. But Malik Nevels, the '94–'95 Public Ally who had worked on the Jackson congressional race, needed to buy a car before starting law school at Urbana-Champaign. He called Barack, who invited him to East View for a test drive. The Saab was in "great condition" and had only sixty thousand miles on it, but Malik, with only $2,500 available, thought it was outside his price range. Barack asked how much he had. "Okay, we've got a deal," Barack responded, and he drafted a handwritten sales contract for the car, the title for which was still in Michelle Robinson's name. Malik eagerly signed, but tried to hand the keys back to Barack because he could see the car was "a complete mess," with "food underneath the seat" and cigarette butts and scraps of paper scattered everywhere. "Aren't you going to clean it up first?" Malik asked. Barack replied, "You're on your way to law school, right?" Malik nodded. "Let me give you your first lesson in contract law," Barack went on, pointing out the words "as is" in the handwritten document. A happy but chagrined Malik cleaned the Saab himself.[18]

Barack also remained engaged with the slow-moving development of the Hope Center. An energetic full-time organizer, Tracy Leary, a two-year veteran of Gamaliel's work in Gary, had been hired in early 1996, and the small initial group of trainees Barack had worked with the previous fall completed their training that spring. By April classes for a second, larger group were under way, and in mid-June Rev. Jeremiah Wright and civil rights organizer Bob Moses each conducted workshops for the group. Tracy had first met Barack in early 1995 through Gamaliel, and by fall 1995, Barack was also informally mentoring another young, somewhat unhappy black Gamaliel organizer, John Eason, whom Hope Center cofounder Sokoni Karanja introduced to Barack. Sokoni and Barack remained committed to "developing African-American leaders who can transform extremely low-income urban communities" where disinvestment had led to "an erosion of hope" that was magnified by "the lack of organization . . . in the African-American community." That community needed "parents who restore standards of civility to the streets and who insist on quality education for their children," but the duo believed that "the current palette of organizing tactics" gave short shrift to "uniquely African-American values and culture." As Barack had articulated previously, "existing schools of organizing have limited visions about leadership identification, community power building, and 'winnable' issue cutting that tend to generate 'big ego' leaders, turfing and divisiveness, and small localized victories." Their hope remained that the Hope Center could "generate a cadre of people who will understand that we can transform deeply impoverished communities if we have persistence, creativity, and the ability to mobilize a critical mass of people."

That summer 1996 program summary stated that "Obama has devoted about 10 percent of his time to the Center," as compared to 25 percent for Sokoni, with Barack doing a two-hour "power analysis" training as one of the twelve weekly sessions for each group of Hope Center "initiates." At a mid-July planning meeting, Barack observed that "we haven't really figured out what our niche is" and "there *has* to be an organizing organization" to "plug" the trainees into. "*Somebody* has to be the organizer of this process" by providing such a group. Tracy Leary persuaded Barack to lead a five-hour workshop on the last Saturday of the month on "Building a Collective Organizing Strategy for Greater Grand Boulevard," the more formal neighborhood name for Bronzeville. Tracy's invitation letter explained how the Hope Center sought "new cadres of citizen leaders who are committed to transforming our community" as well as "transforming extremely low income people."

Barack's interest in the Hope Center reached back to his intense conversations years earlier with first Jeremiah Wright and then Mike Kruglik in which he had envisioned building "a very broad-based, African-American-led black church power base" that could transform Chicago politics citywide. It had a "spiritual dimension to it" yet was "ultimately very pragmatic," and although he had more recently cited the untapped power of unified black churches, as he was entering public office, his political focus shifted to the pitfalls of the electoral system. In early August, the Cleveland *Plain Dealer* published a long, interview-based profile of Barack three weeks before the 1996 Democratic National Convention in Chicago. Barack was decidedly unenthusiastic about both the Republican and Democratic presidential candidates. "Bob Dole seems to me to be a classic example of somebody who had no reason to run," Barack opined. "You're 73 years old, you're already the third-most powerful man in the country. So why? He seems to be drawn by some psychological compulsion. And it's too bad because in a lot of ways, he's an admirable person." President Bill Clinton's campaign was "disturbing to someone who cares about certain issues," and Barack remained just as outspokenly disappointed about Chicago politics as he had been in the *Reader* nine months earlier. "The realization that politics is a business . . . an activity that's designed to advance one's career, accumulate resources and help one's friends," as "opposed to a mission," was "fascinating and disturbing," he said. The upcoming Democratic gathering promised even worse, because "the convention's for sale," Barack realized. "You got these $10,000 a plate dinners and Golden Circle Clubs. I think when the average voter looks at that, they rightly feel they're locked out of the process. They can't attend a $10,000 breakfast, and they know that those who can are going to get the kind of access they can't imagine."[19]

In August 1996, *Dreams From My Father* was published as a $15 paperback, with a new cover featuring a photo of a beaming Barack seated next to Granny Sarah in Kogelo. Random House had been disappointed by the hardcover sales and eager to sell the paperback rights, which Deborah Baker, an editor at Kodansha Globe, had bought under a standard seven-year license for just $4,000. Baker spoke with Barack by phone about the jacket copy and cover design, which used a portion of the Marian Wright Edelman blurb on the front and brief excerpts from newspaper reviews on the back. Baker asked Barack whether he intended to write a second book, and he said that while he might, it could be a novel, just as Scott Turow had shifted to fiction following his memoir. The paperback reprint attracted little attention, although one new reviewer observed that *Dreams* "suffers at times from hints that this public-minded man writes too self-consciously before a reading public."[20]

Labor Day marked the beginning of Barack's pro forma general election campaign. He would face Rosette Caldwell Peyton, a teacher at Hyde Park's Kozminski Academy, the Republican nominee, plus third-party contender David Whitehead. Barack completed another questionnaire for Impact, writing, "I would support passage of a domestic partnership law," and a week later he met with thirteen IVI-IPO members at his supporters Lois and Alan Dobry's home to receive their unanimous endorsement. Barack completed a lengthy IVI-IPO questionnaire, saying that he wanted to "mobilize residents in the district" and that "my top priority is ensuring that our young people are nurtured and supported through fully-funded public schools." The full questionnaire also confronted Barack with thirty-five sometimes multipart specific-issue queries. Number 22 asked, "Will you support a single-payer health plan for Illinois?" and Barack replied, "Yes, with Medicaid incorporated into a single, graduated system of services," though he added that "such a program will probably have to be instituted at a federal level" and that in the interim states "can move more aggressively to expand coverage to the currently uninsured." To question 26, regarding welfare and how the federal reform bill President Clinton had signed into law two weeks earlier would provide block grants to states to fund time-limited assistance in place of the previous Aid to Families with Dependent Children program, Barack responded, "I oppose arbitrary time limit or work requirements that fail to take into account the lack of entry-level jobs." To the final three queries, Barack gave succinct, one word answers: "No" to both "Do you support electronic eavesdropping?" and "Do you support capital punishment?" and "Yes" to "Do you support legislation to (a) ban the manufacture, sale and possession of handguns?" as well as assault weapons.[21]

On the last day of September, Barack began teaching Con Law III as well as Current Issues in Racism and the Law. At a 1L welcoming reception Barack recognized one face: his paternal namesake Barack Echols, one of only two black men in the law school's class of 1999. The race seminar's syllabus remained unchanged, but the twenty-two students were impressed with "how thoughtful his choices" were, 2L African American Felton Newell said. Chapin Cimino, a 3L, remembered him "talking a lot about W. E. B. Du Bois . . . I came away from that course thinking of both Obama and Du Bois as middle-grounders: neither radical nor passivist, neither accommodationist nor separatist." She recalled it as a "very engaged group," and Barack "would play devil's advocate" to ensure that all views were aired.

Liberal and especially minority students sometimes felt uncomfortable in U of C classrooms, both because there were so many conservative students and also because many senior faculty members, personified most outspokenly by law and economics proponent Richard Epstein, never shied from propounding their own views. Epstein was "the defining presence" at the law school in the mid-1990s, his colleagues knew, yet he was far from alone. Prominent federal appellate judge Frank Easterbrook was "so intentionally intimidating" in another seminar that the atmosphere was "horrible" and "hostile." In contrast, Barack was a "great" teacher" and one whose respectful and open-minded demeanor was "refreshing" when compared to the general "lack of politeness at Chicago." Chapin Cimino recalled Barack responding with bemusement when she, an outspoken liberal, said that conservative justice Antonin Scalia was correct about the irrelevance of legislative history to judicial decision-making. Her paper for Barack on the hollowness of class-based affirmative action programs would become her published note in the U of C *Law Review*.

In Con Law III, Barack assigned the new third edition of Geof Stone's *Constitutional Law* casebook, an eighteen-hundred-page behemoth that was used in all of the law school's Con Law courses, to his thirty-seven students. Triste Lieteau, a 3L medical school graduate whose lobbyist dad Mike had met Barack during Alice Palmer's congressional campaign, thought Barack was "a great professor" and "very down to earth." At such "a conservative bastion," liberal professors were often challenged in class by outspoken conservative students, and Triste remembered that Barack handled one such incident "very well." As an African American student, "you do feel this strong sense of isolation and that you're really not fully a part" of the school, she explained. In that context, 3L Hisham Amin welcomed Barack as a black professor who exuded "confidence, presence, and knowledge." At the end of the quarter, the students' evaluations of Barack's teaching registered 9.35 out of 10 for Racism and the Law and 8.71 out of 10 for Con Law III.[22]

On the first Thursday in October, Barack's campaign held a $100-per-person fund-raiser at a West African art gallery. Sponsors included attorneys Jeff Cummings, Allison Davis, Linzey Jones, and Elvin Charity; old Gamaliel colleague David Kindler; former students Jesse Ruiz, Brett Hart, and Jeanne Gills; and family friends Valerie Jarrett, Craig Huffman, and Kathy Stell. The invitation offered attendees the chance to meet "one of the nation's most promising young leaders" and said the proceeds "will help Barack conduct voter registration and outreach efforts to constituents." The next week Barack conducted "an intense discussion on power and self-interest" for the Hope Center, but there was a tension between Sokoni Karanja and Tracy Leary's focus on energizing Bronzeville residents and Barack's vision of a citywide network. "We had some differences of opinion around what it should look like," Sokoni remembered, as Barack "wanted it to be less focused on Bronzeville and more of a citywide thing," perhaps because "I always thought he wanted to be mayor." Given the "dearth of leadership in Bronzeville, I thought this was the place to focus initially and then begin to build out from there," Sokoni explained, while Barack "wanted to take it much more broadly." Sandy O'Donnell and Tracy Leary observed many of their conversations, and young Gamaliel organizer John Eason, who knew how Sokoni and Barack "were thick as thieves," came to realize "how much they differed in terms of how they think about community organizing and building community." Jean Rudd, who knew them both as Woods Fund board members, understood how Barack "became frustrated with the limits of it, that it was going to be limited to a very restricted idea of what leadership development was about and that it was not going to become what he had hoped it would be."

Woods itself, where former UNO Southeast Chicago organizer Todd Dietterle had succeeded Ken Rolling when he became CAC executive director, had moved to make welfare-to-work policies a coequal priority along with the fund's long-standing commitment to organizing. Woods appreciated "how the availability of child care and transportation to work sites affects employment prospects," and with the passage of President Clinton's welfare reform bill, state legislatures would have to devote significant attention to such policies in 1997. Woods, Wieboldt, and MacArthur all cohosted a fall 1996 "Community Organizing Award" ceremony, and Barack's keynote address captured the evolution of his thinking about organizing toward a larger focus. Organizing taught that "we have a collective responsibility to make a better world" and anyone who identified as a victim was "disempowering" themselves. But the focus should not be on "getting concessions out of some power out there," in what John McKnight had called consumer-driven organizing, but upon generating new productivity, just as McKnight had emphasized a decade earlier. "We have to start thinking on a bigger scale," and "organizing on a big scale requires coalition building," Barack argued. "The core idea of organizing—the notion of mutual regard, and mutual responsibility, and collective action," meant that "our individual salvation is tied to our collective salvation" and that "our individual stories are part of a larger story," language Barack would directly echo to a vastly larger audience eight years later.[23]

By mid-October, Michelle was well into her new job at the U of C, organizing an annual volunteerism summit that featured as a speaker her brother Craig's good friend John Rogers Jr. With the general election only three weeks away, Barack had no need to campaign because, as the *Hyde Park Herald* noted, his two opponents "are nowhere to be found." Neither responded to a League of Women Voters invitation to a Saturday candidates' forum, and Barack won some goodwill with a possible new state Senate colleague, Debbie Halvorson, by campaigning for her in the south suburban district where she was challenging longtime Republican incumbent Aldo DeAngelis. Senate Democratic leader Emil Jones Jr., facing a 33–26 Republican majority, hoped to pick up the four seats needed to win control of the chamber. Jones had assigned the Senate Democrats' best staffer, 1987 Georgetown University graduate and former UPI newsman Dan Shomon, to run Halvorson's campaign, and one weekend day, a somewhat overdressed Barack, with Ron Davis and Will Burns in tow, showed up to go door to door for Halvorson.

Both the *Tribune* and the *Sun-Times* endorsed Barack, with the *Trib* calling him "a worthy successor" to Palmer and someone who "has potential as a political leader." Downstate U.S. congressman Dick Durbin, the Democratic candidate to succeed retiring U.S. senator Paul Simon, told reporters he was enjoying reading *Dreams From My Father,* but the only preelection interview request Barack fielded was from the U of C Law School's student newspaper. He said his top priority would be implementing the state's new role under the federal welfare reform measure, and also "the challenge of constructing nontraditional alliances with suburban and downstate legislators."

On Election Day, Barack triumphed with more than 82 percent of the vote, but the overall results were a defeat, because even though Halvorson defeated DeAngelis and Democrat Terry Link picked up another previously Republican seat in Waukegan, Senate Democrats came up two seats short of capturing the majority. The Republicans' reduced margin of 31 to 28 was especially disheartening because two relatively inexperienced Republican candidates, Christine Radogno and recent appointee Dave Luechtefeld, had eked out very narrow victories. Had Emil Jones's campaign staffers turned out another 214 Democratic votes in those two districts, Barack would have been a freshman member of the Senate's majority rather than minority party.

Democrats regained control of the state House of Representatives, 60–58, picking up six seats as part of President Clinton's landslide victory in Illinois that also ensured Dick Durbin's election to the U.S. Senate. The Senate Democrats' narrow failure led to talk about a challenge to Jones's leadership on the part of unhappy colleagues like Quad Cities senator Denny Jacobs, but when Barack went to Springfield for new-members' orientation during the legislature's brief mid-November veto session, his most poignant lesson was how personally popular the departing Alice Palmer was with colleagues, staffers, and progressive lobbyists. "I loved Alice Palmer," assistant staff director Cindy Huebner explained, and although she was prepared to dislike Barack, he was "very nice, very down to earth" once he realized he needed to "mend those fences."[24]

In Chicago, Barack spoke at Jesse Jackson Sr.'s weekly Saturday-morning PUSH rally, brimming with optimism about his new role as a state senator. "If we can rally the community around education, welfare reform, health care, jobs and job training, then we can create a progressive state coalition that can serve not only Chicago's interests but those who lack opportunity everywhere." He said he was creating an issues committee, and he invited district residents to join, calling it a "vehicle for getting people more involved in the political process." Interested citizens should call Cynthia Miller at the soon-to-be senator's district office. "There are big pockets in this district that are underserved," Barack stated, which was very true given how the 13th Senate District, which reached from 47th Street on the north to 81st Street on the south, encompassing Hyde Park and South Shore, also stretched more than five miles westward, encompassing parts of the Englewood, Chicago Lawn, and Marquette Park neighborhoods in a narrow finger between Marquette Road—67th Street—on the north and 75th Street on the south. The district was 20 percent white—largely Hyde Park—and more than 70 percent African American, but that western part of the district, especially Englewood, was a world away from the leafy blocks surrounding the university.

"We plan to organize monthly town hall meetings at rotating locations around the district," Barack announced, and he told the law school's student newspaper that "hopefully I'll be a voice that helps nudge the legislature towards a welfare system that not only recognizes the importance of moving people out of welfare and into work, but also recognizes that there are many barriers to that process," such as the day care and transportation hurdles the Woods Fund's work was highlighting. "We have to address those failures of the market and provide structures of support in those areas," Barack explained, recalling the manuscript he and Rob Fisher had written while at Harvard. He also distinguished himself from his party's president: "on the national level bipartisanship usually means Democrats ignore the needs of the poor and abandon the idea that government can play a role in issues of poverty, race discrimination, sex discrimination, or environmental protection."

Ten days before Christmas Barack got his third traffic ticket in fifteen months when he was pulled over and fined $50 for going sixty-two miles per hour in a forty-five-mile-per-hour zone. His Springfield inauguration would take place on Wednesday, January 8, 1997, but first he had to grade his Con Law III exams and prepare himself to teach Voting Rights and the Democratic Process for the first time when winter quarter began on January 6. Barack knew that an upcoming visiting professor at the law school, Rick Pildes from the University of Michigan, was working on a voting rights casebook along with two other election law scholars, and well before January, Barack had asked Pildes for a copy of their manuscript. Barack's advance description of his course made clear that he would bring to this new class the same relentlessly open-minded style students had experienced in his Racism class. "Does the Voting Rights Act, as amended, promote minority voices, or simply segregate them from the larger political discourse?" he suggestively asked in questioning the overall political utility of so-called minority-majority districts. Indeed, Barack's description asked, "does voting even matter in a complex, modern society where campaigns are dominated by money and issues are framed by lobbyists?"

Barack sounded a similar note regarding districting when the Associated Press asked him to comment on a Supreme Court order returning a challenge to Illinois's oddly drawn, Hispanic-majority 4th Congressional District to the original trial court for reconsideration, a case he knew well because Davis Miner had represented some of the intervenors who were defending that district and three others with black majorities. "It's fair to say that in all these congressional remap cases, the Supreme Court has not yet come up with a clear, manageable, comprehensive approach to determining whether or not a minority-majority district is permissible." Barack again demonstrated a less than worshipful view of the high court in a twelve-page, single-space memo to his Con Law III students, explaining the best type of answers when he returned their mid-December exams. He had posed two questions, one involving a lesbian couple's public employee benefits coverage, the second concerning municipal affirmative action programs, and he limited their answers to a total of fifteen typed double-spaced pages.

For the first question, he cited "the remarkable opacity of Justice Kennedy's opinion" in *Romer v. Evans,* a landmark antidiscrimination decision handed down seven months earlier. On the second, he pointed out how "the type of program at issue in" *Adarand Constructors v. Pena,* an important 1995 affirmative action ruling that applied highly demanding "strict scrutiny" to the federal government's use of explicitly racial classifications, included an "irrebuttable presumption that any contractor who was a member of a minority group fell into this economically disadvantaged category."

Rick Pildes and Barack had dinner one night in Hyde Park, and Pildes got a sense of Barack's approach to law teaching, which was "very different from most" young law professors because of "how independent-minded he was about the issues." Barack "was much deeper and more reflective" than most legal academics. Regarding voting rights, Barack was "open-minded" and "very interested in the facts on the ground, how this stuff was really playing out, rather than ideology." He was such "a wonderful presence" that Rick thought to himself, "I could see this guy as mayor of Chicago."

But another visiting colleague, John R. Lott Jr., had a different sense of Barack. In August 1996, Lott and a graduate student coauthor had presented an evidence-packed paper contending that increased possession of private firearms resulted in less violent crime. That seemingly counterintuitive argument outraged gun control advocates, who responded with angry denunciations of Lott's work. Later that fall, Lott sought to introduce himself to Barack at the law school. "Oh, you're the gun guy," Barack responded coolly. "Yes, I guess so," Lott replied, and he recalled Barack responding, "I don't believe that people should be able to own guns." Lott offered to talk about the issue, but Barack "simply grimaced and turned away, ending the conversation."

Lott's recounting of that exchange matched Barack's frank statement about handguns on his September IVI-IPO questionnaire, but Barack's personal disdain for Lott, exhibited on several subsequent occasions, differed from the warm and approachable professor Barack's twenty-five students encountered in Voting Rights. He assigned photocopied readings from Pildes's manuscript that began with "some early 19th century cases about the denial of the franchise" and continued up through current-day redistricting and campaign finance cases. Adam Bonin, a 3L, recalled that "we learned the law, but we also learned it on the level of real-world impact." David Franklin, also a 3L, remembered that Barack "wasn't interested in high theory at all," and he emphasized that further when students wrote the lengthy papers that were the primary coursework. Bonin addressed racial gerrymandering, and Barack "made me look at this as a real-world issue, and not as a theoretical construct."

Bonin remembered that Barack would push a contrary view if it "wasn't being expressed or students were too complacent in their liberal views." Molly Garhart, another 3L, recalled Barack as "so calm, so level-headed, so fair and so equitable in his treatment of all of us and listening to everyone's ideas." African American 3L Carl Patten had met Barack months earlier in the gym and had lost a one-on-one game to him. In class, Patten was "very impressed" with how Barack "created an environment where people felt free to speak from their viewpoints whether or not they were perceived to be aligned with his." When Carl had seen Barack at the law school after that one-on-one loss, he joshingly remarked, "Next time I'm going to get you." Barack's facial expression immediately said, "This is not the place for me to respond." Patten realized "he's very calculating, extremely calculating," yet "calculating while being genuine." When the quarter ended, the students gave Barack's teaching an 8.71 score on the law school's 10-point scale.[25]

On Wednesday morning, January 8, Barack was sworn into office, signed his oath of office form for his two-year term, and began his true initiation into the world of the handsome Illinois State Capitol building. As a minority party freshman, Barack was assigned to one of the least desirable offices, 105D, a narrow, high-ceilinged ground-floor box in the capitol's north wing. His political home in Springfield would be the Senate's Democratic caucus, an uneasy assemblage whose twenty-eight members included several highly volatile personalities. Democratic leader Emil Jones Jr.'s only meaningful power was hiring staff members and assembling a caucus leadership team to support his role as minority leader. But there were three much more consequential figures in Springfield's political world: newly reinstated Democratic House Speaker Michael J. Madigan, a lifelong politico from Chicago's Southwest Side who had become 13th Ward committeeman at age twenty-seven in 1969 and entered the legislature in 1971; moderate, second-term Republican governor Jim Edgar, who had grown up in the small town of Charleston in eastern Illinois; and Senate Republican president James Peyton "Pate" Philip, from suburban DuPage County.

To outsiders unfamiliar with Springfield, and especially with the capitol's unusual lingo, a description of the city as a land of horseshoes, mushrooms, and lobsters might sound strange. A Springfield "horseshoe" was an often grotesque open-faced sandwich in which a piece of meat was covered first with french fries and then with a cheeselike sauce. Visitors knowledgeable enough to avoid the local delicacy felt rightfully proud. The statement that Springfield was a city of "bad hotels and worse food" was perhaps apocryphal, but there was no shortage of bars, because drinking was state politicos' top recreational activity. Card playing came second, and golf was perhaps third, because the local courses were an everyman's bargain compared to Chicago. But there was little doubt what ranked fourth, as one female lobbyist explained: "there's a lot of people who fucked in Springfield. What else is there to do?"

"Mushrooms" was the name given to rank-and-file members of Mike Madigan's House Democratic caucus, as Madigan's life mission was maintaining a viselike control over legislative decision-making. Madigan was a fervent admirer of former Chicago mayor Richard J. Daley, and his description of Daley was one that fit Madigan himself: "every day of his life, he was about politics." To say that House Democrats were "kept in the dark and fed shit" might slightly exaggerate their experience, but the disdain that senators of both parties exhibited toward their House colleagues was Illinois bipartisanship at its strongest.

"We like to think of ourselves as the House of Lords," African American Chicago senator Donne Trotter explained, and state representatives were political inferiors. "We talk about them like dogs, and we don't really work together." The two chambers sat just yards apart on the Capitol's third floor, separated by a glorious open rotunda that gave the building much of its cathedral-like beauty, but in the capitol's daily life, the two bodies were "very separate worlds," with members interacting almost exclusively with colleagues from their own chamber.

There were, of course, no real lobsters in the nearby Sangamon River, but the capitol's notable "lobsters"—lobbyists—were most often found on "the rail," the brass ring that encircled the rotunda in a building where seats were in short supply. More than legislators, more than legislative staffers, these lobbyists were the crucial players in a state government world where campaign cash and those who could provide it were always the top priority of legislative leaders. Many lobsters were former legislators who had parlayed their knowledge and personal relationships into far more remunerative work than public service. The most experienced lobsters often enjoyed a decisive advantage over part-time legislators, who were often not the sharpest tools in the shed. "Stupid people need representation too, and they're very well represented in the Illinois legislature," one corporate lobbyist acerbically declared. The centrality of lobbyists in Illinois politics was inseparable from the centrality of money in Illinois politics, and in Springfield most of the money flowed to the legislative caucus leaders, "The Four Tops," who hired the staff members and funded the campaigns of those members who, unlike Barack, represented competitive districts.

The power held by the Four Tops, particularly the Senate president and the House Speaker, was the single most defining factor in Illinois politics. Life in the statehouse had changed only in degree, not in kind, from the world that U.S. senator Paul Simon had described three decades earlier in an infamous *Harper's Magazine* essay and that was best personified by Simon's longtime rival, Illinois secretary of state Paul Powell, who died in 1970. Both men were from far downstate Illinois. Powell entered the Illinois House in 1935, at age thirty-three, and Simon in 1953, at age twenty-five. Powell served as Speaker of the House from 1949 to 1951 and again from 1959 to 1963, while Simon moved to the state Senate in 1963. His 1964 article, "The Illinois Legislature: A Study in Corruption," stated that "one third of the members accept payoffs." He quoted one lobbyist recounting that a legislator had said "it will cost you two hundred to five hundred dollars to get the bill out of committee" and another who was told that "for $7,500 I can get you nine votes" from legislative colleagues. Simon never mentioned Powell by name, but he did cite rumors that "under-the-table transactions provide an income of $100,000 a session for

WISCONSIN
Lake Michigan
IOWA
Galena
Rockford
McHENRY
LAKE
COLLAR COUNTIES
Waukegan
Winnetka
Evanston
Elgin
KANE
DUPAGE
COOK
Chicago
Calumet City
Harvey
Gary
Tinley Park
Joliet
WILL
Davenport
Rock Island
Moline
QUAD CITIES
Kankakee
Galesburg
Peoria
Bloomington
Macomb
Champaign-Urbana
ILLINOIS
Quincy
Springfield
Decatur
INDIANA
Mattoon
Terre Haute
Carlinville
MISSOURI
Effingham
Alton
Edwardsville
St. Louis
METRO EAST
Belleville
Mount Vernon
Rend Lake
Du Quion
Carbondale
KENTUCKY
Metropolis
Cairo
0 25 MI.
0 25 KM
43
94
90
88
39
80
74
55
57
65
72
70
64
44
24

one prominent Representative when his party is in power." Simon emphasized that "there are no controls on lobbyists in Springfield . . . they can hand out any amount of money to influence legislators," and "legislators in turn need not account for campaign contributions or disclose their sources."

A year later, Simon wrote a brief follow-up piece, reporting that at the biennial Senate dinner, "I was presented with a 'Benedict Arnold Award.' Previous winners, they said, were Judas Iscariot and Aaron Burr." In 1964, Paul Powell was elected Illinois secretary of state, and he was reelected four years later. In Springfield, he lived at the St. Nicholas Hotel, and when he died at the Mayo Clinic in Rochester, Minnesota, in October 1970, at age sixty-eight, his suite at the St. Nicholas contained $800,000 in cash.

Powell's biographer later contended that some of that money represented "large amounts of dividend income from race-track investments." Even if true, Powell's career in Illinois politics was an indelible lesson in the realities of Springfield: as Barack already sensed, state politics was more concerned with money than it was with the betterment of the lives of Illinois citizens. Of the few people actively interested in changing that state of affairs, Barack already knew the most important one: Joyce Foundation vice president Larry Hansen, a fifty-six-year-old native of Elgin and a former Democratic political operative. At the Joyce board's most recent meeting, Hansen's Money & Politics effort had been elevated to a full-fledged program. But even before that, Hansen had supported University of Illinois at Springfield political science professor Kent Redfield, a former Democratic House staffer, in throwing new light on *Cash Clout* at the Illinois statehouse.

Redfield's 1995 monograph emphasized that "there are no limits on how much an individual, public official, political party, interest group, or corporate entity can contribute to a candidate or spend on behalf of a candidate," even though a 1974 law did require candidates to report most contributions. Redfield noted how "putting campaign reports in an electronic database and making that database easily accessible to everyone would go a long way toward making the law operate as a sunshine law." But Redfield stressed that equally if not more important was "the tremendous growth in the power of the legislative leaders. Their control over both the agenda and the content of policy outcomes has become almost absolute," and went hand in hand with the dominant role they played in the campaigns of many of their caucus members.

In an insightful subsequent essay, "What Keeps the Four Tops on Top?," Redfield wrote that "the power of individual members has continued to decline" as the legislature "has become even more dominated by the legislative leaders." Their "near monopolies on campaign fundraising and the mobilization of legislative staff efforts" allow them "to control the legislative process when the legislature comes into session." Individual members "lack the personal and policy staff necessary to establish political and substantive autonomy apart from the leaders." In Springfield Barack would have one personal secretary—Beverly Criglar, whom he was inheriting from Alice Palmer—and one-fifth of the time of one of the Democratic caucus's Communications and Research staffers, who wrote press releases and such. Barack also had $47,000 to pay for his district office expenses, including Cynthia Miller's part-time salary. As Redfield explained, "strong leaders and relatively weak members . . . mean that interest groups concentrate their campaign contributions on the leaders."

The dominance of the Four Tops and their close relationships with the best-heeled interest groups determined many statehouse decisions. "The 1990s have seen a wholesale retrenchment of the power and autonomy of individual members in the legislative process," Redfield explained, as the Four Tops "control the legislative agenda and determine most of the outcomes." Other observers agreed with Redfield's analysis. "The extraordinary extent to which the Assembly's staff is under the control of party leaders," a University of Illinois report observed, "limits the information available to backbenchers, which forces many of them to look to interest groups for information or cues about policy." With most staffers hired for their "political rather than their policy skills," committees "are to a great degree under leadership control," thus further magnifying "the loss of power by individual legislators."[26]

Those systemic factors would have a huge impact on Barack's experiences as a freshman senator, but equally important, he had entered a body controlled day in, day out by one single man: as one lobbyist put it, in 1997 the Illinois state Senate "was Pate Philip's world." Philip, then sixty-six, was a former marine who never graduated from college and had begun his work life driving a Pepperidge Farm bread truck. Active in the DuPage County Republican Party by his mid-twenties, Philip was elected to the state House in 1966 and moved up to the state Senate in 1974. Seven years later he became the Republican minority leader, and when his party won majority control in 1992, Philip became Senate president. Philip was more than just the Republican Senate leader: since 1970 he had chaired DuPage County's Republican Party, where more than one-third of the 630 precinct committeemen were on public payrolls. That made him the boss of what *Sun-Times* political columnist Steve Neal called "the most effective political unit in the state," and one that in 1988 had single-handedly won Illinois for Republican presidential candidate George H. W. Bush: he carried the state by just 95,000 votes while winning DuPage by more than 123,000.

But Philip was best known for his undisciplined mouth. Neal wrote that "niggers" and "hooknoses" were parts of Philip's regular vocabulary, and in October 1994, he set off a public controversy by telling a suburban newspaper's editorial board that some black state employees at the Department of Children and Family Services "don't have the same work ethics that we have." African American state senator Donne Trotter responded that "these are the kind of comments you expect from the Grand Dragon of the KKK, not the president of the Illinois Senate," and the *Sun-Times* condemned Philip's "racist comments."

Inside the Illinois statehouse, it was universally known that Philip had little respect for Emil Jones Jr., a very dark-skinned African American. Some said that was Philip's partisan nature, but Philip had had an entirely cordial relationship with Phil Rock, Jones's white predecessor. One member of Philip's Republican caucus who tussled with the leader more than most said he was "kind of like a teddy bear" who "said whatever was on his mind," and multiple Republicans cited the amount of authority Philip gave his best committee chairmen, like Steve Rauschenberger on Appropriations and Harvard Law School graduate Carl Hawkinson on Judiciary. One of Philip's deputies acknowledged that the Senate president "was Mr. Politically Incorrect," but the conservative chairman of Public Health and Welfare put it bluntly: Pate Philip was "just an old white racist marine," a view shared by black Democrats.

Philip's counterpart, Emil Jones Jr., grew up in Chicago's Morgan Park neighborhood and got interested in politics at age twenty-five while watching John F. Kennedy's first television debate against Richard M. Nixon. After volunteering in that 1960 presidential race, he became principal aide to 34th Ward alderman Wilson Frost, and in 1972 Jones won election to the state House. A father of four, Jones and his wife Patricia also from 1980 onward raised his war-damaged brother's young son. In 1982 Jones was elevated to the state Senate, and the next year, like most organization loyalists, he backed Jane Byrne for mayor rather than Harold Washington. Jones's husky, guttural voice led many people to seriously underestimate him, because as Governor Jim Edgar later stressed, Emil Jones is "much smarter than he sounds."

Jones recalled that entering the legislature had been an eye-opener. "One of the biggest surprises I found is that outside the city of Chicago, there's a place called Illinois," which "was an entirely different world altogether." Jones said the two books that had had the greatest impact on him were Niccolò Machiavelli's *The Prince*—"government as it is," Jones remarked—and Sun Tzu's *The Art of War*. The legislative process, Jones explained, "is constant war," and he quoted from memory a rough translation of one of Sun Tzu's principles: "do not depend on the enemy not coming, but rather be prepared for the enemy when he does come." When highly popular Senate president Phil Rock retired after Senate Democrats lost their majority in 1992, Jones worked hard to succeed him as Democratic leader. With all seven African American senators and the two Hispanic members in his corner, Jones had nine of the fourteen necessary votes, but only after Chicago mayor Richard M. Daley intervened did Jones win the support of North Side contender Howard Carroll and downstate powerhouse Vince Demuzio. Several Democratic senators remained displeased, most notably Denny Jacobs from Quad Cities, and by early 1997 Jones had served four rocky years as Senate Democratic leader. The narrow 1996 failure to recapture majority control was a huge disappointment to colleagues, and it made them doubt Jones's skills.[27]

If Pate Philip had a distant relationship with Jones, his rapport with his fellow Republican governor Jim Edgar was not much better. "Pate and I were different people," Edgar later recalled. "We were just different cultures, different backgrounds . . . completely different." As everyone in Springfield knew, Jim Edgar was a teetotaler, and he "rarely bothered to develop personal relationships with legislators" in a building that "thrives on personal contact." Philip ran the Senate so that most days ended at 5:00 P.M., and the members adjourned to play cards, smoke a cigar, have a drink or two, and indulge in a little off-color humor. In contrast, the straitlaced governor stayed home with his wife and rarely socialized with fellow politicians. At the time, the outspoken Philip said "that sorry son of a bitch hasn't got a friend in the world," but years later Philip acknowledged that "Jim Edgar was a very good Governor—not much personality," yet a fiscal conservative whose views often but not always matched Philip's.

Edgar realized that Philip was most often influenced by the last person he had spoken with, but Philip was far more attuned to his conservative caucus than he was to the state's moderate Republican governor. Philip's chief of staff, Carter Hendren, who had been Edgar's campaign manager, was usually the glue that held things together. "Carter helped a lot with Pate," Edgar later recalled, and in the eyes of Edgar's top staffers, Hendren was an invaluable presence at the unpredictable Philip's side: "he saved Pate from himself *so* many times," deputy chief of staff Andy Foster explained. Disappointment was widespread among savvy Democrats like Senate Appropriations spokesman Donne Trotter that Edgar was not a more forceful counterbalance to Philip. Trotter saw Edgar as "an absentee governor" who "acquiesced his power to Pate Philip," who "took full advantage of that."

It was not surprising, then, that Philip's closest political partnership was sometimes with the utterly nonideological House Speaker, Mike Madigan. Famous throughout Illinois politics as a man of few words who rarely volunteered his thoughts, Madigan was known as someone whose word was reliable once he finally gave it. Philip and Edgar both remembered Madigan positively, but Emil Jones was the odd man out in Madigan's good working relationship with the two top Republicans. When the Speaker came to the Senate floor to see Philip, members and staffers took note of how Madigan ignored Jones just as Philip did. Jones understandably resented that lack of respect, creating a long-term rift. Donne Trotter recalled that Jones used to acerbically joke that House Democrats carried "a chip in their head" implanted by the Speaker, "and all

Madigan had to do was push a button." The Senate's "lack of respect for House members . . . was in a lot of cases perpetuated by our own leader because he didn't have a real good working relationship with Mike Madigan," Trotter explained. The lack of comity within their chamber and their party left Senate Democrats marginalized in more ways than one.[28]

Four weeks passed between the January 8 swearing-in ceremony and the commencement of legislative business on February 5. On the second weekend in January, Barack held a "community swearing-in" at Kennedy-King College in the Englewood part of his district. He told a weekly newspaper reporter he intended to put "principles before self-interest" in Springfield and that his two top priorities were improving education in Chicago's public schools and job training programs. "We can't just wait on legislation to cause some of the changes," he said. "I'm hoping to participate as part of a broad leadership with political officials, community organizations and ordinary folks to change some of the values that are blocking us from achieving the way we may achieve. Though we may be lobbying for more school funding, it's also important for us to bring education into the homes and ensure parents are checking children's homework, turning off the television, teaching common courtesy and how not to throw trash out windows." Barack said older legislators had advised him to "be quiet and start slow," but he believed that "if you work harder than others, you can still move legislation. I'm going to be the hardest working Senator down in Springfield."

In his annual State of the State address on January 22, Governor Jim Edgar renewed his call for education funding reform and called for expanded campaign contribution disclosures. One week later, the Joyce Foundation–funded Illinois Campaign Finance Project released a two-year study, largely written by Kent Redfield, which aimed "to elevate campaign finance reform onto the legislative agenda." The group's marquee members, cochaired by retiring U.S. senator Paul Simon and former governor William Stratton, also included Barack's Joyce Foundation board colleague Paula Wolff and his longtime proponent and former Harold Washington aide Jacky Grimshaw. The group declared that contrary to what Illinois law still allowed, "campaign funds should be used exclusively for campaign purposes, and not for personal use . . . officials should not be allowed to take campaign contributions with them for personal use after they leave office," and legislators should be prohibited from holding fund-raisers in the Springfield area while the legislature was in session. They stressed that the power of the Four Tops was the crux of the problem: "there is intense pressure to contribute to the leaders, and failure to do so is perceived by many players as a grave mistake," since "not giving money places individuals and groups at a distinct disadvantage."

On the first three days of substantive Senate business, February 5 through 7, Barack introduced *seventeen* different bills. They addressed the redevelopment of blighted areas, fairer credit reporting, better computer technology in public schools, additional job training programs, and the creation of a state Earned Income Tax Credit (EITC) equal to 20 percent of a taxpayer's federal EITC. One called for an additional $50 million for minority business enterprise loans, three dealt with different aspects of welfare reform, and another authorized community colleges to distribute a directory of their graduating students to potential employers. Only three of the seventeen would receive committee consideration in the weeks ahead, but this certainly attested to Barack's claim to be the "hardest working Senator."

So too did remarks Barack offered at a legislative conference convened by AFSCME, the large public employees union. AFSCME lobbyists Ray Harris and Bill Perkins had met Barack in 1992 during Project VOTE!, and at an opening workshop, Harris stressed to new legislators how leadership-dominated the statehouse really was. Barack said that had not been part of the formal orientation, and Perkins remembered being blown away by Barack's lunchtime remarks that elected officials like himself ideally should be answerable to an organized base of voters. Another labor lobbyist, Dan Burkhalter of the Illinois Education Association, one of the state's two major teachers' unions, had dinner with Barack one evening to apologize for how the IEA had ignored him when it seemed Barack would be facing Alice Palmer in an actual primary. "He's nobody's guy," Burkhalter said, invoking a famous old Chicago political saying—"We don't want nobody nobody sent"—so Barack "was a hard guy to figure out." But Barack forthrightly explained his attitude toward the teachers' unions: "Dan, here's the deal. I'll be with you guys. I can be your guy. I'll vote for tax increases, I'll vote for more money for schools, but here's the thing: we've got to figure out how to get bad teachers out of the classroom. You guys should be a part of that instead of blocking it, and if you'll work with me to be a part of it, I'm with you. If you won't, I'm not."

During that first week, Barack also went to Emil Jones Jr. to volunteer for whatever Jones could give him. "You know me," Barack said, alluding to their acquaintance a decade earlier in Roseland. "'You know I like to work hard, so feel free to give me any tough assignments,'" Jones recounted. "I suggested to him that to be successful in this body, you get to know the members, especially those who are not from Chicago," particularly those from downstate Illinois. "Get to know them well" and "Don't hang around with the guys from Chicago. . . . It'll serve you well in this body, and you'll be able to get more things accomplished."

In addition, that week Barack called Judd Miner in Chicago. Both men expected Barack to shift to some part-time role with the firm, but still receiving a regular salary. Allison Davis was transitioning from being a partner in the firm he had cofounded to full-time housing development work in conjunction with Tony Rezko's Rezmar Corp., and by February 1997 both Barack's and Allison's status at the newly named Miner, Barnhill & Galland changed to "Of Counsel," a common law firm title. "We were going to set him up in Springfield," Judd recalled, but Barack quickly realized that would be impossible. "Judd, this is unfair to you guys. It's full-time," so "I don't want to take a draw," Barack explained. "I'm going to be putting in a lot more time than I thought," so "I prefer not to be salaried." George Galland remembered being impressed by how "unusually straightforward" Barack was about the situation, and going forward Barack would be paid only when he could devote time to actual casework.

In early February, Barack summed up his first month's impressions of Springfield in a column for the weekly *Hyde Park Herald,* just as Alice Palmer had during her term. Remarking that he was "somewhat awed by the challenges that lie ahead," Barack confessed that "friends and associates have (only half jokingly) questioned my sanity in entering politics, and warned me of all the corrupting influences lurking in Springfield." But he had run because of "my desire to restore a sense of mission and service to state government," something he believed Illinois's governor lacked. Jim Edgar, he wrote, is "a pleasant enough man who appears incapable of mustering the kind of energy and vision needed to mobilize support around" substantive issues. "But perhaps the biggest impediment to change in Springfield is the lack of meaningful engagement in the process among ordinary citizens, and particularly those persons most vulnerable to government action—or inaction."

The legislature's most "daunting task," Barack wrote, would be coping with the state-level changes required by federal welfare reform. Barack cited the multiple bills he had introduced to "significantly increase and revamp the state's job training and retention services targeted towards unskilled workers" plus support services to help people transition from welfare to work. "It's people power that lifts up issues and frames the debate in new ways," and so he was "organizing citizens' committees" to address "education, economic development, health care, and public safety" issues. "I am extremely humble about what I can accomplish working alone. But I am supremely confident of what we can accomplish together, and look forward to serving you in the years to come."[29]

Barack's organizing experience clearly informed his view of elective office, and it also directly influenced a presentation he delivered on a mid-February Friday morning to Chicago's Local Initiatives Support Corporation (LISC), a well-funded, fifteen-year-old community development organization whose roots lay in a national Ford Foundation effort to revitalize distressed neighborhoods. LISC had created a "Futures Committee" to reshape its ongoing efforts, and several leading members, including Aurie Pennick from the MacArthur Foundation and Howard Stanback, formerly one of Harold Washington's top aides, had known Barack a decade earlier.

Barack shaped his remarks around what he called "eight principles of healthy communities," beginning with "a sound economic base" that offered living-wage jobs and opportunities for advancement. Continuing to echo John McKnight's teachings, his second principle was the need to "organize around production, not consumption." Then came "primacy of the family," inclusively defined, the "central importance of education," an "accessible and transparent public square," emphasizing commonality, "inclusive civic associations and political structures," strong "mediating institutions" such as churches and unions, and lastly a "shared mentality" that values the past while embracing the future. "We must recognize the major tensions that exist in all communities," Barack conceded, but "a healthy community combines a recognition of individuality with a well-defined public sensibility, and emphasizes both personal and collective responsibility." Eight months later, when LISC Chicago ratified its "Futures Committee" findings in a paper entitled "Changing the Way We Do Things," it credited Barack's "incisive observations" as a decisive influence on LISC's conclusions.

Similarly, the Rockefeller Foundation had created a Next Generation Network Program with a half-million dollars a year allotted to "create a corps of 21st-century American leaders with a sense of common purpose and the capability required to build a society committed to fairness and democratic principles, and with the confidence and skills to bring together others in pursuit of those goals." Barack's name came to their attention, and Barack asked Newton Minow at Sidley & Austin to write a recommendation letter. Minow told Rockefeller that Barack "is one of the truly outstanding young men of his generation," as demonstrated by his achievements at Harvard, his authorship of *Dreams From My Father,* and his chairing of the Chicago Annenberg Challenge's board. Minow humorously noted his disappointment that neither Barack nor Michelle had remained at Sidley, but he emphasized that the state Senate was just the start of Barack's public service career. "This is only the beginning for him because I believe one day he will be either Mayor of our City or Governor of our State or a United States Senator. He is going to the top."

Barack's district aide, Cynthia Miller, recalled reading Minow's letter before forwarding it to New York. Rockefeller failed to select Barack, but Cynthia knew Barack's aspirations reached at least as high as Minow's trio of predictions. So did his sister-in-law Janis Robinson, who now was raising two young children while working as the part-time executive director of ABLE, the Alliance of Business Leaders and Entrepreneurs, a small Chicago black executives' group. She and Craig lived within a block of Barack and Michelle, and by 1997

"Michelle and I were very close," Janis recalled. Once Barack was in the state Senate, "Michelle told me at one point that he definitely felt that he could be president," Janis recounted. Michelle had said that Barack "has pretty lofty aspirations," and when Janis asked, "Like what?" Michelle answered frankly: "Like president."

Barack was selected for a somewhat less august program than Rockefeller's, a Harvard "social capital" discussion symposium called the Saguaro Seminar that also included law professor Martha Minow and *Washington Post* columnist E. J. Dionne. Barack's Annenberg involvement continued, even though he often had to participate by phone from Springfield. At one point, Annenberg's efforts were publicly attacked by old ally Patrick Keleher, who after years of intense commitment to Chicago school reform concluded that such efforts were benefiting the reformers more than students. "A decade of waste—waste of public and private resources" had channeled hundreds of thousands of dollars to reform groups, while parents had been left "marginalized and largely voiceless," Keleher argued. "Very little really happens from one wave of Chicago school reform to the next," yet "reformers will fight to the death against giving direct financial control of education to parents" by way of vouchers. In a *Sun-Times* op-ed, Keleher castigated Annenberg's role, and the *Sun-Times* devoted two news stories to how "School Reform Pioneer Decries 'Lost Decade.'"[30]

Barack's principal focus remained the state Senate, and by early March the pace of legislative business picked up appreciably. Judiciary Committee chairman Carl Hawkinson, a 1973 Harvard Law School graduate who had made *Law Review,* was "immediately impressed" when he learned Barack had been president of the *Review*. Fellow Republican Frank Watson remembered standing with Hawkinson on the Senate floor when Barack came up to introduce himself, saying he understood Watson was interested in education. "I'm Barack Obama, and I want to do good things for kids." Watson replied that he would be happy to work together, then expressed amazement once Barack stepped away. "Can you believe that?" Watson remarked about Barack's brimming self-confidence. "That was kind of unique. I'd been around a long time, and that had never happened to me before."

Appropriations chairman Steve Rauschenberger, one of the so-called Fab Five group of reform-minded young Republicans who had entered the Senate together four years earlier, was very intrigued with Barack. "When he first arrived, I thought he was exceptional. He was an intellectual Democrat of color who would engage on public policy." Deno Perdiou, Jim Edgar's Senate lobbyist, gave the governor a similar report: "There's a really bright new African American senator from Chicago who's really good," Edgar remembered being told. Walter Dudycz, Chicago's only Republican senator and a veteran police detective, found Barack "very personable and pleasant," and North Side Democrat Howard Carroll thought Barack was "very impressive." In identical language, both men also saw that he was "very bright." Far downstate Democrat Evelyn Bowles "liked him right off the spot," but both of Chicago's progressive Hispanic senators, Miguel del Valle and Jesus "Chuy" Garcia, thought Barack got a decidedly cool reception from a number of Democrats because of both his University of Chicago affiliation and also because of how he had taken out the much-loved Alice Palmer. On the Senate floor, Barack remained quiet well into March, impressing del Valle as "cautious and careful and very thoughtful." But in Senate Democrats' private caucuses, he spoke up from the outset, often invoking his legal expertise. "He always had something to say," Debbie Halvorson remembered, and older members found Barack's professorial statements a bit much for a newcomer. Southwest Side senator Bobby Molaro, one of the legislature's funniest and most outspoken members, never hesitated to put the overly talkative freshman in his place. "'Barack, shut the fuck up and sit down, you son of a bitch! Who wants to hear from you?' That would happen a lot," Molaro smilingly remembered.

Barack's biggest concern was less his colleagues' reactions than his huge disappointment with the support staff. Initially, as a black Chicago Democrat in a safe seat, he was given a Chicago rather than Springfield-based communications staffer, David Wilson, but Barack quickly asked for a change, and staff director Lori Joyce Cullen gave him Jimmy Treadwell in Springfield. By the beginning of March, Barack again expressed his dissatisfaction, noting that fellow freshman Debbie Halvorson, who had won a competitive seat in the south suburbs, was being staffed by her former campaign manager, Dan Shomon, a thirty-two-year-old Georgetown graduate and former newsman whom everyone regarded as highly energetic and skillful.

Barack had already gone out for a drink with Democratic chief of staff Mike Hoffmann, making clear that "I have ambitions" beyond the Senate, Hoffmann recalled. "He talked about running statewide as a black man with a name like Obama," but "I think he was probably sizing me up as much as anything." Yet in the wake of being doubly disappointed, Barack phoned Hoffmann with a specific request: "'I want Dan Shomon.' I said, 'No way. Shomon's our best guy. We need him for our targeted candidates to make sure they get profile. I can't give him to a safe senator. I just can't afford that.'" Then, "over the course of what seemed to me a half hour's conversation, he wore me down," Hoffmann recounted. Barack claimed it would be "'a light load, just a little extra,' so I finally acquiesced, and I called Shomon."

Dan Shomon had first come to Springfield eight years earlier as United Press International's bureau chief before moving into state government, eventually working for Illinois treasurer Pat Quinn and then shifting to the Senate Democratic staff in early 1995. With Debbie Halvorson's office in the same four-senator suite as his, Barack could easily see how much effort Dan was putting into assisting Debbie. When Hoffmann called to say, "I need you to staff Obama," Shomon responded, "You're kidding. I don't have time for Obama. I don't even know if I like Obama. He wants to change the world in five minutes. I have a lot of other things to do." Hoffmann sweet-talked Shomon, just as Barack had persuaded him. "He just needs a little strategic advice, he doesn't really need you to do any work. He does his own writing. He just needs you to help with strategy."

Soon after Hoffmann's call, Barack caught up with Dan. "I'm overloaded," Shomon insisted, but Barack said, "Let me take you out to dinner," and a night or two later, they met up at Sebastian's, one of Springfield's best restaurants. Barack reiterated Mike's line—"I just want some strategic help. I just want you to give me some advice"—and in the course of dinner, Dan decided, "This is a really nice guy" and agreed to staff Barack.[31]

Soon after the swearing-in, Barack received his committee assignments: Judiciary, following directly from his Harvard and U of C law school pedigree; Public Health and Welfare, which he had specifically requested; and State Government Operations, which was minor in comparison. Senate Judiciary was the least partisan committee in the General Assembly, primarily because independent-minded Republican Carl Hawkinson of Galesburg, who had been chairman since 1993, brought to it an even-handed style. Hawkinson had a good friendship with ranking Democrat John Cullerton, a Loyola Law School graduate who had spent five years as a Cook County public defender. Cullerton, like everyone else, viewed Hawkinson as "very fair, very thoughtful," and "real bright." Judiciary handled both criminal and civil law issues, and lobbyists, such as Illinois Sheriffs Association executive director Greg Sullivan, viewed Hawkinson as "very, very well respected" throughout state government. Other Republican members included razor-sharp former Will County state's attorney Ed Petka, whom a Joliet reporter had nicknamed "Electric Ed" for the number of murderers he had sent to death row.

Dan Cronin and Kirk Dillard were two younger, suburban Republicans. Cronin, a true moderate, saw Judiciary as "a heady committee" with "a nice collegiality" and "a rather significant workload." Dillard, whose warm demeanor masked resolutely conservative views like all-out opposition to abortion, had worked in Jim Edgar's administration before entering the Senate as a protégé of Pate Philip. Among committee Democrats, live wire Bobby Molaro was "a funny guy" who one colleague said provided "a little comic relief," while George Shadid had served twenty-four years as a Peoria police officer before overcoming voter discomfort with his Lebanese ancestry to win election as Peoria County sheriff. After seventeen years, he had entered the Senate in 1993.

Judiciary was significantly better staffed than most committees, with two young attorneys from each caucus assigned to its work: Democrats Gideon Baum and Mike Marrs, and Republicans Matt Jones and John Nicolay, handled the criminal and civil legislation, respectively. Baum realized immediately that Barack was "very smart, very political, very pragmatic," and Nicolay immediately pegged Barack as an "absolutely political, ambitious animal," "one of the most political people I've ever met, from day one." Marrs was impressed when Barack wanted to go over bills prior to the committee's first meeting. But watching Barack on the Senate floor during one of the first days, Marrs thought "he seemed a little bit crestfallen at seeing the process and the fact that there's not these weighty substantive debates going on," that what transpired was "very politically driven and not issue driven." Barack "seemed visibly sort of like 'Is this all there is?'" Marrs remembered. One evening early in the session, "before he really knew anybody," Marrs ran into Barack at a reception. "What's the story on that guy over there?" Barack asked. "Tell me about that person." It was readily apparent that Barack wanted to "soak it all up." Republican Matt Jones had spent four years as a prosecutor in Peoria County before going on Senate staff, and he saw Barack as "trying to fit in and figure things out" in committee. John Nicolay agreed that Barack "was still finding his feet" and "knew his place in the world," yet "from day one" it was clear he was "extremely impressive, very smooth, very bright."

Barack's second committee, Public Health and Welfare, was sometimes called "Nia's committee," because over the previous sixteen years Democratic staffer Nia Odeoti-Hassan had acquired an unequaled reputation for her policy expertise. A Muslim woman who had taught for six years at a small Iowa college before arriving in Springfield in 1981, Nia had become close to Margaret Smith, the aging black Chicago Democrat who was the committee's ranking minority member, as well as Alice Palmer. Nia knew Barack had requested the committee, and with Smith's failing memory, Emil Jones quietly told Nia "to make sure that everything that she knew as minority spokesman that Barack knew too."

At one point, Smith asked Nia, "Why are you down in that young man's office all the time?" but staffers and lobbyists noted that Barack treated Smith with unerring deference and respect. "She had earned her stripes," Republican committee staffer Debbie Lounsberry explained. Barack "was actually functioning as the co-minority spokesman," Nia explained, and while some staffers found Barack's Harvard and U of C credentials intimidating, Nia, who often advised progressive lobbyists like John Bouman and Doug Dobmeyer on how best to handle things, was happy to have such a bright African American senator, one who reminded her of "Dick Newhouse in his early years."

Public Health and Welfare's chairman was Rockford conservative Dave Syverson, who entered the Senate in 1993 as one of the Republicans' Fab Five. Syverson had the committee meet on Tuesday at 8:00 A.M., which meant

Barack had to make the more-than-three-hour drive to Springfield on Monday evenings rather than Tuesday mornings, as he had hoped. As Lounsberry could see, Barack was "not an early morning person, and he was not happy" with the unpleasant schedule. But with Margaret Smith effectively sidelined, Syverson and Barack met "before committee so there wasn't any blindsiding the other person," Syverson recalled, and, according to Lounsberry, they quickly developed "a very cordial professional working relationship." Barack was "pretty quiet when he first started out," she remembered, but gradually he emerged as "the voice of the committee," asking "a lot of questions of the witnesses" when testimony was being heard for and against various bills. But Lounsberry believed the committee was well functioning and collegial because Nia Odeoti-Hassan "knew the issues inside and out."

The chairman of the State Government Operations committee was young Republican Fab Five banker Peter Fitzgerald, who readily admitted that its "housekeeping" tasks made it "one of the least desirable committees" to chair. Barack was the Democratic spokesman, but Fitzgerald, who was distracted by trying to win the 1998 Republican nomination to challenge Democratic U.S. senator Carol Moseley Braun, found Barack overly diligent. "He would make it take up more time than I thought it should," carefully scrutinizing bills even though "there was no partisan material or issues involved." Fitzgerald believed that Barack had "a degree of suspiciousness that I thought was excessive," although he recognized that Barack was "very bright" and "very gifted."[32]

By early March, Emil Jones Jr. and his staff had a clearer sense of how Barack was adjusting. Jones believed "he was naive, totally naive," and communications director Lori Joyce Cullen thought Barack was visibly "pie in the sky: 'I'm going to change the world.'" Her deputy Cindy Huebner remembered that Barack "was really, really wanting to get things done," and Dan Shomon saw that Barack could rub people the wrong way because "he wanted to fix the process too quickly." Citizen Action lobbyist John Cameron thought Barack was "pretentious" and "insufferably ambitious," and recalled how he would "sit in committee and roll his eyes if he thought the testimony was below him."

Dave Gross, Jones's deputy chief of staff, knew that "there were very few opportunities for Democrats to actually participate" in crafting legislation given Pate Philip's control. Shomon's communications colleague Adelmo Marchiori explained that Democrats "had to spend some time eyeing some of the Republicans' bills" to find attractive legislation to cosponsor and claim credit for if it won enactment. Most Democrats' bills were relegated to the Senate's Rules Committee, chaired by Philip's top lieutenant, Urbana undertaker Stan Weaver, where they would quietly die. Mike Hoffmann appreciated that Barack "was quick to understand that in the minority you weren't going to get anything done unless you could partner up with somebody on the other side." Barack spoke frequently with Hoffmann, "probably more so than just about any other member," and clearly understood that his initial months in the capitol were "a learning experience, and he didn't say a lot. I think he tried to keep his ambitions in check," even if others like John Cameron felt otherwise.

A few months later, Barack recounted that "when I first arrived, I anticipated my caucus would meet and discuss what issues we really wanted to focus on and then structure bills for the problems we identified." Instead, he quickly realized that "every legislator is somewhat of an independent contractor, searching out issues that will play well in the district." That produced "a lot of duplicative bills and bills that really hadn't been thought through or thoroughly researched. That was discouraging," yet most Springfield veterans "are resigned to the fact that the work of the legislature is piecemeal and sporadic. I don't think it has to be that way." A few years later, Barack explained that he had been most disappointed by "the degree of control that leadership exerted over the process," which gave Pate Philip, "the original paleoconservative," really "exclusive power to control the flow of legislation and to control the debate." With progressive bills held in the Rules Committee, "you wouldn't even get a debate, much less get a vote," and "that was disappointing and frustrating for someone like myself who was down there to work and discovered that most of my bills were going to get killed before they had even been debated . . . that is not something that I anticipated."

Barack shared his disappointment with Michelle and friends back in Chicago, who could contrast Barack's reaction to the political realities of the capitol with the high expectations for policy innovation he had voiced months before. "He was naive about it," Michelle's young protégé Craig Huffman recalled. "He thought he was getting into the AP class, and he'd really been admitted to the Special Ed class . . . I think it was eye-opening."[33]

On Tuesday, March 4, Barack cast his first notable vote when he and John Cullerton both opposed SB 230, a "partial-birth" abortion ban measure that the Senate Judiciary Committee sent to the floor by a 7–2 margin. The meeting witnessed "spirited testimony from activists but little discussion by lawmakers," with Barack telling reporters that the bill's lack of a health exception meant it was "clearly unconstitutional." Barack remained far more interested in the upcoming welfare law changes that would remove the "safety net" of Aid to Families with Dependent Children, and on March 11 the Public Health and Welfare Committee approved Barack's SB 755, which mandated a study of the impact of those changes on Illinois recipients of public assistance.

The next day, Senate Judiciary approved Barack's SB 574, authorizing administrative hearings for alleged violations of municipal ordinances, but Thursday, March 13, was Barack's rookie debut on the Senate floor, as his SB 837, authorizing the directory of graduating Chicago city college students, came up for passage. The Senate had a long tradition of "roasting" new members when they rose to request passage of their first bill, but Barack suffered a far rougher debut than most. This was due in part to how he had knocked Alice Palmer off the 1996 primary ballot, but to a far greater degree it illustrated the stark dislike several Chicago African American senators had for their new young upstart colleague.

"I come humbly before you on this extremely humble bill," Barack began, explaining that it "authorizes the community colleges to develop and distribute a directory of graduating vocational and technical school students," an idea that someone at Kennedy-King College had suggested to Barack.

Rickey Hendon, a forty-three-year-old West Sider who had entered the Senate four years earlier after serving as 27th Ward alderman, was quick on the attack. "Senator, could you correctly pronounce your name for me? I'm having a little trouble with it," perhaps because Hendon and his buddy Donne Trotter had under their breaths already been mouthing a variety of insulting quips like "Yo Mama," targeting the visibly self-confident, or arrogant, new hotshot.

After Barack answered, Hendon said, "Is that Irish?"

"It will be when I run countywide," Barack replied.

"That was a good joke, but this bill's still going to die. This directory, would that have those 1-800 sex line numbers in this directory?" Hendon queried.

Walter Dudycz, often the presiding officer, called on Barack to answer Hendon's question, and Barack replied, "I apologize. I wasn't paying Senator Hendon any attention."

Hendon responded, "Well, clearly, as poorly as this legislation is drafted, you didn't pay it much attention, either. My question was: Are the 1-800 sex line numbers going to be in this directory?"

Barack stumbled, citing where the idea came from and declaring that those numbers would not be included.

Hendon proceeded. "I seem to recall a very lovely senator by the name of Palmer—much easier to pronounce than Obama—and she always had cookies and nice things to say, and you don't have anything to give us around your desk. How do you expect to get votes? And you don't even wear nice perfume like Senator Palmer did."

Minutes earlier, fellow freshman Debbie Halvorson had presented her first bill, and Hendon had alluded then to how he missed the man she had ousted. Now Hendon said, "The same day that I'm missing Senator Aldo DeAngelis, now I'm missing Senator Palmer because of these weak replacements with these tired bills that make absolutely no sense. I definitely urge a No vote. Whatever your name is."

Friendlier colleagues like Republican Brad Burzynski immediately came to

Barack's aid. "First of all, Senator Obama, I'd like to thank you for not wearing perfume that Senator Hendon likes." He continued with a series of gently needling questions, joined by veteran Democrats Denny Jacobs and Howie Carroll. Then fellow Harvard Law School graduate Carl Hawkinson asked Barack, "How do the quality of these questions compare with those that you received from Professors Dershowitz, Tribe, or Nesson?"

Now Barack was in the spirit. "I must say that they compare very favorably. In fact, this is the toughest grilling that I've ever received. If I survive this event, I will be eternally grateful and consider this a highlight of my legal and legislative career."

Downstate warhorse Vince Demuzio joined in. "Let's not forget, his name ends in a vowel. Senator Obama, let me congratulate you on your first bill. I was sitting here, and as Rickey Hendon was asking you those questions and making those statements, I recalled back to Rickey's first bill. I closed my eyes; I held my nose; I couldn't find the right switch so the staff had to vote me. This time I will keep my eyes open and vote in the affirmative."

Moderate senior Republican Adeline Geo-Karis spoke similarly. "I certainly commend him for his fortitude today, because if you had fortitude at Harvard, you don't know this body yet. But I want to commend you for trying, because you have Mr. Hendon, who never ceases to attack you."

SB 837 then passed unanimously, and five months later Governor Edgar signed it into law.[34]

Hendon's hostility was readily apparent, and the "Yo Mama" comments were "audible" very early on, according to secretary of the Senate Linda Hawker, who remembered Donne Trotter joining in with Hendon. Years later, Hendon cited his "true love for" Alice Palmer to explain his animus, but black lobbyist and former House member Paul Williams was one of many who said the reasons for Hendon and Trotter's performance had deeper and more complex roots. "Everybody wasn't thrilled about Mr. Goody Two-Shoes coming in here and imposing all his great ideas on how we need change, when nobody was begging for it," Williams recalled. "Barack's trying to find his way," but "Barack's a little bit arrogant." Williams believed Barack exhibited an "I'm-smarter-than-these-guys" demeanor, and he recalled that once that spring, Barack told him, "You know, Paul, this forum's a little small for me." Williams thought Barack was "overly intellectual" in an institution that had what another lobbyist called "a very anti-intellectual culture." But to Williams, even more crucial was how "Barack doesn't have the same community experience we do" as black men who had grown up in Chicago.

Early on, Williams introduced Barack to the Senate's only African American member from outside Chicago, thirty-three-year-old James Clayborne, who was from "Metro East," the heavily minority area in southwest Illinois just across the Mississippi River from St. Louis. Williams thought Barack and Clayborne, an attorney who had entered the Senate two years earlier, "had a lot in common," but Clayborne's downstate roots gave him a worldview that was very different from black Chicagoans'. "I can remember specifically Donne Trotter, more so than anybody," Clayborne recalled, "expressing some reluctance, I guess, on the part of accepting Barack and saying that—I'll never forget this—Donne Trotter said to me, 'Can you believe . . . this guy's thirty-some years old [and] he's already written a book about himself?'"

Resentment and jealousy were significant factors, especially toward someone who "had never paid his dues" and who sometimes reminded people of Mel Reynolds. Hyde Park state representative Barbara Flynn Currie recalled that "there was a lot of anger about Mel Reynolds," as well as "a concern that Barack was going to lord it over them in the same kind of way" that the onetime Rhodes Scholar turned convicted felon had. "Here comes another outsider who's got the class and the smarts and the polish that a lot of the rest of us don't," Currie explained. African American AFSCME lobbyist Ray Harris knew Barack "was looking ahead almost from the moment he stepped into the capitol," and he was aware of how Hendon and Trotter "were just senseless in their dislike for the man, because he was everything that they wished they could be." On the Senate floor or off, "they made it their mission in life to make certain people knew that they hated this guy." Southwest Side senator Lou Viverito agreed that for someone so focused on "moving so rapidly," Barack was seen as "an up-and-coming guy that didn't earn his way." Not only was he "more of an outsider," it was also clear "they didn't consider him black enough." Fellow freshman Terry Link, whose office adjoined Barack's and who also sat in the back row in the Senate chamber, agreed that it came down to origins. "Barack wasn't born and raised in Chicago" so "he was considered an outsider more than anything else. . . . There's the part that was more of the problem than anything else."[35]

On March 19 and 20 the Senate passed Barack's administrative hearings bill and then his welfare-study one. Later that day the "partial-birth" abortion ban measure Barack had voted against in committee came to the floor, and Barack remained silent while John Cullerton explained why it was clearly unconstitutional. Then, when the bill received forty-four yes votes and seven nos, Barack joined four other colleagues—Bobby Molaro, Miguel del Valle, Chuy Garcia, and Margaret Smith—in voting "present." Illinois senators actually had three buttons on their desks: red, yellow, and green—and a "present" vote was understood as a no, but it offered senators political cover from criticism that they truly had voted against the measure in question. "What it did was give cover to moderate Democrats who wanted to vote with us but were afraid to do so," Planned Parenthood lobbyist Pam Sutherland explained, and she was happy that Barack chose that option rather than a no. "'If you vote "present," we're going to pick up "present" votes,'" she reassured him, "which we did."

From mid-March to mid-April, the Senate met in only perfunctory sessions. That gave Barack a respite from spending three nights a week in Springfield, where he stayed at the Renaissance Hotel, six blocks from the capitol. Early on Wednesday and Thursday mornings Barack played pickup basketball at the Springfield YMCA. "He had a decent shot, and he liked to talk trash, and he'd throw plenty of elbows," African American lobbyist Vince Williams remembered. "He liked to push and shove," but most of the other regular players, including African American House members Calvin Giles and Art Turner, gave better than they got. Returning to his hotel in sweatclothes, Barack had a cigarette outdoors and got tea, bananas, cereal, and milk from the breakfast buffet to take up to his room before showering and dressing for the day.[36]

Speaking with reporters, Barack willingly voiced disappointment with the Senate. "What I'm surprised at is the lack of an overarching issues strategy" in a legislature where leaders "exert control without a clear road map." Eager to show he was not just a "brainiac," Barack boasted that he could "knock down scotch and tell a dirty joke with the best of them." But he remained focused on the harms that might come from President Clinton's federal welfare reform. "We have to make sure that the families are intact and have some order. If the parents don't have their lives together, a lot of times they'll take it out on their kids." When Aid to Families of Dependent Children ended, "we're going to have real problems placing these folks into jobs," particularly ones that paid a living wage, because "there aren't enough jobs out there."

In late March Barack was shocked to see hundreds of young African American men protesting outside the Dirksen Federal Building in downtown Chicago chanting, "Free Larry." They were supporters of Gangster Disciples leader Larry Hoover, a convicted felon who was again on trial in federal court. In one of his occasional columns for the *Hyde Park Herald* Barack wrote, "Something is terribly wrong" when "an ever growing percentage of our inner-city youth are alienated from mainstream values and institutions, and regard gangs as the sole source of income, protection and community feeling." He realized that "a large proportion are functionally illiterate," thanks to the sorry state of most Chicago public high schools, and since it was mainly "the drug trade that supports the gangs," prosecutors were "locking up these youth in record numbers." Post-incarceration, their chance for gainful employment was "even slimmer than it was when they went in." Barack asserted that "we need to send a strong message to our youth that poverty is never an excuse for violence, individuals are responsible for their behavior, and there are consequences to criminal activity."

But, he argued, Illinois needed to revise its "juvenile justice code so as to balance incarceration with prevention."

In mid-April, Barack flew to Boston for the first meeting of Harvard professor Robert Putnam's thirty-person Saguaro Seminar. Then he was back in Springfield for the final five weeks of the spring session, which would be the busiest, although more for Senate Republicans than for minority party backbenchers. "Springfield is 80 percent after-hours," one experienced female lobbyist explained, but, as progressive Chicago representative Tom Dart recounted, nightlife regulars realized that Barack "definitely was not in the night crowd." African American Commonwealth Edison lobbyist Frank Clark and even Senate president Pate Philip used almost identical phrases: "You didn't see him at night."

The 1997 session was an unusually light one, and the unofficial mayor of Springfield's late-night scene, political newsletter publisher Rich Miller, whose *Capitol Fax* was the weekday bible for state politicos, reported that "Senators spent more time at Springfield's eating and drinking establishments . . . than they did conducting actual business." Tuesday and Wednesday evenings, when everyone was in town, usually featured one or more early-evening receptions that might go on until as late as 8:30 P.M., after which most legislators adjourned to one or another of downtown's many late-night watering holes. Young Democratic Senate staffer John Corrigan described Barack's approach: "I would see him out at these events, and he would squeeze every bit out of Springfield that was good and legitimate for his career. So he would go out, and he'd network, and he would work real hard, and he'd be at all these events. . . . But then, around nine thirty or ten o'clock, you couldn't find him. He was *gone*."

Corrigan explained that "you can get a bad reputation quickly in Springfield" and "Obama understood that." The small world of the statehouse was "a big gossip thing," and Barack "knew that it doesn't matter if he is completely innocent: if it's midnight, and he's talking to a twenty-six-year-old staffer at a bar where everybody's swilling beer, that those kinds of rumors were going to go around about him." Democratic staffer Cindy Huebner had the same view of Barack: "when you would hear the stories from the night before, you never heard any about him."

One progressive female lobbyist was disappointed that Barack "was never at the places," but given how some women lobbyists viewed Barack, that was perhaps wise. "I can clearly remember the first time I saw him," Planned Parenthood's Pam Sutherland recalled. "He's walking down the big stairs" in the capitol. "'Oh my god! He's gorgeous,'" she exclaimed to a female colleague. Barack was "engaging and friendly . . . one of those guys you can't help but like," plus "he's a very attractive man. He had a presence about him . . . his swagger," and "we hit it off right off the bat." One female Republican senator called it "an air of self-confidence, a certain swagger." League of Women Voters lobbyist Cindi Canary put it more bluntly: "women thought Barack was hot." SEIU's Carole Travis, who already knew Barack from their dinners with Bill Ayers, Bernardine Dohrn, and Mona and Rashid Khalidi, saw the same thing: "a lot of women found Barack attractive and were after Barack." But "there was never any sense at all that he was cheating," Carole explained, and it was "very unusual," as one lobbyist euphemistically put it, that Barack "did not participate in Springfield activities."[37]

One Wednesday night in mid-April a trio of business lobbyists—Phil Lackman from the Independent Insurance Agents of Illinois, Dave Manning of the Community Bankers Association, and African American Mike Lieteau of AT&T—invited Barack and fellow freshman Terry Link, plus newly arrived senator Larry Walsh, the Will County Democratic chairman, to join them for dinner, followed by poker at Panther Creek Country Club. Walsh, a lifelong farmer, had arrived in Springfield just a week earlier after appointing himself to a newly vacant seat whose prior occupant had resigned to take a judgeship. Walsh already knew Lake County party chairman Link, a rare legislative teetotaler, and outgoing lobbyists like Lackman and Manning, both of whom were Panther Creek members, were always looking to get to know new senators. Playing cards in the country club's bar, with the Chicago Bulls game against the Miami Heat on the large-screen TV, the group of six was an ideal size for poker, and they had "a great time" until about midnight. They agreed to reconvene the next week, and for the balance of the spring session each Wednesday night they gathered out at Panther Creek.

"There was no shop talk allowed" at these gatherings, Link explained; "nobody talked about their jobs or politics." They played "low-stakes poker: a dollar stake, three-dollar top raise." Someone might win or lose $25 or $30 over the course of a night, but "on the whole it was a thousand laughs," Dave Manning recalled. Smoking was definitely allowed, but teasing humor was the center of the weekly gatherings. Barack was a cautious and careful card player "and would study and study" before making a move. "Barack, it's two bucks. Call, fold, or raise," Phil Lackman would tell him. "Barack enjoyed a couple of beers, he enjoyed a couple of Marlboro Reds," Phil remembered, and that Barack and Mike Lieteau were black was no issue. Lieteau, fourteen years Barack's senior at age fifty, had grown up in Englewood in a family of eight children, attending Catholic schools and then Northern Illinois University. Married to an African American psychiatrist and the father of two daughters, Lieteau lived in South Shore's Jackson Park Highlands enclave. He had begun lobbying for AT&T seven years earlier, and by May 1997 younger daughter Triste, already a graduate of Harvard Medical School, was about to join her sister Michelle as a fellow graduate of the University of Chicago Law School, where she had taken Racism and the Law and Con Law III with Barack.

"It seemed like we were accepted because we were more white" than other statehouse African Americans, Mike Lieteau said in explaining the easygoing biracial poker games. "Barack talked the King's English," as could Mike, but Barack "was never ghetto like me, 'cause I speak Ebonics too." Lieteau had watched all spring as "Trotter and Hendon would just endlessly say, 'yo mama'" to Barack, until eventually Mike asked, "Why don't you say something?" and Barack replied, "I would be stooping to their level if I responded."

Lieteau thought "they resented him because he was a very bright, intelligent person" who "brought his points across objectively rather than emotionally." Barack "was more of an intellectual" and "was more comfortable with people who were of the same intellectual level" as himself. "It was a difference in culture," for Barack "did not fit the stereotype of the typical African American legislator," i.e., a lifelong Chicagoan. "They thought he was arrogant," but Mike believed Barack "was not arrogant, he was confident," even though at times it did appear as if Barack was asking himself, "Why am I here with these idiots?"

Lieteau "felt that Barack was more comfortable with Caucasians, or white people, than he was with ghetto black people," but Lieteau quickly developed an ease of interaction with Barack that was unrivaled not only in Springfield but also with any other black friend. Mike recalled that during the Wednesday-night poker sessions, he humorously sought to knock the other players off their games. "I would try and disturb all of them," he recounted, either by insisting that they "explain the game" when they shifted to some exotic type of poker or simply by holding his cards to irritate them. Barack would demand, "Either call or fold," and Mike responded with ersatz anger: "I'll playing my fucking game and you play yours!" His two favorite targets for teasing were Walsh and Obama. To Larry, "I would make comments about farmers," such as saying "their first experience with sex was with an animal." As expected, "Larry just got very, very angry," declaring that "'I've never fucked a chicken!'" to which Mike replied, "Larry, anybody who responds like that had to fuck a chicken!" With Barack, the obvious target was his ears, which are "like Spock's." Barack "never cursed," and "would never use bad language," but one night Lieteau got him sufficiently riled that Barack called Mike an "anvil head motherfucker," which Mike said was accurate because "my head was like an anvil." But Mike replied, "Barack, you can't even say motherfucker right. . . . You sound just like a fuckin' white boy!"

Mike and Barack's easy friendship had understandable limits. With Lieteau representing AT&T, and with Barack being "very consumer-oriented," he seldom voted with Lieteau. Barack was also a stickler for refusing to accept *anything* of value from lobsters. "You couldn't buy him a Coke," one progressive lobbyist recalled, and the first time Barack had dinner with Mike, he asked, "What do I owe you?" Lieteau responded, "You don't owe me anything," but

Barack objected, knowing that lobbyists filed comprehensive expenditure reports with the secretary of state's office. "You can't report me. What do I owe you? What is my share?" Mike also recalled Barack asking, "How much was that?" even if Mike handed him a can of soda. "This is how bad he was," Mike recounted, but ten years later Barack would boast on national television, "I was famous in Springfield for not letting lobbyists even buy me lunch."[38]

Black lobbyist Ray Harris had warned Barack that as a Democratic rookie, "you're going to have a lot of time on your hands" even during the Senate's busiest weeks, given Republicans' legislative dominance. Many afternoons the day's work, whether in committee or on the floor, was over by 2:30 P.M. Terry Link was "an avid golfer," and Barack already appreciated that in Illinois politics "an awful lot happens on the golf course," as he told his friend Jean Rudd. Springfield offered The Rail, an eighteen-hole public course just north of town that had been designed by the famous Robert Trent Jones Sr., and on weekday afternoons, the fee was just $19 for all you could play, fellow black Democrat James Clayborne remembered. Barack was happy to skip out along with Link, but "when he first started, you couldn't call him a golfer. He left a lot to be desired," Terry recalled. Clayborne and lobbyist Dave Manning sometimes joined them, and Barack also played with Poverty Law Project lobbyist John Bouman, Illinois progressives' lead representative on the federally mandated welfare law changes Barack was so concerned about. He was "pretty much a beginner," Bouman recalled, but by late spring Barack had also recruited Dan Shomon, the high-energy staffer whose assistance he had lobbied for, to take up golf as well.

Shomon's savvy paid off almost immediately when he asked his next-door neighbor, Dave Heckelman, a reporter for the *Chicago Daily Law Bulletin,* the newspaper read by most of the city's lawyers, to write a profile of Barack. Heckelman asked Barack about *Dreams From My Father* and his father, and Barack said, "I suspect that part of my interest in politics, and a particular brand of politics, can be traced back to who he was and what he wanted." For example, Barack believed the tribal rivalries that hindered his father's career in Kenya illustrated that "race and ethnicity can be a destructive force in politics." Barack also said he was very interested in juvenile justice issues, and that tax credits for investors who developed low-income housing, something he had worked on at Miner Barnhill, was "an example of a smart policy." Rejecting the notion that all politics is dirty, Barack mistakenly asserted that "I would argue that Martin Luther King was essentially a politician."

Heckelman asked two of Barack's Judiciary Committee Republican colleagues about him, and both praised him. "He asks very intelligent questions, he has an in-depth understanding of the law, and I think he's going to make a top-notch senator," chairman Carl Hawkinson said. Barack had made "an extremely positive" impression, Kirk Dillard added, and "is a tremendous addition to the Judiciary Committee and to the Senate as a whole." Longtime booster Newton Minow predicted that "he's going places."

Shomon was understandably proud of the story and tried to introduce Barack more widely around town. Ed Wojcicki, the publisher of *Illinois Issues,* a monthly politics and public policy magazine that was a must-read for state politicos, talked with Barack about life in the capitol. "I was expecting to come down here to Springfield and engage in serious debates about the big issues of the day . . . ," Ed recounted Barack complaining to him. "But those discussions aren't happening around here, and nobody seems interested. Am I missing something, or is this just the way it is?" Barack was also exasperated that statehouse reporters displayed little interest in major policy issues. "Ed, why don't we organize a meeting for all the reporters who work down here, and I just want to talk to them about why they're not covering my issues," Barack suggested. Wojcicki replied, "If you try to set up a meeting like that, they're not going to come. They're not going to have that conversation with a public official like you. It's just not what they do." Barack found that "very frustrating," just like he reluctantly grasped Wojcicki's argument that "with leadership like Pate Philip," being in the minority meant "you don't really have much of a chance" to advance any meaningful priorities.

Friends far and wide sensed Barack's disappointment. Democratic campaign consultant Eric Adelstein, who had first met Barack during Project VOTE!, remembered seeing Barack "coming back to the Renaissance" one night looking like "a guy who was not wanting to be in Springfield." Even with legislative license plates, the drive between Chicago and Springfield was a boring three hours plus, and although cell phone coverage was spotty in places, Barack called his district office staffer Cynthia Miller and friends like Rob Fisher, Tom Johnson, and Scott Turow during his Thursday drives back to Chicago and vented about the statehouse. "He was *very* disappointed," Cynthia recalled, especially with "the in-house fighting and the jealousies" he was encountering from colleagues like Hendon and Trotter. Rob remembered Barack explaining that 90 percent of state politics involved fairly technical issues of interest to some small group but to no one else. To Tom, Barack joked about another senator's suggestion about what kind of extracurricular nightlife he might enjoy. Turow recalled Barack saying that the Senate "is different than I thought it would be," and it was clear that "he was having a hard time adjusting to Springfield."[39]

The legislature's annual House-Senate softball game took place on Tuesday evening, April 29. Senate announcer Adeline Geo-Karis said Barack "was fleet of foot, and he looked good in uniform," but other colleagues recalled that Barack's skills on the diamond were minimal, with Senate team manager Frank Watson saying "he was terrible." Back on the floor, the Senate unanimously passed a House bill about drug counseling for which Barack was the lead sponsor, and when another "partial-birth" abortion ban measure returned to the Senate floor, Barack again voted "present," just as he had in committee a week earlier.

That evening Chicago mayor Richard Daley, who was in Springfield to join Republican governor Jim Edgar in seeking more state funding for public schools despite Senate Republicans' reluctance to raise taxes, sponsored a "Taste of Chicago" reception featuring rib sandwiches, ice cream, and cheesecake plus the Lonnie Brooks Blues Band. A line of people gathered to meet and greet Mayor Daley, and Democratic Judiciary Committee staffer Mike Marrs remembered Barack being right behind him. "He introduced himself and posed for a picture with the mayor. I remember hanging back to watch," since Barack indicated he had not previously met Daley.

In the spring session's final days, Barack reluctantly supported a bill authorizing civil commitment hearings for sex offenders whose prison terms were ending. He spoke in favor of a measure allowing state candidates' campaign disclosure reports to be put on the Internet. "There has been a lot of talk about campaign finance reform during the course of this session," with Senate Republicans naming a task force to study the issue rather than take up meaningful House-passed bills, but this "modest" measure, by making the reports "easily accessible," would help "restore people's confidence that in fact we are doing the people's business and not simply the lobbyists' business." Governor Edgar's education funding proposal passed the House and was supported in the Senate by Democrats plus a small handful of Republicans, but with a majority of his Republican caucus firmly opposed to a 25 percent state income tax increase, Pate Philip refused to call the bill for a floor vote.

On Saturday, May 31, the session's final day, the Senate finally moved toward passage of the state-level welfare reform proposal that Edgar had put forward in late April in response to the soon-to-be-implemented changes in federal law. Public Health and Welfare Committee chairman Dave Syverson rose to introduce a lengthy agreed-upon amendment and to "thank Senator Obama for the hours of time that he spent with us in trying to work out a plan that is acceptable and workable." Barack's biggest success was a $30 million increase, to $100 million, in new state funding for subsidized day care, though he failed to win similar support for job training programs. Barack then spoke at greater length than he had at any time that spring. HB 204 was "a huge bill that hasn't received too much attention," but "this may be as important a bill as we pass in this session. It will affect a huge number of people." Barack said "I probably would not have supported the federal legislation" that President Clinton had signed into law nine months earlier, and HB 204 was not a complete response. "I want to emphasize that I continue to have concerns, and I think that those concerns are going to have to be dealt with over the coming months and the coming years." Uppermost was the availability of job training for people lacking employment skills, and a close second was the Illinois program's exclusion of legal immigrants. "I don't like the notion that those people who are here legally, contributing to our society, paying taxes, are not subject to the same benefits, the same social safety net that the rest of us are." Third—as Barack's modest SB 755, now on the governor's desk, would mandate—was a university study of the impact of all the changes on recipients of what would now be called Temporary Assistance to Needy Families.

Barack then recounted a personal experience, confessing that late one night the previous weekend he had been smoking a cigar on his condominium's back porch, which overlooked an alleyway in which men often collected recyclable cans. "What I saw that evening was an entire family, at midnight. A man with a shopping cart, behind him a mother pushing a baby cart, baby inside, at midnight. This was their visible means of support. This is the job that awaited them if they weren't on welfare. We have an obligation to the family. We have an obligation to that child."

Subsequent speakers like moderate Republican Kathy Parker also thanked Barack for his work, and with only Miguel del Valle voting no, the Senate passed the measure by a vote of 56–1–1. Early Sunday morning the House did as well, but in Sunday's news stories, Barack stressed that "it's vital for us to see what we can do within our budget to make sure" the roughly twenty-two thousand legal immigrants in Illinois whose AFDC assistance was ending on July 1 were somehow protected by the state. Welfare recipients "generally are not represented down here in Springfield," Barack drily told one reporter. "They don't have powerful lobbies. They do not contribute to our political candidates."[40]

Unhappy senators voted 49–9, with Barack in the minority, to approve the state's 1998 budget only minutes after receiving copies of the huge final bill. In his *Hyde Park Herald* column, Barack wrote that "this session will largely be remembered for what it failed to accomplish" because legislators "dropped the ball on issues that are critical." Senate Republicans' failure to pass Edgar's Democratic-backed education funding reform was the greatest disappointment, because, Barack noted, the final budget bill "actually increased the percentage of state dollars going to the wealthiest school districts." That was why "I chose to vote against the budget." Barack also complained about Senate Republicans' refusal to call a vote on a utility deregulation bill that would reduce residential rates by up to 15 percent. On welfare reform, "I spent a good deal of time this spring negotiating with the Governor's office and Senate Republicans to come up with a welfare reform plan that would truly move people into jobs. To a large degree we were successful," particularly with increased day-care funding plus $10 million for elderly and disabled legal immigrants who would lose their Supplemental Security Income due to the federal reforms.

On the Judiciary Committee, Barack and Democratic staffer Gideon Baum had spent considerable time working with Republicans on a comprehensive, three-hundred-page Juvenile Justice Reform Act that might pass during the Senate's brief "veto session" in the fall. "The vast majority of juvenile offenders are functional illiterates," Barack told the *Chicago Defender*'s Chinta Strausberg, "high school dropouts who don't feel comfortable in terms of succeeding in the marketplace or in school." That meant "we have to put together prevention programs that work."

Barack told *Herald* readers that "I remain optimistic and enthusiastic," but to another community newspaper he admitted being "disappointed that partisan bickering led by Republicans hurt our school children." His voting record registered 100 percent on "party support" as well as with the American Civil Liberties Union, and his personal scorecard showed that three Senate bills Barack had introduced would be signed into law: his community college student directory, the welfare impact study mandate, and the municipal administrative hearings measure. He told the *Chicago Daily Law Bulletin*'s Dave Heckelman that the latter measure would "make sure that the courts don't get clogged up with a lot of small cases involving quality-of-life issues that can better be handled in an adjudicative setting."

Barack told *Illinois Issues* that the hurried, last-minute passage of a state budget bill no one had had time to read symbolized "the major flaws" in Illinois's legislative process. He cited Wisconsin and Minnesota as superior examples. "Those states have a strong tradition of progressive politics. State government is a leader in making things happen in the state. Illinois is more of a forum for dividing up the spoils and cutting deals. For me, at least, it doesn't make sense to be here just to bring in a few pork projects for my district. The reason we are here should be to develop policies that help people, that make this state a more promising place for our children to live. And if you're not possessed of that passion or vision, then you shouldn't be here."

A few years later, Barack recalled that "I was surprised" during that first spring session "at the degree to which lobbyists, particularly for monied interests, had influence over the details of the legislative process. People probably overestimate the power of lobbyists and special interests on big, well-publicized issues . . . but probably nine-tenths of the legislation that we deal with is highly technical, unspectacular issues that affect people in ways that aren't immediately apparent," just as Barack had told Rob Fisher that spring. "It's those kinds of issues where lobbyists exert a great deal of control," because "they have the capacity to sustain their position down there because they're paying attention to it and nobody else is, and they're better informed about the issue than anyone else is."

Barack had also come to realize "how little the public pays attention to Springfield." Chicagoans focused on city and national politics, "but they have very little awareness of state government." Ironically, that "makes life easier on a lot of legislators here in Chicago because, as long as they don't involve themselves in a scandal, it's very hard for a challenger to mount a challenge against an incumbent legislator." On the other hand, "you don't have the court of public opinion to appeal to because nobody's listening to you . . . and for me that's frustrating I think because there are often times where I'm in debates in Springfield where my opponents will acknowledge that my position is the right one, at least privately, but there's nothing that I can do to change their votes."[41]

Barack's modest campaign fund allowed him to supplement the $47,000 in state Senate funds allocated for his district office and staff with modest additional payments to Cynthia Miller, U of C grad student Will Burns, and his 1995–96 campaign staffers Carol Harwell and Ron Davis for small amounts of their time, but the only donors who contributed as much as $1,000 to him were a trio of labor unions—AFSCME, SEIU, and the Chicago Teachers Union—plus regular benefactor John Schmidt, with Tony Rezko's companies topping the chart at $2,000. Barack was unaware that Rezko simultaneously was failing for weeks at a time to provide heat to tenants of buildings like 7000 South Sangamon Street in Englewood that were owned by his Rezmar Corp. Lawsuits filed against Rezmar by the city drew no public attention.

Prior to the session's end, Dan Shomon told Barack, "You need to see the rest of the state," one that stretched more than two hundred miles farther south from Springfield. When Barack asked why, Dan said, "you have to vote on bills for all of Illinois" and "you need to see how people vote in southern Illinois and what the issues are." The two southernmost state Senate districts were represented by a pair of white Democrats who were sometimes jokingly called "the coal brothers," because coal mining had long been an economic linchpin for the counties that sat just above the Ohio River to the east and the Mississippi River to the west. Senator Jim Rea of Christopher held a fund-raiser late each July at the Rend Lake Golf Course just north of Benton, and once Barack agreed to a weeklong road trip, Shomon told him to pack just khakis and polo shirts and planned an itinerary.

Taking Barack's Jeep Cherokee and their golf clubs, they visited Chester on the Mississippi; Metropolis—population six thousand—on the Ohio; Du Quoin, which hosted the southern Illinois state fair; plus Mt. Vernon and the appropriately named Carbondale. "It was really a great week," Shomon recalled, even with the heat and humidity, and Barack was able to talk with Democratic activists like lawyer John Clemons and farmers Steve and Kappy Scates. When Barack one day asked for Dijon mustard at a TGI Friday's, Shomon lamented his big-city tastes, but Jim Rea remembered that Barack "made a great impression" at the Rend Lake golf tournament. "We had a fabulous turnout," including fellow senators Bruce Farley from Chicago and James Clayborne from Metro East.

Farley, one of the Senate's most popular characters, teased Barack and Clayborne about having to protect them, a joking allusion to the "Little Egypt" region's longtime reputation for virulent racism and "sundown towns." But Shomon recalled that "the whole trip was great," and Barack "had a great time." Barack remembered how white southern Illinoisans seemed "completely familiar to me" because "they were all like my grandparents." Shomon thought "it was really a revelation to him that he had a lot of appeal as a politician" to people in such an unlikely locale.[42]

In mid-June, a lengthy criminal trial got under way in federal court in Springfield, but only at the end of July, when Jim Edgar took the stand, did national newspapers like the *New York Times* pay any attention. On trial were a pair of businessmen who had contributed more than $270,000 to Edgar's gubernatorial campaigns and then had received a $7 million contract for their company, Management Services of Illinois (MSI), from the state Department of Public Aid, whose officials they had plied with extravagant gifts. Edgar testified that he knew nothing about the contract, even though MSI had been his largest contributor during his 1994 reelection race. The *Times* noted what it called Illinois's "regular intertwining of campaign finance and government business," and on Saturday, August 16, both defendants were convicted. The *Chicago Tribune* said the case had "rekindled debate about reforming the state's campaign finance" laws and represented "an ominous warning to the capital city's government and political community."

Then, just four days later, Jim Edgar announced that he would not run for a third term as governor or challenge Senator Carol Moseley Braun, whose time in Washington had been plagued by controversies. The *Washington Post* stated that the MSI verdict had "cast a cloud over his administration," while the *New York Times* cited how the "immensely popular" governor had suffered "a bruising legislative defeat for his education financing reform package" months earlier from Senate Republicans. In Springfield, most Republicans were privately ecstatic because the "cool and aloof" Edgar had endorsed Secretary of State George Ryan, a former House Speaker, as his designated successor. "Ryan's down-to-earth demeanor and willingness to deal and compromise," the *Chicago Tribune* reported, promised to make him a far more effective governor than the standoffish Edgar.

Even before the MSI verdicts, Mike Lawrence, Edgar's former press secretary and a longtime Illinois newsman, had been preparing a proposal for the Joyce Foundation's Larry Hansen to extend the campaign finance reform effort that retired U.S. senator Paul Simon had cochaired six months earlier. Far more influential than his press secretary title might suggest, Lawrence had left the Edgar administration in midsummer to join Simon at Southern Illinois University's Public Policy Institute. The MSI prosecution had begun with a letter to Lawrence from an anonymous whistleblower, and Simon's earlier task force had "highlighted anew the question of whether law-making in Illinois is for sale." Lawrence had already enlisted the support of the League of Women Voters and the newly rejuvenated Illinois chapter of Common Cause, which had named canny, high-energy radio newsman Jim Howard as its new executive director just months earlier.

Lawrence said the new effort would "work with representatives of the governor's office and all four legislative caucuses to . . . help them find common ground" on reform proposals. All five chiefs of staff had agreed to cooperate with the bipartisan Simon-Lawrence duo, and the legislative chiefs pledged to "identify lawmakers who are sincerely interested in reform and enjoy the confidence of their caucus leadership." But Lawrence told Joyce that although "many officeholders and their staffs perceive they must do something to avoid even more animosity toward incumbents" from voters, "legislative leaders are not excited about curbing their ability to fund the campaigns of their members because . . . vital re-election support helps keep their members in line." But with the MSI convictions highlighting "the relationship between state contracts and contributors," Lawrence hoped to build a consensus among the four caucuses' representatives plus "a commitment to pursue legislation embodying that consensus." A few weeks later, even House Speaker Michael Madigan stated that "action should be taken to eliminate the gravy train in Springfield," but campaign finance expert Kent Redfield said, "I just think the chances of something passing the legislature until after the 1998 election is zero."[43]

In late August, Barack spoke at Robeson High School in his district's Englewood neighborhood and praised a restructuring intended to provide "focused, individual attention," since "that's what students need here." He stressed that "we have to see concrete improvement in student achievement, not for the school, not for the principal and the teacher but for the students themselves. If they want to succeed in a global, technological society such as ours, they are going to have to be able to read, write and compute and understand the sciences at a higher level." The Chicago Annenberg Challenge reached the midpoint of its five years in mid-1997, just as the board made its largest allocation of dollars to date. Notable was a grant of $750,000 to Bill Ayers's close friend and UIC colleague Michael Klonsky's Small Schools Network, which had requested only $270,000 and was also receiving $337,000 that the Joyce Foundation's board had approved in late July. As CAC board chair, Barack met with CPS CEO Paul Vallas and school board president Gery Chico to discuss a possible "Public Education Fund" that could boost CAC's efforts by generating additional matching funds, and Joyce allocated $771,000 to the Consortium on Chicago School Research, which was evaluating the impact of CAC's grants.

Barack's Senate post had him speaking to small gatherings such as the Hyde Park Chamber of Commerce and the Park Manor Neighbors Community Council, but he still had time to attend the second Saguaro Seminar gathering in Boston. The next evening he and Michelle had dinner at the home of Northwestern Law School's Bob Bennett, who had hosted them in Italy. Barack and Michelle also made a weekend visit to Washington, D.C., staying with old friend Rob Fisher and his wife Lisa and having an early dinner at a highly regarded restaurant just east of the White House. Michelle and Lisa enjoyed teasing their husbands "about having big heads and that they were lucky that they had us around so they wouldn't get *so* big," Lisa laughingly recalled. Rob remembered that after dinner, Barack "had some really nice cigars," and with both men attired in olive suits and sporting sunglasses, the two couples strolled by the White House, with the women purposely staying behind. "We're *not* walking next to you! You are too ridiculous!" Michelle declared. "Two weirdos in olive suits smoking cigars with sunglasses on!"

Back in Chicago, Barack had breakfast one morning with Charles Halpern, president of the Nathan Cummings Foundation, and his colleague Stephen Heintz, who were launching a new public policy think tank and were looking to recruit a diverse board. Joyce Foundation president Debbie Leff had recommended Barack to Halpern, and he recalled being "extremely impressed" with Barack, who he thought had "this quality of presence—how grounded, focused, and centered he was." Heintz felt similarly and said they had a "fabulous conversation" during which Barack consented to join their nascent board.

That summer, Barack also agreed to become chairman of the Woods Fund's board, as the foundation moved to focus its community organizing grants on "issues that require city-wide, state-wide or national attention," including "reforms in the juvenile justice system." In the *Chicago Defender,* Barack advocated a complete overhaul of political fund-raising. Democrats Bill Clinton and Al Gore have "raised huge sums of money the same way most Republicans did, by appealing to the various special interest lobbyists," Barack observed, but "everyone recognizes that the system is broken and . . . I think the public wants the system fixed, which will require some form of public financing" of campaigns. That "would be an incentive for candidates to run without reaching out to the special interest groups."

On September 29, classes began at the U of C law school. Barack's Racism and the Law drew twenty-five students and remained unchanged from prior years, although Barack did add two new readings at the end of the quarter. Michael Lind's "The End of the Rainbow" appeared in the September–October issue of *Mother Jones* magazine, and "Affirmative Action's Vortex" was a chapter from Paul M. Sniderman and Edward G. Carmines's new *Reaching Beyond Race,* which a *Washington Post* essay had highlighted the weekend Barack and Michelle were visiting Rob and Lisa Fisher. The *Post* said the political scientists' findings were "very bad news for supporters of affirmative action as well as the Democratic party" because they indicated a pressing need to "return to color blind policies that appeal to America's sense of fairness and historical commitment to equality." In the chapter Barack assigned to his students, Sniderman and Carmines argued that affirmative action "represents a dilemma for liberalism," because "even as many liberals reject affirmative action as a violation of fairness, it has come to symbolize the very commitment to fairness for many others. The liberal, Democratic coalition . . . is thus in danger of being politically damned if it continues its commitment to affirmative action, and damned if it does not." They also contended that "the idea that class-based affirmative action provides a way out is an illusion," the same argument Barack's student Chapin Cimino had developed in her now-published paper for him the previous fall.

Barack's Con Law III attracted fifty-eight students, and to accommodate the upcoming Senate veto session in late October and mid-November, Barack arranged for it to meet on most but not all Mondays, Tuesdays, and Fridays. He began the course by asking students "what do we mean by equality?" and threw out a basketball hypothetical including the possibility that "no white men under 6'5" can play," one careful notetaker recorded. Then, beginning with the Constitution, the next eight class meetings traced the U.S. Supreme Court's race cases from *Johnson v. M'Intosh* (1823) and *Dred Scott v. Sandford* (1857), through the post–Civil War era, with Barack telling his students that *Plessy v. Ferguson* (1896) would "prove to be as pernicious as *Dred Scott*." Considering that *McLaurin v. Oklahoma State Regents* (1950) went "most of the way" toward holding that segregated schooling was unconstitutional, Barack asked students whether four years later "is *Brown* anything new?" He characterized *Brown v. Board of Education* (1954) as a "rare case of when the Supreme Court gets out ahead of society," and since "the Court is political and needed support," Barack questioned why the justices confronted school segregation before addressing voting rights. After *Brown II* in 1955, Barack asked "what are courts supposed to do with 'all deliberate speed'?" and "what is the mandate from the *Brown II* Court?" He challenged students to consider Justice Clarence Thomas's powerful critique of *Brown*'s implicit assumption that absent racial integration, black students were incapable of attaining educational parity with whites. By the 1970s, the Court was "looking at equality, not integration."

Then the course's equal protection focus shifted to the familiar standards of "rational basis" and "strict scrutiny" before devoting three classes to the Court's gender discrimination cases. The final third of the quarter tackled substantive due process, with Barack unsurprisingly asking, "Why is travel a fundamental right, but education is not?" When they reached the Court's abortion cases, Barack asked, "should viability be the drawing line?" as the Court repeatedly had held, and how would students apply *Planned Parenthood v. Casey*'s (1992) opaque "undue burden" standard? He also challenged the class to consider William Rehnquist and Antonin Scalia's contention that *Roe v. Wade* (1973) should not have constitutionalized a right to abortion in the first place.

Michael Risch, a careful 3L notetaker (and a future law professor), remembered Barack as "a prepared, engaged teacher who led very good discussions" and made sure all sides were aired. "I would not have been able to guess his personal viewpoints from the way he taught." Andrew Trask, also a 3L, agreed that Barack "was just as likely to probe the logic of an argument he likely agreed with as one he did not," and always gave "students who did not share his likely viewpoint a fair hearing," such as when he "let an outspoken conservative student critique an affirmative action case" and himself pointed out "where the critique had particular strength."

At the end of autumn, student evaluations gave Barack a respectable 8.92 on UCLS's 10-point scale in the Race class, but an unusually low 7.84 in Con Law III, with 7.27 for "guiding and controlling discussion." The Con Law III final exam posed two questions, one about cloning and the other about a school district creating a boys-only Afrocentric academy, and answers could not total more than fifty-two hundred words. After grading the exams, Barack sent students a seventeen-page, single-spaced discussion of the best answers. "It might be argued that the moral judgments at issue with respect to cloning are far more profound than the moral questions involved in consensual sodomy," Barack wrote. "A slim majority of you favored" the boys-only school, he told them, "based on a justifiable skepticism in the prospect of truly integrated schools and an equally justified concern over the desperate condition of many inner city schools."[44]

In October, Auma, her husband Ian Manners, and their five-month-old daughter Akinyi visited Chicago for a week and stayed with Barack and Michelle at East View. Maya, who had finished her master's degree in education at NYU a few months earlier and was working at an experimental middle school on Manhattan's Lower East Side, was there too, and Auma and her family also got to see Craig and Janis Robinson and their two young children. Barack played several rounds of golf with Ian at the nearby Jackson Park course, but the beginning of veto session loomed.

Barack hoped that the long-pending Juvenile Justice Reform Act could be improved, telling reporters, "what we haven't addressed in this bill is the preventative side that prevents kids from getting into the system in the first place . . . we can't just write off these kids at 14, 15, 16." The Senate's consideration of campaign finance reform would wait until the spring, but Barack emphasized that Illinois needed to authorize state matching funds for federally supported health care for uninsured children.

In early November, there was joyful news at East View: after more than a year of trying, Michelle was pregnant. Yvonne Davila, one of Michelle's closest friends from their time together in the mayor's office, knew that it had been "hard for her to get pregnant" and remembered Michelle showing up at her front door with the good news. Barack was busy gathering the petition signatures he needed so he could appear on the uncontested March 1998 primary ballot. He and U of C grad student Will Burns competed to see who could get the most signatures. Barack would face Yesse Yehudah, a little-known Republican candidate, in the subsequent general election. The *Hyde Park Herald* reported that Yehudah was chosen "solely because of his unusual sounding name" in the hope of creating "voter confusion when confronted by the choice of Obama versus Yehudah."

Barack also was distracted by local and statehouse political tussles. In a *Chicago Reader* story chronicling the demise of IVI-IPO, Barack stated that "What it means to be an independent or progressive is less clear today than in the past . . . during the years of the Daley Machine." But when he saw his comments in print, he immediately sent a letter to the editor declaring that "we continue to need strong, independent grassroots organizations." In Springfield, the final week of veto session was marked by intense disagreements over lame-duck governor Edgar's long-stalled education funding reforms. *Capitol Fax*'s Rich Miller reported that "Several Senate Democrats said they couldn't remember a more raucous caucus than yesterday's knock-down-drag-out fight over school funding," noting that the twenty-eight senators split seven yes and eleven no, with ten abstaining, in one private vote. The next day the Senate but not the House passed the funding bill, and Edgar called a one-day special session for December 2 so Chicago Democrats who supported his plan could pressure recalcitrant House members to approve it.

Barack announced that in January he would introduce a proposed amendment to the Illinois Constitution guaranteeing health care coverage for all state residents, and he and Michelle teamed up to organize a University of Chicago panel to discuss the need for juvenile justice reform based on a newly published book by Bill Ayers. Bill had developed a strong interest in the subject as a result of his wife Bernardine Dohrn's work directing Northwestern Law School's Children and Family Justice Center, and Bill's book, *A Kind and Just Parent: The Children of Juvenile Court,* like the Center, focused on the local Cook County Juvenile Court. Written in a conversational style, the book included a description of Hyde Park stating that "our neighbors include Muhammad Ali, former mayor Eugene Sawyer, poets Gwendolyn Brooks and Elizabeth Alexander, and writer Barack Obama. Minister Louis Farrakhan lives a block from our home and adds, we think, a unique dimension to the idea of 'safe neighborhood watch.'"

At the November 20 U of C event, "Juvenile Justice: Predators or Children," Bill read from his book before Barack, his law school colleague Randolph Stone, two teachers from the Cook County Temporary Detention Center, and a former offender who had spent seven years at that Center offered their comments. Bill made clear his stance that "We should call a child a child. A 13-year-old who picks up a gun isn't suddenly an adult." Barack shared Bill's view that the legal system treated too many juveniles as adults, telling the *Defender* that "the current bill unnecessarily incarcerates nonviolent offenders." Barack also publicly recommended Bill's book, telling the *Chicago Tribune* it was "a searing and timely account of the juvenile court system, and the courageous individuals who rescue hope from despair."[45]

For the third Saguaro gathering, Barack had an easy drive to Indianapolis, where Mayor Stephen Goldsmith hosted the event, but in Illinois, momentum began to build to consider campaign finance reform during the legislature's upcoming spring session. The Joyce Foundation and Southern Illinois University, the institutional home of former U.S. senator Paul Simon and his now-deputy Mike Lawrence, agreed to put a combined $100,000 behind "Making Campaign Finance Reform Happen." With that funding, the widely respected Lawrence spoke with the chiefs of staff to all four legislative leaders, starting with Senate Republicans' Carter Hendren, and had them ask that their bosses select a caucus member to take part in such a working group. "I would appreciate it if you didn't give me a hotdog" eager for press coverage, Lawrence said, because he hoped to build consensus behind closed doors before going public with any specifics.

Not by coincidence, Simon's longtime friend Ab Mikva, the now-retired federal appellate judge who had served alongside Simon three decades earlier in Springfield, called Emil Jones and suggested that Jones name Barack as the Senate Democrats' representative. Almost a year had passed since Barack first asked Jones to give him "tough assignments," and Barack immediately said "sure" when Jones followed through on Mikva's suggestion. Senate Republicans named Barack's Judiciary Committee colleague Kirk Dillard, a former Edgar aide who was also close to Pate Philip, the most crucial of the Four Tops, to represent them. House Republicans chose Jack Kubik, a Cook County suburbanite who already had announced he would leave the legislature after 1998 and whom Illinois Common Cause executive director Jim Howard thought was "hugely trusted and respected on both sides of the aisle." House Speaker Michael Madigan selected Gary Hannig, a trusted deputy from rural Montgomery County, south of Springfield. As Howard explained, "Gary *was* Mike Madigan."

Lawrence and Dillard knew each other from the Edgar administration, and a few months later Lawrence would remark that Barack, like Kirk, "was viewed as a rising star," but Dillard and Carter Hendren wondered if Emil Jones's selection of someone as junior as Barack meant "the Democrats really didn't think it was going to go anywhere."

But Barack's interest was sincere, and he devoted his last *Hyde Park Herald* column of 1997 to advocating "a tough gift ban prohibiting legislators and top state officials from receiving anything of value from someone who does business or lobbies with the state." Barack also called for "a shorter campaign season" by moving Illinois's primary elections from March to May or September.

Over the holidays, Barack and Michelle made their annual trip to Honolulu to see Madelyn Dunham, who was now seventy-five years old. During the fall, Madelyn had been one of Barack's top five campaign contributors, with her $1,000 equaling a Teamsters Union local plus two well-known Illinois donors, Irving Harris and Lewis Manilow, and topped only by Tony Rezko, who gave a total of $4,500. Right after New Year's, the law school's winter quarter began, with Barack's Voting Rights class drawing an unusually small enrollment of eleven, perhaps because it met early Monday mornings and late Friday afternoons. It was "such an intimate class," Indira Saladi remembered, and Felton Newell was impressed by "how well he knew the material." Students' grades were based on their twenty-plus-page papers, for which Barack met with them individually, and at the end of term Barack's overall teacher-evaluation score returned to an impressive 9.6 out of 10.[46]

Barack's primary concern in Springfield remained the Juvenile Justice Reform Act. He told the *Defender* that the updated, 310-page draft "still doesn't contain the kind of prevention strategies or mechanisms I think need to be more emphasized." He also explained that he was backing John Schmidt, who had been an important supporter ever since Project VOTE!, over black gubernatorial contender Roland Burris because Schmidt had sought his endorsement before Burris entered the race. Both men also faced conservative southern Illinois congressman Glenn Poshard for the opportunity to take on Republican consensus candidate George Ryan in the November general election. Asked by the *Defender* to address the burgeoning controversy over President Clinton's relationship with White House intern Monica Lewinsky, Obama "said it's too early to tell if Clinton should resign."

On January 29, Senate Judiciary Committee chairman Carl Hawkinson brought the Juvenile Justice Reform Act, on which he had worked for years and which the House had now overwhelmingly passed, to the Senate floor. The committee had approved the bill 9–0, with Barack voting present, the day before, and although Hawkinson expressly thanked Obama "for his work on the bill," he knew Barack would again vote present—a polite no—on the Senate floor. Barack explained his position in the lengthiest remarks he had made since arriving in the Senate a year earlier.

> *I think there are some wonderful things in this bill, and I think the basic principles underlying this bill are terrific. I think the notion that we should catch young people early and make sure that they understand that there are consequences to their actions is vital and critically important, particularly in urban areas. . . . I think the notion that we should get parents responsible for the actions of their children is vital.*

But he had two particular concerns, number one being "the issue of prevention dollars to go with this program" so that young offenders could receive "the support services they need to get on the right track." Barack's second worry was that by authorizing concurrent juvenile and adult sentences, with the latter kicking in if the offender failed to successfully complete the former, the bill "will result in more incarcerations of young people" if insufficient assistance was available. "We should be able to make distinctions between those young people who are serious offenders and do need to be locked up, and those young people who can still be salvaged, still be saved," Barack emphasized. He feared Illinois would "continue down a path where our main strategy for dealing with juvenile crime is to lock kids up. That is an unsustainable, unconscionable approach. It is not smart in terms of crime fighting. It is not the kind of state I want to live in, where we are afraid of our children and we are continually building more prisons as opposed to building more schools." With only Rickey Hendon and Donne Trotter voting no, the Senate passed the bill 50–2, with three other Democrats joining Barack in voting present.[47]

On the campaign finance issue, Mike Lawrence in late January met privately with each of the four caucus designees as well as Jim Edgar's youthful deputy chief of staff, Andy Foster, who would represent the governor, before bringing the group together for its first meeting. Lawrence's initial conversation was with veteran House Democrat Gary Hannig, who recommended they start with the least controversial items, and Lawrence later explained, "I give Representative Hannig credit for the approach that we took." On the eve of that first session, Lawrence told Paul Simon that "Foster and the four legislators have expressed optimism that consensus can be reached on some reforms," such as banning fund-raising while the legislature was in session, strengthening disclosure laws, and prohibiting personal use of campaign funds. However, Lawrence believed "consensus will not be reached" on limiting the overall amounts of contributions or expenditures. In early February the group gathered for an initial discussion, and word of it quickly reached *Capitol Fax*'s Rich Miller, who described the members as "widely respected legislators who should have the ability to convince their respective caucuses to accept their recommendations," including "up and coming state Sen. Barach Obama."

The fourth Saguaro Seminar meeting, on politics and civic engagement, took place in Los Angeles on February's second weekend. In Springfield, Barack introduced nine bills and the proposed constitutional amendment he had announced months earlier. The most notable of the bills would increase the standard individual tax exemption from $1,000 to $2,500 for families with incomes below $25,000, to $2,000 for ones between $25,000 and $50,000 and to $1,500 for those between $50,000 and $75,000. The next Monday Barack used the bill to make a modest press splash, describing his proposal as a modest corrective to how Illinois's state Constitution mandated a flat income tax rate—presently 3 percent—for everyone, regardless of income. Illinois "has one of the most regressive tax systems in the country," one "where a millionaire pays the same tax rate as a single parent who's making a minimum wage," Barack emphasized. The proposed amendment, Senate Joint Resolution 48, would add a new section, Universal Health Care Coverage, to the state Constitution's Bill of Rights: "Health care is an essential safeguard of human life and dignity, and there is an obligation for the State of Illinois to ensure that every person is able to realize this fundamental right. On or before May 31, 2002, the General Assembly by law shall enact a plan for universal health care coverage that permits everyone in Illinois to obtain decent health care on a regular basis."

Life in the capitol offered both levity and tragedy. Emil Jones, understandably displeased with how the Senate's Republican president treated him, found some humorous revenge when a stranger said he looked familiar and asked, "'Are you in politics?' 'I'm in the Senate,'" Jones replied. "What's your name?" the person asked. "Pate Philip," Jones shot back, delighted that at least one Illinoisan thought Pate Philip was a very dark-skinned African American. Far more unforgettable was the death of Decatur Democratic senator Penny Severns from breast cancer at age forty-six. Three years earlier Severns had been the Democratic nominee for lieutenant governor, and she had been hoping to run for secretary of state before her illness became publicly known in December. Although for eleven years she had represented the "conservative, working class, religious community" of Decatur, her constituents had not known that for five years Severns had been in a committed but deeply closeted relationship with a woman fourteen years her junior, former Associated Press Springfield bureau chief Terry Mutchler. Many people in the statehouse knew or suspected as much, and while Severns dreamed of being governor, she and Mutchler had to live "a life that we strove to kept hidden." On February 24 Barack joined Senate colleagues at Penny's funeral in Decatur.[48]

The Wednesday-night poker games from the year before resumed with gusto that spring, although they were now in the kitchen of the simple 7th Street house where Terry Link stayed. Mike Lieteau explained that "some of the senators had reservations about going out to Panther Creek" because they played with chips "right off the bar," which was "a little too public" for anyone worried "that you could be exposed." The weekly gatherings now included Republican senator Tom Walsh, from Chicago's western suburbs; veteran Democratic senator Denny Jacobs, from western Illinois's Quad Cities area; Illinois Manufacturers' Association lobbyist Boro Reljic; and occasionally Mike Lieteau's cousin Fred Fortier, who had become one of Senate Democrats' two Judiciary Committee staffers. Tommy Walsh was a playful jokester, who added "a lot of fun to the group," remembered unrelated Larry Walsh. Denny Jacobs was one of the Senate's most outspoken members. Widely viewed as the father of Illinois's 1990 Riverboat Gambling Act, Jacobs had never resigned himself to Emil Jones's leadership of the Democratic caucus. An old-style politician, Jacobs's blow-ups were legendary. "'They misquoted me in Peoria! They misquoted me in Peoria!'" Democratic staffer Cindy Huebner remembered him once screaming into a phone. "What did they say? 'They said that I said this guy should be hung by his thumbs until he turns blue. I didn't say that! I said he ought to be hung by his *balls* until he turns blue!' He was furious."

Tommy Walsh quickly realized that Barack and Mike were "good friends, and he took liberties with Mike that he didn't take with anybody else." Walsh also recognized what a politically significant trio Barack's Democratic poker buddies were. Waukegan's Terry Link had made the Lake County Democratic Party a significant presence in state politics, and Larry Walsh's Will County Democrats, just like Denny Jacobs's power base in the Quad Cities, represented major areas of Democratic strength outside Chicago's Cook County. Although Barack "never made it real obvious that what he was doing was networking for the future," Walsh explained, "being a friend of Terry Link's was a good thing" for any Democrat who might run statewide in Illinois. Tommy would bring a pack of cigarettes to the game, have one or two, but otherwise make them available to Barack and Denny. Barack "wore the same thing to every card game": a sweatshirt and sweatpants that "looked like pajamas," so his workday clothes would not reek of smoke. Six-packs of beer abounded, and come 10:30 P.M. or so, they ordered two big pizzas. They usually adjourned at midnight because Larry Walsh went to mass every day at 7:00 A.M.

Everyone saw Barack as a conservative card player, who was never afraid to fold, but it "became a standing joke" that he would *never* admit he had lost money over the course of a night, even though he often did. "Maybe I broke even," Barack would say as the others smirked. "The ego was that he could never lose," Mike Lieteau explained, and when Barack did lose a hand, "I always remember him making a point of saying, 'I played that hand right,'" Tommy Walsh recalled. "Yes sir, Senator, you played it like a champ!" Tommy teasingly replied. One night only Barack and Mike Lieteau remained in on a hand, with Barack insistently raising before Mike surprised everyone by winning. Barack "became more agitated" once he realized "he had lost all his money." Barack was "legitimately annoyed," for "Mike should never have been in that hand," but Barack called Lieteau "an idiot" for having stayed in. Mike defused the moment by saying, "I may be an idiot, but I'm not the one leaving mad with no money," forcing even Barack to laugh.

Even at the capitol, Mike could tease Barack in ways no one else did. Telecommunications bills were handled by the Energy and Environment Committee, and one morning Barack was listening to the policy arguments. Mike slid in beside him. "Why are you sittin' in on grown folks' conversations, brother?" he teasingly asked. A serious legislator indeed, "Barack wanted to actually hear the testimony," Lieteau recalled.

By early 1998, the paperback edition of *Dreams From My Father* had been out for eighteen months, a common time frame for publishers to offload their remaining copies of poorly selling titles onto the "remainder" market. Brad Jonas owned three Chicago-area used-book stores, including one in Hyde Park, and he regularly purchased remaindered books. He had met Barack at his daughter's Hyde Park school and had been highly impressed. Early in 1998 Jonas acquired the last four thousand copies of Kodansha Globe's edition of *Dreams* for the grand sum of $900—22 cents a copy—and in early 1998 *Dreams From My Father* was suddenly more visible around Chicago than at any time since its initial publication in hardcover.

One bookstore browser who recognized the author's name and bought a copy was Jim Reynolds Jr. Barack had begun playing pickup basketball every so often at the tony East Bank Club in River North in addition to Hyde Park, and Reynolds, an African American municipal bonds dealer, sometimes played there too. Reynolds had recently left Merrill Lynch to cofound his own firm, Loop Capital Markets. Seven years older than Barack, Reynolds had graduated from a South Side vocational high school before attending college in Wisconsin and taking his first job at Paine Webber. "You're a pretty good writer," Reynolds told Barack the next time he saw him. "I know I'm a good writer," came the self-confident response.

"From that," Reynolds said, "we became friends. We really just started talking a lot. Our careers were mirroring each other's because I had just started Loop Capital" the same year Barack had entered elected office. "We actually started spending a lot of time together. We started playing basketball together on a planned basis. I used to go to Springfield a lot because of my business, and every time I went, I would call him, and I would usually use his office as my home base. So we just became very close friends, very close," playing nine holes of golf every few weeks at the Jackson Park Golf Course just below Hyde Park.

Over the next six years, no one else in Chicago, or anywhere in Illinois, would be a more pivotally helpful figure in Barack's life than Jim Reynolds.[49]

In Springfield, Mike Lawrence's campaign finance reform group met on successive Wednesday mornings. Paul Simon wrote and thanked the participants for their time, adding a handwritten note to Barack: "I also am enthusiastic about your political future!" Lawrence was increasingly confident that his group would agree to "bar use of campaign funds for personal benefit of candidate or his/her family, effective upon the bill becoming law," as he told his participants in a memo asking them to assemble for a four-hour dinner meeting at the Renaissance on Tuesday, March 10. "The goal," he wrote, "will be to come to agreement on a package" and "we then will refer our consensus legislation to staff for drafting." In a subsequent memo to Senate Republican staffers Glenn Hodas and Peg Mosgers, who would write the bill, Lawrence explained that "the majority of those in our group believe the legislation should spell out as explicitly as possible what is not allowed."

The legislature took a one-week break for the March 17 primaries, with Barack receiving all 16,792 Democratic votes cast, far more than upcoming Republican opponent Yesse Yehuda's 401. But other results sent shock waves through Illinois politics. Conservative downstate congressman Glenn Poshard defeated African American former attorney general Roland Burris and Barack's friend John Schmidt for the Democratic gubernatorial nomination with just 38 percent of the vote. In Chicago, progressive Mexican-born state senator Chuy Garcia had faced an all-out challenge mounted by Mayor Richard Daley's increasingly muscular political machine. Daley, furious at Garcia's charges that the city was trying to gentrify the Little Village and Pilsen communities, had his top Latino operative marshal hundreds of organization loyalists against Garcia. A decade earlier Danny Solis had been Barack's UNO organizing colleague, but by 1996, Solis had built UNO into a major political force that backed the mayor. Daley had appointed Solis the 25th Ward's alderman when the seat became vacant, and Solis ally Tony Munoz, an otherwise unknown Chicago police officer, was slated against Garcia. Barack quietly had sent his hardest-working young supporter, Will Burns, and some other U of C youngsters to help Garcia, but on election night, Munoz and the Daley organization came out on top with 54 percent of the vote. The next morning *Capital Fax* observed that "Chicago state legislators will now fear Mayor Daley even more," and Garcia told the *Chicago Reader* that the city now had "a new type of machine, and the mayor is the chairman." Garcia's Senate colleague Miguel del Valle called Solis Daley's "mouthpiece," but Solis insisted, "I don't believe that I've sold out my philosophy of what I'd like to do for my community just because I am part of the Machine." Al Miller, Harold Washington's former press secretary, warned that "anytime anyone raises his head a little bit higher, the Daley folk want to knock it down." Hyde Park's decades of relative independence made it a unique exception to organization influence, but even a Hyde Park legislator had to be wary of drawing Daley's anger.[50]

Barack remained underwhelmed by the impact of the Chicago Annenberg Challenge, asking his board colleagues whether there was evidence of positive trends in student achievement. When legislators returned to Springfield, *Capitol Fax* reported that Lawrence's group was close to agreeing on "a sweeping campaign reform package," including a ban on personal use of campaign funds, but Lawrence knew that the crucial question was whether he and his good friend Carter Hendren, Pate Philip's chief of staff, could convince the Senate president to bring the agreed-upon package to a floor vote with his support. Even though the House had previously passed a broader bill, including a limit on the size of contributions, in private all four caucuses opposed such a ceiling, and everyone knew that passage by the House had occurred only because they were certain Philip would kill it in the Senate.

At the end of March Barack joined U.S. senator Carol Moseley Braun, state representative Barbara Flynn Currie, and 4th Ward alderman Toni Preckwinkle in speaking at a memorial service for Hyde Park political activist Saul Mendelson, who two years earlier had been an outspoken critic of Alice Palmer's attempt to reclaim her Senate seat from Barack. On April 1, Barack resigned from the Senate's State Government Operations Committee because Emil Jones had named him to the slightly more interesting Revenue Committee. Republican chairman Bill Peterson found Barack to be "cordial, conscientious," and "easy to get along with," but poker buddy Denny Jacobs took a dimmer view once Barack joined the committee. "He began asking the witnesses four million questions," Jacobs recalled. "I leaned over and said, 'Hey, enough already! Learn on your own damn time, will you?'" Barack, fellow Democrat Pat Welch realized, "didn't really want to be on Revenue," because the committee's name was "a misnomer—it was a tax break committee, because every bill was to give somebody a tax break." Welch recalled that he and Barack often voted "no" together. When Welch cast a surprising "yes," Barack asked why. "That one helped my district, so I had to vote for it." "Oh," Barack lightheartedly responded. "So that's how this committee works." Jacobs disliked Revenue for a different reason, because "there's no money on Revenue." To "fund your campaigns," he explained, a senator had to be on "committees where you could at least ask for contributions."

In mid-April *The Economist*, the internationally respected British news weekly, published a one-page account of life in Springfield written by an unnamed former capitol aide. Asserting that the statehouse has "a code of ethics that makes Capitol Hill resemble a convent," the magazine reported that "one local journalist responded to a complaint that his newspaper had not exposed a state legislator's activities by saying 'If we cover one mistress, we'd have to cover all of them.' Exposure of all could fill several newspapers each week," *The Economist* stated. It added that while some legislators "are serious and hardworking, a great many others come to Springfield to have a good time." Some drank on the floor, "and at least one has been said to carry cocaine in his briefcase." In the evening, "bars fill up quickly. Sex among the legislators, married or single, is so common as to be invisible. Higher-ranking members can put their girlfriends on the payroll" and "in many offices . . . low-grade sexual byplay is the order of the day." *Capitol Fax* frowned on the story, but then reported that legislators were nervous that the *St. Louis Post-Dispatch* was seeking access to logs of phone calls made from the floor of the House. "Calls to extramarital lovers could be detected," publisher Rich Miller warned.

In mid-April Barack had to write a check for $13,329 to the IRS to cover the balance of the $41,872 in federal taxes he and Michelle owed on a combined 1997 income of $168,903—putting them in a 36 percent tax bracket. With Barack earning only a few thousand dollars now from Miner Barnhill, his $50,000-plus from the U of C law school plus his additional $48,000 from the state Senate and $11,000 from the Joyce Foundation topped Michelle's $60,000-plus salary at the University of Chicago. Only briefly by phone did Barack take part in the inaugural New York meeting of Charles Halpern's nascent new think tank, but he did hold two consecutive Saturday town hall meetings for constituents as well as an evening fund-raiser at a downtown Chicago hotel.

Five days before the fund-raiser, Barack's top contributor, housing developer Tony Rezko, was featured in a *Chicago Tribune* story describing how he and his partner Dan Mahru's Rezmar Corp., in addition to rehabbing mixed-income apartment buildings, was working with Barack's former law firm colleague Allison Davis to begin building medium-priced town houses throughout Kenwood, the neighborhood just north of Hyde Park. Before long, Howard Stanback, whom Barack knew well dating back to the 1987–88 negotiations over the O'Brien Locks site, was working with Davis and Rezko too.[51]

For one of his occasional *Hyde Park Herald* columns, Barack summarized the spring 1998 session by reporting that the education bill approved months earlier "failed to tackle the structural inequities that arise between districts as a result of our state's disproportionate reliance on local property taxes to fund education." He had to reconcile himself with how at each of his town meetings, "constituents cite education as the single most important issue facing our district and the state as a whole." In addition, "one of my top priorities" is "to curb the distorting effects of money on the political process," especially because Illinois has "almost *no rules at all* regarding how much candidates can accept." He predicted that the campaign finance reform proposal "will probably not contain everything I'd like to see . . . but it will contain tough new rules regarding disclosure, personal use of campaign funds, and gifts from lobbyists."

At the end of April, Mike Lawrence told journalists "we do have what I regard as a consensus," and he hoped they could announce it during the first week of May. *Capitol Fax*'s Rich Miller noted that the caucus representatives had only agreed on bullet points, and that at least one of the Four Tops did not want to endorse it until actual statutory language was drafted. Miller said that Common Cause's Jim Howard "has maneuvered himself into a position to effectively assert veto power over the final product," because Carter Hendren and Pate Philip did not want to have Howard publicly lambaste Senate Republicans. Yet Philip was still angling to get a later date for state primary elections included in the package even though Lawrence's group had nixed the idea, an effort that angered House Speaker Mike Madigan.

The Juvenile Justice Reform Act returned to the Senate floor after Governor Jim Edgar's amendatory veto. Illinois's 1970 Constitution gave the governor the authority to put forward "specific recommendations for change" in legislation, and in this instance Senate Judiciary Committee chairman Carl Hawkinson urged the Senate to agree to Edgar's modest changes. The Senate unanimously approved the revised bill, with Barack telling reporters that "the governor's suggestions are sound" and expressing pleasure that the funding issues that led him to vote "present" four months earlier had been remedied. "The vast majority of young people in trouble with the law are young people who aren't getting attention at school, have maybe dropped out of school, and are caught in circumstances of poverty, possibly substance abuse—a host of problems that with effective intervention would be cured," Barack optimistically stated. "Counseling programs for families" and "after school programs for kids" should help teenagers get "back on track."

Four months earlier, Barack's aunt Zeituni Onyango had taken early retirement from Kenya Breweries and had written to Barack asking if she could visit him during a multistop tour of the United States. On Tuesday evening, May 5, she arrived at Chicago O'Hare, called the Obamas' home, and spoke to Michelle, who told her that Barack was in Springfield. She took a taxi to East View, and the next morning Michelle put Zeituni on the train to Springfield, where Barack was waiting for her on the outdoor platform and then took her to the Renaissance Hotel and an evening reception. At the end of the week, she rode back to Chicago in Barack's Jeep Cherokee, staying with Barack and the now seven-months-pregnant Michelle for ten days before boarding a train to Boston to see her brother Omar Onyango Obama.[52]

Back in Springfield the next week, the ongoing campaign finance reform discussions remained almost everyone's top concern, but on May 13 the Senate unanimously passed a bill that received almost no public attention. HB 705 created the Children's Health Insurance Program, or "KidCare," a three-year "Medicaid look-alike" program to provide coverage to children whose low-income, working parents lacked health insurance. Senate Appropriations Committee "budgeteers" Steve Rauschenberger and Donne Trotter played lead roles in shepherding the measure forward, but inside the Public Health and Welfare Committee, which approved the bill on May 7, Barack had helped encourage conservative Republican chairman Dave Syverson to embrace the effort, just as he had with the welfare reform program a year earlier.

For Senate Republicans, it was essential that the program be only a benefit, not an entitlement, but once again expert Democratic staffer Nia Odeoti-Hassan was impressed by the extent to which conservative Republicans could be coaxed into backing a significant if limited expansion of public benefits. Barack also played a behind-the-scenes role in adding a progressive provision that doubled Illinois's personal tax exemption to the budget bill that the Senate would unanimously approve on the final day of the session.

The second and third weeks of May featured day-by-day developments regarding the campaign finance measure. Common Cause's Jim Howard told reporters, "I'd rather get good reform than not get great reform," and on Thursday morning, May 14, Mike Lawrence and his quartet of legislators finally went public with their consensus proposal at a statehouse news conference. Howard jokingly told Barack he needed to speak into the microphones if he wanted to run for higher office, but Barack did not enjoy the joke. "I'm not running for higher office," he coldly responded in what Howard took as "a very arrogant, very condescending kind of putdown." In contrast, Mike Ragen, the veteran Democratic staffer whom Emil Jones had assigned to help Barack on the measure, found him "very nice, very cordial" and "a prince of a guy to deal with." Republican senator Kirk Dillard warned reporters that "this is the toughest of all topics for a legislator," but Senate president Pate Philip made clear that he did not want failure to pass such a measure to be a cudgel Democrats could use against his members, or Republican gubernatorial candidate George Ryan, in the upcoming November election.

Mike Lawrence gave the four legislators a set of talking points to use in privately presenting the proposal to their caucuses. He said that Illinois news media was extremely interested in the issue, and that the consensus "addresses some of the major complaints raised in a *reasonable* way that will *not impact your ability to campaign competitively*." The proposal's three major ingredients—a ban on legislators accepting personal gifts worth more than $100, a requirement that members file their campaign disclosure reports electronically, allowing for immediate journalistic access, and a prohibition on the conversion of campaign funds to personal use—"will be viewed as major reform by the media." Lawrence warned that a failure to move on all three would be seized upon by "campaign opponents and critics. . . . Let's bite the bullet and do it now."

Barack would have gone further and imposed limits on how much contributors could give, but even Lawrence's three points drew intense flak from Democratic colleagues like downstaters Denny Jacobs and Vince Demuzio, as well as African American rivals Rickey Hendon and Donne Trotter. Denny Jacobs understood that Barack had received the assignment because he was "perceived to be Mr. Clean," given his reputation for refusing to let lobbyists buy him any food or drink. "Rickey gave him a hard time, I gave him a hard time, Donne gave him a hard time," Jacobs recalled, and Vince Demuzio really unloaded. "Who the hell are you, Mr. Newbie, Mr. Good Government, trying to change all this?" Barack's fellow freshman Debbie Halvorson recounted. "We did just fine before you got here." Kirk Dillard, Barack's Republican counterpart, remembered Barack calling the experience "brutal" and telling him that Demuzio had condescendingly asked, "Son, how much money do you have in your campaign account? You're negotiating on my behalf" yet would never face meaningful Republican opposition.

Downstate moderate Pat Welch, who held one of the most competitive seats, saw the acrimony as "pretty much an upstate-downstate split," while lame duck Chuy Garcia thought it was more "new school versus the old school." Chief of staff Mike Hoffmann, the only non-senator to attend caucus, recalled that "it got kind of heated, especially Denny," and Emil Jones remembered that Barack "caught pure hell. I actually felt sorry for him," but Jones spoke up firmly for Barack. "It may not be something that we want, but this is the direction that things are going, so we might as well be on the track," just as Mike Lawrence had suggested. Jones remembered that Barack "stuck to his guns" and "he was able to win the majority over." Hoffmann was impressed that "Barack did a nice job of maintaining his composure" and was able to "separate the issue" from the personalities and thus avoid "making it personal." North Side senator Howie Carroll thought Barack "did an excellent job with it," one that "moved him up to the top of the pack" among state Senate Democrats.

Capitol Fax's Rich Miller called the reform proposal "potentially historic," and Paul Simon sent Barack a thank-you letter, predicting that "in the years to come it will be one of the things that you will be proudest of in your legislative years." But Simon and everyone else knew that most of the credit went to Mike Lawrence. House Democrat Gary Hannig, who realized that he, Obama, and Dillard "were seen as traitors by some of the people in our own caucuses," viewed Lawrence as "the guiding light that kept the committee alive" during a process that had begun as "a little bit of a long shot." Hannig sensed that Barack was "very idealistic about how to approach this," but in the House as in the Senate, "there was a lot of skepticism with the Speaker, a lot of skepticism with my caucus" over actually calling the bill for a floor vote. Kirk Dillard knew that Pate Philip "clearly thought that we needed to make changes to protect the institution of the legislature and some members from themselves," but the final consensus "never would have happened without Mike Lawrence." Andy Foster, Governor Edgar's young representative, had found Barack "very collegial, very funny," with "a good sense of humor," and recalled both Barack and Dillard "joking about the beating they're taking in their caucuses." Everyone believed that "if Pate went along with something, everybody was going to be okay with it," and come Monday morning of the session's final week *Capital Fax* reiterated that "the future of this bill is in the hands of Senate President Pate Philip."

On Wednesday evening, Barack and Kirk Dillard appeared on the telecast of *Illinois Lawmakers* with host Bruce DuMont. It was Barack's first time on the show. They acknowledged that a "grandfather clause" had been added to the bill to placate angry old-timers by allowing legislators to convert to personal use whatever sum they had in their campaign accounts as of June 30. Barack explained that "it's difficult to get everybody on board" and that he reluctantly had agreed "to grandfather something in that will soften up the resistance and get on board the majority of members." The crucial decision took place between Mike Lawrence, Pate Philip, and Carter Hendren. "Mike and I had that discussion and I went in and talked to Senator Philip about it. I remember that very vividly, and that was the way to pass the bill," Hendren recalled. "We didn't really like it, but we did it."

On Thursday night, with only twenty-four hours to go, Lawrence believed that everything was almost in order, but he asked Barack how best to word one provision so as to preclude a constitutional challenge. Barack promised to have the language first thing in the morning. On Friday, Barack was late and then apologized for having left the crucial passage in his office. Lawrence recalled that "I blew up at him," but Barack just chuckled and replied, "Gee, Mike, you're really intense."

The Senate convened at noon and Pate Philip's all-powerful Senate Executive Committee sent HB 672 to the floor on a vote of 12–0–1. "The real strategy here was to get it to the floor," Mike Lawrence explained, for "we figured all along that if legislation reached the floor, members would have to vote for it whether they wanted to or not." Philip had been true to his word, and early in the afternoon, Kirk Dillard rose to present the bill, thanking staffers Glenn Hodas, Peg Mosgers, and Mike Ragen for their work on it. "Having spent hundreds and hundreds of hours with Barack Obama on this over the last few months," Dillard said the measure was "the most important piece of campaign finance reform in twenty-five years in the state of Illinois." Then Barack followed, acknowledging that "this is one of the most difficult issues" because "it's something that everybody feels affects them directly." But "this bill moves us in the right direction" and is not "overly onerous." After Dillard was questioned extensively by Denny Jacobs and several other members, the vote was called and HB 672 passed by an overwhelming margin of 52–4–1.

An hour later the Senate adjourned, and the bill moved to the House to await passage there. HB 672 had begun life as a modest piece of legislation, introduced by ranking Republican Tom Cross, who shared a Springfield house with minority leader Lee Daniels, before becoming the vehicle onto which the Lawrence group's measure was appended. Democratic representative Gary Hannig already had House Speaker Mike Madigan's approval, because "you never agree to anything until you ran it by the Speaker," but at 10:00 P.M., Cross told reporters the bill needed more study over the summer. Lawrence knew that meant Daniels "did not want this to become law," but when Lawrence asked Madigan to take the bill away from Cross, as the Speaker could do, Madigan refused. With his apparent triumph now headed for last-minute defeat, and with Jim Howard angrily announcing that Common Cause would declare war on House Republicans, Lawrence called Pate Philip and reminded him that Republican gubernatorial nominee George Ryan, a former House speaker, would be highly vulnerable to Democratic attacks that Republicans had torpedoed campaign reform. "This really gives the Democrats an issue," Lawrence emphasized, and with Carter Hendren agreeing, Philip called Ryan, who readily agreed to call Daniels and lean on him with everything he had.

"Daniels protested, but Ryan insisted," *Capitol Fax*'s Rich Miller reported. Well after midnight, Madigan called HB 672 and recognized Tom Cross, who immediately yielded to Jack Kubik, the House Republican member of Lawrence's group, who brightly remarked, "I thought you'd never ask me," and presented the bill. For close to an hour, Kubik faced an onslaught of objections and remarks from representatives on both sides of aisle. As 2:00 A.M. neared, Madigan closed the debate and called the vote. Just as in the Senate, the number of professed opponents magically shrank and the bill passed by an overwhelming margin of 102–3–13.

Proponents and journalists reacted ambivalently to the victory. The ban on lobbyists' gifts ended up with twenty-three exceptions, including golf and tennis fees as well as "meals or beverages consumed on the premises from which they were purchased." Jim Howard, complaining that "Springfield is a snake pit," admitted "there are parts of it that stink" and acknowledged that "the gift ban has more holes than St. Andrews Golf Course." Pate Philip later remarked that "it was so watered down it was a joke," but Paul Simon and Mike Lawrence called the end result "a very good package," with Lawrence believing that the ban on personal use of campaign contributions was "the most far-reaching aspect of this bill."

Attention quickly turned to the fact that lawmakers had five weeks in which to augment their campaign accounts before the June 30 "grandfather clause" took effect, and *Capitol Fax* noted that "some lobbyists are reporting an unusually heavy number of fundraising calls this month." In a *Chicago Tribune* op-ed, former League of Women Voters director Cindi Canary, now heading the new Joyce Foundation–supported Illinois Campaign for Political Reform, commended Barack and his colleagues for "this landmark legislation." She later reflected that "the central big win, which I don't think any of us realized at the moment, was the electronic disclosure," because it made contribution data readily available on the Web.

Illinois political scientist Kent Redfield, who had assisted the Lawrence group early on, noted that the bill "doesn't address the fundamental question about the role money plays in public policy" and therefore would "force few changes in the real-world relationship between cash and politics," particularly with how the Four Tops funded their vulnerable members' campaigns. Illinois would still have "the least regulated campaign finance system of any state," and the resulting concentration of power "makes most legislators essentially irrelevant to policymaking." Without a limit on how much could be contributed, the wealthiest interest groups would continue to have an outsized impact, with the Illinois State Medical Society, at almost $1.5 million a year, and the Illinois Education Association, at $1.2 million, topping the chart.

Barack was worried about citizens' cynicism about government. "They believe they have no influence on the process since they don't have the money of special interest groups. With the gift ban and the ban on Springfield fundraisers that are contained in this legislation, I think at least some of this confidence will be restored." But he vowed to "continue to push for strict limits on campaign contributions by individuals and political action committees to reduce the amount of money spent on campaigns." Two years later, Barack admitted that "we didn't go as far as I would like" because of the "deep-rooted resistance among a lot of my colleagues." He acknowledged that "many of the lobbyists are very well-informed and do a good job of advocating the interests of the people they represent," yet it remained true that "there is a culture in Springfield where lobbyists buy legislators dinner every night," and the Campaign Finance Reform Act had not changed that.

Overall, *Capitol Fax*'s Rich Miller declared 1998 had been "the most boring legislative session in memory," and with Illinois running an annual budget *surplus* of almost $1 billion, there had been a raft of "new spending plans . . . dreamed up." Legislative leaders had allocated $200 million toward "member initiatives," or what the *Tribune* called "a massive buffet of pork projects," particularly for districts where legislators faced competitive fall races. Barack bragged that he and Hyde Park state representative Barbara Flynn Currie, the number-two House Democrat, had obtained $4.6 million for the rebuilding of the Lake Michigan shorefront between 55th and 79th Streets. Looking toward the fall, Barack's Springfield aide, Dan Shomon, along with Courtney Nottage, Emil Jones's chief legal counsel, prepared to challenge the prospective third-party Senate candidacy of Myra Handy, a South Shore podiatrist. Ron Davis and Tom Johnson, just like two years earlier, filed a challenge before the Chicago Board of Election Commissioners. Handy failed to respond to multiple notices or to appear when a public hearing was convened, and the board barred her from the ballot for not filing the necessary paperwork.[53]

Just after the spring session, Barack and Michelle attended the annual benefit dinner of the Arab American Action Network, a group their friends Mona and Rashid Khalidi had helped found several years earlier. Many of the group's Palestinian and Jordanian American members lived or worked in the westernmost portion of Barack's district. The Obamas wrote a tax-deductible check for $100 and sat at a table with the Khalidis to Michelle's right and keynote speaker Edward Said and his wife on Barack's left. Barack had heard Said speak many times before, at the front of a Columbia University classroom more than fifteen years earlier, when he and Phil Boerner had thought Said was a "flake."

Rashid Khalidi had just written an op-ed in the *Chicago Tribune* stressing that "much of the Palestinian population is today much worse off than when the Oslo accords were signed in 1993," yet Khalidi was in no way anti-Jewish, explicitly noting that the Holocaust was "the modern era's greatest human atrocity." When conversing with Barack about Palestine, Mona Khalidi recalled, "you felt that he understood, and you felt that he was not taking an opposite position," but he always seemed "very, very careful" about what he said. At home in East View, where the Obamas had a German-Jewish neighbor who was a passionate supporter of Israel and an active member of the NRA-affiliated Illinois State Rifle Association, it was much the same. Harry Gendler remembers Barack as a "very polite person" who willingly tolerated "me poking fun at him about environmental issues," particularly Barack driving to and from Springfield all by himself in his Jeep Cherokee. "Barack and I had a running joke" about that not being green, and Gendler also watched as "Barack always hid his smoking," often "smoking in the car." Barack was "happy to talk about the Second Amendment," but whenever Harry raised the subject of Israel, Barack's demeanor changed. "He would turn poker face, turn ice cold" and "told me that we need a more balanced approach—that's the term he used." Barack "did not want to have an exchange of ideas or thoughts on that topic" and indeed "always walked away."

In early June Barack again participated by phone in the second steering committee meeting for Charles Halpern's think tank. Halpern recently had had breakfast with Barack in Chicago, and Barack had "emphasized the importance of not becoming another centrist organization taking safe positions on safe issues." He and Halpern had discussed the "serious possibility" of Barack "giving up law practice & teaching" and devoting "perhaps 75 percent" of his time to head up a Chicago office of the new organization while remaining in the Senate. The next Monday Barack was in Washington, D.C., to speak at a Brookings Institution event addressing "Cities and Economic Revitalization." Barack focused on the "spatial mismatch between job growth and the people who are unemployed" and a parallel "skills mismatch" that was rooted in "an extraordinarily inequitable system of public education finance." He also warned that "unless we . . . fundamentally restructure the public schools . . . it's going to be very difficult to sustain whatever progress has been made" in the economic redevelopment of beleaguered cities. Most problematic were "extraordinarily poor and isolated communities" like Chicago's Englewood, where residents needed access to transportation to locales where jobs were more available. Barack recounted how a recent meeting of the influential Commercial Club of Chicago's Civic Committee had discussed "a fairly bold package of proposals to deal with the regionalism issue." Major corporations had exhibited "some recognition of mutual self-interest" in the plan, but House Speaker Mike Madigan had said "'it sounds Soviet' . . . which indicates that there is going to be some political resistance," Barack drily observed. Later that week in D.C. the Saguaro Seminar's fifth meeting took place, focusing on faith and social capital. Martha Minow remembered that with participants like conservative Republican political operative Ralph Reed, there was "a big fight" featuring "fiery debates about religion and politics." That "was not a happy meeting," she recalled.

In mid-June Barack was reelected chairman of the Chicago Annenberg Challenge board, and he also joined the board of Leadership for Quality Education, the Commercial Club offshoot now headed by his friend John Ayers. In early July Barack completed a lengthy questionnaire for Project Vote Smart detailing his position on a large number of issues. "Abortions should be legally available in accordance with *Roe v. Wade*," and affirmative action was desirable across the board, but Barack endorsed only three out of sixteen principles concerning crime, including "increase state funds for programs which rehabilitate and educate inmates" and "provide funding for military-style 'boot camps' for first-time juvenile felons." On the economy, Barack checked "increase funding for state job-training programs," and on education he endorsed charter schools but not vouchers for students' parents. He also supported "state-funded tuition and fees" for students attending public institutions of higher education who maintain a B average. Barack chose "undecided" on term limits for legislators, on a balanced-budget constitutional amendment, and now, unlike in 1996, on same-sex marriage too. On a list of eight "social issues" such as "increase state funding for Head Start," Barack endorsed all except "favor banning smoking in public places," a reflection of his own ongoing addiction to cigarettes. On state taxes, he did check "greatly increase" for tobacco, and on state spending he chose "greatly increase" for both K–12 education and health care and "slightly increase" for welfare while endorsing a two-year limit on cash assistance.

At 3:00 A.M. on Independence Day, July 4, Michelle woke Barack with the news that she was about to give birth. "Things went fairly smoothly" from there, and a few hours later Michelle gave birth to an eight-pound, fifteen-ounce girl whom they named Malia Ann, her middle name honoring Barack's mother. The baby's midsummer arrival allowed both doting parents to devote large amounts of time to the newborn, and following Michelle's maternity leave the U of C allowed her to shift to part-time status once the new academic year began. A *Chicago Sun-Times* gossip column noted Malia's birth, and among those sending congratulations was former U.S. senator Paul Simon, who reported that his wife was reading *Dreams From My Father* and said, "I want to be supporting you for statewide office one of these days."

Soon after Malia joined the family, Barack's aunt Zeituni arrived back in Chicago after seeing Omar Onyango in Boston and then Abon'go Malik in Washington. Staying now at the nearby Ramada Inn where Barack had first announced his candidacy, Zeituni indulged in an excess of shopping, including a desktop computer. When Barack drove her to O'Hare for her return trip to Nairobi, her luggage was $300 overweight and Barack "used his credit card to pay the extra charges." Zeituni recalled that Barack barely suppressed his anger at the cost, which "did not sit well with him" now that Malia's birth would further stretch the couple's already stressed finances.

Michelle used Malia's arrival to attempt again to get Barack to quit smoking, without success, and the baby's arrival led to a significant decline in their increasingly spotty attendance at Trinity United Church of Christ. Two weeks after Malia's birth, Barack again participated by phone in a third steering committee meeting for the new think tank. When a draft budget was prepared covering the next twelve months of the nascent start-up, Barack and another young steering committee member, David Callahan, were penciled in as committing 50 percent of their time to the new venture.

Barack spent only a little time that summer at Miner Barnhill, but it was a banner season for the small law firm. In mid-June, the Mitsubishi auto company agreed to a $34 million settlement with the EEOC regarding the rampant sexual harassment at its Normal, Illinois, assembly plant, which had first been challenged in the lawsuit that George Galland, Jon Belcaster, and Barack had helped litigate on behalf of several dozen women employees. Mitsubishi had agreed a year earlier to pay $9.5 million to those plaintiffs, but the EEOC settlement benefited about 350 women in what the *Tribune* called "the biggest sexual-harassment case in U.S. history." Then, in early August, Judd Miner finally achieved "a clear-cut win" in the long-running *Barnett* case, alleging that Chicago's city council districts unfairly disadvantaged black voters. A newly hired Miner Barnhill paralegal, Scott Lynch-Giddings, termed it "a big day at the firm" in his diary, and just how big was documented only months later when the federal district judge hearing the case awarded attorneys' fees and tax-free costs to Miner Barnhill. Judge Elaine Bucklo ordered Chicago to pay the firm $5,127,994 for attorneys' time going back to 1992 and an additional $526,472 in documented costs.[54]

In mid-August Barack traveled to Carbondale, in far southern Illinois, to witness Governor Jim Edgar sign the campaign finance reform Gift Ban Act into law at the Southern Illinois University Public Policy Institute that Paul Simon and Mike Lawrence directed. Earlier in the summer, Edgar had been sorely tempted to amendatory veto the legislation in an effort to bar the transfer of money from one campaign fund to another, a move that would defenestrate the Four Tops and require repassage of the bill during mid-November's veto session. That would have guaranteed the easy death of the entire measure, and over the course of the summer, Mike Lawrence "pleaded me out of it," Edgar recalled. He agreed to approve the bill as passed, as "kind of a favor to Mike and Paul Simon," calling it "an important step forward" at the signing ceremony, where he thanked the four lead legislators by name. Barack devoted a *Hyde Park Herald* column to its enactment, noting how "the process was truly bipartisan" and succeeded only because "each of us was willing to take some heat from our respective caucuses." Barack explained that limits "had no chance of passage," but he would "continue to be a strong advocate of contribution limits and public financing of campaigns," because "without such limits and public investment, it's hard to see how we can fully eliminate the influence of big money over the process."

Over the summer, Barack's district staffer, Cynthia Miller, moved his office seven blocks westward from 2152 East 71st Street to 1741, but Cynthia was "completely done and bored" with handling the minutiae of constituent requests that fell to a state legislator's local aide. Barack's district included a good part of the 6th Ward, which earlier that year had acquired an energetic new alderman, Freddrenna Lyle, who "started pushing legislators to come and do things besides legislate." In mid-September Barack joined Lyle and two city Streets and Sanitation Department representatives to listen to and respond to a two-hundred-person crowd concerned about "garbage pickup, tree-trimming, vacant lots and bushes, weed-cutting, alley cleanliness and rodent control," none of which a state senator had anything to do with.

Cynthia recruited her close friend Jennifer Mason to replace her, and both women overlapped for a month in the early fall before Cynthia left for a months-long trip abroad. Both young women received about $2,750 a month in salary, but unknown to Barack, as of August 1998 he on paper acquired an additional district employee, one whom neither he nor anyone else ever met. In fact, "William Higgins" did not exist, but he represented a long-standing Illinois tradition, the "ghost employee." From 1998 to mid-1999 the ever-elusive Mr. Higgins received $780 a month from the state Senate Democratic caucus, with the stipend increasing to $850 a month in 1999–2000 and $950 per month from July 2000 forward until at least February 2003. Who actually pocketed, or divvied up, the grand total of at least $48,950 in state funds that was paid to Mr. Higgins remains unknown, but Barack had no control over or easy access to the caucus payroll records, which were handled entirely by the Springfield-based Democratic leadership staff.[55]

What did concern Democrats statewide were the poor prospects of gubernatorial nominee Glenn Poshard and incumbent U.S. senator Carol Moseley Braun. The southern Illinois congressman's conservative views, especially on abortion, made him unappealing to many progressive Democrats, but his principled refusal to accept campaign contributions from political action committees left him at a huge disadvantage against his Republican opponent. George Ryan was extremely popular among Springfield insiders, who knew him well from his ten years in the House and eight years each as lieutenant governor and now secretary of state. *Capitol Fax*'s Rich Miller praised Ryan as "a genuinely nice man," someone who "has always understood that politics is about interpersonal relationships," and "one of the most loyal politicians you'll ever meet." But a growing scandal in the sprawling bureaucracy that issued commercial driver's licenses threatened to tarnish Ryan's campaign with reports that up to $150,000 in bribes had been paid into Ryan's campaign coffers. Three low-level state employees were soon indicted, but what gave the scandal indelible poignancy were reports that a fiery 1994 accident in which six children had died was caused by an illegally licensed truck driver. Polls still showed Ryan maintaining a lead of at least 10 points over Poshard, while Moseley Braun, who had attracted widespread criticism during her first six years in Washington, was trailing conservative young Republican state senator Peter Fitzgerald by a much narrower margin.

In late September, Barack introduced U.S. representative Jesse Jackson Jr. at the annual dinner of the Chicago chapter of Democratic Leadership for the 21st Century. DL21C, as everyone called it, was a group for aspiring young politicos, so the Jackson-Obama pairing was a natural. Calling Jackson "one of our genuine rising political stars," Barack said he "exemplifies all that's good and right and hopeful about politics." One of DL21C's key figures, young lawyer John Corrigan, found Barack's performance more memorable than the main speaker's, remembering that he "blew me out of the water."

In early October the U of C law school's autumn quarter commenced, with Barack again teaching both Con Law III and Racism and the Law. Con Law was unusually small, with only sixteen students, while the race seminar attracted twenty-eight. It "was packed entirely with people who were liberals," 2L Jay Hines-Shah remembered, and Barack "would really challenge us, pushing us constantly to examine our beliefs." Given the subject, the composition of the class was unsurprising, but a good many U of C law students were astonished to learn that the incoming class of 180 1Ls included only *one* self-identified African American. The number of black law students had been declining for several years, but the composition of the new class of 2001 made it starkly evident that the U of C law school "had a horrible track record for diversity," as 2L Nat Piggee rightly observed. With Barack teaching only upper-class students and spending hardly any additional time in the building, that decline may have escaped his notice, and Barack's own student evaluations remained excellent, with those two autumn classes garnering overall marks of 9.17 and 9.07.

Barack taught both classes while running his own relatively modest reelection campaign against Republican Yesse Yehudah. A host of recently graduated law students, plus young community organizer John Eason, volunteered to do leafleting under the direction of Ron Davis, but when Barack saw a passel of yard signs jointly promoting both Yehudah and Moseley Braun, he overreacted. Dan Shomon was on leave managing a Democratic challenge to an incumbent Republican senator in the southwestern suburbs when Barack called. "You're not going to believe it! There are Yehudah–Moseley Braun signs all over!" "So what?" Dan replied. "He's a Republican." "No, no. I could lose. I really could lose," Barack insisted. "What is wrong with you? You're not going to lose—you're the Democrat!" Dan responded. Barack "was in panic mode," Dan recalled, and Shomon had to call a graphic designer to order up some direct-mail pieces and yard signs so as to assuage Barack's absurd nervousness. As Democratic chief of staff Mike Hoffmann observed, "Dan did *a lot*" for Barack.[56]

In mid-October Barack submitted handwritten responses to questionnaires from two gay rights groups. He now answered "Undecided" to both "Do you support legalizing same-sex marriage?" and "Would you support a bill to repeal Illinois legislation prohibiting same-sex marriage?" Concerning abortion, he volunteered "perhaps for very young teens" when asked about parental notification requirements regarding pregnant minors. The *Chicago Tribune* "enthusiastically endorsed" Barack's reelection, saying he "has emerged from his first term as one of the most impressive members of the General Assembly." The next weekend, Barack headed to Tarrytown, New York, for the sixth Saguaro Seminar, and Martha Minow remembered Barack showing her a picture of three-month-old Malia and saying that "Michelle was not happy that he was away" from home. Participants were so impressed with Barack's ability to synthesize and reframe others' comments in ways that demonstrated their underlying common ground, Minow recalled, that "we started to nickname him 'Governor'" and asked "'When are you running for president?' It became a joke."

Back in Chicago by Monday, one of Barack's new soon-to-be Senate colleagues, Kimberly Lightford, came to Miner Barnhill to introduce herself. A trio of younger, Chicago-area senators—Lightford from the Far West Side and adjoining, mainly black suburbs; Ira Silverstein, an Orthodox Jew from the Far North Side; and Lisa Madigan, daughter of House Speaker Mike Madigan, from the Northwest Side—had won their March primaries and faced little more Republican opposition than Barack. Madigan, just thirty-two years old, had an introductory lunch with Barack over the summer, and Democratic leader Emil Jones Jr. told the thirty-year-old Lightford that she was too young for the Senate but should meet both Barack and James Clayborne, saying they were "the future of the Senate." Barack warned Lightford to temper how much she should expect to accomplish in Springfield, advising her to concentrate on a few particular policy topics. Kim thought Barack was a "very charming, nice guy," and was all the happier when he pulled out his Friends of Barack Obama checkbook and gave her a $250 campaign contribution.

For his general election race against Yehudah, Barack had done a modicum of fund-raising, bringing in a total of $54,000 during the fall, $23,000 of which came from political action committees, including $5,000 from the Illinois Education Association and $3,000 from the Illinois Trial Lawyers Association. His friend John Schmidt, his Joyce Foundation board colleague Carlton Guthrie, and his U of C Law School colleague Martha Nussbaum topped individual contributors with $1,000 each. Old personal friends like Altgeld Gardens school principal Dr. Alma Jones, St. Sabina's Father Mike Pfleger, Sidley's Gerri Alexis, Northwestern's Bob Bennett, and Hope Center colleague Sokoni Karanja all gave from $200 to $500, as did new friends Jim Reynolds and Dr. Eric Whitaker, whom Barack had first met playing pickup basketball at Harvard.

After giving his check to Kim Lightford, Barack headed to Loyola University of Chicago on the city's Far North Side, where progressive lobbyist Doug Dobmeyer had organized a two-day public policy symposium whose plenary speakers included Barack and Aurie Pennick, who a decade earlier had funded Barack's work at DCP. Eager to demystify Illinois government, Barack told the audience that "there is very little policy making that takes place in Springfield that is subject to consistent, ongoing, thoughtful public debate. In fact, the rules and the processes that are set forth in Springfield are primarily designed to constrict debate as much as possible and to localize the decision-making power in as few hands as possible." One example was how "the vast majority of Americans would like to see serious gun control" but "it doesn't pass because there is this huge disconnect between what people think and what legislators think and are willing to act upon."

Equally pressing was "the more fundamental issue of what do we do with this burgeoning number of people who have no health insurance and are one illness away from bankruptcy or worse." Barack's organizing roots were evident as he explained how "the people who are guilty of disempowering the population are not only the bad guys. . . . Sometimes it's also us, sometimes it's the experts or the advocates who are not that much better at advocating on behalf of and with the communities that they purport to represent, so that the lobbyists down in Springfield who represent a host of good causes that I strongly believe in oftentimes have very few troops behind them, have not engaged in conversations with the very communities that they're representing, and the legislators are aware of that." While generally the good guys "lose because they have less money and less resources," to prevail politically "would take more than simply being armed with good facts and making good presentations, it would also have to do with the fact that they had mobilized a constituency around these policy questions."

Barack expressly advocated "public financing for campaigns" and said he was concerned that the Juvenile Justice Reform Act had been overwhelmingly approved "despite the fact that policy research argued against the bill" and without the communities that would be most affected by it knowing what it would do to minority youth. "Policy research for the working poor" should address "issues of economics that diverse populations have in common" as a way to "resuscitate the notion that government action can be effective," because "there has been a systematic . . . propaganda campaign against the possibility of government action and its efficacy, and I think some of it has been deserved. The Chicago Housing Authority has not been a model of good policy-making, and neither necessarily have been the Chicago Public Schools." The most fundamental question was "how do we structure government systems that pool resources and hence facilitate some redistribution, because I actually believe in redistribution, at least at a certain level to make sure that everybody's got a shot," and "regional organizations" bridging urban-suburban boundaries were one promising option.[57]

A few days before the November election, Barack signed two identical, grammatically incorrect four-sentence letters to a pair of state housing officials urging them to support Allison Davis and Tony Rezko's efforts to construct a ninety-seven-unit Kenwood apartment building that would house senior citizens. Barack would later say he did not remember such mundane correspondence, which exemplified the neighborhood-related minutiae his new district aide, Chicago native Jen Mason, dealt with regularly. The next day the *Sun-Times* joined the *Tribune* in endorsing the "highly regarded" Barack for reelec-

purchasing a set of golf clubs at a discount sports store outside Chicago. "I'm sorry, sir, we can't acccept your check," Barack recounted to Dan Shomon. "I think he ended up using a credit card," Shomon recalled, but Barack and Laura each suspected that their zip codes—60615 for East View, but the far more heavily black 60637 for Laura—were subject to redlining in credit scoring just as in home mortgage lending. When Laura proposed writing her course paper on the practice, Barack jumped at the idea, recounting Laura's experience to Shomon as additional support for his idea of introducing a bill to combat such discrimination.

On Wednesday, January 13, Barack was sworn in for his second term in the Illinois Senate—this one, unlike his first, for four years, pursuant to the Senate's staggered, three-terms-per-decade structure. The Chicago newspapers greeted the new session with stories suggesting that the free passes the city's superb museums gave to legislators violated the new gift ban provision. "Of all the things we have to worry about in terms of influence peddling in Springfield, museums fall pretty low down on my list," Barack remarked, saying he had never used his. "The whole intent of the gift ban was to get at the obvious and serious ethical infractions that have occurred on occasion in the past." Newly elected Chicago Democrats Lisa Madigan and Ira Silverstein joined Barack and Terry Link in the junior members' suite of adjoining offices in first-floor room 105, with Barack shifting from 105D to 105B, and just before the swearing-in Barack met the youngest and most junior new Republican, thirty-four-year-old Dave Sullivan, who had been appointed to replace a senator who had passed away just after the November election.

Sullivan had served for six years as an executive assistant to George Ryan, now Illinois's new Republican governor, and in every Springfield bar legislators and lobbyists were celebrating Ryan's election. *Capitol Fax*'s Rich Miller conceded that Edgar was "a pretty decent governor" but was "almost completely incapable of working with the General Assembly." Many members agreed with Debbie Halvorson's description of Edgar as "cold and distant." Miller wrote that Ryan's inauguration felt "like a breath of long-awaited fresh air," and Springfield newspaperman Bernie Schoenburg explained that the former House Speaker "was of the legislature and understood it." Ryan's arrival promised to transform the relationship between the governor's office and the Senate because, as Miller wrote, Ryan and Pate Philip "are great friends" and "are very close." But Ryan's campaign against Glenn Poshard had shown that Ryan was now doing "his best to throw off his reactionary roots," giving Emil Jones an opportunity "to take advantage of the coming clashes by publicly siding with Ryan and taunting Philip to follow the lead of his governor." Edgar recalled that he and Jones had never had much of a relationship because Jones "was looking for jobs" and "I didn't trade a lot on jobs and things like that." But Ryan was a political trader, and Rich Miller predicted that "gun control should be the juiciest opportunity for Jones to create division between Ryan and the gun-loving hunter Philip." Miller likewise warned that Ryan's greatest weakness is that "he trusts his friends too much."

With the Senate in session only occasionally before late February, Barack was able to give a Martin Luther King holiday lecture at Loyola University of Chicago's law school and join his friend Matt Piers and wife Maria Torres for dinner at Bob Bennett and Harriet Trop's home. Michelle was about to return to a heavier schedule at the U of C, and the Obamas were thus looking to hire a full-time babysitter. By late that winter they had hired Glorina Casabal, a friendly woman in her mid-forties, whom they would pay at an annual rate of $16,000. "With a full-time employee suddenly on our payroll," Barack later wrote, family finances became ever more strained.

One piece of financial good fortune came via a lecture agency, Jodi Solomon Speakers Bureau, which for several years had listed Barack as available to talk about "A Story of Race" based on his book. That winter Carleton College, a well-respected liberal arts institution in central Minnesota, agreed to pay $3,500 for Barack to keynote its Black History Month convocation in early February. Barack began his remarks by apologizing for a cold he had caught from seven-month-old Malia, and he spoke informally about how some people questioned his involvement in politics just as others had doubted the value of community organizing. Everyone needed to commit themselves to "something that is bigger than yourself" and to embrace what Martin Luther King Jr., drawing on the Hegelian dialectic, had "called a both/and mentality" rather than the "either/or mentality" that constricted most public debates. Saying that "you have to have some hope to get out of your own private life," Barack implicitly referenced Rev. Jeremiah Wright in declaring that "there's something audacious about hope."

Barack's relatively brief speech left considerable time for questions. The first led Barack into a long discussion of welfare reform and the need for more jobs that paid a living wage. "The minimum wage right now is so low that even people who are working full-time essentially do not get out of poverty." But "political power has shifted in this country to suburban areas that don't see a lot of unemployed people," and job-starved inner-city neighborhoods were witnessing "the disintegration of the fabric of community life." In such areas, churches often were the only "institutions that are still cohesive enough and bring people together," and "the most successful community development organizations . . . are typically ones that are housed in churches or based in an ecumenical effort among a group of churches." In Barack's experience, "if you try to just mobilize people around a purely economic or political agenda, it won't sustain itself and get at some of the very difficult, knotty problems that exist."

Barack also bluntly volunteered that "affirmative action is more important symbolically in some ways than it is as a practical matter. First of all, affirmative action does not function once you get out of college unless you are in a very narrow range of government jobs." Chicago was more than 30 percent black, but "if you go into any major office building you will not see more than five percent black people in there, and most of them are secretaries, receptionists, and delivery people." The federal EEOC "does a miserable job" combating employment discrimination "because it's been starved for funding by Congress," and Barack referenced his experience in *Donnell v. Comdisco Inc.* by telling the audience that he had one client who had been "called nigger by the CEO of a Fortune 500 company in front of the entire office." Barack acknowledged that "there are some legitimate grievances from the majority community in any form of affirmative action," because "some individuals may be marginally harmed."

Finally, Barack said that while "it is not a healthy thing for our civic life that unions have become so beaten down," it was also the case that "in Chicago, the Teachers Union needs to do some work to get out in front of change as opposed to being a barrier to change." Overall, "the union movement has been fairly myopic" in focusing simply on "issues of wages and benefits." Unions need to "think about the people who are unemployed and not only the people who are already in the union." They also should consider "how do we tie our own benefits and wages to performance of a corporation and think of ourselves as shareholders in that company as opposed to simply employees."[59]

Back in Illinois, increased attention focused on the growing number of death row inmates who were being fully exonerated of the murders they previously had been convicted of committing. Northwestern University's Center on Wrongful Convictions was responsible for most of the legwork, and on February 5, Anthony Porter was freed from death row just two hours before his scheduled execution. George Ryan remembered turning to his wife that day and saying, "How the hell does that happen? How does an innocent man sit on death row for fifteen years and gets no relief?" Porter's near-death experience "piqued my interest," Ryan later told the *Chicago Tribune*'s Steve Mills. Six days after Porter's release, Rich Miller wrote in *Capitol Fax* that statehouse denizens "can feel the buzz growing on the death penalty issue" as everyone waited to see what Ryan might do.

In mid-February Emil Jones gave Barack a significant new responsibility by naming him to succeed Donne Trotter as one of Senate Democrats' three members on the powerful Joint Committee on Administrative Rules. JCAR, as everyone in state government called it, oversaw whether executive branch agencies were properly implementing state laws. The committee could impose a ninety-day delay on any agency rule it found questionable, and executive branch officials were required to respond. Barack listened and did not speak at his first JCAR meeting, but within two months he became cochairman of the twelve-member, officially nonpartisan committee. Veteran southern Illinois Democrat Jim Rea, who previously had known Barack primarily socially, was immediately impressed. "Gosh almighty, where did this guy come from?" he remarked to another colleague. "He's sharp, he's asking the right questions, and he wants the answers to those before there's any action taken."

JCAR rarely attracted news coverage, and as February 23's primary elections approached, most eyes were on U.S. representative Bobby Rush's challenge to Chicago mayor Richard Daley, who was approaching ten years in office. The *Tribune* endorsed Daley for a fourth term in a rave editorial praising him for "the transformation of the Chicago Public Schools" under mayoral appointees Paul Vallas and Gery Chico. The *Trib* stated that "Rush has been a good congressman," but *Capitol Fax*'s Rich Miller lambasted Rush's "ineptitude" in running "a less-than-stellar campaign."

Born in southwest Georgia, the fifty-two-year-old Rush had come to Chicago at age seven and enlisted in the U.S. Army when he was seventeen. In the late 1960s, Rush emerged as one of the leading members of the Chicago chapter of the Black Panthers, serving almost six months in prison on a misdemeanor gun charge. After finishing college, in 1975 Rush unsuccessfully challenged 2nd Ward machine alderman William Barnett, but eight years later, with Harold Washington winning the Democratic mayoral nomination, Rush defeated Barnett and joined the City Council, taking an active interest in South Side environmental issues. In 1992 Rush successfully challenged U.S. representative Charlie Hayes for his historic 1st District seat, held by a succession of African Americans ever since Oscar De Priest in 1928 had become the first black member of Congress since 1900 and the first African American ever elected to Congress from outside the South.

Ten days before the primary, polls showed Daley with an overwhelming 65 to 13 percent lead over Rush. Congressional colleague Jesse Jackson Jr. was one of the few elected officials backing Rush, and when votes were tallied on February 23, Daley crushed Rush 72 to 28 percent, winning almost 45 percent of the African American vote citywide and even carrying Rush's own 2nd Ward. A decade earlier, Daley had received 4 percent of the 2nd Ward vote. In defeat, Rush attacked black business leaders "who have aligned themselves as part of the Daley machine," but Chicago journalists marveled at how easily Daley had triumphed in a city that almost every day was earning its title as "the murder capital of the nation." *Washington Post* columnist E. J. Dionne asked his Saguaro Seminar colleague to explain Daley's success. Barack said Daley had benefited from an "aversion to ideology and an emphasis on management." He also explained that political power based on "pinstripe patronage" to wealthy professionals and the campaign contributions that resulted made for a more impregnable mayor than in eras when a payroll patronage machine might split and generate serious challengers. Nowadays, without such a base, "it's harder for folks to build their own independent organizations." *Tribune* columnist Bruce Dold wrote that Chicago's "African-American political leadership seems to be floundering" and quoted an unnamed alderman as predicting "I'm sure that somebody's going to come after Bobby, especially after he encouraged candidates to run against incumbent aldermen." Dold added that "there is talk that state Sen. Barack Obama or [17th Ward] Ald. Terry Peterson would challenge him for his congressional seat."[60]

That February Barack introduced sixteen bills and reintroduced his call for a state constitutional amendment mandating universal health care coverage. His bills sought to lessen low-income workers' tax burdens and to require the state Department of Human Services to establish three pilot job skills enhancement programs. SB 1055, the Check Acceptance Firm Act, sought to remedy what had happened to Barack and Laura Mullens. Companies would be required to provide "information verification services for retail sellers with respect to a consumer's personal check to make available by means of a toll-free telephone number the reasons for not accepting the consumer's check. Requires the reason to be sent to the consumer within 14 days after the request was made."

On March 1 Barack and Jesse Jackson Jr. jointly keynoted Northwestern Law School's Diversity Week. One white 2L thought Barack was the more memorable speaker due to "the level of substance in what he had to say." Rich Miller of *Capitol Fax* noted that "nothing much of substance is happening this year" in Springfield, although legislators were complaining about the campaign finance law because they still viewed "campaign funds as their own private cash." When the chairman of the House Executive Committee showed up audibly drunk at a 10:00 A.M. hearing, newspapers across the state wrote stories about "debating under the influence." Rich Miller protested that the presence of "an allegedly drunken legislator in and of itself is not news. Lots of legislators, executive branch officials and, yes, even journalists have worked day in and day out with the smell of booze on their breath," and he reiterated how "most business conducted in Springfield is done after hours, often over a few drinks." Barack spoke with a reporter investigating the millions Illinois was spending annually on instructional courses for corporate executives at firms such as Chrysler, with overhead payments going to the Illinois Manufacturers' Association and the Illinois Chamber of Commerce, two of the state's top campaign contributors. "I don't think the state should subsidize these folks at the same time we are not able to adequately fund programs for welfare mothers who are trying to move into the work force," Barack complained.

By the second week of March, the George Ryan–Pate Philip friendship was indeed coming apart. Ryan wanted to name House minority leader Lee Daniels, no friend of Philip's, as chairman of the state Republican Party. A bitter argument ensued, and Rich Miller reported that Ryan told "several people afterwards that he had never been spoken to so harshly in his entire life."

By early March Barack was privately pondering whether he should leave the state Senate and Springfield's incestuous political life for more remunerative work and then reenter electoral politics some years later. Barack discreetly spoke with two of the most senior partners at Sidley & Austin: Newton Minow and Eden Martin. Barack had come to know Martin well on the boards of the Woods Fund and John Ayers's Leadership for Quality Education. Martin was just about to be named president of the Commercial Club of Chicago and its highly influential Civic Committee, Chicagoland's most powerful private sector post.

On Friday, March 12, after a Woods Fund board meeting, Barack asked to speak with Eden, and after their conversation, Martin dictated a memo to file memorializing it. Barack was "immensely talented" and "has long term political ambitions," but he "is now thinking about . . . whether he wants to continue to forego the kind of opportunities and income that might be available in the private sector. In particular, he is thinking about the possibility of working for a foundation for several years, after which he would have built up some equity and might have an opportunity to re-enter elective politics. He has talked to Newt and me separately about his thinking" and "I had a long visit with him today." The Joyce Foundation board was in the process of selecting a replacement for Debbie Leff, who four months earlier had been named president and CEO of Second Harvest, a national network of food banks that was the country's sixth largest charity. In a few weeks, Joyce would appoint Cousteau Society executive Paula DiPerna as Leff's successor, and that search plus his experience at Woods had given Barack a good sense of how financially influential foundation leadership roles could be.

They also discussed Barack's joining Sidley to do a very different and more remunerative sort of law than he did at Miner Barnhill. If Barack wants "to return to the practice essentially full time," Martin wrote, "he ought to talk to us about the possibilities." He noted that Barack "has had no significant trial experience. He is also interested in the possibility of a business practice, but recognizes he would have to learn the craft. I think it is very possible that he will call Tom Cole," the head of Sidley's corporate transactional practice, Martin noted.

Two weeks after Barack's conversation with Martin, he and Michelle flew to Boston to attend the wedding of Barack's old Harvard friend Dan Rabinovitz. After the ceremony and reception, they had drinks with Rob and Lisa Fisher, and Barack raised the prospect of leaving the state Senate for a better-paying private-sector job. Michelle made clear her strong desire that Barack leave

Springfield behind so she would not have to continue being a single parent three or four nights each week when the Senate was in session. Rob and Lisa went away with the impression that such a move was just as likely as Eden Martin's memo documented.[61]

On March 17, Illinois's top news was that Governor Ryan had hesitantly signed off on the execution of Andrew Kokoraleis, who seventeen years earlier had participated in several mutilation murders. "The idea of pulling the trigger on a human life had Ryan really freaked out," *Capitol Fax*'s Rich Miller reported, much to the amazement of the governor's longtime friend Pate Philip, who could not fathom Ryan's hesitation. "What the hell's wrong with you, George?" an angry Philip asked Ryan over the phone as fellow senator Ed Petka looked on. Then Pate "literally just slammed the phone down." Petka later said this exchange was "the beginning of the end" of Ryan and Philip's friendship, but Ryan told the *Chicago Tribune* he was confident he "did the right thing" in letting the execution proceed. Meanwhile, the *Tribune* reported from Washington that defeated mayoral candidate Rep. Bobby Rush "hasn't missed a beat" on his congressional work.

Barack's Senate schedule forced him to miss a Chicago Annenberg Challenge meeting where his friend Jim Reynolds was elected to the board, but as chairman Barack already had complained to education researchers Anthony Bryk and Mark Smylie that the board was not happy with the timeliness and usefulness of their evaluation reports, which CAC was liberally funding. Barack sought out coverage from the African American press for his SB 883, the Transportation to Work Act, telling one Chicago weekly that "one of the greatest barriers that welfare recipients face as they return to the workforce is a lack of adequate transportation," because "only about 6 percent of welfare recipients own a car—and most jobs are not located on public transportation routes." As he had argued for three years, "in order for an individual to successfully move from welfare to work, they must first be able to get to their place of employment, and secondly, if they have children, have the means to get them to day care."

On March 18, the Senate passed Barack's modest SB 929, amending the Illinois Enterprise Zone Act to extend such tax abatement zones, of which half a dozen existed in Chicago's low-income areas, from twenty to thirty years. Republican senators Christine Radogno and Steve Rauschenberger spoke against these urban tax breaks, but Barack's bill passed easily, 45–10. Then Barack's Judiciary Committee friend Kirk Dillard, who voted for it, filed a motion to reconsider, and the bill was brought back to the floor for reconsideration on Tuesday, March 23. Barack defended the zones' economic development value, arguing that "they've had a track record of success throughout the state," and that if Republicans were truly "interested in accountability, then we should be looking not just at enterprise zones, but at a whole host of corporate tax breaks that we provide throughout the state." Barack's Democratic colleague Bobby Molaro, a seven-year Senate veteran, questioned the motion to reconsider, saying he could not recall when a bill had been subject to such action, but it was clear that Pate Philip was against the Senate allowing a Democratic bill benefiting inner-city areas to become law. When the new tally was taken, all thirty-two Republicans—eighteen of whom had voted in favor of Barack's bill five days earlier—voted to kill it, with two Democrats, Jimmy DeLeo and John Cullerton, who were no fans of Emil Jones, failing to vote. The outcome "was a shock," Barack's colleague James Clayborne recalled years later, because Barack "thought he had an agreement" and was "greatly disappointed" by the Republicans' turnaround.

Barack had far better success that same day with his check verification measure. "This bill was generated actually from a student of mine at the University of Chicago Law School," he told his colleagues, and it passed unanimously, 58–0. The next day the Senate unanimously approved Barack's job skills enhancement bill and a more modest one amending the Health Care Surrogate Act, but on Thursday, when an uncontroversial Republican bill, the Certified Capital Company Act, was brought to the floor, Barack gave fuller voice to the anger he was still nursing from two days earlier. "I can't help but comment on the fact that the vast majority of these credits the other side of the aisle is going to be voting unanimously on . . . are no different from the enterprise zone legislation" the Republicans killed on Tuesday, and with no greater accountability provisions. "I do not understand why we are getting no objection from various members on the other side of the aisle who objected to my bill because of accountability, but I don't hear a peep out of any of you with respect to" this one. "I would hate to feel cynical and think that it is only because of partisanship that the enterprise zone legislation has stalled and this bill and other bills . . . have not. . . . I would urge all of you on the other side of the aisle to seriously consider . . . treating all of the tax credit bills . . . in the same manner and evaluating them in the same process."

Steve Rauschenberger responded that he felt such "a great deal of discomfort" with this bill that he had met with the sponsors five times to ensure that adequate accountability provisions had been added to the legislation. "I understand Senator Obama's concerned and frustrated with our action on enterprise zones," but Rauschenberger urged passage of the pending bill. Barack replied that had he been given "the courtesy of the opportunity to amend, that would have occurred as well." Barack's Democratic colleague Pat Welch said, "I agree with Senator Obama, but I don't think that agreeing with Senator Obama means that we shouldn't support this bill." Barack was alone in voting present, with the only two no votes cast by Democratic leader Emil Jones and his closest Senate friend, Bill Shaw.

Later that afternoon, when a bill authorizing prosecution as an adult of anyone fifteen or older who was charged with aggravated battery with a firearm in a school zone, Barack again voiced his disappointment. Just a year earlier, this provision had been removed as part of the Juvenile Justice Reform Act because "there is really no proof or indication that automatic transfers and increased penalties and adult penalties for juvenile offenses have, in fact, proven to be more effective in reducing juvenile crime or cutting back on recidivism." But now they were about to "increase penalties further for juveniles and try them further as adults." Barack and four other African American senators all voted present as fifty-two of their colleagues voted yes.[62]

Barack's unhappiness was also manifest in his unusually sour remarks when the Associated Press asked him about new U.S. senator Peter Fitzgerald, his former Springfield colleague. "Based on his record in the state Senate, I would not see Mr. Fitzgerald being a mover and shaker in Washington," Barack said. His Democratic colleagues shared his woe over the session so far, and when Barack wrote his first *Hyde Park Herald* column of the year in early April, he emphasized how "taxpayers are often overpowered by large financial interests in Springfield," such as payday loan companies that "gouge consumers." He also expressed concern about a tuition tax credit proposal for parents with children in private and parochial schools, fearing that its $180 million cost could "potentially drain resources from our chronically underfunded public schools." Barack complained to one reporter that "it's hard to mobilize support for programs that cost money but benefit low-income folks," and he remained disappointed about the lack of support for the Earned Income Tax Credit bill he had again introduced in early February. "I don't understand the logic," Barack told a journalist in a direct echo of John McKnight. "We take this money from people and then spend it on public services they might not need if they had more in their pockets."

Barack's first two years in Springfield had convinced him of the underappreciated importance of state-level policy making, as he told his incipient think-tank colleagues on April 13, but that Tuesday was also runoff primary day, and Barack had become involved in two contentious aldermanic races. In the 15th Ward, he backed independent challenger Ted Thomas, an ACORN colleague of Barack's longtime friends Madeline Talbott and Keith Kelleher, who defeated the Daley machine with 56 percent of the vote. In his home 5th Ward, Barack supported incumbent Barbara Holt, who recalled that he took "me around to speak with his neighbors in East View one Saturday." But Holt's energetic challenger, South Shore attorney Leslie Hairston, emphasized what the *Tribune* termed "questions about Holt's responsiveness to residents' gripes" amid complaints that Holt spent too much time at City Hall and too little in her ward. Holt led Hairston 44 to 30 percent in the initial primary, but the trio of eliminated candidates all endorsed Hairston and evidence grew that Mayor Daley's forces were quietly mustering in support of Holt. Their involvement "may have been her kiss of death" in proudly independent Hyde Park, and on April 13 Hairston easily defeated Holt 55 to 45 percent.

On April 15 Barack and Michelle owed the IRS a whopping $18,401 in additional 1998 taxes. Michelle had earned more than $66,000 from the U of C, and Barack $56,000 from the law school plus another $48,000 from the Senate. His law firm income was barely $10,000, but his board service brought in an additional $9,500 from Joyce and $6,000 from Woods, even though neither foundation yet realized that once Barack had become a legislator two years earlier, a little-known Illinois law adopted in late 1995 with no visible news coverage prohibited him from accepting such honoraria. The Obamas' charitable giving was only $1,100, with $400 going to Trinity United Church of Christ and $100 to the Illinois Coalition Against Handgun Violence. With Malia's babysitter Glorina Casabal costing well over $1,000 a month, family finances remained tight and Barack later said "there were a lot of stresses and strains" between him and Michelle during that time.[63]

That same day, Barack introduced Laura Mullens, the law student who had piqued his interest in discriminatory credit scoring, to the House Financial Institutions Committee. Laura realized that committee members were not expecting a white female with a Georgia accent when Barack said that one of his constituents would testify in support of his Check Acceptance Firm Act. Dan Shomon recalled that Laura impressed the committee, and SB 1055 was unanimously approved. Thanks to Barack, Laura's trip to Springfield was "one of the most educational experiences I had in law school."

As April turned to May, *Capitol Fax*'s Rich Miller wrote that in Springfield "the feeling of bipartisanship and good will towards the governor is still very high," mainly because "Illinois once again has a governor who actually likes people and is willing to rub elbows with them." Barack and his three white Democratic colleagues—Terry Link, Larry Walsh, and Denny Jacobs—had renewed the Wednesday-night poker games, and by early 1999 Terry had refinished the basement of his Springfield house, hosting the games there rather than in his kitchen. Lobbyists Mike Lieteau, Phil Lackman, and Dave Manning remained mainstays, as did Republican senator Tommy Walsh, and in early 1999 the group added another regular, Walsh's good friend and fellow Republican senator Dave Luechtefeld, from Okawville in far southern Illinois.

Often called "coach" for his years leading high school basketball teams, Luechtefeld had noticed Barack right from the start. "He certainly caught your attention when he spoke," and it was obvious "he was looking for something bigger than this from day one," Luechtefeld explained. He sensed the discomfort many urban Democrats felt toward Barack, perhaps because of the "little bit of cockiness, little bit of arrogance" that Barack's self-confident manner exuded. Barack was very respectful toward Luechtefeld, who was more than two decades older, but at his first poker game, Luechtefeld was surprised when Barack "pulls out a cigarette" because he did not expect Barack to be a smoker. Barack "loved to play poker, there was no question about that," and with a growing roster of regulars that soon included Illinois Manufacturers' Association lobbyist Boro Reljic plus occasional participants like Democratic senator Lou Viverito, the Wednesday-night gatherings sometimes split into two adjoining tables. The players also came up with an unofficial code name—a "committee meeting"—in a less-than-successful attempt to shield their pastime from knowledgeable secretaries. With the new players, the monetary stakes crept upward, and after losing one painful hand to tightfisted Barack, Larry Walsh remembered complaining, "Doggone it, Barack, if you were a little more liberal in your card playing and a little more conservative in your politics, we'd get along a lot better!"

In late April Emil Jones officially named Barack as the Democratic spokesman on the Public Health and Welfare Committee, replacing seventy-five-year-old Margaret Smith. At a late April town hall meeting in Woodlawn, Barack and Barbara Flynn Currie were confronted with dozens of people unhappy with their support of a bill that would limit local school councils' power to fire principals. Barack refused to back down, saying principals who clearly were doing a good job deserved protection, and he advised people to get "involved in larger organizations that have a reach beyond this district."

On May 4, Governor Ryan spoke to assembled legislators, proposing a $12 billion public works construction program he called Illinois FIRST—Fund for Infrastructure, Roads, Schools & Transit. The savvy Ryan had gotten the Four Tops, even Pate Philip, to agree to the scope of the program, if not its financing, prior to his announcement, and Ryan later claimed that when he first detailed its pork barrel potential to the quartet, "the drool was running out of their mouths." *Capitol Fax*'s Rich Miller swooned over Ryan's "elegantly written" speech, in which the governor mentioned by name forty-nine legislators whose districts the program would benefit. Barack was busy supporting a House bill to allow pharmacists to substitute generic for more expensive brand-name drugs if the generics had been approved by the federal Food and Drug Administration. Barack told his colleagues, "I actually spoke to FDA officials in Washington to try to get this clarified," and they had confirmed that "the generics are equivalent of the brand-name drugs," as did "physicians that I talked to independently." That effort won overwhelming Senate passage of HB 2256, and that same day the House unanimously approved a somewhat weakened version of Barack's Check Acceptance Firm Act, with SB 1055 then returning to the Senate.

Despite its unanimous approval by both houses of the legislature, within days Barack's Check Acceptance Firm Act was fatally derailed. "I've got bad news for you," former Senate Judiciary Committee staffer Matt Jones told Dan Shomon one morning. "'Your bill is dead.' 'Why?'" Dan asked Matt, who was now a contract lobbyist. "One of the credit scoring companies just hired me. You're done." Dan gave Barack the bad news—"it's dead"—and Barack told Laura what had happened: "the TeleCheck people woke up and killed it in the Senate." Her experience with SB 1055 had been educational indeed, or at least educational in the ways of Springfield. "Lobbyists spend most of their time killing bills," *Capitol Fax* had noted a few weeks earlier, and in early May, Miller observed that spring 1999 "will likely be remembered as the session held for the benefit of rich white guys." That was "not to say that all sessions don't usually benefit rich white guys," Miller quickly added, "but this one just seems to be more blatant."[64]

George Ryan's long immersion in the ways of Springfield was unintentionally highlighted when he named four lobbyists to a seven-member ethics panel he had created following his inauguration. "I was surprised that the governor wasn't more mindful of the possible perceptions that were raised by putting lobbyists on the commission," Barack told the *Chicago Sun-Times*. "I think it would be advisable to put people on the commission that are not perceived to have major interests in Springfield." Barack was not much happier with Ryan's huge Illinois FIRST initiative. It would fund "one of my top priorities," the remodeling of South Lake Shore Drive, the handsome divided highway that linked Hyde Park to downtown Chicago, plus it would pay for the reconstruction of nine miles of the Dan Ryan Expressway south of Bronzeville. But it would be "funded by regressive taxes that will hurt families," because most of the money would come from doubling Illinois's annual vehicle registration fee from $48 to $96 and tripling the title fee for vehicle purchases from $13 to $50. "Regardless of your income or whether you own a Lexus or a Ford Escort, vehicle owners will pay the same fee," exacerbating a state taxation system that was already "one of the most regressive in the United States."

In early May, Senate president Pate Philip made news regarding a bill that would increase criminal penalties for anyone who assaulted a sports referee or umpire. Barack's Republican Health Committee colleague Dave Syverson was presenting the bill when Rickey Hendon asked whether it would cover professional wrestling, where referees were often fair game. Syverson responded, "I don't want to burst your bubble, but when it's part of the act, that would not be included." Hendon replied, "I'm going to let some of my professional wrestling friends know that you think it's just an act, and perhaps when they're here they can lobby you personally and see if you can get out of a choke hold!" Judiciary chairman Carl Hawkinson spoke in support of the bill, but then Philip objected, declaring that "every once in a while, that referee or that umpire ought to get popped, and you ought to pop 'em good!" Then the vote was called, with only thirty-three senators—including Barack—backing the bill and twenty-two bipartisan opponents voting either no or present.

Reporters and columnists jumped all over Philip's remark, but an unrepentant Philip told the *Chicago Tribune,* "They make more mistakes and everybody gets so mad at them, it isn't even funny. . . . I just say maybe they deserve a pop once in a while." *Tribune* columnist Bob Greene labeled Philip "stunningly moronic," and the suburban *Daily Herald,* calling Philip "more inflammatory than usual," editorialized that the state Senate's "level of sportsmanship . . . sometimes sinks pretty low." In contrast, *Capitol Fax*'s Rich Miller defended Philip, arguing that "he embodies older, white, suburban, working class values" and that "his critics ought to take a breath and relax." Even twenty years later Philip continued to stand by his call to arms, saying he was a longtime Chicago Cubs, Chicago White Sox, and Chicago Bears season ticket holder. "Everybody around me bitches about the referees, the calls, like you cannot believe!" Barack had little personal interaction with Philip, but the Senate president reminded him of someone he had known well, as he explained on the floor one day to Republican friend Dave Sullivan. "One of the strangest things for me serving here in the Illinois Senate is that Pate Philip looks just like my grandfather!"[65]

Much Senate attention focused on a pair of protectionist bills aimed at boosting the businesses of liquor distributor William Wirtz and Pepsi bottler Harry Crisp, a pal of Philip's, with Barack voting against both. He also voted against an open-ended deregulation measure sought by Illinois-based banks. "We should have a sense of what exactly this body, this chamber is giving up in terms of its regulatory powers, and I have not heard yet a clear understanding of what exactly those powers are. What legislative regulations and powers do we have that we're potentially sacrificing as a consequence of this bill?" Giving banks carte blanche is "not the way to go," Barack warned, but only moderate suburban Republican Christine Radogno joined him in voting no.

Barack also took to the floor to advocate passage of a House bill requiring that at least two-thirds of the child support payments received by the state go to children, because at present "of the $88 million that we collect in child support, we only distribute $8 million of that to the children who are supposed to be the recipients." Some Republicans thought that too much, and the bill passed with only thirty-four yes votes. The next day Barack became even more peeved when a House bill extending the duration of one particular enterprise zone in Metro East so as to benefit a specific company was called for passage. Taking "a business-by-business approach to the extension of the enterprise zones," in place of his earlier bill that Republicans had killed, would be "highly subject to politics," Barack warned, but even he voted yes.

As the final week of the spring session approached, attention turned to the $65 million that each caucus would receive for so-called member initiatives as part of Governor Ryan's huge Illinois FIRST proposal. One evening Barack joined conservative Republican senator Laura Kent Donahue on Bruce DuMont's *Illinois Lawmakers* telecast, explaining that he believed Illinois had done "as good a job as any state" in aiding welfare recipients' transition to work. Barack expressed guarded support for Illinois FIRST, but emphasized that programs that "benefit the most vulnerable in the state" must not suffer as a result. When DuMont asked if the twenty-year, $50-million-per-year commitment that FIRST entailed might prove problematic during an economic downtown, Donahue agreed, saying it was "too big, too fast." Barack expected to support it but was unhappy about how Illinois was "putting a little bit too much of a bite on working-class families" while the Revenue Committee on which he sat had "voted on about $200 million worth of tax breaks" this year alone.

When Chicago Republican senator Walter Dudycz the next day offered a resolution instructing Illinois's public universities to comply with FOIA requests for their admissions data submitted by a group well known for opposing affirmative action, Senate floor debate dissolved into angry discord. Rickey Hendon and the normally mild-mannered James Clayborne challenged the requester's intent, with Hendon telling Dudycz "don't bring racism and racist legislation to the floor of the Senate." Dudycz was in *no* way racially insensitive, as even Hendon knew, but Barack felt compelled to speak too, stressing that the requesting organization, the Center for Equal Opportunity, "is systematically attempting to dismantle affirmative action in public universities all across the country." Barack readily volunteered that "we know, by definition, that affirmative action means that the minority students may have lower test scores than the white students. . . . That's the essence of affirmative action." But if and when the scores were publicized, then "immediately, essentially every minority student on that campus" would be "under suspicion as being incompetent, unqualified," and "under a cloud that they could not erase." Barack acknowledged that "reasonable minds can differ on the issue of affirmative action" because it was "a difficult, complicated issue," but he urged senators to vote against the resolution, and Dudycz angrily withdrew it.[66]

When the final week began on May 24, the Senate spent all of Monday evening debating a bill to allow riverboat gambling to expand into northwestern Cook County, to permit nonstop gambling on all ten Illinois riverboats, and to underwrite the reopening of a racetrack owned by a well-connected millionaire. The bill had passed the House three days earlier, and veteran senators said they had never seen such intense lobbying on any bill. Emil Jones, who was otherwise pro-gambling, was against a bill he thought offered nothing at all to the parts of Chicagoland that were most in need of economic development. Pate Philip insistently championed the measure, and early in the debate, Barack took the floor. "I am not philosophically very keen on gaming as an economic development strategy," because "oftentimes it tends to be regressive" with regard to the people from whom it sucks money. In addition, "what does disturb me deeply" is how "we are giving away, at least initially, $65 million in tax breaks to people who I think we would all acknowledge are pretty fortunate and doing pretty well." In stark contrast, "during this session, sitting in Public Health and Welfare . . . I watched us, again and again, reject and refuse the requests of people in deep need in this state—personal attendants who are caring for the disabled and are making the minimum wage, and we told them 'You know what? We can't give a raise. We've got a tight budget here.' We had bills that would raise Medicaid eligibility for low-income senior citizens. Said we couldn't do it. Sixteen million dollars. Couldn't find it in the budget. We had a bill that would have shifted the median income of eligibility for subsidized day care. Cost, ten million dollars. We couldn't do it. And many of us sitting here today said to people in desperate need that we couldn't provide those kinds of tax breaks or we couldn't provide those kinds of subsidies because we are fiscally responsible. And yet, here we are today." Barack's disgust was audible. "I've been observing this process for the last three days of people trading their votes, making deals" on the gambling bill, "and I think we should be ashamed of ourselves after saying 'No' to that many people who need some real help in this state, that we are actually willing to go ahead and vote for a piece of legislation that helps people who don't need it."

Later in the debate, Emil Jones warned that the bill's inclusion of an explicit, 20 percent minority set-aside provision would open the measure to easy legal attack because Barack had advised him that the inclusion of a specific quota, rather than a more diffuse goal, was clearly unconstitutional. When the vote was called, with Barack voting no, Philip appeared to have defeated Jones by a one-vote margin, 30–29, with Philip's good friend Ed Petka, who firmly opposed gambling, voting yes. Then anyone who had ever doubted Emil Jones's political savvy quickly learned a lesson. Jones requested a verification of the thirty affirmative votes, in which the presence of every member was to be confirmed. But pro-gaming Democrat Evelyn Bowles had left the chamber, and the result was a 29–29 tie, with the measure's proponents now forced to negotiate with Jones to see what further division of the pie could change his no to a yes.

Eleven Democrats, including some who were not enthused with Jones, had voted with Philip and against their own leader, but the backers of a boat for northwest Cook County now deployed county board president John Stroger's top two campaign contributors, Tony Rezko and Al Johnson, as well-chosen emissaries to Emil Jones. Barack told the *Chicago Defender*'s Chinta Strausberg that the bill would "do very little to develop the South Side, West Side or south suburbs," and that "if we're going to have a boat it should be in an economically distressed area" of Cook County, not in prosperous northwestern Rosemont, "where very few blacks live." *Capitol Fax*'s Rich Miller believed the vote had become "a very personal fight" between Jones and Philip, grounded in part by the perception that Philip "doesn't treat Jones as an equal," but Jones's race-conscious approach to the proposal's shortcomings did not sit well with some white Democrats.

By the time the Senate convened at 1:00 P.M. on Tuesday, Al Johnson and Tony Rezko's entreaties had succeeded: in exchange for $30 million in what Miller called "black pork"—$6 million per year for five years for poor communities—Emil Jones backed the bill. Like Jones, three other African American Chicagoans—Rickey Hendon, Donne Trotter, and Jones's sidekick Bill Shaw—switched their votes as well. After a brief, pro-forma debate, the bill passed 31–27, with Ed Petka and several other anti-gaming Republicans peeling off since their votes no longer were needed.

That night at the governor's mansion, an end-of-session party brought out high spirits all around. The next morning *Capitol Fax* reiterated that 1999 was "The Year of the Rich White Guy." With Illinois benefiting from what one lobbying firm's session summary called a "red-hot economy," the final budget as passed provided for $2 billion more in state spending than Governor Ryan had proposed three months earlier. The firm called the spring session "one of the most exciting and productive in recent legislative history," but Barack firmly disagreed. "It's very hard to separate yourself from the interests of the gaming industry if you're receiving money," he told National Public Radio in an interview excerpt broadcast the same day that *Capitol Fax* reported that Pate Philip had received $257,000 from gambling interests during the spring session and Emil Jones $86,000.

But Barack set off more fireworks on former Chicago alderman Cliff Kelley's WGCI radio show when he told African American listeners that Springfield's legislative black caucus "is not functioning as it should" because "we don't have a unified agenda that's enforced back in the community and is clearly articulated." The next day's *Chicago Defender* put Barack's remarks on its front page under the headline "Obama: Illinois Black Caucus Is Broken." Barack explained that "mistrust between the Black Caucus members in the House and the Black Caucus members in the Senate" dated back several years to disagreements about former governor Jim Edgar's education funding proposal. He told another interviewer that the Wirtz and Crisp special-interest distributorship bills had highlighted the caucus's disunity. "We were struggling to come up with a negotiating position the entire session," he explained. "I don't think we ever had a well-defined agenda, and so we were unable to leverage our votes at the end," as Emil Jones had so successfully done on the hard-fought gaming bill. Black legislators "don't need to be monolithic, but if we're going to be effective, we're going to have to have an agenda and unify around that agenda on at least some of these high profile issues where we potentially do have some leverage."

Barack had expressed similar criticisms of his African American colleagues to fellow senator James Clayborne, but voicing such views on a premier radio program was taking it to another level. Barack "was somewhat impatient," Clayborne remembered, "that we didn't have the unification that we should have, that he wanted to unify us and move an agenda forward." Clayborne had "told him 'Unfortunately, if you jump out there too soon, you'll turn people off. You've gradually got to work your way through the maze to try to get some unification and get us focused on an agenda.'"

When new young African American senator Kimberly Lightford joined the Black Caucus in January 1999, she quickly realized just how poisonous the closed-door dynamics were. Lightford had been surprised by what a cool reception her Democratic colleagues gave her, with southern Illinois veteran Vince Demuzio, Barack, and Miguel del Valle among the few senators who greeted the thirty-year-old warmly, often calling her "kiddo." She knew that some senators saw her "as this little girl" and expected her to perform secretarial tasks, yet at the first Black Caucus meeting after her arrival, Barack nominated her as chairperson to succeed Donne Trotter. Kim quickly learned that the position was more difficult than she imagined because of Trotter and Hendon's dislike for Barack. "We could barely have meetings in caucus because Donne and Rickey would give him hell." Barack "wouldn't argue with them," he "would just leave. He avoided confrontation so well," Kim recalled. "They felt he was arrogant" and believed he "had pretty much tricked Alice out of her seat."

Hendon acknowledged that some meetings of both the Black Caucus and the larger Democratic one "got really heated," with Barack sometimes agreeing with white senators like Denny Jacobs or Terry Link rather than with the Chicago African Americans. "There were some pretty raucous caucuses," Bobby Molaro recalled, and George Shadid would specifically remember Rickey Hendon more than once telling Barack, "you ain't black enough." Hendon and Trotter's personal behavior toward Barack remained just as problematic as it had been two years earlier. "I was so bothered by the both of them, and I was extremely bothered with Senator Hendon because he's loud and he's obnoxious," Kim Lightford recalled. Barack's public criticism of his black colleagues upped the ante, and "Donne more than Rickey took real offense at that," well-connected lobbyist Larry Suffredin remembered. Everyone realized that Trotter was one of the Senate's smartest members, but his sharp tongue was never at a loss for new, insulting put-downs based on Barack's surname. By 1999 Donne was "always referring to him as 'Senator Oh-My-Mama, trying to tell us all what to do,'" Suffredin recounted.[67]

By the end of the spring 1999 session, Barack's disappointment with the Senate was manifest to almost everyone who knew him. "He just felt frustrated," poker buddy Larry Walsh knew. "Springfield was not challenging enough for him." Lobbyist and poker friend Dave Manning saw it similarly. "I think there was some frustration in having his family back home and having to come here all the time, often to accomplish very little" in a minority caucus that suffered under Pate Philip's rule. The two House members with whom Barack had the most contact felt likewise. Barack "began to feel quite frustrated at the inability to make things happen," Hyde Park's Barbara Flynn Currie recalled. Tom Dart, whose largely white House district encompassed much of Chicago's politically influential 19th Ward and who shared Barack's policy interests, used the same adjective as all the others. "I remember him just being frustrated with the fact that so much of it was just personality driven" rather than policy focused. "I remember distinctly those conversations," Dart explained. "Real frustration."

Barack voiced similar feelings to friends who had nothing to do with Springfield. Miner Barnhill colleague John Belcaster would use the same word as Barack's political friends. Barack felt "tremendous frustration" about "being in the minority party. . . . His frustration was palpable, and he would articulate it." Miner Barnhill consultant and occasional golf partner Whit Soule would recall Barack grousing about being "a potted plant" in a system where the lobbyists "were drafting the legislation" and "the party leaders basically cut all the deals." With Miner Barnhill's Paul Strauss Barack "talked about how corrupt it was," and writer friend Scott Turow would remember Barack calling and saying, "Man, I've got to get out of this place." Ed Wojcicki of *Illinois Issues* listened to Barack vent and remembered him saying: "There's really only five jobs I really want, now that I see what it's like. I want to be attorney general, or I want to be U.S. senator; I want to be president of the Senate." None of those were presently available, but to the Springfield denizens to whom he was closest, Barack had broached another possibility, one that *Sun-Times* political columnist Steve Neal had cited immediately after U.S. representative Bobby Rush's loss in the mayoral primary in February: challenging Rush for his congressional seat in the March 2000 Democratic primary.

Barack shared this with Dan Shomon and with Will Burns, who had joined the Senate Democratic staff six months earlier. Burns recalled that "part of it was wanting to get out of Springfield" combined with "a sense that Bobby was vulnerable" because everyone knew Rush had run a "horrible" mayoral campaign. But Barack was also considering that Donne Trotter might challenge Rush. Burns knew "there was a little weird competitive thing between him and Donne," one that had been building for two years now. Trotter was extremely well respected within the Senate, as much on the Republican side as on his own. Trotter's excellent working relationship with Republican budget expert Steve Rauschenberger was recognized by everyone in the capitol. "We could work with Donne," conservative Republican Ed Petka explained, and Trotter's unquestioned status as the best-dressed man in Illinois politics, with a proclivity for colorful bow ties, drew admiration too. Petka saw Trotter as "an African American Paul Simon," who also wore bow ties, and even Pate Philip saw Trotter as "a classy guy." Donne "was a cooperative guy, and you could trust him," Philip explained. "I always liked him."

No matter how handsome and smart Trotter was, the cosmopolitan roots and academic pedigree of his eleven-years-younger junior colleague had led Donne to view Barack with an antipathy that would take years to finally dissolve. Trotter's own undergraduate degree was from humble Chicago State University, and his disdain for everything the U of C represented manifested itself in his interactions with Barack's young acolyte Will Burns, who had now earned two degrees there. Burns recalled, "The first time I met Donne Trotter, he called me . . . a fucked-up motherfucker" because of Burns's ties to the U of C and to Barack. Trotter saw Rush's mayoral run as a sign that "maybe he doesn't want to be" in Congress, and in some quarters resentment still festered about how Rush had taken out Representative Charlie Hayes in the 1992 primary. Some of Trotter's friends thought "Bobby was just not being effective," and Trotter's political godfather, powerful Cook County Board president John Stroger, "didn't have any love for Bobby Rush, so I had that potential support behind me," Donne recalled, especially because Stroger's son Todd could then inherit Trotter's Senate seat. Black lobbyist and former state representative Paul Williams remembered "there was a feeling that Bobby was vulnerable," and Trotter believed that in a one-on-one matchup, all of the disenchantment with Rush could coalesce behind him. Trotter's best friend Rickey Hendon had seen how "from the very beginning, Barack demonstrated ambition," and with Barack now poised to deny Trotter a clean shot at Rush, Donne "clearly resented this new guy trying to jump over him."

As Barack pondered a congressional run, in June he went to Santa Fe for the seventh Saguaro Seminar session and then back to Chicago for meetings of the Chicago Annenberg board and the legislature's Joint Committee on Administrative Rules. Everyone involved in the Annenberg effort now understood that the four-year-old program was not producing dramatic improvements at most of the scores of schools it was assisting. As Northwestern professor Charles Payne told the board, "the moment CAC decided to work with and fund over 220 mostly lower tier schools, we knew there was not going to be a deep impact," in large part because data showed that recipient schools were getting an average of only $49,000, hardly 1 percent of most schools' annual budgets. The board agreed to adopt a "Breakthrough Schools" initiative whereby most of CAC's remaining funds would be concentrated on eighteen particularly strong schools.

At the end of June, Barack flew to New York for the first true board meeting of the incipient think tank that was now calling itself the Network for American Renewal. The advance concept paper envisioned programs addressing "Shared Prosperity and Human Well-Being" and "Renewing American Democracy for the 21st Century." Cofounders Charles Halpern and Stephen Heintz imagined this first gathering as "potentially an historic occasion," and Barack's fellow board members included former Colorado congressman David Skaggs and Public Allies cofounder Vanessa Kirsch. Barack told his colleagues that "part of the opportunity and challenge is to develop norms and values that unify us as Americans but are also inclusive of the norms and values brought by other cultures." He also "stressed that while the Network does not plan to focus on issues of race and culture with specific projects, we must address these issues in all our activities" and must also consider "how we include people who are not sharing in prosperity," because "whatever ails this country ails poor folks worse." One proposed project would focus on "Asset Building," such as "inner-city entrepreneurship." Barack noted that the "concept of 'ownership' is an extremely powerful American notion," and he "expressed special interest in the notion of community assets," again echoing John McKnight.

Another anticipated focus was "Democracy and Devolution," with Barack citing his own past experience in noting how "'motor voter' registration is not really at all significant. The problem is not voter registration but voter participation. How do you engage citizens in their government?" He complained about how in Illinois "hundreds of millions of dollars in tax breaks were approved without knowing what impacts they will have," and no matter what the group might initiate by way of "devolution" at the state level to anticipate "resistance on the part of legislators who like things the way they are." A key question was "Where do citizens collectively assert their rights and interests," which "is central to the current state of democracy"?[68]

Back in Chicago, on July 1 Barack made his first move toward a congressional race by agreeing to Dan Shomon's idea of conducting a small, discreet poll of 1st District voters to explore just how vulnerable Bobby Rush might be. Dan asked his three closest Springfield friends, Senate Democratic staffers John Charles and Adelmo Marchiori and former staffer Jeff Stauter, for their confidential help, explaining that "Barack is thinking about doing this." Adelmo knew that "the relationship between Barack and Shomon was the tightest on staff," and although they thought Barack "was going somewhere, we still questioned Shomon's bromance with Obama," Jeff recalled. "I had the software to do the poll," John explained, "some statistical sampling software that I'd picked up along the way" while polling for state Senate campaigns. Shomon bought "a list of phone numbers from a vendor" there in Springfield for $450, and "we put together this poll," Adelmo remembered. "We were completely on our own personal time," making sure no word of their project seeped back to chief of staff Courtney Nottage. Working out of Stauter's lobbyist office, they "spent night after night calling" residents of the 1st Congressional District, which ran from 26th Street down to 103rd Street on Chicago's South Side, then westward through Englewood to encompass predominantly white Beverly and parts of suburban Evergreen Park, Oak Lawn, Alsip, and Blue Island. "You test your opponent's negatives," and "we would have done a favorability of Barack and Rush," they remembered. Overall they completed about three hundred calls, four young white men asking respondents who were more than 75 percent African American to voice their feelings about former Black Panther leader Rush. The bottom-line question was "Is it doable? Is it worth taking the shot?" and John Charles recalled that after a week's worth of phone calls, "our numbers showed that it was possible."

Barack was encouraged by the amateur poll, but the tallest barrier to launching a challenge was right at home: Michelle. "With every political run that Barack made, my instinct was to talk him out of it," she later recounted. Malia was just celebrating her first birthday, and with Barack in Springfield for so much of the spring, and then out so many evenings at political events like a "Second Thursday" DL21C speaking appearance, Michelle "was in a lot of ways a single mom, and that was not her plan," her close friend Susan Sher explained. This campaign would be far more demanding than either of Barack's state Senate wins, making an already bad situation far worse, no matter what the outcome, and Michelle told Barack she was opposed to it.

As Barack surveyed his closest political friends, he heard far more opposition than support. Carol Harwell declined to be involved, and Terry Link and Denny Jacobs each said such a challenge would be "foolish" and "stupid." So did Mike Lieteau. "I told him he shouldn't run," Mike said, and warned Barack that even he would support Rush. "Bobby was a personal friend of mine" and "was very supportive of me personally and very supportive of telecom issues, so I had no choice." Politically savvy lawyers John Schmidt, Matt Piers, and Newton Minow said the same. "I thought it was a terrible idea," Schmidt recalled, and Piers remembers, "I thought it was crazy." Minow was even blunter: "Are you nuts? Barack, I think this is a mistake." African American acquaintances responded similarly. Sokoni Karanja had long thought Barack eventually would run for mayor but felt this was "a *bad* move. I said, 'Barack, what are you doing? You can't beat Bobby'" and "'folks are going to react badly'" to him trying. Lobbyist and former state representative Paul Williams told Barack, "you aren't the guy for the blackest district in the country," and former Harold Washington aide Tim Wright, who had been Rush's chief of staff before a recent falling-out, advised Barack that "This isn't the race, this isn't the time." When Barack asked 6th Ward alderman Freddrenna Lyle to lunch and revealed his plan, "I explained that I thought the seniors in my area remembered Bobby when he was with the Panthers, and they had a connection to Bobby that I thought was unbeatable," irrespective of how poor a mayoral race Rush had run or how good a state legislator Barack was.

Barack was encouraged by a *Chicago Tribune* Sunday magazine story in which well-known black businessman Dempsey J. Travis, once a top supporter of Harold Washington, said that Barack and Jesse Jackson Jr. were Washington's successors: "there are people we have coming up who have shown they have the class, the intellect and the people skills to do exactly what Harold did—strong people who are bright, sophisticated, sharp. Razor sharp," Travis stressed. "Harold lives on in these people." Barack also took pleasure from a *Hyde Park Herald* interview with him headlined "Hyde Park's Own Renaissance Man." He made no mention of the congressional possibility but cited three top local challenges: "the continuing under-performance of the public schools," whether residents feel "confident that they can raise and educate their children here in safety," and how to "foster economic development and economic integration . . . to include the larger South Side," because Hyde Park was in danger of becoming "an island of prosperity within a sea of poverty."

As Barack continued his private conversations about a congressional run, he did receive some positive feedback. State representative Tom Dart, whose district was part of Rush's, encouraged Barack to run. So did both 4th Ward alderman Toni Preckwinkle, whom Rush had opposed when she defeated two primary challengers back in February, and, more surprisingly, new 5th Ward alderman Leslie Hairston. Barack had not supported Hairston when she took on incumbent Barbara Holt, and Hairston also knew how other Side South politicians disliked Barack because of the perception that "you think you're better than us." A $2,000 contribution from Barack's campaign account to Hairston's was a thank-you for her support.

A trio of independent-minded young black professionals also responded positively. Thirty-two-year-old Tim King was a South Side native and Georgetown University law graduate who in 1995 had become president of a struggling Roman Catholic high school, Hales Franciscan, after finding teaching far more fulfilling than the law. King's work at Hales won rave reviews, with a laudatory *Chicago Reader* profile detailing his unceasing commitment and a *Chicago Tribune* editorial praising him for taking in an orphaned sixteen-year-old who then graduated from Hales and headed to college. Barack had visited Hales to introduce himself, and the two had "hit it off" and would see each other at the East Bank Club and in Hyde Park. When Barack told Tim, "I'm thinking about running for Congress," Tim echoed the thought: "Really? *I'm* thinking about running for Congress!" Barack "looked at me very seriously" and said, "If we both run, there's no chance either one of us will be able to win." Tim concurred: "You're absolutely right, and I defer to you."

Tim agreed to be an active supporter, as did Barack's friend Jim Reynolds, whose thriving Loop Capital firm now employed Barack's brother-in-law Craig Robinson. "Jim, I'm thinking about running for Congress," Barack remarked one summer day while they played golf, and Reynolds eagerly signed on. So did a brand-new acquaintance, Judy Byrd, another Georgetown University law graduate who had met Barack years earlier while working in Chicago city government under Howard Stanback. Like Reynolds, Byrd's friends included a number of well-heeled Chicagoans, and she offered to set up a lunch to introduce Barack to four or five friends who were "all women of some means."

Neither Barack nor Donne Trotter told the other about their plans, but each one spoke privately to Emil Jones, who remembered that Barack "told me that he might run for Congress, so I said, 'I think you'd be good, go ahead and run.' I said, 'You are aware that Donne Trotter's running as well, and that being the case, I cannot be involved . . . I cannot choose between my membership.'" Trotter had the same conversation with Jones. "Emil in my mind was neutral," Donne explained. "His whole thing is he didn't want us using his staff to fight," as Barack already surreptitiously was, and word went out from Jones that no staffers could take leave to work on either campaign: any departure would be permanent.

Thanks to his lobbyist friend Vince Williams, who knew firsthand that "the mayor was very interested in getting to know Barack," Obama also called on Richard Daley, who had badly thrashed Rush just five months earlier. Daley seemed friendly and receptive, reinforcing Barack's assumption that Daley would quietly welcome a challenge to someone who had challenged him. Barack mentioned his meeting with Daley to Madeline Talbott, who along with new 15th Ward alderman Ted Thomas was more than willing to have ACORN back Barack. "Both Barack and I felt like Bobby was very vulnerable," Madeline recalled, even though "no one knew who Barack was." Barack also mentioned his meeting with Daley to former judge and House member Ab Mikva, who believed "Rush was no great shakes as a congressman" and encouraged Barack to challenge him. But Mikva also tried to dissuade Barack from misunderstanding Daley's stance. Barack "thought that he would get Mayor Daley's support," telling Mikva that "at the end . . . the mayor stood up and said, 'Well, good luck to you.' And Barack said, 'Well, I read that that maybe he's open.' And I said, 'No, it's closed, because that's what the old man'—Daley's father—'used to say, "Good luck to you, fella," and that meant that you were on your own.'"

Barack had a similar exchange with John Kelly, a young white Irish lobbyist to whom Emil Jones had taken a liking and who had important family connections in Chicago politics through his uncle, powerful Cook County assessor Jim Houlihan. Kelly was just getting immersed in the world of Springfield, and although Barack "seemed very politically calculating," he was not "a mainstream player in Springfield . . . because he wouldn't go out at night." Kelly thought the congressional race "was a natural progression for him" when Barack mentioned it, asking for his support. But during a summer golf outing at Ravisloe Country Club, in south suburban Homewood, Barack told Kelly that "The key thing is where the mayor's going to be." Kelly replied, "'It's very clear where the mayor's going to be: the mayor's going to be with Bobby Rush.' And he said, 'Why are you so hell-bent on the fact that he's going to support Bobby Rush?' I said, ''Cause it's kind of Politics 101, Barack. He beat Bobby Rush, and he knows that he can beat him again, and he'd rather have him there than you there.' He said 'Why?' And I go, 'Because he doesn't know if he can beat you. Bobby Rush is a known quantity: they know that they can beat him. They don't know where you're going to end up.'" Barack "looked at me kind of like it was shocking to him," and Kelly realized that Barack's "huge" political naiveté stemmed from him not being *from* Chicago.

Chuy Garcia, a 1998 machine victim, said that Daley "had seen Bobby's full potential in the challenge against him," and "didn't see him as a larger threat." But Daley "saw a lot more crossover potential" with Barack, whose "credentials were pretty impressive, so the potential for trouble was much bigger." As Kelly had told Barack, it indeed was Politics 101: "Why should a sitting mayor help someone who can eclipse him down the road: in debate, in credentials, in being an alternative to the mayor?"[69]

One benefit of the huge Illinois FIRST appropriations that Barack had reluctantly backed was the ability it gave senators to direct sizable grants of state funds to quietly chosen recipients, and over the summer, Barack completed the paperwork to channel $462,000 to seven fortunate organizations. He directed almost half of that, a whopping $200,000, to the Citizenship Education Fund's LaSalle Street Project, to assist in "setting up [a] venture capital business for underserved areas." The fund, a nonprofit arm of Rev. Jesse Jackson Sr.'s Rainbow/PUSH, had initiated the project a year earlier as a spinoff of its two-year-old Wall Street Project, which was aimed at drawing investment dollars to poor neighborhoods. Taking its name from Chicago's premier banking street, it was supported by Jim Reynolds as well as the Chicago Urban League. *Crain's Chicago Business* labeled the project "far more of a publicity tool than a power broker" in a story highlighting how local heavyweights were backing the highly similar Social Investment Forum of Chicago. Karin Stanford, the fund's executive director, had just resigned and given birth to a child, and two years later the LaSalle Street Project would be shuttered after news that Jackson Sr. had fathered Stanford's daughter.

Barack also funneled $100,000 to the South Side YMCA for construction of a child development center and $50,000 apiece to two youth service programs for teen mentoring efforts. A trio of smaller grants, including one for a senior day care van and $20,000 for South Shore's Black United Fund, rounded out Barack's summer gift list. In mid-July the weekly *Hyde Park Herald* reported that both Donne Trotter and Barack were considering challenges to Rush. Trotter said "nothing is definite" but he was "leaning in that direction." Local activist Alan Dobry stated that "the congressman is vulnerable," and Trotter said that "my question is what has he done. I think he's a great individual, but if your friends aren't doing the job you expect them to, you don't just leave them there." Barack told the *Herald*, "I haven't made a decision yet" although "it will need to be fairly soon."

By July 25 Barack finally had Michelle's reluctant acquiescence, and he began contacting his oldest and closest friends, asking them to make the maximum $1,000-per-person contribution. Hasan Chandoo immediately wrote a check, as did Gerry Alexis, Rob Fisher, Jeff Cummings, Jean Rudd, Bettylu Saltzman, and Michelle's friend Cindy Moelis. Barack already had $1,000 apiece from Judd Miner and his young friend Robert Blackwell, and before the end of July, Debbie Leff, Douglas Baird, and Newton Minow gave as well. On July 28 Barack mailed both an official "Statement of Candidacy" and a "Statement of Organization" for Obama for Congress 2000 to the Federal Election Commission. Jean Rudd's husband Lionel Bolin was again listed as the campaign's official treasurer, with Barack and Michelle's friend Craig Huffman joining sister-in-law Janis Robinson as two assistant treasurers. Cynthia Miller had just returned to Chicago from abroad, and she plus Al Kindle, 4th Ward alderman Toni Preckwinkle's top political worker, and Amy Szarek, a young fund-raiser who had worked on Carol Moseley Braun's unsuccessful 1998 race, started pulling things together as Dan Shomon and Will Burns, Barack's two most dedicated aides, began planning to leave Senate staff to take charge of the incipient campaign.

On August 1 the *Chicago Tribune* and the *Chicago Sun-Times* caught up to the *Herald*'s report from two weeks earlier. Barack said he was "seriously considering" a race and expected "to make a final decision" in "the next couple of weeks." Four days later the *Defender* followed up, with Barack saying he had been discussing "a possible candidacy" for "the last two months" and would continue "seriously considering" a run "for the next several weeks." Both Toni Preckwinkle and local state representative Barbara Flynn Currie felt strongly that Barack needed to get started as soon as possible if he was going to challenge a well-known incumbent, but ten days into August Barack told the *Herald*, "I will be making a decision before Labor Day." He explained that "it's important for me to discuss the possibility of my candidacy with a wide range of community leaders" and "that process has not yet been completed." The next day he told the *Sun-Times*' Lynn Sweet that "at this point, it looks like we are going to go ahead with this," and Donne Trotter told her that he too was "more than likely" to run.

That same day, August 11, the campaign rented the first of two offices, on 87th Street near Cottage Grove Avenue, arranged phone service, and issued its first payroll checks to Will Burns and Amy Szarek. On Saturday Al Kindle took Barack to march in black Bronzeville's biggest annual event, the *Defender*-sponsored Bud Billiken Day parade, which featured marching band and drill team competitions as well as a full array of South Side politicians. "No one knew who he was," Kindle recounted. "People were asking 'Who is he?'" and Barack "seemed a little embarrassed," Kindle remembered. "You could see how humbling it was in his face." Capitol Hill's *Roll Call* included Bobby Rush in a story on endangered incumbents, saying Rush "has been viewed as vulnerable to a Democratic threat" since his mayoral thrashing, but the *Sun-Times*' Steve Neal warned that Rush "has a solid base of about 40 percent of the vote." Neal predicted that Barack could outpace Rush in fund-raising, collecting at least $500,000 for the race, and the mayoral disaster had shown that Rush's organization "is relatively weak." Donne Trotter "will attempt to portray Obama as a newcomer to the South Side," and a black weekly warned that Barack "is likely to get some unwanted opposition from friends of former state Sen. Alice Palmer," who "has not forgiven Obama."

Neal believed that "Trotter and Obama are genuine threats" to Rush, but former senator Paul Simon was far more cautioning when he had breakfast with Barack one mid-August morning. Simon emphasized that Barack must win public endorsements from prominent African Americans in order to credibly challenge Rush, but Barack had already been told by senior figures like former attorney general Roland Burris that they would not support him. "My instinct is that you will be portrayed as 'the white man's candidate' for this seat," Simon warned in a prescient thank-you letter, "virtually portraying you as an Uncle Tom."[70]

Barack turned his attention toward another round of fund-raising calls, with $1,000 checks coming in from his uncle Charles Payne, Davis Miner colleague Paul Strauss, Sidley & Austin's Eden Martin, Jim Reynolds and his wife Sandra, John Rogers Jr., Pat Graham, and Tony Rezko and his wife Rita. A second and larger campaign office was secured just south of 95th Street on Western Avenue, and by the end of August Cynthia Miller was on the payroll, soon to be followed by campaign manager Dan Shomon. In the world of Springfield, it "was a big deal" when Shomon and Will Burns gave up their Democratic caucus jobs to sign on with Barack, with some colleagues such as Cindy Huebner wondering "why in the hell is Barack doing this?" Paul Simon's sidekick Mike

Lawrence reacted similarly when Barack checked in with him by phone. "Why are you doing that?" Mike asked. "I don't think it's a good idea. I think your future is statewide," not South Side.

Eager for press attention, Barack capitalized on a big mid-August power failure to propose toughening a state law that failed to require utilities to reimburse customers for damages unless an outage lasted more than four hours and losses totaled at least $30,000. Barack devoted a *Hyde Park Herald* column, a *Chicago Tribune* op-ed, and a letter to the editor of the *Chicago Defender* to the problem, his biggest outreach effort to date. He, Shomon, and deputy campaign manager Will Burns began recruiting friends and acquaintances like young organizer John Eason and Democratic activist John Corrigan to devote time to the campaign. Like Will, his good friend Eason thought the race was actually about more than a congressional seat. "You just wanted this guy to go as far as he could, and you wanted to do whatever you could to get him to go as far as he could, and ultimately I think what we're all thinking is 'Oh, we can get another Harold Washington. We'll get him in as mayor,' and even he talked about that." Of course "nobody would say anything publicly against Daley," but "it was the same trajectory as Harold: you run for Congress and you get attention on the national stage and that's just so you can run for mayor. . . . You want to be mayor of Chicago," and Barack "knows that's what I want to hear." But underlying everything, Eason realized, was Barack's intense aversion to Springfield. "I remember him saying he needed to get out of Springfield desperately—he desperately needed to leave Springfield."

Barack also drew public attention, even on the *Defender*'s front page, by decrying how Governor Ryan in mid-August had vetoed his bill mandating that at least two-thirds of child support payments actually go to the children. "One of my biggest concerns in the legislature has been helping people move out of poverty," and HB 1232 was "a pretty common sense" effort in that direction. Barack vowed to seek an override come mid-November's veto session, but a Saturday town hall meeting at a Baptist church in Englewood attracted just twenty people. Barack called for a moratorium on capital punishment and tried to educate the tiny crowd by pointing out how "some of the biggest contributors to Governor Ryan's" and even "Emil Jones's campaigns are gaming interests."

Barack devoted much of mid-September to another burst of fund-raising calls, with $1,000 checks coming in from his brother-in-law Craig Robinson's good friend Marty Nesbitt as well as Nesbitt's boss, wealthy real estate heir Penny Pritzker, John Rogers's Ariel colleague Mellody Hobson, basketball buddy Eric Whitaker, law firm colleagues George Galland and Chuck Barnhill, real estate manager William Moorehead, and Harvard Law School professors Martha Minow, Larry Tribe, and Randall Kennedy as well as new supporter Judy Byrd. Barack prepared to step down as chairman of the Chicago Annenberg board, and in midsummer Howard Stanback had already succeeded him as chairman of the Woods Fund's board, while at the same time Bill Ayers joined the board. On Tuesday, September 21, the campaign reserved a Palmer House ballroom in the downtown Loop for a Sunday-afternoon announcement ceremony five days later, and the next morning Steve Neal reported that Rush was gearing up for a reelection campaign. Former representative Gus Savage's longtime chief of staff, Louanner Peters, would manage it, and while Citizens for Rush had only $94,000 in the bank, Rush had just put his wife Carolyn on its payroll. Rush had met Barack briefly seven years earlier during Project VOTE! and remembered him as "a very *persuasive* young man" with "a sunny disposition" and "a certain charisma about him." That September, reaching out first to political science Ph.D. and Citizen Action veteran Don Wiener and then to political consultant Eric Adelstein, Rush wanted Wiener to research Obama's record. "I don't know much about him. He seems like kind of an Oreo to me."

As of August, Barack believed he had veteran political strategist David Axelrod and his longtime partner John Kupper lined up to be his primary media consultants, but that quickly came unglued. "We actually started doing the race," Kupper recalled, and had "started advising him," but then "we started getting calls saying 'You cannot work against Bobby Rush. If you do, you'll never work for another Democrat again,'" Kupper recounted. "There was pressure from the DNC [Democratic National Committee] and from Washington that caused us to get out of the race," and a direct-mail firm, Crounse Malchow, likewise backed away from Barack. Other well-known Chicagoland consultants like Pete Giangreco and Delmarie Cobb cited their past ties to Rush in declining to get involved, and Axelrod told Dan Shomon to call Chris Sautter, his D.C.-based former partner. Sautter flew to Chicago to have breakfast with Barack and Dan and accepted the handoff from Axelrod and Kupper.

Neal reported that well-known figures like John Schmidt and Newton Minow and prominent black financiers John Rogers and Peter Bynoe were supporting Barack, but warning signs loomed. Even though 17th Ward alderman Terry Peterson, like Schmidt a former aide to Richard Daley, had joined his City Council colleagues Preckwinkle, Hairston, and Thomas in backing Barack, Neal warned that the mayor "might have reason for concern if Obama should emerge as a potential challenger in the 2003 mayoral election. That's why Daley may be neutral for Rush." Neal also noted that even though Jesse Jackson Jr. "is friendly with Obama," he "doesn't want competition as the rising star of the Illinois delegation," and Jackson was "committed to Rush." If Barack won, both youthful Illinois "comptroller Dan Hynes and Rep. Jackson might get eclipsed," and so neither of their fathers, powerful former county assessor Tom Hynes and Jesse Jackson Sr., "want Obama to become the city's rising political star."[71]

Sunday's *Chicago Tribune* called Barack "one of the bright young stars of the local African-American political world" and reported that his campaign "claims"—a bit optimistically—"to have raised more than $100,000" already. That afternoon, an hour before Barack's event at the downtown Palmer House, Rickey Hendon introduced Donne Trotter to several hundred people at Trotter's South Side district office. Trotter had the support of two aldermen, but he did *not* have the support of his longtime mentor, 8th Ward committeeman and Cook County Board president John Stroger, who already had endorsed Rush. Barack's announcement drew a crowd of three hundred, with many of them laughing when Barack used a famous local phrase to declare that "Nobody sent me. I'm not part of some long-standing political organization. I have no fancy sponsors. I'm not even from Chicago. My name is Obama. Despite that fact, somebody sent me . . . the men on the corner in Woodlawn drowning their sorrows in alcohol . . . the women working two jobs . . . They're all telling me we can't wait." Directly attacking Rush, Barack said that "for the past seven years we've been looking for leadership, and it has not been forthcoming. We can't afford to wait another seven years." District residents "deserve to have a congressman who can make a real difference in their lives, and in the life of our community." Local television cameras captured Barack saying that leadership means "framing the debate on the national issues that affect us deeply: gun violence, economic development in the inner city, health care for the uninsured." He warned that the district "is one of the most economically distressed areas in the nation. Our schools still don't make the grade, and our kids are being cheated out of the opportunity to prepare themselves for a high-tech, highly competitive future. Too many of our youth have been lost to the streets. The culture of violence threatens our neighborhoods." Rush "represents a politics that is rooted in the past, a reactive politics that isn't very good at coming up with concrete solutions."

WLS Channel 7's evening newscast told viewers that Trotter and Obama were "serious competition" for Rush and that the primary "could be a close one." Barack was described as a "civil rights attorney and the first black president of the *Harvard Law Review*," while Trotter was called a "Springfield veteran." Viewers were reminded of Rush's "landslide" loss against Daley, and that he "may be vulnerable." Barack told one print reporter, "I don't believe Congressman Rush's voter base is firm," and he dismissed Trotter's candidacy. "I'm not concerned about who else is in the race. By January we'll know who's serious and who's not." Monday's newspapers paid little attention to the dueling announcement ceremonies, but when the *Sun-Times* followed up two days later with a serious look at the race, Barack said "the reason I'm running is not because he did poorly in the mayoral race." Instead, "Congressman Rush's record articulating an agenda for community building has been lacking . . . I think the mayor's race has just underscored his inability to shape that proactive agenda that voters have been looking for." Barack promised "not only a strong message but a lot of energy in our campaign," and declared that "part of what we are talking about is a transition from a politics of protest to a politics of progress."

David Axelrod had walked away from Barack's campaign after the threatening phone calls from Washington, but he was happy to tell the *Hyde Park Herald* what he thought of the race. Increasing Barack's name recognition throughout the entire congressional district should be the campaign's top priority. "It's a challenging task against a long-term incumbent who has universal name recognition. The one thing Barack may have is the ability to raise money and that obviously affects his name recognition." Overall, "I think it's a competitive race," Axelrod said, "but I don't think you should ever underestimate the power of incumbency."[72]

Chapter Eight

FAILURE AND RECOVERY

CHICAGO AND SPRINGFIELD

OCTOBER 1999–JANUARY 2003

Barack's declaration of candidacy attracted little attention, and the next day was an immediate reminder that not until early December could he be a full-time candidate, for autumn classes at the U of C Law School got under way on Monday, September 27. While Voting Rights had been removed from early 2000's winter quarter schedule, in the meantime Barack was again teaching Con Law III and Current Issues in Racism and the Law. Con Law met three mornings a week at 8:30 A.M. and drew thirty-five students, and on the first day, Barack said he would have to reschedule several classes in November because of the veto session. Nate Sutton, a 3L, remembered being surprised by how quickly Barack learned students' names, and it soon became clear that Barack "really had a genuine ability to argue and understand both sides of issues" and "empathize or identify with the bases for arguments on both sides." Barack knew the cases "extraordinarily well," but even on issues like abortion "you would not know what his position was." In the race class, which attracted thirty-two students, 3L Joe Khan also realized how "very approachable" Barack seemed. Both courses remained unchanged from prior years, and 3L Andrew Abrams was impressed by how "enthusiastic and supportive of students" Barack was. When the quarter ended, the two classes gave Barack 8.92 and 8.74 on UCLS's 10-point scale, but a brand-new query was even more instructive. "Would you take a course with this professor again?" with three possible answers: yes, no, and "no influence." Barack's numbers were impressive indeed: a total of forty-three yeses, ten no-influences, and only a single no, just a fraction shy of 80 percent affirmative.

In the political realm, several significant people expressed displeasure about Barack entering the congressional race. Bobby Rush's unhappiness was to be expected. "I was shocked, disappointed, and, you know, frankly somewhat angry," Rush later recalled. "If there are ten issues, we agree on 9.5" of them. "I didn't see the rationale for it." Media consultant Eric Adelstein lined up Washington pollster Diane Feldman to work on Rush's campaign, and research guru Don Wiener began looking into Barack's Springfield record and perusing back issues of the *Hyde Park Herald*. By early October, Rush finally opened a campaign office, but "I don't think Bobby was taking it seriously," field consultant Jerry Morrison explained. Another month would pass before campaign manager Louanner Peters came aboard and spokesperson Maudlyne Ihejirika was hired.

With Cook County Board president and 8th Ward powerhouse John Stroger backing Rush, Donne Trotter's campaign was largely adrift, but Trotter hoped to raise sufficient funds to mount a credible effort, as much against his rival Obama as against Rush. But Barack's biggest problem was at home, because even though Michelle had said yes to the race, once it got started, her unhappiness was visible and sometimes audible. "What I notice about men, all men, is that their order is me, my family," etc., "but me is first," while "for women, me is fourth," Michelle observed. Barack admitted in one joint interview that "we've had our rough patches," and Michelle immediately added, "there were many." Their unceasing financial strains and the need for child care were constant factors. "They had a lot of student loans, and there was a lot of tension around that. Barack's choices made her life harder," Michelle's friend Cindy Moelis recalled. "There was a lot of tension and stress," Michelle agreed, and asked in retrospect whether she was both angry and lonely because of how much time Barack spent on politics, Michelle responded forcefully: "absolutely . . . absolutely." Friends like Kathy Stell knew that Michelle had "an ambivalence about politics in general" that showed in the "skepticism and cynicism" she felt about politicians. Dan Shomon, now Barack's full-time campaign manager, regularly witnessed Michelle's disdain that her super-talented husband was "lowering himself" to electoral politics. Barack and Michelle had dinner every so often with Allison Davis, Bob Bennett, and their wives, and Allison remembered that sometimes "things were unpleasant" because "Michelle was really pissed about the Bobby Rush campaign." That attitude also manifested itself whenever Michelle, with or without Barack, joined Bernardine Dohrn, Bill Ayers, and Mona and Rashid Khalidi at one of those couples' dinners. Being a politician's wife was not something Michelle welcomed, and "I remember lots of conversations" during which Michelle was "really resisting it and fighting, fighting it," Bernardine recalled. Michelle "hates it . . . it's not the life she imagined, not the life she'd prefer." Craig Huffman, Michelle's young protégé, knew that Barack's political life "created enormous tension" at home. Craig remembered calling Barack one fall Saturday to discuss the campaign. "Something had happened, and Michelle, I could hear her in the background. She wasn't happy with him, I'll leave it at that," he recounted. "Hey man, I need to go," Barack apologized as he hung up. Like Kathy Stell and Bernardine Dohrn, Craig realized Michelle was not "enamored with politicians" and was not "the willing political wife." That tension left Barack "chasing something that your partner's not 100 percent on board with."[1]

Eager to draw attention to his campaign, Barack challenged Rush to five debates, just as Rush had done with Mayor Daley. The *Tribune* and the *Defender* reported Barack's challenge, but the incumbent ignored him, just as Daley had Rush. "My assumption was that the race would generate a lot of coverage and I could really focus on making arguments about what I would like to do in Congress," Barack explained, but with the election five months away, Barack realized that "people really weren't paying attention that much," and after the two stories about his debate challenge, more than two weeks passed before any newspaper again mentioned Barack.

Thanks largely to Dan Shomon, as well as many former and present U of C law students, Barack's campaign did draw a good number of weekend volunteers. With Will Burns as deputy campaign manager and Cynthia Miller as Barack's scheduler staffing an office at 87th Street and Cottage Grove, and Shomon and fund-raiser Amy Szarek manning one at 95th and Western Avenue, leafleting and canvassing began. One weekend media consultant Chris Sautter joined Barack, Dan, and 4th Ward alderman Toni Preckwinkle on a daylong series of stops at neighborhood events. Preckwinkle took the lead in introducing Barack, but Sautter realized that "nobody knew who he was." Later in the afternoon, Chris and Barack ended up playing pickup basketball with some teenagers who had no clue that Barack was a congressional candidate. It was "a relaxing day," but even the pickup game underscored that "nobody knew Barack."

Barack called friends and lobbyists to ask for the maximum $1,000-per-person contribution. "You really think you can win?" Scott Turow asked. Barack insisted that Rush was unpopular and vulnerable, and Scott reluctantly wrote a check. City of Chicago lobbyist Bill Luking gave Donne Trotter a $1,000 check at the end of September and soon learned that Barack was keeping an eye on the disclosure reports detailing contributions to his opponents. "Within days, Barack tracked me down and said, 'I trust you will be helpful to all your friends running for Congress!" Luking was surprised by how "very aggressive" Barack was, and some months later he also gave $1,000 to Barack. Jim Reynolds had introduced African American insurance executive Les Coney to Barack, and Coney agreed to contribute, but with young children at home, Les was less than eager to write a $1,000 check. Barack "kept following up to say 'Hey, where's that check, Les?'" and at the end of October, Coney handed it over. By then Barack had also garnered $1,000 apiece from friends like Bernard Loyd and Matt Piers, brother-in-law Craig Robinson, black financier Quintin Primo, black attorney Peter Bynoe and his wife, lobbyist Larry Suffredin, law school colleagues David Strauss and Lawrence Lessig and Lessig's spouse, plus old *Law Review* colleague Julius Genachowski. Newt Minow prevailed upon his friend John Bryan, CEO of Sara Lee Corp., to write a check, and Valerie

Jarrett's boss Daniel Levin gave $1,000 as well. School principal Alma Jones, who had followed Barack's career ever since mentoring him in Altgeld Gardens twelve years earlier, contributed $300.

Keeping up with his other obligations, Barack spoke at a mid-October forum on welfare reform and women of color, where he told listeners how "one of the things I've seen in Springfield is that budget concerns override a lot of moral issues or ethical issues or the desire for consistency." Addressing a sympathetic crowd, Barack explained that "any program dealing with welfare is going to have a disproportionate impact on people of color because we are disproportionately represented. Any failure in the health care system to provide quality health care is going to disproportionately affect people of color because we are disproportionately in need of health care." Barack stressed that "a black face or a brown face is painted on poverty in this country," and as a result "a lot of the policies that are being implemented are targeted towards poor people, and it's politically useful to feed into racist stereotypes to implement the policy." Barack emphasized the importance of distinguishing "between those policies that help women, families, children, teenagers make intelligent choices in their lives and are thereby empowering, versus those policies that are essentially punitive and try to restrict choices." For example, "I can envision policies and programs that offer abstinence as a choice to young women that are not punitive in attempting to reinforce oppressive notions of gender hierarchy because they also target young men and their irresponsibility in terms of how they think about sex." Barack added that "the instinct of a lot of our constituencies out in the neighborhoods is that there is a problem that people take sexuality or child-bearing too lightly and don't consider this a sacred endeavor." When anyone under fifteen years old is sexually active, "the average person thinks that's a problem. We should make sure that we don't just leave that discussion of values or choices to the right wing, in which case they are able to feed on people's general notions to . . . mobilize a political constituency for themselves." He concluded by warning that when progressives oppose abstinence programs, "we get painted into a corner as if we are pro–irresponsible sexuality, pro–teenage pregnancy, and so forth. I think that's always a difficulty."

At JCAR's monthly meeting, Barack challenged a Department of Human Services (DHS) rule change that he believed would harm thirteen families receiving public assistance. DHS's representative said they would happily discuss the details with Barack, who alone among his colleagues voted against the new rule. An effort like that drew no attention, but Barack's campaign staff tried to put him in the public eye. "We had this free cable TV strategy," Dan Shomon explained, to "get Barack on every possible free cable TV" public access program, of which several were aimed at African Americans. On one, with host Joe Green, Barack explained at some length his rationale for challenging Bobby

Rush. In Congress, he, unlike Rush, would provide three types of leadership. First, "leadership at the national as well as local levels in terms of framing the debate and setting an agenda, a specific agenda that people can see, that's clearly articulated." Second, he would offer "leadership in organizing resources in the community." And third, "we should expect out of our political leadership some moral leadership. I think that we need to talk about some of the attitudinal changes that we need to undergo within our communities," including the need to "embrace a notion that academic excellence is important."

Rush's campaign finally got going, and he attacked the Chicago Housing Authority's plan to tear down its infamous high-rises, while Barack was implicitly supportive. "A long-term strategy that involves the creation of mixed-income communities is preferable to the status quo," Barack told the *Hyde Park Herald*, but he knew it would "require a solid commitment from every level of government." The Rush campaign also castigated Barack for holding his announcement ceremony downtown rather than in the district. State representative Lou Jones, who had loathed Barack since their Alice Palmer face-off, proclaimed that Barack "did not know, understand, value or respect the people nor the institutions within the district enough to locate an announcement site among the constituents he says he wants to represent." Barack's campaign responded that "you always start your campaign where you can get the most and best press coverage," and one bimonthly black newspaper commented that Barack "made a good impression at his announcement" and "has been having a strong presence at community events." The *South Street Journal* added that some "still question his loyalty to the black community" because "Obama did not support Rush in the mayoral race" against Richard Daley, but it quoted Aldermen Leslie Hairston and Ted Thomas as praising Barack's record in Springfield.[2]

By mid-October both campaigns scheduled professional polls to see where the race stood with voters, but on October 18 the race took a sudden and tragic turn. At 8:30 P.M., two gunmen pretending to be undercover cops shot and badly wounded twenty-nine-year-old Huey Rich, Rush's out-of-wedlock son who bore his mother's surname and had been raised by his grandmother, a member of Trinity Church. Rich suffered massive blood loss, lost consciousness, and underwent hours of surgery. Chicago newspapers gave the shooting front-page coverage as Rich remained in critical condition for more than seventy-two hours. The young man's kidneys, lungs, and heart were all failing, and the attending physician told Rev. Jeremiah Wright "there is no brain activity." Wright told Rush that Huey was brain dead, and he died without ever regaining consciousness. Newspapers gave front-page headlines to the death, and Rush, speaking at Jesse Jackson's weekly Rainbow/PUSH rally, declared that "we've got to rid our communities of guns . . . guns don't belong in a civilized society."

As police detained one suspect and identified a second, reports emerged that Rich's involvement in drug trafficking had led to the attack. Rush asked Jackson to officiate at the funeral at Trinity, but Wright, fearing the church would be used for a campaign rally, told Rush that he, not Jackson, would preside. Wright was dismayed to see reporters and television in attendance, and Huey's grandmother told Wright to bar the press from the service. "Bobby's not going to make my grandson's death a show and a spectacle. He wasn't there for this kid, and I'm not having him playing the role of grieving father," Wright recounted her telling him. Mayor Daley and former senator Carol Moseley Braun were among the dignitaries who attended, and several years later both killers were convicted and sent to prison, with triggerman Leo Foster sentenced essentially to life without parole.[3]

After Huey Rich's death, the *Chicago Defender* ran a story headlined "Obama Asked to Bow Out of Race Against Rush" after one South Side pastor backing Rush called for Barack to end his campaign. Privately, several savvy observers sympathetic to Barack had the same thought. John Kupper, David Axelrod's partner who had expected to work for Barack, realized that Rush had been recast as a figure of widespread sympathy as opposed to a loser of a mayoral race. "Barack ought to get out," Kupper told Chris Sautter, and Sautter did not disagree, for Huey's death had "washed away any bad feelings that people seemed to have about Bobby." He appreciated Kupper's view that now the race was "completely unwinnable." But with $18,000 spent on polling, and Sautter just two weeks from starting radio advertising and direct mail on Barack's behalf, neither Dan Shomon nor Barack seriously considered standing down.

Four days after Huey's funeral, Bobby Rush formally announced his candidacy for reelection. Cook County Board president John Stroger, state representative Lou Jones, Aldermen Freddrenna Lyle and Arenda Troutman, and former state senator Alice Palmer were among the 150 people in attendance. Lyle, who had a collegial relationship with Barack, told reporters that Obama and Donne Trotter "are fine gentlemen who've down outstanding jobs in Springfield" while stressing that "we shouldn't cannibalize each other." Barack told the *Chicago Tribune* that "an elected official's vision is more important than his seniority" and noted that Rush had challenged and defeated incumbent representative Charlie Hayes eight years earlier. Rush had said "he would bring more energy to the office, and that's what I intend to do."

Although it was not publicly noted at the time, the day after Rush's announcement Barack's campaign received five $1,000 contributions from private citizens who shared the same employer—FORUM, or Fulfilling Our Responsibility Unto Mankind, the South Shore social services agency run by Yesse Yehudah, Barack's 1998 Republican opponent. Yehudah "handed Barack five checks," Dan Shomon remembered. Such an expensive gesture seemed odd, and almost a year would pass before the donation would attract scrutiny. Barack continued to call for the governor to appoint a minority member to the all-white Illinois Commerce Commission and for passage of his "universal health care" state constitutional amendment that would lead to "health insurance for all." At a candidates' forum in Woodlawn, with Bobby Rush represented by a volunteer stand-in, Barack argued that "the test of leadership has to go beyond how you are voting. The question is your ability to mobilize constituencies outside of the 1st Congressional District." In his *Hyde Park Herald* column, Barack did not mention his candidacy and stressed that at the upcoming veto session, "I will be working my hardest to get an override" of the gubernatorial veto that had sidelined his child support reform bill. Donne Trotter, struggling to raise funds and attract attention to his campaign, attacked Barack more than Bobby Rush. Trotter argued that his Springfield experience would make him an excellent congressman, and he asserted that "the only money Barack Obama has brought back is money I've given him as Senate Appropriations spokesman. Has he done better than most? I wouldn't say so. I certainly wouldn't say anything at all based on Obama's two years in office."

Political observers disagreed on how much traction Barack was gaining against Rush. State representative Tom Dart, who represented many of the mostly white neighborhoods that made up the 1st District's western wing, told the *Herald* that given Rush's abysmal showing against Mayor Daley, "you'd have to say he's in trouble." But powerful 19th Ward committeeman Tom Hynes argued that "the mayoral vote was a positive vote for the mayor," not anti-Rush. Then, one week into November, after Dan Shomon had given Chris Sautter more than $20,000 to reserve airtime for upcoming radio ads, the poll results came in. Rush's firm spoke with five hundred 1st District Democratic voters, 90 percent of whom knew Rush's name, compared to only 33 percent for Trotter and just 22 percent for Barack. They also overwhelmingly supported the incumbent over both challengers, with 68 percent choosing Rush, 10 percent Trotter, and only 9 percent Barack, with 14 percent undecided. Among black voters, 72 percent backed Rush, among whites, 52 percent. For Barack, pollster Ron Lester's results were equally dismal. Rush's approval rating was 70 percent, Barack's name was known to just 11 percent, and only 10 percent said they supported him. Lester was struck by how "incredibly popular" Rush was, and especially in the immediate wake of his son's death, Rush's "support ran deep."

Sautter, who was busy producing Barack's first two radio ads, remembered that the poll results were "much worse than anybody imagined," especially given what Shomon and his Springfield buddies thought they had heard from 1st District Democrats four months earlier. Sautter's view of Barack's chances "went way down when we got the poll back," and there was no question that in private "the wind had clearly been let out of the campaign." The painful lesson, Barack later explained, was "do the poll before you announce," because now he had no immediate escape. "Less than halfway into the campaign, I knew in my bones that I was going to lose. Each morning from that point forward I awoke with a vague sense of dread." And the election was still four long months away.[4]

Barack's two sixty-second radio ads debuted Monday, November 15, on black Chicago's two major stations, WVON and WGCI. *Capitol Fax*'s Rich Miller called Barack "a serious challenger" to Rush but mused that Trotter "may help split any anti-incumbent vote that's out there." Rush had been "humiliated" in his loss to Mayor Daley, "but the Daley folks are said to be eyeing Obama as a possible future challenger to hizzoner and may be secretly wishing for a Rush win to stop Obama's rise."

The first of Chris Sautter's ads featured "Diane Johnson," an African American woman with a distinctly down-home voice. "I'm frightened. We can't afford health insurance and some day my family's gonna need a doctor." An announcer followed.

> *For Mrs. Johnson and many others, basic health care today is out of reach. But one new leader is taking on the insurance industry with a bold plan to guarantee that health insurance is available for everyone, regardless of income. That leader is Illinois state senator Barack Obama, candidate for Congress. Barack Obama worked as a community organizer in Roseland and Altgeld Gardens, fighting for the rights of struggling families to receive their fair share of services. After law school, Barack Obama turned down the high-paying corporate jobs to become a civil rights lawyer, and Barack Obama headed up Project Vote, registering over 100,000 minority voters. Now, Barack Obama is running for Congress, a new champion for affordable health care, HMO reform, and better prescription drug benefits for seniors. Barack Obama, Democrat for Congress. New leadership that works for us.*

Barack uttered only the obligatory final sentence: "Paid for by Obama for Congress 2000."

Sautter's goal was to increase Barack's name recognition, and the second ad featured an exchange between a black couple, harking back to August's power failure. "Oh man, there go the lights again." "Another blackout." "I'm tired of this. When's somebody going to do something." "Obama," the woman responded. "Say what?" "State senator Barack Obama. He's fighting for reforms that would force ComEd to refund customers who lose power." Then came the announcer.

> *Barack Obama, Democratic candidate for Congress. As a community organizer, Obama fought to make sure residents in Roseland and Altgeld Gardens received their fair share of services. Barack Obama. As a lawyer, Obama fought for civil rights and headed up Project Vote, registering over 100,000 minority voters. Barack Obama. Elected to the Illinois Senate, Barack Obama pushed to make health insurance available to everyone, regardless of income, and brought millions of dollars into our community for juvenile crime prevention.*

Then the black couple returned. "Here come the lights. ComEd must have heard from that Senator Bama." "That's O-bama. Barack Obama. And they'll be hearing a lot more from him." Then the announcer: "Barack Obama, Democrat for Congress. New leadership that works for us," the same slogan featured in the first ad. Barack again provided just the final words: "Paid for by Obama for Congress 2000."

By November's third week, the campaign had spent $31,000 to broadcast those two ads, and Dan Shomon was working with three young lawyers from the prominent Mayer Brown law firm who had volunteered to help produce a "book" of "Barack's positions on all the issues," particularly "specific economic development ideas" for the 1st District. Barack had to be in Springfield for two veto session weeks before and after Thanksgiving, and during the first he stopped in Decatur, a small city forty-five minutes east of Springfield, to join Rev. Jesse Jackson at an evening protest rally. Two months earlier, a brawl at a high school football game had led to two-year suspensions of seven African American students. The black community was angry about the severity of that penalty, and onstage with Jackson, Barack announced that he would introduce a bill requiring that any suspended student, apart from those who had attacked a teacher or carried a weapon, be provided alternative education. "My attitude is that to the extent that we're dealing with nonviolent offenders or students who haven't struck teachers or used weapons, that probably it makes sense that we have a mandate that in fact these children attend alternative schools," Barack told the *Decatur Herald & Review*. "That's going to be better than having them on the streets," he added. "One of the most difficult things we as a society have to decide is how we deal with young people who have engaged in antisocial behavior. If we simply throw them out, we can anticipate that over time a good number of these people will end up in the prison system." He also worried that Decatur's disciplinary policies "disproportionately affect African-Americans."[5]

On November 21, the *Chicago Sun-Times*' Steve Neal reported the poll results Bobby Rush's campaign had received some days earlier. Neal stressed that in the portion of the 1st District that comprised Barack's state Senate district, Rush led him by only 52 to 24, and with a margin of just 13 points among voters who knew Barack. "The primary might tighten if Obama becomes better known," Neal wrote. A *Crain's Chicago Business* story on the race highlighted Barack's radio ads and noted that Barack had been "perceived as a rising star in Chicago politics since he headed a successful minority voter registration drive in 1992." Barack's campaign had now raised just more than $200,000, and while that could fund the radio ads plus the cost of yard signs, leaflets, office rent, and salaries for Shomon, Will Burns, Amy Szarek, and Cynthia Miller, Barack would not have the funds to run television ads in Chicago's expensive media market. Barack continued to stress that in Rush "we don't have an effective advocate," and he drew coverage in the *Tribune,* the *Defender,* and weekly neighborhood papers like the *Hyde Park Herald* and the *Chicago Citizen* by announcing a plan to spend $50 million over five years to improve computer technology at the 150 public schools in the 1st District. "I believe the role of a congressman is to tap the federal government, the state and private resources to make this possible," with local universities playing key roles. "Our children must have a strong, technology-based curriculum in the schools to be able to compete in the job market," Barack explained. "When kids get excited about computer technology, their parents get excited. They become the catalyst for change in the community."

On the first day of the veto session's second week, Barack's increased child support payments bill returned to the Senate floor. The House had overridden Governor Ryan's veto by a 102 to 15 margin, but Ryan's aides were lobbying fiercely in the Senate to sustain it, arguing that if the bill became law, Illinois would pay out—to the federal government as well as recipients—$6 million more each year than the state took in. On the floor, Barack recounted that of the $88 million Illinois presently collected, half went to Washington, the state retained $35 million, and hardly $8 million went to children. For a successful override, Barack needed at least thirty-six votes in the fifty-nine member Senate, and he described HB 1232 as a "commonsense bill" that would help remedy "how we've screwed up the child support system in this state." On the initial electronic tally, the voting board displayed thirty-seven voting yes, including ten Republicans, with only seven nos and fourteen embarrassed Republicans voting present. But when Public Health and Welfare committee chairman Dave Syverson requested a head-count verification of the two-thirds majority, Barack's victory evaporated. In what *Capitol Fax*'s Rich Miller described as an "obviously scripted move," two yes voters—Democrat John Cullerton and Republican Stan Weaver, Pate Philip's top deputy, had left the floor, and the override failed, 35–7–14.

"I'm very frustrated," Barack told statehouse reporters. "For some reason, the governor dug in his heels on this modest attempt to put more money into the hands of needy children," but Barack vowed to return to the issue in 2000. The short veto session was additionally unpleasant because a number of colleagues questioned the wisdom of both Barack and Donne Trotter remaining in the congressional race against Bobby Rush, and because Trotter's own antipathy toward Barack became increasingly vocal. One staffer overheard, and shared with friends, how Trotter under his breath used two names from *Roots* to refer to Barack as both "Kunta Kinte" and "Toby." To have a fellow senator calling Barack "a derogatory name" was unprecedented, but Trotter's Republican friend Steve Rauschenberger knew that Barack's presence in the congressional race was more of a stimulus for Trotter's candidacy than anything about Bobby Rush. "I'm not sure I ever thought Donne was in that race to win," Rauschenberger explained. Democrat John Cullerton, who was closer to Trotter than to Barack, remembered telling Trotter, "You guys aren't going to win. I don't know why you're doing this. If one of you dropped out, maybe," and Rickey Hendon recalled, "I tried to talk to both of them, but neither one would drop out." Kim Lightford questioned Barack about challenging Rush. "Do you really think you could beat him?" Barack's answer was illuminating. "I don't know if I can beat Bobby, but I'm going to beat Trotter!"

During the veto session, Bev Criglar, the secretary Barack had inherited from Alice Palmer, told him that she was leaving in January for a new job in the comptroller's office. Criglar recommended that he hire Beverly Helm, an African American Springfield native who was well known in the capitol because her late father, Eddie Winfred "Doc" Helm, had served for many years as the Senate's official photographer. Beverly had worked in the Senate years earlier, and one evening Bev Criglar took her to meet Barack for an informal interview. Barack said something about interviewing others, but Bev Criglar disagreed. "No, I don't think so, Barack. She's the one." Barack assented and left it to the two Beverlys to work out their transition.[6]

On December 2, the veto session's final day, the Illinois Supreme Court upended everyone's holiday plans by unanimously striking down the Safe Neighborhoods Act of 1994. Article IV, Section 8(d) of Illinois's 1970 Constitution mandated what was popularly known as the single-subject rule: "Bills, except bills for appropriations and for the codification, revision, or rearrangement of laws, shall be confined to one subject." The framers' goal, as the court stated, was "to prevent the passage of legislation that, standing alone, may not muster the votes necessary for enactment" or, in other words, to prohibit logrolling, which the court in 1974 had held was "both corruptive of the legislator and dangerous to the State." Earlier that year the court had held that "unrelated provisions that by no fair interpretation have any legitimate relation to one another" rendered a bill unconstitutional, and the important 1994 enactment failed that test: "We discern no natural and logical connection between the subject of enhancing neighborhood safety" and the bill's additional "amendments to the civil WIC Vendor Management Act, or the creation of the Secure Residental Youth Care Facilities Licensing Act," because "these purely administrative licensing provisions are not germane to the subject of safe neighborhoods."

The state supreme court's unsurprising holding created controversy because one provision in the 1994 law, upgrading public possession of a firearm from a misdemeanor to a felony, had led to a firestorm two years earlier when savvy National Rifle Association (NRA) lobbyist Todd Vandermyde sought its repeal. Insisting that most members had been unaware of the upgrade's inclusion in the 1994 law, pro-gun legislators successfully championed passage of a bill reinstating the misdemeanor penalty by publicizing cases in which U.S. Marine recruiters, a well-known professional football player, and a pizza deliverywoman who had successfully fended off muggers had all faced felony charges after telling police they had a gun in their car. The Senate had passed that repeal measure 36–14–9, with Barack voting present, but Chicago law enforcement officials reacted with horror, insisting that the felony provision was the "single most significant piece of legislation" for taking gun-toting gang members off the street. Mayor Richard Daley agreed, and three months after the bill's passage, then-governor Jim Edgar had vetoed it, with no override vote attempted.

Four days after the supreme court's decision, Governor Ryan, joined by Daley, state attorney general Jim Ryan, and three of the Four Tops—House Speaker Mike Madigan, House minority leader Lee Daniels, and Senate minority leader Emil Jones Jr., but not Senate president and NRA enthusiast Pate Philip—held a news conference to announce that he was summoning the legislature back to Springfield for a special session beginning on Monday, December 13. All of those officials called for reenactment of the felony provision, with Daley proclaiming that "families are going to be less safe" with that stiffer penalty off the books. *Capitol Fax*'s Rich Miller highlighted that Philip's absence signaled an upcoming tong war between the Senate president and Governor Ryan, with Philip's close sidekick Ed Petka pointing out that Republican senators would refuse to reenact the felony measure they insisted had been snuck into the 1994 bill. Warning that "this session may last a while," with "a long stalemate that lasts until Christmas," Miller predicted that the chances of a compromise between Ryan and Philip over such a binary issue—felony versus misdemeanor—were very slim.[7]

Before Barack returned to Springfield, he held a press conference with ACLU and Mexican American Legal Defense and Education Fund leaders to announce that in January he would introduce a bill that would require police officers statewide to record the race of every driver they stopped. "Racial profiling may explain why incarceration rates are so high among young African-

Americans," Barack asserted. Both the *Tribune* and the *Defender* covered his initiative, and the next weekend Barack appeared on WGCI's *Cliff Kelley Show* and addressed an anti-gun rally in the Greater Grand Crossing neighborhood. On the radio, Barack endorsed the state supreme court's application of the single-subject rule, and he told the *Hyde Park Herald* it was "most important" to retain the felony provision, adding that "there should not be too much difficulty in passing the law in a constitutional fashion" despite "some resistance from the NRA and pro-gun legislators." At Park Manor Christian Church, Barack said he would introduce legislation to restrict gun purchases to one a month and ban most sales at gun shows.

On Monday, December 13, Barack filed more than ten thousand petition signatures to qualify for the March primary ballot, even though only 880 valid ones were required. Collecting that many signatures had taken weeks of work by Will Burns and the campaign's dozens of young volunteers. The *Defender,* the *Tribune,* and one black weekly newspaper reported Barack's challenge that "when he ran for mayor, Rush said that debates are important. It is hypocritical to demand debates in one of his campaigns but hide from debates when he is running for re-election." The petitions deadline also showed that there would be a fourth, long-shot candidate in the race, retired police officer George C. Roby, who contended that Rush, Trotter, and Obama "are from the same tree trunk."[8]

Efforts to work out some sort of compromise between George Ryan and Pate Philip went nowhere. Ryan refused to abandon the felony provision, and although he acceded to one proposal to allow prosecutors to choose between felony and misdemeanor charges on a case-by-case basis, most knowledgeable observers, including Barack, believed that would be held unconstitutional. A *Chicago Tribune* editorial stated the felony provision has "had a dramatic impact in reducing gun injuries and deaths" and proclaimed that Pate Philip "has gone soft on crime." *Capitol Fax*'s Rich Miller observed that "everyone under the dome is feeling serious pressure from their spouses and families to get their rear ends home for the holidays," and that included Barack, who was scheduled to fly to Honolulu with Michelle and Malia on Sunday, December 19, to spend the holidays with Madelyn Dunham. Ryan and Philip remained in regular contact, and each morning when the governor called the Senate president, he said, "This is Scrooge." Pate replied, "Okay, this is the bad guy." On Thursday, December 16, Philip called a vote on the prosecutorial discretion substitute, but no Democrats supported it, leaving it five votes under the three-fifths majority required for approval during a special session.

Barack was content to leave it at that, telling his colleagues that "the public will know where each of us stand," but Ryan ordered the Senate to remain in session. On Friday, December 17, as Ryan tried to recruit Republican senators to his side, Barack told his colleagues that "modest gun control works in making our streets safer," adding, "I believe we have selective enforcement across a whole host of criminal laws." In January, he would introduce "a racial profiling bill," to determine "whether, in fact, stops are being selectively made with respect to race. I'm deeply concerned about that issue, having been the subject of stops that I suspect were selective and based upon my race." Barack added that the legislature should consider allowing broad expungement of criminal convictions, for "it is impossible to find a job if you have a felony record."

Pate Philip, knowing Ryan did not have the thirty-six votes he needed to approve the felony measure, called that bill for a vote, and it fell seven votes under on a 29–18–7 tally. Among the three downstate Democrats supporting Philip was newly arrived Ned Mitchell, who had been appointed to replace the retiring Jim Rea. A friendly, humorous man, Mitchell was lodging in Terry Link's Springfield poker house, and teasingly told Barack he needed to avoid discussing gun bills with him. "You'll have me convinced into voting for that, and it won't be good for me back home," Mitchell warned. "If I were to vote for that bill, I couldn't eat Sunday dinner at my folks' house anymore—they'd break my plate." Barack laughed, but Mitchell reminded him of the fundamental difference between downstate and Chicago: "Where I'm from, we're not shooting one another," and his gun-owning friends were hunters. "It's a *sport* in southern Illinois!"

On Saturday, December 18, senators left Springfield for a forty-eight-hour respite before returning on Monday. Barack postponed his family's travel plans, warning his grandmother that they might have to cancel their trip entirely, and the *Tribune* reported that since the state supreme court's December 2 ruling, 253 people in Chicago, and 60 in the rest of the state, who had been arrested for unlawful possession of a firearm were now facing misdemeanor rather than felony charges. On Tuesday, Barack told senators and spectators that "no one is more eager to get out of here than me. I'm supposed to be on a beach. It's 85 degrees in Honolulu last I checked. I would prefer not being here." He opposed any discretionary solution, explaining that "the notion of creating different punishments for the exact same offense, without any clarity as to why one person would be charged with a felony and another, misdemeanor, would present some constitutional problems."

Barack told one Springfield reporter that the single-subject rule was a good constitutional provision, but that the Illinois Supreme Court was applying it too inclusively, which "really wreaks havoc on the legislative process here." Barack also wrote a *Hyde Park Herald* column blaming the standoff on Republicans who were "listening to extremists in the National Rifle Association" rather than to citizens "who want reasonable gun control." Late Tuesday the Senate adjourned, with senators believing that Ryan would order them back to Springfield by Wednesday, December 29.

On Thursday, December 23, Barack, Michelle, and eighteen-month-old Malia headed to Honolulu for Christmas with Toot. On Monday, Barack learned that the Senate was convening on December 29. He told Michelle they had to leave the next day, but by that evening, Malia had the flu as well as a very severe ear infection. "I tried to get him to come back," Dan Shomon remembered, but maybe he was not "as forceful as I needed to be." Barack also spoke with city of Chicago lobbyist Bill Luking, saying that "he didn't think the votes were there" and that his daughter was more important. "I could not leave my wife alone with my daughter without knowing the seriousness of the baby's condition," Barack later explained, and he remained in Honolulu rather than return to Illinois for the 1:00 P.M. Wednesday vote on SB 224.

By then, even the *New York Times* was covering the Springfield discord, and Illinois newspapers were reporting Ryan's Tuesday claim that "if the Senate calls this tomorrow, it will pass." But when the votes were cast, Ryan came up five votes short, with the tally showing 31–17–2. Democrat Jimmy DeLeo immediately rose to say that his green button had malfunctioned, a claim also made by Republican Wendell Jones. Three absent senators—Tommy Walsh, who said he had stayed home because his pregnant wife was running a fever; Kathy Parker, who was on vacation in New Mexico; and Barack—plus the two claims of button malfunctions had kept Ryan from attaining the thirty-six votes he had anticipated. The governor was livid. "If these folks would have been here, I'd have had my votes," Ryan told statehouse reporters. "We'd have passed this bill today."

Thursday's *Chicago Sun-Times* told readers that "the absences of Obama, Walsh and Parker stunned Ryan," and the *Tribune* stated that Obama and Parker had "decided to remain on vacation instead of returning." Unnamed "aides to Obama, who was in Hawaii, refused to tell the governor's staff how to find him," the *Tribune* added, but Dan Shomon was quoted as saying "his daughter was his first priority." The *Tribune* also featured Bobby Rush's reaction to the news: "This vote was probably the most pivotal vote, one of the most important votes in memory before the General Assembly, and I just can't see any excuse that Mr. Obama could use for missing this vote."[9]

Barack phoned Governor Ryan to apologize for his absence, but by Friday morning editorial denunciations were raining down. "Philip, Criminals Win Again," the *Tribune* declared, calling the absentees "gutless sheep." Obama "who has—had?—aspirations to be a member of Congress, chose a trip to Hawaii over public safety in Illinois." The suburban *Daily Southtown* was no gentler, noting that Barack "was in Hawaii on vacation" when his daughter became ill. "This issue was too important for Obama to take a chance on missing the vote by traveling to the South Pacific in the first place." Barack told Bernard Schoenburg, the political reporter for Springfield's *State Journal-Register,* that "I take my legislative duties really seriously," but "at some point you can't just talk about family values, you actually have to exercise them."

A brief weekend respite allowed Barack to address issues other than his vacation controversy. "Ultimately, if we want to deal with the gun violence taking place in our community, we have to deal with the manufacturing and distribution" of guns and the dealers "who are selling guns to anyone with the cash." A Tuesday-morning *Tribune* gossip column asserted that "those ocean breezes must have blown away the political instincts of state Sen. Barack Obama," who "stayed on vacation in Hawaii instead of returning to Springfield for the post-Christmas vote on tough gun control." Bobby Rush and Donne Trotter "will crucify Obama for his Hawaiian holiday if he persists in his plan to run in the Democratic primary for Rush's congressional seat," and in that day's *Chicago Defender,* Trotter did just that. "I think it is irresponsible for Senator Barack Obama not to come back," Trotter declared. "He and the others dropped the ball for their constituents."

Barack complained to the *Tribune* columnists, and the next morning they ran a follow-up item. "We have a lot of politicians who like to talk about family values. At some point you have to live those family values," Barack argued. The *Sun-Times* ran a prominent story on the controversy, stating that "Obama's absence could cost him dearly." Barack claimed his vote would not have made a difference and that "family has to come first." In the *Hyde Park Herald,* Donne Trotter kept up his assault, declaring that "for anyone not to show up to me is dereliction at the worst and certainly irresponsible," especially because "I was really under the impression that the bill was going to pass." Bobby Rush's campaign spokesperson, Maudlyne Ihejirika, told the media that "while perhaps the most important piece of legislation that could have a significant impact on getting guns off the street was being debated in Illinois, Mr. Obama was on a beach in Hawaii."

Inside Barack's campaign, Cynthia Miller recalled "a tense environment," because "Dan was just so frustrated with Barack" since "once they knew the vote was going to be called, he should have turned around." Media consultant Chris Sautter recognized that the missed vote was "a big deal," and "I wondered whether or not he really wanted this . . . whether or not he appreciated what it took to win a race like this" and "the kind of sacrifice" it required. Chris realized that "Dan was always the taskmaster" for Barack, and Shomon blamed himself for not arguing more forcefully with Barack to come back. "There were other times when he tried to slough out of session . . . and I would really beat him up sometimes, say, 'Look, you've got to show up! This is your job. You need to be there!' But this time I let it slide probably too much. I should have been harder on him."

Bobby Rush's media consultant Eric Adelstein understood that the Barack "on vacation in Hawaii" story played into the perception that the race was between a "dilettante versus the guy from the neighborhood." By January it was clear that Donne Trotter lacked the funding to mount a real campaign. While he had known going in that his mentor John Stroger was backing Rush, over the holidays Stroger's godson Orlando Jones told Donne that word had gone out that members of Stroger's sizable organization should not contribute to Trotter's campaign. "They were fearful," since "many of them worked for the county," and "I felt a little betrayed," Trotter confessed. The "rug was pulled out," but "once you're in it, you're in it." Trotter would remain a candidate, albeit one without a campaign.[10]

By the second week of January, legislators had begun the new spring session with no movement in the standoff between George Ryan and Pate Philip. The Senate president told the suburban *Daily Herald* he did not believe Senator Kathy Parker "deserves any criticism at all" for missing the gun vote because she was in New Mexico. "Our base is pro-guns," Philip explained. "Our districts come first, the state comes second. I mean, people have to get re-elected, so they have to do what they have to do." Concerning unlawful possession, "a first offense with no criminal background shouldn't be a felony."

Barack presided at JCAR's monthly meeting and then introduced his bill calling for a four-year study of traffic stop statistics and training of state police officers in "sensitivity toward racial and ethnic difference." Only the *St. Louis Post-Dispatch*, none of whose Metro East readers lived within 250 miles of the 1st Congressional District, covered Barack's initiative. "Nobody really knows how pervasive the practice of racial profiling is, and that puts a cloud over the able law enforcement officer," Barack told the paper. "There is at least the perception in black communities that 'driving while black' is a crime."

Back in Chicago, Hyde Park's IVI-IPO members voted 28–3–3 to endorse Barack rather than Bobby Rush. The *Herald* publicized the outcome and quoted a spokesman calling Barack "articulate and dedicated" and expressing hope that "Obama will be able to do a better job than the incumbent by bringing more resources into the district." Barack used his *Herald* column to defend his absence for the gun vote, explaining that "my grandmother cannot travel" and that his family's annual visit "is the only means to assure that my grandmother does not spend the holidays alone." A *Herald* editorial robustly stood up for Barack, facetiously writing that he "should have moved his grandmother—and all other family members he might want to visit—to a less glamorous state than Hawaii." A querulous letter mocking Barack as "some progressive, some leader!" was countered by one from former congressman Ab Mikva, whose lengthier letter appeared in the next day's *Chicago Tribune*. Castigating the paper for calling Barack "gutless," Mikva wrote that "the conflict between being a public servant and trying to be a good husband and father . . . is never easy. . . . The media ought not to make it harder by beating up on somebody when family values require priority."

Yet the controversy refused to die. One neighborhood weekly headlined its story "Obama Under Fire for Missing Gun Vote," and another, in a trio of pieces profiling Rush, Trotter, and Obama, mentioned Barack's absence in all three. Bobby Rush told the *Chicago Weekly News*, "I was working on a public service campaign to prevent gun violence over New Year's while Obama was vacationing. We have to keep fighting gun violence and that's where my focus is even if that is not where Senator Obama's focus is." Donne Trotter noted that Barack's "wife was on the trip, and she's a capable individual. It wasn't as if he was abandoning his child in a foreign land." Trotter added that "to use your child as an excuse for not going to work also shows poorly on the individual's character." Barack responded icily. "Perhaps Senator Trotter would feel comfortable leaving his wife behind with a sick child thousands of miles away without any clear idea as to when they would be able to return to Chicago. It's not something I felt comfortable doing," and "for him to try and politicize my child's illness is unfortunate."

In his *Chicago Weekend* profile, Barack said "my ultimate job is to create opportunities through education and employment that prevents youth from getting involved in the criminal justice system," and "we need to devote our talents to keeping youth out of the system." But even in mid-January, the *Tribune* ran a story headlined "Obama Defends Decision to Miss Anti-Crime Vote" and quoted Barack as insisting, "I cannot sacrifice the health or well-being of my daughter for politics." Barack tried to draw attention elsewhere, unveiling a neighborhood beautification plan at Englewood High School and contributing a birthday column on "Why Dr. Martin Luther King, Jr., Is Important to Me" to the *Defender*. "The violence within our community exacts as great a toll as any violence coming from outside our communities," Barack stated. "My commitment to social justice and the sense that serving as an elected official is more than a mere job, but a mission, is largely informed by Dr. King's life."[11]

One evident problem was Barack's lack of entrée into the South Side's rich and complex network of black churches. Neither Dan Shomon nor Cleveland-born Will Burns knew that world. Young Gamaliel organizer John Eason, who was volunteering on weekends, introduced Barack to clergymen he knew, and another regular volunteer, African American UCLS 3L Nat Piggee, was married to Summer Samuels, whose father was a South Side pastor. Selling a Bobby Rush challenger to most churchmen was a daunting task, but by mid-January, Summer became the Obama campaign's fifth full-time staffer. Her godmother,

Lula Ford, was a prominent, recently retired school administrator whose political mentor was Emil Jones Jr., and after getting Jones's approval, Ford hosted a reception at a nightclub called Honeysuckles, where African American educators met Barack. A writer from the weekly *Chicago Reader* listened to Barack's pitch to the black educators. "When Congressman Rush and his allies attack me for going to Harvard and teaching at the University of Chicago, they're sending a signal to black kids that if you're well-educated, somehow you're not 'keeping it real,'" Barack rightly argued.

Like John Eason, Nat and Summer Piggee realized that Barack "wasn't as familiar with some of the customs in the black church" as an African American politician should be. Nat went with Barack to one church appearance and was disappointed when he mangled Martin Luther King's well-known declaration that "injustice anywhere is a threat to justice everywhere." Nat remarked afterward that "You may want to check that King quote," but Barack brushed it off. "Nat, come on, you're the only guy in there who knew it." That dismissiveness reflected arrogance on top of ignorance. "No, dude, you're the only guy in there who *didn't* know it," Nat accurately replied.

On Monday, January 17, Barack held a press conference to declare that "drastic and immediate action is needed to stem the dramatic increases in the cost of prescription drugs." Drug pricing "shows why our entire health care system needs reform" and Barack said he would introduce legislation to require drug companies to sell medications to everyone at the lowest possible price. "This is one step . . . in the movement towards universal health care where all people have insurance." That evening Barack attended a candidates' forum at a field house in the Chatham neighborhood and told the small crowd, "I'm impatient for change. Too often we settle for second best in terms of our political officials. The First Congressional District can do better, but right now we don't have as good a representation as we could have. We don't have a congressman who's effective as a leader at the national as well as the local level."

Bobby Rush did not attend such events, which left plenty of time for Barack to face challenging questions. One man raised the missed gun vote. "If you initiate a lot of ideas, and at the time of a vote you're not there, how can we count on you?" Barack responded curtly. "If you look at my record in Springfield, I don't miss votes. I missed one as a result of my daughter being sick. That's an exceptional situation that doesn't arise often." Afterward the questioner told the *Chicago Reader* reporter, "If you tell me this is one of your issues and then you miss the vote, that concerns me. With that in mind, I'm very reluctant to support him for anything. I think he's biting off a little more than he can chew. He's got some good issues, but he's too green." But when Rashid and Mona Khalidi hosted a small event for Barack, one Palestinian friend, Ali Abunimah, was impressed that Barack was willing to say that the United States needed a more "even-handed" approach toward the Israeli–Palestinian conflict. "He was very aware of the issues," Abunimah remembered, and "critical of U.S. bias toward Israel and lack of sensitivity to Arabs."

Overviews of the Obama–Rush face-off appeared in both highbrow *Illinois Issues* and in *Residents' Journal,* a monthly written and read by Chicago Housing Authority (CHA) tenants. Obama "speaks eloquently for coalition building," the former reported, with Barack arguing that "Rush engages in divisive politics that promotes his name but doesn't deliver the goods. The question here is not policy disagreements, but who can create coalitions." Barack "may be hurt" by the missed vote, and aging black nationalist Lu Palmer declared, "I don't like Obama. He has large Hyde Park liberal backing. He's barely been in the state Senate. Let him get his feet wet." Barack told the CHA magazine, "I'm somebody who believes that politics is a mission," and said Rush had been too passive about CHA's demolition of its notorious high-rises. "If we are providing replacement housing, we need to make sure that public housing residents have a wide range of choices of neighborhoods not just in the city but throughout the region," Barack said. Barack also argued that first-time criminal offenders deserved a second chance. "Too many of our young people think that a gun is a toy and feel that they can intimidate or hurt other people." But "if you haven't been in trouble with the law after a year, then they should take the felony off your record," Barack contended. Yet gun control was imperative. "I think that if you have to take a test to drive a car, a safety test, then there should be no reason it shouldn't be the same way for a gun."[12]

By the third week of January, Barack was back in Springfield at least two evenings a week. Beverly Helm recalled that on their first day together Barack summoned her into his office to make one firm request: "I have people that come in with HIV. I have people come in that have disabilities. I have people come in that are homeless. I never want to see you treat them any differently than you would treat a director or anyone else that walks through these doors." Beverly immediately embraced Barack's point and knew they were "going to get along fine," and "that was day one. We got along fine."

Barack introduced a passel of new bills: higher payments to child support recipients, increased Medicaid eligibility for public aid clients, expanding KidCare into FamilyCare by offering health care coverage to low-income working parents, and stricter regulation of predatory home loan and payday loan vultures. But Barack's primary concern was his congressional campaign, and especially its strained finances. On January 31, all three candidates reported their contributions and expenditures for calendar year 1999: Barack's campaign had raised $266,000 and had spent $250,000, leaving only $16,000 on hand. Staff salaries ate up $4,500 per week, with health benefits and payroll taxes on top of that, and upcoming radio ads would cost another $29,000. But fund-raising had slowed, and on January 25 Barack had loaned the campaign $2,000 of his and Michelle's personal funds. Barack had outraised Bobby Rush, who had taken in $209,000, but a preexisting balance and expenditures of only $147,000 left Rush with $153,000 available. While Barack had received only one single PAC contribution, $3,300 from his poker buddy Dave Manning's Community Bankers Association, a majority of Rush's contributions—$137,000—came from PACs. Donne Trotter had received only $46,000 in contributions and personally loaned his campaign $20,000; with roughly $60,000 in expenses, Trotter had only $8,000 on hand. Barack told reporters, "We're very proud of the fact that we've been competitive with Rush in fund-raising, despite the fact that 99 percent of our fund-raising comes from individuals, while most of Rush's comes from PACs. The fact that we've been competitive indicates the degree of enthusiasm and support that we've gotten in the district." Yet as Barack knew, and as stories in the *Hyde Park Herald* and Washington, D.C.'s *Roll Call* highlighted, his campaign lacked the funds necessary to make any significant push in the race's final six weeks. Illinois political observer Paul Green told *Roll Call* that Barack's challenge "hasn't caught on like I thought it would" and said the contest "looks fairly good for Rush." Progressive former alderman Dick Simpson told the *Herald* that "Obama's done well that he's outraised the incumbent," but needed "at least another $150,000 to $200,000" to make the race competitive. Barack signaled his own realism in telling the *Herald* that "if we can establish more name recognition" before March 21, the outcome would be "very tight."

Media consultant Eric Adelstein had crafted a perfect slogan for Rush's reelection effort: "We're sticking with Bobby." But Rush's campaign was also trying to deploy President Bill Clinton on their candidate's behalf. Rush secured an aisle seat for the president's January 27 State of the Union address, allowing him to hurriedly ask Clinton to cut a radio ad boosting his campaign. On January 31, Governor George Ryan announced he was imposing an open-ended moratorium on the state's death penalty. Citing Illinois's "shameful record of convicting innocent people and putting them on death row," Ryan said there would be no further executions "until I can be sure with moral certainty" that capital defendants are "truly guilty." Even conservative Republican state senator Chris Lauzen hailed Ryan's move, and Barack too praised Ryan for making "a morally correct" decision. Some reacted cynically to Ryan's move, for it had been known for days that one of his closest aides from his time as secretary of state was about to be federally indicted as the commercial-driver's-license bribery scandal crept closer to Ryan himself. Those charges came just twenty-four hours after Ryan's landmark announcement, but those who knew the governor best had no doubt that Ryan's change of heart on capital punishment was utterly sincere and not politically calculated.[13]

Barack had little choice in February but to spend three days each week in Springfield, and he introduced nine additional bills. The first, banning the sale of certain flavored cigarettes, won coverage in the *Defender;* another was the drug pricing measure that had received press attention two weeks earlier. He reintroduced the state earned-income tax credit bill he had first put forward three years earlier, and he also proposed a measure requiring the Department of Human Services to provide domestic violence training for its employees who worked with families on public assistance.

During a long interview in early February with Ted McClelland of the *Chicago Reader,* Barack said the four things of which he was most proud were his 1997 work on welfare reform, his 1998 efforts regarding juvenile justice and campaign finance reform, and his consistent championing of tax justice issues like the earned-income credit, which he now had persuaded Republican Revenue Committee chairman Bill Peterson to cosponsor. They revisited the missed gun bill vote, and Barack admitted that "politically, I took a big hit." He also acknowledged that "any ambivalence that I feel about my life in politics basically revolves around my family." Since entering the congressional race, "I'm out the door by seven, and I'm home at nine, including weekends." He estimated that over the previous six to eight months, he had spent no more than three or four full days with his wife and daughter. If he won, "ultimately I would move my family to Washington," because seeing his daughter only on weekends was "no way to raise a child."

Citing Jesse Jackson Jr. and North Shore progressive Jan Schakowsky as Illinois's two best members of Congress, Barack said that serving in Springfield had taught him that "you can have an enormous influence on the process through the power of your ideas even if you don't get your name as the chief sponsor of a bill." He and Jesse Jr. were the two black elected officials whose credentials would allow them to do well in the private and nonprofit sectors too, so "it makes sense for the community to take advantage of those talents and those gifts," Barack told McClelland. "We need to attract more people like myself and Jesse Jr. into politics in the African-American community." Regarding public service, "I really had to want to be here," for "at every juncture in my life I could have taken a . . . path of least resistance but much higher pay." Barack additionally pointed out that "being president of the *Harvard Law Review* is a big deal."

Finally climbing down off his high horse, Barack said that "before this congressional campaign I was a big pick-up basketball player" and had often enjoyed "some terrific games" at the courts "right off Lake Shore Drive" in Jackson Park. But "right before a big fund-raiser near the beginning of the campaign, I walked in with a big black eye," and so "my staff has banned me from basketball." McClelland asked what books had most influenced Barack's political thinking, and a long pause followed. "My political philosophy is probably more influenced by fiction than it is by . . ." and then pausing again before finally naming "James Baldwin's essays," including *The Fire Next Time; The Autobiography of Malcolm X;* and Toni Morrison's *Song of Solomon*. The only nonfiction work Barack named was *The Power Broker,* Robert Caro's biography of New York urban planner Robert Moses, a "wonderful book" that "really focuses on the intricacies of power in an urban environment." Reaching for a better response, Barack explained that "a lot of times what moves me politically are themes that I find in fiction, themes of people overcoming great obstacles and great odds and creating communities . . . and discovering one's roots and feeling a sense of place," he added.

Barack emphasized to McClelland that he had "raised more money from African Americans than either of the other two," and that "half of my Finance Committee is made up of African American businessmen, most of them under the age of fifty." That wording overlooked Judy Byrd, who, along with Jim Reynolds and Craig Robinson's good friend Marty Nesbitt, was one of Barack's three most dedicated advocates. Barack gave no names, but "I think that part of what they're looking for is a transition in political leadership that reflects their growing sophistication and their involvement in all levels of the economy."[14]

Soon after that, Barack had painful occasion to consider "driving while black" when he was pulled over and given a $75 ticket for going 62 in a 45 mph zone, yet as scheduler Cynthia Miller knew, "Barack does drive fast." On Monday, February 14, media consultant Chris Sautter's new second pair of radio ads started airing on African American stations. The first one began with an announcer introducing "state senator Barack Obama, candidate for Congress, speaking with a group of frustrated South Side seniors." A woman remarked, "Senator Obama, the cost of prescription drugs has gotten so high, sometimes I can't afford the medicine I need." A man added, "You know, I've always backed Bobby Rush, but I don't see him doing anything about these high prices." Then Barack responded: "Let me tell you something. The profits of drug companies are at record levels. Meanwhile we've got seniors who are having to choose between their food, their rent, and their prescription drugs. That's not right. I'm running for Congress to fight for First District families and our seniors on prescription drug and HMO reform." Then the announcer: "Barack Obama. Civil rights lawyer, the head of Project Vote. As our state senator, Barack Obama has taken on the drug and insurance companies with his fight for affordable health care." Barack reentered: "America's health care system locks out too many people. I'm fighting to change that." The woman questioner declared, "Barack, we need a congressman like you who would do more than talk," and a crowd applauded. The announcer returned: "Barack Obama. Democrat for Congress. New leadership that works for us." Barack himself closed with "Paid for by Obama for Congress 2000."

The second new ad began with background noise of a police radio before a white male's voice ordered, "Hand over your driver's license." A black woman replied, "But officer, I wasn't speeding." "Don't talk back to me. Get out of that car," the officer instructed. "But what did I do?" "I'll worry about that. Now open the trunk," and the woman sighed. Then the announcer: "It could happen to you. Or to someone you love. Stopped by the police for no apparent reason, except that you fit a racial profile secretly used by police. It's called racial profiling, and it's an unethical and dangerous practice that needs to end. Now, state senator Barack Obama, candidate for Congress in the First District, is leading the fight to end racial profiling." Then Barack spoke: "This is state senator Barack Obama. Racial profiling is not only wrong and degrading, it's dangerous and can lead to unexpected confrontations. Not only that, it erodes confidence in law enforcement. That's why I've introduced legislation to address the problem of racial profiling and protect you from those who would abuse your rights." Again the announcer, and then Barack, finished as before: "Barack Obama, Democrat for Congress. New leadership that works for us." "Paid for by Obama for Congress 2000."

In a *Hyde Park Herald* column on racial profiling, Barack called it a "growing concern to many minority citizens" and "a serious problem in many Illinois communities." During an interview with the *Chicago Weekly News,* he talked up his welfare reform and juvenile justice efforts in Springfield and said he had helped direct $20 million toward crime prevention. As a congressman, his top priority would be "figuring out how to retool and revamp our public education system to provide the skills to our young people that allow them to compete in the new global economy." In addition, "ultimately we should provide some basic health care to all citizens through programs designed at the federal level, even if some of the implementation takes place at the state level."[15]

In mid-February Barack, Michelle, and Malia attended the seventy-fifth birthday party of his granduncle Charles Payne, who had retired five years earlier from the U of C's library. Maya and Madelyn, who was now seventy-seven years old, flew from Honolulu; Madelyn's sister Arlene and her lifelong companion Margery Duffey came from North Carolina, and their younger brother Jon came from Colorado. It was the first time the four siblings had been together since their mother's funeral thirty-two years earlier. Maya was months away from entering a Ph.D. program at the University of Hawaii, and it was a more relaxed family gathering than the brief and troubled visit Barack, Michelle, and Malia had made to Honolulu seven weeks earlier.

On Thursday, February 17, Bobby Rush finally agreed to appear at several debates and candidates' forums. Rush's press spokesperson, Maudlyne Ihejirika, adopting a demeaning tone, declared that "Obama will get his opportunity to hang out with the congressman—an opportunity he so desperately seeks." She also insisted that Barack's criticisms of Rush's invisibility had played no role in Rush agreeing to debate. "Obama is not even on our radar screen. His constant whining does not even register here," Ihejirika declared.

The next morning Barack, Rush, and Donne Trotter appeared live for two hours on Cliff Kelley's show on WVON radio. Barack expressed opposition to school vouchers because they could "hemorrhage money out of the public school system," but he acknowledged that "when you've just got a monopoly in the public schools without expanding competition, we've got problems. I'm in favor of charters, and I'm in favor of expanding competition within the public school system." During one commercial break, Barack's prescription drug ad aired, and during the next off-air break, Rush expressed outrage. "It's a lie, a fabrication," Rush declared. "It's the worst form of campaign advertising that I've been privy to." Barack responded, "I don't think that it is a distortion of the record for me to say that generally constituents don't have much access or ideas of what the congressman is doing." Rush replied that "maybe you haven't been around the First Congressional District long enough to really see what's going on." Turning to studio onlookers, Rush added that "one of the things that galls me about this guy is that he says one thing about you in your presence" before being cut off as the broadcast resumed.

Rush was scheduled to join his opponents at a Sunday-afternoon forum at Trinity United Church of Christ, but on Saturday Rush left Chicago to comfort his ailing father in southwest Georgia. In Rush's absence, Barack told the Trinity crowd that "if you're only interested in votes, or for that matter bringing money back to the district, stick with the status quo. But if you're concerned about long-term leadership for our community, I think I'm your guy." At a subsequent forum in Hyde Park, Barack stressed that universal health care "has been the centerpiece of my campaign" and told one questioner that "I'm strongly opposed to school prayer, but not prayer. Public schools are not the appropriate place." That issue "goes beyond the separation of church and state and goes to the heart of what we believe in."

On Monday, news spread that Rush's father had died the previous day. More sadly for Barack, former senator Paul Simon's seventy-seven-year-old wife Jeanne had also passed away on Sunday. Barack made plans to drive to Carbondale on Wednesday to attend the visitation and then Thursday morning's funeral service for Simon's wife, but first he loaned his financially struggling campaign an additional $2,000. Dan Shomon and Madelyn Dunham each chipped in the maximum $1,000 apiece that week, but by the end of February, Barack had raised just $89,000 since the first of the year, leaving only $18,000 as well as outstanding bills for three times that amount. Rush had taken in just $103,000, almost half of which came from PACs, and had $72,000 available after spending more than $75,000 on direct mail and another $30,000 on radio ads. Donne Trotter had received only $22,000, had spent $10,000 on one mailing, and had a cash balance of $851.66. Rush told one reporter that Barack's campaign was being fueled by "white money" from outside the district, prompting Barack to reply that "for me to get money from progressive individuals like Judson Miner is far superior to me getting it from Ameritech," although three months later Barack's state campaign committee would accept a $1,500 check from Ameritech.

Rush spent $7,000 for a late-February poll of six hundred 1st District voters that showed him with an overwhelming 71 percent to 10 percent lead over Barack, with Trotter at 8 percent. Seeing Paul Simon in Carbondale was a poignant experience for Barack, and Simon's daughter Sheila recalled that her dad "really made a big deal" of being sure to introduce Sheila to Barack, whom she had heard about but had never met. "It was clear that dad was very fond of him," Sheila remembered. After the funeral Barack returned to Springfield, where one reporter asked about his racial profiling bill, and he explained that "you will be hard pressed to find a black or Hispanic family who does not have at least one member who feels they've been stopped because of their race." The reporter added that Obama "said he has been stopped many times for no apparent traffic violation."[16]

The *Sun-Times* reported Alice Palmer's enthusiastic support of Rush. "African Americans have fought hard for these positions. His seniority is extremely important, and we should not make light of it." Thanks to Summer Samuels-Piggee and John Eason's hard work with South Side pastors, Barack's campaign pulled together a press conference at a Methodist church in Chatham featuring more than a dozen ministers who were publicly endorsing him. Father Mike Pfleger and Rev. Alvin Love, who had known Barack since his time at DCP, were among them; Jeremiah Wright was not. Barack had worked hard for Pfleger's support, meeting at the Pancake House in Hyde Park for a long breakfast, during which Mike had warned him that "you're running against an icon." Mike realized that "ambition was always part of Barack's foundation," yet "there was something about him that I was drawn to" because "when you talked with him you got energized," Mike explained. "I wrestled a whole lot" before agreeing to support Barack, and the backlash after that press conference was intense. "Bobby was very upset with me, very angry with me," and "I got eaten alive on black radio." Mike recalled his stance proved "very costly" because "I had a lot of people turn against me on that."

On Monday, March 6, the *Chicago Tribune* endorsed Barack over Rush, saying, "it's time for a change" and calling Barack "a rising star on the local political scene." Obama "is smart and energetic . . . and he is committed to his community. He has fresh ideas" and "would become a spokesman for African-American concerns nationally and an important voice in shaping urban politics in Chicago and the nation." Barack also announced endorsements by former congressman Ab Mikva, former alderman Leon Despres, and progressive physician Quentin Young, earning a *Sun-Times* story in which Rush's spokeswoman Maudlyne Ihejirika proclaimed that "Obama doesn't know the meaning of leadership." In the *Hyde Park Herald,* Ihejirika reacted to Barack's prescription drugs ad by remarking, "what a fraud Mr. Obama is" and complained about his "constant fabrications." Barack announced a ten-point economic development plan that won coverage in the *Defender,* and on Tuesday night, March 7, Rush appeared—although thirty minutes late—at a candidates' forum at a church in mostly white Beverly.

Barack again resorted to his abstract argument that the contest involved "leadership and who can best articulate a vision for the district." In the suburban *Daily Southtown*'s first article on the race, columnist Phil Kadner said Rush "seemed to have difficulty collecting his thoughts or recalling details," even though the candidates had been given a written list of questions in advance. Barack "has Sidney Poitier good looks and is one of the most articulate candidates you are ever likely to hear," Kadner wrote, but "if Obama said anything memorable, or newsworthy, it slipped my mind." Attendees believed "Obama was the most effective in pleading his case," and Kadner concluded that "Obama would be the more likely to sway opinion in Congress because he's more eloquent."

Kadner later recalled getting a phone call from Barack complaining about the column, but now the primary campaign had less than two weeks to go. On Friday, March 10, Chicago radio stations began airing Chris Sautter's two final ads for Barack. The first began with a ringing school bell followed by a black woman's voice. "As a former teacher and grandparent, I can tell you kids love working on computers. Unfortunately, there aren't enough computers to go around in many South Side schools. Meanwhile, affluent suburban students learn on new computers at home and school. In this new high-tech economy, we can't let our children get caught on the wrong side of the digital divide. That's one reason I'm voting for Barack Obama for Congress on March 21st. As a state senator, Barack has been a leader in bringing quality education for all children." Then Barack spoke: "This is state senator Barack Obama. My school technology plan will give our students the skills they need to compete in the new high-tech job market. But I'll also focus on the basics, by fighting to expand early childhood education programs and by reducing class sizes. Our children deserve the best, and my education plan makes sure they get it." An announcer intoned the campaign slogan—"Barack Obama, Democrat for Congress. New leadership that works for us"—and Barack recited the standard closing: "Paid for by Obama for Congress 2000."

The second ad featured the same African American couple from the ad four months earlier. "I suppose you're not voting in the Democratic primary on March 21?" the woman asked. "Of course I'm voting. You're the one who always forgets to vote," the man replied. "And I suppose you're voting for Bobby Rush again?" she asked. "I guess so. Didn't you vote for Rush last time?" "Not this time," the woman answered. "Rush gets on TV a lot, but I can't think of anything important he's done in Washington. I just found out that Rush has the worst attendance record of any Illinois congressman. He's missed over three hundred votes." "Bad news," the man responded. "So who you for?" "State senator Barack Obama," she answered. "I've seen him speak a couple of times and he's really got a lot on the ball. Civil rights attorney, law professor. Obama's got specific plans to cut prescription drug prices and to stop racial profiling. And Obama can help bring jobs back to the South Side." "Oh yeah, Obama. I've been hearing good things about him lately," the man said. "The *Chicago Tribune* endorsed Obama, calling him an outstanding candidate," she stated. "We need someone who does more than talk," the man offered. "So finally we agree on something?" she asked. "Barack Obama for Congress," he intoned. "New leadership that works for us," she recited. "Paid for by Obama for Congress 2000," Barack concluded.[17]

Barack went into the final ten days with an energized campaign staff if not much money. Dan Shomon, Will Burns, and several dozen regular volunteers placed yard signs on major thoroughfares, and weekend leafleting was especially successful in the predominantly white 19th Ward and in close-in suburbs like Evergreen Park and Blue Island. On most weekday mornings, Barack and a staffer or two shook hands and distributed leaflets at Chicago Transit Authority L—elevated train—stations across the South Side. State representative Tom Dart "definitely got some pushback from people" in Tom Hynes's 19th Ward organization for his active support of Barack, but "everything we were getting back was showing us we were going to do really well" in the western part of the district. Many of the volunteers were involved because of their friendships with Dan Shomon or Will Burns as much as their support for Barack, but a good number of U of C law students pitched in too. Old friend Hasan Chandoo came through Chicago and spent a day with Barack but was not enthused by what he saw. Former *Law Review* colleagues Andy Schapiro and David Goldberg contacted the campaign and asked for a fund-raising letter they could circulate among Harvard Law alumni and were similarly displeased. Chris Sautter, who was in Chicago only occasionally, felt that "down the stretch, Barack was going through the motions," but scheduler Cynthia Miller, who interacted with him multiple times every day, believed "he was still giving it his all."

On Saturday, March 11, the three major candidates appeared at a Chicago Urban League forum. Barack said that for decades federal and city housing policy had tried "to concentrate and segregate poverty," and in 1997, he had "voted against the entire state budget because I thought too much money was going to the prisons." Donne Trotter accused Barack of hijacking his racial profiling bill from state representative Monique Davis, who had introduced one over a year earlier. Barack appeared "stunned" and replied, "Don't lie now. That was my bill. I had it first." With the *Daily Southtown,* Barack abstractly characterized the difference between himself and Rush. "The key issue in this race is who can provide key leadership in framing the issues, organizing resources at a local level to ensure that federal programs are actually benefiting constituents." The *Southtown* endorsed Barack, saying "we have great respect for Rush and have supported him in his previous races . . . but after four terms, we don't believe he has accomplished enough on behalf of his district, particularly the western end." Voters "would be better served by Barack Obama, an energetic young lawyer" who has "emerged as a forceful Democratic spokesman."

Sunday's biggest news was that a thirty-second radio ad featuring President Bill Clinton would begin airing Monday on Rush's behalf. It was 100 percent Clinton: "Illinois has a powerful voice in Congress it can't afford to lose. Bobby Rush's leadership has secured millions of dollars for job training and much needed community development. Bobby Rush has been an active leader in the effort to keep guns away from kids and criminals long before his own family was the victim of senseless gun violence. I'm President Clinton urging you to send Bobby Rush back to Congress where he can continue his fight to prepare our children for the twenty-first century. Illinois and America need Bobby Rush in Congress."

Sunday afternoon Barack marched with Tom Dart and a number of volunteers in the annual South Side Irish Parade, with his name amended to O'Bama in honor of St. Patrick's Day. Monday evening the three major candidates appeared together on a twenty-minute segment of *Chicago Tonight,* public television station WTTW's premier news program. Host Phil Ponce, always a tough questioner, began by asking Rush about the *Tribune* and *Daily Southtown*'s endorsements of Barack. Rush, not the most articulate of speakers, responded by citing President Clinton's endorsement of him. Ponce then asked Barack, "What makes you different from the congressman?" Barack backed into a response, saying, "I don't think there are a lot of ideological differences," because "all three of us are progressive, urban Democrats. I think that we share most views." Barack noted that "Congressman Rush is more hesitant about campaign finance reform. That's a high priority for me . . . but for the most part, I think that we agree ideologically. I think that the issue in this race is really leadership." Then Barack became wordy and repetitive. "The question for the voters in the First Congressional District is who can best articulate and frame the issues that are most important to voters in the district and who can provide the leadership to organize resources, both federal and state and municipal at the local level, to make a difference on issues that count like education, health care, jobs, and economic development, and my assessment is that my experience not only as a legislator but also as a civil rights attorney, as a community organizer, gives me the leadership skills to provide, to really make a difference in this race."

Ponce then asked whether Barack's fund-raising meant he would be "'beholden to interests outside the district,'" as Rush's campaign had insinuated. Barack replied that "99 percent of our money comes from individuals, as opposed to PACs," and his broad support showed "I've got relationships throughout this city," which was valuable because "one of the things that we need in this office is somebody who is able to create coalitions and work across a wide spectrum of interests and I think that's something I can provide." Ponce asked Rush to name his top achievement, and Rush cited protecting a section of the L's Green Line in Englewood. Asked to respond, Barack stated that Englewood had been "festering into other communities" throughout Rush's time in Congress. When Ponce raised Rush's seniority, Barack said, "vision and imagination and hard work is more important" and cited Rush's challenging Charlie Hayes. "I can bring more energy and focus to the job."

Requested to specify his priorities, Barack said "combining education and workforce training to make sure that we've got the human capital in place to compete in a high-tech, high-wage economy." He added that "very few businesses that already exist in that area actually know what kind of federal help they can get," because "one of the arguments I think I have against Congressman Rush in this race is that he has not been particularly accessible in terms of letting businesses know how the federal government can assist them in terms of growing their businesses and how we can market the community to attract new businesses in the community." Ponce challenged Barack that "you simply haven't been in the state Senate very long. You have a limited track record in terms of time. What is your argument based on the one term that you've served in the Senate so far that makes you prepared for the Congress?" Barack corrected Ponce, that he actually was in his second term, and cited his law practice, his Annenberg Challenge role, and his organizing experience as demonstrating that he had "the skill set required not only to deal with government, but also to be able to put government programs together with the not-for-profits and the private sector to make things work for people in the district."

Rush tried to attack Barack, saying, "Senator Barack, he represents a part of Englewood, his district has always been in Englewood. Englewood's always been in his district. What has he done?" Barack replied, "I've brought two million dollars back to the district last year." Trotter sought to dismiss them both by saying, "what we've been hearing here is one who I believe is talking in rhetoric again and someone else is talking in theory." Then Barack volunteered, "I actually thought Congressman Rush did a good thing running against the mayor, because I don't think anybody should have a pass. I don't think Congressman Rush in this situation should have a pass. I don't think the mayor should have a pass. I don't get a pass for my state Senate seat." Barack reiterated, "I think that Congressman Rush did a brave thing running and I don't want it perceived that somehow he is being punished for that race." Yet by mentioning "pass," Barack had left himself open to a closing rebuttal from Rush. "I certainly disagree with the senator, Senator Obama. He did get a pass in his first effort out in terms of running for the Senate. He and others knocked his predecessor, Senator Alice Palmer, off the ballot. He got a free pass on his first time around."[18]

On the campaign trail, Barack held a March 15 press conference to announce that he was channeling some $400,000 in state "member initiative" funding to a job training program at Englewood's Kennedy-King College, where his and Michelle's close friend Kaye Wilson's husband Wellington was president. Barack told the *Chicago Defender*'s Chinta Strausberg that "Rush is fast to issue press releases on economic development but slow or totally absent when it comes to helping real people with their day to day needs and issues in the First District." To another reporter Barack said, "nothing's tougher, I think, than running against an incumbent in a primary." Rush told the *Sun-Times* that Barack's "criticisms are empty" and that he "curses the darkness rather than lighting a candle." Barack met with the *Sun-Times* editorial board, giving each member a paperback copy of *Dreams From My Father* that he took from the remaindered copies he had bought for $1.00 apiece from Chicago book dealer Brad Jonas. But the *Sun-Times* endorsed Rush, saying that Obama and Trotter, while "both impressive figures, have not presented an argument worthy of removing a four-term congressman. . . . Rush deserves re-election for the valuable work he has done in Washington for his district and for Chicago."

On March 17, Ted McClelland's long article "Is Bobby Rush in Trouble?" appeared in the weekly *Chicago Reader*. First it quoted Barack: "Congressman Rush exemplifies a politics that is reactive, that waits for crises to happen and then holds a press conference, and hasn't been particularly effective at building broad-based coalitions." Then Rush was much harsher: "Barack is a person who read about the civil rights protests and thinks he knows all about it. . . . He went to Harvard and became an educated fool" in the university's "ivory tower." Rush added that "we're not impressed with these folks with these eastern elite degrees." McClelland repeated the belief that Mayor Daley "is said to fear Barack Obama's appeal as a citywide candidate," while Barack declared that "Chicago in many ways is the capital of the African-American community in the country."

Some days earlier, in a 43rd Street juice bar, Donne Trotter had uttered a line that would never be forgotten: "Barack is viewed in part to be the white man in blackface in our community." McClelland reported that Trotter "detests" Obama, and aging black nationalist Lu Palmer recounted Barack seeking his support for Project VOTE! seven years earlier. "He came to our office and tried to get us involved, and we were turned off then. We sent him running. We didn't like his arrogance, his air," and "I said, 'Man, you sound like Mel Reynolds.'" Echoing Rush's view, Palmer said that "if they name you head of these elite institutions, the *Harvard Law Review,* that makes one suspect." McClelland wrote that "there are whispers that Obama is being funded by a 'Hyde Park mafia,' a cabal of University of Chicago types, and that there's an 'Obama Project' masterminded by whites who want to push him up the political ladder." Even years later, Rush was utterly unapologetic: "my whole effort was to make sure that people knew that Barack Obama was being used as a tool of the white liberals."

These fevered imaginings of nationalists were less cutting than Rich Miller's statement that Barack "hasn't had a lot of success here, and it could be because he places himself above everybody. He likes people to know he went to Harvard." Neither Barack nor Miller had made any effort to get to know the other because, as one lobbyist wisecracked, "they both saw each other as irrelevant, and one of them was right." Trotter tried to make the Senate's one-vote failure to override Governor Ryan's veto of Barack's child support payments bill an indication of his legislative track record, but moderate Republican Dave Sullivan forcefully rebutted that claim. "No one could have tried any harder than Barack Obama. . . . It would not have gotten out of the Senate in the first place without Barack."[19]

Barack's campaign was less than thrilled with McClelland's article spewing such bad blood to a citywide readership. Everyone understood, as Will Burns explained, that the nationalists' Hyde Park and U of C comments were "code words for both whites and Jews." A more pressing problem, with the election four days away, was how "overextended" the campaign's finances were, as Dan Shomon put it. Their only choice in order to keep going was "deficit spending." On March 15, Shomon, Will Burns, Cynthia Miller, Amy Szarek, and Summer Samuels-Piggee all received their final paychecks, covering the first two weeks of March, but on March 17 a crucial boost arrived in the form of at least five $1,000 checks from Tony Rezko's employees and business partners. The *Chicago Defender* endorsed Rush, declaring that the incumbent has "done much for his constituents and his reelection promises even grander things to come for 1st Congressional District residents." The paper said Obama and Trotter "are both highly qualified . . . but still have much to do" in Springfield.

National coverage of the race accorded Barack significant attention. National Public Radio's *Weekend Edition* featured Barack reiterating the long-term political perspective he had developed more than a decade earlier. "One of the things we need to learn from the Harold Washington era is the importance of institution-building," Barack explained. "When he died, in much the same way as when Martin Luther King was shot, you didn't have strong institutional structures in place and a broad-based set of collective leaders who could carry that work forward." Capitol Hill's *Roll Call* termed Barack "a rising star" while quoting the Rush campaign's claim that "we are not worried in the least." In *USA Today,* longtime Chicago political commentator Don Rose noted that Rush's opponents were "dividing the votes, and they don't have any hard issues against him."

On Sunday, Barack told the *Sun-Times* that "our people are energized, they're interested and I think they will come out to vote." Monday's *Tribune* carried two separate columns both suggesting that even if Barack lost, "Obama has a future," as John Kass put it. Salim Muwakkil further publicized Donne Trotter's "white man in blackface" insult before concluding that Barack "clearly is a man on the way up. Even if he loses Tuesday, it will just be a bump in the road."

Tuesday morning, *Capitol Fax*'s Rich Miller asserted that "the late word is that Sen. Barack Obama may be surging," but no one in the 1st Congressional District, including Barack's own campaign staffers and volunteers, believed any such thing. On 55th Street, a huge billboard featuring the "We're Sticking with Bobby" slogan greeted anyone exiting the Dan Ryan Expressway and turning east toward Hyde Park. Serious money problems continued to plague Barack's campaign, and he wrote an Election Day check for a $5,500 loan, bringing his and Michelle's personal indebtedness in the race to $9,500. Chris Sautter flew in from D.C. to help work polling places and realized that Barack, like Dan Shomon, knew he was not going to win. Barack visited a number of precincts and was struck by how older black women would "shake my hand and pat me on the back and say 'You know, you seem like a really nice young man. You've got a bright future ahead. We just think it's not quite your turn yet.' I knew at that point that I wasn't going to win the race."

Dozens of young volunteers spent Election Day working various precincts and making get-out-the-vote phone calls before heading to an evening gathering at the Hyde Park Ramada Inn once polls closed. Will Burns and several friends had recruited dozens of poor people to distribute leaflets in exchange for a day's pay and evening meal money, but the campaign literally ran out of money as Election Day ended. Ron Davis, who had been so crucial back in 1995–96, chipped in $1,000, but "people are literally emptying their own bank accounts to try and back Barack up," one participant recalled, so that Burns and his friends could avoid serious trouble with the "army of drug users" they had assembled.

Some of the volunteers were shocked by news reports declaring Bobby Rush the winner even before they reached the Ramada. Fund-raiser Amy Szarek had begged her closest girlfriends to come, telling them "I don't want the room to be empty," but while Szarek and Cynthia Miller staffed the Ramada gathering, Dan Shomon stayed at headquarters tabulating returns. Young lawyer Jesse Ruiz remembered a "very somber" scene at the Ramada, with hardly fifty people gathered. For some young volunteers, like enthusiastic Northwestern law student Joe Seliga, the speed with which the race was called, plus Rush's two-to-one margin over Barack, came as "an absolute shock." Michelle Obama was "looking very distraught," and one of Amy Szarek's friends remembered Amy being "so upset" and "devastated" that "she hid herself in the bathroom crying the whole night" with girlfriends "trying to console her."

State representative Tom Dart knew Barack had done well enough in his western portion of the district that he thought Barack might win, until Barack telephoned him to say, "Tom, we got shellacked." At the Ramada, Barack looked "very disappointed," and Craig Huffman, who as assistant treasurer had signed most of the campaign's checks, remembered thinking to himself, "'I wonder if this is it?' because I knew Michelle was very clear" about wanting Barack to abandon electoral politics. Her brother Craig Robinson had the same thought as Huffman: "Wow, I guess that's it."[20]

When Barack spoke to his supporters and a few reporters, he acknowledged the obvious: "I confess to you: winning is better than losing," although "I thought maybe we could pull this off." Barack thanked them for their efforts: "we just had such support with such a diverse base of folks that we can walk away from this feeling proud. What it tells us is that people are hungry for a new kind of politics." Regarding his own future, "I've got to make assessments about where we go from here," Barack volunteered. "We need a new style of politics to deal with the issues that are important to the people. What's not clear to me is whether I should do that as an elected official or by influencing government in ways that actually improve people's lives."

Bobby Rush was astonishingly magnanimous at his own event. "I want to thank my opponents . . . for the way they conducted their races," Rush told the crowd. "The 1st Congressional District is a better district because of Donne Trotter and Barack Obama," he declared. "Reverend Jackson often tells me that only diamonds can sharpen other diamonds," and "my two opponents in this race are diamonds indeed. They've helped me become a better congressman and a better individual." When the final returns came in, Rush had won 61 percent—59,599 out of the 97,664 votes cast, with Barack obtaining 29,649 votes, good for 30.3 percent. Donne Trotter trailed with 6,915, representing 7 percent, and little-known George Roby attracted 1,501.

Some supporters, taking into consideration the sympathy Rush gained from the deaths of his son and father, Barack's missed gun-bill vote, and the earlier polling numbers, felt that Barack had achieved "a moral victory to get 30 percent. I thought we were going to get like 18 percent," young organizer John Eason reflected. Dan Shomon told a reporter "it's difficult to get name recognition" as a challenger in a primary without the $200,000 needed for a television advertising campaign, and Dan expressed regret that Trotter had not cut into Rush's overwhelming support in the heavily black 6th and 8th Wards, where the aldermanic organizations had backed Rush. Barack won a remarkable 72.4 percent of the vote in the mostly white 19th Ward, and 64 percent in the suburban portion of the district, but he received only 17.8 percent in the 6th. In the two most diverse wards, where he was supported by both aldermen, Barack won just 37 percent in Toni Preckwinkle's 4th Ward and 33 percent in Leslie Hairston's 5th. Barack carried Hyde Park's forty-four precincts by only a narrow, eight-hundred-vote margin over Rush.

Thurday's *Sun-Times* featured a warm-hearted editorial commending the 1st District also-rans. "Some stars did not rise, but the efforts and ideas advanced in their losing efforts were impressive." Congressman Rush "called Trotter and Obama 'diamonds.' We concur." More vitally, *Sun-Times* political columnist Steve Neal, reviewing the ward-by-ward percentages, observed that "if Obama is a future candidate for state-wide political office, his impressive totals on the Southwest Side and in the suburbs would indicate that he's got the potential to win." Barack told the *Chicago Defender*'s Chinta Strausberg, "I went into this race knowing it would be an uphill battle," and to another writer he said "the most difficult thing in politics is name recognition," because throughout much of the district "people still didn't know my name."[21]

Looking back four years later, Barack observed that the race "taught me to never overestimate the degree to which voters know who you are." Other knowledgeable observers had different reactions. Toni Preckwinkle and Dan Shomon thought Barack entered the race far too late, and David Axelrod's partner John Kupper, who followed the campaign closely, thought "it didn't seem to be very professionally run or frankly well thought out." In time, Barack agreed. "The problem with that race . . . was in conception. There was no way I was going to beat an incumbent congressman with the limited name recognition that I had."

Hyde Park state representative Barbara Flynn Currie, who believed Barack's 30 percent showing was a "disaster," thought the campaign's fatal flaw was more fundamental, because "the kinds of arguments that he made in his own behalf are pretty subtle." Asserting that Rush was not a policy leader was a difficult argument, for "it's sort of inside baseball," Currie explained. "When you call somebody a hack, you want to be able to identify ways in which hackery is their middle name," something Barack's campaign had failed to do. What's more, "Bobby ran a textbook campaign . . . no mistakes." Will Burns agreed with Shomon that Barack got in too late and that Trotter failed to cut into Rush's base, but he also understood that "it was hard for us to make a persuasive argument to voters as to why they should vote for Barack Obama and effectively take away Bobby Rush's job based on the idea of change" because "it was very abstract, it was not very concrete."

In addition, top fund-raiser Judy Byrd explained, the campaign "miscalculated the strengths and the historical significance of who Bobby Rush was to that community." Furthermore, as Toni Preckwinkle forthrightly said, Barack was "not a very good candidate. He behaves like a U of C professor, doesn't have much of a common touch, and thinks he's smarter than everybody, and that kind of thing doesn't go over very well." Byrd had "worried about that" too and realized, looking back, that Barack had been "learning on the job during that campaign" regarding "how he presented and how he related to the African American community."

Even so, many of Barack's black supporters were outraged by how widely the "white man in blackface" comment had been voiced. Burns recalled that "it got bad. It was real bad," and school director Tim King was one of many young black professionals who were "very offended" that the Rush campaign promoted "this notion that Barack wasn't one of us" because he had graduated from Harvard Law School and was teaching at the U of C. Barack ruefully told Jesse Ruiz, "Me being president of the *Harvard Law Review,* I never thought it could be a liability. But it was a liability in this race." Elvin Charity, who had recruited Barack as a summer associate back in 1989–90, recognized that Rush had succeeded in portraying Barack "as an outsider and an elitist." Carol Harwell, Barack's campaign manager in 1995–96, recalled, "The biggest critique I heard was 'Is he really black,' 'Is he one of us?'" Barack confronted "a lot of skeptical people," Carol realized, who wondered, "Why are you doing this? . . . This is unrealistic," including "in his own household."

Michelle's unhappiness had grown over the course of the campaign and was only heightened further by Barack's landslide loss. They had continued their evening meals with Bill Ayers, Bernardine Dohrn, and the Khalidis, and in the aftermath of Barack's defeat, Michelle's desire that he abandon electoral politics was often verbalized. "Michelle had lost her appetite for it by that point and was ready to give it up. She hated it," Bill remembered.[22]

In subsequent years, Barack blamed his shellacking on "impatience" and "hubris," and acknowledged that it had been "a terrible race . . . everything that could go wrong in that race went wrong." It was "a humbling experience and a useful one," and at another time he admitted, "I was completely mortified and humiliated, and felt terrible." In the course of the campaign, Dan Shomon and Barack had become even closer colleagues, in part because of Michelle's clear disinterest. In the aftermath, Dan saw that Barack "went through kind of a rough period" and was "kind of morose." On one occasion Barack told Shomon that "I don't really know" if politics is "what I want to do" anymore, and Barack later told Cynthia Miller he went through "a dark period" and "did a lot of soul-searching" following his defeat. Barack sounded a similar note to Carol Harwell. "'The hell with this. I could be making money. I've given my all and this is what people think of me,'" Carol recounted Barack telling her. "I think he got so hurt in that election," and while he said "he wanted to get out" of politics, "I think he was just having a super pity party."

Although Barack's campaign had raised more than half a million dollars, the election's immediate aftermath left unpaid bills of more than $50,000, on top of the outstanding $9,500 loan from Barack and Michelle. Barack made up for some of the unpaid salaries by channeling up to $2,000 from his state political fund to Shomon and Will Burns. Dan was quickly hired by a Chicago marketing communications firm, but Will went on unemployment before landing a job at the Metropolitan Planning Council. Among the supportive calls Barack received was one from Mayor Daley, who asked Barack why he had challenged Rush. Barack cited Rush's landslide loss to Daley, but the mayor replied that Rush's defeat in the mayor's race had not meant Rush's congressional seat was vulnerable. "Don't take a defeat as a complete loss. You learn something in defeat," Daley said.

A more surprising call came from someone Barack had never met, but who had recognized his name in press reports, Hawaiian congressman Neil Abercrombie, who forty years earlier had been Obama Sr.'s good friend. Barack responded warmly, but, just as with Barack Echols a few years earlier, he brushed aside Abercrombie's offer to tell him more about his father. "He didn't want to pursue it," Abercrombie explained, and even after they met each other years later, "we've never explored it, not even a little bit." Barack more enthusiastically embraced a letter from former senator Paul Simon, who observed that the outcome was "not a great surprise" but also "shows clearly that you have an ability to reach out to the white community." Simon said, "you should plan to run for state-wide office" rather than Congress, unless Rush stepped down, and added, "I would like to be of help to you if you run for state-wide office."[23]

Around Chicago, other supporters like Simon's old friend Ab Mikva knew Barack was "very dejected." Marty Nesbitt, the third member of Barack's finance committee troika alongside Judy Byrd and Jim Reynolds, realized that he "was disappointed that he'd let so many people around him down." On Wednesday, March 29, one week after his loss, Barack returned to Springfield. Former Senate chief of staff Mike Hoffmann happened by his office and Barack's "not his usual upbeat self." He was "really frustrated," and disconsolately told Hoffmann, "My wife won't let me run again." He could stay in the state Senate, and perhaps aspire to become president if the Democrats won back the majority, or he could leave and "make a lot of money," Hoffmann recounted. "He thought his political career was over."

Kim Lightford encountered Barack on the Senate floor. "I got my butt whipped from Bobby, but I whupped Donne," he told her. Trotter's abysmal performance gave Kim a new put-down to use when presiding at Black Caucus meetings. "Whenever he ended up being a smart mouth in caucuses from there on out, I'll go, 'Mr. 7 Percent,'" Lightford recalled. "The 7 percent man." Emil Jones Jr.'s view was similar to Mayor Daley's. Barack "ran a damned good race," and "it was a good loss for him," because "he learned a heck of a lot." But Jones also realized that the election "hardened the feeling between he and Trotter because he got more than Trotter did in Trotter's senatorial district." That "just added fuel" to "all this hostility between Rickey and Donne Trotter and Barack that existed."

That evening, the regular poker players gathered at Terry Link's house. Link had already spoken with Barack and immediately realized he was at "the lowest" Link had ever seen him. "I think he personally felt humiliated" and he was clearly "depressed. He thought it was the end of the world." That evening no one mentioned the election so Barack took the initiative: "All right. I know you all want to say it. Don't say it!" Everyone laughed and remarked almost in unison: "We told you so!" Recently appointed senator Ned Mitchell, who lived with Link but had just lost his own primary, was not a poker player but came down to the basement to visit "as long as I could stand the smoke." Mitchell did not think Barack was a terribly good poker player, "simply because you could read his facial expressions," but in addition to Barack's "congenial, friendly" manner, he was struck that Barack was "*really* loyal to his family, and that impressed me about him, because you know you see things in Springfield" after dark.

More openly with Kim Lightford than with anyone else, Barack "would talk about temptation in Springfield. And he would say, 'No, no, no, I would never. Michelle would kick my butt. Not only would it not be worth it, but I would not want to have to deal with that.'" To one close acquaintance, Barack boasted about one long-ago conquest. "The only woman he ever talked about screwing was some really hot rich blond chick that he was still proud of." The friend suspected Barack "may have been stretching the story a little bit," although "he was really proud he'd banged some super-hot blonde from a super-rich family," an exaggerated account of Alex McNear's upbringing. Like others, Kim Lightford knew "you'd never see him out much in the evenings" in Springfield because Barack "was very selective about receptions that he went to, how long he was there." Barack "comes and he stands in the doorway, and he glimpses the room, and he makes a decision where he wants to go," Kim recounted. Barack would "work the room and get out," telling Kim "'it's not worth it, going out and hanging out, and getting involved with other women.' He says, 'When I think about it, I just think about . . . how mad Michelle would be at me. I don't want her jumping on me,'" because "'I'm scared of her.'"

Barack skipped the eighth and final Saguaro Seminar session and passed on taking part in reviewing the group's final report. In Springfield the first week of April, Barack hosted five U of C law students, three of whom had volunteered on his congressional race. They had entered the winning bid—$600—for Barack's item, a day in Springfield, in the law school's new Chicago Law Foundation annual fund-raising auction. Joe Khan initiated the bid and found Barack "extremely gracious" during a day that included a committee meeting, a tour of the capitol, and lunch with Barack and Dan Shomon. Heather Sullivan, who had had two classes with Barack, recalled that "he just seemed kind of disengaged" in Springfield, because "it was clear that his ambitions were elsewhere."

On the Senate floor, Barack spoke in favor of a bill to expand free school breakfasts and lunches, saying it "should be a no-brainer" because "we've got hungry kids who are coming to school without enough food." He also reminded his colleagues that "we have a burgeoning number of working poor in this state, who work every day but don't have health insurance," and a few moments later, he rose again to speak on behalf of a bill to provide state aid for the training of day care teachers. "There is a day care crunch all across the state," Barack explained, one that he was personally familiar with because "I've got a two-year-old at home and know firsthand the expense and difficulties of child care." When "there's less turnover, the outcomes are better for the kids." Both bills passed easily and were soon signed into law, but far bigger news, particularly for embattled governor George Ryan, was that Pate Philip finally allowed unanimous Senate passage of a compromise gun bill restoring the felony provision that had caused Barack such grief three months earlier.[24]

In Chicago, Barack joined comedian Dick Gregory at a black women's expo, telling the crowd that "we have a genuine criminal problem in our community. We need to have an agenda on how to rectify some of the things that force our youths to participate in this illegality." A *Hyde Park Herald* reporter asked Barack what he intended to do next, and a lengthy conversation ensued. "Right now my plans are to finish this session and make sure that some priorities that we care about get passed out of the legislature and then to spend some time with my wife and child this summer. I don't have any immediate political plans in the offing. . . . Obviously, after every election you end up needing some time to reflect on where you've been so you know where you're going. Overall, we feel very positive about how the election turned out. I think we laid a terrific foundation for the future, and I think that for a first time out in a large scale, well-publicized race, I think people were favorably impressed. And so I think that options will present themselves as time goes on. I'm in no rush to make any immediate decisions about next steps."

Those comments reflected a significant emotional bounce-back from Barack's state of mind just two weeks earlier, and he also was ready to fully re-engage for the final week of an unusually brief spring session. "What I'm really pushing legislative leaders to look at is a proposal to institute a state Earned Income Tax Credit," a "tax break for the working poor." On April 11, the Four Tops and Governor Ryan reached a budget agreement that included creating a state EITC while also providing another $380 million split among the four caucuses for more "member initiative" pork-barrel projects. The next evening Barack appeared on Bruce DuMont's *Illinois Lawmakers* TV show and said he was happy with the gun-bill outcome but was concerned that Senate Republicans wanted to alter how Illinois disciplined correctional officers who were caught using drugs. Existing law gave prison employees "three strikes" before being fired; a bill being pushed by Pate Philip would reduce that to zero. Prison guards were AFSCME members, and Barack's lobbyist friend Ray Harris had come to him to say that "Pate wants to do this drug testing on our members" and "I want to get it knocked out." Barack reacted dubiously. "Ray, you want me to be against drug testing?" "Not on its merits," Harris responded. "You've been doing this too long," Barack replied.

Barack reluctantly voted yes when the bill passed 45–3–10, since, as he told DuMont, three strikes "was too much." House Democrats killed Philip's measure, but Barack stressed to DuMont that creating an EITC "was something that I had pushed for since I first got down in Springfield." Another of DuMont's guests was *Capitol Fax*'s Rich Miller, and afterward, Barack joined Miller for a drink and told him how unhappy he was at Miller's characterization of him in the *Chicago Reader*. But "he took that criticism the right way," Miller later explained, when "he could have taken it the wrong way," and in the aftermath of his defeat, Barack began to alter how he went about some statehouse interactions.

On the session's final day, only fiscally conservative Republican Chris Lauzen joined Barack in voting against a $6 billion bond authorization bill that senators approved 56–2, with more than a quarter of the funding targeted for state prisons. "We have a tremendous strain on our prison system, partly because of the laws that we pass in this legislature, and we need to build more facilities to house the inmates that we're throwing in," Barack said. Senators should "examine how it is that we're spending such a huge proportion of our budget on corrections," because "we are getting close to the point we are actually spending more money on corrections every year than we are in terms of new school construction, and I think that's something that we need to take note of."

As careful observer Kent Redfield commented, the end-of-session budget negotiations highlighted how "the rank-and-file members of the legislature were largely irrelevant to the process." Barack won a modest 5 percent EITC provision, which meant some 750,000 low-income families earning under $30,000 a year would receive an average credit of $55 at an annual cost to the state of $35 million. It had won approval only because Emil Jones forcefully pushed for its inclusion in a Christmas tree budget made possible by the state's strong economy. Barack told the *Hyde Park Herald* he was "very disappointed" the credit was only 5 percent, and not the 20 percent he had sought, but he hoped to increase that percentage in future years. Barack's only other spring session success was his bill requiring domestic violence training for state employees working with public aid recipients. As legislators left Springfield, eyes turned toward the upcoming general election. Senate Democrats had virtually no chance of erasing Pate Philip's Republican majority, but in 2001, with new census figures, state redistricting might alter those prospects for 2002.[25]

On April 15, the Obamas had to write a check to the IRS for nearly $17,500 to cover the balance of what they owed on their 1999 income of $181,000. Barack had again received board fees from Woods and Joyce on top of his Senate and UCLS salaries, but the congressional race had taken up so much time that he had earned no law firm income. A pair of Steve Neal columns in the *Sun-Times* heralded Barack's future prospects, one headlined "Attorney General May Be Obama's Calling." The second described Barack as a possible mayoral candidate should Richard Daley step down, with Neal declaring that Barack's congressional loss "hasn't diminished his political stock."

But Barack still faced a daunting $60,000 debt from that race, and he turned his attention to reducing it as best he could. Former Democratic National Committee chairman David Wilhelm, who had started a Chicago political consulting firm two years earlier, hosted a "really desultory fund-raiser" on a humble ferry boat on Lake Michigan, with Dan Shomon feeling like "they were going to bring out a casket or something." One attendee, Laura Russell Hunter, a former network news producer who had just moved to Chicago, met Barack and Dan for the first time, and Shomon eagerly recruited her as someone who could spread the word about Barack to other successful white professional women. Dollar-wise, Barack fared better with political colleagues than he had at the marina, receiving $1,000 apiece from the campaign committees of Democratic state senators Jimmy DeLeo and Lou Viverito, $500 each from Democrat Pat Welch and Republican Steve Rauschenberger, plus $1,000 each from lobbyist Al Ronan and Joyce Foundation board colleague Paula Wolff. African American lobbyists Frank Clark and John Hooker saw to it that Obama for Congress received $2,500 from Commonwealth Edison's PAC, and by June Barack's congressional debt had been reduced to $25,000 plus the still-outstanding $9,500 loan from the Obamas.

In May Barack submitted the paperwork for a half dozen "member initiative" grants to the Chicago Park District totaling $1.1 million. Five of them were for improvements ranging from $50,000 to $200,000 at parks across the South Side; the largest by far, $590,000, would create a lovely new park on the north side of 79th Street, less than a block from Father Mike Pfleger's St. Sabina church. Appearing on WTTW's *Chicago Tonight* newscast with his Senate colleague Jimmy DeLeo to spar over state sponsorship of gaming, Barack said that "lower-income and working-class people who can least afford it" bought a disproportionate amount of lottery tickets. It was "troublesome" that Illinois "systematically targets what we know to be lower-income persons as a way of raising revenue" because that was not "the fairest way for us to raise revenue." DeLeo noted that the state lottery contributed $515 million annually for education funding, and Barack leavened his criticism, saying, "I don't want to come off as a scold, because as I mentioned, had I won on my ticket, I would not be here tonight."

But Barack had a more fundamental objection. "The lottery is part of a growing culture that (a) is money obsessed, two, defines success or happiness or even our fantasies are shaped around having lots of money. That's not necessarily a healthy thing." DeLeo stressed that gaming, unlike taxes, was entirely voluntary, and that many people supported riverboat gambling, which led Barack to note that the boats "wield tremendous influence down in Springfield that is probably disproportionate." A few nights later, speaking at a panel discussion on criminal justice issues, Barack rued a political culture that tolerated "the police code of silence" and wrongful prosecutions yet "punished" elected officials who were "perceived as being soft on crime." Again Barack had a more fundamental criticism: "poor folks generally get a raw deal in the criminal justice system," although he also noted that "the only argument for the death penalty is vengeance, and that is a valid emotion."[26]

When the Chicago City Council passed a resolution asking the U.S. Congress to consider paying reparations for slavery to present-day African Americans, Barack appeared briefly on the *CBS Evening News* and firmly dismissed the possibility of that coming to fruition: "generally the Supreme Court has a philosophy that you have to identify a clear wrongdoer and a clear victim" in order to recover damages. In Illinois, almost everyone was astonished when Governor George Ryan, ostensibly a lifelong abortion opponent, vetoed a bill the Senate had passed 33–23–2, with Barack among the nos, ending Medicaid funding of abortions for poor women facing health risks. Ryan's decision, just like his capital punishment moratorium and his unwavering gun-bill stance, left conservatives feeling "completely betrayed."

Capitol Fax's Rich Miller, like Ryan a native of Kankakee, Illinois, was amazed that Ryan had "flip-flopped on so many long-held beliefs and campaign promises," especially given "the reactionary environment he came from."

Kankakee was "one of the most close-minded, bigoted towns I've ever seen," Miller explained. "It's been segregated forever. I know educated Kankakeeans who regularly use the 'N-word' in everyday conversation." Before becoming governor, Ryan had been "the quintessential Kankakee hack," a "warm, genial man" who was "also one of the coldest, most calculating politicians."

Miller's friend Steve Rhodes observed in *Capitol Fax* that Ryan had been so transformed that "our Republican governor is actually a Democrat." In late June the governor called a one-day special session to ask the legislature to approve a six-month suspension of Illinois's 5 percent gasoline tax in the run-up to the fall general election, given rising fuel prices. On the same morning the legislators gathered, Steve Neal published yet another column promoting Barack's political future, again recommending him as the Democratic nominee for state attorney general in two years. Barack reluctantly voted in favor of Ryan's supremely political gas-tax rollback, suggesting Illinois would be better served by not losing that revenue.

The summer of 2000 was the most relaxing time Barack had had in more than a year. Michelle's former Public Allies protégé Malik Nevels, now out of law school but still driving Barack's former Saab, recruited him to join African American broadcaster Tavis Smiley at a Chicago Urban League conference aimed at encouraging one hundred African American teenagers to get "wired and tuned in." In late June Barack attended his first Chicago Annenberg Challenge board meeting in six months. His departure as board chair a year earlier had coincided with the onset of the effort's long-scheduled wind-down. By early 2000, CAC had only $4.5 million left to award after raising the full $59.8 million in required matching funds from Chicago-based foundations and receiving the entire $49.2 million commitment from the Annenberg Foundation.

Only Barack and Pat Graham remained from CAC's original board, and Barack's four and a half years as chair had encompassed a dramatic reining-in of the program's expectations. Initially envisioned as focusing its support on about a hundred schools, the forty-seven networks CAC funded instead included more than 210. "We could have done things more systematically," executive director Ken Rolling confessed to *Catalyst Chicago,* which concluded that CAC "might have had more concrete results if they had invested more money in fewer projects." One friendly critic said "it is extremely difficult to attribute any specific achievements or progress to the work of the Chicago Challenge," and noted how "there was little contact between the Challenge and individual schools." Three and a half million dollars was channeled to support research on the impact of the rest of CAC's grants, a requirement that neither Barack nor Rolling welcomed. Graham repeatedly witnessed how some of the researchers "drove Barack absolutely berserk because they talked in educationese, without clarity," but Pat came to think that "undoubtedly that research was the most lasting contribution of the Chicago Annenberg Challenge." Those researchers concluded that the CAC "provided too few resources and too little support to too many schools" and "had little impact on student outcomes," which all along had been Barack's insistent focus. Fred Hess, who along with Don Moore had the greatest impact on Chicago school reform throughout the two decades predating 2000, nonetheless believed at that time that "the quality of opportunity that kids are starting out with in Chicago today is significantly higher than it was fifteen years ago," when Barack had first arrived in Roseland.

Like Jean Rudd, Ken Rolling had gotten to know Barack well during those early years, but the five years that they worked together were "a time of his life that was I think stressful," Ken recalled. "Money was always a big issue" because "they were in debt" and "living on the edge." More than once Barack vented to Ken. "She's on me. Michelle wants money. 'Why don't you go out and get a good job? You're a lawyer—you can make all the money we need.'" One night Barack got "really upset" when Ken called to say he needed the board chair's input on a pending decision. "God damn it, I'm not getting paid for this. I can't spend my time on this," Barack yelled into the phone. "'Thank you' was very hard to come by," Ken recalled, and after the congressional loss, Ken too heard Barack profess doubt. "I don't know. I think I might just chuck this whole thing, and I'm just going to go get myself a good job and make some money."[27]

Jesse Ruiz heard similar comments at a summer breakfast. Barack "was very demoralized" and was "complaining about Springfield," Ruiz recalled. "A lot of my classmates are partners at law firms now," Barack said, adding that "you probably earn more than I do!" In midsummer Barack, Michelle, and two-year-old Malia traveled to Washington to see Rob and Lisa Fisher. The visit began inauspiciously, with Barack telling their hosts, "'We've forgotten our suitcase,' and Michelle was like, '*We* didn't forget our suitcase! I packed for me and Malia, *you* forgot *your* suitcase!'" Rob and Lisa already knew "there's no question as to who's the boss of that household," and it was readily apparent how unhappy Michelle was with Barack's political life. "Michelle's pushing him hard at that point," Rob recalled. "'Let's take care of our family. This is not worth it.'" During that visit Barack and Michelle also went for brunch to the home of old Harvard friend Julius Genachowski and his wife, and afterward Barack confessed to Julius that Michelle had asked him, "When are we going to have a life like that?"

In late July and again in August, Barack signed off on twenty-eight different "member initiative" projects, committing a total of $1,080,000 in state funds to a wide range of South Side organizations. Twenty-five of the grants ranged in size from $2,500 to $50,000. The three larger ones provided $100,000 each to the South Side YMCA and to Father Mike Pfleger's St. Sabina Church and $450,000—up from the $400,000 announced just before the congressional primary—to Wellington Wilson's Kennedy-King College. Speaking to the Hyde Park Chamber of Commerce, Barack turned from legislative issues to describe his daughter Malia. "She's such a little sponge right now, she's soaking everything in. I know she'll be able to compete at any level because she's had well-educated parents," but "that's not the case for most children." Barack's role as cochair of JCAR led the progressive Woodstock Institute to recruit him to take the lead in pressing Governor Ryan to order the state's Office of Banks and Real Estate, and the Department of Financial Institutions, to strengthen predatory-lending regulations, which a recently passed bill had instructed them to draft. Meeting with Ryan in Chicago on August 10, Barack said the executive branch agencies had not lived up to the legislature's intent, and that JCAR expected to see stronger rules in the near future.

That initiative won Barack front-page attention in both the *Chicago Defender* and the *Hyde Park Herald*, and in the third week of August he flew to Los Angeles, where the Democratic National Convention was about to nominate Vice President Al Gore to take on Texas governor George W. Bush in the November presidential election. Unlike many Illinois politicians, Barack was not a member of the 190-person state delegation, but he thought the trip might be helpful because "I was a little depressed." Barack had reserved a rent-a-car to get from LAX to the Burbank Airport Hilton, where the Illinois delegates were staying, but at the Hertz counter, his credit card was declined and he had to spend "half an hour on the phone" pleading with customer service to increase his credit limit. "Once he got to the hotel he had to increase his debt limit again to get a hotel room," Senate colleague James Clayborne remembered Barack telling him. At a Monday-evening reception at the Los Angeles County Museum of Art, an Associated Press reporter asked Barack to comment on rumors that state Democrats hoped U.S. senator Dick Durbin would run for governor in 2002. "Dick Durbin is not only the most popular Democrat in the state, but the most popular politician in the state," Barack replied. But without a floor pass like official delegates received, Barack and Clayborne ended up taking a bus to the Staples Center and waiting for a day pass together. One night they managed to sneak up to the suite level and watched the proceedings from the box of Donna Brazile, Gore's campaign manager, but Barack found the trip enervating rather than restorative. He was "the guy in the room who nobody knew, but everyone knew didn't belong," he recalled a few years later.

Hasan Chandoo recounted a late-summer conversation: "After the election he's fucked. He's got no money. He's maxed out on credit cards." Michelle would later mention receiving telephone calls from debt collection agents and hesitating to open the mail because of unpaid credit card bills. But with Dan Shomon insistently encouraging him to focus on legislative prospects in the upcoming 2001 session, Barack and Dan revived the "Obama Issues Committee" that had debuted a year earlier at the outset of the congressional race, this time directing volunteers like Will Burns and young attorney Andrew Gruber to "develop bill and research ideas" that Barack could use to introduce legislation in January.[28]

Barack continued to emphasize that "the state's commitment to education is lacking," and in late September he returned to Cambridge for a weekend gathering of six hundred black Harvard Law School alumni. One of his old basketball buddies remembered Barack making some "very poignant" remarks about the challenges of inner-city education, and the conference proceedings recorded him remarking that "what strikes me is how many talented young students come out of law school thinking they have no options." At the end of that month, a board retreat aimed at rejuvenating Bronzeville's Hope Center, where Barack's former aide Cynthia Miller was now program director, showed that Barack's vision of what the center should be had not changed from years earlier. Barack "underscored the need for a citywide space for the development of leadership and issues strategies within the African American community," but just as before, other members wanted to retain a Bronzeville focus. "You cannot be organizing in a local community while trying to develop broader coalitions too," Barack insisted. "There are *no* local organizations in the Black community capable of shaping the *city* agenda," and "what seems to be missing is an overarching strategy for change in the African American community." Barack said that "citywide doesn't mean everything," just that "the conception of what we're trying to build is not confined by geographic boundaries." That meant both "public policy change" and "institution building," but others wanted the Hope Center to remain focused on "changing people's level of awareness in ways that enable them to improve their daily lives," a much more modest mission.

Late September was also the start of the law school's academic year and Barack's return to teaching after nine months away. He once again offered both Constitutional Law III and Current Issues in Racism and the Law, each unchanged from prior iterations and with the Racism syllabus wrongly stating that the class was being offered during the spring quarter, as it had been up through 1996, rather than in autumn. Con Law drew only seventeen students, while Racism attracted thirty-two almost entirely liberal students in a law school that continued to have even fewer African American students than it had a half decade earlier. "It was a school that didn't really understand race relations," one white male 3L explained. It was "stiflingly white male dominated," a 3L female who would become a partner at a huge international law firm recalled, "the most stifling environment I've ever been in."

Once again, as 3L Stacy Monahan Tucker remembered, whenever a student "tried to reduce an issue to a claim that the 'right' result was clear," Barack "would jump in and masterfully argue the other side—not as though he was playing devil's advocate, but as if he truly believed these issues were complex and multilayered, and it would be irresponsible to pretend that there were easy or obvious answers." Barack "seemed to honestly understand and sympathize with the viewpoints of those who took what would be viewed as the more 'bigoted' or less progessive position." Peter Steffen, also a 3L, agreed that Barack "would often argue the more conservative side of the issues" and "was pretty good at keeping a poker face about what he really thought." At one of the course's nine weekly meetings someone asked him directly, "Where do you stand on reparations?" and Barack "paused a long time" before responding, "I think there's a moral case for reparations, but I don't know that in practice we can actually ever do anything about it in any kind of practical way." There was no question that Barack remained a "very effective classroom teacher," but with the final month of classes given over to the students' group presentations, one woman felt "he outsourced about half the class." More importantly, as 3L Rachael Pontikes realized, Barack's syllabus featured 100 percent male authors, causing her to remember that "women's voices were missing from the class."[29]

In the fall, Barack signed off on more "member initiative" grants of state funds to additional South Side groups. In September, $100,000 each went to the Museum of Science and Industry in Hyde Park and to the Chicago Better Housing Association (CBHA) for a six-block-long Englewood Botanical Garden/Education Project that Barack had announced eight months earlier and that was located directly across the street from a CBHA-sponsored housing development. Then, on October 4, he submitted paperwork to award $75,000 to the South Shore Community Partnership Network, a part of Yesse Yehudah's FORUM—Fulfilling Our Responsibility Unto Mankind—"to purchase computer equipment, print literature and training manuals, and to cover administrative costs for the 21st Century 'e'-family.com initiative." Eleven months earlier, Yehudah had handed Obama a bundle of five $1,000 campaign contribution checks from FORUM employees for the congressional race. "Yesse came and kissed his ass," Dan Shomon explained, and then "yes, he got an earmark." Three days after Senate Democrats' chief of staff Courtney Nottage signed off on Barack's $75,000 request, another five $1,000 checks, this time from arm's-length Yehudah friends and acquaintances, arrived to help pay down Barack's outstanding debt. The quid pro quo could not have appeared more explicit, just as was Barack's knowing descent into Illinois's "pay to play" culture: in exchange for $10,000 in campaign contributions to a needy candidate who still needed to recoup $9,500 of his own family's funds from Obama for Congress 2000, $75,000 in state funds had been given to Yehudah's organization to use at his discretion.[30]

In contrast was Illinois's Republican U.S. senator, Peter Fitzgerald. Less than two years into his first term after defeating Carol Moseley Braun, Fitzgerald, on the same day that Barack was submitting the paperwork for Yesse Yehudah's $75,000 grant, took to the floor of the U.S. Senate to filibuster an Interior Department appropriations bill he feared lacked adequate safeguards to prevent Republican insiders' corrupt self-dealing in Springfield. Less than a week earlier, *Capitol Fax*'s Rich Miller had cited Fitzgerald's "super-strong distaste for the pinstripe patronage crowd and the politicians who cater to them" while writing about "how corrupt this state can be." Fitzgerald's fear was that federal funding for Springfield's Abraham Lincoln Presidential Library and Museum would be handled so that "a bunch of political insiders wind up lining their pockets at taxpayer expense." He was particularly concerned about Republican insider Bill Cellini, whose behind-the-scenes influence in state party affairs was infamous and whose wife, along with those of Governor George Ryan and House Republican leader Lee Daniels, sat on the Lincoln Library's board.

Democratic campaign strategist Pete Giangreco praised Fitzgerald, telling *Chicago Magazine* that "he's on target about the Lincoln Library—that thing looks like it's larded up like the proverbial fatted calf." Poll numbers showed Fitzgerald with better than 62 percent approval among Illinois voters, but publicly trashing his own party's top state leaders in the most prominent forum imaginable contributed to the belief that Fitzgerald's reelection in 2004 "is far from a sure thing," *Chicago* reported.

In late October, several weeks before the veto session, Barack traveled to Carbondale to deliver a keynote address at a Youth Government Day event at Southern Illinois University, at the invitation of former U.S. senator Paul Simon. Barack began by highlighting how "pervasive in our culture at this point is a cynicism about not just the political process but public involvement generally. We have a tendency to believe that politics is more of a business than a mission." Barack explained that he held public office because "I feel like meaning in my life depends on me being involved in something bigger than just myself, that unless I hitch my wagon to larger movements of people and the community at large that I'm selling myself short." Barack joked that "having a law degree allows you to pretend that you know what you're talking about even though you don't," and he complained that "change is awfully slow in the political process," especially because "particular groups are so much better organized than the public at large."

Lobbyists, he said, capitalized on a "monopoly of information" in Springfield, where "the degree of partisanship is not based on ideology but simply on the basis of the interests of leadership in maintaining control of their particular chamber." Barack explained that "your state senator or your state representative has ceded much of their individual law-making power to the leadership" of the Four Tops, who focused on "policing their membership and making sure that they don't cross the aisle and work as effectively as they might." Barack was displeased that "our public policy debate isn't as effective as it should be . . . because we frame things as either/or as opposed to both/and," but he referred to his days as an organizer in stressing that "it's important to communicate to people in terms of their self-interest what government and politics" can do for them. Soon after, Paul Simon sent Barack a thank-you letter, commending his "adept handling of the English language" and reiterating that "when the right opportunity presents itself you should be and will be a candidate for state-wide office."[31]

In late October, Michelle discovered that she was again pregnant, happy news but also problematic for the possibility, as the *Sun Times*' Steve Neal repeatedly wrote, that Barack would be an ideal Democratic candidate for state attorney general in 2002. Dan Shomon joined Paul Simon in believing that Barack should look to a statewide race and procured a batch of "Obama Statewide" buttons to hand out at one Democratic gathering. Barack had passingly mentioned the AG possibility to Jim Reynolds, Judy Byrd, and Marty Nesbitt during an early fall conversation in Nesbitt's kitchen, but he told them, "I don't know whether Michelle's going to put up with another race. Maybe one more." Yet having a newborn baby at home just as the 2002 campaign schedule began to ramp up would be difficult, especially given how unhappy Michelle was about Barack's electoral aspirations. In addition, Barack would be up for reelection to the Senate, so any candidacy for another office would mean surrendering the post he already held.

Barack and Dan's informal issues committee focused on "legislative ideas for the 2001" session, and Will Burns recommended that Barack call for an increase in "the number of charter schools operating in Chicago," because such schools enjoyed "broad support . . . within the African American community" and would give Barack "school choice credentials in [a] state-wide race." One week before the presidential election between Al Gore and George W. Bush, Barack devoted a *Hyde Park Herald* column to the third-party candidacy of Ralph Nader. "I fervently believe in Nader's right to run in this election. I think he should have been allowed to participate in debates, and I share a number of his criticisms of the major parties."

Barack kept up a busy, sometimes hectic schedule of speaking appearances around Chicago, with a panel discussion at Northwestern Law School, a dinner address at the awards ceremony of a legal aid hotline service, and an appearance at the Developing Communities Project's annual convention. In the presidential election's confused and uncertain aftermath, with the counting of votes in Florida leaving the winner in doubt, Barack appeared on WTTW-TV's *Chicago Tonight* program and said, "the federal courts have no business being involved in this." While the Republican front-runner's status was still uncertain, Barack said, "I don't think this is going to destroy President Bush's legitimacy in any fashion."

Veto session in Springfield began contentiously with the status of Illinois's payday loan draft regulations directly confronting JCAR. Barack told his colleagues, as well as an unusual crowd of onlookers, that it was "a difficult issue and a controversial one," and he commended the Department of Financial Institutions (DFI) for having done "an admirable job on this." But Barack added that JCAR unanimously believed "we need to make a few further refinements on these rules," because even though "there is a need to institute some significant regulation in this area" in order to curb "the worst abuses," the Illinois Small Loan Association (ISLA) had demonstrated that "some of these provisions will pose an unreasonable economic burden on small lenders," of which Illinois had more than twelve hundred. Barack told reporters, "I'm optimistic that we're going to end up with some rules that will allow these payday loan companies to operate profitably but at the same time curb the egregious abuses that we've been seeing in some neighborhoods." ISLA executive director Steve Brubaker rhetorically asked, "why should a customer of a payday loan business be told by the government how much they can and cannot borrow?" but Barack's Republican Senate JCAR colleague Doris Karpiel explained why the legislators were disagreeing with DFI: "We thought they were going too far, and the industry was upset." In traditional Springfield fashion, JCAR exercised its maximum authority and imposed a six-month delay before DFI's new rules could take effect.

Barack finally won Senate passage of a bill extending the life of tax abatement "enterprise zones" by ten years, an effort that had been narrowly defeated eighteen months earlier. This time, Barack had recruited influential Republican senators Frank Watson and Carl Hawkinson as chief cosponsors, and the measure passed handily, with George Ryan signing it into law early in the new year. More consequentially, on the final day of the session, word spread that Barack's young Senate colleague Lisa Madigan had her eyes on a statewide 2002 race, probably attorney general. During a brief conversation earlier in the session, Barack had heard from House Speaker—and state Democratic Party chairman—Mike Madigan that someone other than Barack was envisioned as the Democratic nominee for attorney general.

In early December, Barack was again on WTTW's *Chicago Tonight* as the battle over Florida's recounting of disputed ballots intensified with a 4–3 state supreme court decision ordering a partial recount. "I was surprised," Barack exclaimed. "I think the Florida state supreme court did the right thing," but the incomplete approach to the recount "makes them a little more vulnerable potentially under Supreme Court review." When a fellow guest observed that "whoever wins this victory is going to be tainted," Barack added, "I think that's true," and several evenings later the U.S. Supreme Court reversed the state court's action, resolving the presidential election in favor of Republican George W. Bush.[32]

Before flying to Honolulu with Michelle and Malia for their annual visit with Toot and Maya, Barack began an effort that would over the following three years bear him very rich fruit. Loop Capital founder Jim Reynolds, who had worked as diligently as anyone at soliciting donations for Barack's congressional race, was about to become president of the Alliance of Business Leaders and Entrepreneurs (ABLE), a low-visibility, thirty-odd-member roundtable of black financial professionals. Barack's sister-in-law Janis Robinson had been ABLE's part-time executive director for several years before taking a job at the U of C's business school, and Barack already knew some members like John Rogers Jr. and Robert Blackwell Jr. After the congressional defeat, Reynolds had begun taking Barack to ABLE's monthly meetings so more members could get to know him. In part the gatherings were gripe sessions, with the energetic business leaders discussing "what the new Black entrepreneur was really up against," publisher Hermene Hartman recalled.

Barack "was usually quiet and more observant than talkative," Hartman said, but "we asked Barack to help us identify business opportunities with the state." Bond attorney Stephen Pugh remembered that Barack "wanted to understand what our needs were," and Barack recalled them in particular telling him that "we are not getting any business from our own state" pension funds regarding asset management as well as equity and fixed-income trading. Barack told Reynolds to reach out to House Speaker Michael Madigan, and both Madigan and Emil Jones Jr. recognized these financial services leaders as potentially valuable new campaign supporters. Shortly before Christmas, Barack accompanied John Rogers and several other ABLE members to an initial meeting with representatives of the Teachers' Retirement System, one of Illinois's three huge public pension investment entities.

Barack also remained highly concerned about his own finances. By the end of 2000, his congressional race debt had been reduced to $23,803, plus the $9,500 that the campaign owed him and Michelle, and Barack told the *Hyde Park Herald* that asking for contributions was "the least attractive part of being an elected official." With a second child on the way, Barack's modest state Senate and law school salaries, plus Michelle's from the U of C, did not make for an easy life, especially since his Miner Barnhill income had dried up entirely during his congressional race. Child care costs topped $1,500 a month, and payments on their student loans were higher than their monthly mortgage tab. Old friend John Schmidt, who was now planning a 2002 gubernatorial bid, tried without success to get Barack some foundation work, but a far better opportunity presented itself thanks to Robert Blackwell Jr., the young ABLE member and IBM veteran who was CEO of Electronic Knowledge Interchange Inc. (EKI), a management and technology consulting firm.

Blackwell offered Barack an $8,000-a-month retainer to serve as EKI's general counsel, reviewing contracts and offering input on "general business strategy." The handsome monthly fee would pass through Miner Barnhill rather than be paid directly to Barack, though Barack often went to EKI's office at 100 South Wacker Drive. Barack was "one of the three smartest people I have ever met," Blackwell recalled, someone who prefers "peaceful relationships" and "nonconfrontational people" while being "hard on the inside and soft on the outside." Raja Krishnamoorthi, a new Harvard Law School graduate who had worked on former New Jersey senator Bill Bradley's unsuccessful presidential campaign, had first met Barack some months earlier and that winter began visiting him at Miner Barnhill. Raja asked Barack why he was now working so hard, and Barack's reply summed up his life as 2000 ended: "I have a lot of bills to pay down."[33]

Before the spring 2001 legislative session got started, the U of C Law School's winter quarter began, with Barack teaching Voting Rights for the first time since 1999. Meeting on Monday mornings and Friday afternoons, the course drew only a dozen students, and Barack asked them to read the U.S. Supreme Court's decision in *Bush v. Gore* before the first class. Each student had to write a long paper, and Barack told them the first day that "no one's getting an A from me whose paper doesn't argue for a new idea . . . your theory, and you argue it," he explained. "You have to have an original idea to get an A." Three 2Ls remembered it as "a really great class" and recalled Barack as a "terrific" professor who "wanted to hear what people thought" and was "so interested and open to competing viewpoints." Nat Edmonds recalled Barack "pushing me, testing me on my ideas and my logic and my arguments," and Kristen Seeger found it "a really rewarding classroom experience." In addition, Barack "always took the time to chat after class," and "you could not miss how much he loved his wife" and family. When another 2L asked Barack to participate in a faculty-hosted weekly law-related film series, Barack agreed and chose *The Verdict,* a 1982 movie starring Paul Newman. "What I like about this movie is that it's about redemption," Barack explained.

In Springfield, a number of issues regarding JCAR required more of Barack's time than usual. He objected to a proposed new rule regarding high-risk loans because it would be "unduly burdensome" to lenders, and he queried whether not enough job-training funds were being spent on actual skills training for the unemployed. JCAR also intervened in a long-running dispute between the Illinois State Board of Education (ISBE) and U.S. District Judge Robert W. Gettleman, who three years earlier, in response to a 1992 lawsuit filed by Chicago special education advocates against the state, had ordered major reforms in how Illinois certified special education teachers. "ISBE continues to deny the undeniable and defend the undefendable," an angry Gettleman had held in 1998, and in 2000 "ISBE's failure to fulfill its responsibilities to ensure that children with disabilities are educated in the least restrictive environment" led Gettleman to order that the board override its normal rule-making process to institute new certification standards. ISBE complied while protesting that it lacked the authority to do so, and now JCAR moved to suspend the new rules, placing the committee in direct confrontation with Judge Gettleman.

On January 10, Barack seconded Emil Jones's nomination for Senate president, a pro forma proceeding given Republicans' unchanged thirty-two-seat majority. Barack told his colleagues that "the reason we're down here" is "to try to make the lives of our constituents, our children, our grandchildren, our parents and grandparents a little bit better." Then, a few moments later, Jones upended the Senate without any forewarning by announcing Senator Lou Viverito as one of his five assistant minority leaders in place of an astonished—and now deposed—Jimmy DeLeo. Only insiders fully appreciated the drama and intrigue because DeLeo, along with his close ally John Cullerton and Barack's poker buddy Denny Jacobs, had never been an enthusiastic supporter of Jones as the Senate Democrats' leader. Only *Capitol Fax* publicized the contretemps, but Jones's unexpected move, like JCAR's confrontation with Gettleman, would roil the statehouse once the spring session picked up steam in late March.[34]

In Chicago, Barack appeared again on WTTW's *Chicago Tonight,* objecting to President Bush's selection of conservative former Missouri senator John Ashcroft as attorney general. Barack noted that Ashcroft's record revealed a bevy of politically questionable associations, including his acceptance of an honorary degree from racially "offensive" Bob Jones University. Barack said he could not imagine former President Clinton nominating someone who had "received an honorary award from the Nation of Islam." The next day Barack joined his law school colleague Dennis Hutchinson on WBEZ Radio's *Odyssey* program for a long discussion of the Supreme Court's record on civil rights. Barack explained that "*Brown v. Board of Education* may be one of those rare circumstances where the court is willing to get slightly beyond conventional opinion," but he added that "very little" actual school desegregation took place over the next decade. During the 1960s, "you've got a cultural transformation that changes how states operate and how states think about the protection of individual rights in ways that didn't exist prior to the Warren Court, and I think that is an important legacy to keep in mind."

Where the court helped African American petitioners, Barack said, "was to vest formal rights in previously dispossessed peoples so that I would now have the right to vote, I would now be able to sit at a lunch counter and order, and as long as I could pay for it, I'd be okay. But the Supreme Court never ventured into the issues of redistribution of wealth and sort of more basic issues of political and economic justice in this society, and to that extent, as radical as I think people try to characterize the Warren Court, it wasn't that radical. It didn't break free from the essential constraints that were placed by the founding fathers in the Constitution, at least as it's been interpreted, and the Warren Court interpreted it in the same way." The Constitution "doesn't say what the federal government or the state governments must do on your behalf, and that hasn't shifted, and I think one of the tragedies of the civil rights movement was because the civil rights movement became so court-focused, I think there was a tendency to lose track of the political and community organizing activities on the ground that are able to put together the actual coalitions of power through which you bring about redistributive change." Barack stressed that "I'm not optimistic about bringing about major redistributive change through the courts," but "organizing's hard. Litigation is hard, but organizing's harder. Part of it is just that it's difficult to mobilize change at the local level."

Aiming to mobilize some tangible financial change, in mid-January, Barack, House Speaker Mike Madigan, Jim Reynolds, and three other black financial executives met with the members and staff of the Illinois State Board of Investment (ISBI), the largest of the state's three major pension investment bodies. The four business leaders had a detailed, written request for ISBI, and the presence of Madigan and Barack demonstrated that Illinois Democrats, black and white, were behind the firms' proposal. Barack later recalled that all he did by way of introduction was say, "Listen to what these folks have to say," and the executives had four specific demands: "1. Immediately require that all current ISBI money managers do 15 percent of their brokerage business . . . with African-American owned broker/dealers." "2. Allocate 15 percent of total ISBI assets to African-American owned investment management firms," which would represent $1.4 billion of ISBI's overall assets of $9.3 billion. ISBI at present had only $107 million, representing 1.2 percent of its assets, invested with one single African American–owned firm. "3. . . . place no less than 50 percent" of those two allocations with Illinois-based firms. "4. Accomplish the above three requests in a 90–120 day timeframe."

As one ISBI staff member noted, the black finance professionals "argued that since the Fund has many African-American beneficiaries, the make up of the fund should be similar to the make up of the beneficiaries." The board members agreed to appoint a committee to "recommend a policy regarding minority investment managers and broker/dealers," and "the guests thanked the Board for their time." Several weeks later, Madigan alone appeared before the Investment Committee of the Teachers' Retirement System to support a similar request to that pension fund.[35]

As spring session geared up, Barack had in hand, thanks to Dan Shomon, Andrew Gruber, and other Obama Issues Committee volunteers, a pair of well-drafted bills that would require state officials to monitor Illinois employers' compliance with the federal WARN—Worker Adjustment and Retraining

Notification—Act. One furniture company had recently laid off more than a hundred workers without providing the sixty days notice required by federal law, and a joint press conference with 4th Ward alderman Toni Preckwinkle announcing the legislation drew coverage in all three of Chicago's major daily newspapers. Barack also had a trio of gun-control bills aimed at depriving anyone with a record of domestic violence allegations from possessing a firearm, and by early February, Barack had introduced ten bills. In tandem with Emil Jones, Barack also championed a Senate bill banning seventeen types of high-powered firearms, limiting handgun purchases to one per month, and providing for handgun registration. "It's hard to find a rationale as to why anyone would want an AK-47 or need more than twelve handguns a year, unless the intent was to distribute," Barack told reporters.

On the political horizon, Barack told the *Chicago Sun-Times,* "I don't think Mayor Daley could be defeated any time soon," even though "African Americans constitute the single largest voting bloc in the city." Steve Neal kept an eye on the Democrats who might run for attorney general in 2002. Obama "has stronger credentials than anyone else in the field," Neal wrote, and though "he has discussed running . . . he may not want to give up his legislative seat for another race that he might not win." Less than two weeks later, Speaker Madigan's daughter Lisa began asking Senate colleagues to support her bid for the office, and with his own possibilities now clearly on hold, Barack devoted much of one mid-February Friday to the longest interview he had ever given.

Julieanna Richardson was a 1980 Harvard Law School graduate who had worked in cable television before deciding in early 2000 to launch a video-interviewing project to record the life experiences of successful African Americans. A year later, *The HistoryMakers* was still a nascent operation, but Richardson already had filmed interviews with Emil Jones, Alice Palmer, and Bobby Rush, among many others, so an up-and-coming state senator was well within her purview. Richardson began with some formulaic personal questions—favorite vacation spot? "Bali," of course, and Barack volunteered that lately he had been tired and sleepy.

He retraced his childhood upbringing and travels, exaggerating his high school misbehavior and his political activism at Oxy and Columbia. He spoke movingly about his work in Roseland before discussing his years at Harvard, his leadership of Project VOTE!, and his frustration with the state Senate. Barack decried the absence of "thoughtful independent voting" and the extent of partisan "jockeying and gamesmanship" by legislative leaders. He also spoke critically about Chicago's African American community. "We don't have a well-defined agenda in terms of what are the issues that we might move forward," and he rued that "we have a personality-driven politics as opposed to an agenda-driven politics. . . . When you look at those groups that are able to accomplish significant things through the political process, it's not because of the charisma of any particular leader, but it's because they have a clear agenda. Certainly that's how the business community operates. They are dogged in pursuit of a particular agenda," which in Springfield often worked to the disadvantage of underrepresented populations because "state government has much more influence on the day-to-day lives of people than the federal government does."

Again echoing John McKnight, Barack stated that African Americans "have to focus on how do we become productive as a community and not just consumers." Finally, Richardson asked Barack about his own legacy and formative influences. "I would like to, first of all, be able to say that I was a good husband and a good father along with being a good elected official." He believed "my career is still largely ahead of me as opposed to behind me," and he intriguingly stated that "most of my influences probably are not so much people that I knew personally as people whose words I've internalized."[36]

In late February the pace of the spring session picked up, and Barack reintroduced his racial profiling bill plus another one to increase the state's earned-income tax credit from 5 to 10 percent. Along with Republican Senate colleague Brad Burzynski and two other JCAR members, Barack met with U.S. district judge Robert W. Gettleman in Chicago to discuss the jurist's unhappiness with the Illinois State Board of Education's continuing opposition to his orders in the *Corey H.* special education case. Legislators like Burzynski wanted Gettleman to back off his insistence that ISBE promulgate new statewide standards, but Barack, the sole lawyer in the group, struck a far more deferential pose, telling Gettleman, "We're just here to listen to you." Afterward, Barack expressed displeasure with Burzynski's forcefulness, saying, "I might have to appear in front of that judge someday!" On February 21, JCAR voted to suspend one of the new rules ISBE had filed pursuant to Gettleman's rulings, and six days later, the judge ordered ISBE to implement the rule immediately, irrespective of JCAR's disapproval. Gettleman did allow ISBE to hold public hearings on the new standards, and Barack commended that move, telling reporters the "judge deserves credit" for that gesture.

Barack introduced another fourteen new bills addressing a wide range of subjects, although none merited a press release, and when Governor George Ryan presented Illinois's proposed budget, Barack expressed disappointment that it failed to expand KidCare to include parents as well as children. At the end of February, Senate Democrats' ugly internal divisions spilled into public view when Jimmy DeLeo and three Democratic friends—John Cullerton, Bobby Molaro, and Tony Munoz—joined unanimous Senate Republicans in voting to change Senate rules to allow President Pate Philip to elevate DeLeo from minority spokesperson to cochairman of the important Executive Appointments Committee. Additional Democrats sympathized with DeLeo, but on the Senate floor, Barack joined Rickey Hendon and Miguel del Valle in challenging the Republicans' power play. "It may be, for example—this is just a hypothetical," Barack facetiously declared, "that a Minority Spokesperson on a committee has gotten into an argument with the Leader and is not happy with the Leader. Maybe he was thinking about running for Leader himself and got upset. And so there's a power struggle taking place within the caucus," one that Republicans were now changing the Senate's rules to take advantage of. "It is a mistake for us to use some power issues that are taking place within this Chamber to fundamentally shift what is already a pretty significant imbalance of power between the majority and the minority and make it even worse," Barack warned. "It's going to produce more recrimination, more rancor."

In early March the Illinois State Library, across the street from the statehouse, hosted Barack for a lunchtime reading of *Dreams From My Father.* He again chose the passage ending with Lolo's lesson that it is "always better to be strong yourself." Barack also cited "the tension between family and ambition, and that is a tension that I continue to struggle with." As JCAR continued to wrestle with the payday loan industry's unhappiness with the state DFI's proposed new regulations, Barack and Lisa Madigan finally dissented from their colleagues' desire to side with the rip-off artists. But in the ongoing dispute between ISBE and Judge Gettleman over his rulings in the *Corey H.* case, JCAR unanimously moved to ask that both houses of the state legislature approve a joint resolution upholding ISBE's stance despite the judge's repeated orders.

The *Chicago Tribune* highlighted Barack's earlier role in directing $200,000 in state funds to Jesse Jackson Sr.'s Citizenship Education Fund, and seeking to defend the expenditure, Barack said that "taxpayers will not be subsidizing a for-profit entity" but instead a venture capital fund that would be "almost like a think-tank about how do we get capital in the community." As Barack had explained to former DNC chairman David Wilhelm, his goal was "asset-based economic development," again echoing John McKnight. "What do we have to build on" so as to create "entrepreneurial enterprises that are indigenous to the community"?[37]

With the Senate rarely in session prior to late March, Barack resumed active lawyering at Miner Barnhill in addition to his highly remunerative consulting for Robert Blackwell Jr.'s EKI. Four years earlier John Belcaster had filed a federal False Claims Act case on behalf of Dr. Janet Chandler, a medical researcher who had been fired by Cook County Hospital after blowing the whistle on fraud and corruption. The district court had dismissed the suit, but with Judd Miner determined to appeal to the Seventh Circuit Court of Appeals, Barack spent considerable time drafting two briefs that were filed with the appellate court in late March. Barack's former law student Marni Willenson, now a young Miner Barnhill partner, remembered them as "beautiful briefs," and when Judd argued the case before a Seventh Circuit panel several months later, Barack's handiwork proved to be "extraordinarily effective." The appellate panel found in Dr. Chandler's favor, and when Cook County appealed that decision to the U.S. Supreme Court, a unanimous affirmance resulted in Cook County and a second defendant paying more than $5.5 million in damages to the federal government and Dr. Chandler.

In Springfield, Barack spoke on the Senate floor in favor of charter schools and also a bill authorizing the sale of clean needles to drug addicts while stressing that "none of us here are seeking to decriminalize drug use." On March 30, three antiabortion bills championed by Republican senator Patrick O'Malley came to the floor, and Barack rose to say the first "is probably not going to survive constitutional scrutiny." He added that "this is an area where potentially we might have compromised and arrived at a bill that dealt with the narrow concerns about how a pre-viable fetus or child was treated by a hospital," but abortion opponents had forsaken that effort. Ten other Democrats as well as Republican Christine Radogno joined Barack in voting present, but SB 1093 was approved anyway, 34–6–12. Barack also voted present as the two subsequent abortion bills passed 33–6–13 and 34–5–13. Speaking to the *Chicago Sun-Times*' Dave McKinney, Barack said the underlying problem with all three bills was that "whenever we define a pre-viable fetus as a person that is protected by the equal protection clause or other elements of the Constitution, we're saying they are persons entitled to the kinds of protections provided to a child, a 9-month-old child delivered to term." That presumption, "if it was accepted by a court, would forbid abortions to take place."

In early April, Barack opposed a bill to allow people with orders of protection to carry concealed firearms, calling it "a Trojan Horse" because of the erroneous claim "that concealed carry will create a safer citizenry." The next day the Senate unanimously passed one of Barack's layoff-notification bills, which soon became law, and he spoke passionately in favor of a measure to increase state education funding. Echoing his Annenberg Challenge experience, Barack professed his "extraordinarily strong commitment to public schools" while emphasizing that "one of the most important things is fostering innovation in the public schools." A bill to require parental notice of minors' abortions produced a 39–7–11 majority reminiscent of the earlier antiabortion trio, with Barack once again voting present. Barack also sought to cosponsor other senators' bills, particularly ones introduced by Republican moderates like Christine Radogno, which he knew stood a better chance of Senate passage than Democrats' measures. Whenever such a bill did pass, the Democratic caucus's press staff churned out a press release. Stories such as a *Chicago Defender* one headlined "Obama Scores Big in Senate Passing of 3 Bills" resulted, with Barack warning that "we can't continue to incarcerate ourselves out of the drug crisis."

By 2001 most progressive lobbyists viewed Barack as a disappointing legislator, despite those bills and press releases. Veterans like John Bouman and Citizen Action's William McNary had befriended Barack early on. The capitol lacked a cloakroom, so lobbyists needed a senator whose office could be their home base, and McNary recalled that Barack's "was the place that a lot of progressive organizations went to hang up our hats and coats. We felt at home with him." Although no one doubted Barack's policy preferences, many thought his pragmatic attitude about what was politically possible in a body dominated by Pate Philip left him unwilling to champion many progressive issues. Doug Dobmeyer, an antipoverty lobbyist, remembered that "there were a number of times when he would say, "'Doug, look, we can't go that far.' That pretty much precluded him from being the point person because he wasn't willing to say 'I can go very far.'" Instead Dobmeyer turned to Hyde Park representative Barbara Flynn Currie, who was "a much stronger legislator than Barack," and in the Senate, expert Public Health and Welfare Committee staffer Nia Odeoti-Hassan was their "go-to person." On Judiciary, hardworking John Cullerton was the most influential Democrat, notwithstanding Barack's desire "to show people that he knew the law," as Republican Kirk Dillard put it. Barack "would ask a lot of more detailed questions, sometimes more theoretical than practical," Dillard remembered. Barack's tendency to "verbalize his thought process" in "the tone of a U of C law professor" could be "a little annoying," moderate Republican Dan Cronin explained.

Al Sharp, the executive director of Protestants for the Common Good, shared Dobmeyer's view. "'I'll do it, Al, but I have to tell you it's not going to happen,'" Sharp remembered Barack saying. "I appreciated the candor, and I appreciated the realism," but Barack was so "very pragmatic" that he was unwilling to fight the good fight. Legal aid veteran Linda Mills recalled that Barack "sponsored a number of bills I wrote," but "I stopped seeking him out as a chief sponsor pretty early on" because Barack was "disengaged" rather than actively pushing the bills. "He was never involved in the legislation," and on many days Barack was simply "unavailable, golfing, playing basketball. He was just out to lunch so often," Mills remembered. "The energetic guy that I had anticipated coming in didn't show up." John Cameron, who worked with McNary at Citizen Action before moving to AFSCME, echoed Mills. "If you were working with him on a bill," Barack had "an attitude generally of 'I shouldn't have to do any work. You guys do the work. . . . It's your bill. Go work for it.'"

During the 2001 spring session, organizing veteran Josh Hoyt worked hand in hand with John Bouman to expand KidCare into FamilyCare. They hoped Barack "would be a leader on this thing, which honestly he was not," Hoyt explained. What "was really disappointing was we went in to meet with Barack, and he gave us this very politically sophisticated explanation of why there was very little room to maneuver in the Senate under Pate Philip, that he had worked very hard to build up the political capital that he had with his fellow Republicans, and he was not going to squander it in a losing fight." Barack's attitude of "sophisticated resignation" meant that "he just wasn't going to fight when it was necessary to fight," Hoyt remembered.

Linda Mills believed Barack could be "exasperatingly rude" at times, with senators as well as lobbyists. On the floor, Barack "put his hands on people's shoulders as though they were his subjects. It was the most patronizing and condescending body language I'd ever seen." John Cameron agreed that Barack "exuded a lot of arrogance in that chamber," and on occasion Barack could be insultingly dismissive to old 1980s acquaintances like Don Moore and Hal Baron. On school reform, Moore said, Barack "never really did much for us," and when Moore took some Local School Council members to see him, Barack's greeting was "Are you people still around?" Baron, Harold Washington's former policy chief, had an almost identical experience, with Barack asking, "Hal, you're still doing this jobs stuff?" Even William McNary experienced Barack being "somewhat dismissive," although one former staffer felt that Barack was someone "you could just forgive being arrogant because he's so clearly smarter than you and just a gifted person." Campaign finance reformer Cindi Canary, who had known Barack since the "motor voter" case that preceded his Senate service, was struck that in Springfield "the people you expect to dislike him," like old-school Democrat Denny Jacobs, "love him," while "the people you expect to like him," like many progressive lobbyists and African American colleagues, "hate him."

The Wednesday-night poker games at Terry Link's house remained a mainstay of Barack's schedule, and he again offered a "day in Springfield" for the law school auction. One onlooker remembered auctioneer Richard Epstein, a conservative law and economics champion, trying to increase the bids by proclaiming that Barack "could be a senator someday! He could be president someday!" The four male 3L winners drove to Springfield on what one recalled was "a pretty quiet day around the Capitol." Barack took them to a committee hearing, onto the Senate floor, and to a late lunch at a small family-run Mexican restaurant near the statehouse. In the capitol's rotunda, Barack paused briefly to chat with a "schlumped-over" Governor George Ryan, who looked like "the saddest sack you'd ever seen" as the commercial-driver's-license scandal continued to roil. Lunch was "a great, very intimate time with him," although Barack was visibly "exhausted." Ted Liazos remembered Barack sounding "pretty reflective" and "telling us that he wasn't sure Michelle was going to let him run statewide" any time in the future. Marty Chester recalled Barack remarking "sort of in jest" that "If I run for something else now, my wife's going to divorce me." As lunch was ending, Barack told the soon-to-be-graduates, "If you have a chance when you get out of law school to make a little money, it's not necessarily a bad thing."[38]

On April 15, Barack and Michelle had to write a check for more than $12,000 to the Internal Revenue Service. Barack told the *Hyde Park Herald* that in 2002 he would run for another term in the state Senate, and an April fund-raiser for which African American attorney Peter Bynoe footed the $6,700 tab brought in $1,000 checks for Barack's state campaign account from African American financial figures John Rogers Jr., Mellody Hobson, Louis Holland, and Lester McKeever, as well as from prominent Chicagoans Penny Pritzker, Abby McCormick O'Neil, and Daniel Levin, plus $200 from Barack's fellow Woods Fund board member Bill Ayers.

In Springfield, JCAR gathered for what *Capitol Fax*'s Rich Miller called its "most closely watched vote in years" on the state's predatory-loan regulations. Thanks to the National Training and Information Center (NTIC), a coalition of neighborhood groups had mobilized in favor of strong rules, counterbalancing the lobbying prowess of the lending industry. "This has probably been as heavily lobbied an issue as I've seen," Barack told one reporter. At the meeting, House Speaker Michael Madigan spoke in favor of the muscular rules, and with all six JCAR Democrats primed to vote yes, and all three Senate Republicans opposed, two moderate House Republicans, Dan Rutherford and Tom Cross, would cast the decisive votes. The highly technical rules prohibited prepayment penalties and other practices that drained homeowners' equity, plus costly single-premium credit insurance.

Barack explained that "what we've been trying to distinguish is the difference between sub-prime and predatory, highly unfavorable loan arrangements that often result in foreclosures. If you're charging 15 or 16 or 17 percent interest, and you're tacking on fees and costs to this each time that loan is refinanced, then potentially you are drawing down thousands of dollars in profits before ultimate foreclosure." Lenders make "a whole lot of money" from such practices, and "in many of these loans, you can't identify any service or value that is being provided to the customer." Given that, "I don't see how as a state we can't step in and say these are essentially practices that we need to restrict and curb." With Cross and Rutherford joining Barack and the other Democrats, JCAR adopted the strong new regulations by a vote of 8–4.

Looming over the end of spring session was the remapping of all Illinois legislative districts pursuant to the 2000 census. The state's 1970 constitution provided for a bipartisan Legislative Redistricting Commission that had been intended to produce partisan compromise, but in each prior decennial iteration, disagreement had led to use of the constitution's final step, where a blind winner-take-all drawing by the secretary of state decides whether a Republican or a Democrat casts the commission's tie-breaking vote. Two young staff members, John Corrigan and Andy Manar, had been working for Emil Jones's Senate Democrats since early that year on a new district map, with the explicit goal of creating "as many minority districts as possible," Corrigan recalled. The affected members did not yet know the new boundaries, but an almost 70 percent increase in Chicago's Hispanic population since 1990 guaranteed the creation of two additional Hispanic-majority districts.

Barack returned to WBEZ's *Odyssey* program to discuss redistricting with Robert Bennett, his good friend from Northwestern University, and Maria Torres of DePaul University, the wife of Barack's legal buddy Matt Piers. "I would like to see more competition for the African American vote. I think it would be a healthy thing," Barack remarked. One fundamental problem with redistricting was exactly the process that was now under way, with the two parties seeking to protect incumbents and maximize the number of safe seats. "We have a tendency in our current system for representatives to choose the people rather than the people choosing the representatives," Barack observed. "What you end up having because of the lack of competition in the majority of these districts is probably an increase in partisanship and a breakdown in terms of the deliberative function of these legislative bodies."

For Barack, much of the session's final weeks was a cascade of disappointments. He criticized a bill regulating telemarketing calls as being "full of holes," and at a Chicago forum, he noted how Pate Philip had blocked consideration of his "driving while black" racial profiling bill. "It's a systemic issue that one person can stifle debate on important issues," Barack complained. "Walk into a room with one hundred black men, and ninety are going to have stories of being pulled over for biased reasons. Even if 50 percent are mistaken, there's a real problem out there." Under the aegis of state attorney general Jim Ryan, Barack and Judiciary Committee colleague Kirk Dillard joined with law enforcement representatives for a series of early-Monday-morning meetings in downtown Chicago to discuss the issue at length, but the conversations came to naught.

Barack unhappily voted present on a bill that would allow Illinois "to commit persons who've committed a sexually violent offense after they've served their time" in prison. This was "steamrolling what are admittedly odious folks," Barack warned, and he also voted present on a bill increasing penalties for drug offenders. "We essentially treat the possession of fifteen grams of cocaine in the same way as we treat a violent rape," Barack argued. "We are currently spending over a billion dollars a year to incarcerate folks, many of whom are nonviolent drug offenders, and we're keeping them for extraordinarily lengthy periods of time." In addition, "a disproportionate number of the youths that are sentenced under these laws are African-American and Latino."

In committee, Barack voted against a bill to expand eligibility for Illinois's on-hold death penalty to anyone convicted of committing a murder "in furtherance of gang activity," and when it reached the floor, he said, "What I'm concerned about is for us to single out 'gang activity' as a standard that is different from activity involving all kinds of other criminal conduct." He also worried "that we use this term 'gang activity' as a mechanism to target particular neighborhoods, particular individuals," with race and ethnicity central to the equation. "If we're going to apply the death penalty, we better make sure that it's absolutely uniform across the board."[39]

That evening the annual capitol skit took place, and Barack's Senate colleague John Cullerton, a truly gifted mimic, performed "some wicked impersonations of Pate Philip," Governor Ryan, Emil Jones, and Mike Madigan. "This has been a rough session," *Capitol Fax*'s Rich Miller wrote the next morning, "a long, boring and difficult session" that "just hasn't been much fun." Tuesday evening was "one laugh riot after another" and a welcome respite. But for Barack the session's final days were thoroughly downbeat. The Senate unanimously passed JCAR's resolution endorsing ISBE's opposition to the new special education rules Judge Gettleman had ordered it to draft. A month later Gettleman mandated that the rules be implemented, and the Seventh Circuit Court of Appeals strongly endorsed his decision, writing that "the district court properly found that the state authorities did not have the power to override an injunctive decree issued by a federal court to remedy a state's violation of standards established by federal law."

On May 25, the Senate voted 46–10–1 to approve a new map of Illinois's congressional districts that representatives of both parties had accepted. Rickey Hendon noted that the boundaries of Bobby Rush and Jesse Jackson Jr.'s districts had been altered so that Donne Trotter and Barack both now lived in the 2nd Congressional District and no longer in the 1st. As *Capitol Fax*'s Rich Miller described, "Jackson's district zips up the lakefront in a narrow band and shoots east a few blocks in Hyde Park to pick up Obama's home." Barack unsurprisingly voted against the new map. He also joined fifty-seven of his colleagues in supporting $3.5 billion in loan guarantees for new coal plants in southern Illinois, stating, "I am a strong supporter, I think, of downstate coal interests and our need to prop up and improve the outputs downstate," environmental worries notwithstanding.

Behind closed doors, the Four Tops and Governor Ryan agreed to the state budget for the upcoming fiscal year, and on the last day of the session, Barack joined in its unanimous passage, although he expressed regret that KidCare had not been expanded to include the children's uninsured parents. "I'm disappointed that 200,000 working moms and dads will have to go another year without health insurance because we couldn't find $7 million this year and $60 million next year out of a $50 billion budget."[40]

With the session over, the big news for the Obamas was the birth on June 10 of Natasha Marian—"Sasha"—her middle name honoring Michelle's mother, as Malia's had Barack's. But a serious family problem loomed nearby, for Michelle's brother Craig and his wife Janis, parents of nine-year-old Avery and five-year-old Leslie, were divorcing. Craig had kept his sister and mother in the dark as he and Janis had grown increasingly apart. Well compensated at Jim Reynolds' Loop Capital Markets and driving a Porsche 944 Turbo and a BMW station wagon, Craig yearned to leave the world of finance for his real passion, basketball.

During the 1999–2000 school year, Craig had coached the University of Chicago Lab School's varsity team, and then Northwestern University basketball coach Bill Carmody invited Craig to join his staff. "Here I was, making all this money, but I really wasn't excited with what I was doing," Craig recalled. The Northwestern job meant a 90 percent cut in pay, but "I had a passion for coaching," and "I didn't have that passion for business." After "a very painful period when we began living separately in the same house," Craig moved back to his childhood home in South Shore, living upstairs over his mother Marian.

Things were tense at East View too. "By the time Sasha was born," Barack later wrote, "my wife's anger toward me seemed barely contained. 'You only think about yourself,' she would tell me. 'I never thought I'd have to raise a family alone,'" Michelle complained. Barack continued, "I found myself subjected to endless negotiations about every detail of managing the house, long lists of things that I needed to do or had forgotten to do, and a generally sour attitude." Michelle's anger at the primacy Barack accorded his political career was understandable. The day after she gave birth to Sasha, Barack hurried downtown for a meeting of an Illinois State Board of Investments (ISBI) subcommittee that had been created several months earlier in response to his and Speaker Madigan's initiative on behalf of African American investment firms.

In late March, Madigan, Jim Reynolds, and other black executives had made a similar request to the State Universities Retirement System's board, and ISBI's staff believed they could increase their use of Illinois-based minority-owned firms without adversely affecting investment earnings. In the interim, Governor Ryan had quietly named two highly sympathetic new members, Robert Newtson and Susan McKeever, to ISBI's board, and at the June 11 meeting, Reynolds took the lead in thanking ISBI for its efforts while firmly requesting that they formally adopt the 15 percent minority-firm investment benchmark that the African American firms were calling for.

By the time of Sasha's birth, Michelle had decided to leave her U of C student affairs job once her maternity leave was over. She was unhappy with her life in manifold ways. "I am sitting there with a new baby, angry, tired, and out of shape. The baby is up for that 4 o'clock feeding. And my husband is lying there, sleeping." She resolved to reorganize her life, taking advantage of her early-to-bed nature whereby she turned in every night at 9:00 P.M., while Barack stayed up past midnight in his small, exceptionally cluttered office at the rear of their condo, which Michelle called "The Hole." Michelle realized "I was pushing Barack to be something I wanted him to be for me. I believed that if only he were around more often, everything would be better." But she realized that if she took the initiative to leave home at 4:30 A.M. for an early workout, Barack would have to assume some basic child care tasks. "I would get home from the gym, and the girls would be up and fed."

Michelle's transition to a better life attitude was gradual, not instantaneous, but it was aided considerably when she received a midsummer phone call from Michael Riordan, who on July 1 had become the new president of the University of Chicago Medical Center. Riordan's general counsel was Susan Sher, Michelle's friend and former colleague from Mayor Richard Daley's office, and Paula Wolff, Barack's Joyce Foundation board colleague, chaired the medical center board that had promoted Riordan to his new post. Riordan wanted Michelle to come speak with him about a new community affairs position at the medical center, but she initially brushed him off, citing newly arrived Sasha. Yet Sher pressed Michelle to at least talk with Riordan, and with no babysitter available, Michelle brought two-month-old Sasha along for a conversation during which Riordan agreed to all of Michelle's requests and convinced her to take up the new post come early fall.

When Maya visited Chicago that summer to see the Obama family's newest member, she found her brother at a low ebb and asked him what was wrong. "I don't know. I'm feeling a little frustrated, like I'm not doing what I'm supposed to be doing, that I could be doing more, that I haven't found my path, my mission. I'm not listening hard enough for my call, and I'm floundering a little bit." Taken aback, "I started laughing at him," Maya recalled. "You're the only guy who could be a state senator, a law professor, and a civil rights lawyer and feel like you're underachieving!" Barack had bounced back from his congressional loss fifteen months earlier, as his more energized legislative performance during the 2001 spring session had reflected. As he explained five years later, he had convinced himself that "I can lose an election and this isn't so bad, and I can still serve and do useful things." But no matter how much effort he had invested from January through May in advancing a plethora of generally modest bills, the meager results had left him "extraordinarily frustrated." Looking back at what were now five spring sessions, Barack summed up his record in an almost plaintive tone: "I'm very proud of the work that I did, even though people didn't know I was doing it."[41]

With a new baby, a discontented wife, and an energetic three-year-old at home, the balance of the summer was relatively quiet for Barack. On July 4 he took Malia along to Hyde Park's annual Independence Day parade, for which local politicians dressed up in patriotic costumes. Two weeks later, Barack was on WTTW's *Chicago Tonight* to weigh in on the city's latest political contretemps. In 1999 white alderman Thomas Murphy had been reelected by the 18th Ward's 85 percent African American population, but now he was being refused membership in the city council's Black Caucus. The dispute drew national coverage in the *New York Times,* and the *Chicago Tribune* supported Murphy, arguing that if a Black Caucus "has any purpose, it is to serve the special needs of Chicago's African-Americans," including "those who elected a white man to be their council voice." The NAACP also backed Murphy, but black political consultant Delmarie Cobb, speaking for the caucus, explained that membership was based on being a "black elected official." On *Chicago Tonight,* Cobb reiterated that position, while Barack took a more nuanced stance. "There's no question that whites can effectively represent blacks, in the same way that blacks can effectively represent whites." The caucus's informal status reminded Barack of black law student associations, where an exclusive group could privately discuss issues. "We all aspire to a vision of a color-blind society," but "what we also have is a reality which is pluralist."

At the end of July, Barack, Dan Shomon, Will Burns, and Laura Hunter held the first new Obama Finance Committee meeting. The short-term focus was an October 11 fund-raiser at the South Shore Cultural Center, where Barack and Michelle's wedding reception had taken place, to generate funds for Barack's state campaign committee. But the real agenda was Barack's political future. Shomon divided his presentation into two headings: "Opportunities" and "Challenges." In the first, Dan cited Barack's "reputation as [a] thoughtful independent," his "demonstrated success at building multi-racial electoral coalitions," his "high visibility," and his "low negatives." Several priorities loomed: "continued consolidation of African-American base," "maintain grass roots event attendance through summer and fall 2001," and "state-wide travel." "Challenges" began with "congressional campaign debt," which had been reduced but was not retired. "Family responsibilities prevent statewide bid in 2002," Shomon wrote, and Barack's "next major campaign will be for statewide office and must be a winner." A "well-funded campaign war chest" would make Barack a "serious contender for emerging opportunities," such as Republican U.S. senator Peter Fitzgerald's expected 2004 race for reelection. "Attorney General 2006" and "Governor 2006," presuming no Democratic incumbents, were more distant possibilities, and another congressional bid would occur "only if there are retirements." But another hurdle, as Barack and Dan well knew from 2000, was "Cost—Examples: 6 weeks of Chicago TV = $2 million," "seven state-wide mail pieces = $500,000." The total cost would be "$5 million or above, depending on [the] office sought."

Barack had raised a little more than $500,000 to challenge Bobby Rush, so he would need a tenfold increase in his fund-raising in order to mount a credible U.S. Senate race. Later that day, at an ACORN leader's eightieth birthday party, Madeline Talbott and Keith Kelleher were struck by how "really depressed" Barack seemed. Given a Republican-controlled Senate, "I don't know if I'm going to stick around with this," he told them, "but I'm thinking of running for something big, real big. I can't tell you what it's about, but I'm going to need your support."

In August, behind-the-scenes tussling about the upcoming state legislative remap intensified, with state representative Lou Jones, Alice Palmer's old friend, showering Democratic staffers with invective when one map showed her House district overlapping with Barack's Senate district. House colleague Barbara Flynn Currie knew Jones was "an incendiary person," and Jones's hatred of Barack was well known. Two years earlier, when state representative Tom Dart backed Barack in his challenge against Rush, Jones went after "me like you wouldn't believe" on the House floor, Dart recalled. "The vehemence was unreal," and the deluge of expletives was "so intense and virulent." Now John Corrigan, who was drawing the Chicagoland districts, became the target of Jones's wrath. "The shit hit the fan. I've never heard anyone so angry," Corrigan recalled, "and I knew it all went back to Alice Palmer."

For Corrigan, the key factor in drawing Cook County's House and Senate districts was "you don't want to waste African American voters if you're a Democrat." The federal Voting Rights Act, with its mandate to maximize the number of black and Hispanic districts, offered a crucial legal assist. "If the end result also helped to produce a Democratic majority, it was not a bad side effect," Corrigan explained. Barack described the same principles to Chinta Strausberg for a front-page story in the *Chicago Defender*. Ten years earlier, when Republicans had won the winner-take-all drawing, the result had been "racial packing," whereby African American influence was diminished by creating heavily black districts. "An incumbent African American legislator with a 90 percent district may feel good about his re-election chances, but we as a community would probably be better off if we had two African American legislators with 60 percent each." Black senators as diverse as Rickey Hendon, Donne Trotter, Barack, and machine warhorse Bill Shaw all grasped that point, but neither Margaret Smith nor a number of black House members endorsed it.[42]

On August 8, Governor George Ryan announced that he would not run for reelection. Saddled with abysmal approval ratings, due primarily to the ongoing federal investigation into the commercial-driver's-license scandal dating from his time as secretary of state, Ryan's decision promised a Republican contest between his moderate lieutenant governor, Corinne Wood, conservative Illinois attorney general Jim Ryan, and ultraconservative state senator Patrick

O'Malley. On the Democratic side, Northwest Side congressman Rod Blagojevich, whose father-in-law, 33rd Ward alderman Richard Mell, was one of the machine's most influential power brokers, had made clear for months that he intended to run. Mayor Richard Daley, who had worked closely with Governor Ryan, loathed Mell, and Barack's friend John Schmidt had considered a gubernatorial bid before word spread that former Chicago school superintendent Paul Vallas would enter the race. Perennial statewide African American candidate Roland Burris was eyeing a run, but even in mid-June, *Capitol Fax*'s Rich Miller put his money on the young congressman. "Blagojevich is wowing Democratic audiences all over Illinois, is raising money by the boatload, has support in the African-American community and has a reliable and sizable Northwest Side base." Blagojevich is "the 'new' face of the Chicago machine," Miller wrote in early August, and "his career has been carefully stage-managed by his father-in-law."

Now officially a lame duck, Ryan vetoed the "gang activity" death penalty bill Barack had opposed. Lisa Madigan formally announced her candidacy for attorney general, soon to be joined by John Schmidt. Joyce Washington, a well-to-do African American health care executive, announced a run for lieutenant governor and impressed Rich Miller as "a real firecracker and about the most personable candidate running for any statewide office." At his frankest moments, Miller loathed the true nature of Chicago and state politics, writing one day about "how brutally greedy and inept the mayor's office is now that all the reformers," like Schmidt and Vallas, were long gone. "When you work for a guy like Mayor Daley (or Speaker Madigan . . . or President Philip) you might as well be in the Mafia," Miller added.

Late that summer television producer David Manilow, whose philanthropist parents Lewis and Susan were regular contributors to Barack's campaigns, recruited him for a pilot episode of *Check, Please!*, a restaurant-review show where a trio of guests dined at three eateries, each guest choosing one. Barack's selection, Hyde Park's Dixie Kitchen and Bait Shop, exemplified just how deeply grounded in Hyde Park Barack and Michelle had become in their eight years there. Barack ate regularly at several local eateries—Bonjour Bakery, Medici, Pizza Capri, and especially Mellow Yellow—and he patronized Hyde Park Hair Salon, Cornell Shoe Service, Golden Touch dry cleaners, plus, on almost every Sunday, he did his grocery shopping at the Hyde Park Co-Op. Many weekday mornings Barack worked out at the Regents Club gym in the Regents Park high-rise condominium before smoking a cigarette outside. One August morning Stephen Heintz, one of the cofounders of the nascent think tank, now named Demos, on whose board Barack had been a usually absent member, and new president Miles Rapoport met Barack in the Loop for lunch. They had flown to Chicago to tell him that "we want this to be a working board," and Barack quickly agreed that he would step down.

At the *Check, Please!* taping, Barack was incredibly relaxed and fabulously personable as he joined a white male firefighter and a white female buyer in discussing Dixie Kitchen and two other restaurants. "I eat there quite a bit," because Dixie Kitchen offered "savory, tasty food for a good price." Firefighter Kevin O'Grady had nominated Zia's Trattoria, in far northwest Chicago, which was "a big surprise" to Barack and his family because it had "terrific" food plus free valet parking. When host Amanda Puck asked Barack what was the worst thing that could happen at a restaurant, he immediately answered "rudeness."[43]

In the run-up to the winner-take-all redistricting drawing on September 5, both parties initiated lawsuits, with Republicans seeking federal court intervention into Illinois's mapmaking and Democrats asking the Democratic-majority state supreme court to exert jurisdiction. One reporter trying to explain the process, whereby the winning party could impose its map to maximize its odds in the 2002 elections, turned to longtime political observer Charles N. Wheeler III. "Whichever side wins, it will be like they've won the lottery. The guys who lose, it's like somebody has just called to tell them their whole family has been wiped out by a serial killer," Wheeler volunteered.

Ten years earlier, when the Republican designee had pulled one of the two red, white, and blue pharmaceutical capsules from a crystal bowl, rumor spread that the Republican one had spent the previous night in a freezer so that the coldest one could be picked. This time one of two small envelopes would be drawn from a Lincolnian stovepipe hat, and the day before the drawing, a dress rehearsal took place in the House of Representatives chamber at the now-ceremonial Old State Capitol. Democratic secretary of state Jesse White and his deputy Jacqueline Price were joined by Democratic designee Tim Mapes, Speaker Madigan's chief of staff, and Republican National Committee member Bob Kjellander.

On September 5, the Republicans insisted that the two slips of paper be weighed to ensure they were identical. Ms. Price read out the relevant constitutional language, then tore open a large envelope into which the two slips of paper and two small envelopes had been placed. White stepped out of the room as Mapes signed the Democratic form and Kjellander the Republican one, with each then placing his slip in one of the small envelopes before licking and sealing it. Into the hat the two envelopes went, with both men and then Price stirring and shaking the hat before White reentered to reach in and select the winner. Tearing open the envelope, he read out the name of Democratic choice Michael A. Bilandic, the former Chicago mayor and state supreme court justice. News spread rapidly, and the *Chicago Defender*'s Chinta Strausberg called

Barack. "One of the most important things that results from this is giving us a real possibility of regaining the Senate. . . . If we can pick up a couple of seats, Emil Jones would be Senate President and that would make a tremendous difference . . . in terms of concrete legislation," Barack emphasized. "If the Democrats controlled the Senate, then all the legislation that we've seen blocked over the last several years like racial profiling, FamilyCare will be passed." Barack's optimism was well grounded, as were the Republicans' fears. Two days later *Capitol Fax*'s Rich Miller reported that while the Republican designee had "licked the edge of the entire flap" on his envelope, the Democratic designee had "licked only a portion of the flap."[44]

When the legislature was not in session, JCAR's monthly meetings took place at midmorning on the first or second Tuesday of each month, on the sixteenth floor of the Thompson Center, in the Chicago Loop. On September's second Tuesday, JCAR members were on their way downtown for the 10:30 A.M. CST meeting. Republican senator Brad Burzynski took a METRA train in from his suburban home, and fellow Republican Doris Karpiel was driving in from Carol Stream. Barack was headed northbound on Lake Shore Drive "when I heard the news on my car radio that a plane had hit the World Trade Center." On the sixteenth floor of the Thompson Center, Emil Jones Jr. and his two top staffers, Courtney Nottage and Dave Gross, were reviewing the proposed map of new districts with John Corrigan.

As soon as Barack reached the sixteenth floor, news spread that a second plane had hit the other tower, and the Thompson Center was evacuated. "We were all outside, and a lot of people were looking up at the Sears Tower," Chicago's tallest building, "thinking that it might be coming down," Barack recalled. He headed to Miner Barnhill's nearby office, where everyone was crowded into the basement conference room to follow the news on television. "We watched that horrific scene of the buildings coming down," and that afternoon Barack drove home. That night Malia and Michelle turned in well before Barack and baby Sasha. "I remember staying up late into the middle of the night burping my child and changing her diapers and wondering what kind of world is she going to be inheriting."

When the *Hyde Park Herald* surveyed local politicians for their comments on the September 11 attacks, Barack replied that "the essence of this tragedy, it seems to me, derives from a fundamental absence of empathy of the part of the attackers: an inability to imagine, or connect with, the humanity and suffering of others." Following Sasha's birth, Barack and Michelle's increasingly spotty Sunday attendance at Trinity United Church of Christ had declined even more, so on September 16 they were not present to hear Rev. Jeremiah Wright deliver a sermon titled "The Day of Jerusalem's Fall." Assailing how the U.S.

government had "supported state terrorism against the Palestinians and black South Africans," Wright echoed Malcolm X's famous remark following John F. Kennedy's assassination to warn that once again "America's chickens . . . are coming home . . . to roost!" Barack was seeing much less of Wright than he had in earlier years, and Wright had no memory of discussing Barack's 1999–2000 congressional race with him. On September 16 Wright was only six days away from his sixtieth birthday, and across black Chicago there was tremendous regard for how he had built Trinity into one of the city's preeminent churches. In an op-ed tribute in the *Sun-Times,* DePaul professor Michael Eric Dyson lauded Wright as "one of the most intellectually sophisticated and scholarly ministers in the land."

In a horrible accident of timing, Barack's good friend Bill Ayers had just published *Fugitive Days,* a memoir of his Weather Underground involvement thirty years earlier. On the morning of September 11, a prominent *New York Times* feature story began with a powerful quote: "I don't regret setting bombs," Ayers said. "I feel we didn't do enough." Bill had previously made similar comments, telling the *Chicago Reader*'s Ben Joravsky in 1990 that "if anything, we didn't go far enough. . . . I wouldn't act any differently. . . . I wouldn't change a thing." But in the aftermath of September 11's almost three thousand deaths, Ayer's embrace of purposeful political violence seemed beyond callous. In a *Chicago Tribune Magazine* profile that same Sunday, Ayers struck an entirely different tone. "We were young and stupid. We allowed our organization to slip into a kind of dogma, a kind of self-righteousness, and there's something dangerous and deadly about that. So, did we make huge mistakes? Yes." But thoughtful remarks like those received little attention as denunciations rained down upon one of Barack and Michelle's ten or so best friends in Hyde Park.

After September 11, Barack "was a wreck," one close acquaintance recalled, and the obvious rhyme—Osama and Obama—now made what was already a highly unusual ballot name an indelible echo of the world's most notorious terrorist. In what may have been his most depressed moment in more than sixteen years, Barack did something he had not done in almost a decade: he wrote "a long beautiful sad letter" to Sheila Jager. She had completed her University of Chicago Ph.D. dissertation—"Narrating the Nation: Students, Romance and the Politics of Resistance in South Korea"—in December 1994, three years after she married U.S. Army intelligence officer Jiyul Kim. Sheila's father Bernd was just as dismissive of Jiyul as he had been of Barack in 1986, but by late 1994, Sheila and Jiyul's first child, Isaac, had arrived, soon to be followed by sister Hannah.

Over the next four years, Sheila published a trio of academic articles drawn from her dissertation while readying the revised manuscript for publication as a book. A recipient of no few fewer than eight notable research grants, by September 2001 Sheila was Henry Luce Assistant Professor of East Asian Studies at Oberlin College in northern Ohio. Jiyul, now an army colonel, had been at work in the Pentagon when the third of the hijacked airliners slammed into the building on the morning of September 11. Hearing the news, Sheila rushed to Isaac and Hannah's elementary school and took the children home. "We didn't hear from him for a couple hours, and my mom was silent," Isaac recalled just before entering the U.S. Military Academy at West Point as a member of the class of 2016.

Barack's long letter made clear that "he was really down and out," Sheila recalled. Barack recounted his unsuccessful congressional race and described the "impact of 9/11" on him. "I remember distinctly how forlorn he sounded," but Barack's real purpose in writing her was to express how much their love still meant to him. He "reconfirmed his thoughts and feelings" even "after ten years of silence," and Sheila was deeply touched that "the nature of the relationship, Barack's thoughts and feelings about it" had "continued post-Chicago" all the way to September 2001. She wrote back, updating him on everything in her life, and Barack replied with "a brief letter acknowledging my response to his first letter. . . . He seemed genuinely happy to know that I was doing well and the tone was far more upbeat."

Seven years later, during a joint interview with Michelle, a questioner asked Barack, "Did you ever have a 'Can this marriage be saved?' moment?" "Sure. I mean, you know, look. We had—" Barack responded, to which Michelle interrupted forcefully by asking, "We did?" causing Barack to backpedal. "Oh, I mean . . . no, no. No, no, no, no. But there's no doubt that there were some strains there" because "we were still paying off student loans, and we'd be short at the end of the month sometimes." In a subsequent joint interview, Barack attempted to walk back his spontaneous honesty, insisting that "there was no point where I was fearful for our marriage," but his outreach to Sheila reflected just how despondent Barack was in the aftermath of 9/11.[45]

On Monday, September 17, Senate Democrats unveiled their new districts map, and it was an over-the-top partisan humdinger, one "that guarantees severe GOP losses in next year's election," the *Chicago Tribune* reported. It placed fourteen of the Senate's thirty-two Republican incumbents into seven districts, with either retirements or primary contests to follow, while creating nine districts without incumbents. Barack told the *Hyde Park Herald*, "it's hard to overestimate the impact the new map will have on policy making in every area." He acknowledged that "redistricting is never fair—it's a partisan process—but the lottery is as good a process as any." With the "continued consolidation of African-American base" uppermost in his mind for whatever race he might next run, on Sunday evening, September 23, Barack drove to Roland Burris's South Side home, where he joined African American congressmen Bobby Rush and Danny Davis plus Senate colleagues Rickey Hendon and Donne Trotter in endorsing Burris, rather than former Chicago schools chief Paul Vallas, in the Democratic race for governor.

The next morning the U of C Law School's autumn quarter got under way, with Barack again teaching Con Law III and Current Issues in Racism and the Law. Con Law drew more than sixty students, with many again impressed that Barack, in contrast to some UCLS professors, had no doctrinal or ideological agenda to propound. One day a self-confident student responded to a query Barack had posed by stating that "the more interesting question" involved something else. Classmates nervously wondered how Barack would respond. He paused and smiled. "The more interesting question, huh?" he said as everyone laughed. Classroom moments like that created "a very comfortable environment" where quieter students were encouraged to participate and "students who wanted to do tons of ranting" were kept in check.

Barack's Racism and the Law drew almost thirty students, and that fall, more than ever before, racial profiling was a front-and-center issue. One Indian American student whose male family members had been repeatedly targeted by Transportation Security Administration saps remembered Barack leading "a very vibrant discussion" of the issue. At a school that even self-conscious white males viewed as a "utopia for white men," the few black male law students were regularly stopped by security guards even after previously introducing themselves to the officers. In class, Barack was "a very fair, even-handed" teacher, "an excellent law professor at a school that really prides itself on the quality of the teaching." Byron Rodriguez, a 3L, had just taken a course with the well-known feminist Catharine MacKinnon, which "was just so painfully one-sided" because "you couldn't disagree with her at all, and Obama was completely in the opposite direction and would really challenge folks of all stripes." Indeed, "I remember him defending, in the way a law professor would, [Robert] Bork's position on the Civil Rights Act" of 1964, which the noted conservative had argued at the time was unconstitutional.

Students found Barack an uncommonly sympathetic professor outside of class as well as in it, even though he spent far fewer hours in the building than any other full-time instructor. One day after class, walking to the parking lot, Barack spoke to 2L Bethany Lampland, who had "taken on incredible loans" and was feeling "tremendous pressure to find a career that was fulfilling." Having recognized "how unhappy I was," Barack gently asked, "Are you happy here?" with "such an earnest, sincere," and "disarming" tone that Bethany unburdened herself. "I'm really unhappy. I'm really miserable, and I'm really scared because I only have a year left, and I don't like this at all, and I'm really afraid that I'm not going to like being a lawyer, and I don't know what else to do with my life." As an experienced law teacher, Barack posed the perfect question: "Did you go straight through?" When Bethany nodded yes, Barack described how Michelle had gone directly from college to law school and had had a much less fulfilling experience both at Harvard and in law practice than he had. But "'going to the University of Chicago Law School is a really big privilege, it's not a set of handcuffs,'" Barack told her. "He talked about how there were many, many ways that I could use the experience of going to law school even if I never practiced law that could be as meaningful or more meaningful than becoming a traditional lawyer, and that he had a lot of confidence that because I was struggling with this so much at such a young age that I would find the right path," Bethany recounted years later. By then she was leading one of New York's largest charity organizations, and that conversation with Barack in the parking lot "really stayed with me."[46]

On WTTW's *Chicago Tonight,* Barack said that new federal detention proposals were "most troubling" because they targeted only noncitizens. "I'm always more concerned about encroachments on civil rights or civil liberties that apply selectively to people. When they apply to everybody, there tends to be a majoritarian check." Barack said he supported slightly broadened wiretapping rules, but emphasized that "the crucial issue here is do we have judicial oversight of some sort" for government surveillance. By late September, Senate Democrats had moderated the political pain their new, now slightly revised map would impose on Republican incumbents, but Republicans believed the new map guaranteed Emil Jones at least twenty-nine of the Senate's fifty-nine seats and perhaps as many as thirty-four. Barack's district would remain centered on Hyde Park, but rather than running westward to encompass so much of deeply poor Englewood, it would instead run northward along the lakefront, taking in not only Kenwood and Oakland but also wealthy neighborhoods like Streeterville and some of the Gold Coast, north of the Chicago River. The underlying rationale, as John Corrigan and Barack both knew, was to make his district less African American so that those voters could prove useful elsewhere, but the remap also gave him "some really interesting new friends," as savvy city lobbyist Bill Luking put it. The redistricting process moved forward, with a three-judge federal court quickly dismissing a Republican challenge to the lottery procedure, but Barack's attention was completely distracted when a sudden medical emergency threatened the life of four-month-old Sasha.

One night Sasha "would not stop crying," Michelle explained, "and she was not a crier, so we knew something was wrong." At dawn they called Sasha's pediatrician, who immediately suspected Sasha had meningitis and told Barack and Michelle to take her right away to the U of C Medical Center's emergency room. There Sasha underwent a spinal tap and was placed on antibiotics. "We didn't know what was wrong, and we were terrified," Barack recalled. "Michelle and I spent three days with our baby in the hospital," with "us not knowing whether or not she was going to emerge okay." But the immediate treatment succeeded, and Sasha was discharged with no harm to her long-term health.

On October 11, Barack's state campaign fund-raiser took place at the South Shore Cultural Center. Barack still confronted his congressional debt, and just a week earlier, he had sweet-talked a brand-new acquaintance, Northwestern University business professor Steven Rogers, into contributing $2,000—in his and his wife's names—to Obama for Congress 2000 as well as $1,000 to Friends of Barack Obama. The congressional donation allowed Barack to pay himself and Michelle back a second $2,000 for the personal funds they had loaned the campaign eighteen months earlier, following an initial repayment in early spring, and by the end of the year, the debt would be down to only $8,830, $5,500 of which was owed to Barack and Michelle. For his state committee fund-raiser, Dan Shomon and Laura Hunter had assembled an almost fifty-person "host committee," and the *Hyde Park Herald* reported that "supporters showed up in droves" at the October 11 event. Many of the host committee members, who gave $1,000 or $1,500 apiece, were old friends and colleagues like Elvin Charity, Jeff Cummings, Valerie Jarrett, and Peter Bynoe, and $1,500 apiece also came from John Rogers at Ariel Capital and Louis Holland of Holland Capital, two of the firms most involved in Barack and Michael Madigan's initiative to reform state pension investment practices. The beneficent Robert Blackwell; old mentors Newton Minow and Abner Mikva; present and past law firm colleagues Judd Miner and Allison Davis; top congressional supporters Jim Reynolds, Judy Byrd, and Marty Nesbitt; school executive Tim King; former student Jesse Ruiz; and fellow Joyce Foundation board members Carlton Guthrie and Paula Wolff were also on the host committee list. Including a total of more than $20,000 from PACs, by the end of the year Barack's state campaign committee boasted a balance of more than $93,000, its healthiest amount ever.[47]

In mid-October, Barack had a strikingly expansive conversation with the *Chicago Defender*'s Chinta Strausberg about the 9/11 attacks. The new profusion of concrete barricades outside government buildings was "the saddest situation and aftereffect of the September 11th tragedy," because "those barricades are a symbol of the fear that people are experiencing. I recognize the need for such precautions, but my strong hope is that over time, we're able to diminish the daily threat of violence and return to the sort of openness and freedom that is the hallmark of our society." While "it's absolutely vital that we pursue a military response and a criminal investigation to dismantle these organizations of violence," Barack said, "we should also examine the foreign policies of the U.S. to make sure that we occupy the moral high ground in these conflicts. In particular, we have to examine some of the root causes of this terrorist activity," because in "nations like Afghanistan, Pakistan, Indonesia, or much of the

Middle East, young men have no opportunities. The only education they are receiving is that provided to them by religious schools that may not provide them with a well-rounded view of the world. They see poverty all around them, and they are angered by that poverty. They may be suffering under oppressive and corrupt regimes, and that kind of environment is a breeding ground for fanaticism and hatred." Rather than hunker down, new initiatives should be launched. "It's absolutely critical that the U.S. is engaged in policies and strategies that will give those young people and these countries hope and make it in their self-interest to participate and create modern, open societies like we have in the U.S."

In early November, Barack keynoted an affordable housing symposium one day and spoke at a lunch for well-heeled female supporters of his own future endeavors the next. At the housing session, Barack reprimanded advocates for failing to put together a coherent policy agenda for poor Chicagoans. For the women's lunch, super-volunteer Laura Hunter pulled together ladies who already knew Barack well, like Judy Byrd, Jean Rudd, and Valerie Jarrett, as well as women of extremely serious means, such as heiress Abby McCormick O'Neil, some of whom would be among Barack's new constituents. Barack said he was "100 percent prochoice" and a supporter of gay and lesbian rights, and he spoke about his efforts on behalf of campaign finance reform, expanded health care coverage, welfare-to-work policies, and tax justice.

In mid-November, veto session got under way in Springfield amid reports showing that state tax revenues were dropping precipitously in the wake of 9/11. It was a desultory six days, with no attempt at overriding Governor Ryan's veto of the gang-related death penalty bill. On the second to last day, the Illinois Supreme Court affirmed the constitutionality of the new legislative map, but the legislature adjourned without taking any action on the state's surging budget deficit. Barack told *Crain's Chicago Business* that "We didn't plan for what happens when all the jobs that were generated evaporate. A lot of the big questions like that we never really tackled." Soon after adjournment, Emil Jones's wife Patricia died of cancer at age sixty-three, and Democratic senators gathered for her funeral.

Politicians' eyes turned increasingly toward the upcoming 2002 elections, with no one filing to oppose Barack in either the March primary or the November general election. In the Democratic gubernatorial race, Rod Blagojevich won the support of the Illinois AFL-CIO and gathered increasing momentum over Paul Vallas and Roland Burris. Just before Christmas, *Crain's Chicago Business* reported that millionaire businessman Blair Hull, who had sold his Hull Trading Company to Goldman Sachs two years earlier for $531 million, had spoken with Mayor Richard Daley about seeking the Democratic nomination to take on Republican U.S. senator Peter Fitzgerald in 2004. Hull had met with Daley at City Hall on November 14, but Hull initially had been interested in challenging Democratic operative Rahm Emanuel for the congressional seat Blagojevich would be vacating. Hull had already spoken with Chicago political hand Jason Erkes, Denver-based consultant Rick Ridder, and Washington-based media expert Anita Dunn about the congressional race, but Daley suggested that instead Hull consider running for the Senate. When *Crain's* reporter Steven Strahler called him for comment, Hull asked, "Who would have told you that? I'm trying to fly under the radar at this point." But Hull confirmed that his discussion with Daley had been "a positive conversation," and Hull also was supporting Blagojevich's gubernatorial campaign.[48]

Michelle was now in her new job at the University of Chicago Medical Center, and her hands were more than full with a protest campaign launched against the hospital by the African American Contractors Association. Unhappy that a construction contract had been awarded to a Hispanic firm, the protesters complained further when university executives shunted them toward "underlings" like Michelle and African American trustee Ralph Moore. "We should not have to be meeting with people who have jobs to protect or don't have the best interests of blacks at heart," protest supporter Rev. Gregory Daniels told the *Defender,* which reported that Daniels "also demanded the removal of Obama and Moore because of their incompetence and inexperience in regards to grassroots community organizations and their assistance in dividing the African American community in these issues." Within less than seventy-two hours, however, the university patched up the dispute, with both the *Defender* and *Chicago Weekend* publishing stories proclaiming the U of C's support for black employment opportunities.

Michelle was not the only one tackling a new opportunity, for throughout December most of Barack's attention was focused on the possibility of becoming the next president of the Joyce Foundation. In November Paula DiPerna, whose two-and-a-half-year tenure had not gone well, submitted her resignation. Former president Debbie Leff, program officers like Unmi Song and Margaret O'Dell, and ranking board figures like chairman Jack Anderson and veterans Chuck Daly and Dick Donahue all saw Barack as a thoughtful and well-informed board member. Barack's legislative interests in campaign finance reform, the earned-income tax credit, and gun violence overlapped extensively with Joyce's policy agenda. Similar Springfield overlap existed with the Woods Fund, and Barack's Annenberg experience also contributed to the Joyce board's discussions of education issues.

When DiPerna resigned, several board members immediately considered Barack before they decided to have Michael Claffey, an old friend of Daly's who had overseen previous presidential searches, handle this one as well. Carlton Guthrie, Barack's fellow African American board member and a Hyde Park neighbor, would run into Barack at the East Bank Club and played golf with him several times a year. Guthrie appreciated that Barack was a politician through and through, but he also knew that Barack "was desperately looking for a way to get out of" Springfield. Guthrie had contributed to Barack's campaigns, but he believed Barack's loss to Bobby Rush had underscored how he "wasn't of the community" and how many people saw him as "a fairly arrogant guy." Guthrie was not on the Joyce board's five-member search committee, but "I didn't see Barack as the next president of the Joyce Foundation" or even "as a serious candidate," because "it didn't seem to be a logical progression for him."

Barack's own view, as he discussed it extensively with Michelle and Dan Shomon, was that the Joyce presidency would allow him to take home a significant salary and accumulate some meaningful savings before reentering the electoral arena some years later. Michelle immediately liked the idea. "My hope was that, okay, enough of this, now let's explore these other avenues for having impact and making a little money so that we could start saving for our future and building up the college fund for our girls." Dan Shomon, who envisioned going with Barack to Joyce as his chief of staff at a higher salary than he could earn in politics, loved the idea too. "We had it all figured out," Dan recalled. "Do it for five years and then run" for a major office. Barack stoked Dan's hopes, leading him to believe that Barack was a hands-down favorite for the job. Barack made similar comments to Cynthia Miller, Newton Minow, Ab Mikva, and Rob Fisher, and they were all left thinking similarly. Like Dan, Cynthia knew "Michelle wants him to take it," but Cynthia, more clearly than Dan, saw that Barack's interest in the job really was just to placate Michelle. Newt Minow enthusiastically supported the idea, as did Mikva, who knew that Michelle was "hoping he would make a firm break with politics" and "get out of the legislature and be a normal person." Rob and Lisa Fisher remembered Barack saying, "I think I'm going to step aside and make money and then go back." They had the impression there "was an offer on the table" and Barack was weighing whether to accept it. "Michelle is saying, 'Are you nuts?'" Lisa recalled, because when the Senate was in session and Barack "wasn't home," Michelle "was a single parent."

Joyce consultant Mike Claffey talked with Barack several times to discuss and then confirm his interest. "Yes, I'm a candidate," Barack told him. Claffey also spoke with mutual acquaintances like Allison Davis and Jean Rudd before Joyce's search committee selected five finalists for formal interviews just after New Year's. Two of the other contenders were Ellen Alberding, Joyce's program officer for culture and the arts, and attorney Mark Real, who since 1981 had directed the Ohio branch of the Children's Defense Fund. Real sensed from preliminary conversations with Claffey and search committee members that "they were looking for somebody who would do real work" and "would show up at the office every day," because "they asked me a lot of questions about work ethic."

Senior board member Dick Donahue tried to dissuade Barack from being a candidate. "For God's sake, Barack, this is a great job. But you don't want it." Underlying Donahue's advice was board members' view that Barack "was insufficiently nonpartisan" and that "it just didn't make any sense to politicize the foundation," as Wisconsin law professor Carin Clauss explained. Board chairman Jack Anderson, who Claffey knew "adored Barack," had evolved like Donahue in his view of Barack's candidacy. The Joyce presidency was "a real full-time job," and it was doubtful "whether you can do it and be active in politics," because they knew Barack wanted to remain a plausible future candidate.

Just before the finalists' formal interviews with the search committee, Mike Claffey briefed each of them. Claffey thought Barack "approached it with some diligence but not a lot," and once the interview got under way, Claffey realized that Barack "was not prepared" and had "ignored my briefing." Later that day, Barack phoned Dan Shomon and confessed that "I was literally shaking with fear that I would get the job because I realized I was deathly fearful." Dan realized that Barack had "walked in there, and he fell apart. He literally just fell apart." In the room, that was what Mike Claffey saw. "This is a thing where he failed," Claffey explained, and Claffey suspected the same thing Dan Shomon, Cynthia Miller, and Rob Fisher all realized: "maybe he didn't want it."

Barack had long believed that "simply pursuing lots of money displays a poverty of ambition, and that there are higher goals and higher aspirations that I could go after," a belief he had held since his 1987 self-transformation. He confessed as much to Cynthia, who knew that Joyce "was never, for him, an option . . . as far as him being true to himself," she explained. "He said, 'This is who Michelle met when she met me.' Now that is a quote. He was saying he's the same person, so he didn't mislead her in terms of where his aspirations are." Yet now, more than a decade later, Michelle's aspirations were far more modest: financial security and a home life where two full-time parents took equal roles in raising their daughters. But when Barack faced a high-pressure situation that ran counter to his deepest beliefs about his destiny, he fell apart and failed.

Mark Real's interview followed Barack's. "I got the sense that they had reservations about other candidates" for whom "this would be sort of their second or third priority. They were looking for somebody for whom this would be their first priority," Real recalled. "They wanted somebody whose principal priority was making the Joyce Foundation work and not somebody who was just a public face." With the interviews complete, the five-member search committee—Jack Anderson, Chuck Daly, Dick Donahue, attorney Roger Fross, and Paula

Wolff—voted. The result was not unanimous, but there was a clear number one, Mark Real, and a runner-up, Ellen Alberding. But Real, who hoped to transform the Ohio Children's Defense Fund into a more influential statewide policy organization with the support of wealthy Columbus patron Abigail Wexner, seemed underwhelmed when he was offered the Joyce presidency at a salary just over $200,000. "He wants $350,000," Mike Claffey recalled, and "Jack Anderson almost falls out of his chair." Real may not have wanted to leave Ohio any more than Barack wanted to abandon electoral politics, and with Joyce unwilling to meet Real's asking price, little time passed before Abigail Wexner and Ohio political leaders announced the creation of a new organization, Kids Ohio, headed by Mark Real. Joyce's search committee turned to Ellen Alberding, who accepted. When Mike Claffey phoned Barack to tell him the news, there was "no moaning and groaning, no anger, no frustration." With that, the Joyce Foundation announced Alberding's appointment and concluded a search in which three finalists all got what they really wanted.[49]

By this time, Barack had already taught the first two classes of his winter quarter Voting Rights course. An embarrassed Barack learned that unbeknownst to him, a new, 2001 second edition of *The Law of Democracy* had been published and stocked at the bookstore, so he quickly updated his syllabus. The early Mondays–late Fridays class drew thirty-five students, including four who had taken either Con Law III or Racism and who realized that Barack's "engagement with the subject matter" made Voting Rights "an even better class" than the other two. "He really came alive, particularly when he was talking about some of the more practical issues," such as "how districting lines were drawn," 2L Karen Schweickart recalled. Barack was also a "very, very nice man" who let Karen bring her baby daughter to class. When someone asked how long it took to drive to and from Springfield, Barack responded, "How long for you, or how long for me, because I have Senate plates?" Byron Rodriguez remembered that when "someone in class once asked who was his favorite circuit judge to argue in front of, he said [Richard] Posner," the senior member of the three-judge panel in *Josephthal,* Barack's only circuit court appearance, "because Posner was smart enough to know when you were right." When Barack casually asked Schweickart what her husband did, and Karen said she was not married, Barack "didn't miss a beat" before saying, "Oh, my mother wasn't married either."

With the Senate's spring session off to a leisurely start, Barack delivered two Martin Luther King Day talks while political eyes all focused on the upcoming November election, with Emil Jones seemingly guaranteed at least a thirty-two-seat Senate majority. Jones had more than $1.6 million in hand to ensure victory. In the Democratic gubernatorial race, Rod Blagojevich had more than $4 million available, former school superintendent Paul Vallas only $300,000, and Roland Burris was being bankrolled by one sole donor, African American TV station owner Joseph Stroud, who had contributed upward of $1 million. By early February Barack had introduced seventeen new bills, only one of which, requiring that state government institute a program to address postpartum depression, would make it to the Senate floor.

When the *Chicago Tribune* ran a story headlined "Philip Urges Review of Pork-Barrel Funds," Barack chimed in to agree with the Senate president that "members' initiatives" had gotten out of hand. "What started off as a good idea to give legislators some control over how money got spent in their districts has become this enormous, unwieldy and unsupervised process that needs to be changed." Three days later a more prominent *Tribune* story, "2 Accused in Scheme at Pork-Rich Charities," reported that the state attorney general had filed suit against Yesse Yehudah, whose FORUM had received more than $750,000 in member initiative funding, beginning with more than $425,000 from Emil Jones plus additional monies from Michael Madigan and Donne Trotter as well as the $75,000 from Barack. In less than forty-eight hours, Obama for Congress 2000 refunded the five $1,000 contributions FORUM employees had made to Barack's congressional campaign in November 1999, although not the five additional $1,000 checks from Yehudah's friends that were received three days after Barack approved the $75,000 grant in October 2000. So as to repay that $5,000, Barack had to loan his federal campaign committee an identical amount, increasing what it owed him and Michelle to $10,500. Even with that embarrassing turn of events, a few days later Barack signed off on an unusually large $500,000 member initiative grant to a Southwest Side women's domestic violence group that within several years would be defunct.

With Barack unopposed in both the March Democratic primary and the general election, he told the *Hyde Park Herald*, "I'm going to spend the next year assessing what my plans might be." As plummeting state revenues led to calls to reduce Illinois's budget, Barack insisted that "the worst thing state government can do during a recession is cut spending." Looking to 2004, a political column in the suburban *Daily Herald* declared that "Illinois is too liberal" for Republican U.S. senator Peter Fitzgerald to be reelected if Democrats nominate "a quality opponent." The *Herald* said that young state comptroller Dan Hynes, progressive U.S. representative Jan Schakowsky, and state legislators Tom Dart, Lisa Madigan, and Barack were among those "already being mentioned" as possible candidates. The next day Barack wrote a check for $5,000, two and a half times his largest political contribution ever, from his state campaign account to his Senate colleague Lisa Madigan's run for attorney general.[50]

In addition to his $8,000-a-month consulting income from Robert Blackwell Jr.'s EKI, Barack in early 2002 invested some time in Miner Barnhill's effort to recover more than $150,000 in fees for its work defending the congressional apportionment that had produced Chicago's first Hispanic-majority district a decade earlier. The state of Illinois had played an entirely passive role in the case, so the successful defense of the map had been handled almost entirely by the defendant-intervenors, represented by Judd Miner and others. But to win a fee award, Miner Barnhill had to argue for "an exception to the general rule that precludes defendants from recovering fees." Barack's colleague John Belcaster could not "see a pathway to success," so Judd asked Barack to tackle the problem. Belcaster recalled that "Barack stays at it and puts together a spell-binding brief," one that persuaded U.S. District Judge David H. Coar to award Miner Barnhill its entire $154,695 fee request. In commending Barack's brief, Coar wrote that "the intervenors offer a thoughtful discussion of the policy implications at play in this case," and "there is no question that the defendant-intervenors' efforts merit a fee award." It was an unprecedented order, but the Seventh Circuit Court of Appeals unanimously affirmed Coar's ruling.

In March, a Senate Judiciary Committee hearing addressed a trio of anti-abortion bills targeting late-term procedures. "It was a very difficult hearing," Barack's Democratic colleague Ira Silverstein recalled, "very emotional, very tough," because of the highly graphic testimony offered by the bills' proponents. "You could have heard a pin drop," Republican Dan Cronin remembered, before a party-line vote of 6–3, with prochoice Republican Adeline Geo-Karis voting present, sent the bills to the Senate floor. On March 19, Rob Blagojevich narrowly won the Democratic gubernatorial nomination over Paul Vallas, whom he had vastly outspent, triumphing by twenty-five thousand votes thanks to winning 55 percent of the vote downstate, where he had advertised heavily on television. African American Roland Burris finished a strong third, with more than 29 percent of the vote. Blagojevich's slogan—"My name is eastern European, my story is American"—went a long way toward alleviating doubts about his unusual surname, but Burris's presence in the race, especially his strong showing in Chicago, had allowed Blagojevich to win a race that otherwise would have gone to Vallas. In the Republican primary, Attorney General Jim Ryan easily outdistanced conservative state senator Patrick O'Malley and moderate lieutenant governor Corinne Wood. Barack received all 30,938 Democratic votes in the 13th Senate District, but wealthy black health executive Joyce Washington piled up more than 362,000 votes statewide while losing the Democratic race for lieutenant governor to gadfly Pat Quinn. Emil Jones had backed Washington strongly, and *Capitol Fax*'s Rich Miller declared that Washington had shown "incredible charisma and potential."[51]

On April 2, federal prosecutors indicted Scott Fawell, George Ryan's chief of staff when the lame-duck governor had been secretary of state, as the commercial-driver's-license scandal probe moved closer to Ryan. *Capitol Fax* termed Fawell's indictment "a political atomic bomb," and *Chicago Sun-Times* columnist Steve Neal called Operation Safe Road, as prosecutors labeled their case, "the worst scandal in Illinois history because there are human casualties," including a half dozen children who had been killed by a truck driver licensed in exchange for a bribe. In the Senate, Barack addressed the importance of helping ex-offenders obtain jobs and voted against the trio of antiabortion bills, noting that they were "really designed simply to burden the original decision of the woman and the physician." The Senate passed two of the three bills but they died in committee in the House, while Barack's postpartum depression bill passed unanimously in both chambers and headed to Governor Ryan's desk. "This condition is very real," Barack stated in a press release. On April 15, a commission Ryan had appointed to study capital punishment in Illinois issued a lengthy report outlining dozens of desirable policy changes. In particular, "we recommend videotaping all questioning of a capital suspect conducted in a police facility, and repeating on tape, in the presence of the prospective defendant, any of his statements alleged to have been made elsewhere."

In mid-April, Barack and Michelle wrote their largest check ever to the Internal Revenue Service because Barack's $98,000 in deduction-free income from Miner Barnhill—$80,000 of which came from Robert Blackwell Jr.'s EKI—meant they owed an additional $44,000. Thanks to the South Shore Cultural Center fund-raiser six months earlier, Barack's state campaign committee was now flush enough to audaciously explore whether Barack could become a serious contender for the 2004 Democratic nomination for U.S. senator. Amy Szarek was brought back on payroll part-time, and Dan Shomon reached out to media consultant Chris Sautter as well as to Eric Adelstein, who had overseen Bobby Rush's successful reelection victory. Meeting Adelstein for lunch, Barack "wants to know what he did wrong" in the congressional race and "really wanted to understand what were the fund-raising benchmarks" to be seen as a serious candidate in a statewide senatorial primary, Adelstein recalled. "You need a million dollars by the fall of '03," Adelstein told him. With painful memories from 1999 in mind, Barack decided they needed to have a top-notch pollster gauge his prospects as a statewide candidate. "We want to do a benchmark to see if the Senate race is do-able," Shomon told Sautter. "Who should I talk to?" Well-known pollster Geoff Garin declined, so Sautter called Paul Harstad, a Colorado-based pollster who had helped U.S. senator Tom Harkin and Governor Tom Vilsack win tough races in Iowa. Harstad had never heard of Barack, so he called John Kupper, David Axelrod's partner, and Kupper said, "You should work for him." He told Harstad that Barack was a promising candidate, and Shomon arranged a conference call with Harstad, Sautter, and Barack. Chris remembered "very distinctly Barack saying 'I don't have to do this. If the numbers aren't good, I can go off and do something else,'" and they agreed to survey both Democratic primary and Illinois general election voters.

Two issues needed to be addressed. If Carol Moseley Braun, who had returned to Chicago months earlier after serving as U.S. ambassador to New Zealand, attempted to reclaim her seat, how would that influence the race? Second, even though Rod Blagojevich had fared surprisingly well outside Chicagoland with an unusual ethnic surname, Shomon "wanted to determine how much of a problem" "Barack Obama" would be as a ballot name. Sautter, Harstad, and Barack all agreed that both samples would be divided, with half of respondents being asked about a potential candidate named "Barack Obama" and the other half presented with one named "Barry Obama."

As Shomon and Harstad worked by phone on a lengthy, comprehensive poll, Barack, in Springfield three days a week, spoke privately with Emil Jones Jr., whose support would be most essential to making him a top-tier contender. Jones's closest staff members knew that Emil "enjoyed conversing with" Barack "and hearing his thoughts on stuff," and with the new legislature map all but certain to elevate Jones to the Senate presidency, Jones in eight months' time would possess the power that Pate Philip had single-handedly exercised throughout Barack's years in Springfield. Jones remembered "a very interesting conversation" that began with Barack stressing how much power Jones would soon have. "In my mind I'm thinking, 'Where's this bullshit coming from? What are you trying to fatten me up for?'" So "I asked Barack, 'What kind of power do you think I'll have?' He says, 'You'll have the power to make a United States senator.' I said, 'Damn, I hadn't thought of that.' He says, 'Well, you've got that power.' And we kept talking, and I said, 'Well, if I have that kind of power, do you know of anyone I could make a United States senator?' I was just sitting there. And he said 'Me.'" Jones said, "I didn't expect that," but he told Barack, "Let me think about that," and the two men continued talking until Jones said, "You know, Barack, that sounds good. Let's go for it!" Jones knew that state comptroller Dan Hynes would be backed by his rival, House Speaker Madigan, should Hynes run, and Jones also realized that Carol Moseley Braun could be a candidate. "Barack says, if she runs, he will not run," Jones remembered, but neither man thought she would. "I said, 'I think this is doable.' He said to me, 'One thing I need from you, early on, is to be out front in the campaign early on.' I told him, 'I have no problem with that.'"

"As I look back at that meeting, it indicated to me that he learned a lot from his congressional race loss," Jones explained. Barack "figured if he had me, his chances would go up 1,000 percent," because "I knew all the players who would be involved." Jones recognized what a difference that would make from two years earlier: "in his loss in the congressional race, he was out there on his own. He had no one with any political influence supporting his candidacy." Barack was appropriately grateful to Jones, and on April 23, Friends of Barack Obama contributed $5,000 to Jones's Illinois Senate Democratic Fund. Following their conversation, Jones mentioned it to John Kelly, the young white lobbyist whose deep family roots in Cook County politics had encouraged Jones to take a shine to him. "I just had a fascinating conversation with Barack," Jones recounted. Barack "came in and said, 'How would you like to make the next U.S. senator?' 'That's great. Who do you have in mind? Who should we support?' He said, 'Me,'" Kelly remembered. Jones also mentioned the conversation to well-known black radio host and former alderman Cliff Kelley. "'Cliff, I'm gonna make me a U.S. senator,'" Kelley recounted. "'Oh, you are? Who might that be?' 'Barack Obama.'"[52]

Barack joined Bill Ayers for a panel discussion at a conference in Chicago titled "Intellectuals: Who Needs Them?" but press reports signaled that Barack quickly needed to expand his outreach about a Senate run. On April 25, *Chicago Sun-Times* gossip columnist Michael Sneed wrote that "Carol Moseley-Braun is angling to reclaim her old job" and that she had called U.S. senator Dick Durbin and former senator Paul Simon to seek their support. Three days later Sneed reported that lawyer and former Chicago Board of Education chairman Gery Chico, a onetime chief of staff to Mayor Daley, "has been making phone calls to see if a run is viable." Sneed added that "financial whiz Blair Hull is also eyeing the job." Two months earlier Hull and his close friend Pat Arbor, a former chairman of the Chicago Board of Trade, had had dinner with Richard Daley at Naha, one of the city's best restaurants. Three months earlier, Daley had encouraged Hull, and on February 27 the mayor told him, "You really should be in the Senate." By late March, Hull had hired two campaign staffers while he and consultant Rick Ridder sought out additional advisers with Illinois experience.

After his conversation with Emil Jones, Barack talked to a highly dubious Michelle about a Senate race and then called their friend Valerie Jarrett. "'I want to come over. Let's invite my closest friends. There's something I want to bounce off of you,'" Jarrett remembered Barack telling her. "Michelle had already told me what it was," Jarrett explained. "She said, 'So we're not in favor of this, right?' 'Absolutely not.' 'That's the right answer!'" Michelle forcefully confirmed. With the congressional race debt still lingering, Michelle recalled that "the big issue around the Senate for me was, how on earth can we afford it?" Jarrett agreed that Barack would "have to raise a ton of cash, and he didn't have it," but Jarrett also knew that underlying all of Michelle's objections was her distaste for Barack's political career. "Michelle hadn't been happy about him going off to Springfield," and a statewide campaign would have Barack away from home even more.

Judy Byrd, the female mainstay of Barack's congressional run, was about to move to New York, so Jarrett invited Barack's two friends whom she knew best, John Rogers and Marty Nesbitt, to join the three of them for brunch that weekend at her apartment. Jarrett had known Rogers for years, and Nesbitt, the chief executive of an airport-parking company that was backed by Chicago heiress Penny Pritzker, had first met Michelle and Barack through his basketball-based acquaintance with Michelle's brother Craig. Rogers's Ariel Capital was a major player in the Illinois investment reform initiative Barack was championing, but Barack's friendship with Nesbitt was rooted in their own very similar family histories. "Up through probably my mid-twenties, the biggest driving motivator for me was to not be like my father," Nesbitt explained.

"So I cooked breakfast," Jarrett recalled. "We were resolved we were going to talk him out of this. No one thought it was a good idea, Michelle being the most clear that it was a bad idea." Then Barack made his pitch. "'I want to run for U.S. Senate. Let me tell you why. I've talked to Emil Jones. He's a huge political force, and he's prepared to support me. When I ran for Congress, I didn't have that kind of support. And if I lose, then Michelle, I'll give up politics.' She's like, 'Yes!'" Jarrett remembered. "'If I can't do it this time, I promise I'll get a normal job in the private sector, so this'll be the last time I ask you to do this—unless I win. And money's a problem, so Valerie, I think you should help me because you're in the business community, you and John. You two should think about helping me do this.'" Jarrett explained that Barack "did this obviously far more eloquently and for ten minutes or so," but "our initial reaction was 'It's too soon. You just lost, and if you lose again, where are you?'" Barack responded, "'If I'm not worried about losing, why are you? If I lose, I lose. But I think I'll win. I think the timing is right. I think I have a lot to offer.'" Barack "didn't convince me he'd win at that moment," but "he convinced me that it was worth a try," Jarrett recounted. "I can't quite explain how it happened," she added. Rogers likewise believed that winning would be "incredibly difficult," but all three friends agreed to help raise money.

Nesbitt's memory was more upbeat: "He convinced us in the room that day that he could pull it off." Most important, Barack's offer to leave politics if he lost was so attractive to Michelle that she reluctantly agreed to countenance one last race. "I said to her that if you are willing to go with me on this ride and if it doesn't work out, then I will step out of politics," Barack recalled. "I think she had come to realize that I would leave politics if she asked me to."

On May 2, Barack's state campaign committee paid Harstad Strategic Research $16,500—the first half of the cost of an extensive, two-sample survey. At the same time, Barack quietly ended his $8,000-a-month consulting arrangement with Robert Blackwell Jr.'s EKI, which over the preceding fourteen months had paid Barack a total of $112,000. In Springfield, he reached out for support from his closest Senate friends. One evening Terry Link walked into Barack's office. "'I've got something to ask you,'" Barack said. "'I'm thinking of running for the U.S. Senate. Would you support me?' And I said, 'Yes.' And he looked at me in a puzzled look, and he said, 'Don't you have to think about it for a while?' And I said, 'No. Now you're ready, and it's the right race.'" Link was the Democratic Party chairman in Lake County, "the third-largest county in the state" after Cook and primarily Republican DuPage, so Link's backing would be outweighed only by Jones's.

Barack also asked Will County Democratic powerhouse Larry Walsh, who hailed from the state's fourth-largest county, if they could have breakfast together. "I'm seriously thinking about running," Barack told Walsh once they sat down. "You represent the kind of district that the state of Illinois is." Larry worried that Barack's congressional loss would be an obstacle, but he too offered his support. Please "keep it quiet," Barack asked, and during a Wednesday poker night, Barack pulled aside Denny Jacobs to make the same request. Denny too agreed, thinking that Barack would do well among suburban voters because "he does have a certain sex appeal for women."

On the Senate floor, Barack quietly told Kim Lightford, "I'm thinking about running for the U.S. Senate." Kim jokingly replied, "I was supposed to replace Carol Moseley Braun, not you!" to which Barack responded, "I'll do two terms, and then I'll hand it over to you." Kim pledged her support, and Barack went by the offices of Metro East's James Clayborne, Southwest Sider Lou Viverito, who had backed him in the congressional race, and the North Side's Miguel del Valle to seek and receive their backing. Downstate Democrat Pat Welch rebuffed Barack, explaining that Barack had not supported him when he had been considering a statewide 2002 race and had instead supported Tom Dart, who had done so much for Barack's congressional run. Barack also got a lukewarm response from lobbyist and regular donor Larry Suffredin, who had just mounted a successful primary challenge for a seat on the Cook County Board. "You cannot afford to lose two races in a row," Suffredin warned Barack.

Then, one day in early May, Barack ran into former senator Carol Moseley Braun and her close lobbyist friend Billie Paige in the capitol's rotunda and invited them to his cramped office. "We discussed the Senate race," Carol remembered, and although Barack did not mention his own interest, he "pissed me off" by asking "How do I know you're not going to get in the same trouble that you've gotten into before?" if she ran again. "When he challenged me like that, I was really taken aback," Carol explained, because she considered it an "insulting" question. "Barack, you don't understand" what "a very hostile environment" the Senate had been for her as an African American woman. "For you not to understand that is pretty shocking to me." The exchange ended there, but in retrospect Moseley Braun thought "I signaled to Barack" that a return to the U.S. Senate "just was not a real appealing prospect" for her.[53]

Barack spoke on the Senate floor about attending the funeral of Dick Newhouse, who had held the Hyde Park seat before Alice Palmer and who had died on May 2 at age seventy-eight. "There's times where all of us get swept up in some of the baser aspects of politics, the gamesmanship, the self-aggrandizement, the vanities," Barack said. "We all have our feet of clay, we all have our weaknesses." On the evening of May 9, Harstad Strategic Research launched a seven-day calling program during which 605 likely Democratic primary voters and 603 likely general election voters, with one-quarter of the former included in the latter, would be asked a series of questions. The pollsters' first question tried to screen out a household where someone worked for the news media, but the next morning the *Daily Southtown* learned that a poll aimed at measuring Barack Obama's potential appeal was under way. Legislative correspondent Kristen McQueary called Barack, who immediately came clean.

On Saturday, May 11, McQueary's story, "State Legislator Considers Bid for Senate," appeared in the *Southtown,* with McQueary introducing Barack as someone "whose name may be his biggest political obstacle." Barack had "said he has approached party leaders and core supporters about running," and agreed that his name might be a hurdle. "I'll have to test to what degree it's a disadvantage in a state-wide race to have an unusual name. It's a legitimate question. But I will say the last election gave me encouragement. If Rod Blagojevich and Rahm Emanuel can win, maybe there's hope for a Barack Obama." The other hurdle was money. "With about $3 to $4 million, you can run about as many television ads as people can stand. Above that, it starts getting on people's nerves." But Barack looked forward to taking on Republican incumbent Peter Fitzgerald. "When I examine his record for some evidence of legislation, projects that help working families, paying for college or managing health care costs, I think his record is pretty thin. The more appropriate role of government is to help working families." On Monday, Rich Miller's *Capitol Fax* alerted everyone to McQueary's scoop: "The *Daily Southtown* reports that Sen. Barack Obama wants to run for the U.S. Senate against Peter Fitzgerald in 2004."

Blair Hull's nascent campaign took interest in Barack, with staffer David Spiegel surveying his biography and unsuccessful race against Bobby Rush. Hull and top consultant Rick Ridder were looking to beef up the millionaire's campaign team, and at 8:00 A.M. on Thursday, May 16, Hull and Ridder met David Axelrod for breakfast at Ceres Cafe, at the Chicago Board of Trade in the Loop. Hull and Axelrod had met previously, when Hull considered challenging Emanuel for Blagojevich's congressional seat before Daley encouraged Hull to think bigger. Hull knew recruiting Axelrod was important because "it would be a sign that we really had a chance if I could get him involved in the campaign," Hull explained. Over breakfast, "they're sort of feeling each other out," Rick Ridder remembered, with Axelrod mentioning Barack's, Gery Chico's, and Dan Hynes's potential candidacies. Ridder knew Hull's background included a domestic violence incident in which charges had not been pursued and research had shown that "unless you know about it, it's almost impossible to find" any record of the incident. But "Blair had this real weakness of telling everybody everything," and "during this conversation, Blair does tell David straight out" that "'I had this domestic violence issue' and I'm going 'Whoa,'" Ridder recounted. "I heard Blair Hull tell him that it was out there." Hull confirmed that "I'm not the kind of guy that hides things," and he too recalled how "there was some discussion of that" at what Ridder labeled that "fateful breakfast with David Axelrod." No commitments were made, and back at his office Axelrod told John Kupper that Hull had left him unimpressed.

With Harstad's poll complete, later that same day Barack's state campaign fund paid the $17,750 balance due on a total tab of $34,250. On May 21 *Chicago Sun-Times* gossip columnist Michael Sneed reported that Obama "is seriously exploring jumping into the U.S. Senate race, even if former USS Carol Moseley-Braun is thinking about getting her old job back." Gery Chico filed his formal statement of candidacy, and African American *Sun-Times* columnist Laura Washington wrote that Moseley Braun "is likely" to run and that Barack "may be" her "most worrisome challenger." When the *Hyde Park Herald* questioned Barack about his plans, he said "we won't know the final slate of candidates until September of 2003," fifteen months away, but he again underscored the financial challenge. "If you're going to be competitive, you need to raise enormous sums of money to get your name and message out, mainly through television advertising."

On May 23 the Illinois Supreme Court upheld the Gift Ban Act that Barack had played such a central role in helping pass four years earlier, dismissing a low-profile challenge that Barack's old-school friend Denny Jacobs had filed against it. The law's grandfather clause loophole had been highlighted when retired senator Jim Rea paid himself $72,500 from his campaign account at the end of 1999, but Barack praised the court for upholding "simple, commonsense rules designed to protect the political process." In the Senate, with Illinois facing a growing budget crisis as state revenues decreased from the previous year for the first time since 1955 and unpaid bills totaled $1 billion, Pate Philip's Republicans moved to reduce state funding of Chicago Public Schools teachers' pensions. "That seems to me to be a deliberate, intentional effort to say 'We don't care about Chicago,'" Barack told reporters. Budget discord delayed the Senate's adjournment until late Sunday night, June 2, the first time in eleven years that spring session had gone beyond May 31, but the budget as passed called for the state to borrow $750 million to fund operating expenses, an unprecedented action.[54]

After spending so much of May in Springfield, Barack awaited Paul Harstad's arrival from Colorado to present him and Dan Shomon with the benchmark poll results. *Sun-Times* political columnist Steve Neal wrote that Barack "would very much like to run for the Senate," but he would be an also-ran if Carol Moseley Braun entered the race. Barack appeared on *Chicago Defender* reporter Chinta Strausberg's cable television show and tried to convince viewers of the importance of state government. "I've been really encouraged by my capacity to influence policy at the state level," even though the new budget was highly disappointing. "Draconian drug laws" meant that prison construction had become "a growing proportion of the state budget," and it had been difficult to win funds for "prevention and intervention programs as opposed to incarceration strategies" because "politics in Springfield is driven by people's perceptions" that "being tough on crime . . . is a winning recipe to win elections." Emphasizing that "it's our youth who are going into the system," Barack rued how "prisons have become an economic development tool. In Illinois it's primarily downstate economies that are being supported by the prison economy" while more dollars were needed to "fund programs for ex-felons to help them transition back into the mainstream." Once Emil Jones became Senate president in 2003, he would have "enormous influence and power" to reshape state priorities.

The bottom line of Paul Harstad's poll results—once one worked through the extremely extensive findings—was astonishingly encouraging. The survey of likely Democratic primary voters began by asking respondents their feelings—very positive, somewhat positive, neutral, somewhat negative, very negative, or don't know—about sixteen different Illinois politicians. The first two names were Jesse Jackson Jr. and Peter Fitzgerald. Jackson's positives versus negative percentages cumulated to 46 and 20, Fitzgerald's to a respectable 37 and 26. The additional fourteen names began with Carol Moseley Braun, Rev. Jesse Jackson Sr., former Chicago schools chief Paul Vallas, and state comptroller Dan Hynes, eventually including Mayor Daley, U.S. representative Jan Schakowsky, Barack, state representative Tom Dart, former Chicago School Board chair Gery Chico, and well-known Cook County treasurer Maria Pappas. On the positive side, Daley, Vallas, and Moseley Braun topped the list at 58, 55, and 50 percent, while the two highest negatives were Jackson Sr. at 33 and Moseley Braun at 26. On "don't know," Barack topped the entire list at 72 percent, with Chico at 62, Hynes at 43, Pappas at 30, and Fitzgerald at just 13. Hynes's positives versus negatives summed to 24 and 4, Chico's to 15 and 3, and Barack's to 12 and 2.

The questionnaire then asked, "How would you rate the performance of Peter Fitzgerald as U.S. Senator?" and Democratic voters registered 39 percent positive and 47 percent negative. Then Harstad honed in on the poll's real purpose. "Thinking about the next Democratic primary election for U.S. Senator, which of these five candidates would you vote for?" The choices were Barack, Moseley Braun, Hynes, Chico, and Pappas, offered in randomized order. Moseley Braun drew 31 percent support, Hynes 14, Pappas 13, Barack 6, and Chico 5, with 30 percent of respondents saying "don't know." Then Democratic voters were asked, "and which candidate for Senate on this list do you find unacceptable?" "None" and "don't know" each got 32 percent, while Moseley Braun received 21, Chico 6, Pappas 4, and Barack and Hynes each 3. After asking about prospective secretary of state and Cook County Board president candidates, including Barack, the poll moved to direct head-to-head face-offs of several Democrats versus Fitzgerald. "Suppose the next election for U.S. Senator were held today, and the candidates were Democrat Barack Obama and Republican Peter Fitzgerald. For whom would you vote?" Barack drew 48 percent of Democrats and Fitzgerald 30 percent, with 22 percent responding "don't know." Using Mayor Daley's lesser-known brother Bill as a representative white Democrat with a well-known surname, Daley drew 64 percent support, 16 points better than Barack.

Finally Harstad focused on his own candidate, using a well-tested device to allay respondents' concern that they were being asked about *their own* racial attitudes. "Ethnic background has long been a factor in Illinois politics, and I'm going to read you the names of a few candidates for U.S. Senator. Thinking about your neighbors' reactions, do you think your neighbors would be more likely or less likely to support a candidate by the name of"—with Gery Chico, Maria Pappas, and either Barack or "Barry" Obama, being posed—"or would this particular name make no difference whatsoever to them?" Of the respondents, 53 percent chose "no difference whatsoever" for Chico, Pappas, and "Barry" Obama. Chico registered 9 percent more likely, 14 percent less likely, and "Barry" drew 7 and 11. However, among the half of the Democratic sample who were asked about "Barack" rather than "Barry," the results were 7 percent more, 23 *percent* "less likely," and only 43 percent "no difference." Thus "Barack" as opposed to "Barry" cost Obama 12 percent of Democratic voters.

Then the poll moved to its most demanding and time-consuming portion, and the caller said they would "read you brief background of four potential candidates. Please tell me which one you would be most likely to vote for." Blair Hull "founded a successful international company that pioneered new technology in stock trading. Hull sold his company to Goldman Sachs for $500 million." Jan Schakowsky's federal and state legislative service was described, as was Gery Chico's on the city school board and with Mayor Daley. "Barack Obama is a state senator and constitutional law professor at the University of Chicago. He was the first African American editor of the *Harvard Law Review*."

After hearing that, 190 respondents—31 percent of the Democratic sample—named Barack as their first choice. Schakowsky, Chico, and Hull all trailed with 23, 21, and 4 percent, respectively. When asked for their second choice, Barack picked up an additional 19 percent of likely Democratic voters, for a grand total of 50 percent, with Chico at 44 and Schakowsky at 39.

The 190 people who had named Barack as their top choice were then asked, "What are the main reasons you support Barack Obama?" With multiple answers allowed, Barack's education—i.e., Harvard—was a clear first, his U of C affiliation a strong second. Then, in a clear give-away of the poll's underlying purpose, all 605 Democrats were told, "let me read you a brief description of Barack Obama," a "possible" Senate candidate. "Barack Obama is a constitutional law professor. He has served six years as state Senator" and "is a long-time member of Trinity Church and is married with two daughters. He was the first African-American editor of the *Harvard Law Review*. As a state Senator, he reduced state income taxes on middle class families and helped pass welfare reform that gets people off welfare and into work. He sponsored a law for after-school programs to keep teens out of trouble and prevent youth crime. He is the lead sponsor of the Cardinal Bernardin Amendment, which provides health care for all Americans. And Obama was one of only four Illinois lawmakers who worked with Senator Paul Simon to pass the toughest ever campaign reform and ethics law in Illinois. Now, based on this description, how much appeal does Barack Obama have as a candidate for U.S. Senator—a great deal, quite a bit, about average, just some, or very little?"

Forty-five percent of likely Democratic voters answered "a great deal," and 27 percent "quite a bit," meaning that 72 percent found Barack highly appealing. Then those 439 respondents were asked "What are the main things you find appealing about Barack Obama as a candidate for U.S. Senator?" Now Barack's issues agenda topped the chart, with welfare reform named by 18 percent of people and 14 percent each citing his interest in youth and his work with Paul Simon. Barack's educational background and professorial status finished a strong second. Then came the final question: "Suppose the next general election for Senator were today," with Peter Fitzgerald against Barack Obama, whom would you vote or lean toward voting for? Barack drew 62 and 10 percent, respectively, now swamping Fitzgerald among Democratic voters by a margin of 72 to 14.

Among Harstad's 603-person sample of likely general election voters, with 148 of the Democratic primary respondents included in this pool, 82 percent admitted they did not know Barack and an additional 13 percent had no feelings pro or con, meaning that 95 percent of Illinois voters were unfamiliar with him. In contrast, 47 percent of respondents felt negatively about Jesse Jackson Sr., 36 percent negatively toward Carol Moseley Braun, and 31 percent negatively toward Jesse Jackson Jr. On the ethnic names question, with half of the sample asked about "Barack" and the other half "Barry," the "less likely" option drew 22 percent with "Barack" and only 12 percent with "Barry." At the end, after Barack was described as a "constitutional law professor" and the "first African-American editor of the *Harvard Law Review,*" and after the Bernardin Amendment and Paul Simon were again cited, the respondents were asked "how much appeal" Barack had. Sixty-two percent of voters said either "a great deal" or "quite a bit." Then, in a final head-to-head choice of Barack versus Peter Fitzgerald, Barack handily triumphed with 50 percent while Fitzgerald drew just 31. In contrast, when Fitzgerald was paired against Carol Moseley Braun, he outpaced her by 10 percent.

The results were exceptionally encouraging. Barack's 72 percent attractiveness to Democratic voters following the candidates' descriptions was "an out-of-the-park score," according to Harstad's partner Mike Kulisheck. Barack's 50 to 31 percent lead over Fitzgerald showed how much potential Barack had if Illinois's statewide electorate knew who he was. That was especially true among female voters, because after the respondents heard Barack's biographical summary, his support among women jumped 37 percent, as compared to 22 percent with men. Barack was not surprised by the results of being "Barry Obama," with Harstad explaining that "you can have one odd name, but two's a lot tougher." But Barack gave no consideration to re-adopting his youthful nickname. "I'm Barack Obama. I may have been raised Barry Obama, but I'm Barack Obama. That's the name I chose and that's who I am, and I'm not changing."[55]

Not long after the Harstad poll, the Washington, D.C., firm of Bennett, Petts & Blumenthal did a slightly smaller poll for Blair Hull. Democratic voters felt favorably about U.S. senator Dick Durbin by a 60 to 12 margin, but Carol Moseley Braun's split was 49 to 28. Only 10 percent of respondents had heard of Hull, with Dan Hynes, Gery Chico, and even Barack coming in measurably higher, at 51, 27, and 21 percent, respectively. Asked their top choice for a Democratic Senate nominee, Moseley Braun led Hynes 47 to 15, with Barack, Chico, and Hull trailing at 6, 5, and 2 percent. Without Moseley Braun, Hynes drew 34 percent, Chico and Barack 12 apiece, and Hull 5, with 55 percent of likely Democratic voters choosing "undecided." In the poll's general election sample, Fitzgerald led Moseley Braun 49 to 42 and Hull 53 to 30. Then the pollsters read out descriptions of the possible Democratic candidates, with Hull characterized as a "successful entrepreneur and businessman" who had risen from "humble roots." Barack was identified as "the first African-American editor of the *Harvard Law Review*" and state comptroller Dan Hynes was described as "the son of a powerful Chicago politician." Those descriptors aided Barack and hindered Hynes, with Democrats giving Moseley Braun 46 percent,

Barack 12, Hynes 11, Chico 10, and Hull 9 following those recitations. Without Moseley Braun, Hynes led with 31 percent and Chico came second with 27, while Barack and Hull trailed with 25 and 19 percent, respectively. Nothing in those results contradicted Harstad's findings, and although the Democratic race was clearly fluid, the evidence seemed clear that Barack enjoyed considerably stronger potential for growth than did Hull.[56]

On Friday, June 7, Illinois governor George Ryan, unhappy with the badly unbalanced budget adopted by the General Assembly the previous Sunday and concerned by the accelerating decline in state revenues, summoned legislators back to Springfield on Monday to confront his amendatory vetos. Barack was furious at some of Ryan's cuts. "He's proposing a $6 million reduction in early childhood block grants that we had specifically restored. He is proposing the total elimination of the HIV-AIDS prevention, outreach and treatment programs for minorities that we had specifically allocated $2 million for," Barack told the *Chicago Defender*. "We should not eliminate programs that deal with children, the aged, the sick and the vulnerable of our society." The root of the problem, Barack explained, "is that most of the time politicians in Springfield operate in obscurity. People don't know what's going on down there so they can cut deals without any fear of voter backlash."

On Tuesday, June 11, the Senate began a tedious series of 234 successive votes on particular line items the governor had deleted from the budget. One-quarter of the way through, Rickey Hendon rose to advocate overturning a Ryan veto that would result in the closing of a Department of Children and Family Services (DCFS) welfare office in his district. "I challenge any of you," he told his colleagues, "to go . . . and see how these children are living in squalor, how these children are dying, how these children are victimized. You wouldn't close this office." The electronic roll was taken, with twenty Democrats plus Republican Kirk Dillard voting to override the veto and thirty-one Republicans and five liberal Democrats—Terry Link, Lisa Madigan, Carol Ronen, Ira Silverstein, and Barack—voting to affirm.

As debate turned to the next line item, Barack spoke, saying he would vote to affirm this veto because it did not concern "core operations that affect those people who need state services the most." He added a lighthearted reference to his mother-in-law, Marian Robinson, explaining how she was "a gold medal winner in the hundred-yard dash in the Senior Olympics. She could whip all of you." Barack teasingly mentioned that he soon would be visiting the Senate president's home district, saying "I hear there are a few Democrats out in DuPage, Pate," before becoming more serious. "We need to open up the appropriations process a little bit more so that every one of the members here, as well as the public and the press, have an opportunity to look at some of these line items and can make some judicious decisions about whether it's worthwhile to fund some of these programs."

Then Rickey Hendon sought the floor. "I just want to say to the last speaker, you got a lot of nerve to talk about being responsible and then you voted for closing the DCFS office on the West Side, when you wouldn't have voted to close it on the South Side. So I apologize to my Republican friends about my bipartisanship comments, 'cause clearly there's some Democrats on this side of the aisle that don't care about the West Side either, especially the last speaker." Barack responded immediately. "I understand Senator Hendon's anger at—actually—the—I was not aware that I had voted no on that last piece of legislation. I would have the Record record that I intended to vote yes. On the other hand, I would appreciate that next time my dear colleague Senator Hendon ask me about a vote before he names me on the floor."

Then, for the first and only time in his entire life, Barack Obama completely fucking lost it. He walked to Hendon's seat, placed a hand on Hendon's shoulder, and as Hendon remembered it, "leaned over and stuck his jagged, strained face into my space and told me in an eerie, dark voice that came from some secret place within the ugly side of him, 'You embarrassed me on the Senate floor, and if you ever do it again, I will kick your ass!'" Hendon replied "'What?'" and Barack said, "'You heard me, motherfucker, and if you come back here by the telephone, where the press can't see it, I will kick your ass right now!'" at the back of the chamber. "I stacked my few papers quietly on my desk in front of me and said, 'Okay, motherfucker, let's go!'"

Pat Welch, seated beside Hendon, was astonished. He knew Hendon was "a volatile character," unlike Barack, and "to have the two attitudes basically switch was very unusual." Also nearby was Debbie Halvorson, who "was shocked" as Barack "just lost it." Hendon walked toward the small telephone room at the left rear of the chamber. "When I got there, I turned and put up my dukes like Muhammad Ali," Hendon recounted. "I begged him to hit me and told him I couldn't go back to the West Side if I let a South Sider punk me and kick my butt. I swelled up my five-foot-seven frame to about six feet and got up to his chest. A little pushing and shoving" ensued, and as more verbal taunts were exchanged, Courtney Nottage, Emil Jones's chief of staff, headed toward them with fellow staffer Jill Rock following behind. "I ran over there," Nottage recalled, in "disbelief that this was happening." Rock had heard their comments on the floor and had watched Barack approach Hendon's seat, "but never in my wildest dreams did I think it was going to go where it went." Nottage physically confronted Barack, placing both his hands on the front of Barack's shoulders. "I was pushing him away from Rickey," with Jill warning, "We are on the Senate floor! There are people watching!" It took Courtney "maybe ten or fifteen seconds" of sustained physical effort to move Barack backward. Hendon recalled that Barack was still taunting him "to come outside where he said he would stomp me into oblivion. I nodded my head and said, 'Let's go.'"

By this time, the confrontation had gotten almost everyone's attention. Donne Trotter remembered that "all of a sudden one of the staff people grabbed me and said, 'Senator, go back and get Rickey.'" Trotter tried to get them to "take this off the floor," and Kim Lightford intervened as well. "I ran over to Rickey, and I was like 'You cut this out!' or 'You hush right now!' I'm in his face. Then I go over to Barack and I'm like 'Hey, brother, what are you doing? Why did you let him get to you?' 'He's gone too far! I've had enough of him!'" Barack told her. "I'm trying to get him off the floor. He would not budge."

Hendon recalled, "I gave Barack a few more choice words and told him I would see him later," before heading back to his seat as "Barack huffed and puffed and sneered at me." Then Barack "walked over to my seat again," saying "I've had enough of you!" but senators and staff finally got Barack to leave the floor. "I never would have expected that from Barack," Jill Rock explained, for it was the most "surprising and shocking" confrontation any of them had seen, and "there's not been an incident like that since." Kim Lightford believed that Barack reacted that way "because Rickey did it on the Senate floor, in front of all the colleagues, the media, the staff," and Jill agreed that was why "'Street Barack' came out." Kim thought that on balance the face-off was a good thing. "That's what Barack needed to do, and it worked." He had endured five years of "the tag team between Trotter and Hendon," and following Barack's eruption "they didn't tease him so much after that. It was like they finally realized that Barack was more than some soft punk to push around."

Soon after the confrontation, Barack's former campaign manager Carol Harwell called him, and she got an earful about Hendon. "'I was getting ready to kick his ass. I'm just about sick of him. He's an idiot, he's a fool, a buffoon. But I'm not a punk, and they need to understand that.'" Not everyone was as understanding as Kim Lightford. Jill Rock remembered that some colleagues "didn't appreciate that he lost his cool," and Emil Jones summoned Barack to his office. "I called Barack in and said, 'Man, come on.' I said, 'Don't let him get under your skin like that. It's not worth it, damaging your image by getting involved with him.'" Barack was still agitated: "'That son of a gun this and blah blah and this,'" Jones recalled. "'Come on. Forget about it. Don't let him get you all riled up.'" Jones knew that "lurking in the back of all this is this whole Alice Palmer thing," but "I told him that 'You're too smart of a man, too intelligent to be involved in this kind of bullshit. Let's forget about it. Rickey's going to be Rickey. You know that. But don't let that damage your image here!'" A sheepish Barack headed back to the floor, confessing to Cindy Huebner, Jones's press secretary, that "Yup, I would have got my ass kicked" had the face-off with Hendon escalated further.[57]

The Senate overrode only twenty-two of Governor Ryan's 234 line-item vetos, but Barack was pleased with a $6 million override regarding early childhood education. Wednesday morning's *Chicago Tribune* noted that "two agitated Chicago lawmakers had a heated exchange on the chamber floor" and said Hendon had "slapped away Obama's extended hand when Obama reached to put his hand on Hendon's shoulder." The rest of the encounter was not reported in the press, and when Barack arrived the next day at a new charter school in his district to deliver its first commencement address, he appeared fully composed. Telling the black eighth graders "to pursue excellence in everything that you do" and to "take responsibility for your actions," he warned that oftentimes African Americans "set the bar too low." Decrying "a certain anti-intellectualism that exists in our communities that says that if you know how to read and write that you're acting white," Barack told them to "commit yourself to something larger than yourself."

That same day one Senate candidacy was publicly confirmed when Blair Hull told the *Chicago Sun-Times* he was prepared to spend up to $40 million on the race after being encouraged by Mayor Daley to pursue it. Hull had previously discussed with consultant Rick Ridder the cost of running, and when Ridder had cited a $30 million price tag, Hull had said that was "not going to change my lifestyle or my children's lifestyles one bit." That evening African American attorney Stephen Pugh's law firm hosted "A Tribute to Senate Democratic Leader Emil Jones Jr.," with Ariel Capital's John Rogers Jr. and several prominent black lobbyists like Paul Williams among the honorary cochairs. Jones had been "bemoaning the fact that the black business community was not really supporting him," Pugh recalled, with contributions to help ensure that Democrats would take control of the state Senate in November.

Four days later, Greg Hinz of *Crain's Chicago Business* caught everyone's attention with a story revealing that legislative payroll records showed that House Republican leader Lee Daniels had staff members working on political campaigns on state time. That was not surprising to anyone employed at the statehouse, but Springfield's *State Journal-Register* soon had a follow-up story headlined "U.S. Attorney to Probe Allegations Against House Republican Staffers." Later that day, Daniels resigned as chairman of Illinois's Republican Party, and as the summer turned to fall, federal authorities would expand their interest in legislative leaders' campaign practices.

Barack was focused on his own campaign funding, and in mid-June Jim Reynolds convened a meeting at his home to ask fellow black financial executives, many of whom were active members of ABLE, to support Barack for the Senate. Reverend Jesse Jackson Sr. was there, and publisher Hermene Hartman remembered that Barack "challenged us to back him." Advertising executive Eugene Morris recalled Hartman saying to Barack, "You want our support, so tell us why we should support you." Barack responded by saying, "I know the folks in this room, I know what a lot of your problems and concerns are, and if I get into office, I will always be there. I'll be your champion." He knew his chances depended largely on raising enough money for a serious television advertising campaign. As Marty Nesbitt said, "it was a matter of having the money to tell his story," and Barack used proportional terms to make his pitch to Chicago's black business leaders: "If you raise $4 million, I have a 40 percent chance of winning. If you raise $5 million, I have a 50 percent chance of winning. If you raise $6 million, I have a 60 percent chance of winning."

Many people in the room felt that Jim Reynolds was the most persuasive. Executive Les Coney remembered Reynolds's appeal to his friends: "Barack wants to run for the Senate, and I really want to know if I can count on your support." Eugene Morris remembered Reynolds insisting, "This guy's for real, we've got to help him." Morris stressed that "Jim was really the guy who spearheaded it," using "his personal capital. Jim was really out there." ComEd's Frank Clark agreed. "Jim took Barack around early, by hand, to meet people who didn't know him. . . . Folks who ultimately fell in love with Barack didn't know him or love him initially." Attorney Martin King, Jesse Jackson Jr.'s best friend, agreed that it was Reynolds who "convinced a lot of us business folks that we should be supportive" of Barack. "Jim was key," King felt. "But not for Jim, no U.S. Senate race" for Barack.

Hermene Hartman was underwhelmed by Barack's presentation and told him so the next day. "You sound very local. You don't sound national. You've got to get a national focus." Barack asked what he should do. "You need to start talking national politics, and the best person is probably for you to go to Jesse" Sr., to absorb some of Rev. Jackson's two decades' worth of experience with national and international issues. Hartman called Jackson, who agreed to talk with Barack on Saturday mornings before Rainbow PUSH's weekly public rallies. Barack also needed to call upon U.S. congressman Jesse Jackson Jr. and ask for his support.

Barack and Jesse Jr. agreed to meet Saturday for breakfast at 312 Chicago, a handsome, well-known Loop restaurant, where they settled into the relatively private southeasternmost booth. "He broaches the question of the Senate," Jackson recalled. "'Michelle wants me to make some money, but I'm thinking about giving this one more shot and running for the U.S. Senate, but Jesse, if you're running, I'm not running,'" Jackson recounted. "There's no chance that I can run," Jackson responded. His father was still active in the Decatur, Illinois, community protests that had begun more than two years earlier with the high school student suspensions. "My dad's practically declared war on downstate," Jackson explained, making a statewide race by anyone with his surname politically impossible. Barack asked if Jesse Jr. would cochair Barack's campaign, and Jackson replied, "I would be honored."

Barack also touched base with staunch supporters like Newton Minow and Ab Mikva, telling them, "I've talked to the Jacksons, and they're going to support me." Thanks to Marty Nesbitt, his financial benefactor Penny Pritzker invited Barack to visit her family at their beachfront home in western Michigan. One morning Jim Reynolds introduced Barack to Andy Davis, a major figure at the Chicago Stock Exchange whose office was in the same building as Loop Capital. Barack asked if he could call Davis, which he did the next day, and Davis agreed to meet Barack for breakfast. Barack explained his interest in running for the Senate, and when Davis asked, "What would it take for you to be a U.S. Senator?" Barack had an immediate answer: "Five million dollars. If I get $5 million I can win." Davis had one crucial question. Explaining, "I'm sick of dealing with guys who I'm going to be embarrassed by later," Davis bluntly asked Barack: "Are there going to be any bimbo eruptions?" "Absolutely not," Barack said. "There will be no bimbo eruptions. There's Michelle and that's it." Davis agreed to join Barack's finance committee and promised to raise $75,000 in the months ahead.[58]

With Barack's outreach intensifying, on June 21 Steve Neal of the *Chicago Sun-Times* reported the full details of Paul Harstad's poll results, clearly aiming to dissuade Moseley Braun from entering the race. Noting that Barack "is exploring a possible run," he wrote that Barack "regards Moseley-Braun as a trailblazer" and "indicates that he would be less inclined to run if she does." Neal emphasized how modest Peter Fitzgerald's ratings were, but Fitzgerald could be reelected if Moseley Braun won the Democratic nomination. Neal's message was clear: "Illinois voters are ready to toss Fitzgerald. But not for Moseley-Braun. Is the Democratic nomination worth having if she can't beat Fitzgerald?"

Toward the end of June, Barack appeared on *Public Affairs with Jeff Berkowitz,* a widely viewed Chicagoland cable television show, and insisted that "right now, my main focus is to make sure that we elect Rod Blagojevich as governor." The acerbic host was taken aback. "You working hard for Rod?" "You betcha!" "Hot Rod?" "That's exactly right. I think that having a Democratic governor will make a big difference. . . . I am working hard to get a Democratic Senate and Emil Jones president, replacing Pate Philip, and once all that clears out in November, then I think we'll be able to make some good decisions about the Senate race." Under Berkowitz's assertive prodding, Barack confirmed that he was a "card-carrying Democrat. I really believe that the core Democratic philosophy is one that . . . really helps working people." Berkowitz was obsessed with school vouchers, and Barack said, "I am not closed-minded on this issue," and that people "shouldn't be didactic or ideological about how" best to provide "the most effective education" for all students. "I am willing to listen to these arguments," but Barack refused to be hectored into explicitly endorsing vouchers.

The next morning Barack had breakfast with Blair Hull at Ceres Cafe. Hull invited Barack because "I was trying to get a sense of who the players were. I wanted to learn as much as I could," and Hull recalled that the two men had "a very pleasant conversation" during which it became "pretty clear that he . . . was very close to declaring." Two days later, Paul Harstad returned to Chicago to present his benchmark poll findings at a well-attended initial meeting of Barack's finance committee. Among Democratic voters, Carol Moseley Braun's positive-negative ratio was 50 to 26 percent, with the split being an even more troubling 39 to 31 percent when asked about Moseley Braun's honesty and integrity. While 59 percent of African Americans identified her as their top choice in the Senate race, 27 percent of white Democrats thought Moseley Braun was unacceptable. Harstad also showed that only 11 percent of white Democratic voters said their "neighbors" would be less likely to vote for someone named "Barry" Obama, whereas 26 percent said that when asked about "Barack." Among general election voters, Moseley Braun's weaknesses were even starker. Asked about her performance when she was a U.S. senator in the 1990s, African Americans felt positively by a 71 to 22 percent margin, but among the entire electorate 59 percent felt negatively and only 32 percent positively. Peter Fitzgerald would defeat Moseley Braun by ten points, and Jesse Jackson Jr. by nineteen, but after voters heard Barack described as a U of C "constitutional law professor" and "the first African-American editor of the *Harvard Law Review*," Fitzgerald's previous 50 to 24 advantage over Obama dissolved, with 50 percent of voters choosing Barack and 31 percent Fitzgerald.

During Harstad's presentation, several of Barack's financial backers requested a written memo summarizing his findings, and Harstad followed up with a barn burner. "Peter Fitzgerald is highly vulnerable," and "Illinois is becoming more Democratic," Harstad stressed. "Carol Moseley-Braun could win a Democratic primary but would have great difficulty winning a general election." However, "without Braun in the race, Obama can win the primary," because "his profile has enormous potential voter appeal." Furthermore, "Obama can beat Fitzgerald in a general election," because "even 27 percent of Republicans choose Barack over Fitzgerald" once they hear his qualifications. "In sum, if Braun opts out of the race, and Obama has the resources to get on television state-wide, Obama can win a primary and go on to win in November."[59]

Steve Neal devoted another column to Barack's potential strength as a statewide candidate, and Barack told the local *Hyde Park Herald* he hoped to create a federal campaign committee by mid-July. Calling the $14 million that Peter Fitzgerald spent to win in 1998 "obscene," Barack said, "Fitzgerald is vulnerable, and I think a strong Democrat could defeat him." He acknowledged that "what Carol does is a factor. It would be silly for me not to take her into account," but Moseley Braun was unwilling to discuss her plans with the *Herald:*

"It's ridiculous. It's an understatement to say it's too early. We haven't even had the '02 elections yet."

As talk of Barack's candidacy spread, even the U of C's student newspaper, the *Chicago Maroon,* sought him out. While a federal campaign filing would happen soon, only in December would he make a final decision. "If this is a race that I do decide to mount, then unfortunately it . . . may cost as much as $3 or $4 million to run in a primary," he told the *Maroon*. Citing the poll results, "if I go forward with the race, then my main liability is my lack of name recognition across the state, which is particularly important given that I have an unusual name. I'm not the kind of person that people will vote for just because they like the sound of my name." Barack said that lots of travel lay ahead too, because he would "have to devote a significant amount of time to meeting with Democratic leaders across the state" to determine if they "think my candidacy would be viable." With Chinta Strausberg of the *Chicago Defender,* Barack sounded more uncertain, saying that 2004 "is still a long way off," and that "I would expect a firm decision sometime early next year." As before, Barack made nice regarding Carol Moseley Braun. "I have been a long-time friend of hers. I have no interest in seeing both myself and Carol run because if we both run, we divide the African-American and progressive voting bases. What Carol does would be a factor in my decision." The *Defender* awarded him a huge, top-of-the-front-page headline—"Barack Obama Explores U.S. Senate Seat"—and he continued to criticize Fitzgerald's record whenever he had the opportunity.

In late July, a number of significant contributions landed in Barack's state committee account. Heiress Abby McCormick O'Neil, whose support Judy Byrd and Laura Hunter had helped nurture, contributed $10,000, a figure that was matched by a trio of gifts from Ariel Capital and its top officers John Rogers and Mellody Hobson. Old friend Greg Dingens added $2,000, but when Barack, Michelle, and their two young girls joined Greg and his wife for dinner one night, they listened as Michelle none too gently argued against going forward with a Senate race. "Michelle started giving Barack some grief, saying 'Barack, if you really want to make a difference, you should go be a high school principal. You can make more difference as a CPS high school principal than as a U.S. senator,' and then the two of them got into this big debate about whether that was true or not," Greg recalled. "I was trying to stay out of it because they were getting pretty animated," with Michelle hesitant because of "the personal costs" of Barack running yet another race.

One day that late summer the family experienced a second frightening medical emergency when four-year-old Malia woke her parents in the middle of the night and "said, 'Daddy, I'm having trouble breathing,'" Barack recounted. "We put her in the car, rushed her to the emergency room, and it turned out that she was having an asthma attack." The crisis passed quickly, and "she's been fine ever since," Barack explained.

Senate duties had him on the road to Peoria and other cities where a Judiciary subcommittee held hearings on the capital punishment reforms proposed in April by Governor Ryan's commission. Even when Barack was at home in Hyde Park, he joined House colleague Barbara Flynn Currie for a panel discussion about instituting public funding for Illinois Supreme Court elections. "Money has come to dominate every aspect of our public lives and our democracy," Barack complained. "At least the judges shouldn't be beholden to special interests," although he knew that "nothing is harder in Springfield than moving forward campaign finance legislation," and there was no choice but "to be patient."[60]

In early August, Steve Neal proclaimed that Blair Hull "is Mayor Daley's choice for the Democratic nomination" for Senate. Hull and Rick Ridder had added well-known Washington-based media consultant Anita Dunn to their campaign team, and although Dunn believed "there was definitely a potential opening" in the race "for somebody outside of the political establishment," her first task was working with Hull to develop "basic candidate skills." Dunn's hiring filled the slot that Hull and Ridder had earlier hoped David Axelrod would take, but over the course of the summer, Axelrod's feelings had evolved. When Barack had first mentioned a Senate run, Axelrod thought Obama was "aiming too high" and suggested he wait until Daley stepped down and run for mayor. But Paul Harstad's poll and the drumbeat of Steve Neal columns made the Democratic race look significantly different in early August than it had in early June.

As a U of C undergraduate in the mid-1970s, David Axelrod had started writing about politics for the *Hyde Park Herald*. Thanks to veteran progressive politico Don Rose, Axelrod got an internship at the *Chicago Tribune*, and by 1979, he was one of the *Trib*'s lead reporters covering city politics. Axelrod's dream was to ascend at the *Trib*, but by 1984, Steve Neal had arrived to write a political column and Axelrod believed he never would become political editor. With his young daughter Lauren suffering from worsening epilepsy, Axelrod was receptive to entreaties by U.S. Senate candidate Paul Simon, who was challenging Republican incumbent Charles Percy. Following Simon's victory, Axelrod and his close friend Forrest Claypool founded a media consulting business, although many Democrats for whom they worked bore no resemblance to the usually principled Simon. Harold Washington, in his last year as mayor, equaled Simon's stature, but machine apparatchiks like Cook County sheriff Richard Elrod were a notch below even state's attorney and mayoral aspirant Richard M. Daley, with whom Axelrod signed on in 1989.

Business was very good, with more than a dozen clients contesting 1990 races, including successful Cook County Board president Richard Phelan. After Phelan's primary win, the *Tribune* credited "$1 million in TV ads, produced by political media adviser David Axelrod," for the "fastest elevation of a political unknown to victory locally, almost exclusively through the use of TV commercials." But Axelrod's usage of little-known court documents to take down African American jurist Eugene Pincham, a top contender, troubled one coworker and led to a *Tribune* story calling the tactic "reminiscent of the Willie Horton controversy during the 1988 presidential campaign," a famously distorted negative ad. Axelrod defended the attack, claiming Pincham "wants his own record obscured" and later stressed that "I believe in partisanship." But his daughter's severe illness forced him to turn down offers to work for Democratic presidential nominees Bill Clinton and Al Gore. By 2002 there was no question that "Axelrod's client list . . . contradicts the self-image" he sought to burnish, and Axelrod later confessed that by August 2002 "Obama offered a path back to the ideals that had drawn me to politics in the first place."

On August 14, Chicago election lawyer Michael Dorf quietly mailed Federal Election Commission Form 1, a "Statement of Organization," signed by Cynthia Miller as treasurer of Obama for Illinois, and FEC Form 2, a "Statement of Candidacy," signed by Barack, to Washington. The next Monday *Capital Fax*'s Rich Miller reported that the Democratic field was rapidly taking shape. Blair Hull was paying for ten campaign workers to aid state Senate Democrats in their upcoming November races, and Gery Chico "is busily learning Spanish and raising money." Barack "is traveling the state, looking for support," but the biggest news was that state comptroller Dan Hynes "is all but certain to run" for the Senate too. "Hynes has a good staff and a strong network. He'll be an instant favorite when he makes the run," Miller stated.

The next day Obama for Illinois sent supporters a "Dear Friend" letter from Barack. "I'm actively exploring" a Senate run, and "we are going forward with fund-raising for this race." A kickoff event was scheduled for October 10. "I wouldn't be considering this race unless our polling data showed we could win," the letter went on, and a copy of Steve Neal's column detailing Harstad's numbers was appended for easy reference. Barack said he was "lining up support from key elected officials," and "we are in the process of assembling a top-flight team to manage my campaign." David Axelrod had "signed on," and "if we can raise the money to put our message on television," prospects would be excellent. Hull's campaign got a copy of the letter, and Axelrod's decision to go with Barack "was clearly seen as significant" for a candidate who previously had seemed like lesser competition than Dan Hynes or even Gery Chico.[61]

As November's elections drew nearer, the odds of a Democratic statewide sweep increased even further. Rod Blagojevich seemed certain to defeat Jim Ryan, and with Emil Jones having more than $3 million to bankroll Democratic Senate races, there seemed to be no doubt that Democrats would seize control of the Senate by at least 32–27. Barack flippantly told the *Hyde Park Herald* that "running unopposed is the way to run," but having a free ride back to Springfield allowed him to focus almost all his attention on raising his profile as a potential U.S. Senate candidate. He expected to be the new chairman of the Senate's Public Health and Welfare Committee if the Democrats took control, so he held a series of ten town hall meetings across Illinois to address health care issues. Approval of his Bernardin Amendment, Barack emphasized, "would change Illinois to a single-payer system where everyone received health benefits." Thanks to new federal funding, Governor Ryan announced that KidCare would expand to provide FamilyCare coverage for almost twenty-nine thousand poor working parents, even though Pate Philip had prevented Barack from winning Senate approval for such an expansion.

In mid-September, Barack flew to Washington to attend the Congressional Black Caucus (CBC)'s annual weekend of meetings. He wanted to call on well-known congressional prognosticator Stuart Rothenberg as well as talk up his candidacy with old Harvard friends like Tom Perrelli and Bruce Spiva. But Barack's primary goal was to use the CBC gathering to get out the word that Illinois had a potential African American U.S. senator whose name was not Carol Moseley Braun. Barack arrived knowing nothing about the long-standing traditions of CBC's annual weekend. Calling it Springfield for African Americans would understate the attractiveness of the surroundings, but when Barack returned to Chicago, he told a bemused Jeremiah Wright he had been astonished by what a "meat market" the gathering was, with multiple women openly propositioning Barack and inviting him up to their hotel rooms. "He came back in shock because he went there to network and to let people know" of his desire to run if Moseley Braun did not. "'Pastor, do you know what they do?'" Wright remembered Barack asking him. "He was in shock," and it left Wright thinking that not only was Barack "a guy with high morals" but also that "he really doesn't know how people are" concerning sex.

Barack continued to talk up his record of tackling "the issues facing working families." Publicity was key. "If people know who I am, meet me, hear me speak or know my track record, I think I can get their votes. The burden is on me to make myself better known." Barack wanted to stress, even to readers of the African American weekly *Chicago Weekend,* that he was seeking more than just black votes. "I have a track record of being able to get crossover votes," as his 2000 race had shown, and he had "an appeal that goes beyond the African-American community. I spent much of the summer traveling across the state," receiving "favorable responses because we're talking about issues that affect everybody." Barack said he had spoken with Carol Moseley Braun, but he was focused "on putting together the resources that would allow me to be an effective candidate," with a fund-raising goal of $3 million. "I've been an effective state Senator, and I can be an even more effective U.S. Senator. I believe there are some issues that can only be effective at the federal level. My running just means I think I can do more in another position."[62]

By mid-September, Barack moved to expand his team beyond Dan Shomon and Cynthia Miller. He asked young attorney Raja Krishnamoorthi, now practicing at Mayer Brown, to serve as his part-time, unpaid policy director. A year earlier, Katrina Emmons, a twenty-year-old graduate of the University of Chicago, had joined Miner Barnhill as a paralegal. She shared an office with Scott Lynch-Giddings, a talented actor holding down a similar day job. As she approached her first anniversary, Katrina told Scott about her interest in politics. He recommended she e-mail Barack, who invited her to lunch at Whole Foods. Katrina anticipated volunteering a few hours a week and was taken aback when Barack was "really direct" in asking if she had "baggage" that could be problematic. He handed her a paperback of *Dreams From My Father,* and said, "You need to read this." A few days later, Barack called and asked, "Do you want to come work for me?" which meant leaving Miner Barnhill. "I can pay you for three months. I can't guarantee you that I'm going to run," but "if it doesn't work out and I don't run, I'll help you find a job somewhere else." Katrina wanted to think about it, but Barack asked her to meet Dan Shomon. Within fifteen minutes, Dan offered twenty-one-year-old Katrina the title of deputy finance director of Obama for Illinois, and she accepted. There was no finance director yet, although Barack and Dan had already asked Claire Serdiuk, the campaign scheduler for U.S. senator Dick Durbin, what her plans were once Durbin's cakewalk reelection race ended in early November. By the end of September, Serdiuk agreed to join Barack's campaign, and a lease was signed for three modest rooms on the fourteenth floor of 310 South Michigan Avenue, a building undergoing renovation. Phone service was set up, and computers and some used office furniture were acquired.

One painful discovery marred the new launch: Obama for Congress 2000 (OFC) had wrongly accepted $5,500 in contributions toward the *general* election, in which Barack of course had never run, and Federal Election Commission rules required immediate refunds. By then, OFC's indebtedness to Barack and Michelle had been whittled down to $5,000, but this new revelation forced Barack to make another personal loan to OFC, leaving him and Michelle $10,500 in the hole. By early October, Cynthia Miller and Katrina Emmons had Obama for Illinois up and running, with Katrina soon purchasing the domain names www.obamaforillinois.com and www.barackobama.com.[63]

In late September, nationwide debate intensified as President George W. Bush warned that Iraqi dictator Saddam Hussein likely possessed weapons of mass destruction and that only a tangible threat of military action against Hussein's regime could force Iraq to allow a thorough international inspection of Hussein's stockpiles. On Saturday night, September 21, progressive Chicago donor Bettylu Saltzman talked about the danger of U.S. military intervention with some friends over dinner, and the next morning she telephoned old activist friend Marilyn Katz to suggest they call together a group of like-minded acquaintances to organize a rally opposing any U.S. use of military force. On Tuesday about a dozen antiwar progressives gathered at Saltzman's home and chose Wednesday, October 2, eight days later, for a lunchtime event outside the U.S. federal building in the downtown Loop.

The day after Saltzman's gathering, Barack told the *Chicago Defender*'s Chinta Strausberg that "the president has not made his case for going into Iraq." Since "we have severe problems here at home" given the huge economic downturn since the 9/11 terror attacks, President Bush's "neglect of the economy does nothing to enhance America's long-term security" and any military action in Iraq would be "a cover-up for a failing economy." Over the weekend, Barack went downstate for a Saturday-night speech to Randolph County Democrats in Sparta, so when Bettylu Saltzman called the Obamas' home to ask Barack to speak at Wednesday's protest, Michelle said she would tell him when he got home. Before returning Saltzman's call, Barack discussed the invitation with Dan Shomon. "What do you think, chief?" Barack asked, using his regular salutation. "He knew, and I knew, that the liberal progressives were key in any Democratic primary," Shomon explained, but Barack worried about the potential political damage if he opposed U.S. military action that most national Democratic politicians were on the cusp of endorsing. "You know I'm going to do this speech," Barack told Shomon, because "I believe it's right," but Barack was "very worried that he was going to eliminate his chance to be in the U.S. Senate." Barack also telephoned David Axelrod, who favored speaking, and a Sunday conference call was set up including Raja Krishnamoorthi and Axelrod's friend Pete Giangreco, a Democratic campaign strategist whom David had been trying to sell on Obama. State comptroller Dan Hynes was a client of Giangreco and his partner Terry Walsh, but Axelrod had been telling Pete that there was "something special" about Barack, whom David believed was "the real deal." Axelrod agreed with Shomon that opposing intervention in Iraq would not harm Barack in a multicandidate Democratic primary, but Giangreco was much more hesitant. On the call, Barack said he intended to call any military action in Iraq a "dumb war," a label that troubled Giangreco. But Barack was firm. "I'm very comfortable saying so," and "my instinct is to do this." On Monday Barack telephoned Bettylu to say that he would speak at the Wednesday rally.

Chicagoans Against War on Iraq, as the organizers dubbed themselves, had secured Jesse Jackson Sr. as their marquee speaker, but otherwise the noontime program featured a lineup that only local progressives would know. Sitting on the podium as the rally got under way, Barack complained to Bettylu about the all-too-familiar musical selections like "Give Peace a Chance." "Couldn't they play something more modern?" he asked. Bettylu and Marilyn Katz got the rally started, reading supportive letters from U.S. senator Dick Durbin and Ohio congressman Dennis Kucinich. Barack's former Senate colleague Chuy Garcia was followed by former MacArthur Foundation president Adele Simmons, and after a musicial interlude, local performer and MC Aaron Freeman asked the crowd of about a thousand to "Please welcome state senator Barack Obama."

Barack began with a reference to the U.S. Civil War before stating, "I don't oppose all wars." Then he mentioned his grandfather's service in World War II before again saying, "I don't oppose all wars." That emphasis surprised many listeners in what organizer Jennifer Amdur Spitz knew was "a peacenik crowd." Musician Peter Cunningham had led off the program and found Barack's theme "a kind of contrarian message that made everybody stop and go, 'What? What is this guy saying?'" Barack repeated his refrain: "I don't oppose all wars. What I am opposed to is a dumb war. What I am opposed to is a rash war," one "based not on reason but on passion, not on principle but on politics." Barack argued that "Saddam poses no imminent and direct threat to the United States, or to his neighbors," and he said that "even a successful war against Iraq will require a U.S. occupation of undetermined length, at undetermined cost, with undetermined consequences." Additionally, intervention would "strengthen the recruitment arm of Al Qaeda. I am not opposed to all wars. I'm opposed to dumb wars." Barack concluded with a second refrain, one he repeated four times: "You want a fight, President Bush?" Then pursue Al Qaeda, pursue UN inspections, get U.S. allies Egypt and Saudi Arabia "to stop oppressing their own people," and "wean ourselves off Middle East oil" at home.

The program closed with Jesse Jackson's remarks and a sing-along, and as a local television news crew interviewed Jackson, Barack appeared alone in the background. Media estimates of the crowd varied from three hundred to three thousand, but Barack had not recognized one supportive couple, a very tall, now eighty-five-year-old black man accompanied by a short white woman: Frank and Bea Lumpkin. Only the suburban *Daily Herald* quoted "Barak Obama" as declaring "I don't oppose all war, I oppose dumb war," but the *Chicago Sun-Times*' brief story devoted as much attention to the views of some nearby construction workers as to the rally's speakers. "I think we should go to war and soon," one ironworker explained.[64]

Two days before the rally, the U of C Law School's autumn quarter had begun, with Barack again teaching Con Law III three mornings a week plus his Racism class on Thursday afternoons. Con Law attracted fifty-two students and Racism thirty-eight, and Barack again faced the challenge of running for office while juggling a heavy teaching schedule. Appearing on WVON's Cliff Kelley radio show, Barack stressed the importance of Democrats winning control of the state Senate while acknowledging that people were "responsive" about "the possibility" of him running for U.S. Senate. "It's important that Democrats at the national level ask serious questions about President Bush's policies."

The next day, after teaching his Con Law III class, Barack addressed three hundred attorneys at the Chicago Council of Lawyers' annual luncheon before heading downstate to Macomb to speak on behalf of Democratic state Senate candidate John Sullivan, who was challenging Republican incumbent Laura Kent Donahue. Dan Shomon was recruiting friends and supporters to drive Barack whenever possible so as to mitigate the demands of his hectic schedule. Will Burns took the wheel on one trip to Champaign and back, and for the Macomb trip, Dan recruited Dave Feller, a former Senate staffer now working in the Cook County Sheriff's Office, to accompany Barack and handle the late-night drive back to Chicago.

At Western Illinois University, Barack told the crowd that "government as it operates and government as it should operate are in conflict." The return trip began inauspiciously when an Illinois state trooper pulled Feller and Barack over not long after leaving Macomb. Feller was at the wheel of Barack's Jeep Cherokee, which had Senate plates, and Barack introduced himself to the trooper, who ran both men's driver's licenses before letting them go with a warning. Back on the road, Barack became worried when Feller began to nod off while driving. "Dave, wake up, wake up!" Feller woke up, but again nodded off a little bit later. An angry Barack insisted on taking the wheel. "He's got the radio up, he's chain smoking," Feller remembered, and when Dave awoke from yet a third nap to suggest a place on Chicago's outskirts where they could get a drink, an exhausted Barack responded simply, "Fuck you!" The two men made it home unscathed, and a few days later, Feller's physician diagnosed him with sleep apnea.

On October 10, the "Obama Issues Committee" met at 7:30 A.M.; that evening would be Barack's long-scheduled kickoff fund-raiser. Dan and Barack both addressed their "game plan/legislative strategy for 2003" with the issues volunteers. Everyone knew Illinois faced an annual budget deficit of $3.4 million, and "the important issue for Senator Obama is to determine which programs should receive further cuts," the agenda explained. With Raja Krishnamoorthi taking the lead, issue assignments were parceled out, with reports due by January 30, before the substantive start of spring session. The fund-raiser that night was a notable success, with old friends like Miner Barnhill's

Jeff Cummings contributing $1,000 and several donors, like Valerie Jarrett's boss Daniel Levin and his wife Fay Hartog-Levin, giving the maximum $2,000 per person.

The next day the U.S. Senate voted 77–23 in favor of a joint resolution authorizing the use of military force against Iraq, with twenty-nine of fifty Senate Democrats siding with Republican president George W. Bush. Barack later observed that "with that vote, Congress became a co-author of a catastrophic war." Barack and Michelle headed to Bloomington, Illinois, to attend the Saturday wedding of Amy Szarek, Barack's congressional campaign fund-raiser. After the ceremony, a number of Barack's current and former campaign aides teased him about his eating habits as everyone enjoyed drinks. Given the amount of time that Barack and Dan spent on the road crisscrossing Illinois beyond Chicagoland, eating in Barack's Jeep Cherokee was almost a daily event. But patronizing drive-through windows had become more problematic as Barack got increasingly picky about what he ate. His strong aversion to cheese exemplified the difficulty. "He had a problem communicating the Whopper order," Dan explained, because Barack asked to skip multiple condiments. "They would always get his order wrong," and in time Dan and Barack gave up on drive-through windows entirely. In Springfield, Barack could always eat his favorite lunch, Cafe Andiamo's vegetarian sandwich, either there or at the capitol, thanks to the young interns who fetched food for busy legislators. Democratic Judiciary Committee staffer Rob Scott remembered Barack cordially thanking the interns by name for such errands, but whenever he could, Barack used Andiamo "as a respite" from the capitol's nonstop glad-handing. Owner Dave Stover, who was also the executive director of the Illinois Association of Rehabilitation Facilities, which represented service providers for people with disabilities, "interacted quite a bit" with Barack on the Public Health and Welfare Committee. Over at Andiamo, Barack "was always, always meticulously worried about getting a parking ticket" from Springfield's aggressive meter attendants. Stover realized that Barack "was always a fairly private man" and "very cautious," and "one of the more conservative liberals that I've ever known." For Barack it was as if "everything had a yellow light on," and not just his "present" button at his Senate desk. Others who saw Barack daily in Springfield, like officemate Terry Link's secretary Bunny Fourez, noted that "he ate so healthy," yet to people in the capitol, just like so many in Chicago, it seemed puzzlingly out of character for someone who was so rigorously health focused and self-controlled to nonetheless be a moderately heavy smoker.

Barack and Dan Shomon had spent so much time together over the previous five years, particularly since 1999–2000, that their relationship had become by far the closest personal bond Barack had formed in his seven years in electoral politics. Dan's friends knew how committed he was to Barack, and although some of them "questioned Shomon's bromance with Obama," no one doubted how devoted Dan was to Barack's best interests, political and personal. As the pace of the nascent U.S. Senate run picked up after Labor Day, Barack asked Dan to sit down with Michelle. Shomon knew Michelle was not enthusiastic about Barack's political career, and he realized things had only gotten worse during and after the 2000 loss to Bobby Rush. "I never sensed his marriage was *really* in trouble," Dan recalled, but there was no question that "there was great tension on the home front," as one Shomon friend remembered him saying. So "I talked to Michelle. She and I had a long conversation about what kind of an impact" a two-year statewide run—late 2002 to potentially late 2004—would have on Barack's family life. Michelle's distaste was at times matched by Barack's guilt about how much he was away from home. "Don't you fucking know I have kids?" Dan recalled Barack asking on more than one occasion when presented with an upcoming travel schedule.

Mulling over Michelle's reluctance and Barack's own regret about missing so much of his daughters' childhoods, Dan took the initiative during one fall drive when they headed south from Springfield to Macoupin County to warn Barack that "if he was running for the U.S. Senate for two years morning, noon, and night, he wasn't going to see his kids, and he was going to feel bad about it. I just knew the personal toll it would take on him," and "I didn't want him to feel guilty." Dan remembered the moment. "We were on the side of the road on Illinois 4 in Carlinville, and I told Barack, 'I don't think you should run.' I said I thought it was a bad idea because of Michelle and the kids." But "Barack just looked at me and said, 'I'm running anyway.'" As close as Barack and Dan were, Barack had never shared with Shomon his sense of destiny as he had with Michelle more than a decade earlier and previously with Sheila and Lena. That unspoken conviction explained why Dan's carefully considered warning about the price to be paid was one Barack could not heed, just as he had been unable to truly pursue the Joyce Foundation presidency nine months earlier.[65]

With that conviction in mind, Barack declined an October invitation to become a candidate for the presidency of Common Cause, for which he had been recommended by former Joyce president Debbie Leff to search consultant and Barack's former Demos board colleague Arnie Miller. Barack's focus was now entirely on the U.S. Senate, and he remained on the road, traveling to places like Elgin to campaign for Democratic state Senate candidates, but Illinois's dire state budget situation dampened the hopes of what could follow from a Democratic sweep. "We're going to be finally driving the bus, and it's got no gas," Barack told one reporter, but he stressed how "it's important that we have Democrats in power to make sure that the state budget is not balanced on the backs of poor people and African-Americans."

On election night, Senate Democrats came away with 33–26 majority control of the chamber, and Rod Blagojevich won the governorship by a margin of more than 230,000 votes. Together with Speaker Michael Madigan's continuing control of the state House, Illinois Democrats now assumed complete dominance of the capitol. But any semblance of unity was less than skin-deep, because "Blagojevich and Madigan despise each other," as *Capitol Fax*'s Rich Miller put it. Appearing the next evening on WTTW's *Chicago Tonight* telecast, Barack emphasized that as Senate president "Emil Jones is now another power center," and one who would "have his own agenda." With everyone on notice about Illinois's gaping budget deficit, which was now projected at $2.2 billion annually, Barack said, "I was at the Blagojevich event last night and spoke to a labor leader who specifically said, 'We are spending a lot of time telling our members not to expect too much on the money side.'" Fellow guest and former Republican governor Jim Edgar laid the blame for Republican losses on the commercial-driver's-license scandal that was still dogging outgoing governor George Ryan.

In contrast, Barack cited the case of the best known of Illinois's thirteen innocent death row inmates. Jim Ryan and unsuccessful Republican attorney general nominee Joe Birkett had continued to defend the faulty conviction of Rolando Cruz, even though DuPage County already had paid Cruz and two codefendants $3.5 million in damages. Weeks earlier, in a *Chicago Tribune* essay entitled "Memo to Voters: Remember the Cruz Case," Barack's friend Scott Turow had denounced "the pattern of prosecutorial misconduct that marked these prosecutions." On television, Barack argued that "the Rolando Cruz case was something that moved a lot of not only Democrats, but independents, to go to the polls in a lot of the swing districts and collar counties" surrounding Chicago. "I think it was a genuine concern. People, even those who support the death penalty, don't like the idea of unfairness," and "I think it did end up becoming an issue in a lot of areas where traditionally you would have seen more Republican votes."[66]

With the election over, Senate Democrats prepared to take charge. With Barack about to become chairman of the Public Health and Welfare Committee, he called lobbyist Jim Duffett, head of the Campaign for Better Health Care, to explain how his Bernardin Amendment aimed at requiring Illinois to move toward universal health care coverage could now be supplanted with actual legislation that Duffett would take the lead in drafting, the Health Care Justice Act. "I want to do hearings around the state" to stimulate public interest in the legislation, Barack told Duffett when they met to discuss a game plan. Barack likewise phoned criminal justice reformer Steven Drizin to inform him that incoming Senate president Emil Jones had tabbed Barack to take the lead on moving forward with legislation to require police interrogations in all potential capital cases be videotaped in full in order to be admissible in court. Drizin remembered that "What struck me was Barack's confidence that he would make sure a bill was passed," even before the session actually got under way.

Bills like those would spotlight a previously obscure, minority-party backbencher as much as any Springfield legislation could, and Barack moved to advance his Senate race prospects on other fronts too. Dan Shomon's friend John Charles, who had quarterbacked the initial 1999 poll that had encouraged Barack's congressional run, turned down a request to join Barack's Senate campaign staff, asking himself "Can he pull this off?" and doubting an affirmative answer. Acutely aware that fund-raising would determine whether Barack could mount a credible campaign including television ads, David Axelrod called Democratic fund-raising consultant Joe McLean, who remembered that "I was skeptical, to say the least, about this young guy with the funny Islamic-sounding name." But he flew to Chicago, "took an immediate liking to Shomon," and "when I met Barack, I knew instantly that he was the real deal. . . . Ax, Dan, and Barack all recognized that fund-raising was going to be key," and the candidate's own willingness and self-discipline to spend hour upon hour calling prospective donors was the most crucial ingredient of all.

"The only resource you have in a political campaign is the candidate's time," McLean believed, and "call time" would determine whether a candidate had the wherewithal "to put yourself in people's living rooms in that three-to-four-week window before the election." McLean hoped to put his own operations director, Susan Shadow, who had just worked an unsuccessful Texas gubernatorial race, on the ground in Illinois, but Barack "wasn't going to pay much" and "it was a crowded primary," so Susan, just like John Charles, decided to give the race a pass. But with Claire Serdiuk joining Katrina Emmons, Barack had two fund-raising staffers ready to go, and the candidate understood McLean's tough-love protocols. "Fund-raising is message delivery to political insiders, and the early game is all about building cred with the insiders," McLean explained. "We wanted to be able to approach the political insiders with a persuasive message" about Barack's potential, as Harstad's poll had demonstrated. "Develop a universe" of "known contributors," "assign a DOLLAR AMOUNT" based on their past giving, and "find phone numbers," McLean's instructions commanded. Most important of all was simply "Make the calls—DO THE WORK." The candidate's own "call room" desk was computer-free to avoid distraction because, McLean explained, "we want somewhere between 25 and 30 hours a week in the call room." Speaking directly to a target, and avoiding voice mail messages, was essential. Then "get to the point . . . the longer the call the less likely you'll get a commit. . . . Tell every prospect why you're running and how you are going to win. . . . Give them the courtesy of asking for a specific amount . . . you can never offend anyone by asking for too much—but you can offend if you ask for too little." Once someone agrees to give, "make it a contract—but gently. 'Thank you for your personal commitment.'" Then hand it off to a finance assistant: "immediate confirmation is the key. E-mail within 5 minutes," and follow up persistently should a check not arrive as promised.

Throughout the fall, Katrina Emmons shepherded Barack through a process he found deeply distasteful. The campaign's office space had three rooms, but almost immediately Barack moved Katrina's desk into his office so their desks faced each other. Barack needed an enforcer because "my incentives are not to make these calls," he admitted. Yet "If I don't make these calls, you don't get paid." But "then he would still try not to make the calls. He hated fund-raising," Katrina explained. Barack would do "just about anything to prevent himself from having to make these calls," and while "he tried to discipline himself to do it, he wasn't doing a very good job of it." Instead, Barack "would talk to me about the Iraq war," and week after week "we talked a lot about the war. . . . I remember arguing about the war, and I remember it being a focus of his attention." Barack also "wanted to gossip about other people" he knew, as if he felt that "I'm going to live vicariously through other people without having to do anything scandalous myself." No matter how many hours Barack spent there, he "wasn't terribly productive," Katrina recalled. "He needed to make certain targets, and he wasn't making them."

Old friends like Judy Byrd, Valerie Jarrett, John Rogers Jr., John Schmidt, Newton Minow, and Ab Mikva all gave the maximum $2,000 per person, as did Al Johnson, Robert Blackwell Jr., and Peter Bynoe, as well as their spouses. African American financiers Louis Holland and Quintin Primo also contributed the maximum, as did heiress Abby McCormick O'Neil and the University of Chicago Law School duo of Martha Nussbaum and Cass Sunstein. Before the end of the year, no less than $8,000 came in from Tony Rezko and his relatives and employees. Longtime supporters like Judd Miner and David Brint and their wives, plus Bettylu Saltzman and Penny Pritzker and their husbands, made significant donations too. Soon after Thanksgiving, election lawyer Michael Dorf closed down Obama for Congress 2000 and assigned the $10,500 debt it owed Barack and Michelle to Obama for Illinois (OFI), which was a "great relief" to Barack. By year's end, more donations from friends like Paul Strauss, Linzey Jones, Les Coney, Greg Dingens and his wife, and Michelle's friend Cindy Moelis and her husband Bob Rivkin pushed OFI's total 2002 income to $290,000. That was modest relative to Barack's goal of no less than $3 million to run a credible race, but with expenses to date totaling only $64,000, one silver lining was that OFI did have a robust cash balance of more than $225,000.[67]

Barack told one reporter he anticipated "a fierce, closely competitive race," and just before Thanksgiving pollster Paul Harstad sent Barack and Dan a strongly worded memo about how to pitch Barack's U.S. Senate candidacy. "I strongly urge you to gear your message toward voters' self-interest because . . . this election will be more about voters and less about candidates," Harstad wrote. Barack's "basic theme" should be "Security for Middle Class Families, Putting PEOPLE First, 'us-versus-them,' putting families above the special interests," Harstad suggested. "People vote their self-interest and typically define it in terms of security." Given Barack's status as a state senator, it would be important to "claim more credit for actually DOING something," and "to the extent that there might be gaps in the Obama legislative record, I urge you to start to fill those gaps," especially concerning retirement security and health care costs. "More than almost any other near-term issue," that latter one, Harstad advised, "has the potential to seriously threaten Republicans' electoral success."

Appearing on Jeff Berkowitz's cable television political talk show, Barack addressed at greater length than ever before his feelings about the looming military attack upon Iraq:

I don't think that there's anybody who imagines that Saddam Hussein is a good guy or somebody that isn't a threat to stability in the region as well as his own people. But I also think that us rushing headlong into a war unilaterally was a mistake and may still be a mistake, and I think that we have to give those inspections a chance. Part of what's going to be difficult to anticipate by the time of the March 2004 primary is whether in fact the United States has invaded Iraq, whether the overthrow of Saddam is complete. If it has happened, then at that point what the debate's really going to be about is what's our long-term commitment there. How much is it going to cost, what does it mean for us to rebuild Iraq, how do we stabilize and make sure that this country doesn't splinter into factions between the Shias and the Kurds and the Sunnis. There's going to be a whole host of critical issues and I think that that's going to be something that whoever the Democratic nominee or those who are seeking the nomination is going to have to be able to grapple with.

Berkowitz pressed Barack on how he would have voted on the resolution authorizing President Bush to use military force had he been a U.S. senator:

If it had come to me in an up or down vote as it came, I think I would have agreed with our senior senator Dick Durbin and voted nay, and the reason is not that I don't think we shouldn't have aggressive inspections. What I would have been concerned about was a carte blanche to the administration for a doctrine of preemptive strikes that I'm not sure sets a good precedent.

Barack also told Berkowitz, "I think you do have to go to a universal health care plan." When Berkowitz asked, "does that mean a single payer?" Barack replied that "we are probably going to have to move towards a single-payer plan of some sort, and I think a major debate is going to be how do you structure that but still retain some of the important components of competitiveness and market-based incentives that are in the existing system."

Barack also had to cope with two later-than-usual veto session weeks in Springfield thanks to November's election. Tongues began wagging when Republican U.S. representative Ray LaHood publicly called for Illinois Republicans to back someone other than incumbent Republican senator Peter Fitzgerald in 2004, and *Capitol Fax*'s Rich Miller wrote that "Fitzgerald is realistically vulnerable" given how "he disappeared from Illinois immediately after winning his Senate seat." The veto session was almost entirely without note, although Barack and ten other liberal Democrats voted against a death penalty reform bill that Barack called "pretty thin gruel." Barack spoke at length before the Senate's 46–11 vote. "I am actually not somebody who is opposed to the death penalty in every single case," he explained, yet he cited his friend Scott Turow's conclusion that "you can't create an error-free system." With "thirteen exonerations from death row in just the past few years," everyone should "acknowledge that that's thirteen innocent people too many to be on death row." Looking ahead to Democratic control come 2003, "we're going to have to do greater reforms than are presented in this bill."

On Wednesday evening, December 4, lame-duck Senate president Pate Philip told his Republican colleagues he would leave office before the spring session began. Three contenders—Kirk Dillard, Steve Rauschenberger, and Frank Watson—announced their candidacies for minority leader in a twenty-four-hour contest that was decided late Thursday after Rauschenbusch threw his support behind the conservative Watson. Philip's departure, after a decade-long vise-grip hold on the Senate, signaled just how dramatic a change was about to transpire with Emil Jones Jr. ascending to the Senate presidency. Barack would get a significant office upgrade, moving from ground-floor suite 105 to the considerably more attractive, though windowless, M114, on the capitol's mezzanine level. Assisted by expert staffer Nia Odeoti-Hassan, Barack would be chairing the Health and Human Services Committee, as he chose to rename it, and superlegislator John Cullerton would succeed the departing Carl Hawkinson as chairman of Judiciary, on which Barack would continue to serve. Illinois's budget woes aside, a slam-bang spring session for Senate Democrats was only weeks away.[68]

Thanks to Raja Krishnamoorthi and other young attorneys like Andrew Gruber, the Obama Issues Committee steamed forward in advance of the spring session. Three subcommittees focused on state spending, state revenue, and Illinois's plethora of corporate tax breaks, all with an eye toward generating "innovative and creative policy positions" so that Barack can "introduce bills related to the ideas that you come up with," Raja and Cynthia Miller told everyone. With the Senate race looming, Barack in mid-December left the boards of both the Joyce Foundation and the Woods Fund. Barack did take time to guest-host Cliff Kelley's popular radio show on WVON, where he addressed the political firestorm created by Senate majority leader Trent Lott's praise for the segregationist record of Senate colleague Strom Thurmond. "The Republican Party itself has to drive out Trent Lott," Barack told listeners. "If they have to stand for something, they have to stand up and say this is not the person we want representing our party." Barack was hardly more charitable toward Illinois's Peter Fitzgerald, declaring on his new but far-from-fancy ObamaForIllinois.com Web page that the Republican incumbent "has represented only narrow concerns, big business and special interests. Meanwhile, corporate and CEO scandals have widened, our health care crisis has gotten worse, and many of us are fighting to get a good-paying job."

But the weightiest factor in the Senate race was still what Carol Moseley Braun would do. Thanks in large part to Valerie Jarrett, Barack had a crucial behind-the-scenes voice working on his behalf. Black financier John Rogers Jr. was one of Moseley Braun's closest friends, and given her painful 1998 loss to Fitzgerald, Rogers explained that "there was just a sense, I think, in the African American community, the business leadership community . . . that Carol should get on with a business life" rather than return to politics. "I was not eager for her to jump back into the game again," and "I had taken Carol over to meet with Jamie Dimon," CEO of Bank One, "to talk about, like, board seats and trying to help her network as she was getting back into Chicago life," following her ambassadorial posting in New Zealand. "I remember vividly the conversations around could I help with Senator Braun, and try to talk with her," Rogers recalled, and "I was doing what I had been asked to by Barack and Valerie to try to sort of feel her out and try and discourage her" from running for Senate.

As Christmas approached, reporters had no success in getting Moseley Braun to reveal her plans. Barack's public comments tried to nudge her toward the sidelines. "Any time you run against somebody who beat you as an incumbent, you have some challenges," Barack remarked. "Her calculation would have to be that people's feelings about Fitzgerald have become so negative, it changes the dynamics from six years ago." When the Associated Press reported that Moseley Braun "said Obama told her earlier this year that if she got in the race, he would get out," Dan Shomon pushed back hard, saying, "We don't know what Senator Moseley-Braun is going to do, but we're in." Privately, Barack's own attitude was far different, as lawyer friend Matt Piers learned when he ran into Barack at U.S. representative Jan Schakowsky's Evanston office. Barack explained that Schakowsky had just told him that she would not run for Senate, "but I'm not going to do it if Carol Braun does." Piers was surprised. "You'd stay out for Carol Braun?" Yes, Barack replied. "Her name recognition is through the roof, and I don't think I could beat her."[69]

Before Barack, Michelle, and their daughters went to Hawaii to spend the holidays with Madelyn Dunham and meet Maya Soetoro's new boyfriend Konrad Ng, Barack for the first time ever sent out Christmas cards to the statehouse press corps. One reporter teasingly gave Barack "the Pressroom Subtle Hint Award," writing that while it was not the first time Springfield journalists had received holiday greetings from a Chicago state senator, "it was the first time during an election cycle in which Peter Fitzgerald is being challenged." Barack had his Con Law III exams to grade, yet his incipient Senate candidacy had in no way harmed the quality of his teaching throughout the fall. His thirty-five Con Law students gave him a strong 6.0 rating on the law school's now seven-point evaluation scale, and thirty-two of them responded "yes" when asked "Would you recommend this class to another student?" Beginning at 8:00 A.M. on Monday, January 6, Barack would again be teaching Voting Rights twice a week, despite the demands of a statewide campaign, but the Obamas' family finances, now without Robert Blackwell Jr.'s EKI beneficence, left no other option.

Rumors spread that Chicago city treasurer Maria Pappas would join the Senate field, but on Monday, January 6, Dan Shomon convened the first weekly conference call of all the political consultants—David Axelrod, John Kupper, and the Strategy Group's Pete Giangreco and Terry Walsh—who would be directing campaign strategy for Obama for Illinois. "Numbers analysis" was the first topic, highlighted by how in a Democratic primary, Cook County African American voters constituted 25 to 27 percent of statewide turnout. "Update on Moseley-Braun's status" led them to conclude that the likelihood of her entering the race was 70 to 80 percent. Per John Rogers's efforts, "Prospective employers?" was touched upon too. An "endorsement press conference with Jesse Jr." could help undercut Moseley Braun's entry, but the most important upcoming date would be January 30, when all campaigns had to publicly file their 2002 fund-raising reports with the Federal Election Commission. Press coverage of those figures would go a long way toward signaling who besides millionaire Blair Hull would be a top-rank contender. John

Kupper jotted down how Barack's goal was "$500K by end of month" with "$400K in [the] bank."

That same morning, Rich Miller's *Capitol Fax* headlined one story "Obama Picks Up Steam." With both Jesse Jackson Jr. and his father backing Barack, some Obama "supporters believe that if he can lock up enough support early, then" Moseley Braun "might be convinced to stay out of the race—or forced out later this year if she has trouble raising money and finding support." Wary of skeptical readers, Miller emphasized, "Don't dismiss Obama. Yes, he has a strange, even Arabic-sounding name (it doesn't take a genius to realize that Obama sounds a lot like Osama) at a time when that attribute is far from ideal . . . but the Harvard Law School grad is raising money and attracting lots of young activists to his campaign."

Later that day, Jesse Jackson Jr. issued a public statement aimed at torpedoing a run by Carol Moseley Braun, saying "it is now time for her to step aside." Reporting Jackson's announcement, *Roll Call* quoted David Axelrod as warning that "she would go into that race with very serious baggage." Further assertions that "she would not be able to beat Fitzgerald" and that "this is a woman who literally needs a job" were not attributed to Axelrod by name, but the purpose of the one-two punch was obvious. That evening in Springfield, Barack's state campaign committee held a "Legislative Inauguration Celebration" at one of the capital's well-trafficked watering holes, Floyd's Thirst Parlor. Former U.S. senator Paul Simon was among those who came to greet Barack, but in the statehouse most attention focused on how Senate president Emil Jones Jr. had promulgated new rules requiring that all amendments to pending bills be referred to the leadership-dominated Rules Committee. "Rank-and-file legislators have given up some of their rapidly disappearing autonomy for the sake of partisan retribution," *Capitol Fax*'s Rich Miller observed.

At the first possible opportunity, Barack introduced a trio of significant but unsurprising bills. One aimed to extend Illinois's earned-income tax credit, a second would require the videotaping of all police interrogations in potential capital cases, and a third would mandate that law enforcement officers record the race or ethnicity of all drivers whom they stopped. Barack also held a press conference to tout an upcoming bill backed by the Service Employees International Union (SEIU) that would require Illinois hospitals to disclose all manner of statistical information, including patient infection rates. "The public has a right to know information that could affect their very ability to survive a hospital stay," Barack declared, information that "should be available to all patients so that they can make responsible health-care choices." Barack also told reporters that it was imperative that incoming governor Rod Blagojevich demonstrate "that he's willing to make tough choices" regarding Illinois's spiraling state budget deficit. "People may grumble and complain, but I think they will respect him making those tough decisions."[70]

Forty-eight hours before leaving office, Governor George Ryan issued a blanket commutation of the death sentences of all 164 Illinois death row inmates. Ryan cited the legislature's 2002 failure to adopt the many reforms recommended by his study commission as one reason for his breathtaking move, and Barack endorsed Ryan's decision. "I think the governor is right that the legislature did a poor job of addressing his proposals," but with Democrats about to take control, "there's going to be a big incentive to carefully examine and hopefully pass some of the governor's reform proposals."

Prior to Rod Blagojevich's Monday, January 13, inauguration, the incoming chief executive "said over the weekend that if he had known how bad the deficit was he might not have run for governor," *Capitol Fax* reported. Miller added that "if he really didn't know, it was a willful ignorance," and statehouse insiders were already worried about the tenor and composition of Blagojevich's new administration. "Anyone who worships secrecy like Team Blago does is headed for trouble," Miller commented. "Paranoid people don't usually govern all that well."

In Chicago following the inauguration, Barack was introduced to two of the city's wealthiest men thanks to longtime cheerleader Newt Minow. Lester Crown and his son James, one of seven siblings, presided over a family-held firm worth $4 billion, and its growth over the years since Lester's father Henry had started a small sand and gravel business in 1919 was the most storied legacy in all of Jewish Chicago. Lester Crown, well known for his hawkish views regarding Israel, responded noncommittally upon meeting Barack, but James, who was only eight years Barack's senior, was extremely impressed. "I pulled him down to my office, and I said, 'Hey, look, I think you should run, and I want you to win,'" James recalled.

On Wednesday night, January 15, John Rogers's phone rang. "I'm not running for the Senate," Carol Moseley Braun told her old friend. "Great!" Rogers responded. "Then she said, 'I'm running for president.'" Rogers was perplexed. "What are you talking about?" Just before calling Rogers, Moseley Braun had telephoned Democratic National Committee chairman Terry McAuliffe to tell him to include her in the lineup for an upcoming Democratic presidential candidates' debate. Moseley Braun later explained that the stimulus for her utterly unexpected decision was a conversation with her ten-year-old niece, who had a social studies text in hand when remarking, "Auntie Carol, all the presidents are boys." Moseley Braun reassuringly replied that "Girls can be president too," but then doubted the truth of her statement. "I was shaken to my core," Carol explained, and her brother asked, "'What's up?' 'I just told Claire a whopping lie,'" Carol responded, to which her brother asked, "What are you going to do about it?"

As Moseley Braun remembered it, she never had any intention of trying to return to the U.S. Senate because her one term there had been "the worst experience of my life," and "it wasn't something I wanted to do" over. Her presidential prospects were beyond long shot, but as soon as John Rogers got off the phone with her, he immediately called Valerie Jarrett with the good news. Valerie in turn phoned Barack, and within less than twenty-four hours, Barack's now five-member campaign staff—with young African American former Senate staffer Randon Gardley having just joined Dan, Cynthia, Claire, and Katrina—began organizing an announcement event for five days later at the Hotel Allegro, in the downtown Loop. The Senate would not be in session on Tuesday, January 21, but Shomon was apologetic for the very short notice when he called Quad Cities senator Denny Jacobs. "I don't care where it is, I don't care when it is, I'm there," Denny told Dan. Terry Link and Larry Walsh responded with similar enthusiasm as Barack's campaign staff worked to draw as many well-known supporters as possible to the 11:00 A.M. event.

By the end of the afternoon on Thursday, January 16, the word was official, and Carol Moseley Braun told the *Chicago Defender,* "I am not running for the Senate." Responding to reporters the next day, Barack said "we weren't planning to drop out immediately" had Moseley Braun announced her candidacy. "We were going to see if we could possibly continue. But realistically, it would have been very tough to stay in against her." Barack explained he had intended to delay any formal declaration of candidacy for another six weeks, but Moseley Braun's heartening move prompted the earlier event. In the *Chicago Tribune,* columnist Eric Zorn praised Barack as the Democrats' "most impressive officeholder" and "the class of the field." In the *Chicago Sun-Times,* African American columnist Laura Washington wrote that Barack's "laid-back charm can be alluring" while noting the congressional race claims that Barack was "not black enough" and warning that his "weakest appeal is to the working class." The local *Hyde Park Herald* reported that Barack's fund-raising had now topped $400,000 and cited how "his squeaky clean image and progressive politics . . . has made him a media darling."[71]

On Tuesday morning, January 21, at the Hotel Allegro, a capacity crowd of several hundred people filled the modest room where Barack would announce his candidacy. Emil Jones Jr. and U.S. representatives Jesse Jackson Jr. and Danny Davis joined white state senators Denny Jacobs, Terry Link, and Larry Walsh on the dais. "I don't have personal wealth or a famous name," Barack told the crowd, "but I have a fire in my belly for fairness and justice." Assailing the incumbent, Barack declared that "four years ago, Peter Fitzgerald bought himself a Senate seat, and he's betrayed Illinois ever since." Saying that he was running to help the state's "working families," Barack argued that "We have a fairness deficit in this country. We have a hope deficit in this country. We have an opportunity deficit in this country." Using as a refrain the phrase "What would Dr. King say?" Barack proclaimed that "we need a politics of hope in this country," not "a politics of division."

A short press conference followed Barack's formal announcement, and Dan Shomon had prepared some "talking points" for the officials supporting Barack. Reminding everyone that more than 25 percent of Democratic primary voters were African Americans, Dan stressed how "Obama did a poll in May 2002 that showed he can win." Barack answered the first question by stating, "I would not have announced if I didn't think I could win," and he expressed hope that he could raise $4 million. "I'm certain we'll have enough resources to get our message out." Jesse Jackson Jr. told the reporters, "I don't think anyone in the entire state has the intellectual capacity of Barack Obama," and conservative white downstater Denny Jacobs called Barack "a gem. He can grasp all the issues, has a knack of being up-front and making friends. He can handle Washington."

A seemingly relaxed Michelle Obama was among those at the Allegro event. Speaking with Mike Marrs, the former Senate staffer who was one of the core members of the Obama Issues Committee, Michelle commiserated about his wife's nursing problems. As the gathering was breaking up, Michelle reminded Marrs about her advice for his wife. "Michael, don't forget to tell Deanna to put lanolin on her nipples!" Mike was bemused that Michelle "had no qualms about shouting this down a hallway," and he believed "she was fantastic—just a wonderful person." Michelle had given her blessing for what she thought would be Barack's final run for office, and in fourteen months, the burden of having a husband who was constantly away from home because of politics would finally end unless Barack somehow triumphed in the March 16, 2004, Democratic primary.

That night's local news on WLS Channel 7, Chicago's ABC affiliate, devoted all of thirty-five seconds to Barack's announcement. The "Harvard-educated lawyer" hailed "from the South Side Hyde Park community" and had lost his 2000 challenge to Bobby Rush. The next morning's *Chicago Defender* was hardly more auspicious, because it quoted a little-known nationalist as declaring that "Obama has no relationship with the Black community and is a product of the white lakefront community. He is a white liberal in blackface." There again was the ghost of 2000, rooted in the events of seven years earlier when Alice Palmer had tried to go back on her word. That evening on WTTW's *Chicago Tonight,* host Phil Ponce had a nine-minute colloquy with Barack, who praised Carol Moseley Braun as a "trailblazer." He said he would have stayed out of the Senate race had she entered because "we share a base." Asked why he was running, Barack replied that "we have a crisis at the national level building, both economically but also in terms of foreign policy." Peter Fitzgerald "has not displayed the sort of leadership on key issues" that was needed, and "the thing that people are most concerned about is their economic security," just as Paul Harstad's memo to Barack had stressed. In closing, Ponce asked Barack if it was helpful to have a governor named Rod Blagojevich if your name was Barack Obama. "Rod's a trailblazer and a hero of mine," Barack gushed, noting how "America is by nature a hybrid culture." Yet there was one ineradicable difference between a Blagojevich and an Obama. "Race, I think, is at the center of the American dilemma, and it always has been."[72]

Chapter Nine

CALCULATION, COINCIDENCE, CORONATION

ILLINOIS AND BOSTON

JANUARY 2003–NOVEMBER 2004

A few hours after Barack's announcement, Blair Hull's campaign notified the Federal Election Commission (FEC) that their candidate's personal contributions had exceeded $1.3 million. Pursuant to the Bipartisan Campaign Reform Act of 2002, popularly known as McCain-Feingold after its Senate sponsors, Arizona Republican John McCain and Wisconsin Democrat Russ Feingold, Hull's filing triggered the so-called millionaires' amendment. When a wealthy candidate like Hull put a huge amount of his own money into a race, the $2,000 limit on personal contributions to competing candidates increased to $6,000 to maintain parity. Once Hull reached $2.1 million in self-funding, a second trigger would allow his primary opponents to accept up to $12,000 from individual contributors. For Barack's campaign, as for Dan Hynes's and Gery Chico's, Hull's investment made available a new pool of wealthy contributors: everyone who already had given them $2,000 toward the primary was now eligible to give an additional $4,000. Fund-raising was uppermost in Barack's mind, because each campaign's final-quarter 2002 FEC reports were due by January 31.

Barack later wrote that he began his formal candidacy "with an energy and joy that I'd thought I had lost." That Sunday Barack accompanied Jesse Jackson Sr. on a visit to Pembroke, Illinois, an almost all-black township an hour south of Chicago where poor farmland had caused dire rural poverty. Monday his campaign team convened to sketch out their game plan for the months ahead. In time both a press spokesperson and an African American "base vote" organizer would be added to the skeletal staff, but four major tasks lay ahead. Number one was of course "raise money." Dan Shomon was hoping to announce having raised $300,000 in 2002 and a total of $400,000 by January 31 in order to demonstrate "that we are competitive," but more crucial would be hitting $1 million by the end of the next FEC reporting period on June 30. A second priority was to secure additional crucial endorsements, particularly Bobby Rush and white progressive U.S. representatives Jan Schakowsky from the North Shore and Lane Evans from Quad Cities. Equally important were major unions: the Service Employees International Union (SEIU), the American Federation of State, County, and Municipal Employees (AFSCME), and the Illinois Federation of Teachers (IFT). With their backing, it might be possible to deny labor favorite Dan Hynes the official endorsement of the Illinois AFL-CIO.

Third was to take advantage of the Democrats' control of the state Senate to bulk up Barack's legislative résumé by passing and publicizing bills he was ready to champion: a renewal and expansion of Illinois's earned-income tax credit (EITC), a hospital "report card" accountability measure that was SEIU's top priority, the racial profiling and death penalty reforms that had been blocked in 2002 as well as health measures that would go through the committee Barack now chaired. Hardworking members of Barack's all-volunteer issues committee were busily pulling together other proposals to demonstrate that "Barack is fiscally responsible and is willing to stand up to powerful special interests." But Barack's political consultants knew they also needed to be prepared to play defense on the "missed gun vote (child was sick)," which had drawn so much public criticism three years earlier, and also the $5 million in low-visibility "member initiative" funds Barack had doled out. Those were "90 percent unimpeachable," John Kupper noted, but the grants that had gone to "Jesse's Citizen Education Project" and to "Yesse Yehudah" could be problematic if opponents focused on them.

Forty-eight hours later, Barack alerted Dan Shomon that his draft of their fund-raising announcement had to be amended. "Chief: The release itself looks very good. Only problem is that the numbers aren't quite right. Our report will show $289,000, not over $300,000, because we can't include some of the checks we got in. In addition, as hard as I'm pushing, the week of the announcement set us back in terms of fundraising, so that at best we'll have close to $400,000 raised by this weekend—and that's only in pledges; actual collected checks will be closer to $350,000. I think we can use the same release, but use 'almost $400,000 raised' and 'close to $300,000'" in hand. "From here on out, it's call time, all the time." Later Barack recounted that "the first $250,000 I raised was like pulling teeth" because beyond Illinois "no major Democratic donors knew me, I had a funny name," and "they wouldn't take my phone calls." Fund-raising consultant Joe McLean had stressed to Barack and finance director Claire Serdiuk that a U.S. Senate race entailed "a high volume, high dollar, candidate-driven call shop," and that for the candidate, "it's hard work that requires real discipline." Claire "understood everything," Joe recalled, and as the legislature's spring 2003 session got under way, Claire joined Barack in his Jeep Cherokee for the three-hour drive to Springfield and back, holding the "call book" and dialing donors on Barack's cell phone as he drove. Their "connect rate" and "commit rate" were fair to middling, but call discipline was much easier to impose on I-55 than it was in the office. By late January, Katrina Emmons had been offered a job at more than double the salary the campaign was paying her, which would allow her to save for law school. Barack "was really not pleased" when she told him she was leaving, "because he's bent over backwards for me" in paying for the health coverage she lost when she left Miner Barnhart. Katrina sensed that Barack's displeasure reflected a belief that she thought he could not win, but young former Senate staffer Randon Gardley stepped into Katrina's shoes as Claire's assistant. When Dan Shomon's mother passed away at the end of January, Dan had to cope with family matters, and Barack's skeletal campaign staff was stretched very thin.

Competitors' FEC filings showed that Gery Chico, with more than $1 million raised, and Dan Hynes, with more than $900,000, were much closer to Blair Hull's total than to Barack's modest $290,000. But Obama for Illinois was spending little compared to the others, and in reporting everyone's numbers, the *Chicago Sun-Times* said Barack "spends between four and five hours a day on the phone doing fund-raising." Senate session days put an increasing crimp on that claim, but Barack proceeded with his plan of demonstrating serious legislative activity as he introduced a passel of new bills, including one to expand KidCare.[1]

In Chicago, John Kupper and David Axelrod prepared an eight-page "Obama for U.S. Senate Strategic Plan." Their uppermost fear was that Barack would not be the sole credible African American candidate in the race, because it appeared that wealthy health care executive Joyce Washington, who had made an impressive but unsuccessful 2002 race for the Democratic nomination for lieutenant governor, intended to run. "Washington's permanent entry in this race would be a significant factor, altering the strategy for achieving an Obama victory in a material way" by lessening the portion Barack could expect of the African American vote. "Anything that can conceivably be done to convince Washington there is no oxygen in the race for her must be undertaken," because "losing a significant chunk of the Obama base may not be fatal but could be terribly harmful." Senate president Emil Jones Jr., who had heartily supported Washington's earlier candidacy, was "livid" that Washington would challenge Barack, and Kupper noted that "blocking off her key funding and political avenues is the greatest thing that can be done to get her to reconsider."

"Obtaining an overwhelming percentage of a strong African-American turnout and a small but significant percentage of the white vote" from "progressive communities in the North Shore suburbs and Lakefront" would be essential for an Obama victory. "The campaign should not invest real resources downstate," the Illinois term for all of the state, not just its southern half, beyond "Chicagoland"—Cook County and the five surrounding "collar counties" that made up Chicago's suburbs. An Obama win would be "much more difficult if the downstate vote" was "gravitating disproportionately to one candidate," as had happened in Blagojevich's 2002 primary win. With Blair Hull and Dan Hynes each expected to follow the governor's playbook, competition for those voters was a certainty.

"It is critical that over the next couple months, Obama uses his leadership position in Springfield to develop a signature health care issue or two," Kupper recommended, so as "to establish early visibility on this central issue for primary voters," as Paul Harstad's earlier polling had shown. Kupper also sketched out three alternative budgets, with spending totals ranging from just under $4.3 million to more than $4.6 million. "The recommended $4.665 million budget attached to this plan is a little more aggressive and expensive than the $4 million figure the campaign currently believes it will raise," Kupper acknowledged, but the higher figure allowed for "earlier and heavier Chicago television" advertising than the lower. The campaign will maintain "a fairly skeletal staff structure for a race of this size" as "every dollar must be preserved for persuasion activities" late in the race. "Obama's solicitation program at this point must be the central thrust of the campaign. The call and meeting time must dictate the schedule—completely. Political and press needs . . . must be scheduled around the 20-to-25 hours a week the candidate must spend on the phone and in solicitation meetings. This must be enforced with maniacal discipline," for January's experience had shown that "there are too few calls being completed each week."

Beyond fund-raising, identifying "a signature health care initiative or two" should be Barack's top goal, so as to "build his reputation as the strongest voice in the race against the GOP's agenda." But the campaign also had to focus on its presumed base, because "any significant African American defections will potentially create an outsized reaction from the insider press. The ongoing effort to secure Congressman Rush's endorsement must continue," and efforts to deter Joyce Washington soon included a public attack by her little-known former campaign manager, Sean Howard. The *Chicago Defender* charitably headlined Howard's denunciation "Activists Say No to Joyce Washington" and quoted him as saying that "there has to be an outside influence for her to make this type of run" when other black political leaders had already pledged their support to Barack.[2]

In Springfield, the first weeks of the new legislative session were dominated by reports of Emil Jones Jr.'s "growing frustration" with Rod Blagojevich. *Capitol Fax*'s Rich Miller wrote that "many insiders" believed that "Jones is flush with his new power and is far too anxious to wield it," but Blagojevich's almost complete absence from the statehouse also left most politicos perplexed. Committee action got under way on SB 15, Barack's bill requiring videotaping of all police interrogations in murder cases, with him acknowledging that "obviously this is a controversial bill" for law enforcement, and he promised "to make sure all the parties concerned have as much input as possible." Barack convened an initial meeting with Illinois Sheriffs' Association executive director Greg Sullivan and other police representatives to lay out his plans. Sullivan said Barack "was very diplomatic about it," but Barack told the fifteen or so attendees, "This is going to happen. Now you can either be part of the solution, or you can be part of the problem, and if you're part of the problem, you're not going to have a seat at the table, so let's sit down, figure this out, figure out what we've got to do, and get this done."

Barack also pushed forward with his "driving while black" racial profiling bill. New Judiciary Committee chairman John Cullerton's top priority was to win approval of his long-sought "primary stop" seat belt bill, but early in the session "Emil says, 'You can't pass that unless you take care of driving while black' . . . so I told Barack, 'I'll help you pass yours, and you help me pass mine.'" Barack agreed, joining Cullerton and John's longtime friend Chuck Hurley, a top official of the National Safety Council (NSC), to plot strategy over dinner along with veteran lobbyist Paul Williams, whose clients included NSC. In prior years Pate Philip had refused to bring a seat belt bill to the floor, telling Hurley, "You know, Chuck, this is a fag bill," but Hurley had won House passage after addressing black legislators' unwillingness to expand police officers' authority to pull over drivers. Hurley enlisted African American former California State Assembly Speaker Willie Brown, who had sponsored such legislation, to address Illinois's Black Caucus by speakerphone. Barack and Donne Trotter posed questions to Brown, with Trotter asking, "Knowing what you know, what would you have done differently?" Brown's response highlighted the deaths that resulted from beltless auto wrecks: "I would have done it sooner. I would have saved more kids."

Barack wondered out loud about linking his racial profiling bill to Cullerton's seat belt measure and whether that might hurt his candidacy outside Chicagoland. "'What about them bubbas downstate, you know? They ain't gonna like this,'" he told Williams. "'Barack, how many goddamn bubbas downstate . . . are voting for someone named Barack Obama in the first place? The second thing is, how many suburban mothers *will* vote for you for that seat belt?'" Williams replied. Barack embraced Williams's analysis, and at dinner the two of them plus Cullerton discussed how Chicago's fifty different Democratic ward organizations would line up among Dan Hynes, Gery Chico, Blair Hull, and Barack. Finally they turned to their legislative package. "Okay, Chuck, tell me about the belt bill," Barack said, and Hurley agreed to use his law enforcement contacts to help move Barack's racial profiling measure forward in tandem with the one authorizing officers to stop any driver not buckled up.

Every Monday Dan Shomon, Cynthia Miller, Claire Serdiuk, and the campaign's five external consultants—David Axelrod, John Kupper, Pete Giangreco, Terry Walsh, and Joe McLean—convened a conference call to agree upon upcoming events and releases. On February 10, they planned a press conference for the next day, at which Barack, as chair of the Senate's Health and Human Services Committee, would announce a series of statewide hearings to address Medicaid shortcomings. Thanks to Hull and the millionaires' amendment, Shomon said the campaign was "confident to hit $1 million by June 30." The next day's press event led to a prominent story in the *Chicago Defender* headlined "Obama: 'Medicaid Cuts Will Kill Hospitals,'" with Barack warning that "the federal government . . . is poised to run up record deficits, effectively borrowing from our children and grandchildren to pay for our own obligations that we don't have the courage to face." A subsequent e-mail from Shomon to pollster Paul Harstad and consultants Kupper and Walsh reiterated that Barack believed health care "is the centerpiece of the campaign."[3]

On February 13, the *Chicago Sun-Times* reported Joyce Washington's formal entry into the race, although the story also noted that Obama "has locked up the major black political leadership." That day's *Chicago Tribune* reported that Dan Hynes's campaign would be cochaired by Mayor Daley's brother John and influential downstate senator Vince Demuzio, but Barack's public presence remained robust with another press conference, this one devoted to his bill to renew and expand Illinois's earned-income tax credit. The *Chicago Defender*'s Chinta Strausberg obliged with another prominent news story, but as the thirteen-month primary campaign officially got started with Hynes, Hull, Chico, and Obama all appearing at a Sunday-night candidates' forum in suburban DuPage County, Barack told one reporter that "it's going to start seeming like 'American Idol.'" Onstage, he targeted his remarks at Republicans, not his Democratic competitors. "We've been making some bad choices at the federal level, and we've also been making some bad choices internationally." Realizing that many questions for prospective U.S. senators would involve federal, not state, policy topics, Obama Issues Committee chairman Raja Krishnamoorthi told his colleagues "we need to get him up to speed on these issues and help formulate his policy positions" by assembling "a briefing book for Barack." The willingness of close to a dozen smart young lawyers like Krishnamoorthi and Andrew Gruber to invest scores of uncompensated hours on his behalf gave his campaign a highly skilled resource that Hynes, Chico, and Hull could not equal, and Barack asked his team to "look for specific initiatives/innovative ideas" he could propose. "People are going to want to know what separates me from the pack, not simply that I toe the progressive line on every issue."

Friends who had known Barack over the previous decade saw some significant changes as he ascended a more public stage. "It was clear to me early on that this was an all-things-otherwise-equal, I'll-do-the-right-thing kind of progressive, not a guy who had either strongly held beliefs or a vision of what he wanted to do with his political career, but rather someone who wanted a political career" and who now had "the politician's annoying ability of telling everybody what they wanted to hear," one progressive attorney who shared dozens of dinners with Barack later reflected. By early 2003 Barack "was already pretty clearly a rising star," but one evening when the conversation turned to newly inaugurated Brazilian president Luiz Inácio Lula da Silva, Barack's ability to turn on a dime was demonstrated all too starkly. Barack said it was "really troubling what's happening down there with all of these anti-American politicians being elected." A woman who knew much more about Latin America than he did objected. "Barack, given what the United States has done to those countries, they're probably all much better off with the kind of leaders that they have ended up with than the kind of leaders they used to have or that our government would want them to have!" Then, "without missing a beat," Barack "switched directions seamlessly" and "was agreeing with her." "Un-fucking-believable," the woman whispered to her equally astonished husband.[4]

As the spring legislative session moved into high gear in the third week of February, Barack introduced twenty-two bills on February 19, followed by an additional *fifty-five* the following day. Several dozen of them were entirely technical measures that fell under his purview as Health Committee chairman, but the breadth of Barack's legislative agenda was impressive and almost overwhelming. Some, like SB 1417, mandating insurance coverage of examinations for colorectal cancer, were brought to Barack by interest groups like the American Cancer Society, but they carried a personal appeal too, in that case reminiscent of his mother's early death from tardily diagnosed cancer. SB 1418, the Ephedra Prohibition Act, was championed by the parents of a young man who had died tragically after ingesting an unregulated dietary supplement. SB 1430, the Health Care Justice Act, would establish a Bipartisan Health Care Reform Commission, and SB 1586 would amend the Open Meetings Act to require all local government bodies to preserve verbatim recordings of any meetings closed to the public. SB 1763, the Victims' Economic Security and Safety Act (VESSA) would require employers to provide leave to employees who had suffered domestic violence, and SB 1765, drafted by Obama Issues Committee volunteer Andrew Gruber, would create a Tax Expenditures Commission to plug corporate tax loopholes.

Behind the scenes, influential Cook County state's attorney Dick Devine signaled his support for Barack's bill mandating videotaping of police interrogations, and *Chicago Tribune* columnist Eric Zorn praised the measure as "smart, cost effective and just public policy." Barack quickly introduced a memorial resolution after twenty-one people died in a nightclub stampede in his district when a security guard used pepper spray in tight quarters. The E2 club had hosted fund-raisers for Carol Moseley Braun and Emil Jones Jr. as well as

Barack, which even the *New York Times* highlighted. *Chicago Tribune* columnist Mary Schmich noted that E2 was completely unknown to white Chicagoans, even though it was well known in the black community, and Barack called a press conference to announce a "commonsense" Nightclub Safety Act outlawing the use of mace, pepper spray, and pyrotechnics like those that had just left one hundred patrons dead in a Rhode Island club fire. "There's something about a criminal law that gets people's attention in a way that a city's building ordinance does not," he told reporters.[5]

With the Senate race now under way for real, Illinois as well as national news outlets began giving it greater attention. U.S. representative Jesse Jackson Jr. told Capitol Hill's *Roll Call* that "I just don't see how Barack Obama loses in a Democratic primary" where African Americans constitute 25 percent of voters. David Axelrod emphasized how Barack is "the progressive candidate in the race," and in the campaign's weekly conference call, Axelrod's partner John Kupper stressed how Barack should emphasize his antiwar stance in challenging his competitors. Bobby Rush told *Roll Call* that "there's just as much potential for the voting black community to support any of the candidates, as it is for them to support one or two of the African-Americans," indicating that his endorsement was up for grabs.

In Illinois, the Bloomington *Pantagraph* commissioned the first poll on the race, with a small sample of 462 voters showing only 27 percent thinking U.S. senator Peter Fitzgerald merited reelection. In a head-to-head matchup against Comptroller Dan Hynes, the Democrat came out on top 34–31, with 35 percent undecided. Privately, Blair Hull's campaign had just conducted their first and much more extensive poll. Among African American Democrats, Barack drew 24 percent, with Cook County treasurer Maria Pappas, whose entry into the race was uncertain, a close second with 18 percent. Both Dan Hynes and Joyce Washington had 12 percent. Among college-educated white Cook County voters, Pappas led Hynes 26 to 24 percent, with Barack at 13 percent. Among all other white voters, downstate as well as in Cook, Hynes topped Pappas 29 to 21, with Barack backed by just 2 percent. Overall, the Hull poll showed Hynes with 29 percent, Pappas 21, Barack 9, and Hull himself barely registering. None of those numbers would have surprised Paul Harstad, Axelrod, or Kupper: Pappas was well known in Chicagoland, only Hynes was known by most downstate Democrats, and Barack's potential among both blacks and white progressives was sizable.[6]

Although Barack's name was not mentioned, on February 24 *Crain's Chicago Business* reporter Greg Hinz published the first story highlighting how Barack's longtime friend Tony Rezko had parlayed his success in raising more than $500,000 for Rod Blagojevich's successful campaign into a highly influential, behind-the-scenes role in the new chief executive's already troubled administration. *Capitol Fax*'s Rich Miller had repeatedly noted that Democratic legislators were angry "with the complete absence of communication with the governor's office," but Hinz highlighted how Rezko was taking a lead role in choosing members of Blagojevich's cabinet and already had placed two of his former employees as the directors of the Department of Commerce and Economic Opportunity and the Illinois Housing Development Authority. Several days later, Blagojevich's top aide, Deputy Governor Doug Schofield, whom *Capitol Fax* called "one of the few good guys over there," quit after less than two months, with an unknown twenty-nine-year-old from New York named as his successor. Unmentioned in Hinz's story was how the leading candidate to become director of the state Department of Public Health, thirty-seven-year-old Dr. Eric Whitaker, had been recommended to Rezko for that job by Whitaker's basketball buddy Barack.

Several weeks later, when Whitaker's appointment was announced, the Associated Press stated that the young internist at Cook County's Stroger Hospital had been "plucked from obscurity." Whitaker recounted that when Blagojevich's staff had called him, "I said, 'Interview for what?'" but Barack told the AP that Eric was "an excellent selection." By then, *Capitol Fax*'s Rich Miller was calling attention to Blagojevich's "tight-knit inner circle of influence-peddling lobbyists and political players" and ruing that the new governor "has alienated so many people in Springfield so quickly," in part thanks to "the growing perception that he doesn't understand his job."

Barack was almost constantly on the road, speaking to local Democrats in downstate counties like McLean and Champaign while juggling his Monday morning–Friday afternoon teaching schedule. Fortunately for him, the law school's winter term classes ended on March 7, and student evaluations showed that the rigors of his campaign had not harmed his classroom performance, with twenty-eight out of twenty-nine students saying they would recommend his Voting Rights class to others. Several remembered how one class session devolved into a long discussion of "the then-imminent invasion of Iraq," with Barack going "out of his way to insure that all different viewpoints were welcome to be heard." Barack knew that one student, 3L Josh Pemstein, had a young daughter who faced serious medical challenges, and in a "very sensitive and empathetic way" Barack regularly asked "how things were going," once mentioning that one of his daughters had been hospitalized. Down in Champaign, Barack focused his remarks on Iraq, just as his strategists had advised, and the headline on the resulting Associated Press story was just what John Kupper wanted: "Obama Challenges Opponents to Speak Out on War."

Kupper and the Obama Issues Committee's volunteers continued to flesh out the campaign's strategies and positions. As the Champaign event exemplified, Kupper wanted at least "one message hit a week" while appreciating that

Barack needed a "minimum of fifteen hours a week fund-raising phone time," a figure that soon needed to ramp up to twenty-five. As Paul Harstad had recommended, they needed a regular mechanism by which to communicate to several thousand "top Democratic opinion leaders," and in April they would launch an "Obama Weekly News" e-mail program. Looking twelve months ahead, Kupper envisioned six weeks of black radio station and four weeks of Chicago television advertising leading up to March 16, 2004. The young issues committee lawyers were churning out impressive seven-to-eight-page papers on topics such as "Homeland Security" and "Bush's Tax Plan" for Barack to study. The latter lambasted how the president's "huge tax cuts, primarily to wealthy individuals . . . will balloon the federal deficit and debt without significantly stimulating the economy."[7]

In Springfield, the Senate Judiciary Committee unanimously approved Barack's videotaping-of-murder-interrogations measure and his racial profiling bill. A *Chicago Tribune* editorial commended Barack for his "serious efforts" to accommodate law enforcement's concerns about the taping protocol, and Barack stressed to reporters that "this is not a bill that is focused on the death penalty. Rather, it is focused on how do we get the right people arrested and convicted." Virtually every newspaper in the state ran stories on one or both bills' progress, with a *Chicago Defender* headline stating "Obama's Anti-Racial Bill Passes Panel." Barack told reporter Chinta Strausberg, "I feel optimistic we'll be able to pass both provisions out of the Senate," and said the profiling measure will "cut down on the potential for discriminatory behavior on the part of law enforcement officers."

A more explicit boost to Barack's Senate candidacy came from his longtime fan Steve Neal at the *Chicago Sun-Times*. In a column headlined "Obama Worthy Heir to Washington Legacy," Neal wrote that Barack was "coming on strong." If he could unite African American voters behind his candidacy, Barack could win the multicandidate primary "with less than a third of the vote. That's why so many Democratic politicians are giving him a solid chance to upset Hynes." As "the most intelligent and articulate contender in a surprisingly strong Democratic field," Barack was "likely to gain the most from televised debates" once the contest reached its final months. Barack "is a coalition builder" with "a record of accomplishment," and "may impress undecided voters as a cut above his competition," Neal predicted. Barack "has the potential to get more than 30 percent of the primary vote," and that might be the tallest hurdle facing him, for among Republicans "there is growing doubt that Fitzgerald can win against any Democrat." Potential donors on Barack's call list could not help be impressed by press clips like that, and in a highly revealing comment to the black newsweekly *N'Digo,* Barack explained his candidacy by declaring, "I see this as a mission, not just as a way of elevating myself individually."

Behind the scenes, Obama for Illinois (OFI) faced two difficult and painful personnel issues. For six years now, no two people had been more dedicated to Barack's political advancement than Cynthia Miller and Dan Shomon. Cynthia had manned his district office before playing a top role in both Barack's 1999–2000 congressional race and now in the first six months of his Senate run. Like her close friend Jen Mason, who had staffed Barack's district office for the last four and half years, Cynthia's religious affiliation was well known to all who worked with her, particularly each fall when she fasted during Ramadan, the Muslim month of prayer and reflection. One colleague remembered that "You walk into his district office, and there are women in there with head wraps on. Nice-looking young ladies with head wraps on. You don't have to question how black he is—just look at his office."

Cynthia and Jen were "both members of the Nation of Islam," and up until the winter of 2002–2003, no one had made an issue of anyone's private religious faith. But then "something happened," one witness recalled. "I think there were some pictures that were a catalyst," and "maybe she had also recently been appointed to a leadership position or something." The bottom line was that "Obama felt that Cynthia's affiliation was something that was going to impact her as well as him" if she remained on staff as the Senate race intensified. "She's starting to have these very difficult conversations with Barack that are really painful for her," and "I had the impression that it was as difficult for him as it was for her." But Cynthia "was more wounded by it," because "she didn't think that it was going to impact her." Yet Barack understandably felt that having a top staffer—indeed, his campaign treasurer—who was a member of Louis Farrakhan's Nation of Islam was politically untenable, and in March 2003 Cynthia shifted to a job in state government. She remained actively involved at arm's length, and several months passed before accountant Harvey Wineberg supplanted her as OFI treasurer. "It was really painful to watch that happen," one fellow staffer explained. "When Cynthia left, I remember feeling that an injustice had occurred," as apparently Barack did too, "maybe because he feels he did something wrong."

Almost simultaneously, super-energetic Dan Shomon was struggling in the weeks following his mother's late January death. It had always been envisioned that as the Senate race ramped up, additional staff would be hired to reduce the burden on Dan, but the combination of Shomon's absences in the wake of his mother's death as well as his own emotional burnout led him to tell Barack they should hire a full-time campaign manager, which would let Dan shift into a less frenetic role as political director. "Dan was fine with it," finance director Claire Serdiuk recalled, and David Axelrod and John Kupper concurred as well. Axelrod "wanted somebody that didn't have as close a relationship to Barack as Shomon," one colleague explained, and by mid-March, everyone was strategizing how to expand Obama for Illinois beyond simply Serdiuk and her deputy Randon Gardley.[8]

In Springfield, Barack's bills mandating hospital "report cards" and the taping of closed government meetings moved forward as he pressed interest group leaders like Illinois Hospital Association executive director Howard Peters and SEIU lobbyist Bill Perkins to come to accord on the union-drafted measure that was SEIU's top legislative priority. Barack continued to win prominent coverage in the *Defender,* with another Chinta Strausberg story featuring his outspoken criticism of President Bush, whose "vision of America's role in the world is fundamentally flawed. Our long-term safety and security is best accomplished by working with our allies, not trying to bully them." Back in Chicago, Claire Serdiuk coordinated Barack's meetings with prominent potential big donors like heiress Penny Pritzker, bank executive Jamie Dimon, *Playboy* publisher Christie Hefner, and arts patron Lewis Manilow.

Democrats' control of the Senate meant that controversial bills involving "infanticide" abortions and expanded gun rights could be defeated in committee, but energetic members of the National Rifle Association's Illinois affiliate still flooded Senate offices with hundreds of phone calls, annoying and at times frightening senators' aides like Barack's Beverly Helm. *Capitol Fax*'s Rich Miller reported that some calls included "personal threats" and "racial epithets," and as soon as Barack learned what was happening he asked Beverly for NRA lobbyist Todd Vandermyde's cell phone number. Beverly remembered that within a half hour the calls ceased, demonstrating Todd's ability to silence the fringe. While Vandermyde knew that Barack would never be with him on any NRA-backed piece of legislation, he recalled that in one office conversation, Barack did volunteer that "I think the Second Amendment means something" when Todd asked him about it.

Beverly Helm had been Barack's Springfield assistant for three years, and with Barack now chairing the Health and Human Services Committee, she and expert staffer Nia Odeoti-Hassan took a lead role in shepherding committee business. Barack moved committee meetings from Tuesdays at 8:00 A.M. to the less burdensome Wednesdays at 9:00 A.M., but Republican staffer Debbie Lounsberry thought that he was "surprised with what it takes to be the chairman of that committee." Given the number of interest groups following the committee's agenda, there was "more pressure" than Barack had expected. Planned Parenthood lobbyist Pam Sutherland wished Barack had exercised a firmer hand because "you'd have the committee hearings on all these bad bills, and they'd go on forever. 'Barack, why are you letting this go on? This is horrible,'" Sutherland would ask. "Pam, I can't do that. They have a right to talk to the committee." Barack "really believed that people could compromise," but on issues like abortion, where that never came to pass, Barack's tolerance "would drive me nuts."

By 2003, Springfield lobbyists knew they would have their hands full whenever they approached Barack about a bill. "You knew you were going to get twenty questions, on any issue," Illinois Federation of Teachers lobbyist Steve Preckwinkle explained. "You always had to be prepared," because "it was almost like giving him a briefing." Utility lobbyist Frank Clark knew "you needed to give him data," and that Barack was "very deliberate," "never" gave an immediate yes-or-no answer, and preferred "policy discussions where he never committed himself." When Mike Lieteau's daughter and former law student Tristé Lieteau visited Barack at Miner Barnhill to tell him she had a job offer from Northwestern University's Memorial Hospital, "we had a long talk about me becoming a lobbyist." Barack wondered "why ever would I want to do that" because he considered it "somewhat of a waste of my talent." When Triste and a colleague called on Barack in Springfield to discuss a hospital provider tax, he "asked several questions that we kind of didn't have the answer for," and Barack's face showed disappointment. "Are you serious? Is this the best you have?" his expression asked. "He was all business," and "I was just so embarrassed."

The new Republican spokesman for the Health Committee, Dale Righter, thought Barack was "genial" and believed "he managed people well" as chairman. Righter quickly realized that Barack was "more interested in reaching a consensus than . . . driving home a particular policy point," and new Democratic vice chair Mattie Hunter saw that Emil Jones was making whatever bills could advance Barack's Senate candidacy a top priority. Hunter thought Barack's style as chairman was "direct," "no nonsense," and at times "domineering." New Evanston Democrat Jeff Schoenberg felt Barack "took a far more Socratic approach" during hearings than most chairs, but just like Righter, he realized that Barack's focus was on bringing "the competing parties to the table" in order to "identify where the commonalities were." New Chicago Latina senator Iris Martinez thought Barack was a "very nurturing, almost teaching kind of person" with a rookie legislator, and, like Mattie Hunter, Martinez agreed to back Barack for U.S. Senate after Barack and Emil Jones asked her to do so. "I didn't realize" that that would infuriate Richard Daley. "The mayor got really mad at me. 'Why are you supporting Barack? Why not somebody else? Why Obama?'" Daley wanted Martinez to support Dan Hynes, just as he had asked 25th Ward alderman Danny Solis, Barack's former organizing colleague, who willingly agreed. Hynes recruited Jeff Schoenberg before Barack or Jones made an approach, and Barack "was pissed," and "it led to some words" when he found out. In addition, Bettylu Saltzman "chewed me out for not being with Barack," Schoenberg recalled.

Emil Jones's support for Barack was visible to all. Inside the statehouse, the phrase "bill-jacking" was used when a leader like Jones mandated that one legislator's bill would now be carried by someone else. "A lot of people were frustrated with the fact that Barack was getting all of this face time, and nobody else," Donne Trotter remembered. "All of a sudden you're the lead, you're getting the press. . . . Why does the light always have to be on you?" Jones was clear with both Trotter and Rickey Hendon that on measures like racial profiling and death penalty reform, Barack would be Senate Democrats' top voice. "Emil was very, very influential in getting other people to go along with him," Democratic freshman Ed Maloney realized, and veteran Denny Jacobs agreed that Emil "helped Barack tremendously." Barack "became the 'box guy.' He had so many bills he had to carry them in a box on the floor," Trotter recalled with exasperation. But by 2003 "you can see the drive in his face" and "an intensity in his eyes" that had not been apparent years earlier. Trotter believed Barack had become "a more committed legislator" and "a more outspoken one," but some fellow senators felt they heard from him more than enough. "Tell that nigger to shut up," one new Hispanic colleague recalled hearing a black senator stage-whisper when Barack spoke on the floor. "He became the fair-haired boy," one friend explained, or what poker buddy Dave Luechtefeld termed "the favored child," and the resulting "resentment" was apparent even to Republicans.

Beverly Helm knew Barack's Springfield routine: in the mornings he might run late, and in midafternoons, once the Senate adjourned, he often snuck away from the capitol if the weather was good. "I'd call him on his cell phone, and I could hear the wind whistling. I said, 'You're on the golf course, aren't you?' He'd say, 'Yes, but don't tell anyone where I am.'" In the evenings "he always talked to his girls every night before they went to bed," and Wednesday's poker games were his only late-night outings. "We never had to worry about whether there was going to be any rumors about him," Beverly remembered. "Barack was a private person," she realized. "He could be very outgoing, but he had a reserved portion about him." In his Springfield office hung a photo of some Hawaiian sea cliffs, and when Beverly asked about it, Barack explained, "that's where we spread my mother's ashes."

Once when Lisa Madigan's secretary told Barack that a Chicago gun enthusiast had bullied Bev after she declined to summon Barack from the Senate floor, Barack firmly reprimanded the fellow when he next appeared. "You will *never* talk to *any* woman like that. . . . You will not bully her again. You need to apologize to her," Beverly recounted Barack saying. "Barack was very quiet, he was very composed. I did not see him raise his voice or anything; he just told the gentleman exactly what he wanted to tell him."

Early that spring Barack asked Beverly, whose unsuccessful school board race he had contributed to, if she could help him "meet some of the leaders here in Springfield in the black community." With her pastor's permission Beverly hosted a "meet and greet" one evening in the fellowship hall and introduced Barack. Bev knew that business lobbyists Phil Lackman and Dave Manning were as close to Barack as anyone in the statehouse, and he still attended the Wednesday-night poker gatherings. When Terry Link moved to a double-wide trailer five miles out, the game moved too, although everyone was nervous about driving back into town post-midnight after having had several beers. When a Sangamon County deputy sheriff pulled Barack over one late night, the need for a new venue was agreed upon, and Illinois Manufacturers' Association lobbyist Dave Manning volunteered his downtown office. "The IMA is basically a Republican organization," Mike Lieteau explained, and that, plus the convenient location, drew additional participants. Conservative Republican Bill Brady, who had entered the Senate a year earlier, found Barack a "nice guy" even with their huge political differences. "I always used to kid him that the only fiscally conservative bone in his body I ever saw was at the poker table with his own money," Brady remembered. "I said to him, 'If you were half as conservative with the taxpayers' dollars as you are with your own, the people of Illinois would be a lot better off!'" By 2003, Barack seemed to be losing less often than he had in earlier years, and Beverly Helm recalled Dave Manning and Terry Link grousing that her boss was winning far too many hands.[9]

On the campaign trail, Barack spoke to a group of young Democratic activists and then addressed a Sunday antiwar rally in the Loop as a U.S. attack on Saddam Hussein's Iraq loomed. "It's not too late," Barack told the crowd of five thousand, but when military action began four days later, Barack's criticism immediately ceased. "Once the president makes the decision to go in, our priority has to be with the safety and success of our troops," Barack told the Associated Press. "Our prayers are with the families, and we have to hope for the best possible outcome in the shortest possible time."

In the state Senate, March 20 was rookie day for Republican Dale Righter and Democrat James Meeks, an African American pastor. Righter's first bill extended the statute of limitations from one year to two for public hospital patients. Barack teasingly complained that "as a fellow lawyer, surely you know that anytime we extend the deadline for lawyers, that doesn't give 'em more time to consider the case, it gives 'em more time to file lawsuits, frivolous lawsuits that are raising health care costs across the board." Meeks's bill prohibited the sale of tobacco from lunch carts near schools, and Barack asked, "I'm curious. Where did you get this bill? Did this come to you in a dream at night, or via some sort of divine intervention?" The American Lung Association and the American Public Health Association, Meeks replied. "I thought you had the direct line on this kind of legislation, that you didn't have to refer to the American Lung Association, that you had higher sources in order to find out what kind of bills needed to be passed." Meeks answered that "I chose not to use the direct line on this one," and Barack responded, "I just wanted to establish that he does have a direct line. So if any of you need, at some point, advice or counsel, that you know where to go."

Later that day, the Senate unanimously approved Barack's two bills banning the dietary supplement ephedra and prohibiting pyrotechnics and pepper spray in nightclubs. Kevin Riggins, whose sixteen-year-old son Sean had died after ingesting ephedra, "deserves enormous commendation," Barack said as Riggins joined him on the Senate floor. "It's the classic example of an ordinary citizen doing extraordinary things, and I really think he deserves all the credit for this legislation." Barack also spoke in favor of a bipartisan resolution expressing support for U.S. military personnel. "The minute that the president makes a decision" to strike, "we all have to unify immediately" and "close ranks."

The next weekend Barack spoke to downstate Democrats in Canton and Macomb at events hosted by progressive U.S. representative Lane Evans, and "they hit it off," Evans's political director Jeremiah Posedel realized. Barack also continued to win good press both downstate and in Chicago, with the *Defender* highlighting his SEIU-backed bill targeting discriminatory hospital pricing. "The health care crisis is enormous and government isn't going to solve this crisis today or tomorrow," Barack acknowledged, but discriminatory pricing is "a fundamental injustice" that "we can solve immediately." At the Macomb event, Mayor Tom Carper praised Barack as "one of the classy people in Springfield," and a Bloomington *Pantagraph* story on Democratic Senate contenders featured Denny Jacobs praising Barack. "He has the right demeanor, he's intelligent, articulate, and presents his arguments well . . . he's been a tremendous state senator, but he'd be an even better U.S. Senator."[10]

Barack took a beautiful Sunday off and went to a nearby playground with his daughters before returning to a busy legislative calendar in Springfield. Barack's SB 1415, to provide public funding for Illinois Supreme Court campaigns, would remedy "the unseemly process where judges have to raise . . . money typically from lawyers," with "at least the appearance, if not the actuality, that this may be impacting the judges' impartiality as they're deciding cases." Worried by "the appearance that the Court can be bought," Barack warned of how "a well-heeled trial lawyer" could "write a $100,000 check," thinking "I've got myself a judge." The Senate passed the bill 39–17, but Speaker Madigan never called it for a vote in the House.

Barack made a brief appearance on CNN's *Newsnight* to talk about the war in Iraq, and he pointedly told the *Chicago Tribune,* "I think there is a bedrock suspicion of George Bush among African-American voters," who "are much more likely to feel that he is engaging in disruptive policies at home and using the war as a means of shielding himself from criticism on his domestic agenda" than white voters. Behind the scenes, negotiations on Barack's "driving while black" racial profiling bill continued because "creating a piece of legislation that actually works in the real world requires that you talk to the people who are operating in the real world," Barack explained. He made a one-night trip to Washington to attend the American Israel Public Affairs Committee (AIPAC) annual dinner, chatting with Harvard basketball buddy Nathan Diament and a reporter for the *Jewish Daily Forward*. Dan Hynes and Peter Fitzgerald attended too, and although AIPAC viewed Fitzgerald as "highly vulnerable," with regard to Barack, Diament realized that "nobody knew who this guy was."

Steve Neal devoted a *Chicago Sun-Times* column to an unnamed campaign's poll showing Cook County treasurer Maria Pappas, a likely but not yet declared candidate, trailing Dan Hynes just 24 to 21 percent, with Barack at 9. This showed "Pappas is a contender," but Neal also wrote that "a third candidate could well emerge as the nominee" and reported that after respondents heard brief descriptions of all the contenders, Hynes led Barack by just 20 to 18, with Pappas now third at 14.[11]

The Senate on April 3 unanimously approved Barack's racial profiling bill and his one requiring the taping of police interrogations in murder cases. On both measures, Barack's many hours of private negotiations with law enforcement representatives had paid off so handsomely that *no* opposition remained to either of the once-controversial bills. New downstate senator Bill Haine, a veteran Madison County state's attorney who had become the Judiciary Committee's most conservative Democrat, was very involved in the videotaping discussions and felt Barack "did an extremely effective job in negotiating the bill with all parties." Judiciary Committee chairman John Cullerton agreed, saying Barack "did a phenomenal job on it." Democratic staff attorney Jim Dodge realized that Barack had focused on "trying to find a middle way through," and Office of the State Appellate Defender representative Kathy Saltmarsh concurred, explaining that Barack's "a negotiator and a compromiser." Even Republican death penalty proponent and former state's attorney Ed Petka termed the videotape bill "tremendous work," admitting that "a year ago, if someone had said I would be supporting this, I would have told them they were bereft of their senses." Barack's hard work paid off politically too: while John Cullerton remained neutral between Barack and Dan Hynes, Bill Haine was so impressed that he readily agreed to support Barack's Senate candidacy despite his conservative downstate district.

Linking the racial profiling bill to Cullerton's seat belt measure, which also was approved that same day, had been smart because the Illinois State Police's strong desire for a provision that would reduce highway traffic deaths led law enforcement groups to agree to Barack's requirement that officers record the apparent ethnic identity of every driver they stopped. During one decisive meeting, Barack articulated that the crux of the issue was officers' perceptions, not census categories. "It doesn't make any difference what race they are. What we're looking for here is what race you think they are. So that's why we want you to put down the race. So you don't have to worry about getting it wrong—it's what's in your mind is what racial profiling is all about." John Cullerton and Jim Dodge remembered the moment. "Whoa, you're right," someone responded. "It just really stopped the whole conversation, and everything sort of turned at that moment," with law enforcement opposition evaporating.

Barack told the *Chicago Defender*'s Chinta Strausberg that approval of both bills on the same day was "historic," and he stressed the value of videotaping to Chicago's suburban *Daily Herald*. "What we're trying to do here is provide as much certainty as possible that when a confession is introduced that in fact it is airtight." Later that afternoon, National Safety Council lobbyist Chuck Hurley, who had mustered crucial support for the seat belt and the profiling bills, was walking away from the capitol when he heard someone call out "Chuck!" He turned to see a black Jeep Cherokee stalled in the middle of South 2nd Street: Barack had run out of gas. Hurley and a friend helped Barack push it to the curb. "I hope your campaign doesn't run out of gas," Chuck teasingly remarked, but the day's successes lingered long after that minor embarrassment. Soon a *Chicago Tribune* editorial heralded the bills' passage, emphasizing that "an enormous amount of credit for the Senate videotaping bill goes to the diligence of state Sen. Barack Obama."[12]

In Chicago, Barack drove to his campaign's first event for gay and lesbian supporters, arriving a bit late and parking illegally. Kevin Thompson, Michelle's onetime City Hall colleague who now lobbied for the U of C Hospitals, had volunteered to organize the gathering at a Boystown bar, and Barack "just knocked it out of the park" when he addressed the fewer than twenty people who attended. Obama for Illinois knew it could announce more than $500,000 in total contributions to date in its mid-April FEC filing for the first quarter of 2003, and U.S. representative Lane Evans had been so impressed by Barack that he offered his endorsement. Equally important, as of April 1 the campaign staff almost doubled in size with the hiring of two new deputy campaign managers, Audra Wilson and Nate Tamarin. Audra, a 1998 law school graduate who had been working for John Bouman at Chicago's National Center on Poverty Law, had offered to volunteer for the campaign before Dan and Barack recruited her to take a full-time post handling the issues coordination and financial tasks Cynthia Miller had overseen before her departure. Nate, a young campaign veteran who had worked under both Pete Giangreco and Terry Walsh, was recommended by them to Shomon, who hired him to handle field operations after Dan's friend Andrew Boron, an issues volunteer who had just gotten married, declined to join the team. Barack and David Axelrod spoke with Peter Coffey, the politically well-connected director of government affairs for the Chicago Botanic Garden, about becoming campaign manager. Coffey declined, telling Barack he wanted to remain at the botanic garden until a bond authorization bill he had been working on became law. But Coffey told Axelrod, "I did not want to be unemployed in March," after the primary, which Coffey expected Dan Hynes would win. "I was pretty blunt with David," because "I didn't believe Obama could raise enough money."

Barack continued to devote serious time to fund-raising, meeting two influential donors, Raghuveer Nayak, a wealthy owner of outpatient surgery centers, and Tony Rezko for lunch on back-to-back days. An evening event hosted by Andy Davis and a keynote speech at a United Auto Workers (UAW) conference were sandwiched in between. Just two days earlier, a *Chicago Tribune* story had reported that Rezko and his business partner Jabir Herbert Muhammad, the son of the late Nation of Islam founder Elijah Muhammad, continued to play major roles in Cook County politics thanks to their prominent alliance with county board president John Stroger. The drive to the UAW event in Ottawa, Illinois, was Nate Tamarin's first real one-on-one exposure to Barack. "Look, when Shomon and I go on road trips, we like to fart, and we like to smoke cigarettes together," Nate remembered Barack telling him. Barack was driving, but as Ottawa drew near, he began reading his notes as well, which "was kind of disconcerting." But "I smoked, and he smoked," Nate explained, "so we bonded over that." Barack gave a "tremendously good speech," and when Nate teased him about playing Jay Z songs on the drive back to Chicago, Barack switched to "Indonesian flute music."

Thanks to Barack's hectic campaign schedule, he and Michelle were rarely attending Trinity church, and Trinity's Sunday-evening services had never been their choice, so they were not present on April 13 for one of Rev. Jeremiah Wright's most impassioned denunciations of the U.S. government. Titled "Confusing God and Government," Wright's stinging sermon culminated in a passage that would stir controversy years later: "The government gives them the drugs, builds bigger prisons, passes a three-strike law, and then wants us to sing 'God Bless America.' No, no, no. Not 'God Bless America!' God damn America! That's in the Bible, for killing innocent people. God damn America for treating her citizens as less than human. God damn America as long as she keeps trying to act like she is God, and she is supreme!"[13]

Obama for Illinois bragged about its $522,000 fund-raising total to date in a news release and in *Obama Weekly News,* an information-packed e-mail that alerted some one thousand recipients to staff hirings and upcoming events. In comparison, the Associated Press reported that Gery Chico had raised $763,000 in the first quarter of 2003 and Dan Hynes $897,000, while Barack's intake totaled just $232,000. Barack's campaign had maintained minimal spending, so it had $356,000 on hand, but Hynes had $801,000 and Chico $1,220,000. Blair Hull reported a balance of only $237,000, but that was after his increasingly staff-heavy effort spent almost $900,000 during the quarter. Barack's list of contributors showed four different members of the Crown family each giving $2,000, and Pritzker relatives likewise totaled $8,000, but thanks to Hull, the millionaires' amendment had allowed John Rogers to chip in an additional $5,000 and both Andy Davis and young banker Alexi Giannoulias to give $4,000 apiece. Valerie Jarrett's boss Dan Levin and his wife Fay Hartog-Levin each contributed $3,000, as did black financier Lou Holland. Old friends Rob Fisher, Al Johnson, Newt Minow, Allison Davis, John Schmidt, and Bettylu Saltzman all added $1,000 or $2,000 apiece.

With April 15 approaching, and Barack and Michelle owing the IRS more than $16,500 in additional federal tax payments on a combined income of $260,000, Obama for Illinois on April 10 finally paid them the $10,500 balance from Barack's loans to his old congressional campaign. Michelle's $98,000 salary at the U of C Medical Center topped Barack's $69,000 at the law school, but his $58,000 as a state senator and $34,000 from Miner Barnhill plus Joyce Foundation board fees went a good way toward helping with the more than $23,000 they had paid to babysitter Sonya Hawes. Home life was a bit more convenient, with Barack having shifted his state Senate district office from East 71st Street to nearby 1013 East 53rd Street in Hyde Park, but every weekday he was in Chicago was spent at the Michigan Avenue campaign office, where he attempted to maximize his "call time."

A "Campaign Prospectus" prepared for potential donors envisioned raising up to $4 million by the March 2004 primary, enough to win the race with a projected 438,000 votes—190,000 from Chicago, 176,000 from suburban Cook and the surrounding collar counties, and 72,000 downstate. The prospectus said that "Barack is on the phone a minimum of four hours daily," and listed three dozen finance committee members, including Robert Blackwell, Marty Nesbitt, Judy Byrd, Allison Davis, Judd Miner, Ab Mikva, Newt Minow, Jim Reynolds, Tony Rezko, John Rogers, Steven Rogers, Bettylu Saltzman, John Schmidt, and Al Johnson; wealthy Chicago worthies John Bryan, Dan and Fay Hartog-Levin, Christie Hefner, Abby O'Neil, Penny Pritzker, and Marjorie Benton; black financial world figures Lou Holland, Susan McKeever, Quintin Primo, Stephen Pugh, André Rice, and Stuart Taylor; plus well-known lobbyist Alfred Ronan.[14]

On Monday, April 14, *Capitol Fax* noted that Peter Fitzgerald had released new poll results showing him leading Dan Hynes by 44 to 38 percent and Barack by 49 to 29 percent in hypothetical 2004 matchups. Then, that evening, word spread that Fitzgerald had begun telling friends he would not run for reelection. Many Republicans were astounded, with state senator Dave Syverson telling the *Chicago Tribune,* "I just can't imagine that," but privately Fitzgerald had long been mulling this decision, irrespective of any poll numbers. Not yet forty-three years old and the father of young children, Fitzgerald was now ambivalent about being a U.S. senator because "what it did to your family life I did not enjoy." He returned to Illinois on more weekends than not to make appearances, ending up in "a Days Inn in Carbondale or Peoria" and missing out on a good portion of his children's young lives. By fall 2002, Fitzgerald had begun questioning whether he had the appetite to run another statewide campaign. "Do I really want to spend the next two years away from my family, raising money, on the road? I thought it was a very hard life," and as he pondered the choice, he increasingly thought that "I want to watch my son's Little League games the next two years." In addition, being a U.S. senator had turned out to be fairly underwhelming. "There were a lot of frustrations in the Senate" because "it seemed like nothing ever got done."

Then a third factor entered Fitzgerald's mind. Five years earlier, when he and Obama overlapped in the state Senate, Peter had found Barack "so cool and passionless and serious." But one winter Sunday morning, Fitzgerald was at an African American church in Roseland when Barack stopped by. "It was clear that he was going around to all the churches on the South Side," but now, unlike in 1997–98, Fitzgerald was "incredibly impressed with how good a speaker" Barack had become, "much better than he had been years earlier." Now Barack "really has the cadence," and listening to him led Fitzgerald to think "what an asset a voice can be in politics," that "an important part of Barack's ability is his voice."

More important, listening to Barack that Sunday morning in Roseland "confirms my impression that Barack will win the Democratic primary. I'm already thinking that I'm not sure I want to run for reelection," but having bested Carol Moseley Braun in 1998, Fitzgerald understood what an advantage a sole African American candidate had in a multicandidate Democratic primary. After that morning, "I knew his talents, and I don't think they were generally known" then. In addition, Barack is "going to get all the editorial board support," so to Peter it seemed "almost certain" that Barack would be his Democratic opponent if he ran for reelection. "I felt that I could beat all the others, but I really had no desire to have a piece of Obama in that campaign. I didn't know how I would run against him. I knew he'd be very tough," since Barack had been "very careful in his years in Springfield" and "I didn't think there was anything to attack." Reflecting on that anticipated matchup, "I thought, 'Boy, this is going to be really hard,'" and why spend two years away from his family if the all-but-certain likelihood was a November 2004 loss to Barack.

When Emil Jones heard that news about Fitzgerald's decision, he said, "I guess he's fearful of losing to Senator Barack Obama." Speculation immediately focused on whether former governor Jim Edgar would run in Fitzgerald's stead, but Barack told the *Hyde Park Herald* he was surprised by the news. "It's rare for a U.S. Senator to step down after one term, but it indicates he had been disengaged from the job," Barack stated. The *Chicago Defender* headlined its story "Fitzgerald's Backing Out Pushes Obama in Front," and Barack voiced an upbeat message to Chinta Strausberg: "If I am effective in presenting a message of change, if I am providing solutions to the health care crises that people are experiencing, the joblessness that exists in our community, the lack of education opportunity, then I can win irrespective of who else is in the race." Barack also said he would have enough money to win: "When you combine those dollars with an energized African American voter base and effective coalition building with other progressive sectors of the population, we think we have a recipe for victory." A few pages further on, a *Defender* editorial told readers that Barack was "the leading force in a battle to move Illinois closer to an enlightened health care system."[15]

Amid reports that President George W. Bush had called Jim Edgar to ask him to enter the Senate race, Barack devoted a *Hyde Park Herald* column to arguing for raising Illinois's minimum wage from $5.15 to $6.50 per hour. When Rod Blagojevich extended his predecessor's moratorium on executions, Barack told the Associated Press "it's a reasonable position for the governor to take," an unsurprising reaction which nonetheless appeared in the *New York Times*. But fund-raising remained Barack's focus. He called on prominent Jewish attorney Alan Solow to pick up a $1,000 check. "We had a long conversation," Solow recalled, and "talked a lot about terrorism," with Barack saying that "the first duty of a government was to protect its own people from physical harm." Barack explained that "I'm against the war in Iraq, but I'm not a pacifist," and over the course of the conversation, he impressed Solow as "the most thoughtful politician and maybe the most thoughtful person about political things that I had ever spent any time with. I was really, really impressed," and when Barack asked Alan to join his finance committee, Solow asked what was expected. When Barack answered simply a pledge to raise at least $10,000, Alan readily signed on: "I was pretty hooked."

The next evening Barack's campaign held an "After Work Fundraising Reception for Young African American Professionals" at "a trendy new jazz and soul venue" in the West Loop featuring a live band and an open bar. Tickets were $100, and Senate colleagues Emil Jones and Jacqueline Collins, a St. Sabina member whom Father Mike Pfleger had encouraged to run for a newly drawn seat, joined Barack along with supportive aldermen Leslie Hairston and Ted Thomas. "I am your instrument potentially to create a new kind of politics," Barack told the five-hundred-person crowd, and the event brought in more than $50,000. The next day Barack joined former U.S. senator Paul Simon on stage at a DeKalb County Democratic fund-raiser before making an overnight trip to Washington to attend the White House Correspondents' Association annual dinner. Back in Chicago, David Axelrod and his wife Susan hosted a Monday-evening, $250-per-person fund-raiser for Barack at their home on the forty-sixth floor of Lake Point Tower, overlooking Lake Michigan. "I'm confident I'll have enough money to develop a message and motivate core voters," Barack told one journalist, while admitting, "I don't have the name of a Kennedy, Daley, or Hynes." He boastfully added that "if I already had $5 million in the bank, this race would be over."[16]

Attention remained focused on Jim Edgar, with *Capitol Fax* reporting that GOP insiders believed the odds of him entering the race were about 60–40. Barack gave "a very moving talk" at the kickoff of the National Community Building Network's tenth annual conference, and as combat operations in Iraq ended, he told the *Chicago Defender* that "winning the peace is going to be the major issue." Raghuveer Nayak hosted a Sunday fund-raising brunch for Barack, after which Barack joined the other Senate candidates at Evanston Democrats' annual meeting. Northwestern University's student newspaper reported that when the contenders were asked to name their top issue, "Obama said he is most concerned with the USA Patriot Act." Dan Hynes and Blair Hull cited the economy, Gery Chico education, and Joyce Washington prioritized universal health care. The next day, with the *Defender*'s Chinta Strausberg in the audience, Barack returned to a similar message at a Chicago hearing of his Health and Human Services Committee. Warning of "draconian cuts in health coverage" at the hands of the Bush administration, Barack argued that a pending reform plan would give states "a strong incentive to drop people from Medicaid to solve a future budget crisis."

While Barack headed downstate to speak to Adams County Democrats in Quincy, his top competitors faced other challenges. Gery Chico had had great fund-raising success early on, but now *Crain's Chicago Business* was warning that Altheimer & Gray, the law firm that Chico headed, was "slashing jobs and salaries" after already having lost thirty-seven attorneys, and a retired managing partner wondered whether the firm might fold. Chico's brother Craig "essentially ran the campaign" in tandem with consultant Eric Adelstein and newly hired manager Michael Golden, a former television newsman, but as Altheimer & Gray's troubles mounted, Chico's call time plummeted. "He was under an intense amount of pressure and stress," finance director Kelly Dietrich remembered, and both Dietrich and operations director Megan Crowhurst, Chico's first salaried staffer, saw their candidate becoming more withdrawn as the law firm's crisis deepened.

After Peter Fitzgerald's withdrawal, comptroller Dan Hynes's pollster Jef Pollock surveyed both Democratic primary and general election voters with mixed results. Hynes led the undeclared Maria Pappas 23–16 among Democrats, with Barack third at 13 percent, but in a general election matchup, Jim Edgar topped Hynes 49–34 and all other Democrats by even more. "We felt good" in terms of the Democratic contest, Hynes's communications director Chris Mather recalled, but in the *Chicago Sun-Times,* Steve Neal wrote that if Edgar accepted Republicans' efforts to draft him, Hynes looked like Democrats' strongest candidate to oppose him.

Blair Hull and his top two consultants, Rick Ridder and Anita Dunn, had had campaign manager Mike Henry in place as 2003 began, but emphasized that candidate training was necessary for the publicly inexperienced millionaire. Given Hull's willingness to fund a top-drawer campaign, "he had the best coaches money could buy," pollster Mark Blumenthal explained, and communications director Mo Elleithee agreed to come on board after "they threw money at me." Blumenthal found Hull "charmingly naive" about what a U.S. Senate candidacy entailed, and Dunn took the lead in doing on-camera training with Hull while Elleithee struggled to do issues Q&A prep. "Mike Henry ran a great operation," Elleithee remembered, but Hull had difficulty articulating why he was running: "there was no core as to why he wanted to do this." Barack made a similar comment when the suburban *Daily Herald* asked him to comment on Hull's candidacy. "I think just by dint of money, he's going to be an important factor in the campaign," but "I think all of us are going to be waiting to see what kinds of proposals he makes and what his vision is for the United States Senate seat."

Hull's spare-no-expense campaign also went overboard on the least publicly discussed aspect of most competitive races: opposition research. Ridder and Dunn engaged Andrew Kennedy's Kennedy Communications to prepare a stunningly thorough research report on Barack, a twenty-five-chapter document totaling several hundred pages that plumbed every knowable corner of Barack's life. Extensive on-the-ground research in Illinois—in-person examination of Chicago and Cook County property, election, and court records, as well as state Senate voting and office expense records in Springfield—was combined with a comprehensive review of online databases and Internet searches plus a criminal records check with the Illinois State Police. An introductory "Executive Summary" included "Top Obama Vulnerabilities," and while many of the twenty-three substantive chapters were topical—Abortion, Education, Guns, Health Care, Taxes, etc.—others reviewed *Dreams From My Father,* "Personal and Court Records," and "Personal Finances."

The Executive Summary, which showed a misunderstanding of the meaning of "present" votes in the state Senate, characterized Barack as "a timid politician often afraid to take a stand on the tough issues," such as failing to vote no on more than half a dozen antiabortion bills. The Campaign Finance chapter noted that in his congressional race, Barack had received at least $6,000 from Tony Rezko and his associates, and the distillation of *Dreams* highlighted that "Obama attended Socialist conferences" in New York. Police records showed that "Obama has been cited for multiple moving violations," but to readers like Ridder and Dunn the almost $19,000 report failed to unearth any notable dirt or powerful political ammunition. Yet pages 3 and 4 of Chapter 19—"Perks, Pay Raises and Office Expenses"—unknowingly contained political dynamite. Senate payroll records showed that "Obama has had several staffers over the years." Anyone close to Barack would recognize Cynthia Miller, Will Burns, and Jennifer Mason's names, but the report also showed that from August 1998 onward, the Senate Democratic caucus had been paying the nonexistent "William Higgins," a supposed Obama staffer, monthly salaries that by February 2003 totaled $48,950. Had anyone in Hull's campaign realized that one of the four payroll recipients was fictional, Barack's political career—through no fault of his own—might have been over as fast as a Chicago newspaper could print the phrase "ghost employee."[17]

In Springfield, House passage of Barack's racial profiling and videotaped-interrogations bills sent both measures to the governor's desk. A *Chicago Tribune* editorial declared that "credit goes to Sen. Barack Obama, who spent months working with every interested constituency and making changes to accommodate their most credible concerns," and a previously hesitant Rod Blagojevich embraced the videotape legislation. "Senator Obama, that was fine work," the *Tribune* declared. "Take a bow."

A parent-teacher conference about five-year-old Malia had Barack and Michelle at the U of C's Lab School early one Friday morning, just a few hours before news spread that former governor Jim Edgar would not run for Peter Fitzgerald's Senate seat. Several considerations led to Edgar's decision. "In Illinois, the governorship is much more important than the U.S. Senate seat," and "after you've been a governor, it's really hard to be a senator," Edgar explained. "I wasn't sure I wanted to be one of a hundred in a debating society" and wondered "would I get bored?" But, just as Peter Fitzgerald had, he saw as his likely opponent the "very articulate, very bright" Barack Obama. "I was always very impressed with him" when they had interacted in prior years, and "I didn't necessarily want to lock horns with Obama. I probably gave Obama more credit that he might be able to emerge as the nominee than most people did at that point in the process" because "I thought he'd appeal to the suburbs" in addition to drawing African American support. With both the incumbent and the heavily favored, widely popular former governor declining to run, Republicans worried about their chances of holding the seat as speculation turned to Elgin state senator Steve Rauschenberger and little-known businessman Andrew McKenna Jr.

Barack mulled the news about Edgar during a Saturday photo shoot intended to get pictures of him in a variety of settings for later use in direct-mail flyers. "Barack should plan on having a few changes of clothes," consultant Terry Walsh had recommended, and the long day started with family scenes at East View before moving to Marty Nesbitt's Kenwood home to visit with a group of senior citizens. As the day progressed, Barack stopped at an East 71st Street barbershop, a hospital, and Kennedy-King College. Several days later, African American communications consultant Pam Smith pulled together ten or so black public relations professionals at Jim Reynolds's Loop Capital Markets office so that Dan Shomon, John Kupper, and Pete Giangreco could hear their advice about how the campaign should present Barack to African American voters. "Integrity of politicians—address head-on. Why *he* is running > make clear," Kupper jotted down. Challenges were flagged immediately: "*Harvard* as a *barrier* to people. *Name*: This is what it means; *proud* of it: okay for A-A audiences. Don't treat it like a liability . . . mixed race/Harvard educated . . . Needs someone recognized as '*street*.' Need to know who he is. Address Harvard education as proud of it, worked for it. More than Harvard grad > was an organizer/community activist." Then Kupper wrote down a phrase someone had mentioned to Pete Giangreco, a phrase that subsequently would resonate far beyond Jim Reynolds's conference room. "It's Time to Believe Again > in what? Don't tell us what to believe . . . Politics of Hope. A *Chance* to Believe Again."[18]

In Springfield, Barack spoke for a House bill that would limit gun purchasers to a maximum of one per month. On Chicago's South Side, "the proliferation of handguns often used by gang members and other criminals is wreaking devastation throughout the community," and "what we are targeting is the straw purchaser who is" fueling "the kinds of violence that we're seeing devastate communities." The bill failed badly, with only twenty-two Democrats—and no Republicans—joining Barack in voting for the restriction.

The next weekend Barack held a press conference and spoke at a West Side Teamsters local before meeting at a coffee shop with Jim Cauley, a thirty-seven-year-old Kentuckian whom Pete Giangreco, Terry Walsh, and David Axelrod hoped to lure to Chicago as Obama for Illinois's new campaign manager. Pete and Terry had worked on a 2001 Jersey City mayoral race Cauley had managed, and Axelrod and Kupper's other two partners, David Plouffe and John Del Cecato, as well as Nate Tamarin, had all previously been colleagues of Cauley at the Democratic Congressional Campaign Committee. By 2003 Cauley was chief of staff to Maryland congressman C. A. "Dutch" Ruppersberger and was interested in managing a large statewide race.

But when Axelrod first cold-called him, Cauley "blew him off," because he had never heard of Barack and had never visited Chicago. Then former Axelrod partner Tom Lindenfeld upbraided Cauley, so he looked at ObamaForIllinois .com, designed by Shomon's friend Bryan Siverly, and found it "less than stellar." But Axelrod called again, as did Pete and Terry, so Jimmy agreed to fly to Chicago to meet Barack, figuring it was a free chance to see the city. By then "I'd already done a lot of research," and "my concerns were whether Obama had the money for a campaign and his name. I didn't think he could win" because "if he didn't get on TV, there's no race." When Cauley and Barack met, Jimmy immediately voiced his doubts. "Your campaign looks like a state Senate race on steroids," he bluntly told Barack. "If you're going to run a traditional African American race and be poor, I've got no time for that." Barack responded that by June 30, he would have raised at least $1 million, a goal that had just become eminently more reachable when word arrived that late on Friday May 9, Blair Hull's campaign had notified the Federal Election Commission that Hull's self-funding had broken through the second trigger provision in the millionaires' amendment, allowing his primary opponents to accept up to $12,000 apiece from individual donors. Chicago newspapers had failed to report that news, but as their conversation stretched past the one-hour mark, Cauley found Barack "so impressive" and "extremely intelligent" that by the time Barack offered him the job, Jimmy had all but changed his mind. They haggled some over what Cauley's monthly salary would be, but he told Barack, "I'm inclined to do this. I'll let you know in the morning." Pete Giangreco picked Cauley up and was overjoyed that Jimmy and Barack had each been impressed with the other. On Monday morning, Cauley formally accepted. "I knew if we put a black in the United States Senate that I was a made man for life," and he told Barack he would arrive in Chicago right after July 4.[19]

A *Daily Southtown* poll found 38 percent of statewide Democratic respondents unsure which U.S. Senate candidate they would support, with Dan Hynes drawing 19 percent support, Maria Pappas 17, and Barack 12. Two Teamster Joint Councils, comprised of dozens of union locals, announced their support for Hynes, while bad news continued for Gery Chico as more lawyers left Altheimer & Gray. The *Chicago Tribune*'s Eric Zorn devoted a column to young Republican Senate hopeful Jack Ryan, a 1985 Harvard Law School graduate who had spent fifteen years at Goldman Sachs before exploring a political career. Seeking to bolster his résumé, Ryan had been hired by Barack's friend Tim King as a teacher at all-black Hales Franciscan High School, and some conservative activists saw Ryan as "the Illinois Republican Party's incarnation of a young Jack Kennedy." Zorn directly compared Ryan and Barack, writing that both have "impressive community service records and huge political futures."

As the spring session moved into its final eight days, Barack applauded Blagojevich's announcement that he would veto any bill expanding gambling in

Illinois. Many legislative Democrats saw increased gaming as a fertile source of much-needed state revenue, but Barack wanted "to take away some of the corporate tax breaks that exist right now," which would "close the budget gap fairly easily." For Barack, gaming was not about revenue, because "I think that the moral and social cost of gambling, particularly in low income communities, could be devastating" if Illinois made gaming devices available in "every tavern and bar across the state."

That same day, Emil Jones Jr. told a private Senate Democratic Caucus something that had already led to statehouse whispers: FBI agents under the direction of Springfield-based assistant U.S. attorney Joseph Hartzler, best known for having sent Oklahoma City terror bomber Timothy McVeigh to the electric chair, were investigating whether Senate Democratic staffers had performed partisan campaign work on state time in 2002. Similar allegations a year earlier had led House minority leader Lee Daniels to step down as chairman of the Illinois Republican Party, and since then the investigation had expanded to House Democrats as well. As Rich Miller soon detailed in *Capital Fax,* early in the 2002 election cycle several now-departed Senate staff members, including Ron Stradt, one of two attorneys assigned to the Judiciary Committee, had been "uncomfortable" with work assignments they received, particularly concerning candidates' ballot petitions. "Morale on the Senate Democratic staff has been abysmal for months," Miller reported, and several months earlier Rob Scott, who handled the committee's criminal legislation and dealt extensively with Barack, had also taken a new job. "The level of griping for Senate Dem workers this spring has exceeded just about anything I've ever experienced," Miller wrote, and federal subpoenas seeking Democratic staffers' time sheets and e-mails reaching back to 1998 were just one factor.

"When the FBI came knocking on my door" to ask about "campaigning being done on state time," one staffer who had witnessed "all the illegal shit they were doing related to campaigns" was not surprised. "You're breaking the fucking law," he had warned more than one superior a year earlier, and as FBI agents fanned out across Springfield, "multiple people met with the feds." Senate Democratic staffers all received a three-page memo stating that "any and all records reflecting or referring to your review of candidate petitions at any state work site" should be given to their supervisor, not destroyed or deleted. But at least one witness said "they came in and wiped everybody's computers." Years later, the story was that some departing staffer had maliciously deleted a huge amount of caucus files, but as FBI agents continued gathering information, their questions shifted to evidence of quid pro quo lawmaking.

Emil Jones casually alerted his members to the probe. "It was almost like 'Oh, by the way,'" one unnamed Democrat said, but the next day, Blagojevich infuriated Jones by heartily endorsing the federal inquiry. "This is part and parcel of investigations that are ongoing here because of a process that has been engulfed in corruption and cynicism for a long, long time," Blagojevich declared. Legislators' feelings about the governor had become increasingly negative as the spring session had progressed, and *Capitol Fax*'s Miller wrote that "the governor has ignored many of his professional responsibilities." Blagojevich met with the Four Tops only twice all spring, and "his duplicity" had been highlighted by his stealthy support for a controversial bill benefiting SBC Communications, one Barack and only three other Democrats had opposed.[20]

Dan Hynes's Senate prospects appeared threatened when millionaire Madison County trial lawyer John Simmons announced he was entering the race. "I am planning to put a lot of money in it," Simmons told the *Chicago Sun-Times*, and the possibility that Simmons might run strongly with downstate white voters "could dramatically reshape the contest," the *Sun-Times* observed. Obama for Illinois's slightly retitled *Obama Weekly Report* all but welcomed the news, saying Simmons's entry "threatens the chances of Dan Hynes and Blair Hull."

Barack accompanied Michelle to a Friday-evening memorial service for her uncle Steve Shields's wife Libby Brewer, who had died at age fifty of cervical cancer, then the next day attended Operation PUSH's weekly Saturday rally. On Sunday, May 25, Rod Blagojevich signed Barack's ephedra-ban bill into law, with Barack telling the *Sun-Times* that "we caught the industry off-guard because we mobilized quickly" thanks to the late Sean Riggins's parents, Kevin and Debbie. Barack also delivered the commencement address at the Chicago-Kent College of Law, uttering many well-worn lines. "There are an awful lot of unhappy lawyers out there," he warned, often because "the only thing that they were concerned about was making it for themselves." An attempt at humor about someone telling him "you seem like a relatively ethical person, for a lawyer," elicited no laughter, but Barack also said that "we need a more equitable way of funding public education in this state."

On Memorial Day, Barack marched in a number of parades before heading back to Springfield for the spring session's final five days. A House bill loosening Temporary Assistance to Needy Families' "family cap" gave Barack an opportunity to praise the bill he helped craft six years earlier as "one of the most effective, least punitive welfare reform policies in the country" while acknowledging that "everything that was said with respect to the value of work over welfare is absolutely correct." On Thursday, May 29, the Senate's agenda included HB 2221, which two months earlier *Capitol Fax* had labeled "a potential nightmare for Democrats" because it addressed an ongoing battle over the organizing of home care workers between SEIU, the service employees union, and AFSCME, the two most powerful pillars on the party's progressive wing.

Blagojevich had already sided with SEIU, and Barack on April 9 had signed on as an alternate chief cosponsor of the bill, which would codify SEIU's right to organize the workers in question.

On the floor, Barack did not speak before it easily passed, 51–2–5, handing AFSCME a painful defeat. Waiting just off the floor was Barack's old friend Keith Kelleher, who had devoted years to building SEIU Local 880 into a potent force for poorly paid home care workers. But AFSCME, which represented most state employees, did not want to surrender jurisdiction over so sizable a workforce, and AFSCME Council 31's deputy director, Roberta Lynch, who two decades earlier had been the Calumet Community Religious Conference (CCRC)'s first staffer, was notoriously feisty for not forgetting or forgiving disloyal politicians.

Leaving the Senate floor that day, "Barack comes out and he gives me this big handshake and hug. 'We did it man, we did it!'" Barack congratulated Keith. A few moments later, veteran SEIU lobbyist Bill Perkins, who had left AFSCME months earlier after a fierce run-in with Roberta, gave Kelleher an astonishing piece of news: "Barack didn't vote for it." Along with four other senators, Barack had voted present. Keith quickly accosted him. "'Barack, Bill just told me that you voted present on this bill?' 'Yeah, man. You know Roberta was really working me, and you know, I'm running'" and did not want to risk losing AFSCME's support. "You've got to be kidding me," Keith replied. "'No.' 'Barack, you're a *sponsor*.' 'Yeah, but you didn't need my vote.'" Kelleher and his leaders were amazed and livid that a chief cosponsor had voted against his own bill because he was afraid it would cost him AFSCME's endorsement.

Behind the scenes, Democratic legislative leaders worked on an ethics act to respond to the federal investigation into campaign work being performed on state time. Barack told the *Chicago Tribune* the bill would be "a much-needed improvement," especially in protecting legislative staffers. "These guys have been hung out to dry because they've been told to go do stuff politically," Barack acknowledged. In private, Barack loyally followed Emil Jones's lead, greatly disappointing advocates like Cindi Canary of the Illinois Campaign for Political Reform, who wanted stronger provisions barring lobbyists from paying for legislators' golf games and the like. On the last day, Barack spoke in favor of HB 3412, stressing that "it's absolutely imperative" to give staff members "a bright line" about "what is not allowable conduct," and the bill passed almost unanimously. Later that day, when a bill modestly raising Illinois's minimum wage came up for a vote, moderate Republican Adeline Geo-Karis noted it did not apply federally. Barack was quick to respond. "Senator Geo-Karis, I promise you, when I get to the federal level, I will make sure that at the federal level we raise the federal minimum wage. But until I get there, this appears to be the best we can do."[21]

While Barack was in Springfield, he received an unexpected gift when the host of WTTW's *Chicago Tonight,* Phil Ponce, asked Jesse Jackson Sr. if he was going to make an endorsement in the U.S. Senate race. "Well, I support Barack Obama," Jackson replied. "Have you already come out?" Ponce queried. "Well, not made an official endorsement, but he brings to the scene the quality of intelligence and freshness that's appealing."

Barack's team continued to prioritize fund-raising, with all eyes focused on the upcoming June 30 FEC reporting date. Attorney Stephen Pugh would host a major event on June 12, and a week later Women for Obama would hold a reception. Barack had been a top sponsor of twenty-six bills that had won passage, a third of which had political appeal: the videotaping and racial profiling measures, significant expansions of both FamilyCare and Illinois's EITC, the ephedra ban and nightclub safety acts, plus the hospital report card and open meetings bills. At a Friday press conference and again on WVON's *Cliff Kelley Show,* Barack called for the governor to sign a newly passed House bill he had championed that would allow ex-offenders to receive state professional licenses necessary for dozens of occupations. "Right now, you can't get a barber's license if you're an ex-offender," Barack stressed. Illinois at present "does not issue a license for a non-violent ex-offender to be a barber, a cosmetologist, to be a nail technologist, a real estate agent, a landscape architect, a court reporter," Barack added. Such a barrier "doesn't make any sense, particularly because for a lot of ex-offenders, one of their best opportunities may be entrepreneurial and starting a small business." Whatever their choice, "it's important that ex-offenders be able to find jobs and housing."

Barack made a two-day fund-raising trip to Manhattan, and one evening prominent attorney Vernon E. Jordan took Barack along to a book party being held in media power-couple Tina Brown and Harold Evans's backyard. Newton Minow's friend John Bryan, the CEO of Sara Lee, had introduced Barack to Jordan, a member of Sara Lee's board, at a Chicago lunch, and Jordan was impressed. Yet at the Upper East Side book party, with former president Bill Clinton in attendance, Barack looked "as awkward and out-of-place as I felt," a young journalist later explained. He was also "one of the few black people in attendance," and "we spoke at length about his campaign" before Barack left. Only then did another guest approach to ask whom she had been chatting with. "Sheepishly he told me he didn't know that Obama was a guest at the party and had asked him to fetch him a drink."

While Barack was in New York, Blair Hull's campaign and Barack's own reviewed their research into his record. "Whether it is because he is ethical or careful, we do not have a smoking gun against Obama," a Hull campaign memo reported. "Much of the analysis of Obama's legislative record has yielded proportionally little in the way of hits," but a Freedom of Information request had been submitted requesting the details on Barack's "member initiatives." The memo noted that "there appears to be a strong connection between Obama and the developers in his district," but that was not pursued. But the Hull campaign's research on Gery Chico "has uncovered lots of negatives and problem areas in Chico's public record."

David Axelrod and John Kupper had asked Joe Sinsheimer, a researcher whom they both respected, to undertake similar work for the Obama campaign. "I remember David calling me" and saying, "This guy is special," after Joe expressed doubt that a black state senator could win. A Duke graduate who had been initiated into "an aggressive style of politics" while working under now-congressman Rahm Emanuel at the Democratic Congressional Campaign Committee, by 2003 Sinsheimer had more than a decade of political research experience. By mid-June, Sinsheimer had completed a two-volume, thirty-one-chapter report on Barack that in form closely mirrored Kennedy Communications' investigation of Barack's record. Many chapters addressed particular issues, like abortion, or eight pages on the economy, and Sinsheimer's "Executive Summary" about Barack as a legislator was decidedly upbeat: "an innovator who understands and works within the system . . . to forge a solution to important problems." Barack's breadth of recently passed bills represented "a record that will allow him to solidify his base vote" as well as "to make the argument to suburban voters that he is a new kind of Democrat—a leader that attacks problems with innovative solutions."

Sinsheimer's defensively oriented chapters cataloged the "E2 Nightclub Stampede," "Obama Finances," "Law Firm," "Forum Inc.," "Citizenship Education Fund," and especially "Present and Missed Votes." Looking at the full record from an opponent's perspective, the argument that "Obama did not have the courage of his convictions on abortion" again represented the top vulnerability. The missed vote on reenacting the Safe Neighborhoods Act in late 1999 also loomed large, but the member initiatives to Yesse Yehudah and Jesse Jackson Sr. just barely made Sinsheimer's top ten. "I don't think that I remember Tony Rezko's name ever being uttered," he later recalled. Kupper also asked Sinsheimer to take a thorough look at Maria Pappas and her husband, Peter Kamberos, and two impressive memos detailed their business activities.

John Kupper immediately read Sinsheimer's report, taking note of potential problems. At the top of his notes, Kupper wrote: "What is Electronic Knowledge Interchange?"—the Robert Blackwell enterprise that had paid Barack $112,000 in 2001 and early 2002. Sinsheimer had vacuumed up Barack's exaggerated March 1997 claim to *Crain's Chicago Business* that he could "knock down scotch and tell a dirty joke with the best of them," but a trio of state Senate actions stood out too: "introduces legislation against zero-tolerance in school expulsions" because of Decatur, "votes to close child welfare office in

Hendon's district," and "votes against extending death penalty to violent gang members" all had "oppo" potential.[22]

When Barack returned from New York, downstate newspapers all reported Congressman Lane Evans's endorsement. "There is one candidate who stands above the rest when it comes to his ability to lead, his knowledge of the issues, his proven record, and his ability to win in November," Evans declared. "We're absolutely thrilled," Barack told the *Quad-City Times,* calling Evans's support "an enormous boost." That night Barack joined Gery Chico at a Democratic gathering in suburban DuPage County, and the next evening featured the big Loop fund-raiser hosted by African American attorney Stephen Pugh. A sustained effort had been made to encourage host committee members to raise "at least $10,000" apiece for the event, and one solicitation letter proclaimed that "very seldom is there an individual like Barack with the ability, will and determination to usher in a new era of progressive politics." With individuals now able to contribute up to $12,000, and the June 30 reporting deadline looming, another letter told potential donors that "Obama is a top tier front runner in this race." More than two hundred people attended, and the reception gave the biggest boost yet to the campaign's goal of having more than $1 million in hand when candidate's FEC reports were filed in mid-July.

With Jimmy Cauley's arrival still a few weeks away, Dan Shomon remained in charge of the campaign's political strategy, and one major concern that remained was how many of Illinois's major unions would support Dan Hynes. Some, like the Teamsters, were already in Hynes's corner. Both SEIU and AFSCME, their internecine warfare aside, were expected to support Barack, but the state AFL-CIO and both teachers' unions, the IFT and the IEA, were still in play. With them, Barack's most powerful ally was Emil Jones, and Shomon knew that having "Jones start calling key labor individuals" was the most essential piece of Obama's labor strategy.

When former 1992 Project VOTE! colleague Bruce Dixon posted a critical essay titled "In Search of the Real Barack Obama" on an African American Web site, Barack wrote a reply and then followed up with a phone call and a second e-mail. Barack explained that his inclusion in a directory published by the avowedly centrist Democratic Leadership Council (DLC), which Dixon had targeted, was simply because he had filled out a brief questionnaire. "I'm proud of the fact that I stood up early and unequivocally in opposition to Bush's foreign policy . . . and I continue to make it a central part of each and every one of my political speeches." In his second message, Barack stressed his progressive stances on divisive issues: "I favor universal health care for all Americans" and "the current NAFTA regime lacks the workers' and environmental protections that are necessary." He declared that "I am not currently, nor have I ever been, a member of the DLC" while also volunteering that he opposed "the misogyny and materialism of much of rap culture." Barack closed by stating, "I've always trusted my moral compass, and have thus far avoided compromising my core values for the sake of ambition or expedience."[23]

With the session over and no law school classes until late September, Barack's schedule was a nonstop mix of public appearances, call time, and meetings with contributors. One June Saturday, Barack spoke at an SEIU rally; the next Saturday he was grand marshal at Bronzeville's annual Juneteenth Parade. One Tuesday evening, Barack was working the crowd at a Clinton Foundation event at the W Hotel in the Loop when he encountered former U of C provost Geof Stone at the shrimp bowl. Geof chided Barack for investing so much time and effort in another political campaign. "The chances of success are so small. Why not become a real, full-time academic," so "you can make a real contribution?" Barack "looked me in the eye," Stone remembered, and said, "Geof, I really appreciate that, but I have a responsibility here, and I think I have an opportunity, and I just have to pursue it." That Friday Barack met Tony Rezko at 8:00 A.M. in preparation for a big fund-raiser that night at Tony's home in Wilmette, one that brought in more than $70,000 for Barack's campaign. Two days later Barack faced a radically different crowd when he sat on the back of a red convertible as two young campaign volunteers drove him through Boystown in the annual Gay Pride Parade.

Blair Hull's campaign got everyone's attention when it launched a two-week, $750,000 radio and television advertising campaign on June 23 in the six non-Chicago markets—Rockford, Quad Cities, Peoria, and Springfield, plus St. Louis, Missouri, and Paducah, Kentucky—that served Illinois residents. The first sixty-second TV spot described Hull as "a successful businessman who built an investment company from scratch." Hull vowed that "we need to create jobs" and "provide affordable health care for our families" before the ad ended with the slogan "he'll work for you." Going "up" on the air nine months before an election was unheard of, but it coincided with Hull's formal announcement of his candidacy on June 25 and visits to Carbondale, Springfield, Decatur, Quad Cities, and Rockford. Barack told the *Chicago Sun-Times* he found Hull's move "pretty remarkable," because "my impression at this stage is that people are not paying close attention to a Senate race that is going to take place in March." But former U.S. senator Paul Simon remarked that given Hull's unlimited funds and his complete lack of name recognition, "I think it's a smart move on his part."

African American health care executive Joyce Washington had retained a well-regarded campaign consultant and a top researcher who examined Barack and Hull. *Sun-Times* political columnist Steve Neal reported that although Barack was "emerging as the consensus choice of the African-American community" and "has a very good chance to win the Democratic nomination," privately Representative Bobby Rush was seeking "payback" by encouraging Washington "to challenge Obama" and helping "recruit allies to her cause." But Washington's challenges paled compared to those facing Gery Chico's campaign. By mid-June operations director Megan Crowhurst had resigned, and Chico's own call time dwindled precipitously even as the FEC's end-of-June reporting deadline drew near. Frustrated finance director Kelly Dietrich gave notice, and then on June 27, Chico as chairman of Altheimer & Gray's executive committee notified the law firm's partners that three days later they would be voting to dissolve the firm. The *Chicago Tribune* reported that the closure "will put 700 administrative staff members out of work," and the *Sun-Times* called the news "a potentially dramatic development" for Chico's Senate candidacy. Campaign staffers heard this from the *Tribune,* not from Chico. "Oh, that's why we haven't raised any money for the past couple of weeks," Dietrich realized. Several weeks later the *Tribune* followed with an embarrassing story detailing how in 2002 Altheimer had booked more than $6 million in uncollected fees as profit while at the same time borrowing more than $22 million into early 2003 to keep the firm afloat. Chico had been paid $800,000 in 2002, but mid-June efforts to merge Altheimer into a stronger firm were rebuffed just days before the collapse.[24]

Barack spent the first two days of July touring downstate cities with Lane Evans to publicize the popular congressman's endorsement. "If every voter in this primary could meet with Barack, he'd win by a landslide," Evans told a Springfield press conference as virtually every downstate newspaper reported his remarks. Shomon and Barack's consultants were also eager to draw attention to the fund-raising prowess that had produced $878,000 in contributions during the second quarter of 2003, giving Obama for Illinois a bank balance of $1,076,000 as July began. "We are going to have enough money to get on television," Barack told downstate reporters, while emphasizing that "I have the track record behind me that doesn't exist for any of the other candidates. I'm the only guy who's ever passed a bill. I'm the only guy that's ever cast a vote." Ceremonies at which Governor Blagojevich signed first the FamilyCare expansion bill and then John Cullerton's seat belt measure put Barack's name in the news too, and on July 4, thanks to Nate Tamarin, Barack appeared at five different Independence Day parades. With Dan Hynes residing in his district, Cullerton explained to Barack over lunch at the Berghoff how he had to remain neutral in the Senate race but volunteered some fund-raising advice. Soon thereafter, when the Cullerton family's 38th Ward Regular Democratic Organization unsurprisingly endorsed Hynes, "I get a call from Obama, saying 'I thought you said you were going to be neutral—your name's on Hynes's Web site.' I said, 'Barack, let me tell you something. I think you're the only person in Illinois who's looked at Dan Hynes's Web site. Okay? But I'll correct it, and I'll call him and get it straight.'"

At an Operation PUSH Saturday forum, Barack told the crowd, "we've got to have health insurance for every family" and "every child. We've got to have national health insurance." At a nearby senior citizens' center, Barack attacked President Bush's prescription drug policies, saying they "create an illusion of providing relief" yet "look like they were written by the pharmaceutical industry." He took the time to write a personal letter to Paul Simon citing his successful downstate swing with Lane Evans, pointedly adding that "your endorsement would mean more than any other." Dan Hynes's breadth of downstate support, with seventy-one out of ninety-six county chairmen already backing him, led wealthy Madison County attorney John Simmons to withdraw from the race and endorse Hynes just seven weeks after having entered.

Absent Hynes, the other Democratic candidates all met for their first joint broadcast appearance on July 13 at a progressive IVI-IPO forum cosponsored by WBEZ radio. Barack devoted his one-minute opening statement to the war in Iraq. "I am the only candidate in this race to stand up and oppose this thing vehemently and vigorously," noting that "I was in the Federal Plaza last fall clearly addressing this issue, and no one else was there." When moderator Laura Washington asked the candidates what they would do to protect civil liberties, Barack responded that "I would introduce a bill that would revoke many of the provisions contained in the first Patriot Act," and "I would absolutely oppose the second Patriot Act," because what's "most threatening are the ability of the Justice Department to obtain wiretaps without a warrant. They basically circumvented the judicial process, and the judiciary is our one check. They are the people that watch the watchers. . . . We need to require proper warrants that specify what the Justice Department is looking for and that is what the Patriot Act has stripped out."

When Washington asked about capital punishment, Barack cited his videotaping bill and said, "the key thing we need at the federal level is to stop all executions," because without a moratorium, "we are going to continue to have people who are on death row who are innocent." When Washington shifted to more personal questions, she asked Barack about his "unusual name." "It's a name that I am very proud of, it's the name that my father gave me, and one that I feel very blessed to have." In a brief television appearance, Barack cited his "wonderfully productive career in the state Senate" and said that "early childhood education is the next big step that we need to take as a society."[25]

In mid-July, Chicago's newspapers gave prominent coverage to all of the Democratic candidates' FEC filings. Most attention focused on Blair Hull, whose campaign had already spent $3.5 million and now employed thirty-three paid staffers. Dan Hynes's fund-raising had slightly outpaced Barack's, leaving Hynes with more than $1.5 million in hand, while Gery Chico's second-quarter contributions badly lagged both Hynes and Barack. The millionaires' amendment was providing their campaigns with a major boost, and a *Sun-Times* analysis showed that twenty-one donors had given Barack between $10,000 and $12,000 per person, while Hynes had twenty-nine such donors. Barack told the *Hyde Park Herald* that Hull had "squandered" his early millions, and a *Chicago Defender* news story stated that Barack "and his staff are looking confident and optimistic."

One weekday evening, Barack spoke at the prominent law firm Jenner & Block's annual "Diversity Dinner" for summer associates. Former *Harvard Law Review* colleague Bruce Spiva had invited him, and Bruce remembered that Barack gave a "very cerebral speech." He tried to joke about how summer associates might meet their future spouses through the program, but the result was "dead silence." The next day brought Barack yet another bevy of publicity as Governor Blagojevich signed a half dozen criminal justice reform bills, including Barack's videotaping, racial profiling, and employment for ex-offenders measures. Barack told the *Chicago Tribune* that the "landmark" videotaping program "can be a very powerful tool to convict the guilty," while a *Chicago Defender* story stated that "the three landmark bills are profoundly important to the community." *Hyde Park Herald* reporter Todd Spivak wrote of Barack's "hugely successful spring session," but in the *Defender* Barack stressed that "this year we had the juice," and "Emil Jones was the juice."

On three Friday mornings in a row—July 11, 18, and 25—Barack met fund-raising dynamo Tony Rezko for 7:30 A.M. breakfasts. An entirely unexpected new candidate entered the Senate race on June 20 when Nancy Skinner, an articulate, white, thirty-eight-year-old progressive talk-radio host, jumped into the contest and was put on leave by WLS-AM. *Chicago Tribune* columnist Eric Zorn gave Skinner a warm welcome to electoral politics, but much greater reaction ensued the next day when Blair Hull's campaign announced that Bobby Rush had agreed to chair Hull's effort. The campaign's press release cited the millionaire's and the former Black Panther's "remarkably similar backgrounds" as a reason for Rush's endorsement. "Rush will play an integral role in campaign strategy," the release stated, but Rush quickly sought to rebut most observers' suspicions by telling the *Sun-Times* that "there has not been a red cent—not one red cent—exchanged between Blair Hull and myself with regards to this campaign." Asked for comment, Barack responded, "Why would I stir up something unnecessarily? You can draw your own conclusions," before magnanimously telling the *Hyde Park Herald* that "I have great respect for Congressman Rush and look forward to his support in the general election." A bit later, Barack said, "I was not surprised by Congressman Rush's unwillingness to endorse me" while pointedly adding, "I am surprised at his endorsement of an individual who has no track record of community service" and "has never been involved in the political process." In a column headlined "Bobby Rush a Sellout," the *Sun-Times*'s Steve Neal declared that Rush "has exposed himself as a small-time hack and petty grudge-holder" while reminding readers that Rush's 2002 support of Republican state treasurer Judy Baar Topinka had netted his wife Carolyn a job in Topinka's office.[26]

Inside Hull's campaign, lead consultant Rick Ridder critiqued their existing game plan and its downstate focus. The Chicago media market reached 73 percent of Illinois voters, while St. Louis accounted for just 7 percent, but the plan envisioned spending 20 percent of their advertising budget in that "Metro East" market and only 43 percent in Chicago. "Our winning scenario involves garnering a huge percentage of downstate votes, but it also requires being competitive in Chicago and running up pretty big numbers in the collar counties." Yet Ridder appreciated that the campaign faced a more existential challenge. "In order for Blair to win, he must ride a wave of change. We have to define him as someone who will change Washington in ways that will benefit the average Illinois family. The advantages that Obama (black & progressive vote) and Hynes (state-wide & machine) have over us cannot be overcome simply by having a better run, better funded campaign." They must "define Blair as someone who is 'above' politics, yet can get things done . . . in a way that sets him apart from all other politicians."

Barack's campaign consultants were likewise strategizing, and two successive meetings, first with AFSCME, then with SEIU, were held so that they could detail just how Barack could win the primary. At the July 22 AFSCME session, Barack intended to lead off with his "stump speech and overview," but before he could, Roberta Lynch confronted him about his May "present" vote on SEIU's HB 2221. Barack parried. "Maybe somebody hit my switch," he claimed, and "Roberta lost it," AFSCME's Ray Harris remembered. "She goes insane," and Barack "took it." Then the meeting got back on track, with David Axelrod explaining "how we win with a name like Barack Obama." Pete Giangreco addressed "targeting—how we get to 35 percent." Newly arrived campaign manager Jim Cauley described how funds would be husbanded for 2004 advertising, and Audra Wilson detailed outreach efforts toward the campaign's African American base. Two days later Barack, John Kupper, Pete, Jim, and Audra did a similar presentation for top SEIU staffers.

The centerpiece of both meetings was Pete Giangreco's finely honed analysis based upon prior elections' vote totals. Barack began with a fundamental advantage in a multicandidate field: "no major African American candidate in a major state-wide campaign has won less than 29 percent" of the statewide primary vote. In Chicago, "Obama will be able to win at least 43 percent of the vote," or about 204,000 votes—a total derived from Roland Burris's 2002 numbers in African American wards and John Schmidt's 1998 ones in white ones, each discounted somewhat in light of Joyce Washington, Nancy Skinner, and imaginably Maria Pappas's presence on the ballot. In the balance of Cook County, Barack's target would be about 75,000 votes, in the surrounding "collar" counties, some 47,000, and in the balance of Illinois—"downstate"—a further 58,000, for a raw vote total of approximately 384,000—33 percent of the anticipated primary turnout. "Measurable portions of the downstate vote are African American," so "Obama's tactical goal for downstate is to win the majority of African American voters and leave it to the other candidates to compete for the rest." Giangreco stressed that all of these numbers were "conservative, realistic estimates," tallied with an eye toward convincing AFSCME and SEIU that they should see Barack as the likely winner.

The evening after the SEIU presentation, Barack was interviewed by suburban cable-TV host Jeff Berkowitz, who began by asking Barack about the Iraq war. "Saddam Hussein was a genuinely dangerous despot," but "my analysis said that Saddam Hussein was not an imminent threat" and that "vigorous inspections" targeting Saddam's suspected weapons programs could have contained the danger. Stressing that he had spoken publicly when others had not, Barack argued that "what absolutely we can't have out of our U.S. senator from Illinois is somebody who waffles on the issue and somebody who ducks on the issue and puts their finger out to the wind and waits to find out how the wind is blowing before they make a statement."

Berkowitz asked whether Barack supported Senate Democrats' filibuster of President Bush's nomination of a highly credentialed Hispanic attorney for a circuit court judgeship. "It's perfectly appropriate for a senator to oppose a nominee even if they're smart, even if they're nice, if their fundamental vision of the Constitution is flawed. If somebody doesn't believe in the protection of civil liberties, if somebody does not believe in the promotion of civil rights, then that I think is a rationale. . . . Look at the way that the federal bench has generally caved in the face of John Ashcroft and the Executive Branch and the Patriot Act. It's a disgrace. And that is the bulwark that we have to protect ourselves from an overreaching—" Berkowitz interrupted before Barack could say "presidency." "So you are opposed to the Patriot Act?" he asked. "Absolutely. Absolutely. Would have voted against it. And I think it is a shame that we had only one U.S. senator in the entire U.S. Senate who voted against it."

Then Berkowitz shifted to school vouchers, which he had previously pressed Barack on. "Can't we put the old tape on?" Barack replied. "We have to consider every possibility of improving what admittedly is an intolerable school system for a lot of inner-city kids. I do not believe in vouchers. I am a strong supporter of charter schools" because "we do have to innovate and experiment to encourage competition in the school systems." In contrast, over time vouchers would "reduce the options available particularly for the hardest-to-teach kids, because a private market system will not ultimately try to reach the toughest-to-teach kids."[27]

In August, the Hull campaign launched additional television advertising downstate, with their candidate declaring that "every American deserves quality, affordable health care." In Chicago, the campaign debuted on seven black-oriented stations a sixty-second radio ad featuring Bobby Rush. "My friend Blair Hull," Rush proclaimed, "is an independent voice who will make sure that we get our fair share." Barack acerbically told the *Sun-Times* that "the nice thing about actually having a track record of service in the community is that you don't have to pay for all of it." Skepticism was in order, for "whether the message is coming from Bobby Rush or anybody else, one would be hard pressed to believe that an individual who has never worked on issues important to the African American community during the first sixty years of his life has suddenly discovered these issues."

Capitol Fax's Rich Miller reported that "all the talk now on the Democratic side is about Barack Obama and Dan Hynes." Like Rick Ridder, Miller believed Democratic voters wanted candidates "who will shake up Washington." Dan Hynes "can be painted as a cautious careerist, which is the kiss of death in the current atmosphere. Obama could be seen as too cerebral and maybe even unelectable because of his race and his different name. Democratic primary voters don't trust unknown millionaires . . . like Blair Hull. Gery Chico is mostly unknown" and faced dire questions about the collapse of Altheimer & Gray. Treasurer Maria Pappas "is well-liked and well-known in Cook County," but "she hasn't raised any money, she has no campaign infrastructure," and "insiders generally dismiss her."

Barack continued his day-in, day-out rounds of small public appearances and donor solicitation meetings. A few months earlier his good friend Rashid Khalidi had accepted Columbia University's offer of a prestigious academic chair. As he and Mona prepared to leave Chicago for New York, Barack spoke at a Friday-night good-bye party that doubled as a fund-raiser for Mona's Arab American Action Network. Rashid, like Mona, had remained a public political voice, telling one U of C forum that George W. Bush "uses 'terrorism' to justify measures which are blatantly unconstitutional" and observing that "the office of the President is the costliest to buy."

As Barack's Senate candidacy had intensified over the previous twelve months, Barack and Michelle's attendance at the almost nightly dinners at the Khalidis' or Bill Ayers and Bernardine Dohrn's home declined. "Sometimes it was just her," Mona recalled, because "we would call her when she was alone." Michelle always seemed "more natural, less reserved" than Barack, Mona thought, but "all our kids were very taken by him." Malia and then Sasha too had often come along, especially to Bill and Bernardine's home, because "our children were great with younger children," Bernardine explained, "and so it just worked for them to come for dinner because the children would be carried around and would be busy and would be happy." Bill remembered Barack from those years as "a decent, compassionate, lovely person: pragmatic, middle-of-the-road, and ambitious." Bernardine recalled that by 2003 it was "so clear that he's driven, he's a politician . . . a politician, heart and soul." At the Khalidis' August 1 good-bye party, Mona remembered that in his remarks, Barack "never said the word Arab, he never said the word Palestine," which at the time, given the setting, seemed "very strange."[28]

By mid-July, Barack's team of consultants had prepared a second comprehensive poll that Paul Harstad and his partner Mike Kulisheck conducted later that month. Sixteen months had passed since Harstad's exceptionally encouraging first "benchmark," and the 806 likely Democratic primary voters who responded to this one showed that potential candidate Maria Pappas was notably popular among black Democrats, with 50 percent viewing her positively, only 7 percent negatively, and 24 percent saying "don't know." Dan Hynes's numbers came in at 33–5–34, while Barack trailed at 20–2–64. With the other 14 percent of respondents responding "neutral," 78 percent of African Americans remained unfamiliar with Barack, only slightly better than the 84 percent who were unfamiliar with Blair Hull. Among all respondents, Dan Hynes polled 24 percent support, and Barack and Pappas each 15, when voters were asked to choose a candidate. Among African Americans, Barack led Pappas 34 to 16 percent.

The poll then proceeded to candidate descriptions, with half of respondents being told that Barack was "a proven progressive voice" who had "passed landmark laws to help working families" and the others hearing that "he brings fresh thinking" and "innovative, new ideas." Following the biographical recitations, Hynes led Barack among all respondents 26–17, while among African Americans Barack led Joyce Washington 39–16. When a question about Barack's name was posed, responses turned from negative to mildly positive after "named for his late father" was added. Barack's status as a U of C law professor outdid even "first African American president of the *Harvard Law Review*" as a positive, and once Barack's many legislative achievements were cited, respondents' support for him grew to 46 percent, including 71 percent of blacks. Additionally, the 203 African American respondents in the Chicago media market were asked the positive-neutral-negative-don't-know question about a host of mainly African American names, with Father Mike Pfleger coming in tops at 70–0, Rev. Clay Evans 69–3, former judge Eugene Pincham 69–5, and Emil Jones Jr. 54–5.

In preparation for an all-day August 4 meeting of Barack's team of consultants, Harstad prepared a graphical presentation highlighting particular findings. In response to the initial vote-choice question, 42 percent of black men, but only 28 percent of black women, chose Barack, and while 44 percent of South Side African Americans named Barack, only 20 percent of other blacks did so. But, after all of the biographical information had been offered, not only had Barack's overall support risen from 15 percent to 46, among African American men it had jumped from 42 to 73, and among black women from 28 percent to 69. Looking at the entire sample, "the Most Movable Voters to Obama," once his background had been fleshed out, were white women in relatively prosperous households. Another summary chart highlighted how both Dan Hynes and Maria Pappas had 66 percent name recognition statewide, with Pappas at 86 percent and Hynes 69 in the Chicago media market, while Barack was 36 and 45 percent. Interestingly, among Democratic voters who knew all three candidates, Barack led both Hynes and Pappas 39–21–13.

Obviously "Obama has major room for growth" as voters learned more about him. A cover memo that Paul gave his colleagues in advance of the August 4 meeting stated that Barack "is already fully competitive with" Hynes and Pappas, "even though Obama is only known to half as many voters." Barack "is by far the preferred favorite of African Americans," but also "has tremendous room for growth once he becomes better known in this community." Barack "also has a lot of room for growth in downstate Illinois," where at present only 15 percent of voters had heard of him. Overall, the poll documented "that to know Barack Obama is to like him," for Barack "has broad appeal and a lot of room for growth." Harstad rendered his bottom-line finding in headline fashion: "Obama Has Potential to Reach as High as 46% in Multi-Candidate Primary."

In the wake of Harstad's upbeat presentation, Cauley circulated a memo to the consultants and the staff that he said was "the beginnings of the campaign plan." In his first four weeks in Chicago, Cauley initially was stunned that "everybody in the campaign" had a credit card. "I went in there, and all those kids had credit cards, and I cut them all up" as one way of husbanding precious funds. "All those kids" was a bit of an overstatement because in addition to Dan Shomon, Claire Serdiuk, and deputy campaign managers Audra Wilson and Nate Tamarin, only on August 1 had Obama for Illinois's tiny staff grown from six to seven when intern Liz Drew went on salary to join Randon Gardley as another finance assistant for Claire. A recent graduate of Brown University, Liz had been recommended to Barack by Tom Hazen, a family friend whom Judd Miner had successfully represented in an age-discrimination case. Along with University of Illinois junior Lauren Kidwell, who was about to return to school, Liz had supported Claire and Randon in staffing events, to which Barack insisted upon driving himself rather than having a driver.

Cauley expanded the campaign's tight quarters by moving from the fourteenth floor at 310 South Michigan to a larger, unfinished space on the seventeenth. Barack initially was perturbed by all the exposed wiring, but when Jimmy deftly explained that encasing them would cost several hours' worth of Barack's call time, Barack immediately replied that they looked just fine. Nate Tamarin knew that Barack "just hated" call time, but Cauley's arrival meant increased discipline, and "the campaign really ramped up," Audra Wilson recalled. The campaign had yet to hire an African American outreach director, and when Barack asked longtime AFSCME lobbyist Ray Harris to take on the role, he instead recommended AFSCME organizer Kevin Watson, whom Dan Shomon also knew. Cauley had the experience to be campaign manager, but as a newcomer to Chicago and Illinois politics, "the care and feeding of the local animals," as Jimmy put it, remained Shomon's responsibility. After some brief "uneasiness at the beginning," Cauley realized how "immensely useful" Dan was for anything in Illinois politics, and the two became an effective team.

Dan and Ray Harris arranged to introduce Barack to Kevin Watson at Mellow Yellow in Hyde Park, and as often happened when Barack drove himself, he arrived more than an hour late. "You've got regular time, and you've got Barack Obama time," Watson came to learn. Barack apologized, and almost immediately asked Kevin to join the campaign as base vote director, saying that Ray and Dan's judgment was conclusive. Cauley's campaign plan called for expanding the staff as slowly as possible, although both another finance assistant and a press person could not be delayed much longer. Cauley's memo also recommended some "media training" for Barack to "Work on the Passion Gap" the consultants saw in Barack's overly professorial public speaking style. By the end of September, campaign scheduling would be complicated by Barack's four-days-a-week class schedule at the U of C, but he could not afford to take leave for the autumn quarter, although he would have no choice but to do so for the winter 2004 one. Some initial appearances on "urban" radio during the fall could boost Barack's name recognition among black voters, but the veto session would occupy seven days in November.

Once candidate petitions were filed, Joyce Washington's would be reviewed with an eye toward whether Barack could become the only African American candidate. In mid-January a final eight weeks of radio ads would commence, with the extent of Chicagoland television advertising dependent on how the next six months of fund-raising went: hopefully five weeks at a frequency of 4,000 "gross rating points," or in the poorer alternative three and a half weeks at 3,050 GRPs. The bottom line was crystal clear: "Barack Obama goal of 35 percent on election day," a 2 percent increase from Pete Giangreco and Terry Walsh's initial targeting figure, but one they had agreed to revise slightly upward. Their updated "Integrated Strategic Plan" hoped for the Democratic presidential race to remain competitive up through Illinois's March 16 primary because "turnout will be enhanced, and Obama will be the likely beneficiary" as more black voters went to the polls. "To become the established African American candidate on the ballot is to build votes from a rock-solid base," Giangreco wrote as he increased Barack's projected vote share to just under 50 percent in Chicago and just more than 40 in the balance of Cook County. A statewide goal of 413,000 votes would mean victory with just more than 35 percent.[29]

The morning after the consultants' daylong meeting with Harstad, Barack made a three-hour drive northwestward to Galena, where the Illinois Federation of Teachers' thirty-eight-member state board was interviewing the candidates as part of an annual retreat. "We'd had a good relationship with Dan Hynes for many years," IFT political director Steve Preckwinkle recalled, and there was only a "pretty slim" chance that the union's endorsement might go to anyone else. Some board members had questioned the utility of even interviewing Hynes's opponents, but Preckwinkle's experience with Barack in Springfield led him to say, "there's at least one more serious candidate for us to interview." Just before Barack's 11:00 A.M. session, Preckwinkle cautioned him, "I can't promise you that the outcome of this is going to be favorable to you." Barack "puts his hand on my shoulder. 'Steve, that's fine. I didn't ask for any guarantees'" and said he appreciated the opportunity. "'Now it's up to me.'"

The session lasted only thirty minutes, but Barack "literally just blew them away," answering "every question" on issues like No Child Left Behind with "total command of the facts." As a result, "our board was probably ready to make an endorsement decision that day," but Preckwinkle recommended that they hold off for three months to see how Barack's fund-raising progressed. "I told Barack I thought he did an excellent job," that "he helped himself a great deal and probably made himself a contender with the board when he had not been prior to that." When Dan Hynes asked how he had done, Preckwinkle answered honestly: "Frankly, probably a little flat, at least compared to one of the other candidates."

The next Saturday was Bronzeville's annual Bud Billiken Parade, with Jim Cauley reprimanding Barack for walking "the whole damn parade with his shades on. 'You gotta look people in the eye!'" he scolded. A Monday-night event in Elgin hosted by the African American Ministerial Alliance drew only twenty-five people to hear about "a chance to believe again," Barack's campaign theme, but his entire staff plus young volunteers Juleigh Nowinski and Tarak Shah pulled out all the stops to have as many supporters as possible present for the Illinois State Fair's annual Democrat Day on August 13. Barack led off his remarks by describing a conversation he had had only an hour or so earlier. "Walking into the hotel this afternoon . . . I was stopped by a woman who actually works in the statehouse" and who told Barack that her mother had suffered a stroke and her Medicare coverage was expiring. Now the secretary might have to quit work to stay home and take care of her. "Ultimately, the only reason that we should be involved in politics is to give this woman a fighting chance," Barack told the crowd as Hynes, Hull, and Chico waited their turns. The question "is not whether you're going to vote the right way as a Democrat, it's whether you're going to fight as a Democrat. I'm tired of watching Democrats go to Washington and fail in the face of the Republican onslaught. I'm tired of Republicans taking a mile and then we take back an inch and call it a victory." Democrats needed a Senate candidate "who is battle-tested, has been in the trenches, fighting in the state Senate" for programs such as the FamilyCare expansion, thanks to which fifty-five thousand Illinois families "are going to have health insurance because of a bill I passed."

Six months later, Barack would use that hotel lobby conversation again. "'Senator, I know how busy you are, but I want to talk to you for a moment,'" Barack recounted. "'Sure, what's going on?' 'My mother, who's aging, fell down and broke her hip. And she went into the hospital, and it looked like it was supposed to be a simple operation but something went wrong . . . she is now paralyzed. . . . She has two choices. She can get long-term care up in Chicago 200 miles away from me, or she can come home, and I will have to care for her 24 hours a day, which would mean that I would have to give up my job. And I don't know what to do.'" Then "she started crying," and Barack added, "I always think back to that story."

Rich Miller reported that at the fairgrounds that Wednesday, "Barack Obama gave the best speech, hushing the often chatty crowd with a stirring account of a family overwhelmed with medical bills." Thursday was Republican Day, with Miller reporting that women staffing the Democrats' tent were distracted by Republican Senate candidate Jack Ryan's good looks. "Ryan has something about him which attracts attention and interest," Miller explained. Barack cited the hotel conversation during a cable-TV interview with host Frank Avila, saying such a "typical" story showed why the U.S. needed to "move in the direction of a universal health care plan." Barack also called for "universal preschool" and emphasized how as a U.S. senator, he would "be a check against overreaching by the executive branch," such as when a U.S. attorney general "decides that 'I want to be able to read Frank Avila's emails' without having to go to a judge'" for a warrant. Asked how he would best describe himself, Barack replied, "I think I'm a populist."

On Sunday evening, Barack and Republican state senator Steve Rauschenberger jointly appeared on host Bruce DuMont's two-hour, nationally syndicated *Beyond the Beltway* radio show. Also in Chicago but on the phone was Democratic presidential candidate Howard Dean. "It's great to talk to you, Governor. Congratulations on the terrific work you've done so far," Barack gushed. Dean returned the compliment. "I see a lot of people around with Obama buttons on. I didn't know who you were until I saw those buttons and asked." Barack was exultant. "I like that. I like that," he told Dean. "I actually share many of your views." After Dean signed off, Barack explained that "I like Dean a lot" because "there is an enormous hunger for plain-speaking Democrats" such as Dean. "He takes a clear stand, he speaks his mind."

Asked why he was running, Barack said, "I think that George Bush is making the wrong choices, and Democrats need somebody who can stand up to him. . . . The problem with the Democrats in Washington sometimes is they get rolled, or they don't stand up for issues, and then they carp at Republicans who do have oftentimes the courage of their convictions, and I give George Bush credit for that." Too often "Democrats seem timid when it comes to very commonsense issues that I think ultimately could generate support not only from Democrats but also from Republicans." When a caller asked about illegal aliens, Barack responded that "immigration is a net plus to the economy, and that includes illegal and legal," because "we end up benefiting from the work that's done and the taxes that are paid by undocumented workers." Throughout the show, the give-and-take between Barack and Rauschenberger was very genial, and as the program was wrapping up, Steve said in the Illinois Senate race Barack is "the Republicans' number-one fear."[30]

The next morning the *Chicago Sun-Times*' Steve Neal embraced the results of Paul Harstad's late July poll. Obama is in "a very strong position to win" and a growing number of politicians "regard Obama as the most likely winner. . . . If the black community rallies behind Obama, he will be very difficult to stop" and "would be the instant favorite over any Republican now in the field." In a subsequent column, Neal stated that Barack was "consolidating his base in the African-American community and among lakefront and suburban independents," although Neal also said that Barack's campaign was benefiting from a "finance committee that includes real estate developers Penny Pritzker and Tony Rezko" as well as Newton Minow, Abner Mikva, John Rogers, and Bettylu Saltzman.

Dan Hynes's polling showed him with a strong 26–15–12 percent lead over Maria Pappas and Barack, but with Blagojevich signing Barack's nightclub safety, EITC expansion, and domestic violence protection bills into law, commendatory press clips ensued almost daily. John Kupper joked about "riding the wave of Barack bill-signings and Steve Neal columns" in a memo to Barack's other aides while calling for an event featuring "jobs and/or pensions," because "fighting to protect pensions was the number one argument for Barack" in Harstad's detailed poll findings. The consultants turned their attention to selecting a press secretary for the campaign, eventually recommending Pam Smith, an African American public relations consultant who had worked on

Jesse Jackson's 1988 presidential campaign. Barack took the lead in interviewing Kaleshia Page, a Smith College and Vanderbilt Law School graduate whom Audra Wilson knew and who had just finished a fellowship in the governor's office. "You went to law school? Why aren't you practicing?" Barack demanded. "He just hammers me," because "he did not understand why I was not using the skills and experience I had." She thought "this man's attacking me," and "it was just a little paternalistic," but as soon as Kaleshia left, Barack called her cell phone to apologize and invited her to join Randon Gardley as a finance assistant to Claire Serdiuk. "Sorry I was a little rough on you," she remembered him saying.

Jim Cauley brought on Madhuri Kommareddi, an impressive Northwestern University senior who had volunteered in the spring before devoting the summer to her senior thesis, as operations director. Fellow undergraduate Lauren Kidwell, back at the University of Illinois for her senior year, began staffing a new Obama for Illinois Springfield office two days a week, and a young Rock Island woman, Andrea Hance, began working in Lane Evans's Quad Cities area. Before long both Juleigh Nowinski and Adam Stolorow, a Brown graduate and army veteran whose best friend had met Cauley in a D.C. bar, went on staff as well. As Jimmy put it with only slight exaggeration, Barack's campaign staff "was me and sixteen kids and Shomon."

Barack got a notable piece of good news when he was endorsed by well-respected Whiteside County Democratic Party chairman Lowell Jacobs, breaking Dan Hynes's phalanx of downstate support, on account of Barack's strong stance against the war in Iraq. Old friend Scott Turow remembered that it took Barack three phone calls before Scott agreed to host a late-August fund-raiser. "Why won't you support me?" Barack pressed. "Look, you told me that you were going to beat Bobby Rush, and he kicked your ass, and now you're telling me that you're going to win the Senate race?" Scott responded. "Okay, tell me one of the people who's either announced or who's thinking about this that you'd rather see as U.S. senator," Barack replied. "No one." "Then you have to support me," Barack insisted, and Scott gave in. "I thought it was pretty adept," Turow said.

That same day Barack called on U.S. representative Jan Schakowsky, whom Dan Hynes also "was really courting," Jan recalled. "What both of them wanted from me was the progressive stamp of approval," but she was very troubled by Hynes's position on the Iraq war: "that was huge for me." In contrast, Barack's "war speech was very important," but she was concerned about how his unusual name would fare on a statewide general election ballot. Attorney Alan Solow, who had been so impressed with Barack months earlier, had spent much of a weeklong summer trip to Israel trying to convince Jan to endorse Barack, arguing that "he stands for everything you believe in." But someone familiar with Schakowsky's thinking explained that when Barack visited her office to ask for her endorsement, "there was such an arrogance that it was off-putting." She held back from making a decision, telling Pete Giangreco that Barack had been "kind of disrespectful" and complaining to AFCSME's John Cameron that Barack "came to see me, and he was kind of a jerk." Cameron, who had witnessed Roberta Lynch's angry confrontation with Barack a month earlier, fully understood: "he came to see us, and he *was* a jerk!"

In contrast, Emil Jones talked up Barack's candidacy during a long, late-summer interview with *Capitol Fax*'s Rich Miller. "We as a caucus had a great session," and "I think our greatest accomplishment was reform of the criminal justice system," Jones explained. Barack's "chances are excellent. He's a very, very bright, articulate individual. I've known him long before he went to law school. We've been friends many, many years." Jones predicted Barack "will pick up substantial support. He has a great mixture of supporters," and "I think he will do an excellent job because he is able to attract voters from outside his base." Would Joyce Washington's presence hurt Barack among black voters? "I don't think her candidacy is going anywhere. I told her she shouldn't be running." What about Blair Hull? "I don't think you can buy an election."[31]

By early August, Hull's consultants and top campaign staffers realized they knew less about their candidate than they should. Some Hull staffers knew that his 1998 divorce from former wife Brenda Sexton had been contentious. One of Hull's earliest hires recalled that "I heard about it from one of Axelrod's people." Then the opposition researcher whom Joyce Washington had retained to look into both Barack's and Hull's backgrounds gave a friend who was leaving Hull's staff a more specific tip, and Kennedy Communications soon located a police incident report that described how on February 9, 1998, "Hull Was Arrested for Domestic Violence." While Hull had been "arrested and taken into custody," the charges were dropped before any court appearance after Sexton filed for divorce, and a significant financial settlement was quickly reached. The incident report was available only at the Chicago Police Department's 18th District Near North Station if someone knew enough to go there and request it.

Rick Ridder convened a Saturday-morning conference call to discuss the problem. "There's some stuff he hasn't told us," media consultant Anita Dunn warned pollster Mark Blumenthal, but of the more than fifteen people on the call, only Hull and his close friend Pat Arbor agreed with veteran Chicago political hand Fred Lebed that the campaign should inoculate itself by quickly giving the story to *Sun-Times* political columnist Steve Neal. "We would have weathered that," Lebed explained, because whatever controversy ensued would be old news by March 2004. Dunn and others disagreed, and pollster Blumenthal later admitted to *The New Yorker*'s David Remnick that "the amount of money people were making was a factor" in the failure of anyone who was being paid by the campaign to advocate the disclosure of something that potentially could cut it short.

Soon after the call, Ridder asked Hull's secretary for Hull's complete divorce file, and as soon as Ridder reviewed it, he called Hull at his vacation home in Sun Valley, Idaho. By far the most explosive document was a seven-page affidavit Brenda Sexton had executed on March 12, 1998, and filed as part of her divorce suit, one that Hull had never laid eyes on: "he never saw the allegations," Ridder explained. But given the affidavit's contents, "this campaign's over if this ever comes out," Ridder told Hull. He flew to Sun Valley to go over everything with his candidate, and the good news from Hull's lawyers was that the case file in *Hull v. Hull* was sealed, making the documents immune from public request.

Discovery of Sexton's affidavit convinced Ridder there was no point in leaking news of Hull's arrest to the *Sun-Times* because "it only leads to the sealed document, and we can't survive really the sealed document." An in-person meeting of all the campaign's consultants was quickly set for Saturday, September 6, at Anita Dunn's Washington office. Almost twenty people attended, because, as pollster Mark Blumenthal put it, by that point the Hull campaign was "like the Noah's Ark of consulting: two of everything." Ridder led the meeting, describing the arrest and the incident report, and alluding to the further allegations in Sexton's sealed affidavit. There was some discussion regarding how vulnerable a sealed Chicago court file was, but Hull and Sexton were now back on friendly terms. Five months earlier, Sexton had been named director of the Illinois Film Office by Governor Rod Blagojevich, whose gubernatorial campaign Hull had supported with more than $450,000, and Ridder was cautiously optimistic that Sexton would cause no harm. Meeting her at a Chicago Starbucks, Ridder gently asked whether some of the allegations in the sworn affidavit might in retrospect seem overstated, but Brenda "breaks down in tears," explaining that "I can't recant" because that "would be perjury."

Hull's background was such that "I'm able to estimate probabilities pretty well," and he asked Ridder: "What are the chances of this story getting out?" Fifty percent, Rick answered. Hull then asked Ridder to poll all the consultants for their estimates of his chances in either instance. "Blair wants your odds on winning if the affidavit becomes public and if the affidavit does not become public," Ridder told them. Ridder's own estimate was "5 to 10 percent at most" in the first scenario, but about 40 percent in the latter. Blumenthal had not seen the document, but he knew Barack and Hynes had stronger political bases than did Hull. "If the affidavit becomes public: 20 percent," he wrote Ridder. "If the affidavit does not become public: 25 percent." Ridder recalled that these numbers "didn't faze" Hull, and Hull had a similar memory of it. Most estimates were "25 or 30 percent," and "that was probably higher than my number," because Hull believed victory was "a long shot," although to him that "was a wager that I felt was worth taking."[32]

On Labor Day, Barack was in Rockford and then Rock Island. As John Kupper had recommended, a press release drew five Chicago TV stations to a news conference and garnered a *Defender* headline Kupper could have written: "Obama Seeks to Protect Pensions." Steve Neal reported that Emil Jones "has bluntly told union leaders that labor should support Obama," and those efforts had so far blocked Dan Hynes from getting the two-thirds support necessary for the Illinois AFL-CIO's endorsement. Rich Miller described Hynes as "the man to beat" because of his "big-time labor support" and "the backing of just about every county Democratic chairman." While Hynes "has put together a pretty good organization," his name recognition remained "pretty low" for someone who twice had won statewide races for comptroller. But Hynes "has yet to say why he's running," and "his campaign is bereft of issues." In addition, Hynes's "overly cautious, conservative manner" showed that he "has no anger about him," despite many Democrats' strong antipathy toward the Bush Administration. Miller saw Hull "more as a spoiler than a potential victor," someone who could draw enough votes away from Hynes to help Obama. As "a thoughtful progressive," Barack "would probably be the best U.S. Senator of all the candidates in either party" if he could overcome "his name and his race." Miller dismissed Joyce Washington and Gery Chico while commending Nancy Skinner and Maria Pappas's potential, though the undeclared Pappas "has no organization, no campaign infrastructure," and "no policy ideas." Miller wrote that if Barack could "somehow portray" to voters that "by far" he is "the most impressive potential Senator," he would have "a good shot" at prevailing.

In Springfield, almost everyone knew Steve Rauschenberger "was always the smartest guy in the room." But Rauschenberger's desire for the Republican nomination to replace Peter Fitzgerald was hobbled by uncertainty about "whether we can pull together a fund-raising plan that we can execute." *Capitol Fax* noted that insiders considered a past DUI arrest as a stain on Rauschenberger's record, but Miller saw potential in little-known, pro-choice former air force general John Borling because he had been "blown away" by Borling's "candor and forthright style." But Illinois Republicans were most focused on Jack Ryan, who "has the good looks, the money and the personal bio to wow voters," especially suburban women. Given that challenge, soon after Labor Day, Rauschenberger publicly proposed that he and Barack mount their own "Lincoln-Douglas style" series of two-man, bipartisan statewide debates. But Barack, with his campaign on firmer footing than Rauschenberger's, saw no reason to go toe-to-toe with a whip-smart conservative.

At a West Side birthday event for Congressman Danny Davis, Barack reached back to his Roseland roots and proclaimed that "corporations have an obligation to their workers and to their communities." Barack's press clips continued to glow, with *Sun-Times* columnist Laura Washington echoing Rich Miller in declaring that "Obama is by far the most qualified candidate," even though "he may be too smart, too reserved, and perceived as too elitist for regular black folk." Washington explained that "it's the Uncle Leland problem. My uncle says that low-income and working-class blacks don't think Obama is 'down' enough. It's a cultural phenomenon, and it's rooted in an unfortunate strain of anti-intellectualism and distrust of those with close associations with the white power structure." Steve Neal was more upbeat, writing that "there is growing momentum for Obama in the black community." Washington contended that Barack "needs a better slogan than his lame 'a chance to believe again,'" and downstate political reporter Mark Samuels noted that Barack's "last name rhymes with 'Osama'" while praising him as "a really bright guy." Barack told him, "I probably enjoy campaigning more downstate than I do in Chicago," and oftentimes southern Illinois white voters "remind me of my grandparents." But Samuels warned that "the backdoor whispering campaign surrounding Obama promises to give a whole new definition to the word 'ugly.'"[33]

Crain's Chicago Business, the weekday newspaper of the city's commercial elite, looked at Barack's second-quarter FEC filing. "Obama's surprisingly strong showing" negated the question of "whether he could raise enough money" to be a top-tier candidate. "I think a lot of people are surprised," Barack told *Crain's.* "It has exceeded my expectations, and it's very heartening. I think we're going to be able to keep pace and be competitive with the other candidates," so long as Barack continued to devote more than twenty hours a week to call time.

One recurring challenge that Barack had faced since 1995 was keeping up with the extensive questionnaires presented to him by various interest groups. A thirty-nine-item one from Illinois NOW proved especially taxing. Audra Wilson took charge of drafting responses to questions that included, "Would you vote to repeal the U.S. Patriot Act?" "Yes. The Patriot Act has the potential to infringe on personal freedoms of many Americans, which is wrong." Reviewing her answers after unsalaried policy director Raja Krishnamoorthi e-mailed him the draft, Barack rewrote that response. "Yes, I would vote to repeal the U.S. Patriot Act, although I would consider replacing that shoddy and dangerous law with a new, carefully crafted proposal that addressed in a much more limited fashion the legitimate needs of law enforcement in combating terrorism."

Audra initially answered "yes" to two questions that asked if Barack supported the elimination of sexual orientation discrimination in the military and repeal of the federal Defense of Marriage Act. Barack changed the first response to say, "I would oppose policies that fail to advance equal rights in the military," and suggested "I believe changing the federal definition of 'marriage' would unnecessarily detract from efforts to pass a domestic partnership law" as a reply to the second. Barack asked Raja "whether people think it sounds too squirrelly," and in the end his answer stated, "I support laws recognizing domestic partnerships and providing benefits to domestic partners. However, I do not support legislation to repeal the Defense of Marriage Act." In the affirmative, Barack emphasized that "universal access to health care is one of my top priorities," but addressing a question asking "Would you oppose legislation that criminalizes transporting a minor across state lines by anyone other than a parent for the purpose of obtaining an abortion?" Barack e-mailed Raja to say, "I'm still not clear on this. . . . Is this an issue the choice community feels strongly about?" After Krishnamoorthi's team of policy volunteers weighed in, the final answer was "I would oppose any legislation . . . which fails to include an exception for minors who have been victims of sexual abuse, neglect, or physical abuse by a parent."

Barack won more Chicagoland news coverage with a press release that generated a September 11 *Defender* headline touting "Obama's Security Plan Aims to Stop Terrorists." Claiming that "our homeland security efforts are flagging because this administration has failed to provide adequate support to the state and local agencies on whom our security depends," what Barack actually called for was "$3 billion to hire, train, and equip" seventy-five thousand new firefighters, a wonderful gift horse for organized labor.[34]

With Dan Hynes finally formally declaring his Senate candidacy, Barack's campaign team refined their "Campaign Prospectus" and "Integrated Strategic Plan" for presentation to donors. "We think 35 percent will win the election" and "fighting for middle class families" will be the central theme. After summarizing Paul Harstad's highly encouraging July poll results, the prospectus made the campaign's potential clear: "if Obama has the resources to get on television state-wide, he will win the Democratic primary and go on to win in November." Harstad outlined a plan for four to six early 2004 tracking polls leading up to March 16, plus a pair of January "focus groups" tilted toward "high income women" to refine the campaign's message points before television ads were written and produced. Depending on fund-raising success, the campaign's overall broadcast media budget could range from $2 million up to $3 million.

Raja Krishnamoorthi's issues and policy team volunteers continued to churn out succinct, high-quality briefing papers for Barack to absorb on topics ranging from "Bush's Economic Stimulus Plan" to "Health Insurance Coverage Proposals and Costs." Attorney Andrew Gruber remained a mainstay of Raja's team, but throughout 2003 Dino Christenson, a young political science Ph.D. who was working for former DNC chairman David Wilhelm, became another invaluable contributor. Barack often e-mailed his policy advisers late at night after returning home from that day's donor appointments, call time, and evening appearances. One Sunday former UNO organizer Bruce Orenstein and his wife Nancy MacLean hosted a fund-raiser at their Rogers Park home, and the next weekend Barack attended multiple events in Springfield, Champaign, and Decatur. Barack was still resisting a regular driver, but Dan Shomon recruited his friend Andrew Boron, another policy team volunteer, to drive Barack from Champaign back to Chicago. Boron had drafted a position paper on Israel, and during the drive, he told Barack about Harvard law professor Alan Dershowitz's just-published book, *The Case for Israel*. Boron was taken aback when Barack "said to me point blank that 'The problems in the Middle East were caused by the Jewish Americans' community's unwillingness,' or 'inflexibility,' I think was the exact word, and I was sort of floored" that Barack had said this.

Randon Gardley, Kaleshia Page, and Liz Drew staffed Barack's Chicagoland events, but finance director Claire Serdiuk and consultant Joe McLean tried to expand the campaign's fund-raising efforts to other big U.S. cities where Barack had old friends or new acquaintances. Hasan Chandoo recruited close to a dozen people for a late-September New York event. Hasan, like Rob Fisher, was one of the few old friends whom Barack regularly telephoned, but his enthusiastic support of the war in Iraq had created an awkwardness in their relationship. From New York Barack headed to Washington, joining Congressman Jesse Jackson Jr.'s table at the annual Congressional Black Caucus dinner and attending a fund-raiser at the Kalorama home of former *Harvard Law Review* colleague Jon Molot. One attendee at Molot's home remembered that Barack "spoke forever. He went on for an hour," answering questions interminably. "Barack's been a law professor too long . . . it really was not good."[35]

Barack got far better reviews back in Illinois for several weekend events in Carbondale and a Sunday reception at Raghuveer Nayak's Oak Brook home. On Monday, September 29, the U of C Law School's autumn quarter began, with Barack's Con Law III meeting Monday, Tuesday, and Friday mornings at 8:30 A.M., and his Racism seminar Thursdays from 4:00 to 6:00 P.M. Gery Chico's struggling campaign launched a $500,000 television advertising campaign featuring four thirty-second ads, which media consultant Eric Adelstein knew was "a high-risk strategy" to increase Chico's name recognition. A pair of prominent Sunday *Chicago Sun-Times* stories drew attention to Chico, but none that he welcomed. "Gery Chico and the Firm That Failed" said that "hundreds of employees were left without jobs" when Altheimer & Gray imploded in late June. One former partner blamed "a combination of mismanagement and malfeasance," and another said, "Chico pretty much abandoned his post." A sidebar piece titled "Polls Show Scant Support" noted that Chico was drawing only 3 or 4 percent support even though he had been the first Democrat to declare, and "some of Chico's rivals predict he'll pull out in the next few months." With a double-barreled assault like that, Chico's donations were in danger of declining even further.

Like Eric Adelstein, David Axelrod was thinking about the Obama campaign's upcoming television ads, and on October 2, he sat down with Barack to film a conversation aimed at producing air-worthy snippets. "I'm not that different from many young African American men who don't know their fathers," Barack asserted, "and . . . I think that helped motivate me to get into the kind of work I do." He spoke to a checklist of his proudest Springfield achievements: KidCare, Illinois's EITC, and being "the point person for Democrats in the Senate for protecting and preserving civil liberties and civil rights" by "making sure that we have got some sort of privacy and security from government." Then he turned to health care reform, his signature issue: "We lack the political will to fight the special interests that are preventing significant health care reform. Drug companies are making record profits, insurance companies are making enormous amounts of money . . . we lack the political will to fight back against companies that are driving up these health care costs. . . . If in fact we wanted to provide health care to all citizens, we could do it with the amount of money that we have." Democrats were part of the problem, because too often "people get elected and then they're so afraid to lose" that "they're not willing to fight for the people that sent them there in the first place. Too often Democrats lack the courage of their convictions."

Regarding the Iraq war, "the price tag on it should have been apparent to anybody who was reading the newspapers. . . . We should have understood that that was going to be a decade-long commitment that was going to cost billions of dollars precisely at a time when we have enormous needs here at home." Last, Barack turned to religion and Trinity UCC, saying that Jeremiah Wright "is a wonderful minister and it's a wonderful congregation. . . . My church life informs me and provides great support not only to me, but to my family."[36]

That evening, former U.S. senator and Democratic presidential candidate Bill Bradley publicly endorsed Barack. Raja Krishnamoorthi, who had worked for Bradley, had arranged an introductory meeting a year earlier after telling Bradley that Barack was someone he should meet. "That was the only time I've ever known Barack to be on time for a meeting," Raja recalled. "It went really well," and "Obama was on cloud nine" afterward. Bradley's Chicago appearance failed to generate any notable local headlines, but just forty-eight hours later was the campaign's first televised debate, sponsored by the Illinois NAACP. All six Democrats—Barack, Gery Chico, Blair Hull, Dan Hynes, Nancy Skinner, and Joyce Washington—were joined on stage by five Republicans.

Barack appeared quite subdued even before hearing three references to Osama bin Laden early in the telecast. When Barack's turn finally came, he stressed that "I was the only candidate who spoke out publicly against this war" because "we had no proof that there was an imminent danger, an imminent threat." He stated that "we have to support Israel, the only democracy in the Middle East," while backing "a two-state solution that allows Palestinians their legitimate aspirations." He called George W. Bush "the most fiscally irresponsible president we have had in the history of our nation," citing the need to "roll back those Bush tax cuts that go to the wealthiest Americans." Republican Jack Ryan proclaimed that "the biggest civil rights issue of our generation" is the hundreds of thousands of mainly minority children trapped in "failing schools." Blair Hull emphasized that "universal health care is a right," and Nancy Skinner observed that "there could be nothing more unpatriotic than the Patriot Act." Barack complained that "we don't have a voice in Washington for working families," also saying that "we have no African Americans in the U.S. Senate." He advocated "full federal funding of campaigns in order to take the special interests out," and in his closing statement he asserted that "manufacturing has vanished from the Illinois landscape." In his final remarks, Barack said that "our government seems more intent on rebuilding Iraq than rebuilding here at home" and forcefully stressed that "I have a track record of delivering."

The debate received little attention, and the *Chicago Tribune* ignored it entirely. *Sun-Times* columnist Steve Neal wrote that Dan Hynes "is maintaining a steady lead" in the Democratic race while saying that Barack and Chico had turned in the two best debate performances. The next Sunday Barack campaigned in suburban Oak Park before attending a fund-raiser hosted by Oak Park acquaintances Susan Himmelfarb and Mike and Susan Klonsky. Tuesday was Barack's SEIU endorsement interview, where he again apologized for his present vote five months earlier. At 7:00 A.M. on Wednesday, Barack did his first rush-hour campaigning at a Chicago L station, at the busy North Side Fullerton stop. A campaign press release announced "Obama Breaks $2 Million Fundraising Barrier" thanks to raising $775,000 between July and September. His campaign had almost $1.5 million in hand, and the next day, Barack again had breakfast with Tony Rezko before heading to Metro East's St. Clair County and rescheduling his Racism class from Thursday to Friday. On Sunday morning, Barack attended Trinity's 7:30 A.M. service, and early on Monday, he was at the South Side 95th Street Red Line terminus before meeting with Rev. Jeremiah Wright.[37]

By mid-October, even while on vacation in Mexico, David Axelrod was drafting a script for the first of Barack's two initial sixty-second radio ads, scheduled to debut on WVON 103 FM and WGRB 1390 AM the week of October 20. He sent John Kupper an e-mail suggesting "State Senator Barack Obama, candidate for U.S. Senate, talks about his family, his faith and his mission." His unusual name was explained immediately—"My father was from Kenya"—just as Paul Harstad's poll results had instructed. "I've been a member of Trinity United Church of Christ for fifteen years" established Barack's Christian bona fides. Axelrod's suggested closing declared, "Join the movement. Barack Obama for the U.S. Senate."

The final version, which would air for six weeks, began with a black female announcer using Axelrod's opening line—"his family, his faith, and his mission"—word for word as David had drafted it. That was followed by Barack saying: "My father was from Kenya, and he was a part of that first generation who came to the United States right after independence. I teach constitutional law at the University of Chicago. I've been a member of Trinity United Church of Christ for the last fifteen years, and my faith as a Christian is deeply important to me; it provides me with the values that inform my politics." Then back to the announcer: "In Springfield, Barack Obama won battles to extend health coverage to tens of thousands of uninsured children, tax relief for the working poor, a ban on racial profiling, and historic death penalty reforms." Barack again: "Every person in this country should have basic health care coverage, every child should have a chance to have a decent education, every able-bodied person who wants to work should be able to get a job." The announcer proclaimed Axelrod's slogan—"Join the movement. Barack Obama, Democrat for the U.S. Senate"—before the required closing: Barack said, "This is Barack Obama, and I approve this message," and the announcer said, "Paid for by Obama for Illinois Inc."

Two overlapping versions of the second ad featured four figures well known throughout black Chicago: Father Michael Pfleger, Operation PUSH's female Rev. Willie Barrow, Far South Side state senator Rev. James Meeks, and West Side congressman Danny K. Davis. In one, Rev. Barrow called Barack "a proven leader" and "the most qualified candidate." She declared that "Barack Obama stands up for *us*," and Barack said, "On health care, the war in Iraq, and so much more, our voices haven't been heard," before the announcer again invited listeners to "Join the movement."

The second began with the announcer saying, "Respected voices within our community agree. State senator Barack Obama for the United States Senate. Here's Reverend and Senator James Meeks." "Barack Obama is a leader among leaders. We look to him in the state Senate when there is a tough and a difficult decision," Meeks asserted. "Barack Obama will be a voice for the little guy, for the common person, a senator who's concerned about those who are unemployed, about those with no health care, about those who are paying these high prices for prescription drugs." "Here's Congressman Danny K. Davis," the announcer said. "All of us will benefit from a Barack Obama in the Senate. As Barack begins to talk, America will say, 'Here is another great voice, another great possibility, another great potential to move this country forward,'" Davis intoned. "Join the movement. Barack Obama, Democrat for U.S. Senate," the announcer declared.

Barack's campaign began organizing a rally for Saturday, November 1, at Pleasant Ridge Missionary Baptist Church on the West Side, where they hoped to garner footage for future ads. Ten days earlier, Barack was at an evening planning meeting at the popular MacArthur's Restaurant when "some young men came in, sort of crashed the meeting," Barack recounted. "They came in, and they said, 'You know what, Senator Obama, you should have respected us in coming here. You should have contacted us first.' I said 'Well, I didn't know about you.' And one of them I did know, and I said 'I'll give you the number.' And they were kind of confused, and they're frustrated and angry . . . 'We feel as if we've been ignored too long . . . that all the politicians . . . y'all don't care, and y'all have failed us.' That's what they said: 'Y'all have failed us,'" and "I told them, I said, 'You're right. I agree. We have failed. We failed because somewhere along the line we decided that it was standard operating procedure for half of the young men in Chicago to be out of work and out of school.'" Barack promised them, "Just because we've failed in the past does not mean that we have to fail in the future," and the gang members left.[38]

The next morning, Barack flew to Washington for a fund-raising event at the home of African American power broker Vernon E. Jordan. About fifty Democratic insiders, including fund-raising consultant Joe McLean, old *Law Review* friend Julius Genachowski, and well-known attorney Greg Craig listened as Barack spoke from Jordan's staircase. For some, it was their first exposure to Barack, and Craig was hugely impressed. Back in Illinois, Emil Jones Jr. and Jill Rock, Jones's staff operations director, reenergized Democrats' public pension investment reform efforts, which had lain largely dormant since 2001. The Illinois State Board of Investment (ISBI) had a new executive director, William Atwood, and several new board members, including Barack's close friend and former law firm colleague Allison Davis, whom Tony Rezko had proposed. The ISBI board moved to allocate a minimum of 5 percent of its funds to minority-owned firms, and Jones's staff pressed the board of the State Universities Retirement System (SURS) to detail its use of minority fund managers. SURS reported that 7.1 percent of its broker commissions were going to African American firms, but Jones moved to add further oversight, creating a special Senate Select Committee on Public Pension Investments and naming Barack as one of four Democratic members. Bill Atwood remembered Barack asking, "How are we doing with emerging managers?" whenever they ran into each other, and John Rogers of Ariel Capital recalled that Barack "did a great job of coming and helping us as African American business leaders connect into Springfield."

On Saturday morning, Barack and his five Democratic opponents all addressed Citizen Action of Illinois's eighty-nine-member policy council. "You need a U.S. senator who's not going to be thinking 'How's this going to play?' but one who leads," Barack told the sympathetic group, and he won their Senate endorsement with 88 percent support. The *Chicago Tribune*'s first Senate poll found the yet-to-declare Maria Pappas drawing 16 percent support. Dan Hynes and Barack trailed with 12 and 9 percent, respectively, but 45 percent of Democrats said they were undecided. Hynes launched his first television ad that same day, telling viewers that "if building schools, roads, and hospitals is right for Iraq, then it's right for America too." The diversity of Barack's campaign outreach was highlighted on two back-to-back days when he delivered the keynote address at the annual meeting of the Illinois State Baptist Convention and then held a press conference to publicize his support in the gay and lesbian community. Michelle's longtime friend Kevin Thompson had taken the lead in pulling the group together, and although only one reporter, Tracy Baim from the LGBT *Windy City Times,* attended, "Obama addressed the room as if it was full," Baim remembered.

On Friday, October 31, Steve Neal reported that Maria Pappas would finally enter the race the next week. While Dan Hynes "has already raised more than $2 million, has locked in the endorsements of key Northwest and Southwest side Democratic organizations, and has lined up significant Downstate support," Pappas's high name recognition and personal popularity throughout the Chicago media market immediately made her a top-tier contender.

Barack's big Saturday-morning rally drew a bevy of West Side politicians, but Pleasant Ridge Missionary Baptist Church was only half full for the Obama campaign's cameras. Cook County commissioner Earlean Collins told the crowd, "We have a separate set of laws and justice for African American people in this country," while praising Barack's fortitude. "I know that he will deal with the terrorists in the White House because we got a lot of them up there. We talk about terrorists and all of this stuff. We better look at the White House first when we talk about terrorism around this country." When Barack took the podium, he ran through his Springfield achievements before turning to the Senate race. Referencing how Blair Hull's wealth could manifest itself on Election Day in what Chicagoans called "street money," Barack cited the late Mayor Washington: "I remember what Harold said: 'Take the money, then vote for me.'" Saying "we've got five months left in this race," he promised that "I will go to Washington, and we will start taking this country back."[39]

Forty miles to the north, the Illinois Federation of Teachers' forty-one-member executive board was meeting in Gurnee to decide the powerful union's Senate race endorsement. Barack's seven-page IFT questionnaire, recounting his positions during "six long and sometimes lonely years" in Springfield, had raised no hackles and bore several personal touches. "My experiences with my five- and two-year-olds confirm that their brains are sponges," and Barack rued "teachers who have left the profession out of frustration and enter law school." A "slam-dunk endorsement" for Barack passed overwhelmingly. Media outlets across the state noted that the IFT's endorsement undercut both Dan Hynes's stature as labor's favorite son and Gery Chico's efforts to stress education. "It's a tremendous boost to my campaign," Barack told *State Journal-Register* political reporter Bernard Schoenburg. *Capital Fax* described the IFT's decision as "potentially huge," with Miller soon adding that further major union endorsements of Barack's candidacy—namely SEIU and AFSCME—loomed as well.

As the IFT press conference ended, Bernie Schoenburg asked Barack about his history of illegal drug use, and Barack said that during high school and college, he had smoked marijuana, and "I did inhale." Schoenburg followed up, asking about anything else, and Barack responded, "I haven't done anything since I was twenty years old." Three days later, Schoenburg reported that conversation at the end of a catch-all piece, but within a day or two, someone told him he should look at *Dreams From My Father.* Schoenburg went to the library and quickly realized that Barack's abrupt ending of their chat—"That'll suffice"—cloaked how his response contradicted how *Dreams* said he had done "a little blow." Schoenburg reached Barack on his cell phone seeking clarification, and Barack quickly apologized, saying that "I had caught him off guard," and he "didn't want to step on" the IFT endorsement by creating other news. "My life is literally an open book," Barack told Schoenburg, and the next Monday a follow-up story headlined "Frank Talk About Drug Use in Obama's 'Open Book'" appeared on page 17 of the *Journal-Register.*

In a taped interview with journalist Terry Martin that ran statewide on the Illinois Channel, Barack emphasized that the federal Patriot Act had caused "fundamental principles" to be violated thanks to the widespread, warrantless electronic surveillances it allowed. Asked about his *Harvard Law Review* editing experience, Barack said, "I haven't read a law review article probably in quite some time." His hardworking issues team had prepared a sophisticated tax incentive program that would reward "REAL USA Corporations"—REAL as in "Responsible, Accountable, Loyal"—for creating new U.S. manufacturing jobs, provided those companies "limit top management compensation to 50 times the lowest-paid full-time worker." That press release drew no attention at all, but at a Saturday South Loop candidates' forum for one hundred top Chicago high school students, Barack's "outreach director" Randon Gardley was swamped with interested teenagers following Barack's remarks. "I don't like to base my decisions on auras of people," fifteen-year-old Andrew Miller told the *Chicago Tribune,* but Barack "could be president in 15 years."[40]

Capitol Hill's *Roll Call* profiled the Illinois race, calling Barack "poised and charismatic" while reporting that Dan Hynes is "widely regarded as the favorite." *Crain's Chicago Business* predicted that in twelve months "Barack Obama narrowly beats Jack Ryan," but a quick fund-raising trip to Boston was a huge disappointment for Barack. Harvard law professor David Wilkins, who already had donated $5,000, hosted a fund-raiser, but Barack's former teacher Martha Minow recalled that hardly five people attended. Her law school colleague Elizabeth Warren was one of them, writing a $250 check, and several weeks later David's wife Ann Marie, a successful music manager, contributed $10,000 to Barack's campaign too.

Back in Illinois, Barack continued to resist having a driver for his often-hectic schedule of appearances. When Lyndell Luster, a Chicago State University police officer and part-time student, turned up to volunteer after hearing about Barack from a professor close to Emil Jones, both Dan Shomon and Jim Cauley took notice. With the primary race moving into its last four months and more attention focusing on the candidates, neither Dan, Jimmy, nor especially Michelle Obama were keen on Barack being on the road by himself. Randon Gardley often accompanied Barack to African American events, and in late October AFSCME veteran Kevin Watson arrived on staff as base vote director, but Luster's CSU job meant he routinely carried a sidearm. When Lyndell drove Barack a time or two, mentioning that "I never knew my father" and had experienced some earlier troubles, the two developed a good rapport. Both men smoked, and when Lyndell drew Barack out a bit, seeking assurance that no "secret stuff" would torpedo his campaign, Barack reassured him: "Look, brother, I promise you, none of that stuff will ever surface about me." Lyndell thought Barack shared his view that "I never have a bad day," because "I never saw him lose it. He never lost his cool." Over time Lyndell realized that Barack "had this self-checking mechanism" on his emotions, because "I never saw him get upset."

No one except Michelle knew Barack better than Dan Shomon, and as the campaign intensified, Dan saw that the increased stress was taking a toll. The most evident indication was that Barack was smoking even more. Almost a decade had passed since Michelle had first made Barack promise to quit smoking once *Dreams From My Father* was finished, but as Dan knew, Michelle "wasn't successful as an enforcer." Barack "would smoke in front of Michelle," but he especially smoked while driving, and he was "smoking a lot in the Senate campaign."

Compared to years earlier, Barack also had become more finicky about what he ate. On one downstate drive, Barack "just blew up" because "he wanted some healthy food, and there wasn't healthy food" anywhere to be found. "We were tired and stressed," Dan explained, but young Randon Gardley experienced the same thing. Randon too knew that Barack was "pretty even-keeled," but "the weight of the campaign" plus how "at home it was just a real challenge for him financially" led to "a lot of cigarettes." Under stress, Barack could "be snappy" and "not a very pleasant person to be around." Some staffers thought Barack barked more at African American men than at others. "There were some of us that he was a lot more unpleasant with than others," one put it, and "I noticed the difference and didn't appreciate it."

With the veto session's final week about to start, a *Chicago Tribune* profile lambasted Bobby Rush for his "suspiciously punitive" support of Blair Hull over Barack. Noting that Rush's brother Marlon was drawing a handsome salary as Hull's deputy campaign manager, African American journalist Don Wycliff wrote that Rush's behavior "has inspired bitter anger and deep suspicion among many opinion leaders in Chicago's black community." Barack tried to persuade veteran Citizen Action leader William McNary to join his campaign to give it further African American heft, but McNary demurred, telling Barack he needed to be less professorial in front of black audiences.

On Monday, November 17, the second big union endorsement fell into place, with SEIU announcing its widely expected endorsement of Barack. House Speaker and state Democratic Party chairman Michael Madigan quickly made official his support for Dan Hynes, but *Rockford Register Star* political editor Chuck Sweeny warned that "when Hynes' ward bosses battle Hull's dollars, the two men could cancel each other out," allowing Barack to win. Barack joined Jesse Jackson Sr. and other black political leaders at a small pre-Thanksgiving rally in the Loop protesting the Bush administration's record on job creation, but Jackson was shouted down by the same young black men who had interrupted Barack's meeting at MacArthur's a month earlier. Kevin Watson was with Barack and watched as one protester "snatched a microphone" from Congressman Danny Davis while several others were "tearing up people's signs." They "were really shouting at Barack," and Kevin thought "it was such a scary scene," but Barack was "calm and cool and he's talking to them." A *Tribune* reporter heard Barack say, "I understand you're angry, brother, and I want to talk to you—please call me" as he handed one young man his card. Barack "was the only one that they allowed to go up there and speak that day," Kevin remembered.

On Sunday morning, Barack and Kevin drove two hours northwest to Rockford, for appearances at five black churches. Introducing Barack to "base vote" Democrats outside of Chicagoland was a key element in Pete Giangreco's statewide targeting strategy, and Rockford, just like Decatur and Springfield in central Illinois and Madison and St. Clair counties down in Metro East, was a community where African American outreach was essential. Barack also took time to speak with *Register Star* political editor Chuck Sweeny, whose story "Obama Leaves Good Impression" told readers that "if resumes were dollars,

Obama would be worth a few billion." Revisiting his theme from two weeks earlier, Sweeny wrote that "Democratic establishment types tell me the race is between Hull and Hynes. I say if he gets enough money and troops to market himself in all Democratic strongholds, the charismatic academic from Hyde Park can win the nomination."[41]

Sweeny was not alone. Suburban *Daily Herald* columnist Jack Mabley complained that "opponents try to imply he is a Muslim," but predicted that "if Illinois votes Democratic as usual next year, Barack Obama will make history." Barack's campaign continued to churn out almost daily policy-specific press releases, most of which continued to win prominent coverage in the *Chicago Defender.* One story highlighted Barack's plan for "providing health insurance to all," and the West Side's *Austin Weekly News* offered an enthusiastic endorsement of Barack's candidacy, complaining that "it's perverse for black nationalists to reject the son of a Kenyan for not being black enough." *Chicago Tribune* reporter David Mendell asserted that Barack has "so far campaigned almost exclusively in the black community," but more accurately added that Hynes's union support "is not as broad and solid as had been expected." Barack's fund-raising remained strong, with members of Jim Crown's extended family donating tens of thousands of dollars after meeting Barack at a December 1 reception.

Friday, December 5, was the end of the law school's autumn quarter. Barack's Con Law III had attracted twenty-six students, and Racism and the Law had drawn twenty-eight. Barack had been late several times, with students willingly waiting for him, and in both courses his teacher evaluation scores held up fairly well even with the scheduling issues. In Con Law only one respondent said they would not recommend the class to others, while all but two in Racism said they would recommend it. In the race class, which included a number of left-wing students from the university's school of social work, 3L Nathan Smith remembered that Barack "often actually took kind of the conservative position in class in response to their comments." Barack "knew all of our names" and "was incredibly engaged as a professor." He no longer had time to chat with students after class, but his comments led Smith to conclude that Barack believed that "legislative action is more effective than judicial action in bringing about actual results in structurally oppressed minority communities." During the last class, Barack cited Derrick Bell's critique of *Brown v. Board of Education* and Justice Clarence Thomas's powerful 1995 opinion in *Missouri v. Jenkins* arguing that there was nothing necessarily inferior about all-black schools. "When you have two people on the left and the right that come together on a point of analysis so outrageously controversial, namely that *Brown v. Board* wasn't the panacea that all American lawyers are required to say that it is, I would suggest for your consideration that it is a proposition worthy of thinking about," Smith would recount Barack telling his students.

Early Monday morning, December 8, Barack and Dan Shomon filed ten thousand petition signatures in Springfield to secure Barack's place on the March 16 ballot. Nate Tamarin and longtime Barack ally Al Kindle had used the petition effort to begin building up the Chicago field operation they would need for the primary. That same day, in a move that attracted no attention, Barack resigned from the Senate Select Committee on Public Pension Investments to which Emil Jones had appointed him only six weeks earlier. From Springfield, Dan and Barack began a downstate swing encompassing Decatur, Edwardsville, Marion, and Rock Island.

Late on Tuesday, sad and politically horrible news arrived: former U.S. senator Paul Simon, who earlier in the fall had told David Axelrod he would endorse Barack, had died at age seventy-five after undergoing eight hours of unsuccessful heart surgery. Axelrod had favored postponing Simon's endorsement until early 2004, but now he was gone. A *Chicago Tribune* headline proclaimed "A Legacy of Honesty and Dignity," and Barack told the Associated Press, "Paul Simon was an inspiration to every American who believes that government can be a force for good, a tool for empowering people and righting injustice." With a Saturday visitation and a Sunday service in Carbondale, Barack altered his schedule to join a who's who of Illinois politics plus such Washington eminences as Senator Edward M. Kennedy in honoring Simon's life. Comparing the Senate candidates to Simon's career, *Capitol Fax*'s Rich Miller wrote that "nobody stacks up, although Barack Obama has the potential to at least step into Simon's shoes, if not actually fill them, and Steve Rauschenberger is in the ballpark." In Carbondale, Simon's daughter Sheila, who knew of her dad's intent to endorse, offered her condolences to Barack: "I am so sorry this really worked out horribly for you."[42]

On December 15, the last day on which to file petitions for the March primary, Joyce Washington and Nancy Skinner submitted their signatures. Barack's campaign was ready and waiting to take a fine-toothed comb to a copy of Washington's submission, but the deadline for filing an objection was just seven days away: Monday, December 22, at 5:00 P.M. On Tuesday morning, it was announced that U.S. representative Jan Schakowsky would hold a press conference later that day to endorse Barack. Her husband Bob Creamer recalled that Jan's earlier hesitancy had dissolved "once it became increasingly clear" that Barack "was putting a real good campaign together." Evanston Democrat Larry Suffredin explained that "Jan's endorsement of Barack was a breakthrough for him" because among North Side and North Shore progressives, Schakowsky's support was "a big deal."

Barack was disappointed by the media turnout at the press conference, with the *Chicago Tribune* not attending. Afterward, a young journalist asked Barack to address the challenge of fund-raising "without having to prostitute yourself." Barack conceded that "certainly I believe as you advance up the political ladder, it becomes more difficult, which is why I think ultimately we will fully recapture our democracy when we provide free airtime for candidates." Barack reiterated his belief that "every single person should have a basic right to health care," and in the wake of Schakowsky's endorsement, observers like Rich Miller and Steve Neal concluded that "Obama has the momentum" among the Senate contestants.

Inside Blair Hull's heavily staffed, no-expense-spared campaign, the results of Mark Blumenthal and his partner Anna Bennett's early December poll led Hull's consultants to think much the same thing. Barack's support among African American Democrats was now up to 29 percent, thanks to his increased name identification, and his support among liberal, college-educated Cook County white Democrats had risen to 18 percent. In the overall statewide numbers, Maria Pappas narrowly led Dan Hynes, 19 to 16 percent, with Barack and Hull trailing with 13 and 12 percent, respectively. The downstate numbers were worrisome for Hynes because over the previous ten months, his support had slipped from 37 to 23 percent as Hull's had risen from virtually nothing to 23 percent thanks to his television advertising. As Hull's consultants prepared for a December 9 meeting to discuss the numbers, pollsters Bennett and Blumenthal realized that "for Obama, getting to 30 percent was relatively easy." Barack could win the race with about 35 percent if he captured 80 percent among African Americans, 50 percent of Cook County white liberals, and a smattering of white liberals elsewhere.

Lead consultant Rick Ridder understood the challenge: "the big issue in the campaign was could we cut Barack Obama's numbers in the black community with Bobby Rush, with cash, whatever to close to 75 percent," rather than anything higher, so as to keep Barack at 30 percent and allow Hull to win a narrow plurality. Mark Blumenthal thought the hurdle was insurmountable because "there was no scenario that I could get Hull over 30 percent." The realization that "I just didn't see how we could win this" led Blumenthal to wonder about the propriety of going forward. "People were getting paid a lot," and when Mark raised the question with his partner, Anna's response was "It's not our job to talk him out of running. It's our job to help him win. If he wants to run, it's our job to help him." Bobby Rush and his brother Marlon attended the December 9 conclave, and with the congressman insisting that Barack could not win more than 60 percent of the African American vote, Hull's consultants agreed to push forward.[43]

The next Saturday, the top African American women on Barack's campaign, Audra Wilson and Pam Smith, convened two focus group discussions at Malcolm X College on the West Side. Pam moderated the first group, composed of eight black women, asking them what they knew about the Democratic Senate candidates. One of two female pastors in the group immediately said Barack "is a member of Trinity Church, Pastor Jeremiah Wright, and he lets his biblical teaching inform his politics." The second clergywoman noted that "Obama has experience" thanks to "all these bills that he has passed, and he worked with the devil himself, Pate Philip, and got some legislation passed." She had been at the PUSH rally and had watched as Barack dealt with the young men trying to disrupt it. "Obama said, 'Have your piece.' Man, this brother is smooth. I liked the way he handled them. . . . He knows what the concerns of the people are. I see young white kids and older white people working for this young man, and they respect him. I think he has a sense of fairness about him. He is concerned about employment."

An older woman said she would vote for Barack. "The name kind of struck me, but when I found out that his dad was African," that explained it. "He kind of reminds me of Martin Luther King" in "his fairness, his calmness." A second lady agreed. "He does have good spirit. He is friendly and warm. He is not flirtatious. He is a young man full of wisdom, but he is settled. I think he genuinely cares for all people. . . . He has something about him that makes everyone around him want to help him. And he tells you that his name means 'blessed by the Lord,' and that is the way he acts. I mean, this brother has his act together." A third woman said she had attended a small fund-raiser and had witnessed "this conversation he had with his daughter and her little friend" after "one was running around and one got mad at the other one. And the mother was there too, but he took the lead, and I said, 'Let me check out his negotiating skills. If he could negotiate with these two kids, then he could go to Capitol Hill.' . . . So I was just watching him negotiate with these two little girls, and that sold me. I could tell a relationship was there."

Then Pam read a brief summary of Barack's biography and top issues to the group and asked what appealed. "Civil rights attorney," one woman responded. "That really stood out to me." "Health care for all children," said a second. A third lady agreed: "Health care. When I think of health care, I think that should be available to everyone." One of the pastors said "what stands out to me with Obama is the fact that he is a law professor, and that he is biracial. Because to me, it would take something like that for him to advance, even after he gets into the Senate, to maybe run for president." She explained that "the fact that he is biracial should let them know that he understands both cultures, and that he is a law professor, so he is certainly intelligent enough to do the job. . . . And the fact that he passed legislation for racial profiling lets his own people know that he is really aware of things." Pam asked about Barack's Kenyan heritage, and several women in succession said, "I love it." "This man has a global vision," one explained. "If your parent is from another country . . . your view is larger than the United States."

Audra Wilson moderated the second group, of eleven black men. One said Barack is "very articulate, very neat looking," with "a good clean reputation." While "driving through a white neighborhood" he had seen some Obama yard signs, and "I searched the Web," and "I found that he's a very active and effective political leader." A second man was impressed by Barack's work to expand job opportunities for ex-offenders because "the record keeps most people unemployed when they come back to the community." A third said, "here's a guy who is Harvard educated," who "has *all* this education and is trying to do something for the community. That's a positive." Another volunteered that "considering all that he has done in the community, I think it's cool that he has a white woman for a wife." Others quickly corrected him, saying mother, not wife, and Audra noted "lots of laughter—group is happy his wife is black." The one fellow explained, "I thought his wife was white, and I said, 'Wow, he has a white trophy, and he's still out doing all this in the community.'" Another man noted that Barack had been "consistently in and out of issues in the *Defender*," and the group also compared notes about what benefits Bobby Rush may have received in exchange for backing Blair Hull.

Those two discussions went a long way toward indicating that Barack's base was indeed solidifying, but Jim Cauley still had his campaign workers examine Joyce Washington's petition signatures to see if she, like Alice Palmer, might have submitted too few valid ones to make the March 16 ballot. On Sunday evening, December 21, the Obama team convened for a two-hour conference call, and election lawyer Michael Dorf, who had reviewed the findings, told the group that an objection had a 50–50 chance of success. "What the hell does that mean?" Jim Cauley asked. Under pressure, Dorf increased the odds to 55–45, but he said he was not certain that "this was a winning challenge." Almost everyone on the call spoke in favor of filing. Given their goal of winning with a 33 to 35 percent plurality, Washington might attract just enough black voters to deny Barack the margin he needed. As Cauley argued, "if she clips off three points, you can't get there," but David Axelrod was concerned that targeting Washington "might really crash our brand." Barack felt similarly, asking, "How would it look? Will I get flak?" Everyone appreciated Barack's "history with Alice Palmer," which Audra Wilson thought "so haunted him," and Dorf understood that "the worst thing was you challenging a black woman and you lose." Barack sided with Axelrod. "Yeah, I don't want to do this," but Barack agreed to sleep on it until the next day. So as to keep their options open until the last minute, Dorf on Monday was "standing in the lobby of the Board of Elections at five to five with two suitcases" and "my cell phone . . . waiting for Nate Tamarin to tell me 'Go file it' or 'Don't file it,'" Dorf recalled. But Barack did not change his mind. "We're not going to do that," and Nate's 4:55 P.M. call to Dorf was brief: "Go back downstairs."[44]

As the Christmas holidays approached, the big news was the federal criminal indictment of former governor George Ryan in the bribes-for-licenses scandal from his time as secretary of state. Barack, Michelle, and their two daughters were leaving for Honolulu on December 23, but when Jim Cauley first found out about that plan, "I went ass stomping in there . . . ready to chew his ass" and "a big fight" ensued. Jimmy knew about Barack's missed vote on the 1999 gun bill but did not realize that Christmas in Hawaii was an annual pilgrimage. Jimmy wanted as large a fund-raising total as possible when the FEC quarterly reporting closed on December 31, and his candidate was about to take a week off. "Jim, my sister's getting married" and her father was deceased, Barack explained. Maya and Konrad Ng were indeed having a wedding ceremony on December 28 at Oahu's Paliku Gardens, but they had already legally wed eleven months earlier in Las Vegas. Jimmy told Barack, "Fuck you!" and walked out. "What pissed me off was his fuck-you attitude about it," Cauley explained.

In Honolulu, Maya was teaching at the U of H's Educational Laboratory School and working toward a Ph.D. in education. At the ceremony, Barack gave a "poetic introductory speech" that his father's old friend Congressman Neil Abercrombie found "mesmerizing." Konrad and Maya read love letters to each other and exchanged rings, after which everyone shared a potluck meal.

On December 31 Barack and his family flew back to Chicago. His consultants had been busy projecting that with $1,750,000 cash on hand, and hopes for raising another $1,500,000, the campaign could afford three to four weeks of Chicago TV ads right before the primary. Jesse Jackson Jr. joined Barack at the opening of a south suburban Obama campaign office, and the next morning Barack was on V103, Chicago's top R&B radio station, for a "Battle of the Best" face-off against DJ Herb Kent. Choosing the Ohio Players, his favorite group during his youth, Barack came out on top. Barack's campaign schedule had him at Chicago L stations some mornings from 7:00 to 8:00 A.M., with weekend evenings at South Side nightspots like the Sandpiper Lounge and Sunday mornings in multiple black churches. A downstate strategy memo laid out a three-point game plan: capturing 70 to 75 percent of the African American vote outside Chicago by mounting a serious campaign effort in Metro East's Madison and St. Clair counties, pursuing support in conservative white areas where Barack enjoyed the backing of popular local politicians such as Lane Evans and state senators Denny Jacobs and Bill Haine, and mobilizing youthful supporters in university communities like Champaign, DeKalb, Carbondale, and Macomb.

Two groups of fifteen suburban North Shore white women were invited to meet with Paul Harstad on the night of January 8. David Axelrod had stressed to Harstad that "our goal should be, first and foremost, how to sell Barack to white progressive voters." He wanted "a great deal of probing about the war . . . to find out if it is, indeed, the trump card we think it is." Harstad should also "explore the name issues, and see if it engenders any reaction," because "the mission here is to focus on how to present Barack." The first group of women were fifty-five and older, and featured four who intended to support Gery Chico, two apiece who backed Dan Hynes and Maria Pappas, one each favoring Barack and Joyce Washington, with the balance undecided. Harstad began by asking them about the state of the country and the upcoming Senate vacancy. One remarked that Peter Fitzgerald was "probably too honest" for Illinois politics, and a second singled out Barack. "I like Obama. He is a constitutional lawyer with very high credentials."

When Harstad zeroed in on the Senate race, one woman cited Jan Schakowsky's endorsement of Barack and a profusion of Obama yard signs. "Young married man with family. Harvard." Another said, "his name is so interesting. What is he? . . . Is he black?" Someone cited Gery Chico's Altheimer & Gray debacle. "He took a multimillion business and threw it down the toilet." When Harstad focused on what the women knew about Barack, several cited Schakowsky's support, his educational background, and his U of C affiliation. Then Harstad showed the group video excerpts from the NAACP debate and sought their reactions. Various women said Barack sounded "very confident," "knowledgeable," "honest and real," and that he "is easy on the eyes." "Hynes sounded stiff" and Blair Hull "sounded embalmed," one remarked. When Harstad asked whose endorsements mattered the most to them, Jan Schakowsky, Abner Mikva, and the Sierra Club topped the list. Did Barack remind them of any celebrities? "Denzel Washington, Harry Belafonte, Sidney Poitier," came the replies. "We can't deny the fact that we are impressed by people's looks," one woman explained.

Then Harstad probed for Barack's negatives. "Many Democrats . . . are worried that a candidate named Barack Obama cannot win a statewide general election," he asserted, eliciting little reaction. There was also little response when Paul mentioned Barack's present votes. Harstad recited Barack's evasive initial denial to Bernie Schoenburg about drug usage beyond marijuana, and one woman remarked, "it just seems like a weird thing to protect." Finally Paul read a long, praiseful description of Barack's background and issue positions, including that "Obama has pledged to repeal aspects of the Patriot Act that threaten our civil liberties" and asked for reactions. "Be still my heart." "He sounds perfect." "The first African American president of the *Harvard Law Review*. That impressed me." "I like the fact that he is a constitutional law scholar." Lastly, Harstad asked what advice they would give Barack, and the responses were "Answer the discrepancies on the drug thing," "Answer the discrepancies on voting present," and "He should explain how he got his name."

After a half-hour break, the evening's second group was composed of women in their twenties to early fifties. Three supported Chico, with two each for Barack and Maria Pappas, one for Dan Hynes, and the balance unsure. For the Senate race, one immediately cited "Obama. He was in the Skokie parade. . . . Paul Simon was his mentor . . . professor of constitutional law." When Harstad followed up, other participants embraced Barack because of the first speaker's comments. "Obama, because I just learned here that he was associated with Paul Simon." "Obama . . . I knew about Paul Simon, and I was always incredibly impressed with him." "I have always liked Paul Simon, and if he is his protégé, I am interested." "Obama . . . Aside from the Paul Simon connection, I also like that he was well educated, taught constitutional law." "Obama. He sounds electable. . . . Look at what happened around this table: he can garner a coalition!"

Then Harstad screened the debate excerpts. "Obama was very articulate," one commented. "Hynes made me uncomfortable. Just another politician," a second observed. "Do you think Obama's race makes him more or less electable?" Harstad asked. "More" came the answer. "Hynes was so stiff," another stated. "Hynes seemed like Dan Quayle," someone added. At the Skokie parade, Barack "was as cool as a cucumber," one recalled, but a second complained that Barack sounded "a little bit too street" on the video. Asked what endorsements mattered most, the women chose Planned Parenthood and IFT over Mikva and Schakowsky. Harstad closed by requesting their feelings about Barack, and one woman said "we all fell in love with him here tonight as Paul Simon's protégé."

Both groups of women were asked what items in Barack's biography most stood out to them, and two points concerning expanded health coverage scored even higher than first black president of the *Review*. Schakowsky's endorsement proved the most significant, and when the women's numerical ratings of the video excerpts on a 1-to-10 scale were averaged, Barack scored 8.4, with Chico, Hynes, Pappas, and Hull trailing with numbers ranging from 6.6 to 4.2. Harstad and partner Mike Kulisheck found it "incredibly promising" that the North Shore women had been so "extremely receptive and supportive" regarding Barack's biography. John Kupper, who along with Pete Giangreco had joined Kulisheck to watch the two groups from behind a one-way window, found the women's reactions "really exciting," with that "be still my heart" response the most memorable of all. The evening's comments would go a long way toward informing how Kupper and Axelrod would prepare Barack's broadcast ads, and how Giangreco and his partner Terry Walsh would write and design Barack's direct-mail flyers.[45]

The next morning's papers carried news that Dan Hynes had finally been endorsed by the state AFL-CIO. Hynes told reporters, "this is by far the most significant endorsement of the campaign," and in Washington, *The Hill* said that it "could lock up the race for him." The conservative *National Review* wrote that "Hynes appears well on his way to winning his party's nomination" and that Illinois Republicans' "best hope would be for state Sen. Barack Obama to upset him in the primary." Far more knowledgeably, *Capital Fax*'s Rich Miller noted that Hynes "has been struggling to break out of the pack," and "his unrepentant support for the Iraq war will probably be a big issue" as more Democratic voters focused on the campaign. With purposeful timing, the next day AFSCME announced its support for Barack. While the AFL-CIO endorsement would aid Hynes, the three big unions backing Barack—the IFT, SEIU, and AFSCME—"are incredibly active, organized and wealthy, and can field thousands of volunteers," Miller stressed. Hynes had better use the AFL announcement "to create some much-needed momentum," Miller warned, because "so far, his campaign has been on cruise control. The guy has held statewide office for 5 years and still hasn't broken out of the pack. Not good."

With only eight weeks until the election, the blizzard of questionnaires and endorsement interviews increased significantly. The progressive IVI-IPO questionnaire ran sixteen pages, and in it Barack called "the so-called Patriot Act" "one of the most dangerous pieces of legislation to pass Congress in decades." Barack reiterated his opposition to the Iraq war and called for "normalization of relations with Cuba" because U.S. "policies toward Cuba have been a miserable failure." He noted that "our government is running an unsustainable budget deficit. We must move away from this perilous path of deficits." Barack favored "a national moratorium" on the death penalty, even though "in theory, I support capital punishment for a very narrow band of heinous crimes, such as serial killing" and terrorist attacks. "What is occurring today in Guantanamo . . . is an affront to our Constitution. We cannot allow terrorists to cause us to sacrifice the very rights that make our nation special."

Barack told his staff and consultants to take particular care with the *Chicago Tribune*'s ten-item questionnaire because the paper might publish his responses. Both a NOW representative as well as Jan Schakowsky had upbraided Barack for his "squirrelly" answers regarding gay rights on their earlier questionnaire, and Barack now realized that "we actually fudged more than Hynes, and a lot more than any of the other candidates." Although he wanted "to make our position clear and unequivocal," the next day Barack again faced the same troublesome questions when he met with the lesbian and gay editors of the *Windy City Times*. Barack started out safely by saying that "one of my top priorities is moving in the direction of universal health care." When the questions moved to gay equality, Barack declared, "I am a fierce supporter of domestic partnership and civil union laws. I am not a supporter of gay marriage . . . primarily just as a strategic issue. I think that marriage, in the minds of a lot of voters, has a religious connotation. I know that's true in the African American community." Editor Tracy Baim remembered Barack asking her to turn the tape recorder off so that "we could have an off-the-record conversation about same-sex marriage" and what was "realistic and pragmatic." Back on the record, "strategically, I think we can get civil unions passed," Barack explained, and "to the extent that we can get the rights, I'm less concerned about the name." Barack said, "I've walked the walk on every single issue that's been important to the LGBT community" and believed he had "an ability to translate my passion for equality and justice into a language that a broad audience can relate to and understand." But afterward Barack called Baim "to clarify that he opposes" any new measures to prohibit same-sex marriages.

On January 15, the *Chicago Tribune* published the results of a small-sample poll that had been conducted a week earlier. It showed Dan Hynes, Maria Pappas, and Barack all tied with 14 percent apiece, Blair Hull at 10 percent, and 38 percent of Democratic voters undecided. *Capitol Fax*'s Rich Miller was stunned. "How can Dan Hynes possibly be polling just 14 percent in this race? And how is a relative unknown with a stunningly awful ballot name like Barack Obama staying even with him?" Although "nobody seems to be paying attention to the campaign . . . on name recognition alone, Hynes ought to be far ahead of the pack." Hynes had run strong campaigns in 1998 and 2002, when he was elected and then reelected as comptroller, and he "is well liked and respected by the party regulars," Miller wrote. "You'd think that all of that ought to be worth a benchmark of at least 20 to 25 points. Maybe when people finally figure out that a Senate race is on the horizon and they're reminded who Dan Hynes is, his numbers will start to rise. But there's something missing from his campaign. Back in 1998, Hynes put together a remarkable, vibrant organization. . . . But so far, he's been running this campaign like a cautious old man." Criticizing Hynes's "ultra-traditional, low-key" effort, Miller highlighted how "so far, almost nobody knows that Hynes supported the war in Iraq," which is "a huge no-no for most Democratic primary voters." Noting Blair Hull's 10 percent showing, "you can bet good money that he is taking most of those votes away from Hynes," Miller observed. While Hynes's Web site declared that "elections are about inspiring people and bringing them together," Miller stated that "the only candidate who seems to be inspiring Democratic primary voters these days is Barack Obama."[46]

Neither Hynes's pollster Jef Pollock nor his communications director Chris Mather was worried. "The AFL was huge for us," Mather recalled, and the campaign was untroubled that Barack was backed by IFT, SEIU, and AFSCME "because we got the AFL-CIO," Mather emphasized. "We were premising a lot of what we had on a downstate strategy" featuring television advertising, Pollock explained, and he was more worried about competing with Blair Hull's pocketbook there, not Obama. "We thought that downstate was where we would win it," Mather agreed, "and we spent a lot of time down there," for "downstate was absolutely key." In Chicagoland, where Democratic ward and township organizations were at their strongest, "we were going to run more of a field campaign," Mather remembered, and Pollock expected they would reap the benefits of "an amazing political infrastructure."

Dan Hynes was similarly optimistic. "I felt like I had a lot of advantages . . . statewide name recognition, a support base, some of the traditional Democratic organizational support," so "everybody viewed me as the front-runner." Hull's millions concerned him, but he thought Maria Pappas would further fractionalize the Chicagoland vote. Given Hull's big media effort, "I had decided I was going to spend a ton of money downstate" to "keep pace with Blair Hull." Hynes's campaign had just gone up on downstate television, with Dan's campaign manager, his brother Matt, telling the Associated Press that while "you might have 25 percent undecided in Chicago," that could well be "50 percent downstate." Dan realized that "the unions that went with Barack brought a lot to the table in terms of producing," and he appreciated too how "that one legislative session was defining for him," because Barack's long list of 2003 bills enabled him to "show that he could get things done."

Barack made just that argument in a campaign e-mail that went out the day the 2004 spring session got under way. The Senate race "will not be won by saturating the airwaves with commercials, nor will it be delivered by party insiders," Barack wrote. "Instead, it will be won by communicating a record of real achievement on critical issues." An astonishing three hundred people showed up at campaign headquarters for a Saturday volunteer rally, and Monday's Martin Luther King holiday marked the opening of one Obama for Illinois office on the West Side and another on the South Side, just off the Dan Ryan Expressway at 5401 South Wentworth. A pair of College Democrats debates had Barack in Carbondale on January 20 and the next night at Northwestern University in Evanston. "I think the war on drugs has been a failure, and I think we need to rethink and decriminalize our marijuana laws," Barack told the Northwestern crowd, adding, "I'm not somebody who believes in legalization of marijuana." Friday morning at 7:00 A.M. Barack and Randon Gardley spent two hours greeting potential voters at the 87th and 79th Street South Side Red Line L stations.[47]

One by one, the *Chicago Tribune* had been running Sunday profiles of each Democratic candidate, and the early "bulldog" edition of Barack's first appeared on Saturday morning, January 24. *Tribune* reporter David Mendell had had a long, pre-Christmas interview with Barack at 310 South Michigan and had watched him at a good many campaign events. In their first conversation, Mendell asked Barack about the millions of dollars Mayor Daley's administration had devoted to the glitzy Millennium Park project just across Michigan Avenue rather than to struggling residential neighborhoods. Mendell was struck by the candor of Barack's response. "How do you really expect me to answer that? If I told you how I really felt, I'd be committing political suicide right here in front of you." While several of Barack's closest confidants, particularly Valerie Jarrett and David Axelrod, were well-known Daley loyalists—or "stone-cold machine operatives," as one longtime organizer put it—Barack kept his distance from the mayor "without ever opposing Daley's wishes."

Barack might mouth off to young aides about challenging Daley for mayor should he lose the Senate race, but Barack "never took on any issues about the functioning of Chicago," school reform champion Don Moore explained. Moore saw Barack as "a very brilliant pragmatist" who had "worked his way up through the system." Campaign reform advocate Cindi Canary, who had been disappointed by Barack's complete deference to Emil Jones when the 2003 ethics bill was being weakened, agreed that Barack was "a very pragmatic politician." That was true in Barack's arm's-length relationship with Daley's "new machine" and in his close relationship with old-school Senate president Emil Jones, whom one top Illinois progressive termed "a very transactional guy." As Canary said of Barack's relationship to the Daley organization, "he was never really of it, or against it. He worked it, and it worked him. It was very mutually beneficial."

The same was true of the alliance between Barack and Emil Jones. By January 2004, Jones's investment in Barack's Senate candidacy was obvious to all who knew him. Everyone in Illinois Democratic politics realized that Jones was "a very, very smart man," and he was also seen as "warm and likable," a stark contrast to his Springfield bête noire, the "icily calculating" Michael Madigan. Jones's seething anger at the disrespect the House Speaker had shown him for years energized Jones in the legislature and in his fervent support for Barack over Madigan-backed Dan Hynes. "I think he wanted to send the message that there are multiple sources of power here in Illinois besides the traditional white Irish enclave of votes that had run Cook County politics for so long," Jones's number-two staffer Dave Gross explained. Unlike Madigan, Jones had some profound policy commitments, especially regarding increased education funding, and only Jones's steadfast backing had allowed Barack to make the headway he had with his racial profiling and videotaping bills a year earlier. Some of Jones's commitments, like his devotion to Chicago State University and his support for black business interests seeking increased public patronage, were indeed to differing degrees "transactional" relationships, but Jones's newfound devotion to Barack's advancement seemed heartfelt to everyone upon whom he leaned.

Within the Senate Democratic caucus, Jones's message was crystal clear: "You need to get behind your colleague," and "I want everybody to support Barack." While Terry Link, Denny Jacobs, Larry Walsh, and Bill Haine, as well as African Americans James Clayborne, Kim Lightford, and Jacqueline Collins had endorsed Barack, some liberal senators like Jeff Schoenberg and Susan Garrett had backed Dan Hynes. For organization loyalists like Martin Sandoval and Tony Munoz, the question of endorsing Barack never crossed their minds, but three white Southwest Side and south suburban senators who had warm relations with him—Lou Viverito, Ed Maloney, and Debbie Halvorson—felt they had to back Hynes. Maloney's ties to Hynes's father Tom were well known, and Viverito, who had supported Barack's congressional run, was Mike Madigan's state senator. "I understood my obligations," Viverito explained, and "I tried to explain" to Barack "that it was not a personal decision," but Barack took offense. "He was kind of bitter at the people that didn't support him," Barack's friend Mike Lieteau remembered, and with the good-natured Viverito, "Barack treated him like shit," Mike recalled. "Lou was very hurt." For Debbie Halvorson, who had taken on lots of new, mostly white territory in the 2002 remap, it was even worse. "I had no choice" because several local officials "threatened me" if she did not back Hynes, Halvorson explained. Dan Shomon had managed Debbie's 1996 upset victory, but once her name appeared on Hynes's Web site, Dan "comes over to me and says, 'You are now the enemy,'" Debbie recalled. "Dan was the worst. Barack took it better than Dan did."

"There were people in the caucus who went with Dan Hynes, and Emil stuck it to them every day," Jones's press secretary Cindy Huebner remembered. Jones made a special effort to neutralize the duo who had been Barack's worst detractors since he arrived in Springfield. "There's only two members that I had conversations with, and that was Hendon and Trotter," Jones explained. "'It wouldn't be wise for me as the leader to put you guys in the position of power and authority if you cannot support this man over some silly, petty jealousy,'" Jones told them, directly threatening their Senate positions. "I said, 'On my watch, it's not going to happen.'" Hendon buckled. "I knew the president would like me a whole lot more if I went along with his wishes."

Jones was equally direct with lobbyists. ComEd's Frank Clark and John Hooker heard Jones's message clearly. "Look, Frank, he's absolutely my guy. I totally support him," Jones told Clark, and Clark responded accordingly. "I did everything I could to get corporate support for him from both ComEd and Exelon and others" in order to comply with Jones's request. Lobbyist John Kelly was as close to Jones as almost anyone, yet as a member of Chicago's white

Irish tribe, Kelly's support for Hynes was unquestionable. Jones, however, "made people think twice about supporting other people," Kelly explained. "He did that with lobbyists, he did that with people in Springfield. . . . It became clear that almost everything he was going to do, it was going to have some angle on getting Obama elected." Whenever someone spoke to Emil about a pending bill, Jones asked, "Where are you on the U.S. Senate race? I want to support people's initiatives who support my initiatives." Concerning Barack, "to say that he was invested I think is an understatement," Kelly observed.

Jones was not shy about acknowledging the forceful role he played. Jones felt obliged to live up to Barack's observation that as Senate president he possessed "the power to make a United States senator." Jones described it as "reverse psychology." Barack "put the onus on me to prove that I had the power to do it. . . . He put it on me to prove that I could do it." Barack "needed someone who had enough influence to be able to help clear the path. Otherwise he would never have gotten out of the backfield."

By late January 2004, with the end-of-the-month FEC reporting deadline for fourth-quarter 2003 contributions approaching, Barack's campaign knew that it would have the funds necessary for Chicagoland TV ads in the run-up to March 16. Obama for Illinois's quarterly total of $827,000 had been helped by a last-minute $10,000 check from Tony Rezko's friend Elie "Lee" Maloof, a sum that Rezko secretly—and illegally—reimbursed, but that $827,000 figure eclipsed supposed front-runner Dan Hynes's $708,000 total by more than $100,000. More important, Jim Cauley's insistent spending discipline meant that Obama for Illinois now had $1,789,000 on hand.

Barack's long-standing friendship with Rezko reflected a comfort level that was not shared by campaign staffers who dealt with Tony. "I was told to meet with him" right after arriving in Chicago, Jim Cauley remembered, and "I met him more than a few" times. "Rezko was a little weird," Cauley thought, and he knew that "some of that . . . money is a little dirty," but Barack pressed him. "I was told to meet with him more, and I told him I wouldn't, because I had a very keen sense of people and that guy made me nervous from day one." If anyone "says he wasn't around" Barack's campaign, "he's full of shit," Cauley emphasized.

Dan Shomon acknowledged that Rezko "was very, very important on fundraising in '04, putting checks together," and finance director Claire Serdiuk realized that Rezko did not understand how different the FEC's requirements were from Illinois's lax standards. Tony "was coming from a system that had no laws, no limits" to one "where there were a lot of laws and a lot of limits," Claire explained. "We were very, very fortunate, because we were running a federal campaign, and there were very tight rules. There were several times when I had to go back and tell Tony, 'I'm not accepting this check.'" For example, the millionaires' amendment's increased limits did not include corporate or partnership funds. "I told Barack, 'We have to deny this check,' and Barack would call Tony and say, 'We have to deny this check—we cannot accept this on the federal,'" Claire recalled. "It happened a few times in the primary," and as Maloof's $10,000 illustrated, it could have happened more.

David Mendell's *Tribune* profile lauded Barack as "without a doubt the most dynamic speaker" and "the most commanding presence" in the Democratic field, but those plaudits were not what generated angry reactions that Saturday morning. Instead it was a vignette Mendell used at the end of his article:

> *There's no doubt Obama can draw attention. Shortly after signing autographs at the recent forum, Obama grabbed the hand of Christina Hynes, the wife of one of his opponents, Illinois Comptroller Dan Hynes, and then kissed her cheek, prompting her to flush and smile broadly.*
>
> *"He has a smooth personality, sometimes a little too smooth," said his campaign manager, Jim Cauley. "He's still young, and we have a ways to go, but he has the potential to be something very special in this business."*

Barack was furious at Jimmy for calling him "a little too smooth," as Cauley heard from his "thin-skinned" candidate in a 6:00 A.M. telephone call. But Barack's anger was mild compared to the reaction in Dan Hynes's camp. "We went apeshit," communications director Chris Mather remembered, over Mendell's explicit insinuation that Christina Hynes "might have the hots for another candidate." Christina blushed easily and was "not a public person," and campaign manager Matt Hynes called Mendell at home to complain vociferously about the *Tribune*'s "total cheap shot," threatening to end Mendell's access to Dan unless the passage was removed from later editions. Mendell hesitated. "It happened. I was standing right there," and "if anything, I had toned down Mrs. Hynes's overt response to Obama." Indeed, Mendell had expressed surprise. "I said to Senator Obama, 'Do you really think it's wise to kiss the other candidates' wives?' He said, 'We've known each other a long time. It's been a long trail.'" Mendell called *Tribune* political editor Bob Secter, and they agreed to remove Christina Hynes's name while retaining the description, with Barack now kissing "a supporter of one of his opponents."[48]

On Sunday afternoon, Barack was back in Evanston to speak to about three hundred people at city Democrats' endorsement meeting. Dan Hynes and Gery Chico spoke as well, but Barack "just blew everybody away," his old Miner Barnhill golf partner Whit Soule thought, and a remarkable 90 percent of the crowd voted to support Barack. The next afternoon at 1:00 P.M., Barack had his meeting with the *Chicago Tribune*'s editorial board, and David Axelrod, again in Mexico, e-mailed Barack to reiterate "the overall arguments to stress." Number one, "you offer more than words. You offer a record of legislative accomplishment which uniquely prepares you for this job. You are a proven, skilled legislator." David also wanted Barack to present himself as "a Paul Simon Democrat" who represented independence and change. "Look for opportunities to demonstrate your willingness to buck party orthodoxy," Axelrod advised. "Health care is our best issue, and you should hit it hard," stressing that "real health care reform requires the ability to stand up to the power of the insurance and drug lobbies."

On Monday afternoon, Barack told the *Tribune* journalists that "what I bring to this race is a track record of accomplishment, the ability to work with others and the ability to frame the debate in a way that does not simply polarize but brings people together around a set of common goals." On Wednesday, Barack was set to appear at the City Club of Chicago's noontime candidates' forum, and ahead of a Tuesday-afternoon practice session, Axelrod and John Kupper prepared a memo calling for Barack to reiterate "the independence/fighting special interests/pragmatic accomplishments theme." They wanted Barack to emphasize "our overarching message that he has a seven-year legislative record of standing up to the special interests in Springfield." Barack "needs to be forceful and passionate," displaying "some outrage and indignation" while using "short, declarative sentences rather than professorial soliloquies." The City Club event turned out to be a largely sedate affair, but Barack did declare that "we have a crisis in our democracy right now, and so much of it has to do with the use and abuse of money in the political process." Republican *Sun-Times* columnist Tom Roeser called Barack "the winner" as he "trumped his opponents with an understated charisma," and on the same page, Steve Neal wrote that the late Paul Simon had viewed Barack "as the rising star of the Illinois Democratic Party."[49]

Political insiders were highly attentive as the primary race entered its final six weeks, but that was not true for most citizens, as Chicagoland's major suburban paper documented. "A random survey last week by *Daily Herald* reporters and editors found 87 percent of 162 suburban residents could not name even one Senate candidate from either party." Among Democrats, "millionaire businessman Blair Hull of Chicago netted 11 responses," thanks to his robust TV advertising, and "nine people answered state Senator Barack Obama of Chicago, with full credit given to two who could remember only that his name rhymed with Osama." *Capitol Fax*'s Rich Miller reported that a new WBBM automated poll showed Dan Hynes leading Hull and Barack by one point each, at 20–19–19, with Maria Pappas at 18, again reflecting an uptick for Hull. By late January, Paul Harstad for Barack's campaign and Hull's own pollsters had results of their first 2004 tracking surveys. Harstad's results showed Hynes at 22 percent and Barack at 16, with Pappas and Hull trailing at 15 and 11 percent, respectively. However, 66 percent of Democratic voters still did not know

Barack, a surprising *increase* of 2 points since Harstad's July 2003 poll. Among black voters, Barack's don't-knows remained at 46 percent, and the 34 percent supporting him was static as well, even though Barack's favorability total had risen from 37 to 45 percent. After Barack's issue positions and legislative achievements were presented, he picked up an additional 23 percent support, showing that his existing 16 percent share had the potential to jump to 39 once more voters knew about him.

Anna Bennett and Mark Blumenthal's late January poll, the first of *ten* tracking surveys they would conduct for Hull's campaign prior to mid-March, showed Barack with 35 percent overall name recognition and about 45 percent African American support, ten points higher than Harstad's number from a week earlier. That gave Barack a narrow 19–18–16–13 lead over Hynes, Hull, and Pappas, with Hynes and Hull in a tight duel among nonliberal white voters. But after respondents heard long descriptions of each of the candidates, Barack's support jumped to 29 percent while Hull rose to 19 and Hynes slipped a point to 17.

Barack was "extraordinarily saddened" and "heartbroken" to learn that Beverly Criglar, his Springfield secretary from 1997 through 1999, had died suddenly of a heart attack at age sixty-three. Barack's focus turned toward preparing for a February 4 statewide radio debate, but he told the Associated Press that "the undocumented workers currently in this country and woven into the fabric of our lives are not simply strangers to be deported." At the radio debate, Barack stressed his opposition to the U.S. invasion of Iraq as "a dumb war" and pressed Dan Hynes over his support of it while highlighting his state Senate experience. "The legislature is full of tough calls. It's not like an administrative job—it requires tough calls." But Hynes's campaign faced a greater challenge than radio-only criticism from Barack. As *Capitol Fax*'s Rich Miller highlighted, Hynes's focus on promoting his candidacy via downstate television ads was going nowhere. Whenever a Hynes ad debuted, Blair Hull was "up in a couple of days with his own TV spot on the same issue, but with four or five times the money—effectively drowning out Hynes' message."[50]

Building good footage from which to craft Barack's TV ads, David Axelrod and John Kupper booked studio interviews with several people who knew Barack well and scheduled a variety of events where they could film him on Saturday, February 7. Mike Pfleger assured David that "Barack is the kind of character that ten years from now, people will be just as proud with, because he is consistent with who he is." Jan Schakowsky said her endorsement of Barack was based on his opposition to the war and his insistence that he "would have been a no vote on the Patriot Act," because "we need someone who's going to stand up and defend civil liberties."

On Friday, February 6, John Kupper sat in a Near North Side studio with first Jeremiah Wright and then Michelle Obama. Wright told Kupper that Barack "is a dreamer," a "combination of intelligence . . . and soul." From "almost twenty years of knowing him," Wright was certain that Barack "is very much interested . . . in health care for all" and especially "child health care, because his daughter has asthma." Jerry endorsed Barack as fervently as he knew how. "There are only a few people in life, and Barack is one of them, that you feel privileged to have been blessed to share space and time with in your lifetime."

Michelle spoke of her husband even more personally. "I think there is a part of him that's trying to live up to the image that he has of what the father he didn't know would have wanted him to be," she explained. "I think Barack grew up hearing these wonderful stories about his father and how imposing and committed and serious he was, and I think there is an element to his determination and passion that is trying to live up to the reputation of the father he didn't know . . . to live up to this dream."

Michelle volunteered that "Barack's role in government is our sacrifice to the bigger picture," and that "our spiritual life and our spiritual development is key. . . . We are fortunate to come from such a strong church home, with a wonderful pastor like Reverend Wright, who not only has the spiritual message, but he ties that into the bigger picture: politics and just sort of civil rights, and there's a whole range of education and awareness that happens at our church that sort of speaks to us."

Michelle emphasized the importance of Trinity. "Church, religion, family—those are the things that ground us and help Barack and I make it through what will be a challenging several months and a challenging lifetime if we continue to do—and we plan to continue to do—this work in the future." Michelle said that Barack "is truly gifted for this work," and "we're doing this because, as I talked to him before we entered this race, that this race has to be about telling the truth," as could be done by "somebody with his gifts."

Some of what Michelle told Kupper was not fit for any television ad. "I think down in Springfield there are frustrations with the lack of leadership that we see in our own party," because "the party's kind of lost its way. I hate to say it, but we've talked about this on many occasions." Michelle knew that the U.S. Senate is "a millionaire's club," but "the bottom line is that we're just not moving in the right direction morally as a country, and I think that he's very passionate about playing a role." Her regard for her husband was astonishing. "It's almost like he's too good to be true, and people don't believe he can be that perfect." It was true that "he doesn't put his dishes in the sink, and I have to remind him to take the garbage out, and he doesn't put his shoes up, and I could go on about his imperfections. . . . So he is a man, but for a man, he's pretty right on point."

Barack's long Saturday, February 7, began with a 10:00 A.M. rally at Rogers

Park's Heartland Cafe. His remarks recycled familiar themes, such as how "particularly here in Illinois . . . politics operates as a business instead of a mission." Several hours of video and still photo shoots with a variety of people—young and old, black and white, male and female—at an Evanston nonprofit preceded another rally at nearby Mount Zion Missionary Baptist Church. "There are very few people in Springfield who actually take the time to study the issues, to craft legislation, who hold hearings, who get community input," Barack told the crowd. More relaxed than earlier, he complained that "I'm having to put up with consultants, and I'm having to put up with not having any lunch," but he vowed that in January 2005, "we're gonna start changing this country."

A brief fund-raiser was followed by Barack sitting down on camera with David Axelrod at a donor's lakefront home. To an earlier interviewer, Barack had seconded Michelle's statement that "I can't imagine a better church home than Trinity," but with Axelrod, a somewhat tired Barack spoke almost exclusively about his Springfield résumé, like KidCare and Illinois's EITC. "I'm enough of a pragmatist that I know how to reach across the aisle, and I know how to work on a bipartisan basis to make things happen." He said that "somebody who's able to speak out with honesty and conviction over time can gather allies and eventually persuade a majority, and so that's a role that I welcome, and it's a role that I've always felt comfortable with down in Springfield." Barack similarly declared, "I haven't been afraid to stand up to some of the most powerful interests, whether it was the drug companies or utility companies or gaming interests down in Springfield, and say, 'This doesn't make sense.'" Only at the end did the interview turn personal. Barack regretted "all the time that I'm spending away from" his daughters, but "thanks largely to my wife, they are thriving and happy." Barack mentioned that "I struggle with my wife during tax time to figure out how we're going to meet our obligations," but he was deeply troubled that the Senate campaign had all but removed him from his daughters' daily lives. He tried to avoid thinking about it, but when Valerie Jarrett asked him how he was holding up as the race intensified, Barack responded with rare emotion: "I don't want to be the kind of father I had."[51]

On Sunday morning, Barack started the day at Grand Boulevard's Liberty Baptist Church. Fourth Ward politico Al Kindle and journalist Ted McClelland, who had written a profile of Barack for the *Chicago Reader* four years earlier, were struck by how much more fluid and conversational a public speaker Barack was than he had been during his unsuccessful congressional race. McClelland wondered what to make of "a campaigner whose persona changed so drastically in four years," but even one-on-one, Barack seemed different from four years earlier. McClelland recalled that "the aloofness was gone" and "he didn't mention Harvard once. . . . I was impressed that he finally seemed to believe in something more than the fact that being president of the *Harvard Law Review* is a pretty big deal."

Others who had heard Barack speak over the years also noted how much more relaxed and self-confident he was now. "The more he campaigned, the more natural he got at it and the more disciplined he got at it," AFSCME and Citizen Action veteran John Cameron explained. "This whole 'I'm the guy, I'm Mr. Cool, you gotta be with me' kind of stuff all disappeared." Dan Hynes saw it too. As the race progressed, "the more confidence he had, and it showed in his speaking style and the way he interacted with people." Kevin Watson, who was squiring Barack to countless South Side barbershops, beauty salons, and nightclubs, watched Barack change before his very eyes. "You could see him more and more relaxing," Kevin explained. "I can see it happening, and I can *feel* it happening. . . . It was almost like he was turning into something other than what he was . . . something that is bigger than him."

U of C graduate student Warren Chain had volunteered a lot during Barack's congressional run, believing he was a "good guy" though a "very poor" public speaker. Chain kept in touch briefly, visiting Barack at Miner Barnhill, and then "didn't see him for two or three years" before running into him at the Walgreens Pharmacy on 55th Street, just a few blocks from East View, during the Senate campaign. He thought Barack "had a different look in his eye, a different energy" than three years earlier. "He was just so intense." People who had known Barack more than a decade earlier also noticed a change. Old organizing colleague David Kindler attended an Arab American–hosted reception for Barack during the Senate race and found him "kind of disappointing" because he "wasn't the candid Barack" of years past and was now "a little bit more cagey." Among themselves, Barack's organizer friends mourned how different he was from years earlier, when "the wall was down" and "when he was far less guarded and before the drawbridge to his feelings went up." Ali Abunimah, Mona and Rashid Khalidi's Arab American Action Network colleague, felt similarly after Barack encouraged him, "Keep up the good work!" while offering an apology. "I'm sorry I haven't said more about Palestine right now," Abunimah recalled Barack telling him, "but we are in a tough primary race. I'm hoping when things calm down, I can be more up front."[52]

As a full-time candidate, Barack's schedule was filled with editorial board meetings and media interviews. He told the *Daily Herald* "that for lesbians and gays, it's not just the marriage issue, it's also the moral issue of recognition and acceptance," but he did not want to "shove the issue" forward rather than stress "the principle of non-discrimination." To WBEZ's Steve Edwards, Barack bragged about his basketball talent while admitting that "my coat is in shreds. It's an embarrassment to my wife." WBBM's weekly automated SurveyUSA poll showed Blair Hull with a significant 29 to 19 percent lead over Barack and Dan Hynes, thanks to 44 percent support downstate. Barack "has just 46 percent of black voters," *Capitol Fax*'s Rich Miller reported with amazement. That number would surely rise, and Barack's "new yard signs in black neighborhoods feature his photo." Barack "is the only candidate in this race with any real charisma, which could help give his campaign a late spark" once he started airing TV ads. But "Hull's coming deluge means that Hynes almost has to go negative on the guy," though "Hynes has limited funds. . . . If Hynes doesn't do something about Hull's ascension, then Hynes is toast."

By the second week of February, Barack's campaign was broadcasting several new radio ads. In one, Barack was the only voice:

This is state senator Barack Obama. I'm running for the United States Senate, and I want to tell you why. We're spending billions of dollars in Iraq while millions in our own country are out of work. Our health care system's collapsing. Our schools don't have the support they need. But too many in Congress have been afraid to stand up to George Bush and his destructive policies. Too many have let him have his way, and now our country and our community are paying the price. I'm a Democrat fighting for an America where anyone who wants to work can find a job, where everyone gets the health care they need, and every single child gets the chance they deserve. I want to make our voices heard in the United States Senate. This is Barack Obama and I approve this message because I hope you'll join me in this movement. Please register to vote. If you want to volunteer, call my headquarters at 312-427-6300. Together, we can make history and send a message of hope and change.

On black radio stations, Barack's campaign deployed a pair of new ads. In one, entitled "Champion," Operation PUSH's Rev. Willie Barrow had "an important message for our community. February 17th is the last day to register for the primary election. And on that primary ballot is a champion for our community . . . state senator Barack Obama . . . a husband, a father, and a leader in his church." Barrow signed off "urging you to join the movement." A second featured Congressman Jesse Jackson Jr.:

A generation ago, our parents marched, bled and died to secure our civil rights. But equal opportunity, it's still an unfulfilled promise. Now we can send a powerful message of change to Washington by electing state senator Barack Obama to the United States Senate. Obama has spent his life knocking down barriers. He made history as the first African American president of the Harvard Law Review, but instead of cashing in, he has chosen a life of public service. In Chicago, he led a registration drive that helped elect Bill Clinton and Carol Moseley Braun. In Springfield, Obama's passed laws extending health care to uninsured children, for the videotaping of police interrogations and new curbs on racial profiling. What a difference it would make to have Barack Obama in the United States Senate. He'll hold America to its promise for all of us.

"This is Barack Obama, and I approve this message to ask for your vote on March 16th." Jackson then urged listeners to "Join the movement! Barack Obama for Senate" before Barack said, "Paid for by Obama for Illinois Inc."

Barack's own staff was becoming aware that his prospects were looking up. "Things start to build, and it really did start to feel like a campaign and not a job," events coordinator Kaleshia Page recalled. Jim Cauley was pushing to have the funds to make as big a late TV buy as possible, but among the campaign's African American staffers, unspoken disquiet slowly increased. Barack had pressed Cauley not to allow the staff to become too white, but even so, strategy discussions with Barack's all-white, all-male team of consultants often left Audra Wilson and Pam Smith, the campaign's two senior black women, feeling slighted. "If this is considered the base, you've got to listen to us," Audra finally remarked. The disagreements reached a crescendo over whether the campaign should devote upward of $35,000 to billboard advertising aimed at African American drivers. Kevin Watson argued that it was essential to show Barack and Michelle in order to send the message that "this is a light-skinned African American marrying somebody darker." Barack agreed. "We need to do billboards," he told Cauley, but Jimmy resisted: "No we don't." "Jim, I'm just telling you, if they think I've got a white wife, the boogie shit sticks," Barack replied. "He'd just get up in my face," Jimmy remembered, but Axelrod and the other consultants sided with Cauley. "Barack, there are two people in this room who have won major African American campaigns, and you're not one of them," Joe McLean recounted Axelrod stating.

Similar tensions arose when top supporters Jim Reynolds, Marty Nesbitt, and Valerie Jarrett pressed Barack about why the campaign was not yet running television ads. "We've got to go up on TV, Jim. Even if it's a hundred grand, we've got to go up," Barack told Cauley. Again the consultants supported Cauley. "Barack is getting impatient," John Kupper recounted, and "we had to continue to persuade him. . . . 'We have a strategy here . . . we have to hold this money till the end, when it matters.'" Barack conceded on the television issue, but in the end Kevin Watson did prevail on billboards. "Jimmy was a hero," Kupper explained. "We needed somebody who was strong enough to stand up" to pressure, "and sometimes stand up to the candidate," and "to a very large extent, Jimmy did."

By February 16, Barack had been endorsed by two more important north suburban Democratic Party groups, drawing 78 percent support in New Trier Township and 60 percent in Northfield. David Axelrod scheduled video photo shoots with Barack, his family, Jesse Jackson Jr., and Jan Schakowsky at Marty Nesbitt's Kenwood home on February 17 as the timeline for producing Barack's initial TV spots closed in. Axelrod had a draft script in hand for their debut ad, but Barack hesitated after his first read-through, caviling at its final phrase, "Yes we can." "Is that too corny?" Barack asked. David explained his thinking, but Barack sought Michelle's reaction. "Miche, what do you think?" "Not corny," she replied, and the taping proceeded.

Meanwhile, attention in the race shifted to questions about new front-runner Blair Hull. In the *Chicago Tribune*'s February 15 profile, headlined "Political Novice Hull Uneasy in Spotlight," reporter David Mendell portrayed Hull's efforts to make himself into a political candidate. "I forgot to tell them I served in the army and taught school. I have to remember these things," Hull told himself after his endorsement interview at the *Rockford Register Star*. As an "introverted personality, Hull is anything but a natural," Mendell stated, despite Hull's investment of almost $24 million of his own money. Only at the end did Mendell mention that Hull, who had divorced his first wife of almost thirty years in 1994, had then *twice* married and been divorced from Brenda Sexton. Indeed, "in 1998, Sexton filed [for] an order of protection against him, public records show," Mendell revealed. Chris Mather from Dan Hynes's campaign later acknowledged that she had "worked for months to make that happen," but when Mendell asked Hull about what had occurred, he said only that the second divorce was "contentious," but that he and Sexton were now friends.

Two days later, *Chicago Sun-Times* gossip columnist Michael Sneed reported that Hull and Sexton had shown up together at an evening fund-raiser, news that did not surprise Hull's campaign staff, who believed they were weathering the storm. Then an Eric Zorn column in Thursday's *Tribune* headlined "Public Entitled to Know About Hull's History" quoted the candidate as saying, "I won't talk about it. I'm not available to talk about my personal life." Zorn argued that "Hull's candidacy has given us a right to know," and that evening on CBS Channel 2's local newscast, reporter Mike Flannery stated that an order of protection "usually requires evidence of domestic abuse." As with Mendell and Zorn, Hull had turned aside Flannery's questions too, even when Flannery probingly asked whether Hull had ever threatened Sexton with violence. "This really doesn't play a role in the Senate race" because "it's a private matter," Hull responded. Flannery got Sexton on the phone, who said, "I'll leave it in Blair's hands. . . . He's the one running for office." Flannery again suggestively asked about any threats of violence, but Sexton replied, "I've said all I'm going to say."[53]

David Mendell upped the pressure, reporting in Saturday's paper that Sexton had "obtained two orders of protection" against Hull. With top consultant Rick Ridder wondering "how long we can stonewall this," Hull volunteered what the one publicly obtainable document would indicate. "The police report from that night says that I hit her shin." Hull added, "I regret those months, and that night in particular, as well as the legal posturing that resulted. Most importantly, Brenda and I are now friends."

Barack's first TV ad was scheduled to begin running in the Chicago media market on Monday, February 23, a week earlier than initially planned, thanks to his campaign's better-than-expected fund-raising. Barack lost his greatest press champion when *Sun-Times* political columnist Steve Neal committed suicide at age fifty-four, but on Sunday, Barack received another highly valued union endorsement when Teamsters Local 705, whose twenty-one thousand members made it the state's largest, backed his candidacy. With new poll numbers landing almost daily, there now was no question that Blair Hull's massive paid media campaign had vaulted him into the lead. An automated *Daily Southtown* poll showed Hull with 27 percent, and Barack and Dan Hynes trailing badly with 17 percent apiece. A Roosevelt University poll had Hull at 27, Hynes at 21, and Barack at 14 percent, while the *Tribune* showed Hull with 24, Barack with 15, and Hynes with only 11. Hull's name recognition was now at 63 percent, Hynes at only 54, and Barack a dismal 32, although 34 percent of likely Democratic voters remained undecided.

Capitol Fax's Rich Miller warned that many top Democrats "worry that a Hull win could be a disaster this fall" if damaging information emerged post-primary. Like Rick Ridder, Hull media consultant Anita Dunn and campaign manager Mike Henry knew how widespread the rumors were about Hull's 1998 behavior toward Brenda Sexton. The trio had long believed, as Dunn put it, that "nobody would feel they had to use those against us unless they thought that Blair was a real threat to win the primary." But now "Blair had gone into the lead in public polling, so it was tick, tick, tick," and "people are going to come after us," Henry agreed.

A little before 3:00 P.M. on Sunday afternoon, February 22, Carbondale City Council member Sheila Simon arrived at the Akkineni family's home at 506 Midwest Club Parkway in suburban Oakbrook. Sheila's day job as a law teacher at Southern Illinois University had her in Chicagoland for a domestic violence conference, and weeks earlier David Axelrod had telephoned her to say, "We'd like to do something where you would talk about how your dad was going to endorse Barack." Sheila was noncommittal. "My hesitancy was I didn't want to get involved in saying what would my dad have done," and "David understood that," but in a second call, Axelrod asked, "How about you just talk about what your dad and Barack worked on?" Sheila agreed, and David sent her a draft script that Sheila was 99 percent happy with. "The language that David proposed was so really on target," and "the only thing I needed to change" was deleting one word calling her dad "excellent" because it felt "a little boastful for me to say that."

Roopa Akkineni was a good friend of Madhuri Kommareddi, the Northwestern senior who was doing most of the campaign scheduling, and Axelrod had a full crew plus a teleprompter on hand, awaiting Sheila's arrival. "You have great brows. I'm just removing the distractions," the makeup artist told Sheila. The taping "took a long time," because "we did it so many times," Sheila remembered. "David would say, 'Well, this time emphasize this word a little bit more,'" or "'be a little more cheery when you talk about the death penalty.' 'David, give me a break,'" Sheila responded, but by afternoon's end Barack's second TV ad was ready for final editing and production.

On Monday, Barack's first thirty-second spot, entitled "Yes," was broadcast. It began with Barack in a suit and tie directly addressing the camera. "They said an African American had never led the *Harvard Law Review*," as the video shifted first to a 1990 photo of Barack and then to an image of the February 6, 1990, *New York Times* story: "First Black Elected to Head Harvard's Law Review." Then back to Barack: "In the state Senate, they said we couldn't force insurance companies to cover routine mammograms, but we did," in a 2001 bill that won unanimous passage in the Senate and the House but for which Barack was not among its thirteen senatorial sponsors. Barack reappeared. "They said we couldn't find the money to cover uninsured children, or give tax relief to the working poor, or pass new laws to stop wrongful executions, but I have," as the video flashed through a trio of graphics citing to Barack's 2003 KidCare, EITC, and videotaping bills. Then Barack returned. "Now they say we can't change Washington? I'm Barack Obama. I'm running for the United States Senate, and I approve this message to say Yes We Can."[54]

On Monday evening, all of the Democratic Senate candidates assembled in Springfield for a second, one-hour radio debate. When Blair Hull called himself "an outspoken opponent" of the Iraq war, Barack objected. "The fact of the matter is, Blair, that you were silent when these decisions were being made. And you didn't end up being outspoken about it until well after the war had been completed, to the extent that it's been completed. . . . And now you're sending out mailings saying you were a strong opponent of the war. You were AWOL on this issue. And that's important . . . because Illinois voters deserve to know whether or not their U.S. senator is going to duck issues or whether he's going to be up front on issues."

Hull responded by saying that Joyce Washington, Nancy Skinner, and Gery Chico had also opposed the war, but Barack pressed his attack. "At what stage did you make a statement suggesting you were opposed to it? On your Web site, there was no mention of it. We made a joint appearance in front of the Champaign County Democrats in which you didn't utter a peep about it. I give credit to people like Dan Hynes and Maria [Pappas], who were consistent in saying they were supportive of it. . . . I disagree with them . . . but I admire the fact that they did take a consistent position on it." That backhanded compliment did Hynes no favor, but Barack's energetic critique gave Pappas an opening. "Excuse me, you want to talk about ducking issues, Mr. Obama. Where were you in Springfield when there were six prochoice votes called? . . . But you were not there to vote. Let's not talk about who ducked issues." Barack said he expected voters to look at "the totality of your record and your life," and after the debate, reporters sought further answers from Blair Hull. "Unfortunately, there are contentious divorces, and I don't believe this really affects the kind of senator I'm going to be," Hull said.

From Springfield, Barack headed south to Edwardsville and Marion. State senator Bill Haine, his retired predecessor Evelyn Bowles, and Sheila Simon all praised Barack at a series of campaign events. "Barack and my dad were cut from the same cloth," Sheila told a Williamson County crowd, using a phrase that tens of thousands of TV viewers would soon hear her use in Barack's second ad. Paul Simon "was a mentor and role model for me," Barack told reporters. "I can't say what he was going to say, but it had been scheduled. He was somebody who I think felt confident that I would carry on the work that he did."

In Chicago, the impact of Barack's first television ad was immediately apparent. An automated SurveyUSA poll of fifteen hundred Democratic voters conducted Sunday, Monday, and Tuesday, February 22–24, showed Barack jumping ahead of Blair Hull by 27 to 25 percent, with Dan Hynes at 18. Then SurveyUSA separated out the respondents by day. Among those polled on Sunday and Monday, Hull led Barack 32 to 23 percent, but on Tuesday, the numbers flipped dramatically, with Barack ahead of Hull 32 to 22 percent and now backed by 59 percent of African Americans. Privately, the Hull campaign's twice-weekly tracking polls told much the same story: Barack now had better than 50 percent support among Cook County white liberals, enough to win the primary with a 35 percent or better plurality. In Thursday morning's *Chicago Tribune,* metro columnist Eric Zorn kept up his attack on the now-fading former front-runner, asserting that Blair Hull has "made a mess of his first big test of character." At midmorning, Hull's campaign decided to reverse course, reluctantly issuing a pair of statements from Hull and Brenda Sexton announcing that they would jointly ask Cook County Circuit Court to immediately unseal their 1998 divorce file.

In the midst of the Hull firestorm, the *Chicago Sun-Times* on Friday made the first major newspaper endorsement in the race, praising Barack as "a rising star with impressive political skills and a keen intellect." *Sun-Times* gossip columnist Michael Sneed told readers that three months earlier, a young woman had been found dead from carbon monoxide poisoning in a property owned by Blair Hull, but that story passed almost unnoticed as all attention focused on the divorce file. Only on Thursday had Jason Erkes, Hull's latest communications director, become privy to Brenda Sexton's 1998 affidavit that Rick Ridder had read six months earlier. "Wow—this is sensational," Erkes realized, and he immediately began calling Chicago journalists, asking to speak to them off the record in advance of Friday's formal release of the affidavit. "There's some sensational words there," Erkes explained, and "it read much more sensational" than Hull's frank comments to his aides had led them to anticipate.

Brenda Sexton also told several of her closest friends about what was coming, and longtime Obama supporter Laura Hunter immediately phoned Dan Shomon, who was on the road with Barack northwest of Chicago. "Holy fuck, Blair is going down. He doesn't survive this," Laura told an astonished Shomon, who switched his cell to speakerphone. The most explosive revelation was one single four-letter word: Hull had called Sexton a "cunt" while allegedly threatening her with violence. Shomon immediately called David Axelrod, who was amazed by the specifics. "That's impossible. Nobody would ever run for Senate with that in their record." Barack had a more pressing concern. "Call Fred Lebed," he told Dan. "Find out if Hull's dropping out, because if he drops out, we're screwed." Six years earlier, in 1998, Barack had held off on backing Dan Hynes for comptroller when Lebed was mulling the race, and Shomon referenced that history as soon as Fred answered his phone. "I have a message for you from Barack Obama: You fucking owe us! You better keep Blair in this race!"[55]

"Barack's big concern was that the stuff on Hull was going to come out too soon," deputy campaign manager Nate Tamarin recalled. "He spent a lot of time worrying about when it was going to come out, and the timing of it and what it would mean for Hynes" if Hull's departure from the race allowed Hynes to run up a huge margin among white downstate voters. Michelle Obama put it similarly. "As the candidate, you know more about the potential stuff than the public does . . . so you go into this race knowing all your opponents' issues . . . whether their house is made of bricks or glass. The things that came to pass were completely known to everybody except the public. These were things we talked about at the outset," even if Sexton's specific allegations were more explosive than Axelrod, Barack, or most of Hull's own campaign team had anticipated.

In her March 12, 1998, affidavit, Brenda Sexton had declared, "Blair is a violent man with an ungovernable temper," and "I fear for my emotional and physical well-being." One night in September 1997, "Blair threw a remote control across the room and called me a 'fucking cunt.'" In early December, he had called her an "evil bitch" and "'a fucking cunt,' repeatedly," rhetorically asking "'Do you want to die? I am going to kill you, you fucking bitch.' I was deathly afraid of him." In early February, she had ordered Blair out of bed, and he "held one of my legs and punched me extremely hard in the left shin. After that, he swung at my face with his fists a couple of times in a menacing manner just missing me." She had called the police, who arrested Hull, but after Sexton obtained an initial order of protection, prosecutors dismissed it after reviewing the February police report detailing how she had first kicked Hull.

Friday evening's newscasts and Saturday morning's headlines were predictable, if demure on the specific alleged language: "Hull's Dirty Laundry on Line," blared the *Sun-Times*. The *Tribune*'s Eric Zorn was not alone in wondering why Hull had not made "a full and apologetic disclosure back last summer," but Hull told the suburban *Daily Herald* that he had raised that option with his children as well as Sexton. "We felt that the benefit of getting somebody like me in the United States Senate was so much greater" that the political dangers "paled in comparison, and I had to take that risk." The *Sun-Times* wrote that Sexton had dropped the protection order within forty-eight hours after a divorce agreement gave her "more than $3 million in cash and property," and the female friend who had first introduced her to Hull told the paper "it was all about money for Brenda."

Sunday morning the *Tribune* joined the *Sun-Times* in endorsing Barack as an "outstanding candidate" who "has a proven record of spirited, principled and effective leadership in the legislature." Even more important, by Sunday Barack's second TV ad, featuring Sheila Simon, had been on the air since Friday. Entitled "Mentor," the first two-thirds of the spot featured Sheila's unidentified voice and old footage of her father. "For half a century, Paul Simon stood for something very special in public life: integrity, principle, and a commitment to fight for those who most needed a voice. State senator Barack Obama is cut from that same cloth. With Paul Simon, Barack led the fight to stop wrongful executions and to pass new ethics and campaign finance laws to clean up our politics." Then Sheila appeared on camera for the final ten seconds. "I know Barack Obama will be a U.S. senator in the Paul Simon tradition. You see, Paul Simon was my dad." The FEC-required closing helped too. "I'm Barack Obama, and I proudly approve this message."

"It was really lovely the way they put it together," Sheila thought. Barack's consultants all realized the importance of the ad. It "may have been a more effective endorsement than had Paul done it himself, because of the emotion involved," John Kupper thought. Dan Shomon agreed, explaining that Barack "got something better: he got him speaking from the grave." Crucial to the spot's power was that Sheila "looks just like him," Kupper noted. Pete Giangreco understood "how important Paul Simon's brand was to the emerging Obama brand," as the two early January focus groups with white suburban women had demonstrated: "that's what they wanted: they wanted another Paul Simon." Young African American campaign aide Randon Gardley recalled that "at that point, he was a different kind of candidate. He was no longer a black candidate because he had Sheila Simon's face and images of her father."

Reporters praised Barack's first two ads. The suburban *Daily Herald* said the first one "highlights his accomplishments in Springfield in a quick-hit, easy-to-understand way that makes Obama seem like a candidate of accomplishment." *Capitol Fax*'s Rich Miller commended Barack's "well-produced TV ads," especially "his 'beyond the grave' TV endorsements by the sainted Paul Simon." For Barack, the depiction of him to tens of thousands of strangers produced a palpable change in his life. Just before the ads started, Barack had told Jim Cauley, "Don't leave me with any debt" on March 17. "That was his lesson from the congressional" race, Cauley realized: "'Don't wake up the day afterwards, if we should lose, and have me owing any money,'" Barack insisted. But the debut of Barack's first ad, followed just four days later by the affidavit that seemed to torpedo Hull's chance of victory, led Barack to realize that he might win. "It starts at that moment," Jim Cauley thought, when Laura Hunter first called Shomon and Barack with the Hull details, but as increasing numbers of passersby now recognized him, Barack began to grasp just what winning would entail. "All of a sudden folks recognize him," Nate Tamarin remembered, and "he said he couldn't go to the store without people noticing him," Cauley added. "He seemed stunned by it," and Jimmy was "amazed at his incredulousness." Cauley had come to think that Barack was "more naive than people would care to know," but "it was quite stunning to watch" Barack comprehend that he was now a public figure. "I've been running around this damn state for two years and ain't nobody known who I am, and now all of a sudden everybody knows who I am." Jimmy responded with sarcasm: "Well, it's called TV, son."[56]

The Hull campaign quickly conducted a pair of female focus groups that suggested Sexton's allegations were not fatal, although Hull feared they were. Rick Ridder, Anita Dunn, and Mike Henry met with Hull and his adult daughters to discuss whether to cut back or suspend the campaign. Dunn and Henry were reluctant to go forward, yet Hull's daughters felt strongly that their father should not drop out, and Hull agreed, believing that the additional money the campaign would spend would make no difference in his or his children's future lifestyles. But by early the next week, the Hull campaign's daily tracking polls showed Barack's support among both white educated Cook County Democrats and African Americans was above 60 percent. Even among non-Cook whites, Barack's support moved up to about 25 percent, and in overall statewide tallies, Barack moved upward to 36 percent, then 38, and then 40, with Hull and Hynes mired in the upper teens.

Paul Harstad completed his second tracking poll for Barack's campaign on March 1, and his numbers confirmed what Hull's pollsters were seeing. Barack's TV ads had dramatically increased both his name recognition and his positive-regard scores. Only 20 percent of black Democrats remained unfamiliar with him, and Joyce Washington's support among African Americans, once as high as 10 percent, had plummeted to just 1, while Maria Pappas had dropped from 19 to 7. Blair Hull's negative-regard tally had jumped from 6 to 26 percent, and black voters appeared to be moving to support Barack almost en masse. Sixth Ward alderman Freddrenna Lyle explained that once they learned his life story, "he made my seniors proud. They talked about that. He was everything that they had wanted for their children."

Maria Pappas, Dan Hynes, and even Gery Chico now joined Hull and Barack in running TV ads, with a new Hynes spot featuring his wife Christina, a physician. "I know Dan understands the problems of our health care system that I see every day," she said before an announcer stated that "Hynes has a detailed plan to lower health care costs, insure every child, and protect your insurance coverage when you change jobs." The ad closed with Dan saying, "I believe we can fix our broken health care system," but relations between Hynes's campaign staff and their media consultants were on the verge of breakup too, with observers believing that the quality of Hynes's excellent 2003 debut ad had not been equaled by subsequent spots. "It became a vacuum of guidance, and that was problematic," Hynes's communications director Chris Mather recalled. All the campaigns had to submit a March 1, preprimary FEC filing, and reporters noticed that while Barack still had $1,277,000 cash on hand even after laying out almost $1.8 million since January 1, Hynes's campaign had only $635,000 available. Hynes's downstate strategy meant that he had never budgeted for a late TV splurge in the Chicago media market, yet Hull's collapse now left Hynes unable to respond. "When that vacuum happened, I didn't have the money to go up on TV in Chicago," Dan recalled.

Barack's campaign had taken in $1,274,000 during the first eight weeks of 2004, with a bevy of $12,000 donors jumping out from Barack's FEC filing. Some were old friends like John Rogers Jr. and prominent Chicago donor Lewis Manilow, and three more members of the Crown family had each contributed the maximum. Texas junk-bond salesman John Gorman, whom Vernon Jordan had introduced to Barack, and his wife each donated $12,000, as did Sara Lee CEO John Bryan's wife Neville, financier Sam Zell's wife Helen, and Abby McCormick O'Neil's son Conor. Numerous friends like Bettylu Saltzman and Stephen Pugh fell just shy of the ceiling, and two business associates of Tony

Rezko, M. A. Dabbouseh and Joseph Aramanda, each kicked in $10,000. In Aramanda's case, the entire donation actually came from Rezko, another undisputed violation of federal law. The $10,000 apiece from Dabbouseh and Aramanda attracted no attention whatsoever, but a March 2 check in a similar amount from famed basketball star Michael Jordan, a friend of top Barack backer Peter Bynoe, led to a *Chicago Sun-Times* headline as well as an editorial.[57]

Barack continued to attract plaudits. Suburban *Daily Herald* columnist Burt Constable wrote that Barack "has the best qualifications and the worst name" among the Senate contenders. "Given a chance on the national scene, he'll soon be on a short list to become our nation's first African-American president." In the *Sun-Times,* African American columnist Mary Mitchell expressed disgust that former U.S. representative Cardiss Collins had joined Bobby Rush in endorsing Blair Hull. "When a black person has a real chance of going to the U.S. Senate, and a former Black Panther campaigns for the other guy, you have to wonder whether black politics has really made much progress." Mitchell noted that "a candidate who is embraced by many as a rising star on the political scene has to prove to some voters that he is not too black and to others that he is black enough."

In the wake of Hull's divorce allegations, journalists began asking why court documents concerning Republican Senate front-runner Jack Ryan's 1999 divorce from his ex-wife Jeri were also sealed. Ryan cited issues involving the couple's young son, but Jeri, an actress and former Miss Illinois better known for her roles in *Star Trek: Voyager* and *Boston Public* than for TV films like *Co-Ed Call Girl,* was even more of a celebrity than Jack. Behind the scenes, journalistic curiosity was piqued because Rod McCulloch, the campaign manager for Republican long shot John Borling, had quietly begun spreading the word about what the documents contained. Three years earlier, when Ryan had been considering a 2002 Senate run, McCulloch had spoken with him about that possibility. "What if my ex-wife comes to Illinois and accuses me of being a sexual deviant?" Ryan had asked a startled McCulloch. "What the hell does that mean?" McCulloch wondered, but Ryan pulled back. At the outset of the current race, Ryan had required his consultants and top staffers to sign confidentiality agreements, but one day McCulloch received a phone call from a consultant who asked, "Do you want to see what Jack Ryan's hiding?" McCulloch said yes, and "they come to my office . . . they lay it out and open it up to the pages in question, and there it is." Given the agreement, McCulloch was allowed to read quickly but not copy the documents, and he immediately called his candidate, who then discussed the situation with at least two other Republican contenders. A March 4 *Sun-Times* headline signaled what was going on: "Senate Rivals Urge Ryan to Unseal Divorce Records." Borling stated that "Jack

Ryan is going to have to own up with respect to some of those questions about sealed court records," and *Tribune* divorce specialist Eric Zorn took Ryan aside to ask whether Ryan had forced his ex-wife "to go to sex clubs." Ryan denied it, but when Sangamon County Republican Party chairman Irv Smith queried the candidate about rumors of "kinky sex," Ryan had responded: "How do you define 'kinky sex'?" Smith remembered. "He didn't deny anything."

On Thursday evening, March 4, all the Democratic candidates assembled for a televised debate at Chicago's WTTW. Everyone had been asked to arrive by 6:30 P.M. for the 7:00 P.M. live start, but Barack arrived several minutes late following a drive from Springfield. Dan Hynes reached over to smooth Barack's jacket collar just as host Phil Ponce began by immediately confronting Blair Hull about what Hull acknowledged was a "messy divorce" from a "financially motivated" spouse. Ponce followed up twice before asking Maria Pappas whether Hull should quit. Pappas demurred, saying that one of Hull's daughters had just spoken to her in the ladies' room in defense of her father. When Ponce asked Barack to address domestic violence, Barack cited his recently passed Victims' Economic Security and Safety Act. Then Ponce shifted to gay marriage, with Barack evading the first question. When Ponce followed up, Barack said, "we have to make sure that people have the rights that they deserve: to transfer property, to visit hospitals" before adding that "I'm very mindful of the moral implications of this—my parents were of different races" and could not have married in southern states, "so this is not something that is a distant abstraction for me. I recognize its importance." Commending 1960s civil rights proponents for prioritizing the landmark 1964 and 1965 statutes, "I'm not necessarily going to lead the issues with a repeal of the anti-miscegenation act because we are trying to consolidate rights for those people who need them right now."

When Ponce turned to the other candidates, only Gery Chico embraced gay marriage. "It is a basic matter of civil rights. It is the civil rights issue of our era. Separate but equal has been ruled unconstitutional a long time ago. To me, this is no different than giving women the right to vote in 1920." Ponce shifted to individual questions, and when Blair Hull cited his complete independence from donors, Barack chimed in. "Blair identifies a genuine problem" because "we have a system of financing campaigns in this country that tends to be corrupting, and I don't think there's any doubt about that." When Hull asserted that both Barack and Dan Hynes had accepted contributions from pharmaceutical interests, Hynes responded that "neither of us are beholden to special interests."

Ponce asked each of the candidates what book they currently were reading, and Barack immediately cited *The 2% Solution,* "a terrific book" whose sub-title, *Fixing America's Problems in Ways Liberals and Conservatives Can Love* foretold its argument that "American politics urgently needs . . . a new center." Author Matthew Miller decried that "our two major political parties are organized around ideologies and interest groups that systematically ban commonsense, well-funded policies that blend liberal and conservative ideas." In a less combative nation, "politicians might sit down quietly with thoughtful colleagues in the other party and . . . develop a bipartisan caucus that refines an agenda informed by [a] problem-solving . . . mind-set."

Seeking further personal insights, Ponce asked the candidates to describe the last time they had used public transit. Maria Pappas drew laughter by asking, "Does that include a cab?" Barack said several months earlier in New York City, and "probably last year some time" after Ponce specified Chicago. As the debate moved along, Blair Hull returned to the attack, accusing Barack and Dan Hynes of taking a total of $130,000 from the U.S.'s top-ten outsourcing companies. Barack responded, "I've got a track record of fighting on behalf of working people," but Dan Hynes angrily took offense. "This is the second time Blair Hull has suggested that Barack and I are beholden to special interests. I resent that. I mean I think that is offensive. You don't know me, you don't know Barack. I know Barack to be an honest person, and I know I'm an honest person. Just because we don't have millions of dollars, and we have to fund our campaigns through the generosity of others, does not give you a right to suggest that we are dishonest or beholden to anyone, and I think that's wrong."

Hull again said "we need to separate the money from the decision-making process. That's why we need public funding of campaigns." Barack cited working with Paul Simon to pass the 1998 campaign reform bill and then seconded Hynes. "We can't self-finance, and the reason Dan and I can't self-finance is because when we graduated from school, we made a decision, that my wife has to deal with and my two small children have to deal with, which was we were not going to pursue wealth because we were going to engage in public service." As the hour drew to a close, Ponce asked Barack why Bobby Rush and other black figures had not backed him, and Barack did not respond directly but again cited his "track record" regarding racial profiling and job opportunities for ex-offenders.[58]

Brenda Sexton did not like what she heard during the debate, and she issued a statement saying she was "shocked and hurt" by Hull's comments about her. Hull's campaign returned the fire with a press conference featuring Terry Forrest, the Sexton friend who had first introduced her to Hull, as well as Hull's three daughters and his former wife Kathy. Forrest recounted Sexton's comments to her in 1998 following the tussle with Hull. "It was all about 'I've got to get a lawyer and get some money.' I believed it was all an embellishment and a positioning in an effort to get more money."

With only nine days left, Dan Hynes spent the weekend campaigning downstate while Barack told the *Chicago Defender*, "We have enough money to stay on TV until the end of the campaign. The main task is to see that our organization is adequate to capture the enthusiasm we see on the ground." For months Jim Cauley, Nate Tamarin, and Kevin Watson had been preparing an extensive field operation for the run-up to Election Day. Staffers like David LeBreton were organizing phone banks and get-out-the-vote operations in collar counties like Lake, McHenry, and DuPage, taking advantage of the scores of volunteers who were offering their free time to Barack's campaign.

An even more extensive Election Day operation was planned for the city of Chicago and suburban African American areas. Michelle's former colleague Kevin Thompson had joined the campaign to handle both the North Side and outreach to gay and lesbian supporters, but it was the three offices in the largely black West and South Sides and south suburbs that saw the most activity. On the West Side, Barack was astounded to learn that longtime nemesis Rickey Hendon's significant organization was going all-out on his behalf. "Hendon got so enthusiastically involved that it was a puzzle to Barack," Emil Jones Jr. remembered. "Barack called me and said, 'What did you do to Hendon? He's so overwhelmingly involved. How'd you get him?' I jokingly said to Barack, 'I made him an offer, and you don't want to know.' That's how it went down."

Jones also put the hammer down on lobbyists who had no choice but to back Dan Hynes but who still needed to stay in the Senate president's good graces. As the race entered the home stretch, Jones summoned "some of Hynes's financial backers" and told them, "'I want you to raise me some money for Barack.' Of course they would say to me, 'You know we're with Dan Hynes.' I say, 'I know that, but that doesn't preclude you from raising some money for Barack.' And they would go out there and raise me the money." One young Irish lobbyist experienced "Emil's old school, brass-knuckle politics" at Petros Diner in the Loop. "'You need to get off of the Hynes bandwagon and come with Obama, and don't do it because you want to, do it because I'm asking you. Don't do it for Barack, do it for me,'" Jones told him. Barack's increasing lead in the polls lent further weight to Jones's demand. "This is going to happen. You've got to be with this guy. . . . We've got to do this for him. . . . If you needed something from him, I would ask him to do the same thing for you, and he would do it, not for you but for me.'" As the young lobbyist realized, "Emil was holding all the cards at the time as Senate president," and that gave Jones decisive leverage over many Springfield players. "He did that, I think, with a lot of people that Barack didn't even know about," as Emil "just totally took it upon himself" to do it. "That's the power of the presidency," Jones proudly but matter-of-factly explained.

The near-disaster of Barack being late for the debate caused David Axelrod and Jim Cauley to insist on greater discipline in Barack's travel. With Lyndell Luster unable to drive Barack full-time, Kevin Watson recruited Mike Signator, a forty-six-year-old south suburban police officer, as another driver, and Tom Crane, a supportive donor who owned Tri-State Auto Auction, provided a Denali SUV. In several South Side wards, Watson could count on Al Kindle to run Barack's Election Day ground game, but to cover the balance of the huge South Side, Jim Cauley reached out to David Axelrod's Washington-based former partner, Tom Lindenfeld, who was well known to national Democrats as the best GOTV operative in the country. "Get your ass out here," Cauley told Lindenfeld, and with $150,000 budgeted for Election Day efforts, Lindenfeld was able to bring several dozen experienced operatives from around the country for the campaign's final week. His top deputy, Greg Naylor, was from eastern Pennsylvania, where a local magazine said he "has attained near-mythic status in Philadelphia politics." Lindenfeld housed his crew downtown, and first he established a working relationship with Kindle and other South Side honchos. "Al was very helpful," Naylor recalled, and Lindenfeld began implementing his ground game. A large parking lot on the west side of Stony Island Avenue at 76th Street, across from Jackson Park Hospital, was secured, and Lindenfeld began renting scores of fifteen-person vans. On March 16, the vans would spread out across the South Side with paid day workers going block by block, knocking on doors, and if no one was home, leaving an Obama door hanger on the knob.[59]

On Sunday, March 7, Barack completed his Chicagoland newspaper trifecta when the suburban *Daily Herald* joined the *Sun-Times* and *Tribune* in endorsing him. While praising Dan Hynes as "a sincere, thoughtful candidate," the *Herald* was bowled over by Barack's "evident sense of decency and justice" as well as "his intellect, confidence, ease in accepting challenges and compassion. Very few candidates for public office have impressed us in this way. Paul Simon comes to mind." The *Joliet Herald-News* joined the Bloomington *Pantagraph* in endorsing Dan Hynes, and the *Aurora Beacon News* surprisingly backed Gery Chico. That Sunday Barack also received the support of the *Peoria Journal Star* and the *St. Louis Post-Dispatch,* which many in Metro East read.

A new *Tribune* poll showed Barack leading Dan Hynes 33 to 19 percent, with Blair Hull dropping to 16. Sixty-two percent of African Americans now supported Barack, and in Paul Harstad's third tracking poll for Barack's campaign, the numbers were even better: Barack led Hynes and Hull 36–21–13, with 67 percent of black Democrats backing Barack and only 12 percent still unfamiliar with him. Two-thirds of downstate voters still did not know Barack, but Blair Hull's negative rating had risen to 30 percent statewide. Hull's own tracking polls were even more dire: Hull still led downstate with 35 percent, but Barack's black support was more than 70 percent, boosting his statewide share to better than 40.

Campaigns' direct-mail efforts usually flew beneath journalists' radar, but the Hull campaign had three new mail pieces out. One featured Hull's issue stances and said the divorce allegations were "only about money . . . Brenda wanted a ten million dollar settlement." The two others targeted Barack. One rebutted Barack's claim that "he ALONE opposed Bush's war in Iraq," quoting the late Steve Neal calling Hull "an early and outspoken critic" of the war. The second lambasted Barack's "present" votes on antiabortion bills. "When Barack Obama had the chance, he refused to take a position on a woman's right to choose . . . seven times!" That flyer then stated inaccurately that those votes involved "legislation pushed by pro-choice extremists," but the *Tribune*'s Eric Zorn quickly knocked down the Hull campaign's misrepresentation of what a "present" vote meant. Illinois Planned Parenthood's Pam Sutherland explained that "the idea is to recruit a group to vote 'present'" that would include "Democrats who were shaky on the issue in an effort to convince them not to vote 'yes.'" Barack stated that "no one was more active to beat back those bills than I was."

Barack's direct-mail program, handled by Pete Giangreco and Terry Walsh's Strategy Group, featured different pieces targeted at different voters. The first flyer, aimed at white progressives, featured Pete's "Finally, a chance to believe again" phrase and quoted the *Sun Times*' report that Paul Simon had viewed Barack as a "rising star." Inside, headlines described Barack's "courage," "unmatched qualifications," and "real record." A second piece featured a photo of Paul Simon and the headline "Illinois Had a Senator Who Stood *for Something Special*." Only upon opening it did the recipient realize it was promoting Barack, as the left page featured a photo of him and Simon in early 2003. On the right was a photo of Sheila Simon, and her quote that "I know Barack Obama will be a U.S. Senator in the Paul Simon tradition." A third flyer, designed for black voters, stated "Our Cause Is Progress. Our Voice Is Barack Obama." His bills "requiring videotaping" and "ending racial profiling" were highlighted, as was his 1992 Project VOTE! work that "helped elect Carol Moseley Braun and Bill Clinton." A family photo caption said the Obamas "are active members of Trinity United Church of Christ," and another tag line said, "On March 16, we have the power to make Barack Obama the only African-American in the U.S. Senate."

Barack's urban radio ads also ramped up. In a sixty-second one entitled "Faith," Rev. Jeremiah Wright declared, "now more than ever, we need those rare leaders who will stand up for what's right, what's just and good. Barack Obama is such a leader." In the state Senate, Barack had "led the fight to bring health care to our children, tax relief for low-income workers, and new laws to stop racial profiling and the execution of the wrongly accused," Wright stated. "I've known Barack as a parishioner in my church. I've seen his devotion to his wife and two young daughters. I've been witness to his faith. Imagine what it would mean for our state and our nation to have Barack Obama bring that sense of moral mission to the floor of the U.S. Senate!" An announcer recited the closing refrain: "Join the movement. On March 16th, vote Barack Obama, Democrat, for U.S. Senate."

Another, entitled "Assault," featured two of black Chicago's best-known lawyers. "This is attorney R. Eugene Pincham with an urgent message for our community. From the Bush White House to the local courthouse, our rights are under assault, but we can fight back by electing Barack Obama to the United States Senate. In the state Senate, Barack Obama led the battle for a law requiring the videotaping of interrogations in murder cases to stop the railroading of the innocent. Obama will bring a voice to Washington they need to hear, a voice for change, hope, and justice." Then "This is James Montgomery, city corporation counsel under Mayor Harold Washington. And like Mayor Washington, Barack Obama's a fighter for fairness" who "passed a new law to stop racial profiling by police," was "the first African American president of the *Harvard Law Review,*" and "is a constitutional law professor at the University of Chicago." A female announcer ended with the now-familiar phrase: "Join the movement. On March 16th, vote Barack Obama."[60]

As Barack's poll numbers strengthened across the board, campaign contributions skyrocketed, including big dollars from out-of-state Democratic donors like billionaire George Soros. Marty Nesbitt, Jim Reynolds, Peter Bynoe, and John Rogers remained the mainstays of Barack's rock-solid support in Chicago's black business community, but Valerie Jarrett also became increasingly involved in the campaign. "A week, two weeks before primary election night, then she became a constant in my life," one finance staffer recalled. One finance committee colleague believed that once Valerie "realized that this could happen, she got more involved." Another perceptive staffer thought that with Jarrett, it was "the mayor releasing her to be more engaged."

As money poured in, Jim Cauley and the consultants realized that they could expand the television advertising to downstate markets like the Quad Cities; Paducah, Kentucky; and even St. Louis. The *Moline Dispatch* endorsed Dan Hynes, saying it wished Barack "had more first-hand knowledge of the Quad-Cities area," but Barack racked up endorsements from several African American Chicago weeklies and from a suburban chain of more than two dozen local papers. Those publications praised Barack's "reputation for intelligence and effectiveness," and a few days later both the *Rockford Register Star* and the *Chicago Defender* added their backing. Barack's "exemplary personal qualifications" plus his "character and demonstrated ability as a legislator make it a privilege to endorse him," a front-page *Defender* editorial declared.

The *Chicago Tribune* zinged Barack with a story detailing how his Senate staff had mailed out more than seventy thousand copies of a "Legislative Update" newsletter to constituents just forty-eight hours before a deadline banned such state-funded preelection advertisements. In a telling reflection of how the race had turned, *Capitol Fax*'s Rich Miller noted "persistent rumors that Mayor Daley's folks may be leaning towards Obama now." Daley's inner circle was "extremely paranoid" and "Obama is doing so well in the polls and on the campaign trail that he is being seen as a threat to hizzoner's hold on the city and Cook County. Better to send him to D.C. than have him looking for another office to seek if he loses."

Wednesday night, March 10, was the campaign's final debate, with now-underdog Dan Hynes contrasting his own role as state comptroller to bills Barack had backed in the state Senate. "When George Ryan and the General Assembly refused to curb their pork spending, I froze the pork. I often stood alone—that's what leaders do," Hynes declared. "Barack Obama chose a different course. He stayed silent. He didn't do anything." Hynes asserted that "for four years there was a feeding frenzy in Springfield," and "Barack Obama played a part in it." Instead of a direct rebuttal, Barack said, "Dan has been a fine comptroller, but the role of the comptroller is to stand on the sidelines and comment." Only afterward did Barack complain that "it's a little disingenuous" for Hynes "to suggest he somehow was this warrior" during Springfield's spendthrift years.

That evening Barack's campaign got an additional boost when the League of Conservation Voters (LCV) launched its own thirty-second television ad that endorsed his candidacy. Entitled "Rising Star," it featured LCV president Deb Callahan praising Barack for one bill he had introduced and a second he cosponsored at the behest of the Sierra Club, neither of which ever reached the floor. "There's one candidate for Senate who's won the tough fights for a cleaner and healthier Illinois—Barack Obama," Callahan declared. "Obama took on the polluters and President Bush for cleaner power plants. And he led the fight to reduce Illinois's record rates of asthma. The *Sun-Times* calls him a rising star, and the *Tribune* endorses him too. And the League of Conservation Voters is responsible for this message because we believe Barack Obama is the best candidate to protect our environment."

The League announced it was investing more than $100,000 into airing the ad, and Barack's campaign now had out its third and final ad, entitled "Hope." Images of both Paul Simon and Harold Washington invoked two strong legacies. "There have been moments in our history when hope defeated cynicism, and the power of people triumphed over money and machines. The *Tribune* says Barack Obama rises above the field with 'a proven record of spirited, principled and effective leadership.' The *Sun-Times* calls Obama 'the man for this time . . . and place.' The *Daily Herald* backs Obama for his 'sense of decency and justice.' On Tuesday, let's make this moment count: Barack Obama for the

U.S. Senate." Barack uttered only the required closing: "I'm Barack Obama, and I approve this message."

Barack's campaign also produced its last flyer, which cited the *Daily Herald, St. Louis Post-Dispatch,* and *Tribune* endorsements. "We need to rise above partisan politics and powerful interests," and "Barack Obama says Yes We Can!" it declared. "Barack Obama offers Illinois hope, opportunity, and a chance to believe again." Little came of an Associated Press query asking Barack to detail his history of drug use. Reporters were excited by Hull's admission of extensive cocaine use as an adult, and Barack said he had used cocaine "a few times" and "primarily smoked pot" during high school and his first two years of college. "Drugs and divorces. That's what it has turned into. It's very depressing," Barack told the AP.

In Thursday's *Chicago Defender,* Emil Jones Jr. called on all African Americans to unite behind Barack. "It's important for our children, our nation, and it disturbs me that you've got so-called black elected officials," like Bobby Rush, not supporting Barack. "I'm calling on the community to repudiate those individuals and vote strongly for Barack Obama." Jones especially called out five black state representatives who, like Speaker Madigan, were backing Dan Hynes. "They are all concerned about Tom Hynes' son. Well, Barack Obama is my son. . . . When do we get a chance? When are they going to make up their mind? He's been endorsed by practically every major newspaper in this state, and he's the most qualified. We need to integrate the U.S. Senate. There's not a black face in the U.S. Senate, and I think a black male going to the Senate will inspire a lot of young black people in this state and across the nation with some pride and hope. . . . I'm calling on the people in those respective districts and wards to come out strong and support the most qualified person for the office and that is Barack Obama."

Axelrod warned Barack's campaign staff that "it's important that we have balance in our schedule for the final few days. I know we have to get our base going, but we really have TWO bases, and we can ill-afford stories about how we are concentrating only in the black community." They should consider "what pictures do we want?" on Sunday and Monday's local TV news programs. "We need events in a variety of communities each day to reflect the broad coalition we are building," Axelrod stressed. "If the story is merely us trying to rally the black community, it doesn't help us." That night, a South Side campaign rally at Liberty Baptist Church drew a largely black crowd of more than fifteen hundred. "Nothing like this has happened to us since Harold Washington," U.S. representative Danny Davis proclaimed.[61]

Just a few hours before that rally, Rod McCulloch, John Borling's campaign manager, went public about the Jack Ryan divorce records he had seen a week earlier. McCulloch had spoken with state treasurer and Republican Party chair

Judy Baar Topinka, who told him: "If you believe what you saw is true, you have to go forward." In a written statement faxed to news organizations from a suburban Kinko's, McCulloch stated that the documents "have nothing to do with his son" but instead detailed ex-wife Jeri's allegations that Jack had had an affair with a colleague and "took her to various sex clubs" in New York, New Orleans, and Paris. McCulloch told the *Sun-Times,* "My conscience tells me that this must be aired before the primary next Tuesday" and that "my motivation is limited to my belief that voters deserve full disclosure."

Friday's newspapers all carried the story, omitting the most lascivious details, but Barack's campaign quickly learned the contents of McCulloch's statement. "Everybody's hearing for the first time that he's taken his wife to swingers' clubs, and they're saying, 'This is the thing that's going to come out,'" and that it would be "a nail in the coffin," one young volunteer recalled. "What I distinctly remember is me not knowing what a swingers' club is, and everybody's like, 'Oooh.' I remember feeling like a little kid at the dinner table" when she asked "'What's a swingers club?' . . . and I remember Obama saying, 'We'll tell you after the meeting.'" A *Tribune* editorial revealed that four days earlier the paper had petitioned the court to unseal the records. Noting how one prior filing suggested the documents might contain "inflammatory, inappropriate and embarrassing material," the *Tribune* called upon Ryan to "share it with voters now," because "this issue won't disappear" if he won the Republican nomination.

Ryan refused opponents' requests that a neutral figure like former governor Jim Edgar privately review the file and reassure Illinois Republicans. Sunday's newspapers featured poll results showing Ryan and Barack leading their respective races, with Barack at more than double Dan Hynes's 18 percent. *Tribune* reporter Rick Pearson upbraided Barack's opponents for not highlighting his missed gun vote in 1999, but a whirlwind day awaited the candidate. A trio of Far South Side church visits preceded an appearance at the Southwest Side's St. Patrick's Day Parade. With Hynes supporters in charge, Barack's cohort had been relegated to the rear of the parade, but Barack quickly improved his place upon seeing Senate Republican colleague Dave Sullivan in the front rank. "Barack put his arm around me and said, 'Okay, Sully, let's go!'" Sullivan was amazed at the onlookers who cheered Barack. "This election's over. Barack's going to win," Dave told his wife.

From there, Barack hopped a chartered two-engine plane to fly first to Peoria and then East St. Louis for more appearances at African American churches. Aware of how many American politicians had died in single-engine plane crashes, Barack had always refused free trips to or from Springfield on the state aircraft available to the governor and the Four Tops. When Barack returned to Chicago late that evening, he met with Jim Cauley, who had just seen new polling numbers showing Barack's support surging close to 50 percent thanks to a big jump among downstate voters now seeing his TV ads. Cauley was astonished. Given Barack's name, "I thought he had a ceiling," and "I thought 45 would be a huge number for him," Cauley explained. "Barack, I don't think you can break 50," Cauley told his candidate. "Oh yes I can," Barack replied. "Before that, it was me and him. After that, he became Barack," Cauley remembered.

For months Cauley had headed to Garrett Ripley's bar at 712 North Clark Street, close to his Chicago apartment. Field director Nate Tamarin often joined him, as did SEIU politico Jerry Morrison, consultant Pete Giangreco, young Obama campaign staffers, and occasionally Mike Henry, Blair Hull's campaign manager. "That was Jimmy's sort of home away from home," Pete explained. "When I think of Jim, I think of Garrett Ripley's," volunteer coordinator Adam Stolorow remembered. "Late nights with Jim and Nate." As Tuesday neared, all the regulars wrote down their Senate race predictions, with the most optimistic participant giving Barack a 7-point win over Dan Hynes.

Hynes was worried. "Obama had all the buzz" and "the trends were very, very disturbing." Barack's "numbers were moving, Blair Hull's were plummeting, and mine were flatlined," Hynes remembered. Monday morning his concern intensified. "I opened the newspaper and looked at the picture from the St. Patrick's Day parade. I mean, St. Patrick's Day, that's my day. And there was Barack Obama surrounded by every single Irish politician in town. I'm cropped out of the picture. And I thought to myself, 'That's not good!'"

In a room at the Schwab Rehabilitation Center, eighty-five-year-old columnist Vernon Jarrett filled out his absentee ballot while struggling to recover from major cancer surgery. His son Bill, the former husband of Barack and Michelle's friend Valerie Jarrett, had died ten years earlier at age forty. No one had championed Harold Washington more fervently during the 1980s than Vernon Jarrett. In Monday's *Chicago Defender,* in the last column he would write before dying several weeks later, Jarrett invoked the cherished mayor's legacy to instruct black Chicago to embrace Barack. "Obama is one of the highest qualified young men that this country can present regardless of race, creed or color. He is a superior Harold Washington at an early age. All those fine qualities that we have saluted in Harold Washington are magnified in Obama, including his intellectual, academic and moral conduct," Jarrett wrote. "I am here to tell you that you have another Harold Washington in Obama. Only he's younger, brighter and equally committed."[62]

Barack told reporters "we are in the position to potentially win this thing," and Monday morning he began by greeting potential voters at the South Side's 87th Street L station. Tom Lindenfeld and Greg Naylor's robust Election Day effort aimed to have more than two thousand paid workers hit the streets on Barack's behalf, while Dan Hynes's campaign boasted that it would deploy at least eight thousand experienced volunteers. As any politically experienced Chicagoan knew, Election Day was a homeless person's favorite day of the year, because anyone able to walk could win a cash return for a day's work from one or another campaign. Blair Hull's operation had signed up fifty homeless black men at a Bronzeville shelter, promising $75 plus meals on Tuesday.

On the West Side, Barack's campaign had quietly channeled tens of thousands of dollars to ward-level political organizations to help mobilize black voters there. In total, several hundred thousand dollars would be devoted to maximizing the African American percentage of the overall Democratic vote, an expenditure Jim Cauley and Tom Lindenfeld had persuaded David Axelrod, Pete Giangreco, and the other media-focused consultants to agree to at a meeting with top finance committee members Valerie Jarrett and Marty Nesbitt. Cauley remembered that after eight months of fiscal discipline, all of a sudden "I was spending money like a drunken sailor," and most of all "a fortune on the turnout operation." On the vast South Side, Tom Lindenfeld's massive GOTV effort was logistically mind-boggling. In addition to the 175 or so fifteen-passenger rental vans that on Tuesday would fan out across scores of neighborhoods, port-o-potties and a big RV command center supported the all-outdoors operation. With temperatures below freezing and snow in the forecast, ponchos and hand warmers for the army of day workers were last-minute additions to a budget that also included thousands of dollars in fast-food coupons. On Monday night, Lindenfeld and his out-of-town veterans struggled to stock the vans with campaign literature and other supplies, and as midnight approached, Lindenfeld realized, "I don't have enough people on the lot to be able to get the work done by 8:00 A.M." Wandering out onto Stony Island Avenue, he accosted a homeless man. "Want to make $20?" met with eager acceptance. "'You got some other buddies who want to make some money?' 'Yeah, but we've got to go get them.'" So "we drive to an underpass," and "he jumps out of the van, he disappears into the dark, he comes back with different people," more than enough people to fill the fifteen-person van. Working through the early-morning hours, Lindenfeld's homeless army got the vans fully stocked, but Tom needed additional cash to pay his recruits. "There are no ATMs where there's poor people," so Tom drove northward to Hyde Park to get the cash. "That was just part of life on the South Side."

As Tuesday dawned, Lindenfeld's operation swung into action. The van drivers, all recruited from among South Side Obama supporters, quickly trained and then supervised the day workers assigned to their vehicle. The canvassers would do a midday sweep, then one in the late afternoon, and finally a third one just before the polls closed. Anyone needing a ride would be picked up by another vehicle. Only at the end of the day, when the vans reported back to base, would the drivers receive their $100 in cash and the other workers $75 each.

It was a hectic as well as expensive day, with multiple sites, plus Al Kindle's similar effort and the various West Side operations. Operations director Madhuri Kommareddi oversaw the funding for the Election Day effort. With tens of thousands of dollars in cash on hand, Madhuri, a Lindenfeld aide, and an armed bodyguard worked out of Valerie Jarrett's ninth-floor lakefront apartment at 4950 South Chicago Beach Drive. "We were doing the cash distribution that would go to the different sites out of Valerie's apartment," Madhuri explained. Jarrett had advanced $8,160 for Lindenfeld's troops' hand warmers, plus an additional $16,869 for more van rentals, but as the polls closed and Barack's top supporters gathered in the ballroom at the downtown Hyatt Regency, both Lindenfeld's command post and Al Kindle's South Side office made emergency requests for more cash to cope with higher-than-expected costs. Frantic phone calls were made, with Lindenfeld telling Cauley that chaos would erupt if another $10,000 was not delivered to his site. "They were just being dicks," Tom thought, and although a quick delivery did arrive, a similar crisis also played out at 54th and Wentworth. Kindle's "was a mob scene," Greg Naylor recalled, but "some additional cash" was delivered there too, and the hundreds of workers all headed home happy.[63]

At the Hyatt Regency, Barack's consultants had carefully planned the evening's schedule. Former Democratic National Committee chairman David Wilhelm, Emil Jones Jr., Rev. Jesse Jackson Sr., and U.S. representatives Jesse Jackson Jr., Jan Schakowsky, Lane Evans, and Danny Davis would speak to the crowd before Barack made his first appearance, hopefully a little before 10:00 P.M. David Axelrod arranged for camera crews to film the event for use in general election ads, but strict orders were issued about who should appear in camera shots with Barack and his family members and who must not. Sheila Simon and Rev. Jeremiah Wright joined Barack, Michelle, their daughters, Maya Soetoro, and close family friend Kaye Wilson onstage, but Axelrod was adamant that Jackson Sr. not appear alongside Barack. "The one thing Ax didn't want was him in the shot," so there were "*explicit* orders to tackle Jesse Jackson if he got near the stage," Jim Cauley recounted. "There's no less than ten staffers that were told to take him out, and Valerie Jarrett was assigned to stand next to him and hold his hand and not let him on the stage."

Anticipating victory, Barack had spent a good portion of the day committing to memory his prepared remarks once he and Michelle, accompanied by their daughters, had voted early that morning in Hyde Park. While Dan Shomon hosted a Springfield party at Floyd's Thirst Parlor, Barack's campaign staffers from around Chicagoland headed for the Hyatt as soon as their Election Day tasks ended. David LeBreton, who had been coordinating phone bank efforts in a trio of collar counties, asked his parents to meet him at the Hyatt with a clean pair of socks so that the pair he had worn nonstop for the previous three days could go into the men's room trash can.

Amid the hustle of election night, one donor made demands. "Tony Rezko wanted a suite" there at the Hyatt, Jim Cauley remembered, but the hotel was sold out. Rezko called finance assistant Kaleshia Page to reiterate his request, but Kaleshia said no. Then her phone rang again: "Find him a room," Barack ordered. "'Kaleshia, I know you're busy, but see what you can do.' So I screamed at a whole bunch of hotel people and found him a room," Kaleshia recalled. "I might have handed him the keys." On what would be one of the most momentous nights of his life, Barack took the time to make sure his aides did everything they could to make Tony Rezko happy.

Polls closed at 7:00 P.M., but none of the campaigns were prepared for as quick a verdict as TV newscasters delivered. "It was over in the blink of an eye," a dejected Dan Hynes remembered. "Three minutes after the polls closed. That was a bit disappointing." Chris Mather, Hynes's communications director, got worried when Barack's TV ads went up on downstate stations, but on Election Day, "I thought we still had a good chance to win." But in the hour after 7:00 P.M., Mather's reaction was just like her candidate's: "I can't believe what I'm seeing."

At the Hyatt, Barack's warm-up speakers paused when first Blair Hull publicly conceded the race and called Barack shortly before 9:00 P.M., and then Hynes did the same twenty minutes later. "All I can say is I will never read the newspapers the same way again," Hull told reporters, while a magnanimous Hynes praised Barack. "He ran a great campaign. He's an unbelievably talented individual." Other congratulatory calls came in, including one from presumptive Democratic presidential nominee John Kerry. As 10:00 P.M. approached, Sheila Simon presented Barack with one of her father's trademark bow ties as everyone prepared for Barack to speak to the big crowd that had gathered. The numbers on the TV newscasts were astonishing: Barack was winning a *majority* of the vote, with projections showing him winning as much as 53 percent, despite the presence of six other candidates on the ballot. In the corridor outside the ballroom, Barack encountered pollster Paul Harstad. "Did you poll this?" Barack asked in amazement. "I told David 47 percent," Paul responded. "Thank you so much. I really appreciate it. I really do," Barack replied.

Following Sheila Simon's introduction, Barack took the stage at 10:14 P.M. "Sixteen months ago"—actually fourteen—"I stood in the Allegro Hotel with some of you and announced my candidacy for the United States Senate. And although the announcement was respectfully received, I think it's fair to say that the conventional wisdom was we could not win. We didn't have enough money, we didn't have enough organization. There was no way that a skinny guy from the South Side with a funny name like Barack Obama could ever win a statewide race." Barack took a long pause. "Sixteen"—again—"months later we are here as Democrats from all across Illinois to say, 'Yes we can. Yes we can. Yes we can,'" as the crowd joined in, picking up the chant. "We are all connected as one people," Barack stated, and victory must result "in some concrete good for ordinary people out there. . . . That is our end goal. It's not just winning an election, it's making a change in this country." Saying, "I am fired up," Barack called for "a politics of hope," invoked "the audacity of hope," and declared that "I am absolutely positive that we can change the course of this country."

A few weeks later, Barack said the results "surprised everybody but me," yet several months further on he stated, "I was as surprised as anybody in terms of the margin of victory. . . . I didn't expect to win as well as we did." David Axelrod told Barack that "Harold is smiling down on us tonight," and Axelrod stressed to Chicago newsmen that "one of the great disciplines of the campaign was not to spend money early and waste those resources." Some observers were surprised that Barack was not more visibly excited, and Michelle told one columnist, "he's basically a calm guy. It takes a lot to push his buttons. He has incredibly low blood pressure." WBBM's late-evening newscast told viewers that "the size of his triumph in the suburbs is truly historic," and WMAQ's anchor said, "it's as if once everyone realized just who Barack Obama was, that's the person they tended to turn to." At the Hyatt, David Wilhelm told local stations that Barack "ran an unfailingly positive campaign where he told a story about himself, he told a story about his experience, he told a story about his track record."[64]

Barack and his family spent the night at the Hyatt so he was more easily available for early Wednesday morning television news shows. Most of Barack's campaign staff went to Garrett Ripley's, where hardworking intern Alex Okrent "literally falls asleep at the bar," student volunteer Tarak Shah remembered. "You couldn't rouse him." Wednesday's *Tribune* declared Barack's 52.8 percent victory a "landslide." With 655,923 votes statewide, he easily outdistanced second-place Dan Hynes, who ended up with 23.7 percent. Blair Hull finished third, with 10.8 percent, having spent more than $200 for each vote he received. Maria Pappas, Gery Chico, Nancy Skinner, and Joyce Washington all trailed badly, receiving 6.0, 4.3, 1.3, and 1.1. percent, respectively.

One national reporter predicted that "Obama is certain to be an overnight sensation in national Democratic circles," and *Sun-Times* columnist Mark Brown declared that "Obama has the potential to be the most significant political figure Illinois has sent to Washington since Abraham Lincoln." Retired jurist R. Eugene Pincham told one news service that Barack "has this tone that distinguishes him from the typical black, and yet he's black." A lady from Chicago's 29th Ward put it differently to the *Tribune*: "for lack of a better word, it's like he's multi-colored." Most Illinois commentators focused on the remarkable breadth of Barack's support. In Chicago, black voter turnout was 41.4 percent, the highest in a Democratic primary since Carol Moseley Braun's historic 1992 win. In primarily black wards, Barack received 88 percent of the vote, plus 78 percent in predominantly black areas of suburban Cook County. In county board president John Stroger's 8th Ward, where the Democratic organization backed Dan Hynes, Barack received 15,684 votes—more than 91 percent—and Hynes just 649. Overall, Barack won forty-one of the city's fifty wards and received more than 66.5 percent of the vote. Dan Hynes carried only five wards, and Gery Chico four, but even in the Hynes family's home 19th Ward, Barack received more than 41 percent of the vote.

Beyond Chicago's city limits, Barack won 89 percent of the vote in Evanston and 87 percent in Oak Park. He carried both Cook County and Terry Link's Lake County with more than 64 percent of the vote, and also won the other four collar counties, including a 57 percent majority in DuPage. Thanks to strong support in university communities like Champaign, Barack carried seven additional counties across Illinois, winning an astonishing 25 percent of the downstate vote. As Rich Miller wrote in *Capitol Fax,* "it looks like Obama won wherever people saw his TV ads." Jack Ryan won the Republican nomination with 36 percent of the vote, but in the traditional Republican stronghold of DuPage, the raw vote comparison was striking: Barack received 35,770 votes, Ryan only 25,276. Yet Ryan's victory threatened to prove hollow, for as Miller observed in Wednesday's *Capitol Fax,* "if the *Chicago Tribune* wins its case and forces open Ryan's divorce records, and the rumors are true, he could find himself off the ticket in a matter of weeks."

Even prior to Tuesday night's resounding triumph, Barack and his consultants had turned their minds to a trio of concerns. First, Illinois tradition called for victorious candidates to do a day-after fly-around of the state, and Peter Coffey, who a year earlier had rebuffed Barack's request to come on board the campaign, volunteered to organize that tour. Then the candidate and his family sought a much-deserved vacation. "They wanted to go to Florida," finance director Claire Serdiuk remembered, but "we could not get them booked on a flight to Florida" because of spring break week. Claire instead recommended Scottsdale, Arizona, and particularly its famous Biltmore Resort. "Ooh, Michelle would love to stay at the Biltmore," Barack replied, but that too was sold out, and a fallback was found. Third, the candidate and especially his consultants realized that once Barack was the Democratic nominee for a Republican-held U.S. Senate seat, his young, bare-bones campaign staff would require a major transfusion in the form of nationally experienced Democratic operatives.

After Wednesday's early TV shows and a quick call-in to WVON's Cliff Kelley program, Barack was the star attraction at an Illinois Democratic Unity press conference. Both his vanquished primary opponents and even U.S. representative Bobby Rush attended, with Barack giving special thanks to his "political godfather," Emil Jones Jr. Speaking with Chicago reporters before heading to Springfield and then four other downstate destinations, Barack said, "I have been chasing this same goal my entire adult career, and that is creating an America that is fairer, more compassionate and has a greater understanding between its various peoples." Asked about Jack Ryan's divorce, Barack expressed disinterest. "I think it's important to define character issues broadly. Character is not just a matter of whom you're sleeping with or what you did when you were a teenager," and he instead targeted "Ryan's embrace of George Bush's agenda." When reporters questioned him about his own admissions in *Dreams,* Barack explained that young men "engage in self-destructive behavior because they don't have a clear sense of direction" and that he was able "to pull out of that and refocus" by "tying myself to something much larger than myself."

A young woman told the *Chicago Tribune*'s David Mendell that Barack's campaign "feels like a movement," and "for people of our generation, we haven't been a part of something like this before." Mendell thought Barack was "on the verge of becoming a significant voice for the black American, and perhaps even for the common American," but Barack told the *New York Times* he believed voters "are more interested in the message than the color of the messenger." Under a headline declaring that "Overnight, a Democratic Star Is Born," the *Times* said Barack was now "a treasured commodity in the Democratic Party nationally." *USA Today* agreed, titling a story "Dems See a Rising Star in Illinois Senate Candidate." Some people "whisper about a presidential future," the paper added, and the *Boston Globe* said that with his primary win, "Obama seemed to step into history."[65]

Amid all the kudos, one Wednesday-afternoon e-mail stood out. Dr. Farr A. Curlin, a young, quite religious U of C medical internist, had heard Barack speak and had voted for him, but he wrote to complain that ObamaForIllinois.com, in its section on "Protecting Choice and Achieving Gender Equity," promised that Barack would oppose "right-wing ideologues who seek to overturn *Roe v. Wade*." As someone who opposed "the systematic killing of human fetuses," Curlin believed that Barack was "fair-minded and wise enough to know that the ad hominem, slanderous rhetoric that marks your website . . . demeans you even as it demeans those with whom you disagree. . . . I sense that you are above that." Because "I believe you will win and may go on to be the President one day," Curlin hoped that in the future Barack would address abortion "with plain reason and fairness."

At 1:14 A.M. Thursday, Barack replied, thanking Curlin for his "thoughtful letter" and agreeing "that the language in my website has fallen victim to some of the short-hand and jargon of the Democratic party," which is "not always a healthy thing." Barack said, "I do not believe that the question of abortion is an easy one" and indeed was "a profoundly difficult" one. "I think it should be the woman, and not others, that acts as the final decision-maker," because "the overwhelming majority of women do not make these decisions with the casualness that opponents of abortions sometimes portray." Barack added that "I personally view the ethical issues involved in early term abortion as being fundamentally different from the issues raised by late term abortions, although I see no clear dividing line provided by science." Two years later, Barack recounted that exchange, saying he had felt "a pang of shame" at Curlin's complaint but incorrectly added that he immediately had the website language altered—it remained unchanged for at least seven weeks.

On Friday Barack, Michelle, and their daughters flew to Arizona for a four-day vacation. "Scottsdale's beautiful," Barack told Claire Serdiuk in a thank-you phone call. It was his first brief respite in more than fourteen months, and it allowed him time to reflect on how much his life had changed since his initial television ad introduced him to hundreds of thousands of people. Asked four years later when he first thought about being president of the United States, Barack said, "it wasn't until I won my Senate primary" that he began "believing . . . in a very concrete way" that "being president was something I would pursue." Since "we had gotten a pretty powerful response while I was running in the primary . . . I had a sense that the message I was delivering might resonate with a broad cross section of the American people."

Barack's aides understood that "things just changed overnight" for him after his victory, as Kevin Thompson put it. Finance assistant Liz Drew recalled that "his whole world just changed like the next day," and "nothing is going to be the same again." Barack's consultants knew that a major ramp-up would quickly have to take place, and by the time Barack and his family returned to Chicago, two long memos were waiting. The first, written by the Strategy Group's Pete Giangreco and Terry Walsh, analyzed the primary triumph. Their targeting had always anticipated that black voters would rally to Barack's candidacy once they knew who he was, so their analysis argued that the scale of his victory was not within Chicago but outside the city's borders. Pete and Terry were astonished by Dan Hynes's "dramatic inability to capture undecided votes or other votes that came into play" when Blair Hull's candidacy imploded and Maria Pappas failed to run a serious campaign. "Obama became the default candidate for uncommitted voters," especially in suburban Cook and the surrounding collar counties. Their goal for suburban Cook County had been 42 percent, but Barack captured 61, and his overall 54 percent in the five collar counties, including an "extraordinary" 64 percent in Lake, outperformed the targeting by

15 percent. In rural counties in both northern Illinois and in true downstate, Barack ran only 4 to 6 percent ahead of expectations.

Barack's top staffers and outside observers all agreed that Hull's triggering of the McCain-Feingold "millionaires' amendment" had given Barack's campaign an important although not decisive boost. Jim Reynolds recalled that "the thing that we were most afraid of in the Senate race"—whether Barack would be able to raise enough money to be a top-tier candidate—turned out to be no obstacle at all. "A huge amount of the black business community wrote the biggest checks they had ever written to a politician," and thanks to the entrée Barack gained to other major donors like the Pritzker, Crown, and Soros families, by March 16, scores of donors had given five-figure amounts. Several slightly divergent analyses agreed that between $1.7 and $2 million represented donations above the $2,000-per-person normal limit. John Kupper felt that money was "hugely important," for "otherwise we would never have gotten to the level of funding and thus media advertising that we were able to achieve." Jim Cauley agreed: "If we didn't have the millionaires' amendment, we would have been on TV at ten days instead of three weeks" before March 16, "and maybe we wouldn't have accelerated enough to win." Yet far less attention was devoted to an even more important achievement: 95 percent of Obama for Illinois's total expenditures occurred after December 31, 2003.

John Kupper believed Barack benefited hugely from the absence of a truly strong female candidate, but Dan Hynes was not alone in thinking that Barack's opposition to the Iraq war had proven decisive. "I think it was highly significant and had a huge impact," Hynes explained, for "it defined him and his candidacy with some very key, important constituencies." In progressive white suburbs like Evanston and Oak Park, Barack's antiwar stance was "a major factor" in having "a whole lot of folks fall in love with him very quickly," state senator Don Harmon believed. As North Side political commentator Russ Stewart cogently concluded, "Obama won because, unlike Hynes, he crafted an image that appealed to a liberal base."

After the primary, Hynes's camp was left wondering "how did the Blair Hull voters in the polls all go to Barack?" once Hull's candidacy imploded. "Maybe in our messaging we just weren't getting things to stick enough," Chris Mather conjectured. "I'm not sure that we could have done anything differently," pollster Jef Pollock thought. "Obama picks up the Hull collapse, and we didn't," for Barack became "the agent of change that people were looking for." As Mather put it, Barack's campaign "had a narrative that was perfect." Blair Hull realized that his "biggest mistake"—indeed a "fatal" one—was not pushing his consultants much harder to make the details of his 1998 divorce public six months before the primary. "It was clearly the right thing to do and we should have done it very early." As Mather noted, Hull "had all the money in the world to rewrite the narrative" once that was old news. Hull walked away from the race with equanimity, even after one of Barack's Springfield poker buddies ran into him in Mexico soon after the primary. "Blair! Blair Hull! You just spent tens of millions of dollars for someone to recognize you in a bar in Cabo San Lucas!" "Life is not always fair," Blair Hull explained. "You don't regret the things you did, you regret the things you didn't do. I don't regret that I ran for the Senate."

Consultants, staffers, and journalists all recognized that "a series of excellent, well-conceived television spots," particularly the "enormously effective" Sheila Simon one, had boosted Barack's appeal. "The turning point was the Sheila Simon commercial," Raja Krishnamoorthi, Barack's uber-volunteer policy director, believed. "People took notice right away." Speaking with CBS Chicago reporter Mike Flannery, Barack's Republican Senate colleague Kirk Dillard called it "a great ad," and Flannery added, "which many people credit as one of the turning points of the campaign." Anita Dunn explained that "what that ad accomplished for him downstate was that he became the alternative to Blair." *Capitol Fax* agreed: "The Obama ads gave former Hull supporters and leaners a place to go right away." Dunn understood "the phenomenon of people feeling better about themselves for supporting an African American," and Axelrod's "advertising was pitch-perfect for Obama."[66]

The second memo to Barack, titled "Finishing the Job," critically analyzed his campaign to date and specified the changes needed for the general election race. Written by David Axelrod, his D.C.-based partner David Plouffe, and John Kupper, it warned that while Barack had won by "a margin that was inconceivable," there would be no "free pass" for "an African-American nominee with a funny name and a seven-year legislative record." Axelrod and Kupper's involvement with the campaign over the previous fourteen months had been almost daily, while Plouffe had had just one long breakfast conversation with Barack a year earlier when he was struggling to do his necessary call time. "You have to be the candidate. Not the campaign manager, scheduler, or driver," Plouffe told Barack, who replied, "I understand that intellectually, but this is my life and career. And I think I could probably do every job on the campaign better than the people I'll hire to do it. It's hard to give up control when that's all I've known in my political life. But I hear you and will try to do better."

The ensuing months had repeatedly demonstrated Barack's difficulty in surrendering control over his daily life and schedule, a struggle highlighted by how long it took him to transition from driver to passenger. Now "the level of scrutiny facing you and your campaign will enter another magnitude," because the Axelrod team assumed that Jack Ryan's "campaign will be taken over by skilled national operatives." Four main staffing challenges loomed. First, Barack should bring "on board a senior communications strategist to oversee the entire press operation." During the primary, "the press relationships of your consultants helped obscure the deficiencies of the campaign's press operation," and "a lot of talent" was now available. In place of Raja Krishnamoorthi's band of volunteers, Barack would also need a full-time "professional research director."

Although "Joe McLean has done great work" and "Claire and your fundraising team obviously exceeded expectations," Barack now had "close to a limitless fundraising potential." Events in major cities across the nation needed to be scheduled, and going forward "your fundraising operation . . . should damn near be presidential." The good news was that "with certain exceptions, your days of asking for individual contributions should be over. (We know you are unhappy to hear this)," but "a campaign fundraiser with serious national experience" should immediately be hired. Lastly, although Northwestern senior Madhuri Kommareddi "has done a terrific job" overseeing scheduling and operations, Barack needed both "a senior scheduling and advance director" and "a full-time 'body' person" in addition to whoever was driving Barack when he was on the road. "The pace and breakneck speed of the general will dwarf the primary," Axelrod's team warned.

For several of the new roles, Barack and his consultants knew almost immediately whom they wanted. Even before flying to Scottsdale, Barack asked Peter Coffey, who a year earlier had declined to become campaign manager, to take charge of scheduling just as he had done for Barack's postprimary swing. Coffey agreed, as did Scott Kennedy, a former David Wilhelm aide whom Giangreco and Cauley asked to take charge of the FEC compliance work that Madhuri and Claire's young assistant Liz Drew had been trying to stay on top of. Joe McLean asked his colleague Susan Shadow, who likewise had refused Barack's initial offer a year earlier, to move to Chicago to join Claire in overseeing national fund-raising. "Can you be here Monday?" Barack asked Susan by phone, and with the campaign now able to pay senior staff $10,000 a month, Susan signed on too. For Barack's personal "body man," Michelle recommended Kevin Thompson, who accepted Barack's invitation as well. The consultants also had someone in mind for the research job, plus several serious contenders to interview for the communications post.

For the tight-knit group that had worked closely with Jimmy, Nate, and Claire to bring about such a stunning primary win, the transition was jarring. "Nothing will ever be like the closeness and camaraderie of that primary, and the chaos," Liz Drew remembered. Adam Stolorow agreed that "it was such a fun staff," but the arrival of so many higher-ranking, experienced newcomers made for hurt feelings. "I want people that have done this before," Barack instructed. "Those are the exact words," Dan Shomon recounted. Audra Wilson recalled that it was "almost as though there wasn't appreciation or recognition of all that we had done to get to this point." Lyndell Luster agreed that "it was a little rough" and "kind of cold." Before the primary "it was like a family," but it became "a business" afterward. In the ensuing weeks, "a lot of people got pushed aside" and "it was kind of heart-wrenching for me," Lyndell confessed. Virtually all of the staff veterans who felt slighted were African American, and longtime AFSCME lobbyist Ray Harris complained to Barack about his treatment of Kevin Watson, who had been "with him day in, day out." Barack replied, "You know, Ray, Kevin's had me in an awful lot of bars," not at all surprising for South Side campaigning. "Kevin was devastated," Ray thought. "A lot of the staff wasn't being properly utilized, so people started to quit," Audra Wilson explained. "I had a long talk with Barack, and he was a little taken aback."[67]

As the Axelrod memo reflected, Barack's aides expected a demanding general election campaign against Jack Ryan. John Kupper thought "he could be problematic," because his well-publicized inner-city teaching stint at Hales Franciscan would make Ryan "more palatable to suburban whites" than other conservative Republicans. Nate Tamarin said, "We took him very seriously," because with "a guy named Barack Obama . . . we anticipated that this thing was going to be a challenge." Claire Serdiuk concurred. Ryan was "a very, very competitive candidate," for with a "white Irishman" up against "Barack Hussein Obama," the contest was "not at all a walk-away." Many Republicans agreed. Barack's lobbyist friend John Nicolay thought "Jack Ryan was probably by far the most dangerous opponent for Barack." Senate minority leader Frank Watson had assumed Dan Hynes would be the Democratic nominee, but once Barack won, a lot of Republicans started thinking, "We've got a real chance at this." In contrast, former governor Jim Edgar told reporters that Barack's victory demonstrated that in Illinois, "race is no longer a hindrance. In fact, it's a plus."

Barack and his family had returned from Scotsdale directly to Springfield for a special event on the Senate floor. "I am proud that today, I, I believe, in front of my five-year-old and two-year-old, I can justify my work in public service because I have the honor of introducing on the floor one of the biggest stars of stage and screen," a character from the Chicago Children's Museum. "I know the theme song. I know most of the plots of Clifford the Big Red Dog, who is here. Give it up for Clifford." Many senators remembered Barack from that first day back, but not for his rousing introduction of Clifford. Poker buddy Tommy Walsh remembered that "he was struttin' more than you've ever seen a guy." Walsh was unperturbed, but senators who were not fans took umbrage.

One month earlier a reporter had noted that Barack had "a slightly stiff manner of walking that some people close to him suspect is an affectation of John F. Kennedy." Judiciary Committee Republican Ed Petka remembered that Barack "had a gait in his walk," a "type of gait that was sending off a signal of cockiness." A foul-mouthed Chicago senator recalled Barack's "pimp-ass fuckin' gait" that Wednesday. "Too bad you couldn't be with me," Barack remarked. When the colleague offered pro forma congratulations, Barack told him to "save it" and uttered an ethnic put-down. "Barack, go fuck yourself," the senator replied, explaining that "Barack was an arrogant motherfucker." Barack's postprimary demeanor led senators to ask "Who the fuck made him king?" or, as one African American colleague remarked to a black lobbyist, "Who this nigger think he gonna be? President?"

The Democratic Senatorial Campaign Committee (DSCC) had sent Barack's campaign their opposition research report on Jack Ryan, whom it labeled "an extreme right wing ideologue." Ryan "is virtually incapable of answering a question in a succinct yet comprehensive fashion," the DSCC believed, and "his facts . . . are often wrong." That Wednesday in Springfield, Barack's top staff and consultants assembled for a long meeting to discuss policy stances as well as a proposed general election media budget of more than $8 million.

The next day *Illinois Times,* Springfield's alternative newsweekly, posted online a piece by Todd Spivak, who previously had covered Barack for the *Hyde Park Herald*. Spivak noted that African American House Democrats like Lou Jones and Monique Davis had refused to back Barack in the primary. But Spivak's lede—"It can be painful to hear Ivy League–bred Barack Obama talk jive," using words like "homeboy" while being "perhaps more vanilla than chocolate"—easily overshadowed the House members' put-downs. Early the next morning Spivak's phone rang. It was Barack, "screaming at me" while denying he had ever used "homeboy." Spivak recounted that "it seemed so silly; I thought for sure he was joking. He wasn't." Another reporter had heard Barack call Spivak an "asshole" at a fund-raiser, but Spivak was left "stunned" and "trembling" as Barack "shouted me down" when Spivak tried to respond. "I asked if there was anything factually inaccurate," but Barack "cut me off" when he tried to say more. Barack had previously told Spivak not to call his cell phone, but now Barack insisted that he should have after Barack's staff dropped the ball on Spivak's interview request.

A landslide victory had not made Barack any less thin-skinned, but later that day on the Senate floor he acknowledged how conscious he was of his new status. Speaking in favor of a bill to allow for the sealing of former drug offenders' criminal records in order to improve their employment chances, Barack explained that "nobody, obviously, right now in this room, probably is more mindful of the politics of issues and how he or she votes than I am right now. Nevertheless, I feel obliged to support this bill, knowing that it could cause me problems in a mailer come November." Given "the enormous numbers of people who are now going through the criminal justice system," Barack cast a crucial thirtieth vote to "give people a second chance" as three other Democrats joined with unanimous Republicans in voting no.[68]

On Saturday afternoon, Barack met *Chicago Sun-Times* religion reporter Cathleen Falsani at the coffee shop a few doors south of his campaign office. "I have a deep faith," and "I'm rooted in the Christian tradition. I believe that there are many paths to the same place, and that is a belief that there is a higher power," Barack explained. He spoke about his 1985–88 experiences in Roseland, erroneously saying he had joined Trinity United Church of Christ in "1987 or '88." Asked if he still attended Trinity, Barack answered "Every week. Eleven o'clock service," though four years later, he accurately acknowledged that "there was quite a big chunk of time, especially during the Senate race, where we might not have gone to Trinity for two, three months at a time."

Barack told Falsani that he believed that "religion at its best comes with a big dose of doubt," especially given that "there's an enormous amount of damage done around the world in the name of religion and certainty." Barack said, "I think I have an ongoing conversation with God," and that "the biggest challenge, I think, is always maintaining your moral compass. Those are the conversations I'm having internally. I'm measuring my actions against that inner voice that for me at least is audible, is active, it tells me where I think I'm on track, and where I think I'm off track."

Asked who in his life he looked to for guidance, Barack immediately cited Rev. Jeremiah Wright as "certainly someone I have an enormous amount of respect for," adding that Father Michael Pfleger is "a dear friend and somebody I interact with closely." Barack volunteered that "I think there is an enormous danger on the part of public figures to rationalize or justify their actions by claiming God's mandate," and he confessed that "the nature of politics is that you want to have everybody like you and project the best possible traits onto you. Oftentimes that's by being as vague as possible." He added that "nothing is more powerful than the black church experience," and said that "in my own sort of mental library, the civil rights movement has a powerful hold on me." In closing, Falsani asked Barack what he remembered about his decision to join Trinity. Barack said it was "a gradual process," not "an epiphany," because "there is a certain self-consciousness that I possess as somebody with probably too much book learning and also a very polyglot background."

Asked several years later whether he prayed, Barack said, "Yes, I do . . . every day." Asked what he prayed for, Barack answered, "Forgiveness for my sins and flaws, which are many, the protection of my family, and that I'm carrying out God's will, and not in a grandiose way, but simply that there is an alignment between my actions and what he would want." Michelle said the family prayed at every meal, but among the scores of people who knew Barack well over the years, very, very few believed that religious faith played any significant role in his life. One who did was state Senate colleague Ira Silverstein, an Orthodox Jew whose Springfield office for four years adjoined Barack's and whose daughters attended the same preschool as Barack's. "We used to talk about Israel all the time," Silverstein recalled. "We talked about religion a lot. He is a very religious person." Another was attorney and author Scott Turow, with whom Barack had discussed capital punishment in such "a very spiritual way" that Scott was certain Barack's religious faith "was really sincere."[69]

On Monday, Barack, Jim Cauley, and Claire Serdiuk flew to Washington to meet with DSCC officials and attend a Congressional Black Caucus (CBC) fund-raiser. They also met Robert Gibbs, Axelrod's preferred candidate for communications director, a white Alabaman who until four months earlier had worked on John Kerry's presidential campaign, and Amanda Fuchs, a Northwestern Law School graduate whom the consultants were recommending as the new policy director. Like Peter Coffey, Susan Shadow, and Kevin Thompson, Gibbs and Fuchs were both white—"nobody was whiter than Gibbs," Pete Giangreco explained—and Barack's insistence upon maintaining a racially balanced staff was teetering badly with the departures of Pam Smith, Kevin Watson, and Audra Wilson, and the influx of white newcomers. Barack "was very insistent that the staff be diverse," and Giangreco recalled that agreeing to hire Gibbs "was a little bit of a pill for Barack to swallow," especially because another candidate, Jamal Simmons, was African American. In addition, Axelrod recruited Nora Moreno, a young woman who knew Chicago well, as Gibbs's deputy.

"Barack's really pounding for black operatives with some experience," Cauley explained, and he, Joe McLean, and Tom Lindenfeld recommended Darrel Thompson, an African American who had served as a top aide to House Democrat Richard Gephardt. Barack remembered Thompson from Alice Palmer's congressional campaign nine years earlier, but Darrel sought assurance of Barack's political toughness when he and Cauley asked Darrel to join the campaign. "Well, if you're asking me if I'm a punk, the answer is no," Barack replied. "I intend to run this race and win it." Pleased by that, Darrel agreed to come on board as chief of staff under Cauley. African American Vera Baker, who had staffed the CBC fund-raiser, agreed to move to Chicago to join Barack's finance team. Susan Shadow added two more young fund-raisers, Jenny Yeager and Jordan Kaplan, to her and Claire's growing team. When Gibbs asked Cauley to hire another young white press assistant, Tommy Vietor, Cauley did so knowing there would be blowback. "Barack chewed me out one afternoon and told me if I hired another white person he was going to kick my ass. 'Jimmy, you've got too many white folk in here,'" Barack complained. Cauley replied by saying Vietor was only a volunteer. "I just lied to Barack," he recounted.

While in Washington, Barack also taped a radio program and a TV show with African American broadcaster Tavis Smiley. "I'm certainly black enough to have trouble catching a cab in New York City," Barack assured Tavis, and on

TV Barack remarked that viewers could get confused by telecasts. "Sometimes they're watching Fox News. You *know* that's gonna get 'em confused." In an article for Salon.com entitled "The New Face of the Democratic Party—and America," Scott Turow wrote that "Obama is the very face of American diversity." Comparing him to prominent African American Republicans Colin Powell and Condoleezza Rice, Turow predicted that Barack "may become the first black Democrat able to rise above race."

One reader of Turow's essay was Rachel Klayman, an editor at Crown Publishing who dimly remembered that Barack had once published a book. Searching online, Klayman discovered that not only was the paperback edition of *Dreams From My Father* long out of print, but the rights to issue a reprint were held by Crown itself. Neither the company nor agent Jane Dystel still had a copy of *Dreams,* but Barack still had copies of the original paperback, and sent several to Klayman. Reading *Dreams,* Klayman quickly resolved to reissue the book and asked Dystel if Barack could write a preface to the new edition. Within a few weeks, Barack did just that in the space of an afternoon. Briefly surveying his life since 1995, Barack wrote that his legislative work had proven "satisfying" because "the scale of state politics allows for concrete results." Explaining how his primary win had led to *Dreams'* republication, Barack predicted, "I have a tough general election coming up." Rereading it "for the first time in many years . . . I have the urge to cut the book by fifty pages or so," without saying how. Citing 9/11, Barack wrote that "the bombs of Al Qaeda have marked, with eerie precision, some of the landscapes of my life . . . Nairobi, Bali, Manhattan," but demonstrated how "the embrace of fundamentalism and tribe dooms us all."

Not only were Honolulu newspapers trumpeting Barack's new fame, Columbia University's student newspaper also wrote that the 1983 alumnus was "one of the rising stars of the Democratic party." Barack's campaign sent out a fund-raising e-mail warning that "our precious civil liberties are threatened by the Patriot Act," and his policy team volunteers turned their attention to Jack Ryan's record. Andrew Gruber surveyed the issues content on Ryan's campaign Web site and wrote Raja Krishnamoorthi to express amazement: "it has a section on the 2nd Amendment, but does not have a section on health care." Illinois Republicans remained privately worried about how the *Tribune*'s effort to unseal Ryan's divorce file would play out, with *Capitol Fax* reporting that Illinois congressman and U.S. House Speaker Dennis Hastert had told Ryan he would be unable to raise funds in Washington until the problem was resolved. Given Ryan's problems and the strength of Barack's primary win, "unless Ryan unearths some very damaging political stances, legislative votes or personal ethical lapses of Obama's, Obama will win in a walk," Chicago columnist Russ Stewart predicted.

A serious Republican effort was under way to investigate Barack's background and record. Matt Tallmer, a former Republican National Committee operative, was tasked with much of the work, and he began by reading *Dreams From My Father*. "Most of the potential problems involve very minor issues," and the book "omits whole portions of his life," Tallmer reported. He concluded that Barack "is not certain whether he is black or white," but resolved his racial conflict "by joining the white world." A review of Miner Barnhill's legal clients was also unproductive, and two memos examining Barack's major donors were badly off-target. Only contributors to his Senate campaign, but not his earlier races, were reviewed, and although Tony Rezko's close relationship with Governor Blagojevich was noted, neither Rezko nor any of his associates were included in a list highlighting possible "bad actors." Instead Valerie Jarrett, an old friend who had fought an expressway, and a close friend of Michelle's who opposed corporal punishment in schools were all oddly prioritized. An initial effort to examine Barack's Springfield record floundered badly, but several weeks later Debbie Lounsberry, who had previously staffed the Public Health and Welfare Committee for Republicans, compiled a highly competent analysis of Barack's potentially problematic Senate votes.[70]

Barack's campaign received encouraging news from Paul Harstad's first general election poll. A plurality of Illinois voters felt negatively about President George W. Bush, and Democrat John Kerry led Bush 52 to 38 percent. Barack registered 85–1 in favorability among black voters, 74–3 among Democratic women, a surprising 35–13 among Republican women, and 17–28 among Republican men. Burrowing down into positive and negative themes, "represents change" was a clear plus for Barack, with health care and retirement security ranked high as well. Initially voters chose Barack over Jack Ryan by 52 to 33 percent. After both candidates' issue positions and biographies were detailed, Barack's lead increased to 53–30, with Ryan weakened more by his views on policies than he was strengthened by his résumé. In a summary memo to Barack's campaign team, Harstad highlighted that even with a 19-point lead, Barack's overall name recognition was only 73 percent, as compared to an ostensible 84 percent for Ryan. "Thus Obama has more room for growth in support than does Ryan."

Barack continued to say Ryan's divorce was not "an appropriate topic for debate." A Saturday campaign appearance at a College Democrats conference was followed by an evening party at Tony Rezko's Wilmette mansion. Asked about Iraq on WTTW's *Chicago Tonight*, Barack emphasized that "we've got to be sure we get the job done" and stressed that "our support of Israel is unequivocal." On April 8 John Kerry arrived in Chicago for a gala evening fund-raiser at the Hyatt Regency and a Friday-morning appearance with Barack at a West Side job-training center. Barack spoke at the gala and gave Kerry a "stirring introduction" at the Friday event. That evening Barack and Michelle joined Kerry and his wife Teresa for a private dinner. A *Washington Post* story on their joint appearance called Barack one of the Democrats' "rising stars," and two days later the *Chicago Sun-Times* commented on how many such references to Barack were being made.

On Monday, April 12, Paul Harstad presented his poll results to Barack's team, followed by an all-day "consultants retreat" on Tuesday to plan the fall campaign. Harstad's numbers were clear: "Target voters: women & Downstate" with "Women 70 percent of target" and emphasize "Pocketbook/Kitchen-table issues." There was a "Big advantage in 'change,'" and voters were concerned about "healthcare affordability." "Need inoculation on crime & drugs. Need inoculation on raising taxes," although John Kupper noted that "Ryan has terrific vulnerabilities on taxes & health care: we *want* those debates." At Tuesday's large gathering, there were "all these new faces," with many newcomers being introduced for the first time. Nora Moreno, Gibbs's deputy, recalled that Barack asked of his new communications director, "Is it Robert or Bob?" "We're all trying to figure out what our roles are," she explained. Gibbs "comes on board and he's like 'Who are you? Why you?'" Mostly the answer was "we all had relationships with David."

Paul Harstad and partner Mike Kulisheck reprised their poll presentation, Pete Giangreco and Terry Walsh outlined their general election targeting, and Axelrod and John Kupper addressed messaging. Susan Shadow spent an hour on fund-raising, warning everyone that "I didn't know where the money was going to come from" because unless wealthy Jack Ryan began self-funding, individual general election donors would be limited to $2,000 apiece, not the five-figure contributions allowed in the primary. That meant a new emphasis on out-of-state fund-raising, rather than Barack's Chicagoland acquaintances, which meant Axelrod's promise of no more call time went out the window big time. Jimmy Cauley sketched out an overall budget of $13 million, with $8 million for broadcast media, $1.7 million for direct mail, and $1.8 million for salaries. Scheduler Peter Coffey recalled that throughout all the discussions, Barack "didn't really say much. . . . He would sit there and let it play out. That was his management style," just as it had been at the *Harvard Law Review*.

At one point, Axelrod mentioned the upcoming Democratic National Convention (DNC) and the desirability of using it to showcase Barack to a wider audience. "We should really try to get a prime-time speaking slot." Pete Giangreco spoke up. "I know Jack Corrigan. I can call him," Pete volunteered. Corrigan, whom John Kerry's campaign had named as its DNC manager three months earlier, had been Giangreco's boss in 1988 as national field director of Michael Dukakis's presidential campaign. Pete telephoned Corrigan. "Yeah, we know all about Obama," Jack responded. Months earlier AFSCME's Henry

Bayer had told Corrigan that Abner Mikva, whom Corrigan had worked for a quarter century earlier, believed Barack was "the most talented politician he's met in fifty years." That had gotten Corrigan's attention, and he called law school classmate Elena Kagan, who had clerked for Mikva and overlapped at the U of C with Barack from 1991 to 1995, to ask what she thought of him. "He's great," Kagan said, and Corrigan made a note to ask Barack to join Kerry's campaign staff once "this kid" lost his March primary. Corrigan responded positively when Giangreco said, "It would be really great if we got a prime-time speaking slot" for Barack.[71]

Tax day was less painful for Barack and Michelle than in earlier years: rather than owing the Internal Revenue Service a five-figure sum, they were due a $1,500 refund. But the good financial news ended there. The University of Chicago had paid Barack half of his normal $64,000 annual salary even though he was not teaching in early 2004, but his primary win meant he would have to take a leave for the autumn as well. Michelle's salary was $121,000, but with child care costs of more than $2,000 a month, family finances were about to get tighter than ever, with their 2004 income headed well below 2003's $238,000. Michelle had long feared exactly this scenario, as she told the *Tribune*'s David Mendell. If Barack won in November, "now we're going to have two households to fund, one here and one in Washington." Her sarcasm could be cutting, if not demeaning. "Even if you do win, how are you going to afford this wonderful next step in your life?" By April they were discussing taking out a second mortgage on their East View condominium to make up for the loss of Barack's law school salary.

Over the past decade-plus, Barack had taught twenty-six courses: Racism and the Law twelve times, Con Law III eight times, and Voting Rights six. Among the roughly 750 students he taught, their cumulative evaluations documented the consensus expressed by Douglas Baird and other faculty colleagues: Barack had been a "splendid," "sensational," and "tremendously gifted teacher" who was "off-the-charts popular with students." Given Barack's consistent absence from the law school except for actual classroom time, most law faculty felt that "being here had zero effect on Barack's thoughts." Baird disagreed. "Being a teacher in a University of Chicago classroom makes a big difference. I think the kinds of conversations that he had to have had as a teacher of constitutional law here would be completely different than if he was doing it at Harvard or a place like that because he's getting conservative students, and people are going to press him on issues," and "that's going to affect you," Baird believed. "That kind of serious intellectual engagement that he had here makes a huge difference" in how Barack understood disagreements over constitutional interpretation.[72]

Barack joined U.S. senator Dick Durbin for a three-day, twelve-stop downstate swing, spending one night at a Super 8 motel in Benton. While Barack was visiting Cairo, Metropolis, and Harrisburg, Jack Ryan held a Springfield news conference to demonstrate that as a state senator "Blank Check Barack" had backed no fewer than 428 tax and fee increases. But reporters realized something was amiss when Ryan displayed a chart showing that Illinois had 846,000 state employees, a vast exaggeration, and things got even worse when they sought documentation for the 428 claim. "We'll get it to you later," Ryan replied, but when Bernard Schoenburg of the *State Journal-Register* cornered Ryan aide Dan Proft, it immediately became clear that "Ryan didn't know what bills he was talking about": of the supposed increases, 146 dated from a single 1999 bill that seventeen Republicans, including Pate Philip and Frank Watson, had joined Barack and most Democrats in approving 42–17, and the balance of 282 came from a 2003 bill that Barack had voted *against*. The resulting headlines—"Ryan's Chart Is Off" in the *Chicago Sun-Times*, "Jack Ryan Woefully Unprepared for Attack on Obama" in the *Journal-Register*—were "more proof that Ryan's campaign is not yet ready for prime time," *Capitol Fax*'s Rich Miller observed. Barack responded that "Mr. Ryan obviously hasn't served in any policy-making position, so it's not surprising he doesn't have a very good grasp of the statistics and the issues involved." Ryan's campaign replied by sending out an e-mail claiming that Barack "has earned a reputation to the left of Mao Tse-Tung."

"I don't recall ever covering a candidate who accused his opponent of being for something he had voted against," Bernie Schoenburg wrote in the *Journal-Register*. While a court-appointed referee weighed what to do with Ryan's divorce file, the candidate assured Republican Party chair Judy Baar Topinka and former governor Jim Edgar "that there was nothing in there that would be embarrassing to the party." Edgar shared those reassurances with others while Ryan's campaign turned up the pressure on Barack by assigning a twenty-four-year-old staffer to follow Barack and videotape all of his public comments. "Initially, I tried to talk with him. I said, 'Listen, I don't mind you following me, but please be fifteen feet away. I'm on the phone with my wife,'" Barack recalled. A *New Yorker* correspondent also spent several days with Barack, who at the end of a long one said, "I'd rather be doing what you're doing: sitting in the corner listening, watching everybody, taking notes. That comes more naturally to me than this does."

On Wednesday, May 19, Barack brought to the Senate floor the final version of his longest-standing legislative commitment, the Health Care Justice Act. As reintroduced early in the 2003 session by House counterpart Willie Delgado, HB 2268 mandated that by 2007, the state "shall implement a health care access plan" that would provide "quality health care for all residents of Illinois." A thirty-member "Bipartisan Health Care Reform Commission" would submit a report that "shall make recommendations that shall be the basis for a health care access plan or plans" that would be passed into law. The House had narrowly approved the bill on April 1, 2003, but soon thereafter insurance industry opponents pointed out that the Illinois Supreme Court had long held that "the General Assembly cannot bind its successors unless authorized by the Constitution." Barack amended the bill to say that the commission "shall make recommendations that shall be *considered by the General Assembly as* the basis" for a health plan, but insurance industry entreaties and more to Senate president Emil Jones Jr. halted the bill's progress. "We had to keep on meeting with the insurers," top proponent Jim Duffett of the Campaign for Better Health Care (CBHC) explained.

Duffett offered to find another chief sponsor, telling Barack he realized "you're getting beat up on this issue pretty bad," but Barack said, "No, absolutely not. I'm really committed, and I'm going to do this." At a 2003 AFL-CIO conference that Duffett and Barack both addressed, Barack had spoken out strongly for the bill. "I happen to be a proponent of a single-payer universal health care program," he told the audience. "I see no reason why the United States of America, the wealthiest country in the history of the world, spending 14 percent of its gross national product on health care, cannot provide basic health insurance to everybody. . . . And that's what Jim's talking about when he says 'everybody in, nobody out.' A single-payer health care plan, a universal health care plan. That's what I'd like to see. But as all of you know, we may not get there immediately. Because first we've got to take back the White House, we've got to take back the Senate, and we've got to take back the House."

In Springfield, a number of Democratic senators "did not want to oppose the bill but they did everything they could to try to water it down," Duffett explained, because they, just like Emil Jones, benefited from insurance industry support. In early March, Barack told the *Chicago Defender* that "the resistance of special interests means it will not be possible to craft a new national program," and the same held true in Springfield. Encouraged by David Wilhelm, a CBHC board member, Duffett endorsed Barack's view that the bill would have to be significantly weakened to pass the Senate. As Illinois Planned Parenthood's Pam Sutherland put it, "you always knew with Barack you were going to compromise, whether you liked it or not." Fervent single-payer proponents like Hyde Park's Dr. Quentin Young, who practiced alongside Barack's personal physician, David Scheiner, strongly rued "Barack's penchant for compromise," and when on May 11 Barack filed a thoroughgoing rewrite of HB 2268, Health Committee ranking Republican Dale Righter realized it now looked "absolutely nothing like what had been introduced" fifteen months earlier.

In place of the original mandate that "On or before January 1, 2007, the State of Illinois shall implement a health care access plan," the revised bill said, "Illinois is strongly encouraged to implement a health care access plan." What had been the "Bipartisan Health Care Reform Commission" was now renamed the "Adequate Health Care Task Force." The already-amended language instructing that the commission's final report "shall be considered by the General Assembly as the basis for a health care access plan" was supplanted by a statement that "the final report by the Task Force shall make recommendations for a health care access plan." Insurance industry opposition weakened because, as Barack's poker buddy and insurance lobbyist Phil Lackman explained, "it's hard to oppose a task force." Ranking Republican Dale Righter saw that single-payer proponents were "very unhappy" with Barack, because "they believed that they had a champion there." Quentin Young agreed: Barack "abandoned" HB 2268's original call for universal access and "sold out to the insurance companies."

When Barack brought HB 2268 to the Senate floor on May 19, he said, "we have a full-blown health care crisis" and explained that "what this bill does is to create a task force that over the next years will examine mechanisms by which we can expand affordability and accessibility of health care." The Four Tops would appoint six task force members each, and the governor an additional five. In contrast to months earlier, Barack emphasized, "I want to say on record that I am not in favor of a single-payer plan. I don't think we can set up that kind of plan, and if we were going to even attempt to some sort of national health care, that would have to obviously be done at the federal level." Stressing that "we can't expand accessibility and affordability unless the baseline costs of health care are curbed," Barack also said that "what we don't have right now is any kind of sense of urgency about the extraordinary difficulties that our constituents are experiencing with respect to health care costs." Oak Park Democrat Don Harmon commended Barack for his work. "I know this has been no easy task for Senator Obama, but I applaud him for navigating this bill and building a truly impressive list of organizations and entities that are supporting the bill."

When conservative Republican Bill Brady questioned Barack about the insurance industry's view, Barack explained that "we radically changed" HB 2268 "in response to concerns that were raised by the insurance industry. The assurance that I received from the insurance industry was that if we took out the mandate, that the legislature would have to implement the bill, then, in fact, their objections would be lifted, and I got repeated commitments from the insurance industry to that effect. What then happened is, after we removed the provisions that initially had been the source of criticism and made it a bill that studies the problem and does not mandate that the legislature act on it, I got a reversal from industry which said 'No, we're still not interested in it.' And it was in the face of that obstinance" that Barack had the Health Committee send it to the Senate floor.

Brady conceded, "I don't disagree with the fact that you did everything you could to try to get an agreement," and Barack deplored how "insurance lobbyists here in Springfield had been engaging in such fear-mongering." Minority floor spokesman Peter Roskam was already Barack's least-favorite Republican, and after Roskam invoked the phrase "socialized medicine" in calling HB 2268 "a bill with an agenda," Barack objected, saying, "the notion that you would blatantly characterize this bill as being that is dishonest." "Don't lie about it," Barack told Roskam. Veteran Democrat Miguel del Valle thought that exchange was "the angriest I've ever seen Barack," and del Valle commended him, saying, "you've got to get angrier more often!" On the floor, Evanston Democrat Jeff Schoenberg rose to commend "Senator Obama's hard work, his perseverance, and perhaps most importantly, his integrity and sense of fairness." Objecting to Roskam's words, Schoenberg found it "very distressing that when all else fails, we have to try to discredit the personal integrity of a member of this chamber because we may have a political or ideological difference. I don't think it should ever reach that point."

Barack agreed, and on an almost straight party-line vote, the Senate approved the Health Care Justice Act 31–26–1. Barack's good friend Denny Jacobs, who realized Barack "has that ability to be a little bit of all things to all people," was the only Democrat to vote no. Press coverage was limited to a single, page 10 story in the *State Journal-Register*. One week later the House ratified the bill, and in late summer Governor Blagojevich signed it into law. Two and a half years would pass before the task force submitted its final report, with eleven of its twenty-nine members supporting a single-payer plan. Republican Dale Righter felt the enactment was "meaningless by the time it was over," producing only a report that "gathers dust and is really never to be seen or heard from again." Fellow Health Committee Republican Christine Radogno, who had voted present rather than no, agreed. The Health Care Justice Act had allowed Barack to "claim success even though it was nowhere near what he initially started with." But HB 2268 as it was passed "gave him a fig leaf and gave him something to claim as a win."[73]

Meanwhile David Axelrod had been crafting a six-page fall campaign memo to "Team Obama" entitled "Message: Yes We Can!" "The remarkable strength of the Obama candidacy" demonstrates that "Barack inspires a vision of an honest and uplifting politics, and fires hope for real change," Axelrod wrote. "Barack stands apart" by "preaching a politics of civility and community, of mutual respect and responsibility." Voters "respond to his character and sincerity. Our challenge is to maintain that tone, protect that special character and sincerity and always bear in mind that the brain dead politics of Washington is as much our target as Jack Ryan." Axelrod warned that "if we begin to operate in much the same fashion, singing from the partisan hymn book, slinging mud . . . and sliding around on issues in an overtly political way, we will destroy what people like best about our man." The campaign needed to emphasize "strengthening the embattled middle class" through "making health care affordable," "creating quality jobs," and "improving the quality and affordability of education."

The day after the Health Care Justice Act passed, the presence of Justin Warfel, the Ryan campaign videographer, in Barack's life became a top statewide story. Warfel "has stalked him for ten days," Chicago's local ABC news telecast told viewers. Barack wanted Warfel to "keep a respectful distance" so that "you can actually have a quiet conversation," but astonishingly Warfel usually remained "literally a foot away" until a bevy of press questions led him to disappear. "He has a tape recorder, so if I'm calling my wife at home, I've got a guy a foot and a half away," Barack told reporters. "I think an expectation of some basic personal space that when I call my wife or I'm calling my kids to wish them good night" was not unreasonable. "It's this kind of incivility in politics that turns people off to politics." The *Chicago Sun-Times* told readers that Warfel "interrupted Obama several times with heckling questions," but only *Capitol Fax*'s Rich Miller reported that Warfel was "tailing the candidate in his car." Barack told reporters that so far "he hasn't come into the restroom," but Senate Republican leader Frank Watson and state party chair Judy Baar Topinka publicly criticized Warfel's behavior. Rich Miller termed it "truly creepy" and a *Sun-Times* editorial proclaimed "Ryan Takes Campaign to New Low."

When Ryan's team issued a press release calling Barack "the criminals' good friend," Barack said that claim was "a little wacky," and the *Journal-Register*'s Bernie Schoenburg observed that "Ryan needs to get control of the people working for him." Before long, Republican congressman Ray LaHood publicly criticized Ryan for running "a boneheaded campaign" and singled out Warfel's conduct as "the stupidest thing I've ever seen in a high-profile campaign."[74]

In Springfield, Barack was far less involved in the spring session's legislative work than he had been in his seven previous years. Colleagues and staffers remember that "he was always on the phone," and while campaign events sometimes kept him away from Wednesday night's poker games, on some afternoons, he still managed to escape to the golf course. Playing in a foursome with lobbyist friends Phil Lackman and Dave Manning plus young campaign photographer David Katz, Barack's cell phone rang. It was Michelle, so he had to answer, but he pretended he was outside the statehouse. "She'd just kill me if she knew I was playing golf," he explained after hanging up.

The legislature failed to agree on a state budget by May 31 even though all three central players—Governor Blagojevich, House Speaker Mike Madigan, and Senate president Jones—were Democrats. Two deep fissures were responsible, one of which everyone understood: Jones's long-held anger over the disrespect Madigan exhibited toward him. The second was the even stronger animus between Blagojevich and Madigan. *Capitol Fax*'s Rich Miller rued

Blagojevich's "complete refusal to engage in even minimal governance," and even one of the governor's top supporters spoke about Blagojevich's "immaturity" and "lack of seriousness." Madigan had concluded that he could not trust Blagojevich, and the governor's top aides had come to believe he was a "selfish, insecure narcissist" who "had some sort of learning disability that made him incapable of focusing." On May 31, the last day of session before a constitutional supermajority would be required to adopt a tardy budget, Emil Jones allowed Barack to be the only Democrat who opposed his and the governor's spending plan, which Madigan's House would not approve. Barack told reporters, "My instinct is we are close enough that if we eliminated the egos and the politics, we could come up with a deal relatively quickly." Rich Miller rightly observed that "ultimately, it's the governor's fault," but he also castigated Jones for "embracing Blagojevich far too closely and leading his members over the cliff" and into overtime. Miller reported that Blagojevich was hankering for a televised speaking appearance at the Democratic National Convention, but predicted that "the Washington folks will want to showcase their newest rock star, Barack Obama."

Barack's new status as the heavily favored candidate for a U.S. Senate seat had not altered his self-discipline issues. "He was late everywhere," scheduler Peter Coffey remembered, and sometimes almost a no-show, even for meetings with prospective donors. Dan Shomon and Jim Cauley had tried to instill greater discipline, but as Nate Tamarin had learned, Barack "doesn't respond to enforcer behavior" and regularly put off necessary tasks until the last possible moment. With Mike Signator replacing Lyndell Luster as Barack's regular driver, matters went from bad to worse. Darrel Thompson explained that "Mike was Barack's back door" during times when "Barack didn't want to be found." Signator quickly became "Barack's buddy" and "biggest enabler," and "with Mike in the car" the campaign "was a lot less able to keep Barack doing the things that he was supposed to be doing." The "Mike and Barack show" became a particular problem for Cauley. One day the fallout was explosive when Barack was a no-show for Malia's end-of-the-year parent-teacher conference. Cauley remembered that Michelle "lit into me," "really reaming me out," because "in his infinite wisdom" Barack had "told her it was my fault. But it was another one of those things where him and fucking Signator were out dicking around, so he throws me under the bus." Cauley realized that Barack "was balancing a lot of crap in his life," but he objected to Barack interjecting him into his marital tensions. "Dude, you got problems there?" Jimmy asked.

Kevin Thompson, who on campaign trips often served as the candidate's "body man," thought Barack now seemed "a little more pensive" than in previous years. In part that was because he still had to do call time during the drives to and from Springfield, seeking out-of-state donors for a race promising a Democratic Senate pickup. But it also reflected how much his life had changed since his campaign first went up on television four months earlier. "The level of coverage that he was getting," with national reporters now joining an increased cast of Illinois journalists, was "a little stunning to him," Kevin realized. "It certainly had an impact on him, and it was noticeable." Articles in *The New Yorker* and *The New Republic* discussed Barack's "miraculous" primary triumph, with *TNR*'s Noam Scheiber citing academic findings that white Americans regard non-American blacks more favorably than native African Americans. "Obama's name—or at least its provenance—may have actually helped him, by distinguishing him from other African Americans," Scheiber suggested. "The power of Obama's exotic background to neutralize race as an issue, combined with his elite education and his credential as the first African American *Harvard Law Review* president, made him an African American candidate who was not *stereotypically* African American," Scheiber argued.

The number of complete strangers who recognized Barack now made his desire for downtime more insistent. "His favorite little hideout was going to the East Bank Club," Kevin explained, because he could get a daily workout there away from public view. "He'd get violent with me if he didn't get his two hours at the East Bank Club, or an hour," Jim Cauley remembered. "There would be huge fights in scheduling meetings about his workout time," which Barack, though not his campaign team, insisted was a priority over call time. The impact of the Senate race was also visible in how Barack now "smoked a lot," Kevin recalled, and Dan Shomon agreed that Barack "became a chain smoker in '04." When Barack was at his campaign headquarters, he smoked outside the building, often alongside a new young African American finance staffer, Clinton Latimore Jr., who was struck by "how down to earth" and "very personable" Barack was. "You couldn't get him away from his BlackBerry," Latimore remembered, a device Cauley had first introduced Barack to ten months earlier.

The arrival of all the new, mainly white, nationally experienced staffers meant that relationships at campaign headquarters "changed overnight," Kevin Thompson remembered. There was "a cultural shift," indeed "a cultural war going on between the national people and the local Illinois people," and the veterans who had signed on when Barack was a little-known long shot were the casualties. When Cauley first arrived, he had developed a good accord with Dan Shomon, and Cauley's "very strong personality" had enabled him to ride herd on Barack just as the consultants had hoped. But Robert Gibbs's arrival was entirely different. He sought to exert authority from the start despite being a newcomer to Illinois. The contrast—and conflict—was sharpest with Shomon, who everyone realized "knows everyone you're supposed to know" in Illinois politics. But Gibbs and Shomon "didn't get along" and "it was kind of a pissing match," one staffer recalled. "Nobody works harder, nobody brings more energy, nobody knows more people" than Dan Shomon, Peter Coffey said, and everyone who respected Shomon took offense when "Gibbs disdains him." Unlike Dan and Jimmy, Gibbs indulged Barack rather than challenged him, but as Barack's shift to a beckoning national stage accelerated, his need for Shomon's expertise dissipated. "Dan was one of the only people who truly cared about what was best for Barack," one supporter explained, and "I became very disenchanted with Barack" over his failure to appreciate how absolutely essential Shomon had been to his political success. As Jimmy put it, Dan "took him a long ways on his journey."

The tensions between Gibbs and Shomon were mild compared to the problems generated by newly hired African American fund-raiser Vera Baker. As soon as she arrived in Chicago, it became clear she was "very, very high maintenance," because whether it was her apartment or a health club membership, "nothing was good enough," one supervisor recalled. "It was so unpleasant," and "the person who had to deal with it day to day," Joe McLean explained, was new national fund-raising director Susan Shadow. "It was extremely difficult to work with a prima donna," and even worse with someone who "didn't seem to have any common sense" and whose ability to organize events fell woefully short of what DSCC officials had said she could do. Two events Baker had responsibility for "were disasters," with the far more able duo of Jenny Yeager and Jordan Kaplan picking up the slack. "It was just a frigging nightmare constantly," and "Barack was very upset with one of the D.C. events she did." At least a trio of superiors urged that Baker be terminated, but fears of offending the DSCC, where Baker had once been deputy political director, stood in the way.

But more than Baker's professional competence was at issue. "It was a disaster on a lot of levels," one consultant explained, and "we knew immediately" that there was an additional problem "that was not fixable" and that posed a risk to the candidate. The campaign's most inimitable voice put it bluntly: "she liked hanging off his elbow and being in the rooms" when Barack met with famous donors like George Soros. "It was touchy," one witness explained, and "it took us a long time to fix it" given the DSCC's role. Then one remark by an unfamiliar stranger sealed it. Cauley recalled that "Vera was off his elbow and somebody went up to Valerie Jarrett and said, 'Wow, Michelle looks great tonight.' That was the beginning of the end of Vera Baker." A trio of top campaign members all attested to Jarrett's decisive role in the "hard decision" to relocate Baker from Chicago back to D.C. as soon as possible. As Joe McLean phrased it, "we eliminated the risk fairly early, but the problem was still there."[75]

In public, Barack's campaign was going incredibly well. A *Chicago Tribune* poll showed him with a commanding 52–30 percent lead over Jack Ryan. Barack's favorability ratio was 46–9, with 20 percent of voters saying they had not heard of him, while Ryan's favorable-to-unfavorable score was a worrisome 29–25. *New York Times* columnist Bob Herbert instructed readers to "remember the name Barack Obama," and a Manhattan fund-raiser hosted by George Soros was soon followed by one in Washington at Hillary Rodham Clinton's home. Simultaneously, Jim Cauley and Darrel Thompson called on top figures in John Kerry's campaign to lobby for a prime-time speaking slot at the convention on one of the evenings when network television would broadcast the proceedings. They left a videotape of Barack's three TV spots plus footage of Barack speaking at his March 16 victory celebration. "We were really making a strong case and a strong pitch," Darrel recalled.

On June 13 Barack publicly challenged Ryan to six debates between August and October, and Ryan responded by calling for ten. Ryan also asserted that Barack favored universal health care, and Barack inaccurately replied, "I have never advocated" for what he called "a much larger national program," saying, "there are a whole host of ways to do it." Barack's campaign team was "eager to do some qualitative research downstate," as John Kupper explained, to learn how best to present their oddly named biracial candidate to white voters who remained largely unfamiliar with him and his record. "We were kind of astounded that Barack did so well with downstate voters in the primary" and "wanted to find out if that possibility extended to general election voters as well," Kupper recalled. Paul Harstad and Mike Kulisheck organized two focus groups in Peoria for the evening of June 16. Among thirteen women in the first, eight professed support for Jack Ryan and only one for Barack, with another asking, "Isn't he Iranian?" But once Harstad played video excerpts of the two candidates, seven expressed support for Barack, with Ryan drawing five. When one woman compared "Obama" to "Osama," Paul asked if Barack's name was a real barrier, and multiple women said yes.

Among the ten men in the second group, three expressed support for Democrat John Kerry and five for President George Bush, with one volunteering that "it's unfortunate that Senator Fitzgerald . . . is leaving. . . . I think he did a good job. He took stands on issues." Turning to Barack and Jack Ryan, one man said, "the fact that he won't let Obama go to a restroom without a cameraman tells me he's an ass, frankly. That's why I won't vote for him." When Harstad played the videos, two participants liked how Barack spoke of "we" rather than "I," with one calling Barack "very charismatic" and a second terming him "a strong presence." In the end, six of the men favored Ryan and four Barack. Reviewing the evening's full transcripts and the participants' numerical score sheets, Harstad and Kulisheck highlighted participants' "longing for candidates who bring people together" and Barack's high marks for seeking "common ground" and avoiding "negative politics." Barack's greatest vulnerability was the charge that he "favors a Hillary Clinton–style universal health" care program, and among other dangers, "the guns issue was totally eclipsed by the gay marriage issue in these groups." Although "it did not take too much information to swing" a majority of the women "to support Obama at the end," such a shift did not occur among the men.[76]

Early in June, a court-appointed referee recommended unsealing only a small portion of Jack Ryan's divorce file. Ryan told reporters that the documents would cause "no problems for the campaign. Would there be something that might be embarrassing to me? Maybe." But when the *Chicago Tribune* challenged that recommendation and sought a full unsealing, Los Angeles Superior Court judge Robert A. Schnider sided with the newspaper rather than the referee. "In the end, the balance tips slightly to the public," Schnider held, because "protection from embarrassment cannot be a basis for keeping from the public what's put in public courts." Schnider accepted Jack Ryan's argument that release of the file would harm his young son, acknowledging that "the nature of the publicity generated will become known to the child and have a deleterious effect on the child," but he set release for twelve days later. The *Tribune* gave its victory front-page coverage and reported that Ryan privately had told top Republicans that the file would be embarrassing but he could "weather" the disclosures. Ryan announced he would not appeal the decision and sought the judge's approval for quicker release of the file the following Monday.

On June 21, reporters were notified that copies of the documents would be made available at a 4:00 P.M. Ryan press conference at the Chicago Hilton. That hour came and went with no candidate and no copies, but at 5:00 P.M. waiters brought in flowers and food. "Was this a party?" columnist Carol Marin wondered. Two more hours passed before the documents arrived, accompanied by a raft of Ryan supporters attesting to his virtues before the candidate appeared at 8:00 P.M. Inarticulate and unwilling to admit the documents were highly embarrassing, Ryan's hour-long appearance devolved into what the suburban *Daily Herald* called a "marathon, often surreal news conference."

The next morning's *Tribune* declared "Ryan File a Bombshell—Ex-Wife Alleges GOP Candidate Took Her to Sex Clubs." In New York, "respondent wanted me to have sex with him there with another couple watching. I refused," Jeri Lynn Ryan had stated in a sworn declaration. "Respondent asked me to perform a sexual activity upon him, and he specifically asked other people to watch. I was very upset." At the Paris club Ryan took her to, "people were having sex everywhere. I cried. I was physically ill. Respondent became very upset with me and said it was not a 'turn-on' for me to cry."

On Monday, Barack had been on a "Downstate Jobs Tour," with appearances in Springfield and East Alton preceding a dinner in far south Carbondale, where Dan Hynes introduced him as "one of the greatest United States senators this country will ever see." Early in the day, Barack had rebuffed reporters' questions about Ryan's divorce by saying it was not among the issues impacting downstate families. In Carbondale, Barack went out of his way to insist, "I've said clearly and unequivocally, I'm not in favor of gay marriages." When reporters then pressed him on the evening's new Ryan revelations, Barack again emphasized that "people are not talking to me about Jack Ryan's personal life."

Top Illinois Republicans faced the same questions as Barack, and some were eager to respond. Ray LaHood said he was "shocked" and that Ryan "needs to immediately withdraw from the race. There is no way Republicans in Illinois will vote for somebody with this kind of activity in their background." Tuesday's *Chicago Sun-Times* reported that state party chair Judy Baar Topinka "believes Ryan lied to her," and former governor Jim Edgar was "stunned" and "furious." Until that evening, Edgar had been defending Ryan, but Edgar said that what came out "was far worse than what he had told me." Ryan had "misrepresented what was in there to me" and "what he had told me was completely wrong."

A wire service correspondent asserted that "a married man trying to have kinky sex with his wife seems tame," but an Illinois Republican told *Roll Call* that "it's not so much what's in the files, it's the fact that he lied about it" that left GOP leaders so angry. *Capitol Fax*'s Rich Miller wrote that "most everyone has figured for weeks now that Ryan couldn't beat Barack Obama. The big fear now, though, is that Ryan might bring down the rest of the Republican ticket." Miller predicted that "this is going to get very ugly, very fast," and by late Tuesday, Ryan and Topinka were publicly disagreeing about their prior conversations. "I remember her asking me whether there was anything in the documents that would preclude me from being a U.S. Senator, and no, there isn't," Ryan asserted. "What's in those documents at its worst is that I propositioned my wife in an inappropriate place. That's the worst." Topinka's version was that "I said, very specifically, twice, so that I would hear it twice, 'Is there anything in your divorce document that would be personally embarrassing to you or to the Republican Party?' He said 'no' both times."

Wednesday's *Sun-Times* featured an unnamed Illinois Republican saying, "I don't think anyone can overstate the impact of the lying and misleading statements he made. It's not just to Judy. It's to a whole host of people. That's severely damaged his credibility." The *Tribune*'s editorial page declared that "Ryan was not honest with Republican primary voters," and the *Sun-Times* explicitly called for "Ryan to Hit the Road," castigating "his unprincipled behavior." Downstate papers like the Bloomington *Pantagraph* chimed in similarly, saying "Ryan should withdraw." At midmorning Wednesday, Illinois Republican U.S. House Speaker Dennis Hastert canceled a Ryan fund-raising event scheduled for that evening, and on *The Tonight Show* host Jay Leno joked that Ryan was "going after the 'swing vote.'"

By Thursday, some Republicans were speaking up on Ryan's behalf. A column in the conservative *Illinois Leader* asserted that "the real problem the ILGOP had with Jack Ryan is he wants to keep U.S. Attorney Patrick Fitzgerald," the federal prosecutor who had unraveled former governor George Ryan's bribe-filled tenure as secretary of state. Then lame-duck U.S. senator Peter Fitzgerald said he had told Ryan to stay in the race and that the lack of support from party leaders like Topinka was doing Ryan more harm than the divorce documents. Fitzgerald complained that "she was so supportive of George Ryan and was never offended by the indictments or the corruption there, but she is so offended" by the "unproven allegations" in Jack Ryan's divorce. "It does not add up to me."

In the *Tribune,* an op-ed contributor asked why shouldn't Ryan "be judged on the quality of his ideas and the positions he would advocate in the Senate, rather than on divorce negotiations?" Blair Hull also had a second question: "why under the circumstances would anyone who hasn't spent his or her entire life planning a political career and cautiously behaving accordingly . . . ever run for office?" Instead of pursuing divorce pleadings, "the Chicago press corps should be hounding Jack Ryan with questions" about "why he supports a Bush White House policy in Iraq that is so clearly a costly failure" and "why he supports a Medicare prescription drug bill that was a taxpayer giveaway to the pharmaceutical special interests." Just like Barack, Hull also said that questions about Ryan's divorce were not among "the issues that make a difference to Illinois families," and he criticized "a press obsession with personal lives when so many public policy issues demand a thorough debate."

In Springfield, where the legislature was still in session thanks to the ongoing state budget standoff, Barack remarked that it was "probably not surprising" that the press had fixated on Jeri Ryan's claims. "So much of our culture is caught up in celebrity and sensationalism. It's an unfortunate aspect of our culture generally, and our politics ends up taking on that same flavor." Barack said, "I do regret the personalization of politics of this kind. It's not something that I wish on anybody." Ryan's salacious problem drew new national attention to the Illinois contest, with *Washington Post* columnist E. J. Dionne noting speculation that his former Saguaro Seminar colleague "may be the first African-American president of the United States." Similarly, a *New York Times* news story stated that Barack "is widely regarded as a rising Democratic star" who in a few months' time would be a U.S. senator.

Ryan's campaign was polling to learn how much damage had been done. A new *Daily Southtown* poll showed Ryan trailing Barack 54–30, and while 49 percent of respondents viewed Ryan unfavorably, 57 percent said he should remain in the race. In hypothetical matchups, Barack led former governor Edgar by 45–42 percent and Senator Fitzgerald 47–40. Republicans publicly compared notes on who could replace Ryan, but neither Edgar, Fitzgerald, nor Topinka expressed interest. Conservative state senators Kirk Dillard and Dave Syverson supported Steve Rauschenberger, whose underfinanced primary campaign had led him to a third-place finish behind Ryan. Neither Topinka nor Republican national committeeman Robert Kjellander had any love for Rauschenberger, who, like Fitzgerald, had publicly attacked Republican insiders for profiteering off state government business.

On Friday morning, the Ryan campaign's poll mirrored the *Southtown* numbers, with Barack holding a better than 20-point lead. Ryan summoned his top staff to his Gold Coast apartment to discuss the results. "This can be done, but it's going to get ugly," one participant argued, advocating that if they "go nuclear on Obama for four months" it could become a competitive race. "There is enough good opposition research to defeat Barack Obama, to make clear to the voters of Illinois that he should not be a member of the United States Senate." But Ryan rejected that course. "This is not the way I saw this race," he explained. "I don't want to be remembered as the guy that ran that kind of campaign." Peter Fitzgerald called, and Ryan told him he would soon issue a statement withdrawing from the race. Ryan assembled his full staff to tell them his decision. "I believe that one man, living for purposes larger than himself, can make a difference," Ryan told the press. He rebuked the *Tribune* for its "truly outrageous" pursuit of the divorce file and said he had no interest in mounting "a brutal, scorched-earth campaign." Jim Edgar stated that Ryan had "made the right decision," and Topinka announced that Republicans would select a replacement within three weeks. Steve Rauschenberger declared that "if the circumstances are right, and we can organize a campaign to win, I'm very interested," but the *Daily Herald* reported that Peter Fitzgerald "hopes Rauschenberger is too smart to enter a race that is already lost. 'I would not encourage anyone who is my friend to accept the nomination.'"[77]

Barack offered compassionate words. "I can only imagine what Jack and his family have gone through over the last several days. I admire Jack's commitment to public service," and "I deeply admire the work that he has done as a teacher" at Hales Franciscan High School. Although Barack was now without an opponent, his campaign team plowed ahead, with policy director Amanda Fuchs relying on long-standing volunteers like Raja Krishnamoorthi and new ones like law professors Cass Sunstein and Elizabeth Warren to prepare position papers on everything from homeland security to consumer protection policies. On Saturday, June 26, Barack delivered the Democratic response to President Bush's weekly radio address, recording a text about job loss that had been written by David Axelrod and Robert Gibbs. His delivery was flat and uninspired, and Gibbs recalled that "it was kind of obvious that he was recording the words of somebody else."

Michelle Obama had begun making occasional campaign appearances by herself, extolling her husband's virtues to small audiences in downstate towns like Charleston. Darrel Thompson had taken the lead in organizing more than

130 "Ba-Rock the House" fund-raising parties all across Illinois for Tuesday evening, June 29, and that morning John Kerry let slip that Barack would address the Democratic National Convention, telling reporters, "I cannot wait to hear his voice." Late that afternoon as Barack, Darrel, Robert Gibbs and Mike Signator headed west to the suburban Wheaton house party Barack would attend, Gibbs received a message that within moments, Kerry campaign manager Mary Beth Cahill would be calling to speak with Barack.

"Jack Corrigan probably mentioned him first to me," Cahill remembered, after first Pete Giangreco's mid-April call to talk up Barack as a prime-time DNC speaker and then an early June conversation Jack had with another old friend from the 1988 Dukakis presidential campaign—Lisa Hay, who had then entered Harvard Law School and become the *Law Review*'s treasurer. Corrigan had asked Hay to support John Kerry, but Lisa told Jack she wanted to contribute to her old friend Barack Obama. "You know him?" Corrigan responded. "I think he's great," Lisa replied, and she recounted how Barack's remarks at the *Law Review*'s April 1990 banquet had been so powerful that even the waiters at Boston's Harvard Club had paused to listen. "It was the most moving speech about the promise of America that I had ever heard. And I remember it struck me that Barack touched everyone in the room," Hay told Corrigan. "That's when I put him on the list for keynote," Corrigan remembered. Michigan governor Jennifer Granholm had been Corrigan's top pick, but Lisa's account altered Corrigan's plan. "I'm thinking about making a pretty unusual recommendation for the keynote speaker, and I could really be stepping in it here," Jack told Vicky Rideout, a lifelong friend whom he had recruited "to come and be the head of speechwriting for the convention."

Together they watched the video of Barack that Jim Cauley and Darrel Thompson had given the Kerry team, and Corrigan shared his thinking with campaign manager Mary Beth Cahill, who spoke with Kerry every day. "I told him that I was leaning toward this," Cahill remembered, and Kerry responded affirmatively. "He was happy with the recommendation of Obama" because Barack had "definitely made an impression upon him" back in April when they had campaigned together. Shortly before June 29, Pete Giangreco again called Corrigan to lobby for Barack, but missed the point when Corrigan said something like, "I think you're going to be pretty pleased." Jack's "being opaque, and I don't get it," Pete explained.

In the SUV headed to Wheaton, Barack's end of the conversation with Cahill was simple. "Thank you . . . Thank you . . . Thank you." But Barack was overjoyed. "I know exactly what I want to say. I really want to talk about my story as part of the larger American story," and he immediately told his aides that unlike the radio address, this one he would write himself "in a way that was personal," Robert Gibbs recalled. That evening, speaking to the one-hundred-plus house parties all across Illinois via conference call, Barack told his supporters, "I didn't get into this race to run against anyone. I got into this race to run *for* something: a set of ideals, a set of values."

Barack was back in Springfield for most of the rest of the week because of the ongoing state budget deadlock. "After I'd scribbled some notes, I wrote it in about three nights" on a yellow legal pad in his room on the twelfth floor of the Renaissance Hotel. "I just sat in a hotel room watching a basketball game and wrote it up, most of it, sort of in one sitting," Barack recalled. His initial draft was about twenty-four hundred words long, and with little of substance taking place on the Senate floor, Barack could sit at his desk, tinkering with the language. On Friday, he and a camera crew headed forty-five miles south to Carlinville to film a conversation David Axelrod hoped to use in fall television ads. Barack boasted about his skill at poker, saying, "I'm using it to bankroll my kids' education, because I'm a pretty good player so I usually win." But he also spoke with unusual frankness about his family's financial struggles. "My wife . . . if we could afford it, she would stay at home until the three-year-old is in first grade, but we had to make sure that she was still working just because I'm in public service, so I don't make a lot of money, and we just couldn't afford a situation where she wasn't working. But what that means is that we are paying someone to come in the house and look after the kids. It's like having someone on payroll—it's like you're paying somebody, but you're not making a profit on it. You've got to pay their taxes, their Social Security, you've got to pay workers' compensation. So it's a huge challenge, a huge struggle."

Shifting to policy issues, Barack noted that "our child support system in Illinois is terrible and it's been terrible for like a decade now." The state was not "going after these fathers and making sure that they take responsibility for their kids. My mother was a single mother . . . and so I know how hard it is for a single parent to try to raise kids. It's a struggle." Changing tack, Barack stated that the Patriot Act "requires you speaking out and voting no. We only had one U.S. senator, Russ Feingold from Wisconsin, vote against the Patriot Act out of one hundred. And that really is a shame." Barack's remarks were of mixed value for advertising purposes, but with his campaign having taken in $4 million from more than eighty-five hundred donors between April and June, his fall media bankroll was hefty indeed. The new issue of *Time* magazine described him as "the hottest property in this year's Senate races," and a *Chicago Tribune* story reporting his fund-raising windfall observed that "Obama's apparent strategy is to ward off a deep-pocketed or popular Republican from seeking the GOP nomination."[78]

With state senator Kirk Dillard saying that Steve Rauschenberger is the "overwhelming choice of suburban Republicans," public discussion focused on the Elgin legislator. Yet Rauschenberger wisely hesitated, seeking assurances from national Republican campaign committees that funding would be available for a serious race. "I did not have the financial resources to get the campaign going," and given Barack's huge head start, it soon became clear that it would be "impossible to raise the kind of money needed." Rauschenberger declined, and while some young Republican activists called for Jack Ryan to reenter the race, speculation next turned to fiery former Chicago Bears football coach Mike Ditka.

Some of Barack's staff were initially "really, really worried" about facing an opponent who was universally known, but with the *Tribune* reporting that Ditka was not registered to vote and columnists highlighting his explosive temper, he too stood aside. Ditka later said that not challenging Barack was the "biggest mistake I've ever made," and Barack professed disappointment, saying, "I was looking forward to what was sure to be a fun, exciting race with the coach." Even the *New York Times* covered Ditka's withdrawal, describing Barack as "highly popular and lavishly financed." With Illinois Republicans becoming desperate, Barack teasingly asked his Senate friend Dave Sullivan to enter the race. "'Sully, why don't you run against me?' 'Sure, Barack. The first thing your campaign will do is produce the check I wrote to you when you ran against Bobby Rush.'"

Eager to burnish his slim record on foreign policy, Barack delivered an address to the Chicago Council on Foreign Relations. He singled out "Iran, which the Bush Administration has correctly targeted as a dangerous cheater in the nuclear game," as a top concern, and stated that "our first and immutable commitment must be to the security of Israel, our only true ally in the Middle East and the only democracy." Journalists covering John Kerry's presidential campaign reported that Barack, "a rising African-American star," would deliver a prime-time DNC speech, but only on July 14 did the Kerry campaign publicly announce that Barack would deliver the convention's much-coveted keynote address on Tuesday evening, July 27. Kerry termed Barack "an optimistic voice for America," and Mary Beth Cahill, noting that the three major networks were not planning live coverage on Tuesday, told reporters she hoped Barack's selection would change their minds. Barack called the selection "an enormous honor and enormous responsibility," admitting "it's not something that I would have ever anticipated this early in my career. As my wife said, I better not screw it up." Barack wanted to "move past the politics of division toward the politics of hope," and he distinguished himself from President Bush by saying that "my philosophy is that there's nobody you shouldn't talk to." He complained to reporters that "there's been a corporate takeover of my life," and the next morning, on NBC's *Today Show,* host Katie Couric introduced Barack to viewers as "a rising star."

By then, David Axelrod and especially John Kupper had been hard at work trying to shorten the draft Barack had shared with them via a 1:00 A.M. e-mail. Axelrod and his wife were vacationing in Florence, Italy, and everyone was concerned that Barack's text was longer than what DNC speechwriting chief Vicki Rideout was looking forward to receiving. Kupper recalled that "I would cut out parts of the speech and send him the revised draft, and it would come back with almost all of them restored. This was a painstaking process. Eventually we got it down to seventeen minutes." Barack wanted to emphasize the resonant phrase "the audacity of hope," and in one exchange, Axelrod passed along word that Barack had to reference Kerry's military service. Almost daily, new, slightly tweaked drafts were created, with shifting word counts—"2,370 words"—noted on many of the *thirteen* successive drafts Kupper reviewed. At the convention in Boston, Barack would deliver the speech using a teleprompter, and his campaign leased one so Barack could begin practicing with the unfamiliar device. An initial Sunday-morning run-through in the big volunteer room at campaign headquarters in Chicago was "very good," Darrel Thompson thought, and then the teleprompter was moved to a conference room in the building housing Axelrod and Kupper's offices, so that Barack could continue working on his delivery in quiet quarters.[79]

While Barack was busy practicing, invaluable contributor Tony Rezko was drawing unfavorable press attention. *Capitol Fax*'s Rich Miller had warned months earlier about "the sort of influence peddlers who surround the governor," but only in midsummer 2004 did the *Chicago Sun-Times* and then the *Tribune* begin to cover Rezko's behind-the-scenes role in choosing gubernatorial appointees for such obscure but powerful entities as the Illinois Health Facilities Planning Board (IHFPB). Rich Miller observed that if Blagojevich's "buddies are somehow mixing campaign contributions with appointments . . . all he has to do is look at George Ryan to see how his future will end up." In mid-July, the *Sun-Times* reported that Rezko had gone "weeks without paying employees" at a fast-food chain he owned, and two days later, the *Tribune* detailed how a podiatrist and Rezko business partner who had contributed $25,000 to Blagojevich a year earlier had been rewarded with a seat on the IHFPB.

With Republicans still lacking a candidate to oppose Barack, state Senate friend Kirk Dillard toyed with entering the race. "Barack Obama is beatable," Dillard declared. "Somebody in the Republican Party needs to do some research on Barack's record," because Barack "is far to the left of where most Illinoisans are." In a radio interview, Dillard revealed that "Barack and I kidded yesterday that it would probably be the most gentlemanly" face-off ever, but it "would be good for politics, not only in Illinois but nationally, to have people who genuinely admire each other and their families to wage a totally above-the-board, on-the-issues race." Barack "is a wonderful man" and "has a great future in whatever he does," but "Barack Obama is beatable," Dillard insisted. Asked about Dillard, Barack responded that "with someone of his caliber, the race tightens immediately." In a decidedly odd gesture, a *New York Times* editorial called upon Dillard to enter the race. "Run, Kirk, run. Illinois needs you—and so does Mr. Obama," because "Illinois voters deserve to see a capable opponent force him to answer tough questions and defend his positions." But just like Rauschenberger and Ditka before him, Dillard said he would give the race a pass, citing his two young daughters.

In a cable-television interview, Jeff Berkowitz asked Barack how his recently professed opposition to a single-payer health care plan contradicted what he had said on Berkowitz's show in November 2002. "The problem we have is what kinds of time frames are we talking about," Barack responded. "Do I think that over time we are essentially going to have a system in which you have got a market system but people are getting basic health care of some sort and a market system superimposed on that for discretionary things—cosmetic surgery or elective surgery?" Barack added before Berkowitz interrupted. "There is going to be some sort of coverage that we want to provide everybody," Barack explained, "minimum coverage. . . . Some people are going to be getting that through their jobs," but "there are going to be other people that don't have that." As Berkowitz accused him of "backing off from single-payer," Barack insisted, "I have been consistent in saying what I am in favor of is universal health care . . . we have to move in a direction of universal health care, because if we don't, the question I have for people who argue against me is, 'Who is it that you think shouldn't be covered?'" From there, the conversation devolved into cross talk as school voucher enthusiast Berkowitz sought to interest Barack in health care vouchers.

Early in the week of July 19, Barack finally sent his speech to Vicky Rideout in Boston. Stuck in Springfield as the Four Tops struggled to reach a budget accord with Blagojevich, Barack spoke with Associated Press reporter Chris Wills, who had covered the statehouse since 1998. Editors titled Wills's story "Rising Democratic Star Readies for His Moment in the Lights," and when a Honolulu-based AP writer followed up, Barack sounded every bit a native in anticipating his next visit. "I look forward to getting up in the morning, driving to Sandy Beach and doing some bodysurfing and then getting a shave ice and plate lunch." *Capitol Fax*'s Rich Miller expressed astonishment at "Obama's superstar status," but Barack told the *Journal-Register*'s Bernie Schoenburg that his daughters preferred to stay in Chicago and attend day camp rather than accompany their parents to Boston.

Barack had to be in Boston no later than first thing Sunday, July 25, for an early-morning taping of NBC's *Meet the Press*. With the legislature hoping to complete its business and adjourn sometime on Saturday, NBC sent a chartered jet to Springfield to fly Barack directly to Boston. Only a little after 8:00

P.M. did the state Senate finally adjourn. Barack was pleased that an additional $55 million was allocated to expand FamilyCare, but the late hour meant that only at 1:30 A.M. did he and Michelle arrive at Boston's Back Bay Hilton.

At much the same hour, a long-delayed rental truck pulled up at the Fleet Center's loading dock in Boston's North End. Ten days earlier, just after the public announcement of Barack's keynote role, GOTV mastermind Tom Lindenfeld had called Jim Cauley and firmly advised him that if Barack's address was going to have maximum impact on television, the campaign needed to ensure that the crowd on the convention floor had readily visible "Obama" placards in hand. Tom had long been a DNC floor manager, and he warned Jimmy, "You've got to get these signs printed up. If you don't have the signs, your speech isn't going to be the same." Cauley hesitated, because printing the signs—with a huge white "Obama" on a blue background plus a new, shorter Web address, www.obama2004.com—came with a $20,000 price tag. Cauley wanted the DNC or Kerry's campaign to pick up the tab, but Lindenfeld was insistent. "I like beat the shit out of them over this," and Jimmy finally relented. The signs needed to arrive in Boston by Friday. Late that week campaign staffer Adam Stolorow helped twenty-one-year-old intern Alex Okrent load the newly printed placards into a rental truck before Okrent set out on the fifteen-hour drive to Boston. Then, somewhere in Ohio, the truck broke down, and only in the wee hours of Sunday morning did Alex call Lindenfeld with tardy good news: "We're here."

Barack was up before 6:00 A.M. to get ready for *Meet the Press,* then CBS's *Face the Nation,* ABC's *World News Tonight,* and finally CNN's *Late Edition.* On *Meet the Press,* Barack told host Tim Russert that he had been fortunate to grow up in Hawaii and endorsed comedian Bill Cosby's call for African Americans to demand more of each other. "He's right. . . . There's got to be an element of individual responsibility and communal responsibility for the uplift of the people in inner-city communities." On *Face the Nation,* Bob Schieffer called Barack "kind of a rock star of Democratic politics," and on ABC's *World News Tonight,* anchor Terry Moran told Barack that one Democrat had said that "most people who know Barack Obama believe that he will be a presidential candidate in the very near future." Barack demurred, saying, "I may be flavor of the month this week," but CNN's Wolf Blitzer raised the same question on *Late Edition,* asking, "Do you want to be president?" Barack replied that he wanted to be the best U.S. senator "that I can be," and when Blitzer followed up, saying "there's people talking about this," Barack again ducked. "That's silly talk. Talk to my wife. She'll tell me I need to learn to just put my socks in the hamper."[80]

In addition to the television tapings, Barack also met with Michael Sheehan, a prominent speaking coach who had tutored every top Democratic National Convention speaker since 1988. David Axelrod already had called Sheehan to sing Barack's praises. "This guy's special. This guy could be the one." Sheehan recalled that in two decades of knowing Axelrod, "I never heard him that complimentary" about anyone, and Michael knew "they're always looking for . . . a rising star." The practice podium and teleprompter were located in a hastily converted locker room in the bowels of the Fleet Center, and on Sunday afternoon Barack had his first one-hour session with Sheehan. The essential thing for a first-time convention speaker to appreciate was "the difference between what you hear in the hall and what is heard at home on television," Sheehan stressed. "You hear everything in the hall. You don't hear everything at home," so a skillful speaker should "surf the applause" rather than pause for too long. "As soon as you hear the volume start to drop, start talking again," Sheehan explained. "You can talk on top of it, and we still hear you." No matter how loud the hall became, "don't yell," because "you don't have to scream" for television viewers to hear you.

Kerry aides Vicky Rideout and John Corrigan watched silently as Barack worked on the pacing of his remarks under Sheehan's guidance. By now, Barack barely needed the teleprompter because he had his text almost perfectly memorized, yet given the small, windowless practice room, "you can't really orate in that circumstance," so Barack's delivery seemed flat. "First rehearsals are often underwhelming, and this one was no exception," Rideout recalled. "I remember conversing with Corrigan after that and being like, 'Well, I don't know. I'm not sure if he's got what it takes or not.'"

A second rehearsal session was set for Monday, and on Sunday night an exhausted Barack went to bed at 9:30 P.M. His schedule was filled with almost nonstop media interviews, and both the *New York Times* and the *Chicago Tribune* published front-page profiles. "Political Phenomenon Obama Vaults into National Spotlight," read the *Tribune* headline, with reporter David Mendell noting that Barack "has been riding a wave of adoration by the national media." Barack calmly observed that "the hype obviously has reached a fever pitch, and I think the fever will break." Congressman Jesse Jackson Jr. confessed, "I pray for him every day," and warned that "we can't allow him to be overburdened with too many expectations." A *Chicago Defender* editorial echoed Jackson's sentiments. "There needs to be a tempering of expectations" because "it is vital that Obama not be considered the next John F. Kennedy or the next Martin Luther King Jr. or the next Jesse Jackson Sr. He should be given the time and opportunity to stand on his own two feet and put forth a legislative agenda and record that will do us proud."

At a Monday lunch with *Tribune* journalists, Barack conceded that on Iraq, "there's not that much difference between my position and George Bush's position at this stage." Conservative columnist John Kass confessed, "I couldn't help but be impressed by the man," calling Barack "the real thing." The reporters described how "nearly everywhere he goes" in Boston, Barack "is mobbed by well-wishers, reporters and others who want a moment of his time. 'It's getting incrementally harder to move him through crowds,'" Jim Cauley told them. Late that afternoon Barack's buddy Marty Nesbitt witnessed the same phenomenon. "We were walking down the street in Boston, and this crowd was growing behind us. . . . I turned to Barack, and I said, 'This is incredible. You're like a rock star.' And he looked at me and said, 'If you think it's bad today, wait till tomorrow.' And I said, 'What do you mean?' and he said, 'My speech is pretty good.'"

In Honolulu, Madelyn Dunham told a caller that she was "a little overwhelmed," because "this has all come on the national level really fast." In a recent phone call, "I told him to smile when he's on TV," and "of course, I'm proud of him." At a sunny, early-evening outdoor reception hosted by Rod Blagojevich, Barack ran into former law student Lisa Ellman, an Illinois delegate, who recalled him saying, "I'm excited but I'm not nervous," explaining that giving the keynote speech would not be as tough as teaching a U of C law class because there would no hard questions from smart students.

On Tuesday morning, newspapers on both sides of the Atlantic all featured the same phrase. New Jersey senator Jon Corzine told the *Boston Globe* that Barack was "a rising star," and the *Philadelphia Inquirer,* Toronto's *Globe and Mail,* and Britain's *Guardian* all applied that label too. In Chicago's suburban *Daily Herald,* DuPage County Democratic chairman Gayl Ferraro remarked that "it would be great for Illinois if he was to become the first African-American president," and a column in Britain's *Independent* asked, "Can Obama Be First Black President?" and emphasized that by 2012, Barack would have eight years experience as a U.S. senator.

Barack was up before 6:00 A.M. on Tuesday for appearances on ABC's, CBS's, CNN's, and NBC's morning shows, plus PBS and NPR tapings. ABC's Diane Sawyer introduced Barack as a "rising star" on *Good Morning America,* as did Hannah Storm on CBS's *Early Show.* Welcoming Barack to NBC's *Today,* Katie Couric told viewers that "some are already saying he could be the first African American president." ABC and NBC interviewed Michelle as well. "Don't screw it up" was her advice for tonight, and "I still remind him, 'You still have to win my vote, buddy.' And when I get mad at him, 'I'm not voting for you.'" When NPR's Melissa Block asked Barack about his late father, Barack responded, "I think in some ways I still chase after his ghost a bit, but also I think I try to balance the importance of family with my career in ways that he wasn't able to accomplish."

Barack spent the rest of the morning and early afternoon doing remote interviews with *ten* Illinois TV stations as well as speaking at a League of Conservation Voters rally. For lunch he got a packaged sandwich from a convenience store, and in search of personal privacy, he happily chose a port-a-potty. "When I go into the regular restroom, all these people want to shake my hand, and that's not the place I want to be shaking hands," he told David Mendell.

A little before 3:00 P.M., Barack headed back to the Fleet Center for his final rehearsal session with Michael Sheehan. On the way, word arrived that John Kerry's speechwriting team wanted Barack to remove five sentences from his text because they too closely echoed a line in the acceptance speech Kerry would deliver on Thursday night. "Obama was furious," David Axelrod recalled. Cauley remembered that "Axelrod was in the van, I was in the van when we thought John Kerry was stealing a piece of the speech, and Barack said, 'Fuck him. That fucker is trying to steal a line from my speech. They didn't have that in Kerry's speech. They saw it, they liked it, and now they're stealing it.' Barack was *real* frustrated with that," and Jimmy had never seen Barack so angry.

The Kerry team was divided, with speechwriting chief Vicky Rideout believing that Kerry's personal speechwriters were overreaching about what they wanted Barack to delete. "It was the emotional peak of the speech," Rideout realized, and when Barack called her "and made the case of why it was important to him to keep it in there" and pressed as to whether Kerry himself was requesting the change, Rideout was entirely sympathetic. "There was really not a very good reason for them wanting him to take it out," and Vicky pushed back. "Do you really understand what you're asking of Obama here?" she told Kerry speechwriter Josh Gottheimer. Rideout called Barack back. "It's Kerry's convention," Vicki explained, but she thought Barack could deal with the perceived overlap by rewording one sentence in his text rather than removing the entire paragraph.

Soon Barack was back in the Fleet Center's "blue room," almost directly underneath the stage where six hours later he would speak to a national audience. A dour David Axelrod, a pensive Robert Gibbs, and a glum Michelle Obama sat along the wall and watched as Barack went through his final rehearsal with Sheehan. Then the juniormost member of Kerry's speechwriting team, Jon Favreau, arrived to ask if the offending sentence—"We're not red states and blue states; we're all Americans, standing up for the red, white, and blue"—had indeed been dropped. Favreau remembered that Barack "kind of looked at me, kind of confused, like 'Who is this kid?'" before evenly asking, "Are you telling me I have to cut this line out?" Sheehan recalled that Favreau told him, "We'd like you to take it out," and Barack replied, "No. It's staying in."

Axelrod asked Favreau to join him in the hallway to see if they could work out a compromise. Kerry's text stated, "Maybe some just see us divided into those red states and blue states, but I see us as one America: red, white, and blue." Looking at Barack's "not red states and blue states," Axelrod instead wrote, "We are one people, all of us pledging allegiance to the stars and stripes, all of us defending the United States of America." Favreau concurred, remaining in the hallway with Rideout as Barack's session continued. Then Gottheimer called to say that he wanted Barack's entire paragraph removed. Rideout recalled, "I was standing outside of the rehearsal room" when Gottheimer called Favreau. "He was on the phone with them when they're saying, 'No, actually it has to be the whole paragraph that comes out.'" Then Rideout instructed Favreau, "Tell him you can't hear him," and then she took the phone herself. "I just did an 'I can't hear you,'" she said. "'Are you there? Hanging up now.'" The battle was over, and thanks to Rideout, Barack had won. "It stayed in," she self-effacingly explained.

Barack was due back at the Fleet Center that evening for a 9:45 P.M. curtain call. Illinois senator Dick Durbin would introduce him, but DNC officials had caviled when Barack said that he wanted Michelle with him backstage. DNC officials also resisted when Darrel Thompson from Barack's campaign had protested the DNC's initial provision of only four floor passes. With Darrel, Joe McLean, and Robert Gibbs all working the phones, a podium pass for Michelle materialized along with a DNC apology and the offer of John Kerry's personal skybox for that evening for Barack's friends and donors. On the way back to the Fleet Center, Barack called Madelyn Dunham in Honolulu. Backstage, Chicago political consultant Kevin Lampe, the Podium Operations Team's "speaker tracker" assigned to Barack, was ready with a large-type hard copy of Barack's text in case the teleprompter failed. Debating with Michelle and several aides which of several ties to wear, Michelle's dislike of the entire lot led someone to suggest the one Robert Gibbs was wearing. Everyone agreed, and Gibbs removed his tie and Barack put it on. Only seven years later would he ceremonially return it. As Durbin delivered his introduction, Michelle hugged her husband and invoked her favorite line: "Just don't screw it up, buddy!"[81]

Walking out onto the thrust stage that almost put speakers among the crowd, Barack shook Durbin's hand and hugged him before turning to the podium. He repeatedly thanked the audience of five thousand for their cheers and applause before the venue finally quieted. In each cutaway shot, white-on-blue Obama signs bobbed everywhere on the convention floor. Barack began by telling the story of his parents. Jack Corrigan and a "really deathly nervous" Vicky Rideout stood just offstage. She thought Barack "seemed nervous to me at first," and up in a skybox, Mary Beth Cahill recalled feeling that Barack "seemed quite nervous," and "he started out a little shaky." But when Barack mentioned his mother's Kansas birth, that delegation's audible reaction made Barack pause. He looked their way, gestured and smiled, and appeared to relax. "I saw a moment where he took a breath, and his shoulders went back," Vicki recalled. "Then he just hit his stride and got going," Cahill explained. Up in the Obama team's temporary box, Terry Link and Darrel Thompson realized it too. "After that, it's another speech. That's when the speech takes off," Darrel remembered.

Barack explained his African name, asserting that "in a tolerant America, your name is no barrier to success." The crowd reaction built, and to television viewers, Obama signs were everywhere. "It was awesome," Tom Lindenfeld remembered. "That's what we're doing this whole thing for," this "big orchestration," because "without the signs, he would not have had the punch." Barack declared that "my story is part of the larger American story, that I owe a debt to all of those who came before me, and that in no other country on earth is my story even possible." The crowd erupted in applause, with the TV cameras again showcasing the Obama signs. The crowd's energy and the speaker "feed off each other," as Lindenfeld put it, and Mary Beth Cahill recalled that in the arena "it was riveting" with "the people in the hall reacting so viscerally to him." Barack described Americans' faith "that we can say what we think, write what we think, without hearing a sudden knock on the door," that "we can participate in the political process without fear of retribution."

Barack spoke about Illinoisans both downstate and in the collar counties before declaring that in inner-city neighborhoods, "children can't achieve unless we raise their expectations and turn off the television sets and eradicate the slander that says a black youth with a book is acting white." Then he turned to praise the character and courage of John Kerry, stating that Kerry "believes in the constitutional freedoms that have made our country the envy of the world." The broadcast camera cut away to show Jesse Jackson Sr. standing and applauding. Barack told of meeting a young marine enlistee in East Moline and praised America's troops, with the next cutaway shot showing Hillary Rodham Clinton standing and clapping. Asserting that Americans are "all connected as one people," Barack said that "if there's a child on the South Side of Chicago who can't read, that matters to me, even if it's not my child," and that "if there's an Arab American family being rounded up without benefit of an attorney or due process, that threatens my civil liberties."

Barack warned of "those who are preparing to divide us, the spin masters, the negative ad peddlers who embrace the politics of anything goes. Well, I say to them tonight, there's not a liberal America and a conservative America, there's the United States of America." As the crowd erupted, Barack's pace accelerated, and his voice strengthened. "There's not a black America and a white America and Latino America and Asian America, there's the United States of America." The television broadcast again cut away to Jesse Jackson Sr. rising and applauding. "The pundits like to slice and dice our country into red states and blue states: red states for Republicans, blue states for Democrats. But I've got news for them too: we worship an *awesome* God in the blue states, and we don't like federal agents poking around in our libraries in the red states. We coach Little League in the blue states and, yes, we've got some gay friends in the red states." Barack's voice had intensified, and the power of his words crescendoed. On the convention floor, Robert Gibbs turned to David Axelrod and asked, "Are you seeing what I'm seeing?" and Axelrod, standing just behind a pair of broadcasters, heard CNN's Jeff Greenfield tell ABC's George Stephanopoulos, "This is a great fucking speech!"

When Barack rhetorically asked, "Do we participate in a politics of cynicism, or do we participate in a politics of hope?" the crowd immediately answered, "Hope." Barack sustained that theme, invoking the phrase he first had heard from Jeremiah Wright: "the audacity of hope. In the end, that is God's greatest gift to us, the bedrock of this nation, a belief in things not seen, a belief that there are better days ahead." He closed by declaring that if John Kerry was elected president, "out of this long political darkness, a brighter day will come." The hall erupted as Michelle hugged him and cutaway TV shots again pictured Hillary Rodham Clinton and Jesse Jackson Sr. Dick Durbin and his wife Loretta joined Barack and Michelle onstage as cheers and applause continued.

On the convention floor, tears flowed freely in the Illinois delegation. "Some of the most hardened politicians that I ever met, people who wouldn't hesitate to knock their mothers over to get whatever it is they wanted, had tears streaming down their cheeks," state senator Jeff Schoenberg recounted. "I had tears in my eyes," Senate president Emil Jones Jr. confirmed. "It was electrifying." To comptroller Dan Hynes, "it was surreal."

Up in the Obama team's skybox, "almost everybody in the room was crying," state senator Terry Link remembered, including Valerie Jarrett and Penny Pritzker. When Link's cell phone rang, it was Republican Tommy Walsh: "Our buddy hit a home run tonight." Backstage, everyone was ecstatic. For Jack Corrigan, "it exceeded my wildest expectations," and "right after the speech," Vicky Rideout turned to Jack to say, "'We just elected our first black president.' 'Yeah, you might be right in twelve years or so,'" Corrigan replied. Then Barack appeared, whom Corrigan had never met. Vicky took the initiative. "I introduced him to Corrigan, and I said, 'This is the person who selected you for this.'"

On MSNBC's live telecast, host Chris Matthews said, "I have to tell you, a little chill in my legs right now. That is an amazing moment in history right there. A keynoter like I have never heard." Matthews's guest, Representative Dick Gephardt, agreed. "A star is born." Matthews concurred: "A star is born," highlighting Barack's "strong language" about how black families needed to take responsibility for their children's education. On CNN, Jeff Greenfield praised Barack's remarks as "one of the really great keynote speeches of the last quarter century."

On the convention floor, a Latina delegate from Nevada spoke to veteran

Washington Post reporter David S. Broder. "Look at the energy he brought to this room," she remarked. "He is definitely a rising star." Watching Barack's performance from his nearby home, John Kerry "was rightly blown away," one aide recalled. Barack's own team fully appreciated what had just occurred. "I realized at that moment that his life would never be the same," David Axelrod recalled. Jim Cauley put it similarly: "he was born that night," as it was "the transformational moment."

Backstage, NBC's Brian Williams asked both Barack and Michelle to join him for a live appearance on Matthews's MSNBC show. "You are, after all, a state senator from Illinois," Williams began, "so it's fair to ask you, when you get up there and see all those signs with your name on them and hear the chant that happens to match your name, what must that feeling be like?" Barack replied that "it's an enormous honor to be able to address a convention like this, to have the opportunity to speak to the country about the values that I care about . . . and it's especially nice to do it in front of my wife." Williams sought Michelle's reaction. "I was incredibly proud. And I am tough on him. And all I have to say is, honey, you didn't screw up, so good job!" Michelle said, "I was the bigger ball of nerves, even though I tried to act like I wasn't," but Barack "was terrific" and "he brought me to tears. . . . He was fabulous." Williams asked Barack what came next, saying he had "just heard a pundit behind me refer to the White House." Barack cited a multiday downstate swing his campaign had scheduled. "As soon as this convention is over, we are loading up our kids in an RV. We are traveling around to county fairs, eating ice cream . . . and meeting voters along the way." The media "is fickle," and voters want to know "is this a guy who delivers," who can "actually help my life in some concrete way?" Then Michelle said, "I am just hoping that our kids watched the whole thing." She had told her mother Marian "they can only stay up as long as they keep it on this channel, and that if they want to see Mom, they have to wait until Dad finishes."

Minutes later, NBC's Andrea Mitchell declared, "the real breakout tonight is Obama," that "Obama is a rock star," with Chris Matthews adding, "I've just seen the first black president." Across the nation, any number of TV viewers who had known Barack—or "Barry"—years earlier had trouble grasping what they had just seen. One Haines Annex friend from twenty-four years earlier thought "maybe that's Barry's brother." Phil Boerner's cousin Pern Beckman, who had shared beers with Barack in Morningside Heights, "watched the entire 2004 convention speech and never made the association. I'd said to my wife, 'Why isn't this guy running for president?'" and only long after did Phil tell Pern whom he had seen. "About six months later it turns out that I've known him all along. That's how much of a transformation took place from this shy kid" he had known in the early 1980s. In New York, Business International-al's Barry Rutizer was "completely dumbfounded" when he realized who the speaker was. "I practically fell out of my chair." In Honolulu, Madelyn Dunham called Barack's cell phone and left a message. "That was a very nice speech, Barack. You did well." State Senate colleague James Clayborne left an archly congratulatory message recalling their unhappy sojourn to Los Angeles. "Something's wrong. Four years ago, we were at the convention together, and four years later I'm sitting on my couch watching you give the keynote. Something's wrong." In Chicago, most of Barack's campaign staffers who had not gone to Boston gathered at Trinity United Church of Christ to watch the speech. In Springfield, the man who had done more than any other person to bring Barack to this point in his political career, Dan Shomon, watched the speech at the VFW post on Old Jacksonville Road.[82]

Barack's campaign team had booked a three-hour after-party at a large downtown restaurant, Vinalia, and worries that too few people would show up were proven terribly wrong when a crowd of thousands descended on the venue. "It was mayhem," Claire Serdiuk recalled, with Jim Cauley and Darrel Thompson having trouble getting Barack inside. "He made some short remarks at that event, even better than on the convention floor," Illinois delegate David Munar remembered. "He said something like 'If you're expecting another speech, I am going to disappoint you. But I am glad I moved you, inspired you, I'm happy with it. I just want to remind you that they're just words. It is our job to make those words reality. The hard work is not making a speech, the hard work is living up to the speech.' He was so eloquent and so on point. . . . It moved me more than anything, that he had that self-awareness. He checked himself immediately."

Tuesday's *Chicago Tribune* commended Barack for "a brilliant, passionate and heartening speech," and *USA Today* christened him "a rising Democratic star." At campaign headquarters, things were "crazy," with hundreds of people calling to volunteer and tens of thousands of dollars arriving in Internet contributions. "I don't think any of us had any idea that it would be like it was," national finance director Susan Shadow remembered. "We were just overwhelmed." A number of staffers had seen an advance text of Barack's speech, "but none of us was prepared for the reaction it got," young press assistant Tommy Vietor recalled. On paper, Barack's words were "nowhere close to the way he delivered it," as Vicky Rideout also appreciated. "The words are not amazing in any way. They're somewhat routine." What had captivated television viewers was the passionate intensity of Barack's delivery as he responded to the crowd in the hall. His performance had been rousing, but it was not because of "the words on the page."

In Boston, plaudits rained down. Chicago mayor Rich Daley told reporters that Barack "hit a grand slam home run. He had passion and emotion, and that's what it's all about." Congressman Lane Evans told the *Journal-Register*'s Bernie Schoenburg that "there's already people talking about Barack being a presidential candidate," and *USA Today* quoted a Pennsylvania delegate saying, "I think he could be our first black president." Even a Republican National Committee spokeswoman acknowledged that "Obama is a rising star," and she was echoed by Salim Muwakkil, who years earlier had publicized black nationalists' hostility toward Barack. Now he was "a genuine rising star" who "literally embodies our multicultural future."

Just like the night before, Barack took the acclaim in stride while weighing other concerns. "Malia is six, and I can't believe it, but a third of her childhood is over already," Barack told David Mendell. How much of her young life had he missed because of politics? "Too much," Barack answered. By midday Wednesday, Barack was running forty minutes behind schedule, but he took the time to speak to a Massachusetts state capitol gathering of 165 high school students. "We've got to call an audible and skip some of these things," journalists heard him say. "I've got to see my wife." That evening, when the national networks commenced their live coverage, NBC anchor Tom Brokaw told viewers he wondered "who will be remembered most" once the convention ended, "Kerry or Obama?" Thursday's *Christian Science Monitor* said Barack "could be one of the most exciting and important voices in American politics in the next half century," and the *Boston Globe* understandably noted that he "seemed to have the term 'rising star' formally appended to his name."

At a *Science Monitor*–hosted breakfast, Barack again endorsed Bill Cosby's critique of black America, saying his "underlying premise was right on target," that Cosby's "underlying idea . . . that we have young people who aren't disciplined, that we have a strain of anti-intellectualism . . . is absolutely true." Barack also cited his own family's financial worries, saying they had recently taken out a second mortgage on their East View condominium and that the challenge of living in both Chicago and Washington on a U.S. senator's salary of $158,100 "is a source of concern."

Filming an interview with African American broadcaster Tavis Smiley, Barack stressed that "I'm very clear that I've got to stay focused on Illinois, and that . . . the worst thing that I could do is start taking all this hype and all this press seriously." While magazine editor Tina Brown mused in the *Washington Post* about "Obama-mania" and Barack's "genderless air," the drumbeat of "rising star" accolades continued apace, with Friday's *Boston Globe,* the *Harvard Crimson,* and Massachusetts's black weekly, the *Bay State Banner,* all adopting the label. On Friday Barack and Jim Cauley headed to Boston's Logan Airport because the candidate's five-day downstate tour would get under way first thing on Saturday. At security, the new rock star was selected for an additional pat-down, and then "we had to hide him in the lounge because we couldn't move him through the airport. It was a little bit Beatle-ish," Cauley recalled.

One by-product of Barack's newfound national fame was an intense spurt of interest in his soon-to-be-reissued *Dreams From My Father*. Less than forty-eight hours after Barack's keynote speech, *Dreams* ranked number 10 on Amazon's bestseller list, and Crown quickly accelerated its schedule, aiming to get forty thousand copies into stores by August 10 rather than at the end of the month. Editor Rachel Klayman of course called Barack "a rising political star," and *New York Post* gossip columnist Cindy Adams and *Crain's New York Business* both reported that Jane Dystel was talking with publishers about what *Crain's* called "a mid-six-figure advance" for a new book about Barack's "ideas and beliefs." Dystel confirmed "we are already talking about it," adding drily that "this one should make some money." Within a month's time, the new paperback of *Dreams* would appear on the *New York Times'* paperback nonfiction bestseller list, remaining there for fourteen of the next eighteen weeks.[83]

Only on Friday evening, back home in Hyde Park, did Barack realize just how demanding this downstate tour would be. *Thirty-nine* stops were scheduled over the next five days. After the rented RV pulled up at East View Saturday morning, the first stop was a Hyde Park event, which would be followed by seven more, with the long day ending four hundred miles later in DeKalb after an evening cookout in the Moline backyard of downstate campaign director Jeremiah Posedel, who remembered that Barack "called my cell phone the night before the tour started and he was very pissed." Barack's staff had little sympathy for the candidate's unhappiness. Chief scheduler Peter Coffey recalled that "Jeremiah put together a great schedule," one that now was all the more important after Barack's five-day celebrity circuit in Boston. "He needed to see these towns, he needed to be in these communities," Peter knew, "to remind him that this is the job he's running for" on behalf of regular Illinoisans, not broadcast interviewers.

With Michelle, Malia, and Sasha on board and a fifteen-passenger van filled with reporters following behind, the RV left Hyde Park for an eighty-mile drive to Ottawa, in Democrat-rich LaSalle County. Posedel had expected the rapid-fire small-town stops to draw approximately one hundred people each, but in Ottawa a crowd of more than five hundred welcomed Barack. Then it was on to Granville, then DePue, and there were another five hundred people in Kewanee, followed by Monmouth and Oquawka. The crowd of hundreds there was "electrified and energized" by the late-afternoon appearance of "a rising star," who "was swamped by those seeking autographs." In Moline, more than three hundred people crowded into Posedel and girlfriend Anita Decker's backyard before the RV headed back eastward to overnight in DeKalb. At 9:00 A.M.

Sunday morning, fourteen hundred people awaited Barack in a DeKalb park, even in ninety-degree heat. Then it was on to Marengo, Belvidere, and Rockford, where a crowd of a thousand gathered in a riverfront park. Barack took the time to greet hundreds of people, which meant that by Freeport the RV was running an hour late. In Galena, Malia and Sasha finally got their promised ice cream before the tour made Sunday's seventh and last stop in Mount Carroll.

Tribune reporter David Mendell felt that "Obama fever" had given the tour "the feel of a coronation." Barack hesitated to articulate his own feelings about it. "This is all so, well, interesting. But it's all so ephemeral," he told Mendell. "I don't know how this plays out, but there is definitely a novelty aspect to it all." But "it can't stay white hot like it is right now." Both the Boston experience and now people's reactions across central Illinois left Michelle thinking about Barack's security. "You have to achieve a balance between looking out for his safety and not looking like he is afraid of the community he is serving." They began Monday in Rock Falls, then Dixon, Rochelle, Lacon, Pekin, Normal, and Clinton. "Barack seems to become energized at every stop by the people that he meets," young press assistant Tommy Vietor blogged, but by Monday's final stop at a $100-per-person Danville fund-raiser that drew seven hundred, Barack was ninety minutes late.

Tuesday, day 4, again featured what Vietor called a "fast and furious" pace, beginning at 8:30 A.M. in Champaign. Local state representative Naomi Jakobsson introduced Barack as "Illinois's rising star," and at the day's second stop, in Tuscola, retired farmer Boyd Stenger told Barack, "You need to run for president in the next election." When *Chicago Sun-Times* reporter Dave McKinney remarked to Barack about "the stunning amount of positive press" he was getting, Barack's cool reply was "Don't jinx me." They forged on to Mattoon, Neoga, Mount Vernon, Fairfield, Mount Carmel, Lawrenceville, and Olney, where five hundred people greeted Barack even though he was more than an hour late. Tuesday's tenth and final stop was Salem, where another crowd of five hundred awaited. Wednesday was Barack's forty-third birthday, and it featured a stop in Taylorville before the RV turned northeastward and headed back toward Chicagoland. A $50-per-person evening birthday party at the Matteson Holiday Inn featured Representative Jesse Jackson Jr. serenading Barack with "Happy Birthday" before Barack praised Jackson and Senate president Emil Jones Jr. as his "two rocks of Gibraltar." As the celebration wore down, Barack turned to tour architect Jeremiah Posedel. "You did a great job, but don't ever fucking do that to me again!"[84]

While Barack was traversing central Illinois's small towns, David Axelrod and John Kupper had been revising the general election plan. "Barack could spend not a penny on advertising and still win," they noted, but "Barack remains relatively unknown," so his newfound national fame created a new hurdle. "We need to meet and exceed expectations. A 55–45 win in November will now be read as a disappointing showing. . . . We need to drive up Barack's margin as high as we reasonably can to cement his standing—both state-wide and nationally—and help discourage a more credible challenger six years from now." But given how long Barack had gone without an opponent, the earlier game plan of spending more than $8 million on paid media could be scaled back to "about half that amount, with the focus on Downstate."

The weeks since Jack Ryan's withdrawal had been quiet for those Obama staffers who did not go to Boston or have to cope with the postconvention deluge, and policy director Amanda Fuchs had used the time to produce a comprehensive "Defensive Research Book" on Barack's Springfield record. "Soft on crime" was the top vulnerability, but Amanda's desire to have a long, comprehensive interview with Barack about his life's record beyond the state legislature was repeatedly brushed aside by the candidate and David Axelrod.

By the first week of August, Illinois Republicans' desperation to find a plausible stand-in candidate reached new depths as the nineteen-member State Central Committee began interviewing a motley collection of applicants. Rockford state senator Dave Syverson had the odd idea of asking Alan Keyes, an outspoken African American abortion opponent who in 1988 and 1992 had lost Senate races in his home state of Maryland before launching quixotic presidential runs in 1996 and 2000. Keyes agreed to apply, and flew to Chicago on August 3. Calling Barack "a radical ideologue," Keyes interviewed with the Republican committee for ninety minutes on August 4 and garnered two-thirds support. Keyes requested several days to decide, during which time Barack told the *Tribune,* "I think he'll need to explain how he can best represent the people of Illinois, not having ever lived here." *Capitol Fax*'s Rich Miller dismissed Keyes as "an arrogant blowhard," and the *Tribune* discovered that Keyes had $524,000 in unpaid debts from his two presidential quests plus a $7,400 income tax lien he quickly settled.

Barack and his family were headed to Canada to see his sister Maya, her husband Konrad, and their new baby Suhaila, who were visiting Konrad's parents in Burlington, Ontario. A visit to nearby Niagara Falls was de rigueur, and while Barack was enjoying a weekend away from politics, Alan Keyes returned to Chicago to accept the Republican nomination before a small crowd at a suburban restaurant. Reading Monday's news stories on his return, Barack told the *Tribune* that Keyes "did not say a word about jobs, he didn't say a word about health care—the two issues that when I travel around the state people seem to be talking about all the time."

Barack joked that his old offer of six debates with Jack Ryan was a "special for in-state residents," but his campaign's insistence that three face-offs with Keyes in the fall would be more than enough met with criticism from Chicago editorialists. With a CBS Chicago poll showing Barack leading Keyes 67–28, and the *Chicago Defender* denouncing Keyes as "a political hit man" intent on "a character assassination campaign," Barack's desire to minimize his personal exposure made sense. *Capitol Fax*'s Rich Miller wrote that Republicans hoped to keep Barack "pinned down in Illinois" for the fall, but by refusing six debates, Barack had been "knocked off his game for the first time this year." Miller noted that "Obama isn't accustomed to negative publicity, to put it mildly," and wondered how well Barack would cope with the rhetorical onslaught everyone expected from Keyes, who already had made "the 'rock star' frontrunner look like a mope."

Suburban *Daily Herald* political writer Eric Krol noted that Barack "has gone out of his way to sound more like a moderate since the March primary." Appearing on NPR's *Fresh Air,* Barack predicted that "people are just weary of . . . slash-and-burn politics that demonizes whoever doesn't agree with you." Yet "ruthless" and "cynical" political professionals stoked such tactics, because they have "a monetary stake in the outcomes." At Saturday's annual Bud Billiken Parade in Bronzeville, Barack, Michelle, and Congressman Jesse Jackson Jr. rode on top of a float. "Obama was treated to a king's welcome" on account of his "super celebrity status," wrote the *Tribune*'s David Mendell, while the *Defender* declared that a "serene" Barack "looked more like a film star than a senatorial candidate." In contrast, when Alan Keyes appeared, he was met with cries of "Go back to Maryland!" and he "was loudly booed by the crowd." When the parade MC reprimanded the crowd, "the disapproval grew louder and was taken over by chanting Obama's name."

The next morning Barack and Keyes appeared separately on ABC's *This Week,* with host George Stephanopoulos stating that at the DNC, Barack "convinced a lot of those delegates that he's got the makings of a future president." Barack predicted that "there are going to be some trying moments" in facing off against Keyes, and he understood Republicans wanted to "bloody me up a little bit before I got to Washington." When Stephanopoulos asked Barack about Keyes's position "that life begins at conception," Barack parried. "Well, I as a Christian might agree with that, but if I agree with that it's based on a religious premise, and not one that I think is subject to scientific proof." Barack added that Democrats err if they believe "that only secularism could express tolerance," because "part of my job as a Christian is to recognize that I may not always be right, that God doesn't speak to me alone."

At midday Sunday, Barack headed to Rogers Park for the annual Indian Independence Day Parade. Also on the northwest corner of Devon and Western Avenues was Alan Keyes, whose wife Jocelyn was Indian American. Meeting for the first time, Barack approached his opponent to shake hands. Keyes made some remark about the number of debates, and Barack shot back, "I guarantee we're going to debate, because you've been talking a lot." Keyes replied that "I have the very bad habit of telling the truth," then adding that "the sad truth of the matter is, one test of a candidate, I believe, is that you say what you mean and mean what you say." Barack replied, "I always do."

Sun-Times reporter Natasha Korecki stood close by as the two men "frequently interrupted each other, continuing to grasp hands as they exchanged barbs only inches from the other's face." Barack told Keyes, "Don't just go and keep on talking. You've got to do a little listening. You're not a very good listener." Finally they parted, with Barack telling a *Tribune* journalist that Keyes's "entire premise is that people that don't share his convictions are bad people." *Capitol Fax*'s Rich Miller wrote that "the alleged rock star lost his famously unflappable cool" when confronted by Keyes in the flesh. "Both men shouting at each other and pointing fingers" was "par for the course for Keyes" but "not a good thing at all for Senator Suave," who "needs to get his act together." Privately, David Axelrod agreed, reprimanding Barack for a scene that left Axelrod "shocked" by his candidate's behavior. Barack replied, "I just wasn't going to let him punk me."

Following campaign stops in Peoria and Decatur, Barack "was greeted with wild enthusiasm at every appearance" when he arrived at the huge Illinois State Fair in Springfield on what was nominally "Governor's Day." "The star of the day was Obama, not Gov. Rod Blagojevich," the *Tribune* stated, with Barack's Senate buddy Denny Jacobs declaring, "it's an Obamafest!" As if any further evidence was needed, several days later a *Tribune* poll showed Barack with a 65 to 24 percent statewide lead over Keyes. Barack's favorable-to-unfavorable margin was 62 to 14, with 24 percent uncertain. Regarding Keyes's signature issue, only 28 percent of voters wanted greater restrictions on abortion. In contrast, 56 percent said Jack Ryan should not have dropped out, and, of all people in Illinois, Rod McCulloch now agreed with them. "Honestly, if I'd have known they were going to bring in Alan Keyes, I might have kept my mouth shut, because an October surprise [about Ryan] would have been better than Alan Keyes."[85]

Barack and his family slipped out of Chicago for a late-August vacation on Martha's Vineyard, staying with Valerie Jarrett at her home in Oak Bluffs, long a summer destination for privileged African Americans. Barack made a surprise appearance at a forum in an Edgartown church, but otherwise relaxed, playing golf with Chicago friend Allison Davis. After going out to jog one morning, he returned to Jarrett's with what she called "a look of complete disbelief on his face. 'You aren't going to believe what happened to me,' he said. 'A guy took my picture as I jogged by.'"

While Barack was away, his campaign team debuted his first general election TV ad in downstate markets from Rockford to Springfield. Focused on his

REAL USA Corporations bill, it featured Barack stating, "I came to Illinois nineteen years ago to work in a community torn apart by the closing of a steel plant." Saying, "we need some common sense," Barack called for ending "tax breaks for corporations who move job overseas," giving them instead to "companies who create jobs here." At the same time, Paul Harstad conducted the campaign's second general election poll, focusing particularly on downstate voters and Republican women in the Chicago media market. Only 62 percent of respondents knew about Barack's DNC speech, and while his lead over Keyes was a whopping 64–20, a hypothetical pairing of Barack versus former governor Jim Edgar showed Barack with just a modest 49–40 percent advantage.

Once Barack returned, he made a quick trip to Birmingham, Alabama, for a $1,000-per-person fund-raiser at the Civil Rights Institute, and afterward he walked across 6th Avenue North to visit Sixteenth Street Baptist Church. "Let's say a little prayer, because I believe in the power of prayer," Barack told pastor Rev. Arthur Price. Back in Chicago, his consultants met to absorb Harstad's polling results and to discuss how Barack could be deployed nationally to help raise money for Democratic Senate candidates who were facing tougher races than his, particularly minority leader Tom Daschle. At the same time, David Axelrod, John Kupper, and David Plouffe gave Barack a twelve-page "Transition" memo outlining what he should do after he won election to the Senate on November 2. "You are permitted to hire two transition staff to work from the time of your election until your swearing-in," and they expressly recommended Robert Gibbs as "a solid choice" to remain as Barack's communications director. In addition, Daschle's chief of staff, Pete Rouse, was "someone with whom you should develop a peer relationship." Noting "the galvanizing national presence you created through your keynote address," Barack's team said he needed to formulate his own policy agenda, and "one issue we'd suggest you focus on is healthcare reform."

Making his first campaign swing through Metro East since the DNC, Barack told a mostly black crowd at East St. Louis's Mount Zion Baptist Church that visiting the Birmingham church where four young girls had been killed in a 1963 Klan bombing had put his newfound fame in perspective. "It's easy to get swept up in the hoopla, to read your name in the papers and to see yourself on TV. And all of that is fun. But standing in that church in Birmingham, Alabama, I realized and I reminded myself that the reason you get into public service is not for yourself. It's not about your family. It's not about your vanity. It's not about your ambition. If you want to be first, you have to be first in service."

At his next appearance, before an almost entirely white crowd at a Waterloo VFW hall, Barack joked about his DNC speech. "My wife, I asked her for a little support. She said, 'Don't screw it up.'" That afternoon in Benton, Barack spoke frankly about his opponent. "I don't just want to win, I want to give this guy who is running against me a spanking. The reason I do is because he exemplifies the kind of scorched-earth, slash-and-burn negative campaign that has become the custom in Washington, and it is the reason why we can't get anything done. Ordinary people just don't act like that, they don't call people names all the time."

Responding to downstate reporters' questions, Barack measured his remarks accordingly. "On some issues I'm quite conservative. I believe in the death penalty. I believe that there is an important place for the Second Amendment in this state. I believe that it is important that we don't think we can solve all our problems with government programs. I've said that publicly and repeatedly, and I've voted in that fashion." Barack called abortion "a deeply difficult moral issue" and declared, "I'm not in favor of gay marriage." He remained firmly critical of the Patriot Act, stating that "we have to make sure that we don't get so swept up in our fears that we throw overboard those constitutional protections that make this country so special."

Appearing before the Rock Island County NAACP on Labor Day, Barack was greeted "more like he was a rock star than a politician," the *Quad-City Times* reported. Barack stressed that "I wouldn't be standing here today if it were not for Lane Evans," the local congressman who early on had endorsed Barack. In Chicago, Alan Keyes called a press conference to criticize Barack's "spanking" remark, but made bigger headlines by proclaiming that "Christ would not vote for Barack Obama" because of abortion. *Capitol Fax*'s Rich Miller noted Keyes's "continuing bizarre behavior," but Barack calmly told Springfield reporters that "I leave it up to God to judge how good of a Christian I'm going to be. I leave it up to the voters to judge how good of a U.S. senator I'm going to be. I don't concern myself too much with Mr. Keyes's judgment on either matter," and "I will leave Mr. Keyes to the theological speculations. My job is to focus on the issues the voters care about: jobs, health care, education."[86]

Chicago Sun-Times reporter Lynn Sweet publicly castigated Barack's "manipulative, cagey campaign" for "hiding practically all of Obama's out-of-state treks" to fund-raisers from the press. Barack also was attending a ramped-up schedule of Chicagoland fund-raisers, such as a Tuesday-evening one at Raghuveer Nayak's home in Oak Brook. Michael Parham, Jeff Cummings's attorney friend who had attended one of Barack's earliest state Senate fund-raisers before leaving Chicago for Seattle, offered to have his RealNetworks colleagues host a fund-raiser, and Barack called to confirm. "When are we going to get you back to Chicago?" he asked. "You know, your governor wants me to come right now," Parham replied. "For what?" Barack asked. "To be his general counsel," Michael said. "Without skipping a beat," Barack forcefully declared, "Don't do it. Don't take that job. Stay away from him. You don't need him. Don't take that job." Parham was taken aback by Barack's "instinctual reaction," but followed his advice.

Meeting with newspaper editorial boards prior to their endorsement decisions, Barack mentioned to *Daily Herald* journalists his disappointment that Republican colleague Steve Rauschenberger had criticized him over the summer. Called for comment, Steve praised Barack personally while asserting that in Springfield he had been "a show horse, not a workhorse." Barack is "brilliant, his potential is unlimited," and his ambition knew no bounds. "I don't think even the U.S. Senate will be enough for Barack," Rauschenberger exclaimed. Speaking with *Tribune* editors about foreign policy dangers, Barack said that "launching some missile strikes into Iran is not the optimal position for us," but "on the other hand, having a radical Muslim theocracy in possession of nuclear weapons is worse." Barack felt similarly about Pakistan's nuclear arsenal. If "certain elements" gained power, "I think we would have to consider going in and taking those bombs out." Yet Barack said that health care reform, not foreign policy, would be his top Senate priority.

A justifiably proud Emil Jones Jr. told *Tribune* reporters that "Obama's the future" and that Barack "embodies all that I dream for and work for." An essay in the leftist weekly *In These Times* said that Barack's Senate "campaign points toward a new era . . . of progressive politics in search of a new majority," and when Barack appeared at a Saturday event in suburban Bolingbrook, the *Joliet Herald News* reported that he was "greeted like a film star." Campaigning in Philadelphia for Pennsylvania Democratic Senate candidate Joe Hoeffel, Barack attracted a fervent crowd. "I am quite confident he could be the first African American president," one college student told a *Philadelphia Inquirer* reporter, and an undergraduate journalist for the *Daily Pennsylvanian* was struck by how Barack "stayed and mingled with the crowd for more than 30 minutes." Back in Illinois, a *St. Louis Post-Dispatch* poll showed Barack leading Keyes 68–23, with 54 percent of Illinois voters viewing Keyes unfavorably. A *Tribune* poll gave Barack a 51-point lead, 68–17, with 39 percent of Republicans viewing Barack favorably and 38 percent calling Keyes "too extreme." Rich Miller wrote that at one otherwise-unreported event in Quincy, "Keyes led the crowd in a chant of 'Obama been lyin',' which sounded to some ears like 'Osama bin Laden.'" Miller believed that "'arrogant' is perhaps too soft a word to describe Mr. Keyes," and former governor Jim Edgar recounted a similar reaction that month from the president of the United States. "I ran into Bush, and he said, 'I'm not coming to Illinois. You guys are nuts out there. Alan Keyes?'"[87]

Barack continued to receive more fawning profiles than tough questions. *Black Enterprise* magazine put him on its cover, heralding "the birth of a new political star" who may "one day become America's first African American president." *Chicago Sun-Times* reporter Scott Fornek got both Michelle and Barack's sister Maya to describe Barack's behavior during their annual Christmastime games of Scrabble. Michelle called him "a big trash talker," and Maya termed her brother "an indelicate winner" who "would crow like a rooster and flap his wings and make slam-dunk motions." Barack confessed that "when it comes to Scrabble I just can't help myself. . . . I've got a competitive nature."

Television host Oprah Winfrey interviewed Barack and Michelle, telling him that when he had called out "the slander that says a black youth with a book is acting white" in his DNC speech, "I stood up and cheered." Michelle told Oprah that if she had met the Ryan campaign's video stalker, "I would've punched him," and in the face of Barack's many accolades, she emphasized that "we're clear on the fact that we have to stay humble and prayerful." Barack lamented "the people I meet in these little towns who have lost their jobs" and "are struggling and occasionally slipping into bitterness," but he also revealed that "I've had ten days off in the last three years, and that includes weekends. My workdays are often sixteen hours." Barack told Winfrey, "I think the biggest mistake politicians make is being inauthentic."

Traveling north to Madison to praise Wisconsin's Russ Feingold, the sole U.S. senator to vote against the Patriot Act, Barack said, "I like to think that, had I been in the Senate, I would have cast the second vote against the Patriot Act," although "I wasn't there in the pressure of that moment." A Cleveland reporter who had first met Barack eight years earlier was now struck by Barack's "detached skepticism." Barack told him that "there is value to considering the possibility that you may be wrong," but in his *Tribune* questionnaire, Barack bluntly declared that "the global community cannot tolerate nuclear technology in the hands of a radical theocracy."

Barack continued to spend a disproportionate amount of time downstate, appearing at events like Macoupin County's thirtieth annual Country Fun Day on a farm just west of Gillespie. Top Democrats like Dick Durbin, Rod Blagojevich, and Emil Jones Jr. were all there that Sunday, and Jones recalled an elderly white woman, in what was an almost all-white corner of Illinois, turning to him after Barack finished speaking and saying, "This young man is going to be president of the United States some day. I just hope I live long enough to vote for him." On Monday, speaking at the University of Illinois College of Law in Champaign a day before his first debate with Alan Keyes, Barack vented some of the anger he felt toward his challenger. "When they're not being bombarded by messages from Michael Moore or Fox News, citizens today have to deal with politicians who habitually lie, blatantly mischaracterize their opponents, and do other things that would not be accepted at all in any other aspect of society" beyond politics.

The fifty-three-minute, radio-only debate turned out to be a major letdown for Illinois journalists. "Not only were there no fireworks, there weren't even many sparks," wrote *Sun-Times* reporter Scott Fornek. *Capitol Fax*'s Rich Miller lambasted the questioners for posing "almost nothing but softballs" and failing to confront Keyes with any of his outlandish prior statements. A *Los Angeles Times* correspondent found Barack's performance "calm and analytical," and his most notable statement was that "with terrorists I have a very hawkish position. I think we should hunt them down, kill them, dismantle their operations." Miller reported that "several of Obama's advisors privately admitted disappointment with their guy's performance," and by now David Axelrod and John Kupper knew all too well that Barack was "a guy who just never really took debates seriously."

National and Chicago journalists marveled that "Obama Extends Reach Beyond Illinois Race," as a *USA Today* headline put it. The paper said Barack had campaigned in fourteen states since the DNC, raising more than $1.2 million for the Democratic Senatorial Campaign Committee. A *Tribune* analysis showed that Barack had donated at least $268,000 from Obama for Illinois to other Democratic candidates, including $53,000 to embattled Senate minority leader Tom Daschle, but with contributions *to* Barack's campaign now totaling more than $14 million, such generosity came easily.

Michelle now was making campaign appearances of her own, despite having just lost valued babysitter Sonja Hawes, with a young African American U of C student, Marlease Bushnell, taking her place. Michelle told an Edwardsville audience that "I grew up as a poor girl" and described how she had to escape black anti-intellectualism as a youngster. "I confronted it growing up. I had to duck and dodge to cover up the fact that I enjoyed school and excelled in it so I didn't get my butt kicked on the way home from school." In a Springfield conversation, Michelle again emphasized the centrality of religion in her and her husband's lives. "We've been guided by faith throughout the course of this political campaign, a faith that says to whom much is given, much is expected . . . a faith that says we're all put on this earth for a higher purpose, a faith that tells us that we can always do better." In place of the disdain with which Michelle had viewed Barack's political aspirations in earlier years, she now seemed to embrace a loftier mission. "Winning is just a small part," she told the *Journal-Register*. "This campaign is about so much more—it's about changing the tone of politics."[88]

Since his primary win, Barack also exhibited a greater religious faith than close acquaintances had ever previously sensed. After the DNC, *Tribune* reporter David Mendell covered Barack almost everywhere in Illinois, noticing that "along his Senate campaign trail, Obama would never fail to carry his Christian Bible. He would place it right beside him, in the small compartment in the passenger-side door of the SUV, so he could refer to it often." When Mendell asked Barack about his "ever-present Bible," he was "a bit hesitant to answer" and was "uncharacteristically short" in saying he read it a "couple times a week." "It's a great book and contains a lot of wisdom," Barack added.

Campaigning in Denver for Democratic Senate candidate Ken Salazar, Barack was christened a "rising star" by the *Denver Post,* the same phrase Soledad O'Brien used to introduce him to viewers of CNN's *American Morning.* Barack told O'Brien that all of the attention is "a little over the top" because "I know how hard it is to actually get stuff done in politics," but in the Denver crowd, one African American woman told her children, "Look hard, honey. That man might be your president someday."

The *Chicago Tribune* commented on what an artfully balanced position Barack took on affirmative action. "Promoting diversity is a compelling national interest, but it has to be done in a way that is not a back-door use of quotas and takes into account the full record of the students, not just race and test scores," Barack stated. Barack's increasing fame attracted a number of attempts to plumb the impact of his racial identity. *St. Louis Post-Dispatch* correspondent Kevin McDermott observed that Barack "often seems like a different man from setting to setting, depending on his audience," but McDermott took special note when Barack told a group of black students, "You've got to have a higher standard of excellence than the scumbag folks back in the neighborhood expect of you."

Even before his DNC speech, the Springfield *State Journal-Register* had run an editorial observing that Barack "could just as easily be considered white" as black, and as his election as the first-ever black male Democrat in the U.S. Senate neared, additional commentators weighed in. Political figures like Colin Powell and Barack "give off the sense that they have transcended traditional racial categories" by virtue of "their speech and demeanor, their personal narratives and career achievements," Benjamin Wallace-Wells wrote in the *Washington Monthly.* For them, "race had been an advantage because whites see in them confirmation that America, finally, is working." An idiosyncratic African American commentator, emphasizing what he believed was Barack's "substantially more dominant white heritage," argued that "Obama has adopted a bizarre practice that might be most accurately described as *passing in reverse*" by asserting his African American identity.

Veteran black journalist Don Terry interviewed Barack and a number of other Chicagoans before concluding that Barack "is a Rorschach test. What you see is what you want to see." Barack understood that dynamic. "I'd describe myself as coming from several cultures, and feel I belong to all of them," he told Terry. "I think people project a lot of their racial questions onto me." Barack nicely captured the complexity of American blackness when telling an *Ebony* magazine writer that "All of us had parents or grandparents who were the subject of much greater discrimination. That's true whether they were from the Caribbean, or Africa or Mississippi." When Don Terry asked Chicago black activist Bamani Obadele about Barack, he answered that "to black people he's black. To some whites, they don't see him as a black man. They see him almost as one of them." A middle-aged white female told Terry that Barack could well be the first "minority" president. Why that rather than black, Terry asked. "I guess it's because I don't think of him as black."[89]

All across Illinois, newspapers endorsed Barack for election to the U.S. Senate. The *Chicago Sun-Times* lauded his "meteoric rise" as "a walking advertisement for the American dream," and Bloomington's *Pantagraph,* which had backed Dan Hynes in the primary, praised Barack's "pragmatic, bipartisan approach to problem solving." The *Chicago Tribune* embraced him as "a rare example of someone who is able to rise above ethnic and racial divides and political partisanship," and the *Chicago Defender* proclaimed that "Obama Will Do Us Proud" in the U.S. Senate. Speaking to voters east of the Mississippi River, the *St. Louis Post-Dispatch* commended Barack's "nuanced understanding of the issues facing the nation" and said that "over the course of the long senatorial campaign, Mr. Obama has skyrocketed from an obscure northern Illinois legislator to national celebrity."

Only on October 19 did Barack's campaign begin running its first Chicagoland TV ad, a thirty-second spot featuring his DNC keynote declaration that "there is not a liberal America and a conservative America, there is the United States of America." The prepaid $2 million expenditure demonstrated "we are not taking any vote for granted," David Axelrod told the *Sun-Times.* On Thursday night, October 21, Barack and Alan Keyes met for their first televised debate, though the viewing audience was modest given that the St. Louis Cardinals, whom most downstate baseball fans rooted for, were hosting the Houston Astros for Game 7 of the National League Championship Series at the same time. Early on, Barack emphasized his legislative goal of "setting partisanship aside and seeking common ground." After Keyes at almost endless length spoke about Christianity and the Lord, Barack expressed exasperation in what journalists unanimously thought was the night's best statement. "I'm not running to be the minister of Illinois. I'm running to be its United States senator." Barack also declared that "everyone who lives in the United States knows that race still matters. It matters powerfully. I think that we still suffer from the legacies of slavery and Jim Crow, and this has been the single biggest blight on the history of this nation."

In a long preelection profile in the *Tribune,* David Mendell called Barack "the voice of a new generation of politicians" and potentially "the country's first black president." David Axelrod acknowledged that "Barack has been shot out of a cannon" and that "I think Barack understands that he has been on an incredible ride." In a similar piece in the *Daily Herald,* Barack joked about *Dreams From My Father*'s presence on the *New York Times* bestseller list. "I was feeling pretty good about that until I read that actually the fastest-growing best seller on the list was Jenna Jameson's *How to Make Love Like a Porn Star*," which showed that "there's not necessarily any correlation between celebrity and accomplishment."

Associated Press correspondent Chris Wills, citing Michelle's "unusually high-profile role" in the fall campaign, devoted a story to her as well. "Barack was not as cynical about politics as I was. Unlike me, Barack always had faith in the American people," Michelle told him. Speaking at Illinois State University, Michelle reprised her Springfield remarks that "this race means much more to us than winning," because "it's about changing the tone of politics." Barack "wasn't going to do this if he couldn't tell the truth, if he couldn't operate with dignity, if he couldn't treat his opponents with some respect and if he couldn't change the discourse." Sounding multiple notes, Michelle stated that "the only thing I'm telling people in Illinois is that Barack is not our savior. I want to tell it to the whole country, and I will if I get the opportunity." She added that "Barack can't do a thing if Bush is in office," but that if the Democrats won the presidency, and "Barack gets in under Kerry, the possibilities are endless."

On CNN, Barack praised Michelle as "tougher" than him. "I love to write," he told host Carlos Watson. "I love fiction. I love to read fiction, but I'm not sure I have enough talent to write fiction," and he attributed *Dreams*' phoenixlike rebirth to "luck, happenstance, God's blessings." Barack said, "I think there's a direct line between Harold Washington, to Carol Moseley Braun, to myself," emphasizing that "I stand on the shoulders of Harold Washington. I got the Ivy League degrees because somebody in Selma, Alabama, or Birmingham was willing to make enormous sacrifices."

In Kenya, a *New York Times* correspondent visited the "rising star's" relatives in Nyang'oma, and with just ten days to go before the November 2 election, a *Tribune* poll showed Barack leading Keyes 66 to 19 percent. Among Republicans, 32 percent said they would vote for Barack, and among African Americans only 3 percent favored Keyes. Given such an overwhelming lead, Barack spent that Sunday speaking first in Orangeburg, South Carolina, and then stoking Democratic GOTV efforts in Daytona Beach, Florida, one of the most tightly contested states in John Kerry's close race against George Bush.[90]

WTTW's assertive Phil Ponce moderated the third and final debate between Barack and Keyes on Tuesday evening, October 26. Whenever Barack spoke, Keyes exhibited a smirky smile that many viewers might have found inappropriate. Ponce put Barack on the defensive by asking if as a public official, his children would attend public school. Barack responded by saying they attended the Lab School at the U of C, "where I teach, and my wife works, and we get a good deal for it," which was witty enough to draw laughter from the studio audience. "We're gonna choose the best possible education for our children," Barack replied. Asked about gay equality, Barack said, "we have an obligation to make sure that gays and lesbians have the rights of citizenship," the "same set of basic rights." On an easier note, Ponce asked Barack where his family would live after his election. "We haven't decided yet," Barack replied. "My priority is gonna be making sure I've got as much time as possible with my wife and children," and he lauded Michelle as "an extraordinary wife who carries more than her load."

After a brief postdebate press conference, Barack appeared on Fox News's *Hannity & Colmes* with Sean Hannity, who introduced him as "the new rock star of the Democratic Party." Barack said, "we can disagree without being disagreeable," and declared that "the most important thing that we have to do in improving our politics is not assuming bad faith on the part of other people." With opponents of legal abortion, for example, "I don't assume that they're automatically trying to oppress women." The next day, in an implicit reflection of just how uncompetitive the Senate race was, Barack spoke at two Naperville high schools, where the proportion of eligible voters among his listeners was very modest. A *St. Louis Post-Dispatch* poll showed a lead of 67 to 25 percent, and on the final Friday before election day, Barack enjoyed a leisurely lunch with Chicago mayor Richard Daley at Manny's Deli on the Near South Side, David Axelrod's favorite eatery.

Barack campaigned for real on Saturday and Sunday before returning home to join his daughters for Halloween trick-or-treating in Hyde Park. Meanwhile, Alan Keyes had declared that any Roman Catholic voting for Barack would be committing "a mortal sin." Barack responded, "I think everyone is accustomed now to Mr. Keyes's histrionics, and I don't think it is going to have much of an impact on this election." On Monday, Barack headed downstate for a Springfield rally at state AFL-CIO headquarters and another stop in Madison County's Granite City before returning home. On election morning, the whole family was up early, with Barack and Michelle arriving to vote at 7:58 A.M. at Hyde Park's Catholic Theological Union, two blocks west of their home. Then all eyes turned toward that evening's victory celebration at the downtown Hyatt Regency Chicago.[91]

Close to two thousand people, including 167 journalists, filled the Hyatt's Grand Ballroom once Illinois's polls closed at 7:00 P.M. Only the precise extent of Barack's victory was in question, and as results came in, it was clear that Barack's landslide win reached from Chicago's suburbs to all but a few small, heavily Republican downstate counties. In Chicago, Barack received 88 percent of the vote, winning at least 75 percent in all fifty wards. In nearby Will County, Barack's huge ninety-thousand-vote margin helped Senate colleague Larry Walsh win a "stunning upset" in his race for county executive. Across downstate, Barack carried 61 percent of the vote, with Keyes winning only ten of Illinois's 102 counties. In the final statewide tally, Barack trounced Keyes by a margin of 3,597,456 to 1,390,690, or 70 to 27 percent, with two minor candidates receiving the balance. It was, unsurprisingly, the most overwhelming U.S. Senate victory in Illinois history.

At the Hyatt Regency, an exasperated Robert Gibbs received a series of apologetic phone calls from Keyes aide Dan Proft, who was trying without success to persuade his candidate to call Barack and concede defeat. Keyes never called. "I'm supposed to make a call that represents the congratulations toward that which I believe ultimately stands for and will stand for a culture evil enough to destroy the very soul and heart of my country?" Keyes told a Christian radio station. "I can't do this. And I will not make a false gesture."

Realizing there was no point in waiting for Keyes to concede, Barack appeared onstage at the Hyatt to thank "the best political staff that has been put together in this state." Singling out only "my pastor Jeremiah A. Wright Jr. of Trinity United Church of Christ" for thanks by name, Barack "most of all" thanked his family members who had joined him onstage, including sister Maya and brother-in-law Craig Robinson. In remarks that the suburban *Daily Herald* called "eloquent," Barack made no reference to his defeated opponent, but he did praise retiring senator Peter Fitzgerald for serving "ably and with integrity."

It had been a long and grueling campaign since Barack had formally announced his candidacy in a far smaller room at the Hotel Allegro twenty-one months earlier, on January 21, 2003—so grueling that Barack got the math wrong, saying that was "656 days ago," when actually it was 650. "This is really just the end of the beginning," Barack stressed, emphasizing that "in the ultimate equation we will not be measured by the margin of our victory" but by "whether we are able to deliver concrete improvements to the lives of so many" Illinoisans who were struggling. Barack's answer to that challenge was the same slogan he had trumpeted in his first television ad: "Yes we can!"

Barack's 70 percent landslide certainly lived up to journalists' expectations, but across the rest of the United States, election night results were not what he had hoped for. In South Dakota, Senate Democratic leader Tom Daschle lost his seat to Republican John Thune by forty-five hundred votes. Five other previously Democratic seats also went to Republicans, and with Colorado's Ken Salazar representing the only other Democratic pickup, Republican control of the Senate jumped from 51–49 to a healthy 55–45. Even more crucially, only well after midnight did it become clear that President George W. Bush had narrowly carried Ohio, thereby defeating John Kerry, who formally conceded the race at 11:00 A.M. Wednesday.

By then Barack, after only two hours' sleep, had already appeared on all three networks' morning news shows. Accepting congratulations, he told *Today Show* host Matt Lauer that as a politician "the most difficult thing is dealing with the lack of time for my family." A bit later, Barack told broadcaster Tavis

Smiley that "I am taking my wife out on a date this weekend, and I'm going to take the kids to the park." Gently rebuffing Smiley's suggestion that as the only African American senator, he would be representing black America, Barack stressed that "my first obligation is to my family, my second obligation is to the people of Illinois." Once "I get more experience," Barack added, "then hopefully I can also be an important voice at the national level." Citing the efforts he had made on behalf of Democratic Senate candidates nationwide, Barack believed he would be "getting some good committee assignments," but Democrats' failure to capture the presidency showed that "we haven't perfected a language that describes our values as effectively as the Republicans have."

President Bush called Barack to congratulate him on his election, and Barack thanked him, saying, "I'm not rooting for your failure. I'm rooting for your success" in Iraq and elsewhere. The president said he was eager to collaborate on tax reform, and Barack assured him, "I hope we can find common ground."

On Wednesday afternoon Barack met with reporters at his campaign office. "I want to be seen as a workhorse and not a show horse," Barack told them, because "ultimately what is most important is your ability to shape the world around you." He became "slightly testy" when Lynn Sweet of the *Chicago Sun-Times* pestered him repeatedly about his presidential possibilities. "It's a silly question," Barack insisted. "I'm a state Senator. I was elected yesterday. I have never set foot in the U.S. Senate. I have never worked in Washington. And the notion that somehow I am going to start running for higher office, it just doesn't make sense." When Sweet persisted, Barack declared, "I can unequivocally say I will not be running for national office in four years, and my entire focus is making sure that I'm the best possible Senator on behalf of the people of Illinois." For Sweet's benefit, Barack repeated his insistence three times. "I am not running for president. I am not running for president in four years. I am not running for president in 2008."[92]

Chapter Ten

DISAPPOINTMENT AND DESTINY

THE U.S. SENATE

NOVEMBER 2004–FEBRUARY 2007

On Thursday, November 4, less than forty-eight hours after his election victory, Barack resigned his Illinois Senate seat, ended his affiliation with Miner, Barnhill & Galland, and went on indefinite leave from the U of C Law School. He also formally retained prominent Washington lawyer Robert Barnett as his new literary representative, ending his fourteen-year relationship with Jane Dystel. Weeks earlier, Dystel had secured a two-book contract for Barack, one of which would lay out "What I Believe," but Washington attorney Vernon E. Jordan had told Barack to consider Barnett prior to signing. Eleven years earlier, Dystel had all but saved Barack's life when she succeeded in reselling "Journeys in Black and White" following Poseidon's cancellation of his initial contract, but Barnett was Washington's superagent for political stars, charging an hourly rate for his services rather than taking a traditional 10 or 15 percent agency commission on a contract's total value.

On Saturday, the Democratic ward committeemen from the 13th Senatorial District met to elect Barack's successor. Barack's young protégé Will Burns, now back on Senate staff, was one of several highly qualified applicants, and Barack and Senate president Emil Jones Jr. both supported his candidacy. Fourth and Fifth Ward aldermen Toni Preckwinkle and Leslie Hairston effectively controlled the choice, and neither felt that Barack or Jones had gone out of their way to be helpful to their wards. Forty-year-old attorney Kwame Raoul, a Haitian American Hyde Park native and Trinity UCC member who had twice challenged Preckwinkle and possessed a $60,000 campaign bankroll, was Burns's top competitor. "I was Barack's guy," Will recalled, but Barack did not press Preckwinkle or Hairston to select Burns. "Barack stayed out of the competition," Kwame confirmed, and when the committeemen cast their votes, Raoul was their almost unanimous choice. Barack's disinterest infuriated Burns's closest friends, and when Kwame called Barack on his cell phone a few hours later to ask his advice, Barack semiseriously responded, "Stay out of jail!" Barack added that Raoul should be very careful about who worked in his district office, but then quickly said he was at a theater and ended the call by saying, "Hey, man, this is family time. I have to see a movie with my daughters, so I have to let you go." Many in the audience applauded when they saw Barack with Malia and Sasha.

Jennifer Mason handled the district office transition to Kwame. Within days she and Cynthia Miller discarded all of Barack's old files, but on Tuesday, Barack returned to Springfield for several send-off events with his ex-colleagues. Emil Jones introduced a resolution congratulating Barack, saying that his DNC speech represented "my proudest moment in politics." Then Barack took the podium, acknowledging that it was "a bittersweet moment" and saying "I have to give special credit to President Jones for his friendship and support, because if it weren't for him, I don't think I would be standing here today."

Those in the know throughout the statehouse and all of Illinois politics unanimously concurred. "Emil was more instrumental in him winning that U.S. Senate race than anybody," lobbyist and poker buddy Mike Lieteau stressed. Chicago civic powerhouse Eden Martin said, "I think Emil played a much bigger role in his career than anybody knows." Dan Shomon emphasized that Emil "was critical," and even top donors acknowledged that Jones was "actually the most critical person to Barack's success." Barack happily acknowledged that Emil was his "political godfather," but on the floor of the state Senate, even Republican leader Frank Watson warmly wished Barack "good luck" while humorously adding that "if you need any humility, you know where you can find it, right back here." After an evening reception, his closest Springfield friends held a special poker game to drive home Watson's point. "We brought him down to earth real quick," explained Terry Link, describing how they worked together so that Barack lost every hand. By night's end he had "lost everything he had," Mike Lieteau remembered, and had to write someone a personal check. "He didn't throw his cards or take a swing at anybody, but he wasn't a happy person," Terry recalled.

Barack told reporters that the Senate committees he was most interested in were Finance, Foreign Relations, and Agriculture, and senior Illinois senator Dick Durbin stressed that Barack had "really helped himself immeasurably" with his out-of-state fund-raising work. "He's contributed to every single Senate candidate, and that goes a long way. People really appreciate it." Barack had already called Republican Foreign Relations Committee chairman Richard Lugar to express interest, and old Harvard Law School friend Cassandra Butts took charge of setting up Barack's Capitol Hill office while other former classmates warned him against taking a seat on the often-contentious Judiciary Committee. Postponing for the moment any decision about whether the whole family would move to Washington, Barack said he would commute back and forth from Chicago at least until the end of the school year.[1]

On Sunday morning, November 7, Barack appeared on both ABC's *This Week* and NBC's *Meet the Press*. Saying that "George Bush and I shared a million voters" across Illinois, Barack called for people to "disagree without being disagreeable." Asked if he would serve his full six-year Senate term, Barack answered, "Absolutely . . . I expect to be in the Senate for quite some time." Barack devoted the next week to a five-day "Thank You, Illinois" tour, holding town hall meetings in Peoria, Moline, Belleville, and other cities. In Peoria, Barack met with Republican congressman Ray LaHood to talk about "how we would work together for Illinois." In Moline, *Quad-City Times* reporter Ed Tibbetts listened as one man told another, "You better get his autograph. He may be president one day." A woman interjected, "He will be president someday." In Belleville Barack declared, "it's irresponsible to add more debt onto future generations," while in Bloomington, the *Pantagraph* advised him to "stay humble" and "don't forget you were sent to Washington, D.C. to represent Illinois." *Time* magazine ran an eight-page Obama profile that described him as "charismatic" and "meticulously self-aware." George Galland from Miner Barnhill said, "There aren't many blindingly talented people, and most of them are pains in the ass. Barack is the whole package." Former Virginia governor Doug Wilder said, "Obama could be president. There's nothing to stop him."

Later on Sunday, Barack flew to Washington for the Senate's four-day orientation for new senators. A temporary basement office with two desks and two telephones was an inauspicious start for a U.S. senator, and Barack told one reporter he and Michelle had "a lot of big decisions to make on how to protect our kids and keep our marriage strong." On a more mundane level, Barack also had to decide how to allocate the significant staff budget he would have once he was officially sworn into office on January 4, 2005. "I'm tapping into friends and colleagues who have worked in Washington before to give me some sense of how to shape the office." Most important was Cassandra Butts's outreach to Pete Rouse, Tom Daschle's longtime chief of staff. "I had thirty years in the federal retirement system, so I figured this was a time to go," Rouse explained, but Cassandra had other plans, and asked Pete to meet her and Barack one evening in the Mandarin Oriental Hotel restaurant, southwest of the Capitol.

"We talked for the first hour about how he should approach getting organized and getting established and getting set up," Rouse recalled. Then Barack asked "would I be interested in being his chief of staff." Pete first said no, but Barack and Cassandra persisted. "It was very much us selling Pete on why he'd want to work for a freshman senator who was like ninety-ninth in seniority," Cassandra remembered. Barack explained how invaluable Pete could be. "I know what I'm good at. I know what I'm not good at. I know what I know, and I know what I don't know," Barack told Pete. "I can give a good speech. . . . I know policy. I know retail politics in Illinois. I don't have any idea what it's like to come into the Senate and get established in the Senate. . . . I want to get established and work with my colleagues and develop a reputation as a good senator." Also, "I don't know how to build a large staff and negotiate the potential pitfalls of being a relatively high-profile newcomer to the Senate. I have no intention of running for president in 2008." Barack stressed, "I don't want to be a black senator, I want to be a senator who happens to be black." Given Pete's unique level of respect within the Senate, Barack wanted him "to partner with me to make sure I get off on the right foot." As Rouse warmed toward Barack's request, Barack told Pete he wanted to hire a staff that was diverse in multiple ways, and would leave most of the D.C. hiring decisions to him. "It was really important to bring Pete on board," Cassandra explained, because with Rouse as chief of staff, it all but eliminated Barack's learning curve as a newcomer to Capitol Hill. Pete "knew the Senate. He knew what it took to be a good senator," especially how representing one's home state well was crucial to building a great reputation.

To oversee what would be multiple offices across Illinois, Barack asked first Paul Williams and then Ray Harris to take charge, but both turned him down. "Barack, this only makes sense to me if you're going to be president in the next few years," Paul responded, and Barack said that would not be so. Before flying back to Chicago to take his daughters to the circus, Barack met with Peter Fitzgerald and joined him for lunch in the Senate Dining Room. Fitzgerald had given up his seat by choice, and he told Barack about "all the downsides of life in the Senate." Barack asked, "'Do you ever have time to think?' and I said 'No,'" whereas in Springfield, "in the state Senate, you did have time to think."[2]

Barack stressed to journalists that "I don't intend to be a major spokesman for the Democrats," but a two-day visit to Manhattan to help promote *Dreams From My Father*'s paperback sales turned into a media blitz. Speaking on WNYC's *Leonard Lopate Show,* Barack noted, "I've got to set up a good constituent service office" for Illinoisans and said that New York senator "Hillary Clinton has been a terrific model for me." On CBS's *The Early Show,* host Hannah Storm told Barack, "you are such a big star. Everyone's calling you the future, the savior of the Democratic Party," and said that "you've been compared to everyone from Tiger Woods to Abe Lincoln." Barack told Charlie Rose, "I was surprised by the degree to which" his DNC speech had resonated but admitted having "what my wife considers probably too strong a sense of self-esteem." On *The Late Show,* Barack joked to host David Letterman that "the main reason my wife married me was I still had family in Hawaii," but in a subsequent radio interview Barack said, "I've been broke for the last ten years." Billionaire investor Warren Buffett invited Barack to have lunch with him and his daughter Susan, and Barack chartered a private jet to fly from Chicago to Omaha for the luncheon and then back.

In Washington the next weekend, Barack joined Massachusetts governor Mitt Romney as one of the two keynote speakers for the exclusive Gridiron

Club's humorous winter dinner. Barack joked that his 70 percent statewide victory had included "102 percent in Chicago," and offered an update on his DNC speech. "Since the election, that gay couple I knew in the red states? They've moved back to the blue states." A low-key announcement indicated that Barack's committee assignment hopes had been met only in part: Foreign Relations and Environment and Public Works were his two major ones, with Veterans Affairs as a third. But the media and Gridiron hoopla produced some pushback, with congressional prognosticator Stuart Rothenberg writing in *Roll Call* that he was "a bit tired of the fawning" over Barack. Noting that "I've even heard his name floated as a potential Democratic presidential nominee in 2008," Rothenberg exclaimed, "C'mon, let's get real. It's silly to be talking about him as a national figure—or for higher office—at this point."

A few days later, Barack and his family flew to Honolulu for a three-week vacation. Madelyn Dunham was hospitalized but was soon released, and a week passed before Barack made the first of two public appearances, addressing an enthusiastic crowd of 850 at a $100-per-person Democratic fund-raiser in Waikiki following a tearful introduction by his father's old friend U.S. representative Neil Abercrombie. Barack autographed scores of copies of *Dreams From My Father,* telling reporters, "this is a place where I always replenish my spirits." The next day he spoke to four hundred students at Punahou, telling them to "dream big" and recalling his eighth-grade ethics class. At much the same time, Crown Publishers in New York issued a press release announcing that Barack was well on his way to becoming a millionaire. The new paperback of *Dreams* had more than three hundred thousand copies in print, and Crown had now signed Barack to a new contract that would pay him $850,000 apiece for two future books, the first of which would appear in 2006. The proceeds of a third volume, for children, would go to charity.

On Christmas Day, Michelle and Maya Soetoro-Ng cooked breakfast for all the family members at Madelyn's tenth-floor apartment at 1617 South Beretania Street, where Barack had lived throughout high school. Following their annual game of Scrabble, everyone except Madelyn then went to the beach, followed by dinner at a nice restaurant. All the downtime allowed Barack and Michelle to discuss the challenges that lay ahead, particularly whether the family should remain in Chicago once the school year ended or move to Washington. The new January issue of *Vanity Fair* had a two-page photo spread featuring Barack and Michelle, and on the Monday after Christmas, Barack graced the cover of *Newsweek*. Michelle told reporter Jonathan Alter that Barack "is not a politician first and foremost. He's a community activist exploring the viability of politics to make change." Michelle also said she believed her husband had no hidden skeletons. "We've been married thirteen years, and I'd be shocked if there was some deep, dark secret." Barack declared that Democrats need to "reassert in very explicit language that our best ideas rise out of communal values," and dismissed talk of being on the 2008 Democratic ticket. "This speculation is almost comically premature for an incoming senator."[3]

On the day after New Year's, Barack and his family were in Washington preparing for his official swearing in on Tuesday, January 4. Half a dozen Illinois reporters were in town to cover the ceremony, and when Barb Ickes of the *Quad-City Times* asked Barack to autograph her copy of *Dreams From My Father,* she reminded him, "You wrote in here that Congress is 'compliant and corrupt.' What if you were right?" Wielding his pen, Barack replied, "I suspect it is." On Monday morning, he signed a six-month lease on an unfurnished one-bedroom apartment at 300 Massachusetts Avenue NW, a building populated mostly by Georgetown University law students, and asked a staffer to help him buy a mattress. Meeting with thirty journalists in his temporary basement office, Barack said again, "I will not be running for president in 2008." When one reporter asked him what he thought his place in history was, Barack caviled. "I don't think I have a place in history yet. I've just been elected. I haven't done anything yet. Making a speech and getting elected isn't historical. When you start talking about history, that's measured over decades, over a lifetime of accomplishment."

Expected at the White House for a meeting between President Bush and the newly elected legislators, Barack's unfamiliarity with Capitol Hill led him to miss the bus that would take them down Pennsylvania Avenue. "I've got to figure out how to get to the White House. Should I be driving on my own? I don't want Bush to think I'm blowing him off," he told a staffer before an SUV materialized. Smoking a cigarette en route, Barack arrived just in time. Except for Madelyn Dunham, all of Barack's family was arriving in Washington for Tuesday's events, and Monday evening, they all had dinner together. Malik Abon'go and Granny Sarah flew in from Nairobi, Auma from London, and Maya and Konrad from Honolulu. "When you have relatives all over the planet, and you can't afford to get together on a regular basis, this is an excellent excuse to get them together," Michelle observed. Representative Jesse Jackson Jr. told the *Chicago Tribune* that "if we keep faith and keep hope alive, they will be the First Family of the United States," and Tuesday's *Tribune* described Barack as "an almost unprecedented national celebrity," someone "akin to an African-American version of JFK" and representing "a new Camelot."

Emil Jones Jr., Jeremiah Wright, and several relatives joined the Obama family for Tuesday's festivities. At midday Barack took the formal oath of office from Vice President Dick Cheney on the Senate floor. As family and friends then walked across the Capitol grounds to a reception at the Library of Congress, Malia asked her father, "Are you going to try to be president? Shouldn't you try to be the vice president first?" Taking a far more dyspeptic view was

Michelle, who remarked within journalists' earshot that "Maybe one day he will do something to warrant all this attention." As *Chicago Sun-Times* reporter Lynn Sweet wrote that evening, "on Google, enter 'Barack Obama' and 'rising star' and out will pop about 15,500 entries."

Before heading back to Chicago, Barack and Michelle looked at several homes for sale in the posh Maryland suburb of Chevy Chase and spoke with a school principal or two, but Michelle was strongly inclined to stay in Hyde Park rather than move to Washington. Her mother Marian, still living in nearby South Shore, enjoyed helping out with Malia and Sasha, and both girls were doing well at the U of C Lab School. Barack, who initially had wanted his family to join him in D.C. in the summer, now agreed. "This is a healthier environment to raise kids in," he explained. "Everybody in the neighborhood knows us, and they knew us before, and they knew the girls before, and they're not going to start acting different with them."[4]

With $330,000 soon coming in royalties from *Dreams*' republication, the Obamas could afford to move from their East View condominium to a larger family home. Along with Miriam Zeltzerman, the real estate agent who had sold them the condo twelve years earlier, Michelle looked at almost a dozen houses that were for sale in Hyde Park and in neighboring Kenwood. Narrowing that group to four, she asked Barack to look at them, including her favorite, a grand house at 5046 South Greenwood Avenue whose asking price, $1.95 million, "was above what we'd originally intended to pay," Barack recalled. "So I went with Miriam to take a look at the house. It was a wonderful house."

Originally built in 1910, 5046 South Greenwood had once been the home of the founder of Chess Records, the legendary blues label. By the late 1990s, when former hospital executive F. Scott Winslow bought it, the huge home was in "terrible, unlivable disrepair," its third floor full of old vinyl records and the autographed master discs used to produce them. Winslow oversaw a painstaking, yearlong, $1 million restoration, subdivided the large property into two separate lots, and late in 1999 put the 6,199-square-foot house on the market for $1.8 million. "The restored mansion," the *Chicago Tribune* reported, "has six bedrooms, a 28-foot-long glass conservatory, three fireplaces, a formal library with built-in Honduran mahogany glass-door bookcases, double red oak pocket doors, a 30-foot-by-40-foot entertainment loft with a 15-foot-high cathedral ceiling and several large stained glass windows." It also featured "brand new windows, a whole-house stereo and video system . . . oak floors, a 600-bottle wine and cigar cellar, a new four-car garage with an indoor, half-court basketball court and an 800-square-foot roof deck." On the new southerly lot, which buffered the big home from busy 51st Street, a second house could eventually be built.

In the summer of 2000, Fredric Wondisford and Sally Radovick, two Harvard endocrinologists who were joining the U of C's medical faculty, purchased both properties, paying $1,650,000 for the mansion and $621,000 for the adjoining lot. By January 2005, Wondisford and Radovick had accepted new appointments at Johns Hopkins University, beginning in the fall, and put both up for sale, at $1.95 million and $625,000, specifying simultaneous closings in June. Listing agent Donna Schwan Jackson was well acquainted with Tony Rezko, who maintained an apartment just across 51st Street at 5109 South Ellis, and Barack asked Tony to join him and Zeltzerman on a second tour of the home. "He thought it was a good house," Barack recalled, and Rezko asked about the adjoining lot, which Barack said was not worth its $625,000 asking price to him and Michelle. Tony said he might be interested in acquiring it, and Barack replied, "That would be fine." As Barack later explained, "my basic view at that time was having somebody who I knew, a friend of mine, who would be developing the lot if he could, would be great. It would be somebody who we know."

"We were eager to make a purchase, make a decision," and with Miriam Zeltzerman believing that the $1.95 million asking price was too high, Barack and Michelle on January 15 made an initial lowball offer of $1.3 million, agreeing to a delayed June closing. Rezko wanted to involve another close friend's wife, Patti Blagojevich, a part-time real estate agent, in the transaction, her husband Rod remembered. "Rezko was talking to Patti about a property, a friend of his, who was buying a house in Hyde Park," and "actually almost had Patti do the real estate deal for Obama's house," but was unable to interject her into the sale. The sellers responded to Barack and Michelle's initial offer with a counteroffer, and on January 21, the Obamas increased their bid to $1.5 million. The sellers immediately replied, and on January 23, the Obamas again increased their offer, this time to $1.65 million, the same price Wondisford and Radovick had paid five years earlier. The sellers accepted, and soon after they accepted a list-price offer of $625,000 for the adjoining lot from Rita Rezko, Tony's wife. Neither Barack nor the sellers knew how deeply troubled Rezko's finances were, with GE Commercial Finance Corporation having obtained an unreported $3.5 million judgment against him just two months earlier. The $1.65 million price was significantly more than the Obamas had intended to pay, but Michelle's enthusiasm for 5046 South Greenwood carried the day. After the previous eight years, and especially the past two, "Barack felt he owed Michelle the home," David Axelrod explained.[5]

With Senate business getting off to a slow start, Barack spent much of January in Illinois. Opening a Springfield office, Barack also visited the state capitol, telling House members that "the Democrats are better-looking here than they are in Washington. The Republicans are tougher here than they are in Washington." At a town hall meeting in Lockport, Barack received a greeting "usually reserved for movie and rock stars," the *Tribune* reported, and a *Daily Herald* journalist who watched a second one in Waukegan wrote that the "sharp, engaging and often witty" yet "humble" senator "hasn't lost any of the charisma and energy he exuded" during the 2004 campaign. Returning to Washington for his first Senate hearing, Barack said "how grateful I am" to join the Foreign Relations Committee given its "wonderful reputation for bipartisanship."

In Chicago, Oprah Winfrey took a camera crew to East View to spend a day with the Obamas. When the show was broadcast, Winfrey introduced Barack as "a man for our time" and "our electrifying new senator" who "is fast becoming America's favorite son." Oprah told Barack that his DNC speech had "felt monumental and momentous to me," but Barack was self-deprecating, worrying "if you don't have people around you who can remind you that actually 'What you just said makes no sense,' and fortunately I have my wife to do that continually." Michelle joined in, telling Winfrey, "Barack is absolutely the messiest person in the household," and reminding him, "You had dirty clothes on top of the basket this morning. And I'm just like, 'There's a basket with a lid. Lift it up, put it in.'" Michelle reminisced that "Friday night was always date night," but "we're too tired now to do a dinner and a movie, so it's usually just dinner." Barack agreed. "Yeah. We're getting too old to do both."

The next day *Tribune* columnist Eric Zorn, the lead voice in the takedowns of Blair Hull and Jack Ryan, published a piece predicting that "Obama Will Run for President" in 2008 and would win. Dismissing Hillary Clinton as "a poisonously polarizing figure who will build a bridge back to the 20th century and those dreadful Clinton Wars," Zorn noted that "among average Democrats, Obama's is the only name that doesn't tend to provoke either a yawn, a puzzled look or an anguished cry of 'Please, God, not again!'" On the Republican side, "John McCain's time will have passed." While Barack's communications director Robert Gibbs "denied again Wednesday that Obama will run in 2008," Zorn told readers, "Don't you believe it."

After holding additional town hall meetings at VFW posts in Evergreen Park and Springfield, Barack was soon back in Washington, where he used his seat on the Veterans Affairs Committee to grill a Bush administration nominee about a seeming disparity in how much funding Illinois veterans were receiving. When the nomination of Condoleezza Rice for secretary of state came to the Senate floor, Barack joined thirty-two other Democrats in voting to confirm her. Thirteen Democrats, including California's outspoken Barbara Boxer, Illinois's Dick Durbin, and Massachusetts's Ted Kennedy and John Kerry, voted against Rice. When progressive Illinois representative Jan Schakowsky asked Barack about his vote, citing Rice's role in instigating the war in Iraq, Barack explained, "I want to be able to cross party lines. . . . I don't want to just be a Barbara Boxer."[6]

In late January, Barack launched his own individual PAC, Hopefund, as a vehicle for political fund-raising and contributing to other Democrats' campaigns. Former Senate campaign fund-raisers Jenny Yeager and Jordan Kaplan, plus staffers Nate Tamarin and Tarak Shah, came on board, with Valerie Jarrett serving as the PAC's treasurer. On February 1 Barack had a long private conversation about life as a famous yet junior U.S. senator with New York's Hillary Rodham Clinton, and two days later Barack joined Clinton and thirty-four other Democrats in voting against the nomination of Alberto Gonzales to be U.S. attorney general. "If we are willing to rationalize torture through legalisms and semantics," no longer would the U.S. represent "a higher moral standard" in the world. The next week Barack was one of nineteen Democrats joining unanimous Senate Republicans in approving President Bush's Class Action Fairness Act by a 72–26 margin.

In Barack's Senate office, chief of staff Pete Rouse sketched out a "Strategic Plan" for Barack's first year as a U.S. senator. The top priority "was demonstrating that he was serious about being a senator for Illinois and delivering for Illinois." Most weeks Barack flew back to Chicago on Thursday afternoons to hold regular town hall meetings across the state while also spending at least Sundays at home with his family. At one session in Naperville, Republican state senator Kirk Dillard introduced Barack, telling the crowd that working with him in Springfield had been "one of the highlights of my life."

Rouse and Barack were in firm agreement that he should modulate his national profile by limiting his national speechmaking and declining interviews to appear on television talk shows. Rouse's second focus was for Barack "to fit into the Senate and to . . . show that he was a team player, that he was not a headline hunter, that he deserved better committee assignments, that he could help the leadership." Barack "was very aware of the importance of being a team player and not raising people's hackles." As Barack told his aides, "I think it's important to take it slow. I want to be liked." Only a year or more in would Barack begin to seek a national stage, but in reviewing Rouse's game plan with other aides, Barack complained that his busy weekly schedule already allowed him insufficient time to study the major issues coming before the Senate.

Barack's newfound wealth led him to dip a toe into the stock market, with wealthy black investor George W. Haywood recommending a broker who purchased about $90,000 worth of stock in SkyTerra Communications and about $10,000 in AVI BioPharma for Barack. Haywood and his wife Cheryl were major investors in both firms, as were two other men, John J. Gorman of Texas and Jared Abbruzzese of New York, both of whom had contributed generously to Barack's 2004 Senate race. In Chicago, it took Barack and Michelle less than two weeks to obtain an attractive rate commitment from Northern Trust for the 80 percent, $1.32 million mortgage their purchase of 5046 South Greenwood would require, but *Tribune* stories busily detailed how Barack's friend Tony Rezko was the beneficiary of insider dealings that had led the Illinois Tollway to award its fast-food franchises to two chains closely linked to Tony and his friend Chris Kelly, Rod Blagojevich's top fund-raiser.[7]

Barack made a special exception to his low-national-profile stance to fly to Atlanta to address a $500-per-person sixty-fifth-birthday fund-raiser for Congressman John Lewis, but Barack's next three speaking appearances were at town hall meetings in Columbus, Rock Island, and Bloomington, Illinois. A *Washington Post* Style Section profile of the "rising star" proclaimed that Barack's "signature quality is the ease with which he inhabits his charisma," but it was senior senator Ted Kennedy's comment in the story—"He seems to be keeping his head down and doing everything right"—that most pleased Pete Rouse and Barack's other aides, who now finally moved into his permanent offices on the top floor of the Hart Senate Office Building. Further good news came when U of C Hospitals' president Michael Riordan promoted Michelle to a new post as one of seventeen vice presidents. With Michelle now ready to resume full-time work, and her mother Marian reducing her work hours to pick Malia and Sasha up from school each weekday, the promotion promised to raise Michelle's annual salary from $120,000 to upward of $300,000.

In mid-March a further deluge of critical news coverage hit Tony Rezko. The *Tribune* revealed that Tony operated two O'Hare Airport fast-food outlets that a minority-set-aside program had awarded to Rezko's longtime business partner Jabir Herbert Muhammad, the seventy-six-year-old son of Nation of Islam founder Elijah Muhammad. Those outlets had grossed $15.7 million from 1999 through 2002, but as the *Chicago Sun-Times* underscored, the city ordinance governing the set-aside program "requires that a minority not just own, but actually control and operate the business day-to-day." Tony's close relationship to Cook County Board president John Stroger and his championing of his large extended family were both reflected by the presence of "six Rezkos on Cook County payrolls." Tony and Muhammad also owned another minority-set-aside firm that operated one thousand pay phones controlled by Cook County, but the *Tribune* discovered that Deloris Wade, its supposed African American chief executive, had actually died fourteen months earlier.[8]

In a long *Tribune* profile, Barack told reporter Jeff Zeleny that in the Senate, "I feel very humble about what I don't know" and said that his voting record "won't be as easy to categorize as many people expect." Barack expressed open ambivalence, saying that "it's more fun being a governor or president" because in the Senate "one of the difficulties is that you're constantly confronted with these votes that may not reflect your views." At the end of March, Barack conducted a three-day statewide tour of college campuses to promote his first bill, the Higher Education Opportunity Through Pell Grant Expansion or "HOPE"

Act. Highlighting how "college tuition is rising at a stunning rate of almost 10 percent a year," the bill would increase need-based Pell Grants from a maximum of $4,050 a year to $5,100, a figure President Bush had embraced five years earlier. At SIU Edwardsville students "mobbed" Barack, "asking him for autographs," and when his Hopefund team used his name to e-mail members of the progressive MoveOn.org network seeking donations to conservative West Virginia Democrat Robert Byrd, the results were astonishing. "In less than 24 hours, more than 15,000 contributors gave $634,000 to Byrd's campaign," a West Virginia newspaper reported.

On April 15, with their modest, pre-royalties 2004 income totaling only $207,000, Barack and Michelle received a $6,200 IRS refund. Four days later Barack joined President Bush and other political luminaries to celebrate the opening of Springfield's Abraham Lincoln Presidential Library and Museum. At a grand dinner the evening before the public ceremony, Barack introduced old friends to Tony Rezko, and in his own remarks the next day, Barack complained that "image all too often trumps substance" while lauding Lincoln as "this man who was the real thing." After hitching a ride back to Washington with President Bush on Air Force One, Barack told a National Press Club audience he already had held nineteen town hall meetings across Illinois and said that "Democrats have to be able to talk about faith and family and community in ways that resonate with the American people but also embrace tolerance and diversity." He warned that "if the Democrats are dismissive or patronizing toward people's genuine concerns about faith, family and community, we will lose."

Introducing his second and third bills, Barack demonstrated his desire to address important issues that had not received widespread attention. The E-85 Fuel Utilization and Infrastructure Development Incentives Act aimed to increase the small number of gas stations available for flexible fuel vehicles, and the Attacking Viral Influenza Across Nations (AVIAN) Act sought to prevent an avian flu pandemic. Sitting down with WTTW's Phil Ponce in Chicago, Barack explained that he was receiving "about three hundred invitations a week" but was rigorously limiting his speaking appearances outside Illinois. "Most of the time that I'm spending right now is on policy," he explained, and asked about the 2008 presidential race, Barack replied, "I've ruled that out."[9]

In the *Chicago Tribune* and *Chicago Sun-Times,* Tony Rezko was receiving far more coverage than Barack. Governor Rod Blagojevich's estranged father-in-law, 33rd Ward alderman Dick Mell, had publicly alleged that the administration was trading state appointments for campaign donations, triggering investigations that led to grand jury subpoenas to gubernatorial fund-raisers like Chris Kelly and Rezko. One top Democratic lawmaker told the *Tribune* that Tony "seems to have a hand in everything Rod Blagojevich does," and a front-page profile of Rezko noted that he had raised funds for Barack and Cook County Board president John Stroger as well as for the governor. "Tony's the type of guy who is always helping host fundraisers and getting his friends to give you money," Stroger told the *Tribune*. "But he's real upfront with you, and I've never known him to ask for favors, which is unusual in this business."

Keeping up a busy speaking schedule all across Illinois, Barack expressed shock when he learned at a Thornton Township youth summit that south suburban high schools ended classes at 1:30 P.M. Such an abbreviated school day could spur youthful criminality, and Barack warned that a criminal record was often "an economic death sentence" for young adults' job prospects. At Knox College's commencement in Galesburg, Barack told graduates that "focusing your life solely on making a buck shows a poverty of ambition. . . . It's only when you hitch your wagon to something larger than yourself that you will realize your true potential." Recycling some of those remarks at the University of Chicago's Pritzker School of Medicine's commencement, Barack used that setting to bemoan how "the crushing cost of health care in America" as well as the nation's "lack of collective will to ensure that every single American has access to effective, affordable health care" represented "a moral shame," "an economic disaster," and "a national crisis."

Barack and Republican Foreign Relations Committee chairman Richard Lugar coauthored a *New York Times* op-ed warning of the danger of an avian flu epidemic, and speaking at Jesse Jackson Sr.'s Operation PUSH, Barack decried what he had learned in Thornton Township. "You've got high schools in the south suburbs that are letting children out at 1:30 P.M. because they can't afford to keep them until 3:00 P.M. These young people were saying that they know they are being shortchanged and no one seems to be concerned about them."

When *USA Today* asked Barack about his life as a freshman senator, he expressed disappointment that "there very rarely is real debate" in the Senate. "Each of us is speaking to an empty floor and to C-SPAN and giving stock speeches." Barack later explained, "I think fairly quickly I realized that it was going to be hard to get things done in the minority. But that wasn't a surprise. That was true when I was in the state legislature." But Washington nonetheless was a disappointment as compared to Springfield. "I was surprised by the slow pace of the Senate. In the state legislature, we could get a hundred bills passed during the course of a session. In the Senate, it was maybe twenty. And that I think made me realize how resistant to change Washington is."

With the Senate badly divided over several conservative appellate court nominees put forward by President Bush, Barack voiced disappointment that "this argument we have been having over the last several weeks about judicial nominations has been an enormous distraction from some of the work that is most important to the American people." Highlighting that he was "speaking to an empty chamber for the benefit of C-SPAN," Barack said, "the last thing I would like to be spending my time on right now is talking about judges." Barack made clear his limited interest in the composition of the federal bench, but he did "express, in the strongest terms, my opposition to the nomination of Janice Rogers Brown," a California African American whom Barack termed "a political activist who happens to be a judge." When Barack's Illinois seatmate Dick Durbin drew Republican criticism for comparing interrogation techniques used upon U.S. detainees at Guantánamo Bay, Cuba, to those employed by the Nazis, Barack rose in support after Durbin returned to the Senate floor to apologize. "We have a tendency, perhaps because we don't share as much time on the floor as we should, perhaps because our politics seemed to be ginned up by interest groups and blogs and the Internet, we have a tendency to demonize and jump on and make a mockery of each other across the aisle. That is particularly pronounced when we make mistakes."

In late June, a front-page *Tribune* story examined Barack's role as the Senate's sole African American. "One of the things that I'm trying to be mindful of is not starting to get so comfortable or risk-averse that I end up sounding like everyone else," Barack told reporter Jeff Zeleny. "One of the scripts for black politicians is that for them to be authentically black they have to somehow offend white people," which Barack rejected. Zeleny believed that Barack "has purposely refrained from being a leading voice" on racial issues, and "he doesn't want to be seen as the black conscience of the Senate." When Zeleny asked Barack whether he believed the United States was ready for a black president, Barack said yes. "I think an African American candidate—if he's the best candidate—can be president."[10]

In midsummer Barack rented a new D.C. apartment, this one on the second floor of a town house at Massachusetts Avenue and 6th Street NE. When his Hopefund PAC's finance report covering the first half of 2005 was released, it showed that more than $851,000 had been raised, of which $406,000 had been spent, leaving a healthy balance of $445,000. About $234,000 in contributions came from within Illinois, and Hopefund had contributed $2,000 toward Dan Hynes's 2004 Illinois campaign debt plus $10,000 to Virginia gubernatorial candidate Tim Kaine. Barack also gave $4,200 to every Democratic senator who was up for reelection in 2006, including Florida Democrat Bill Nelson. In early July, Barack flew to Orlando to join Nelson at a black Baptist church, and his Senate colleague introduced Barack as "a rock star who carries himself with dignity and humility and is so smart." Back in Washington before the Senate's August recess, Barack outspokenly attacked "the misplaced priorities of the Senate leadership" on the floor. Warning that "we are not focusing on the problems that truly matter," Barack asked "is giving liability protection to gun manufacturers really more important than passing the Department of Defense authorization bill during a time of war?" Barack sounded a similar theme the next day, highlighting "one of the most pressing problems of our time, our dependence on foreign oil" and calling "energy independence" a national priority.

By August, Barack and Michelle had sold their East View condo to jazz musician Kurt Elling for $415,000—a nice return on their $277,500 1993 investment—closed on their $1.6 million purchase of 5046 South Greenwood, and moved to the spaciously luxurious new home. Michelle had further burnished her income by joining the board of TreeHouse Foods, which, along with its parent company, Bay Valley Foods, would pay her $45,000 annually in director's fees. In mid-August, Barack joined former president Bill Clinton and other dignitaries in speaking at a lengthy memorial service for longtime *Ebony* publisher John H. Johnson and then sandwiched two days of downstate town hall meetings around Governor's Day at the Illinois State Fair in Springfield. Most citizens asked about soaring gasoline prices and rising health care costs, but in Champaign a group of antiwar protesters were waiting in the parking lot with signs criticizing Barack's unwillingness to call for an immediate withdrawal of U.S. troops from Iraq. At the forum, Barack's responses left one critic complaining that "he dodged a lot of questions by giving noncommittal answers," but when Barack remarked, "I am not the president—yet," the crowd responded with "loud cheers."

Foreign Relations Committee chairman Richard Lugar had long pursued the decommissioning of obsolete Soviet weapons of mass destruction, annually visiting Soviet bloc sites to monitor progress. Soon after joining the committee, Barack expressed interest in nuclear proliferation challenges, and Lugar readily agreed when Barack asked if he could join the chairman's 2005 ten-day trip. On August 24 their fourteen-member group, including Barack's new foreign policy staffer Mark Lippert and communications director Robert Gibbs, boarded a government DC-9 to fly to Moscow. From there they flew first to Saratov and then on to Perm, where aging nuclear warheads were being destroyed. Russian space agency officials greeted Lugar's party with vodka toasts, but when the time came for them to fly to Ukraine, Russian border guards sought to search the U.S. aircraft.

A standoff ensued, with Lugar, Barack, and their aides, minus their passports, locked in an airport lounge while frantic phone calls bounced between Washington, Moscow, and Perm. Barack napped on a sofa while Lugar remarked to Jeff Zeleny of the *Chicago Tribune,* the only journalist on the trip, that "it makes you wonder who really is running the country." Diplomatic niceties were restored when the "Rod Blagojevich of Perm," as Barack described the regional governor, took charge, and the U.S. group flew on to Kiev, where the senators toured a "dilapidated building" where bioterror viruses "were locked behind thin padlocks or not at all." From Kiev, they headed to Donetsk, then southward to Baku, Azerbaijan, before making a final stop in London. Lugar and Barack called upon British prime minister Tony Blair at 10 Downing Street and then flew back to Washington.[11]

By the time of their return, Hurricane Katrina had caused what Barack called a "crisis of biblical proportions" along the Gulf Coast, with massive flooding in low-lying areas of New Orleans leading to a disproportionately poor and African American death toll that topped eighteen hundred. On September 5 Barack joined Oprah Winfrey and Jesse Jackson Sr. to travel to Houston, where former presidents George H. W. Bush and Bill Clinton, along with Barack's Senate colleague Hillary Clinton, greeted evacuees who were being housed at the Astrodome. As commentators debated the extent to which the federal government's tardy response to the disaster reflected disinterest in poor communities of color, Barack took to the Senate floor to denounce the "unconscionable ineptitude." He said, "I think the ineptitude was colorblind," and he quickly introduced a trio of bills aimed at helping refugees and those assisting them. Launching a series of podcasts, Barack spoke of "communities that had been abandoned before the hurricane" and called for the country to "close this divide between the wealthy and the poor." Accepting a Sunday TV show invitation for the first time since his swearing-in, Barack told ABC's George Stephanopoulos that the Bush administration's response was "so detached from the realities of inner-city life" that it reflected the government's "historic indifference" toward poor communities that "are disproportionately African American." Federal officials exhibited "a terrific spin operation, but not the kind of soul-searching that I think you'd want to see from any administration, Democrat or Republican."

Barack emphasized to the *Tribune*'s Jeff Zeleny that "it is way too simplistic just to say this administration doesn't care about black people," but "I think it is entirely accurate to say that this administration's policies don't take into account the plight of poor people in poor communities." Visiting Harvard to deliver keynote remarks at the law school's "Celebration of Black Alumni," Barack chided the criticism directed at President Bush. "We haven't displayed the kind of cool, focused outrage that Charles Hamilton Houston displayed" when launching the NAACP's legal campaign against school segregation. "In fact, our anger at Bush and the administration lets us off the hook. It allows us to say, 'Well, I didn't vote for him. I wrote John Kerry a check, so it's not my problem.' But of course it is our problem." Barack's friends joked about how Harvard president Larry Summers twice mispronounced his name as "Bare-ack" when introducing him, but Barack joked too about his colleagues in the U.S. Senate. "There are some folks there who you're wondering, 'How'd you get here?'"

Out of public view, private fund-raisers were a staple of Barack's schedule. A Philadelphia one featuring Barack produced more than $500,000 for Pennsylvania Democratic Senate candidate Bob Casey, and during Barack's Harvard visit, a $1,000-per-person event increased Hopefund's coffers. David Axelrod and his wife Susan were deeply devoted to an epilepsy research organization their daughter Lauren's illness had led them to found, and for the second time since his election to the Senate, Barack headlined a CURE fund-raiser, this one a roast of Chicago congressman Rahm Emanuel, a onetime ballet dancer who had lost part of a finger in a youthful accident. Rahm was "the first to adapt Machiavelli's *The Prince* for dance," Barack joked. "As you can imagine, there were a lot of kicks below the waist." The loss of much of Emanuel's middle finger "rendered him practically mute," Barack asserted. "Has he ever flashed that little stubby thing at you?"[12]

But Katrina's aftermath remained the foremost national issue, with Barack telling NPR listeners that "most of the effects of race have to do with institutional and structural barriers to opportunity." He explained that "we've got to see this as a twenty-, thirty-, forty-year project," but "the last presidential election" had not included any "conversations about poverty." Just as in his foreign affairs partnership with Indiana Republican Richard Lugar, so too on Katrina Barack partnered with conservative Oklahoma Republican freshman Tom Coburn to introduce a bill requiring careful oversight of federal recovery spending. Barack and Coburn had met during Senate orientation, and "our wives hit it off, and we became friends," Coburn explained. "We both share a philosophy that whatever money is spent by the federal government should be well spent," and Barack struck Coburn as "a very honest and sincere man."

When the *Chicago Tribune*'s Jeff Zeleny asked Barack how work on his forthcoming book was progressing, Barack answered "slowly," saying it would be very different from the still-bestselling *Dreams*. "It will be less autobiographical and be more focused on public policy and the direction I think the country should be going." Following the September 3 death of Chief Justice William H. Rehnquist and President Bush's September 5 nomination of conservative circuit judge John G. Roberts Jr. as Rehnquist's successor, Roberts met privately with Barack and other senators. Impressed with Roberts's sterling credentials and judicial demeanor, Barack was "sorely tempted" to support his confirmation, telling David Axelrod that "If I become president someday, I don't want to see my own, qualified nominees for the Court shot down because of ideology."

Chief of staff Pete Rouse vehemently disagreed, eventually changing Barack's mind. "I will be voting against John Roberts's nomination. I do so with considerable reticence," Barack said on the Senate floor before Roberts was confirmed by a 78–22 margin, with Democrats splitting evenly for and against. "It was a close call," Barack told broadcaster Tavis Smiley, because "I do not think that John Roberts is Antonin Scalia or Clarence Thomas," the two most conservative justices. Yet Barack objected strenuously to "the sort of broad-brush dogmatic attacks" that interest groups launched against Senate

Democrats like Vermont's Patrick Leahy who voted to confirm Roberts. In a long essay on the progressive Daily Kos Web site, Barack criticized "vilifying good allies" and said he was "convinced that . . . the strategy driving much of Democratic advocacy, and the tone of much of our rhetoric, is an impediment to creating a workable progressive majority in this country." Most Americans "don't think George Bush is mean-spirited or prejudiced," but they are "aware that his administration is irresponsible and often incompetent." John Roberts was "a conservative judge who is (like it or not) within the mainstream of American jurisprudence, a judge appointed by a conservative president who could have done much worse (and probably, I fear, may do worse with the next nominee)." Attacks on Democrats who supported Roberts "make no sense," because "to the degree that we brook no dissent within the Democratic party, and demand fealty to the one, 'true' progressive vision for the country, we risk the very thoughtfulness and openness to new ideas that are required to move this country forward." Barack warned that "by applying such tests, we are hamstringing our ability to build a majority. We won't be able to transform the country with such a polarized electorate. . . . Whenever we exaggerate or demonize, or oversimplify or overstate our case, we lose. Whenever we dumb down the political debate, we lose. A polarized electorate that is turned off of politics, and easily dismisses both parties because of the nasty, dishonest tone of the debate, works perfectly well for those who seek to chip away at the very idea of government because, in the end, a cynical electorate is a selfish electorate." But Barack was no centrist, because "on issues like health care, energy, education and tackling poverty, I don't think Democrats have been bold enough," and boldness will "require us to admit that some existing programs and policies don't work very well" while also appreciating that "the tone we take matters."

In early October, Barack spoke on the Senate floor in fervent support of an amendment offered by Arizona Republican John McCain to prohibit torture. "Torture is morally reprehensible" and "the use of torture does not enhance our national security," Barack declared. "We must make it absolutely clear . . . that the United States does not and will not condone this practice." In the heavy schedule of often ponderous committee hearings convened by the trio of panels on which he sat, Barack usually exhibited an easygoing manner. While Foreign Relations was far more interesting than Environment and Public Works, Barack also enjoyed a highly active role on Veterans Affairs, partly because of the friendly bipartisan style that chairman Larry Craig, an Idaho Republican, and Hawaii Democrat Danny Akaka, the ranking minority member, jointly cultivated. Bantering with one witness, Barack referred to the self-confident Robert Gibbs in joking that "my communications director would insist that he is the most important person in my office." After Secretary of State Condoleezza Rice refused to give a timeline for U.S. withdrawal from Iraq during a testy

Foreign Relations Committee hearing, an unhappy Barack called her responses "entirely unsatisfactory." When Centers for Disease Control and Prevention director Julie Gerberding offered bureaucratic doublespeak when questioned about avian flu preparedness, Barack expressed irritation—"that doesn't make sense to me"—before becoming even more exasperated by Gerberding's obfuscatory answers: "I have to say I'm now confused again."[13]

With Barack's post-Katrina emergence on the national stage drawing no discernible criticism in either Illinois or the Senate, he and Pete Rouse decided that the rigorous self-restraint that had characterized Barack's first eight months in Washington could be relaxed. But with Barack in D.C. no more than three evenings a week, Rouse realized that Barack was not able to "spend a lot of time building relationships" on Capitol Hill and around town. Barack "was off in Chicago most of the time because of his family," and while Barack "could have developed much deeper personal relationships" if he was spending "more time in D.C.," that was difficult "when you're living in Chicago and that's not your natural inclination."

With Michelle, Malia, and Sasha ensconced in the grand mansion on South Greenwood, Barack was "home more now than when he was running" for the U.S. Senate in 2002–2004, Michelle explained. Yet with town hall meetings and other events all over Illinois, Barack remained "a weekend dad. They see him on Sundays and Saturdays. That's been what they know." Michelle recounted that "it's been all of their lives that it's rare to have Dad at home for dinner, to see him in the mornings before you go to school, to have him there when you go to bed at night." Most days she remained a single parent, although now one with more extensive household help and far greater financial resources. When Michelle was in Washington, she refused to stay in Barack's modest apartment on Massachusetts Avenue NE. "When she came, I had to get a hotel room," Barack explained.

But there was no denying that Barack was deeply disappointed by life in the U.S. Senate. Pete Rouse called those feelings "the freshman blues," but David Axelrod remembered Barack musing that perhaps he, just like Peter Fitzgerald, should give up on the Senate after one term and run for governor of Illinois in 2010. Visiting Springfield to speak to the Illinois Press Association, Barack decried the tenor of D.C. "What you have in Washington all too often are a Fox News version of the world and a *New York Times* version of the world talking past each other," he told the *Journal-Register*'s Bernie Schoenburg. In by far the most revealing of the weekly podcasts he was recording, Barack described life as a U.S. senator to whoever listened in on his Obama.Senate.gov Web site. "Being away from your family" remained "the most difficult" challenge, but in other ways Capitol Hill was a letdown compared to Springfield. "You don't really have genuine debate in the U.S. Senate in the ways that I had become accustomed to" in the statehouse, where "maybe one out of five, one out of ten bills that came up would actually be modified as a consequence of the debates" on the state Senate floor. "People would start asking the sponsor of a bill questions and maybe as a consequence of those questions the sponsor might withdraw the bill or he might promise to amend it." Outcomes could be altered, and "it happened frequently enough to get a sense that . . . people were still listening to the merits" of whatever was being debated. In contrast, "here in Washington that doesn't happen very much," Barack observed. "I've seen a lack of independence on the part of my colleagues . . . on a lot of major issues."[14]

Barack's Hopefund PAC held an all-day "policy workshop" at the Hyatt Regency Chicago to allow over one hundred donors who had given $2,500 or more to interact with Barack. Two days later Barack flew to Omaha for a private fund-raiser hosted by billionaire investor Warren Buffett and his daughter Susie. Buffett had been impressed with Barack ever since watching his 2004 DNC speech and told the *Omaha World-Herald*, "I'm as enthusiastic about him as I am about anybody in political life." Barack's "very smart and articulate" and "a natural leader." Susie Buffett added that "I want him to be president," and Barack told the *Tribune*'s Jeff Zeleny that "the wonderful thing about Warren Buffett—similar to my relationship with Oprah—it's somebody who doesn't need anything from me."

In late October, Barack traveled to Virginia to campaign for Democratic gubernatorial nominee Tim Kaine, and two weeks later a family vacation in Phoenix enabled him to appear with a Democratic Senate candidate before a large crowd at Arizona State University. An invitation to speak at civil rights heroine Rosa Parks's funeral in Detroit preceded an appearance on the Comedy Channel's *Daily Show*, where Barack told host Jon Stewart that the most pressing foreign affairs challenge was "how fast can we get our troops home without causing all-out chaos in Iraq."

A Kaiser Family Foundation report decrying the extent of "Sex on TV" occasioned an unusual broadside in which Barack decried "a mass media culture that saturates our airwaves with a steady stream of sex, violence, and materialism. . . . As we're spending more free time immersed in this media culture, the amount of questionable content spilling across our screens is growing by the year." Barack believed that "mass media is contributing to an overall coarsening of our culture," and for children "the content of their viewing is not enriching their minds, but numbing them" by "trivializing the important and desensitizing us to the tragic." Since "mindless violence and macho aggression on television begets the same behavior in our kids," everyone should "start by turning off the TV altogether." Particularly "as parents, we have an obligation to our children to turn off the TV, pick up a book, and read to them more often," as

Barack regularly did with his daughters. Yet he wished for a culture "that has a higher calling than simply peddling indecency and materialism for profit."

One search for profit that instead produced a $13,000 loss came to a quick end when Barack sold all of his SkyTerra Communications and AVI BioPharma stock, liquidating the "Freedom Trust" that attorney Bob Bauer had created six months earlier to oversee Barack's holdings. Speaking at a commemoration of Robert F. Kennedy's eightieth birthday, his widow Ethel called Barack "our next president," and two days before Thanksgiving, Barack for the first time since his election formally addressed the Iraq war in a lengthy speech to the Chicago Council on Foreign Relations. Looking back to 2003, "at the very least, the administration shaded, exaggerated and selectively used the intelligence available in order to make the case for invasion." Now "we have to manage our exit in a responsible way," starting with "a limited drawdown of U.S. troops" and adopting "a time-frame" for "a phased withdrawal." At bottom, "we need to say that there will be no bases in Iraq a decade from now."[15]

Two days before Thanksgiving, Barack devoted his eleventh weekly podcast to describing how this would be the first time his family would celebrate the day at their new home rather than at Marian Robinson's. His mother-in-law would still take charge of the turkey, mashed potatoes, string beans, and macaroni and cheese. "I try to sneak off with my brother-in-law to watch the football game," and that evening Barack and his daughters visited old family friend Allison Davis's home, just two blocks northward. Malia did an impromptu demonstration of what she had learned at ballet class, "and poor Barack just started crying," Allison remembered. "'I'm never at home,' and 'They're growing up, and I'm missing out,'" he tearfully complained. When Michelle a few days later consented to a telephone interview with her alma mater's undergraduate newspaper, she brushed aside a question about Barack's next step. "Some day, he may take on a more influential role in politics. For now, our future is our kids."

Following a quick trip to Montgomery, Alabama, to attend a fiftieth-anniversary commemoration of the famous bus boycott, Barack sat down with first the *Chicago Tribune*'s editorial board and then the suburban *Daily Herald*'s. He expressed disquiet about Judge Samuel Alito, President Bush's newest Supreme Court nominee. "There's an amazing consistency in which he is ruling for the more powerful against the less powerful, across the board. And that concerns me. That makes me suspicious." With the *Daily Herald* journalists, Barack explained that in early January he would make a brief visit to Baghdad, with additional stops in Doha, Amman, Jerusalem, and Ramallah. Looking back on his first eleven months as a U.S. senator, "I got more done than I expected," so "I think this year exceeded expectations." He had held thirty-nine town hall meetings across Illinois, and his vote against Chief Justice John Roberts was clearly his toughest. "I did not want to put my imprimatur on a set of decisions he'll be making that I think won't be good for the American people." Asked what had changed in his life, "I guess the biggest change is that I'm no longer a private person," because "I can't go to a restaurant now with my wife or go to a movie without a majority of the people knowing who you are and then a large portion of the people coming up to you. And everybody has cameras now." Barack admitted that he was still smoking and unable to quit, but he explained that he thought his remarkable popularity was because people "think that I'm not overly calculating."

In *The New Republic*, Washington correspondent Ryan Lizza published a piece entitled "Why Barack Obama Should Run for President in 2008." Obama "is the most promising politician in America, and eventually he is going to run for president." Asserting that "experience is an overrated asset in presidential politics," Lizza argued that additional years in the U.S. Senate would not benefit Barack. "Obama must strike while he is hot." Two days later, *Daily Herald* columnist Burt Constable, citing his own March 2, 2004, column as the first published item touting Barack as a future presidential candidate, seconded Lizza's argument. Barack "should be president," and Constable wished that George W. Bush would "outsource the next three years of presidential duties to Obama." Asserting that "he's ready now," Constable recounted that when someone at the *Herald* had cited the downside of accumulating a lengthier Senate voting record prior to a presidential run, Barack had replied, "I'm not unmindful of that." Constable concluded, "Let's hope Obama changes his mind about running for president in 2008. The country needs him."[16]

On December 8, Barack's celebrity got a further unexpected boost when it was announced that he was one of five nominees for a Grammy Award—"best spoken word album"—for his audiobook edition of *Dreams*. His competition was stiff: George Carlin, Al Franken, Garrison Keillor, and Sean Penn, and by chance Barack already was scheduled to appear on Franken's syndicated radio show the next day. Barack joked, "I told my wife this morning that I wanted to be referred to from now on as 'The Artist Formerly Known as Barack Obama,'" and "she told me to go shovel snow." Barack was looking forward to two weeks in Hawaii before heading to Iraq, and when Franken told him "you would be the strongest nominee in 2008" for Democrats, Barack responded that "I'm not there yet." Franken jokingly riffed on Ryan Lizza's line: "There's a time when you're hot, and remember President Cuomo."

At the behest of Senator Bill Nelson, Barack flew to Orlando to speak to the Florida Democratic Party's annual convention. Back in D.C., Barack coauthored a *Wall Street Journal* op-ed on immigration reform with Florida Republican senator Mel Martinez and a few days later coauthored a *Washington Post* one on Darfur with extremely conservative Kansas Republican senator Sam Brownback. As the Senate moved toward adjournment for the holidays, Barack joined with almost all other Democrats to block a renewal of the Patriot Act amid press reports that government electronic surveillance programs illegally exceeded what even that statute generously authorized. "Before I ever arrived in the Senate, I began hearing concerns from people of every background and political leaning that this law" was "threatening to violate our rights and freedoms as Americans," Barack said on the Senate floor. The Patriot Act gave government "powers it didn't need to invade our privacy," and the pending renewal was "legislation that puts our own Justice Department above the law" by allowing it to "go on a fishing expedition through every personal record," including phone calls and e-mails. Saying "this is just plain wrong," Barack warned that "doing it without any real oversight seriously jeopardizes the rights of all Americans and the ideals America stands for."

The Senate's ongoing business delayed Barack's departure to Honolulu, but on a podcast version of his floor statement opposing the Patriot Act, Barack explained that the rest of his family was already en route. "My wife basically said, 'Well, I hope you can make it, buddy,' and took off." Several journalists used the extra time to prepare end-of-the-year stories on Barack's freshman experience, with Oklahoma Republican Tom Coburn telling the *St. Louis Post-Dispatch* that Barack was a "phenomenal young man who will go to great heights." Arizona Republican John McCain added that Barack is "very impressive, he's thoughtful, he's centrist." When the *Chicago Tribune*'s Jeff Zeleny interviewed him, Barack confessed, "I'm subject like everyone else to vanity and what Dr. King called 'the drum major instinct' of wanting to lead the parade." He explained that "one sort of measure of my own wisdom is the degree to which I can clear my mind of ego and focus on what's useful, and I'm not always successful at that." Asked about his smoking, Barack admitted that "it's an ongoing battle" because "the flesh is weak," and when Zeleny inquired about how work on his second book was going, Barack responded, "Don't ask." He added that "it's not coming as fast as I would like. It needs to be done by March. I feel like I do have to write it myself. I would feel very uncomfortable putting my name to something that was written by somebody else or co-written or dictated. If my name is on it, it belongs to me."

Zeleny revealed that in Barack's "early months in Washington, a handful of friends and close advisers wondered whether Obama actually enjoyed being a senator." When Zeleny spoke with Michelle, she said, "it's a tough choice between 'Do you stay for Malia's basketball game on Sunday or do you go to New Jersey and campaign for [Jon] Corzine,'" who in November had won the governorship there. "It's a constant pull to say, 'Hey guys, you have a family here,'" but her "hope is that that is going to change, and we're going to go back to our normal schedule of keeping Sundays pretty sacred." On balance, "this has been a good year, but it's still overhyped," Michelle remarked. "If we don't mature,

Barack Obama is going to fall and fall hard because he's going to have to make some decisions that people will not agree with. That's the nature of politics."[17]

On Wednesday, January 4, Barack, Senators Evan Bayh (D-IN) and Kit Bond (R-MO), and Representative Harold Ford (D-TN) departed Washington en route to Kuwait City. After playing basketball and eating with some Illinois troops at Camp Arifjan and a brief visit to Qatar, Barack and his colleagues proceeded on to Baghdad on a U.S. C-130 that performed evasive maneuvers before landing. Fitted out in Kevlar vests and helmets, the group climbed into a Black Hawk helicopter to fly to the U.S.-controlled Green Zone. There Barack lunched with additional Illinois soldiers before the delegation dined with Iraqi president Jalal Talabani. Housed in a former Saddam Hussein pool house, Barack spoke by conference call with a trio of Illinois reporters. The next morning, again clad in vests and helmets, the delegation boarded a Black Hawk and flew to first Fallujah and then Kirkuk. Hours earlier another army Black Hawk had crashed in northwestern Iraq, killing all twelve Americans on board.

Returning to Kuwait City, Barack proceeded on to Amman, Jordan, recording by phone another semiweekly podcast in which he praised the U.S. military as "probably the most capable institution on the planet." From Amman Barack traveled to Israel, where the Jewish Federation of Metropolitan Chicago hosted his five-day visit. A planned meeting with Israeli prime minister Ariel Sharon was canceled after Sharon suffered a serious stroke, and Barack met instead with foreign minister Silvan Shalom. The next day Barack boarded an Israeli Black Hawk for a visit to the remote Arab Catholic village of Fassouta on the Lebanese border, where he was welcomed at the century-old Mar Elias Church. "Thank you so much. I extend greetings from my pastor. He's like my priest. His name is Jeremiah Wright," Barack remarked as Chicago newsman Chuck Goudie stood nearby. On Thursday Barack crossed into the West Bank to meet both Palestinian president Mahmoud Abbas and a group of Palestinian students in Ramallah. On Friday, Barack met with AIPAC officials in Jerusalem before visiting the Yad Vashem Holocaust memorial. "In stark silence, Obama placed a wreath next to the eternal flame," Goudie reported, and on Saturday Barack headed home to Chicago.

Michelle had been especially on edge during Barack's time in Iraq, but back at 5046 South Greenwood for three days with his family, Barack turned his attention to a lingering property concern. An old swing set sat astride the boundary between their home and Tony Rezko's adjoining lot, but construction of a new one for Malia and Sasha would require a bit more space than where the property line now fell. "That's what triggered my thinking that it would be nice to widen the lot," Barack explained, and he called Tony to ask if he would be amenable to such a deal as well as the construction of a fence between the two properties. Rezko quickly agreed to sell Barack a ten-foot strip of land and to pay for erecting a fence. Tony offered to convey the land at the appraised value of $40,000, but Barack, leery of accepting a large monetary favor, instead offered to pay $104,500, or about one-sixth of the $625,000 that Rezko and his wife had paid for the entire lot, and Tony agreed.[18]

Returning to Washington, Barack accepted Senate minority leader Harry Reid's offer to be Democrats' point man on a new push for stronger congressional ethics reform. Also awaiting him was a memo from Pete Rouse. "It makes sense for you to consider now whether you want to use 2006 to position yourself to run in 2008 if a 'perfect storm' of personal and political factors emerges in 2007," Rouse recommended. "If making a run in 2008 is at all a possibility, no matter how remote, it makes sense to begin talking and making decisions about what you should be doing 'below the radar' in 2006 to maximize your ability to get in front of this wave should it emerge and should you and your family decide it is worth riding."

David Axelrod and Robert Gibbs shared Rouse's view, and Axelrod remembered Barack jotting "This makes sense" on Pete's memo. As fund-raising totals for 2005 emerged into public view, Barack's fiscal stature among his colleagues was again underscored. Filings revealed that he had raised a total of $6.55 million: $2.1 million for his own Hopefund, $300,000 of which he had donated to other Democrats, plus $4.4 million that he had raised for others. At least twenty-three flights on private jets had helped him rack up those totals. *Chicago Sun-Times* correspondent Lynn Sweet tallied those numbers and surveyed reactions to Barack's first year in Washington. Progressives both nationally and in Chicago expressed disappointment that Barack had not been a more outspoken figure, with veteran activist Marilyn Katz telling Sweet that "people project on him their own politics." But New York senator Hillary Rodham Clinton gave Sweet a very different verdict. "He and I have talked often about how to get off on the right foot in the Senate, and he has done a superb job this year. He is a careful, thoughtful, effective person, and he is doing what is right for him."

Speaking with Fox News about his Iraq trip, Barack confessed that "I am probably less optimistic" now than he was before. "There is no military solution to the problems in Iraq" and "only political solutions are going to bring about some semblance of peace." Appearing two days later on NBC's *Meet the Press,* Barack explained that thanks to the continuing presence of U.S. forces, "we help spur the insurgency." When host Tim Russert challenged him about his future plans, Barack vowed, "I will serve out my full six-year term" in the Senate. "So you will not run for president or vice president in 2008?" Russert asked. "I will not," Barack answered.

Barack announced that he indeed would vote against Judge Samuel Alito's nomination to the Supreme Court, and when he and other senators visited the

White House to discuss Iraq with President Bush, what Barack called "a frank exchange of views" ensued. "We need to bring our troops home as quickly as possible, but to do so in a way that does not precipitate all-out civil war in Iraq," Barack stated in a same-day podcast. As Barack's ethics reform role drew greater attention, he told a lobbying reform summit that "how extensively money influences politics is the original sin of everyone who's ever run for office, myself included."

Returning to Illinois for another round of town hall meetings, Barack sat down for a lengthy conversation with the *Quad-City Times*' editorial board. "I'm actually surprised at how slow things move in the Senate," Barack explained, but "I've gotten more done than I expected" and "the things I think I'm proudest of" were "some excellent work on veterans" and improving the VA's responsiveness. With the Patriot Act, "the key principle in my mind is making sure that there's somebody watching the watchers," but ethics reform took up much of the discussion. "Over the last five years" congressional "corruption has gotten worse," Barack told the editorial board, "and I don't think, by the way, that the Democratic leadership is strong enough on this front." What was needed was "some sort of independent commission that examines these issues, an Office of Public Integrity that is empowered to field complaints and investigate them," Barack explained. He admitted that "Democrats aren't without sin," and said that "sometimes what my party does is uncomfortable to me, but I know that it has to be defended because it's part of the job of being a member of a party."

Appearing two days later on ABC's *This Week,* Barack on national television refused to acknowledge that Senate Democratic leader Harry Reid had insisted that Barack take a more partisan stance on ethics reform than he wanted to. "The only way we're going to get something passed is to make sure that we work with Republicans who are interested in serious reform," Barack acknowledged, but "I do think that the Democrats can create a contrast by making sure that the ethics bill we present is big and meaningful and bold and not just tinkering around the edges."[19]

On Wednesday, February 1, following a phone invitation from Arizona Republican John McCain, Barack met with nine other senators—seven Republicans and two Democrats—to discuss possible ways forward on lobbying reform legislation. Republican majority leader Bill Frist of Tennessee had suggested that a task force, rather than Senate committees, might be a way to proceed, but on Thursday a letter was drafted over Barack's signature, addressed to McCain, citing "the culture of corruption that has permeated the nation's capital" as reason for moving ahead with legislation rather than forming a task force. McCain had left for a quick trip to Germany, but on Friday Democratic minority leader Harry Reid's staff e-mailed the Obama letter to reporters. McCain learned of the missive over the weekend during a phone conversation with

Mark Salter, his speechwriter and closest staffer, and he told Salter to compose a response. "Brush him back," McCain instructed.

Monday afternoon, February 6, the Salter-McCain reply arrived in Barack's office as a political letter bomb.

> *I would like to apologize to you for assuming that your private assurances to me regarding your desire to cooperate in our efforts to negotiate bipartisan lobbying reform legislation were sincere. When you approached me and insisted that despite your leadership's preference to use the issue to gain a political advantage in the 2006 elections, you were personally committed to achieving a result that would reflect credit on the entire Senate and offered the country a better example of political leadership, I concluded your professed concern for the institution and the public interest was genuine and admirable. Thank you for disabusing me of such notions with your letter to me dated February 2, 2006, which explained your decision to withdraw from our bipartisan discussions. I'm embarrassed to admit that after all these years in politics I failed to interpret your previous assurances as typical rhetorical gloss routinely used in politics to make self-interested partisan posturing appear more noble. Again, sorry for the confusion, but please be assured I won't make the same mistake again.*

Three paragraphs then summarized McCain's long-standing support for bipartisan reform legislation. It closed with this:

> *I understand how important the opportunity to lead your party's effort to exploit this issue must seem to a freshman Senator, and I hold no hard feelings over your earlier disingenuousness. Again, I have been around long enough to appreciate that in politics the public interest isn't always a priority for every one of us. Good luck to you, Senator.*

Barack was astonished. "The perception in our office was that this was a very innocuous boilerplate letter," Barack said of his initial missive. He immediately called McCain's office, but McCain was not back from Germany. Barack and Robert Gibbs quickly set to work drafting a defensive response that sought to dial down the temperature on what was already a Capitol Hill e-mail sensation. "I am puzzled by your response," and "I have no idea what has prompted your response," Barack's Monday-evening reply stated. "The fact that you have now questioned my sincerity and my desire to put aside politics for the public interest is regrettable." By Tuesday, reporters were all over the story, with Barack explaining to the *Sun-Times*' Lynn Sweet that McCain had misunderstood his initial letter. "I think there was confusion over the reference to a task force. We were specifically referring to the proposal that Bill Frist had had, to set up a formal task force to do this," not McCain's more informal bipartisan meetings. Tuesday afternoon Barack and McCain spoke briefly by phone, and Barack told the *Chicago Tribune*'s Jeff Zeleny that while the tone of the Arizonan's letter "was a little over the top . . . John McCain's been an American hero and has served here in Washington for twenty years, so if he wants to get cranky once in a while, that's his prerogative." In private, Barack was more scathing: "I'm not interested in being bitch-slapped by John McCain."

On Wednesday Barack and McCain both testified about lobbying reform before the Senate Rules Committee. "I'm particularly pleased to be sharing this panel with my pen pal, John McCain," Barack joked. Republicans declined to embrace Barack's call for an independent "Congressional Ethics Enforcement Commission," and afterward Barack told Sheryl Gay Stolberg of the *New York Times* that since Congress is "a clubby institution . . . the idea that there might be a watchdog process outside of it . . . will be tough to achieve." That evening Barack got an emotional lift when he was awarded the spoken word Grammy, and Thursday on his semiweekly podcast he praised McCain as "a good and decent man." The news cycle quickly moved on, with McCain aide Mark Salter realizing he had overdone his boss's "brush him back" instruction. "I guess I beaned him instead."[20]

The McCain tiff aside, Barack continued to win plaudits. Vermont Democrat Patrick Leahy told the Associated Press, "I've been here 31 years and seen a small handful of people that have made as much of an impression as he has, and he has done it by working hard." In a five-page profile in *Time* magazine, Barack remarked that "I probably always feel on some level I can persuade anybody I talk to," but reporter Perry Bacon Jr. noted that "liberals in particular have often projected onto him views he doesn't have." The *Chicago Tribune*'s Jeff Zeleny agreed that "Obama remains largely undefined on a broad spectrum of issues," but he realized that "Obama is purposely increasing his visibility as he steps beyond an early strategy of political caution." Barack joined thirty-three other Democrats in supporting a compromise extension of the Patriot Act that won Senate approval by a vote of 89–10, but well-known liberal Democrats like Patrick Leahy, Tom Harkin, and Ron Wyden, plus conservative Robert Byrd, joined longtime opponent Russ Feingold in voting no.

A front-page *USA Today* headline proclaimed that "Democrats See Obama as Face of 'Reform and Change,'" and Barack explained to the *New York Times*' Sheryl Stolberg his recent decision to stop flying on corporate jets. "This is an example where appearances matter," because "very few of my constituents have a chance to travel on a corporate jet." Under Senate rules, Barack had been reimbursing the planes' owners the normal first-class fare, rather than actual cost, and "I said to my staff, 'We may be following the rules but it's hard for me to reconcile this.'" On March 11, Barack was the Democratic headliner at the annual Gridiron Club dinner, turning in what *New York Times* columnist Maureen Dowd called "a smooth, funny performance" that included singing the lyrics "If I only had McCain" to the *Wizard of Oz* tune "If I Only Had a Brain." The next morning on CBS's *Face the Nation,* Bob Schieffer gushed to Barack that "you were the absolute star of last night's Gridiron show." Barack replied that "in my family, my wife is the funny one," and that if asked, Michelle "could've done a rip on me that would've lasted twenty minutes." Barack cited energy independence, health care, and education as his top issues beyond drawing down U.S. forces in Iraq, but he also addressed abortion. "I think the Democrats historically have made a mistake in just trying to avoid the issue, or pretend that there's not a moral component to it. There is," as he had known going back eighteen years. "I also think that it's important, even as I indicate that I'm prochoice, to say this is not a trivial issue."

African American syndicated columnist Leonard Pitts Jr. bluntly warned readers that "Barack Obama is not Jesus." Pitts explained, "I feel the need" to say that given how "seemingly every exhalation of his name" is "accompanied by angels singing hosannas and sighs of adoration from a congregation of Democrats looking to him for political salvation. Or, if you prefer, resurrection." Pitts plaintively wondered, "Is it asking too much that people wait until he actually does something before they start chasing his name with a hallelujah chorus?" The *Times'* Maureen Dowd, so impressed by Barack's Gridiron performance, told Democrats that instead of nominating Hillary Rodham Clinton for president in 2008, they "should find someone captivating with an intensely American success story—someone like Senator Obama," but whoever wrote Dowd's headline—"What's Better? His Empty Suit or Her Baggage?"—inserted an insult where none was intended.[21]

Barack consented to two long interviews with the *National Journal*'s Kirk Victor for what became a ten-page profile in the widely read D.C. weekly. "I am surprised by the lack of deliberation in the world's greatest deliberative body," Barack again stressed, drawing a contrast to Springfield, "where every bill had to be defended and subject to questions. Here, there is a lot more of competing press releases, and I think that contributes to some of the partisanship and lack of serious negotiation" he had witnessed in the Senate. Barack again denied that minority leader Harry Reid had made him take a more partisan position on ethics reform than he wanted. "Is there some tension there between my role as a member of the Democratic Caucus and how I might operate as an entirely free agent? Yes . . . that is the role that I have been placed in," but "I was very explicit that I wouldn't do something like this if the ultimate objective was not to actually get a bill passed that would move the country forward," rather than score political points. Barack again cited his work on the Veterans Affairs

Committee as his most consequential work to date, and said that the 2008 Democratic presidential ticket was "not something that I am spending a lot of time thinking about." He also obliquely highlighted how out of place he felt in the Senate because most of his colleagues "are significantly older" and "have raised their families here in Washington." In contrast, "I am not a part of the Washington social set. A lot of the interactions . . . I just can't participate in because I want to get home to see my wife and two daughters."

Back in Chicago for a weekend with his family, Barack joined Congressman Jesse Jackson Jr. at an antiviolence service at Englewood's Fellowship Missionary Baptist Church that drew more than two thousand people. Two young neighborhood girls, one aged ten and the other fourteen, had recently been killed by stray bullets from gang members' shootouts. The gunmen "don't have a sense of self-respect," and "if we don't change how we raise our children," gang violence would continue, Barack warned. "There's a reason they shoot each other, because they don't love themselves, and the reason they don't love themselves is we are not loving them, we're not paying attention to them, we're not guiding them, we're not disciplining them. We've got work to do."[22]

The next day was Illinois's 2006 primary election. Back in December, Barack had publicly endorsed twenty-nine-year-old basketball buddy Alexi Giannoulias for state treasurer. Early in Barack's U.S. Senate race, Giannoulias had begun introducing Barack to potential contributors in Chicagoland's vibrant Greek community, and Dan Shomon had agreed to move Obama for Illinois's bank account to Alexi's well-regarded, family-owned Broadway Bank. State Democratic Party chairman and House Speaker Michael Madigan had tapped Paul Mangieri, a little-known state's attorney from Galesburg, for the Treasurer's race, but Barack told Greg Hinz of *Crain's Chicago Business* that "if someone has stepped out on my behalf, I think it's important to reciprocate." Asked if Giannoulias had sufficient experience to be treasurer, Barack said, "Alexi is qualified," but "it's important for you and others to put him through his paces and ask tough questions." By early March, Barack was starring in a thirty-second Giannoulias TV ad, telling viewers "he's one of the most outstanding young men that I could ever hope to meet. He's somebody who cares deeply about people. He got that from his family. They really exemplify and embody the American dream. . . . Alexi Giannoulias—he's going to be an outstanding treasurer."

Then, eight days before the primary, Hinz revealed that Broadway Bank had made loans to a man named Michael Giorango, whom the *Miami Herald* in 2002 had labeled an "alleged Chicago organized crime figure" and who in early 2004 had been convicted in Miami federal court of running a multistate prostitution ring. Two days later the *Tribune* reported that Giorango, a "Chicago crime figure" nicknamed "Jaws," had also been convicted of bookmaking and gambling violations in 1989 and 1991. Still, Giorango had received more than $6 million in loans from Broadway Bank since the 1990s. Thanks to his TV campaign, Giannoulias easily defeated Mangieri, 62 to 38 percent. In another race, David Axelrod's close friend and former partner Forrest Claypool had challenged Cook County Board president John Stroger.

Axelrod had pressed Barack to endorse Claypool, who had volunteered his help during Barack's 2004 Senate campaign, whereas Stroger had backed Dan Hynes. But Claypool's past role as a top aide to Mayor Richard Daley made many black aldermen leery, and even progressives like the 5th Ward's Leslie Hairston supported Stroger, whom everyone in black Chicago politics viewed as "a good soldier." Fully aware of the racial divide, Barack had rebuffed Axelrod's insistent pleading with a "flash of anger," David recalled. A week before the primary Stroger suffered a massive stroke, and on election eve Barack told a newscaster he would vote for Claypool. On Tuesday the incapacitated Stroger eked out a 53–47 victory, and the contrast between Barack's kingmaker role in Giannoulias's victory and his hands-off stance in the Claypool-Stroger contest offended many progressives. One Claypool friend, Sunil Garg, wrote a strongly worded *Tribune* op-ed asserting that Barack "has weakened his moral authority by seemingly safeguarding his political career" rather than endorsing Claypool. "I—and many others—cannot understand how someone can be considered presidential when he refuses to take a stand on the most important race in recent memory in his own backyard."[23]

Back in Washington, Barack took to the Senate floor to say that he "didn't anticipate the deafening silence" that had greeted his bill to create an independent congressional ethics watchdog. When a far weaker bill was called for a vote, Barack was one of eight senators—including Democrats Russ Feingold and John Kerry as well as Republicans John McCain and Tom Coburn—who voted no in protest. Every night when he was in Washington, Barack returned to his modest apartment to put in as much time as he could drafting chapters for his forthcoming book. His deadline for submitting a complete manuscript was March 31, but Barack was going to miss that by at least a month. Each time he completed a chapter, the draft was passed off to staff members and a few old friends for fact-checking and political vetting. Rob Fisher, who had contributed so much editorial help on *Dreams,* was among them, and Rob remembered that "we got into a deep back and forth about global warming." After Rob left Barack a voice mail warning him that one chapter was "a mess," Rob's wife Lisa was the first to hear the message Barack left in response: "You don't call a U.S. senator who has a deadline for a book and tell him his book's a mess!"

Barack's hefty initial advance for that book, plus robust royalties from *Dreams*' strong paperback sales, skyrocketed the Obamas' annual income for 2005 to $1,670,000, more than they had earned in the previous seven years combined. In addition to Barack's $1.2 million in author's royalties, Michelle's U of C Hospitals salary had jumped from $121,000 to $316,000, and she took in an additional $45,000 from her service on TreeHouse Foods' two boards. The Obamas had upped their charitable contributions to $77,000, including $5,000 to Trinity Church, and were in the process of contributing a further $22,500 to Trinity in 2006, but in mid-April the Obamas' 2005 windfall had Barack writing a check for a whopping $430,000 to the IRS.

Barack remained often on the road, speaking at a Hartford dinner for Senate colleague Joseph Lieberman, who was facing a progressive primary challenger, and defending Lieberman even though he had supported the Iraq war. "I know that some in the party have differences with Joe. I'm going to go ahead and say it. It's the elephant in the room. And Joe and I don't agree on everything. But what I know is that Joe Lieberman's a man with a good heart, with a keen intellect, who cares about the working families of America." A week later, traveling to Minnesota to speak on behalf of Democratic Senate candidate Amy Klobuchar, a Minneapolis *Star Tribune* reporter covering Barack's appearances wrote that some attendees were "talking about an Obama presidency as if it were a sure thing."

Barack's full-throated support for Alexi Giannoulias's candidacy reemerged as an issue when the *Chicago Tribune* reported that his family's Broadway Bank had made additional loans totaling $11.8 million to convicted felon Michael Giorango *in 2005*. One of them, for $3.6 million, went to Giorango and a *second* convicted felon, Demitri Stavropoulos, for the purchase of a floating casino in Myrtle Beach, South Carolina. In 1994 Stavropoulos had been convicted of explosives possession, and a 2004 bookmaking conviction had had him *in federal prison* since that time. The "senior loan officer" and Democratic candidate for treasurer told reporters, "It's a loan that I don't know the details of," as "I don't cultivate the relationships. I don't bring these deals in." A *Tribune* editorial asked whether Giannoulias's defense "is that he was clueless as to what his bank was doing?" and at a town hall meeting in Elmhurst, Barack told reporters, "I'm going to ask Alexi directly what is happening." Several days later Barack said he had told Giannoulias that "appearances matter," but "so long as these loans were legal" and "were not financing illegal activities . . . I'm not going to pass judgment on how the bank handled its loan portfolio." Alexi finally told reporters that "I probably should have looked into it more," and although journalists doubted the viability of Giannoulias's candidacy, Barack's support for him dipped from public view.[24]

Across Illinois, Barack continued to hold town hall meetings that often drew overflow crowds. At Loyola University in Chicago he warned more than two thousand people that because of the country's dependence on Middle Eastern oil, "we're financing both sides of the war on terrorism through our SUVs." He told cable-TV interviewer Jeff Berkowitz, "I don't think there's any doubt that Iran is seeking nuclear weapons." Illinois seatmate Dick Durbin told the *State Journal-Register*'s Bernie Schoenburg that he liked Barack's judgment, style, and sense of humor. "He's a bright guy. Being mixed-race, people feel comfortable with him," and "he gets along well not only with our side of the aisle but with the other side as well." In contrast, a *Time* magazine item asserted that Barack "has reached so often across the aisle . . . that some Democrats complain he won't be their firebrand," but Durbin had privately already told friends he believed Barack should run for president in 2008. To Schoenburg, Durbin hinted at just that. "Hanging around the Senate for an extra term may not make him a change agent." Instead, "it may make him a person with a long voting record," and Barack "shouldn't rule out any possibility at this point." When Harvard law professor Martha Minow ran into Barack at a May 5 Chicago dinner, she asked him directly, "Are you going to run?" and Barack's response was anything but discouraging. "He said, 'There's one answer: It's up to Michelle. Go talk to her.'" Minow did just that. "I said, 'So, is Barack going to run?' and she said, 'We're having serious family discussions about it.'" Minow replied, "It would be so great for the country," and Rob Fisher recalled a long telephone conversation about that possibility while they were still wrestling with Barack's book manuscript. Rob recalled that "when he made the decision to run for the presidency, he called me up" and "we talked for two or three hours. He was mostly concerned about his family," but "we both came to the conclusion at the end of it that we thought he could win."

The morning after his exchange with Minow, Barack flew to Omaha to speak at the Nebraska Democratic Party's annual dinner. Before that event, Barack addressed a cheering crowd of a thousand people at Salem Baptist Church, declaring that the Bush administration's largesse in Iraq had left the United States with a national debt now totaling "$30,000 from every man, woman, and child in America" as well as inner-city "neighborhoods where rats outnumber computers." At the evening dinner, which conservative Nebraska senator Ben Nelson had asked Barack to keynote, Nelson introduced Barack as "one of the most effective members of the U.S. Senate," and Barack returned the compliment by telling the crowd that Nelson was the body's most popular member. Pete Rouse was struck by Nelson's private comment that Barack was the only national Democrat Nelson had considered inviting, and Barack told a local interviewer that Democrats "tend to talk in bullet points as opposed to telling the story of where we want to take the country and what our vision is for this country's future."

Back in Washington, Barack continued to criticize the ballooning national debt, complaining on the Senate floor that the federal government's behavior "simply passes the burden to our children and grandchildren" and spontaneously adding that "this place never ceases to amaze me." On May 20 Barack told the *Chicago Sun-Times*' Lynn Sweet that he finally had completed his book manuscript and now had time to work on it further. "The copyeditor needs it at the end of June. So I'm good. I have a month and a half to tool around with it and make changes." Then a quick flight to Springfield allowed Barack to address the SIU School of Medicine's commencement, and he warned the new doctors that "there is something fundamentally broken about our health care system." The good news was that "Massachusetts just signed into law a groundbreaking plan that would cover most of its citizens," and it was incumbent upon politicians and physicians alike to "ensure that every American has routine checkups and screenings and information about how to live a healthy lifestyle."[25]

In late May, when Barack hired former Blair Hull strategist Anita Dunn to head up his Hopefund PAC, journalists wondered what that said about his political plans, with a front-page *Chicago Tribune* headline asking "Obama in '08?" Later that same day, Dick Durbin invited reporters to his Springfield home to tell them he hoped Barack would run. "People like Barack don't come along very often," and "he should seriously consider it. I think he could bring a great deal to the national race for the presidency." Durbin's statement drew little press attention, but other top Senate Democrats told Barack they felt the same. Minority leader Harry Reid had been impressed with Barack since soon after he first arrived on Capitol Hill. Reid recalled that after an early address on the Senate floor, Reid approached the new freshman and told him, "'That speech was phenomenal, Barack.'" Then, "without the barest hint of braggadocio or conceit, and with what I would describe as deep humility, he said quietly: 'I have a gift, Harry.'"

After Durbin's announcement, Reid and New York senator Chuck Schumer each told Barack he should run. "It was interesting that they felt as strongly as they do," Barack told David Axelrod. In Chicago, Axelrod's friend Eric Zorn, the *Tribune* columnist who had predicted an Obama presidential candidacy sixteen months earlier, echoed Durbin in saying, "if he's ever going to run for president, 2008 is his year." Zorn believed Barack was "bored and disillusioned by the Senate," and the longer he remained a senator the more opportunities he would have "to alienate his core supporters with clumsy political moves such as turning his back on former close ally Forrest Claypool in the Democratic primary for the presidency of the Cook County Board, playing both sides of the gay marriage issue while more courageous Democrats in the Senate come out in support, and endorsing the lightweight, ethically obtuse Alexi Giannoulias in the Democratic primary for state treasurer as apparent payback for Giannoulias' earlier financial help."

The Bloomington *Pantagraph* advised differently, asserting in an editorial headlined "Obama Presidential Talk Quite Premature" that "it takes more than an engaging smile and oratorical gifts to make a good president." *Time* columnist Joe Klein reported that "close friends of Obama's say he really doesn't know yet what he's going to do in 2008," but "the discussions have grown more serious in recent months." Klein believed that "the best reason for Obama to run is . . . that he is young and everybody else seems so old," but in the left-wing *Nation,* blogger David Sirota warned that "because the media have not looked as closely at his political positions, Obama has taken on the quality of a blank screen on which people can project whatever they like." Sirota understood that Barack "hasn't discouraged this," but a formal interview left Sirota admitting that "Obama has an impressive control of the issues and a mesmerizing ability to connect with people." Barack also demonstrated "a remarkable ability to convince you that his positions are motivated purely by principles, not tactical considerations," and Barack revealingly stressed to Sirota that "I don't think in ideological terms. I never have."[26]

Barack seemed to exhibit how profoundly he was reflecting on his life's course in a remarkably self-revealing commencement address he delivered at the University of Massachusetts Boston on June 2. He returned to the summer of 1985 and his two-day drive toward his new future in Chicago, and he recalled how he had "stopped for the night at a small town in Pennsylvania whose name I can't remember anymore." Barack proceeded to describe how the motel owner—Bob Elia—had told him to go into broadcasting rather than community organizing because "You can't change the world, and people will not appreciate you trying." Barack told his Boston audience that "objectively speaking, he made some sense . . . but I knew that there was something in me that wanted to try for something bigger." What that "something bigger" would entail reached all the way back to Barack's intimately revelatory comments first to Sheila and then to Lena in 1987–88, but what he had first envisioned as his destiny two decades earlier was now a reality that even long-experienced politicians like Dick Durbin and Harry Reid were telling him it was time to pursue.

A few days later, Barack took to the Senate floor to oppose a constitutional amendment aimed at prohibiting gay marriages. "Decisions about marriage," Barack stated, "should be left to the states," but he stressed that "personally, I do believe that marriage is between a man and a woman," period. In mid-June, Michelle's brother Craig, who had just been named head basketball coach at Brown University, married his longtime girlfriend Kelly McCrum in a ceremony at Chicago's Hotel Allegro that Barack, Michelle, and their daughters happily attended. The next day a front-page *Washington Post* story announced that "Obama's Profile Has Democrats Taking Notice." Dick Durbin said, "I don't believe there is another candidate I've seen, or an elected official, who really has the appeal that he does." Durbin revealed he recently had told Barack, with reference to 2008's first state, "'Why don't you just kind of move around Iowa and watch what happens?' I know what's going to happen," Durbin asserted, "and I think it's going to rewrite the game plans of a lot of presidential candidates if he makes that decision." New York senator Chuck Schumer concurred. "I haven't seen a phenomenon like this, where someone comes in so new and is so dazzling." Barack told the *Post* that "at this stage, I haven't changed my mind from previous demurrals," although he confirmed that "we've visited 25 states since taking office."

Even a conservative *Wall Street Journal* contributor argued that Barack's running for president in 2008 might be "the smartest thing he ever does," because his "political stock may never be higher than it is right now." At present, Barack was "part Clinton, part Roosevelt, part JFK and MLK, but in eight or twelve or sixteen years, he might be John Kerry." In the Senate, Barack declined to join Kerry, Russ Feingold, and eleven other liberal Democrats in voting to withdraw U.S. forces from Iraq within twelve months' time, instead supporting a more widely backed measure calling for a withdrawal to *begin* before the end of 2006. In late June, Barack used a chapter on religious faith from his forthcoming book as the basis for a keynote address he delivered to a progressive believers' symposium. It recounted Barack's 2004 discomfort when Dr. Farr Curlin had e-mailed him to complain about how Obama for Illinois's Web site dismissed abortion opponents as "right-wing ideologues." Barack's "Call to Renewal" speech attracted more press attention than anything he had done since his 2004 DNC keynote, with *Washington Post* columnist and fellow Saguaro Seminar alumnus E. J. Dionne heralding it as "the most important pronouncement by a Democrat on faith and politics since John F. Kennedy's Houston speech in 1960," in which Kennedy had discussed his own Roman Catholicism. Fellow *Post* columnist Dana Milbank warned that Barack "is enormously charismatic—and utterly undefined." In the *Chicago Defender,* a left-wing contributor upbraided Barack for his Iraqi war stance, claiming "he was against it before he was for it," and stating that "it is amazing that Obama is the recipient of so much praise when he says so little."[27]

The day after his "Call to Renewal" address, Barack sat down with journalist Jacob Weisberg for an interview that was as remarkably revealing as his commencement speech in Boston four weeks earlier. Barack mentioned that "I kept a journal basically from my junior year in college until I went to law school," and when former president Lyndon B. Johnson's name came up, Barack remarkably stated that "there's a piece of him in me—that kind of hungry, desperate to win, please, succeed, dominate—I don't know any politician who doesn't have some of that reptilian side to him. Or any person for that matter. But that's not the dominant part of me." Barack blamed his Hawaii childhood for how "there's a big part of me that's pretty lazy," and when Weisberg asked whether it was a belief in God that had led Barack to join Trinity Church, Barack replied that "it wasn't an epiphany. I didn't have a revelation. It was a very quiet process where I came to understand many of the things I cared about in a religious way and as part of the tradition of the Christian church." Barack went on to explain that "there was a voice inside me that I could hear that was true—a kernel of myself—this is from a Borges poem—sort of a love poem to a mistress of his," one that Jorge Luis Borges had written in 1934. "He says 'I offer you that kernel of myself that is basically untouched, doesn't traffic in the trivial or the petty or is just there'" (a close rendition of Borges's declaration, "I offer you that kernel of myself that I have saved, somehow—the central heart that deals not in words, traffics not with dreams, and is untouched by time, by joy, by adversities"). Barack told Weisberg, "I felt that. That sounds like god to me," and "to me god is connection . . . making connections, and in that connection, that's where I find god." Barack admitted that "I don't know what happens to us after we die," but "when I tuck in my kids, I have a little piece of heaven . . . I genuinely mean that in a religious way. When I look at them at night sleeping, it is a religious experience. It's miraculous to me."

Barack mentioned that Jeremiah Wright "was just here two weeks ago" and cited his own "frustrations of being here as opposed to being in the state legislature," although "I actually find Congress identical to the state legislature." When Weisberg asked about the presidency, Barack said, "my attitude about something like the presidency is that you don't want to just *be* the president. You want to change the country. You want to make a unique contribution. You want to be a *great* president," Barack emphasized. "There are what, maybe ten presidents in our history out of forty-something who you can truly say led the country? And then there are thirty-odd who just kind of did their best. And so I guess my point is, just being the president is not a good way of thinking about it."

Weisberg also spoke with Michelle Obama, who said, "politics is a completely unappealing way to live your life. There's nothing that makes this attractive to go through as a family. But I also know very deeply and much more intimately than anyone out there how truly gifted Barack is. Anyone wondering, 'Is he the real deal?' and I know it." She also said that "Barack would walk away from it tomorrow if I said so," but "part of me looks at my children and the world that I want my kids and grandkids to live in and says, 'How can I stand in the way?' But I struggle with it every day."[28]

In early July, Barack's office announced that he would take a two-week trip to Africa, including several stops in Kenya, in late August. On July 20, it was announced that Barack would attend Iowa senator Tom Harkin's twenty-ninth annual "steak fry" on September 17, one of the state's premier political events. At the invitation of Louisiana senator Mary Landrieu, Barack joined her and other senators on a brief trip to New Orleans, which was still struggling to rebuild from Hurricane Katrina almost a year earlier. Barack said he "was astonished by what I saw" and by "the terrible reality and scope of the destruction."

Sitting down at a D.C. event with NBC's Tim Russert, Barack called for the "creation of nonpartisan maps" in federal and state legislative districting so as to "encourage and reward more bipartisanship." Barack voiced disappointment that most members of Congress were so attached to "the status of office" that "they don't want to take any risks" for fear of being defeated. "A good 50 percent of what's wrong with Washington has to do with fear," Barack explained. Being an elected official "puts a severe strain on your family," and "the biggest challenge and the biggest strain on me has been the fact that I didn't move my family to Washington, and so I'm away from my wife and kids three days, four days a week, and that's very difficult, and it's much more difficult for my wife." In a similar conversation with John Patterson of suburban Chicago's *Daily Herald,* Barack stressed that as a senator representing a large state "the schedule is relentless" and now "people recognize me" all the time. "The only time when it's tough is when I'm with my kids," because "a lot of times people will come up, and it kind of intrudes a little bit on our time together." Barack believed he remained unchanged by all the public and media attention. "I really credit . . . the seven years of laboring in almost total obscurity in Springfield for a healthy attitude and healthy skepticism about this." When Patterson mentioned the presidency, Barack said, "I think what you look for in presidents is good judgment and knowledge of the issues and the ability to make decisions." "Do you think you could do a better job than the current president?" Patterson asked. "No comment," Barack replied.

After taking several days in early August to record the audio version of his new book in a Chicago studio, Barack held eight town hall meetings during a two-day downstate swing before attending Governor's Day at the Illinois State Fair. House Speaker and state Democratic chairman Michael Madigan remained unwilling to support treasurer nominee Alexi Gianoullias, and communications director Robert Gibbs told Kristen McQueary of the *Daily Southtown* that "Barack cautioned the speaker that he might be viewed as petty and vindictive if he's not united behind the party's ticket in the fall." In response, Madigan sarcastically dismissed Barack by telling McQueary, "There's been no word from the messiah." *Capitol Fax*'s Rich Miller later wrote that "sources close to the usually easy-going Obama say he has an enmity for Madigan unlike that for anyone else."[29]

On Friday, August 18, Barack, accompanied by Gibbs, foreign policy aide Mark Lippert, and three military officers, flew from Washington to Amsterdam's Schiphol Airport. Barack had specifically invited air force major general Scott Gration, who had spent much of his childhood in Africa and was fluent in Swahili. Also on the plane was a trio of reporters—Jeff Zeleny of the *Tribune,* Lynn Sweet of the *Sun-Times,* and Bill Lambrecht of the *St. Louis Post-Dispatch,* as well as a pair of documentary filmmakers who had begun filming Barack off and on three months earlier. At Schiphol, Barack visited the meditation room before the whole party boarded a second KLM flight en route to Cape Town, South Africa, where they arrived on Saturday night. Barack told the reporters that "my goal is less personal and more about public policy, and my hope is that my visit can shine a spotlight on the enormous challenges Africa faces," particularly the AIDS crisis and how its spread was fostered by "the lack of control women have over sex" in Africa. On Sunday Barack visited Nelson Mandela's former prison cell on Robben Island, and on Monday he toured an AIDS treatment facility. "There needs to be a sense of urgency and an almost clinical truth-telling about AIDS," Barack warned. "If it is not addressed in an unambiguous fashion, the percentage of people who are infected is going off the charts." That afternoon, Barack met Archbishop Desmond Tutu, who told him "you're going to be a very credible presidential candidate." On Tuesday Barack toured Soweto and one of its anti-apartheid museums, telling journalists, "if it wasn't for some of the activities that happened here, I might not be involved in politics and might not be doing what I am doing in the United States."

On Thursday, August 24, Barack and his party flew to Nairobi, Kenya, where Michelle and their daughters joined him for a six-day stay. Barack's group got an immediate introduction to Kenya's endemic corruption when airport customs officials extorted more than $1,700 from the party's videographers. The next morning Barack mentioned the rip-off to Kenyan president Mwai Kibaki, telling reporters afterward that "at every level, the people of Kenya suffer through corruption of political offices." That evening, a Kenyan official gave Barack's aide Mark Lippert "wads of cash in brown envelopes" to repay the camera crews.

Barack drew huge crowds and ecstatic greetings everywhere he went in Nairobi, and on Saturday he, his family, and their traveling press corps boarded an East African Airlines flight to Kisumu on their way to Granny Sarah's family homestead in Kogelo. In Kisumu, Barack and Michelle visited a mobile AIDS center to get tested for HIV, with Barack telling onlookers, "I just want everybody to remember that if a senator from the United States can get tested, and his wife can get tested, then everybody in this crowd can get tested, and everyone in the city can get tested." Michelle was stunned by the mass enthusiasm for her husband. "It's rare that I'm speechless, but it's hard to interpret what all this means to me and means to us," she told the *Tribune*'s Jeff Zeleny, who described Kenyans greeting Barack "more like a prophet than a politician." The nineteen-car motorcade made the one-hour drive to Kogelo, where crowds forced the planned two-hour visit to be cut to hardly forty-five minutes before the return trip to Nairobi began. Villagers presented Barack with a well-fed three-year-old goat, but Barack declined the gift, explaining, "I don't think they'd let me take it on the plane."

Back in Nairobi, Barack visited the huge Kibera slum, where an estimated 20 percent of the 700,000 residents were HIV-positive. On Monday Barack delivered an outspoken anticorruption speech at the University of Nairobi that was broadcast live by Kenya's largest television station. "Where Kenya is failing is in its ability to create a government that is transparent and accountable, one that serves its people and is free from corruption," Barack declared. Citing his own hometown, Barack said that "while corruption is a problem we all share, here in Kenya it is a crisis—a crisis that's robbing an honest people of the opportunities they have fought for, the opportunity they deserve." Terming corruption "one of the great struggles of our time," Barack called for the Kenyan government to "downsize the bureaucracy" and channel the savings to higher salaries for the remaining officials. He also said that "ethnic-based tribal politics has to stop. It is rooted in the bankrupt idea that the goal of politics or business is to funnel as much of the pie as possible to one's family, tribe, or circle with little regard for the public good." In contrast, "when people are judged by merit, not connections, then the best and brightest can lead the country."[30]

Sitting down with two journalists from the *Nation,* Kenya's largest newspaper, Barack answered a question about President Kibaki by saying that "sometimes in an environment where there's a lot of pressure . . . maybe you forget what exactly you were trying to do in the first place, which is something all of us in politics have to contend with." When they asked how could Kenyans "push our leaders toward being more accountable," Barack returned to his Roseland roots in responding that "ultimately it's going to come from the ground up, not the top down. . . . There's got to be a very systematic organizing at the grassroots level in order for politicians to ultimately respond." Barack stressed that "one of the dangers is that those who start out idealistic get pulled into the system," and that in any multiethnic society, "the only way to have a strong grassroots movement is if it cuts across the different communities." With 56 percent of Kenyans living below the poverty line, "Kenya can't afford to be divided" along tribal lines, and knowingly or unknowingly, Barack invoked his father's legacy by highlighting the mistreatment of Kenyan women. "An entire segment of the population is continuing to suffer sexual violence and abuse," and Kenyans needed to appreciate that "the single biggest indicator of whether a country develops or not is how well it treats and educates its women."

On Tuesday Barack and his party flew from Nairobi to the Masai Mara National Reserve in southern Kenya for a safari tour, during which they saw a lion devouring a wildebeest. The next evening, minus Barack's family, the group left Nairobi on a U.S. military aircraft bound for Djibouti. Barack had the page proofs of his book to read and correct, and at Crown Books in New York, an assistant editor answered her phone to hear Barack report, "I'm calling from Djibouti" and was wondering where he could find a fax machine. On Thursday, August 31, Barack visited a tent camp in eastern Ethiopia filled with people displaced by flooding. On Friday he flew to N'Djamena to meet with Chadian president Idriss Déby, and the next morning Barack's group flew to a dirt airstrip in eastern Chad to join a UN convoy for a one-hour drive to a Darfur refugee camp just inside Chad's border with Sudan. Barack's visit to the sprawling, fifteen-thousand-person camp, one of a dozen in Chad housing some 235,000 Sudanese refugees, lasted only ninety minutes, but he told accompanying journalists that "it is important that these folks here are not forgotten." On Sunday morning, Barack's military aircraft departed N'Djamena for Frankfurt, from where Barack and his aides returned to the U.S. A *Chicago Sun-Times* editorial praised his trip as "admirable and honorable," and Barack told the paper's Lynn Sweet that meeting Desmond Tutu and fellow Noble Peace Prize winner Wangari Maathai had been two high points of the journey. "There is something about when people serve; somehow it enriches them in all sorts of ways. They just seemed like happy, fulfilled people, even though they are not particularly wealthy."[31]

Two days after Barack returned to Washington, the Senate without recorded objection passed his and Tom Coburn's Federal Funding Accountability and Transparency Act, sending the measure aimed at ending no-bid federal contracting to the House. Recording a podcast, Barack welcomed that move in seeming amazement, remarking that "every once in a while, you actually get something done around here." Three days before Barack was due at Iowa senator Tom Harkin's much-heralded steak fry, former Illinois competitor Dan Hynes called to give ten minutes' advance notice that he was about to tell the press he believed Barack should run for the presidency. "Well, Dan, I'm flattered," Barack replied, but "Michelle will never forgive you." Hynes told reporters "what we need to do is both create a real movement for those of us who believe he's the right man for the presidency and to give him a better understanding of just how broad and deep and emotional the support is for him across the state."

Sun-Times reporter Lynn Sweet observed that "the charismatic Obama is now on a pedestal," and she believed Barack already was giving "answers that suggest the thought of running for president has crossed his mind." Indeed, within days of his return from Africa, Barack told David Axelrod that "with so many folks talking to me about running, I feel like I have an obligation to at least think about this in a serious, informed way." In *Newsweek,* African American journalist Ellis Cose called Barack "a political phenomenon unlike any previously seen" and described him as "the perfect mirror for a country that craves to see itself as beyond race, beyond boundaries, beyond the ugly parts of its past." On the same morning that Barack arrived in Iowa for the Harkin event, the nearby *St. Louis Post-Dispatch* reported Illinois poll results showing Barack with "an unheard-of 70 percent approval rating" and cited Cose's essay as a prime example of the "runaway train of gushing publicity" Barack was receiving.

Pete Rouse had recruited veteran Democratic operative Steve Hildebrand, who had managed Al Gore's 2000 Iowa campaign, to squire Barack to the Harkin steak fry, at the Warren County Fairgrounds in Indianola. "I thought, let's have a little fun with this. I wanted to create a little buzz," Rouse explained, and political reporters immediately took notice. Tom Harkin introduced Barack by citing the U2 musician Bono: "I couldn't get him, so I settled for the second biggest rock star in America today!" as the huge turnout of thirty-five hundred attested. But "Obama turned in a rare mediocre performance," David Mendell of the *Tribune* reported, offering only what *Salon*'s Walter Shapiro called "an artful pastiche of earlier rhetoric" and speaking for far too long—thirty-eight minutes. Still, Barack received what influential *Des Moines Register* political columnist David Yepsen called "a nearly unprecedented response," and as he was "mobbed by fans," another *Register* reporter caught Barack's self-satisfied reaction: "I'm going to have to come back to Iowa. This is all right!" When Shapiro asked whether this meant he would change his thinking about 2008, Barack said, "nothing has changed in my mind. But you never know." Shapiro immediately interrupted: "You mean things can change?" "Yeah, things can change," Barack replied.

On the drive back to Des Moines's airport, Steve Hildebrand told Barack how remarkable the Iowans' excitement had been. "What does all this mean?" Steve asked. "I am not sure," Barack answered. But Hildebrand, who had not previously met him, was now convinced that Barack was the Democrats' best possible presidential candidate for 2008, and he began e-mailing friends to promote an Obama candidacy. Several days later, David Yepsen wrote that the steak fry crowd had included "a lot of people I'd never seen at a Democratic event. Young people. People of color. Most old pros will tell you that attracting new people to your campaign is one sign of a winner." Yepsen believed "Obama's got to look at the reception he got and think there just might be some chemistry happening" with Iowa Democrats. "Maybe I didn't want to make the run this soon in life," Yepsen imagined Barack thinking, "but the presidency is up for grabs, and I've got to take my chances now, not later. As the old saying goes: He who hesitates—is lost."

As Barack's remarks to David Axelrod even before his Iowa reception indicated, Barack thought there should be a serious discussion once the November 7 midterms took place. "We ought to see how this goes and after the election just talk about whether or not we ought to consider running," Barack told his aides. Within a few days Pete Rouse sent a memo to Barack's closest political advisers recommending they all assemble in Chicago on November 8.[32]

Barack resumed his jam-packed schedule of speaking appearances and congressional business. A weekend return to Illinois saw him hold his fifty-seventh town hall meeting, drawing a Friday-night crowd of more than thirteen hundred in Joliet. On September 26 Barack and Tom Coburn visited the White House to watch President Bush sign their Transparency Act into law. When journalists the next day asked Chicago mayor Richard Daley about Barack running for president, Daley responded that "everybody wants him. If he wants to run, he should run. I mean, why not?" The day after that, Oprah Winfrey, who was reading an advance copy of Barack's book, chimed in to say, "I do wish he would run. And if he would run, I would do everything in my power to campaign for him," because Barack's "sense of hope and optimism for this country and what is possible for the United States is the kind of thing that I would like to get behind."

In Washington, Barack spoke up in support of an amendment aimed at curtailing military tribunal proceedings for captured terrorists. "We don't know when this war against terrorism might end" but "the fundamental human rights of the accused should be bigger than politics." The procedure the U.S. was using for captives held at Guantánamo Bay, Cuba, "allows them no real chance to prove their innocence" and thus represented a "betrayal of American values." Instead, Barack called for "a real military system of justice that would sort out the suspected terrorists from the accidentally accused" who had ended up in U.S. custody.

A pair of weekend town hall meetings back in Illinois also allowed Barack to hop across the Mississippi River and campaign for a Democratic congressional candidate in Davenport, Iowa. National journalists like *Time* magazine's Joe Klein were in tow, and in Rockford, *Register Star* political editor Chuck Sweeny told Klein, "Obama is reaching out. He's saying the other side isn't evil. You can't imagine how powerful a message that is for an audience like this." In the Quad Cities, longtime Democratic activist Bill Gluba told Klein that people's reactions to Barack reminded him of when he had driven New York senator Robert F. Kennedy around Davenport in May 1968. "I'll never forget the way people reacted to Kennedy. Never seen anything like it since—until this guy."

When a local reporter asked Barack if he would run in 2008, he responded, "I'll let you know." Speaking privately with Klein, Barack said that once his upcoming book tour and the November 7 election were over, "I will think about how I can be most useful to the country and how I can reconcile that with being a good dad and a good husband . . . I haven't completely decided or unraveled that puzzle yet." Klein was struck by "the elaborate intellectual balancing mechanism" that Barack "applies to every statement and gesture, to every public moment of his life." Barack, like fellow African Americans Colin Powell, Tiger Woods, Oprah Winfrey, and Michael Jordan, all "transcend racial stereotypes," Klein wrote. Black academic John McWhorter believed that race was the very core of Barack's appeal. "The key factor that galvanizes people around the idea of Obama for president is, quite simply, that he is black," McWhorter stated. Absent his blackness, Barack would just be "some relatively anonymous rookie," and thus "Obama is being considered as presidential timber not despite his race, but because of it."

But to Joe Klein, watching people's reactions to Barack in Rockford, Rock Island, and Davenport, Barack was "an American political phenomenon," who was "the political equivalent of a rainbow—a sudden preternatural event inspiring both awe and ecstasy." Klein's *Time* cover story stressed that Barack was "not a screechy partisan. Indeed, he seems obsessively eager to find common ground with conservatives." Oklahoma's Tom Coburn agreed, telling another journalist that "Barack's got the capability . . . and the pizzazz and the charisma to be a leader of America, not a leader of Democrats."[33]

Sitting down with Oprah Winfrey to tape a long interview for broadcast during his book tour, Barack expressed dismay at "this celebrity culture" that "gobbles you up." Barack worried about saying "what they want to hear, as opposed to you trying to stay in touch with that deepest part of you, that kernel of truth inside," invoking again the Borges phrase he had recited to Jacob Weisberg three months earlier. "I think what's happened is that we are so interested in spin and we're less interested in facts," but "the country is not as divided as Washington . . . not as divided as those cable news shows make it out to be." Barack complained about seeing erectile dysfunction ads during hours when children could be watching television and noted that "one of the tragedies of Africa is that the relationship between men and women, I think, has broken down. There's a lot of sexual violence, a lot of AIDS is caused by women whose husbands are bringing it home to them." Michelle told Oprah that the Kenya "trip was overwhelming for me," but that their daughters "just sort of took it all in stride." The girls were so "patient" that "we have to be careful to really structure boundaries for them" because if not "they'd let their time be eaten up by other things. So it's really up to us to protect that time, to make sure that we demand for them what they don't demand for themselves." But Michelle saw her family's greatest challenge as "how do you make sure . . . that you don't . . . get swept up in this celebrity, that you don't get caught up in the hype . . . that you remain centered and focused."

In Illinois, headlines like "'Obama Fever' Grips Peoria" attested to Barack's popularity as he campaigned for local candidates. The *Peoria Journal Star* lauded

Barack as "a veritable national celebrity," and the *Champaign News-Gazette* complimented him as "an attractive, intelligent guy who conducts his politics in a gentlemanly fashion" and "shows great promise." But the *News-Gazette* cited questions concerning Alexi Giannoulias's "family links to organized crime figures" as an example of Barack's "lack of experience and seasoned judgment" and said it would be "the height of folly" for Barack to run for president. When Barack campaigned in Kansas City for Democratic Senate candidate Claire McCaskill, a local columnist praised him as "an icon of the American Dream," but back in Metro East the *Belleville News-Democrat* took note of that coverage and complained that "Obama always seems to be making headlines in places other than Illinois."

The next day, following a media advisory that U.S. attorney Patrick Fitzgerald would be holding a 1:00 P.M. press conference, Illinois headlines focused not on Barack but on his friend Tony Rezko: "Top Blagojevich Adviser Indicted." Six days earlier a federal grand jury had returned a twenty-four-count, sixty-five-page criminal indictment against Rezko, charging him with trying to extort more than $7.5 million in contract kickbacks and campaign contributions to Illinois's governor. With Tony out of the country and his whereabouts unknown, the indictment initially was issued under seal, but the U.S. attorney labeled the crimes committed by Rezko a "pay-to-play scheme on steroids."

Chicago journalists had expected this shoe to drop ever since Tony's pal Stuart Levine had been indicted months earlier, but the charges focused white-hot attention on Rezko's political patrons, especially Rod Blagojevich. Tony's former business partner Dan Mahru told the *Tribune* that the man he first met in 1989 had for years "worked so hard" to help his extended family. "But the Tony Rezko I knew after the governor got elected was not the same person. He changed," Mahru explained. A quick *Tribune* tally showed that Rezko and his companies had contributed at least $385,000 to Illinois politicians since 1994, and Barack's communications director Robert Gibbs immediately announced that Barack's federal campaign fund would divest itself of the $11,500 that Tony had donated to Barack's 2004 Senate race.

A *Sun-Times* gossip columnist wrote that "Rezko is hurt by Obama's lack of support since reports surfaced" of the federal probe, and competing newspapers struggled to collate just how much Tony had raised for Barack from friends and business associates, with reliable tallies reaching as high as $200,000. Eight days later, FBI agents took Rezko into custody at Chicago's O'Hare Airport as soon as his plane taxied to the gate. One Democratic legislator noted that even two years earlier "it wasn't a secret that Tony Rezko was a very connected moneyman in the Blagojevich administration who was coming under increased scrutiny." Asked about Barack's friendship with Tony, Illinois campaign reform advocate Cindi Canary expressed dismay. "It surprised me that late in the game he continued to take contributions from somebody who was under a rather dark cloud in the state."[34]

The same morning that the Rezko headlines dominated Chicago newspapers' front pages, the *Tribune* ran a page-3 story previewing Barack's new book, *The Audacity of Hope*—its title echoing a Jeremiah Wright sermon first preached years earlier. An advance review in *Publishers Weekly* had dismissed the book as a "sonorous manifesto" offering "muddled, uninspiring proposals" that represented only "tepid Clintonism." With its official publication date set for Tuesday, October 17, Barack on Monday made a quick trip to Indianapolis for a fund-raising luncheon before returning to Chicago for an evening book party at the home of Valerie Jarrett's parents, Jim and Barbara Bowman—just like eleven years earlier with *Dreams*. When radio executive Melody Spann Cooper asked Barack about a presidential race, he said, "I don't think Michelle is going to let me do this." Jarrett remembered that when the time came for Barack to offer some brief remarks, he "started to say he was sorry to have been away from his family so much . . . and began crying so hard he couldn't go on."

In Tuesday morning's *New York Times,* Michiko Kakutani wrote that portions of *Audacity* "read like outtakes from a stump speech" and that the book "occasionally slips into . . . flabby platitudes" that were "little more than fuzzy statements of the obvious." A *Los Angeles Times* review was somewhat kinder, calling *Audacity* "an easy-reading, congenial book that is halfway successful," but adding that too many of Barack's proposals "are incremental, timid, tangential—anything but audacious." But early reviewers often missed those brief occasions when *Audacity* offered refreshing political honesty—"much of what ails the inner city involves a breakdown in culture that will not be cured by money alone"—and bracing self-criticism: "as a consequence of my fund-raising I became more like the wealthy donors I met." Barack acknowledged that "I serve as a blank screen on which people of vastly different political stripes project their own views," and "for the broad public at least, I am who the media says I am. I say what they say I say. I become who they say I've become." Barack conceded that "I don't consider George Bush a bad man" and admitted that "it is my obligation . . . to remain open to the possibility that my unwillingness to support gay marriage is misguided."

But without a doubt by far the most outspoken of *Audacity*'s nine topical chapters was the one on race. "The collapse of the two-parent black household," with 54 percent of African American children now living in single-parent homes, "is occurring at such an alarming rate" that black America's divergence from the rest of the United States "has become a difference in kind, a phenomenon that reflects a casualness toward sex and child rearing among black men." Warning that "conditions in the heart of the inner city are spinning out of control," Barack wrote that "liberal policy makers and civil rights leaders" have "tended to downplay or ignore evidence that entrenched behavioral patterns among the black poor really were contributing to intergenerational poverty." Barack suggested that "perhaps the single biggest thing we could do to reduce such poverty is to encourage teenage girls to finish high school and avoid having children out of wedlock." More broadly, *Audacity* also insisted that "proposals that solely benefit minorities . . . can't serve as the basis for the kinds of sustained, broad-based political coalitions needed to transform America."[35]

As Barack's book tour commenced, Illinois seatmate Dick Durbin renewed his call for Barack to run. "I said to him, 'Do you really think sticking around the Senate for four more years and casting a thousand more votes will make you more qualified for president?'" Barack responded that "it's a family decision," but other voices echoed Durbin. A *Chicago Sun-Times* reporter wrote, "Why wait? Obama's stock has never been higher," even though "he is very untested. He has never even had a tough, adversarial press conference." Springfield's *State Journal-Register* agreed, citing Barack's "ability to inspire and touch people," and advising "Run, Sen. Obama, run." The *Chicago Tribune,* praising Barack for "beautifully" handling his first nineteen months as a U.S. senator, singled out how he had "not strayed from his early message that America must rise above angry partisanship."

In the national press, conservative *New York Times* columnist David Brooks echoed the *Journal-Register* in a column headlined "Run, Barack, Run." Declaring that "Obama is a new kind of politician" and "a mega-hyped phenomenon that lives up to the hype," Brooks asserted that "the times will never again so completely require the gifts that he possesses." Fellow *Times* columnist Frank Rich concurred, saying "of course he should run," because Barack "is Bill Clinton without the baggage." African American columnist Clarence Page advised "Seize the day . . . I hope you run . . . you may never again see this many people who are this eager for you to run. Just don't expect us to be nice to you after you decide to do it."

On October 19, Barack blanketed the airwaves from dawn until midnight, beginning with NBC's *Today* show. Barack rued the impact of the nation's capital. "The minute you arrive in Washington, suddenly there are all these forces, whether it's the media or parties or the need to raise money, that kind of tamp down those basic human responses that you have towards other people." On NPR's *All Things Considered,* Barack told host Michele Norris that "there's a strong impulse when you're in public life to try to control your image as much as possible." Regarding *Audacity,* Barack said, "I'm sure that if you asked my wife whether I adequately listed my failings in the book she would say no, that she could supplement it substantially." Fund-raising was a never-ending problem. "My preference would be that we've got public financing of campaigns, and nobody has to raise money whatsoever." Failing that, "the question then you constantly have to ask yourself is, are the means that you're using to make sure you're competitive in elections in any way undermining those core values that brought you into politics in the first place? I feel confident that hasn't happened, but it's something you constantly have to monitor." A presidential run has "got to be based on you feeling that somehow you can be useful, that you can offer something that is unique and will help create a better life for the people you seek to represent. And those are questions that I'm constantly asking myself."

On CNN's *Larry King Live,* Barack explained that "I've got a wife at home who is more interested in whether I rinsed out the dishes and put them in the dishwasher" than in his public acclaim. "This has been a very unproductive Congress since I've arrived," but Barack thought his Africa trip and public HIV test had been highly valuable. "I thought that was probably the best investment of fifteen minutes that I'd ever make," because across Africa "men continue to oftentimes abuse woman, have multiple partners, and so there's a whole situation in terms of how men treat women that has to be dealt with." Barack told King that "faith is very important in my life on a daily basis," and that after November 7 he would focus on 2008. "I love that idea of deciding what will be most useful."

With PBS late-night TV host Charlie Rose, Barack explained that "what I'm really trying to do here is to see, can I change the political culture? . . . What we have now is a surplus amount of conflict that is manufactured. It's manufactured in television ads, it's manufactured in terms of how the parties portray each other." Regarding 2008, "when you decide to run for president . . . you are saying to the American people, 'I am giving my life to you' . . . and that's not a decision that I think you can or should make solely on the basis of ambition. It transcends ambition."

The next day, when a questioner asked Barack what his greatest fear was as a politician, he answered, "my greatest fear, I think, is that I lose track of that voice: who I am and what my values are and what I care most deeply about." At a Kennedy Library Q&A in Boston, *New York Times* columnist Bob Herbert cited Barack's previous night appearances with King and Rose. "It's a bit much, isn't it?" Barack replied as the audience laughed. "That's what my wife says, anyway: 'I am fed up with reading about you.'" Barack explained, "I actually find the attention and seeing my name in the papers and the stuff that feeds your ego less satisfying as time goes on." Concerning the presidency, "that office is so different from any other office on the planet that you have to understand that if you seek that office, you have to be prepared to give your life to it. . . . The bargain that any president, I think, strikes with the American people is 'You give me this office, and in turn my fears, doubts, insecurities, foibles, need for sleep, family life, vacations, leisure is gone. I am giving myself to you,'" and "you don't make that decision unless you are prepared to make that sacrifice, that trade-off, that bargain."

As he had before, Barack explained to Herbert how the U.S. Senate was less satisfying than his years in Springfield. "When I was in the state legislature in Illinois, every senator had to be on the floor on every bill. And the sponsor would stand up and present it, and anybody could ask questions . . . so you'd actually have a sense of deliberation." On Capitol Hill, "that almost never happens on the floor of the Senate" and that was a "disappointment."[36]

As Barack traveled the country for book signings and Democratic campaign rallies, two Philadelphia events on October 21 produced an *Inquirer* news story telling readers that Barack was "as handsome as John F. Kennedy and as charismatic as Bill Clinton." He offered "a schoolboy's charm and a movie star's smile," and reporter Christine Schiavo quoted a fifty-three-year-old man as explaining, "You feel like you're in the presence of greatness when you're around him." Scheduled to appear the next morning on NBC's *Meet the Press,* Barack spoke by conference call with Robert Gibbs, David Axelrod, and Axelrod's partner David Plouffe about the likelihood that host Tim Russert would confront him with his January declaration that he would not run in 2008. Barack told his aides he considered such a race "unlikely," but "Why don't I just tell the truth? Say I had no intention of even thinking about running when I was on the show in January, but things have changed, and I will give it some thought after the 2006 election."

On the air with Russert, Barack said it was "important not to buy into your own hype," and "I've got a wife who knocks me down a peg any time I start thinking what they're writing about me is true." When Russert raised the presidency, Barack explained that "most of the time it seems that the president has maybe 10 percent of his agenda set by himself and 90 percent of it set by circumstances." Asked about greatness, Barack replied that "when I think about great presidents, I think about those who transformed how we think about ourselves as a country in fundamental ways," who "transformed the culture and not simply promoted one or two particular issues." The presidency "can't be something that you pursue on the basis of vanity and ambition. I think there's a certain soberness and seriousness required when you think about that office that is unique."

As he had with Bob Herbert, Barack reiterated that as a candidate, "the bargain you're making with the American people is that 'You put me in this office and my problems are not relevant. My job is to think about your problems.'" Just as expected, Russert presented Barack with the videotape of his January denial. "Given the responses that I've been getting over the last several months, I have thought about the possibility, but I have not thought about it with the seriousness and depth that I think is required. . . . After November 7, I'll sit down and consider it." Russert pressed the point. "It's fair to say you're thinking about running for president in 2008?" "It's fair, yes," Barack agreed. "I am still at the point where I have not made a decision to pursue higher office, but it is true that I have thought about it over the last several months." "So it sounds as if the door has opened a bit," Russert observed. "A bit," Barack replied.

On Monday morning every major newspaper headlined Barack's comments to Russert. The *New York Times* quoted Steve Hildebrand recounting how "the reaction that Obama got in Iowa was like nothing I've ever seen before with another politician." *Times* columnist Bob Herbert, reflecting on his Boston Q&A with Barack, warned he would have to "develop the kind of toughness and savvy that are essential in the ugly and brutal combat of a presidential campaign." Beginning a West Coast swing that would take him from Phoenix to Denver, San Francisco, Seattle, and Los Angeles, Barack told a magazine editors' conference that his family might well not be prepared for a presidential race. "It's not clear that they are ready for it, or that I even want to put them in a position where they've got to make that decision." Michelle "cares more about whether I'm a good father and a good husband than she does about whether I'm a U.S. senator."

In a lengthy television taping, Barack told host Tavis Smiley that on *Meet the Press,* it would have been "foolish of me to pretend that somehow I hadn't thought about it." Yet "the presidency is a unique position," one "that consumes you, it consumes your family," and "you have to feel that you are prepared in a very internal conversation between you and your maker, and your family." Barack confessed that "I haven't even thought through the process to think it through. . . . I need to take a look at what message do I have that would be unique, and am I the right messenger for it? Most profoundly, I'd have to talk to Michelle and my two little girls and find out whether this is something that they're signed up for." On the issues, "the most important area where we have not made serious efforts is when it comes to serious inner-city poverty," especially with regard to opportunities for ex-offenders. Barack twice cited "the capacity to disagree without being disagreeable," and while he stressed avoiding "the demonization of the other side," he nonetheless attacked the "radically ideological Bush administration."[37]

In Denver, more than five hundred people began lining up as early as 5:00 A.M. to have Barack autograph their copies of *Audacity*. "We need healing, and this man can bring it about," one customer told a reporter. Back in Chicago, longtime cheerleader Newton Minow penned a *Tribune* op-ed headlined "Why Obama Should Run for President." Up until Minow happened to see a rebroadcast of Barack's Iowa steak fry appearance on C-SPAN, "I did not think he should run for president," Minow explained. But watching Iowans' enthusiasm for Barack had changed his mind, and Minow called him, saying, "You ought to go for it now."

In the *Washington Post,* conservative columnist Charles Krauthammer also called for Barack to run, predicting that while he would not win in 2008—"the country will simply not elect a novice in wartime"—a loss would "put him irrevocably on a path to the presidency," in part because "there are more Americans who would take special pride in a black president than there are those who would reject one because of racism." Writing in the *Financial Times,* Jacob Weisberg, who had so memorably interviewed Barack four months earlier, asserted that the nationwide acclaim that had greeted Barack's book tour "has overthrown much of the conventional wisdom about what is likely to happen in the 2008 U.S. presidential campaign." If Barack announced his candidacy, he "will rapidly become the de facto Democratic frontrunner," eclipsing Senate colleague Hillary Rodham Clinton and former vice presidential nominee John Edwards.

In Marin County, north of San Francisco, eleven hundred people paid $125 per person for a copy of *Audacity* and the opportunity to hear Barack address a noontime charity fund-raiser. The following morning in Seattle, more than two thousand people heard Barack tell a political rally, "We've got to have hope. We've got to have a belief in things not seen." Later that day, a book signing attracted a sellout crowd of twenty-five hundred. On Saturday morning in Austin, Texas, people began lining up at 6:00 A.M. for a Barack book signing in the Texas capitol. Barack made it back to Chicago in time to tardily join Michelle at a gala sixty-fifth-birthday celebration for Rev. Jesse Jackson at the South Shore Cultural Center. CNN correspondent Don Lemon buttonholed an exuberant Michelle. "I love this man. I grew up in this man's house. I've seen it all," exclaimed Michelle in recalling her childhood friendship with Jackson's daughter Santita. Then Lemon asked, "Are you ready to be first lady?" and Michelle clammed up. "No comment."

Barack complained to *Chicago Tribune* books editor Elizabeth Taylor that "I never get a chance to read anymore," and when Taylor asked what books had most impacted him, Barack cited *Gandhi's Truth,* calling Erik Erikson's psychoanalytic biography "a great book." On Monday morning, Barack was at a Borders in downtown Minneapolis, where the store quickly sold out of its one thousand copies of *Audacity.* "I saw him on Oprah," one student told a reporter. "I'm not really into politics, but I like inspiring people, and he inspires me." At a subsequent rally in suburban Rochester, Minnesota, Barack was "mobbed for autographs," and on Tuesday in Milwaukee fifteen hundred people braved a chilly wind at a riverfront park to hear Barack speak on behalf of Democratic Wisconsin governor Jim Doyle. U.S. senator Herb Kohl told the crowd "we might just have a future president in town," and Milwaukee mayor Tom Barrett said the enthusiasm was "something we haven't seen for decades." An older woman told a journalist that Barack was "our Jack Kennedy," and a

New York Times correspondent called Barack "the prize catch of the midterm campaign."[38]

But trouble was brewing back in Chicago as *Tribune* reporters discovered that the vacant lot adjoining Barack's new home was owned by newly indicted Tony Rezko. Barack later recounted, "I called him to let him know that 'Look, you may be getting inquiries about this, and so it's important for you to be able to talk to folks about your intentions in terms of development and so forth,'" but now Tony was not taking anyone's questions. Chicago columnists came down hard. Revisiting the purchase prices of the two properties, the *Tribune*'s John Kass concluded that "Obama bought his home at a $300,000 discount. Rezko bought the adjoining lot from the same sellers at full price. One got a juicy bargain. The other overpaid."

Appearing on NPR's *Talk of the Nation,* Barack alluded to the criticism, saying, "there're going to be some days where" as a politician "you get knocked around a little bit," and the next morning a *Tribune* editorial called for Barack "to explain, fully and quickly" whether Rezko "in effect subsidized Obama's purchase of the opulent house and also provided an abutting private preserve that adds to its ambience." Late Friday Barack told the *Sun-Times* he had not spoken with Tony for more than six months, and that neither he nor Michelle had ever represented Rezko or his various companies. In a written statement, Barack said "it was a mistake to have been engaged with him at all in this or any other personal business dealing that would allow him, or anyone else, to believe that he had done me a favor." Barack added that "I consider this a mistake on my part, and I regret it," but a *Crain's Chicago Business* editorial criticized the conjoined purchases as "a colossally stupid move."

But Barack's hectic schedule of Democratic campaign rallies, especially in states where Senate challengers were waging competitive races against incumbent Republicans, left him little time to ponder the Rezko criticism. In Richmond, Virginia, he "received a rock star's welcome" and "thunderous cheers" when speaking on behalf of Democrat Jim Webb. A *State Journal-Register* poll of Illinois voters showed that 59 percent would vote for Barack for president, although only 50 percent believed he should run in 2008. Appearing in Cleveland with Ohio Senate candidate Sherrod Brown on Saturday, Barack told Brown's wife Connie Schultz that a 2008 race was doubtful. "You know, Michelle really does not want me to do this." Indeed, when Barack first told Michelle about the upcoming Wednesday meeting to discuss a run, "I said 'No way. Absolutely not.' The last thing I wanted was for my girls to have their worlds turned upside down. It broke my heart just to think about it," Michelle recalled. "Let's not do this now," she told her husband.

But on Sunday, November 5, two days before Election Day, Barack made his third trip that fall to Iowa, giving "a boisterous speech" on behalf of Democratic gubernatorial candidate Chet Culver. "After the election, I'm going to sit down and give the possibility the serious consideration it deserves," Barack told reporters, and discuss it "with my family, with my pastor." A *Washington Post* story surveying the fall campaign efforts of possible 2008 contenders stated that "Obama has generated rapturous enthusiasm among Democrats," and that "his success on the campaign trail in recent weeks has added to his cachet." The *Post* also calculated that Barack's fund-raising efforts had raised a total of $3.6 million for Democratic candidates. On Tuesday, Barack spent the evening watching election returns and making congratulatory calls to Democratic winners. As the hours passed, it became clear that Democrats had won control of the Senate, defeating six Republican incumbents to take a 51–49 majority. Democrats also won control of the House, seizing thirty Republican seats, and in Illinois young Alexi Giannoulias won election as state treasurer. Savoring how his party now controlled both houses of Congress, Barack told the *New York Times,* "Democrats have a wonderful opportunity to show that we have an agenda for change."[39]

On Wednesday Barack headed to David Axelrod's office in River North to have lunch with mayoral brother and former commerce secretary Bill Daley, who told him, "You gotta run." But the day's main event was still ahead: an afternoon meeting in Axelrod's conference room that Pete Rouse had begun pulling together weeks earlier. In attendance were Axelrod's partner David Plouffe, the enthusiastic Steve Hildebrand, Robert Gibbs, former Senate staffer Alyssa Mastromonaco, who was now at Hopefund, close friends Marty Nesbitt and Valerie Jarrett, and Michelle. Rouse had prepared a six-point memo to guide the discussion, but Barack had three uppermost questions. Number one, "Could I win?" especially against Hillary Rodham Clinton. "How would you organize a campaign against the best brand name in Democratic politics?" Second, what would be the impact on his family? "What would the schedule look like? How much money would I have to raise?" Michelle asked whether Barack could get home to see his daughters each weekend. Hildebrand eagerly said yes, but Plouffe, who thought that running was very unlikely, interjected a firm no. "If you run, you'll never see your family, you'll be under pressure the likes of which you can't imagine, and it will be absolutely miserable from a personal standpoint." Marty Nesbitt reacted jokingly to that litany: "I'm so glad I'm not running for president." Barack immediately corrected him: "Oh, but you are. This is going to affect your life, too."

But Barack's third question was the toughest: "Should I win? Is there some unique message, something distinct enough from the other candidates, that would justify me running this soon?" Over the past three weeks, Barack had come to believe that the answer to this question was yes. From the Iowa steak fry onward, throughout all of his book tour and campaign events, "there was just this remarkable, visceral response" crowds had shown toward him. "What it told me was that people really were looking for something different." Yet Michelle worried that even if she could be convinced that a campaign would still allow Barack to be the father she wanted her daughters to have, "can we actually chart a course to victory? I want you to show me how you're going to do this. You need to show me that this is not going to be a bullshit, fly-by-night campaign," one participant recalled. Barack spoke up, but Michelle cut him off. "We're talking about you right now." Barack finally said that the question going forward was "whether we can build not a winning campaign, but a credible one," yet his tone was measured. "Well, I think it's highly unlikely that I'm going to do this, but we should go and do due diligence on this and have another meeting in a month."

Barack, Michelle, Nesbitt, and Jarrett headed to a nearby Italian restaurant for dinner. Plouffe left the meeting thinking that a race remained improbable, but Barack "was more serious about running than I had anticipated." David Axelrod agreed: "I think he wants to run," but "Michelle is the wild card." At dinner, Michelle voiced her doubts and worries to Jarrett and Nesbitt, especially regarding Barack's physical safety plus the impact on their daughters. Eventually Valerie pushed back. "Let's try this from a different perspective. Michelle, let's say Barack answers all your questions to your full satisfaction," that "he's got an answer for every one of them. Are you in?" Michelle's reply surprised her husband: "I'm in a hundred and ten percent," but she quickly turned to Barack. "You're going to be really specific with me. You're going to tell me exactly how we're going to work it out" before she would give her blessing for a race.[40]

The next morning the *New York Times* reported that *The Audacity of Hope* would be the number 1 hardback nonfiction bestseller in the newspaper's next Sunday rankings. Calling the book "something of a publishing stunner," the paper reported that to date *Audacity* had sold 182,000 copies. With 860,000 copies in print, the book's success promised to increase Barack's royalty income even further. But his focus was elsewhere. Buoyed by the previous afternoon's discussion, Barack was on the phone to Democratic activists in Iowa and New Hampshire, asking what John Norris, who had run John Kerry's 2004 Iowa campaign, called "earnest" questions about the Democratic race. In New Hampshire, state party chairman Kathy Sullivan and Jim Demers, Dick Gephardt's former New Hampshire campaign chair, received calls. "Kathy, this is Barack Obama. I might be coming to New Hampshire." Sullivan had been thinking about holding a December event to celebrate Democrats' November victories, and she immediately invited Barack to address it.

Returning to Washington, Democrats' new Senate majority allowed Barack to trade in his membership on the Environment and Public Works Committee while adding two more appealing ones—Health, Education, Labor and Pensions, and Homeland Security and Governmental Affairs—while retaining Foreign Relations. Meeting privately with a dozen or so D.C. friends, many of whom he had known since Harvard, Barack found the group hesitant about his presidential prospects. One African American expressed doubt that the country was ready to elect a black president, and several comments obliquely touched on Barack's safety. Speaking with the Associated Press, Barack remarked that "the people who are most hesitant about this oftentimes are African-Americans because they feel protective of me."

Appearing on CNBC with Tim Russert, Barack said that "the way we gerrymander congressional districts now has a big impact on the inability of the parties to get together, because if every district is drawn 70 percent Republican or 70 percent Democratic, then the congressional representatives in those districts don't really feel rewarded by working with the other side" and instead fear a primary challenge. Barack favored "nonpartisan line-drawing" that would "make all these districts competitive." Similarly, "reducing the impact of money on politics would have an impact because I think that part of the incentive here in Washington is to please the best-organized and most vocal interest groups. And they oftentimes aren't representing the common good," most strongly backing legislators "who are most orthodox in how they approach problems."

When Russert asked about his family, Barack explained "this is probably the one area of my life where I feel the most nagging doubt because I'm gone from home a lot. I like to tell myself that it's worth the sacrifice," that "the good that I'm doing in public office offsets" missing his daughters' ballet recitals and soccer games. Describing Michelle as "a tough woman" who "constantly keeps me in line," Barack said, "she's not somebody who's naturally political" but "tolerates it." With "as much guilt as I already feel sometimes about being away, running for president" promised to be "extraordinarily stressful" for his family, "so that's at the center of my considerations." As he had told his aides in Chicago, seeking to be Americans' president would require "a particular message" as well as "a belief in how you can improve their lives that is unique" and that he could "execute better than anybody else out there." His newfound status, Barack emphasized, is "something that has happened very rapidly. It's not something that's gone according to my own internal clock and timetable." Russert asked, "Has it changed you?" "So far it hasn't, I don't think," Barack replied. "I ask my wife and I ask my close friends whether it has, and their estimation is that no, I'm still recognizably me."[41]

David Axelrod told an AP reporter that "if he decides to run," Barack "can put the money together, and he can attract the talent." In fact, Pete Rouse and Steve Hildebrand already had already reached out to Julianna Smoot, the finance director for Tom Daschle's unsuccessful 2004 reelection race, to ask her to head up Barack's fund-raising if he indeed ran. In a major speech to the Chicago Council on Global Affairs, Barack addressed "a way forward in Iraq," "a conflict that grows more deadly and more chaotic with each passing day. A conflict that has only increased the terrorist threat it was supposed to help contain." Noting that "al Qaeda is successfully using the war in Iraq to recruit a new generation of terrorists," Barack called for "a strategy no longer driven by ideology or by politics, but one that's based on a realistic assessment of the sobering facts on the ground." Any belief that Iraq would become a "flourishing democracy" was "an ideological fantasy," and "my deepest suspicions about this war's inception have been confirmed and exacerbated."

With Iraq now "quickly spiraling out of control," there are "no good options left in this war," and Barack called for a withdrawal of U.S. forces, to begin within four to six months. He also warned that Afghanistan was "backsliding . . . toward chaos" on account of "our lack of focus and commitment." Appearing live on CNN right after his address, Barack asserted that "almost every problem that we've confronted is one that I anticipated" in his October 2002 antiwar speech. "I seem to have gotten it right," Barack told correspondent Don Lemon, who then asked about 2008. "I don't have a particular timetable," Barack replied, but when Lemon said "Very soon?" Barack responded, "absolutely."

Eight days later David Axelrod gave Barack a twelve-page memo making the case for why he should run. George W. Bush had been a "hyper-partisan, ideological, and unyielding" president, and voters were yearning for the opposite:

> *You are uniquely suited for these times. No one among the potential candidates within our party is as well positioned to rekindle our lost idealism as Americans and pick up the mantle of change. No one better represents a new generation of leadership, more focused on practical solutions to today's challenges than old dogmas of the left and right. That is why your convention speech resonated so beautifully. And it remains the touchstone for our campaign moving forward.*

All told, "this is a splendid time to be an outsider. That's one of the principal reasons to run now." Hillary Clinton "is a formidable candidate," but "she is not a healing figure" and "will have a hard time escaping the well-formulated perceptions of her among swing voters as a left-wing ideologue." Then Axelrod turned up the volume. "History is replete with potential candidates for the presidency who waited too long rather than examples of people who ran too soon. . . . You will never be hotter than you are right now" and "there are many reasons to believe that if you are ever to run for the presidency, this is the time."

Axelrod warned that with *Dreams,* "the disarming admissions of weakness in your book will become fodder for unflattering, irritating inquiries," such as

"How *many* times did you use cocaine . . . When did you stop?" Such questions will be

> *more than an unpleasant inconvenience. It goes to your willingness and ability to put up with something you have never experienced on a sustained basis: criticism. At the risk of triggering the very reaction that concerns me, I don't know if you are Muhammad Ali or Floyd Patterson when it comes to taking a punch.*
>
> *You care far too much what is written and said about you. You don't relish combat when it becomes personal and nasty. When the largely irrelevant Alan Keyes attacked you, you flinched.*

In conclusion, Axelrod contended that "all of this may be worth enduring for the chance to change the world. And many, many people who believe in you are ready to march because we think the world so badly needs the change and leadership you have to offer. . . . If you pull the trigger, I am confident that we can put together a great campaign and campaign message of which we can all be proud."[42]

The next day, New Hampshire Democrats announced that Barack would visit the state to speak at their December 10 rally. Evangelical pastor Rick Warren invited Barack to join him and ultraconservative Kansas senator Sam Brownback for a World AIDS Day summit at his Orange County megachurch. When Brownback greeted Barack by saying, "Welcome to my house," Barack demonstrated his bona fides to the crowd: "Sam, this is my house too. This is God's house." Telling the audience, "I don't demonize other folks," Barack exclaimed that "faith is not just something you have, it's something you do." The crowd was impressed, with one Christian journalist explaining that Barack "almost speaks here like a pastor. That's why he gets a standing ovation from an ardently, ardently pro-life audience" despite Barack's prochoice position.

In a front-page *New York Times* story headlined "Early 'Maybe' from Obama Jolts '08 Field," Republican political consultant Mark McKinnon called Barack "the most interesting persona to appear on the political radar screen in decades." Barack was "a walking, talking hope machine, and he may reshape American politics." That evening liberal billionaire George Soros hosted a Manhattan session at which Barack met with a dozen top Democratic donors. In an editorial "Obama Should Run," Barack's hometown *Chicago Tribune* joined the chorus. "When a leader evokes the enthusiasm that Obama does, he should recognize that he has something special to offer, not in 2012 or 2016, but right now." Stating that Barack has "an approach that transcends party, ideology and geography," the *Tribune* declared, "no one else has shown a comparable talent for appealing to the centrist instincts of the American people." With "a voice that celebrates our common values instead of exaggerating our differences," Barack's "magnetic style and optimism would draw many disenchanted Americans back into the political process. . . . He and the nation have little to lose and much to gain from his candidacy."

Returning to Washington, Barack met privately with former secretary of state Colin Powell, who a decade earlier had refused to run for the presidency because of his wife's opposition. Barack also met former senator Tom Daschle for dinner at his favorite restaurant, Tosca, where they "took the kitchen table in the back where nobody could see us. We had a bottle of wine and a great meal and what was supposed to be a conversation that lasted about an hour I think went over three," Daschle recalled. "I told him that I thought his lack of Washington experience was one of his greatest assets" and "that windows of opportunity for running for the presidency close quickly, and that he shouldn't assume, if he passes up this window, that there will be another. I had that experience" and Daschle did not want "to see the same thing happen to him."[43]

As Barack prepared to fly to New Hampshire for his Sunday Democratic event, he told the Associated Press "the whole prospect of a presidential race for me is not something I've engineered. I was on a different internal clock," and "it's only been in the last couple of months that the amount of interest in a potential candidacy reached the point where I had to consider seriously" actually running. Barack told the state's largest newspaper that he believed people "are looking for somebody who is authentic," for "a pragmatic politician rather than an ideologue." David Axelrod told another journalist that Barack was doing "a lot of soul searching" in wrestling with multiple questions. "Some are very personal. Some are political. And some are the largest questions, about the contribution he thinks he can make." Just as he had in his memo, Axelrod stressed that "I think this country is very hungry for new leadership that will take us past the kind of hyper-partisanship and hyper-ideological kind of politics we've seen" over the last decade.

On Saturday evening, December 9, Barack stayed in Chicago to attend Malia's ballet recital before flying to Manchester and arriving close to midnight. A bookstore appearance in Portsmouth and a private reception at the state's top law firm were on Sunday's schedule in addition to the sold-out, two-thousand-person Democratic rally. New Hampshire governor John Lynch introduced Barack, telling the cheering crowd that "we originally scheduled the Rolling Stones . . . but then we canceled them when we realized Senator Obama would sell more tickets." More than 150 journalists and twenty-two camera crews covered Barack's remarks, with former Dover mayor Jack Buckley telling the *Boston Globe* that "I have never seen anything like this in my forty years of being active in politics." He added that "if I were Hillary, I would be more than a little concerned," and in Monday's *New York Times,* reporter Adam Nagourney wrote that Barack had received "the kind of reception typically afforded a movie star," one that was "nothing short of a spectacle." Taking questions afterward, Barack said, "I want to take my time" on making a decision and brushed aside an inquiry about "Hussein": "the American people are not concerned with middle names." One journalist observed that Barack was "a politician who happens to be black, not a black politician," and a conservative columnist warned that Barack was "an uncommonly opaque rock-star politician." But veteran political reporter Walter Shapiro stated "something is happening around Obama that we have not seen in American politics for decades."[44]

On Monday morning, David Axelrod's warning about "unflattering, irritating inquiries" proved prescient when *Crain's Chicago Business* ran a story headlined "Sen. Obama Sees No Hypocrisy in His Wife's Post at a Firm That Does Business with Wal-Mart." Three weeks earlier, in a gesture covered only by the Associated Press, Barack had endorsed the efforts of a union-backed group called Wake Up Wal-Mart, which was attacking the giant employer's salaries and health benefits for its workers. TreeHouse Foods, whose board Michelle served on, sold more of its products to Wal-Mart than to any other retailer, and also had recently closed a pickle plant in La Junta, Colorado, costing 153 workers their jobs. *Crain's* reporter Greg Hinz, noting Michelle's $45,000-a-year board fee and the CEO's $26 million annual salary, called TreeHouse a "company that pays its executives a very hefty amount of money while laying off mostly minority workers in an economically-deprived area." Contacted by Hinz, Michelle claimed that "my income is pretty low compared to my peers," and told Hinz "you wouldn't ask that question if, like some people in politics, we had trust funds and were rich." Four months later, the Obama's IRS Form 1040 would show that Michelle's 2006 income had totaled $324,000—$273,00 from the U of C Hospitals and an additional $51,000 from TreeHouse. La Junta mayor Don Rizzuto told Hinz that "if she and her husband are the champions of the little guy, it's amazing what they're doing."

With a second and potentially decisive meeting of Barack's closest advisers scheduled for Wednesday, December 13, Barack sat down with his brother-in-law Craig Robinson to talk about the pending decision. "Barack asked if I minded talking to Michelle about how this window of opportunity might not ever be available again," Craig recounted. "He didn't expect me to convince her, since he himself hadn't been able to do that yet," but Craig realized that a second, preliminary obstacle loomed too: Marian Robinson, who was among the many people who worried about Barack surviving a presidential run. Speaking first to his mother, Craig won her over to reluctant acceptance, but "you'll never get Michelle to agree to it," she told her son. Then Craig spoke with Michelle. "Just let him take his shot. You can't deprive him of that. He wouldn't hold you back if the roles were reversed."

Over the course of more than an hour, Craig gradually wore down his younger sister's opposition. As Michelle remembered her change of heart, she explained, "I would have felt guilty not doing it. I would have felt I was being selfish." In the end she came to believe "I had no choice," and Craig privately exulted over what he remembered as "that long but ultimately successful phone call."[45]

On Monday Barack taped a special surprise appearance for that evening's *Monday Night Football* telecast:

> *Good evening. I'm Senator Barack Obama. I'm here tonight to answer some questions about a very important contest that's been weighing on the minds of the American people. This is a contest about the future, a contest between two very different philosophies, a contest that will ultimately be decided in America's heartland. In Chicago they're asking, 'Does the new guy have the experience to lead us to victory?' In St. Louis they're wondering, 'Are we facing a record that's really so formidable, or is it all just a bunch of hype?' Let me tell you, I'm all too familiar with these questions, so tonight I'd like to put all the doubts to rest. I would like to announce to my hometown of Chicago and all of America that I am ready for the Bears to go all the way, baby!*

Barack put a Chicago Bears cap on his head during the last sentence. On Tuesday morning the segment was replayed on NBC's *Today* show, with host Matt Lauer saying it "shows he's got a sense of humor."

The next day Barack's closest advisers reconvened in David Axelrod's conference room. Valerie Jarrett already had given Barack her verdict: "I think it's a go. I think it's your moment." Pete Rouse recalled that Barack told the group, "I'm still inclined not to do this, but I've talked to Michelle about it, and if we're going to do this at some point, this may be the best time. I'm worried about my daughters, how this will affect them." Michelle concurred, but said that the family issues were manageable, and David Plouffe presented a draft budget, an early strategy outline, and a list of issues that would have to be tackled before any actual kickoff. Then Michelle turned to her husband to pose a bigger question. "You need to ask yourself *why* do you want to do this? What are you hoping to uniquely accomplish by getting the presidency?" Barack had a ready answer. "There are a lot of things I think I can accomplish, but two things I know. The first is, when I raise my hand and take that oath of office, there are millions of kids around this country who don't believe that it would ever be possible for them to be president of the United States, and for them the world would change on that day. And the second thing is, I think the world would look at us differently the day I got elected. . . . I think I can help repair the damage that's been done" to America's international reputation during George Bush's presidency. Some of Barack's aides were more hesitant than Axelrod, Jarrett, and even Barack himself, but over the course of the discussion, it became clear that in Barack's mind the odds were shifting modestly but perhaps decisively toward an affirmative decision. As the meeting ended, Barack said, "I think I've moved past the 50-50 mark and I'm inclined to do it." His aides should continue planning, and "I'm going to go to Hawaii to think about it" during his family's annual winter vacation.[46]

In Thursday morning's *Washington Post,* conservative columnist George Will joined the chorus: "Run Now, Obama." Will termed Hillary Rodham Clinton "the optimal opponent" for a challenger who "offers a tone of sweet reasonableness," but in the conservative *Washington Times,* chief political correspondent Don Lambro warned that Barack seems "deeply risk-averse to getting into a principled fight about anything larger than himself." In Chicago, Barack met with Mayor Richard Daley and then the editorial boards of the *Tribune* and the *Sun-Times*. At the *Trib,* Barack voiced the same question that Michelle had posed to him a day earlier: "Do I have something that is sufficiently unique to offer the country that it is worth putting my family through a presidential campaign?" Politically, "I think I would be a viable candidate. So that's a threshold question, and I wouldn't run if I didn't think I could win." Asked about Tony Rezko's adjoining lot, Barack called the purchase of the ten-foot strip "stupid," "a boneheaded move," but the discussion focused on the questions he was pondering. "Do I have a particular ability to bring the country together around a pragmatic, commonsense agenda for change that probably has a generational element to it as well?"

African American columnist Clarence Page came away from the conversation reminding readers that "hardly anyone knows much about" Barack, but there was no question he was "sounding very much like a candidate." Columnist John Kass saw Barack sneak outside to smoke and realized "I'd never see Obama alone again" if he ran. At the *Sun-Times,* Barack said he would have "a pretty good chance of winning the nomination" over Hillary Clinton and John Edwards, but conceded that his safety was "something that is on Michelle's mind, and the minds of many of my friends." Barack drolly added that "being shot, obviously . . . is the least attractive option."

Barack also met with ABLE's group of black entrepreneurs and with éminences grises Newton Minow and Abner Mikva at Minow's office. "What Barack wanted to talk about was his concern about the impact of a run on Michelle and the girls. He was worried that he would not be able to spend time with them, and that the campaign could be disturbing for them. That was the biggest question in his mind," Minow recounted. He and Mikva had helped raise three daughters apiece, and both men told Barack they thought he should run. "We both said to him that, as fathers, we thought that if he was going to run for president, this was the best time," because it was "better to be away when your daughters are very young than when they're teenagers, because our experience is that when a daughter really needs a father is when they're teenagers." Barack welcomed their advice. "That's very interesting. I'm going to tell that to Michelle." Minow added, "Besides, if you're elected, you'll be with your family all the time, because you'll all be living in the same place." In retrospect, "I think he'd made up his mind before he ever asked us," Minow explained. "I just think he wanted some reassurance," as "I could see he wanted to do it."[47]

Illinois seatmate Dick Durbin told reporters he was optimistic about Barack running, and *Newsweek*'s Jonathan Alter, a Chicagoland native, heard reliable chatter that there were "certainly no red flags" from Michelle. Barack invited David Plouffe to his South Greenwood home one night to ask him to become campaign manager. "I won't make a final decision until the Hawaii trip. But I think it is all but a certainty that I'll run," Barack told him. Plouffe needed to mull the impact that job would have on his own young family, and the day before the Obamas headed to Honolulu, Barack met with Plouffe, Axelrod, and Gibbs at Axelrod's office. "I am 90 percent certain I am running. Maybe even higher. You guys should proceed quietly as if I am, and keep making progress on planning and sizing up potential personnel. . . . I want to mull it over a bit more while I'm away" and "when I get back in early January, I'll give you the final green light."

As a *Wall Street Journal* story noted, "the grass roots excitement" about an Obama candidacy, exemplified by Barack's "rock-star-like reception last week in New Hampshire," forced Hillary Rodham Clinton's campaign to consider how to respond if Barack did run. Even two months earlier, top strategist Mark Penn had called Barack "the biggest" potential surprise in the race, and now he gave Clinton an eleven-page memo, warning that "the national press is relatively hostile," "want to be king makers," and are looking for "someone 'new' who can be their own." Thus "Obama represents a serious challenge because at least for the moment he represents something big—an inspirational movement. But the more you analyze what he says, the more you wonder what is behind the hype. No big original ideas. No incredible accomplishments for others, only for himself. No one could fill the expectations that have been built up for him" and "all politics is an expectations game." Clinton's campaign should "research his flaws, hold our powder, see if he fades or does not run. Attacking him directly would backfire. His weakness is that if voters think about him [for] five minutes, they get that he was just a state senator and that he would be trounced by the big Republicans," because "Obama has not the experience to be a serious challenger to" Republican John McCain.

Calling for Obama to run but echoing Penn, *New York* magazine political writer John Heilemann observed that "for all his promise, Obama is basically an empty vessel." In the Senate, "the legislation he has offered has been uniformly mundane, marginal, and provincial." Barack's second bill to pass Congress—providing $104 million to promote democracy in the Congo—would be signed into law three days before Christmas without ceremony and with hardly a mention in the press. "How many times has he used his megaphone to advance a bold initiative or champion a controversial cause?" Heilemann asked. "Zero." The excitement about Barack "isn't issue-based: It's stylistic. His popularity is rooted in his calm, consensus-seeking deliberative demeanor and in his calls to common purpose."[48]

In Honolulu, Barack, Michelle, and their daughters, plus Chicago buddies Marty Nesbitt and Eric Whitaker, stayed at the Hyatt Regency in Waikiki. On December 19 Maya Soetoro-Ng told an AP reporter that Barack is "going to make his decision here and announce it to us." Playing almost daily at the Olomana Golf Links in Waimanalo, "it's a much needed time for reflection" for Barack, Maya explained. "He's got to figure out what he's going to do," and "I think he's trying to reconnect with family and get away from the excessive demands of his schedule." Maya said that she had "discussed the pros and cons of running for president" with her brother, "but he hasn't indicated his decision yet." On December 24 an unwelcome *Chicago Tribune* story highlighted how a prized internship in Barack's Senate office had gone to a daughter of Joseph Aramanda, who in 2004 had contributed $10,000 of Tony Rezko's money to Barack's Senate campaign. Other reports showed Barack tied for the lead with other Democratic contenders in early polls in both Iowa and New Hampshire. David Plouffe called Barack to accept his offer to manage the campaign, but over the phone "I thought I was hearing a guy experiencing huge second thoughts," Plouffe remembered. "He said he was still much more likely than not but would need to spend more time thinking about it during the remaining days of his vacation." Staying in a huge Waikiki hotel had turned out to be a mixed blessing, for now scores of strangers recognized Barack, even in Honolulu. "I really can't go down the street. This feels different," Barack told Plouffe, who replied that if Barack ran, "your life is going to change forever. You'll lose all privacy."

Michelle got her husband to once again promise he would quit smoking if he ran, but Barack spent as much time golfing as focusing on his big decision. Playing at Waialae Country Club on December 27, Barack "was very relaxed and he had a great time on the golf course," club president and state appellate judge James Burns told the *Advertiser* for a story headlined "Obama's Visit All About Golfing, Not Presidency." "He had some good shots and some bad shots. He's very friendly and everyone wanted to say hello to him, meet him and take pictures with him." By phone with Valerie Jarrett in Chicago, Barack signaled that he was leaning strongly toward a run, saying "this is pretty much done," but

Maya now declined to tell the AP that any decision would be made while Barack remained in Hawaii, explaining that "the world will find out soon enough."[49]

By New Year's Day, 2007, Barack and his family were back in Chicago. In the *Sun-Times,* African American columnist Laura Washington advocated "run, Barack, run," and on Tuesday, January 2, Barack turned up unannounced at David Axelrod's office. "I think I want to do this," but he had still not decided for sure: "I'm going to give it a few more days." David said, "My main concern is that you're not obsessive enough to run for president," that "you may be too normal to run." Leaving Axelrod's office, Barack ran into Forrest Claypool. "It may not be exactly the time I would pick," Barack remarked, "but sometimes the times pick you." Speaking with Plouffe, Axelrod thought a final decision remained uncertain, but during a January 3 podcast in which Barack called any "surge" of U.S. forces in Iraq "a chilling prospect," he said his two weeks in Hawaii had left him feeling "renewed and refreshed."

On January 4 Barack returned to Washington, telling one journalist he believed "there is a weariness with the ideological battles and cultural wars of the past," especially given how George W. Bush's presidency "has been the most ideological administration in my lifetime, even more so than" Ronald Reagan's. Along with other senators, Barack met with Bush at the White House to express strong opposition to any "surge" in Iraq. "I said definitively that I thought it was a bad idea," Barack told reporters. To Pete Rouse, Barack said he was close to a final decision. "I've decided to do it, you can plan that we're going to do it, but don't do anything yet because I want to go home to Chicago this weekend and make sure I don't have buyer's remorse." In his own mind, "the way I thought about it was more of a sense of duty" to run, Barack explained. "It was more the sense of, given what's been given to me, I should probably just give it a shot and see whether in fact there's something real there." In retrospect, "I gave myself 25 percent odds, maybe 30," but absent a run he would face continued boredom in the Senate. A conference call was scheduled with all of Barack's closest family members—Maya, Auma, Craig and Kelly Robinson—plus Marty Nesbitt, to alert them to the upcoming news. "Okay. I think we're going to do this. It looks like this thing is going forward. But before we go for sure, I feel obligated to let you guys know how considerably hard this is going to be on all of you." No one replied until Auma broke the silence with "I think this is wonderful," and everyone agreed.

Late that Saturday night Barack phoned David Plouffe: "It's a go." Then he called Axelrod: "Axe, I just called Plouffe and told him it's a go." On Sunday Barack called Tom Daschle to give him the confidential news, and plans were made for Barack and Michelle to make a quick, unpublicized trip to Nashville to speak with former vice president Al Gore and his wife Tipper about the rigors of a race.[50]

In Washington, reporters who showed up for a press conference on ethics legislation instead sought Barack's reaction to a *People* magazine photo spread titled "Beach Babes" that included a picture of a shirtless, clearly fit Barack taken on some Hawaiian beach. "I really appreciate you toting that around," an "uncharacteristically flustered" Obama told them. "It's embarrassing," he added. "It's paparazzi. Stop looking at it." Robert Gibbs wisecracked that next on Barack's schedule was "photo shoot on South Beach," but instead Barack made a quick trip to Manhattan to speak at the annual dinner of Rev. Jesse Jackson's Wall Street Project. "I owe him a great debt," Barack told the crowd. "I would not be here had it not been for 1984, had it not been for 1988," the two times Jackson had run for president.

Meeting privately in Washington with his top aides, Barack stressed that "We're all in this together. We're going to rise or fall together. No sharp elbows. No big egos. I want us to be a team," and "I want a campaign that is buttoned up like a business," with "no drama" and "no leaks." Reporters gleaned that David Plouffe would be the campaign manager if Barack went forward, but no word that the actual decision to run already had been made reached reporters. Barack met privately with Secretary of State Condoleezza Rice prior to her testimony before the Foreign Relations Committee to defend President Bush's "surge" in Iraq. At the hearing, Barack bluntly challenged her. "Essentially, the administration repeatedly has said, 'We're doubling down. We're going to keep on going. You know, maybe we lost that bet, but we're going to put a little more money in, and because now we've got a lot in the pot, and we can't afford to lose what we've put in the pot.' And the fundamental question that the American people have and I think every senator on this panel, Republican and Democrat, are having to face now is at what point do we say 'enough'?" Barack asked.

Appearing on CBS's *Face the Nation,* Barack said, "we'll be making an announcement fairly soon" about 2008, and the next morning, prior to an MLK holiday event in south suburban Harvey, Barack updated it to "very soon" while complaining about the *People* picture. "Finding out that there was a photographer lurking in the bushes while I was playing on the beach with my kids is a source of concern." With even the local *Daily Southtown* calling Harvey a "corruption-plagued, crime-ridden town," Barack took the opportunity to lecture city officials that public office was not supposed to be "a place where they can help their family and their friends instead of helping the people who elected them. We don't need that kind of leadership. You want to make a lot of money, go start a business. Don't run for office."[51]

Twenty-four hours later, on January 16, Barack finally declared his candidacy via a three-minute Web video, with a formal announcement set for February 10 in Springfield. "I certainly didn't expect to find myself in this position a year ago," Barack admitted. Emphasizing "how hungry we all are for a different kind of politics," he noted that "today our leaders in Washington seem incapable of working together in a practical, commonsense way. Politics has become so bitter and partisan, so gummed up by money and influence, that we can't tackle the big problems that demand solutions." Above all, "we have to change our politics and come together around our common interests and concerns as Americans." The *New York Times* called the video "a blue-sky plan of optimism, offering few specifics," but Barack told the *Times*' Jeff Zeleny that "one thing that I'm convinced of is that people want something new." Speaking with Zeleny again the next day, Barack said that "there are moments in American history where there are opportunities to change the language of politics or set the country's sights in a different place, and I think we're in one of those moments." Barack paused. "Whether I'm the person to help move that forward or somebody else is, is not for me to determine."

The *Washington Post* welcomed Barack's candidacy by commending "his willingness to try to understand rather than to dismiss the arguments of the other side." In Chicago, the *Tribune*'s John Kass warned that Barack was an "empty vessel," and when Barack fulsomely endorsed Mayor Richard M. Daley for reelection, one opinionated reporter said the move "puts to rest any doubt that Obama is anything but at the center of the same old machine." *Capitol Fax*'s Rich Miller expressed amazement at what was happening. "This Obama phenomenon is not rational in any form. It is, in fact, almost completely irrational . . . the average Obama supporter knows very little about the man he or she adores," but "the more exposure he gets, the more people swoon over him." Several national commentators noted how Barack, like retired general Colin Powell a decade earlier, "is a child of immigration" who "doesn't sound or look too black" and had "succeeded at a prestigious white institution." Indeed, unreported November 1996 nationwide exit polling data showed that had Powell, rather than Kansas senator Bob Dole, been the Republican nominee challenging President Bill Clinton, Powell's 50 percent of the vote would have easily defeated Clinton. Thus "America has for some time been ready to elect a black president," and Eric Krol of the suburban *Daily Herald* warned that Barack's "opponents would be foolish to underestimate him. His ability to inspire a popular groundswell among regular people could well carry him to his party's nomination."[52]

Attending Sunday's NFC championship game, in which the Chicago Bears defeated the New Orleans Saints 39–14, Barack told friends that if he could win the first-in-the-nation Iowa caucuses on January 3, 2008, then "I'm credible." That was David Plouffe's view too. "We thought we had a path to the nomination, but it was a very narrow path," Plouffe explained. "We had to win Iowa." Yet Barack's four-person campaign team was starting totally from scratch. Working from a temporary office on Connecticut Avenue NW before securing permanent office space in Chicago, staffers purchased a wireless router and then a printer soon after Barack's Web video went live. "It was definitely seat of the pants," Plouffe recalled. "It was a start-up in every sense of the word," and "getting the campaign up and running was a huge challenge."

For Barack, President Bush's decision to ramp up U.S. military efforts in Iraq remained a top target. "I originally opposed this war precisely because I thought that, once we were in, it would become a morass," he remarked on CBS. At a D.C. symposium, Barack also vowed that "universal health care for every single American must not be a question of whether, it must be a question of how." He declared, "I am absolutely determined that by the end of the first term of the next president we should have universal health care in this country." Considering that "one out of four health care dollars is spent on nonmedical costs, mostly bills and paperwork," eliminating that morass was simply "a problem of political will."

On January 29 Barack joined other senators for a quick trip to New Orleans to hold a hearing on reconstruction since Katrina. The next day, he introduced his own bill to address the Iraq war, calling for the withdrawal of U.S. combat forces to begin on May 1 and to be fully complete by March 31, 2008. In a lengthy interview with a young blogger, Barack cited his Iraq proposal, his health care stance, and the need to eliminate partisan gerrymandering, "which discourages the kind of robust debate that we need to have. If people feel like this is a 90 percent Democratic district or a 90 percent Republican district, then at a certain point folks start opting out of the process." But health coverage was uppermost in his mind. "You have to run on the notion that by the end of your first term you're going to have health care for all. That then would give you, should you win, a mandate, and you have to use that mandate quickly in the first one hundred days before the corrosive process of Washington starts setting in."[53]

On February 1, Barack's nascent campaign notified the Federal Election Commission that both major parties' presidential nominees should commit to accept public funding and limit their general election spending. This would "facilitate the conduct of campaigns freed from any dependence on private fundraising," but *New York Times* reporter David K. Kirkpatrick predicted that "a candidate with a much bigger bank account at the start of the general election would be reluctant to relinquish that advantage." The next day Barack spoke at a student-organized rally at George Mason University in Fairfax, Virginia, that drew thirty-five hundred young people. That weekend Barack's staff moved from D.C. to Chicago, with David Plouffe explaining that "it was brutal to try to get this set up and running in a matter of weeks. We had to find office space, we had to get accounts up and running, we had to raise some initial seed money, we had to get staff hired." An initial fund-raiser was scheduled at

Chicago's Hyatt Regency for the day after Barack's Springfield announcement, and a second was set for Beverly Hills nine days later. New staffers were told that discipline was paramount. "The culture of the organization from the get-go was, a leak could not be tolerated," and if one occurred, Plouffe warned, "corrective action" would be taken.

Michelle Obama took part-time leave from her U of C job, and Barack told the *Chicago Tribune*'s Christi Parsons that "I've got an ironclad demand from my wife" to stop smoking. "I've been chewing Nicorette strenuously," but while Barack claimed, "I've never been a heavy smoker," his long-standing inability to quit, plus his repeated denials of how much he smoked, reflected his embarrassment about being a smoker and the strength of his addiction. Norman Mailer once confessed that "Smoking cigarettes insulates one from one's life, one does not feel as much, often happily so," and an insightful observer explained that Barack tried to hide and minimize his smoking so as "to deny . . . that he relies on a . . . nicotine addiction to manage his fears."

David Axelrod told the *Washington Post*'s Anne Kornblut that Barack "is very focused on the fact that he doesn't want to lose his essential self in this process, and if he does—if what he projects and delivers is just more of the kind of politics people have become accustomed to—it would be a disappointment to him." In a pair of February 7 interviews, Barack said that "what I dread most is being away from my kids," but he believed "there is this enormous hunger for a new kind of politics . . . a politics that brings us together." "I think we are in a moment where there is a possibility" that his campaign "could reshape the political landscape." He emphasized that he was running "to win. But it's also to transform the country."[54]

At 4:00 A.M. Thursday, Barack e-mailed a final draft of his Saturday announcement speech to his advisers. On Friday afternoon, he and his campaign team climbed into multiple vans for the drive to Springfield, where, at 10:00 A.M. Saturday, Jeremiah Wright was scheduled to deliver an invocation before Barack took the podium. During the trip southward, David Axelrod got a call alerting him to a forthcoming *Rolling Stone* cover story headlined "The Radical Roots of Barack Obama." Those "radical roots" were Trinity Church, which author Ben Wallace-Wells labeled "a leftover vision from the sixties of what a black nationalist future might look like." Calling Jeremiah Wright "a sprawling, profane bear of a preacher," one who "has a cadence and power that make Obama sound like John Kerry," Wallace-Wells quoted at length from a Wright sermon that actually had been delivered in 1993 at Washington's Howard University, not at Trinity. "We are deeply involved in the importing of drugs, the exporting of guns and the training of professional killers," Wright declared. "We care nothing about human life if the ends justify the means," and "God has *got* to be *sick of this shit*!"

Wallace-Wells asserted that Trinity represented "as openly radical a background as any significant American political figure has ever emerged from, as much Malcolm X as Martin Luther King Jr." Wallace-Wells quoted a characterization Barack had given the *Tribune,* that Wright was "a sounding board for me to make sure that I'm not losing myself in some of the hype and hoopla and stress that's involved in national politics." Wallace-Wells added that "when you read" *Dreams,* "the surprising thing—for such a measured politician—is the depth of radical feeling that seeps through, the amount of Jeremiah Wright that's packed in there."

In the van headed southbound on I-55, Axelrod "looked stricken" as he read the article, David Plouffe recalled. "'This is a fucking disaster,' he said. 'If Wright goes up on that stage, that's the story. Our announcement will be an asterisk.'" Axelrod, Plouffe, and Robert Gibbs called Barack in his van, and Barack asked them to forward him the story. Barack soon called back, expressing agreement: the article featured "some pretty incendiary language" from Wright, and "we can't afford to let this story hijack the day. I'll call him and tell him it will overshadow everything. I still want him to come; maybe he can do a private prayer with my family before I go out to speak."

In Amherst, Massachusetts, Jeremiah Wright's cell phone rang. Barack cited the *Rolling Stone* article, telling Wright, "You can kind of go over the top at times" and "get kind of rough in the sermons, so it's the feeling of our people that perhaps you'd better not be out in the spotlight, because they will make you the focus, and not my announcement. Now, Michelle and I still want you to have a prayer with us. Can you still come and have prayer, before we go up?" Wright was puzzled at what sermon *Rolling Stone* had unearthed, but he already was scheduled to catch an early-morning flight out of Boston to O'Hare and then on to Springfield. Two years earlier, Wright had said Barack "is not a person to compromise," so he was disappointed by this turn of events, but he quickly agreed and also concurred with Barack's suggestion that Wright's associate pastor and designated successor, Otis Moss III, be asked to give the invocation instead. Jerry gave Barack Moss's cell number and immediately called Moss to tell him he had done so. To Jerry's astonishment, Moss said that Barack's buddy Eric Whitaker, also a Trinity member, had *already* called to invite him to Springfield. Moss told Wright he would decline, as he barely knew Barack, and after a brief night's sleep, Saturday morning Wright flew to Springfield.[55]

Saturday dawned exceptionally cold in Springfield. Michelle warned that "there's no way anyone is going to come out and stand in the bitter cold to hear you," but Obama's aides brushed aside her plea that they hold the event indoors rather than in front of the picturesque Old State Capitol where Abraham Lincoln had served 150 years earlier. Anticipating the weather, Alyssa Mastromonaco's advance team had a strong heater hidden inside Barack's outdoor podium. The temperature was just 12 degrees, but by 10:00 A.M. a crowd of up to seventeen thousand people filled the streets and plaza surrounding the Old State Capitol. Inside the historic structure, Jeremiah Wright led a prayer before Barack headed outside. Standing nearby, David Axelrod "caught Wright's withering glare as he walked by." After being introduced by Illinois seatmate Dick Durbin, Barack confessed, "it's humbling to see a crowd like this." Some of Barack's early stanzas echoed his 2004 DNC speech, and he said his three years in Roseland had sent him first to Harvard and then to Springfield. Pushing back against any suggestion that his Senate service was too brief to qualify him for the presidency, Barack declared, "I've been there long enough to know that the ways of Washington must change," a line crafted by Axelrod and one that won the crowd's loudest cheers. Barack blamed "the failure of leadership" for "our chronic avoidance of tough decisions, our preference for scoring cheap political points instead." President after president had campaigned on promises of change, but time and again, "after the election is over, and the confetti is swept away, all those promises fade from memory, and the lobbyists and special interests move in, and people turn away, disappointed as before." Barack vowed he would be different, and asked his listeners to join him "if you feel destiny calling."

After his twenty-minute speech, Barack's party headed to Springfield's airport, where a chartered aircraft waited to fly them to Cedar Rapids, Iowa. There, more than two thousand people packed Barack's first campaign event in 2008's first state. "I want to win," Barack told them, "but I don't just want to win. I want to transform this country."[56]

Epilogue

THE PRESIDENT DID NOT ATTEND, AS HE WAS GOLFING

In Chicago, Barack's growing campaign staff set up shop in February 2007 on the eleventh floor at 233 North Michigan Avenue. Barack's first campaign swing took him to South Carolina and Virginia. "Body man" Reggie Love, a former Duke University basketball player who had been on Barack's Senate staff, and trip director Marvin Nicholson, who had been on Senator John Kerry's staff, became Barack's two closest traveling aides. Michelle was now campaigning too, having reduced her U of C job to one day a week, but her introduction of Barack at a Hollywood fund-raiser drew critical attention after *New York Times* columnist Maureen Dowd brought Michelle's remarks to a wide readership. "I have some difficulty reconciling the two images I have of Barack Obama," Michelle said. "There's Barack Obama the phenomenon. He's an amazing orator . . . best-selling author, Grammy winner. Pretty amazing, right? And then there's the Barack Obama that lives with me in my house, and that guy's a little less impressive. For some reason this guy still can't manage to put the butter up when he makes toast, secure the bread so that it doesn't get stale, and his 5-year-old is still better at making the bed than he is." Dowd suggested that Michelle's "chiding was emasculating, casting her husband . . . as an undisciplined child." But Barack was unoffended. "I think creating a life for children that is stable and in which they have reliable, regular adult figures in their lives that they can look up to is important," he told *Times* columnist Nick Kristof.[1]

Speaking at the annual commemoration of Selma, Alabama's 1965 voting rights campaign, Barack said that because of those protests, his parents "got together and Barack Obama Jr. was born." Reporters quickly noted, as the *New York Times* put it, "that he was born in 1961, four years before the confrontations at Selma took place." Barack responded that "I meant to say the whole civil rights movement," not Selma. The focus on Barack's biographical misstatement led reporters to ignore his comment that "we've still got a lot of economic rights that have to be dealt with." A *Washington Post* columnist who read *Dreams* warned that Barack's "tendency to manipulate facts may bear watching," since "we hardly know him."

To another *Post* columnist, Barack said that beyond winning the presidency, "there's the possibility" he could "also transform the country in the process, that the language and the approach I take to politics is sufficiently different that I could bring diverse parts of this country together in a way that hasn't been done in some time." One possible example, at least in Barack's mind, came in a CNN appearance. "I think that 'marriage' has a religious connotation in this society, in our culture, that makes it very difficult to disentangle from the civil aspects of marriage." Thus for gay Americans Barack supported "civil unions that provide all the civil rights that marriage entails."

On March 6, the *New York Times* reported how Barack had disinvited Jeremiah Wright from taking the stage at his February 10 announcement. "When his enemies find out that in 1984 I went to Tripoli" with Nation of Islam leader Louis Farrakhan, Wright told reporter Jodi Kantor, "a lot of his Jewish support will dry up quicker than a snowball in hell." Five days earlier, Wright had appeared on Fox News' *Hannity & Colmes,* and when Sean Hannity had peppered him with aggressive questions, Wright responded with insistent anger. "How many of Cone's books have you read? How many of Cone's books have you read?" he repeated, citing the well-known founder of black liberation theology, James H. Cone. "How many books of Dwight Hopkins's have you read?"—a Cone student and Trinity member.

In Chicago, African American *Sun-Times* columnist Mary Mitchell cited Wright's comment to the *Times* about Farrakhan in arguing that "Obama took some bad advice from campaign staff who underestimated the impact such a slight could have" in producing "hurt feelings." Wright prophetically told a PBS interviewer that if "you think it's ugly now, it's going to get worse. It's going to get much worse," and *Sun-Times* columnist Laura Washington observed that "hell hath no fury like a preacher scorned."

Asked by one CNN interviewer, Barack said he and Wright had spoken, and Wright told Trinity's congregation that Barack had apologized for the disinvitation, explaining that he had gotten "some bad advice from his own campaign people." On another CNN show, Barack claimed he had stopped smoking because of "how scared I am of my wife," but when the *New York Times*' Jodi Kantor continued to pursue the Wright story, Barack disingenuously told her that "I've never had a thorough conversation with him about all aspects of politics." Kantor observed that "it is hard to imagine" how "Obama can truly distance himself from" his pastor, but Wright told her that "if Barack gets past the primary, he might have to publicly distance himself from me," and that Barack had said, "yeah, that might have to happen."[2]

In the midst of the Wright contretemps, Barack received an unexpected call from a long-forgotten schoolmate: Keith Kakugawa, his closest friend when Barry was a Punahou tenth grader, had just been released from a California prison. Barack "was utterly amazed," Keith said, and "I told him how proud

I am of him." Keith had endured multiple convictions for cocaine possession with intent to sell, plus one for auto theft. "I'm really sorry," Barack said. "How bad is it for you?" Keith knew all too well that as a felon "you can't get a decent job, college education or not, once you have a record," that "you're not eligible for almost any kind of state assistance, and you have very few places to turn." Barack told Keith, "Hey, I'd rather you not talk to reporters," but within days first the *Wall Street Journal* and then the *Chicago Tribune* ran front-page stories about him. "Please don't put Senator Obama in a bad light for knowing me," Keith e-mailed the *Journal*'s Jackie Calmes. Barack was not enthused by the stories. "This is something I'm not enjoying about the presidential race," he told the *Trib*. "Me getting screwed I'm fine with. Suddenly everybody who's ever touched my life is subject to a colonoscopy on the front page of the newspaper."

In Chicago, Barack's former friendship with Tony Rezko was drawing sustained press attention. The *Sun-Times* was doggedly pursuing the details, with columnist Carol Marin writing that "this gleaming presidential hopeful and paragon of new politics behaves just like any other dissembling, dismissive Chicago pol" after Barack's aides physically shielded him from insistent *Sun-Times* reporter Tim Novak as his SUV pulled away. "Maybe it was the image of that getaway, played on the 5 o'clock news, that finally persuaded Obama to hastily call a news conference to which Novak was not invited," Marin explained. The *Sun-Times* editorialized that Barack's efforts to dodge Novak "make it appear as if he has something to hide," and Barack refused to discuss Tony with the *New York Times*. In an ensuing front-page story, the *Times* accurately concluded that "the two men were closer than the senator has indicated."

When *Times* columnist David Brooks caught him in a Senate hallway, Barack responded eagerly when Brooks asked if he had ever read Reinhold Niebuhr. "I love him. He's one of my favorite philosophers." What had Barack learned from Niebuhr? "I take away the compelling idea that there's serious evil in the world," that "we should be humble and modest in our belief that we can eliminate those things." Brooks was impressed, but he was also frustrated. Barack "loves to have conversations about conversations. You have to ask him every question twice, the first time to allow him to talk about how he would talk about the subject, and the second time so you can pin him down to the practical issues at hand." The result was "Bromide Obama, filled with grand but usually evasive eloquences about bringing people together and showing respect."

All of the journalistic interest in Barack's past led new readers to absorb *Dreams*. Many erroneously presumed that the book predated any interest in a political career, but *Washington Monthly* blogger Kevin Drum offered a remarkably perceptive reading. A striking detachment between the author and his retrospective self was "clearest in the disconnect between emotions and events: Obama routinely describes himself feeling the deepest and most painful emotions imaginable . . . but these feelings seem to be all out of proportion to the actual events of his life, which are generally pretty pedestrian." Readers "get very little sense of what motivates him," for "there's very little insight into what he believes and what really makes him tick." Drum concluded that "by the time I was done, I felt like I knew less about him than before." Likewise, in the *Weekly Standard* Andrew Ferguson discussed how in *Dreams* Barack "was making it up, to an alarming extent," while creating "a fable . . . far bigger and more consequential" than Barack's actual life. In private, Barack told his aides that "there's a certain ambivalence in my character that I like about myself. It's part of what makes me a good writer," but "it's not necessarily useful in a presidential campaign."[3]

In early May, responding to entreaties from Senators Dick Durbin and Harry Reid, the Secret Service assigned agents to Barack and his family, replacing the private security guards his campaign had hired. A front-page *Washington Post* story quoted Barack as telling black officials in South Carolina that sometimes in Chicago "when I talk to the black chambers of commerce, I say 'You know what would be a good economic development plan for our community would be if we make sure folks weren't throwing their garbage out of their cars.'" In response, a *Chicago Defender* contributor wrote that "all this sounds very derogatory, à la Amos 'n' Andy or Stepin Fetchit."

Garnering more attention was a front-page *New York Times* profile of Michelle, which quoted her brother Craig as explaining that "everyone in the family is afraid of her." The *Times* also quoted Barack saying that "she's a little meaner than I am," and a nationwide Associated Press story revisited Michelle's fund-raiser comments describing how the man with whom she lived differed so much from Barack's media image. Only the Chicago papers covered Michelle's resignation from the board of TreeHouse Foods, but Michelle told ABC's *Good Morning America* that Barack "is very able to deal with a strong woman, which is one of the reasons why he can be president: because he can deal with me." When the Capitol Hill town house where Barack stayed suffered a minor fire, Michelle's reaction was almost gleeful. "I was like, I *told* you it was a dump!" Then *Esquire* magazine named Barack as one of the "Best Dressed Men in the World," citing him for "sharply tailored two-piece suits offset by a peerless collection of light-blue ties." Michelle was astonished. "He's got like five white shirts and three black suits," and "he's like 'Mr. GQ' all of a sudden."

Barack found the relentless schedule of a nationwide presidential campaign daunting. "I like a certain brand of green tea" and "I'm a big trail mix guy," he told the *Today* show. "But the main thing is to get my workouts in . . . you've got to block out an hour somewhere." In another interview, he confessed that "you never get over the sacrifice of being away from your family. I was home one day this week and rode bikes with my daughters, went to the dentist last week, and took them to dinner." Asked if he dreamed of the White House, Barack said no. "I don't dream of the White House. I dream of the beach. I dream of playing basketball. I dream about my kids."[4]

At a forum televised by CNN, Barack volunteered that "I am where I am today because of the education that I received," and he appeared to have Keith Kakugawa in mind when he cited how "we have ex-offenders who are coming out of prisons constantly—thousands each and every day. We're going to have to make a commitment to provide them a second chance," which "will require a government investment in transitional jobs because" ex-offenders were such anathema to the private sector.

In the Senate, on June 6 Barack offered a controversial floor amendment to a delicately crafted bipartisan immigration reform bill. The pending measure included a "points" system to prioritize visas for skilled immigrants over those with family ties for the next fourteen years. Barack wanted to "sunset" that factor, which liberals viewed with distaste, at five years. "I am willing to defer to those Senators who negotiated this provision and say we should give it a try, but I am not willing to say this untested system should be made virtually permanent." That made South Carolina Republican Lindsey Graham, one of the architects of the bipartisan compromise, "very, very mad," Graham recalled, and his floor response reflected that. "Some people, when it comes to tough decisions, back away, because when you talk about bipartisanship, some Americans on the left and right consider it heresy, and we are giving in if we adopt this amendment," Graham warned. "It means that everybody over here who has walked the plank and told our base you are wrong" in opposing compromise would be shafted. "You are going to destroy this deal," for "some people don't want to say to the loud folks 'No, you can't have your way all the time.'" Then Graham directly addressed Barack. "So when you are out on the campaign trail, my friend, telling about why we can't come together, this is why."

Obama "briefly appeared stunned" by Graham's comment, the Associated Press reported, and replied meekly. The amendment, he said, says that "we will go forward with the proposal that has been advanced by this bipartisan group. It simply says we should examine after five years whether the program is working." Barack asked senators to "consider the nature of the actual amendment that is on the floor as opposed to the discussion that preceded mine." Just before the vote, Barack spoke again. "The authors of this legislation deserve credit for working diligently and coming up with a carefully balanced bill," but the new system would be "a radical departure from the one we have had in the past" and his amendment "simply says that after five years, we will reexamine the bill." Graham disagreed. "If we sunset the merit-based system at five years, there is no vehicle left," he warned. Barack's amendment was defeated 42–55, with eight Democrats, including lead negotiators Dianne Feinstein and Ted

Kennedy, joining all but one Republican in voting against it. Immediately after the vote, Barack and Graham got into a "heated exchange" just off the Senate floor. "They were going at it," Florida Republican Mel Martinez told the AP. "We could hear them inside." Graham told the AP, "I said, 'I'm very disappointed in you,'" explaining that Barack "gave in to pressure from the left, and the pressure from the left and right was enormous." Barack said "it's a matter of too much coffee and people being on the floor too long," but Graham later remarked that Barack had "folded like a cheap suit."[5]

In mid-June, Barack reacted with unusual fury when the *New York Times* revealed that his campaign had given journalists a not-for-attribution document attacking Hillary Clinton's close ties to the Indian American community and mocking her as "(D-Punjab)." Close friends like Hasan Chandoo had purposely kept a low profile once Barack announced. "We all disappeared. The last thing he needs is a bunch of Paki friends," Hasan explained, but following the *Times* story Barack recounted that "I had to go and call some of my best friends and explain why my campaign was engaging in xenophobia. That was the most angry I've been in this campaign." The following morning, a *Chicago Tribune* story headlined "Obama Team Can Play Rough" cited the *Times*' "Punjab" story as one reflection of how "Obama has surrounded himself with operatives skilled in the old-school art of the political back-stab."

The *Tribune* also previewed a forthcoming biography of Barack by *Tribune* reporter David Mendell, who had covered his 2004 Senate campaign. The *Trib* said Mendell portrayed Barack as "a far more calculating politician than his most ardent supporters might imagine," citing how Barack had given his October 2002 antiwar speech in part "to curry favor" with rally organizer and wealthy donor Bettylu Saltzman. "Barack was not happy," Saltzman recalled, and when she told him she was submitting a letter of rebuttal, Barack's response was "You better." Early in the book Mendell highlighted Barack's "imperious, mercurial, self-righteous and sometimes prickly nature" and said that after compiling an "aggressively liberal" record in Springfield, Barack in Washington had taken "a dramatic turn toward calculation and caution."[6]

On the campaign trail, Barack vowed that "I will sign a universal health care plan that covers every American by the end of my first term as president." In Iowa, Barack cited how his childhood years in Indonesia and his Kenyan roots informed his opposition to the Iraq war, because they had taught him "how powerful tribal and ethnic sentiments are." Journalists were astonished when Barack's first Iowa television ad featured Illinois Republican state senator Kirk Dillard. "Senator Obama worked on some of the deepest issues we had and was successful in a bipartisan way. Republican legislators respected Senator Obama. His negotiation skills and an ability to understand both sides would serve the country well." Dillard told reporters he supported Republican John

McCain for the presidency, but said that Barack's candidacy, "whether he wins or loses, is good for Illinois and it's good for the United States."

Barack told an NPR host that "it's important to make sure that I don't lose my core honesty with myself and that I don't start trimming my sails or biting my tongue in order to get elected." Finishing a "downright tentative" speech to a predominantly black female audience at New Orleans's Superdome, Barack had trouble finding an exit through the stage curtain as his lapel mike captured him telling himself, "If I can just find which way to go." Back in Chicago, speaking to yet another South Side antiviolence rally, Barack noted that "in this last school year, thirty-two Chicago public school students were killed," more deaths "than the number of soldiers from this whole state who were killed in Iraq. Think about that. At a time when we're spending $275 million a day on a war overseas, we're neglecting the war that's being fought in our own streets." To a black audience in Washington's most forlorn neighborhood, Barack cited the millions of Americans who "cannot write thousand-dollar campaign checks to make their voices heard. They suffer most from a politics that has been tipped in favor of those with the most money."[7]

Appearing before the National Association of Black Journalists, Barack made fun of the number of commentators who had wondered whether he was "black enough" by arriving twenty minutes late. "I want to apologize for being a little late, but you guys keep on asking whether I'm black enough," Barack began as the audience convulsed in laughter. "Uh-huh, that's right—so I figured I'd stroll in about ten minutes after deadline. I've been holding that in my pocket for a while." Then Barack turned serious. "The day I'm inaugurated, the racial dynamics in this country will change to some degree. If you've got Michelle as first lady, and Malia and Sasha running around on the South Lawn, that changes how America looks at itself. It changes how white children think about black children, and it changes how black children think about black children."

Michelle too was now an almost full-time public figure. "As the campaign has moved along, her speeches have become stronger, funnier and more personable," a *Chicago Sun-Times* correspondent wrote. "She speaks with more emotion than her husband; you feel she is the power propelling him, that she has the psychological mettle, the tough skin, the searing ambition." As for campaigning, "the little sacrifice we have to make is nothing compared to the possibility of what we could do if this catches on," Michelle now believed. Barack likewise said that "Michelle and I would not be doing this if we didn't think that at some level it's worth it. . . . You're doing this because you genuinely think that you can bring the country together. . . . If I didn't think I was going to be able to do that, then this wouldn't be worth it. It exacts a high toll."

Reporter Ryan Lizza, who had begun covering Barack in 2004, expressed his "realization that Obama . . . is more of an old-fashioned pol than you think," that "underneath the inspirational leader who wants to change politics . . . is an ambitious, prickly, and occasionally ruthless politician." Judd Miner confessed to Lizza that "his biggest disappointment was that Obama hired as his top strategist David Axelrod," whom Judd and other Harold Washington–era progressives rightly viewed as a Daley machine operative pretending to be something better. *Washington Post* columnist David Ignatius wrote that Barack's "almost eerie self-confidence" reflected a persona that is "closer to a rock star than a typical politician." Without naming Axelrod, Ignatius quoted "a top Obama adviser" as explaining that Barack "is totally pragmatic. He asks what would work and what wouldn't."

An enterprising Denver reporter visited the Hyde Park Hair Salon to ask barber Abdul Karim about Barack. "He's changed a little bit. I'll be honest with you. He's not as vocal. He's not as natural as he used to be. There's two Baracks now. There's the Barack who comes in on a Sunday when we're all busy and he can't be himself. He has to be the politician, shake all the hands. And then there's the Barack who comes in here late on Saturday, nobody here. That's the guy we always knew. Talk a little basketball, talk about the Bulls, the fight that was on last night. You know what I'm saying? I know that Barack. I don't necessarily know the other one."

Old friend Rob Fisher explained that Barack was coming to realize "that the public image of who he is is not who he actually is." Another unnamed friend made the same point to *The New Yorker*'s Larissa MacFarquhar: "I think sometimes he feels phony to himself." Barack has "to struggle with . . . being a regular guy who has become a persona named Barack Obama. The persona is going to get . . . more and more distant from him and the way he used to live his life." Barack himself later acknowledged how "there's me, and then there's this character named 'Barack Obama.'" A wide range of longtime acquaintances now saw a "veneer" that had not existed before 2004. "There's a little bit of a wall there. You don't really get to know him," explained a *Law Review* colleague who continued to host fund-raisers for Barack. There is a "wall that's up," or a "drawbridge" marking "as close as Barack lets you get to him," explained another old friend. Prior to his fame, Barack "was much more accessible emotionally," former Chicago aide Cynthia Miller explained. "To everyone."[8]

Barack's campaign had outdone all other candidates in both parties by raising over $58 million during the first half of 2007, but by mid-September, with national polls showing Hillary Clinton maintaining a commanding lead, nervousness among donors led Barack and Michelle to ask Valerie Jarrett to assume an active role in his campaign. Although Jarrett had known Barack since 1991, she had played almost no role in his 2000 congressional race and became an active presence in his U.S. Senate campaign only in early 2004. With no campaign experience, to staffers and reporters alike Jarrett seemed like an odd addition, but Valerie's regard for Barack knew no bounds. "He's always wanted to be president," she told *The New Yorker*. "He didn't always admit it, but oh, absolutely. The first time he said it to me, he said 'I just think I have some special qualities, and wouldn't it be a shame to waste them.'" To *New Yorker* editor David Remnick, Jarrett went even further. "I think Barack knew that he had God-given talents that were extraordinary. He knows exactly how smart he is. . . . He knows how perceptive he is. He knows what a good reader of people he is. And he knows he has that ability—the extraordinary, uncanny ability—to take a thousand different perspectives, digest them, and make sense of them, and I think that he has never really been challenged intellectually." Jarrett told another interviewer that Barack was "really by far smarter than anybody I know," and to Remnick she emphasized that Barack was "somebody with such extraordinary talents . . . he's just too talented to do what ordinary people do." In another conversation, Valerie emphatically stated that Barack "is a man who is devoted to his wife. There aren't going to be any skeletons in his closet in terms of his personal life at all. Period."

Following a pair of mid-September news stories, a more critical evaluation came from someone who had once known Barack far better than even Jarrett. On September 20, reports quoted Jesse Jackson Sr. as privately complaining that Barack was "acting like he's white" given his disinterest in a racial controversy in Jena, Louisiana. Later that day the Senate debated a resolution denouncing a MoveOn.org ad that had mocked General David Petraeus as "General Betray Us." Barack declined to vote for or against it, saying that "the focus of the United States Senate should be on ending this war, not on criticizing newspaper advertisements. . . . By not casting a vote, I registered my protest against this empty politics." A *Washington Post* politics blog ran a story on the contretemps, and by 11:00 A.M. the following morning logged-in readers had posted eleven comments on the report. Then came a twelfth:

> *I can't believe the excuses the Obamamanics give their gutless leader. He should have taken a stand—yea or nay—on the Moveon.org ad. He claims he's above such politics but his decision not to vote was motivated by politics: he doesn't want to alienate the Moveon.org people since he needs to compete in the Democratic primaries, but he still wants to be seen as a moderate in the general elections. It was a political move, not a courageous one. Leadership is about courage. If he's too pure for dirty politics, then what does he want to be President for? And where was he in Jena, by the way??*

The commenter signed off as "Recovered Obamamanic," but her Web log-in showed that it was "Posted by: sheila.jager." Six years had passed since she and Barack had last been in touch; Barack had launched his presidential campaign without contacting her. Sheila later explained, "I remember being really, really upset at this time over the 'Betray Us' ad and really mad at Barack for not being courageous enough to smack them down." Courage had been "a big issue between us"; "it was a passing fury, though. I've forgiven him since."[9]

Barack's campaign manager David Plouffe admitted to one reporter that winning "Iowa—that's the whole shebang!" before adding, "I guess I'm not supposed to say that." On October 11, all of Barack's top aides gathered in Chicago for a comprehensive review of strategy and messaging. Voters wanted a president "who can unite the country and restore our sense of purpose," an overview memo said. Barack's "change you can believe in" slogan was aimed at creating a character contrast with Hillary Clinton, who was "driven by political calculation, not conviction." Appearing on Tavis Smiley's television show, Barack acknowledged that for many people his appeal "has to do less with the positions I'm taking than the tone I'm taking."

On October 30, Barack turned in his best debate performance to date. That same day, however, Sheila Jager weighed in with yet another biting critique, this time in response to a *Washington Post* story detailing how Barack "has changed his position" on Social Security since last May. "Obama had moved from being open to a solution that might include raising the retirement age or indexing benefits to prices rather than wages . . . to one of making the protection of benefits one of his three core principles," the *Post* stated. "Obama's decision to wall off benefits might be interpreted as a political calculation," and "Obama now owes it to voters to explain his own evolution." Sheila was similarly disappointed:

> *One cannot help but feel Obama's desperation, especially when his stance on the issues are now presented in stark contrast to what he said before (e.g. on Iran and now Social Security). Does he think that the electorate is too dumb to remember his earlier pronouncements or that we cannot see political maneuvering when he does it in our face? Is this the politics of hope? What happened to the Obama of old? Where did he go????*[10]

Barack took the next evening off to go trick-or-treating with his daughters in Hyde Park, a Halloween mask disguising his identity from most passersby. Old CCRC colleague Bob Klonowski nonetheless recognized him, and Barack jokingly explained, "it's the only way I can go out now." The *Chicago Sun-Times*' Lynn Sweet mounted an insistent effort to find out what had become of Barack's state Senate records that had been discarded three years earlier, but a far bigger milestone was Barack's energetic performance at Iowa Democrats' all-important Jefferson-Jackson Dinner on November 10. "Fucking home run," David Axelrod told David Plouffe.

On *Meet the Press* the following morning, Barack seemingly bragged about his friendship with billionaire investor Warren Buffett, who "made $46 million last year" while acknowledging that "money is the original sin in politics and I am not sinless." Host Tim Russert asked him about the discarded state Senate records and about Tony Rezko, whom Barack said he had not spoken with since his indictment. Talking with a writer for Oprah Winfrey's *O Magazine* who was profiling Michelle, Barack explained, "I'm more easygoing than she is" and "I don't get as tense or stressed. I'm probably more comfortable with uncertainty and risk." He also expressed pleasant surprise at how much campaigning Michelle was doing. "She has been much more enthusiastic than I expected, much more engaged and involved." Michelle's friend Cindy Moelis agreed, noting what a change it was from Barack's previous races. "This is the first time in all these elections that I've seen her be as passionate and committed as she has been."

In a long interview on MSNBC, Michelle talked about how well her daughters were doing thanks to their grandmother Marian Robinson. "This wouldn't be possible without her," and the girls were happy "to stay in their world, in their lives. They want to be with their friends, and we allow that to happen." For Michelle and Barack, "we always wonder whether politics changes people, and one of the things we've desperately tried to do is not to allow our political lives to change who we are fundamentally." Michelle also stressed that "the point isn't winning, it's changing the country, it's changing America, it's changing the way you live. It's throwing this game out, shaking it up and throwing it out the window, because it's not just playing it better than the people who played it before."

Michelle also spoke about the pervasiveness of black self-doubt. "There's always that doubt in the back of the minds of people of color . . . deep down inside you doubt that you can really do this because that's all you've been told is 'No, wait.' That's all you hear and you hear it from people who love you, not because they don't care about you and because they're afraid. . . . I would not be where I am" and "neither would Barack if we listened to that doubt. . . . There are a lot of kids who I know aren't pushing themselves or going for what they know they can do because of that doubt." But "Barack has been doing stuff he's not supposed to, I'm doing stuff that people told me I wasn't supposed to do—that's my whole life!"[11]

A few days before Christmas, Barack answered a question from CBS's Harry Smith by explaining that the "last thing I do at night, I say a prayer. I say a prayer. I ask that my family is protected and that I'm an instrument of God's will." With the all-important Iowa caucuses less than two weeks away, on January 3, Barack felt confident that David Plouffe's all-Iowa strategy was building toward an upset victory. Barack stepped away from the campaign trail for just thirty-six hours to celebrate Christmas at home in Chicago, with his sisters Maya and Auma and their daughters plus Maya's husband joining the Obamas for a non-Hawaii gathering. On Christmas morning they invited Barack's Secret Service detail in as the girls opened presents, and following dinner at a downtown hotel they played their annual game of Scrabble. Maya described to NPR how Barack won a come-from-behind victory, "which was maddening, and then he gloats."

Barack returned to Iowa on December 26 with all of his relatives in tow. On New Year's Eve, Michelle told a crowd in Grinnell that "Barack is one of the smartest people you will ever encounter who will deign to enter this messy thing called politics," and on Thursday evening, January 3, an "extraordinary turnout" of Iowa Democrats—239,000 voters as compared to 124,000 in 2004—carried Barack to a decisive 38 percent victory. John Edwards barely edged Hillary Clinton for second place, 29.8 to 29.5 percent. Barack gloried in his triumph. "Thank you, Iowa . . . they said this day would never come. . . . But on this January night, at this defining moment in history, you have done what the cynics said we couldn't do. . . . We are one nation. We are one people. And our time for change has come . . . to move beyond the bitterness and pettiness and anger that's consumed Washington." Barack vowed to be "a president who finally makes health care affordable and available to every single American, the same way I expanded health care in Illinois, by bringing Democrats and Republicans together to get the job done," a claim that prompted double takes by Illinois reporters. "My journey began on the streets of Chicago" and "sometimes, just sometimes, there are nights like this, a night that, years from now," when America is "a nation less divided and more united, you'll be able to look back with pride and say that this was the moment when it all began. . . . This was the moment. Years from now, you'll look back and you'll say that this was the moment, this was the place" for people "who are not content to settle for the world as it is, who have the courage to remake the world as it should be."[12]

Forty-eight hours later, during a Democratic debate, Barack told Hillary Clinton, "You're likable enough," a choice of words that led many commentators to criticize what one called Barack's "dismissive condescension" toward the female front-runner. Two days later, on ABC's *Good Morning America,* Barack told Diane Sawyer, "I'm doing this stuff not for me, but for something bigger than me. There's an aspect of politics that's about me, but then there's an aspect of politics that is larger than me. And you have to have a certain amount of megalomania to think that you should be president of the United States, but I think you have to cross a certain threshold when you say this is not about my ambition and it's about something bigger." Responding with irritation to Clinton's assertions that he was too inexperienced to be president, Barack used a basketball analogy in telling *Newsweek*'s Richard Wolffe that "at some point people have to stop asserting that because I haven't been in the league long enough I can't play. It's sort of like Magic Johnson or LeBron James—[they] keep on scoring thirty, and their team wins, but people say they can't lead their team because they're too young."

Nonetheless, when New Hampshire Democrats went to the polls on January 8, Clinton squeaked by with a surprising seventy-five-hundred-vote win over Barack, defeating him 39.1 to 36.5 percent, with Edwards a distant third. Clinton likewise won a narrow victory in Nevada's caucuses, while Barack triumphed in South Carolina, and with Edwards's exit from the race the Democratic contest moved toward the twenty-three states that would select 1,681 Democratic convention delegates on "Super Tuesday," February 5. In a pair of interviews with religious periodicals, Barack spoke of "the sacredness of sex" while distancing himself from Jeremiah Wright, who was scheduled to give his final pre-retirement sermon at Trinity on February 10. "Sometimes he's provocative in ways that I'm not always comfortable with and in ways that I deeply disagree with occasionally." Yet "I am proud of Reverend Wright and what he's done in his life." Michelle spoke similarly, saying "there are plenty of things he says that I don't agree with, that Barack doesn't agree with." But "your pastor is like your grandfather," she added. "You can't disown yourself from your family because they've got things wrong."

After Hillary Clinton brought up Barack's long relationship with Tony Rezko during a televised debate, Barack faced a new gauntlet of questions about their friendship. "Nobody had any indications that he was engaged in wrongdoing" while he was still interacting with Tony, Barack claimed, yet both Chicago newspapers had begun publicizing Rezko's dubious dealings in mid-July 2004, months before the Obamas' purchase of 5046 South Greenwood. In late January the *Tribune* endorsed Barack for the Democratic presidential nomination while nonetheless stating that Barack's account concerning Rezko "strains credulity."

On Super Tuesday, the two Democratic finalists battled to a tie, with Barack and Clinton each winning over eight hundred delegates. Throughout a dozen mid-February primaries and caucuses, Barack won modest but consistent delegate gains over Clinton. *New York Times* columnist Paul Krugman warned that "the Obama campaign seems dangerously close to becoming a cult of personality," and in the *Washington Post* Robert Samuelson observed that Barack "seems to have hypnotized much of the media and the public with his eloquence and the symbolism of his life story. The result is a mass delusion that Obama is forthrightly engaging the nation's major problems," for in actuality "the contrast between his broad rhetoric and his narrow agenda is stark." When Tony Rezko's criminal trial began on March 3, questions again intensified, and Barack ended one tense press conference that campaign manager David Plouffe called "a disaster" by saying, "we're running late."[13]

On March 4, Hillary Clinton bested Barack in both Ohio and Texas, and the two candidates remained neck-and-neck in the delegate race with the next big primary, in Pennsylvania, seven weeks away. Then, early on March 13, ABC News reported that Jeremiah Wright's videotaped sermons that were available for purchase at Trinity Church included both his September 16, 2001, one in which he had said that the 9/11 attacks reflected how "America's chickens are coming home to roost," and his April 13, 2003, one declaring "God damn America." Barack first learned of that quotation during a phone interview with a Pittsburgh reporter. "I haven't seen the line" and "I profoundly disagree," but this is "what happens when you just cherry-pick statements from a guy who had a forty-year career as a pastor." Back in Chicago, campaign manager David Plouffe was aghast. "We were seeing the clips for the first time" and the campaign's failure to research what the public might learn about Wright "was unforgivable," Plouffe realized. By Friday morning, more and more attention was focusing on the Wright tapes, but Barack was scheduled to have a pair of hour-plus meetings that afternoon with first the *Tribune* and then the *Sun-Times* to finally answer each and every question the newspapers wanted to put to him about Tony Rezko. Barack gave each paper a set of documents detailing the purchase of 5046 South Greenwood, and he told the *Tribune* group that his campaign has "a mentality of let's be very protective of information." With the *Sun-Times* journalists, Barack recalled Tony as "a very gracious individual," but Barack's willingness to respond to every question the two papers asked about Rezko represented a major turn in the road. "Obama should have had Friday's discussion 16 months ago," the *Tribune* editorialized, and *Sun-Times* columnist Mark Brown observed that "Obama can be too obstinate when he feels under attack."

Soon after Barack left the *Sun-Times,* his campaign had a comment, "On My Faith and My Church," up on *Huffington Post*. "I vehemently disagree and strongly condemn" Wright's statements, and "I categorically denounce any statement that disparages our great country." The two video clips "were not statements I personally heard him preach" or "utter in private conversation." That evening, Barack spoke with MSNBC's Keith Olbermann, Fox News' Major Garrett, and CNN's Anderson Cooper, emphasizing that same point: "they were not statements that I heard when I was in church." But "I would not repudiate the man," and "one thing that I hope to do is to use some of these issues to talk more fully about the question of race in our society." Barack had long been wanting to deliver a major speech about race, and with campaign manager David Plouffe realizing that the I-wasn't-there-that-day explanation "was a woefully inadequate response," he and David Axelrod reluctantly concurred when Barack later that evening said the time for that speech had come. "Wright will consume our campaign if I can't put it into a broader context," Barack told Plouffe, who responded that time was of the essence. With Barack campaigning in Pennsylvania in advance of that state's April 22 primary, Plouffe recommended Philadelphia's National Constitution Center, on Tuesday morning—just four days hence. Barack would have to devote every offstage moment to crafting the speech.[14]

In front of a small audience of supporters and against a backdrop of eight American flags, Barack stressed that Wright "has been like family to me," saying "I can no more disown him than I can disown my white grandmother . . . a woman who loves me as much as she loves anything in this world, but a woman who once confessed her fear of black men who passed her by on the street, and who on more than one occasion has uttered racial or ethnic stereotypes that made me cringe." Seeking to place both Wright and Madelyn Dunham in the broader context of America's painful racial history, Barack painted a broad canvas of resentment and reconciliation. Immediate praise came from former Democratic presidential candidate Senator Joe Biden, who termed the address "an important step forward in race relations." Laudatory editorials quickly followed, with the *Washington Post* calling the speech "compelling," "eloquent," and "an extraordinary moment of truth-telling." The *New York Times* labeled it "powerful and frank," reflecting "an honesty seldom heard in public life." Obama's "eloquent speech should end the debate over his ties to Mr. Wright," the *Times* opined, and "it is hard to imagine how he could have handled it better."

The political impact of Barack's address suggested that the speech had remedied the controversy that had threatened to torpedo his campaign, but listeners in Honolulu did not share the *Post* and *Times*' enthusiasm. Maya Soetoro-Ng declined comment to a reporter, but a family friend explained that "she was furious" over Barack's comments about their grandmother. His characterization of her "seemed completely gratuitous, even cruel," one critic rightly thought, and one of the late Stanley Dunham's close friends acidly remarked, "I have a far better opinion of both Stan and Madelyn than I have heard from Barry's speeches." His anger was shared by "the people who worked for Madelyn" during her years at the Bank of Hawaii, who thought Barack's "disgusting" remarks represented "absolute pandering." To them, Barack represented the utter fulfillment of Stan and Madelyn's dreams. "When folks look at Barack Obama, most people see a black man running for president. I see Stan Dunham," his friend Rolf Nordahl explained.[15]

Barack created a minor controversy when he called Madelyn "a typical white person" during a Philadelphia interview while denying that "my grandmother harbors any racial animosity." Then remarks that Barack made at a private fund-raiser in San Francisco drew criticism because he had cited "bitter" Pennsylvanians who "cling to guns or religion." In Pittsburgh, Barack asserted that as a community organizer, "block by block, we helped turn those neighborhoods around," a claim that no one in Roseland or Altgeld Gardens would affirm. To ABC's Terry Moran, Barack confessed that "during the course of campaigning" "sometimes you lose" your "core truthfulness" and "it becomes a performance," as it had that day in Pittsburgh. Now half empty, Altgeld "reeks of decay, fear and despair," one journalist wrote, and Hazel Johnson's daughter Cheryl stated that "Obama left no real legacy behind."

In early April, Barack reached out to Jeremiah Wright and got a blistering reception. "I cussed him out. 'You pronounced judgment on my sermon and you never heard the sermon. You don't know what the sermon was about 'cause you didn't hear it.'" Barack asked to speak with Wright in person at a "secure location," and on Saturday, April 12, Barack went to Wright's home in suburban Tinley Park. It was "just Barack and me. I don't know if he had a wire on him," but the reason for Barack's visit soon became clear: he told Wright, "I really wish you wouldn't do any more public speaking until after the November election." Wright had upcoming engagements in Detroit and Washington, and he dismissed the request: "How am I supposed to support my family? I have a daughter and a granddaughter in college whose tuition I pay. I've got to earn money." Barack was insistent. "I really wish you wouldn't. The press is going to eat you alive." Barack left empty-handed, but before long Wright received an e-mail from Barack's close friend Eric Whitaker, also a Trinity member, offering Wright $150,000 "not to preach at all" in the months ahead. Wright refused and mentioned the offer to his friend Father Mike Pfleger, and "when Father Mike went off on Eric, Eric went back and told Barack," Wright explained.

On April 16, just hours before the final televised debate preceding Pennsylvania's April 22 vote, Barack's campaign released his and Michelle's 2007 Form 1040, showing that he had received more than $4 million in royalty income the previous year. The Obamas had upped their charitable giving to a total of $240,000, including $26,270 to Trinity Church, and owed the IRS $1,059,000. At that night's debate, ABC News moderators Charles Gibson and George Stephanopoulos pressed Barack about his "bitter" and "cling" remark, and repeatedly about Reverend Wright, before asking Barack what his relationship was with long-ago Weather Underground bomber turned education professor Bill Ayers. Barack called Bill "a guy who lives in my neighborhood . . . who I know," who's "not somebody who I exchange ideas from on a regular basis." Ayers had "engaged in detestable acts forty years ago, when I was eight years old," but the implication that that "somehow reflects on me and my values, doesn't make much sense, George. The fact is, is that I'm also friendly with Tom Coburn, one of the most conservative Republicans in the United States Senate, who during his campaign once said that it might be appropriate to apply the death penalty to those who carried out abortions. Do I need to apologize for Mr. Coburn's statements? Because I certainly don't agree with those either."[16]

Fortunately for Barack, his comparison of Senate colleague Coburn to the now-controversial Ayers, similar to his pairing of Madelyn and Wright in his Philadelphia speech, attracted little press comment, and Barack called Coburn to apologize. "I said, 'Fine with me,'" Coburn recalled, but he still wondered, "Why answer a question by throwing a friend under a bus?" Come April 22, Hillary Clinton easily won Pennsylvania, defeating Barack 55 to 45 percent, but a more ominous problem loomed when Jeremiah Wright returned to the public stage on April 25. Speaking with sympathetic television host Bill Moyers, Wright argued that "we have supported state terrorism against the Palestinians" and complained that the media "paint me as some sort of fanatic" who is "un-American" and has "a cult at Trinity United Church of Christ. . . . I felt it was unfair. I felt it was unjust. I felt it was untrue," but it was being disseminated to millions of Americans "who know nothing about the prophetic theology of the African American experience." When Moyers asked about Barack's Philadelphia speech, Wright answered that Barack "does what politicians do," a comment that Barack felt impugned his motives. "How could he say that about me?" Barack asked Valerie Jarrett. "He knows that's not true. He knows I wasn't being a politician."

Speaking with Fox News' Chris Wallace, Barack mentioned his private conversation with Wright and his apology to Coburn, "who I consider a close friend." Later that day, speaking to the Detroit NAACP, Wright contended that African American and European American children learn differently, with the former having a "right-brained, subject-oriented" learning style, and the latter "a left-brained, cognitive object-oriented learning style." Then, the next morning, Wright, his wife, and his eldest daughter joined a trio of black journalists for an appearance at the National Press Club.

"This is not an attack on Jeremiah Wright, it is an attack on the black church," Wright asserted. "Black preaching is different from European and European American preaching. It is not deficient. It is just different. It is not bombastic. It is not controversial. It's different." Wright reiterated his notion that "black learning styles are different from European and European American learning styles," but far more usefully mentioned that Trinity has "sent thirty-five men and women through accredited seminaries." When the session turned to Q&A, Wright brushed aside some inquiries. "You haven't heard the whole sermon? That nullifies the question." He noted, "I served six years in the military. Does that make me patriotic?" Returning to his main point, Wright insisted that "this is not an attack on Jeremiah Wright. It has nothing to do with Senator Obama. This is an attack on the black church." He recklessly declared that Louis Farrakhan "is one of the most important voices in the twentieth and twenty-first century," adding that "I said to Barack Obama last year, 'If you get elected, November the 5th I'm coming after you, because you'll be representing a government whose policies grind under people." Then Wright made reference to two books, *Emerging Viruses,* by Leonard Horowitz, whom serious scholars viewed as an "AIDS conspiracy theorist," and Harriet Washington's *Medical Apartheid,* a credible, mainstream work of history detailing how government programs had victimized black patients.

By nightfall, excerpts of Wright's comments flooded cable news. In Chapel Hill, North Carolina, Barack knew he would have to denounce Wright far more bluntly than before if his candidacy was to survive. "The person that I saw yesterday was not the person I met twenty years ago," Barack told a Winston-Salem press conference on Tuesday. Still smarting from Wright's characterization of his Philadelphia speech, Barack said "he doesn't know me very well," and, given his Monday remarks, "I may not know him as well as I thought, either." Wright "has done enormous good in the church. He has built a wonderful congregation. The people of Trinity are wonderful people, and what attracted me has always been their ministries' reach beyond the church walls." But Wright's embrace of AIDS conspiracies and Louis Farrakhan are "ridiculous propositions" which "should be denounced. And that's what I'm doing very clearly and unequivocally here today. . . . When I say that I find these comments appalling, I mean it. It contradicts everything that I am about and who I am."

Replying to a question, Barack explained that "upon watching it, what became clear to me was that it was more than just defending himself. What became clear to me was that he was presenting a worldview that contradicts who I am and what I stand for. And what I think particularly angered me was his suggestion somehow that my previous denunciation of his remarks were somehow political posturing." Furthermore, "the insensitivity and the outrageousness of his statements and his performance in the question and answer period yesterday, I think, shocked me. . . . In some ways what Reverend Wright said yesterday directly contradicts everything that I've done during my life." Uttering "a bunch of rants that aren't grounded in truth" was "a show of disrespect to me," and given how Wright "caricaturized himself . . . that made me angry, but also made me saddened."[17]

The forcefulness of Barack's statements allayed the burgeoning controversy. The *Washington Post* thought "the whole sorry episode raises legitimate questions about his judgment," but the *New York Times* saluted Barack's "powerful, unambiguous denunciation" as "the most forthright repudiation of an out-of-control supporter that we can remember." Barack and his aides were deeply fearful of the impact Wright's behavior would have in the decisive Indiana and North Carolina primaries on May 6, and on NBC's *Meet the Press* Barack again addressed the issue. By employing "divisive, hateful language," Wright demonstrated "that he did not share my fundamental belief and my fundamental values in terms of bringing the country together and moving forward." Barack confessed, "I feel a great loyalty to that church," yet with Wright he insisted that "I never sought his counsel when it came to politics. . . . My commitments are to the values of that church, my commitment is to Christ, it's not to Reverend Wright." To *Newsweek,* Barack even claimed that "I cannot recall a time where he and I sat down and talked about theology or we had long discussions about my faith."

The night before Indiana and North Carolina voted, Barack was as despondent as Valerie Jarrett had ever seen him. "How could someone I knew, sometime I trusted, do this to me?" Barack vented, fearing that Wright might have sunk his candidacy. When David Axelrod reported that polling showed Barack losing Indiana and neck-and-neck in North Carolina, Barack responded acerbically. "Get Axelrod out of here. He's a downer." But twenty-four hours later, the actual vote totals were vastly better: Barack won North Carolina by 15 points, and Hillary Clinton's margin over him in Indiana was just a single point. The overall delegate race remained tight, but now it looked virtually certain that Barack could narrowly clinch the Democratic nomination in four weeks' time, following the last six small-state primaries.

In private, Jeremiah Wright reassured old friend Tim Black, "I support Barack one hundred percent. I believe in his vision, and I believe that he is the best qualified candidate to offer us the hope of going in a different direction as a country." Across black Chicago, scores of people who knew both men were pained beyond words, for as one of the city's wisest African American voices emphatically emphasized, "Jeremiah Wright was the black male father figure for Barack," someone whose long-term influence upon Barack should not be underestimated. Then, on the last Sunday in May, Father Mike Pfleger mocked Hillary Clinton while guest-preaching at Trinity, and Barack's campaign was forced to disassociate him from those remarks too. Two days later, Barack announced that "with some sadness," he and Michelle were withdrawing as members of Trinity. "This was one I didn't see coming," Barack told reporters.[18]

On June 3, as expected, delegates won in the two final Democratic primaries, in Montana and South Dakota, allowed Barack to clinch the Democratic nomination. The final lines of Barack's victory speech were remarkable:

> *I am absolutely certain that generations from now, we will be able to look back and tell our children that this was the moment when we began to provide care for the sick and good jobs to the jobless; this was the moment when the rise of the oceans began to slow and our planet began to heal; this was the moment when we ended a war and secured our nation and restored our image as the last, best hope on Earth. This was the moment—this was*

the time—when we came together to remake this great nation so that it may always reflect our very best selves, and our highest ideals

Very few U.S. newspapers took note of those words, but Barack also took the time to call Madelyn Dunham in Honolulu. "She just said she was really proud," Barack told ABC's Charles Gibson the next morning. "She's going blind, so to hear in her voice what this meant to her, that was a pretty powerful moment." Madelyn "is a very steady person," Barack told another interviewer. "She never gets too high, and never gets too low. And I think a lot of my temperament comes from her." In a speech discussing how he had grown up fatherless, Barack said, "I resolved many years ago that it was my obligation to break the cycle—that if I could be anything in life, I would be a good father to my children."

In a mid-June videotaped message, Barack announced that his campaign would reject public financing, which "was set up to reduce the influence of private donations in the political process," a front-page *New York Times* story noted. "His decision to break an earlier pledge to take public money will quite likely transform the landscape of presidential campaigns, injecting hundreds of millions of dollars into the race and raising doubts about the future of public financing for national races." The *Times* added that "the decision means that Mr. Obama will have to spend considerably more time raising money . . . at the expense of spending time meeting voters." *Times* columnist David Brooks eviscerated Barack, writing that "Obama didn't just sell out . . . the primary cause of his professional life" to win "a tiny political advantage" over Republican John McCain, but did so in "a video so risibly insincere that somewhere down in the shadow world" the notorious late Republican hatchet man "Lee Atwater is gaping and applauding."[19]

In early July, Barack told a religious magazine that "we can prohibit late-term abortions" so long as "there is a strict, well-defined exception for the health of the mother. Now, I don't think that 'mental distress' qualifies as health of the mother. I think it has to be a serious physical issue that arises in pregnancy, where there are real, significant problems to the mother carrying that child to term." Citing both that and the campaign finance U-turn, *Chicago Tribune* columnist John Kass reminded readers that "Obama is a Chicago politician" who "has flip-flopped again and again." Progressive Chicago author Paul Street stressed how in Springfield, "Obama was more interested in having his name associated with resume-padding legislative victories than with attaining 'progressive' victories."

When a pair of interviewers from a women's magazine asked Barack about Michelle earning more than him thanks to her high-paying U of C Medical Center job, Barack replied that "that wasn't much of an issue, because the truth is, for eleven out of the thirteen years we were married, I was making substantially more than her, and during the two years that I wasn't, I was running for the United States Senate, which had its own gratifications. So it wasn't as if I felt inadequate. . . . I didn't feel threatened by that at all," Barack insisted.

The weeks leading up to the Democratic National Convention in Denver at the end of August featured scores of news features examining various aspects of Barack's life, but journalists found Barack's campaign far from helpful. A *New York Times* reporter complained that "they attacked me like I'm a political opponent" when one of his stories drew the campaign's ire, and *The New Republic* noted that communications director Robert Gibbs "has built a particularly large reservoir of ill will" among journalists. "Reporters who have covered Obama's biography or his problems with certain voter blocs have been challenged the most aggressively," Gabriel Sherman wrote. "'They're terrified of people poking around Obama's life,' one reporter says. 'The whole Obama narrative is built around this narrative that Obama and David Axelrod built, and, like all stories, it's not entirely true. So they have to be protective of the crown jewels.'"[20]

As Barack's nomination drew near, more and more journalists attempted to gauge his character. *New York* magazine noted that thanks to Barack's "calm, collected nature . . . he has been able to completely defang the stereotype of the angry black male." In *The New Republic,* John B. Judis, who had interviewed Barack's community organizing mentors, warned that Obama is "a politician of vision, not interests" and has "fashioned a political identity in near-total opposition to the core principles of his one-time profession." Former U of C Law School colleague Cass Sunstein pronounced Barack "a minimalist" whose "pragmatism is heavily empirical." In one of the most perceptive readings ever of *Dreams,* the well-known writer David Samuels pinpointed Barack's "distanced and writerly view of a self as something that is summoned through a creative act of will." Noting "the painful process of self-formation" that Barack went through, Samuels highlighted "Obama's uncharitable treatment of his white family" and "his distaste for his mother" while pointing out the "many structural parallels" that *Dreams* shared with Ralph Ellison's *Invisible Man.* Barack claimed to *Time* that as an undergraduate, "I was reading Milton Friedman and Friedrich Hayek," famous conservative economists, and Barack told the *Tribune* that "I had to scratch and claw my way to the point I am now, and I think I've done so without cutting corners or compromising my integrity."[21]

On August 23, Barack announced his selection of Senate Foreign Relations Committee colleague Joe Biden as his vice presidential running mate, and five days later Barack accepted the Democratic Party's nomination for president of the United States. As his campaign against the Republican ticket of Senator John McCain and Alaska governor Sarah Palin got under way, Barack spoke haltingly with ABC's Diane Sawyer about the absence of his father from his life. What would Barack ask him? Sawyer queried. "I'd like to ask him what he felt about his wives. There was no sense of real commitment there. Not just me, but many of his subsequent wives and their children. And I'd like to know what it was that led him to miss that aspect of what a marriage should be about."

With grandmother Madelyn Dunham dying of cancer and requiring surgery for a broken hip, Barack broke off from campaigning on October 23 to make the long flight to Honolulu to see her in the same apartment where he had grown up. "I'm still not sure whether she makes it to Election Day," Barack told ABC's *Good Morning America*. "I wanted to make sure . . . that I had a chance to sit down with her and talk to her. She's still alert and she's still got all her faculties, and I want to make sure that I don't miss that opportunity right now," as he had thirteen years earlier with his mother.

Ten days later, early on Monday morning, November 3, Madelyn died in her apartment at age eighty-six, with Maya Soetoro-Ng by her bedside. Maya called Barack with the news, and at a campaign rally in Charlotte, North Carolina, Barack spoke somberly about her role in his life. Madelyn's body was cremated, and the next evening, Maya watched the election returns with Madelyn's urn by her side. At 11:00 P.M. the networks declared Barack the next president of the United States, and the final tally showed him defeating John McCain by almost 10 million votes, 53 to 46 percent. Barack carried twenty-eight of the fifty states, with his Electoral College margin over McCain a whopping 365 to 173.[22]

In New York, former Business International Corporation vice president Lou Celi spoke for everyone who had known Barack prior to 1985. "There was no clue. . . . You would never think in a million years that this was going to be the president of the United States." In Chicago, Emil Jones Jr. thought back to how Alice Palmer's supporters had filed such inept petition signatures in December 1995. "That was a fatal mistake which would change the course of history." Across Illinois, multiple Democrats thought about someone who was absent from election night celebrations. "Barack Obama is not where he is today without Dan Shomon, and I don't think anyone can or will dispute that." Another Obama supporter put it more pungently. "If it wasn't for Dan Shomon, Barack Obama wouldn't be shit." In San Francisco, DNC speechwriting coordinator Vicky Rideout recalled July 27, 2004: "if it weren't for that moment, he wouldn't be president."

In Chicago, David Plouffe and David Axelrod knew that January 3, 2008, had been crucial. If Clinton "had been able to prevent us from winning Iowa, she would have been the nominee," Plouffe explained. The Iowa victory had been essential in a second way too: at the campaign's outset, "the biggest race problem we had to start was not with the white voters," Axelrod felt, "but with African American voters" who had "a deep sense of skepticism" that a black candidate could win. Among black observers, there was little disagreement about why Barack had won the support of 43 percent of white voters. Barack's "racial idealism" touched "a deep longing . . . on the part of whites to escape the stigma of racism," and many whites "literally longed for ways to disprove the stigma" by supporting Barack. "Voting for Obama was like reparations on the cheap," African American law professor Richard Thompson Ford cogently explained. Yet even conservative commentator Shelby Steele predicted that "the profound disparity between black and white Americans . . . will persist even under the glow of an Obama presidency."

Barack immediately was faced with a major economic crisis he would be inheriting from George W. Bush, as well as the need to make scores of crucial personnel decisions. Appearing on CBS's *60 Minutes* alongside Michelle, Barack pledged, "I intend to close Guantánamo, and I will follow through on that"; in a subsequent interview, Barack also emphasized "my broad commitment: no torture." On December 20, the Obama family flew to Honolulu for their annual holiday getaway, and three days later, following a memorial service, Barack and Maya traveled to Lanai Lookout to scatter Madelyn Dunham's ashes into the Pacific at the same spot where they had scattered Ann's in 1995.[23]

In advance of Barack's inauguration on January 20, 2009, former Harvard Law classmate Frank Hill Harper wrote in *Essence* that "President Obama has sparked a transformation in the psyche, self-esteem and aspiration of young black males. His positive impact will be seen for generations to come," but on inauguration day itself, some of Barack's most important black supporters, like Chicago's Jim Reynolds, were all but left in the lurch. "Do you realize that Jim took a bus to the inauguration?" another unhappy black donor complained. Some of Barack's old friends and advisers, like Berkeley Law dean Chris Edley, had strongly opposed his decision to name Chicago congressman Rahm Emanuel as White House chief of staff, and even in the first week of Barack's presidency it became starkly apparent that "Rahm wanted Obama out in public constantly," as David Axelrod put it. Hardly a week passed before *Wall Street Journal* columnist Peggy Noonan nailed the problem. "When the office is omnipresent, it is demystified. Constant exposure deflates the presidency, subtly robbing it of power and making it more common." Robert M. Gates, whom Barack had asked to stay on as secretary of defense, soon came to agree. "When it comes to the media, often less is more. . . . If one appears infrequently, then people pay more attention when you do appear." In time Axelrod concurred. "We used him way too much. It robbed him of his power as the narrator of a larger story."

By March, close observers worried that "Obama's White House was slipping into a kind of dysfunction," with Emanuel and National Economic Council director Larry Summers, the former president of Harvard University, playing more dominant roles than Barack. "The president was concerned about showing his uncertainty, or his lack of acquired knowledge on lots of these policies, to his own advisers," one of them told Pulitzer Prize–winning journalist Ron Suskind. On March 11, Barack reluctantly signed into law the $410 billion Omnibus Appropriations Act of 2009, which included more than eighty-five hundred congressional "earmarks" totaling over $7.7 billion. The earmarks had a richly deserved reputation for reflecting what the *Washington Post* called "a connection between campaign contributions and spending programs," and the *Post*'s front-page news story called Barack's signing of the bill "an early blow to his attempt to change how business is done in Washington." The *Wall Street Journal* reported that David Axelrod had strongly but unsuccessfully argued for a veto, with Emanuel and House Speaker Nancy Pelosi prevailing. The *Post* detailed how "Obama backed away from bolder proposals" for earmarks reform advocated by Senate reformer Russ Feingold, and NBC News' Chuck Todd rued how "one of Obama's core principles was jettisoned for the sake of expediency."[24]

Speaking with Steve Kroft of CBS's *60 Minutes,* Barack revealed that "the hardest thing about the job is staying focused." He was very concerned about "making sure that . . . we start getting a handle on our long-term structural deficit," a point he reiterated to C-SPAN's Steve Scully. "The long-term problem is Medicaid and Medicare. If we don't reduce long-term health care inflation significantly, we can't get control of the deficit." On a brighter note, living above the office could not be beat. "The White House has been terrific for our family life compared to some of our other previous situations like campaigns, because we are all in the same place," and each evening Barack could go upstairs at six thirty for dinner with his daughters before retiring to his study.

On Monday night, May 25, Barack made one of his most consequential decisions when he called federal appellate judge Sonia Sotomayor to tell her that on Tuesday he would nominate her for the Supreme Court seat from which Justice David H. Souter was retiring. "He asked me to make him two promises. The first was to remain the person I was, and the second was to stay committed to my community," Sotomayor recalled a few months later, after winning Senate confirmation and being sworn into office.

By mid-July Barack had turned his attention toward his long-envisioned health care reform measure, telling ABC's Terry Moran that "we can get this done by the fall" and that "we also do want to reform the insurance industry." Polls showed that over the course of Barack's first six months in the White House, his disapproval rating had risen from an initial 20 percent to 39, but journalists continued to be impressed—and worried—by the sort of support that Barack attracted. *The Economist*'s Lexington column stated that "Obama has inspired more passionate devotion than any modern president," but warned of how "the personality cult that surrounds him . . . has stoked expectations among its devotees to such unprecedented heights" that disappointment was certain.

Inside the West Wing, decision-making remained troubled, with a *Wall Street Journal* headline calling Barack "a micromanager." The former law professor would ask aides "to do justice to a view other than their own," saying "I want more than what's in this room" when they could not do so. But "the president had lost control of his White House," Ron Suskind learned from his interviewing. Obama "had almost no process to translate his will into policy on the occasions when he could decide on a coherent path. But such decisions were rare," and the tenor of debate was often fouled by chief of staff Rahm Emanuel, who decried liberals as "fucking retards" in one large meeting. "We are treated as though we are children," a leader of one progressive group told a reporter.[25]

On Thursday morning, August 6, in Oberlin, Ohio, Sheila Miyoshi Jager opened her e-mail. Sheila had long worried that her identity could prove "politically explosive given the nature of our relationship," but now she was happy to reminisce about Barack. "So you see, this is really, really sensitive stuff" since "we did not go our separate ways after 1988" and "I do not think Michelle knows."[26]

In mid-September, the administration's health reform plan, the Affordable Care Act, was introduced into Congress, and Barack told CBS's Steve Kroft that "once this bill passes, I own it. And if people look up and say, 'You know what? This hasn't reduced my costs. My premiums are still going up 25 percent, insurance companies are still jerking me around,' I'm the one who's going to be held responsible." Barack again noted that "the only way I can get medium- and long-term federal spending under control is if we do something about health care." On November 7, the Affordable Care Act was approved by the House of Representatives, but passage came on a vote of 220 to 215, with only one Republican in favor and 39 Democrats opposed.

Writing in *New York* magazine, John Heilemann lamented how the White House had "badly botched the job of presenting reform to the country" and seemed "more interested in passing something, anything, no matter how ineffectual, that could be labeled reform than in making sound policy." As one Democratic senator told Heilemann, "We'll get a health care bill, but we've squandered the ability to change America."

On December 1 Barack approved a "surge" of thirty thousand U.S. troops in Afghanistan. Barack told Steve Kroft that authorizing that deployment was "absolutely" the most difficult decision he had made as president. In Afghanistan, his hope was "to scale this down to a point where it is manageable and we are protecting American lives. . . . That is my ultimate job."

When Oprah Winfrey asked him in a pre-Christmas interview how he would grade his first year in office, Barack gave himself a "good, solid B+." Stating he had "inherited the biggest set of challenges of any president since Franklin Delano Roosevelt," Barack said he had "stabilized the economy" and prevented "a significant financial meltdown. . . . We are on our way out of Iraq. I think we've got the best possible plan for Afghanistan. . . . We have achieved an international consensus around the need for Iran and North Korea to disable their nuclear weapons. And I think that we're going to pass the most significant piece of social legislation since Social Security, and that's health insurance for every American. . . . If I get health care passed," Barack would up his grade to A-, and on December 24 the Senate passed a bill slightly different from the House one with sixty affirmative votes, the minimum number needed to foil a filibuster. One day before the first anniversary of Barack's inauguration, Massachusetts voters upended his political game plan for final approval of the Affordable Care Act by electing Republican Scott Brown to fill the seat of the late Edward M. Kennedy, thereby reducing to fifty-nine the Democrats' Senate majority.

Barack granted a trio of first-anniversary interviews. He expressed pleasure in how well his daughters had adjusted to life in the White House, but disappointment that "what I haven't been able to do . . . is bring the country together. . . . That's what's been lost this year—that whole sense of changing how Washington works." Barack told *Time*'s Joe Klein he had hoped "that health care wouldn't take this long" and that "if there's one thing I have learned" it's that things "always take longer than you think." To ABC's Diane Sawyer, Barack said, "one thing I'm clear about is that I'd rather be a really good one-term president than a mediocre two-term president. . . . The easiest thing for me to do . . . would be to go small bore . . . just make sure that everybody's comfortable and we only propose things that don't threaten any special interests. . . . I don't want to look back on my time here and say to myself, all I was concerned about was nurturing my own popularity."

Barack's one-year anniversary also offered commentators a chance to assay his presidency to date. CBS News' assiduous Mark Knoller reported there had been only twenty-one days when Barack "did not have a public or press appearance," and that he had golfed twenty-nine times. With Barack relying heavily upon the quartet of Rahm Emanuel, David Axelrod, Valerie Jarrett, and Robert Gibbs, the *Financial Times*' Edward Luce rued how not since Richard Nixon's presidency had there been "an administration that has been so dominated by such a small inner circle." Given the all-but-complete partisan divide over health reform, Luce quoted an academic as observing that Barack "totally lost control of the narrative in his first year in office."[27]

On March 21, by a vote of 219 to 212, the House of Representatives passed the Senate's previously approved version of the Affordable Care Act, with no Republicans in favor and thirty-four Democrats voting nay. Two days later, Barack signed it into law, and the White House celebrated what it viewed as the greatest achievement of Barack's presidency to date. As Neera Tanden, one of Barack's top aides on the measure, emphasized, "there was never an issue that was in the bill that was more important than just passing the bill." Yet as NBC's Chuck Todd stated, "the Affordable Care Act can hardly qualify as the universal health care Obama promised," and he believed it exemplified "a familiar Obama pattern: grandly proclaim a bold new direction, then move to the establishment middle."

In April, David Remnick's *The Bridge,* the first serious biography of Barack, was published to mixed reviews. Some critics were disappointed by Remnick's "idolatrous" and "all but starry-eyed" view of Obama, ruing how in Remnick's portrait "Obama emerges as nearly flawless." Others objected to Remnick's "defining Mr. Obama largely through the prism of race," with veteran *New York Times* reporter Howard W. French writing that Remnick's "insistence on race" represented "an approach that all but declares the man's race is the most interesting thing about him." Another reaction was equally fundamental. *New York Times* critic Michiko Kakutani lamented that *The Bridge* does "not really explain how this young, rootless outsider acquired the self-confidence that fueled his ascent in national politics." In the *New York Review of Books,* former *Times* executive editor Joseph Lelyveld complained that "no one can pin down exactly when the young Obama first thought seriously of running for president." *Salon* editor in chief Joan Walsh zeroed in on a lengthy comment Barack had made to Remnick about his 1980s transformation, which he dated to moving to New York City rather than to Chicago. "Obama's answer is weirdly unsatisfying, disembodied and impersonal, like he's narrating someone else's story."

The Bridge also drew the attention of Sheila Jager, who took issue with how it spoke of Barack's unnamed girlfriend as being "a white University of Chicago student." "I don't consider myself exclusively white, as I am half Asian," and "what Remnick wrote is wrong. Race and identity being a central theme of the book, I think Remnick completely missed what I believe was the most important factor that led Barack to resolve his torment over this central issue of his life." Sheila felt that Remnick "seems to take things too much at face value," especially *Dreams.*

But for Sheila *The Bridge* reopened a deep emotional wound. "The nature of the relationship" and how "it continued post-Chicago . . . is the real problem," for "in fact there is a lot to hide." Indeed, "our relationship was a tragedy that has weighed and haunted my life," such that "I hoped so much for Clinton to be the nominee and then that McCain would win not for politics or policy positions, but only because I feared this would happen. And now, what I had feared more than anything, feared because I didn't know exactly how it would affect me, only that I knew it would affect me in a manner that would bring out all the pain that I had worked so hard to put past me" had indeed come to pass.[28]

As a trio of valuable books emerged covering the first year of Barack's presidency—Jonathan Alter's *The Promise,* Bob Woodward's *Obama's Wars,* and Richard Wolffe's *Revival*—press commentary on how to understand Barack continued. Writing that "the president seems to stand foursquare for nothing much," *Washington Post* columnist Richard Cohen asked, "Who is this guy? What are his core beliefs?" A month later, Cohen repeated his plaintive call: "Who Is Barack Obama?" Given how the president "has opined on virtually everything in hundreds of public statements," as the *Los Angeles Times* put it, many analysts' remained puzzled about Barack's core beliefs.

On August 4, Senate Republican minority leader Mitch McConnell visited the Oval Office for his first one-on-one conversation with Barack, more than eighteen months into his presidency. Reporters learned that the invitation had been extended only after former Senate Republican leader Trent Lott asked former Senate Democratic leader Tom Daschle to complain to the White House about Barack's disinterest in developing personal relationships with congressional leaders. The following day, with only five Republicans voting yes, the Senate confirmed Barack's second Supreme Court nominee, Solicitor General Elena Kagan, to replace the retiring John Paul Stevens by a tally of 63–37.

While journalistic commentary on his presidency grew increasingly critical, one problem Barack faced was beyond his control. A July CNN poll found that 27 percent of Americans believed Barack likely was born outside the United States, and an early August *Time* one showed that 24 percent of respondents, and 46 percent of Republicans, thought Barack was a Muslim. Speaking with *Rolling Stone*'s Jann Wenner, Barack boasted that "here I am, halfway through my first term, and we've probably accomplished 70 percent of the things that we said we were going to do." Granted "Afghanistan is harder than Iraq," but "we have been very successful in recruiting and beginning to train Afghan security forces." Barack also emphasized that "we are not going to use a shroud of secrecy to excuse illegal behavior on our part." He repeated his "70 percent" claim to the *New York Times*' Peter Baker while explaining that "we probably spent much more time trying to get the policy right than trying to get the politics right." That would remain Barack's view going forward. "In those first two years, I think a certain arrogance crept in, in the sense of thinking as long as we get the policy ready, we didn't have to sell it," he recounted in 2015.[29]

But Washington's best-connected journalists offered portraits of Obama's presidency far more critical than Barack's own rosy image. In the style he had mastered for more than three decades, Bob Woodward described how top Democratic insider John Podesta believed that "Obama's approach was so intellectual" and that "sometimes a person's great strength, in this case Obama's capacity to intellectualize, was also an Achilles' heel." The national security adviser, General Jim Jones, who was leaving the White House, found Barack "cerebral and distant," Woodward wrote, and Defense Secretary Bob Gates would recount that "one quality I missed in Obama was passion." But "the only military matter, apart from leaks, about which I ever sensed deep passion on his part was 'Don't Ask, Don't Tell,'" concerning gay service members. As someone whose government service reached back forty years, Gates also appreciated how Barack's "White House was by far the most centralized and controlling in national security of any I had seen since Richard Nixon and Henry Kissinger."

Liberal commentators took some solace when Rahm Emanuel resigned as chief of staff in order to run for mayor of Chicago. Pete Rouse temporarily replaced Emanuel before former commerce secretary Bill Daley took up the post. In a long and comprehensive *New York Times Magazine* cover story titled "The Education of President Obama," Peter Baker wrote that Barack "rarely reaches outside the tight group of advisers" led by Jarrett; "he's opaque even to us," a top White House aide told Baker. "Except maybe for a few people in the inner circle, he's a closed book." Baker revealingly described how "on long Air Force One flights," Barack "retreats to the conference room and plays spades for hours, maintaining a trash-talking contest all the while, with the same three aides," Reggie Love, Marvin Nicholson, and presidential photographer Pete Souza.

One of the president's top economic aides described to Ron Suskind "the Barack Obama he first met in 2007. He felt there was a clarity of thought and purpose to that earlier version that was increasingly difficult to find in the years he was the president . . . somehow the president had lost ownership of his words and, eventually, his deeds." The *New York Times*' Sheryl Stolberg, building on her colleague Peter Baker's story, quoted Democratic policy maven William Galston: "I can't believe that he wants to go down in history as the president who promised to overcome polarization and ended up intensifying it." Nebraska Republican senator Mike Johanns told Stolberg that Barack "needs to build friendships and he needs to build trust" in order to work more successfully with Congress.[30]

When the midterm elections took place on November 2, Republicans gained six seats in the U.S. Senate and sixty-three in the House, seizing control of the latter chamber. In *Newsweek,* Eleanor Clift wrote that Barack's problem was that "there's too much hero worship around him" and that "if his aides weren't so in love with him and wrapped up in the idea of him as a transformational president, they might have seen this coming."

Speaking with CBS's Steve Kroft, Barack expressed regret over having signed the earmark-laden 2009 appropriations bill. "That's an example of where I was so concerned about getting things done that I lost track of the reason I got elected, which was we were to change how business was done here." Barack said that "in terms of setting the tone and how this town operates . . . that's an aspect of leadership that I didn't pay enough attention to." In addition, passing the Affordable Care Act was "actually a little more costly than we expected, politically," particularly one-offs like the "Cornhusker kickback," necessary for Nebraska senator Ben Nelson's vote, which "really hurt us politically." Acknowledging that "part of my promise to the American people when I was elected was to maintain the kind of tone that says we can disagree without being disagreeable," Barack admitted that "over the course of two years, there have been times where I've slipped on that commitment."

Following New Year's, first David Axelrod and then Robert Gibbs exited the White House. Axelrod "was pushed out" in what wired-in journalist Richard Wolffe called "a wrenching expression of disaffection from the president he had fallen for," and Wolffe purposely used the exact same words to characterize Gibbs's departure, too. "The president broke Axe's and Robert's hearts," a "member of Obama's inner circle" told Wolffe. "Axe and Gibbs were effectively fired," that person explained, and David Plouffe moved into Axelrod's former office while former *Time* reporter Jay Carney succeeded Gibbs.

In Chicago, unhappiness with the Obama White House was widespread. African American publisher Hermene Hartman sarcastically congratulated Emanuel on taking the first black president and making him non-black, and retiring mayor Richard Daley complained that "we've got a Chicago guy in Washington and he's forgotten Chicago." In March 2011, Hartman went public about black Chicago's disappointment with Barack, telling *Chicago Magazine*'s Carol Felsenthal that Barack "has not been loyal to some of the people who were there for him from day one," like Jim Reynolds. "Did you outgrow us, or did you forget?" Hartman asked.

Sentiments like Hartman's could also be found in Roseland's modest bungalows. Barack "forgot where he came from," longtime Reformation Lutheran custodian John Webster said, his voice full of tears. "I don't forget people who help me," but "he lost his soul," and "it hurts me so bad." On Roseland's decrepit South Michigan Avenue shopping strip, one young woman, asked what she would tell Barack, had an easy reply: "Just take all the guns away. The guns are ridiculous." Pastors whose churches abutted Palmer Park, across from what once had been Barack's office at Holy Rosary, warned that the park was entirely unsafe after dark. Data showed that Chicago's public schools were serving black students no better than they had twenty years earlier, with only one in two graduating high school. That of course helped explain why black unemployment in Chicago was 24.2 percent; in 1987 it had been 17.2. "Chicago remains the most racially segregated city in the country," the *Tribune* noted, but across the South Side everyone's uppermost fear was the "escalating violence." Visiting Roseland, a journalist described "the barren, empty feel of the streets," and a thirty-nine-year-old native said that "things in Roseland were pretty bad when I was growing up . . . but they are much worse today."[31]

In Washington, the president refused to embrace the proposals of a bipartisan panel he had appointed to address the United States' burgeoning deficit. The following day, in a budget speech to which Republican lawmakers had been specially invited, Barack used what the *Washington Post* called "a sharp, partisan tone" to accuse Republicans of seeking to "end Medicare as we know it." Furious House Budget Committee chairman Paul Ryan warned a presidential aide that "You just poisoned the well," and dismayed congressional Republicans told the *Post* that "Obama's partisan tone made no sense." Panel cochairs Erskine Bowles and Alan Simpson were similarly upset. Bowles called the speech harsh, and Simpson said, "I thought it was like inviting a guy to his own hanging."

Visiting 5046 South Greenwood for only the third time in two years, Barack discovered a Kapiolani Hospital booklet documenting his August 4, 1961, birth among his late mother's papers. Charmed by the keepsake, Barack directed his attorneys to ask Hawaiian officials to release his long-withheld "long form" birth certificate. Five days later, in a far more consequential surprise, U.S. special operations troops confronted and killed Osama bin Laden in what promised to be the signature accomplishment of Barack's presidency. Speaking again with Steve Kroft, Barack said the time that U.S. forces were inside bin Laden's compound in Abbottabad, Pakistan, was "the longest forty minutes of my life with the possible exception of when Sasha got meningitis . . . and I was waiting for the doctor to tell me that she was all right."

In midsummer, in the aftermath of a golf game with Republican House Speaker John Boehner, Barack appeared to be "within a handshake" of a bipartisan "grand bargain" on deficit reduction before suddenly altering his stance. "There appeared to be a very different president in attendance" than three days earlier, a superb *Washington Post* "tick tock" reported, and Boehner said no, but forty-eight hours later Barack reverted to his prior position. "His message: I'll take your last offer," the *Post* wrote, but by then the deal was dead. "White House advisers conceded that the collapse of the debt talks was a disaster from a policy perspective," the *Post* reported. In the wake of the debt deal collapse, one of Barack's "closest friends" told Richard Wolffe that "I have never seen him this mad," as Barack "cursed the media and what he saw as its gross inadequacies and failings."

A scathing verdict came from distinguished African American Harvard

Law professor Randall Kennedy. Barack's "lists of retreats from progressive commitments is considerable," Kennedy wrote, and "Obama's much-vaunted pragmatism degenerates at key moments into mere expediency." Barack's less than persuasive fence-straddling on gay marriage "is a sad spectacle: the prevarication of a decent politician impelled by his perception of electoral realities to adopt an indecent position with which he disagrees." There now was no question that "Obama is a professional politician first and last. For the sake of attaining and retaining power, he is willing to adopt, jettison, or manipulate positions as evolving circumstances require."[32]

In late September, after a U.S. drone strike ordered by Barack killed Anwar al-Awlaki, an American-born al Qaeda leader in Yemen, the president met with his top aides. "Turns out I'm really good at killing people. Didn't know that was gonna be a strong suit of mine," Barack remarked before launching into an extended self-critique, the first one any of them had ever heard him voice. Barack questioned his carefully calibrated political stances on many issues, particularly gay marriage, but chief of staff Bill Daley was worried that word of Barack's comments would leak out. Six weeks later, just before another meeting, David Plouffe informed Barack that *New York* magazine's John Heilemann knew about Barack's earlier comments. "How could someone do this to me?" Barack asked, telling his assembled aides, "I can't trust anybody here anymore," as Heilemann again soon learned. Barack demanded that the guilty party apologize and stalked out. Vice President Joe Biden echoed Barack's anger before likewise leaving. The aides all agreed that Barack's presidency was under siege.

In his annual round of pre-Christmas television interviews, a visibly defensive Barack insisted to CBS's Steve Kroft that "our deficit problems are completely manageable." Barack boastfully claimed, "I would put our legislative and foreign policy accomplishments in our first two years against any president" except perhaps Lyndon Johnson, Franklin Roosevelt, and Abraham Lincoln, although he conceded that "when it comes to the economy, we've got a lot more work to do." When ABC's Barbara Walters asked him, "What's the trait you most deplore in yourself?" Barack answered "laziness," explaining that "deep down, underneath all the work I do, I think there's a laziness in me" that he attributed to "growing up in Hawaii."

As 2011 headed toward its close, Barack's closest personal aide, Reggie Love, left the White House to finish his MBA at the Wharton School of Business. Reporting Love's departure, the *New York Times*' Jodi Kantor, whose forthcoming book *The Obamas* had White House aides worried, called Barack "a private, solitary president who allows few people to grow close to him," but Kantor was unaware that in the wake of the Affordable Care Act's passage, Barack had finally, after years of failure, completely given up smoking. Her book's portrait of Barack was far from warm, and she wrote that Barack's "self-confidence often made him lazy." Aides believed "there was nothing spontaneous or vulnerable about the president," and Kantor emphasized how even after three years "they could not fully read their boss." The most notable controversy generated by *The Obamas* concerned a quotation that Barack's friend Allison Davis had given Kantor, which he said was off-the-record, but she nonetheless published: "When I leave office there are only two things I want. I want a plane and I want a valet," Barack had told Davis.

In the *Atlantic Monthly,* James Fallows quoted an unnamed administration official as explaining how "Obama's extra-high intellectual capacity is simply not matched by his emotional capacity . . . he does not seem to have the ability to connect with people." Noting how multiple tomes all detailed Barack's "emotional distance from all but a handful of longtime friends and advisers," Fallows described Barack as an "incrementalist operator" whose interpersonal demeanor was "cool to the point of chilly."[33]

In late January, Office of Management and Budget director Jack Lew was named chief of staff, replacing Bill Daley, whose one-year tenure Richard Wolffe termed "a walking disaster." Both the *New York Times* and then the *Washington Post* published long stories detailing how the previous summer's debt deal had failed, with the *Times*' Jackie Calmes saying that Barack's "partisan turn undercuts a central promise of his 2008 campaign, to rise above the rancor," and that "Obama arguably failed to show leadership on perhaps the country's biggest problem."

Barack had known since November that a forthcoming book on his parents and his own young adulthood would "out" both Alex McNear and Genevieve Cook. Author David Maraniss had learned about Alex's private life from an old Oxy acquaintance, and at the behest of a mutual friend she spoke with Maraniss by phone, telling him that "he could not read the letters" Barack had written her in 1982–83 while giving him "a handful of quotes . . . that I didn't think would in any way embarrass" Barack. Maraniss had also gotten Genevieve Cook on the phone, and Genevieve later explained that her "main motivation in my desire to talk to Maraniss was anger/indignation: I thought it was exclusively about having a venue to respond to Barack's decision to conflate (and thus misrepresent) the girlfriend in NYC" in *Dreams*. Genevieve shared with Maraniss some excerpts from her journals, and one photo of her and Barack, but she made no mention of the letters Barack had written her.

With *Vanity Fair* scheduled to release in early May an excerpt from the book featuring Alex and Genevieve, Barack's deep fear about a vastly more explosive revelation led him to pick up an aide's phone to do something that he had not done in over twenty years. As Sheila Jager related, "Barack called me out of the blue a few weeks ago. Of course it was a surprise," and Barack said he was "disgusted" by Maraniss's interest in his sexual history. "I told him that Maraniss had contacted me," as he had another 1980s U of C anthropology student also named Sheila, "but that I had . . . cut him off. Barack seemed relieved." Sheila was ecstatic that Barack had contacted her, for that showed "it never ended badly" and reflected "the warm feelings we still have for one another."

The *Vanity Fair* excerpt previewed the only two newsworthy nuggets in Maraniss's book, and reviewers panned the volume's shortcomings. "Obama does not really emerge in Maraniss's biography," complained Darryl Pinckney in the *New York Review of Books,* and "Maraniss can't find in Obama's story the scene when he revealed to someone that he possessed a sense of having an extraordinary destiny." In *The New Republic,* Nicholas Lemann dismissed Maraniss's book as "raw and under-processed." Lemann appreciated that "Obama's childhood circumstances were more emotionally difficult than he has made them out to be," but he found it "remarkable and unexplained" that Barack "wound up with both an unstoppable drive to power and complete self-control."[34]

On May 9, three days after Vice President Joe Biden publicly voiced his support for same-sex marriage, Barack tardily admitted that he agreed. "I've just concluded that for me personally it is important for me to go ahead and affirm that I think same-sex couples should be able to get married," he told ABC's Robin Roberts in a hastily scheduled interview. It was now all but official that former Massachusetts governor Mitt Romney would be Barack's Republican opponent come November, yet America's increasingly widespread acceptance of gay equality made Barack's long-standing political pusillanimity appear embarrassingly outdated. Among onetime supporters, expressions of profound disappointment with Barack's presidency continued apace. "Obama must be defeated in the coming election. He has failed to advance the progressive cause in the United States," declared Roberto Mangabeira Unger, once Barack's favorite teacher.

In the African American community, Columbia University professor Fredrick Harris observed that "Obama stopped talking about racial inequality once he solidified support from black voters." Similarly, "campaign ads highlighting Obama's white heritage . . . constantly reminded voters of the candidate's not-completely-black identity." Equally bad, "instead of pressuring Obama to move on certain policies, black leaders and black voters act more as cheerleaders than players."

Speaking with Barack's best-known black supporter, Michelle Obama agreed with alacrity when Oprah Winfrey asked Barack, "So you can compartmentalize?" "Oh, he's very good at that," Michelle immediately volunteered. Barack's younger brother Mark believed Barack "compartmentalizes his life into the private and public spheres," but Mark explained that Barack "is so impenetrable" that "he remains an enigma." Chicago friends acknowledged Barack's "deep compartmentalizing," but to Winfrey Barack said, "I'm in constant conversation with God and that voice that is true about doing the right thing."[35]

Barack's renomination for a second term on September 6 offered journalists and insiders an opportunity to judge his first. "He doesn't feel he needs to thank his friends," Valerie Jarrett admitted to Jonathan Alter, and an unnamed big donor complained to *The New Yorker*'s Jane Mayer that Jarrett herself "seemed to think she was blessing me by breathing in the same space." *Politico*'s John Harris and Jonathan Martin judged Barack "a man of conventional instincts, practicing conventional politics," and people who were disappointed failed to "understand the arc of Obama's career before the presidency." Barack "has presented no set of ideas that collectively represent anything that might last beyond his term," and in addition, "Obama's cultural impact has been virtually nil."

The day after Barack's renomination, Bob Woodward's second book on the Obama presidency appeared, with former National Economic Council director Larry Summers saying of Barack, "I don't think anybody had a sense of his deep feelings about things. I don't think anybody has a sense of his deep feelings about people. I don't think people have a sense of his deep feelings around his public philosophy." In his autobiography, the late Paul Simon had emphasized that "a strong president has to have strong convictions," but in Summers's eyes Barack's fundamental problem was his "excessive pragmatism."

Speaking once again with Steve Kroft, Barack admitted that "the spirit that I brought to Washington, that I wanted to see instituted . . . I haven't fully accomplished that. Haven't even come close in some cases . . . If you ask me what's my biggest disappointment, it's that we haven't changed the tone in Washington as much as I would have liked." Unlike Winfrey, Kroft posed some challenging questions. "The national debt has gone up 60 percent in the four years that you've been in office," he accurately told Barack, who responded that that was 90 percent George W. Bush's fault. "I said I'd end the war in Iraq. I did," Barack declared.

In New York in mid-September, Barack attended a $40,000-per-person fund-raiser hosted by hip-hop stars Jay Z and Beyoncé that raised $4 million for his campaign. "Let me just begin by saying to Jay and Bey, thank you so much for your friendship," Barack told the one hundred guests as they imbibed "$800-per-bottle Armand de Brignac champagne." The White House declined to schedule a New York meeting with visiting Israeli prime minister Benjamin Netanyahu, "telling Netanyahu that Obama was too busy to see him," *USA Today* reported. In response, Chicago's Susan Crown, once a major Obama donor, switched her support to Mitt Romney. "I am so embarrassed that our current President, instead of taking a meeting with Netanyahu, is going to New York and hanging out with Jay Z and Beyoncé," Crown stated. "Why Jay Z and Beyoncé are your priority?"

Similarly unhappy was veteran Industrial Areas Foundation organizer Arnie

Graf, whom Barack twenty-six years earlier had quizzed about the realities of interracial marriage during his IAF training. In 2003 Graf had been favorably impressed with Mitt Romney when he dealt with him in Massachusetts, but now both presidential contenders "are not the equals of their own former selves," Graf wrote. Citing "the errors and lapses that have made his first term a disappointment," Graf said, "I'd like to hear the Obama of 1986 comment on the Obama of 2012." With both Romney and Obama, "their ambition to be president has overwhelmed their core beliefs," and "the earlier version of each man would never have supported the current version—or even recognized him."[36]

When the tightly contested presidential race ended late on November 6, Barack narrowly won reelection, receiving 51 percent of the popular vote and carrying 26 states. Overall Barack topped Romney by almost 5 million votes, and his Electoral College margin was a healthy 332 to 206. Reacting to Barack's victory, Sheila Jager recounted how he "always said to me that becoming the president wasn't the goal in and of itself," that "he wanted to be a great one."

Speaking with *Time* magazine, Barack stressed how as president "the amount of power you have is overstated." In addition, "anything we do is going to be somewhat imperfect . . . so what we try to do is just tack in the right direction." Yet in a remarkable implicit comparison, Barack also stated that as president one was no longer who one had been previously. "The Lincoln who is a lawyer in Springfield, Illinois isn't the same Lincoln as the one who addresses Gettysburg. For that matter, the Lincoln who's elected President is not the same as the Lincoln who delivers the second inaugural. They're different people."

Soon after New Year's, an unusual poll reported that many Americans believed that Barack Obama was a different person too. Indeed, *36 percent* thought that Obama is "hiding important information about his background and early life," New Jersey's Fairleigh Dickinson University reported. Since "among Republicans . . . more knowledge leads to greater belief in political conspiracies," that 36 percent figure masked how *64 percent of Republicans,* and *42 percent of whites,* but only 17 percent of blacks, agreed that it was "probably true" that Barack was hiding something crucial.[37]

As Barack prepared to take the oath of office for a second time, CBS News' Mark Knoller summarized the statistical record of Barack's first term. Unemployment had remained static at 7.8 percent, but the national debt had risen from $10.6 to $16.4 trillion, "the largest increase in the national debt under one president." Over the course of four years, Barack had delivered 1,852 public speeches and had granted 591 media interviews, 104 of which had been with national television networks. Barack had taken thirteen vacations, totaling 83 days, and had played 113 rounds of golf. Some months earlier Barack had told Oprah Winfrey that in 2008 "I promised I wouldn't be a perfect president, but

I'd wake up every single day working as hard as I could . . . and that promise I've kept."

NBC's Chuck Todd believed that even before his second inauguration, Barack "set himself on a path to failure" by embracing a conflictual policy agenda on top of his lack of outreach to members of Congress. Without expressly naming his source, Todd reported that moderate West Virginia Democratic senator Joe Manchin, elected to office in 2010, received his first telephone call from Barack more than two years later. But "Obama didn't call many senators, period."

Speaking in unusually personal terms at Morehouse College's commencement, Barack said, "I sure wish I had had a father who was . . . involved" in his life, and "I've tried to be for Michelle and my girls what my father was not for my mother and me." He described what he called "the special obligation I felt, as a black man . . . to help those who need it most, people who didn't have the opportunities that I had."

In early June, Edward Snowden's astonishing revelations about the depth and extent of U.S. government electronic surveillance programs shook Obama's administration and all of American politics. European anger at the U.S. executive branch was especially acute, and as an aura of failure began to settle over Barack's presidency, old friends struggled to come to terms with why things had gone so wrong. "Barack is a tragic figure: so much potential, such critical times," but "such failure to perform . . . like he is an empty shell," one longtime Hyde Park friend lamented. "Maybe the flaw is hubris, deep and abiding hubris, as if his touch, his deep understanding, will clear the path. . . . How else could he possibly be so . . . unruffled" despite what was happening? Yet no one alive brought deeper insight into the tragedy of Barack Obama than Sheila Jager.

> *I think the seeds of his future failings were always present in Chicago. He made a series of calculated decisions when he began to map out his life at that time and they involved some deep compromises. There is a familiar echo in the language he uses now to talk about the compromises he's always forced to make and the way he explained his future to me back then, saying, in effect, I *wish* I could do this, but pragmatism and the reality of the world has forced me to do that. From the bailout to NSA to Egypt, it is always the same. The problem is that "pragmatism" can very much look like what works best for the moment. Hence, the constant criticism that there is no strategic vision behind his decisions. Perhaps this pragmatism and need to just "get along in the world" (by accepting the world as it is instead of really trying to change it) stems from his deep-seated need to be loved and admired which has ultimately led him on the path to conformism and not down the path to greatness which I had hoped for him.*[38]

In late August, Barack unexpectedly reversed course about launching military strikes against the Syrian regime of Bashar al-Assad, which was continuing to murder thousands of Syrian civilians. His sudden change of heart, which occurred following a one-on-one conversation with Denis McDonough, who seven months earlier had succeeded Jack Lew as White House chief of staff, stunned other top aides and world leaders alike. Jeffrey Goldberg, Obama's most-favored foreign policy commentator, wrote that National Security Adviser Susan Rice was "shocked" and believed that "the damage to America's credibility would be serious and lasting. Others had difficulty fathoming how the president could reverse himself." One Middle Eastern head of state told Goldberg that Obama was "untrustworthy," and even Barack's former CIA director and defense secretary Leon Panetta spoke disparagingly of him. "Once the commander in chief draws that red line, then I think the credibility of the commander in chief and this nation is at stake if he doesn't enforce it." Former Clinton administration aide William Galston rued how Barack's "allergy to the use, or even the threat, of force has rendered U.S. diplomacy all but toothless."

At the end of September, as the Affordable Care Act's new Web site, HealthCare.gov, prepared to debut, Barack promised that "any American out there who does not currently have health insurance can get high-quality health insurance." But as it soon became clear that the Web site was a dysfunctional disaster, journalists like NBC's Chuck Todd wondered "how could a president so powerfully make a law so central to his legacy and then fail so miserably to make it real?" *Politico*'s Michael Grunwald observed that "the crash of the Obamacare website" was "a debacle that seemed to confirm every attack ever leveled at Obama's competence." Simultaneously, press anger at Barack's White House increasingly spilled into public view. Veteran *New York Times* reporter David Sanger said that "this is the most closed, control-freak administration I've ever covered," and longtime CBS newsman Bob Schieffer concurred: "Whenever I'm asked what is the most manipulative and secretive administration I've covered, I always say it's the one in office now."

In mid-October, in his third book on Obama's campaigns and presidency, Richard Wolffe zinged Barack's "lack of preparation and disengagement at vital times," a failing that a *Washington Post* story soon highlighted. "Former Obama administration officials said the president's inattention to detail has been a frequent source of frustration," and even after almost five years in the White House, "Obama has yet to master the management of information within his administration."

The negative commentary kept on coming. A front-page *New York Times* story on Democratic fund-raising stated that "more than a dozen Obama supporters . . . described the president as an introvert," and in a new book on the 2012 campaign, Mark Halperin and John Heilemann wrote that stories of Barack's "aloofness and inattentiveness to his donors were legion." In the *Wall Street Journal*, the distinguished foreign affairs commentator Fouad Ajami wrote that what energized opposition to Barack "wasn't his race" but "his exalted view of himself" as "above things, a man alone and anointed."[39]

A seeming culmination of all the months of negative commentary came in early November when *Vanity Fair* published former *New York Times* reporter Todd S. Purdum's profile of Barack, "The Lonely Guy." Purdum believed "Obama has always been alone," since "he was abandoned not only by his black African father . . . but also by his white American mother." Longtime Chicago friends agreed that Barack "is the functional equivalent of an orphan," someone who "craved acceptance" as a result of his abandonment. Purdum thought that "Obama's resolute solitude—his isolation and alienation from the other players . . . has emerged as the defining trait of his time in office," for "self-possession is the core of his being."

Purdum highlighted "Obama's persistent assumption that supporters (and staffers, for that matter) don't need to be thanked." In Chicago, that behavior was well known, but as Barack's time in the presidency lengthened with few former aides and friends hearing anything at all from the White House, feelings of disappointment hardened into anger. Barack "doesn't feel indebted to people," a once-close assistant explained, but among friends who had known Michelle as well as or even better than Barack, the surprise and pain at the extent to which both Obamas had "gone Hollywood," as with Jay Z and Beyoncé, was pronounced. Friends had believed that "Michelle will always be there to keep him in check," because "she grounded him," but now the evidence was undeniable that "she's more caught up" in celebrity hobnobbing than Barack was. "She's an even bigger disappointment than he is," one black Chicagoan remarked. In New York too friends of many years expressed anger over how "they just fucked all these people who were so loyal to them." Yet conversations about "all the people that Barack has burned" were most common in Chicago, with once-fervent campaign volunteers explaining how they no longer supported Barack "because of how he was treating all of those people." One former protégé had realized "that's what he's about," but "you don't treat people like that," he exclaimed, his voice full of emotion. "I don't think people know who politicians *are* versus who they project." Todd Purdum wondered "how someone wired the way Obama is got so far in politics," yet Purdum concluded that the absence of any "strong intellectual rivals" in Barack's inner circle reflected "a basic insecurity" on the part of the president.[40]

As 2013 ended, *Politico* quoted longtime Washington observer David Gergen criticizing how, as with the Affordable Care Act and HealthCare.gov, "there's enormous value placed, within the White House, upon winning" but "there's less and less value placed upon executing the law itself." Barack's team is "much, much better at campaigning . . . than they are in governing and in management," Gergen explained. Speaking with *New Yorker* editor David Remnick, Barack once again minimized his role, saying that "as president . . . you are essentially a relay swimmer in a river full of rapids, and that river is history." The president, he emphasized, "cannot remake our society," but congressional expert Charlie Cook reported that "on Capitol Hill, few Democrats love, fear, or even respect their president." In part that was because "this administration has delivered very little," and the *New York Times*' Peter Baker concurred that Democrats also blamed Barack "for not doing more to work across the aisle."

By late February 2014, progressive Afro-English *Guardian* journalist Gary Younge wondered if any purpose whatsoever remained in Barack's presidency. Health care reform is "not likely to be remembered as transformational," Younge rightly pointed out, and if Barack "can't reunite a divided political culture . . . then what is the point of his presidency?" Barack's "presence in power . . . lacks purpose," Younge believed. A front-page *New York Times* story about Barack's impact documented "the gap between what his supporters expected and what they now see." As a young arts activist explained, "there are a lot of people being let down by a president they were very enthusiastic about. There's a big sense of betrayal." A subsequent *Politico* profile labeled Barack's presidency "a remarkable descent," contrasting his "lack of engagement" with Congress to how "Obama spent 46 days on the golf course in 2013" and "follows obsessively" the NBA playoffs. A friend described Barack as "fatigued," but the piece powerfully conveyed people's "disappointment for what might have been."[41]

On April 7, Barack's aunt Zeituni Onyango, the closest living link to his late father, died in Boston at age sixty-one. Her sudden passing surprised friends, for just weeks earlier she had still been sending e-mails whose upbeat tone—"With fondest joy and love, Auntie Zeituni"—belied her penurious circumstances and made no mention of her health problems. Omar Onyango and Malik Abon'go hosted a wake, with the *New York Times*' Jason Horowitz reporting that Barack "helped pay funeral expenses and sent a condolence note" but "did not attend, as he was golfing."

Aboard Air Force One in late April, a president peeved at news stories highlighting what one journalist called his "flaccid responses" to "global turmoil" told his traveling press corps that "I can sum up my foreign policy in one phrase. Don't do stupid shit." *New York Times* reporter Mark Landler felt Obama's remark "seemed crude, almost juvenile," but it quickly threatened to become "perhaps the signature slogan of his presidency" as it was "roundly mocked for its blinkered message about American power." The *Times*' editorial board protested Barack's "maddeningly bland demeanor" and his "frustratingly cautious" foreign policy. Picking up on his paper's editorial, Landler wrote of Barack's "shrunken vision" for U.S. foreign policy and observed that "Obama's instinct when he gets into a difficult situation has always been to deliver a speech."

In *Foreign Policy,* former Defense Department counselor Rosa Brooks apologized to Hillary Clinton for supporting Barack in 2008. "In the Obama White House, innovation became reactiveness, discipline became rigidity, and a tight circle of campaign aides and Chicago pals tried to micro-manage the entire executive branch," Brooks explained. It was readily apparent that "Obama's an introvert. . . . Look at his tight body language; listen to the undertone of irritation in his voice. Barack Obama is a man who almost always looks and sounds like he'd prefer to be somewhere else. He's a politician who hates politics."[42]

The chorus continued. In an op-ed headlined "Barack Obama's Presidency Is Spiraling Downward," U of C political science professor Charles Lipson told *Chicago Tribune* readers that Barack "is a poor manager. He doesn't pay attention to crucial details, surrounds himself with sycophants and doesn't hold anyone accountable." In *Politico,* Harvard law professor Randall Kennedy asked, "Did Obama Fail Black America?" Kennedy's answer was that "for many African-Americans, he has been a hero—but also a disappointment. On critical matters of racial justice, he has posited no agenda, unveiled no vision," and seemed "virtually mute." Although "blacks as a whole have fared badly during Obama's tenure," the latest Gallup Poll showed Barack with 87 percent approval among blacks as compared to only 35 percent among whites. With his "serious" and "dignified" demeanor, plus his marriage to a black woman, for African Americans Barack was "the antithesis of The Black Man as Failure," Kennedy wrote. In contrast, Cornel West, another well-known black voice, intensified his long-standing criticism of Barack. "He posed as a progressive and turned out to be a counterfeit. We ended up with a Wall Street presidency, a drone presidency, a national security presidency . . . we ended up with a brown-faced Clinton: another opportunist."

Back in Chicago, libertarian *Tribune* columnist Steve Chapman complained about how "Obama led Americans to believe that he would be far more sensitive to privacy and civil liberties than George W. Bush was. But more often than not, he reflexively indulges the demands of law enforcement agencies. . . . In clashes between government and the individual, the president almost invariably sides with the former." Similarly, the *New York Times*' Maureen Dowd quoted her courageous colleague James Risen calling Barack "the greatest enemy to press freedom in a generation."

In the *Wall Street Journal,* Peggy Noonan complained that Barack "seems unserious, frivolous, shallow. He hangs with celebrities, plays golf," and "seems disinterested, disengaged almost to the point of disembodied. He is fatalistic, passive, minimalist," the former Reagan speechwriter lamented. Presidents are "not supposed to check out psychologically," and "it is weird to have a president who has given up. . . . It is unprecedented and deeply strange" plus "unbelievably dangerous." A *Politico* tally showed that as of midsummer, Barack had played golf on more than 180 days as president, about two-thirds of the time with White House "body man" Marvin Nicholson. "His lonely golfing sojourns should be seen as a reflection of the man as well as the president," wrote *Politico* editor Michael Hirsh.

Even *Washington Post* columnist Ruth Marcus, a consistently supportive voice, admitted that "this has been a disappointing presidency" and wondered "whether a more experienced hand in the Oval Office might have done a better job," especially on foreign policy. Marcus warned that if Barack "leaves office with the country in more danger than it was eight years earlier, the rest of the legacy will be for naught."[43]

In Oberlin, Ohio, Sheila Jager worried about the man whom she still loved. "My past with Barack still seems very present and I think that's true for both of us. I still care for him very deeply as you already know; our relationship was so complex and deep and filled with so many contradictions." Seven years earlier, when she had posted those critical Web comments, "I was angry that he hadn't contacted me and had left me, so to speak, in the lurch . . . that he wrote me completely out of his memoir, which I thought was a deliberate effort to avoid confronting some painful truths. I was really angry about that." Sheila's "feelings have evolved" since 2007 "because some of the issues between us have since been resolved" once Barack began calling her. In retrospect, Sheila wished Barack had read history, rather than literature. "I think if he had, he'd have much less faith in the 'arc of history' or '21st century behavior and norms' and a better understanding that we lead blessed lives not because it is the natural order of things but precisely because it is so very unnatural, and that people have had to fight and die to attain and maintain it. I think that history teaches us that evil is much more a prevalent force in the world than goodness and peace."

In retrospect, Sheila believed "that something changed in him after we went our separate ways after Harvard, as if the part of him that was so vulnerable and open (and sensual?) went underground and something else—raging ambition, quest for greatness, whatever just took over instead." But her feelings for Barack remained unchanged. "You of course know how deeply I loved him, and will always love him, and . . . I feel a sense of protectiveness toward him." Barack "is almost always the brightest person in the room, and he knows that." Yet "he is also stubborn and prone to think he is always right."[44]

Appearing on NBC's *Meet the Press* and CBS's *60 Minutes* in September, Barack was visibly defensive when first Chuck Todd and then Steve Kroft posed tough questions about U.S. strategy for combating Daesh's expanding presence in Syria and Iraq. When former CIA director and defense secretary

Leon Panetta published a memoir of his time in government, coverage immediately focused on his statement that Barack sometimes "avoids the battle, complains, and misses opportunities." In a interview with *USA Today*'s Susan Page, Panetta contended that Barack as president had "kind of lost his way."

Most eyes focused on November 4's upcoming midterm elections, with Republicans appearing poised to win enough Democratic Senate seats to easily seize control of the upper chamber. Come election night, the Republican sweep was overwhelming: thirteen Democratic House seats and a decisive seven Senate seats were lost, a number that soon grew to nine as ones in Alaska and Louisiana also switched hands. Mitch McConnell replaced Harry Reid as majority leader, and a *Washington Post* headline stated "Midterm Disaster Rips Apart Awkward Ties Between Obama and Senate Democrats." *Post* reporter Paul Kane wrote that Barack's "approach to social engagement with lawmakers is almost nonexistent."

Chicago Magazine columnist Carol Felsenthal took to *Politico* with a call that many Obama friends would echo: "Fire Valerie Jarrett!" Felsenthal's sources detailed how "Jarrett micromanages guest lists . . . hangs out in the private quarters and often joins the Obamas for dinner" upstairs at the White House. Felsenthal knew that the reason for Jarrett's seemingly all-powerful role lay in how "it was important to Michelle that Valerie be in the room." Yet Jarrett "seems to isolate" Barack "from people who might help him," Felsenthal wrote. A *New Republic* profile of Valerie by Noam Scheiber emphasized "Jarrett's obsessiveness about control." The piece described how Jarrett rebuffed helpful criticism. "She just cuts off. It's stone cold," one source told Scheiber, who thought Jarrett was as responsible as anyone for how "Obama has become even more persuaded of his righteousness as the years have gone on."[45]

On CBS's *Face the Nation,* host Bob Schieffer asked Barack the most astonishing questions. "Do you like politicians? Do you like politics? Do you like this job?" and "Is it what you thought it would be?" Barack's response was unchanged from years prior. "There are times," he explained, when "we have not been successful in going out there and letting people know what it is that we are trying to do and why this is the right direction. So there is a failure of politics." The following week, *Foreign Policy* publisher David Rothkopf's book *National Insecurity* targeted Barack's relationships with foreign leaders. An unnamed Latin American president contrasted Barack to George W. Bush, who "was always direct and a man of his word. I cannot say the same for Obama." Indeed, out of "three dozen or so" heads of state, "not a single one of those foreign leaders with whom I spoke preferred Obama to Bush," a devastating verdict. An even more telling blow was landed by General Michael Hayden, CIA director under George W. Bush, who was pleased that "there was surprising continuity between the 43rd and 44th presidents of the United States . . .

despite the campaign rhetoric" Barack had used in 2008. "He ran against Bush 43.1 and governed like Bush 43.2 in terms of security issues," Hayden stated. Even Barack's signature accomplishment—killing Osama bin Laden—"did not actually reduce the terrorist threat in any material way," Rothkopf noted.[46]

Public criticism of Barack's all-out support of U.S. intelligence agencies burgeoned. *New York Times* reporter James Risen lamented how six years earlier Barack "had a mandate to do something different, and he didn't do it." As president, Barack "surrounded himself with a lot of the Bush people," like now-CIA director John Brennan, and "he normalized the war on terror. He took what Bush and Cheney kind of had started on an emergency, ad hoc basis and turned it into a permanent state and allowed it to grow much more dramatically than it ever had under Bush or Cheney, and part of that . . . was his attack on whistleblowers and journalists."

Retiring Democrat Jay Rockefeller of West Virginia, a member of the Senate Intelligence Committee, forcefully criticized "the White House's strong deference to the CIA. . . . While aspiring to be the most transparent administration in history, the White House continues to quietly withhold from the committee more than 9,000 documents" detailing the CIA's use of torture during the Bush years. "I am simply disappointed, rather than surprised, that even when the CIA inexplicably conducted an unauthorized search of the committee's computer files and e-mails at an offsite facility," Rockefeller added, that "the White House's support for the CIA's leadership was unflinching." His Intelligence Committee colleague, California Democrat Dianne Feinstein, likewise denounced the CIA's surveillance of the committee as "a separation of powers violation." Lame-duck Colorado Democrat Mark Udall, who had lost in November, took to the Senate floor to protest "the unprecedented actions that some in the intelligence community and administration have taken in order to cover up the truth. . . . One would think this administration is leading the efforts to right the wrongs of the past and ensure the American people learn the truth about the CIA's torture program. Not so." Instead, Barack's White House "continues to try to cover up the truth" by supporting "the Agency leadership's persistent and entrenched culture of misrepresenting the truth to Congress and the American people." Decrying "the White House's willingness to let the CIA do whatever it likes," even when "barbaric programs" in which "real actual people engaged in torture" were involved, Udall issued a personal challenge. "The President needs to purge his administration of high-level officials who were instrumental to the development and running of this program. He needs to force a cultural change at the CIA" and take "real action to live up to the pledges he made" when first campaigning for the presidency.

A front-page *New York Times* story endorsed Udall's attack and focused on CIA director John Brennan. "In the 67 years since the CIA was founded, few presidents have had as close a bond with their intelligence chiefs as Mr. Obama has forged with Mr. Brennan," Peter Baker and Mark Mazzetti wrote. "The result is a president who denounces torture but not the people accused of inflicting it." Retiring Michigan Democratic senator Carl Levin joined the chorus of critics, telling the *Times* that "Brennan has gotten away with frustrating congressional oversight. He shouldn't have gotten away with it." Writing in *Politico,* Lindsay Moran, a former undercover officer who had resigned in disgust, termed the CIA "a morally bereft wasteland," explaining that even in 2003 she knew "that to stay with the Agency would be to end up on the wrong side of history." In retrospect, she realized that "our enemies' principal triumph may" be "in what they prompted us to become." As 2014 came to a close, another front-page *New York Times* story said the CIA was "an agency that has been ascendant since President Obama came into office." One of Barack's former cabinet members explained that "presidents tend to be smitten with the instruments of the intelligence community. I think Obama was more smitten than most," for "this has been an intelligence presidency in a way we haven't seen maybe since Eisenhower."[47]

As 2015 dawned, public colloquy about Barack's shortcomings continued apace. Arizona Republican John McCain, now chairman of the Senate Armed Services Committee, recounted to the *New York Times* how sixteen months earlier he had endorsed Barack's plan to intervene militarily in Syria before the president reversed course. "When somebody looks you in the eye in the Oval Office and says they're going to do something, don't you take their word for it? I did. I took his word for it. And obviously I shouldn't have." In an even more remarkable statement, former presidential adviser David Axelrod told the *New York Times* that the Obama presidency had failed to accomplish its uppermost goal. "We didn't achieve what we set out to achieve. We clearly haven't changed the tone in Washington."

Inside the White House, stasis reigned, with *Washington Post* columnist Dana Milbank ruing how "for far too long, this president has surrounded himself with yes men, living in a self-congratulatory world of affirmation." Milbank cited what he called "the twin pillars of detachment that have underpinned his presidency: insularity and secrecy." He noted how after Barack had offered up "some contemptuous remarks about Democratic critics" on Capitol Hill, "Obama went to play golf . . . not with lawmakers but with old friends and staffers." A *Politico* survey of the White House press corps found almost 75 percent agreement that "President Obama dislikes the press," and when asked which administration was "the least press-friendly," Barack outpointed George W. Bush 65 to 11 percent. Looking back to 2008, CBS's Bob Schieffer struck an apologetic note, remarking that "maybe we were not skeptical enough" about Barack's candidacy.[48]

More than a continent away, Genevieve Cook pondered why Barack's presidency had gone so badly wrong. Just like Sheila Jager, Genevieve too remembered how the Barack whom she once loved "was afraid of consigning himself to a life where he cut off his emotional side in the name of his political ambitions." Now she grasped "what a political animal Barack is," so "ruthlessly ambitious," papering over "a great deal of self-doubt" with "arrogance . . . some degree of self-glorification, and a strong dislike for being embarrassed, losing status, or having his reputation tarnished." Underlying everything was "a willingness to be insincere in order to bolster his need to be on top and in control . . . a personality where the need to win . . . trumps all the other stuff."

In midsummer, soon after a white gunman killed nine black people inside Charleston's Emanuel AME Church, Barack told an interviewer that gun violence "has been the one area where I feel that I've been most frustrated and most stymied." But Barack understood that "in the absence of a movement politically in which people say 'enough is enough,' we're going to continue to see, unfortunately, these tragedies take place." He rightly pointed out that "Americans are not more violent than people in other developed countries. But they have more deadly weapons to act out their rage, and that's the only main variable that you see between the U.S. and these other countries."

In late July, with almost eighteen months left in his second term, Barack astonishingly admitted, "I'll be honest with you—I'm looking forward to life after being president." Asked whether his daughters felt likewise, Barack intriguingly answered "not as much as Michelle, but certainly ready," without explaining his wife's eagerness to leave the White House. Barack revealingly confessed that his favorite television program was a Golf Channel reality show, *Big Break,* in which the winner gets to join the professional tour, and reporters highlighted how golf had become "something of an obsession" for him. Barack "now lets few weekends go by without hitting the links," and by early fall CBS's careful tally showed that "Obama has played 256 rounds since becoming president." As the *New York Times*' Mark Landler noted, Barack's meetings with foreign leaders at various resorts rather than in the White House "have seemed an elaborate excuse to hit the links."[49]

Behind the scenes, many Democrats were just as eager for Barack to exit the White House as he himself now seemed. "No president in modern times has presided over so disastrous a stretch for his party, at almost every level of politics," wrote the historically savvy journalist Jeff Greenfield. When Barack took office, Democrats held sixty seats in the U.S. Senate; they now held forty-six. In the House of Representatives, there were sixty-nine fewer Democrats than there had been in January 2009. "The decimation of the Democratic Party nationally" had begun in 2010 with the loss of Ted Kennedy's Massachusetts Senate seat, and "when you look at the states, the collapse of the party's fortunes are worse," with nine—and soon ten—more Republican governors in office than in 2009. Democrats had then controlled twenty-seven state legislatures; now that number was reduced to eleven, with Republicans up to thirty from fourteen in 2009.

Following Greenfield's lead, a lengthy *New York Times* story cataloged how Barack "today presides over a shrinking party whose control of elected offices at the state and local levels has declined precipitously" since he took office. "Democratic losses in state legislatures under Mr. Obama rank among the worst in the last 115 years, with 816 Democratic lawmakers losing their jobs" and Republicans controlling more legislative chambers that at any time in American history.

With interviewers like Steve Kroft, Barack's defensiveness was now greater than ever. "I feel like I'm being filibustered," an exasperated Kroft remarked during an early October taping, and Barack continued to minimize how much influence a president had. "What I didn't fully appreciate, and nobody can appreciate until they're in the position, is how decentralized power is in this system," Barack claimed. When asked about his signature domestic achievement, Barack asserted that the Affordable Care Act "is working better than even I thought it was going to work," a statement that a trio of *New York Times* stories shredded. The headlines depicted Obamacare as underwhelming: "Many Say High Deductibles Make Their Health Care All but Useless," "Many See IRS Fines as More Affordable Than Insurance," and "Lost Jobs, Houses, Savings: Even Insured Often Face Crushing Medical Debt."

One New Jersey man told the *Times* that "we have insurance, but can't afford to use it," since "the deductible, $3,000 a year, makes it impossible to actually go to the doctor." Many people were "paying for insurance I could not afford to use," like an Illinois family who were paying $1,200 a month in premiums for coverage whose annual deductible was $12,700. A Kaiser Family Foundation study found that "more than seven million people who are eligible for exchange coverage would pay less in penalties than for the least expensive insurance available to them," such as a couple who were paying $455 a month but faced a $6,000 deductible. "It literally covered zero medical expenses," the wife explained. "I do not believe it serves the public good to entrench private insurance programs that put actual care out of reach for those they purport to serve."

Journalists reacted with dismay but not surprise when Marilyn Tavenner, who had been "chiefly responsible for" the administration's rollout of HealthCare.gov, was named "the top lobbyist for the nation's health insurance industry," as the *Times* worded its story. Tavenner's appointment "highlights how federal health programs have become a priority for insurers, which increasingly depend on revenues from Medicare and Medicaid and the new public insurance marketplaces," reporter Robert Pear wrote. *Politico* confirmed that "the insurance industry has largely made peace with Obamacare," and the *Times* highlighted how the United States remained "the most expensive place in the world to get sick."[50]

Overseas, Barack's foreign policy legacy was faring worse than his domestic one. The *Washington Post*'s Jackson Diehl wrote that thanks to Barack, all around the world "the dictators are winning." *Post* columnist Richard Cohen mocked Barack's "ringing call to do as little as possible" in Syria, and Cohen's colleague David Ignatius highlighted how "Obama's failure to develop a coherent strategy left the field open" for Russian dictator Vladimir Putin to wreak bloody havoc. Barack's "policy of minimalism" was one that he defended "with an off-putting petulance," Cohen complained, yet Cohen understood how Syria was "a foreign policy debacle in which the measure of Obama has been taken. He's been bullied off the playground," or as someone might once have put it in Springfield, punked by Putin.

In early December, the weightiest and most widely respected voice of all joined the chorus of critics, albeit without using Barack's name. Decrying "self-serving politicians more concerned with getting reelected than with the nation's future," former defense secretary Robert Gates wrote that

> *the next president must be resolute. He or she must be very cautious about drawing red lines in foreign policy, but other leaders must know that crossing a red line drawn by the president of the United States will have serious—even fatal—consequences. The public, members of Congress and foreign leaders alike must know that the president's word is his or her bond, and that promises and commitments will be kept and threats will be carried out. The next president must hold people in government accountable; when programs or initiatives are bungled, senior leaders should be fired. He or she needs to have the courage to act in defiance of public opinion and polls when the national interest requires it.*

Gates also wrote that "we desperately need a president who will strive tirelessly to identify and work with members of both parties in Congress interested in finding practical solutions to our manifold problems," someone who would be a "true unifier of Americans." Gates added that "we need a president who is restrained . . . in rhetoric, avoiding unrealistic promises, exaggerated claims of success and dire consequences if his or her initiatives are not adopted exactly as proposed. Restrained in expanding government when so much of what we have works so poorly . . . Restrained from questioning the motives of those who disagree and treating them as enemies with no redeeming qualities." It was a devastating verdict on Barack's presidency from a figure of unchallenged international stature who for over two years had worked alongside him for better and for worse.

As 2015 turned to 2016, denunciations of the United States' glaring failure in Syria increased even further. The *Post*'s Jackson Diehl thought Barack's stubbornness had reached the point of being "untethered from reality." The *Post*'s editorial board decried how "Putin is reveling in the geopolitical victory handed to him by the Obama administration," whose "passivity" allowed Putin "to act as he chooses." Lamenting "the absence of American leadership," former Democratic senator Joe Lieberman dramatically disclosed what one European head of government had told him the United States needed to do: "Elect a president who understands the importance of American leadership in the world."[51]

As if Bob Gates's *Washington Post* op-ed was not enough, Barack's other former Republican cabinet member, four-year secretary of transportation Ray LaHood, weighed in as well. In a memoir of his years in politics, LaHood wrote, "I do not believe the White House ever committed fully to a genuine bipartisan approach to policy making." As president, "Obama depended almost exclusively on a handful of folks" inside the White House, and Barack "rarely sought counsel outside that group. . . . As time passed, the president seemed to me to become more isolated, more insulated from those outside the in-group, less engaged with others." In *Politico,* another once-supportive Republican rued "the ugly partisanship" that had permeated "these long, bitter, brutal years" of Barack's presidency. But "I don't think he really tried all that hard" to remedy that, "and even if he did, he failed spectacularly."

On February 10, 2016, the ninth anniversary of his announcement for the presidency outside the Old State Capitol, Barack returned to Springfield to address the Illinois General Assembly and visit briefly with old friends of both parties from years past. "When he saw me, he gave me a big hug," former Republican state senator Dave Sullivan recounted, and perhaps only in Springfield could Barack find a Republican whom he would hug. Speaking to the legislature, Barack warmly recalled former Senate president James "Pate" Philip and commended still-sitting Republican senators Christine Radogno and Dave Syverson for how they had all worked well together years earlier. Barack conceded "my inability to reduce the polarization and meanness in our politics," but when old friends Kirk Dillard, Denny Jacobs, and Larry Walsh joined Barack for a public Q&A at a nearby arts center, Barack disclaimed responsibility for why the nation's politics were now so much more vile than the Illinois statehouse's fifteen years earlier. "It's not like I changed. I'm the same guy now that I was then," Barack asserted. "The interpersonal stuff makes a huge difference," Barack rightly noted. "I've gotten to know George W. Bush quite well" and "he's a good man." Even in the U.S. Senate, "I had very good relationships and friendships with" Republican senators "who now can't take a picture with me. It wasn't like I changed," Barack insisted, yet his years in the White House had witnessed virtually none of the bipartisan outreach he had practiced once upon a time with Sullivan and Dillard, nor any of the non-ideological Wednesday-night poker games he had played with conservative Democrats Jacobs and Walsh.

More than five decades earlier, Norman Mailer had pinpointed the "elusiveness" of another president, "the detachment of a man who was not quite real to himself," one who would come to epitomize presidential toughness while achieving what success he did thanks to a rich cast of senatorial friends and journalistic buddies. In Washington, Richard Wolffe, who had by far the best camaraderie with Barack of any reporter, had come to grasp what Genevieve and Sheila had realized about Barack years earlier, "that he willed himself into being." In Springfield too a perceptive woman understood how Barack "is an invention of himself." But it was essential to appreciate that while the crucible of self-creation had produced an ironclad will, the vessel was hollow at its core. "You didn't let anyone sneak up behind you to see emotions—like hurt or fear—you didn't want them to see," Barack long ago had taught himself, yet hand in hand with that resolute self-discipline came a profound emptiness. "I had no idea who my own self was," Barack had realized at Punahou, and while he had indeed "willed himself into being"—as an African American man, as a loving father, and as a successful politician—eight years in the White House had revealed all too clearly that it is easy to forget who you once were if you have never really known who you are.[52]

ACKNOWLEDGMENTS

More than twenty years ago, in the acknowledgments for *Liberty and Sexuality,* I wrote that books like that, and this, fundamentally depend on "the kindness of strangers." Many of the more than one thousand people who took the time to speak with me for *Rising Star* are no longer strangers, and a number of the most important and most helpful have become personal friends, yet there is no gainsaying how a book as long and detailed as this one comes into existence only because of the assistance of hundreds and hundreds of people—interviewees, research assistants, and especially librarians and archivists worldwide.

Unlike some authors, I do one hundred percent of my own interviewing, reading, and note taking, and I will not repeat here the names of everyone listed in the bibliography who was kind enough to speak with me, whether in person, by telephone, or by e-mail. Early on in this nine-year project, thanks to Chris Stansell and Jane Dailey, Alix Lerner became my first Chicago-based research assistant, contributing invaluably—never more so than when she perused the old phone directories sitting on the shelves of Regenstein Library. When Alix left Hyde Park to begin graduate study at Princeton, Chris thankfully introduced me to U of C Ph.D. candidate Ali Lefkovitz, now an assistant professor of history at Rutgers University–Newark, who suffered through hours and hours of microfilm review in obscure libraries but who also accompanied me on an utterly unforgettable interviewing trip southward from Cook County.

Thanks again to Jane, after Ali completed her doctorate, U of C Ph.D. candidate Sarah Miller-Davenport stepped in in Ali's stead, and after Sarah accepted a lectureship at the University of Sheffield, my dear friend Dennis Hutchinson recruited Italia Patti—a 2014 U of C Law School J.D. who is now clerking on the U.S. Court of Appeals for the Sixth Circuit—to complete my Chicago research work. Very early on, Mike Klarman kindly lent me his Harvard Law School research assistant, Jenn Schultz (now a senior associate at Goodwin Procter LLP), for Harvard library work, and some years later wonderful Beryl Satter recruited Christopher Witrak to do some very savvy library work for me in New York City. Very late in the game, superb Steve Smith from Montreal performed yeoman picture-taking in central Harlem.

Many journalists, writers, and scholars whose earlier works have touched on one or another part of this huge story shared information and advice. Liza Mundy, Janny Scott, and Sally Jacobs top this list, but it is long indeed: Jim Kloppenberg, Rachel Swarns, Jim Meriwether, Tom Shachtman, Tenisha

Armstrong, Edgar Tidwell, Phil Dougherty, Verica Jokic, John Conroy, Nancy Hewitt, Alan Brinkley, Dick Powers, Evan Gahr, Jim Sleeper, Nan Rubin, Al Brophy, Scott Helman, Larry Gordon, Serge Kovaleski, Bob Secter, Hank DeZutter, David Moberg, Ted McClelland, Eleanor Kerlow, Kurt Andersen, Randy Kryn, Becca Williams, Mark Johnson, Nancy Benac, Peter Wallsten, Lawrence Hurley, Will Saletan, Jacob Weisberg, David Shribman, Ken Gormley, Mike Leahy, Stanley Kurtz, and Jerome Corsi. Jamie Weinstein shared with me some invaluable files, and Robert Draper kindly gave me his notes from some pioneering, early interviews.

From Amsterdam, Arlette Strijland at Atria sent me copies of Ann Dunham's unpublished writings; in Hawaii, Carlyn Tani and Kylee Mar at Punahou were especially helpful, as were Bron Solyom and Ellen Chapman at the University of Hawaii at Manoa. Ian Mattoch, Emme Tomimbang, Joanne Corpuz, Andrew Walden, and multiple staff members at the Hawaii State Library all aided me too as did Mark Davis and Fred Whitehead. Susan Corley, Alec Williamson, Tom Topolinski, Mark Hebing, Keith Peterson, and Pete Vayda all provided instructive items, and archivists at a trio of institutions—Nicole Dittrich and Nicolette Dobrowski at Syracuse University; Peter Berg and Peter Limb at Michigan State University; and Steven Fullwood, Mary Yearwood, and Chris McKay at the Schomburg Center—all helped provide invaluable documents concerning Barack H. Obama Sr.

Of all the people who devoted scores of hours to aiding and coaching me on this huge project, none revealed a more pure spirit than the late Zeituni Onyango Obama, who welcomed me to her modest Boston abode and then patiently showered me with scores of e-mails, explaining the intricacies of her family's history in Kenya and beyond, prior to her untimely passing in April 2014 at age sixty-one. I recall her voice fondly, and with tears.

In Eagle Rock, the kindness of Jean Paule and the efforts of Jim Jacobs were especially helpful, as were the contributions of Dale Stieber, Vanessa Zendejas, Debra Plummer, and Kyle Herrara. Adam Sherman, Susan Keselenko Coll, and John Drew all shared items from long ago, but my greatest Occidental debts without question are owed to Margot Mifflin and Phil Boerner, both of whom time and again shared with me memories and reached out to others on my behalf, easing my path with people who had learned all too well that many callers came with hidden agendas either political or pecuniary.

Beginning in May 2010, Alex McNear extended to me her trust, welcoming me into her home and sharing with me almost in full the letters she had retained from 1982 to 1984. In September 2012, Genevieve Cook (whose present-day surname is omitted from this book) similarly entrusted me, hosting me and my spouse at her and her partner's home, and then sharing in toto the letters she had retained from 1985 to 1986. Andy Roth and Keith Patchel were invaluably helpful, as were Jeremy Feinberg and Greg Poe. Jocelyn Wilks and Tara Craig at Columbia University, Sydney Van Nort and Samuel Sanchez at City College, and Derick and Jeremy Schilling all aided my research as well. If there was a gold medal for reference librarians, it would go to Theresa McDevitt at Indiana University of Pennsylvania, who found the rarest thing of all, and Glee Murray then unearthed similar items.

In Chicago, my innumerable debts stretch literally from Altgeld Gardens and Hegewisch to the Evanston border. Greg Galluzzo, Mike Kruglik, Jerry Kellman, Paul Scully, and especially Ken Rolling all willingly gave me hours of their time on multiple occasions, and no single comment has more enriched this book than Greg's recommendation that I contact Mary-Ann Wilson, whose file folder from 1986 to 1987 introduced me to a new world of wonderful people. Roberta Lynch, Dick Poethig, Robin Rich, Adrienne Jackson, Eva Sturgies, Rick Williams, David Kindler, Ed Grossman, Joan Keleher, and Gabriela Franco at St. Victor Parish all unearthed items from years past. Jeremiah Wright, Mike Pfleger, and terrific Hermene Hartman all kindly vouched for me to parishioners and colleagues, and when Jeremiah first referred to me as "Brother Garrow," I knew once again who was still blessing me. Ivey Matute-Brooks and Charles Lofton likewise helped me, as did Anita Beard, Gregory Callaway, Pandwe Gibson, Lisette Spraggins, Mary Pat White, Father Robert Cooper, Rachel Mikva, Dorothy Shipps, Niaz Dorry, Charlie Cray, Ben Gordon, and Robert T. Gannett. Glenn Humphreys and especially Teresa Yoder were wise guides at the Harold Washington Library, as were Michael Flug and Beverly Cook at the Harsh Collection on 95th Street. Valerie Harris does a superb job at the University of Illinois at Chicago, as did Debbie Vaughan at the Chicago Historical Museum. May the ghost of Archie Motley take revenge. Tom Joyce and Malachy McCarthy showed me the riches at the Claretian Missionaries Archives, as did Rod Sellers at the Southeast Side Historical Society, and a very special thank you goes out to now-Justice Mary Yu. Kathryn DeGraff and Kelly Gosa aided me at DePaul University, as did Harry Miller at the State Historical Society of Wisconsin; Ashley Howdeshell at Loyola University Chicago; Nicole Gotthelf at the Center for Neighborhood Technology; Lisa Jacobson, David Koch, and Quincina Jackson at the Presbyterian Historical Society; Nancy Carroll at Wartburg Theological Seminary; Julie Satzik and Jac Treanor at the Archdiocese of Chicago; and Wm. Kevin Cawley at the University of Notre Dame. Regina McGraw and Carmen Prieto valuably helped me at the Wieboldt Foundation, as did Deborah Harrington at the Woods Funds, Barbara Denemark Long at the Chicago Community Trust, Richard Kaplan at the MacArthur Foundation, and Katherine Litwin at the Donors Forum. It is unfortunate how the present-day Joyce Foundation stands apart from the helpful traditions of Chicago's philanthropic community.

For years and years forward from August 2009, Sheila Miyoshi Jager responded helpfully and often enthusiastically to my many queries, notwithstanding how some of them caused her to revisit or experience deep emotional pain. As a dedicated scholar herself, Sheila understood from day one the importance of historical accuracy, even when she understandably rued how it eventually would intrude on her and her family's privacy. I believe that in keeping Sheila's many confidences from 2009 until 2017, the best possible balance between a rightful desire to live one's life free from paparazzi scum while accommodating the demands of history has been struck.

From January 2010 onward, no one has been a more encouraging and buoyant historical companion than Rob Fisher. From the discovery of the first half of a long-forgotten book manuscript to his mother's serendipitous unearthing of the second half, Rob has been tremendously helpful again and again and again. Mark Kozlowski, Gina Torrielli, Jackie Fuchs, Radhika Rao, Brad Berenson, John Parry, Adam Charnes, Trent Norris, William Terry Fisher, Joel Freid, and my Pitt Law colleague Debbie Brake all likewise unearthed valuable items from years earlier. At Harvard Law School, Nicola Seaholm, Margaret Peachy, and especially Lesley Schoenfeld were very helpful, as was Fran O'Donnell at Harvard Divinity School. Karen Cariani and Keith Luf at WGBH; Jen Christensen at CNN; Peter Filardo, Alison Lotto, and Janet Bunde at New York University; and Jeff Flannery at the Library of Congress all helped uncover previously unknown materials. Special thanks also go to David Wigdor, Steve Wermiel, Laura Demanski, Dale Carpenter, and Celeste Moses.

My longtime friend and academic colleague Dennis Hutchinson warmly welcomed me to the University of Chicago Law School again and again, and I thank many members of the law school's support staff for their assistance when I was working in one or another office in that wonderful building. Great Geof Stone and Tom Rosenblum both aided my Chicago research, as did Adam Gross, Michael Risch, Marty Chester, Lisa Ellman, and Peter Steffen, as well as Julia Gardner and Barbara Gilbert at the U of C Library. Jesse Jackson Jr. and Martin King were exceptionally helpful and forthcoming, and John Levi, Judd Miner, Jeff Cummings, Scott Lynch-Giddings, Marilyn Katz, Katrina Browne, Stephen Heintz, Sandy O'Donnell, Tracy Leary, Ellen Schumer, Tom Johnson, Alan and Lois Dobry, and Rob Mitchell all located valuable items from years past, as did Lindsay Booker at Echoing Green, Maggie Bertke at the Appleseed Network, and Andrew Mooney at LISC.

No one could have been a more relentlessly helpful and instructive guide to the world of Illinois politics than Dan Shomon, who scores and scores of times introduced me to people who had never before recounted their memories. Likewise, Dave Sullivan, Carter Hendren, Jo Johnson, and Patty Schuh all invaluably aided my outreach to others. In Springfield, Kathleen Bloomberg at the Illinois State Library enthusiastically arranged the greatest interlibrary loan feat of all time, and Mike Lawrence dug through old computer files to discover invaluable memos. Vicki Thomas, Claire Eberle, and Rita Messinger at JCAR; Walter Ray and Matt Baughman at Southern Illinois University; Dave Joens at the Illinois State Archives; Terry Martin at the Illinois Channel; Yvonne Davis at WTTW; Bruce DuMont and John Gieger at the Museum of Broadcast Communication; and Angela White at Indiana University–Purdue University Indianapolis all helped unearth valuable old materials. William Atwood, Tim Blair, Karen Maggio, and Linsey Schoemehl all helped with my many FOIA requests to different Illinois state government entities. Judy Byrd, Ellen Chube, Jai Winston, Tara Zavagnin, and Donna Anderson all deserve special thanks, as does Arthur Miller at Lake Forest College.

Chris Sautter, Andrew Gruber, Laura Hunter, Eden Martin, Fred Lebed, Ed Wojcicki, Jonathan Goldman, Stephen Pugh, Don Wiener, and Marv Hoffman all searched through old files to locate important items. John Kupper shared a large box of invaluable files, and Rick Ridder proved that gold can be found in Colorado basements. Terry Walsh, Pete Giangreco, Terrie Pickerill, Paul Harstad, Joe McLean, Jim Cauley, Mark Blumenthal, and Ann Bertell all likewise shared instructive items from 2002 to 2004, and Peter Fitzgerald was especially helpful.

When this book first began in 2008–2009, Kate Pretty and John Gray at Homerton College at Cambridge University extended tangible and crucial support. My affection for Homerton remains unequaled, and I thank so many of my former Fellows, especially Peter Raby, Louise Joy, and Rich Williams for their deeply supportive friendship. Stuart Marcelle has eased my work as much as anyone, and both the Cambridge Blue and the Live and Let Live remain wonderful real ale freehouses. At the University of Pittsburgh School of Law, Dean Chip Carter supported my work on this book to a degree one would find in few places in academia. Law librarian extraordinaire Marc Silverman has contributed as much to this book as Alix and Ali, and never more so than when he unearthed Bob Elia's identity from old property records. At Pitt, I have benefited again and again from the wonderful assistance of Patty Blake, and from the help of Sue Leroy, Amy Change, LuAnn Driscoll, Sarah Lynn Barca, Kim Getz, and Mike Dabrishus. Tom Ross, Jessie Allen, Larry Frolik, and Doug Branson have been especially supportive colleagues, as have Ed McCord and Pat Gallagher; and Tony Infanti, Nancy Burkoff, Tony Novosel, and John Stoner have greatly aided my teaching.

Many longtime friends—my wonderful former spouse, Susan Newcomer; Kim Fellner; Charles Kaiser; Ellen Chesler; Steve Fayer; Keith Miller; Jim Cone; Rachelle Horowitz; JoBeth McDaniel; and Marland Buckner; as well as Pete and Matt Kurzweg, and Carole and Tom Horowitz in Pittsburgh, Catherine O'Brien in D.C., and Patrick Parr—have all been wonderfully supportive in manifold ways. Gary Pomerantz, Laura Hunter, Charles and Brenda Eagles, and Mark Bradley and Liza Mundy all willingly devoted hours of their time to reading this book in final manuscript form, and I remain undyingly grateful to them all.

Andrew Wylie has been unstinting in his support for this complicated endeavor, and Jackie Ko and Lauren Rogoff have been superb as well. Steve Wasserman and especially Scott Moyers provided valuable advice for which I remain deeply indebted. At HarperCollins, Henry Ferris and Nick Amphlett have been repeatedly helpful, and assistant general counsel Trina Hunn made the legal review as pleasant as possible. Throughout the final months of this herculean endeavor, senior production editor Dale Rohrbaugh oversaw a process whose complexity was exceeded only by its scale. Greg Villepique performed an excellent copyedit; Sarajane Herman conducted an astonishingly impressive proofread; and Cynthia Crippen produced a far better index than the author could have. Publicist Sharyn Rosenblum demonstrated her savvy even months prior to publication.

Beginning in July 2013, Bob Bauer has striven impressively to balance an appreciation of the scholarly mission against the professional demands of representing the world's busiest client, and both Judy Casey and Ferial Govashiri have been unstintingly helpful. Barack Obama devoted dozens of hours to reading the first ten chapters of this manuscript, and his understandable remaining disagreements—some strong indeed—with multiple characterizations and interpretations contained herein do not lessen my deep thankfulness for his appreciation of the scholarly seriousness with which I have pursued this project and for what became eight full hours of always-intense "off-the-record" conversations.

No one has sacrificed more of their own personal happiness on account of this nine-year undertaking than has Darleen Opfer. There were scores of enjoyable experiences along the way, whether in New Zealand or Chicago, but there were also hours alone in a rental car in a CBS Studios parking lot and evenings when talkative interviewees made dinner a notional concept. Her support, from early on in Cambridge through her painstaking edit of the epilogue, has again and again made this a better book.

NOTES

CHAPTER ONE: THE END OF THE WORLD AS THEY KNEW IT

1. This account of Wisconsin Steel's downfall draws from scores of sources, some of which contradict others regarding particular details and timing. As a general rule, more contemporaneous accounts are relied upon more than subsequent recollections years later. The three best sources for Frank Lumpkin's experiences on March 28 are Larry Galica, "Wis. Workers Mark Closing," *DC,* 29 March 1984, p. A1; Bill Gleason, "Steelyard Blues," *CT Magazine,* 13 June 1982, pp. 13–21; and Michael Flannery, "Plant Closing Wrecks Plans," *CST,* 25 May 1980, p. 66. Barber Tony Mastronardi is quoted in C. D. Matthews, "Stream of Customers Now Only a Trickle," *DC,* 28 March 1985, p. A5. Other sources on Frank and Bea Lumpkin are copious, including Frank Lumpkin, *Keep Strong,* August 1980, p. 25, Richard Poethig Papers [RPP] Box 25; Lisa Newman, "So. Chicago's Man of Steel," *DC,* 11 April 1988, p. 1; Studs Terkel, "I'm Saying Racism Is Unnatural," *The Nation,* 6 April 1992, pp. 448–52 (an interview with Frank which later appeared in Terkel's book *Race*); Laurent Belsie, "Southeast Chicago," *CSM,* 20 October 1993, p. 10; R. C. Longworth, "Man of Steel—Frank Lumpkin's Boxing Days Are Over, but There's Plenty of Fight—and Heart—Left in Him," *CT,* 31 May 1995, p. T1; Cheryl Ross, "Love Is a Battlefield," *CR,* 4 May 2000; Beatice Lumpkin's two books, *"Always Bring a Crowd!"—The Story of Frank Lumpkin, Steelworker* (International Publishers, 1999) and *Joy in the Struggle: My Life and Love* (International Publishers, 2013); and memorial items from the time of Frank's death at age ninety-three: Trevor Jensen, "Frank Lumpkin, 1916–2010, Fought for Rights of Workers at Wisconsin Steel," *CT,* 6 March 2010; Phil Ponce on *Chicago Tonight,* WTTW, 9 March 2010; and John Bachtell, "Frank Lumpkin, 'Saint of Chicago,' Dies at 93," *People's World,* 10 March 2010, which acknowledges that Frank was "a longtime member of the CPUSA's National Committee."

Sources on the history of the Southeast Side's neighborhoods and the history of the steel industry there are copious, too, but anyone seeking a rich sense of that history must begin with Rod Sellers and Dominic A. Pacyga, *Chicago's Southeast Side* (Arcadia Publishing, 1998). Also crucial are John Conroy's impressively rich, five-part series of articles, all titled "Mill Town," *CM,* November 1976, pp. 164–82, December 1976, pp. 162–63, 210–19, 304, January 1977, pp. 114–25, 189, February 1977, pp. 106–15, 185, and March 1977, pp. 114–15, 132–46; the late Jerry Sullivan's beautifully written "Paradise Doomed," *CM,* August 1985, pp. 142–56, 160–61; Peter T. Alter, "Mexicans and Serbs in Southeast Chicago: Racial Group Formation During the Twentieth Century," *Journal of the Illinois State Historical Society* 94 (Winter 2001–2002): 403–19; and David Bensman, "South Deering," *Encyclopedia of Chicago,* 2004. Essential books include Paul A. Tiffany, *The Decline of American Steel: How Management, Labor, and Government Went Wrong* (Oxford University Press, 1988); William Kornblum, *Blue Collar Community* (University of Chicago Press, 1974); Richard M. Dorson, *Land of the Millrats* (Harvard University Press, 1981), a wonderful folklore ethnography of Indiana's steel towns and plants; and, first and foremost, David Bensman and Roberta Lynch's powerful and pioneering *Rusted Dreams: Hard Times in a Steel Community* (McGraw-Hill, 1987), esp. p. 69 (Wisconsin's shutdown "was the beginning of the end of an era"). See also Lance Trusty, "End of an Era: The 1980s in the Calumet," in James B. Lane, ed., "Life in the Calumet Region During the 1980s," *Steel Shavings* 21 (1992), pp. 1ff. On Edward R. Vrdolyak's career up through the mid-1980s, the essential sources are the November 1976, February 1977, and March 1977 installments of John Conroy's "Mill Town" articles, Robert J. McClory, "Is There Political Gold in Vrdolyak?," *II,* October 1979, pp. 9–12, and David L. Protess, "The Vrdolyak Chronicles," *CL,* June 1988, pp. 1, 16–20; see also Michael Miner, "Fast Eddie's Editor: An Unabashed Confession," *CR,* 16 June 1988.

The crucial, prescient story predicting Wisconsin's closing was Tom O'Boyle, "Steel Firm Faces Bankruptcy, Crippled by Harvester Strike," *CCB,* 10 March 1980, pp. 1, 28. Richard Longworth's subsequent pair of stories, "Wisconsin Steel Scrambles for Financial Aid to Stay Alive," *CT,* 17 March 1980, p. D10, and "Wisconsin Steel Seeks New Funds to Avoid Bankruptcy," *CT,* 27 March 1980, p. C7, are likewise essential, as are Charles Storch and Pam Sebastian, "Wisconsin Steel Fights to Survive," *CT,* 6 April 1980, p. D5; R. C. Longworth, "Steel Union Understated Dues Income in Reports to U.S.," *CT,* 27 April 1980, p. 6; R. C. Longworth, "Shutdown of Bankrupt Steel Mill Leaves U.S., Creditors Holding Bag," *CT,* 4 May 1980, p. M3; and Longworth's invaluable later essay, "Winners and Losers in Steel Firm Collapse," *CT,* 8 June 1980, pp. M1–2.

Essential sources for understanding Envirodyne's acquisition of Wisconsin Steel from International Harvester are William Gruber, "Fearless Buying: Tiny Firm Tackles Ailing Steel Outfit," *CT,* 7 August 1977, pp. B1, B14; Carol J. Loomis, "A Lot of Losers at Wisconsin Steel," *Forbes,* 19 May 1980, p. 94; Barbara Marsh and Sally Saville's stunningly impressive duo of stories, "International Harvester's Story: How a Great Company Lost Its Way," *CCB,* 8 November 1982, pp. 21–43, and "A McCormick's Mission: Revive the Sleeping Giant," *CCB,* 15 November 1982, pp. 19–42; Gordon L. Clark's extremely valuable "Piercing the Corporate Veil: The Closure of Wisconsin Steel in South Chicago," *Regional Studies* 24 (October 1990): 405–20 (which observes that "Harvester sold its steel division hoping that the new owner would be held responsible in whole or in part for the pension liabilities"); Circuit Judge Walter J. Cummings Jr.'s opinion in *Lumpkin v. Envirodyne Industries,* 933 F. 2d 449 (7th Cir.), 17 May 1991; and David C. Ranney, "The Closing of Wisconsin Steel," in Charles Craypo and Bruce Nissen, eds., *Grand Designs: The Impact of Corporate Strategies on Workers, Unions, and Communities* (ILR Press, 1993), pp. 65–91. See also John F. Wasik, "Waiting for the Brimstone," *Progressive,* March 1984, pp. 28–30; Patricia Moore, "Once-Proud Harvester Plowed Deep Furrows," *CST,* 2 December 1984, p. B1; James Fallows, "America's Changing Economic Landscape," *Atlantic,* March 1985, pp. 47ff. (agreeing that for Wisconsin's workforce March 28 "was the end of the world"); Gregory D. Squires et al., *Chicago: Race, Class, and the Response to Urban Decline* (Temple University Press, 1987); and Judith Stein, *Running Steel, Running America* (University of North Carolina Press, 1998), esp. pp. 249–50.

George Harper's recollections of March 28 appear in Bonita Brodt, "Steel Firm's Collapse Was Quick, Painful," *CT,* 21 August 1980, p. M2, and workers' recollections of that day appear in *News from Calumet Community Religious Conference* Vol. 1, #1, 9 November 1980 ("We got no warning"), and "Five Years and Fighting Hard," *Labor Today,* April 1985, p. 4 (FLP Box 5 Fld. 1). Contemporaneous stories include Storer Rowley, "Steel Mill Closes; 3,400 Idled," *CT,* 29 March 1980, p. M2; Dave Schneiderman, "Agency That OK'd Loan to Study Plant Closing," *CT,* 30 March 1980, p. 3; and John Wasik, "Wisconsin Steel Closes," *DC,* 31 March 1980, p. 1. See also Maury Richards, "Wisc. Steel Bankrupt," *1033 News & Views,* April 1980, pp. 1, 3. And last, but most certainly not least, the most emotionally powerful commentary of all on March 28—and the morning of March 29—is Christine J. Walley's account in "Deindustrializing Chicago: A Daughter's Story," in Hugh Gusterson and Catherine Besteman, eds., *The Insecure American* (University of California Press, 2010), pp. 113–39, esp. pp. 113 and 124, and *Exit Zero: Family and Class in Postindustrial Chicago* (University of Chicago Press, 2013), esp. pp. 1–2. *Exit Zero* can bring tears to your eyes each time you open it. This account benefits as well from DJG's interviews with Alma Avalos, Tom Geoghegan, Ken Jania, and Beatrice Lumpkin.

2. Longworth, "Winners and Losers in Steel Firm Collapse," *CT,* 8 June 1980, p. M2; *CT,* 1 April 1980, p. C6; *DC,* 1 April 1980, p. 1; John Wasik, "Court to Decide Wisconsin Fate," *DC,* 2 April 1980, p. 1; Editorial, "Wisconsin Mess A Danger to Us All," *DC,* 2 April 1980, p. 4; John Wasik, "Wisconsin Woes in Bankruptcy Court," *DC,* 3 April 1980, p. 1; Wasik, "Order Loan Release to Protect Coke Ovens," *DC,* 4 April 1980, p. 1; Wasik, "Wisconsin Lenders Seek Payments" and Wasik, "Layoff, Forms Mar Jim's Easter Weekend," *DC,* 5 April 1980, p. 1; Charles Storch and Pam Sebastian, "Wisconsin Steel Fights to Survive," *CT,* 6 April 1980, p. D5; John Wasik, "Insurance Benefits, Certainty Elude Wisconsin Workers," *DC,* 8 April 1980, p. 1; Editorial, "Save Wisconsin Bill Needs Support," *DC,* 8 April 1980, p. 4. Only come spring 1981 were workers allowed in to clean out their old lockers. Lynn Emmerman, "Lives Built on Steel Rust with Idleness," *CT,* 14 September 1986, p. T1.

3. John Wasik, "Wisconsin Workers Prompt Investigation into PSW Union," *DC,* 17 April

1980, p. 1; Amelia Forglia, "Unemployment Compensation Wanted Now," *CT,* 18 April 1980, p. D2 (letter to the editor); Leonard Roque to Dear Member, "Progressive Steel Workers Union Membership Meeting," 18 April 1980, FLP Box 2 Fld. 4; R. C. Longworth, "Plan to Reopen Steel Firm Studied," *CT,* 23 April 1980, p. C1; R. C. Longworth, "Harvester Will Pay Wisconsin Steel Benefits," *CT,* 26 April 1980, p. 6; Longworth, "Steel Union Understated Dues Income in Reports to Feds," *CT,* 27 April 1980, p. 6; Longworth, "Plan to Sell Steel Mill Inventory Is Approved," *CT,* 29 April 1980, p. B6; Longworth, "Strike May Kill Wisconsin Steel," *CT,* 2 May 1980, p. C8; Longworth, "Shutdown of Bankrupt Steel Mill Leaves U.S., Creditors Holding Bag," *CT,* 4 May 1980, p. M3; Longworth, "Firm Owning Wisconsin Steel Asks $750,000 'to Protect' Its Assets," *CT,* 7 May 1980, p. E3; John McCarron, "Bleak Future Seen for Wisconsin Steel," *CT,* 17 November 1980, p. 3.

4. DJG interviews with Tom Joyce and Dick Poethig; Richard P. Poethig, "Telling the Story: The Role of Information-Sharing in Urban and Industrial Mission," *International Review of Mission* 87 (January 1998): 113–22; National Conference on Religion and Labor, 13–14 May 1980, St. Thomas More College, Covington, KT, pp. 3–5, RPP Box 20; and Chuck Rawlings e-mail to DJG.

Sources on the Youngstown shutdown and its immediate aftermath are copious, but Staughton Lynd's excellent *The Fight Against Shutdowns* (Singlejack Books, 1982), Thomas G. Fuechtmann, *Steeples and Stacks: Religion and Steel Crisis in Youngstown* (Cambridge University Press, 1989), and Charles W. Rawlings's own subsequent "Steel Shutdown in Youngstown: Ecumenical Response to the Opening Hand of Globalization," *Church & Society,* January–February 2003, pp. 71–91, are the richest historical overviews. Also valuable are Lynd's contemporary essay, "The Fight to Save the Steel Mills," *NYRB,* 19 April 1979; Roger T. Wolcott, "The Church and Social Action," *Journal for the Scientific Study of Religion* 21 (March 1982): 71–79; Terry F. Buss and F. Stevens Redburn, "Religious Leaders as Policy Advocates: The Youngstown Steel Mill Closing," *Policy Studies Journal* 11 (June 1983): 640–47; and Lynd's later article, "The Genesis of the Idea of a Community Right to Industrial Property in Youngstown and Pittsburgh," *Journal of American History* 74 (December 1987): 926–58.

Charles W. Rawlings's initial memo was "The Steel Crisis and Growing Unemployment," 29 September 1977 (ICUIS 3700); his subsequent writings on Youngstown include "The Religous Community and Economic Justice," in John C. Raines et al., eds., *Community and Capital in Conflict: Plant Closings and Job Loss* (Temple University Press, 1982), pp. 136–51; "New Challenges for the Church and Labor," *Radical Religion* 5 (June 1981): 69–73; and "U.S. Steel vs. the People," *Christianity & Crisis,* 29 March 1982, pp. 75–80. NCEA's first two contributions were Gar Alperovitz, "Preliminary Observations: Youngstown, Ohio," 9 October 1977 (ICUIS 3698), and Alperovitz and Jeff Faux, "Youngstown Lessons," *NYT,* 3 November 1977, p. 25. ECMV's "A Religious Response to the Mahoning Valley Steel Crisis," 29 November 1977 (ICUIS 3796), is most easily available in Feuchtmann, *Steeples and Stacks,* pp. 137–45; Alperovitz summarized his preliminary report in "The Youngstown Experience," *Challenge,* July–August 1978, pp. 21–24, and his subsequent view is presented in Alperovitz and Faux, "When Steel Goes Cold," *NYT,* 12 December 1979, p. A31. Three oral histories conducted by Philip Bracy in early 1981 with ECMV members—Robert Campbell, John Sharick, and Richard Speicher—are available online, and the Ohio Historical Society possesses two small collections (MSS 793 and YHC MSS 0008) of ECMV Papers.

Contemporaneous journalism includes Gene Smith, "Youngstown Steel to Pare Operations," *NYT,* 20 September 1977, p. 1; William K. Stevens, "Shutdown of Steel Works Stuns Youngstown," *NYT,* 21 September 1977, p. A1; Robert A. Dobkin (AP), "Steel Workers Urge U.S. Restrictions on Imports," *WP,* 24 September 1977, p. D9; Agis Salpukas, "Steel Towns Hit by Layoffs Face Hard Times," *NYT,* 13 October 1977, p. D1; Salpukas, "U.S. Funds Available to Reopen Plants, Steel Group Is Told," *NYT,* 26 November 1977, p. 38; William H. Jones, "Steel Plant Reopening Is Possible, Report Says," *WP,* 16 December 1977, p. E9; AP, "Coalition Unveils Proposal to Reopen Idled Steel Plant," *WP,* 18 December 1977; Robert Reinhold, "H.U.D. Rushing to Revive an Ohio Steel Works," *NYT,* 31 December 1977, p. 27; Bill Peterson, "Closed Steel Plant Model for HUD Recovery Effort," *WP,* 31 December 1977, p. A2; Paula L. Cizmar, "Steelyard Blues," *Mother Jones,* April 1978, pp. 36–42; Edward Cowan, "Steel Merger? Debate Grows in Youngstown,"

NYT, 10 April 1978, pp. D1–2; Agis Salpukas, "Hope Glimmers for Reopening Campbell Steel Works," *NYT,* 12 April 1978, p. D1; Susanna McBee, "Desperate Jobless City Seeks to Buy Failing Steel Plant," *WP,* 9 May 1978, p. A2; Reginald Stuart, "Youngstown Seeks a Grasp on Its Fading Steel Industry," *NYT,* 20 June 1978, p. B10; Edward Cowan, "U.S. Approves Merger of LTV with Lykes, Creating Steel Giant," *NYT,* 22 June 1978, pp. A1, D13; Agis Salpukas, "At Youngstown Steel, Relief but No Joy," *NYT,* 22 June 1978, p. D13; Richard Brookhiser, "Mr. A Goes to Youngstown," *NR,* 7 July 1978, p. 835; Susanna McBee, "Youngstown Group Presses to Reopen Steel Mill," *WP,* 19 September 1978, p. A3; Marjorie Hyer, "Religous Leaders Backed in Plan to Reopen Steel Mill," *WP,* 28 September 1978, p. A28; Ivar Peterson, "Public Money and Private Ambition Clash over Future of Steel in Ohio's Mahoning Valley," *NYT,* 8 December 1978, p. A18; Agis Salpukas, "Steel Union, Making Concession, Backs Plant's Reopening in Ohio," *NYT,* 30 March 1979, p. A10; Roger Wilkins, "Carter Rejects a Plan to Reopen Steel Plant in Ohio," *NYT,* 31 March 1979, p. 46; Wilkins, "Reopening a Steel Plant in Ohio: Reaction to the Federal Rejection"; Robert Howard, "Going Bust in Youngstown," *Commonweal,* 25 May 1979, pp. 301–5; Nick Kotz's excellent overview, "Youngstown's Tragedies: A Legacy for Other Cities," *WP,* 17 June 1979, pp. B1, B4–B5; and John Logue, "When They Close the Factory Gates," *Progressive,* August 1980, pp. 16–23. Note as well Judge George C. Edwards's opinion in *Local 1330 United Steelworkers v. United States Steel Corp.*, 631 F. 2d 1264 (6th Cir.), 25 July 1980.

5. On Leo T. Mahon, Leo's own wonderful and impressive autobiography, coauthored with Nancy Davis, *Fire Under My Feet: A Memoir of God's Power in Panama* (Orbis Books, 2007), is an essential starting point; Thomas G. Kelliher Jr.'s rich dissertation, "Hispanic Catholics and the Archdiocese of Chicago, 1923–1970" (University of Notre Dame, December 1996), esp. pp. 131–35, 181–84, 197, 203, 311–12, 339, and Robin Rich's unpublished paper, "Explaining Historical Change: The Development of Christian Base Communites in Latin America," March 1991 (RRP), which is based upon a May 1990 interview with Leo, are essential, too. DJG's interviews with Leo Mahon, Bill Stenzel, Bernie Pietrzak, Jan Poledziewski, Fred Simari, and Christine Gervais, as well as Greg Sakowicz's 2008 interview with Leo, all inform this account; Leo's support of the ERA appears in "From the Pastor's Desk," *St. Victor Parish Sunday Bulletin,* 4 May 1980, p. 2. See also Leo Mahon, "Law and Creativity," *Upturn,* February–March 1985, p. 13, Leo T. Mahon, *Jesus and His Message: An Introduction to the Good News* (ACTA Publications, 2000), and his Chicago archdiocesan obituary, "Archdiocesan Priest, Msgr. Leo T. Mahon, Dies," 22 May 2013.

On Saul Alinsky, and von Hoffman's work with him as well, the definitive biography is Sanford D. Horwitt, *Let Them Call Me Rebel: Saul Alinsky, His Life and Legacy* (Alfred A. Knopf, 1989); see also Lawrence J. Engel, "Saul D. Alinsky and the Chicago School," *Journal of Speculative Philosophy* 16 (2002): 50–66. On John Cardinal Cody, Charles W. Dahm's authoritative *Power and Authority in the Catholic Church: Cardinal Cody in Chicago* (University of Notre Dame Press, 1981); Andrew M. Greeley, *Confessions of a Parish Priest: An Autobiography* (Simon & Schuster, 1986), pp. 133, 406–19; John Conroy's remarkable "Cardinal Sins," *CR,* 4 June 1987; Andrew M. Greeley's "The Fall of an Archdiocese," *CM,* September 1987, pp. 128–31, 190–92; Bill Clements, "Uncovering the Cardinal," *CM,* December 2002, pp. 66–87; and Roy Larson, "In the 1980's, A Chicago Newspaper Investigated Cardinal Cody," *Nieman Reports,* Spring 2003, are essential; see also Alexander L. Taylor III, "God and Mammon in Chicago," *Time,* 21 September 1981, and Linda Witt and John McGuire, "A Deepening Scandal over Church Funds Rocks a Cardinal and His Controversial Cousin," *People,* 28 September 1981. Roy Larson, Bill Clements, and Gene Mustain's pioneering *CST* stories on Cody, which debuted on September 10, 1981, are accessible only on microfilm.

6. DJG interviews with Tom Joyce, Dick Poethig, and Leo Mahon; "Present at the Meeting at St. Victor's, Calumet City, on the Closing of the Area Steel Mills," [6 June 1980], Tom Joyce to the Committee to Follow Up on the St. Victor's Meeting Concerning the Closings in the Steel Mills, [7 June 1980], "Steering Committee Minutes," 23 June 1980, Ensign Leininger to Dick Poethig, 23 June 1980, "Meeting of the Steering Committee on the Steel Crisis in the Calumet Region," 7 July 1980, all RPP Box 20.

7. "Wisconsin Workers Meeting," 7 July 1980, FLP Box 2 Fld. 4; R. C. Longworth, "Union May Try to Block Steel Inventory Removal," *CT,* 19 July 1980, p. 6; Leo T. Mahon, "From the Pastor's Desk," *St. Victor Parish Sun-*

day Bulletin, 27 July 1980, p. 2; Bob Gehring et al., to Dear Pastor & Church Member, 28 July 1980, RPP Box 19; "Planning Committee Meeting . . . ," 4 August 1980, RPP Box 20. See also Larry Galica, "Wisconsin Steel Five Years Later, Closure Still Haunts Community," *DC,* 28 March 1985, p. A5.

8. Mary Gonzales Weekly Schedule, Latino Institute Papers (LIP), Box 4; DJG interviews with Mary Gonzales, Greg Galluzzo, Bob Creamer, Tom Cima, and Bea Lumpkin; Constantine Angelos, "Feisty Activist Group Disbanding—The Power-Tweaking Sesco Has Long List of Accomplishments," *Seattle Times,* 17 September 1991; Minutes, "Administrative Staff Meeting," Latino Institute, 17 January 1980 and 5 March 1980, LIP Box 5; Mary Gonzales, "A Proposal Presented by United Neighborhood Organization of Southeast Chicago," June 1980, LIP Box 103 Fld. 15; Minutes, "Administrative Staff Meeting," Latino Institute, 3 July 1980, LIP Box 5; Donald C. Reitzes and Dietrich S. Reitzes, "Alinsky in the 1980s: Two Contemporary Chicago Community Organizations," *Sociological Quarterly* 28 (1987): 265–83; Wilfredo Cruz, "The Nature of Alinsky-Style Community Organizing in the Mexican-American Community of Chicago: United Neighborhood Organization," Ph.D. dissertation, University of Chicago, December 1987; Wilfredo Cruz, "UNO: Organizing at the Grass Roots," *II,* April 1988, pp. 18–22; Danny Westneat, "Obama Won Race from Ground Up," *Seattle Times,* 4 June 2008, p. B1.

9. Leo T. Mahon, "From the Pastor's Desk," *St. Victor Parish Sunday Bulletin,* 17 August 1980, p. 2, and 24 August 1980, p. 2; "Clergy Focus on Steel Layoffs," *CT,* 23 August 1980, p. N10; Mark Kiesling, "Churches Fight Steel Pullout," [Lansing] *Sun Journal,* 26 August 1980 (RPP Box 20); [Dick Poethig], "Meeting of the Calumet Religious Community," 23 August 1980, and "Organizing Committee for the Calumet Religious Community Conference," 26 August 1980, both RPP Box 20; and DJG interviews with Leo Mahon, Dick Poethig, Tom Joyce, and Roberta Lynch.

10. R. C. Longworth, "Steel Workers Reject Chase Back Pay Offer," *CT,* 5 August 1980, p. C4; Pam Sebastian, "Wis. Steel Gets $1.6 Million," *CT,* 26 August 1980, p. C7; "Marchers Protest Wis. Steel Pay Deal," *CT,* 30 August 1980, p. 5. On Walt Palmer, see Pam Sebastian, "Wis. Steel Hopes Rise Just a Bit," *CT,* 31 August 1980, p. M5; "Black Activist Trying to Revive Steel Concern," *WSJ,* 23 September 1980, pp. 29, 32; Frank Lumpkin to Anthony Roque, 1 October 1980, FLP Box 2 Fld. 4; R. C. Longworth, "Aid Coming to Reopen Mill: Byrne," *CT,* 7 October 1980, p. M1; John Wasik, "Byrne Vows November Mill Opening," *DC,* 7 October 1980, p. 1 (FLP); R. C. Longworth and Pam Sebastian, "Wisconsin Steel Won't Reopen Soon," *CT,* 12 October 1980, p. M5; John McCarron and Richard Longworth, "Byrne Tells Plan for Wisconsin Steel," *CT,* 1 November 1980, p. M3.

11. Minutes, CCRC, 30 September 1980, "Toward a Founding Convention of the CCRC," 30 September 1980, Minutes, CCRC, 14 October 1980, all RPP Box 20; Leo T. Mahon to Dear Sirs (Inland Steel), Richard P. Poethig to Sarah Cunningham, 17 October 1980, both RPP Box 24; Minutes, CCRC, 28 October 1980, RPP Box 20; "We Face Today a Crisis of Unprecedented Dimensions," CCRC trifold flyer, n.d., Roberta Lynch Papers (RLP); *News from Calumet Community Religious Conference,* Vol. 1, #1, 9 November 1980, RLP; [Roberta Lynch], "Director's Report," 10 November 1980 and "Agenda," CCRC, 11 November 1980, RPP Box 20; Richard P. Poethig, "The Church's Response to Plant Closings," 12 November 1980, RPP Box 19 and ICUIS Box 74 Fld. 1295; Richard P. Poethig, "A Strategy Paper for the Participation of Protestant Congregations in the CCRC," 17 November 1980, RPP Box 19; Minutes, CCRC, 25 November 1980, *News from Calumet Community Religious Conference* Vol. 1 #2, 28 November 1980, RLP, "Steering Committee and Who It Represents," n.d., "1981 January–December CCRC Budget Proposal," n.d., Minutes, CCRC, 16 December 1980, all RPP Box 20; *News from Calumet Community Religious Conference* Vol. 1, #3, 19 December 1980, RLP; Roberta Lynch to Dick Poethig et al., 29 December 1980, Poethig Papers; "CCRC 1981 Program," n.d., RPP Box 20; Richard Poethig, "Crisis in Steel Industry," *A. D.,* January 1981.

12. Minutes, Administrative Staff Meeting, 31 October 1980, LIP Box 5; DJG interviews with Mary Gonzales, Greg Galluzzo, Tom Cima, and Ted Aranda; Lynn Emmerman, "Picketing Steelworkers Claim Byrne Broke Vow," *CT,* 28 November 1980, p. B6; [Tom Joyce], "Claretian Social Development Grants for 1981," December 1980, CSDF Papers; Wieboldt Foundation, 1980 Annual Report, p. 14; Woods Charitable Fund, Report for the Years 1979 and 1980, pp. 3, 12.

13. CCRC Research Committee Meeting, 5 January 1981, RPP Box 20; Samuel Ayyub

Bilal, "Look at Steel Employee's Crisis," *Bilalian News,* 9 January 1981, pp. 7, 28; [Roberta Lynch], "Organizing in 1981," [13 January 1981]; Minutes, CCRC Steering Committee, 13 January 1981, both RPP Box 20; *News from Calumet Community Religious Conference* Vol. 1, #4, 16 January 1981, RLP; Roberta Lynch and John Beckman to CCRC Steering Committee, 19 January 1981, RPP Box 20; Leo Mahon to Jack Egan, 26 January 1981, Egan Papers Box 151 Fld. 21; *News from Calumet Community Religious Council* Vol. 1, #5, 6 February 1981; and Richard P. Poethig, "The Calumet Steel Economy and the Church's Response," 9 February 1981, both RPP Box 19; Richard P. Poethig, ed., "Plant Closings: The Church's Response," *Justice Ministries: Resources for Urban Mission* 10 (Fall 1980) [actually post–6 February 1981], 54pp., esp. p. 16; CCRC, "Training Session for Local Congregation Seminars," [16 February 1981], 8pp., RLP and RPP Box 19; *News from Calumet Community Religious Conference* Vol. 1, #6, 20 February 1981, RLP; Roberta Lynch to Steering Committee, "CCRC Progress," 26 February 1981; *News from Calumet Community Religious Conference* Vol. 1, #7, 6 March 1981; Minutes, CCRC Steering Committee, 10 March 1981; [Dick Poethig Notes], 10 March 1981; Leo Mahon to Dear Pastor, 19 March 1981, all RPP Box 20; *News from Calumet Community Religious Conference* Vol. 1, #8, 20 March 1981, RLP; *News from Calumet Community Religious Conference* Vol. 1, #9, 3 April 1981, RPP Box 20. See also Robert McClory, "Alinsky Lives," CR, 3 April 1981, pp. 1, 28–36; and James Balanoff, "U. S. Steel Cons the Congress and Union," *ITT,* 20–26 January 1982, p. 11.

14. Pam Sebastian, "U.S. Agency Buys Wisconsin Steel," *CT,* 21 January 1981, p. C3; "Relief Office to Assist Idle Steelworkers," *CT,* 22 January 1981, p. B1; Pam Sebastian, "No Fast Revival Seen at Wisconsin Steel Despite Role of U.S.," *CT,* 25 January 1981, p. M5; Mitchell Locin, "200 March on CTA, Win Concessions," *CT,* 30 January 1981, p. M6; "How to Organize Friends and Influence People—*Salt* Interviews Greg Galluzzo," *Salt,* June 1983, pp. 21–25; Pam Sebastian, "New Cloud Clings to Wisconsin Steel," *CT,* 22 February 1981, p. M5; Sebastian, "Pritzkers May Help Revive Steel Mill," *CT,* 15 March 1981, pp. 1, 4; Eileen Ogintz, "Laid-Off Workers Plan Lobby Effort," *CT,* 16 March 1981, p. C6; *CRec,* 9 April 1981, pp. HR7172–75. Unemployment benefits ended after thirty-nine weeks. Lynn Emmerman, "Lives Built on Steel Rust with Idleness," *CT,* 14 September 1986, p. T1. On mini-mills, see Christopher Scherrer, "Mini-Mills: A New Growth Path for the U.S. Steel Industry?," *Journal of Economic Issues* 22 (December 1988): 1179–1200, and Breandan O'Huallachain's excellent "The Restructuring of the U.S. Steel Industry," *Environment and Planning A* 25 (September 1993): 1339–59.

15. Minutes, CCRC Steering Committee, 14 April 1981; Roberta Lynch to CCRC Steering Committee, 24 April 1981, *News from Calumet Community Religious Conference* Vol. 1, #10, 1 May 1981; Roberta Lynch to File, "Illinois Presbyterians," 1 May 1981; Roberta Lynch to File, "Other Denominations," 5 May 1981; Minutes, CCRC Steering Committee, 12 May 1981; Minutes, CCRC Steering Committee, 9 June 1981; Minutes, CCRC Steering Committee, 12 August 1981, all RPP Box 20; *Calumet Reporter,* September 1981, RLP; *Calumet Reporter,* October 1981, Poethig Papers; Minutes, CCRC Steering Committee, 13 October 1981, RPP Box 20; *Calumet Reporter,* January 1982, RLP and Poethig Papers; Minutes, CCRC Steering Committee, 12 January 1982, RPP Box 20; "Latinos Get Child-Care Grant," *CT,* 30 April 1981, p. M4; Minutes, Administrative Staff Meeting, Latino Institute, 15 May 1981, LIP Box 5; Chicago Community Trust, Annual Report Fiscal Year 1981; Wieboldt Foundation, 1981 Annual Report; Father John F. Moriarty to Cardinal Cody, 24 March 1981 and 6 May 1981, CHD Files, Cody Papers; *Chicago Catholic,* 20 November 1981, p. 18; Mary Gonzales Weekly Schedule, LIP Box 4 (12 November 1981 "Breakfast w/ Jean Rudd"); Minutes, UNO Convention Committee, 18 November 1981, 1 and 15 December 1981, 5 and 19 January 1982, LIP Box 101; Latino Institute, "Community-Based Advocacy Leadership: A Proposal to the Ford Foundation," 24 December 1981, LIP Box 102; Leon M. Despres to Frank Lumpkin, 4 May 1981, FLP Box 3 Fld. 5; R. C. Longworth and Pam Sebastian, "Workers Ask Day in Court vs. IH," *CT,* 6 September 1981, p. V1; Frank Lumpkin to Crossroads Fund, "Grant Proposal for Wisconsin Steelworkers Save Our Jobs Committee," 15 February 1982, Crossroads Fund Papers (CFP), Box 38 Fld. 476; Jackie Schad to Greg Galluzzo, 8 June, 27 September, and 1 November 1982, all CFP Box 38 Fld. 476; Thomas Geoghegan, *Which Side Are You On?* (Farrar, Straus & Giroux, 1991), pp. 91–121; DJG interview with Tom Geoghegan. Steve Gorecki, "CHD Funds Help Hispanics Help Themselves," *Chicago Catholic,*

16 December 1983, pp. 42, 44, 45, reports that UNO of Southeast Chicago received $34,000 in 1981 and $24,000 in 1982 from national CHD funds. See also "Some Peacemakers—Working for Justice," *Chicago Catholic,* 18 November 1983, p. 11, and William T. Poole and Thomas W. Pauken, *The Campaign for Human Development* (Capital Research Center, 1988), p. 123.

On CHD, also see *Chicago Catholic,* 10 July and 17 July 1981, p. 1; Robert T. Reilly, "CHD: 'Perhaps the Best Thing the Church Has Done in Decades,'" *Salt,* November–December 1981, pp. 12–18; Joseph Cardinal Bernardin, "The Story of the Campaign for Human Development," 25 August 1995 (reprinted in *CRec,* 11 September 1995, pp. S13174–76); Jim Castelli, "How the Church Passes the Buck to the Poor," *U.S. Catholic,* October 1996, pp. 21–26; Lawrence J. Engel, "The Influence of Saul Alinsky on the Campaign for Human Development," *Theological Studies* 59 (December 1998): 636–61; and Joel R. Schom, "Catholic Campaign for Human Development Faith-Based Grantmaking," Philanthropy & Nonprofit Sector Program, Loyola University Chicago, 2007.

16. DJG interviews with Greg Galluzzo, Mary Gonzales, Jerry Kellman, and Leo Mahon; Donald C. Reitzes and Dietrich S. Reitzes, "Alinsky in the 1980s: Two Contemporary Chicago Community Organizations," *Sociological Quarterly* 28 (1987): 265–83, esp. 271; Phil Davidson, "Obama's Mentor," *II,* March 2009; Don Walton, "Obama Mentor Was Here," *Lincoln Journal Star,* 26 April 2010; Connie Bolt to CCRC Steering Committee Members, 1 February 1982, Minutes, CCRC Steering Committee, 9 February 1982, CCRC Financial Report, 9 March 1982, all RPP Box 20; Jerry Kellman, Mary Gonzales, and Greg Galluzzo to Leo Mahon, "Our Suggestions for a Church-Based Organization in the Calumet Region," 13 April 1982, Poethig Papers and RPP Box 20.

17. Leo Mahon to Dear CCRC Member, 28 May 1982, RPP Box 20; Leo T. Mahon, "From the Pastor's Desk," *St. Victor Parish Sunday Bulletin,* 18 April 1982, p. 2; R. C. Longworth, "Working to Save What's Theirs," *CT,* 30 June 1982, p. B1; Tom Cima, "From the Shadows to New Spirit," *Upturn,* March–April 1983, p. 3; Monroe Anderson, "100 Jobless March in Loop, Protest Lack of Job Creation," *CT,* 29 July 1982, p. 17. In the absence of any true biography of Joseph Bernardin, the best existing source is Eugene Kennedy, *My Brother Joseph: The Spirit of a Cardinal and the Story of a Friendship* (St. Martin's Press, 1997). The full extent to which the economic crisis of the Calumet region did radicalize Leo Mahon's political views is most starkly revealed in Mahon, "From the Pastor's Desk," *St. Victor Parish Sunday Bulletin,* 30 January 1983, p. 2, where he began by asking "is it possible that there is something wrong with our system?" and queried of people who live off of investment income, "do they not tend to be useless and unproductive parasites?" Questioning "the basic tenet of Capitalism that says that the one real owner of an industry is the investor and that the money invested has prior rights," Leo declared that "we should consider changing our economic system so that the workers who manufacture the product will also be considered owners, owners with rights equal or even superior to the stockholders who do not work. . . . I am well aware that I shall be called a socialist or even a communist" on account of those comments.

18. DJG interviews with Bob Moriarty and John McKnight; McKnight, "The Medicalization of Politics," *Christian Century,* 17 September 1975, pp. 785–87 (reprinted in McKnight, *The Careless Society* [Basic Books, 1995], pp. 55–62); McKnight, "Professionalized Service and Disabling Help," in Ivan Illich et al., *Disabling Professions* (Marion Boyars, 1977), pp. 69–91 (reprinted in *Careless Society,* pp. 36–52); McKnight, "The Need for Oldness, *St. Croix Review* 10 (February 1979): 22–31 (reprinted in *Careless Society,* pp. 26–35), McKnight, "The Professional Problem," *Institutions* 2 (September 1979) (reprinted in *Careless Society,* pp. 16–25), McKnight, "A Nation of Clients?," *Public Welfare* 38 (Fall 1980): 15–19 (reprinted in *Careless Society,* pp. 91–100). See also John L. McKnight, "Service Growth Changing Society," *CT,* 2 January 1981, p. B4. Powerful evidence of McKnight's influence appears in Stanley Hallett, "The Limits of a Model Community Bank," *TNW,* August 1983, pp. 10–16: "political action must be taken to transform the flows of resources from maintaining people in a welfare trap to enabling them to get the tools and techniques needed to become productive." A harmful institution, Hallett wrote, is one that "convinces people that they should be passive recipients of its services" and thereby "inhibits the formation of strong community organizations"; "such an institution stands in the way of needed political change."

19. DJG interviews with Bob Moriarty, Kevin Limbeck, Mary Ellen Montes, Ellen Schumer, Danny Solis, Phil Mullins, George Schopp,

Alma Avalos, and Bob Ginsburg. Moriarty warmly recalled Lena and Alma: "Those two were the stars. They were real leaders. They had and they could produce a following, and they had nerve." Also Robert Ginsburg, "The Dirt Comes Out from Under the Carpet," *CBE Environmental Review*, March–April 1983, pp. 3–5ff.; Gayle Peterson, "We Are the People: The Fight Against Toxic Waste," *Health & Medicine* 2 (Winter 1983–84): 13–22; Jerry Sullivan, "Of Dumps, Chicago Politics & Herons," *Audubon*, March 1987, pp. 122–26; Casey Bukro, "'Health Hazard' Waste Heads for Chicago," *CT*, 8 September 1982, pp. 1, 8; Bukro, "Toxic Waste Destroyer to Get a Trial by Fire," *CT*, 12 September 1982, p. 4. On Foster Milhouse, see his 30 July 2012 *CST* obituary.

On the Trumbull Park riots, see Robert Gruenberg, "Chicago Fiddles While Trumbull Park Burns," *Nation*, 22 May 1954, pp. 441–43, and "Trumbull Park: Act II," *Nation*, 18 September 1954, pp. 230–32; Ben Joravsky and Eduardo Camacho, *Race and Politics in Chicago* (Community Renewal Society, 1987), pp. 21–29; Arnold R. Hirsch, "Massive Resistance in the Urban North: Trumbull Park, Chicago, 1953–1966," *Journal of American History* 82 (September 1995): 522–50; Edwina Leona Jones, "From Steel Town to 'Ghost Town': A Qualitative Study of Community Change in Southeast Chicago" (M.A. thesis, Loyola University Chicago, May 1998), pp. 82–84; Alter, "Mexicans and Serbs in Southeast Chicago," and D. Bradford Hunt, *Blueprint for Disaster: The Unraveling of Chicago Public Housing* (University of Chicago Press, 2009), pp. 102–3.

20. Hank DeZutter, "Battle Cry of the Southeast Side: Don't Dump on Us," *CR*, 11 March 1983, pp. 3, 34; Thomas J. Murphy to Harold Washington, 26 August 1983, Harold Washington Papers (HWP) Public Safety Box 12 Fld. 11. The *Tribune* failed to cover Washington's visit to Bright School. See Thom Shanker and David Axelrod, "Mayoral Candidates Gain More Support," *CT*, 30 March 1983, p. 9. On Washington's election, the richest sources are Paul Kleppner, *Chicago Divided: The Making of a Black Mayor* (Northern Illinois University Press, 1985), esp. p. 217; Dempsey J. Travis, *An Autobiography of Black Politics* (Urban Research Press, 1987), pp. 489–610; Travis, *Harold: The People's Mayor* (Urban Research Press, 1989), pp. 153–97; and James Hawking, "Political Education in the Harold Washington Movement," D.Ed. dissertation, Northern Illinois University, August 1991, pp. 52–139.

21. R. C. Longworth, "Union Chief Enters Steel Saga," *CT*, 18 October 1982, pp. 10–11; John Kass, "Roque Counters Save Our Jobs' Charges," *DC*, 19 October 1982, p. 1; R. C. Longworth, "System Foundered on Lies, Selfishness," *CT*, 17 February 1988, p. B3; David Bensman and Roberta Lynch, "The Workers Who Wouldn't Go Away," *Nation*, 30 April 1988, p. 606; Geoghegan, *Which Side Are You On?* p. 96; Wisconsin Steelworkers Save Our Jobs Committee to Crossroads Fund, "Grantee Fiscal and Progress Report," 27 January 1983, CF Box 38 Fld. 476; Thomas Hardy, "Hanging Tough in Tough Times," *CT*, 10 April 1983, p. K3; Jackie Schad to Gregory Galluzzo, 20 May 1983, CFP Box 38 Fld. 475; CCRC Research Committee Meeting, 5 January 1981, RPP Box 20; Mary Schmich, "A Union Leader Who Wasn't Made of Steel," *CT*, 25 August 1985, Terry Atlas, "U.S. Steel Gives Go-Ahead to New S. Side Rail Mill," *CT*, 18 September 1982, p. M1; Robert Kearns, "Steel Woes Show on South Side," *CT*, 20 September 1982, p. B10; Michael Tackett, "Steelworkers Accept Contract: 'We're Working,'" *CT*, 2 March 1983, p. 5; Robert Kearns, "South Works May Close, U.S. Steel Chief Says," *CT*, 3 May 1983, p. C1; Jesse Auerbach, "Jobs for Displaced Workers," *TNW*, May 1983, pp. 1, 8–12; Linnet Myers, "A Saga of Hard Times Set in Cold Steel," *CT*, 29 August 1983, p. T1. See also Elizabeth Balanoff's 1977 interview with then-USW Local 65 president Alice Peurala.

22. DJG interviews with Bob Moriarty, Mary Ellen Montes, Alma Avalos, Ed Grossman, George Schopp, Bob Ginsburg, Phil Mullins, and Danny Solis; Hank Greenberg and Steve Kerch, "Protesters Rally at Meeting of Waste Company," *CT*, 21 May 1983, p. 1; Jose L. Rodriguez, "17 Arrested at South Side Landfill Rally," *CT*, 17 June 1983, p. B3; Kenan Heise, "Churches Battle in Behalf of Jobless," *CT*, 10 June 1983, p. 3; "The Future of Calumet City," *St. Victor Parish Sunday Bulletin*, 26 June 1983, p. 4; Casey Bukro, "Mayor to Block 2 Dump Projects, Pending Review," *CT*, 24 August 1983, p. B1; Thomas J. Murphy to Harold Washington, 26 August 1983, HWP Public Safety Box 12 Fld. 11; Thom Shanker and David Axelrod, "Washington Invades Ald. Vrdolyak's 10th Ward Turf," *CT*, 25 August 1985, p. B1; Scott Buckner, "Mayor Vows 'No More Dumps'—600 Pack St. Kevin's Meeting," *DC*, 25 August 1983, p. A1; John Kass and Mark Eissman, "Washington Tells Crowd Waste Management Permits

'Wrong,'" *DC,* 25 August 1983, p. A1; Kass and Eissman, "Washington Promises a Zoning Board Revamp," *DC,* 25 August 1983, p. A2; Dennis Geaney, "From the Editor's Notebook," *Upturn,* October–November 1983, p. 7; DeZutter, "Battle Cry of the Southeast Side"; Lee Botts, "Women Form Backbone of City Environmental Groups," *TNW,* July 1984, p. 17; Sullivan, "Of Dumps, Chicago Politics & Herons"; Jerry Sullivan, "Field & Street," *CR,* 1 March 1990; Jim Schwab, *Deeper Shades of Green* (Sierra Club Books, 1994), p. 176. Benny Scheid's 1981 letter is quoted in John Conroy, "Cardinal Sins," *CR,* 4 June 1987, and Scheid's brief obituary appears in *CST,* 2 May 1987 and *CT,* 3 May 1997; also see Steve Gorecki, "Through Good Times and Bad St. Kevin Still 'Family,'" *Chicago Catholic,* 30 October 1981, p. 23, and George Schopp, "Tying Together the Telephone Lines," *Upturn,* March–April 1983, p. 2. Harold Washington indeed had served thirty-six days in the Cook County jail in 1972 for failure to file tax returns for at least four years. See James L. Merriner, *Grafters and Goo Goos* (Southern Illinois University Press, 2004), pp. 219–20. The most thorough review of WMI's history is Edwin L. Miller Jr., "Waste Management, Inc.," Final Report, March 1992, San Diego (CA) County District Attorney.

23. DJG interview with George Schopp; Benny [Scheid], "An Open Letter to Dennis Geaney," *Upturn,* January–February 1984, p. 14; Joravsky, "Dumpers Swamp City's Southeast Side"; Steve Kerch and Monroe Anderson, "Dumping Ban Called Threat to Business," *CT,* 21 September 1983, pp. B1, B4; Michael Arndt, "Firm Permitted to Burn PCBs on Southeast Side," *CT,* 5 October 1983, p. 1.

24. William T. Hogan, *Steel in the United States: Restructuring to Compete* (D. C. Heath & Co., 1984), esp. pp. 25–26, Hogan, "The Steel Industry Today," *Iron and Steel Engineer,* April 1987, pp. 50–52, John P. Hoerr, *And the Wolf Finally Came: The Decline of the American Steel Industry* (University of Pittsburgh Press, 1988), esp. pp. 100, 425–26, and Gordon L. Clark, "Corporate Restructuring in the Steel Industry: Adjustment Strategies and Local Labor Relations," in George Sternlieb and James W. Hughes, eds., *America's New Market Geography* (Rutgers Center for Urban Policy Research, 1988), pp. 179–214, primarily inform the initial summary of the steel industry's decline; Robert Kearns, "Plan Offered for New South Works," *CT,* 12 August 1983, p. D1; Kearns, "U.S. Steel Puts Off Decision on Rail Mill," *CT,* 13 August 1983, p. B6; Linnet Myers, "Hopes Vanish Along with Steel Mills," *CT,* 23 October 1983, p. C1; Steven Greenhouse, "U.S. Steel May Keep Plant Open," *NYT,* 9 November 1983, p. D1; Robert Kearns, "South Works Talks 'At a Standstill,'" *CT,* 10 November 1983, p. 1; Kearns, "U.S. Steel Plan Irks Union," *CT,* 11 November 1983, p. 1; Kearns, "The South Works Political Football," *CT,* 14 November 1983, p. C1; Kearns, "U.S. Steel Irks Workers with New Demands," *CT,* 8 December 1983, p. 1; Linnet Myers, "South Works Concessions Rejected," *CT,* 22 December 1983, p. 1; Myers, "Unions at Mills No Longer Strong as Steel," *CT,* 25 December 1983, p. C1; Myers, "South Works Dream Dies," *CT,* 28 December 1983; Sam Smith, "U.S. Steel Reneged, Rostenkowski Says," *CT,* 30 December 1983, p. 1; Linnet Myers, "Steelworkers Find Unity in Defeat," *CT,* 1 January 1984, p. 1; Editorial, "South Works Fate Heads Things to Watch For as 1984 Rolls In," *DC,* 1 January 1984, p. A4; Larry Galica, "U.S. Sues for Rail Mill," *DC,* 11 January 1984, p. A1; R. C. Longworth, "A Betrayal on South Works," *CT,* 11 January 1984, p. P19; Larry Galica, "USS to Demolish 2 Blast Furnaces," *DC,* 26 January 1984, p. A1; "Keep South Works Open, Bernardin Asks U.S. Steel," *CST,* 31 January 1984, p. 6; Fr. Dennis J. Geaney, "Loss of Steel Plant Jobs," *CT,* 3 February 1984, p. 18; Larry Galica, "So. Works Story Is Tale of Woe," *DC,* 6 February 1984, p. A2; Linda Wolohan (UPI), "South Works Employees Are Struggling in the Face of a Very Uncertain Future," *DC,* 6 February 1984, p. B2; Larry Galica, "Reaction Mixed to USS Decision," *DC,* 13 March 1984, p. A2; Joseph Cardinal Bernardin, "The Need for Collaboration," *CT,* 21 March 1984, p. 15; Larry Galica, "Local 65 May Sell Its Union Hall," *DC,* 29 March 1984, p. A2; Galica, "Rumors of South Works Closing Denied by Firm," *DC,* 10 May 1984, p. B3; Galica, "Unemployment Rate Drops," *DC,* 2 June 1984, p. A1; Galica, "U.S. District Court to Hear South Works Demolition Suit," *DC,* 28 June 1984, p. A1; Galica, "U.S. Court Rejects Suit," *DC,* 30 June 1984, p. A1; Galica, "City May Join Suit to Stop Dismantling at South Works," *DC,* 17 July 1984, p. A2.

25. Ann Grimes, "UNO: Catholic Clergy Up Front in Back of the Yards Organizing Effort," *CR,* 1 April 1983, pp. 3, 31; Latino Institute 1984 Annual Report, LIP Box II-4; Untitled selective list of 1983 Chicago CHD Applications, n.d. [spring 1983], Joseph Cardinal Bernardin Papers; "Claretian Social Development Grants

Awarded," July 1983, "1984 Claretian Social Development Grants Awarded," June 1984, p. 93 (renumbered p. 61), Claretian Social Development Fund 1974–1993, Report to the Eastern Province Assembly from the Peace and Justice Committee, June 1993, p. 12, all TJP; Julie Kemble Borths, "Five Local Projects Awarded National CHD Grants," *Chicago Catholic,* 30 September 1983, p. 2; Woods Charitable Fund, *Report for the Year 1984,* p. 11 (noting a $15,000 1984 grant to CCRC "to expand staff"); [Draft] "By Laws of the Calumet Community Religious Conference," n.d. [1983], RPP Box 20; Paul Burke, "Time for Twelve," *Upturn,* October–November 1983, pp. 6, 8; Leo T. Mahon, "From the Pastor's Desk," *St. Victor Parish Sunday Bulletin,* 12 February 1984, pp. 2–3; Suzanne Carter, "Eight Local Projects Get Development Funds," *Chicago Catholic,* 6 and 13 July 1984, p. 22; Wilfredo Cruz, "UNO: Organizing at the Grass Roots," *Illinois Issues (II),* April 1988, pp. 18–22; M. W. Newman and Lillian Williams, "People Power: Chicago's Real Clout," *CST,* 8 April 1990, pp. 12–14; Graciela Kenig, "For Veteran Organizer, Education Is Power," *CST,* 2 October 1991, p. 30; Helen Sundman, "UNO: Taking Organizing to a New Level, or Leaving the Community Behind?," *Chicago Reporter,* May 1994; Carol Felsenthal, "Danny Solis," *CM,* 29 March 2011; DJG interviews with Jerry Kellman, Greg Galluzzo, Mary Gonzales, Ellen Schumer, Kevin Limbeck, Josh Hoyt, Peter Martinez, Leo Mahon, Tom Joyce, Bernie Pietrzak, Fred Simari, Jan Poledziewski, Christine Gervais, Bill Stenzel, Bonnie Nitsche, Wally Nitsche, Ken Jania, Alma Avalos, Tom Kaminski, and Betty Garrett; "History of Hegewisch," *HN,* 25 January 1984; Linnet Myers, "Grim Reality Hits Steel Mill Community," *CT,* 15 June 1984, p. B8; Robert Bergsvik, "Film Captures True Spirit of Southeast Side," *DC,* 15 October 1984, p. A1; Walley, "Deindustrializing Chicago," pp. 128–30, and Walley, *Exit Zero,* pp. 57, 63–70, 74, 117, 153; Silvio DeAntoni et al., "Holy Rosary—The Last 26 Years," in *Holy Rosary Parish 1882– 2008,* pp. 19–21. On Bill Stenzel's timeline, see *St. Victor Parish Sunday Bulletin,* 7 June 1981, p. 2, William Stenzel, "Revival Week at St. Helena's," *Upturn,* November–December 1982, p. 5, and Bill Stenzel, "1990 in Roseland," *Upturn,* April–May 1985, p. 4. On Fred Simari, see Leo T. Mahon, "From the Pastor's Desk," *St. Victor Parish Sunday Bulletin,* 6 May 1984, p. 2. On Raymond P. Nugent, see his 28 October 1995 *CST* and *CT* obituaries.

26. This brief history of Greater Roseland draws on dozens of sources, beginning first and foremost with Janice L. Reiff's rich and authoritative "Rethinking Pullman: Urban Space and Working-Class Activism," *Social Science History* 24 (Spring 2000): 7–32, as well as her "Roseland" and "West Pullman" essays in the *Encyclopedia of Chicago* (2004) and Clinton E. Stockwell's "Washington Heights" essay in that same project; also Arnold R. Hirsch's *Making the Second Ghetto: Race and Housing in Chicago, 1940– 1960* (University of Chicago Press, 1983), esp. pp. 3–4, 46–47, 54–55; Melaniphy & Associates' *Chicago Comprehensive Neighborhood Needs Analysis* [CCNNA]: *Volume I: Citywide Needs Analysis,* January 1982, and particularly their incredibly rich *CCNNA: Volume II: Roseland Community Area,* 1982, *CCNNA: Volume II: West Pullman Community Area,* 1982, and *CCNNA: Volume II: Washington Heights Community Area;* and *Local Community Fact Book, Chicago Metropolitan Area* (Chicago Review Press, 1984), p. 129. Lilydale's history is best presented in Michael W. Homel, "The Lilydale School Campaign of 1936: Direct Action in the Verbal Protest Era," *Journal of Negro History* 59 (July 1974): 228–41; see also Homel's *Down from Equality: Black Chicagoans and the Public Schools, 1920– 41* (University of Illinois Press, 1984), esp. p. 166. The richest account of the 1947 riots is John Bartlow Martin, "Incident at Fernwood," *Harper's,* October 1949, pp. 86–98. Mark T. Mulder's "Mobility and the (In) Significance of Place in an Evangelical Church: A Case Study from the South Side of Chicago," *Geographies of Religions and Belief Systems* 3 (2009): 16–43, and especially his superb "Evangelical Church Polity and the Nuances of White Flight: A Case Study from the Roseland and Englewood Neighborhoods in Chicago," *Journal of Urban History* 38 (January 2012): 16–38, esp. pp. 17, 24, dramatically outclass Robert P. Swierenga, *Dutch Chicago* (Eerdmans Publishing, 2002), pp. 35–38, 295–349. A fifth CRC congregation, Pullman Christian Reformed Church, established as a mission church in late 1972 and by 1977 located at 103rd Street and South Vernon Avenue, eight blocks east of South State Street, is not addressed by Mulder but is very much located in Roseland. Anthony Van Zanten, "Worship in Pullman CRC: An Interview with Rick Williams," *Reformed Worship,* March 1988; Sonya Jongsma Knauss, "Van Zanten Eases Out of Decades in Inner City Ministry, *The* [Dordt College] *Voice,* Winter 2004; "Have You Heard of the City?—A Roseland Christian Ministries Story" (DVD, 2005); DJG interviews with Donna and Tony Van Zanten and Rick Williams.

27. Calvin P. Bradford and Leonard S. Rubinowitz, "The Urban-Suburban Investment-Divestment Process: Consequences for Older Neighborhoods," *The Annals* 422 (November 1975): 77–87, esp. 83–84; Bradford, "Financing Home Ownership: The Federal Role in Neighborhood Decline," *Urban Affairs Quarterly* 14 (March 1979): 313–35, esp. 326, 328; Bradford, "Never Call Retreat: The Fight Against Lending Discrimination," in *Credit by Color: Mortgage Market Discrimination in Chicagoland* (Chicago Area Fair Housing Alliance, 1991), pp. 5–18, esp. pp. 9–10; Bradford and Anne B. Shlay, "Assuming a Can Opener: Economic Theory's Failure to Explain Discrimination in FHA Lending Markets," *Cityscape* 2 (February 1996): 77–87, esp. 80–81. Also see Judith D. Feins, "Urban Housing Disinvestment and Neighborhood Decline: A Study of Public Policy Outcomes," Ph.D. dissertation, University of Chicago, March 1977, esp. p. 278 regarding FHA Commissioner Letter 68–8, 2 August 1968; Gregory D. Squires, "Community Reinvestment: An Emerging Social Movement," in Squires, ed., *From Redlining to Reinvestment: Community Responses to Urban Disinvestment* (Temple University Press, 1992), pp. 1–37, esp. 6, and Squires, "Friend or Foe? The Federal Government and Community Reinvestment," in W. Dennis Keating et al., eds., *Revitalizing Urban Neighborhoods* (University Press of Kansas, 1996), pp. 222–34, esp. 223.

28. *Greater Roseland Newsletter,* Fall 1970, CHM; "U.S. to Renovate 100 Homes Abandoned in Roseland," *CT,* 9 September 1979, p. XIV-1; Greater Roseland Organization, Constitution of Umbrella Organization, 17 September 1982, CRF Box 17 Fld. 207; Father John F. Moriarty to Cardinal Cody, 19 and 27 March 1980, 12 May 1980, 23 February 1981, and 2 March 1981, CHD Files, Cody Papers; William Mullen, "Teen-Aged Terror Stalks Fenger High," *CT,* 7 June 1981, p. III-1; David Ibata, "Formulas Traced for Urban Revival," *CT,* 25 October 1981, p. 1; Steve Gorecki, "Pilsen Benefits from CHD," *Chicago Catholic,* 20 November 1981, p. 31; Chicago Community Trust, *Annual Report FY 1982,* pp. 24–25; Melaniphy & Associates, *CCNNA, Volume I,* pp. 55–56, 398, *CCNNA, Volume II: Roseland Community Area,* pp. 19, 23, 26, 46, 82, 92, 94, 133, *CCNNA, Volume II: West Pullman Community Area,* pp. 29–31; Julie S. Putterman, "Chicago Steelworkers: The Cost of Unemployment" (Steelworkers Research Project, January 1985), esp. pp. 15, 45; John Williams Jr., "GRO Moves into Housing Area," *SEC,* 15 April 1982, p. 1; Kathleen Myler, "Jobs, Gangs Major Roseland Problems," *CT,* 18 April 1982, p. M2; GRO, "Report for Chicago Community Trust 2nd Quarter," ca. 5 January 1983, CRF Box 17 Fld. 207; Lenora Rodgers to Crossroads Fund, 1 February 1983, CRF Box 17 Fld. 207; "West Pullman . . .Called Home By 40,900 People," *CT,* 22 February 1983, p. B1; Stanley Ziemba, "Federal Mortgage Aid Asked," *CT,* 18 March 1983, p. B1; *GRO Community Memo,* May 1983, Jackie Schad to Lenora Rodgers, 9 May 1983, CRF Box 17 Fld. 207; Stanley Ziemba, "FHA Mortgage Defaults in Black Neighborhoods Top 81%," *CT,* 4 November 1983, p. A19; Jackie Schad to Ronald Carter, 3 April 1984, CFP Box 17 Fld. 207; Bernard C. Taylor, "The Study of Community Development Block Grant Funds of the City of Chicago," 16 April 1984, HWP Central Office Files Box 9 Fld. 91; Rev. Robert C. Behnke and Mary Margaret Moran to Superintendent Fred Rice, 27 August 1984, HWP Central Office Files Box 42 Fld. 6. Also see Robert A. Slayton, "Foreclosure in Chicago: The Differing Impact on Race," Chicago Urban League, May 1987, esp. pp. 4–5; David Moberg, "Can Chicago Be Saved?," *Inc. Magazine,* March 1988; Isabel Wilkerson, "Small Inner-City Hospitals in U.S. Face Threat of Financial Failure," *NYT,* 21 August 1988; Don DeBat, "Roseland, Pullman Get Second Wind," *CST,* 28 June 1991, p. 3; "Congratulations Extended to Mr. & Mrs. Theodis Rodgers Sr. on Their Fiftieth Wedding Anniversary," *Chicago City Council Journal,* 11 September 1991, p. 5362; Isabel Wilkerson, "Black Neighborhood Faces White World with Bitterness," *NYT,* 22 June 1992; *South Street Journal* [*SSJ*], 29 September–12 October 1994, p. 12; Kristin Ostberg, "Can This Neighborhood Be Changed?," *CR,* 9 October 1997; Judith McCray, "One Man's Vision Puts Economic Goals in Focus for Community," *CST,* 9 July 1993, p. BE3; Ray Quintanilla, "Black Expo's Key Man," *Chicago Reporter,* September 1993; "The Rev. Bernard C. Taylor, 46, Black Expo Chairman," *CST,* 23 November 1994, p. 75 (obituary); Chuck McWhinnie, "Mary Bates, Community Activist," *CST,* 22 February 1999, p. 51 (obituary at age eighty); *SSJ,* 25 March–7 April 1999, p. 2; Grant Pick, "Stayin' Alive," *CR,* 16 November 2000; Mark Konkol, "Roseland's Only Steak House Survives Tough Times," DNAInfo.com, 19 October 2015.

29. Thomas Maier, "Study Finds Little S. E. Side Toxic Hazard," *CST,* 9 January 1984, p. 19; Diana Strzalka, "Residents Protest Landfill Request," *DC,* 14 January 1984, p. A1; Scott Buckner, "Waste Ban Vote Next Week," *DC,* 25 January 1984, p. A2; Buckner, "Council OKs Waste Ban," *DC,* 31 January 1984, p. A1; Buckner, "Waste Ban Prompts Debate," *DC,* 31 January 1984, p. A2; Harry Golden Jr., "Panel Urges Ban on New Dumps," *CST,* 31 January 1984, p. 6; Mark Kiesling, "Simon Visits South Deering Landfill," *DC,* 14 February 1984, p. A1; "Waste Dumps Hurt Area, Simon Says," *CT,* 14 February 1984, p. A4; Mark Kiesling, "Landfill Moratorium Passed," *DC,* 16 February 1984, p. A1; "Waste Fight Not Over," *DC,* 17 February 1984, p. A4; Scott Buckner, "South Deering Landfill Expansion Denied by ZBA," *DC,* 25 February 1984, p. A1; Mark Kiesling, "Right-to-Know Bill Garners Support," *DC,* 2 March 1984, p. A1; Bob Kostanczuk, "Nature Park Meetings Set," *DC,* 3 March 1984, p. A1; "More News in Waste War," *DC,* 5 March 1984, p. A4; John Kass and Mark Eissman, "Waste Hill in Plan for Southeast Side," *CT,* 13 March 1984, p. B2; Kass and Eissman, "Landfill Proposal Thorn in Lake Calumet's Side," *CT,* 23 March 1984, p. B3; Bob Kostanczuk, "Critics Rip Landfill Plan," *DC,* 28 March 1984, p. A1; "Reject Waste Management's Offer," *DC,* 29 March 1984, p. A4; Robert Bergsvik, "SE Side Residents Skeptical of Dump," *Columbia College Chronicle,* 2 April 1984, p. 1; Bill Bero, "Waste Firm's Official Clarifies Position on Proposed Nature Park," *DC,* 9 April 1984, p. A5; Steve Kerch, "Home-Grown Experts Crusade Against Waste Dumps," *CT,* 15 April 1984, p. B2; Illinois EPA, "The Southeast Chicago Study: An Assessment of Environmental Pollution and Public Health Impacts," HWP DS Box 5 Fld. 8; Harlan Dragor, "Far S. Side Pollution No Threat, Says Study," *CST,* 25 April 1984, p. 12; Casey Bukro, "S. Side's High Cancer Rate Cited," *CT,* 25 April 1984, p. D1; Bukro, "Study Finds Cancer High, Downplays Risk," *CT,* 26 April 1984, p. D14; Mark Kiesling, "South Deering Landfills Must Obtain Permits or Shut Down," *DC,* 26 April 1984, p. A1; Kiesling, "Confusion Surrounds SD Landfill," *DC,* 1 May 1984, p. A1; Kiesling, "Right-to-Know Bill Is in Trouble," *DC,* 2 May 1984, p. A3; Bob Kostanczuk, "Waste Management Calls for Environmental Impact Study," *DC,* 15 May 1984, p. A2; Mark Brown, "Calumet Landfill Foes Win Delay," *CST,* 15 May 1984, p. 15; Mark Kiesling, "Community Right-to-Know Bill Fails to Pass," *DC,* 1 June 1984, p. A3; Kiesling, "Waste Firm Merger Questioned," 14 June 1984, p. A1; Patrick Barry, "The Trash Clash," *CST,* 15 June 1984, p. N1; Mark Kiesling, "PCB Monitoring Station Is to Be Installed Soon," *DC,* 16 June 1984, p. A3; Judy Freeman, "Beating Down the Waste Dumps," *TNW,* July 1984, pp. 1, 15–19; Mark Kiesling, "Environmental Legislation Has Some Success," *DC,* 6 July 1984, p. A3; Larry Galica, "Study of Area Pollution Will Come Under Scrutiny," *DC,* 13 July 1984, p. A1; Casey Bukro, "U.S. Study Urged for Lake Calumet," *CT,* 14 July 1984, p. 5; Mark Kiesling, "Pollution Monitors to Be Installed," *DC,* 14 July 1984, p. A3; Harlan Dragor and Clem Richardson, "SE Siders Ask EPA Help on Toxins," *CST,* 14 July 1984, p. 18; Mark Kiesling, "More Pollution Monitoring Approved for Lake Calumet," *DC,* 21 July 1984, p. A1; Casey Bukro, "Firm Seeks to Ship PCBs to S. Side," *CT,* 23 July 1984, pp. 11–12; Mike Nolan, "Waste Firm Has No Plans to Bring PCBs to Chicago," *DC,* 24 July 1984, p. A1; Judy Freeman, "Bulging Landfills Breed New Trash Crisis," *TNW,* August 1984, pp. 1, 15–18; Mark Kiesling, "Decision on SCA Operation Is Not Expected Until Fall," *DC,* 15 August 1984, p. A2; Kiesling, "Lake Calumet Monitors Due," *DC,* 30 August 1984, p. A1; Kiesling, "Paxton Landfill Will Ask for Contempt of Court Order," *DC,* 1 September 1984, p. A1.

30. Larry Galica, "Wis. Steel to Be Scrapped," *DC,* 26 January 1984, p. A1; Helene McEntee, "Pact Signed to Scrap Most of Wis. Steel," *CST,* 26 January 1984, p. 75; Larry Galica, "Bids Sought to Develop Wis. Steel's No. 6 Mill," *DC,* 27 January 1984, p. A1; Galica, "Ex-Wis. Workers Hope Demolition Will Provide Jobs," *DC,* 6 February 1984, p. A1; Larry Galica, "Mayor to Rap Reagan, Thompson at Rally for Wisconsin Steel Works," *DC,* 15 March 1984, p. A1; Galica, "Tempers Flare at Wis. Steel Rally," *DC,* 16 March 1984, p. A1; Galica, "Wis. Workers Mark Mill Closing," *DC,* 28 March 1984, p. A1; Galica, "Wis. Workers Mark Closing," *DC,* 29 March 1984, p. A1; U.S. House of Representatives, Committee on Agriculture, Subcommittee on Domestic Marketing et al., *Hearings—Hunger in the United States and Related Issues,* Serial No. 98-63, pp. 62–64, 70–71, 2 March 1984, Chicago; Larry Galica, "2 SC Businesses Closing," *DC,* 5 May 1984, p. A1; James Strong and Linnet Myers, "Wrecker Crushing Hopes at Idle Wisconsin Steel," *CT,* 11 May 1984, p. 1; Larry Galica, "UNO Meeting on Jobs Draws 500 to

So. Chicago Church," *DC,* 12 May 1984, p. A2; "UNO Is Sponsoring Registration Drive," *DC,* 21 August 1984, p. A6. Also see Frank Lumpkin to Crossroads Fund, 20 December 1983, Lumpkin, "Grantee Fiscal and Progess Report," 4 January 1984, Jackie Schad to Greg Galluzzo, 19 April 1984, and "Statement of Frank Lumpkin," 13 August 1984, all CRF Box 38 Fld. 474.

31. Mark Kiesling, "Thornton Township Will Get Share of Job Retraining $," *DC,* 1 August 1984, p. A2; "1984 Claretian Social Development Grants Awarded," June 1984, p. 93 (renumbered p. 61), TJP; DJG interviews with Jerry Kellman, Fred Simari, Jan Poledziewski, Christine Gervais, Bill Stenzel, Bonnie and Wally Nitsche, Betty Garrett, Paul Burak, Cathy Askew, Dan Lee, Tommy West, Tom Kaminski, Eva Sturgies, Joe Bennett, Adrienne Jackson, Marlene Dillard, John Calicott, Stanley Farier, Dominic Carmon, Loretta Augustine-Herron, Yvonne Lloyd, Len Dubi, Tom Knutson, Ken Jania, and Mike Kruglik; Edward McClelland, *Young Mr. Obama* (Bloomsbury Press, 2010), p. 9; Donald Ehr, "The Midwest District Story, 1925–2012," in *Communities of the Word—Stories of the Chicago Province* (Society of the Divine Word, 2012), available at www.divineword.org; Donna Ulanowski, "Calumet College Adopts Elementary School," *DC,* 15 December 1984, p. A5, clearly places the late Father Elmer Powell, who died in 1989, at Our Lady in 1984, but the Ehr essay makes no mention of Father Powell, instead indicating a direct succession from Farier to Carmon. On Tony Vader, see Dahm, *Power and Authority,* pp. 180–82, also see *Catholic New World,* 16 February 2003; on Joe Bennett, see Jon Yates and Charles Sheehan, "Parishioners Reeling Amid Abuse Inquiries" and "Priest Investigated in Sex Abuse," both *CT,* 6 February 2006, and Manya Brachear et al., "Cardinal Lifts the Veil on Abuses," *CT,* 13 August 2008.

32. By far the richest source on the history of Altgeld Gardens is J. S. Fuerst's superb and absolutely wonderful *When Public Housing Was Paradise: Building a Community in Chicago* (Praeger, 2003), esp. pp. 34–35, 57–60, 78, 92–93, 114–16, 137–38, 153. Also informing this account are Altgeld-Carver Alumni Association, *History of Altgeld Gardens, 1944–1960* (Taylor Publishing, 1993); Devereux Bowly Jr., *The Poorhouse: Subsidized Housing in Chicago, 1895–1976* (Southern Illinois University Press, 1978), esp. pp. 42–45, 85, 175; Janice L. Reiff, "Riverdale," *Encyclopedia of Chicago,* (2004); Hunt, *Blueprint for Disaster,* pp. 3–4, 58, 61, 62–63, 127; Jeffrey Helgeson, *Crucibles of Black Empowerment* (University of Chicago Press, 2014), pp. 3–6; Craig E. Colten, "Industrial Wastes of the Calumet Area, 1869–1970: An Historical Geography" (Illinois Department of Energy and Natural Resources, September 1985), esp. p. 71; "CHA Management Offices & Developments," n.d., HWP Financial & Administration Box 2 Fld. 19; Julianna Richardson's History Makers interview with Gloria Jackson Bacon; DJG's interviews with Alma Jones, Gloria Jackson Bacon, Loretta Augustine-Herron, Yvonne Lloyd, Martha Kindred, Dan Lee, Salim al-Nurridin, Hazel Johnson, Cheryl Johnson, and James V. Jordan; Larry Hawkins' *CT* obituary, 3 February 2009; Pam Belluck, "End of a Ghetto: Razing the Slums to Rescue the Residents," *NYT,* 6 September 1998 ("In 1969 . . . Congress changed public housing policy with disastrous results for projects nationally. Instead of a fixed rent, tenants would pay a percentage of their income"), and D. Bradford Hunt's account of a similar decline at the CHA's infamous Robert Taylor Homes: "Between 1967 and 1974, the percentage of working-class families fell from 50% to 10%, while reliance on ADC [Aid to Families with Dependent Children] shot up from 36% to 83%. The mass exodus of two-parent, working-class familes and their replacement with non-working, female head families caused the bulk of the change." Hunt, "What Went Wrong with Public Housing in Chicago? A History of the Robert Taylor Homes," *Journal of the Illinois State Historical Society* 94 (Spring 2001): 96–123, at 109.

The contemporaneous journalism is "Calumet Area House Project to Open Soon," *CT,* 6 August 1944, p. S1; "Altgeld Gardens Housing Dedicated Before 5,000," *CT,* 27 August 1945, p. 13; Albert G. Barnett, "Chicago Negroes Fight Housing Shortage," *CD,* 7 June 1947, p. 13; "Cover Drainage Ditch, Altgeld Residents Ask," *CT,* 12 August 1951, p. S1; Gladys Priddy, "Figures Tell Tale of Change in Riverdale," *CT,* 3 April 1955, p. S4; Frank Starr, "Poverty 'Inner Ring' Grips South Side," *CT,* 13 August 1964, p. 1; Gail Stockholm, "Hearings Continue on Altgeld Gardens Bus Fare Increase," *CT,* 14 March 1968, p. 3A; "Altgeld Obtains New Health Facility," *Greater Roseland Newsletter,* Fall 1970, p. 5, CHM; Donald Yabush, "Embattled Church Will Be 25," *CT,* 20 August 1972, p. S5; Constance Lauerman, "Altgeld Gardens Wins Dope Fight, Faces Bigger One," *CT,* 19 November 1982, p. S7; James Jackson, "Altgeld Dreams Reverting to Nightmares," *CT,* 22 March 1973, p. S1.

Regarding Altgeld, also see Judy Freeman, "Where Have All the Toxins Gone?," *TNW,* June 1985, pp. 1, 6–9; Craig E. Colten, "Industrial Wastes in Southeast Chicago: Production and Disposal 1870–1970," *Environmental Review* 10 (Summer 1986): 93–105; Colten, "Chicago's Waste Lands: Disposal and Urban Growth, 1840–1990," *Journal of Historical Geography* 20 (April 1994): 124–42; Don Coursey et al., "Environmental Racism in the City of Chicago: The History of EPA Hazardous Waste Sites in African-American Neighborhoods" (Harris School Working Paper, University of Chicago, October 1994); and David N. Pellow, *Garbage Wars: The Struggle for Environmental Justice in Chicago* (MIT Press, 2002).

33. Emmett George and Dave Schneiderman, "Thousands Flee Acid Fumes," *CT,* 27 April 1974, p. 1; George and Joseph Sjostrom, "'Man, This Stuff Is Poison; The People Are All Scared,'" *CT,* 27 April 1974, p. W1; Dave Schneiderman and Jack Fuller, "Workers Draining Acid Tank; Vapors Gone," *CT,* 28 April 1974, p. 1; Jon Van, "Gas Victims Help Each Other; Kept in Dark, They Assert," *CT,* 28 April 1974, p. 2; "250 Altgeld Residents Meet," *CT,* 3 May 1974, p. 7; Frank Zahour, "Fume Victims Still Feeling Symptoms," *CT,* 5 May 1974, p. 22; John O'Brien, "Acid Leak Starts in Confusion, Ends in Probe," *CT,* 6 May 1974, p. 3; Casey Bukro, "Acid Leak: Could It Reoccur?," *CT,* 26 May 1984, p. 31; Melaniphy & Associates, *CCNNA, Volume II: Riverdale Community Area,* 1982, esp. pp. 17, 25, 30; Michael Arndt, "Mom's Dream: Moving 2 Sons from Gang Terror," *CT,* 13 December 1982, p. A19; Monroe Anderson & Casey Bukro, "An Air of Mystery at Carver," *CT,* 23 January 1983, p. 4; DJG interviews with Dan Lee, Salim al-Nurridin, and Alma Jones.

Sharon Pines, "Public Housing Residents Fight for Their Health," *TNW,* March 1985, pp. 14–17; Hazel Johnson to Jackie Schad, 6 June 1986, CRF Box 30 Fld. 368; Deborah Nelson, "'Hotspot' Air Smells, Kills," *CST,* 3 June 1987, p. 10; Virginia Mullery, "Hazel Johnson," *Salt,* July–August 1988, pp. 4–6; and DJG's interviews with Hazel Johnson, Cheryl Johnson, Martha Kindred, and Bob Ginsburg most inform this brief sketch of PCR, whose surviving records are now part of the Harsh Collection at the Chicago Public Library's very rich 95th Street branch. The April 1984 draft report and the ensuing stories were Illinois EPA, "The Southeast Chicago Study: An Assessment of Environmental Pollution amd Public Health Impacts," HWP DS Box 5 Fld. 8; Harlan Dragor, "Far S. Side Pollution No Threat, Says Study," *CST,* 25 April 1984, p. 12; Casey Bukro, "S. Side's High Cancer Rate Cited," *CT,* 25 April 1984, p. D1; and Bukro, "Study Finds Cancer High, Downplays Risk," *CT,* 26 April 1984, p. D14. Also see Robert Ginsburg, "Multiple Exposure—A New Approach to Pollution Control," *CBE Environmental Review,* Winter 1984, pp. 3–4, 15.

Other accounts discussing Johnson's work include Michael H. Brown, *The Toxic Cloud* (Harper & Row, 1987), pp. 65–92; Jim Ritter, "A Nose for Woes on S. E. Side," *CST,* 22 April 1990, p. 1; John Elson, "Dumping on the Poor," *Time,* 13 August 1990, p. 46; Leslie Ansley, "Homesick—From the Fumes," *USA Weekend,* 17–19 April 1991, pp. 17, 19; Linc Cohen, "Waste Dumps Toxic Traps for Minorities," *Chicago Reporter,* April 1992; Josh Getlin's excellent "Fighting Her Good Fight," *LAT,* 18 February 1993; Michael Lipske, "Hazel Johnson: Protecting a Community's Health," *National Wildlife,* October–November 1993; Heather M. Little, "Toxin Shock," *CT,* 15 January 1995, p. WN3; Jennifer Peltz, "Living at CHA Site Called Health Hazard," *CT,* 24 October 1999, p. 1; Fuerst, *When Public Housing Was Paradise,* pp. 189–90; Frank Sennett, "An Urban Refusal," *TNW,* October—November 1993, pp. 9–13; three obituaries: Robert D. Bullard, "Environmental Justice Movement Loses Southside Chicago Icon Hazel Johnson," OpEdNews.com, 14 January 2011, Margaret Ramirez, "Hazel M. Johnson, 1935–2011," *CT,* 16 January 2011, Mary Mitchell, "Champion of Environment a Quiet Hero," *CST,* 16 January 2011; and Dawn Turner Trice, "Final Tribute Sought for Environmental Activist," *CT,* 18 July 2011. An impressive account of a CHA Local Advisory Council that *did* represent tenants, at the Wentworth Gardens project, is Roberta M. Feldman and Susan Stall, *The Dignity of Resistance: Women Residents' Activism in Chicago Public Housing* (Cambridge University Press, 2004).

34. DJG interviews with Madeline Talbott, Keith Kelleher, Grant Williams, and Steuart Pittman; Illinois ACORN to Dear Neighbors, September 1983, and *ACORN News,* October 1983, both Illinois ACORN Papers [IAP] Box 3, Fld. 76; Cynthia-Val Jones, "Fear CTA Service Cut for Altgeld Gardens," *CD,* 8 March 1984, p. 1; Jones, "Altgeld Claims Victory in CTA Fight," *CD,* 3 April 1984, p. 3; Illinois ACORN, South Side Toxics Project Budget," ca. August

1984, IAP Box 3, Fld. 47; "South Side United Neighbors—Transition from Grant Williams to Steuart Pittman," 20 September 1984, IAF Box 2, Fld. 33. On ACORN's history, see Gary Delgado, *Organizing the Movement: The Roots and Growth of ACORN* (Temple University Press, 1986), and particularly John Atlas, *Seeds of Change: The Story of ACORN, America's Most Controversial Antipoverty Community Organizing Group* (Vanderbilt University Press, 2010), esp. pp. 48–53, 99–101. On Eden Green, see Bowly, *The Poorhouse,* p. 175; Melaniphy & Associates, *CCNNA Analysis, Volume II: Riverdale Community Area,* p. 29; Reiff, "Riverdale," *Encyclopedia of Chicago;* and Rebekah Levin et al., "Community Organizing in Three South Side Chicago Communities: Leadership, Activities, and Prospects," Center for Impact Research, September 2004, pp. 19–23.

35. DJG interviews with Jerry Kellman, Bob Klonowski, Tom Knutson, Loretta Augustine-Herron, Len Dubi, Jan Poledziewski, and Fred Simari; Robert Bergsvik, "Jobless Group Making Comeback," *DC,* 2 August 1984, p. A2; Mark Kiesling, "Thornton Township's Share of Jobs Grant Is Still Not Decided," *DC,* 3 August 1984, p. A2; Kiesling, "2 Named to Task Force," *DC,* 10 August 1984, p. A1; Kiesling, "Job Retraining Classes to Be Conducted at TCC," *DC,* 6 September 1984, p. A3; C. D. Matthews, "Aid for Jobless Is Available," *DC,* 1 January 1985, p. A2; Naomi Schreier, "Job Retraining Program Offered," *DC,* 11 January 1985, p. A2. On Father Len Dubi, also see Dolores Madlener, "The Lord's Maverick," *Catholic New World,* 17 January 2010. On what was officially called "Prairie State 2000," see Duane E. Leigh, "Possible Alternative Uses of UI Trust Funds: A Survey of Recent State Initiatives," Washington State Institute for Public Policy, October 1986, p. 15; Robert G. Sheets and Yuan Ting, "Determinants of Business Response to Government Tax Incentives for Privately Financed Termination Benefit Programs," *Policy Studies Review* 11 (Autumn/Winter 1992): 222–40; and Paul Osterman and Rosemary Batt, "Employer-Centered Training for International Competitiveness: Lessons from State Programs," *Journal of Policy Analysis and Management* 12 (Summer 1993): 456–77, esp. pp. 472–73.

36. Casey Bukro, "Southeast Siders Reject State Health Assurances," *CT,* 21 September 1984, p. A4; Leslie Baldacci, "S. E. Siders Confront Public Health Officials," *CST,* 21 September 1984, p. 6; Charlotte Chun to Robert Buchanan, "Wells in the 9th Ward," 10 August 1984, HWP Community Services Box 16 Fld. 18; Press Release, "Cyanide Found in Southeast Side Water Wells," TNW, 19 September 1984, HWP Development Box 5 Fld. 8 & Community Box 16 Fld. 18; Juanita Salvador Burris to Benjamin Reyes, "Concerns About Water in the 9th Ward," 20 September 1984, HWP Development Box 16 Fld. 31; Harlan Draeger, "Cyanide in Well Water Here," *CST,* 21 September 1984, p. 6; Casey Bukro, "Fear Flows over Tainted Wells," *CT,* 24 September 1984, p. II-1; Benjamin Reyes to Commissioners John Corey and Jesse Madison et al., "Contaminated Water Wells in Maryland Manor (134th Place)," 25 September 1984, HWP Development Box 5 Fld. 8; Judy Freeman, "We've Got da Sulfur Water Blues," *TNW,* October 1984, pp. 1, 17–20; Thom Clark, "From the Editor," *TNW,* October 1984, p. 2; Judy Freeman, "Residents Challenge EPA Findings," *TNW,* November 1984, pp. 8–9; Pines, "Public Housing Residents Fight," *TNW,* March 1985, pp. 14–17; "Percy Skips S.E. Side Meeting, Riles Crowd," *CST,* 24 October 1984, p. 7; Robert Bergsvik, "Percy Absence Angers UNO," *DC,* 25 October 1984, p. A3; DJG interviews with Hazel and Cheryl Johnson, Mary Gonzales, Danny Solis, Phil Mullins, and Mary Ellen Montes. See also Robert Bergsvik, "Voter Drives Head into Last Busy Weekend," *DC,* 27 September 1984, p. A2.

On Marian Byrnes, see Hank DeZutter, "War of the Wetlands," *CR,* 5 February 1982, pp. 3, 30; Michael Abramowitz, "Chicago's 'Toxic Wasteland' Breeds Blue-Collar Environmentalism," *WP,* 8 November 1992, p. A3; Rod Sellers's class presentation interview with Byrnes from February 2001; Arthur M. Pearson, "Marian Byrnes: Conscience of the Calumet," *Chicago Wilderness Magazine,* Fall 2001; Illinois House Joint Resolution 16, 21 February 2003, honoring Byrnes's work; and her 30 May 2010 *CT* obituary: Duaa Eldeib, "Southeast Side Activist Fought for Environment."

37. Naomi Schreier, "State EPA Reviewing Hazardous Waste Sites," *DC,* 15 October 1984, p. A5; Manuel Galvan, "Teamwork Urged to Counter Ground-Water Contamination," *CT,* 29 October 1984, p. 17; Susan Salter, "S. Side Well Tests Urged Following Cyanide Find," *CST,* 29 October 1984, p. 10; Mark Kiesling, "EPA Denies Petition Requesting Probe into Environmental Risks," *DC,* 31 October 1984, p. A2; JSB [Juanita Salvador-Burris] to Ben [Reyes], 1 November 1984, HWP Development Box 16

Fld. 31; Mark Kiesling, "Cleanup of S. Deering Waste Site Begins," *DC,* 16 November 1984, p. B2; Casey Bukro, "State to Clean Up South Side Dump," *CT,* 21 November 1984, p. A4; Robert Bergsvik, "South Deering Residents Leery About Bright School Adoption," *DC,* 22 November 1984, p. A2; Bergsvik, "Adoption of Bright School Is Halted," *DC,* 26 November 1984, p. A1; Bergsvik, "Bright School Adoption Nixed," *DC,* 1 December 1984, p. A2; Bergsvik, "Community Group Renews Its Push for Jobs Center," *DC,* 1 December 1984, p. A1; "Jobs Center Closer," *DC,* 6 December 1984, p. A1; Wisconsin Steelworkers SOJC, "Progress Report," 9 December 1984, FLP Box 4 Fld. 9 and CFP Box 38 Fld. 473; Larry Galica, "Ex-Wis. Steel Workers Still Hoping," *DC,* 28 January 1985, p. A2.

38. "South Side United Neighbors—Transition from Grant Williams to Steuart Pittman," 20 September 1984, IA Box 2 Fld 33; "ACORN Organizing Committee Meeting," 20 September [1984], "Altgeld Tenants United (ATU)–ACORN Second Organizing Committee Meeting," 4 October 1984, both IA Box 3 Fld. 58; Press Release, "Altgeld Tenants Demand Sewer Repairs," 10 October 1984, IA Box 3 Fld. 59; Esther Wheeler to ATU/ACORN, 11 October 1984, IA Box 4 Fld. 60; "Odors Protested by CHA Residents," *CT,* 12 October 1984; "Last Organizing Committee Meeting, Altgeld Tenants Union–ACORN," 29 October 1984, "First Neighborhood Meeting of Altgeld Tenants United–ACORN," 30 October 1984, "What Is ACORN?," ca. 1 November 1984, all IAP Box 2 Fld. 32 and Box 3 Fld. 58; Press Release, "ACORN Members in Altgeld Gardens Demand Sewer Repairs," 8 November 1984, Carl Taylor, "ACORN Members Protest Sewers in Altgeld," 12 November 1984, ACORN–Altgeld Tenants United, "Facts About Our Campaign to Get the Sewers Fixed," [14 November 1984], ACORN "Press Release," 14 November 1984, all IA Box 3 Fld. 59; Altgeld Tenants United News, n.d., IA Box 2 Fld. 32; Juanita Bratcher, "CHA Hit on Sewer Crisis at Altgeld," *CD,* 15 November 1984, p. 1; Bratcher, "Cite $6 Mil Altgeld Repairs," *CD,* 20 November 1984, p. 1; Agenda, ACORN–ATU Planning Meeting, 26 November 1984, IA Box 2 Fld. 32; Perry H. Hutchinson to Superintendent Fred Rice, 28 November 1984, HWP Central Office Files Box 42 Fld. 6; Agenda, ACORN Altgeld Tenants United Neighborhood Meeting, 29 November 1984, IA Box 2 Fld. 32; Steuart Pittman, "Status Report," 17 December 1984, IA Box 2 Fld. 62; ACORN Altgeld Tenants United Monthly Meeting, 20 December 1984, IA Box 2 Fld. 58; Madeline Talbott, "Year End–Year Begin Report and Plan, Illinois ACORN," 1 January 1985, IA Box 2 Fld. 63; Leon Despres to William Walls, 16 January 1985, and Seborn E. Blackburn to Leon Despres, 21 February 1985, both HWP Central Office Files Box 1 Fld. 20; "ACORN Planning Meeting for Tour of Waste Management, Inc.," 21 January [1985], IA Box 3 Fld. 56; Illinois ACORN, "Notes on March 26 [1985] Board Meeting," IA Box 1 Fld. 10; ACORN–Altgeld Tenants United Meeting, 27 March 1985, IA Box 2 Fld. 72; Steuart Pittman, "Transition Notes," Altgeld Tenants United–ACORN, March 1985, IA Box 2 Fld. 33; Steuart Pittman to Wade Rathke, 27 October 1984, IA 2003-072 Box 1; DJG interviews with Steuart Pittman, Madeline Talbott, and Keith Kelleher. On Esther Wheeler, also see Cheryl W. Thompson, "Altgeld Gardens, City and Parks Join to Give Children a Place of Their Own," *CT,* 6 June 1993, Flynn McRoberts, "CHA Tries a Blend of Management," *CT,* 16 November 1994, and Obama, *Dreams From My Father,* p. 229 ("Mrs. Reece").

39. PCR Press Release, "Southeast Side Resident Release Health Survey and Call on City to Ban New Landfills on the Southeast Side," 16 January 1985, CRF Box 30 Fld. 368; "Southeast Siders Back Waste Ban," *CT,* 17 January 1985, p. C6; "Statement by Mayor Harold Washington," 23 January 1985, HWP Development Box 15 Fld. 26; Casey Bukro, "Mayor Urges Banning New Dumps, Tighter Rules," *CT,* 24 January 1985, p. A3; Harry Golden Jr., "Mayor Asks New Restrictions on Waste Disposal Facilities," *CST,* 24 January 1985, p. 34; Mark Kiesling and Robert Bergsvik, "City Council Extends Ban on Landfill Development," *DC,* 24 January 1985, p. A1; Solid Waste Management Task Force list, 6 February 1985, HWP Press Secretary Box 4 Fld. 28; "Waste Management Earns $142 Million," *DC,* 22 February 1985, p. B2; "City Extends Landfill Moratorium," *TNW,* March 1985, p. 16; Juanita Salvador-Burris to Benjamin Reyes et al., "Request for Meeting by Coalition for Appropriate Waste Disposal," 11 March 1985, and "Briefing Summary," ca. 17 April 1985, both HWP Development Box 16 Fld. 40; Robert Bergsvik, "Elimination of Landfill Is Sought," *DC,* 26 April 1985, p. A2; "Tough Talk," *DC,* 29 April 1985, p. A3; Robert Bergsvik, "Timing Not Right for UNO Protest," *DC,* 9 May 1985, p. A3; "MSD Takes Dump Off Its Agenda," *HN,* 16 May 1985, p. 1; James E. Landing, "Things You Should Know,"

HN, 16 May 1985, pp. 1, 5; Casey Bukro, "U.S. to Sniff Out Calumet," *CT,* 10 June 1985, p. 1; "Take Urgent Action Against New Dump," *HN,* 13 June 1985, p. 1; Violet T. Czachorski, "We Stand Alone in MSD Fight," *HN,* 20 June 1985, p. 1; Robert Bergsvik, "Sludge: Mounting Problem Forces MSD Land Action," *DC,* 20 June 1985, p. A3; Bergsvik, "MSD Agrees to Open Talks with Waste Management," *DC,* 28 June 1985, p. A1; Vi Czachorski, "MSD to Begin Dump Talks," *HN,* 4 July 1985, p. 1; and Hazel Johnson to Jackie Schad, 6 June 1986, CRF Box 30 Fld. 368. On the O'Brien Lock & Dam, see especially John Kass and Mark Eissman, "Waste Hill in Plan for Southeast Side," *CT,* 13 March 1984, p. B2; and also Mark J. Bouman, "A Mirror Cracked: Ten Keys to the Landscape of the Calumet Region," *Journal of Geography* 100 (May–June 2001): 104–10; on *HN* publisher Vi Czachorski, see her 11 July 2013 *CT* obituary: Joan Kates, "Violet Czachorski, Publisher of Hegewisch News."

40. DJG interviews with Ellen Benjamin and Roberta Lynch; "Agenda—Meeting with Mayor Washington, February 5, 1985," HWP Central Office Files Box 9 Fld. 41; Robert Mier to Benjamin Reyes, "Monthly Management Plan Status Report," 13 March 1985, HWP DS Box 2 Fld. 18; "Roseland Community" Notes, ca. 28–29 April 1985 and "Roseland" 3x5 card, n.d., both HWP DS Box 8 Fld. 9; John Betancur et al., "Roseland: Needs Assessment and Organizational Infrastructure, An Introduction," Center for Urban Economic Development, University of Illinois at Chicago, June 1985.

On Ronald Nelson's murder, see Jerry Thornton, "Gunman Kills Iowa Professor Outside Church," *CT,* 18 March 1985, p. A7; "Iowan Dies in Robbery on S. Side," *CST,* 18 March 1985, p. 22; "Police Issue Sketch of Professor's Killer," *CT,* 21 March 1985, p. A4; William Recktenwald, "Suspect Held in Church Lot Killing," *CT,* 15 April 1985, p. 6; Leon Pitt, "Sketch in Sun-Times Leads to Murder Suspect's Arrest [sic]," *CST,* 15 April 1985, p. 54; "Ex-Convict Charged in Slaying at Church," *CT,* 16 April 1985, p. A4; Tom Fitzpatrick, "Senseless Murder Haunts Courtroom," *CT,* 11 September 1986, p. 7; Linnet Myers, "Robber Convicted in Prof's Slaying," *CT,* 12 September 1986, p. A4; Rosalind Rossi, "Ex-Con Guilty of Professor's Murder," *CST,* 12 September 1986, p. 16; Myers, "Robber Gets Death in Slaying," *CT,* 9 October 1986, p. B2; Rossi, "Ex-Con Due to Die for Killing Prof," *CST,* 9 October 1986, p. 34. *People v. Hayes,* 564 N. E. 2d 803 (Ill. S. Ct.), 21 November 1990, offers a detailed recounting of the crime and Hayes's trial; subsequent affirmances of his conviction are *Hayes v. Illinois,* 499 U.S. 967 (cert. denied), 15 April 1991, *Hayes v. Carter,* U.S.D.C.N.D. Ill., 22 May 2003 (2003 USD Lexis 8650), *Hayes v. Battaglia,* 403 F. 3d 935 (7th Cir.), 13 April 2005, and *People v. Hayes,* Ill. App. 1st Dist., 4 November 2013 (2013 WL 5940308). Also see Illinois Department of Corrections Inmate Search (also providing current, full-color photos of Hayes).

41. "EDA Approves Way to Sell Wis. Number 6 Mill," *DC,* 7 March 1985, p. A1; Larry Galica, "Wisconsin Steel: Five Years Later, Closure Still Haunts Community," *DC,* 28 March 1985, p. A5; C. D. Matthews, "Ex-Employees Remember the Painful Past," *DC,* 28 March 1985, p. A7; Linnet Myers, "Good Times Ceased Rolling with the Steel," *CT,* 28 March 1985, p. C1; Frank Lumpkin to Crossroads Fund, 31 March 1985 and 4 April 1985, Lumpkin, "Grant Agreement," 26 June 1985, all CRF Box 38 Fld. 473; Tom Geoghegan to Josh Gotbaum, 5 April 1985, FLP Box 4 Fld. 12; Susan Gallagher, "Steelworkers Seek Justice," *TNW* May 1985, pp. 1, 6–7; R. C. Longworth, "It's Almost Over for Wisconsin Steel Plant," *CT,* 24 June 1985, p. C1; "Meeting Planned to Save Steel Mill," *DC,* 25 June 1985, p. A3; Larry Galica, "Ex-Wis. Steel Workers Told of Efforts to Save No. 6 Mill," *DC,* 27 June 1985, p. A1; Galica, "Wis. Steel Vigil is Planned," *DC,* 29 June 1985, p. A1; R. C. Longworth, "Final Days Close in at Wisconsin Steel," *CT,* 5 July 1985, p. II-1; "Reaction Is Mixed to GM Plant Loss," *DC,* 11 July 1985, p. A1; and Ranney, "The Closing of Wisconsin Steel," p. 80.

On South Works, see Larry Galica, "Union Will Try to Stop Demolition," *DC,* 25 October 1984, p. A1; Robert Bong, "South Works Is Finished," *DC,* 29 October 1984, p. A4; Larry Galica, "Demolition Begins at South Works," *DC,* 15 December 1984, p. A1; "So. Works Closed Until Jan. 2," *DC,* 27 December 1984, p. A2; Helene McEntee, "S. Works Planning Beam Output Boost," *CST,* 24 January 1985, p. 83; "Quit Foot-Dragging in USS Suit," *DC,* 18 February 1985, p. A4; Larry Galica, "Action Expected on So. Works Suit," *DC,* 22 February 1985, p. A1; James Warren, "Alice Peurala Regains Reins of Steel Union Local," *CT,* 1 May 1985, p. 3; "USS Demolition Ban Continued," *DC,* 13 June 1985, p. A1; Larry Galica, "USS Demolition Delayed," *DC,* 19 July 1985, p. A1; Robert Mier to Father

George Schopp, 24 July 1985, Schopp to Frank [Lumpkin], 30 July 1985, and Paul W. Bateman (EDA) to Schopp, 2 August 1985, all FLP Box 4 Fld. 12; and Hoerr, *And the Wolf Finally Came,* pp. 425–26.

On Republic, see DJG interview with Maury Richards; Conroy, "Mill Town," November 1986, p. 173; *1033 News & Views,* May 1982, p. 1; Linnet Myers, "Hopes Vanish Along with Steel Mills," *CT,* 23 October 1983, p. C1; Larry Galica, "150 to Lose Jobs at Republic Steel," *DC,* 6 January 1984, p. 1; Mark Kiesling, "Local 1033 Rejects Richards," *DC,* 26 January 1984, p. A2; Kiesling, "Local 1033 Votes to Back Richards," *DC,* 2 February 1984, p. A1; Larry Galica, "Republic–J&L Merger Rejected," *DC,* 17 February 1984, p. A8; Mark Kiesling, "Unemployment, Dumps Concerning Richards," *DC,* 6 March 1984, p. A2; "Richards Not Down," *DC,* 23 March 1984, p. A2; Larry Galica, "Union Leader Not Surprised by Republic–LTV Merger OK," *DC,* 23 March 1984, p. B3; Galica, "Local 1033 Leader Optimistic About Republic–LTV Merger," *DC,* 2 June 1984, p. A2; "Merger Combines Major Bar Producing Facilities," *LTV Steel Reporter,* July 1984, p. 1; Robert Bergsvik, "LTV Won't Deny Layoff Rumors for Mill," *DC,* 6 September 1984, p. A1; Larry Galica, "Union Griever Tosses Hat into Local 1033 Ring," *DC,* 26 October 1984, p. A8; Frank Guzzo, "Outlook for South Chicago Plant Dim," *1033 News & Views,* November 1984, p. 1; Larry Galica, "Guzzo, LTV Officials Meet," *DC,* 27 November 1984, p. A1; Galica, "Union Membership Angry over Death at LTV Plant," *DC,* 30 November 1984, p. A1; Galica, "LTV Meeting Rocky but 'Constructive,'" *DC,* 1 December 1984, p. A1; Raul Hernandez et al., "Work for Less Dues," *DC,* 19 December 1984, p. A4; Hogan, *Steel in the United States,* p. 63; Larry Galica, "Protesters Demand Jobs," *DC,* 19 January 1985, p. A2; Galica, "Local 1033 Members Reject Job Cuts," *DC,* 28 January 1985, p. A1; Frank Guzzo, "Are We Satisfied with the Progress," *Labor Management Plan Talk,* April 1985, p. 6; Larry Galica, "Richards Ousts Guzzo in USW Local 1033 Election," *DC,* 29 April 1985, p. 1; *1033 News & Views,* May 1985; Larry Galica, "Narrow 1033 Races Settled," *DC,* 4 May 1985, p. A2; "New Leader," *DC,* 31 May 1985, p. A2; Robert Bong, "LTV Seeks State Bail-Out," *DC,* 3 June 1985; Larry Galica, "LTV Tax Relief Request Rapped," *DC,* 11 June 1985, p. A1; *1033 News & Views,* July 1985; and Hoerr, *And the Wolf Finally Came,* pp. 496–502, 507–12.

42. DJG interviews with Jerry Kellman, Ken Jania, Fred Simari, Len Dubi, Mike Kruglik, Tom Kaminski, Nadyne Griffin, Betty Garrett, Marlene Dillard, Adrienne Jackson, Dan Lee, Loretta Augustine-Herron, Yvonne Lloyd, John Calicott, Jan Poledziewski, Bill Stenzel, Bonnie Nitsche, Cathy Askew, Bob Klonowski, and Glee Murray; Robert L. Johnston, "In Defense of CHD," *Chicago Catholic,* 3 and 10 August 1984, p. 8; Marianne Comfort, "New CHD Grants Announced; Fund Effort Defended," *Chicago Catholic,* 14 September 1984, p. 3; Mary Claire Gart, "5 Local Projects Get National CHD Grants," *Chicago Catholic,* 19 October 1984, p. 7 (stating the amount was $50,000); Robert L. Johnston, "Deanery Project Links Parishes, Community," *Chicago Catholic,* 14 December 1984, pp. 1, 14 (citing $42,000 rather than $50,000); "Time for 'Time,'" *Chicago Catholic,* 14 December 1984, p. 8; "Time for XII Lay Ministry Training Program," *St. Victor Parish Sunday Bulletin,* 10 February 1985, p. 3; Thomas Casaletto, "CHD Attacks Poverty's Causes," *Chicago Catholic,* 17 May 1985, p. 15; Renee Brereton, "CHD Proposal Evaluation Form—The Developing Communities Project," #86-D-02-02, 7 March [1986], Joseph Cardinal Bernardin Papers (citing $42,000 rather than $50,000); Poole and Pauken, *The Campaign for Human Development,* p. 95 (citing $40,000); Woods Charitable Fund, Report for the Year 1985, pp. 3, 15; Marty [Adams] to Ben [Reyes], 16 April [1985], HWP DS Box 16 Fld. 47; "Woods Fund Supports Citizen Involvement," *TNW,* May 1985, p. 19; Woods Charitable Fund 1985 Form 990-PF, Part XVI, 14 May 1986; *Mother Jones,* March 1983, p. 47, April 1983, p. 43, and May 1983, p. 52; and most crucially, of course, "Two Minority Jobs, Chicago," *Community Jobs,* June 1985, p. 3. The assertion that "half" of CCRC's budget came from church dues was an aspiration rather than reality. Kellman had advertised in *Community Jobs* ten months earlier, before he hired Mike Kruglik, but that ad—identical in both the July–August and September 1984 issues—had simply sought an "Organizer" or "Trainee," with no references *at all* to "Minority," "95 percent black," "experience in the black community," or "Affirmative action position." "Organizer Chicago," *Community Jobs,* July–August 1984, p. 7, and September 1984, p. 4. Multiple published assertions circa 2008 that Kellman also advertised in the *New York Times* are incorrect, as is easily confirmable: the only "community organizer" ads that appear in the *Times* during the first six months of 1985 are on 17 and 24 February for South Bronx People for Change and on 31 March for the Community Family Planning Council of New York City; no CCRC ad whatsoever appeared in the *Times*.

On San Antonio, see Donald C. Reitzes and Dietrich S. Reitzes, *The Alinsky Legacy: Alive and Kicking* (JAI Press, 1987), pp. 117–26. On Father John W. Calicott, see his essays "Black Priestly Vocations," *Upturn*, December 1980–January 1981, pp. 12, 14; "A Practical and Real Necessity," *Upturn*, November–December 1982, pp. 6, 8; "Another Beginning," *Upturn*, December 1984–January 1985, p. 8; and "Church and Young People," *Upturn*, August–September 1986, p. 14. In 1994, Calicott was removed from ministry on account of two allegations of sexual misconduct dating from 1976. He was reinstated in 1995, then removed again in 2002 solely on account of a change in U.S. Conference of Catholic Bishops policy regarding prior allegations. See Scott Fornek and Philip Franchine, "Pastor of Holy Angels Accused of Sex Abuse," *CST*, 12 April 1994, p. 4; Ingrid E. Bridges, "Controversial Priest Makes Official Return to Holy Angels," *CD*, 16 October 1995, p. 3; Cathleen Falsani and Ana Mendieta, "Accused Priest Lecturing Kids About Sex," *CST*, 22 January 2004, p. 8; Falsani, "Priest Ran Afoul of Policy on Abusers, Archdiocese Says," *CST*, 23 January 2004, p. 9; and Cheryl V. Jackson, "Church Defrocks Popular Ex-Priest Accused of Abuse," *CST*, 28 October 2009, p. 8.

CHAPTER TWO: A PLACE IN THE WORLD

1. Accounts of the Obama family's history in Kenya are replete with uncertain dates, conflicting memories, and a plethora of relatives with multiple names or nicknames. In sorting through the history from Hussein Onyango forward, the most assiduously researched account is Peter Firstbrook's *The Obamas: The Untold Story of an African Family* (Preface Publishing, 2010), esp. pp. 174, 184, and 201. Between April 2010 and December 2013, the late Zeituni Onyango, both in person and in scores of e-mails, kindly explained countless intricacies of her family's history to me, and her autobiography, *Tears of Abuse* (Afripress Publishing, 2012), esp. pp. 7–8, is usually the best guide to the extended family's genealogy. Her untimely passing is hugely mourned. See Katharine Q. Seelye, "Zeituni Onyango, Obama's Aunt from Kenya, Dies at 61," *NYT*, 9 April 2014, p. B19. Abon'go Malik Obama and Frank Koyoo's *Barack Obama Sr.: The Rise and Life of a True African Scholar* (Xlibris, 2012), esp. p. 118, offers some valuable specifics but must be used with caution. See also Barack Obama, *Dreams From My Father* (Times Books, 1995), pp. 394–427, esp. 413, and Mark Obama Ndesandjo, *An Obama's Journey* (Lyons Press, 2014), pp. 23, 277. Sally H. Jacobs's *The Other Barack* (Public Affairs, 2011) is a generally reliable journalistic account. Countless newspaper stories, especially from 2007–2009, must be read with extreme caution. Stories that *do* merit attention are John Oywa, "Sleepy Little Village Where Obama Traces His Own Roots," *Nation*, 15 August 2004; Laurie Goering, "Obama Campaign Closely Watched—in Kenya," *CT*, 10 October 2004, p. 1; John Oywa, "Tracing Obama Snr.'s Steps as a Student at Maseno School," *Standard*, 4 November 2008; Joe Ombuor, "Obama's Father and the Origin of the Muslim Name," *Standard*, 4 November 2008; John Oywa, "Keziah Obama: My Life with Obama Senior," *Standard*, 11 November 2008; and Stephen Kinzer, "The Man Who Made Obama," *Guardian*, 8 February 2009. Subsequent journalistic books have recycled materials from these stories, sometimes with citations, sometimes without. "Resume, Barack H. Obama," Tom Mboya Papers Box 47 Fld. 5, which Betty Mooney Kirk prepared for him, notes the law firm and engineering firm employment; Obama, "Work Experience After Leaving School," [18 May 1959], Phelps Stokes Fund Papers (PSFP) Box 214, describes multiple jobs without naming any employers. In a similarly dated "Application for Financial Assistance" to AAI, in which Obama listed his date of birth as 18 June 1934, he asserted, "I was forced to leave school in my last year due to financial difficulties at home . . . due to the fact that my father got very ill." He also wrote that after college, "I hope to open my own firm of Civil Engineering and Architecture."

On Tom Mboya, see particularly his own *Freedom and After* (Andre Deutsch, 1963), esp. p. 59, and David Goldsworthy, *Tom Mboya: The Man Kenya Wanted to Forget* (Heineman, 1982), esp. pp. 61–62; on Bill Scheinman, see Joseph B. Treaster, "W. X. Scheinman, 72, Broker and Friend of Kenyan Leader," *NYT*, 25 July 1999. Obama arrived in New York more than a month prior to the well-publicized educational "airlift" of Kenyan students that began in 1959 and continued through 1961. Organized under the aegis of the African-American Students Foundation, which Mboya and Scheinman created in the spring of 1959, the initial airlift plane delivered eighty-one young Kenyans to the U.S. to begin collegiate studies. See "81 Kenyans to Hold U.S. Scholarships," *NYT,* 6 August 1959, p. 7; "81 Youths in Kenya Off to School in the U.S.," *NYT,* 8 September 1959, p. 7; Gay Talese, "81 in From Kenya to Go to College," *NYT,* 10 September 1959, p. 8; William X. Scheinman, "Higher Education in Kenya," *NYT,* 10 September 1959, p. 34; Tom Mboya, "To Aid African Students," *NYT,* 24 November 1959, p. 36; Albert G. Sims, "Africans Beat on Our College Doors," *Harper's,* April 1961, pp. 53–58; and Tom Mboya, "African Higher Education: A Challenge to America," *Atlantic,* July 1961, pp. 23–26. Subsequent sources on the airlift include Alan Rake, *Tom Mboya: Young Man of New Africa* (Doubleday, 1962), pp. 172–79; Mboya, *Freedom and After,* pp. 137–38 and 143–46; Mansfield I. Smith, "The East African Airlifts of 1959, 1960, and 1961," Ph.D. dissertation, Syracuse University, June 1966; Goldsworthy, *Tom Mboya,* pp. 116–19; Michael Dobbs's excellent article, "Obama Overstates Kennedy's Role in Helping His Father," *WP,* 30 March 2008, pp. A1, A8; James H. Meriweather, "'Worth a Lot of Negro Votes': Black Voters, Africa, and the 1960 Presidential Campaign," *Journal of American History* 95 (December 2008): 737–63, esp. pp. 745–48; Tom Shachtman, *Airlift to America* (St. Martin's Press, 2009); Cora Weiss, "From Kenya to America," *NYT Book Review,* 9 May 2010, p. 6; Robert F. Stephens, *Kenyan Student Airlifts to America, 1959–1961: An Educational Odyssey* (Kenway Publications, 2013), esp. p. 62; and Ndesandjo, *An Obama's Journey,* p. 25.

2. Betty Mooney to Frank Laubach, 5 June 1958, 4 and 17 August 1958, Laubach Papers Box 7; Obama to John M. Livingstone (AAI), 20 October 1958, and Obama to Gordon P. Hagberg (AAI), 30 October 1958, PSFP Box 214; Mooney to Laubach, 21 December 1958, Frank Laubach to Floyd et al., 30 December 1958, Betty Mooney to Mrs. Laubach, 11 January 1959, Laubach Papers Box 7; Hagberg to Obama, 20 January 1959, and Mooney to Hagberg, 28 January 1959, PSFP Box 214; *The Key: Kenya Adult Literacy News* #5, February 1959, p. 5, Laubach Papers Box 237; Betty Mooney to Frank Laubach, 16 and 24 February 1959, Laubach Papers Box 7; Obama to Hagberg, 4 March 1959, PSFP Box 214; Mooney to Laubach, 2 March 1959, Laubach Papers Box 8; Frank J. Taylor, "Colorful Campus of the Islands," *Saturday Evening Post,* 24 May 1958, pp. 38–39, 95–98; Edward T. White, Director of Admissions, University of Hawaii, "Certificate of Eligibility," 19 February 1959, Obama INS File and PSFP Box 214; Betty Mooney to Frank Laubach, 23 March 1959, Mooney and Helen Roberts to Laubach, 28 March 1959, Laubach to Mooney, 30 March 1959, Laubach to Registrar, University of Hawaii, 30 March 1959, Mooney to Laubach, 26 April 1959, Laubach Papers Box 8; Hagberg to Obama, 14 May 1959, Hagberg to Mooney, 14 May 1959, Obama to Hagberg, 18 May 1959, Mooney to Hagberg, 20 May 1959, and Mooney to AAI, 20 May 1959, PSFP Box 214; Mooney to Laubach, 1 and 4 June 1959, Laubach Papers Box 8; Hagberg to Loyd Steere, 4 June 1959, PSFP Box 214; Mooney to Laubach, 9 June 1959, Laubach Papers Box 8; Steere to Hagberg, 12 June 1959, Mooney to Hagberg, 6 July 1959, Hagberg to Obama, n.d. [pre–13 July 1959], Hagberg to Obama, 13 July 1959, Hagberg to Mooney, 14 July 1959, Mooney to Hagberg, 21 July 1959, Hagberg to Mooney, n.d., Obama to Hagberg, 21 July 1959, Harry Heintzen to Bruce McGavren, "Air Ticket for Barack Obama," 23 July 1959, McGavren to Heintzen, "Air Ticket—Barack Obama," 28 July 1959, PSFP Box 214; Barack H. Obama to Dr. Frank Laubach, 28 July 1959, Laubach Papers Box 8; Barack Hussein Obama, "Statement to Be Signed by Applicant for Nonimmigrant Visa," 29 July 1959, Obama INS File; Betty Mooney to Frank Laubach, 5 August 1959 ("Barack left last night"), Laubach Papers Box 8; Heintzen to File, "Transportation and Hotel Arrangements for Mr. Barack Obama," 6 August 1959, PSFP Box 214; "Arrival-Departure Record," Obama INS File; Heintzen to File, "Barack Obama," 11 August 1959, and Anna Trimiar to Heintzen, "Barack Obama," 28 August 1959, PSFP Box 214; Betty Mooney to Frank Laubach, 30 August 1959 (reporting that "Barack is at Koinonia now"), Laubach Papers Box 8. The Hawaii admissions certificate records Obama's birth year as 1934.

On Koinonia, see Timothy Miller, *The Quest for Utopia in Twentieth-Century America: 1900–1960* (Syracuse University Press, 1998), p. 178. Helen M. Roberts's biography is *Champion of the Silent Billion: The Story of Frank C. Laubach, Apostle of Literacy* (Macalester Park Publishing, 1961). Obama expressly credits the *Saturday Evening Post* story in a conversation with a student newspaper reporter six months later. See "First African Enrolled in Hawaii Studied

Two Years by Mail," *Ka Leo O Hawaii,* 8 October 1959, p. 3. Jacobs, *The Other Barack,* p. 92, dates the *Ramogi* ad to July 7, 1959; an impressive academic article, Matthew Carotenuto and Katherine Luongo, "*Dala* or Diaspora? Obama and the Luo Community of Kenya," *African Affairs* 108 (April 2009): 197–219, at 204n29, dates it to April 7; both could well be correct. Of the three Luo primers, only the second, at least as recorded in the comprehensive worldwide OCLC database, officially lists Obama as the author. See Barack H. Obama, *Otieno: The Wise Man. Book 2. Wise Ways of Farming [Otieno Jarieko]* (East African Literature Bureau, 1959). See also Dana Seidenberg, "Cold Warrior for Racial Equality," *East African,* 7 February 2009.

3. Shurei Hirozawa, "Young Men from Kenya, Jordan and Iran Here to Study at U.H.," *Honolulu Star-Bulletin* [*HSB*], 19 September 1959, p. 5; Johnny Brannon, "Hawaii's Imperfect Melting Pot a Big Influence on Young Obama," *Honolulu Advertiser* [*HA*], 10 February 2007; University of Hawaii to INS, "Report of Initial Registration," 21 September 1959, Obama INS File; DJG interview with [Andy] Pake Zane, Abercrombie speaking in "Barack Obama Revealed," CNN, 20 August 2008; to Dan Boykin, "'08: Year of Obama," *Midweek,* 2 January 2008; in Jacobs, *The Other Barack,* pp. 107–8; on "A Childhood of Loss and Love," ABC's *20/20,* 26 September 2008; and to David Nather, "Obama's Quick Rise on a Non-Traditional Career Path," *CQ Weekly,* 21 August 2008. The UH registration report to INS, like the *Star-Bulletin* story, gave Obama's age as twenty-five, with an 18 June 1934 birth date. Sometime in the fall of 1959 Bill Scheinman's African-American Students Foundation paid $143 to Obama, or to UH on his behalf, but no specific detail is recorded. "African-American Students Foundation, Inc., Obligations and Bills Due as of November 18, 1959," AASF Papers Box A Fld. WXS-Bills & November 1959 Expenses; "From: W. X. S. Re: Four Year Jackie Robinson Scholarships," n.d., AASF Papers Box A Fld. WXS-Bills, Mboya. See also "Kenya Students in the United States—May 11, 1960," p. 6, AASF Papers Box B Fld. Lists—Airlift Students & Scheinman Papers Box 20 Fld. 2, listing "Barrack H. Obama," and an undated document from sometime after March 1961, "IIE Grants $100,000," AASF papers Box A Fld. IIE, p. 28, listing "Obama, B. H." but not indicating any sum of money.

4. "First African Enrolled in Hawaii Studied Two Years by Mail," *Ka Leo O Hawaii* [*KLOH*], 8 October 1959, p. 3, *KLOH,* 5 November 1959, p. 1; "Isle Inter-Racial Attitude Impresses Kenya Student," *HSB,* 28 November 1959, p. 5; Betty Mooney to Frank Laubach, 22 October 1959, Laubach Papers Box 8; Obama (1810 University Avenue) to Gordon Hagberg, 8 December 1959, PSFP Box 214; "Application by Alien Student for Permission to Accept Employment," 8 December 1959, Obama INS File; Harry Heintzen to Obama, 4 February 1960, Obama (1648A Tenth Avenue) to Heintzen, 10 February 1960, Sumie F. McCabe to AAI, "Barack H. Obama," 12 February 1960, Harold M. Bitner to AAI, 15 February 1960, Elizabeth Mooney to Heintzen, 26 February 1960, PSFP Box 214. Marjorie Yoshioka, "Model UNers to Debate Race," *KLOH,* 7 April 1960, p. 4; "VJ" (AAI) to File, "Barack Obama," 19 April 1960, PSFP Box 214; Barack H. Obama (1648-A Tenth Avenue), "Terror in the Congo," *HSB,* 8 June 1960, p. 8; Obama UH Transcript, 22 June 1960 (625 Eleventh Avenue), Heintzen to Obama, 11 July 1960, Obama (2036 Round Top Terrace) to Heintzen, 24 July 1960, PSFP Box 214; Sumie McCabe, "Certificate of Eligibility," 24 July 1960, and Obama (2036 Round Top Terrace), "Application to Extend Time of Temporary Stay," 28 July 1960, Obama INS File; Robert L. Sherman (AAI) to UH, 7 September 1960, Edward T. White to Sherman, 22 September 1960, Elizabeth Mooney Kirk to Heintzen, 24 March 1961, PSFP Box 214; Obama, "Application to Extend Time of Temporary Stay," 31 August 1961, Obama INS File; Dorothy Heckman Gregor quoted in Jacobs, *The Other Barack,* p. 111; Judy Ware in Janny Scott, *A Singular Woman* (Riverhead Books, 2011), p. 84; Philip Ochieng, "The Pride of a People: Barack Obama, the Luo," *Nation,* 17 January 2009. See also Philip Ochieng, "It's About America's Humanity, Not Black Power," *East African,* 7 November 2008. On Dunham's virginal status prior to fall 1960, see Scott, *A Singular Woman,* p. 84.

The 1960–61 and 1961–62 Polk City Directories for Honolulu both list Barack H. Obama as living at 625 Eleventh Avenue.

At least two previous authors, including Jacobs, *The Other Barack,* p. 115, and a subsequent one who subcontracted out much of his research and interviewing, including to a published expert on UFOs, name Ella Wiswell as the instructor for Russian 101. In their absence of more extensive footnoting, however, *KLOH,* 19 January 1961, p. 5, giving a 28 January 1961 final exam date for Russian 101, and "Schedule of Courses," *KLOH,* 26 January 1961, p. 8, offer a potentially decisive corrective. Four fall 1960 Russian courses were taught—101, 151, 153, 201—with 102 (which required three additional language laboratory sessions per week) and again three more advanced ones offered for spring 1961. Wiswell is identified as teaching only the advanced courses, while two sections of the introductory course were taught by Isabella Troupiansky, a 1930 graduate of the University of Paris who, born in 1907, was fifty-three years old in 1960. Troupiansky taught at UH at least from 1960 through 1961. See *Modern Language Journal* 46 (March 1962): 124, and Patricia A. Polansky, "Who Created Us? . . . ," *Slavic & East European Language Resources* 9 (2008): 174–225. Troupiansky died in 2002 at age ninety-five. Ann Dunham's subsequent University of Washington transcript confirms that she took a fall 1960 UH Russian course, but lists it as Russian 100, presumably a typo.

5. The starting point for any examination of Ann Dunham's early life and those of her parents is Janny Scott's rich and valuable *A Singular Woman,* esp. pp. 20–23, 25–26, 37–38, 41–42. Also informing this account are Stephen MacDonogh, *Pioneers: The Frontier Family of Barack Obama* (Brandon, 2010); two excellent Associated Press stories by a pair of superb reporters, Allen G. Breed, "'Toot': Obama Grandmother a Force That Shaped Him," 23 August 2008, and Nancy Benac, "Obama's Gramps: Gazing Skyward on D-Day in England," 30 May 2009; and DJG's interviews with Ralph Dunham and Charles Payne. See also David Nitkin and Harry Merritt, "A New Twist to an Intriguing Family History," *Baltimore Sun,* 2 March 2007.

6. Scott, *A Singular Woman,* esp. pp. 56, 58; DJG interviews with Ralph Dunham and Charles Payne; Hugh Pickens, "Barack Obama's Grandfather Stanley Dunham," peacecorpsonline.typepad.com/poncacityweloveyou, 6 February 2009; Beverly Bryant, "Obama's Family Once Lived in PC," *Ponca City News,* 8 February 2009, p. A1; Ann Arnold, "Vernon's Obama Connection," TexomasHomePage.com, 14 February 2009; Fred Mann, "Kansas Roots Show in Obama, Say Relatives," *Wichita Eagle,* 2 February 2008; Rick Montgomery, "Barack Obama's Mother Wasn't Just a Girl from Kansas," *Kansas City Star,* 26 May 2008, p. A1; Susan Peters, "President Obama: From Kansas to the Capital," KAKE.com (Wichita), 27 January 2009. Ann's other childhood books, all in Box 7 of the Obama Family Papers at UH and all in editions dating from between 1946 and 1950, are almost all well-known classics, including *Aesop's Fables, Pinocchio,* Robert Louis Stevenson's *Treasure Island,* and L. Frank Baum's *The Wizard of Oz.*

7. Scott, *A Singular Woman,* pp. 43–72, esp. p. 47; Tim Jones, "Obama's Mom: Not Just a Girl from Kansas," *CT,* 27 March 2007; Nicole Brodeur, "Memories of Obama's Mother," *Seattle Times,* 5 February 2008; Jonathan Martin, "Obama's Mother Remembered as 'Uncommon,'" *Seattle Times,* 8 April 2008, p. A1; Amanda Ripley, "The Story of Barack Obama's Mother," *Time,* 21 April 2008, pp. 36ff; Susan Essoyan, "Strong Women Led Obama," *HSB,* 10 August 2008; most especially Phil Dougherty, "Stanley Ann Dunham, Mother of Barack Obama, Graduates from Mercer Island High School in 1960," HistoryLink.org, 22 January 2009, which corrects errors in earlier stories; Lisa Fairbanks-Rossi, "Like Mother, Like Son: Before Barack Became a Kid in Hawaii," *Pacific Northwest Inlander,* 27 January 2009; and DJG interviews with Ralph Dunham, Charles Payne, and Susan Botkin Blake. Obama, *DFMF,* p. 123, indicates Ann first saw *Black Orpheus* in Chicago while working there as an au pair during the summer of 1959, but no other evidence or recollection, including most significantly Scott's *A Singular Woman,* testifies to Ann spending a summer in Chicago.

David Maraniss, *Barack Obama: The Story* (Simon & Schuster, 2012), pp. 135–37, 154 [hereinafter *BOTS*] describes Isle-Wide. Given the group project nature of its research and interviewing, specifics presented in *BOTS* must be treated with considerable caution. See, for example, in a publication that always must be read with the *greatest possible* caution, Don Wilkie, "What David Maraniss Left Out of *the Story,*" *American Thinker,* 11 July 2012. However, Ralph Dunham's excellent memory independently confirms the Isle-Wide story.

Most loony musings about the Obamas and the Dunhams merit no citation or comment. One regrettable exception, given both the author's vita and the essay's venue, is Angelo M. Codevilla, "The Chosen One," *Claremont Review of Books,* Summer 2011, pp. 52–58, esp. p. 54. Obama Sr. was not selected by the CIA to study in the U.S., nor was Stanley Dunham a CIA operative. First arriving in Hawaii in the summer of 1960, neither was Dunham on hand to greet Obama Sr. upon his arrival in 1959. The fascinating photo in question, of uncertain date and provenance, may well picture Stanley Dunham with Obama Sr., but if so it almost certainly dates from June 1962, when Obama was leaving Hawaii.

8. Stanley Ann Dunham Obama's University of Washington Transcript; Barack Obama to Genevieve Cook, 1 January [1985], Genevieve Cook Papers; *Stanley Ann D. Obama v. Barack H. Obama,* Libel for Divorce, #57972, Domestic Relations Division, First Judicial Circuit Court, Hawaii, 25 January 1964; DJG interviews with Zeituni Onyango, Ralph Dunham, and Charles Payne; Auma Obama, *And Then Life Happens* (St. Martin's Press, 2012), p. 35 (recalling later being told that "Her father is insisting that the two of them get married"); Kadi Warner in Scott, *A Singular Woman,* p. 91; Suzanne Roig, "Hawaii Friends, Former Co-Workers Remember Madelyn Dunham," *HA,* 15 November 2008; Herbert A. Sample, "Barack Obama's Late Grandmother Remembered," AP, 15 November 2008.

Ann's son would later write that "how and when the marriage occurred remains a bit murky" (Obama, *DFMF,* p. 22) and would tell a subsequent interviewer, "I never probed my mother about the details. Did they decide to get married because she was already pregnant?" (Ripley, "The Story of Barack Obama's Mother," *Time,* 21 April 2008, pp. 36ff.). On occasion, the younger Obama has offered comments regarding his parents' marriage that are patently inaccurate. See Obama, Remarks at Selma Voting Rights March Commemoration, 4 March 2007; and Patrick Healey and Jeff Zeleny, "Clinton and Obama Unite, Briefly, in Pleas to Blacks," *NYT,* 5 March 2007, p. A14: "Obama relayed a story of how his Kenyan father and his Kansas mother fell in love because of the tumult of Selma, but he was born in 1961, four years before the confrontation at Selma took place." In one of the few interviews Madelyn Dunham ever gave prior to her death in 2008, she said in October 2004 that her daughter "was sometimes startling." Asked how, she replied, "she married Barry's father." David Mendell, *Obama: From Promise to Power* (HarperCollins, 2007), p. 27. See also Mendell, "The Teachings of Toot," *CM,* 21 October 2008. That comment rightly stresses that Ann did not marry against her own will, above and beyond her father's insistence.

9. "Phi Kap, Phi Beta Kappa Present Awards Tonight," *KLOH,* 12 January 1961, p. 1; Robert L. Sherman to UH, 27 January 1961, Obama (1704 Punahou Street, Apt. 15) to Hagberg, 8 February 1961, Obama to Hagberg, 25 February 1961, Obama, "Application for Domestic Scholarship," 1 March 1961, Obama to AAI, 4 March 1961, Allan F. Saunders to AAI, 8 March 1961, PSFP Box 214; "248 Students Make Dean's List, *KLOH,* 9 March 1961, p. 4; Hagberg to Obama, 10 March 1961, James K. Lowers to Hagberg, 14 March 1961, Lee E. Winters to Hagberg, 2 April 1961, Sherman to Obama, 7 June 1961, and Obama to Sherman, 18 June 1961, PSFP Box 214; Lucy Ikeda, "Petition Urges Repeal of Georgia Act," *KLOH,* 19 January 1961, p. 1; E. W. Putman (president, NAACP Honolulu Branch), "Georgia Petition," *KLOH,* 16 February 1961, p. 4; "Integration Pickets Greet Alabama Governor," *HSB,* 25 June 1961, p. 1; "Bias Foes Greet 'Bama Governor," *HA,* 25 June 1961, p. A8; Leo Egan, "Governors Asked to Back Kennedy," *NYT,* 27 June 1961, p. 13; Neil Abercrombie, "Racial Equality Now!" *KLOH,* 30 June 1961, p. 2; Hal Abercrombie in Jacobs, *The Other Barack,* p. 109, confirming Obama's presence.

One of the other leading Patterson pickets, John M. Kelly Jr., became locally famous in a very different context after founding Save Our Surf that very same year. See Catherine E. Toth, "Hawaii Surf Activist John Kelly Dies," *HA,* 5 October 2007; Diana Leon, "His Save Our Surf Gave Waves a Break," *HSB,* 5 October 2007; "John Kelly: Tireless Savior of the Surf," *HSB,* 7 October 2007; Suzanne Roig, "Hawaii Surf Legend John Kelly Laid to Rest," *HA,* 19 November 2007; and Brett Thomas and Chris Evans, "The Aloha of John Kelly," *Surfer's Journal* 16 (Summer–Fall 2007): 6, 8.

10. Obama, "Application by Alien Student for Permission to Accept Employment," 3 March 1961, Ralph H. Holton (INS District Director) to Obama, 9 March 1961, and Lyle H. Dahlin, "Memo for File," n.d. [10–11 April 1961], all Obama INS File; Sally Jacobs, "Father Spoke of Having Obama Adopted," *BG,* 7 July 2011, p. A1; Zeituni Onyango e-mail to DJG, 7 July 2011; DJG interviews with Charles Payne, Ralph Dunham, and Virginia Dunham Goeldner; Jacobs, *The Other Barack,* pp. 122–24; Scott, *A Singular Woman,* p. 25; "Certificate of Live Birth," 8 August 1961, Hawaii State Department of Health # 61-10641 (issued 25 April 2011); "Births, Marriages, Deaths," *HA,* 13 August 1961, p. B6; "Births," *HSB,* 14 August 1961, p. 24. Obama's 1704 Punahou Street #15 address also appears in "East African Students in the United States as of December, 1961," p. 8, AASF Papers Box B Fld. Lists—Airlift Students, and Scheinman Papers Box 20 Fld. 2. On Dr. Sinclair, see his 24 August 2003 *HSB* obituary notice; Dan Nakaso, "Born in the USA, Certificate Proves," *Honolulu Star-Advertiser,* 28 April 2011; and Corky Siemaszko's 29 April 2011 *New York Daily News* interview with his widow. One self-proclaimed secondhand witness to the 4 August birth can be entirely discounted. See Paula Voell, "Teacher from Kenmore Recalls Obama Was a Focused Student," *Buffalo News,* 20 January 2009 (former Punahou teacher Barbara Nelson describing well-known Honolulu obstetrician Dr. Rodney T. West as the attending physician rather than Dr. Sinclair); see also Maraniss, *BOTS,* pp. 166, 276. Dr. West died in 2008. See Dan Nakaso, "Pearl Harbor Survivor a Hero to Visitors," *HA,* 11 November 2007, and "Rodney T. West," *Physician Executive,* May 2008.

11. Obama, "Application to Extend Time of Temporary Stay," 31 August 1961, Obama, "Statement to Be Signed by Applicant for Nonimmigrant Student Visa," n.d. [31 August 1961], William T. Wood II, "Memo: Obama, Barack H.," 31 August 1961, Ralph H. Holton to Obama, 6 September 1961, all Obama INS File; *News from East-West Center #6,* July 1961, p. 5, Obama to AAI, 19 September 1961, and Robert L. Sherman to Obama, 19 September 1961, PSFP Box 214; "Zeigler Gets Adviser Post," *KLOH,* 29 September 1961, p. 1; Obama to AAI, 3 October 1961, Obama, "Application for Domestic Scholarship," 27 October 1961, A. Lee Ziegler to Sherman, 9 November 1961, Allan F. Saunders to AAI, 15 November 1961 (stating that Obama "has free relations with a variety of students"), Sherman to Saunders, 28 November 1961, and Saunders to Sherman, 6 December 1961, PSFP Box 214; Stanley Ann Dunham Obama Transcript, University of Washington; University of Washington Catalogue, 1961–62, p. 5; Jonathan Martin, "Obama's Mother Remembered as 'Uncommon,'" *Seattle Times,* 8 April 2008, p. A1; Michael Patrick Leahy, *What Does Barack Obama Believe?* (Harpeth River Press, [1 September] 2008), esp. pp. 5–6, 31–32 (based upon 23 August 2008 phone interviews with Susan Botkin Blake, Maxine Hanson Box, and Barbara Cannon Rusk); Charlette LeFevre, "Barack Obama: From Capitol Hill to Capitol Hill," *Capitol Hill Times,* 9 January 2009; Patti Payne, "Obama's Mother Went to Mercer Island High School," *Puget Sound Business Journal,* 11 January 2009; Jenny Neyman, "Obama Baby-Sitter Awaits New Era," *Redoubt Reporter* [Soldotna, AK], 20 January 2009; Charlette LeFevre and Philip Lipson, "Baby Sitting Barack Obama on Seattle's Capitol Hill," Seattle Museum of the Mysteries, 28 January 2009 (reprinted in *Seattle Gay News,* 6 February 2009); Phil Dougherty, "Barack Obama Moves to Seattle in August or Early September 1961," HistoryLink.org, 10 February 2009; Scott, *A Singular Woman,* pp. 87–88; Jacobs, *The Other Barack,* p. 127; DJG interview with Susan Botkin Blake; and, with the *greatest reluctance* (and only because it recounts a phone interview with Mary Toutonghi), Jerome Corsi, "New Doubts Revealed in Obama Nativity Story," WorldNetDaily, 28 July 2009. A one-minute video interview with Susan Botkin Blake entitled "She Changed His Diapers" was posted on www.chicagotribune.com on 27 March 2007 but has since vanished from the Web. Maraniss, *BOTS,* p. 176, states regarding Ann that "her mother encouraged her escape, and apparently abetted it. She called a friend on Mercer Island to arrange temporary shelter." This is a very plausible *guess,* but no source whatsoever is cited to support the claim that Madelyn actively aided Ann's move to Seattle.

12. Jacobs, *The Other Barack,* p. 126; Robert M. Ruenitz, "I Called Him 'Obama,'" in Naranhkiri Tith et al., "Remembering My Friend Barack Obama," Cambodiana.org, January 2006 (now defunct yet preserved in an otherwise often unreliable Website, www.theobamafile.com/_family/senior.htm); "Hitosubashi, UH Combine for Economic Seminar," *KLOH,* 17 October 1961, p. 1; "Nuclear Stalemate Views to Be Aired at Symposium" and "Enrollment at New High," both *KLOH,* 20 October 1961, p. 1; "Foreign Scholars Discuss Nuclear Arms Stalemate," *KLOH,* 28 November 1961, p. 3; Kiri Tith e-mails to DJG; Neil Abercrombie, "He Protests," *KLOH,* 22 December 1961, p. 4; "NAACP Group to Hear Obama," *KLOH,* 12 January 1962, p. 3; Obama to Robert L. Sherman, 31 January 1962, PSFP Box 214; Ruth H. Bunche to Dear Friend, 8 February 1962, AASF Papers Box B Fld. Financial Items Bills; Sherman to Obama, 9 February 1962, and A. Lee Ziegler to AAI, 21 February 1962, PSFP Box 214; "Liberal Arts Society Elects 26 Students to Membership," *KLOH,* 27 April 1962, p. 3; "A Call to All the People of Hawaii," *KLOH,* 8 May 1962, p. 3; "'Peace Rally' Set for 3PM Today," *HSB,* 13 May 1962, p. 12; George Eagle, "Peace Rally Hears World Amity Pleas," *HA,* 14 May 1962, pp. A1, A2; "Marchers Make Plea for

Peace," *HSB*, 14 May 1962, p. 20; "Community Leaders Call for Action to Prevent War," *Voice of the ILWU*, 18 May 1962, p. 5; "Barack Obama Senior in 1962," ILWULocal142.org, 12 March 2009.

13. Betty Mooney Kirk to Harry Heintzen, 22 March 1962, PSFP Box 214; Kirk to Tom Mboya, 8 May 1962, Mboya to Kirk, 14 May 1962, Mboya Papers Box 47 Fld. 5; Helen Roberts letters of 15 May and 4 July 1962 as quoted in Jacobs, *The Other Barack*, p. 130; Auma Obama, *And Then Life Happens*, p. 32; Zeituni Onyango, *Tears of Abuse*, pp. 26–27; Barack Obama to "Dear Tom," 29 May 1962, Mboya Papers Box 41 Fld. 5. Zeituni wrote that Kezia "left Kogelo for her parent's home in Kendu Bay leaving the children behind" when "Auma was just a baby" after hearing that Barack "had a girlfriend" in the U.S. "Kezia returned to check on them but never stayed." *Tears*, pp. 26–27.

14. Jacobs, *The Other Barack*, p. 128; Abon'go Malik Obama and Frank Koyoo, *Barack Obama Sr.*, p. 183; "1,201 U. of H. Students Get Diplomas, Degrees," *HSB*, 18 June 1962, p. 3; "1200 Students Get Degrees at UH Commencement," *HA*, 18 June 1962, p. A5; "Kenya Student Wins Fellowship," *HSB*, 20 June 1962, p. 7; John Griffin, "First UH African Graduate Gives View on East-West Center," *HA*, 22 June 1962, p. B3 (reporting that Obama "leaves today"); Neil Abercrombie in Tim Jones, "Obama's Mom," *CT*, 27 March 2007; Jacobs, *The Other Barack*, p. 131; Barbara Cannon Rusk's 23 August 2008 interview as quoted in Leahy, *What Does Barack Obama Believe?* p. 36; Rusk in Phil Dougherty, "Barack Obama Moves to Seattle in August or Early September 1961," HistoryLink.org, 10 February 2009; Scott, *A Singular Woman*, p. 84; Janny Scott e-mail to DJG, 22 September 2011; Dinesh Sharma, *Barack Obama in Hawaii and Indonesia* (Praeger, 2011), p. 220; Auma Obama, *And Then Life Happens*, p. 37.

15. Obama to Robert L. Sherman, 13 June 1962, Sherman to Obama, 21 June 1962, Betty Kirk to Emory Ross (AAI), 22 June 1962, Robert S. Laubach to Ross, 29 June 1962, Ross to Laubach, 9 July 1962, Ross to Kirk, 13 July 1962, Obama to Sherman, 14 July 1962, Obama, Phelps Stokes Fund Application, 14 July 1962, William Cullen Bryant II to Sherman, 18 July 1962, and Sherman to Kirk, 24 July 1962, PSFP Box 214; Maraniss, *BOTS*, p. 183 (quoting from the Kirks' guestbook); Obama, "Certificate of Eligibility," 10 August 1962, Obama, "Certificate by Nonimmigrant Student," 17 August 1962, Obama, "Application to Extend Time of Temporary Stay," 17 August 1962, Alien's Change of Address Card, #A11 938 537, 26 September 1962, all Obama INS File; Obama Transcript, UH, 20 September 1962, PSFP Box 214; Stanley Ann Dunham Obama Transcript, University of Washington (showing a copy was sent to UH on 30 December 1962); Johnny Brannon, "Hawaii's Imperfect Melting Pot a Big Influence on Young Obama," *HA*, 10 February 2007; Will Hoover, "Obama's Hawaii Boyhood Homes Drawing Gawkers," *HA*, 9 November 2008; "President Barack Obama's Many Dwellings Over the Years," BergProperties.com, 4 February 2009; Neil Abercrombie in Kirsten Scharnberg and Kim Barker, "The Not-So-Simple Story of Barack Obama's Youth," *CT*, 25 March 2007, in Jon Meacham, "On His Own," *Newsweek*, 1 September 2008, p. 26, on ABC's *20/20*, "A Childhood of Loss and Love," 26 September 2008, in Dan Nakaso, "Army Veteran Grandfather Was Obama's Boyhood Pal in Hawaii," *HA*, 14 November 2008, in Gloria Borland and Kris Anderson, "An American Boyhood: Barack Obama in Hawaii," 2008, and in Sharma, *Barack Obama in Hawaii and Indonesia*, p. 28; Cindy Pratt Holtz in Maraniss, *BOTS*, p. 187; Yatushiro in Gloria Borland, "Barack Obama: Made in Hawaii," Demo Reel, 2011, and in Claudine San Nicolas, "Retired Teachers on Maui Recall Young, 'Cute' Student Barry," *Maui News*, 21 January 2009; and Barack Obama's very rich and revealing 2001 interview with Julieanna Richardson.

Obama has recalled witnessing the astronauts' arrival in a 22 October 2008 statement on India's lunar launch, archived at SpaceRef.com, in Remarks at the White House, 20 July 2009, *Public Papers of the Presidents* [*Public Papers*] 2009 Vol. II, p. 1133, and in Commencement Address at Miami Dade College, 29 April 2011, *Public Papers* 2011 Vol. I, p. 476. Obama has incorrectly said Apollo rather than Gemini, for a review of all Apollo and Gemini splashdowns reveals that only Gemini 8 astronauts Neil Armstrong and David Scott's 18 March 1966 arrival could have been attended by Stanley and Barry. See AP Honolulu, "Spacemen on Way to Cape Kennedy," *LAT*, 19 March 1966, p. 1. By the time of the Apollo 8 astronauts' arrival in Hawaii on 27 December 1968, Obama was in Indonesia.

On the political obstruction Borland encountered in attempting to complete her planned film "Barack Obama: Made in Hawaii," see Emily Wax, "Obama's Link to Hawaii Not Ignored by Islanders," *WP,* 1 May 2013, and especially Stewart Lawrence, "Why Did Top Democrats Try to Kill a Pro-Obama Documentary?," *Daily Caller,* 9 September 2013.

16. Scott, *A Singular Woman,* pp. 97–100; Obama to Robert L. Sherman, 28 November 1962, Sherman to Obama, 30 November 1962, Sherman to Obama, 16 January [1963], Obama to Sherman, 2 April 1963, Gordon Hagberg to IIE, "Barack Obama," 18 April 1963, Hagberg to Sherman, 19 April 1963, Sherman to Hagberg, 22 April 1963, Richard C. Raymond to Sherman, "Barack Obama," 2 May 1963, Obama Transcript, Harvard University, 6 May 1963, Obama to Raymond, 13 May 1963, Elizabeth V. Murrell (IIE) to Obama, 20 May 1963, Sherman to Obama, 27 May 1963, PSFP Box 214; *Stanley Ann D. Obama v. Barack H. Obama,* Libel for Divorce, #57972, Domestic Relations Division, First Judicial Circuit Court, 20 January 1964; Jacobs, *The Other Barack,* pp. 135, 138–39, 144–61; Shachtman, *Airlift to America,* p. 221; Obama, "Application to Extend Time of Temporary Stay," 6 June 1963 (approved 10 June 1963), Obama INS File; Obama to Sherman, 20 June 1963, Sherman to Obama, 23 July 1963, Obama to Sherman, 29 August 1963, Obama, "Partial Scholarship Program," 30 August 1963, Obama, "Information Sheet," n.d. (falsely asserting that "I passed my Cambridge School Certificate" at Maseno School), Hendrik S. Houthakker to Sherman, 30 August 1963, Edward S. Mason to Sherman, 3 September 1963 (Obama "has a lot of native intelligence . . . I find him an impressive fellow and I think he will go far"), Sherman to Obama, 16 September 1963 (with a handwritten "P.S. No more aid unless grades improve"), and Obama to Sherman, 16 September 1963, PSFP Box 214; Dana E. Klotzle, Associate Director, Unitarian Universalist Service Committee, to INS, 21 January 1964, Inspector K. D. MacDonald to District Director J. A. Hamilton Jr., 31 January 1964, Obama INS File; Alice B. Hall to Sherman, 7 February 1964, and Sherman to Obama, 11 February 1964, PSFP Box 214; Hamilton (by Deputy District Director Robert L. Suddath) to American Consul, London, 2 March 1964, Obama, "Application to Extend Time of Temporary Stay," 21 April 1064, M. F. McKeon, "Re: A11 938 537—Barack Hussein Obama," 28 April 1964 and 19 May 1964, David D. Henry to Barack H. Obama, 27 May 1964, M. F. McKeon, "Re: A11 938 537—Barack Hussein Obama," 8 June 1964, District Director to Barack H. Obama, 9 June 1964, Obama INS File; Sherman to Obama, 17 June 1964, PSFP Box 214; (FNU) Mulrean, "Memo for File: A11 938 537," 18 June 1964, "Non-Citizen Departure Information," Harvard University International Students Office, 21 July 1964, Joseph M. O'Connell, "Report of Investigation," 22 July 1964, and "Arrival-Departure Record," Obama INS File. See also Sally Jacobs, "The Trials of Omar, Obama's Uncle," *BG,* 8 January 2012. Jacobs also reports, per a subsequent Obama CV that she located in Nairobi, that his proposed dissertation was entitled "An Econometric Model of Staple Theory of Development," *The Other Barack,* p. 160.

Several second- or thirdhand witnesses remember hearing that Ann filed for divorce upon learning about Kezia and Obama's first two Kenyan children. When Ann first learned of Kezia is unrecorded. By January 1964 Obama had been gone from Honolulu for over a year and a half; if any new knowledge came to Ann during that time, it too is unrecorded. See Arlene Payne in Scott, *A Singular Woman,* p. 92, and DJG interview with Jan-Michelle Lemon Kearney, who recalled her Harvard Law School classmate the younger Barack recounting that "when his mom found out about his father having this other family, that kind of ended it." Ann's subsequent UH graduate program mentor, Alice Dewey, recounted her understanding from a decade later of Ann's earlier view: "When Barack Sr. goes off to Harvard, he promises to send funds to take care of her and Barry, and he doesn't do it, and that's why she's angry. 'I don't care who he's sleeping with, I want the money because I'm trying to feed the two of us,'" Dewey recounted to DJG.

17. Dr. Tori Nishigawa, "To Whom It May Concern: Re: Mrs. Stanley Ann Dunham Soetoro," 29 April 1965, Frank R. Porter, "Officer's Review and Action Sheet," A14 128 294, 24 May and 7 June 1965, John F. O'Shea (INS District Director), "Deportation Docket Control Action Slip," 7 June 1965, Gary Fujiwara to Travel Control, "Lolo Soetoro," 30 June 1965, O'Shea to Mr. and Mrs. Lolo Soetoro, 6 July 1965, Robert Wooster to Robert Zumwinkle, "Visa Status of East-West Alumnus from Indonesia, Mr. Lolo Soetoro," 7 July 1965, Zumwinkle to O'Shea, "Mr. Lolo Soetoro," 13 July 1965, Lolo Soetoro, untitled statement, n.d., 3pp., Frank R. Potter, "Memo to File: A14 128 294," 19 July 1965, Potter, "Operations Supervisory Review," 21 July 1965, Robert Aitkin (EWC) to R. E. Soehardi (Indonesian

Consulate, San Francisco), 17 September 1965, Aitkin to Soehardi, 27 September 1965, S. Ann Soetoro, "Affidavit," 30 November 1965, O'Shea to Assistant Regional Commissioner, 7 December 1965, L. W. Gilman to Deputy Associate Commissioner, 14 December 1965, J. P. Sharon to District Director, Honolulu, 12 January 1966, and Robert R. Schultz, "Memorandum for File," 22 November 1966, all Soetoro INS file.

Throughout the first six months of 1965, superb *New York Times* correspondent Neil Sheehan filed dozens of stories from Indonesia reporting increased anti-Americanism on the part of Communist activists and allies there. See particularly "Indonesia Seizes Third U.S. Library," *NYT*, 16 February 1965, p. 1, and "Moslems on Java Clash with Reds," *NYT*, 17 March 1965, p. 18. For contemporaneous U.S. press coverage of the violence that commenced on 30 September, see Seth S. King, "Indonesia Says Plot to Depose Sukarno Is Foiled by Army Chief," *NYT*, 2 October 1965, p. 1; Ian Stewart, "Army-Red Clash Stirs in Jakarta," *NYT*, 6 October 1965, p. 1; Seth S. King, "Indonesian Army Battles Rebels in Key Java City," *NYT*, 7 October 1965, p. 1; Max Frankel, "U.S. Is Heartened by Red Setback in Indonesia Coup," *NYT*, 11 October 1965, p. 1; Seth S. King, "Indonesia Coup Is Still a Mystery," *NYT*, 17 October 1965, p. 1; King, "Indonesia Orders Red Units Curbed," *NYT*, 22 October 1965, p. 1; Stanley Karnow, "First Report on Horror in Indonesia," *WP*, 17 April 1966, pp. 1, 20; King, "The Great Purge in Indonesia," *NYT Magazine*, 8 May 1966, pp. 25ff.; and Seymour Topping, "Slaughter of Reds Gives Indonesia a Grim Legacy," *NYT*, 24 August 1966, pp. 1, 16. Scholarly understanding of the mass killings was very slow to develop. See Robert Cribb, "Genocide in Indonesia," *Journal of Genocide Research* 3 (2001): 219–39, esp. 231–37; Cribb, "Unresolved Problems in the Indonesian Killings of 1965–1966," *Asian Survey* 42 (July–August 2002): 550–63; Mary S. Zurbuchen, "History, Memory, and the '1965 Incident' in Indonesia," *Asian Survey* 42 (July–August 2002): 564–81; John Roosa, "Violence and the Suharto Regime's Wonderland," *Critical Asian Studies* 35 (2003): 315–23; Roosa's superbly rich and thoughtful *Pretext for Mass Murder: The September 30th Movement and Suharto's Coup d'Etat in Indonesia* (University of Wisconsin Press, 2006), esp. pp. 31, 33, 91, 177, 207–8, 214, and 218; Roosa, "President Sukarno and the September 30th Movement," *Critical Asian Studies* 40 (2008): 143–59; and Douglas Kammen and Katharine McGregor, eds., *The Contours of Mass Violence in Indonesia, 1965–68* (University of Hawaii Press, 2012), esp. pp. 3–4 and 8.

18. Untitled INS note, 31 October 1966, Robert R. Schultz, "Memorandum for File, 22 November 1966, John F. O'Shea to Lolo Soetoro, 13 December 1966, O'Shea to Assistant Regional Commissioner, 18 December 1966, L. W. Gilman to Deputy Associate Commissioner, 19 December 1966, Assistant Commissioner to Regional Commissioner, 5 January 1967, J. P. Sharon to District Director Honolulu, 13 January 1967, Robert R. Schultz, "Adjudicator's Basis for Decision," 24 May 1967, O'Shea to Assistant Regional Commissioner, 6 June 1967, Sharon to District Director Honolulu, 13 June 1967, all Soetoro INS file; Stanley Ann Soetoro, "Application for Amendment of Passport," 29 June 1967, DOS FOIA Release 200807238; Assistant Commissioner to Regional Commissioner, 21 August 1967, Sharon to District Director Honolulu, 25 August 1967, W. I. Mix, "Memorandum to File," 4 September 1967, O'Shea to Assistant Regional Commissioner, 25 September 1967, Gilman to Deputy Associate Commissioner, 29 September 1967, O'Shea to Chief, FSS, BECA, DOS, 6 October 1967, O'Shea to Assistant Regional Commissioner, 11 July 1968, E. J. Strapp to District Director Honolulu, 29 July 1968, O'Shea to Assistant Regional Commissioner, 1 August 1968, O'Shea to Ann D. Soetoro, 1 and 22 August 1968, all Soetoro INS file; Stanley Ann Dunham Soetoro, "Request by United States National for and Report of Exception to Section 53.1 . . . ," 21 October 1971, DOS FOIA; Kelli Abe Trifonovitch, "Being Local, Barry and Bryan," *Hawaii Business*, October 2008 (first publishing of Obama's Noelani Elementary School kindergarten class photo); Barack Obama, Remarks in Tokyo, 14 November 2009, *Public Papers* 2009 Vol. II, p. 1675; Press Conference in Canberra and Remarks in Canberra, 16 November 2011, *Public Papers* 2011 Vol. II, pp. 1433, 1440; Jackie Calmes, "President Hits His Stride," *NYT*, 21 November 2011. A photograph clearly picturing young Barry and Noelani classmate Scott Inoue, with apparent Christmas drawings in the background, is said to date from their third-grade year, i.e., December 1969. Ann Dunham, and Barry, may have returned to Honolulu several weeks prior to Christmas; Ann's 1993 résumé states she was in Jakarta until December 1969, and then again from January 1970 forward—with an implicit interruption. See Pat

Gee, "Third-Grade Photo Captures Obama's Grin," *HSB*, 28 December 2009, and Scott Inoue e-mails to DJG.

19. Journalistic sources on Obama's less than four years in Jakarta are very extensive, especially from 2007 to '08, but almost always present memories of asserted playmates, neighbors, fellow students, and former teachers with no apparent hesitation about what clear recollections such individuals all had more than thirty-five years later. Sources critically reviewed and relied upon in this account include, in generally chronological order, Stanley Ann Dunham Soetoro, Application for Passport Renewal, 13 August 1968, DOS FOIA (stating with regard to residing overseas "Indefinite—Married to an Indonesian Citizen"); Phil Mayer, "Ex-Islander Gets Prestigious Harvard Post," *HSB*, 9 February 1990 (noting Obama's attendance of first grade at Noelani Elementary School); Obama, *DFMF*, esp. pp. 29–31, 36–52; Christopher Wills (AP), "Obama Mixes Exotic Background with Street-Level Experience," 16 October 2004; John Vause's contributions to CNN broadcasts and CNN.com on 22 and 23 January 2007; Obama on *The Early Show*, CBS, 24 January 2007; Nedra Pickler (AP), "Obama Challenges Allegation About Islamic School," 24 January 2007 (quoting Darmawan); Prodita Sabarini, "Impish Obama Couldn't Sit Still, Says School Pal," *Jakarta Post*, 31 January 2007; Paul Watson, "As a Child, Obama Crossed a Cultural Divide in Indonesia," *LAT*, 15 March 2007; Kirsten Scharnberg and Kim Barker, "The Not-So-Simple Story of Barack Obama's Youth," *CT*, 25 March 2007; Barker, "Obama Madrassa Myth Debunked," *CT*, 25 March 2007; Haroon Siddiqui, "Obama's Muslim Heritage," *Toronto Star*, 14 June 2007; Trish Anderton, "Obama's Jakarta Trail," *Jakarta Post*, 26 June 2007; Kim Chipman and Wahyudi Soeriaatmadja, "Obama's Jakarta Friends Recall a Would-Be Leader," Bloomberg, 31 December 2007; Muhammad Cohen, "The Indonesian Candidate," *Asia Times*, 20 February 2008; "Obama's Former School Friends Form Fan Club," *Jakarta Post*, 2 March 2008; Ed Davies, "Indonesia Left Deep Imprint on Obama Family," Reuters, 22 March 2008 (quoting Hendro); Richard Wolffe and Michael Hirsh, "A Man at Home in the World," *Newsweek*, 12 April 2008; Devi Asmarani, "Indonesian Roots for Obama in U.S. Election," *Singapore Straits Times*, 14 April 2008; Roger Cohen, "Obama's Indonesian Lessons," *NYT*, 14 April 2008, Ripley, "The Story of Barack Obama's Mother," *Time*, 21 April 2008, pp. 36ff.; Christine Oelrich, "Obama the 'Curly-Haired' One," Deutsche Presse-Agentur, 29 April 2008 (quoting Darmawan saying the essay declared, "I will become president"); Obama on CNN, 13 July 2008; Margaret Conley, "Obama's Early Days in Jakarta," ABCNews. com, 25 September 2008 (Sinaga stating the paper was in Indonesian); Mark Forbes, "Obama, AKA Fat Little Barry, Remembered," *Sydney Morning Herald*, 1 October 2008; "Obama as We Knew Him," *Observer* (UK), 26 October 2008, p. 4; Robin McDowell (AP), "Obama's Childhood Home in Indonesia Up for Sale," 28 October 2008; Obama in Richard Wolffe, *Renegade: The Making of a President* (Crown Publishers, 2009), p. 250; "Obama Family Rejects Liah's Claims, Calls Her a Fraud," *Jakarta Post*, 16 January 2009; Endy M. Bayuni, "Obama's Indonesian Classroom," *NYT*, 18 January 2009; Obama, Remarks to the United States Hispanic Chamber of Commerce, 10 March 2009, *Public Papers* 2009 Vol. I, p. 208; Obama, Remarks to Wakefield High School Students and Faculty, 8 September 2009, *Public Papers* 2009 Vol. II, p. 1355; Chris Brummitt (AP), "Indonesia Remembers Young Obama," 16 March 2010; "Childhood Photo of Barack Obama in Indonesia Found," *Telegraph* (UK), 17 March 2010; Andrew Higgins, "Catholic School in Indonesia Seeks Recognition for Its Role in Obama's Life," *WP*, 9 April 2010, p. A1; Kelly Heffernan-Taylor, "Indonesia: Obama's Childhood Friends and Teachers Share Memories," CBSNews.com, 7 November 2010; Norimitsu Onishi, "Obama Visits a Nation That Knew Him as Barry," *NYT*, 8 November 2010; Haryo "Pongky" Soetendro, Sonny Trisulo et al., *A Gift from Your Family: For Barack, Michelle, Malia and Sasha on the Occasion of Your Visit to Indonesia During November 2010* (Saraswati Papers, 2010), esp. pp. 6, 9, 15; Obama, Remarks at the University of Indonesia, Jakarta, 10 November 2010, *Public Papers* 2010 Vol. II, p. 1787, Obama, Remarks Honoring the 2011 National and State Teachers of the Year, 3 May 2011, *Public Papers* 2011 Vol. I, p. 488 (attending Noelani School for the outset of first grade); Scott, *A Singular Woman*, pp. 113–16, 126–28; DJG interview with Bronwen and Garrett Solyom, "Obama's Elementary School Teacher Passes Away," *Jakarta Post*, 10 December 2011 (Israella Darmawan); Niniek Karmini (AP), "Obama's Transgender Ex-Nanny Outcast," 5 March 2012; Karmini (AP), "Obama's Former Nanny Overwhelmed by New Celebrity," 8 March 2012; and Maraniss, *BOTS*, pp. 213–24, 229–43, esp. p. 220 ("no reason not to trust"). Alone among all journalistic commentators, only one writer correctly doubted the unbelievable clarity and precision of the Jakarta teachers' supposed recollections four decades later. See Amy Hollyfield, "Yep, Young Obama Had Big Goals. So?," *St. Petersburg Times,* 7 December 2007, p. A1. Likewise, claims by one Menteng Dalam neighbor and one Matraman servant that Lolo on multiple occasions physically assaulted Ann, once leaving her bleeding, have been put in print by several journalists but are simply far too distant in time and uncorroborated to be accepted as part of a true historical record. See also Scott, *A Singular Woman,* p. 128. Also particularly meriting dismissal is the claim that "religion permeated Barry's years in Indonesia," made in Stephen Mansfield, *The Faith of Barack Obama* (Thomas Nelson, 2008), p. 19.

20. Obama, *DFMF,* pp. 29–30, 51–52; Obama in Monica Mitchell, "Son Finds Inspiration in the Dreams of His Father," *Hyde Park Herald* [*HPH*], 23 August 1995, p. 10; Kay Ikranagara in Judith Kampfner, "Dreams from My Mother," BBC Radio World Service, 16 September 2009 (and, similarly, in Ben Barber, "Obama's Mother Worked for USAID, World Bank, in Indonesia," *FrontLines,* March 2009); "Oprah Talks to Barack Obama," *O, The Oprah Magazine,* November 2004; Obama in Charles Barkley, *Who's Afraid of a Large Black Man?* (Penguin Press, 2005), pp. 22–23 (printing a fall 2004 interview with Obama conducted by Barkley and Michael Wilbon); Obama in Scharnberg and Barker, "The Not-So-Simple Story," *CT,* 25 March 2007; Obama on ABC's *20/20,* 26 September 2008; "I Wish I Were Black—Again," *Ebony,* December 1968, pp. 119–20, 122, 124; John Howard Griffin, *Black Like Me* (Houghton Mifflin, 1961); Jonathan Yardley's luminous "John Howard Griffin Took Race All the Way to the Finish," *WP,* 17 March 2007; Bruce Watson, "*Black Like Me* 50 Years Later," *Smithsonian Magazine,* October 2011; Phillip L. Hammack, "The Political Psychology of Personal Narrative: The Case of Barack Obama," *Analyses of Social Issues and Public Policy* 10 (December 2010): 182–206, pp. 190, 193; Obama Remarks at Cornell College, Mt. Vernon, IA, 5 December 2007. See also especially Gayle Wald, *Crossing the Line: Racial Passing in Twentieth-Century U.S. Literature and Culture* (Duke University Press, 2000), who notes that in earlier decades *Ebony* had published at least three illustrated stories about the downsides of passing: "Case History of an Ex-White Man," December 1946, pp. 11–16, "I'm Through with Passing," March 1951, pp. 22–27, and "The Curse of Passing," December 1955, pp. 50–56.

21. John F. O'Shea to Ann D. Soetoro, 22 August 1968, Soetoro INS File (letter addressed to 2234 University Avenue returned to sender); Obama, *DFMF,* pp. 54, 58; Will Hoover, "Obama's Hawaii Boyhood Homes Drawing Gawkers," *HA,* 9 November 2008; Ron Jacobs, *Obamaland* (Trade Publishing, 2009), p. 30; Scott, *A Singular Woman,* p. 97; DJG interviews with Ralph Dunham, Virginia Dunham Goeldner, Charles Payne, Alec Williamson, Susan Williamson Corley, and Rolf Nordahl. See also Obama in Jackie Calmes, "President Hits His Stride," *NYT,* 21 November 2011 (perhaps referencing another transit through Sydney at age eight). Bob Pratt would apparently operate a furniture store in Kailua, Kona, on the big island of Hawaii, during at least the late 1970s and early 1980s: see *In the Matter of the Petition of T.S.K., Associates,* Hawaii Land Use Commission #A80-482, 14 May 1981, p. 9. Per his 17 February 2005 *Honolulu Advertiser* obituary, Albert Robert Pratt was born in 1927 and died in Kailua at age seventy-seven. On Punahou, see especially Nelson Foster, ed., *Punahou: The History and Promise of a School of the Islands* (Punahou School, 1991), a beautifully rich pictorial history, and Lawrence Downes, "For Obama, Estranged in a Strange Land, Aloha Had Its Limits," *NYT,* 9 April 2007, p. A16.

22. The literature on Frank Marshall Davis is extensive, but anyone interested in his life story must begin with his exceptionally rich and revealing autobiography, *Livin' the Blues: Memoirs of a Black Journalist and Poet,* ed. John Edgar Tidwell (University of Wisconsin Press, 1992), esp. pp. 312, 323, and 327, and then proceed to "Bob Greene," *Sex Rebel: Black* (Greenleaf Classics, 1968), which of course is difficult to obtain. Tidwell's works are the starting point for any critical appreciation of Davis. See Tidwell, "An Interview with Frank Marshall Davis," *Black American Literature Forum* 19 (Autumn 1985): 105–8; "Frank Marshall Davis," *Kansas History* 18 (Winter 1995–1996): 270–83; Tidwell, "'I Was a Weaver of Jagged Words': Social Function in the Poetry of Frank Marshall Davis," *Langston Hughes Review* 14 (Spring–Fall 1996): 65–78, esp. p. 65 ("did not survive"); Tidwell, "Alternative Constructions to Black Arts Autobiography: Frank Marshall Davis and

1960s Counterculture," *CLA Journal* 41 (December 1997): 147–60; Frank Marshall Davis, *Black Moods—Collected Poems,* ed. Tidwell (University of Illinois Press, 2002); Tidwell, ed., *Writings of Frank Marshall Davis: A Voice of the Black Press* (University Press of Mississippi, 2007). See also Bill V. Mullen, Review of *Livin' the Blues, Black Moods,* and *Writings of Frank Marshall Davis, African American Review* 42 (Fall 2008): 768–770 ("among the best"); James E. Smethurst, *The New Red Negro: The Literary Left and African American Poetry, 1930–1946* (Oxford University Press, 1999), pp. 135–41 ("a certain distance"); Bill V. Mullen, *Popular Fronts: Chicago and African American Cultural Politics, 1935–1946* (University of Illinois Press, 1999); Stacy I. Morgan, *Rethinking Social Realism: African American Art and Literature, 1930–1953* (University of Georgia Press, 2004), pp. 24–25, 28–29, 184–206; Dudley Randall, "An Interview with Frank Marshall Davis—'Mystery' Poet," *Black World* 23 (January 1974): 37–48; Davis to Ron Welburn, 31 August 1982, in Ronald G. Welburn, "American Jazz Criticism, 1914–1940," Ph.D. dissertation, New York University, October 1983, pp. 189–90; Davis's May 1987 oral history interview with Chris Conybeare and Joy Chong of UH's CLER; Kathryn Waddell Takara, "Frank Marshall Davis," in Steven C. Tracy, ed., *Writers of the Black Chicago Renaissance* (University of Illinois Press, 2011), pp. 161–84; and Takara's *Frank Marshall Davis: The Fire and the Phoenix—A Critical Biography* (Pacific Raven Press, 2012). Takara's 2011 essay and 2012 book supplant her earlier "Rage and Passion in the Poetry of Frank Marshall Davis," *Black Scholar* 26 (Summer 1996): 17–26, "Frank Marshall Davis in Hawaii: Outsider Journalist Looking In," *Social Process in Hawaii* 39 (1999): 126–44, and "Frank Marshall Davis: A Forgotten Voice in the Chicago Black Renaissance," *Western Journal of Black Studies* 26 (Winter 2002): 215–27. Regarding *Sex Rebel,* see Davis's 1 November 1968 and 27 October 1982 letters to Margaret Burroughs in the Burroughs Papers, DuSable Museum, Chicago. See also William Grimes, "Margaret T. Burroughs, Archivist of Black History, Dies at 95," *NYT,* 27 November 2010. The DuSable Museum also houses a small Frank Marshall Davis archival collection.

On Frank and Helen's December 1948 move to Oahu, see especially Davis, "Inventory After a Year," *Honolulu Record,* 8 December 1949, pp. 8, 6; see also "Helen Peck's Marriage to Lt. Kline Told," *CT,* 16 February 1944, p. 17. On Frank's CPUSA affiliation and involvements, anyone must begin with Frank's FBI headquarters main file, 100-328955, whose first serial dates from July 1944 and its last from September 1963. Other sources referencing or discussing Frank in that vein are U.S. House of Representatives, Committee on Un-American Activities, *Hearings Regarding Communist Activities in the Territory of Hawaii—Part 3,* 17–19 April 1950, 81st Cong., 2nd Sess. (U.S. GPO, 1950), pp. 2065–68; Hawaii Residents' Association, *Communism in Hawaii: A Summary of the 1955 Report of the Territorial Commission on Subversive Activities,* p. 44; U.S. Senate, Committee on the Judiciary, *Scope of Soviet Activity in the United States—Hearings Before the Internal Security Subcommittee, Part 41,* 5 and 6 December 1956, 84th Cong., 2nd Sess. (U.S. GPO, 1957), pp. 2518–19, *Part 41-A, Appendix II,* pp. 2697–98, and *Appendix III,* pp. 2782–83; Gerald Horne, *Fighting in Paradise: Labor Unions, Racism, and Communists in the Making of Modern Hawaii* (University of Hawaii Press, 2011), pp. 36–38, 129–33, 198–202, 269–70, 298–99, and esp. 330–31; and Paul Kengor, *The Communist—Frank Marshall Davis* (Simon & Schuster, 2012). On Jack Hall, see Sanford Zalburg, *A Spark Is Struck! Jack Hall and the ILWU in Hawaii* (University of Hawaii Press, 1979).

The Koa Cottages were located within the rectangle formed by Kuhio, Liluokalani, Kalakaua, and Kealohilani Avenues. Mark Kaleoluaoha Davis's "glee" remark appeared in a 3 April 2009 comment on neoneocon.org. Stan and Barry's August 1970 visit there is informed by Obama, *DFMF,* pp. 76–77, DJG interviews with Charles Payne and Dawna Weatherly-Williams, Sudhin Thanawala (AP), "Writer Offered a Young Barack Obama Advice on Life," 2 August 2008 (quoting Maya Soetoro-Ng); Toby Harnden, "Barack Obama's True Colours" and "Frank Marshall Davis, Alleged Communist, Was Early Influence on Barack Obama," *Telegraph* [Sunday] *Magazine* (UK), 23 August 2008, pp. 26ff.; David Remnick, *The Bridge* (Alfred A. Knopf, 2010), p. 96; Kathryn Takara interview quotation ("His grandfather was one of Frank's closest friends") in Cliff Kincaid and Herbert Romerstein, *Communism in Hawaii and the Obama Connection* (America's Survival, May 2008), p. 5; and Maraniss, *BOTS,* p. 270 (apparently quoting Obama). Davis was first publicly identified as the "Frank" in *DFMF* by Gerald Horne, "Rethinking the History and Future of the Communist Party," *Political Affairs,* 28 March 2007.

23. Scott, *A Singular Woman,* photo section following p. 120; Ed Davies, "Indonesia Left Deep Imprint on Obama Family," Reuters, 22 March 2008; Soetendro, Trisulo et al., *A Gift from Your Family,* p. 9; Obama, *DFMF,* pp. 31, 38–41; Obama, Illinois Center for the Book Presentation, Illinois State Library, Springfield, 7 March 2001; Obama, *The Audacity of Hope* (Crown Publishers, 2006), p. 346; Obama in Todd Purdum, "Raising Obama," *Vanity Fair,* March 2008; Obama on C-Span Book TV, 23 November 2004; Ian Buruma, "A Free Spirit," *NYRB,* 26 May 2011; Justin A. Frank, *Obama on the Couch* (Free Press, 2011), pp. 126, 205; Tammerlin Drummond, "Barack Obama's Law," *LAT,* 12 March 1990, pp. E1, E2; Allison J. Pugh (AP), "First Black President of School's Law Review Uninterested in a Cushy Job," 15 April 1990; Obama, Book Channel interview with Marc Strassman, Studio City, CA, 11 August 1995; Obama, "A Life's Calling to Public Service," *Punahou Bulletin,* Fall 1999, pp. 28–29; Julieanna Richardson's wonderfully rich 2001 HistoryMakers.com interview with Obama; Obama on *Fresh Air,* NPR, 12 August 2004; Obama in Stephanie Griffith, "White House Prospect Obama Forever Changed by Asia Sojourn," Agence France-Presse, 9 January 2007; Obama interview with Nicholas D. Kristof, NYTimes.com, 5 March 2007; Wolffe, *Renegade,* p. 235. See also Obama in Mitchell, "Son Finds Inspiration on the Dreams of His Father," *HPH,* 23 August 1995, p. 10 ("My mother gave me a positive self-image of being a black person. She raised me on stories about my father" and indeed "I think to some extent, she romanticized black life") and in Q&A with Wakefield High School Students, Arlington, VA, 8 September 2009, *Public Papers* 2009 Vol. II, p. 1352 ("she never spoke badly about him"). See as well Sharma, *Barack Obama in Hawaii and Indonesia,* p. xxiii: "The early formative experiences in Jakarta shaped Obama indelibly."

24. Obama, *DFMF,* pp. 53–62; Will Hoover, "Obama's Hawaii Boyhood Homes Drawing Gawkers," *HA,* 9 November 2008; Punahou Catalog 1971–1972, Punahou School Archives; Jacobs, *Obamaland,* pp. 30, 39; Dan Nakaso, "Obama's Tutu a Hawaii Banking Female Pioneer," *HA,* 30 March 2008; DJG interviews with Pal Eldredge, Susan Williamson Corley, Alec Williamson, and Mark Hebing; Joella Edwards, "Buff 'N Blue & Black," in Jacobs, *Obamaland,* pp. 60–63; Ronald Loui and Joella Edwards on *Real Talk with Jack McAdoo,* JMacRadio.com, Shows 9, 10, and 12, (1, 4, and 10 November 2010), and Joella Edwards, "Growing Up with Obama," *Lowcountry View with Earl Yates,* 15 and 23 January 2011, Youtube.com.

Incorrect perceptions of Obama's financial aid status at Punahou are legion. John Rowehl, "Financial Aid Explained," *Ka Punahou,* 17 December 1976, p. 1, is an almost definitive guide. Punahou did not use "the term 'scholarship' because" that "implies that monetary assistance is given for special talents or abilities," which was not the case. "Punahou gives financial aid only to those who cannot afford to pay full tuition," and during the initial admissions process "a financial status form" is filled out by parents. "Usually if a family's net income is $12,000 or more, they won't be eligible." In 1975–76, only 240 Punahou students were receiving aid, with only two being awarded the top sum of $1,500: "No full scholarships exist." If a student is "granted aid, those in grades seven and up are placed in work service" totaling 100 to 140 hours per year. Not one of Obama's many friends and classmates, nor any Punahou teacher, has ever recounted Obama working at Punahou.

On Mabel Hefty, see Susan Essoyan, "A Teacher's Hefty Influence," *HSB,* 29 July 2007 (reprinted in *Punahou Bulletin,* Fall 2007); Maraniss, *BOTS,* p. 269; and Obama's "Remarks Honoring the 2011 National and State Teachers of the Year," 3 May 2011, *Public Papers* 2011 Vol. I, p. 488, "Remarks at Campaign Rally," 22 August 2012, North Las Vegas, Nevada, and "Email from President Obama: 'My Fifth-Grade Teacher,'" White House Blog, 29 April 2015. On the fifth-grade year, also see Jason Hagey, "Man Who Grew Up with Barack Obama . . . ," *Tacoma News Tribune,* 23 August 2008, p. A1; Chris McGann, "Basketball Buddy Remembers Years with Obama," *Seattle Post-Intelligencer,* 25 August 2008, p. B1; Woody Paige, "A Special Boy Called Barry," *Denver Post,* 28 August 2008, p. P6; Matt Roper, "Our School Pal Barack," *Mirror,* 15 November 2008, p. 14; Ronald P. Loui in Constance F. Ramos, ed., *Our Friend Barry: Classmates' Recollections of Barack Obama and Punahou School* (Lulu.com, September 2008), pp. 83, 88–89; and Jackie Calmes, "On Campus, Obama and Memories," *NYT,* 3 January 2009.

25. Stanley Ann Dunham Soetoro, "Request by U.S. National for and Report of Exception to Section 53.1 . . . ," 21 October 1971, Dunham Passport File; Obama, *DFMF,* pp. 62–71; Scott, *A Singular Woman,* pp. 90–91, 143–45;

P. Huddait [?] to Mr. [Patrick F.] Coomey, "A11 938 537, Barack H. Obama," 17 July 1964, and E. Golden, untitled memo, 28 August 1964, Obama INS File; Mark Obama Ndesandjo, *Nairobi to Shenzhen* (Aventine Press, 2009), pp. 110–15; Ndesandjo, *An Obama's Journey,* pp. 14–20, 27, and esp. 359; Edmund Sanders, "Obama Not Quite His Father's Son," *LAT,* 17 July 2008; Robert Shenton (Harvard Registrar) to Obama, 16 November 1965, and Alice B. Hall to INS District Director, 17 November 1965, both Obama INS File; DJG interviews with Zeituni Onyango; Onyango, *Tears of Abuse,* pp. 26–41; Malik Obama, *Barack Obama Sr.*, pp. 217, 228–35; Jacobs, *The Other Barack,* pp. 165–90; Firstbrook, *The Obamas,* pp. 205–6, 213–15. Maraniss, *BOTS,* pp. 206–7, appears to present the most accurate account of Obama's mid-1965 auto accident, and it is relied upon here. Photographic evidence suggests that Lolo Soetoro and one-year-old Maya, as well as Lolo's parents Tik and Titi, also traveled to Honolulu in the fall of 1971. See the three photos each dated 1971 and clearly taken in Honolulu in Soetendro, Trisulo, et al., *A Gift from Your Family,* pp. 13–15.

As of 2014–15, a full and accurate PDF of Sessional Paper No. 10 is easily accessible on the Web page of the World Bank. Barack H. Obama, "Problems Facing Our Socialism," *East Africa Journal,* July 1965, pp. 26–33, might best be read after surveying Goldsworthy, *Tom Mboya,* pp. 234–38, and David William Cohen, "Perils and Pragmatics of Critique: Reading Barack Obama Sr.'s 1965 Review of Kenya's Development Plan," 24 March 2010 (unpublished seminar paper), a PDF of which is readily available on the Web. See also William Cohen and E. S. Atieno Odhiambo, *The Risks of Knowledge: Investigations into the Death of the Hon. Minister John Robert Ouko in Kenya, 1990* (Ohio University Press, 2004), p. 182, and Ben Smith and Jeffrey Ressner, "Long-Lost Article By Obama's Dad Surfaces," *Politico,* 15 April 2008. The essential monograph for tracing Kenya's 1966–67 political upheaval is Cherry Gertzel, *The Politics of Independent Kenya 1963–8* (Northwestern University Press, 1970), esp. pp. 32, 49–50, 73–124, 145–46, 152, 175–76; also central are *Jet,* 30 June 1966, p. 50 (reporting the expulsion of Ernestine Hammond Kiano), D. Pal Ahluwalia, *Post-Colonialism and the Politics of Kenya* (Nova Science, 1996), pp. 38–59, Barack Echols e-mails to DJG, and DJG interview with Barack Echols (with Echols quoting an e-mail to him from his mother). Echols was born in August 1968; Hansen's forced departure from Nairobi thus was circa November 1967. See also William R. Ochieng, "Structural & Political Change," in B. A. Ogot and Ochieng, eds., *Decolonization & Independence in Kenya 1940–93* (James Currey, 1995), pp. 83–109, esp. pp. 98, 107; Kiano's 1989 interview with Harry Kreisler; a biographical citation to Dr. Kiano upon the conferment of an honorary doctorate from the University of Nairobi on 1 December 1997, available at www.uonbi.ac.ke/sites/default/files/Kiano.pdf; and Stephens, *Kenyan Student Airlifts to America,* pp. 8–12.

26. Jacobs, *The Other Barack,* pp. 177, 190–228; Firstbrook, *The Obamas,* p. 214; Ndesandjo, *Nairobi to Shenzhen,* pp. 131–33; Ndesandjo, *An Obama's Journey,* pp. 30–31, 131; DJG interviews with and e-mails from Zeituni Onyango, especally 22 May 2012 (reporting that Malik Obama "who has the birth certificate of David" told her by phone that it shows David was born on 11 September 1968); Onyango, *Tears of Abuse,* pp. 39, 79, 217; Malik Obama, *Barack Obama Sr.*, pp. 236, 257, 261, 265; Auma Obama, *And Then Life Happens,* pp. 59–60; Andrea Sachs, "Auma Obama on Her Famous Brother," *Time,* 4 May 2012; Angella Johnson, "My Little Brother Barack Obama," *Daily Mail,* 9 March 2013.

Both Jacobs, *The Other Barack,* pp. 208, 271, and Maraniss, *BOTS,* pp. 246, 592, quote from an article titled "Tourism Officer on Drinks Charge" in the 4 November 1967, *Daily Nation* newspaper; *BOTS* also presents a partial photo of the article. Neither book cites a particular page number, and a careful review of the microfilm edition of the 1967 *Daily Nation* available in United States libraries fails to show any such article in that edition of the 4 November paper.

On Abercrombie and Zane's late 1968 visit to Nairobi, see Tim Jones, "Obama's Mother: Not Just a Girl from Kansas," *CT,* 27 March 2007; Kevin Merida, "The Ghost of a Father," *WP,* 14 December 2007, p. A12; Lee Cataluna, "Glimpse of Obama Sr. in Photos," *HA,* 8 August 2008; David Maraniss, "Though Obama Had to Leave to Find Himself, It Is Hawaii That Made His Rise Possible," *WP Magazine,* 24 August 2008; and DJG interview with Pake Zane.

On Mboya's assassination and Obama's testimony, see especially Goldsworthy, *Tom Mboya,* pp. 279–81; "Njenga Defence Case Today," *East African Standard,* 9 September 1969, p. 1; and Malik Obama, *Barack Obama Sr.*, p. 257.

27. Scott, *A Singular Woman,* pp. 90–91, 143–44; Obama, *DFMF,* pp. 62–71, 126; Jacobs, *The Other Barack,* p. 229; Obama in Mitchell, "Son Finds Inspiration in the Dreams of His Father," *HPH,* 23 August 1995, p. 10; Obama's 2001 oral history interview with Julieanna Richardson; Obama on *Fresh Air,* NPR, 12 August 2004, Obama in Barkley, *Who's Afraid of a Large Black Man?,* p. 23; Obama on *The Oprah Winfrey Show,* 19 January 2005; Obama on "Barack Obama Revealed," CNN, 20 August 2008; Obama, Remarks at the White House, 19 June 2009, *Public Papers* 2009 Vol. I, p. 859; Obama, Q&A with Wakefield High School Students, Arlington, Virginia, 8 September 2009, *Public Papers* 2009 Vol. II, p. 1350; and Obama, Remarks at the Kennedy Center, *Public Papers* 2009 Vol. II, p. 1780. On Arlene Payne and Margery Duffey, see Arlene's obituary, Andrew Kenney, "UNC Researcher, a Pioneering Academic, Kept Kinship to Obama to Herself," *News & Observer,* 23 June 2014.

On Obama Sr.'s visit to Hefty and Eldredge's fifth-grade classes, also see B. J. Reyes, "Punahou Left Lasting Impression on Obama," *HSB,* 8 February 2007; Eldredge in Meacham, "On His Own," *Newsweek,* 1 September 2008, p. 26; Eldredge in Remnick, *The Bridge,* p. 74; and DJG interviews with with Pal Eldredge and Mark Hebing. Eldredge recalls Ann coming to the second, midyear student evaluation conference; Hefty's invitation to Obama could have stemmed from that meeting. On Brubeck's 19 and 21 December concerts, see *HSB,* 17 December 1971, and Brubeck Papers, University of the Pacific, Section 1.F.4.2, particularly the February 1972 Aida program, p. 34. From Honolulu, Obama went to stay with his brother Omar at 17 Perry Street in Cambridge, Massachusetts, where Ruth wrote to him on 4 January 1972 to complain that she and his Kenyan children "Haven't heard from you in ages—perhaps you have forgotten us?" Ruth B. Obama to Obama Sr., 4 January 1972, Abon'go Malik Obama Papers.

28. Stanley Ann Dunham Soetoro, "Application for Passport by Mail," 4 January 1972, Dunham Passport File; Mark Hebing e-mail to Serge Kovaleski, 25 January 2008, Hebing Papers; DJG interviews with Mark Hebing and Pal Eldredge; Eldredge in Tani, "A Kid Called Barry," *Punahou Bulletin,* Spring 2007, pp. 16–21, on ABC *Nightline,* 26 April 2007, and in Glauberman and Burris, *The Dream Begins,* p. 13; Jacobs, *Obamaland,* p. 78; Austin Murphy, "Obama Discusses His Hoops Memories at Punahou High," SportsIllustrated.com, 21 May 2008; Abercrombie in Dan Nakaso, "Army Veteran Grandfather Was Obama's Boyhood Pal in Hawaii," *HA,* 14 November 2008; Punahou Catalog 1972–73; "Obama's Ex-Teacher Expects Successful Visit," *Japan Times,* 14 November 2009; Jayne Arakawa Kim in Ramos, ed., *Our Friend Barry,* pp. 45; Barton George, "Obama in '73," blogs.sun.com, 7 February 2007; Punahou's middle school yearbook *Na Opio;* "Obama on Zionism and Hamas," Jeffrey Goldberg Blog, The Atlantic.com, 12 May 2008. On Abercrombie's eventual 1974 UH Ph.D. dissertation, "Mumford, Mailer and Machines: Staking a Claim for Man," see Richard Borreca, "Life Lessons from Gov.-Elect's Thesis," *Honolulu Star-Advertiser,* 28 November 2010. Either prior to admission, in 1970, or sometime during the fifth-grade year, Punahou reportedly administered IQ tests to students. Pal Eldredge, who has given all of his evaluative materials to Punahou's archives, recalls of Obama that "he was really up there. His IQ was up there."

On the 1973 summer trip, see Amanda Griscom Little, "Obama on the Record," Grist .org, 30 July 2007; Jeff Zeleny, "As America Learns About Obama, He Returns the Favor," *NYT,* 10 July 2008; Obama on *The Late Show,* 10 September 2008; Wolffe, *Renegade,* p. 147; Obama, Remarks on the 160th Anniversary of the Department of the Interior, 3 March 2009, *Public Papers* 2009 Vol. I, p. 179; Scott, *A Singular Woman,* pp. 139–42; Obama in David Remnick, "Going the Distance," *New Yorker,* 27 January 2014; and Obama's remarks at Senator Daniel K. Inouye's funeral, Washington, D.C., 21 December 2012. Obama, *DFMF,* pp. 144–45, incorrectly dates the trip as 1972 rather than 1973.

On Obama's seventh- and eighth-grade years, the 1973–74 and 1974–75 Punahou Catalogs as well as Kelli Furushima's rich collection of yearbooks and other materials are most instructive, as well as DJG interviews with Furushima, Mark, Joann, and Philip Hebing, Laurie Tom, Bob and Kent Torrey, Pal Eldredge, and Alan Lum; Cheryl Lister in Ramos, ed., *Our Friend Barry,* p. 55; Laurel Bowers Husain and Laurie Uemoto Chang, "Obama Encourages Students to 'Dream Big,'" *Punahou Bulletin,* Spring 2005, pp. 14–17; Carlyn Tani, "Facts About Barack Obama's Ties to Punahou School," Punahou, n.d.; and Obama, Remarks at Benjamin Banneker High School, 28 September 2011, *Public Papers* 2011 Vol. II, p. 1180. See as well S. Ann

Soetoro to John F. O'Shea, 1 May 1974, O'Shea to Lolo Soetoro, 11 May 1974, and Douglas H. Brehm, "Memorandum for File," 23 May 1974, Soetoro INS File. On the tennis incident, see Caldwell in Ramos, ed., *Our Friend Barry,* pp. 51–52, and Caldwell's 2012 Jim Gilmore interview for *Frontline.* Ronald Loui on *Real Talk with Jack McAdoo,* JMacRadio.com, Show 10, 4 November 2010, endorsed Caldwell's memories ("that certainly was the kind of environment that was going on in the tennis court"). Three DJG attempts in 2014 to ask Mauch about his recollections were unavailing. On Mauch see also "Oakland Man Gets Tennis Job," *Oakland Tribune,* 26 December 1966, p. A22. Obama cited the incident in *DFMF,* p. 80, and may have alluded to it—"coaches"—in his 19 January 2005 appearance on *The Oprah Winfrey Show.* Obama on occasion would mention playing football in the ninth grade. See Obama, "Community Revitalization," Nebraska Wesleyan University Forum, 9 September 1994; also see Dave Reardon, "Coaching Teammates Eldredge, McLachlin Retire from Punahou," *HSB,* 20 June 2007. Punahou records indicate it was eighth grade.

29. Scott, *A Singular Woman,* pp. 156–57, Maya Soetoro-Ng in Emme Tomimbang, "Barack Obama: Hawaii Roots," 8 February 2008; Ann Dunham to Alice Dewey, 8 January 1976 and 24 February 1976, and Ann Dunham Sutoro to Ph.D. Committee, "Review and Update on Activities Relating to Ph.D.," September 1984, Dunham Papers; *Stanley Ann Soetoro v. Lolo Soetoro,* "Information Concerning Conciliation and Child Care," Hawaii First Circuit Family Court #117619, 15 June 1980; Dan Nakaso, "Obama's Mother's Work Focus of UH Seminar," *HA,* 12 September 2008; Sara Lin, "Obama Slept Here," *WSJ,* 7 November 2008; Will Hoover, "Obama's Hawaii Boyhood Homes Drawing Gawkers," *HA,* 9 November 2008; Punahou Catalog 1975–1976 (among Kelli Furushima's Punahou materials but absent from Punahou's own archives), *Ka Punahou* Volume 60, #1, 12 September 1975, through #24, 21 May 1976 (one of the two editors-in-chief was Steve Case), esp. Andrew Steele, "Freshpersons: A Social Guide to the Academy," 12 September 1975; *Oahuan* 1976; Catherine Black, "A Tradition of Innovation," *Punahou Bulletin,* Fall 2014; Tavares in Brian Charlton (AP), "In Honolulu, Young Obama Was Part of a Multiethnic Existence," 6 February 2007; Ann Sanner (AP), "Personal Side: The '08 Candidates Name a Few of Their Favorite and Not So Favorite Things," 20 December 2007; DJG interviews with Eric Kusunoki, Paula Miyashiro Kurashige, Larry Tavares, Greg Ramos, Tony Peterson, Keith Peterson, David Craven, Laurie Tom, Andrea Dolan Owen, and Kelli Furushima; Kusunoki in Richard Wolffe, "When Barry Became Barack," *Newsweek,* 31 March 2008, pp. 24ff.; Whitey Kahoohanohano in Dan Boylan, "'08: Year of Obama," *Midweek,* 2 January 2008; Sharon Yanagi in Gloria Borland and Kris Anderson, "An American Boyhood: Barack Obama in Hawaii," Video, 2008, Rik Smith in Scharnberg and Barker, "The Not-So-Simple Story of Barack Obama's Youth," *CT,* 25 March 2007, and in Sudhin Thanawala (AP), "In Multiracial Hawaii, Obama Faced Discrimination," 19 May 2008; Tony Peterson, "Obama's Depth of Mind, Ability to Grow Don't Surprise High School Friend," *Nashville Tennessean,* 6 February 2008 (reprinted in *HSB,* 15 February 2008, and *Salem Statesman Journal,* 12 May 2008); Peterson in David James Smith, "The Ascent of Mr. Charisma," *Sunday Times Magazine* (UK), 23 March 2008, pp. 58ff., in "Obama as We Knew Him," *Observer* (UK), 26 October 2008, pp. 4–7, and in Kevin Simpson, "Identity Questions Shaped Obama into New Kind of Politician," *Denver Post,* 28 August 2008 (Peterson noting that with regard to *DFMF* "the year I knew him isn't recorded in his book"). See also Kenji Kakugawa, "Young Mr. Obama," my.bo.com/kenjikakugawa, 6 November 2008, and Dennis Bader, Facebook posting, 27 September 2012. Obama would later say, with apparent reference to those Poki Street years, that "my mother was on food stamps while she was getting her Ph.D." Joe Klein, "The Fresh Face," *Time,* 23 October 2006, pp. 44ff. No other sources or recollections speak to this issue.

30. Ann Dunham to Alice Dewey, 8 January 1976 and 24 February 1976, Dunham Papers; Scott, *A Singular Woman,* pp. 153, 157, 167–68; Stanley Ann Dunham Soetoro, "Application for Passport," 2 June 1976, Dunham Passport File; Maya Soetoro-Ng on MetroTVNews.com, 12 December 2009; DJG interviews with Alec Williamson, Rolf Nordahl, Pal Eldredge, Alice Dewey, Ralph Dunham, and Charles Payne; Williamson e-mail to DJG; Nordahl, "Gramp's Dream," in Jacobs, *Obamaland,* p. 130; Eldredge on *Nightline,* ABC, 26 April 2007; Obama, *DFMF,* pp. 76–77; Dan Nakaso, "Obama's Tutu a Hawaii Banking Female Pioneer," *HA,* 20 March 2008; Susan Essoyan, "Strong Women Led Obama," *HSB,* 10 August 2008;

Allen G. Breed (AP), "'Toot': Obama Grandmother a Force That Shaped Him," 23 August 2008; Suzanne Roig, "Hawaii Friends, Former Co-Workers Remember Madelyn Dunham," *HA,* 15 November 2008; Dunham herself in Scott Fornek, "'I've Got a Competitive Nature,'" *CST,* 3 October 2004, p. 12, and in Amy Rice and Alicia Sams, *By the People* [film], Pivotal Pictures, 2009; Maraniss, *BOTS,* p. 270. On "Guess Who's Coming to Dinner?" see also Don Terry, "The Skin Game," *CT Magazine,* 24 October 2004, pp. 14ff. On Bob's Soul Food Place and the Family Inn, see *HA,* 16 October 2003. Another bar they frequented was the Black Cat, at Hotel and Richards Streets. Obama would later assert, incorrectly at least with regard to Stan, that by 1972 his grandparents were "voting for Nixon and concerned with law and order." Book Channel Interview with Marc Strassman, 11 August 1995. Nine years later he would say that Madelyn was a Republican. David Mendell, "Running as If He's Got a Rival," *CT,* 13 July 2004, pp. 1, 20.

31. Punahou Catalog 1976–1977, *Ka Punahou* Volume 61 #1, 10 September 1976, through #18, 20 May 1977, Punahou Archives, (#5 is missing), esp. Darwin Sen, "JV Basketball and Soccer Conclude Winning Seasons," 18 March 1977, p. 7; *Oahuan* 1977, esp. p. 143, "Junior Varsity Basketball Schedule 1976–77," Punahou Archives; DJG interviews with Tony Peterson, Keith Kakugawa, Mark Haine, Dan Hale, Pal Eldredge, David Craven, Mark Hebing, Mike Ramos, and Kelli Furushima (including Kelli's 1977 yearbook inscription); Jacobs, *Obamaland,* p. 46; Dave Burgess, *A Tale of Two Brothers: The Keith Kakugawa Story* (Lulu Press, 2009), esp. pp. 2, 23, 29–34, 47, 52, 183–84, 187, 191; Obama, *DFMF,* pp. 72–73, 83–86; Kakugawa in Jackie Calmes, "From Obama's Past: An Old Classmate, a Surprising Call," *WSJ,* 23 March 2007, p. A1; Kakugawa and Orme in Scharnberg and Barker, "The Not-So-Simple Story of Barack Obama's Youth," *CT,* 25 March 2007; Kakugawa on Jake Tapper, "Tale of Two Men: Senator Obama's Best Friend Ray," *Good Morning America,* ABC, 30 March 2007; Kakugawa in Peter Wallsten, "Obama Defined by Contrasts," *LAT,* 24 March 2008, in Richard Wolffe et al., "When Barry Became Barack," *Newsweek,* 31 March 2008, pp. 24ff., on *Anderson Cooper 360,* CNN, 15 April 2008; and on *Real Talk with Jack McAdoo,* Shows 7 and 10, 28 October and 4 November 2010; Ramos in Tomimbang, "Barack Obama: Hawaii Roots"; Orme in Rosemarie Bernado, "Punahou Alumni Celebrate Pride," *HSB,* 20 January 2009; Ann Dunham Soetoro, "Review and Update on Activities Relating to Ph.D.," September 1984, Dunham Papers; Obama on *The Oprah Winfrey Show,* 18 October 2006, Obama Remarks at Cornell College, Mt. Vernon, Iowa, 5 December 2007; Furushima in B. J. Reyes, "Punahou Left Lasting Impression on Obama," *HSB,* 8 February 2007, and in Minna Sugimoto, "Obama's Sweeties Quietly Campaign for Former Punahou Classmate," KHNL.com, 5 November 2008. On Obama's early interest in jazz, see also Russ Cunningham in Brian Charlton (AP), "In Honolulu, Young Obama Was Part of a Multiethnic Existence," 6 February 2007, Dean Ando in Reyes, "Punahou Left Lasting Impression on Obama," and Jann S. Wenner, "A Conversation with Barack Obama," *Rolling Stone* #1056-57, 10 July 2008 ("When I was in high school, probably my sophomore or junior year, I started getting into jazz"). Kakugawa's and Ramos's memories disagree about when the Schofield party occurred; Ramos's dating of it fits better with all other evidence. Regarding his feelings concerning his mother's absence, Obama in 2006 would say, "I'm somebody who did not get a lot of attention as a very young person" and in 2008 he would allow, "in retrospect, it was probably harder on me than I cared to admit" and "I suspect it had more of an impact than I know." Jimmie Briggs, "A Man of Vision," *Essence,* October 2006, p. 160; *20/20,* ABC, 26 September 2008; Remnick, *The Bridge,* p. 83. Apropos of Kakugawa's insistence, Obama on one occasion would explicitly acknowledge, "the trouble I had didn't have to do with being black. It had to do with the fact that I had this funny name, Barack Obama . . . You wish your name was Tim Smith." *Fresh Air,* NPR, 12 August 2004. Almost alone among subsequent commentators, Nicholas Lemann, "The Cipher," *TNR,* 25 October 2012, pp. 28–31, bluntly confronts Ann's behavior and its impact: "she consistently decided, from the time he was about ten, to structure her life so that she spent almost no time with him," and Barry "sensed this and resented it deeply." See also Paul H. Elovitz, "A Comparative Psychohistory of McCain and Obama," *Journal of Psychohistory* 36 (Fall 2008): 98–143, at 130 ("his sense of abandonment by his parents"). Observers who speak of "how carefully his mother sheltered him from the harsh reality of paternal abandonment" are missing at least half of the picture. David Lauter, "Review," *LAT,* 17 June 2012. Obama could more easily acknowledge his father's absence, once remarking, "I know the toll it took on me, not having a father in the house." Julie Bosman, "Obama Calls for More Responsibility from Black Fathers," *NYT,* 16 June 2008.

32. "Who Is the Real Barack Obama?," *Good Morning America,* ABC, 9 February 2007; Scott Fornek, "'Blessed by God,' Rooted in Two Continents," *CST,* 22 January 2003, p. 8; Burgess, *A Tale of Two Brothers,* pp. 183–84; DJG interviews with Keith Kakugawa, Mark Haine, Pal Eldredge, Bob Torrey, Jim Iams, Jay Seidenstein, Tom Topolinski, Mike Ramos, Greg Ramos, Dan Hale, and Mark, Joann, and Philip Hebing; Wayne Weightman e-mail to DJG; Kakugawa in Serge F. Kovaleski, "Old Friends Say Drugs Played Only Bit Part in Obama's Young Life," *NYT,* 9 February 2008, p. A1; Topolinski's interview with Jim Gilmore for *Frontline;* Topolinski comments of 25 November and 12 December 2007 on politicalticker.blogs.cnn.com; Weightman in Doug Ward, "U.S. Democrats Hope Vancouver Convention Can Help Change the World," *Vancouver Sun,* 14 April 2008; Punahou Catalog 1977–78 and 1978–79, esp. pp. 60, 86; Bill Messer in David James, "Meet Barack Obama's Welsh Teacher," *Western Mail,* 15 December 2008; and "Ex-Teacher on 'Oddly Quiet' Obama," BBCNews.com, 20 January 2009; Alicia Eler, "President Obama Pens Personal Apology to Art Historian," Hyperallergic.com, 18 February 2014; Linne Nickelsen-Willis, "The Narc Squad," in Ramos, ed., *Our Friend Barry,* pp. 60–62; *Ka Punahou* Vol. 62 #1, 9 September 1977, through #18, 19 May 1978, especially Steve Yamane, "Cagers Win Four in Row," 16 December 1977, p. 8, Natalie Steele, "Vars. AA Cagers Upset First Place University High," 17 February 1978, p. 7, "Vars. A Cagers Close Out ILH Season Finishing Fifth," 3 March 1978, p. 7; *Oahuan* 1978, esp. p. 163; Titcomb in Tani, "A Kid Called Barry," *Punahou Bulletin,* Spring 2007, pp. 16–21. The third member of "The Narc Squad" was Tom Krieger. On the widespread presence of pakalolo in 1970s Honolulu, see Michael Corcoran, "Obama's Oahu," *Austin American Statesman,* 9 November 2008, p. G1: "The pungent odor of marijuana could be smelled everywhere: movie theaters, Waikiki sidewalks, the Kam Drive In swap meets. . . . Everybody in Hawaii was getting high in the '70s." On the Bendix family, see Mark's younger sister Dyno in David Keyes, "Remembering a Kid Named 'Barry,'" *Bonner County Daily Bee,* 27 December 2007. Obama, *DFMF,* p. 80, describes "our assistant basketball coach" once using the word "niggers," but none of Obama's teammates or friends has ever made reference to any such incident.

33. Ann Dunham Sutoro to Ph.D. Committee, "Review and Update on Activities Relating to Ph.D," September 1984, Dunham to Alice Dewey, 28 July 1978, and Dunham résumé, ca. 1993, Dunham Papers; Judith Kampner, "The Untold Story of Obama's Mother," *Independent,* 16 September 2009; Christine Finn and Tony Allen-Mills, "Jungle Angel Was Barack Obama's Mother," *Times,* 8 November 2009; S. Ann Dunham, *Surviving Against the Odds: Village Industry in Indonesia* (Duke University Press, 2009), pp. xxxi and 299; Titcomb in Amy Rice and Alicia Sams, *By the People,* Pivotal Pictures, 2009; DJG interviews with Alice Dewey, Kelli Furushima, Kent Torrey, Annette Yee, Clyde Higa, Dawna Weatherly-Williams, Mark Hebing, Ian Mattoch, David Craven, and Tom Topolinski; Kelli Furushima in Hans Nichols, "In Hawaii, Media Surfs Obama's Past," *Politico,* 14 March 2007, and on *Anderson Cooper 360,* CNN, 15 April 2008; Scott Fornek, "Barack Obama," *CST,* 1 March 2004, p. 6; Obama on CNBC, 18 November 2006; Ann Sanner (AP), "Personal Side: The '08 Candidates Name a Few of Their Favorite and Not So Favorite Things," 20 December 2007; Sudhin Thanawala (AP), "Obama Worked to Fit In at Elite School," 26 March 2008; Obama in Kenneth T. Walsh, "Becoming Barack Obama," *U.S. News,* 9 June 2008, p. 17; Obama, "Here's the Scoop," LinkedIn.com, 25 February 2016; Punahou Catalog 1978–79, esp. pp. 28, 60, 63–80; Arlene Kishi, "Mattoch Explores 'Law and Society,'" *Ka Punahou,* 2 April 1976; *Ka Punahou* Vol. 63 #1, 8 September 1978, through #17, 24 May 1979; Law and Society syllabus from 1980 and other materials in the Ian Mattoch Papers, Honolulu; Samuel Mermin, *Law and the Legal System: An Introduction* (Little Brown, 1973); Connie Ramos in Ramos, ed., *Our Friend Barry,* p. 14; Barry Obama, "The Old Man," *Ka Punahou,* 15 December 1978, p. 5 (reprinted in *Ka Wai Ola,* 26th ed., May 1979, p. 65; the first comma in the fifth line is omitted in the newspaper copy); Mariko Gordon in *Daruma Newsletter,* November 2008, and Bernice Glenn Bowers in Dan Nakaso, "Hawaii Democrats Plan Parties Hoping to Celebrate Obama Win," *HA,* 3 November 2008. On Mermin, a 1936 graduate of Yale Law School, see "Memorial Resolution of the Faculty of the University of Wisconsin–Madison on the

Death of Professor Emeritus Samuel Mermin," 7 April 2008.

In contrast to Scott, *A Singular Woman,* pp. 192–93, Dunham's 1984 and 1993 timeline summaries indicate she did not return to Honolulu again in 1978 after departing for Indonesia in late May. On Obama at Baskin-Robbins, also most guardedly see Matt Pearl, "Kenmore Native Is Obama's Former Teacher," WGRZ.com, 21 January 2009. In the 26 March 2006 AP story, Obama also mentions working at "a burger chain," with the apparent implication that this also was summer 1978.

34. Tyler Dacey and Darwin Sen, "McLachlin Ends Sabbatical," *Ka Punahou,* 22 April 1977, p. 6; Leslyn Leu, "AA Captures Tournament," *KP,* 15 December 1978, p. 11; Punahou School Varsity AA Roster 1978–79 and 1978–79 Varsity AA Basketball Schedule (annotated with scores), Punahou Archives; Stephen Greaney, "Basketball Proves Power," *KP,* 26 January 1979, p. 8; Layne Yamada and Heather Murakami, "Basketball Team Shoots for More Wins," *KP,* 9 February 1979, p. 7; Sean Brennan, "Varsity Shows Spunk," *KP,* 23 February 1979, p. 4; Clyde Mizumoto, "At Last! Puns Nab ILH Title," *HA,* 6 March 1979, pp. E1, E2; "State Prep Meet Today," *HA,* 8 March 1979, p. D1; Troy Egami, "Chris Teaches Champ 'Psychos'" and Jim Ashford, "Basketball Wins ILH Title," *KP,* 9 March 1979, p. 11; "Punahou 77, Kapaa 29," *HA,* 9 March 1979, p. D3; Clyde Mizumoto and Steve Kimura, "Moanalua, Puns in Finals Tonight," *HA,* 10 March 1979, pp. D1, D6; Mizumoto, "Punahou Takes State Title," *HA,* 11 March 1979, pp. K1, K4; Troy Egami, "Basketball Triumphs After 3 Year Drought," *KP,* 16 April 1979, pp. 4–5; Carita Zimmerman, "McLachlin Quits Coaching," *KP,* 24 May 1979, p. 2; Barry Obama, "Winner," *Oahuan* 1979, p. 104; DJG interviews with Chris McLachlin, Tom Topolinski, Mike Ramos, Larry Tavares, Dan Hale, Alan Lum, and Greg Ramos; Obama in Mendell, *Obama,* p. 48; Lum in Brian Charlton (AP), "In Honolulu, Young Obama Was Part of a Multiethnic Existence," 6 February 2007; McLachlin in Tani, "A Kid Called Barry"; Lum and Obama in Todd Purdum, "Raising Obama," *Vanity Fair,* March 2008; Obama in Bryant Gumbel, "The Audacity of Hoops," HBO Sports, 15 April 2008, and Ann Sanner (AP), "Obama Says He Looked to Basketball in His Youth," 16 April 2008 ("improvisation"); Austin Murphy, "Obama Discusses His Hoops Memories at Punahou High," SportsIllustrated.com, 21 May 2008; Jaymes Song (AP), "Obama Finds Refuge, Identity in Basketball," 16 June 2008, McLachlin in Harnden, "Barack Obama's True Colours," and in Remnick, *The Bridge,* p. 92; Allison Schaefers, "Punahou Rejoices in Success of Obama," *HSB,* 5 November 2008. Also see David Chircop, "Young Obama Impressed Teacher in Hawaii," *Everett Herald,* 18 January 2009.

35. Obama in Barkley, *Who's Afraid of a Large Black Man?,* pp. 23, 26; Obama, *DFMF,* pp. 93–96; Obama on *The Friday Night Show,* WTTW.com, 3 December 2004; Obama's 2001 interview with Julieanna Richardson; DJG interviews with Annette Yee, Kent Torrey, Tom Topolinski, Dan Hale, Mark Hebing, and Mike Ramos; McLachlin on Orme in Hans Nichols, "In Hawaii, Media Surfs Obama's Past," *Politico,* 14 March 2007; Orme and Titcomb in Tani, "A Kid Named Barry"; Topolinski interview with Jim Gilmore; "KP's Illustrated Guide to Punahou Sapiens," *Ka Punahou,* 26 January 1979, p. 5; Madelyn in Scott Fornek, "'I've Got a Competitive Nature,'" *CST,* 3 October 2004, p. 12; Obama in David Remnick, "Going the Distance," *New Yorker,* 27 January 2014; Titcomb in Tomimbang, "Barack Obama: Hawaii Roots." See also Jessica Curry, "Barack Obama: Under the Lights," *Chicago Life,* Fall 2004 ("during my teenage years, I engaged in a lot of antisocial behavior"), and Obama, *TAOH,* p. 30 ("my rejection of authority spilled into self-indulgence and self-destructiveness"). With regard to six-packs, Obama in 2001 would remark, "I might have drunk a six-pack in an hour before going back to class," but seven years later said, "I would have two, three beers probably a day for certain periods of time" during high school. Obama's 2001 interview with Julieanna Richardson and *20/20,* ABC, 26 September 2008. On Titcomb's father and family, see Curtis Lum, "Frederick Titcomb, Retired Judge," *HA,* 14 March 2000, Lori Tighe, "Colorful Magistrate, Isle Actor Frederick Titcomb Dies," *HSB,* 14 March 2000, and a 9 May 2010 obituary item at titcombfami lyhistory.blogspot.com.

36. Obama, *DFMF,* p. 270; DJG interviews with Paula Kurashige, Eric Kusunoki, Keith Peterson, and Kim Jones Nelson; Kusunoki in Will Hoover, "Obama's Declaration Stirs Thrills at Punahou," *HA,* 11 February 2007, in Richard Serrano, "Obama Classmates Saw a Smile, But No Racial Turmoil," *LAT,* 11 March 2007, in Hans Nichols, "Angry Obama the Pothead Is Not How They Remember Him in Hawaii,"

Telegraph (UK), 25 March 2007, on "The Candidates: Barack Obama," MSNBC, 20 February 2008, in Borland and Anderson, "An American Boyhood," in Remnick, *The Bridge,* p. 78, in Sasha Abramsky, *Inside Obama's Brain* (Portfolio, 2010), p. 25, and on Martin Kaste, "Hawaii Prep School Gave Obama Window to Success," NPR, 12 October 2012; Bart Burford, "Barry and Me," in Jacobs, *Obamaland,* pp. 58–59; Furushima in Nichols, "Angry Obama"; Titcomb in Tomimbang, "Barack Obama"; Hebing in Serrano, "Obama Classmates"; Ramos and Haworth in Ramos, ed., *Our Friend Barry,* pp. 11, 78; Kolivas in Heidi Chang, "Hawaii Helped Shape Barack Obama," VOANews.com, 12 August 2008; Charles Payne on *Chicago Tonight,* WTTW, 12 February 2007; Dan Nakaso, "Family Precedent," *USA Today,* 8 April 2008, p. 7A. Come 2008, Obama would acknowledge, "on the surface, I remained very polite," but in previous years he repeatedly had insisted that "for much of my high school years" he had taken on a "macho African-American image of a basketball player talking trash" and "had this divided identity—one inside the home, one for the outside world." No one within the Punahou portion of that "outside world" ever saw any such thing. "Barack Obama Revealed," CNN, 20 August 2008; Obama on *This Week,* ABC, 15 August 2004; "Oprah Talks to Barack Obama," *O, The Oprah Magazine,* November 2004. See also Obama with Tim Russert on CNBC, 18 November 2006 (a "real macho guise"). As of Barry's final 1978–79 school year, Richard Haenisch, who was half black, half German, and a future college basketball star, would also arrive at Punahou, but he did not play basketball that year.

37. Obama, "A Life's Calling to Public Service," *Punahou Bulletin,* Fall 1999, pp. 28–29; Austin Murphy, "Obama Discusses His Hoops Memories at Punahou High," SportsIllustrated.com, 21 May 2008; Dwight N. Hopkins, "The Hawaiian President: Beyond Black and White," *Christian Century,* 10 February 2009, pp. 10–11.

38. 1979 *Oahuan,* esp. p. 271; "Dead Man Identified," *HSB,* 10 January 1986, p. A2; "Homicide Probe," *HSB,* 11 January 1986, p. B4; "Police Seek Help," *HSB,* 12 January 1986, p. A5; "Arraignment Set," *HSB,* 16 January 1986, p. A14; Jim Borg, "Man, 20, Ordered Held for Trial in Bludgeon Death," *HA,* 21 January 1986; "Killing Blamed on 'Jilted' Lover," *HSB,* 11 December 1986; "Man Convicted of Murder in Bludgeon Slaying," *HA,* 19 December 1986; "Man Found Guilty in Ex-Lover's Death," *HSB,* 19 December 1986; "Life Term in Hammer Slaying," *HA,* 24 January 1987; "Claw-Hammer Killer Sentenced," *HSB,* 24 January 1987; *Devere v. Falk,* 951 F. 2d 359 (9th Cir.), 20 December 1991 (1991 WL 275356 full text); Richard Alleyne, "Obama's High School Pot Dealer . . . ," *Daily Mail,* 29 January 2014. Killer Andrew Devere served over twenty years in prison for Ray's murder before being paroled in 2007. When Obama's yearbook page was first brought to public attention in a blog posting by a 1980 Punahou graduate, Mark Bendix objected and the post disappeared. Eve Maler, "Barry Obama and the Gang," *Pushing String Blog,* www.xmlgrrl.com, 16 January 2007. On January 31 Bendix commented: "just keep things on the down low and be cool. The choom gang had a name for people like you. Goober!" Years later Mike Ramos would incorrectly assert that unlike Greg Orme, neither he nor Obama were Choom Gang members. Jackie Calmes, "Visits with School Pals Are a Touchstone on President's Trips to Hawaii," *NYT,* 4 January 2014. Ramos must have momentarily forgotten that the most famous public photo of the Choom Gang includes a smiling Obama plus a glimpse of his own forehead, glasses, and left arm.

39. Ann Dunham Suturo to Alice Dewey, 2 February 1979 and 13 May 1979, Sutoro to Ph.D. Committee, "Review and Update on Activities Relating to Ph.D.," September 1984, Dunham Papers; Scott, *A Singular Woman,* pp. 6, 196, 214; Dunham, *Surviving Against the Odds,* p. 299; DJG interview with Alice Dewey, Adam Sorensen, "President Obama's 1979 Prom Photos," Swampland.time.com, 23 May 2013; Sorensen, "Obama's Grand Old Party," *Time,* 3 June 2013, pp. 12–13; "Double Date from President Obama's High School Prom Reveals Details," InsideEdition.com, 23 May 2013; "President Obama's Prom Date," InsideEdition.com, 24 September 2013.

40. Punahou 1979 Commencement film, YouTube; "Dave Donnelly's Hawaii," *HSB,* 4 June 1979, p. B2; "Where They Are Now Class of '79," *Punahou Bulletin,* November 1979, pp. 22–23; Jacobs, *Obamaland,* p. 51; DJG interviews with Kent Torrey, Greg Ramos, Tom Topolinski, Mike Ramos, Dan Hale, Mark Hebing, Kelli Furushima, and Kraig King; Topolinski interview with Jim Gilmore; Eli Saslow and Philip Rucker, "Obama Looks to Future with a Nod to His Past," *WP,* 20 January 2009, p. A1; Obama in Jaymes Song (AP), "Obama Returns to School," 13 August 2008, Wolffe, *Renegade,* p. 187, Obama in Maggie Murphy and Lynn Sherr, "The President and Michelle Obama on Work, Family, and Juggling It All," *Parade,* 20 June 2014; Neena Cherayil, "Did Obama Really Get Rejected from Swarthmore?," *Swarthmore Daily Gazette,* 8 November 2008 (quoting two students to whom Obama each spoke); Obama, *DFMF,* p. 96–98; Ann Dunham Sutoro to Alice Dewey, 13 May 1979, Dunham Papers. On Kenji Salz's family, see his father's obituary, "Andrew J. Salz," *San Francisco Chronicle,* 9 October 1995, p. A20. Like Bobby Titcomb's father, Andrew Salz, a graduate of Yale Law School, also became a Honolulu district court judge.

CHAPTER THREE: SEARCHING FOR HOME

1. Occidental College 1979–80 Catalog, esp. pp. 5–6, 28, Occidental College Student Directory 1979–80, Occidental College Student Handbooks ("Lookbook") 1976–77, 1977–78, 1978–79, 1979–80, Jean Paule to Richard Anderson, "Fall Term 1979," 18 November 2008, and Jean Paule, "Racial Diversity at Occidental College: A History," April 2004, Occidental College Archives; Judy Strong, "Student Surplus Jams Dormitories," *The Occidental* [*TO*], 27 September 1979, pp. 1, 10; Bill Hollingsworth, "Gilman Cites Improved Quality as Goal of '78–'79 Freshman Class Reduction," *TO,* 24 February 1978, pp. 1, 7; Alison Bell, "Sophomore Exodus: Why Are They Leaving?," *TO,* 12 May 1978, pp. 1, 10; Tom Hammitt, "ASOC Schedules Minority Recruitment Day for March," *TO,* 16 February 1979, pp. 1, 7; Tyler Kearn and Aidan Lewis, "Retracing Obama's Legacy," *Occidental Weekly,* 25 November 2008, p. 6; IRS to Stanley Ann Dunham Sutoro, "Notice of Federal Tax Lien Under Internal Revenue Laws," 12 November 1991 (indicating that in October 1985 assessments totaling over $17,000 had been levied against Ann regarding her 1979 and 1980 taxes), IRS FOIA; Andy Torres, "Torres on Noon Ball," *TO,* 15 February 1980, p. 13; "CORE," *TO,* 9 May 1980, p. 2; Pete Hisey, "Center Stage," *Occidental Magazine,* Fall 2004, pp. 24–29, esp. p. 26 ("full scholarship"); Larry Gordon, "Occidental Recalls 'Barry' Obama," *LAT,* 29 January 2007 ("full scholarship"); Lynn Sweet, "The Obama We Don't Know," *CST,* 19 July 2008 (supposedly a full scholarship only for his freshman year); Scott Helman, "Small College Awakened Future Senator to Service," *BG,* 25 August 2008, p. A1; Jerry Crowe, "Obama Honed His Skills at Occidental," *LAT,* 10 November 2008, p. D2; Huell Howser, "Barack Obama's Occidental College Days," *California Gold,* KCET, 13 November 2008; Lyle James Slack, "Obama at Occidental," *Verdugo Monthly,* January 2009, pp. 15–16 (originally posted on *Arroyo Monthly,* 25 November 2008); Colleen Sharkey, "Friends of Barry," *Occidental Magazine,* Winter 2009, pp. 12–17; Ben Bergman, "Obama's Other College Hopes for Presidential Boost," NPR, 16 January 2009; Larry Gordon, "An Early Influence," *LAT,* 18 January 2009, Maraniss, *BOTS,* pp. 337, 343–44, 346; Obama on *20/20,* ABC, 23 December 2011; Obama Q&A with Elizabeth Taylor, *CT,* 29 October 2006, p. B3; DJG interviews with Kraig King, Paul Carpenter, Andy Roth, Roger Boesche, Kent Goss, Romeo Garcia, Eric Newhall, Tom Topolinski, Phil Boerner, Paul Anderson, Ken Sulzer, John Boyer, Sim Heninger, Adam Sherman, Bill Snider, Tom Moyes, Samuel Yaw "Kofi" Manu, Michael Schwartz, and Paul Herrmannsfeldt; Adam Sherman, "Back to the Annex," Sherman Papers. Obama himself would assert that "I received a scholarship to Occidental" in his 2001 interview with Julieanna Richardson. As of 1979–80, 52 percent of Oxy undergraduates received some amount of nonloan student aid. Jan Klunder, "Oxy Backs Arts in Coming Crisis," *LAT,* 10 February 1980, pp. 1, 6–7. Per Phil Boerner's retrospective 15 March 1983 diary entry, Obama most definitely had a stereo in his Haines Annex room. On Richard Reath, see Cindy Zedallis, "'Sophisticated' Reath Speaks in Thorne," *TO,* 27 September 1979, pp. 1, 10. Obama's textbook from the Boesche and Reath course, Kenneth Prewitt and Sidney Verba, *Introduction to American Government,* 3rd ed. (Harper & Row, 1979), would be passed down to 1984 Occidental graduate Asad Jumabhoy. On Phil Boerner's father and family, see Michael Boerner's *WP* obituary, 9 May 2012. Boerner and Boyer recall attending what they thought was the premier showing of *Apocalypse Now,* but that had taken place over a month earlier, on 15 August 1979. See *LAT,* 12 August 1979.

2. DJG interviews with John Boyer, Paul Carpenter, Mike Ramos, Roger Boesche, Lawrence Goldyn, Ken Sulzer, Larry Hogue, and Alex McNear; Occidental College 1979–80 Catalog, esp. pp. 5, 28; 1980–1981 Catalog, esp. p. 158; Lawrence Goldyn, "ERA/Draft Fusion Threatens Women's Progress," *TO,* 15 February 1980, pp. 5, 9; ad, *TO,* 22 February 1980, p. 5; Cindy Zedalis, "Goldyn Notes Hatred Towards Homosexuals," *TO,* 29 February 1980, pp. 1, 2, 11; Goldyn, "Goldyn" (letter), *TO,* 7 March 1980, p. 11; BO to McNear, 20 November 1982; Tracy Baim, "Obama Seeks U.S. Senate Seat," *Windy City Times,* 4 February 2004; Kerry Eleveld, "Obama Talks All Things LGBT with *The Advocate,*" *The Advocate,* 10 April 2008; Lou Chibbaro Jr., "Nominee's Gay Mentor Speaks Out," *Washington Blade,* 12 September 2008; Brandon Hamilton, "Obama's Gay Mentor: Lawrence Goldyn '73," *Reed Magazine,* December 2010; Howser, "Obama's Occidental College Days."

3. Rick Cole, "Nader's Challenge: Are You a Bystander?," *TO,* 10 February 1978, pp. 3, 7; Andrew Roth, "Corporate Apartheid: Occidental Commits Itself to an Investment in Racism," *TO,* 17 February 1978, pp. 3, 6; Gary Chapman, "Democratic Socialism: Breaking the Chains of Corporate Power," *TO,* 17 February 1978, pp. 4–5; Bill Ellis, "Students Call for College Divestment," *TO,* 24 February 1978, pp. 1, 7; Bill Hollingsworth, "Divestment Rally to Greet Trustees," *TO,* 7 April 1978, p. 1; Martin Burns, "Students Protest Trustee Inaction," *TO,* 14 April 1978, p. 1; Bill Ellis, "Student Protestors Rally at Bank of America Stockholders Meeting," *TO,* 28 April 1978, pp. 1, 11; "SCC Endorsements" and "SCC Candidate Statements," *TO,* 5 May 1978, pp. 2, 5–8; Andy Roth, Doyle Van Fossen, and Gary Chapman, "Mary Jane," *TO,* 12 May 1978, p. 2; Bill Hollingsworth, "Faculty Asks Trustees to Divest," *TO,* 19 May 1978, pp. 1, 7; 1978 *La Encina* Yearbook, p. 113; "Trustees Consider South African Investments," *Occidental College Magazine,* August 1978, pp. 18–21; Tom Hammitt and Bill Davis, "Affirmative Action Committee Promises Minority Profs," *TO,* 9 February 1979, pp. 1, 11; "News '78–'79: Recapping the Issues and Events," *TO,* 15 September 1979, pp. 1, 4; Michael Boerner and Dorothy Boerner to Phil Boerner, 29 and 30 January 1980, as excerpted in 19 May 2009 Phil Boerner memo, Boerner Papers; Cindy Zedalis, "Quad Lights Up at Anti-Draft Rally," *TO,* 1 February 1980, pp. 1, 11; Karen Aidem and Nancy Eisenberg, "Anti-Registration Agitators Begin Campaign for Student Education," *TO,* 8 February 1980, pp. 1, 7; Cindy Zedalis, "Profs' Perspectives Offered at Teach-In," *TO,* 15 February 1980, pp. 1, 14; Roger Boesche, "New Cold War Mentality Critiqued," *TO,* 15 February 1980, pp. 4, 6–7; Cindy Zedalis, "Honor Court and Writing Reviewed by Professors," *TO,* 22 February 1980, pp. 1, 11; Susan Keselenko, "Anti-Draft Activism Fades with Finals," *TO,* 4 April 1980, p. 3; Student Coalition Against Registration and the Draft letter to the editor, *TO,* 18 April 1980, pp. 2, 10; Larry Hogue, "Jamming Draft Gears by Massive Resistance," *TO,* 16 May 1980, p. 3; DJG interviews with Andy Roth, John (Doyle) Van Fossen, Stephanie Fulks Westerman, and Larry Hogue; Brad Weber e-mail to DJG; Katie Hafner, "Gary Chapman, Internet Ethicist, Dies at 58," *NYT,* 17 December 2010; Valerie J. Nelson, "Gary Chapman Dies at 58," *LAT,* 24 December 2010. On Oxy DSA, also see Susan Keselenko, "[Dorothy] Healey Condemns U.S. Response to Soviet Action," *TO,* 29 February 1980, pp. 1, 11.

4. Bill Davis and Tom Hammitt, "Minority Enrollment Plummets," *TO,* 19 January 1979, pp. 1, 7; Earl Chew et al., letter to the editor, *TO,* 18 January 1980, p. 2; Sue Paterno, "Just How Much Do YOU Know," *TO,* 1 February 1980, p. 4; Bill Mullen, "Oxy Lacrosse: They Bring 'Em Back Alive," *TO,* 1 February 1980, p. 9; Susan Keselenko, "Minority Fraternity Proposal Approved," *TO,* 7 March 1980, p. 1; Bill Knutson, "Assistant Dean Resigns," *TO,* 4 April 1980, pp. 1, 11; Anne Ball, "Rowe Reinstated After Resignation," *TO,* 10 October 1980, p. 1; 1980 *La Encina* Yearbook, esp. pp. 95, 101 (a Haines residents' group picture not including Obama), 174 (misidentifying Richard Casey as Eric Moore), 186, 190, 211, 216; Jean Paule e-mail to DJG, 27 July 2009; DJG interviews with Eric Moore, Romelle Rowe Ecung, Judith Pinn Carlisle, Hasan Chandoo, Amiekoleh Usafi, Margot Mifflin, and Tom Grauman; Usafi in Sharkey, "Friends of Barry," p. 14. On Delanney, see Yann Masson, "Presidentielle Americaine: Avec Barack, on Regardait les Matchs de Basket," *Nice Matin,* 6 November 2008, Edward Rangsi, "Laurent Delanney," ISportConnect.com, 6 February 2012, and "All in the Game of Love," *CEO Magazine,* January 2015, pp. 44–51. Per Obama, *DFMF,* p. 99–100, Oxy contemporaries such as Mifflin and Grauman are convinced that Carlisle was the basis for the character "Joyce," though Judith herself understandably doubts that: "I did not know Barack Obama. As far as I know, I never said a word to him."

5. Occidental College 1979–80 Catalog, p. 5; "Dorm Closing," *TO,* 15 February 1980, p. 16; DJG interviews with Susan Keselenko Coll, Caroline Boss, Eric Moore, Romeo Garcia, Margot Mifflin, Dina Silva, Alex McNear, and John Boyer; Group Y, "The MX Missile: Bigger Is

Not Better," Political Science 94, 12 May 1980, Group A, "Critique on the MX Missile," n.d., Caroline Boss et al., "The MX Missile: Superiority Not Worth the Price: A Rebuttal by Group Y," 16 May 1980, "Critique of Group A by Group Y," n.d., Susan Keselenko Coll Papers; Dick Anderson, "O Triumphe!" *Occidental Magazine,* Winter 2009, pp. 10–11; Monica Corcoran, "Barack Obama Went Hawaiian Casual at Occidental College in L.A.," *LAT,* 18 January 2009; Sue Paterno, "Pakistan's Ambassador to Speak," *TO,* 2 May 1980, p. 4; Tom Cotrel, "Mirza Speaks on the Middle East" and Bob Mitrovich, "Ambassador Mirza Meets Interviewers," *TO,* 9 May 1980, pp. 1, 11; "Dictionary," *TO,* 15 September 1979, p. 3 (Cooler); field hockey ad, *TO,* 5 October 1979, p. 9; Bill Mullen, "Hasan Chandoo and His $163.00 All-Stars," *TO,* 26 October 1979, p. 9; "Women Can-Do for Chandoo," *TO,* 16 November 1979, p. 8; Lisa Jack, "Recollections of President-Elect Barack Obama," 17 December 2008, Augsburg.edu; Laura Fitzpatrick, "The Long-Lost Negatives," *Time,* 17 December 2008; Mike Boehm, "Photos Show President Barack Obama as Barry the Freshman," *LAT,* 27 May 2009, p. D1; Guy Trebay, "When He Was Barry," *NYT,* 18 June 2009, p. E4; Erin Carlyle, "Minneapolis Photographer Captures Obama as a Young Man," *CityPages,* 24 June 2009. Nine months later, Jack would win first prize in an Oxy photo contest. *TO,* 6 February 1981, p. 6. Two subsequent photos Tom Grauman would take of Obama, one picturing his left-handed removal of a volume from a shelf in the library's basement, the other showing him with Hasan at the 8 February 1981 Ujima dinner, both capture the ring on his left index finger as well. See also Paul Herrmannsfeldt, "Profile: Romeo Garcia an Occidental Legend," *TO,* 16 May 1980, p. 5.

6. Serge F. Kovaleski, "Old Friends Say Drugs Played Only Bit Part in Obama's Young Life," *NYT,* 9 February 2008, p. A1; Sharon Cohen (AP), "Barack Obama: Finding Common Bonds in Different Worlds," 3 June 2008; Chidanand Rajghatta, "McCain vs. Obama: Who's Better for India?," *Times of India,* 8 June 2008; Andrea Brown, "To Couple, He's Barry in a Leather Jacket," *Colorado Springs Gazette,* 20 January 2009; DJG interview with Alice Dewey; Barack Hussein Obama, Selective Service Registration Form, 29 July 1980, FOIA; Maya Soetoro 1979–1984 Passport, Obama Family Papers; Stanley Ann Soetoro v. Lolo Soetoro, "Complaint for Divorce" and "Information Concerning Conciliation and Child Care," both 15 June 1980, and "Decree," 5 November 1980, Hawaii First Circuit Family Court #117619; Obama, *DFMF,* pp. 87–91; Maraniss, *BOTS,* pp. 317, 362, 382; Frank Marshall Davis to Margaret Burroughs, 27 October 1982, Burroughs Papers. Obama believes that his grandparents' argument occurred while he was still in high school, but the balance of evidence suggests otherwise. Also see Bob Owens, "Did Obama Actually Register for Selective Service?," PajamasMedia.com, 12 August 2008 (quoting the Selective Service System as stating that Obama's registration number was 61-112559-1). Years later one news story would have Obama working "to sell island trinkets in a gift shop" and "making sandwiches" at a deli in Honolulu that summer, but the story contains other obvious errors and attributes its information to campaign staffers, not Obama. Sweet, "The Obama We Don't Know," *CST,* 19 July 2008. Sometime in 1980, Ann completed work on a monograph entitled *Women's Work in Village Industries on Java,* a 157-page volume published in Jakarta that no more than half a dozen libraries around the world would catalog and retain.

7. Occidental College 1980–81 Catalog; DJG interviews with Hasan Chandoo, Asad Jumabhoy, Margot Mifflin, Dina Silva, Caroline Boss, Tom Grauman, Larry Hogue, Paul Carpenter, Bill Snider, Sim Heninger, Eric Moore, Chris Welton, and Paul Herrmannfeldt; Welton, "Yes He Can," January 2009 (e-mailed to DJG 14 May 2014); Mifflin e-mails to DJG, 2 April 2013 and 22 November 2014; Caroline Boss and Larry Hogue, "Socialists," *TO,* 3 October 1980, p. 2; Debbie Rush, "Women's Hockey Doing Well," *TO,* 10 October 1980, p. 9; Seth Borenstein (AP), "Obama to Boldly Go Where No Geek Has Gone Before," 24 December 2008; Boss in Remnick, *The Bridge,* p. 109; Jonathan Alter, *The Promise* (Simon & Schuster, 2010), p. 179; Larry Wilson, "When Barry Obama, Oxy Student, Lived in Pasadena," *Pasadena Star-News,* 10 November 2015. In perhaps his most direct subsequent acknowledgment, Obama would state, "I did all my drinking in high school and college. I was a wild man. I did drugs and drank and partied." Amanda Griscom Little, "Barack Obama," *Rolling Stone,* 30 December 2004, p. 88. See also "Barack Obama Revealed," CNN, 20 August 2008 ("My first two years of college, I was continuing some bad habits from high school").

8. Cindy Zedalis, "Faculty Tables Divestment Resolution," "Divestiture" editorial, and Todd

Gold, "Richard Gilman: Interview," *TO,* 17 October 1980, pp. 1, 2, 4–6; "Activate" editorial, *TO,* 24 October 1980, p. 2; Dan McCormac, "Student Coalition Is Revived to Fight Apartheid," and "South Africa," *TO,* 31 October 1980, pp. 21, 24; Anne Ball, "SCC Encourages Ethnic Awareness," and Diana Takata, "U.S. Multinationals Bankroll Racism," *TO,* 7 November 1980, pp. 1, 3; Debbie Levi, "Morality Above and Beyond Financial Concerns," *TO,* 14 November 1980, p. 5; Lee Cauble, "College Council Discusses Club Sports, Apartheid Speakers," and "S. African Teach-In," *TO,* 21 November 1980, pp. 15, 16; Anthony Russo, "Faculty Demands Divestment Action," Arpie Balekjian, "Speaker Cleaves to Corporations," Julie Evans, "Opinions Differ on Apartheid Issue," and Debbie Levi, "Namibian People's War on Apartheid," *TO,* 5 December 1980, pp. 1, 4; DJG interviews with Caroline Boss, Larry Hogue, Andy Roth, Dario Longhi, Eric Moore, Asad Jumabhoy, Hasan Chandoo, Margot Mifflin, Bill Snider, and Paul Carpenter, "Crossroads Africa," *TO,* 8 February 1980, p. 8; *TO,* 3 October 1980, p. 11; Obama in Richard Wolffe et al., "When Barry Became Barack," *Newsweek,* 31 March 2008, pp. 24ff.; Wolffe, *Renegade,* pp. 31–32; Huell Howser, "Barack Obama's Days at Occidental," KCET TV, 25 November 2008; Sharkey, "Friends of Barry," *Occidental Magazine,* Winter 2009, p. 13; Jim Schlosser, "Downtown Baker Knew Obama Way Back When," *Greensboro News & Record,* 19 April 2010; Eric Moore interview with Jim Gilmore.

9. DJG interviews with Caroline Boss, Hasan Chandoo, John Drew, Tom Grauman, Roger Boesche, and Andy Roth; Sohale Siddiqi interview with Jim Gilmore; Drew's 1979 Occidental College transcript, Boss to Drew, 24 January 1980, Drew to Erik V., 10 April 1980, Drew Papers. Drew's recollections of Barack and Hasan's visit to Caroline's family home are best read in chronological order, particularly in light of the conclusions of Herbert L. Packer, Ex-Communist *Witnesses: Four Studies in Fact Finding* (Stanford University Press, 1962). John C. Drew, "Face-to-Face with Young Marxist Obama: Remembering My Days as an Anti-Apartheid Student Activist," Anonymous PoliticalScientist.blogspot.com, 18 November 2009, updated on 7 February 2010, the same date as a 59-second Youtube video; Ronald Kessler, "Obama Espoused Radical Views in College," Newsmax.com, 8 February 2010; Drew, "Was Obama a Committed Marxist in College?," Breitbart.TV, 12 February 2010; Drew, "Glorious Leader Gap," AnonymousPoliticalScientist.blogspot.com, 5 July 2010; Drew, "Obama at Occidental," DirectorBlue.blogspot.com, 23 October 2010; Paul Kengor, "Obama's 'Missing Link,'" *American Thinker,* 10 December 2010; Drew, "Meeting Young Obama," *American Thinker,* 24 February 2011; Drew, "Even Republicans Rejected Info About Obama's Past," *American Thinker,* 14 September 2011; Drew, "My White Girlfriend Inspired Obama's Big, Dark Regina in *Dreams From My Father,*" *American Thinker,* 17 July 2012; Jamie Glazov, "Young Obama's Dreams of a Communist Revolution in America," *FrontPage Magazine,* 8 August 2012 (quoting Drew as acknowledging "I only remember bits and fragments of the actual conversation"); Kengor, *The Communist,* pp. 249–58; Drew, "White Like Me," *American Thinker,* 27 May 2013. Fourteen years later Obama would mention having read "philosophers like Sartre and Fanon" while at Oxy. "Community Revitalization," Nebraska Wesleyan University Forum, 9 September 1994, Lincoln, NE. Contrary to one published conclusion of Drew's, Roger Boesche's gradebooks indicate Obama was not in his fall term 1980 PS 132 II. In line with Packer's analysis, Drew in "Glorious Leader Gap" would come to further recall that Obama had seemed "strikingly effeminate." On Hamid and Obama's road trip, see Wolffe, *Renegade,* p. 238; on New Year's Eve see Maraniss, *BOTS,* p. 367.

10. Occidental College 1980–81 Catalog, esp. pp. 73, 122, 158; Siddiqi interview with Gilmore; DJG interviews with Anne Howells, David James, Margot Mifflin, Mark Dery, Bill Snider, Larry Hogue, Alex McNear, Roger Boesche, Ken Sulzer, John Boyer, Caroline Boss, Tom Grauman, and Asad Jumabhoy; Mark Dery, "Dancing with Words," *Occidental Magazine,* Fall 2005, p. 56 (memorializing Jensvold, who died in an accidental fall in June 2005 at age fifty-one); David Mendell, *Obama: From Promise to Power* (Amistad, 2007), pp. 60–62; Larry Gordon, "Occidental Recalls 'Barry' Obama," *LAT,* 29 January 2007; John Schwada, "A Look at Barack Obama's Time at Occidental College," FoxLA11, 25 February 2008; Boesche in Wolffe et al., "When Barry Became Barack," *Newsweek,* 31 March 2008, pp. 24ff.; Oliver Haydock, "Meet Roger Boesche, Who Knew 'Barry Obama' in Passing at Occidental," *New York Observer,* 3 April 2008; Howser, "Barack Obama's Occidental College Days," *California Gold,* KCET, 13 November 2008; Kearn and Lewis,

"Retracing Obama's Legacy," *Occidental Weekly,* 25 November 2008; Slack, "Obama at Occidental," *Verdugo Monthly,* January 2009, pp. 15–16; Sharkey, "Friends of Barry," p. 15; Gordon, "An Early Influence," *LAT,* 18 January 2009; Jeremy Oberstein, "Obama's Ascent to the Top," *Glendale News Press,* 20 January 2009; Gordon, "Obama's Oxy Professor Learns the President Has a Long Memory," *LAT,* 15 August 2009; Gabriel Bernadett-Shapiro, "Boesche Makes a White House Call," *Occidental Weekly,* 23 September 2009; "12 Minutes with Obama," *Occidental Magazine,* Fall 2009; Margot Mifflin, "Obama at Occidental," NewYorker.com, 3 October 2012; Eric Krol, "Students Put Candidates on the Spot," *CDH,* 30 September 2004, p. 10; Obama on *The Friday Night Show,* WTTW.com, 3 December 2004; Obama in Barkley, *Who's Afraid of a Large Black Man?* pp. 23–24; Obama on *20/20,* ABC, 26 September 2008.

11. Anne Ball, "Students Lack Interest on Draft Issue," Ball, "SCC Supports Faculty," Arpie Balekjian, "Gregory Captivates and Keeps Large Audience," Anthony Russo, "Forum Explores Prejudices," and "African Speaker," *TO,* 16 January 1981, pp. 1, 13, 14, 16; "Calendar of Events," *LAT,* 15 January 1981, p. WS17; "Bread Not Bombs Martin Luther King Vigil" flyer, 17 January 1981, McNear Papers; Henry Weinstein and Erwin Baker, "S. Africa to Move Consulate to L.A.," *LAT,* 22 February 1980, p. C5; 6 October 1980 flyer in the Carol B. Thompson and Bud Day Papers, Michigan State University Archives; "Protest Staged Against South African Consulate," *LAT,* 7 October 1980, p. B19; "The Land of the Free (Trips)," *LAT,* 5 October 1980, p. F4; Carol B. Thompson, "Land of the Free (Trips)," *LAT,* 11 October 1980, p. C3; John L. Mitchell, "South African Consul Sits Tight in B.H. Office," *LAT,* 30 October 1980, pp. 1, 15; "The Southland," *LAT,* 18 January 1981, p. A2; Anthony Russo, "Speaker, Film Denote Tragedy of Apartheid," *TO,* 23 January 1981, pp. 1, 5; Obama in Barkley, *Who's Afraid of a Large Black Man?* p. 28; Stewart Ikeda and Alexia Robinson, "U.S. Senator Barack Obama on Black Student Politics," *Black Collegian,* 1st Semester 2006, pp. 48–51 (but misdating it to his freshman rather than sophomore year at Oxy); "Tim Ngubeni: The Bringer of Hope," YouTube, 11 June 2007 (two parts); DJG interviews with Hasan Chandoo, Caroline Boss, Alex McNear, Eric Moore, and Paul Anderson. A Chandoo letter in *TO,* 30 January 1981, p. 12, voiced a fear that Pakistan "may be the next underdeveloped country which becomes an annihilated sideshow to the super power hegemonic games." On the Gathering, see Janet Clayton and John L. Mitchell, "Black Ministers in Political Move," *LAT,* 7 May 1979, pp. 1, 26–28, and Sandy Banks, "Watts: Many Churches But How Helpful?," *LAT,* 19 August 1980, pp. 1, 14–15.

12. Debbie Levi, "Minority Recruitment Decline Reflects Priorities," *TO,* 23 January 1981 [I], p. 1, and 30 January 1981 [II], p. 1; Dan Karasic, "MEC Will Receive Tentative Funding," *TO,* 30 January 1981, p. 1; *Third World Voice* 1, #1, January 1981 (date stamped 21 January), Occidental College Archives; Anne Ball, "SCC Supports Faculty," *TO,* 16 January 1981, p. 1; "Divestiture," *TO,* 30 January 1981, p. 16; Michael Abelson, "SCC Discusses Budget, Divestment at Meeting," and "Gay Awareness Week," *TO,* 6 February 1981, p. 1; Larry Hogue to Alex McNear, 7 February 1981, McNear Papers; *Occidental Alumni Magazine,* Spring 1981, p. 37; Debbie Levi, "Gays Present Alternative Lifestyle," *TO,* 13 February 1981, pp. 1, 11, 16; "Rally," *TO,* 13 February 1981, p. 2; DJG interviews with Hasan Chandoo, Caroline Boss, and Chris Welton ("the three of us went there and heckled"); Obama himself is uncertain as to whether he joined in the heckling.

13. Anthony Russo, "Students Demand Divestment," *TO,* 20 February 1981, pp. 1, 11; "Divestment," *TO,* 20 February 1981, p. 2; "Strictly Opinion" and Brad Weber and Lupe Nolasco, "Affirmative Action and Divestiture," *TO,* 27 February 1981, pp. 4, 12; *La Encina* 1981, pp. 137, 185; Obama, *DFMF,* pp. 105–7; Rivera in Margaret Talev, "For Obama, a Tale of Two Speeches," KRDC McClatchy, 16 November 2007, and in Helman, "Small College Awakened Future Senator to Service," *BG,* 25 August 2008, p. A1; Mifflin in "Obama as We Knew Him," *Observer,* 26 October 2008, pp. 4–7; Mifflin, "The Occidental Tourist," *NYT,* 18 January 2009, p. IV–14; Mifflin, "Obama at Occidental," NewYorker.com, 3 October 2012; Chris Welton, "Yes He Can," January 2009, and Welton, "That Troublemaker Barry Obama," *Financial Times,* 7 May 2014, p. 8; DJG interviews with Caroline Boss, Tom Grauman, Hasan Chandoo, Chris Welton, Rebecca Rivera, Margot Mifflin, Eric Moore, Amiekoleh Usafi, Sim Heninger, Paul Carpenter, Larry Hogue, and Tom Moyes. On Sarah-Etta Harris, see Tom Hammitt, "ASOC Schedules Minority Recruitment Day for March," *TO,* 16 February 1979, p. 1 ("From what I've seen in my three years here,

Occidental couldn't give a damn if it returned to all Anglo"), "Richter Fellowship Recipients Announced," *TO,* 22 February 1980, p. 10, and "Possibilities & Realities," 1981–1982, p. 64, Occidental College Archives. Justin A. Frank's characterization of the rally as "in retrospect, a crucial juncture in the history of our nation" is too hilarious not to quote. *Obama on the Couch* (Free Press, 2011), p. 99.

14. DJG interviews with Hasan Chandoo, Caroline Boss, Tom Grauman, and Barack Obama; Obama, *DFMF,* pp. 103–4, 107–12, 116; Obama, "Community Revitalization," Nebraska Wesleyan University Forum, 9 September 1994, Lincoln, NE; Obama, Northwestern University Commencement Address, 16 June 2006; Obama interview with Katie Couric, *CBS Evening News,* 5 December 2007; Obama, "A New Era of Service," University of Colorado at Colorado Springs, 2 July 2008; Obama on *20/20,* ABC, 26 September 2008; "Possibilities & Realities," 1981–82, p. 64, Occidental College Archives. Regarding abandonment, one psychologist would later assert that "abandonment was the primary issue of Barry's life." Avner Falk, *The Riddle of Barack Obama* (Praeger, 2010), p. 84. Years later, and well prior to any public suggestion that "Regina" was based upon Boss, John Drew would reflect on that passage in *DFMF* and observe "that really sounds like Caroline." DJG interview with Drew, 14 December 2011. In an interesting chronological coincidence, the Oxy newspaper would report that following an excessive night of partying on Friday, February 20, as part of a student event called "Da Getaway," a special allocation of student government funds was necessary to pay four people for "removing trash, mopping floors, etc.," as part of "the cleanup of several dorms." Alan Tong, "SCC Allocates Funds for 'Getaway' Workers," *TO,* 6 March 1981, p. 11. Oxy employed thirty-seven unionized housekeepers, who worked—rather unhappily—thirty hours per week at an hourly wage of $3.87. See "Support," *TO,* 29 May 1981, p. 2, a letter to the editor from twelve students noting that "unfortunately most people take them for granted."

15. Karla Olson, "Rising Above Oxy Through Columbia," *TO,* 20 February 1981, p. 5; DJG interviews with Eric Moore, Phil Boerner, Anne Howells, Romelle Rowe Ecung, Hasan Chandoo, Asad Jumabhoy, Caroline Boss, Alex McNear, Sim Heninger, Bill Snider, and Paul Carpenter; Boerner in Sharkey, "Friends of Barry," p. 16; Howells in Gordon, "Occidental Recalls 'Barry' Obama," in Kovaleski, "Old Friends Say Drugs Played Only a Bit Part," and in Mary Anne Ostrom, "Democrats Get Down to Business," *San Jose Mercury News,* 25 August 2008; Obama in Tammerlin Drummond, "Barack Obama's Law," *LAT,* 12 March 1990, pp. E1, E2; Obama, *DFMF,* pp. 115–16 (erroneously imagining "a transfer program that Occidental had arranged with Columbia"); Obama in Gordon, "Occidental Recalls"; Obama's 2001 interview with Julieanna Richardson ("a lot of my friends . . . had graduated, so I transferred"); Obama in Pete Hisey, "Center Stage," *Occidental Magazine,* Fall 2004, pp. 24–29 ("I really wanted to see New York"); Obama, Q&A at My Brother's Keep Townhall, 21 July 2014, YouTube.com; Obama in Wolffe, "When Barry Became Barack"; "Barack Obama Revealed," CNN, 20 August 2008; Chris Welton, "Yes He Can," January 2009; Snider in Jim Schlosser, "Downtown Baker Knew Obama Way Back When," *Greensboro News & Record,* 19 April 2010; Phil Boerner to grandparents, 24 March 1981, Boerner Papers (mentioning submission of transfer application to Columbia). The application deadline was April 15. *Columbia University Bulletin 1981–1983,* p. 32, Columbia Archives. To one early questioner, Obama would assert, "I wanted to be around more black folks in big cities." Linda Matchan, "A Law Review Breakthrough," *BG,* 15 February 1990, p. 29. Two decades later Obama would say, "I think I had a hunger to shape the world in some way, to make the world a better place, that was triggered around the time that I transferred." Remnick, *The Bridge,* p. 114.

16. DJG interviews with Paul Anderson, Alex McNear, Tom Grauman, David James, Margot Mifflin, Mark Dery, Dina Silva, and Barack Obama; Mifflin in Kevin Fagan, "Obama: Transformations," *San Francisco Chronicle,* 14 September 2008, and in Sharkey, "Friends of Barry"; Alex McNear and Tom Grauman, "Feast," *TO,* 6 March 1981, pp. 2, 11; "Feast," *TO,* 13 March 1981, p. 12; *Feast,* Spring 1981, esp. pp. 11–12; Michael Dobson, "New Literary Journal Puts Only the Best on Display," *TO,* 22 May 1981, pp. 6–7. "Underground" was also published in classmate Mark Dery's *Plastic Laughter #2,* 1981, p. 6. On Obama's poems, also see Kevin Batton, "Ode to Obama," *Occidental Weekly,* 21 March 2007, Ian McMillan, "The Lyrical Democrat," *Guardian,* 28 March 2007, Rebecca Mead, "Obama, Poet," *New Yorker,* 2 July 2007, Jack Cashill, "Race, Fantasy, and the *New York-*

er's Editor," *American Thinker,* 30 April 2011 (someone who is cited with the greatest reluctance), Maraniss, *BOTS,* pp. 382–84, Kengor, *The Communist,* pp. 264–66. On McNear, also see Alex McNear and Wendy Clark, "SCC Lacks Initiative in Cultural Funding," *TO,* 22 May 1981, pp. 3, 11.

17. Andy Roth postcards to Alex McNear, n.d. March and 22 September 1981, McNear Papers; Phil Boerner diary entries, 18 April and 22 May 1981, Boerner Papers; DJG interviews with Andy Roth, Phil Boerner, Tom Grauman, Larry Hogue, Susan Keselenko, Alex McNear, Margot Mifflin, Hasan Chandoo, Caroline Boss, Eric Moore, and Bill Snider; I. M. Anal, "Visionaries Get Down Get Latent at Colloquium" and "Dorm Damage," *TO,* 24 April 1981, pp. 3, 12; Laura Pelegrin, "Campus Response to Sex Questionnaire," *TO,* 1 May 1981, p. 5,

18. "Apartheid," *TO,* 27 February 1981, p. 16; "SCAA," *TO,* 6 March 1981, p. 12; Debbie Levi, "Minority Exodus Creates Concern" and Arpie Balekjian, "Faculty Re-establishes GPAs on Transcripts," *TO,* 13 March 1981, pp. 1, 11; "Housing," *TO,* 10 April 1981, p. 16; "Confluence to Explore Oxy's Level of Tolerance," *TO,* 17 April 1981, p. 7; *La Encina* 1981, p. 77; *Black Student Directory,* n.d. (ca. March 1981), Occidental College Archives; Kevin Terpstra, "Racism?," *TO,* 13 March 1981, p. 2; "SCC Statement," *TO,* 4 June 1981, p. 4; Stephanie O'Neill, "Occidental College Rejects Student Petition for Divestiture," *LAT,* 16 April 1987; Blanca Labunog Araujo e-mail to DJG. See also "Southern Africa Resource Project," n.d. (ca. June 1981), Carol B. Thompson and Bud Day Papers, Michigan State University, detailing a metropolitan Los Angeles effort in the spring of 1981 in which the two Oxy participants were Ghanaian undergraduate Yaw Manu and sociology professor Dario Longhi.

19. "Free U. by DSA and Urban Studies" flyer, 3 March 1981, and "Oxy CISPES Chapter" flyer, 4 March 1981, Alex McNear Papers; DJG interviews with Paul Carpenter, Hasan Chandoo, and Paul Anderson; Bill Billiter, "3,000 Protest U.S. Role in El Salvador," *LAT,* 19 April 1981, pp. B1, B4; Debbie Levi, "CISPES Symposium Slated," *TO,* 8 May 1981, p. 4; Levi and Susan Keselenko, "Panel Ponders El Salvador Crisis," *TO,* 15 May 1981, pp. 1, 9, 11; "A Day for El Salvador" flyer, 24 May 1981, "The Liberation of El Salvador" flyer, 31 May 1981, and CISPES "Dear Friend" letter, n.d. (early June 1981), McNear Papers; Pete Hamill, "What Does Lou Grant Know About El Salvador?," *New York Magazine,* 15 March 1982, pp. 24–30; Chandoo in Hisey, "Center Stage," *Occidental Magazine,* Fall 2004, pp. 24–29. On CISPES, see also Van Gosse, "'The North American Front': Central American Solidarity in the Reagan Era," in Mike Davis and Michael Sprinker, eds., *Reshaping the U.S. Left: Popular Struggles in the 1980s* (Verso, 1988), pp. 11–50, at 23–26, U.S. Senate, Select Committee on Intelligence, *The FBI and CISPES,* S. Rpt. 101-46, July 1989, esp. pp. 20–21, and Nora Hamilton and Norma Stoltz Chinchilla, *Seeking Community in a Global City: Guatemalans and Salvadorans in Los Angeles* (Temple University Press, 2001), pp. 130–31. On Romero's assassination, see especially Tom Gibb, "The Killing of Archbishop Oscar Romero," *Guardian,* 22 March 2000. Carpenter would recall Obama encouraging him to take courses taught by Egan, who would die at age forty in a Honduran airplane crash. Suzanne P. Kelly, "Carleton Professor Believed to Be Among Honduran Crash Victims," *Minneapolis Star Tribune,* 24 October 1989.

20. "Forum on Pakistan Scheduled Sunday," *TO,* 8 May 1981, p. 14; Steve Tulchin and Chris Welton, "Forum Focuses on Pakistani Oppression," *TO,* 15 May 1981, pp. 1, 10; "Dissident Judge Steps Down," *Sydney Morning Herald,* 20 October 1980, p. 3; Feroz Ahmed, "Afzal Bangash: A Life Dedicated to Militant Struggle," *Economic & Political Weekly,* 20 December 1986, p. 2219; Jose Salcedo and Linda Duffy, "Support," *TO,* 15 May 1981, p. 13; Susan Keselenko, "Lawrence Goldyn: Reflections on an Abrupt Occidental Experience," *TO,* 29 May 1981, pp. 4–5; Lawrence Goldyn e-mail to DJG.

21. DJG interviews with Tom Grauman, Caroline Boss, Alex McNear, Phil Boerner, Eric Moore, Sim Heninger, and Kent Goss; Phil Boerner diary entries, 29 May and 5 and 6 June 1981, Boerner Papers; Jeremy Feldman, "Tight Housing Discourages Transfer Applications to CC," *Columbia Spectator* [CS], 18 November 1981, pp. 1, 2; Moore in Sharkey, "Friends of Barry," in Wolffe et al., "When Barry Became Barack," in Howser, "Barack Obama's Days at Occidental," and in his interview with Jim Gilmore; Goss in Helman, "Small College Awakened Future Senator," and in Remnick, *The Bridge,* p. 105. The second *Feast* would not appear until mid-December 1981; Occidental College Archives. See also Jeremy Feldman, "Going After Obama," *Columbia Spectator,* 5 June 2012, a wonderful response to an Obama detractor who distorted Feldman's story from thirty-one years earlier in order to demean: Charles C. Johnson, "Did Obama Have Lower SAT Scores Than George W. Bush?," Brietbart.com, 22 May 2012.

22. DJG interviews with Dina Silva, Caroline Boss, John Drew, Phil Boerner, Paul Carpenter, Hasan Chandoo, Keith Kakugawa, and Asad Jumabhoy; Drew in Kessler, "Obama Expressed Radical Views"; Drew, "Was Obama a Committed Marxist in College?"; Drew, "Meeting Young Obama"; John Drew to Alex McNear, 13 July 1981, McNear Papers; Paul Carpenter to Phil Boerner, 13 June 1981, Boerner Papers; Madelyn Dunham in Mendell, *Obama,* p. 58; Stanley Ann Dunham, "Application for Passport," 27 April 1981, Dunham Passport File; Ann Dunham Sutoro résumés circa 1981 and 1993, Dunham Papers; Andra Wisnu, "Friends Reflect on Legacy of Obama's Mother," *Jakarta Post,* 14 November 2008; Mary S. Zurbuchen, "Ann Dunham Sutoro and the Ford Foundation's International Philanthropy," American Anthropological Association Annual Meeting, Philadelphia, December 2009; Dunham, *Surviving Against the Odds,* p. 300; Scott, *A Singular Woman,* p. 220, 254–55; Larry Rohter, "Obama Says Real Life Experience Trumps Rivals' Foreign Policy Credits," *NYT,* 10 April 2008, p. A18; Richard Wolffe and Michael Hirsh, "A Man at Home in the World," *Newsweek,* 12 April 2008; Adam Goldman and Robert Tanner (AP), "Old Friends Recall Obama's Years in L.A., N.Y.," 15 May 2008; Rajghatta, "McCain vs. Obama," *Times of India,* 8 June 2008; Azhar Masood, "Obama's Larkana Connection," *Pakistan Times,* 10 July 2008; Hamid in Remnick, *The Bridge,* p. 112; Wolffe, *Renegade,* pp. 238–40. Years later, Obama would tell a French questioner that "I have traveled through the south of France when I was in college," perhaps to visit Occidental friend Laurent Delanney. Interview with Laurence Haim, Canal Plus, 1 June 2009, *Public Papers of the Presidents* 2009, Vol. I, p. 747.

23. Obama, *DFMF,* pp. 113–14, 118–19; Sohale Siddiqi in Goldman and Tanner, "Old Friends Recall Obama's Years"; Siddiqi interview with Jim Gilmore; Phil Boerner diary entries, 16 and 29 June and 8 September 1981, Phil Boerner to Karen McCraw, 8, 10, 20, and 25 August and 1 September 1981, Boerner Papers; Boerner, "Barack Obama '83, My Columbia College Roommate," *Columbia College Today,* January–February 2009; Alison Leigh Cowan, "Recollections of Obama's Ex-Roommate," NYTimes.com, 20 January 2009; Kim Clark, "Obama's Lessons for Transfer Students," *U.S. News,* 16 January 2009; Lydia DePillis, "Obama Slept Here," *New York Observer,* 1 February 2009; DJG interviews with Boerner and Paul Carpenter; Bradley Gallo, "Meet the Roommate-in-Chief," Northattan.org, 20 October 2009; Boerner in Marie-Joelle Parent, "Obama's Guide to New York," *Ottawa Sun,* 15 November 2009; Elizabeth A. Harris, "President Obama Studied Here," *NYT,* 13 June 2010; Krisanne Alcantara, "Inside President Barack Obama's Former NYC Apartment," realestate.aol.com, 13 August 2012.

On Emmett W. Bassett, who graduated from Tuskegee Institute in 1942 under the tutelage of George Washington Carver before earning a Ph.D. at Ohio State University in 1956 and teaching first at Columbia and then at the New Jersey College of Medicine, see Nathan Mayberg, "Black History Month Celebrated at College," *Sullivan County Democrat,* 24 February 2004; David Colman, "Director's Corner," *Neuro News* (McGill University), April 2008; Bassett and his wife Priscilla's 2011 oral history interview with Joseph Mosnier, Grahamsville, NY; his 4 October 2013 obituary in the *Sullivan County Democrat;* and the program from his memorial service, Boerner Papers. On Obama's cooking, see also Kenneth T. Walsh, "No Silver Spoon for Obama," *U.S. News,* 9 June 2008, p. 18. On Earl Chew, see Laurie Becklund, "Jurors Haunted by Guilty Verdict Want It Set Aside," *LAT,* 1 July 1992, p. A1; Becklund, "Judge Heeds Jurors' Pleas, Orders Retrial," *LAT,* 2 July 1992, p. A1; "California Case Puts Spotlight on Jury Coercion and Peer Pressure," *NYT,* 17 July 1992; and especially John Schwada, "Looking for Obama," MyFoxLA.com, 29 February 2008.

24. Columbia University Bulletin 1981–83, esp. pp. 2–8, 32–33, 35, 45–47, 53, and Columbia College Courses of Instruction 1981–82, Columbia University Archives; Phil Boerner Columbia transcript and Phil Boerner diary entry, 8 September 1981, Boerner Papers; Obama to Ann Sutoro, 21 September 1981, Dunham/Solyom Papers; Tammerlin Drummond, "Barack Obama's Law," *LAT,* 12 March 1990, pp. E1–E2 (stating that Obama paid off his undergraduate student loans during 1984); Obama interview with David Axelrod, 7 February 2004, JKP; DJG interviews with Phil Boerner, Pern Beckman, Gerrard Bushell, Darwin Malloy, Ron Sunshine, and William Araiza; Rifka Rosenwein, "Alienation Is Common for Minority Students," *CS,* 25 September 1981, p. 1; Joseph Walsh,

"Students Label CU Life Depressing," *CS,* 13 October 1981, p. 1; Danny Ly, "Striking Tenants Demand Front-Door Locks," *CS,* 16 November 1981, p. 1; Tom Todreas in *Columbia College Today,* May–June 2009, pp. 56–57; Obama to Boerner, 21 October 1986, Boerner Papers.

On Frank Ayala, who ironically would end his career as dean of students at Occidental College, see Jon Elsen, "Princeton Dean Named to Columbia Position," *CS,* 20 September 1979, p. 1; Richard Froehlich, "Ayala Goes West—Back to School," *CS,* 22 April 1982, p. 4; and *Occidental Weekly,* 26 November 2007.

25. "A Forum on South Africa—Including the Film *The Rising Tide*" flyer, 21 October 1981, Boerner Papers; LaVerne McDowell, "Sen. Urges Protest of Reaganomics," *CS,* 12 November 1981, p. 1; DJG interviews with Phil Boerner, Pern Beckman, and Ron Sunshine; Dorothy Boerner to Phil Boerner, 1 November 1981, Boerner diary entry, 16 November 1981, Boerner to Karen McCraw, 5 and 15 December 1981, Boerner to grandparents, 28 December 1981, Boerner diary entries, 20, 21, and 22 January 1982, Boerner to Karen McCraw, 23 January 1982, Boerner diary entries, 24, 29, and 30 (2) January and 1, 5, and 6 February 1982, Boerner to McCraw, 7 February 1982, Boerner diary entries, 12 and 13 February 1982, Boerner to McCraw, 2 April 1982, Boerner Papers; Cathie M. Currie, "A Columbia Classmate Remembers Obama," FactCheck.org, 23 February 2010. On Ndaba/Gulabe, see Susan Benesch, "Fighting for Divestment: David Ndaba of the African National Congress," *CS,* 25 May 1983, p. 3; Bill Cecil, "George Harrison," *Workers World,* 4 November 2004; and William Minter's 2004 interview with Jennifer Davis. Neither Boerner nor Obama recorded or remember Ron's last name, and his identity remains unknown. In mid-February 1982 Phil Boerner jotted down that Barack's studio was in a building numbered 932 and that his phone number was 866-4417. Neither Obama's name nor that number appear in the 1982 *Cole's Cross Reference Directory* for Manhattan, but Boerner's reference to a nearby record store strongly suggests that Obama's studio apartment was at 932 Amsterdam Avenue rather than 932 Columbus Avenue. No fewer than three record stores—at 906, 963, and 964—were located very close to the Amsterdam building.

26. Obama, *DFMF,* pp. 4, 119–27; Obama's 2001 interview with Julieanna Richardson; Kenneth Meeks, "Favorite Son," *Black Enterprise,* October 2004, pp. 88–95; Shica Boss-Bicak, "Barack Obama '83: Is He the New Face of the Democratic Party?," *Columbia College Today,* January 2005; David Saltonstall and Michael Saul, "Obama: Oh Boy, the Rent!" *NYDN,* 8 November 2005; Obama in Jacob Weisberg, "The Path to Power," *Men's Vogue,* September 2006, pp. 218–23; Mendell, *Obama,* pp. 16, 59, 62; Maya in Jennifer Steinhauer, "Charisma and a Search for Self," *NYT,* 17 March 2007, p. A1; Jeff Chang, "Maya Soetoro Ng: Q&A," *Vibe,* September 2007; Susan Page, "Obama Is Still Seeking Traction," *USA Today,* 2 October 2007, p. 1A; Jennifer 8. Lee, "Where Obama Lived in 1980s New York," NYT Cityroom blog, 30 January 2008; Todd Purdum, "Raising Obama," *Vanity Fair,* March 2008; Wolffe et al., "When Barry Became Barack," *Newsweek,* 31 March 2008, pp. 24ff.; Lisa Miller and Wolffe, "Finding His Faith," *Newsweek,* 21 July 2008, pp. 26ff.; Wolffe, *Renegade,* pp. 30–31; "Barack Obama Revealed," CNN, 20 August 2008; Obama in Remnick, *The Bridge,* p. 114; Siddiqi in Goldman and Tanner, "Old Friends Recall Obama's Years in L.A., N.Y."; Michael Daly, "Young Barack Obama Came to the City to Find Himself," *NYDN,* 11 November 2008; Sheryl Gay Stolberg, "Obama Conveys Principle to Students," *NYT,* 8 June 2010; Siddiqi interview with Jim Gilmore; Maraniss, *BOTS,* p. 444; Scott, *A Singular Woman,* p. 254; DJG interviews with Andy Roth and Mahboob Mahmood; Mahboob Mahmood, "Readings with Barack," 30 September 2008. *Cole's Cross Reference Directory* for Manhattan, 1983, p. 765, and 1984, p. 771, confirm "B Obama 410-2857" at 339 East 94th with that phone number beginning in 1982. Lynn Sweet, "The Obama We Don't Know," *CST,* 19 July 2008, contains factual errors and asserts secondhand that "one summer Obama worked for a private company holding a contract to process health records" for city employees and also "was a telemarketer in midtown Manhattan selling *New York Times* subscriptions over the phone, wearing a headset." Obama himself has never mentioned either such job, nor has he ever referenced the construction site job in any extemporaneous remarks.

27. DJG interviews with Andy Roth and Alex McNear.

28. Phil Boerner to Karen McCraw, 31 August 1982, Boerner Papers; Columbia University Bulletin 1981–83, pp. 2–8, 191–93, Columbia College, Courses of Instruction 1982–83, esp. pp. 38, 85, 90, Columbia Archives; Warwick

Daw in *Columbia College Today,* January–February 2009, p. 61, and in Maraniss, *BOTS*, p. 456; DJG interviews with Michael Ackerman, Pern Beckman, Phil Boerner, Robert Y. Shapiro, Michael Baron, William Araiza, and Jeremy Feinberg; Michael Waldman, "The Question of Edward Said," *CS*, 4 March 1982, pp. B6–B7, B9; Michael L. Baron, "Tug of War: The Battle over American Policy Toward China, 1946–49," Ph.D. dissertation, Columbia University 1980; Maurice Possley, "Activism Blossomed in College," *CT*, 30 March 2007; Janny Scott, "Obama's Account of New York Years Often Differs from What Others Say," *NYT*, 30 October 2007, p. B1; Baron in Naftali Bendavid, ed., *Obama: The Essential Guide to the Democratic Nominee* (Triumph Books, 2008), p. 36, in Jim Popkin, "Obama and the Case of the Missing Thesis," MSNBC, 24 July 2008, in Betsy Morais, "After Long Absence, Obama, CC '83, Speaks at Alma Mater," *CS*, 11 September 2008; William J. Broad and David E. Sanger, "Obama's Youthful Ideals Shaped the Long Arc of His Nuclear-Free Vision," *NYT*, 5 July 2009, p. A1; "Gerald Feinberg, 58, Physicist; Taught at Columbia University," *NYT*, 23 April 1992; Gerald Feinberg, *What Is the World Made of? Atoms, Leptons, Quarks, and Other Tantalizing Particles* (Doubleday, 1977), esp. pp. xii, xiii, xv; Gerald Feinberg, Elementary Physics 1002Y Syllabus, Spring 1991, Feinberg Papers. Per the latter, Feinberg in 1982 may well also have assigned some portions of Robert H. March, *Physics for Poets*, 2nd ed. (McGraw-Hill, 1978).

29. Barack Obama to Alex McNear, 26 September 1982, McNear Papers; DJG interviews with Alex McNear and Jeremy Feinberg; Anatole Broyard, "Revising the Heroine," *NYT*, 11 September 1982.

30. Gerald Feinberg, "Physics C1001X, First Exam," 12 October 1982, Feinberg Papers; DJG interview with Gregory Poe ("physics for poets was a famous gut"); Aaron Freiwald, "South African: Americans Should Denounce Apartheid," *CS*, 3 November 1982, p. 2; Cynthia Gelper, "Book, Students Say CC Could Be Better for Blacks," *CS*, 19 November 1982, pp. 1, 7; Barack Obama to Alex McNear, 20 November 1982, McNear Papers; Friedrich Nietzsche, *Basic Writings of Nietzsche* (Modern Library, 1992), p. 229.

31. Obama, *DFMF*, pp. 3–5, 114, 127–28; Scott, *A Singular Woman*, p. 256; William Onyango, "Economic Planning Man Dies in Crash," *Nairobi Times*, 30 November 1982, p. 5; DJG interviews with Zeituni Onyango, Pake Zane, and Alba De Souza; Philip Ochieng, "From Home Squared to the U.S. Senate," *Nation*, 1 November 2004; Edmund Sanders, "Obama Not Quite His Father's Son," *LAT*, 17 July 2008 ("I can't stand it" and "unrecognizable"); Zane in Maraniss, "Though Obama Had to Leave to Find Himself," *WP Magazine*, 24 August 2008; Kenneth Ogosia and Michael Mugwang'a, "New Links in Dreams From Obama's Father," *Nation*, 25 August 2008; Zane in Sally Jacobs, "A Father's Charm, Absence," *BG*, 21 September 2008; Joe Ombuor, "Obama's Father and the Origin of the Muslim Name," *Standard*, 4 November 2008; John Oywa and George Olwenya, "Obama's Dad and His Many Loves," *Standard*, 14 November 2008 (referencing Sebastian Peter Okoda); Jenn Jagire, "Keziah: First and Last [sic] Wife of Obama Sr.," *New Vision* (Uganda), 2 January 2009; Zane in Remnick, *The Bridge*, pp. 67–68; Mark Obama Ndesandjo, *Nairobi to Shenzhen*, pp. 6, 174; William Foreman (AP), "Obama's Half-Brother Recalls Their Abusive Father," 3 November 2009; Andrew Jacobs, "An Obama Relative Living in China Tells of His Own Journey of Self-Discovery," *NYT*, 4 November 2009; Keith Richburg, "Obama Half Brother Steps Into Spotlight," *WP*, 5 November 2009; David Eimer, "Our Alcoholic Father Beat Me," *Telegraph*, 14 November 2009; Tristan McConnell and Andrew Rice, "The Obama Diaspora," *NY Magazine*, 23 November 2009; Zeituni Onyango, *Tears of Abuse*, esp. pp. 39, 41–42, 90–91, 101, 217, 275, 283; Auma Obama, *And Then Life Happens*, esp. pp. 55, 58, 62, 68, 82, 89, 119–20, 131–35; Ndesandjo, *An Obama's Journey*, pp. 33, 44, 51, 59; Abon'go Malik Obama, *Barack Obama Sr.*, esp. pp. 258–59, 265, 28, 325 (noting with regard to Zeituni that "the bond that existed between the two was extraordinary"), 373; Firstbrook, *The Obamas*, pp. 225–32, 247; Jacobs, *The Other Barack*, esp. pp. 210, 227, 230, 235–43, 247–54; Maraniss, *BOTS*, pp. 277–78, 414. Multiple sources state that Hussein Onyango died in 1979, not 1975, but the three who knew best—Zeituni Onyango, Abon'go Malik Obama, and the very thorough Peter Firstbrook—all convincingly agree upon 1975. Likewise, some later sources offer differing exact dates for Obama Sr.'s death, but the most contemporaneous source, the November 30, 1982, *Nairobi Times* story, is on balance the most likely to be correct.

32. Obama on *The Friday Night Show*, WTTW, 3 December 2004; Obama, *TAOH*, p.

205; Jeff Zeleny, "Obama Talks About Growing Up Without His Father," NYT Caucus Blog, 3 October 2007; Obama in Kevin Merida, "The Ghost of a Father," *WP,* 14 December 2007, p. A12; Obama in Purdum, "Raising Obama," *Vanity Fair,* March 2008. See also Obama's 2 October 2003 interview with David Axelrod, Chicago. Instances where Obama would incorrectly state that he did not adopt "Barack" until late 1982 include Bernard Schoenburg, "Issues of Race a Painful Part of Obama Family History," *SJR,* 11 January 2004, p. 15; Obama interview with David Axelrod, 7 February 2004; and Obama on *The Oprah Winfrey Show,* 19 January 2005. Obama would once say that "some of my drive comes from wanting to prove that he should have stuck around," but Barack was indeed deeply fortunate to have grown up half a world away from a man who was "little short of a monster" and "suffered from a severe narcissistic personality disorder." "Who Is Barack Obama?," *ABC World News Tonight,* 1 November 2007; Toby Harnden, "A Life of President Obama's Father," *Telegraph,* 5 August 2011; Falk, *The Riddle of Barack Obama,* p. 28.

33. Siddiqi interview with Jim Gilmore; Kakugawa in Burgess, *A Tale of Two Brothers,* p. 94; McNear in Maraniss, *BOTS,* p. 455; DJG interviews with Alex McNear, Michael Schwartz, Mike Ramos, and Asad Jumabhoy; Barack Obama to Phil Boerner, 3 March 1983, and Katy Budge to Phil Boerner, 15 March 1983, Boerner Papers; Columbia University Bulletin 1981–83, pp. 2–8.

34. Obama to Alex McNear, 10 February 1983, McNear Papers (mentioning Spanish); DJG interviews with Michael Baron, Pern Beckman, Len Davis, and Judy Clain; Columbia College, Courses of Instruction 1982–83, p. 41; Lennard J. Davis, *Factual Fictions: The Origins of the English Novel* (Columbia University Press, 1983), esp. pp. 157–58, 213, 221–22; Lennard J. Davis Class Recording, June 1983, Davis Papers; Lennard J. Davis, *Resisting Novels: Ideology and Fiction* (Methuen, 1987), esp. pp. ix, 24, 224; Davis, "Barack Obama's Professor Gives Him Mid-Term Grades," *Huffington Post,* 21 January 2010. Internal evidence makes it clear that the Davis class recording dates from early June 1983 during a summer session iteration of his seminar, and not from the spring one that had ended just four weeks earlier. Jan Hoffman, "Making Up Is Hard to Do," *Village Voice,* 31 May 1983, pp. 14–18; Gary Arnold, "Film Notes," *WP,* 20 May 1983, p. A25; *NYT* advertisement, 29 May 1983, p. H14; *Newsweek,* 30 May 1983, p. 96; and *NYT,* 3 June 1983, pp. C8, C17.

35. Columbia College, Courses of Instruction 1982–83, pp. 34, 103; Maurice Obstfeld e-mails to DJG; Jonathan Zimmerman, "Playbooks for the White House," *CT,* 30 January 2007; Thomas Vinciguerra, "Dreams From My Mater," *Columbia Magazine,* Summer 2012; DJG interviews with Jonathan Zimmerman and Gregory Poe; Andrew G. Walder, Sociology W3229y, "State Socialist Societies," Spring 1983, Poe Papers; Andrew Walder e-mails to DJG; Alex Nove, *Stalinism and After,* 2nd ed. (George Allen & Unwin, 1981), esp. p. 40; Timothy E. O'Connor, untitled review of *Stalinism and After,* 3rd ed., *Studies in Soviet Thought* 41 (March 1991): 157–61, esp. pp. 159 and 160; Bill Wallace, "Obituary: Professor Alex Nove," *Independent,* 20 May 1994 ("when Mikhail Gorbachev raised the cry of perestroika, he seemed almost to be quoting" Nove); Elizabeth Pond, *From the Yaroslavsky Station: Russia Perceived* (Universe Books, 1981); John Leonard, "Review," *NYT,* 6 October 1981, p. C9; Walter Goodman, *NYTBR,* 29 November 1981, p. 11; Susan Sherer Osnos, "Conversations on the Trans-Siberian Express," *WPBW,* 27 December 1981, p. 6; Donald Bremner, "A View of Soviet Life from Siberia," *LAT,* 15 July 1982, p. G36; David Lane, *The End of Social Inequality? Class, Status and Power Under State Socialism* (George Allen & Unwin, 1982), esp. p. 159; Ivan Szelenyi, "The End of Social Inequality," *Acta Sociologica* 26 (July 1983): 313–16; T. Anthony Jones, untitled review of *The End of Social Inequality? Soviet Studies* 36 (January 1984): 146–47; Milovan Djilas, *The New Class: An Analysis of the Communist System* (Praeger, 1957), esp. pp. 47, 213; Serge Schmemann, "Milovan Djilas, Yugoslav Critic of Communism, Dies at 83," *NYT,* 21 April 1995.

36. Obama to Alex McNear, 10 February 1983, McNear Papers; DJG interviews with McNear; Raymond Williams, *Marxism and Literature* (Oxford University Press, 1977). Twelve years later, Obama would describe his job search quite differently: "in the months leading up to my graduation, I wrote to every civil rights organization I could think of, to any black elected official in the country with a progressive agenda, to neighborhood councils and tenants rights groups. When no one wrote back, I wasn't discouraged." Obama, *DFMF,* p. 135.

37. Obama to Boerner, 3 March 1983, Boerner Papers; Obama, "Breaking the War Mentality," *Sundial: The Weekly Newsmagazine,* 10 March

1983, pp. 2–3, 5; Remnick, *The Bridge,* p. 117. The *Sundial* article was first discovered by Evan Gahr, "Young Obama Wrote Article Blaming America for 'Growing Threat of War,'" *Human Events,* 9 January 2009; see also Broad and Sanger, "Obama's Youthful Ideals Shaped the Long Arc of His Nuclear-Free Vision," *NYT,* 5 July 2009.

38. 1983 Socialist Scholars Conference, Cooper Union, 1 and 2 April 1983, Democratic Socialists of America Papers Box 81; Obama to McNear, 1 April 1983, McNear Papers; Obama, *DFMF,* p. 122; Stanley Kurtz, *Radical-in-Chief: Barack Obama and the Untold Story of American Socialism* (Simon & Schuster, 2010), pp. 1, 25–26, 30, 62–63; Angelo M. Codevilla, "The Chosen One," *Claremont Review of Books,* Summer 2011, pp. 52–58, at 58; DJG interviews with Andrew Roth and Alex McNear.

Kurtz insists that mailing labels generated from a list of attendees who signed in at that first April 1983 conference proves that Obama also attended the second such conference, which was held April 19 through 21, 1984, and that he had the especially dangerous experience of hearing this author and the seminal, wonderful James H. Cone speak about Dr. King on Friday morning, April 20. The mailing labels list Obama's address as 339 E. 94th Street #6A, the apartment in which he lived with Sohale up through mid-May 1983. By mid-November 1983, and continuing through April and indeed most of 1984, Obama was living at 622 W. 114th Street #43, and on April 20 he would have been at work at One Dag Hammarskjold Plaza on Manhattan's East Side. Kurtz, *Radical-in-Chief,* pp. 69–71.

39. Obama to McNear, 1 April 1983, McNear Papers; Corey Kilgannon, "Keyboard Virtuoso, *NYT,* 19 November 1995; Pamela Dorian, "Everyone Is a Winner!" UrbanMysticMusing.com, 14 April 2009; DJG interviews with William Araiza and Jeremy Feinberg. SSDI indicates that Diane Dee was born in June 1924, and died in December 2009, age eighty-five.

40. Julius Genachowski, "Senate to Trustees: Divest!" *CS,* 28 March 1983, p. 1; Jeff Adler, "Divest Rally Outside Trustee Meeting Draws 100," *CS,* 5 April 1983, p. 1; Julius Genachowski, "Students Keep Up Vigil to Force CU Divestment," *CS,* 21 April 1983, p. 2; Leslie Dreyfous, "Danny Armstrong," *CS,* 17 May 1983, p. 19; Julius Genachowski, "Columbia Will Not Divest," *CS,* 13 July 1983, p. 1; Tony Vellela, *New Voices: Student Activism in the '80s and '90s* (South End Press, 1988), pp. 19–38; Siddiqi interview with Jim Gilmore; DJG interviews with Alex McNear, Verna Bigger Myers, Janis Hardiman Robinson, Wayne P. Weddington III, Darwin Malloy, Gerrard Bushell, Michael Ackerman, Charles V. Hamilton, Robert Y. Shapiro, and Michael Baron; Daniel Armstrong, Barbara Ransby, Kieth Cockrell, Derek Hawkins, Joseph H. Zwicker, Tim Todreas, and John G. Ruggie e-mails to DJG; Hollis R. Lynch, "After Bad Start, CU Must Be More Representative," *CS,* 22 February 1983, pp. 3, 6; Susan Scheiner, "Blacks Have Rough Time with CU's White Climate," *CS,* 25 April 1983, pp. 5, 13; Helen Kennedy, "Obama's Quiet Yrs. in NYC," *NYDN,* 14 January 2007, p. 26; Ross Goldberg, "Obama's Years at Columbia Are a Mystery," *New York Sun,* 2 September 2008; "Obama's Lost Years," *WSJ,* 11 September 2008, p. A14; Maraniss, *BOTS,* pp. 436–37, 459; Obama's 2001 interview with Julieanna Richardson; Obama's 2002 television appearance with Chinta Strausberg; Obama's 21 August 2006 remarks at the South African Institute for International Affairs, available at southafrica.usembassy.gov; Jim Popkin, "Obama and the Case of the Missing Thesis," MSNBC, 24 July 2008; Betsy Morais, "After Long Absence," *CS,* 11 September 2008; Broad and Sanger, "Obama's Youthful Ideals," *NYT,* 5 July 2009. See also Obama's unique reference in John Corr, "From Mean Streets to Hallowed Halls," *Philadelphia Inquirer,* 27 February 1990, pp. C1, C4: Columbia "was on a hill. . . . A lot of my education happened when I walked down the hill and into Harlem." On the wonderful Chuck Hamilton, see Wilbur C. Rich, "From Muskogee to Morningside Heights: Political Scientist Charles V. Hamilton," *Columbia Magazine,* Spring 2004.

41. Obama to Alex McNear, 27 April 1983, McNear Papers; McNear to Andrew Roth, 12 May 1983, Roth Papers; DJG interviews with Alex McNear, Hasan Chandoo, Asad Jumabhoy, and Tim Jessup; Richard Bernstein, "For 7,300 at Columbia, a Day Full of Balloons and Diplomas," *NYT,* 18 May 1983, pp. B1, B4; Richard Pollack, "7,329 Graduate CU in a Festive Mood," *CS,* 25 May 1983, pp. 1, 6; John R. Pulliam, "Obama Big on Ceremonies," *GRM,* 5 June 2005; Scott, *A Singular Woman,* pp. 243, 245, 249, 254–55; Obama to Alex McNear, 27 June (postmarked 5 July) 1983, McNear Papers; Obama to Phil Boerner, n.d., ca. 27 June 1983, Boerner Papers; Ann Dunham Sutoro, "Civil Rights of Working Indonesian Women," Institute for Legal Aid, Jakarta, 10 December 1982, 5pp., and Sutoro, "The Effects of Industrialization on Women Workers in Indonesia," n.d., ca. 1982, 18pp., Atria Institute Archives; Ann Dunham Soetoro Passport, April 1981–April 1986, Obama Family Papers, UH; Ann D. Sutoro to the Files, "Women's Economic Activities in North Coast Fishing Communities," Ford Foundation, 22 May 1983, 10pp., and Sutoro, "II. Program Statement: Women and Employment (Indonesia)," n.d., 1983, 41pp., esp. pp. 14, 17, Atria Institute Archives; Alice Dewey and Geoffrey White, "Ann Dunham: A Personal Reflection," *Anthropology News,* November 2008, p. 20. Apropos of Dewey's comment, Ann's son would later recount, "I actually remember her saying to me once, 'I don't feel white. That's not my identity.'" Wolffe, *Renegade,* p. 178. One news story would report that Dewey accompanied Dunham to Kenya. Dan Nakaso, "Obama's Mother's Work Focus of UH Seminar," *HA,* 12 September 2008.

42. Obama to Alex McNear, 1 September and 15 November 1983, McNear Papers; DJG interviews with Alex McNear and Hasan Chandoo. Obama would later write, "I decided to find more conventional work for a year, to pay off my student loans and maybe even save a little bit." *DFMF,* p. 135. Per *Cole's Cross Reference Directory for Manhattan*, 1984, p. 825 and 1985 p. 839, listing "D Reilly 866-8172," the first public report of the 622 West 114th Street address was Beth J. Harpaz (AP), "Obama Slept Here: Roots in 3 Nations, 6 Time Zones," 11 November 2008. When very limited selections from some but not all of Obama's letters to McNear were publicized in 2012 (Maraniss, "Becoming Obama," *Vanity Fair,* June 2012, and *BOTS,* pp. 450–52, 454–55, 464–65, 468–70, erroneously calling them "passionate"), one far more perceptive reporter labeled them "terribly pretentious . . . every discussion of the Great Existential Questions ends up being, basically, all about him and his superior philosophical mind and how he is on a higher plane than anyone else." With "a slightly patronizing" attitude toward his girlfriend, "Who did this guy think he was, with all these delusions of grandeur?" Noreen Malone, "Barack Obama's Old Girlfriends Get Dishy," *New York Magazine,* 2 May 2012.

43. Obama in John Corr, "From Mean Streets to Hallowed Halls," *Philadelphia Inquirier,* 27 February 1990, pp. C1, C4; DJG interviews with Cathy Lazere, Lou Celi, Barry Rutizer, Beth Noymer Levine, Dan Armstrong, Lisa Shachtman Hennessey, Peggy Mendelow, Brenda Vinson, Michael Williams, and Steve Delaney; Susan Arterian Chang e-mail to DJG; Dan Armstrong, "Barack Obama Embellishes His Resume," analyzethis.net, 9 July 2005, and ensuing comments by Bill Millar and Beth Noymer Levine; *Financing Foreign Operations—Interest Rate & Foreign Exchange Updater,* 15 December 1983, 15 January 1984, 15 and 31 March 1984, 15 April 1984; *Financing Foreign Operations Financial Update Bulletin,* 15 December 1983, 2pp.; *Business International Money Report* [*BIMR*], 6, 13, and 20 January 1984, 3 February 1984, esp. pp. 33–35, 9 March 1984; *Business International Weekly Report to Managers of Worldwide Operations,* 6, 13, and 20 January 1984, 3 and 10 February 1984, 26 March 1984; Janny Scott, "Obama's Account," *NYT,* 30 October 2007; Scott Horsley, "Obama's Early Brush with Financial Markets," NPR, 9 July 2008; Noymer in Lisa Miller and Richard Wolffe, "Finding His Faith," *Newsweek,* 21 July 2008, pp. 26ff.; Jessica Ravitz, "Utahn Recounts Sharing Job with Obama in New York," *Salt Lake Tribune,* 27 August 2008.

44. DJG interviews with Andrew Roth, Genevieve Cook, Bill Ayers, Bernardine Dohrn, and John Ayers; John Ayers e-mail to DJG, 1 May 2012 ("Bill does not remember Genevieve Cook from Bank Street. 'Hey, that was almost thirty years ago'"); Genevieve Cook, "Dancing in Doorways," B.A. thesis, Swarthmore College, May 1981, esp. pp. 1–2, 95; "Barack Obama" and "Recipes from Bill Ayers," Cook Papers; Cook in Maraniss, *BOTS,* p. 472. On Helen Ibbitson Jessup, see her book *Court Arts of Indonesia* (Harry N. Abrams, 1990), and Paul Richard, "Under Java's Spell," *WP,* 19 May 1991. On Michael J. Cook, see Pilita Clark, "Ambassador on the Warpath," *Sydney Morning Herald,* 11 March 1993, p. 13. On Philip C. Jessup's work in Indonesia, see Henry Kamm, "Indonesia Nickel Project Reflects 2 Worlds," *NYT,* 14 April 1978, pp. D1, D14. Jessup died in August 2013 at age eighty-seven. See Evi Mariani, "In Memoriam: Mining Executive Philip Jessup's Deep Connection to Indonesia," *Jakarta Post,* 23 October 2013.

Bill Ayers and Bernardine Dohrn had lived underground for over a decade following their April 2, 1970, indictment on federal criminal charges. Those charges were dropped in early 1974 because they involved unconstitutional warrantless electronic surveillance, but both Ayers and Dohrn remained underground until Dohrn surrendered on December 3, 1980, in Chicago to face state charges still pending against her—but not Ayers—from the violent October 1969 "Days of Rage." A front-page November 27 *New York Times* story had reported they had been living at 520 W. 123rd and that more than a dozen New Yorkers had identified Dohrn. All three major television networks covered Dohrn's December 3 surrender on their evening news broadcasts. On January 13, 1981, Dohrn pled guilty to reduced charges, was fined $1,500, and was granted probation that allowed her and her family to return to 520 West 123rd Street.

Sixteen months later, however, in the wake of an October 20, 1981, armored car robbery in Nanuet, New York, that left one Brink's employee and two police officers dead, a federal grand jury subpoenaed Dohrn to supply particular examples of her handwriting. Rental vehicles used in both the Nanuet holdup and an earlier April 1980 one had been rented by women using counterfeit drivers' licenses of customers who had patronized Broadway Baby, an Upper West Side store where Dohrn had worked under a pseudonym for six months prior to the February 1980 birth of her second son, Malik. Dohrn refused to submit the handwriting samples and was jailed for contempt. Seven months later, concluding that Dohrn's handwriting samples were superfluous to the prosecution of the Nanuet killers, Judge Gerard L. Goettel let Dorhn go free. *In re Dohrn,* 560 F. Supp. 179 (S. D. N. Y.), 5 January 1983. See also "Indictments Voided for 12 Weathermen," *NYT,* 4 January 1974, p. 41; Nathaniel Sheppard Jr., "2 in Weather Underground Are Bargaining to Surrender," *NYT,* 24 November 1980, p. A18; Josh Barbanel, "Bernardine Dohrn Reportedly Seen on the West Side," *NYT,* 27 November 1980, pp. A1, D12; Benita Brodt and Ronald Koziol, "Defiant Dohrn Surrenders," *CT,* 4 December 1980, p. 1; Nathaniel Sheppard Jr., "Bernardine Dohrn Gives Up to Authorities in Chicago," *NYT,* 4 December 1980, p. A18; Nathaniel Sheppard Jr., "Chicago Home of Friend Was Refuge for Miss Dohrn," *NYT,* 5 December 1980, p. A22; Douglas E. Kneeland, "Ex-Radical Leader Gets Probation and Fine in Chicago," *NYT,* 14 January 1981, p. A12; M. A. Farber, "Behind the Brink's Case: Return of the Radical Left," *NYT,* 16 February 1982, pp. A1, B4; "Miss Dohrn Refusing to Aid Brink's Case," *NYT,* 18 May 1982, p. B3; "Miss Dohrn Jailed for Contempt," *NYT,* 20 May 1982, p. B12; "Bernardine Dohrn Freed by Judge," *NYT,* 6 January 1983, p. B3; Eileen Putman (AP), "Radical Dohrn Released, Not Sorry," *Palm Beach Post,* 15 February 1983, pp. A7, A8; Sam Roberts, "Bar Panel to Consider Dohrn's Fitness," *NYT,* 26 August 1985, p. B4; Peter Howell, "Hippie Terrorists Reincarnated," *Toronto Star,* 18 October 2003; Patrick Healy, "A Playwright's Glimmers of a Fugitive Childhood," *NYT,* 3 September 2009. On "smash monogamy," see Bernardine Dohrn, Bill Ayers, and Jeff Jones, *Sing a Battle Song* (Seven Stories Press, 2006), pp. 136, 204.

45. Genevieve Cook Journal, 10, 12, 22, 26, and 29 January 1984, Cook to Obama, 10 January 1984, Cook poem beginning "saya rasa," 29 January 1984, Cook journal, 5 and 11 February 1984, Cook alphabetical poem for Obama, n.d., Cook Papers; Dunham to Alice Dewey, 13 February 1984, Dunham Papers; DJG interviews with Genevieve Cook, Hasan Chandoo, Cathy Lazere, Beth Noymer Levine, Lou Celi, Barry Rutizer, Lisa Shachtman Hennessey, Dan Armstrong, and Peggy Mendelow; Celi in Issenberg, *BG,* 6 August 2008.

46. Genevieve Cook journal, 18, 19, 20, 22, 24, 25, and 27 February 1984, 5, 9, 17, and 20 March 1984; DJG interviews with Genevieve Cook and Phil Boerner; Michael Isbell e-mail to DJG.

47. Genevieve Cook journal, 22, 25, and 29 March 1984, 1 April 1984; Richard F. Shepard, "Going Out Guide," *NYT,* 9 February 1984, p. C32; Mel Gussow, "How Billie Whitelaw Interprets Beckett," *NYT,* 14 February 1984, p. C13; Frank Rich, "A Whitelaw Beckett," *NYT,* 17 February 1984, p. C3; Bruce Weber, "Billie Whitelaw, 82, Longtime Beckett Muse," *NYT,* 23 December 2014, p. A20; DJG interviews with Genevieve Cook, Hasan Chandoo, and Beenu Mahmood; Sohale Siddiqi interview with Jim Gilmore; Howell Raines, "Hart and Mondale Clash Repeatedly in Sixth Debate," *NYT,* 29 March 1984, p. A1; Mike Shanahan (AP), "Debate: Most Personal, Acrimonious Exchanges," 29 March 1984; Fred Rothenberg (AP), "Close Contact Makes for Lively Debate," 29 March 1984; Gerald M. Boyd, "Thousands Cheer Jackson at Rally in Harlem," *NYT,* 1 April 1984, p. 29; Chandoo in Pete Hisey, "Center Stage," *Occidental Magazine,* Fall 2004, pp. 24–29; Mahmood, "Readings with Barack," 30 September 2008. In 2008 Obama's campaign would acknowledge that in 1984 he was registered to vote at 622 West 114th Street, although the site where that acknowledgment appeared, www.commongroundpolitics.net, has since disappeared from the Web.

Over two decades later, Jackson would assert that "it was in a presidential debate with Walter Mondale at Columbia University in 1984" when he first met Obama. "After the debate he introduced himself and said 'This is doable, we can do this, and a black man can run and win the White House.'" Robert T. Starks, "Jackson's Tears from All the Years," *N'Digo Profiles,* December 2008, pp. 16, 18. Remnick, *The Bridge,* p. 490, quotes Jackson as saying that years later Obama told him he had seen the debate.

48. Genevieve Cook journal, 3, 8, and 9 April 1984; Obama to Alex McNear, 14 April 1984, McNear Papers.

49. Susan Arterian Chang e-mail to DJG; DJG interviews with Peggy Mendelow and Dan Armstrong (who also recounted Tom Ehrbar's comment concerning Dan Kobal); Bill Millar's 2007 comments on Dan Armstrong's www.analyzethis.net blog; Issenberg, "Obama Shows Hints of His Year in Global Finance," *BG,* 6 August 2008; *Financing Foreign Operations—Interest Rate & Foreign Exchange Rate Updater,* 30 April, 15 and 31 May, 15 and 30 June, and 15 and 31 July 1984; *Business International Money Report,* 20 April, 18 May, and 15 June 1984; *Business International Weekly Report,* 20 April, 5 May, and 8 June 1984. Public reports of ties between the CIA and BI's network of foreign-based correspondents predated Obama's time there. John M. Crewdson, "CIA Established Many Links to Journalists in U.S. and Abroad," *NYT,* 27 December 1977, pp. 1, 40, reported that BI, "a widely respected business information service . . . had provided cover" for at least four CIA agents twenty years earlier. See also Angelo M. Codevilla, "The Chosen One," *Claremont Review of Books,* Summer 2011, pp. 52–58, at 55.

50. Genevieve Cook journal, 16 and 30 April 1984; Sim Heninger to Phil Boerner, 2 April 1984, Boerner Papers; DJG interviews with Genevieve Cook, Sim Heninger, and Phil Boerner; Genevieve to Barack unsent letter, 6 May 1984, Genevieve Cook journal, 8, 9, 16, and 26 May 1984, Cook Papers; Ann Dunham to Alice Dewey, 13 February 1984, Dunham to Leatrice Mirikitani, 26 March 1984, Dunham to Alan Howard, 26 March 1984, Ann Dunham Sutoro to Ph.D. Committee, "Review and Update on Activities Relating to Ph.D.," September 1984, Dunham Papers; Scott, *A Singular Woman,* pp. 257–62. See also Christine Finn and Sarah Baxter, "Long-Range Love of Barack's Absent Mother," *Sunday Times,* 23 November 2008, p. 24.

Years later, Obama would offer conflicting comments about his attitude toward his mother at this time. In 1995, citing her and his grandparents, he would say, "I never pushed them away"; in 2008 he would acknowledge "that there were tensions" and that he was "rebelling or pushing away from her and my grandparents." Monica Mitchell, "Son Finds Inspiration in the Dreams of His Father," *HPH,* 23 August 1995, p. 10; Wolffe, *Renegade,* pp. 31 and 149.

51. Genevieve Cook journal, 1, 4, 8, 10, 11, 19, 20, 24, 25, 27, and 29 June 1984; DJG interviews with Genevieve Cook, Hasan and Raazia Chandoo, and Beenu Mahmood; Sohale Siddiqi interview with Jim Gilmore. On 640 2nd Street, see James Barron and Peter Baker, "In Brooklyn Brownstone, President Found a Home on the Top Floor," *NYT,* 2 May 2012, and Barron, "Obama? Just the Forgettable 1980s Boyfriend of a Landlord's Tenant," *NYT,* 7 May 2012.

52. Genevieve Cook journal, 30 June 1984, 1, 15, and 25 July 1984, 5, 19 and 22 August 1984, trio of small photo booth pictures, summer 1984, Cook Papers; DJG interviews with Genevieve Cook and Hasan and Raazia Chandoo; Obama, *DFMF,* pp. 128–29. On the film, see Janet Maslin, "Redgrave in James's 'Bostonians,'" *NYT,* 2 August 1984. On the San Ysidro mass murder, see "Coast Man Kills 20 in Rampage at a Restaurant," *NYT,* 19 July 1984, p. A1.

53. Genevieve Cook journal, 22, 24, 27, 30, and 31 August 1984, 1 and 5 September 1984; DJG interviews with Alex McNear, Genevieve Cook, Mike Ramos, Phil Boerner, and Paul Herrmannsfeldt. On Alex's subsequent life, her eventual ex-husband Bob Bozic, and her daughter Vesna, born in 1991, see Nick Paumgarten, "The Ring and the Bar," *New Yorker,* 30 January 2012, pp. 28–34, and Alex Vadukul, "After Last Call," *NYT,* 24 January 2016.

54. *Business International Money Report,* 5 October 1984, p. 314; DJG interviews with Cathy Lazere, Michael Williams, Jeanne Reynolds Schmidt, Maria Stathis Batty, Gary Seidman, Dan Armstrong, Lou Celi, Brent Feigenbaum, Beth Noymer Levine, Lisa Shachtman Hennessey, and Genevieve Cook; Susan Arterian and John Geanuracos e-mails to DJG; Jeanne Reynolds's comments on analyzethis.net, 9 July 2005; Obama, *DFMF,* pp. 135–38, 210–11; Obama in Mendell, *Obama,* p. 62, and in John Corr, "From Mean Streets to Hallowed Halls," *Philadelphia Inquirer,* 27 February 1990, pp. C1, C4; Genevieve Cook journal, 2, 10, and 13 December 1984; Janet Maslin, "Murphy in 'Beverly

Hills Cop,'" *NYT,* 5 December 1984, p. C25. In 2001 Obama would tell Julieanna Richardson that "for a year I worked as a financial journalist to pay off my student loans and as soon as I had those paid off" he sought more appealing work. Brent Feigenbaum would recall beginning at BI with a salary of $19,500 after rejecting Lou Celi's initial offer of $17,500.

Also note "How a Master Plays the Game: The World Bank Approach to FX, Interest Rate Swaps," *BIMR,* 16 November 1984, pp. 361–63, based on an interview with World Bank "swap operations" manager Cyrus Ardalan. The only one of the Financial Action Report series to appear in OCLC is the subsequent one by Cathy A. Lazere and Brent H. Feigenbaum, *Improving Financial Management in France* (Business International, 1985), 128 pp. Also see Business International Research Report, *Kenya: Foreign Investment in a Developing Economy* [Geneva]: Business International S.A., April 1980, 166pp., a document not produced by BI's New York office nor updated in 1984. See as well *Business International Weekly Report,* 27 July, 10 August, 14 September, and 2 and 16 November 1984; *Financing Foreign Operations—Interest Rate & Foreign Exchange Rate Updater,* 15 and 31 August 1984, 14 and 28 September 1984, 15 and 31 October 1984, 27 November 1984, and 11 December 1984; and *The BIMR Handbook on Global Treasury Management* (Business International Corporation, August 1984), a lengthy (227 pp.) topical reprinting of *BIMR*'s weekly articles. No one recalls Barack attending the "BO Annual Picnique August 1984," whose custom T-shirt featured a number of quasi-humorous lines including "If I don't get my own Wang . . ." and "They're no longer with the company."

On David Opiyo's death, see also Auma Obama, *And Then Life Happens,* p. 145; Zeituni Onyango e-mails to DJG; and especially Mark Obama Ndesandjo, *An Obama's Journey,* p. 81.

55. Barack Obama to Genevieve Cook, 1 January 1985, Cook Papers; Ann Dunham notebook, esp. pp. 57, 61, 93, 103, Dunham/Solyom Papers; Scott, *A Singular Woman,* pp. 263–65. On the Terkel book, see Loudon Wainwright, "I Can Remember Every Hour," *NYTBR,* 7 October 1984, pp. 7, 9. Bobby Titcomb would later volunteer that "after high school . . . I had struggles of my own and Barack supported me through those tough times." Emme Tomimbang, "Barack Obama: Hawaii Roots," 8 February 2008. On Adi Sasono and Ann's personal life, see Julia Suryakusuma in Judith Kampfner, "Dreams From My Mother," BBC Radio World Service, 16 September 2009; Scott, *A Singular Woman,* pp. 235–41; Ian Buruma, "A Free Spirit," *NYRB,* 26 May 2011 ("her liaisons with Indonesian men"); and Barack's reference to his mother displaying "a certain recklessness" in Amanda Ripley, "The Story of Barack Obama's Mother," *Time,* 21 April 2008, pp. 36ff.

56. Genevieve Cook journal, 25 January 1985; Obama, *DFMF,* pp. 138–39; Ralph Nader et al., *Action for a Change: A Student's Manual for Public Interest Organizing* (Grossman, 1971), Bob Belfort to TVAJ Campaign Students and Staff, "Campaign Update," 24 January 1985, NYPIRG Papers; DJG interviews with Barack Obama, Arthur Barnes (Obama's description "fits me to a tee"), Eileen Hershenov, Chris Meyer, Tom Wathen, Diana Mitsu Klos, Alison Kelley, Jay Halfon, Neal Rosenstein, Rebecca Weber, and Michael Gecan; Ann F. Lopez e-mail to DJG; Ben Smith, "Becoming Obama," *Politico,* 30 January 2007; Hershenov in Patrick Whelan, *The Catholic Case for Obama* (Catholic Democrats, 2008), p. 51n4; Jason Fink, "Obama's Years in New York Left Lasting Impression on Colleagues," *AM New York,* 9 November 2008; [Rosenstein], "President Barack Obama's Work History as an Organizer with the New York Public Interest Research Group," [January 2010], NYPIRG Papers; course and departmental records, CCNY, spring semester 1985. NYPIRG's 1985 CCNY telephone number, 212-234-1628, remained unchanged even thirty years later. On NYPIRG's TVAJ legislation, see also Robin Topping, "Trying to Quell the DES Nightmare," *Newsday,* 20 January 1985. On Jeffries, see particularly Richard M. Benjamin, "The Bizarre Classroom of Dr. Leonard Jeffries," *Journal of Blacks in Higher Education,* Winter 1993–1994, pp. 91–96, and Kwame Okoampa-Ahoofe Jr., *The New Scapegoats: Colored-On-Black Racism* (iUniverse, 2005), pp. 58–60, 68–71. This author has no memory of such a February 1985 request, but he knows what his response would have been.

57. Genevieve Cook journal, 4 and 5 February, 15 and 31 March, and 22, 23, and 28 April 1985; DJG interviews with Genevieve Cook, Andrew Roth, Keith Patchel, Tim Jessup, and Pete Vayda. The resulting Richard Lloyd album was *Field of Fire.* Phil Boerner attended a reading by well-known poet Allen Ginsberg at Columbia on Tuesday, March 26, and believes it very likely that Barack went with him. See also Ted Kenney, "Beat Poet Ginsberg Returns to Columbia," *CS,* 27 March 1985, pp. 1, 4.

Obama's comments in later years about where he lived in New York City were often highly inaccurate. On one occasion in 2007 he said he once lived in Park Slope, implicitly referencing Genevieve's 2nd Street apartment, and "a little in Brooklyn Heights," perhaps a reference to his stays with Hasan Chandoo. "I don't remember the exact address" but "I subletted for about three months in Brooklyn Heights, near the Promenade." That is inaccurate, but memories of buying bagels and the Sunday *New York Times* near the Clark Street IRT station almost certainly date from times when he stayed with Hasan either at his Eagle Warehouse apartment or later in Brooklyn Heights proper. See Dana Rubenstein, "Barack and Roll: Obamamania Hits Brooklyn Heights," *The Brooklyn Paper,* 28 July 2007; Michael Saul, "He Wants Army of Bams," *NYDN,* 23 August 2007; and "Barack Obama at Cooper Union," audacityofparkslope.org, 27 March 2008. A 2008 Obama campaign statement had Barack in April–May 1985 living in Park Slope and June–July 1985 in Brooklyn Heights, perhaps a mislocating of Genevieve's Boerum Hill apartment on Warren Street. Referencing 622 West 114th, that statement acknowledged "we actually only have 1 actual address, otherwise just streets or neighborhoods," but the site where that information appeared in a thread dated 18 August 2008, www.commongroundpolitics.net, has since disappeared from the Web.

58. Bob Belfort to TVAJ Students and Staff, "Campaign Update," 7 March 1985, Tom Wathen to All Staff, "Staff Meeting," 26 March 1985, Belfort to TVAJ Students and Staff, "Campaign Update," 11 April 1985, NYPIRG Papers; "NYPIRG Opposes Reagan Budget Cuts," *The Paper* [CCNY], 3 May 1985, p. 7; Genevieve Cook journal, 6 May 1985; S.B. 5494, 1 May 1985, NYPIRG Papers; Chris Meyer, "Higher Education in 1985" flyer, Frances Fox Piven Papers, Smith College, Box 85 Fld. 4; Bob Belfort to TVAJ Students and Staff, "Campaign Update," 9 May 1985, NYPIRG Papers; Jeffrey Schmalz, "Filing-Period Extension Expected in Toxic Suits," *NYT,* 8 May 1985, p. B2; NYPIRG CCNY, "A Community Forum: Federal Budget Cuts '85–'86," 8 May 1985, Piven Papers, Box 11 Fld. 5; Jeffrey Schmalz, "Longer Period for Toxic Suits Wins Backing in State Senate," *NYT,* 23 May 1985, p. B2; Gene Russianoff and Peter Skeie, *Back to Go: A Study of Subway Service* (NYPIRG, August 1985), 31pp. (thanking Barack Obama, Alison Kelley, and Diana Klos among many others); Schmalz, "Cuomo Shifts on Toxic-Substance Bill," *NYT,* 10 December 2005, p. B3; "Cuomo Cites Two Bills as Key for Legislature," *NYT,* 7 May 1986, p. B2; Schmalz, "New York Officials Reach Liability Insurance Accord," *NYT,* 20 June 1986, p. B3; Cuomo's 30 July 1986 signing remarks in *Product Safety & Liability Reporter,* 1 August 1986, p. 528; Martha H. Cotiaux, "A Crucial Deadline for New York Toxic Victims," *NYT,* 5 June 1987; Ben Whitford, "Coal and Clear Skies: Obama's Balancing Act," *Plenty Magazine,* 31 October 2008; Fink, "Obama's Years in New York"; [Rosenstein], "Obama's Work History"; DJG interviews with Genevieve Cook, Alison Kelley, Eileen Hershenov, Chris Meyer, and Tom Wathen; Obama, "Community Revitalization, Nebraska Wesleyan University Forum, 9 September 1994; Obama, *DFMF,* p. 139; Obama's 20–21 October 1999 cable television interview with Joe Green; Obama's June 2002 interview with Chinta Strausberg; and Obama's 16–17 August 2003 interview with Frank Avila and Morgan Carter on CAN-TV Channel 19, Chicago.

Obama would later write of traveling to Washington to deliver the CCNY students' letters to New York members of Congress, although no contemporary documents or other possible participants attest to such a trip. See Obama, *TAOH,* p. 43 (misdating it as 1984), and Remnick, *The Bridge,* p. 121. In May 1985 NYPIRG was also mounting a campaign against milk prices in New York City. See "NYPIRG Survey Shows New York City Shoppers Are Paying Premium Milk Prices," 8 May 1985, "Milking Us Dry: A Survey of Milk Prices in New York City," May 1985 (noting that "New York is the only state that limits competition amongst milk distributors"), and Paul Herrrick to NYC Outreach Staff, "Milk Licensing," 13 June 1985, NYPIRG Papers.

59. Genevieve Cook poem, 9 May 1985, Genevieve Cook journal, 13, 14, and 23 May 1985, Cook poem written on PS 133 "Dear Parents" notice, 11 June 1985, on the back side of which is an unsent "Dear Ida" letter, 19 June 1985, Genevieve Cook journal, 20 and 28 June 1985, Cook Papers; DJG interviews with Genevieve Cook.

60. DJG interviews with Genevieve Cook, Andy Roth, Keith Patchel, Jerry Kellman, Hasan and Raazia Chandoo, and Beenu Mahmood; Obama's 2001 interview with Julieanna Richardson; Obama's 7 February 2004 interview with David Axelrod; Obama's 25 June 2005 remarks in "Bound to the Word," *American Libraries,* August 2005, pp. 48–52, at 51; Obama, "A Politics of Conscience," United Church of Christ, Hartford, 23 June 2007; "Barack Obama Revealed," CNN, 20 August 2008; Kellman in Byron York, "The Organizer," *National Review,* 30 June 2008; Obama, *DFMF,* pp. 140–43; Obama in Erin Meyer, "Discovering Hyde Park," *HPH,* 14 February 2007, p. 4; Obama, University of Massachusetts at Boston Commencement Speech, 2 June 2006 ($1,000 to purchase a car for which he paid $800); Kellman in Bob Secter and John McCormick, "Portrait of a Pragmatist," *CT,* 30 March 2007, in Mike Litwin, "Obama's 'Change' Could Be More Than a Coined Phrase," *Rocky Mountain News,* 29 August 2007, in Serge Kovaleski, "In Organizing, Obama Led While Finding His Place," *NYT,* 7 July 2008; Kellman's 24 July 2008 interview with Jim Gilmore; Kellman in Tim Harper, "The Making of a President," *Toronto Star,* 16 August 2008, p. A1; Kellman's 25 August 2008 DNC Speech, Denver; "Barack Obama Tribute" video, DNC, 28 August 2008; Kellman in Jennifer Liberto, "Origin of Obama's Run Is on the South Side," *St. Petersburg Times,* 26 October 2008, p. A1; Kellman on *The Takeaway,* WNYC, 5 November 2008; Genevieve Cook journal, 16 and 26 July 1985.

Eighteen years later, Obama would assert, "By the time I was 20, I was no longer engaged in any of this stuff" and "By the time I was 20, I don't think I indulged again." An accurate answer would have been twenty-three rather than twenty. Bernard Schoenburg, "Painting of Ex-Gov. Ryan," *SJR,* 9 November 2003, p. 17, and Schoenburg, "Frank Talk About Drug Use in Obama's 'Open Book,'" *SJR,* 16 November 2003, p. 17. Four months later Obama would say that he used cocaine "a few times" and in high school and at Oxy "primarily smoked pot"; three months further on he would say "when I was a teenager, 16, 17, 18, I experimented with drugs." "Candidates for U.S. Senate Make Last Push," (AP), *CDH,* 12 March 2004, p. I-12; and *NBC Nightly News,* 28 July 2004. Ryan Lizza would notably write that "While Obama was in search of an authentic African American experience, Kellman was simply in search of an authentic African American." "The Agitator," *TNR,* 19 March 2007, pp. 22ff.

61. Barack Obama to Genevieve Cook, 14 August 1985, Cook Papers; DJG interviews with Genevieve Cook and Robert Elia; Obama, "Why Organize? Problems and Promise in the Inner City," *Illinois Issues,* August–September 1988, pp. 40–42; Obama, *DFMF,* pp. 135–36; Obama Commencement Addresses at the University of Massachusetts at Boston, 2 June 2006, and Northwestern University, 16 June 2006; Obama remarks to College Democrats of America, Columbia, SC, 26 July 2007; Obama remarks at Cornell College, Mt. Vernon, IA, 5 December 2007; Obama Commencement Address, Wesleyan University, 25 May 2008. Cf. Willie Morris, *North Toward Home* (Houghton Mifflin, 1967). Avner Falk, *The Riddle of Barack Obama,* p. 149, terms Obama's move to Chicago "the turning point in his" life.

On Robert Elia, also see Joe Pinchot, "Property Being Stolen, Hermitage Man Says," *Sharon Herald,* 22 November 2006, and his sister's obituary, "Rose Elizabeth O'Hare," *Sharon Herald,* 29 March 2009. On Dr. King's 27 January 1956 experience, see Garrow, *Bearing the Cross,* pp. 56–58. On Coffin Point, see Henry Adams, *The Education of Henry Adams* (Houghton Mifflin, 1918), p. 290; Willie Lee Rose, *Rehearsal for Reconstruction: The Port Royal Experiment* (Bobbs-Merrill, 1964), passim; Michael O'Brien, *Henry Adams & the Southern Question* (University of Georgia Press, 2005), pp. 118–19; Charles R. Babcock, "Soviet Secrets Fed to FBI for More Than 25 Years; Infiltration by Two Brothers Detailed in Book on Dr. King," *WP,* 17 September 1981, p. A1; and DJG, *The FBI and Martin Luther King Jr.: From "SOLO" to Memphis* (W. W. Norton, 1981).

CHAPTER FOUR: TRANSFORMATION AND IDENTITY

1. Barack Obama to Genevieve Cook, 14 August 1985, Cook Papers; Obama in Erin Meyer, "Discovering Hyde Park," *HPH,* 14 February 2007; DJG interviews with Genevieve Cook, Beenu Mahmood, Bill Stenzel, Jerry Kellman, Mike Kruglik, Adrienne Jackson, Asif Agha, Joe Bennett, Tom Kaminski, and John Calicott; Edward T. Chambers with Michael A. Cowan, *Roots for Radicals: Organizing for Power, Action, and Justice* (Continuum, 2005), Chapter 2; Jim Capraro in "Mr. Obama's Neighborhood," Hudson Institute, 1 October 2008, p. 5; Kellman in Richard Wolffe, *Renegade* (Crown, 2009), p. 63, in Jonathan Kaufman, "For Obama, Chicago Days Honed Tactics," *WSJ,* 21 April 2008, p. A1, Kellman, "Training Obama," *Organizing,* 11 September 2009, Kellman interview with Jim Gilmore, in Shira Schoenberg, "Obama Keeps

Making the Rounds," *Concord Monitor,* 20 December 2007, on CNN's "American Morning," 4 February 2008, in Tim Harper, "The Making of a President," *Toronto Star,* 16 August 2008, p. A1, in Sarah Kliff, "Service Changes People's Character," Newsweek.com, 5 September 2008, in "Obama as We Knew Him," *Observer,* 26 October 2008, pp. 4–7, and in Kelly E. Carter, "Leadership Lessons from the Top," *Black Enterprise,* March 2009, p. 105; Obama, *DFMF,* pp. 146–49; Obama, remarks to the Alliance for American Manufacturing, Pittsburgh, PA, 14 April 2008; Roger Simon, "Obama: 'I Have the Potential of Bringing People Together," *Politico,* 8 February 2007; Ryan Lizza, "The Agitator: Barack Obama's Unlikely Political Education," *TNR,* 19 March 2007, pp. 22ff.; Byron York, "The Organizer: What Did Barack Obama Really Do in Chicago?," *National Review,* 30 June 2008; Schopp in Edwina Jones, "From Steel Town to 'Ghost Town,'" M.A. thesis, Loyola University of Chicago, May 1998, p. 45, and also in Angela Bradbery, "Mill Area Knew Good Times, Now It Knows Bad," *CT,* 31 January 1992; Jerry Sullivan, "Paradise Doomed," *Chicago Magazine,* August 1985, pp. 142–45, 160–61; Bill Granger, "Requiem for the Southeast Side," *CT,* 12 April 1992, Ying-Kuang Hsu et al., "Locating and Quantifying PCB Sources in Chicago: Receptor Modeling and Field Sampling," *Environmental Science & Technology* 37 (2003): 681–90; Seung-Muk Yi et al., "Emissions of Polychlorinated Biphenyls (PCBs) from Sludge Drying Beds to the Atmosphere in Chicago," *Chemosphere* 71 (2008): 1028–34; Robert Bergsvik, "SCA Pleads Not Guilty to PCB Exposure Charges," *DC,* 6 June 1985, p. A1; Bergsvik, "Third PCB Monitor Ordered," *DC,* 16 July 1985, p. A1; Bergsvik, "CID Sewage Allowance Vote Due," *DC,* 17 July 1985, p. A1; Bergsvik, "Higher South Deering Cancer Rate Found," *DC,* 18 July 1985, p. A3; Pat Wingert, "Specialist Called in Dump Fire," *CT,* 19 August 1985, p. M3; Ernest G. Barefield to Mike Holewinski, "Landfill at 122nd Street," 20 August 1985, HWP COF Box 9 Fld. 34; Robert Bergsvik, "Toxic Waste Fire Finally Capped," and C. D. Matthews, "Waste Fire Concerns Residents," *DC,* 21 August 1985, pp. A1 and A2; Don B. Gallay to Jesse D. Madison, "122nd & Cottage Grove," 23 August 1985, HWP COF Box 9 Fld. 34; "UNO Schedules Conference," *DC,* 24 August 1985, p. A5; Casey Bukro, "Stubborn Fire Puts Heat on Dump Owner," *CT,* 26 August 1985, p. A1; Bukro, "Fire Fuels Toxic Clean-Up Fight," *CT,* 28 August 1985, p. A3; Robert Bergsvik, "Toxic Waste Cleanup Urged," and "EPA Monitoring Landfill Fire," *DC,* 28 August 1985, p. A1; Jim Quinlan, "U.S. Aid on Dump Fire Asked," *CST,* 28 August 1985, p. 22; Henry Locke, "Increase Efforts to Extinguish Underground Fire," *CD,* 29 August 1985, p. 8; Debbie Nelson, "S.E. Side Dump Cited for Hazards Since '71," *CST,* 4 September 1985, p. 25; Jesse M. Madison to Ernest Barefield, "Status Report—122nd & Cottage Grove Hazardous Waste Site," 11 September 1985, HWP COF Box 9 Fld. 34; *Citizens for a Better Environment, United Neighborhood Organization of Southeast Chicago & Mary Ellen Montes v. William K. Reilly,* 1991 WL 95040 (N. D. Ill.), 24 May 1991 (filed 16 September 1985); "UNO to Meet with Simon," *DC,* 19 September 1985, p. A2; Casey Bukro and Lisa Frazier, "Wastes Buried at Dump May Be Blown Up," *CT,* 11 October 1985, p. 7; "SE Sides Demand EPA Control Toxics," *TNW,* November 1985, p. 11; Harold Henderson, "Don't Dump on Us," *CR,* 23 May 1986, pp. 1, 20–29.

2. "Ice Cream Vendor Slain in Front of Son," *CT,* 7 July 1985, p. A2; William Recktenwald, "Police Hunt Killers of Ice Cream Vendor," *CT,* 8 July 1985, p. 15; Clem Richardson, "Children Mourn Slain Ice Cream Vendor," *CST,* 8 July 1985, pp. 1, 6; William Recktenwald, "Shooting Creates Fear in Vendors," *CT,* 9 July 1985, p. A3; "Obituaries," *CT,* 12 July 1985, p. A7; Wes Smith, "Mortician Promotes His Lively Undertaking," *CT,* 14 July 1985, p. B3; "Suspect Charged in Vendor's Slaying," *CT,* 31 August 1985, p. 5; Linnet Myers, "Son's Testimony Avenges Ice Cream Man's Killing," *CT,* 13 November 1987, p. 10; Rosalind Rossi, "Man Guilty of Killing Ice Cream Driver," *CST,* 13 November 1987, p. 24; Rossi, "Killer of 'Model Citizen' Gets Life Term," *CST,* 12 March 1988, p. 4; *People v. L. C. Riley & Willie Dixon,* 596 N. E. 2d 122 (Ill. App. Ct. 1st Dst.), 22 June 1992; *People v. Riley,* 606 N. E. 2d 1233 (Ill. S.Ct.), 2 December 1992; *People v. Riley,* 622 N. E. 2d 1222 (Ill. S.Ct.), 6 October 1993; *U.S. ex rel. Willie Dixon v. Godinez,* 1994 WL 411374 (N. D. Ill.), 3 August 1994.

Madeline Talbott 1985 and 1986 appointment books, Illinois ACORN Papers (IAP) Box 2 Flds. 25 and 26; Illinois ACORN, "Notes from the Board Meeting," 24 September 1985, IAP Box 1 Fld. 10; Madeline Talbott, "Chicago ACORN Year End/Year Begin Report and Plan," 4 January 1986, IAP Box 2 Fld. 62; Chicago ACORN Board List, January 1986, IAP Box 3 Fld. 54; DJG interviews with Madeline Talbott and Ted

Aranda; Juanita Bratcher, "Ald. Hutchinson Boasts of 'Go-Getting' Ways in 9th," *CD,* 22 June 1985; Perry H. Hutchinson to Maurice Parrish, 12 August 1985, both in HWP FAS Box 13 Fld. 23; Henry Locke, "Official Charges: Prostitution Allowed to Run Rampantly on Michigan Ave. in Roseland," *CD,* 5 September 1985, p. 8; "Hutchinson Clarifies Resolution," *CD,* 26 September 1985, p. 4.

3. DJG interviews with Jerry Kellman, Bob Klonowski, Loretta Augustine-Herron, Yvonne Lloyd, Dan Lee, Cathy Askew, Marlene Dillard, Tom Kaminski, and Eva Sturgies; Augustine-Herron in Nicole Duran, "The New Kid on the Block," *Campaigns & Elections,* June 2007, pp. 52ff., in Mike Littwin, "Obama's 'Change' Could Be More Than a Coined Phrase," *Rocky Mountain News,* 29 August 2007, in Markus Ziener, "Barack Obama in Chicago," *Handelsblatt,* 25 June 2008, in "They Met Obama When," *CT,* 16 November 2008, p. 19, in *Becoming Barack: Evolution of a Leader,* Little Dizzy Home Video, 2009, and in Dahleen Glanton and Katherine Skiba, "Community Organizers Who Taught Obama Get an Invitation to the Inauguration," *CT,* 20 January 2013; Lloyd in Michael Cass, "Obama's Rise Is No Surprise to Nashville Woman Who Worked with Him," *Tennessean,* 5 October 2012; Obama to Genevieve Cook, 14 August 1985, Cook Papers.

4. Auma Obama, *And Then Life Happens* (St. Martin's Press, 2012), pp. 153–71; Auma Obama in *Becoming Barack: Evolution of a Leader,* Little Dizzy Home Video, 2009, in "Obama as We Knew Him," *Observer,* 26 October 2008, pp. 4–7, on *Today,* NBC, 30 April 2012, on *Piers Morgan Tonight,* CNN, 2 May 2012, and on *Starting Point,* CNN, 3 May 2012; DJG interview with Asif Agha; Obama, *DFMF,* pp. 207–10, 211–22, 227; Obama on *All Things Considered,* NPR, 27 July 2004, on *Fresh Air,* 12 August 2004, on *The Oprah Winfrey Show,* 19 January 2005, in Hill Harper, *Letters to a Young Brother* (Gotham Books, 2006), pp. 156–57, in Todd Purdum, "Raising Obama," *Vanity Fair,* March 2008, in Jon Meacham, "Obama on His Parents' Influence" and "What Barack Obama Learned from His Father," *Newsweek,* 22 August 2008, and on *20/20,* ABC, 26 September 2008. Referring to his father's 1971 appearance in Honolulu, Obama told Davies that "Later in life . . . I would realize how much that brief visit and interaction with him altered me in all sorts of ways."

5. Jim Duffy, "Strike: City Teachers Walk Out Again," *Southtown Economist* [SE], 3 September 1985, pp. 1, 2; Linda Lenz and Tim Padgett, "Teachers Strike," *CST,* 3 September 1985, pp. 1, 4; Lenz and Donald M. Schwartz, "Teachers Still Out," *CST,* 4 September 1985, p. 3; "Unconsionable Strike," *CST,* 4 September 1985, p. 33; Dave Roeder, "Politicians Quickly Bicker Over Strike," *SE,* 4 September 1985, pp. 1, 2; Casey Banas and Philip Lenz, "Governor Enters School Fray," *CT,* 4 September 1985, pp. 1, 2; Banas and Jean Davidson, "City Schools Back in Business," *CT,* 5 September 1985, pp. 1, 2; Linda Lenz and Clem Richardson, "Pupils Head Back," *CST,* 5 September 1985, p. 6.

Illinois House of Representatives, 9 April 1985, p. 39 (Giglio's introduction of House Bills 1167 and 1168); Bob Kostanczuk, "4 Giglio Bills Clear First Hurdle," *DC,* 30 April 1985, p. 2; "Support Urged for Jobs Bank," *DC,* 2 July 1985, p. A3; "Computer Job Bank Bill Due," *DC,* 12 July 1985, p. A1; Illinois Senate, 3 July 1985, pp. 37–38; Illinois Public Act 84-0050, signed 19 July 1985; Peggy Shaw, "Gov. Thompson Approves Area Job Program Funds," *The Star* [*TS*] (Chicago Heights), 1 August 1985, p. 1; "Plugging the Gaps," *TS,* 4 August 1985, p. A16; Juanita Bratcher, "Deny Jobs Plan a Political Plum," *CD,* 6 August 1985, p. 3; "Job Bank Office to Open Oct.1 at GSU," *TS,* 19 September 1985, p. 1; "New Regional Office to Help Jobless Find Work Will be Based at Governors State," *TS,* 19 September 1985, Adrienne Bitoy Jackson Papers (ABJP); "State, Religious Leaders to Help Launch Network," *TS,* 26 September 1985, p. A3; "Leaders to Dedicate Job Bank Sept. 30," *Chicago Catholic* [CC], 27 September 1985, p. 5; "Employment 'Network' Starts Tomorrow," *TS,* 29 September 1985, p. A1; Leo T. Mahon, "From the Pastor's Desk," *St. Victor Parish Sunday Bulletin* [*SVPSB*] 29 September and 6 October 1985, p. 2; C. D. Matthews, "Mood Upbeat at Jobs Rally," *DC,* 1 October 1985, p. A1; Ken Brucks and Sharon Jacobson to All CHD Funded Groups, "Memorandum," 1 October 1985, IAP Box 3 Fld. 55; Thom Gibbons, "Leaders Launch Employment Service," *TS,* 3 October 1985, pp. A1, A4; "Jobless Effort Gets Top Church Support, Praise," *CC,* 4 October 1985, p. 6; "Funds to be Available," *DC,* 5 October 1985, p. A7; Sharon Jacobson to Voice of the People et al., "The Bridge to a Better Future Program," 7 October 1985, IAP Box 3 Fld. 55; "Area Gets New Job-Assessment Program," *Kankakee Sunday Journal,* 13 October 1985, p. 23; "GSU Prof to Direct Job Finding Network," *TS,* 17 October 1985, p. A5; "Employment Network Update," *SVPSB,* 20 and 27 October 1985, p. 4; David A. Gorak, "New Jobs Network to Target Hard-Hit Southern Suburbs," *CCB,* 21 October 1985; "GSU Prof Heads New Jobs Network," *DC,* 23 October 1985, p. A11; "6 Self-Help Groups Get Grants," *CT,* 28 October 1985, p. 7 (Obama); "Skills Bank Receives Grant," *DC,* 29 October 1985, p. A1; George H. Ryan, "Build Illinois: Abundant Opportunities for Local Governments," *Illinois Municipal Review,* November 1985, pp. 11–12; Timothy Auer, "Local Self-Help Groups Get $185,000 from CHD," *CC,* 1 November 1985, pp. 1, 28; Bob Kostanczuk, "Unemployed May Find Hope Through Network," *DC,* 4 November 1985, p. A1; "Research Associates Join GSU Staff," *DC,* 7 November 1985, p. B3; "Data Bank," *SVPSB,* 10 November 1985, p. 5; Juanita Bratcher, "Rice Raps Libby on Plan to Close S. Side Plant," *CD,* 18 November 1985, p. 13; "New Job Program Draws 86 Hopefuls," *Hammond Times,* 18 November 1985; Bob Kostanczuk, "Assessment Program Successful," *DC,* 22 November 1985, p. A3; "HACO Receives Grant from Thornton Township," *Standard Newspaper,* 23 November–7 December 1985, ABJP; "Church Leads in Bringing New Hope to Steel Depressed Calumet Area," *Illlinois Lutheran,* November–December 1985, p. 4; Violet T. Czachorski, "Background on CCRC," *Hegewisch News,* 12 December 1985, p. 1; Tom Joyce, "1985 Social Development Grants Completed," n.d., p. 67, TJP; "The Spotlight Is On . . . Gloria Boyda!" *SVPSB,* 16 February 1986, p. 6; Obama, *DFMF,* pp. 151–55; Thomas C. Knutson, "In Memoriam: Joseph Cardinal Bernardin," *Let's Talk* 2 #2, Easter 1997; Trevor Jensen, "Rev. Paul E. Erickson, Ex-Lutheran Bishop," *CT,* 5 April 2007; Mike Kruglik on *All Things Considered,* NPR, 6 September 2008; Julie Ratey, "Interview with Senator Barack Obama," *Catholic Digest,* October 2008; Obama, Notre Dame Commencement Address, 17 May 2009, *Public Papers of the Presidents* 2009 Vol. I, p. 661 (recalling Bernardin "speaking at one of the first organizing meetings I attended"); Patricia Zapor, "Obama Cites Influence of Cardinal Bernardin," Catholic News Service, 2 July 2009; DJG interviews with Fred Simari, Jerry Kellman, Mary Bernstein, Mike Kruglik, Tom Knutson, Loretta Augustine-Herron, Bob Klonowski, Jan Poledziewski, Len Dubi, Maury Richards, John Calicott, Dan Lee, Tommy West, Paul Burak, Tom Joyce, Ken Brucks, Mary Yu, and Sharon Jacobson. See also "6 Community Groups Get Self-Help Grants," *CT,* 23 May 1985, p. B12; Mary Yu, "The Center Stage," *Upturn,* June–July 1985, pp. 13–14; "CHD, $88 Million Later, to Mark 15th Anniversary," *CC,* 26 July–2 August 1985, p. 13; Suzanne Carter, "CHD-Backed Project Shows Upbeat Side of City Housing," and Peter Kelly, "One Effective Self-Help Effort Very Often Leads to Another," *CC,* 9–16 August 1985, pp. 20, 21; and Robert L. Johnston, "CHD Arrives!" *CC,* 23–30 August 1985, pp. 3, 28. Within two years of Bernardin's remarks at the rally, two of the most progressive priests of the Chicago archdiocese would publicly express their disappointment with him. Leo's friend Jack Egan remarked that "I've never met a more calculating man" and Andrew Greeley would decry Bernardin's "timorous and equivocal leadership style." Egan in Roy Larson, "The Pope's Man in the Middle," and Greeley, "The Fall of an Archdiocese," *Chicago Magazine,* September 1987, pp. 122–27, at 124, and 128–31, 190–92 at 192.

6. Frank Lumpkin et al., "A Plan to 'Nationalize' Steel," *CT,* 11 August 1985, p. P14; Larry Galica, "Peurala Chides Parton 'Giveaways,'" *DC,* 20 August 1985, p. A3; Mary Schmich, "A Union Leader Who Wasn't Made of Steel," *CT,* 25 August 1985; Frank Lumpkin in U. S. House of Representatives, Committee on Education and Labor, Subcommittee on Employment Opportunities, *Hearing: The Income and Jobs Action Act of 1985,* 99th Cong., 1st sess., Serial 99-26, 4 September 1985, pp. 94–96; "Atty. Gen. Files Suit Against USS for No Rail Mill," *DC,* 21 September 1985, p. A3; Larry Galica, "Local 1033 President No Supporter of 1984 Steel Co. Merger," *DC,* 27 September 1985, p. A1; Maury Richards, "We Gave LTV Every Chance," *1033 News & Views,* October 1985, p. 3; Maria Cerda to Alton Miller, "Monthly Press Activity Reports," 1 October 1985, HWP DS Box 5 Fld. 16; Larry Galica, "Ex-Wisconsin Employees Picket at Downtown Sites," *DC,* 5 October 1985, p. 1, FLP Box 5 Fld. 1; Juanita Bratcher, "Coalition to Fight Foreclosures," *CD,* 29 October 1985, p. 3 (including Frank Lumpkin and Alice Peurala); "Dedication Scheduled," *DC,* 14 November 1985, p. A1; Larry Galica, "LTV Pension Probe Requested," *DC,* 25 November 1985, p. B5; *1033 News & Views,* December 1985, p. 1; Robert Mier et al., "Strategic Planning and the Pursuit of Reform, Economic Development, and Equity," *Journal of the American Planning Association* 52 (Summer 1986): 299–309. Identical copies of Lumpkin's *CT* letter previously had appeared as "Nationalized Steel Would Help

Nation and the Community," *CST,* 15 July 1985, p. 22, and "Nationalize Steel Mills," *DC,* 24 July 1985, p. A4.

7. Barack Obama to Genevieve Cook, 3 September [sic] 1985 ("9/3" but postmarked 3 October 1985), Cook Papers; John Rockwell, "Bonnie Raitt Sings at the Ritz," *NYT,* 18 August 1985, p. 61; Bill Stenzel, "Rights and Reality," *Upturn,* February–March 1987, p. 7; Bill Stenzel, "Dean's Role—Pro-Active," *Upturn,* December 1987–January 1988, pp. 10, 14; Bonnie Nitsche in *Holy Rosary Parish 1992–2008,* pp. 41–42; Loretta Augustine-Herron in Serge Kovaleski, "In Organizing, Obama Led While Finding His Place," *NYT,* 7 July 2008; Bill Stenzel in Patrick Whelan, *The Catholic Case for Obama* (Catholic Democrats, October 2008), pp. 9–11; Stenzel and Jerry Kellman in Eli Saslow, "Obama's Path to Faith Was Eclectic," *WP,* 18 January 2009, p. A18; DJG interviews with Genevieve Cook, Fred Simari, Jerry Kellman, Mike Kruglik, Mary Bernstein, Bill Stenzel, Bonnie and Wally Nitsche, Betty and Selmo "Lucky" Garrett Jr., Jan Poledziewski, Bob Klonowski, Loretta Augustine-Herron, Nadyne Griffin, and Tom Kaminski; Bob Rakow, "Retired Priest Made All Feel at Ease," *CST,* 20 February 2011; Patrick Whelen and William Stenzel, "Chicago Priest, Hired Barack Obama as Organizer, Dies at 70," *NCR,* 28 February 2011; Byron York, "The Organizer," *National Review,* 30 June 2008.

8. "'Council Wars' at Mann Park," *Hegewisch News,* 17 October 1985, p. 1; "UNO Banquet Set," *DC,* 29 October 1985, p. A1; Larry Galica, "Reverend Subject of Protest Rally" and "UNO Issues Awards," *DC,* 2 November 1985, p. A3; Robert Bergsvik, "Vet to Oppose Panayotovich in 35th District," *DC,* 12 November 1985, p. A1; Judy Keane, "Alinsky Leftovers," *DC,* 14 November 1985, p. A4; Larry Galica, "Radical Tactics Chided," *DC,* 21 November 1985, p. A1; Galica, "UNO Criticized at Public Forum," *DC,* 23 November 1985, p. A3; Judy Keane, "The Truth Finally," *DC,* 4 December 1985, p. A4; Don Terry and Andrew Bagnato, "Thompson to Release Cash for Schools," *CT,* 27 November 1985; "Public Officials Join Dump Fight" and Violet Czachorski, "Mayor Is Silent," *Hegewisch News,* 22 August 1985, p. 1; Czachorski, "Groups Unite Against Plan," *HN,* 26 September 1985, p. 1; Robert Bergsvik, "Protest Fails to Obtain Commitment Against More New Landfill Locations" and "Sanitary Landfill Permits Approved," *DC,* 5 October 1985, pp. A1, A3; "Residents on the Stump Against More Dumps," *HN,* 10 October 1985, p. 1; James E. Landing, "Landing Tallies Objections," *HN,* 14 November 1985, pp. 1, 5; Mary Ellen Montes, "Hazardous Dumps," *Save Our Jobs News* [#2], December 1985, p. 3, FLP Box 4 Fld. 2; Robert Bergsvik, "Ban on Landfills May Be Extended," *DC,* 4 December 1985, p. A3; Gertrude T. Smith to Harold Washington, 7 December 1985, HWP COF Box 22 Fld. 62; Casey Bukro, "Agency Cools Its Efforts on Overly Warm Landfill," *CT,* 13 December 1985, p. A3; "Simon Plans Meeting with EPA," *DC,* 21 December 1985, p. A3; John B. W. Corey to Benjamin Reyes, "Estimate of Cost for a Water Main to Serve Homes on 134th Place," 3 October 1984, HWP DS Box 16 Fld. 31; Lester S. Dickinson to Kari Moe et al., "Meeting Concerning Environmental Health and Safety Problems," 17 September 1985, HWP COF Box 9 Fld. 34; Jean Davidson, "Families So Close Yet So Far from Solution to Tainted Water," *CT,* 1 December 1985, p. C1; Juanita Bratcher, "Residents Fight Water Blues," *CD,* 3 December 1985, pp. 1, 18; Mayor's Press Office News Release, 24 December 1986, HWP Box 53.

9. Obama to Phil Boerner, 20 November 1985, Boerner Papers, Obama, *DFMF,* pp. 159–62, 170–71, 175–76, 178–79 (calling Cathy "Mary"); Obama Interview with Cathleen Falsani, 27 March 2004; Obama Commencement Address, Northwestern University, 16 June 2006; Obama, remarks at James C. Wright Middle School, Madison, WI, 4 November 2009, *Public Papers of the Presidents* 2009 Vol. II, p. 1636; Peter Slevin, "For Clinton and Obama, a Common Ideological Touchstone," *WP,* 25 March 2007, p. A1; DJG interviews with Asif Agha, Jerry Kellman, Loretta Augustine-Herron, Yvonne Lloyd, Dan Lee, Cathy Askew, Stephanie Askew, Betty Garrett, Tom Kaminski, and Marlene Dillard.

10. "Mayor Appoints Task Force on Steel," 1 June 1984, and "Chicago's Task Force on Steel Industry," 30 October 1984, HWP POS Box 15, Fld. 109; "Steel and Southeast Chicago: Reasons and Opportunities for Industrial Renewal," Northwestern University Center for Urban Affairs, November 1985, 354 pp.; Robert Bergsvik, "Chicago Steel Report Is Released," *DC,* 4 December 1985, p. A3; "Good News, Bad News on Steel Job Scene," *TNW,* January–February 1986, p. 6; Ann R. Markusen, "National Policies Cause Local Steel Decline," *TNW,* March 1986, pp. 3–4; Markusen, "Planning for Industrial Decline: Lessons from Steel Communities," *Journal of Planning Education and Research* 7

(1988): 173–84, esp. 181; Markusen, "City on the Skids," *CR*, 23 November 1989; David Ranney, *Global Decisions, Local Collisions* (Temple University Press, 2003), pp. 108–20, esp. 111 and 118; Pierre Clavel and Sara O'Neill-Kohl, "Losing Out on Industrial Policy: The Chicago Case," City & Regional Planning Working Paper, Cornell University, January 2010, esp. pp. 21–24. See also Robert Giloth and Josh Lerner to Kari Moe, "Task Force on Steel and Southeast Chicago, 23 May 1986, HWP CSS Box 25 Fld. 51; Rob Mier to Harold Washington, "Task Force on Steel and Southeast Chicago," 27 May 1986, Rob Mier to Al Miller, "Release of Steel Task Force Pre-Publication Report," 18 July 1986, and Rob Mier to Brenda Gaines, "Steel Task Force Pre-Publication Report," 25 July 1986, all HWP DS Box 8 Fld. 29; and Wim Wiewel and Nicholas C. Rieser, "The Limits of Progressive Municipal Economic Development: Job Creation in Chicago," *Community Development Journal* 24 (April 1989): 111–19.

11. Gerald S. Kellman and Maurice Richards to Harold Washington, 19 December 1985, HWP FAS Box 13 Fld. 69; Richard Foster, "Local 1033 Joining with Community & Religious Organizations to Stop Plant Closings," *1033 News & Views*, January 1986, p. 1; Rob Mier to Harold Washington, "LTV Steel Potential Closure" and "LTV Steel Fact Sheet," 2 January 1986, HWP DS Box 8 Fld. 29 and FAS Box 13 Fld. 69; Laurie P. Cohen and Thomas F. O'Boyle, "Ill Fated Merger: LTV, Dragged Down by Steel Subsidiary, Struggles to Survive," *WSJ*, 6 January 1986, p. 1; Merrill Goozner, "Ailing LTV May Close Huge South Side Mill," *CCB*, 13 January 1986, p. 1; Rob Mier's 1989 interview with Betty Brown-Chappell; Obama, *DFMF*, pp. 150, 168–69; DJG interviews with Maury Richards, Jerry Kellman, Loretta Augustine-Herron, and Marlene Dillard. On Rob Mier, see also Mier et al., *Social Justice and Local Policy Development* (Sage Publications, 1993), his interview in Norman Krumholz and Pierre Clavel, eds., *Reinventing Cities* (Temple University Press, 1994), pp. 68–82; and Robert Giloth and Wim Wiewel, "Equity Development in Chicago: Robert Mier's Ideas and Practice," *Economic Development Quarterly* 10 (August 1996): 204–16.

12. Obama to Phil Boerner, 20 November 1985 (noting he planned to be in D.C. December 20–24), Boerner Papers; Obama to Genevieve Cook, 5 January 1986, Cook Papers; Obama, *DFMF*, p. 204, 261–67; Tom Maliti (AP), "Obama's Brother Chooses Life in the Slow Lane," 26 October 2004; DJG interviews with Tom Kaminski, Zeituni Onyango, Beenu Mahmood, Hasan and Raazia Chandoo, and Genevieve Cook. On Abon'go Malik, see also Marshall Sella, "The Audacity of Bro," *GQ*, July 2013.

13. Thomas M. Burton and Dean Baquet, "FBI Mole Gave Alderman Cash Campaign Donation," *CT*, 26 December 1985; Manuel Galvan and Burton, "Aldermen, Contractor Feel Heat," *CT*, 28 December 1985; E. R. Shipp, "'Council Wars' at an End, Says Chicago Mayor," *NYT*, 7 January 1986; Burton, "FBI Mole Gave Ald. Kelley Cash," *CT*, 14 January 1986; Maurice Possley and Douglas Frantz, "FBI Phoned in Plays to Mole at Meetings," *CT*, 16 January 1986; Frantz and John O'Brien, "Waste Firm Lobbyist Opened Doors for FBI Mole," *CT*, 22 January 1986, pp. 1, 2; Irene Benjamin to Harold Washington, 27 January 1986, HWP COF Box 22 Fld. 62; Frantz, "FBI Mole Says He Tried to Buy 'Veto-Proof' Council," *CT*, 2 February 1986, p. 16; Frantz and Mark Eissman, "City Counsel Disqualifies Himself as Payoff Prober," *CT*, 6 February 1986, pp. 1, 7; James Strong and Burton, "Washington Aide Coffey Quits City Hall Post," *CT*, 8 February 1986; John Camper and John Kass, "Scandal Gets Too Big for Mayor to Ignore," *CT*, 9 February 1986, pp. 1, 18; Camper and Strong, "Probe Puts Buffer on Council Wars," *CT*, 19 February 1986; Baquet and Strong, "City's Counsel Quits," *CT*, 20 February 1986, p. 1; John Camper, "City Hall Probe Puts Squeeze on Washington's Inner Circle," *CT*, 21 February 1986, p. 1; Ray Gibson, "City Garbage Disposal Contract Probed by Federal Grand Jury," *CT*, 3 March 1986; Bob Wiedrich and Baquet, "Lobbyist Paid for City Aide's Vacation," *CT*, 6 April 1986, p. 1; Baquet and Wiedrich, "Claim of 'Finder's Fee' for Ald. Kelley Probed," *CT*, 11 April 1986, p. 1; Fran Spielman, "'Dumb' Mole Gave Me $28,500—Alderman," *CST*, 2 May 1986, pp. 1, 14; Ray Gibson, "Alderman Tells of $28,500 Gift," *CT*, 3 May 1986, p. 5; Baquet and William Gaines, "Waste Firm Courted Clout," *CT*, 18 May 1986, p. 1; Baquet and Gaines, "Waste Management Hauls Loads of Clout," *CT*, 16 June 1986, p. 1; Possley, "Mole Bribes of Officials Confirmed," *CT*, 7 October 1986, p. 1; Burton and O'Brien, "City Hall Bribe Probers to Name Names," *CT*, 21 November 1986, p. 1; Burton and Possley, "7 Indicted in City Hall Probe," *CT*, 22 November 1986, p. 1; Andrew H. Malcolm, "2 City Council Members

Among 7 Indicted in Chicago Inquiry," *NYT,* 22 November 1986; Baquet and Gaines, "Chicago Clout Feeds Waste Empire," *CT,* 23 November 1986, p. 2; Possley, "Mole's Taping of Ald. Kelley Disclosed," *CT,* 27 November 1986, p. 1; Possley, "Ex-Lobbyist Claims He Bribed Aldermen," *CT,* 4 February 1987; Baquet, "Probe Looks Higher in Waste Firm Bribe," *CT,* 9 February 1987; Possley and Robert Davis, "Ex-Ald. Kelley to Plead Guilty," *CT,* 24 April 1987; William B. Crawford Jr., "Ex-Ald. Kelley Gets 1-Year Prison Term," *CT,* 12 June 1987; Crawford, "Jury Sees How Mole Ran Sting," *CT,* 22 September 1987; Crawford, "$5,000 Bribe Shown on Tape," *CT,* 23 September 1987; Possley, "Incubator Figure Caught in Middle," *CT,* 28 October 1987; Crawford, "Ex-Lobbyist Sentenced in Bribes to Former Aldermen," *CT,* 15 March 1988; Joseph P. Fried, "Michael Burnett, 67, Criminal Who Exposed City Corruption," *NYT,* 2 November 1996.

Brian Kelly, "Harold Washington's Balancing Act," *Chicago Magazine,* April 1985, pp. 180–83, 200–207; Melvin G. Holli and Paul M. Green, *Bashing Chicago Traditions: Harold Washington's Last Campaign* (Wm. B. Eerdmans, 1989), pp. vii–viii, 19–22, 27–32, esp. 22; *Ketchum v. Byrne,* 740 F. 2d 1398, 7th Cir., 1984, cert. denied, 3 June 1985, 471 U.S. 1135; James Strong, "Council Factions Huddle to Draw Ward Boundaries," *CT,* 25 October 1985, pp. 1, 12; Maurice Possley and Manuel Garcia, "Ward Remap Deal Could Put Minorities in Control," *CT,* 30 October 1985, p. 1; Possley and Galvan, "Ward Remappers Draw Marzullo's Wrath," *CT,* 31 October 1985, p. 1; Possley, "Ward Remap in Judge's Hands," *CT,* 1 November 1985, p. 2; Strong and William B. Crawford Jr., "Final Draft of Ward Remap Drawn," *CT,* 8 November 1985, p. 1; Hanke Gratteau, "Negotiations on Remap Boil Down to 3 Wards," *CT,* 4 November 1985, p. 2; *Ketchum v. City Council,* 630 F. Supp. 551 (N. D. Ill.), 30 December 1985; Robert Davis and Galvan, "'86 Vote Ordered for Remap Wards," *CT,* 31 December 1985, p. 1; Strong and Davis, "Remap Crusader New City Counsel," *CT,* 1 March 1986, p. 1; Jeffrey D. Colman and Michael T. Brody, "*Ketchum v. Byrne:* The Hard Lessons of Discriminatory Redistricting in Chicago," *Chicago-Kent Law Review* 64 (1988): 497–530; and Jud Miner's 1990 interview with Betty Brown-Chappell. See also Robert J. McClory, "Washington at Midterm: A Look at the Record," *Illinois Issues,* January 1985, pp. 20–24.

14. Cohen and O'Boyle, "Ill-Fated Merger," *WSJ,* 6 January 1986, p. 1; Goozner, "Ailing LTV May Close Huge South Side Mill," *CCB,* 13 January 1986, p. 1; "Report: LTV Closing?," *DC,* 13 January 1986, p. A2; Larry Galica, "1986: The Year They Loved to Hate," *DC,* 31 December 1986, p. A2; "Churches Ask Steel Study," *CT,* 15 January 1986, p. B1; "Rep. Stresses Reinvestment for LTV Steel," *DC,* 15 January 1986, p. A3; IAF to CHD, "Minority Organizer Training Program," 12 July 1985, IAFP Box 162 Fld. 1686; "Minority Organizers Training," 14–16 January 1986, IAFP Box 163 Fld. 1693; Sharon A. Jacobson to Renee Brereton, 13 May 1986, IAFP Box 162 Fld. 1678; Matt O'Connor, "LTV to Idle 775 in Chicago," *CT,* 22 January 1986, p. B1; Barry Cronin, "LTV Laying Off 775 Here," *CST,* 22 January 1986, p. 4; "LTV Closing Unit," *NYT,* 22 January 1986, p. D5; Larry Galica, "LTV Announces 775 Layoffs," *DC,* 22 January 1986, p. A1; Galica, "Laid-Off LTV Employees to Be Given Job Assistance," *DC,* 23 January 1986; Frank Lumpkin to Crossroads Fund, 23 January 1986, CFP Box 38 Fld. 472; "LTV Rally Postponed," *DC,* 25 January 1986, p. A1; Mark Hornung, "LTV Slashes 775 Jobs at Huge Chicago Works," *CCB,* 27 January 1986, p. 49; Galica, "Caucus Hears LTV's Woes," *DC,* 30 January 1986, FLP Box 5 Fld. 6; Mike Parker, "Keep Union United in Quitting QWL Program," *Labor Notes,* February 1986, pp. 10, 15; Janis Parker, "Job Seekers to Receive Assistance Via Network," *DC,* 1 February 1986, p. A5; "Skills Assessment Offered at College," *South Suburban Citizen,* 7 February 1986, ABJP; "1033 Announces Candidate Endorsements," *DC,* 14 February 1986, p. A4; Galica, "Jobless Help Available," and "LTV-Local 1033 Contract Talks to Resume," *DC,* 15 February 1986, p. A1; "Reaction Mixed to Job Network," *DC,* 19 February 1986; John Conroy, "The Silence in Steel City," *CR,* 21 February 1986, pp. 1, 18–30; "Plant Closing Law Is Discussed," *DC,* 27 February 1986, p. A1; "Save LTV Jobs! Public Ownership the Only Way," *Save Our Jobs News,* March 1986, p. 3, FLP Box 5 Fld. 5; *1033 News & Views,* March 1986; Galica, "Wis. Steel's Sudden Closing Spurs Proposal," *DC,* 3 March 1986, p. A1; Galica, "Bill Designed to Regulate Closures Opposed by Some," *DC,* 4 March 1986, p. A3.

Obama, *DFMF,* p. 172, 231–33; Gloria Jackson Bacon's 2002 interview with Julieanna Richardson; DJG interviews with Loretta Augustine-Herron, Yvonne Lloyd, Dominic Carmon, Alma Jones, Hazel and Cheryl Johnson, Gloria Jackson Bacon, James V. Jordan, Jerry

Kellman, and Mike Kruglik; Loretta Augustine-Herron in Anthony Painter, *Barack Obama: The Movement for Change* (Arcadia Books, 2009), p. 101; Altgeld Gardens Residents to Harold Washington, 20 July 1985, HWP COF Box 1 Fld. 20; Jean Davidson, "Child-Parent Centers Helping Needy Families Learn Together," *CT,* 13 October 1985, p. 1; Sandra Crockett, "Altgeld Fears Toxic Threat," *CD,* 22 January 1986, p. 1; Crockett, "Move to Ease Altgeld's Waste Scare," *CD,* 13 February 1986, p. 3; Illinois Environmental Protection Agency, "The Southeast Chicago Study: An Assessment of Environmental Pollution and Public Health Impacts," March 1986; Casey Bukro, "Southeast Side Cancer Death Rates High," *CT,* 15 March 1986, p. 5; Dean Congbalay, "Health Study's Bad News Is No News to Southeast Side Groups," *CT,* 17 March 1986, p. C5; Sharon Pines, "Minorities, Others Organize Against Toxic Threat," *TNW,* April 1986, pp. 5–6; Harold Henderson, "Don't Dump on Us," *CR,* 23 May 1986, pp. 1, 20–29. See also Lisa Newman, "Pioneer CHA Clinic Gets $500,000 Grant," *CT,* 7 October 1991; Deborah Pinkney, "Altgeld Tenants Sick of Bad Air," *CST,* 10 July 1993, p. 3; Joyce Kelly, "CHA Project House Call," *CT,* 25 June 1995; and Kim Geiger, "Obama Tour Covers President's South Side Past," *CT,* 11 November 2013.

15. Obama to Andrew Roth, 27 February 1986, Roth Papers; Obama to Roy and Mary Obama, 17 March 1986, Abon'go Malik Obama Papers; Obama to Phil Boerner, 19 March 1986, Boerner Papers; Obama, *DFMF,* pp. 187–88; Obama on *The Friday Night Show* with Bob Sirott, WTTW.com, 3 December 2004; Loretta Augustine in Bob Secter and John McCormick, "Portrait of a Pragmatist," *CT,* 30 March 2007, and in Sasha Abramsky, *Inside Obama's Brain* (Portfolio, 2010), p. 213; Yvonne Lloyd in Janice Malone, "Nashville Resident Recalls Senator Obama Always Destined for Success," *Tennessee Tribune,* 17–23 January 2008, p. A1; DJG interviews with Asif Agha, Jerry Kellman, Mary Bernstein, Loretta Augustine-Herron, Yvonne Lloyd, Marlene Dillard, Nadyne Griffin, Tom Kaminski, Andrew Roth, and Phil Boerner.

16. Madeline Talbott 1986 appointment book, IAP 2011-111 Box 2 Fld. 26; Renee Brereton, "CHD Proposal Evaluation Form—The Developing Communities Project," #86-D-02-02, 7 March 1986, Bernardin Papers; "CHD Allocation Process Rolling," *CC,* 21 March 1986, p. 5; Mary Yu to Cardinal Bernardin, "Recommendations for Funding—Nationally Funded Proposals," 16 April 1986, Sharon Jacobson and Father Ken Brucks to Bishop Rodriguez and Rev. Frank Kane, "Recommendations for Funding—Nationally Funded Grants," 16 April 1986, Jacobson to CHD Allocation Committee of Washington Funds, "Our Recommendations to National CHD," 17 April 1986, Jacobson to Renee Brereton, "Diocesan Recommendations for National CHD Funding of Chicago Community Organizations," 17 April 1986, Cardinal Bernardin to Mary Yu, "Recommendations for Nationally Funded Proposals," 22 April 1986, Rev. Alfred LoPinto to Joseph Cardinal Bernardin, 29 May 1986, all Bernardin Papers; Industrial Areas Foundation, "Notice: July 1986 National Training," 3 April 1986, IAFP Box 73 Fld. 882; Viki Cucciardo (CHD D.C.) to Patricia DeMaria (IAF), "Recruits for IAF Minority Training—Los Angeles," 29 April 1986, and Sharon A. Jacobson to Renee Brereton, 13 May 1986, IAF Box 162 Fld. 1678; IAF to CHD, "Minority Training Program," n.d., IAFP Box 162 Fld. 1689; Rey Flores, "The Catholic Campaign for Human Development: Reform or Bust," *Crisis Magazine,* 9 November 2011; Billy Hallowell, "New Book's Ironic Claim: Catholic Church Paid to Send Obama to an Alinsky-Founded Group's Community Organizing Training," The Blaze.com, 23 July 2012; Phyllis Schlafly and George Neumayr, *No Higher Power: Obama's War on Religious Freedom* (Regnery Publishing, 2012), pp. 194–97, 200–201, 204, and esp. 80 (citing "a source who once had access to copies of documents" from the Bernardin Papers) and 205 (thanking Rey Flores); William S. McKersie, "Fostering Community Participation to Influence Public Policy: Lessons from the Woods Fund of Chicago, 1987–1993," *Nonprofit and Voluntary Sector Quarterly* 26 (March 1997): 11–26, esp. 15–17; Jean Rudd in McKersie, "Strategic Philanthropy and Local Public Policy," Ph.D. dissertation, University of Chicago, March 1998, esp. pp. 343 and 349; DJG interviews with Jean Rudd, Ken Rolling, Rochelle Davis, Madeline Talbott, Renee Brereton, Sharon Jacobson, Mary Yu, and Ken Brucks. See also Jean Rudd, "How One Foundation Came to Support Community Organizing," in Sandy O'Donnell et al., *Promising Practices in Revenue Generation for Community Organizing* (Center for Community Change, October 2005).

17. Manuel Galvan and Robert Davis, "26th Ward Race Mirrors State of City Politics," *CT,* 12 March 1986; Davis and Galvan, "Mayor's Council Gains May Be Hollow Victory," John

Camper, "Mayor, Vrdolyak Show Their Muscle," and Tim Franklin, "5 Chicago Legislators Rejected," *CT*, 20 March 1986, pp. 1, 2; Galvan and Mark Eissman, "Mayor's Man Takes Lead in 26th," *CT*, 21 March 1986; *Torres v. Board of Election Commissioners*, 492 N.E. 2d 539 (Ill. App. 1st Dst.), 14 April 1986; Andy Knott and Eissman, "Back to the Streets in the 26th Ward," *CT*, 18 April 1986, p. 1; Davis and Eissman, "Council Balance Hinges on 2 Runoff Elections," and "Heavies to Turn Out for 26th Ward Rematch," *CT*, 20 April 1986; Davis and Galvan, "Vote Theft Charged in 26th Ward Race," *CT*, 22 April 1986; Davis and Mark Zambrano, "Mayor Hopes 15, 26 Add Up to Power," *CT*, 27 April 1986; Eissman, "Campaigns Go Door-to-Door as Dust Settles in 26th Ward," *CT*, 28 April 1986; Eissman, "Races May Hinge on Absentees," *CT*, 29 April 1986; Davis and Galvan, "Mayor Feels the Power Shifting His Way," and Jack Houston and Zambrano, "Election Watchers Marshal Forces as Ward Runoffs Go to the Voters," *CT*, 30 April 1986; Davis and Galvan, "Mayor the Winner in Aldermanic Elections," *CT*, 1 May 1986; Cecilia Cummings, "Mayor Explores Vrdolyak Territory," *CST*, 24 June 1986; Hawking, "Political Education in the Harold Washington Movement," pp. 164–68.

Kevin Reese, "Superfund Urged for Jobless," *DC*, 22 March 1986, p. A2; Matt O'Connor, "Union's Pay Cuts in LTV Pact Are Either Too Much or Too Little," *CT*, 24 March 1986, p. 4; Delbert C. Augustson, "Thanks to Officials Who Didn't Help Out," *DC*, 26 March 1986, p. A4; Larry Galica, "Vote Deadline on LTV Pact Near," *DC*, 28 March 1986, p. A1; Jim Masters, "Wis. Steel Workers Hear Encouraging Words," *DC*, 1 April 1986, p. A3; James Warren, "LTV Pact Ratified; Pay, Benefits to Be Slashed," *CT*, 5 April 1986, p. 6; Maury Richards, "New Contract Ratified," *1033 News & Views*, April 1986; Tim Offutt to Frank Lumpkin, 15 April 1986, CFP Box 38 Fld. 472; *1033 News & Views*, May 1986, p. 8.

18. Sandra Crockett, "A Tale of Two Parks: The White One Green, The Black One Dirty," *CD*, 18 December 1985, p. 18; Juanita Bratcher, "S. Side Residents Charge Favoritism in Parks, *CD*, 4 March 1986, p. 7; [Friends of the Parks], untitled list, n.d., HWP COS Box 24A Fld. 4; Michael Orenstein, "Park Projects Bring Communities Together," *TNW*, April–May 1989, pp. 3–4, 21; John Owens in Linda Matchan, "A Law Review Breakthrough," *BG*, 15 February 1990, p. 29, in Letta Tayler and Keith Herbert, "Obama Forged Path as Chicago Community Organizer," *Newsday*, 2 March 2008, p. A6, on *Tell Me More*, NPR, 23 July 2008, and in Bill Glauber, "Chicago Neighborhood Watching President-Elect Closely," *MJS*, 12 January 2009; DJG interviews with Eva Sturgies, Nadyne Griffin, Aletha Strong (Gibson), and John Owens; Owens in Lisa Lerer, "Obama Decades-Old Shooting Scare Guides Anti-Poverty Plan," Bloomberg, 19 February 2013; Maraniss, *BOTS*, pp. 550–51. See also Tom Brune, ed., *Neglected Neighborhoods: Patterns of Discrimination in Chicago City Services* (Chicago Reporter, 1981). One Saturday evening in April or early May, Barack, apparently accompanied by one of the DCP ladies ("Ruby"), attended a performance of Ntozake Shange's *For Colored Girls Who Have Considered Suicide/When the Rainbow Is Enuf* at the Edgewater Presbyterian Church in far-north Chicago. See Richard Christiansen, "Cast Lends a Fierce Beauty to Shange's 'Colored Girls,'" *CT*, 3 April 1986, Obama, *DFMF*, pp. 204–6, and Philip Weiss, "Obama's Dark Side," MondoWeiss.net, 22 April 2008. No surviving DCP member has any recollection of accompanying Barack to the play.

19. Philip J. O'Connor, "$3.9 Million Joyce Gifts to 84 Groups," *CST*, 8 May 1986, p. 54; Obama to Genevieve Cook, 14 August 1985, Cook Papers; [Tom Joyce], "Report of the Justice & Peace Committee—1980–1986," 9 May 1986, and Rev. Thomas Joyce to Rev. Martin Kirk, "Recommendations for Grants from the Claretian Social Development Fund," 10 May 1986, TJP; "Draft of Letter from Bishop Gaughan to Pastors," n.d., Bernardin Papers; Jerry Rodell, "South Suburbs—Alive and Well," *Upturn*, October–November 1987, p. 7; Claretian Social Development Fund 1974–1993: Report to the Eastern Province Assembly from the Peace and Justice Committee, June 1993, pp. 13–14, TJP; Kellman in John B. Judis, "Creation Myth," *TNR*, 10 September 2008, pp. 18ff.; Gregory A. Galluzzo, "Gamaliel and the Barack Obama Connection," *Organizing* #1, December 2008, p. 3 (updated early 2009); DJG interviews with Jerry Kellman, Tom Joyce, Leo Mahon, Greg Galluzzo, Tom Knutson, Len Dubi, Bob Klonowski, Mike Kruglik, Adrienne Jackson, Mary Bernstein, Mary Gonzales, Peter Martinez, Lynn Feekin, Nancy Jones, Kerry Taylor, and Fred Simari. See also Greg LeRoy, "Report on the Activities of the Calumet Project for Industrial Jobs, April 1–September 30, 1984," RPPP Box 21; Bruce Nissen, "Union Battles Against Plant Closings," *Policy Studies Journal*

18 (Winter 1989–90): 382–95; Lynn Feekin and Nissen, "Early Warnings of Plant Closings," *Labor Studies Journal* 16 (Winter 1991): 20–33; and Nissen, *Fighting for Jobs: Case Studies of Labor*-Community *Coalitions Confronting Plant Closings* (State University of New York Press, 1995), esp. p. 7. By happenstance, a very low-visibility "South Suburban Action Coalition on Health, Unemployment and Low Income" had briefly existed eighteen months earlier. See Bob Kostanczuk, "Unemployment More Serious in Cal City," *DC*, 8 October 1984, p. A1.

"College Sponsors Luncheon for Local Business Leaders," *DC*, 6 March 1986, p. A9; "Skills Assessment Sessions Slated," *DC*, 13 March 1986, p. A3; "University Data Bank User Recommends Service Highly," *DC*, 26 March 1986, p. A10; "Jobs Skills Assessment Regional Employment Network at St. Victor Jubilee House April 28th and 30th," *SVPSB*, 13, 20, and 27 April 1986, pp. 9, 6, 5; *Inscapes* (GSU Office of University Relations), 9 May 1986; "Sessions Set on Job Skills," *DC*, 9 May 1986, p. A1; "Skill Assessment Session for Jobless," *The Star*, 15 May 1986, p. A2; "Open Skill Assessment Centers," *The Star*, 18 May 1986, p. A7; "Regional Employment Network Reports Initial Success," *The Star*, 22 May 1986, p. C1; C. D. Matthews, "Area Employment Network Gathering Data," *DC*, 22 May 1986, p. A8; "Job Skills Site Opens," *DC*, 29 May 1986, p. A3; "Designate Three Sites for Skill Assessment Tests," *The Star*, 29 May 1986, p. A8; *Inscapes*, 4 June 1986; "Job Skill Interviews Are Slated," *DC*, 1 July 1986, p. A1; "Job Help," *DC*, 16 July 1986, p. A2; "GSU Hosts Job Interviews," *DC*, 23 July 1986; *Inscapes*, 23 November 1987, p. 3; Obama, *DFMF*, pp. 167–68. See also Julie S. Putterman, *Chicago Steelworkers: The Cost of Unemployment* (Steelworkers Research Project, January 1985), pp. 48, 53, 58.

20. Chicago Housing Authority Ad, "Specification No. 8632," *CST*, 14 April 1986, HWP DS Box 1 Fld. 10; Mrs. Callie Smith et al. to Zirl Smith, 17 May 1986, Mrs. Callie Smith et al. to Harold Washington, 20 May 1986, J. D. Stubblefield (Air Quality Testing) to James W. Wesley (CHA), "Altgeld Gardens/Murray Homes . . . Identification & Evaluation of Asbestos," 27 May 1986, all HWP DS Box 1 Fld. 10; "Altgeld Tenants Suspect Asbestos in CHA Units," *CD*, 28 May 1986, p. 8; "Asbestos at Altgeld Gardens," "Possibility of Asbestos," and "Conduct Tests for Presence of Asbestos," WBBM Newsradio, 28 May 1986, 6:35 A.M., 3:25 P.M., and 4:35 P.M. transcripts, HWP COS Box 22 Fld. 9; John Blake, "Asbestos Test Worries Altgeld Gardens Residents," *CT*, 29 May 1986, p. C7; Marilynn Marchione, "Asbestos Peril Reported in Ida B. Wells Homes," *CST*, 29 May 1986, p. 12; "Presence of Asbestos," WBBM TV News, 29 May 1986, 10 P.M. transcript, HWP COS Box 22 Fld. 9; CHA Press Statement, 30 May 1986, HWP DS Box 1 Fld. 10; "Independent Firm to Do Testing," WBBM TV News, 30 May 1986, 10 P.M. transcript, HWP COS Box 22 Fld. 9; Martha Allen, "Asbestos in CHA Apartments Poses Possible Health Hazards," *Chicago Reporter*, June 1986, pp. 1–4; "CHA Hiring Firms to Test for Asbestos," *CT*, 2 June 1986, p. C5; "Asbestos Danger," WBBM TV News, 2 June 1986, 6 P.M. transcript, and "Emergency Asbestos Investigation," WBBM TV News, 2 June 1986, 10 P.M. transcript, HWP COS Box 22 Fld. 9; Cheryl Devall, "Asbestos Tests Slated for Wells Homes," *CT*, 3 June 1986, p. C3; "Smith Takes the Heat" and "Asbestos a Health Hazard," WBBM Newsradio, 3 June 1986, 5:58 A.M. and 6:30 A.M. transcripts, "Testing Begun" and "Lab Testing Begins," WBBM TV News, 4 June 1986, 5 P.M. and 9 P.M. transcripts, all HWP COS Box 22 Fld.; Harry Golden Jr., "Mayor's Aid Asked to Remove Asbestos," *CST*, 5 June 1986, p. 24; Chinta Strausberg, "Altgeld Residents Seek Meeting with Mayor on Asbestos," *CD*, 5 June 1986, p. 3; Rob Karwath and Dave Schneiderman, "Poisonings Prompt Tests Near Plant," *CT*, 5 June 1986, p. 1; Zirl S. Smith to Renault A. Robinson et al., "Status of Asbestos, Lead Based Paint, PCB and Other Toxic Substances," 5 June 1986, HPW DS Box 1 Fld. 10 and COS Box 22 Fld. 9; Obama, *DFMF*, pp. 235–42; Callie Smith in Peter Wallsten, "Fellow Activists Say Obama's Memoir Has Too Many I's," *LAT*, 19 February 2007; Patrick Reardon, "Obama's Chicago," *CT*, 25 June 2008, p. T1; Linda Randle in Jen Christensen and Matt Hoye, "Slow Change Frustrated Young Obama, Friends Say," CNN .com, 18 August 2008, and in "Barack Obama Revealed," CNN, 20 August 2008, Adrienne P. Samuels, "Barack's Chicago," *Ebony*, January 2009; *Becoming Barack: Evolution of a Leader*, Little Dizzy Home Video, 2009 (brief footage of the 4 June meeting with Evans); "Evangeline Katherine Irving," *CT*, 26 January 2012; DJG interviews with Linda Randle, Mike Kruglik, Adrienne Jackson, Alma Jones, Hazel and Cheryl Johnson, Martha Kindred, Callie Smith, Loretta Augustine-Herron, Yvonne Lloyd, Ben Reyes, Luz Martinez, Hal Baron, Jane Ramsey, Robert Ginsburg, and Tim Evans.

21. "Asbestos in Third CHA Building," "Public Housing Asbestos Contamination," and "CHA Has Known About Asbestos," WBBM TV News, 5 June 1986, 5 P.M., 6 P.M., and 10 P.M. transcripts, "CHA Residents Joining Forces," WGN TV News, 5 June 1986, 10:30 P.M. transcript, HWP COS Box 22 Fld. 9; "CHA's Chief Flushed Out" and Marilynn Marchione, "Asbestos Tests Set at 2 CHA Properties," *CST,* 6 June 1986, pp. 10 and 20; Rob Karwath, "Lead Poison Tests Set for Firefighters," *CT,* 6 June 1986, p. A3; "Asbestos Removal," WBBM TV News, 6 June 1986, 6 P.M. transcript, HWP COS Box 22 Fld. 9; Rogers Worthington, "Students to Be Tested for Lead," *CT,* 8 June 1986, p. A3; Henry Locke, "Free Health Examinations," *CD,* 9 June 1986, p. 3; "Grave Concerns About Asbestos," and "Crowd Keeps Smith from Speaking," WBBM TV News, 9 June 1986, 6 P.M. and 10 P.M. transcripts; "Stormy Meeting," WLS TV News, 9 June 1986, 10 P.M. transcript; "Residents Demand Asbestos Removal," WMAD TV News, 9 June 1986, 10 P.M. transcript, HWP COS Box 22 Fld.9; *Becoming Barack* footage, 9 June 1986; Marilynn Marchione, "CHA Asbestos Meeting Fizzles," *CST,* 10 June 1986, p. 12; Cheryl Devall, "CHA Director Booed from Talks on Asbestos," *CT,* 10 June 1986, p. C2; John McCarron, "CHA Moves on Asbestos Clean-Up," and John Blake, "300 Line Up for Lead Testing, *CT,* 11 June 1986, pp. C3, A3; Benjamin Reyes to CCRC, 12 June 1986, and Artensa Randolph to Renault A. Robinson et al., "Ad Hoc Asbestos Abatement Committee Meeting June 16, 1986," 18 June 1986, both HWP DS Box 1 Fld. 10; Jess Bravin, "City Reveals Factory Lead Poisoning Cases," *CT,* 18 June 1986, p. A3; Martha Allen, "CHA Begins Asbestos Cleanup at Ida B. Wells," *Chicago Reporter,* July 1986, pp. 6–7; "CHA Removes Asbestos," *DC,* 8 July 1986, p. A2; "CHA Steps Up Efforts to Remove Asbestos," *CST,* 12 July 1986, p. 6; Marilynn Marchione, "Funds Asked in Asbestos Fight," *CST,* 30 July 1986, p. 16; Patrick Berry, "Toxics and Communities' Right to Know," *TNW,* September 1986, pp. 1, 13–15; Obama, *DFMF,* pp. 243–46; Patrick Reardon, "Obama's Chicago," *CT,* 25 June 2008, p. T1; Edward McClelland, *Young Mr. Obama* (Bloomsbury Press, 2010), pp. 45–46, DJG interviews with Loretta Augustine-Herron, Yvonne Lloyd, Alma Jones, Callie Smith, Linda Randle, Hazel and Cheryl Johnson, Martha Kindred, Jerry Kellman, Mike Kruglik, and John Owens. The existing microfilm copy of the *Daily Calumet* omits the entire month of June 1986 as well as other scattered dates throughout 1984–1987.

22. CCRC Invoice to CHD Chicago, 15 May 1986, Obama to Sharon [Jacobson], 20 May 1986, and Sharon Jacobson to Obama, 30 June 1986, Bernardin Papers; Obama to Phil Boerner, 2[9]? May 1986, Boerner Papers; Obama to Genevieve Cook, n.d. (ca. 3 June 1986), Cook Papers; DJG interviews with Sharon Jacobson, Phil Boerner, and Genevieve Cook.

23. DJG interviews with Asif Agha, Sheila Jager, Michael Dees, Loretta Augustine-Herron, Yvonne Lloyd, Marlene Dillard, Maria Cerda, Eva Sturgies, Aletha Strong, Nadyne Griffin, George Galland, Mary Yu, Sharon Jacobson, and Ken Brucks; Obama, *DFMF,* p. 210; Bernd Jager, "Memories and Myths of Evil: A Reflection on the Fall from Paradise," *Collection du CIRP* 1 (2007): 211–30, esp. 215–16; Christian Thiboutot, "Brief Biography of Bernd Jager," *Collection du CIRP* 1 (2007): 288–90; Penny Hastings, "Sonoma County Artists Open Their Houses This Weekend," *San Francisco Chronicle,* 12 October 2001.

Developing Communities Project, "A Proposal to Expand Grass-Roots Organizing Around Employment and Education Issues in the Far South Side of Chicago," n.d. (ca. early October 1986), MAWP; John Camper and Robert Davis, "Mayor's Allies Push Aside City Parks Boss Kelly," *CT,* 17 June 1986, p. 1; Chinta Strausberg, "Mayor Seizes Control of Park Board," *CD,* 17 June 1986, p. 1; Ann Marie Lipinski and Camper, "Kelly Forces Mobilize to Resist Parks Coup," *CT,* 19 June 1986, p. 1; Donald M. Schwartz, "Organization Invited to Air Parks Concerns," *CST,* 3 July 1986, p. 11; Cheryl Devall, "Neglected Parks Get a Look-See," *CT,* 3 July 1986, p. M11; Elaine M. Johnson to Jesse Madison, "Gately Stadium," 3 July 1986, HWP L&I Box 5 Fld. 16; Tom McNamee, "Netsch Takes His Wine and Cheese to the Parks," *CST,* 6 July 1986, p. 1; Camper, "Kelly Ally May Quit Parks Board," *CT,* 10 July 1986, p. M10; "Status of Work Done on List of Requests Submitted by Developing Communities Project as of July 7th, 1986," and "Presentation to the Board of Commissioners of the Chicago Park District," 10 July 1986, Sturgies Papers; Camper, "Vrdolyak's Park Joins Waiting List," *CT,* 11 July 1986, p. 1; Barry Cronin, "Field House OK Delayed," *CST,* 11 July 1986, p. 16; Jerry Schay, "Will Gately Swing Open?," *CT,* 14 July 1986, p. B11; Camper, "Court Reaffirms Parks Takeover," *CT,* 15 July 1986, p. 1; Schwartz, "Gately

Stadium Resurfacing Set," *CST*, 15 July 1986, p. 2; John Hoellen, "Athletic Directors Cheer Gately Resurface Plan," *CT*, 16 July 1986, p. C5; Camper, "Vrdolyak's Ward to Get Fieldhouse," *CT*, 19 July 1986, p. M5; George Papajohn and Steve Neal, "Kelly Surrenders Parks Job," *CT*, 21 July 1986, p. 1; Ann Marie Lipinski, "Rights Lawyer Named Parks Counsel," *CT*, 9 August 1986, p. 5; Johnson to Ernest Barefield et al., "Gately Stadium," 22 July 1986, HWP Legislative Series Box 5 Fld. 16; Abdul Alkalimat, "Mayor Washington's Bid for Re-Election," *Black Scholar*, November–December 1986, pp. 2–13, esp. 8–9; Basil Talbott Jr., "Jesse Madison's Big Bout," *Chicago Magazine*, November 1987, pp. 145, 184–96; David Moberg, "One Year Without Washington," *CR*, 24 November 1988; Obama, *DFMF*, pp. 189, 226–27. See also Megan Pellegrini, "Eva Sturgies, UIC Cornea Service Patient," n.d., www.uic.edu/com/eye/PatientCare/PatientSuccessStories/ and Noreen S. Ahmed-Ullah, "Historic Gately Stadium Re-Opens After Facelift," *CT*, 14 October 2011.

CHD, "Working for Justice . . . Restoring Human Dignity," 23 June 1986, and "CHD Funded Projects 1985–1986," Bernardin Papers; Madeline Talbott 1986 appointment book, IA 2001-111 Box 2 Fld. 26; "Self-Help for Poor Takes in $300,000," *CT*, 24 June 1986, p. C4; "CHD Local Grants to Community Units Total Over $300,000," *CC*, 27 June–4 July 1986, pp. 1, 26; Jim Masters, "Self-Help Groups Given Grants," *DC*, 3 July 1986, p. A2; Obama Southwest Airlines ticket, Bernardin Papers.

24. Ed Chambers, *Organizing for Family and Congregation* (Industrial Areas Foundation, March 1978), esp. pp. 4, 14, 18–19, 21, 32; Donald C. Reitzes and Dietrich C. Rietzes, "Alinsky's Legacy: Current Applications and Extensions of His Principles and Strategies," *Research in Social Movements, Conflict and Change* 6 (1984): 31–55; Gregory F. Augustine Pierce, *Activism That Makes Sense: Congregations and Community Organization* (ACTA Publications, 1984); "10-Day National Training, July 12 to 22, 1983," and "IAF National Training July 1985," IAFP Box 74 Fld. 902; IAF to CHD Committee, "Minority Organizers Training Program," 12 July 1985, IAFP Box 162 Fld. 1686; IAF to CHD, "Minority Training Program," 31 July 1986, IAFP Box 163 Fld. 1694; Patricia DeMaria (IAF) to Jim Jennings (CHD), 5 September 1986, IAFP 162-1689; Stephanie Chavez, "Activists Seek to Organize Church-Based Political Coalition," *LAT*, 29 November 1987; "IAF 10-Day Training Schedule," n.d. (ca. 1990–91), Gecan Papers; Mary Beth Rogers, *Cold Anger* (University of North Texas Press, 1990), pp. 180–82; Dolly Merritt, "Dancer Pirouettes Her Way to Famed New York School," *Baltimore Sun*, 14 June 1992; Larry B. McNeil, "The Soft Arts of Organizing," *Social Policy* 26 (Winter 1995): 16–22; Tonya Jameson, "Teenager Dances onto Bigger Stage," *Baltimore Sun*, 18 November 1996; Holly Selby, "On Her Toes," *Baltimore Sun*, 6 April 1997; Michael Gecan, *Going Public: An Organizer's Guide to Citizen Action* (Beacon Press, 2002), esp. pp. 9, 19–20, 36; Edward T. Chambers with Michael A. Cowan, *Roots for Radicals: Organizing for Power, Action, and Justice* (Continuum, 2005), esp. pp. 21, 65, 84; Lonnie Parker O'Neal, "Alicia Graf, Rising to a Greater Height," *WP*, 11 February 2006; Will Englund, "How Would Obama Govern?," *National Journal*, 25 October 2008; Charlotte Allen, "From Little ACORNs, Big Scandals Grow," *Weekly Standard*, 3–10 November 2008; Gecan, *After America's Midlife Crisis* (MIT Press, 2009), p. 9; Ralph Benko, "Dear President Obama: You're No Saul Alinsky," Forbes.com, 16 January 2012; Arnie Graf, "Barack and Mitt, As I Knew Them," *NYDN*, 28 September 2012; Rowenna Davis, "Arnie Graf: The Man Ed Miliband Asked to Rebuild Labour," *Guardian*, 21 November 2012; Owens in Lisa Lerer, "Obama Decades-Old Shooting Scare Guides Anti-Poverty Plan," Bloomberg, 19 February 2013; Mary Riddell, "Power to the People," *Telegraph*, 1 August 2013; Jane Merrick, "Arnie Graf," *Independent*, 10 November 2013; Heather Wood Rudulph, "Get That Life: How I Became a Professional Ballerina," *Cosmopolitan*, 29 December 2014; Aaron Schutz and Mike Miller, eds., *People Power* (Vanderbilt University Press, 2015), pp. 201–14; DJG interviews with Johnnie Owens, Mike Gecan, Charles Payne, Arnie Graf, Greg Galluzzo, Peter Martinez, and Ken Rolling. Joseph Heathcott, "Urban Activism in a Downsizing World: Neighborhood Organizing in Postindustrial Chicago," *City & Community* 4 (September 2005): 277–94, is an especially insightful analysis of Alinsky-style organizing, and Tom Streithorst, "In Search of Obama," *Prospect Magazine*, 21 June 2012, is remarkably acute in its articulation of how Barack's "appearance was different to his identity."

25. Obama airline ticket, Bernardin Papers; Harry Boyte, *The Backyard Revolution: Understanding the New Citizen Movement* (Temple University Press, 1980), pp. 49–51; Robert

Fisher, *Let the People Decide: Neighborhood Organizing in America* (Twayne Publishers, 1984), pp. 48–49; Galica, "USW Loses Rail Mill Suit," *DC,* 29 May 1986, p. A1; Maury Richards, "LTV Files Bankruptcy," *1033 News & Views,* July 1986; Matt O'Connor, "LTV Files for Bankruptcy," *CT,* 18 July 1986, p. 1; Larry Galica, "LTV Files Bankruptcy Petition," *DC,* 18 July 1986, p. A1; Brian Bremner, "Chi. Works' Chances Dim as LTV Files Ch. 11," *CCB,* 21 July 1986, p. 4; Galica, "Union to Fight LTV Benefit Cuts," *DC,* 22 July 1986, p. A1; Galica, "Hospitals Will Treat LTV Retirees," and C. D. Matthews, "Banks Are Cashing LTV Checks," *DC,* 23 July 1986, pp. A1, A2; Galica, "Strike Threat Sparks South Works Idling," and "Union, LTV Eye End to Insurance Problem," *DC,* 24 July 1986, p. A1; Galica, "U.S. Steel Preparing for Strike," and "Employees to Picket LTV," *DC,* 25 July 1986, pp. A1, A2; Galica, "Protesters Arrested at LTV," *DC,* 26 July 1986, p. A1; Galica, "Steel Struggles Make for Long, Hot Summer," *DC,* 28 July 1986, p. A1; Galica, "Effects of USX Strike Would Be Hard to Feel," *DC,* 28 July 1986, p. B3; Galica, "300 Picket LTV Plant," *DC,* 29 July 1986, p. A1; "LTV Strike Is Over," *DC,* 31 July 1986, p. A1; Galica, "Union Charges USS With 'Lockout,'" and "USW Celebrates LTV Victory," *DC,* 1 August 1986, p. A1; Galica, "Wait Begins for USW," and "Picket Line Spirits High," *DC,* 2 August 1986, pp. A1, A8; Robert Greene (AP), "LTV Announces Layoffs in Chicago, Hammond, Youngstown," 7 August 1986; Robert Mier to Harold Washington, "LTV Chicago Plant Layoffs," 7 August 1986, HWP DS Box 8 Fld. 29; O'Connor, "New Blow to Chicago Steel Mills," *CT,* 8 August 1986, p. 1; David Greising, "Steel Firm to Lay Off 1,650 Here," *CST,* 8 August 1986, p. 1; Galica, "LTV Will Idle 1,650," *DC,* 8 August 1986, p. A1; "S. Side Group Weighs Chi. Works Options," *CCB,* 11 August 1986, p. 70; "Layoffs Far-Reaching," *DC,* 11 August 1986, p. A2; Galica, "Pickets Block Tracks; Police Escort Train," *DC,* 12 August 1986, p. A1; Galica, "Steelworkers Block Tracks," *DC,* 14 August 1986, p. A1; Maria B. Cerda to Benjamin Reyes, "Letter to LTV Steelworkers from Mayor Washington," and Harold Washington, "Dear LTV Worker," 19 August 1986, HWP DS Box 5 Fld. 12; Galica, "Steel Woes Move to Acme," *DC,* 22 August 1986, p. A3; Galica, "LTV Filing Hurts Small Business," *DC,* 25 August 1986, p. A1; USWA Local 1033, Monthly Membership Meeting Minutes, USWA Local 1033 Papers, Box 18, esp. 30 September 1986, p. 4; Maury Richards, "A Day That Will Live in Infamy," *1033 News & Views,* September 1986, p. 1; Jane Slaughter, "Steelworkers Locked Out at USX," *Labor Notes,* September 1986, pp. 1, 14–15; M. W. Newman, "Broken Hearts of Steel," *Chicago Magazine,* September 1986, pp. 188, 191; Galica, "LTV Layoffs to Begin," *DC,* 12 September 1986, p. A1; Lynn Emmerman, "Lives Built on Steel Rust with Idleness," *CT,* 14 September 1986, p. T1; Brian Bremner, "LTV to Upgrade Indiana Harbor, Eyes Chi. Works Sale," *CCB,* 15 September 1986, p. 1; Cerda to Reyes, "Update on LTV Activities," 16 September 1986, HWP DS Box 5 Fld. 12; Galica, "'Super Wednesday' to Help Laid Off Workers," *DC,* 17 September 1986, p. A3; Galica, "LTV Union Kicks Off 'Super Wednesday,'" *DC,* 19 September 1986, p. A3; "LTV Bid for Concessions Rapped," *DC,* 26 September 1986, p. A1; Galica, "Once Proud So. Works Faces Struggle to Avoid Knock-Out," *DC,* 26 September 1986, p. A2; Galica, "So. Works Has Rich History," *DC,* 27 September 1986, p. A2; Galica, "So. Works in Its Twilight," *DC,* 29 September 1986, p. A2; Galica, "No End Seen in USX-USW Struggle," *DC,* 15 October 1986, p. A3; Galica, "1986: The Year They Loved to Hate," and "Labor Leader's Loss Still Is Felt," *DC,* 31 December 1986, p. A2; Gordon L. Clark, "Regulating the Restructuring of the U.S. Steel Industry: Chapter 11 of the Bankruptcy Code and Pension Obligations," *Regional Studies* 25 (April 1991): 135–53, esp. 138–39; James B. Lane, ed., "The Uncertainty of Everyday Life: A Social History of the Calumet Region During the 1980s," *Steel Shavings* 38 (2007): 181–88; DJG interviews with Maury Richards and George Schopp.

26. Robert Bergsvik, "Environmental Report Delayed," *DC,* 18 January 1986, p. A3; Bergsvik, "Air Pollution Monitor at Heg. School Linked into Program," *DC,* 27 January 1986, p. A10; Bergsvik, "Council to Vote on Waste Ban Extension," *DC,* 29 January 1986, p. A3; Bergsvik, "CBE Prepares for Suit," *DC,* 13 February 1986, p. A5; Bergsvik, "Landfill Countdown Is On," *DC,* 29 May 1986, p. A3; *A New Direction for Solid Waste Management in Chicago. Final Report of the Mayor's Task Force on Solid Waste Management,* April 1986, esp. p. 62; Bergsvik, "CID Operations Heightened," *DC,* [5 June 1986], p. 1; Hazel Johnson to Jackie Schad, 6 June 1986, CRP Box 30 Fld. 368; James W. Collum to Hubert W. Holton, "Neighborhood Forums—District 4," 9 June 1986, Tina Vicini to Office of the Mayor's Press Secretary, "Mayor's Forum," 13 June 1986, and Greg Longhini to Dinorah Marquez, "Briefing Material for Mayor Washington's Neighborhood Forum in the South Chicago/South Deering Community," 18 June 1986, all HWP DS Box 1 Fld. 10; Marquez, "Briefing Notes," 24 June 1986, HWP COS Box 44 Fld. 4; News Release, "Mayor Washington Meets Residents from South Chicago, East Side, South Deering, and Hegewisch," 24 June 1986, HWP DS Box 13 Fld. 23; John Kass, "Ailment Makes Mayor Sing Praise Faintly," *CT,* 14 August 1986, p. 3; Gail Weisberg to Anthony C. Gibbs Jr., "Neighborhood Forums," 15 August 1986, HWP CSS Box 22 Fld. 23; Jerry Sullivan, "Of Dumps, Chicago Politics & Herons," *Audubon,* March 1987, pp. 122–26.

27. John Camper and Dean Baquet, "Patronage Accusations Stall Mayor's Job Referral Program," *CT,* 15 December 1985, p. 1; "Developing Communities Project: A Proposal to Expand Grass-Roots Organizing Around Employment and Education Issues in the Far South Side of Chicago," n.d. [ca. early October 1986], MAWP, pp. 3–4; Gamaliel Institute, "Community Participation and Leadership Program," n.d. [ca. March–April 1986], 12pp., and [Gamaliel], "Chicago's Greatest Challenge: Lost Minds, Lost Hopes," n.d. [May 1986], 21pp., Gamaliel Papers; Phil Mullins, "The Name of the Game Is Power," *TNW,* May 1986, pp. 7–8; Divine Word International to Gamaliel Foundation, Statement, n.d. [August 1986], Gamaliel Foundation: An Organizing Institute, "Report," September 1986, 5pp., and Gamaliel Foundation, "Community Participation and Leadership Program," n.d. [ca. September–October 1986], 10pp., Gamaliel Papers; Obama, *DFMF,* pp. 184–86; Yvonne Lloyd in "A Rusty Toyota," *WP,* 11 February 2007, p. B3; David Moberg, "Obama's Third Way," *NHI Shelterforce* #149, Spring 2007; Christine Lagoria, "Where Obama's Journey Began," CBSNews.com, 27 April 2007; David James Smith, "The Ascent of Mr. Charisma," *Sunday Times Magazine,* 23 March 2008, pp. 58ff.; Byron York, "The Organizer," *National Review,* 30 June 2008; "Barack Obama Revealed," CNN, 20 August 2008; McClelland, *Young Mr. Obama,* p. 20; DJG interviews with Loretta Augustine-Herron, Yvonne Lloyd, Tommy West, John Ayers, Mary Ellen Montes, and Barack Obama. On Gamaliel's growth, see Kathryn I. Coughlin's tough-minded "A Fund Development Strategic Plan for the Gamaliel Foundation," M.S. thesis, DePaul University, February 1999, esp. pp. 6–7; Dennis A. Jacobson, *Doing Justice: Congregations and Community Organizing* (Fortress Press, 2001); and Stephen Hart's excellent *Cultural Dilemmas of Progressive Politics: Styles of Engagement Among Grassroots Activists* (University of Chicago Press, 2001), esp. pp. 27–120. On Mary Ellen's role at Fiesta Educativa, later the National Center for Latinos with Disabilities, see "LeFevour Appointed President," *CST,* 10 December 1986; Mike Ervin, "Kill My Program," *CR,* 8 May 1988; Susy Schultz, "City Colleges Name Brady as Chancellor," *CST,* 23 December 1988; and Graciela Kenig, "Hispanic Women Gain Power," *CST,* 4 October 1991.

28. Chicago Phone Directory, 1986, 1987, and 1988; University of Chicago Student Directory, 1986, 1987, and esp. 1988, p. 69; Hyde Park Neighborhood Phone Book, July 1988, esp. p. 47; Abigail Foerstner, "Exhibit Is a 'Who's Who' of 150 Years of Masterworks," *CT,* 3 October 1986, p. 83; Obama to Phil Boerner, 21 October 1986, Boerner Papers; Obama, *DFMF,* pp. 227–28; Ben Calhoun, "The Neighborhood That Launched Barack Obama," WBEZ Radio, 3 July 2008; Kellman in Moberg, "Obama's Third Way," and in John B. Judis, "Creation Myth," *TNR,* 10 September 2008, pp. 18ff.; McClelland, *Young Mr. Obama,* pp. 49–50; DJG interviews with Jerry Kellman, Mike Kruglik, Bill Stenzel, Tom Kaminski, Loretta Augustine-Herron, Dan Lee, Cathy Askew, Asif Agha, Johnnie Owens, Sheila Jager, John Morillo, Andrea Atkin, and Phil Boerner.

29. Leo T. Mahon, "From the Pastor's Desk," *SVPSB,* 2 February 1986, p. 2; *SVPSB,* 15 June 1986, p. 2; Mahon, "From the Pastor's Desk," *SVPSB,* 17 August 1986, pp. 2–3; "Mahon Leaves St. Victor," *DC,* 30 August 1986, p. A8; Mahon, "A Last Letter," *SVPSB,* 21 September 1986, pp. 2–3; "Pastor to Speak," *DC,* 10 December 1986, p. A3; Eloise Valadez, "New Pastor Named for St. Victor," *DC,* 13 December 1986, p. A5; Mahon, "St. Victor Parish," Dan Tomich, "The Caretaker," Gloria Boyda, "Transition," and Tom Cima, "A Tale of Three Sundays," *Upturn,* December 1986–January 1987, pp. 4–8; Karen Galvin, "Communal Reconciliation," *Upturn,* February–March 1987, p. 10; DJG interviews with Leo Mahon, Bill Stenzel, Bernie Pietrzak, Tom Cima, George Schopp, Len Dubi, Fred Simari, Jan Poledziewski, and Jerry Kellman.

On SSAC, see Robert Bergsvik, "Coalition Eyes Loan Practices," *DC,* 18 December 1985, p. A3; "South Suburban Action Conference," *SVPSB,* 24 August 1986, p. 8; John Gallagher, "SSAC Calls for Action on HUD Woes," *DC,*

24 September 1986, p. A1; "Senators' Support Sought," *DC*, 7 October 1986, p. A1; C. D. Matthews, "Cal City Group Addressing Housing Issues," *DC*, 3 November 1986, p. A3; Bergsvik, "Housing Group Protests Lack of HUD Office," *DC*, 18 November 1986, p. A1; Gerald Kellman, SSAC, "CHD Pre-Application Form," 30 November 1986, Bernardin Papers; Andrew Fegelman, "Suburban Group Seeks U.S. Housing Changes," *CT*, 14 January 1987, p. C8; Bergsvik, "Housing Talks 'No Blockbuster,'" *DC*, 16 January 1987, p. A2; Bergsvik, "SSAC Shuns HUD," *DC*, 6 February 1987; Nick Manetas, "Business Group Becoming a Political One," *DC*, 17 March 1987, p. A4; Deborah Snow, "End to 'Dual Housing Market' Urged," *DC*, 15 June 1987, p. A1; Leslie Morse, "SSAC Urges Open Real Estate Lists," *DC*, 30 July 1987, p. A2; John Gallagher, "SSAC Makes Housing Plans Top Priority," *DC*, 30 October 1987, p. 1.

30. "Articles of Incorporation," Developing Communities Project, 26 September 1986, Mary-Ann Wilson, Notes of Meeting with Barack Obama, n.d. [ca. 1 October 1986], 3pp., Obama to Wilson, 6 October 1986, "By Laws of the Developing Communities Project," [6 October 1986], 6pp., "Board of Directors—Developing Communities Project," n.d., "Developing Communities Project: A Proposal to Expand Grass-Roots Organizing Around Employment and Education Issues in the Far South Side of Chicago," n.d. [ca. early October 1986], 10pp., Wilson to Obama, 13 November 1986, Obama to Wilson, 22 November 1986, Form SS-4, Form 2848, "Registration Statement—Charitable Organization," and "Financial Information Form," November 1986, Form 1023, "Application for Recognition of Exemption," 25 November 1986, Wilson to Marlene Dillard, 25 November 1986, Wilson to IRS, 25 November 1986, Wilson to Charitable Trust Division, Illinois Attorney General's Office, 2 December 1986, IRS to DCP, 15 December 1986, Wilson to Obama, 24 December 1986, IRS to Wilson, 16 January 1987, Obama to Wilson, 20 January 1987, Wilson to Obama, 22 January 1987, all Mary-Ann Wilson Papers; Juanita Bratcher, "Grant $7.6 Million to 151 Agencies," *CD*, 23 December 1986, p. 1; Woods Charitable Fund, Report for the Year 1986 [ca. March 1987], esp. pp. 2–5, 20–22; Obama to Anne Hallett, 30 December 1986, Wieboldt Foundation, Report for 1984 and 1985, 23 April 1986, and 1986 Annual Report [March 1987], p. 2, Wieboldt Foundation Papers; Mike Flannery, "Obama's Community Service Called Under Question," WBBM Radio, 20 February 2007; DJG interviews with Greg Galluzzo, Mary Gonzales, Bob Moriarty, Danny Solis, Phil Mullins, Peter Martinez, Mary-Ann Wilson, Jerry Kellman, Mary Bernstein, Marlene Dillard, Dan Lee, Betty Garrett, Loretta Augustine-Herron, Tom Kaminski, Cathy Askew, Yvonne Lloyd, Nadyne Griffin, Aletha Strong, Ann West, Isabella Waller, Deloris Burnam, Ernest Powell, Kathy Kish, Ken Rolling, Jean Rudd, John Owens, and Anne Hallett.

31. Larry Galica, "Waste Management Given OK to Buy SCA Chemical," *DC*, 13 September 1984, p. A1; Thomas Petzinger and Matt Moffett, "Plants That Incinerate Poisonous Wastes Run into a Host of Problems," *WSJ*, 26 August 1985, pp. 1, 12; Linda Lenz, "Bright Elementary Shines Through," *CST*, 20 April 1986; Vi Czachorski, "Get Ready to Fight Dump," *HN*, 2 October 1986, p. 1; "Landfill Battlers Plan Celebration," *DC*, 3 October 1986, p. A3; Casey Bukro, "South Side Facility Burning 'Superwaste,'" *CT*, 25 October 1986, p. 1; Gail Weisberg to Department Heads, "Far South Neighborhood Forum," 27 October 1986, WHP CSS Box 11 Fld. 8 and DS Box 6 Fld. 14; Czachorski, "ERV Gives Hope to HOPE," *HN*, 30 October 1986, p. 1; Bukro, "Industrial Free Increase Urged to Fight Toxins," *CT*, 31 October 1986, p. 1; Bukro, "Waste Fight Turning to Ballot Box," *CT*, 3 November 1986, p. C1; Czachorski, "From the Editor's Scratch Pad," *HN*, 6 November 1986, p. 1; United Neighborhood Organization of Southeast Chicago, Annual Awards Banquet, 7 November 1986, FLP Box 5 Fld. 6; Dean Baquet and William Grimes, "Firm Goes Far with Vrdolyak Name," *CT*, 9 November 1986; p. 1, "Hispanic Hope," *DC*, 11 November 1986, p. A5; Galica, "Steelworkers Face Dim Holiday, and "LTV Pensions in Danger," *DC*, 26 November 1986, p. A1; Weisberg to Department Heads, "South Deering/East Side/Hegewisch Forum," 28 November 1986, HWP CSS Box 22 Fld. 16; "UNO Honors Ortega and Lumpkin," *Save Our Jobs News* 2 #2 (December 1986): 3, FLP Box 5 Fld. 5; Czachorski, "From the Editor's Scratch Pad," *HN*, 4 December 1986, p. 1; Galica, "Wis. Steel Workers Picket Navistar Meeting," *DC*, 10 December 1986, p. A5; Czachorski, "Anti-Dump Petition May End Up in Trash Heap," *HN*, 25 December 1986, p. 1; Galica, "After Five Months, Labor Troubles Move into 1987," *DC*, 30 December 1986, p. A2; Jim Vallette, *Waste Management Inc.: The Greenpeace Report* (Greenpeace USA, May

1987), pp. 15–16; Helen Sundman, "UNO: Taking Organizing to a New Level, or Leaving the Community Behind?," *Chicago Reporter,* May 1994; DJG interviews with Danny Solis, Phil Mullins, Bruce Orenstein, Mary Ellen Montes, Hazel Johnson, and Bob Ginsburg.

32. John McKnight, "John Deare and the Bereavement Counselor," fourth annual E. F. Schumacher Lecture, October 1984, in McKnight, *The Careless Society* (Basic Books, 1995), pp. 3–15; McKnight and John Kretzmann, "Community Organizing in the '80s: Toward a Post-Alinsky Agenda," *Social Policy,* Winter 1984, pp. 15–17; McKnight, "A Reconsideration of the Crisis of the Welfare State," *Social Policy,* Summer 1985, pp. 27–30; Christopher Chandler, "John McKnight," *Chicago Magazine,* November 1985, pp. 202–7; McKnight, "Capturing Local Health Dollars," *CT,* 17 February 1986, p. 11; "Things Go Better with Neighbors—Mary O'Connell Interviews John McKnight," *Salt,* June 1986, pp. 4–11; McKnight, "Looking at Capacity, Not Deficiency," in Marc Lipsitz, ed., *Revitalizing Our Cities: New Approaches to Solving Urban Problems* (Fund for an American Renaissance, 1986), pp. 101–6; McKnight, "Well-Being: The New Threshold to the Old Medicine," *Health Promotion* 1 (1986): 77–80; McKnight, "Regenerating Community," *Social Policy,* Winter 1987, pp. 54–58; Wieboldt Foundation, 1987 Annual Report [March 1988], p. 7; DJG interviews with John McKnight and Ellen Schumer.

33. [Donald R. Moore], *The Bottom Line: Chicago's Failing Schools and How to Save Them* (Designs for Change, January 1985), 126pp.; Jean Latz Griffin and Casey Banas, "Dropout Ills More Severe in Recount," *CT,* 7 February 1985, p. 1; "Dismal School Dropout Study," *CT,* 11 February 1985, p. 18; Banas, "Failures Climb at City Schools," *CT,* 12 March 1985, p. 1; G. Alfred Hess, Jr., *Dropouts from the Chicago Public Schools: An Analysis of the Classes of 1982–1983–1984* (Chicago Panel on Public School Finances, 24 April 1985), 121pp.; "School Dropout Rate Nearly 50%!" *CST,* 24 April 1985, p. 1; Patrick Reardon, "Board Policy Blamed for Dropout Epidemic," *CT,* 25 April 1985, p. 1; Banas, "'Average' High School Below U.S. Par," *CT,* 30 April 1985; Jean Davidson, "Push Must Come from Outside of Schools to Be Felt Within," *CT,* 30 June 1985, pp. D1, D6; "Criminal Behavior Is Reinforced," *CT,* 17 Septenber 1985, p. 10; Reardon and Davidson, "Schools Outclassed in Fight with Failure," *CT,* 30 September 1985, p. 1; Banas, "School Reform Attacked as 'Minimal,'" *CT,* 14 November 1985, p. 1; Banas, "Dropout Theory Takes New Course," *CT,* 20 March 1986, p. 1; Julie Whitmore, "A Year After: Why Reform Failed in State Education," *CCB,* 21 July 1986, p. 3; Hess, "Power to the People," *CCB,* 21 July 1986, p. 12; Griffin, "Parents Often the Pushers," *CT,* 23 September 1986; Hess Jr. et al., "'Where's Room 185?': How Schools Can Reduce Their Dropout Problem," *Education and Urban Society* 19 (May 1987): 330–55; Banas, "8 Schools Found Cutting Class Time," *CT,* 3 December 1986, p. 1; Linda Lenz, "Panel Charges 3 Schools Fake 'Study Halls,'" *CST,* 3 December 1986, p. 3; Juanita Bratcher, "Average Student 'Shortchanged,' Study Reports," *CD,* 3 December 1986, p. 1; William Recktenwald, "Classroom Hours Don't Add Up at City Schools," *CT,* 4 December 1986, p. 1; Bratcher, "Bad Report Irks Byrd," *CD,* 4 December 1986, p. 3; Jack Houston, "Class Time Audit Set," *CT,* 5 December 1986, p. 3; Bratcher, "Dropout Rate Irks Parents," *CD,* 11 December 1986, p. 3; "Panel Investigates High School Drop Out Rates," *CD,* 16 December 1986, p. 11; Banas, "Dropout Crisis Called Even Worse," *CT,* 17 December 1986, p. 1; "How High Schools Cheat Students," *CT,* 17 December 1987, p. 22; "Byrd Defends School Anti-Dropout Role," *CT,* 18 January 1987, p. 2; Robert McClory, "Manford Byrd's Report Card," *Chicago Magazine,* February 1987, pp. 125–31; Donald R. Moore, "Voice and Choice in Chicago," in William H. Clune and John F. Witte, eds., *Choice and Control in American Education, Volume* 2 (Falmer Press, 1990), pp. 153–98, esp. 161; G. Alfred Hess Jr., *School Restructuring, Chicago Style* (Corwin Press, 1991), pp. 10–14; Charles L. Kyle and Edward R. Kantowicz, *Kids First—Primero Los Ninos: Chicago School Reform in the 1980s* (Illinois Issues, 1992), pp. 149–50, 215–16; Hess, "Race and the Liberal Perspective in Chicago School Reform," in Catherine Marshall, ed., *The New Politics of Race and Gender* (Falmer Press, 1994), pp. 85–96, esp. 88; Jon Yates, "G. Alfred Hess Jr.: 1938–2006—Educator Challenged Inequity of Resources in Chicago Schools," *CT,* 30 January 2006; Marilyn Sherman, "Fred Hess Led the Way in School Reform," *Inquiry,* Spring 2006; DJG interview with Don Moore; Linda Lenz, "Don Moore: Reform Leader, LSC Champion," *Catalyst Chicago,* 4 September 2012; Graydon Megan, "Donald Moore, 1942–2012," *CT,* 9 September 2012. See also Hess, "Global Development Training for Village Residents: The Maharashtra Village Development Project," Ph.D. dissertation, 1980, Northwestern University.

34. Chinta Strausberg, "Teachers Leave Pre-School Fearing Asbestos," *CD,* 4 December 1986, p. 3; Donald M. Schwartz, "Schools to Act on Crowding," *CST,* 4 December 1986, p. 64; "Parents Threaten School Boycott," *CT,* 14 December 1986, p. C19; Robert Davis, "Boycott over Asbestos Fears Thins Out Attendance at Preschool," *CT,* 16 December 1986, p. C8; Roger Flaherty, "Hazard Spurs School Boycott," *CST,* 16 December 1986, p. 2; Jean Davidson, "Prove School Safe or Else, Parents Say," *CT,* 17 December 1986, p. C3; Casey Banas, "Alternate Wheatley Site Sought," *CT,* 18 December 1986, p. C2.

Patrick Reardon and Bonita Brodt, "CHA Finances Falling Apart as Quickly as Its Buildings," *CT,* 5 December 1986, p. 1; "A Neighborhood No More" and "Any Day Would Make a Great Moving Day," *CT,* 9 December 1986, p. 18; Stanley Ziemba and R. C. Longworth, "Leadership Battles Keep CHA in the Basement," *CT,* 11 December 1986, p. 1; Reardon and Ziemba, "'New Day' for CHA as Bad as Old One," *CT,* 12 December 1986, p. 1; Ziemba and Davis, "Smith Resigns in CHA Feud," *CT,* 8 January 1987, p. 1; Reardon and Ziemba, "Robinson Quits as CHA Chief," *CT,* 17 January 1987, p. 1; Leslie Baldacci, "New CHA Team Vows to Improve Conditions," *CST,* 30 January 1987, p. 20; Baldacci, "CHA Nearly Loses New $9 Million in Funding," *CST,* 5 February 1987, p. 14; Jerry Thornton, "Asbestos Is Latest Problem for CHA," *CT,* 20 April 1987, p. 7; Joel Kaplan and Ziemba, "Ex-Chief Smith Targeted by HUD," *CT,* 1 May 1987, p. 1; Reardon, "HUD to Deny $82 Million Request, CHA Says," *CT,* 28 May 1987, p. 1; "8.9 Million Cleanup Slated at 5 Projects," *CST,* 14 June 1987, p. 8; Ziemba, "CHA Asbestos Removal Stalled," *CT,* 8 August 1987, p. 5; Jack Houston, "CHA Is Again Seeking Asbestos Bids," *CT,* 9 August 1987, p. C3; Reardon and Thornton, "CHA Asbestos Plan in Danger," *CT,* 13 August 1987, p. 1; Reardon, "CHA Gets $2 Million in Program to Abate Asbestos," *CT,* 25 August 1987, p. 7; Reardon and Ziemba, "Rallying Cry Turns into a Whisper," *CT,* 1 September 1987, p. C1; Reardon and Ziemba, "CHA Tenant Control Project OK'd," *CT,* 23 September 1987, p. C6; Baldacci, "U.S. Extends CHA Asbestos-Pact Deadline," *CST,* 1 January 1988, p. 24; Baldacci, "CHA Asbestos Removal to Start," *CST,* 10 March 1988, p. 10; William E. Schmidt, "Bold Plans for Curing Sick Housing," *NYT,* 11 October 1988; Alton Miller, *Harold Washington: the Mayor, the Man* (Bonus Books, 1989), pp. 306–15, esp. 307. See also Andrew H. Malcolm, "Study Portrays Chicago as a City That's Divided," *NYT,* 5 October 1986, p. 38, and John McCormick, "Chicago Bounces Back," *Newsweek,* 6 October 1986, pp. 42–43.

35. DJG interviews with Asif Agha, Chin See Ming, Sheila Quinlan, Sheila Jager, and Jerry Kellman; Sheila Miyoshi Jager, "Narrating the Nation: Students, Romance and the Politics of Resistance in South Korea," Ph.D. dissertation, University of Chicago, December 1994, p. viii; Florence Hamlish Levinsohn, *Harold Washington: A Political Biography* (Chicago Review Press, 1983); *Kirkus Reviews,* 1 November 1983; Bechetta Jackson, "Florence Hamlish Levinsohn, Writer," *CT,* 30 October 1998; Kellman in Jonathan Kaufman, "For Obama, Chicago Days Honed Tactics," *WSJ,* 21 April 2008, p. A1.

36. DJG interviews with Sheila Jager and Michael Dees. Years later, Barack disagreed with Sheila's account: "it's not accurate that I asked her to marry me." Sheila's younger brother David would later write that their father Bernd believed that humans "live in a dual cosmos in which one's own perspective is always countered and complemented by the different perspective of one's neighbor" and that life was "a never-ending cultural task of building bridges." David Jager, "Conversing with My Father," *Collection du CIRP* 1 (2007): 12–16, at 15–16.

37. "How to Organize Friends and Influence People—Salt Interviews Greg Galluzzo," *Salt,* June 1983, pp. 21–25, esp. 24; Greg Galluzzo 1987 planning diary, Galluzzo Papers [GGP]; Gamaliel Foundation, "Community Participation and Leadership Program," [9 January 1987], Greg Galluzzo to Stewart Wagner, 21 January 1987, Gamaliel Foundation Annual Report: 1986–1987, 10 June 1987, Gamaliel Papers; Galluzzo in Patti Wolter, "Consumer's Guide to Organizer Training," *TNW,* October–November 1991, pp. 16–21, esp. 17, and in Ben Whitford, "Coal and Clear Skies," *Plenty Magazine,* 31 October 2008; Galluzzo, "Gamaliel and the Barack Obama Connection," *Organizing* #1, December 2008, p. 3 (updated 2009); Galluzzo interview with Don Elmer, January 2009; DJG interviews with Greg Galluzzo, Mary Gonzales, Peter Martinez, Kevin Limbeck, Ellen Schumer, and Todd Dietterle; Tina Sfondeles, "Activist Who Taught Obama to Be Community Organizer Is Retiring," *CST,* 2 June 2013. See also Galluzzo, "Church-Based Model of Community Organization," *Upturn,* October–November 1983, pp. 3–4, and Galluzzo, "What Is a Community Organization," *SVPSB,* 9 September 1984, pp. 3–4.

38. Linda Lenz, "Wheatley Parents Resume Boycott," *CST,* 6 January 1987, p. 52; Jon Jeter, "Parent Pullout Set at Wheatley in Asbestos Fight," *CST,* 9 January 1987, p. 28; Lenz, "Wheatley Boycott Ends," *CST,* 21 January 1987, p. 10; untitled photo, *CT,* 22 January 1987, p. C3; People for Community Recovery, January and February 1987 Activity Reports, esp. 19 February, CFP Box 30 Fld. 368; Lucille R. Dobbins to Olga Davidson, 6 February 1987, HWP DS Box 2 Fld. 2; Casey Bukro, "38 Carcinogens Found in Lake Calumet Air," *CT,* 12 February 1987, p. C3; "Waste Management's Earnings Up for Quarter," *DC,* 20 February 1987, p. A2; DJG interviews with Bruce Orenstein, Mary Ellen Montes, and Hazel Johnson.

39. "Julian Bond Narrates 'Eyes on the Prize' with Dr. King as Pivotal Force on WTTW," *CD,* 13 January 1987, p. 15; Jack E. Wilkinson, "An Unvarnished Study of Dr. King," *CD,* 15 January 1987, p. 40; Clifford Terry, "'Prize' Series a Masterful Telling of Painful Times," *CT,* 21 January 1987; DJG, *Bearing the Cross: Martin Luther King, Jr., and the Southern Christian Leadership Conference* (Morrow, 1986), esp. pp. 431–525; Basil Talbott Jr., "Raby's Experience Helps Washington," *CST,* 13 March 1983, p. 20; Robert McClory, "The Activist," *CT Magazine,* 17 April 1983, pp. 26ff.; Manuel Galvan, "Al Raby to Head City Commission on Civil Rights," *CT,* 16 January 1985, p. A3; Wes Smith, "Rush Street Area Trying to Survive Its Popularity," *CT,* 3 October 1985, p. 1; Patrick Reardon, "Too Proud to Trade Welfare for Minimum Wage," *CT,* 1 December 1985, p. D3; Galvan, "An Insider Looks Out for Rights," *CT,* 2 January 1986, p. 1; Chinta Strausberg, "Mayor Names Grimshaw to IGA Position," *CD,* 21 July 1986, p. 20; Wieboldt Foundation 1987 Annual Report [March 1988], p. 15; Wieboldt Foundation 1987 Form 990-PF, Part XVI, 13 May 1988; Grimshaw's 1989 interview with Betty Brown-Chappell; Steve Perkins, "A Personal Remembrance of the Remarkable Life of My Dear Friend Al Raby," 19 June 2004, esp. pp. 5, 23, 29, 32; McClelland, *Young Mr. Obama,* p. 47; John P. Martin, "Sarmina Described as Smart, Fair, Ambitious," *Philadelphia Inquirer,* 19 February 2012; Pauline Dubkin Yearwood, "Long and Winding Road," *Chicago Jewish News,* 16 March 2012; DJG interviews with John McKnight, Gwendolyn LaRoche Rogers, Steve Perkins, Patty Novick, Judy Stevens, Wayne Whalen, Jane Ramsey, and Jacky Grimshaw.

40. Maria B. Cerda, "Update on LTV Activities," 16 September 1986, HWP DS Box 5 Fld. 12; Larry Galica, "LTV Pension Takeover Set," *DC,* 13 January 1987, p. A1; Galica, "LTV Pension Plan Takeover Reaction Mixed," *DC,* 16 January 1987, p. A3; Robert Bergsvik, "LTV Pension Hearing Set for Tomorrow," *DC,* 22 January 1987, p. A3; C. D. Matthews, "Reaction Is Mixed to New USX Pact," *DC,* 24 January 1987, p. A3; Galica, "Labor Dispute Ends," *DC,* 2 February 1987, p. A1; Galica, "Steel Poured at So. Works," *DC,* 17 February 1987, p. A1; Galica, "Steel Supply Workers Waiting," *DC,* 23 February 1987, p. A1; USWA Local 1033, Monthly Membership Meeting Minutes, esp. 25 February 1987, p. 2, and 26 May 1987, p. 6; Galica, "LTV Workers Fight for Pensions," *DC,* 3 March 1987, p. A1; Galica, "Protection of Benefits Eyed," *DC,* 30 March 1987, p. A2; David Moberg, "The Trouble in Steel City," *CR,* 9 April 1987; Galica, "Tentative LTV Agreement Is Reached on Lost Benefits," *DC,* 10 April 1987; "LTV Workers to Seek Federal Aid," 21 April 1987, p. A1; Galica, "LTV Pension Aid Pledged," *DC,* 12 May 1987, p. A1; Galica, "LTV Workers Feel Hassled Over Pensions," *DC,* 15 May 1987, p. A3; "LTV Won't Seek Court Help to Avoid Current Contract," and Galica, "LTV to Honor Current Contract," *DC,* 18 May 1987, p. A1; "LTV Pension Protection Extended 4 Months," *DC,* 21 May 1987, p. A3; Galica, "LTV Talks Shift to Home Front," *DC,* 2 June 1987, p. A1; Galica, "LTV Pact Lacks Local Support," *DC,* 27 June 1987, p. A2; "LTV Pact OK'd," *DC,* 7 August 1987, p. A1; DJG interviews with Maury Richards and Jerry Kellman.

Raymond Coffey and John Camper, "Washington Wins at Wire," and Jon Margolis, "Mayor's Tight Win Just Half the Battle," *CT,* 25 February 1987, p. 1; Gary Rivlin, "Mr. Machine," *CR,* 2 April 1987.

41. DJG interviews with Dan Lee, Adrienne Jackson, and Emil Jones Jr.; Juanita Bratcher, "Educational Potpourri," *CD,* 7 May 1986, p. 4; Manuel Galvan, "Frost Successor Wins Approval," *CT,* 22 November 1986; Galvan and Charles Mount, "Successor to Frost Is In—For Now," and Mount, "Frost Clear to Vote on Successor," *CT,* 27 November 1986; Galvan, "Frost's Successor Sworn In," *CT,* 2 December 1986; Fred Marc Biddle, "10 Vie to Replace Frost as Alderman," *CT,* 19 February 1987; Biddle,

"Mayor May Decide Race in 34th Ward," *CT,* 6 April 1987; Obama, *DFMF,* p. 223; Bernard Schoenberg, "Don't Expect Many Niceties in Keyes-Obama Contest," *SJR,* 22 August 2004, p. 15; Jones in 93rd Illinois General Assembly, 8 November 2004, pp. 11–13, in Christopher Wills and Deanna Bellandi (AP), "Obama Learned Politics Helping the Poor," 20 February 2008, on "The Making of Obama," *Nightline,* ABC, 25 February 2008, in Letta Tayler and Keith Herbert, "Obama Forged Path as Chicago Community Organizer," *Newsday,* 2 March 2008, p. A6, in Abdon M. Pallasch, "Grooming a Politician: Jones Took Obama, the 'Pushy' Organizer, Under His Wing," *CST,* 24 August 2008, p. A14, in "They Met Obama When," *CT,* 16 November 2008, p. 19, and in David Smallwood, "Two Guys Who Laid It on the Line," *N'Digo Profiles,* December 2008, pp. 14–15.

42. Anthony Van Zanten, "Worship at Pullman CRC: An Interview with Rick Williams," *Reformed Worship,* March 1988; Kenan Heise, "Cecil Partee, Pioneer in Politics," *CT,* 17 August 1994; "Cecil Partee, 73, Illinois Politician," *NYT,* 19 August 1994; Williams in Kalari Girtley, "How Obama Got His Start," *HPH,* 14 February 2007, p. 14; DJG interviews with Rick Williams, Adrienne Jackson, Tyrone Partee, John Webster, and Alonzo C. Pruitt. See also "Asque Named New Roseland Director," *CD,* 8 April 1986, p. 2, and Patty Pensa, "Former Roseland YMCA Finds New Life as Affordable Housing," *CT,* 15 January 2010.

43. Larry Galica, "Steelworkers Face Dim Holiday," *DC,* 26 November 1986, p. A1; Beth Donaldson (Housing Commissioner) to Harold Washington, "Reverend Bernard Taylor and the Trinity Roseland Organizations," 13 December 1985, DS Box 8 Fld. 9; Bruce A. Gottschall (NHS) to Benjamin Reyes, "Meeting with City Departments Regarding NHS Roseland Development," 21 October 1985, HWP DSBox 6 Fld. 15; Benjamin Reyes to Harold Washington, "Departmental Monthly Reports," 14 March 1986, HWP DS Box 2 Fld. 18; Minutes, NHS Central Board Meeting, 20 May 1986, and "Calendar of NHS Events," 4 June 1986, HWP DS Box 13 Fld. 27; "Housing Group Gets $200,000 Lift," *CST,* 9 June 1986, p. 16; Celeste Busk, "Roseland Ready for Revitalization," *CST,* 13 June 1986, p. H26; Steve Kerch, "Housing Rehab Plan Set for Troubled Roseland," *CT,* 29 June 1986, p. M1; Lorenzo Ealy (Acting President, GRO) to Jackie Shad, 17 July 1986, CFP Box 17 Fld. 207; Minutes, NHS Central Board Meeting, 16 September and 18 November 1986, NHS Chicago, *Annual Report 1986,* and "NHS Resource Development Campaign '87," 12 March 1987, all HWP DS Box 13, Flds. 26 and 27; M. W. Newman, "Execs Risk Millions on Housing," *CST,* 12 April 1987, pp. 1, 34; DJG interviews with Ellen Benjamin and Salim Al Nurridin. See also William Peterman, *Neighborhood Planning and Community-Based Development* (Sage Publications, 2000), pp. 129–51.

44. Henry Locke, "Petitions Seeks Bus Shelters in Roseland," *CD,* 30 April 1986, p. 8; Madeline Talbott, "Organizing Plan . . . Chicago ACORN, 1 November 1986–30 June 1987," IAP 2001-111 Box 2 Fld. 32; Talbott, "Chicago ACORN Report & Plan, YE/YB '88," 2–3 January 1988, IAP Box 2 Fld. 62; Henry Locke, "City Colleges Open Doors to Train Journeymen Here," *CD,* 30 September 1985, p. 6; Juanita Bratcher, "Ministers Tackle Homeless Problem in Roseland Area," *CD,* 28 January 1987, p. 3; Bratcher, "Ministers Don't Bank on Beverly," *CD,* 31 January 1987, p. 3; Bratcher, "Clerics Picket Beverly Bank," *CD,* 3 February 1987, p. 3; Bratcher, "Ministers Will Bank on Beverly," *CD,* 28 February 1987, p. 1; Bratcher, "Demand Jobs for Youth at Evergreen," *CD,* 7 April 1987; Susan Chandler, "Battling the Color of Money," *CST,* 23 July 1989, p. 55; Chandler, "Bankers Paint Two Pictures of Roseland," *CST,* 24 July 1989, p. 37; Obama, *DFMF,* pp. 159–62; Sampson interview with Julieanne Richardson, Maraniss, *BOTS,* pp. 527–29, 603; DJG interviews with Alonzo Pruitt, Nadyne Griffin, Al Sampson, Alvin Love, and Barack Obama.

45. Richard Philbrick, "Religious News Notes," *CT,* 6 April 1972, p. W10; Jeremiah A. Wright Jr., "Buying Co-Op Food," *CT,* 8 February 1981, p. A2; Sam Smith, "Food Co-ops: Faster, Fresher, Cheaper," *CT,* 14 May 1981, p. A2; Wright, "Black Church Renewal," in Trinity United Church of Christ, *Perspectives: A View from Within. A Compendium Text for Churchwide Study* (Trinity United Church of Christ, [September] 1982), 187pp., pp. 32–42, esp. 35–37; Dempsey J. Travis, *An Autobiography of Black Politics* (Urban Research Press, 1987), pp. 581–83; Roger Wilkins, "Keeping the Faith," *Frontline,* WGBH, 16 June 1987; Daniel Ruth, "Chicago Minister Inspires 'Faith,'" *CST,* 16 June 1987, p. 50; James McBride, "Ghetto," *WP,* 16 June 1987, p. F2; Tracey Robinson, "Community Links Pay Off," *CST,* 6 May 1988, pp. 7–8; Wright, "The Significance of Harold Washington," in Henry J. Young, ed., *The Black Church* *and the Harold Washington Story* (Wyndham Hall Press, 1988), pp. 1–9, esp. p. 2; Audra D. Strong, "Message Remade to Foster Black Faith," *CT,* 26 May 1989; Shirley L. Malcom, "Reclaiming Our Past," *Journal of Negro Education* 59 (Summer 1990): 246–59, at 257; Wright, "A Black Congregation in a White Church: Trinity United Church of Christ, Chicago," in Robert L. Burt, ed., *Good News in Growing Churches* (Pilgrim Press, 1990), pp. 37–63, esp. 43–45; Wright, "An Underground Theology," in Dwight N. Hopkins, ed., *Black Faith and Public Talk: Critical Essays on James H. Cone's Black Theology and Black Power* (Baylor University Press, 1991), pp. 96–102, esp. 96–97, 99–100; "Chicago's New School Boss Starts March 25," *Jet,* 29 January 1981, p. 15; Jini Kilgore Ross, ed., *What Makes You So Strong? Sermons of Joy and Strength from Jeremiah A. Wright, Jr.* (Judson Press, 1993), esp. pp. 97–109; Obama, *DFMF,* pp. 272–74, 280–87; Julia M. Speller, "Unashamedly Black and Unapologetically Christian: One Congregation's Quest for Meaning and Belonging," Ph.D. dissertation, University of Chicago, December 1996, esp. pp. 2, 16–17, 68–69, 77, 151, 157n13, 162, 197–99; Wright's interviews with Julieanna Richardson, 2002, John Kupper, 2004, and Arlen Passa, 2009; Obama in E. Janet Wright, "Obama for Illinois," *Trumpet,* February 2004, pp. 30–32; Julia M. Speller, *Walkin' the Talk: Keeping Faith in Africentric Congregations* (Pilgrim Press, 2005), pp. 72–102; Wright, "Growing the African American Church Through Worship and Preaching," in Carlyle F. Stewart, ed., *Growing the African American Church* (Abingdon Press, 2006), pp. 63–81, esp. 72; Manya A. Brachear and Bob Secter, "Race Is Sensitive Subtext in Campaign," *CT,* 6 February 2007, p. 1; Wright in Erin Meyer, "Where Obama Developed His Audacity of Hope," *HPH,* 14 February 2007, p. 20, and in Ryan Lizza, "The Agitator," *TNR,* 19 March 2007; Obama's 2004 interview with Cathleen Falsani; Obama in Mendell, *Obama: From Promise to Power,* p. 77; Obama in Marni Goldberg, "Obama Urges Democrats to Embrace Faith," *CT,* 29 June 2006, p. 3, in Bob Secter and John McCormick, "Portrait of a Pragmatist," *CT,* 30 March 2007, and in Ariel Sabar, "Barack Obama: Putting Faith Out Front," *CSM,* 16 July 2007, p. 1; Wright on *Bill Moyers Journal,* 25 April 2008; Wright, *A Sankofa Moment: The History of Trinity United Church of Christ* (Saint Paul Press, 2010), esp. pp. 43, 48–55, 61, 64, 72, 91, 94, 144–45, 171; DJG interviews with Adrienne Jackson, Alvin Love, Lacey K. Curry, Jeremiah A. Wright Jr., Donita Powell Greene, and Rick Williams; Carl A. Grant and Shelby J. Grant, *The Moment: Barack Obama, Jeremiah Wright, and the Firestorm at Trinity United Church of Christ* (Rowman & Littlefield, 2013), esp. pp. 94–99, 104–9.

Also see Trevor Jensen, "Rev. Kenneth B. Smith Sr.: 1931–2008," *CT,* 23 January 2008, and Bruce Lambert, "Reuben A. Sheares, 58, a Pastor and a Leader in Church of Christ," *NYT,* 16 July 1992. On the 1981 appointment of Ruth Love rather than Manford Byrd, see Robert McClory, "A Look at Love's Labors," *Illinois Issues,* April 1983, pp. 27–31, Ben Joravsky, "A Kind of Death," *CT,* 9 February 1992, Frances Ratliff Stewart, "The Urban School Superintendent: A Case Study of Ruth Love in Chicago," D.Ed. dissertation, Northern Illinois University, December 1996, and especially her three interviews with Love, Byrd, and *Tribune* education reporter Casey Banas at pp. 217–73 of that volume. On J. H. Jackson and Olivet church, see Clarence Page, "Powerful, Paradoxical," *CT,* 11 November 1983, p. 45, and Jerry Thornton, "Influential Minister Is Honored," *CT,* 19 February 1990, pp. II-1, 4.

46. Robert McClory, "The Holy Terror of St. Sabina's," *CR,* 16 November 1989; McClory, "Black and Catholic Are Joint Ventures at Chicago Parish," *NCR,* 13 March 1998, p. 5; Rob Csillag, "Chicago's Renegade Priest," *Toronto Star,* 8 November 2003; McClory, *Radical Disciple* (Lawrence Hill Books, 2010), esp. p. 139; Evan Osnos, "Father Mike," *TNY,* 29 February 2016; DJG, *Bearing the Cross,* pp. 498–99; DJG interviews with Mike Pfleger, Jeremiah Wright, and Tommy West.

47. Janny Scott, *A Singular Woman* (Riverhead Books, 2011), p. 273, Greg Galluzzo 1987 planning diary, GGP; Perry H. Hutchinson to Harold Washington, 1 December 1986, HWP COF Box 29 Fld. 53; Larry J. B. Houston, "Hits Ald. Hutchinson," *CD,* 19 January 1987, p. 15; Juanita Bratcher, "Ald. Hutchinson Pulls for School Custodian," *CD,* 31 January 1987, p. 6; Hutchinson to Jesse L. Jackson, 11 March 1987, HWP COF Box 29 Fld. 53; Robert Davis, "Hurdles Face 9th Ward Incumbent," *CT,* 16 March 1987; Chinta Strausberg, "Mayor to Dedicate Roseland Job Center," *CD,* 23 March 1987, p. 3; "Mayor's Schedule—Monday, March 23, 1987," "Briefing Notes, Roseland Community Visit," 7pp., "Mayor Washington Remarks," Roseland Vocational Training/Job Placement Assistance

Center Ribboncutting, 23 March 1987, and Rich Redmond, "Roseland Community Visit and Tour," 25 March 1987, all HWP SES Box 11 Fld. 9; People for Community Recovery, "March [1987] Activity Report," CFP Box 30 Fld. 368; "Break Ground in Roseland," *CD,* 1 April 1987, p. 4; Ben Joravsky, "Mystery Zoning," *CR,* 5 May 1988; Obama, *DFMF,* pp. 180–81, 223–26; David L. Scheiner, Hyde Park Associates in Medicine, "To Whom It May Concern," 29 May 2008 (attesting that he has been Obama's physician "since March 23, 1987"); DJG interviews with Sheila Jager, Greg Galluzzo, Johnnie Owens, Loretta Augustine-Herron, Salim Al Nurridin, Marlene Dillard, Adrienne Jackson, Betty Garrett, Eva Sturgies, and Tommy West. See also Joanne Gabbin et al., "Black Rap," *Negro American Literature Forum* 5 (Autumn 1971): 80–84, 108–14 (meaningless student chatter including Salim Al Nurridin); an April 1984 Chicago ACORN sign-in sheet including "Dr. Omar Faruq" as well as Imam S. T. Ibrahim of 33 E. 111th Place, IAP Box 3 Fld. 79; and especially Monica Copeland, "Gangs' Spread Sends Chill Through Far South Side," *CT,* 1 June 1992, citing "Sheikh Al-Faruqi Muhammad, who has lived in the area 20 years" and has fathered twelve children.

48. Larry Galica, "Former Wisc. Steel Workers Planning Rally," *DC,* 26 March 1987, p. A3; Merrill Goozner, "Steel Mill's Ghost Haunts Navistar," *CT,* 29 March 1987, p. B1; Galica, "Wisc. Steel Workers Rally," *DC,* 30 March 1987, p. A1; Galica, "Lawsuit Could Begin in Summer," *DC,* 31 March 1987, p. A3; "Wisc. Steel Wait Is Continuing," *DC,* 14 September 1987, p. A2; "Two Groups Fight New Landfills," *CST,* 3 April 1987, p. 38; Robert Bergsvik, "Waste Permit Applications Almost Ready," *DC,* 13 April 1987, p. A1; Patrick Barry, "Can Garbage Bring Back Steel City Jobs?," *TNW,* May 1987, pp. 15–17; C. D. Matthews, "Waste Reactor to Be Built," *DC,* 2 May 1987, p. A1, Greg Galluzzo 1987 planning piary, GGP; "Harold Washington for Mayor," *CT,* 29 March 1987; "Mayor's Schedule—Tuesday, April 7, 1987," and Rich Redmond, "Altgeld Gardens Campaign Tour—Election Day," 13 April 1987, HWP SES Box 13 Fld. 3; Robert Davis and Terry Wilson, "Washington Sweeps to 2nd Term," and "Washington's Showdown Victory," *CT,* 8 April 1987; Dirk Johnson, "With Victory in Chicago Mayor Finally in Control," *NYT,* 9 April 1987; Gary Rivlin, "Enemies of the People," *CR,* 7 May 1987; Ray Hanania, "Shaw Learns, Listens to Voters," *CST,* 7 September 1987, p. 7; Melvin G. Holli and Paul M. Green, *Bashing Chicago Traditions: Harold Washington's Last Campaign* (Wm. B. Eerdmans, 1989), esp. pp. 111–14; William B. Crawford and Maurice Possley, "Alderman, 19 Others Indicted," *CT,* 14 April 1987; Crawford and Possley, "Ald. Hutchinson Pleads Innocent to Swindle Charges," *CT,* 15 April 1987; Crawford, "U.S. Outlines Its Case Against Ex-Alderman," *CT,* 15 March 1988; Linnet Myers, "Former Alderman Convicted of Fraud," *CT,* 20 May 1988; John Gorman, "Ex-Alderman Gets 11 Years for Fraud," *CT,* 20 August 1988; Gorman, "Ex-Alderman Is Told to Pay Back Bribes," *CT,* 29 September 1989; George Papajohn, "Former Ald. Perry Hutchinson Dies While Serving Prison Term," *CT,* 2 January 1992. On Robert Shaw and his brother William, see especially Ben Joravsky, "By Any Means Necessary," *CR,* 31 October 2002. On Cuyahoga Wrecking, also see Scott Glover and Luisa Yanez, "Wrecking Company Owner to Surrender," *Sun Sentinel,* 21 May 1997.

49. Cheryl Devall, "Status of Black Catholics Slowly Moving Up," *CT,* 20 June 1986; "87 People to Watch in 1987," *CT,* 7 January 1987; Nancy Ryan, "Black Catholic Conference to Study 10 Key Issues," *CT,* 23 January 1987; Devall, "Black Catholics Convene to Reassess Role, Needs," *CT,* 13 March 1987; Devall, "Black Catholics Discuss Role," *CT,* 15 March 1987; Devall, "Black Catholics Hold Congress," *CT,* 22 May 1987; Devall, "Blacks Seek a Role in Catholic Church," *CT,* 29 May 1987; Cathy Green and Monique Irvin, "The National Black Catholic Congress," *In a Word* 5 (July–August 1987): 2–11; Edward K. Braxton, "The National Black Catholic Congress: An Event of the Century," *Josephite Herald,* Summer 1987, as reprinted in *U.S. Catholic Historian* 7 (Spring–Summer 1988): 301–6; Carol Norris Green, "A Postcard from Barack," *Catholic San Francisco,* 30 January 2009, p. 13; Jason Horowitz, "The Catholic Roots of Obama's Activism," *NYT,* 23 March 2014, p. A1; DJG interview with Cynthia Norris.

Gregory A. Galluzzo to Margo Dunlap (Borg-Warner Foundation), 9 February 1987, Galluzzo to Gwendolyn Jordan, 5 May 1987, Galluzzo to Ellen J. Benjamin (Borg-Warner), 16 May 1987, Techny Towers to Gamaliel Foundation, Invoice, 27 May 1987, Gamaliel Foundation, "Support for Community Lead-Organizers-in-Training," n.d. [mid 1987], 13pp., Gamaliel Papers; David T. Kindler, "My Awesome Inaugural Adventure," January 2009,

DTKindler.com; DJG interviews with Margaret Bagby, Linda Randle, Renee Brereton, Greg Galluzzo, Mary Gonzales, Peter Martinez, Phil Mullins, Mike Kruglik, Mary Ellen Montes, and David Kindler.

50. Daniel Lee and Barack H. Obama to Harold Washington, 4 May 1987 (page 2 of the proposal is missing), and Obama to Luz Martinez, 7 May 1987, HWP CSS Box 10 Fld. 17; G. Alfred Hess Jr., "Schools Need Parent Involvement," *CT,* 16 March 1987, p. 10; Jack Houston, "City Class of '85 Nearly Cut in Half by Dropouts," *CT,* 30 April 1987, p. 1; Tracey Robinson and Tim Padgett, "School Boss Sees 'Disaster' in Plans to Scrap Board," and Linda Lenz, "'85 Dropout Rate Topped 50% at 29 City High Schools," *CST,* 22 May 1987, pp. 8 and 22; Casey Banas, "Early Marking Deadline Spurs School Absenteeism," *CT,* 2 June 1987; Hess and James L. Greer, "Bending the Twig: The Elementary Years and Dropout Rates in the Chicago Public Schools," Chicago Panel, 30 July 1987, 70pp.; Jean Latz Griffin, "Dropout Rate Tied to Early Failures," *CT,* 31 July 1987, p. 1; Daniel Lee et al. to Harold Washington, 12 June 1987, HWP COF Box 27 Fld. 12; Delores T. Woods to Jane Ramsey, 29 June 1987, HWP COF Box 27 Fld. 27; Kyle and Kantowicz, *Kids First,* pp. 70, 155, 216–18; DJG interviews with Bruce Orenstein, Danny Solis, Nadyne Griffin, Ann West, Rosa Thomas, Eddie Knox, Luz Martinez, Jacky Grimshaw, Jane Ramsey, and Hal Baron. A quarter century later, Luz Martinez would have no recall of Obama, though she did remember DCP's name. On Kari Moe, see her 1990 interview with Betty Brown-Chappell and her 1994 interview in Krumholz and Clavel, eds., *Reinventing Cities,* pp. 95–100.

On the initial 21 October 1986 Education Summit, which the *Tribune* did not cover, see William S. McKersie, "Philanthropy's Paradox: Chicago School Reform," *Educational Evaluation and Policy Analysis* 15 (Summer 1993): 109–28, at 123, Dan A. Lewis and Kathryn Nakagawa, *Race and Educational Reform in the American Metropolis* (SUNY Press, 1995), p. 82, and Jim Carl, "Harold Washington and Chicago's Schools Between Civil Rights and the Decline of the New Deal Consensus," *History of Education Quarterly* 41 (Fall 2001): 311–43, at 332. The three additional signatories on the 12 June letter were Father Alan Scheible (not "Schieble" as in the letter) of St. Willibrord, who died in 2000, Rev. Samuel Fluker of the House of Inspiration Church of God in Christ, who died in 2006, and a Rev. "John Murphy" of True Right Missionary Baptist Church on 95th Street, whose correct name may have been Rev. Willie J. Murphy and who is also deceased. See Shereice Garrett, *My Life Is in God's Hands* (Trafford, 2012), Chapter One.

51. DJG interviews with Cathy Askew, Stephanie Askew, Dan Lee, Marlene Dillard, Yvonne Lloyd, Loretta Augustine-Herron, Aletha Strong Gibson, Ann West, Betty Garrett, Eva Sturgies, Aurie Pennick, Margaret Bagby, Johnnie Owens, Rosa Thomas, Carolyn Wortham, Ernest Powell, and Emil Jones; Richard J. Kaplan (MacArthur) e-mail to DJG; Gamaliel Foundation Annual Report: 1986–1987, 10 June 1987, p. 5, Gamaliel Papers; Obama to Anne Hallett, 26 January 1988, Wieboldt Foundation Papers; Carolyn Wortham YouTube video, 21 May 2007; Lloyd in Sharon Cohen (AP), "Barack Obama Straddles Different Worlds," 14 December 2007, in Janice Malone, "Nashville Resident Recalls Senator Obama Always Destined for Success," *Tennessee Tribune,* 17–23 January 2008, p. A1, and in Michael Cass, "Obama's Rise Is No Surprise to Nashville Woman Who Worked with Him," *Tennessean,* 5 October 2012; Bagby in Dahleen Glanton and Katherine Skiba, "Community Organizers Who Taught Obama the Ropes," *CT,* 20 January 2013; Edward McClelland, "When Barack Obama Partied," NBCChicago.com, 15 January 2013; Obama, remarks at Brooks College Preparatory Academy, Chicago, 19 February 2015 ("the car didn't heat up real well"). See also Isabel Wilkerson, "The Tallest Fence," *NYT,* 21 June 1992. On Chicago philanthropy, the essential overview source is Robert Matthews Johnson, *The First Charity: How Philanthropy Can Contribute to Democracy in America* (Seven Locks Press, July 2008), esp. p. 84 ("in Chicago . . . there is more understanding and support among funders about organizing than anywhere else").

52. Greg Galluzzo, 1987 planning diary, GGP; Gwendolyn J. Jordan (CRS) to Galluzzo, 1 June 1987, Gamaliel Papers; Obama in Marc Strassman's August 1995 Book Channel interview, in Monica Mitchell, "Son Finds Inspiration in the Dreams of His Father," *HPH,* 23 August 1995, in his 2001 interview with Julieanna Richardson, in Scott Fornek, "'I've Got a Competitive Nature,'" *CST,* 3 October 2004, p. 12, in Ronald Roach, "Obama Rising," *BIHE,* 7 October 2004, pp. 20–23, and on *The Friday Night Show* with Bob Sirott, WTTW.com, 3 December 2004; Obama, *TAOH,* p. 206 ("my resolve to lead a public life"); Obama on CNBC with Tim Russert, 18 November 2006; Obama, remarks at the National Prayer Breakfast, 5 February 2009, *Public Papers of the Presidents* 2009 Vol. I, p. 46; Stanley Ann Dunham, "Application for Passport by Mail," 27 March 1986, DOS FOIA Release; Ann Dunham to Alice Dewey, 3 November 1987, and Ann Dunham résumé, n.d. (ca. 1993), Dunham Papers; S. Ann Dunham, *Surviving Against the Odds* (Duke University Press, 2009), pp. 280, 300; Julia Suryakusuma in Prodita Sabarini, "Ann Dunham Soetoro: Love for Indonesia," *Jakarta Post,* 17 March 2010; Scott, *A Singular Woman,* pp. 267–70; Michael R. Dove, "Dreams from His Mother," *NYT,* 11 August 2009; Dove, book review of *Surviving Against the Odds, Anthropological Quarterly* 83 (Spring 2010): 449–54, esp. 451; Jeff Chang, "Maya Soetoro Ng: Q & A," *Vibe,* September 2007; Dan Boylan, "Obama's Sister Goes Campaigning," *Midweek,* 12 September 2007; Sudhin Thanawala (AP), "Sister: Hawaii's Cultural Mix Laid Basis for Political Career," 14 February 2008; Anna Scott, "Senator's Sister Fills in the Blanks," *Sarasota Herald-Tribune,* 18 July 2008, p. A1; Stuart Coleman, "Tea with Barack Obama's Sister," *Salon,* 23 October 2008; Evan Drellich, "BU Prof Recalls College Road Trips with Obama," *Pipe Dream,* 4 November 2008; John Nichols, "Friend of Barack," *Madison Capital Times,* 14 January 2009; Obama, Q&A with Student Journalists, 27 September 2010, and remarks at DNC Rally, 28 September 2010, *Public Papers of the Presidents* 2010 Volume II, pp. 1435, 1455; Meredith Turits, "Maya Soetoro-Ng," Glamour.com, 20 January 2013; DJG interviews with Greg Galluzzo, Mary Gonzales, Kevin Limbeck, Nancy Jones, Lynn Feekin, Sheila Jager, Tyrone Partee, Asif Agha, Charles Payne, Mike Dees, and Douglas Glick; Maureen O'Donnell and Jon Seidel, "Barack Obama's Great-Uncle Dies at 89," *CST,* 12 August 2014. "Obama's Mother Stayed in Pakistan for 5 Years," *Daily Waqt* (Lahore), 31 August 2008, badly misstates the duration of Ann's 1986–87 work but is nonetheless instructive.

53. People for Community Recovery (PCR), April, May, and June 1987 Activity Reports, and "Grantee Fiscal and Progress Report," n.d. (ca. 30 June 1987), CFP Box 30 Fld. 368; Jim Vallette, *Waste Management, Inc.: The Greenpeace Report* (Greenpeace USA, May 1987); Casey Bukro, "Firm Buying Time for Landfill," *CT,* 28 May 1987, p. C6; Deborah Nelson, "Our Toxic Trap: Crisis on the Far South Side," *CST,* 31 May 1987, p. 1; Nelson, "Waste Pits Poisoning Air, Water," *CST,* 1 June 1987, p. 1; Nelson, "San. Dist. Cancer Rate Is Ignored," and "They Can't Escape Sludge Stench," *CST,* 2 June 1987, pp. 1, 11; Nelson, "'Hotspot' Air Smells, Kills," *CST,* 3 June 1987, p. 10; Nelson, "No Haste to Clean Up Waste," "Pollution Suits Often Caught in Legal Muck," and John Krukowski, "Illinois Senate Panel Asks Toxic Waste Probe," *CST,* 4 June 1987, pp. 10–11; Nelson, "Tackling the Pollution Scourge," and Bob Olmstead, "4 Area Residents Sue San. Dist.," *CST,* 5 June 1987, p. 47; "Sanitary District Hit with Lawsuit," *CT,* 5 June 1987, p. D2; People for Community Recovery, "COME!!! PROTEST!!!" flyer, 11 June 1987, CFP Box 30 Fld. 368; Olmstead, "San. Dist. Pledges to Probe Pollution," *CST,* 12 June 1987, p. 3; Bukro, "Southeast Side Fights to Shut Toxic Dumps," *CT,* 14 June 1987, p. C1; Roger Flaherty, "McHenry County Villages Battle Plans for Landfill," *CST,* 17 June 1987, p. 7; Bukro, "Toxic Sites Called Not That Bad," *CT,* 23 June 1987, p. C3; Fred Marc Biddle, "City Begins Clean-Up of Huge Illegal Dump," *CT,* 25 June 1987, p. C3; Jacquelyn Heard, "Study Aims to Halt Ill Effects of Plant," *CT,* 26 June 1987, p. C8; "Presentation by Waste Management of Illinois to City of Chicago," 30 June 1987, HWP COS Box 71 Fld. 2; [James Landing,] Lake Calumet Study Committee, "Update," 30 June 1987, Landing Papers (JLP) Box 39; Marian Byrnes and Hazel Johnson to Richard Carlson (IEPA), 30 June 1987, Robert Ginsburg Papers (RGP); Robert Bergsvik, "Region Waste Wars Rage On," *DC,* 6 July 1987, p. A1; Bergsvik, "Waste Incidents Spark Fears," *DC,* 7 July 1987, p. A1; Bergsvik, "Waste Worries Head for 21st Century," *DC,* 8 July 1987, p. A2, Richard Carlson to Marian Byrnes and Hazel Johnson, 8 July 1987, RGP; Tim Padgett and Nelson, "Waste Site Health Review Demanded," *CST,* 9 July 1987, p. 42; Bukro, "Calumet City Residents Urge Freeze on New Permit for Landfill," *CT,* 9 July 1987, p. C7; Bergsvik, "Landfill Complaints Aired," *DC,* 9 July 1987, p. A1; Bergsvik, "2 State Lawmakers Back Permit Freeze for SCA Burner," *DC,* 11 July 1987, p. A3; Leslie Morse, "Committee Tackles Area Waste Woes," *DC,* 14 July 1987, p. A3; Nelson and Padgett, "Can David Trash Goliath? Far South Siders Fight Permits for Giant Waste Management," *CST,* 19 July 1987, p. 38; Ben Gordon and Jim Vallette, "Action Alert: Rally for Environmental Rights Chicago, July 29, 1987," and "Say 'No!' to Waste Management" flyer, 29 July 1987, JLP Box 37

Fld. SCA; Bergsvik, "CID Protesters Arrested," and "CID Target of Protest," *DC,* 30 July 1987, pp. A1, A4; "Protesters Halt Trucks at Landfill," *CT,* 30 July 1987, p. 1; Leon Pitt, "Protesters Block Far S. Side Landfill," *CST,* 30 July 1987, p. 32; PCR, "Summary" and "Introduction," n.d. (ca. 1 August 1987), CFP Box 30 Fld. 367; Dennis Geaney, "Confessions of a Street Walker," *SVPSB,* 2 August 1987, p. 7; "Protesters Make Not Guilty Plea," *DC,* 1 September 1987, p. A3; "Protesters Sentenced," *DC,* 16 September 1987, p. A1; Bergsvik, "Garbage Crisis Dumps Big in '87," *DC,* 28 December 1987, p. 2; Virginia Mullery, "Hazel Johnson," *Salt,* July–August 1988, pp. 4–6; DJG interviews with Hazel Johnson, Dan Lee, Cathy Askew, Margaret Bagby, Loretta Augustine-Herron, Betty Garrett, Scott Sederstrom, and Leonard Lamkin.

54. "CID Protest Staged," *DC,* 6 August 1987, p. A1; "CID Is Target of Another Protest," *DC,* 18 August 1987, p. A1; Henry Locke, "Protesters Razz Dump Site," *CD,* 18 August 1987, p. 3; "Activities of the Joint Committee," 25 August–13 October 1987, RGP; Deborah Nelson, "Agency to Run S. Side Cleanup Urged," *CST,* 26 August 1987, p. 6; Jean Davidson and Wes Smith, "Southeast Siders Irked Over Pollution-Health Report," *CT,* 26 August 1987, p. C3; Robert Bergsvik, "Waste Cleanup Tax Eyed," *DC,* 26 August 1987, p. A1; Nicholas J. Melas (Metropolitan Sanitary District President) to Emil Jones Jr., 26 August 1987, RGP; Howard Stanback to Harold Washington, "Solid Waste Management Strategy Options," 28 August 1987, HWP COS Box 42 Fld. 6; Thomas J. Murphy to Maurice Parrish, 31 August 1987, HWP PSS Box 12 Fld. 1; Terry Ayers to Bill Child, "Cook Co.-Dutch Boy," 4 September 1987, RGP; "CID Landfill Protested Again," *DC,* 9 September 1987, p. A1; Henry Locke, "Waste Site Tour Set for So. Side," *CD,* 10 September 1987, p. 4; Nelson, "Dump-Site Tour an Eye-Opener for Legislators," *CST,* 11 September 1987, p. 36; James A. Fitch to Sharon Gist Gilliam, 11 September 1987, HWP COS Box 42 Fld. 6; Bergsvik, "Toxic Waste Sites Focus of Hearing," *DC,* 14 September 1987, p. A1; Gilliam to Fitch, 23 September 1987, HWP COS Box 42 Fld. 6; "State Sen. Jones to Challenge Savage," *CT,* 16 November 1987, p. C2; Howard J. Stanback C.V. in HWP COS Box 42 Fld. 6; Gene Maeroff, "The New School Pushes to Revive Ph.D. Program," *NYT,* 18 May 1986; Stanback's 1989 interview with Betty Brown-Chappell; Stanback in Krumholtz and Clavel, eds., *Reinventing Cities,* pp. 176–77; Lynne M. Cunningham, "Local Initiatives to Catalyze Economic Development," *Planning News* (American Planning Association Illinois Chapter), Spring 1988, pp. 4–5; DJG interviews with Hazel Johnson, Howard Stanback, Hal Baron, Mary Ryan, and Bruce Orenstein.

55. Greg Galluzzo 1987 planning diary, GGP; Scott, *A Singular Woman,* photo insert and pp. 243, 245, 268–71; "Frank Davis Dead at 81," *CD,* 13 August 1987, p. 5; Obama's 2001 interview with Julieanne Richardson; DJG interviews with Greg Galluzzo, Johnnie Owens, Genevieve Cook, Tim Jessup, Pete Vayda, Hasan and Raazia Chandoo, Genevieve Cook, Sheila Jager, Mike Dees, Doug Glick, Asif Agha, and Cathy Askew. Sheila accompanied Barack on one trip to New York City, but not this summer 1987 one. "I only met the Paki crowd once in NY and don't remember them or the visit very well," and Hasan does not believe he ever met Sheila.

56. Karen Thomas and Jean Latz Griffin, "As Different as City, Suburbs," *CT,* 2 September 1987, p. 1; Casey Banas, "Teachers Want 10% Pay Raise," and Griffin and Michele Norris, "Money's There, But It's Wasted," *CT,* 3 September 1987, pp. 1, 6; Linda Lenz, "Teachers OK Strike," *CST,* 5 September 1987, p. 1; Lenz, "No Quick End Seen to Strike," *CST,* 9 September 1987, p. 1; Jack Houston and Griffin, "Teachers Strike Closes Schools," and Banas and Griffin, "School Board Counting Days Till Strike Pays," *CT,* 10 September 1987, p. 1; G. Alfred Hess Jr., "Need School Reform, Not a Showdown," *CT,* 12 September 1987, p. 8; Griffin and Fred Marc Biddle, "No Progress in School Strike," *CT,* 13 September 1987; Banas, "Plan Would Give Teachers a Raise," *CT,* 15 September 1987, p. 1; Banas, "Cuts to Fund School Raises Proposed," *CT,* 16 September 1987, p. 1; Nancy Ryan and Blair Kamin, "Mayor Wants Strike Ended by Monday," and Ryan and Banas, "Parents Urge Mayor to Act in School Crisis," *CT,* 17 September 1987, p. 1; Ryan and Kamin, "Pickets at Mayor's Home Demand an End to Teachers Strike," *CT,* 18 September 1987, p. 7; Banas and R. Bruce Dold, "Board Doesn't Want to End Strike Soon, Vaughn Says," *CT,* 19 September 1987, p. 5; Manford Byrd Jr., "Byrd Defends School System Reforms," *CT,* 24 September 1987, p. 26; Banas and Andrew Martin, "Teachers' Raises Cost Jobs: Board," *CT,* 25 September 1987; Norris, "School Board May OK Learning with Lunch," *CT,* 26 September 1987, p. 7; Banas, "Outcry Rises for Reform of Schools," *CT,* 2 October 1987, p. 1; Lenz and Jim Merriner, "Board,

Union Study 3% Raise as Compromise," *CST,* 2 October 1987, p. 3; Banas, "Teachers Cool to Latest Offer," *CT,* 3 October 1987, p. 1; Banas, "Union, Schools Reach Pact," and "End of Strike Could Be a New Beginning," *CT,* 4 October 1987, pp. 1, 14; Banas, "Mayor Tilts Spotlight to Reform," and "Strike's Wake Leaves Winners and Losers," *CT,* 5 October 1987, pp. 1, 2; Banas, "Call Is Out to School Reformers," *CT,* 7 October 1987, p. 1; Bill McKersie, "And a New Test for Mayor Washington," *CT,* 8 October 1987, p. 27; C. D. Matthews, "School Reform Gains Support," *DC,* 29 December 1987, pp. 3, 6; Howard Stanback and Robert Mier, "Economic Development for Whom? The Chicago Model," *NYU Review of Law & Social Change* 15 (1987): 11–22, 37–41; Thom Clark, "Chicago Parents Organize for Better Schools," *TNW,* January–February 1988, pp. 3–6; Ben Joravsky, "Mad as Hell: School Reformers Declare War on the Central Bureaucracy," *CR,* 1 March 1990; Alex Poinsett, "School Reform, Black Leaders: Their Impact on Each Other," *Catalyst,* May 1990, pp. 7–11, 43 (which is unreliably hostile toward both Hess and Moore); Hess, *School Restructuring, Chicago Style* (Corwin Press, 1991), pp. 25–27; Kyle and Kantowicz, *Kids First,* esp. pp. 178–81, 209–10, 219–24; David Moberg, "Can Democracy Save Chicago's Schools?," *American Prospect,* January 1992, pp. 98–108; Jeffrey Mirel, "School Reform, Chicago Style: Educational Innovation in a Changing Urban Context, 1986–1991," *Urban Education* 28 (July 1993): 116–49, esp. p. 121; William Ayers, "Chicago: A Restless Sea of Social Forces," in Charles T. Kerchner and Julia E. Koppich, eds., *A Union of Professionals* (Teachers College Press, 1993), pp. 177–93, esp. 181–83; Ayers and Michael Klonsky, "Navigating a Restless Sea: The Continuing Struggle to Achieve a Decent Education for African American Youngsters in Chicago," *Journal of Negro Education* 63 (Winter 1994): 5–18; Maribeth Vander Weele, *Reclaiming Our Schools* (Loyola University Press, 1994), p. 10; Hess, *Restructuring Urban Schools,* pp. 14–15, 21–22, 194; Carl, "Harold Washington and Chicago's Schools," esp. pp. 316, 334–36. See also Michael B. Katz, "Chicago School Reform as History," *Teachers College Record* 94 (Fall 1992): 56–72, Linda Lenz, "Missing in Action: The Chicago Teachers Union," in Alexander Russo, ed., *School Reform in Chicago: Lessons on Policy and Practice* (Harvard Education Press, 2004), pp. 125–31, and Jim Carl, "'Good Politics Is Good Government': The Troubling History of Mayoral Control of the Public Schools in Twentieth-Century Chicago," *American Journal of Education* 115 (February 2009): 305–36.

57. Delores T. Wood to Jane Ramsey, "Close: Developing Communities Project," 29 June 1987, HWP COF Box 27 Fld. 27 ("Hal Baron and Joe Washington handling matter . . . . Mtg. set next Monday Sept. 28"); David Giffey, "Mulling Change with a Young Obama in Spring Green," *Madison Capital Times,* 21 July 2009; Kellman in David Moberg, "Obama's Community Roots," *Nation,* 16 April 2007, and in Sharon Cohen (AP), "Barack Obama Straddles Different Worlds," 14 December 2007; McKnight, "The Future of Low-Income Neighborhoods and the People Who Reside There," Center for Urban Affairs and Policy Research, Northwestern University, June 1987, 18pp., esp. pp. 3, 5, 7–8, 12; McKnight, "Do No Harm," *Social Policy,* Summer 1989, pp. 5–14, reprinted in *The Careless Society,* pp. 101–14; Beryl Satter, *Family Properties* (Metropolitan Books, 2009), p. 128; Boyte in "Mr. Obama's Neighborhood," Bradley Center, Hudson Institute, Washington, D.C., 1 October 2008, p. 13; William Upski Wimsatt, "Anonymous Benefactor," *CR,* 26 March 1998; Anthony Burke Boylan, "Scholar Stanley J. Hallett, 68, Mixed Theology, Urban Planning," *CT,* 26 November 1998; John McCarron, "Stan Hallett's Quiet Crusade for Simple Solutions Will Live On," *CT,* 30 November 1998; Linda Randle in David Jones, "Obama the Chameleon," *Daily Mail,* 1 November 2008, p. 46 (recalling Sheila as "super intelligent" and having "a lot of ambition"); DJG interviews with Hal Baron, John McKnight, Jerry Kellman, Greg Galluzzo, Mike Kruglik, Johnnie Owens, Bruce Orenstein, Linda Randle, Sheila Jager, and Harvey Lyon. See also John P. Kretzmann and McKnight, *Building Communities from the Inside Out: A Path Toward Finding and Mobilizing a Community's Assets* (ACTA Publications, 1993).

58. Barbara Marsh, "Strike, Learning Goals Stall School-Biz Pact," *CCB,* 21 September 1987, p. 3; Andrew Martin and Casey Banas, "Parents Rally to Reform Schools," *CT,* 9 October 1987, p. 7; Banas, "Schools Risk Funding Loss," *CT,* 11 October 1987, p. 1; Marsh, "School Reform Gets Tepid Biz Response," *CCB,* 12 October 1987, p. 3; Banas and Cheryll Devall, "A 1st Step to Reform in Schools," *CT,* 12 October 1987, p. 1; Linda Lenz, "Mayor Vows Unified School Reform Plan," *CST,* 12 October 1987, p. 5; Chinta Strausberg, "Mayor Tabs School Summit a 'Success,'" *CD,* 13 October 1987, p. 1; William

Snider, "Chicago's 'Unprecedented' Populist Revolt," *Education Week,* 28 October 1987, pp. 1, 20–21; Robert Davis and Jan Crawford, "Reform Must Be Tied to Aid, Schools Told," *CT,* 3 November 1987, p. 3; Banas and Devonda Byers, "Chicago's Schools Hit as Worst," *CT,* 7 November 1987, p. 1; "Schools in Chicago Are Called the Worst by Education Chief," *NYT,* 8 November 1987; Jack Houston, "Mayor: We'll Clean Up Schools," *CT,* 8 November 1987, p. 3; Marsh, "School Reform a Must: Biz Group," *CCB,* 9 November 1987, p. 3; Banas, "School Plan Calls for Local Control," *CT,* 10 November 1987; William S. McKersie, "Reforming Chicago's Public Schools: Philanthropic Persistence, 1987–1993," *Advances in Educational Policy* 2 (1996): 141–57; H. Gregory Meyer, "Taking a Novel Approach Toward Harold Washington's Reign," *CT,* 24 September 2004.

59. Pullman Christian Reformed Church Bulletins, 6 and 27 September and 4 and 11 October 1987; "Ill. Economy to Be Explored," *DC,* 15 September 1987, p. A5; *Inscapes* (GSU), 30 September 1987, p. 2; Board of Governors News, "First Annual Board of Governors Public Policy Conference to Be Held at GSU," 5 October 1987; Mark Bouman, "The Place of Place in Community Economic Development," 16 October 1987, "Developing Illinois' Economy: A Statewide Perspective," 16 October 1987 (pp. 15–16), Mark Bouman Papers; Richard Ringer, "Chicago Bank Agrees to Plan for Local Loans," *American Banker,* 26 August 1987, p. 3; Leon Pitt, "Bank, Blacks in $20 Million Pact," *CST,* 18 September 1987, p. 42; John Camper, "Community Pressure Keeps Bank Funds Right at Home," *CT,* 22 September 1987, p. C1; Ben Joravsky, "South Side Hardball: Residents and White Bankers Team Up in $20 Million Investment Agreement," *CR,* 15 October 1987; R. Bruce Dold, "Minority Bank Faces Surprise Foe," *CT,* 5 May 1988, p. C1; Joravsky, "Roseland's Thorny Problem," *CR,* 7 July 1988; Susan Chandler, "Bankers Paint Two Pictures of Roseland Activist Lomax," *CST,* 24 July 1989, p. 37; Maggie Garb, "Foreclosures Plague Austin, Roseland," *CST,* 7 August 1988, p. H1; Garb, "Community Leaders Working to Put Hope Back in Roseland," *CST,* 29 July 1990, p. 16; Brian Edwards, "Dressed for Success," *CT,* 11 October 1991; "Chicago Man Opens Mall," *Jet,* 10 May 1993, pp. 24–25; Satter, *Family Properties,* pp. 371–75; Richard Mertens, "Dream Deferred," *University of Chicago Magazine,* May–June 2011; Darryl E. Getter, "The Effectiveness of the Community Reinvestment Act," Congressional Research Service, 7 January 2015; DJG interviews with Mark Bouman, Richard Longworth, Josh Hoyt, and Ed Grossman.

60. Nina Totenberg, "Justice Brennan," *All Things Considered,* NPR, 29 and 30 January 1987, AT-87-0129.01/01-C; Obama to Brennan, 7 December 1990, WJBP Box II-94; "The Disadvantaged Among the Disadvantaged: Responsibility of the Black Churches to the Underclass," Harvard Divinity School, 23–25 October 1987; Luix Overbea, "Black Clergy Look for Ways to Stem the Spread of Urban Poverty," *CSM,* 6 November 1987, p. 6; "The Return of America's Conscience," *Harvard Divinity Bulletin,* February 1988, p. 4; Obama to Phil Boerner, 26 October 1987, Boerner Papers; Titcomb in Peter Serafin, "Punahou Grad Stirs Up Illinois Politics," *HSB,* 21 March 2004, and in Mendell, *Obama,* p. 82; Kellman in Bob Secter and John McCormick, "Portrait of a Pragmatist," *CT,* 30 March 2007, in Liza Mundy, "A Series of Fortunate Events," *WP Magazine,* 12 August 2007, in Smith, "The Ascent of Mr. Charisma," *Sunday Times Magazine,* 23 March 2008, on *Anderson Cooper 360,* CNN, 15 April 2008, in Rodric J. Bradford, "Hiring Barack Obama," BustedHalo .com, 19 April 2008, in Byron York, "The Organizer," *National Review,* 30 June 2008, in Serge Kovaleski, "In Organizing, Obama Led While Finding His Place," *NYT,* 7 July 2008; Kellman's July 2008 interview with Jim Gilmore; Kellman in Jon Meacham, "On His Own," *Newsweek,* 1 September 2008, p. 26, in John B. Judis, "Creation Myth," *TNR,* 18 September 2008, on *All Things Considered,* NPR, 17 October 2008, and in "Obama as We Knew Him," *Observer,* 26 October 2008, pp. 4–7; Obama in John Corr, "From Mean Street to Hallowed Halls," *Philadelphia Inquirer,* 27 February 1990, pp. C1, C4, in David Heckelman, "Lawyer-Legislator Breathes Life into the Dreams of His Father," *CDLB,* 26 April 1997, p. 9, in his 2001 interview with Julieanna Richardson, on *The Strausberg Report* in 2002, in "Introducing Barack Obama" in 2003, and on *Charlie Rose,* 23 November 2004; Obama's 2 and 16 June 2006 commencement speeches at the University of Massachusetts at Boston and Northwestern University; Obama's remarks at Campus Progress Annual Conference, D.C., 12 July 2006, and remarks to College Democrats of America, Columbia, SC, 26 July 2007; Obama's 5 December 2007 remarks at Cornell College, Mt. Vernon, Iowa, and 25 May 2008 Wesleyan University commencement address; Obama, "A

New Era of Service," University of Colorado, Colorado Springs, 2 July 2008; Obama in Richard Stengel et al., "The Interview," *Time,* 17 December 2008; Robert D. McFadden, "Rev. Peter J. Gomes Is Dead at 68; A Leading Voice Against Intolerance," *NYT,* 1 March 2011; DJG interviews with Arnie Graf, Asif Agha, Sheila Jager, Charles Payne, and Jerry Kellman.

61. Obama to Phil Boerner, 26 October 1987, Boerner Papers; Greg Galluzzo 1987 planning diary, GGP; Galluzzo to Ellen Benjamin (Borg-Warner Foundation), 16 May 1987, and Galluzzo to Mariita Conley (Beatrice Foundation), 2 December 1987, Gamaliel Papers; Kruglik in Secter and McCormick, "Portrait of a Pragmatist," *CT,* 30 March 2007, in Remnick, *The Bridge,* p. 161, and in McClelland, *Young Mr. Obama,* p. 175; DJG interviews with David Kindler, Kevin Jokisch, Danny Solis, Phil Mullins, Mike Kruglik, Paul Scully, Mike Pfleger, Mary Yu, Ken Brucks, and Regina Foran Thibeau. On Foran, see Jan Crawford, "10 Groups Granted $148,000," *CT,* 26 July 1987, and Foran, "Glimpses of Discipleship," *Upturn,* October–November 1987, p. 9. In September 1987, Ken Rolling of Woods wrote that UNO's groups "together represent perhaps the strongest organizing neighborhood constituency in the city outside of the political machines." McKersie, "Strategic Philanthropy and Local Public Policy," p. 370.

62. Greg Galluzzo 1987 planning diary, GGP; Patrick J. Keleher 1987 pocket Day-Timers, 17 September 1987 ("Nice chat with Barack"), 13 October 1987, 13 and 17 November 1987 (7 P.M. Reformation Lutheran in the latter), PJKP; "Classes to Begin on College Testing," *CT,* 22 September 1987; "College Entrance Exam Forms Available," *CT,* 25 September 1987; "Al Raby to Leave City Rights Post," *CT,* 19 October 1987, p. 2; Nat Krieger, "DCP Seeks to Improve Public Education," *CD,* 23 November 1987, p. 8; Larry Finley, "Agencies Get Help to Help Others," *CST,* 29 December 1987, p. 9; Jean Franczyk, "School Reform Pressure Continues," *Chicago Reporter,* January 1988, pp. 1, 8–9, 11; Obama to Anne Hallett, Developing Communities Project 1988 Proposal to the Wieboldt Foundation, 26 January 1988, Wieboldt Foundation 1988 Annual Report, p. 15, Wieboldt Papers; "New School Reform Effort in Roseland," *TNW,* March–April 1988, p. 11; Edith R. Sims, "Successful Programs, Policies, and Practices Employed at Corliss High School," *Journal of Negro Education* 57 (Summer 1988): 394–407, esp. 394–97; "A Guide to Regional and Local Grantmakers," [1988], esp. pp. 203, 205, Human SERVE Papers Box 16, Fld. 742; Woods Charitable Fund, Report for the Year 1987, esp. p. 20, and Report for the Year 1988, esp. p. 26 (showing $25,500 for "operating support" to DCP); Robert Foulkes and William Peterman, "Roseland Community Credit Needs Assessment," Voorhees Center, University of Illinois at Chicago, June 1992, esp. parts G1 and G2; Obama, *DFMF,* pp. 256–61, 267–68, 279–80; Jerry Kellman, AriseChicago Breakfast Talk, 24 November 2009; Kim Geiger, "Obama Tour Covers President's South Side Past," *CT,* 11 November 2013; DJG interviews with Greg Galluzzo, Johnnie Owens, Aletha Strong Gibson, Ann West, Carolyn Wortham, Homer Franklin, Shirley Chappell, Sokoni Karanja, Steve Perkins, Jerry Kellman, Mary Bernstein, Tyrone Partee, John Webster, and Jean Rudd. On Robert Healey, see Ana Mendieta, "Labor Leader Robert Healey Dies at 72," *CST,* 24 July 2002, p. 16, James Janega, "Stalwart Led Chicago Labor Unions," *CT,* 24 July 2002, and "Robert M. Healey: A Legacy of Vision and Strength," *Insight* (IFT), Winter 2003. On Carver High School, where the late Marcellus Stamps Jr. succeeded the late Alexander Whitfield as principal in mid-1987, see "Carver's Walters: More Than a Coach," *CST,* 26 October 1986, Bonita Brodt, "That 'Mean' Lady," *CT,* 3 March 1987, and Vernon Jarrett, "Academic 'All-Star' Inspires Carver Kids," *CST,* 23 November 1987, p. 28; on Fenger High School, whose principal was the late Leo Dillon, see Robin Downing, "Gangs Lead to Closed Campus," *New Expression,* September 1987, p. 5; on Julian High School, whose Dr. Oliver is also deceased, see Kimberly Ward, "Social Clubs Don't 'Clique' with All Students" and "Social Clubs Banned at Julian," *New Expression,* January 1988, p. 12, and February 1988, p. 4.

63. Thomas J. Murphy to Lawrence E. Kennon (ZBA), 18 September 1987, HWP PSS Box 12 Fld. 11; Casey Bukro, "South Side Smog Report Singes Wood Stoves, Traffic," *CT,* 25 September 1987, p. 3; Bill Richards, "Waste Management Faces More Inquiries," *WSJ,* 28 September 1987, p. 6; Murphy to Michael Holewinski, 7 October 1987, HWP PSS Box 12 Fld. 11; Deborah Nelson, "U.S. EPA Refuses to Testify in S. South Pollution Probe," *CST,* 14 October 1987, p. 22; Nelson, "S. Side Landfill in Use Without Permits Since '83," *CST,* 16 October 1987, p. 5; Robert Bergsvik, "Paxton Court Date Is Set,"

DC, [17 October 1987] (HWP PSS Box 12 Fld. 11); Bukro, "Garbage Piles Up as City Loses Ground," *CT*, 26 October 1987, p. 1; "Burying Our Heads in the Garbage," *CT*, 28 October 1987, p. 18; Howard Stanback to Harold Washington, "Proposed Solid Waste Management Plan: An Outline," 28 October 1987, HWP COS Box 49 Fld. 6; Bergsvik, "Toxic Panel Wraps Up Public Hearings," *DC*, 29 October 1987, p. A2; Jean Davidson, "State EPA Calls for Crackdown on Southeast Side Pollution," *CT*, 30 October 1987, p. C6; Don B. Gallay to Lee Botts, "Paxton Landfill," 3 November 1987, HWP PSS Box 12 Fld. 11; Botts to Stanback, "Comments on Proposal on Toxic Waste Incineration," 12 November 1987, HWP CSS Box 49 Fld. 6; "Waste Proposal Meeting Agenda," 16 November 1987, JLP Box 41 Fld. SAMP; James E. Landing to LCSC Members, 20 November 1987, JLP Box 29 Fld. LCSC.

64. Casey Banas, "948 School Jobs Axed for Teachers' Raises," *CT*, 25 November 1987, p. 4; Thomas Hardy and R. Bruce Dold, "Mayor's Death Stuns City," Hardy and Dold, "Washington Suffers Heart Attack," Robert Davis, "Black Leader, 65, on Verge of Dream," and John Camper and John Kass, "Private Man with a Public Life," *CT*, 26 November 1987, p. 1; William Braden, "Heart Attack Kills Mayor," Philip Franchine, "Stunned City Mourns," and "The Mayor and His Rich Legacy," *CST*, 26 November 1987; "Washington's Showdown Victory," *CT*, 8 April 1987; "Harold Washington and Chicago," *CT*, 26 November 1987, p. 30; Davis and Manuel Galvan, "Mayor Will Lie in State in City Hall," *CT*, 27 November 1987; Hardy and Mitchell Locin, "Mayor's Coalition in Peril," *CT*, 29 November 1987; John Kass, "Who Will Wear the Mantle?," *CT*, 28 February 1988; Holli and Green, *Bashing Chicago Traditions*, p. vii; Robert Mier et al., "Decentralization of Policy Making Under Mayor Harold Washington," *Research in Urban Policy* 4 (1992): 79–102, at 95; Orr in Camper and Kass, "Scandal Gets Too Big for Mayor to Ignore," *CT*, 9 February 1986; William J. Grimshaw, *Bitter Fruit: Black Politics and the Chicago Machine, 1931–1991* (University of Chicago Press, 1992), pp. 184, 195–97; Kass and Michele Norris, "Service Becomes Rally for Evans," Dold and Hardy, "Sawyer's Mayoral Bid Gathers Steam," Dold and James Strong, "Sawyer Says He Has Enough Votes to Win," *CT*, 1 December 1987; Galvan, "Memorial Gives Way to Politics," William Recktenwald and Galvan, "Madhouse Inside and Outside," and Kass and Camper, "Council Elects Sawyer Mayor," *CT*, 2 December 1987; Strong and Hardy, "Sawyer Elected After a Raucous Night," Hardy, "Succession Fight Split Blacks, United Whites," Camper, "Mayor Makes Pitch for Unity," and Kass, "Key Washington Ally Watches Glumly as Her Political World Comes to an End," *CT*, 3 December 1987; Patrick Reardon, "Fallout Burns, Soothes Sawyer's Backers," *CT*, 7 December 1987; Vernon Jarrett, "Ald. Shaw: Our City's Leading Phony," *CST*, 8 January 1988, p. 41; Gary Rivlin, "Seven Days as Mayor," *CR*, 10 March 1988; Sawyer's December 1989 interview with Betty Brown-Chappell; Hawking, "Political Education in the Harold Washington Movement," pp. 209–23. See also Larry Bennett, "Harold Washington's Chicago: Placing a Progessive City Administration in Context," *Social Policy*, Fall 1988, pp. 22–28, and Bennett, "Harold Washington and the Black Urban Regime," *Urban Affairs Quarterly* 28 (March 1993): 423–40.

65. Obama on *The Friday Night Show* with Bob Sirott, WTTW.com, 3 December 2004; Obama in Gretchen Reynolds, "Vote of Confidence," *Chicago Magazine*, January 1993; Obama, *DFMF*, pp. 287–89; Obama in Hank DeZutter, "What Makes Obama Run?," *CR*, 8 December 1995; Obama's 2001 interview with Julieanna Richardson; Obama in Caroline Daniel, "Political Rookie Traverses the Ethnic Divide," *FT*, 24 June 2004, p. 8; Obama in Barkley, *Who's Afraid of a Large Black Man?* p. 36; Obama, *TAOH*, p. 360; Obama in Maraniss, *BOTS*, p. 562; Obama's 30 July 2013 interview with David Blum; McKnight in Judis, "Creation Myth," *TNR*, 10 September 2008; McKnight YouTube video, 1 August 2008; McClelland, *Young Mr. Obama*, p. 178; Greg Galluzzo in Daniel Libit, "The End of Community Organizing in Chicago?," *Chicago Magazine*, April 2011; Peter Slevin, *Michelle Obama* (Knopf, 2015), p. 163; DJG interviews with Steve Perkins, Patty Novick, Judy Stevens, Michael Baron, John McKnight, Greg Galluzzo, and Mary Gonzales.

66. Greg Galluzzo 1987 planning diary, GGP; Patrick J. Keleher 1987 pocket Day-Timers, 8 and 11 December 1987, PJKP; Jack Houston, "School Reform Efforts Move Ahead," *CT*, 30 November 1987, p. 1; Casey Banas, "Forum Asks Local Power for Schools," *CT*, 2 December 1987; Harry Golden Jr., "Mayor to 'Reach Out' to Heal Racial Rifts," *CST*, 4 December 1987, p. 7; Barbara Marsh, "Report Rips Chicago United Record," *CCB*, 7 December 1987, p. 1; Houston, "Teachers' Rejection of Reform Draws Ire of Watchdog Groups," *CT*, 13 December 1987, p. 3; Patrick Reardon, "City School Integration Spending Hit," *CT*, 16 December 1987; Manford Byrd Jr., "Help Schools Before Pushing Jobs Plan," *CT*, 21 December 1987; Jean Franczyk, "School Reform Pressure Continues," *Chicago Reporter*, January 1988, pp. 1, 8–9, 11; Howard Stanback to Ernest Barefield, "Solid Waste Management Planning," 7 December 1987, HWP COS Box 49 Fld. 6; Landing to Lake Calumet Study Committee Members, 7 December 1987, and Fitch to Landing, 7 December 1987, JLP Box 42 Fld. Fitch; Casey Bukro, "City's Garbage Disposal Crisis Major Issue Facing New Mayor," *CT*, 8 December 1987, p. C4; Report of the Joint Committee of Hazardous Waste in the Lake Calumet Area, December 1987, esp. pp. 16, 20, RGP; Robert Bergsvik, "Pollution Report Urges Health Studies," *DC*, 9 December 1987, p. 1; Lake Calumet Area Health Study Urged," *CT*, 9 December 1987; Barefield to Stanback, "Solid Waste Management Planning," 11 December 1987, HWP COS Box 49 Fld. 6; Bergsvik, "2 Landfill Permits Expected," *DC*, 18 December 1987, p. 1; Larry Gross, "5 Teens Charged as Adults in Altgeld Garden Slaying," *CD*, 8 December 1987, p. 3; Dan Nakaso, "Obama's Tutu a Hawaii Banking Female Pioneer," *HA*, 30 March 2008; Mary Zurbuchen in Amanda Ripley, "The Story of Barack Obama's Mother," *Time*, 21 April 2008; Allen G. Breed (AP), "'Toot': Obama Grandmother a Force That Shaped Him," 23 August 2008; DJG interviews with Johnnie Owens, Alice Dewey, Asif Agha, and Sheila Jager.

67. Greg Galluzzo 1987 planning diary (1988 pages), GGP; Patrick J. Keleher 1988 Pocket Day-Timers (18 January 1988), PJKP; Liz Sly, "Schools Seek Big Role for Parents," *CT*, 8 January 1988, p. 2; Jack Houston, "Parents' Panel Urged for Schools," *CT*, 10 January 1988, p. 1; Karen M. Thomas, "Parents Vote for Local School Control," *CT*, 21 January 1988, p. 1; Thomas, "Parents Make Push for Schools," *CT*, 24 January 1988, p. 3; Thomas, "Reform Finds Common Ground," *CT*, 5 February 1988, p. 7; Thomas, "Community Role Urged for Schools," *CT*, 9 February 1988, p. 3; Obama to Phil and Karen Boerner, n.d. (sometime soon after 16 January 1988), Boerner Papers; Obama to Anne Hallett, 26 January 1988, enclosing "Developing Communities Project's 1988 Proposal to the Wieboldt Foundation," Wieboldt Papers; DJG interviews with Phil Boerner and Anne Hallett.

68. Alvin Love in Kalari Girtley, "How Obama Got His Start," *HPH*, 14 February 2007, p. 14, in Kovaleski, "In Organizing, Obama Led While Finding His Place," *NYT*, 7 July 2008, in York, "The Organizer," *NR*, 30 July 2008, in "They Met Obama When," *CT*, 16 November 2008, p. 19, in Robert Beckford, "God Bless You Barack Obama?," BBC 2, 25 January 2010, and in McClelland, *Young Mr. Obama*, p. 169; Robert McClory, "The Holy Terror of Saint Sabina's," *CR*, 16 November 1989; McClory, "Black and Catholic Are Joint Ventures at Chicago Parish," *NCR, 13 March 1998, p.* 5; McClory, *Radical Disciple*, esp. p. 163; Obama in Angela Burt-Murray et al., "America's Teachable Moment," *Essence*, March 2010, pp. 121–26, at 125; Patrick T. Reardon, "Obama's Chicago," *CT*, 25 June 2008, p. T1; Maudlyne Ihejirika, "Old Obama Haunts Becoming Popular Tourist Attractions," *CST*, 23 May 2009; DJG interviews with Alvin Love, Jeremiah Wright, Donita Powell Greene, Sokoni Karanja, Tommy West, Cathy Askew, Nadyne Griffin, Rosa Thomas, Loretta Augustine-Herron, Mike Pfleger, Sheila Jager, Asif Agha, Doug Glick, John Morillo, and Andrea Atkin.

69. Casey Bukro, "Southeast Side Targeted for EPA Cleanup," *CT*, 31 January 1988, p. 1; Robert Bergsvik, "O'Brien Locks Eyed for Landfill," *DC*, 31 January 1988, pp. 1, 10; C. D. Matthews, "Environmentalists Are Fuming Over Proposed Landfill," *DC*, 2 February 1988, pp. 1, 6; Bukro, "Southeast Siders See Some Green in Landfill Plans," *CT*, 2 February 1988, p. 1; James A. Fitch to James E. Landing, 2 February 1988, Mayor's Office Press Release, "City Reaffirms Landfill Moratorium Policy," 8 February 1988, and United Neighborhood Organization of Southeast and Developing Communities Project to the People of the Southeast Side of Chicago, 8 February 1988, JLP Box 42 Flds. Fitch and Fitch Blue; Bergsvik, "Landfill Foes 'Ambush' Local Leaders," and "New Landfill Proposal Told for O'Brien Locks," *DC*, 9 February 1988, pp. 1, 7; Bukro, "80 Southeast Side Residents Reject Firm's Offer on Landfill Use," *CT*, 9 February 1988, p. 3; Bergsvik, "Leaders Disgaree on Landfill," *DC*, 10 February 1988, pp. 1, 13; Bukro, "Sawyer Pushes Mandatory Recycling," *CT*, 10 February 1988, p. 1; Vi Czachorski, "O'Brien Locks Proposal," *HN*, 11 February 1988, p. 1; Bergsvik, "Vrdolyak Ripped by Pucinski," *DC*, 11 February 1988, pp. 1, 6; Bergsvik, "Chamber Takes Stand Against O'Brien Landfill," *DC*, 16 February 1988, pp. 1,

8; Marian Byrnes to Eugene Sawyer, 16 February 1988, JLP Box 40 Fld. CURE; Czachorski, "From the Editor's Scratch Pad," *HN,* 18 February 1988, pp. 1, 5; Bergsvik, "Groups Hope Protest Is Cure for Landfill 'Doom,'" *DC,* 19 February 1988, pp. 1, 16; Bergsvik, "Fitch Finds Pitching Landfill Tough Job," *DC,* 22 February 1988, pp. 1, 6; Fitch to Landing, 23 February 1988, JLP Box 42 Fld. Fitch; "CURE Holds Meeting," *HN,* 25 February 1988, pp. 1, 5; "Fitch Cancels" and "CURE at City Hall," *HN,* 3 March 1988, p. 1; Bergsvik, "Avalon Trails Residents Say No to Landfill," *DC,* 6 March 1988, pp. 1, 14; Wilfredo Cruz, "UNO: Organizing at the Grass Roots," *Illinois Issues,* April 1988, pp. 18–22; Tayler and Herbert, "Obama Forged Path as Chicago Community Organizer," *Newsday,* 2 March 2008, p. A6; McClelland, *Young Mr. Obama,* p. 58; DJG interviews with Bruce Orenstein, Mary Ellen Montes, Ed Grossman, Johnnie Owens, Loretta Augustine-Herron, Bob Klonowski, Howard Stanback, Judy Byrd, and Mary Ryan.

70. Dave Kehr, "'Unbearable Lightness of Being' Takes Heavy-Handed View," *CT,* 5 February 1988; Vincent Canby, "Film 'Lightness of Being,'" *NYT,* 5 February 1988; Pat Watters, *Down to Now: Reflections on the Southern Civil Rights Movement* (Pantheon, 1971), p. 381; DJG interviews with Sheila Jager, Mary Ellen Montes, Phil Mullins, Danny Solis, Bob Moriarty, and Bruce Orenstein.

71. C. D. Matthews, "Wisconsin Steel Settlement Offered," *DC,* 11 February 1988, p. 1; Cindy Richards, "$15 Mill. to Wis. Steel Workers," *CST,* 11 February 1988; Matt O'Connor, "Mill Workers Settle with Wisconsin Steel," *CT,* 11 February 1988, p. C1; Robert Bergsvik, "Ex-Wisconsin Steel Workers to Vote on Pact," *DC,* 12 February 1988, pp. 1, 9; Richards, "Navistar Settlement Far Less—and Far More—Than Hoped," *CST,* 12 February 1988; Timothy P. Vick, "Workers Accept Navistar Offer," *DC,* 15 February 1988, pp. 1, 5; George Papajohn, "Sad Victory for Ex-Steel Workers," *CT,* 15 February 1988, p. 1; Lynn Sweet, "Wis. Steel Workers Vote to OK Deal," *CST,* 15 February 1988; R. C. Longworth, "System Foundered on Lies, Selfishness," *CT,* 17 February 1988, p. B3; Susan Moran, "Mill a Fond Memory," *DC,* 20 March 1988, p. 3; "Wisconsin Steel Accord to Be Discussed Sunday," *DC,* 25 March 1988, p. 2; M. W. Newman, "Rusty Dreams," *CST,* 27 March 1988; Bill Clements, "Emotion Runs High at Reunion," *DC,* 28 March 1988, pp. 1, 3; John F. Wasik, "End of the Line at Wisconsin Steel," *Progressive,* October 1988, p. 15; Thomas Geoghegan, *Which Side Are You On?* (Farrar, Straus, 1991), pp. 101, 104; Beatrice Lumpkin, *Always Bring a Crowd* (International Publishers, 1999), pp. 175, 196; DJG interviews with Bea Lumpkin and Tom Geoghegan.

On South Works, see Larry Galica, "USX Nets Loss of $97 Million," *DC,* 2 May 1987, p. A3, Galica, "Local USW Leaders Seek to Strike Down Proposed Pact," *DC,* 7 July 1987, p. A3, and "After-Effects of Steel Shutdowns," *TNW,* August–September 1989, p. 18. On Republic LTV, see Bergsvik, "Steelworkers Back Panayotovich," *DC,* 25 February 1988, p. 3, Bergsvik, "United Steelworkers Union Locals Gear for Elections," *DC,* 28 February 1988, p. 3, Bergsvik, "LTV Workers to Share Profit," *DC,* 10 April 1988, p. 3, Bergsvik, "President Re-Elected in USW Local 1033," *DC,* 24 April 1988, p. 18, "LTV Union OKs Pact, ESOP Management, Workers to Purchase Bar Division," *Gary Post-Tribune,* 6 October 1989, "LTV Concludes Sale of Bar Division," PR Newswire, 29 November 1989, and DJG interview with Maury Richards. South Works would close for good in April 1992. See Bill Granger, "Requiem for the Southeast Side," *CT,* 12 April 1992, and Cara Jepsen, "Retooling South Works," *TNW,* February–March 1995, pp. 16–20.

72. Neighborhood Schools Coalition, "The Neighborhood Schoolhouse That Works: Education Reform Proposal for the Chicago Public Schools," February 1988, 29pp., Arthur L. Berman Papers Box 6 Fld. 6; Linda Lenz, "Big Increase Urged in School Personnel," *CST,* 18 February 1988, p. 3; Constanza Montana, "Meeting on School Reform Halted," *CT,* 19 February 1988, p. C3; "Parents Rebuffed at Hearing," *CST,* 19 February 1988, p. 3; Karen M. Thomas, "Push Coming to Shove on School Reform," *CT,* 21 February 1988, p. 1; Montana, "Educators Urge Shared School Decision-Making," *CT,* 24 February 1988, p. 3; James Deanes, "The Need for Reform," *DC,* 26 February 1988, p. 5; Patrick J. Keleher 1988 pocket Day-Timers (29 February 1988), PJKP; Jean Franczyk, "Parents Continue Push for Local School Control," *Chicago Reporter,* March 1988, p. 2; Rudolph Unger and Steve Johnson, "City Principals Have Their Say," *CT,* 11 March 1988; Thomas, "Reformers Reject Elected School Board," *CT,* 15 March 1988, p. 1; Lenz, "Education Summit Groups Reject Elected School Bd.," *CST,* 15 March 1988, p. 16; Thomas, "Parents the Key in School Plan,"

CT, 22 March 1988, "New School Reform Effort in Roseland," *TNW,* March–April 1988, p. 11; Mary O'Connell, "Groups Organize Legislative Push on School Reform," and "School Reform: Don't Lose the Golden Moment," *TNW,* March–April 1988, pp. 7–8 and 22; Ben Joravsky, "Too Many Bosses," *Chicago Magazine,* April 1988, pp. 104, 137–44; Joravsky, "The Chicago School Mess," *Illinois Issues,* April 1988, pp. 12–15; Paul M. Green, "SON/SOC: Organizing in White Ethnic Neighborhoods," *Illinois Issues,* May 1988, pp. 24–28; O'Connell, *School Reform Chicago Style,* pp. 11–16; Kyle and Kantowicz, *Kids First,* pp. 212–15; Don Moore, "Earl Durham, a Long-Distance Runner," 3 November 2007; DJG interviews with Danny Solis, Peter Martinez, Phil Mullins, Steve Perkins, Bill Ayers, Johnnie Owens, Dan Lee, Loretta Augustine-Herron, Yvonne Lloyd, Aletha Strong Gibson, Tyrone Partee, and John Webster. On Lourdes Monteagudo, see Isabel Wilkerson, "Schools Chief Learns About Politics," *NYT,* 23 May 1990, and Leigh Behrens, "Lourdes Monteagudo," *CT,* 2 September 1990.

73. Obama, *DFMF,* p. 275–76, 289; Meg McSherry Breslin, "George Kelm, Advocate for Kids, Families," *CT,* 19 December 1998; Owens in Smith, "Ascent of Mr. Charisma," *Sunday Times Magazine,* 23 March 2008, in McClelland, *Young Mr. Obama,* pp. 59–60, and in Maraniss, *BOTS,* pp. 563–64; Auma Obama, *And Then Life Happens,* p. 189; DJG interviews with Jean Rudd, Ken Rolling, Lionel Bolin, Robert Bennett, and Johnnie Owens. Obama may have received a similar scholarship offer from NYU's law school, and he almost certainly was admitted to Stanford. Richard A. Serrano and David G. Savage, "Obama's Harvard Law School Days Marked by Bridge-Building," *LAT,* 27 January 2007, and *Harvard Law Revue,* 20 April 1990, p. 8 n. 8: "Regret Often Turning Down Stanford (Winters, 1988–1990)."

74. Casey Bukro, "Waste Firm Dismisses 2 After Probe," *CT,* 5 March 1988; Robert Bergsvik, "Protesters Demand Closing of Area Chemical Incinerator," *DC,* 8 March 1988, pp. 1, 11; Juanita Bratcher, "S. Siders Want to CURE Toxic Waste Incinerator," *CD,* 8 March 1988, pp. 1, 18, Jim Ritter, "Toxic Waste Furnace Gauge Off; Execs Fired," *CST,* 8 March 1988, p. 45; "WM Violation at SCA," and Jim Landing, "The Waste Assault on Hegewisch Continues," *HN,* 10 March 1988, pp. 1, 6; Harlan Draeger, "S.E. Siders Block Sludge-Dumping Plan," *CST,* 11 March 1988, p. 21; Howard Stanback to Solid Waste Advisory Committee, "O'Brien Locks," 14 March 1988, JLP Box 42 Fld. Negotiations Fitch; Kathy O'Malley and Hanke Gratteau, "The Voting Booth," *CT,* 3 February 1988; Vernon Jarrett, "Grumbling Not Enough," *CST,* 21 February 1988, p. 13; Fran Spielman, "Pro-Sayer Blacks Face Test at Polls," *CST,* 10 March 1988, p. 3; R. Bruce Dold, "Savage Faces Toughest Congressional Fight," *CT,* 11 March 1988; Bergsvik, "A Local Newspaper Sells Out, but Is Anybody Reading?," *DC,* 13 March 1988, p. 9; Ann Marie Lipinski and Jorge Caruso, "Sawyer Suffers Blow as 4 Allies Lose Ward Race," *CT,* 16 March 1988, p. C1; Don Hayner and Ray Hanania, "4 Sawyer Allies Lose Committee Races," *CST,* 16 March 1988, p. 3; James Warren, "East Day for Savage, Other Congressional Incumbents," *CT,* 17 March 1988, p. 4; Jarrett, "'Harold's People' Deliver the Message," *CST,* 20 March 1988, p. 15; Tom Gorman, "Superior Organization Key to Trotter Victory," *HPH,* 23 March 1988, pp. 1, 2; Gorman, "Braun Resigns; Trotter Takes His 25th District Seat Early," *HPH,* 7 December 1988, p. 1; David K. Fremon, *Chicago Politics Ward by Ward* (Indiana University Press, 1988), pp. 358–64; Palmer's June 2000 interview with Julieanna Richardson; DJG interviews with Alice Palmer and Donne Trotter.

75. Shawnelle Richie, "Mayor Eugene Sawyer Briefing Notes," 17 March 1988, and Sawyer, "Remarks," Southeast Chicago Community Meeting, 17 March 1988, Sawyer Papers; "UNO/DCP Meeting Fact Sheet: The Land-Fill Issue in South-East and Far South Chicago" [17 March 1988], CURE, "Keep Your Eyes on the Prize! Hold On!" 17 March 1988, and Marian Byrnes to Father Dominic Carmon, 21 March 1988, all JLP Box 40 Fld. CURE; Robert Bergsvik, "Sawyer Promises Task Force on Area Landfill Expansion," *DC,* 18 March 1988, pp. 1, 7; Casey Bukro, "Sawyer Backs Dumping Foes," *CT,* 18 March 1988, p. C3; Tim Padgett and Deborah Nelson, "Sawyer Assures S.E. Side of Landfill Voice," *CST,* 18 March 1988, p. 4; Bergsvik, "3rd Task Force the Charm in Landfill Fight," *DC,* 20 March 1988, pp. 1, 14; Edward R. Vrdolyak to Eugene Sawyer, 22 March 1988, JLP Box 40 Fld. CURE; Vi Czachorski, "From the Editor's Scratch Pad," *HN,* 24 March 1988, pp. 1, 5; Bergsvik, "Vrdolyak Seeks Local Representation of Landfill Panel," *DC,* 24 March 1988, p. 3; Bergsvik, "Chamber Rep Drops Out of Talks," *DC,* 25 March 1988, p. 3; Bergsvik, "Expert Speaks Against Landfill," *DC,* 30 March 1988,

p. 3; Bergsvik, "Sawyer Names Landfill Panel," *DC,* 31 March 1988, p. 3; Mike Wilkinson, "Deals, Dollars, and Dumps on the Southeast Side," *TNW,* May–June 1988, pp. 3–5; Jonathan Kaufman, "Black Power Brokers Ready to Rise in Tandem with a New President," *WSJ,* 6 November 2008; DJG interviews with Bruce Orenstein, Howard Stanback, Judy Byrd, Mary Ellen Montes, Loretta Augustine-Herron, Hazel Johnson, Phil Mullins, Dan Lee, George Schopp, Alma Avalos, Marlene Dillard, Margaret Bagby, Dominic Carmon, and Bob Klonowski.

76. Obama, *DFMF,* pp. 227, 290–91; Robert Matthews Johnson, *The First Charity* (Seven Locks Press, July 1988), p. 171; Emil Jones Jr., Illinois State Senate, 93rd General Assembly, 8 November 2004, pp. 11–13; Yvonne Lloyd in "A Rusty Toyota . . . ," *WP,* 11 February 2007, p. B3; DJG interviews with Johnnie Owens, Loretta Augustine-Herron, Eddie Knox, Rick Williams, Dan Lee, Cathy Askew, Tommy West, Adrienne Jackson, Marlene Dillard, Alma Jones, Ernest Powell, Betty Garrett, Tyrone Partee, Alvin Love, Yvonne Lloyd, Margaret Bagby, Nadyne Griffin, Howard Stanback, Emil Jones Jr., Renee Brereton, and Linda Randle.

77. Vi Czachorski, "From the Editor's Scratchpad," *HN,* 7 April 1988, p. 1; Robert Bergsvik, "Mayor's Landfill Panel Ripped by Coalition," *DC,* 7 April 1988, pp. 1, 9; Bergsvik, "Landfill Option Reopened," *DC,* 8 April 1988, p. 3; "No More Landfills," *DC,* 10 April 1988, p. 4; Bergsvik, "Dump Referendum Sought for SE Side Wards," *DC,* 11 April 1988, p. 3; "Vote on Landfills," *DC,* 13 April 1988, p. 4; Casey Bukro, "Dump Urged for Southeast Side," *CT,* 15 April 1988, p. C4; Bergsvik, "Sawyer Aide Pushes Landfill Moratorium," *DC,* 15 April 1988, pp. 1, 12; Bergsvik, "Paxton Files Second Lawsuit," *DC,* 17 April 1988, p. 3; Phil Kadner, "Some People Treated Worse Than Garbage," *DC,* 18 April 1988, p. 3; Mike Nolan, "Hegewisch Residents Urged to Speak Out on Landfills," *DC,* 19 April 1988, p. 5; Czachorski, "Task Force Meets Again," *HN,* 21 April 1988, pp. 1, 6; Bukro, "Illinois EPA Wants to Shut Southeast Side Incinerator," *CT,* 22 April 1988; Bergsvik, "Delay Urged for Landfill Decision," *DC,* 24 April 1988, p. 2; Nolan, "Landfill Expansion Battle May Go to Court," *DC,* 26 April 1988, pp. 1, 5; "Landfill Meeting Changed," *DC,* 27 April 1988, p. 7; John Jeter and Harlan Draeger, "Corps Report Outlines S. Side Wetlands Uses," *CST,* 27 April 1988, p. 25; Czachorski, "From the Editor's Scratchpad," and "IEPA Requests Shutdown," *HN,* 28 April 1988, p. 1; Czachorski, "From the Editor's Scratch Pad," *HN,* 5 May 1988, pp. 1, 5; Bergsvik, "Pollution Study Proposed," *DC,* 5 May 1988, pp. 1, 12; Harlan Draeger, "Landfill Idea Sparks 'War,'" *CST,* 5 May 1988, p. 32; Mark Brown, "Calumet Cleanup Panel Voted," *CST,* 7 May 1988, p. 7; Draeger, "Trash Squeeze Heats Landfill Fight," *CST,* 8 May 1988, p. 36; Bergsvik, "Cap Tops Landfill Activist List," *DC,* 9 May 1988, pp. 1, 12; Czachorski, "A War Against O'Brien Locks," *HN,* 12 May 1988, pp. 1, 5; Bergsvik, "Steel Firms Challenge Suit Against EPA," *DC,* 15 May 1988, pp. 1, 18; John Gallagher, "Dolton to Say No to Landfill," *DC,* 18 May 1988, pp. 1, 9; Bergsvik, "Sawyer Response Sought on Landfill," *DC,* 19 May 1988, p. 14; "Our Petition Drive," *DC,* 26 May 1988, p. 4; Wilkinson, "Deals, Dollars, and Dumps on the Southeast Side," DJG interviews with Mary Ellen Montes, Howard Stanback, Bruce Orenstein, Loretta Augustine-Herron, Dominic Carmon, and Bob Klonowski.

78. Greg Galluzzo 1987 planning diary, GGP; Galluzzo to Mariita Conley, 2 December 1987, Gamaliel Papers; Woods Fund Report for the Year 1987, pp. 3, 23; Obama, "Why Organize?," *Illinois Issues,* August–September 1988, pp. 40–42; Woods Fund Report for the Year 1989, p. 23; Wayne Klatt, untitled review, *Illinois Historical Journal* 84 (Autumn 1991): 211–12; Obama, *DFMF,* pp. 104, 115; David Moberg, "Obama's Third Way," *NHI Shelterforce* #149, Spring 2007; Obama in Ariel Sabar, "For Obama, Bipartisan Aims, Party-Line Votes," *CSM,* 17 April 2008, p. 1; Jeffrey Goldberg, "Obama on Zionism and Hamas," TheAtlantic.com, 12 May 2008; Obama in Ryan Lizza, "Making It: How Chicago Shaped Obama," *New Yorker,* 21 July 2008; Obama on "Barack Obama Revealed," CNN, 20 August 2008; Wolffe, *Renegade,* p. 64; Obama Remarks to Campaign Staff, Chicago, 7 November 2012; DJG interviews with Greg Galluzzo, Mary Gonzales, David Kindler, Kevin Jokisch, Mike Kruglik, Jean Rudd, Fred Simari, Sheila Jager, and Ted Aranda. See also Greg Galluzzo, "Community Organizing Through Faith-Based Networks," and Mike Kruglik, "Faith-Based Organizing and Core Values," in M. Paloma Pavel, ed., *Breakthrough Communities* (MIT Press, 2009), pp. 231–33 and 234–35, and Wade Rathke, "Greg Galluzzo of Gamaliel Begins to Look Back," *Social Policy,* Spring 2011, pp. 55–56. On Mike Kruglik and SSAC's focus on housing rehabilitation, see David Greising, "Housing Re-Starts," *CST,* 5 October 1988, p. 48.

79. Lawrence Bommer, "'Ma Rainey' Sings Again in City Premiere," *CT,* 8 January 1988; Richard Christiansen, "'Ma Rainey' Bursts Out in a Vivid Spectacle," *CT,* 14 January 1988; Christiansen, "Pegasus' 'Ma Rainey' Packs Both Power, Pity," *CT,* 28 February 1988; Jay Pridmore, "Spertus Exhibit Centers on Impact of the Eichmann Trial," *CT,* 8 April 1988, p. 56; Ann Marie Lipinski and Dean Baquet, "Sawyer Aide's Ethnic Slurs Stir Uproar," *CT,* 1 May 1988; Andrew Herrmann and Phillip J. O'Connor, "Mayor Aide's Ouster Sought," and "Excerpts from Lectures by Cokely," *CST,* 1 May 1988; Cheryl Devall, "Sawyer Stands by Aide Who Made Ethnic Slurs," *CT,* 2 May 1988; Philip Franchine and Tim Gerber, "Mayor Rejects Calls to Punish Aide for Racial, Religious Slurs," *CST,* 2 May 1988; Lipinski and Devall, "Cokely's 'Sorry' Fails to End Fury," Devall, "Pressure Builds to Fire Mayoral Aide," and "Mayor Sawyer's Strange Tolerance," *CT,* 3 May 1988; Lynn Sweet and Ray Hanania, "Mayoral Aide Issues Apology as Furor Grows," *CST,* 3 May 1988; Lipinski and Devall, "Cokely Apology Fails to End Fury," Lipinski and James Strong, "Sawyer, Top Aides Huddle on Cokely," and Bruce Dold, "Why Mayor Treads Lightly on Cokely," *CT,* 4 May 1988; Sweet and Hanania, "Sawyer Hedges on Cokely," *CST,* 4 May 1988; Sweet and Hanania, "Cokely Decision Due Today," and Vernon Jarrett, "A Primer on Buffoonery," *CST,* 5 May 1988; Lipinski and Baquet, "Sawyer Fires Cokely," and David McCracken, "Spertus Exhibit a Different Look at Tradition," *CT,* 6 May 1988; Sweet, "Cokely Is Fired," *CST,* 6 May 1988; Doug Cassel, "Is Tim Evans for Real?," *CR,* 16 March 1989; Obama, *DFMF,* p. 211; DJG interviews with Sheila Jager and Asif Agha. On Davis's background, see his 2010 interview with Terence Sims.

80. William Snider, "Chicago's 'Summit': A Populist Blueprint to Reshape Schools," *Education Week,* 30 March 1988, pp. 1, 19; Karen M. Thomas, "$210 Million Price Tag Put on School Reform," *CT,* 30 March 1988, p. 1; Thomas, "Lawmakers Cool to School Plan," *CT,* 31 March 1988, p. 1; Thomas, "Critics Say School Reform Falls Short," *CT,* 1 April 1988, p. 1; Robert Davis, "Criticism Mounts on Reform Plan," *CT,* 7 April 1988, p. 8; Linda Lenz, "School Summit Has Blowup," *CST,* 8 April 1988, p. 12; Chad Carlton, "Bill Seeks Reform of City Schools," *CT,* 14 April 1988, p. 2; Patrick J. Keleher 1988 pocket Day-Timers (18 ["Call Barack"] and 19 April 1988 [showing Barack at "LU"—Loyola University—at 6 P.M.]), PJKP; Rudolph Unger, "Some School Reformers Hit for Bypassing Rules," *CT,* 20 April 1988, p. 6; Snider, "State Board Calls for Chicago Reform Monitor," *Education Week,* 27 April 1988, p. 8; Daniel Egler and Tim Franklin, "Sawyer Backs State Income-Tax Hike," *CT,* 27 April 1988, p. 1; Lynn Sweet, "Sawyer Pushes Hike in State Income Tax," *CST,* 27 April 1988, p. 3; Thomas, "School Summit Rebuffs Legislators," *CT,* 29 April 1988, p. 1; Lenz, "School Summit Wants Outsiders Kept Out," *CST,* 29 April 1988, p. 16; Manford Byrd Jr., "School Board Working Hard for Reform," *CT,* 30 April 1988, p. 10; Lenz, "Scaling the Summit of City School Reform," *CST,* 1 May 1988, pp. 4–5, 32; Tracey Robinson, "New Aim: 'Local' Schools," *CST,* 2 May 1988, pp. 7–8; Lenz, "School Reform Role for France," and Neil Steinberg, "Can School 'Choice' Work?," *CST,* 3 May 1988, pp. 1, 6, 7; Lenz, "Schools Try to Boost Odds," *CST,* 4 May 1988, pp. 6–7; Alf Siewers, "Bad Teachers a Sour Spot," *CST,* 5 May 1988, pp. 7–8; Lenz, "School Reform Unit Urged," *CST,* 7 May 1988, p. 8; Cindy Richards, "Bottom Line: Education," *CST,* 8 May 1988, p. 11; Ezra Bowen, "A New Battle Over School Reform," *Time,* 9 May 1988; Bonita Brodt, "At Goudy, the Future Dies Early," and "Reforming the Nation's Worst Schools," *CT,* 15 May 1988, pp. 1, 2; Patrick Reardon and R. Bruce Dold, "No Clout and No Concern Add Up to No Education," *CT,* 17 May 1988, p. 1; Brodt, "For Many Pupils, Learning Never Starts at Home," *CT,* 18 May 1988, p. 1; Michele Norris and Jean Latz Griffin, "Pupils May Move On, but Inept Teachers Stay On," *CT,* 19 May 1988, p. 1; Egler and Thomas, "Legislators Take Lead on School Reform," *CT,* 20 May 1988, p. 1; Griffin and Jack Houston, "Bureaucrats Bungle, Students Pay the Price," *CT,* 22 May 1988, p. 1; Reardon, "Manford Byrd's View from the Top," *CT,* 22 May 1988, p. 11; Griffin and Norris, "Teachers Union Has Power to Run System," *CT,* 23 May 1988, p. 1; Reardon and Owen Youngman, "City Brings Up the Rear in Taxes for Education," *CT,* 24 May 1988, p. 1; Reardon and Merrill Goozner, "City's Willing to Pay for Good Schools, Poll Says," *CT,* 24 May 1988, p. 11; "How to Fix America's Worst Schools," *CT,* 29 May 1988, p. 2; Michael D. Klemens, "The Emerging Form of Chicago School Reform," *Illinois Issues,* June 1988, pp. 10–12; "Education Summit Challenges System, Community," [Chicago Community] *Trust Quarterly,* Summer 1988, pp. 24–26; Kyle and

Kantowicz, *Kids First,* pp. 243–47, 251–58, 265; Loretta Augustine-Herron in Abdon M. Pallasch, "Taught Residents to Lobby," *CST,* 24 August 2008, p. A15; DJG interviews with Loretta Augustine-Herron, Rosa Thomas, Aletha Strong, Ann West, Gwendolyn LaRoche Rogers, Danny Solis, Phil Mullins, Don Moore, Robert Gannett, Miguel del Valle, and Bill Ayers. Kyle and Kantowicz's superb *Kids First,* p. 265n12, cites "Key Hispanic and Black Community Organizations Endorse HB #3707," 16 May 1988, CURE-Haymarket Papers, which were later given to the Chicago Historical Society. The remade Chicago History Museum has failed to process those papers, reports that its inventory of them makes particular subject files impossible to locate, and is unable to retrieve them from its own remote storage facility. Debbie Vaughan e-mails to DJG, 6 and 8 April 2011.

81. 1989 University of Chicago Student Directory, p. 58; Invitation, Dick Longworth to Frank Lumpkin, 2 May 1988, Roberta Lynch to Lumpkin, n.d., and Program, "A Tribute to Frank Lumpkin," 22 May 1988, all FLP Box 6 Fld. 3; Pat Harper, "Lumpkin Honored at Dinner," *DC,* 23 May 1988, p. 3; "Hanging Tough," *DC,* 26 May 1988, p. 4; Obama, *DFMF,* p. 302; Obama in Jacob Weisberg, "The Path to Power," *Men's Vogue,* September 2006, pp. 218–23; Loretta Augustine-Herron in Christopher Andersen, *Barack and Michelle* (William Morrow, 2009), pp. 107, 113; Bruce Orenstein in McClelland, *Young Mr. Obama,* p. 178; Maraniss, *BOTS,* p. 570; DJG interviews with Sheila Jager, Mary Ellen Montes, Cynthia Norris, Loretta Augustine-Herron, Bruce Orenstein, Greg Galluzzo, Yvonne Lloyd, Dan Lee, Betty Garrett, Margaret Bagby, John Webster, Tom Kaminski, Len Dubi, and Cathy Askew.

82. Barack Obama to Sheila Jager, 29 May 1988; Obama to Cynthia Norris, n.d., Norris Papers; Chris Peachment, "Gently Does It, Bruno," *Times* (London), 22 June 1988; David Robinson, "Engagement Is Broken," *Times* (London), 23 June 1988; Nigel Andrews, "At Odds with the Angels," *FT,* 24 June 1988, p. 17; George Perry, "Angel Who Came in from the Cold," *Sunday Times* (London), 26 June 1988; Obama, *DFMF,* pp. 299–436, esp. 299–301, 306, 311, 328–29, 338–47, 381, 431–34, 436; Maina wa Kinyatti, *Kenya's Freedom Struggle: The Dedan Kimathi Papers* (Zed Books, April 1988); Obama on *Fresh Air* with Dave Davies, NPR, 12 August 2004; Obama on *The Friday Night Show* with Bob Sirott, WTTW, 3 December 2004; Laurie Goering, "Obama Campaign Closely Watched—in Kenya," *CT,* 10 October 2004, p. 1; Obama Commencement Address, University of Massachusetts at Boston, 2 June 2006; Obama, "An Honest Government, a Hopeful Future," University of Nairobi, 28 August 2006; Macharia Gaitho and Julie Gichuru, "Obama: 'I Speak What I Think Is True and Say It Best,'" *Nation* (Nairobi), 1 September 1986; Carole Norris Green, "A Postcard from Barack," *Catholic San Francisco,* 30 January 2009, p. 13; Auma Obama in "Obama as We Knew Him," *Observer,* 26 October 2008, pp. 4–7; Auma Obama, *And Then Life Happens,* pp. 189–204; George Obama with Damien Lewis, *Homeland* (Simon & Schuster, 2010), pp. 17–18; Mike W. Thomas, "Technology Is Helping Secure Firm's Future," *San Antonio Business Journal,* 8 August 2010; Zeituni Onyango, *Tears of Abuse* (Afripress, 2012), p. 196; Ezra Ogosa Obama, aka Obeid, in Malik Obama, *Barack Obama Sr.* (Xlibris, 2012), p. 320; Kelvin Chann (AP), "Obama's Brother Writes About Abuse," 18 December 2013; Jenni Marsh, "The Other Obama," *South China Morning Post,* 22 March 2014; Mark Obama Ndesandjo, *An Obama's Journey* (Lyons Press, 2014), pp. 144–58, 365–72; Jason Horowitz, "The Catholic Roots of Obama's Activism," *NYT,* 23 March 2014, p. A1; DJG interviews with Sheila Jager, Cynthia Norris, Cathy Askew, Hasan and Raazia Chandoo, and Zeituni Onyango. The book about Africa Barack read while in Europe was probably either David Lamb's *The Africans* (Random House, 1983) or, perhaps more likely, Sanford J. Ungar's *Africa: The People and Politics of an Emerging Continent* (Simon & Schuster, 1985). See Alan Cowell, "Inside Africa 30 Years Later," *NYTBR,* 6 February 1983, pp. 7, 14, Xan Smiley, "Sub-Saharan Balance Sheet," *NYTBR,* 1 September 1985, pp. 4–5, and Gilbert Khadiagala's untitled review of both books in *SAIS Review of International Affairs* 6 (Winter–Spring 1986): 255–57. Note as well Elspeth Huxley, *Out in the Midday Sun: My Kenya* (Viking, 1987), and Mort Rosenblum and Doug Williamson, *Squandering Eden: Africa at the Edge* (Harcourt Brace Jovanovich, 1987). See Louisa Dawkins, "Steaming to Mombasa," *NYTBR,* 22 March 1987, p. 20, and James Brooke, "A Continent Left Behind," *NYTBR,* 29 November 1987, pp. 7, 9.

83. Johnnie Owens in Linda Matchan, "A Law Review Breakthrough," *BG,* 15 February 1990, p. 29; Obama, "Community Revitalization," Nebraska Wesleyan University Forum, 9 September 1994; Obama's 7 February 2004 interview with David Axelrod; Michael Levenson and Jonathan Saltzman, "At Harvard Law, A Unifying Voice," *BG*, 28 January 2007, p. A1; Secter and McCormick, "Portrait of a Pragmatist," *CT*, 30 March 2007; Cassandra Butts in "A Rusty Toyota, a Mean Jump Shot, Good Ears," *WP*, 11 February 2007, p. B3; Owens in Smith, "Ascent of Mr. Charisma," *Sunday Times Magazine*, 23 March 2008, in Indira A. R. Lakshmanan, "Obama Draws on Lessons from Chicago Streets to Propel Campaign," Bloomberg, 3 July 2008, on *Tell Me More*, NPR, 23 July 2008, in McClelland, *Young Mr. Obama*, p. 181, and in Lisa Lerer, "Obama Decades-Old Shooting Scare Guides Anti-Poverty Plan," Bloomberg, 19 February 2013; Kellman on "Obama Presidential Announcement Reaction," C-Span, 10 February 2007, in Brian DuBose, "Obama's Early Near-Miss," *WT*, 26 July 2007, p. A1, and in Liza Mundy, "A Series of Fortunate Events," *WP Magazine*, 12 August 2007, pp. 10ff.; Jeremiah Wright in Erin Meyer, "Where Obama Developed His Audacity of Hope," *HPH*, 14 February 2007, p. 20; Martin Peretz, "Standing by His Man," *TNR*, 23 April 2008 (Wright as "his father's vagrant presence"); Wright on Robert Beckford, "God Bless You Barack Obama?," BBC 2, 25 January 2010; DJG interviews with Sheila Jager, Johnnie Owens, Mike Kruglik, Jean Rudd, Ken Rolling, Richard J. Kaplan, Bruce Orenstein, Mary Ellen Montes, Alma Avalos, Marlene Dillard, Dominic Carmon, Margaret Bagby, Hazel Johnson, Bob Klonowski, Yvonne Lloyd, George Schopp, Mary Ryan, Cathy Askew, Loretta Augustine-Herron, Dan Lee, Cynthia Norris, Jerry Kellman, Zeituni Onyango, Hermene Hartman, and Jeremiah Wright.

On Chicago's Southeast Side landfill developments, see Robert Bergsvik, "Health Official Warns of Leaks," *DC*, 10 June 1988, pp. 1, 14; Bergsvik, "Officials Mum on Extension of Landfill Ban," and "Extend Moratorium," *DC*, 12 June 1988, pp. 1, 4, 14; Vi Czachorski, "From the Editor's Scratch Pad," *HN*, 16 June 1988, p. 1, and 28 July 1988, pp. 1, 5; Czachorski, "Waste Management Holds Closed Meeting," *HN*, 11 August 1988, pp. 1, 4; Final Report of Mayor's Southeast Chicago Community Task Force on the Environment and Solid Waste Disposal, 24 August 1988, RGP; Press Release, "Mayor Sawyer Supports Solid Waste Task Force Recommendations," 21 September 1988, JLP Box 38 Fld. Chicago Works Together; Cheryll Devall, "Southeast Side Group Wants End to New Dumps," *CT*, 22 September 1988, p. C2; Ray Hanania, "Bar 134th St. Dump, S.E. Siders Ask Sawyer," *CST*, 22 September 1988; "SE Side Task Force Demands Landfill Tradeoffs," *TNW*, October–November 1988, pp. 25, 28; People for Community Recovery, "Crossroads Fund's Grantee Fiscal and Progress Report," n.d. [ca. November 1988], CFP Box 30 Fld. 367; *Citizens for a Better Environment, United Neighborhood Organization of Southeast Chicago & Mary Ellen Montes v. Thomas*, 704 F. Supp. 149 (N.D. Ill.), 10 January 1989; Stevenson Swanson, "City Toughens Its Landfill Rules," *CT*, 9 March 1989, p. C3; Steven A. Salzman to Jean Anne Kingrey (DOJ), 31 October 1989, RGP; Swanson, "Waste Firm Deals for Landfill Gap," *CT*, 24 November 1989, p. 8; Salzman to Harlee Strauss, 12 December 1989, RGP; Larry Galica, "Residents Protest Landfill Proposal," *Northwest Indiana Times*, 24 May 1994; AP, "3 Ill. Bank Executives Indicted," 6 February 1998; Matt O'Connor, "2 Banks Admit to Looting Funds for Political Use," *CT*, 24 June 1999; "Former Bank Chief Pleads Guilty in Thefts of Unclaimed Cash," *CT*, 8 October 1999; "18-Month Sentence for Ex-Bank President," *CT*, 21 January 2000; Casey Sanchez, "Building Power," *Chicago Reporter*, January 2006; Dan Mihalopoulos and Azam Ahmed, "A Rising Force in Hispanic Chicago," *CT*, 22 June 2009; Mihalopoulos, "A Lifetime of Close Ties and Growing Influence," *NYT*, 14 January 2012; DJG interviews with Howard Stanback, Bruce Orenstein, and Mary Ryan.

On school reform developments, see C. D. Matthews, "UNO Wants More Parent Participation in Schools," *DC*, 27 May 1988, pp. 1, 11; Daniel Egler and Jean Latz Griffin, "House Expected to Hear Reform Proposals," *CT*, 1 June 1988, p. 1; Karen M. Thomas, "House Democrats Wait to Hear From Chicago in School Bills," *CT*, 2 June 1988, p. 7; "How Lawmakers Can Help the Schools," *CT*, 5 June 1988, p. 2; Thomas and Constanza Montana, "School Board Seeks Reform Now," *CT*, 7 June 1988, p. 6; Maudlyne Ihejirika, "2,500 at Rally Demand School Reform Measure," *CST*, 7 June 1988, p. 2; William Snider, "Reform Measure for Chicago Schools Unveiled," *Education Week*, 8 June 1988, p. 9; Devall and Thomas, "Oversight Board Urged for Schools," *CT*, 9 June 1988, p. 1; Thomas and Griffin, "Teachers May Bend for School Reform," *CT*, 15 June 1988, p. 1; Peter S. Willmott, "Essentials of School Reform," *DC*, 22 June 1988, p. 5; Jack Houston, "School

Board Exec Quits in Frustration," *CT,* 23 June 1988, p. 1; Griffin and Thomas, "Democrats Offer School Plan," *CT,* 24 June 1988, p. 1; Thomas and Griffin, "General Assembly Limping into Home Stretch," *CT,* 26 June 1988, p. 1; Griffin and Thomas, "Democrats' School Plan Under Fire," *CT,* 28 June 1988, p. 1; Dave Roeder, "School Reform Consensus Reached," *DC,* 29 June 1988, pp. 1, 16; Roeder, "School Reform Bill Faces GOP Hurdle," *DC,* 30 June 1988, pp. 1, 12; Lynn Sweet, "Plan Gives Governor Bigger Role on Schools," *CST,* 2 July 1988, p. 6; Sweet, "School Bill OK'd," *CST,* 3 July 1988, pp. 1, 10; Griffin, "School Plan Now Faces the Real Test," *CT,* 3 July 1988, p. 12; Devall and Thomas, "City School Reforms Add Spark but Lack Funding Fuel," *CT,* 4 July 1988, p. 8; Sweet, "School Funding Level Hit," *CST,* 4 July 1988; Sweet, "Praise, Caution Greet School Reforms," *CST,* 5 July 1988, p. 20; Barbara Marsh, "Small School Win Leaves Reformers with Tough Job," *CCB,* 11 July 1988, p. 3; Patrick Reardon, "School Reform Bill Hurts Poor Children, Byrd Says," *CT,* 16 July 1988, p. 5; Sweet, "Byrd Rips School Bill," *CST,* 16 July 1988, p. 12; Al Raby and Kale Williams, "School Reform a Triumph of People," *CT,* 27 July 1988, p. 18; Mary O'Connell, "Organizing Effort Pays Off for Schools," *TNW,* August–September 1988, pp. 3–6; Michael D. Klemens, "Chicago Schools: Reform to Come but No Money," *Illinois Issues,* August–September 1988, pp. 43–44; William Snider, "Illinois Awaiting Governor's Pen on Chicago Plan," *Education Week,* 3 August 1988; Florence Hamlish Levinsohn, "School Revolt," *CR,* 26 May 1989; Moore, "Voice and Choice," pp. 167–171; Mary O'Connell, *School Reform Chicago Style,* pp. 16–21 and esp. 40, which lists DCP as a June 1988 member of the ABCs Coalition and "Prof. William Ayers" as ABCs contact person; Hess, *School Restructuring,* pp. 59–78; Kyle and Kantowicz, *Kids First,* pp. 257, 262–64, 267–79; Dan A. Lewis and Kathryn Nakagawa, *Race and Educational Reform in the American Metropolis* (SUNY Press, 1995), pp. 85–91; Shipps, "Big Business and School Reform," pp. 273–75; Shipps, *School Reform, Corporate Style,* pp. 115–27; DJG interviews with Danny Solis, Phil Mullins, and Bill Ayers.

CHAPTER FIVE: EMERGENCE AND ACHIEVEMENT

1. Obama Onyango, "Soul to Soul," *Ebony,* August 1971, p. 14; "Notice of Federal Tax Lien," #49017340, 11 April 1990 (Omar O. Obama) & #49245786, 10 June 1992 (Obama O. Onyango); Michael Levenson and Jonathan Saltzman, "At Harvard Law, A Unifying Voice," *BG,* 28 January 2007, p. A1; Rachel Rose-Sandow, "Obama's Somerville Apartment a Source of Pride," GateHouse News Service, 14 October 2008; James Bone, "Found in a Rundown Boston Estate," *Times* (London), 30 October 2008; Alan Wirzbicki, "Obama's Old Digs," *BG,* 7 March 2011; Denise Lavoie (AP), "Obama Uncle Held in Mass. by Immigration Officials," 29 August 2011; Billy Baker and Glen Johnson, "Obama Kin Arrested on DUI Charge," *BG,* 30 August 2011; Maria Sacchetti and Dan Adams, "Obama's Uncle Is Called a Fugitive," *BG,* 31 August 2011; Sacchetti and John R. Ellement, "Obama's Uncle Set to Fight Deportation," *BG,* 1 September 2011; Sally Jacobs, *The Other Barack* (Public Affairs, 2011), pp. 151–53; Jacobs, "The Trials of Omar, Obama's Uncle," *BG,* 8 January 2012; Clennon L. King, "Few Neighbors Recall Obama During Harvard Years," *Somerville Times,* 19 May 2013; Matt Byrne, "Obama's Stay in Somerville Explored by Students," *BG,* 21 September 2013; Sacchetti, "Judge Says Obama's Uncle Can Stay in U.S.," *BG,* 3 December 2013; Sacchetti, "In Reversal, Obama Says He Lived with Uncle," *BG,* 5 December 2013; DJG interviews with Zeituni Onyango.

Harvard Law School 1988–89 Catalog, esp. pp. 4–5, 16, 65–67; Dan Kroll, "Bell Questions Minority Hiring," David Snouffer, "Denied Again, Dalton Waits," and Andrew Pollis and Jeff Gershowitz, "Jackson Fleeing Faculty Turmoil?," *Harvard Law Record* [*HLRec*], 15 April 1988, pp. 1, 4, 5, 7, and 14; "Vorenberg Steps Down from Deanship," *HLRec,* 29 April 1988, pp. 1, 4, 14, and 16; "Black Law Students Occupy Harvard Law School Dean's Office Protesting Faculty Hiring Practices," BLSA Press Release, 10 May 1988, p. 25 of *Special Edition—BLSA Memo,* Vol. 1, No. 1, Spring/Fall 1988, 50 pp., Harvard Law School Archives [HLSA] (also esp. p. 7); Michelle Robinson, "Minority and Women Law Professors: A Comparison of Teaching Styles," in *Special Edition—BLSA Memo,* pp. 30–33; Jonathan S. Cohn, "Bok: New Law Dean by Spring," *Harvard Crimson* [*HC*], 14 September 1988; Erick Hachenburg, "One L Class Is Select & Diverse," *HLRec,* 16 September 1988, pp. 1, 9; "Dean Search Gets Student Input," *HC,* 28 September 1988; Dan Kroll and Hachenburg, "Interesting First Years Arrive at HLS," *HLRec,* 7

October 1988, p. 3; Tara A. Nayak, "A Confident Vision in Turbulent Times," *HC,* 8 June 1989; Allan R. Gold, "Departing Dean Looks Back at Dream and Reality," *NYT,* 23 June 1989, p. B10; Linda Matchan, "Harvard Law Dean Leaving with Sadness," *BG,* 28 June 1989, p. 29; Richard D. Kahlenberg, *Broken Contract: A Memoir of Harvard Law School* (Hill & Wang, 1992), esp. pp. 5–7, 17, 120; Eleanor Kerlow, *Poisoned Ivy* (St. Martin's Press, 1994), p. 51; William Glaberson, "James Vorenberg, Watergate Prosecutor's Right-Hand Man, Dies at 72," *NYT,* 13 April 2000; Harvard Law School, *Report on the State of Black Alumni, 1869–2000* (HLS, 2002), esp. pp. 18, 21; George W. Hicks Jr., "The Conservative Influence of the Federalist Society on the Harvard Law School Student Body," *Harvard Journal of Law & Public Policy* 29 (Spring 2006): 623–712, esp. pp. 684, 701; David B. Wilkins, "The New Social Engineers in the Age of Obama," *Howard Law Journal* 53 (Spring 2010): 557–644, esp. p. 560; Kevin K. Washburn, "Elena Kagan and the Miracle at Harvard," *Journal of Legal Education* 61 (August 2011): 67–75, esp. p. 67; Robert Granfield, "Legal Education as Corporate Ideology: Student Adjustment to the Law School Experience," *Sociological Forum* 1 (June 1986): 514–23; Granfield and Thomas Koenig, "From Activism to Pro Bono: The Redirection of Working Class Altruism at Harvard Law School," *Critical Sociology* 17 (April 1990): 57–80, esp. pp. 58, 71, 74, 78; Granfield, "Making It by Faking It: Working-Class Students in an Elite Academic Environment," *Journal of Contemporary Ethnography* 20 (October 1991): 331–51; Granfield and Koenig, "The Fate of Elite Idealism: Accomodation and Ideological Work at Harvard Law School," *Social Problems* 39 (November 1992): 315–31, esp. 318, 320, 327; Granfield and Koenig, "Learning Collective Eminence: Harvard Law School and the Social Production of Elite Lawyers," *Sociological Quarterly* 33 (Winter 1992): 503–20; Granfield, *Making Elite Lawyers: Visions of Law at Harvard and Beyond* (Routledge, 1992); DJG interviews with Linda Singer and Verna Myers.

The literature plumbing the substance and impact of Roberto Unger's scholarly writings is truly vast. From the perspective of 1988, the best frame of reference for grasping Unger's import is offered by the fourteen articles that comprise the Summer 1987 issue (Vol. 81, No. 4) of *Northwestern University Law Review.* See also Eyal Press, "The Passion of Roberto Unger," *Lingua Franca,* March 1999. Regarding BLSA, no other issues of *BLSA Memo* are held by the HLS Archives, and no African American law student from 1988–91 recalls any BLSA newsletter.

2. *HLS Adviser,* Vol. 20, #1, 1 September 1988, p. 5; Harvard Law School 1988–89 Catalog, pp. 22–24; Harvard Law School First Year Students 1988–1989, Rob Fisher Papers [RFP]; Section 3 List (incomplete—99 names), Gina Torielli Papers [GTP]; Robert M. Fisher, *The Logic of Economic Discovery: Neoclassical Economics and the Marginal Revolution* (NYU Press, 1986), esp. chapter 2; A. W. Coats, review of *The Logic of Economic Discovery, Kyklos* 40 (March 1987): 118–19; Philip Mirowski, review of *The Logic of Economic Discovery, Journal of Economic History* 47 (March 1987): 295–96; P. J. Eijgelshoven, review, *De Economist* 135 (July 1987): 257–59 ("clearly written"); DJG interviews with Rob Fisher, Richard Cloobeck, Eric Collins, Diana Derycz-Kessler, Mark Kozlowski, Jennifer (Radding) Gardner Trulson, Sarah Leah Whitson, and Scott Becker.

On Lakatos, see especially his own posthumous *Proofs and Refutations: The Logic of Mathematical Discovery* (Cambridge University Press, 1976), Jancis Long, "Lakatos in Hungary," *Philosophy of the Social Sciences* 28 (June 1998): 244–311, Matteo Motterlini, ed., *For and Against Method* (University of Chicago Press, 1999), John Kadvany, *Imre Lakatos and the Guises of Reason* (Duke University Press, 2001), Alex Bandy, *Chocolate and Chess: Unlocking Lakatos* (Akademiai Kiado, 2010), and Kadvany, "Chocolate and Chess (Unlocking Lakatos)," *Philosophy of the Social Sciences* 42 (June 2012): 276–86.

3. HLS Schedule of Courses 1988–89, HLSA; *HLS Adviser,* #1, 1 September 1988, pp. 6–7; Ian R. Macneil, *Contracts: Exchange Transactions and Relations,* 2nd ed. (Foundation Press, 1987), pp. xvii, xix; DJG interviews with Alicia Rubin Yamin, Michelle Jacobs DeLong, David Attisani, Eric Collins, Richard Cloobeck, Amy Christian McCormick, Lisa Hay, Martin Siegel, Mark Kozlowski, Morris Ratner, David Troutt, Paolo Di Rosa, Jennifer (Radding) Gardner Trulson, Greg Sater, Sarah Leah Whitson, Rob Fisher, Tim Driscoll, Gina Torielli, Jackie Fuchs, Shannon Schmoyer, Steve Berkow, Diana Derycz-Kessler, Ken Mack, Lauren Ezrol Klein, Sima Sarrafan, Erin Edmonds, David Smail, Leonard Feldman, Lisa Paget-Kahn, Jonathan Z. King, and Yi-Fun Hsueh.

4. Melanie Eversley, "Best Supporting Actor," *The Root,* 1 October 2008; Obama in Meg

Vaillancourt, "Derrick Bell Threatens to Leave Harvard," *Ten O'Clock News,* WGBH, 24 April 1990, bostonlocaltv.org; Dawn Ross, "Program Offers Minority Survival Kit," *HLRec,* 30 September 1988, p. 3; "Obama Mentor Charles Ogletree Shares Insight on Our President," Essence.com, 16 December 2009; DJG interviews with (Frank) Hill Harper, Kevin Little, Leon Bechet, Kenny Smith, David Hill, Jeffrey Selbin, Paolo Di Rosa, Jan-Michele Lemon Kearney, Ursula Dudley Oglesby, Cassandra Butts, Sheryll Cashin, Christine Lee, Alan Jenkins, and Erin Edmonds.

5. DJG interviews with Sheila Jager, Asif Agha, Rob Fisher, and Cassandra Butts.

6. *Three Speech* #1, 26 September 1988, #2, 2 October 1988, #3, 11 October 1988, #4, 17 October 1988, #5, 25 October 1988, #6, 31 October 1988, #7, 7 November 1988, #8, 15 November 1988, #9, 22 November 1988, #10, 6 December 1988, #11, 13 December 1988, #12, 14 December 1988, Mark Kozlowski Papers [MKP]; King, "Few Neighbors Recall Obama," *Somerville Times,* 19 May 2013; George P. Hassett, "Obama Finally Pays Local Parking Tickets," *Somerville News,* 7 March 2007; David Abel, "Obama Paid Late Parking Tickets," *BG,* 8 March 2007, p. B1; Rudy Rodriguez, "Federalists Cry Foul Over Coalition Seat Assignments," *HLRec,* 21 October 1988; Tara A. Nayak, "Sitting In and Speaking Out in a Search for Change," *HLRec,* 8 June 1989; Andrew Blake, "Jackson, in N.E., Stresses Priorities Over Passion," *BG,* 25 October 1988, p. 12; Joanne Ball, "Report Urges More Blacks on Faculty at Harvard," *BG,* 25 October 1988, p. 17; Erick Hachenburg, "HLS Students Not Afraid of 'L' Word," *HLRec,* 4 November 1988, p. 1; Jonathan S. Cohn, "Law Profs Debate Dean's Role," *HC,* 17 November 1988; Allan R. Gold, "Harvard Awaits Appointment of New Law Dean," *NYT,* 2 December 1988, p. B9; Kahlenberg, *Broken Contract,* pp. 166–67; Ken Mack in "Barack Obama Before He Was a Rising Political Star," *Journal of Blacks in Higher Education,* Autumn 2004, pp. 99–101; Mack, remarks at John F. Kennedy School of Government, 9 April 2007; Mack, "Even at Harvard, Obama Had a Knack for Bonding with Diverse People," *Harrisburg Patriot-News,* 17 February 2008; Mack on *Anderson Cooper 360,* CNN, 15 April 2008; Mack's 2012 interview with Jim Gilmore; Joe Fernandez and Emily Maranjian in M. Charles Bakst, "Brown Coach Robinson a Strong Voice for Brother-in-Law Obama," *Providence Journal,* 20 May 2007, p. D1; Lisa Hay on Neal Conan, "Who Is Barack Obama?," *Talk of the Nation,* NPR, 19 August 2008; Ayelet Waldman in Mark Matthews, "Classmate Speaks Fondly of Obama," KGO-TV, San Francisco, 25 August 2008; Jackie Fuchs, "Why Barack Obama Reminds Me of Joan Jett," Huffington Post, 26 August 2008; Jerry Sorkin, "The Great Obama," NouvelleBlogger, 8 November 2008; Donald Gaffney in Don J. DeBenedictis, "Judicial Profile: Hon. Donald F. Gaffney," *Los Angeles Daily Journal,* 16 March 2009; Edward Felsenthal in Richard Alley, "Edward Felsenthal Tames Web," *Memphis Commercial Appeal,* 8 November 2009; Clifford Krauss, "Lawyer Who Beat Chevron in Ecuador Faces Trial of His Own," *NYT,* 31 July 2013; DJG interviews with David Shapiro, Morris Ratner, Edward Felsenthal, Jackie Fuchs, Brad Wiegmann, Ken Mack, Jan-Michele Lemon Kearney, Jeffrey Ellman, Steven Heinen, Paolo Di Rosa, Sarah Leah Whitson, Richard Cloobeck, Rob Fisher, Amy Christian McCormick, Sherry Colb, Roger Boord, Jonathan Z. King, Sima Sarrafan, Alicia Rubin Yamin, David Troutt, Lisa Paget-Kahn, Martin Siegel, Scott Sherman, Steven Berkow, Greg Sater, David Attisani, Timothy Driscoll, Jennifer (Radding) Gardner Trulson, Erin Edmonds, Lauren Ezrol Klein, John "Vince" Eagan, Jackie Fuchs, Gina Torielli, Roger Boord, David Smail, Leonard Feldman, Lisa Hay, Diana Derycz-Kessler, Yi-Fun Hsueh, and Rachel Cano. For a professorial discussion of gunners and "turkey bingo," see Robert M. Lloyd, "Why Every Student Should Be a Gunner," University of Tennessee College of Law, 16 September 2011.

7. *Three Speech* #3, 11 October 1988, #6, 31 October 1988, MKP; David Rosenberg, "The Causal Connection in Mass Exposure Cases: A 'Public Law' Vision of the Tort System," *HLR* 97 (February 1984): 849–929, esp. 928; Rosenberg, "The Dusting of America: A Story of Asbestos—Carnage, Cover-Up, and Litigation," *HLR* 99 (May 1986): 1693–1706; Rosenberg, "Of End Games and Openings in Mass Tort Cases: Lessons From a Special Master," *Boston University Law Review* 69 (May 1989): 695–730, esp. 695 and 698; DJG interviews with Rob Fisher, David Rosenberg, Mark Kozlowski, Richard Cloobeck, Amy Christian McCormick, Eric Collins, David Attisani, Alicia Rubin Yamin, Steven Berkow, Diana Derycz-Kessler, and Jennifer Gardner Trulson.

8. Harvard Law School 1988–89 Catalog, esp. p. 5; *HLS Adviser* #2, 2 September 1988,

p. 2; Harvard Law School *Yearbook 1989,* pp. 14, 145 (Obama fails to appear in the alphabetical presentation of the 1991 class); HLS Schedule of Courses 1988–89, HLSA; Mary Mitchell, "Memoir of a 21st Century History Maker," *Black Issues Book Review,* January–February 2005, pp. 18–21; Cassandra Butts in David Mendell, *From Promise to Power* (Amistad, 2007), p. 92, on NPR's "Tell Me More," 22 February 2007, in Larissa MacFarquhar's superb "The Conciliator: Where Is Barack Obama Coming From?," *New Yorker,* 7 May 2007, pp. 48ff., and in John Heilemann, "When They Were Young," *New York Magazine,* 22 October 2007; DJG, *BTC,* p. 7; Butts's July 2008 interview with Jim Gilmore; Gina Torielli in David Jackson and Ray Long, "Obama Knows His Way Around a Ballot," *CT,* 4 April 2007; Jackie Fuchs, "Why Barack Obama Reminds Me of Joan Jett," Huffington Post, 26 August 2008; Maina Kiai, "Kenya: Learning from the Obama Campaign," *Nairobi Star,* 9 September 2008; Richard Mauer, "Path Cleared in Ice, Whales Swim Free," *NYT,* 27 October 1988; Cassandra Butts in Linda Kramer Jenning, "Meet the Women Who Can Handle Anything!," *Glamour,* February 2009; DJG interviews with Kenny Smith, Paolo Di Rosa, Brad Wiegmann, Martin Siegel, Steven Berkow, Tim Driscoll, Greg Sater, Rob Fisher, Ken Rothwell, Cassandra Butts, Gina Torielli, David Troutt, Scott Sherman, Lisa Paget-Kahn, Ursula Dudley Oglesby, David Attisani, Sherry Colb, Jennifer Gardner Trulson, Mark Kozlowski, Sarah Leah Whitson, Rachel Cano, Diana Derycz-Kessler, Erin Edmonds, David Hill, Jackie Fuchs, Alan Jenkins, James Esseks, Scott Becker, Julius Genachowski, Dan Rabinovitz, and Thom Thacker.

9. Harvard Law School 1988–89 Catalog, esp. pp. 67–68 and 170; First Year Ames Committee to All Students in Sections I, III, and IV, "Dates for 1989 First-Year Ames," 23 September 1988, and Board of Student Advisers First-Year Ames Committee to All Students in Sections I, III, and IV, "First-Year Ames Moot Court Program," 31 October 1988, GTP; *HLS Adviser* #15, 8 December 1988, p. 4; Mahmood in Maraniss, *BOTS,* p. 497; "Civil Rights Organizer Al Raby, 55," *CT,* 24 November 1988; Don Terry, "Albert Raby, Civil Rights Leader in Chicago with King, Dies at 55," *NYT;* John Camper, "Al Raby's Friends Tell of His Work," *CT,* 30 November 1988; Judith S. Kleinfeld and Suzanne Yerian, eds., *Gender Tales: Tensions in the Schools* (St. Martin's Press, 1995), pp. 179–92, at 182; *Three Speech* #10, 6 December 1988; DeBenedictis, "Judicial Profile: Hon. Donald F. Gaffney," *LADJ,* 16 March 2009; Exams, Civil Procedure III and IV, 9 January 1989, Criminal Law III, 11 January 1989, and Torts IIIa, 13 January 1989, *Harvard Law School Annual Examinations in Law 1988–89,* pp. 19–23, 125–30, and 186–91; DJG interviews with Rob Fisher, Mark Kozlowski, Lisa Hertzer Schertler, Scott Becker, Beenu Mahmood, Asif Agha, Michelle Jacobs DeLong, Tim Driscoll, David Troutt, Scott Sherman, Gina Torielli, John Levi, and Geraldine Alexis.

10. "A Guide to Regional & Local Grantmakers," [1988], p. 205 (noting a $25,000 Woods Fund grant to LIFT on 21 June 1988), Human SERVE Papers Box 16 Fld. 742; Stephen Rynkiewicz, "Back-to-Roots Look Could Spread Further," *CST,* 6 November 1988; "Group Targets Abandoned, Run-Down Buildings," *Gary Post-Tribune* [*GPT*], 5 February 1989; "Gary Gets a LIFT," *GPT,* 8 February 1989; Woods Charitable Fund, Report for the Year 1989, esp. pp. 20–21; Wieboldt Foundation, Annual Report for 1989, p. 12; Peg Spindler in Jeff Manes, "'Sister Peg' Gives Her All for Those with Nothing," *GPT,* 11 January 2009; Thomas D. Rush, *Reality's Pen: Reflections on Family, History & Culture* (Mill City Press, 2012), pp. 55–56, 96–103, 231; Jerry Kellman's 2008 inteview with Jim Gilmore; Greg Galluzzo's 2009 interview with Don Elmer; Obama, *TAOH,* p. 328; DJG interviews with Jerry Kellman, Mike Kruglik, John Owens, Mary Gonzales, Greg Galluzzo, Danny Solis, Phil Mullins, Todd Dietterle, David Kindler, Betty Garrett, Aletha Strong Gibson, Dan Lee, Nadyne Griffin, Alvin Love, Desta Houston, Jean Rudd, Ken Rolling, Mary Ellen Montes, Thomas D. Rush, John Levi, and Geraldine Alexis.

Lena remarried in 1992, with George Schopp performing the ceremony. "Before I got married I felt it was important to throw away things," namely "letters from Barack and postcards," she explained in 2010. See also Ben Joravsky, "Trouble in Paradise," *CR,* 25 July 1991, and Maria Elena Montes, "New Airport Offers South Side Hope," *CT,* 14 February 1992. Additional spring 1989 *GPT* stories mentioning LIFT appeared on 19 February, 15, 21, and 27 March, 21, 25, and 27 April, and 10 and 13 May 1989.

On school reform developments during the last four months of 1988, see Karen M. Thomas, "School Reform Bill Is Amended," *CT,* 27 September 1988, p. 1; Lynn Sweet and Linda Lenz,

"Thompson Alters and Inks School Reform Bill," *CST*, 27 September 1988, p. 7; Thomas and Jack Houston, "Teacher Union Vows to Fight Reform Bill," *CT*, 28 September 1988, p. 6; Michael D. Klemens, "Governor Steps into Chicago School Reform," *Illinois Issues*, November 1988, p. 25; Rob Karwath, "School Law Deal May Be Near," *CT*, 24 November 1988, p. 6; Patricia Smith, "Kids Lobby for School Reform," *CST*, 29 November 1988, p. 26; Daniel Egler and Jan Crawford, "School Reform Bill Approved," *CT*, 2 December 1988, p. 1; Thomas, "City School Reform Is Signed into Law," *CT*, 13 December 1988, p. 1; William Ayers, "Chicago's Schools: The Real Work Can Get Underway," *CT*, 19 December 1988, p. 19; Klemens, "Chicago School Reform Signed into Law," *Illinois Issues*, January 1989, p. 28; David Bednarek, "Supporting Leaders for Tomorrow: Chicago Business Leadership and School Reform," Institute for Educational Leadership, 1989, esp. pp. 21, 23; Donald R. Moore, "Voice and Choice in Chicago," in Clune and Witte, eds., *Choice and Control in American Education, Volume 2* (Falmer Press, 1990), pp. 171–72; Kyle and Kantowicz, *Kids First*, pp. 281–83; Jeffrey Mirel, "School Reform, Chicago Style: Educational Innovation in a Changing Urban Context, 1976–1991," *Urban Education* 28 (July 1993): 116–49, esp. p. 137; and Rosetta Vasquez, *Reforming Chicago Schools* (LEPS Press, 1994), pp. 13–58. In context, see also John McCarron's seven-part series of *Chicago Tribune* articles, "Chicago on Hold: The New Politics of Poverty," which ran from 28 August 1988 through 4 September 1988, with the first ("'Reform' Takes Cost Toll," 28 August) and seventh (Patrick Reardon and McCarron, "City Leaders Blew Schools Reform Chances," 4 September 1988) addressing schools. See as well McCarron, "The Timid City," *CT*, 4 September 1988, p. 1; "Exploiting the Housing Shortage," *CT*, 7 September 1988, p. 20; Robert Mier, "City Official Rebuts Tribune's Development Series," *CT*, 16 October 1988, p. P1; and McCarron's response to Mier, "It's Past Time We Saw Some Results," *CT*, 16 October 1988, p. P6.

11. Harvard Law School 1988–89 Catalog, esp. p. 71; *HLS Adviser* #19, 12 January 1989, p. 3; Syllabus, American Legal History: 1760–1900, Spring 1989, William "Terry" Fisher Papers; *Three Speech* #13, 25 January 1989, *Three Speech* #14, 2 February 1989, MKP; *HLS Adviser* #23, 9 February 1989, pp. 3–4; *Three Speech* #15, 14 February 1989, MKP; *HLS Adviser* #24, 16 February 1989, p. 3; *HLS Adviser* #25, 23 February 1989, p. 2; *Three Speech* #16, 3 March 1989, *Three Speech* #17, 27 April 1989, *Three Speech* #18, 9 May 1989, *Three Screech* #666 Abort Section 3 (by Jackie Fuchs), MKP; Paul Hutcheon, "The Scottish Professor Who Moulded the Young Obama," *Scotland Sunday Herald*, 8 June 2008; DJG interviews with Rob Fisher, Mark Kozlowski, Cassandra Butts, David Attisani, Leonard Feldman, David Troutt, Jackie Fuchs, Jennifer Gardner Trulson, Martin Siegel, Sherry Colb, Paolo Di Rosa, Rob Fisher, Ken Mack, Richard Cloobeck, Edward Felsenthal, Tim Driscoll, Shannon Schmoyer, Roger Boord, Jan-Michele Lemon Kearney, Brad Wiegmann, and Steven Heinen. Glendon refuses to respond to requests to comment about her recollections of Obama. See Glendon's 27 April 2009 letter to University of Notre Dame president Father John Jenkins, declining to receive an honorary medal because her former student, whom she called "a prominent and uncompromising opponent of the Church's position on issues involving fundamental principles of justice," would be delivering the commencement address on the same platform. "Declining Notre Dame," FirstThings.com.

12. Troy Morgan, "Minority President, Supervising Editor Elected to Review," *HLRec*, 17 February 1989, pp. 1, 12; Tara A. Nayak, "Law Review Chooses Masthead," *HC*, 8 March 1989; Nayak, "Law School Dean Search Near Finish," *HC*, 1 February 1989; Jonathan S. Cohn, "Committee Was Wary of Clark," *HC*, 17 February 1989; Linda Matchan, "Conservative to Head Harvard Law," *BG*, 18 February 1989, p. 1; Allan Gold, "Traditionalist Is Named as Harvard Law Dean," *NYT*, 18 February 1989; Cohn and Nayak, "Clark Appointment Made Official," and Cohn, "Bok Opts for 'Character' Over Consensus," *HC*, 18 February 1989; "Lux et Veritas Redux?," *WSJ*, 23 February 1989, p. 15; "Bok Taps Clark as New Dean," "Faculty Reactions Range From Praise to Scorn," and "Clark Responds to Charges of Faculty Divisiveness," *HLRec*, 24 February 1989, pp. 1, 5, 8–10, 16; "Students Meet with Clark," *HC*, 1 March 1989; Lisa Green Markoff, "Harvard's New Leader," *Legal Times* [*LT*], 6 March 1989, p. 4; Markoff, "Can New Harvard Dean Bring On a Calmer, Gentler Atmosphere?," *LT*, 20 March 1989, p. 4; Nayak, "HLS Students Protest Lack of Hispanic Profs," *HC*, 23 March 1989; Cohn, "Divided Law Faculty Finishes a Chapter," *HC*, 24 March 1989.

13. *HLS Adviser* #25, 23 February 1989; Robert M. Fisher HLS Transcript; Kahlenberg, *Broken Contract*, p. 51, First-Year Ames Committee to All Students in Sections I, III, and IV, "Dates for 1989 First-Year Ames," 23 September 1988, and Board of Student Advisers First-Year Ames Committee to All Students in Sections I, III, and IV, "First-Year Ames Moot Court Program," 31 October 1988, GTP; "1988–89 Ames Case Descriptions," n.d., "Indictment for Securities Fraud, United States of America v. Janine Egan," 22 pp., Barack Obama, "Issues Analysis—Hertzer Team—Insider Trading Cases—BSA Advisor Becker," n.d., 11 pp., and Barack Obama and Mark Kozlowski, U.S. Attorney's Office—Appellee, "Brief for the Prosecution—Appellee," Egan v. United States of America, Argument: March 23, 1989, Ames Courtroom, 7:30 P.M., 13 pp., Mark Kozlowski Papers; DJG interviews with Rob Fisher, Gina Torielli, Mark Kozlowski, Lisa Hertzer Schertler, and Scott Becker.

14. Macneil, *Contracts,* p. 963; Robert Lowe, "At Harvard: Sexism in the Paper Chase?," *BG,* 10 April 1989, pp. 33, 35; Tom Bilello, "Visiting Professor Responds to WLA Sexism Charges," and "Macneil Threatens Disciplinary Action for Absenteeism," *HLRec,* 14 April 1989, pp. 1, 3, 12; Erin Edmonds, "Macneil Reveals Hypocrisy," and Raymond D. Jasen, "Macneil Article Misleads," *HLRec,* 21 April 1989, pp. 4, 7; Ian R. Macneil, "Macneil Rebuts Sexism Charges," *HLRec,* 5 May 1989, pp. 4, 6; Chester E. Finn Jr., "The Campus: 'An Island of Repression in a Sea of Freedom,'" *Commentary,* September 1989, pp. 17–23; Richard Bernstein, "On Campus, How Free Should Free Speech Be?," *NYT,* 10 September 1989, p. IV-5; Ian R. Macneil, "Harvard Law School," *Commentary,* March 1990, pp. 10–11; Dinesh D'Souza, *Illiberal Education* (Free Press, 1991), pp. 197–99; Mark Tushnet, "Political Correctness, the Law, and the Legal Academy," *Yale Journal of Law and the Humanities* 4 (Winter 1992): 127–63, at 131–43 and 149–52; Kleinfeld and Yerian, eds., *Gender Tales,* pp. 179–92; Hutcheon, "The Scottish Professor Who Moulded the Young Obama"; DJG interviews with Martin Siegel, Sherry Colb, Richard Cloobeck, Erin Edmonds, Leonard Feldman, David Smail, Steven Heinen, Sarah Leah Whitson, Shannon Schmoyer, Jan-Michele Lemon Kearney, Jackie Fuchs, David Troutt, Lisa Hay, Ken Mack, Paolo Di Rosa, Rob Fisher, David Attisani, Michele Jacobs DeLong, Ali Rubin Yamin, Diana Derycz-Kessler, and Lauren Ezrol Klein. *HLS Adviser* #8, 13 October 1988, listed Section III's two official Law School Council representatives as Lauren Ezrol and Kevin Mohan. Macneil died on 16 February 2010, at age eighty. Trevor Jensen, "Northwestern University Law Professor Ian Macneil," *CT,* 21 February 2010, "Ian Macneil," *Telegraph,* 23 February 2010.

15. *HLS Adviser* #30, 6 April 1989, pp. 3, 6; #31, 13 April 1989, p. 13; #33, 27 April 1989, p. 3; #35, 11 May 1989, pp. 1, 5; masthead pages, *Harvard Civil Rights–Civil Liberties Law Review,* Vol. 24, No. 1, Winter 1989 (date stamped 27 March 1989), and Vol. 24, No. 2, Spring 1989 (date stamped 7 August 1989) (the masthead pages appear in the hard copy issues, not in digitized versions); Exams, Legal History 100, Property III, Contracts III, and Civil Procedure III, *Harvard Law School Annual Examinations in Law 1988–89,* pp. 217–19, 160–68, 76–100, 24–31, Laurence Tribe on *20/20,* ABC, 26 September 2008, in "Obama as We Knew Him," *Observer,* 26 October 2008, pp. 4–7, and in Seth Stern, "A Commander in Chief," *Harvard Law Bulletin,* Fall 2008; Newton Minow in Pauline Dubkin Yearwood, "Obama and the Jews," *Chicago Jewish News,* 12 September 2008, and in Peter F. Zhu, "Obama's Quiet Harvard Roots," *HC,* 2 November 2008; Tribe, "Morning-After Pride," Forbes.com, 5 November 2008; Tribe on "Obama Made a Strong First Impression at Harvard," NPR, 22 May 2012; Tribe's June 2012 interview with Jim Gilmore; Newton Minow's June 2008 interview with Jim Gilmore; DJG interviews with Rob Fisher, David Rosenberg, Laurence Tribe, Martha Minow, Newton Minow, John Levi, Alan Jenkins, Wendy Pollack, David Goldberg, David Rosenberg, Sung-Hee Suh, Cassandra Butts, Michele Jacobs DeLong, David Attisani, Brad Wiegmann, Greg Sater, Lauren Ezrol Klein, Mark Kozlowski, Steven Heinen, and Paolo Di Rosa.

16. David S. Hilzenrath, "Hallowed Be Its Name," *HC,* 14 March 1984; *Hazelwood School District v. Kuhlmeier,* 484 U.S. 260 (13 January 1988); *DeShaney v. Winnebago County,* 489 U.S. 189 (22 February 1989); Tara A. Nayak, "Law Review Chooses Masthead," *HC,* 8 March 1989; *HLS Adviser* #35, 11 May 1989, pp. 1, 5; Kristen Swartz, "Law Review Adds New Faces," *HLRec,* 29 September 1989, p. 10; Mary Flood, "At Law Review, Number of Women Members Falls Sharply," *HLRec,* 12 October 1990, pp. 1, 16; Kerlow, *Poisoned Ivy,* pp. 21–23; Rob Fisher in Levenson and Saltzman, "At Harvard Law, A Unifying Voice," *BG,* 28 January 2007, p. A1; DJG interviews with Alicia Rubin Yamin, Tom Perrelli, Radhika Rao, Amy F. Kett, Susan Higgins, Dan Bromberg, Radhika Rao, Carl Coleman, Rob

Fisher, Gordon Whitman, and Anne Toker. On *The Bluebook,* see A. Darby Dickerson's impressive "An Un-Uniform System of Citation: Surviving with the New *Bluebook,*" *Stetson Law Review* 26 (Fall 1996): 53–104, at 55–65, Christine Hurt, "*The Bluebook* at Eighteen: Reflecting and Ratifying Current Trends in Legal Scholarship," *Indiana Law Journal* 82 (Winter 2007): 49–68, and Fred R. Shapiro and Julie Graves Krishnaswami, "The Secret History of the *Bluebook,*" *Minnesota Law Review* 100 (2016). On *DeShaney,* see Lynne Curry, *The DeShaney Case: Child Abuse, Family Rights, and the Dilemma of State Intervention* (University Press of Kansas, 2007), and Lynda G. Dodd's excellent "*DeShaney v. Winnebago County*: Governmental Neglect and 'The Blessings of Liberty,'" in Myriam Gilles and Risa Goluboff, eds., *Civil Rights Stories* (Foundation Press, 2007), pp. 185–210.

17. Obama, *TAOH,* pp. 327–32; Obama, "My First Date with Michelle," *O Magazine,* February 2007 (from *TAOH*); Gene Siskel, "Spike Lee's Mission," and Clarence Page, "Spike Lee's Warning," *CT,* 25 June 1989; Dave Kehr, "'Do the Right Thing' at Least Takes a Stab at Complexity," *CT,* 30 June 1989; Michele Obama in January 2004 in Mendell, *From Promise to Power,* pp. 93–97; Michele Obama's February 2004 interview with John Kupper; Michelle Obama Form Letter to Timuel Black, 4 May 2007, TBP Box 129 Fld. 2; Michelle in Krista Lewin, "Obama Brings Husband's Campaign to Cole County," *Mattoon Journal Gazette* and *Charleston Times-Courier,* 28 June 2004, p. A1, and in Cassandra West, "Her Plan Went Awry," *CT,* 1 September 2004, pp. W1–W2; Debra Pickett, "'My Parents Weren't College-Educated Folks,'" *CST,* 19 September 2004, p. 10; Karen Springen, "First Lady in Waiting," *Chicago Magazine,* October 2004; both Obamas on *The Oprah Winfrey Show,* 19 January 2005; Obama on CNBC, 18 November 2006; Springen and Jonathan Darman, "Ground Support," *Newsweek,* 29 January 2007, p. 40; Michelle in Lynn Norment, "The Hottest Couple in America," *Ebony,* February 2007, pp. 52ff.; Michelle on *Chicago Tonight,* WTTW, 12 February 2007; Obama in Erin Meyer, "Discovering Hyde Park," and Michelle in Kalari Girtley and Brian Wellner, "Michelle Obama Is Hyde Park's Career Mom," *HPH,* 14 February 2007, pp. 4, 7–8, 16–17; Michelle in Brian DeBose, "Obama's Wife 'Commands a Room,'" *Washington Times,* 30 April 2007, p. A1; Anne E. Kornblut, "Michelle Obama's Career Timeout," *WP,* 11 May 2007, p. A1; Michelle on *Good Morning America,* ABC, 23 May 2007; Mark Konkol, "Mr. Obama's Neighborhood," *CST,* 25 May 2007, p. 12; Richard Wolffe and Daren Briscoe, "Across the Divide," *Newsweek,* 16 July 2007, pp. 22ff.; Gwen Ifill, "Beside Barack," *Essence,* September 2007, pp. 200ff.; Scott Fornek, "'He Swept Me Off My Feet,'" *CST,* 3 October 2007, p. 6; Melinda Henneberger, "The Obama Marriage," *Slate,* 26 October 2007; Peter Slevin, "Her Heart's in the Race," *WP,* 28 November 2007, p. C1; Monica Langley, "Michelle Obama Solidifies Her Role in the Election," *WSJ,* 11 February 2008, p. A1; "Michelle Obama on Love, Family & Politics," CBSNews.com, 15 February 2008; Patrick T. Reardon, "Obama's Chicago," *CT,* 25 June 2008, p. T1; Michelle's remarks to DNC's Gay and Lesbian Leadership Committee, New York City, 26 June 2008; Obama and Michele Obama on "Barack Obama Revealed," CNN, 20 August 2008; "Michelle Obama: South Side Girl," video, Democratic National Convention, 25 August 2008; "Barack Obama Tribute" video, Democratic National Convention, 28 August 2008; Harriette Cole, "The Real Michelle Obama," *Ebony,* September 2008; "Michelle Obama," *20/20,* ABC, 3 October 2008; Abigail Pesta, "Michelle Obama Keeps It Real," *Marie Claire,* 22 October 2008; Rosemary Ellis, "A Conversation with Michelle Obama," *Good Housekeeping,* November 2008; Obama and Michelle on *60 Minutes,* CBS, 16 November 2008; Stacy St. Clair and Dahleen Glanton, "Scenes from Obamas' Love Story," *CT,* 30 November 2008; Carol Felsenthal's superb "The Making of a First Lady," *Chicago Magazine,* February 2009, pp. 50–56, 72–78; Michelle in Wolffe, *Renegade,* p. 60; Thomas Reed in Jodi Kantor, *The Obamas* (Little, Brown, 2012), p. 16; Michelle on *106 & Park,* BET, 19 November 2013, "President Barack Obama, Michelle Obama Praise First Date Movie *Do the Right Thing,*" *US Magazine,* 30 June 2014; Peter Slevin, *Michelle Obama* (Knopf, 2015), pp. 121–31; DJG interviews with Geraldine Alexis, Linzey Jones, Rob Fisher, Kelly Jo MacArthur, Thomas Reed, Evie Shockley, Newton Minow, Eden Martin, Jerry Kellman, and Jean Rudd.

18. Fraser Robinson III Employee Work History, City of Chicago, Marian Robinson Voter Registration Card, 7436 S. Euclid Avenue, 25 January 1971, Robbie S. Terry to Fraser Robinson and Marion [sic] L. Robinson, Quit Claim Deed, 17 January 1980 (conveying ownership of 7436 S. Euclid Avenue to the Robinsons), and

Michelle L. Robinson Voter Registration Card, 7436 S. Euclid Avenue, 11 October 1988, all Liza Mundy Papers; Michelle LaVaughn Robinson, "Princeton-Educated Blacks and the Black Community," Princeton University, 1985, esp. pp. 2–3, 12, 14, 26, 55; Dave Kehr, "Black on Black," *CT,* 23 July 1989; Dean Golemis, "International Festival Showcases Black Film Alternatives," *CT,* 28 July 1989; Obama in "Organizing in the 1990s: Excerpts from a Roundtable Discussion," in Peg Knoepfle, ed., *After Alinsky* (Illinois Issues, 1990), p. 134; Ted Shen, "Ruby Oliver Goes to Hollywood," *CR,* 26 July 1990; Dave Kehr, "'Mama' Has Strength, On Screen and Off," *CT,* 5 March 1993; Hugh Hart, "Beating the Odds," *CT,* 10 March 1993; Craig Robinson in Courtney McCarty, "High 'Hoops' for Brother-in-Law," *Daily Northwestern,* 27–28 September 2004; Sarah Brown, "Obama '85 Masters Balancing Act," *Daily Princetonian,* 7 December 2005; Rosalind Rossi, "Obama's Anchor," *CST,* 21 January 2007, p. 9; Bill Reynolds, "Welcome to Obama's Family," *Providence Journal* [*PJ*], 15 February 2007, p. C1; Pete Thamel, "Coach with a Link to Obama Has Hope for Brown's Future," *NYT,* 16 February 2007, p. D1; Jonathan Tilove, "At Princeton, Michelle Obama Worried About Being on the Periphery," *Newark Star-Ledger,* 18 February 2007, p. P1; DeBose, "Obama's Wife 'Commands a Room,'" *Washington Times,* 30 April 2007, p. A1; Craig in M. Charles Bakst, "Brown Coach Robinson a Strong Voice for Brother-in-Law Obama," *PJ,* 20 May 2007, p. D1; W. Gardner Selby, "In Austin, Michelle Obama Hails Husband's Background," *Austin American-Statesman,* 12 July 2007, p. B3; Michelle in Mary Mitchell, "Makeup's Too Much Work for Michelle," *CST,* 7 August 2007, p. 12; Maria L. LaGanga, "It's All About Priorities for Michelle Obama," *LAT,* 22 August 2007; Rebecca Johnson, "The Natural," *Vogue,* September 2007, pp. 774ff. ("I'm smarter than him"); Andy Katz, "Brown Coach Robinson Coaching Brother-in-Law Obama, Too," ESPN.com, 13 September 2007; Michelle in Scott Helman, "Early Defeat Launched a Rapid Political Climb," *BG,* 12 October 2007, p. A1; Craig in Melinda Henneberger, "The Obama Marriage," *Slate,* 26 October 2007; Holly Yeager, "The Heart and Mind of Michelle Obama," *O Magazine,* November 2007, pp. 286ff.; "Michelle Obama Speaks with MSNBC's Mika Brzezinski," MSNBC.com, 13 November 2007; Slevin, "Her Heart's in the Race," *WP,* 28 November 2007, p. C1; Leslie Bennetts, "First Lady in Waiting," VanityFair.com, 27 December 2007; Liza Mundy, *Michelle: A Biography* (Simon & Schuster, 2008), esp. pp. 1–2, 35, 78, 81–82, 89–92; Michelle in Richard Wolffe, "Inside Obama's Dream Machine," *Newsweek,* 14 January 2008; Liz Halloran, "Q&A: Michelle Obama," USNews.com, 1 February 2008, Craig in Stu Woo, "For His Candidate-in-Law, Brown Hoops Coach Robinson Records an Assist," *Brown Daily Herald,* 1 February 2008, and in Bill Reynolds, "Yes, He's Much More Than Obama's Brother-in-Law," *PJ,* 10 February 2008, p. C1; Richard Wolffe, "Barack's Rock," *Newsweek,* 25 February 2008; Craig in Desmond Conner, "Coach Has His Own Campaign," *Hartford Courant,* 28 February 2008, and in Mark Patinkin, "Obama's Got Game," *PJ,* 1 March 2008, p. D1; Lauren Collins, "The Other Obama," *New Yorker,* 10 March 2008, pp. 88ff.; Craig Robinson on *Anderson Cooper 360,* CNN, and on "Barack Obama Revealed," 13 March, 4 July, and 20 August 2008; Kristen Gelineau (AP), "Would-Be First Lady Drifts into Rock Star Territory," 30 March 2008; Brian Feagins, "Georgian Recalls Rooming with Michelle Obama," *AJC,* 12 April 2008; Sally Jacobs, "Learning to Be Michelle Obama," *BG,* 15 June 2008, p. A1; Michael Powell and Jodi Kantor, "Michelle Obama Is Ready for Her Close-Up," *NYT,* 18 June 2008; Lawrence Aaron, "A Look at Michelle Obama During Her Jersey Years," *Bergen Record,* 22 June 2008, p. O4; Lynne Marek, "The 'Other Obama' Honed Her Skills at Sidley Austin," *NLJ,* 23 June 2008, p. 6; Craig in Anne C. Mulkern, "Husband Held to 'High Standard,'" *Denver Post,* 25 August 2008; Amanda Paulson, "Michelle Obama's Story," *CSM,* 25 August 2008; "Michelle Obama: South Side Girl," video, Democratic National Convention, 25 August 2008; Kelly Eleveld, "It's Not Just About the Hair," *The Advocate,* 27 August 2008; Harriette Cole, "From a Mother's Eyes," *Ebony,* September 2008; Gwen Ifill, "The Obamas," *Essence,* September 2008, pp. 150–57, 214; Melinda Henneberger, "Michelle Obama Interview: Her Father's Daughter," *Reader's Digest,* October 2008, pp. 196–203; Geraldine Brooks, "Michelle Obama: Camelot 2.0?," *More,* October 2008; Matt Hutchins, "Ogletree on Obama," *HLRec,* 30 October 2008; Esther Breger, "All Eyes Turn to Michelle Obama '85," *Daily Princetonian,* 5 November 2008; Robin Abcarian, "Michelle Obama Speaks from Her Heart," *LAT,* 16 November 2008; Robinson, "B-Ball with Barack," *Time,* 17 December 2008; Wolffe, *Renegade,* p.

35; Felsenthal, "The Making of a First Lady," *CM,* February 2009; Eli Saslow, "From the Second City, An Extended First Family," *WP,* 1 February 2009; Chuck Klosterman, "Craig Robinson: America's First Coach," *Esquire,* February 2009; "Augie Alum Holds Court with Obama," Augustana.edu, 13 February 2009; Rachel L. Swarns, "Friendship Born at Harvard Goes to White House," *NYT,* 9 March 2009; Slevin, "Mrs. Obama Goes to Washington," *Princeton Alumni Weekly,* 18 March 2009; "How We Keep Our Love Alive," *In Touch Weekly,* 27 April 2009, pp. 32–35; Michael Scherer and Nancy Gibbs, "Interview with the First Lady," *Time,* 21 May 2009; Craig Robinson, *A Game of Character* (Gotham Books, 2010), esp. pp. xviii–xix, xxiii, 5, 11, 58, 95–97, 112–15, 143–44, 147–49; Katherine Skiba, "First Grandma Keeps Low Profile," *CT,* 8 March 2010; Kantor, *The Obamas,* p. 39; Tamara Jones, "Michelle Obama Gets Personal," *More,* February 2012; Theodore Ford, "Kindergarten with a Girl Named Michelle," *CST,* 24 November 2012; Michelle on *106 and Park,* BET, 19 November 2013; Slevin, *Michelle Obama,* esp. pp. 32, 42, 57–58, 61, 87–88, 118–22; Jennifer Connelly e-mail to Newton Minow and John Levi, 16 September 2008; DJG interviews with John Rogers, Janis Robinson, Jan Anne Dubin, and Paul Carpenter. On Dukes Happy Holiday Resort, see Kenan Heise, "Rufus Dukes, Pioneer Black Automobile Dealer," *CT,* 25 September 1987. In Ann Dunham's life, Barack's uncle Charles Payne would later remark, "everything seemed temporary," and Barack would later volunteer that "My choosing to put down roots in Chicago and marry a woman who is very rooted in one place probably indicates a desire for stability that maybe I was missing" throughout his earlier life. Amanda Ripley, "The Story of Barack Obama's Mother," *Time,* 21 April 2008, pp. 36ff.

On the extended Robinson family and its earlier generations, see Anthony Weiss, "Michelle Obama Has a Rabbi in Her Family," *Forward,* 2 September 2008; Shailagh Murray, "A Family Tree Rooted in American Soil," *WP,* 2 October 2008, p. C1; Zev Chafets, "Obama's Rabbi," *NYT Magazine,* 5 April 2009; Rachel L. Swarns and Jodi Kantor, "In First Lady's Roots, a Complex Path from Slavery," *NYT,* 8 October 2009; Swarns, "Meet Your Cousin, the First Lady," *NYT,* 17 June 2012; and Swarns, *American Tapestry: The Story of the Black, White, and Multiracial Ancestors of Michelle Obama* (Amistad, 2012), esp. pp. 94–95.

19. Erwin N. Griswold, "The Harvard Law Review—Glimpses of Its History as Seen by an Aficionado," in *Harvard Law Review: Centennial Album* i (1987); Griswold's January 1992 interview with Victoria Radd, pp. 6–7; Griswold, *Ould Fields, New Corne: The Personal Memoirs of a Twentieth Century Lawyer* (West Publishing, 1992); Robert E. Keeton, "Griswold," *HLR* 105 (April 1992): 1386–1401; Dennis Hevesi, "Erwin Griswold Is Dead at 90," *NYT,* 21 November 1994; "Griswold, Law School Legend, Dies at 90," *HC,* 21 November 1994; "Editors Directory," Harvard Law Review, 1989–90, Deborah Brake Papers; Mark Tushnet and Timothy Lynch, "The Project of the Harvard *Forewords:* A Social and Intellectual Inquiry," *Constitutional Commentary* 11 (Winter 1994): 463–500; DJG interviews with Tom Krause, Susan Freiwald, David Goldberg, John Parry, Ken Mack, Michael Guzman, Michael Weinberger, Jacqueline Scott Corley, Brad Wiegmann, Barry Perlstein, Brad Berenson, Scott Collins, Diane Ring, Frank Cooper, Tracy Higgins, Lisa Hay, Scott Siff, Julie Cohen, Kevin Downey, Gordon Whitman, Christine Lee, Susan Higgins, and Dodie Hajra.

20. Harvard Law School 1989–90 Catalog, esp. pp. 4–5, 16, 74, and 84; *HLS Adviser* Vol. 21, #1, 31 August 1989, esp. p. 5; Robert M. Fisher HLS Transcript; Kahlenberg, *Broken Contract,* p. 53; Tribe in Shira Schoenberg, "Law Expert: Obama Will Preserve Constitution," *Concord Monitor,* 14 November 2007; Carrie Budoff Brown, "School Buds," *Politico,* 5 December 2008; DJG interviews with Rob Fisher, Chris Edley, Mark Kozlowski, Scott Scheper, Laura Jehl, Jennifer Gardner Trulson, Sarah Leah Whitson, Kevin Downey, Larry Tribe, Paolo Di Rosa, David Troutt, Tom Perrelli, Julius Genachowski, and Roger Boord.

21. Terry Carter, "At Harvard Law, a New Era Dawns," *NLJ,* 7 August 1989, pp. 1, 44–45; Greg Herbert, "Accused Rapist, Barred from School, Sues Harvard," and Patrick Miles, "Clark Cuts Public Interest Position," *HLRec,* 8 September 1989, pp. 1, 2, 7, 8; Carter, "Harvard Dean Eliminates Office for Public Interest Counseling," *Legal Times,* 11 September 1989, p. 4; "Student Groups Meet with Clark, Rally Planned for Tuesday," and Herbert, "Clark's First Step Sends Wrong Signal," *HLRec,* 15 September 1989, pp. 1, 3, 6; Tara A. Nayak, "Law Students Sign Petition, Plan Rally," *HC,* 15 September 1989; George Paul, "Students, Professors Rally in Support of Public Interest," and Herbert,

"Watkins Indicted on Rape, Assault Charges," *HLRec,* 29 September 1989, pp. 1, 16; Nayak, "Public Interest Squabble," *HC,* 30 September 1989; Paul, "Student Demands Halt Interviews of Placement Advisor Candidates," and Steven Donziger, "Clark Must Broaden Perspective in Public Interest Choices," *HLRec,* 6 October 1989, pp. 1, 7, 13; Nayak, "Law School Gets $1.5M for Public Service," *HC,* 11 October 1989; Herbert, "Watkins Pleads Not Guilty," and Paul Tarr, "Clark Announces $1 Million Endowment," *HLRec,* 20 October 1989, pp. 1, 13, 16; Anthony Flint, "Harvard Group Decries Loss of Law Counselor," *BG,* 25 October 1989, p. 91; Nayak, "Clark Defends Move to Cut Counseling," *HC,* 28 October 1989; "The Abandonment of Public Interest Law," *Legal Times,* 30 October 1989, p. 21; Felicia Ravago, "Students Rally for Interview Boycott of Baker and McKenzie," and Julia Gordon, "Harvard Watch Report Critical of Clark's Public Interest Reorganization," *HLRec,* 3 November 1989, pp. 1, 3, 4, 12; Nayak, "A Law Dean with a New 'Mission,'" *HC,* 4 November 1989; Nayak, "Commitment Often Ends After Graduation," *HC,* 6 November 1989; Greg Dingens, "Kennedy Visit Prompts 'Awareness Rally' on Abortion, Civil Rights Issues," *HLRec,* 17 November 1989, p. 12; Herbert, "Students Demand Immediate Action at Public Interest Forum," *HLRec,* 1 December 1989, pp. 1, 9; Flint, "Students Dump Protest Letters Outside Harvard Law Dean's Office," *BG,* 5 December 1989, p. 31; Nayak, "Letters Protest Clark Public Interest Policy," *HC,* 5 December 1989; DJG interview with Chris Edley. Accused rapist Kevin Watkins would be tried, convicted, and sentenced to five years in prison, with his conviction repeatedly affirmed. John Thornton, "Watkins on Trial in Middlesex Courthouse," *HLRec,* 4 May 1990, pp. 1, 16; George Paul, "Watkins Found Guilty; Sentenced to 5 Years," *HLRec,* 14 September 1990, p. 3; "Watkins to Appeal Rape Conviction," *HLRec,* 19 April 1991, p. 1; *Commonwealth v. Watkins,* 33 Mass. App. Ct. (14 July 1992), appeal denied 413 Mass. 1105 (3 September 1992), *Watkins v. DiPaolo,* 37 F.3d 1481 (1st Cir.) (14 October 1994) (affirming denial of habeas corpus), 513 U.S. 1194 (27 February 1995) (cert. denied). Apropos of the *Record*'s 3 November story about an 18 October protest against Baker and McKenzie, a photo in the 1990 *Harvard Law School Yearbook,* p. 14, would picture Barack, in glasses, attending some event related to that protest.

22. "Organizing in the 1990s: Excerpts from a Roundtable Discussion," in Peg Knoepfle, ed., *After Alinsky: Community Organizing in Illinois* (Illinois Issues, 1990), pp. 123–52, esp. pp. 133–35, 143, 148–51; John B. Judis, "Creation Myth," *TNR,* 10 September 2008, pp. 18ff.; Ben Joravsky, "When the Next Derrick Rose Is Right Under Your Nose," *CR,* 15 January 2009; DJG interviews with Jean Rudd and Ken Rolling. See also William E. Schmidt, "Chicago Nears Vote on Choice for Mayor as Race Flares," *NYT,* 27 February 1989, Dirk Johnson, "Daley Wins Primary in Chicago," *NYT,* 1 March 1989, and Johnson, "Daley Wins as Mayor of Chicago, Ending Six Years of Black Control," *NYT,* 5 April 1989.

23. Richard H. Stewart, "Organizing Is His Vocation," *BG,* 8 August 1988, p. 19 (Finfer); Obama University of Chicago Law School Faculty Web Page, 2001, web.archive.org; *HLS Adviser #8,* 12 October 1989, p. 9 (Kozinski speaking on 16 October); Peter M. Cicchino, "An Activist at Harvard Law School," *American University Law Review* 50 (2001): 551–65; Glen Johnson (AP), "At Harvard Law School, Obama Foreshadowed Presidential Appeal," 26 January 2007; Greg Dingens on "Obama Presidential Announcement Reaction," C-SPAN, 10 February 2007; Butts in "A Rusty Toyota, a Mean Jump Shot," *WP,* 11 February 2007, p. B3; Harper in Jodi Kantor, "One Place Where Obama Goes Elbow to Elbow," *NYT,* 1 June 2007, p. A1; Finfer in Irene Sege, "Community Organizers Fault Comments at GOP Gathering," *BG,* 6 September 2008, p. A9; Harper in Melanie Eversley, "Best Supporting Actor," *The Root,* 1 October 2008; Finfer, "The Obama Experience," *Social Policy,* Fall 2008, pp. 11–12; Kenny Smith in Jim Morrill, "Remembering Obama," *Charlotte Observer,* 13 January 2009; Eric Fingerhut, "Be It Bush or Obama, White House Door Open for O.U.'s Diament," JTA.org, 27 January 2009; Finfer in Sara Rimer, "Community Organizing Never Looked So Good," *NYT,* 12 April 2009; Abramsky, *Inside Obama's Brain* (Portfolio, 2010), p. 53; A. J. Buckley, "Before the Scene: Hill Harper," *Scene Louisiana Magazine,* December 2012–January 2013, pp. 18–19; DJG interviews with Rob Fisher, Cassandra Butts, Laura Jehl, Christine Lee, Karla Martin, Morris Ratner, James Esseks, Keith Boykin, Ken Mack, Gordon Whitman, Lew Finfer, Tom Krause, Kenny Smith, Hill Harper, Tom Lininger, Ursula Dudley Oglesby, Tynia Richards, David Hill, Nathan Diament, Greg Dingens, and Tom Wathen. See also Sondra

Hemeryck, Cassandra Butts, Laura Jehl, et al., "Reconstruction, Deconstruction and Legislative Response: The 1988 Supreme Court Term and the Civil Rights Act of 1990," *Harvard Civil Rights–Civil Liberties Law Review* 25 (Summer 1990): 475–590. On Walpole, see Kelsey Kauffman, *Prison Officers and Their World* (Harvard University Press, 1988), esp. pp. 23–37.

24. Laurence H. Tribe, "The Curvative of Constitutional Space: What Lawyers Can Learn From Modern Physics," *HLR* 103 (November 1989): 1–39; Harvard Law School 1989–90 Catalog, p. 105; Rob Fisher to Laurence Tribe, 1 November 1989, RFP; Peter M. Yu to Erwin N. Griswold, 9 June and 9 November 1989, Griswold to Yu, 13 and 29 November 1989, Griswold Papers [EGP] Box 133 Fld. 2; Patrick Philbin et al., to The Body, "The Frug Piece," 15 January 1992, Trent Norris Papers [TNP]; Christine Lee in Debra McCown, "So What Was President Obama Like in College?," *Bristol Herald Courier,* 27 January 2009; Cass R. Sunstein, "Interpreting Statutes in the Regulatory State," *HLR* 103 (December 1989): 405–509; [Thomas J. Perrelli], "Search and Seizure—Suspicionless Drug Testing—Seventh Circuit Upholds Drug Testing of Student Athletes in the Public Schools," *HLR* 103 (December 1989): 591–97; Yu to Griswold, 5 December 1989, and Griswold to Yu, 21 December 1989, EGP Box 133 Fld. 2; Ann Dunham to Alice Dewey, 3 November 1989, Dunham résumé, n.d. (ca. 1993), Dunham Papers; Julia Suryakusuma, "Obama for President . . . of Indonesia," *Jakarta Post,* 29 November 2006, and "Barack Obama: Mum's the Word," *Jakarta Post,* 4 November 2008; Paula Bender, "Legacy of the President's Mother," *Malamalama: The Magazine of the University of Hawaii,* January 2009; Dunham, *Surviving Against the Odds* (Duke University Press, 2009), p. 301; Judith Kampfner, "Dreams from My Mother," BBC Radio, 16 September 2009; Suryakusuma, "'Ibu' Ann's Legacy to Indonesia," *Jakarta Post,* 21 March 2010; Janny Scott, *A Singular Woman* (Riverhead, 2011), pp. 265–66, 285–91; *Stallman v. Youngquist,* 125 Ill. 2d 267, 531 N.E. 3d 355 (21 November 1988); [Barack Obama], "Tort Law—Prenatal Injuries—Supreme Court of Illinois Refuses to Recognize Cause of Action Brought by Fetus Against Its Mother for Unintentional Infliction of Prenatal Injuries," *HLR* 103 (January 1990): 823–28; DJG interviews with Rob Fisher, Larry Tribe, Mike Dorf, Barbara Schneider, Pauline Wan, Andy Schapiro, Chad Oldfather, Brad Berenson, Susan Freiwald, Kevin Downey, Frank Cooper, Gordon Whitman, Carl Coleman, Radhika Rao, Tracy Higgins, Jennifer Borum Bechet, John Parry, Christine Lee, and Tom Krause. The *Harvard Law Review*'s own bound, anonymously annotated master copy of Volume 103, housed in the president's office, attests to Obama's authorship of the January 1990 case note and lists "P. editors C. Lee, B. Obama, J. Parry, G. Whitman, S. Freiwald" on the first page of the March 1990 Cook article. DJG interview with then-*HLR* president Mitchell Reich, 18 April 2011.

25. Harvard Law School 1989–90 Catalog, esp. pp. 4–5; *HLS Adviser* #14, 30 November 1989, p. 5; *Harvard Law School Annual Examinations in Law 1989–90,* pp. 78–81, 17–20, 35–38; Cassandra Butts's interview with Jim Gilmore; DJG interviews with Elvin Charity, Jan-Michele Lemon Kearney, Radhika Rao, Ken Mack, Chris Edley, Larry Tribe, and Rob Fisher.

26. Dunham to Suryakusuma in Margaret Conley, "Obama's Early Days in Jakarta," ABCNews.com, 25 September 2008, on *20/20,* ABC, 3 October 2008, and in Scott, *A Singular Woman,* pp. 296–97; Obama, *TAOH,* p. 327; Sheila Miyoshi Jager, "Narrating the Nation: Students, Romance and the Politics of Resistance in South Korea," Ph.D. dissertation, University of Chicago, December 1994, pp. vii–viii; Sheila Jager e-mails to DJG.

27. Harvard Law School 1989–90 Catalog, pp. 4–5, 114; Harvard Law School Schedule of Courses 1989–90, HLSA; *Harvard Law School Annual Examinations in Law 1989–90,* pp. 336–50; "Memorandum to Barak [sic] Obama and Rob Fisher From Larry Tribe Re Your February 1 Memo," 2 February 1990, RFP; Paul Tarr and John Thornton, "Obama Election Lauded Coast to Coast," *HLRec,* 9 February 1990, pp. 1, 12; Linda Matchan, "A Law Review Breakthrough," *BG,* 15 February 1990, p. 29; Tammerlin Drummond, "Barack Obama's Law," *LAT,* 12 March 1990, pp. E1–E2; Elise O'Shaughnessy, "Harvard Law Reviewed," *Vanity Fair,* June 1990, p. 106; Granfield, "Making It by Faking It," p. 342; Mack in "Barack Obama Before He Was a Rising Political Star," *Journal of Blacks in Higher Education,* Autumn 2004, pp. 99–101; Richard A. Serrano and David G. Savage, "Obama's Harvard Law School Days Marked by Bridge-Building," *LAT,* 27 January 2007; Levenson and Saltzman, "At Harvard Law, a Unifying Voice," *BG,* 28 January 2007, p. A1; Cassandra Butts on *Tell Me More,* NPR, 22 February 2007; Mack in John Heilemann, "When They Were Young,"

New York Magazine, 22 October 2007; Crystal Nix Hines, "Testimonial," undated 2008 video, on BarackObama.com; Goldberg in Ed Pilkington, "Obama Dreams Big, Achieves Bigger," *Guardian,* 8 January 2008; Ken Mack, "Even at Harvard, Obama Had a Knack for Bonding with Diverse People," *Harrisburg Patriot-News,* 17 February 2008; Marian Robinson in Richard Wolffe, "Barack's Rock," *Newsweek,* 25 February 2008, and in Melinda Henneberger, "Michelle Obama Interview," *Reader's Digest,* October 2008, pp. 196–203; Michael Cohen in David Nather, "Obama's Quick Rise on a Non-Traditional Career Path," *CQ Weekly,* 21 August 2008; Ken Mack on "The Choice 2008," *Frontline,* PBS, 14 October 2008; Jodi Kantor, "Barack Obama, Forever Sizing Up," *NYT,* 26 October 2008, p. IV-1; Jennifer Borum Bechet in Jonathan Tilove, "Few in Louisiana Have Close Ties to Obama," *New Orleans Times-Picayune,* 24 November 2008; Jim Chen in Jack Brammer, "Two Kentuckians Reflect on Their Time with Barack Obama," *Lexington Herald-Leader,* 19 January 2009; Stephanie Schorow, "Obama's Narrative," *Harvard Gazette,* 16 November 2011; Fatima Mirza, "HLS Professor Offers Look into President Obama the Student," *HC,* 16 November 2011; DJG interviews with Rob Fisher, Gina Torielli, Jennifer Gardner Trulson, Jonathan King, Susan Freiwald, David Rosenberg, Larry Tribe, Chris Edley, Cassandra Butts, John "Vince" Eagan, Ken Mack, Frank Cooper, Jennifer Borum Bechet, Christine Lee, Tom Krause, Monica Harris, Debbie Brake, Scott Collins, Radhika Rao, Diane Ring, Andy Schapiro, Pauline Wan, Gordon Whitman, Brad Berenson, Marisa Chun, Michael Cohen, Kevin Downey, Michael Guzman, Lourdes Lopez-Isa, David Nahmias, Jackie Scott Corley, Daniel Slifkin, Anne Toker, Michael Weinberger, Brad Wiegmann, Dan Bromberg, Jane Catler Nober, Micki Chen, Jack Chorowsky, Carl Coleman, Barbara Eyman, Tracy Higgins, Mark S. Martins, Patrick O'Brien, Jorge Ramírez, Barbara Schneider, Frank Amanat, Adam Charnes, Julie Cohen, Jennifer Collins, Sarah Eaton Stuart, Julius Genachowski, John Parry, Scott Siff, Tom Perrelli, Lisa Hay, Amy Kett, David Goldberg, Chad Oldfather, Monica Harris, Ryan Stoll, Lars Noah, Nancy Kao, and Barry Perlstein. Jennifer Gardner Trulson believes Barack also took Trial Advocacy Workshop C, "a good one," alongside her in January 1990 with several instructors, but this is unconfirmed. Based almost entirely on individuals' own recollections of how they faired, the ten candidates eliminated after the first round were Frank Amanat, Adam Charnes, Jim Chen, Julie Cohen, Jennifer Collins, Sarah Eaton, Ken Mack, John Parry, Scott Siff, and Andrew Schlafly. All told, fifty-three editors who were present on 4 February 1990 have voiced their recollections of that day.

28. Paul Tarr and John Thornton, "Obama Election Lauded Coast to Coast," *HLRec,* 9 February 1990, pp. 1, 12; Jim Morrill, "Remembering Obama," *Charlotte Observer,* 13 January 2009; "First Black President of Harvard Law Review Elected," AP, 5 February 1990, and in *USA Today,* 6 February 1990, p. 2A, *St. Petersburg Times,* 6 February 1990, p. A4, *WP,* 8 February 1990, p. B3, *AJC,* 9 February 1990, p. A2, and *HA,* 11 February 1990, p. A32; undated television interview footage in Becoming Barack, [February 1990]; Fox Butterfield, "First Black Elected to Head Harvard's Law Review," *NYT,* 6 February 1990, p. A20; Anthony Flint, "First Black Chosen Head of Harvard Law Journal," *BG,* 6 February 1990, p. 17; Philip M. Rubin, "Obama Named New Law Review President," *HC,* 6 February 1990; Michael J. Ybarra, "Activist in Chicago Now Heads Harvard Law Review," *CT,* 7 February 1990, p. C3; Phil Mayer, "Ex-Islander Gets Prestigious Harvard Post," *HSB,* 9 February 1990; Linda Matchan, "A Law Review Breakthrough," *BG,* 15 February 1990, p. 29; J. D. Reed, "People," *Time,* 19 February 1990, p. 93; John Corr, "From Mean Streets to Hallowed Halls," *Philadephia Inquirer,* 27 February 1990, pp. C1, C4; *HLS Adviser* #27, 8 March 1990, p. 10 (Mikva speaking on 8 March); Laurence H. Tribe, *Abortion: The Clash of Absolutes* (W.W. Norton, 1990), pp. xv–xvi; Obama in "The First Black President of the *Harvard Law Review,*" *Journal of Blacks in Higher Education,* Winter 2000–2001, pp. 22–25; David Wilkins in Lauren A. E. Schuker, "Obama Stars at Convention," *HC,* 30 July 2004, in Marie C. Kodama, "Obama Left His Mark on HLS," *HC,* 19 January 2007, and in Jon Meacham, "On His Own," *Newsweek,* 23 August 2008; Tony Mauro, "Obama's Road Not Taken," *Legal Times,* 13 December 2007; Shelby Steele, *A Bound Man* (Free Press, 2007), p. 14; McClelland, *Young Mr. Obama,* p. 79; Michelle in Liza Mundy, *Michelle* (Simon & Schuster, 2008), pp. 108–9; Mikva in Ryan Lizza, "Making It: How Chicago Shaped Obama," *TNR,* 21 July 2008; DJG interviews with Frank Cooper, David Hill, Hill Harper, Nicole Lamb-Hale, Lori-Christina Webb, Susan Higgins, Debbie

Brake, Jackie Scott Corley, Sarah Eaton Stuart, Sheryll Cashin, Ab Mikva, Julius Genachowski, Mark Kozlowski, and Larry Tribe, and Tribe's 2012 interview with Jim Gilmore. A few weeks later, *L.A. Law* star Blair Underwood visited the law school and had dinner with Barack and huge fan Jennifer Borum at a Chinese restaurant. DJG interviews with Jennifer Borum Bechet and Christine Lee. See also Brian McIver, "In Treatment Star Blair Underwood on Being Barack Obama's Predecessor," *Daily Record* (Glasgow), 14 October 2009. Cashin's and Mikva's clear and consistent memories—Barack "never visited our chambers"—suggest that specific details in Claire Cooper, "Jeff Bleich—Finding Common Ground," *San Francisco Bar Journal,* Fall 2008, pp. 48–50, and Jess Bravin, "Obama's New Man Down Under," *WSJ,* 22 December 2009, must be discounted. The *Record* story stated that Obama had made "significant contributions to the final two chapters" of Tribe's book, which, if accurate, would mean Chapter 9, "In Search of Compromise," pp. 197–228, which addressed the idea of "common ground," and Chapter 10, "Beyond the Clash of Absolutes," pp. 229–42, which addressed "What Our Views of Abortion Can Teach Us About Ourselves," "Fetal Endangerment: Child Abuse?," similar to Barack's January 1990 case comment, and "How Can We See and Talk to One Another." Tammerlin Drummond, "Barack Obama's Law," *LAT,* 12 March 1990, pp. E1–E2, says, "Obama wrote an insightful research article showing how contrasting views in the abortion debate are a direct result of cultural and sociological differences" for Tribe. Rob Fisher, an anti-abortion Roman Catholic, recalls that "Barack and I discussed abortion quite a bit," particularly "what a sad thing it is." Tribe has been unable to locate any abortion-related work Obama did for him. DJG interviews with Rob Fisher, Larry Tribe, Mike Dorf, and Peter Rubin.

29. George Paul, "Obama Selects Review Editors," *HLRec,* 16 February 1990, pp. 1, 11; Perrelli and Berenson in Levenson and Saltzman, "At Harvard Law, a Unifying Voice," *BG,* 28 January 2007, p. A1; Berenson in Richard Wolffe and Daren Briscoe, "Across the Divide," *Newsweek,* 16 July 2007, pp. 22ff.; Christine Lee on "The Choice 2008," *Frontline,* PBS, 14 October 2008; Obama, "Remarks on the Death of Nelson Mandela," 5 December 2013; DJG interviews with David Goldberg, Lisa Hay, Tom Perrelli, Julie Cohen, Christine Lee, Michael Cohen, Kevin Downey, David Nahmias, Susan Freiwald, Adam Charnes, Jennifer Collins, Brad Berenson, Amy Kett, Radhika Rao, Tom Krause, Bruce Spiva, Jennifer Borum Bechet, Monica Harris, Scott Siff, Michael Weinberger, and John Parry.

30. *HLR* 103 (January 1990): 601–828; Peter M. Yu to Erwin M. Griswold, 4 January 1990, EGP Box 133 Fld. 1; Charles Rothfeld, "Minority Critic Stirs Debate on Minority Writing," *NYT,* 5 January 1990, p. B6; Griswold to Yu, 12 January 1990, Barbara Schneider to Griswold, 23 January 1990, and Griswold to Schneider, 25 January 1990, EGP Box 133 Fld 1; Katharine T. Bartlett, "Feminist Legal Methods," *HLR* 103 (February 1990): 829–889, esp. 829; Kenneth Lasson, "Scholarship Amok: Excesses in the Pursuit of Truth and Tenure," *HLR* 103 (February 1990): 926–50, esp. 928, 930; Yu to Griswold, 7 February 1990, and Griswold to Yu, 8 February 1990, EGP Box 133 Fld. 1; Anthony E. Cook, "Beyond Critical Legal Studies: The Reconstructive Theology of Dr. Martin Luther King, Jr.," *HLR* 103 (March 1990): 985–1045; Floyd Abrams, "A Worthy Tradition: The Scholar and the First Amendment," *HLR* 103 (March 1990): 1162–1173, esp. 1162; Yu to Griswold, 7 March 1990, and Griswold to Yu, 13 March 1990, EGP Box 133 Fld. 1; DJG interviews with Gordon Whitman, Frank Amanat, Adam Charnes, Lisa Hay, John Parry, Scott Siff, and Brad Wiegmann.

31. Harvard Law School 1989–90 Catalog, esp. pp. 75, 132, 147; Gerald E. Frug, "The Ideology of Bureaucracy in American Law," *HLR* 97 (April 1984): 1276–1388; Frug, *Local Government Law* (West Publishing, 1988); Joan Chalmers Williams, "The City, the Hope of Democracy: The Casebook as Moral Act," *HLR* 103 (March 1990): 1174–86, esp. 1183, 1185–86; [Roberto Unger, Readings Packet], Jurisprudence, Harvard Law School, Spring Term 1990, esp. pp. 2 and 468, Radhika Rao Papers; Unger, *False Necessity: Anti-Necessitarian Social Theory in the Service of Radical Democracy* (Cambridge University Press, 1987); Unger, *Passion: An Essay on Personality* (Free Press, 1984); Rob Fisher—Discussant: Barack Obama, "Class Presentation, Tribe's Post-Newtonian Seminar," 15 February 1990, RFP; Holmes in *Towne v. Eisner,* 245 U.S. 418, 425 (1918); Laurence H. Tribe and Michael C. Dorf, "Levels of Generality in the Definition of Rights," *University of Chicago Law Review* 57 (Fall 1990): 1057–1108, at 1077n83 (citing Rob's *The Logic of Economic Discovery*); Tribe and Dorf, *On Reading the Con-*

stitution (Harvard University Press, 1991), esp. pp. iii, 125, and 132 (again citing Rob's *Logic*); Harry N. Scheiber, "What the Framers Didn't Say," *NYTBR*, 17 March 1991; Cassandra Butts in Larissa MacFarquhar, "The Conciliator," *New Yorker*, 7 May 2007, pp. 48ff.; Joel Freid, "It's California's Turn to Elect a President Like John Kennedy," *California Progress Report*, 28 January 2008; Freid in "Bay Area Residents in D.C. to Witness History," *Oakland Tribune*, 18 January 2009, and in *Nuts & Boalts*, 16 March 2009; Noam Scheiber, "Crimson Tide," *TNR*, 4 February 2009; Butts in Remnick, *The Bridge*, p. 189; DJG interviews with Rob Fisher, Chris Edley, Dan Rabinovitz, Cassandra Butts, David Troutt, Joel Freid, David Smail, Jennifer Gardner Trulson, John Parry, Tynia Richard, Jerry Frug, Roberto Unger, Radhika Rao, Susan Freiwald, Julie Cohen, Steve Ganis, Larry Tribe, Jack Chorowsky, Mike Dorf, Peter Dolotta, Andy Schapiro, and Tom Perrelli.

32. Tammerlin Drummond, "Barack Obama's Law," *LAT*, 12 March 1990, pp. E1–E2 (reprinted in *Toronto Star*, 18 March 1990, *San Francisco Chronicle*, 1 April 1990, and *Punahou Bulletin*, Fall 1990, pp. 40–42); Diana Quinn Rose, "Why Not Call Him 'Interracial'?," *LAT*, 30 March 1990, p. E10; Caroline V. Clarke, "A First Choice at Harvard Law Review," *American Lawyer*, April 1990, p. 92; "Law Review Firsts," *National Law Journal*, 2 April 1990, p. 4; Peter M. Yu and Barack H. Obama to Erwin N. Griswold, 10 April 1990, and Griswold to Yu and Obama, 12 April 1990, EGP Box 200 Fld. 2; Allison J. Pugh (AP), "First Black President of School's Law Review Uninterested in a Cushy Job," 15 April 1990 (and in *Miami Herald*, 18 April 1990, p. C1, *Philadelphia Tribune*, 20 April 1990, p. A6 [more complete text], and *Chicago Daily Herald*, 3 May 1990, p. 2); "Alumni Album," *Punahou Bulletin*, Spring 1990, p. 56; "Movers and Shakers," *Emerge*, May 1990, p. 62; "Illinois Organizer Heads Harvard Law Review," *Illinois Issues*, May 1990, p. 33; "A Landmark Election at the Harvard Law Review," *Harvard Magazine*, May–June 1990, p. 86; Elise O'Shaughnessy, "Harvard Law Reviewed," *Vanity Fair*, June 1990, p. 106; "Law Review Election Reflects Growing Diversity," *Harvard Law Bulletin*, Summer 1990, p. 26; "HLS in Perspective: Obama Elected," and "Law Review," *Harvard Law School Yearbook* 1990, pp. 74 and 182; Roxanne Brown, "In Pursuit of Excellence," *Ebony*, August 1990, pp. 112ff.; Drummond, "Sen. Obama Was Real Deal Even Way Back Then," *Contra Costa Times*, 5 November 2006, p. F4; Lee in Remnick, *The Bridge*, p. 209; DJG interviews with Tammerlin Drummond and Christine Lee. Notwithstanding that *Illinois Issues* references a 9 February 1990 Obama interview with PBS, Barack did not appear on that evening's *PBS NewsHour* and no other PBS video, audio, or transcript of Barack from early 1990 has surfaced.

33. *HLS Advisor* #23, 8 February 1990, p. 5, #25, 22 February 1990, p. 9, #27, 8 March 1990, p. 6; Maggie S. Tucker, "Affirmative Action Debated," *HC*, 9 March 1990; Sharon Stone, "BLSA Spring Conference Raises Issues for Professionals in '90s," *HLRec*, 16 March 1990, p. 15; Schuker, "Obama Stars at Convention," *HC*, 30 July 2004; Serrano and Savage, "Obama's Harvard Law School Days," *LAT*, 27 January 2007; Levenson and Saltzman, "At Harvard Law, a Unifying Voice," *BG*, 28 January 2007, p. A1; Heilemann, "When They Were Young," *NYM*, 22 October 2007; Kennedy in "The TNR Primary," *TNR*, 13 February 2008, p. 16; Jodi Kantor, "For a New Political Age, a Self-Made Man," *NYT*, 28 August 2008; Mack on "Obama: Professor President," BBC Radio 4, 6 January 2009; Randall Kennedy, *The Persistence of the Color Line* (Pantheon Books, 2011), p. 43; DJG interviews with Tynia Richard, Alan Jenkins, Nicole Lamb-Hale, and Ken Mack.

34. Harvard Law School Schedule of Courses 1989–90, HLSA; Obama to Andrew Roth, 19 March 1990, ARP; Paul Simon to Obama, 7 March 1990, PSP; O'Shaughnessy, "Harvard Law Reviewed," *Vanity Fair*, June 1990, p. 106; Obama Remarks at the Illinois State Library, 7 March 2001; J. Michael Kennedy, "Speech Gives a Boost to Obama's 'Dreams,'" *LAT*, 6 August 2004; Charles Leroux, "The Buzz Around Obama's Book," *CT*, 6 August 2004; Cindy Adams, "Obama's Memoir Really 18 Years Old," *New York Post*, 9 June 2008; Jane Dystel e-mail to Robert Draper, 2009; Obama on *Anderson Cooper 360*, CNN, 14 March 2008; DJG interviews with Ken Mack, Rob Fisher, Dan Rabinovitz, Bernard Loyd, Bruce Orenstein, David Troutt, Scott Siff, and Cassandra Butts.

35. Greg Galluzzo to Ed Shurna, 17 April 1989, Gamaliel Papers; "SSAC Rally to Launch New Cities Community Development Corp.," 27 April 1989, Flossmoor, Tom Knutson Papers; Darren Hillock, "Cardinal Joins SSAC to Initiate New Housing Rehab Program," *Star*, 30 April 1989, pp. A1–A2; Bill Kemp, "Community Organizing in South Suburban Islands of Poverty," *Illinois Issues*, May 1989, pp. 23–27;

Florence Hamlish Levinsohn, "School Revolt," *CR,* 26 May 1989; Mary O'Connell, "School Reform," *TNW,* August–September 1989, p. 3; Isabel Wilkerson, "New School Term in Chicago Puts Parents in Seat of Power," *NYT,* 3 September 1989; Barbara B. Buchholz, "Hard-Hit NW Indiana to Get Housing Grant," *CCB,* 2 October 1989, p. 7; "LIFT Obtains Housing Grants," *GPT,* 10 August 1989; "LIFT Raises $1 Million for Rehabbing," *GPT,* 25 September 1989; "LIFT Battles Blight," *GPT,* 12 November 1989; "Elementary School Council Election Results," *CT,* 23 October 1989; William Ayers, "Reforming Schools & Rethinking Classrooms: A Chicago Chronicle," *Rethinking Schools,* October–November 1989, pp. 6–7; Ayers, "Chicago, Laboratory for Schools," *NYT,* 18 December 1989, p. A19; Ayers, *The Good Preschool Teacher* (Teachers College Press, 1989); Ben Joravsky, "Mad as Hell," *CR,* 1 March 1990; Wilkerson, "Fate of Principals Splits Some Chicago Schools," *NYT,* 2 March 1990; Patrick J. Keleher Jr., "Business Leadership and Education Reform: The Next Frontier," Heritage Foundation Forum, 28 April 1990; "School Reform Marches On," *TNW,* June–July 1990, p. 11; "Coalition Pushes to Add 45,000 Hispanic Voters," *CST,* 14 September 1990; Kyle and Kantowicz, *Kids First,* pp. 219–20; Wieboldt Foundation Annual Report 1990, p. 14; Alex Poinsett, "Corporate Chicago Weighs in with Clout, Money, Time," *Catalyst,* March 1991, pp. 2–5; G. Alfred Hess Jr., *School Restructuring, Chicago Style* (Corwin Press, 1991), pp. 162–91; Joe Reed, "Grass Roots School Governance in Chicago," *National Civic Review* 80 (Winter 1991): 41–45; James G. Cibulka, "Local School Reform: The Changing Shape of Educational Politics in Chicago," *Research in Urban Policy* 4 (1992): 145–73; Hess, "City Schools Can't Wait for Poverty Cure," *Catalyst,* May 1992, pp. 17–18; William S. McKersie, "Philanthropy's Paradox: Chicago School Reform," *Educational Evaluation and Policy Analysis* 15 (Summer 1993): 109–28; Wilfredo Cruz, "From Blue Jeans to Pin Stripes," *Illinois Issues,* December 1993, pp. 20–23; Sharon G. Rollow and Anthony S. Bryk, "Democratic Politics and School Improvement: The Potential of Chicago Reform," in Catherine Marshall, ed., *The New Politics of Race and Gender* (Falmer Press, 1994), pp. 97–106; Hess, "School-Based Management as a Vehicle for School Reform," and "The Changing Role of Teachers," *Education and Urban Society* 26 (May 1994): 203–19 and 248–63; Dorothy Shipps, "Big Business and School Reform," pp. 55, 101, Anthony S. Bryk et al., *Charting Chicago School Reform: Democratic Localism as a Lever for Change* (Westview Press, 1998); Madeline Talbott, "Parents as School Reformers," in Alexander Russo, ed., *School Reform in Chicago* (Harvard Education Press, 2004), pp. 55–62; Lynne Mock, "The Personal Vision of African American Community Leaders," Ph.D. dissertation, University of Illinois at Chicago, 1999, pp. 29–30; James G. Kelly, "Contexts and Community Leadership: Inquiry as an Ecological Expedition," *American Psychologist* 54 (November 1999): 953–61, at 955; L. Sean Azleton, "Bounday Spanning and Community Leadership: African American Leaders in the Greater Roseland Area," M.A. thesis, University of Illinois at Chicago, 1996, p. 16; William Rohe et al., *Sustainable Nonprofit Housing Development: An Analysis of Maxwell Award Winners* (FannieMae Foundation, 1998), pp. 123–30; Kelly et al., "On Community Leadership: Stories About Collaboration in Action Research," *American Journal of Community Psychology* 33 (June 2004): 205–16, at 207; DJG interviews with Bill Ayers, Greg Galluzzo, Mary Gonzales, Phil Mullins, Danny Solis, Josh Hoyt, Bruce Orenstein, Harvey Lyon, Todd Dietterle, Mary Ellen Montes, Mike Kruglik, Johnnie Owens, Ken Rolling, Alvin Love, Aletha Strong Gibson, Desta Houston, Len Dubi, Tom Knutson, Kathy Kish, Jerry Kellman, Paul Scully, David Kindler, and Kerry Taylor. Many additional 1989–90 *Gary Post Tribune* stories mention LIFT's development, and Gamaliel's 1990–91 success in Milwaukee is superby analyzed in Stephen Hart, *Cultural Dilemmas of Progressive Politics* (University of Chicago Press, 2001), pp. 27–120, esp. pp. 33. On Gamaliel since the early 1990s, see especially Robert Kleidman, "Community Organizing and Regionalism," *City & Community* 3 (December 2004): 403–21.

36. R. Bruce Dold, "Opposition Enlists a Top Fundraiser for Bid to Unseat Savage," *CT,* 26 June 1989; Dold, "Savage's Rival Hit by Allegation," *CT,* 20 August 1989; Dold, "Prospective Savage Rival Says Allegations Are Lies," *CT,* 22 August 1989; Dold, "Ald. Shaw Landed Job for Reynolds' Accuser," *CT,* 26 August 1989; Dold, "Savage Challenger Opens Campaign," *CT,* 13 September 1989; Dold, "Savage's Foe Acquitted of Sex Charges," *CT,* 7 October 1989; Gilbert Jimenez, "Reynolds Cleared on Sex Charges," *CST,* 7 October 1989; "Educator's Office Robbed, Petitions Taken," *CT,* 3 December 1989; Weston Kosova, "Savage Gus," *TNR,* 29

January 1990, pp. 21–22; Paul Merrion, "Reynolds Groundswell Poses Savage Threat," *CCB*, 29 January 1990, p. 4; David C. Rudd, "Sawyer Backs Reynolds Over Savage," *CT*, 19 February 1990; Florence Hamlish Levinsohn, "The Street Scrapper and the Rhodes Scholar," *CR*, 8 March 1990; Dirk Johnson, "Challenge at Home for Adept Player of Racial Politics," *NYT*, 11 March 1990; Vernon Jarrett, "Defeat of Savage Would Be a Blow to Black Independence," *CST*, 13 March 1990; Jarrett, "Reynolds' Support Isn't Homegrown," *CST*, 15 March 1990, p. 37; William E. Schmidt, "Rep. Savage Claims Victory in Illinois," *NYT*, 21 March 1990; John Camper and Andrew Fegelman, "Savage Overcomes Tough Challenge," *CT*, 21 March 1990; Camper, "Everything Falls in Place for Savage," *CT*, 22 March 1990; Jarrett, "Why Gus Savage Keeps Winning," *CST*, 29 March 1990; Michael Miner, "Savage Appeal," *CR*, 29 March 1990; Robert Davis, "Shaw Finds Cheers Quickly Turn to Jeers," *CT*, 29 June 1990, p. M1; DJG interview with Andy Schapiro.

37. *HLS Advisor* #31, 12 April 1990, p. 7, #33, 26 April 1990, p. 5; Kerlow, *Poisoned Ivy*, pp. 170, 182–83; Susan Higgins on travelingmanrick.blogspot.com, 16 May 2008; Lisa Hay in Neal Conan, "Who Is Barack Obama," *Talk of the Nation*, NPR, 5 August 2008; DJG interviews with Lee Hwang, Andy Schapiro, Lisa Hay, Scott Smith, Jorge Ramírez, Kevin Downey, Cassandra Butts, Susan Higgins, and Chad Oldfather. See also Frank J. Tipler, "The Obama-Tribe 'Curvative of Constitutional Space' Paper Is Crackpot Physics," unpublished paper, SSRN.com, 26 October 2008, esp. p. 3 ("Tribe's paper demonstrates an appalling ignorance of elementary mathematics").

38. David Warsh, "Rebuilding Beirut: A Liberal Vision for Harvard Law," *BG*, 11 February 1990, p. A1; Jay K. Varma, "Law School Hires New Prof," *HC*, 15 February 1990; Philip M. Rubin, "Despite Advances, Hiring Plans Criticized," *HC*, 17 February 1990; Rubin, "Law Students Seek Curriculum Change," *HC*, 2 March 1990; Patrick Miles Jr., "Public Interest Committee Prepares Position Paper," *HLRec*, 2 March 1990, pp. 1, 12; Daniel Golden, "An Unconventional Traditionalist," *BG Magazine*, 4 March 1990, pp. 12ff.; Rubin, "Clark Holds Forum on Public Interest," *HC*, 8 March 1990; Jim Houpt, "Clark: HLS Should Further Public Interest," *HLRec*, 16 March 1990, pp. 1, 16; Brian Timmons, "That's No Okie, That's My Torts Professor," *WSJ*, 3 April 1990, p. A20; Rubin, "Harvard Law Students to Boycott Classes," *HC*, 4 April 1990; Linda Popejoy, "Students Protest Dean on Diversity," *HLRec*, 11 April 1990, pp. 1, 12; Popejoy and John Thornton, "Clark and Students Talk at Forum," *HLRec*, 20 April 1990, pp. 1, 12; Ken Emerson, "When Legal Titans Clash," *NYT Magazine*, 22 April 1990, pp. 26ff.; Diane Bartz(AP), "Harvard Law Professor May Request Leave to Protest Lack of Diversity," 23 April 1990; Fox Butterfield, "Harvard Law Professor Quits Until Black Woman Is Named," *NYT*, 24 April 1990, p. A1; John H. Kennedy, "Harvard Black Eyes a Leave in Tenure Dispute," *BG*, 24 April 1990, p. 1; WGBH News footage, 24 April 1990, 11:39, Item 7232, WGBH Archives; Kennedy, "Harvard Students Rally in Support of Bell's Vow," *BG*, 25 April 1990, p. M1; "Students Rally in Support of Harvard Professor's Protest," *NYT*, 25 April 1990, p. A16; Tynia Richard on *CBS This Morning*, 25 April 1990; Butterfield, "Harvard Law School Torn by Race Issue," *NYT*, 26 April 1990, p. A20; Derrick Bell, "Why We Need More Black Professors in Law School," *BG*, 29 April 1990, pp. A1, A4; Robert Zafft, "Bell's Message Is Repugnant," *HC*, 30 April 1990, and *HLRec*, 4 May 1990, p. 9; Anthony Flint, "Jackson Offers to Mediate Harvard Dispute," *BG*, 3 May 1990, p. 27; Popejoy, "Clark Offers Cool Response to Bell's Protest," Houpt, "Timmons Attracts National Attention for CLS-Bashing," and Matt Kairis, "In Defense of Randall Kennedy," *HLRec*, 4 May 1990, pp. 1, 3, 4, 8, 16; Flint, "Bell at Harvard: A Unique Activisim," *BG*, 7 May 1990, p. M1; Flint, "Harvard Dean Says No to Jackson Parley," *BG*, 8 May 1990, p. 21; "Harvard Law Declines Jackson Offer of Help," *NYT*, 8 May 1990, p. A20; Flint, "Jackson Says Eyes Are on Harvard," *BG*, 9 May 1990, p. 41; Meg Vaillancourt, "Jesse Jackson at Harvard Law School," *Ten O'Clock News*, WGBH, 9 May 1990, 5:43, WGBH Archives; Butterfield, "At Rally, Jackson Assails Harvard Law School," *NYT*, 10 May 1990, p. A14; Flint, "Rev. Jackson, at Harvard, Urges National Conference on Racism," *BG*,10 May 1990, p. 1; David Troutt, "The Challenge of Diversification," *LAT*, 13 May 1990, and *HLRec*, 14 September 1990, p. 9; Richard Cohen, "The Question of Merit at Harvard Law," *WP*, 15 May 1990, p. A23; George F. Will, "Academic Set-Asides," *WP*, 17 May 1990, p. A27; "For Whom the Bell Tolls," *TNR*, 21 May 1990, p. 8; Butterfield, "Old Rights Campaigner Leads a Harvard Battle," *NYT*, 21 May 1990, p. A18; Alfred Edmond Jr., "Derrick Bell Lays

Down the Law at Harvard," *Black Enterprise,* July 1990, p. 15; Timmons, "Fraudulent Diversity," *Newsweek,* 12 November 1990, p. 8; Sharon Stone, "Law Student's Column Heats Up Diversity Debate," *HLRec,* 30 November 1990, pp. 3, 7; Ian Haney López, "Community Ties, Race, and Faculty Hiring: The Case for Professors Who Don't Think White," *Reconstruction* 1, #3 [August 1991]: 46–62, at 49; Derrick Bell, *Confronting Authority* (Beacon Press, 1994), esp. pp. 63, 110–11; David A. Hill, "Thoughts on the Diversity Movement at HLS," Keith Boykin, "The Initial Meeting: A Reflection," and Marie Lott Pharoah, "Reflections on the Diversity Movement," in *HLS Diversity: A Celebration of the Movement,* 30 October 1998; Keith Boykin, "Barack Obama, Blacks and Harvard Law," KeithBoykin.com, 26 January 2007; Ian Haney López, *Dog Whistle Politics* (Oxford University Press, 2014), esp. p. x; DJG interviews with David Hill, Hill Harper, Tynia Richard, Mark Kozlowski, Cassandra Butts, Keith Boykin, Linda Singer, Ken Mack, Morris Ratner, Steve Ganis, Derrick Bell, Ursula Dudley Oglesby, Paolo Di Rosa, Micki Chen, Michael Weinberger, Nathan Diament, Ian Haney López, Gerry Singsen, and Rob Fisher. See also Randall Kennedy, "Racial Critiques of Legal Academia," *Harvard Law Review* 102 (June 1989): 1745–1819, and Derrick Bell, Tracy Higgins, and Sung-Hee Suh, eds., "Racial Reflections: Dialogues in the Direction of Liberation," *UCLA Law Review* 37 (August 1990): 1037–1100.

39. Robin West, "Jurisprudence and Gender," *University of Chicago Law Review* 55 (Winter 1988): 1–72; Tushnet and Lynch, "The Project of the Harvard *Forewords:* A Social and Intellectual Inquiry," *Constitutional Commentary* 11 (Winter 1994): 463–500; John Parry to Brad Berenson, "Results," and Berenson reply comments, 2 May 1990, "To: Two-L Editors, 'Results of Faculty Case Comment Mailbox Vote,'" 2 May 1990, and Parry to Berenson, 3 May 1990, JPP; *HLR* 103 (May 1990): McConnell, "The Origins and Historical Understanding of Free Exercise of Religion," 1409–1519, "Developments in the Law—Medical Technology and the Law," 1520–1676; Obama to Griswold, 10 May 1990, and Griswold to Obama, 17 May 1990, EGP Box 133 Fld. 1; *HLR* 103 (June 1990): Vicki Schultz, "Telling Stories About Women and Work," 1749–1843, "Colloquy—Responses to Randall Kennedy's *Racial Critiques of Legal Academia,*" 1844–87, [Radhika Rao,] "The Priest Who Kept His Faith But Lost His Job," 2074–81, "Scholarship Admired: Responses to Professor Lasson," 2085–86; Obama to Griswold, 10 [June] 1990, Susan M. Higgins to Griswold, 13 June 1990, and Griswold to Obama, 18 July 1990, EGP Box 133 Fld. 1; Jeffrey Ressner and Ben Smith, "Obama Kept Law Review Balanced," *Politico,* 23 June 2008; DJG interviews with Michael Weinberger, David Goldberg, Lourdes Lopez-Isa, Adam Charnes, Jennifer Collins, John Parry, Brad Berenson, Robin West, Gordon Whitman, Scott Siff, Susan Freiwald, Michael Guzman, Lisa Hay, Susan Higgins, Brad Wiegmann, Chad Oldfather, Kevin Downey, Jacqueline Scott Corley, Michael Cohen, David Nahmias, Chris Sipes, David Wilkins, Christine Lee, Radhika Rao, Barbara Schneider, Debbie Brake, Brian Bertha, and Ken Mack.

40. *Harvard Law School Annual Examinations in Law 1989–90,* pp. 66–69, 572–77; Obama remarks at the Cambridge (MA) Public Library, 20 September 1995, and at the Illinois State Library, 7 March 2001; Dystel in Cindy Adams, "Obama's Memoir Really 18 Years Old," *NYP,* 9 June 2008; Dystel's 2009 e-mail to Robert Draper; Dystel in Jeremy Greenfield, "Agents Unwilling to Adapt Won't Last," DigitalBook World.com, 3 January 2013; DJG interviews with Dan Rabinovitz, Gerry Frug, Rob Fisher, and David Troutt.

41. *HLS Adviser* #36, 17 May 1990, p. 2; PR Newswire, 13 June 1990; David Rubinstein and Jennifer Juarez Robles, "Law Firms Still Lag in Minority Hiring," and Rubinstein, "Top Student: What Kind of Minorities Do Firms Want?," *Chicago Reporter,* July–August 1990, pp. 1, 3–7; Louis Aguilar, "Survey: Law Firms Slow to Add Minority Partners," *CT,* 11 July 1990, p. B1; *Roy Abon'go Obama v. Mary K. Cole,* #90-16597, Prince George's County Circuit Court, 16 July 1990 (judgment signed 9 October 1991); *Austin v. Michigan Chamber of Commerce,* 494 U.S. 652, 660–61 (27 March 1990); Obama to Trent Norris, 23 July 1990, TNP; Mary Flood, "At Law Review, Number of Women Members Falls Sharply," *HLRec,* 12 October 1990, pp. 1, 16; "Suzanne Alele '85," *Princeton Alumni Weekly,* 24 October 1990; Kenan Heise, "Albert Maule, 40, President of Police Board," *CT,* 12 October 1995; Tom Maliti (AP), "Obama's Brother Chooses Life in Slow Lane," 26 October 2004; Obama's June 2006 interview with Jacob Weisberg; Obama, *TAOH,* p. 331; Laura Washington, "What Has Brown Done?," *CST,* 23 October 2006, p. 39; Ray Gibson and David Jackson, "Rezko Owns

Vacant Lot Next to Obama's Home," *CT,* 1 November 2006, p. 1; Dave McKinney and Chris Fusco, "Obama Spells Out 'Regret' After Land Deal with Rezko," *CST,* 5 November 2006, p. 4; Tim Novak, "Obama and His Rezko Ties," *CST,* 23 April 2007, p. 22; Novak, "Broken Promises, Broken Homes," *CST,* 24 April 2007, p. 20; James L. Merriner's essential "Mr. Inside Out," *Chicago Magazine,* November 2007; Edward McClelland, "How Close Were Barack Obama and Tony Rezko?," *Salon,* 1 February 2008; Martin Fricker and Graham Brough, "Barack Obama's Stepmother in a Bingo-Lover from Bracknell," *Daily Mirror,* 9 February 2008; "Complete Transcript of the *Sun-Times* Interview with Barack Obama," *CST,* 15 March 2008; Transcript of Obama's *Chicago Tribune* Interview, *CT,* 16 March 2008, p. 22; Reardon, "Obama's Chicago," *CT,* 25 June 2008, p. T1; Wolffe, *Renegade,* pp. 34–35; Obama remarks at the White House, 19 June 2009, and Q&A session with students, 8 September 2009, *Public Papers of the Presidents,* 2009, Vol. 1, p. 858, and Vol. 2, p. 1352; Yang Suwan in Judith Kempfner, "The Untold Story of Obama's Mother," *Independent,* 16 September 2009; Scott, *A Singular Woman,* pp. 290–91, 295–96; Auma Obama, *And Then Life Happens* (St. Martin's, 2012), pp. 39–40, 210; Slevin, *Michelle Obama,* p. 132; Newton Minow's 2008 interview with Jim Gilmore; DJG interviews with Elvin Charity, Alan Solow, Geraldine Alexis, John Levi, Newton Minow, Linzey Jones, Greg Dingens, Eden Martin, Evie Shockley, Jerry Kellman, David Brint, Lisa Hay, Susan Higgins, Sean Lev, Charlie Robb, Marisa Chun, Brad Berenson, John Parry, Frank Amanat, Leonard Feldman, Darin McAtee, Sondra Hemeryck, Chris Sipes, Trent Norris, Lori-Christina Webb, Bruce Spiva, Lee Hwang, Nancy Cooper, and Zeituni Onyango. Sheree and Abon'go's union would last less than six years. See *Sheree A. Obama v. Abon'go M. Obama (c/o Nyangema Kogelo Primary School, Siaya),* #96-04351, Prince George's County Circuit Court, 4 March 1966, granted 11 July 1997. See also *Si Van Lee v. Maya K. Soetoro,* Hawaii 1CC90-0-002339 (motor vehicle tort), 19 September 1990, dismissed 4 February 1991.

On Tony Rezko's life prior to 1990, see especially Sam Lesner, "Hyde Parkers, All!," *HPH,* 12 June 1985, p. 43; see also Tom Gorman, "Crucial Concessions Gets Park Contract," *HPH,* 24 July 1985, pp. 3, 28; Gorman, "Caterer Changes Subcontractor," *HPH,* 25 February 1987, p. 3; Gorman, "Crucial May Get New South Shore Center Catering Contract," *HPH,* 29 March 1989, p. 1; Carolyn Hirschman, "Politicians Join Community in Protest Over Crucial Contract," *HPH,* 2 August 1989, p. 1; Dan Mihalopoulos and Laurie Cohen, "City Knew O'Hare Vendor Wasn't Minority-Run Firm," *CT,* 18 March 2005, p. 1; John Chase, "Tony Rezko," *CT,* 26 May 2005, p. 1; and David Jackson and Chase, "Rezko's Life a Story of Pizza and Politics," *CT,* 12 October 2006, p. 1. On Herbert Muhammad, see his 1989 interview with Blackside's Sam Pollard, and Richard Goldstein, "Jabir Herbert Muhammad, Who Managed Muhammad Ali, Dies at 79," *NYT,* 28 August 2008.

42. Michael Cohen in David Nather, "Obama's Quick Rise," *CQ Weekly,* 21 August 2008; DJG interviews with Carol Platt Liebau, Lori-Christina Webb, Sean Lev, Charlie Robb, Christine Lee, Brad Berenson, Michael Cohen, David Nahmias, Adam Charnes, Daniel Slifkin, Julie Cohen, Tom Perrelli, Susan Freiwald, Chris Sipes, Kevin Downey, Orin Sellstrom, and Howard Ullman.

43. Carol Platt Liebau, "The Barack I Knew," TownHall.com, 5 March 2007; Artur Davis in Nicole Duran, "New Kid on the Block," *Campaigns & Elections,* June 2007, pp. 52ff., and Seth Stern, "A Commander in Chief," *Harvard Law Bulletin,* Fall 2008; DJG interviews with Paul Massari, Lois Leveen, Robin West, Bruce Spiva, Jeff Hoberman, Daniel Slifkin, Brad Berenson, John Parry, Kevin Downey, Adam Charnes, Christine Lee, Susan Freiwald, Trent Norris, Earl Martin Phelan, Tom Perrelli, Julie Cohen, Marisa Chun, Jonathan Putnam, and Carol Platt Liebau.

44. Harvard Law School 1990–91 Catalog, esp. pp. 4–5, 115, 138–39; *HLS Adviser* Vol. 22, #1, 30 August 1990, pp. 5, 11; Glendon, Minow, Rakoff, "Law and Society Seminar Fall 1990: Civil Society," RFP; Kahlenberg, *Broken Contract,* p. 64; Granfield and Koenig, "Learning Collective Eminence," pp. 505, 515; Minow in Lauren A. E. Schuker, "Obama Stars at Convention," *HC,* 30 July 2004; Minow's May 2007 interview with Liza Mundy; Adam Charnes in Morrill, "Remembering Obama," *Charlotte Observer,* 13 January 2009; Gary Orren in "Obama 'A Great Persuader' Says U.S. Expert," University of Ulster Press Release, 25 March 2010; DJG interviews with Rob Fisher, David Hill, Michelle Jacobs DeLong, Lourdes Lopez-Isa, David Nahmias, Brad Berenson, Chris Sipes, Scott Siff, Christine Lee, Marisa Chun, Sean Lev, Tom Perrelli, Jennifer Collins, Adam Charnes,

Brad Wiegmann, John Parry, Daniel Slifkin, Jon Molot, Michael Cohen, Susan Higgins, Gary Orren, Martha Minow, and Eric Posner.

45. *HLR* 104 (November 1990): West, "Foreword: Taking Freedom Seriously," 43–106, Fried, "Comment: *Metro Broadcasting, Inc. v. FCC:* Two Concepts of Equality," 107–127, [Kevin M. Downey], "Right to Die—Incompetents' Right to Refuse Life-Saving Medical Treatment," 257–66, at 265–66n74, [Darin McAtee], "Death Penalty—Aggravating and Mitigating Circumstances," 139–49, [Frank Amanat], "Political Patronage in Promotion, Hiring, Transfer, and Recall Decisions," 227–37; Charlie Robb on "Political Animal," WashingtonMonthly.com, 8 October 2008; DJG interviews with Adam Charnes, John Parry, Brad Berenson, Lee Hwang, Kevin Downey, Darin McAtee, Charlie Robb, and Frank Amanat.

46. Jonathan M. Berlin, "HLS Students Hold Vigil Outside Faculty Meeting," *HC,* 10 September 1990; Linda Popejoy, "Professor Derrick Bell, 1990–91: The Spirit of Protest Continues," and Kristen Swartz, "Placement Office Cautions Third Years on Their Job Opportunities," *HLRec,* 14 September 1990, pp. 1, 3, 16; George Paul, "Public Interest Office Swamped," "Rights Group Plans to Rally for Diversity," and Karen Levy, "Student Groups Demand Boycott of Baker & McKenzie," *HLRec,* 21 September 1990, pp. 1, 3, 15, 16; Scott M. Finn, "Law Students Demand More Minority Faculty," *HC,* 28 September 1990; Peter S. Canellos, "Harvard Law School Reinstates Public Interest Career Counseling," *BG,* 30 September 1990, p. 34; Berlin, "Students Form New Pro Bono Group," *HC,* 5 October 1990; Malcolm E. Harrison, "After Rally, CCR Begins to Rethink Strategy on Diversity," and Felicia Ravago, "At Forum, Trouble Over Clinicals," *HLRec,* 5 October 1990, pp. 1, 13, 16; Ivan Oransky, "Law Students Protest Recruiting Interviews," and Philip M. Rubin, "Law Students Picket Interviews," *HC,* 10 October 1990; Mary Flood, "Students Confront Firm on Fifth Floor of Charles Hotel," and "At Law Review, Number of Women Members Falls Sharply," *HLRec,* 12 October 1990, pp. 1, 16; Sean P. Murphy, "Students Honor Law Professor as He Begins His Protest Leave," *BG,* 21 October 1990, p. 34; Jim Chen, "WLA's Misguided Call for Affirmative Action," *HLRec,* 26 October 1990, pp. 4, 7; Anthony Flint, "Flexibility, Determination at Harvard Law," *BG,* 30 October 1990, p. 21; Berlin, "Flyers Allege Ethical Violations," *HC,* 1 November 1990; Patricia Arzuaga and Michelle Benecke, "WLA Blasts Record for Article on Law Review," *HLRec,* 2 November 1990, pp. 4–5; Obama, "Review President Explains Affirmative Action Policy," *HLRec,* 16 November 1990, pp. 4, 7; Berenson in Levenson and Saltzman, "At Harvard Law, a Unifying Voice," *BG,* 28 January 2007, p. A1; McCullough in Jodi Kantor, "In Law School, Obama Found Political Voice," *NYT,* 28 January 2007, p. 1, and in Todd Purdum, "Raising Obama," *Vanity Fair,* March 2008; Berenson in Mary Vallis, "Unlikely Presence," *National Post,* 7 June 2008, p. A17; Rachel L. Swarns, "Obama Walks a Delicate Path on Class and Race Preferences," *NYT,* 2 August 2008, p. A1; Marisa Chun in Nather, "Obama's Quick Rise on a Non-Traditional Career Path," *CQ Weekly,* 21 August 2008; Mack on "The Choice 2008," *Frontline,* PBS, 14 October 2008; 3L grade-on John Bu in Joe L. Stanganelli, "Young Obama Draws Praise from Law School Classmates from Erie," *Erie Times-News,* 30 October 2008; Chen in Jack Brammer, "Two Kentuckians Reflect on Their Time with Barack Obama," *Lexington Herald-Leader,* 19 January 2009; DJG interviews with Brad Berenson, Jackie Scott Corley, Frank Amanat, Ken Mack, Julie Cohen, John Parry, Michael Cohen, Tom Perrelli, Michael Guzman, Edith Ramirez, Janis Kestenbaum, Bruce Spiva, Lourdes Lopez-Isa, and Jonathan Putnam. Although the exact date is unknown, Barack also participated in a panel discussion on free speech on campuses, moderated by Charles Nesson and briefly memorialized in the 1991 Harvard Law School *Yearbook:* "I don't see a lot of conservatives getting upset if minorities feel silenced," Barack said. It was not covered by either the *Record* or the *Crimson.* See also Charles C. Johnson, "At Harvard Law School in 1991, Obama Approved of Restricting Speech to Protect Minorities," *Daily Caller,* 8 October 2012. See as well Cassandra Butts's negative review of Shelby Steele, *The Content of Our Character, HCRCLLR* 26 (Winter 1991): 306–15.

47. David S. Broder, "Jesse Helms, White Racist," *WP,* 29 August 2001; Phalen in Levenson and Saltzman, "At Harvard Law, a Unifying Voice," *BG,* 28 January 2007, p. A1; Ann Sanner (AP), "Obama Says He Looked to Basketball in His Youth," 16 April 2008; Berenson in Zhu, "Obama's Quiet Harvard Roots," *HC,* 2 November 2008; Dave DeWitt, "Harvey Gantt Leaves a Legacy in NC," WUNC.org, 5 September 2012; DJG interviews with Rob Fisher, Julius Genachowski, Brad Berenson, and Earl Martin Phelan. Also see DJG, "The Helms Attack on King," *Southern Exposure* 12 (March–April 1984): 12–15.

48. Trent [Norris] to Barack, "Meltzer's Letter," 5 October 1990, TNP; Griswold to Obama, 8 October 1990, EGP Box 133 Fld. 1; "Celebrating Brennan," *Legal Times,* 19 November 1990, p. 11; Griswold to Obama, 26 November 1990, EGP Box 199 Fld. 8; Obama to Griswold, 30 November 1990, EGP Box 133 Fld. 1; *HLR* 104 (December 1990): Charny, "Nonlegal Sanctions in Commercial Relationships," 373–467, Wilkins, "Legal Realism for Lawyers," 468–524, Williams, "*Metro Broadcasting, Inc. v. FCC*: Regrouping in Singular Times," 525–46, [Jonathan Putnam], "First Amendment—Voters' Speech Rights—Federal District Courts Mandate Availability of Write-in Voting," 657–64; Obama to William J. Brennan, 7 December 1990, and Brennan to Obama, 13 December 1990, WJBP Box II-94; Griswold to Obama, 9 January 1991, EGP Box 132 Fld. 13; "Summary of Revenue and Expenses," HLR, June 1990, and "Agreement Among the Holders of the Copyright in the Uniform System of Citation," May 1991, TNP; Nancy Day, "The Battle for Illinois," *Harvard Magazine,* July–August 2004, pp. 75–78; Adam Charnes in Dahlia Lithwick, "A Complicated Record on Race," *Newsweek,* 7 April 2008, p. 34; Alex M. Azar II in Shapiro and Krishnaswami, "The Secret History of the *Bluebook,*" *Minnesota Law Review* 100 (2016); DJG interviews with Frank Amanat, David Goldberg, Michael Weinberger, Adam Charnes, David Wilkins, John Parry, Amy Kett, Jonathan Putnam, Lisa Hay, Ken Mack, Dodie Hajra, Susan Higgins, Richard Parker, and Trent Norris. See also Tushnet and Lynch, "The Project of the Harvard *Forewords,*" p. 470 ("*Forewords* are systematically likely to be disappointing, in the sense that they are likely to be less interesting than the most interesting work their authors have previously written"). In 1990–91 the *Review*'s graduate treasurer was the late Ernest J. "Ernie" Sargeant, managing partner of the premier Boston law firm Ropes and Gray, who died in January 2008; the other faculty trustee was William D. Andrews. See *ALI Reporter,* Spring and Summer 2008, and *HC,* 19 October 1992.

49. "Agreement Among the Holders of the Copyright in the Uniform System of Citation," May 1991, TNP; [Laura Demanski] to Elaine [Pfefferblit], "Journeys in Black and White by Barack Obama from Jane Dystel," 25 October 1990, LDP; Esther B. Fein, "Book Notes: A Duchy Turns 12," *NYT,* 20 May 1992, p. C20; Obama at Illinois State Library, 7 March 2001; Peter Osnos, "Barack Obama and the Book Business," Century Foundation, 30 October 2006; Janny Scott, "Obama's Story, Written by Obama," *NYT,* 18 May 2008, p. A1; Adams, "Obama's Memoir Really 18 Years Old," *NYP,* 9 June 2008; Weinberger in Tracy Connor, "See, I Did Too Know Bam," *NYDN,* 16 November 2008; Osnos, "The Making of the Book That Made Obama," *Daily Beast,* 27 January 2009; Dystel e-mail to Robert Draper, 2009; Draper, "Barack Obama's Work in Progess," *GQ,* November 2009; Maraniss, *BOTS,* pp. 362–63, 527–28, and 596n; DJG interviews with Lisa Hay, Brad Berenson, Ann Patty, Laura Demanski, and Michael Weinberger.

50. [Christine M. Lee], "Invisible Man: Black and Male Under Title VII," *HLR* 104 (January 1991): 749–68; Griswold to Obama, 9 January 1991, EGP Box 132 Fld. 13; Barack Obama and Robert Fisher, "Outline—Transformative Politics," Paper for Law and Society, n.d., 10pp., RFP; Seth Gitell, "Obama and the 'Invisible Man,'" *New York Sun,* 12 August 2008; DJG interviews with Christine Lee, Rob Fisher, Mark Kozlowski, and Martha Minow.

51. Anthony Flint, "Student Lawyers Sue Harvard on Hiring," *BG,* 21 November 1990; "A Class Sends Message to Harvard Law School," *NYT,* 21 November 1990, p. B11; Dan Greaney, "Students Sue HLS Over Faculty Hiring," *HLRec,* 30 November 1990, pp. 1, 12; Seth S. Karkness, "Taking the Law School to Court," *HC,* 1 December 1990; Ian Haney López, "Community Ties, Race, and Faculty Hiring," *Reconstruction* Vol. 1, #3 [August 1991]: 46–62; Kerlow, *Poisoned Ivy,* pp. 104–5; Maggie S. Tucker, "Alumni Association Candidates Sweep Overseers Election," *HC,* 7 June 1990; Donald M. Solomon, "Social Responsibility Should Be Top Priority," *HC,* 6 December 1990; Chester Hartman to HRAAA Executive Committee, 10 December 1990, and Solomon to Dear Friend of HRAAA, 31 December 1990, HRAAAP; Tucker, "HAA and HRAAA Name Candidates," *HC,* 31 January 1991; Solomon, "Press Advisory," 31 January 1991, HRAAAP; Gady A. Epstein, "The Last Hurrah?," *HC,* 4 June 1992; Alexander Reid and Adrian Walker, "Recent Parolee Is Charged in Assault-Weapon Killing," *BG,* 7 December 1990, p. 1; John Ellement, "Board Could Have Kept Rifle Suspect Behind Bars," *BG,* 8 December 1990, p. 1; Matthew Brelis, "Johnston Criticizes Parole of Suspect," *BG,* 9 December

1990, p. 33; Ellement, "Slaying Suspect Paroled in Error, Probe Concludes," *BG,* 3 January 1991, p. 17; Doris Sue Wong, "Parolee Gets Split Verdict in Fatal Spree," *BG,* 12 August 1992, p. 24; *Harvard Law School Annual Examinations in Law 1990–91,* pp. 324–26; Jacobs, *Obamaland,* p. 123; *Punahou School Alumni Directory 1841–1991* (Harris Publishing, 1991), p. 226; Harvard Law School 1990–91 Catalog, pp. 143–44; Eric F. Saltzman and Tom Bywaters, *The Shooting of Big Man: Anatomy of a Criminal Case,* ABC, 8 June 1979; Jeffrey Kaye, "Documentary Draws Fire from Public Defenders," *American Lawyer,* October 1979, p. 10; Fred Kobrick e-mail to Jeff Kobrick, 4 August 2008; Jeff Kobrick, "Notes on Barack Obama," 21 November 2008, Jeff Kobrick Papers; DJG interviews with Linda Singer, Keith Boykin, Derrick Bell, Donald Solomon, Chester Hartman, Robert Paul Wolff, Hill Harper, Rob Fisher, Gina Torielli, and Jeff Kobrick.

On Ronald Stokes, also see "Prisoner Attacks Officers," *BG,* 31 October 1980; *Stokes v. Superintendent,* 452 N.E. 2d 1123 (Mass. SJC), 15 August 1983; *Stokes v. Fair,* 795 F. 2d 235 (1st Cir.), 16 July 1986; *Stokes v. Commissioner of Correction,* 530 N.E. 2d 801 (Mass. Ct. Apps.), 16 November 1988; *Commonwealth v. Stokes,* 653 N.E. 2d 190 (Mass. Ct. Apps.), 19 July 1995; appeal denied, 653 N.E. 2d 1277 (Mass. SJC), 27 September 1995; *U.S. v. Stokes,* 947 F. Supp. 546 (D. Mass.), 18 November 1996, 124 F. 3d 39 (1st Cir.), 22 August 1997; Paul Langner, "Career Criminal Faces Life in Prison," *BG,* 23 March 1999, p. B3; *U.S. v. Stokes,* 2000 WL 246478 (D. Mass.), 28 February 2000, 388 F. 2d 21 (1st Cir.), 5 November 2004, 544 U.S. 917, 21 March 2005, and 2005 WL 2170091 (1st Cir.), 8 September 2005.

52. Ian Ayres, "Fair Driving: Gender and Race Discrimination in Retail Car Negotiations," *HLR* 104 (February 1991): 817–73; William E. Schmidt, "White Men Get Better Deals on Cars, Study Finds," *NYT,* 13 December 1990; Merrill Goozner, "Study Faults Car Sales Tactics," *CT,* 13 December 1990; AP, "Study of New-Car Sales Finds That White Men Get Best Deals," *LAT,* 13 December 1990; Rhonda Mahony, "Blacks Pay More for Cars," *Black Enterprise,* April 1991, p. 14; Obama to Griswold, 8 February 1991, and Griswold to Obama, 8 February 1991, EGP Box 132 Fld. 13; David Margolick, "At the Bar," *NYT,* 11 January 1991, p. B5; John Parry to Griswold, 15 January 1991, Griswold to Parry, 18 January 1991, Amy Folsom Kett to Griswold, 23 and 28 January 1991, Griswold to Kett, 29 January 1991, EGP Box 132 Fld. 13; Michael W. McConnell, "The Selective Funding Problem: Abortions and Public Schools," *HLR* 104 (March 1991): 989–1050; [A. Marisa Chun], "Unfulfilled Promises: School Finance Remedies and State Courts," *HLR* 104 (March 1991): 1072–92; [Monica Harris], "Cleaning Up the Debris After 'Fleet Factors': Lender Liability and CERCLA's Security Interest Exemption," *HLR* 104 (April 1991): 1249–68; [Scott Siff], "Judicial Enforcement of International Law Against the Federal and State Governments," *HLR* 104 (April 1991): 1269–88; Dorothy E. Roberts, "Punishing Drug Addicts Who Have Babies: Women of Color, Equality, and the Right of Privacy," *HLR* 104 (May 1991): 1419–83; [Trenton H. Norris], "Developments in the Law—International Environmental Law—Institutional Arrangements," *HLR* 104 (May 1991): 1580–1608; Obama, "Community Revitalization," Nebraska Wesleyan University Forum, Lincoln, 9 September 1994; Lisa Hay in Neal Conan, "Who is Barack Obama?," *Talk of the Nation,* NPR, 19 August 2008; Ken Mack in "The Choice 2008," Frontline, PBS, 14 October 2008; Adam Charnes in Morrill, "Remembering Obama," *Charlotte Observer,* 13 January 2009; Scott Siff in Noam Scheiber, "Crimson Tide," *TNR,* 4 February 2009; Jamshid Ghazi Askar, "Scholar Michael W. McConnell Mixes Law, Religion," *Deseret News,* 9 January 2011; DJG interviews with David Goldberg, Scott Siff, Kevin Downey, Lisa Hay, David Troutt, David Hill, Karla Martin, Mark Kozlowski, John Parry, Julie Cohen, Anne Toker, Susan Freiwald, Brad Berenson, Charlie Robb, Trent Norris, Tom Perrelli, Jon Molot, Lori-Christina Webb, Bruce Spiva, Adam Charnes, Michael Guzman, Sondra Hemeryck, Howard Ullman, Orin Sellstrom, Jeff Hoberman, Chris Sipes, Michael Cohen, Lee Hwang, Sean Lev, Dodie Hajra, Darin McAtee, and Marisa Chun.

53. *Summit for the '90's,* WTBS, 28 January 1991 [Tape 136], CNN; Gerry Yandel, "Talk Won't End Troubles, Black Summit Cautions," *AJC,* 29 January 1991, p. B1; "Editor, Harvard Law Review, Black History Minutes," TBS Public Affairs, [February 1991]; Marie [Milnes], Trent [Norris], and Jon [Molot] to Governance Committee, "Report on Our Three Areas," 5 January 1991, TNP; "Law Review Elects New Leader," *HLRec,* 8 February 1991, p. 3; Obama to Griswold, 8 February 1991, EGP Box 132 Fld. 13; Kerlow, *Poisoned Ivy,* pp. 10–11, 15–16;

Butts and Ellen in Remnick, *The Bridge,* pp. 191, 210; DJG interviews with Cassandra Butts, Rob Fisher, Randall Kennedy, Mark Kozlowski, Roger Boord, Trent Norris, David Goldberg, John Parry, Carol Platt Liebau, Bruce Spiva, Jon Molot, Sean Lev, Howard Ullman, Charlie Robb, Jonathan Putnam, Sondra Hemeryck, Mark Rosen, and Tom Perrelli.

54. Bob Arnold, "Obama Opens Up: Looks Back on Historic Year," *HLRec,* 15 February 1991, pp. 1, 8; Daniel J. Greaney, "Reviewing the *Review,*" *HLRec,* 1 March 1991, p. 4; Griswold to Obama, 15 March 1991, Ellen to Griswold, 4 April 1991, Griswold to Ellen, 12 April 1991, Ellen to Griswold, 6 May 1991, Griswold to Ellen, 10 May 1991, EGP Box 132 Fld. 13; Trent Norris and Ken Mack to The Body, "*Bluebook* Changes," 7 March 1991, Alan Diefenbach to David Ellen, "Proposal for a Permanent Advisory Board for the *Bluebook,*" 25 April 1991, Norris to David Ellen et al., "*Bluebook* Process," 9 September 1991, Norris, "Proposal for a Uniform System of Citation Advisory Board," 9 September 1991, Pat [Philbin], Eric [Reifschneider], Rob [Niewyck], and Simone [Francis] to David [Ellen], Bruce [Spiva], and Trent [Norris], "Bluebook Process," 24 September 1991, all TNP; Jim C. Chen, "Something Old, Something New, Something Borrowed, Something Blue," *University of Chicago Law Review* 58 (Fall 1991): 1527–40, esp. 1536, 1537; Norris, "Report of the *Bluebook* Editor," 4 February 1992, TNP; James W. Paulsen, "An Uninformed System of Citation," *HLR* 105 (May 1992): 1780–94, at 1792; James D. Gordon, "Oh No! A New Bluebook!," *Michigan Law Review* 90 (May 1992): 1698–1704, at 1700; Scheiber, "Crimson Tide," *TNR,* 4 February 2009; DJG interviews with Trent Norris, Frank Amanat, Ken Mack, Carol Platt Liebau, Monica Harris, Susan Freiwald, Brad Wiegmann, Brad Berenson, Lori-Christina Webb, Lourdes Lopez-Isa, Edith Ramirez, Janis Kestenbaum, Mark Kozlowski, Sean Lev, Tom Perrelli, and Karla Martin.

55. Fraser Robinson III Employee Work History, City of Chicago; Obama, *TAOH,* p. 332; Michelle Obama in Pete Thamel, "Coach with a Link to Obama Has Hope for Brown's Future," *NYT,* 16 Feruary 2007, p. D1; Felsenthal, "The Making of a First Lady," *CM,* February 2009; Craig Robinson, *A Game of Character,* p. 153–55; Jodi Kantor, "The Obamas' Marriage," *NYT,* 1 November 2009; Kantor, *The Obamas,* p. 40; Slevin, *Michelle Obama,* pp. 132–33; *HLS Adviser #25,* 21 February 1991, p. 5; Paul Tarr, "Bell Stuns BLSA Conference," *HLRec,* 15 March 1991, pp. 1, 7; Meredith Ebbin, "Ombudsman Arlene Brock Happy for Barack, Her Harvard Law School Chum," *Bermuda Sun,* 6 June 2008; Obama to Bell, 27 March 1991, DBP Box 136 Fld. 1; George Paul, "Students Intervene in CCR Suit," *HLRec,* 8 February 1991, p. 1; Daniel J. Greaney, "The Radical Fringe Must Be Stopped," *HLRec,* 15 February 1991, p. 4; "Judge Rejects Suit on Bias in Harvard's Hiring," *NYT,* 26 February 1991, p. A18; Anthony Flint, "Harvard Law Wins in Bias Suit Ruling," *BG,* 26 February 1991, p. 24; Robert Arnold, "Discrimination Suit Dismissed," *HLRec,* 1 March 1991, pp. 1, 8; Ken Myers, "Students Take Diversity Fight Against Harvard to State Court," *NLJ,* 4 March 1991, p. 4; *Harvard Law School Coalition for Civil Rights v. President and Fellows of Harvard College,* 412 Mass. 66 (9 July 1992).

56. Harvard Law School 1990–91 Catalog, p. 157; Unger, *Social Theory: Its Situation and Its Task* (Cambridge University Press, 1987); Unger, *False Necessity: Anti-Necessitarian Social Theory in the Service of Radical Democracy* (Cambridge University Press, 1987); Unger, *Plasticity into Power: Comparative Historical Studies in the Institutional Conditions of Economic and Military Success* (Cambridge University Press, 1987); Unger, *Democracy Realized: The Progressive Alternative* (Verso, 1998); Unger, *What Should the Left Propose?* (Verso, 2006); Unger, *The Left Alternative* (Verso, 2009); Unger e-mails to David Remnick, 1 and 6 March 2009; Chris Szabla, "After Rocky but Influential Tenure, Brazil's 'Minister of Ideas' Returns to HLS," *HLRec,* 4 October 2009; Gary Dorrien, *The Obama Question* (Rowman & Littlefield, 2012), p. 3 (echoing Unger in highlighting how "Obama's blend of extroverted charm . . . and personal guardedness epitomizes the style of sociability prized by American professional and business culture"); [Obama and Fisher], "Plant Closings: Creative Destruction and the Viability of the Regulated Market," n.d. [post 5 March 1991], RFP; DJG interviews with Roberto Unger, Gerry Frug, Rob Fisher, Steve Ganis, and Ian Haney López.

57. [Obama and Fisher], "Race and Rights Rhetoric," n.d. [ca. April 1991], RFP; Aldon D. Morris, *The Origins of the Civil Rights Movement* (Free Press, 1984), DJG, *Bearing the Cross: Martin Luther King Jr. and the Southern Christian Leadership Conference* (Morrow, 1986), DJG, *The FBI and Martin Luther King, Jr.: From*

'SOLO' to Memphis (W. W. Norton, 1981); Martha Minow to Rob Fisher and Barack Obama, "Re: Your Manuscript," n.d., RFP; Obama's 2001 interview with Julieanna Richardson; Martha Minow's 2007 interview with Liza Mundy; Jager, "Narrating the Nation: Students, Romance and the Politics of Resistance in South Korea," Ph.D. dissertation, University of Chicago, December 1994; DJG interviews with Rob Fisher, Martha Minow, Scott Siff, and Sheila Jager. On Jiyul Kim, see Kristin Wilson, "Prof Named Fulbright Scholar," *Cumberland County Sentinel,* 31 January 2005, Jiyul Kim, "Self Introduction," Korea History Group Blog, 21 January 2006, and Jiyul Kim, "Cultural Dimensions of Strategy and Policy," Strategic Studies Institute, U.S. Army War College, May 2009, p. vi.

58. Douglas Baird in Daniel J. Yovich, "Obama's Star Power in a Bright Legal Galaxy," *HPH,* 14 February 2007, p. 15, in Scott, "Obama's Story, Written by Obama," *NYT,* 18 May 2008, p, A1, in Jason Zengerle, "Con Law," *TNR,* 30 July 2008, p. 7, in Michael Lipkin, "The Professor and the President," *Chicago Maroon,* 7 November 2008, in *Becoming Barack,* Little Dizzy Home Video, 2009, and on CNBC, 12 January 2009; Jud Miner in Jodi Enda, "Great Expectations," *American Prospect,* February 2006, pp. 22ff., in Margaret Talev, "For Obama, a Tale of Two Speeches," KRDC McClatchy, 16 November 2007, in Jo Becker and Christopher Drew, "Pragmatic Politics, Forged on the South Side," *NYT,* 11 May 2008, p. A1, in Tim Harper, "The Making of a President," *Toronto Star,* 16 August 2008, p. A1, in Yearwood, "Obama and the Jews," *Chicago Jewish News,* 12 September 2008, in Hilary Leila Krieger, "Mr. Obama's Neighborhood," *Jerusalem Post,* 23 October 2008, in Olivia Clarke, "A Firm with a Misson," *Chicago Lawyer,* January 2009, and in McClelland, *Young Mr. Obama,* p. 64; DJG interviews with Kelly Jo MacArthur, Geraldine Alexis, John Levi, Charles Payne, Douglas Baird, Geof Stone, Judd Miner, and Allison Davis. On Payne's library work, see Charles Payne et al., "The University of Chicago Library Data Management System," *The Library Quarterly* 47 (January 1977): 1–22; Maureen O'Donnell and Jon Seidel, "Barack Obama's Great-Uncle Dies at 89," *CST,* 12 August 2014; Graydon Megan, "Obama's Great-Uncle, Innovative Librarian," *CT,* 13 August 2014; and "Charles T. Payne, Leader in Library Automation, 1925–2014," University of Chicago Library *News,* 15 August 2014.

59. Dolly Smith, "Law Students Rally at Harvard, Demand Diverse Faculty," *BG,* 5 April 1991, p. 19; Toyia R. Battle, "Law Students End Overnight Sit-In," *HC,* 6 April 1991; Sharon Stone, "Students Strike for Diversity," Robert Arnold, "Students Storm Dean Clark's Office," and Betsy Fishman, "CCR Intimidates, Insults and Condescends," *HLRec,* 12 April 1991, pp. 1, 2, 4, 11, 12; Dan Greaney, "Double Standard for Student Dissidents," *HLRec,* 19 April 1991, p. 4; Ira E. Stoll, "Law Dean Clark Slams Protesters," *HC,* 25 April 1991; "Law Students Rally, Decry Dean's Letter," *HC,* 26 April 1991; Arnold, "'What I Do Is Actually Quite a Bit of Fun," "Diversity Protesters Picket Griswold Hall," Charissee Carney and Nicole Lamb, "Record: A Fascist Enquirer!?," and Andrew Thomas, "An Open Letter to Dean Clark," *HLRec,* 3 May 1991, pp. 1, 3, 4, 9, 11–12; Mark N. Templeton, "The Last Laugh Is Dean Clark's," *HC,* 6 June 1991; Eric Felten, "Freedom to Think at Harvard Law," *Washington Times,* 17 July 1991, p. E1; Matthew S. Bromberg, "Harvard Law School's War Over Faculty Diversity," *Journal of Blacks in Higher Education* 1 (Autumn 1993): 75–82; *Harvard Law Revue,* 13 April 1991; DJG interviews with David Goldberg, Scott Siff, John Parry, Kevin Downey, Bruce Spiva, Sondra Hemeryck, Trent Norris, Mark Rosen, and Chad Oldfather. On Mary Joe Frug's never-solved murder, Alice McQuillan, "The Professor and the Murder," *Boston Herald Magazine,* 29 March 1992, pp. 7ff., is vastly superior to Peter Collier, "Blood on the Charles," *Vanity Fair,* October 1992, pp. 144–64.

60. Harvard Law School 1990–91 Catalog, p. 5; "Student Profile: Barack Obama," Harvard Law School 1991 *Yearbook,* pp. 109, 184; Harvard University, *The Three Hundred and Fortieth Commencement,* 6 June 1991, esp. p. 37 (seventy-two 1991 magna cum laude J.D. graduates), HUA; Emma Coleman Jordan, "Degrees of Hatred," *WP,* 16 June 1991, pp. B1–B2; Kerlow, *Poisoned Ivy,* p. 126; Obama, *DFMF,* p. 437; Obama's 2001 interview with Julieanna Richardson; Obama's 2004 interview with David Axelrod; Obama on NPR with Tavis Smiley, 29 March 2004; *HLS Adviser* #32, 18 April 1991, p. 4, #36, 16 May 1991, p. 2; "Summer Addresses—3Ls," n.d. [May 1991], Adam Charnes Papers; Tucker, "HAA and HRAAA Name Candidates," *HC,* 31 January 1991; Donald M. Solomon to "Dear Friend," April 1991, HRAAAP; "Slates for Harvard's Spring Elections," *Harvard Mag-*

azine, May–June 1991, p. 79; Tucker, "Overseers Redefine Role," *HC,* 18 November 1991; Gady A. Epstein, "HRAAA Offers No Board Nominees," *HC,* 16 April 1992; Scott, *A Singular Woman,* p. 297; Levenson and Saltzman, "At Harvard Law, a Unifying Voice," *BG,* 28 January 2007; Obama, remarks at Hampton University Annual Ministers' Conference, 5 June 2007; Obama on *All Things Considered,* NPR, 10 June 2008 ($60,000); Obama, Remarks at a Question-and-Answer Session, 6 July 2011, *Public Papers,* 2011, Vol. II, p. 828 ($60,000); DJG interviews with Jonathan King, Roberto Unger, Rob Fisher, Ken Mack, Gordon Whitman, Christine Lee, Jan-Michele Lemon Kearney, and Mark Kozlowski. Lynn Sweet, "The Obama We Don't Know," *CST,* 19 July 2008, which contains a number of errors, declares that "Obama took out $42,753 in loans to pay for Harvard" but gives no indication of the source for so specific and unique an amount.

CHAPTER SIX: BUILDING A FUTURE

1. Steve Kerch, "South Shore: Country Club Symbolizes Rebirth of Neighborhood," *CT,* 25 November 1984, p. 1; Jan Ferris and Henri Cauvin, "Roseland Tries to Ride Out the Shock Wave," *CT,* 6 September 1994; Obama, *DFMF,* p. 438–39; David Moberg, "Danny Davis: A Sharecropper's Son Searches for Common Ground," *CR,* 7 February 1991; Ben Joravsky, "Aldermania," *CR,* 21 February 1991; Thomas Hardy and Tim Jones, "It's a Cakewalk for Daley Slate," *CT,* 27 February 1991; Hardy, "Daley Flattens His Opposition," *CT,* 3 April 1991; Paul M. Green, "Chicago's 1991 Mayoral Elections: Richard M. Daley Wins Second Term," *Illinois Issues,* June 1991, pp. 17–25; Fremon, *Chicago Politics Ward by Ward,* p. 49; Tom Gorman, "Braun Resigns," *HPH,* 7 December 1988, p. 1; Steve Neal, "As Dems Drift, Dunne Feels Draft," *CST,* 2 April 1990, p. 1; Neal and Jim Merriner, "Dem Pick Lyons," *CST,* 3 April 1990, p. 1; Neal, "Evans Backers Target Sawyer Council Allies," *CST,* 13 May 1990, p. 4; John Kass, "Evans Battles Longtime Foe in 4th Ward," *CT,* 6 March 1991; Hardy, "4th Ward Runoff May Settle Old Score," *CT,* 25 March 1991; Vernon Jarrett, "Sleazy Tactic Mars Race in 4th Ward," *CST,* 2 April 1991, p. 25; "Dobry Should Resign Committeeman Post," *CST,* 3 April 1991, p. 29; Jarrett, "Racial Slander Pays Off in Preckwinkle's Ward," *CST,* 4 April 1991, p. 46; Hardy, "5th Ward Democrat Holds Firm," *CT,* 5 April 1991; Lynn Sweet, "Dobry Spurns Call to Resign Over Hate Flyers," *CST,* 5 April 1991, p. 26; Neal, "Sleazy Race-Baiting Sets Dobry Apart," *CST,* 5 April 1991, p. 35; Kass, "What's Next for Chicago?," *CT,* 7 April 1991, p. P1; Carole Ashkinaze, "Dobry 'Logic' Sick," *CST,* 7 April 1991, p. 39; Neal, "LaPaille Moves to Heal Racial Rift," *CST,* 8 April 1991, p. 19; Joravsky, "Dobry's Folly," *CR,* 25 April 1991; Green, "A Tale of Two Wards," *II,* May 1991, pp. 38–39; Neal, "Dobry Defies Calls for His Resignation," *CST,* 13 May 1991, p. 25; P. Davis Szymczak and Kass, "Judge Rules Evans' Loss Is Official," *CT,* 15 May 1991; William J. Grimshaw, *Bitter Fruit: Black Politics and the Chicago Machine, 1931–1991* (University of Chicago Press, 1992), pp. 203–5; Neal, "Dobry Tries to Salvage Battered Image," *CST,* 17 May 1991, p. 37; Gretchen Reynolds, "Is Mayor Daley Really a Republican?," *Chicago Magazine,* December 1991; Robert Howard, "Victory Goes to Newhouse Over Kinnally," *CT,* 15 June 1966, p. 5; Robert McClory, "Invisible Man," *CR,* 13 September 1990; Hardy, "Newhouse Leaving an Impressive Record," *CT,* 31 May 1991; Sweet, "Political Briefing," *CST,* 31 May 1991, p. 26; Aamer Madhani, "Richard Newhouse Jr.," *CT,* 2 May 2002; Alice J. Palmer, "Concepts and Trends in Work-Experience Education in the Soviet Union and the United States," Ph.D. dissertation, Northwestern University, June 1979, esp. pp. 158–60; "Erskine W. Roberts," *CST,* 7 March 2013; Lisa Y. Henderson's exceptionally impressive genealogical work on Dr. Joseph H. Ward at Scuffalong.com; Sheryl Fitzgerald, "Black Police League Has Tough Job Ahead," *CD,* 14 September 1968, p. 1; "Palmer Loses Court Battle; No Longer a Cop," *CD,* 2 January 1971, p. 3; Edward L. Palmer ("President, Black Press Institute"), "South Africa, Reagan and 'Child Torture,'" *CT,* 4 June 1986; Black Press Institute, "An Afro-American Journalist in the USSR," *People's Daily World,* 19 June 1986, p. 18A; Alice Palmer with Mike Giocondo, "A Peaceful World, a Just World, Requires Outreach," *People's Daily World,* 24 December 1986, pp. 10A, 15A; Cliff Kincaid and Herbert Romerstein, "Communism in Chicago and the Obama Connection," www.usasurvival.org, 22 May 2008; Carter L. Clews, "Identity Thrall," *Washington Times,* 30 July 2008; Trevor Loudon, "The Pro-Soviet Agent of Influence Who Gave Barack Obama His First Job in Politics," www.usasurvival.org, 2012;

Chuck O'Bannon and David Robinson, "Buzz and Alice Palmer," 2012 (which provides a very richly instructive view indeed of Buzz Palmer); Grimshaw, *Bitter Fruit,* p. 133 (Stroger); Michael Sneed, "Sneed," *CST,* 4 June 1991, p. 2; Sweet, "Palmer in Line for Senate Seat," *CST,* 4 June 1991, p. 63; Hardy, "Educator Seen as Successor to Newhouse," *CT,* 5 June 1991, p. C7; Sneed, "Sneed," *CST,* 5 June 1991, p. 2; Paula Corrigan, "Newhouse Seat Fight Looms," *HPH,* 5 June 1991, pp. 1, 2; Monica Copeland, "Independent Palmer to Replace Newhouse," *CT,* 7 June 1991, p. C3; "Palmer Wins Newhouse Senate Seat," *CST,* 7 June 1991, p. 24; Jarrett, "Bad News, Good News," *CST,* 9 June 1991, p. 41; Jarrett, "Palmer Remains the Right Choice to Succeed Sen. Newhouse," *CST,* 11 June 1991, p. 27; Corrigan, "Palmer Wins Senator's Seat," and "Senator Resigns at Critical Time," *HPH,* 12 June 1991, pp. 1, 2, 4; Sneed, "Sneed," *CST,* 18 June 1991, p. 2; Corrigan, "New State Senator Prepares for a Long Career," *HPH,* 26 June 1991, p. 3; Kathleen Furore, "New State Senator Seeks Wedding of Politics, Education," *CT,* 30 June 1991, p. W11; Corrigan, "Area Legislative Boundaries May Change with Remap," *HPH,* 3 July 1991, p. 3; Alice Palmer, "School Funding Needs Attention," *HPH,* 25 September 1991, p. 4; "Names," *Illinois Issues,* October 1991, p. 22; "State Senator Wins Environment Award," *HPH,* 30 October 1991, p. 2; Alice Palmer's interviews with Julieanna Richardson and Kate McAuliff; DJG's interviews with Alan Dobry, Tim Evans, Alice Palmer, Donne Trotter, and Ab Mikva.

2. Obama on WTTW's *Chicago Tonight,* 26 October 2004; Jesse Jackson Jr., *A More Perfect Union* (Welcome Rain, 2001), pp. 36–37; Michelle Obama, "How I Got Unstuck," *O Magazine,* September 2005, pp. 22ff.; Michelle in Kalari Girtley and Brian Wellner, "Michelle Obama Is Hyde Park's Career Mom," *HPH,* 14 February 2007, pp. 7–8, 16–17; Michelle and Jarrett in Richard Wolffe, "Barack's Rock," *Newsweek,* 25 February 2008; Michelle on CNN's "Barack Obama Revealed," 20 August 2008; Quincy White in Liza Mundy, *Michelle* (Simon & Schuster, 2008), p. 92; Trevor Jensen, "Charles E. Lomax, 1924–2009: Longtime Lawyer for Don King," *CT,* 25 September 2009; Chinta Strausberg, "Local Man Makes Good with Money Management Firm," *CD,* 7 July 1987, p. 18; Lauren Young, "Mr. Rogers' Neighborhood," *Smart Money,* March 2002; John W. Rogers Jr.'s 2002 interview with Julieanna Richardson; Lynn Norment, "John Rogers: Setting New Standards at the Top of the Money Market," *Ebony,* August 2004, pp. 140ff.; Phil Vettel, "Despite Changes in the Kitchen, Stylish Gordon Has Aged Gracefully," *CT,* 6 October 1989; Vettel, "The Best of a Good-Eating City," *CT,* 2 June 1991; Scott Fornek, "'He Swept Me Off My Feet': Obamas Recall First Date," *CST,* 3 October 2007, p. 6; Bob Dart, "Obama: Trying to Take His Inspiring Vision to the White House," Cox News Service, 11 January 2008; Michelle on *20/20,* ABC, 3 October 2008; Katherine Skiba, "First Lady's Former Chief of Staff 'Blissfully' Unemployed," *CT,* 29 January 2011; Sher's 2009 interview with Robert Draper; Fran Spielman, "Daley Selects Mosena as Chief-of-Staff," *CST,* 1 June 1991, p. 5; Jarrett in Kenneth Meeks, "Favorite Son," *Black Enterprise,* October 2004, pp. 88–95, in Christopher Benson, "Camelot Rising: Barack & Michelle Obama Begin Their Storied Journey," *Savoy,* February 2005, pp. 60–69, 103–6, in Sylvester Monroe et al., "Anatomy of a Moment," *Ebony,* March 2008, pp. 67ff., in Douglas Belkin, "For Obama, Advice Straight Up," *WSJ,* 12 May 2008, p. A5, and in Kathryn Knight, "Is She the U.S. Cherie?," *Daily Mail,* 26 June 2008, p. 56; Don Terry's excellent "In the Path of Lightning," *CT Magazine,* 27 July 2008; Michelle Cottle, "The Woman to See," *TNR,* 27 August 2008, pp. 14ff.; Jonathan Van Meter's superb "Barack's Rock," *Vogue,* October 2008, pp. 336ff.; Lynn Sweet, "Ms. Jarrett Goes to Washington," DailyBeast. com, 24 November 2008; Isabel Wilkerson, "The Closer," *Essence,* April 2009, pp. 106–9, 150; Christi Parsons, "The President's Right-Hand Woman," *CT,* 19 February 2011; Levi in Melinda Henneberger, "The Obama Marriage," *Slate,* 26 October 2007; Minow in James L. Merriner, "The Friends of O," *Chicago Magazine,* June 2008, and in Jay Newton-Small, "Michelle Obama's Savvy Sacrifice," *Time,* 25 August 2008; Miner in Ryan Lizza, "Making It," *New Yorker,* 21 July 2008; George Galland in Dean Barnett, "Would You Hire Barack Obama?," *Weekly Standard,* 1 September 2008; Clarke, "A Firm with a Mission," *Chicago Lawyer,* January 2009; John W. Rogers Jr.'s 2002 interviews with Julieanna Richardson; Jarrett's 2012 interview with Jim Gilmore; Minow's 2008 interview with Gilmore; DJG interviews with Sondra Hemeryck, Jesse Jackson Jr., John Levi, Kelly Jo MacArthur, Gwendolyn LaRoche Rogers, John W. Rogers Jr., Geraldine Alexis, Newton Minow, Linzey Jones, Eden Martin, Judd Miner, Allison Davis, Paul

Strauss, George Galland, Elvin Charity, and Rob and Lisa Fisher; Bob Goldsborough, "John W. Rogers Sr., 1918–2014," *CT,* 22 January 2014. On Valerie Jarrett, also see Maurice Lee, "When the Going Gets Tough, Jarrett Steps Up," *HPH,* 21 May 2003, p. 7, Christi Parsons et al., "Barack's Rock," *CT,* 22 April 2007, and Jarrett on *Tell Me More,* NPR, 24 July 2008. Paul Kengor, "Letting Obama Be Obama," *American Spectator,* July–August 2011, pp. 12–19, is a relentlessly hostile Jarrett profile. Her former spouse, Dr. William Robert Jarrett, whom she married in 1983, left in 1987, and divorced in 1988, died of a heart attack in 1993 at age forty. Kenan Heise, "Dr. William Jarrett of Jackson Park Hospital," *CT,* 23 November 1993. On Jarrett's father, see Dawn Rhodes, "Dr. James E. Bowman, 1923–2011," *CT,* 30 September 2011. On her mother, see Barbara Mahany, "Bard of Education," *CT,* 20 August 1997, and Kimberly Sweet, "Behind Every Child," *University of Chicago Magazine,* February 1999. Timuel D. Black Jr. interviewed both Bowmans at length in 1994. See Black, *Bridges of Memory* (Northwestern University Press, 2003), pp. 568–99, and the full transcripts in the Black Papers, CPL.

3. Obama in Mary Mitchell, "Memoir of a 21st Century History Maker," *Black Issues Book Review,* January–February 2005, pp. 18–21 ("I started writing it right after law school in 1991. I took about four months off to get it started"); Kennedy Communications, "State Senator Barack Obama, Candidate for U.S. Senate: Background Report," April 2003, RRP; "Academic Non-Faculty New Hire Form," 3 September 1991, and Margaret Fallers to Obama, 30 September 1991, Provost Office Records, Faculty Files, Accession #2004-150, Box 20, UC Special Collections; UCLS, *The Glass Menagerie 1991–92,* esp. p. 40; Mark Megalli, "Bellwether Conservative: The University of Chicago Law School," *Journal of Blacks in Higher Education,* Autumn 1993, p. 82; Stone in Merriner, "The Friends of O," *CM,* June 2008; Douglas Baird on CNBC, 12 January 2009; Stone in Tom Hundley, "Ivory Tower of Power," *CT Magazine,* 22 March 2009; DJG interviews with Jean Rudd, Lionel Bolin, Douglas Baird, Geof Stone, and Kathryn Stell; Joel B. Pollak, "Obama's Literary Agent in 1991 Booklet," Breitbart.com, 17 May 2012.

4. No contemporaneous sources or subsequent memories date with any precision Barack and Michelle's trip to Kenya, though all contextual evidence suggests it took place sometime between late August 1991 and January 1992. By far the richest source on it is Auma Obama's *And Then Life Happens* (St. Martin's, 2012), pp. 212–16. See also Michelle L. Robinson Employee Work History, City of Chicago (first day of work 16 September 1991); Obama, *DFMF,* pp. 439–40 ("after our engagement"); Obama's 2001 interview with Julieanna Richardson; John Oywa, "Sleepy Little Village Where Obama Traces His Own Roots," *Nation* (Nairobi), 15 August 2004; Laurie Goering, "Obama Campaign Closely Watched—in Kenya," *CT,* 10 October 2004, p. 1; Obama, *TAOH,* p. 53 ("shortly before" their October 1992 wedding); DJG interviews with Kevin Thompson and Zeituni Onyango. In late 1991 Barack's brother formally petitioned to change his name from Roy Abon'go Obama to Abon'go Malik Obama. *In re Obama,* #91-22380, Prince George's County Circuit Court, 29 November 1991 (granted 7 March 1992).

5. Barack and Michelle on *60 Minutes,* CBS, 16 November 2008; National Center for Careers in Public Life, "Taking Responsibility for Our Future," 1–3 November 1991, Wingspread Conference Center, KBP; Katrina Browne, "Program Designed to Help Young People Help Others," *Wingspread The Journal,* Winter 1992, pp. 1, 9; Todd Savage, "Driving for Change," *Community Jobs,* March 1992, pp. 7, 10; Public Allies Board of Directors, 31 March 1992, Illinois ACORN Papers (98-059) Box 6 Fld. 2; "Mayor, Youth Organization to Honor 100 Volunteers in Public Service," U.S. Newswire, 24 April 1992; Public Allies, Benefit Committee, D.C. Awards Ceremony program, 28 April 1992, IAP Box 6 Fld. 2; Kris Worrell, "Rock the Capital," *AJC,* 15 January 1993, pp. D1, D4; Karen E. Klages, "At Your Service," *CT,* 31 March 1993; Paul Schmitz, *Everyone Leads: Building Leadership from the Community Up* (Jossey-Bass, 2012), pp. 14–20; DJG interviews with Katrina Browne, Jacky Grimshaw, Jason Scott, and Paul Schmitz.

6. Jeremiah A. Wright Jr., "The Audacity of Hope," in Jini Kilgore Ross, ed., *What Makes You So Strong? Sermons of Joy and Strength from Jeremiah A. Wright, Jr.* (Judson Press, 1993), pp. 97–109 (as preached at Houston's Wheeler Avenue Baptist Church in January 1991); Obama, *DFMF,* pp. 291–95, 440; Julia M. Speller, "Unashamedly Black and Unapologetically Christian: One Congregation's Quest for Meaning and Belonging," Ph.D. dissertation, University of Chicago, December 1996, p. 163; Speller, *Walkin' the Talk* (Pilgrim Press, 2005), p. 92; Barack in E. Janet Wright, "Obama for

Illinois," *Trumpet,* February 2004, pp. 30–32; Barack on *The Tavis Smiley Show,* 23 October 2006 (believing he heard "The Audacity of Hope" in 1988); Elizabeth Taylor, "Q&A with Barack Obama," *CT,* 29 October 2006, p. B3 ("a sermon that I heard probably 15 [sic] years ago when I was a community organizer"); Jodi Kantor, "A Candidate, His Minister and the Search for Faith," *NYT,* 30 April 2007, p. A1; Ariel Sabar, "Barack Obama: Putting Faith Out Front," *CSM,* 16 July 2007, p. 1; Dan Gilgoff, "Barack Obama: Praying to Be 'An Instrument of God's Will,'" Beliefnet.com, 21 January 2008; Dwight N. Hopkins, "The Wright Neighborhood," *Religion in the News,* Spring 2008; Scott Helman, "Obama's Odyssey on Race," *BG,* 20 March 2008; Stanley Kurtz, "The Audacity of the Real 'Audacity,'" NationalReview.com, 21 April 2008; Obama's 29 April 2008 press conference in Winston-Salem, NC; Cone in Christi Parsons and Manya A. Brachear, "What Led Obama to Wright's Church," *CT,* 4 May 2008, p. 1; Obama's 31 May 2008 press availability in Aberdeen, SD; Kurtz, "'Context,' You Say?," *National Review,* 19 May 2008, pp. 28–36; Barack in Lisa Miller and Richard Wolffe, "'I Am a Big Believer in Not Just Words, but Deeds and Works,'" *Newsweek,* 12 July 2008 (print issue 21 July); Barack in Toby Harnden, "Barack Obama's True Colours," *Telegraph Magazine,* 23 August 2008, pp. 26ff.; Michelle in Melinda Henneberger, "Michelle Obama Interview," *Reader's Digest,* October 2008, pp. 196–203 ("Barack knew the reverend from his work in the city, and we joined together"); Salim Muwakkil on "The Choice 2008," *Frontline,* PBS, 14 October 2008; Toni Preckwinkle in Lizza, "Making It," *TNR,* 21 July 2008, and in Martin Fletcher, "Before He Was Famous: The Truth Behind Obama's Meteoric Ascent," *Times* (London), 14 October 2008; Robert T. Starks, "How He Did It," *N'Digo Profiles,* December 2008, p. 34; Melissa Harris-Lacewell in Rebecca Traister, *Big Girls Don't Cry* (Free Press, 2010), p. 48; Janny Scott, *A Singular Woman* (Riverhead, 2011), pp. 291–92, 298, 301; Maya in Maraniss, *BOTS,* p. 547; "Notice of Federal Tax Lien Under Internal Revenue Laws," #91-155273, 12 November 1991; Dunham, *Against All Odds,* pp. v, 99, 254; DJG interviews with Jeremiah Wright, Kevin Tyson, Deloris Burnam, Carolyn Wortham, Patricia Novick, Kathryn Stell, Greg Galluzzo, Sheila Jager, Ralph Dunham, Charles Payne, and Barack Obama. On Trinity, see also Jeremiah A. Wright Jr., "Doing Black Theology in the Black Church," in Linda E. Thomas, ed., *Living Stones in the Household of God* (Fortress Press, 2004), pp. 13–23; Jason Byassee, "Africentric Church: A Visit to Chicago's Trinity UCC," *Christian Century,* 29 May 2007, pp. 18–23; Kareem Crayton, "You May Not Get There with Me: Obama and the Black Political Establishment," in Manning Marable, ed., *Barack Obama and African American Empowerment* (Palgrave Macmillan, 2009), pp. 195–207; and Dwight N. Hopkins, "Race, Religion, and the Race for the White House," in Charles P. Henry et al., eds., *The Obama Phenomenon* (University of Illinois Press, 2011), pp. 181–99. The assertion in Allison Samuels, "Why Oprah Winfrey Left Rev. Jeremiah Wright's Church," *Newsweek,* 3 May 2008 (print issue 12 May), that "Winfrey was a member of Trinity" is incorrect.

7. Kathleen Best, "Dixon's Challengers Growing," *SLPD,* 20 November 1991, p. A1; Glen C. Thompson, "New Remap Puts Currie Back in Hyde Park-Kenwood," *HPH,* 22 January 1992, pp. 1, 2; Rob Karwath, "Remap Alters the Boundaries and the Rules in 2nd District," *CT,* 26 January 1992; Florence C. Goold, "IVI/IPO Sets Primary Slate," *HPH,* 12 February 1992, pp. 1, 2; Steve Johnson, "Savage Ahead in 2nd, Reynolds Has Hope," *CT,* 2 March 1992; Johnson, "Reynolds Questions Savage's Lead," *CT,* 3 March 1992; Florence Hamlish Levinsohn, "Carol Moseley Braun," *CR,* 5 March 1992; Ted Gregory and Rick Pearson, "Jewish Groups Hit Savage Attack Against Reynolds," *CT,* 9 March 1992; Lynn Sweet, "Dixon Pulls Away in Senate Contest," *CST,* 9 March 1992, p. 1; Sweet, "Braun Capitalizes on TV Debate Exposure," *CST,* 10 March 1992, p. 9; "Local Candidates Gear Up for Next Week's Primary Election," and "Support for Local Candidates," *HPH,* 11 March 1992, pp. 1, 2, 4; Robert Davis, "Not Anti-Israel, Just Pro-U.S., Savage Says," *CT,* 12 March 1992; Janine Poe and Helaine Olen, "Reynolds Is Injured in Shooting," *CT,* 13 March 1992; "Savage Plays the Race Card Again," *CT,* 14 March 1992; Sweet and Bill Braden, "Poll Puts Braun at Dixon's Heels as Hofeld Stalls," *CST,* 15 March 1992, p. 17; Johnson, "Reynolds, Lipinski Win," *CT,* 18 March 1992; Sweet, "A Braun Upset," *CST,* 18 March 1992, p. 1; Edward Walsh, "Sen. Dixon Loses in Stunning Upset," *WP,* 18 March 1992, p. A1; Isabel Wilkerson, "Illinois Senator Is Defeated by County Politician," *NYT,* 18 March 1992, p. A19; John Kass and Monica Copeland, "Braun's Win a Chance to Forge New Black Power," and Karwath, "Remap,

Blacks Keys to Reynolds' Victory," *CT,* 19 March 1992; Walsh, "Illinois Opens Mean Season for Veteran Officeholders," *WP,* 19 March 1992, p. A1; Goold, "Women Sweep Election Slate Here as in Rest of the State," *HPH,* 25 March 1992, pp. 1, 2; Goold, "Local Election Results Don't Reveal Surprises," *HPH,* 1 April 1992, p. 3; Manuel Galvan, "Bobby Rush and Mel Reynolds Defeat Incumbent Congressmen," *Illinois Issues,* May 1992, pp. 14–17; "Herald Endorsements in Local Races," *HPH,* 28 October 1992, p. 4; Gretchen Reynolds, "Vote of Confidence," *Chicago Magazine,* January 1993; Barack's 27 October 1993 interview with Zeke Gonzalez, his 20 September 1995 remarks at the Cambridge (MA) Public Library, his 1 July 2002 interview with Chinta Strausberg, and his 7 February 2004 one with David Axelrod; Paul Simon, *P.S.: The Autobiography of Paul Simon* (Bonus Books, 1999), p. 177; Peter Slevin, "For Obama, a Handsome Payoff in Political Gambles," *WP,* 13 November 2007, p. A3; "Barack Obama and the History of Project VOTE!," YouTube, 21 March 2008; Sanford A. Newman, "ACORN Didn't Employ Obama in '92," *WSJ,* 22 July 2008; Allison Davis in David Smallwood, "Veteran Attorneys Recall Obama," *N'Digo Profiles,* December 2008, p. 24; Remnick, *The Bridge,* pp. 221–22; McClelland, *Young Mr. Obama,* pp. 65–72; Fredrick C. Harris, *The Price of the Ticket* (Oxford University Press, 2012), pp. 57–58; DJG interviews with Sandy Newman, Jacky Grimshaw, John Schmidt, John Rogers, Judd Miner, Allison Davis, Paul Strauss, Carol Harwell, and Brian Banks.

8. Yvonne D. Delk, Joseph Gardner, and Barack Obama to Keith Kelleher, 28 April 1992, Illinois ACORN Papers (IAP) Box 6 Fld. 28; [Agenda], "Project Vote Steering Committee Meeting, 4 May 1992," IAP (2001-111) Box 2 Fld. 55; "Project Vote Chicago Coalition Current as of 5/11/92," Gwen Jordan (CRS), Obama, and Gardner to Steering Committee Members, Joint Voter Registration Project, "Meeting—May 18, 1992," 15 May 1992, Project Vote, "Agenda Illinois Project Vote Steering Committee," 19 May 1992, "Voter Registration Time-Line," n.d., Gardner and Obama to Project Vote Steering Committee, "Re: Press Conference on Wednesday, June 3rd," 28 May 1992, [Draft], "Coalition Pledges to Increase Minority Voter Registration by Over 100,000," 1 June 1992, "Agenda—Project Vote Steering Committee," n.d. [3 June 1992?], Obama to Project Vote Steering Committee, "Re: Press Conference Wednesday, June 10th, 10AM–11AM Daley Plaza," n.d., "Voter Registration Priorities Illinois, Fall 1992," 10 June 1992, Keith Kelleher, Local 880, to Barack Obama, Project Vote, "Request for Project Vote Funds," 23 June 1992, all IAP Box 6 Fld. 28; Michael Sneed, "Sneed," *CST,* 2 August 1992, p. 2; Vernon Jarrett, "Voter Registration Is Key to Respect," *CST,* 4 August 1992, p. 23; Bruce Dixon to Deputy Registrars and Affiliated Organizations, "Re: Bud Billiken Day Parade & Picnic," n.d. [early August 1992], IAP Box 6 Fld. 29; Jarrett, "'Project Vote' Brings Power to the People," *CST,* 11 August 1992, p. 23; "V.O.T.E. Community & Project Vote 'Voice of the Ethnic Community,'" 26 August 1992, IAP Box 6 Fld. 29; Sandy Newman and Linda Dickinson Kendrix to Steve Cullen, 4 September 1992, AFSCME Council 31 Papers; Zach [Pollett] to Craig [St. Louis], Madeline/Caroline/Keith [Chicago], Amy [Detroit], "Re: VR/GOTV Funding Meetings in Your Cities with Citizen Vote Initiative," 4 September 1992, IAP Box 7 Fld. 41; "Register to Vote to Help Yourself," *CD,* 8 September 1992, p. 9; Sandy Newman to Al Leuin (Coulterville, IL), 14 September 1992, AFSCME 31 Papers (A31P); Keith Kelleher to Robin Leeds, "Voter Registration, GOTV, Election Day Proposal for SEIU Local 880 and ACORN in Chicago and Southwest Illinois," 14 September 1992, IAP Box 7 Fld. 41; Kendrix to Cullen, 16 September 1992, A31P; Chinta Strausberg, "Clinton Gets Big Welcome at Regal," *CD,* 21 September 1992, pp. 1, 26; Carol Jouzaitis, "Clinton Hammers Away at Bush," *CT,* 21 September 1992; Tim Gerber, "Clinton Says Registration Vital for Win in November," *CST,* 21 September 1992, p. 6; Michael K. Frisby, "In Mich., Clinton Assails Bush on Economy," *BG,* 21 September 1992, p. 11; Joseph E. Gardner to All Interested Parties, "Voter Registration/Ward Analysis," 21 September 1992, IAP Box 7 Fld. 41 and IAP (162) Box 6 Fld. 29; Frisby, "Business Executives Give Clinton a Boost," *BG,* 22 September 1992, p. 12; "Project VOTE Adds 50,000 Voters," *CD,* 22 September 1992, p. 22; Strausberg, "Voter Registration Up in 19 Key Wards," and "Stars to Fire Up Registration Drive," *CD,* 23 September 1992, pp. 3, 5; Basil Talbott, "Jackson Brings Flair to Rush's Voter Drive," *CST,* 25 September 1992, p. 4; Ray Long, "Most New Voters from Black Wards: Study," *CT,* 26 September 1992, p. 5; Jarrett, "Need for a Black Summit Is Urgent," *CST,* 1 October 1992, p. 39; Mary A. Johnson and Philip Franchine, "Mail Drives, House Calls Fuel Voter Roll Surge," *CST,* 3 October

1992, p. 4; Karen Ball (AP), "Voter Registration: Setting Records in Some Regions," 4 October 1992; Reynolds, "Vote of Confidence," *CM*, January 1993; Ben Joravsky, "Lou Pardo, the Czar of Voter Registration, Nears 100,000," *CR*, 23 September 1993; Chicago DSA's 36th Annual Debs-Thomas-Harrington Dinner Program, 14 May 1994; Obama, Illinois State Senate Transcript, 9 May 2002, p. 90 (recalling Newhouse's help); Toni Foulkes, "Case Study: Chicago—The Barack Obama Campaign," *Social Policy*, Winter 2003/Spring 2004, pp. 49–52, esp. p. 50; Keith Kelleher, "Growth of a Modern Union Local: A People's History of SEIU Local 880," *Just Labour* 12 (Spring 2008): 1–15, esp. p. 10 ("for" Braun); Stanley Kurtz, *Radical-in-Chief* (Simon & Schuster, 2010), pp. 223–27; John Atlas, *Seeds of Change* (Vanderbilt University Press, 2010), p. 101; Eileen Boris and Jennifer Klein, *Caring for America* (Oxford University Press, 2012), pp. 164–76; UCLS *Announcements 1991–92*, 6 September 1991, p. 27; "Academic Non-Faculty Termination or Transfer Form," 30 October 1992, Provost Office Records, Faculty Files, #2004-150, Box 20 (Barack's appointment as a Fellow ended 30 June); UCLS *Announcements 1992–93*, 27 August 1992, p. 27; UCLS, *The Glass Menagerie 1992–93*, esp. p. 44; Chicago Video Project, "Introducing Barack Obama," Obama for Illinois, 2003; Obama, transcript of remarks at Heartland Cafe Rally, Chicago, 7 February 2004, JKP and TPP ("to get Bill Clinton and Carol Moseley Braun elected"); Obama, remarks at the Take Back America Forum, Washington, D.C., 19 June 2007 ("to help Bill Clinton get elected"); Monique Garcia, "Election Volunteer, Extraordinaire," *CT*, 11 July 2007; Barack's 1993 interview with Zeke Gonzalez, his 2001 one with Julieanna Richardson, and his 2002 one with Chinta Strausberg; Brian Banks and Rita Whitfield in "Barack Obama and the History of Project Vote," BarackObama.com, 21 March 2008; Mark S. Allen, "From Adversary to Longtime Ally," BlackVoices.com, 19 January 2009; DJG interviews with Carol Harwell, Brian Banks, Keith Kelleher, Madeline Talbott, Al Kindle, Mike Pfleger, Jeremiah Wright, Kevin Tyson, Tondra L. Loder-Jackson, Miguel del Valle, Sandy Newman, John Schmidt, Yvonne Delk, Gwendolyn Jordan, Jeff Schoenberg, Marv Dyson, James D. Holzhauer, Bill Perkins, Roberta Lynch, Ray Harris, Margaret Bagby, Alvin Love, Desta Houston, Eric Adelstein, Peter Coffey, and Carol Moseley Braun. Edward Gardner's long-standing interest in voter registration in readily evident in his 1989 interview with Blackside's Dave Lacy.

On DCP as of 1991–92, see especially Ben Joravsky, "A Confederacy of Churches," *CR*, 21 February 1992, which is alone in noting DCP's huge 20 November 1991 "convention"; Lynne Owens Mock, "Validation of an Instrument to Assess the Personal Values of African-American Community Leaders," M.A. thesis, University of Illinois at Chicago, 1994; L. Sean Azelton, "Boundary Spanning and Community Leadership: African American Leaders in the Greater Roseland Area," M.A. thesis, University of Illinois at Chicago, 1996; John C. Glidewell et al., "Natural Development of Community Leadership," in R. Scott Tindale et al., eds., *Theory and Research on Small Groups* (Plenum Press, 1998), pp. 61–86; S. Darius Tandon et al., "Constructing a Tree for Community Leaders: Contexts and Processes in Collaborative Inquiry," *American Journal of Community Psychology* 26 (August 1998): 669–96; Mock, "The Personal Vision of African American Community Leaders," Ph.D. dissertation, University of Illinois at Chicago, 1999, esp. pp. 127, 130; James G. Kelly, "Contexts and Community Leadership: Inquiry as an Ecological Expedition," *American Psychologist* 54 (November 1999): 953–61; Kelly et al., "Collaborative Inquiry with African-American Community Leaders: Comments on a Participatory Action Research Process," in Peter Reason and Hilary Bradbury, eds., *Handbook of Action Research: Participative Inquiry and Practice* (Sage Publications, 2001), pp. 348–55; Tandon et al., "Participatory Action Research as a Resource for Developing African American Community Leadership," in Deborah L. Tolman and Mary Brydon-Miller, eds., *From Subjects to Subjectivities: A Handbook of Interpretive and Participatory Methods* (New York University Press, 2001), pp. 200–17; Kelly et al., "On Community Leadership: Stories About Collaborations in Action Research," *American Journal of Community Psychology* 33 (June 2004): 205–16. Johnnie Owens left DCP in August 1993 and was succeeded by Cassandra Lowe. DJG interviews with Johnnie Owens, Mike Kruglik, Greg Galluzzo, Mary Gonzales, and Paul Scully.

9. Obama's August 1995 interview with Connie Martinson; Obama's September 1995 remarks at the Cambridge (MA) Public Library; Tom Maliti (AP), "Obama's Brother Chooses Life in Slow Lane," 26 October 2004; Loretta Augustine-Herron in Philip Sherwell, "The Big Hitter," *Sunday Telegraph*, 28 January 2007, p.

18, and in Sherwell, "Obama: The Secret of His Success," *ST,* 8 June 2008, p. 17; Kirsten Scharnberg and Kim Barker, "The Not-So-Simple Story of Barack Obama's Youth," *CT,* 25 March 2007; Jeff Chang, "Maya Soetoro Ng: Q&A," *Vibe,* September 2007; Janny Scott, "A Biracial Candidate Walks His Own Fine Line," *NYT,* 29 December 2007, p. A1; Jacobs, *Obamaland,* p. 123; Emme Tomimbang, "Barack Obama: Hawaii Roots," 8 February 2008; Loretta in David James Smith, "The Ascent of Mr. Charisma," *Sunday Times Magazine (London),* 23 March 2008, pp. 58ff.; Adam Goldman and Robert Tanner (AP), "Old Friends Recall Obama's Years in LA, NY," 15 May 2008; Stefano Esposito, "2 People Who Love Each Other," *CST,* 13 July 2008, p. A2; Yann Masson, "Presidentielle Americaine: Avec Barack, on Regardait Les Matchs de Basket," *Nice Matin,* 6 November 2008; Andrea Brown, "To Couple, He's Barry in a Leather Jacket," *Colorado Springs Gazette,* 20 January 2009; "Obamas Credit Stevie Wonder," *NYT,* 27 February 2009; Allison Davis's 2009 interview with Robert Draper; Kantor, *The Obamas,* p. 176; Auma Obama, *And Then Life Happens,* p. 221–22; Carol Felsenthal, "Yvonne Davila," ChicagoMag.com, 14 December 2012; Obama, remarks at Solyndra, Fremont, CA, 26 May 2010, *Public Papers* 2010 Vol. 1, p. 714; Tracy Baim, *Obama and the Gays* (Prairie Avenue Productions, 2010), pp. 536–37; Michael Cass, "Obama's Rise Is No Surprise to Nashville Woman Who Worked with Him," *Tennessean,* 5 October 2012; DJG interviews with Mike Ramos, Kelly Jo MacArthur, Hasan Chandoo, Jeremiah Wright, Jerry Kellman, Greg Galluzzo, Johnnie Owens, David Kindler, Sokoni Karanja, Ken Rolling, Jean Rudd, Rob Fisher, Mark Kozlowski, Dan Rabinovitz, Tom Perrelli, Jan-Michele Lemon Kearney, Ken Mack, Geraldine Alexis, Tom Reed, Allison Davis, Paul Strauss and Marlies Carruth, Laura Tilly, Loretta Augustine-Herron, Yvonne Lloyd, Margaret Bagby, John Schmidt, John Rogers Jr., Carol Harwell, Brian Banks, Kevin Thompson, Charles Payne, Janis Robinson, Bernard Loyd, and Jesse Jackson Jr.

10. Strausberg, "Voter Registration Drive Gets Push on Final Day," *CD,* 5 October 1992, pp. 1, 22; "Voter Registration: Project Vote! Targets Urban Swing Areas," *The Hotline,* 8 October 1992; Linda M. Harrington and John C. Patterson, "Issues Spark Registration Jump," *CT,* 10 October 1992, p. 1; Strausberg, "VOTE Spokesman Warns of Polling Place Confusion," *CD,* 14 October 1992, p. 8; David Shribman and James M. Perry, "Perot's Support Slips, but Impact in Race Remains," *WSJ,* 30 October 1992, p. A1; Eric Zorn, "Better Not to Vote Than Vote Stupidly," *CT,* 1 November 1992, p. C1; Newman to Cullen, 2 December 1992, A31P; "Breaking Records in Illinois," *Project Vote!* [newsletter], n.d., pp. 1, 3; Reynolds, "Vote of Confidence," *CM,* January 1993; Frank James, "25 Chicagoans on the Road to Making a Difference," *CT,* 10 February 1993, p. 5; Obama UCLS Faculty Curriculum Vitae, ca. 2000, web.archive.org, 2001 ($200,000); Evans in Sherwell, "The Big Hitter," *Sunday Telegraph,* 28 January 2007, p. 18, and in Daniel J. Yovich, "Favorite Son," *HPH,* 14 February 2007, pp. 2, 14; Erika George in Deanie Wimmer, "Utah Delegate's Life Touched by Obama," KSL.com, 24 August 2008, and in Matt Canham, "Utah Prof Remembers Obama," *Salt Lake Tribune,* 26 August 2008; Toni Preckwinkle in Mark Karlin, "Who Is Obama's Alderwoman [sic] in Chicago?," Buzzflash.com, 15 July 2008, and in "A Salute to Barack Obama," *Hyde Park History* 31 (Winter 2008–2009), p. 4; Saltzman in Terry McDermott, "What Is It About Obama?," *LAT,* 24 December 2006, in Lizza, "Making It," *TNR,* 21 July 2008, in Yearwood, "Obama and the Jews," *Chicago Jewish News,* 12 September 2008, and in Abramsky, *Inside Obama's Brain,* p. 207; Axelrod in Maria L. La Ganga, "The Man Behind Obama's Message," *LAT,* 15 February 2008; Axelrod's June 2008 interview with Jim Gilmore; Axelrod in Remnick, *The Bridge,* p. 225; Axelrod, *Believer,* p. 119; Leslie Hairston, "Tribute to Late Mr. Ronald L. Davis," *Chicago City Council Journal,* 14 May 2008, pp. 15–16; Obama's 2001 interview with Julieanna Richardson and his 2002 one with Chinta Strausberg; DJG interviews with Carol Harwell, Brian Banks, Sandy Newman, John Schmidt, Bill Perkins, Toni Preckwinkle, Laura Washington, Mike Pfleger, Phil Mullins, Roberta Lynch, and Bettylu Saltzman.

11. Robert Davis, "Daley Merges Planning, Economic Development Staffs," *CT,* 11 October 1991; Alf Siewers, "Commission Has Big Plans for Neighborhoods," *CST,* 21 October 1991, p. 4; Wilma Randle, "Aide Brings Own Vision to City Post," *CT,* 8 December 1991; John Kass, "Daley's New City Hall Excludes Blacks, Suit Charges," *CT,* 25 June 1992; Fran Spielman, "Daley's Chief of Staff to Oversee Aviation," *CST,* 16 September 1992, p. 14; Michael Sneed, "Sneed," *CST,* 15 October 1992, p. 2; Michelle L. Robinson Employee Work History, City of Chicago;

Ann Dunham résumé, n.d. (post-February 1993), Dunham Papers; Andrew J. Salz obituary, *San Francisco Chronicle,* 9 October 1995, p. A20; Obama's 2001 interview with Julieanna Richardson; Charles Leroux, "The Buzz Around Obama's Book," *CT,* 6 August 2004; Peter Osnos, "Barack Obama and the Book Business," The Century Foundation, 30 October 2006; Robert Draper's 2009 interviews with Allison Davis, Michelle Obama, Valerie Jarrett, and Peter Osnos; Jane Dystel's 2009 e-mail to Draper; Draper, "Barack Obama's Work in Progress," *GQ,* November 2009; Remnick, *The Bridge,* pp. 227–28; Scott, *A Singular Woman,* p. 301–4; Diana Darling, *Tandjung Sari: A Magical Door to Bali* (Editions Didier Millet, 2012), esp. pp. 17–28; Obama in Jonathan Van Meter, "Leading by Example," *Vogue,* April 2013, pp. 248–55, 318–19; Mark Obama Ndesnadjo, *An Obama's Journey* (Lyons Press, 2014), pp. 281–84; DJG interviews with Jean Rudd, John Schmidt, Kelly Jo MacArthur, Jerry Kellman, Sandy Newman, Dan Hale, Bronwyn Solyom, Allison Davis, Kevin Thompson, Janis Robinson, Laura Demanski, and Ann Patty.

On Patty and the death of Poseidon, see Esther B. Fein, "Book Notes: A Duchy Turns 12," *NYT,* 20 May 1992, p. C20; *Estate of Virgina C. Andrews v. U.S.*, 804 F. Supp. 820 (E. D. Va.), 27 October 1992; David Streitfeld, "A Novelist's Tales from the Crypt," *WP,* 7 May 1993, p. A1; Streitfeld, "Dead Writer's 'Ghost' Says He Paid Large Fee," *WP,* 27 May 1993, p. D1; Maureen O'Brien, "What's Happening at Poseidon," *Publishers' Weekly,* 5 July 1993, p. 18; Sarah Lyall, "Book Notes: Absorbing Poseidon," *NYT,* 7 July 1993, p. C13; Lyall, "Book Notes: Leaving After All," *NYT,* 4 August 1993, p. C13; *Estate of Virginia C. Andrews v. U.S.*, 850 F. Supp. 1279 (E. D. Va.), 10 May 1994; Nina Shengold, "Mother of Invention," *Chronogram,* March 2005, pp. 72–73; Patty, "The Future for Book Editors: Royalties?," *Publishing Perspectives,* 9 April 2010.

12. *Barnett v. Daley,* 835 F. Supp. 1063 (N. D. Ill.), 8 October 1993; *African American Voting Rights Legal Defense Fund v. Villa,* 999 F. 2d 1301 (8th Cir.), 4 August and 1 November 1993; *Tyus v. Schoemehl,* 93 F. 3d 449 (8th Cir.), 16 August 1996; *Baravati v. Josephthal Lyon & Ross,* 834 F. Supp. 1023 (N. D. Ill.), 4 October 1993; "People," *CT,* 27 June 1993, p. B9; "Business Appointments," *CST,* 5 July 1993, p. 40; *Chicago Lawyer,* September 1993, p. 66; Judson Miner, "Barack Obama's Career at Miner, Barnhill & Galland," n.d. [2007], 4pp., Miner Papers; Abdon M. Pallasch, "Obama's Legal Career," *CST,* 17 December 2007; David Smallwood, "Veteran Attorneys Recall Obama," *N'Digo Profiles,* December 2008, p. 24; Allison Davis's 2009 interview with Robert Draper; DJG interviews with Judd Miner, Allison Davis, Chuck Barnhill, George Galland, Paul Strauss, Laura Tilly, Jeff Cummings, Mark Kende, Sarah Siskind, and Whit Soule. On Miner and the firm's history, also see Michael Powell, "Anatomy of a Counter-Bar Association: The Chicago Council of Lawyers," *American Bar Foundation Research Journal* 4 (Summer 1979): 501–41, esp. 506, 511–13, Olivia Clarke, "A Firm with a Mission," *Chicago Lawyer,* January 2009, and Don Rose, "The Lawyers Who Reformed Chicago," *The Week Behind,* 23 March 2011.

13. Jessica Bassett, "Barack's Students Gabe and Nicole Gore Reflect on 'Professor Obama,'" *St. Louis American,* 30 October 2008; Jesse Ruiz in "Obama's Chicago Roots," Time.com, 17 December 2008; DJG interviews with Douglas Baird, Jesse Ruiz, Barack Obama, Julie Fernandes, Brett Hart, Jeanne Gills, Julia Bronson, and Elaine Horn. See also Joan Giangrasse Kates, "Jesus Ruiz, 1917–2013," *CT,* 24 April 2013. Per Marsha Ferziger Nagorsky's 31 December 2009 e-mail to Dennis J. Hutchinson, no UCLS course evaluations survive: "the entire Spring quarter of 1993 is missing." No surviving copy of Barack's 1993 syllabus has yet been discovered, but all former students' recollections suggest the 1993 class proceeded just as it did in 1994, and that Spring Term 1994 Syllabus (Adam Gross Papers) is relied upon here. Stacy Marlise Gahagan and Alfred L. Brophy, "Reading Professor Obama: Race and the American Constitutional Tradition," *University of Pittsburgh Law Review* 75 (2014): 495–581, reprints the syllabus in full and offers a careful and thorough analysis of it while noting (522) that "the readings are remarkably non-theoretical."

14. Michelle L. Robinson Employee Work History, City of Chicago (last day of work 30 April 1993); Woods Charitable Fund, *Report for the Year 1992,* p. 24; *Public Allies Chicago,* Winter 1993; Alphonso Myers, "Student Awarded for Leadership and Service," *Columbia College Chronicle,* 3 May 1993, pp. 1–2; Jane Meredith Adams, "Youths Vow War on 'Me' Generation," *CT,* 21 June 1993; "Comings and Goings," *CCB,* 12 July 1993, p. 25 ("Public Allies: Michelle Obama, 29, to regional director from assistant commissioner of the city's Department of Planning and Development"); Public Allies: The

National Center for Careers in Public Life, "An Action Plan by Young People," n.d., KBP; Ed Cohen, "Echoing Green: Building Institutions, Impacting Policy, Creating a Community," 13 July 1995, esp. p. 18 ("provided assistance with student loans enabling Michelle Obama to work at Public Allies"), EGP; Michelle in Sarah Brown, "Obama '85 Masters Balancing Act," *Daily Princetonian,* 7 December 2005; Schmitz in Melinda Henneberger, "The Obama Marriage," *Slate,* 26 October 2007; Richard Wolffe, "Barack's Rock," *Newsweek,* 25 February 2008; Suzanne Perry, "Fired Up and Ready to Grow," *Chronicle of Philanthropy,* 17 April 2008; Michelle in Geraldine Brooks, "Michelle Obama: Camelot 2.0?," *More,* October 2008; Cindy Moelis in Carol Felsenthal, "The Making of a First Lady," *Chicago Magazine,* February 2009, pp. 50–56, 72–78; Paul Schmitz, *Everyone Leads* (Jossey-Bass, 2012), pp. 23–24; DJG interviews with Katrina Browne, Jason Scott, Sunny Fischer, Julianne Alfred Sullivan, Bethann Hester, Julian Posada, and Paul Schmitz.

15. Obama's 27 October 1993 interview with Zeke Gonzalez; Obama's 7 March 2001 remarks at the Illinois State Library; Charles Leroux, "The Buzz Around Obama's Book," *CT,* 6 August 2004, p. T1; "Oprah Talks to Barack Obama," *O Magazine,* November 2004; Obama, U.S. Senate Financial Disclosure Report, 15 June 2005; Osnos, "Barack Obama and the Book Business," The Century Foundation, 30 October 2006; Scott, "Obama's Story, Written by Obama," *NYT,* 18 May 2008, p. A1; Kevin Simpson, "Identity Questions Shaped Obama into New Kind of Politician," *Denver Post,* 28 August 2008, p. P8; D. D. Guttenplan, "The Washington Insider Who Made Obama Rich," *FT,* 25 October 2008; Osnos, "The Making of the Book That Made Obama," DailyBeast.com, 27 January 2009; Osnos's 2009 interview with Robert Draper; Dystel's 2009 e-mail to Draper; Draper, "Barack Obama's Work in Progress," *GQ,* November 2009; James Rosen, "Obama's 'American Story' Faces Fresh Scrutiny," FoxNews.com, 5 July 2012; DJG interviews with Rob Fisher, Henry Ferris, and Alex McNear.

16. Warranty Deed #93-608056, Stephen D. and Helen N. Anderson to Barack and Michelle Obama, 5450-1 S. East View Park, 28 July 1993; Francine Knowles, "Video Store in Hyde Park Finds Niche Among Giants," *CST,* 23 July 1995; "Five Minutes with: Sen. Barack Obama," CampusProgress.org, ca. 22 March 2005; Kathy Chaney, "Time for Fun," and "Who Is Barack Obama?," *HPH,* 14 February 2007, pp. 4, 6, 15; Mark Konkol, "Mr. Obama's Neighborhood," *CST,* 25 May 2007, p. 12; Patrick T. Reardon, "Obama's Chicago," *CT,* 25 June 2008, p. T1; Obama on MSNBC's "Hardball," 2 April 2008; Ray Gibson et al., "How Broke Were Obamas?," *CT,* 20 April 2008; "Obama's Chicago Roots," Time.com, 17 December 2008; Gayle Keck, "Touring Obama's Chicago," *San Francisco Chronicle,* 18 January 2009; Christopher Benson, "The President's Neighborhood," *Chicago Magazine,* June 2009; Dennis Rodkin, "Renters Snap Up Obamas' Former Condo," ChicagoMagazine.com, 4 August 2010; Alford A. Young Jr., "The Black Masculinities of Barack Obama," *Daedalus,* Spring 2011, pp. 206–14; Maggie Murphy and Lynn Sherr, "The President and Michelle Obama on Work, Family, and Juggling It All," *Parade,* 20 June 2014; Craig Robinson in Bill Reynolds, "Welcome to Obama's Family," *Providence Journal,* 15 February 2007, p. C1, in Eric Tucker (AP), "Family Ties: Brown Coach, Barack Obama," 1 March 2007, and in Frank Fitzpatrick, "Barack Obama's Closest Opponent," *Philadelphia Inquirer,* 2 March 2007; Obama on *This Week,* ABC, 13 May 2007; Craig in M. Charles Bakst, "Brown Coach Robinson a Strong Voice for Brother-in-Law Obama," *PJ,* 20 May 2007, p. D1, in Liza Mundy, "A Series of Fortunate Events," *WP Magazine,* 12 August 2007, pp. 10ff., in Mundy, *Michelle,* p. 100, and in John Feinstein, "Obama's Brother-in-Law Is a Court Authority," *WP,* 18 January 2009; *Gary W. E. Forth v. Maya K. Soetoro,* #1DV94-0-001619, 9 May 1994 (decree entered 12 May 1995); Obama on the *CBS Evening News,* 16 January 2008; Jeff Chang, "Maya Soetoro Ng: Q&A," *Vibe,* September 2007; Maya in Scott, *A Singular Woman,* p. 296; DJG interviews with John Belcaster, Charles Payne, Harry Gendler, Julian Posada, Jim Reed, Greg Dingens, Bernard Loyd, Tom Reed, Barack Echols, Janis Robinson, and Verna Myers. John Belcaster recalls with precise clarity that his starting salary when Davis Miner hired him in late 1993 was $73,000.

17. Gary Wisby, "Youths Sought to Honor as New Leaders," *CST,* 28 March 1994, p. 13; Public Allies Chicago, *Check-In,* Winter 1995, p. 5; DJG interviews with Bethann Hester, Julie Alfred Sullivan, Julian Posada, Jeremiah Wright, Kevin Tyson, Carlton Guthrie, Deloris Burnam, Keith Kelleher, Madeline Talbott, Sarah Siskind, and Joel Rogers; "The Fifteen Greatest Black Preachers," *Ebony,* November 1993, p. 156; Speller, *Walkin' the Talk,* p. 92; Keith Kelleher, "Jacky Grimshaw," 5 August 1992, IAP Box 10 Fld. 91; "Contacts We Need to Make for New Party/Progressive Chicago," 11 February 1993, IAP Box 10 Flds. 40 and 77; Kelleher, "Mtg. with Jacky Grimshaw," 12 May 1993, and [Kelleher], "Mtg. w/ Barak Obama," 27 July 1993, IAP (162) Box 6 Fld. 49; Daniel Cantor to Obama, 12 August 1993, and Zach Pollett to Madeline [Talbott] and Keith [Kelleher], "Agenda for Phone Conversation on Illinois Political Odds 'n Ends," 21 August 1993, IAP Box 6 Fld. 48; Joe Gardner et al. to Joe Gardner, IAP Box 6 Fld. 44 and Box 10 Fld. 58; Joel Rogers, "Is It Third Party Time?," *In These Times,* 4 October 1993, pp. 28–29; Kelleher to Cantor, "New Party Progress and Plans Since June IEC," 30 September 1993, IAP (2003-066) Box 1 Fld. New Party 1993; Keith [Kelleher] to Talbott et al., "Conversation with Steve Saltzman," 2 November 1993, IAP Box 6 Flds. 47, 48, and 49; Gardner et al. to Gardner, 5 November 1993, IAP Box 6 Fld. 44; Gardner et al. to Obama, 31 December 1993, IAP Box 6 Fld. 49 and Box 12 Fld. 4; [Kelleher], "PRO Chicago Calls," 14 February 1994, IAP Box 11 Fld. 71; Gardner et al. to Talbott, 28 March 1994 and 22 April 1994, IAP Box 11 Flds. 69 and 64, Lou Pardo to Kelleher, 29 April 1994, IAP Box 11 Fld. 63; Gardner et al. to Talbott, 27 May 1994, and 5 July 1994, IAP Box 11 Flds. 62 and 54; *New Party News* ("For a Progressive Chicago"), Vol. 1, #1, September 1994, 4pp., IAP Box 11 Fld. 45; Steve Mills, "Looking for the Left," *CT,* 1 December 1994 (the New Party's "Chicago organization is not yet fully formed"); Stanley Kurtz, *Radical-in-Chief* (Simon & Schuster, 2010), p. 244; Kurtz, "Obama on the Fringe," *National Review,* 25 June 2012, pp. 25–27.

18. UCLS *Announcements 1993–94,* pp. 37–38; UCLS, *The Glass Menagerie 1993–94,* esp. p. 45; "40 Under Forty," *CCB,* 27 September 1993, pp. 27ff.; Obama's 27 October 1993 interview with Zeke Gonzalez; Woods Charitable Fund, *Report for the Year 1993,* esp. pp. 3, 6, 26; William S. McKersie, "Watchful Waiting, and Other Evaluation Lessons Learned: A Chicago Case Study," *Foundation News & Commentary* 38 (September–October 1997): 32–38 ("the Woods board gets full copies of all proposals along with the staff's recommendations"); Gil Dietz, "He's Still Waiting for His Own Beer Summit," *Muscatine Journal,* 3 May 2010; Mundy, *Michelle,* pp. 119–20; Gary Wisby, "Youths Sought to Honor as New Leaders," *CST,* 28 March 1994, p. 13; Eric Krol, "Service to Community Helps Pay Way Toward College Education," *CT,* 29 April 1994; Public Allies Chicago, *Check-In,* Winter 1995; Obama to Phil and Karen Boerner, n.d. [circa mid-March 1994, as Laura Boerner was born March 4]; DJG interviews with Robert Bennett, Allison Davis, David Kindler, Kevin Jokisch, Tom Knutson, Steve Perkins, Jean Rudd, Ken Rolling, Howard Stanback, Julie Alfred Sullivan, Bethann Hester, Julian Posada, Bernard Loyd, Sunny Fischer, Ian Larkin, and Jan Anne Dubin.

19. Obama, "Current Issues in Racism and the Law," Spring Term 1994 Syllabus, "Reading Packet #1," for 29 March and 5 April 1994, and "Reading Packet #2," for 12 and 19 April 1994, Adam Gross Papers; Cornel West, "Learning to Talk of Race," *New York Times Magazine,* 2 August 1992, pp. 24–25; UCLS Course Evaluations, Racism and the Law, Spring 1994 (14 respondents); Susan Epstein, "Obama 10 Years Ago," DailyKos.com, 1 August 2004; Jodi Kantor, "Teaching Law, Testing Ideas, Obama Stood Apart," *NYT,* 30 July 2008, p. A1; Jesse Ruiz in Michael Lipkin, "The Professor and the President," *Chicago Maroon,* 7 November 2008, and in Robin I. Mordfin, "From the Green Lounge to the White House," *University of Chicago Law School Record,* Spring 2009, pp. 2–9; Susan Epstein, "In the Classroom with Barack Obama and Elena Kagan," *Santa Barbara Independent,* 8 July 2010; DJG interviews with David Strauss, Adam Gross, Paul Strauss, Susan Epstein, Jesse Ruiz, Carolyn Shapiro, Robert Mahnke, and Salil Mehra. On Cose's late 1993 *The Rage of a Privileged Class,* see Arnold Rampersad, "Another Day, Another Humiliation," *NYTBR,* 9 January 1994, p. 6.

20. Audio recording, oral argument in *Baravati v. Josephthal, Lyon & Ross,* #93-3647, 7th Cir., 7 April 1994, http://web.archive.org/web/20081009195106/http://www.suntimes.com/images/cds/MP3/obama.mp3; *Baravati v. Josephthal, Lyon & Ross,* 28 F. 3d 704 (7th Cir.), 1 July 1994; John Flynn Rooney, "Arbitration Panel Can Award Punitive Damages: Court," *Chicago Daily Law Bulletin,* 5 July 1994, p. 3; "Appeal Is Filed on Ward Boundaries," *CT,* 21 April 1994; *Barnett v. Daley,* 32 F. 3d 1196 (7th Cir.), 23 August 1994 (argued 20 April 1994); "Barack Obama, Of Counsel," www.lawmbg.com, 30 August 2000 ("real estate financing, acquisition, construction and/or redevelopment of low and moderate income housing"); "Barack Obama CV," www.law.uchicago.edu, 4 June 2004 ("helped to structure and finance efforts to construct mixed-income housing to replace public housing in and around Cabrini Green"); Pallasch, "Obama's Legal Career," *CST,* 17 December 2007, p. 4; Dan Morain, "Obama's Lawyer Days Were Effective but Brief," *LAT,* 6 April 2008, p. A14; David Smallwood, "Veteran Attorneys Recall Obama," *N'Digo Profiles,* December 2008, p. 24; Sony Kassam and Bushra Kabir, "Mr. Belcaster: 'Integrity, Intelligence, and Energy,'" *The Hoofbeat,* 7 December 2009; Obama in Wolffe, *Renegade,* p. 190; Michael Tapscott and Richard Pollock, "The Obama You Don't Know," *Washington Examiner,* 19 September 2012; DJG interviews with Rob Fisher, Ken Mack, John Belcaster, Paul Strauss, Jeffrey Cummings, Judd Miner, George Galland, Laura Tilly, Matt Piers, and Scott Lynch-Giddings. On Arthur Brazier, see his 1995 interview in Black, *Bridges of Memory,* pp. 547–67. Woodlawn Preservation and Investment Corp. was founded in 1987, acquired the 504-unit Woodlawn Gardens from HUD in 1988 and renamed it Grove Parc Plaza. On WPIC, see Patrick T. Reardon, "Woodlawn Project Gives Hope For Future," *CT,* 29 October 1993, Tom Andreoli, "Redevelopment and Redemption: Woodlawn, Kenwood Try . . . Again," *CCB,* 4 June 1994, R. C. Longworth, "Inner-City Areas Get 2nd Chance," *CT,* 27 December 1994, Cal McAllister, "Building for the Future," *CT,* 15 April 1995, and especially Binyamin Appelbaum, "Grim Proving Ground for Obama's Housing Policy," *BG,* 27 June 2008.

21. Obama's September 1995 remarks at the Cambridge (MA) Public Library ("I had to take some leaves of absence for a month or two once the first draft was written to try to polish it"); Obama in Mary Mitchell, "Memoirs of a 21st-Century History Maker," *Black Issues Book Review,* January–February 2005, pp. 18–21; Michelle in Kevin Merida, "The Ghost of a Father," *WP,* 14 December 2007; Peter Osnos, "The Making of the Book That Made Obama," DailyBeast.com, 27 January 2009; Osnos's 2009 interview with Robert Draper; Draper, "Barack Obama's Work in Progress," *GQ,* November 2009; Maraniss, *BOTS,* p. 566; DJG interviews with John Belcaster, Sheila Jager, Rob Fisher, Henry Ferris, Peter Osnos, and Ruth Fecych. See also Ellen Schumer phone conversation notes, "Barack Obama," 2 May 1994, ESP ("Following a theme, touching on being biracial—looks at organizing—family—ancestry"), and Schumer notes, "5-18-94—Barack Obama 12:30PM–2PM," ESP ("discussed Hawaii"). Portions of an early-stage typescript, one reflecting Ferris's editing and Barack's handwritten alterations yet preceding Ruth Fecych's involvement, were put up for commercial sale by Abon'go Malik Obama in 2015 (ms. pp. 6–7, 17, 22–24, 43–46, 48, 70, and 20, corresponding to *DFMF* pp. ix–x, 14–15, 18–20, 32–35, 36, 52 and 403). Barack's changes were stylistic rather than substantive, with his childhood Jakarta home offering a "crocodile" pond rather than an "alligator" one (ms. p. 36, *DFMF* p. 36).

22. Keith [Kelleher] to Interested Parties, "Voter Registration, Politics Update, 20 July 1994, IAP Box 11 Fld. 55; Sandy [O'Donnell] to Ken [Rolling] and Barack, "Panel on the Future of Organizing," 9 June 1994, Sandy to WFC Evaluation Team, "Interviews, Case Studies, Records," 14 June 1994, [O'Donnell], "Minutes of the First Meeting of the Panel of the Future of Organizing," 29 June 1994, Sandy to Ken and Barack, "Next Panel Meetings," 16 July 1994, Sandy to Evaluation Team, "Ideas for Focus Group of Cross-City Leaders," 17 July 1994, Sandy to Ken, "Organizing Evaluation," 20 July 1994, O'Donnell to Obama, "August 29 Organizing Discussion," 12 August 1994, O'Donnell to Obama, "Questions re Public Policy & Organizing," 19 August 1994, Obama to Gonzales et al., "August 29 Meeting of Discussion Group on Community Organizing," 22 August 1994, [O'Donnell], "Minutes of the Second Panel Discussion on Organizing," 29 August 1994, "Suggested Questions for the Panel Discussion on Community Organizing," 25 October 1994, [O'Donnell], "Minutes of the Third Panel Discussion on Organizing," 26 October 1994, "Panel Discussion on Community Organizing: Advice to the Woods Fund of Chicago," 13 January 1995, [O'Donnell], "Minutes of the Fourth Panel Discussion on Organizing," 13 January 1995, all SOP; Alex Rodriguez and Scott Fornek, "Gangs Besiege Once-Peaceful Neighborhood," *CST,* 2 September 1994, p. 6; Mary A. Johnson, "A Gang's Violence Often Strikes Its Own," "Community Spirit Wanes," and "Ministry Offers 'Neutral' Zone," *CST,* 4 September 1994, p. 7; Jan Ferris and Henri Cauvin, "Roseland Tries to Ride Out the Shock Wave," *CST,* 6 September 1994; Obama and Ken Rolling to "Dear Colleagues," 25 April 1995, CACP; Sandra O'Donnell et al., *Evaluation of the Fund's Community Organizing Grant Program* (Woods Fund of Chicago, April 1995), esp. pp. 15–16; Woods Fund of Chicago, *1994 Annual Report,* esp. pp. 1–7, 17, 20; Woods Fund of Chicago, 1994 Form 990-PF, Part XV, 29 June 1995; Jean

Rudd, "How One Foundation Came to Support Community Organizing," in Sandy O'Donnell et al., *Promising Practices in Revenue Generation for Community Organizing* (Center for Community Change, October 2005); DJG interviews with Keith Kelleher, Jean Rudd, Ken Rolling, Mary Gonzales, Jacky Grimshaw, Madeline Talbott, Sokoni Karanja, and Sandy O'Donnell. On Keith's 1994 efforts, with repeated references to Barack, also see Kelleher to Carol Harwell, "List of Positions That Need Resolution," 8 June 1994, Leslie Watson-Davis to Marvin Randolph and Zach Pollett, "Chicago Trip—For Your Information," 13 June 1994, and Keith to Madeline Talbott et al., "Conversations with Carol Harwell," 17 June 1994, all IAP (2003-072) Box 1 Fld. 1994, and Keith to Interested Parties, "Research on Voter Registration and 24th Ward Candidates," 7 July 1994, IAP Box 11 Fld. 57.

23. Nina Burleigh, "The Great Black Hope," *Spy,* March 1993, pp. 20–21; Elaine S. Povich, "The Housewarming—It Has Not Been One Big Welcome Party for Freshman Rep. Reynolds," *CT,* 17 August 1993; Steve Neal, "Weak Field Puts Reynolds in Front," *CST,* 28 December 1993, p. 17; John Kass, "2nd District Campaign Not as Loud, Still Heated," *CT,* 20 February 1994; David Elsner and Lou Carlozo, "Weller, Giglio Win Bids in 11th District," *CT,* 16 March 1994; Neal, "Last-Minute Cash Netted Stroger Win," *CST,* 25 March 1994, p. 39; Chinta Strausberg, "Alice Palmer Pushed for Mayor's Post," *CD,* 11 July 1994, p. 3; Kass, "Anti-Daley Forces Start to Beat Campaign Drums," *CT,* 29 July 1994, p. 1; Ray Gibson, "Edgar Has 5–1 Lead in Cash Race," *CT,* 2 August 1994, p. C1; Fred Lebed datebook for 9 August 1994; Kevin Knapp, "Kenwood Developers Are Donors to Ald. Preckwinkle's Campaign," *HPH,* 10 August 1994, pp. 1–2; Knapp, "Alderman Explains Campaign Donations," *HPH,* 17 August 1994, pp. 1–2; Knapp, "Campaign Finances Examined," *HPH,* 7 September 1994, pp.1–2; Knapp, "Campaigns Driven by Finances," *HPH,* 26 October 1994, pp. 1–2; Knapp, "Building Rehabbed with Help from County Program," *HPH,* 2 November 1994, p. 4; Andrew Fegelman, "Stroger's Patience Is Rewarded," *CT,* 9 November 1994; Fegelman, "Stroger's Win Built on Voter, Party Unity," *CT,* 10 November 1994; Knapp, "Alderman's Contributions Drop Off as Election Comes Closer," *HPH,* 8 February 1995, pp. 1, 6; Knapp, "Incumbent Gets Financial Boost," *HPH,* 22 February 1995, pp. 1–2; Knapp, "Local Aldermanic Races Hit New High in Cost," *HPH,* 9 August 1995, p. 1; Knapp, "Tough Races Marked '95 Elections," *HPH,* 3 January 1996, pp. 3, 9; Knapp, "Campaign Finances Examined," *HPH,* 27 March 1996, pp. 1–2; DJG interviews with Toni Preckwinkle, Laura Tilly, David Brint, Fred Lebed, Roland Burris, Paul Strauss, George Galland, and John Schmidt.

24. Ray Gibson, "Reynolds Blasts Cops, O'Malley Over Probe," *CT,* 12 August 1994, p. 4; Christopher Drew and Gibson, "Reynolds' Handling of Campaign Funds Focus of U.S. Probe," *CT,* 14 August 1994, p. 1; Timothy J. Burger, "Frosh Rep. Reynolds Is Target of Probe," *RC,* 15 August 1994; Edward Walsh, "Lawmaker Accused of Affair with Teen," *WP,* 16 August 1994, p. A4; Gibson and Drew, "Aide Unveils More Reynolds Campaign Accounts," *CT,* 17 August 1994, p. 1; "The Troubles of Mel Reynolds," *CT,* 18 August 1994; Andrew Fegelman and Robert Becker, "Reynolds Indicted in Teen Sex Scandal," *CT,* 20 August 1994, p. 1; Judy Pasternak, "Congressman Charged with Sex Crimes," *LAT,* 20 August 1994; Don Terry, "Illinois Indicts a Congressman in a Sex Case Involving 2 Girls," *NYT,* 20 August 1994, p. 7; James Webb (AP), "Reynolds Promise Conflicts with Controversies," 20 August 1994; Steve Neal, "Palmer Could Be Contender in 2nd," *CST,* 29 August 1994, p. 23; "Mel Reynolds: The Absentee Congressman," *Illinois Politics,* September 1994, pp. 1, 3; John Kass, "Gardner Takes a Risk in Rush to Face Daley," *CT,* 4 September 1994, p. P1; Michael Sneed, "Sneed," *CST,* 15 September 1994, p. 2. On Alice Palmer's work in Springfield, see Palmer, "King's Legacy of Change, Not Dreams," *HPH,* 24 February 1993, p. 4; Senator Alice Palmer, "Springfield Report" #1, February–March 1993, 8pp.; Palmer, "Edgar's Budget Hurts City," *HPH,* 17 March 1993, p. 4; Palmer et al., "Asks Edgar Change on College Boards," *CD,* 1 May 1993, p. 19; Palmer, "Springfield Must Make Law, Not Haste," *HPH,* 26 May 1993, p. 4; Palmer, "Hails Article," *CD,* 16 June 1993, p. 11; Ethan Michaeli, "Recycling Could Come to Low-Income Areas," *CD,* 17 June 1993, p. 3; Palmer, "Failure on Schools Will Haunt Assembly," *HPH,* 28 July 1993, p. 4; Palmer, "Springfield Report" #2, Summer 1993, 8pp.; "Palmer Lauds NAACP's Anti-NAFTA Position," *CD,* 5 August 1993, p. 6; Palmer, "Why I Oppose NAFTA," *Illinois Politics,* September 1993, pp. 7, 10; Palmer, "Tax Codes Need a Careful Revision," *HPH,* 17 November 1993, p. 4; Palmer, "Side Issues Delay Riverboat Gambling," *HPH,* 27 April 1994, p. 4;

Chinta Strausberg, "Sen. Palmer Calls State of Higher Education Dismal," *CD,* 14 May 1994, p. 5; Palmer, "Medicaid Spurs State Budget Problems," *HPH,* 29 June 1994, p. 4; "Senator Alice Palmer's 13th District Education Committee Meeting Highlights," 16 July 1994, Timuel Black Papers Box 135 Fld. 6; Citizens for Senator Alice J. Palmer, flyer for 4 August 1994 BBQ, TBP Box 135 Fld. 26; Palmer, "Downstate Answers to City Questions," *HPH,* 21 September 1994, p. 4; Sheila Washington, "Palmer Gets Down to Nitty-Gritty," *CD,* 3 October 1994, p. 9.

25. *Barnett v. City of Chicago,* 32 F. 3d 1196 (7th Cir.), 23 August 1994; Jan Crawford Greenburg, "Suit Contesting City Ward Map Gets New Life," *CT,* 25 August 1994; "Forum Offers New Thoughts at Wesleyan," *Omaha World Herald,* 6 September 1994, p. 18; Obama, "Community Revitalization," Nebraska Wesleyan University Forum, 9 September 1994, NWU Archives; Michael Kelly, "Obama's 1994 Speech Telling," *OWH,* 9 November 2008, p. 1B.

26. Kennedy Communications, "State Senator Barack Obama, Candidate for U.S. Senate: Background Report," April 2003, RRP; Obama, "Charles Murray's Political Expediency Denounced," *All Things Considered,* NPR, 28 October 1994; Susan Locke to David Plotkin, 30 September 1994, Human SERVE Papers (HSP) Box 16 Fld. 747 and Box 35 Fld. 1497; Zach Pollett to ACORN Offices, "National Voter Registration Act Implementation Campaign," 23 October 1994, IAP (98-059) Box 5 Fld. 42; Doug Hess to Madeline Talbott, "Re: ACORN Notice to IL Head Election Official," 3 November 1994, ACORN to Ronald D. Michaelson, 14 November 1994, and David [Plotkin] to All, "Illinois Implementation (or lack thereof)," 2 December 1994, HSP Box 16 Fld. 747; DJG interview with Rob Fisher.

27. Chinta Strausberg, "Palmer Throws Hat into 2nd District Race," *CD,* 18 October 1994, p. 3; "Announcement of Congressional Exploratory Committee State Senator Alice Palmer (D-13)," 21 November 1994, TBP Box 135 Fld. 26; Susan Kuczka, "State Sen. Palmer Ponders Bid for Reynolds' 2nd District Post," *CT,* 22 November 1994; Steve Neal, "Palmer Beats Jackson Jr. to Punch," *CST,* 25 November 1994, p. 49; Judy Pasternak, "Following in His Father's Footsteps," *LAT,* 12 December 1995; "Barack Obama—12/1/94—Noon–1:30PM," Ellen Schumer Papers; Monica Copeland, "Agency's Emphasis on Life Success," *CT,* 12 June 1992; Lynne Duke, "Mixed Legacies on Boulevards of the 'Dream,'" *WP,* 4 April 1993, p. A1; George Papajohn, "Honoree a 'Genius' at Helping," *CT,* 15 June 1993; Allan Johnson, "South Side Agency Honors Donors with a Feast," *CT,* 25 June 1993; David Moberg, "Bold Vision: Sokoni Karanja Nurtures the Restoration of Bronzeville," *CT,* 19 June 1995; Karanja's 2005 HistoryMakers interview with Larry Crowe; Michael I. J. Bennett, "The Rebirth of Bronzeville: Contested Space and Contrasting Visions," in John P. Koval et al., eds., *The New Chicago: A Social and Cultural Analysis* (Temple University Press, 2006), pp. 213–20; "Bronzeville Wants City's Empty Lots," *CT,* 23 November 2006; Michelle Robinson, "Motion to Transfer to Inactive Status Pursuant to Supreme Court Rule 770," *In Re Michelle Lavaughn Robinson,* Ill. S.Ct., # 94-CH-456, 8 June 1994; Althea J. Kuller to Michelle Obama, and Kuller to Juleann Hornyak, 23 June 1994 (filing said motion); Jeremy Mindich, "Americorps: Young, Spirited and Controversial," *CT Magazine,* 9 April 1995; Carol Kleiman, "'Xers' Don't Fit the Stereotypes," *CT,* 9 April 1995; Randee Fenner, "Rivkin and Moelis: Together in Public Service," *Stanford Lawyer* 87, 12 November 2012; "Business Appointments," *CST,* 28 November 1994, p. 52; Joyce Foundation, *1995 Annual Report,* esp. p. 83; Barbara Sullivan, "The Power of Paula," *CT,* 12 September 1993; Lew Butler's 2009 interview with Ann Lage, pp. 351 and 466–67; Carole Feldman (AP), "Tycoon Hopes $500 Million Gift Will Stem School Violence," CT, 18 December 1993, p. 3; "School Violence and Annenberg's Gift," *CT,* 21 December 1993; Charles Storch and V. Dion Haynes, "Schools Go After Windfall," *CT,* 23 October 1994, p. C1; Storch and Haynes, "Philanthropist Puts His Money on City Schools," *CT,* 21 January 1995; Storch, "School Reformers Getting Wish," *CT,* 23 January 1995; "Annenberg Grant Is More Than Money," *CT,* 26 January 1995; William Ayers and Michael Klonsky, "Navigating a Restless Sea: The Continuing Struggle to Achieve a Decent Education for African American Youngsters in Chicago," *Journal of Negro Education* 63 (Winter 1994): 5–18; Ben Joravsky, "The Long, Strange Trip of Bill Ayers," *CR,* 8 November 1990; Jon Anderson, "Weathering Change," *CT,* 8 July 1993, p. T1; Ayers, Chapman, and Hallett, "A Booster Shot for Chicago's Public Schools," *CT,* 31 January 1995, p. 15; [Patricia Graham], CAC Board of Directors, "Minutes," 15 March 1995, CACP; Ellen Schumer notes, "Barack Obama," 22 March 1995, ESP; CAC Board of Directors, "Minutes," March 31,

1995, CACP; William C. Ayers, "Minutes," CAC Board of Directors, 13 April 1995, CACP; Peter Applebome, "Annenberg School Grants Raise Hopes, and Questions on Extent of Change," *NYT,* 30 April 1995, p. 30; CAC, "Minutes, Organizational Meeting of the Board of Directors," 11 May 1995, CACP (only Obama, Graham, and Hallett present, formally authorizing the Collaborative's advisory role); Lynn Olson, "Annenberg Institute Seeks to Find Voice in 1st Year," *Education Week,* 21 June 1995; Obama to Abon'go Malik Obama, 23 July 1995, Abon'go Malik Obama Papers; Ken Rolling to Vartan Gregorian et al., "Chicago Annenberg Challenge Program Report," 8 May 1996, CACP; Steve Diamond, "That 'Guy Who Lives in My Neighborhood': Behind the Ayers-Obama Relationship," Global Labor Blog, 23 June 2008; Stanley Kurtz, "Obama and Ayers Pushed Radicalism on Schools," *WSJ,* 23 September 2008, p. A29; Michael Bechek, "In Debate Over Obama's Past, U. History Revealed," *Brown Daily Herald,* 24 September 2008; Scott Shane, "Obama and '60s Bomber: A Look into Crossed Paths," *NYT,* 4 October 2008; Dakarai I. Aarons, "Backers Says Chicago Project Not 'Radical,'" *Education Week,* 15 October 2008, pp. 1, 14–16; Ellen Goodman, "Sharing Dreams from His Mother," *BG,* 9 May 2008; Scott, *A Singular Woman,* pp. 304–33; DJG interviews with Alice Palmer, Jesse Jackson Jr., Ellen Schumer, Sokoni Karanja, Julian Posada, Bethann Hester, Jan Anne Dubin, Jacky Grimshaw, Sunny Fischer, Ian Larkin, Craig Huffman, Debbie Leff, Chuck Daly, Jack Anderson, Paula Wolff, Carin Clauss, Carlton Guthrie, Anne Hallett, Bill Ayers, Pat Graham, Adele Simmons, and Charles Payne. Apropos of Barack's comments to Pat Graham over dinner, also see Obama on University of Illinois at Springfield's "Inside Illinois Government," 3 December 1998 ("how difficult it is to get a book published when you finally write it"). On COFI, see Sandy O'Donnell and Ellen Schumer, "Community Building & Community Organizing," *NHI Shelterforce* #85, January–February 1996, and Schumer and O'Donnell, "Turning Parents into Strong Community Leaders," *NHI Shelterforce* #114, November–December 2000.

28. Obama to Bell, 3 February 1995, Bell to Obama, 27 March 1995, and Obama to Bell, 27 March 1995, Bell Papers; Obama, "Complaint for Injunctive and Declaratory Relief," *ACORN v. Edgar,* #95-C-174, N. D. Ill., 11 January 1995; Juan Cartagena to NVRA Implementation Attorney Network, "Status of Litigation," 1 February 1995, Human SERVE Papers Box 35 Fld. 1495; *ACORN v. Edgar,* 880 F. Supp. 1215, esp. 1219, 1222 (N. D. Ill.), 28 March 1995; Bernie Mixon and Peter Kendall, "Illinois Told to Move on 'Motor-Voter Law,'" *CT,* 29 March 1995, p. C1; Obama to David R. Melton, 29 March 1995, HSP Box 16 Fld. 748; *ACORN v. Edgar,* 56 F. 3d 791 (7th Cir.), 5 June 1995; *ACORN v. Edgar,* 1995 WL 359900 (N. D. Ill.), 13 June 1995; *ACORN v. Edgar* Docket, 26 June 1995; 51st Annual IVI/IPO Independents' Day Dinner, 22 July 1995, HSP Box 16 Fld. 749; *ACORN v. Edgar,* 1995 WL 532120 (N. D. Ill.), 7 September 1995; McClelland, *Young Mr. Obama,* pp. 78–79; *African American Voting Rights Legal Defense Fund v. Villa,* 54 F. 3d 1345 (8th Cir.), 12 May 1995; cert. denied, 516 U.S. 1113, 20 February 1996; Barnaby J. Feder, "Kenneth Pontikes, 54, Founder of Computer Leasing Company," *NYT,* 25 June 1994; Henri Cauvin, "Kenneth N. Pontikes, 54, Founder of Comdisco Inc.," *CT,* 25 June 1994; *Donnell v. Comdisco,* #95-C-512 (N. D. Ill.), 26 January 1995; Ronald E. Yates, "Ex-Comdisco Worker Sues, Alleging Bias," *CT,* 7 February 1996; Matthew Schifrin, "Jack Slevin's Ordeal," *Forbes,* 14 August 1995, pp. 82–83, 86; *Donnell v. Comdisco,* 1995 U. S. Dist. Lexis 13214 (N. D. Ill.), 8 September 1995; Obama's 20 September 1995 remarks at the Cambridge (MA) Public Library; Barbara Sullivan, "Now Is She a Household Name?," *CT,* 31 January 1994; *Buycks-Roberson v. Citibank,* #94-C-4094 (N. D. Ill.), 6 July 1994; Buycks-Roberson *v. Citibank,* 162 F.R.D. 322 and 338 (N. D. Ill.), 29 June 1995; "Settlement Agreement," 9 January 1998, "Order," 7 May 1998, and "Order," 12 May 1998, *Buycks-Roberson v. Citibank,* #94-C-4094; Mike Robinson (AP), "Some Cases Obama Worked on in His Career as an Attorney," 20 February 2007; Pallasch, "Obama's Legal Career," *CST,* 17 December 2007, p. 4; Dan Morain, "Obama's Lawyer Days Were Effective but Brief," *LAT,* 6 April 2008, p. A14; Pallasch, "Judge Ruben Castillo Makes Top 10 List of Obama's Supreme Court Possibilities," *CST,* 13 May 2009; Matt O'Connor, "Federal Judge Bua Hanging Up the Robe," *CT,* 11 September 1991; Bill Lueders, "They Knew Him When," *Isthmus,* 21 February 2008; Miner, "Barack Obama's Career at Miner, Barnhill & Galland"; Jeff Cummings, Speech Outline, [29 April 2009]; Dick Lovell, "Obama Is 'An Extraordinary Person,' Friend Jeff Cummings Tells Rotarians," *Madison Rotary News* #44, 1 May 2009; Neil Munro, "With Landmark Lawsuit,

Barack Obama Pushed Banks . . . ," and Stephen Elliott, "Obama's African-American Clients Got Coupons, Not Cash," *Daily Caller,* 3 September 2012; Matt Gertz, "*Daily Caller* Tries and Fails to Read Legal Documents," MediaMatters.org, 4 September 2012; DJG interviews with Allison Davis, Alex Kotlowitz, Derrick Bell, Madeline Talbott, Judd Miner, Jeff Cummings, Cindi Canary, David Melton, Jeanne Gills, Whit Soule, and Chuck Barnhill.

29. UCLS *Announcements 1994–95,* pp. 44–45; UCLS *Directory 1994–95,* p. 34; UCLS Course Evaluation, Current Issues in Racism and the Law, Spring 1995; Noni-Ellison Southall, "When the President Was My Professor," CNN.com, 25 February 2013; DJG interviews with Marni Willenson, Tiffanie Cason De Liberty, Susannah Baruch, Linda Simon, Liisa Thomas, and Deborah Burnet. By 1995 Barack was contributing "an outstanding lecture" on the legal roots of American racism each time UC medical professor Dr. Deborah Burnet, the spouse of CCRC veteran Rev. Bob Klonowski, taught her interdisciplinary "Inequality and Health" course.

30. Times Books catalog, page 23, Rob Mitchell Papers (and listing *Dreams* as a June title); *Publishers Weekly,* 10 April 1995, p. 46; *Kirkus Reviews,* 15 April 1995; Thomas Hardy and John Kass, "Daley Hits the Campaign Running," *CT,* 9 December 1994; Bernie Mixon, "Gardner Says He'll Stay in Mayoral Race," *CT,* 12 December 1994; Kass, "Daley Calls for End to City Racial Politics," *CT,* 15 December 1994; Kass, "Moseley-Braun Steps into Daley's Corner," *CT,* 21 December 1994; Burney Simpson, "Voter Registration: Too Good to Be True," *Chicago Reporter,* February 1995; Hardy, "Ho-Hum Mayoral Campaign Is Good News for Daley," *CT,* 12 February 1995; Joseph A. Kirby and Kass, "Gardner Tries Tossing a GOP Label at Daley," *CT,* 14 February 1995; Kass and Theresa Puente, "Gardner Expresses Bitterness," *CT,* 16 February 1995; Kirby, "Jackson's Words Lift Up Gardner," *CT,* 19 February 1995; Hardy, "Daley's Machine Humming," *CT,* 26 February 1995, Hardy and Kass, "Daley Rolls Just Like Dad," *CT,* 1 March 1995; Hardy and Kass, "Daley Rests While Burris Maps Strategy," *CT,* 2 March 1995; Hardy, "Shades of Resentment Evident in Burris Campaign Against Daley," *CT,* 26 March 1995; Don Terry, "In a GOP World, a Democrat Rules Chicago," *NYT,* 1 April 1995; Hardy, "Daley Again in a Cakewalk," *CT,* 5 April 1995; Hardy, "Daley Keeps Minority Inroad," *CT,* 7 April 1995; Dick Simpson et al., "The New Daley Machine: 1989–2004," The City's Future Conference, Chicago, July 2004, esp. p. 7; Dan Conley, "Look Homeward, Obama," *Salon,* 23 May 2008 ("Daley stalwarts like Valerie Jarrett . . . and John Rogers"); Draper's 2009 interviews with Robin Schiff and Benjamin Dreyer; DJG interviews with Rob Fisher, Ruth Fecych, and Roland Burris.

31. Steve Neal, "Emil Jones Considers Run for Treasurer," *CST,* 12 December 1994, p. 23; Janan Hanna, "Case Against Reynolds Dealt Blow," *CT,* 10 January 1995; "Jack O'Malley's Burden of Proof," *CT,* 11 January 1995; Neal, "Palmer Clouds Reynolds' Political Future," *CST,* 11 January 1995, p. 29; Alice Palmer, "Legislature Seeks to Change Higher Education," *HPH,* 1 March 1995, p. 4; Maurice Possley, "Reynolds Accused of Sex with 2nd Teen," *CT,* 5 April 1995; Ellen Schumer notes, 7 April 1995, ESP; "Twice Burned in the 2nd District," *CT,* 21 April 1995; Gabriel Kahn, "Legal Woes Mounting, Reynolds Now Pursued by Political Rivals," *Roll Call,* 24 April 1995; Schumer notes, 1 May 1995, ESP; Possley, "Reynolds Is Socked with New Indictments," *CT,* 4 May 1995; Ellen Warren, "Amid Hopes, Always Doubt," *CT,* 7 May 1995; Schumer notes, 11 May 1995, ESP; Illinois State Senate Floor Transcript, 26 May 1995, esp. pp. 66–68 (del Valle), 71–73 (Garcia), and 77 (Palmer); Thomas Hardy and Rick Pearson, "State GOP Had Its Way All the Way," and Suzy Frisch, "State Lawmakers Adjourn with GOP Flaunting Majority," *CT,* 28 May 1995; Ray Gibson, "Tapes May Decide Reynolds' Fate," *CT,* 25 June 1995; Friends of Alice Palmer, FEC Form 3, 31 January 1995–30 June 1995 (as amended 28 June 1996), FEC; Obama to Abon'go Malik Obama, 23 July 1995, Abon'go Malik Obama Papers; Obama in Joe Frolik, "A Newcomer to the Business of Politics," *Cleveland Plain Dealer,* 3 August 1996, p. A1; Obama's 2001 interview with Julieanna Richardson; Obama's remarks at the Heartland Cafe, Chicago, 7 February 2004, JKP and TPP, and on *The Charlie Rose Show,* 23 November 2004; "Complete Transcript of the *Sun-Times* Interview with Barack Obama," Sun-Times.com, 15 March 2008 ("we were both active in Alice Palmer's campaign, so I got to know him at that point"); Obama interview transcript, *CT,* 16 March 2008, p. 22 (Rezko introducing Obama to aldermen); Jean Rudd in Ryan Lizza, "The Agitator," *TNR,* 19 March 2007, pp. 22ff., and in Michael Powell, "Calm in the Swirl of History," *NYT,* 4 June 2008; Michelle in Suzanne Bell, "Michelle Obama Speaks at ISU," *Daily Vidette,* 26 October 2004, and in Sarah Brown, "Obama '85 Masters Balancing Act," *Daily Princetonian,* 7 December 2005; Rosalind Rossi, "Obama's Anchor," *CST,* 21 January 2007, p. A9; Michelle in Eric Tucker (AP), "Family Ties: Brown Coach, Barack Obama," 1 March 2007, on *All Things Considered,* NPR, 9 July 2007, in Scott Helman, "Early Defeat Launched a Rapid Political Climb," *BG,* 12 October 2007, p. A1, in Rosemary Ellis, "A Conversation with Michelle Obama," *Good Housekeeping,* November 2008, in Liza Mundy, *Michelle,* p. 124, and in Sandra Sobieraj Westfall, "The Obamas Get Personal," *People,* 4 August 2008; McClelland, *Young Mr. Obama,* p. 106; McClelland, "Mel Reynolds's Horniness Opened Door for Obama," NBCChicago.com, 28 November 2012; DJG interviews with Alice Palmer, Hal Baron, Brian Banks, Kitty Kurth, Darrel Thompson, Ellen Schumer, Andy Foster, Judd Miner, Allison Davis, Jeff Schoenberg, John Hooker, Frank Clark, Ray Harris, Jean Rudd, Larry Suffredin, Steve Derks, Marilyn Katz, Matt Piers, Chuy Garcia, Toni Preckwinkle, Barbara Holt, and Barbara Flynn Currie. On Barbara Holt's election, see Steve Neal, "Holt, Oliver-Hill Neck and Neck in South Side Aldermanic Race," *CST,* 29 March 1995, p. 33, Jorge Oclander, "Loss of Clout Dims IVI-IPO's Fate," *CST,* 2 April 1995, p. 22, and Ben Joravsky, "Mud Flap," *CR,* 20 April 1995. On Chuy Garcia, also see his 1990 interview with Betty Brown Chappell, Steve Bogira, "What Pilsen Taught Chuy Garcia," *CR,* 12 December 2014, and Bogira, "Jesus 'Chuy' Garcia's Journey," *CR,* 21 January 2015. See also "The Barack Obama *We* Know," *Lake & Prairie,* 4th Quarter 2008, p. 3.

32. "Elementary School Council Election Results," *CT,* 13 October 1989; Michael Hirsley, "Islam Pilgrimage Goal of a Lifetime," *CT,* 20 May 1994, p. C9; Judy Hevrdejs and Mike Conklin, "Changing of Guard in Schools Puts Daley Staff at Head of Class," *CT,* 16 June 1995, p. 18; Kevin Knapp, "Names Begin to Surface for Committeeman Races," *HPH,* 23 August 1995, p. 3; Michael Sneed, "Sneed," *CST,* 25 June 1995, p. 4; Chinta Strausberg, "St. Sen. Palmer Makes Bid for Reynolds Seat," *CD,* 27 June 1995, p. 3; Strausberg, "Palmer Ready to Take on Reynolds," *CD,* 28 June 1995, p. 1; Thomas Hardy, "Palmer Seeks to Replace Reynolds," *CT,* 28 June 1995, p. C3; Scott Fornek, "Reynolds' 1st Challenger Announces," *CST,* 28 June 1995, p. 14; Knapp, "Alice Palmer to Run for Reynolds Seat," *HPH,* 5 July 1995, pp. 1–2; Hevrdejs and Conklin, "Something Different," *CT,* 7 July 1995, p. 20; "The Friends of Alice J. Palmer (in formation)," n.d., TBP Box 103 Fld. 21 (including Barbara Flynn Currie, Al Johnson, Antoin Rezko, Allison Davis, Toni Preckwinkle, Jean Rudd, Barack Obama, Hal Baron, Brian Banks, and Alan and Lois Dobry); Friends of Barack Obama, Report of Campaign Contributions and Expenditures, Form D-2, 1 July 1995–31 December 1995, 2 February 1996; Bruce Bentley, "Chicago New Party Update," *New Ground* 42, September–October 1995; Al Johnson's 1989 interview with Betty Brown-Chappell; Robert Mier and Kari J. Moe, "Decentralized Development: From Theory to Practice," in Pierre Clavel and Wim Wiewel, *Harold Washington and the Neighborhoods: Progressive Development in Chicago, 1983–1987* (Rutgers University Press, 1991), p. 87; Cheryl V. Jackson, "Albert W. Johnson: GM's 1st Black Franchise Owner," *CST,* 14 January 2010; Trevor Jensen, "Albert W. Johnson: 1920–2010," *CT,* 21 January 2010; "Former Neighbor Al Johnson, First Black GM Dealer, Dies at 89," RanchoMurieta.com, 29 January 2010; DJG interviews with Hermene Hartman, Abner Mikva, Don Johnson, and Willie Delgado.

33. Obama [by Hallett] to Bruce Newman, 31 May 1995, CACP Box 130 Fld. 924; CAC Board of Directors "Minutes," 5 June 1995, CACP; Jacquelyn Heard and V. Dion Haynes, "GOP Changes the Rules for City Schools," *CT,* 13 November 1994; Rick Pearson and Joseph A. Kirby, "School Power Lobbed to Daley," *CT,* 26 April 1995; John Kass and Pearson, "Daley Quickly Flexes School Muscle," *CT,* 25 May 1995; Haynes, "Schools Get Ready for Annenberg Challenge on Innovative Education," *CT,* 23 June 1995, p. 5; Kass, "Daley Names School Team," *CT,* 30 June 1995; Kass, "City Schools Get Chief Who Rejects Failure," *CT,* 2 July 1995; Jacquelyn Heard, "New School Officials Plan Coordinated Effort," *CT,* 4 July 1995; Joanne Esters-Brown, "School Cluster Must Meet Aug. 1 Deadline," *HPH,* 5 July 1995, pp. 1–2; Haynes, "School Team Tells Principals of Plans," *CT,* 14 July 1995; "Rising to the Annenberg Challenge," *CT,* 17 July 1995; Heard, "Schools Get New Whiz Kid," *CT,* 23 July 1995; Haynes, "Proposed Budget Has Pupil Focus," *CT,* 14 August 1995; "Vallas' Amazing Balancing Act," *CT,* 15 August 1995; R. Bruce Dold, "It's Too Late to Cry Over Spilled Milk," *CT,* 18 August 1995; Linda Lenz, "The New Law," and Grant Pick, "179 Partnerships Bid for First-Round Annenberg Funds," *Catalyst Chicago,* September 1995; Lenz and Lynnette Richardson, "89 Groups Get Annenberg Go-Ahead," *Catalyst Chicago,* October 1995; CAC Board of Directors, "Minutes," 12 October 1995, CACP; Ken Rolling to Vartan Gregorian et al., "Chicago Annenberg Program Report," 8 May 1996, CACP; Anthony S. Bryk et al., "Chicago School Reform," in Diane Ravitch and Joseph P. Viteritti, eds., *New Schools for a New Century* (Yale University Press, 1997), pp. 164–200; Dorothy Shipps, "The Invisible Hand: Big Business and Chicago School Reform," and Michael B Katz et al., "Poking Around: Outsiders View Chicago School Reform," *Teachers College Record* 99 (Fall 1997): 73–116 and 117–57; David Moberg, "Chicago's 4 R's: Reading, 'Ritin, 'Rithmetic and Reform," *Examining Education,* 20 September 1998, p. 10; Shipps and Karin Sconzert, *The Chicago Annenberg Challenge: The First Three Years* (Consortium on Chicago School Research, March 1999), pp. 15–17; Paul G. Vallas, "Saving Public Schools," *Civic Bulletin* 16, March 1999, pp. 1–9; William S. McKersie and Anthony Markward, "Lessons for the Future of Philanthropy: Local Foundations and Urban School Reform," in Charles T. Clotfelter and Thomas Ehrlich, eds., *Philanthropy and the Nonprofit Sector in a Changing America* (Indiana University Press, 1999), pp. 385–412; G. Alfred Hess Jr., "Expectations, Opportunity, Capacity and Will: The Four Essential Components of Chicago School Reform," and Shipps et al., "The Politics of Urban School Reform: Legitimacy, City Growth, and School Improvement in Chicago," *Educational Policy* 13 (September 1999): 494–517 and 518–45; William Rau et al., "The Chicago School Reforms: Are They Working?," *Sociological Quarterly* 40 (Autumn 1999): 641–61; Hess, "School Reform Struggles in an Era of Accountability," *PRAGmatics* 2 (Fall 1999): 3–5; Shipps, "Regime Change: Mayoral Takeover of the Chicago Public Schools," in James G. Cibulka and William L. Boyd, eds., *A Race Against Time: The Crisis in Urban Schooling* (Praeger, 2003), pp. 106–28; Shipps, "The Businessman's Educator: Mayoral Takeover and Nontraditional Leadership in Chicago," in Larry Cuban and Michael Usdan, eds., *Powerful Reforms with Shallow Roots* (Teachers College Press, 2003), pp. 16–37; Shipps, "Pulling Together: Civic Capacity and Urban School Reform," *American Educational Research Journal* 40 (Winter 2003): 841–78; Shipps, "Chicago: The National 'Model' Reexamined," in Jeffrey R. Henig and Wilbur C. Rich, eds., *Mayors in the Middle* (Princeton University Press, 2004), pp. 59–95; Christine C. George, "The Chicago Annenberg Challenge: The Messiness and Uncertainty of Systems Change" (Loyola University of Chicago), 2007, esp. pp. 14–17; Shipps, "Updating Tradition: The Institutional Underpinnings of Modern Mayoral Control in Chicago's Public Schools," in Viteritti, ed., *When Mayors Take Charge: School Governance in the City* (Brookings Institution Press, 2009), pp. 117–47; DJG interviews with Pat Graham, Ken Rolling, Anne Hallett, and Bill Ayers.

34. [Sandy O'Donnell], "Minutes of Meeting of Hope Center Planning Team & WAALIMU," 22 July 1995, SOP; DJG interviews with Sokoni Karanja and Sandy O'Donnell; Maurice Possley, "Reynolds' Trial Delayed 2 Weeks," *CT,* 30 June 1995; Possley, "Cast Is Ready for Reynolds Sex-Trial Drama," *CT,* 16 July 1995; Possley, "Reynolds Jury Has Balance," *CT,* 22 July 1995; Thomas Hardy, "Courtroom Foes Reynolds, O'Malley Both at Crossroads in Their Careers," *CT,* 23 July 1995; Possley and Peter Kendall, "Reynolds' Graphic Phone Calls Detailed," *CT,* 25 July 1995; Janita Poe and Graeme Zielinski, "Among Reynolds' Constituents, Frustration Runs Deep," *CT,* 26 July 1995; Hardy, "In Tapes and Trial, Reynolds Is Projecting a Profile in Shame," *CT,* 30 July 1995; "Sexual Assault Trial Puts a Chicago Congressman's Career on the Brink," *NYT,* 30 July 1995; Stephen Braun, "Lawmaker's Trial Rivets Politically Jaded Chicago," *LAT,* 1 August 1995; Steve Neal, "2nd District Race May Get Crowded," *CST,* 30 July 1995, p. 27; Jerry Thomas, "The Son Is Rising," *CT,* 26 May 1995; Ray Gibson, "Reynolds Foe Already Has Raised $51,000 Toward Her '96 Campaign," *CT,* 3 August 1995; Hardy, "Even in August, We're Still Dogged with Politics," *CT,* 6 August 1995; Richard L. Berke, "Behind-the-Scenes Role for a 'Shadow Senator,'" *NYT,* 27 March 1991; Steve Rhodes, "What Does Junior Want?," *Chicago Magazine,* May 2005; Christopher Wills, "Jackson: Trailblazer, Not Obama's Mentor," AP, 7 March 2008; Jason Zengerle, "Jr.: Who Thwarted the Ambitions of Jesse Jackson's Son?," *New York Magazine,* 4 November 2012; DJG interviews with Hal Baron, Brian Banks, Alice Palmer, Kitty Kurth, Darrel Thompson, Delmarie Cobb, Jesse Jackson Jr., and Marty King,

35. Friends of Barack Obama, Form D-2, 1 July 1995–31 December 1996, 2 February 1996; John Chase, "Tony Rezko," *CT,* 26 May 2005, p. 1; Remnick, *The Bridge,* p. 277; DJG interviews with David Brint, Kevin Thompson, Carol Harwell, and Janis Robinson. A month later, on September 6, Barack would inscribe a copy of *Dreams* to Jean Rudd: "You have been such an important person in my life—as a friend, a mentor, a colleague and a sounding board. I am grateful for everything and look forward to our friendship only deepening in the years to come. Love, Barack Obama."

36. *Dreams From My Father* (Times Books, 1995), esp. pp. v, xiii, xv–xvii, 5, 12, 16, 21–22, 29–31, 40, 51, 70, 72, 79, 86, 93–95, 104–5, 135–39, 151–53, 167–80, 189, 195–98, 207–53, 269, 279–85, 291–95, 340–45, 427–30, 434, 437, 440–42 (page citations are to the widely available, 442-page 2004 paperback edition, rather to the very rare 403-page Times Books hardback original); Jeffrey S. Siker, "President Obama, the Bible, and Political Rhetoric," *Political Theology* 13 (September 2012): 586–609, at 588 (highlighting the epigraph and observing that "Obama clearly viewed this passage as an important commentary on his life"); Mike Littwin, "Obama's 'Change' Could Be More Than a Coined Phrase," *Rocky Mountain News,* 29 August 2007; Rebecca Janowitz, *Culture of Opportunity—Obama's Chicago: The People, Politics, and Ideas of Hyde Park* (Ivan R. Dee, 2010), p. 5; Michael Cass, "Obama's Rise Is No Surprise to Nashville Woman Who Worked with Him," *Tennessean,* 5 October 2012; Peter Osnos, "The Making of the Book That Made Obama," DailyBeast.com, 27 January 2009; Robert Draper's interviews with Mary Beth Roche and Jessica Reighard; Draper, "Barack Obama's Work in Progress," *GQ,* November 2009; Barbara Foley, "Rhetoric and Silence in Barack Obama's *Dreams From My Father,*" *Cultural Logic* 2009, pp. 1–46, at 23; Jefferson A. Singer and Peter Salovey, *The Remembered Self: Emotion and Memory in Personality* (Free Press, 1993), p. 157 ("identity may be as"); David B. Pillemer, *Momentous Events, Vivid Memories* (Harvard University Press, 1998), pp. 10, 55, 59, 70, 73, 76, 83; Dan P. McAdams, "What Psychobiographers Might Learn from Personality Psychology," in William T. Schultz, ed., *Handbook of Psychobiography* (Oxford University Press, 2005), pp. 64–83, at 74–75; Pierre-Marie Loizeau, "Barack Obama's Autobiography: A Quest for Personal and Political Identity?," in Benaouda Lebdai, ed., *Autobiography as a Writing Strategy in Postcolonial Literature* (Cambridge Scholars Publishing, 2015), pp. 53–65 at 63–64; Mitchell Aboulafia, *Transcendence* (Stanford University Press, 2010), p. 434; Philip Weiss, "Obama's Dark Side," MondoWeiss.net, 22 April 2008; Lynn Sweet, "Obama's Book: What's Real, What's Not," *CST,* 8 August 2004; Roger Cohen, "The Obamas of the World," *NYT,* 6 March 2008; DJG interviews with Loretta Augustine-Herron, Yvonne Lloyd, Dan Lee, Ralph Dunham, Cathy Askew, Margaret Bagby, Eva Sturgies, Maury Richards, Jerry Kellman, Mike Kruglik, Greg Galluzzo, Alex McNear, Genevieve Cook, Sheila Jager, Mary Ellen Montes, Johnnie Owens, Jeremiah Wright, and Zeituni Onyango.

37. *NYTBR,* 6 August 1995, pp. 21, 17; "Off the Shelf Memories," *Bergen Record,* 21 May 1995, p. E3; *Booklist,* July 1995, p. 1844; "One Man's Story of Race and Inheritance," *California Voice,* 23 July 1995, p. S5 (which also appears under the byline Don Thomas in the *New York Beacon,* 9 October 1995, p. 29, which indicates that it earlier may also have appeared in the *Big Red News*); "Young Writer Chronicles His Search for Better Meaning in His Life," *Sacramento Observer,* 2 August 1995, p. E2; Michael Harris, "Man on a Mission Comes to Terms with His African Roots," *LAT,* 7 August 1995, p. E4; Susan Bickelhaupt and Maureen Dezell, "Living," *BG,* 8 August 1995, p. 62; Malaika Brown, "Mid-Summer Reading Primer," *Los Angeles Sentinel,* 9 August 1995, p. C4; "August at Barnes & Noble," *New York Magazine,* 7 August 1995, p. 7; Obama on *Bill Thompson's Eye on Books,* 9 August 1995, on *Chicago Tonight* with Phil Ponce, WTTW, 22 January 2003, on *Connie Martinson Talks Books,* [11] August 1995, and on *Book Channel* with Marc Strassman, 11 August 1995; Robert Draper's 2009 interviews with Connie Martinson, James Fugate, and Capril Bonner Thomas (whose signed copy, "To Capril: Thanks for your welcoming spirit. Best wishes for the future" is dated 11 August 1995); Draper, "Barack Obama's Work in Progress," *GQ,* November 2009; DJG interview with Pete Vayda and Hasan Chandoo (whose inscribed copy of *Dreams* is dated 8 August 1995).

38. "Kup's Column," *CST,* 11 August 1995, p. 46; Sher in Jonathan Van Meter, "Barack's Rock," *Vogue,* October 2008, pp. 336ff.; Dorothy Collin, "For Ex-Council Wars General Vrdolyak, Talk's a Good Game," *CT,* 6 February 1995; Hermene D. Hartman, "Magic Is in the Air," *N'Digo Profiles,* December 2008, p. 4; Robert Draper's 2009 interviews with Susan Sher, Valerie Jarrett, and Eric Whitaker; Draper, "Barack Obama's Work in Progess," *GQ,* November

2009; Hartman in Janny Scott, "Obama's Story, Written by Obama," *NYT,* 18 May 2008, p. A1; Monice Mitchell, "Son Finds Inspiration in the Dreams of His Father," *HPH,* 23 August 1995, pp. 10, 14; Yvonne Dennis, "Mixed Emotions over Dual Heritage," *Philadelphia Daily News,* 14 August 1995, p. 36; Zachary R. Dowdy, "Across 3 Continents, He Explores Who He Is," *BG,* 14 August 1995, p. 31; Paul Ruffins, "Racism or Attitude?," *WPBW,* 20 August 1995, p. 6; "At Lincoln Library," *State Journal-Register,* 20 August 1995, p. M1; Nich Charles, "Continental Divide," *Emerge,* September 1995, p. 74; Benjamin D. Berry, "Search for Racial Identity Speaks to Many," *Virginian Pilot,* 6 September 1995, p. E3; Eric Clark, "Author Barack Obama Comes to Grips with His Mixed-Race Heritage," *Crisis,* October 1995, pp. 6ff.; Ida Peters, "Books for Christmas Giving and Pleasure," *Baltimore Afro-American,* 23–30 December 1995, p. B4; Kevin Shawn Sealey, *The Black Book Review,* 31 December 1995, p. 9; Dori J. Maynard in About . . . *Time* [Rochester, NY], 28 February 1996, p. 33; Orme in Kirsten Scharnberg and Kim Barker, "The Not-So-Simple Story of Barack Obama's Youth," *CT,* 25 March 2007; Titcomb in Emme Tomimbang, "Barack Obama: Hawaii Roots," 8 February 2008; Furushima in Richard A. Serrano, "Obama Classmates Saw a Smile, but No Racial Turmoil," *LAT,* 11 March 2007, and in Hans Nichols, "In Hawaii, Media Surfs Obama's Past," *Politico,* 14 March 2007; Larissa MacFarquhar, "The Conciliator: Where Is Barack Obama Coming From?," *New Yorker,* 7 May 2007, pp. 48ff.; Obama in Bernard Schoenberg, "Don't Expect Many Niceties in Keyes-Obama Contest," *SJ-R,* 22 August 2004, p. 15, and on *News & Notes,* NPR, 8 February 2005; Glauberman and Burris, *The Dream Begins,* p. 105; Naftali Bendavid, ed., *Obama: The Essential Guide to the Democratic Nominee* (Triumph Books, 2008), p. 32; Kevin Drum, "Getting to Know Obama," *Washington Monthly,* 5 May 2007; Jonathan Raban, "All the President's Literature," *WSJ,* 10 January 2009; David Greenberg, "Hope, Change, Nietzsche," *TNR,* 26 May 2011; David Mendell, "Obama Lets Opponent Do Talking," *CT,* 24 June 2004; Dan Armstrong, "Barack Obama Embellishes His Resume," AnalyzeThis.net, 9 July 2005, and subsequent comment by Jeanne [Reynolds Schmidt]; John Drew, "My White Girlfriend Inspired Obama's Big, Dark Regina in *Dreams From My Father,*" *American Thinker,* 17 July 2012; Gary Kamiya, "Biracial, but Not Like Me," *Salon,* 5 February 2008; Peter Wallsten, "Obama Defined by Contrasts," *LAT,* 24 March 2008, p. A1; Obama on *Off Topic with Carlos Watson,* CNN, 24 October 2004; DJG interviews with Judd Miner, Gwen LaRoche, Barbara Holt, Craig Huffman, Richard Steele, Barack Echols, Ty Wansley, Monice Mitchell Simms, Hermene Hartman, Mike Ramos, Tom Topolinski, Alan Lum, Kelli Furushima, Pal Eldredge, Dan Hale, Cassandra Butts, Cathy Lazere, Susan Arterian Chang, and Greg Galluzzo. See also Raban, "Diary," *London Review of Books,* 20 March 2008, pp. 46–47.

39. Michael Lipkin, "The Professor and the President," *Chicago Maroon,* 7 November 2008; Maurice Possley and Peter Kendall, "Reynolds' Defense Facing Key Decisions," *CT,* 11 August 1995; Bruce Bentley, "Chicago DSA Executive Committee Meeting Minutes," 12 August 1995, DSA Papers Box 116; Thomas Hardy, "Palmer Gains Endorsements in Bid for Reynolds' Seat," *CT,* 14 August 1995; Chinta Strausberg, "Palmer Earns Support for Reynolds' Seat," *CD,* 15 August 1995, p. 7; Possley and Kendall, "Reynolds: 'I Am Not Guilty,'" *CT,* 15 August 1995; Gabriel Kahn, "Reynolds Sticks with 'Phone Sex' Defense," *Roll Call,* 17 August 1995; Possley and Kendall, "Reynolds' Fury Erupts on Stand," *CT,* 17 August 1995; Steve Neal, "Jesse Jackson Jr. in Bid for Congress," *CST,* 21 August 1995, p. 25; Possley and Kendall, "War of Words a Fitting End to Reynolds Trial," *CT,* 22 August 1995; Possley and Kendall, "Reynolds Guilty on All Counts," Nancy Ryan and Laurie Cohen, "Constituents Feel Sad Over Destroyed Career," and Hardy, "Candidates Line Up for Reynolds' Spot," *CT,* 23 August 1995; "Congressman Convicted of Sexual Assault," *NYT,* 23 August 1995; "Hubris, Nemesis and Mel Reynolds," *CT,* 24 August 1995; R. Bruce Dold, "Shattered Dreams: First Gus, Then Mel," *CT,* 25 August 1995; Strausberg, "Reynolds Foes Raring to Go," *CD,* 24 August 1995, p. 3; Dirk Johnson, "In Congressman's District, Conviction Evokes Regret," *NYT,* 24 August 1995; Edward Walsh, "Moseley-Braun Urges Reynolds to Quit," *WP,* 24 August 1995, p. A4; Scott Fornek, "Crush of Candidates Keeps the 2nd Hopping," *CST,* 27 August 1995, p. 20; Steven R. Strahler, "Young and Restless: Meet GOP's Fab 5," *CCB,* 20 May 1995; Strausberg, "2nd District Storm," *CD,* 28 August 1995, pp. 1, 10; Strausberg, "Streeter Weighs Run for Reynolds Seat," *CD,* 29 August 1995, p. 3; Neal, "Palmer Might Have the Right Salve for 2nd District Wounds," *CST,*

30 August 1995, p. 33; Russ Stewart, "Winter Election Makes Jones the Favorite in 2nd District," *Illinois Politics*, September 1995, pp. 8–9; Neal, "Gardner to Help Jones' Bid for Reynolds' Seat," *CST*, 1 September 1995, p. 35; Patricia Callahan and V. Dion Haynes, "Line to Succeed Reynolds Is Running Out the Door," *CT*, 3 September 1995; Strausberg, "Jackson Jr. Wants to Aid Reynolds Family," *CD*, 5 September 1995, p. 5; "Jesse Jackson Jr. Announces Bid for 2nd District Congressional Seat," *South Street Journal*, 7 September–4 October 1995, p. 4; Hardy, "Reynolds Gets Around to Making Resignation Official," *CT*, 9 September 1995; Cornelia Grumman, "Jackson Wears Name Proudly," *CT*, 10 September 1995; "Jesse Jackson's Son to Run for House Seat," *NYT*, 10 September 1995; Strausberg, "Jesse Jr. Throws Hat in Ring," *CD*, 11 September 1995, p. 4; Rick Pearson and Mitchell Locin, "Voting Set on Successor to Reynolds," *CT*, 12 September 1995; Michael Gillis and Fornek, "Primary and Election Dates Set to Fill 2nd District Seat," *CST*, 12 September 1995, p. 18; Strausberg, "Date Set for 2nd District Race," *CD*, 12 September 1995, p. 3; Friends of Alice Palmer, "FEC Form 3, 1 July 1995–8 November 1995," 28 June 1996; Dold, "Paying a Visit to Mel Reynolds—Both of Them," *CT*, 20 December 1996; DJG interviews with Jesse Ruiz, Alice Palmer, Steve Rauschenberger, William McNary, and Tim Wright. Joe Gardner died at age fifty on 16 May 1996. See Jacquelyn Heard and Andrew Fegelman, "Gardner Loses Fight with Cancer," *CT*, 17 May 1996, Kevin Knapp, "Committeeman's Spot Remains Up for Grabs," *HPH*, 22 May 1996, pp. 1–2, and Art Golab, "Admirers Pay Tribute to Joseph Gardner," *CST*, 24 May 1996.

40. Kevin Knapp, "Politicians Scramble in Wake of Reynolds Resignation," *HPH*, 13 September 1995, pp. 1–2; Strausberg, "Group Set to Tab Jones for Reynolds' Seat," *CD*, 13 September 1995, p. 3; Agenda, Chicago New Party Progressive Candidate Forum, 14 September 1995, IAP Box 12 Fld. 13; Obama on *The Oprah Winfrey Show*, 19 January 2005; Palmer in David Jackson and Ray Long, "Obama Knows His Way Around a Ballot," *CT*, 4 April 2007; Obama, *DFMF*, pp. 123–25; Bron Solyom in Dinesh Sharma, *Barack Obama in Hawaii and Indonesia* (Praeger, 2011), pp. 36–37; Robin Toner, "An Issue That Hits Home on the Campaign Trail," *NYT*, 6 July 2007; John McCormick, "Obama's Mother in New Ad," *CT*, 21 September 2007; Ceci Connolly, "President Gets Personal at Forum on Health Care," *WP*, 20 July 2009; Obama, "Remarks at Opening Session of Bipartisan Meeting on Health Care Reform," 25 February 2010, *Public Papers* Vol. 1, p. 256; Scott, *A Singular Woman*, pp. 231, 333–41; Kevin Sack, "Book Challenges Obama on Mother's Deathbed Fight," *NYT*, 13 July 2011; Jane Hall, "Moyers Named President of Two Funding Groups," *LAT*, 8 March 1990; "Florence F. Schumann, A Philanthropist, 99," *NYT*, 26 February 1991; DJG interviews with Carol Harwell, Alice Palmer, Andrew "Pete" Veyda, and Bill Moyers. Barack one week later would call Moyers to decline the board membership, citing his just-launched state Senate candidacy.

41. John Kass and L. Linn Allen, "Daley's Development Chief Leaving to Become Housing Firm Executive," *CT*, 15 September 1995; Larry Hartstein, "Jarrett to Get Nod as CTA Chairman," *CT*, 18 September 1995; Jodi Kantor, *The Obamas* (Little, Brown, 2012), p. 18; Scott Fornek, "Sen. Jones Is Joining the Pack," *CST*, 15 September 1995, p. 6; Veronica Anderson et al., "Who'll Be Standing When the Music Stops Playing?," *CCB*, 18 September 1995, p. 8; Chinta Strausberg, "Barrow Lobbies for Jesse Jr.," and "Harvard Lawyer Eyes Palmer Seat," *CD*, 19 September 1995, p. 3; Monice Mitchell, "Hyde Parker Announces Run for State Senate Seat," *HPH*, 4 October 1995, p. 3; Kennedy Communications, "State Senator Barack Obama, Candidate for U.S. Senate: Background Report," April 2003, RRP; Will Burns's 2008 interview with Jim Gilmore; Adam Doster, "A Thoughtful Politician," *CR*, 16 February 2011; Carol Felsenthal, "Barack Obama Protege Will Burns Begins His Ascent," *CM*, May 2011; DJG interviews with Carol Harwell, Kevin Thompson, Alice Palmer, Alan and Lois Dobry, Toni Preckwinkle, Barbara Holt, David Kindler, John Rogers, Kathy Stell, and Will Burns.

CHAPTER SEVEN: INTO THE ARENA

1. Robert E. Mitchell to Jessica Reighard, 8 May 1995, Concord Festival of Authors Brochure, 21–23 September 1995, Rob Mitchell Papers; Robert Taylor, "Bookmaking," *BG*, 17 September 1995, p. B42; Matthew Brelis, "Colin Powell Book Tour Hits Hub," *BG*, 20 September 1995, p. 32; Obama, remarks at the Cambridge Public Library, 20 September 1995, Cambridge Municipal Television; Joseph P. Kahn, "Powell Fever Grips Hub; Book Signing Draws 2,000,"

BG, 21 September 1995, p. 63; Kennedy Communications, "State Senator Barack Obama, Candidate for the U.S. Senate: Background Report," April 2003, RRP; Steve Neal, "Jones May Deal His Way into Win," *CST,* 20 September 1995, p. 41; Ethan Michaeli, "Palmer States Case for 2nd Dist.," *CD,* 20 September 1995, p. 3; "2nd District Hopefuls," *CD,* 21 September 1995, p. 3; "Candidates for the 2nd Congressional District Speak," *Chicago Weekend,* 24 September 1995, p. 1; Barbara Holt, "New School System Responds to Community's Needs," *HPH,* 27 September 1995, p. 4; Maurice Possley and Peter Kendall, "Congressman Gets 5 Years," *CT,* 29 September 1995: "Candidates for the 2nd Congressional District Speak: Part II," *CW,* 1 October 1995, p. 1; Janita Poe and Thomas Hardy, "Jones Sets Sights on Congress," *CT,* 1 October 1995; Hardy, "Electing Replacement for Reynolds May Touch Off Tussle in Springfield," *CT,* 1 October 1995; Maudlyne Ihejirika, "Senate Leader Runs in 2nd District," *CST,* 1 October 1995, p. 7; Chinta Strausberg, "Jones Officially Enters into 2nd District Race," *CD,* 2 October 1995, p. 5; Neal, "GOP May Give Covert Aid to Jones," *CST,* 3 October 1995, p. 31; Possley and Kendall, "Reynolds Starts Prison Term," *CT,* 6 October 1995; Sunya Walls, "Sen. Emil Jones Announces Run for Congress," *CW,* 8 October 1995, p. 2; Strausberg, "Ghost of Politics Past," *CD,* 9 October 1995, p. 1; Rick Pearson, "14 Candidates Petition to Run for Seat Reynolds Vacated," *CT,* 10 October 1995; Michaeli, "Dead Presidents," *CD,* 11 October 1995, p. 1; Michaeli, "Jones Runs on Record," *CD,* 12 October 1995, p. 3; "The Illinois New Party Is Proud to Endorse Alice Palmer and Willie Delgado," n.d. ("at the October 12 meeting"), IAP Box 12 Fld. 14; "Candidates for the Second Congressional District Speak," *CW,* 15 October 1995, p. 1; Kevin Knapp, "Local Senate Race Heats Up," *HPH,* 25 October 1995, pp. 1, 2; Strausberg, "Palmer: Keep AMA Out of Campaign," *CD,* 25 October 1995, p. 5; Neal, "Palmer Drops to 3rd in 2nd District Race," *CST,* 27 October 1995, p. 37; Neal, "Jones and Jackson Tied for Lead in 2nd District," *CST,* 30 October 1995, p. 27; Strausberg, "Jackson Jr. Claims Registration Drive Netted 5,000 New Voters," *CD,* 31 October 1995, p. 4; Fornek, "Palmer Says She's Not Worried About Polls," *CST,* 1 November 1995, p. 6; Michaeli, "Palmer Gets Boost in 2nd Dist.," *CD,* 1 November 1995, p. 3; Bonnie Miller Rubin, "Jackson's Name, Age Stand Out," *CT,* 3 November 1995; Neal, "Is Jones Making Right Move in 2nd District?," *CST,* 6 November 1995, p. 29; Vladimire Herard, "Candidates Hit the Issues," *CD,* 7 November 1995, p. 3; "IVI-IPO Endorses Palmer for 2nd District Seat," *HPH,* 8 November 1995, p. 3; Salim Muwakkil, "Palmer's Thoughtful Vision Is Salve for Wounds in 2nd," *CST,* 8 November 1995, p. 49; Francis X. Clines, "Black Men Fill Capital's Mall in Display of Unity and Pride," *NYT,* 17 October 1995, p. A1; Clark McPhail and John McCarthy, "Who Counts and How: Estimating the Size of Protests," *Contexts* 3 (August 2004): 12–18; Obama's 19 March 2008 interview with ABC News' Terry Moran; DJG interviews with Rob Mitchell, Jesse Jackson Jr., Marty King, Malik Nevels, Mike Lieteau, Don Rose, Alan and Lois Dobry, Lou Viverito, Hal Baron, Sue Purrington, Kitty Kurth, Paul Williams, Willie Delgado, Janis Robinson, and Carol Harwell. On Don Rose, see Dennis L. Breo, "Confessions of a Political Gadfly," *CT Magazine,* 13 December 1987.

2. Friends of Barack Obama, Report of Campaign Contributions and Expenditures, Form D-2, 1 July–31 December 1995, 2 February 1996; Robert Davis, "Bit Subdued, Radical Reemerges," *CT,* 12 August 1988; Patricia Lear, "Rebel Without a Pause," *CM,* May 1993, pp. 65–69, 96–100; Susan Chira, "At Home with Bernardine Dohrn," *NYT,* 18 November 1993; Tom Johnson to Obama, 27 October 1995, TEJP; Scott Fornek, "Foot Soldiers for the '96 Elections," *CST,* 29 October 1995, p. 14; Chinta Strausberg, "Palmer Challenger Says He Won't Step Aside in Race," *CD,* 21 December 1995, p. 3; Obama's 2001 interview with Julieanna Richardson; Maria Warren, "Get to Know Barack Obama," WarrenPeaceMuse.blogspot.com, 27 January 2005; David Jackson and Ray Long, "Obama Knows His Way Around a Ballot," *CT,* 4 April 2007; Deanna Bellandi (AP), "Obama's Neighborhood Rich in Diversity," 23 July 2007; Ben Smith, "Obama Once Visited '60s Radicals," *Politico,* 22 February 2008; Peter Slevin, "Former '60s Radical Is Now Considered Mainstream in Chicago," *WP,* 18 April 2008; Joanna Weiss, "How Obama and the Radical Became News," *BG,* 18 April 2008; Christopher Wills (AP), "Obama Faced Tough Choice in First Legislative Race," 24 April 2008; Ron Grossman, "Family Ties Proved Ayers' Point," *CT,* 18 May 2008; Andrew Ferguson, "Mr. Obama's Neighborhood," *Weekly Standard,* 16 June 2008; Carol Felsenthal, "Barack Obambi?," *Huffington Post,* 1 July 2008; Ryan Lizza, "Making It: How Chicago Shaped Obama," *New Yorker,* 21 July 2008; Norman Kelley, "The '2 A.M. Booty Call': Q&A with Adolph Reed Re Obama and American Politics," DevilsAdvocateDivision.blogspot.com, 9 August 2008; Tim Harper, "Civil Rights Elders Aided Obama's Political Rise," *Toronto Star,* 18 August 2008, p. A1; Pauline Yearwood, "Obama and the Jews," *Chicago Jewish News,* 12 September 2008; Drew Griffin and Kathleen Johnston, "Ayers and Obama Crossed Paths on Boards, Records Show," CNN.com, 7 October 2008; Lynn Sweet, "McCain Misleading Public on Role Ayers Played in Obama Political Career," *CST,* 15 October 2008; Hilary Leila Krieger, "Mr. Obama's Neighborhood," *Jerusalem Post,* 23 October 2008; "William Ayers Sets the Record Straight," *Good Morning America,* ABC, 14 November 2008; Margalit Fox, "Arnold Jacob Wolf, a Leading Reform Rabbi, Is Dead at 84," *NYT,* 29 December 2008; Marc S. Allen, "From Adversary to Longtime Ally," BlackVoices.com, 19 January 2009; Dick Lovell, "Obama Is 'An Extraordinary Person,' Friend Jeffrey Cummings Tells Rotarians," *Madison Rotary News* #44, 1 May 2009, p. 1; Ayers in Remnick, *The Bridge,* p. 280–81; McClelland, *Young Mr. Obama,* pp. 114–16; Bernardine Dohrn, "Seize the Little Moment: Justice for the Child—20 Years at the Children and Family Justice Center," *Northwestern Journal of Law and Social Policy* 6 (Spring 2011): 334–57; Carol Felsenthal, "Obama Protege Will Burns Begins His Ascent," *CM,* May 2011; Will Burns's 2008 interview with Jim Gilmore; DJG interviews with Bernard Loyd, Carol Harwell, Janis Robinson, Sunny Fischer, Ellen Benjamin, Jesse Ruiz, Jeanne Gills, Brett Hart, Eric Pelander, Jeff Cummings, John Belcaster, Greg Dingens, Steve Derks, Michael Parham, Alice Palmer, Allison Davis, Tim Evans, Kitty Kurth, Bill Ayers, Bernardine Dohrn, Rosellen Brown, Mona Khalidi, Carole Travis, Quentin Young, Ken Warren, Judd Miner, Tom Johnson, George Galland, Will Burns, Kathy Stell, Barbara Holt, Al Kindle, Tim Black, and Adolph Reed. On Leon Finney and Lincoln South Central Real Estate, see Angela Caputo, "Following Finney," *Chicago Reporter,* January 2012; also see Antonio Olivo, "Questions Raised About Leon Finley's Woodlawn Organization," *CT,* 6 January 2012.

3. Mark PoKemper photos, ca. 6–7 November 1995, MPP; Hank DeZutter, "What Makes Obama Run?," *CR,* 8 December 1995, pp. 1ff.; Ellen Schumer notes, 8 November 1995, and Schumer to Obama, 10 November 1995, ESP; Scott Fornek, "Barack Obama," *CST,* 1 March 2004, p. 6; Notice of Federal Tax Lien Under Internal Revenue Laws, #91-155273, 12 November 1991; Obama, *DFMF,* pp. xi–xii, 124–25; Ron Jacobs, *Obamaland* (Trade Publishing, 2008), p. 79; Obama on *Situation Room,* CNN, 8 May 2008; David Maraniss, "Though Obama Had to Leave to Find Himself, It Is Hawaii That Made His Rise Possible," *WP Magazine,* 24 August 2008; Remnick, *The Bridge,* pp. 286–88; Abramsky, *Inside Obama's Brain,* p. 104; Dinesh Sharma, "She Had a Dream," *Asia Times,* 16 January 2010; Scott, *A Singular Woman,* pp. 343–46; DJG interviews with Mark PoKemper, Gerry Alexis, Ellen Schumer, Charles Payne, and Bronwen and Garrett Solyom.

4. Ellen Schumer to Obama, 10 November 1995 (with further 21 November annotation), ESP; Lisa Ely, "Candidates Move Ahead as Election Draws Near," *Chicago Citizen,* 9 November 1995, p. 1; Chinta Strausberg, "Latest Support Is Music to Ears of Jackson Jr.," *CD,* 9 November 1995, p. 3; John Kass, "Mob-Linked Union Paid Jackson," *CT,* 10 November 1995, p. C1; Scott Fornek, "2nd District Candidates Rip into Each Other," *CST,* 10 November 1995, p. 17; Thomas Hardy, "Jones Happy to Run as Political Insider," *CT,* 12 November 1995, p. C1; Strausberg, "Jones Builds a Bigger Base," *CD,* 13 November 1995, p. 3; Jesse Jackson Jr. TV Ad, 13 November 1995, JJJP; Steve Neal, "Slogans Fly as 2nd District Race Picks Up Steam," *CST,* 14 November 1995, p. 31; "Orr Backs Palmer for Reynolds' Seat," and "Another Chance in the 2nd District," *CT,* 14 November 1995; Bonnie Miller Rubin, "Palmer Defined by, and Penalized for, Independence," *CT,* 15 November 1995, p. C1; Strausberg, "Palmer Defends Wife Against Davis Attacks," *CD,* 15 November 1995, p. 21; Fornek, "2nd District Candidates Talk Taxes," *CST,* 16 November 1995, p. 22; Mitchell Locin, "Davis Runs Against All Odds Again," *CT,* 17 November 1995, p. C1; Hardy, "Jackson Jr. Leads in Race to Replace Reynolds," *CT,* 19 November 1995; Hardy, "Jackson Dismisses Youth Issue," *CT,* 20 November 1995; Fornek, "Charges Fly in 2nd Dist. Race," *CST,* 20 November 1995, p. 9; Strausberg, "Second Dist. Candidates Battle Over Polls, Issues," *CD,* 20 November 1995, p. 5; Ethan Michaeli, "Jackson Leads Money Race," *CD,* 21 November 1995, p. 1; Joanne Esters-Brown, "2nd District Debate Turns into Mudslinger," *HPH,* 22 November 1995, p. 1; Strausberg, "Jackson Hits Jones Campaign Literature in 2nd Dist. Race," *CD,*

22 November 1995, p. 3; Gary Mays and Hardy, "Jackson's Bid Clouded by New Support," *CT,* 23 November 1995; Hardy, "2nd District Primary May Be Turkey," *CT,* 24 November 1995; "Emil Jones Deserves Election to Congress," *CST,* 24 November 1995, p. 53; Don Terry, "In House Election, a Familiar Name," *NYT,* 24 November 1995; Kass, "Ex-Gang Member Adds Fuel to Campaign Fire in 2nd," *CT,* 25 November 1995; Hardy, "Variety Is Theme in 2nd District," *CT,* 26 November 1995; Art Golab and Michelle Campbell, "2nd District Candidates Trade Accusations," *CST,* 26 November 1995, p. 9; Benjamin Sheffner, "On the Special Election Trail," *Roll Call,* 27 November 1995; Michaeli, 'Palmer Takes Campaign into Home Stretch," and Strausberg, "Palmer Disputes 'Soft on Crime' Charge," *CD,* 27 November 1995, pp. 1, 5; Teresa Puente and Patricia Callahan, "2nd District Foes Preach Hope," *CT,* 27 November 1995; Strausberg, "Candidates: Jobs at Stake," *CD,* 28 November 1995, p. 1; Fornek, "Weather May Be a Deciding Factor Today at the Polls," Michael Briggs, "Jones Gets Big Boost of Money at the End," and Steve Neal, "For 2nd District Voters, It's Youth vs. Experience," *CST,* 28 November 1995, pp. 9 and 31.

5. Jesse Jackson Jr. with Frank E. Watkins, *A More Perfect Union: Advancing New American Rights* (Welcome Rain, 2001), p. 39; Burney Simpson, "Computers and Cash Boot Up Jackson's New Machine," *Chicago Reporter,* January 1996; Minutes, CAC Board of Directors, 28 November 1995, 3:15 P.M., CACP; Sabrina Miller, "Schools Get Big Bucks to Fund Reform," *CT,* 20 December 1995; Lorraine Forte, "35 Networks Get First Annenberg Funds," *Catalyst Chicago,* February 1996; Deanna Bellandi (AP), "U of I at Chicago Opens Ayers Records," 27 August 2008; Thomas Hardy and Bonnie Miller Rubin, "Jesse Jackson Jr. Rolls Over Veteran Opponents," *CT,* 29 November 1995; Scott Fornek, "Jesse Jr. Wins," *CST,* 29 November 1995, p. 1; "Jesse Wins," and Ethan Michaeli, "Palmer Concedes in Dem Primary," *CD,* 29 November 1995, pp. 1 and 3; "Jesse Jackson Jr. Wins Primary in Chicago," *NYT,* 29 November 1995; Kevin Knapp, "Another Entry in State Senate Race, *HPH,* 29 November 1995, p. 1; Benjamin Sheffner, "His Last Name Proves Golden for Jesse Jackson Jr.," *Roll Call,* 30 November 1995; Hardy, "Jackson Names Just Spells Victory," *CT,* 30 November 1995; Rick Bryant, "Jesse Jr. Comes Up Aces," *Southtown,* 30 November 1995; Chinta Strausberg, "Somer, Jesse to Do Battle," *CD,* 30 November 1995, p. 1; Lisa Ely et al., "Jackson Wins Primary Bid to Congress," *Chicago Citizen,* 30 November 1995, p. 1; Russ Stewart, "Daley Has Big Stake in 7th District," *Illinois Politics,* December 1995, pp. 9–10; Strausberg, "Jackson Jr. Campaign Heats Up," *CD,* 4 December 1995, p. 3; Strausberg, "Draft Palmer Campaign Launched," *CD,* 5 December 1995, p. 4; David K. Fremon, *Chicago Politics Ward by Ward* (Indiana University Press, 1988), pp. 260, 267; Ellen Schumer note, "12/5—Barack Obama—751-1170," 5 December 1995, ESP; Rick Bryant, "Palmer Will Seek Re-Election After All," *Daily Southtown,* 9 December 1995, p. A7; Kevin Knapp, "Palmer May Re-Enter State Senate Race," *HPH,* 13 December 1995, p. 1; Scott Fornek, "'I've Got a Competitive Nature,'" *CST,* 3 October 2004, p. 12; Adolph Reed Jr., "Obama No," *Progressive,* May 2008; DJG interviews with Kitty Kurth, Alan Dobry, John Charles, Dan Shomon, Carol Harwell, Emil Jones Jr., Lula Ford, Jesse Jackson Jr., Marty King, Hal Baron, Paul Williams, Jesse Ruiz, Howard Stanback, Bob Klonowski, Madeline Talbott, Sunny Fischer, Alice Palmer, Don Wiener, Barbara Holt, Tim Black, Adolph Reed, Michael Dawson, Willie Delgado, William McNary, and Nia Odeoti-Hassan. "My husband to this day," Palmer volunteered in 2011, "he's never forgiven me" for promising to give up her seat.

6. Hank DeZutter, "What Makes Obama Run?," *CR,* 8 December 1995, pp. 1ff.; John H. Sibley, "Obama's Myopia," *CR,* 22 December 1995; Brent Watters, "Accomplished Black Chicago Author Explores the Subject of Race," *Hyde Park Citizen,* 28 December 1995, p. 3; DJG interview with Hank DeZutter.

7. Bonnie Miller Rubin, "Democracy Costs Time and Money," *CT,* 12 December 1995; Thomas Hardy and William Presecky, "Win for Both," *CT,* 13 December 1995; Steve Daley, "Jackson Jr. Out of the Shadows, into Spotlight," *CT,* 15 December 1995; Kevin Knapp, "List of Next Year's Candidates Is Sparse," *HPH,* 13 December 1995, pp. 1–2; Tom Johnson fax to Ron Davis or Carol Harwell, 28 November 1995, TEJP; Friends of Barack Obama, Report of Campaign Contributions and Expenditures, Form D-2, 1 July 1995–31 December 1995, 2 February 1996; Chinta Strausberg, "Jackson Jr. to File Petitions Today," *CD,* 18 December 1995, p. 22; Adam Lashinsky et al., "One Good Thing," *CCB,* 18 December 1995, p. 8; Strausberg, "Palmer OKs Draft to Run for Re-Election," *CD,*

19 December 1995, p. 3; Thomas Hardy, "Jackson Foe Now Wants Old Job Back—Palmer Must Now Battle Own Endorsee," *CT,* 19 December 1995; "The Unzip" and "State Senator Alice J. Palmer Announces Run for Re-Election," *South Street Journal,* 19 December 1995–4 January 1996, pp. 2, 9; Knapp, "Palmer Caught in Campaign Draft," and "Candidates File Petitions for Local Political Offices," *HPH,* 20 December 1995, pp. 1–2; Ellen Schumer notes (on lunch with Barack), 20 December 1995, ESP; Strausberg, "Palmer Challenger Says He Won't Step Aside in Race," *CD,* 21 December 1995, p. 3; Ronald Davis, Notarized Statement, 22 December 1995, TEJP; "Candidates Prepare to Wage Battles Over Nominating Petitions," *CT,* 24 December 1995; Sunya Walls, "Alice Palmer Decides to Run for Re-Election," *Chicago Weekend,* 25 December 1995, p. 2; Judy Hevrdejs and Mike Conkling, "Early Christmas for Bill Melton," *CT,* 25 December 1995; Mathias W. Delort, Odelson and Sterk, "Appearance," *Davis v. Palmer,* #96-EB-SS-09, 28 December 1995; Thomas E. Johnson, "Appearance," *Davis v. Ewell,* #96-EB-SS-11, 2 January 1996, TEJP; Kevin Knapp, "Candidates Face Petition Challenges," *HPH,* 3 January 1996, p. 3; Raymond Ewell fax to Thomas E. Johnson, and including Marc Ewell, "Motion to Strike Objector's Petition," 3 January 1996, TEJP; Saul Mendelson, "Independents Defend Their Pre-Election Work," *HPH,* 7 February 1996, pp. 4, 14; David Jackson and Ray Long, "Obama Knows His Way Around a Ballot," *CT,* 4 April 2007; Leslie Hairston, "Tribute to Late Mr. Ronald L. Davis," *Chicago City Council Journal,* 14 May 2008, pp. 15–16; Drew Griffin, *Newsroom,* CNN, 29 May 2008; David Freddoso, "Obama Played by Chicago Rules," *WSJ,* 20 August 2008, p. A19; Michelle Norris, "How Chicago Politics Shaped Obama," *All Things Considered,* NPR, 16 October 2008; Will Burns's 2008 interview with Jim Gilmore; Rickey Hendon, *Black Enough/White Enough: The Obama Dilemma* (Third World Press, 2009), p. 10; Remnick, *The Bridge,* pp. 291–92; McClelland, *Young Mr. Obama,* p. 118; George E. Condon Jr., "A Sense of Self," *National Journal,* 22 September 2012, pp. 16–19; DJG interviews with Delmarie Cobb, Jesse Jackson Jr., Alice Palmer, Will Burns, Carol Harwell, Emil Jones, Mike Hoffmann, Dan Shomon, John Charles, Tim Black, Paul Williams, Alan and Lois Dobry, Ray Harris, Barbara Flynn Currie, Ellen Schumer, Mike Pfleger, Jacky Grimshaw, Jan Schakowsky, Dave Gross, Courtney Nottage, Tom Johnson, and Kathy Stell. Also see Steven R. Strahler, "Jackson's PUSH Comes to Shove," *CCB,* 8 April 1996, p. 1, and Dale Eastman, "The Rise of Jesse Jackson Jr. and the First Family of Black America," *Chicago Magazine,* May 1996.

8. *Lumpkin v. Envirodyne Industries,* 933 F.2d 933 (7th Cir.), 17 May 1991, cert. denied, 502 U.S. 939, 4 November 1991; Paul Galloway, "Steel Workers' Labor of Love," *CT,* 4 April 1994, p. T1; R. C. Longworth, "Forged in Steel, Soon Forgotten," *CT,* 30 March 1995, p. 1; Longworth, "Man of Steel," *CT,* 31 May 1995, p. T1; Longworth, "Wisconsin Steel Deal May End Ex-Workers' Agony," *CT,* 9 December 1995, p. 1; "Envirodyne, Steel Workers OK Settlement," *CST,* 10 December 1995, p. 10; "Wisconsin Steel," *Daily Southtown,* 9 December 1995, pp. A1, A7; Tamara Kerrill, "Wisconsin Steel Veterans Accept 10% of Their Due," *CST,* 3 January 1996, p. 6; DJG interviews with Tom Geoghegan, Tom Johnson, and Bea Lumpkin.

9. Saul Mendelson and Lois Friedberg-Dobry, "Independents Defend Their Pre-Election Work," *HPH,* 7 February 1996, pp. 4, 14; Ellen Schumer note, "1-5-96—Barack Obama," ESP; Adolph Reed Jr., "The Curse of Community," *Village Voice,* 16 January 1996; Saul [Mendelson] to Alice [Palmer], 25 January 1996, Dobry Papers and TEJP; Kevin Knapp, "Local Independent Voters Still Divided," *HPH,* 31 January 1996, pp. 1, 3; David Jackson and Ray Long, "Obama Knows His Way Around a Ballot," *CT,* 4 April 2007; McClelland, *Young Mr. Obama,* p. 119; DJG interviews with Alan and Lois Dobry, Kathy Stell, Tom Johnson, and Carol Harwell.

10. Barack Obama, Impact Endorsement Questionnaire, 7 January 1996, 3pp., and Obama to Trudy Ring, Outlines, 15 February 1996, both Web-posted in conjunction with Tracy Baim, "Obama Changed Views on Gay Marriage," *Windy City Times,* 14 January 2009, p. 6, and both reprinted in Baim, *Obama and the Gays: A Political Marriage* (Prairie Avenue Productions, 2010), pp. 190, 192–93.

11. Tom Johnson notes, 8 January 1996, 4 P.M., TEJP; Kevin Knapp, "Petition Challenges Shape Political Ballot," *HPH,* 10 January 1996, pp. 1–2; Michael Sneed, "Sneed," *CST,* 11 January 1996, p. 4; Madeline Talbott, "Response to 'What Kind of Democracy Are We Building in the New Party?,'" 6 January 1996, IAP Box 11 Fld. 4; "Chicago New Party 1996 Goals," n.d., IAP Box 10 Fld. 23; Agenda, New Party Membership Meeting, 11 January 1996, IAP Box 10

Fld. 24; "Minutes, New Party Meeting (Draft)," 11 January 1996, IAP Box 10 Fld. 23; *New Party News* Vol. 5, #2, Spring 1996 ("New Party members" who won spring primaries include Barack); Bruce Bentley, "New Party Update," *New Ground* 47, July–August 1996 (noting Barack's attendance at an 11 April New Party membership meeting, where he "encouraged NPers to join his task forces on Voter Education and Voter Registration"); Jim Cullen, "Editorial: The Next Campaign," *Progressive Populist,* November 1996 ("New Party member Barack Obama"); [Madeline Talbott], "Proposal for Maintaining a City-Wide Structure," November 1996, IAP Box 12 Fld. 11; [Madeline Talbott], "Chicago New Party Chairperson's Report Year End '96/ Year Begin '97," n.d., IAP Box 10 Fld. 103; Illinois New Party Membership List, 3 January 1997, IAP Box 10 Fld. 118 (no listing of Obama, esp. per pp. 11, 17, 20, and 22–23); "Targeting Resources to Electoral Districts," n.d. [ca. February 1997], IAP Box 11 Fld. 1; "New Party—Not Paid Up," 17 April 1997, IAP (2010-080) Box 10 Fld. 50, p. 8 ("Obama Barack Date Join 1/11/96"); Carl Davidson, "Carl Davidson on Barack O'bomb'em," 21 January 2007, mailman .lbo-talk.org; Carl Davidson Web comment per Ron Radosh, 18 July 2012, pjmedia.com; Jackson and Long, "Obama Knows His Way Around a Ballot," *CT,* 4 April 2007; John Nichols, "How to Push Obama," *The Progressive,* January 2009; Stanley Kurtz, "Obama's Third-Party History," NationalReview.com, 7 June 2012; Ben Smith, "Obama and the New Party," Buzzfeed.com, 8 June 2012; Rosie Gray and Ben Smith, "In Chicago, No Memory of Obama New Party Membership," Buzzfeed.com, 12 June 2012; Kurtz, "Obama on the Fringe," *National Review,* 25 June 2012, pp. 25–27; DJG interviews with Madeline Talbott, Keith Kelleher, Joel Rogers, and Sarah Siskind.

12. Kevin Knapp, "Primary Ballot Gets Face-Lift from Board," *HPH,* 17 January 1996, p. 1; Thomas E. Johnson, "Notice of Filing" and "Objector's Motion to Strike Candidate's Rule 8 Motion," *Davis v. Palmer,* #96-EB-SS-09, 17 January 1996, TEJP; Nancy Ryan and Thomas Hardy, "Sen. Palmer Ends Bid for Re-Election," *CT,* 18 January 1996; "Palmer Out Again," *CST,* 18 January 1996, p. 14; Chinta Strausberg, "Palmer Throws in Towel," *CD,* 18 January 1996, p. 3; Rick Bryant, "Palmer Pulls Her Hat from 13th District Ring," *Daily Southtown,* 18 January 1996, p. A4; Johnson notes, 20 January 1996, TEJP; Sunya Walls, "Alice Palmer Withdraws from Race for Re-Election," *Chicago Weekend,* 21 January 1996, p. 3; Cook County Electoral Board, "Decision," *Davis v. Palmer,* #96-EB-SS-09, 23 January 1996, 2pp., TEJP; Knapp, "Final Primary Ballot Taking Shape," *HPH,* 24 January 1996, p. 1; *Ulmer D. Lynch Jr. v. State Board of Election, Chicago Board of Election Commissioners, Thomas E. Johnson & Barack Obama,* #96-C-0270, N. D. Ill., 24 January 1996; Cook County Electoral Board, "Remand to Hearing Examiner," *Davis v. Ewell,* #96-EB-SS-11, 31 January 1996; Johnson to Obama, 30 January 1996 ("enclosed is the Election Board's decisions removing Alice Palmer and Ulmer Lynch from the ballot"), 30 January 1996; James M. Scanlon to Johnson, 5 February 1996, Johnson to Obama, 5 and 6 February 1996, Johnson, "Answer to Amended Complaint," *Lynch v. State Board of Elections,* #96-C-0270, N. D. Ill., 7 February 1996, Cook County Electoral Board, "Decision," *Davis v. Ewell,* #96-EB-SS-11, 9 February 1996, Johnson to Obama, 12 February 1996 ("enclosed are the decisions knocking Ewell and Askia off the ballot"), all TEJP; *Marc Ewell v. Board of Election Commissioners,* #96-C-0823, N. D. Ill., 13 February 1996; Charles R. Norgle Sr., "Minute Order," *Lynch v. State Board of Elections,* #96-C-0270, N. D. Ill., 16 February 1996; Johnson to Ron Davis, 19 February 1996, TEJP; Knapp, "Ex-Candidates Sue Election Board," *HPH,* 21 February 1996, p. 1; Raymond W. Ewell, "Memorandum of Law in Support of Petition for Temporary Restraining Order," James M. Scanlon, "Answer," Thomas E. Johnson, "Barack Obama and Ronald Davis' Motion to Intervene as Defendants," Johnson, "Intervening Defendants' Answer and Defense," and Ruben Castillo, "Minute Order," *Ewell v. Board of Election Commissioners,* #96-C-0823, 21 February 1996, TEJP; Starks in Remnick, *The Bridge,* p. 292; Hendon, *Black Enough/White Enough,* p. 10; Hendon's 2010 interview with Peter Slen; DJG interviews with Howard Carroll and Don Wiener.

13. Kevin Knapp, "Local Independent Voters Still Divided," *HPH,* 31 January 1996, pp. 1, 3; Senate Floor Transcript, 8 February 1996, p. 2 (SB 1554); Salim Muwakkil, "Candidate Not What He Seems, Foes Insist," *CST,* 12 February 1996, p. 29; Minutes, CAC Board of Directors Executive Committee, 22 February 1996, CACP; Ellen Schumer notes, 5 January 1996, ESP; Bruce Bentley, "Chicago DSA Executive Committee Meeting Minutes," 20 January

1996, DSAP Box 116; Byron White and William Mullen, "U. of C. Is Losing Its Top Urban Sociologist," *CT,* 8 February 1996; Flyer, "Chicago DSA Membership Meeting," and Bruce Bentley, "Chicago DSA Membership Meeting Minutes," 8 February 1996, DSAP Box 116; "Hyde Park Happenings," *HPH,* 21 February 1996, p. 21; Kate Olson, "Town Meeting Addresses Economic Problems," *Chicago Maroon,* 27 February 1996, pp. 1, 4; Bob Roman, "A Town Meeting on Economic Insecurity: Employment and Survival in Urban America," and "Chicago DSA Endorsements in the March 19th Primary Election," *New Ground,* March–April 1996; Joel Bleifuss, "Red-Boating Obama," *In These Times,* 11 March 2008; Henry Louis Gates Jr., "A Conversation with William Julius Wilson on the Election of Barack Obama," *Du Bois Review* 6 (March 2008): 15–23, at 18 (Barack was "very, very engaging" when Wilson first met him at the home of Tracy Meares); DJG interviews with Tim Wright, John Schmidt, and Katie Romich.

14. Raymond W. Ewell, "Memorandum of Law in Support of Plaintiff's Standing to Bring Instant Action," James M. Scanlon, "Defendant's Motion for Summary Judgment," and Thomas E. Johnson, "Intervening Defendants' Motion for Summary Judgment," *Ewell v. Board of Election Commissioners,* #96-C-0823, N. D. Ill., 26 February 1996, TEJP; Knapp, "Court Rules Candidate Off Ballot," *HPH,* 28 February 1996, pp. 1–2; Ewell, "Plaintiff's Response to Defendants' and Intervenor's Motion for Summary Judgment," and Scanlon, "Defendant's Response to Plaintiff's Memorandum of Law in Support of Plaintiff's Standing to Bring Instant Action," *Ewell v. Board of Election Commissioners,* #96-C-0823, N. D. Ill., 29 February 1996, *Ewell v. Board of Election Commissioners,* #96-C-0823, N. D. Ill., 4 March 1996, Johnson to Obama, 4 March 1996, Ulmer D. Lynch, "Plaintiffs' Second Amended Complaint," *Lynch v. State Board of Elections,* #96-C-0270, N. D. Ill., 26 February 1996, Charles R. Norgle Sr., "Minute Order," *Lynch v. State Board of Elections,* #96-C-0270, N. D. Ill., 11 March 1996, Johnson to Obama, "Ewell and Lynch," 11 March 1996, all TEJP; "The Unzip" and "South Street Journal Endorsements," *South Street Journal,* 14–27 March 1996, pp. 2, 13; Kennedy Communications, "State Senator Barack Obama, Candidate for U.S. Senate: Background Report," April 2003, RRP; Chinta Strausberg, "Several Candidates Run Unopposed," *CD,* 18 March 1996, p. 19; Illinois State Board of Elections, *Official Vote Cast at the Primary Election, General Primary, March 19, 1996,* p. 109; Knapp, "Election Day Yields a Few Surprises," *HPH,* 27 March 1996, pp. 1–2; "Senate Hopeful Still Has Plenty to Spend," *HPH,* 3 April 1996, pp. 1–2; Friends of Barack Obama, Report of Campaign Contributions and Expenditures, 1 January 1996–30 June 1996, 31 July 1996; DJG interviews with Tom Johnson, Janis Robinson, and Carol Harwell.

15. UCLS *Announcements 1995–96,* p. 38; UCLS *Directory 1995–96;* UCLS Course Evaluations, Current Issues in Racism and the Law, Spring 1996; Maya in Emme Tomimbang, "Barack Obama: Hawaii Roots," 8 February 2008, DVD; Delanney in Yann Masson, "Presidentielle Amercaine: Avec Barack, on Regardait Les Matchs de Basket," *Nice Matin,* 6 November 2008; *Tyus v. Schoemehl,* 93 F.3d 449 (8th Cir.), 16 August 1996 (argued 9 April 1996), cert. denied, 520 U.S. 1166 (1997); *ACORN v. Edgar,* 75 F.3d 304 (7th Cir.), 26 January 1996; *ACORN v. Edgar,* 1996 WL 284959, N. D. Ill., 28 May 1996; David [Plotkin] to Staff, "Re: Legal," 29 May 1996, HSP Box 35 Fld. 1494; *ACORN v. Edgar,* 1996 WL 406652, N. D. Ill., 17 July 1996; *ACORN v. Edgar,* 1996 WL 447256, N. D. Ill., 1 August 1996; *ACORN v. Edgar,* 99 F. 3d 261, 262 (7th Cir.), 31 October 1996 (Illinois's behavior in the case "borders on the frivolous"); Paul Jaskunas, "EEOC v. Mistubishi," *American Lawyer,* July–August 1996, pp. 95–96; Phil Vettel, "Park Avenue Cafe Deserves Attention," *CT,* 19 July 1996; Baird in Alexandra Starr, "Case Study," *NYT Magazine,* 21 September 2008, on "Obama: Professor President," BBC Radio 4, 6 January 2009, and on CNBC video interview, 12 January 2009; DJG interviews with Mary Ellen Callahan, Hisham Amin, Triste Lieteau Smith, Bob Bennett, Judd Miner, George Galland, Paul Strauss, Jeff Cummings, Laura Tilly, John Belcaster, Douglas Baird, Geof Stone, David Strauss, and Dennis Hutchinson.

16. Public Allies Chicago, *Check-In,* Fall 1995; "Obama Appointed to New University of Chicago Job," *HPH,* 5 June 1996, p. 5; "Obama Named First Associate Dean of Student Services," *University of Chicago Chronicle,* 6 June 1996; "Any Volunteers?," *University of Chicago Magazine,* August 1996; Hilary F. Ryder, "New Faculty Arrive at U of C," *Chicago Maroon,* 23 September 1995, p. B15; Public Allies Chicago, *Check-In,* Fall 1996 and Spring 1997; Maudlyne Ihejirika, "Program Turning Xers Into Leaders," *CST,* 27 May 1997, p. 46; Michelle in Holly Yeager, "The Heart and Mind of Michelle Obama," *O Magazine,* November 2007, pp. 286ff., in Geraldine Brooks, "Michelle Obama: Camelot 2.0?," *More,* October 2008, and in Michael Scherer and Nancy Gibbs, "Interview with the First Lady," *Time,* 21 May 2009; Paul Schmitz, *Everyone Leads* (Jossey-Bass, 2012), pp. 51–53, 243–45; Carl A. and Shelby J. Grant, *The Moment* (Rowman & Littlefield, 2013), p. 150; Mariana Cook, "A Couple in Chicago," *New Yorker,* 19 January 2009; Slevin, *Michelle Obama,* pp. 162, 171; DJG interviews with Pat Graham, Julian Posada, Bethann Hester, Jan Anne Dubin, Ian Larkin, Sunny Fischer, Craig Huffman, Chuck Supple, Leif Elsmo, Paul Schmitz, Kathy Stell, Kevin Tyson, Carlton Guthrie, and Deloris Burnam.

17. David Moberg, "Last Train Out: GM's Electro-Motive Plant Has Assembled Its Final Locomotive," *CT,* 24 January 1993; Jon Anderson "Weathering Change," *CT,* 8 July 1993, p. T1; Tara Gruzen, "'60s Radicals Try to Activate Students," *CT,* 17 April 1996; Scott Turow, "The New Face of the Democratic Party—and America," *Salon,* 30 March 2004; Peter Wallsten, "Allies of Palestinians See a Friend in Barack Obama," *LAT,* 10 April 2008; Marc Santora and Elissa Goctman, "Political Storm Finds a Columbia Professor," *NYT,* 31 October 2008; Scott Turow's 2009 interview with Robert Draper; Carol Felsenthal, "Bill Ayers Talks Obama," *Chicago Magazine,* December 2013; DJG interviews with Bill Ayers, Bernardine Dohrn, Mona Khalidi, Carole Travis, John Ayers, Scott Turow, and Tom Geoghegan. On Rebecca Jumper Coleman Shields, who died 24 January 1988, see Rachel L. Swarns, *American Tapestry* (Amistad, 2012), esp. pp. 87–109.

18. Minutes, CAC Board of Directors, 2 May 1996, Ken Rolling to Vartan Gregorian et al., "Chicago Annenberg Challenge Program Report," 8 May 1996, Gregorian to Obama, 28 May 1996, Box 114 Fld. 765, Minutes, CAC Board of Directors, 18 June 1996, CAC, "Program Report for the Period Ending December 31, 1996," CACP; Shipps and Sconzert, *The Chicago Annenberg Challenge: The First Three Years,* March 1996, esp. p. 19; Joyce Foundation, *Work in Progress,* January, May, and September 1996; Kenneth P. Vogel, "Obama Linked to Gun Control Efforts," *Politico,* 20 April 2008; Abramsky, *Inside Obama's Brain,* p. 235; Adrian Butler, "Barack's British Background," *Sunday Mirror,* 13 January 2008; Paul Burnell, "Why Obama Came to My Stag Do," *The One Show,* BBC, 29 January 2008; Martin Fricker and Graham Brough, "Barack Obama's Stepmother Is a Bingo Lover from Bracknell," *Daily Mirror,* 9 February 2008; Oliver Harvey, "Obama's Brother Is in Bracknell," *The Sun,* 26 July 2008; Auma Obama, *And Then Life Happens,* p. 233; DJG interviews with Ken Rolling, Pat Graham, Carin Clauss, Paula Wolff, Carlton Guthrie, Jack Anderson, Chuck Daly, Debbie Leff, Margaret O'Dell, Unmi Song, Rob and Lisa Fisher, Mark Kozlowski, John Charles, Jill Rock, Cynthia Miller, Carole Travis, Michael McGann, Jeff Stauter, Pat Welch, Carol Harwell, and Malik Nevels. At some point in the 1990s, "Michelle and I . . . decided we were going to lease a car," Barack later told Elizabeth Warren. "When I returned the car four years later . . . it was the first time I realized how expensive that car was." Ron Suskind, *Confidence Men* (HarperCollins, 2011), p. 445.

19. Tracy P. Leary to Rev. Vincent McCutcheon, 12 February 1996, "Hope Center Profiles Spring Session 1996," n.d., TLP; "Hope Institute Receives Grant to Organize Local School Election," *South Street Journal,* 14–27 March 1996, p. 2; "Hope Center Spring Session Practicums," n.d., "Hope Center Spring Session Overview of Participation," n.d., Tracy P. Leary to Mrs. Nicholson, 24 May 1996, Leary, "Weekly Report (June 11th–June 18th)," 19 June 1996, "The Lugenia Burns Hope Center Presents a Community Forum Guest Speaker: Robert 'Bob' Moses," [14 June 1996], Leary, "Weekly Report (June 19th–25th)," TLP; [Sandy O'Donnell], "The Lugenia Burns Hope Center: Purposes, Description, First Year Progress, Second Year Plans," n.d. [ca. late June 1996], SOP; Tracy Leary 1996 appointments book, esp. 2 July and 23 August 1996, Leary, "Weekly Report (June 26th–July 2nd)," 3 July 1996, Leary, "Weekly Report (July 3rd–9th)," 16 July 1996, Leary to "Dear ____" Form Letter, 11 July 1996, Leary, "Weekly Report (July 10th–16th)," 17 July 1996, TLP; [Sandy O'Donnell], "Hope Center Planning Group Notes," 18 July [1996], SOP; Leary, "Weekly Report (July 31st–August 6th)," 7 August 1996, Leary, "4120 [S. Prairie] Weekly Report (August 21st–26th)," Leary, "Weekly Report (August 21st–28th)," 28 August 1996, The Lugenia Burns Hope Center for the Study and Development of New World Community, "Building a Collective Organizing Strategy for Grand Boulevard," 29 August 1996 (draft agenda?), Leary, "Weekly Report (August 28th–September 3rd)," 4 September 1996, TLP; Joe Frolik, "A Newcomer to the Business of Politics," *Plain Dealer,* 3 August 1996, p. A1; Beverly Reed, "Senator Obama's Bronzeville Connection," *South Street Journal,* 17 January 2008, p. 14; DJG interviews with Tracy Leary, Paul Scully, Sokoni Karanja, and Mike Kruglik.

20. *Dreams From My Father* (Kodansha Globe), 1996; Tim Engles, *Magill Book Reviews,* March 1997; Janny Scott, "Obama's Story, Written by Obama," *NYT,* 18 May 2008, p. A1; Philip Turner, "'Dreams From My Father' and Kodansha Globe, 1995–96," philipsturner.com, 11 March 2012; DJG interviews with Philip Turner, Deborah Baker, and Rob and Lisa Fisher. Two subsequent reviews are Dorothy S. Latiak, "Reflections on the Life of Obama," *HPH,* 23 June 1999, p. 4, and E. J. Graff, "Being Black and White," *American Prospect,* 10 September 2001, pp. 42ff. (dismissing Obama's narrative as "unfocused meandering" and "boring writing").

21. "IMPACT Questionnaire," 3 September 1996, in Baim, *Obama and the Gays,* pp. 9–11; Tracy Ring, "Elections Set for Illinois House, Senate," *Outlines,* November 1996; "IVI-IPO General Candidate Questionnaire," Dobry Papers and DWP; IVI-IPO, "Candidate Report Sheet for Interviews, Barack Obama," 9 September 1996, DWP; "Two Candidates to Challenge Obama for State Senate Seat," *HPH,* 25 September 1996, p. 2; Mike Allen and Ben Smith, "Liberal Views Could Haunt Obama," *Politico,* 11 December 2007; Lynn Sweet, "The Past Haunts Obama?," *CST,* 12 December 2007, p. 26; Smith, "Chicago Activists Doubt Obama's Explanation on Questionnaire," *Politico,* 19 December 2007; Kenneth P. Vogel, "Obama Had Greater Role on Liberal Survey," *Politico,* 31 March 2008; DJG interviews with Alan and Lois Dobry and Don Wiener. The existence of incomplete versions of the IVI-IPO questionnaire on the Web has in the past led to confusion.

22. UCLS *Directory 1996–1997,* esp. p. 20; UCLS *Announcements, 1996–1997,* esp. pp. 37 and 40; Chapin Cimino, "Class-Based Preferences in Affirmative Action Programs After *Miller v. Johnson:* A Race-Neutral Option, or Subterfuge?," *University of Chicago Law Review* 64 (Fall 1997): 1289–1310; Geoffrey R. Stone et al., *Constitutional Law,* 3rd ed. (Little Brown, 1996); UCLS Course Evaluations, Autumn 1996; "Candidates as Academics: I Got a Professor Named Obama," 8 February 2008, and "U of C '97" comment, 2:28 P.M., 11 February 2008, AbovetheLaw.com; John K. Wilson, *Barack Obama: The Improbable Quest* (Paradigm, 2008), pp. iv, 80; DJG interviews with Barack Echols, David Strauss, Felton Newell, Chapin Cimino, Dennis Hutchinson, Trieste Lieteau Smith, and Hisham Amin.

23. "Please Join Us for an Evening of Art, Jazz & Politics with Barack Obama," 3 October 1996, DTKP; Friends of Barack Obama, Report of Campaign Contributions and Expenditures, 1 July 1996–31 December 1996, 31 January 1997; Harvard Law School, Class of 1991 Fifth Anniversary Report, 1996, p. 73 (Obama entry dated 3 October 1996); Tracy Leary 1996 appointments book, esp. 24 September 1996; Leary to "Dear Friend," 27 September 1996, Leary, "Weekly Report (Nov. 6th–12th)," 12 November 1996, Leary to "Dear Friend," 13 November 1996, TLP; Regina McGraw, "Loud Way Sometimes the Best Way," *CT,* 1 October 1996; Woods Fund of Chicago, 1995 and 1996 Annual Reports; DJG interviews with David Kindler, Sokoni Karanja, Tracy Leary, Sandy O'Donnell, John Eason, Jean Rudd, Todd Dietterle, and Josh Hoyt. As Hoyt rightly states, "church-based organizing was incredibly successful" and "became the basis of a huge amount of empowerment in the Latino community, and it never really took root in the black community."

24. "Chance to Learn About Volunteer Opportunities," *UCC,* 10 October 1996; Brenda Warner Rotzoll, "Lessons in Volunteerism," *CST,* 13 October 1996, p. 30; Kevin Knapp, "Volunteering Becomes a Full-Time Job for Obama," *HPH,* 8 January 1997, p. 2; "Public-Service Internships Available Through Summer Links," *UCC,* 20 March 1997; Lisa Fingeret, "University to Fund Volunteerism," *HPH,* 2 April 1997, p. 2; "Volunteerism Thriving Among Students," *UCC,* 12 June 1997; Charlotte Snow, "Connecting Communities: New Campus Volunteer Efforts Show Promise of Broadening Undergraduate Perspectives and Improving Town-Gown Relations," *UCM,* June 1998; Snow, "Student Activists Raise Signs Over Sweatshops," *UCM,* June 2000; Russ Stewart, "The Future of the Republican Senate," *Illinois Politics,* October 1995, pp. 9, 14; Stewart, "Big Democratic 'Wave' Needed to Win State Senate," *Illinois Politics,* October 1996, pp. 11–12; Jennifer Halperin, "On Target: State Senate Democrats Count on Shifts in the South Suburbs," *II,* October 1996, pp. 22–25; "Illinois Senate Endorsements," *CT,* 16 October 1996; Kevin Knapp, "Two 'Stealth Candidates' on Ballot in State Senate Race," *HPH,* 23 October 1996, p. 3; "Our Endorsements for Illinois Senate," *CST,* 27 October 1996, p. 39

(and repeated 3 and 5 November); Paul Krawzak, "Two Candidates, One Senator," *SJ-R,* 27 October 1996, p. 1; John Drumwright, "U of C Professor Closes In on Illinois State Senate Seat," *The Phoenix,* 4 November 1996, p. 1; Illinois State Board of Elections, Official Vote Cast at the General Election, 5 November 1996, p. 32 (Barack 48,592, David Whitehead 7,461, Rosette Caldwell Peyton 3,091); Michael Tackett, "Except in Illinois," *CT,* 10 November 1996; Judy Hevrdejs and Mike Conklin, "Devine's Key Asset in Victory," *CT,* 12 November 1996; "House Changes Hands in Election," *Cook-Witter Report* 12, #10, 22 November 1996; *CF,* 9 April 1999 (Halvorson's victory was "the single greatest political accomplishment by the Senate Democrats" during the 1990s); "Rosette Caldwell Peyton, 78," *HPH,* 2 May 2007, p. 19; Dan Shomon's 2008 interview with CNN's Jen Christensen; Debbie Halvorson, *Playing Ball with the Big Boys* (Solutions Unlimited, 2012), p. 26; DJG interviews with Dan Shomon, Debbie Halvorson, Terry Link, Tom Walsh, Denny Jacobs, Cindi Canary, Glenn Hodas, John Charles, Linda Hawker, Lori Joyce Cullen, and Cindy Huebner Davidsmeyer.

25. Maudlyne Ihejirika, "Black Officials Push for State-Level Activism," *CST,* 10 November 1996, p. 6; Kevin Knapp, "Election Holds No Surprises for Local Candidates," *HPH,* 13 November 1996, p. 2; Craig A. Roberts, Paul Kleppner, et al., *Almanac of Illinois Politics—1996* (University of Illinois at Springfield, 1996), pp. 76–77 (13th District); Melanie Carr, "Law School Professor Barack Obama Wins State Senate Seat," *The Phoenix,* 18 November 1996, p. 2; Minutes, CAC Board of Directors, 2 December 1996 (Obama absent); Kennedy Communications, "State Senator Barack Obama, Candidate for U.S. Senate: Background Report," April 2003, RRP; "Tentative Legislative Schedule for 1997 Spring Session," *CWR* 12 #11, 18 December 1996; UCLS *Announcements 1996–97,* p. 65; *King v. State Board of Elections,* 979 F. Supp. 582 (N. D. Ill.), 7 March 1996, remanded, 519 U.S. 978, 12 November 1996; Sarah Nordgren (AP), "Lower Court Ordered to Take Another Look at Hispanic District," 12 November 1996; *King v. State Board of Elections,* 979 F. Supp. 619 (N. D. Ill.), 1 August 1997, affirmed 522 U.S. 1087, 26 January 1998; Samuel Issachaoff, Pamela S. Karlan, and Richard H. Pildes, *The Law of Democracy: Legal Structure of the Political Process* (Foundation Press, 1998), esp. p. viii; Obama, Constitutional Law III Final Examination, 12 December 1996; Obama to Students in Con Law III, "The Exam," n.d., law professors.typepad.com; Pildes in Jeffrey Toobin, "Bench Press," *New Yorker,* 21 September 2009; John R. Lott Jr. and David B. Mustard, "Crime, Deterrence, and Right-to-Carry Concealed Handguns," UCLS Law and Economics Working Paper #41, August 1996; "Concealed Weapons Deter Crime, Study Says," *CT,* 10 August 1996; Stephen Chapman, "Taking Aim," *CT,* 15 August 1996; Lott, "More Guns, Less Violent Crime," *WSJ,* 28 August 1996; Lott, "Obama and Guns: Two Different Views," Fox News.com, 7 April 2008; Grover G. Norquist and Lott, *Debacle* (John Wiley, 2012), pp. 1–2; Jamie Weinstein, "Gun Rights Advocate and Former Obama Colleague: President Used to Treat Him as 'Evil,'" *Daily Caller,* 8 April 2012; Lott, *At the Brink* (Regnery Publishing, 2013), pp. xv, 126; Adam Bonin, "Professor Obama and Me," DailyKos.com, 20 December 2007; Bonin in Lydialyle Gibson, "Elemental Obama," *University of Chicago Magazine,* September–October 2008, p. 45; Alexandra Starr, "Case Study," *NYT Magazine,* 21 September 2008; David Franklin in Jodi Kantor, "As a Professor, Obama Held Pragmatic Views on Court," *NYT,* 3 May 2009, and in Jeffrey Rosen, "Race to the Top," *TNR,* 6 May 2009; UCLS Course Evaluations, Voting Rights, Winter 1997; DJG interviews with Rick Pildes, John Lott, Molly Garhart, Carl Patten, and Adam Bonin.

26. Senate Transcript, 90th General Assembly, 8 January 1997, pp. 3–4, 18; Doug Finke and Bill Bush, "Power Sharing Begins," *SJ-R,* 9 January 1997, p. 1; Jean Latz Griffin, "Senate Rookies Soon Learn to Go with the Flow," *CT,* 26 January 1995; Michael Madigan's richly revealing 2009 interview with Robert V. Remini, p. 13; Paul Simon (as told to Alfred Balk), "The Illinois Legislature: A Study in Corruption," *Harper's,* September 1964, pp. 74–78; Simon, "Cleaning Up the Illinois Legislature: A Follow-Up Report," *Harper's,* September 1965, p. 125; Paul Simon, *P.S.: The Autobiography of Paul Simon* (Bonus Books, 1999), pp. 41, 62–69, 104–5, 232, 307; Robert E. Hartley, *Paul Powell of Illinois: A Lifelong Democrat* (Southern Illinois University Press, 1999), esp. pp. xi, 19, 34, 134, 140, 143, 149, 185, 193; Margaret Ramirez, "Lawrence N. Hansen, 1940–2010," *CT,* 16 November 2010; Joyce Foundation, *Work in Progress,* January 1997; Kent D. Redfield, *Cash Clout: Political Money in Illinois Legislative Elections* (University of Illinois at Springfield,

1995), esp. pp. 5, 11, 150–51; Doug Finke et al., *Illinois for Sale: Do Campaign Contributions Buy Influence?* (University of Illinois at Springfield, 1997), esp. pp. 121, 188; Redfield and Paul J. Quick, *Reforming Campaign Finance in Illinois: Issues and Prospects* (University of Illinois, July 1998), pp. 2, 6–7; Redfield, *Money Counts* (University of Illinois, 2001), pp. vii–xi; James L. Merriner, "The Four Tops," *Illinois Issues,* November 1996, pp. 16–20; Michael Caldwell, "The New Pattern of Power in Illinois: Effective Government, Deadlock, or Tyranny?," (IGPA) *Policy Forum* 11 (1997): 1–6; Redfield, "What Keeps the Four Tops on Top," in David A. Joens and Paul Kleppner et al., *Almanac of Illinois Politics—1998* (University of Illinois at Springfield, 1998), pp. 1–8; Gary MacDougal, *Make a Difference* (St. Martin's Press, 2000), p. 235 ("Springfield is full of people who care a lot more about making money from government one way or another than they care about good government"); Obama phone number card, 1997, Timuel Black Papers Box 129 Fld. 1 (listing Cynthia's hours as M–Th 12:30–4:30 P.M., Fridays 9 A.M. to noon, and Saturdays 10 A.M. to 2 P.M.); David Kidwell and John Chase, "Favorable Legislation Flows to Private Clients of House Speaker Madigan," *CT,* 3 June 2012; Chase, Kidwell, and Ray Long, "Madigan Consolidates Power by Holding Sway in Legislative Races," *CT,* 5 June 2012; Will Caskey, "Why Are Illinois Democrats So Awful?," *CCB,* 31 January 2013 ("we suck *because* of strong party leadership"); DJG interviews with Jim Collins, Carole Travis, Dave Joens, Mike Ragen, Boro Reljic, Jan Schakowsky, Barbara Flynn Currie, Willie Delgado, Tom Dart, Donne Trotter, Lisa Madigan, and Cynthia Miller.

27. Michael D. Klemens, "Pate Philip: Leader of Senate Minority and DuPage's GOP Machine," *Illinois Issues,* March 1989, pp. 13–16; Steve Neal, "Few Are Neutral About GOP Leader Philip," *CST,* 19 January 1990, p. 29; Neal, "Philip Makes Name as Backward Bully," *CST,* 13 July 1992, p. 21; Ted Gregory, "Philip Poised to Lead State Senate," *CT,* 4 November 1992; Jennifer Halpern, "Pate Philip," *II,* March 1993, pp. 13–15; Robert Becker and Susan Kuczka, "'Pate' Philip Knocks Minorities," *CT,* 6 October 1996; Ray Long and Michael Gillis, "Racial Remarks by Philip Spark Political Storm," *CST,* 7 October 1994, p. 4; "Pate Philip's Racism and the Work Ethic," *CST,* 7 October 1994, p. 41; John K. Wilson, "The Bigot Who Runs Illinois," *Chicago Ink,* August 1997; Rick Pearson, "Pate," *II,* October 1997, pp. 14–17; *Capitol Fax,* 17 May 1999, p. 2; Carter Hendren's 2009 interview with Mark DePue; Thomas Hardy, "Sen. Jones Ahead in Race to Be Democratic Leader," *CT,* 16 December 1992; Rick Pearson, "Daley Settles Senate Fight—For Now," and Pearson and John Kass, "Jones Started Schmoozing City Hall Early On," *CT,* 17 December 1992; Pearson, "Democratic Senate Squabble Goes Along with Jones' Selection," *CT,* 18 December 1992; Pearson, "Democrats in Senate Rebuff Chief," *CT,* 27 January 1993; Jennifer Halpern, "Emil Jones Passes Muster as New Senate Democratic Leader," *II,* August–September 1993, pp. 26–29; Jones's 2000 interview with Julieanna Richardson and his 2006 one with Bernard Sieracki; Paul Davis, "State Senator Emil Jones, Jr.," *N'Digo,* 6–12 February 2003, pp. 6–7; Tracey Robinson-English, "Statehouse Clout," *Ebony,* April 2005, pp. 150ff.; Ben Smith, "Emil Jones Blowback," *Politico,* 5 February 2007; Jones's 2008–2009 interviews with Erma Brooks-Williams; Bernie Schoenburg's 2009 interview with Mark DePue, p. 98; Jim Edgar's 28 May 2010 (#15, p. 678) interview with DePue; DJG interviews with John Hooker, James "Pate" Philip, Emil Jones Jr., Linda Hawker, Larry Bomke, Steve Rauschenberger, Carl Hawkinson, Dan Cronin, John Nicolay, Walter Dudycz, Dave Syverson, Donne Trotter, Howard Carroll, Pat Welch, John Cullerton, Robert Molaro, Miguel del Valle, Chuy Garcia, and William Luking. Translations of *The Art of War* 8.10 differ widely on the second phrase of Jones's quote; some render it as "but rely on your own preparations."

28. *Capitol Fax,* 28 December 1998 and 27 April 1999; MacDougal, *Make a Difference,* p. 255; Bernie Schoenburg's 2009 interview with Mark DePue, p. 65; Jim Edgar's 28 May (#15, pp. 676–78) and 2 September (#20, p. 878) 2010 interviews with DePue; "Edgar Talks About Madigan," *Capitol Fax,* 27 November 2012; Halvorson, *Playing Ball with the Big Boys,* pp. 43–44 (describing Edgar as "a remote, aloof figure" even toward Republican senators); DJG interviews with Pate Philip, Jim Edgar, Kirk Dillard, Carter Hendren, Scott Kaiser, Deno Perdiou, Andy Foster, Kurt DeWeese, Linda Hawker, and Donne Trotter. On Madigan's long-term power base, see Rick Pearson, "What Is Mike Madigan Up To?," *II,* April 1997, pp. 12–17; Madigan's 2009 interview with Remini; David Kidwell and John Chase, "Favorable Legislation Flows to Private Clients of House Speaker Madigan,"

CT, 3 June 2012; Chase, Kidwell, and Ray Long, "Madigan Consolidates Power by Holding Sway in Legislative Races," *CT,* 5 June 2012; James Ylisela Jr., "Michael Madigan Is the King of Illinois," *Chicago Magazine,* December 2013; and Kidwell et al., "How Madigan Built His Patronage Army," *CT,* 5 January 2014.

29. Brent Watters, "Obama Pledges to Be Hardest Working Senator in Springfield," *Chicago Citizen,* 16 January 1997, p. 4; "Barack Obama, Senior Lecturer," www.law.uchicago.edu, 24 January 1997; Jennifer Davis, "State Politicians Have Promised to Work Together, Get Things Done," *Illinois Issues,* February 1997, pp. 6–7; "Edgar's Seventh State of the State," *CWR* 12 #12, 7 February 1997 (22 January); Illinois Campaign Finance Project, *Tainted Democracy: How Money Distorts the Election Process in Illinois and What Must Be Done to Reform the Campaign Finance System* (University of Illinois at Springfield), 29 January 1997, pp. 4, 8, 11, 14, 36; Joyce Foundation *Annual Report 1996,* esp. pp. 26–29; Ed Wojcicki, "Still the Wild West?: A 10-Year Look at Campaign Finance Reform in Illinois," Southern Illinois University, September 2006, pp. 5–6; Senate Floor Transcripts, 90th General Assembly, 5 February 1997, pp. 9, 14 (SBs 398, 399 and 472), 6 February 1997, p. 7 (SBs 574 575, 576, 577, 578, and 579), 7 February 1997; Peter Slevin, "Obama Forged Political Mettle in Illinois Capitol," *WP,* 9 February 2007, p. A1; Janny Scott, "In Illinois, Obama Proved Pragmatic and Shrewd," *NYT,* 30 July 2007, p. A1; Christopher Wills (AP), "Old-School Politician Mentored Obama," 31 March 2008; Eli Saslow, "From Outsider to Politician," *WP,* 9 October 2008, p. A1; Kevin Knapp, "Developers Make No Small Plans in North Kenwood," *HPH,* 5 February 1997, pp. 1–2; Lisa Fingeret, "Connected Developers Strike Out," *HPH,* 12 March 1997, pp. 1–2; Carol McHugh Sanders, "Former Law Firm Partner Takes Firm Stands About Housing," *Chicago Lawyer,* April 1997, pp. 73ff.; Tim Novak, "Obama's Ex-Boss a Rezko Partner," *CST,* 23 April 2007, p. 25; Miner, Barnhill & Galland letterhead, 5 February 1997, IAP Box 11 Fld. 3; Miner in Mike Robinson (AP), "Obama Got Start in Civil Rights Practice," 20 February 2007, in Abdon M. Pallasch, "Obama's Legal Career," *CST,* 17 December 2007, p. 4, and in Pauline Yearwood, "Obama and the Jews," *Chicago Jewish News,* 12 September 2008; Obama, "Help Needed to Change Springfield," *HPH,* 19 February 1997, pp. 4, 9; Maudlyne Ihejirika, "City Ranks Low in a Survey on Kids' Well-Being," *CST,* 19 February 1997, p. 20; DJG interviews with Dave Manning, Ed Wojcicki, Cindi Canary, Ray Harris, Bill Perkins, Sharon Corrigan, Dan Burkhalter, Emil Jones Jr., Judd Miner, Chuck Barnhill, George Galland, and Paul Strauss.

30. Rebecca Riley et al., Futures Committee, to Dear Colleague, 23 January 1997, "Futures Committee Moves Forward," *Looking Forward,* February 1997, Futures Committee, "Healthy Communities: How Do We Know Them When We See Them?," 14 February 1997, "Kemmis, Obama Provides Food for Thought at First Futures Committee Symposium," *Looking Forward,* March 1997, Andrew Mooney, "Changing the Way We Do Things: Recommendations and Findings of the Futures Committee, October 1997, all LISC Chicago Papers; Patrick Barry, "History: Our Quarter-Century Journey," LISC *2005 Donors Report;* Newton N. Minow to JoAnn Benini, 26 February 1997, NMP; Miller in Dan Morain, "Obama: A Fresh Face or an Old-School Tactician?," *LAT,* 8 September 2007; CAC, Program Report for the Period Ending December 31, 1996, CACP; Patrick Keleher, "River of School Grants Won't Wash Away Failure," *CST,* 19 February 1997, p. 41; Raymond R. Coffey, "School Reform Pioneer Decries 'Lost Decade,'" *CST,* 20 February 1997, p. 6; Coffey, "Daley, Vallas as Reform's Best Hope," *CST,* 21 February 1997, p. 6; Minutes, CAC Board of Directors Executive Committee, 24 February 1997, CACP Box 114 Fld. 770; CAC Board of Directors Minutes, 2 April 1997; Ann Bradley, "Staying in the Game," *Education Week,* 30 April 1997, pp. 23ff.; Dirk Johnson, "Chicago Schools Set Standard in Insisting Students Perform," *NYT,* 6 June 1997, p. A1; Shipps and Sconzert, *The Chicago Annenberg Challenge: The First Three Years,* pp. 25–26; DJG interviews with Marilyn Katz, Cynthia Miller, Newton Minow, Janis Robinson, and Ken Rolling.

31. Rauschenberger in Chris Bailey, "Contemplating Steve and Barack, and Oh, What Might Have Been," *CDH,* 19 September 2004, p. 18; Eli Saslow, "From Outsider to Politician," *WP,* 9 October 2008, p. A1; Jim Edgar's 14 December 2010 (#26) interview with Mark DePue, p. 1087; Halvorson, *Playing Ball with the Big Boys,* p. 32; Dan Shomon in Philip Sherwell, "The Big Hitter," *Telegraph,* 28 January 2007, Hoffmann in Rick Pearson and Ray Long, "Careful Steps, Looking Ahead," *CT,* 3 May 2007; Shomon in Kevin McDermott, "In Illinois, Obama Made Waves from the Start," *SLPD,* 10 December

2007; Shomon's July 2008 interview with Jen Christensen; Ben Calhoun, "1997 Trip Sparked Obama's Rural-Voter Courtship," NPR, 10 August 2008 (and WBEZ, 8 August); Shomon on "Barack Obama Revealed," CNN, 20 August 2008; "Dan Shomon—First Person," BBC News, 26 August 2008; Shomon in Sherwell, "Revealed: Barack Obama, the Boring Speaker," *Telegraph,* 26 October 2008; Remnick, *The Bridge,* pp. 317–18; DJG interviews with Carl Hawkinson, Frank Watson, Deno Perdiou, Jim Edgar, Walter Dudycz, Patty Schuh, Steve Rauschenberger, Debbie Halvorson, Howard Carroll, Denny Jacobs, Art Berman, Evelyn Bowles, Miguel del Valle, Chuy Garcia, Robert Molaro, Lori Joyce Cullen, Linda Hawker, Jill Rock, Mike Hoffmann, Dave Gross, Cindy Huebner Davidsmeyer, Dave Joens, Adelmo Marchiori, John Charles, Dan Shomon, and Cynthia Miller.

32. DJG interviews with Carl Hawkinson, John Cullerton, Ed Petka, Kirk Dillard, Dan Cronin, Robert Molaro, George Shadid, Greg Sullivan, Jim Collins, Boro Reljic, Mike Marrs, Gideon Baum, John Nicolay, Matt Jones, Nia Odeoti-Hassan, Dave Syverson, Christine Radogno, Kathy Parker, Debbie Lounsberry, Linda Hawker, John Bouman, William McNary, Doug Dobmeyer, Dave Stover, David Weisbaum, Dave Joens, Kurt DeWeese, and Peter Fitzgerald. See also Bethany Krajelis, "Lobbyist's Rise to Top 'Speaks Volumes,'" *CDLB,* 27 April 2010 (Nicolay), and Rick Pearson, "Kirk Dillard Reloads," *CT,* 21 February 2014.

33. Emil Jones Jr. in George E. Condon Jr., "A Sense of Self," *National Journal,* 22 September 2012, pp. 16–19; John Cameron in Abramsky, *Inside Obama's Brain,* pp. 89–94; Shomon in John F. Harris and Glenn Thrush, "The Ego Factor: Can Obama Change?," *Politico,* 5 November 2010; Ray Long, "Stanley B. Weaver, 78," *CT,* 13 November 2003; Obama in Jennifer Davis, "Freshmen Ponder Spring Lessons as They Return for the Fall Session," *II,* October 1997, pp. 6–7; Obama's 2001 interview with Juliette Richardson and his 23 October 2006 American Magazine Conference one with David Remnick; DJG interviews with Emil Jones Jr., Lori Joyce Cullen, Dave Gross, Mike Hoffmann, Adelmo Marchiori, Dan Shomon, and Craig Huffman.

34. Christi Parsons and Courtney Challos, "Ban on Abortion Procedure Wins a Senate OK," *CT,* 5 March 1997; David Hein, "Abortion Bill Heads to Vote in Senate," *BP,* 5 March 1997, p. A5; Heather Ryndak, "Ban on Partial-Birth Abortion Clears Panel," *CST,* 5 March 1997, p. 70; Research Department, National Republican Senatorial Committee, "Barack Obama (D-IL): On the Record," 22 July 2004, p. 10, Jack Ryan for U.S. Senate Opposition Research Files (JRORF); Senate Floor Transcript, 7 March 1997, p. 4 (Barack introducing SB 1183 regarding job training); Cathy D. Gardner, "Sen. Obama Takes Aim at Edgar's Budget," *CD,* 11 March 1997, p. 5; Senate Floor Transcript, 13 March 1997, pp. 112–23; [Debbie] Lounsberry, "Barack Obama—The Record: A Summary," 10 May 2004, p. 40 (noting Barack's 13 March "present" vote on SB 853, which passed 54–0), JRORF; David Heckelman, "Bill Would Direct Elder Jurors to Court Nearest Home," *CDLB,* 14 March 1997, pp. 1, 18; Rickey Hendon, *Black Enough/White Enough: The Obama Dilemma* (Third World Press, 2009), p. 179. Illinois State Senate floor proceedings are easily available in high-quality daily transcripts at www.ilga.gov, and for the 91st General Assembly (1999–2000) onward, Senate Journals, containing the texts of amendments, conference committee reports, and motions, plus floor votes, are also available at www.ilga.gov. "The Senate does not tape its committee hearings," although caucus staff members sometimes did so informally. See Richard C. Edwards, "Researching Legislative History," Illinois Legislative Reference Bureau, October 2008.

35. Kirk Dillard in Sharon Cohen (AP), "Barack Obama Straddles Different Worlds," 19 October 2007; Paul Williams in Ariel Sabar, "For Obama, Bipartisan Aims, Party-Line Votes," *CSM,* 17 April 2008; Hendon on the Illinois Channel with Terry Martin, 13 January 2009; DJG interviews with Linda Hawker, Debbie Halvorson, Pat Welch, Carol Harwell, Cynthia Miller, Paul Williams, Cindi Canary, James Clayborne, Nia Odeoti-Hassan, Miguel del Valle, Chuy Garcia, Barbara Flynn Currie, Vince Williams, Ray Harris, Lou Viverito, and Terry Link.

36. Senate Floor Transcript, 19 March 1997, p. 190 (Barack stating that he erroneously voted no rather than yes on SB 1000 on 18 March); 20 March 1997, pp. 50–66 and 210 (Barack noting on the latter that earlier he erroneously voted present rather than yes on SB 700); Daniel T. Zanoza, "Senator Shows Courage," *CDH,* 29 March 1997, p. I-7; "Edgar's Budget Proposal," *CWR* 13 #2, 4 April 1997; "Senator Obama's Bills Pass Senate," *South Street Journal,* 4–24 April 1997, p. 4; Nathan Gonzales, "The

Ever-'Present' Obama," *WSJ,* 14 February 2007; Sam Youngman, "Abortion Foes Target Obama Because of His Vote Record on Illinois Legislation," *The Hill,* 15 February 2007; "Obama Abortion Dodges Blessed by Planned Parenthood," ABCNews.com, 17 July 2007; Dave McKinney, "Hillary Slams Obama 'Present' Votes on Abortion, Gun Laws," *CST,* 4 December 2007; Raymond Hernandez and Christopher Drew, "It's Not Just 'Ayes' and 'Nays,'" *NYT,* 20 December 2007; Richard Wolffe and Karen Springen, "The Incremental Revolutionary," *Newsweek,* 21 January 2008, pp. 38ff.; David Schaper, "Examing Obama's 'Present' Votes in Illinois," *All Things Considered,* NPR, 23 January 2008; Peter Wallsten, "Obama Said Oops Six Times on Votes as a State Senator," *LAT,* 24 January 2008 (noting 18 and 20 March 1997); Daniel C. Vock, "'Present' Votes Defended by Illinois Lawmakers," Stateline.org, 25 January 2008; Eli Saslow, "From Outsider to Politician," *WP,* 9 October 2008; DJG interviews with Pat Welch, Vince Williams, Mike Lieteau, and Willie Delgado.

37. Steven R. Strahler, "Chicago Observer," *CCB,* 24 March 1997, p. 42; Lisa Fingeret, "Political Organization Celebrates 30 Years of Spaghetti," *HPH,* 2 April 1997, p. 3; Chinta Strausberg, "Obama Says Jobs Could End Child Abuse," *CD,* 8 April 1997, p. 22; Jeff Zeleny, "Hoover Trial Resonates in Many Ways," *CT,* 25 March 1997; Obama, "Rethinking Approach to Juvenile Crime," *HPH,* 16 April 1997, p. 4; Liza Mundy, "A Series of Fortunate Events," *WP Magazine,* 12 August 2007, pp. 10ff.; E. J. Dionne Jr., "Full Faith," *TNR,* 9 April 2008; *Capitol Fax,* 5 February 1998 ("last spring"); DJG interviews with Tom Dart, Dave Manning, Frank Clark, Mike Lieteau, Pate Philip, John Corrigan, Cindy Huebner Davidsmeyer, Linda Mills, Pam Sutherland, Christine Radogno, Cindi Canary, and Carole Travis.

38. Bob Merrifield, "Senate Job Is Up to Walsh," *CT,* 12 January 1997; Merrifield, "Party Boss Looks at Senator Post," *CT,* 4 April 1997; Merrifield, "Democratic Chief Named State Senator," *CT,* 8 April 1997; Nancy Ryan, "Whites at Ease in Minority Role," *CT,* 8 June 1997; *Capitol Fax,* 21 January 1998 ("most lobbying expenses must be reported to the state, which creates a written record"); Rick Pearson and Ray Long, "Careful Steps, Looking Ahead," *CT,* 3 May 2007; Obama on *This Week,* ABC, 13 May 2007; Christopher Wills (AP), "Clues to How Obama Would Play His Hand as President Can Be Found in Poker Style," 24 September 2007; Wills (AP), "Obama's Complex History with Lobbyists," 1 December 2007; James McManus, "Aces," *New Yorker,* 4 February 2008; Eli Saslow, "From Outsider to Politician," *WP,* 9 October 2008; Link in "Obama as We Knew Him," *Observer,* 26 October 2008, pp. 4–7; Dennis Yohnka, "President Beckons Friend Walsh to Washington for Inauguration," *KDJ,* 22 January 2009; John Sullivan et al., "Public Officials Found Helping Clients of Family," *NYT,* 7 January 2012; DJG interviews with Larry Walsh, Mike Lieteau, Phil Lackman, Dave Manning, Terry Link, Triste Lieteau Smith, and John Bouman.

39. Rudd in Janny Scott, "In Illinois, Obama Proved Pragmatic and Shrewd," *NYT,* 30 July 2007, p. A1; Link in "Obama as We Knew Him," *Observer,* 26 October 2008, pp. 4–7; Saslow, "From Outsider to Politician," *WP,* 9 October 2008; Link in Don Van Natta Jr., "The New First Golfer," *Golf Digest,* February 2009, pp. 80ff.; David Heckelman, "Lawyer-Legislator Breathes Life into the Dreams of His Father," *CDLB,* 26 April 1997, p. 9; Turow's 2009 interview with Robert Draper; DJG interviews with Terry Link, James Clayborne, Dave Manning, John Bouman, Dan Shomon, Ed Wojcicki, Eric Adelstein, Rob Fisher, Tom Johnson, and Scott Turow.

40. Senate Floor Transcript, 30 April 1997, p. 13; Jennifer Davis, "The Missing Piece," *II,* May 1997, pp. 16–19; Senate Floor Transcript, 9 May 1997, p. 35 (House Bill 1080, which becomes PA 90-0249, signed 29 July 1997); R. Bruce Dold, "Happy Ending for Keystone Kids—and the Court," *CT,* 21 March 1997; House Floor Transcript, 17 April 1997, pp. 315–16; "Statehouse Glance," AP, 10 May 1997; Research Department, NRSC, "Barack Obama (D-IL): On the Record," 22 July 2004, JRORF, pp. 6 (Obama voted "present" on HB 1558, mandating consecutive prison sentences, in Senate Judiciary on 6 May, and did not vote when the Senate unanimously passed it on 9 May) and 10 (HB 382); Senate Floor Transcript, 9 May 1997, p. 99 (Obama as Senate sponsor of HB 1619, allowing nonresidents as administrators and guardians of estates and trusts, signed into law 17 August 1997 as PA 90-0472); Senate Floor Transcript, 13 May 1997, pp. 23–24; House Floor Transcript, 13 May 1997, p. 113; "Number of Bills Down in General Assembly," *CWR* 13 #3, 14 May 1997; Janota and Allen, "Daley Gives Lawmakers a Taste of Chicago During Springfield

Visit," *CDH,* 18 May 1997, p. I-10; "Mayor Daley and Representative Kenner Meet in Springfield 'Taste,'" *SSJ,* 27 June–18 July 1997, p. 2; Senate Floor Transcript, 21 May 1997, pp. 24–25 (passage of SB 6 regarding sex offenders) and 67 (unanimous passage of Obama's Senate Resolution 80, memorializing former Chicago police commander Mildred B. Stewart); John C. Patterson, "Sex Offender Law Could Face Court Challenge," *CDH,* 22 May 1997, p. A6; Senate Floor Transcript, 22 May 1997, pp. 53, 61 (passage of HB 729, which was signed into law 18 August 1997 as PA 90-0495); Senate Floor Transcript, 31 May 1997, pp. 21–22 (final passage of Obama's SB 574), 40–63 (HB 204), and 92–93 (SB 1129, the 1998 fiscal year budget, passed 49–9); Russ Stewart, "Philip Makes Sure GOP Remains 'Anti-Tax' Party," *Illinois Politics,* June 1997, pp. 10–11; Dave McKinney, "State Welfare Reform OK'd; Proposal Won't Protect Immigrants," *CST,* 1 June 1997, p. 2; Kristi O'Connor, "Welfare Overhaul Bill Clears Both Houses," *SJ-R,* 1 June 1997, p. 1; Rick Pearson and Christi Parsons, "School Funding Reform Plans Left for Dead," *CT,* 1 June 1997; Parsons, "State Welfare Revamp Approved," *CT,* 1 June 1997; Parsons, "State Set to Take Work-to-Welfare Step," *CT,* 3 June 1997; Senate President James 'Pate' Philip, *End of Session Report 1997* (Senate Republican Caucus, 1997), pp. 74–84, 106–9; "Spring Session Ends," *CWR* 13 #4, 23 June 1997; *Capitol Fax,* 24 November 1997 and 23 April 1999; MacDougal, *Make a Difference,* pp. 17, 263–67; Dan A. Lewis et al., "Welfare Reform Effects in Illinois," in Lawrence B. Joseph, ed., *Families, Poverty, and Welfare Reform* (Harris School, University of Chicago, 1999), pp. 126–29; David Glenn, "What the Data Actually Show About Welfare Reform," *Chronicle of Higher Education,* 21 June 2002, pp. 14ff.; Lewis, *Gaining Ground in Illinois: Welfare Reform and Person-Centered Policy Analysis* (Northern Illinois University Press, 2010), esp. pp. 5, 20, 131; Gonzales, "The Ever-'Present' Obama," *WSJ,* 14 February 2007; Remnick, *The Bridge,* p. 300; Jim Edgar's 2010 interviews with Mark DePue; DJG interviews with Frank Watson, Dan Shomon, Mike Marrs, Scott Kaiser, Deno Perdiou, Howard Peters, John Bouman, Wendy Pollack, Dave Syverson, Kathy Parker, Debbie Lounsberry, and Nia Odeoti-Hassan. See also Obama's remarks at the Saddleback Civil Forum on the Presidency, 16 August 2008, Lake Forest, California: "I always believed that welfare had to be changed. I was much more concerned . . . when President Clinton initially signed the bill that this could have disastrous results" but "it worked better than . . . a lot of people anticipated."

41. Obama, "Legislature Squandered Opportunity," *HPH,* 18 June 1997, p. 4; "Obama Disappointed with Defeat of Landmark School Funding and Electric Deregulation Plans," *South Street Journal,* 27 June–18 July 1997, p. 3; Jennifer Davis, "Illinois Gears Up to Retool Its Social Welfare System," *II,* July–August 1997, pp. 6–7; "Governor Signs New Laws," *Illinois Politics,* July–August 1997, pp. 7–8; Senator Barack Obama [by Illinois Department of Human Services], "Federal and State Welfare Reform: How Does It Effect You," 6pp. flyer, August 1997, TBP Box 129 Fld. 1; "Some Death Row Inmates to Get DNA Tests," *Chicago Independent Bulletin,* 7 August 1997, p. 7 (regarding HB 2138, PA 90-0141, of which Obama became a chief cosponsor on 9 May 1997, the same day it unanimously passed the Senate); Dalia Dangerfield, "Obama Cites Environmental Danger," *CD,* 16 August 1997, p. 1; Samuel J. Pernacciaro and Jack R. Van Der Slik, "Roll Call Voting in the 90th General Assembly," in Joens, Kleppner et al., *Almanac of Illinois* Politics—*1998,* pp. 9–16 and 80–81; David Heckelman, "Chief Judge Lauds New Jury Commission Act," *CDLB,* 18 August 1997, pp. 1, 24; "Heed Obama in Lake's Environmental Danger," *CD,* 19 August 1997, p. 11; Heckelman, "City Courts to the Rescue," *CDLB,* 26 August 1997, pp. 1, 22; "Governor Edgar Signs New Laws," *Illinois Politics,* September 1997, pp. 6–7; Davis, "Freshmen Ponder Spring Lessons as They Return for the Fall Session," *II,* October 1997, pp. 6–7; Obama's 2001 interview with Julieanne Richardson; DJG interviews with Gideon Baum and Dan Shomon.

42. Friends of Barack Obama, Report of Campaign Contributions and Expenditures, 1 January 1997–30 June 1997, 28 July 1997; Kevin Knapp, "Obama Spends Campaign Cash," *HPH,* 20 August 1997, p. 3; Adam Doster, "A Thoughtful Politician," *CR,* 16 February 2011; Tim Novak's excellent "Obama and His Rezko Ties," *CST,* 23 April 2007, p. 22; Novak, "Broken Promises, Broken Homes," and "Obama: I Didn't Know About Rezko Problems," *CST,* 24 April 2007, pp. 8 and 20; Larry Bell, "Obama Kick-Back Cronyism," Forbes.com, 25 October 2011; Garance Franke-Ruta, "The Next Generation," *American Prospect,* August 2004; Obama, *TAOH,* p. 49; Mike Ramsey, "Backing Barack: Early Obama Supporters Find Recent Events Affirm Faith," *SJ-R,* 31 December 2006; Alex MacGillis and Steve Mufson, "Coal Fuels a Debate over Obama," *WP,* 24 June 2007, p. A1; MacGillis, "In Illinois, Clues to Obama's Electability," *WP,* 15 June 2008, p. A1; Shomon's 2008 interview with Jen Christensen; Ben Calhoun, "1997 Trip Sparked Obama's Rural-Voter Courtship," NPR, 10 August 2008, "Dan Shomon—First Person," BBC News, 26 August 2008; Obama in "Barack Obama Tribute" video, DNC, 28 August 2008; Shomon in Van Natta Jr., "The New First Golfer"; Remnick, *The Bridge,* pp. 318–19; McClelland, *Young Mr. Obama,* p. 137; DJG interviews with Dan Shomon, Jim Rea, and John Charles.

43. Pam Belluck, "Illinois Governor Testifies in Federal Bribery Trial," *NYT,* 30 July 1997; Rick Pearson and Christi Parsons, "MSI Verdict Jolts Springfield," and Parsons, "Trial Confirms Cynic's View of Government," *CT,* 17 August 1997; Pearson and Bob Kemper, "MSI Case Puts Favor Seekers on Notice," *CT,* 18 August 1997; Belluck, "Illinois Governor Surprises by Retiring from Politics," *NYT,* 21 August 1997; Edward Walsh, "Illinois Republican Gov. Edgar to Retire," *WP,* 21 August 1997; John Kass, "One Vacancy, Many Opportunities," *CT,* 21 August 1997; Kass, "Legislature May Pluck Edgar's Feathers," *CT,* 22 August 1997; Bill Bush, "Edgar Orders Agency Gift Ban," *SJ-R,* 27 August 1997, p. 1; Parsons, "Gov. Edgar Sets Limits for Gifts," *CT,* 27 August 1997; Parsons, "Friends Like These," and Charles N. Wheeler III, "Sometimes You Just Have to Kill the Alligators," *II,* September 1997, pp. 24–25 and 38; *United States v. Martin,* 195 F. 3d 961 (7th Cir.), 1 November 1999, cert. denied, 530 U.S. 1263, 26 June 2000; David Bernstein, "Lawmakers Gone Wild," *Chicago Magazine,* January 2013; "People," *II,* April 1997, p. 34; Harold Henderson, "Eight Easy Pieces?," *CR,* 24 July 1997; Joyce Foundation, *Work in Progress,* August 1997, p. 14; Mike Lawrence, proposal to the Joyce Foundation, ca. mid-August 1997, MLP and PSP; Don Thompson, "Scandals Bring Pressure for Campaign Finance Reforn," *CDH,* 25 September 1997, p. 1; Dave Gosnell, "Passing the Bucks," *BND,* 5 October 1997; David Rheingold, "U.S. Campaign Finance Limits Stiffer Than State's," *BND,* 6 October 1997; Gosnell, "Deregulation Push Gets Utility Boost," *BND,* 13 October 1997; Lawrence, "Mission Impossible," *II,* September 1998, pp. 38–41; Kirk Dillard's 9 November 2009 interview with Mark DePue, p. 47 ("Mike Lawrence was a major policy advisor to Governor Edgar"); Howard Peters's 2009 interview with DePue, p. 95 ("Mike functioned in the administration as much more than a press secretary . . . Mike was very much involved in policy"); Mike Lawrence's 2009 interviews with DePue, pp. 161, 183–84; DJG interviews with Jim Howard, Cindi Canary, and Mike Lawrence.

44. Nathaniel K. Wilkes, "Robeson High School Reinvents Itself for New School Year," *Chicago Weekend,* 28 August 1997, p. 6; Minutes, CAC Board of Directors, 25 June 1997, Minutes, CAC Board of Directors Executive Committee, 29 July 1997, Minutes, CAC Board of Directors, 30 September 1997; Ken Rolling, "Chicago Annenberg Challenge 1997 Interim Report," 2 October 1997; "Chicago Annenberg Challenge 1998 Annual Report" (reporting that $2.7 million was awarded in both 1995 and 1996, and $19.8 million in 1997); William Ayers, Michael Klonsky, and Gabrielle H. Lyon, eds., *A Simple Justice: The Challenge of Small Schools* (Teachers College Press, 2000), esp. p. 189; Michael Klonsky and Susan Klonsky, *Small Schools* (Routledge, 2008), pp. 25–26; Michael Klonsky's 2014 interview with Mark Larson; Joyce Foundation, *1997 Annual Report,* p. 42; Mark A. Smylie et al., *Getting Started: A First Look at Chicago Annenberg Schools and Networks* (Consortium on Chicago School Research), June 1998, pp. 15, 97; G. Alfred Hess Jr., "Community Participation or Control: From New York to Chicago," *Theory into Practice* 38 (Autumn 1999): 217–24; Dorothy Shipps, "The Businessman's Educator: Mayoral Takeover and Nontraditional Leadership in Chicago," in Larry Cuban and Michael Usdan, eds., *Powerful Reforms with Shallow Roots* (Teachers College Press, 2003), pp. 16–37, at 33 (referencing a 10 October 1997 interview with a Democratic state legislator who almost certainly was Barack, although the CCSR most unpersuasively claims it cannot locate Shipps's interviews); Minutes, CAC Board of Directors Executive Committee, 21 November 1997, Minutes, CAC Board of Directors, 10 December 1997; Obama, "Genetic Testing Could Help Save Lives," *HPH,* 10 September 1997, p. 4; Phyllis C. Richman, "Red Hot Red Sage," *WP,* 5 April 1992; "Schedule of Century Project Meetings," 22–23 September [1997], Chicago, SHP; Abramsky, *Inside Obama's Brain,* p. 100; Woods Fund of Chicago, *1997 Annual Report,* pp. 1, 20 (a $30,000 grant to the Hope Center); Woods Fund of Chicago, 1997 Form 990-PF, Part VIII, 14 May 1998

(Obama formally became board chairman 10 December 1997); Chinta Strausberg, "Obama Wary of Probe of Gore," *CD,* 23 September 1997, p. 7; Brett Schaefer, "University Not Helpful, Say Local Business Leaders," *HPH,* 24 September 1997, p. 3; "PMNCC Welcomes Senator Obama," *Chicago Independent Bulletin,* 9 October 1997, p. 6; UCLS *Announcements 1997–98,* pp. 34–35, 37–38; UCLS *Directory '97–'98,* p. 20; Michael Lind, "The End of the Rainbow," *Mother Jones,* September–October 1997; Sniderman and Carmines, *Reaching Beyond Race* (Harvard University Press, 1997), pp. 15–58, esp. pp. 53–58; Richard Morin, "The Hidden Truth About Liberals and Affirmative Action," *WP,* 21 September 1997, p. C5; Michael Risch, Con Law III Notes, MRP; UCLS Course Evaluations, Autumn 1997; Obama, Constitutional Law III Final Examination, 13 December 1997; Obama to Students in Con Law III, "The Exam," n.d. [ca. March 1998], esp. pp. 7 and 17; Abdon M. Pallasch, "Law Students Gave Obama Big Thumbs-Up," *CST,* 18 December 2007, p. 7; Jason Zengerle, "Con Law," *TNR,* 30 July 2008, p. 7; Michael Risch in "Obama as a Constitutional Law Professor," Volokh.com, 3 May 2008; Elysia Solomon in Robin I. Mordfin, "From the Green Lounge to the White House," *University of Chicago Law School Record,* Spring 2009, pp. 2–9; DJG interviews with Bob Bennett, Rob and Lisa Fisher, Charles Halpern, Stephen Heintz, Peter Steffen, Michael Risch, Elysia Solomon, and Andrew Trask.

45. Adrian Butler, "Barack's British Background," *Sunday Mirror,* 13 January 2008; Oliver Harvey, "Obama's Brother Is in Bracknell," *The Sun,* 26 July 2008; Don Van Natta Jr., "The New First Golfer," *Golf Digest,* February 2009, pp. 80ff.; Jacobs, *The Other Barack,* p. 5; Auma Obama, *And Then Life Happens,* pp. 243–44; Sasha Abramsky, "Obama's Sister Has Big Ideas for Education," *The Nation,* 23 March 2011; Lorraine Forte, "Legislators Divided on 'Get Tough' Reforms," *CST,* 28 September 1997, p. 21; M. A. Stapleton, "Devine Backs Bill to Update Juvenile Delinquency Laws," *CDLB,* 30 September 1997, pp. 1, 22; *Capitol Fax,* 13 October 1997, p. 2; Obama, "Hot Topics for the Fall Veto Session," *HPH,* 15 October 1997, p. 4; *CF,* 15 and 16 October 1997, Senate Floor Transcript, 16 October 1997, p. 6 (Obama introducing S. Res. 220, seeking to declare 1 November as South Shore Islamic Community Center Day, which languished in the Rules Committee); Doug Finke, "Edgar, GOP Propose Gift Restrictions," *SJ-R,* 17 October 1997, p. 9; *CF,* 17 October 1997; "Busy Veto Session Expected," *CWR* 13 #7, 20 October 1997; *CF,* 27, 28, 29, 30, and 31 October 1997; Davila in Kevin O'Leary, "Why Barack Loves Michelle," *Us Weekly,* 30 June 2008, pp. 52–57; Christopher Andersen on *The Early Show,* CBS, 22 September 2009 (asserting, without any apparent support, that the Obamas "discussed adoption with some of their closest friends"); "Obama Says He'll Run Again," *HPH,* 5 November 1997, p. 3; Burns in David Maraniss, "The Roots of Two Presidents," *WP,* 25 March 2012, p. A17; Kevin Knapp, "A Few Races Worth Watching in March," *HPH,* 24 December 1997, pp. 1–2; *CF,* 3, 5, 7, 11, 13, 14, 17, 19, 21, 24, and 25 November 1997; Paul Krawzak, "Campaign Finance Reform Task Force Optimistic on Procurement Changes," *SJ-R,* 5 November 1997, p. 5; Jeffrey Felshman, "Fighting Over Scraps," *CR,* 7 November 1997; Obama, "Endorsing the IVI-IPO," *CR,* 14 November 1997; Melita Marie Garza, "Universal Health Care Proposed," *CT,* 17 November 1997; "Fall Veto Session Overview," *CWR* 13 #8, 1 December 1997; Rick Pearson, "Legislators Get Message, Pass School-Fund Bill," *CT,* 3 December 1997; *CF,* 3 and 8 December 1997; "Governor Edgar Signs New Laws," *Illinois Politics,* January–February 1998, p. 13; [Debbie] Lounsberry, "Barack Obama—The Record: A Summary," 10 May 2004, JRORF, p. 23 (Obama recorded as not voting on HB 452, the school funding measure, although per Senate Floor Transcript, 14 November 1997, p. 142, Barack supported it); Peter Wallsten, "Obama Said Oops Six Times on Votes as a State Senator," *LAT,* 24 January 2008 (Barack erroneously voting yes rather than no on SB 493, per Senate Floor Transcript, 14 November 1997, p. 120); Jennifer Vanasco, "Close-Up on Juvenile Justice," *University of Chicago Chronicle,* 6 November 1997; "Juvenile Justice: Predators or Children," *HPH,* 12 November 1997, p. 12 and 19 November 1997, p. 5; Ayers, *A Kind and Just Parent* (Beacon Press, 1997), p. 82; Chinta Strausberg, "Obama: Juvenile Justice System Flawed," *CD,* 3 December 1997, p. 3; "Mark My Word," *CT,* 21 December 1997, p. B5; Dan Morain and Bob Drogin, "Obama and the Former Radicals," *LAT,* 18 April 2008; Stanley Kurtz, "Barack Obama's Lost Years," *Weekly Standard,* 11 August 2008; Dohrn, "Seize the Little Moment: Justice for the Child—20 Years at the Children and Family Justice Center," *Northwestern Journal of Law and Social Policy* 6 (Spring 2011): 334–57; DJG interviews with Scott Kaiser, Cynthia Miller, Will Burns, Bernardine Dohrn, Bill Ayers, Carole Travis, and Mona Khalidi.

46. Larry Hansen to Mike Lawrence, 21 November 1997, PSP; David Rheingold, "It's All Legal: Campaign Fund Pays Bills, Buys Tickets, Etc.," *BND,* 30 November 1997; *CF,* 2, 4, 5, and 12 December 1997; SIU Acceptance Form, "Making Campaign Finance Reform Happen," 5 December 1997; Mike Lawrence to James "Pate" Philip (Attn. Carter Hendren), Emil Jones (Attn. Mike Hoffmann), Michael J. Madigan (Attn. Tim Mapes), and Lee Daniel (Attn. Mona Martin), 12 December 1996, MLP and PSP; "Rep. Kubik to Return to Newspaper Roots," *CT,* 9 September 1997; *CF,* 23 and 31 December 1997; Obama, "Ethics Reform Comes to Springfield," *HPH,* 31 December 1997, p. 4; Hansen to Lawrence, 31 December 1997, PSP; Friends of Barack Obama, Report of Campaign Contributions and Expenditures, 1 July 1997–31 December 1997, 30 January 1998; Charity Crouse, "Fundraising Is Looking Up for Local Politicians," *HPH,* 11 February 1998, p. 3; UCLS *Announcements 1997–98,* p. 63; UCLS Course Evaluations, Winter 1998; Mike Lawrence, "Mission Impossible," *II,* September 1998, pp. 38–41; Ed Wojcicki, "Still the Wild West?," pp. 9–11; Jo Becker and Christopher Drew, "Pragmatic Politics, Forged on the South Side," *NYT,* 11 May 2008, p. A1; DJG interviews with Mike Lawrence, Jim Howard, David Starrett, Cindy Canary, Pate Philip, Carter Hendren, Emil Jones, Mike Hoffmann, Jill Rock, Gary Hannig, Tom Dart, Andrew Trask, Indira Saladi, and Felton Newell. Rezko also contributed $457 worth of food for a 22 January 1998 party. Friends of Barack Obama, Report of Campaign Contributions and Expenditures, 1 January 1998–30 June 1998, 27 July 1998.

47. *CF,* 3 December 1997; Chinta Strausberg, "Juvenile Justice Bill in Jeopardy?," *CD,* 12 January 1998, pp. 1, 7; *CF,* 12, 13, 15, 27, 28, 29, and 30 January 1998; "Kids, Crime and the Law," *CT,* 13 January 1998; David Heckelman, "Juvenile Justice Among First Topics of New Legislative Session," *CDLB,* 14 January 1998, pp. 1, 18; Annie Sweeney, "Grievances Against UIC Aired," *CT,* 25 January 1998; Rev. Gordon McLean, "Re-evaluate Juvenile Justice Act," *CT,* 28 January 1998; Christi Parsons and Rick Pearson, "Edgar Vows a New Push on School Tax Reform," *CT,* 29 January 1998; Strausberg, "Obama Backs Schmidt; Hints at Trouble for Democrats," *CD,* 29 January 1998, p. 8; Senate Floor Transcript, 29 January 1998, pp. 4–5, 10–15; Ray Serati, "Bill Would Revamp Illinois Juvenile Law," *PJS,* 30 January 1998; Christi Parsons, "Edgar Plan Faces Early Resistance," *CT,* 30 January 1998; "Edgar Delivers Final State of the State," *CWR* 13 #11, 13 February 1998; DJG interviews with Matt Jones and Gideon Baum.

48. Mike Lawrence to Paul Simon, "Campaign Finance Reform Meeting," n.d., PSP; *CF,* 2, 3, 4, 11, 12, 13, 19, 20, 23, 24, 25, 26, and 27 February 1998; Senate Floor Transcript, 18 February 1998, pp. 2 (SJRCA 48), 7, 19 February 1998, pp. 14, 16 (SB 1545), 21, 20 February 1998, pp. 2, 4–5, 16–17, 22; Jason Piscia, "Bigger State Income-Tax Relief Proposed," SJ-R, 24 February 1998, p. 9; Chinta Strausberg, "Obama Unveils 'Fair' Tax Plan," *CD,* 24 February 1998, p. 5; Obama, "Illinois Needs a Fairer Tax System," *HPH,* 25 February 1998, p. 4; Jennifer Davis, "Penny Severns Remembered," *II,* March 1998, p. 35; Lawrence, "Mission Impossible," *II,* September 1998, pp. 39–40; Edward McClelland, "Illinois's Pioneering Gay Legislator," NBCChicago.com, 11 May 2013; Bernard Schoenburg, "Severns Would Be Proud of Colleagues, Former Partner Says," *SJ-R,* 10 November 2013; Mutchler, *Under This Beautiful Dome* (Seal Press, 2014), pp. 3, 43, 149, 233; DJG interviews with Mike Lawrence, Andy Foster, Gary Hannig, Kirk Dillard, Kent Redfield, and Linda Hawker.

49. DJG interviews with Denny Jacobs, Mike Lieteau, Fred Fortier, Boro Reljic, Terry Link, Phil Lackman, Dave Manning, Larry Walsh, Cindy Huebner Davidsmeyer, and Brad Jonas; Jim Reynolds in David Smallwood, "Genesis of the Grassroots Movement," *N'Digo Profiles,* December 2008, pp. 10–11, 44; Becky Yerak, "Straddling the Worlds of High Finance and Community Activism," *CT,* 1 April 2013; Cheryl L. Reed, "The Used Car Lot of the Book World," *CST,* 21 January 2007, p. B12. See also Jacobs in Steven Thomma, "Barack Obama, Not Your Typical Politician," KRDC, 16 February 2007, and Julie A. Dressler, "The Establishment of Illinois Riverboat Gambling and Its Impact on Alton, Peoria and Joliet," Senior Honors Thesis, Illinois Wesleyan University, 10 May 1994.

50. [Mike] Lawrence to Dillard, Foster, Hannig, Kubik, Obama, ca. 20 February 1998, MLP and PSP; Paul Simon to Obama, 23 February 1998, PSP; Lawrence to Dillard, Foster, Hannig, Kubik, Obama, 27 February 1998, MLP and PSP; Adriana Colindres, "New Political

Web Site Right on the Money," *SJ-R* (and *PJS*) 4 March 1998, p. 1; *CF*, 4, 5, 6, 9, 10, 11, 12, 13, 16, 17, 18, 19, and 24 March 1998; Lawrence to Dillard, Foster, Hannig, Kubik, Obama, 6 March 1998, [Lawrence], "[Legislative Staff on] Elements of Tentative Consensus," n.d., MLP; John Kass, "Latino Leader Tapped to Run the 25th Ward," *CT*, 2 March 1996; Joseph N. Boyce, "Hispanic Influx Spurs Legislator to Practice Multiethnic Diplomacy," *WSJ*, 7 May 1997, p. A1; Christi Parsons and Ray Long, "Star Is Rising for a 2nd Madigan," *CT*, 18 March 1998; "A Message from Latino Voters," *CT*, 19 March 1998; Linda Lutton, "War on Independents," *CR*, 3 September 1998; Solis in Carlos Hernandez Gomez, "Population Soars but Political Power Lags," *Chicago Reporter*, September 2001; Obama, "Getting the Lead Out of Our Children," *HPH*, 25 March 1998, p. 4; DJG interviews with Mike Lawrence, Chuy Garcia, Will Burns, and Danny Solis.

51. Minutes, CAC Board of Directors Executive Committee, 24 February 1998, Minutes, CAC Board of Directors, 19 March 1998, p. 3; *CF*, 26, 27, and 30 March 1998; Lynn Sweet and Dave McKinney, "Ethics Law Reveals Too Little, Group Says," *CST*, 29 March 1998, p. 4; Obama to Mr. Falk, "Saul Mendelson Website," 3 May 1998; Carl Shier, "So Long, Saul," *New Ground* 58, May–June 1998; Bill Dedman, "Study Criticizes Illinois' Rules for Legislators," *NYT*, 30 March 1998; "Governor Delivers Final Message," *CWR* 14 #1, 31 March 1998; James Evans, "Study Cites 'Anything Goes' Ethics in Legislature," *CDH*, 1 April 1998, p. 15; Senate Floor Transcript, 1 April 1998, pp. 104–5; *CF*, 6, 14, 16, 17, 20, 21, 22, 23, 27, and 29 April 1998; Chinta Strausberg, "Pols Rev Up for Rainy Day Funding Battle," *CD*, 9 April 1998, p. 6; "Little Local Difficulties," *Economist*, 11 April 1998, p. 21; James Merriner (AP), "Simon Seeks Legislative Reform," *PJS*, *MJG* and *CTC*, 11 April 1998; Barack H. Obama and Michelle Robinson-Obama, 1997–1998 Tax Summary, appended to Obama and Robinson-Obama, 1998 Income Tax Return, Vicki L. Hudson and Co., 9 April 1999; Joyce Foundation, 1996 Form 990-PF, Part VIII(1), 15 May 1997, and 1997 Form 990-PF, Part VIII(1), 14 May 1998 ($11,000 in director's fees paid to Obama each year); "Sen. Obama to Hold Town Meeting," *HPH*, 15 April 1998, p. 2; Charlie Halpern to Barack O'Bama et al., "Century Project Meeting," 24 March 1998, Halpern to "Dear Colleagues," 10 April 1998, and "Meeting Summary, New Institute Planning Group," 20 April 1998, SHP; Jeanette Almada, "Reverse Commute: North Kenwood/Oakland Enjoying an Influx of the Middle Class," *CT*, 26 April 1998, p. RE1; Almada, "Turning the Corner," *CT*, 2 May 1999, p. RE1; Jacobs in Jonathan Kaufman, "For Obama, Chicago Days Honed Tactics," *WSJ*, 21 April 2008, p. A1, and in Eli Saslow, "From Outsider to Politician," *WP*, 9 October 2008, p. A1; DJG interviews with Bill Peterson, Denny Jacobs, and Pat Welch.

52. Obama, "Education Most Important Town Hall Issue," *HPH*, 29 April 1998, p. 4; Kate Clements, "Legislative Group in Final Stages on Campaign Finance Reform Plan," *CDH*, 29 April 1998, p. 10; Ray Long, "Philip Hoping to Give Campaign Reform a Shot," *CT*, 30 April 1998; *CF*, 30 April and 1, 5, 6, 7, and 8 May 1998; Christi Parsons, "Juvenile Reform Act Gets Edgar Backing," *CT*, 25 April 1998; Senate Floor Transcript, 5 May 1998, pp. 10–15 (SB 363); Jason Piscia, "Senate Approves Edgar's Changes to Juvenile Justice Bill," Copley News Service, 5 May 1998; David Heckelman, "Revamped Juvenile Justice Bill Nears Passage," *CDLB*, [6] May 1998, pp. 1, 22; Obama in Mary Wisniewski Holden, "The Juvenile Justice Reform Act: Walking a Fine Line," *Chicago Lawyer*, June 1998, p. 1; Senate President James "Pate" Philip, *End of Session Report 1998*, pp. 61–63 (terming SB 363 "a major, desperately-needed step forward in revolutionizing the juvenile justice system"); "The Juvenile Justice Reform Act of 1998" and Daniel Dighton, "Balanced and Restorative Justice in Illinois," *The Compiler* 18 (Winter 1999): 1, 4–5, 9–15, and 20; Daniel Dighton, "Three Years After Sweeping Reform, What's Different in the Juvenile Justice System?," *The Compiler* 21 (Spring 2002): 1–5; Christine Kraly, "Finding Justice and Peace for Illinois Youth," *Chicago Lawyer*, April 2012; Zeituni Onyango, *Tears of Abuse* (Afripress Publishing, 2012), pp. 120–38; "State Senator to Speak at Annual ACLU Banquet," *SJ-R*, 6 May 1998, p. 32; DJG interviews with Mike Lawrence, Jim Howard, Carter Hendren, and Zeituni Onyango. See also John Nelson Walters, "The Illinois Amendatory Veto," *John Marshall Journal of Practice and Procedure* 11 (1978): 415–40.

53. Ray Long, "Edgar Sees Some Merit in Tax-Relief Proposal," *CT*, 9 May 1998; Senate Floor Transcript, 13 May 1998, pp. 6–19 (HB 705, PA 90-736); Tom Schafer, *Meeting the Challenge: The Edgar Administration, 1991–1999* (Jim Edgar, 1998), pp. 77–79; Nia

Odeoti-Hassan, "Senate Democratic Staff House Bill Analysis 91st General Assembly, HB 1732," 27 April 1999, Berman Papers Box 4 Fld. 31; David Moberg, "Back to Its Roots: The Industrial Areas Foundation and United Power for Action and Justice," in John P. Koval et al., eds., *The New Chicago* (Temple University Press, 2006), pp. 239–47, at 243–44; Paul Merrion, "One Tiny Step for Campaign Reform," *CCB,* 11 May 1998, p. 4; *CF,* 11, 14, 18, 19, 20, 21, 22, 26, and 27 May 1998; Christi Parsons and Mike Cetera, "Small Tax Break Is Seen Likely," *CT,* 14 May 1998; [Mike Lawrence], "Talking Points," n.d. [ca. 14 May 1998], MLP and PSP; SIU Public Policy Institute Press Release, "Bipartisan Group Reaches Consensus on Campaign Finance Reform," n.d. [14 May 1998], PSP; Rick Pearson, "Campaign Fund Reform Talk Wears on Officials," and "Some Common Sense Campaign Rules," *CT,* 15 May 1998; Doug Finke, "Simon: Even Chance of Ethics Reform," *SJ-R,* 15 May 1998, p. 13; Don Thompson, "Skeptics Question Legislators' Resolve for Campaign Reform," *CDH,* 15 May 1998, p. 10; John C. Patterson, "Rule Changes Will Center on Lobbyists," *DH&R,* 15 May 1998, p. A3; Lawrence to Carter Hendren, "Ethics Legislation," 18 May 1998, MLP; Simon to Obama, 18 May 1998, PSP; "Illinois Lawmakers," 20 May 1998, MBC; Bruce Dold, "Why Pols, Porches and Lucrative Campaign Chests Are a Bad Mix," *CT,* 22 May 1998; Senate Floor Transcript, 22 May 1998, pp. 55–57, 79–80, 114–17 (HB 455); House Floor Transcript, 22 May 1998, pp. 124, 159–211; Ray Long and Parsons, "Pork Tops Menu for State Lawmakers," *CT,* 23 May 1998; Dave McKinney, "State Senate OKs Pay Raises, Tougher Ethics Rules," *CST,* 23 May 1998, p. 4; James Merriner (AP), "House GOP Trying to Block Reforms," *PJS,* 23 May 1998, p. B6; Thompson, "State Lawmakers End Session That Would Have Made Perfect 'Seinfeld' Fodder," *CDH,* 24 May 1998, p. 10; Rick Pearson, "No Crises, No Fights to Detain Legislators," *CT,* 24 May 1998; "Illinois Legislature OKs Most Restructive Ethics Bill Since Watergate," *SLPD,* 24 May 1998, p. D10; John McCarron, "General Assembly Made Everyone a Winner, Sort Of," *CT,* 25 May 1998; Lawrence to Cindi Canary, "Campaign Reform Legislation," 27 May 1998, MLP; Madeleine Doubek, "Campaign Finance Reform, with Loopholes," *CDH,* 28 May 1998, p. 14; Steven R. Strahler, "Chicago Observer," *CCB,* 1 June 1998; *CF,* 3, 5, 12, and 20 June 1998; "Senator Obama Co-Sponsors Reform Bill," *CIB,* 4 June 1998, p. 7; Parsons, "Ethics Bill Lets Lawmakers Squeeze Between Loopholes," *CT,* 4 June 1998; Obama, "Important Lakeshore Funds Allocated," *HPH,* 10 June 1998, p. 4; Steve Neal, "Reforming Elections," *CST,* 15 June 1998, p. 27; "State Campaign Finance Reform," *Morning Edition,* NPR, 16 June 1998; Paul Simon to Pate Philip, Emil Jones Jr., Michael J. Madigan, and Lee Daniels, 16 June 1998, PSP; Cynthia Canary and James Kales, "Voters Must Demand Clean Elections," *CT,* 20 June 1998; Dennis Conrad (AP), "Campaign Trips Take Exotic Turn," *PJS,* 21 June 1998, p. A1; *CF,* 24 June 1998; "Spring Session Ends," *CWR* 14 #3, 25 June 1998; "Pate" Philip, *End of Session Report 1998,* pp. 87–89; Stephen S. Morrill, "Client Report for 1998 End of Session," Morrill and Associates, June 1998, esp. p. 4; Kent D. Redfield and Paul J. Quick, "Reforming Campaign Finance in Illinois: Issues and Prospects," Institute of Government and Public Affairs, University of Illinois, July 1998, pp. 2–3, 9; "Edgar Signs New Laws," *Illinois Politics* 7 #5 (ca. 31 July 1998), p. 18; Jennifer Davis, "Campaign Finance Reform Is a Matter of Style Over Substance," *II,* July–August 1998, pp. 6–7; Dan Shomon to Thomas E. Johnson, 11 May 1998, Larry O'Brien to Courtney Nottage, 20 May 1998, Shomon to Johnson, 22 May 1998, *Davis v. Handy,* #98-EB-RES-003, filed 22 May 1998, Herman W. Baker Jr. "Appearance," *Davis v. Handy,* 1 June 1998, Chicago Board of Election Commissioners, "Findings and Decision," *Davis v. Handy,* 11 June 1998, and Johnson to Obama, 19 June 1998 (enclosing a requested bill for $225), TEJP; "Governor Edgar Signs New Laws," *Illinois Politics,* August 1998, pp. 12–13; Joyce Foundation, "Campaign Reform Roundup," *Work in Progress,* September 1998, p. 16; Lawrence, "Mission Impossible," *II,* September 1998, pp. 38–41; Lawrence, "Cooperative Approach Forges Campaign Finance Reform," *Illinois Country Living,* October 1998, p. 4; Simon, *P.S.: The Autobiography of Paul Simon,* p. 375; Joyce Foundation, *1999 Annual Report,* esp. p. 68 ($547,000 to the Illinois Campaign for Political Reform); Obama's February 2000 interview with Ted McClelland; Redfield, *Money Counts: How Dollars Dominate Illinois Politics and What We Can Do About It* (Institute for Public Affairs, University of Illinois, 2001), pp. 5, 37–40; Shomon in Philip Sherwell, "The Big Hitter," *Sunday Telegraph,* 28 January 2007; Jones in Jackie Calmes, "Statehouse Yields Clues to Obama," *WSJ,* 23 February 2007; Lawrence in Peter Slevin, "Obama Forged Political Mettle in Illinois Capitol," *WP,* 9 February 2007, p. A1; Jones in Janny Scott, "In Illinois, Obama Proved Pragmatic and Shrewd," *NYT,* 30 July 2007, p. A1; Jones on "The Candidates: Barack Obama," MSNBC, 20 February 2008; Obama at the Saddleback Civil Forum on the Presidency, 16 August 2008, Lake Forest, CA ("I remember one of my colleagues . . . who said, 'Where do you expect us to eat, McDonald's?'"); Dillard in David Nather, "Obama's Quick Rise on a Non-Traditional Career Path," *CQ Weekly,* 21 August 2008; Conrad and Christopher Wills (AP), "Obama's Role in Ill. Ethics Bill Was Complicated," 6 October 2008; Jones in Kelly E. Carter, "Leadership Lessons from the Top," *Black Enterprise,* March 2009, p. 105; Edgar's 9 September, 8 November, and 14 December 2010 interviews with Mark DePue, pp. 912–14, 981, and 1087; Lawrence in John F. Harris and Glenn Thrush, "The Ego Factor: Can Obama Change?," *Politico,* 5 November 2010; Shomon in George E. Condon Jr., "A Sense of Self," *National Journal,* 22 September 2012; DJG interviews with Dave Syverson, Nia Odeoti-Hassan, Debbie Lounsberry, John Bouman, Steve Rauschenberger, Kurt DeWeese, Mike Lawrence, Jim Howard, David Starrett, Cindi Canary, Gary Hannig, Glenn Hodas, Carter Hendren, Patty Schuh, Pate Philip, Emil Jones, Denny Jacobs, Terry Link, Larry Walsh, Donne Trotter, Pat Welch, Chuy Garcia, Howie Carroll, Debbie Halvorson, Miguel del Valle, Lou Viverito, Jim Rea, Evelyn Bowles, Mike Ragen, Mike Hoffmann, Jill Rock, Courtney Nottage, Dan Shomon, and Jim Edgar. On Currie, see Jennifer Davis, "Designated Hitter," *II,* September 1997, pp. 21–23. Also see Senate Floor Transcript, 13 May 1998, pp. 68–72 (HB 3063) and 96–97 (Obama on HB 3575, which passes 58–0). Research Department, National Republican Senatorial Committee, "Barack Obama (D-IL): On the Record," 22 July 2004, JRORF, p. 5, identifies Barack as voting no on HB 3063 when the Senate Judiciary Committee on 6 May approved it 7–1–1, but on the Senate floor on 13 May, when it is passed 55–1–1, Barack's committee colleague John Cullerton expressly states his opposition. On 14 May, the Senate approved HB 3129, a bill condemning the Kyoto global warming treaty, 54–3, with Barack in the majority. Senate Floor Transcript, p. 64. Ken Dilanian, "Obama Shifts Stance on Environmental Issues," *USAT,* 18 July 2008, and Ben Whitford, "Coal and Clear Skies: Obama's Balancing Act," *Plenty Magazine,* 31 October 2008. On 21 May, Senate Judiciary approved 6–3 the House-amended SB 1846, which now addressed "unlawful contact with street gang members" by convicted criminals, and later that day the Senate passed it 54–3–1, with Barack voting no. See [Debbie] Lounsberry, "Barack Obama—The Record: A Summary," 10 May 2004, p. 4, ["State Senate"], "Obama Votes—Lonely Votes, 1997–2003," n.d., JRORF, Sam Youngman and Aaron Blake, "Obama's Crime Votes Are Fodder for Rivals," *The Hill,* 14 March 2007, p. 1.

54. Rashid Khalidi, "Israel's Misstep, *CT,* 6 March 1997; Jon Anderson, "'Bridge of Voices' Links Lives Divided by Ignorance," *CT,* 26 September 1997; Khalidi, "The Catastrophe," *CT,* 10 May 1998; Louise Cainkar, "The Arab American Action Network: Meeting Community Needs, Building on Community Strengths," June 1998, tesa.leb.net; Teresa Puente, "Arabs Built Solid Base in Chicago, Study Says," *CT,* 18 June 1998; Barack H. Obama and Michelle Robinson-Obama, 1998 Income Tax Return, Schedule A, p. 1, Vicki L. Hudson & Co., 9 April 1999; Ali Abunimah, "How Barack Obama Learned to Love Israel," Electronic Intifada, 4 March 2007; Philip Klein, "Obama Rising," *American Spectator,* July–August 2007, pp. 24–31; Peter Wallsten, "Allies of Palestinians See a Friend in Barack Obama," *LAT,* 10 April 2008; Hilary Leila Krieger, "Mr. Obama's Neighborhood," *Jerusalem Post,* 23 October 2008; David A. Miller, "Meeting Minutes," Steering Committee of the New Institute, 2 June 1998, and Stephen Heintz, "June 2" [1998] notes, SHP; National Issues Forum, "Cities and Economic Revitalization," Brookings Institution, 8 June 1998, C-SPAN Video, 14 June 1998, at 63:00–71:30, 84:30, 96:00, 103:00, 110:00, 113:30, 122:00, 166:00, and 181:00; Minutes, CAC Board of Directors, 18 June 1998 and 16 September 1998; Barbara Rose, "School Watchdog Loses Some Bark," *CCB,* 29 November 1993, p. 1; Michael Klonsky, "Business Leaders Switching Sides?," *Catalyst,* February 1994, pp. 1, 4–6; Ann Bradley, "Business Group's New Agenda Worries Reformers in Chicago," *Education Week,* 16 February 1994; Rose, "Reformer to Lead School Watchdog," *CCB,* 7–13 March 1994, p. 36; Northwestern News Release, "Rep. Jesse Jackson Jr. and Sen. Barack Obama Featured at Diversity Week," 24 February 1999 (Leadership for Quality Education is among Barack's board memberships); Obama, Illinois State Legislative Election 1998 National Political Awareness Test [Project Vote Smart], 2 July 1998; Obama, "Remarks at the White House," 19 June 2009, *Public Papers of the Presidents* 2009 Vol. 1, p. 863; Obama, *TAOH,* p. 339; Anne E. Kornblut, "Michelle Obama's Career Timeout," *WP,* 11 May 2007, p. A1; Obama in Lisa Miller and Richard Wolffe, "I Am a Big Believer in Not Just Words, but Deeds and Works," *Newsweek,* 12 July 2008; Geraldine Brooks, "Michelle Obama: Camelot 2.0?," *More,* October 2008; Christopher Andersen, *Barack and Michelle* (William Morrow, 2009), pp. 185–86; Remnick, *The Bridge,* p. 302; Slevin, *Michelle Obama,* p. 166; Obama, "How the Presidency Made Me a Better Father," *More,* July–August 2015, pp. 68–69; Zeituni Onyango, *Tears of Abuse,* pp. 135–38; Grant and Grant, *The Moment,* pp. 150–52; Scott Fornek, "'I've Got a Competitive Nature,'" *CST,* 3 October 2004, p. 12; Michael Sneed, "Sneed," *CST,* 7 July 1998, p. 4; Simon to Barack and Michelle Obama, 14 July 1998, PSP; Agenda, New Institute Steering Committee, 19–20 July 1998, Stony Point, NY; "Century Project Retreat," 19–20 July 1998; Stephen Heintz notes, "NCF Retreat," 19–20 July [1998], Mary Nakashian to Stephen Heintz and David Callahan, "Budget and Narrative for the New Institute," 31 July 1998, Jennifer Paradise to Alan Morrison, and "Proposal for a New Institute," 13 August 1998, all SHP; "Mitsubishi Settles with Women in Sexual Harassment Lawsuit," *NYT,* 29 August 1997; "Mitsubishi Settlement Said to Total $9.5 Million," *WP,* 30 August 1997; Greg Burns, "Mitsubishi, EEOC Settle Suit," *CT,* 11 June 1998; Kathy Bergen and Carol Kleiman, "Mitsubishi Will Pay $34 Million," *CT,* 12 June 1998; *Barnett v. City of Chicago,* 969 F. Supp. 1359 (N. D. Ill.), 9 June 1997, vacated and remanded, 141 F. 3d 699 (7th Cir.), 1 April 1998, cert. denied, 524 U.S. 954, 26 June 1998, on remand, 17 F. Supp. 2d 753 (N. D. Ill.), 7 August 1998; Scott Lynch-Giddings diary, 7 August 1998, SLGP; Andrew Martin and Abdon M. Pallasch, "Judge Favors City Remap Making 18th a Black Ward," *CT,* 8 August 1998; *Barnett v. City of Chicago,* 122 F. Supp. 2d 915, 25 February 2000; DJG interviews with Mona Khalidi, Harry Gendler, Charles Halpern, Stephen Heintz, Sheryll Cashin, Martha Minow, Zeituni Onyango, Kevin Tyson, Kathy Stell, Judd Miner, Paul Strauss, George Galland, Laura Tilly, Steven Mange, and Scott Lynch-Giddings.

55. *CF,* 3 June, 15 July, 10 and 17 August 1998; Jim Edgar news release, "Governor Signs Legislation Installing Campaign Finance and Ethics Reforms," 12 August 1998; Finke, "Edgar to Sign Ethics Reform Package," SJ-R, 12 August 1998, p. 1; Long and Parsons, "Campaign Finance Reform Not Without Its Loopholes," *CT,* 13 August 1998; Finke, "Edgar OKs Campaign, Ethics Reform, but Urges Changes," *SJ-R,* 13 August 1998, p. 1; Ray Serati, "Campaign Reform Moves at Least One Step Forward," Copley News Service, 13 August 1998; Obama, "Progress on Campaign Finance Reform," *HPH,* 26 August 1998, p. 4; Kennedy Communications, "State Senator Barack Obama, Candidate for U.S. Senate: Background Report," April 2003, Part 19, pp. 3–4, RRP; Mark Johnson, "New Kind of Mayor Is Emerging," *Richmond Times-Dispatch,* 14 September 1998, p. A10; "6th Warders Discuss Local Concerns," *Chicago Independent Bulletin,* 17 September 1998, p. 3; Michael Miner, "Saul Alinsky, Poster Child," *CR,* 9 February 2012 (recounting Barack's attendance at a Sunday 27 September 1998 panel discussion of a play titled *The Love Song of Saul Alinsky);* "Edgar Vetoes 24 Bills," *CWR* 14 #6, 15 September 1998, p. 4; Simon to Obama, 29 September 1998, PSP; "Governor Edgar Signs New Laws," *Illinois Politics* 7 #7, n.d. [ca. September 1998], pp. 17–18; Christi Parsons, "New Laws Are Even Tough on Lawmakers," *CT,* 1 January 1999 (Jim Howard calling the campaign finance reform bill "a very, very significant step forward" that "will make a difference in how business is conducted"); DJG interviews with Mike Lawrence, Jim Edgar, Cynthia Miller, Will Burns, and Freddrenna Lyle.

56. *CF,* 3, 14, 21, 26, and 31 August, 4, 8, 15, 16, 18, and 29 September, 2, 6, 8, 9, 14 October 1998; Matt O'Connor and Ray Long, "U.S. Says Ryan Fund Got Money in License Scam," *CT,* 7 October 1998; Cornelia Grumman and Ted Gregory, "Ryan Setting Himself Apart from Scandal," *CT,* 8 October 1998; Gail Mansfield, "Congressman Jackson Impresses Crowd at Bow Tie Event," *DL21C Report,* January 1999, pp. 1–2; "Member Profile: Activism, Law Mesh for John Corrigan," *DL21C Report,* Summer 1999, p. 2; UCLS *Announcements 1998–99,* pp. 36–37, 58; UCLS *Directory '98–'99,* esp. p. 18; UCLS Course Evaluations, Autumn 1998; Obama, Constitutional Law III Final Examination, December 1998; DJG interviews with Glenn Poshard, John Corrigan, Jesse Ruiz, Laura Mullens Nolen, Jay Hines-Shah, Nat Piggee, Dan Sokol, Triste Lieteau Smith, John Eason, Will Burns, Dan Shomon, and Mike Hoffmann.

57. Outlines/Blacklines Candidate Questionnaire '98, 9 October 1998, in Baim, *Obama and the Gays,* p. 191; "Endorsements for the Illinois Senate," *CT,* 12 October 1998; Minow in Lauren A. E. Schuker, "Obama Stars at Convention," *HC,* 30 July 2004, in Carrie Budoff Brown, "Obama: The Journey of a Confident Man," *Politico,* 28 August 2008, and in Seth Stern, "A Commander in Chief," *Harvard Law Bulletin,* Fall 2008; Jodi Kantor, "For a New Political Age, a Self-Made Man," *NYT,* 28 August 2008; James T. Kloppenberg, *Reading Obama* (Princeton University Press, 2011), p. 74; *CF,* 16, 19, 20, and 22 October 1998; Citizens for Lightford, Report of Campaign Contributions and Expenditures, 1 July 1998–31 December 1998, 1 February 1999, p. 4; Lightford in Ron Fournier (AP), "Obama's Life and Record in Springfield, Ill., Hint at the President He Would Be," 17 June 2007, in Eli Saslow, "From Outsider to Politician," *WP,* 9 October 2008, p. A1, and in Remnick, *The Bridge,* p. 302; Lightford's 2000 interview with Adele Hodge; Friends of Barack Obama, Report of Campaign Contributions and Expenditures, 1 July 1998–31 December 1998, 28 January 1999; Obama, "Public Policy in the 21st Century," Loyola University of Chicago, 19 October 1998; Doug Dobmeyer and Asma Ali, "Research Conference Links Universities and Community Efforts in Chicago," *PRAGmatics* 2 #1 (Winter 1999), p. 8; "Edgar's Last Veto Session," *CWR* 14 #8, 21 October 1998; DJG interviews with Martha Minow, Ira Silverstein, Lisa Madigan, Kim Lightford, and Doug Dobmeyer.
58. Obama to Julia Stasch and to Sondra C. Ford, 28 October 1998; Tim Novak, "Obama's Letters for Rezko," *CST,* 13 June 2007, p. 8; Mike Robinson (AP), "Obama's Relationship with Alleged Fixer," 22 January 2008; "Complete Transcript of the Sun-Times Interview with Barack Obama," SunTimes.com, 14 March 2008; "State Senate Picks," *CST,* 29 October 1998, p. 29; *CF,* 29 October, 3, 4, 5, 6, 9, 12, 13, 18, 19, and 20 November, 1, 2, 4, 7, 9, 11, 14, 18, 21, 23, 24, and 30 December 1998, 19 and 27 January 1999; Michael Sneed, "Sneed," *CST,* 3 November 1998, p. 4; Brett Schaeffer, "Elections Expected to Yield Few Surprises," *HPH,* 4 November 1998, p. 3; "Illinois Senate," and Steve Kloehn, "Care Amendment Has Way to Go," *CT,* 5 November 1998; Jennifer Paradise, "Steering Committee Notes," 9 November 1998, SHP; "Local Incumbents Re-Elected in Landslide," *HPH,* 11 November 1998, p. 2; Bernard Schoenburg, *SJ-R,* 12 November 1998, p. 7; "Incumbency Important in '98 Elections," *CWR* 14 #9, 3 December 1998; Obama, "Inside Illinois Government," UIS TV, 3 December 1998; Minutes, CAC Board of Directors Meeting, 10 December 1998; Scott Lynch-Giddings diary, 11 December 1998; Stephen S. Morrill, "Client Report for 1998 Veto Session," Morrill and Associates, December 1998; "Veto Session Provides Sparks, No Fire," *CWR* 14 #10, 18 December 1998, p. 2; Steve Neal, "Black Candidates Flex More Political Muscle," *CST,* 23 December 1998, p. 9; Mark Brown, "New Kid on the Block," *II,* April 1999, pp. 14–20 (Dan Hynes); Redfield, *Money Counts,* p. 77; Ron Jacobs, *Obamaland* (Trade Publishing, 2008), p. 122; "Native Moberlyan Served in Senate with Obama," *Moberly Monitor-Index,* 24 October 2008; Halvorson, *Playing Ball with the Big Boys,* p. 32; DJG interviews with Terry Link, Debbie Halvorson, Pat Welch, John Charles, Dan Shomon, Dave Joens, and Scott Lynch-Giddings.
59. UCLS *Announcements 1998–99,* p. 67; UCLS Course Evaluations, Winter 1999 (showing a 9.7 mark for Barack from a total of ten respondents); *CF,* 28 December 1998, 12 and 14 January, 3, 4 February 1999; Tim Novak, "Pols' Museum Perk May Violate State's Gift Ban," *CST,* 11 January 1999, p. 10; Christi Parsons and Rick Pearson, "Museums Rethinking Political Free Passes," *CT,* 12 January 1999; Senate Floor Transcript, 13 January 1999, pp. 3, 5, 16; "Sens. Jones, Obama Vow to Continue Fairness Fight," *CD,* 14 January 1999, p. 3; "In the Law Schools," *CDLB,* 18 January 1999, p. 3; "New Governor, Executive Officers Sworn In," *CWR* 14 #11, 4 February 1999; "A Story of Race," Jodi Solomon Speakers Bureau, Boston, n.d.; Kennedy Communications, "State Senator Barack Obama, Candidate for the U.S. Senate: Background Report," April 2003, RRP; Obama, "Politics, Race and the Common Good," Carleton College, Northfield, MN, 5 February 1999; Shuchi Anand, "Speaker Ushers in Black History Month," *Carletonian,* 12 February 1999, pp. 1, 3; Obama, *TAOH,* p. 340; Schoenburg's 2009 interview with Mark DePue, p. 89; Jim Edgar's 28 May 2010 interview with DePue, pp. 676–78; Halvorson, *Playing Ball with the Big Boys,* p. 45; DJG interviews with Heather Sullivan, Laura Mullens Nolen, Dan Shomon, Lisa Madigan, Ira Silverstein, Dave Sullivan, and Bob Bennett.
60. *CF,* 8, 11, 12, 16, 18, 24, 25, 26 February 1999; "Richard Daley for Mayor," *CT,* 7 February 1999; Rick Pearson, "Rush Can't Make Dent in Daley's Huge Lead," *CT,* 14 February 1999;

James Hill and Cornelia Grumman, "Rush Fires Up Backers as Daley Plays It Cool," *CT,* 14 February 1999; Flynn McRoberts and Laurie Cohen, "Challenger Sees Self as Man of the People," *CT,* 16 February 1999; JCAR Minutes, 17 February 1999; Pearson, "Daley Triumphs in Landslide," *CT,* 24 February 1999; Hill, "Rush Sees Gains, but Few Agree," *CT,* 25 February 1999; "Committee Assignments," *CWR* 14 #12, 26 February 1999; E. J. Dionne, "A Mayoral Confession," *WP,* 26 February 1999, p. A27; R. Bruce Dold, "Politics of Race Almost Ignored," *CT,* 26 February 1999; Adam Cohen, "The Frame Game," *Time,* 21 March 1999; Steve Mills, "What Killed Illinois' Death Penalty," *CT,* 10 March 2011; Rush's 1988 interview with Blackside's Terry Rockefeller, his 2000 and 2001 ones with Julieanna Richardson, and his 2005 one with Julian Bond; Dirk Johnson, "A Politician's Life, from Militant to Mainstream," *NYT,* 3 June 1990; Don Wycliff, "Soul Survivor," *CT Magazine,* 16 November 2003; Jennifer Yachnin, "Lawsuits, Debts Plague Rep. Bobby Rush," *Roll Call,* 1 March 2011; Katherine Skiba, "Numerous Battles Haven't Slowed Rep. Bobby Rush," *CT,* 19 September 2011; DJG interviews with Claire Eberle, Jim Rea, and Brad Burzynski.

61. Senate Floor Transcripts, 2 February 1999, p. 5 (introducing SB 166, his EITC bill), 17 February 1999, p. 8, 24 February 1999, pp. 8–9, 16–18, 29, 25 February 1999, pp. 20, 23, 32, 26 February 1999, p. 25; Northwestern News Release, "Rep. Jesse Jackson Jr. and Sen. Barack Obama Featured at Diversity Week," 24 February 1999; "NU Launches Diversity Week," *CDLB,* 26 February 1999, pp. 3, 20; *CF,* 3, 4, 5, 9, 10, 11, and 12 March 1999; Ray Long and Michelle Brutlag, "State Legislator Talks as Witnesses and Bills Wait," *CT,* 4 March 1999; Kurt Erickson, "DUI: Debating Under the Influence?," *BP,* 4 March 1999, p. A3; George Pawlaczyk (AP), "Meeting in Secrecy, Illinois Agency Gives Tax Dollars for Executive Perks," *BND,* 8 March 1999; Senate Floor Transcript, 11 March 1999, pp. 63–64 (58–0–0 Senate passage of Barack's entirely technical SB 565, with 117–0–0 House passage on 5 May and signed into law 29 July 1999 as PA 91-354) and 74 (Obama correcting his vote on SB 485 from no to yes); R. Eden Martin, "Memo to File Re: Barack Obama," 12 March 1999, REMP; Andy Davis, "Influential Civic Group Picks New Leadership," *CT,* 23 June 1999; PR Newswire, "Second Harvest Announces New President and CEO," 18 December 1999; Joyce Foundation, *1998 Annual Report,* esp. pp. 2–3; Joyce Foundation, *Work in Progress,* January 1999; Jeff Borden, "In New Role, Leff Out to Reap Gains for Second Harvest," *CCB,* 25 January 1999; James Janega, "New Leader to Take Over Joyce Foundation Control," *CT,* 28 April 1999; Joyce Foundation, *Work in Progress,* May 1999, p. 22; Jennifer Cassell, "Joyce Foundation's New Chief to Focus on Jobs, Environment," *CST,* 11 June 1999, p. 20; DJG interviews with Joe Seliga, Eden Martin, Dan Rabinovitz, Rob and Lisa Fisher, and Thom Thacker. See also David Freddoso, *The Case Against Barack Obama* (Regnery, 2008), pp. 169–71, reporting Republican state senator Patrick O'Malley's recollection that on 25 February 1999 Barack in the Judiciary Committee initially voted present on O'Malley's Senate Resolution 26, opposing U.S. participation in the International Criminal Court, before switching to yes, and then not voting when the Senate on 3 March passed it 55–1–1. Senate Floor Transcript, 3 March 1999, pp. 11–12.

62. Joe Mahr, "Criminal Code Due for a Cleanup," *SJ-R,* 14 March 1999, p. 1 (and *PJS,* 15 March and *ECN,* 29 March); Ron Eckstein, "Rush Refocuses on Congress, Working with Daley," *CT,* 14 March 1999; *CF,* 15, 16, 17, 18, and 29 March 1999; JCAR Minutes, 16 March 1999; Senate Floor Transcript, 16 March 1999, p. 29 (Obama introducing Senate Joint Resolution 24 to name I-57 from Cairo to Chicago the Thurgood Marshall Memorial Freeway); "From Welfare to Work," *Chicago Independent Bulletin,* 18 March 1999, p. 15; Cornelia Grumman and Rick Pearson, "Ryan Agonized, but Confident He 'Did the Right Thing,'" *CT,* 18 March 1999; Senate Floor Transcript, 18 March 1999, pp. 3, 127–32; Minutes, CAC Board of Directors, 18 March 1999; "Kreiter & Tejeda," *CDH,* 20 March 1999, p. I-8; Senate Floor Transcript, 23 March 1999, pp. 43–44 (SB 1055) and 273–78 (SB 929); Senate Floor Transcript, 24 March 1999, pp. 64–65 (passage of Obama's SB 561 59–0–0) and 73–74 (passage of Obama's SB 680 56–0–0); Senate Floor Transcript, 25 March 1999, pp. 8–12 (SB 11) and 210–11 (SB 759); *CF,* 4 and 5 December 2000 (reprinting Ryan's 30 November 2000 remarks at Northwestern Law School); Bruce Shapiro, "A Talk with Governor George Ryan," *Nation,* 8–15 January 2001, p. 17; Maurice Possley and John Chase, "Cruz, 2 Others Pardoned," *CT,* 20 December 2002; Ryan, "The Role of the Executive in Administering the Death Penalty," 2003 *University of Illinois Law Review* 1077–86; DJG interviews with Mark

Warnsing, Ed Petka, and James Clayborne. On Bill Shaw and his twin brother Bob, see Ben Joravsku, "By Any Means Necessary," *CR*, 31 October 2002, and Gregory Tejeda, "Shaws Used Local Government to Build Allies," *Chicago Argus*, 28 November 2008.

63. Jennifer Loven (AP), "Born to Privilege, Steeped in Caution, Fitzgerald Takes His Place in the Senate," 27 March 1999; Obama, "Keep Church and State Separate," *HPH*, 7 April 1999, p. 4; Cornelia Grumman, "State Taxes Block Road Off Welfare," *CT*, 10 April 1999; John McCarron, "Welfare State: Illinois Ranks at Top in Soaking the Poor," *CT*, 12 April 1999; Stephen Heintz, "Steering Committee" Notes, 13 April 1999, SHP; Cindy Richards, "South Side Has Many Aldermanic Choices," *CT*, 11 February 1999; Andrew Martin and Monica Davey, "Murphy Top Aldermanic Surprise," *CT*, 24 February 1999; Curtis Lawrence, "Independence Is Hot Issue in 5th Ward Race," *CST*, 25 March 1999, p. 32; Steve Neal, "Independence Is at Issue in 5th Ward," *CST*, 29 March 1999, p. 8; Megan O'Matz and Monica Davey, "5th Ward Race Paved with Questions," *CT*, 8 April 1999; Scott Fornek and John Carpenter, "Daley Doesn't Take the 5th," *CST*, 14 April 1999, p. 37; Gary Washburn and Melita Marie Garza, "Daley Double Blow," *CT*, 14 April 1999; Brett Schaeffer, "Hairston Upsets Holt to Win Tight Fifth Ward Race," *HPH*, 14 April 1999, pp. 1–2; Scott Fornek, "5th Ward Reasserts Its Identity," *CST*, 16 April 1999, p. 6; Schaeffer, "Polls Show Win Was a Ward-Wide Effort," *HPH*, 21 April 1999, p. 1; Keith Kelleher and Madeline Talbott, "The People Shall Rule," *NHI Shelterforce* #114, November–December 2000; Atlas, *Seeds of Change*, pp. 110–11; Barack H. Obama and Michelle Robinson-Obama, 1998 Income Tax Return, Vicki L. Hudson & Co., 9 April 1999; Kennedy Communications, "State Senator Barack Obama, Candidate for the U.S. Senate: Background Report," April 2003, RRP; Senate Floor Transcript, 3 November 1995, pp. 28–30 (HB 1498), Illinois Governmental Ethics Act, 5 ILCS 420 Section 2-110, PA 89-405, 8 November 1995; Barack in Lynn Norment, "The Hottest Couple in America," *Ebony*, February 2007, pp. 52ff.; DJG interviews with Madeline Talbott, Keith Kelleher, Barbara Holt, Leslie Hairston, Jim Reed, and Laura Washington.

64. Status of SB 1055, 91st General Assembly (House Financial Institutions Committee Hearing, 15 April 1999); *CF*, 10 March, 15 April and 4, 5, 6, 7 May 1999; JCAR Minutes, 20 April 1999; "Domestic Battery," Copley News Service, 20 April 1999 (and *SJ-R*, 21 April 1999); Caitlin Devitt, "Parents Say School Bill Hits LSCs Too Hard," *HPH*, 28 April 1999, p. 3; Joe Mahr, "Bill Would Allow Victims of Sex Crimes to Have Criminal Case Files Sealed," *SJ-R*, 28 April 1999, p. 38; "State Sen. Obama Gets Top Post, Seeks Funds for Freeways," *South Street Journal*, 29 April–12 May 1999, p. 4; "Session Clears Half Way Point," *CWR* 15 #2, 4 May 1999; Ray Long and Christi Parsons, "Ryan Puts $12 Billion Works Bill on Table," *CT*, 5 May 1999; Ryan in James L. Merriner, *The Man Who Emptied Death Row* (Southern Illinois University Press, 2008), p. 97; Senate Floor Transcript, 6 May 1999, pp. 108–9, 122 (HB 2256, amendatorily vetoed on 2 August 1999); House Floor Transcript, 6 May 1999, pp. 58, 103–7 (114–0–0 passage of SB 1055); Erika Slife, "Margaret Smith, 82, Longtime State Senator a Health Care Advocate," *CT*, 17 May 2005; Kurt Erickson, "Poker Buddies: Obama's Decision Was in the Cards," *QCT*, 17 October 2006 (and *BP*, 24 October 2006, p. A3); Terry McDermott, "What Is It About Obama?," *LAT*, 24 December 2006; Peter Slevin, "Obama Forged Political Mettle in Illinois Capitol," *WP*, 9 February 2007, p. A1; Michael Scherer and Michael Weisskopf, "Candidates' Vices: Craps and Poker," *Time*, 2 July 2008; "Obama as We Knew Him," *Observer*, 26 October 2008; DJG interviews with Laura Mullens Nolen, Dan Shomon, Terry Link, Larry Walsh, Denny Jacobs, Mike Lieteau, Phil Lackman, Dave Manning, Tommy Walsh, Dave Luechtefeld, Boro Reljic, and Lou Viverito.

65. Dave McKinney and Fran Spielman, "Ryan Backs Lobbyists on His Ethics Panel," *CST*, 7 May 1999, p. 8; *CF*, 10, 11, 12, and 13 May 1999; Obama, "The Effects of Gov. Ryan's Plan," *HPH*, 12 May 1999, p. 4; Senate Floor Transcript, 7 May 1999, pp. 30–32 (passage of HB 41 by 52–3–3, with Barack, John Cullerton, and Lisa Madigan voting present) and 55–63 (HB 448 passes 33–16–6); Michelle Brutlag, "Senator's Jab at Referees Is Outvoted by Colleagues," *CT*, 8 May 1999; Dave McKinney and Matt Adrian, "Foul Called on Philip for Quip on 'Popping' Referees," *CST*, 8 May 1999, p. 9; Aaron Chambers, "Measure to Widen Hearsay Evidence Heads to Ryan," *CDLB* [8] May 1999, pp. 1, 24; Bob Greene, "What Would You Call a Person Who Thinks Like This?," *CT*, 12 May 1999; "Poor Sportsmanship," *CDH*, 14 May 1999, p. 12; DJG interviews with Pate Philip and Dave Sullivan.

66. Senate Floor Transcript, 11 May 1999, pp. 17 (Senate passage of HB 854 regarding the sealing of trial records, which the Judiciary Committee had approved 7–0–3 on 28 April, by 58–0–1 with Barack voting present), 26–34 (defeat of HB 1061 by 23–36, with Barack voting no), and 104–16 (passage of HB 2204); Senate Floor Transcript, 12 May 1999, pp. 30–31 (passage of HB 402, lowering the sales tax on car leases, by 57–1 with Barack the sole no); Senate Floor Transcript, 13 May 1999, pp. 54–65 (passage of HB 1232), 102–8 (passage of HB 1812 by 38–18–3, with Barack voting no), and pp. 179–81 (reconsideration of HB 1961 by 34–23, with Barack not voting, and then passage by 31–26); Research Department, NRSC, "Barack Obama (D-IL): On the Record," 22 July 2004, p. 2, JRORF; Senate Floor Transcript, 14 May 1999, pp. 38–55 (passage by 33–26 of HB 152 with Barack voting no) and 120–28 (passage by 56–2 of HB 2320); Ray Long and Michelle Brutlag, "Wirtz Liquor Business Gets Senate Boost," *CT,* 15 May 1999; "How They Voted on Liquor Licensing and Soft Drink Distribution Bills," *Illinois Politics* Vol. 8 #3 [2 August 1999], pp. 6–7; Redfield, *Money Counts,* pp. 125–31; JCAR Minutes, 18 May 1999; *CF,* 19 and 24 May 1999; Obama on *Illinois Lawmakers* with Bruce DuMont, 19 May 1999, MBC; Rick Pearson and Long, "Bargaining Begins as Lawmakers Deliver Pork Wish Lists to Ryan," *CT,* 20 May 1999; Senate Floor Transcript, 20 May 1999, pp. 22–39 (SR 115); "Affirmative Action," *CDH,* 21 May 1999, p. 9; Jeffrey K. Finley, "Senators Debate Minority Information Resolution," *DH&R,* 21 May 1999, p. A3; Long and Pearson, "Black Legislators Play Trump to Block Rosemont Casino," *CT,* 21 May 1999; Long and Pearson, "Ryan's Building Plan Wins Funding," *CT,* 22 May 1999. On Walter Dudycz, see Bryan Miller, "Mr. Dudycz Goes for Washington," *CR,* 18 October 1990.

67. Senate Floor Transcript, 24 May 1999, pp. 16–62 (SB 1017 fails verification 29–29); Rick Pearson and Ray Long, "Tie Vote Sinks Gambling Bill," *CT,* 25 May 1999; Chinta Strausberg, "Black Senators Blast Casino Bill," *CD,* 25 May 1999, pp. 1, 6; *CF,* 25, 26, 27, and 28 May and 23 June 1999; Senate Floor Transcript, 25 May 1999, pp. 6–10 (SB 1017 passes 31–27); Pearson and Long, "Reversal of Fortune," *CT,* 26 May 1999; Dean Olsen, "Senate OKs Gambling Expansion," *ECN,* 26 May 1999; Senate Floor Transcript, 27 May 1999, pp. 73–80 (passage of SB 652 by 56–2); Alexander Russo, "Chicago School Days," *Slate,* 2 April 2008; Jennifer Davis, "Is Welfare-to-Work Working?," *II,* September 1999, pp. 25–28; Senate President James 'Pate' Philip, *End of Session Report 1999,* esp. pp. 53–72, 85–90; Stephen S. Morrill, "Client Report for 1999 End of Session," Morrill and Associates, June 1999; "Session Puts Illinois FIRST," *CWR* 15 #3, 18 June 1999, and 15 #7, 10 November 1999; Emil Jones in *CF,* 6 August 1999; Strausberg, "Obama: Illinois Black Caucus Is Broken," *CD,* 1 June 1999, pp. 1, 8; "Black Caucus Divides on Private Interest Legislation," *Illinois Politics* Vol. 8 #3 [2 August 1999], p. 6; David Welna, "Growing Concerns Regarding Campaign Contributions to Political Candidates from Gambling Industries," *All Things Considered,* NPR, 23 June 1999; Lightford in Kevin McDermott, "Beyond Black," *II,* February 2006, and in Niall Stanage, *Redemption Song* (Liberties Press, 2009), p. 28; Hendon, *Black Enough/White Enough,* p. 11; DJG interviews with Larry Walsh, James Clayborne, Kim Lightford, George Shadid, Robert Molaro, and Larry Suffredin.

68. Hendon, *White Enough/Black Enough,* pp. 11–12; Saguaro Seminars Schedule; Obama and Barbara Flynn Currie, "State Allocates Drive Funding," *HPH,* 16 June 1999, p. 4; Pat Ford to CAC Board of Directors, "Breakthrough Schools," June 1999; Minutes, CAC Board of Directors, 17 June 1999; Mark A. Smylie and Stacy A. Wenzel et al., *The Chicago Annenberg Challenge: Successes, Failres, and Lesson for the Future* (Consortium on Chicago School Research, August 2003), pp. 114, 136–37; JCAR Minutes, 22 June 1999; "Chronology," n.d. [1999], [David Callahan], "Concept Paper," The Network for American Renewal, June 1999, Agenda, First Meeting of the Board of Trustees, Network for American Renewal, 29–30 June 1999, all SHP; David A. Miller, Minutes, Network for American Renewal Board of Trustees, 29–30 June 1999, Demos Papers; Stephen Heintz notes, Network for American Renewal First Board of Trustees Meeting, 29–30 June 1999, SHP; DJG interviews with Larry Walsh, Dave Manning, Tom Dart, Barbara Flynn Currie, John Belcaster, Whit Soule, Paul Strauss, Scott Turow, Ed Wojcicki, Dan Shomon, Will Burns, Steve Rauschenberger, Ed Petka, Pate Philip, Paul Williams, Donne Trotter, Charles Halpern, Stephen Heintz, and David Callahan.

69. Obama for Congress 2000, FEC Report of Receipts and Disbursements, 1 July 1999–31 December 1999, 31 January 2000, p. 59; Greg Downs, "Campaign Dollars Flow Through 4th and 5th Wards," *HPH,* 11 August 1999, pp. 1–2; "Successful 'Second Thursday' Season Wraps Up," *DL21C Report,* Fall 1999, p. 3; Janny Scott, "In 2000, a Streetwise Veteran Schooled a Bold Young Obama," *NYT,* 9 September 2007, p. A1; Minow in Tim Harper, "Civil Rights Elders Aided Obama's Political Rise," *Toronto Star,* 18 August 2008, p. A1; Mikva on PBS NewsHour, 25 September 2008; Rosemary Ellis, "A Conversation with Michelle Obama," *Good Housekeeping,* November 2008; Sher in Jodi Kantor, "The Obamas' Marriage," *NYT Magazine,* 1 November 2009; Schmidt in Remnick, *The Bridge,* p. 314; Travis in Flynn McRoberts, "Chicago's Black Political Movement: What Happened?," *CT Magazine,* 4 July 1999, pp. 11ff.; Deborah Bayliss, "Hyde Park's Own Renaissance Man," *HPH,* 7 July 1999, p. 8; Friends of Barack Obama, Report of Campaign Contributions and Expenditures, 1 January 1999–30 June 1999, 2 August 1999; Cheryl Ross, "Hales Angel," *CR,* 30 October 1997; Steve Kloehn, "Love Binds Grad and 'Dad,'" *CT,* 1 June 1998; "A Model for Making a Difference," *CT,* 2 June 1998; Reynolds in David Smallwood, "Genesis of the Grassroots Movement," *N'Digo Profiles,* December 2008, pp. 10–11, 44; Robinson, *A Game of Character,* p. 159; DJG interviews with Dan Shomon, John Charles, Adelmo Marchiori, Jeff Stauter, Carol Harwell, Terry Link, Denny Jacobs, Mike Lieteau, John Schmidt, Matt Piers, Newton Minow, Sokoni Karanja, Paul Williams, Tim Wright, Bernard Loyd, Tim Black, Freddrenna Lyle, Tom Dart, Toni Preckwinkle, Leslie Hairston, Tim King, Judy Byrd, Emil Jones, Donne Trotter, Vince Williams, Madeline Talbott, Keith Kelleher, Ab Mikva, John Kelly, and Chuy Garcia.

70. Illinois State Senate Democratic Caucus, Department of Commerce and Community Affairs Project Authorization Forms, 5 August 1999, pp. 14–20, per SB 630, Illinois Department of Commerce & Economic Opportunity FOIA Release; Steven R. Strahler, "Rev. Jackson's New Pulpit Is LaSalle St.," *CCB,* 13 July 1998; Eric L. Smith, "How Jesse Jackson's Focus on the Financial Markets Could Make a Difference," *Black Enterprise,* October 1998, pp. 111ff.; Kevin Knapp, "Connecting Biz with Black Community," *CCB,* 13 March 1999; James Hill and Rick Pearson, "Jackson Sets Sights Close to Heart, Home," *CT,* 25 March 1999; Strahler, "Exchanges Give Jesse a Risky Cold Shoulder," *CCB,* 31 May 1999; Julie Johnson, "Another Social Agenda for LaSalle Street," *CCB,* 13 September 1999; Pam Belluck, "Jackson Says He Fathered Child in Affair with Aide," *NYT,* 19 January 2001; Belluck, "Questioned About Finances, Jackson Will Amend Taxes," *NYT,* 8 March 2001; Sabrina L. Miller, "Jackson Faces Suit for Child Support," *CT,* 29 April 2001; Monica Davey, "Jackson Education Fund Amends Disclosure Forms," *CT,* 29 April 2001; Davey, "Jackson Donor List Has Errors," *CT,* 1 May 2001; Strahler, "Detour Ahead for Rev. Jackson's Diversity Push on LaSalle Street," *CCB,* 15 October 2001; Ray Gibson, "Obama Has Long Backed Faith Charities," *CT,* 12 July 2008; Greg Downs, "Challengers Lining Up to Take On Rep. Rush," *HPH,* 14 July 1999, p. 3; "State Senator Trotter to Challenge Rush," *South Street Journal,* 15–28 July 1999, p. 2; JCAR Minutes, 20 July 1999; Friends of Barack Obama, Report of Campaign Contributions and Expenditures, 1 July 1999–31 December 1999, 31 January 2000; Obama for Congress 2000, FEC Candidate Summary Report, 31 December 2000; Obama, "Statement of Candidacy," 28 July 1999 (received by FEC 4 August 1999); Lionel E. Bolin, "Statement of Organization," Obama for Congress 2000, 28 July 1999; AP, "Great for the State," *CDH,* 30 July 1999, p. I-9; Pearson, "2 May Challenge Rush for Congressional Seat," *CT,* 1 August 1999; Steve Neal, "Obama Set to Take on Rush," *CST,* 1 August 1999, p. 9; "News: The City," *CDH,* 1 August 1999, p. 9; Chinta Strausberg, "Obama Set to Take on Rush for Congress," *CD,* 5 August 1999, p. 6; Greg Downs, "Sen. Obama Files Candidacy," and Mitchell A. Pravatiner, "Area Politicians Should Remain in Their Seats," *HPH,* 11 August 1999, pp. 1, 5; Lynn Sweet, "Rush May Face Tough 2000 Bid," *CST,* 12 August 1999, p. 35; "Bud Billiken's 70th Year," *CD,* 14 August 1999, p. 1; Olivia Ridgell, "Drill Team and Drum Corps Competitions Created Excitement," and "And a Good Time Was Had by All," *CD,* 16 August 1999, pp. 5, 11; John Mercurio and Rachel Van Dongen, "Safe Seats? At Least 14 House Members Face Serious Intraparty Foes," *Roll Call,* 16 August 1999; Neal, "Rush's Record an Asset in 2000 Re-Election Fight," *CST,* 16 August 1999, p. 8; JCAR Minutes, 17 August 1999; Kindle in Jo Becker and Christopher Drew, "Pragmatic Politics, Forged on the South Side," *NYT,* 11 May 2008, p. A1, and in Remnick, *The Bridge,* pp. 319–20; Simon to Obama, 18 August 1999, PSP; Neal, "Trotter Getting Ready to Make a Run at Rush," *CST,* 23 August 1999, p. 8; Buvan Nathan, "Looking at Politics," *Chi-*

cago Independent Bulletin, 26 August 1999, p. 2; Neal, "New African-American Leaders Are Emerging," *CST,* 1 September 1999, p. 8; DJG interviews with David Wilhelm, Lionel Bolin, Janis Robinson, Craig Huffman, Al Kindle, Dan Shomon, Will Burns, Toni Preckwinkle, Barbara Flynn Currie, and Roland Burris.

71. Friends of Barack Obama, Report of Campaign Contributions and Expenditures, 1 July 1999–31 December 1999, 31 January 2000; Obama for Congress 2000, FEC Candidate Summary Report, 31 December 2000; Obama, "Customers Need Better Protection," *HPH,* 1 September 1999, p. 4; "Democratic Senators Propose Legislation to Increase Protections for Utility Customers," *Chicago Citizen,* 2 September 1999, p. 18; Harriet Trop notes (dinner with the Obamas and the Miners, 5 September 1999); Obama, "Reimbursing Customers for Blackouts," *CT,* 7 September 1999, p. 14; Obama, "ComEd: A Senator's Solution," *CD,* 9 September 1999, p. 9; Chinta Strausberg, "Pols Seek Override on Vetoed Child Support," *CD,* 14 September 1999, pp. 1, 6; JCAR Minutes, 14 September 1999; Greg Downs, "StreetWise Exec May Bid for Rush's Seat," *HPH,* 15 September 1999, p. 13; "The Rush Race for the 1st District," *South Street Journal,* 16–29 September 1999, p. 5; Ken Rolling, "1999 Mid-Year Report to the Annenberg Foundation," 2 August 1999 ("Obama is seriously contemplating running"); Obama to Members of the [CAC] Board of Directors, 20 September 1999 (resigning as board president effective 30 September); Woods Fund of Chicago, *1998 Annual Report,* p. 1, and *1999 Annual Report,* pp. 2, 6; Steve Neal, "Rep. Rush Enlists Re-Election Muscle," *CST,* 22 September 1999, p. 7; Citizens for Rush, FEC Form 3, Report of Receipts and Disbursements, 1 January–30 June 1999, 5 August 1999, and 1 July–31 December 1999, 30 July 2000; Corey Hall, "State Senator Obama Discusses Past, Future Legislation at Town Hall Meeting," *Chicago Weekend,* 23 September 1999, p. 4; Neal, "2 Challengers Take Aim at Ousting Rush," *CST,* 26 September 1999, p. 10; Lauren W. Whittington, "When Obama Wasn't a Star," *Roll Call,* 12 February 2007; Robert Frank and Mark Maremont, "Money Maven Penny Pritzker Raises Record Amounts of Cash for Obama," *WSJ,* 15 March 2008; Dan Morain and Bob Drogin, "Obama and the Former Radicals," *LAT,* 18 April 2008; Ari Berman, "Obama Under the Weather," *Nation,* 19 May 2008; John Lippert, "Penny Pritzker Shows Why She Convinced Buffett to Support Obama," Bloomberg, 20 August 2008; Jason Zengerle, "The Message Keeper," *TNR,* 5 November 2008; Rush in Remnick, *The Bridge,* p. 315; Evan Osnos, "The Daley Show: Dynastic Rule in Obama's Political Birthplace," *New Yorker,* 8 March 2010, pp. 38ff.; Axelrod, *Believer,* pp. 118, 122; DJG interviews with Cynthia Miller, Dan Shomon, Will Burns, Adelmo Marchiori, Cindy Huebner Davidsmeyer, Jill Rock, Mike Lawrence, John Eason, John Corrigan, Eden Martin, Jean Rudd, Bill Ayers, Howard Stanback, Don Wiener, Eric Adelstein, John Kupper, Pete Giangreco, Delmarie Cobb, and Chris Sautter.

72. Mike Dorning, "Farms, Cuba Form Capitol Crisis," *CT,* 26 September 1999; Obama on WLS Channel 7 TV, 26 September 1999, MBC; Michael Ko, "2 Challengers Seek to Replace Rush in Washington," *CT,* 27 September 1999; Chinta Strausberg, "Trotter, Obama Seek to Oust Rush," *CD,* 27 September 1999, p. 3; "2 Launch Bids to Unseat Rush," *CST,* 27 September 1999, p. 14; "News: The City," *CDH,* 27 September 1999, p. 9; Obama to Jeanne and Paul Simon, n.d. (received 28 September 1999) ("our prayers are with you during your recovery"), PSP; Greg Downs, "Obama Announces Bid for Rush's Congressional Seat," *HPH,* 29 September 1999, pp. 1–2; Scott Fornek, "Running Against Rush," *CST,* 29 September 1999, p. 6; Corey Hall, "State Sen. Donne Trotter Announces for Congress to 'Make a Visible Difference,'" *Chicago Citizen,* 30 September 1999, p. 1; Hall, "State Sen. Barack Obama Announces for Congress," *Hyde Park Citizen,* 30 September 1999, p. 2; Frank Ottman, "1st District Congressional Race, Historical," *South Street Journal,* 21 October–3 November 1999, pp. 4–5; DJG interviews with Dan Shomon, Will Burns, Carolyn Shapiro, and John Rogers Jr.

CHAPTER EIGHT: FAILURE AND RECOVERY

1. UCLS *Announcements 1999–2000,* pp. 39, 43, 75–76; UCLS, *The Glass Menagerie 1999–2000,* esp. p. 20; UCLS Course Evaluations, Autumn 1999; Obama, Constitutional Law III Final Examination, [8–12] December 1999 (again two questions with a 5,200-word limit); Obama to Phil Boerner, n.d. [September 1999], PBP ("I'm running for Congress. Life is hectic but good!"); Rush on *Morning Edition,* NPR, 27 July 2004, and in Scott Helman, "Early

Defeat Launched a Rapid Political Climb," *BG*, 12 October 2007; Citizens for Rush, FEC Form 3, Report of Receipts and Disbursements, 1 July–31 December 1999, 30 July 2000; Michelle in Cassandra West, "Her Plan Went Awry," *CT*, 1 September 2004; "Oprah Talks to Barack Obama," *O Magazine*, November 2004; Obama, *TAOH*, p. 340; Sandra Sobieraj Westfall, "Michelle Obama: 'This Is Who I Am,'" *People*, 18 June 2007, p. 118; Shira Schoenberg, "Obama Keeps Making the Rounds," *Concord Monitor*, 20 December 2007; Moelis in Leslie Bennetts, "First Lady in Waiting," VanityFair.com, 27 December 2007, in Judy Keen, "Michelle Obama: Campaigning Her Way," *USAT*, 11 May 2007, and on *20/20*, ABC, 3 October 2008; Shomon in Carol Felsenthal, "The Making of a First Lady," *CM*, February 2009; Obama, Remarks at Workplace Flexibility Forum, 31 March 2010, *Public Papers of the Presidents* 2010 Vol. 1, p. 433; Kantor, *The Obamas*, pp. 142–43; DJG interviews with Jay Hines-Shah, Nathan Sutton, Michael Edney, Ted Liazos, Heather Sullivan, Joe Khan, Andrew Abrams, Don Wiener, Eric Adelstein, Jerry Morrison, Kathy Stell, Allison Davis, Bernardine Dohrn, and Craig Huffman. On Wiener, see David Moberg, "Grass-Roots Activist for a New World Order," *CT*, 18 August 1992.

2. Minutes, CAC Board of Directors, 16 and 30 (at 6500 S. Champlain) December 1999; "State Senator Challenges Rush to Debate Him," *CT*, 5 October 1999; Chinta Strausberg, "Obama Challenges Rush to Debate," *CD*, 5 October 1999, p. 3; Greg Downs, "Congressional Candidates Angle for Hyde Park Voters," *HPH*, 6 October 1999, pp. 1–2; Barack in John Patterson, "Betting on Name Recognition," *CDH*, 15 February 2000, p. 15; Sautter in Lauren W. Whittington, "When Obama Wasn't a Star," *Roll Call*, 12 February 2007; Obama for Congress 2000, FEC Candidate Summary Report, 31 December 2000; NOW LDEF and Illinois Caucus for Adolescent Health, "Caught in the Crossfire: Illinois Roundtable on Welfare Reform, Women of Color and Reproductive Health," 13 October 1999, Chicago Foundation for Women; Downs, "Hyde Park Activists Launch Campaign Salvo," "Rush: A Collective Vision," and "Rep. Rush Denounces CHA Rehab Plans," *HPH*, 20 October 1999, pp. 2, 6, 9, 11; JCAR Minutes, 19 October 1999; Joe Green, "Barack Obama: An Agenda for the 21st Century," ca. 20–21 October 1999, YouTube; Strausberg, "Obama, Pols, File Bill to Protect Utility Customers," *CD*, 21 October 1999, p. 10; Frank Ottman, "1st District Congressional Race, Historical," *SSJ*, 21 October–3 November 1999, pp. 4–5; "Five Democratic Senators File Legislation to Increase Utility Customers' Protections," *HPC*, 28 October 1999, p. 3; Carol Felsenthal, "Les Coney, Top Obama Bundler and Networker Extraordinaire," *Chicago Magazine*, 21 July 2011; DJG interviews with Will Burns, Cynthia Miller, Dan Shomon, Susan Pierson Lewers, Leslie Corbett, Terrie Pickerill, Chris Sautter, Toni Preckwinkle, Scott Turow, Bill Luking, and Les Coney.

3. Frank Main, "Bobby Rush's Son Is Shot on S. Side," *CST*, 19 October 1999, p. 1; Evan Osnos and Shawn Taylor, "2 Sought in Shooting of Rep. Rush's Son," *CT*, 19 October 1999; Molly Sullivan and Main, "Rush Son Critically Injured in Shooting," *CST*, 20 October 1999, p. 1; Osnos and Jennifer Peltz, "2 Sought in Rush's Son's Shooting," *CT*, 20 October 1999; Gene O'Shea, "Gunmen Shoot Rush's Son," *DS*, 20 October 1999; Walter S. Mitchell III, "Congressman's Son in Critical Condition," *CD*, 20 October 1999, p. 4; Main and Ana Mendieta, "Clues Sought in Rush Son Shooting," *CST*, 21 October 1999, p. 18; "Congressman's Son Remains in Critical Condition," *CT*, 21 October 1999; Main, "Rush's Son Dies of Wounds from Monday Attack," *CST*, 22 October 1999, p. 1; "Rep. Rush's Son Dies of Wounds," *CT*, 22 October 1999; Main, "Rep. Rush's Son Dies 4 Days After Attack," *CST*, 23 October 1999, p. 1; Terry Wilson and Aamer Madhani, "Rush's Son Dies After Organs Fail," *CT*, 23 October 1999; O'Shea and Sean D. Hamill, "Rush's Son Dies of Wounds," *DS*, 23 October 1999, pp. 1–2; Main, "Man Quizzed in Slaying of Rush's Son," and John Carpenter and Robert C. Herguth, "Rush Speaks Out on Death of His Son," *CST*, 24 October 1999, pp. 3, 20; Wilson and Madhani, "Rush's Son Dies," *CT*, 24 October 1999; Bradley Keoun, "Rush Mourns Son," *CT*, 24 October 1999; O'Shea, "One Held in Death of Rush's Son," *DS*, 24 October 1999, p. B3; Chinta Strausberg, "Rush Speaks Out on Son's Tragic Shooting," *CD*, 25 October 1999, p. 4; Osnos and Wilson, "Charges Expected in Fatal Shooting of Rush's Son," *CT*, 25 October 1999; Main and Michael Sneed, "Police Cite 'Confession' in Rich Slaying," *CST*, 26 October 1999, p. 10; Wilson and James Hill, "Suspect Charged in Slaying of Rush's Son," *CT*, 26 October 1999; Anne Bowhay, "Man Charged in Rich's Murder," *DS*, 26 October 1999, pp. 1–2; "Man Charged in Rush Son Killing," *CD*, 26 October 1999, p. 1; Main, "Cops Seek Boss in Rich Slaying," *CST*,

27 October 1999, p. 24; Hill, "Suspect Held Without Bail in Slaying of Rush's Son," *CT*, 27 October 1999; O'Shea, "Huey Rich Murder Motive Disputed," and Jamie Parker, "Funeral Scheduled Thursday for Rich," *DS*, 27 October 1999, pp. B3, B4; Madhani, "Friends Old and New Bid Rush's Son Farewell," *CT*, 29 October 1999; Carpenter, "Goal Emerges from Grief," *CST*, 31 October 1999, p. 2; John McCormick, "A Father's Anguished Journey," *Newsweek*, 28 November 1999; Madhani, "2 Found Guilty of Murdering Rush's Son," *CT*, 15 March 2002; Jeff Coen, "Judge Sentences Murderer of Rush's Son to 90 Years," *CT*, 22 June 2002; *People v. Prince*, 840 N. E. 2d 1240 (Ill. Ct. Apps.), 8 December 2005; DJG interview with Jeremiah Wright.

4. "Obama Asked to Bow Out of Race Against Rush," *CD*, 26 October 1999, p. 8; Paul Simon to Obama, 26 October 1999, PSP; Obama for Congress 2000, FEC Report of Receipts and Disbursements, 1 July–31 December 1999, 31 January 2000, esp. pp. 69 and 75; Citizens for Rush, FEC Report of Receipts and Disbursements, 1 July–31 December 1999, 30 July 2000; Greg Downs, "Candidates Seeking Votes in Unlikely Places," *HPH*, 27 October 1999, p. 9; John H. White, "Rush Announces Campaign for a 5th House Term," *CST*, 1 November 1999, p. 8; Carl Kozlowski, "Democrat Rush Going for 5th Term in Congress," *CT*, 2 November 1999; Downs, "Obama Gains an Endorsement, Rush Announces Re-Election Bid," *HPH*, 3 November 1999, pp. 1–2; Buvan Nathan, "Looking at Politics," and "Rush Seeks Fifth Term on Record," *CIB*, 4 November 1999, pp. 2 and 8; Corey Hart, "Cong. Rush Officially Declares Candidacy for Reelection" and "Senator Trotter Unveils 'Economic Plan' for 1st Congressional," *CW*, 4 November 1999, p. 1; Jennifer Leovy, "Community Tours Encourage Students to Explore Beyond Home," *UCC*, 4 November 1999; Senate Floor Transcript, 4 November 1999, p. 5 (Obama introducing SBs 1256 and 1257); "Week of Lectures on Public Interest Law," *CDLB*, 5 November 1999, p. 2; Rick Pearson, "Democrats Hope to Build a Youthful Base for Future," *CT*, 8 November 1999; Beverly A. Reed, "Obama Seeks Minority Appointee for ICC," *CD*, 9 November 1999, p. 3; *Communications Daily*, 9 November 1999; Downs, "Candidates to Appear at Health Care Forum" and "Trotter: 1st District Short of Washington Cash," Obama, "State Health Plan on the Table," and Don Robinson, "Rush Should Take Obama's Debate Challenge," *HPH*, 10 November 1999, pp. 3, 4, 10, and 17; Downs, "Candidates Spar Over Health Care," *HPH*, 17 November 1999, pp. 1–2, Steve Neal, "Poll Gives Rush Big Lead," *CST*, 21 November 1999, p. 4; Sheldon Lane, "Appoint Minority to ICC," *HPH*, 8 December 1999, p. 9; Rachel Van Dongen, "Serious Primary Threats Abound This Year," *RC*, 21 February 2000; Sautter in Don Gonyea, "Obama's Loss May Have Aided White House Bid," NPR, 19 September 2007, and in Janny Scott, "In 2000, a Streetwise Veteran Schooled a Bold Young Obama," *NYT*, 9 September 2007; Dan Morain, "Obama: A Fresh Face or an Old-School Tactician?," *LAT*, 8 September 2007; Lauren W. Whittington, "When Obama Wasn't a Star," *RC*, 12 February 2007; Obama, *TAOH*, p. 106; Wolffe, *Renegade*, p. 120; DJG interviews with John Kupper, Chris Sautter, Dan Shomon, Will Burns, and Jerry Morrison.

5. *CF*, 12 November 1999, p. 3; Chinta Strausberg, "Obama Launches Congressional Campaign Ads," *CD*, 15 November 1999, p. 3; "Champion," BO-01, and "Blackout," BO-02, Cassette #101287, 11 November 1999, CSP; Andrew Gruber e-mail to Michael Cabonargi, 11 November 1999, Shomon e-mail to Joe Seliga, Gruber, and Don Harmon, 15 November 1999, Gruber e-mail to Harmon et al., 8 December 1999, Seliga to Obama and Shomon, "Obama 2000 Jobs and Economic Development Platform," 20 December 1999, 6pp., [Seliga, Minutes], "Obama 2000 Economic Development Committee," 20 December 1999, 2pp., and Seliga e-mail to Gruber et al., 20 December 1999, all AGP; JCAR Minutes, 16 November 1999; Dirk Johnson, "7 Students Charged in a Brawl That Divides Decatur, Ill.," *NYT*, 10 November 1999; Flynn McRoberts and Janan Hanna, "Embattled School Board Issues Defiant Statement," *CT*, 17 November 1999; Joe Mahr, "Bill Calls for State's Expelled Students to Be Offered Alternate Form of Education," *SJ-R*, 17 November 1999, p. 10 (and *PJS* and *ECN*); John C. Patterson, "State Senator Calls for Changes to School Expulsion Laws in Wake of Decatur Dispute," *DHR*, 17 November 1999, p. A3; Senate Floor Transcript, 18 November 1999, p. 2 (Obama introducing SB 1280); Anthony Man, "Mitchell Sees Politics Behind Black Caucus," *DHR*, 22 November 1999, p. A3; Greg Downs, "New Candidate Enters Race," *HPH*, 17 November 1999, p. 2; DJG interviews with Chris Sautter, Dan Shomon, Andrew Gruber, Don Harmon, and Joe Seliga.

6. Steve Neal, "Poll Gives Rush Big Lead," *CST,* 21 November 1999, p. 4; Paul Merrion, "Corporate Backers Take Sides in a Tough District Race," *CCB,* 22 November 1999, p. 3; "Rush Challenger Has Plan for Computers in Schools," *CT,* 24 November 1999; Chinta Strausberg, "Obama Aims to Wipe Out 'Digital Divide,'" *CD,* 29 November 1999, p. 5; Greg Downs, "Obama: More Tech Funds for Schools," *HPH,* 1 December 1999, p. 8; J. Coyden Palmer, "Senator Obama Looking to Decrease 'Digital Divide,'" *CC,* 2 December 1999, p. 1; "Quiet Fall Veto Session Expected," *C-WR* 15 #7, 10 November 1999; Ray Long and Douglas Holt, "Ryan Veto of Child-Support Break Resoundingly Rebuffed by House," *CT,* 18 November 1999; Senate Floor Transcript, 30 November 1999, pp. 11–26 (HB 1232); Christopher Wills, "Move to Give Welfare Families More Child Support Fails in Senate," AP, 30 November 1999 (and *DS,* 1 December 1999, p. A3); Adriana Colindres, "Senate Nixes Child-Support Bill," *SJ-R,* 1 December 1999, p. 7 (and *PJS* and *ECN*); Long, "Veto Stands Barring More Cash for Kids on Welfare," *CT,* 1 December 1999; *CF,* 1 December 1999, "Few Surprises in Fall Veto Session," *C-WR* 15 #8, 15 December 1999, pp. 2–3; Obama, "Improve Child Support System," *HPH,* 15 December 1999, p. 4; Obama's 4 February 2000 interview with Ted McClelland; Hendon, *Black Enough/White Enough,* p. 12; Lisa Kernek, "School Board Election Takes Shape," *SJ-R,* 11 November 2000, p. 7; "Beverly Ann Criglar, 63," *CT,* 4 February 2004; Bernard Schoenburg, "Former Obama Assistant Remembers Him as Friendly Boss," *SJ-R,* 8 November 2008; Beverly Helm-Renfro's 2010 interview with Mark DePue, pp. 50–52; DJG interviews with Jeremy Flynn, Adelmo Marchiori, Jeff Stauter, Steve Rauschenberger, John Cullerton, Kim Lightford, and Beverly Helm-Renfro.

7. *People v. Cervantes,* 723 N.E. 2d 265 (Ill. S.Ct.), 2 December 1999; Ken Armstrong and Ray Long, "High Court Invalidates Anti-Crime Law," *CT,* 3 December 1999; *People v. Reedy,* 708 N.E. 2d 1114 (Ill. S.Ct.), 22 January 1999; *Fuehrmeyer v. City of Chicago,* 311 N.E. 2d 116 (Ill. S.Ct.), 29 March 1974, quoting *People ex rel. Drake v. Mahaney,* 13 Mich. 481, 494–95 (1865), *CF,* 6, 7, 8 10, and 13 December 1999; Douglas Holt, "Ryan Calls Legislature Back to Fix Crim Laws," *CT,* 7 December 1999; Mark Konkol, "State Lawmakers Back in Springfield," *DS,* 7 December 1999, p. A6; Senate Floor Transcript, 21 May 1997, pp. 25–33 (SB 71); Christi Parsons and Andrew Martin, "Bead Drawn on Gun Law," *CT,* 22 May 1997; Gary Washburn, "Daley Fears More Bloodshed If Gun Law Is Watered Down," *CT,* 5 June 1997; Teresa Puente, "Gov. Edgar Vetoes Measure to Relax Gun Penalties," *CT,* 17 August 1997; Rick Pearson, "Philip vs. Ryan Showdown Looms on Gun Law," *CT,* 10 December 1999; Pearson, "Ryan to Go to the Mat on Gun Issue," *CT,* 11 December 1999; Dave McKinney and Fran Spielman, "Ryan, Philip at Odds Over Gun-Law Penalty," *CST,* 11 December 1999; Long, "Anti-Crime Law Rejection Could Have Ripple Effect," *CT,* 12 December 1999; Jeremy Manier, "Hillard, Devine Out in Front," *CT,* 13 December 1999; Long and Holt, "Standoff Stalls Vote on Revamp of Gun Law," *CT,* 14 December 1999; Christopher Wills, "Great Gun Debate," AP, *DS,* 14 December 1999, pp. 1–2; Holt and Long, "Philip Flip-Flops on Gun Penalty He Once Backed," *CT,* 15 December 1999; DJG interviews with Gideon Baum, Mark Warnsing, Ed Petka, and Pate Philip.

8. Evan Osnos, "Bobby Rush," *CT,* 5 December 1999, p. P3; Osnos, "'Racial Profiling' Target of Bill," *CT,* 7 December 1999; Chinta Strausberg, "Coalitions' Bill Hits Racial Profiling Tiff," *CD,* 7 December 1999, p. 3; Strausberg, "Daley, Devine Fight Battle for Tough Laws" and "Obama Unveils Federal Gun Bill," *CD,* 13 December 1999, p. 3; Hurley Green III, "Sen. Obama Hosts Anti-Violence Rally" and "Sen. Obama Challenges Cong. Rush in 1st," *CIB,* 9 [sic] December 1999, pp. 3 and 7; Caitlin Devitt, "Legislators Back in Session to Pass Gun Law," *HPH,* 15 December 1999, p. 2; Jennifer Loven (AP), "Man on a Mission," *DS,* 13 December 1999, pp. 1–2 (and as "Son's Slaying Transforms Congressman's Priorities," *LAT,* 19 December 1999); Greg Downs, "New Candidate Enters Race," *HPH,* 17 November 1999, p. 2; Douglas Holt, "Early Bird Challenger Gets Jump on Campaign Filing," *CT,* 14 December 1999; Beverly A. Reed, "Obama Demands Debate," *CD,* 15 December 1999, p. 3; JCAR Minutes, 14 December 1999.

9. *CF,* 15, 16, 17, 18, 20, 21, 22, and 30 December 1999; Ray Long and Douglas Holt, "Stalemate Over Gun Law," and "Hang Tough on Gun Law," *CT,* 16 December 1999; Don Thompson, "Governor Backs Misdemeanor Option for Gun Crimes," *CDH,* 16 December 1999, p. 14; Senate Floor Transcript, 16 December 1999, pp. 6–37; Rick Pearson and Long, "The Showdown in Springfield," *CT,* 17 December

1999; Senate Floor Transcript, 17 December 1999, pp. 19–34; Long and Pearson, "Compromise Crime Bill Collapses," *CT,* 18 December 1999; Long and Pearson, "Ryan Takes Gun-Law Fight to Public," and Pearson, "More Than a Bill Riding on Line," *CT,* 19 December 1999; Pearson and Long, "Ryan Seeks GOP Support on Crime Bill," *CT,* 20 December 1999; John O'Connor, "Legislative Leaders Continue Temporary Solution to Gun Debate," AP, 20 December 1999 (and *Evansville Courier,* 21 December 1999, p. B7); Long and Pearson, "Philip's Plan Fails to Move Gun Talks," *CT,* 21 December 1999; Phil Kadner, "Dishonest Legislators Concoct Bad Laws," *DS,* 21 December 1999, p. A3; Senate Floor Transcript, 21 December 1999, pp. 20–22; Pearson and Long, "Ryan Going After Philip's Loyalists," *CT,* 22 December 1999; Holt, "Ryan Tries to Downplay Gun-Bill Rift," *CT,* 23 December 1999; Holt, "Philip Forces Not Bending for Ryan," *CT,* 24 December 1999; Pearson, "GOP Revives Debate on Concealed Weapons," *CT,* 25 December 1999; Joe Mahr, "Dart Tightens Single-Subject Clause," *SJ-R,* 26 December 1999, p. 17, Rich Miller, "GOP Senators Play the 'Integrity' Card in Gun Battle," *DS,* 26 December 1999, p. A13; Lisa Song, "Top Health Official Campaigns for Ryan on Anti-Crime Legislation," *CT,* 27 December 1999; Long and Pearson, "Gun Bill No Threat, Ryan Tells Hunters," *CT,* 28 December 1999; "Ryan Trying Again," *DS,* 28 December 1999, p. A10; Obama, "Ideologues Frustrate Gun Law," *HPH,* 29 December 1999, p. 5; Dirk Johnson, "Republicans in Illinois Feud Over Gun Control," *NYT,* 29 December 1999; Pearson and Long, "Ryan Likes Chances, but Philip in Control," *CT,* 29 December 1999; Mark J. Konkol, "Ryan: I Have Votes for Gun Bill," and Phil Kadner, "Making Streets Safe Is a Nasty Business," *DS,* 29 December 1999, p. A3; Pearson and Long, "Ryan Comes Up Short," *CT,* 30 December 1999; Dave McKinney, "Ryan Loses on Gun Bill," *CST,* 30 December 1999, p. 1; Thompson, "Anti-Crime Bill Falls Again," *CDH,* 30 December 1999, p. 1; Obama, "Family Duties Took Precedence," *HPH,* 12 January 2000, p. 4; Obama's 4 February 2000 interview with Ted McClelland, *CF,* 9 January 2001 (Wendell Jones); DJG interviews with Ed Petka, Mark Warnsing, Todd Vandermyde, Ned Mitchell, John Charles, Dan Shomon, and Bill Luking.

10. Don Thompson, "Crime Bill Trouble Causes Worry About GOP Control in Senate," *CDH,* 31 December 1999, p. 1; "Philip, Criminals Win Again," *CT,* 31 December 1999; "Pols Take a Pass," *DS,* 31 December 1999, p. A10; Steve Neal, "Weak-Kneed Senators Won't Hobble Ryan," *CST,* 31 December 1999, p. 16; Bernard Schoenburg, "One Odd Thing After Another as Senate Session Dragged On," *SJ-R,* 2 January 2000, p. 15; Rich Miller, "Ryan, Like Edgar, Learns He Can't Win Pate's Game," *DS,* 2 January 2000, p. A17; Chinta Strausberg, "Clergy Continue Anti-Shooting Battle," *CD,* 3 January 2000, p. 4; Ellen Warren and Terry Armour, "The Inc. Column," *CT,* 4 January 2000; Strausberg, "Gov. Ryan, Trotter Say 'Gun Bill Is Not Dead,'" *CD,* 4 January 2000, p. 6; Warren and Armour, "The Inc. Column," *CT,* 5 January 2000; Dave McKinney, "Senators Explain Gun Vote Absence," *CST,* 5 January 2000, p. 8; Greg Downs and Caitlin Devitt, "Obama Misses Gun Law Vote," and Devitt, "Candidates Jockey for Ballot Position," *HPH,* 5 January 2000, pp. 1–2; Dana Dubriwny, "State Senator Who Missed Gun Vote Says She's Sorry," *CDH,* 6 January 2000, p. 4; Ihejirika in Corey Hall, "Congressional Concerns: State Senator Barack Obama," *CW,* 13 January 2000, p. 16; Remnick, *The Bridge,* p. 321; DJG interviews with Cynthia Miller, Chris Sautter, Dan Shomon, and Donne Trotter.

11. Don Thompson, "Philip Says Ryan May Be Getting Bad Advice on Gun Issue," *CDH,* 9 January 2000, p. 1; *CF,* 11 January 2000; JCAR Minutes, 12 January 2000; Senate Floor Transcript, 12 January 2000, p. 6; Lisa Snedeker and Kevin McDermott, "Lenders Criticize Bill," *SLPD,* 13 January 2000, p. A1; Steve Warmbir, "Trial Sets Off Debate on Racial Profiling," *CDH,* 21 January 2000, p. 1; Senate Floor Transcript, 13 January 2000, p. 3; "Eye on Legislation," *CDLB,* 18 January 2000, p. 3; Karen Shields, "Independent Organization Favors Obama," Obama, "Family Duties Took Precedence," "Truth and Tales in Gun Law Vote," Mikva, "Obama Not to Blame," and James Cracraft, "Where Was Obama?," *HPH,* 12 January 2000, pp. 2, 4, 7; Mikva, "Family Values," *CT,* 13 January 2000, p. 22; Adam Harrington, "Obama Under Fire for Missing Gun Vote," *CWN,* 13 January 2000; Corey Hall, "Congressional Concerns: Congressman Bobby Rush," "Congressional Concerns: State Senator Donne Trotter," and "Congressional Concerns: State Senator Barack Obama," *CW,* 13 January 2000, pp. 4 and 16; David Mendell, "Obama Defends Decision to Miss Anti-Crime Vote," *CT,* 17 January 2000; "Senator Obama Unveils Englewood Beautification Plan," *CD*, 24 January 2000, p. 14; Obama, "Why Dr. Martin Luther King, Jr., Is Important to Me," *CD*, 15 January 2000, p. 18; "Race Profiling," *II*, March 2000, p. 13; Heather Nickel, "Driving While Black," *II*, June 2000, pp. 18–21.

12. Chinta Strausberg, "Democrats Urge Congress to Help Seniors, Patients Meet Rx Costs," *CD*, 19 January 2000, p. 5; Beverly A. Reed, "Obama Calls for Drop in Drug Costs," *CD*, 20 January 2000, p. 50; Hall, "Congressional Candidates Answer Questions at Forum," *CW*, 20 January 2000, p. 4; Ted Kleine, "Is Bobby Rush in Trouble?," *CR*, 17 March 2000; McClelland, "How Obama Learned to Be a Natural," *Salon*, 12 February 2007; McClelland, *Young Mr. Obama*, pp. 158–59; Kathleen Lavey, "Book Traces Obama's Political Roots in City," *CST*, 23 October 2010; Burney Simpson, "The 1st Congressional District Encompasses Extreme Contrasts," *II*, February 2000, pp. 6–7; Lu Palmer's 1989 interview with Blackside's Dave Lacy; H. Gregory Meyer, "Lu Palmer, 82," *CT*, 14 September 2004; Beauty Turner, "Race for Congress," *Residents' Journal*, February 2000, p. 9; Ali Abunimah, "How Barack Obama Learned to Love Israel," *Electronic Intifada*, 4 March 2007; Michael McAuliff, "Hil Bam Gear Up for Battle over Jewish Vote," *NYDN*, 6 March 2007, p. 32; Larry Cohler-Esses, "Obama Pivots Away from Dovish Past," *Jewish Week*, 9 March 2007; Edward McClelland, "The Crazy Uncles in Obama's Attic," *Salon*, 18 March 2008; Peter Wallsten, "Allies of Palestinians See a Friend in Barack Obama," *LAT*, 10 April 2008; DJG interviews with Will Burns, John Eason, Nat Piggee, and Lula Ford.

13. Senate Floor Transcripts, 20 January 2000, p. 2 (introducing SBs 1407, 1408, and 1413), 26 January 2000, pp. 9 and 11 (introducing SBs 1489 and 1506), 27 January 2000, p. 6 (introducing SJR 51); Obama for Congress 2000, FEC Form 3, Report of Receipts and Disbursements, 1 July 1999–31 December 1999, 31 January 2000, and 1 January 2000–1 March 2000, 22 November 2002; Citizens for Rush, FEC Form 3, Report of Receipts and Disbursements, 1 July 1999–31 December 1999, 30 July 2000; Trotter for Congress, FEC Form 3, Report of Receipts and Disbursements, 1 July 1999–31 December 1999, 1 February 2000; Rush campaign ad, *Residents' Journal*, February 2000, p. 17 ("We're sticking with Bobby"); Karen Shields, "Congressional Candidates Win Endorsements," and Ann Cadge, Letter to the Editor, *HPH*, 2 February 2000, pp. 7 and 4; Shields, "Obama Surpasses Incumbent in Money-Raising Campaign," *HPH*, 9 February 2000, pp. 1–2; Adam Harrington, "Rush Ahead in Cash Race," *CWN*, 10 February 2000; Rachel Van Dongen, "Serious Primary Threats Abound This Year," *RC*, 21 February 2000; William Hatfield, "Democrats Seek Tax Credits for Day Care, Tuition, Rents," *CST*, 28 January 2000, p. 14; Curtis Lawrence, "Gay Activists Savor Success at Fund-Raiser," *CST*, 30 January 2000, p. 3 (Obama attending Equality Illinois event); Chinta Strausberg, "Obama Calls for End to 'Star Wars' Funding," *CD*, 31 January 2000, p. 5; Ken Armstrong and Steve Mills, "Ryan: 'Until I Can Be Sure,'" *CT*, 1 February 2000; Pat Karlak, "Death Penalty Put on Hold," *CDH*, 1 February 2000, p. 1; Dirk Johnson, "Illinois, Citing Faulty Verdicts, Bars Executions," *NYT*, 1 February 2000; Johnson, "Illinois Governor Hopes to Fix a 'Broken' Justice System," *NYT*, 19 February 2000; Obama, "State Should Curtail All Executions," *HPH*, 23 February 2000, p. 4; *CF*, 27 January and 1 and 2 February 2000; Matt O'Connor, "Charges Reach Ryan's Inner Circle," *CT*, 2 February 2000; Dirk Johnson, "No Executions in Illinois Until System Is Repaired," *NYT*, 21 May 2000; Ted Gregory, "DuPage Ready to Settle Cruz Prosecution Suits," *CT*, 22 September 2000; Bernard Schoenburg, "Former Obama Assistant Remembers Him as Friendly Boss," *SJ-R*, 8 November 2008; Beverly Helm-Renfro's 13 July 2010 interview with Mark DePue, p. 53; DJG interviews with Beverly Helm-Renfro, Eric Adelstein, Jim Covington, Kathy Saltmarsh, and Mark Warnsing.

14. Senate Floor Transcript, 1 February 2000, pp. 4–5 (Obama introducing SBs 1583, 1594 and 1606), 2 February 2000, pp. 6–7 (Obama introducing SBs 1644 and 1663), 3 February 2000, p. 1 (Obama introducing SBs 1710, 1711, 1712, and 1713); AP, "State Senator Introduces Bill to Ban Flavored Cigarettes," 4 February 2000; Chinta Strausberg, "Sen. Obama Seeks Ban on Sale of Bidi Cigarettes," *CD*, 5 February 2000, p. 4; Obama's 4 February 2000 interview with Ted McClelland.

15. Kennedy Communications, "State Senator Barack Obama, Candidate for U.S. Senate: Background Report," April 2003, RRP; Beverly A. Reed, "State Senator Obama Joins Efforts to Rebuild Englewood," *CD*, 7 February 2000, p. 5; "Durbin Endorses Rush for Fund-Raising Efforts," *CDH*, 7 February 2000, p. I-9; "Race-Reporting Bill Is Good Legislation," *MD*, 8 February 2000; JCAR Minutes, 8 February

2000; Sydney G. Bild, "Rush Leads Way in Prescription Drug Benefits," *HPH,* 9 February 2000, p. 4; Jennifer Nelson, "Pair of Measures Would Discount Drugs for Seniors," *SJ-R,* 9 February 2000, p. 11; Corey Hall, "Rush, Trotter and Obama Present Themselves to CBA," *CW,* 10 February 2000, p. 2; Obama to Keith Peterson, 10 February 2000, KPP; Reed, "Barack Obama Promises Vision, Leadership in 1st Congressional Seat," *CD,* 12 February 2000, p. 3; "Afford," BO-03, and "Profile," BO-04, Cassette #102436, 14 February 2000, CSP; "IVI-IPO Backs Moore, Obama in Contests," *CT,* 15 February 2000; Chinta Strausberg, "IVI/IPO Backs Cousins, Obama for March 21st Race," *CD,* 15 February 2000, p. 3; Obama, "Putting a Stop to Racial Profiling," *HPH,* 16 February 2000, p. 4; Adam Harrington, "Mr. Obama Tries to Go to Washington," *CWN,* 17 February 2000.

16. Scott, *A Singular Woman,* p. 10; Andrew Kenney, "UNC Researcher, a Pioneering Academic, Kept Kinship to Obama to Herself," *N&O,* 23 June 2014; Dean Olsen, "Plan to Let Schools Exceed Tax Cap Fails," *SJ-R,* 17 February 2000, p. 36 (Obama in Revenue Committee not voting on SB 1878); Scott Fornek, "Rush Agrees to Debate Dem Opponents," *CST,* 18 February 2000, p. 12; "News: The City," *CDH,* 18 February 2000, p. I-13, and 19 February 2000, p. 8; Beverly A. Reed, "Earned Income Tax Credit Near," *CD,* 19 February 2000, p. 1; Curtis Lawrence, "Rush, Opponents Clash Off the Air," *CST,* 19 February 2000, p. 4; Lawrence, "Rush Disputes Ad Portrayals at Debate," *CST,* 20 February 2000, p. 12; Evan Osnos, "Rush Could Face His Toughest Test," *CT,* 20 February 2000; Chinta Strausberg, "Congressional Hopefuls Battle Over Vouchers, Health," *CD,* 21 February 2000, p. 3; Osnos, "Challengers Campaign Without Rush," and John Chase, "Jeanne Simon, 77, Wife of Ex-Senator," *CT,* 21 February 2000; Senate Floor Transcript, 22 February 2000, p. 2 (Obama introducing SB 1943); "Rush's Father Dies After Falling Ill," *CT,* 22 February 2000; "Ryan Looks Ahead with Budget Address," *C-WR* 15 #10, 23 February 2000; Karen Shields, "Candidates Struggle to Articulate Differences," *HPH,* 23 February 2000, p. 3; Senate Floor Transcripts, 24 February 2000, pp. 43–44 (Obama apologizing for wrongly voting no on SB 649), 93–96 (SB 1532), 25 February 2000, pp. 80–87 (SB 1885); Jennifer Nelson, "Senate Wants No Tobacco Money for Lawyers," *SJ-R,* 26 February 2000, p. 5; Kellie Gormly, "Several Bills Target Racial Profiling," *SJ-R,* 27 February 2000, p. 11 (and *PJS* 27 February and *ECN* 28 February 2000); Diane Lewis, "Attorneys' Tobacco Pay Debated," *BP,* 29 February 2000, p. A5; Alysia Tate, "Outside Dollars Drive Campaigns in 1st District Race," *Chicago Reporter,* March 2000; Steve Neal, "Rush Should Practice What He Once Preached," *CST,* 3 March 2000, p. 9; Jennifer Loven, "Panther Turned Congressman Faces Tough Re-Election Battle," AP, 15 March 2000 (20 February Trinity forum); Rachel Van Dongen, "Rush Confident He'll Be Back," *RC,* 20 March 2000; Obama for Congress 2000, FEC Form 3, Report of Receipts and Disbursements, 1 January 2000–1 March 2000 and 2 March 2000–31 March 2000, 22 November 2002; Citizens for Rush, FEC Form 3, Report of Receipts and Disbursements, 1 January 2000–1 March 2000, 14 February 2001; Trotter for Congress, FEC Form 3, Report of Receipts and Disbursements, 1 January 2000–1 March 2000, 14 March 2000; Friends of Barack Obama, Form D-2, Report of Campaign Contributions and Expenditures, 1 January–30 June 2000, 31 July 2000; Lauren W. Whittington, "When Obama Wasn't a Star," *RC,* 12 February 2007; Obama, *TAOH,* p. 233 ("police cars pulling me over for no apparent reason"); DJG interviews with Charles Payne and Sheila Simon.

17. Lorraine Forte, "Female Leaders Support Rush," *CST,* 28 February 2000, p. 10; Chinta Strausberg, "Obama Picks Up Clergy Support for Congressional Bid," *CD,* 29 February 2000, p. 3; Curtis Lawrence, "Pastors Turn to Obama Over Rush in Race," *CST,* 29 February 2000, p. 8; "Group of Ministers Refuse to Back Rush," *CDH,* 29 February 2000, p. 9; Robert Roman, "Chicago DSA Recommendations for the March Primary Election," *New Ground* #45, March–April 2000 ("recommending both Bobby Rush and Barak Obama"); R. Eugene Pincham, "Rush Challengers Play Follow the Leader," *HPH,* 1 March 2000, p. 4; "Local Ministers Endorse Obama for Congress," *CD,* 4 March 2000, p. 1; "Barack Obama for Congress," *CT,* 6 March 2000; "News: The City," *CDH,* 6 March 2000, p. 9; Susan Dodge, "Mikva Endorses Obama in 1st District," *CST,* 6 March 2000, p. 9; "Obama Announces Help for Goldblatt's Workers," Press Release, 6 March 2000, AGP; Strausberg, "Obama Tries to Rescue Laid Off Goldblatt Employees," *CD,* 7 March 2000, p. 5; JCAR Minutes, 7 March 2000; Karen Shields, "Congressional Race Heats Up," and Obama, "Fight for Equal Pay Is Vitally Important,"

HPH, 8 March 2000, pp. 1–2, 4; Forte, "Better Late Than Never: Rush, Rivals Debate," *CST,* 8 March 2000, p. 24; Phil Kadner, "There's No 'Bulworth' in Congressional Debate," *DS,* 9 March 2000, p. A3; "Divide," BO-05, and "Agree," BO-06, Cassette #102848, 10 March 2000, CSP; John Presta, *Mr. and Mrs. Grassroots* (Elevator Group, 2010), pp. 1–3, 24–25, 35, 48; Steve Metsch, "Beverly Man Writes Book About Obama Campaigns," *SS,* 22 March 2010; DJG interview with Mike Pfleger.

18. Michelle McFarland-McDaniels, "Rep. Rush's Glass House Seems to Be Crumbling," *CD,* 11 March 2000, p. 14; Evan Osnos, "Challengers Debate Rush on His Voting," *CT,* 12 March 2000; Steve Neal, "Rush Picks Up Clinton Support," *CST,* 12 March 2000, p. 11; Mark J. Konkol, "Rush Takes Challengers Seriously," and "U. S. House Choices," *DS,* 12 March 2000, pp. A1, A4, A12; "Clinton Endorses Rush's House Campaign," *CDH,* 12 March 2000, p. I-11; Karen Shields, "Congress Campaign Moves into Home Stretch," *HPH,* 15 March 2000, pp. 1, 8; Corey Hall, "Disagreements, Differences on Display at 1st Congressional District Candidates Forum," *CW,* 16 March 2000, p. 1; Bill Clinton TV and radio ads for Bobby Rush, 13 March 2000; *Chicago Tonight,* WTTW, 13 March 2000; Curtis Lawrence, "Deeds, Not Issues, in 1st Dist. Spotlight," *CST,* 14 March 2000, p. 12; AP, "Quest to Lure Voters," *CDH,* 21 March 2000, p. 1; Christopher Wills, "Obama Friend Sells Candidate," AP, 16 December 2007; Presta, *Mr. and Mrs. Grassroots,* pp. 2, 58; DJG interviews with Dan Shomon, Will Burns, Cynthia Miller, Tom Dart, Bill Fuhry, Gideon Baum, John Corrigan, Joe Seliga, Andrew Boron, Nat Piggee, Warren Chain, John Eason, Tasneem Goodman, Dan Johnson-Weinberger, Joe Khan, Hasan Chandoo, Andy Schapiro, and Jesse Ruiz. Davis introduced HB 3911 on 17 February 1999; Barack introduced SB 1324 on 12 January 2000. Dave McKinney, "Videotape Bill Fails in House," *CST,* 26 March 1999, p. 18; Maurice Possley and Ray Long, "Bill Seeks Statewide Taping of Suspects," *CT,* 8 February 2000; Long and Possley, "Videotaping of Suspects Gets Push," *CT,* 18 February 2000; "Racial Profiling," *II,* April 2000, p. 12; Steven A. Drizin and Beth A. Colgan, "Let the Cameras Roll: Mandatory Videotaping of Interrogations Is the Solution to Illinois' Problem of False Confessions," *Loyola University of Chicago Law Journal* 32 (Winter 2001): 337–424, esp. 344–48, 391, 404, 412–19.

19. Curtis Lawrence, "Rush Defending Seat Against 3 Dem Challengers," *CST,* 14 March 2000, p. 4; Michelle McFarland-McDaniels, "Rush Campaign Should Stop Throwing Stones," Lois Friedberg-Dobry, "Despite Trotter Claims, IVI-IPO Backs Obama," and Obama Campaign Ad, *HPH,* 15 March 2000, pp. 4 and 9; Gregory L. Giroux, "Chicago Voters Will Apply Judgment to Rush in Tuesday's Primary," *CQ Daily Monitor,* 15 March 2000; "State Senator Obama Delivers $400,000 to Kennedy-King College for Jobs Training Program," Business Wire, 15 March 2000; Chinta Strausberg, "Obama Ends Job Training Plan at Kennedy-King College," *CD,* 16 March 2000, p. 14; "Another Term for Rush," *CST,* 16 March 2000, p. 35; Stanley Ziemba, "Rush's Rival Backed by Suburban Officials," *CT,* 16 March 2000; Ted Kleine [McClelland], "Is Bobby Rush in Trouble?," *CR,* 17 March 2000; Wayne D. Watson, "Opening Doors Changing Lives Program," Kennedy-King College, 4 May 2000; Mary Mitchell, "Memoir of a 21st-Century History Maker," *Black Issues Book Review,* January–February 2005, pp. 18–21; Cheryl L. Reed, "The Used Car Lot of the Book World," *CST,* 21 January 2007, p. B12; Rush in Remnick, *The Bridge,* p. 331; McClelland, *Young Man Obama,* p. 153; McClelland, "When Obama Loved Reporters," NBCChicago.com, 29 May 2013; Katherine Skiba and Lolly Bowean, "Behind the Scenes at the White House," *CT,* 6 April 2011; Rich Miller, "Obama's Roots a Reminder That Ambition Can Be a Grand Thing," *CCB,* 20 January 2017; DJG interviews with Brad Jonas and Sharon Corrigan. The Watson statement asserts that the newly announced $400,000 would come on top of a prior $500,000, but subsequent state documents reflect only one single $450,000 Obama-initiated grant to Kennedy-King College.

20. Obama for Congress 2000, FEC Form 3, Report of Receipts and Disbursements, 2 March–31 March 2000, 22 November 2002; "Our Endorsements," and Chinta Strausberg, "2nd [sic] Congressional District Voters Set to Clear Smoke," *CD,* 18 March 2000, pp. 1, 17; "Hotly Contested Race for the Democratic Nomination for Representative of the 1st District Based on the South Side of Chicago," *Weekend Edition,* NPR, 18 March 2000; Abdon M. Pallasch, "NRA Lends Support to Zwick Campaign," *CST,* 19 March 2000, p. 17; Rachel Van Dongen, "Rush Confident He'll Be Back," *RC,* 20 March 2000; Kevin Davis, "Former '60s Militant Working to Change 'Culture of Violence,'" *USAT,* 20 March 2000, p. 26A; Scott Fornek, "GOP Crying Foul Over Phony Ballot," *CST,* 20 March 2000, p. 4; John Kass, "Let Election Day Become Holiday from Political Stiffs," and Salim Muwakkil, "Ironies Abound in 1st District," *CT,* 20 March 2000; *CF,* 21 March 2000; Rick Pearson, "It's Over for Big Guys," *CT,* 21 March 2000; Obama in Pat Guinane, "Star Power," *II,* October 2004; Scott Helman, "Michelle Obama Revels in Family Role," *BG,* 28 October 2007; Craig Robinson on *Anderson Cooper 360,* CNN, 15 April 2008; Jodi Kantor, "Michelle Obama, Reluctant No More," *NYT,* 26 August 2008; Presta, *Mr. and Mrs. Grassroots,* p. 63; Beverly Helm-Renfro's interview with Mark DePue, p. 59; DJG interviews with Will Burns, Dan Shomon, Beverly Helm-Renfro, Barbara Flynn Currie, Chris Sautter, Craig Huffman, John Eason, Warren Chain, Tasneem Goodman, Joe Seliga, Jesse Ruiz, Terrie Pickerill, Susan Pierson Lewers, Leslie Corbett, Paul Strauss, Marlies Carruth, Tim King, and Tom Dart. The five 17 March Rezko contributors were Dan Mahru, Joseph Aramanda, Michael Sreenan, Deloris Wade, and Jennifer Shaxted Arons. Elie "Lee" Maloof and developer Mark Temple and his wife Janette all likewise contributed $1,000 each on 17 March. Patrick J. Fitzgerald, "Government's Motion for Admission of Other Acts of Evidence," *U.S. v. Antoin Rezko,* U. S. D. C. N. D. Ill. #05-CR-691, 11 January 2008, p. 18n.9; Kenneth P. Vogel, "Rezko Role Bigger Than Admitted," *Politico,* 14 March 2008; Vogel, "Obama Releases Names of Rezko-Linked Donors," *Politico,* 15 March 2008; and Don Wiener e-mail to DJG. See also Thomas A. Corfman, "Developer Agrees to Buy 62-Acre Site on Near South Side," *CT,* 11 April 2001, and Corfman, "Housing Planned for Near South—Rezmar Buys Site for $67.5 Million," *CT,* 22 March 2002.

21. Susan Kuczka and Flynn McRoberts, "Kirk, Rush, Fend Off Strong Ballot Challengers," *CT,* 22 March 2000; Curtis Lawrence, "Rush Wins in 1st," *CST,* 22 March 2000, p. 2; Mark Konkol, "Rush Holds Off Fierce Competition," *DS,* 22 March 2000, p. B2; "Low Voter Turnout," *CD,* 22 March 2000, p. 1; Corey Hall, "Cong. Rush Wins Democratic Nomination for Fifth Term," *CC,* 23 March 2000, p. 4; "Dismay and Hope," *CST,* 23 March 2000, p. 39; Chinta Strausberg, "Rush Thanks Voters," *CD,* 23 March 2000, p. 1; Konkol, "Rush Rebounds for Easy Primary Win," *DS,* 23 March 2000, p. B6; Adam Harrington, "Rush to Victory," *CWN,* 27 March 2000; Strausberg, "Rush, Obama and Trotter Return to Teamwork," *CD,* 27 March 2000, p. 5; Karen Shields, "Rush Re-Elected," *HPH,* 29 March 2000, pp. 1–2; Steve Neal, "Rush's Solid Win Proves His Worth in 1st District," *CST,* 3 April 2000, p. 8; Caitlin Devitt, "March Primary Reverberates Through Political Year," *HPH,* 27 December 2000, pp. 6, 8 (Obama "did much worse than expected in the race"); Jim Reynolds in David Smallwood, "Genesis of the Grassroots Movement," *N'Digo Profiles,* December 2008, pp. 10–11, 44; DJG interviews with Terrie Pickerill, John Eason, and Dan Shomon.

22. Obama in John Patterson, "Betting on Name Recognition," *CDH,* 15 February 2004, p. 15; Shomon in Lauren W. Whittington, "When Obama Wasn't a Star," *RC,* 12 February 2007; Obama in Wolffe, *Renegade,* p. 120; Ruiz in McClelland, *Young Mr. Obama,* p. 166; Rebecca Traister, *Big Girls Don't Cry* (Free Press, 2010), p. 50; DJG interviews with Toni Preckwinkle, Dan Shomon, John Kupper, Barbara Flynn Currie, Will Burns, Judy Byrd, Elvin Charity, Carol Harwell, and Bill Ayers.

23. Obama in Garance Franke-Ruta, "The Next Generation," *American Prospect,* August 2004, pp. 13ff., in Scott Fornek, "'I've Got a Competitive Nature,'" *CST,* 3 October 2004, p. 12, on *Off Topic with Carlos Watson,* CNN, 24 October 2004, on *The Tavis Smiley Show,* 3 November 2004, at the American Magazine Conference, Phoenix, AZ, 23 October 2006, in Christopher Wills, "Obama Learned from Failed Congress Run," AP, 24 October 2007, and in Mendell, *From Promise to Power,* p. 146; Shomon in Carol Felsenthal, "What If: Would Obama Be Living in the White House Today If . . . ," *Huffington Post,* 30 January 2009, in Felsenthal, "The Making of a First Lady," *CM,* February 2009, pp. 50ff., and in James Warren, "Obama After His First Shellacking," *NYDN,* 9 November 2014; Obama for Congress 2000, FEC Form 3, Report of Receipts and Disbursements, 2 March–31 March 2000, 22 November 2002; Friends of Barack Obama, Report of Campaign Contributions and Expenditures, 1 January–30 June 2000, 31 July 2000; Bernard Schoenburg, *SJ-R,* 18 May 2000, p. 5 (Shomon); Greg Hinz, "Profile: Launching a New Campaign to Overhaul Education Funding," *CCB,* 2 July 2001 (Burns); Daley in David Smallwood, "Mayor Daley Smiles on Barack," *N'Digo Profiles,* December 2008, p. 12, and in Evan Osnos, "The Daley Show," *New Yorker,* 8 March

2010, pp. 38ff.; Abercrombie in Kevin Merida, "The Ghost of a Father," *WP,* 14 December 2007, p. A12; Mary Orndorff, "Davis-Obama Alliance Has Deep Roots," *Birmingham News,* 18 January 2007, p. B1; Paul Simon to Obama, 12 April 2000 (dictated 6 April) and 13 April 2000, PSP; DJG interviews with Dan Shomon, Cynthia Miller, Carol Harwell, Will Burns, John Schmidt, Ab Mikva, and Bruce Spiva.

24. Mikva in Michael Weisskopf, "Obama: How He Learned to Win," *Time,* 8 May 2008; Nesbitt in Wolffe, *Renegade,* p. 121; Research Department, National Republican Senatorial Committee, "Barack Obama (D-IL): On the Record," 22 July 2004, p. 5, JRORF (Obama absent for 22 and 28 March 2000 Judiciary Committee votes); HB 665 Senate Vote Tally, 29 March 2000; Senate Floor Transcript, 31 March 2000, pp. 44–45; Link in Scott Helman, "Early Defeat Launched a Rapid Political Climb," *BG,* 12 October 2007, and in "The Candidates: Barack Obama," MSNBC, 20 February 2008; Kristin Goss, *Better Together* (Saguaro Seminar on Civic Engagement in America, Harvard University), December 2000, esp. p. 104; Rick Pearson and Ray Long, "Philip Offers 2 Gun Compromises," *CT,* 5 April 2000; Senate Floor Transcript, 5 April 2000, pp. 13–33 (HB 2379) and 40–42 (HB 4021); Pearson and Ryan Keith, "Ryan Spurns 1 Philip Plan for a Deal on Firearms," *CT,* 6 April 2000; Pearson and Long, "Governor, GOP Near Deal for Gun Law," *CT,* 7 April 2000; Pearson and Long, "Senate Gun Deal Restores Felony Status," *CT,* 8 April 2000; Pearson, "Ryan Ultimatum Propelled Compromise on Gun Bill," *CT,* 9 April 2000; Rick Hepp, "Ryan Signs Modified Safe Neighborhoods Act," *CT,* 13 April 2000; *CF,* 14 April 2000; "2000 Legislative Session Summary," *LRU First Reading* 14 (June 2000), p. 7; Lightford in Melinda Henneberger, "The Obama Marriage," *Slate,* 29 October 2007; Diane Salvatore, "Barack and Michelle Obama: The Full Interview," *Ladies' Home Journal,* August 2008 ("I still remember when . . . Michelle called close to tears because our wonderful babysitter had quit and decided to go back to nursing." Michelle "had depended so heavily on this person to kind of hold it all together. And she was, frankly, mad at me, because she felt as if she was all alone in this process"); DJG interviews with Dan Shomon, Mike Hoffmann, Kim Lightford, Emil Jones Jr., Debbie Halvorson, Jill Rock, Cindy Huebner Davidsmeyer, John Hooker, Christine Radogno, Terry Link, Tommy Walsh, Denny Jacobs, Ned Mitchell, Dan Johnson-Weinberger, Joe Khan, Nat Piggee, and Heather Sullivan.

25. Joseph Omoremi, "Dick Gregory Zings Gangs and Drugs," *CD,* 10 April 2000, p. 5; JCAR Minutes, 11 April 2000 (with Obama and Lisa Madigan dissenting from an 8–2 vote objecting to state agencies' collection of permit requesters' Social Security numbers); Senate Floor Transcript, 11 April 2000, pp. 7–22 (HB 2855); Karen Shields, "Obama Focuses on Senate, Mulls Over Next Move," *HPH,* 12 April 2000, p. 10; Rick Pearson and Ryan Keith, "Budget Deal Serves Tax Cuts, Pork," *CT,* 12 April 2000; Jaime Ingle, "Lawmakers Tone Down Expulsion Legislation," *DHR,* 12 April 2000, p. A1; Obama on *Illinois Lawmakers* with Bruce DuMont, 12 April 2000, MBC; Senate Floor Transcript, 9 January 2001 (Obama votes present as HB 4659 passes 35–10–13); Rich Miller, "Maybe It's Right Time for Obama to Run," *CST,* 8 December 2006, p. 43; Ted McClelland, "How Obama Learned to Be a Natural," *Salon,* 12 February 2007; Senate Floor Transcript, 15 April 2000, pp. 33–36; Christi Parsons and Ryan Keith, "Generous Budget Reflects Economy," *CT,* 16 April 2000; Stephen S. Morrill, "Client Report for 2000 End of Session," Morrill and Associates, April 2000; "Session Ends in April," *C-WR* 15 #12, 4 May 2000; Martha Irvine, "Working Poor to Get 5 Percent Worth of State Income Tax Relief," AP, 11 May 2000; Monique Smith, "Low-Income Families to Receive Extra Tax Credit," and Obama, "Earned Income Tax Credit Highlights a Short Session," *HPH,* 24 May 2000, pp. 3, 4; State Senate President James "Pate" Philip, *End of Session Report 2000,* June 2000, esp. p. 63 (SB 1712 signed into law 2 June as PA 91-0759); "Obama Sponsored Law Helps Domestic Violence Victims," *HPC,* 26 October 2000, p. 6; "Obama Strikes Out at Domestic Violence," *HPH,* 1 November 2000, p. 11; "Welfare Workers Get Training," *SSJ,* 1–16 November 2000, p. 2; Redfield, *Money Counts,* p. 7; *CF,* 16 July (Paul Green) and 6 August 1999; Tom Waldron, *Supporting Working Families When State Coffers Are Empty: Lessons from Illinois' Successful Campaign for a Better State EITC* (Hatcher Group, November 2003), pp. 3–7; Christopher Wills, "Obama Record Shows a Liberal Open to Compromise," AP, 25 June 2008; DJG interview with Emil Jones Jr.

26. Barack H. and Michelle R. Obama 1999 Form 1040, n.d.; Kennedy Communications, "State Senator Barack Obama, Candidate for U.S. Senate: Background Report," April 2003,

RRP; Steve Neal, "Attorney General May Be Obama's Calling," *CST,* 19 April 2000, p. 8; Neal, "A Dozen to Consider If Mayor Skips Race," *CST,* 8 May 2000, p. 8; Arthur Fournier, "Distinguished Group to Lead Meditations at Holy Week Performance in Rockefeller," *UCC,* 13 April 2000 (Haydn's *The Seven Last Words of Christ* on 19 April); "Senate Democrat Member Project Authorization Forms," 9 November 1999 and 3 May 2000, Illinois DCEO FOIA; Obama for Congress 2000, FEC Form 3, Report of Receipts and Disbursements, 1 April–30 June 2000, 22 November 2002; *Chicago Tonight,* WTTW, 10 May 2000; Maggie Bertke e-mails to DJG, 4 and 8 April 2011 (Obama speaking about midwestern poverty at a 12 May 2000 Appleseed Network board meeting); "Law School to Co-Sponsor Forum on State's Death Penalty May 15," *UCC,* 11 May 2000; Elizabeth J. Ptacek, "Execution Moratorium Prompts Debate at U. Chicago Law School," *CM,* 17 May 2000; JCAR Minutes, 16 May 2000; Shomon in Scott Helman, "Early Defeat Launched a Rapid Political Climb," *BG,* 12 October 2007; DJG interviews with Robert Bennett, David Wilhelm, Dan Shomon, and Laura Russell Hunter. The Woods Fund board's 2000 decision to invest $1 million into Allison Davis's Neighborhood Rejuvenation Partners would later attract critical queries. See Tim Novak and Fran Spielman, "Obama Helped Ex-Boss Get $1 Mil. from Charity," *CST,* 29 November 2007, p. 10, Binyamin Appelbaum, "Grim Proving Ground for Obama's Housing Policy," *BG,* 27 June 2008, p. A1, and Larry Bell, "Obama Kick-Back Cronyism, Part 1," Forbes.com, 25 October 2011.

27. *CBS Evening News,* 17 May 2000; Senate Floor Transcript, 7 April 2000, pp. 126–36 (HB 709); Ryan Keith and Ray Long, "Abortion Funding Ban Dropped in Ryan's Lap," *CT,* 8 April 2000; Rick Pearson, "Ryan Vetoes Abortion Curb," *CT,* 10 June 2000; *CF,* 12 and 28 June and 9 August 2000; "Honors," *CDLB,* 1 June 2000, p. 3 (10 June IVI-IPO annual dinner); JCAR Minutes, 13 June 2000; Dave McKinney, "Treasurer's Deal Stirs Debate," *CST,* 19 June 2000, p. 8; Steve Neal, "Once More, with Feeling for Hartigan," *CST,* 28 June 2000, p. 43; Chinta Strausberg, "Daley to Monitor Tax Relief at Pumps," *CD,* 1 July 2000, p. 3; Monique Smith, "Currie: Gas Tax Cut Promises a 'Pig in a Poke,'" *HPH,* 5 July 2000, p. 2; "Obama Says General Assembly Suspends Sales Tax on Gas," *CC,* 6 July 2000, p. 2; "A Fourth Special Session Recap," *C-WR* 16 #2, 26 July 2000, p. 4; Tommiea Jackson, "Tavis Smiley Gets Chicago's Black Youth 'Wired,'" *CD,* 6 July 2000, p. 5; "Tavis Smiley Foundation Works with Chicago Urban League to Present Conference for 100 African American Youth," PR Newswire, 6 July 2000; "Teen Leadership Conference," *Jet,* 7 August 2000, p. 40; CAC Board of Directors Minutes, 16 December 1999; CAC, 1999 Annual Report to the Annenberg Foundation, 31 January 2000; Cindy Richards, "Annenberg Looks for Lessons as Program Winds Down," *Catalyst Chicago,* March 2000; CAC Board of Directors Minutes, 27 March 2000 (Obama absent); Michael Martinez, "Education Fund Aims to Boost City Schools," *CT,* 29 March 2000; Alexander Russo, "From Frontline Leader to Rearguard Action: The Chicago Annenberg Challenge," 1 April 2000; CAC Board of Directors Minutes, 26 June 2000; Ken Rolling, "Year 2000 Mid-Year Report to the Annenberg Foundation," [16] August 2000; CAC, "Summary of Private Matching Grant Contributors," 22 February 2001; CAC, Year 2000 Annual Report to the Annenberg Foundation, 28 February 2001; Stacy A. Wentzel et al., "Development of Chicago Annenberg Schools: 1996–1999," Consortium on Chicago School Research (CCSR), July 2001; Mark A. Smylie, Wentzel et al., *The Chicago Annenberg Challenge: Successes, Failures, and Lessons for the Future* (CCSR, August 2003), esp. pp. vii, 113; Ken Rolling, "Reflections on the Chicago Annenberg Challenge," in Russo, ed., *School Reform in Chicago: Lessons on Policy and Practice* (Harvard Education Press, 2004), pp. 23–28; Linda Lenz, "A Conversation with Fred Hess," *Catalyst Chicago,* February 2006; Christine C. George, "The Chicago Annenberg Challenge: The Messiness and Uncertainty of Systems Change," Loyola University of Chicago, 2007; David J. Hoff, "Obama's Annenberg Stint Informs White House Bid," *Education Week,* 7 March 2007, p. 10; Sam Dillon, "Obama Looks to Lessons from Chicago in His National Education Plan," *NYT,* 10 September 2008; Will Englund, "How Would Obama Govern?," *National Journal,* 25 October 2008; Carol Felsenthal, "Susan Crown on Why She Supports Mitt Romney," ChicagoMagazine.com, 20 March 2012; Shia Kapos, "Susan Crown's Juggling Act," *CCB,* 26 May 2012; "On the Air," *CDLB,* 17 July 2000, p. 3 (Obama on the *Public Affairs with Jeff Berkowitz* cable television interview show today); JCAR Minutes, 18 July 2000; DJG interviews with Barbara Flynn Currie, Malik Nevels, Pat Graham, and Ken Rolling.

28. "Senate Democratic FY '01 Member Project Authorization Form," 28 July and 9, 17, and 31 August 2000, Illinois DCEO FOIA; "$2 Million in Illinois FIRST Projects for Chicago," Illinois Governor's Office Press Release, 30 January 2001 ($450,000 for Kennedy-King); Leon Tripplett, "Obama Talks Education and Politics with Chamber," *HPH*, 2 August 2000, p. 6; Chinta Strausberg, "Predatory Lending Reform Near," *CD*, 10 August 2000, p. 1; Joe Ruklick, "Obama Pushes to Eliminate Mortgage Loan Sharks," *CD*, 14 August 2000, p. 3; Billy Poorten, "State Sen. Obama Targets Predatory Lending Rules," *HPH*, 16 August 2000, p. 2; "Assembly Urged to Aid Affordable Housing," *JHN*, 16 August 2000; Monique Smith, "City and State Reps. Go After Predatory Mortgage Loans," *HPH*, 6 September 2000, pp. 1–2; Kyriaki Venetis, "Senator Seeks New Rules," *Origination News*, October 2000, p. 12; Smith, "Ald. Hairston Lobbies for Drug Bill in Washington," *HPH*, 4 October 2000, p. 2; Malcolm Bush and Daniel Immergluck, "Research, Advocacy, and Community Reinvestment," in Gregory D. Squires, ed., *Organizing Access to Capital* (Temple University Press, 2003), pp. 154–68; Jennifer Loven, "Illinois Democrats Wishing for Durbin Gubernatorial Bid," AP, 14 August 2000; JCAR Minutes, 15 August 2000 (Obama absent); Monique Smith, "Fresh Off Election Year, State Legislators Ring It Up," *HPH*, 16 August 2000, p. 3; Obama, commencement address at the University of Massachusetts at Boston, 2 June 2006; Obama, *TAOH*, pp. 354–55; Christi Parsons, "Michelle Obama's Mission," *CT*, 20 April 2008; "Senator Barack Obama Issues Committee Meeting," 21 August 2000; Dan Shomon e-mail to Andrew Gruber et al., "Obama Issues Committee—Update!," 10 September 2000, Will Burns e-mail to Shomon, 11 September 2000, Gruber e-mail to Shomon, "Issues Committee," 22 September 2000, and Shomon e-mail to Gruber, 29 September 2000, all AGP; Friends of Barack Obama, Form D-2, Report of Campaign Contributions and Expenditures, 1 July–31 December 2000, 31 January 2001; Ruiz in Robin I. Mordfin, "From the Green Lounge to the White House," *UCLS Record*, Spring 2009, pp. 2–9; DJG interviews with Jesse Ruiz, John Rogers Jr., Rob and Lisa Fisher, Julius Genachowski, Hasan Chandoo, James Clayborne, Dan Shomon, Will Burns, Andrew Gruber, Mike Marrs, and Bill Fuhry.

29. JCAR Minutes, 19 September 2000; Obama, "Education First; Funding Key to Property Tax Reform," *HPH*, 27 September 2000, p. 4; Robin Washington, "Civil Rights Lawyers Honored," *Boston Herald*, 24 September 2000, p. 7; Lewis Rice, "School Hosts First Celebration of Black Alumni," *Harvard Law Bulletin*, Fall 2000; [Sandy O'Donnell], "Hope Center Retreat: Notes on Discussion on Purposes, Identity, Niche, and the Mission of the Hope Center Going Forward," 30 September 2000, "Proposal for the 'New' Hope Center's Identity, Niche, Purposes," n.d. [post 30 September 2000], SOP; UCLS *Announcements 2000–01*, pp. 40, 45; UCLS, *The Glass Menagerie 2000–01*, esp. p. 19; "Constitutional Law III Prof. Obama—Autumn 2000 Syllabus," Martin Chester Papers; Obama, Constitutional Law III Final Examination, December 2000; "Current Issues in Racism and the Law—Fall 2000," Peter Steffen Papers; Jodi Kantor, "Barack Obama, Forever Sizing Up," *NYT*, 26 October 2008; DJG interviews with Kenny Smith, Cynthia Miller, Will Burns, John Eason, Sokoni Karanja, Marty Chester, Daniel Lin, Stacy Monahan Tucker, Peter Steffen, Dan Sokol, and Rachael Pontikes.

30. "Senate Democrat FY 01 Member Project Authorization Form," 19 September and 4 October 2000, Illinois DCEO FOIA; "Ryan: $215,000 in Illinois FIRST Projects for Chicago's Southeast Side," 13 October 2000, "Ryan: $1.4 Million in Illinois FIRST Projects for Chicago South," 31 October 2000, "$1.5 Million in Illinois FIRST Projects for South Chicago," 20 November 2000, "Ryan: $1.4 Million in Illinois FIRST Projects for Chicago," 29 November 2000, "$1.7 Million in Illinois FIRST Projects for Chicago," 15 December 2000, Illinois Governor's Office Press Releases; "Currie, Obama Bring It Home," *HPH*, 29 November 2000, p. 3; "Senate Democrat FY 01 Member Project Authorization Form," 19 December 2000, Illinois DCEO FOIA; "$2 Million in Illinois FIRST Projects for Chicago," Illinois Governor's Office Press Release, 30 January 2001 ($100,000 for the Englewood Botanical Garden); Kenny B. Smith to Carla Needham, 31 December 2002, Smith to Paula Vehovic, 14 October 2003, Illinois DCEO FOIA; Obama for Congress 2000, FEC Form 3, Report of Receipts and Disbursements, 1 October–31 December 2000, 22 November 2002 (five 7 October donors: Margaret A. Davis, J. Archie Hargraves, Fred Morrison, Shirley Newman, and Alvin Shields); "$5.6 Million in Illinois FIRST Projects for Chicago," Illinois Governor's Office Press Release, 28 February 2001 (including $75,000 to FORUM "to provide infrastructure, training and support maintenance for computer technology and internet access for inner city, low income families"); Ray Gibson and

David Jackson, "State Pork to Obama's District Included Allies, Donors," *CT,* 3 May 2007; Dan Morain, "Obama: A Fresh Face or an Old-School Tactician," *LAT,* 8 September 2007; Christopher Wills, "Obama Says Illinois Earmarks Are Public," AP, 13 March 2008; Chris Fusco and Dave McKinney, "Obama's $100,000 Garden Grant Wasted," *CST,* 11 July 2008; Bob Secter and Gibson, "Obama Has Long Backed Faith Charities," *CT,* 12 July 2008; Chris Fusco et al., "Obama's Goodie Bag," *CST,* 17 July 2008, p. 9; Fusco and McKinney, "Obama Grant Being Probed," *CST,* 25 September 2008; DJG interview with Dan Shomon.

31. *CF,* 29 September 2000, *Congressional Record,* 4 and 5 October 2000, pp. S9800–08, 9812, 9814–15, 9818–20, 9896–98; Mike Dorning, "Library Feud Spurs Fitzgerald Filibuster," *CT,* 5 October 2000; Steve Rhodes, "Party Pooper," *Chicago Magazine,* May 2001, pp. 98–101, 120–26; Geoffrey Johnson, "The Lincoln Crusade," *Chicago Magazine,* April 2005; Obama, Youth Government Day address, Southern Illinois University, 21 October 2000; Simon to Obama, 26 October 2000, PSP.

32. JCAR Minutes, 17 October 2000; [Will Burns] to Senator Obama Issues Committee, "Education Policy," 17 October 2000, "Obama Issues Committee Meeting," 18 October 2000, Fran Tobin e-mail to Leslie Corbett, "Obama Issues Committee," 19 October 2000, Tracy Occomy e-mail to Corbett, "Obama Issues Committee," 23 October 2000, all AGP; Obama, "By George, Gore and Bush Are Different," *HPH,* 1 November 2000; "Around Town," *CDLB,* 19 October 2000, p. 3; "In the Law Schools," *CDLB,* 1 November 2000, p. 3; CARPLS Annual Meeting 2000, 2 November 2000, CARPLS Papers; Bonnie McGrath, "After Hours," *CDLB,* 14 November 2000, p. 3; Bill Fuhry e-mail to Dan Shomon and Obama, "Issues," 3 November 2000, AGP; "DCP 14th Annual Victory Convention" sign-in sheet, 3 November 2000; *Chicago Tonight,* WTTW, 6 November 2000; John R. Weinberger e-mail to Shomon, "Obama Issues," 9 November 2000, AGP; *Chicago Tonight,* WTTW, 13 November 2000; Andrew Gruber e-mail to Shomon, "Obama Issues Committee Meeting," 13 November 2000, Burns e-mail to Shomon, 14 November 2000, AGP; JCAR Minutes, 14 November 2000; Dean Olsen, "Proposal to Halt Payday Loan Abuses Blocked," *SJ-R,* 15 November 2000, p. 28; "Obama Issues Committee Meeting," 15 November 2000, AGP; Senate Floor Transcript, 15 November 2000, pp. 2 (SJR 75), 5–28 (SB 1426), 51–61 (SB 1867); Woods Fund of Chicago, *1999 Annual Report,* pp. 4–6; Paul Merrion, "Taxpayer Champion Emerges," *CCB,* 27 November 1999; Senate Floor Transcript, 30 November 2000, pp. 7–8 (HB 1511), 8–14 (HB 1991 passes 51–3–1 and is signed into law 11 January 2001 as PA-91-0937); Joe Biesk, "New Law Extends Life of Economic Enterprise Zones," *CT,* 12 January 2001; *CF,* 1 December 2000; Stephen S. Morrill, "Client Report for 2000 Veto Session," Morrill and Associates, December 2000; Billy Poorten, "Local Officials Vote 'Yea' on Soldier Field Plan," *HPH,* 6 December 2000, p. 3; "General Assembly Maintains Status Quo," *C-WR* 16 #6, 8 December 2000, p. 2; *Chicago Tonight,* WTTW, 8 December 2000; "Loan Battle Brews Anew," *CCB,* 11 December 2000, p. 4; JCAR Minutes, 12 December 2000; Shomon e-mail to Gruber et al., "Obama Issues Committee Meeting," 12 December 2000, AGP; CAC Board of Directors Minutes, 14 December 2000; "Veto Session Wrap Up," *C-WR* 16 #7, 15 December 2000, p. 2; Melissa Allison, "State Takes on Predatory Lending," *CT,* 15 December 2000; Randy Blaser, "Payday Loan Rules Go Back to Legislature," *Lake Zurich Courier,* 21 December 2000 (and eleven additional newspapers, 21 December 2000 through 4 January 2001); JCAR 1997–2001 Annual Reports, ca. 2002, p. 69; [Debbie] Lounsberry, "Barack Obama—The Record: A Summary," 10 May 2004, p. 4, JRORF (highlighting above all other Senate floor votes Obama's voting present on HB 1511, a tougher-sentencing measure, which passed 54–2–2, on 30 November 2000); Nesbitt in Kevin McDermott, "Unlikely Competitors: Years as Outsider Gave Obama the Ability to Relate, Friends Say," *SLPD,* 24 October 2004, p. A1, in Mendell, *Obama,* pp. 151–54, and in Wolffe, *Renegade,* pp. 122, 330; DJG interviews with Dan Shomon, Andrew Boron, Judy Byrd, Bernardine Dohrn, James Clayborne, Ray Harris, Steven Mange, Will Burns, Warren Chain, Leslie Corbett, Mike Marrs, Bill Fuhry, Desta Houston, and Lisa Madigan.

33. Scott Lynch-Giddings diary entry, 15 December 2000; Melissa M. Bernardoni, "Robinson to Lead Diversity Affairs," *GSB Magazine,* Winter 2000; James B. Arndorfer, "Minority Funds Seek More Pension Work," *CCB,* 26 March 2001, p. 58; Jim Reynolds to Robert Newtson, 11 June 2001, ISBI Papers; Hermene D. Hartman, "Barack and Black Enterprise," *N'Digo,* 7–13 June 2007, p. 3; Obama, remarks at National Urban League Annual Conference,

St. Louis, MO, 27 July 2007; Christopher Drew and Raymond Hernandez, "Loyal Network Backs Obama After His Help," *NYT*, 1 October 2007; Hartman, "Barack's Friends Were ABLE!," and Reynolds in David Smallwood, "Genesis of the Grassroots Movement," *N'Digo Profiles*, December 2008, pp. 10–11, 44; Obama for Congress 2000, FEC Candidate Summary Report, 31 December 2000; Todd Spivak, "No Money? State Reps Say It's No Sweat," *HPH*, 21 February 2001, pp. 1–2; Chuck Neubauer and Tom Hamburger, "Obama Donor Received a State Grant," *LAT*, 27 April 2008; Peoria Public Schools, *Staff News Weekly*, 4 February 2009, p. 1; Raoul V. Mowatt, "Sport of Table Tennis Gets a Local Spin," *CT*, 2 January 2003; Robert Blackwell Jr.'s 2012 interview with John Page; DJG interviews with Scott Lynch-Giddings, David Wilhelm, David Spiegel, Hermene Hartman, Janis Robinson, John Rogers Jr., Stephen Pugh, Eugene Morris, Les Coney, Martin King, Dan Shomon, Mike Lieteau, Carol Harwell, John Schmidt, Robert Blackwell Jr., and Raja Krishnamoorthi.

34. Dave McKinney, "Partisan Rumble," *II*, January 2001, pp. 12–28; UCLS *Announcements 2000–01*, p. 84; JCAR Minutes, 9 January 2001; Dean Olsen, "Lawmakers Slow Down Ruling for Special Education Changes," *SJ-R*, 10 January 2001, p. 6; *Corey H. v. Board of Education*, 995 F. Supp. 900, 903, 912 (N. D. Ill.), 19 February 1998; Michael Martinez, "State Board Ripped Over Disabled Kids," *CT*, 24 February 1998; Stephanie Banchero, "State Board OKs Special-Ed Change," *CT*, 17 March 2000; Banchero, "State Appeals U.S. Ruling on Special Education," *CT*, 13 October 2000; Order, *Corey H. v. Board of Education, 27 February 2001; Reid L. v. Illinois State Board of Education*, 289 F. 3d 1009 (7th Cir.), 13 May 2002; *CF*, 10 (noting a $10,000 annual increase in Senate district office allotments), 11 and 12 January 2001; Senate Floor Transcript, 10 January 2001, pp. 11–12 and 24; John S. Sharp, "New Rules for DUI Blood Test Approved by Legislative Panel," *CDH*, 11 January 2001, p. 10; JCAR Minutes, 11 January 2001; James Fuller, "Bill Would Inform Voters How to Fix Ballot Mistakes," *CST*, 15 January 2001, p. 12; Obama, "How to Keep Cool Over the Rising Heating Costs," *HPH*, 17 January 2001, p. 4; Barbara Sherlock, "Legislators, Groups Praise Fight Against Predatory Lending," *CT*, 11 March 2001; "Leadership and Committee Assignments," *C-WR* 16 #8, 19 March 2001; Jodi Kantor, "As Professor, Obama Held Pragmatic Views on Court," *NYT*, 3 May 2009; DJG interviews with Kristen Seeger, Nathaniel Edmonds, Alleza Strubel, Mark Warnick, Denny Jacobs, John Cullerton, and Miguel del Valle.

35. *Chicago Tonight*, WTTW, 17 January 2001; Gretchen Helfrich, "The Court and Civil Rights," *Odyssey*, WBEZ Radio, 18 January 2001; Jane R. Patterson and Michael L. Mory to Governor George H. Ryan and Members of the General Assembly, "Report on Emerging Investment Managers Fiscal Year 2000," 31 August 2000, "Presentation of Requests to Jane Patterson, Executive Director," 20 December 2000, "Presentation of Requests to the Illinois State Board of Investment by African-American Owned Investment Management and Broker/Dealer Securities Firms," [19 January 2001], Joseph P. Cacciatore, Minutes, Illinois State Board of Investment, 19 January 2001, Keith Cardoza to Board Members, "Minority Brokers and Money Managers," 21 February 2001, ISBI Papers; Keith Bozarth, "Minutes of the Investment Committee," and Bozarth, "Minutes of the Regular Meeting of the Board of Trustees," Teachers' Retirement System, 15–16 February 2001, TRSI Papers; Obama's remarks at the National Urban League Annual Conference, St. Louis, MO, 27 July 2007. On WTTW, Barack also volunteered that "I, for example, don't agree with a missile defense system."

36. Chinta Strausberg, "Vouchers Will Hurt Public Schools," *CD*, 24 January 2001, p. 1; Dan Shomon e-mail to Andrew Gruber, "Draft Release," 26 January 2001, AGP; Curtis Lawrence and Scott Fornek, "After Washington, Several Figures Fill Political Voice," *CST*, 28 January 2001, p. 11; "Obama, Aldermen Announce State Legislation to Guarantee Faster Assistance for Laid-Off Workers," Press Release, 28 January 2001, AGP; Strausberg, "Sen. Obama Bill to Aid Laid-Off Workers," *CD*, 29 January 2001, p. 3; Courtney Challos, "Legislators Pushing Plan to Aid Unemployed," *CT*, 29 January 2001; "Bills Would Aid Laid-Off Workers," *CST*, 29 January 2001, p. 16; Shomon to Cheryl Howard (Illinois Coalition Against Domestic Violence) and Tom Mannard and Chris Boyster (Illinois Council Against Handgun Violence), "Legislation on Gun Control and Domestic Violence," 30 January 2001, AGP; Senate Floor Transcript, 31 January 2001, pp. 6–7 (Obama introducing SBs 33, 41, 51, 59, 61, and 62); AP, "Reactions to Ryan's Address," 31 January 2001; Senate Floor Transcript, 1 February 2001, p.

3 (Obama introducing SBs 123, 124, 125, and 126); Pam Belluck, "Illinois Governor Dogged by Inquiry into Corruption," *NYT,* 3 February 2001; Steve Neal, "Battle for Attorney General's Office Already Taking Shape," *CST,* 5 February 2001, p. 29; Barbara Flynn Currie, Leslie Hairston, Barack Obama, and Toni Preckwinkle, "No Secret Meetings Here!," *HPH,* 7 February 2001, p. 4; Henry Bayer (AFSCME) to Illinois State Board of Investment, 9 February 2001, ISBIP; Shomon e-mail to Gruber and Gruber e-mail to Shomon, "Layoffs," 13 February 2001, AGP; Obama and Preckwinkle, "State Government Must Protect Laid-Off Workers," *HPH,* 14 February 2001, p. 4; Gruber e-mail to Shomon, "WARN," 28 February 2001, AGP; Lisa B. Song, "Senate Democrats Seek to Close Gun Loopholes," *CT,* 14 February 2001; Sean Noble, "Senate Democrats' Proposals Take Aim at Weapons Control," *SJ-R,* 15 February 2001, p. 7; Richard Goldstein, "Florida Election Fallout Still Being Heard," *QCT,* 19 February 2001 and as "Funding May Derail State Election Reform," *DHR,* 20 February 2001, p. A1; Strausberg, "Jones, Obama Dare Pate to Bring Gun Bill to Floor," *CD,* 20 February 2001, p. 6; *CF,* 21 February and 7 March 2001; Obama's February 2001 interview with Julieanna Richardson; Baim, *Obama and the Gays,* pp. 13, 194. To Richardson, Barack further asserted, "I think that the civil rights movement probably had the biggest influence on my life, the participants in that. Not just Dr. King or Malcolm X, but Bob Moses and Fannie Lou Hamer and Rosa Parks and E. D. Nixon."

37. "Obama, Davis, Leading Pastors Announce Legislation to Study Racial Profiling in Illinois," Obama Staff Press Release, 19 February 2001; Chinta Strausberg, "End Racial Profiling," *CD,* 20 February 2001, p. 1; Senate Floor Transcript, 20 February 2001, pp. 10, 17 (Obama introducing SBs 266 and 375); JCAR Minutes, 21 February 2001; Order, *Corey H. v. Board of Education,* 27 February 2001; Sean Noble and Dean Olsen, "Judge Allows Hearings on Changes in Training, Work of Special-Ed Teachers," *SJ-R,* 28 February 2001, p. 26; *Reid L. v. Illinois State Board of Education,* 289 F. 3d. 1009 (7th Cir.), 13 May 2002; Senate Floor Transcript, 21 February 2001, pp. 8, 11, 22 (Obama introducing SBs 578, 619, 620, 621, 622, 623, 624, and 801); Senate Floor Transcript, 22 February 2001, pp. 2, 16, 31–32 (Obama introducing SBs 893, 894, 1111, 1112, 1322, and 1338); Bridget M. Kuhn, "KidCare Becomes a Healthy Success," *DS,* 26 January 2000, p. A3; Strausberg, "Ryan Urges Lawmakers to OK Health, Heating Bill," *CD,* 27 February 2001, p. 8; Gretchen Helfrich, "The Right to Vote," *Odyssey,* WBEZ Radio, 27 February 2001; Todd Spivak, "Obama Lobbies for State Racial Profiling Legislation," *HPH,* 28 February 2001, pp. 1–2; Strausberg, "No Gambling at O'Hare," *CD,* 28 February 2001, p. 3; Senate Floor Transcript, 28 February 2001, pp. 11–35 (SR 63 adopted 35–22) and 35 (Obama introducing SJR 14); *CF,* 1 March 2001; Rick Pearson, "Some See a Hidden Motive in Philip's Bipartisan Move," *CT,* 1 March 2001; "Senate Democrat FY 01 Member Project Authorization Form," 2 March 2001, DCEO Papers FOIA; "Obama 'Booked for Lunch,'" *SJ-R,* 4 March 2001, p. 45; Obama remarks, Illinois State Library, 7 March 2001; Daniel C. Vock, "Panel Votes to End Defense of Intoxication," *CDLB,* [8] March 2001, pp. 1, 22; Monica Davey and E. A. Torriero, "Jackson Group Defends Its Links to State Grant," *CT,* 17 March 2001; Pearson, "State Hopefuls Test Waters for 2002 Primary," *CT,* 19 March 2001; "On the Air," *CDLB,* 19 March 2001, p. 3; JCAR Minutes, 20 March 2001; DJG interviews with Brad Burzynski, Beverly Helm-Renfro, David Wilhelm, and Scott Kennedy.

38. Briefs filed 21 March 2001, #00-4110 and #01-1810, in *U.S. ex rel Chandler v. Cook County,* 277 F. 3d 969 (7th Cir.), 22 January 2002, *Cook County v. U.S. ex rel Chandler,* 538 U. S. 119, 10 March 2003; Gabriel Piemonte, "Sen. Obama Introduces [sic] State Affordable Housing Bill," *HPH,* 21 March 2001, pp. 1–2; "$16.1 Million in Illinois FIRST Projects for Chicago," Illinois Governor's Office Press Release, 23 March 2001; Adriana Colindres, "Capital Beat," *PJS,* 25 March 2001, p. B5; "Senate Committee Action Report," Judiciary, 27 March 2001 (SB 1095); Senate Floor Transcript, 29 March 2001, pp. 48 (Obama commending Revenue Committee chairman Bill Peterson on his SB 1135), 93–99 (SB 78), and 105–25 (SB 155); John S. Sharp, "House Advances Plan Addressing Race Profiling," *CDH,* 30 March 2001, p. 6; Senate Floor Transcript, 30 March 2001, pp. 84–90 (SBs 1093, 1094, and 1095); Dave McKinney, "Bill Proposes Care for Fetus After Abortion," *CST,* 31 March 2001, p. 1; Jeannine Koranda, "Senate OKs Limits on Abortion," *DHR,* 31 March 2001, p. 1; Doug Finke, "Dollars for Dialing," *II,* April 2001, pp. 17–18; "Obama's Tax Credit Bill Moves to Senate Floor," and "Obama's Housing Tax Credit Bill Passes Through Illinois Senate," Obama Staff Press Releases, 2 April 2001; "Obama Passes Bill to Expedite Order of Protection Process," Obama Staff Press Release, 3 April 2001; Piemonte, "Obama's Housing Bill Passes Senate," *HPH,* 4 April 2001, p. 10; Minutes of the Regular Meeting of the Board of Trustees, General Assembly Retirement System of Illinois, 4 April 2001 (#219), GARS Papers, pp. 1839–44 (Obama's first meeting as a newly appointed member of the board, per Tim Blair e-mail to DJG, 17 December 2013); Senate Floor Transcript, 4 April 2001, pp. 94–118, esp. 106 (SB 604 fails 29–27); Ray Long and Joe Biesk, "Gun Show Checks OK'd but House Rejects Firearms Database," *CT,* 5 April 2001; "Obama Helps Defeat Concealed-Carry Legislation," Obama Staff Press Release, 5 April 2001; Senate Floor Transcript, 5 April 2001, pp. 68–70 (Obama's SB 62, which on 18 July 2001 became PA 92-0087), 91–99, esp. 91 (SB 636), and 175–91, esp. 175–76 (SB 663); Rick Pearson, "2nd Chance Vote System Under Fire," *CT,* 6 April 2001; Senate Floor Transcript, 6 April 2001, pp. 18–26; "Obama: Drug Offender Rehabilitation Initiative OK'd by Senate," "Obama Co-Sponsors Bill Providing Vital Information to Rape Victims," and "Obama Health Care Measures Gain Senate Approval," Obama Staff Press Releases, 6 April 2001; Chinta Strausberg, "Obama Scores Big in Senate Passing of 3 Bills," *CD,* 9 April 2001, p. 5; Todd Spivak, "Sen. Obama Helps Defeat a Concealed Firearm Bill," *HPH,* 11 April 2001, p. 3; "$4.4 Million in Illinois FIRST Projects for Chicago," Illinois Governor's Office Press Release, 11 April 2001; Strausberg, "Sen. Obama's Bill Passes; Helps Laid-Off Workers," *CD,* 12 April 2001, p. 5; "Obama: Legislation to Help Laid-Off Workers Passes Senate" and "Obama Legislation to Prevent Telemarketers from Calling Certain Illinois Residents," Obama Staff Press Releases, 13 April 2001; Ryan Keith, "Obama's Past Offers Ammo for Critics," AP, 17 January 2007; Mike Robinson (AP), "Obama Got Start in Civil Rights Practice," *WP,* 20 February 2007; Abdon M. Pallasch, "Obama's Legal Career," *CST,* 17 December 2007, p. 4; Christopher Wills, "Obama's Thin But Varied Record," AP, 16 January 2008; Adrienne P. Samuels, "Barack's Chicago," *Ebony,* January 2009; DJG interviews with Judd Miner, Jeffrey Cummings, John Belcaster, Marni Willenson, Carolyn Shapiro, Rob Scott, Beverly Helm-Renfro, John Bouman, William McNary, Doug Dobmeyer, Kirk Dillard, Dan Cronin, Al Sharp, Linda Mills, Nancy Shier, Randy Wells, Josh Hoyt, Don Moore, Hal Baron, Mike Marrs, Cindi Canary, Dan Shomon, Peter Steffen, Marty Chester, and Ted Liazos.

39. Barack H. and Michelle L. Obama, 2000 Form 1040, 15 April 2001; Obama, "Statement of Economic Interests," IL-0221-2001, 27 April 2001; Friends of Barack Obama, Report of Campaign Contributions and Expenditures, 1 January–30 June 2001, 9 August 2001; "People," *CT,* 29 April 2001; Woods Fund of Chicago, *2000 Annual Report* (ca. May 2001); Todd Spivak, "Currie, Obama Team During Key Springfield Session," *HPH,* 18 April 2001, p. 6; Joseph Omoremi, "Coalition Mobilizes Support for Predatory Lending Rule," *CD,* 5 April 2001, p. 3; *CF,* 17 April 2001; JCAR Minutes, 17 April 2001; Ryan Keith, "Legislative Committee Approves Rules on Predatory Lending," AP, 17 April 2001; Abdon M. Pallasch, "State Committee Passes 'Predatory Lending' Law," *CST,* 18 April 2001, p. 66; Ray Long and Melissa Allison, "Illinois OKs Predatory Loan Curbs," *CT,* 18 April 2001; Dean Olsen, "Regulations to Stop Predatory Lending OK'd," *SJ-R,* 18 April 2001, p. 9; "Obama Leads Way to Predatory Lending Reform," Obama Staff Press Release, 18 April 2001; Chinta Strausberg, "Ryan Says Anti-Predatory Lending Bill Is Progress," *CD,* 19 April 2001, p. 26; Spivak, "Obama a Key Leader on Predatory Lending Reforms," *HPH,* 25 April 2001, p. 8; "Illinois Groups Win Sweeping Anti-Predatory Lending Regulations," and Gale Cincotta, "State Anti-Preddatory Lending Regulations Show Power of Community Groups," *Disclosure,* April–May 2001; Bush and Immergluck, "Research, Advocacy, and Community Reinvestment," in Squires, ed., *Organizing Access to Capital,* p. 165; *CF,* 22 March 2001; Strausberg, "City State Pols Say Remap Cannot Take from Blacks," *CD,* 19 April 2001, p. 3; Gretchen Helfrich, "Redistricting," *Odyssey,* WBEZ Radio, 23 April 2001; Joseph Omoremi, "Community Organization 'Butts Heads' with Aldermen," *CD,* 24 April 2001, p. 6; "Seniors Rally for Bill to Provide Discount Medications to Elderly," *SJ-R,* 26 April 2001, p. 23; Richard Goldstein, "Illinois Senate Approves Sexual Offender Bill," *QCT,* 1 May 2001; Spivak, "Reps but No Residents at Redistricting Hearings," and "Currie, Obama: Tight Budget Dictates New Legislation," *HPH,* 2 May 2001, pp. 3 and 8; Kevin McDermott, "Senate Panel Waters Down Telephone Solicitation Bill,"

SLPD, 2 May 2001, p. B1; Lesley R. Chinn, "St. Mark Church Hosts Forum on Racial Profiling," *CC*, 3 May 2001, p. 12; "Obama Legislation Moves to Governor's Desk," Obama Staff Press Release, 3 May 2001; "$5.5 Million in Illinois FIRST Projects for Chicago," Illinois Governor's Office Press Release, 8 May 2001; Spivak, "Racial Profiling Bill Blocked by Senate Again" and "Obama Bill Passes Senate," *HPH*, 9 May 2001, pp. 1, 2, 8, and 12; Goldstein, "Senate to Hear Sex Offender Bill," *DHR*, 9 May 2001, p. A6; Senate Floor Transcript, 9 May 2001, pp. 21–24 (HB 313); "Obama Passes Bill to Strengthen Nursing Home Inspections," Obama Staff Press Release, 10 May 2001; Senate Floor Transcript, 10 May 2001, pp. 50–59 (HB 2088) and 95–106 (HB 126); Ben Kieckhefer, "Bill Would Increase Penalties for Dealing Ecstasy," *SJ-R*, 11 May 2001, p. 16 (and *PJS* p. B1); Obama e-mails to Ronald Loui, 11 November 2000, 15 February 2001, and 12 May 2001 (withdrawing from a tentatively scheduled 23 May banquet speech in St. Louis); Goldstein, "Senate to Hear Telemarketing Bill," *DHR*, 15 May 2001, p. D1; "Obama, Currie: District Receives Illinois First Money for Projects in Hyde Park," Obama Staff Press Release, 15 May 2001; JCAR Minutes, 15 May 2001; Senate Floor Transcript, 15 May 2001, pp. 44–66, esp. 46–50 (HB 1812); Jeff Zeleny and Ray Long, "Senate OKs Expanded Death Penalty," *CT*, 16 May 2001; Dave McKinney and James Fuller, "Bill Seeks Death Penalty for Gang Killings," *CST*, 16 May 2001, p. 12; Obama, "Give Nonviolent Female Offenders 2nd Chance," *HPH*, 16 May 2001, p. 4; Charles N. Wheeler III, "Remap 2000," in David A. Joens et al., *Almanac of Illinois Politics—2000* (University of Illinois at Springfield, 2000), pp. 13–22; Samuel K. Gove, "Origins, Mechanics, and Politics," in *Redistricting Illinois 2001* (University of Illinois, 2001), pp. 1–5; "Census 2000 and Redistricting in Illinois," *CW-R* 16 #9, 23 May 2001; Spivak, "Gang Affiliation a Death Penalty Determinant," and Gabriel Piemonte, "Tax Credit Bill at Final Hurdle," *HPH*, 13 June 2001, pp. 6 and 11; Senate President James "Pate" Philip, *End of Session Report 2001*, esp. p. 91; Carlos Hernandez Gomez, "Population Soars but Political Power Lags," *Chicago Reporter*, September 2001; "Major Bills Passed by the Illinois General Assembly," *LRU First Reading* 15 (September 2001): 8–15; "Legislative Highlights," *C-WR* 16 #11, 19 September 2001; Charles N. Wheeler III, "Remap 2001," in Barbara Van Dyke-Brown et al., *Almanac of Illinois Politics—2002* (University of Illinois at Springfield, 2002), pp. 3–13; Sam Youngman and Aaron Blake, "Obama's Crime Votes Are Fodder for Rivals," *The Hill*, 14 March 2007; Sara Just, "How Republicans Plan on Ganging Up on Obama for Being 'Weak' on Street Gangs," ABCNews.com, 23 April 2008; Sam Dillon, "Obama Pledge Stirs Hope in Early Education," *NYT*, 17 December 2008; Dillard in Abramsky, *Inside Obama's Brain*, p. 54; Shia Kapos, "Abby McCormick O'Neil Steps into Chicago's Cultural Spotlight," *CCB*, 23 October 2010; DJG interviews with Sokoni Karanja, Todd Dietterle, Howard Stanback, John Corrigan, Dave Gross, Kirk Dillard, and Mark Donahue.

40. *CF*, 16, 21, 22, and 29 May and 8 June 2001; Senate Floor Transcript, 16 May 2001, pp. 35–43 (HB 1900); Senate Floor Transcript, 17 May 2001, p. 55 (HB 1887); "Obama Legislation Removes Juveniles from Paying Non-Emergency Medical or Dental Services," Obama Staff Press Release, 18 May 2001; Senate Floor Transcript, 21 May 2001, pp. 9–10 (SJR 26); JCAR, *Annual Report 2001*; *Reid L. v. Illinois State Board of Education*, 289 F. 3d 1009, 1016, 1020 (7th Cir.), 13 May 2002; Chinta Strausberg, "Illinois Black Caucus Bills Pass," *CD*, 23 May 2001, p. 10; Corey Hall, "Legislation Seeks to Protect Children from Lead Poisoning Passes," *CC*, 24 May 2001, p. 8; Senate Floor Transcript, 25 May 2001, pp. 7–28 (HB 2917); Anthony Man, "State Senate Passes Remap," *SJ-R*, 25 May 2001, p. 14 (and *DHR*, 26 May, p. A1, and *QCT*, 26 May); "$1.2 Million in Illinois FIRST Projects for Chicago," Illinois Governor's Office Press Release, 25 May 2001; Senate Floor Transcript, 30 May 2001, pp. 15–30 (HB 1599); "Obama Legislation to Reduce Lead Poisoning Exposure," Obama Staff Press Release, 30 May 2001; Senate Floor Transcript, 31 May 2001, pp. 56–64 (SR 153) and 106–9 (HB 3440); Doug Finke, "Lawmakers OK Budget," *SJ-R*, 1 June 2001; Dan Mihalopoulos, "Political Opponents Cast Out by Remap," *CT*, 27 June 2001; John Mercurio, "Between the Lines," *RC*, 2 July 2001; Todd Spivak, "Rush Opponents Are Drawn Out of First District," and Obama, "Tallying Wins and Losses in Springfield," *HPH*, 4 July 2001, pp. 3, 4, 14; Woods Fund of Chicago, *2001 Annual Report*, pp. 8–11; Corey Hall, "Senator Obama Assesses State Legislature's Failures and Triumphs," *HPC*, 12 July 2001, p. 5; Ken Dilanian, "Obama Shifts Stance on Environmental Issues," *USAT*, 18 July 2008; Ben Whitford, "Coal and Clear Skies: Obama's Balancing Act," *Plenty Magazine*, 31 October 2008.

41. Jodi Wilgoren, "Chief Executive of Chicago Schools Resigns," *NYT,* 7 June 2001; Irv Kupcinet, "Kup's Column," *CST,* 24 June 2001, p. 16; Pete Thamel, "Coach with a Link to Obama Has Hope for Brown's Future," *NYT,* 16 February 2007; Paul Buker, "Trading It All for Basketball," *Oregonian,* 13 April 2008; Ginny Graves, "Back to Life in Midlife," *More,* July–August 2008, pp. 132ff.; Robinson, *A Game of Character,* pp. 157–63; Obama, *TAOH,* pp. 4, 340; Wolffe, *Renegade,* pp. 52, 120; Keith Cardoza to Board Members, "Minority Broker and Money Managers," 21 February 2001, Ron Schmitz to Board Members, "Manager Search Process—Public Security Managers," 21 February 2001, Joseph P. Cacciatore, "Minutes of the Minority Broker/Manager Policy Committee," 23 February 2001, Cardoza to Board Members, "Minority Brokers and Money Managers," 18 April 2001, Cacciatore, "Minutes of the Illinois State Board of Investment Sub-Committee for Minority Broker/Manager Policy," 23 April 2001, ISBI Papers; Robin Goldwyn Blumenthal, "Quid Pro Quota? How Jesse Jackson Aims to Get More Pension Fund Business for His Contributors," *Barron's,* 21 May 2001; Jane R. Patterson to Board Members, "June 11, 2001 Sub-Committee Meeting," 4 June 2001, Cardoza to Board Members, "Barron's Article," 6 June 2001, Jim Reynolds to Robert Newtson, 11 June 2001, Cacciatore, "Minutes of the Illinois State Board of Investment Sub-Committee for Minority Broker/Manager Policy," 11 June 2001, ISBIP; J. Fred Giertz, "Minutes," Investment Committee, Board of Trustees of the State Universities Retirement System, 22 March 2001, James M. Hacking, "Minutes, Quarterly Meeting of the Board of Trustees of the State Universities Retirement System," 23 March 2001, John R. Krimmel and Dan M. Slack to Members of the SURS Investment Committee, "Report from the Working Group on Minority-Owned and Female-Owned Investment Managers," 23 April 2001, SURSI Papers; "Michael Riordan to Become CEO of University of Chicago Hospitals," UCH Press Release, 11 June 2001; Michelle on *The Oprah Winfrey Show,* 19 January 2005, and in Cassandra West, "Her Plan Went Awry," *CT,* 1 September 2004; Michelle Obama, "How I Got Unstuck," *O Magazine,* September 2005, pp. 22ff.; Obama, *TAOH,* p. 338; Michelle in Anne E. Kornblut, "Michelle Obama's Career Timeout," *WP,* 11 May 2007, in Judy Keen, "Michelle Obama: Campaigning Her Way," *USAT,* 11 May 2007, in Sandra Sobieraj Westfall, "Michelle Obama: 'This Is Who I Am,'" *People,* 18 June 2007, p. 118, in Maria L. La Ganga, "It's All About Priorities for Michelle Obama," *LAT,* 22 August 2007, in Holly Yeager, "The Heart and Mind of Michelle Obama," *O Magazine,* November 2007, pp. 286ff., in Kristen Gelineau, "Would-Be First Lady Drifts into Rock-Star Territory," AP, 30 March 2008, in Stefano Esposito, "'2 People Who Love Each Other,'" *CST,* 13 July 2008, p. A2, and in Geraldine Brooks, "Michelle Obama: Camelot 2.0?," *More,* October 2008; Joe Stephens, "Obama Camp Has Many Ties to Wife's Employer," *WP,* 22 August 2008; Katherine Skiba, "First Lady's Former Chief of Staff 'Blissfully Unemployed,'" *CT,* 29 January 2011; Maya in Scott Helman, "Early Defeat Launched a Rapid Political Climb," *BG,* 12 October 2007, on *American Morning,* CNN, 26 August 2008, in Stuart Coleman, "Tea with Barack Obama's Sister," *Salon,* 23 October 2008, and in Fran Korten, "Obama's Sister: What Our Mother Taught Us," *Yes Magazine,* Summer 2011; Obama on *Hannity & Colmes,* FOX, 24 October 2006, at the 23 October 2006 American Magazine Conference, and on *Off Topic with Carlos Watson,* CNN, 24 October 2004; DJG interviews with Janis Robinson and Kevin Thompson.

42. JCAR Minutes, 12 June 2001; Minutes, Board of Trustees Meeting, Demos: A Network for Ideas and Action, 19–20 June 2001, SHP; Gary Washburn, "Murphy Eyes Role in Black Caucus," *CT,* 20 June 2001; John W. Fountain, "A White Alderman Who Wants to Join a Black Caucus," *NYT,* 23 June 2001; Washburn, "Caucus Mulling Murphy's Plea," *CT,* 28 June 2001; "Ryan: $1 Million in Illinois FIRST Projects for Chicago," Illinois Governor's Office Press Release, 28 June 2001; "The Black Caucus' Race Card," *CT,* 4 July 2001; Chinta Strausberg, "Obama and Trotter Take Aim at IRA for Waging Unjust Campaign," *CD,* 5 July 2001, p. 22 (Obama remains "in favor of handgun registration requirements and licensing requirements"); JCAR Minutes, 10 July 2001; Washburn, "Murphy's Bid to Join Black Panel Postponed," *CT,* 13 July 2001; Tim Novak, "Murphy May Join Black Caucus Today," *CST,* 15 July 2001, p. 10; Fran Spielman, "Daley: Let Murphy in Black Caucus," *CST,* 17 July 2001, p. 1; Sarah E. Richards, "Murphy Drops His Caucus Bid," *CT,* 17 July 2001; *Chicago Tonight,* WTTW, 18 July 2001; B. Drummond Ayres, "White Drops Bid to Join Black Caucus," *NYT,* 22 July 2001; Michael Eric Dyson, "Murphy's Efforts Misguided," *CST,* 24 July 2001, p. 29; "$3.1 Million in Illinois FIRST

Projects for Chicago," Illinois Governor's Office Press Release, 17 July 2001; "Agenda for Finance Committee Meeting," 26 July 2001, "What Will a Statewide Campaign Cost? (Dan Shomon)," "Obama 2002 and Beyond," Finance Committee Presentation, 26 July 2001, LRHP; Chinta Strausberg, "Civil Rights Group Calls for End to Racial Districting," *CD*, 2 August 2001, p. 1; "Governor Signs Bill Requiring Vital Information Be Given to Rape Victims" and "Obama: State Treasurer to Publish List of Unclaimed Properties," Obama Staff Press Releases, 7 August 2001; Todd Spivak, "Unclaimed Property List," *HPH*, 8 August 2001, p. 27; Spivak, "Obama Bill Increases Rights for Rape Victims," *HPH*, 15 August 2001, p. 2; Spivak, "Local Legislators Build Election Campaign Warchests," *HPH*, 5 September 2001, p. 3; DJG interviews with Leslie Hairston, Laura Russell Hunter, Dan Shomon, Will Burns, Madeline Talbott, Keith Kelleher, John Corrigan, Barbara Flynn Currie, Mike Lieteau, Ray Harris, and Tom Dart.

43. Rick Pearson and Ray Long, "Ryan Won't Seek Second Term," *CT*, 9 August 2001; John W. Fountain, "Illinois Governor, Unpopular in Polls, Won't Run Again," *NYT*, 9 August 2001; Russ Stewart, "Kosovo Fame Fuels Blagojevich's Ambitions," *Illinois Politics* Vol. 8 #3 [2 August 1999], pp. 16–17; *CF*, 19 December 2000, 15 June, 19, 20, and 30 July, 1, 6 11, and 17 August, 29 October ("this is Illinois . . . land of the fix"), and 21 December (Madigan "has too much power for his own good") 2001; Pearson, "Hynes Backs Out of Governor Field," *CT*, 29 June 2001; Christi Parsons, "Ryan Uses Veto to Push Gay Rights," *CT*, 11 August 2001; Amanda Puck, *Check, Please!*, 14 August 2001; Gabriel Piemonte, "Lobbying for Living Wage," *HPH*, 15 August 2001, p. 6; Pearson, "Madigan Gives His Blessing to Daley," *CT*, 17 August 2001; Parsons, "Ryan Vetoes Expanded Death Penalty," *CT*, 18 August 2001; Art Barnum, "Birkett Opens Door to Attorney General Bid," *CT*, 21 August 2001; "Ryan: $1.3 Million in Illinois FIRST Projects for Chicago," Illinois Governor's Office Press Release, 21 August 2001; "Governor Signs Bills to Increase Penalties for Use of Falsified FOID Card and Possession of Illegal Ammunition" and "Governor Signs Bills Improving Rights of Crime Victims," Obama Staff Press Releases, 23 August 2001; "Leaders Promote 1st Day Attendance," *CT*, 24 August 2001; Gary Wisby, "Welfare Reform Brings Dramatic Drop in Cases," *CST*, 26 August 2001, p. 8; Strausberg, "Obama, Smith Join Chicago Urban League School Push," *CD*, 27 August 2001, p. 7; Piemonte, "Obama Tax Credit Bill Signed by Governor Ryan," and Todd Spivak, "Cook County Voters Will Get a Second Chance," *HPH*, 5 September 2001, p. 6; Gretchen Helfrich, "Slavery and the Constitution," *Odyssey*, WBEZ Radio, 6 September 2001 (Obama one of three guests); Steve Neal, *CST*, 10 September 2001, p. 1; Pearson, "Schmidt Bows Out of One Race, Enters Another," *CT*, 11 September 2001; Steve Rhodes, "The Case Against Daley," *CM*, December 2002 ("the most consistently troubling aspect of the Daley administration is the way the mayor's friends have loaded up on city business"); Terry Armour, "Salute to the Cutting Room Floor," *CT*, 16 September 2002; *HPH*, 11 December 2002, p. 2; Lenore T. Adkins, "U. of C. Holds Barbershop, History in Its Hands," *HPH*, 2 April 2003, p. 6; Johnathan E. Briggs, "Hyde Park Haircut Hub on the Move," *CT*, 27 December 2006; Erin Meyer, "Discovering Hyde Park," and "Who Is Barack Obama?," *HPH*, 14 February 2007, pp. 4, 6, 15; Mark Konkol, "Mr. Obama's Neighborhood," *CST*, 25 May 2007, p. 12; Todd Spivak, "Barack Obama and Me," *Houston Press*, 28 February 2008; Steve Johnson, "'Check, Please' Reveals Another Side of Obama," *CT*, 15 January 2009; Christopher Benson, "The President's Neighborhood," *Chicago Magazine*, June 2009; DJG interviews with John Schmidt, Dan Shomon, Will Burns, David Kindler, Stephen Heintz, and Miles Rapoport.

44. Christopher Wills, "Lawmakers Want Redistricting Process Tossed Out," AP, 18 July 2001; Christi Parsons, "Democrats Make Their Own Move for Remap," *CT*, 26 July 2001; Rick Pearson, "Jim Ryan Takes Heat on Remap," *CT*, 20 August 2001; Daniel C. Vock, "State Legislative Remap Spawns More Litigation," *CDLB*, 20 August 2001, p. 1; Vock, "Bilandic or Miller May Decide State Remap," *CDLB*, [23] August 2001, p. 1; Vock, "James Ryan Takes Redistricting Case to Federal Court," *CDLB*, 24 August 2001, p. 1; Patricia Manson, "Judge: Consolidation Ruling Comes First in Remap Flap," *CDLB*, [29] August 2001, p. 1; John Patterson, "Political Power for the Next Decade Is All in a Drawing," *CDH*, 4 September 2001, p. 4; Vock, "Dems Mount Final Attack on Remap Lottery," *CDLB*, [5] September 2001, p. 1; Pearson, "Lottery on Remap Being Challenged," *CT*, 5 September 2001; Pearson and Hugh Dellios, "Republicans Hit Jackpot in Legislative Remap," *CT*, 6 September 1991; *CF Extra*, 5 September 2001; Chinta

Strausberg, "Jesse White Pulls Democratic Majority for Remap," *CD,* 6 September 2001, p. 6; Pearson, "Democrats Win Lottery for Remap," *CT,* 6 September 2001; Vock, "Luck Sides with Dems for Remap Panel Majority," *CDLB,* [6] September 2001, p. 1; Dave McKinney, "Dems Win Control of Legislative Remap," *CST,* 6 September 2001, p. 5; Doug Finke, "Democrats Win Remap Draw," *SJ-R,* 6 September 2001, p. 1; Patterson, "Democrats Gain Redistricting Advantage," *CDH,* 6 September 2001, p. 8; *CF,* 6, 7, and 10 September 2001; Bernard Schoenburg, "The Envelope, Please," *SJ-R,* 6 September 2001, p. 7; Vock, "State Remap Battle Moves to Federal Court," *CDLB,* [7] September 2001, p. 1; David Maraniss, "The Luck That Propelled President Obama's Rise," *WP,* 11 March 2012; DJG interviews with John Charles, Will Burns, Carter Hendren, Patty Schuh, Carter Hendren, and Dave Sullivan.

45. JCAR Minutes, 7 August 2001; "Our Politicians Weigh In on Attack," *HPH,* 19 September 2001, p. 4; Obama, remarks at the Woodrow Wilson Center, Washington, D.C., 1 August 2007; Obama's 17 August 2011 interview with CBS News' Anthony Mason; Sam Hudzik, "Where Was President Obama on September 11th, 2001?," WBEZ Radio, 2 September 2011; Jeremiah Wright, "The Day of Jerusalem's Fall," 16 September 2001, blakfacts.blogspot.com/2008/03/day-of-jerusalems-fall.html; Grant and Grant, *The Moment,* pp. 150–52; Michael Eric Dyson, "Fight for Equality Doesn't Diminish with Age," *CST,* 25 September 2001, p. 39; Bill Ayers, *Fugitive Days: A Memoir* (Beacon Press, 2001); Dinitia Smith, "No Regrets for a Love of Explosives," *NYT,* 11 September 2001, pp. E1, E3; Don Terry, "The Calm After the Storm," *CT Magazine,* 16 September 2001, pp. 10ff.; Ben Joravsky, "The Long, Strange Trip of Bill Ayers," *CR,* 8 November 1990; Sheila Miyoshi Jager, "Narrating the Nation: Students, Romance and the Politics of Resistance in South Korea," Ph.D. dissertation, Department of Anthropology, University of Chicago, December 1994; Jager, "Women, Resistance and the Divided Nation: The Romantic Rhetoric of Korean Reunification," *Journal of Asian Studies* 55 (February 1996): 3–21; Jager, "A Vision for the Future, or Making Family History in Contemporary South Korea," *Positions* 4 (Spring 1996): 31–58; Jager, "Woman and the Promise of Modernity: Signs of Love for the Nation in Korea," *New Literary History* 29 (Winter 1998): 121–34; Jager, *Narratives of Nation Building in Korea: A Genealogy of Patriotism* (M. E. Sharpe, 2003); Cindy Leise, "Oberlin High Holds Signing Day for 84 Seniors Heading to College," [Lorain County] *Chronicle-Telegram,* 2 May 2012; Diane Salvatore, "Barack and Michelle Obama: The Full Interview," *Ladies' Home Journal,* August 2008; Barack in Jodi Kantor, "The Obamas' Marriage," *NYT Magazine,* 1 November 2009; DJG interviews with Doris Karpiel, Brad Burzynski, John Corrigan, Jeremiah Wright, Bill Ayers, Bernardine Dohrn, Mona Khalidi, Carol Travis, Madeline Talbott, Sheila Jager, and Mike Dees.

46. "Dems Seek Own Counsel in Remap Case," *CDLB,* 14 September 2001, p. 1; *CF,* 17 and 19 September 2001; *Barnow v. Ryan,* N. D. Ill. #01-C-6566, 2001 WL 1104729, 17 September 2001; Rick Pearson, "Democrats' Remap Could Tilt Senate," *CT,* 18 September 2001; Doug Finke, "Legislative Map Unveiled," *SJ-R,* 18 September 2001; "Ryan: $1.4 Million in Illinois FIRST Projects for Chicago," Illinois Governor's Press Release, 18 September 2001; Todd Spivak, "Lack of Opponents and Money for U.S. Rep. Rush," *HPH,* 19 September 2001, p. 10; Finke, "Republican vs. Republican in Six Districts in Democrats' Map," *SJ-R,* 19 September 2001; Pearson, "Republicans Lament Remap," *CT,* 19 September 2001; *Currie v. Ryan,* Ill. S. Ct. #92341, 19 September 2001; Pearson, "Democrats Set Remap Vote, GOP Looks to Court," *CT,* 20 September 2001; Pearson, "GOP Takes Wraps Off Rival Remap Proposal," *CT,* 21 September 2001; Daniel C. Vock, "Republicans Release Proposed Legislative Remap," *CDLB,* 21 September 2001, p. 1; Christi Parsons and Kevin Lynch, "Burris Says 3rd Time Is Charm in Governor Run," *CT,* 24 September 2001; Chinta Strausberg, "Burris Vows to Extend the Death Moratorium," *CD,* 26 September 2001, p. 7; Spivak, "Democrats Would Be King Under New State Remap," *HPH,* 26 September 2001, pp. 1–2; UCLS *Announcements 2001–02,* esp. pp. 42, 48; UCLS, *The Glass Menagerie 2001–02,* esp. pp. 20, 59; Eli Saslow, "Obama Worked to Distance Self from Blagojevich Early On," *WP,* 12 December 2008; David Bernstein, "Chicago Straight," *Chicago Magazine,* June 2009; Alexandra Starr, "Case Study," *NYT Magazine,* 21 September 2008; Sam Hudzik, "Where Was President Obama on September 11th, 2001?," WBEZ Radio, 2 September 2011; DJG interviews with Debbie Halvorson, Roland Burris, Delmarie Cobb, Fred Lebed, Dan Shomon, Jesse Jackson Jr., Stephen Feldman, Martha

Pacold, David Bird, Aaron Solomon, Mark Warnick, Tasneem Goodman, Karen Schweickart, Denise Walia Levin, Byron Rodriguez, Bethany Lampland, Alex Montgomery, Nathaniel Edmonds, and Beth DeLisle.

47. Rick Pearson, "Democrat Remap Clears Panel Over GOP Protests," *CT,* 26 September 2001; *CF,* 26, 27, and 28 September and 1, 9, 10 and 15 October 2001; Paul Simon to Obama, 26 September 2001, PSP; *Winters v. Illinois State Board of Elections,* 197 F. Supp. 1110 (N. D. Ill.), 20 November 2001; Pearson, "Democrats Score Another Victory—Remap Challenge by GOP Rejected," *CT,* 29 September 2001; Patricia Manson, "Legislative Redistricting Squabble Making for Strange Bedfellows in Deed," *CDLB,* [29] September 2001; *Chicago Tonight,* WTTW, 2 October 2001; William Grady and John Chase, "DuPage Clout Threatened," *CT,* 11 October 2001; Obama, *TAOH,* p. 186; Obama, "Remarks at a Virtual Town Hall Meeting," 26 March 2009, *PPotP* 2009 Vol. 1, p. 361; Michelle Obama remarks, 18 September 2009, Washington, D.C.; Obama, "Remarks at Opening Session of Bipartisan Meeting on Health Care Reform," 25 February 2010, *PPotP* 2010 Vol. 1, p. 256; Byron Tau, "Obama: Hardest Moment in My Life Was Sasha's Illness," *Politico,* 30 March 2012; Carrie Healey, "Obama Recalls Daughter Sasha's Medical Scare," *The Grio,* 26 September 2013; Healey, "Michelle Obama Discusses Sasha's Meningitis Scare," *The Grio,* 20 December 2013; "Host Committee," Friends of Barack Obama fall fund-raiser, 11 October 2001, LRHP; Todd Spivak, "Obama Raises Money for Next Election Season," *HPH,* 17 October 2001, p. 3; Jennifer Mason e-mail to Laura Hunter, "Names," 23 October 2001, LRHP; Spivak, "Developers Contribute to Alderman's Election Fund," *HPH,* 24 October 2001, p. 8; Obama to Timuel Black, 24 October 2001, TBP Box 103 Fld. 23; Steven R. Strahler, "Chicago Observer," *CCB,* 29 October 2001, p. 50; Obama for Congress 2000, FEC Form 3, Report of Receipts and Disbursements, 1 October–31 December 2001, 22 November 2002; Friends of Barack Obama, Report of Campaign Contributions and Expenditures, 1 July–31 December 2001, 30 January 2002; Christopher Drew and Mike McIntire, "After 2000 Loss, Obama Built Donor Network from Roots Up," *NYT,* 3 April 2007; DJG interviews with Bill Luking, Dan Shomon, and Laura Russell Hunter.

48. David Mendell and Gary Washburn, "Race Shifts Seen in Remap Effort," *CT,* 14 October 2001; "Remap Foes," *CDLB,* 15 October 2001; JCAR Minutes, 16 October 2001; Chinta Strausberg, "Sen. Obama: Barriers 'Sad, Symbols of Fear,'" *CD,* 17 October 2001, p. 8; Daniel C. Vock, "Legislative Remap Case Before State's High Court," *CDLB,* [23] October 2001, p. 1; *Illinois Legislative Redistricting Commission v. White,* Ill. S. Ct. #92454, 23 October 2001; Vock, "Justices Deny Dem Request for a 3rd Time in Remap Case," *CDLB,* [24] October 2001, p. 1; Vock, "GOP Senators Ask High Court to Redraw Remap," *CDLB,* 30 October 2001, p. 1; "Ryan: $4.5 Million in Illinois FIRST Projects for Chicago," Illinois Governor's Office Press Release, 30 October 2001; Annah Dumas-Mitchell, "Metamorphosis of Woodlawn Nearly Complete," *CD,* 5 November 2001, p. 5; Vock, "Dems, GOP Face to-Face over Remap," *CDLB,* 8 November 2001, p. 1; Adam Gross notes, Business and Professional People for the Public Interest, "New Avenues for Affordable Housing," 8 November 2001, AGP; "Anita Braver, Judy Byrd and Abby O'Neil Invite You to a Lunch and Discussion with Senator Barack Obama," 9 November 2001, "Obama 2002 and Beyond," Presentation for Women's Lunch, 9 November 2001, "Women's Luncheon Friday, November 9, 2001," Sign-In Sheets, "Obama Women's Luncheon, 11-9-01," LRHP; Crystal Yednak, "Foes of Death Penalty Bill Threaten Political Revenge," *CT,* 25 October 2001; Todd Spivak, "South Side Politicians, Leaders Battle Gang Bill," *HPH,* 31 October 2001, p. 3; "Some Vetoes Controversial but Little Action Expected," *C-WR* Vol. 17 #1, 13 November 2001; Strausberg, "Jesse Connecting Black Statewide Coalition for Unity," *CD,* 13 November 2001, p. 5; JCAR Minutes, 13 November 2001; Gabriel Piemonte, "Veto Session Has Full Plate Following Sept. 11 Attacks," *HPH,* 14 November 2001, p. 2; Minutes of the Regular Meeting of the Board of Trustees, General Assembly Retirement System of Illinois, 14 November 2001, #220, pp. 1845–48; Vock, "Minorities Fail in Bid to Join Remap Case," *CDLB,* 16 November 2001, p. 1; "Ryan: $11.9 Million in Illinois FIRST Projects for Chicago," Illinois Governor's Office Press Release, 16 November 2001; Greg Hinz, "Lobbying Bid Falls Short for Bank One," *CCB,* 19 November 2001, p. 4; Strausberg, "Final Burial of Death Penalty Big Win for Coalition," *CD,* 20 November 2001, p. 1; Spivak, "Gang Bill Veto Upheld," *HPH,* 21 November 2001, p. 7; Vock, "Redistricting Arguments Seem Like Deja Vu All Over Again," *CDLB,* [22] November 2001,

p. 1; Vock, "Court to Examine Redistricting Complaints," *CDLB,* 26 November 2001, p. 1; "Obama Favors 'Economic Security' Package in Effort to Help Working Families, Businesses," Obama Staff Press Release, 27 November 2001; Piemonte, "HP Legislators Ponder Bills' Threat to Civil Liberties," and Obama, "Budget Knife Shouldn't Cut the Homeless," *HPH,* 28 November 2001, pp. 3, 4; Senate Floor Transcript, 28 November 2001, pp. 84–97 (HB 2299); John Patterson, "Anti-Terrorism Bill Clears One Hurdle in Legislature," *CDH,* 29 November 2001, p. 8; *Cole-Randazzo v. Ryan,* 762 N.E. 2d 485 (Ill. S. Ct.), 28 November 2001; Rick Pearson and Gary Washburn, "High Court's Democrats Keep Party Remap Lines," *CT,* 29 November 2001; Vock, "Justices Vote Party Lines to Uphold Remap," *CDLB,* 29 November 2001, p. 1; Stephen S. Morrill, "Client Report for 2001 Veto Session," Morrill and Associates, December 2001; James Janega, "Patricia A. Jones, 63," *CT,* 4 December 2001; *CF,* 5, 12, and 21 December 2001; Vock, "GOP Goes for Broke in U.S. Court in Remap Fight," *CDLB,* 10 December 2001, p. 1; "Veto Session Highlights," *CW-R* 17 #2, 17 December 2001, pp. 2–3; Steven R. Strahler, "2001: The Year in Review," *CCB,* 17 December 2001, p. 18; JCAR Minutes, 18 December 2001; Spivak, "Filing: Hyde Park Pols Face Few Challengers in 2002," *HPH,* 19 December 2001, p. 1; Spivak, "City, State Remap Bickering Dominates Year," *HPH,* 26 December 2001, p. 11; *Beaubien v. Ryan,* 762 N. E. 2d 501 (Ill. S. Ct.), 27 December 2001; John Flynn Rooney, "GOP Suffers Shutout in Remap Battle," *CDLB,* [28] December 2001, p. 1; *Campuzano v. Illinois State Board of Elections,* 200 F. Supp. 2d 905 (N. D. Ill.), 3 May 2002; Strahler, "Dollar for Dollar: Senate Race Could Be One for the (W)ages," *CCB,* 24 December 2001, p. 22; Kate A. Kane, "Risky Business, Sound Thinking," *Fast Company,* April 1997; Blair Hull, "The Future of Trading," *Futures Industry,* December 2000–January 2001; Joshua Green, "A Gambling Man," *Atlantic Monthly,* January–February 2004, pp. 34–38; Mark Jannot, "A Rahm for the Money," *Chicago Magazine,* August 1992; William Neikirk, "Mr. Fixit: Rahm Emanuel's Chicago Street Smarts Saved His Career," *CT,* 23 November 1997; DJG interviews with Jason Erkes, Rick Ridder, Blair Hull, and Anita Dunn; Ann Bertell e-mail to DJG, 9 May 2014. Zev Chafets, "Obama's Rabbi," *NYT Magazine,* 5 April 2009, asserts that Capers Funnye's Blue Gargoyle received $75,000 in state funds thanks to Obama, but state records document only $25,000 authorized by Barack on 9 August 2000 and announced by Governor Ryan's office on 16 November 2001.

49. Maurice Lee, "Black Contractors Protest U. of C. Hospital Project," *HPH,* 14 November 2001, pp. 1–2; Annah Dumas-Mitchell, "Officials to Contractors: Blacks Won't Be Cheated," *CD,* 29 November 2001, p. 1; Dumas-Mitchell, "U of C Commits to Economic Advancement of Minorities," *CD,* 5 December 2001, p. 5; LaRisa Lynch, "AACA and University of Chicago Hospitals Reach Agreement to Increase Black Participation in Construction, Employment Opportunities," *CW,* 6 December 2001, p. 1; Slevin, *Michelle Obama,* pp. 174–77; Obama, Constitutional Law Final Examination, [ca. 5–10] December 2001; Joyce Foundation, *Work in Progress,* January 2002; "Joyce Foundation Names New President," *CT,* 10 January 2002; "Ellen Alberding Named Joyce Foundation President," Joyce Foundation News Release, 10 January 2002; "Joyce Foundation Appoints New President," *Philanthropy News Digest,* 13 January 2002; "KidsOhio.org Organized to Advocate for Ohio Children," PR Newswire, 21 April 2002; Obama's February 2004 interview with David Axelrod; Scott Helman, "Early Defeat Launched a Rapid Political Climb," *BG,* 12 October 2007; Kenneth P. Vogel, "Obama Linked to Gun Control Efforts," *Politico,* 20 April 2008; Mundy, *Michelle,* p. 136; Carol Felsenthal, "What If: Would Obama Be Living in the White House Today If . . . ?," *Huffington Post,* 30 January 2009; Newton Minow's 2008 interview with Jim Gilmore; DJG interviews with Debbie Leff, Unmi Song, Margaret O'Dell, Paula Wolff, Carlton Guthrie, Carin Clauss, Jack Anderson, Charles Daly, Michael Claffey, Mark Real, Dan Shomon, Abner Mikva, John Schmidt, Newton Minow, Rob Fisher, and Linda Hawker.

50. UCLS *Announcements 2001–02,* p. 87; Obama, Voting Rights Syllabus, Winter 2002; *The Solon,* January 2002; *CF,* 9, 10, 25, 28, and 31 January, and 1, 5, 7, 8, 13, 18, 25, and 26 February 2002; JCAR Minutes, 9 January 2002; Carrie Golus, "Obama to Speak for MLK Day," *UCC,* 10 January 2002; *CDH,* 21 January 2002, p. N1; B. Drummond Ayres Jr., "A Governor in the Middle of a Race He Won't Run," *NYT,* 25 January 2002; Senate Floor Transcript, 30 January 2002, p. 3 (Obama introducing SBs 1652 and 1653); Min Lee, "Public Funding Sought for Supreme Court Candidates," *CDLB,* 31 January 2002, pp. 1, 22; Aaron Chambers, "Political Kaleidoscope," *II,* February 2002;

Rick Pearson, "Blagojevich War Chest Dwarfs Opponents," *CT,* 1 February 2002; JCAR Minutes, 5 February 2002; Senate Floor Transcript, 5 February 2002, pp. 6–7 (Obama introducing SBs 1781, 1782, 1783, 1788, and 1789); Ray Long and Adam Kovac, "Philip Urges Review of Pork-Barrel Funds," *CT,* 6 February 2002; Senate Floor Transcript, 6 February 2002, p. 20; Senate Floor Transcript, 7 February 2002, pp. 1, 5, 7, 11, 13, and 17; Douglas Holt and Ray Gibson, "2 Accused in Scheme at Pork-Rich Charities," *CT,* 9 February 2002; Chinta Strausberg, "ACORN, Pols Unveil Bill to Aid Heat Bills," *CD,* 13 February 2002, p. 3; "Senate Democrat FY 02 Member Project Authorization Form," 13 February 2002 ("Southwest Women Working Together" filed its final IRS Form 990 in 2006), Illinois DCEO FOIA; Dan Mihalopoulos, "Burris Faces Toughest Step," *CT,* 17 February 2002; Shamus Toomey, "Are Fitzgerald's Enron Blasts Real Anger or Opportunism?," *CDH,* 17 February 2002, p. 1; Todd Spivak "Despite Breezy Primaries, Banks Big for State Reps," and Obama, "Human Alternatives to State Budget Cuts," *HPH,* 20 February 2002, pp. 1–2, 4; Long and Christi Parsons, "Ryan Budget Puts State on Crash Diet," *CT,* 21 February 2002; "Governor Ryan's Final Budget Address," *C-WR* 17 #4, 27 February 2002; Spivak, "Preckwinkle Bank Bulges, Hairston Trailing Behind," *HPH,* 27 February 2002, pp. 1–2; Obama for Congress 2000, FEC Form 3, Report of Receipts and Disbursements, 1 January–31 March 2002, 4 December 2002; Christopher Drew and Mike McIntire, "After 2000 Loss, Obama Built Donor Network from Roots Up," *NYT,* 3 April 2007; Ben Protess and Steve Mills, "Roland Burris Received About $1.2 Million in Loans from Businessman," *CT,* 9 February 2009; Frank, *Obama on the Couch,* pp. 203–4, 218; DJG interviews with Karen Schweickart, David Bird, Byron Rodriguez, and Tasneem Goodman.

51. *King v. State Board of Elections,* 2002 WL 356383 (N. D. Ill.), 5 March 2002, and 2003 WL 22019357 (N. D. Ill.), 19 August 2003; *King v. Illinois State Board of Elections,* 410 F. 3d 404 (7th Cir.), 13 June 2005 (argued 28 October 2004 by Jeffrey Cummings); JCAR Minutes, 5 March 2002; "Senate Committee Action Report," Judiciary, 5 March 2002 (SB 1662); *CF,* 6, 8, 14, 15, 19, 20, 21, 22, and 29 March 2002 and 19 August 2013; "Off-Road Track Noise Bill Clears Committee," *DHR,* 12 March 2002, p. A3; "Illinois Primary Election Set for March 19," *C-WR* 17 #5, 13 March 2002; Rick Pearson and Jeff Zeleny, "Bush Drops In for City's Green Party," *CT,* 17 March 2002; Jon Van, "New Telemarketer Twist on Computerized Calls," *CT,* 18 March 2002; Pearson, "Ryan to Face Blagojevich," *CT,* 20 March 2002; Mike Dorning, "Blagojevich Rode to Win by Way of Downstate," *CT,* 21 March 2002; Todd Spivak, "Jesse Jr. Wins Big, Currie and Obama Rev Up for Nov.," and "Bill Would Protect Activists," *HPH,* 27 March 2002, pp. 1–2, 7; Chinta Strausberg, "Pols to Ryan: Don't Mess with Workers," *CD,* 27 March 2002, p. 5; [Debbie] Lounsberry, "Barack Obama—The Record: A Summary," 10 May 2004, JRORF, pp. 12–13; Amanda B. Carpenter, "Obama More Pro-Choice Than NARAL," HumanEvents.com, 26 December 2006; Obama's 16 August 2008 interview with David Brody on CBN.com; DJG interviews with Judd Miner, John Belcaster, Jeff Cummings, Geoffrey Rapp, Ira Silverstein, and Dan Cronin.

52. Chinta Strausberg, "Howard Pushes Expungement Bills," *CD,* 1 April 2002, p. 4; *CF,* 19 January 2001, 4 March 2002 ("George Ryan has been one of the most honest governors this state has ever had"), and 3, 4, 11, 12, 16, and 17 April 2002; Obama, "Laws Help Ex-Offenders Re-Enter Society," *HPH,* 3 April 2002, p. 4; Matt O'Connor, "U.S. Accuses Ryan Fund, 2 Ex-Aides of Corruption," *CT,* 3 April 2002; Senate Floor Transcript, 3 April 2002, pp. 6 (Obama introducing SR 374) and 139 (Obama amending SB 1782); Rick Pearson and O'Connor, "U.S. Probe Indicts 'Political Culture,'" *CT,* 4 April 2002; Senate Floor Transcript, 4 April 2002, pp. 28–35 (SBs 1661, 1662, and 1663), 45 (SB 1782), 104–5 (SB 2098), and 153–56 (SB 2303); "Obama's Postpartum Depression Bill Is Sent to Governor Ryan," Obama Staff Press Release, 19 June 2002; Steve Neal, "Bloody Scandal Tops Them All," *CST,* 5 April 2002, p. 41; Steve Rhodes, "The Fawell Affair," *Chicago Magazine,* May 2004; Adriana Colindres, "Arguments Made for Cigarette Tax Hike," *SJ-R,* 5 April 2002, p. 4; Daniel C. Vock, "Senators Scorch Fee Claim in Tobacco Case," *CDLB,* 5 April 2002, pp. 1, 22; Strausberg, "Medicaid Cuts Will Cost Millions in Lost Federal Funds," *CD,* 6 April 2002, p. 1; Todd Spivak, "Critics Call the Remaps Partisan, Undemocratic," *HPH,* 10 April 2002, p. 2 (Obama admitting that "incumbents drawing their own maps will inevitably try to advantage themselves"); JCAR Minutes, 10 April 2002; Senate Floor Transcript, 10 April 2002, pp. 9–20 (SJR 18, seeking to remove the Illinois Supreme Court's criminal jurisdiction);

Report of the Governor's Commission on Capital Punishment, 15 April 2002; John O'Connor, "Ryan to Keep Moratorium While Death Penalty Overhaul Faces Uncertain Future," AP, 16 April 2002; Steve Mills and Maurice Possley, "'We're Talking About Life and Death,'" *CT,* 16 April 2002; Jodi Wilgoren, "Panel in Illinois Seeks to Reform Death Sentence," *NYT,* 16 April 2002; Thomas P. Sullivan, "Repair or Repeal—Report of the Governor's Commision on Capital Punishment," *Illinois Bar Journal* 90 (June 2002): 304–7, 325; Sullivan, "Preventing Wrongful Convictions," *Judicature* 86 (September–October 2002): 106–9, 120; Scott Turow, *Ultimate Punishment* (Farrar, Straus and Giroux, 2003); Barack H. Obama and Michelle L. Obama, IRS Form 1040, 13 April 2002; Woods Fund of Chicago, *2001 Annual Report,* p. 34 (noting a $6,000 grant to Trinity United Church of Christ "in recognition of Barack Obama's contribution of services," pursuant to 5 ILCS 420 §2-110 [b]); Dorothy A. Brown, "Lessons from Barack and Michelle Obama's Tax Returns," *Tax Notes,* 10 March 2014, pp. 1109–13; Friends of Barack Obama, Report of Campaign Contributions and Expenditures, 1 January–30 June 2002, 31 July 2002; Obama, *TAOH,* p. 3; Adelstein in Andersen, *Barack and Michelle,* p. 195, on Sam Hudzik, "Where Was President Obama on September 11th, 2001?," WBEZ.org, 2 September 2011, and in Kantor, *The Obamas,* p. 276; Jones in "Jones Recalls Request for Help with Running for U.S. Senate," *SJ-R,* 31 December 2006, p. 5, in Ryan Lizza, "The Agitator," *TNR,* 19 March 2007, in Ron Fournier, "Obama's Life and Record in Springfield, Ill., Hint at the President He Would Be," AP, 17 June 2007, in Janny Scott, "In Illinois, Obama Proved Pragmatic and Shrewd," *NYT,* 30 July 2007, on *Nightline,* ABC, 25 February 2008, in Todd Purdum, "Raising Obama," *Vanity Fair,* March 2008, on "Barack Obama Revealed," CNN, 20 August 2008, in Abdon M. Pallasch, "Grooming a Politician," *CST,* 24 August 2008, p. A14, on *All Things Considered,* NPR, 16 October 2008, in David Smallwood, "Two Guys Who Laid It on the Line," *N'Digo Profiles,* December 2008, pp. 14–15, in Adrienne P. Samuels, "Barack's Chicago," *Ebony,* January 2009, and in David Bernstein, "Chicago Straight," *CM,* June 2009; Cliff Kelley in Todd Spivak, "Barack Obama and Me," *Houston Press,* 28 February 2008; DJG interviews with Dan Shomon, Chris Sautter, Paul Harstad, Pete Giangreco, Eric Adelstein, Jill Rock, Emil Jones Jr., and John Kelly.

53. "Resolution Calls Attention to Health," *HPH,* 17 April 2002, p. 2; GARS Board of Trustees Minutes, 17 April 2002, #221, pp. 1849–53; Dan Shomon e-mail to Obama, "Press Conference on WARN," 18 April 2002, AGP; "Intellectuals: Who Needs Them?," Center for Public Intellectuals, UIC, 19–20 April 2002 ("Intellectuals in Times of Crisis: Experiences and Applications of Intellectual Work in Urgent Situations"); Michael Sneed, "Hmmm . . . ," *CST,* 25 April 2002, p. 4; *CF,* 26 April and 1, 2, 3, 6, and 7 May 2002; Sneed, "Join the Crowd: Chico May Run for Senate," *CST,* 28 April 2002, p. 12; Thomas A. Corfman, "Chico Takes Top Spot at Altheimer," *CCB,* 8 May 1999; Lisa Brennan, "Altheimer and Gray's Bold New Leader," *NLJ,* 7 June 1999; Fred Lebed e-mail to DJG, 3 May 2011 (lunch with Obama 29 April 2002); Michelle and Nesbitt in Mendell, *Obama,* pp. 151–53; Jarrett in Christopher Drew and Mike McIntire, "After 2000 Loss, Obama Built Donor Network from Roots Up," *NYT,* 3 April 2007, in Scott Helman, "Early Defeat Launched a Rapid Political Climb," *BG,* 12 October 2007, in Peter Slevin, "For Obama, a Handsome Payoff in Political Gambles," *WP,* 13 November 2007, in Slevin, "Her Heart's in the Race," *WP,* 28 November 2007, in Shira Schoenberg, "17 Minutes to Stardom," *Concord Monitor,* 6 December 2007, in Richard Wolffe, "Inside Obama's Dream Machine," *Newsweek,* 14 January 2008, in Roger O. Crockett, "Obama's Executive Sounding Board," *Business Week,* 19 May 2008, and most especially in Robert Draper, "The Ultimate Obama Insider," *NYT Magazine,* 26 July 2009; Remnick, *The Bridge,* pp. 357–58 and 368; Ryan Lizza, "Making It: How Chicago Shaped Obama," *New Yorker,* 21 July 2008; Jodi Kantor, "Obama's Friends Form Strategy to Stay Close," *NYT,* 13 December 2008; Kantor, *The Obamas,* pp. 197–98; Melissa Harris, "Martin Nesbitt, the First Friend," *CT,* 21 January 2013; Friends of Barack Obama, Report of Campaign Contributions and Expenditures, 1 January–30 June 2002, 31 July 2002; Chuck Neubauer and Tom Hamburger, "Obama Donor Received a State Grant," *LAT,* 27 April 2008; Link on "The Candidates: Barack Obama," MSNBC, 20 February 2008; Bryan Smith, "Cook County Sheriff Tom Dart," *Chicago Magazine,* January 2009; Jarrett's 2012 interview with Jim Gilmore; DJG interviews with Blair Hull, Rick Ridder, Fred Lebed, David Spiegel, Robert Blackwell Jr., Larry Walsh, Terry Link, Denny Jacobs, Kim Lightford, James Clayborne, Miguel del Valle,

Lou Viverito, Pat Welch, Larry Suffredin, Elvin Charity, Craig Huffman, Carlton Guthrie, Paul Strauss, and Carol Moseley Braun.

54. Senate Floor Transcript, 8 May 2002, pp. 76–95 (HB 5646); Aamer Madhani, "Richard Newhouse Jr., 78," *CT,* 2 May 2002; Senate Floor Transcript, 9 May 2002, pp. 23–25 (HB 3212, the Technology Development Act, passed 56–0–0 and signed into law 26 August 2002 as PA 92-0851) and 90 (SJR 78); Dave Lundy, "Governor Must Lead on Tech, Capital," *CST,* 23 January 2003, p. 52; Harstad Strategic Research, "Illinois Statewide Survey—Democratic Primary Voters, May 2002," and "General Election Survey—Illinois Statewide Survey—May 2002," 9–15 May 2002, PHP and JKP; David [Spiegel] to Rick [Ridder], "Barack Obama," 10 May 2002, RRP; Kristen McQueary, "State Legislator Considers Bid for Senate," *DS,* 11 May 2002, p. A3; *CF,* 13, 15, 20, and 22 May and 1, 2 and 3 June 2002; JCAR Minutes, 14 May 2002; Michael Sneed, "Sneed," *CST,* 21 May 2002, p. 4; Kip Kolkmeier to Secretary of the Senate, 21 May 2002, FEC; *Flynn v. Ryan,* 771 N. E. 2d 414 (Ill. S. Ct.), 23 May 2002; *CF,* 1 February 2000; Christi Parsons, "Justices Revive Ethics Law," *CT,* 24 May 2002; Adriana Colindres, "Justices Put Gift Ban Act Back in Effect," *SJ-R,* 24 May 2002, p. 1; "Obama Praises Supreme Court's Ethics Ruling," Obama Staff Press Release, 28 May 2002; Laura Washington, "Moseley-Braun Looking for Rerun," *CST,* 27 May 2002, p. 21; Senate Floor Transcript, 28 May 2002, pp. 24–70, esp. 47 (SB 2390); Todd Spivak, "Two HP Politicians Consider Bid for U.S. Senate," *HPH,* 29 May 2002, p. 3; "Budget Crisis Tops Spring Session Agenda," *C-WR* 17 #6, 23 April 2002, p. 2; Rick Pearson and Christi Parsons, "Illinois Senate Passes $54 Billion Budget Measure," *CT,* 29 May 2002; Daniel C. Vock, "Appeals Switch Encounters Budget Hitch," *CDLB,* 29 May 2002, pp. 1, 24; Senate Floor Transcript, 31 May 2002, pp. 28–49 (HB 4680); Ray Long and Christi Parsons, "Legislators End Session with Pork, Tax Hikes," *CT,* 3 June 2002; "Obama Says 92nd General Assembly Produces Some Progress, Some Disappointment," Obama Staff Press Release, n.d. [ca. 3 June 2002]; Spivak, "Budget Deemed Reasonable," Obama, "Laws Address Illinois Sentencing Disparities," and Spivak, "Supreme Court Upholds Campaign Finance Law," *HPH,* 5 June 2002, pp. 3, 4, 6; Senate President James "Pate" Philip, *End of Session Report 2002,* esp. pp. 62–63, 82; "Major Bills Passed by the Illinois General Assembly," *LRU First Reading* 16 (July 2002): 6–11; Remnick, *The Bridge,* p. 359; Axelrod, *Believer,* p. 124; DJG interviews with Chris Sautter, Paul Harstad, Mike Kulisheck, Kristen McQueary, David Spiegel, Blair Hull, Rick Ridder, and John Kupper. Earlier proceedings in the case challenging the Gift Ban Act are noted in Ryan Keith, "Judge Throws Out State Gift-Ban Law," *SJ-R,* 4 August 2000, p. 1, Charles N. Wheeler III, "A Judge's Ruling Left Ethics and Campaign Finance Reforms in Limbo," *II,* September 2000, pp. 42–43, Dave McKinney, "Ruling Lets Pols Off the Hook," *CST,* 19 September 2000, p. 10, and Doug Finke, "House Approves New Ethics Package," *SJ-R,* 6 April 2001, p. 11.

55. Steve Neal, "Dem Senate Race to Be Free-for-All," *CST,* 3 June 2002, p. 33; "The Strausberg Report," n.d. [ca. 6 June 2002]; Russ Stewart, "'Roadblock Strategy' Is Daley Goal for 2003 Race," *NSP,* 12 June 2002; Harstad Strategic Research, "Illinois Statewide Survey—Democratic Primary Voters, May 2002," and "General Election Survey—Illinois Statewide Survey—May 2002," 9–15 May 2002, PHP and JKP; Harstad in Noam Scheiber, "Cruel Intentions," *TNR,* 28 May 2008, p. 7; DJG interviews with Chinta Strausberg, Paul Harstad, Chris Sautter, and Dan Shomon.

56. Bennett, Petts & Blumenthal Poll #1427, Final Illinois Senate, June 2002, MBP; DJG interviews with Mark Blumenthal, Anita Dunn, and Blair Hull.

57. Christi Parsons and Adam Kovac, "Ryan Orders Legislators to Return," *CT,* 8 June 2002; *CF,* 10, 11 June 2002; *PBS NewsHour,* 10 June 2002; Chinta Strausberg, "Budget Cuts Severe in State's HIV/AIDS Program," *CD,* 11 June 2002, p. 1; JCAR Minutes, 11 June 2002; Senate Floor Transcript, 11 June 2002, pp. 77–82, 301; Strausberg, "Jones, Howard, Obama Bristle at Gov. Ryan's Cuts," *CD,* 12 June 2002, p. 5; Rick Pearson and Ray Long, "Careful Steps, Looking Ahead," *CT,* 3 May 2007; Obama on the *CBS Evening News,* 17 December 2007 (asked "When was the last time you lost your temper?," Obama avoided a direct answer and replied "I don't get mad too often"); David Freddoso, *The Case Against Barack Obama* (Regnery, 2008), pp. 34–36; Eli Saslow, "From Outsider to Politician," *WP,* 9 October 2008; Rickey Hendon, *Black Enough/White Enough: The Obama Dilemma* (Third World Press, 2009), pp. 29–34; Bernard Schoenburg, "Not All Sweetness and Light in Hendon's Obama Book," *SJ-R,* 28 December 2008; Hendon's 13 January 2009 Illinois Channel interview with Terry Martin and his June 2010 one with Peter Slen; Terry Link in Remnick, *The Bridge,* p. 339; McClelland, *Young Mr. Obama,* pp. 189–92; Halvorson, *Playing Ball with the Big Boys,* p. 39; DJG interviews with Courtney Nottage, Jill Rock, John Charles, Cindy Huebner Davidsmeyer, Emil Jones, Donne Trotter, Kim Lightford, Carol Ronen, Pat Welch, Debbie Halvorson, Terry Link, Denny Jacobs, Larry Walsh, Walter Dudycz, and Carol Harwell.

58. Ray Long and Rick Pearson, "Legislators Back Most Ryan Cuts," *CT,* 12 June 2002; "To the Contrary Notwithstanding," *C-WR* 17 #8, 24 June 2002; Obama, "Special Session Brings Some Hard Choices," *HPH,* 3 July 2002, p. 4; Lynn Sweet, "Mega-Donor Gives Senate a Go," *CST,* 13 June 2002, p. 41; Obama, commencement address, North Kenwood/Oakland Charter School, 13 June 2002; Pugh, Jones, and Johnson, "A Tribute to Senate Democratic Leader Emil Jones Jr.," 13 June 2002, SPP; Greg Hinz, "GOP Leader Doled Out Office Help," *CCB,* 17 June 2002; Amanda York, "Legislators Pony Up Millions of Dollars for Homeland Security," AP, 17 June 2002; Rick Pearson, "Daniels Staffers Are Targeted in Probe," *CT,* 21 June 2002; Adriana Colindres, "U.S. Attorney to Probe Allegations Against House Republican Staffers," *SJ-R,* 28 June 2002, p. 11; Ray Long, "U.S. Will Probe Daniels Staffers, Jim Ryan Says," *CT,* 28 June 2002; Pearson, "Daniels Quits GOP Post," *CT,* 29 June 2002; Bernard Schoenburg, "Daniels Quits as GOP Chairman," *SJ-R,* 29 June 2002, p. 1; *CF,* 2 July 2002; Hermene D. Hartman, "Barack and Black Enterprise," *N'Digo,* 7–13 June 2007, p. 3; Mendell, *Obama,* pp. 155–56; Nesbitt in Michael Weisskopf, "Obama: How He Learned to Win," *Time,* 8 May 2008, and in Ryan Lizza, "Making It: How Chicago Shaped Obama," *New Yorker,* 21 July 2008; Jim Reynolds in David Smallwood, "Genesis of the Grassroots Movement," and Robert T. Starks, "Jackson's Tears from All the Years," *N'Digo Profiles,* December 2008, pp. 10–11, 16, 18, 44; Hartman in Klein, *The Amateur,* p. 31; Jason Zengerle, "Operation Push: Obama's Artful Corralling of the Jesse Jacksons," *TNR,* 13 February 2008, pp. 7–8; Jo Becker and Christopher Drew, "Pragmatic Politics, Forged on the South Side," *NYT,* 11 May 2008; Keith D. Picher, "Steve Pugh," *Leading Lawyers Network Magazine,* January 2009; Carol Felsenthal, "Les Coney, Top Obama Bundler and Networker Extraordinaire," *Chicago Magazine,* 21 July 2011; Zengerle, "Jr.: Who Thwarted the Ambitions of Jesse Jackson's Son?," *New York Magazine,* 4 November 2012; Andrew Davis in Jodi S. Cohen and David Jackson, "Blagojevich Allies Help State Sell Off Student Loans," *CT,* 11 March 2007; John McCormick, "Pritzker Blazes Campaign Trail," *CT,* 30 September 2007; James L. Merriner, "Network Ties," *Chicago Magazine,* June 2008; Newton Minow's interview with Jim Gilmore; DJG interviews with Stephen Pugh, Preston Pugh, Hermene Hartman, Eugene Morris, Les Coney, Martin King, Newton Minow, Abner Mikva, Jesse Jackson Jr., and Andy Davis.

59. Steve Neal, "Fitzgerald Could Face an Old Foe," *CST,* 21 June 2002, p. 45; *Public Affairs with Jeff Berkowitz,* 27 June 2002; Ann Bertell e-mail to DJG, 9 May 2014; Harstad Strategic Research, "Statewide Survey Among Democratic Primary Voters in Illinois" and "Statewide Survey Among General Election Voters in Illinois," 9–15 May 2002; Harstad to Obama Finance Committee, "Summary of Poll Results," 11 July 2002, PHP; DJG interviews with Dan Shomon, Blair Hull, Paul Harstad, and Bettylu Saltzman.

60. Steve Neal, "Obama Could Add Drama to Senate Race," *CST,* 3 July 2002, p. 41; JCAR Minutes, 9 July 2002; Todd Spivak, "Obama Begins Raising Funds for U.S. Senate Bid," *HPH,* 10 July 2002, pp. 1–2; Russ Stewart, "Large Democratic Field Targets Fitzgerald in 2004," *NSP,* 10 July 2002; Jennifer Bussell, "Obama May Seek U.S. Senate Seat in '04 race," *Chicago Maroon,* 12 July 2002; Chinta Strausberg, "Barack Obama Explores U.S. Senate Seat," *CD,* 17 July 2002, pp. 1, 3; "On the Air," *CDLB,* 19 July 2002, p. 3; John Kass, "In the Interest of Full Disclosure, Start at O'Hare," *CT,* 23 July 2002; *Chicago Tonight,* WTTW, 23 July 2002 (Obama stating that "I think it's absolutely vital not to think in terms of the single black leader"); Eric Krol, "Eyes Already on 2004 Senate Race," *CDH,* 31 July 2002, p. 1; Friends of Barack Obama, Report of Campaign Contributions and Expenditures, 1 July–31 December 2002, 31 January 2003; Obama's 7 February 2004 interview with David Axelrod, p. 19; Obama in USS, EPW, "S. 131, 'The Clear Skies Act of 2005,'" S. Hrg. 109-867, 2 February 2005, p. 28; Obama on *Good Morning America,* ABC, 8 April 2015; Valerie Jarrett's 2012 interview with Jim Gilmore; Min Lee, "Senate Panel to Review Death Penalty Reforms," *CDLB,* 31 May 2002, pp. 3, 24; Jennifer Davis, "Death Penalty Reforms

Aired at Peoria Hearing," *PJS*, 1 August 2002, p. B1; Spivak, "Currie and Obama Bills Seek to Clean Up Courts," *HPH*, 7 August 2002, p. 7; DJG interviews with Greg Dingens and Matt Jones.

61. Steve Neal, "Sky's the Limit in Senate Race," *CST*, 7 August 2002, p. 43; Russ Stewart, "Hynes' Job as 'Fiscal Grouch' Will Be Taxing," *NSP*, 7 August 2002; Dennis Byrne, "Feeling Like a Political Nitwit?," *CT*, 12 August 2002; JCAR Minutes, 13 August 2002; Michael C. Dorf to Secretary of the Senate (2), 14 August 2002, FEC; "Decision 2002," *Lake and Prairie*, 4th Quarter [August] 2002, p. 2; *CF*, 19 August 2002; David Mendell and Darnell Little, "Rich '90s Failed to Lift All," *CT*, 20 August 2002; Obama to "Dear Friend," Obama for Illinois, 20 August 2002, RRP; Grant Pick, "Hatchet Man: The Rise of David Axelrod," *Chicago Magazine*, December 1987; Doug Cassel, "Is Tim Evans for Real?," *CR*, 16 March 1989; Bryan Miller, "The Art of the Campaign," *CR*, 11 July 1991; Donald Sevener, "Political Pitchman," *II*, December 1993, pp. 15–19; Patrick T. Reardon, "The Agony and the Agony," *CT Magazine*, 24 June 2007; Robert G. Kaiser, "The Player at Bat," *WP*, 2 May 2008; Jim Merriner, "The Champion of Cronyism: David Axelrod," *Chicago Daily Observer*, 7 April 2009; David Plouffe, *The Audacity to Win* (Viking, 2009), pp. 7–8; Evan Osnos, "The Daley Show," *New Yorker*, 8 March 2010, pp. 38ff.; Paul Kengor, "David Axelrod, Leftie Lumberjack," *American Spectator*, March 2012, pp. 16–29; Axelrod, *Believer*, pp. 123, 125; DJG interviews with Anita Dunn, Fred Lebed, Blair Hull, Rick Ridder, John Kupper, Don Wiener, Michael Dorf, Cynthia Miller, and Judy Byrd. On the Phalen–Pincham race, see Thomas Hardy, "Phelan and Pincham Punch It Up," *CT*, 13 March 1990, Hardy, "Democratic Rivals Pull Out All the Stops," *CT*, 14 March 1990, Hardy, "Race Seeps into County Campaign," *CT*, 16 March 1990, Hardy and Robert Davis, "Phelan Edges Pincham," *CT*, 21 March 1990, Rob Karwath and Robert Davis, "TV Ads Put Phelan in the Driver's Seat," *CT*, 22 March 1990, and Hardy and Constanza Montana, "Pincham Keeps Heat on Phelan," *CT*, 1 May 1990; David Elsner, "Phelan, Sheahan Win," *CT*, 7 November 1990. See also *People v. Ruiz*, 396 N.E. 2d 1314 (Ill. App. 1st Dst. 2nd Div.), 13 November 1979, Robert Enstad, "2 Gang Members Guilty of Slayings," *CT*, 22 March 1980, Jane Fritsch, "2d Man Gets Death in Slayings," *CT*, 25 April 1980, *People v. Ruiz*, 447 N.E. 2d 148 (Ill. S. Ct.), 17 December 1982, Trevor Jensen and Steve Mills, "R. Eugene Pincham 1925–2008," *CT*, 4 April 2008, and James Janega, "R. Eugene Pincham Remembered for Contributions to Law, Community," *CT*, 13 April 2008.

62. Todd Spivak, "HP Pols Working Toward Democratic State Control," *HPH*, 28 August 2002, pp. 1–2; Memorandum Opinion and Order, *U S. ex rel King v. F. E. Moran, Inc.*, N. D. Ill. #00-C-38777, 29 August 2002, 2002 WL 2003219 (Obama appearance listings of 22 August and 26 September 2002); *CF*, 1 August and 4, 5 27 September 2002; John Patterson, "The Heat Is On," and Kurt Erickson, "Snoozeville?," *II*, September 2002; Bernard Schoenburg, "No Shortage of Candidates for U.S. Senate Run in '04," *SJ-R*, 1 September 2002, p. 15; "Town Meeting on Health Care," *CDH*, 10 September 2002, p. 3; JCAR Minutes, 10 September 2002; Douglas Holt, "U.S., State Expand Health Insurance KidCare Plan to Add Parents," *CT*, 13 September 2002; "The Family Care Campaign," in Woods Fund of Chicago, *2001 Annual Report*, pp. 8–13; Eric Krol and John Patterson, *CDH*, 16 September 2002, p. 5; Daniel C. Vock, "ABA Chief: End Private Funding of Judge Races," *CDLB*, 18 September 2002, p. 1; Lynn Sweet, *CST*, 19 September 2002, p. 8; Photos of Obama and Shomon at Illinois Democratic Women, Bloomington, IL, 21 September 2002, ObamaForIllinois.com; Spivak, "Obama Sponsors West Nile Virus Legislation," *HPH*, 25 September 2002, p. 10; Corey Hall, "Obama to Set Sights on U.S. Senate Seat," *CW*, 26 September 2002, p. 1; Jodi Wilgoren, "GOP Death Penalty Feud Sinks to First-Name Calling," *NYT*, 26 September 2002; Dave McKinney, "Clean Sweep? Illinois Democrats Pin Their Fall Hopes on a Big Broom," *II*, October 2002; Obama, "State-Wide Meetings Focus on Uninsured," *HPH*, 2 October 2002, p. 4; Obama to Stuart Rothenberg, 10 October 2002; Friends of Barack Obama, Report of Campaign Contributions and Expenditures, 1 July–31 December 2002, 31 January 2003; Jeremiah Wright's 6 February 2004 interview with John Kupper; Wright in Mendell, *Obama*, p. 159–60; Artur Davis in Nicole Duran, "The New Kid on the Block," *Campaigns and Elections*, June 2007, pp. 52ff.; DJG interviews with Scott Kennedy, David Wilhelm, Josh Hoyt, Todd Dietterle, John Bouman, Sheryll Cashin, Tom Perrelli, Bruce Spiva, and Jeremiah Wright.

63. Obama for Illinois, FEC Form 3, Report of Receipts and Disbursements, 1 July–31 December 2002, 29 January 2003; Obama for Congress 2000, FEC Form 3, Report of Receipts and Disbursements, 1 July–30 September 2002, 1 October–27 November 2002, and 28 November–31 December 2002, 12 March 2003; Ryan Ori, "Taste: Presidential Approval," *PJS,* 14 January 2009; DJG interviews with Raja Krishnamoorthi, Scott Lynch-Giddings, Katrina Emmons Mulligan, Dan Shomon, and Cynthia Miller.

64. Chinta Strausberg, "Opposition to War Mounts," *CD,* 26 September 2002, p. 1; Marilyn Katz, "Chicago Residents Form Coalition Opposed to War in Iraq," n.d., MKP; Stevenson Swanson and Jill Zuckman, "Iraq OKs Inspection Deal," *CT,* 2 October 2002; Strausberg, "Leaders Rip Cost of Pending War," *CD,* 2 October 2002, p. 3; Chicagoans Against War on Iraq, "Don't Give President Bush a Blank Check to Wage War on Iraq," 2 October 2002, Chicagoans Against War on Iraq, "Program," [2 October 2002], MKP; Bill Glauber, "War Protesters Gentler, but Passion Still Burns," *CT,* 3 October 2002; Annie Sweeney, "Lunch-Hour Crowd Debates War Issue," *CST,* 3 October 2002, p. 22; Strausberg, "War with Iraq Undermines U.N.," *CD,* 3 October 2002, p. 1; Greg Bryant and Jane B. Vaughn, "300 Attend Rally Against Iraq War," *CDH,* 3 October 2002, p. 8; Bettylu Saltzman, "Going Back a Few Years with Obama," *CT,* 3 July 2007; Obama, "Weighing the Costs of Waging War in Iraq," *HPH,* 30 October 2002, pp. 5–6; Mendell, *Obama,* pp. 173–75; "Carl Davidson on Barack O-bomb'em," beyondchron.org, 21 January 2007; Perry Bacon Jr., "War Critics Question Obama's Fervor," *WP,* 15 September 2007; Jennifer Hunter, "Obama's Touted Anti-War Stance Not So Audacious," *CST,* 2 October 2007, p. 18; Obama, "A New Beginning," DePaul University, 2 October 2007; Marilyn Katz, "CAWI's Start, Oct. 2 and Obama's Future," www.noiraqwar-chicago.org, 2 October 2007; John McCormick, "Obama Marks '02 War Speech," *CT,* 3 October 2007; Scott Helman, "Early Defeat Launched a Rapid Political Climb," *BG,* 12 October 2007; Jim Rutenberg, "Finding Archives Lacking, Obama Returns to 2002," *NYT,* 12 October 2007; Michael Crowley, "Is Obama's Iraq Record Really a Fairy Tale?," *TNR,* 27 February 2008; Don Gonyea, "Obama Still Stumps on 2002 Anti-War Declaration," *Morning Edition,* NPR, 25 March 2008; Jo Becker and Christopher Drew, "Pragmatic Politics, Forged on the South Side," *NYT,* 11 May 2008; Paul Street, *Barack Obama and the Future of American Politics* (Paradigm Publishers, 2008), p. 137; Groundswell Films, "What Changed?," YouTube, 23 December 2009; Remnick, *The Bridge,* pp. 343–49; McClelland, *Young Mr. Obama,* pp. 202, 230; David Maraniss, "Obama's Military Connection," *WP,* 6 May 2012; Beatrice Lumpkin, *Joy in the Struggle* (International Publishers, 2013), pp. 244, 331; Axelrod, *Believer,* pp. 129–30; David Axelrod's 2009 interview with Robert Draper, DJG interviews with Bettylu Saltzman, Marilyn Katz, Jennifer Amdur Spitz, Dan Shomon, Raja Krishnamoorthi, Pete Giangreco, Chris Sautter, Chuy Garcia, and Beatrice Lumpkin.

65. UCLS *Announcements 2002–03,* esp. pp. 39, 45; UCLS, *The Glass Menagerie 2002–03,* esp. p. 21; "Black Faculty Member at the University of Chicago Law School Is Running for the U.S. Senate," *Journal of Blacks in Higher Education,* Autumn 2002, p. 26; *CF,* 1, 4, 9, 16, 17, and 18 October 2002; Chinta Strausberg, "Obama Ponders Run for U.S. Senate," *CD,* 10 October 2002, p. 6; Greg Hinz, "Jones Powers Up Senate Role," *CCB,* 7 October 2002, p. 3; JCAR Minutes, 8 October 2002; Michael Meiferdt, "Sen. Obama Visits WIU Campus," *MJ,* 9 October 2002; "Obama Issues Committee," 10 October 2002, AGP; H. J. Res. 114, Authorization for Use of Military Force Against Iraq Resolution of 2002, 11 October 2002; *BP,* 13 October 2002, p. 39 (Szarek-Martin wedding); Obama for Illinois, FEC Form 3, Report of Receipts and Disbursements, 1 July–31 December 2002, 29 January 2003; Obama, *TAOH,* p. 293; Obama's Remarks at the Woodrow Wilson Center for International Studies, Washington, D.C., 1 August 2007; Shomon in David Mendell, "Barack Obama—Democrat for U.S. Senate," *CT,* 22 October 2004, in Mendell, *Obama,* pp. 148–49, in Liza Mundy, *Michelle,* p. 137, in Carol Felsenthal, "The Making of a First Lady," *Chicago Magazine,* February 2009, and in Remnick, *The Bridge,* pp. 336, 353–54; "Office Hours with Professor Obama?," *WSJ* Law Blog, 6 February 2008; "Candidates as Academics: I Got a Professor Named Obama," AbovetheLaw.com, 8 February 2008; DJG interviews with Jonathan Yi, Lisa Ellman, Ilisabeth Bornstein, Josh Pemstein, Dan Shomon, Will Burns, Dave Feller, John Sullivan, Andrew Gruber, Mike Marrs, Raja Krishnamoorthi, Susan Pierson Lewers, Leslie Corbett, Rob Scott, Jeff Stauter, Dave Stover, Bunny Fourez, Chris Sautter, Preston Pugh, and Cindy Huebner Davidsmeyer.

66. Doug Finke, "Democrats Try to Sweep State Elections," *SJ-R,* 21 October 2002, p. 1; *CF,* 22, 24, 28, and 31 October and 6, 7, 8, 14, 15, and 19 November 2002; "Steffen to Discuss Health Care," *CDH,* 28 and 29 October 2002, p. I-3; John O'Connor, "Budget Problems Could Dampen Democrat Control of Senate," AP, 29 October 2002; Strausberg, "Daley, Jesse Hold Mass GOTV Rally," *CD,* 2 November 2002, p. 3; Ray Long and Christi Parsons, "Democrats Win Senate and Lock on Legislature," *CT,* 6 November 2002; Finke, "Democats Take Charge of General Assembly," *SJ-R,* 6 November 2002, p. 5; *Chicago Tonight,* WTTW, 6 November 2002; Scott Turow, "Memo to Voters: Remember the Cruz Case," *CT,* 6 October 2002; Long and James Janega, "Illinois Democrats United—For Now," and Parsons, "GOP Scoffs at Dealing on Budget," *CT,* 7 November 2002; Ellen Warren, "How He Made It to the Mansion," *CT,* 8 November 2002; Dave McKinney, "The 93rd Unpacks Its Agenda," *II,* January 2003; Karen Wilcox e-mail to Arnie Miller, and Miller e-mail to DJG, 2 February 2010; DJG interviews with Debbie Leff, Arnie Miller, and David Spiegel.

67. *CDH,* 17 November 2002, p. J5 (Obama hiring Claire Serdiuk); Michael Dorf to Obama, 26 November 2002, FEC; Obama for Illinois, FEC Form 3, Report of Receipts and Disbursements, 1 July–31 December 2002, 29 January 2003; Obama for Congress 2000, FEC Form 3, Report of Receipts and Disbursements, 1 July–30 September 2002, 1 October–27 November 2002, and 28 November–31 December 2002, 12 March 2003; McLean-Clark, "Fundraising: Planning, Prospecting and the Importance of Money," n.d., 21 pp., JMP; Michael D. Shear and Ceci Connolly, "In Illinois, a Similar Fight Tested a Future President," *WP,* 9 September 2009; McClelland, *Young Mr. Obama,* p. 221; Peter Wallsten, "Ruling Secures for Obama a Larger Place in History," *WP,* 29 June 2012; Dean Olsen, "Seeds of Federal Health Care Law Were Planted in Illinois Senate," *SJ-R,* 15 July 2012; DJG interviews with Jim Duffett, Willie Delgado, Steve Drizin, John Charles, Joe McLean, Susan Shadow Ransone, Katrina Emmons Mulligan, Claire Serdiuk, Cynthia Miller, Michael Dorf, and Dan Shomon.

68. Dennis Conrad, "Democrats Eyeing Fitzgerald in 2004 Senate Race," AP, 16 November 2002; JCAR Minutes, 19 November 2002; *CF,* 20 and 21 November and 2, 4, 5, 6, 10, 19, and 23 December 2002; Minutes, General Assembly Retirement System Board of Trustees, 20 November 2002, #222, pp. 1854–58; Senate Floor Transcript, 20 November 2002, p. 9 (Obama introducing SBs 2432 and 2433); Harstad to Obama and Shomon, "Initial Stream of Conscious Thoughts on Issues and Message for the 2003–2004 Senate Campaign," 21 November 2002, JKP; Ryan Keith, "Lawmakers Alter Ryan's Reform Package But Say Issue Isn't Over," AP, 22 November 2002, and *CDH,* 24 November 2002, p. I-14; *Public Affairs with Jeff Berkowitz,* 25 November 2002 and 18 July 2004; Bernard Schoenburg, "LaHood, Fitzgerald Trade Jabs," *SJ-R,* 29 November 2002, p. 1; "Fall Veto Session Begins," *C-WR* 18 #1, 1 December 2002, p. 3; Lauren W. Whittington, "Fitzgerald Vulnerable; Democrats and Even Some in GOP Ponder Senate Challenge," *RC,* 2 December 2002; Obama on *Newsnight,* CNN, 2 December 2002; Richard Goldstein, "Court Examines Death-Sentence Decision," *QCT* and *MJG,* 3 and 4 December 2002; Ryan Keith, "Senate Committee Approves Death Penalty Package Despite Criticism," AP, 3 December 2002; Senate Floor Transcript, 3 December 2002, p. 6 (Obama introducing SB 2434); Senate Floor Transcript, 4 December 2002, pp. 39–67 (HB 5657); Adriana Colindres, "Senate OKs Death Penalty Plan," and Doug Finke, "Departing Lawmakers Get New Jobs," *SJ-R,* 5 December 2002, p. 13; Rick Pearson, "Philip Will Resign as Senate's GOP Chief," *CT,* 5 December 2002; Finke, "Philip to Retire as Senate GOP Leader," *SJ-R,* 5 December 2002, p. 10; John Patterson, "What Will Philip's Legacy Be?," *CDH,* 6 December 2002, p. 8; Russ Stewart, "Fitzgerald's Prospects in '04 Are Not That Bleak," *NSP,* 11 December 2002; "Quiet Veto Session," *C-WR* 18 #2, 20 December 2002, pp. 2–3; Flynn McRoberts, "Small Town Guy Steps Up," *CT,* 29 December 2002; Stephen S. Morrill, "Client Report for 2002 Veto Session," Morrill and Associates, December 2002; DJG interviews with Frank Watson, Ed Petka, Steve Rauschenberger, and Nia Odeoti-Hasan.

69. "Upcoming Events," ObamaForIllinois.com, 10 December 2002; Jennifer Holuj (Center for Tax and Budget Accountability) e-mail to Raja Krishnamoorthi and Andrew Gruber, "Documents," 10 December 2002, Cynthia Miller e-mail to Andrew Boron et al., "Follow Up," 11 December 2002, Krishnamoorthi e-mail to Miller, "Issues Committee Database," 12 December 2002, Gruber e-mail to Corporate Tax Breaks Subcommittee, "Preliminary Materials on Tax Expenditures," 12 December 2002, Dan

Shomon e-mail to Gruber et al., "Obama: Tax Breaks," 16 December 2002, Gruber e-mail to David Adams et al., "Obama Corporate Tax Breaks," 19 December 2002, Gruber e-mail to Joe Seliga et al., "Obama Corporate Tax Breaks," 20 December 2002, Seliga e-mail to Gruber et al., "Obama Corporate Tax Breaks," 24 December 2002, John Kerr, "Individual Income Tax Credit Analysis," 29 December 2002, 5pp., Gruber e-mail to Adams et al., "Obama Corporate Tax Breaks," 31 December 2002, Krishnamoorthi e-mail to Gruber, 2 January 2003, all AGP; Chinta Strausberg, "Trent Lott Agrees to Meet with Black Caucus," *CD,* 12 December 2002, p. 3; "Four New Directors Join Joyce Foundation Board," Press Release, 12 December 2002; Joyce Foundation, *2002 Annual Report;* Woods Fund of Chicago, *2002 Annual Report,* p. 4; Carol Marin, "Talking Some Sense into Moseley-Braun," *CT,* 13 December 2002; Obamafor Illinois.com, 17 December 2002 (first Internet Archive capture); JCAR Minutes, 17 December 2002; JCAR, *2002 Annual Report;* Eric Krol, "Some Not Sure Moseley-Braun Should Return," *CDH,* 22 December 2002, p. 1; Dennis Conrad, "Moseley-Braun Says She Could Win Her Old Senate Seat Back," AP, 23 December 2002; Laura Washington, "Lott Is No Friend," *CST,* 23 December 2002, p. 29; *CF,* 31 December 2002; Hartman and Melody Spann-Cooper in Don Terry, "Raising the Voice," *CT Magazine,* 19 March 2006, pp. 14ff.; Obama in Mundy, *Michelle,* p. 143 ("I probably would have stepped out of politics for a while" had Moseley Braun run); Remnick, *The Bridge,* p. 359; McClelland, *Young Mr. Obama,* pp. 195–96; DJG interviews with Carol Moseley Braun, John Rogers Jr., Abner Mikva, David Wilhelm, Joe McLean, Claire Serdiuk, Jan Schakowsky, and Matt Piers. All told, Barack received a total of $70,000 for his service on the Joyce Foundation board. Kenneth P. Vogel, "Obama Linked to Gun Control Efforts," *Politico,* 20 April 2008.

70. Kurt Erickson, "Ryan, Gadhafi, Arafat Have a Friend at U of I," *BP,* 12 January 2003, p. C1; Jacobs, *Obamaland,* p. 88; "They Met Obama When," *CT,* 16 November 2008; Obama, Constitutional Law III Final Examination, [11–16] December 2002; UCLS Course Evaluations, Fall 2002 (Current Issues in Racism is missing); UCLS *Announcements 2002–03,* p. 81; *CF,* 3, 6, 7, 8, 9, 10 January 2003; "Pappas Eyes Run for U.S. Senate," *CT,* 4 January 2003; "Agenda, Obama for Illinois, Strategy Conference Call, Monday, January 6th [2003], 1PM CT," John Kupper notes, "Obama 1-6 [2003]," and John Kupper 2003 daybook entries, JKP; Andrew Gruber e-mail to Bill Fuhry, 6 January 2003, David J. Adams e-mail to Gruber, 8 January 2003, Gruber e-mail to Adams et al., and John Kerr e-mail to Gruber, "Another Suggestion," 10 January 2003, all AGP; Senate Floor Transcript, 8 January 2003, p. 4; "Legislative Inauguration Celebration," 8 January 2003, AGP; Lauren W. Whittington, "Waiting for Carol," *RC,* 9 January 2003; Chinta Strausberg, "Senate Elects Black President," *CD,* 9 January 2003, p. 1; "Nice Words Abound as Senate Convenes to Pick Leader," *SJ-R,* 9 January 2003, p. 5; JCAR Minutes, 9 January 2003; Senate Floor Transcript, 9 January 2003, pp. 2–3 (Obama introducing SBs 4, 8, 9, 15, and 30); Mary Massengale, "Certain Hospital Data Could Be Made Available to Public," *SJ-R,* 10 January 2003, p. 8; Matthew Kemeny, "Law Would Require Hospital Disclosures," *BP,* 10 January 2003, p. A1; Andrew Binion, "Hospital Bill Might Create 'Report Card,'" *The Southern,* 17 January 2003, *QCT,* 18 January 2003, and *DHR,* 26 January 2003; Christopher Wills, "Blagojevich Faces Budget Crisis with Big Promises," AP, 10 January 2003; Strausberg, "Blagojevich Reveals $5 Billion Deficit," *CD,* 14 January 2003, p. 1; Paul Simon to Obama, 15 January 2003, PSP; "Blagojevich Leads Democrats into Power," *C-WR* 18 #4, 24 March 2003; DJG interviews with Raja Krishnamoorthi, Andrew Gruber, Dan Shomon, and Claire Serdiuk.

71. Maurice Possley and Steve Mills, "Clemency for All," *CT,* 12 January 2003; Mary Massengale, "Ryan's Actions May Spark Reforms," *SJ-R,* 12 January 2003, p. 6; Eric Krol, "Legislators Defend Themselves After Ryan Puts Some Blame on Them," *CDH,* 12 January 2003, p. 5; Jodi Wilgoren, "Citing Issues of Fairness, Governor Clears Out Death Row in Illlinois," *NYT,* 12 January 2003; Bruce Shapiro, "Ryan's Courage," *The Nation,* 16 January [print edition 3 February] 2003; Christi Parsons and Karen Mellen, "House Bill Seeks Limit on Blanket Clemency," *CT,* 17 January 2003; *CF,* 13, 14, 16, and 22 January 2003; Timothy J. Gilfoyle, "Chicago Fortunes: Interviews with Lester Crown and John H. Johnson," *Chicago History,* Fall 2000, pp. 58–72; Louise Daly, "Crown," *CCB,* 15 October 2005; Christopher Drew and Mike McIntire, "After 2000 Loss, Obama Built Donor Network from Roots Up," *NYT,* 3 April 2007; Jane Ammeson, "The Busy Life of Billionaire Lester Crown," *Chicago Life,* February 2008; Jo Becker and Drew, "Pragmatic Politics, Forged on the South Side," *NYT,* 11 May 2008; Pauline Yearwood, "Obama and the Jews," *Chicago Jewish News,* 12 September 2008; Hilary Leila Krieger, "Mr. Obama's Neighborhood," *Jerusalem Post,* 23 October 2008; Steve Neal and Lynn Sweet, "Ex-Senator Won't Try Again," *CST,* 17 January 2003, p. 3; Jeff Zeleny, "Eyes on the Presidency," *CT,* 17 January 2003; Dan Balz, "Moseley-Braun May Join the Crowd," *WP,* 17 January 2003; Chinta Strausberg, "Moseley-Braun Won't Run for U.S. Senate," *CD,* 18 January 2003, p. 33; Krol, "Ex-Senator Doesn't Want Rematch with Fitzgerald," *CDH,* 18 January 2003, p. 11; Eric Zorn, "Moseley-Braun Gives Democrats Reason for Hope," *CT,* 18 January 2003; Andrew Gruber e-mail to David Adams et al., "Obama Corporate Tax Breaks," 19 January 2003, AGP; Laura Washington, "Obama's Got Perfect Pedigree to Take Over Senate Race," *CST,* 20 January 2003, p. 31; Todd Spivak, "Moseley-Braun Backs Out, Obama in for Senate Race," *HPH,* 22 January 2003, pp. 1–2; Dan Mihalopolous, "Presidential Contender Braun Makes Bid Official," *CT,* 23 September 2003; DJG interviews with Newton Minow, Alan Solow, John Rogers Jr., Carol Moseley Braun, John Kupper, Dan Shomon, Randon Gardley, Katrina Emmons Mulligan, Cynthia Miller, Claire Serdiuk, Denny Jacobs, Terry Link, Larry Walsh, and John Eason.

72. "Obama Kicks Off U.S. Senate Run," Press Release, 21 January 2003; "Talking Points for Obama Press Conference, January 21, 2003," JKP; Herbert G. McCann, "State Senator Declares U.S. Senate Bid," AP, 21 January 2003; WLS ABC Chicago, 21 January 2003, MBC; Gregory Tejeda, "Chico's Senate Chances Rise," UPI, 21 January 2003; Rick Pearson and John Chase, "Legislator in Race to Unseat Fitzgerald," *CT,* 22 January 2003; Scott Fornek, "Obama Takes Jab at Fitzgerald as He Starts Run," and "'Blessed by God,' Rooted in Two Continents," *CST,* 22 January 2003, p. 8; Chinta Strausberg, "Obama to Challenge Sen. Fitzgerald," *CD,* 22 January 2003, p. 5; Eric Krol, "Democratic Candidate Says Fitzgerald 'Betrayed' State," *CDH,* 22 January 2003, p. 11; *Chicago Tonight,* WTTW, 22 January 2003; Todd Spivak, "Powerful Pals Promise Votes for Obama in 2004," *HPH,* 29 January 2003, pp. 1–2; Krol, "Obama's Debut Gave Little Hint of What Was to Come," *CDH,* 19 January 2007, p. 16; DJG interviews with John Eason, Tim King, Dan Johnson-Weinberger, Jesse Jackson Jr., Emil Jones, Denny Jacobs, Terry Link, Larry Walsh, Jacqueline Collins, Freddrenna Lyle, Newton Minow, Andy Davis, Stephen Pugh, Preston Pugh, Linzey Jones, Joe Seliga, David Narefsky, Al Sharp, Tommy Walsh, Bill Fuhry, and Mike Marrs.

CHAPTER NINE: CALCULATION, COINCIDENCE, CORONATION

1. Edward B. Chez, Hull for Senate, to Secretary of the Senate, 21 January 2003, FEC; "Big Spender Triggers Higher Donation Limits," *CST,* 1 February 2003, p. 4; Lauren W. Whittington, "Millionaires on Notice," *RC,* 3 February 2003; Rick Pearson and Ray Gibson, "Campaign Fund Law Has Giant Loophole," *CT,* 5 February 2003; Jennifer A. Steen, "Self-Financed Candidates and the 'Millionaires' Amendment,'" in Michael J. Malbin, ed., *Election After Reform: Money, Politics, and the Bipartisan Campaign Reform Act* (Rowman & Littlefield, 2006), pp. 204–18; Obama, *TAOH,* p. 5; Senate Floor Transcript, 22 January 2003, p. 1 (Obama introducing SB 59); Robert Themer, "U.S. Senate Hopeful: Rural Strategy May Grow in Pembroke," *KDJ,* 28 January 2003; Whet Moser, "The Ongoing Poverty of Pembroke, Illinois," *CM,* 21 September 2011; "Topics to Discuss at Axelrod Meeting, January 27, 2003, 3:45PM," Shomon e-mail to Kupper, 27 January 2003, "Barack 1-27," John Kupper conference call notes, JKP; Andrew Gruber e-mail to Shomon and Raja Krishnamoorthi, "Obama Corporate Tax Breaks—Summary Memo," 23 January 2003, Gruber e-mail to Joe Seliga et al., "Obama," 27 January 2003, Gruber e-mail to Krishnamoorthi and Shomon, "Obama Corporate Tax Breaks," 30 January 2003, AGP; Shomon e-mail to Kupper, "Press Release on FEC Report," 28 January 2003, Obama e-mail to Shomon, "Re: Press Release on FEC Report," 29 January 2003, JKP; OFI, FEC Form 3, Report of Receipts and Disbursements, 1 July–31 December 2002, 29 January 2003; Senate Floor Transcript, 29 January 2003, pp. 3–4 (Obama introducing SBs 125, 126, 127, 128, 129, 130, and 131); "Fully Implement FamilyCare Budget Initiative for FY04," NCPL, 5 February 2003; Obama in Ken Silverstein, "Barack Obama Inc.: The Birth of a Washington Machine," *Harper's,* November 2006, pp. 31–40; Martin Dupuis and Keith Boeckelman, *Barack Obama, the New Face of America* (Praeger, 2009), pp. 46–48; DJG interviews with Michael Dorf, Claire Serdiuk,

Kelly Dietrich, Dan Shomon, John Kupper, Paul Harstad, Raja Krishnamoorthi, Andrew Gruber, Joe Seliga, and Bill Fuhry.

2. "Obama for U.S. Senate Strategic Plan," n.d., 9pp., JKP, and a subsequent, updated but highly similar 8pp. version, TWP; Kupper Campaign Expenditure Projections, n.d., ca. January 2003, JKP; Chinta Strausberg, "Activists Say No to Joyce Washington," *CD,* 29 January 2003, p. 3.

3. *CF,* 23, 24, 27, and 28 January and 4, 5, 7, and 14 February 2003; JCAR Minutes, 4 February 2003; Senate Floor Transcript, 5 February 2003, p. 5 (Obama introducing SBs 263 and 264); Lynn Sweet, "Fitzgerald Needs to Shed Cloak," *CST,* 6 February 2003, p. 37; Matthew Kameny, "Democrats Support Taped Interrogation," *BP,* 6 February 2003, p. 3; Shomon e-mail to Kupper et al., "Monday's Strategy Conference Call," 10 February 2003, OFI, "Obama to Hold State Hearings on 'Dangerous' Bush Medicaid Plan," 11 February 2003, Press Conference Transcript, 11 February 2003, JKP; Chinta Strausberg, "Obama: 'Medicaid Cuts Will Kill Hospitals,'" *CD,* 12 February 2003, p. 3; Steve Neal, "Fitzgerald's Fading Fast," *CST,* 12 February 2003, p. 45; Shomon e-mail to Paul Harstad et al., "Conference Call on Obama Issues Committee," 13 February 2003, JKP; Andrew Gruber e-mail to Raja Krishnamoorthi et al., "Obama Bill," 14 February 2003, AGP; Christopher Wills, "Obama Record Shows a Liberal Open to Compromise," AP, 25 June 2008; Steven Thomma, "Obama an Effective Mediator in Illinois Senate," KRDC, 27 March 2008; DJG interviews with Rob Scott, Greg Sullivan, Jim Dodge, Matt Jones, Bill Haine, Kathy Saltmarsh, John Cullerton, Paul Williams, Chuck Hurley, Dan Shomon, and John Kupper.

4. Lynn Sweet, "Vallas Backing Chico for Senate," *CST,* 13 February 2003, p. 33; "Hynes Readies for Senate Race," *CT,* 13 February 2003; "Democrat Announces Bid for Fitzgerald Seat," *CT,* 14 February 2003; OFI, "Obama Pushes for Renewal of Illinois Earned Income Tax Credit," 16 February 2003, JKP; Chinta Strausberg, "Obama Unveils New Tax Credit Bill," *CD,* 17 February 2003, p. 3; Beth Sneller, "DuPage Democats Turn Attention to Beating Fitzgerald," *CDH,* 17 February 2003, p. 1; H. Gregory Meyer and Sean D. Hamill, "4 Democrats Take Aim at GOP, Bush Stances," *CT,* 17 February 2003; Krishnamoorthi e-mail to Cynthia Miller et al., Krishnamoorthi e-mail to Andrew Gruber et al., "Obama Issues Committee," 18 February 2003, AGP; Jeremiah Posedel e-mail, "Meet & Greet—Blair Hull," 18 February 2003, JKP; Senate Floor Transcript, 18 February 2003, p. 14 (Obama introducing SB 552); Obama Block Schedule, 18–28 February 2003, JKP; Krishnamoorthi e-mail to Miller, "Pls. Forward to Issues Committee," 21 February 2003, AGP; Tom Waldron, "Supporting Working Families When State Coffers Are Empty: Lessons from Illinois' Successful Campaign for a Better EITC," Hatcher Group, November 2003, pp. 3–4; DJG interviews with John Bouman, Sean Noble, Raja Krishnamoorthi, Andrew Gruber, Joe Seliga, Dino Christenson, and Matt Piers.

5. *CF,* 19, 20, and 21 February 2003; Senate Floor Transcript, 19 February 2003, pp. 2, 18–32 (Obama introducing SBs 777, 780, 781, 782, 783, 784, 860, 861, 862, 863, 864, 890, 891, 986, 987, 988, 989, 990, 994, 995, and 997, and SR 43); Andrew Gruber e-mail to Obama, "Obama Bill," 19 February 2003, AGP; Mary Massingale, "Democrats Pushing Plan to Cut Drug Prices," *SJR,* 20 February 2003, p. 1; Jodi Wilgoren, "Jesse Jackson, a Club Owner and Lasting Ties," *NYT,* 20 February 2003; "Devine's Intervention," *CT,* 20 February 2003; Senate Floor Transcript, 20 February 2003, passim (Obama introducing fifty-five bills: SBs 1183, 1407–30, 1432–33, 1437, 1493 (altered March 5), 1524, 1554–58, 1582–88, and 1618–22); Eric Zorn, "Numbers Add Up to Suffering and Heartbreak," *CT,* 22 February 2003; Mary Schmich, "Tragedy Pulls Racial Divisions into Spotlight," *CT,* 23 February 2003; Sean D. Hamill, "Nightclub Safety Laws Mapped by Legislator," *CT,* 27 February 2003; Chinta Strausberg, "Obama, Club Owners Push Anti-Panic Bill," *CD,* 27 February 2003, p. 5; Senate Floor Transcript, 27 February 2003, p. 6; "Obama Weighs In on E2," *HPH,* 5 March 2003, p. 6; Andrew Gruber e-mail to David Adams et al., "Obama Corporate Tax Breaks," 12 March 2003, AGP; DJG interview with Steve Derks.

6. Lauren W. Whittington, "Illinois Black Turnout Key," *RC,* 24 February 2003; *CF,* 25 and 28 February 2003; Dan Shomon e-mail to Terry Walsh et al., "Obama Strategy Conference Call," 25 February 2003, JKP; Senate Floor Transcript, 26 February 2003, p. 2 (Obama introducing SJRs 16 and 22); Kirk Erickson, "Poll: Few Voters Back Fitzgerald," *BP,* 28 February 2003; Bennett, Petts & Blumenthal, "Recap of Surveys of Illinois Democratic Voters," 4 May 2004, MBP; DJG interview with Mark Blumenthal.

7. Greg Hinz, "Deals Cloud Ill. Housing Agency Bid," *CCB,* 24 February 2003, p. 3; *CF,* 24, 26, and 27 February and 3, 4, 6, 10, 12, and 21 March 2003; "Stroger Hospital Doctor Named to Top Health Post," *CT,* 26 March 2003; Lindsey Tanner, "Bleak Public Health Scene Confront's New Chief's Idealism," AP, 10 April 2003; Stephanie Zimmerman, "The Advocate Is In," *II,* June 2003; Sarah A. Klein, "Eric E. Whitaker," *CCB,* 3 November 2003; Scott T. Shepherd, "The Politics of Health," *The New Physician,* March 2004; "Complete Transcript of the *Sun-Times* Interview with Barack Obama," 14 March 2008; Obama Block Schedule, 1–7 March 2003, JKP; UCLS Course Evaluations, Voting Rights, Winter 2003; "Bill Would Block Tech Investments," *CCB,* 1 March 2003; *The May Report,* 3, 5, and 14 March 2003; John O'Connor, "Obama Challenges Opponents to Speak Out on War," AP, 3 March 2003; OFI, "Planning Timeline" (2 versions), n.d., 3pp., JKP; John J. Koenigsknecht, "Homeland Security," 25 February 2003, 7pp., Andrew Gruber e-mail to Raja Krishnamoorthi, "Issue Paper: Bush's Tax Plan," 28 February 2003, Krishnamoorthi e-mail to Gruber, "Tax Cuts," 2 March 2003, Krishnamoorthi e-mail to Issues Committee, "Sample Paper," 3 March 2003, AGP; Korecki, *Only in Chicago,* p. 160; DJG interviews with John Kupper, Raja Krishnamoorthi, Andrew Gruber, Josh Pemstein, H. Ron Davidson, and Beth DeLisle.

8. "Will the Senate Seek Justice?," *CT,* 4 March 2003; Christopher Wills, "Senate Committee Approves Death Penalty Safeguards, Taping Requirement," AP, 4 March 2003; Ray Long and Christi Parsons, "Senate Begins Assault on Crime," *CT,* 5 March 2003; Anne Marie Tavella, "Taping Murder Interrogations Wins Early Senate Support," *CDH,* 5 March 2003, p. 11; Chinta Strausberg, "Obama's Anti-Racial Bill Passes Panel," *CD,* 5 March 2003, p. 3; "Death Penalty Bill Headed for Senate Vote," *CDLB,* 5 March 2003, pp. 1, 24; Adriana Colindres, "Death Penalty Reforms Advance to Senate," and "Bill Targeting Racial Profiling OK'd by Committee," *SJ-R,* 5 March 2003, pp. 9, 15; Matthew Kemeny, "Death-Penalty Reforms Win Panel's Support," *BP,* 5 March 2003, p. A4; Dan Campana, "Taping Murder Suspects Gains Support in County," *DDC,* 13 March 2003; Steve Neal, "Obama a Worthy Heir to Washington Legacy," *CST,* 5 March 2003, p. 55; Neal, "Hynes Could Defeat a Weakened Fitzgerald," *CST,* 12 March 2003, p. 49; Senate Floor Transcript, 5 March 2003, pp. 25–26 (Obama colloquy with Peter Roskam); *CF,* 6 March 2003, p. 2 (detailing the controversy); Kristy Hessman, "Senate Committee OKs Public Funding of Supreme Court Campaigns," AP, 5 March 2003 (approval of Barack's SB 1415); Paul Davis, "Barack Obama Seeks to Shape National Debate for Illinois," *N'Digo,* 6–12 March 2003, pp. 6, 18; Michael C. Dorf to FEC, 20 August 2003, FEC; Mendell, *Obama,* p. 182; DJG interviews with Dan Shomon, Chris Sautter, Cynthia Miller, Will Burns, Kevin Thompson, Beverly Helm-Renfro, John Eason, Claire Serdiuk, Paul Harstad, John Kupper, Forrest Claypool, Joe McLean, Michael Dorf, Katrina Emmons Mulligan, Randon Gardley, Andrew Boron, Terry Walsh, and Pete Giangreco.

9. *CF,* 6, 12, 13, 17, 18, 19, and 20 March 2003; John Patterson, "Why Local Leaders May Need to Record Closed-Door Talks," *CDH,* 7 March 2003, p. 8; Obama Block Schedule, 8–31 March 2003, JKP; OFI, "Obama Says $1.35 Minimum Wage Increase Will Help Families," 9 March 2003, OFI; Chinta Strausberg, "Blacks, Whites Unite in Anti-War Efforts," *CD,* 10 March 2003, p. 3; *CF,* 11 and 12 March 2003; JCAR Minutes, 11 March 2003; "Blagojevich Delivers His First State of the State Address," *C-WR* 18 #5, 14 April 2003 (12 March); Senate Committee Action Report, Health and Human Services, 12 March 2003 (SB 1082); Brock Willeford, "Senate Republican Staff Analysis," SB 1082, 13 March 2003; Kristy Hessman, "Lawmakers Approve Abortion-Friendly Legislation," AP, 13 March 2003; Ray Long and Kate McCann, "Club Safety Bill Heads to Senate," *CT,* 13 March 2003; "Leave This Legal Reform Alone," *CT,* 14 March 2003; OFI, "Obama Bill to Stop Racial Profiling in Illinois Passes Senate for First Time," 15 March 2003; Andrew Binion, "Former Army Scout Loves His Role as NRA Lobbyist in Springfield," *SI,* 4 May 2003; Terence P. Jeffrey, "More on Obama and Babies Born Alive," *Human Events,* 16 January 2008; Larry Rohter, "Obama's 2003 Stand on Abortion Draws New Criticism in 2008," *NYT,* 20 August 2008; Jess Honig, "Obama and 'Infanticide,'" Factcheck.org, 25 August 2008; Trotter in David Nather, "Obama's Quick Rise on a Non-Traditional Career Path," *CQ Weekly,* 21 August 2008, and on *PBS Newshour,* 25 September 2008; Rickey Hendon, *White Enough/Black Enough,* pp. 37–38; Hendon's 2010 interview with Peter Slen; Beverly Helm-Renfro's 2012 interview with Mark DePue, esp. pp. 55,

58, 61, 65; Christopher Wills, "Clues to How Obama Would Play His Hand as President Can Be Found in Poker Style," AP, 24 September 2007; Peter Hitchens, "The Black Kennedy: But Does Anyone Know the Real Barack Obama?," *Daily Mail,* 2 February 2008; Will Englund, "When Is Obama Bluffing?," *NJ,* 5 December 2009; DJG interviews with Howard Peters, Bill Perkins, Dale Righter, Jo Johnson, Pam Sutherland, Beverly Helm-Renfro, Todd Vandermyde, Richard Pearson, Nia Odeoti-Hassan, Debbie Lounsberry, Glenn Hodas, Christine Radogno, Steve Preckwinkle, Frank Clark, Tristé Lieteau Smith, Mattie Hunter, Jeff Schoenberg, Iris Martinez, Jill Rock, Jim Dodge, Emil Jones Jr., Donne Trotter, Kim Lightford, Jacqueline Collins, Martin Sandoval, Bobby Molaro, Denny Jacobs, Ed Maloney, Dave Luechtefeld, Terry Link, Larry Walsh, Bill Brady, Mike Lieteau, Boro Reljic, Dave Manning, Phil Lackman, and Pete Baroni.

10. "DL21C's March 2003 Second Thursday Event" and "Candidates Greet 2nd Thursday Crowds," DL21C-Chicago.org; Senate Floor Transcript, 19 March 2003, pp. 1 (Obama introduces SR 89) and 6 (Senate passage of Obama's SB 59); OFI, "Obama Challenges Democratic U.S. Senate Candidates to Stand Up, Speak Out and Oppose Iraq War," 15 March 2003; Jim Ritter, "Anti-War Rally Here Draws Thousands," *CST,* 17 March 2003, p. 3; H. Gregory Meyer and Grace Aduroja, "Demonstrators Say No to War," *CT,* 17 March 2003; Megan Reichgott, "Some Illinoisans Cheer, Others Protest U.S. Attack on Iraq," AP, 20 March 2003; Senate Floor Transcript, 20 March 2003, pp. 56–58, 106–12, 197–200, 204–5, 219–24; Rich Frederick, "Ephedra Bill Passed by Senate," *SJR,* 21 March 2003, p. 21; "State Sen. Obama to Speak at Evans Reception," *CDLB,* 21 March 2003; Chinta Strausberg, "Uninsured Black Patients Gouged: SEIU Study," *CD,* 22 March 2003, p. 20; Matthew Kemeny, "Three Democrats Hope to Unseat Fitzgerald," *BP,* 23 March 2003, p. B3; Tammie Leigh Brown, "Democrats Meet with Community Saturday," *MJ,* 24 March 2003; Todd Spivak, "E2 Tragedy Prompts New State, City Safety Laws," *HPH,* 9 April 2003, p. 6; Jan Dennis, "Son's Death Spurs Crusade Against Herbal Stimulant," AP, 25 April 2003; DJG interview with Jeremiah Posedel. On James Meeks, see Monica Davey, "At the Crossroads: A Neighborhood, a Church and a Pastor," *CT,* 6 April 1999, Greg Hinz, "State Senator, TV Host and Megachurch Pastor," *CCB,* 4 July 2005, p. 2, and Mick Dumke, "The Church of Clout," *CR,* 25 January 2007.

11. Senate Floor Transcript, 24 March 2003, pp. 77–89, esp. 87; Chris Wetterich, "Senate Votes to Allow Needle Purchases," *CST,* 25 March 2003, p. 19; OFI, "Obama: Governor's Budget Protects Crucial Services," 25 March 2003; Senate Floor Transcript, 25 March 2003, pp. 69–71 (SB 4) and 77–84 (SB 130); Senate Floor Transcript, 26 March 2003, pp. 86–90 (SB 125), 90–93 (SB 130), and 187–97 (SB 1415); AP, "Senate Approves Public Funding of Supreme Court Campaigns," 26 March 2003; Jermain Griffin, "Senate Approves Public Funding for High Court Campaigns," *CDLB,* 27 March 2003, pp. 1, 24; *CF,* 27, 28, and 31 March 2003; *Newsnight,* CNN, 28 March 2003; Tim Jones, "Poll: 66% of Blacks Oppose War," *CT,* 29 March 2003; Christopher Wills, "Civil Rights Activists Seizing Opportunity Under Democrats," AP, 30 March 2003; Obama Block Schedule, 8–31 March 2003 and April 2003, JKP; E. J. Kessler, "Campaign Confidential," *JDF,* 4 April 2003; Todd Spivak, "Controversial Obama Bill Would 'Open Government,'" *HPH,* 2 April 2003, p. 3; Steve Neal, "Pappas Nears Hynes, But It's Still Early," *CST,* 2 April 2003, p. 53; DJG interview with Nathan Diament.

12. Senate Floor Transcript, 3 April 2003, pp. 16–18 (SB 15), 21–23 (SB 30), and pp. 23–34 (SB 50); Christopher Wills, "Senate Approves Broad Overhaul of Death Penalty System," AP, 3 April 2003; Anne Marie Tavella, "Profiling, Taping Plans Pass Senate," *CDH,* 4 April 2003, p. 17; Christi Parsons and Kate McCann, "Senate OKs Death Penalty Bills," *CT,* 4 April 2003; *CF,* 4 April 2003; Senate Floor Transcript, 4 April 2003, pp. 18–33 (SB 1035) and pp. 126–29 (Obama and Steve Rauschenberger colloquy on SB 1414); Chinta Strausberg, "Senator Obama's Racial Profiling, Videotaping Bills Clear Senate," *CD,* 5 April 2003, p. 4; "A Remarkable Week for Justice," *CT,* 7 April 2003; Christopher Wills, "Obama Cites Death Penalty Reforms," AP, 12 November 2007; Kevin McDermott, "In Illinois, Obama Made Waves from the Start," *SLPD,* 10 December 2007; Obama, remarks on Trayvon Martin, 19 July 2013, Washington, D.C.; DJG interviews with Greg Sullivan, Matt Jones, Jim Covington, Kathy Saltmash, Pete Baroni, Bill Haine, Jim Dodge, Ed Petka, Dan Cronin, Don Harmon, Mark Donohue, Jim Finley, Steve Drizin, Lawrence Marshall, David Starrett, and Jacqueline Collins.

13. Obama Block Schedule, April 2003, JKP;

OFI, "Join Senator Barack Obama for a Community Discussion," 6 April 2003; John Kupper conference call notes, "Obama," 7 April [2003], JKP; JCAR Minutes, 8 April 2003; Senate Floor Transcript, 8 April 2003, pp. 15–20 (SB 1586), 134–35 (SB 1414), 136–37 (SB 1417); AP, "Senate Approves Taping of Closed Meetings," 8 April 2003; "In Springfield," *CDLB*, 9 April 2003, p. 3; *CF*, 9, 10, and 11 April 2003; "Sen. Obama's Minimum Wage Bill [sic] Passes Illinois Senate," *CD*, 9 April 2003, p. 8 (SB 600, sponsored by Kim Lightford and seven additional senators, not including Obama); Mickey Ciokajlo, "Calling Cards Eyed for County Inmates," *CT*, 10 April 2003; Bruce Japsen, "Illinois May Demand Report Cards from Hospitals," *CT*, 13 April 2003; *Obama Weekly News*, 13–17 April 2003, TWP; Cleve Mesidor e-mail to John Kupper, 15 April 2003, JKP; Jeremiah Wright, "Confusing God and Government," 13 April 2003, www.blackpast.org/2008-rev-jeremiah-wright-confusing-god-and-government; Chinta Strausberg, "Obama to Bush: 'Fix Up U.S. Budget First,'" *CD*, 15 April 2003, p. 3; "Governor Sets Stage for Budget Dance," *C-WR* 18 #6, 17 April 2003; "Democratic State Senator Visits CMS," *Chicago Medicine*, May 2003, p. 11 ("early April"); "FY 2004 Budget Review," *C-WR* 18 #8, 8 September 2003; Baim, *Obama and the Gays*, pp. 15, 521, 538; Gayle Kaiser, "Up Close with Obama," *The Paper* [Galesburg], 12 January 2005, pp. 1, 3; Brian Ross and Rehab El-Buri, "Obama's Pastor," ABC News, 13 March 2008; Roland S. Martin, "The Full Story Behind Wright's 'God Damn America' Sermon," CNN.com, 21 March 2008; Audra Wilson, "Witnessing History from the Very Beginning," *Chicago Reporter*, 25 December 2008; Claire Bushey, "How Blagojevich Go-Between Built Surgery Center Business," *CCB*, 8 June 2013; Marc Solomon, *Winning Marriage* (ForeEdge, 2014), pp. 268–69; DJG interviews with Kevin Thompson, Jeremiah Posedel, Jim Collins, Audra Wilson, John Bouman, Nate Tamarin, Andrew Boron, Pete Giangreco, Terry Walsh, Dan Shomon, and Peter Coffey.

14. Dan Shomon e-mail to John Kupper et al., "Draft Weekly Report for Week 2," 13 April 2003, JKP; *Obama Weekly News*, 13–17 April 2003, TWP; OFI, FEC Report of Receipts and Disbursements, 1 January–31 March 2003, 14 April 2003; OFI, "Obama U.S. Senate Campaign Raises Half-Million," 15 April 2003; Dennis Conrad, "Illinois Democrats Busy Raising Money for U.S. Senate Race," AP, 16 April 2003; "New Digs," *HPH*, 16 April 2003, p. 4; Peter Savodnik, "Democrats See Big Opportunity in Open Illinois Senate Race," *TH*, 7 May 2003; Dupuis and Boeckelman, *Barack Obama*, p. 48; Lawrence A. Horwich, Barack H. and Michelle L. Obama, 2002 Form 1040, 24 March 2003; OFI, "Campaign Prospectus," n.d. [ca. soon after 16 April 2003], TWP; Bob Secter, "For Obama, Charity Really Began in the U.S. Senate," *CT*, 25 April 2007; Dorothy A. Brown, "Lessons from Barack and Michelle Obama's Tax Returns," *Tax Notes*, 10 March 2014, pp. 1109–13; Steven R. Strahler, "Lou Holland . . . Dies at 74," *CCB*, 29 February 2016.

15. "2004 Campaign Polling: Illinois Senate," NationalJournal.com, 8–11 April 2003; *CF*, 14, 15, and 16 April 2003; Rick Pearson, "Fitzgerald Not Going to Pursue Re-Election," *CT*, 15 April 2003; Lynn Sweet, "Fitzgerald Quits Race," *CST*, 15 April 2003, p. 1; Jodi Wilgoren, "Illinois Senator Announces He Won't Seek Re-Election," *NYT*, 16 April 2003; Nick Anderson, "GOP's Fitzgerald Won't Try to Retain Senate Seat," *LAT*, 16 April 2003; Pearson and Ray Long, "Edgar Leads GOP's List to Replace Fitzgerald," *CT*, 16 April 2003; Sweet, "Doubts Force Fitzgerald Out," Scott Fornek and Dave McKinney, "White House Choice Is Edgar, but Field Could Get Crowded," and Steve Neal, "Writing Was on the Wall After Latest Fitzgerald Polls," *CST*, 16 April 2003, pp. 9, 55; Peter Savodnik, "Fitzgerald Bows Out," *TH*, 16 April 2003; Chinta Strausberg, "Fitzgerald's Backing Out Pushes Obama in Front," and "An Enlightened Health Care Policy," *CD*, 16 April 2003, pp. 3, 9; Mike Ramsey, "Fitzgerald Won't Run Again for Senator," *SJR*, 16 April 2003, p. 1; Sweet, "Edgar Watch Puts Poll on Hold," *CST*, 17 April 2003, p. 33; Corey Hall, "Union Accuses Hospitals of Price-Gouging Poor and Minorities," *HPC*, 17 April 2003, p. 4; Shia Kapos, "HIV Compromise Wins Support," *CT*, 20 April 2003; Lauren W. Whittington, "Waiting for Edgar," *RC*, 21 April 2003; Richard Goldstein, "Republicans Ponder Senate Chances," *MJG* and *SI*, 22 April 2003; Todd Spivak, "Obama Raises $600,000," *HPH*, 23 April 2003, p. 2.

16. Obama Block Schedule, April 2003, JKP; Dan Shomon e-mail to John Kupper et al., Kupper to Shomon, and Raja Krishnamoorthi to Shomon et al., "Obama Weekly Report Draft," [20] and 21 April 2003, *Obama Weekly News*, 21–27 April 2003, TWP; Obama, "Increasing Minimum Wage to Lift Up the Poor," *HPH*, 23 April 2003, p. 4; Rick Pearson, "President

Asks Edgar to Consider Senate Run," *CT,* 23 April 2003; Dennis Conrad, "Bush Urges Ex-Ill. Gov on Senate Run," AP, 23 April 2003; Chinta Strausberg, "Obama Joins Hunger Group in Fighting to Remove Red Tape for Thousands of Hungry," *CD,* 24 April 2003, p. 2; "Illinois: GOP Got Ryan-ed All the Way Down the Ticket," *The Hotline,* 24 April 2003; John O'Connor, "Governor Says Death Penalty Moratorium Will Continue," AP, 24 April 2003; "Illinois Keeps a Moratorium on Executions," *NYT,* 25 April 2003; Obama to Stephen Pugh, 3 April 2003, SPP; OFI, "Black Young Professionals Raise $65,000 for Obama," 24 April 2003; *Obama Weekly News,* 28 April–3 May 2003, TWP; Lenore T. Adkins, "Obama Hosts Swank Party, Nets $50K for Senate Race," *HPH,* 30 April 2003, p. 7; "Young Black Professionals Raise $61,000 for Senate Candidate Obama," *HPC,* 1 May 2003, p. 2; "Black Professionals Raise $61,000 for Obama," *CD,* 10 May 2003, p. 6; Chris Rickert, "Simon Calls on U.S. to Be More Sensitive," *DDC,* 26 April 2003; Bernard Schoenburg, "Democrats Lining Up Money, Staffers for Senate Run," *SJR,* 27 April 2003, p. 15; Susan and David Axelrod, Nancy and Ed Burke, Sara Paretsky to Dear Friend, n.d., JKP; E. J. Kessler, "Illinois Senate Candidates Eyeing State's Jewish Voters," *JDF,* 9 May 2003, p. 1; Axelrod, *Believer,* pp. 154–55; DJG interviews with Alan Solow, John Kupper, and Tom Geoghegan.

17. "Democratic Poll Overview," ILSenate.com, Global Strategy Group, 25–30 April 2003; *CF,* 29 and 30 April and 2 and 6 May 2003; National Community Building Network 10th Anniversary 2003 Annual Conference, 30 April–3 May 2003; Russ Stewart, "Recruitment of Edgar Is GOP's Top Priority," *NSP,* 30 April 2003; Obama Block Schedule, May 2003, JKP; Bernard Schoenburg, "It's Not a Great Shock," *SJR,* 1 May 2003, p. 7; Daniel C. Vock, "House Vote to Come on Taped Confessions," *CDLB,* 2 May 2003, pp. 1, 24; Chinta Strausberg, "Bush: Combat Ends, but War Must Go On," *CD,* 3 May 2003, p. 3; Jaime Griesgraber, "Candidates Introduced at Annual Meeting of City Democrats," *DN,* 5 May 2003; Steven R. Strahler, "Wheels Wobbling at Chico's Law Firm," *CCB,* 5 May 2003, p. 3; Alison Neumer, "Michael Golden," *CT,* 28 May 2003; OFI, "Obama, Medicaid Advocates Criticize Bush Plan at Senate Hearing," 5 May 2003; Strausberg, "Obama, Activists Rip Bush Over Medicaid Reform Plan," *CD,* 6 May 2003, p. 2; Thomas P. Sullivan and Scott Turow, "Taping Confessions Is a Much-Needed Reform," *CT,* 6 May 2003; Doug Wilson, "U.S. Senate Candidate Brings Message to Quincy," *QHW,* 7 May 2003, p. B1; Steve Neal, "Don't Count Comptroller Out," *CST,* 7 May 2003, p. 59; Eric Krol, "High Roller Set to Spend Big to Unseat GOP Senator," *CDH,* 13 April 2003, p. 1; Kennedy Communications, "State Senator Barack Obama, Candidate for U.S. Senate: Background Report," April 2003, RRP; Hull for Senate, FEC Form 3, Report of Receipts and Disbursements, 1 April–30 June 2003, 3 July 2003, p. 173 ($18,840.26 payment to Kennedy Communications on 21 May 2003 for "Research"); DJG interviews with Paul Schmitz, Eric Adelstein, Megan Crowhurst, Kelly Dietrich, Dan Hynes, Chris Mather, Jef Pollock, Rick Ridder, Fred Lebed, Anita Dunn, Mark Blumenthal, David Spiegel, Kitty Kurth, Mike Henry, Mo Elleithee, Don Wiener, and Blair Hull.

18. Obama Block Schedule, May 2003, JKP; *CF,* 7, 8, 9, 12, and 13 May 2003; Senate Floor Transcript, 7 May 2003, pp. 11–24 (HB 60, PA 93-0007, 20 May 2003); Christi Parsons and John Chase, "Women's Equal Pay Advances," *CT,* 8 May 2003; John O'Connor, "Bill Requiring Taped Interrogations Heads to Governor," AP, 8 May 2003; Parsons and Kate McCann, "Taped Confessions Bill Passes," and "First Light of Justice Reform," *CT,* 9 May 2003; Anne Marie Tavella, "Legislators Want Police to Roll Tape," *CDH,* 9 May 2003, p. 8; Jermain Griffin, "Legislators to Negotiate Scope of Proposed 'Sunshine in Litigation' Law," *CDLB,* 9 May 2003, pp. 1, 24; Rick Pearson, "GOP Senate Floodgates Open as Edgar Says No," *CT,* 10 May 2003; Bernard Schoenburg, "Edgar Won't Run," *SJR,* 10 May 2003, p. 1; Shamus Toomey, "No Party for Republicans Without Edgar," *CDH,* 10 May 2003, p. 5; Terry Walsh to Dan Shomon, "Photo Shoot," 25 April 2003, JKP; "Photo Shoot Schedule," 10 May 2003, TWP; Dan Shomon e-mail to John Kupper, "PR Brainstorming on Wednesday, May 14," 12 May 2003, JKP; Ryan Keith, "Labor Support, Money Helped SBC Win Big at Capitol," AP, 12 May 2003; Rich Miller, "The Floodgates," *IT,* 13 May 2003; Vikas Bajaj, "SBC Finds Allies in State Legislatures," *DMN,* 13 May 2003; Rick Pearson, "Illinois Governor Defends Controversial Measure Benefiting SBC Communications," *CT,* 13 May 2003; JCAR Minutes, 13 May 2003; Senate Floor Transcript, 13 May 2003, pp. 55–66 (HB 3316) and 122–31 (HB 1574); Pat Guinane, "Lawmakers

Vote to Hold Slowpokes to the Right," *SJR,* 14 May 2003, p. 1; Anne Marie Tavella, "Slow Driver Bill Passes Senate," *CDH,* 14 May 2003, p. 7; Marni Pyke, "Shining a Light on Racial Profiling," *CDH,* 14 May 2003, p. 1; Chinta Strausberg, "Obama Urges Congress to Resist $300 Billion Deficit," *CD,* 14 May 2003, p. 3; "Agenda, PR Brainstorming for Senator Barack Obama," 14 May 2003, John Kupper meeting notes, "Obama 5/14," JKP; Senate Floor Transcript, 15 May 2003, p. 89 (Obama remarking concerning HB 515 that "I know that Pete Baroni did not draft this because it's not as clear as I would like"); Senate Floor Transcript, 16 May 2003, pp. 118–31, esp. 120–21 (HB 2579); Edwin Colfax, "Taped Confessions," *CT,* 18 May 2003; Obama YouTube Video, 16 June 2003 (HB 60 "is a major victory that we achieved this year" and "there is no logical reason why we would not provide health care plans to immigrant youth"); Mark K. Matthews, "States Tell Police to Turn on the Camera," Stateline.org, 4 May 2005; Jerry Crimmins, "Videotaped Interrogations Get High Marks," *CDLB,* 3 November 2005, p. 1; Rob Warden's excellent "Illinois Death Penalty Reform: How It Happened, What It Promises," *Journal of Criminal Law and Criminology* 95 (Winter 2005): 381–426; "A Chronicle Q&A with Barack Obama," Chronicle.com, 12 November 2007 (HB 60); Jim Edgar's 2010 interviews with Mark DePue, pp. 1050–51, 1087, 1089; DJG interviews with Kathy Saltmarsh, Laura Russell Hunter, Deanne Benos, Jim Edgar, John Millner, John Kupper, Pete Giangreco, and Terry Walsh.

19. Obama Block Schedule, May 2003; OFI, "Obama Announces State Hearings on Bush Plan to 'Download' Key Parts of Head Start," 18 May 2003; *Obama Weekly Report,* 19–26 May, 20 May 2003, TWP; Hull for Senate, FEC Form 10, 9 May 2003, 4:55 P.M. (re-sent 12 May 2003, 3:42 P.M.), FEC; Jack Brammer, "Two Kentuckians Reflect on Their Time with Barack Obama," *Lexington Herald-Leader,* 19 January 2009; DJG interviews with Pete Giangreco, Terry Walsh, John Kupper, Dan Shomon, Nate Tamarin, Jim Cauley, and Rob Fisher.

20. Obama Block Schedule, May 2003, JKP; "Democratic Poll Overview," IlSenate.com, 19 May 2003; Greg Hinz, "Gov. Plays Favorite in Business," *CCB,* 19 May 2003; *CF,* 9 April, 19, 20, 21, 22, 23, and 27 May and 11 and 18 September 2003; Kelly Quigley, "Teamsters Back Hynes for Senate," *CCB,* 20 May 2003; Eric Zorn, "Candidate Ryan Is Not Confused About Who He Is," *CT,* 20 May 2003; Adam Clymer, "Two Parties on the Prowl to Claim the Senate in '02," *NYT,* 23 June 2001; Jill Stanek, "Jack Ryan: From Harvard to Hales," *IL,* 30 September 2002; Daniel K. Proft, "Putting Conservative Principles into Action," *IL,* 29 May 2003; Fran Eaton, "Jack Ryan, GOP Candidate for U.S. Senate," *IL,* 14, 15, and 16 October 2003; Lynn Sweet, "Jack Ryan Holds Back in Biography," *CST,* 27 November 2003, p. 45; Ray Long and Christi Parsons, "Scolding for SBC Law," *CT,* 20 May 2003; "WomanView: Illinois Victims' Economic Security and Safety Act (SB 1763)," NCPL, 25 March 2003; Senate Floor Transcript, 20 May 2003, pp. 27–32 (unanimous passage of HB 3486, the successor to Obama's SB 1763); Charlie Crain, "Teamsters Back Hynes for Senate," *NIT,* 21 May 2003; Peter Savodnik, "Pols vs. Businessmen in Illinois Race," *TH,* 21 May 2003; Chinta Strausberg, "Governor: No Gambling!," *CD,* 21 May 2003, p. 3; Ray Long and Rick Pearson, "U.S. Turns Probe to State Senate Democrats," *CT,* 21 May 2003; Dearbail Jordan, "Altheimer's International Chairman Quits After Internal Shake-Up," TheLawyer.com, 21 May 2003; Tom Gradel to Kelly Steel et al., 21 May 2003, HORP; Rick Pearson and Ray Long, "Capitol Collision Course," *CT,* 22 May 2003; Scott Fornek, "Downstate Dem Enters Senate Fray," *CST,* 23 May 2003; Bernard Schoenburg, "McVeigh Prosecutor," *SJR,* 19 April 2010; DJG interviews with Tim King, Wendy Pollack, Rob Scott, Dan Shomon, and Jim Dodge.

21. Obama Block Schedule, May 2003, JKP; "Notes," *CDLB,* 16 May 2003, p. 24; Libby Brewer Notice, *CT,* 18 May 2003; *CF,* 26 March and 21, 23, 28, 29, and 30 May 2003; *Obama Weekly Report,* 19–26 May 2003 and 27 May–1 June 2003, TWP; OFI to Team Captain, "June 12th Event," 20 May 2003; Claire Serdiuk e-mail to Marty Nesbitt et al., 22 May 2003, SPP; Scott Fornek, "Downstate Dem Enters Senate Fray," *CST,* 23 May 2003, p. 4; Obama, Commencement Address, Chicago-Kent College of Law, 25 May 2003; Gubernatorial Press Release, "Governor Calls for Ban on Sale of Dietary Supplement Ephedra," 14 May 2003; Brandon Loomis, "Ill. Gov. Signs Ephedra Ban," AP, 25 May 2003, and *CDH,* 26 May 2003, p. I-9; Jim Ritter, "Ephedra Sales Banned in Illinois," *CST,* 26 May 2003, p. 6; Senate Floor Transcript, 27 May 2003, pp. 29–42, esp. 34–35 (HB 3023); John Kupper notes, "Barack 5/27," 27 May 2003, JKP; Minutes, GARS Board of Trustees, 29 May

2003, #223, pp. 1859–63; Senate Floor Transcript, 29 May 2003, pp. 82–88 (HB 2221); Senate Journal, 29 May 2003, p. 46; Ray Long and Rick Pearson, "State Set to Clean Its House," and Christi Parsons and Long, "Death Penalty Reform Goes to Blagojevich," *CT,* 30 May 2003; Senate Floor Transcript, 30 May 2003, pp. 42–43 (HB 569), 140–42 (SB 1493) and 160–61 (SR 89); AP, "Nightclub Safety Bills Heads to Governor," 31 May 2003; Senate Floor Transcript, 31 May 2003, pp. 27–39 (HB 3412), 181–86 (SB 600), and 201–3 (SB 4); Long and Parsons, "O'Hare Plan Passes Amid Budget Rush," *CT,* 1 June 2003; Pearson and Long, "Blagojevich Lauds His Budget," and Kate McCann, "Democrats Take Charge," *CT,* 2 June 2003; Jermain Griffin, "Campaign Finance Reform Still on Group's Docket," *CDLB,* 3 June 2003, pp. 1, 24; Stephen S. Morrill, "Client Report for 2003 End of Session," Morrill & Associates, June 2003, esp. pp. 6–9; Senate Republican Leader Frank Watson, *End of Session Report 2003,* esp. pp. 15, 136–200; Illinois Environmental Council, "2003 Environmental Scorecard," and "End of Session Report," 2 June 2003, JGP; "Major Bills Passed by the Illinois General Assembly," *LRU First Reading* 17 (August 2003): 7; "Spring Legislative Session Mostly Mild," *C-WR* 18 #7, 8 August 2003; Barbara Van Dyke-Brown et al., *Almanac of Illinois Politics—2004* (University of Illinois at Springfield, 2004), pp. 150–51 (crediting Obama with nineteen bills passed); Tom Waldron, "Supporting Working Families When State Coffers Are Empty: Lessons from Illinois' Successful Campaign for a Better EITC, Hatcher Group, November 2003, pp. 8–10; Linda Mills, *Illinois Prisoner Reentry: Building a Second Chance Agenda* (Annie E. Casey Foundation, July 2004); Carrie Cox et al., "Certificates of Relief from Disabilities Implementation and Tracking," Safer Foundation, November 2006, esp. p. 5; David Moberg, "Union Blues Lift in Chicago," *Nation,* 9 April 2005; Ed Wojcicki, "Still the Wild West?: A 10-Year Look at Campaign Finance Reform in Illinois," Simon Institute, Southern Illinois University, September 2006, p. 17; Cindi Canary in Jackie Calmes, "Statehouse Yields Clues to Obama," *WSJ,* 23 February 2007; Olivia Clarke, "Choosing Sides," *CL,* January 2008; Brad Spirrison, "Barack Obama Has an Affinity for Venture Capital That Dates Back to His Days as a State Senator," *Venture Capital Journal,* 1 February 2009; Maura O'Hara, "IVCA Mourns a Key Contributor," IllinoisVC.org, 17 August 2011; John Sexton, "Obama Praised Private Equity When He Needed Cash," Breitbart.com, 2 July 2012; Alejandra Cancino, "Executive Profile: Keith Kelleher," *CT,* 2 December 2013; DJG inteviews with Larry Walsh, Miguel del Valle, Paula Wolff, Will Burns, Pete Baroni, Linda Mills, Sean Noble, John Nicolay, Keith Kelleher, Madeline Talbott, Roberta Lynch, Bill Perkins, Ray Harris, Michael McGann, Jerry Morrison, Cindi Canary, and Ed Wojcicki.

22. *Chicago Tonight,* WTTW, 29 May 2003; David Mark, "'04 Illinois Democratic Senate Candidates," *Campaigns & Elections,* June 2003, p. 40; Obama Block Schedule, June 2003, Dan Shomon e-mail to John Kupper et al., "Agenda for Today's Conference Call/Meeting," 2 June 2003, JKP; *Obama Weekly Report,* 2–8 June 2003, JKP; Scott Fornek, "Poshard Never Heard of Downstater, Leans Toward Hynes," *CST,* 4 June 2003, p. 12; Chinta Strausberg, "Pols Urge Governor to Sign 2nd Chance Bill," *CD,* 12 June 2003, p. 3; Lesley R. Chinn, "Ex-Offender Legislation Passes; Awaits Governor's Signature," *HPC,* 12 June 2003, p. 2; Hull Research to Hull Communications et al., "Current/Additional Research on Barack Obama," 9 June 2003, HORP and RRP; Hull Research to Hull Communications et al., "Current/Additional Research on Dan Hynes," 9 June 2003, HORP; Sean Cartwright, Kennedy Communications, to Hull for Senate, "Current/Additional Research on Gery Chico," 9 June 2003, HORP; Jeremiah Posedel e-mail to Dear Friend, "Rep. Evans Endorses Obama!," 9 June 2003, TWP; *Voices for Choices v. Illinois Bell Telephone,* #03-C-3290 (N. D. Ill.), 9 June 2003 (Judge Charles P. Kocoras granting preliminary injunction); Jon Van and Christi Parsons, "Judge Blocks SBC Rate Law Ruling," *CT,* 10 June 2003; *CF,* 13 June 2003; Peter Savodnik, "Illinois Lawyer Simmons Garners Support," *TH,* 10 June 2003; JCAR Minutes, 10 June 2003; *Obama Weekly Report,* 10–16 June 2003, TWP; [Joe Sinsheimer], "Research—State Sen. Barack Obama," 2 vols., 12 June 2003, esp. pp. 2, 4, 51, 90, 167, TWP; John Kupper notes, "Obama Research," n.d., 2pp., "Horizon/CMS," n.d., 11pp., and "First Health Group Corp.," n.d., 18pp., JKP; Mark Schreiner, "Passion for Politics Keeps Drawing Him In," *Wilmington Star News,* 25 March 2007; James L. Merriner, "Obama 2008?," *CM,* March 2006; Howard Fineman, "The Broker: Vernon Jordan," *Newsweek,* 5 April 2008; Vernon Jordan, *Make It Plain* (Public Affairs, 2008), p. 207; Eric Alterman, "Life of the Party,"

Guardian, 6 November 2008; Katherine Rosman, "Before He Was President," *WSJ*, 7 November 2008; DJG interviews with John Kupper and Joe Sinsheimer.

23. *The Demo Memo* (Milton Township, DuPage County), 3 June 2003; Bernard Schoenburg, *SJR*, 12 June 2003, p. 7; Ed Tibbetts, "Evans Endorses Obama in Senate Race," *QCT*, 12 June 2003; Stephen Pugh and Marty Nesbitt to Pugh, 9 May 2003, and Bettylu K. Saltzman (OFI) to Stephen Pugh, 2 June 2003, SPP; "Fundraising Reception," 12 June 2003, Dan Shomon, "Obama Labor Strategy," 10 June 2003, TWP; Bruce Dixon, "In Search of the Real Barack Obama," *Black Commentator*, 5 June 2003; "Not 'Corrupted' by DLC, Says Obama," *Black Commentator*, 19 June 2003; "Obama to Have Name Removed from DLC List," *Black Commentator*, 26 June 2003; DJG interviews with John Kupper, Joe McLean, Claire Serdiuk, Bettylu Saltzman, Stephen Pugh, and Pete Giangreco.

24. "Upcoming Events," ObamaForIllinois.com, 14 June 2003; Steven R. Strahler, "Center Not Holding at Altheimer," *CCB*, 16 June 2003, p. 30; *Obama Weekly Report*, 16–23 June 2003, TWP; Russ Stewart, "'Black Is Beatable' Gives Hope to State GOP," *NSP*, 18 June 2003; Bill Zwecker, "Looks Like the Party's Over," *CST*, 19 June 2003, p. 40; "Armour & Co.," *CT*, 22 June 2003; "Hynes Receives Backing," *CDH*, 23 June 2003, p. I-3; OFI, "Obama Hails Supreme Court Decision Upholding Affirmative Action," 23 June 2003; www.blairhull.com, Internet Archive (TV and radio ads); Scott Fornek, "Dem Senate Hopeful Starts Earliest Illinois TV Blitz Ever," *CST*, 24 June 2003, p. 16; Eric Krol, "Senate Hopeful Already Thinking About March," *CDH*, 24 June 2003, p. 11; Meg Kinnard, "Hull Gets an Early Start for Illinois Senate," NationalJournal.com, 24 June 2003; Steve Neal, "Rush Senate Endorsement Puts Payback Before Prudence," *CST*, 25 June 2003, p. 47; Rick Pearson, "Top-Dollar Senate Race Looms," *CT*, 25 June 2003; Chinta Strausberg, "Maynard Jackson's Death at 65 Leaves 'Big Hole,'" *CD*, 25 June 2003, p. 8; Russ Stewart, "Flood of 'RWGs' Mean Avalanche of Senate Ads," *NSP*, 25 June 2003; Eric Krol, "Unorthodox Candidate Enters U.S. Senate Race," *CDH*, 26 June 2003, p. 6; Bernard Schoenburg, "Blair Hull Announces Bid for U.S. Senate," *SJR*, 26 June 2003, p. 13; Patrick J. Powers, "Senate Hopeful First to Place Metro-East Ads," *BND*, 27 June 2003, p. B1; AP, "Hull's Early Push Reflects His Wealth, Lack of Name Recognition," 27 June 2003; "The Long, Hot Political Days of Summer Have Arrived," *SI*, 1 July 2003; Meg Kinnard, "To Hull, Business Savvy Would Help Senate," NationalJournal.com, 10 July 2003; OFI, "Obama Asks Hastert to Act Immediately to Make Sure 378,000 Illinois Families Receive Tax Credit in Washington," 26 June 2003; Michael C. Dorf to Dan Shomon and Terrie Pickerill, "New Disclaimer Requirements for Radio and Televised Communications," 27 June 2003, JKP; Ameet Sachdev, "Altheimer & Gray to Close," *CT*, 28 June 2003; Lynn Sweet, "Chico Could Be In for Bumpy Ride," *CST*, 3 July 2003, p. 33; Dearbail Jordan, "The Final Dark Days of the Firm That Wanted Too Much," TheLawyer.com, 7 July 2003; Sachdev, "Altheimer's Lawyers Migrate to New Firms," *CT*, 17 July 2003; Sachdev, "Altheimer Practices Questioned," *CT*, 20 July 2003; Chris Fusco and Tim Novak, "Rezko Cash Triple What Obama Says," *CST*, 18 June 2007, p. 3; Bob Secter and David Jackson, "Funds Tough to Figure for Rezko Aid," *CT*, 31 January 2008; Kenneth P. Vogel, "Obama Releases Names of Rezko-Linked Donors," *Politico*, 15 March 2008; Baim, *Obama and the Gays*, pp. 15, 201, 521; Stone in James L. Merriner, "The Friends of O," *CM*, June 2008; DJG interviews with Geof Stone, Susan Pierson Lewers, Leslie Corbett, Kevin Thompson, Nate Tamarin, Alan Reger, Megan Crowhurst, and Kelly Dietrich.

25. V. Dion Haynes, "Fight Racial Profiling at Local Level, Lawmaker Says," *CT*, 29 June 2003; Dan Shomon e-mail to John Kupper et al., "Obama Weekly Meeting Agenda, Monday, June 30," 29 June 2003, Nate Tamarin e-mail to Kupper, 30 June 2003, Kupper notes, "Obama 6/30," Obama Block Schedule, July 2003, JKP; *Obama Weekly News*, 1–7 July 2003, TWP; OFI, "U.S. Rep. Lane Evans Endorses Barack Obama," 1 July 2003; Jennifer Davis, "Evans Endorses Illinois Senator," *PJS*, 2 July 2003, p. B3; Matt Adrian, "Rep. Evans Endorses Obama in 2004 U.S. Senate Race," *DHR*, 2 July 2003, p. A3 (and as "Evans Gives Senate Race Endorsement to Obama," *QCT*, and "Evans Endorses Obama for Senate," *SI*, 2 July 2003); Bernard Schoenburg, "Lane Evans Endorses Barack Obama for Senate," *SJR*, 2 July 2003, p. 12; Doug Wilson, "Evans Endorses Obama for Senate Race," *QHW*, 2 July 2003, p. A9; Dave Dorsett, "Senate Hopeful Makes Macomb Stop," *MJ*, 6 July 2003; Daniel Duggan, "Hull's Senate Bid: No Experience Necessary?," *ECN*, 2 July 2003; *CF*,

2, 7, and 10 July 2003; Shamus Toomey, "Health Program Expanded," *HPH,* 2 July 2003, p. 8; OFI, "Obama Reports $1 Million on Hand for U.S. Senate Bid, Has Raised Over $1.4 Million for Campaign So Far," 2 July 2003, TWP; V. Dion Haynes, "Racial Profiling Ban Leaves Some with Mixed Emotions," *CT,* 3 July 2003; NTSB, "NTSB and Illinois Make History When Governor Signs Three Most Wanted Improvements in One Day," 3 July 2003; gubernatorial press release, "Governor Strengthens Traffic Safety Laws in Time for Summer Travel," 3 July 2003; Matt Adrian and Richard Goldstein, "Senate Candidates Stack Up Money," *DHR,* 6 July 2003, p. B4 (and also *QCT,* 5 July and *SI,* 6 July); *Obama Weekly News,* 8–15 July 2003, TWP; Paul Harstad e-mail to John Kupper et al., "Draft Q'aire for Our Dem Primary Survey," 8 July 2003, and Pete Giangreco e-mail to Harstad, 11 July 2003, JKP; Obama to Paul Simon, received 11 July 2003, and Simon to Obama, 15 July 2003, PSP; Corey Hall, "Senator Barack Obama Celebrates Healthcare Legislation at PUSH Forum," *HPC,* 10 July 2003, p. 2; "Peace Group Will Discuss Patriot Act," *Glencoe News,* 26 June 2003 (9 July); OFI, FEC Report of Receipts and Disbursements, 1 April–30 June 2003, 10 July 2003; Friends of Barack Obama, Report of Campaign Contributions and Expenditures, 1 January–30 June 2003, 31 July 2003; Scott Fornek, "Downstate Dem Quits Senate Race," *CST,* 10 July 2003, p. 3; OFI, "Obama Calls on Congress to Declare Independence from Drug Lobby," 10 July 2003; Chinta Strausberg, "Obama Seeks Creation of Federal Drug Czar," *CD,* 12 July 2003, p. 2; Todd Spivak, "Obama Shoots Down Federal Prescription Drug Plans," *HPH,* 16 July 2003, p. 6; "Baggage Ex-Cons Don't Need," *CST,* 13 July 2003, p. 31; Laura S. Washington and Leslie Gryce Sturino to Candidates and Campaigns, "July 13th Forum Format and Ground Rules," 10 July 2003, JKP; Antonio Mora, *Eye on Chicago,* CBS 2 TV, 13 July 2003; "IVI-IPO Partial Transcript," WBEZ, 13 July 2003, JKP; John McCormick, "Senate Hopefuls Abound for '04," *CT,* 14 July 2003; Carol Felsenthal, "Governor Sunshine," *CM,* November 2003; David Bernstein, "Mr. Un-Popularity," *CM,* February 2008; Bernstein, "Chicago Straight," *CM,* June 2009; John Presta, *Mr. and Mrs. Grassroots* (Elevator Group, 2010), pp. 111–12; DJG interviews with Chuck Hurley and John Cullerton.

26. Obama Block Schedule, July 2003, JKP; *Obama Weekly Report,* 16–22 July 2003, TWP; Peter Savodnik, "Illinois Senate Democratic Primary Heats Up," *TH,* 16 July 2003, p. 6; Ray Gibson and Rick Pearson, "Candidate Hull Spends at Record Pace," *CT,* 16 July 2003; Scott Fornek, "Blackjack King Outspends Dem Rival 4–1 in Senate Bid," *CST,* 16 July 2003, p. 1; Lynn Sweet, "Senate Rivals Find Silver Lining in Hull's Spending," *CST,* 17 July 2003, p. 34; "Diversity Committee Hosts Summer Program Reception," *Equal Time* (Jenner & Block), Summer 2003, p. 11; Jenner & Block press release, "State Senator Barack Obama Addresses Summer Associates at Annual Diversity Dinner," 17 July 2003; gubernatorial press release, "Governor Signs Six-Bill Package Addressing Serious Problems in Illinois' Criminal Justice System," 17 July 2003; OFI, "Ground-Breaking Obama Criminal Justice Package Signed into Law," 17 July 2003; Pearson, "Taped Confessions to Be Law," *CT,* 17 July 2003; Monica Davey, "Illinois Will Require Taping of Homicide Investigations," *NYT,* 17 July 2003; Chinta Strausberg, "Jones Credited with Criminal Justice Bill Signing," *CD,* 17 July 2003, p. 2; Mike Ramsey, "Taped-Confessions Bill Signed into Law," *SJR,* 18 July 2003, p. 1; Kevin Pang, "Long-Distance Service Has a Conservative Ring to It," *CT,* 18 July 2003; Joe Ruklick, "Hail Passage of Anti-Crime Bills," *CD,* 19 July 2003, p. 1; Chad Anderson, "You Might Be a Candidate for U.S. Senate If," *RRS,* 19 July 2003, p. A10; *CDH,* 20 July 2003, p. I-3; Obama and Fred Tsao, "Seek to Improve Working Conditions for All," *CT,* 21 July 2003; *CF,* 21 July 2003; Eric Zorn, "Radio Host Sees Senate Bid with a Citizen's Eye," *CT,* 22 July 2003; Robert Feder, "Talk Show Hosts Aim for U.S. Senate," *CST,* 22 July 2003, p. 47; Todd Spivak, "Obama Passes $1M Mark in Bid for U.S. Senate Seat," and Obama, "Gov. Okays Obama's Cutting-Edge Legislation," *HPH,* 23 July 2003, pp. 1–2 and 4 (and 30 July 2003, p. 2); Corey Hall, "Gov. Blagojevich Signs Six Criminal Justice Reform Bills," *HPC,* 24 July 2003, p. 1; John Cullerton, Kirk Dillard, and Peter G. Baroni, "Capital Punishment Reform in Illinois: A Model for the Nation," *DCBA Brief: Journal of the DuPage County Bar Association,* April 2004; Thomas P. Sullivan, "Police Experiences with Recording Custodial Interrogations," Northwestern University School of Law, Summer 2004, esp. pp. C1–C3; Joe Ruklick, "Fund-Raising Success Gives Obama Momentum," *CD,* 26 July 2003, p. 5; Strausberg, "Trotter, Washington Butt Heads Over Obama," *CD,* 22 July 2003, p. 3; Hull for Senate press release,

21 July 2003, BlairHull.com; "South Side Congressman Bobby Rush to Chair Blair Hull's U.S. Senate Campaign," WBBM Radio, 21 July 2003; Pearson, "Rush Endorses Hull for U.S. Senate," *CT,* 22 July 2003; Fornek, "Rush Chairing Hull's Senate Campaign," and Steve Neal, "Bobby Rush a Sellout," *CST,* 22 July 2003, pp. 15, 25; Spivak, "Rush Snubs Obama, Endorses Hull," *HPH,* 23 July 2003, p. 2; Strausberg, "Rush Backing of Blair 'Divides Rep's Base,'" *CD,* 24 July 2003, p. 2; La Risa Lynch, "U.S. Rep. Rush Backs Hull for Senate," *CC,* 24 July 2003, p. 1; "Seeking Higher Elected Office," *City Journal,* 24 July–7 August 2003, pp. 3–4; *Obama Weekly Report,* 23–29 July 2003, TWP; Presta, *Mr. and Mrs. Grassroots,* pp. 122–26; DJG interview with Bruce Spiva.

27. Rick Ridder and Craig Hughes to Mike Henry, "Media Plan," 22 July 2003, RRP; Dan Shomon e-mail to David Axelrod et al., "Barack's Meeting with AFSCME on Tuesday, July 22," 16 July 2003, Jim Cauley e-mail to Axelrod et al., "AFSCME Tomorrow," Nate Tamarin e-mail to Pete Giangeco et al., "SEIU," 21 July 2003, "2004: How Obama Wins by the Numbers," n.d., 9pp., JKP; "Obama for Illinois: Vote Model & Targeting," n.d., 27pp., TWP; *Public Affairs with Jeff Berkowitz,* 24 July 2003; DJG interviews with John Cameron, Ray Harris, John Kupper, Pete Giangreco, Terry Walsh, Jim Cauley, and Audra Wilson.

28. Scott Fornek, "Web Site Draws Few Senate Hopefuls So Far," *CST,* 25 July 2003, p. 16; *CF,* 28 July 2003; John Kupper fax to Deb Schommer, 25 July 2003, JKP; Joe Ruklick, "Obama's Stand on Issues," *CD,* 29 July 2003, p. 6; Peter Savodnik, "Illinois Senate Candidate Compared to Moseley Braun," *TH,* 30 July 2003, p. 9; Charla Brautigan, "Democrats Endorse Hynes for U.S. Senate," *JHN,* 30 July 2003; Mark Rodeffer, "Hull Takes on Health Care Industry," NationalJournal.com, 31 July 2003; Fornek, "Senate Hopefuls Vie for Black Vote," *CST,* 5 August 2003, p. 8; Rick Pearson, "Rauschenberger Weighs Senate Run," *CT,* 5 August 2003; Meg Kinnard, "Hull Vies for Black Vote with Rush Support," National Journal.com, 6 August 2003; Chinta Strausberg, "Obama: 'Blacks Too Smart for PR Divide,'" *CD,* 7 August 2003, p. 4; Rashid Khalidi, "Israel Blocks Path to Peace," *CT,* 11 February 2002; Khalidi, "Basic Truths from Both Sides of the Conflict," *CT,* 3 April 2002; Isaac Wolf, "Khalidi Accepts Chair Offer from Columbia," *Chicago Maroon,* 31 January 2003; Peter Wallsten, "Allies of Palestinians See a Friend in Barack Obama," *LAT,* 10 April 2008; Terry Gross, "Which Way the Wind Blows: Bill Ayers on Obama," *Fresh Air,* NPR, 18 November 2008; William Ayers, "The Real Bill Ayers," *NYT,* 6 December 2008; "An Interview with Bill Ayers," *The Point* #5, Spring 2012; Rachel DeWoskin, "The Sunday Rumpus Interview: Bill Ayers," TheRumpus.net, 1 December 2013; Presta, *Mr. and Mrs. Grassroots,* pp. 132–34; Steve Metsch, "Beverly Man Writes Book About Obama Campaigns," *SS,* 22 March 2010; Scott Lynch-Giddings, "Why Vote for Obama? Because of Who He Is," 29 October 2012, SLGP; DJG interviews with Scott Lynch-Giddings, Mona Khalidi, Bernardine Dohrn, and Bill Ayers.

29. Harstad Strategic Research, "Illinois Statewide Democratic Party Survey, 16–22 July 2003," 34pp., graphical supplement 24pp., Senate Summary, 4pp., 30 July 2003, JKP and PHP; Harstad to Obama for Senate Campaign, "Survey Shows Barack Obama Among Leaders in Democratic Nomination for U.S. Senator in Illinois," 1 August 2003, PHP; John Kupper daybook entries, August 2003, JKP; Cauley to Obama Team, "The Consultant Meeting Follow-Up," 7 August 2003, TWP; "Obama Earned Media Plan," 8 August 2003, JKP; OFI, "Integrated Strategic Plan," n.d., JCP; Ben Austen, "In the Loop," *Harper's,* June 2010, pp. 29–37; DJG interviews with John Kupper, Pete Giangreco, Terry Walsh, Paul Harstad, Mike Kulisheck, Jim Cauley, Dan Shomon, Claire Serdiuk, Randon Gardley, Nate Tamarin, Audra Wilson, Madhuri Kommareddi, Liz Drew, Tom Hazen, Lauren Kidwell, Pam Smith, and Kevin Watson.

30. Obama Block Schedule, August 2003, JKP; "Democratic Poll Overview," IlSenate.com, 10–12 August 2003; Paul Merrion, "In Senate Race, Give a Big Edge to the Officeholders," *CCB,* 11 August 2003, p. 9; *CDH,* 11 August 2003, p. I-3; Anne Marie Apollo, "U.S. Senate Candidate Obama Introduces Himself to Elgin," *ECN,* 12 August 2003, p. A1; Todd Spivak, "State Politicians Put Their Fundraising on Cruise Control," *HPH,* 13 August 2003, p. 2; Christi Parsons, "Closed-Door Meetings to Be Taped," *CT,* 13 August 2003; OFI, "Obama Law Strengthens Open Government in Illinois," 13 August 2003; Spivak, "Gov. Signs Obama-Currie Bill for Open Meetings," *HPH,* 20 August 2003, p. 6; Obama [State Fair] transcript, [13 August 2003], TWP; Frank Avila and Morgan Carter, "Election 2004," CAN-TV

Channel 19, 16–17 August 2003; Obama, transcript of remarks, Evanston, IL, 7 February 2004, JKP and TPP; *CF,* 14 and 15 August 2003; Chinta Strausberg, "Starks, Jesse, Obama Say Recall Vote 'Orchestrated,'" *CD,* 16 August 2003, p. 2; Bernard Schoenburg, "Senate Race Poll," *SJR,* 17 August 2003, p. 17; Angela Rozas and Joshua Howes, "2 Juries Dubious over Confession Tapes' Merits," *CT,* 17 August 2003; Bruce DuMont, *Beyond the Beltway,* 17 August 2003, MBC; Ben Joravsky, "When Obama Needed Public Access TV to Reach Voters," *CR,* 14 January 2014; DJG interviews with Steve Preckwinkle, Margaret Blackshere, Jim Cauley, Audra Wilson, Randon Gardley, Jesse Jackson Jr., Juleigh Nowinski, and Tarak Shah.

31. Steve Neal, "New Poll Shows Obama Gaining Among Dems in Senate Race," *CST,* 18 August 2003, p. 41; "Democratic Poll Overview," IlSenate.com, 14–20 August 2003; Paul Simon to Obama, 18 August 2003, PSP; Obama Block Schedule, August 2003, JKP; *Obama Weekly Report,* 19 August 2003, TWP; Todd Spivak, "HP Pols Blast Veto Against Death Penalty Reform Bill," *HPH,* 20 August 2003, p. 7; Neal, "Favorite Sons Call in Big Guns in High-Priced Race for Senate," *CST,* 20 August 2003, p. 53; Ray Long and John Chase, "State Acts to Bar Future E2s," *CT,* 20 August 2003; John Kupper to Obama Team, "Free Press Hits," 22 August 2003, Kupper notes, "Obama 8/25," 25 August 2003, Rachel Dahan e-mail to Nate Tamarin, "IWHC's 2003 Legislative Voting Report Card—Obama 100%," 25 August 2003, JKP; Todd Spivak, "Bobby Rush Explains Endorsement of Hull," *HPH,* 27 August 2003, pp. 12–13; Chinta Strausberg, "Davis, Obama: Bush's $500 Billion Runaway Deficit Train Wrecking Economy," *CD,* 27 August 2003, p. 3; Kate Clements, "Low-Income Families Will Benefit from Laws," *CNG,* 27 August 2003; Strausberg, "Jones to Blagojevich: 'We Will Override Veto,'" *CD,* 28 August 2003, p. 3; "WomanView: Blagojevich Signs Two Bills Promoting Women's Economic Security," National Center on Poverty Law, 29 August 2003; Mary Massingale, "State Alters Earned Income Tax Credit," *SJR,* 6 September 2003, p. 17; *CF,* 11, 16, and 18 September 2003; Emily M. Shaules, "New Illinois Law Protects Victims of Domestic Violence," *The Brief,* February 2004; Julie A. Iotrowski-Govreau, "The Illinois Victims' Economic Security and Safety Act of 2003: A Comparative Analysis," ISBA, May 2005; DJG interviews with Wendy Pollack, Jeremiah Posedel, Jeff Stauter, Scott Turow, Jan Schakowsky, Bob Creamer, Alan Solow, John Cameron, and William McNary. On Rezko, also see Thomas A. Corfman, "Giant South Side Project Planned," *CT,* 31 July 2003, and David Roeder and Fran Spielman, "Up to 5,000 Residential Units in Plans for Near S. Side Development," *CST,* 31 July 2003, p. 55.

32. *CF,* 13 August 2003; Kennedy Communications to Hull Campaign, "Top Hull Negatives," n.d., RRP; "Affidavit of Brenda Sexton Hull," *Hull v. Hull,* #35363, Cook County Circuit Court, 12 March 1998, 7pp., JKP and TWP; Terry Armour, "Sexton Appointed," *CT,* 8 April 2003; Armour, "Sexton Stands By Her Business Background," *CT,* 9 April 2003; Rick Ridder e-mail to Mark Blumenthal and Blumenthal e-mail to Ridder, 5:44 P.M. and 11:08 P.M., 10 September 2003, MBP; Jason Horowitz, "Newly Out in Front for White House," *WP,* 15 October 2009; Blumenthal in Remnick, *The Bridge,* p. 372; DJG interviews with Alan Reger, Mo Elleithee, David Spiegel, Rick Ridder, Joanie Braden, Anita Dunn, Mark Blumenthal, Mike Henry, Blair Hull, Fred Lebed, Kitty Kurth, and Don Wiener.

33. Obama Block Schedule, September 2003, JKP; Deidre Cox Baker, "Activists, Candidates Decry Weak National Economy," *QCT,* 1 September 2003; Steve Neal, "Dems Scuffle for AFL-CIO Backing," *CST,* 1 September 2003, p. 21; *CF,* 11 and 28 August and 2 September 2003; OFI, "Obama Offers Strong Steps to Honor Labor by Protecting Worker Pensions," 2 September 2003; Chinta Strausberg, "Obama Seeks to Protect Pensions," *CD,* 4 September 2003, p. 5; Terry Walsh e-mail to Jim Cauley et al., "Pension Redux," 5 September 2003, JKP; *Obama Weekly Report,* 8–14 September 2003, TWP; Thomas Roeser, "Rauschenberger's Gambit," *CST,* 6 September 2003, p. 12; Scott Fornek, "Poshard Never Heard of Downstater," *CST,* 4 June 2003, p. 12; Obama YouTube Clip with Danny Davis, 1:30, 7 September 2003; Chicago Video Project, "Introducing Barack Obama," 5:18, 2003; Laura Washington, "If He Can Turn Out His Black Base and Build a Coalition," *CST,* 8 September 2003, p. 39; Mark Samuels, "Senate Candidate Should Be Judged on Ideas, Not Ethnicity," *MJG* and *CTC,* 8 September 2003; Lauren W. Whittington, "Populist Pappas," *RC,* 10 September 2003; Neal, "Obama, Daley Would Make Formidable Political Team," *CST,* 12 September 2003, p. 45; Washington, "Uproar Over Top Cop," *CST,* 15 September 2003, p. 41; Samuels, "Obama

Works to Overcome Name Confusion," *MJG* and *CTC*, 15 September 2003; Rich Miller, "Illinois Comptroller Leads the Democratic Field for Senate," *RCR, MD*, and *IT*, 16, 17, and 18 September 2003; Obama, "Legislation Would Protect Worker Pensions," *HPH*, 17 September 2003, p. 4; Bernard Schoenburg, "Rauschenberger Not Shy in Commenting About Foes," *SJR*, 28 September 2003, p. 15 (Obama "is certainly brilliant and a great speaker"); Lynn Sweet, "Running to the Right," *II*, January 2004; Obama on *The Leonard Lopate Show*, NPR, 22 November 2004; Michael Weisskopf, "Obama: How He Learned to Win," *Time*, 8 May 2008; DJG interviews with Emil Jones Jr., Dave Gross, Margaret Blackshere, Steve Rauschenberger, Linda Mills, Claire Eberle, and Laura Washington.

34. Paul Merrion, "Obama's Appeal Drives Cash Flow," *CCB*, 15 September 2003, p. 3; Dan Shomon e-mail to Kupper, "Illinois NOW Questionnaire," 25 August 2003, "Final Draft, Illinois NOW Questionnaire for Senator Barack Obama," 3 September 2003, Raja Krishnamoorthi e-mail to Obama and Obama to Krishnamoorthi, "UAW and NOW Questionnaires," 6 and 7 September 2003, Fredrick Vars e-mail to Krishnamoorthi and Krishnamoorthi e-mail to Kupper, "Repro Rights," 8 September 2003, JKP; OFI, "Illinois NOW Questionnaire for Senator Barack Obama," 10 September 2003, pp. 2–4; OFI, "Obama Releases Comprehensive Plan to Strengthen Homeland Security," 10 September 2003; Chinta Strausberg, "Obama's Security Plan Aims to Stop Terrorists," *CD*, 11 September 2003, p. 3.

35. Rick Pearson and Brett McNeil, "Hynes' Senate Bid Leans on Labor," *CT*, 15 September 2003; Jim Ritter, "Hynes Joins Race for U.S. Senate," *CST*, 15 September 2003, p. 12; Eric Krol, "Comptroller Running for U.S. Senate," *CDH*, 15 September 2003, p. 7; Mike Ramsey, "Hynes Announces Bid for U.S. Senate," *SJR*, 1 September 2003, p. 7; Obama Block Schedule, September 2003, OFI, "Campaign Prospectus," 15 September 2003, 16pp., Paul Harstad, "Primary Polling Budget and Options for Obama Campaign," 17 September 2003, "Agenda, Media Planning Meeting," 19 September 2003, "Barack Obama, Democrat for U.S. Senate," n.d. (mid 2003), "Barack Obama—Democrat for U.S. Senate: A Chance to *Believe* Again," n.d. (post–18 August 2003), 13pp., OFI, "Integrated Strategic Plan," n.d. (post 19 September 2003), JKP; Dino P. Christenson, "Bush's Economic Stimulus Plan," 6 June 2003, 8pp., AGP; N.A., "Prescription Drugs," 9 June 2003, JKP; Gruber e-mail to Krishnamoorthi, "Tax Cuts Issues Paper," 16 June 2003, Christenson, "Economic Stimulus Initiatives," 1 August 2003, 15pp., AGP; Tom Staunton, "Health Insurance Coverage Proposals and Costs," 1 September 2003, 9pp., JKP; Bill Edley e-mail to Jim Cauley and Krishnamoorthi e-mails to Gruber and Christenson, "Patriot Corp Overview" (2), 19 September 2003, Christenson e-mail to Gruber and Gruber e-mail to Krishnamoorthi, "Top 1%," 19 and 23 September 2003, Christenson e-mail to Krishnamoorthi and David Adams e-mail to Krishnamoortki, "Patriot," 23 and 24 September 2003, AGP; Nancy MacLean and Bruce Orenstein to "Dear Friends," 18 August 2003, BOP; OFI, "Obama Calls on Conference Committee to Protect Seniors, Not Drug Companies," 21 September 2003; Dan Mihalopolous, "Presidential Contender Braun Makes Bid Official," *CT*, 23 September 2003; Bill Myers, "Rallying Cry: Defend Civil Liberties," *CDLB*, 24 September 2003, p. 1; Peter Savodnik, "Ex-Panther Rush Setting Race Aside in Backing Hull," *TH*, 24 September 2003, p. 14; *Obama Weekly Report*, 24 September 2003, TWP; Mike Pittman, "From the Editor," *CCC*, February 2007, p. 2; DJG interviews with John Kupper, Paul Harstad, Pete Giangreco, Terry Walsh, Raja Krishnamoorthi, Andrew Gruber, Dino Christenson, Dan Shomon, Bruce Orenstein, Nancy MacLean, Andrew Boron, Hasan and Raazia Chandoo, Jesse Jackson Jr., and Jon Molot.

36. Obama Block Schedule, September 2003 and October 2003, JKP; "Democratic Poll Overview," IlSenate.com, 22–24 September 2003; Shamus Toomey, "Half of Voters Undecided on Senate Pick," *CST*, 26 September 2003, p. 9; OFI, "Upcoming Events," 27 September 2003; Leah Williams, "Senator Obama Speaks at Delta Sigma Theta's Anniversary Banquet," *DE*, 29 September 2003; Lynn Okamoto, "Braun Paves Way to White House for Women in Future," *DMR*, 28 September 2003 (Obama praising Carol Moseley Braun); OFI, "Obama: 'No Blank Check on Iraq,'" 28 September 2003; UCLS *Announcements 2003–04*, esp. pp. 41, 46, 173; UCLS, *The Glass Menagerie 2003–04*, esp. pp. 33, 100; David Axelrod's 29 and 30 September 2003 interviews with Danny Davis and James Meeks; Lauren W. Whittington, "Chico Is First on Air in Crowded Senate Race," *RC*, 30 September 2003; Meg Kinnard, "Chico Debuts with Plans for Reform" and "Fourth Chico

Ad Highlights Latino Roots," NationalJournal.com, 1 and 3 October 2003; Abdon M. Pallasch, "Gery Chico and the Firm That Failed," and "Polls Show Scant Support a Year into Race," *CST*, 5 October 2003, pp. 12–13; Obama's 2 October 2003 interview with David Axelrod, JKP; DJG interviews with Mike Lawrence and Eric Adelstein. On Chico, see also Ben Joravsky, "Chico and the Man," *CR*, 20 January 2011, and Achy Obejas, "Why Gery Chico Is the White Candidate for Mayor," WBEZ, 18 February 2011.

37. OFI, "U.S. Senate Candidate Barack Obama Welcomes Very Special Guest Former U.S. Senator Bill Bradley" and "Obama Endorsed by Former U.S. Senator Bill Bradley," 2 October 2003; Raja Krishnamoorthi e-mail to John Kupper et al., "NAACP Debate Questions," 2 October 2003, JKP; *CDH*, 3 October 2003, p. I-5; "NAACP Debate," 4 October 2003, Illinois Channel DVD; Dave Newbart, "11 Senate Hopefuls Focus on Bush, War," *CST*, 5 October 2003, p. 20; Lesley R. Chinn, "Eleven Senate Candidates Debate Issues at NAACP Forum," *HPC*, 9 October 2003, p. 1; Eric Krol, "Oberweis Criticized for Raising Money at Ritzy Club," *CDH*, 6 October 2003, p. 1; *Obama Weekly Report—October 6–12*," 9 October 2003, TWP; Steve Neal, "Downstate Could Push Hynes Over the Top," *CST*, 8 October 2003; Neal, "Senate Debate Yields 4 Standouts," *CST*, 17 October 2003, p. 47; OFI, "Obama: Protect Seniors, Not Drug Companies, HMOs," *CW*, 9 October 2003, p. 4; Obama to Timuel Black, 10 October 2003, TBP Box 129 Fld. 1; Stuart Rothenberg, "There's No Sure Thing When It Comes to the Illinois Senate Race," *RC*, 14 October 2003; JCAR Minutes, 14 October 2003; OFI, FEC Form 3, Report of Receipts and Disbursements, 1 July–30 September 2003, 14 October 2003; OFI, "Obama Breaks $2 Million Fundraising Barrier," 15 October 2003; Chris Cillizza, "Senate Money Rolling In," *RC*, 16 October 2003; Rick Pearson, "Hull Off to Fast Start in Senate Campaign Spending," *CT*, 16 October 2003; OFI, "Obama Leading Legislation for METRA Improvements," 16 October 2003; Glen Justice, "In Races with One Deep Pocket, the Law Tries to Tailor a Second," *NYT*, 17 October 2003 ("Barak"); Christopher Wills, "Some Senate Candidates Already on the Air for 2004," AP, 17 October 2003; James Muhammad, "Obama, Collins Attack Hospital Collection Practices," *CD*, 21 October 2003, p. 2; Todd Spivak, "Obama Passes $2M Mark in Expensive Senate Race," *HPH*, 22 October 2003, pp. 1–2 (ten $12,000 donors as of 30 September, including Joseph Stroud and Andy Davis's wife Cynthia Wheeler); DJG interviews with Keith Kelleher, Bill Perkins, Jerry Morrison, Raja Krishnamoorthi, John Kupper, Marni Willenson, Carolyn Shapiro, and Robert Bennett.

38. David Axelrod e-mail to John Kupper, "What About This?," 15 October 2003, JKP; *Obama Weekly Update—October 13*," 17 October 2003, TWP; OFI, "Join," www.ilsenate.com/News/Media/join.mp3, Internet Archive; "Rolled" and "Promise" 60 Second Urban Radio Scripts, Obama Block Schedule, October 2003, JKP; Maudlyne Ihejirika, "Nigerian Building Professionals Host Annual Scholarship Gala," *HPC*, 23 October 2003, p. 1; *Obama Weekly Update—October 20*," 23 October 2003, TWP; Lauren W. Whittington, "Anti-Obama Site Shut Down," *RC*, 23 October 2003; AP, "Web Site Targeting Senate Candidate Withdrawn," *CDH*, 24 October 2003, p. 17; Transcript, West Side Rally to Elect Barack Obama, 1 November 2003, John Kupper daybook entry, 5 November 2003, JKP; DJG interviews with John Kupper. No audio of the second radio ad has been located.

39. Obama Block Schedule, October 2003 and November 2003, JKP; Ken Silverstein, "Barack Obama Inc.: The Birth of a Washington Machine," *Harper's*, November 2006, pp. 31–40; Robert Novak, "Clinton Ally, a Washington Super-Lawyer, Switches Allegiance to Back Obama," *CST*, 4 March 2007, p. B6; Vernon Jordan, *Make It Plain* (Public Affairs, 2008), p. 207; Tim Novak, "How Reform-Minded City Hall Critic Became a Cozy Insider," *CST*, 11 November 2007, p. A14; Allison S. Davis, "Minutes of the ISBI Committee for Emerging Manager Policy," 8 October 2003, ISBIP; James M. Hacking to Antonio Munoz, c/o Jillayne Rock, 23 October 2003, SURSP; Emil Jones Jr. to Linda Hawker, Senate Floor Transcript, 23 October 2003, pp. 1–2; Allison S. Davis, "Minutes, Meeting of the Illinois State Board of Investments," 24 October 2003, ISBIP; "Citizen Action/Illinois Endorses State Senator Barack Obama in U.S. Senate Democratic Primary Race," 25 October 2003; OFI, "Obama Endorsed by Citizen Action," 25 and 27 October 2003; Brian Wallheimer, "Policy Council Endorses Obama," *SJR*, 26 October 2003, p. 13; Rick Pearson, "Illinois Undecided on Bush Challenger," *CT*, 27 October 2003; Jennifer Koons, "Hynes Wants Nation Building at Home," NationalJournal.com, 4 November

2003; OFI, "Obama Charges Bush with Undercutting Support for Sexual Assault Victims," 28 October 2003; OFI, "Obama Announces Committee for LGBT Rights," 29 October 2003; Tracy Baim, "GLBT Group Forms for Candidate Obama," *WCT,* 5 November 2003; Dennis R. Wheeler, "Barack Obama, Running Hard for Senator," *The Star,* 30 October 2003; Mike Wiser, "Police Get Preview of New Traffic Stop Requirements," *RRS,* 30 October 2003, p. A9; *Obama Update—Week of October 26,* 31 October 2003, TWP; Steve Neal, "Pappas Has Potential to Take Hynes to Wire," *CST,* 31 October 2003, p. 39; Eric Krol, "Cook County Treasurer Joins U.S. Senate Race," *CDH,* 1 November 2003, p. 1; Scott Fornek, "Pappas Tosses Her Baton into Senate Ring," *CST,* 10 November 2003, p. 14; Obama remarks at National Urban League Annual Conference, St. Louis, MO, 27 July 2007; William Atwood in Christopher Drew and Raymond Hernandez, "Loyal Network Backs Obama After His Help," *NYT,* 1 October 2007, and in Michael Tapscott and Richard Pollock, "The Obama You Don't Know," *Washington Examiner,* 19 September 2012; Baim, *Obama and the Gays,* p. 16; transcript, "West Side Rally to Elect Barack Obama," 1 November 2003, JKP; "Introducing Barack Obama," Chicago Video Project, 2003; DJG interviews with Joe McLean, Jill Rock, William Atwood, John Rogers Jr., John Cameron, William McNary, Alvin Love, Kevin Thompson, and Bruce Orenstein. Prior pension systems' developments are reflected in Joseph P. Cacciatore, "Minutes of the Illinois State Board of Investment for Minority Broker/Manager Policy," 27 July 2001, Jane R. Patterson to Minority Sub-Committee, "Proposed Policy," 17 August 2001, Cacciatore, "Minutes," 20 August 2001, ISBIP; Jon Bauman, "Minutes, Meeting of the Investment Committee," Teachers' Retirement System of Illinois, 9 August 2001, TRSIP; Keith Cardoza to Board Members, "Proposed Minority Policy," 14 September 2001, and Cacciatore, "Minutes," 21 September 2001, p. 3 (unanimously approving the policy), ISBIP; Bauman, "Minutes," 22 October 2001, TRSIP; Laurel Martin et al. to Investment Committee, "Minority Manager Due Diligence," 24 October 2001, and Emma E. McIntosh, "Minutes," 14 November 2001, SURSP; Bauman, "Minutes," 17 December 2001 (authorizing allocation of up to $450 million to minority managers plus an additional $107.6 million to Capri Capital), Bauman, "Minutes" (2), 14 February 2002, Bauman, "Minutes" (2), 11 April 2002, TRSIP. On Maria Pappas, see especially Bryan Miller, "What Does Maria Pappas Want?," *CR,* 13 August 1992.

40. "2003–2004 IFT Questionnaire for State Senator Barack Obama," n.d., 7pp. IFTP; IFT, "Illinois Federation of Teaches Backs Obama," 5 November 2003; Chinta Strausberg, "Pushes to Veto Death Penalty Reform Bill," *CD,* 3 November 2003, p. 1; *CF,* 4 and 7 November 2003; Obama, "Obama Sets Legislative Sights on Metra Neglect," *HPH,* 5 November 2003, p. 4; Maura Kelly, "U.S. Senate Candidates Seek Younger Voters," AP, 4 November 2003; Kate Clements, "Prison Employees Seek $10.6 Million to Increase Staffing," *CNG,* 5 November 2003; Strausberg, "Blagojevich, Jones Reach Accord on Lying Cops Bill," *CD,* 6 November 2003, p. 1; Sarah Okeson, "Lawmaker Trying to Be '1 Out of 100,'" *PJS,* 6 November 2003, p. B3; Senate Floor Transcript, 6 November 2003, pp. 2–19 (SB 67) and 20 (Obama introduces SB 2120); Christopher Wills, "Senate Rejects Drivers Licenses for Illegal Immigrants," AP, 6 November 2003; Obama Interview with Terry Martin, 6 November 2003, IllinoisChannel.org; Scott Fornek, "Teachers Union Backs Obama in Senate Bid," *CST,* 7 November 2003, p. 20; Eric Krol, "Senate Hopefuls Get Key Endorsements," *CDH,* 7 November 2003, p. 19; Bernard Schoenburg, "Teachers' Union Backs Obama for U.S. Senate," *SJR,* 7 November 2003, p. 8; Mark Samuels, "Illinois Federation of Teachers Endorses Obama for U.S. Senate," *DHR,* 7 November 2003, p. A5; Schoenburg, "Painting of Ex-Gov. Ryan 'Straightforward, Simple,'" *SJR,* 9 November 2003, p. 17; Schoenburg, "Frank Talk About Drug Use in Obama's 'Open Book,'" *SJR,* 16 November 2003, p. 17; *CF,* 10 and 12 November 2003; Raja Krishnamoorthi e-mail to Andrew Gruber and Dino Christenson, 8 October 2003, Christenson, "Economic Stimulus Initiative: Offshore Tax Haven, 10 October 2003, Christenson e-mail to Krishnamoorthi, 13 October 2003, Krishnamoorthi e-mail to Christenson, 15 October 2003, Krishnamoorthi e-mail to Gruber and Christenson, "Latest," 22 October 2003, Pam Smith e-mail to Krishnamoorthi, "Draft Real Release," 5 November 2003, AGP; OFI, "Obama Offers Innovative Plan to Boost U.S. Manufacturing Jobs," 7 November 2003; Krishnamoorthi e-mail to Issues Committee, "Issues Update—Employment," 10 November 2003, Krishnamoorthi e-mail to Gruber, Tara Thompson e-mail to Gruber, "Obama Economic Plan," and Kupper e-mail to

Krishnamoorthi, "Early Education Initiative," 19 November 2003, AGP; Manya A. Brachear, "Program Provides Living Civics Lessons to High Schoolers," *CT,* 9 November 2003; Lauren W. Whittington, "Unions Split in Illinois Primary," *RC,* 13 November 2003; Schoenburg, "Obama's Meteoric Rise Hard to Predict 5 Years Ago," *SJR,* 6 November 2008; Terry Martin, "Reflections on Barack Obama," *Illinois Municipal Review,* January 2009, pp. 7–8; Schoenburg's 9 March 2009 interview with Mark DePue, pp. 94–96; DJG interviews with Steve Preckwinkle, Raja Krishnamoorthi, Andrew Gruber, and Dino Christenson.

41. Obama Block Schedule, November 2003, JKP; *Obama Update—Week of November* 5, 8 November 2003, TWP; "Illinois Is Ground-Zero in Battle for the Senate," *RC,* 10 November 2003, Paul Merrion, "Turn On, Log In, Find Out," *CCB,* 10 November 2003, p. 11; Steve Neal, "Hull Ups the Ante in Senate Bid," *CST,* 12 November 2003, p. 55; Meg Kinnard, "Chico Touts Education Credentials," NationalJournal.com, 12 November 2003; Don Wycliff, "Soul Survivor," *CT Magazine,* 16 November 2003, pp. 12ff.; OFI, "Obama Says Pending Medicare Bill Protects Drug Companies, Not Seniors," 16 November 2003; David Mendell, "Obama Lures Some Unions from Hynes," *CT,* 17 November 2003; SEIU, "SEIU Illinois Endorses Obama for U.S. Senate," 17 November 2003; Bernard Schoenburg, "Madigan Backing Hynes for U.S. Senate," *SJR,* 18 November 2003, p. 1; *CF,* 18, 20, and 21 November 2003; JCAR Minutes, 18 November 2003; JCAR, *2003 Annual Report;* Peter Savodnik, "Obama Vying for Key Union Support to Turn Back Hynes," *TH,* 19 November 2003, p. 13; Mendell, "Madigan Gives Boost to Hynes' Senate Campaign," *CT,* 19 November 2003; Scott Fornek, "Speaker Madigan Backs Hynes for Senate," *CST,* 19 November 2003, p. 17; Doug Pokorski, "Senate Candidates Gather for Forum," *SJR,* 19 November 2003, p. 14; Minutes, GARS Board of Trustees, 19 November 2003; Senate Floor Transcript, 19 November 2003, pp. 73–74 (SB 783) and 77–78 (SB 1935); OFI, "Obama Blasts Federal Energy Bill as 'Giveaway' to Big Polluters," 19 November 2003; *Obama Weekly Report,* 19 November 2003, TWP; Chuck Sweeney, "Democratic Senate Primary Winner May Surprise," *RRS,* 20 November 2003, p. A9; Ray Long and John Chase, "Gay-Rights Proposal Shelved in Legislature," *CT,* 20 November 2003; Senate Floor Transcript, 20 November 2003, pp. 43–45 (commending Pete Baroni) and 54–68 (SB 702); Long and Chase, "Landmark Ethics Law Approved," *CT,* 21 November 2003; Long and Chase, "Legislature Lurches to Wild Finish," *CT,* 22 November 2003; Travis Morse, "Police Prepare for Racial Profiling Study," *FJS,* 23 November 2003; Chinta Strausberg, "Rejected City Plan to Lead to 800 Layoffs," *CD,* 24 November 2003, p. 3; Todd Spivak, "Key Unions Back Obama," *HPH,* 26 November 2003, p. 3; Mendell, "Protesters Use Bullhorns to Give Jackson an Earful," *CT,* 26 November 2003; Strausberg, "Jesse Rally Disrupted," *CD,* 26 November 2003, p. 2; "Jackson Leads Rally for Jobs," *CST,* 26 November 2003, p. 11; Daniel Duggan, "Party Bigwigs Slow to Endorse," *ECN,* 27 November 2003, p. A1; "Church Tour," *RRS,* 30 November 2003; Sweeny, "Obama Leaves Good Impression in Forest City," *RRS,* 2 December 2003, p. A7; Stephen S. Morrill, "Client Report for 2003 Veto Session," Morrill and Associates, November 2003; "Veto Session Offers Landmark Legislation," *C-WR* 19 #1, 20 January 2004; Obama for Illinois, FEC Form 3, Report of Receipts and Disbursements, 1 July–31 December 2003, 30 January 2004, esp. pp. 246, 252; Virgil Wright, "Upstart Illinois Senator a Hit with Delegates, News Media," *BSB,* 5 August 2004, p. 2; Ray A. Coleman, *The Obama Phenomenon* (Prioritybooks Publications, 2009), pp. 20–22, 31–37; McClelland, *Young Mr. Obama,* pp. 237–39; Christopher Drew and Mike McIntire, "After 2000 Loss," *NYT,* 3 April 2007; Abramsky, *Inside Obama's Brain,* p. 87; David B. Wilkins, "The New Social Engineers in the Age of Obama," *Howard Law Journal* 53 (Spring 2010): 557–644, at 585; Ron Suskind, *Confidence Men* (HarperCollins, 2011), p. 79; DJG interviews with Martha Minow, David Wilkins, Claire Serdiuk, Peter Coffey, Pete Baroni, Sean Noble, John Cameron, Cindi Canary, Lyndell Luster, Dan Shomon, Jim Cauley, Randon Gardley, Raja Krishnamoorthi, Madhuri Kommareddi, Kaleshia Page, and William McNary.

42. Lynn Sweet, "Downstate Dem to Quite Senate Race, Back Chico," *CST,* 1 December 2003, p. 13; Kelly Quigley, "O'Shea Backs Out of Senate Race," *CCB,* 1 December 200; OFI, "Obama Announces Plan to Address AIDS Crisis in Minority Communities," 1 December 2003; Chinta Strausberg, "Obama Seeks $1 Bill to Combat AIDS," *CD,* 3 December 2003, p. 7; Jack Mabley, "Politicians Who Do What's Right—Like Fitzgerald—Are Rare," *CDH,* 3 December 2003, p. 15; OFI, "Obama Announces

Health Care Plan to Cover All Children," 3 December 2003; Strausberg, "Obama Announces $68 Billion Health Plan," *CD,* 4 December 2003, p. 2; "Back Barack, the Best of Bunch for U.S. Senate," *AWN,* 4 December 2003; *Obama Weekly Update,* 4 December 2003, TWP; Dan Shomon e-mail to Pam Smith, 4 December 2003, JKP; David Mendell, "Key Race May Tip Balance in Senate," *CT,* 7 December 2003; Obama Block Schedule, December 2003, JKP; Drew and McIntire, "After 2000 Loss," *NYT,* 3 April 2007; Bob Secter, "Obama's Fundraising, Rhetoric Collide," *CT,* 1 February 2008; UCLS course evaluations, Autumn 2003 (5.85 for Con Law and 5.73 for Race on UCLS's 7-point scale); Obama, Constitutional Law Final Examination, [10–15] December 2003; Salim Muwakkil, "Worthy of the Land of Lincoln," *ITT,* 8 December 2003, p. 11; *CDH,* 8 December 2003, p. I-9; OFI, "Obama Files 10,000 Petition Signatures," 8 December 2003; Obama to Emil Jones Jr., 8 December 2003 ("effective today"), in *Senate Journal,* 14 January 2004, p. 4; David Jackson and John McCormick, "Building Obama's Money Machine," *CT,* 13 April 2007; Antonio Munoz to Stanley Rives, 22 December 2003, Rives to J. Fred Giertz et al., "Re-establishment of the Minority and Emerging Manager Subcommittee," 29 December 2003, John R. Krimmel to Members of the Minority and Emerging Manager Subcommittee, "Agenda for the January 19, 2004 Meeting," 8 January 2004, "Minutes," 19 January 2004, Krimmel to Investment Committee Members, "Report from the Minority and Female-Owned Business Enterprise Committee," 29 January 2004, "Minutes," 17 and 18 February 2004, Krimmel, "Report and Minutes," 1 March 2004, "Minutes," 18 March 2004, SURSP; Ray Long, "Candidates Line Up for Next Run," *CT,* 9 December 2003; Strausberg, "Leaders: Medicare a 'Horror' for Seniors," *CD,* 9 December 2003, p. 3; "Illinois Lawmaker Announces," *BND,* 9 December 2003, p. B3; OFI, "Statement on the Passing of Honorable Paul Simon," 9 December 2003; AP, "Reaction to the Death of Former Sen. Paul Simon," 9 December 2003; Diane Wilkins, "Obama Unveils Job Plan While Announcing Candidacy," *MDR,* 10 December 2003; Ray Long, "A Legacy of Honesty and Dignity," *CT,* 10 December 2003; *CF,* 10 December 2003; *Obama Weekly Update,* 11 December 2003, TWP; "Simon Services Set for Weekend," *CT,* 11 December 2003; Long, "Clinton to Be at Simon Rites," *CT,* 12 December 2003; "2004 Tentative Session Calendar," *C-WR* 18 #10, 12 December 2003, p. 4; Eric Krol, "Succeeding Fitzgerald," *CDH,* 13 December 2003, p. 1; Long, "Simon Recalled as Moral Crusader," *CT,* 15 December 2003; Richard Goldstein, "Funeral Draws Well-Known Crowd," *SI,* 15 December 2003; Abdon M. Pallasch, "Professor Obama Was a Listener, Students Say," *CST,* 12 February 2007, p. 5; Pallasch, "Law Students Gave Obama Big Thumbs Up," *CST,* 18 December 2007, p. 7; "Fired Up and Ready to . . . Teach!," AbovetheLaw.com, 21 February 2008; "Brian"'s 12:59 A.M. 31 July 2008 comment (#356) on Jodi Kantor, "Teaching Law, Testing Ideas," *NYT,* 30 July 2008; Alexandra Starr, "Case Study," *NYT Magazine,* 21 September 2008; Axelrod, *Believer,* p. 143; DJG interviews with Kaleshia Page, Claire Serdiuk, Nathan Smith, Thomas FitzGibbon, Jonathan Yi, Dan Shomon, Michael Dorf, Dave Feller, Andrew Boron, Juleigh Nowinski, Jill Rock, John Kupper, Terrie Pickerill, and Sheila Simon.

43. David Mendell and Liam Ford, "Candidate Filings Find 1 Surprise Hat in Ring," *CT,* 16 December 2003; Lynn Sweet, "Schakowsky Backs Obama for U.S. Senate," *CST,* 16 December 2003, p. 20; Eric Krol, "Schakowsky Endorses Obama," *CDH,* 16 December 2003, p. 7; JCAR Minutes, 16 December 2003; OFI, "Schakowsky Endorses Barack Obama for U.S. Senate," 16 December 2003; Rebecca Ephaim, "Looking for a Knockout," *Conscious Choice,* February 2004, pp. 26–31; *Obama Weekly Update,* 17 December 2003; *CF,* 17, 19, and 22 December 2003; Steve Neal, "Obama's Endorsements Stacking Up," *CST,* 31 December 2003, p. 33; Eric Krol, "Building from the Base," *II,* January 2004; Joshua Green, "A Gambling Man," *Atlantic,* January–February 2004, pp. 34–38; Bennett, Petts & Blumenthal, "Recap of Surveys of Illinois Democratic Primary Voters," 4 May 2003, MBP; Jan Schakowsky's 2004 interview with David Axelrod; McClelland, *Young Mr. Obama,* p. 240; DJG interviews with Dan Shomon, Nate Tamarin, Michael Dorf, Jan Schakowsky, Bob Creamer, John Cameron, William McNary, Larry Suffredin, Mark Blumenthal, Rick Ridder, Anita Dunn, Fred Lebed, and Mike Henry.

44. "Focus Group—Women," and "Focus Group—Men," 13 December 2003, JKP; Cauley in Mendell, *Obama,* p. 205; Noam Scheiber, "Cruel Intentions," *TNR,* 28 May 2008; DJG interviews with Audra Wilson, Pam Smith, Jim Cauley, Randon Gardley, Pete Giangreco, Terry Walsh, John Kupper, Joe Sinsheimer, Forrest Claypool, Dan Shomon, and Michael Dorf.

45. Matt O'Connor and Ray Gibson, "Ryan Indicted," *CT,* 18 December 2003; Monica Davey, "Former Gov. Ryan of Illinois Indicted on Graft Charges," *NYT,* 18 December 2003; *CF,* 19 December 2003; Obama Block Schedule, December 2003, JKP; "Affidavit of Application for Marriage License," D520068, Clark County, NV, 21 January 2003 (performed 23 January); Bennett Guira, "Wedding of the Year," *The Rainbow Edition,* January 2004 (Educational Laboratory School); Neil Abercrombie in Stu Glauberman and Jerry Burris, *The Dream Begins* (Watermark Publishing, 2008), p. vi; Untitled budget projection sheet, n.d., ca. 2 January 2004, "4.5 million Budget," n.d., ca. 6 January 2004, "Media Discussion," 2 January [2004], David Axelrod e-mail to Paul Harstad, 2 January 2004, JKP; UCLS *Announcements 2003–04,* pp. 83, 173; Scott Fornek, "Firefighters Support Hynes for Senate," *CST,* 6 January 2004, p. 5; Paige Wiser, "The 'It' Factor," *CST,* 7 January 2004, p. 44; *Obama Weekly Update—January 7, 2004,* TWP; Roland Klose, "The Changing Face of Voters," *IT,* 8 January 2004; Axelrod e-mail to Kupper, 8 January 2004, OFI, "Downstate Strategy Memo," 9 January 2004, and "Obama for Illinois Strategy Meeting," 9 January 2004, John Kupper daybook entries, January 2004, Harstad Strategic Research, "Focus Group Outline," 8 January 2004, JKP; "Primary Election Focus Group, Older Women," "Primary Election Focus Group, Younger Women," and "Primary Election Focus Group Summary," 8 January 2004, PHP; "Raw counts of the number . . ." and "Focus Group Observations," n.d., JKP; Remnick, *The Bridge,* p. 375; Axelrod, *Believer,* p. 144; DJG interviews with Jim Cauley, Bron Solyom, Kevin Watson, Jesse Jackson Jr., John Kupper, Pete Giangreco, Paul Harstad, and Mike Kulisheck.

46. Peter Savodnik, "Ill. Labor Leaders to Decide on a Senate Endorsement," *TH,* 7 January 2004, p. 9; Steve Neal, "Hynes Poised to Snare Key Labor Backing," *CST,* 7 January 2004, p. 37; *CF,* 7 13, and 15 January 2004; Pat Guinane, "Hynes Wins Coalition Backing," *QCT,* 8 January 2004 (and *SI,* 9 January); David Mendell, "Hynes Garners AFL-CIO Backing," *CT,* 9 January 2004; Scott Fornek, "Hynes Lands AFL-CIO Endorsement," *CST,* 9 January 2004, p. 5; Eric Krol, "AFL-CIO Supports Hynes in Senate Run," *CDH,* 9 January 2004, p. I-11; OFI, "AFSCME Picks Obama," 10 January 2004; Dave McKinney and Leslie Griffy, "Obama Scores Backing from AFSCME," *CST,* 11 January 2004, p. 3; "Union Endorses Obama," *CDH,* 11 January 2004, p. I-5; Mendell, "Hynes Takes Low-Key Road in Senate Bid," *CT,* 11 January 2004; Paul Merrion, "Riches Alone Can't Win Political Office," *CCB,* 12 January 2004, p. 1; John J. Miller, "Aiming for the Hill," *NR,* 14 January 2004; Russ Stewart, "Here's New Year's Political Forecasts," *NSP,* 14 January 2004; Ramsin Canon, "Profiles of Candidates for Senate: Barack Obama, Democrat," GapersBlock.com, 14 January 2004; "House Divided," *IT,* 15 January 2004; "Illinois AFL-CIO Endorses Dan Hynes in Illinois Senate Race," *LL,* January 2004, pp. 1–2, IVI-IPO *Action Bulletin,* December 2003, p. 3; Sara Schmitt e-mail to Audra Wilson, and Raja Krishnamoorthi e-mail to Andrew Gruber and David Adams, "Questionnaire & Platform," 9 and 10 December 2003, Megan Handley e-mails to Gruber, Wilson e-mail to Gruber, David Munar e-mail to Handley, and Adams e-mail to Gruber and Wilson, "AIDS Foundation Q&A," 16 and 17 December 2003, AGP; AIDS Foundation of Chicago's Candidate Questionnaire, 12 February 2004, Leslie Gryce Sturino to Dear Candidate, 15 December 2003, pp. 3, 6, JKP; Obama, IVI-IPO 2004 U.S. Senate Questionnaire, 5 January 2004, 16pp.; Obama e-mail to John Kupper et al., 12 January 2004, Harstad e-mail to Kupper et al., "A Few Suggestions on Trib Questionnaire," 22 January 2004, JKP; Tracy Baim, "Obama Seeks U.S. Senate Seat," *WCT,* 4 February 2004; "Letters: Obama on Marriage," *WCT,* 11 February 2004; John Chase, "Senate Chase Remains Murky," *CT,* 15 January 2004; Rich Miller, "Dan Hynes Squandering His Political Capital," *GZ, MD,* and (as "Seeking Inspiration") *IT,* 22 January 2004; Baim, *Obama and the Gays,* pp. 19–26; Sheryl Gay Stolberg, "Obama's Views on Gay Marriage 'Evolving,'" *NYT,* 18 June 2011; DJG interviews with Ray Harris, Steve Preckwinkle, Margaret Blackshere, John Cameron, and Bill Fuhry.

47. *The Solon,* GARS, January 2004; Senate Floor Transcript, 15 January 2004, p. 18 (Obama introducing SB 2269); OFI, "Obama Pushes Automatic Enrollment for KidCare," 15 January 2004; Obama e-mail prefacing *Obama Weekly Update,* 15 January 2004, Teamster.net; *CNG,* 15 January 2004; [Audra Wilson], "Debate Preparation," n.d. [ca. 15 January 2004], and Wilson to John Kupper et al., "Subject Matter for Northwestern Debate on Wednesday 1/21," 15 January 2004, JKP; *CDH,* 16 January 2004, p. I-3; Tona Kunz, "Senate Hopefuls Debate Finer Points," *CDH,* 18 January 2004, p.

I-6; AP, "Chicago Democrats Take Senate Campaigns on the Road," 18 January 2004; "Upcoming Events," ObamaForIllinois.com, 19 January 2004; Daniel Duggan, "Union Rallies Around Local Candidate," *ECN,* 20 January 2004; Caleb Hale, "Democratic Candidates for U.S. Senate Debate at SIUC," *SI,* 21 January 2004, p. A1; AP, *CDH,* 21 January 2004, p. I-6; "Senate Candidates Propose Solutions to National Issues," *DE,* 21 January 2004; David More, "Going to the Candidates' Debate," *Carbondale Dispatch,* 21 January 2004; Seth Freedland, "Democratic Senate Contenders View for Voter Confidence at Tech," *DN,* 22 January 2004; George Olwenya, "Siaya Ecstatic Over Would-Be Illinois Senator," *EAS,* 22 January 2004; *Obama Weekly Report,* 22 January 2004, TWP; Jen Haberkorn, "Obama '04 at Odds with Obama '08," *Washington Times,* 1 February 2008, p. A1; DJG interviews with Jef Pollock, Chris Mather, and Dan Hynes.

48. David Mendell, "Obama Banks on Credentials, Charisma," *CT,* 25 January 2004; Mendell, *Obama,* pp. 198–203, 219–23; OFI, FEC Form 3, Report of Receipts and Disbursements, 1 October–31 December 2003, 29 January 2004; Friends of Barack Obama, Report of Campaign Contributions and Expenditures, 1 July–31 December 2003, 2 February 2004; Mendell and John Chase, "Pappas Trails in Race for Funds," *CT,* 1 February 2004; Tamara E. Holmes, "Will Mr. Obama Go to Washington?," *BE,* February 2004, p. 26; Michael Miner, "The Kiss and the Cover-Up," *CR,* 19 March 2004; Hendon in Todd Spivak, "In the Black," *IT,* 25 March 2004; Christi Parsons and Ray Long, "Wheeler-Dealer with a Cause," *CT Magazine,* 19 September 2004; Tracy Robinson-English, "Statehouse Clout," *Ebony,* April 2005, pp. 150ff.; Aaron Chambers, "Unofficial Advisors Create a Stir," *RRS,* 11 January 2004; Patrick J. Fitzgerald, "Government's Motion for Admission of Other Acts of Evidence," *U.S. v. Antoin Rezko,* N. D. Ill. #05-CR-691, 11 January 2008, p. 18 (Maloof cooperated with prosecutors); Matt Baron, "Covering a Phenomenon," *Wednesday Journal* and *Forest Park Review,* 25 March 2008; Ryan Lizza, "Making It," *New Yorker,* 21 July 2008; Clark in David Smallwood, "Two Guys Who Laid It on the Line," *N'Digo Profiles,* December 2008, pp. 14–15; Hendon, *White Enough/Black Enough,* p. 63; Dupuis and Boeckelman, *Barack Obama,* p. 48; Rebecca Janowitz, *Culture of Opportunity—Obama's Chicago* (Ivan R. Dee, 2010), p. 218; Eric Lipton, "Ties to Obama Aided in Access for Big Utility," *NYT,* 23 August 2012; DJG interviews with John Rogers Jr., Mike Gecan, Randon Gardley, Don Moore, Cindi Canary, Jan Schakowsky, Emil Jones Jr., Frank Clark, John Hooker, John Kelly, Courtney Nottage, Dave Gross, Jill Rock, Cindy Huebner Davidsmeyer, Debbie Halvorson, Lou Viverito, Mike Lieteau, John Cameron, Bill Perkins, Jim Cauley, Dan Shomon, Claire Serdiuk, and Chris Mather.

49. "Upcoming Events," ObamaForIllinois .com, 19 January 2004; OFI, "Obama Endorsed by Democratic Party of Evanston," 26 January 2004; Anne Broache, "Dems Give Support to Obama," and Jerome C. Pandell, "Candidates Still Confident After Endorsement Choice," *DN,* 26 January 2004; Richard Goldstein, "Most Senate Candidates Keep Campaign Offices Close to Home," *SI,* 26 January 2004; John Kupper daybook entries, January 2004, David Axelrod e-mail to Obama, "A Few Thoughts for Trib," 25 January 2004, JKP; John Chase and David Mendell, "Candidates Find Their Foe in Bush," *CT,* 27 January 2004; Audra Wilson e-mail to John Kupper et al., "Debate Prep for 2004 City Club of Chicago," and Axelrod and Kupper to the Obama Team, "Strategy for City Club Debate," 27 January 2004, JKP; City Club of Chicago Forum, 28 January 2004, CAN-TV, Illinois Channel DVD; Bob Seidenberg, "Evanston Democrats Back Obama in U.S. Senate Race," *ER,* 29 January 2004; Scott Fornek, "Democrats Tout Experience in Cordial Senate Debate," *CST,* 29 January 2004, p. 22; Mendell, "Hynes Says Charge That He Laundered Funds Ridiculous," *CT,* 29 January 2004; Eric Krol, "Hynes Defends Himself After Questions About Finances," *CDH,* 29 January 2004, p. I-11; Steve Neal, "Obama Has Good Shot at Senate Seat," and Thomas Roeser, "If Dems Pick Best Candidate," *CST,* 30 January 2004, p. 43; DJG interview with Whit Soule.

50. Stacy St. Clair, "Do You Know These People?," *CDH,* 27 January 2004, p. 1; Terry Walsh e-mail to Mike Kulisheck et al., "Questionnaire Drafting for Tracking Poll," 13 January 2004, JKP; Harstad Strategic Research, "Track 1: Illinois Statewide Democratic Primary Survey," 14–18 January 2004, JKP and PHP; *CF,* 19 January 2004; Bennett, Petts & Blumenthal, "Final Illinois Statewide," BPB 1643, January 2004, RRP; Bennett, Petts & Blumenthal, "Recap of Surveys of Illinois Democratic Primary Voters," 4 May 2004, MBP; *CDH,* 26 January 2004, p. IV-2, 30 January 2004, p. I-3, and 1 February

2004, p. III-2; "Tentative Media Schedule," 31 January 2004, JKP; Mick Dumke, "Beyond the Base," "Conceding No Demographic," and "Slim Chances," *Chicago Reporter,* February 2004, pp. 8ff., 18–19; Peter Schuler, "A Senate Story," *University of Chicago Magazine,* February 2004; Salim Muwakkil, "Obama's Legacy Is Likely to Ignite a Progressive Firestorm," *Trumpet,* February 2004, pp. 28–29; Eric Krol, "Where Candidates Stand on Iraq War, Security," *CDH,* 1 February 2004, p. 1, John Kupper daybook entries, February 2004, OFI Block Schedule, February 2004, JKP; Kevin McDeermott, "Illinois Senate Race Remains Up for Grabs," *SLPD,* 2 February 2004, p. A1; AP, "Democrats in Senate Race Want Options for Immigrants," *SI,* 2 February 2004; Lauren W. Whittington, "Land of Mystery," *RC,* 3 February 2004; Bernard Schoenburg, "Condolences," *SJR,* 1 February 2004, p. 31; "Beverly Ann Criglar, 63," *CT,* 4 February 2004; Senate Floor Transcript, 3 February 2004, p. 2 (Obama introduces SR 394); Senate Floor Transcript, 4 February 2004, pp. 3, 5 (Obama introduces SBs 2579, 2581, 2608, 2609, 2610, and 2611); John Chase and David Mendell, "Opponents Put Hynes on Defensive Over Fundraising, War Stand," *CT,* 5 February 2004; Scott Fornek, "Dem Rivals Bash Hynes' Support for Iraq War," *CST,* 5 February 2004, p. 6; Eric Krol, "Democrats Battle Over Iraq War," *CDH,* 5 February 2004, p. 1; *CF,* 5 February 2004; Senate Floor Transcript, 6 February 2004, pp. 14, 18 (Obama introduces SBs 3088 and 3147); Dick Durbin to "Dear Friend," 6 February 2004, JKP; Senate Floor Transcript, 10 February 2004, p. 5; "Governor Blagojevich Makes Education Focus of Second State of the State Address," *C-WR* 19 #2, 29 February 2004, pp. 3–4. No complete recording or transcript of the 4 February WBEZ debate seems to survive.

51. Terrie Pickerill e-mail to John Kupper, "Obama Shoot," 21 January 2004, JKP; Pickerill to Jim Cauley and Juleigh Nowinski, "REV Shoot Details," 4 February 2004, TPP; Axelrod's 2004 interviews with Mike Pfleger and Jan Schakowsky; John Kupper's 2004 interviews with Jeremiah Wright and Michelle Obama; "Schedule, February 7, 2004," TPP; Obama, transcript of remarks, Heartland Cafe, Rogers Park, and transcript of remarks, Mount Zion Missionary Baptist Church, Evanston, 7 February 2004, JKP and TPP; E. Janet Wright, "Obama for Illinois," *Trumpet,* February 2004, pp. 30–32; David Axelrod's 7 February 2004 interview with Obama, JKP and TPP; John Del Cecato e-mail to Axelrod and Kupper, "Potential Script from Heartland Cafe?," 9 February 2004, JKP; Jarrett in Wolffe, *Renegade,* p. 123.

52. "Tentative Media Schedule," 31 January 2004, OFI Block Schedule, February 2004, JKP; Dennis Duggan, "Searching for Dem Votes in Kane County," *ECN,* 9 February 2004, p. A6; Chuck Sweeny, "Senate Hopefuls Blast Bush, Republicans," *RRS,* 9 February 2004, p. B3; David Mendell, "Hull Proves Money No Object in Bid for Senate," *CT,* 10 February 2004; *CF,* 10 February 2004; Joe Sinsheimer e-mail to David Axelrod, "Situational Analysis," and Sinsheimer e-mail to John Kupper, "Hull Fines," 10 February 2004, JKP; Rich Miller, "Maybe It's Right Time for Obama to Run," *CST,* 8 December 2006, p. 43; Ted McClelland, "How Obama Learned to Be a Natural," *Salon,* 12 February 2007, and *Young Mr. Obama,* pp. 227–29; Ali Abunimah, "How Barack Obama Learned to Love Israel," *Electronic Intifada,* 4 March 2007; Coleman, *The Obama Phenomenon,* p. 30; Presta, *Mr. and Mrs. Grassroots,* p. 152; DJG interviews with Ab Mikva, Liz Drew, John Cameron, Dan Hynes, Kevin Watson, Warren Chain, David Kindler, Mike Kruglik, and Paul Scully.

53. "Tentative Media Schedule," 31 January 2004, OFI Block Schedule, February 2004, Pam Smith e-mail to John Kupper, "Press," 29 January 2004, JKP; SurveyUSA, "IL Senate Primaries," IlSenate.com, 8 February 2004; OFI, "Obama Gains Support of New Trier Dems," 9 February 2004; John Kupper notes, "Celinda [Lake]," 10 February [2004], JKP; Russ Stewart, "Hull, Obama Surge in Democrats' Senate Race," *NSP,* 11 February 2004; OFI, "Obama Picks Up Four Key Latino Endorsements," 11 February 2004; Elaine Hopkins, "Obama: Parents Must Be Involved in Schools," *PJS,* 11 February 2004, p. B5; Joshua Kolar e-mail to Andrew Gruber et al., "Obama," 11 February 2004, AGP; Lauren W. Whittington, "The Great Black Hope," *RC,* 12 February 2004; Andrew Schroeder, "New Trier Democrats Endorse Kerry, Obama," *WL,* 12 February 2004; Terrie Pickerill memo to Jim Cauley and Juleigh Nowinski, "2/17/04 Shoot Details," 12 February 2004, TPP; *CF,* 13 February 2004; Steve Edwards, "Democratic Senate Candidate Barack Obama," *Eight Forty-Eight,* WBEZ, 13 February 2004; OFI, "U.S. Rep. John Conyers Stumps for Obama," 13 February 2004; Kristen McQueary, "Black Caucus Leader Backs Obama for Senate," *DS,* 14 February 2004; OFI, "Champion" and "March" Urban Radio Ads, n.d., JKP and

TWP; John Patterson, "Betting on Name Recognition," *CDH,* 15 February 2004, p. 15; David Mendell, "Political Novice Hull Uneasy in Spotlight," *CT,* 15 February 2004; Laura Washington, *CST,* 16 February 2004, p. 41; Jeff Berkowitz, "Making It a Hat Trick," Blogspot.com, 16 February 2004; Michael Sneed, "Sneed," *CST,* 17 February 2004, p. 4; "Shooting Schedule," 17 February 2004, JKP; Michelle Obama to "Dear Friends," ObamaForIllinois.com, 17 February 2004; Jeremy Adragna, "Big Dollars Boost Obama's Chances as Primary Looms," *HPH,* 18 February 2004, pp. 1–2; "Analysis of the Governor's Budget—Part I," *C-WR* 19 #3, 14 April 2004; SB 2800 Bill Status, 5, 19–20, and 26 February and 26 and 28 April 2004; Senate Floor Transcripts, 24 February 2004, p. 17, and 20 May 2004, p. 6; Eric Zorn, "Public Entitled to Know About Hull's History," and Mendell, "Hull Infuses Campaign Chest with $5 Million," *CT,* 19 February 2004; *Obama Weekly Update,* 19 February 2004, TWP; Mike Flannery, "Hull Faces Questions of Domestic Abuse," CBS2Chicago.com, 19 February 2004, Internet Archive; Eric Krol, "Candidate Refuses to 'Clear the Air,'" *CDH,* 20 February 2004, p. 15; Kate Clements, "Legislation Seeks Charity Care Changes," *CNG,* 20 February 2004 (SB 2579); Pam Smith e-mail to John Kupper, 20 February 2004, JKP; Eric Krol, "Then and Now: Obama's Views on Gay Rights," *CDH,* 26 March 2007, p. 1; Remnick, *The Bridge,* pp. 371, 374; McClelland, *Young Mr. Obama,* p. 253; Axelrod, *Believer,* p. 143; Axelrod in Amy Chozick, "When He Walks Out of That Building," *NYT Magazine,* 15 February 2015; DJG interviews with Bill Haine, Chris Mather, Jim Cauley, Audra Wilson, Pam Smith, Kevin Watson, Randon Gardley, Pete Giangreco, Terry Walsh, John Kupper, Joe McLean, and Terrie Pickerill.

54. OFI, "State's Largest Teamsters Local Backs Obama in U.S. Senate Race," 15 February 2004; *Obama Weekly Update,* 19 February 2004, TWP; Rick Pearson, "Steve Neal, 54," *CT,* 20 February 2004; David Mendell, "Blair Hull Defends Self on Divorce Incident," *CT,* 21 February 2004; OFI, "Young Voters Rally for Obama," 21 February 2004; Debra Schommer Klein e-mail to Kupper, 21 February 2004, JKP; Terrie Pickerill scheduling e-mails and "TV Scripts," n.d., ca. 20 February 2004, TPP; "2004 Campaign Polling: Illinois Senate," NationalJournal.com, 21–22 February 2004; Dave Orrick, "Hull Responds to Report of Domestic Violence," *CDH,* 22 February 2004, p. 17; Bernard Schoenburg, "Special Interest Foe Hull Helps Ex-Wife Land State Job," *SJR,* 22 February 2004, p. 15; Andrew Herrmann and Scott Fornek, "Wealthy Hopefuls Lead Senate Race," *CST,* 22 February 2004, p. 3; John Chase, "Pappas Is Picking Up 1 Vote at a Time," *CT,* 22 February 2004; OFI, "Former Clinton Campaign Manager Endorses Obama," 22 February 2004; John Chase, "TV Spots Pay Off in Ryan, Hull Senate Bids," *CT,* 23 February 2003; *CF,* 23 February 2004; Axelrod & Associates, "Yes," 23 February 2003, JKP and YouTube; SB 866, 3 July 2001, PA 92-0048; SB 130, 30 June 2003, PA 93-0063; SB 4, 18 August 2003, PA 93-0534; SB 15, 12 August 2003, PA 93-0517; Greg Hinz, "Dems Hit Local Airwaves," *CCB,* 23 February 2004; Adriana Colindres, "Obama Sees Himself as Fresh Voice on Issues," Copley News Service, 23 February 2004; Madhuri Kommareddi e-mail to Terry Walsh et al., "Obama Consultant Call Tuesday," 23 February 2004, JKP; "Obama Hits Airwaves on Road to Victory," *CD,* 24 February 2004, p. 2; Lauren W. Whittington, "Hull Goes for Broke," *RC,* 25 February 2004; John Bachtell, "Illinois Senate Race Offers Historic Opportunity," *PWW,* 28 February 2004; John S. Jackson, "The Making of a Senator," Southern Illinois University, August 2006; Presta, *Mr. and Mrs. Grassroots,* p. 154, 156; McClelland, *Young Mr. Obama,* p. 245; DJG interviews with Rick Ridder, Anita Dunn, Mike Henry, Jason Erkes, Don Wiener, Sheila Simon, and Madhuri Kommareddi.

55. OFI Block Schedule, February 2004, JKP; *Obama Weekly Update,* 19 February 2004, TWP; David Mendell and Molly Parker, "Opponents Take Aim at Hull on His Divorce, Other Issues," *CT,* 24 February 2004; Bernard Schoenburg, "Senate Candidates Spar in Springfield," *SJR,* 24 February 2004, p. 9; Scott Fornek, "Hull Was Late Opposing War, Obama Says," and "Hull Says He's OK with Releasing Divorce Record," *CST,* 24 February 2004, p. 6; Christopher Wills, "Abuse Accusations Against Hull Take Center Stage," AP, 24 February 2004; John Patterson, "Democrats Press Hull for Details on Protection Order," *CDH,* 24 February 2004, p. 5; *CF,* 24, 25, and 27 February 2004; Dorothy Schneider, "Officials: Clean Air Act Haunts Illinois," *SJR,* 25 February 2004, p. 12; OFI, "Sheila Simon, Daughter of the Late Sen. Paul Simon, Endorses Obama for U.S. Senate," 25 February 2004; Steve Horrell, "Simon Backing Obama," *EI,* 26 February 2004; Diane Wilkins, "Simon Endorses Obama," *MDR,*

26 February 2004; Schoenburg, "Obama Gets Endorsement from Simon's Daughter," *SJR,* 26 February 2004, p. 12; "2004 Campaign Polling: Illinois Senate," NationalJournal.com, 29 February 2004 (Hynes' Global Strategy Group, polling 800 likely voters February 22–24, had Hull leading Hynes and Obama 21–19–19); "Obama, Post-Debate, Overtakes Hull in IL Democrat Sen Primary," SurveyUSA, 24 February 2004, 8pp.; Schoenburg, *SJR,* 26 February 2006, p. 7; Andrew Herrmann, "New Poll Shows Obama Pulling Ahead of Hull," *CST,* 26 February 2004, p. 30; Eric Zorn, "The Real Hull Slips Out from Behind the Ads," *CT,* 26 February 2004; Senate Floor Transcript, 26 February 2004, p. 4 (Obama introduces SR 441); Terry Walsh e-mail to Obama et al., "Revisiting Topline Vote Goals," and John Del Cecato e-mail to David Axelrod et al., "New Poll," 26 February 2004, JKP; Zorn, "Hull Relents: 'Read My Files!,'" Change of Subject Blog, ChicagoTribune.com, 26 February 2004, 1:10 P.M.; Mendell, "Hull, Ex-Wife Ask Judge to Unseal Divorce Files," *CT,* 27 February 2004; Michael Sneed, "Sneed," Frank Main, "Police Report: Wife Said She Kicked Hull, He Hit Her," and "This Democratic Senate Pick Can Hit the Ground Running," *CST,* 27 February 2004, pp. 4, 10, 55; Eric Krol, "Spenders Lead Race for Senate," and "Hull to Release Divorce Records," *CDH,* 27 February 2004, pp. 1, 5; Schoenburg, "Hull, Wife Want Divorce File Unsealed," *SJR,* 27 February 2004, p. 15; Chris Rickert, "Democrats Obama, Hynes Take Similar Stands on the Issues," *DDC,* 27 February 2004; United Power for Action and Justice, "Citizens' Report on Candidates for U.S. Senate," March 2004, pp. 5–6 (26 February 2004); Bennett, Petts & Blumenthal, "Recap of Surveys of Illinois Democratic Voters," 4 May 2004, MBP; Remnick, *The Bridge,* p. 375; Coleman, *The Obama Phenomenon,* pp. 43, 85; Axelrod, *Believer,* p. 145; DJG interviews with Bill Haine, Evelyn Bowles, Rick Ridder, Anita Dunn, Mike Henry, Jason Erkes, Laura Russell Hunter, Dan Shomon, Fred Lebed, Claire Serdiuk, and Dan Hynes. No complete recording or transcript of the 23 February debate seems to survive.

56. "Affidavit of Brenda Sexton Hull," *Hull v. Hull,* #35363, Cook County Circuit Court, 12 March 1998, 7pp., JKP and TWP; Eric Zorn, "The Hull Files," Change of Subject Blog, ChicagoTribune.com, 27 February 2004, 1:10 P.M.; Rick Ridder to Kathy Hull, "Current Controversy," [27 February 2004], RRP; David Mendell, "Hull's Ex-Wife Called Him Violent Man in Divorce File," and Zorn, "Hull's Biggest Mistake Isn't in Divorce File," *CT,* 28 February 2004; Frank Main, "Hull's Dirty Laundry on the Line," *CST,* 28 February 2004, p. 4; Rob Olmstead and Eric Krol, "Papers Show Hull Threats, Profanity," *CDH,* 28 February 2004, p. 1; Daniel Duggan, "One-Liners Show Deep Differences with President," *ECN,* 28 February 2004, p. A1; OFI, "Barack Obama Endorsed by Sun-Times AND Tribune," 28 February 2004; "For the Democrats: Obama," and John Kass, "Hull Learning Nothing Fair in Illinois Politics," *CT,* 29 February 2004; Michael Sneed, "Deciphering Blair Hull," *CST,* 29 February 2004, p. 4; Krol, "Hull Campaign Not Ruined, Experts Say," *CDH,* 29 February 2004, p. 1; Bernard Schoenburg, "Costly Primary," and "U.S. Senate Democratic Candidates: Barack Obama," *SJR,* 29 February 2004, pp. 1, 4; Mike Ramsey, "U.S. Senate Hopefuls Make Pitch Outside Chicago," *ODT,* 29 February 2004; Chuck Sweeny, "Democratic Candidates for U.S. Senate," *RRS,* 29 February 2004, p. H4; "Primary Vote an Important One," and "Barack Obama," *JHN,* 29 February 2004; OFI, "Star Newspapers Endorse Obama," and "Service Union Rally for U.S. Senate Candidate Barack Obama Draws 1,000," 29 February 2004; "Mentor," JKP, TWP, and JCP; LGBT Committee for Obama, Speakeasy Supper Club Brunch, 29 February 2004; Nicole Saunders, "The Next Black Senator?," *Essence,* March 2004, p. 32; Dan Johnson-Weinberger, "Picking the Good Ones," *TCP,* March 2004, p. 3; Michael Ibrahem, "U.S. Senate Candidate Barack Obama," *Residents' Journal,* March–April 2004; Krol and John Patterson, "Oberweis Not Backing Down on Immigration," *CDH,* 1 March 2004, p. 9; "Illinois Needs Senator with Track Record, Experience," *BP,* 1 March 2004, p. A8; *CF,* 1 March 2004; Scott Fornek, "Barack Obama," *CST,* 1 March 2004, p. 6; Mendell, "Hull Criticizes Drugmakers, but Heavily Invests in Them," *CT,* 1 March 2004; Mundy, *Michelle,* p. 147; Shomon in Remnick, *The Bridge,* p. 377; McClelland, *Young Mr. Obama,* p. 245; DJG interviews with Nate Tamarin, Paul Harstad, John Kupper, Randon Gardley, and Jim Cauley. Subsequent speculation seeking to credit David Axelrod with a decisive role in the public release of Sexton's deposition reflected both exaggeration and implications of false credit-claiming. *CF,* 4 March 2004 ("It's highly doubtful that any of his opponents had much, if anything to do with the story's genesis"), Mendell, "Obama Lets Opponent Do Talking,"

CT, 24 June 2004 ("the Democrat's campaign worked aggressively behind the scenes to fuel controversy about Hull's filings"), *CF,* 30 August 2004 ("some prominent media types wrongly credited Dan Hynes with the kill"), Mendell in Ben Smith, "Obama's Style: Go Negative, Stay Clean," *Politico,* 3 November 2011 ("Axelrod and the Obama Senate campaign played no public role in unsealing the Hull or Ryan divorce records. But behind-the-scenes machinations were more complicated. . . . You really have to connect the dots to pin things on Axelrod"), and DJG interviews with Audra Wilson, Don Wiener, John Kupper, Terrie Pickerill, Pete Giangreco, Terry Walsh, and Toni Preckwinkle. Thanks to opposition researcher Don Wiener, Blair Hull's campaign was directly responsible for the FEC imposing a $275,000 fine on James Chao, an executive who reimbursed his APEX Healthcare employees for $69,500 in donations to Dan Hynes, and a $76,500 penalty on Hynes's campaign. Mendell, "Exec Fined $275,000 over Donations to Hynes Campaign," *CT,* 29 April 2005, and Rick Pearson, "Hynes' Senate Campaign Will Pay FEC to Settle Complaints," *CT,* 20 December 2006.

57. Bennett, Petts & Blumenthal, "Recap of Surveys of Illinois Democratic Primary Voters," 4 May 2004, MBP; Harstad Strategic Research, "Track 2: Illinois Statewide Democratic Primary Survey," 29 February–1 March 2004, JKP and PHP; "Left to Spend as of March 1," Obama Block Schedule, March 2004, John Kupper daybook entries, March 2004, JKP; John Chase and Molly Parker, "NOW Puts Hull on the Defensive Over Divorce File," *CT,* 2 March 2004; *CF,* 2 and 3 March 2004; Jennifer Koons, "Chico, Pappas Focus on Education & Jobs," NationalJournal.com, 2 March 2004; "2004 Campaign Polling: Illinois Senate," National Journal.com, 1–3 March 2004; OFI, "Obama Raises Another $1.3 Million Since Jan. 1st," 2 March 2004; David Mendell and Liam Ford, "Hull Calls in Press, Then Takes a Hike," *CT,* 3 March 2004; Christopher Wills, "In Senate Race, Government Experience Comes in All Flavors," AP, 3 March 2004; Rich Miller, "Blair Hull Has a Really Big Problem," *GZ,* 4 March 2004; Lauren W. Whittington, "Man and Machine," *RC,* 4 March 2004; Peter Savodnik, "Obama-Ryan Match a 'GQ Race to Watch,'" *TH,* 4 March 2004; "Mixed Messages," *IT,* 4 March 2004; Chase and Mendell, "Divorce Is Now an Issue for GOP," *CT,* 4 March 2004; OFI, FEC Form 3, Report of Receipts and Disbursements, 1 January–25 February 2004, 1 March 2004; Koons, "Hynes Details Plans for Health Care, Jobs," NationalJournal.com, 4 March 2004; Channel 7 ABC TV, 4 March 2004, MBC; Scott Fornek, "Contribution from MJ Helps Obama Campaign Fund Soar," *CST,* 5 March 2004, p. 9; "Michael Likes Obama Too," *CST,* 8 March 2004, p. 31; Chris Fusco and Robert C. Herguth, "Gov Adviser Quietly Held Casino Hand," *CST,* 7 March 2004, p. 7; Fusco, "Gov Rejects Request to Stay Out of Gaming Regulation," *CST,* 23 March 2004, p. 9; Fusco et al., "Gov Taps D.C. Lawyer to Probe Casino Ruling," *CST,* 25 March 2004, p. 12; Dave McKinney, "Gov. Fund-Raiser Linked to Man in Indictment," *CST,* 7 August 2005, p. A10; Robert Elder, "Brokerage Leader Helping Others Invest in Obama," *AAS,* 23 February 2007, p. B6; Sandra Sobieraj Westfall and Maureen Harrington, "No Ordinary Contender," *People,* 19 February 2007, p. 83; Patrick J. Fitzgerald, "Government's Motion for Admission of Other Acts of Evidence," *U.S. v. Antoin Rezko,* N. D. Ill. #05-CR-691, 11 January 2008, pp. 10 and 17 (Aramanda cooperated with prosecutors); Korecki, *Only in Chicago,* p. 144; DJG interviews with Rick Ridder, Mark Blumenthal, Anita Dunn, Mike Henry, Blair Hull, Chris Mather, Jef Pollock, Dan Hynes, Freddrenna Lyle, and Rob Fisher.

58. Burt Constable, "Best Candidate Has Worst Name," *CDH,* 2 March 2004, p. 9; Scott Fornek and Andrew Herrmann, "Senate Rivals Urge Ryan to Unseal Divorce Records," and Mary Mitchell, "Don't Black Politicians Know Who Their People Are?," *CST,* 4 March 2004, pp. 9 and 14; Phil Ponce, City Club Debate, *Chicago Tonight,* WTTW, 4 March 2004; Matthew Miller, *The 2% Solution* (Public Affairs, 2003), esp. pp. xi, 53, 262; Ron Ingram, "Obama States His Case in Decatur for Senate Nomination," *DHR,* 5 March 2004, p. A3; Rick Pearson, "Divorce Turns into Albatross of Senate Race," *CT,* 5 March 2004; Herrmann and Fornek, "Obama Leads Pack for Dem Senate Nod," Curtis Lawrence and Fornek, "Hull Says He Struck Wife to Defend Himself," and Dave McKinney, "Obama Pals Block Bill Helpful to Rival Hynes," *CST,* 5 March 2004, pp. 9, 10, 32; John Patterson, "Messy Divorce Tale Told by Hull," *CDH,* 5 March 2004, p. 1; Kurt Erickson, "Poll: Hull Slips; Ryan Far Ahead," *BP,* 5 March 2004, p. A1; Mike Robinson, "Democrat Obama Grabs Lead in Fresh Poll," AP, 5 March 2004; Kristen McQueary, "Obama Surges to Lead in Southtown Poll," *ECN,* 6 March 2004; Eric Krol, "Your Senate Choice: Millionaire or Fund Raiser," *CDH,* 7 March 2004, p. 1; "Handshakes Aren't Like Rubbing Elbows," *CT,* 7 March 2004; Lauren W. Whittington, "Final Days for Fightin' Illini," *RC,* 9 March 2004; Ellen Warren, "Scenes from a Senate Campaign," *CT,* 12 March 2004; "Rod McCulloch's Response to Jeff Berkowitz," ArchPundit.com, 3 July 2004; DJG interviews with Rod McCulloch, Andy Davis, and Anita Dunn.

59. Channel 7 ABC TV, 5 and 6 March 2004, MBC; John Chase and John Bebow, "Ex-Wife Refuses to Back Down from Hull," *CT,* 6 March 2004; Andrew Herrmann and Scott Fornek, "Hull's Ex-Wife Tells Her Side," *CST,* 6 March 2004, p. 2; Eric Krol, "Hull Responds in Ad About His Divorce," *CDH,* 6 March 2004, p. 4; Monica Davey, "Closely Watched Illinois Senate Race Attracts 7 Candidates in Millionaire Range," *NYT,* 7 March 2004; Chuck Sweeny, "Obama, Hynes Talk Jobs in Race for Senate," *RRS,* 7 March 2004, p. B1; David Mendell, "Proud of 'Battle Scars,' Chico Still in the Fight," *CT,* 7 March 2004; Dorothy Schneider, "Hynes Makes Campaign Stop in City," *SJR,* 7 March 2004, p. 11; John J. Miller, "Racing for the Senate," *NR,* 8 March 2004; Chinta Strausberg, "Obama, Davis, Others Remember 'Bloody Sunday,'" and Joe Ruklick, "Obama Holding Slight Lead in Democratic Senate Nomination Race," *CD,* 8 March 2004, pp. 2 and 4; Krol, "Obama Says Autobiography Teaches Lessons," *CDH,* 8 March 2004, p. 11; Mark Samuels, "Fight for Clean Air Hits Illinois Floor," *QCT,* 8 March 2004; *CF,* 8 March 2004; John Bebow, "Hull's Ex-Wife Wants a Different Spotlight," *CT,* 10 March 2004; Carol Slezak, "Heel or a Hero?," *CST,* 10 March 2004, p. 51; "Manpower vs. Machine," *Philadelphia Magazine,* March 2007; Marcia Gelbart, "Obama Helps Fattah Seek Some Late Funds," *PI,* 8 May 2007; Kenneth P. Vogel, "'Mystery' Man Lends Support to Obama," *Politico,* 19 October 2008; Paul Schwartzman, "A Political Strategist with D.C. on His Mind," *WP,* 8 April 2014; DJG interviews with Jim Cauley, Nate Tamarin, Kevin Watson, David LeBreton, Kevin Thompson, Lyndell Luster, Emil Jones Jr., John Kelly, Al Kindle, Tom Lindenfeld, and Greg Naylor.

60. "Choice for U.S. Senate," *CDH,* 7 March 2004, p. 16; "United States Senate—Democrat: Barack Obama," *PJS,* 7 March 2004, p. A4; "For U.S. Senate," *SLPD,* 7 March 2004, p. B2; "Endorsement: U.S. Senate, Democrats," *ABN,* 7 March 2004, p. D2; OFI, "Obama Picks Up Endorsements from the *Daily Herald, St. Louis Post-Dispatch, Champaign News-Gazette,* and the *Peoria Journal Star,*" 7 March 2004; Harstad Strategic Research, "Track 3: Statewide Democratic Primary Survey," 7–8 March 2004, JKP and PHP; Rick Pearson, "Obama, Ryan Out Front," Eric Zorn, "Disparagment of Obama Votes Doesn't Hold Up," and Joseph Kellman, "2 Hopefuls Are Primarily Long Shots," *CT,* 9 March 2004; *CF,* 9 March 2004; John Patterson, "Pro-Choice Advocates Defend Obama Votes," *CDH,* 10 March 2004, p. 15; Bennett, Petts & Blumenthal, "Recap of Surveys of Illinois Democratic Primary Voters," 4 May 2004, MBP; Blair Hull direct mail, "WDM," "MSHC," and "HUL0469," "Joyce Washington Democrat for U.S. Senate," SM&M, Obama direct mail, "BO_PRO1," "BO_PRO3," and "BO_A1cr," JKP, PGP, and RRP; "Faith" and "Assault" urban radio ads, JKP, TWP, and TPP; DJG interviews with Paul Harstad, Mike Kulisheck, Mark Blumenthal, Rick Ridder, Pete Giangreco, Terry Walsh, John Kupper, and Terrie Pickerill.

61. Eric Fidler, "Obama Takes Lead in Polls, Donations" and "Barack Obama Moves Front and Center in Illinois Democratic Senate Race," AP, 9 and 10 March 2004; Russ Stewart, "Hull's 'Collapse' Has Senatorial Precedent," *NSP,* 10 March 2004; "For U.S. Senate," *MD,* 10 March 2004, p. A4; "Hyde Park Citizen's Endorsements,"*HPC,* 10 March 2004, p. 2; John Chase and David Mendell, "Candidate Without Faults Is a Rarity," *CT,* 10 March 2004; Audra Wilson e-mail to John Kupper et al., "ABC 7 Debate," 20 January 2004, JKP; ABC 7 News, 10 March 2004, 10 P.M.; Ron Carter, "Barack Obama for U.S. Senator," *SSJ,* 11 March 2004, p. 11; "Obama for Democrats," *DPT,* 11 March 2004; Lauren W. Whittington, "Rivals Race to Catch Obama," *RC,* 11 March 2004; Harold Henderson et al., "15 Candidates! We Can Help," Michael Miner, "The View from Downstate," and Ben Joravsky, "Why Won't He Commit?," *CR,* 11 March 2004; Todd Spivak, "Head of the Class," *IT,* 11 March 2004; Chinta Strausberg, "Pols: 'Kerry's the Man to Defeat Bush,'" *CD,* 11 March 2004, p. 2; Chase and John McCormick, "Hull Concedes He Took Cocaine in Early '80s," *CT,* 11 March 2004; Abdon M. Pallasch, "Hynes Pounces on Obama at Last Debate," *CST,* 11 March 2004, p. 10; Eric Krol, "Hynes Criticizes Obama for Votes on Spending," *CDH,* 11 March 2004, p. 7; Mike Ramsey, "Hynes, Obama Clash Over Budget at Debate," *SJR,* 11 March 2004,

p. 11; "Hynes Takes Shot at Obama at Forum," *PJS,* 11 March 2004; Mike Robinson, "Democratic Senate Hopefuls, in Home Stretch, Exchange Shots," AP, 11 March 2004; OFI, "Sierra Club and LCV Endorse Obama," 5 February 2004; "LCV Mounts Independent Campaign on Behalf of Barack Obama for U.S. Senate," 9 March 2004; Public Citizen's "New Stealth PACs" archive at www.stealthpacs.org; Jennifer Koons, "LCV Praises Obama's Enviro Record," NationalJournal.com, 11 March 2004; OFI, "Obama Places Major Ad Buy in Metro East," 11 March 2004; "Hope," BUO-04-17, JCP and TPP; "Faith" (DNA: Did Not Air), TWP and TPP; "Barack OBAMA U.S. Senate Yes, We Can!," PGP; David Axelrod e-mail to Audra Wilson et al., 11 March 2004, JKP; "The Five Questions," *CJN,* 12 March 2004; Harold Meyerson, "A Bright Hope in Illinois," *WP,* 12 March 2004; AP, "Candidates for U.S. Senate Make Last Push," *CDH,* 12 March 2004, p. I-12; Kurt Erickson, "Obama, Ryan Pulling Away in Senate Races," *BP,* 12 March 2004, p. A1; Jeff Smyth, "Hynes Says U.S. Senate Primary Down to Two," *SI,* 12 March 2004; Matt O'Connor et al., "U.S. Indicts Consumer Advocate," *CT,* 12 March 2004; Dave Newbart, "Schakowsky's Husband Indicted in Bank Fraud," *CST,* 12 March 2004, p. 6; Laurie Cohen, "For a Public Watchdog, Questions of Politics, Principles," *CT,* 6 June 1993; Jeff Berkowitz, "Buzzzzz," Blogspot.com, 12 March 2004; Jennifer Koons, "Obama Touts Past Work with Simon," NationalJournal.com, 12 March 2004; OFI, "The Final Countdown to Victory," Bob Chinn's Crabhouse, 12 March 2004, 6 P.M.; "Barack Obama Is Our Choice," "Our Endorsements," and "Barack Obama," *CD,* 13 March 2004, p. 1, 19, 25; "Democrat: Barack Obama," *RRS,* 14 March 2004, p. H1; John Chase and David Mendell, "Senate Rivals Struggle to Wash Off Mud Stains," *CT,* 14 March 2004; Jonathan Goldman, Executive Director, Illinois Environmental Council, "2004 Environmental Scorecard," n.d.; Rick Pearson and Ray Long, "Careful Steps, Looking Ahead," *CT,* 3 May 2007; Mendell, *Obama,* p. 240; Presta, *Mr. and Mrs. Grassroots,* p. 156; DJG interviews with Claire Serdiuk, Kaleshia Page, Andy Davis, Randon Gardley, Jim Cauley, Pete Giangreco, Terry Walsh, John Kupper, Terrie Pickerill, Jeremy Flynn, Jack Darin, Pat Welch, Howard Learner, and Jonathan Goldman. No audio or video recording or transcript of the 10 March WLS and League of Women Voters debate seems to survive, despite MBC having copies of most WLS news broadcasts. The Internet Archive did not save either WLS's or the Illinois League's Web pages during March 2004.

62. *CF,* 11 and 12 March 2004, "Statement by Rod McCulloch," 11 March 2004, 3:13 P.M., IllinoisLeader.com; John Chase and Liam Ford, "Bid Aimed at Derailing Ryan," and "Mr. Ryan, Clear the Air," *CT,* 12 March 2004; Scott Fornek and Andrew Herrmann, "Borling Fires Aide Who Leaked Claims About Ryan," *CST,* 12 March 2004, p. 11; "Money and Brains," *Economist,* 13 March 2004, p. 34; Dan Stumpf, "Obama Captures Student Support at NU," *DN,* 14 March 2004; P. J. Huffstutter, "Sordid Pasts, Hefty Pockets at Fore of Illinois Senate Race," *LAT,* 14 March 2004; John Chase and Liam Ford, "Candidates Scramble for 11th Hour Support," and Rick Pearson, "Not Why, Who," *CT,* 14 March 2004; Abdon M. Pallasch, "Can Machine Hand It to Hynes with Obama in Lead?," *CST,* 14 March 2004, p. 2, Eric Krol, "No Negative TV Ads, but Race Got Nasty Anyway," *CDH,* 14 March 2004, p. 1; Bernard Schoenburg, "Obama, Ryan Pace Race for Senate," *SJR,* 14 March 2004, p. 1; "Poll: Ryan, Obama Hold Senate Leads," *PJS,* 14 March 2004; "Barack Obama Runs for U.S. Senate Seat," *Jet,* 15 March 2004, p. 6; Eric Ferkenhoff, "In Ill. Senate Race, Character Is the Issue," *BG,* 15 March 2004; Ofelia Casillas and H. Gregory Meyer, "Even a Dog Gets in Act as Senate Race Ends," and Dawn Turner Trice, "Obama Unfazed by Foes' Doubts on Race Question," *CT,* 15 and 16 March 2004; Scott Fornek, "Dems in Final Sprint for Senate Race," *CST,* 15 March 2004, p. 7; Eric Krol, "Senate Candidates Make Final Sweep of the State," *CDH,* 15 March 2004, p. 1; Vernon Jarrett, "This Is the Odyssey of One Man's Vote for Barack Obama for U.S. Senate," *CD,* 15 March 2004, p. 2; Maura Kelly, "Jackson Makes Appeal for Obama," AP, 15 March 2004; *CF,* 15 March 2004; Bennett, Petts & Blumenthal, "Recap of Surveys of Illinois Democratic Primary Voters," 4 May 2004, MBP; OFI, "Statement from Barack Obama on the Passing of Vernon Jarrett," 24 May 2004; Ferman Beckless, "Reflections on the Life of Vernon Jarrett," *CD,* 25 May 2004, p. 3; Jeff Berkowitz, "Fair and Balanced with Mark Brown?" and "Correction," Blogspot.com, 29 and 30 June 2004; Debra Pickett, "Sunday Lunch with Dan Hynes," *CST,* 26 December 2004, p. 18; Cauley in Remnick, *The Bridge,* p. 381; DJG interviews with Dave Sullivan, Katrina Emmons Mulligan, Jim Cauley, Dan Hynes, Nate Tamarin, Adam

Stolorow, Pete Giangreco, Jerry Morrison, and Mike Henry.

63. *CF,* 15 March 2004; Lauren W. Whittington, "Obama, Ryan Lead Late Polls," *RC,* 16 March 2004; Rick Pearson and James Janega, "Candidates Hopeful to the End," and Eric Zorn, "Election Day Tidbits," *CT,* 16 March 2004; Fornek, "Precinct Workers Hit the Streets," and Mark Brown, "Election Day Gives Homeless Chance to Earn a Buck," *CST,* 16 March 2004, pp. 13, 2; Chinta Strausberg, "Leaders: 'Turnout Key to Victory,'" *CD,* 16 March 2004, p. 3; OFI, FEC Form 3, Report of Receipts and Disbursements, 25 February–31 March 2004, 14 April 2004, esp. pp. 591–92, and 1 April–30 June 2004, 14 July 2004, esp. p. 1737; DJG interviews with Jim Cauley, Madhuri Kommareddi, Tom Lindenfeld, Greg Naylor, Pete Giangreco, Al Kindle, and Adam Stolorow.

64. Madhuri Kommareddi e-mail to Terry Walsh et al., "Obama Consultants Call Tuesday 11AM," 15 March 2004, JKP; Obama remarks and transcript, JCP, JKP, and PHP; WGN, Fox, CBS, and NBC Chicago TV news excerpts, 16 March 2004, JKP; David Mendell, "Obama Routs Democratic Foes," Eric Zorn, "Victory Party Puts Tiny Crack in Obama Calm," and "The Senate Race of a Generation," *CT,* 17 March 2004; Scott Fornek, "Obama Defeats Hull's Millions, Hynes' Name," and Mary Mitchell, "As the Machine Sputters, the People Are Heard," *CST,* 17 March 2004, pp. 2, 3; Eric Krol, "Suburban Voters Boost Ryan, Obama to Big Wins," *CDH,* 17 March 2004, p. 1; Chinta Strausberg, "OBAMA WINS BIG," *CD,* 17 March 2004, p. 3; Doug Finke, "Two Emerge from Pack of Senate Candidates," and Bernard Schoenburg, "Obama vs. Ryan," *SJR,* 17 March 2004, p. 1; Doug Wilson, "State Senator to Face Former Teacher in Race for U.S. Senate," *QHW,* 17 March 2004, p. A11; David S. Broder, "In Illinois, a Contest of Contrasts," *WP,* 17 March 2004; Fornek, "Obama's Appeal Spans Racial Lines," *CST,* 18 March 2004, p. 9; Doug Wilson, "Democratic U.S. Senate Candidate Obama Makes a Stop in Quincy," *QHW,* 18 March 2004, p. A7; OFI, "Obama Takes Downstate Unity Tour," 18 March 2004; Jeremy Adragna, "Obama Wins Rush's Support After Landslide Victory," *HPH,* 24 March 2004, p. 3; Tracy Douglas, "Democratic Senate Candidate Rallies for Younger Voters in C-U," *DI,* 5 April 2004; Jeff Smyth, "Democrats Rally for Obama," *SI,* 22 June 2004; Obama remarks transcript, Carlinville, IL, 2 July 2004, JKP; Elizabeth Brackett, "Rising Star," *PBS NewsHour,* 27 July 2004; Carol Felsenthal, "My Neighbor's Surprise Meeting with Obama," *CM,* 13 August 2012; David Axelrod's 2009 interview with Robert Draper; DJG interviews with Dan Shomon, David LeBreton, Jim Cauley, Claire Serdiuk, Kaleshia Page, Liz Drew, Dan Hynes, Chris Mather, Paul Harstad, and Sheila Simon.

65. David Mendell, "Obama Routs Democratic Foes," and Glenn Jeffers and Rex W. Huppke, "Blacks United Behind Obama," *CT,* 17 March 2004; Mark Brown, "Voters Warmed to Obama, the Next Hot Politician," *CST,* 17 March 2004, p. 2; Jonathan Tilove, "That's Obama with a 'B,'" Newhouse, 17 March 2004; *CF,* 17 and 19 March 2004, "It's Kerry and Obama," *LL,* March 2004, pp. 1–2; Chinta Strausberg, "Now, Obama to Take on Bush," *CD,* 18 March 2004, p. 2; Mendell, "Ryan, Obama Enter New Ring," and Gary Washburn and H. Gregory Meyer, "Hynes' Loss Puts Machine in Doubt," *CT,* 18 March 2004; Eric Krol, "Campaign Starts Early for Ryan, Obama," and "Stage Set for Good Senate Campaign," *CDH,* 18 March 2004, pp. 1, 14; Tom Polansek, "No Rest for the Winners," *SJR,* 18 March 2004, p. 7; Kurt Erickson, "Nominee Obama Takes Message Downstate," *BP,* 18 March 2004, p. A1; Todd Spivak, "The Contenders," *IT,* 18 March 2004; Jeff Smyth, "Obama Stops in Region on Victory Tour of State," *SI,* 18 March 2004; Shawn Clubb, "Obama Stops in Area to Thank Supporters," *AT,* 18 March 2004; Monica Davey, "As Quickly as Overnight, a Democratic Star Is Born," *NYT,* 18 March 2004; Lauren W. Whittington, "Gentlemen from Illinois," *RC,* 18 March 2004; Chuck Sweeny, "Jobs Talk of Senate Candidates," *RRS,* 19 March 2004; Debbie Howlett, "Dems See a Rising Star in Illinois Senate Candidate," *USAT,* 19 March 2004; Aaron Chambers, "Theories on Support Accrue in Senate Race," *RRS,* 20 March 2004, p. A7; Lynda Arakawa, "U.S. Senate Candidate Has Hawaii Roots," *HA,* 19 March 2004; Al Swanson, "Analysis: Obama New Kind of Democrat?" UPI, 19 March 2004; Peter Serafin, "Punahou Grad Stirs Up Illinois Politics," *HSB,* 21 March 2004; Kurt Erickson, "Bracket Creep Symptom of Hullacious Election," *BP,* 21 March 2004, p. C1; Eric Krol, "Jack Ryan Has Uphill Battle for Black Voters," *CDH,* 22 March 2004, p. 1; Russ Stewart, "'Democratic Wing' Fuels Obama's Senate Triumph," *NSP,* 24 March 2004; Rich Miller, "Barack Obama Wins Big," *GZ,* 25 March 2004; Mick Dumke, "Crossing Over," *Chicago Reporter,* April 2004, pp. 19ff.; Dan Johnson-Weinberger, "Primary Opportunities," *TCP,* April 2004, p. 3; Eric Ferkenhoff, "Democrats Pinning High Expectations on an Illinois Senator," *BG,* 3 April 2004; Stewart, "Hynes' White House Plan Dimmed by Loss," *NSP,* 28 April 2004; Dumke, "Running on Race," *Colorlines Magazine,* 22 September 2004, pp. 17ff.; DJG interviews with Pam Smith, Tarak Shah, Peter Coffey, Emil Jones Jr., Dan Shomon, and David LeBreton.

66. "Protecting Choice and Achieving Gender Equity," ObamaForIllinois.com (26 March, 9 May, and 9 September 2004); Farr Curlin e-mail to Obama, "Congratulations and a Concern," 17 March 2004, and Obama to Curlin, 18 March 2004; Obama, "'Call to Renewal' Keynote Address," 28 June 2006; Obama, *TAOH,* pp. 195–98; Obama, Commencement Address at the University of Notre Dame, 17 May 2009, *Public Papers of the Presidents* 2009, Vol. 1, p. 660; Greg Jaffe, "The Quiet Impact of Obama's Christian Faith," *WP,* 22 December 2015; Jann S. Wenner, "A Conversation with Barack Obama," *Rolling Stone* 1056/57, 10 July 2008; The Strategy Group to Barack Obama and Jim Cauley, "Turnout and Performance Review," 22 March 2004, JKP; Christopher Hayes, "Check Bounce," *TNR,* 17 March 2004; Steen, "Self-Financed Candidates and the 'Millionaires' Amendment,'" in Malbin, *Election After Reform,* pp. 204–18; Andrew Malcolm, "Sen. Obama Might Be Just Obama Without Law Written by Sen. McCain," LATimesBlog, 27 June 2008; Dupuis and Boeckelman, *Barack Obama,* pp. 22–23, 49; Reynolds in David Smallwood, "Genesis of the Grassroots Movement," *N'Digo Profiles,* December 2008, pp. 10–11, 44; Dunn in Remnick, *The Bridge,* p. 380; Eric Fidler, "Barack Obama Moves Front and Center," AP, 10 March 2004; *CF,* 19 March 2004; Russ Stewart, "Hynes' White House Plan Dimmed by Loss," *NSP,* 28 April 2004; Don Rose, "Beyond Race—or Not," *CT,* 2 May 2004; Kirk Dillard and Mike Flannery, "At Issue," WBBM Newsradio 780, 18 July 2004, JKP; McClelland, *Young Mr. Obama,* p. 255; DJG interviews with Claire Serdiuk, Kevin Thompson, Liz Drew, Dan Hynes, Don Harmon, Chris Mather, Jef Pollock, Blair Hull, Amita Dunn, Raja Krishnamoorthi, Rick Ridder, and Mark Blumenthal.

67. Axelrod & Associates to Obama, "Finishing the Job," 16 March 2004, JKP; David Plouffe, *The Audacity to Win* (Viking, 2009), p. 8; DJG interviews with John Kupper, Forrest Claypool, Dan Shomon, Joe McLean, Pete Giangreco, Terry Walsh, Jim Cauley, Peter Coffey, Claire Serdiuk, Nate Tamarin, Pam Smith, Audra Wilson, Kaleshia Page, Liz Drew, Madhuri Kommareddi, Susan Shadow Ransone, Scott Kennedy, Adam Stolorow, Randon Gardley, Lyndell Luster, Kevin Watson, and Ray Harris.

68. Fran Eaton, "GOP Tops Meet at 'Morning After' Breakfast Wednesday," *IL,* 17 March 2004; Liam Ford and Gary Washburn, "Ryan Quits 2 Corporate Boards," *CT,* 19 March 2004; DSCC to Democratic U.S. Senate Nominee and Staff, "Jack Ryan Research Summary and Recommendations," 19 March 2004, John Kupper to Paul Harstad and Mike Kulisheck, "Thoughts on March Poll," 19 March 2004, JKP; Thomas Roeser, "Think Kerry Is Liberal? Get a Load of Obama," *CST,* 20 March 2004, p. 18; Kurt Erickson and Scott Miller, "The Race Is On," *BP,* 21 March 2004, p. B1; Kevin McDermott, "Ryan's Divorce Woes Just Won't Die," *SLPD,* 21 March 2004, p. B1; Trevor Jensen, "In Illinois, It's Still Pay to Play," *Adweek,* 22 March 2004, p. 14; Chinta Strausberg, "Obama to Bush: 'Coalition of the Willing Is a Paper Wish,'" *CD,* 22 March 2004, p. 2; Lauren W. Whittington, "Down on the Farm," *RC,* 23 March 2004; John Kupper daybook entries, March 2004, Kupper e-mail to Harstad, "Qualities," 23 March 2004, JKP; Senate Floor Transcript, 24 March 2004, p. 8; Mick Dumke, "Beyond the Base," *Chicago Reporter,* February 2004, pp. 8ff.; Senate Floor Transcript, 25 March 2004, pp. 225–26 (SB 2579); "Preliminary Media Budget for Obama for Senate—General Election," n.d. [24 March 2004], JKP; Tara Malone, "Fighting for Funding," *CDH,* 25 March 2004, p. 1; Todd Spivak, "In the Black," *IT,* 25 March 2004; Senate Floor Transcript, 26 March 2004, pp. 126–27 (SB 3007); Ray Long and Molly Parker, "Gun Curbs Suffer Setbacks," *CT,* 27 March 2004; Channel 7 ABC TV, 27 March 2004, MBC; OFI, "Obama and SEIU Pass Landmark Legislation to Guarantee Free and Low Cost Health Care for the Uninsured," 28 March 2004; David Mendell, "Obama Has Center in His Sights," *CT,* 27 April 2004; Spivak, "Barack Obama and Me," *Houston Press,* 28 February 2008; Harstad in Abramsky, *Inside Obama's Brain,* p. 39; Axelrod, *Believer,* p. 154; DJG interviews with John Kupper, Nate Tamarin, Claire Serdiuk, John Nicolay, Frank Watson, Tommy Walsh, Ed Petka, Martin Sandoval, Paul Williams, Terry Walsh, Joe McLean, Paul Harstad, and Todd Vandermyde.

69. Obama's 27 March 2004 interview at CathleenFalsani.com; Falsani, "'I Have a Deep

Faith,'" *CST,* 5 April 2004, p. 14; Falsani, *The God Factor* (Farrar, Straus, Giroux, 2006), pp. 45–51; Silverstein in Pauline Yearwood, "Obama and the Jews," *CJN,* 12 September 2008; Obama in Lisa Miller and Richard Wolffe, "'I Am a Big Believer in Not Just Words, but Deeds and Works,'" *Newsweek,* 12 July 2008; Rosemary Ellis, "GH Talks with Michelle Obama," *Good Housekeeping,* November 2012; DJG interviews with Ira Silverstein and Scott Turow.

70. Erin P. Billings, "CBC Boosts Obama," *RC,* 29 March 2004; Eric Krol and John Patterson, "GOP, Ryan Begin to Take Aim at Obama," *CDH,* 29 March 2004, p. I-15; Obama on Tavis Smiley's TV show and Smiley's NPR radio show, 29 March 2004; Chinta Strausberg, "Obama Takes Aim at President Bush Again," *CD,* 30 March 2004, p. 2; Liam Ford, "Some Ryan Divorce Files Should Be Unsealed," *CT,* 30 March 2004; *CF,* 30 and 31 March 2004; Scott Turow, "The New Face of the Democratic Party—and America," *Salon,* 30 March 2004; J. Michael Kennedy, "Speech Gives a Boost to Obama's 'Dreams,'" *LAT,* 6 August 2004; Charles Leroux, "The Buzz Around Obama's Book," *CT,* 6 August 2004; Janny Scott, "Obama's Story, Written by Obama," *NYT,* 18 May 2008; David Axelrod's 2009 interview with Robert Draper; Obama, *DFMF* (Three Rivers Press, 2004), pp. vii–xii; Russ Stewart, "Ryan's Weak Showing Bodes Ill Versus Obama," *NSP,* 31 March 2004; John Harwood, "Presidential Politics Overshadows Rise of State-Level Stars," *WSJ,* 31 March 2004; Nick Klagge, "Barack Obama, CC '83, Up for Senate," *CS,* 31 March 2004; "Obama Tells Supporters 'Distinctions Could Not Be Clearer,'" *IL,* 31 March 2004; Eddie Pont e-mail to Andrew Gruber, "Jack Ryan," 31 March 2004, AGP; Senate Floor Transcript, 31 March 2004, pp. 25–31 (SB 1645, PA 93-0672, 2 April 2004); *KDJ,* 1 April 2004; "Illinois Sends a Strong Message to Bush—NO Overtime Pay Cuts," *LL,* March 2004, pp. 1–2; Errol Louis, "A Democratic Star Is Born," *NYS,* 2 April 2004; Nate Tamarin e-mail to Paul Harstad et al., "Scheduling Issues," 2 April 2004, JKP; Gruber to Raja Krishnamoorthi, "Ryan Etc.," 5 April 2004, AGP; Matt Tallmer, "Thoughts on Obama's Memoir" and "Potential Factual Problems in Obama's Memoir," 30 March 2004, "Obama's Memoir—The Race Issue" and "Obama & Project Vote," 7 April 2004, "Obama's Law Clients," 3 April 2004, "Review of Obama's Law Clients," 6 April 2004, "Major Obama Donors," 8 April 2004, "Obama Corporate & Union Donors," 12 April 2004, "Obama's Institutional Backers: A Who's Who of Corporate and Union Malfeasance, and Companies That Cost Illinois Jobs," 16 April 2004, "Major Donors to Obama's 2004 Senatorial Campaign," 18 April 2004, "Obama Connections to E2 Nightclub," 24 April 2004, [Gary Maloney], "Obama Legislation—92nd G.A.," 36 pp., n.d., and "Obama Legislation—93rd G.A.," 81 pp., n.d., [Brian McNeill], "Barack Obama: Media Report," 29 April 2004, 253 pp., [Debbie] Lounsberry, "Barack Obama—The Record: A Summary," 10 May 2004, ["State Senate"], "Obama Votes—Lonely Votes, 1997–2003," n.d., JRORF; Bernard Schoenburg, "Obama's Team," *SJR,* 2 May 2004, p. 23; Michael Futch, "Fayetteville Woman Is On Board with Obama," *Fayetteville Observer,* 3 February 2008 (Jenny Yeager); Mark Leibovich, "Between Obama and the Press," *NYT Magazine,* 21 December 2008 (Gibbs); Will Rahn, "The Jack Ryan Files," *Daily Caller,* 25 September 2012; Vince Coglianese, "The Jack Ryan Files," *Daily Caller,* 26 September 2012; DJG interviews with Jim Cauley, Claire Serdiuk, John Kupper, Pete Giangreco, Terry Walsh, Amanda Fuchs Miller, Joe McLean, Scott Turow, and Debbie Lounsberry.

71. Harstad Strategic Research, "Illinois Statewide General Election Survey," 25–31 March 2004, JKP; "Illinois Statewide General Election Survey," 25 pp. pie charts, PHP; "Target Voters: Illinois Statewide General Election Survey," JKP; "2004 Campaign Polling: Illinois Senate," NationalJournal.com, 25–31 March 2004; Harstad to Obama for U.S. Senate Campaign, "Barack Obama Starts Off with a 19% Lead Over Jack Ryan," 2 April 2004, PHP; Scott Fornek, "Obama: Back Off Divorce Files," *CST,* 3 April 2004, p. 4; Tracy Douglas, "Democratic Senate Candidate Rallies for Younger Voters in C-U," *DI,* 5 April 2004; Obama, *Chicago Tonight,* WTTW, 5 April 2004; Bill Burton and Audra Wilson, "John Kerry Joins Barack Obama to Talk Jobs and Economy in Chicago," 8 April 2004, "Talking Points for Job Training Tour with John Kerry," n.d., and Kim Molstre to Senator Kerry, "Conversation on Jobs/Economy with Barack Obama," 8 April 2004, JKP; Jill Zuckman, "Illinois Democrats Ring Up at Least $2 Million for Kerry Run," *CT,* 9 April 2004; Eric Krol, "Kerry Collects $2 Million at Chicago Dinner," *CDH,* 9 April 2004, p. 13; "Kerry, Obama Outline New Direction for America," U.S. Newswire, 9 April 2004; Fornek, "Obama, Kerry Plug Each Other at Chicago Job-Training Center," *CST,* 10 April 2004, p. 10; Zuckman, "Kerry Gets Questions on War, Not Jobs," *CT,* 10 April 2004; Lois Romano, "Kerry Sprinkles Jobs Message with Attacks on Iraq Policy," *WP,* 10 April 2004; Glen Johnson, "Kerry Calls for Larger Role for Allies," *BG,* 10 April 2004; Fornek, "Obama's Poll Puts Him Far Ahead of Ryan," *CST,* 12 April 2004, p. 7; Eric Krol and John Patterson, "Kerry Visit a 'Tease' to Durbin," *CDH,* 12 April 2004, p. 11; *CF,* 12 April 2004; OFI, "Upcoming Downstate Events," 12 April 2004; John Kupper daybook entries, April 2004, Kupper notes, "Barack 4-12," "Barack-Downstate," n.d., "Barack poll," n.d., "Obama for Illinois Consultant Retreat Agenda," 13 April 2004, JKP; Salim Muwakkil, "Shades of 1983," *ITT,* 26 April 2004, p. 13 ("rising star"); Krol and Patterson, *CDH,* 26 April 2004, p. 11; Harstad fax to Kupper, 29 April 2004, Cauley e-mail to Vera Baker et al., "Monday Strategy Call at 7:30PM," 7 May 2004, JKP; Pete Hisey, "Center Stage," *Occidental Magazine,* Fall 2004, pp. 24–29; Christopher Drew and Mike McIntire, "An Obama Patron and Friend Until an Indictment," *NYT,* 14 June 2007; Bob Secter, "The Rezko Trial—Levine: Obama a Guest at '04 Party for Tycoon," *CT,* 15 April 2008; James A. Barnes, "Word of Mouth Fueled Obama's Star Turn," *NJ,* 25 August 2008; Corrigan in Remnick, *The Bridge,* pp. 384–85; Corrigan's 2013 interview with Daniel Urman; DJG interviews with Robert Bennett, Paul Harstad, John Kupper, Pete Giangreco, Peter Coffey, Amanda Fuchs Miller, Nora Moreno Cargie, Kevin Thompson, Jeremiah Posedel, and Jack Corrigan.

72. AP, *CDH,* 14 June 2003, p. I-11; Lawrence A. Horwich, Barack H. Obama, and Michelle L. Obama, 2003 Form 1040, 15 April 2004; William Finnegan, "The Candidate," *New Yorker,* 31 May 2004; Dori Meinert, "Senators Release Financial Records," *SJR,* 15 June 2004, p. 17 (stating Obama listed "no liabilities"); Obama, U.S. Senate Financial Disclosure Report, 15 June 2005; Dorothy A. Brown, "Lessons from Barack and Michelle Obama's Tax Returns," *Tax Notes,* 10 March 2014, pp. 1109–13; Joel Lanceta, "Obama Prepares to Run for U.S. Senate," *Chicago Maroon,* 17 February 2004; UCLS, *The Glass Menagerie 2003–04,* pp. 33, 100, *2004–05,* pp. 27–28; UCLS, Obama courses summary sheet, n.d., ca. 2007; Mendell, *Obama,* pp. 151–52; Michael Lipkin, "The Professor and the President," *Chicago Maroon,* 7 November 2008; Jeffrey Rosen, "Obama: The Teacher President," *Washingtonian,* January 2009; Kwame Anthony Appiah, "Obama: Professor President," BBC Radio 4, January 2009; Carol Felsenthal, "The Making of a First Lady," *CM,* February 2009; Tom Hundley, "The Ivory Tower of Power," *CT Magazine,* 22 March 2009; DJG interviews with Dennis Hutchinson, Geof Stone, Eric Posner, David Strauss, Douglas Baird, and Saul Levmore. Bart Schultz, "Obama's Political Philosophy," *Philosophy of the Social Sciences* 39 (June 2009): 127–73, can be discounted in its entirety, and multiple UCLS faculty concur that Cass Sunstein "was not particularly close to him."

73. Sara Nelson to Christina Angarola et al., "Updated Schedule," 12 April 2004, JKP; Paul Hampel, "Senate Candidate Obama Says He Will Campaign More in Metro East," *SLPD,* 15 April 2004; "Durbin, Senate Candidate Obama Tout Ethanol Bill," *BND,* 15 April 2004; Eric Fodor, "Candidate Barack Obama: President Bush Has Created a Big Mess," *HDR,* 16 April 2004; Caleb Hale, "Defending Coal Throughout Illinois," *SI,* 16 April 2004; OFI, "Upcoming Events," 16–17 April 2004; Tom Polansek, "Senate Candidates Wage War of Words," *SJR,* 16 April 2004, p. 14; Dave McKinney, "Ryan's Chart Is Off," *CST,* 16 April 2004, p. 6; Liam Ford and David Mendell, "Ryan, Obama Side by Side in Race for Election Funds," *CT,* 16 April 2004; Bernard Schoenburg, "Jack Ryan Woefully Unprepared for Attack on Obama," *SJR,* 18 April 2004, p. 21; Eric Krol and John Patterson, "Candidate's Numbers Don't Add Up at Press Conference," *CDH,* 19 April 2004; SB 1028, 21 May 1999, and SB 1903, 31 May 2003; *CF,* 19, 20, 21, and 28 April and 3 and 13 May 2004; Dana Jay, "Teaching Experience Helps Shape Ryan's Political Career," *CCC,* 19 April 2004; Jim Muir, "Southern Illinois Needs Jobs, Not Political Rallies," *SI,* 20 April 2004; Obama to Timuel Black, 21 April 2004, TBP Box 129 Fld. 1; Todd Spivak, "Ready, Aim, Misfire," *IT,* 23 April 2004; Mick Dumke, "A Lesson in Politics," *Chicago Reporter,* April 2004; Minutes, GARS Board of Trustees, 28 April 2004; Obama to Jack Lavin, 28 April 2004, ILDCEO FOIA; Ford, "Stand on Labor Sets Ryan Apart," *CT,* 30 April 2004; Vera Baker e-mail to Jim Cauley et al., "Text for Bundling Site," 30 April 2004, JKP; Tom Polansek, "Winning Strategies Differ Among Black Politicians," *SJR,* 3 May 2004, p. 1; Dan Shomon e-mail to Robert Gibbs et al., "Legislative Update for This Week—Controversial Bills," 3 May 2004, JKP; *Inside Politics,* CNN, 4 May 2004, Schoenburg, "U.S. Senate Candidates Present Plans to Improve State Business," *SJR,* 5 May 2004, p. 11, Schoenburg, "Jack Ryan Sticking with Creative Tax-Increase Story," *SJR,* 6 May 2004, p. 7, Rich Miller, "This Isn't Dixie," *IT,* 6 May 2004, Dan Proft Memorandum to Political Reporters, "Outside the Mainstream: Barack Obama Truth Squad Report #4," 7 May 2004, JKP (no #s 1–3 seem to survive); "A GLBT Salute to Senator Hillary Rodham Clinton, Senator Jon Corzine, Senator Richard J. Durbin and State Senator Barack Obama," 7 May 2004; Lisa Ellman and Rafi Jafri, "DNC Young Professionals' Event," *DLC21C Report,* Summer 2004, p. 5; Muir, "Durbin Says Defense Secretary Should Resign," *SI,* 8 May 2004; Krol and Patterson, "Campaign Notebook," *CDH,* 10 May 2004, p. 11; Michael Martinez, "Ryan's Ex-Wife Signs on to Keep Divorce Files Sealed," *CT,* 11 May 2004; Lauren Shepherd, "Obama Takes 16-Point Lead in New Poll," *TH,* 12 May 2004, p. 3; Scott Richardson, "Obama Stops in County," *BP,* 12 May 2004, p. A3; Senate Floor Transcript, 13 May 2004, pp. 118 (HB 4371) and 135–40 (HB 4566); "Bush Campaign Names Edgar as State Chief," *CT,* 15 May 2004; Mark Samuels, "Classic Matchup," *SI,* 16 May 2004; Ron Ingram, "Democratic U.S. Senate Candidate Obama Speaks in Decatur to Trade Group," *DHR,* 16 May 2004; "Illinois Senate Primary Victors Prepare for November," *RC,* 17 May 2004; OFI, "On 50th Anniversary of Landmark Decision, Obama Proposes Adding 25,000 Teachers to High-Need Schools," 17 May 2004; "Obama Wants 25,000 New Teachers for High-Need Schools," *CD,* 18 May 2004, p. 2; Kristen McQueary, "Poll: Obama Ahead of Ryan by 8 Points in Senate Race," *CST,* 18 May 2004, p. 4; ObamaBlog.com, 18, 19, and 20 May 2004; HB 2268, 19 February 2003; *Routt v. Barrett,* 396 Ill. 322, 337 (22 January 1947); HB 2268, Senate Committee Amendment No. 1, 24 April 2003; Obama remarks to AFL-CIO Civil, Human and Women's Rights Conference, YouTube, 2003 (the exact date is uncertain and remains unconfirmed); Joe Ruklick, "Obama Holding Slight Lead in Democratic Senate Nomination Race," *CD,* 8 March 2004, p. 4; HB 2268, Senate Floor Amendment No. 2, 11 May 2004; Senate Floor Transcript, 12 May 2004, p. 67; HB 2268, Senate Floor Amendment No. 4, 13 May 2004; Senate Floor Transcript, 19 May 2004, pp. 31–49 (HB 2268, PA 93-0973, 20 August 2004) and 64 (Obama introduces SR 565); Rodney Hart, "Candidate Prefers Campaign with Personal Touch," *QHW,* 20 May 2004, p. A9; "Health Care Study," *SJR,* 21 May 2004, p. 10; OFI, "Obama Celebrates Passage of Illinois Citizen Soldier Initiative," 28 May 2004; Hisey, "Center Stage," *Occidental Magazine,* Fall 2004, pp. 24–29; "Oprah Talks to Barack Obama," *O Magazine,* November 2004; Scott Helman, "In Illinois, Obama Dealt with Lobbyists," *BG,* 23 September 2007; Jonathan Cohn, "Obama's Clean Bill of Health," *TNR,* 30 January 2008, pp. 15–17; Amy Goodman, "Dr. Quentin Young, Longtime Obama Confidante," DemocracyNow.org, 11 March 2009; Angie Drobnic Holan, "Obama Statements on Single-Payer Have Changed a Bit," Politifact.com, 16 July 2009; Michael D. Shear and Ceci Connolly, "In Illinois, a Similar Fight Tested a Future President," *WP,* 9 September 2009; William Finnegan, "One Year: Dreams from Another Life," *New Yorker,* 20 January 2010; John Heilemann and Mark Halperin, *Game Change* (HarperCollins, 2010), pp. 13–14; McClelland, *Young Mr. Obama,* pp. 222–25; Beverly Helm-Renfro's 2010 interview with Mark DePue, p. 63; Jim Edgar's 2010 interviews with DePue, pp. 1052–53; Dean Olsen, "Seeds of Federal Health Care Law Were Planted in Illinois Senate," *SJR,* 15 July 2012; Debra Pressey, "Duffett Leaving After Three Decades with Health Care Campaign," *CNG,* 12 June 2014; DJG interviews with Dan Shomon, Jim Rea, Glenn Hodas, Jeremy Flynn, Jim Duffett, Quentin Young, John Bouman, Willie Delgado, Phil Lackman, Dale Righter, Howard Peters, Nia Odeoti-Hassan, Mattie Hunter, Miguel del Valle, Denny Jacobs, Dave Syverson, Christine Radogno, Mark Warnsing, John Cameron, William McNary, Doug Dobmeyer, and Kurt DeWeese. The Adequate Health Care Task Force's 26 January 2007 report is available at www.idph.state.il.us/hcja/index.htm. On Dr. David Scheiner, who felt Barack was "not someone I could get to know," see David Whelan, "Obama's Doctor Knocks ObamaCare," *Forbes,* 18 June 2009, Mike Dorning, "Obama's Doctor Says President's Health-Care Reform Plans Are Too Timid," *CT,* 29 July 2009, Sam Stein, "Obama's Doctor: President's Vision for Health Care Bound to Fail," HuffingtonPost. com, 29 July 2009, Kennedy Elliott, "Dr. Scheiner," *Medill Reports,* 9 December 2009, and Carol Felsenthal, "Obama's Former Doc on the President . . . ," *CM,* 25 May 2012.

74. David Axelrod to Team Obama, "Message: Yes We Can!," n.d. [ca. 22 May 2004], 6pp., JKP; Axelrod, *Believer,* p. 162; Channel 7 ABC TV, 20 May 2004, MBC; Ray Long, "Shadow Keeps Eye on Obama," *CT,* 21 May 2004; Dave

McKinney, "Obama Admits He Dislikes His Most Loyal Follower," *CST,* 21 May 2004, p. 8; John Patterson and Sara Hooker, "Obama Takes Political Stalker in Stride," *CDH,* 21 May 2004, p. 15; Doug Finke and Mary Massingale, "Ryan's Cameras Bugging Obama," *SJR,* 21 May 2004, p. 1; Kevin McDermott, "Ryan Has Cameraman Tailing Obama," *SLPD,* 21 May 2004, p. B2; *CF,* 21, 25, and 26 May 2004; OFI, "Obama Calls on Senate to Pass Law Allowing Prescription Drug Re-Importation," 21 May 2004; ObamaBlog.com, 21, 23, 24, 25, and 27 May 2004; "On the Street," *QHW,* 23 May 2004, p. A9; David Mendell, "Ryan Aide to Give Obama More Space," *CT,* 23 May 2004; Amy Taylor, "Obama Inspires Local Democrat Rally," *JHN,* 23 May 2004; "Jack Ryan Should Know Better," *SJR,* 23 May 2004, p. 26; "Ryan Takes Campaign to New Low," *CST,* 24 May 2004, p. 41; James Kuhnhenn, "With Seven Retirements, Control of Senate Is at Stake in Election," *PI,* 24 May 2004, p. A2; Bruce DuMont, *Illinois Lawmakers,* WTTW, 26 May 2004; Gary Washburn and Christi Parsons, "Burke Asks Ban on Video Stalking," *CT,* 27 May 2004; Senate Floor Transcript, 30 May 2004, p. 7 (Obama introduces SJR 84); Bernard Schoenburg, "Jack Ryan's Staffers May Be a Little Too 'Passionate,'" *SJR,* 3 June 2004, p. 7; Peter Savodnik, "LaHood Rips Ryan over Senate Race," *TH,* 8 June 2004; Dori Meinert, "LaHood Blasts Ryan's 'Bonehead Campaign,'" *SJR,* 9 June 2004, p. 14; Beverly Helm-Renfro's 2010 interview with DePue, p. 75; Baim, *Obama and the Gays,* p. 27; DJG interviews with John Sullivan, Beverly Helm-Renfro, Dale Righter, Frank Watson, Phil Lackman, Jim Dodge, Dan Cronin, Jeff Schoenberg, Pat Welch, Boro Reljic, Jim Collins, Debbie Halvorson, John Kelly, and Glenn Hodas.

75. *CF,* 25, 28, and 31 May 2004, 1, 2, 4, 9, 10, 11, and 14 June 2004, and 27 November 2012; Senate Floor Transcript, 31 May 2004, p. 99 (HB 2721, 32–26) and 100 (HB 864); Christi Parsons and Ray Long, "11th-Hour Budget Push Seeks Malpractice Deal," *CT,* 31 May 2004; Long and Parsons, "Democrats Stalemated Over Budget," *CT,* 1 June 2004; Rick Pearson and Long, "3 Kings' Egos Entangle Budget," *CT,* 2 June 2004; Doug Finke, "Assembly Heads Home; Negotiators Will Call Them Back When Budget Is Reached," *SJR,* 2 June 2004, p. 1; Ryan Keith, "Obama Walks Fine Line with Legislative Votes in U.S. Senate Race," AP, 13 June 2004; Pat Guinane, "Under Lock and Key," *II,* September 2004; Ed Murnane, "Senator Obama: Quite a Contrast with Senator Durbin," *MCR,* 20 February 2005; Paul Merrion, "Obama's Star Rating," *CCB,* 22 January 2007, p. 3; Jeff Coen and John Chase, *Golden* (Chicago Review Press, 2012), pp. 148, 150; Noam Scheiber, "Race Against History—Barack Obama's Miraculous Campaign," *TNR,* 31 May 2004, pp. 21–26; Marc Solomon, *Winning Marriage* (ForeEdge, 2014), pp. 269–71; DJG interviews with Emil Jones Jr., Debbie Halvorson, Pat Welch, Jeff Schoenberg, Glenn Poshard, Dave Gross, Jill Rock, Frank Watson, Dale Righter, Dan Cronin, Jim Dodge, Phil Lackman, Jim Cauley, David Wilhelm, Scott Kennedy, Deanne Benos, Peter Coffey, Darrel Thompson, Kevin Thompson, David LeBreton, Nate Tamarin, Dan Shomon, Nora Moreno Cargie, Tarak Shah, Kaleshia Page, Liz Drew, Claire Serdiuk, Clinton Latimore Jr., Jeremiah Posedel, Andrew Boron, Laura Russell Hunter, Joe McLean, and Susan Shadow Ransone.

76. ObamaBlog.com, 31 May and 1, 2, 3, 4, 5, 7, 8, 9, 10, 11, and 13 June 2004; National Journal.com, 21–24 May 2004; John Chase, "Obama Gets Early Boost from Voters," *CT,* 31 May 2004; Lauren W. Whittington, "Illinois: Democratic Heavies to Fete Obama in N.Y., D.C.," *RC,* 3 June 2004; Ted McClelland, "The Obama Juice," *CR,* 4 June 2004; Bob Herbert, "A Leap of Faith," *NYT,* 4 June 2004; Liam Ford and David Mendell, "Keep Ryan Records Closed, Court Referee Says," *CT,* 4 June 2004; Chase, "Cheney Goes to Bat for Ryan," *CT,* 5 June 2004; OFI, "Obama Unveils Energy Independence Plan for Illinois," 6 June 2004; Obama to Jack Ryan, 13 June 2004, ObamaBlog.com; OFI, "Obama Proposes Series of Debates," 13 June 2004; Mendell, "Obama, Ryan Debate About Debates," *CT,* 14 June 2004; "Hellenes for Obama," 14 June 2004, SPP; Ford and Mendell, "Ryan Calls 6 Debates and Raises by 4," *CT,* 15 June 2004; Bernard Schoenburg, "Jack Ryan Blasts Obama on Universal Health Care," *SJR,* 15 June 2004, p. 10; OFI, "Obama Joins Bi-Partisan Lawmakers in Calling for Passage of Stem Cell Research Bill," 16 June 2004; John Kupper daybook entries, June 2004, [Paul Harstad], "Focus Group Outline," 16 June 2004, JKP; "General Election Focus Group, Women," and "General Election Focus Group, Men," 16 June 2004, PHP; Harstad and Mike Kulisheck to Obama for Senate Campaign, "Participant Ratings from Two June 16, 2004 Focus Groups in Peoria, Illinois," 9pp., and "Summary of Key Findings," JKP; Todd Spivak, "Little Big Man," *IT,* 17 June 2004; OFI, "Illinois AFL-CIO Endorses Barack Obama for Senate," and "Associated Fire Fighters of Illinois Endorses Barack Obama for Senate," 17 June 2004; Gregg Walker, "Reminiscing—Obama in Manhattan—June [8] 2004," ManhattanViewpoint.blogspot.com, 4 November 2008; DJG interviews with Joe Sinsheimer, John Kupper, Paul Harstad, and Mike Kulisheck.

77. Liam Ford and David Mendell, "Keep Ryan Records Closed, Court Referee Says," *CT,* 4 June 2004; Ford and Mendell, "Ryan Calls 6 Debates and Raises by 4," *CT,* 15 June 2004; "Judge to Decide Fate of Sealed Docs, Conservatives Brace for Worst," *IL,* 17 June 2004; Michael Martinez and Rick Pearson, "Court Sets Release of Ryan's Divorce File," *CT,* 18 June 2004; Ford and Pearson, "Ryan Won't Appeal Ruling on Files," *CT,* 19 June 2004; "Declaration of Jeri Lynn Ryan," 9 June 2000, pp. 21–23, "Declaration of John C. Ryan," n.d., p. 17, and "Declaration of Sharon Zimmerman," n.d., *In re Marriage of Ryan,* Los Angeles Superior Court #BD 290 382; John Chase and Ford, "Ryan File a Bombshell—Ex-Wife Alleges GOP Candidate Took Her to Sex Clubs," and Eric Zorn, "Ryan's Effort to 'Protect' Son Was Self-Serving," *CT,* 22 June 2004; Scott Fornek and Dave McKinney, "Ex-Wife Says Ryan Took Her to Sex Clubs," *CST,* 22 June 2004, p. 6; Eric Krol, "Ryan Denies Sex Club Claim," *CDH,* 22 June 2004, p. 1; Mike Ramsey, "Ryan Denies Ex-Wife's Claims," *SJR,* 22 June 2004, p. 1; Al Swanson, "Analysis: Embarrassment or Sex Scandal?" UPI, 22 June 2004; *CF,* 22, 23 (2), 24 (2), and 25 June 2004; OFI, "Obama Says Companies Creating Jobs in American Should Be Rewarded with Tax Cuts, 21 June 2004; ObamaBlog.com, 21, 22, and 23 June 2004; Andrea Zimmerman, "Obama Offers Plan to Improve Climate for Workers," *AT,* 22 June 2004; Jeff Smyth, "Democrats Rally for Obama," *SI,* 22 June 2004; "Obama Rallies for Election at SIU," *DE,* 22 June 2004; Lauren W. Whittington, "Republicans Quiet on Ryan—For Now," *RC,* 23 June 2004; Pearson and Ford, "GOP Leaders Say They Felt Misled on Ryan File," Carol Marin, "Ryan's Trouble with the Truth," and "Sex, Lies and Politics," *CT,* 23 June 2004; Fornek and McKinney, "Ryan Insists He Has No Plans to Withdraw from Senate Race," and "Time for Ryan to Hit the Road," *CST,* 23 June 2004, pp. 6, 53; Krol, "GOP Mostly Silent on Ryan," *CDH,* 23 June 2004, p. 1; Bernard Schoenburg, "Ryan Still Running," *SJR,* 23 June 2004, p. 1; Stephen Kinzer, "Illinois Senate Campaign Thrown into Prurient Turmoil," *NYT,* 23 June 2004; Dan Balz, "GOP Nominee Fights Calls to Exit Contest," *WP,* 23 June 2004; Jennifer DeWitt, "Obama Plan Would Reward Firms for Creating U.S. Jobs," and "Obama Says He Wants to Keep Focus on Real Issues," *QCT,* 23 June 2004; Chuck Sweeny, "Tax Breaks, Creating Jobs Focus for Obama," *RRS,* 23 June 2004; *USAT,* 23 June 2004, p. A7; Todd Spivak, "Ryan's Hope," *IT,* 24 June 2004; David Mendell, "Obama Lets Opponent Do Talking," and John Kass, "Ryan Should Quit Senate Race That He Lost Long Ago," *CT,* 24 June 2004; Fornek, "Republicans Band Together Against Ryan," *CST,* 24 June 2004, p. 3; Krol, "Ryan Event Canceled When Hastert Bails Out," *CDH,* 24 June 2004, p. 1; Schoenburg, "Obama Reaps Benefits from Jack Ryan's Troubles," *SJR,* 24 June 2004, p. 7; Kevin McDermott and Joel Currier, "Illinois GOP Committee Is Divided on Ryan's Future," *SLPD,* 24 June 2004; "Ryan Should Step Down from U.S. Senate Race," *BP,* 24 June 2004, p. A10; Caroline Daniel, "Political Rookie Traverses the Ethnic Divide," *FT,* 24 June 2004, p. 8; Jill Stanek, "Lies and Sex Clubs," *IL,* 24 June 2004; "2004 Campaign Polling: Illinois Senate," NationalJournal.com, 24 June 2004; Pearson and Rudolph Bush, "With Successor in Mind, GOP Plots Ryan's Exit," and Blair Hull, "Press Hijacks Senate Race Again," *CT,* 25 June 2004; Lynn Sweet and Scott Fornek, "GOP Lawmakers Agree Ryan Must Go," and Sweet, "Fitzgerald Steps Up as Party Leaders Ditch Ryan," *CST,* 25 June 2004, p. 6; Krol, "Ryan Gets the Cold Shoulder from GOP," and "Jack Ryan Made His Choice, Now Is Living with It," *CDH,* 25 June 2004, pp. 1, 14; Dori Meinert, "Republicans Question Ryan's Future," *SJR,* 25 June 2004, p. 13; E. J. Dionne, "In Illinois, a Star Prepares," *WP,* 25 June 2004; "Ryan: Some Ears Didn't Want to Listen," *IL,* 25 June 2004; Ford and Bush, "Ryan Quits Race," and "Jack Ryan's Painful Exit," *CT,* 26 June 2004; Fornek and Stephanie Zimmerman, "Sex Scandal Drives Ryan from Race," and Rich Miller, "Ryan Just Couldn't Tell the Truth," *CST,* 26 June 2004, pp. 4, 12; Krol, "Top Prospects Won't Take Ryan's Place," *CDH,* 26 June 2004, p. 1; Stephen Kinzer, "Candidate, Under Pressure, Quits Senate Race in Illinois," *NYT,* 26 June 2004; Chase, "Shaken Again, GOP Forced to Control Damage," and Pearson, "Ryan's Ruin," *CT,* 27 June 2004; Fornek, "Ryan's Final Campaign Meeting," and McKinney, "GOP Replacement Likely in 3 Weeks," *CST,* 27 June 2004, p. 3; Mark Brown, "It's Never Too Late to Say 'I Told You So,'" *CST,* 28 June 2004,

p. 2; Christopher Hayes, "Original Sin," *TNR*, 29 June 2004; Clarence Page, "The 'No-Sex' Sex Scandal," *CT*, 30 June 2004; Patrick Rucker, "Senator Blasts GOP on Ryan Exit," and Zorn, "Rising Stars Probably Won't Take On Obama," *CT*, 1 July 2004; Mike Ramsey, "Fitzgerald Criticizes Ryan Replacement Process," *SJR*, 1 July 2004, p. 9; *CF*, 1 July 2004; Pearson and Long, "Gidwitz Rejects a Senate Bid," *CT*, 2 July 2004; Pearson, "Candidates Line Up to Replace Ryan," *CT*, 3 July 2004; Aricka Flowers, "What Ever Happened to the Republican Party in Illinois?," *Chicago Life*, November 2004; Robert Feder, "All Good News for Media Mogul Ryan," Robert Feder.com, 12 December 2013; Michael Miner, "Former Senate Hopeful Jack Ryan Is Back in Politics—As a Publisher," *CR*, 25 November 2014; Jim Edgar's 2010 interviews with Mark DePue, pp. 1053–54; Heilemann and Halperin, *Game Change*, pp. 20–21; DJG interviews with Terry Walsh, Peter Coffey, Glenn Hodas, Jim Edgar, Kirk Dillard, Steve Rauschenberger, Dan Cronin, and Peter Fitzgerald.

78. OFI, "Obama Statement on Jack Ryan's Decision to Withdraw from U.S. Senate Race," 25 June 2004; Eric Krol, "Facing 'Brutal' Campaign, Ryan Quits," *CDH*, 26 June 2004, p. 1; Bernard Schoenburg, "Ryan Out of the Running," *SJR*, 26 June 2004, p. 1; Amanda Fuchs to Obama Team, "Policy Development Schedule," 24 June 2004, JKP; Obama, "Democratic Response to the President's Weekly Radio Address," 26 June 2004; OFI, "In Radio Address, Obama Says Middle Class Being Squeezed," 26 June 2004; AP, "Democrat Economy Squeezing Middle Class," 26 June 2004; Stefan C. Friedman, "Dems Boost Rising Black Star," *NYP*, 27 June 2004, p. 13; Krista Lewin, "Obama Brings Husband's Campaign to Coles County," *MJG* and *CTC*, 28 June 2004, p. A1; "Fundraiser for Barack Obama," DL21C-Chicago.org, Summer 2004 (28 June); Jodi Wilgoren, "Who Will Give the Dems' Keynote? Kerry Lets a Hint Slip," *NYT*, 29 June 2004, 5:27 P.M.; "Ba-Rock the House Party for Barack Obama," www.sebby.org, 29 June 2004; OFI, "More Than 3,000 Gather to Raise $75,000-Plus for Barack Obama at House Parties Across Illinois," 29 June 2004; "In the Law Firms," *CDLB*, 29 June 2004, p. 3; Obama to "Dear Table Tennis Friends," 30 June 2004; OFI, FEC Form 3, Report of Receipts and Disbursements, 1 April–30 June 2004, 14 July 2004; Friends of Barack Obama, Report of Campaign Contributions and Expenses, 1 January–30 June 2004, 2 August 2004; Senate Floor Transcript, 30 June 2004, p. 1 (3rd Day of 5th Special Session); Senate Floor Transcript, 2 July 2004, p. 1 (1st Day of 6th Special Session); OFI, "Obama Statement Commemorating 40th Anniversary of the Signing of the Civil Rights Act of 1964," 2 July 2004; "Obama in Diner," 2 July 2004, Carlinville, IL, JKP and TPP; OFI, "Obama Statement on Independence Day Weekend," 4 July 2004; Douglas Waller, "Dreaming About the Senate," *Time*, 5 July 2004, pp. 34–35; *CF*, 6 July 2004; OFI, "Obama Raises $4 Million in Three Months," 6 July 2004; David Mendell, "Fundraising Has Set Record, Obama Says," *CT*, 7 July 2004; OFI, "Obama's Retirement Security Plan," 7 July 2004; John Froehling, "Obama Brings Positive Campaign to Canton," *CDL*, 8 July 2004, pp. 1, 12; Paul Merrion, "Obama Lead Brings Bucks from Biz PACs," *CCB*, 12 July 2004, p. 3; Peronet Despeignes, "Dem Star Taking Convention Debut in Stride," *USAT*, 27 July 2004; "Oprah Talks to Barack Obama," *O Magazine*, November 2004; Obama, *TAOH*, p. 354; Mike Dorning and Christi Parsons, "Inside Obama's Inner Circle," *CT*, 14 January 2007; David Bernstein, "The Speech," *CM*, June 2007; Shira Schoenberg, "17 Minutes to Stardom," *Concord Monitor*, 6 December 2007; James Pitkin, "Obama Girl," *Willamette Week*, 12 March 2008; John McCormick, "1-on-1 with Obama," *CT*, 23 August 2008; James A. Barnes, "Word of Mouth Fueled Obama's Star Turn," *NJ*, 25 August 2008; Eli Saslow, "The 17 Minutes That Launched a Political Star," *WP*, 25 August 2008; Dayo Olopade, "Barack's Big Night," *TNR*, 25 August 2008; Bernard Schoenburg, "Obama's Meteoric Rise Hard to Predict 5 Years Ago," *SJR*, 6 November 2008; Remnick, *The Bridge*, p. 386; Mary Frances Berry and Josh Gottheimer, eds., *Power in Words* (Beacon Press, 2010), p. 11; Heilemann and Halperin, *Game Change*, pp. 20–21; Axelrod's 2008 interview with Jim Gilmore; Cahill's and Axelrod's 2009 interviews with Robert Draper; Corrigan's 2013 interview with Daniel Urman; DJG interviews with Amanda Fuchs Miller, Darrel Thompson, Claire Serdiuk, Jim Cauley, Peter Coffey, Juleigh Nowinski, Pete Giangreco, Jack Corrigan, Lisa Hay, Mary Beth Cahill, Josh Gottheimer, Vicky Rideout, John Kupper, Jill Rock, Matt Jones, and Terrie Pickerill.

79. Peter Savodnik, "Ill. GOPers Question Party's Senate Plans," *TH*, 7 July 2004, p. 12; "Our View," *JHN*, 7 July 2004; "Groups Want Ryan Back in Illinois Race," *WT*, 8 July 2004; Rick Pearson, "GOP Told Thanks, but No Thanks,"

CT, 9 July 2004; Bernard Schoenburg, "Rauschenburger Won't Run," *SJR*, 9 July 2004, p. 1; Kevin McDermott, "Field for Republican Replacement Is Wide Open," *SLPD*, 11 July 2004, p. B1; Pearson, "Ditka Tells GOP He's Game," *CT*, 13 July 2004; "The GOP's Hail Mary Play?," and John Kass, "Ditka Doesn't Have Makeup to Be Senator," *CT*, 14 July 2004; Pearson and John Chase, "Ditka Punts on Possible GOP Run for U.S. Senate," *CT*, 15 July 2004; Mike Ramsey, "Ditka Won't Run for Senate," *SJR*, 15 July 2004, p. 1; Stephen Kinzer, "Ex-Coach Won't Run for Senate," *NYT*, 15 July 2004; Pearson and Rudolph Bush, "Deflated GOP Goes Back to Its Search," *CT*, 16 July 2004; Rahul Sangwan, "Jack Ryan '81: The Conservative Idealist," *Dartmouth Independent*, 1 October 2004; Aricka Flowers, "What Ever Happened to the Republican Party in Illinois?," *Chicago Life*, November 2004; Ann Dalrymple, "Ditka Impressed with Ambition of the Bakken," *Dickinson Press*, 3 October 2013; Obama, Speech to the Chicago Council on Foreign Relations, 12 July 2004; OFI, "Obama Calls for Renewal of American Leadership in the World," 12 July 2004; David Mendell, "Running as If He's Got a Rival," *CT*, 13 July 2004; Eric Krol, "Obama Favors Keeping Troops in Iraq," *CDH*, 13 July 2004, p. 6; Patrick Healy, "Kerry Hones Campaign Themes," *BG*, 13 July 2004, p. A3; Axelrod e-mail to Kupper, 13 July 2004, JKP; Kerry-Edwards campaign press release, "Barack Obama to Deliver Keynote Address at 2004 Democratic National Convention," 14 July 2004; Healy, "Kerry Takes Steps to Attract African-American Voters," *BG*, 15 July 2004, p. A3; Jill Zuckman and Mendell, "Obama to Give Keynote Address," *CT*, 15 July 2004; Scott Fornek, "Obama to Keynote Dem Convention," *CST*, 15 July 2004, p. 3; Krol, "Convention Spotlight to Shine on Obama," *CDH*, 15 July 2004, p. 15; Bill Lambrecht, "Democrats Give Obama Keynote Slot at Convention," *SLPD*, 15 July 2004, p. A4; Lei-Ling Hopgood, "Obama the Democrats' Next Big Thing," Cox News Service, 15 July 2004; Jeremy Berkowitz, "Obama to Give Keynote in Boston," *TH*, 15 July 2004; Obama on *Today*, NBC, 15 July 2004; "Obama Heads for the Convention," *Economist*, 17 July 2004, p. 32; Shira Schoenberg, "17 Minutes to Stardom," *Concord Monitor*, 6 December 2007; Eli Saslow, "The 17 Minutes That Launched a Political Star," *WP*, 25 August 2008; Axelrod's 2008 interview with Jim Gilmore; Axelrod's 2009 interview with Robert Draper; Remnick, *The Bridge*, pp. 393–94; Berry and Gottheimer, *Power in Words*, p. 12; Axelrod, *Believer*, pp. 156–57; DJG interviews with Steve Rauschenburger, Dave Sullivan, Dale Righter, Dan Cronin, Glenn Hodas, John Nicolay, Claire Serdiuk, John Kupper, Jim Cauley, Darrel Thompson, and Vicky Rideout.

80. *CF*, 28 April 2003 and 7, 8, 9, 16, 20, 21, 23, and 26 July 2004, Dave McKinney and Chris Fusco, "Madigan: Fire Hospital Board," *CST*, 30 June 2004, p. 10; Eric Herman, "Businessman Looking to Sell Chain of Pizza Places," *CST*, 16 July 2004, p. 55; Ray Gibson and Crystal Yednak, "Blagojevich Adviser Tied to Appointee," *CT*, 18 July 2004; Kirk Dillard, *At Issue*, Newsradio 780 WBBM, 18 July 2004, JKP; Obama, *Public Affairs with Jeff Berkowitz*, 18 July 2004; Rick Pearson, "Hinsdale Legislator May Face Obama," *CT*, 19 July 2004; "Obama on the Right Track," *HPH*, 21 July 2004, p. 4; Carol Marin, "Illinois Republicans May Deserve Spanky the Clown," *CT*, 21 July 2004; Scott Fornek, "Dillard Says He Can Beat Obama," *CST*, 21 July 2004, p. 11; "Chase the Guy with the Ball," *NYT*, 21 July 2004; John Chase and Liam Ford, "Does Anyone Want This Job?," *CT*, 22 July 2004; OFI, "Barack Obama Statement on the 9/11 Commission Report," 23 July 2004; Peronet Despeignes, "Dem Star Taking Convention Debut in Stride, *USAT*, 27 July 2004, p. 6A; Christopher Wills, "Rising Democratic Star Readies for His Moment in Lights," AP, 23 July 2004; Ron Staton, "Illinois Senate Candidate Says Lucky to Have Been Raised in Hawaii," AP, 24 July 2004; Ray Long and Christi Parsons, "8 Weeks Late, Lawmakers OK $46 Billion Plan," *CT*, 25 July 2004; Bernard Schoenburg, "Obama's Children Have More Important Things in Mind," *SJR*, 25 July 2004, p. 15; Pearson, "Party Unity," *CT*, 26 July 2004; Stephen S. Morrill, "Client Report for 2004 End of Session," esp. pp. 5, 15, Morrill & Associates, July 2004; "Major Bills Passed by the Illinois General Assembly," *LRU First Reading*, September 2004; *C-WR* 19 #7, 5 October 2004; Eric Krol, "Obama Tries to Keep Cool as Star Rises," *CDH*, 27 July 2004, p. 1; Jo Mannies, "Charming, Brainy Obama Creates Party Sensation," *SLPD*, 27 July 2004, p. A1; Mark Leibovich, "The Other Man of the Hour," *WP*, 27 July 2004; John Patterson, "What's Changed for Obama in the Last Two Years," *CDH*, 27 July 2006, p. 7; Obama, *TAOH*, p. 357; David Bernstein, "The Speech," *CM*, June 2007; Eli Saslow, "The 17 Minutes That Launched a Political Star," *WP*, 25 August 2008; Jeff Zeleny, "Former Outsider Navigated a Path to the Marquee,"

NYT, 27 August 2008; Obama, *Meet the Press,* NBC, 25 July 2004; Obama, *Face the Nation,* CBS, 25 July 2004; Obama, *World News Tonight,* ABC, 25 July 2004; Obama, *Late Edition,* CNN, 25 July 2004; Jim Edgar's 2010 interviews with Mark DePue, p. 1054; DJG interviews with Kirk Dillard, Vicky Rideout, Jeff Schoenberg, Tom Lindenfeld, Jim Cauley, David LeBreton, and Adam Stolorow.

81. David Mendell, "Political Phenomenon Obama Vaults into National Spotlight," *CT,* 26 July 2004; Christopher Wills, "Obama Riding Wave to Keynote Speech and Star Status," AP, 26 July 2004; Scott Fornek, "Former Obama Adversaries Look on Without Regret," *CST,* 27 July 2004, p. 6; Eric Krol, "Obama Tries to Keep Cool as Star Rises," *CDH,* 27 July 2004, p. 1; "Let Barack Obama Be the Next Barack Obama," *CD,* 27 July 2004, p. 9; John Kass, "Obama's a Star Who Doesn't Stick to the Script," and Mendell and Jeff Zeleny, "Obama Says War to Decide Election," *CT,* 27 July 2004; "Convention Speaker Says Hawaii Shaped His Life," *HA,* 27 July 2004; Bernard Schoenburg, "Obama Fights Image of Stardom," *SJR,* 27 July 2004, p. 1; Susan Milligan, "In Obama, Democrats See Their Future," *BG,* 27 July 2004; Carl Chancellor, "A Rising Star Gets a Key Role Tonight," *PI,* 27 July 2004; P. J. Huffstutter, "Obama Making a Name for Himself," *LAT,* 27 July 2004; Lisa Ellman, "The Audacity of Hope," *UCLSR,* Fall 2004, pp. 20–21; Shawn McCarthy, "Minorities Looking for Gains in Battle for the Presidency," *G&M,* 27 July 2004, p. A3; Julian Borger, "Rising Star to Woo Voters with Upbeat Keynote Speech," *Guardian,* 27 July 2004, p. 10; Rupert Cornwell, "An Unknown Rookie, but Can Obama Be First Black President?," *Independent,* 27 July 2004; Mary Carey, "Local Eyes Look at National Figures, Up Close," *Daily Hampshire Gazette,* 27 July 2004; Obama on *Good Morning America,* ABC, 27 July 2004, on *Today,* NBC, 27 July 2004, on *The Early Show,* CBS, 27 July 2004, and on *American Morning,* CNN, 27 July 2004; "Rising Star," *PBS NewsHour,* 27 July 2004; Obama on *All Things Considered,* NPR, 27 July 2004; Mendell, "Obama Finding Himself Flush with Media Attention," *CT,* 28 July 2004; Scott Fornek, "Obama Delivers," *CST,* 28 July 2004, p. 1; Amanda Griscom, "Barack Star," Grist.com, 4 August 2004 (and *Salon,* 6 August 2004); Peter Newbould, "Boston Journal: An Inside Look at the 2004 Democratic National Convention," *AAP Advance,* Fall 2004, pp. 1, 12; Obama, *TAOH,* pp. 358–59; Eric Krol and John Patterson, "Obama: The Path to Today," *CDH,* 10 February 2007, p. 1; David Bernstein, "The Speech," *CM,* June 2007; Favreau in Ashley Parker, "What Would Obama Say?," *NYT,* 20 January 2008; Nesbitt in Todd Purdum, "Raising Obama," *Vanity Fair,* March 2008, in Ryan Lizza, "Making It," *TNY,* 21 July 2008, and on "Barack Obama Revealed," CNN, 20 August 2008; Eli Saslow, "The 17 Minutes That Launched a Political Star," *WP,* 25 August 2008; Saslow, "Helping to Write History," *WP,* 18 December 2008; Favreau in Charles P. Pierce, "The Speechwriter in Chief," *BG Magazine,* 21 December 2008; Daren Briscoe, "An Interview with Barack Obama," *Newsweek,* 8 January 2009 (recorded ca. 18 May 2008); Remnick, *The Bridge,* p. 395; Berry and Gottheimer, *Power in Words,* p. 13; Obama, "Remarks," 11 February 2011, *Public Papers of the Presidents* 2011 Vol. 1, p. 104; Axelrod, *Believer,* p. 158; Elizabeth Segrin, "Obama's Former Law Student Is Shaping the Nation's Policy on Drones," *Fortune,* 4 February 2015; Marty Nesbitt's 2009 interview with Robert Draper; DJG interviews with Michael Sheehan, Vicky Rideout, Josh Gottheimer, Jim Cauley, Lisa Ellman, Joe McLean, Kitty Kurth, and Raja Krishnamoorthi.

82. Obama, DNC Keynote Address, C-Span Video, 27 July 2004; *Hardball with Chris Matthews,* MSNBC, 27 July 2004, 10 P.M.; Gwen Ifill, "Rising Star," PBS NewsHour, 27 July 2004; David S. Broder, "Democrats Focus on Healing Divisions," *WP,* 28 July 2004; Katharine Q. Seelye, "Senate Nominee Speaks of Encompassing Unity," *NYT,* 28 July 2004; Michael Tackett, "Old, New Guards Rebuke Bush for Drive to War," *CT,* 28 July 2004; Mary Mitchell, "Obama Embodies the Hopes of the People," *CST,* 28 July 2004, p. 7; Eric Krol and John Patterson, "Obama Shares His 'American Story,'" and Judy Baar Topinka, "Beyond the Glare of the Spotlight, Obama Facts Tell Different Story," *CDH,* 28 July 2004, pp. 1, 16; Bernard Schoenburg, "Preaching Hope: Delegates Cheer Obama's Keynote Address," *SJR,* 28 July 2004, p. 1; Robert Bianco, "What Happened to Conventional Wisdom?," *USAT,* 28 July 2004; Brent Baker, "MSNBC and CNN Lionize 'Rock Star' Obama, See a Future President," Media Research Center, 28 July 2004; Scott Fornek, "Obama Has Arrived, Illinois Dems Say," *CST,* 29 July 2004, p. 7; Michele Steinbacher, "ISU Student Trustee Working for Obama," *BP,* 29 July 2004; Mike Thomas, "What's Behind Barack's Celebrity?," *CST,* 9 August 2004, p.

44; George Vradenburg, "Here's to the Skinny Kid with the Funny Name," *Tikkun*, September–October 2004, p. 6; Senate Floor Transcript, 8 November 2004, p. 12; Obama on *Today*, NBC, 19 October 2006; Jones in Terry McDermott, "What Is It About Obama?," *LAT*, 24 December 2006; Krol, "Practice and Passion: How Obama Went from 'Uh, Uh' to a Polished Speaker," *CDH*, 25 February 2007, p. 1; Fornek, "Madelyn Payne Dunham," *CST*, 9 September 2007; Scott Helman, "Early Defeat Launched a Rapid Political Climb," *BG*, 12 October 2007; Eli Saslow, "The 17 Minutes That Launched a Political Star," *WP*, 25 August 2008; Niall Stanage, *Redemption Song* (Liberties Press, 2009), p. 23; Presta, *Mr. and Mrs. Grassroots*, p. 182; Remnick, *The Bridge*, p. 401; Axelrod's 2009 interview with Robert Draper; DJG interviews with Vicky Rideout, Mary Beth Cahill, Tom Lindenfeld, Terry Link, Darrel Thompson, Jeff Schoenberg, Emil Jones Jr., Forrest Claypool, Kitty Kurth, Andy Schapiro, Dan Hynes, Jim Cauley, Josh Gottheimer, Randon Gardley, Nate Tamarin, Dan Shomon, Carolyn Shapiro, Adam Sherman, Paul Anderson, Pern Beckman, Barry Rutizer, and James Clayborne. By far the best academic discussion of the speech is Robert C. Rowland and John M. Jones's "Recasting the American Dream and American Politics: Barack Obama's Keynote Address to the 2004 Democratic National Convention," *Quarterly Journal of Speech* 93 (November 2007): 425–48, which notes that "in many ways the message was quite unremarkable" (433) and, like Chris Matthews, how Barack had emphasized the need for "personal responsibility values" (439). See also David A. Frank and Mark Lawrence McPhail's truly awful "Barack Obama's Address to the 2004 Democratic National Convention: Trauma, Compromise, Consilience, and the (Im)Possibility of Racial Reconciliation," *Rhetoric & Public Affairs* 8 (Winter 2005): 571–99, which states that a speech by Al Sharpton was "far superior to Obama's" (585), Babak Elahi and Grant Cos, "An Immigrant's Dream and the Audacity of Hope," *American Behavioral Scientist* 49 (November 2005): 454–65, and Deborah F. Atwater, "Senator Barack Obama: The Rhetoric of Hope and the American Dream," *Journal of Black Studies* 38 (November 2007): 121–29.

83. "The Phenom," *CT*, 28 July 2004; Bill Nichols, "Second Night's Speakers Stoke the Party Fires," *USAT*, 28 July 2004, p. 4A; Salim Muwakkil, "Barack Obama Made Smashing National Debut," *Progressive*, 28 July 2004; Obama on *NBC Nightly News*, 28 July 2004; Randal C. Archibold, "Day After, Keynote Speaker Finds Admirers Everywhere," and Monica Davey, "In a Star's Shadow, Republicans Strain to Find Opponent," *NYT*, 29 July 2004; Richard Benedetto, "Address Throws Illinois' Obama into Whirlwind of Political Hope," *USAT*, 29 July 2004, p. 6A; Mark Jurkowitz, "With One Speech, Neophyte's Political Stock Soars," *BG*, 29 July 2004; Jeremy Dauber, "A Star Is Born," *CSM*, 29 July 2004; Rick Pearson and David Mendell, "Democrats: We've Got a Winner in Obama," *CT*, 29 July 2004; Mary Mitchell, "Democrats Need More Than Speech from Obama," *CST*, 29 July 2004, p. 7; Bernard Schoenburg, "Illinois Delegates Foresee Bright Future for Obama," *SJR*, 29 July 2004, p. 1; Michele Steinbacher, "ISU Student Trustee Working for Obama," *BP*, 29 July 2004; Sumana Chatterkee (KR), "Democrats Buzzing About Obama's Future," *The State*, 29 July 2004 ("rising star"); Brian A. DeBose, "Obama Emerges as Major Party Player," *WT*, 29 July 2004, p. A1; Obama on *The Tavis Smiley Show*, 29 July 2004; Allan M. Winkler, "Catch a Rising Democratic Star," *BG*, 30 July 2004; Liam Ford and John Chase, "With Ryan Out, GOP Moves On," and Rick Pearson, "Obama: Those Feuding Should Take a 'Timeout,'" *CT*, 30 July 2004; Dave McKinney and Scott Fornek, "Ryan Formally Exits Senate Race vs. Obama," *CST*, 30 July 2004, p. 10; Eric Krol, "From New Heights, Any Obama Fall Could Be a Long One," *CDH*, 30 July 2004, p. 1; Schoenburg, "Obama Details His Stands on the Issues," and "Obama Suggests a 'Timeout' for Bickering Democrats," *SJR*, 30 July 2004, pp. 1, 4; Michael Cader, "Publishers Eyeing Obama," *NYS*, 30 July 2004, p. 14; Tom Zucco, "Speech Pulls Obama Book into Limelight," *SPT*, 30 July 2004, p. A10; Mark Wood, "Obama Has Strength of His Own," *FTU*, 30 July 2004, p. A6; Lauren A. E. Schuker, "Obama Stars at Convention," *HC*, 30 July 2004; Christopher Caldwell, "A Democrat Star Is Born," *FT*, 31 July 2004; Scott L. Malcolmson, "An Appeal Beyond Race," *NYT*, 1 August 2004; Jennifer Skalka, "Foes Out to Prove Obama Not Alone," and Clarence Page, "Obama's Drama and Our Dreams," *CT*, 1 August 2004; John Aloysius Farrell, "Obama Revives MLK's Dream," *Denver Post*, 1 August 2004; Terence Samuel, "A Shining Star Named Obama," *U.S. News*, 2 August 2004, p. 25; Paul Merrion, "The Siren Call of Clout," *CCB*, 2 August 2004, p. 3; Matthew Continetti, "The Anti-Obama," *Weekly Standard*, 2 August 2004; "Who Is Barack Obama?,"

WT, 2 August 2004 ("rising star"); Tina Brown, "Barack Obama, Shaking Up the Sound-Bite Culture," *WP,* 5 August 2004; "A Rising Star," *BSB,* 5 August 2004, p. 4; Cindy Adams, "Dems Rising Star Barack a Writer Too," *NYP,* 5 August 2004, p. 14; Lynn Sweet, "Obama's Book," *CST,* 8 August 2004, p. 32; "At Deadline," *CNYB,* 9 August 2004, p. 1; Kevin Chappell, "Barack Obama," *Jet,* 16 August 2004, pp. 4–10, 59–64; *NYTBR,* 29 August 2004 through 26 December 2004; "Sales of Obama's Memoirs Go Up as Candidate's Star Rises," *CT,* 5 September 2004; Janet Maslin, "Crowd Pleasers," *NYT,* 10 September 2004; David Mendell, "Barack Obama," *CT,* 22 October 2004; Dan Holly, "Onward and Upward," *Black Issues Book Review,* November–December 2004, p. 63; Lynn Sweet, "Obama Enlists D.C. Superagent," *CST,* 8 November 2004, p. 26; Shira Schoenberg, "17 Minutes to Stardom," *Concord Monitor,* 6 December 2007; Jeff Zeleny, "Former Outsider Navigated a Path to the Marquee," *NYT,* 27 August 2008; Munar in Baim, *Obama and the Gays,* p. 519; Edward-Isaac Dovere and David Nather, "10 Years Later: Obama's Hits and Blunders," *Politico,* 27 July 2014; DJG interviews with Claire Serdiuk, Jim Cauley, Darrel Thompson, Kaleshia Page, Susan Shadow Ransone, and Liz Drew.

84. Obamablog.com, 28, 30, and 31 July and 6 August 2004; "Obama Will Be in Kewanee Saturday," *KSC,* 28 July 2004; Mary Rae Bragg, "Galena Democrats Set to Welcome Rising Star Obama," *TH,* 29 July 2004, p. A1; Dan Churney, "Obama Blitz Begins Here," *ODT,* 1 August 2004; Brandon Coutre, "Obama Brings Message to Illinois," *PJS,* 1 August 2004; Greg Stanmar, "Obama Hits the Trail," *BP,* 1 August 2004; Carol Clark, "Obama in Oquawka," *GRM,* 1 August 2004; Thomas Geyer, "Democrats Line Up for Evans," *QCT,* 1 August 2004; Dustin Lemmon, "Obama Brings Star Power to Evans Fundraiser," *MD,* 1 August 2004; Chuck Sweeny, "Obama 'Voice of the Democratic Future' in Illinois," *RRS,* 1 August 2004; Travis Morse, "Obama to Visit Freeport Library," *FJS,* 1 August 2004; Christopher Wills, "Obama Returns to Illinois Carrying Big Hopes on His Shoulders," AP, 1 August 2004; Matt Smolensky, "Large Crowd Attend Obama Event," *MRA,* 2 August 2004; Joan E. Snyder, "Joy and Hope on the Campaign Trail," Parkinson's Information Exchange Network, 2 August 2004; Chris Rickert, "Obama Mania," *DDC,* 2 August 2004; Matt Hanley, "Obama: Ba-rock Star Impact in DeKalb," *AB,* 2 August 2004; Jeff Kolkey, "Obama Stumps in Marengo," *NH,* 2 August 2004; Brad Broders, "From Boston to a Bus," WFIR, 2 August 2004; Mike Wiser, "Q&A with Barack Obama," and "Rockford Shows Its Obama Spirit," *RRS,* 2 August 2004; Melanie M. Schroeder, "Obama Makes Freeport History," *FJS,* 2 August 2004; Scott T. Holland, "Obama Makes Speech in Front of Gracious Crowd," *CH,* 2 August 2004; David Mendell, "Heady Week Yields to Hard Work," *CT,* 2 August 2004; Paul Gale, "Obama Wants to Carry on Traditions of Paul Simon," *SVN,* 2 August 2004; Jennifer Speiser, "Obama in the Heart of Illinois," WHOI, 2 August 2004; "Obama Campaigns in Central Illinois," WEEK, 2 August 2004; "Under Gaze of Spotlight, Obama Is Praised for Bridging Racial Divide," JTA, 3 August 2004; Mike Berry, "Obama Wows the Crowd in Kewanee," *KSC,* 3 August 2004; Michelle Gibbons, "Obama Points Locals to His Side at Rally," [NIU] *Northern Star,* 3 August 2004; Barb Kromphardt, "Barack Obama Visits Bureau County," *BCR,* 3 August 2004; Dawn Turner Trice, "Obama's Appeal May Be the Life of Both Parties," *CT,* 3 August 2004; Scott Richardson, "Obama: I'll Fight for Workers," *BP,* 3 August 2004, p. A1; Amy Hoak, "Obama Creates a Stir During Campaign Swing Through Central Illinois," *DHR,* 3 August 2004; Edith Brady-Lunny, "Obama Draws Crowd of More Than 300 in Clinton," *BP,* 3 August 2004; Stephanie Miller, "Obama Visit," WTHI, 3 August 2004; Pat Phillips, "Obama Greets Area Voters," *CNG,* 3 August 2004; Amber Griswold, "New Fame for Chicago Senator," WFIE, 3 August 2004; Molly Stephey, "Obama Swings Through C-U," *DI,* 4 August 2004; Dave McKinney, "Obama Can't Help but Shine," *CST,* 4 August 2004, p. 6; Herb Meeker, "Obama Takes U.S. Senate Campaign to Lake Land College in Mattoon," *DHR,* 4 August 2004; Greg Cima, "'Skinny Guy from the South Side' Tells Voters Here What He Believes," *MVRN,* 4 August 2004 ("rising star"); Nicole Sack, "U.S. Senate Candidate Talks About Platform in Mount Vernon," *SI,* 4 August 2004; Kevin Ryden, "Obama Makes Campaign Stop in Olney," and "Obama Crowd Includes Happy Democrats," *ODN,* 4 August 2004; Jennifer Weakley, "Obama Energizes Local Democrats," *DCN,* 5 August 2004; Bob Livingston, "Obama: When You Pay Attention to Politicians They Act Right," *MCR,* 5 August 2004; Bernard Schoenburg, "Obama Tells Taylorville Crowd Fame Won't Go to His Head," *SJR,* 5 August 2004, p. 1; Lauren Fitz-

Patrick and Alice Hohl, "Obama Celebrates 43rd at Matteson Fundraiser," *DS,* 5 August 2004; Mundy, *Michelle,* pp. 152–53; Mendell's 2008 interview with Jim Gilmore; DJG interviews with Jeremiah Posedel, Peter Coffey, Jim Cauley, Lauren Kidwell, Matt Jones, Kelly Jo MacArthur, and Darrel Thompson.

85. David Axelrod and John Kupper to Obama Campaign, "Revised Media Plan," 4 August 2004; Amanda [Fuchs] to Obama Team, "Executive Summary—Defensive Research Book," 6 August 2004, 7pp., JKP; Liam Ford and John Chase, "GOP Nears Choice for Senate," *CT,* 3 August 2004; Scott Fornek, "GOP Wooing Keyes to Take On Obama," *CST,* 3 August 2004, p. 6; *CF,* 3, 6, 10, 12, and 20 August 2004; Chase and Ford, "GOP Narrows Senate List to 2," *CT,* 4 August 2004; Fornek, "Keyes, GOP Discuss Taking On Obama," and Fornek and Lynn Sweet, "State GOP Narrows Senate Hopefuls to 2," *CST,* 4 August 2004, pp. 1, 3; Ford and Chase, "GOP Wants Keyes," *CT,* 5 August 2004; Fornek, "GOP Asks Keyes to Run," *CST,* 5 August 2004, p. 12; Fran Eaton, "Peeks into IL GOP's Executive Session Reveal Struggle for Party Control," *IL,* 5 August 2004; Rick Pearson and Jon Yates, "Keyes Reportedly Set to Accept GOP Bid," *CT,* 6 August 2004; Fornek, "Keyes Promises Republicans He'll Run," *CST,* 6 August 2004, p. 3; Ford and Rudolph Bush, "Debt, Tax Questions Surface for Keyes," *CT,* 7 August 2004; Ford, Bush, and Chase, "Conservative Heart, Soul Make Keyes Tick," *CT,* 8 August 2004; Fornek, "Keyes a Hot Property," *CST,* 8 August 2004, p. 3; OFI, "Statement from Barack Obama on Alan Keyes' Entry into the Illinois U.S. Senate Race," 8 August 2004; Juliet Eilperin, "Keyes to Face Obama," *WP,* 9 August 2004; Ford and Chase, "Keyes Guarantees Fight, If Not Victory," *CT,* 9 August 2004; Maureen O'Donnell and Fornek, "Keyes Fires Up GOP Faithful," *CST,* 9 August 2004, p. 3; Ford and Chase, "On Day 1, New Foes Already Getting Licks In," *CT,* 10 August 2004; Fornek, "Keyes Says Obama Afraid to Face Him," *CST,* 10 August 2004, p. 18; Ford and David Mendell, "Obama Runs Away, Keyes Says," and John Kass, "Debate Strategy Is a Winner, But Voters Lose Out," *CT,* 11 August 2004; Mark Brown, "Obama's Backpedaling Is a Disappointment," and Fornek, "Keyes Says He Wants to Rumble with Obama," *CST,* 11 August 2004, pp. 2, 24; "Keyes Guilty of Black-on-Black Crime," *CD,* 11 August 2004, p. 9; "When You Propose Six Debates . . . ," *CT,* 12 August 2004; Michael Van Winkle and Greg Blankenship, "The Obama Myth," *American Spectator,* 12 August 2004; Obama on *Fresh Air,* NPR, 12 August 2004; OFI, "Illinois Fraternal Order of Police Endorse Barack Obama for Senate," 12 August 2004; Ford and Mendell, "Keyes Sets Up House in Cal City," *CT,* 13 August 2004; Fornek, "Obama Ads May Start Next Week," *CST,* 13 August 2004, p. 5; Eric Krol, "Who Will Scare Soccer Moms More, Keyes or Obama?," *CDH,* 13 August 2004, p. 14; Fornek, "Obama to Debut TV Ads Next Week—But Not Here," and Thomas Roeser, "How the GOP Ended Up Picking Keyes," *CST,* 14 August 2004, pp. 6, 18; OFI, "Obama Celebrates Signing of the Illinois Citizen Soldier Initiative," 14 August 2004; Mendell, "Billiken Crowd Jeers Keyes, Cheers Obama," *CT,* 15 August 2004; Maudlyne Ihejirika, "Bud Billiken Parade a Mix of Pompons and Politics," *CST,* 15 August 2004, p. 11; Karen E. Pride, "More Than 1 Million Attend 75th Annual Billiken Parade," *CD,* 16 August 2004, p. 3; La Risa Lynch, "Obama Greets Voters at Bud Billiken Parade," *HPC,* 18 August 2004, p. 1; Obama on *This Week,* ABC, 15 August 2004; Glenn Jeffers, "Parade Heralds a Hearty Rivalry," "Obama, Meet Keyes," and Ford and Mendell, "Senate Race to Hit Airwaves," *CT,* 16 August 2004; Natasha Korecki, "How Do You Do?," *CST,* 16 August 2004, p. 8; OFI, "Obama Calls for Increased Funding for Agricultural Resarch and Development," 16 August 2004; Eric Zorn, "Keyes Saddles Wrong Horse for Senate Race," *CT,* 17 August 2004; Steve Tarter, "Senate Candidate Obama Tours Peoria Agricultural Lab," *PJS,* 17 August 2004; OFI, "Obama Calls for Doubling of Ethanol Production for Gasoline by 2012," 17 August 2004; Korecki and Fornek, "Keyes Likens Abortion to Terrorism," *CST,* 17 August 2004, p. 6; H. Gregory Meyer, "Keyes Says State Needs an Outsider," *CT,* 18 August 2004; Obama, "Expanded O'Hare, or 3rd Airport? Chicago Area Needs Both," *CST,* 18 August 2004, p. 59 (and *HPH,* 25 August 2004, p. 4); Rich Miller, "Deal with It," *IT,* 19 August 2004; Ray Long and Christi Parsons, "Democrats Try to Be Friends," *CT,* 19 August 2004; McKinney, "With Obama Around, Dems Put Aside Bickering," *CST,* 19 August 2004, p. 20; Doug Finke, "Obama Stars on Governor's Day," *SJR,* 19 August 2004, p. 1; Jeff Berkowitz, "Chairman Topinka Sticks the Shiv in Jack Ryan One More Time," Blogspot.com, 19 August 2004; OFI, "Obama Calls on Administration to Increase Intelligence Sharing Between Local and Federal Law Enforcement,"

19 August 2004; Colleen Mastony, "Cops Give Obama Subdued Reception," *CT,* 20 August 2004; McKinney, "GOP Conservatives Rally Behind Keyes at State Fair," *CST,* 20 August 2004, p. 20; Bob Secter, "Obama Holds 41-Point Lead Over Keyes," and Clarence Page, "A Man for All Racial Reasons," *CT,* 22 August 2004; Bernard Schoenburg, "Don't Expect Many Niceties in Keyes-Obama Contest," *SJR,* 22 August 2004, p. 15; Lynch, "Obama Stumps for Votes in South Suburbs," *HPC,* 25 August 2004, p. 1; Rich Miller, *GZ,* 26 August 2004; Daniel Nolan, "Relative: Obama's Got 'A Good Handle on Canada,'" *Hamilton Spectator,* 11 June 2008; Jason Misner, "Barack Obama Was Here," *Burlington Post,* 20 June 2008; Axelrod, *Believer,* p. 161; DJG interviews with Amanda Fuchs Miller, Nora Moreno Cargie, Dave Syverson, Ed Petka, Steve Rauschenberger, and Rod McCulloch. A comprehensive collection of all of Alan Keyes's public comments from 8 August 2004 through the balance of the fall campaign is available at www.keyesarchives.com.

86. OFI, "Obama Releases New Campaign Ad Focusing on Creating Quality Jobs in Illinois," 25 August 2004; Todd Spivak, "Judging Judy," *IT,* 26 August 2004; OFI, "Illinois Families Lose More in Income Than Any Other State Over Last Three Years," 26 August 2004; John Chase and John McCormick, "Race for Senate Turns from Odd to Kind of Wacky," *CT,* 28 August 2004; Harstad Strategic Research, "Illinois Statewide Survey—August 2004," 22–29 August 2004, PHP; Rudolph Bush, "Keyes Hits Obama's Gay Marriage Stance," *CT,* 29 August 2004; Obamablog.com, 30 August 2004; Katherine Bouma, "Obama Encouraged by City's Racial Progress," *BN,* 31 August 2004, p. B1; Lynn Sweet, "Keyes Symbolizes Deep Rift Between State GOP Factions," *CST,* 31 August 2004, p. 6; "Decision 2004: Sierra Club Endorsed Candidates," *Lake & Prairie,* 4th Quarter [August] 2004, p. 3; Ryan Lizza, "The Natural," *Atlantic,* September 2004; Allison Horton, "Obama Releases Television Ad Focusing on Job Creation," *HPC,* 1 September 2004, p. 1; "Block Schedule for Obama for Illinois," 2 September 2004, "Obama for Illinois Consultants' Retreat Agenda," 2 September 2004, AKP Message and Media to Barack Obama, "Transition," 2 September 2004, 12pp., AKP; "Obama for U.S. Senate—Illinois Media/General Election Planning 2004 (Revision C)," 7 September–2 November 2004, JKP; *CF,* 3 and 8 September 2004; OFI, "Obama Fights for Horizon Miners, Calls for Bankruptcy Reform," 3 September 2004; Scott Reeder, "Kenya Believe It?—Obama's Rise in Politics," *ODT,* 4 September 2004 (and *KDJ,* 11 September 2004); Nicole Sack, "Obama: Time for Reform," *SI,* 4 September 2004; Matt Adrian and Mark Samuels, "A Look at Barack Obama One-on-One," *QCT,* 4 September 2004 (and *DHR,* 5 September, p. B1); Rick Pearson, "Keyes, State GOP Gearing Up Blame Campaign," and Clarence Page, "Keyes Becomes GOP's Burden," *CT,* 5 September 2004; Kevin J. Kelley, "A Homecoming for Obama, 'If I Win,'" *The Nation* (Nairobi), 5 September 2004; Liam Ford and David Mendell, "Candidates Rally Their Strongholds," and Eric Zorn, "To Put It Bluntly, Tactless Keyes Isn't Senatorial," *CT,* 7 September 2004; Tory Brecht, "Obama Fires Up RI Crowd," *QCT,* 7 September 2004; Mike Ramsey, "Keyes Questions Obama Spanking Remark," Copley, 7 September 2004; OFI, "Obama Outlines Plan to Help Small Businesses Pay for Health Insurance," 7 September 2004; Lesley R. Chinn, "Obama Speaks to AFL-CIO Members," *HPC,* 8 September 2004, p. 1; Ford and Mendell, "Jesus Wouldn't Vote for Obama, Keyes Says," *CT,* 8 September 2004; Stephanie Sievers, "Obama Says He's Not Worried About Getting Jesus' Vote," *ODT,* 9 September 2004; Christopher Wills, "Obama, Keyes Make Pitches to Farm Group," AP, 9 September 2004; Kurt Erickson, "Obama Wants to Focus on Issues, Not Theology," *BP,* 9 September 2004; "Obama: U.S. in Health-Care Crisis," *PJS,* 9 September 2004; Debra Pickett, "My Parents Weren't College-Educated Folks," *CST,* 19 September 2004, p. 10; Pat Guinane, "Star Power," *II,* October 2004; "A Rusty Toyota, a Mean Jump Shot, Good Ears," *WP,* 11 February 2007; Lauren Martin, "Welcome, Mr. President!!!," *Martha's Vineyard Magazine,* August 2009; Tom Dresser, "Encountering Barack Obama Amid Vineyard's Black History," *Vineyard Gazette,* 19 August 2010; Matt Viser, "For Obamas, Vineyard Has Enduring Attraction," *BG,* 17 July 2013; DJG interviews with Sheryll Cashin and Allison Davis. On Davis, also see Mark Konkol, "We Still Live in a Slum," *CST Neighborhoods* blog, 29 June 2007.

87. Lynn Sweet, "What's Up with Obama Secrecy?," *CST,* 9 September 2004, p. 37; OFI, "Statement from Barack Obama on Darfur, Sudan," [10 September 2004, not 7 October]; David Mendell, "Obama Urges Push on Sudan," *CT,* 11 September 2004; Ryan Keith, "Legislative Record Puts Obama at Heart of

Philosophical Debate," AP, 11 September 2004; Mendell and Liam Ford, "Keyes Derails Obama from Traditional Track," *CT,* 13 September 2004; Steve Edwards, "Guns and Crime," *Eight Forty-Eight,* WBEZ, 13 September 2004; Obama on *Paula Zahn Now,* CNN, 13 September 2004; Block Schedule for Obama for Illinois, 2 September 2004, JKP; Rick Pearson, "Keyes Says Game Plan Is Controversy," *CT,* 14 September 2004; OFI, "Obama Outlines Plan to Make College More Affordable," 15 September 2004; AP, "Obama Calls for Increase in Pell Grants," 15 September 2004; John Chase, "Keyes, Obama Are Far Apart on Guns," *CT,* 15 September 2004; Esther Githui, "Democratic Senatorial Candidate Barack Obama Stresses Importance of Roots," VOA, 15 September 2004; Eric Krol, "Obama Offers Mild Retort to Keyes," *CDH,* 16 September 2004, p. 7; *CF,* 16, 20, 27, 29, and 30 September 2004; Dave McKinney, "Obama Suggests Closing Beach to Study Asbestos," *CST,* 17 September 2004, p. 8; Molly Parker, "Obama Wants to Make Pell Grants More Accessible," *PJS,* 17 September 2004; Chris Bailey, "Contemplating Steve and Barack," *CDH,* 19 September 2004, p. 18; Christi Parsons and Ray Long, "Wheeler-Dealer with a Cause," *CT Magazine,* 19 September 2004, pp. 12ff.; Adriana Colindres, "Obama, Keyes Give Opinions on War, Health Care," *SJR,* 19 September 2004, p. 17; David Moberg, "Audacious and Hopeful," *ITT,* 20 September 2004, p. 22; Chase, "Keyes, Obama Differ on Tactics," *CT,* 20 September 2004; Kevin McDermott, "Obama May Trounce Keyes for Senate," *SLPD,* 20 September 2004, p. A1; Edwards, "Immigration," *Eight Forty-Eight,* WBEZ, 20 September 2004; Obama, "Springfield Report," *HPH,* 22 September 2004, p. 4; R. Eden Martin, "Introduction of Barack Obama," Commercial Club, 22 September 2004, REMP; OFI, "Women for Obama Luncheon," 23 September 2004; Mendell, "Obama Would Consider Missile Strikes in Iran," *CT,* 25 September 2004; Nicole Ziegler Dizon, "Obama Says His Beliefs Say No to Gay Marriage," *CST,* 25 September 2004; Tamara Sharman, "Obama Delivers Message of Hope," *KDJ,* 25 September 2004; Louise Brass, "Obama Touches Base in Bolingbrook," *JHN,* 26 September 2004; Chase, "State GOP Wrestling for Identity," *CT,* 26 September 2004; Edwards, "Education," *Eight Forty-Eight,* WBEZ, 27 September 2004; Mendell, "Obama Takes Show on Road," *CT,* 28 September 2004; Sweet, "Kerry Taps Obama to Court African-American Vote," *CST,* 28 September 2004, p. 20; Carrie Budoff, "Obama Draws a Crowd at Hoeffel Event," *PI,* 28 September 2004; Kali Backer, "Dem. Party Star Electrifies Crowd at Pa. Senate Race Rally," *Daily Pennsylvanian,* 28 September 2004; Krol, "Students Put Candidates on the Spot," *CDH,* 30 September 2004, p. 10; Mike Christianson, "Obama Comes to Baltimore," *Afro-American,* 2–8 October 2004, p. A1; Jim Edgar's 2010 interviews with Mark DePue, p. 1055; Jason Meisner, "Blagojevich Fundraiser Gets 2 Years in Bribe Case," *CT,* 10 February 2014; DJG interview with Michael Parham.

88. "The Obama Record," *LL,* September–October 2004, p. 8; Kenneth Meeks, "Favorite Son," *BE,* October 2004, pp. 88–95; Barbara Ransby, "Barack Obama," *The Progressive,* October 2004, pp. 35–38; Scott Fornek, "'I've Got a Competitive Nature,'" *CST,* 3 October 2004, p. 12; John Chase, "A Big Split Over Abortion, Stem Cells," *CT,* 4 October 2004; Ed Tibbetts, "Obama Boosts Fellow Candidates," *QCT,* 5 October 2004; "Barack Obama Visits Quad Cities Area to Help Fellow Democrats," *WCFC,* 5 October 2004; Kurt Allemeier, "Obama: People Are Struggling Out There," *RIA,* 5 October 2004; Lesley R. Chinn, "Obama Reminds Supporters Not to Take Election for Granted," *HPC,* 6 October 2004, p. 2; Nathaniel Zimmer, "Obama Rejects Liberal Label," *JHN,* 6 October 2004; Eric Krol, "Obama Clarifies Sex Ed Views at Benedictine," *CDH,* 6 October 2004, p. 17; Ronald Roach, "Obama Rising," *BIHE,* 7 October 2004, pp. 20–23; Lauren W. Whittington, "Obama Endearing Himself with Cash," *RC,* 7 October 2004; Lynn Sweet, "Obama's Stealth," *TH,* 7 October 2004, p. 19; Peter Slevin, "Ill. GOP Watches Take-No-Prisoners Campaign," *WP,* 7 October 2004; Sweet, "Obama's Sharing the Wealth with Fellow Dems," *CST,* 7 October 2004, p. 47; David Mendell, "Obama's Record a Plus, a Minus," and Kathryn Masterson, "Truly a World Apart," *CT,* 8 October 2004; Slevin, "Obama Lending Star Power to Other Democrats," *WP,* 11 October 2004; Chase, "Obama, Keyes Clash on Terrorism," *CT,* 11 October 2004; Steve Edwards, "Foreign Policy," *Eight Forty-Eight,* WBEZ, 11 October 2004; Joe Frolik, "A Sense of Mission," *Plain Dealer,* 12 October 2004, p. B9; Hetal Bhatt, "Obama Enthralls Spirited Crowd," *DI,* 12 October 2004; Jodi Heckel, "Democratic Senate Hopeful Denounces Negative Campaigns," *CNG,* 12 October 2004; Obama-Keyes debate transcript, 12 October 2004; P. J. Huffstutter, "Candidates

Miles Apart in History-Making Illinois Election," *LAT,* 13 October 2004; Chase and Liam Ford, "Obama, Keyes Put on Kid Gloves," *CT,* 13 October 2004; Fornek, "Keyes, Obama Spar Over War in Iraq," *CST,* 13 October 2004, p. 18; John Patterson, "A Restrained First Meeting," *CDH,* 13 October 2004, p. 1; Bernard Schoenburg, "Debate Sticks to the Issues," *SJR,* 13 October 2004, p. 1; Kurt Erickson, "Keyes, Obama Stake Positions," *BP,* 13 October 2004, p. A1; Corey Hall, "Obama and Keyes Discuss Issues at First Joint Appearance," *HPC,* 13 October 2004, p. 1; *CF,* 13 October 2004; Eric Zorn, "Senate Debate Hints at What Might Have Been," *CT,* 14 October 2004; Debbie Howlett, "Obama Extends Reach Beyond Illinois Race," *USAT,* 14 October 2004, p. 14A; Andrew Stern, "Illinois Democrat Set to Win All-Black Senate Race," Reuters, 14 October 2004; Mendell and Rudolph Bush, "Obama Spreads His Influence," *CT,* 15 October 2004; Krol, "In Senate Race, Moderates Finally Coming into Style," *CDH,* 15 October 2004, p. 16; Kurt Erickson, "Obama Boosts Renner as He Campaigns for 'Team,'" *BP,* 15 October 2004, p. A1; Elaine Hopkins, "Michelle Obama in Peoria," *PJS,* 15 October 2004; Mendell and Ford, "Keyes Manages to Rake in Cash," *CT,* 16 October 2004; Norma Mendoza, "Obama's Wife Makes Local Stop," *EI,* 16 October 2004; Mary Massingale, "Michelle Obama Says Faith Guides Her Husband," *SJR,* 16 October 2004, p. 4; Christopher Wills, "Obama Mixes Exotic Background with Street-Level Experience," AP, 16 October 2004; Erickson, "Obama-Keyes Debate Lacked Spark, Fire," *BP,* 17 October 2004, p. C1; Yussuf Simmonds and Kathy Williamson, "National Leaders Converge on L.A.," *LAS,* 21–27 October 2004, p. A1; Rich Miller, "Obamarama," *IT,* 28 October 2004 (and *MD,* 29 October); "Oprah Talks to Barack Obama," *O Magazine,* November 2004; Lawrence A. Horwich, Barack H. Obama, and Michelle L. Obama, 2004 Form 1040, 2 March 2005; Jones in Terry McDermott, "What Is It About Obama?," *LAT,* 24 December 2006; Obama in Jonathan Van Meter, "Leading by Example," *Vogue,* April 2013, pp. 248–55, 318–19; "Meryl Streep's Interview with Michelle Obama," *More,* July–August 2015, pp. 84–85, 140; DJG interviews with Emil Jones Jr. and John Kupper.

89. Mendell, *Obama,* pp. 76–77; Jeffrey S. Siker, "President Obama, the Bible, and Political Rhetoric," *Political Theology* 13 (September 2012): 586–609; Claire Martin, "Dems' Rising Star Shines for Salazar," *DP,* 17 October 2004, p. C1; Obama on *American Morning,* CNN, 19 October 2004; John Chase, "Obama, Keyes Concur on Little," *CT,* 18 October 2004; Kevin McDermott, "Unlikely Competitors: Years as Outsider Gave Obama the Ability to Relate, Friends Say," *SLPD,* 24 October 2004, p. A1; "Look Past the Race Issue," *SJR,* 18 July 2004, p. 20; Benjamin Wallace-Wells, "The Great Black Hope: What's Riding on Barack Obama?," *Washington Monthly,* November 2004, pp. 30–36; Amos N. Jones, "Obama's Oversimplified Identity," *NYAN,* 5–11 August 2004, p. 13; Jones, "Black Like Obama," *Thurgood Marshall Law Review* 31 (2005–2006) 79–100, esp. 80, 95; Don Terry, "The Skin Game—Do White Voters Like Barack Obama Because 'He's Not Really Black?,'" *CT Magazine,* 24 October 2004, pp. 14ff.; Joy Bennett Kinnon, "Barack Obama," *Ebony,* November 2004, pp. 196ff. That fall, University of Chicago political scientist Melissa Harris-Lacewell surveyed some 567 Illinoisans, presenting Barack as multiracial to one half of them and black to the other. There was no difference in how *white* Illinoisans evaluated him, yet among *African American* respondents, "when Obama was framed as a multiracial candidate, he was assessed more positively than when he was framed as black." Harris-Lacewell and graduate student Jane Junn rightly observed that "most black respondents understand themselves as being multiracial in some sense." Harris-Lacewell and Junn, "Old Friends and New Alliances," *Journal of Black Studies* 38 (September 2007): 30–50, esp. 41; see also David J. Anderson and Junn, "Deracializing Obama: White Voters and the 2004 Illinois U.S. Senate Race," *American Politics Research* 38 (May 2010): 443–70.

90. "Obama Still Looks Like a Winner to Us," *CST,* 15 October 2004, p. 45; "Obama Shines Through Senate Circus," *JHN,* 17 October 2004; "For U.S. Senate: Barack Obama," *CNG,* 18 October 2004; John Patterson, "Foreign Policy Also Splits Keyes, Obama," *CDH,* 18 October 2004, p. 1; Steve Edwards, "Health Care," *Eight Forty-Eight,* WBEZ, 18 October 2004; "Obama Can Best Serve Illinois in U.S. Senate," *BP,* 19 October 2004; Scott Fornek, "Is Obama Overconfident?," *CST,* 20 October 2004, p. 22; Jay Field, "Barack Obama Profile," *Eight Forty-Eight,* WBEZ, 20 October 2004; John Chase and Liam Ford, "GOP Leaves Keyes Off Its Team Ad," *CT,* 21 October 2004; "An Easy Decision," *QCT,* 21 October 2004; "Obama

Certain to Glide to Nov. 2 Victory," *SI,* 21 October 2004; Obama-Keyes debate transcript, 21 October 2004; Chase and Ford, "Senate Debate Gets Personal," and Mendell, "Barack Obama," *CT,* 22 October 2004; Fornek, "Rivals Take Gloves Off in First Televised Senate Debate," *CST,* 22 October 2004, p. 12; Eric Krol, "Fire and Brimstone Keyes, Obama Hit on Spiritual Issues in Heated Debate," *CDH,* 22 October 2004, p. 1; *CF,* 22 October 2004; Christopher Wills, "Obama's Wife Takes Big Role in Campaign," AP, 22 October 2004; Bernard Schoenburg, "Obama Speaks at Local Rally, Endorses Auditor Candidate," *SJR,* 23 October 2004, p. 13; Rick Pearson and Chase, "Democrats Grip on State Firm," and "Obama for U.S. Senate," *CT,* 24 October 2004; "Obama for Illinois," *SLPD,* 24 October 2004, p. B2; "Senate Hopeful Obama to Rally Voters," *DBNJ,* 24 October 2004, p. C7; Obama on *Off Topic with Carlos Watson,* CNN, 24 October 2004; Marc Lacey, "Illinois Democrat Wins Kenyan Hearts," *NYT,* 25 October 2004; Chase, "The Issues: Taxes/ Education," *CT,* 25 October 2004; Krol, "'Politics of Hope' Appeals to Wide Range of Supporters," *CDH,* 25 October 2004, p. 1; "U.S. Senate: Barack Obama," *PJS,* 25 October 2004; "Obama Is the Better Candidate to Represent Ill.," *CCB,* 25 October 2004; Thomas Brown, "Ill. Senate Hopeful Rouses S.C. Voters," *Orangeburg Times & Democrat,* 25 October 2004; Virginia Smith, "Religious Zeal Flows During Obama Visit," *DBNJ,* 25 October 2004, p. C1; Steve Edwards, "Jobs and the Economy," *Eight Forty-Eight,* WBEZ, 25 October 2004; Suzanne Bell, "Michelle Obama Speaks at ISU," *DV,* 26 October 2004; "Obama for U.S. Senate," *SJR,* 29 October 2004, p. 8; "Obama Will Do Us Proud in U.S. Senate," *CD,* 29–31 October 2004, p. 15.

91. John Chase, "Senate Hopefuls Far Apart on War," *CT,* 26 October 2004; Obama-Keyes debate, 26 October 2004, WTTW; Obama on *Hannity & Colmes,* Fox News, 26 October 2004; Monica Davey, "Where to Catch a Rising Political Star?," *NYT,* 27 October 2004; Chase and Courtney Flynn, "Keyes, Obama Disagree Sharply," *CT,* 27 October 2004; Scott Fornek, "Final Debate Offers Spirit, Substance," *CST,* 27 October 2004, p. 6; Eric Krol, "Last Shots at Each Other," *CDH,* 27 October 2004, p. 1; Mike Ramsey, "Keyes, Obama Trade Jabs Over Moral Issues," *SJR,* 27 October 2004, p. 13; Liam Ford and Aamer Madhani, "Keyes Commercial Has Plenty of Praise," *CT,* 28 October 2004; Fornek, "Keyes' First Ad Focuses on Positive Spin," *CST,* 28 October 2004, p. 18; Kevin McDermott, "Obama Keeps Huge Lead over Keyes in Senate Race," *SLPD,* 28 October 2004, p. A1; Kari Allen, "Obama Impresses Naperville Students," *CDH,* 28 October 2004, p. 1; Bernard Schoenburg, "U.S. Senate: Obama, Keyes Offer Clear Choices," *SJR,* 28 October 2004, p. B2; *CF,* 28 October 2004; Ted McClelland, "Keyes and Friends Wind Up Down at the Pokey," *CR,* 29 October 2004; David Mendell and Ford, "Obama, Daley Enjoy Lunch," *CT,* 30 October 2004; Rudolph Bush and Liam Ford, "Hopefuls Making Last Dash for Votes," and Chase, "Obama's Message Has Edge on Keyes," *CT,* 31 October 2004; Krol, "Keyes, Obama Sound Off About Bin Laden Tape," *CDH,* 31 October 2004, p. 17; Kristyna C. Ryan, "Illinois State Senator Barack Obama," *Chicago Bar Association Record* 18 (November 2004): 46–48; Chase and Ford, "Keyes Presses Catholic Voters," *CT,* 1 November 2004; Fornek, "Keyes Rips Obama," *CST,* 1 November 2004, p. 3; Rick Pearson and Chase, "Unusual Match Nears Wire," *CT,* 2 November 2004; Bernard Schoenburg, "A Final Scramble for Votes in Illinois," *SJR,* 2 November 2004, p. 1; Kevin McDermott, "A Tale of Two Campaigns," *SLPD,* 2 November 2004; Christopher Benson, "Camelot Rising," *Savoy,* February 2005, pp. 60–69, 103–6; Marc Solomon, *Winning Marriage* (ForeEdge, 2014), pp. 271–73. David Airne and William L. Benoit's vapid "2004 Illinois U.S. Senate Debates," *American Behavioral Scientist* 49 (October 2005): 343–52, can be ignored.

92. OFI, Obama victory remarks, 2 November 2004; John Chase and David Mendell, "Obama Sails to Senate Win," Mendell and Chase, "Huge Victory Sets Stage for National Role," and Mendell, "Obama to Arrive with Status," *CT,* 3 November 2004; Scott Fornek, "Obama Takes Senate Seat in a Landslide," *CST,* 3 November 2004, p. 6; Eric Krol, "Obama Wins Landslide," and "The Election That Wasn't," *CDH,* 3 November 2004, pp. 1, 10; Karen E. Pride, "Barack Supporters Gather to Watch History Come Alive," *CD,* 3 November 2004, p. 16; Kevin McDermott, "Democrats' Rising Star Has Rousing Victory," *SLPD,* 3 November 2004, p. A1; Cindy Wojdyla Cain, "Walsh Wins County Exec Job," *JHN,* 3 November 2004; *CF,* 3 November 2004; Obama on *Today,* NBC, on *Good Morning America,* ABC, and on *The Early Show,* CBS, 3 November 2004; Obama on *The Tavis Smiley Show,* 3 November 2004; Sharon Cohen, "Obama's Incredible Year," AP, 3 November 2004; Todd

Spivak, "Rising Son," *IT,* 4 November 2004; Mendell, "After Reaching Heights, Obama Lowering His Sites," *CT,* 4 November 2004; Fornek, "Obama for President? That's 'Silly,'" *CST,* 4 November 2004, p. 17; Krol, "Obama Silences Presidential Talk," *CDH,* 4 November 2004, p. 7; Mike Ramsey, "'I Am Not Running for President in 2008,' Obama Says," *SJR,* 4 November 2004, p. 1; Christopher Wills (AP), "Obama Wants to Start by Learning," *BP,* 4 November 2004; Liam Ford, "From Keyes, No Congratulations," *CT,* 5 November 2004; William Lamb, "Don't Expect Too Much at First, Obama Tells Belleville Backers," *SLPD,* 13 November 2004, p. 10; Obama on *The Leonard Lopate Show,* NPR, 22 November 2004; Christopher Benson, "Camelot Rising," *Savoy,* February 2005, pp. 60–69, 103–6; Wills, "Obama Puts Rural Illinois Experience to Work in Effort to Win Iowa," AP, 13 October 2007.

CHAPTER TEN: DISAPPOINTMENT AND DESTINY

1. "Drew" to Cassandra [Butts], "Possible Committee Assignments," n.d., JKP; Eric Krol, "Obama Can Hit the Ground Running," *CDH,* 3 November 2004, p. 10; John Chase, "8 Vie for Obama's Legislature Seat," *CT,* 6 November 2004; Jodi S. Cohen, "Obama's Springfield Seat Goes to Lawyer," *CT,* 7 November 2004; Cheryl L. Reed, "Dems Name Obama Replacement," *CST,* 7 November 2004, p. 8; Lynn Sweet, "Obama Enlists D.C. Superagent," *CST,* 8 November 2004, p. 26; Karen E. Pride, "Obama's Senate Replacement Chosen," *CD,* 8 November 2004, p. 5; Senate Floor Transcript, 8 November 2004, pp. 1–2, 11–15; SR 716, 8 November 2004; "Obama Says 'Thank You' to Area Volunteers," *LL,* November–December 2004, p. 3; David Mendell, "Obama in Bipartisan Goodbye," and "Obama Says Goodbye to Illinois," *CT,* 9 November 2004; Dave McKinney, "Springfield Lovefest," *CST,* 9 November 2004, p. 60; Bernard Schoenburg, "Colleagues Congratulate Obama," *SJR,* 9 November 2004, p. 9; Kevin McDermott, "Obama Discusses Writing Book," *SLPD,* 9 November 2004, p. B1; *CF,* 9 November 2004; Cindy Adams, "Barack's Dreams Beyond His Agents," *NYP,* 10 November 2004, p. 14; Jonathan Alter, "The Audacity of Hope," *Newsweek,* 27 December 2004; Sweet, "Be-Bop, Barack and Bucks from Book," *CST,* 17 March 2005, p. 39; Jeff Zeleny, "New Man on the Hill," *CT,* 20 March 2005; Tracey Robinson-English, "Statehouse Clout," *Ebony,* April 2005, pp. 150ff.; UCLS, *The Glass Menagerie, 2005–06,* p. 22, *2006–07,* pp. 21–22, *2007–08,* pp. 22–23; Martin Schram, "Senate Duo Exception to Partisan Bickering," Scripps Howard, 6 December 2005; Christina Larson, "Hoosier Daddy," *Washington Monthly,* September 2006, pp. 9ff.; Peter Osnos, "Barack Obama and the Book Business," The Century Foundation, 30 October 2006; Mike Dorning and Christi Parsons, "Inside Obama's Inner Circle," *CT,* 14 January 2007; Kwame Raoul, "'Stay Out of Jail,'" *HPH,* 14 February 2007, p. 18; Christopher Wills, "Clues to How Obama Would Play His Hand as President Can Be Found in Poker Style," AP, 24 September 2007; Dan Morain, "Obama's Lawyer Days Were Effective but Brief," *LAT,* 6 April 2008 (Obama totaled 3,723 billable hours at Miner Barnhill, mostly prior to 1997); Bernard Schoenburg, "Former Obama Assistant Remembers Him as Friendly Boss," *SJR,* 8 November 2008; Jim Merriner, "Barack Obama and the Parallel to Jimmy Carter," *CDO,* 24 March 2009; Edward McClelland, "The Ones Obama Left Behind," NBCChicago.com, 23 March 2011; Monique Garcia, "Hyde Park–Kenwood Area Fosters Another Set of Democratic Leaders with Clout," *CT,* 13 April 2011; James Warren, "Nobel Laureates Here, Grinding Poverty There," *NYT,* 12 August 2011; DJG interviews with Toni Preckwinkle, Leslie Hairston, Emil Jones Jr., Will Burns, John Eason, Cynthia Miller, Dan Shomon, Mike Lieteau, Eden Martin, Andy Davis, Terry Link, Tommy Walsh, Cassandra Butts, and Nathan Diament. On Obama's state legislative record, see John Patterson, "Obama's State Record No Mystery," *CDH,* 19 February 2007, p. 1, Randolph Burnside and Kami Whitehurst, "From the Statehouse to the White House?," *Journal of Black Studies* 38 (September 2007): 75–89, and Boris Shor, "How Liberal Was Obama as a State Senator in Illinois?," RedBlueRichPoor.com, 6 October 2008. Charles Peters, "Judge Him by His Laws," *WP,* 4 January 2008, cogently argued that "the media have not devoted enough attention to Obama's bills and the effort required to pass them," indicative of a presumption "that state legislatures are not to be taken seriously." On the contrary, Obama's record "is revealing enough to merit far more attention than it has received."

2. Obama on *This Week,* ABC, 7 November 2004, and on *Meet the Press,* NBC, 7 November 2004; Rudolph Bush and David Mendell,

"Obama Makes Case for 'Common Ground,'" *CT,* 8 November 2004; "Obama to Host Meeting at Peoria Library," *PJS,* 9 November 2004; Ed Tibbetts, "Obama Fires-Up Q-C Crowd," *QCT,* 11 November 2004; Tibbetts, "Obama Gets Immediate Star Status," *QCT,* 12 November 2004; William Lamb, "Don't Expect Too Much at First, Obama Tells Belleville Backers," *SLPD,* 13 November 2004, p. 10; "Obama Must Keep Balance While Working for Illinois," *BP,* 13 November 2004, p. A6; "LaHood: It's a Good Time for Illinoisans," *PJS,* 14 November 2004; Bush and Mendell, "Obama Off to Washington," *CT,* 14 November 2004; Amanda Ripley et al., "Obama's Ascent," *Time,* 15 November 2004, pp. 74–81; "A Star Is Born," *U.S. News,* 15 November 2004, p. 53; Sarah Frank and Bush, "New Lawmakers Learn the Ropes in Washington," *CT,* 15 November 2004; Nicole Ziegler Dizon (AP), "Obama Works to Balance Fame, Illinois Interests," *BP,* 15 November 2004, p. B1; Dori Meinert, "Obama Gets Oriented in the Senate," *SJR,* 16 November 2004, p. 26; Minutes, GARS Board of Trustees, 17 November 2004, #226, pp. 1876–81; Deirdre Shesgreen, "Obama Arrives, and Fitzgerald Says Goodbye," *SLPD,* 19 November 2004, p. A1; Rudolph Bush, "Fitzgerald Stood Up," *CT,* 19 December 2004; Bernard Schoenburg, "Durbin, LaHood Say Some Good Things About Obama," *SJR,* 17 December 2006, p. 39; Tim Dickinson, "Obama's Brain Trust," *Rolling Stone* 1056/57, 10 July 2008; Pete Rouse's 2008 interview with Jim Gilmore; Ceci Connolly, "Daschle's Woes Test an Insider's Insider," *WP,* 3 February 2009; Richard Wolffe, *Renegade* (Crown, 2009), p. 39; John Heilemann and Mark Halperin, *Game Change* (HarperCollins, 2010), p. 26; Cassandra Butts's 2012 interview with Gilmore; Ray LaHood, *Seeking Bipartisanship* (Cambria Press, 2015), pp. 203–4; DJG interviews with Cassandra Butts, Peter Fitzgerald, Paul Williams, Ray Harris, and Rob Fisher.

3. Edward Felker, "Obama: Pledges to Avoid Spotlight," *KDJ,* 17 November 2004; Obama on *The Leonard Lopate Show,* NPR, 22 November 2004; Rudolph Bush, "Obama Greets New York on 2-Day Publicity Blitz," *CT,* 23 November 2004; Obama on *The Early Show,* CBS, 23 November 2004, on *Charlie Rose,* PBS, 23 November 2004, and on C-Span Book TV at Barnes & Noble, 23 November 2004; "Obama on 'Letterman,'" *CT,* 26 November 2004; Robyn Tysver, "Democratic Star Dines with Buffetts," *OWH,* 30 November 2004, p. B2; Rick Perlstein, "Conviction Politics," *American Prospect,* December 2004, p. 26; Obama on *The Friday Night Show,* WTTW, 3 December 2004; William Neikirk, "Obama Shows Media He's a Stand-Up Guy," *CT,* 6 December 2004; Lynn Sweet, "Obama Takes Secretive Flight to Visit Buffett," *CST,* 6 December 2004, p. 12; Stuart Rothenberg, "A Reality Check as Obama-Mania Hits the Media," *RC,* 6 December 2004; Maura Kelly Lannan, "Obama Gets Committee Assignments," AP, 6 December 2004; Rudolph Bush, "Obama Named to Key Panels," *CT,* 7 December 2004; Finlay Lewis, "Obama Jokes About Being a Celebrity," *SJR,* 7 December 2004, p. 11; Mary Vorsino, "Obama Pushes Message of Diversity," *HSB,* 17 December 2004; Gordon Y. K. Pang, "Democrats Call Obama Hawaii's 'Third Senator,'" *HA,* 17 December 2004; Jaymes Song, "Obama Gets Celebrity Reception as Hawaii's 'Third Senator,'" AP, 17 December 2004; Press Release, "Barack Obama . . . Signs Book Contract with Crown Publishers," 17 December 2004; Rudolph Bush, "Obama's $1.9 Million, 3-Book Deal Includes 1 Children's Book," *CT,* 18 December 2004; Sara Nelson, "Obama Inks 2-Book Deal," *NYP,* 18 December 2004, p. 22; "Obama Urges Kids to 'Dream Big,'" *HSB,* 18 December 2004; "Obama Gets $1.9 Million Deal for Books," *NYT,* 20 December 2004; Greg Hinz, "Barack on a Roll," *CCB,* 20 December 2004; Maura Webber Sadovi, "An Offbeat Christmas Stirs Up Harmony," *CST,* 22 December 2004, p. F1; *American Morning,* CNN, 24 December 2004; Jonathan Alter, "The Audacity of Hope," *Newsweek,* 27 December 2004, pp. 74ff.; Amanda Griscom Little, "Barack Obama," *Rolling Stone* 964/965, 30 December 2004, p. 88; "Best Rookie," *Vanity Fair,* January 2005, pp. 86–87; Laurel Bowers Husain and Laurie Uemoto Chang, "Obama Encourages Students to 'Dream Big,'" *Punahou Bulletin,* Spring 2005, pp. 14–17; Godfrey Sperling, "Why Not an Obama-Romney Match Up in 2008?," *CSM,* 5 July 2005; Sweet, "When Obama and Romney First Met," *CST,* 30 September 2012; DJG interview with Rob Fisher.

4. Barb Ickes, "Face-to-Face with the Freshman Senator," *QCT,* 3 January 2005; Obama, *DFMF,* p. 133; Eric Krol, "Obama's Green but Full of Steam," *CDH,* 3 January 2005, p. 1; *CRec,* 4 January 2005, p. S4; Maureen Groppe, "Obama Tells Reporters to Slow Down," *USAT,* 4 January 2005; Rudolph Bush and David Mendell, "Great Expectations for Junior Senator," *CT,* 4 January 2005; Lynn Sweet, "Obama,

Bean, Lipinski Sworn In," *CST,* 4 January 2005; Ickes, "Obama Humble in New Role," *QCT,* 4 January 2005; Nicole Sack, "Ready to Serve," *SI,* 4 January 2005; Edward Felker, "Obama Begins Senate Term," *ODT,* 4 January 2005; Faye Fiore, "He's the Hill's King for a Day," *LAT,* 5 January 2005; Sweet, "While Obama Basks, Durbin Rises," and Carol Marin, "Obama Knows Best, Worst Is Yet to Come," *CST,* 5 January 2004; Bush and Mendell, "State Fields Lineup of Stars in Congress," *CT,* 5 January 2005; Krol, "Obama Takes the Oath," *CDH,* 5 January 2005, p. 13; Dori Meinert, "Obama Takes Capitol by Storm," *SJR,* 5 January 2005, p. 11; Sack, "Obama Officially Sworn In as Senator," *SI,* 5 January 2005; Obama, *CRec,* 6 January 2005, p. S53; "Obama Unlikely to Visit Kenya Until 2006," *Nation,* 8 January 2005; Kevin Chappell, "Barack Obama Takes His Seat in U.S. Senate," *Jet,* 24 January 2005, p. 4; Jeremiah A. Wright Jr., "Senator Barack Obama's Swearing-In Ceremony," *Trumpet,* February 2005, pp. 42–43; Jeff Zeleny, "New Man on the Hill," *CT,* 20 March 2005; Bernard Schoenburg, "Obama Maintains Chicago Home," *SJR,* 19 April 2005, p. 12; Joy Bennett Kinnon, "Michelle Obama," *Ebony,* March 2006, pp. 58ff.; Michelle's July 2006 interview with Jacob Weisberg; Zeleny, "Testing the Water," *NYT,* 24 December 2006; Kalari Girtley and Brian Wellner, "Michelle Obama Is Hyde Park's Career Mom," *HPH,* 14 February 2007, pp. 7–8, 16–17; Leslie Bennetts, "First Lady in Waiting," *Vanity Fair,* December 2007; Scott Helman, "Holding Down the Obama Family Fort," *BG,* 30 March 2008; Jonathan Darman, "Mr. Obama's Washington," *Newsweek,* 30 June 2008; Eli Saslow, "A Rising Political Star Adopts a Low-Key Strategy," *WP,* 17 October 2008; Saslow, "Helping to Write History," *WP,* 18 December 2008; Auma Obama, *And Then Life Happens,* p. 289; Malik Obama, *Barack Obama Sr.,* pp. 320–21.

5. Bob Goldsborough, "On the Ball," *CT,* 21 November 1999; Cook County Assessment Records, 1 July and 25 August 2000, #661660 ($1,650,000); Cook County Mortgage Records, 8 August 2000, Documents # 603541 and 603542 (Emigrant Mortgage Co., $1,237,500 and University of Chicago, $412,500); Cook County Mortgage Records, 23 August 2000, Documents # 650071 and 650073 (Mid Town Bank & Trust, $414,000, and University of Chicago, $207,000); "New Faculty Appointments Made Across University," *UCC,* 21 September 2000; Obama, U.S. Senate Financial Disclosure Report, 15 June 2005; Ray Gibson and David Jackson, "Rezko Owns Vacant Lot Next to Obama's Home," *CT,* 1 November 2006; Lynn Sweet, "The Obama Chronicles: Waukegan Interview Transcript," *CST,* 6 November 2006; Sweet, "Obama's Sticky Switch from Media Darling to Media-Hounded," *CST,* 7 November 2006, p. 2; Edward McClelland, "How Close Were Barack Obama and Tony Rezko?," *Salon,* 1 February 2008; Fredric Wondisford e-mail to Bob Bauer, 7 February 2008; Timothy J. Burger, "Obama Bought Home Without Rezko Discount, Seller Says," Bloomberg, 18 February 2008; Jackson and Bob Secter, "Obama: I Toured Home with Rezko," *CT,* 19 February 2008; Tim Novak, "Gov's 'Clout Lists' Surface in Rezko Trial," *CST,* 22 February 2008; Mike McIntire and Christopher Drew, "As Developer Heads to Trial, Questions Linger Over a Deal with Obama," *NYT,* 2 March 2008; James Bone, "Boneheaded Deal Haunts House Obama Built," *TOL,* 7 March 2008; "Complete Transcript of the *Sun-Times* March 14 Interview with Barack Obama," *CST,* 15 March 2008; Obama, Transcript of March 14 *Tribune* Interview, *CT,* 16 March 2008; Binyamin Appelbaum, "Obama Haunted by Friend's Help Securing Dream House," *BG,* 16 March 2008; Coen and Chase, *Golden,* p. 273; Axelrod, *Believer,* p. 165; DJG interviews with F. Scott Winslow.

6. House Floor Transcript, 10 January 2005, pp. 9–10; Bernard Schoenburg, "Obama Opens Local Office," *SJR,* 11 January 2005, p. 1; Stanley Ziemba, "Obama a Big Hit in Lockport," *CT,* 12 January 2005; Russell Lissau, "Obama Meets, Greets in Suburbs," *CDH,* 14 January 2005, p. 1; Brian DuBose, "Obama Sets Out to Build Legacy of Achievement," *WT,* 17 January 2005, p. A1; USS, CFR, "The Nomination of Condoleezza Rice to Be Secretary of State," S. Hrg. 109-151, 18–19 January 2005, pp. 86–90, 141–45, 189–92; *The Oprah Winfrey Show,* 19 January 2005; Eric Zorn, "'08 Reasons Why Obama Will Run for President," *CT,* 20 January 2005; Daniel Pike, "Obama, Vets Talk," *SJR,* 23 January 2005, p. 13; Matt Adrian, "Obama Hears Vets Concerns in Town Hall Meeting," *SI,* 24 January 2005; USS, CVA, "Nomination of Hon. R. James Nicholson to Be Secretary, Department of Veterans Affairs," S. Hrg. 109-14, 24 January 2005, pp. 5–6, 29–38, 47–48, 50–51; Rudolph Bush, "Obama Grills Vet Nominee on Aid," *CT,* 25 January 2005; USS, EPW, "The Need for Multi-Emissions Legislation," S. Hrg. 109-876, 26 January 2005, pp. 23–24,

40–41; Charles Babington, "Rice Is Confirmed Amid Criticism," *WP,* 27 January 2005; Sweet, "Senator May Be Exhausting Progressive Goodwill Very Early," *CST,* 27 January 2005, p. 37; Obama's remarks at Families USA Annual Health Conference, 27 January 2005; Jennifer Skalka, "Obama Rips Bush on Vets' Funds," *CT,* 15 February 2005; Cheryl L. Reed, "Obama Says Bush 'Bull-Headed' About Veterans," *CST,* 15 February 2005, p. 3; DJG interview with Jan Schakowsky.

7. Lynn Sweet, "Obama Using His Fame to Help Other Dems," *CST,* 26 January 2005, p. 24; Charles Keeshan, "Obama Gets Warm Welcome at Town Hall Meeting in Republican Stronghold," *CDH,* 1 February 2005, p. 1; Brian Slupski, "Obama Inspires; Can He Succeed?," *NH,* 1 February 2005; USS, CFR, "Strategies for Reshaping U.S. Policy in Iraq and the Middle East," S. Hrg. 109-21, 1 February 2005, pp. 106–9; USS, EPW, "S. 131, 'The Clear Skies Act of 2005,'" S. Hrg. 109-867, 2 February 2005, pp. 28–29, 90–91; "Bozeman-Evans Cultivated UCSC's Commitment to Community Service," *UCC,* 3 February 2005; USS, CVA, "Benefits for Survivors," S. Hrg. 109-32, 3 February 2005, pp. 7, 49–50; Obama, *CRec,* 3 February 2005, pp. S927–28; Charles Babington and Dan Eggen, "Senate Confirms Gonzales, 60 to 36," *WP,* 4 February 2005; Andrew Zajac, "Senate OKs Gonzales for Cabinet Post," *CT,* 4 February 2005; Sweet, "Anti-War Obama Opposes Gonzales," *CST,* 4 February 2005, p. 29; Obama on *News & Notes,* NPR, 4 and 8 February 2005; Nicole Sack, "Ready for Business," *SI,* 7 February 2005; Emily Udell, "Keeping the Faith," *ITT,* 8 February 2005, p. 10; Amy Boerema, "Obama Jabs at Bush, Himself," *CDH,* 8 February 2005, p. 3; Gary Washburn and Rick Pearson, "A Fracas in the Family," *CT,* 9 January 2005; James Kimberly and Ray Long, "Governor Assigns His Aide to Probe Mell's Charge," *CT,* 13 January 2005; John Chase and Pearson, "Feud Sparks 2nd Probe," *CT,* 14 January 2005; Chase and Washburn, "About-Face: Mell Recants Job-Selling Charge," *CT,* 21 January 2005; Long and John McCormick, "State Contracts, Donations Tied," *CT,* 31 January 2005; USS, EPW, "Environmental Protection Agency's Fiscal Year 2006 Budget," S. Hrg. 109-843, 9 February 2005, pp. 14–15, 33–34; USS, CFR, "Tsunami Response: Lessons Learned," S. Hrg. 109-153, 10 February 2005, pp. 76–81; Virginia Groark and John Chase, "Tollway Oasis Pact Rich with Links to Governor's Allies," *CT,* 13 February 2005; Obama, *CRec,* 14 February 2005, p. S1333; Chase and Ray Long, "Critics Call for Tollway Inquiry," *CT,* 15 February 2005; USS, CVA, "Proposed Fiscal Year 2006 Budget for Department of Veterans Affairs Programs," S.Hrg. 109-72, 15 February 2005, pp. 47, 67–69; USS, CFR, "The President's Budget for Foreign Affairs," S. Hrg. 109-98, 16 February 2005, pp. 43–46; USS, CFR, "Democracy in Retreat in Russia," S. Hrg. 109-83, 17 February 2005, p. 40; Groark and Chase, "Tollway Firm Has Link to Governor," *CT,* 18 February 2005; Groark and Chase, "Oases Developer Won't List Vendors," *CT,* 24 February 2005, Groark, "State to Get Vendor Details," *CT,* 25 February 2005, Jeff Zeleny, "New Man on the Hill," *CT,* 20 March 2005, Zeleny, "Q&A with Michelle Obama," *CT,* 25 December 2005, Sweet, "After Cautious, Bipartisan Year, Obama Opens New Chapter," *CST,* 22 January 2006, p. A12, Mike McIntire and Christopher Drew, "In '05 Investing, Obama Took Same Path as Donors," *NYT,* 7 March 2007; Zeleny and Drew, "Obama Says His Investments Presented No Conflicts of Interest," *NYT,* 8 March 2007; Christi Parsons and Jill Zuckman, "Obama Says He Was Unaware of Stocks in Trust Fund," *CT,* 8 March 2007; Sweet, "Stock Answers," *CST,* 8 March 2007, p. 16; Mike Dorning and Parsons, "Carefully Crafting the Obama Brand," *CT,* 12 June 2007; Perry Bacon Jr., "The Outsider's Insider," *WP,* 27 August 2007; Marc Ambinder, "Teacher and Apprentice," *Atlantic,* December 2007; Joe Stephens, "Obama Got Discount on Home Loan," *WP,* 2 July 2008; Justin Peters, "Behind Barack's 'Suspicious' Mortgage," *CJR,* 2 July 2008; Deborah Howell, "More Story Than a Loan Merited," *WP,* 13 July 2008; Pete Rouse's 2008 interview with Jim Gilmore; Eli Saslow, "A Rising Political Star Adopts a Low-Key Strategy," *WP,* 17 October 2008; Wolffe, *Renegade,* pp. 39–40, 43; Heilemann and Halperin, *Game Change,* pp. 23–25; Suskind, *Confidence Men,* p. 50; Cassandra Butts's 2012 interview with Gilmore; DJG interviews with Nate Tamarin and Tarak Shah.

8. Mark Preston, "Obama, Thune in Demand," *RC,* 15 February 2005; Obama, remarks at John Lewis's 65th-birthday gala, 21 February 2005; Errin Haines, "Hundreds Gather to Celebrate Congressman's 65th Birthday," AP, 21 February 2005; Tom Baxter and Charles Yoo, "Luminaries Applaud Congressman's Legacy," *AJC,* 22 February 2005, p. B1; "U.S. Senator Brings Powerful and Moving Message to

Columbus," *Columbus Times,* 24 February–2 March 2005, p. 2; Deirdre Cox Baker, "Obama Draws Hundreds in Q-C," *QCT,* 24 February 2005; Mark Leibovich, "The Senator's Humble Beginning," *WP,* 24 February 2005, p. C1; Paul Ayars, "Obama Stresses Benefits of Education," Copley, 25 February 2005; Steve Silverman, "Obama Pledges to Improve Schools, Health-Care Access," *BP,* 25 February 2005, p. A3; Obama, remarks to the American Legion Legislative Rally, 1 March 2005; USS, CFR, "Foreign Assistance Oversight," S. Hrg. 109-141, 2 March 2005, pp. 72–73, 166–67; Obama, *CRec,* 2 March 2005, pp. S1904–5; "BarackObama .com Rescued from the Deep End by Pool.com," Canadian Corporate Newswire, 8 March 2005; La Risa Lynch, "Obama Discusses Budget Cuts, Veterans Issues at Town Hall Meeting," *HPC,* 9 March 2005, p. 1; Mary Curtius and Tom Hamburger, "Bush's Clear Skies Act Stalls in the Senate," *LAT,* 10 March 2005; Obama on *All Things Considered,* NPR, 10 March 2005; Kathleen Day, "Senate Passes Bill to Restrict Bankruptcy," *WP,* 11 March 2011; Jeff Zeleny, "Obama Slams Bush," *CT,* 11 March 2005; Lynn Sweet, "Obama Finds Bush's Pitch 'Offensive,'" *CST,* 11 March 2005, p. 3; Obama's CURE keynote address, 11 March 2005; Sandy Banks, "He's Comfortable in His Skin," *LAT,* 13 March 2005; Obama, *CRec,* 15 March 2005, pp. S2727–28; Dan Mihalopoulos and John Chase, "O'Hare Vendor Called Minority Front," *CT,* 16 March 2005; Fran Spielman, "City Accuses O'Hare Firm of Using Minority Front," *CST,* 16 March 2005, p. 20; Obama, *CRec,* 16 March 2005, pp. S2789–90, 2793; Mihalopoulos and Chase, "Governor Shuns O'Hare Scandal," *CT,* 17 March 2005; Chris Fusco et al., "City Ignored Possible O'Hare 'Front' for Years," *CST,* 17 March 2005, p. 8; USS, CVA, "Back from the Battlefield," S. Hrg. 109-109, 17 March 2005, pp. 4–5, 22–24; Obama, *CRec,* 17 March 2005, pp. S2885–86; Mihalopoulos and Laurie Cohen, "City Knew O'Hare Vendor Wasn't Minority-Run Firm," *CT,* 18 March 2005; Spielman et al., "Daley: I Don't Know Why City Slow to Act," *CST,* 18 March 2005, p. 11; Dori Meinert, "Obama Marches with Democrats, but Not in Lock-step," and "Obama's Staff Combines Experience, Enthusiasm," Copley, 18 March 2005; "Five Minutes with: Sen. Barack Obama," CampusProgress.org, ca. 23 March 2005; Mickey Ciokajlo and Mihalopoulos, "Minority Firm Led by a Dead Woman," *CT,* 26 March 2005; Rummana Hussain, "Dead Woman Listed as Leading Firm," *CST,* 27 March 2005; Ciokajlo and Chase, "Stroger Taps Blagojevich Aide," *CT,* 8 April 2005; Mike Dorning, "In Big Jump, Obamas Earn $1.67 Million," *CT,* 26 September 2006; Dorning, "Employer: Michelle Obama's Raise Well-Earned," *CT,* 27 September 2006; Obama, *TAOH,* p. 342; Joe Stephens, "Contracts Went to a Longtime Donor," *WP,* 22 August 2008; Ben Whitford, "Coal and Clear Skies," *Plenty Magazine,* 31 October 2008.

9. Jeff Zeleny, "New Man on the Hill," *CT,* 20 March 2005; David Mendell, "Obama Quietly Helps Draw Fine Line," *CT,* 27 March 2005; Press Release, "Obama Unveils Bill to Make College More Affordable," 28 March 2005, Obama.Senate.gov; Lynn Sweet, "Obama's 1st Bill: Raising Pell Grants," *CST,* 29 March 2005, p. 28; Obama to "Dear MoveOn Member," 29 March 2005; Alexa Aguilar, "First on Obama's Agenda: A More Affordable Education," *SLPD,* 30 March 2005, p. B1; Patrick J. Powers, "Obama Tackles Issues in Town Hall Meeting," *BND,* 30 March 2005; Paul J. Nyden, "Senator Gives Big Boost to Byrd," *Charleston Gazette,* 31 March 2005, p. 1A; Nate Sandstrom, "Obama Pushes Pell Grant Bill," *DI,* 31 March 2005; Ed Tibbetts, "Silvis Marine Greets Obama at Black Hawk," *QCT,* 1 April 2005; Kurt Allemeier, "Obama, Q-C Marine Reunite," *MD,* 1 April 2005; Obama, *CRec,* 5 April 2005, p. S3205 (introducing S. 697, the Higher Education Opportunity Through Pell Grant Expansion [HOPE] Act); USS, EPW, "Pending Nominations," S. Hrg. 109-646, 6 April 2005, pp. 9, 139–41; Obama, opening statement at confirmation hearing of John Bolton, 11 April 2005; Obama, remarks at the Herblock Foundation annual lecture, 11 April 2005; Steven R. Weisman, "Nominee to U.N. Defends Record at Senate Panel," *NYT,* 12 April 2005; Obama Press Release, "My Visit to Walter Reed Army Medical Center," 12 April 2005, Obama.Senate.gov; Allison Horton, "Senator Obama Hosts Townhall Meeting," *HPC,* 13 April 2005, p. 1; Obama, *CRec,* 13 April 2005, pp. S3511–12; Obama, *CRec,* 14 April 2005, pp. S3641–43; Lawrence A. Horwich, Barack H. Obama and Michelle L. Obama 2004 Form 1040, 5 March 2005; Obama, "Pell Grant Expansion Offers Students Hope," *BP,* 17 April 2005, p. C3; Perry Bacon Jr., "Barack Obama: The Future of the Democratic Party?," *Time,* 18 April 2005; Bernard Schoenburg, "Obama Maintains Chicago Home," *SJR,* 19 April 2005, p. 12; Doug Wilson, "Obama Speaks Out," *QHW,* 20 April 2005; Obama,

"Remarks at the Opening of the Abraham Lincoln Presidential Library and Museum," 20 April 2005; Schoenburg, "Each Speaker Finds Lessons in Lincoln," *SJR,* 20 April 2005, p. 1; USS, EPW, "Pending Nominations," S. Hrg. 109-652, 20 April 2005, pp. 17, 21–23; Obama, *CRec,* 20 April 2005, p. S3996; Jon Sawyer and Deirdre Shesgreen, "Obama Is Reluctant to Defend Filibuster in Judicial Battle," *SLPD,* 21 April 2005, p. A4; Caleb Hale, "Obama Stands Firm in His Opposition to Clear Skies Act," *SI,* 24 April 2005; Obama, *CRec,* 21 April 2005, pp. S4082 and 4087–88; USS, CFR, "The Millennium Challenge Corporation's Global Impact," S. Hrg. 109-187, 26 April 2005, pp. 18–21; Obama, "Remarks at the National Press Club," 26 April 2005; Obama, *CRec,* 27 April 2005, pp. S4405–6 (introducing S. 918, the E-85 Fuel Utilization and Infrastructure Development Incentives Act); Obama, *CRec,* 28 April 2005, pp. S4502–3, S4581–82 (introducing S. 969, the Attacking Viral Influenza Across Nations [AVIAN] Act); Obama, remarks at NAACP Fight for Freedom Fund dinner, 2 May 2005; Tom Collins, "Obama Addresses Veterans Care, Abortion, Social Security," *LNT,* 6 May 2005; Jonathan Bilyk, "Obama Blasts Bush, GOP in Ottawa Speech," *ODT,* 6 May 2005; Jo Ann Hustis, "Q&A: Obama Answers Questions at Town Meeting," *MDH,* 6 May 2005; Obama, remarks at the *Rockford Register Star* Young American Awards, 7 May 2005; Aracely Hernandez, "Crowd of 400 Cheers Obama in Sycamore," *DDC,* 9 May 2005; Obama on *Chicago Sunday,* WTTW, 8 May 2005; Obama, *CRec,* 10 May 2005, pp. S4830–31; USS, CFR, 12 May 2005 (John Bolton); USS, CVA, "An Open Discussion: Planning, Providing, and Paying for Veterans' Long-Term Care," S. Hrg. 109-240, 12 May 2005, pp. 48–49; Obama, *CRec,* 12 May 2005, pp. S5025–26 and S5058–59; "Many View Obama as He Does Lincoln," *BP,* 4 July 2005, p. A3; Mara Liasson, "Barack Obama, Still on Rise," *All Things Considered,* NPR, 8 December 2006; DJG interview with Bethann Hester.

10. Abdon M. Pallasch, "Jail's Phone Contractor Has Links to Stroger," *CST,* 2 May 2005, p. 27; Rick Pearson, "Governor's Ratings Tank," *CT,* 15 May 2005; Dave McKinney and Frank Main, "Mell's Claims Spur Grand Jury Probe of Gov's Inner Circle," *CST,* 15 May 2005; Fran Spielman, "Company to Give Up Disputed O'Hare Restaurants," *CST,* 16 May 2005, p. 12; "Oprah Thows a Party Honoring Black Women Who Inspired Her," *SJMN,* 17 May 2005, p. A2 (14 May); John Chase and Jeff Coen, "Blagojevich Blames Probe on Family Feud," *CT,* 17 May 2005; USS, CFR, "The Commission for Africa," S. Hrg. 109-203, 17 May 2005, pp. 31–34, 39–42; McKinney, "2 State Officials Say Gov's Fund-Raiser Helped Get Jobs," *CST,* 19 May 2005; USS, CFR, "Iran Weapons Proliferation, Terrorism, and Democracy," S. Hrg. 109–211, 19 May 2005, p. 1; John Chase, "First Lady of Illinois Linked to Developer," *CT,* 20 May 2005; transcript of Veterans Town Hall Meeting with Secretary Nicholson and Senators Obama & Durbin, Wright Junior College, Chicago, 20 May 2005; David Usborne, "The Contender," *Independent,* 25 May 2005; Chase, "Tony Rezko," *CT,* 26 May 2005; Jackie Kucinich, "Obama and CBC Split on Cloture," *TH,* 26 May 2005, p. 1; USS, CVA, "Battling the Backlog," S. Hrg. 109-216, 26 May 2005, pp. 2–3, 61–63; Obama, *CRec,* 26 May 2005, pp. S5965–67; Henry Stuttley, "Obama Comes to Hear from DuPage," *CDH,* 28 May 2005, p. 3; Obama, remarks at the Abraham Lincoln National Cemetery, 30 May 2005; Allison Horton, "Obama Hosts Youth Summit," *HPC,* 1 June 2005, p. 1; Obama, remarks at Knox College commencement, 4 June 2005; John R. Pulliam, "Obama Big on Ceremonies," *GRM,* 5 June 2005; Kevin Sampier, "Providing Inspiration," *PJS,* 5 June 2005; David Kusnet, "Stumped," *TNR,* 20 June 2005; Obama and Lugar, "Grounding a Pandemic," *NYT,* 6 June 2005; Kathy Kiely, "Fresh Faces in Congress Stress Cooperation," *USAT,* 6 June 2005, p. 10A; USS, CFR, "The Emergence of China Throughout Asia," S. Hrg. 109-239, 7 June 2005, pp. 20–22, 28–30; Obama, *CRec,* 7 June 2005, p. S6159 (introducing S. 1180, the Shelter All Veterans Everywhere [SAVE] Reauthorization Act); Jackie Kucinich, "This Time, Obama and CBC Agree in Close Cloture Vote," *TH,* 8 June 2005, p. 4; Obama, *CRec,* 8 June 2005, pp. S6178–80 and S6228 (introducing S. 1194, the Spent Nuclear Fuel Tracking and Accountability Act); USS, CVA, "Pending Health Care Related Legislation," S. Hrg. 109-217, 9 June 2005, pp. 7–8; Jeff Zeleny, "Obama Grills Nominee on '87 Racial Remarks," *CT,* 10 June 2005; Obama, commencement address at the Pritzker School of Medicine, 10 June 2005; Obama, *CRec,* 13 June 2005, p. S6375; USS, CFR, "North Korea," S. Hrg. 109-226, 14 June 2005, pp. 32–35; Lesley R. Chinn, "Obama Wants Black People to Have Courage to Lead," *CC,* 15 June 2005, p. 3; Obama, *CRec,* 15 June 2005, pp. S6609, S6647; Obama, *CRec,* 16 June

2005, pp. S6690, 6698; Scott Fornek, "Obama Is No. 1 Most Popular Senator," *CST,* 17 June 2005, p. 4; Carl Hulse, "Senator Seeks Details on Nominee's '87 Speech," *NYT,* 18 June 2005; Liam Ford, "Obama's Church Sermon to Black Dads: Grow Up," *CT,* 20 June 2005; USS, CFR, "U.S. Policy Toward Russia," S. Hrg. 109-250, 21 June 2005, pp. 40–45; Obama, *CRec,* 21 June 2005, p. S6897; Obama, "Here's What It Takes to Be a Bona Fide 'Full-Grown' Man," and "Obama's Colorblind Words," *CT,* 22 June 2005; Obama, *CRec,* 22 June 2005, p. S7008; Zeleny, "Obama Wants to See FBI File on Nominee," *CT,* 23 June 2005; Rodney D. Sieh, "Obama: Debt Relief Should Be Contingent on Taylor Turn Over," *NYB,* 23–29 June 2005, p. 10; Mary Ann Ford, "Obama Fields Questions in B-N," *BP,* 25 June 2005, p. A1; Carrie Kepple, "Obama Finds Job 'Exhilarating,'" *PJS,* 25 June 2005; Obama, address to the American Library Association, 25 June 2005; Obama, "Bound to the Word," *American Libraries,* August 2005, pp. 48–52; Rummana Hussain, "Obama Fears 'Big Brother' Over Our Shoulders," *CST,* 26 June 2005, p. 31; Zeleny, "When It Comes to Race, Obama Makes His Point—with Subtlety," *CT,* 26 June 2005; USS, CVA, "Emergency Hearing to Examine the Shortfall in VA's Medical Care Budget," S. Hrg. 109-263, 28 June 2005, pp. 11–12, 28–29; Obama, *CRec,* 28 June 2005, p. S7460; La Risa Lynch, "Obama Addresses 21st Ward Town Hall Meeting," *HPC,* 29 June 2005, p. 1; Peggy Noonan, "Conceit of Government," *WSJ,* 29 June 2005; Zeleny, "Obama Reviews FBI File on Stalled State Nominee," *CT,* 29 June 2005; Obama, *CRec,* 29 June 2005, pp. 7558–59; Obama, "Why I Oppose CAFTA," *CT,* 30 June 2005; USS, CFR, "The Challenge to the Middle East Road Map," S. Hrg. 109-268, 30 June 2005, pp. 26–29; Obama, *CRec,* 30 June 2005, pp. 7697–98; Obama, "What I See in Lincoln's Eyes," *Time,* 4 July 2005, p. 74; Molly Parker, "Up Close and Personal," *PJS,* 6 July 2005; USS, CVA, "Is the VA Prepared to Meet the Needs of Our Returning Vets?," S. Hrg. 109-218, 6 July 2005, pp. 1–43; Sharon Woods Harris, "Obama Listens to Residents' Concerns," *PDT,* 7 July 2005; Obama in Wolffe, *Renegade,* p. 41.

11. Lynn Sweet, "Book Deals Put Obama in Millionaires' Club," *CST,* 16 June 2005; Warranty Deed 517233010, Northern Trust Co. Trust #10209, 21 June 2005; Federal Election Commission, "Factual and Legal Analysis," *In re Barack Obama,* MUR 6035; Sandra Mathers, "Obama Lends Star Power to Nelson," *Orlando Sentinel,* 10 July 2005; USS, CFR, "North American Cooperation on the Border," S. Hrg. 109-423, 12 July 2005, pp. 37–40; USS, EPW, "Nominations of Marcus Peacock et al.," S. Hrg. 109-725, 14 July 2005, pp. 31–35, 57–59; Sweet, "Obama Draws on African Roots as He Steps onto Global Stage," *CST,* 18 July 2005, p. 30; USS, CFR, "Policy Options for Iraq," S. Hrg. 109-312, 18–20 July 2005, p. 147; Jill Zuckman, "State Department Nominee to Go Forward After Obama Relents," *CT,* 19 July 2005; Obama, *CRec,* 19 July 2005, pp. 8447–50 and S 8490 (introducing S. 1426, the Drinking Water Security Act); Chris L. Jenkins, "Obama Rallies N. Va. Democrats Behind Kaine," *WP,* 21 July 2005; USS, CFR, "United Nations Reform," S. Hrg. 109-356, 21 July 2005, pp. 32–33; Obama Press Release, "Darfur and the U.N.," 22 July 2005, Obama.Senate.gov; Karen E. Pride, "Obama Discusses Sudan Crisis," *CD,* 22–24 July 2005, p. 2; Sharon Cohen, "Obama Keeps Perspective as Freshman," AP, 24 July 2005; Obama, remarks to the AFL-CIO national convention, Chicago, 25 July 2005; Obama, "Upgrading Health-Care Technology Would Save Many Lives, Much Money," *CDH,* 26 July 2005, p. 10; USS, CFR, "Energy Trends in China and India," S. Hrg. 109-326, 26 July 2005, pp. 26–27; Obama, *CRec,* 28 July 2005, p. S9248; Brendan R. Linn, "Power to Advise Obama for Year," *HC,* 29 July 2005; Obama, *CRec,* 29 July 2005, pp. S9338–39; Tedd Carrison, "Point Rehab Future in Sen. Obama's Hands," and "Dear Sen. Obama, Please Save the Point," *HPH,* 3 August 2005, pp. 1, 4, 7; Gary Washburn, "Obama Coy on Backing Daley," *CT,* 5 August 2005; Fran Spielman, "Obama Won't Commit to Daley in '07," *CST,* 5 August 2007, p. 3; Kari Lydersen, "Thousands Pay Tribute to Publisher of Ebony and Jet," *WP,* 16 August 2005; James Janega and Charles Storch, "Saluting Pioneer Publisher John H. Johnson," *CT,* 16 August 2005; Steve Horrell, "Talking to the Senator," *EI,* 17 August 2005; Paul Hampel, "Obama Tells Crowd in Godfrey He Opposes Tapping Oil Reserve," *SLPD,* 17 August 2005, p. 1; Doug Wilson, "Meeting the Masses," *QHW,* 17 August 2005; Bernard Schoenburg, "Democrats Rally for Blagojevich," *SJR,* 18 August 2005, p. 1; Mike Frazier, "Economic Issues Fill Sen. Obama's Ears," *DHR,* 19 August 2005; Paul Wood, "Lawmaker Expresses Reservations About War," *CNG,* 19 August 2005; Joe Parrino, "Obama Talks to Voters," *DI,* 22 August 2005; Stanley Ziemba, "Illinois' Sen.

Obama Steps into Peotone Airport Feud," *CT,* 22 August 2005; Ziemba, "Airport Compromise on Table," *CT,* 23 August 2005; Obama, *CRec,* 25 May 2005, pp. S5862–64; Press Release, "Obama to Visit Nuclear, Biological Weapons Destruction Facilities in Former Soviet Union," 23 August 2005, Obama.Senate.gov; David Holley, "U.S.-Russian Efforts to Protect Arsenal Gain Steam," *LAT,* 27 August 2005; Jeff Zeleny, "U.S. Focuses on Russian WMD," *CT,* 27 August 2005; Zeleny, "Wasn't the Cold War Supposed to Be Over?," *CT,* 29 August 2005; Zeleny, "U.S., Ukraine Sign Pact on Germ Threat," *CT,* 30 August 2005; Sweet, "Obama Part of Group Locked Up at Russian Airport," *CST,* 29 August 2005, p. 2; Sweet, "Russia Plane 'Delay' Wasn't a 'Detention,'" *CST,* 30 August 2005, p. 20; "Lugar and Obama Urge Destruction of Conventional Weapons Stockpiles," 30 August 2005; Zeleny, "U.S. Gets Pathogens from Ex-Soviet Republic," *CT,* 3 September 2005; Zeleny, "A Foreign Classroom for Junior Senator," *CT,* 23 September 2005; Mike Dorning, "In Big Jump, Obamas Earn $1.67 Million," *CT,* 26 September 2006; Sholto Byrnes, "This Cat Can Scat!," *The New Review,* 14 October 2007; Joe Stephens, "Obama Got Discount on Home Loan," *WP,* 2 July 2008; David Nather, "Obama's Quick Rise on a Non-Traditional Career Path," *CQ Weekly,* 21 August 2008.

12. Stephanie Zimmerman and Scott Fornek, "Daley 'Shocked' as Feds Reject Aid," *CST,* 3 September 2005, p. 3; Lynn Sweet, "To Obama, Tragedy More About Class Than Race," *CST,* 5 September 2005, p. 8; Mike Colias, "Obama Says Hurricane Victims Will Need Long-Term Support," AP, 5 September 2005; Lisa Rein, "Clinton and Bush in Charity Reprise," *WP,* 6 September 2005; Ron Harris, "Big Names Converge on Houston to Greet Evacuees," *SLPD,* 6 September 2005, p. A10; Obama, *CRec,* 6 September 2005, pp. S9628–29; Obama, *CRec,* 7 September 2005, p. S9737 (introducing S. 1630, the National Emergency Family Locator Act); Obama, *CRec,* 8 September 2005, pp. S9781–82, S9786, and S9844–45 (introducing S. 1638, the Hurricane Katrina Emergency Health Workforce Act); Obama Podcast #1, "Hurricane Katrina," 8 September 2005, Obama.Senate.gov; Obama on *This Week,* ABC. 11 September 2005; Jeff Zeleny, "Judicious Obama Turns Up Volume," *CT,* 12 September 2005; Obama, *CRec,* 12 September 2005, p. S9940 (introducing S. 1685, the Evacuation of Individuals with Special Needs Act); Allison Samuels and Richard Wolffe, "When Oprah Met Obama," *Newsweek,* 10 December 2007, p. 12; Carrie Budoff, "Santorum Strikes Back at Casey," *PI,* 13 September 2005, p. B4; Clarence Page, "We're Still Two Nations," *CT,* 14 September 2005; Obama Podcast #2, "Katrina Relief Spending Accountability," 14 September 2005, Obama.Senate.gov; transcript, Coburn-Obama news conference, 14 September 2004; USS, CFR, "U.S.-Indonesia Relations," S. Hrg. 109-407, 15 September 2005, pp. 3–4, 67–70; Obama, "Securing Our Energy Future," Resources for the Future, Washington, D.C., 15 September 2005; Obama, response to president's speech about Gulf Coast rebuilding effort, 15 September 2005; Dori Meinert, "Sen. Obama Broadening His Scope," Copley, 16 September 2005; Zeleny, "Obama on Bush," *CT,* 18 September 2005; Tracy Jan, "Obama Urges Alumni to Help Fight Poverty," *BG,* 18 September 2005; Obama on *Face the Nation,* CBS, 18 September 2005; Javier C. Hernandez, "Obama Talks Katrina Relief," *HC,* 19 September 2005, Carol Beggy and Mark Shanahan, "Obama Draws a Fund-Raising Crowd," *BG,* 19 September 2005, p. B8; Alvin Powell, "Rights, Equality Center Stage at HLS Events," *Harvard Gazette,* 22 September 2005; Marcella Bombardieri and Sarah Schweitzer, "Talking About U.S. Senator," *BG,* 25 September 2005; Dick Dahl, "Obama Leads Focus on Katrina Aftermath During Celebration of Black Alumni," *Harvard Law Today,* November 2005; Obama, remarks at Rahm Emanuel roast for CURE, 20 September 2005.

13. Jeff Zeleny, "Please Move Over, Dear Oprah," *CT,* 19 September 2005; Obama on *News & Notes,* NPR, 21 September 2005; Obama Podcast #3, "Poverty in America," 21 September 2005, Obama.Senate.gov; Terry Neal, "Race, Class Re-Enter Politics After Katrina," *WP,* 22 September 2005; Steve Ivey, "Obama: White House Blind to Poverty in U.S.," *CT,* 22 September 2005; Obama, *CRec,* 22 September 2005, pp. 10358–59 and S10365–66; Tom Coburn and Obama, "Rebuilding with Accountability," *WT,* 23 September 2005, p. A20; Carol Marin, "Obama Dissent Gets Thumbs Up," *CST,* 25 September 2005; Karen E. Pride, "Obama Voices Opposition to Photo Voting Requirement," *CD,* 26–27 September 2005, p. 3; Obama, *CRec,* 26 September 2005, p. S10435 (introducing S. 1770, the Hurricane Katrina Fast-Track Refunds for Working Families Act); Obama on *The Tavis Smiley Show,* 27 September 2005; Perry Bacon Jr., "Barack Obama Steps (Carefully) into the Spotlight," *Time,* 28

September 2005; Tedd Carrison, "Obama Last Hope to Preserve Point," and Obama, "The Point Means a Lot to the Community," *HPH,* 28 September 2005, pp. 1–2, 4; USS, CFR, "Darfur Revisited," S. Hrg. 109-868, 28 September 2005, pp. 42–44, 65–71, 86–89; Obama, *CRec,* 28 September 2005, pp. S10628–29; Obama Podcast #4, "Judge John Roberts Supreme Court Nomination," 28 September 2005, Obama .Senate.gov; Emily Pierce, "Coburn, Obama Latest Odd Couple," *RC,* 29 September 2005; Deirdre Shesgreen, "Measure Would Reduce Medical Mistakes and Lawsuits," *SLPD,* 29 September 2005, p. A2; USS, CVA, "Pending Nominations," S. Hrg. 109-394, 29 September 2005, pp. 7, 81–83; Obama, *CRec,* 29 September 2005, pp. S10654–55; Obama, "Tone, Truth, and the Democratic Party," DailyKos.com, 30 September 2005; USS, CFR, "Iraq in U.S. Foreign Policy," S. Hrg. 109-449, 19 October 2005, pp. 63–67; Steven R. Weisman, "Rice, in Testy Hearing, Cites Progress in Iraq," *NYT,* 20 October 2005; Robin Wright, "Rice Declines to Give Senators Timeline for Iraq Withdrawal," *WP,* 20 October 2005; USS, CFR, "Avian Influenza: Are We Prepared?," S. Hrg. 109-628, 9 November 2005, pp. 55–60, 116–20; Bacon, "The Outsider's Insider," *WP,* 27 August 2007; Rouse's 2008 interview with Gilmore; Wolffe, *Renegade,* p. 42; Axelrod, *Believer,* p. 170.

14. Tom Philpott, "Obama Fights Review of Stress Cases," *Military Update,* 3 October 2005; Tedd Carrison, "Obama Calls Emergency Point Meeting," *HPH,* 5 October 2005, p. 1; USS, EPW, "Kyoto Protocol," S. Hrg. 109-1016, 5 October 2005, pp. 20, 40–42, 79–80, 217; Obama, *CRec,* 5 October 2005, pp. S11069–70 and 11138–39 (introducing S. 1821, the Pandemic Preparedness and Response Act); Obama Podcast #5, "Avian Flu Preparedness," 5 October 2005, Obama.Senate.gov; USS, EPW, "The Roles of the Environmental Protection Agency et al.," S. Hrg. 109-978, 6 October 2005, p. 1; Judy Ogutu, "Senator Obama to Visit Kenya Next Year," *The Standard* (Nairobi), 10 October 2005; "Sen. Obama Deserves Kudos," *HPH,* 12 October 2005, p. 4; Obama Podcast #6, "From the Road," 14 October 2005, Obama.Senate.gov; Christopher Wills, "Obama Pushes for Mix of Taxes, Spending Cuts for Katrina," AP, 14 October 2005; Caleb Hale, "Obama Sees Progress, Frustration as Senator," *SI,* 15 October 2005; Jeff Zeleny, "Voters Give Obama, Durbin Good Marks," *CT,* 16 October 2005; Bernard Schoenburg, "Sen. Obama Sees Little 'Genuine Debate' in New Job," *SJR,* 16 October 2005, p. 15; Tara Fasol, "Barack Obama on Hand to Answer Questions," *WFDA,* 17 October 2005; USS, EPW, "S. 1772, The Gas Petroleum Refiner Improvement and Community Empowerment Act of 2005," S. Hrg. 109-1001, 18 October 2005, pp. 8–9, 25–26; Obama, *CRec,* 18 October 2005, pp. S11451–52; Obama Podcast #7, "Impressions of the Senate After the First 10 Months," 19 October 2005, Obama.Senate.gov; Obama, *CRec,* 20 October 2005, p. S11647; Obama, "Thanks for the Feedback," DailyKos.com, 20 October 2005; Joy Bennett Kinnon, "Michelle Obama," *Ebony,* March 2006, pp. 58ff.; Kalari Girtley and Brian Wellner, "Michelle Obama Is Hyde Park's Career Mom," *HPH,* 14 February 2007, pp. 7–8, 16–17; Jonathan Darman, "Mr. Obama's Washington," *Newsweek,* 30 June 2008; Eli Saslow, "A Rising Political Star Adopts a Low-Key Strategy," *WP,* 17 October 2008; Ashley Fantz, "Getting Inside Obama's Brain," CNN.com, 13 November 2008; Rouse's 2008 interview with Gilmore; Paul Schwartzman, "Mr. Obama's (Giddy) Neighborhood," *WP,* 18 Janaury 2009; Michael Scherer and Nancy Gibbs, "Interview with the First Lady," *Time,* 21 May 2009; Wolffe, *Renegade,* p. 42; Jonathan Van Meter, "Leading by Example," *Vogue,* April 2013, pp. 248–55, 318–19; Axelrod, *Believer,* pp. 168, 171.

15. Michael Kelly, "Buffetts: Barack a Name to Note," *OWH,* 25 October 2005, p. B1; Obama, *CRec,* 25 October 2005, pp. S11814–15; Obama, "Teaching Our Kids in a 21st Century Economy," Center for American Progess, 25 October 2005; Obama, statement on the death of Rosa Parks, 25 October 2005; Obama Podcast #8, "Remembering Rosa Parks," 26 October 2005, Obama.Senate.gov; Obama, *CRec,* 26 October 2005, pp. S11909–10; Obama on *Talk of the Nation,* NPR, 26 October 2005; Lynn Sweet, "Opening Soon: Obama's School," *CST,* 27 October 2005, p. 45; Obama, *CRec,* 27 October 2005, pp. S11972–73; USS, CVA, "The Rising Number of Disabled Veterans Deemed Unemployable," S. Hrg. 109-425, 27 October 2005, pp. 28–31; Christina Nuckols, "Ill. Senator Gives Kaine a Hand on Stump," *Virginian Pilot,* 31 October 2005; Jeff E. Shapiro, "Kaine: 'We've Got the Momentum,'" *RTD,* 31 October 2005, p. B1; Obama, "The Political Movement in Black America," *Ebony,* November 2005, pp. 116ff.; Obama and Richard Lugar, "Challenges Ahead for Cooperative Threat Reduction," Council on Foreign Relations, *CRec,* 1 November 2005, pp. S12130–31; Jeff Zeleny,

"Obama-Lugar Proposal Targets Stockpiles of Conventional Weapons," *CT,* 2 November 2005; AP, "Seven-Hour Funeral Pays Tribute to Parks," 2 November 2005; Tom Curry, "Obama Builds Foreign Policy Credentials," MSNBC.com, 3 November 2005; Obama, *CRec,* 3 November 2005, p. S12312; Ben Roberts, "Obama Leads Democrats in Efforts on Bird Flu," *SLPD,* 6 November 2005, p. B4; Zeleny, "Obama Dabbles in a Moment of Comedy," *CT,* 8 November 2005; Obama, *CRec,* 8 November 2005, pp. S12521–22 and 12531 (introducing S. 1975, the Deceptive Practices and Voter Intimidation Prevention Act); Obama, remarks regarding the Kaiser Family Foundation's "Sex on TV" report, 9 November 2005; Lisa De Moraes, "Television More Oversexed Than Ever," *WP,* 10 November 2005; Zeleny, "Study Twice as Much Sex Aired on TV," *CT,* 10 November 2005; Obama, remarks at the National Women's Law Center, 10 November 2005; Obama Podcast #9, "Protecting Voters Against Deceptive Practices," 10 November 2005, Obama.Senate.gov; James A. Barnes and Peter Bell, "In the Year 2025," *NJ,* 12 November 2005, pp. 3516–17; Obama, "Cutting 'Pork' to Rebuild Coast," *CT,* 12 November 2005; Dan Lavoie, "A World of Thanks," *DS,* 12 November 2005; Eugene Scott, "Sen. Obama Urges Arizonans to Vote for Pederson," *Arizona Republic,* 13 November 2005, p. B5; Emilia Arnold, "Obama, Pederson Call for Lower Education Costs," *ASU Web Devil,* 14 November 2005; Obama, *CRec,* 15 November 2005, p. S12805, 12817–18; Obama, *CRec,* 16 November 2005, p. S12943; USS, EPW, "Oversight to Examine Transportation Fuels of the Future," S. Hrg. 109-1012, 16 November 2005, pp. 8–9, 36–37, 70, 93; Obama Podcast #10, "Fiscal Responsibility and Gulf Coast Reconstruction," 16 November 2005, Obama.Senate.gov; Obama, remarks at the Robert F. Kennedy Human Rights Award ceremony, 16 November 2005; USS, CFR, "African Organizations and Institutions," S. Hrg. 109-900, 17 November 2005, pp. 18, 22–25; Obama, *CRec,* 17 November 2005, p. S13090–92, 13121–22, 13124, 13131–32, and 13174–75 (introducing S. 2047, the Healthy Communities Act, and S. 2048, the Lead Free Toys Act); Zeleny, "Obama's National Appeal Rallies an Army of Backers," *CT,* 20 November 2005; Dennis Conrad, "Obama Turns to Speeches to Spread the Word," AP, 21 November 2005; Obama, "Moving Forward in Iraq," Chicago Council on Foreign Relations, 22 November 2005; Peter Slevin, "Obama Calls on Bush to Admit Iraq Errors," *WP,* 23 November 2005; Zeleny, "Obama: Pull GIs from Iraq Gradually," *CT,* 23 November 2005; "Wise Words from Obama," *SJR,* 23 November 2005, p. 6; Katherine Hutt Scott, "Obama Picked as Most Likely to Become President," Gannett, 23 November 2005; Glen Ford and Peter Gamble, "Obama Mouths Mush on War," *Black Commentator* 161, 1 December 2005; Obama, U.S. Senate Financial Disclosure Report, 15 May 2006; Christi Parsons and Jill Zuckman, "Obama Says He Was Unaware of Stocks in Trust Fund," *CT,* 8 March 2007; Lynn Sweet, "Stock Answers," *CST,* 8 March 2007, p. 16; Mike Dorning, "Questions Stalk Obama's Portfolio," *CT,* 9 March 2007.

16. Obama, Podcast #11, "Happy Thanksgiving," 22 November 2005, Obama.Senate.gov; Sarah Brown, "Obama '85 Masters Balancing Act," *Daily Princetonian,* 7 December 2005; Richard G. Lugar and Obama, "Junkyard Dogs of War," *WP,* 3 December 2005; Jeff Zeleny and Rick Pearson, "Obama: Iraq War Splits Democrats," *CT,* 6 December 2005; Martin Schram, "Senate Duo Exception to Partisan Bickering," Scripps Howard, 6 December 2005; Ryan Lizza, "Why Barack Obama Should Run for President in 2008," *TNR,* 6 December 2005; Eric Krol, "Obama Heads to Iraq in January to Assess Progess," *CDH,* 7 December 2005, p. 7; Burt Constable, "Obama's Ready to Be President If We Can Convince Him to Run," *CDH,* 8 December 2005, p. 15; Krol, "Better Than Expected," *CDH,* 2 January 2006, p. 1; DJG interview with Allison Davis.

17. Jeff Zeleny, "Sen. Obama Gets Grammy Nomination in Narration Category," *CT,* 9 December 2005; Obama, *The Al Franken Show,* Air America, 9 December 2005; Wes Allison, "State Party Hitched to Rising Star," *SPT,* 10 December 2005, p. A1; Steve Bousquet, "Obama Rocks the House at Florida's Convention," *SPT,* 11 December 2005, p. B4; Obama, *CRec,* 14 December 2005, pp. S13528–29; Obama and Mel Martinez, "Coming to America," *WSJ,* 15 December 2005; transcript of Obama and Martinez press conference, 15 December 2005; Obama, *CRec,* 15 December 2005, pp. S13629–30; Obama, *CRec,* 16 December 2005, pp. S13773, 13788 (introducing S. 2125, the Democratic Republic of the Congo Relief, Security and Democracy Promotion Act); Obama Podcast #12, "The Reauthorization of the Patriot Act," 16 December 2005, Obama.Senate.gov; Deirdre Shesgreen, "Democrats' Rising Star Takes First Year Carefully," *SLPD,* 16 December 2005; Sheryl Gay

Stolberg and Eric Lichtblau, "Senators Thwart Bush Bid to Renew Law on Terrorism," *NYT,* 17 December 2005; Kurt Erickson, "Durbin Helps Kill Patriot Act Renewal," *BP,* 17 December 2005, p. A1; Dori Meinert, "Democrats Assail Wiretaps," *PJS,* 20 December 2005; Obama, *CRec,* 20 December 2005, pp. 14149–50 and S14187–88 (introducing S. 2149, the Summer Term Education Programs for Upward Performance [STEP UP] Act); Obama, *CRec,* 21 December 2005, p. 14248; Zeleny, "The First Time Around: Sen. Obama's Freshman Year," and "Q&A with Sen. Barack Obama," *CT,* 25 December 2005; Obama and Sam Brownback, "Policy Adrift on Darfur," *WP,* 27 December 2005; Heilemann and Halperin, *Game Change,* pp. 29–30.

18. Jeff Zeleny, "Obama Making 1st Trip to Iraq," *CT,* 5 January 2006; Jason Straziuso, "Obama: Involve More Minorities in Iraq Government," AP, 7 January 2006; Borzou Daragahi, "Congressmen Warn Iraqis to Shape Up," *CT,* 8 January 2006; Lynn Sweet, "Obama Sees No Military Solution in Iraq," *CST,* 8 January 2006, p. A8; Eric Krol, "Obama: 'Everything's Up for Grabs' in Iraq," *CDH,* 8 January 2006, p. 7; Dori Meinert, "Obama: Obstacles Still in Iraq," *SJR,* 8 January 2006; "U.S. Helicopter Crash Kills 12 in Iraq," CNN.com, 9 January 2006; Obama Podcast #13, "From the Road: Speaking with American Troops in Iraq," 9 January 2006, Obama.Senate.gov; Herb Keinon, "Rising Star Illinois Senator Obama on First Visit Here," *JP,* 10 January 2006, p. 5; Chuck Goudie, "Obama Visits Remote Israeli Town with Chicago Ties," ABC7 Chicago, 11 January 2006; "JUF Leaders Join Obama on Israel Visit," *JUF News,* [11 January 2006]; Goudie, "Obama Meets with Arafat's Successor," ABC7 Chicago, 12 January 2006; Goudie, "Obama Wraps Up Middle East Trip," ABC7 Chicago, 13 January 2006; Obama Podcast #14, "From the Road: Israel and the Palestinian Territories," 14 January 2006, Obama.Senate.gov; Flynn Murphy, "Obama Townhall," *NS,* 18 January 2006; USS, CFR, "New Initiatives in Cooperative Threat Reduction," S. Hrg. 109-880, 9 February 2006, pp. 5–6, 23–29, 44–46; Ray Gibson and David Jackson, "Rezko Owns Vacant Lot Next to Obama's Home," *CT,* 1 November 2006; Dave McKinney and Chris Fusco, "Obama Spells Out 'Regret' After Land Deal with Rezko," *CST,* 5 November 2006, p. A4; "How Obama Won Over a Tiny Arab Village," *JP,* 22 February 2008; Obama's *CT* and *CST* interviews, 14 March 2008; Goudie, "Obama's Foreign Mission #2," ABC7 Chicago, 17 July 2008; Goudie, "Obama World Trip Leaves Personal Controversies Behind," *CDH,* 28 July 2008.

19. Jeff Zeleny, "Obama Is Democrats' Point Man on Ethics," *CT,* 18 January 2006; Obama on *Good Morning America,* ABC, and *American Morning,* CNN, 18 January 2006; Obama Press Release, "Honest Leadership and Open Government," 18 January 2006, Obama.Senate.gov; Obama, *CRec,* 18 January 2006, p. S14 (introducing S. 2179, the CLEAN UP Act); Mike Dorning and Zeleny, "Democrats Put Their Ethics Reform Package on the Table," and Zeleny, "Obama: 'Plantation' Remark Warranted," *CT,* 19 January 2006; Obama on *On the Record,* Fox News, 20 January 2006; Lynn Sweet, "Obama to Pay Full Fare on Private Jet," *CST,* 21 January 2006, p. 6; John Kass, "Obama Might Be Walking on Hot Water," *CT,* 22 January 2006; Sweet, "After Cautious, Bipartisan Year, Obama Opens New Chapter," *CST,* 22 January 2006, p. A12; Obama on *Meet the Press,* NBC, 22 January 2006; Sweet, "Obama Makes It Very Clear: No White House Bid in '08," *CST,* 23 January 2006, p. 4; "Obama to Vote No on the Nomination of Judge Alito to the Supreme Court," 24 January 2006; Obama Podcast #15, "Today's Meeting on Iraq with President Bush," 25 January 2006, Obama.Senate.gov; Obama, *CRec,* 26 January 2006, pp. S190–91 and S228–29 (introducing S. 2201, the FAA Fair Labor Management Dispute Resolution Act); Obama, remarks at the Lobbying Reform Summit, 26 January 2006; Jamie Francisco, "Obama Packs 'Em In, Riffs on Iraq, Alito," *CT,* 31 January 2006; Obama Podcast #16, "Supreme Court Nomination of Samuel Alito," 31 January 2006, Obama.Senate.gov; Obama on the *CBS Evening News,* 31 January 2006; Jodi Enda, "Great Expectations," *American Prospect,* February 2006, pp. 22ff.; Obama on *Good Morning America,* ABC, 1 February 2006; Obama, *CRec,* 1 February 2006, p. S431; Sweet, "Obama in Flight," *TH,* 2 February 2006, p. 15; USS, CVA, "Jobs for Veterans Act Three Years Later," S. Hrg. 109-632, 2 February 2006, pp. 9–10; Rachelle Treiber, "Sen. Obama Hosts Town Meeting in South Holland," *DS,* 4 February 2006; Chas Reilly, "Obama Visits South Suburban College," *NIT,* 4 February 2006; Dan Balz and Haynes Johnson, *The Battle for America 2008* (Viking, 2009), p. 26; Heilemann and Halperin, *Game Change,* pp. 30–31; Axelrod, *Believer,* pp. 173–75.

20. Obama to John McCain, 2 February 2006;

McCain to Obama, 6 February 2006; Obama to McCain, 6 February 2006; Elana Schor, "McCain Counters Obama with Sharp Rebuttal on Lobbying Reform," *TH,* 7 February 2006, p. 3; Jeff Zeleny, "McCain Slams 'Posturing' Obama," *CT,* 7 February 2006; Lynn Sweet, "McCain: Obama Is Insincere," *CST,* 7 February 2006, p. 8; Eric Krol, "McCain, Obama Sparring in Exchange of Letters," *CDH,* 7 February 2006, p. 13; Josh Gerstein, "McCain Harangues Obama Over Bipartisan Lobbying Reform Effort," *NYS,* 7 February 2006, p. 5; Zeleny, "McCain, Obama Bury the Hatchet," *CT,* 8 February 2006; Sweet, "Obama on McCain: 'I Think His Feelings Got Bruised,'" *CST,* 8 February 2006, p. 3; Obama and Jay Inslee, "Salvaging the Auto Industry," *BG,* 8 February 2006; Obama, *CRec,* 8 February 2006, pp. S864, S867–68 (introducing S. 2257, the Hurricane Katrina Working Family Tax Relief Act, S. 2259, the Congressional Ethics Enforcement Commission Act, and S. 2261, the Transparency and Integrity in Earmarks Act); Obama, "Testimony Before the Senate Rules Committee," 8 February 2006; Sheryl Gay Stolberg, "Differences Emerge Over Efforts to Restrict Lobbying," *NYT,* 9 February 2006; "McCain v. Obama," and Sweet, "Obama and Clarity," *TH,* 9 February 2006, pp. 26, 27; Tory Newmyer, "Feud Settled, but Paths Split," *RC,* 9 February 2006; Zeleny, "'Pen Pals' McCain, Obama Talk Pork," *CT,* 9 February 2006; Sweet, "'Pen Pals' Obama, McCain Call a Truce," *CST,* 9 February 2006, p. 3; Obama Podcast #17, "Lobbying and Ethics Reform," 9 February 2006, Obama.Senate.gov; Eric Pfeiffer, "Senate's Golden Boy Shines Not So Brightly," *WT,* 10 February 2006, p. A5; Howard Fineman, "Mark Salter: McCain's Closest Aide," *Newsweek,* 18 July 2008; Pete Rouse's July 2008 interview with Jim Gilmore; Heilemann and Halperin, *Game Change,* pp. 323–25.

21. Christina Chapman, "Morris Greets Obama," *JHN,* 13 February 2006; Obama, *CRec,* 13 February 2006, pp. S1107–8; Obama, *CRec,* 14 February 2006, pp. S1177–78, 1179 (introducing S. 2280, the Stopping Transactions Which Operate to Promote Fraud, Risk and Under-Development [STOP FRAUD] Act and S. 2286, the Equality for Two-Parent Families Act); David Jackson, "Obama, Durbin Propose Federal Mortgage Reform," *CT,* 15 February 2006; Dennis Conrad, "Obama: The Senator with the Midas Touch," AP, 15 February 2006; USS, CFR, "The President's Budget for Foreign Affairs," S. Hrg. 109-769, 15 February 2006, pp. 44–46; USS, EPW, "Environmental Protection Agency's Fiscal Year 2007 Budget," S. Hrg. 109-1022, 15 February 2006, p. 1; Obama Podcast #18, "Darfur: Current Policy Not Enough," 15 February 2006, Obama.Senate.gov; USS, CVA, "Hearing on Proposed Fiscal Year 2007 Budget for Department of Veterans Affairs Programs," S. Hrg. 109-627, 16 February 2006, pp. 6–7; Obama, *CRec,* 16 February 2006, pp. S1401 and S1426–27; Obama, remarks at Ethics Commission press conference, 16 February 2006; Obama on *PBS NewsHour,* 16 February 2006; Perry Bacon Jr., "The Exquisite Dilemma of Being Obama," *Time,* 20 February 2006, pp. 24–28; Jeff Zeleny, "Stepping Off the Sidelines, into the Spotlight," *CT,* 26 February 2006; Alexander Bolton, "Dems Cool on Obama Bill," *TH,* 28 February 2006, p. 1; Obama Press Release, "Energy Security Is National Security," Governors' Ethanol Coalition, 28 February 2006, Obama .Senate.gov; David Bernstein, "Battle Lines," *CM,* March 2006; Seth Borenstein, "Obama's Fuel-Efficiency Swap," KRDC, 1 March 2006; Obama, *CRec,* 1 March 2006, pp. S1581–82 (introducing S. 2348, the Nuclear Release Notice Act); Obama Podcast #19, "Energy Security Is National Security," 1 March 2006, Obama .Senate.gov; Obama, *CRec,* 2 March 2006, p. S1651 (introducing S. 2358, the VA Hospital Quality Report Act, and S. 2359, the Hospital Quality Report Card Act); USS, CFR, "A Nuclear Iran: Challenges and Responses," S. Hrg. 109-679, 2 March 2006, pp. 39–43 ("We have generally, across the board, said that a nuclear Iran is unacceptable. I happen to share that view"); Sheryl Gay Stolberg, "Senate Passes Legislation to Renew Patriot Act," *NYT,* 3 March 2006; Obama, *CRec,* 3 March 2006, p. S1678; Kathy Kiely, "Democrats See Obama as Face of 'Reform and Change,'" *USAT,* 6 March 2006; Obama, *CRec,* 7 March 2006, pp. S1822–23; Stolberg, "Fight Looms on Lawmakers' Use of Corporate Jets," *NYT,* 8 March 2006; Obama, *CRec,* 8 March 2006, p. S1868; Deirdre Shesgreen, "Obama Moves into Leadership Role on Ethics Legislation," *SLPD,* 9 March 2006, p. A5; USS, EPW, "Oversight on the Nuclear Regulatory Commission," S. Hrg. 109-1027, 9 March 2006, pp. 9–10, 26–30; Obama, *CRec,* 9 March 2006, p. S1953; Dori Meinert, "Obama Carries a Tune," *SJR,* 12 March 2006, p. 11; Obama on *Face the Nation,* CBS, 12 March 2006; Jeff Zeleny, "A Tuneful Obama Leads Gridiron Potshots at Cheney," *CT,* 13 March 2006; Lynn Sweet, "Obama in Prime Form at

Gridiron Club Roast," *CST,* 13 March 2006, p. 28; Leonard Pitts Jr., "Don't Assume Obama Will Be a Modern Savior," *MH,* 13 March 2006; Obama Press Release, "21st Century Schools for a 21st Century Economy," 13 March 2006, Obama.Senate.gov; Obama, *CRec,* 14 March 2006, pp. S2082–83, S2110–11, and S2128; Maureen Dowd, "What's Better? His Empty Suit or Her Baggage?," *NYT,* 15 March 2006; Harold W. Anderson, "Obama, Others Served Up Laughs at Annual Gridiron Club Dinner," *OWH,* 19 March 2006, p. B9; "An Evening of Laughs with Obama at Gridiron Dinner," *BSB,* 13 April 2006, p. 7.

22. USS, CFR, "Post-Palestinian Election Challenges in the Middle East," S. Hrg. 109-903, 15 March 2006, p. 1; USS, CVA, "Looking at Our Homeless Veterans Programs," S. Hrg. 109-533, 16 March 2006, pp. 36–38; Obama, *CRec,* 16 March 2006, pp. S2227, 2237–38, 2289–90, and S2335 (introducing S. 2441, the Innovation Districts for School Improvement Act); Obama Podcast #20, "21st Century Schools for a 21st Century Economy," 16 March 2006, Obama.Senate.gov; Kirk Victor, "Reason to Smile," and "In His Own Words: Barack Obama," *NJ,* 18 March 2006, pp. 18–27; Anne E. Kornblut, "But Will They Love Him Tomorrow?," *NYT,* 19 March 2006; Obama, remarks at Fellowship Baptist Church, Chicago, 20 March 2006; Demetrius Patterson, "Englewood Community Rallies Against Violence," *CD,* 22 March 2006, p. 3; Bacon, "Obama Reaches Out with Tough Love," *WP,* 3 May 2007.

23. Rosalind Rossi, "4 Indicted in S. Suburb Gambling," *CST,* 18 June 1992, p. 74; Michael Perlstein, "Brothel Case Expands into Miami," *NOTP,* 7 August 2002, p. 1; Larry Lebowitz, "Multicity Brothel Network Smashed," *MH,* 7 August 2002; "Federal Jury Convicts Defendant in 'Circuit' Prostitution Case," States News Service, 9 January 2004; John T. Slania, "On Broadway: City's Most Profitable Bank," *CCB,* 25 October 2004; Rick Pearson, "Parties Get Set for Busy Primary," *CT,* 5 December 2005; Greg Hinz, "Halo Tarnish," *CCB,* 10 December 2005, Hinz, "Sen. Obama, Elevated to Saintly Status, Pulls an Earthly Political Deal," *CCB,* 12 December 2005; Eric Zorn, "For His Sake, Stroger's Talk Better Be Phony," *CT,* 2 February 2006; Mickey Ciokajlo, "Ex-Aide to Daley Forges Own Path," *CT,* 10 March 2006; Hinz, "Gun Shop Owner, Felon Got Loans from State Treasurer Hopeful," *CCB,* 13 March 2006; Ciokajlo, "Stroger Holding Off Claypool," *CT,* 14 March 2006; Ciokajlo, "Stroger Suffers Stroke," Courtney Flynn and David Jackson, "Loans to Crime Figure Haunt State Treasurer Hopeful," and "For the Democrats: Mangieri," *CT,* 15 March 2006; Janet Rausa Fuller, "Giannoulias Secures Dem Treasurer's Nod," *CST,* 22 March 2006; Jeff Berkowitz, "Another 'Just Another Pol' Day for Senator Obama," Blogspot.com, 22 March 2006; Ciokajlo and Robert Becker, "Stroger Wins. Can He Serve?," and Eric Zorn, "Obama on the Fence Is a Fat Target," *CT,* 23 March 2006; Sunil Garg, "A Profile in Discouragement—Why Was Obama MIA in the Cook County Board Race?," *CT,* 24 March 2006; ". . . And Obama Is Confounding," *CT,* 28 April 2006; Bob Secter and John McCormick, "Sweet—and Sour—Home Chicago," *CT,* 23 June 2008; Mick Dumke, "Alexi's Albatross," *CR,* 3 December 2009; Axelrod, *Believer,* p. 180; DJG interviews with Dan Shomon, Forrest Claypool, Leslie Hairston, Freddrenna Lyle, and John Rogers Jr.

24. Alexander Bolton, "Sens. Obama, Coburn Make Unlikely Duo," *TH,* 28 March 2006, p. 1; Obama, *CRec,* 28 March 2006, pp. S2434–35, S2441–43; USS, EPW, "The Impact of the Elimination of MTBE," S. Hrg. 109-1026, 29 March 2006, pp. 35–36; Obama Podcast #21, "Improving Chemical Plant Security," 29 March 2006, Obama.Senate.gov; Sheryl Gay Stolberg, "Senate Passes Pared-Down Lobbying Bill," *NYT,* 30 March 2006; Jeffrey H. Birnbaum, "Senate Passes Lobbying Bill," *WP,* 30 March 2006; USS, CVA, "Legislative Presentations of the National Association of State Directors of Veterans Affairs et al.," S. Hrg. 109-564, 30 March 2006, p. 34; Obama, *CRec,* 30 March 2006, pp. S2611–12; Jeff Zeleny, "Writing Bug Bites Obama Again," *CT,* 31 March 2006; Mark Pazniokas, "A Rising Star Visits Hartford," *HC,* 31 March 2006, p. A1; Obama, "Fueling the Future," *American Prospect,* April 2006, p. 12; William Yardley, "Support for Iraq War Strains Good Will as Lieberman Seeks Re-Election," *NYT,* 2 April 2006; Obama on *This Week,* ABC, 2 April 2006; Obama, *CRec,* 3 April 2006, pp. S2722–23; Obama Press Release, "Energy Independence and the Safety of Our Planet," 3 April 2006, Obama.Senate.gov; John Chase, "Obama Scolds Bush on Environment, Energy Proposals," *CT,* 4 April 2006; Obama, *CRec,* 4 April 2006, p. S2805 (introducing S. 2506, the Healthy Places Act); Obama, "Restoring America's Promise of Opportunity, Prosperity and Growth," Brookings Institution, 5 April 2006;

Obama Podcast #22, "Immigration Reform," 5 April 2006, Obama.Senate.gov; Obama, *CRec,* 6 April 2006, pp. S3178–79; Rob Hotakainen, "Riding High, Obama Hits the Road for His Party," *MST,* 8 April 2006, p. A1; Deborah Caulfield Rybak, "Obama Gets DFLers on Their Feet," *MST,* 9 April 2006, p. B3; David Jackson, "Loans Cast Pall Over Candidate," *CT,* 9 April 2006; "Deposit Some Answers, Please," *CT,* 12 April 2006; Rick Pearson and Jackson, "Obama Leans on Treasurer Nominee," *CT,* 13 April 2006; Harvey Wineberg, Barack H. and Michelle L. Obama, 2005 Form 1040, 15 April 2006; Jackson, "Giannoulias Speaks Up on Loans," *CT,* 27 April 2006; Scott Fornek, "Giannoulias: I Take It Back," *CST,* 27 April 2006, p. 6; "Giannoulias Is Elusive," *CT,* 28 April 2006; Eric Krol, "Treasurer Candidate Squeezed by Catch-22 Over Loans," *CDH,* 5 May 2006, p. 16; Obama, U.S. Senate Financial Disclosure Report, 15 May 2006; Travis Tritten, "Execs Bet on Development," (Myrtle Beach) *Sun News,* 21 May 2006; "SunCruz Sunshine," *Sun News,* 23 May 2006; Patrick Danner, "Boulis Kin, 2 Felons Own Martha's Site," *MH,* 18 June 2006; Garrison Wells, "SunCruz Chief Sees Calculated Risk in Casinos," *Sun News,* 25 June 2006; Tritten, "Casinos Offer Fee to County," *Sun News,* 28 September 2006; Jacob Weisberg, "The Path to Power," *Men's Vogue,* September 2006, pp. 218–23; Mike Dorning, "In Big Jump, Obamas Earn $1.67 Million," *CT,* 26 September 2006; Dorning, "Employer: Michelle Obama's Raise Well-Earned," *CT,* 27 September 2006; Tritten, "Casino Deal Under Attack," *Sun News,* 17 October 2006; Tritten, "Panel Reviews Casino Tax Talks," *Sun News,* 24 February 2007; Bob Secter, "For Obama, Charity Really Began in the U.S. Senate," *CT,* 25 April 2007; Mike Dorning and Christi Parsons, "Carefully Crafting the Obama 'Brand,'" *CT,* 12 June 2007; Bacon, "The Outsider's Insider," *WP,* 27 August 2007; Steve Warmbir and Frank Main, "The Face of the New Mafia," *CST,* 24 February 2008, p. 2; James Bone, "Boneheaded Deal Haunts House Obama Built," *TOL,* 7 March 2008; Lauren Collins, "The Other Obama," *New Yorker,* 10 March 2008; Michael Crowley, "Mutual Contempt," *TNR,* 26 March 2008; David Jackson et al., "$20 Million in Loans to Felons," *CT,* 2 April 2010; DJG interviews with Rob and Lisa Fisher and Madhuri Kommareddi.

25. Obama, "Town Hall Meeting at Loyola University," 10 April 2006, Obama.Senate.gov; Laura Burns, "Obama Talks Politics, Future," *Loyola Phoenix,* 12 April 2006; John Barrett, "York Visit Senator's 46th Town Hall Forum," *EP,* 14 April 2006; Caleb Hale, "Obama Discusses State, National Issues in Carbondale Meeting," *SI,* 15 April 2006; Obama on *Public Affairs with Jeff Berkowitz,* 17 April 2006; Tom Johnson, "Senator Obama Tells BHS Students Change Is Needed in Government," *BCR,* 20 April 2006; Bernard Schoenburg, "Durbin: Obama Shouldn't Rule Out National Office," *SJR,* 20 April 2006, p. 5; Perry Bacon Jr. and Massimo Calabresi, "The Up-and-Comers," *Time,* 24 April 2006, p. 38; Obama on *Charlie Rose,* PBS, with Nicholas Kristof, 25 April 2006; USS, CVA, "VA Research," S. Hrg. 109-598, 27 April 2006, p. 1; Obama Podcast #23, "Update on Darfur," 27 April 2006, Obama.Senate.gov; Deanna Belandi, "Obama Uses Town Hall Meetings to Connect as a Regular Guy," AP, 30 April 2006; Holli Chmela, "Rally Urges U.S. Role in Sudan," *CT,* 1 May 2006; Obama, *CRec,* 1 May 2006, pp. S3779–80; Antonio Olivo and Oscar Avila, "United They March," *CT,* 2 May 2006; Obama, *CRec,* 2 May 2006, pp. S3879–80; Jeff Zeleny, "Senate OKs Plan to Open Storm Bids," *CT,* 3 May 2006; Obama, *CRec,* 4 May 2006, pp. S4014, S4029; USS, CFR, "Housing and Urbanization Issues in Africa," S. Hrg. 109-946, 4 May 2006, pp. 4–5; Obama Podcast #24, "Immigration Rallies," 4 May 2006, Obama.Senate.gov; Obama, remarks at Chicago benefit dinner for Facing History and Ourselves, 5 May 2006; Don Walton, "Obama: Door Opens for Democrats," *Lincoln Journal Star,* 7 May 2006, p. C1; Jason Kuiper, "Sen. Obama Lauds Nelson During Visit to Omaha," *OWH,* 7 May 2006, p. B1; Kyle Michaels, "My Interview with Barack Obama," Blogspot.com, 7 May 2006; Josh Noel, "Illinois Vets Can Reapply for Benefits," *CT,* 9 May 2006; Hal Dardick, "Obama Still to Press Spill Bill," *CT,* 10 May 2006; Obama, *CRec,* 10 May 2006, pp. S4324–25; Obama, *CRec,* 11 May 2006, p. S4425; Obama, remarks at Emily's List annual luncheon, 11 May 2006; Dennis Conrad, "Obama Mocks Bush's 'Subliminal Messages' of Iraq Victory," AP, 11 May 2006; Obama Podcast #25, "A Real Solution for High Gas Prices," 11 May 2006, Obama.Senate.gov; Obama on *Late Night with Conan O'Brien,* 12 May 2006; Megan Reichgott, "Obama Jokes About Possible Presidential Bid in '08," AP, 13 May 2006; Obama, *CRec,* 17 May 2006, pp. S4674–75; USS, EPW, "Nominations of Molly O'Neill and Dale Klein," S. Hrg. 109-1036, 17

May 2006, pp. 10–11; Obama Podcast #26, "Honoring Our Commitment to Veterans," 18 May 2006, Obama.Senate.gov; USS, CFR, "Iran's Political/Nuclear Ambitions and U.S. Policy Options," S. Hrg. 109-763, 17–18 May 2006, pp. 38–40, 117–20; Jeff Zeleny, "Obama Writes of 'Spooky' Good Luck," *CT,* 20 May 2006; Lynn Sweet, "In New Book, Obama Muses on Politics," *CST,* 20 May 2006, p. 6; Obama, prepared remarks at SIU School of Medicine commencement, Springfield, 20 May 2006; Sweet, "Obama Touts 2nd Volume," *CST,* 21 May 2006, p. A9; Daniel Pike, "Obama Talks to SIU Medical Graduates," *SJR,* 21 May 2006, p. 1; Obama, *CRec,* 23 May 2006, pp. S4941–42, S4953; Obama, *CRec,* 24 May 2006, pp. S5055, S5106–7; Marc Ambinder, "Teacher and Apprentice," *Atlantic Monthly,* December 2007; Mundy, *Michelle,* p. 165; Wolffe, *Renegade,* p. 45; DJG interviews with Larry Suffredin, Martha Minow, and Rob Fisher.

26. Hillary Rodham Clinton and Obama, "Making Patient Safety the Centerpiece of Medical Liability Reform," *New England Journal of Medicine* 354 (25 May 2006): 2205; Obama, *CRec,* 25 May 2006, pp. S5177–78, S5296; USS, CVA, "Veterans Affairs Data Privacy Breach," S. Hrg. 109-577, 25 May 2006, pp. 13–14; Jeff Zeleny, "Obama in '08?," *CT,* 28 May 2006; Lynn Sweet, "Chatter About Obama's 2008 Plans Picks Up," *CST,* 26 May 2006, p. 32; Amanda Reavy, "Durbin: Obama Should Run for President," *SJR,* 29 May 2006, p. 8; Eric Zorn, "Better '08 Than Never?," *CT,* 31 May 2006; "Obama Presidential Talk Quite Premature," *BP,* 2 June 2006, p. A6; Scott Helman and Lisa Wangsness, "Candidates Rally for the Homestretch," *BG,* 2 June 2006; John J. Monahan, "Patrick Basks in Obama's Glow," *Worcester Telegram & Gazette,* 2 June 2006, p. A6; Joe Klein, "Barack Obama Isn't Not Running for President," *Time,* 5 June 2006, p. 28; David Sirota, "Mr. Obama Goes to Washington," *Nation,* 26 June 2006, pp. 20–23 (Web posted 8 June); Harry Reid, *The Good Fight* (Berkley Books, 2009), p. 299; Heilemann and Halperin, *Game Change,* pp. 33–37; Axelrod, *Believer,* p. 181.

27. Obama, commencement address at the University of Massachusetts at Boston, 2 June 2006; Cristina Silva, "Speakers Brighten Ceremony at UMass-Boston," *BG,* 3 June 2006; Neil Mirochnick, "Sen. Obama Tells UMass-Boston Graduates to 'Do What Is Hard,'" *Quincy Patriot-Ledger,* 3 June 2006, p. 23; William Hershey, "Obama Says Ohio at Heart of Democrats' National Strategy," *Dayton Daily News,* 4 June 2006, p. A4; Obama, *CRec,* 7 June 2006, pp. S5523–24, S5576, S5601–2 (introducing S. 3475, the Homes for Heroes Act); Obama, *CRec,* 8 June 2006, pp. S5616–17; Obama Podcast #27, "Network Neutrality," 8 June 2006, Obama.Senate.gov; USS, CFR, "Counterterrorism: The Changing Face of Terror," S. Hrg. 109-859, 13 June 2006, pp. 38–43; Obama, "Remarks at 'Take Back America,'" Washington, D.C., 14 June 2006; USS, EPW, "Oversight of the Superfund Program," S. Hrg. 109-1042, 15 June 2006, pp. 12–13, 24–26; Obama Podcast #28, "Katrina Reconstruction Contracts," 15 June 2006, Obama.Senate.gov; Obama, Northwestern University commencement address, 16 June 2006; Charles Babington, "Obama's Profile Has Democrats Taking Notice," *WP,* 18 June 2006; Ross Douthat, "The Great Democratic Hope," *WSJ,* 21 June 2006; Obama, *CRec,* 21 June 2006, pp. 6233–34, S6288 (introducing S. 3554, the Alternative Diesel Standard Act); USS, EPW, "Inherently Safer Technology in the Context of Chemical Site Security," S. Hrg. 109-1044, 21 June 2006, pp. 12–14; David Stout, "Senate Rejects Democratic Efforts on Iraq," *NYT,* 22 June 2006; "Obama's Supporters Should Encourage Him to Run," *SI,* 22 June 2006; Obama, *CRec,* 22 June 2006, p. S6364; Obama Podcast #29, "Senate Iraq Resolutions," 22 June 2006, Obama.Senate.gov; Obama, remarks at Book-Expo America author breakfast, 23 June 2006; Jeffrey H. Birnbaum and Jim VandeHei, "Call for Lobbying Changes Is a Fading Cry, Lawmakers Say," *WP,* 26 June 2006; Obama, *CRec,* 27 June 2006, pp. S6523–24; USS, EPW, "EPA Regional Inconsistencies," S. Hrg. 109-1046, 28 June 2006, pp. 25–26; Obama, "Call to Renewal Keynote Address," Washington, D.C., 28 June 2006; Obama on *Hannity & Colmes,* Fox News, 28 June 2006; *CRec,* 29 June 2006, pp. S7178–80 (unanimous consent passage of Obama's S. 2125, as amended); Dana Milbank, "Will Democrats Put Their Faith in Obama?," *WP,* 29 June 2006; Marni Goldberg, "Obama Urges Democrats to Embrace Faith," *CT,* 29 June 2006; Lynn Sweet, "Obama Puts His Faith in Spotlight," *CST,* 29 June 2006, p. 45; Obama on *Good Morning America,* ABC, 29 June 2006; "UCC Member Sen. Barack Obama Discusses Faith and Politics," *UCC News,* 29 June 2006; Obama, *CRec,* 29 June 2006, pp. S6759–60, S6793–94 (introducing S. 3607, the Responsible Fatherhood and Healthy Families Act); E. J. Dionne, "Obama's Eloquent Faith," *WP,* 30 June

2006; Michelle Goldberg, "What's the Matter with Barack Obama?," *Huffington Post,* 30 June 2006; Eric Zorn, "Change of Subject," *CT,* 2 July 2006; Amy Sullivan, "In Good Faith," *Slate,* 3 July 2006; Paul Wood, "Obama's Prayer: Wooing Evangelicals," *NR,* 6 July 2006; PastorDan, "Interview with Barack Obama," StreetProphets.com, 7 July 2006; Margaret Kimberley, "Obama Gets Religion," *CD,* 7–9 July 2006, p. 11; "Faith, Race and Barack Obama," *Economist,* 8 July 2006, p. 52; Zorn, "Change of Subject," *CT,* 9 July 2006; Obama, "Politicians Need Not Abandon Religion," *USAT,* 10 July 2006; Clarence Page, "Democrats Need Old-Time Religion," *CT,* 12 July 2006; Obama on *Morning Edition,* NPR, 14 July 2006; Obama, "The Religious Divide," *CT,* 18 July 2006; Robinson, *A Game of Character,* pp. 182–86; "Kelly Robinson," *Mom Magazine,* February 2011, pp. 10–15.

28. Obama's 29 June 2006 interview with Jacob Weisberg; Jorge Luis Borges, "Two English Poems (II)," 1934; Michelle Obama's July 2006 interview with Weisberg; Weisberg, "The Path to Power," *Men's Vogue,* September 2006, pp. 218–23. The Borges stanza Barack could approximately quote from memory follows a line that says, "I offer you the loyalty of a man who has never been loyal."

29. Lynn Sweet, "Obama Trip to Highlight U.S. Interests in Africa," *CST,* 10 July 2006, p. 34; Obama, remarks at Campus Progress annual conference, Washington, D.C., 12 July 2006; USS, CFR, "An Iraq Update," S. Hrg. 109-869, 13 July 2006, pp. 39–41; Obama, *CRec,* 13 July 2006, pp. S7502–3; Obama Podcast #30, "Homeland Security Appropriations," 13 July 2006, Obama.Senate.gov; Obama, *CRec,* 17 July 2006, pp. S7598–99; Obama, "Statement Before the Senate Subcommittee on Federal Financial Management," 18 July 2006; Obama, *CRec,* 19 July 2006, pp. S7854 and S7926–27 (introducing S. 3694, the Fuel Economy Reform Act); Jeff Zeleny, "Democrats Hold Bush Accountable," *CT,* 20 July 2006; Ray McAllister, "Obama's Plan," *RTD,* 20 July 2006, p. B1; Obama, *CRec,* 20 July 2006, pp. S7984–85; Obama Podcast #31, "Fuel Economy Reform Act," 20 July 2006, Obama.Senate.gov; Zeleny, "Obama Trip to Iowa Ups Buzz on '08," *CT,* 21 July 2006; Sweet, "Obama Plan to Visit Iowa Steps Up White House Buzz," *CST,* 21 July 2006, p. 36; Thomas Beaumont and Jane Norman, "Illinois' Obama to Attend Harkin Event," *DMR,* 21 July 2006; Charlotte Eby, "Obama to Headline Harkin's Annual Steak Fry," *QCT,* 21 July 2006; John Patterson, "Obama in '08? Not So, Senator Says," *CDH,* 21 July 2006, p. 6; Zeleny, "Obama Joins Parade to See New Orleans," *CT,* 22 July 2006; Obama on *Public Affairs with Jeff Berkowitz,* 22 July 2006; Dennis Conrad, "Obama Takes Political Spotlight on Return to Africa," AP, 23 July 2006; Mary Mitchell, "Obama Reassures Ministers He's Still on Their Side," *CST,* 25 July 2006, p. 14; Obama, *CRec,* 25 July 2006, p. S8181; Obama, "Support Fuel Economy Reform," *HPH,* 26 July 2006, p. 4; Terry Dean, "Obama Addresses Small Business Concerns," *AWN,* 26 July 2006; Obama, remarks at Partnership for Public Service, Washington, D.C., 26 July 2006; John Patterson, "What's Changed for Obama in Last 2 Years," and "Senator Says He's Still Willing to Help Blagojevich Despite Hiring Concerns," *CDH,* 27 July 2006, p. 7; USS, CFR, "Nomination of Hon. John R. Bolton to Be U.S. Representative to the United Nations," S. Hrg. 109-935, 27 July 2006, pp. 64–69; USS, EPW, "The Stafford Act: A Plan for the Nation's Emergency Preparedness and Response System," S. Hrg. 109-1066, 27 July 2006, pp. 6–7, 22–24; Obama, *CRec,* 27 July 2006, pp. S8379–80 (introducing S. 3757, to name a post office); Obama, *CRec,* 1 August 2006, pp. S8502–3; Perry Bacon Jr., "The Outsider's Insider," *WP,* 27 August 2007; Obama, *CRec,* 3 August 2006, pp. S8840–41 (introducing S. 3822, the Genomics and Personalized Medicine Act); Obama, remarks to AFSCME national convention," Chicago, 7 August 2006; "Obama Addresses Public Service Union," *CT,* 8 August 2006; Robin Givhan, "Mussed for Success," *WP,* 11 August 2006; Obama, Xavier University commencement address, 12 August 2006; AP, "Sen. Obama Addresses Xavier Graduates," 13 August 2006; "Obama to Take Seven-County Downstate Town Hall Tour," 31 July 2006, Obama.Senate.gov; Kristen McQueary, "Sore Spot Remains on Statewide Ticket," *DS,* 14 August 2006; Bill Bartleman, "Rising Star: Obama Draws Crowd in Massac," *Paducah Sun,* 15 August 2006; Nathaniel West, "Obama Addresses Packed House During Town Hall Meeting," *DHR,* 16 August 2006; Rich Miller, "Olive Branch," *IT,* 18 January 2007.

30. Lynn Sweet, "Media, Sudan Watching Trip Closely," *CST,* 18 August 2006, p. 14; Jeff Zeleny, "Obama Returns to Africa as Celebrity," *CT,* 20 August 2006; Sweet, "1st Up: Mandela's 27-Year Prison," *CST,* 20 August 2006, p. A5; Bill Lambrecht, "As Obama Heads to Africa, Terrorism Is a Top Priority," *SLPD,* 20 August

2006, p. A8; Christopher Wills, "Obama Hopes Fame, Family Combine to Produce Results for Africa Trip," AP, 20 August 2006; Sweet, "A Humbling Visit to Mandela's Cell," *CST,* 21 August 2006, p. 26; Clare Nulis, "Obama Visits Former Mandela Prison," AP, 21 August 2006; Obama, remarks at the South African Institute for International Affairs, Cape Town, 21 August 2006; Jeff Zeleny and Laurie Goering, "Obama Challenges South Africa to Face AIDS Crisis," *CT,* 22 August 2006; Sweet, "Confronting Africa's Scourge: AIDS," *CST,* 22 August 2006, p. 6; Lambrecht, "Obama Tackles South Africa's Response to AIDS," *SLPD,* 22 August 2006, p. A1; Edmund Sanders, "Senator's Kenya Visit Inspires Obama-Mania," *LAT,* 22 August 2006; Zeleny, "Congo's Unrest Blocks Obama Visit," *CT,* 23 August 2006; Sweet, "Sen. Expects to Disappoint in Kenya," *CST,* 23 August 2006, p. 13; Michelle Faul, "Sen. Barack Obama Says South Africans' Fight for Freedom Inspired His Political Career," AP, 23 August 2006; Sweet, "Kenya Eager to See Its 'Great Senator,'" *CST,* 24 August 2006, p. 16; AP, "Sen. Obama to Take HIV Test in Kenya," 24 August 2006; Zeleny, "'Our Blood Is in His,'" *CT,* 25 August 2006; Sweet, "In Kenya, a Hero's Welcome Awaits," *CST,* 25 August 2006, p. 6; Lambrecht, "Visit Puts Kenya in Grip of Obama-Mania," *SLPD,* 25 August 2006, p. A6; Christopher Wills, "Obama Meets Kenyan President," AP, 25 August 2006; Jeffrey Gettleman, "Obama Gets a Warm Welcome in Kenya," *NYT,* 26 August 2006; Sweet, "Senator Mobbed in Kenya," *CST,* 26 August 2006, p. 3; Lambrecht, "Obama Says Culture of Corruption Continues to Be Obstacle for Kenya," *SLPD,* 26 August 2006, p. A29; Wills, "Obama Returns to Father's Village with New Fame, New Message," AP, 26 August 2006; Zeleny, "Kenyans' Welcome Is Heavy with Hope," *CT,* 27 August 2006; Sweet, "For Kenyans, Senator from Illinois Has Come Home," *CST,* 27 August 2006, p. A24; Lambrecht, "Village Welcomes Obama Home," *SLPD,* 27 August 2006, p. A3; Wills, "Obama Promises Help During Visit to One of Africa's Biggest Slums," AP, 27 August 2006; Sweet, "Taxpayers, Campaign Fund, Obama's Own Money Pay for Trip," *CST,* 28 August 2006, p. 7; Obama, "An Honest Government, a Hopeful Future," University of Nairobi, 28 August 2006; Gettleman, "Obama Urges Kenyans to Get Tough on Corruption," *NYT,* 29 August 2006; Zeleny, "Kenya Chided Over Corruption," *CT,* 29 August 2006; Sweet, "Senator Rebukes Kenya's Corruption," *CST,* 29 August 2006, p. 7; Wills, "Obama Urges Kenyans to Oppose Corruption," AP, 29 August 2006; "The Imperfections of Man," *CT,* 30 August 2006; Auma Obama, *And Then Life Happens,* pp. 291–95.

31. Lynn Sweet, "Senator Has a Front-Row Seat as Lion Devours Wildebeest," *CST,* 31 August 2006, p. 20; AP, "Obama Visits Flood-Displaced Ethiopians," 31 August 2006; Macharia Gaitho and Julie Gichuru, "Obama: 'I Speak What I Think Is True and Say It Best,'" *Nation,* 1 September 2006; Jeff Zeleny, "Sudan Refugees Plead for World's Help," *CT,* 3 September 2006; Sweet, "Visit to Camp Spotlights Ordeal of Darfur Refugees," and "Africa Trip Leads to Many Valuable Successes," *CST,* 3 September 2006, pp. A24, B6; Christopher Wills, "Sudanese Refugees Tell U.S. Senator Darfur Needs U.N. Force," AP, 3 September 2006; Sweet, "Next Step Pressuring U.S. on Sudan," *CST,* 4 September 2006, p. 20; Wills, "Obama Visit Highlights Precarious Situation of Darfur Refugees," AP, 4 September 2006; Greg Hinz, "Obama on Oil," *CCB,* 4 September 2006; James Kirchick, "In Africa, Obama Was Both VIP and Persona Non Grata," *TH,* 5 September 2006; Wills, "Sen. Obama in Africa, Polishing His Resume," AP, 6 September 2006; Obama Podcast #32, "Africa Trip Recap," 6 September 2006, Obama.Senate.gov; Choire Sicha, "Obama in Orbit," *NYO,* 6 November 2006, p. 1; Laurie Abraham, "Mr. Obama Goes to Washington (and Cape Town, and Nairobi, and . . .)," *Elle,* December 2006; Nicholas D. Kristof, "Obama: Man of the World," *NYT,* 6 March 2007.

32. USS, EPW, "Federal Renewable Fuels Programs," S. Hrg. 109-1079, 6 September 2006, pp. 30–32; Lynn Sweet, "What Obama Needs to Reach the Next Level," *CST,* 7 September 2006, p. 41; Obama, *CRec,* 7 September 2006, pp. S9211, S9297; Dennis Conrad, "Obama Says U.S. Has Plenty of Reasons to Consider Africa," AP, 8 September 2006; William Neikirk, "Bloggers Help Obama Pass Senate Pork Bill," *CT,* 9 September 2006; Ellis Cose, "Walking the World Stage," *Newsweek,* 11 September 2006, p. 26; David Yepsen, "Now-or-Never Fears Push Presidential Wannabees," *DMR,* 12 September 2006, p. A9; Obama, *CRec,* 12 September 2006, pp. S9354–55; Richard Clough, "Efforts to Rein In Lobbyists Stall in Congress," *CT,* 13 September 2006; Obama, *CRec,* 13 September 2006, pp. S9493–94; USS, CFR, "Lebanon: Securing a Permanent Cease-Fire," S. Hrg. 109-655, 13 September 2006, pp. 30–34; USS, EPW, "Nominations of Roger Romulus Martella Jr. et al.," S. Hrg. 109-1080, 13 September 2006, pp. 15–17; USS, EPW, "Nuclear Release Notice Act," S. Rpt. 109-347, Calendar No. 637, 25 September 2006, esp. pp. 2–3, 5–6 (13 September 2006); Mike McIntire, "Nuclear Leaks and Response Tested Obama in Senate," *NYT,* 3 February 2008; Ben Whitford, "Coal and Clear Skies: Obama's Balancing Act," *Plenty Magazine,* 31 October 2008; Elana Schor, "Obama and Coburn Revive Effort to Stop No-Bid FEMA Contracts," *TH,* 14 September 2006, p. 18; Obama Podcast #33, "Coburn-Obama Transparency Bill," 14 September 2006, Obama.Senate.gov; Obama on *The Situation Room,* CNN, 14 September 2006; Crystal Yednak, "Ex-Rival Hynes Tries to Start Obama National Bandwagon," *CT,* 15 September 2006; Eric Herman, "Hynes to Obama: Run for President," *CST,* 15 September 2006, p. 12; Kurt Erickson, "Comptroller Urges Obama to Run for President," *QCT,* 15 September 2006; Thomas Beaumont, "Obama Coming to Harkin Steak Fry," *DMR,* 15 September 2006, p. B5; Obama, *CRec,* 15 September 2006, p. S9657; Kevin McDermott, "Obama: Could He Be Too Popular?," *SLPD,* 17 September 2006, p. C1; Anne E. Kornblut, "For This Red Meat Crowd, Obama's '08 Choice Is Clear," *NYT,* 18 September 2006; Beaumont, "Obama Insists He's Focusing on Party in '06," *DMR,* 18 September 2006, p. B1; Tim Jones, "Iowa Democrats See Contender in Obama," *CT,* 18 September 2006; Mike Glover (AP), "Obama to Dems: Get Tougher on Security," and Scott Fornek, "Poll: Obama for President—but When?," *CST,* 18 September 2006, pp. 18, 22; Charlotte Eby, "Obama's Fans Flock to Iowa," *QCT,* 18 September 2006; Elizabeth Ahlin, "Embrace Idealism, Obama Tells Dems," *OWH,* 18 September 2006, p. A1; Walter Shapiro, "Obama in '08?," *Salon,* 18 September 2006; David Yepsen, "What Triggers the Big Buzz Around Obama? It's Hope," *DMR,* 21 September 2006, p. A19; Ed Tibbetts, "Obama's Visit Includes Both Sides of River," *QCT,* 23 September 2006; David Mendell, "Looking Beyond Obama-Mania: Is He Ready Yet?," *CT,* 24 September 2006; Perry Bacon Jr., "The Outsider's Insider," *WP,* 27 August 2007; Roger Simon, "Relentless," *Politico,* 25 August 2008; Carol Felsenthal, "The Making of a First Lady," *CM,* February 2009; Balz and Johnson, *The Battle for America 2008,* p. 27; Wolffe, *Renegade,* pp. 46–48; Heilemann and Halperin, *Game Change,* pp. 56–58; Axelrod, *Believer,* p. 183.

33. USS, CFR, "Responding to Iran's Nuclear Ambitions: Next Steps," S. Hrg. 109-648, 19 September 2006, pp. 34–38; USS, CVA, "The Legislative Presentation of the American Legion," S. Hrg. 109-771, 20 September 2006, p. 1; Obama, "Energy Independence: A Call for Leadership," MoveOn Progressive Vision, Georgetown University, 20 September 2006; Tim Craig, "Obama Boosts Webb with Alexandria Rally," *WP,* 21 September 2006; John McWhorter, "The Color of His Skin," *NYS,* 21 September 2006, p. 11; Obama, *CRec,* 21 September 2006, pp. S9879–80; Mary Baskerville, "Sen. Obama at Joliet Town Hall," *KDJ,* 23 September 2006; Clarence Page, "Obama's Audacious Aura of Hope," *CT,* 24 September 2004; "President Bush Signs Federal Funding Accountability and Transparency Act," 26 September 2006; Obama Podcast #34, "Transparency Act," 26 September 2006, Obama.Senate .gov; Mike Flannery, "Obama's First Law," CBS2 Chicago, 26 September 2006; Obama, *CRec,* 27 September 2006, pp. S10346–48; Fran Spielman, "Daley on Obama: 'Everybody Wants Him,'" *CST,* 28 September 2006, p. 14; Scott Fornek, "If Obama Runs, I'll Campaign for Him, Winfrey Says," *CST,* 28 September 2006, p. 3; David Montgomery, "The Democrats' Charisma Doctor," *WP,* 28 September 2006; Obama, *CRec,* 28 September 2006, pp. S10388–89, S10471–72 (introducing S. 3969, the Lead Poisoning Reduction Act); Obama, *CRec,* 29 September 2006, pp. S10642–43; Mike Glover, "Obama: Democrats Should Embrace National Security Debate," AP, 30 September 2006; Chuck Sweeny, "Barack Obama Assails Bush's Policies," *RRS,* 1 October 2006, p. B1; Tina Stein, "Obama Comes to Rockford," WIFR .com, 1 October 2006; Kay Luna, "Obama Visits Q-C to Stump for Democrats," *QCT,* 1 October 2006; Jennifer Senior, "Dreaming of Obama," *NYM,* 2 October 2006; Joe Klein, "The Fresh Face," *Time,* 23 October 2006, pp. 44ff. (Web posted 16 October).

34. Obama and Michelle on *The Oprah Winfrey Show,* 18 October 2006 (taped 3 October); Clarence Page, "Oprah, Obama, and the Leadership Gap," *CT,* 1 October 2006; "Obama Is Riding a Wave of Celebrity," *CNG,* 4 October 2006; Frank Radosevich II, "'Obama Fever' Grips Peoria," *PJS,* 5 October 2006; Jenee Osterheldt, "Sen. Barack Obama Gets My Vote," *KC Star,* 7 October 2006; Ezra Klein, "Too Soon for Obama," *LAT,* 8 October 2006; "Everywhere but Illinois," *BND,* 10 October 2006; Eric Zorn,

"It's Simple Scandals That Shock," *CT,* 17 September 2006; Carol Marin, "'Man Behind the Curtain' Faces Lots of Questions," *CST,* 27 September 2006, p. 55; *U.S. v. Stuart Levine and Antoin Rezko,* "Superseding Indictment," No. 05-CR-691, 5 October 2006; Rick Pearson et al., "Top Blagojevich Adviser Indicted," and David Jackson and John Chase, "Rezko's Life a Story of Pizza and Politics," *CT,* 12 October 2006; Natasha Korecki et al., "It's 'Pay to Play' . . . on Steroids," Michael Sneed, "The Rezko Scandal," and Korecki and Dave McKinney, "Where Is He?," *CST,* 12 October 2006; Jeff Coen and Matt O'Connor, "U.S. Doesn't Know Where Rezko Is," *CT,* 13 October 2006; Obama on the *CBS Evening News,* 13 October 2006; Michael Tackett, "Sen. Obama, Take a Chance in 2008 Race," *CT,* 15 October 2006; Kathy Chaney and Erin Meyer, "Rezko Gives to Local Pols," *HPH,* 18 October 2006, pp. 1–2; Coen and O'Connor, "Rezko Is Expected to Show Up," *CT,* 19 October 2006; Coen, "Rezko Pleads Not Guilty," *CT,* 20 October 2006; Chaney, "Obama Regrets Land Deal He Made with Tony Rezko," *HPH,* 8 November 2006, pp. 1–2; Brian Ross and Rhonda Schwartz, "The Rezko Connection," ABCNews.com, 10 January 2008; Mike Robinson, "Obama's Relationship with Alleged Fixer," AP, 22 January 2008; Bob Secter et al., "Obama, Rezko Ties Again at Issue," *CT,* 23 January 2008; Secter and David Jackson, "Funds Tough to Figure for Rezko Aid," *CT,* 31 January 2008; Dan Morain, "Obama to Be in Background of Trial," *LAT,* 3 March 2008.

35. *Publishers Weekly,* 2 October 2006, p. 52; Jennifer Senior, "Dreaming of Obama," *NYM,* 2 October 2006; Mike Dorning, "Obama's New Book Laments 'Empathy Deficit,'" *CT,* 12 October 2006; James L. Merriner, "Barack Obama for Prez," *CST,* 15 October 2006, p. B12; James Wensits, "'Red to Blue' Bolsters Donnelly," *South Bend Tribune,* 17 October 2006, p. B1; Kurt Erickson, "Obama Durbin's '08 Pick," *BP,* 17 October 2006, p. A1; Michiko Kakutani, "Foursquare Politics, with a Dab of Dijon," *NYT,* 17 October 2006; John Balzar, "Obama Finds Common Ground on Political Plain," *LAT,* 17 October 2006; Eric Krol, "Obama Is Not a Morning Person," *CDH,* 17 October 2006, p. 6; Obama, *TAOH,* pp. 11, 114, 121, 47, 223, 245, 333, 251, 254, 256, 248; Lynn Sweet, "Obama Riding Momentum," *CST,* 19 October 2006, p. 47; Glenn Thrush, "Is He Waiting in the Wings?," *Newsday,* 19 October 2006; Obama, "My Spiritual Journey," *Time,* 23 October 2006, pp. 52ff. (Web posted 16 October); Jodi Kantor, "For a New Political Age, a Self-Made Man," *NYT,* 28 August 2008; Evan Thomas et al., "Barack Obama: How He Did It," *Newsweek,* 17 November 2008 (Web posted 4 November); Melody Spann-Cooper in *Becoming Barack,* Little Dizze Home Video, 2009; Heilemann and Halperin, *Game Change,* pp. 58–59.

36. Jill Zuckman, "Obama Dips Toe in 2008 Waters," *CT,* 18 October 2006; Dennis Conrad, "Obama's Book Hints That Presidential Run Could Be Family Decision," AP, 18 October 2006; Lynn Sweet, "Obama Riding Momentum," *CST,* 19 October 2006, p. 47; David Brooks, "Run, Barack, Run," *NYT,* 19 October 2006; "Charismatic Obama Looks Presidential," *SJR,* 19 October 2006, p. 6; Obama on *Today,* NBC, on *All Things Considered,* NPR, on *Larry King Live,* CNN, and on *Charlie Rose,* PBS, 19 October 2006; John Kass, "Politicians Don't Get Much Past Readers," and "Is This Obama's Time?," *CT,* 20 October 2006; Margaret Talev, "A Conversation with Sen. Obama," KRDC, 20 October 2006; Obama at the John F. Kennedy Library Forum, 20 October 2006; Maureen Dowd, "Obama's Project Runway," *NYT,* 21 October 2006; Frank Rich, "Obama Is Not a Miracle Elixir," *NYT,* 22 October 2006; Clarence Page, "Enjoy, Senator, It'll Get Worse," *CT,* 22 October 2006.

37. Christine Schiavo, "'Presence of Greatness': Sen. Obama Gets a Rock-Star Reception in Phila.," and "Verbatim: Sen. Obama on 'The Audacity of Hope,' Iraq and More," *PI,* 22 October 2006, pp. B1, C2; Michael Kazin, "Rising Star," *WP,* 22 October 2006; Obama on *Meet the Press,* NBC, 22 October 2006, Adam Nagourney and Jeff Zeleny, "Crowd-Pleaser from Illinois Considers White House Run," and Bob Herbert, "The Obama Bandwagon," *NYT,* 23 October 2006; Dan Balz, "Obama Says He'll Consider a 2008 Bid for the Presidency," *WP,* 23 October 2006, p. A1; Chuck Neubauer, "Obama Admits He's Thinking '08," *LAT,* 23 October 2006; William Neikirk, "Audacity of a Hopeful," *CT,* 23 October 2006; Lynn Sweet, "Obama Mulling '08 Presidential Bid," *CST,* 23 October 2006; Edward Luce, "A Subtle but Effective Political Soft Sell," *FT,* 23 October 2006; Obama's 23 October 2006 American Magazine Conference interview with David Remnick, Pheonix, AZ; Obama on *The Tavis Smiley Show,* 23 October 2006; Ken Silverstein, "Barack Obama Inc. The Birth of a Washington Machine," *Harper's,* November 2006, pp.

31–40; Press Release, "Senator Obama's Office Responds to Misleading Harper's Magazine Story," 23 October 2006, Obama.Senate.gov; Katharine Q. Seelye, "Obama Offers More Variations from the Norm," *NYT,* 24 October 2006; Richard Cohen, "Why Not Obama?," *WP,* 24 October 2006; "Obama Needs to Keep Feet on Ground," *BP,* 24 October 2006, p. A6; "Iowa a Good Place to Make a Tough Decision," *QCT,* 25 October 2006; Jon Friedman, "Obama Fails to Wow a Magazine Conference," Marketwatch .com, 25 October 2006; Plouffe, *The Audacity to Win,* p. 6.

38. Obama on *Hannity & Colmes,* Fox News, 24 October 2006; Kevin Simpson, "Dems' 'Rock Star' Rocks Colorado," *DP,* 25 October 2006, p. A1; Newton N. Minow, "Why Obama Should Run for President," *CT,* 26 October 2006; Richard Halstead, "Obama Mania Visits Marin," *Marin Independent Journal,* 26 October 2006; Charles Krauthammer, "Winning by Losing," *WP,* 27 October 2006; Jacob Weisberg, "The Audacity of the Democratic Hopeful," *FT,* 27 October 2006; Doug Ward, "'Obamamania' Taking Off in U.S.," *Vancouver Sun,* 28 October 2006, p. A15; Carla Hall, "A Rising Star from Illinois Makes Some Noise in L.A.," *LAT,* 28 October 2006; Danny Westneat, "Obama Doesn't Dazzle, but He Does Intrigue," *Seattle Times,* 29 October 2006, p. B1; Isadora Vail, "Senator, Star: Obama Greets Frenzied Crowd," *AAS,* 29 October 2006, p. B1; Elizabeth Taylor, "Q&A with Barack Obama," *CT,* 29 October 2006; Bill Lambrecht, "Obama's Book Tells What He Thinks," *SLPD,* 29 October 2006, p. F8; Bill Maxwell, "Senator, You Better Walk Before You Run," *SPT,* 29 October 2006, p. P1; Howard Kurtz, "Obama? So Handsome, and Probably Delicious," *WP,* 30 October 2006; Meggen Lindsay, "Obama Inspires His Minnesota Fans," *SPPP,* 31 October 2006, p. A1; Bob von Sternberg, "Obama Knows How to Attract a Crowd," *MST,* 31 October 2006, p. B3; Graeme Zielinski, "'He's Our Jack Kennedy': Obama Excites Milwaukee as He Stumps for State Dems," *MCT,* 31 October 2006, p. A1; Garrett M. Graff, "The Legend of Barack Obama," *Washingtonian,* November 2006, pp. 68–71, 122–24; Anne E. Kornblut, "A Senate Newcomer, Helping Fellow Democrats," *NYT,* 1 November 2006; Art Sims, "Rev. Jesse Jackson's Bash a Celebration Fit for a King," *CD,* 1 November 2006, p. 15; *CNN Newsroom,* 1 November 2006, 2 P.M.; Christine G. Sabathia, "Sen. Barack Obama Talks One-on-One," *LAS,* 2–8 November 2006, p. A1; Joel McNally, "No Denying That History Is Calling Upon Obama," *MCT,* 4 November 2006; David Mendell, "Backers' Mission Is Political, Personal," *CT,* 17 January 2007; DJG interview with Newton Minow.

39. Ray Gibson and David Jackson, "Rezko Owns Vacant Lot Next to Obama's Home," *CT,* 1 November 2006; Stanley Crouch and Alan Wolfe, "Barack Obama's The Audacity of Hope," *Slate,* 1 November 2006; Crouch, "What Obama Isn't: Black Like Me," *NYDN,* 2 November 2006, p. 35; John Kass, "Obama Fuzzy on Fence That Tony Built," and Eric Zorn, "Letter to Voters a Letdown for Obama Idealists," *CT,* 2 November 2006; Mark Brown, "Obama's Dealings with Rezko Buy a Parcel of Questions," *CST,* 2 November 2006, p. 2; Obama on *Talk of the Nation,* NPR, 2 November 2006; "Obama, Rezko," and "And the Inevitable Backlash," *CT,* 3 November 2006; Pamela Stallsmith and Kiran Krishnamurthy, "Big Names Stump for Webb," *RTD,* 3 November 2006, p. A6; Dana Heupel, "Backed for President," *SJR,* 4 November 2006, p. 1; Sarah Baxter, "Obama Mania," *Sunday Times Magazine* (London), 5 November 2006, pp. 26ff.; Jackson and Gibson, "Obama: I Regret Deals with Rezko," *CT,* 5 November 2006; Dave McKinney and Chris Fusco, "Obama Spells Out 'Regret' After Land Deal with Rezko," and Carol Marin, "Machine's Shadow Creeps Over Obama," *CST,* 5 November 2006, pp. A4, B6; Rebecca Traister, "Obama on Tour," *Salon,* 5 November 2006; William Hershey and Laura Bischoff, "Democrats Optimistic Tuesday Will Bring Victory," *DDN,* 5 November 2006, p. A31; Sarah Hollander and Rena A. Koontz, "Democrats Urge Voters to Raise State Wage," *CPD,* 5 November 2006, p. B3; Gloria Borger, "Does Barack Really Rock?" *U.S. News,* 6 November 2006; "Obama's Star Power Dims in Rezko's Orbit," *CCB,* 6 November 2006, p. 28; Lee Hermiston, "Obama Pumps Up Iowa Crowd," *DMR,* 6 November 2006, p. A6; Mike Glover, "Obama Keeps Mind on 2006 Vote, for Now," AP, 6 November 2006; Lynn Sweet, "The Obama Chronicles: Waukegan Interview Transcript," *CST,* 6 November 2006; Obama on *Nightline,* ABC, 6 November 2008; Shailagh Murray, "Eyeing 2008, Political Stars Ran Hard for Others," *WP,* 7 November 2006; Jeff Zeleny and Kate Zernike, "For Democrats, Time to Savor Victory at Last," *NYT,* 8 November 2006; John Dickerson, "Barackwater," *Slate,* 14 December 2006; Peter Slevin, "Obama Says He Regrets Land Deal with Fundraiser," *WP,* 17 December 2006;

Tim Novak, "Lot Next to Obama Can Be Yours for $1.5M," *CST,* 10 October 2007, p. 28; James L. Merriner, "Mr. Inside Out," *CM,* November 2007; Obama's 14 March 2008 *CST* interview; Carol Felsenthal, "The Making of a First Lady," *CM,* February 2009.

40. Bill Reynolds, "Welcome to Obama's Family," *PJ,* 15 February 2007, p. C1; Jodi Kantor and Jeff Zeleny, "Michelle Obama Adds New Role to Balancing Act," *NYT,* 18 May 2007; Mike Dorning and Christi Parsons, "Carefully Crafting the Obama 'Brand,'" *CT,* 12 June 2007; Perry Bacon Jr., "The Outsider's Insider," *WP,* 27 August 2007; Obama on *The Tavis Smiley Show,* 18 October 2007; Obama, news conference with Google employees, Mountain View, CA, 14 November 2007; Audie Cornish, "Rare National Buzz Tipped Obama's Decision to Run," *All Things Considered,* NPR, 19 November 2007; Marc Ambinder, "Teacher and Apprentice," *Atlantic Monthly,* December 2007; Leslie Bennetts, "First Lady in Waiting," *Vanity Fair,* December 2007; Daren Briscoe, "An Interview with Barack Obama," *Newsweek,* 8 January 2009 (18–19 May 2008); Matt Schofield, "Michelle Obama Says She Didn't Want Husband to Run," *KC Star,* 10 July 2008; Pete Rouse's July 2008 interview with Jim Gilmore; Roger Simon, "Relentless," *Politico,* 25 August 2008; Evan Thomas et al., "Barack Obama: How He Did It," *Newsweek,* 17 November 2008; Wolffe, *Renegade,* pp. 49, 52–54; Heilemann and Halperin, *Game Change,* pp. 61–63; Plouffe, *The Audacity to Win,* p. 13; Axelrod, *Believer,* p. 190.

41. Julie Bosman, "Obama's New Book Is Surprise Best Seller," *NYT,* 9 November 2006; Chris Lehmann, "Disciplined Politician Dispenses Tough Love," *NYO,* 13 November 2006, p. 9; Charles Babington, "Time in Senate May Be Irrelevant If Obama Runs," *WP,* 13 November 2006; Glenn Thrush, "Democratic Surge Boosts Obama," *QCT,* 13 November 2006; Obama on *Good Morning America,* ABC, 13 November 2006; Obama, remarks at the King National Memorial groundbreaking ceremony, 13 November 2006; Anne E. Kornblut, "Purpose Driven Candidate?," *NYT,* 14 November 2006; "Obama Announces New Committee Assignments," 14 November 2006, Obama.Senate.gov; Christi Parsons, "'Purpose' Pastor Has Pulpit for Obama," *CT,* 15 November 2006; Obama Podcast #35, "Democratic Senate Majority," 15 November 2006, Obama.Senate.gov; Obama, *CRec,* 16 November 2006, pp. S11021–22; Leonard Pitts Jr., "Obama: Not Your Father's Politician," *MH,* 18 November 2006; "Excerpts from Obama's Interview with AP," AP, 18 November 2006; Obama on CNBC with Tim Russert, 18 November 2006; Nicholas Spangler, "Packed Theater Greets Obama," *MH,* 19 November 2006; Thomas Beaumont, "Obama Talks with Top Advisers in Iowa," *DMR,* 26 November 2006, p. A1; Deirdre Shesgreen, "On the Way to New Hampshire, Obama Tests Presidential Waters," *SLPD,* 10 December 2006, p. A4; Walter Shapiro, "Obama's Magic," *Salon,* 12 December 2006; Ginger Thompson, "Seeking Unity, Obama Feels Pull of Racial Divide," *NYT,* 12 February 2008; Pete Rouse's July 2008 interview with Jim Gilmore; Evan Thomas et al., "Barack Obama: How He Did It," *Newsweek,* 17 November 2008; Plouffe, *The Audacity to Win,* p. 14; Heilemann and Halperin, *Game Change,* p. 71.

42. Jim Kuhnhenn, "Obama Rides Personal, National Political Wave and Ponders His Future," AP, 19 November 2006; Obama, "A Way Forward in Iraq," Chicago Council on Global Affairs, 20 November 2006; Obama on *CNN Newsroom,* 20 November 2006; Christi Parsons and David Mendell, "Obama: Time to Stop 'Coddling' Iraq," *CT,* 21 November 2006; Scott Fornek, "'No Good Options Left: Obama," and Lynn Sweet, "Obama, Dems Now Setting Agenda," *CST,* 21 November 2006, pp. 8, 9; Wally Sipers, "Obama Apologizes to Reporter After Potential Love Is Lost," *BND,* 21 November 2006; Kathy Kiely, "Returning Rookies Reflect on Lessons Learned," *USAT,* 27 November 2007; Chuck Goudie, "Question Whispered About Obama Now Asked in Public," *CDH,* 27 November 2006, p. 13; Sweet, "Durbin's Online Petition: Run, Obama, Run," *CST,* 28 November 2006, p. 20; Alexander Cockburn, "Meet Senator Slither," Counterpunch.org, 9 December 2006; Matthew Mosk, "The $75 Million Woman," *WP,* 8 October 2007; Ryan Lizza, "Battle Plans: How Obama Won," *New Yorker,* 17 November 2008; Balz and Johnson, "Obama Campaign Offers Lessons for Present Governing Challenges," *WP,* 31 July 2009, and *The Battle for America 2008,* pp. 29–30; Axelrod, *Believer,* pp. 194–98.

43. Mike Dorning, "Obama Reaches Out in Key 2008 States," *CT,* 29 November 2006; Lynn Sweet, "Obama Bound for New Hampshire," *CST,* 29 November 2006, p. 26; Michael Tomasky, "The Phenomenon," *NYRB,* 30 November 2006; Sweet, "Obama Is Going to Go for It," *CST,* 30 November 2006, p. 43; Seema

Mehta, "Obama an Unlikely Guest at O.C. Church," *LAT,* 30 November 2006; Dorning, "Obama's Mega-Church Visit Spotlights Waning 'God Gap,'" *CT,* 1 December 2006; Obama, World AIDS Day speech, Saddleback Church, 1 December 2006; Martin Wisckol, "Obama Warmly Welcomed by Evangelical Christians," *OCR,* 1 December 2006; Michael Finnegan, "AIDS Fight Needs Churches, Obama Says," *LAT,* 2 December 2006; Dorning, "Evangelicals' Open Arms, Wary Hearts Greet Obama," *CT,* 2 December 2006; Tim Grieve, "Left Turn at Saddleback Church," *Salon,* 2 December 2006; Maureen Dowd, "What's in a Name, Barry?," *NYT,* 2 December 2006; Adam Nagourney, "Early 'Maybe' from Obama Jolts '08 Field," *NYT,* 4 December 2006; Patrick Healy, "Obama Meets Party Donors in New York," *NYT,* 5 December 2006; "Obama Should Run," *CT,* 6 December 2006; Obama, *CRec,* 6 December 2006, pp. S11291–92; Obama on *The Early Show,* CBS, 7 December 2006; Obama, *CRec,* 7 December 2006, p. S11501 (introducing S. 4102, the Election Jamming Prevention Act); Obama Podcast #36, "Iraq Study Group," 7 December 2006, Obama.Senate.gov; Charles Babington and Shailagh Murray, "For Now, an Unofficial Rivalry," *WP,* 8 December 2006; Mara Liasson, "Barack Obama, Still on Rise," *All Things Considered,* NPR, 8 December 2006; Terry McDermott, "What Is It About Obama?," *LAT,* 24 December 2006; Marc Ambinder, "Teacher and Apprentice," *Atlantic Monthly,* December 2007; Tom Daschle's June 2008 interview with Jim Gilmore; Evan Thomas et al., "Barack Obama: How He Did It," *Newsweek,* 17 November 2008; Wolffe, *Renegade,* pp. 55–56; Heilemann and Halperin, *Game Change,* pp. 69–70.

44. Holly Ramer, "Obama Says 'Internal Clock' Kept Him from N.H. Until Now," AP, 8 December 2006; John DiStasio, "Obama Heads to NH," *Union Leader,* 9 December 2006, p. A1; Adam Nagourney, "The Pattern May Change, If," *NYT,* 10 December 2006; Deirdre Shesgreen, "On Way to New Hampshire, Obama Tests Presidential Waters," *SLPD,* 10 December 2006, p. A4; Nagourney, "Obama, Visiting New Hampshire, Offers Flavor of a Campaign That Might Be," *NYT,* 11 December 2006; Dan Balz, "Obama Takes First Steps in N.H.," *WP,* 11 December 2006; Terry McDermott and Mark Z. Barabak, "Crowds Adore Obama," *LAT,* 11 December 2006; Jill Lawrence, "Obama Draws Crowds as He Tours New Hampshire," *USAT,* 11 December 2006, p. 9A; Susan Milligan, "Obama's Star Power Shows on N.H. Visit," *BG,* 11 December 2006; Scott Brooks, "Obama Fever Grips NH," *Union Leader,* 11 December 2006, p. A1; Christi Parsons, "N.H. Crowds Warm to Obama," *CT,* 11 December 2006; Lynn Sweet, "Hamp-Sure? N.H. Seems to Love Obama," *CST,* 11 December 2006, p. 6; Craig Gilbert, "Obama Dips His Toe Deeper into Presidential Waters," *MJS,* 11 December 2006, p. 1; John Dickerson, "Barack Star," *Slate,* 11 December 2006; Scott Lehigh, "Obama Bandwagon Is Filling Up Fast," *BG,* 12 December 2006; Gail Russell Chaddock, "What's Driving Obama-Mania?," *CSM,* 12 December 2006; John Podhoretz, "Obama: Rorschach Candidate," *NYP,* 12 December 2006; Steven Thomma, "If America Is Ready for a Black President, It Might Be Someone Like Obama," KRDC, 12 December 2006; Walter Shapiro, "Obama's Magic," *Salon,* 12 December 2006; Maureen Dowd, "Will Hillzilla Crush Obambi?," *NYT,* 13 December 2006.

45. Marcus Kabel, "Obama Throws Weight Behind Wal-Mart Critics," AP, 16 November 2006; Greg Hinz, "Sen. Obama Sees No Hypocrisy in His Wife's Post at a Firm That Does Business with Wal-Mart," *CCB,* 11 December 2006, p. 2; Barack H. and Michelle L. Obama, 2006 Form 1040, 15 April 2007; Audie Cornish, "Rare National Buzz Tipped Obama's Decision to Run," *All Things Considered,* NPR, 19 November 2007; Leslie Bennetts, "First Lady in Waiting," *Vanity Fair,* December 2007; Michelle Obama on *The Situation Room,* CNN, 1 February 2008; Scott Helman, "Holding Down the Obama Family Fort," *BG,* 30 March 2008; Daren Briscoe, "A Conversation with Barack Obama," *Newsweek,* 9 January 2009 (18–19 May 2008); Matt Schofield, "Michelle Obama Says She Didn't Want Husband to Run," *KCS,* 10 July 2008; Rachel L. Swarns, "Family Mainstay to Move into White House," *NYT,* 10 January 2009; Robinson, *A Game of Character,* pp. 186–92. Journalists would later note how in December 2006 and January 2007, Michelle's employer, the University of Chicago Medical Center, hired ASK Public Strategies, the decidedly low-profile partner firm of David Axelrod's AKPD political consulting shop, to examine UCMC's Urban Health Initiative. ASK—Axelrod, Eric Sedler, and John Kupper—was best known for creating corporate front groups such as "Consumers Organized for Reliable Electricity," on behalf of ComEd, and the "New York Association for Better Choices," on behalf of Cablevision. Howard Wolinsky, "The Secret Side of David Axelrod," *Business Week,* 14 March 2008, Joe Stephens, "Obama Camp Has Many Ties to Wife's Employer," *WP,* 22 August 2008, and Mike McIntire, "For Obama Strategist, Ties to Corporations and Non-Profits as a Consultant," *NYT,* 18 October 2008.

46. *Today,* NBC, 12 December 2006; Zachary A. Goldfarb, "On Monday Night Football, An Announcement from Obama," *WP,* 17 December 2006; Marc Ambinder, "Teacher and Apprentice," *Atlantic Monthly,* December 2007; Daren Briscoe, "An Interview with Barack Obama," *Newsweek,* 8 January 2009 (18–19 May 2008); Pete Rouse's July 2008 interview with Jim Gilmore; Wolffe, *Renegade,* pp. 20, 56–57; Heilemann and Halperin, *Game Change,* pp. 62, 72; Plouffe, *The Audacity to Win,* p. 16; Suskind, *Confidence Men,* pp. 125–26; Axelrod, *Believer,* p. 199.

47. George Will, "Run Now, Obama," *WP,* 14 December 2006; Mike Dorning, "Obama Donors Signal His Grass-Roots Support," *CT,* 14 December 2008; Donald Lambro, "The Obama Factor," and "The Other Barack Obama," *WT,* 14 December 2006, pp. A21, A22; Rick Pearson, "Obama on Obama," *CT,* 15 December 2006; Lynn Sweet, "Trail of Fears," *CST,* 15 December 2006, p. 3; Rosa Brooks, "Barack's Ready," *LAT,* 15 December 2006; Peggy Noonan, "The Man from Nowhere," *WSJ,* 16 December 2006; Joe Garcia, "Barack Obama: The Guise of Change," DailyKos.com, 16 December 2006; Clarence Page, "What's in a Middle Name?," *CT,* 17 December 2006; David Jackson and Ray Gibson, "Obama Intern Had Ties to Rezko," *CT,* 24 December 2006; John Kass, "Obama All Pecs and Peccadilloes, Now," *CT,* 17 January 2007; Carol Marin, "Frank Discussions About Family Helped Obama Decide," *CST,* 17 January 2007, p. 37; Cathleen Falsani, "Evangelical? Obama's Faith Too Complex for Simple Label," *CST,* 19 January 2007, p. 28; Philip Sherwell, "The Big Hitter," *Sunday Telegraph,* 28 January 2007, p. 18; David Jackson and Ray Gibson, "Rezko Sells Lot Next to Obama," *CT,* 24 February 2007 (on 28 December 2006); Chris Fusco and Dave McKinney, "Rezko Lawyer Wants to Build Kenwood Condos," *CST,* 24 February 2007, p. 2; Kathy Chaney, "Developer Desires to Be Obama's Neighbor," *HPH,* 28 February 2007, pp. 1–2; Hermene D. Hartman, "Barack and Black Enterprise," *N'Digo,* 7–13 June 2007; Mundy, *Michelle,* p. 169; Newton Minow's 2008 interview with Jim Gilmore; DJG interviews with Newton Minow, Abner Mikva, and Hermene Hartman.

48. Bernard Schoenburg, "Durbin, LaHood Say Some Good Things About Obama," *SJR,* 17 December 2006, p. 39; Gregory Rodriguez, "Is Obama the New 'Black'?," *LAT,* 17 December 2006; Jonathan Alter, "2008: The Contenders," *Newsweek,* 18 December 2006, p. 7; Jackie Calmes, "Obama Forces Shuffle in Early '08 Lineup," and John Fund, "Not So Fast," *WSJ,* 18 December 2006; Eric Zorn, "Obama Critics Build Cases on Faulty Premises," *CT,* 19 December 2006; "Barack Obama, U.S. Senator," *WT,* 19 December 2006, p. A20; [Mark Penn], "Launch Strategy Thoughts," [21 December 2006], 11pp.; *CRec,* 8 December 2006, pp. S11836–38 (S. 2125; P. L. 109-456, 22 December 2006); John Heilemann, "The Chicago Cipher," *NYM,* 25 December 2006; Josh Gerstein, "Obama Shapes an Agenda Beyond Iraq War," *NYS,* 26 December 2006, p. 1; Joshua Green, "'The Plan,' October 2006," and "Penn's 'Launch Strategy' Ideas, December 21, 2006," TheAtlantic.com, 11 August 2008; Balz and Johnson, *The Battle for America 2008,* pp. 50–52; Plouffe, *The Audacity to Win,* pp. 23–25.

49. Mary Clare Jalonick, "Daschle: Obama Has 'Unlimited Potential,'" AP, 20 December 2006; Brian Charlton (AP), "Obama Run in '08 Will Be Decided Here, Sister Says," *HA* (and *HSB*), 20 December 2006; Conor Clarke, "Barack Obama's Non-Scandal," *TNR,* 20 December 2006 (and *CST,* 24 December 2006, p. B3); Gary Hart, "American Idol," *NYRB,* 24 December 2006; Jeff Zeleny, "Testing the Water, Obama Tests His Own Limits," *NYT,* 24 December 2006; David Jackson and Ray Gibson, "Obama Intern Had Ties to Rezko," and Christi Parsons, "Obama's No 'Smith,' but Is That Really Bad?," *CT,* 24 December 2006; Frank Main, "Internship Also Links Obama, Rezko," *CST,* 24 December 2006, p. A5; Jonathan Alter, "Is America Ready?," *Newsweek,* 25 December 2006, pp. 28ff.; Michael Barone, "The Experience Factor," *U.S. News,* 25 December 2006; Lynn Sweet, "Obama Eyes Iowa in Putting '08 HQ Here," *CST,* 25 December 2006, p. 3; "Poll Shows Obama Pulling Even with Clinton in NH," *Union Leader,* 26 December 2006, p. A2; Johnny Brannon, "Obama's Visit All About Golfing, Not Presidency," *HA,* 28 December 2006; Charlton, "Obama Stays Out of Public Eye in Hawaii," AP, 29 December 2006; "Obama Needs to Walk, Not Run," *BND,* 30 December 2006; Evan Thomas et al., "Barack Obama:

How He Did It," *Newsweek,* 17 November 2008; Mendell, *Obama,* p. 380; Wolffe, *Renegade,* pp. 22–25; Plouffe, *The Audacity to Win,* pp. 25–26; Heilemann and Halperin, *Game Change,* p. 73.

50. Laura Washington, "Whites May Embrace Obama, but Do 'Regular' Black Folks?," *CST,* 1 January 2007, p. 23; Lois Romano, "Effect of Obama's Candor Remains to Be Seen," and Ruth Marcus, "The Audacity of Nope," *WP,* 3 January 2007; Obama Podcast #37, "Against an Escalation in Iraq," 3 January 2007, Obama.Senate.gov; Patrick Healy and Adam Nagourney, "In the Back Room and Over Drinks, Senator Clinton Plans for 2008," *NYT,* 4 January 2007; Obama, "A Chance to Change the Game," *WP,* 4 January 2007; Lynn Sweet, "Past Drug Use May Test Obama," *CST,* 4 January 2007, p. 7; Margaret Carlson, "For Obama, It's Public Character That Counts," Bloomberg, 4 January 2007; Mark Silva, "Democrats Warn Bush Against Iraq Buildup," *CT,* 6 January 2007; Kenneth T. Walsh, "'A Great Hunger for Change," *U.S. News,* 8 January 2007; Paul Merrion, "Obama Taps Local Fundraising Bigwigs," *CCB,* 8 January 2007; Ambinder, "Teacher and Apprentice," *AM,* December 2007; David Axelrod's and Pete Rouse's 2008 interviews with Jim Gilmore; Harriette Cole, "The Real Michelle Obama," *Ebony,* September 2008, pp. 72–79; Wolffe, *Renegade,* pp. 57–58; Robinson, *A Game of Character,* pp. 195–96; Balz and Johnson, "A Political Odyssey: How Obama's Team Forged a Path," *WP,* 2 August 2009, and *The Battle for America 2008,* p. 19; Plouffe, *The Audacity to Win,* p. 27; Heilemann and Halperin, *Game Change,* pp. 73–74; Robert M. Gates, *Duty* (Knopf, 2014), p. 300; Axelrod, *Believer,* pp. 201–4.

51. Jim Abrams, "Senate Begins Work on Ethics Bill," AP, 8 January 2007; Obama Podcast #38, "Ethics Legislation," 8 January 2007, Obama.Senate.gov; Jeff Zeleny and Patrick Healy, "Decisions on the Horizon," *NYT,* 9 January 2007; Dana Milbank, "At Newsstands Everywhere, the Honorable Beach Babe from Illinois," *WP,* 9 January 2007; Christi Parsons, "Democrats Target Ethics," and Eric Zorn, "Honesty Helps Obama Separate Past from Future," *CT,* 9 January 2007; Lynn Sweet, "Fit to Be President?," *CST,* 9 January 2007, p. 3; Helen Kennedy, "Jesse's Bid to Be Prez Paved Way, Says Obama," *NYDN,* 9 January 2007, p. 4; Annie Karni, "Obama Seeks Support on Clinton Turf," *NYS,* 9 January 2007, p. 2; USS, CHS, "Ensuring Full Implementation of the 9/11 Commission's Recommendations," S. Hrg. 110-865, 9 January 2007, pp. 37–40; Elizabeth Williamson, "The Green Gripe with Obama," *WP,* 10 January 2007; USS, HEL, "Health Care Coverage and Access," S. Hrg. 110-24, 10 January 2007, p. 1; "Obama Statement on President's Call for Troop Escalation in Iraq," 10 January 2007; Sweet, "Clinton Locks Up Key Endorsement," *CST,* 11 January 2007, p. 29; Obama on *Today,* NBC, 11 January 2007; *CRec,* 8 December 2006, p. S11830 (H. R. 6060, P. L. 109-472, 11 January 2007); "Lugar-Obama Nonproliferation Legislation Signed into Law by the President," 11 January 2007; USS, CFR, "Securing America's Interest in Iraq: The Remaining Options," S. Hrg. 110-153, 11 January 2007, pp. 139–41; Anne E. Kornblut, "Devastating Criticism by Both Parties," *NYT,* 12 January 2007; Milbank, "The Secretary vs. the Senators," *WP,* 12 January 2007; Obama on *Face the Nation,* CBS, 14 January 2007; Mark Silva, "Obama, McCain: 2 War Views," *CT,* 15 January 2007; "Beach Babes," *People,* 15 January 2007, pp. 10–11; Mark Z. Barabak, "How Obama Went from Underdog to Alpha," *LAT,* 4 June 2008; Tim Dickinson, "Obama's Barin Trust," *Rolling Stone* 1056/57, 10 July 2008; Shailagh Murray, "In Obama's Circle, Chicago Remains the Tie That Binds," *WP,* 14 July 2008.

52. Obama, Announcement for President, 16 January 2007, YouTube; Jeff Zeleny, "Obama Starts Bid, Reshaping Democratic Field," *NYT,* 17 January 2007; Shailagh Murray and Chris Cillizza, "Obama Jumps into Presidential Fray," *WP,* 17 January 2007; Judy Keen, "The Big Question About Barack Obama: Does He Have Enough Experience to Be President?," *USAT,* 17 January 2007; Mike Dorning and Christi Parsons, "He Sets His Sights on the White House," and John Kass, "Obama All Pecs and Peccadilloes, Now," *CT,* 17 January 2007; Bernard Schoenburg, "Why Springfield," *SJR,* 17 January 2007, p. 1; "Obama Statement on Iraq," 17 January 2007; Zeleny, "Obama Knows Novelty Goes Only So Far as an Agent of Change," *NYT,* 18 January 2007; "He's In," *WP,* 18 January 2007; Eric Zorn, "Why Obama Just Might Click with the Voters," *CT,* 18 January 2007; Scott Fornek, "Obama Sizing Up Abe's Old Digs," *CST,* 18 January 2007, p. 16; Obama, *CRec,* 18 January 2007, pp. S722–23; Christi Parsons, "Senate OKs Tougher Ethics Bill 96-2," *CT,* 19 January 2007; Eric Krol, "Obama's Debut Gave Little Hint of What Was to Come," *CDH,* 19 January 2007, p. 16; Bernard Schoenburg, "Site of Obama's Announcement Still a Mystery," *SJR,* 19 January

2007, p. 1; Leonard Pitts Jr., "America's Baggage Is Obama's Burden," *MH,* 19 January 2007; Dori Meinert, "Obama's Senate Record Shows High Party Loyalty," Copley, 19 January 2007; Kevin McDermott, "Site of Obama Speech Likely Will Evoke Lincoln," *SLPD,* 21 January 2007, p. C7; Kurt Erickson and Ed Tibbetts, "Obama's Records Reveal His Liberal Side," *BP* (and *QCT*), 21 January 2007; Debra J. Dickerson, "Colorblind," *Salon,* 22 January 2007; Fran Spielman, "Obama Endorses Daley," *CST,* 22 January 2007; Gary Washburn, "Daley Gets Backing of Obama," *CT,* 23 January 2007; Steve Rhodes, "Barack Obama (D-Daley)," *BR,* 23 January 2007; Dante Chinni, "Obama-Mania May Backfire," *CSM,* 23 January 2007; Rich Miller, "Love Is Blind," *IT,* 25 January 2007; George Scialabba, "The Work Cut Out for Us," *Nation,* 29 January 2007, pp. 23–27; Salim Muwakkil, "Barack's Black Dilemma," *ITT,* February 2007, p. 16; Kevin Alexander Gray, "The Packaging of Obama," *Progressive,* February 2007, pp. 41–44; Paul Street, "The Obama Illusion," *Z Magazine,* February 2007; James Crabtree, "Barack Obama," *Prospect Magazine* (UK) 131, February 2007; Ta-Nehisi Coates, "Is Obama Black Enough?," *Time,* 1 February 2007; Peter Beinart, "Black Like Me," *TNR,* 5 February 2007, p. 6; Martin Plissner, "Ready for Obama Already," *NYT,* 7 February 2007.

53. Obama on MSNBC, 23 January 2007; USS, CVA, "DOD/VA Collabortion and Cooperation," S. Hrg. 110-1, 23 January 2007, pp. 1, 56, 63; Obama on *Good Morning America,* ABC, and *The Early Show,* CBS, 24 January 2007; Obama, remarks at Families USA Conference, 25 January 2007; Mike Dorning, "Obama Joins Ranks of Democrats Supporting Universal Health Care," *CT,* 26 January 2007; Scott Fornek, "Clinton Pal Bobby Rush: I'm Supporting Obama," *CST,* 27 January 2007, p. 5; USS, CHS, "Hurricanes Katrina and Rita," S. Hrg. 110-33, 29 January 2007, pp. 10–12, 31–35, 38–41, 60–61; Adam Nagourney, "Senators at Louisiana Hearing Criticize Federal Recovery Aid," *NYT,* 30 January 2007; USS, CFR, "Nominations of the 110th Congress," S. Hrg. 110-777, 30 January 2007, pp. 23–25, 60–61; Obama, *CRec,* 30 January 2007, p. S1322 (introducing S. 433, the Iraq War De-Escalation Act); Shailagh Murray, "Obama Bill Sets Date for Troop Withdrawal," *WP,* 31 January 2007; Obama Podcast #39, "Iraq War De-Escalation Act," 31 January 2007, Obama.Senate.gov; Jason Horowitz, "Biden on the Biden Story," *NYO,* 31 January 2007; Nagourney, "Biden Unwraps His Bid for '08 with an Oops!," *NYT,* 1 February 2007; Dan Balz, "Biden Stumbles at the Starting Gate," *WP,* 1 February 2007; Elana Schor, "Sen. Barack Obama Tries to Flesh Out His Record with an Election-Fraud Bill," *TH,* 1 February 2007; Rachel L. Swarns, "So Far, Obama Can't Take Black Vote for Granted," *NYT,* 2 February 2007; Obama, "Remarks at the DNC Winter Meeting," 2 February 2007; Jonathan Singer, "MyDD Interview with Barack Obama," 2 February 2007; Ben Smith, "Black Rift Widens over Obama," *Politico,* 3 February 2007; Horowitz, "Biden Unbound," *NYO,* 5 February 2007; Anne E. Kornblut, "Obama Confronts 'Outsider' Dilemma," *WP,* 5 February 2007; David Remnick, "The Joshua Generation," and Ryan Lizza, "Battle Plans: How Obama Won," *TNY,* 17 November 2008; Lloyd Grove, "The World According to David Plouffe," Portfolio.com, 11 December 2008; Wolffe, *Renegade,* pp. 68–69; Balz and Johnson, *The Battle for America 2008,* p. 71.

54. Lynn Sweet, "Obama Begins Fund-Raising Drive," *CST,* 1 February 2007, p. 31; Zachary A. Goldfarb, "Mobilized Online, Thousands Gather to Hear Obama," *WP,* 3 February 2007; Sweet, "Obama in Campaign Mode on a Theme of 'Hope,'" *CST,* 4 February 2007, p. A19; Anne E. Kornblut, "Obama Confronts 'Outsider' Dilemma," *WP,* 5 February 2007; Bernard Schoenburg, "Cold Weather Won't Move Obama's Announcement," *SJR,* 5 February 2007, p. 1; Christi Parsons and Manya A. Brachear, "Obama Launches an '07 Campaign—to Quit Smoking," *CT,* 6 February 2007; Norman Mailer, "Superman Comes to the Supermarket," *Esquire,* November 1960; Frank, *Obama on the Couch,* p. 130; Roger Simon, "Obama: 'I Have the Potential of Bringing People Together,'" *Politico,* 8 February 2007; USS, CFR, "The President's Foreign Affairs Budget," S. Hrg. 110-389, 8 February 2007, pp. 57–60; USS, HEL, "NCLB Reauthorization," S. Hrg. 110-295, 8 February 2007, pp. 46–47; Judy Keen, "Obama Touts Non-Political Resume," and "Dem Cites 'Unique' Skills as Key to His Campaign," *USAT,* 9 February 2007; Mike Allen, "Undoing Obama," *Politico,* 9 February 2007; Peter Wallstein, "Blacks Wonder If Obama's Multiculturalism Is a Problem," *LAT,* 10 February 2007; "Michelle Obama to Work Part Time During Campaign," *UCC,* 15 February 2007; Lloyd Grove, "The World According to David Plouffe," Portfolio.com, 11 December 2008.

55. Sara Karp, "Sen. Barack Obama's Pastor Frames Progressive Issues Through Lens of Faith," Religion News Service, 10 March 2005; Manya A. Brachear, "Pastor Inspires 'Audacity to Hope,'" *CT,* 21 January 2007; Ben Wallace-Wells, "The Radical Roots of Barack Obama," 14 February 2007, and "Destiny's Child," *Rolling Stone,* 22 February 2007, pp. 48–57; Jodi Kantor, "Disinvitation by Obama Is Criticized," *NYT,* 6 March 2007; Michael Powell and Kantor, "A Strained Wright-Obama Bond Finally Snaps," *NYT,* 1 May 2008; Daren Briscoe, "An Interview with Barack Obama," *Newsweek,* 8 January 2009 (18–19 May 2008); Evan Thomas et al., "Barack Obama: How He Did It," *Newsweek,* 17 November 2008; Heilemann and Halperin, *Game Change,* p. 75; Balz and Johnson, *The Battle for America 2008,* pp. 33, 201; Remnick, *The Bridge,* pp. 468–72; Plouffe, *The Audacity to Win,* pp. 28, 40–41; Axelrod, *Believer,* pp. 209–12; Jeremiah Wright's interviews with DJG and Edward Klein; Klein, *The Amateur,* p. 46.

56. Obama, "Remarks Announcing His Candidacy for President," 10 February 2007, YouTube; Adam Nagourney and Jeff Zeleny, "Obama Formally Enters Presidential Race with Calls for Generational Change," and Brent Staples, "Decoding the Debate Over the Blackness of Barack Obama," *NYT,* 11 February 2007; Dan Balz and Anne E. Kornblut, "Obama Joins Race with Goals Set High," *WP,* 11 February 2007; Rick Pearson and Ray Long, "Obama's Kickoff Is Steeped in Symbolism," Christi Parsons, "Obama: 'Destiny Calling,'" and Michael Tackett, "Obama Entry Heavy in Symbolism," *CT,* 11 February 2007; Scott Fornek and Dave McKinney, "Obama Makes It Official," *CST,* 11 February 2007, p. A1; Daniel Pike, "A Stirring Start," and Bernard Schoenburg, "Obama Has No Problem Writing His Own Speeches," *SJR,* 11 February 2007, pp. 1, 17; Kevin McDermott, "Obama, Off and Running," *SLPD,* 11 February 2007, p. A1; Kurt Erickson, "All In: Obama Enters '08 Race," *BP,* 11 February 2007, p. A1; Jennifer Jacobs, "Obama Enters Race for President," *DMR,* 11 February 2007, p. A1; David Yepsen, "Can a Candidate Be Too Candid?," *DMR,* 12 February 2007, p. B3; Maureen Dowd, "Obama, Legally Blonde?," *NYT,* 14 February 2007; David Montgomery, "Barack Obama's On-Point Message Man," *WP,* 15 February 2007; Anna Schneider-Mayerson, "It's Obamalot!," *NYO,* 12 March 2007, p. 1; Matthew Mosk, "The $75 Million Woman," *WP,* 8 October 2007; M. Charles Bakst, "Up Close: Michelle Obama," *PJ,* 21 February 2008, p. B1; Plouffe, *The Audacity to Win,* pp. 41–42; Axelrod, *Believer,* pp. 212–15.

EPILOGUE: THE PRESIDENT DID NOT ATTEND, AS HE WAS GOLFING

1. Matt Taibbi, "Between Barack and a Hard Place," *Rolling Stone,* 15 February 2007 (and Alternet.org); Curtis Lawrence, "I Want a Black President," *CST,* 17 February 2007; Louis Chude-Sokei, "Shades of Black," *LAT,* 18 February 2007; Michael D. Shear, "Obama Takes First Campaign Trip South," *WP,* 19 February 2007; Eugene Robinson, "Authentic Obama," *WP,* 20 February 2007; Rich Miller, "Why Obama Will Surprise National Political Pundits," *CST,* 23 February 2007; Jeff Zeleny, "As Candidate, Obama Carves Antiwar Stance," *NYT,* 26 February 2007; Obama, "Foreword," in National Urban League, *The State of Black America 2007* (Beckham Publications, 2007), pp. 9–12; Maureen Dowd, "Where's His Right Hook?," *NYT,* 3 March 2007; Nicholas D. Kristof, "Obama on the Issues," *NYT,* 5 March 2007; Jason Horowitz, "Barack Attack," *NYO,* 26 March 2007, p. 1; Dowd, "She's Not Buttering Him Up," *NYT,* 25 April 2007; Lynn Sweet, "Getting Ready to Roll," *CST,* 3 May 2007, p. 37; John McCormick, "Chicago Is Heart, Brain Center of Obama Campaign," *CT,* 11 June 2007; McCormick, "Obama's 'Body Man' on His Game," *CT,* 15 November 2007; "Obama's Right Hands," *Glamour,* December 2007; Jacob Dagger, "Q&A: Inseparable from Obama," *Duke Magazine,* September–October 2008; Axelrod and Plouffe et al. on *60 Minutes,* CBS, 9 November 2008.

2. Jeremiah Wright on *Hannity & Colmes,* Fox News, 1 March 2007; Obama, remarks at the Selma voting rights march commemoration, 4 March 2007; Patrick Healy and Jeff Zeleny, "Clinton and Obama Unite, Briefly, in Pleas to Blacks," *NYT,* 5 March 2007; Jodi Kantor, "Disinvitation by Obama Is Criticized," *NYT,* 6 March 2007; Brian DeBose, "Obama's Short Tenure Seen as a Plus," *WT,* 7 March 2007; Jonathan Tilove, "Who Is Obama's 'Cousin Pookie'?," Newhouse, 8 March 2007; Mary Mitchell, "Obama Slights His Own Pastor," *CST,* 8 March 2007, p. 14; Bob Abernethy, "Link Between Senator Barack Obama and Reverend Jeremiah Wright of Trinity United Church of Christ Issue in Presidential Campaign," *Religion & Ethics NewsWeekly,* 9 March 2007; Dawn Turner Trice, "Is Obama Black Enough?," *CT,* 11

March 2007; Henry C. Jackson, "AP Interview: Obama Doesn't Take 'Great Offense' Over Word Play," AP, 11 March 2007; Laura Washington, "National Media Can't Get Enough of Obama," *CST,* 12 March 2007, p. 33; Eugene Robinson, "The Moment for This Messenger?," *WP,* 13 March 2007; Marc Hujer, "Spiegel Interview with Barack Obama's Pastor," *Spiegel,* 13 March 2007; Henry Bodget, "Barack Obama and His Money," *Slate,* 13 March 2007; Obama on *Anderson Cooper 360,* CNN, 14 March 2007; Leon Wieseltier, "Audacities," *TNR,* 19 March 2007, p. 64; David Ehrenstein, "Obama the 'Magic Negro,'" *LAT,* 19 March 2007; Obama on *The Larry King Show,* CNN, 19 March 2007; Michael Tarm, "Activist Obama Church Enters Campaign Spotlight," AP, 20 March 2007; Rich Miller, "Will 3rd Ward Politics Affect Obama's Presidential Race?," *CST,* 23 March 2007, p. 31; Richard Cohen, "Obama's Back Story," *WP,* 27 March 2007; Obama on The *Situation Room,* CNN, 28 March 2007; Jodi Kantor, "A Candidate, His Minister and the Search for Faith," *NYT,* 30 April 2007.

3. Jackie Calmes, "From Obama's Past: An Old Classmate, a Surprising Call," *WSJ,* 23 March 2007; Maurice Possley et al., "The Troublesome Return of a Long-Lost Classmate," *CT,* 25 March 2007; Jake Tapper, "Tale of Two Men: Obama Mentor Now Homeless," *Good Morning America,* ABC, 30 March 2007; Peter Hart, "Obamamania," *Extra!,* March–April 2007; Adam Nagourney, "2 Years After Big Speech, A Lower Key for Obama," *NYT,* 8 April 2007; Obama on *The Late Show with David Letterman,* 11 April 2007; John Heilemann, "Money Chooses Sides," *NYM,* 15 April 2007; Tim Novak, "Obama and His Rezko Ties," *CST,* 23 April 2007, p. 22; Novak, "Obama's Ex-Boss a Rezko Partner," *CST,* 23 April 2007, p. 25; Rick Pearson and Ray Long, "Obama Still Answers for Old Ties to Rezko," *CT,* 24 April 2007; Novak, "Broken Promises, Broken Homes," *CST,* 24 April 2007, p. 20; Novak, "Obama: I Didn't Know About Rezko Problems," *CST,* 24 April 2007, p. 8; Carol Marin, "Obama Ducks the Questions," *CST,* 25 April 2007, p. 43; David Brooks, "Obama, Gospel and Verse," *NYT,* 26 April 2007; "Answers Are Needed About Rezko's Deals," *CST,* 27 April 2007, p. 38; Katie Couric, "Questions About Barack Obama's Dealings with Chicago Business Developer," *CBS Evening News,* 27 April 2007; Kevin Drum, "Getting to Know Obama," *Washington Monthly,* 5 May 2007; Andrew Ferguson, "Self-Made Man: Barack Obama's Autobiographical Fictions," *Weekly Standard,* 18 June 2012; Sally Quinn, "A Welcome Face—and a Question," *WP,* 5 May 2007; Lynn Sweet, "Rezko Question Dogs Obama," *CST,* 17 May 2007, p. 31; Marin, "Out of Touch with Illinois?," *CST,* 6 June 2007, p. 41; Ray Gibson and David Jackson, "Obama Campaign Distances Itself from Indicted Fundraiser Rezko," *CT,* 7 June 2007; Jackson and John McCormick, "Critic: Obama Endorsements Counter Calls for Clean Government," *CT,* 12 June 2007; Novak, "Obama's Letters for Rezko," *CST,* 13 June 2007, p. 8; Christopher Drew and Mike McIntire, "An Obama Patron and Friend Until an Indictment in Illinois," *NYT,* 14 June 2007; Chris Fusco and Tim Novak, "Rezko Cash Triple What Obama Says," *CST,* 18 June 2007, p. 3; "The Drip, Drip, Drip on Obama," *CT,* 19 June 2007; Oona King, "Barack Obama's Dreams From My Father," *Times (London),* 15 September 2007; Jess Bravin, "Barack Obama: The Present Is Prologue," *WSJ,* 7 October 2008; Evan Thomas et al., "Barack Obama: How He Did It," *Newsweek,* 17 November 2008; Liam Julian, "Niebuhr and Obama," *Policy Review,* April–May 2009, pp. 19–33; Dave Burgess, *A Tale of Two Brothers: The Keith Kakugawa Story* (Lulu Press, 2009), p. 171; DJG interview with Keith Kakugawa.

4. Perry Bacon Jr., "Obama Reaches Out with Tough Love," *WP,* 3 May 2007; Jeff Zeleny, "Secret Service Guards Obama," *NYT,* 4 May 2007; Christi Parsons and Mike Dorning, "Obama Given Secret Service Protection," *CT,* 4 May 2007; Lynn Sweet, "Agents to Shield Obama," *CST,* 4 May 2007, p. 3; Obama, "Remarks to the National Conference of Black Mayors," Baton Rouge, LA, 5 May 2007; Lynne Duke, "How Big a Stretch?," *WP,* 7 May 2007; John McCormick, "Obama Downplays Security Bubble," *CT,* 7 May 2007; Harry C. Alford, "Did Barack Obama Lie?," *CD,* 9 May 2007, p. 7; Obama on *This Week,* ABC, 13 May 2007; Dorning, "Obama: President Needs Judgment with Toughness," *CT,* 14 May 2007; Joe Klein, "Barack Obama," *Time,* 14 May 2007, p. 56; McCormick, "Her Money Is on Obama," *CT,* 14 May 2007; Parsons, "Obama Doesn't Toe Line at Labor Forum," *CT,* 15 May 2007; Jodi Kantor and Jeff Zeleny, "Michelle Obama Adds New Role to Balancing Act," *NYT,* 18 May 2007; "Michelle Obama One-on-One," *Good Morning America,* ABC, 22 and 23 May 2007; Parsons, "Obama's Wife Cuts Ties to Wal-Mart," *CT,* 23 May 2007; Sweet, "Ties to Wal-Mart Cut," *CST,* 23 May

2007, p. 36; Gary Dorrien, "Hope or Hype?," *Christian Century,* 29 May 2007, pp. 24ff.; Deanna Bellandi, "Michelle Obama Razzes Her Husband, but Says He Doesn't Mind," AP, 29 May 2007; Obama on *Today,* NBC, 29 May 2007; Walter Shapiro, "Barack Obama's Quiet Rebellion," *Salon,* 30 May 2007; Donovan Taylor, "Sen. Barack Obama Does Support Local Chambers of Commerce," *CD,* 6 June 2007, p. 7; AP, "House Where Obama Rents Is Hit by Blaze," *WP, CT,* and *CST,* 18 June 2007; Margaret Talev, "Michelle Obama Fits Her Public Role," KRDC, 8 July 2007; Richard Wolffe and Daren Briscoe, "Across the Divide," *Newsweek,* 16 July 2007, pp. 22ff.; Megan Reichgott, "Obama's 2 Daughters Question Mom," AP, 20 July 2007; Mike McIntire, "Blogger Well Known in Politics Turns His Attention to Attack on Obama Campaign," *NYT,* 30 August 2007; Jeff Chang, "Barack Obama Q&A," *Vibe,* September 2007 [23 May 2007]; Tonya Lewis Lee, "Your Next First Lady?," *Glamour,* September 2007; "The Best Dressed Men in the World," *Esquire,* September 2007; Dana Slagle, "Michelle Obama Juggles Marriage, Motherhood & Work on the Campaign Trail," *Jet,* 10 September 2007, pp. 14ff.; Obama and Michelle Obama on *60 Minutes,* CBS, 16 November 2008; Jonathan Van Meter, "Leading by Example," *Vogue,* April 2013, pp. 248–55, 318–19.

5. Obama at the Sojourners Presidential Forum, 4 June 2007, CNN; Obama, remarks at Hampton University annual Ministers' Conference, 5 June 2007; Mike Dorning, "Obama Warns of Black 'Quiet Riot,'" *CT,* 6 June 2007; Obama, *CRec,* 6 June 2007, pp. S7155–56, S7162; Charles Babington, "Obama, Graham Trade Barbs in Senate," AP, 7 June 2007; Dena Bunis, "Second Thoughts by Bill's Backers," *OCR,* 7 June 2007; Kate Zernike and Jeff Zeleny, "Obama in Senate," *NYT,* 9 March 2008; Graham on "The Choice 2008," *Frontline,* PBS, 14 October 2008; Margaret E. Dorsey and Miguel Diaz-Barriga, "Senator Barack Obama and Immigration Reform," *Journal of Black Studies* 38 (September 2007): 90–104.

6. Jeff Zeleny, "A New Kind of Politics Closely Resembles the Old," *NYT,* 16 June 2007; Mike Dorning, "Obama Urges Fathers to Be Responsible," *CT,* 16 June 2007; John McCormick, "Obama Bio Cites Hidden '02 Agenda," *CT,* 21 June 2007; Mendell, *Obama,* pp. 7, 12; John McCormick, "Obama Team Can Play Rough," *CT,* 22 June 2007; Bettylu Saltzman, "Going Back a Few Years with Obama," *CT,* 3 July 2007; Richard Wolffe, "Inside Obama's Dream Machine," *Newsweek,* 14 January 2008, pp. 30ff.; Wolffe, *Renegade,* p. 127; DJG interviews with Hasan Chandoo and Bettylu Saltzman.

7. Obama, remarks at the Take Back America Forum, 19 June 2007; Eugene Robinson, "A Father's Absence and a Son's Message," *WP,* 19 June 2007; Carol Hunter, "Thoughtful Obama Says He's Ready to Lead," *DMR,* 20 June 2007, p. A11; Obama, "A Politics of Conscience," UCC 50th anniversary convention, Hartford, 23 June 2007; Jeff Zeleny, "Obama, in New TV Ads, Focuses on His Pre-Senate Years," *NYT,* 26 June 2007; Rick Pearson and John McCormick, "GOP Stalwart Lauds Obama in TV Ads," *CT,* 26 June 2007; Larry Blumenfeld, "Barack Obama in New Orleans," *Salon,* 6 July 2007; Amina Luqman, "Obama's Tightrope," *WP,* 6 July 2007; Ruth Marcus, "From Barack Obama, Two Dangerous Words," *WP,* 11 July 2007; Obama on *News & Notes,* NPR, 13 July 2007; Obama, remarks on Chicago Violence at the Vernon Park Church of God, 15 July 2007; McCormick, "Obama Attacks Violence in Chicago," *CT,* 16 July 2007; Obama, remarks to the Planned Parenthood Action Fund, 17 July 2007; Obama, remarks at Changing the Odds for Urban America, Anacostia, D.C., 18 July 2007; Obama, remarks to the College Democrats of America, Columbia, SC, 26 July 2007; Obama, remarks at the National Urban League Annual Conference, St. Louis, MO, 27 July 2007; Obama, "Renewing American Leadership," *Foreign Affairs* 86 (July–August 2007): 2–16; Salim Muwakkil, "The Squandering of Obama," *ITT,* August 2007, p. 15; Obama, remarks at the Woodrow Wilson Center, 1 August 2007; Eric Zorn, "Obama Missing Chance to Cut Bad Local Ties," *CT,* 2 August 2007; Scott Helman, "PACs and Lobbyists Aided Obama's Rise," *BG,* 9 August 2007; Mike Dorning, "In Caregiver's Shoes," *CT,* 9 August 2007; Andrew Sullivan, "Goodbye to All That," *Atlantic,* December 2007; Jeff Zeleny, "Reprising a Career," *NYT,* 22 January 2008.

8. MacFarquhar, "The Conciliator," *TNY,* 7 May 2007, pp. 48ff.; Obama, remarks to the NABJ Convention, Las Vegas, 10 August 2007; Gromer Jeffers Jr., "Is He Black Enough?," *DMN,* 11 August 2007; Brian DeBose, "Obama Confronts 'Blackness' Issue," *WT,* 11 August 2007; Perry Bacon Jr., "The Obamas Are Tired of the Blackness Question," *WP,* 14 August 2007; Leonard Pitts, "Concentrate on Obama's Record, Not His Color," *MH,* 15 August 2007;

John McCormick, "Buffett Is Obama Meal Ticket," *CT,* 16 August 2007; Les Payne, "Sen. Barack Obama: In America, a Dual Audience," *Newsday,* 19 August 2007; Steve Penn, "Barack Obama Exudes Confidence," *KC Star,* 20 August 2007; Obama, "Our Main Goal: Freedom in Cuba," *MH,* 21 August 2007; Jennifer Hunter, "Michelle Gets Stronger All the Time," *CST,* 21 August 2007, p. 18; Obama on *The Daily Show,* Comedy Central, 22 August 2007; David Ignatius, "The Pragmatic Obama," *WP,* 23 August 2007; Mike Littwin, "Obama's 'Change' Could Be More Than a Coined Phrase," *RMN,* 29 August 2007; Obama, "Hit Iran Where It Hurts," *NYDN,* 30 August 2007; Gwen Ifill, "Beside Barack," *Essence,* September 2007, pp. 200ff.; Ryan Lizza, "Above the Fray," *GQ,* September 2007; Maureen Dowd, "The 46-Year-Old-Virgin," *NYT,* 5 September 2007; McCormick and Christi Parsons, "Winfrey Draws Rich, Famous to Obama Bash," *CT,* 9 September 2007; David Baumann, "Is Obama as Bipartisan as His Campaign Says?" *SPT,* 10 September 2007; Obama, remarks at Ashford University, Clinton, IA, 12 September 2007; Obama, "Slate-Yahoo-Huffington Post Presidential Forum," 12 September 2007; Jeff Zeleny and Michael R. Gordon, "Obama Offers Most Extensive Plan Yet for Winding Down War," *NYT,* 13 September 2007; Obama on MSNBC, 17 September 2007; Gwen Ifill, "The Candidate," *Essence,* October 2007, pp. 224ff.; Obama in Philip Galanes, "Barack Obama and Bryan Cranston on the Roles of a Lifetime," *NYT,* 6 May 2016; DJG interviews with Rob Fisher, Jon Molot, George Galland, Paul Scully, and Cynthia Miller.

9. MacFarquhar, "The Conciliator," *TNY,* 7 May 2007, pp. 48ff.; Christi Parsons and Mike Dorning, "Jackson Softens Criticism of Obama in Jena 6 Case," *CT,* 20 September 2007; Alec MacGillis, "Obama Skips Vote, Iowa Debate," *WP,* 20 September 2007, and twelfth of twenty-one comments, 11:05 A.M., 21 September; Ryan Grim, "Senate Votes to Do PR for MoveOn," *Politico,* 20 September 2007; "Campaign Digest," *CT,* 21 September 2007; Nedra Pickler, "Democrat Barack Obama Brings in Friend and Long-Term Adviser Amid Concerns About Campaign," AP, 21 September 2007; Perry Bacon Jr. and Matthew Mosk, "Obama's Challenge," *WP,* 23 September 2007; Obama, remarks at Howard University convocation, 28 September 2007; Obama, "Foreword," *Charleston Law Review,* Fall 2007; David Moberg, "Obama's in the Eye of the Beholder," *ITT,* October 2007, pp. 29ff.; Obama, remarks at DePaul University, 2 October 2007; Jeff Zeleny, "Obama Talks About Growing Up Without His Father," *NYT,* 3 October 2007; Marc Ambinder, "Teacher and Apprentice," *Atlantic,* December 2007; Robert Frank and Mark Maremont, "Money Maven Penny Pritzker Raises Record Amounts of Cash for Obama," *WSJ,* 15 March 2008; Noam Scheiber, "Plouffe Piece," *TNR,* 7 May 2008; Douglas Belkin, "For Obama, Advice Straight Up," *WSJ,* 12 May 2008; Tim Dickinson, "Obama's Brain Trust," *Rolling Stone* 1056/57, 10 July 2008; Valerie Jarrett's 2009 interview with Robert Draper; Jarrett in Remnick, *The Bridge,* p. 274; Plouffe, *The Audacity to Win,* pp. 91–92, 95–99; Suskind, *Confidence Men,* pp. 10–11; Noam Scheiber, "The Partner," *TNR,* 7 June 2012, pp. 4–6; Cal Fussman, "Valerie Jarrett: What I've Learned," *Esquire,* May 2013; Jarrett, "Unconditional Love and a Legacy of Early Education," *Politico,* 5 December 2013; Axelrod, *Believer,* p. 232; DJG interviews with Judy Byrd, Claire Serdiuk, and Sheila Jager.

10. Jim Davenport, "Obama Says Faith 'Plays Every Role' in His Life," AP, 7 October 2007; Michael Crowley, "Hope Sinks," *TNR,* 8 October 2007, p. 20; Obama on *The Tonight Show,* NBC, 17 October 2007; Joan Walsh et al., "All the Candidates' Books," and "Reading Barack Obama," *Salon,* 18 and 19 October 2007; Obama on *The Tavis Smiley Show,* 18 October 2007; Steve Chapman, "Obama vs. the Anger Mongers," *CT,* 21 October 2007; David S. Broder, "Trying Times for the Obama Faithful," *WP,* 25 October 2007; Christine G. Sabathia, "Meet Michelle Obama," *LAS,* 25–31 October 2007, p. A1; Linda Douglass, "Michelle Obama on the Family and the Campaign Trail," *National Journal,* 26 October 2007; Melinda Henneberger, "The Obama Marriage," *Slate,* 26 October 2007; Adam Nagourney and Jeff Zeleny, "Obama Promises a Forceful Stand Against Clinton," *NYT,* 28 October 2007; Dan Balz, "What's on the Table for Obama and Social Security?," *WP,* 30 October 2007, and third comment, 2:58 P.M.; Ryan Lizza, "The Relaunch," *TNY,* 26 November 2007, pp. 72ff.; Richard Wolffe, "Inside Obama's Dream Machine," *Newsweek,* 14 January 2008, pp. 30ff.; Lizza, "How to Beat Hillary Clinton," *TNY,* 13 October 2015.

11. Holly Yeager, "The Heart and Mind of Michelle Obama," *O Magazine,* November 2007, pp. 286ff.; Michael Gordon and Jeff Zeleny, "Interview with Barack Obama," *NYT,* 1 November

2007; Gordon and Zeleny, "Obama Envisions New Relationship with Iran," *NYT,* 2 November 2007; James Traub, "Is (His) Biography (Our) Destiny?," *NYT Magazine,* 4 November 2007, pp. 50ff.; Lynn Sweet, "Living in a Glass House," *CST,* 5 November 2007, p. 26; "Barack Obama Attends Campaign Fund Raiser on St. Thomas," *St. John Source,* 5 November 2007; Sweet, "Where Are Obama State Senate Files?," *CST,* 10 November 2007, p. 10; Obama on *Meet the Press,* NBC, 11 November 2007; Sweet, "Successor Got Key Obama Files," *CST,* 12 November 2007, p. 26; "Michelle Obama Speaks with MSNBC's Mika Brzezinski," MSNBC .com, 13 November 2007; Pew Research Center, "Blacks See Growing Values Gap Between Poor and Middle Class," 13 November 2007; Mike Baker and Christopher Wills, "Obama, Who Rapped Clinton on Records, Says He Has None From Ill. State Senate," AP, 14 November 2007; Obama, "News Conference with Google Employees," Mountain View, CA, 14 November 2007; Margaret Talev, "Q&A with Michelle Obama," KRDC, 21 November 2007; Obama, "Former Prisoners Deserve a Second Chance," *LAS,* 22–28 November 2007, p. A6 (and *CD,* 5 December 2007, p. 8); Andrea Billups, "Mrs. Obama, Savvy Surrogate," *WT,* 25 November 2007, p. A1; David S. Broder, "Obama v. the Icon," *WP,* 25 November 2007; Obama on Nightline, ABC, 26 November 2007; Rebecca Traister, "Michelle Obama Gets Real," *Salon,* 28 November 2007; Perry Bacon Jr., "Foes Use Obama's Muslim Ties to Fuel Rumors About Him," *WP,* 29 November 2007; Juan Williams, "Obama's Color Line," *NYT,* 30 November 2007; Leslie Bennetts, "First Lady in Waiting," *Vanity Fair,* December 2007; Richard Wolffe, "'I Can Only Be Who I Can Be,'" *Newsweek,* 16 February 2008; Plouffe, *The Audacity to Win,* pp. 109–15; Axelrod, *Believer,* pp. 236–40; DJG interview with Bob Klonowski.

12. Obama on the *CBS Evening News,* 6 December 2007; Noam Scheiber, "The Closer," *TNR,* 10 December 2007; Roger Cohen, "Obama's American Idea," *NYT,* 10 December 2007; Daniel Heim, "Sending Mixed Messages?," *RC,* 11 December 2007; Obama on the *CBS Evening News,* 17 December 2007, on *The Early Show,* CBS, 18 December 2007, on the *CBS Evening News,* 19 December 2007, on *The Early Show,* CBS, 20 December 2007, on *Good Morning America,* ABC, 20 December 2007, and on *Today,* NBC, 21 December 2007; Shailagh Murray and Peter Slevin, "Message, Method Are Behind Obama's Climb," *WP,* 22 December 2007; Obama on *Face the Nation,* CBS, 23 December 2007; Scott Helman, "'Law Professor' Obama Embraces Nuance on Trail," *BG,* 25 December 2007; Bob Dean, "Candidates Brake for Christmas," *AJC,* 25 December 2007; Maya Soetoro-Ng on *Tell Me More,* NPR, 27 December 2007; Obama on *Meet the Press,* NBC, 30 December 2007; Dan Boylan, "'08: Year of Obama," *Midweek,* 2 January 2008; Maureen Dowd, "Deign or Reign?," *NYT,* 2 January 2008; "Barack Obama Interview," 2 January 2008, ChicagoTribune.com; John McCormick and Christi Parsons, "Obama Feeling Like a Winner," *CT,* 3 January 2008; Obama on *Good Morning America,* ABC, *The Early Show,* CBS, and *Today,* NBC, 3 January 2008; "Obama's Caucus Speech," *NYT,* 3 January 2008; Dan Balz et al., "Obama Wins Iowa's Democratic Caucuses," *WP,* 4 January 2008; Yudhijit Bhattacharje, "Barack Obama," *Science,* 4 January 2008, pp. 28–29; Nick Shields, "Obama's Iowa Victory Speech Brings Double Takes," *CDH,* 5 January 2005; Deborah Solomon, "All in the Family," *NYT Magazine,* 20 January 2008; Amy Rice and Alicia Sams, *By the People,* Pivotal Pictures, 2009.

13. Richard Wolffe, "'Hungry for Change,'" *Newsweek,* 4 January 2008; Ben Smith, "Likable Enough?," *Politico,* 5 January 2008; Stanage, *Redemption Song,* p. 82; Obama on *Good Morning America,* ABC, 7 January 2008; Christopher Hitchens, "Identity Crisis," *Slate,* 7 January 2008; Obama on *Good Morning America,* ABC, *The Early Show,* CBS, *Today,* NBC, *Morning Edition,* NPR, and *Fox and Friends,* Fox News, 9 January 2008; William Jelani Cobb, "As He Rises, the Old Guard Scowls," and David Greenberg, "Why Obamamania?," *WP,* 13 January 2008; Wolffe, "Inside Obama's Dream Machine," *Newsweek,* 14 January 2008, pp. 30ff.; Obama on the *CBS Evening News,* 16 January 2008; David Ignatius, "The Obama of 'Dreams,'" *WP,* 17 January 2008; Dan Gilgoff, "Barack Obama," Beliefnet.com, 21 January 2008; Sarah Pulliam and Ted Olsen, "Q&A: Barack Obama," *Christianity Today,* 21 January 2008; Dan Morain and Tom Hamburger, "Obama Dogged by Ties to Donor," *LAT,* 23 January 2008; Bob Secter et al., "Obama, Rezko Ties Again at Issue," *CT,* 23 January 2008; Obama on *Good Morning America,* ABC, *The Early Show,* CBS, *Today,* NBC, and *Morning Edition,* NPR, 23 January 2008; Jodi Kantor, "Obama's Christian Campaign," *NYT,* 25 January 2008;

Jay Hancock, "Obama's Exelon Ties Merit Close Look," *Baltimore Sun,* 25 January 2008; Obama on *Today,* NBC, 26 January 2008; John Kass, "If You Look Closely, It's Plain: Rezko Is Obama's Problem," *CT,* 27 January 2008; "For the Democrats: Obama," *CT,* 27 January 2008; Obama on *This Week,* ABC, 27 January 2008; Bob Secter and David Jackson, "Funds Tough to Figure for Rezko Aid," and Eric Zorn, "Tony Rezko, the Early Warnings," *CT,* 31 January 2008; Liz Halloran, "Q&A: Michelle Obama," *U.S. News,* 1 February 2008; Obama on *World News Tonight,* ABC, 2 February 2008; Valerie Jarrett, "Barack Obama," *CT,* 4 February 2008; Obama on *The Early Show,* CBS, and *The Situation Room,* CNN, 4 February 2008, on *Good Morning America,* ABC, *The Early Show,* CBS, *Today,* NBC, and *Fox and Friends,* Fox News, 5 February 2008, and on *60 Minutes,* CBS, 10 February 2008; Paul Krugman, "Hate Springs Eternal," *NYT,* 11 February 2008; Margaret Ramirez, "Obama Pastor Gives Last Trinity Sermon," *CT,* 11 February 2008; David Mark, "Full Text: Obama Interview," *Politico,* 12 February 2008; Susan Saulny, "Michelle Obama Thrives in Campaign Trenches," *NYT,* 14 February 2008; Jack Shafer, "How Obama Does That Thing He Does," *Slate,* 14 February 2008; Ramirez, "Pastor Remixes Church Service," *CT,* 17 February 2008; Obama on *Today,* NBC, 19 February 2008; Robert Samuelson, "The Obama Delusion," *WP,* 20 February 2008; Robin Abcarian, "Michelle Obama in Spotlight's Glare," *LAT,* 21 February 2008; Christi Parsons and John McCormick, "Analyzing Obama's Stump Speeches," *CT,* 24 February 2008; Kenneth T. Walsh, "One-on-One with Barack Obama," *U.S. News,* 25 February 2008, p. 44; Sandra Sobieraj Westfall, "25 Questions for Barack Obama," *People,* 25 February 2008, p. 96; Lynn Sweet, "Before Jewish Group, Obama Distances Himself," *CST,* 25 February 2008; Claudia Feldman, "An Interview with Michelle Obama," *Houston Chronicle,* 28 February 2008; Obama, "'I Will Be a President Who Draws Upon the Energy and Expertise of the Indian-American Community,'" *India Abroad,* 29 February 2008, p. A10; David A. Hollinger, "Obama, Blackness, and Postethnic America," *CHE,* 29 February 2008; David Ignatius, "Obama: A Thin Record for a Bridge Builder," *WP,* 2 March 2008; Howard Kurtz, "'Soft' Press Sharpens Its Focus on Obama," *WP,* 3 March 2008; Obama on *Good Morning America,* ABC, 3 March 2008; Rick Pearson and McCormick, "Barack Obama Takes Heat Over NAFTA Memo, Rezko," *CT,* 4 March 2008; Lynn Sweet, "Obama Talks a Lot, But Answers Little," *CST,* 4 March 2008; Dana Milbank, "Ask Tough Questions? Yes, They Can!," *WP,* 4 March 2008; Lauren Collins, "The Other Obama," *TNY,* 10 March 2008, pp. 88ff.; Hollinger, "Obama, the Instability of Color Lines, and the Promise of a Postethnic Future," *Callaloo* 31 (Fall 2008): 1033–37; Plouffe, *The Audacity to Win,* pp. 197, 203.

14. Obama on *Good Morning America,* ABC, *The Early Show,* CBS, *Today,* NBC, and *Fox and Friends,* Fox News, 5 March 2008; Darryl Pinckney, "Dreams from Obama," *NYRB,* 6 March 2008; Lynn Sweet, "Obama's Inner Circle," *CST,* 9 March 2008; Christopher Wills, "An Obama-Rezko Primer," AP, 10 March 2008; Obama on *Hardball with Chris Matthews,* MSNBC, 11 March 2008; Obama on *Today,* NBC, *The Early Show,* CBS, and *Good Morning America,* ABC, 12 March 2008; Brian Ross on *Good Morning America,* ABC, 13 March 2008, 7:07 A.M.; Ross and Rehab El-Buri, "Obama's Pastor: God Damn America, U.S. to Blame for 9/11," ABCNews.com, 13 March 2008; Paul Beatty, "Cool We Can Believe In," *TNR,* 14 March 2008; Christopher Drew and Jo Becker, "Obama Lists His Earmarks, Asking Clinton for Hers," *NYT,* 14 March 2008; Cass R. Sunstein, "The Obama I Know," *CT,* 14 March 2008; Lynn Sweet, "Obama Lists Pet Projects, at Last," *CST,* 14 March 2008, p. 4; David Brown, "Obama Addresses PA's Importance, Pastor's Remarks," *PTR,* 14 March 2008; Jodi Kantor, "Pastor's Words Still Draw Fire," *NYT,* 14 March 2008; Obama, "Transcript of *Tribune* Interview," *CT,* 16 March 2008 [14 March]; "Complete Transcript of the *Sun-Times* Interview with Barack Obama," *CST,* 15 March 2008 [14 March]; Obama, "On My Faith and My Church," *Huffington Post,* 14 March 2008, 4:28 P.M.; Obama on *Countdown,* MSNBC, *Fox on the Record,* Fox News, and *Anderson Cooper 360,* CNN, 14 March 2008; David Jackson, "Obama: I Trusted Rezko," *CT,* 15 March 2008; Tim Novak et al., "More Rezko Dough Found," *CST,* 15 March 2008, p. 2; Jodi Kantor, "Obama Denounces Statements of His Pastor as 'Inflammatory,'" *NYT,* 15 March 2008; Peter Slevin, "Outspoken Minister Out of Obama Campaign," *WP,* 15 March 2008; "Obama's Rezko Narrative," and John Kass, "Obama Opens Up on Rezko," *CT,* 16 March 2008; Chris Fusco et al., "Obama Explains Rezko Relationship," and Mark Brown, "Obama Cleared the Air on Rezko," *CST,* 16

March 2008, p. 2; Obama on *PBS NewsHour*, 17 March 2008; Esther Kang, "The Rezko Refund," *CM*, April 2008, p. 20; Peter Nicholas, "Obama Relies on Close-Knit Inner Circle," *LAT*, 8 November 2008; Rice and Sams, *By the People*, Pivotal Pictures, 2009; Heilemann and Halperin, *Game Change*, p. 235; Plouffe, *The Audacity to Win*, pp. 41, 206–12; Alby Gallun, "Rezko Creditors Come Up Short in Bankruptcy Case," *CCB*, 9 June 2015.

15. Kantor and Jeff Zeleny, "On Defensive, Obama Plans Talk on Race," *NYT*, 18 March 2008; Eli Saslow, "Congregation Defends Obama's Ex-Pastor," and "The Wright Question," *WP*, 18 March 2008; Ron Fournier, "The Trouble with Obama's Arrogance," AP, 18 March 2008; Obama, "A More Perfect Union," Philadelphia, 18 March 2008; Michael Crowley, "Obama's Speech Was Brilliant, But—," *TNR*, 18 March 2008; Zeleny, "Obama Urges U.S. to Grapple with Race Issue," and "Mr. Obama's Profile in Courage," *NYT*, 19 March 2008; Shaleigh Murray and Dan Balz, "Obama Urges U.S.: 'Move Beyond Our Old Racial Wounds,'" and "Moment of Truth," *WP*, 19 March 2008; Ben Calhoun, "Chicagoans: Reports Misrepresent Obama's Church," NPR, 19 March 2008; Obama on *Anderson Cooper 360*, CNN, 19 March 2008; Arnold Jacob Wolf, "My Neighbor Barack," Jews4Obama 2008, 20 March 2008; Obama on *Larry King Live*, CNN, 20 March 2008; Peter Wallstein, "Obama Defined by Contrasts," *LAT*, 24 March 2008; Patricia Novick, "Prophets Don't Speak the Language of Politics: A Voice from the Pews," New America Media, 25 March 2008; William A. Von Hoene Jr., "Reverend Wright in a Different Light," *CT*, 26 March 2008; Dan Nakaso, "Obama's Tutu a Hawaii Banking Female Pioneer," *HA*, 30 March 2008; Kelefa Sanneh, "Project Trinity: The Perilous Mission of Obama's Church," *TNY*, 7 April 2008; Julianna Goldman, "Obama Confidante Jarrett Wields Clout of Campaign Inner Circle," Bloomberg, 8 April 2008; Adam Clark, "Wright's Teachings Are Part of African-American Opposition to Empire," *Cincinnati Enquirer*, 8 April 2008; Obama on *Today*, NBC, 8 April 2008; Dayo Olopade, "Far Wright," *TNR*, 9 April 2008; Freddoso, *The Case Against Barack Obama*, p. 166; Rolf Nordahl, "Gramp's Dream," in Jacobs, *ObamaLand*, p. 130; David A. Frank, "The Prophetic Voice and the Face of the Other in Barack Obama's 'A More Perfect Union' Address, March 18, 2008," *Rhetoric & Public Affairs* 12 (Summer 2009): 167–94; Robert E. Terrill, "Unity and Duality in Barack Obama's 'A More Perfect Union,'" *Quarterly Journal of Speech* 95 (November 2009): 363–86; G. Reginald Daniel, "Race, Multiraciality, and Barack Obama: Toward a More Perfect Union?," *The Black Scholar* 39 (Fall–Winter 2009): 51–59; T. Denean Sharpley-Whiting, ed., *The Speech: Race and Barack Obama's 'A More Perfect Union'* (Bloomsbury, 2009); Todd, *The Stranger*, p. 113; DJG interviews with Alec Williamson and Rolf Nordahl.

16. Obama interview with ABC News' Terry Moran, 19 March 2008; Dan Gross, "Obama on WIP," Philly.com, 21 March 2008; Pamela Reid Bussard, "Barack Obama Back on St. Thomas for Golf, Relaxation," *St. John Source*, 24 March 2008; Leon Wieseltier, "Oybama," *TNR*, 26 March 2008; Kenneth P. Vogel, "Taxes: Obama Prospered as He Soared," *Politico*, 27 March 2008; Scott Helman, "Holding Down the Obama Family Fort," *BG*, 30 March 2008; Spencer Ackerman, "The Obama Doctrine," *American Prospect*, April 2008, pp. 12ff.; Obama on *Today*, NBC, 1 April 2008, *The Early Show*, CBS, and *Hardball with Chris Matthews*, MSNBC, 2 April 2008; Kerry Eleveld, "Obama Talks All Things LGBT," *The Advocate*, 10 April 2008; Mayhill Fowler, "Obama: No Surprise That Hard-Pressed Pennsylvanians Turn Bitter," *Huffington Post*, 11 April 2008; Obama, remarks to the Alliance for American Manufacturing, Pittsburgh, 14 April 2008; Curtis Lawrence, "Clinic That Filed Bankruptcy Getting Officials' Support," *CST*, 1 April 2002, p. 4; Helena Safron and Kelly Forman, "Field Guide for Altgeld Gardens in Chicago IL," unpublished paper, Geology Department, University of Wisconsin, June 2003; Clemolyn 'Pennie' Brinson, "Altgeld Gardens Lawsuit Settlement," *Residents' Journal*, January–February 2004; Susan Dosemagen, "Are There Enough Students to Fill New Altgeld Gardens School?," *Medill Reports*, 25 November 2008; Tom Baldwin, "Anger and Despair Rule the Mean Streets That Set Obama on Road to Power," *Times* (London), 16 January 2009; Layton Ehmke et al., "In Altgeld Gardens, Problems Run Deeper Than Fenger Violence," *Medill Reports*, 27 October 2009; Linda Paul, "Altgeld Gardens: The Wall of Death," WBEZ Chicago, 5 November 2009; Christopher Drew, "At Developer's Trial, Witness Recalls Seeing Obamas at 2004 Party for Investor," *NYT*, 15 April 2008; Bob Secter, "The Rezko Trial—Levine: Obama a Guest at '04 Party for Tycoon," *CT*, 15 April 2008; Barack H. and Michelle

L. Obama, 2007 Form 1040, 13 April 2008; Obama, Senate Financial Disclosure Report, 15 May 2008; "Transcript of Democratic Debate in Philadelphia," *NYT,* 16 April 2008; Obama on *All Things Considered,* NPR, and *The Daily Show,* Comedy Central, 21 April 2008, on The *Early Show,* CBS, *Today,* NBC, and *Good Morning America,* ABC, 22 April 2008; Heilemann and Halperin, *Game Change,* p. 245; Remnick, *The Bridge,* pp. 528–29; DJG interview with Jeremiah Wright; Wright's November 2011 interview with Ed Klein; Klein, *The Amateur,* pp. 50–51.

17. Wright on *Bill Moyers Journal,* PBS, 25 April 2008; Wright, address to the Detroit branch of the NAACP, 27 April 2008; Obama on *Fox News Sunday,* 27 April 2008; Abdon M. Pallasch, "Obama's Subprime Pal Penny Pritzker," *CST,* 28 April 2008; Wright, remarks at the National Press Club, 28 April 2008; Nicoli Nattrass, *The AIDS Conspiracy: Science Fights Back* (Columbia University Press, 2013), p. 25; Alessandra Stanley, "Not Speaking for Obama, Pastor Speaks for Himself, at Length," *NYT,* 29 April 2008; Obama, press conference on Jeremiah Wright, Winston-Salem, NC, 29 April 2008; Jeff Zeleny and Adam Nagourney, "An Angry Obama Renounces Ties to His Ex-Pastor," *NYT,* 30 April 2008; Tom Coburn on *The Sean Hannity Show,* Fox News, 28 May 2008; Clarence E. Walker and Gregory D. Smithers, *The Preacher and the Politician* (University of Virginia Press, 2009), pp. 46–50; Wolffe, *Renegade,* pp. 171, 174, 181–84; Heilemann and Halperin, *Game Change,* pp. 245–49; Remnick, *The Bridge,* pp. 529–31.

18. "Mr. Obama and Rev. Wright," *NYT,* 30 April 2008; "Parting with the Pastor," *WP,* 30 April 2008; Adolph Reed Jr., "Obama No," *Progressive,* May 2008; Michael Powell and Jodi Kantor, "A Strained Wright-Obama Bond Finally Snaps," *NYT,* 1 May 2008; Michelle Obama on *Today,* NBC, 1 May 2008; Obama on *Meet the Press,* NBC, 4 May 2008; Obama on *Today,* NBC, *American Morning,* CNN, *The Early Show,* CBS, *Morning Joe,* MSNBC, and *Fox and Friends,* Fox News, 5 May 2008; Fred Siegel, "The Obama Way," *NR,* 5 May 2008; Obama on the *NBC Nightly News,* 8 May 2008; Jeremiah A. Wright to Timuel Black, 13 May 2008; Ta-Nehisi Coates, "A Deeper Black," *The Nation,* 19 May 2008; Obama and Michelle Obama on *Good Morning America,* ABC, 19 May 2008; Stanley Kurtz, "Wright's Trumpet," *Weekly Standard,* 19 May 2008; Kurtz, "'Context,' You Say?," and "Left in Church: Deep Inside the Wright Trumpet," *NR,* 19 and 20 May 2008; Cinque Henderson, "Maybe We Can't," and John B. Judis, "The Big Race: Obama and the Psychology of the Color Barrier," *TNR,* 28 May 2008, pp. 16–18 and 21–24; Perry Bacon Jr., "On Policy, Obama Breaks Little New Ground," *WP,* 29 May 2008; Jeff Zeleny, "A Pulpit Matter Again," *NYT,* 30 May 2008; Susan Saulny, "Mocking of Clinton at Obama's Church Reverberates," *NYT,* 31 May 2008; Obama, remarks in Aberdeen, SD, 31 May 2008; Ben Smith and Mike Allen, "Obama Quits Trinity UCC," *Politico,* 31 May 2008; Charles P. Pierce, "The Cynic and Senator," *Esquire,* June 2008; Laura S. Washington, "Obama Not Feelin' the Love from Smiley," *ITT,* June 2008, p. 17; Michael Powell, "Following Months of Criticism, Obama Quits His Church," *NYT,* 1 June 2008; Derek Kravitz and Keith Richburg, "Obama Quits Longtime Church Over Inflammatory Comments," *WP,* 1 June 2008; Jon Meacham, "What Barack Obama Learned from His Father," *Newsweek,* 22 August 2008; Daren Briscoe, "An Interview with Barack Obama," *Newsweek,* 8 January 2009 [18–19 May 2008]; Dan Balz and Haynes Johnson, "A Political Odyssey," *WP,* 2 August 2009; Balz and Johnson, *The Battle for America 2008,* p. 211–14; Heilemann and Halperin, *Game Change,* pp. 248–49; Wolffe, *Renegade,* p. 184; DJG interviews with Timuel Black and Hermene Hartman. Two years later, Wright would write that "no one in the Obama administration will respond to me, listen to me, talk to me or read anything that I write to them. I am 'toxic'" and "'radioactive.' . . . When Obama threw me under the bus, he threw me under the bus literally!" Wright to Joseph Prischak, 18 February 2010.

19. Obama, remarks in St. Paul, MN, 3 June 2008; Obama's interview with ABC News' Charles Gibson, 4 June 2008; John McCormick, "A Wizard of Odds Steers Obama Run," *CT,* 9 June 2008; Julie Bosman, "Obama Calls for More Responsibility from Black Fathers," *NYT,* 16 June 2008; Obama on *World News Tonight,* ABC, 16 June 2008; Michael Powell and Jodi Kantor, "After Attacks, Michelle Obama Looks for a New Introduction," *NYT,* 18 June 2008; Michael Eric Dyson, "Obama's Rebuke of Absentee Black Fathers," *Time,* 19 June 2008; Michael Luo and Jeff Zeleny, "Obama, in Shift, Says He'll Reject Public Financing," and David Brooks, "The Two Obamas," *NYT,* 20 June 2008; Diane Salvatore, "Barack and Michelle

Obama: The Full Interview," *Ladies' Home Journal,* August 2008; Rice and Sams, *By the People,* Pivotal Pictures, 2009; Jonathan V. Last, "American Narcissus," *Weekly Standard,* 22 November 2010.

20. Howard Kurtz, "Applying a Personal Touch to the Campaign," *WP,* 23 June 2008; Cameron Strang, "Q&A with Barack Obama," *Relevant Magazine,* 1 July 2008; Frank James, "Obama Backs Late, Mental-Health Abortion," *CT,* 5 July 2008; Laura Miller, "Barack by the Books," *Salon,* 7 July 2008; Andrew Delbanco, "Deconstructing Barry," *TNR,* 9 July 2008, pp. 20ff.; Dwight D. Murphey, *DFMF* review, *Journal of Social, Political and Economic Studies* 33 (Summer 2008): 271–77; David S. Broder, "Obama's Enigma," *WP,* 13 July 2008; John Kass, "Obama Backers on the Left Are Doing the Wincing Now," *CT,* 13 July 2008; Obama, "My Plan for Iraq," *NYT,* 14 July 2008; Adolph Reed Jr., "Where Obamaism Seems to Be Going," *Black Agenda Report,* 15 July 2008; Obama on *Larry King Live,* CNN, 15 July 2008; Darryl Pinckney, "Obama & the Black Church," *NYRB,* 17 July 2008, pp. 18–21; Harry C. Alford, "Beyond the Rhetoric," *Pittsburgh Courier,* 17 July 2008; Paul Street, "Statehouse Days," Znet, 20 July 2008; Tyrone Simpson, "Barack Obama and the Abuse of Black Fathers," *Black Agenda Report,* 22 July 2008; Sandra Sobieraj Westfall, "Barack Obama Gives Daughter $1 Allowance a Week," *People,* 23 July 2008; Gabriel Sherman, "Barack Obama and the Press Break Up," *TNR,* 24 July 2008; "Barack Obama Interview," *CT,* 26 July 2008; Obama Q&A, Unity Journalism Conference, Chicago, 27 July 2008; Westfall, "Michelle Obama on Date Nights with Her Husband," *People,* 30 July 2008; Joanna Coles and Lucy Kaylin, "Barack in the Saddle," *Marie Claire,* 15 August 2008; Damon Linker, "How Obama Found Himself in Trouble This Time," *TNR,* 21 August 2008; Ben Smith and Jeffrey Ressner, "Barack Obama's Lost Law Review Article," *Politico,* 22 August 2008.

21. David Brooks, "Where's the Landslide?," *NYT,* 5 August 2008; Chris Bailey, "It's Still Obama-rama in Hawaii," *HM,* 12 August 2008; Michelle Cottle, "Be Not Cool," *TNR,* 15 August 2008; Obama at the Saddleback Civil Forum on the Presidency, 16 August 2008; Vanessa Grigoriadis, "Black & Blacker," *NYM,* 18 August 2008; Karen Tumulty and David Von Drehle, "Obama on His Veep Thinking," *Time,* 20 August 2008; John Lippert, "Penny Pritzker Shows Why She Convinced Buffett to Support Obama," Bloomberg, 20 August 2008; "Barack Obama Interview," *CT,* 21 August 2008; Mike Pflanz, "Barack Obama Is My Inspiration, Says Lost Brother," *Telegraph,* 21 August 2008; Von Drehle, "The Five Faces of Barack Obama," *Time,* 21 August 2008; John McCormick, "1-On-1 with Obama," *CT,* 23 August 2008; Jodi Kantor, "For a New Political Age, a Self-Made Man," *NYT,* 28 August 2008; Carrie Budoff Brown, "Obama: The Journey of a Confident Man," *Politico,* 28 August 2008; David Moberg, "Loving Obama Left," *ITT,* September 2008; John B. Judis, "Creation Myth," and Cass Sunstein, "The Empiricist Strikes Back," *TNR,* 10 September 2008; David Samuels, "Invisible Man—How Ralph Ellison Explains Barack Obama," *TNR,* 22 October 2008. On *Dreams,* see also especially Kelly Bulkeley, "Dreams Shed Light on Obama's Values," *SFC,* 17 August 2008, p. G9 (focusing on two dreams featuring loincloths, Barack's "feelings of hostility toward his father," and "personal qualities he deplores in his paternal ancestors yet fears may dwell in him too"), Barbara Foley, "Rhetoric and Silence in Barack Obama's *Dreams From My Father,*" *Cultural Logic* 2009, pp. 1–46, at 4 ("the negation of leftist and revolutionary ideas . . . is the intent guiding Obama's text"), James T. Kloppenberg, *Reading Obama* (Princeton University Press, 2010), esp. pp. 9, 14, 20, and 251 (calling *Dreams* "a fable or an allegory" and noting how "Obama's debts to Ellison run particularly deep. Many borrowed images from *Invisible Man* . . . pop up in passages in *Dreams*"), and Glenda R. Carpio, "Race & Inheritance in Barack Obama's *Dreams From My Father,*" *Daedalus,* Winter 2011, pp. 79–89.

22. Obama on *The O'Reilly Factor,* Fox News, 4, 8, 9, and 10 September 2008, and on *20/20,* ABC, 26 September 2008; Obama Interview Transcript, *DFP,* 3 October 2008; Michelle Obama on *20/20,* ABC, 3 October 2008; John Kass, "A Presidential Debate, the Chicago Way," *CT,* 5 October 2008; Deanna Bellandi, "Family, Friends Help Obamas Juggle Kids, Campaign," AP, 13 October 2008; Liza Mundy, "Michelle and Me," *Slate,* 14 October 2008; David Brooks, "Thinking About Obama," *NYT,* 17 October 2008; Jasmin Sandelson, "This Sterile Presidency," *Varsity,* 17 October 2008, p. 28; Richard Epstein, "The Obama I (Don't) Know," *Forbes,* 21 October 2008; Michael Powell, "After a Year on the Road, Obama Is Changing His Tempo," *NYT,* 22 October 2008; Michael Gerson, "The Irony of Obama," *WP,* 22 October 2008; Colm

Toibín, "James Baldwin & Barack Obama," *NYRB,* 23 October 2008; Robert Shikina, "Somber Obama Returns Home," *HSB,* 24 October 2008; Obama on *Good Morning America,* ABC, 24 October 2008; Jeff Zeleny, "On Perhaps a Final Visit to a Most Beloved Supporter," *NYT,* 25 October 2008; Robert Barnes, "Obama Visits Grandma Who Was His 'Rock,'" *WP,* 25 October 2008; Jodi Kantor, "Barack Obama, Forever Sizing Up," *NYT,* 26 October 2008; Keith Richburg, "America Is Showing Europe the Way Again," *Observer,* 26 October 2008; Jeff Zeleny, "Long by Obama's Side," *NYT,* 27 October 2008; Eric Bates, "Obama's Moment," *Rolling Stone* 1064, 30 October 2008, pp. 74–81; John Heilemann, "Obama Sheds Tears for Grandma in North Carolina," *NYM,* 3 November 2008; Dan Nakaso, "Obama's Sister 'Wept Tears of Joy' Over His Win," *HA,* 12 November 2008; Dalton Tanonaka, "We're All Coming to the Ball," *JP,* 30 November 2008; Christopher Hitchens, "Cool Cat," *Atlantic,* January–February 2009; Frank Rudy Cooper, "Our First Unisex President?: Black Masculinity and Obama's Feminine Side," *Denver University Law Review* 86 (2009): 633–61; Rice and Sams, *By the People,* Pivotal Pictures, 2009.

23. Adam Nagourney et al., "Near-Flawless Run Is Credited in Victory," *NYT,* 5 November 2008; Shelby Steele, "Obama's Post-Racial Promise," *LAT,* 5 November 2008; John W. Dean, "Predicting the Nature of Obama's Presidency," FindLaw, 14 November 2008; Robert Fitch, "The Change They Believe In," speech to Harlem Tenants' Association, 14 November 2008; Kevin Merida, "Letting the Big Win Sink In," *WP,* 15 November 2008; Obama on *60 Minutes,* CBS, 16 November 2008; Evan Thomas et al., "Barack Obama: How He Did It," *Newsweek,* 17 November 2008; Jodi Kantor, "An Old Hometown Mentor, and Still at Obama's Side," *NYT,* 24 November 2008; Marie Arana, "He's Not Black," *WP,* 30 November 2008; Kantor, "Obama's Friends Form Strategy to Stay Close," *NYT,* 13 December 2008; Richard Stengel et al., "The Interview: Person of the Year Barack Obama," *Time,* 17 December 2008; Chris Bailey, "Obama Bids Farewell to Grandmother on Oahu Coast," *HM,* 24 December 2008; Jeff Zeleny, "Obama's Zen State," *NYT,* 25 December 2008; Eli Saslow, "As Duties Weigh Obama Down, His Faith in Fitness Only Increases," *WP,* 25 December 2008; Obama's interview with John Harwood, CNBC, 7 January 2009; Richard Thompson Ford, "Barack Is the New Black," *DuBois Review* 6 (March 2009): 37–48, at 40; Michelle C. Bligh and Jeffrey C. Kohles, "The Enduring Allure of Charisma: How Barack Obama Won the Historic 2008 Presidential Election," *Leadership Quarterly* 20 (June 2009): 483–92; DJG interviews with Lou Celi, Emil Jones Jr., Peter Coffey, Dave Feller, and Vicky Rideout.

24. Hill Harper, "I Want to Be Like Him," *Essence,* January 2009, p. 110; Obama, "What I Want for You—and Every Child in America," *Parade Magazine,* 14 January 2009; Michael D. Shear, "Obama Pledges Entitlement Reform," *WP,* 16 January 2009; Peggy Noonan, "Look at the Time," *WSJ,* 30 January 2009; David Brooks, "A Moderate Manifesto," *NYT,* 3 March 2009; Carol E. Lee and Jonathan Martin, "Obama: 'I Am a New Democrat,'" *Politico,* 10 March 2009; Obama, remarks on earmark reform, 11 March 2009; Peter Baker and David M. Herszenhorn, "Obama Signs Spending Bill but Criticizes Earmarks," *NYT,* 12 March 2009; Paul Kane and Scott Wilson, "Obama Signs Spending Bill, Vowing to Battle Earmarks," *WP,* 12 March 2009; Jonathan Weisman and Greg Hitt, "Obama Outlines Plan to Curb Earmarks," *WSJ,* 12 March 2009; Noam Scheiber, "What's Eating David Axelrod?," *TNR,* 14 October 2010; Edward McClelland, "The Ones Obama Left Behind," NBCChicago.com, 23 March 2011; Kantor, *The Obamas,* p. 33; Suskind, *Confidence Men,* pp. 195–96, 239; Gates, *Duty,* p. 96; Todd, *The Stranger,* pp. 75–79; Axelrod, *Believer,* p. 346; Axelrod in Michael Grunwald, "The Selling of Obama," *Politico Magazine,* May–June 2016.

25. Obama on *60 Minutes,* CBS, 22 March 2009; Jim Merriner, "Barack Obama and the Parallel to Jimmy Carter," *CDO,* 24 March 2009; Oprah Winfrey, "Oprah Talks to Michelle Obama," *O Magazine,* April 2009; Dale Singer, "The Same Forces Shaped Obama and Blagojevich, Scott Simon Says," *St. Louis Beacon,* 6 April 2009; Angela Burt-Murray, "A Mother's Love," *Essence,* May 2009, pp. 104–13; Obama's interview with Steve Scully, C-SPAN, 22 May 2009; Jon Meacham, "What He's Like Now," *Newsweek,* 25 May 2009, pp. 36–42; Peter Baker and Jeff Zeleny, "Obama Chooses Hispanic Judge for Supreme Court Seat," *NYT,* 27 May 2009; Robert J. Samuelson, "The Obama Infatuation," *WP,* 1 June 2009; Ben Smith, "A Sheep in Wolffe's Clothing," *Politico,* 3 June 2009; Kevin Baker, "Barack Hoover Obama: The Best and Brightest Blow It Again," *Harp-*

er's, July 2009, pp. 29–37; Obama's 23 July 2009 interview with ABC's Terry Moran; Lexington, "The Obama Cult," *Economist*, 25 July 2009, p. 42; Glenn C. Loury, "Obama, Gates and the American Black Man," *NYT*, 26 July 2009; Neil King Jr. and Jonathan Weisman, "A President as Micromanager," *WSJ*, 12 August 2009; Sonia Sotomayor's interview with Susan Swain, C-SPAN, 16 September 2009; Edward Luce, "A Fearsome Foursome," *FT*, 3 February 2010; Suskind, *Confidence Men*, p. 323; Philip Galanes, "Barack Obama and Bryan Cranston on the Roles of a Lifetime," *NYT*, 6 May 2016.

26. Sheila Jager e-mails to DJG, 6 (2), 7 (3), 8 (2), and 9 (2) August 2009. Sheila was relieved to be contacted by someone she viewed as "a first-rate historian" rather than a journalist, and sharing her memories unburdened her. "It has been thrilling speaking to you about him."

27. Obama on *60 Minutes*, CBS, 13 September 2009; Obama on *Face the Nation*, CBS, 20 September 2009; Richard Cohen, "Obama's Identity Crisis," *WP*, 20 October 2009; Obama's 27 October 2009 interview with *U.S. News'* Kenneth T. Walsh; John Heilemann, "Obama Lost, Obama Found," *NYM*, 29 November 2009; Peter Spiegel et al., "Obama Bets Big on Troop Surge," *WSJ*, 2 December 2009; Obama on *60 Minutes*, CBS, 7 December 2009; Owen Fiss, "Obama's Betrayal," *Slate*, 4 December 2009; Obama's 11 December 2009 interview with Will Smith; Jake Tapper, "President Obama Grades Self," ABCNews.com, 14 December 2009; Christ Frates, "Payoffs for States Get Reid to 60," *Politico*, 19 December 2009; Josef Joffe, "Who Is This Guy?," *American Interest*, January 2010; Christi Parsons, "A Look at Obama's First Year in Presidency," *LAT*, 17 January 2010; Mark Knoller, "Obama's First Year: By the Numbers," CBSNews.com, 20 January 2010; Joe Klein, "Q&A: Obama on His First Year in Office," *Time*, 21 January 2010; John Heilemann, "Mr. Cool Gets Hot," *NYM*, 22 January 2010; Larry Hackett and Sandra Sobieraj Westfall, "Our First Year," *People*, 25 January 2010, pp. 60ff.; James W. Ceasar, "The Roots of Obama Worship," *Weekly Standard*, 25 January 2010; Obama's 25 January 2010 interview with ABC's Diane Sawyer; Byron York, "Has Obama Become Bored with Being President?," *Washington Examiner*, 29 January 2010; Edward Luce, "A Fearsome Foursome," *FT*, 3 February 2010; Lawrence Lessig, "How to Get Our Democracy Back," *The Nation*, 22 February 2010; Jackson Diehl, "Where Are Obama's Foreign Confidants?," *WP*, 8 March 2010; David Brooks, "Getting Obama Right," *NYT*, 12 March 2010.

28. Douglas Brinkley, "The Bridge," *LAT*, 28 March 2010; Frank Rich, "It's a Bird, It's a Plane, It's Obama," *NYT*, 4 April 2010; Ginger Adams Otis, "The Bridge," *NYP*, 4 April 2010; Mike Wilson, "Success Crafted by Timely Adaptation," *SPT*, 5 April 2010; Erik Spanberg, "The Bridge," *CSM*, 5 April 2010; Joan Walsh, "Barack Obama: The Opacity of Hope," *Salon*, 5 April 2010; Michiko Kakutani, "Seeking Identity, Shaping a Nation's," *NYT*, 6 April 2010; "His Big Moment Begins," *Economist*, 8 April 2010; Christopher Borelli, "New Yorker Editor David Remnick Pens New Obama Biography," *CT*, 8 April 2010; Amy Black, "Taking the Measure of Barack Obama," *Books & Culture*, 9 April 2010; David W. Blight, "The Bridge," *SFC*, 10 April 2010; Garry Wills, "Behind Obama's Cool," *NYTBR*, 11 April 2010; Jon Meacham, "The New Book on Barack Obama," *Newsweek*, 12 April 2010; Darryl L. Wellington, "Autobiographical Fire and Obama's Creation of Self," *The Common Review*, 14 April 2010; Howard W. French, "Dirt Off His Shoulders," *The National*, 16 April 2010; Joseph Lelyveld, "Who Is Barack Obama?," *NYRB*, 19 April 2010; John R. MacArthur, "Under False Colours," *Spectator*, 8 May 2010, p. 33; Mona Gable, "Chronicling President Barack Obama," *LAT*, 19 May 2010; Todd, *The Stranger*, pp. 86, 126; Sheila Jager e-mails to DJG, 8 (2), 9, 12, and 13 April 2010.

29. Richard Cohen, "President Obama's Enigmatic Intellectualism," *WP*, 22 June 2010; Kathleen Parker, "Obama: Our First Female President," *WP*, 30 June 2010; Cohen, "Who Is Barack Obama?," *WP*, 20 July 2010; David Ignatius, "A President Tripped Up by the Spontaneous," *WP*, 25 July 2010; Peter Nicholas and Janet Hook, "Obama the Velcro President," *LAT*, 30 July 2010; Mitch McConnell on *Fox on the Record*, Fox News, 4 August 2010; Obama's interview with Brian Lamb, C-SPAN, 12 August 2010; Frank Rich, "Why Has He Fallen Short?," *NYRB*, 19 August 2010; Josh Gerstein, "Should Obama Show His Faith?," *Politico*, 19 August 2010; James Hohmann and John F. Harris, "Dems Urge Obama to Take a Stand," *Politico*, 23 August 2010; Obama's interview with Peter Baker, 27 September 2010, *NYT*, 12 October 2010; Jann S. Wenner, "Obama in Command," *Rolling Stone*, 14 October 2010; Obama's interview with Bill Simmons, *GQ*, 17 November 2015.

30. Woodward, *Obama's Wars*, pp. 38–39, 207,

338, 344; Gates, *Duty,* pp. 298–99, 569, 585, 587; Peter Baker, "The Education of President Obama," *NYT Magazine,* 17 October 2010; Hodge, *The Mendacity of Hope* (HarperCollins, 2010), pp. 1, 20, 130; Johnny E. Williams, "'Change You Can Believe In': You Better Not Believe It," *Critical Sociology* 38 (September 2012): 747–68, at 761; Suskind, *Confidence Men,* pp. 451, 459; Sheryl Gay Stolberg, "Obama's Playbook After Nov. 2," *NYT,* 25 October 2010.

31. Joe Klein, "Where Obama Goes from Here," *Time,* 4 November 2010; John F. Harris and Glenn Thrush, "The Ego Factor: Can Obama Change?," *Politico,* 5 November 2010; Obama on *60 Minutes,* CBS, 7 November 2010; Eleanor Clift, "The Problem with the Cult of Obama," *Newsweek,* 21 November 2010; Jonathan V. Last, "American Narcissus," *Weekly Standard,* 22 November 2010; Alan Brinkley, "The Philosopher President," *Democracy Journal,* Winter 2011, pp. 80–85; Jim Sleeper, "An Unlikely Pragmatist," *Dissent,* Spring 2011, pp. 106–9; Wolffe, *The Message,* pp. 48–55, 64; Halperin and Heilemann, *Double Down,* pp. 15–16; Bill Glauber, "Chicago Neighborhood Watching President-Elect Closely," *MJS,* 12 January 2009; Carol Felsenthal, "N'Digo Publisher Hermene Hartman: Loyalty Not One of Obama's Qualities," *CM,* 21 March 2011; Edward McClelland, "The Ones Obama Left Behind," NBCChicago.com, 23 March 2011; Chris Hedges, "The Obama Deception: Why Cornel West Went Ballistic," TruthDig.com, 16 May 2011; Don Terry, "Fighting Crime and Pressuring the Mayor," *NYT,* 4 August 2011; Joel Hood, "CPS Fails to Close Performance Gap," *CT,* 14 November 2011; Michael Leahy, "Old Obama Acquaintance Voices South Side's Disillusionment with His Former Ally," *WP,* 17 August 2012; Paul Tough, "Obama vs. Poverty," *NYT Magazine,* 19 August 2012; Dawn Turner Trice, "Harold Washington Made Us Feel Proud," *CT,* 17 November 2012; Ben Austen, "The Death and Life of Chicago," *NYT Magazine,* 2 June 2013; Dahleen Glanton and Lolly Bowean, "Trouble in Pill Hill," *CT,* 24 November 2013; Glenn Thrush, "Obama's Obama," *Politico Magazine,* January–February 2016; DJG interviews with Hermene Hartman, Stephen Pugh, and John Webster.

32. Paul Krugman, "The President Is Missing," *NYT,* 11 April 2011; Perry Bacon Jr., "Obama's Budget Speech Carries Partisan Tone," *WP,* 13 April 2011; Lori Montgomery, "Obama Address Was Surprise Attack, GOP Lawmakers Say," *WP,* 15 April 2011; Ryan Lizza, "The Consequentialist," *TNY,* 2 May 2011; Obama's 4 May 2011 interview with CBS's Steve Kroft; Jacob Weisberg, "Obama Mama," *Slate,* 9 May 2011; David Greenberg, "Hope, Change, Nietzsche," *TNR,* 26 May 2011; Walter Russell Mead, "Can This Presidency Be Saved?," *American Interest,* 17 June 2011; Toby Harnden, "A Life of President Obama's Father," *Telegraph,* 5 August 2011; Drew Westen, "What Happened to Obama?," *NYT,* 7 August 2011; Scott Wilson, "Obama, the Loner President," *WP,* 7 October 2011; Kennedy, *The Persistence of the Color Line,* pp. 22–24, 273; Jackie Calmes, "Obama's Deficit Dilemma," *NYT,* 27 February 2012; Peter Wallsten et al., "Obama's Evolution," *WP,* 17 March 2012; Wolffe, *The Message,* p. 17; Halperin and Heilemann, *Double Down,* pp. 11–12, 18–20.

33. Obama, remarks on ending the war in Iraq, 21 October 2011 ("the tide of war is receding"); Jodi Kantor, "Leaving Obama's Shadow," *NYT,* 10 November 2011; Obama's 9 December interview with CBS' Steve Kroft; Obama on *20/20,* ABC, 23 December 2011; Jackson Lears, "A History of Disappointment," *London Review of Books,* 5 January 2012, pp. 10–13; Kantor, *The Obamas,* pp. 76, 178, 185, 192, 213, 251, 275, 329; Connie Schultz, "Partners in Love and the Presidency," *NYT,* 8 January 2012; Keach Hagey, "W.H. vs. Kantor: Overkill?," *Politico,* 13 January 2012; Renshon, *Barack Obama,* pp. 181, 187; James Fallows, "Obama, Explained," *Atlantic,* March 2012; Halperin and Heilemann, *Double Down,* pp. 54–58, 63–65; Todd, *The Stranger,* pp. 285–86.

34. Wolffe, *The Message,* p. 19; Jackie Calmes, "Obama's Deficit Dilemma," *NYT,* 27 February 2012; James Fallows, "Obama, Explained," *Atlantic,* March 2012, pp. 54–70; Obama's 1 March 2012 interview with Bill Simmons; Peter Wallsten et al., "Obama's Evolution," *WP,* 17 March 2012; David Maraniss, "Becoming Obama," VanityFair.com, 2 May 2012, 9:30 A.M.; James Fallows, "The Making of the President," *NYT,* 14 June 2012; Mike Godwin, "From the Choom Gang to the White House," *Reason,* 18 June 2012; Darryl Pinckney, "Young Barry Wins," *NYRB,* 16 August 2012; Nicholas Lemann, "The Cipher," *TNR,* 25 October 2012, pp. 28–31; DJG interviews with Alex McNear, Genevieve Cook, Sheila Jager, and Sheila Quinlan.

35. Obama's 9 May 2012 interview with ABC's Robin Roberts; Eleveld, *Don't Tell Me to Wait,* pp. 158, 254; Wear, *Reclaiming Hope,* pp. 142–56; Roberto Mangabeira Unger, "Beyond Obama," YouTube, 22 May 2012; Fredrick Harris, "Still Waiting for Our First Black President," *WP,* 3 June 2012; Harris, *The Price of the Ticket,* pp. 139, 154, 167; Jodi Kantor and Nicholas Confessore, "Leading Role in Obama '08, But Backstage in '12," *NYT,* 15 July 2012; Harris, "The Price of a Black President," *NYT,* 28 October 2012; "Oprah Talks to the Obamas," *O Magazine,* November 2012 (July 2012); Mark Obama Ndesandjo, *An Obama's Journey,* p. 299; DJG interview with Hermene Hartman.

36. Jane Mayer, "Schmooze or Lose," *TNY,* 27 August 2012; Jo Becker, "The Other Power in the West Wing," *NYT,* 2 September 2012; Dan Balz, "Obama Did Not Change Washington," *WP,* 2 September 2012; Ryan Grim and Sam Stein, "Barack Obama Promised a New Kind of Politics, but Played the Same Old Game," *Huffington Post,* 2 September 2012; John F. Harris and Jonathan Martin, "Barack Obama, the Conventional President," *Politico,* 3 September 2012; Peter Baker, "4 Years Later, Scarred But Still Confident," *NYT,* 6 September 2012; Harris and Alexander Burns, "Verdict: Obama Levels More Attacks," *Politico,* 6 September 2012; Woodward, *The Price of Politics,* p. 81; Simon, *P.S.*, p. 138; Ryan Lizza, "Let's Be Friends," *TNY,* 10 September 2012; Bob Herbert, "No More Excuses," *Huffington Post,* 4 October 2012; Obama's 12–13 September 2012 interview with CBS' Steve Kroft; DeWayne Wickham, "Netanyahu Hedges His Political Bets," *USAT,* 20 September 2012; Hunter Walker, "President Obama Parties with Jay Z and Beyoncé," *NYO,* 18 September 2012; Ann Gearhart and Jason Horowitz, "Behind the Big-Dollar Fundraisers," *WP,* 20 September 2012; Geoff Earle, "When O Met Z and Beyoncé Outdazzled Them Both," *NYP,* 20 September 2012; Arnie Graf, "Barack and Mitt as I Knew Them," *NYDN,* 28 September 2012; Jasmine Velasco, "Chicago's Susan Crown," Breitbart.com, 3 October 2012; David Maraniss, "Obama's Way with Words," *WP,* 14 October 2012; Alter, *The Center Holds,* p. 141,

37. Michael Lewis, "Obama's Way," *Vanity Fair,* October 2012; Glenn Thrush, "Barack Obama the Not-So-Happy Warrior," *Politico,* 5 November 2012; Richard Stengel et al., "Setting the Stage for a Second Term," *Time,* 19 December 2012; "Conspiracy Theories Prosper: 25% of Americans Are 'Truthers,'" Fairleigh Dickinson University Public Mind Poll, 17 January 2013; Sheila Jager e-mail to DJG, 7 November 2012.

38. Mark Knoller, "Obama's First Term: By the Numbers," CBSNews.com, 19 January 2013; "Oprah Talks to the Obamas," *O Magazine,* November 2012; Christopher H. Pyle, "Barack Obama and Civil Liberties," *Presidential Studies Quarterly* 42 (December 2012): 867–80; David K. Shipler, "Will Obama the Constitutional Lawyer Please Stand Up?," *Nation,* 23 January 2013; Franklin Foer and Chris Hughes, "Barack Obama Is Not Pleased," *TNR,* 27 January 2013; Obama on *60 Minutes,* CBS, 27 January 2013; Peter Baker, "Obama's Turn in Bush's Bind," *NYT,* 9 February 2013; Jim VandeHei and Mike Allen, "Obama, the Puppet Master," *Politico,* 18 February 2013; John Podesta, "Obama Should Lift Secrecy on Drones," *WP,* 13 March 2013 (Obama "is acting in opposition to the democratic principles we hold most important"); Obama, commencement speech at Morehouse College, 19 May 2013; Carol Felsenthal, "Obama's Buddy Eric Whitaker: Is He Clueless or What?," *CM,* 12 August 2013; Todd, *The Stranger,* pp. 383, 398, 403, 413; DJG interview with Carole Travis; Sheila Jager e-mail to DJG, 20 August 2013.

39. John F. Harris and Todd S. Purdum, "What's Wrong with President Obama?," *Politico,* 18 September 2013; Obama's 30 September 2013 interview with NPR's Steve Inskeep; Leonard Downie Jr., "In Obama's War on Leaks, Reporters Fight Back," *WP,* 6 October 2013; Wolffe, *The Message,* pp. 31, 206–7; Nicholas Confessore, "How He Won," *NYTBR,* 20 October 2013; Scott Wilson, "Controversies Show How Obama's Inattention to Detail May Hurt His Presidential Legacy," *WP,* 30 October 2013; Amy Chozick, "In Clinton Fund-Raising, Expect a Full Embrace," *NYT,* 22 October 2013; Halperin and Heilemann, *Double Down,* pp. 30, 47; Fouad Ajami, "When the Obama Magic Died," *WSJ,* 15 November 2013; Todd, *The Stranger,* 451–52, 466; Gideon Rose, "What Obama Gets Right," *Foreign Affairs,* September–October 2015, pp. 2–12; Jeffrey Goldberg, "The Obama Doctrine," *Atlantic,* April 2016, pp. 70–90; Grunwald, "The Selling of Obama," *Politico Magazine,* May–June 2016; William A. Galston, "Obama's Toothless Foreign Policy," *WSJ,* 7 September 2016.

40. George E. Vaillant, *Adaptation to Life* (Little, Brown, 1977), p. 29; Dan P. McAdams, "What Psychobiographers Might Learn from Personality Psychology," in William T. Schultz, ed., *Handbook of Psychobiography* (Oxford University Press, 2005), pp. 64–83; Steven Berglas, "Why Does Barack Keep Making Weird Jokes About Michelle?," *Politico,* 19 November 2013; the Obamas on *20/20,* ABC, 29 November 2013;

Todd S. Purdum, "The Lonely Guy," Vanity *Fair,* December 2013; Obama's 20 December 2013 interview with Steve Harvey; Jonathan Van Meter, "Michelle Obama: A Candid Conversation," *Vogue,* December 2016 ("living in the White House is isolating . . . you slowly start losing touch"); DJG interviews with Matt Piers, Katrina Emmons Mulligan, Eugene Morris, Jason Scott, Dan Shomon, Andrew Boron, and John Eason.

41. Todd, *The Stranger,* p. 466; Elizabeth Titus and John F. Harris, "Management Experts Knock Obama," *Politico,* 31 December 2013; Frank Bruni, "The Obama-Bush Nexus," *NYT,* 21 January 2014; Michael Gerson, "Our Complex President," *WP,* 24 January 2014; David Remnick, "Going the Distance," *TNY,* 27 January 2014; Michael Hirsh, "Kerry Unbound," and Charlie Cook, "Moving Right Along," *NJ,* 1 February 2014; Gary Younge, "What the Hell Is Barack Obama's Presidency For?," *Guardian,* 23 February 2014; Fred Hiatt, "Is There Change President Obama Can Believe In?," *WP,* 23 February 2014; Fred Kaplan, "The Realist," *Politico Magazine,* 27 February 2014; Hirsh and James Oliphant, "Obama Will Never End the War on Terror," *NJ,* 27 February 2014; Adolph Reed Jr., "Nothing Left," *Harper's,* March 2014; Jonathan Chait, "The Color of His Presidency," *NYM,* 6 April 2014; Peter Baker, "For Obama Presidency, Lyndon Johnson Looms Large," *NYT,* 9 April 2014; Dan Balz, "Obama and LBJ," *WP,* 13 April 2014; Jason Horowitz, "Obama Effect Inspiring Few to Seek Office," *NYT,* 14 April 2014; Leon Wieseltier, "The Inconvenience of History," *TNR,* 23 April 2014; Carrie Budoff Brown and Jennifer Epstein, "The Obama Paradox," *Politico,* 1 June 2014.

42. Zeituni Onyango e-mail to DJG, 6 December 2013; Katharine Q. Seelye, "Zeituni Onyango, Obama's Aunt from Kenya, Dies at 61," *NYT,* 8 April 2014; Jason Horowitz, "Amid Politics, Obama Drifted Away from Kin," *NYT,* 23 April 2014; Michael D. Shear, "The Rise of the Drone Master," *NYT,* 30 April 2014; Maureen Dowd, "Is Barry Whiffing?," *NYT,* 30 April 2014; Todd S. Purdum, "Barack Obama Laughs At but Not With," *Politico,* 2 May 2014; "President Obama and the World," *NYT,* 4 May 2014; Mike Dorning, "Obamas Diverge with Bidens on Mortgage Refinancing," Bloomberg, 15 May 2014; Kathleen Parker, "VA Scandal Shows Obama Is Out of the Loop—Again," *WP,* 24 May 2014; Obama's 28 May 2014 interview with NPR's Steve Inskeep; Mark Landler, "In Obama's Speeches, a Shifting Tone on Terror," *NYT,* 1 June 2014; Albert R. Hunt, "Obama's Destructive War on the Media," Bloomberg, 8 June 2014; Ron Fournier, "'I've Had Enough': When Democrats Quit on Obama," *NJ,* 9 June 2014; Rosa Brooks, "Sorry, Hillary: I Should Have Voted for You," *Foreign Policy,* 11 June 2014; Michael Grunwald, "The Selling of Obama," *Politico Magazine,* May–June 2016; Mark Landler, *Alter Egos* (Random House, 2016), pp. xii–xiv.

43. Charles Lipson, "Barack Obama's Presidency Is Spiraling Downward," *CT,* 17 June 2014; Ruth Marcus, "A Cautious Obama Misreads History," *WP,* 18 June 2014; Randall Kennedy, "Did Obama Fail Black America?," *Politico,* 27 June 2014; Steve Chapman "Obama: Wrong on Cellphone Searches and More," *CT,* 29 June 2014; Peggy Noonan, "The Daydream and the Nightmare," *WSJ,* 5 July 2014; Matt Lewis, "The Obama Years: The Trailers Were Great, the Movie Was Horrible," *Telegraph,* 5 July 2014; Michael Hirsh, "Our Lonely First Duffer," *Politico,* 7 August 2014; Maureen Dowd, "Where's the Justice at Justice?," *WP,* 17 August 2004; Michael Kazin, "Obama's Biggest Problem: Political ADD," *TNR,* 19 August 2014; Carl Hulse et al., "Obama Is Seen as Frustrating His Own Party," *NYT,* 19 August 2014; Thomas Frank, "Cornel West," *Salon,* 24 August 2014; Ruth Marcus, "A Threat Puts Obama's Legacy in Question," *WP,* 24 August 2014.

44. Sheila Jager e-mails to DJG, 1 September and 10 and 14 November, 2014; David A. Graham, "The Wrong Side of 'the Right Side of History,'" *Atlantic,* 21 December 2015.

45. Obama on *Meet the Press,* NBC, 7 September 2014; Obama on *60 Minutes,* CBS, 26 September 2014; Leon Panetta, *Worthy Fights* (Penguin, 2014); Susan Page, "Panetta: '30-Year War,'" *USAT,* 6 October 2014; Ron Fournier, "What a Real White House Shake-Up Looks Like," *NJ,* 23 October 2014; Paul Kane, "Midterm Disaster Rips Apart Awkward Ties Between Obama and Senate Democrats," *WP,* 6 November 2014; Carol Felsenthal, "Fire Valerie Jarrett," *Politico,* 7 November 2014; Todd, *The Stranger,* p. 273; Robert Draper, "The Enigma of Barack Obama," *WSJ,* 8 November 2014; Noam Scheiber, "The Obama Whisperer," *TNR,* 9 November 2014.

46. Obama on *Face the Nation,* CBS, 9 November 2014; David J. Rothkopf, *National Insecurity,* pp. 85, 204, 207, 318, 324, 358; Jeffrey Goldberg, "A Withering Critique of Obama's National Security Council," *Atlantic,* 12 November

2014; Dana Milbank, "Obama Is Turning into George W. Bush," *WP*, 24 November 2014.

47. Glenn Greenwald, "Talking to James Risen About *Pay Any Price*," *The Intercept*, 25 November 2014; Chris Ariens, "Ann Compton," Media Bistro.com, 8 December 2014; Josh Gerstein, "President Obama's Mixed Signals," *Politico*, 10 December 2014; Mark Udall, *CRec*, 10 December 2014, pp. S6474–78; Lindsay Moran, "The CIA Book I Wish I'd Written," *Politico*, 14 December 2014; Peter Baker and Mark Mazzetti, "Brennan Draws on Bond with Obama in Backing CIA," *NYT*, 15 December 2014; Obama's 16 December 2014 interview with NPR's Steve Inskeep; Mazzetti, "After Scrutiny, CIA Mandate Is Untouched," *NYT*, 27 December 2014; Hadas Gold, "Risen," *Politico*, 17 February 2015; James Risen, "If Donald Trump Targets Journalists, Thank Obama," *NYT*, 30 December 2016.

48. Elizabeth Drew, "The Unmaking of the President," ProjectSyndicate.org, 6 January 2015; Sheryl Gay Stolberg, "At 78, McCain Savors a Dream Job in the Senate," *NYT*, 14 January 2015; Kenneth Roth, "Obama & Counterterror: The Ignored Record," *NYRB*, 5 February 2015; Ben Smith, "Full Transcript of Buzzfeed News' Interview with President Barack Obama," Buzzfeed.com, 10 February 2015; "A True Believer Meets Reality," *Economist*, 14 February 2015; Amy Chozick, "When He Walks Out of That Building, I Don't Think He's Gonna Look Back," *NYT Magazine*, 15 February 2015; Dana Milbank, "Is This the End of the White House's Insufferable Insularity?," *WP*, 29 March 2015; Alan J. Kuperman, "Obama's Libya Debacle," *Foreign Affairs*, March–April 2015; Obama's 6 April 2015 interview with NPR's Steve Inskeep; "The Truth About Covering Obama," *Politico*, 22 April 2015; Milbank, "Obama's Trade Deal," *WP*, 11 May 2015; Jack Shafer, "The Rise and Fall of the Obama-Media Romance," and Nick Gass, "Schieffer on Obama," *Politico*, 1 June 2015.

49. Genevieve Cook e-mails to DJG, 28 September 2012 and 26 and 28 February 2015; Obama on *WTF with Marc Maron* podcast, 18 June 2015; Obama's 24 July 2015 interview with the BBC's Jon Sopel; Obama, remarks to the people of Africa, 28 July 2015; Obama's 6 August 2015 interview with NPR's Steve Inskeep; Gardiner Harris, "After 1600 Pennsylvania Avenue, Where To?," *NYT*, 12 October 2015; Obama's 17 November 2015 interview with GQ's Bill Simmons; Mark Landler, "Obama Revives the Republican Glory Years of Sunnylands Estate," *NYT*, 8 February 2016. Note must be taken of Mohamed Moustafa, "The Other Face of Obama!," YouTube, 7 May 2012, Moustafa, "Obama and His Ex-Girlfriend's Husband," YouTube, 10 May 2012, and two subsequent YouTube postings on 11 June and 28 July 2012, all offered up by Genevieve's once-abusive former spouse.

50. Jeff Greenfield, "Democratic Blues," *Politico*, 20 August 2015; Sheryl Gay Stolberg et al., "In Obama Era, GOP Bolsters Grip in the States," *NYT*, 13 November 2015; Gerald F. Seib, "The President Remains a Paradox," *WSJ*, 12 January 2016; Juan Williams, "The Carolina Pander for Black Votes," *WSJ*, 13–14 February 2016; Marilynne Robinson, "A Conversation in Iowa," *NYRB*, 5 and 19 November 2015 (14 September 2015); Obama's 6 October 2015 interview with CBS' Steve Kroft; Obama's 12 November 2015 interview with ABC's George Stephanopoulos; Obama's 17 November 2015 GQ Interview with Bill Simmons; Obama's 20 March 2015 interview with the *Huffington Post*'s Sam Stein; Robert Pear, "Head of Obama's Health Care Rollout to Lobby for Insurers," *NYT*, 16 July 2015; Lucia Graves, "Marilyn Tavenner," *NJ*, 24 October 2015, pp. 24–25; Pear, "Many Say High Deductibles Make Their Health Care Insurance All but Useless," *NYT*, 15 November 2015; Liz Hamel et al., "The Burden of Medical Debt," Kaiser Family Foundation, December 2015; Abby Goodnough, "Many See I.R.S. Fines as More Affordable Than Insurance," *NYT*, 3 January 2016; Margot Sanger-Katz, "Lost Jobs, Houses, Savings: Even Insured Often Face Crushing Medical Debt," *NYT*, 5 January 2016; Josh Gerstein, "How Obama Failed to Shut Washington's Revolving Door," and Michael Grunwald, "The Nation He Built," *Politico Magazine*, January–February 2016; Sarah Kliff, "Is Obamacare Failing?," *Vox*, 24 August 2016; Amy Goldstein, "In North Carolina, ACA Insurer Defections Leave Little Choice for Many Consumers," *WP*, 14 October 2016; Scott Rasmussen, "GOP Hangs On to Historic Gains in State Legislatures," Rasmussen Media Group, 8 December 2016; Dan Balz, "Democrats Search for a Path Back into Rural America's Good Graces," *WP*, 10 December 2016.

51. Karen DeYoung, "How the Obama White House Runs Foreign Policy," *WP*, 5 August 2015; David Ignatius, "Russia and 'The Facts on the Ground' in Syria," *WP*, 2 October 2015; Jackson Diehl, "Obama's Olive Branches are Lifelines for Authoritarian Regimes," *WP*, 9 November 2015; Richard Cohen, "The President Who

Lost His Voice," *WP,* 1 December 2015; Robert M. Gates, "The Kind of President We Need," *WP,* 4 December 2015; Obama's 21 December 2015 interview with NPR's Steve Inskeep; Michael Crowley, "'We Caved,'" *Politico Magazine,* January–February 2016; Obama's 22 January 2016 interview with *Politico*'s Glenn Thrush; Diehl, "Can Obama Let Go of His Wished-For Legacy?," *WP,* 8 February 2016; Joseph I. Lieberman, "The Absence of U.S. Leadership," *WP,* 25 February 2016; "Mr. Putin Continues to Call the Shots in Syria," *WP,* 26 February 2016; "Grozny Rules in Aleppo," *Economist,* 1 October 2016, pp. 12–13 ("the agony of Syria is the biggest moral stain on Barack Obama's presidency" and "his greatest geopolitical failure"); Greg Jaffe, "The Problem with Obama's Account of the Syrian Red-Line Incident," *WP,* 4 October 2016; Nicholas Kristof, "The Blot on Obama's Legacy," *NYT,* 6 October 2016 (Syria is "a huge blot on his legacy"); Jaffe, "Washington's Foreign Policy Elite Breaks with Obama over Syrian Bloodshed," *WP,* 20 October 2016.

52. Ray LaHood, *Seeking Bipartisanship* (Cambria Press, 2015), pp. 209, 215, 232; Peter Baker, "Promised Bipartisanship, Obama Adviser Found Disappointment," *NYT,* 12 November 2015; Darren Samuelson, "Obama's Vanishing Administration," *Politico,* 5 January 2016; Matt Latimer, "True-Blue Obama," *Politico,* 6 January 2016; Jeff Jacoby, "Obama Regrets Polarized Rancor. He Should," *BG,* 24 January 2016; David Brooks, "I Miss Barack Obama," *NYT,* 9 February 2016; Obama, address to the Illinois General Assembly, Springfield, 10 February 2016; Christi Parsons's interview with Obama, Kirk Dillard, Denny Jacobs, and Larry Walsh, Springfield, 10 February 2016; Mark Landler, "Obama Revisits Springfield, and His Vow to Bridge a Partisan Divide," *NYT,* 11 February 2016; Mike Riopell, "President Obama Greets Old Friends During Springfield Visit," *CDH,* 11 February 2016; Tamara Coffman Wittes, "The Slipperiest Slope of Them All," *Atlantic,* 12 March 2016; Jackson Diehl, "The Costs of Obama's Syrian Policy Are Apparent to Everyone but Him," *WP,* 21 March 2016; Fred Hiatt, "U.S. Leadership Matters," *WP,* 28 March 2016; Obama, remarks at the Toner Prize ceremony, Washington, DC, 28 March 2016; David Luban, "Has Obama Upheld the Law?," *NYRB,* 21 April 2016; William A. Galston, "How Obama's Economy Spawned Trump," *WSJ,* 4 May 2016; Dan Corey, "President Obama Sits Down with the *Daily Targum,*" *Daily Targum,* 12 May 2016; Margaret Sullivan, "Obama Promised Transparency," *WP,* 24 May 2016; Jeffrey Goldberg, "What Obama Actually Thinks About Radical Islam," *Atlantic,* 15 June 2016; "Obama Retreats from Putin in Syria—Again," *WP,* 3 July 2016; Michael D. Shear, "A Classified Matter at the White House," *NYT,* 8 August 2016; Nicholas Kristof, "Obama's Worst Mistake," *NYT,* 11 August 2016; David Cole, "The Drone Presidency," *NYRB,* 18 August 2016 (observing that "the only person to win the Nobel Peace Prize based on wishful thinking" then "pioneered a dramatically dangerous and ethically dubious form of warfare"); Charles Lane, "Obama Criticizes Past Presidents' Foreign Policies," *WP,* 15 September 2016; Jackson Diehl, "Putin's Lesson for Obama in Syria," *WP,* 19 September 2016; "Barack Obama and Doris Kearns Goodwin: The Ultimate Exit Interview," *Vanity Fair,* 21 September 2016; Jonathan Chait, "Five Days That Shaped a Presidency," *NYM,* 3 October 2016; Vanessa K. De Luca et al., "The Obama Legacy," *Essence,* October 2016, pp. 100–111; Obama, "The Way Ahead," *Economist,* 8 October 2016, pp. 22–24; Obama, "Now Is the Greatest Time to Be Alive," *Wired,* November 2016; Josh Kraushaar, "How Obama Inadvertently Set the Stage for Trump's Presidency," *National Journal,* 9 November 2016; Jann S. Wenner, "The Day After," *Rolling Stone,* 29 November 2016; Obama on *The Daily Show with Trevor Noah,* Comedy Central, 12 December 2016; Ta-Nehesi Coates, "My President Was Black," *The Atlantic,* January–February 2017; William A. Darity Jr., "How Barack Obama Failed Black Americans," *The Atlantic,* 22 December 2016; Obama on David Axelrod's *The Axe Files,* CNN.com, 26 December 2016; Obama, "The President's Role in Advancing Criminal Justice Reform," *HLR* 130 (January 2017): 811–66; Obama, "Repealing the ACA Without a Replacement: The Risks to American Healthcare," *NEJM,* 6 January 2017, pp. 1–3; Obama on ABC's *This Week,* 8 January 2017; Obama, "The Irreversible Momentum of Clean Energy," *Science,* 9 January 2017, pp. 1–4; Obama, "Farewell Address," Chicago, 10 January 2017; Obama on *Dateline NBC,* 13 January 2017; Obama on CBS's *60 Minutes,* 15 January 2017; Obama's 13 January 2017 interview with Michiko Kakutani, *NYT,* 16 January 2017; Norman Mailer, "Superman Comes to the Supermarket," *Esquire,* November 1960; Wolffe in Maureen Dowd, "The Ungrateful President," *NYT,* 7 August 2012; DJG interview with Cindi Canary; Obama, *DFMF,* pp. 79, 82.

BIBLIOGRAPHY

SELECTED BOOKS

Abramsky, Sasha. *Inside Obama's Brain*. Portfolio, 2010.

Almanac of Illinois Politics. Institute of Public Affairs, University of Illinois at Springfield, 1996, 1998, 2000, 2002, 2004.

Alter, Jonathan. *The Promise: President Obama, Year One*. Simon & Schuster, 2010.

Alter, Jonathan. *The Center Holds: Obama and His Enemies*. Simon & Schuster, 2013.

Altgeld-Carver Alumni Association. *History of Altgeld Gardens, 1944–1960*. Taylor Publishing, 1993.

Andersen, Christopher. *Barack and Michelle: Portrait of an American Marriage*. William Morrow, 2009.

Atlas, John. *Seeds of Change: The Story of ACORN, America's Most Controversial Antipoverty Community Organizing Group*. Vanderbilt University Press, 2010.

Axelrod, David. *Believer: My Forty Years in Politics*. Penguin Press, 2015.

Baim, Tracy, *Obama and the Gays*. Prairie Avenue Productions, 2010.

Balz, Dan, and Haynes Johnson. *The Battle for America 2008*. Viking, 2009.

Barkley, Charles. *Who's Afraid of a Large Black Man?* Penguin Press, 2005.

Bell, Derrick. *Confronting Authority: Reflections of an Ardent Protester*. Beacon Press, 1994.

Bensman, David, and Roberta Lynch. *Rusted Dreams: Hard Times in a Steel Community*. McGraw-Hill, 1987.

Berry, Mary Frances, & Josh Gottheimer, eds. *Power in Words: The Stories Behind Barack Obama's Speeches*. Beacon Press, 2010.

Black, Timuel D., Jr. *Bridges of Memory*. Northwestern University Press, 2003.

Burgess, Dave. *A Tale of Two Brothers: The Keith Kakugawa Story*. Lulu Press, 2009.

Chait, Jonathan. *Audacity*. HarperCollins, 2017.

Chambers, Edward T., with Michael A. Cowan. *Roots for Radicals: Organizing for Power, Action, and Justice*. Continuum, 2005.

Clavel, Pierre, and Wim Wiewel. *Harold Washington and the Neighborhoods: Progressive City Government in Chicago, 1983–1987*. Rutgers University Press, 1991.

Cobb, William Jelani. *The Substance of Hope: Barack Obama and the Paradox of Progress*. Walker & Co., 2010.

Coleman, Ray A. *The Obama Phenomenon*. Prioritybooks Publications, 2009.

Dahm, Charles W. *Power and Authority in the Catholic Church: Cardinal Cody in Chicago*. University of Notre Dame Press, 1981.

D'Antonio, Michael. *A Consequential President*. St. Martin's Press, 2017.

Darling, Diana. *Tandjung Sari: A Magical Door to Bali*. Editions Didier Millet, 2012.

Davis, Frank Marshall. *Livin' the Blues: Memoirs of a Black Journalist and Poet*, ed. John Edgar Tidwell. University of Wisconsin Press, 1992.

[Davis, Frank Marshall] "Bob Greene." *Sex Rebel: Black*. Greenleaf Classics, 1968.

Dorrien, Gary. *The Obama Question: A Progressive Perspective*. Rowman & Littlefield, 2012.

Dorson, Richard M. *Land of the Millrats*. Harvard University Press, 1981.

Dunham, S. Ann. *Surviving Against the Odds*. Duke University Press, 2009.

Eleveld, Kerry. *Don't Tell Me to Wait*. Basic Books, 2015.

Falk, Avner. *The Riddle of Barack Obama: A Psychobiography*. Praeger, 2010.

Feinberg, Gerald. *What Is the World Made Of? Atoms, Leptons, Quarks, and Other Tantalizing Particles*. Doubleday, 1977.

Firstbrook, Peter. *The Obamas: The Untold Story of an African Family*. Preface Publishing, 2010.

Fisher, Robert M. *The Logic of Economic Discovery: Neoclassical Economics and the Marginal Revolution*. New York University Press, 1986.

Fisher, Robert M., and Bella the Dog. *Field of Gourds: A Guide to Intellectual Rebellion*. CreateSpace, 2012.

Foster, Nelson, ed. *Punahou: The History and Promise of a School of the Islands*. Punahou School, 1991.

Frank, Justin A. *Obama on the Couch*. Free Press, 2011.

Freddoso, David. *The Case Against Barack Obama*. Regnery Publishing, 2008.

Fremon, David K. *Chicago Politics Ward by Ward*. Indiana University Press, 1988.

Fuechtmann, Thomas G. *Steeples and Stacks: Religion and the Steel Crisis in Youngstown*. Cambridge University Press, 1989.

Fuerst, J. S. *When Public Housing Was Paradise: Building a Community in Chicago*. Praeger, 2003.

Gates, Robert M. *Duty*. Knopf, 2014.

Gecan, Michael. *Going Public: An Organizer's Guide to Citizen Action*. Beacon Press, 2002.

Geoghegan, Thomas. *Which Side Are You On?* Farrar, Straus and Giroux, 1991.

Gertzel, Cherry. *The Politics of Independent Kenya 1963–8*. Northwestern University Press, 1970.

Glauberman, Stu, and Jerry Burris. *The Dream Begins: How Hawaii Shaped Barack Obama*. Watermark Publishing, 2008.

Go, Julian, ed. *Rethinking Obama*. Emerald Group, 2011.

Goldsworthy, David. *Tom Mboya: The Man Kenya Wanted to Forget*. Heineman, 1982.

Grant, Carl A., and Shelby J. Grant. *The Moment: Barack Obama, Jeremiah Wright, and the Firestorm at Trinity United Church of Christ*. Rowman & Littlefield, 2013.

Grimshaw, William J. *Bitter Fruit: Black Politics and the Chicago Machine, 1931–1991*. University of Chicago Press, 1992.

Haas, Michael, ed. *Barack Obama, The Aloha Zen President*. Praeger, 2011.

Halperin, Mark, and John Heilemann, *Double Down: Game Change 2012*. Penguin Press, 2013.

Halvorson, Debbie. *Playing Ball with the Big Boys*. Solutions Unlimited, 2012.

Harris, Fredrick C. *The Price of the Ticket: Barack Obama and the Rise and Decline of Black Politics*. Oxford University Press, 2012.

Haskins, Ron, and Greg Margolis. *Show Me the Evidence: Obama's Fight for Rigor and Results in Social Policy*. Brookings Institution Press, 2015.

Hart, Stephen. *Cultural Dilemmas of Progressive Politics: Styles of Engagement Among Grassroots Activists*. University of Chicago Press, 2001.

Hartley, Robert E. *Paul Powell of Illinois: A Lifelong Democrat*. Southern Illinois University Press, 1999.

Heilemann, John, and Mark Halperin, *Game Change*. HarperCollins 2010.

Helgeson, Jeffrey. *Crucibles of Black Empowerment: Chicago's Neighborhood Politics from the New Deal to Harold Washington*. University of Chicago Press, 2014.

Hendon, Rickey. *Black Enough/White Enough: The Obama Dilemma*. Third World Press, 2009.

Hess, G. Alfred, Jr. *School Restructuring, Chicago Style*. Corwin Press, 1991.

Hess, G. Alfred Jr., *Restructuring Urban Schools: A Chicago Perspective*. Teachers College Press, 1995.

Hirsch, Arnold R. *Making the Second Ghetto: Race and Housing in Chicago, 1940–1960*. University of Chicago Press, 1983.

Hoerr, John P. *And the Wolf Finally Came: The Decline of the American Steel Industry*. University of Pittsburgh Press, 1988.

Hogan, William T. *Steel in the United States: Restructuring to Compete*. D. C. Heath & Co., 1984.

Holder, R. Ward, and Peter B. Josephson. *The Irony of Barack Obama: Barack Obama, Reinhold Niebuhr and the Problem of Christian Statecraft*. Ashgate Publishing, 2012.

Holli, Melvin G., and Paul M. Green. *Bashing Chicago Traditions: Harold Washington's Last Campaign*. Wm. B. Eerdmans, 1989.

Horne, Gerald. *Fighting in Paradise: Labor Unions, Racism, and Communists in the Making of Modern Hawaii*. University of Hawaii Press, 2011.

Horwitt, Sanford D. *Let Them Call Me Rebel: Saul Alinsky, His Life and Legacy*. Alfred A. Knopf, 1989.

Hunt, D. Bradford. *Blueprint for Disaster: The Unraveling of Chicago Public Housing*. University of Chicago Press, 2009.

Jackson, Jesse L., Jr., with Frank E. Watkins. *A More Perfect Union: Advancing New American Rights*. Welcome Rain Publishers, 2001.

Jacobs, Ron. *Obamaland: Who Is Barack Obama?* Trade Publishing, 2009.

Jacobs, Sally H. *The Other Barack*. Public Affairs, 2011.

Janowitz, Rebecca. *Culture of Opportunity—Obama's Chicago: The People, Politics, and Ideas of Hyde Park*. Ivan R. Dee, 2010.

Joravsky, Ben, and Eduardo Camacho. *Race and Politics in Chicago*. Community Renewal Society, 1987.

Kahlenberg, Richard D. *Broken Contract: A Memoir of Harvard Law School*. Hill & Wang, 1992.

Kammen, Douglas, and Katharine McGregor, eds. *The Contours of Mass Violence in Indonesia, 1965–68*. University of Hawaii Press, 2012.

Kantor, Jodi. *The Obamas*. Little, Brown, 2012.
Kengor, Paul. *The Communist—Frank Marshall Davis*. Simon & Schuster, 2012.
Kennedy, Randall. *The Persistence of the Color Line: Racial Politics and the Obama Presidency*. Pantheon Books, 2011.
Kerlow, Eleanor. *Poisoned Ivy: How Egos, Ideology, and Power Politics Almost Ruined Harvard Law School*. St. Martin's Press, 1994.
Klein, Edward. *The Amateur: Barack Obama in the White House*. Regnery, 2012.
Kleppner, Paul. *Chicago Divided: The Making of a Black Mayor*. Northern Illinois University Press, 1985.
Kloppenberg, James T. *Reading Obama: Dreams, Hope, and the American Political Tradition*, 2nd ed. Princeton University Press, 2012.
Knoepfle, Peg, ed. *After Alinsky: Community Organizing in Illinois*. Illinois Issues, 1990.
Korecki, Natasha. *Only in Chicago*. Midway, 2013.
Kornblum, William. *Blue Collar Community*. University of Chicago Press, 1974.
Kurtz, Stanley. *Radical-in-Chief: Barack Obama and the Untold Story of American Socialism*. Simon & Schuster, 2010.
Kyle, Charles L., and Edward R. Kantowicz. *Kids First—Primero Los Ninos: Chicago School Reform in the 1980s*. Illinois Issues, 1992.
LaHood, Ray. *Seeking Bipartisanship*. Cambria Press, 2015.
Landler, Mark. *Alter Egos*. Random House, 2016.
Leahy, Michael Patrick. *What Does Barack Obama Believe?* Harpeth River Press, 2008.
Leeman, Richard W. *The Teleological Discourse of Barack Obama*. Lexington Books, 2012.
Loudon, Trevor. *Barack Obama and the Enemies Within*. Pacific Freedom Foundation, 2011.
Lumpkin, Beatrice. *"Always Bring A Crowd!" The Story of Frank Lumpkin, Steelworker*. International Publishers, 1999.
Lumpkin, Beatrice. *Joy in the Struggle: My Life and Love*. International Publishers, 2013.
Lynd, Staughton. *The Fight Against Shutdowns: Youngtown's Steel Mill Closings*. Singlejack Books, 1982.
MacDonogh, Stephen. *Pioneers: The Frontier Family of Barack Obama*. Brandon, 2010.
MacDougal, Gary. *Make a Difference: How One Man Helped Solve America's Poverty Problem*. St. Martin's Press, 2000.
Macneil, Ian R. *Contracts: Exchange Transactions and Relations*, 2nd ed. Foundation Press, 1978.
Mahon, Leo T., with Nancy Davis. *Fire Under My Feet: A Memoir of God's Power in Panama*. Orbis Books, 2007.
Mann, James. *The Obamians*. Viking, 2012.
Maraniss, David. *Barack Obama: The Story*. Simon & Schuster, 2012.
McClelland, Edward. *Young Mr. Obama: Chicago and the Making of a Black President*. Bloomsbury Press, 2010.
McKnight, John. *The Careless Society: Community and Its Counterfeits*. Basic Books, 1995.
Mendell, David. *Obama: From Promise to Power*. HarperCollins, 2007.
Merriner, James L. *The Man Who Emptied Death Row: Governor George Ryan and the Politics of Crime*. Southern Illinois University Press, 2008.
Miller, Alton. *Harold Washington: The Mayor, the Man*. Bonus Books, 1989.
Miller, Matthew. *The 2% Solution: Fixing America's Problems in Ways Liberals and Conservatives Can Love*. Public Affairs, 2003.
Mundy, Liza. *Michelle*. Simon & Schuster, 2008.
Ndesandjo, Mark Obama. *Nairobi to Shenzhen*. Aventine Press, 2009.
Ndesandjo, Mark Obama. *An Obama's Journey*. Lyons Press, 2014.
Obama, Abon'go Malik, and Frank Koyoo. *Barack Obama Sr.: The Rise and Life of a True African Scholar*. Xlibris, 2012.
Obama, Auma. *And Then Life Happens*. St. Martin's Press, 2012.
Obama, Barack. *Dreams From My Father*. Times Books, 1995.
Obama, Barack. *The Audacity of Hope*. Crown Books, 2007.
Obama, Barack. *Public Papers of the Presidents, 2009–*. USGPO, 2010–.
Obama, Barack H. *Otieno: The Wise Man. Book 2. Wise Ways of Farming*. East African Literature Bureau, 1959.
Obama, George, with Damien Lewis. *Homeland: An Extraordinary Story of Hope and Survival*. Simon & Schuster, 2010.
O'Connell, Mary. *School Reform Chicago Style: How Citizens Organized to Change Public Policy*. Center for Neighborhood Technology, 1991.
Onyango, Zeituni. *Tears of Abuse*. Afripress Publishing, 2012.
Pellow, David N. *Garbage Wars: The Struggle for Environmental Justice in Chicago*. MIT Press, 2002.
Pillemer, David B. *Momentous Events, Vivid Memories*. Harvard University Press, 1998.

Plouffe, David. *The Audacity to Win*. Viking, 2009.

Presta, John. *Mr. and Mrs. Grassroots: How Barack Obama, Two Bookstore Owners, and 300 Volunteers Did It*. Elevator Group, 2010.

Rahming, Melvin B., ed. *Critical Essays on Barack Obama*. Cambridge Scholars Publishing, 2012.

Raines, John C., et al., eds. *Community and Capital in Conflict: Plant Closings and Job Loss*. Temple University Press, 1982.

Ramos, Constance F., ed. *Our Friend Barry: Classmates' Recollections of Barack Obama and Punahou School*. Lulu Press, 2008.

Ranney, David. *Global Decisions, Local Collisions*. Temple University Press, 2003.

Redfield, Kent D. *Money Counts: How Dollars Dominate Illinois Politics and What We Can Do About It*. Institute for Public Affairs, University of Illinois at Springfield, 2001.

Reitzes, Donald C., and Dietrich S. Reitzes. *The Alinsky Legacy: Alive and Kicking*. JAI Press, 1987.

Remnick, David. *The Bridge: The Life and Rise of Barack Obama*. Alfred A. Knopf, 2010.

Renshon, Stanley A. *Barack Obama and the Politics of Redemption*. Routledge, 2012.

Rivlin, Gary. *Fire on the Prairie: Chicago's Harold Washington and the Politics of Race*. Henry Holt & Co., 1992.

Robinson, Craig, with Mim Eichler Rivas. *A Game of Character*. Gotham Books, 2010.

Rock, Philip J., with Ed Wojcicki. *Nobody Calls Just to Say Hello: Reflections on Twenty-Two Years in the Illinois Senate*. Southern Illinois University Press, 2012.

Roosa, John. *Pretext for Mass Murder: The September 30th Movement & Suharto's Coup d'État in Indonesia*. University of Wisconsin Press, 2006.

Rosen, Louis. *The South Side: The Racial Transformation of an American Neighborhood*. Ivan R. Dee, 1998.

Ross, Jini Kilgore, ed. *What Makes You So Strong? Sermons of Joy and Strength from Jeremiah A. Wright, Jr.* Judson Press, 1993.

Rothkopf, David J. *National Insecurity*. PublicAffairs, 2014.

Satter, Beryl. *Family Properties: Race, Real Estate, and the Exploitation of Black Urban America*. Metropolitan Books, 2009.

Savage, Charlie. *Power Wars*. Little, Brown, 2015.

Schmitz, Paul. *Everyone Leads*. Jossey-Bass, 2012.

Schutz, Aaron, and Mike Miller, eds. *People Power: The Community Organizing Tradition of Saul Alinsky*. Vanderbilt University Press, 2015.

Scott, Janny. *A Singular Woman: The Untold Story of Barack Obama's Mother*. Riverhead Books, 2011.

Sellers, Rod, and Dominic A. Pacyga. *Chicago's Southeast Side*. Arcadia Publishing, 1998.

Shane, Scott. *Objective Troy*. Tim Duggan Books, 2015.

Sharma, Dinesh. *Barack Obama in Hawaii and Indonesia: The Making of a Global President*. Praeger, 2011.

Shipps, Dorothy. *School Reform, Corporate Style: Chicago, 1880–2000*. University Press of Kansas, 2006.

Slevin, Peter. *Michelle Obama: A Life*. Alfred A. Knopf, 2015.

Smith, Susan Williams. *The Book of Jeremiah: The Life and Ministry of Jeremiah A. Wright Jr.* Pilgrim Press, 2013.

Soetendro, Haryo "Pongky," Sonny Trisulo, et al. *A Gift from Your Family: For Barack, Michelle, Malia and Sasha on the Occasion of Your Visit to Indonesia During November 2010*. Saraswati Papers, 2010.

Speller, Julia M. *Walkin' the Talk: Keepin' the Faith in Africentric Congregations*. Pilgrim Press, 2005.

Squires, Gregory D., ed. *From Redlining to Reinvestment: Community Responses to Urban Disinvestment*. Temple University Press, 1992.

Stanage, Niall. *Redemption Song—Barack Obama: From Hope to Reality*. Liberties Press, 2009.

Stephens, Robert F. *Kenyan Student Airlifts to America, 1959–1961: An Educational Odyssey*. Kenway Publications, 2013.

Suskind, Ron. *Confidence Men: Wall Street, Washington, and the Education of a President*. HarperCollins, 2011.

Swarns, Rachel. *American Tapestry*. Amistad, 2012.

Takara, Kathryn Waddell. *Frank Marshall Davis: The Fire and the Phoenix—A Critical Biography*. Pacific Raven Press, 2012.

Todd, Chuck. *The Stranger: Barack Obama in the White House*. Little, Brown, 2014.

Travis, Dempsey J. *An Autobiography of Black Politics*. Urban Research Press, 1987.

Travis, Dempsey J. *Harold: The People's Mayor*. Urban Research Press, 1989.

Turow, Scott. *Ultimate Punishment: A Lawyer's*

Reflections on Dealing with the Death Penalty. Farrar, Straus and Giroux, 2003.

United States Senate, Committee on Environment and Public Works, *Hearings*, 2005–2006. (USS, EPW)

United States Senate, Committee on Foreign Relations, *Hearings*, 2005–2007. (USS, CFR)

United States Senate, Committee on Health, Education, Labor and Pensions, *Hearings*, 2007. (USS, HEL)

United States Senate, Committee on Homeland Security and Governmental Affairs, *Hearings*, 2007. (USS, HSGA)

United States Senate, Committee on Veterans' Affairs, *Hearings*, 2005–2007. (USS, CVA)

Vaillant, George E. *Adaptation to Life*. Little, Brown, 1977.

Walley, Christine J. *Exit Zero: Family and Class in Postindustrial Chicago*. University of Chicago Press, 2013.

Wear, Michael. *Reclaiming Hope*. Thomas Nelson, 2017.

Whelan, Patrick. *The Catholic Case for Obama*. Catholic Democrats, 2008.

Wolffe, Richard. *Renegade: The Making of a President*. Crown Publishers, 2009.

Wolffe, Richard. *Revival: The Struggle for Survival Inside the Obama White House*. Crown, 2010.

Wolffe, Richard. *The Message: The Reselling of President Obama*. Twelve, 2013.

Woodward, Bob. *Obama's Wars*. Simon & Schuster, 2010.

Woodward, Bob. *The Price of Politics*. Simon & Schuster, 2012.

Wright, Jeremiah A. Jr. *A Sankofa Moment: The History of Trinity United Church of Christ*. Saint Paul Press, 2010.

Yanes, Nicholas A., and Derrais Carter, eds. *The Iconic Obama, 2007–2009*. McFarland & Co., 2012.

Zeigler, James. *Red Scare Racism and Cold War Black Radicalism*. University Press of Mississippi, 2015.

SELECTED ARTICLES AND BOOK CHAPTERS

Alter, Peter T. "Mexicans and Serbs in Southeast Chicago: Racial Group Formation During the Twentieth Century." *Journal of the Illinois State Historical Society* 94 (Winter 2001–02): 403–19.

Bradford, Calvin. "Financing Home Ownership: The Federal Role in Neighborhood Decline." *Urban Affairs Quarterly* 14 (March 1979): 313–35.

Bradford, Calvin. "Never Call Retreat: The Fight Against Lending Discrimination." In *Credit by Color: Mortgage Market Discrimination in Chicagoland*, 5–18. Chicago Fair Housing Alliance, January 1991.

Bradford, Calvin, and Leonard S. Rubinowitz. "The Urban-Suburban Investment-Disinvestment Process: Consequences for Older Neighborhoods." *The Annals* 422 (November 1975): 77–87.

Bradford, Calvin, and Anne B. Shlay. "Assuming a Can Opener: Economic Theory's Failure to Explain Discrimination in FHA Lending Markets," *Cityscape* 2 (February 1996): 77–87.

Carl, Jim. "Harold Washington and Chicago's Schools: Between Civil Rights and the Decline of the New Deal Consensus, 1955–1987." *History of Education Quarterly* 41 (Fall 2001): 311–343.

Carotenuto, Matthew, and Katherne Luongo. "Dala or Diaspora? Obama and the Luo Community of Kenya." *African Affairs* 108 (April 2009): 197–219.

Clark, Gordon L. "Corporate Restructuring in the Steel Industry: Adjustment Strategies and Local Labor Relations." In *America's New Market Geography*, edited by George Sternlieb and James W. Hughes, 179–214. Rutgers Center for Urban Policy Research, 1988.

Clark, Gordon L. "Piercing the Corporate Veil: The Closure of Wisconsin Steel in South Chicago." *Regional Studies* 24 (October 1990): 405–20.

Colman, Jeffrey D., and Michael T. Brody. "Ketchum v. Byrne: The Hard Lessons of Discriminatory Redistricting in Chicago." *Chicago-Kent Law Review* 64 (1988): 497–530.

Conroy, John. "Mill Town." *Chicago,* November 1976, 164–82; December 1976, 162–63, 210–19, 304; January 1977, 114–25, 189; February 1977, 106–15, 185; March 1977, 114–15, 132–46.

Conroy, John. "Cardinal Sins." *Chicago Reader*, 4 June 1987.

Cruz, Wilfredo. "UNO: Organizing at the Grass Roots." *Illinois Issues*, April 1988, 18–22.

DeZutter, Hank. "What Makes Obama Run?" *Chicago Reader*, 8 December 1995, 1, 12, 14, 16, 18, 20, 22, 24.

Draper, Robert. "Barack Obama's Work in Progress." *GQ*, November 2009, 154ff.

Drizin, Steven A., and Beth A. Colgan. "Let the Cameras Roll: Mandatory Videotaping of Interrogations Is the Solution to Illinois' Problem of False Confessions." *Loyola University Chicago Law Journal* 32 (Winter 2001): 337–424.

Drummond, Tammerlin. "Barack Obama's Law—Harvard Law Review's First Black President Plans a Life of Public Service." *Los Angeles Times*, 12 March 1990, E1–2.

Engel, Lawrence J. "The Influence of Saul Alinsky on the Campaign for Human Development." *Theological Studies* 59 (December 1998): 636–61.

Gahagan, Stacey Marlise, and Alfred L. Brophy. "Reading Professor Obama: Race and the American Constitutional Tradition." *University of Pittsburgh Law Review* 75 (2014): 495–581.

Giloth, Robert, and Kari Moe. "Jobs, Equity, and the Mayoral Administration of Harold Washington in Chicago (1983–87)." *Policy Studies Journal* 27 (February 1999): 129–46.

Goldberg, Jeffrey. "The Obama Doctrine." *The Atlantic,* April 2016, 70–90.

Granfield, Robert, and Thomas Koenig. "From Activism to Pro Bono: The Redirection of Working Class Altruism at Harvard Law School." *Critical Sociology* 17 (April 1990): 57–80.

Granfield, Robert, and Thomas Koenig. "Learning Collective Eminence: Harvard Law School and the Social Production of Elite Lawyers." *Sociological Quarterly* 33 (Winter 1992): 503–20.

Granfield, Robert, and Thomas Koenig. "The Fate of Elite Idealism: Accommodation and Ideological Work at Harvard Law School." *Social Problems* 39 (November 1992): 315–31.

Greeley, Andrew M. "The Fall of an Archdiocese." *Chicago,* September 1987, 128–31, 190–92.

Hammack, Phillip L. "The Political Psychology of Personal Narrative: The Case of Barack Obama." *Analyses of Social Issues and Public Policy* 10 (December 2010): 182–206.

Heathcott, Joseph. "Urban Activism in a Downsizing World: Neighborhood Organizing in Postindustrial Chicago." *City & Community* 4 (September 2005): 277–94.

Hirsch, Arnold R. "Massive Resistance in the Urban North: Trumbull Park, Chicago, 1953–1966." *Journal of American History* 82 (September 1995): 522–50.

Homel, Michael W. "The Lilydale School Campaign of 1936: Direct Action in the Verbal Protest Era." *Journal of Negro History* 59 (July 1974): 228–41.

Hopkins, Dwight N. "Race, Religion, and the Race for the White House." In *The Obama Phenomenon,* edited by Charles P. Henry et al., 181–99. University of Illinois Press, 2011.

Hunt, D. Bradford. "What Went Wrong with Public Housing in Chicago? A History of the Robert Taylor Homes." *Journal of the Illinois State Historical Society* 94 (Spring 2001): 96–123.

Jager, Bernd. "Memories and Myths of Evil: A Reflection on the Fall from Paradise." *Collection du Cirp* 1 (2007): 211–230.

Jager, David. "Conversing with My Father." *Collection du Cirp* 1 (2007): 12–16.

Joravsky, Ben. "Dumpers Swamp City's Southeast Side with Noxious, Toxic Waste." *Chicago Reporter,* August 1983, 1–5.

Kelleher, Keith. "ACORN Organizing & Chicago Homecare Workers." *Labor Research Review* 8 (April 1986): 33–45.

Kelleher, Keith. "Growth of a Modern Union Local: A People's History of SEIU Local 880." *Just Labour* 12 (Spring 2008): 1–15.

Kelly, James G. "Contexts and Community Leadership: Inquiry as an Ecological Expedition." *American Psychologist* 54 (November 1999): 953–61.

Kelly, James G., L. Sean Azelton, et al. "On Community Leadership: Stories About Collaboration in Action Research." *American Journal of Community Psychology* 33 (June 2004): 205–16.

King, Richard H. "Becoming Black, Becoming President." *Patterns of Prejudice* 45 (2011): 62–85.

Kloppenberg, James T. "Barack Obama and Progressive Democracy." In *Making the American Century,* edited by Bruce J. Schulman, 267–87. Oxford University Press, 2014.

Lawrence, Mike. "Mission Impossible." *Illinois Issues*, September 1998, 38–41.

Levinsohn, Florence Hamlish. "The Street Scrapper and the Rhodes Scholar." *Chicago Reader,* 8 March 1990.

Lewis, George. "Barack Hussein Obama: The Use of History in the Creation of an 'American' President." *Patterns of Prejudice* 45 (2011): 43–61.

Loizeau, Pierre-Marie. "Barack Obama's Autobiography: A Quest for Personal and Political Identity?" In *Autobiography as a Writing Strategy in Postcolonial Literature,* edited by Benaouda Lebdai, 53–65. Cambridge Scholars Publishing, 2015.

Lynd, Staughton. "The Genesis of the Idea of a Community Right to Industrial Property in Youngstown and Pittsburgh, 1977–1987." *Jour-*

nal of American History 74 (December 1987): 926–58.

McAdams, Dan P. "What Psychobiographers Might Learn from Personality Psychology." In *Handbook of Psychobiography,* edited by William T. Schultz, 64–83. Oxford University Press, 2005.

McKersie, William S. "Fostering Community Participation to Influence Public Policy: Lessons from the Woods Fund of Chicago, 1987–1993." *Nonprofit and Voluntary Sector Quarterly* 26 (March 1997): 11–26.

Mirel, Jeffrey. "School Reform, Chicago Style: Educational Innovation in a Changing Urban Context, 1976–1991." *Urban Education* 28 (July 1993): 116–49.

Moore, Donald R. "Voice and Choice in Chicago." In *Choice and Control in American Education, Volume* 2, edited by William H. Clune and John F. Witte, 153–98. Falmer Press, 1990.

Mulder, Mark T. "Mobility and the (In)Significance of Place in an Evangelical Church: A Case Study from the South Side of Chicago." *Geographies of Religions and Belief Systems* 3 (2009): 16–43.

Mulder, Mark T. "Evangelical Church Polity and the Nuances of White Flight: A Case Study from the Roseland and Englewood Neighborhoods in Chicago." *Journal of Urban History* 38 (January 2012): 16–38.

Murphy, John M. "Barack Obama, the Exodus Tradition, and the Joshua Generation." *Quarterly Journal of Speech* 97 (November 2011): 387–410.

Obama, Barry. "The Old Man." *Ka Punahou*, 15 December 1978, 5, and reprinted in *Ka Wai Ola*, May 1979, 65.

Obama, Barry. "Winner." *Oahuan* 1979, 104.

Obama, Barack. "Breaking the War Mentality." *Sundial: The Weekly Newsmagazine*, 10 March 1983, 2–3, 5.

Obama, Barack. "Why Organize? Problems and Promise in the Inner City." *Illinois Issues*, August–September 1988, 40–42.

[Obama, Barack]. "Tort Law—Prenatal Injuries—Supreme Court of Illinois Refuses to Recognize Cause of Action Brought by Fetus Against Its Mother for Unintentional Infliction of Prenatal Injuries." *Harvard Law Review* (January 1990): 823–28.

Obama, Barack. "Review President Explains Affirmative Action Policy." *Harvard Law Record*, 16 November 1990, 4, 7.

Obama, Barack. "Help Needed to Change Springfield." *Hyde Park Herald*, 19 February 1997, 4, 9.

Obama, Barack. "Rethinking Approach to Juvenile Crime." *Hyde Park Herald*, 16 April 1997, 4.

Obama, Barack. "Legislature Squandered Opportunity." *Hyde Park Herald*, 18 June 1997, 4, 15.

Obama, Barack. "Illinois Needs a Fairer Tax System." *Hyde Park Herald*, 25 February 1998, 4.

Obama, Barack. "Education Most Important Town Hall Issue." *Hyde Park Herald*, 29 April 1998, 4.

Obama, Barack. "Progress on Campaign Finance Reform." *Hyde Park Herald*, 26 August 1998, 4.

Obama, Barack. "Keep Church and State Separate." *Hyde Park Herald*, 7 April 1999, 4.

Obama, Barack. "A Life's Calling to Public Service." *Punahou Bulletin*, Fall 1999, 28–29.

Obama, Barack. "Why Dr. Martin Luther King, Jr. is Important to Me." *Chicago Defender*, 15 January 2000, 18.

Obama, Barack. "State-Wide Meetings Focus on Uninsured." *Hyde Park Herald*, 2 October 2002, 4.

Obama, Barack. "Weighing the Costs of Waging War in Iraq." *Hyde Park Herald*, 30 October 2002, 5, 6.

Obama, Barack H. "Terror in the Congo." *Honolulu Star-Bulletin*, 8 June 1960, 8.

Obama, Barak [sic] H. "Problems Facing Our Socialism." *East Africa Journal*, July 1965, 26–33.

Peterson, Gayle. "We Are the People: The Fight Against Toxic Waste." *Health & Medicine* 2 (Winter 1983–84): 13–22.

Pines, Sharon. "Public Housing Residents Fight for Their Health." *The Neighborhood Works*, March 1985, 14–17.

"Principales Publications du Professeur Bernd Jager." *Collection du Cirp* 1 (2007): 291–301.

Protess, David L. "The Vrdolyak Chronicles." *Chicago Lawyer*, June 1988, 1, 16–20.

Purdum, Todd S. "Raising Obama." *Vanity Fair*, March 2008, 314ff.

Ranney, David C. "The Closing of Wisconsin Steel." In *Grand Designs: The Impact of Corporate Strategies on Workers, Unions, and Communities*, edited by Charles Craypo and Bruce Nissen, 65–91. ILR Press, 1993.

Rawlings, Charles W. "Steel Shutdown in Youngstown: Ecumenical Response to the Opening Hand of Globalization." *Church & Society*, January–February 2003, 71–91.

Reiff, Janice L. "Rethinking Pullman: Urban Space and Working-Class Activism," *Social Science History* 24 (Spring 2000): 7–32.

Reitzes, Donald C., and Dietrich S. Reitzes. "Alinsky in the 1980s: Two Contemporary

Chicago Community Organizations." *Sociological Quarterly* 28 (1987): 265–83.

Robinson, Michelle. "Minority and Women Law Professors: A Comparison of Teaching Styles." *BLSA Memo Special Edition*, Spring–Fall 1988, 30–33.

Rolling, Ken. "Reflections on the Chicago Annenberg Challenge." In *School Reform in Chicago: Lessons on Policy and Practice,* edited by Alexander Russo, 23–28. Harvard Education Press, 2004.

Siker, Jeffrey S. "President Obama, the Bible, and Political Rhetoric." *Political Theology* 13 (September 2012): 586–609.

Sullivan, Jerry. "Paradise Doomed." *Chicago,* August 1985, 142–45, 160–61.

Sullivan, Jerry. "Of Dumps, Chicago Politics & Herons." *Audubon*, March 1987, pp. 122–26.

Tandon, S. Darius, L. Sean Azelton, James G. Kelly, and Debra A. Strickland. "Constructing a Tree for Community Leaders: Contexts and Processes in Collaborative Inquiry." *American Journal of Community Psychology* 26 (August 1998): 669–96.

Walley, Christine J. "Deindustrializing Chicago: A Daughter's Story." In *The Insecure American*, edited by Hugh Gusterson and Catherine Besteman, 113–39. University of California Press, 2010.

Wright, Jeremiah A., Jr. "The Significance of Harold Washington." In *The Black Church and the Harold Washington Story,* edited by Henry J. Young, 1–9. Wyndham Hall Press, 1988.

Wright, Jeremiah A., Jr. "A Black Congregation in a White Church: Trinity United Church of Christ, Chicago." In *Good News in Growing Churches*, edited by Robert L. Burt, 37–63. Pilgrim Press, 1990.

Wright, Jeremiah A., Jr. "An Underground Theology." In *Black Faith and Public Talk: Critical Essays on James H. Cone's Black Theology and Black Power,* edited by Dwight N. Hopkins, 96–102. Baylor University Press, 1991.

Wright, Jeremiah A., Jr. "Doing Black Theology in the Black Church." In *Living Stones in the Household of God,* edited by Linda E. Thomas, 13–23. Fortress Press, 2004.

SELECTED MOVIES, VIDEOS, AND TAPE RECORDINGS

An American Boyhood: Barack Obama in Hawaii. Gloria Borland and Kris Anderson, 2008.

Barack Obama, Oral Argument in *Baravati v. Josephthal, Lyon & Ross, Inc.* #93-3647 (7th Cir.), 7 April 1994. http://web.archive.org/web/20081009195106/. http://www.suntimes.com/images/cds/MP3/obama.mp3.

Barack Obama, "Community Revitalization." Nebraska Wesleyan University Forum, Lincoln, NE, 9 September 1994.

Barack Obama, remarks at the Cambridge (MA) Public Library. 20 September 1995, Cambridge Municipal Television.

Barack Obama, "Cities and Economic Revitalization." National Issues Forum, Brookings Institution, Washington, D.C., 8 June 1998.

Barack Obama, "Public Policy in the 21st Century." Loyola University Chicago, 19 October 1998.

Barack Obama, "Inside Illinois Government." University of Illinois at Springfield TV, 3 December 1998.

Barack Obama, "Politics, Race and the Common Good." Black History Month Convocation, Carleton College, MN, 5 February 1999.

Barack Obama, Youth Government Day Remarks. Simon Institute, Southern Illinois University, Carbondale, 21 October 2000.

Barack Obama, Illinois Center for the Book presentation. Illinois State Library, Springfield, 7 March 2001.

Barack Obama, *Check, Please!* WTTW, 14 August 2001.

Barack Obama, Commencement Address, North Kenwood/Oakland Charter School, 13 June 2002.

Barack Obama, Commencement Address, Chicago-Kent College of Law, 25 May 2003.

Barack Obama, Commencement Address, Pritzker School of Medicine, University of Chicago, 10 June 2005.

Barack Obama, Commencement Address, University of Massachusetts at Boston, 2 June 2006.

Barack Obama, Commencement Address, Northwestern University, Evanston, 16 June 2006.

Barack Obama, Commencement Address, Xavier University, New Orleans, 11 August 2006.

Barack Obama, Commencement Address, Wesleyan University, Middletown, 25 May 2008.

"Barack Obama: An Agenda for the 21st Century." Joe Green, YouTube, 20–21 October 1999.

Barack Obama: Hawaii Roots. Emme Tomimbang, 8 February 2008.

Barack Obama: Made in Hawaii. Demo reel. Gloria Borland, 2011.

"Barack Obama's Days at Occidental." Huell Howser, KCET, 25 November 2008.

"Barack Obama's Occidental College Days." Huell Howser, *California Gold,* 13 November 2008.

Becoming Barack: Evolution of a Leader. [Zeke Gonzalez, originally recorded 27 October 1993], 2009.

"Buzz and Alice Palmer," Chuck O'Bannon and David Robinson, YouTube, 2012.

By the People: The Election of Barack Obama. Amy Rice and Alicia Sams, Pivotal Pictures, 2009.

Chicago Tonight Archive. WTTW, Chicago.

Eight Forty-Eight Archive, WBEZ Radio, Chicago.

Have You Heard of the City? A Roseland Christian Ministries Story. 2005.

Illinois Lawmakers Archive. Museum of Broadcast Communications, Chicago.

Introducing Barack Obama. Chicago Video Project, 2003.

"Keeping the Faith" [Trinity United Church of Christ], Roger Wilkins, *Frontline,* WGBH, 16 June 1987.

Museum of Broadcast Communications Archives (MBC), Chicago.

"The Novel and Ideology." Class recording, Columbia University. Lennard J. Davis, [June 1983].

Odyssey Archive, WBEZ Radio, Chicago.

Punahou School Commencement, 2 June 1979, YouTube.

Turner Broadcasting System (WTBS) Archive, Atlanta.

WGBH Archives, Boston.

Wrapped in Steel—Neighborhoods of Southeast Chicago. Jim Martin, 1985.

SELECTED UNPUBLISHED WORKS

Azelton, L. Sean. "Boundary Spanning and Community Leadership: African American Leaders in the Greater Roseland Area." M.A. thesis, University of Illinois at Chicago, 1996.

Baron, Michael. "Tug of War: The Battle Over American Policy Toward China." Ph.D. dissertation, Columbia University, 1980.

Betancur, John, et al. "Roseland: Needs Assessment and Organizational Infrastructure, An Introduction." Center for Urban Economic Development, University of Illinois at Chicago, June 1985.

Carey, Curtis D. "Barack Obama's Transnational Translation: A Rhetoric of Postmodern Unity." Ph.D. dissertation, Howard University, August 2008.

Carriere, Michael H. "Building a New House of Hope: The Rise of the African-American Megachurch in Postindustrial Chicago." Breslauer Symposium Paper, University of California, 2006.

Chicago Foundation for Women. "Caught in the Crossfire: Illinois Roundtable on Welfare Reform, Women of Color and Reproductive Health," 13 October 1999.

Clavel, Pierre, and Sara O'Neill-Kohl. "Losing Out on Industrial Policy: The Chicago Case." Cornell University City and Regional Planning working paper, January 2010.

Cohen, David William. "Perils and Pragmatics of Critique: Reading Barack Obama Sr.'s 1965 Review of Kenya's Development Plan." Seminar paper, 24 March 2010.

Colten, Craig E. "Industrial Wastes in the Calumet Area, 1869–1970: An Historical Geography." Illinois Department of Energy and Natural Resources, September 1985.

Cook, Genevieve. "Dancing in Doorways." B.A. thesis, Swarthmore College, May 1981.

Cook, Genevieve. "Third Culture Kids." Personal essay, December 2014.

Coughlin, Kathryn I. "A Fund Development Strategic Plan for the Gamaliel Foundation." M.S. thesis, DePaul University, February 1999.

Coursey, Don, et al. "Environmental Racism in the City of Chicago: The History of EPA Hazardous Waste Sites in African-American Neighborhoods." Harris School working paper, University of Chicago, October 1994.

Cruz, Wilfredo. "The Nature of Alinsky-Style Community Organizing in the Mexican-American Community of Chicago: United Neighborhood Organization." Ph.D. dissertation, University of Chicago, December 1987.

Ehr, Rev. Donald, et al. "Communities of the Word—Stories of the Chicago Province: The Midwest District Story, 1925–2012." Society of the Divine Word, 2012.

Faraone, Dominic E. "Urban Rifts and Religious Reciprocity: Chicago and the Catholic Church, 1965–1996." Ph.D. dissertation, Marquette University, May 2013.

Feins, Judith D. "Urban Housing Disinvestment and Neighborhood Decline: A Study of Public Policy Outcomes." Ph.D. dissertation, University of Chicago, March 1977.

George, Christine C. "The Chicago Annenberg Challenge: The Messiness and Uncertainty of Systems Change." Philanthropy & Non-Profit

Sector Program, Loyola University Chicago, 2007.

Greetham, David. "Chicago's Wall: Race, Segregation and the Chicago Housing Authority." Senior thesis, College of Wooster, Spring 2013.

Hall, Dorothy I., and Judy Meima. "Economic Development in Southeast Chicago." Voorhees Center for Neighborhood and Community Improvement, University of Illinois at Chicago, 1983.

Haney, LeViis A. "The 1995 Chicago School Reform Amendatory Act and the CPS CEO: A Historical Examination of the Administration of CEOs Paul Vallas and Arne Duncan." D.Ed. dissertation, Loyola University Chicago, May 2011.

Hawking, James. "Political Education in the Harold Washington Movement." D.Ed. dissertation, Northern Illinois University, August 1991.

Illinois Campaign Finance Project. "Tainted Democracy: How Money Distorts the Election Process in Illinois and What Must Be Done to Reform the Campaign Finance System." Institute for Public Affairs, University of Illinois at Springfield, 29 January 1997.

Jackson, John S. "The Making of a Senator: Barack Obama and the 2004 Illinois Senate Race." Paul Simon Public Policy Institute, Southern Illinois University, August 2006.

Jager, Sheila Miyoshi. "Narrating the Nation: Students, Romance and the Politics of Resistance in South Korea." Ph.D. dissertation, Department of Anthropology, University of Chicago, December 1994.

Jones, Edwina Leona. "From Steel Town to 'Ghost Town': A Qualitative Study of Community Change in Southeast Chicago." M.A. thesis, Loyola University Chicago, May 1998.

Kelliher, Thomas G., Jr. "Hispanic Catholics and the Archdiocese of Chicago, 1923–1970." Ph.D. dissertation, University of Notre Dame, December 1996.

Kennedy Communications. "State Senator Barack Obama, Candidate for U.S. Senate: Background Report, Analysis Only." April 2003.

Levin, Rebekah, et al. "Community Organizing in Three South Side Chicago Communities: Leadership, Activities, and Prospects." Center for Impact Research, September 2004.

Mahmood, Mahboob. "Readings with Barack." 30 September 2008.

Mahmood, Mir Mahboob. "Mahatma Gandhi: Saint vs. Visionary." B.A. thesis, Princeton University, 1981.

Mangcu, Xolela MacPherson. "Harold Washington and the Cultural Transformation of Local Government in Chicago, 1983–1987." Ph.D. dissertation, Cornell University, January 1997.

McKersie, William S. "Strategic Philanthropy and Local Public Policy: Lessons from Chicago School Reform, 1987–1993." Ph.D. dissertation, University of Chicago, March 1998.

McKnight, John L. "The Future of Low-Income Neighborhoods and the People Who Reside There." Center for Urban Affairs and Policy Research, Northwestern University, June 1987.

Melaniphy & Associates. *Chicago Comprehensive Neighborhood Needs Analysis, Volume I: Citywide Needs Analysis*; *Volume II: Roseland Community Area, Volume II: Washington Heights Community Area, Volume II: West Pullman Community Area*. January 1982.

Miller, Edwin L., Jr. "Waste Management, Inc." Final report, San Diego County (CA) District Attorney, March 1992.

Mills, Linda. "Illinois Prisoner Reentry: Building a Second Chance Agenda." Annie E. Casey Foundation, July 2004.

Mock, Lynne Owens. "Validation of an Instrument to Assess the Personal Values of African-American Community Leaders." M.A. thesis, University of Illinois at Chicago, 1994.

Mock, Lynne. "The Personal Vision of African American Community Leaders." Ph.D. dissertation, University of Illinois at Chicago, 1999.

Obama, Barack, and Robert Fisher. "Plant Closings: Creative Destruction and the Viability of the Regulated Market," March 1991, 103 pp., Rob Fisher Papers.

Obama, Barack, and Robert Fisher. "Race and Rights Rhetoric," April 1991, 144 pp., Rob Fisher Papers.

O'Donnell, Sandra, Yvonne Jeffries, Frank Sanchez, and Pat Selmi. "Evaluation of the Fund's Community Organizing Grant Program." Woods Fund of Chicago, April 1995.

Palmer, Alice J. "Concepts and Trends in Work-Experience Education in the Soviet Union and the United States." Ph.D. dissertation, Northwestern University, June 1979.

Putterman, Julie S. "Chicago Steelworkers: The Cost of Unemployment." Steelworkers Research Project, January 1985.

Reardon, Kenneth M. "Local Economic Development in Chicago 1983–1987: The Reform Efforts of Mayor Harold Washington." Ph.D. dissertation, Cornell University, 1990.

Rich, Robin. "Explaining Historical Change: The Development of Christian Base Communities in Latin America." March 1991, Robin Rich Papers.

Robinson, Michelle LaVaughn. "Princeton-Educated Blacks and the Black Community." B.A. thesis, Princeton University, 1985.

Sadlowski, Edward A. "Deindustrialization and Displacement: Southeast Chicago's Basic Steel Neighborhoods and the Community Response." Ca. 1991, Frank Lumpkin Papers, Chicago History Museum.

Shipps, Dorothy. "Big Business and School Reform: The Case of Chicago 1988." Ph.D. dissertation, Stanford University, February 1995.

Shipps, Dorothy, and Karin Sconzert with Holly Swyers. "The Chicago Annenberg Challenge: The First Three Years." Consortium on Chicago School Research, March 1999.

Similly, Leslie E. "Black Vernacular English and the Rhetoric of Jeremiah Wright." M.A. thesis, University of Central Oklahoma, 2008.

Simpson, Dick, et al. "The New Daley Machine: 1989–2004." Paper presented at The City's Future Conference, Chicago, July 2004.

Smith, Mansfield I. "The East African Airlifts of 1959, 1960, and 1961." Ph.D. dissertation, Syracuse University, June 1966.

Smylie, Mark A., Stacy A. Wenzel, et al. "The Chicago Annenberg Challenge: Successes, Failures, and Lessons for the Future." Consortium on Chicago School Research, August 2003.

Speller, Julia M. "Unashamedly Black and Unapologetically Christian: One Congregation's Quest for Meaning and Belonging." Ph.D. dissertation, University of Chicago, December 1996.

Stewart, Felicia Ratliff, "The Urban School Superintendency: A Case Study of Ruth Love in Chicago." Ph.D. dissertation, Northern Illinois University, December 1996.

[Sunflower Management]. "Research—State Sen. Barack Obama." 2 vols., 12 June 2003.

Sutoro, Ann Dunham. "Civil Rights of Working Indonesian Women." Institute for Legal Aid, Jakarta, 10 December 1982. Atria Institute.

Sutoro, Ann Dunham. "The Effects of Industrialization on Women Workers in Indonesia." Lecture, Indonesian Society of Development, [1982]. Atria Institute.

Sutoro, Ann D. "Program Statement: Women and Employment (Indonesia)." [1983]. Atria Institute.

Sutoro, Ann D. "Women's Economic Activities in North Coast Fishing Communities: Background for a Proposal from PPA." Ford Foundation, 22 May 1983. Atria Institute.

Thompson, Keriann B. "Explaining the Audacity of Hope: The Rhetoric of Illinois Senator Barack Obama." M.A. thesis, Pepperdine University, July 2007.

Wojcicki, Ed. "Still the Wild West?: A 10-Year Look at Campaign Finance Reform in Illinois." Paul Simon Public Policy Institute, Southern Illinois University, September 2006.

Wright, Jeremiah A., Jr. "Black Church Renewal," in *Perspectives: A View from Within. A Compendium Text for Churchwide Study*. Trinity United Church of Christ, September 1982, pp. 32–42.

Wright, Jeremiah A., Jr. "The Day of Jerusalem's Fall," Trinity United Church of Christ, Chicago, 16 September 2001, http://blakfacts.blogspot.com/2008/03/day-of-jerusalems-fall.html.

Wright, Jeremiah A., Jr. "Confusing God and Government," Trinity United Church of Christ, 13 April 2003, http://www.blackpast.org/2008-rev-jeremiah-wright-confusing-god-and-government.

Zurbuchen, Mary S. "Ann Dunham Sutoro and the Ford Foundation's International Philanthropy." Paper presented at American Anthropology Association annual meeting, December 2009.

ARCHIVAL COLLECTIONS AND PERSONAL PAPERS

Abraham Lincoln Presidential Library Oral History Collections, Springfield, IL (ALPL)

African American Students Foundation Papers, Michigan State University, East Lansing, MI (AASF)

AFSCME Council 31 Papers, Chicago

Atria Institute Archives, Amsterdam, NL

Derrick Bell Papers, New York University Archives, New York, NY

Brad Berenson Papers, Fairfield, CT

Arthur L. Berman Papers, Loyola University of Chicago Archives

Joseph Cardinal Bernardin Papers, Archives of the Archdiocese of Chicago

Timuel D. Black Papers, Harsh Collection, Chicago Public Library (TBP)

Mark Blumenthal Papers, Washington, DC (MBP)

Phil Boerner Papers, Sacramento, CA

Mark Bouman Papers, Chicago

Deborah Brake Papers, Pittsburgh, PA

William J. Brennan Papers, Manuscript Division, Library of Congress, Washington, DC
Betty Brown-Chappell Papers, Chicago History Museum (BBCP)
Katrina Browne Papers, Cambridge, MA
Dave Brubeck Papers, University of the Pacific, Stockton, CA
Margaret Burroughs Papers, DuSable Museum, Chicago
CARPLS (Coordinated Advice & Referrals Program for Legal Services) Papers, Chicago
Jim Cauley Papers, Louisville, KT
Adam Charnes Papers, Dallas, TX
Martin Chester Papers, Minneapolis
Chicago Annenberg Challenge Papers, University of Illinois at Chicago Archives (CACP)
Chicago Community Trust Papers, Chicago
Citizens for Rush, Federal Election Commission (FEC) Filings
City College of New York Archives, New York, NY
Claretian Social Development Fund Papers, Claretian Missionaries Archives, Chicago (CSDF)
John Cardinal Cody Papers, Archives of the Archdiocese of Chicago
Susan Keselenko Coll Papers, Washington, DC
Columbia University Archives, New York, NY
Genevieve Cook Papers, "Norfolk, VT"
Crossroads Fund Papers, University of Illinois at Chicago Archives
Frank Marshall Davis FBI HQ Main File 100-328955, FOIA
Frank Marshall Davis Papers, DuSable Museum, Chicago
Laura Demanski Papers, Chicago
Democratic Socialists of America Papers, Tamiment Library, New York University, New York, NY
Alan and Lois Dobry Papers, Chicago
John Drew Papers, Laguna Niguel, CA
Ann Dunham Papers, Bronwyn Solyom, Honolulu
Stanley Ann Dunham State Department Passport File, FOIA
Echoing Green Papers, New York, NY
John "Jack" Egan Papers, University of Notre Dame Archives
Jeremy Feinberg Papers, New York, NY
Rob Fisher Papers, Arlington, VA (RFP)
William "Terry" Fisher Papers, Cambridge, MA
Joel Freid Papers, Oakland, CA
Friends of Alice Palmer, Federal Election Commission (FEC) Filings
Friends of Barack Obama, Illinois State Board of Election (ISBE) Filings
Jackie Fuchs Papers, Los Angeles, CA
Greg Galluzzo Papers, Chicago (GGP)
Gamaliel Foundation Papers, Chicago
Gery Chico for Senate, Federal Election Commission (FEC) Filings
Jonathan Goldman Papers, Chicago
Erwin N. Griswold Papers, Harvard Law School Archives (EGP)
Adam Gross Papers, Chicago
Ed Grossman Papers, Chicago
Andrew Gruber Papers, Salt Lake City (AGP)
Paul Harstad Papers, Boulder, CO (PHP)
Harvard Law School Archives (HLSA), Cambridge, MA
Harvard-Radcliffe Alumni Against Apartheid Papers, Michigan State University Archives, East Lansing, MI (HRAAA)
Harvard University Archives (HUA), Cambridge, MA
Mark Hebing Papers, Mill City, OR
Stephen Heintz Papers, New York, NY (SHP)
[Blair] Hull for Senate Federal Election Commission (FEC) Filings
Hull Opposition Research Papers (HORP), "East Middleton," WI
Human SERVE Papers, Columbia University Archives, New York, NY
Laura Russell Hunter Papers, Washington, DC (LHP)
[Dan] Hynes for Senate Federal Election Commission (FEC) Filings
Illinois ACORN Papers, State Historical Society of Wisconsin (IAP)
Illinois Department of Commerce and Economic Opportunity Files (ILDCEO), FOIA
Illinois General Assembly, State Senate Floor Transcripts, 90th, 91st, 92nd & 93rd Sessions, 1997–2004
Illinois State Archives Materials, Springfield, IL
Illinois State Board of Investment Papers (ISBI), Chicago, FOIA
Illinois State Senate Republican Caucus Files, Springfield, IL
Industrial Areas Foundation (IAF) Papers, University of Illinois at Chicago Archives
Institute on the Church in Urban Industrial Society Papers (ICUIS), University of Illinois at Chicago Archives
Adrienne Bitoy Jackson Papers, Chicago (ABJP)
Jesse Jackson Jr. Papers, Chicago (JJJP)
Sally Jacobs Papers, Boston
Thomas E. Johnson Papers, Chicago
Joint Committee on Administrative Rules (JCAR) Papers, Springfield, IL

Tom Joyce Papers, Claretian Missionaries Archives, Chicago (TJP)
Marilyn Katz Papers, Chicago
Patrick J. Keleher Papers, Wilmette, IL
David T. Kindler Papers, Oak Park, IL
Tom Knutson Papers, Harvey, IL
Jeff Kobrick Papers, San Mateo, CA
Mark Kozlowski Papers, New York, NY (MKP)
John Kupper Papers, Wilmette, IL (JKP)
James Landing Papers, University of Illinois at Chicago Archives (JLP)
Latino Institute Papers, DePaul University Archives, Chicago (LIP)
Frank Laubach Papers, Syracuse University Archives, Syracuse, NY (FLP)
LawMBG.com, Internet Archive Wayback Machine, 2000–2004
Mike Lawrence Papers, Carbondale, IL (MLP)
Law.UChicago.edu, Internet Archive Wayback Machine, 1997–2004
Tracy Leary Papers, New York, NY
LISC (Local Initiatives Support Corporation) Papers, Chicago
Frank Lumpkin Papers, Chicago History Museum (FLP)
Roberta Lynch Papers, Chicago (RLP)
Scott Lynch-Giddings Papers, Blue Island, IL
Mahboob Mahmood Papers, Singapore
Eden Martin Papers, Chicago
Ian Mattoch Papers, Honolulu
Tom Mboya Papers, Hoover Institution Archives, Palo Alto, CA
Joe McLean Papers, Washington (JMP)
Alex McNear Papers, Sag Harbor, NY (AMP)
Judson Miner Papers, Chicago
Newton Minow Papers, Chicago History Museum
Liza Mundy Papers, Arlington, VA
Glee Murray Papers, Washington, DC
Trent Norris Papers, San Francisco, CA
NYPIRG Papers, New York, NY
Barack H. Obama [Sr.] INS File, FOIA
Obama Family Papers, University of Hawaii at Manoa, Honolulu.
ObamaBlog.com, Internet Archive Wayback Machine, 2004
ObamaForIllinois.com, Internet Archive Wayback Machine, 2002–2004
Obama for Illinois, Federal Election Commission (FEC) Filings
Obama.Senate.gov, Internet Archive Wayback Machine, 2005–2007
Obama 2000, Federal Election Commission (FEC) Filings
Occidental College Archives, Los Angeles
Sandy O'Donnell Papers, Chicago (SOP)
Alice Palmer Papers, Harsh Collection, Chicago Public Library
Maria Pappas for U.S. Senate Federal Election Commission (FEC) Filings
John Parry Papers, Portland, OR
Keith Peterson Papers, Denver, CO
Phelps Stokes Fund Papers, Schomburg Center, New York, NY (PSFP)
Terrie Pickerill Papers, Chicago (TPP)
Frances Fox Piven Papers, Smith College Archives, Northampton, MA
Gregory Poe Papers, Washington, DC. (GPP)
Richard Poethig Papers (RPP), University of Illinois at Chicago Archives
Richard Poethig Papers, Chicago
Stephen Pugh Papers, Chicago
Pullman Christian Reformed Church Papers, Chicago (PCRC)
Punahou School Archives, Honolulu
Radhika Rao Papers, San Francisco
Robin Rich Papers, Gary, IN
Rick Ridder Papers, Denver, CO (RRP)
Michael Risch Papers, Villanova, PA
Andrew Roth Papers, San Francisco
Jack Ryan for U.S. Senate Opposition Research Files, Washington, DC (JRORF)
St. Victor Parish Records, Calumet City, IL
Chris Sautter Papers, Washington, DC
Eugene Sawyer Papers, Chicago Public Library Archives
William X. Scheinman Papers, Hoover Institution Archives, Palo Alto, CA
Ellen Schumer Papers, Chicago (ESP)
Paul Simon Papers, Southern Illinois University Archives, Carbondale, IL (PSP)
Lolo Soetoro INS File, FOIA
State Universities Retirement System of Illinois Papers (SURSIP), Champaign, IL, FOIA
Peter Steffen Papers, Chicago
Eva Sturgies Papers, Chicago
Teachers Retirement System of Illinois Papers (TRSIP), Springfield, IL, FOIA
Carol B. Thompson and Bud Day Papers, Michigan State University Archives, East Lansing, MI
Gina Torielli Papers, Lansing, MI (GTP)
Trotter for Congress, Federal Election Commission (FEC) Filings
United Steelworkers Local 1033 Papers, Chicago History Museum
University of Chicago Law School Files, Chicago
University of Chicago Library Special Collections, Chicago

University of Washington Files, Seattle, WA
Terry Walsh Papers, Evanston, IL (TWP)
Harold Washington Papers, Chicago Public Library Archives (HWP)
Washington Administration Neighborhood & Economic Development Policy Papers, Chicago History Museum
[Joyce] Washington for Senate Federal Election Commission (FEC) Filings
Wieboldt Foundation Papers, Chicago
Mary-Ann Wilson Papers, Chicago
Ed Wojcicki Papers, Springfield, IL
Woods Fund Papers, Chicago

PERIODICALS

Alton Telegraph (*AT*)
Atlanta Journal Constitution (*AJC*)
Atlantic
Aurora Beacon (*AB*)
Austin Weekly News (*AWN*)
Barrington Courier-Review (*BCR*)
Bay State Banner (*BSB*)
Beachwood Reporter (*BR*)
Belleville News-Democrat (*BND*)
Birmingham News (*BN*)
Black Enterprise (*BE*)
Black Issues in Higher Education (*BIHE*)
Bloomington Pantagraph (*BP*)
Boston Globe (*BG*)
Bureau County Republican (*BCR*)
Business International Money Report (*BIMR*)
Business International Weekly Report (*BIWR*)
Canton Daily Ledger (*CDL*)
Capital City Courier (*CCC*)
Capitol Fax (*CF*)
Catalyst Chicago
Catholic New World (*CNW*)
Champaign News-Gazette (*CNG*)
Charleston Times-Courier (*CTC*)
Check-In [Public Allies Chicago]
Chicago Bar Association Record
Chicago Catholic
Chicago Citizen (*CC*)
Chicago Daily Herald (*CDH*)
Chicago Daily Law Bulletin (*CDLB*)
Chicago Daily Observer (*CDO*)
Chicago Defender (*CD*)
Chicago Independent Bulletin (*CIB*)
Chicago Jewish News (*CJN*)
Chicago Lawyer (*CL*)
Chicago Life
Chicago Magazine (*CM*)
Chicago Maroon
Chicago Reader (*CR*)
Chicago Reporter
Chicago Sun Times (*CST*)
Chicago Tribune (*CT*)
Chicago Weekend (*CW*)
Chicago Weekly News (*CWN*)
Christian Science Monitor (*CSM*)
Clinton Herald (*CH*)
Columbia College Chronicle [Chicago]
Columbia Spectator (*CU*)
Community Jobs
Congressional Record (*CRec*)
Cook-Witter Report [Springfield, IL] (*CWR*)
Crain's Chicago Business (*CCB*)
Daily Calumet (*DC*)
Daily Egyptian [SIU] (*DE*)
Daily Illini [UICU] (*DI*)
Daily Northwestern [NU] (*DN*)
Daily Southtown (*DS*)
Daily Vidette [ISU] (*DV*)
Dallas Morning News (*DMN*)
Danville Commercial News (*DCN*)
Daytona Beach News-Journal (*DBNJ*)
Decatur Herald & Review (*DHR*)
DeKalb Daily Chronicle (*DDC*)
Denver Post (*DP*)
Des Plaines Times (*DPT*)
DL21C Report [Chicago]
East African Standard [Nairobi] (*EAS*)
Edwardsville Intelligencer (*EI*)
Elgin Courier-News (*ECN*)
Elmhurst Press (*EP*)
Evanston Review (*ER*)
Feast [Occidental College]
Financial Times (*FT*)
Financing Foreign Operations (*FFO*)
Freeport Journal Standard (*FJS*)
Galesburg Register-Mail (*GRM*)
Galesburg Zephyr (*GZ*)
Harrisburg Daily Register (*HDR*)
Harvard Crimson (*HC*)
Harvard Law Bulletin (*HLB*)
Harvard Law Record (*HLRec*)
Harvard Law Review (*HLR*)
Harvard Law Revue
Harvard Law School Adviser (*HLSA*)
Harvard Magazine
Hawaii Business
Hawaii Magazine (*HM*)
Health & Medicine
Hegewisch News (*HN*)
Honolulu Advertiser (*HA*)
Honolulu Record
Honolulu Star Bulletin (*HSB*)

Hyde Park Citizen (HPC)
Hyde Park Herald (HPH)
Illinois Issues (II)
Illinois Leader (IL)
Illinois Lutheran
Illinois Politics
Illinois Times (IT)
In These Times (ITT)
Jakarta Post (JP)
Jersualem Post
Jewish Daily Forward (JDF)
Joliet Herald News (JHN)
Journal of Blacks in Higher Education (JBHE)
Ka Leo O Hawaii [UHM] *(KLOH)*
Ka Punahou (KP)
Kankakee Daily Journal (KDJ)
Kansas City Star (KCS)
Kewanee Star Courier (KSC)
Labor Letter [Illinois AFL-CIO] *(LL)*
LaSalle News Tribune (LNT)
Los Angeles Sentinel (LAS)
Los Angeles Times (LAT)
LRU First Reading [Legislative Research Unit, ILGA]
Macomb Journal (MJ)
Madison Capital Times (MCT)
Madison County Record (MCR)
Marion Daily Republican (MDR)
Mattoon Journal Gazette (MJG)
Miami Herald (MH)
Midweek [Honolulu]
Minneapolis Star Tribune (MST)
Moline Dispatch (MD)
Monmouth Review Atlas (MRA)
Morris Daily Herald (MDH)
Mount Vernon Register News (MVRN)
Mt. Carmel Register (MCR)
Nairobi Times
Naperville Sun (NS)
Nation [Nairobi]
National Catholic Reporter (NCR)
National Journal (NJ)
National Review (NR)
N'Digo
New Ground [Chicago DSA]
New York Amsterdam News (NYAN)
New York Beacon (NYB)
New York Daily News (NYDN)
New York Magazine (NYM)
New York Observer (NYO)
New York Post (NYP)
New York Sun (NYS)
New York Times (NYT)
NHI Shelterforce
Northwest Herald (NH)
Northwest Indiana Times (NIT)
Northwest Side Press [Chicago] *(NSP)*
Occidental Magazine
Olney Dail Mail (ODM)
Oprah Magazine
Orange County Register (OCR)
Ottawa Daily Times (ODT)
Pekin Daily Times (PDT)
People's Weekly World (PWW)
Peoria Journal Star (PJS)
Philadelphia Inquirer (PI)
Pittsburgh Tribune Review (PTR)
Politico
Progressive
Providence Journal (PJ)
Punahou Bulletin
Quad-City Times (QCT)
Quincy Herald Whig (QHW)
Reconstruction
Residents' Journal [Chicago]
River Cities Reader (RCR)
Rock Island Argus (RIA)
Rockford Register Star (RRS)
Roll Call (RC)
Salt Magazine
Sauk Valley Newspapers (SVN)
South End Citizen (SEC)
South Street Journal (SSJ)
Southern Illinoisan (SI)
Southtown Economist (SE)
Southtown Star (SS)
State Journal-Register [Springfield, IL] *(SJR)*
St. Louis Post-Dispatch (SLPD)
St. Paul Pioneer Press (SPPP)
Standard (Nairobi)
Steel Shavings
Telegraph Herald (Dubuque, IA) (TH)
1033 News and Views [USW Local]
The Hill (TH)
The May Report (TMR)
The Nation (TN)
The Neighborhood Works (TNW)
The New Republic (TNR)
The Occidental (TO)
The Paper [CCNY]
The Phoenix [UCLS]
The Solon [Illinois General Assembly Retirement System]
The Star [South Suburban]
Third Coast Press [Chicago]
Third World Voice [Occidental College]
Three Speech [HLS] *(TS)*
Trumpet [Trinity United Church of Christ]
University of Chicago Chronicle (UCC)

University of Chicago Law School Record (*UCLSR*)
University of Chicago Magazine (*UCM*)
Upturn [Association of Chicago Priests]
USA Today (*USAT*)
Wall Street Journal (*WSJ*)
Washington Post (*WP*)
Washington Times (*WT*)
Waterloo-Cedar Falls Courier (*WCFC*)
Waukegan News Sun (*WNS*)
West Frankfort Daily American (*WFDA*)
Wilmette Life (*WL*)
Windy City Times (*WCT*)
Work in Progress [Joyce Foundation]

INTERVIEWS AND ORAL HISTORIES

Andrew Abrams (DJG), 7 March 2014, Chicago (T)
Michael Ackerman (DJG), 8 and 13 December 2011 (E); 14 December 2011 (T); 20 February 2012 (E), Los Angeles
Eric Adelstein (DJG), 10 November 2011, Chicago
Asif Agha (DJG), 24 January 2010, Philadelphia
Geraldine Alexis (DJG), 30 April 2013, San Francisco
Salim Al Nurridin (DJG), 14 January 2010, 10 November 2010, Chicago
Frank Amanat (DJG), 19 October 2010, Washington, DC
Hisham Amin (DJG), 6 March 2014, Baltimore (T)
John T. "Jack" Anderson (DJG), 13 March 2014, Naples, FL (T)
Paul Anderson (DJG), 10 February 2013 (E); 15 February 2013 (T), Minneapolis
Arthe Anthony (Oby Okpalanmma), 21 March 2014, Los Angeles, Occidental College
William Araiza (DJG), 11 October 2010, Brooklyn, NY
Ted Aranda (DJG), 14 March 2013, Chicago
Blanca (Labunog) Araujo (DJG), 4 January 2010, Poway, CA (E)
Dan Armstrong (DJG), 29 January 2010, New York, NY
Daniel Armstrong (DJG), 6 December 2011, Los Angeles (E)
Susan Arterian (DJG), 28 September 2010, White Plains, NY (E)
Catherine Askew (DJG), 17 September 2009, Chicago
Stephanie Askew (DJG), 17 September 2009, Chicago
Andrea Atkin (DJG), 22 October 2010, Raleigh, NC
David Attisani (DJG), 25 February 2013, Boston (T)
William Atwood (DJG), 25 June 2013, Chicago
Loretta Augustine-Herron (DJG), 9 May 2009; 17 May 2010 (T) Calumet City, IL
Alma Avalos (DJG), 14 November 2010, Chicago
David Axelrod (Jim Gilmore), 20 June 2008, Chicago (*Frontline*)
David Axelrod (Robert Draper), 2009, Washington, DC
Bill Ayers (DJG), 10 September 2009, Chicago
John Ayers (DJG), 9 December 2011; 1 May 2012 (E), Palo Alto, CA
Gloria Jackson Bacon (Julieanna Richardson), 10 July 2002, Chicago (TheHistoryMakers.com)
Gloria Jackson Bacon (DJG), 10 March 2014, New Orleans (T)
Margaret Bagby (DJG), 20 May 2010, South Holland, IL
Douglas Baird (DJG), 18 September 2009, Chicago
Deborah Baker, 15 (2) June 2015, Brooklyn, NY (E)
Casey Banas (Felicia Ratliff Stewart), 26 June 1996, Chicago
Brian Banks (DJG), 16 January 2010; 12 December 2010 (E), Chicago
Arthur Barnes (DJG), 18 October 2016, New York (T)
Chuck Barnhill (DJG), 24 July 2012, Madison, WI
Hal Baron (DJG), 3 May 2011; 11 June 2011, 6 June 2014 (E), Chicago
Michael Baron (DJG), 14 March 2014, Sarasota, FL (T)
Peter Baroni (DJG), 1 June 2010, Western Springs, IL
Susannah Baruch (DJG), 19 March 2014, Washington, DC (E)
Emmett and Priscilla Bassett (Joseph Mosnier), 21 July 2011, Grahamsville, NY
Maria (Stathis) Batty (DJG), 20 March 2013, Atlanta (T)
Gideon Baum (DJG), 29 June 2013, Chicago
Jennifer Borum Bechet (DJG), 28 October 2010; 28 October 2010 (E), New Orleans
Leon Bechet (DJG), 29 October 2010, New Orleans (T)
Scott Becker (DJG), 11 May 2010, Chicago
Peregrine "Pern" Beckman (DJG), 8 January 2010, Burbank, CA (T)
John Belcaster (DJG), 18 September 2009; 22 July 2010 and 25 June 2015 (E), Chicago

Derrick Bell (DJG), 26 January 2010 (E); 28 January 2010, New York, NY
Ellen Benjamin (DJG), 11 January 2010, Chicago
Joseph R. Bennett (DJG), 20 March 2013, Hazel Crest, IL (T)
Robert Bennett (DJG), 2 November 2010; 3 November 2010 (E), Chicago
Deanne Benos (DJG), 20 November 2015, Chicago (T)
Brad Berenson (DJG), 22 January 2010, 25 April 2011, Washington, DC
Steven Berkow (DJG), 22 April 2013, Washington, DC
Arthur L. Berman (DJG), 12 November 2011, Chicago
Mary Bernstein (DJG), 5 December 2011, Oakland, CA
Brian Bertha (DJG), 2 December 2011, San Francisco
Maggie Bertke (DJG), 4 and 8 April 2011, Washington, DC (E)
David Bird (DJG), 6 May 2014, Pittsburgh, PA
Timuel Black (DJG), 6 May 2009, Chicago
Margaret Blackshere (DJG), 12 November 2011, Niles, IL
Robert Blackwell Jr. (DJG), 7 May 2009, Chicago
Robert Blackwell Jr. (John Page), 2012, Chicago
Susan (Botkin) Blake (DJG), 3 January 2012, Mercer Island, WA (T)
Mark Blumenthal (DJG), 22 July 2013; 7 August 2013 (E), Washington, DC
Phil Boerner (DJG), 26 (2) and 28 (2) March 2009 and 5 April 2009 (E); 17 April 2009; 18 (2), 19, 20, 21, 22 (2), 23, and 24 (2) April 2009, 1, 20 (2), 22, 27, 28, 29, and 31 May 2009, 2, 28 (3), and 29 (2) June 2009, 12, 28 (3), 29 (2), and 30 (2) July 2009, 24, 25, 30, and 31 (2) August 2009, 2 and 3 September 2009, 13 and 21 December 2009, 7, 8, 11, 22 (3), 23, 24 (3), and 26 February 2010, 30 (2) March 2010, 1, 2, and 26 April 2010, 11 June 2010, 19, 20, 21, and 22 July 2010, 14 August 2010, 12 January 2011, 9, 10, and 12 February 2011, 29 March 2011, 16 May 2011, 5 June 2011, 1 May 2012, 15 and 16 August 2012, 16 September 2012, 16 October 2012, 13 (2), 17, 20, 23, 24, and 25 November 2014, 3 (2), 6, and 31 December 2014 (E), Sacramento
Roger Boesche (DJG), 8 April 2009, Los Angeles
Lionel Bolin (DJG), 28 July 2012, Beverly Shores, IN
Larry Bomke (DJG), 9 May 2011, Springfield, IL
Adam C. Bonin (DJG), 28 February 2014, Philadelphia (E)
Roger Boord (DJG), 16 April 2013, Chicago (T)
Ilisabeth Bornstein (DJG), 4 April 2014, Pittsburgh, PA
Andrew Boron (DJG), 13 August 2013, Chicago
Caroline Boss (DJG), 4 January 2012, Friday Harbor, WA
John Bouman (DJG), 3 April 2009, Chicago
Mark Bouman (DJG), 19 February 2010 (E), Chicago
Evelyn Bowles (DJG), 14 March 2013, Glen Carbon, IL (T)
John Boyer (DJG), 21 December 2011, Honolulu
Keith Boykin (DJG), 13 October 2010, New York, NY
Chris Boyster (DJG), 7 November 2011, Springfield, IL
Joannie Braden (DJG), 21 April 2014, Denver, CO
Bill Brady (DJG), 18 November 2010, Springfield, IL
Deborah Brake (DJG), 26 March 2013, Pittsburgh, PA (T)
Carol Moseley Braun (Julian Bond), 16 March 2005, Charlottesville, VA (Institute for Public History, UVA)
Carol Moseley Braun (DJG), 14 May 2010, Chicago
Arthur Brazier (David Axelrod), 28 January 2004, Chicago (TPP)
Renee Brereton (DJG), 9 and 15 September 2009, Washington, DC (T)
David Brint (DJG), 30 March 2014, Northbrook, IL (T)
Rudy Broad (DJG), 14 May 2011, Springfield, IL (T)
Dan Bromberg (DJG), 9 December 2011; 9 December 2011 (E), Redwood Shores, CA
Julia Bronson (DJG), 26 March 2014, New York, NY (T)
Rosellen Brown (DJG), 8 July 2015 (T), 15 July 2015 (E), Chicago
Katrina Browne (DJG), 19 April 2011; 29 May 2011 (E), Newton, MA
Kenneth Brucks (DJG), 11 July 2012, Chicago
Paul Burak (DJG), 13 January 2010, Orland Park, IL
Dan Burkhalter (DJG), 16 July 2012, Madison, WI
Deloris Burnam (DJG), 17 September 2009, Chicago
Deborah Burnet (DJG), 4 April 2009, Chicago
Will Burns (Jim Gilmore), 25 July 2008, Chicago (*Frontline*)
Will Burns (DJG), 8 September 2009, Chicago
Roland W. Burris (DJG), 16 March 2014, Chicago (T)

Brad Burzynski (DJG), 24 March 2014, Clare, IL (T)
Gerrard Bushell (DJG), 28 and 30 January 2010, New York, NY
Lewis Butler (Ann Lage), 16 April 2009, Stinson Beach, CA (Bancroft Library, University of California, Berkeley)
Cassandra Butts (Jim Gilmore), 10 July 2008, Washington (*Frontline*)
Cassandra Butts (DJG), 28 April 2011, Washington, DC
Judy Byrd (DJG), 3 September 2013; 18 September 2013 (T), New York, NY
Manford Byrd Jr. (Felicia Ratliff Stewart), 8 May 1995, Chicago
Marian Byrnes (Rod Sellers), 14 February 2001, Chicago (Mark Bouman)
Mary Beth Cahill (Robert Draper), 2009, Washington, DC
Mary Beth Cahill (DJG), 6 February 2014, Washington, DC
Kristen B. Caldwell (Jim Gilmore), 27 June 2012, Sunnyvale, CA (*Frontline*)
John Calicott (DJG), 13 January 2010, Country Club Hills, IL
David Callahan (DJG), 14 October 2010, New York, NY
Mary Ellen Callahan (DJG), 13 March 2014, Washington, DC (T)
John Cameron (DJG), 8 November 2010; 8 November 2010 (E), Chicago
Robert Campbell (Philip Bracy), 28 April 1981, Youngstown, OH (Youngstown State University)
Cindi Canary (DJG), 4 May 2011, Chicago
Rachel Cano (DJG), 20 February 2013, San Diego (E)
Nora Moreno Cargie (DJG), 15 March 2013, Chicago
Judith Pinn Carlisle (DJG), 12 February 2013 (E); 13 February 2013 (T), Flint, MI
Dominic Carmon (DJG), 30 October 2010, New Orleans
Paul Carpenter (DJG), 16 December 2011, Santa Monica, CA
Howard Carroll (DJG), 22 March 2013, Lincolnwood, IL (T)
Marlies Carruth (DJG), 15 January 2010, Chicago
Sheryll Cashin (DJG), 22 January 2010; 30 March 2010 (E), Washington, DC
Jim Cauley (DJG), 26 October 2010, Louisville, KY
Maria B. Cerda (DJG), 12 May 2010, Chicago (T)
Lou Celi (DJG), 28 January 2010, New York, NY
Warren Chain (DJG), 24 March 2014, Lumberton, NJ (T)
Hasan Chandoo (DJG), 19 September 2013, Armonk, NY
Raazia Chandoo (DJG), 19 September 2013, 18 December 2014 (E), Armonk, NY
Shirley Chappell (DJG), 5 February 2015 (E), Chicago
Elvin Charity (DJG), 15 September 2009, Chicago
John Charles (DJG), 9 May 2011, Springfield, IL
Adam Charnes (DJG), 23 October 2010, Winston-Salem, NC
Micki Chen (DJG), 20 October 2010, Arlington, VA
Marty Chester (DJG), 6 March 2014, Minneapolis (T)
Jack Chorowsky (DJG), 20 April 2011, New York, NY
Dino Christenson (DJG), 18 December 2013, Boston (T)
A. Marisa Chun (DJG), 26 April 2011, Washington, DC
Thomas Cima (DJG), 13 May 2010, Chicago
Chapin Cimino (DJG), 30 April 2014 (T); 30 April 2014 (E), Philadelphia
Michael Claffey (DJG), 26 June 2013 (T); 27 June 2013 (E), Urbana, IL
Judith Clain (DJG), 9 December 2014, New York, NY (E)
Frank Clark (DJG), 8 August 2013, Chicago
Carin Clauss (DJG), 16 July 2012, Madison, WI
James Clayborne (DJG), 17 November 2010, Springfield, IL
Forrest Claypool (DJG), 27 June 2013, Chicago
Richard Cloobeck (DJG), 23 February 2013 (E); 24 February 2013 (T), Las Vegas
Delmarie Cobb (DJG), 13 August 2013, Chicago (T)
Kieth Cockrell (DJG), 11 February 2013, Detroit (E)
Peter Coffey (DJG), 26 June 2013, Chicago
Tom Coffey (DJG), 2 March 2015, Chicago (E)
Julie E. Cohen (DJG), 29 April 2010, Cambridge, MA
Michael L. Cohen (DJG), 15 December 2011, Los Angeles
Sherry Colb (DJG), 27 February 2013, Ithaca, NY (T)
Carl Coleman (DJG), 7 January 2011, Lancy, France (T)
Susan Keselenko Coll (DJG), 9 October 2013, Washington, DC
Eric Collins (DJG), 9 September 2014, London, UK (T)

Jacqueline Y. Collins (DJG), 11 May 2011, Springfield, IL
James M. Collins (DJG), 10 May 2011, Springfield, IL
Jennifer Collins (DJG), 23 October 2010, Winston-Salem, NC
Scott Collins (DJG), 7 December 2010, London, UK
Les Coney (DJG), 24 June 2013, Chicago
Danna Cook (DJG), 14 April 2009, Los Angeles (E)
Genevieve Cook (DJG), 4 and 5 September 2012 (E); 5 September 2012 (T); 6, 7, 8, 19, 22, 23, 26, 27, 28, and 30 September 2012, 15 November 2012 (E); 9, 10, and 11 December 2012; 6 February 2013, 3, 5, and 17 October 2013, 21, 22, and 23 October 2014, 12 (5), 13 (3), 15 (2), 19 (4), and 20 (4) December 2014, 20 (2) January 2015, 24, 26, and 28 February 2015 (E), "Norfolk, VT"
Frank Cooper (DJG), 10 December 2010, Armonk, NY (T)
Nancy Cooper (DJG), 21 December 2011, Honolulu
Leslie Corbett (DJG), 30 April 2014, Chicago (T)
Jacqueline Scott Corley (DJG), 2 December 2011, San Francisco
Susan (Williamson) Corley (DJG), 12 May 2010, Scottsdale, AZ (T)
Jack Corrigan (Daniel Urman), 2013, Brookline, MA (Northeastern University)
Jack Corrigan (DJG), 18 March 2014, Brookline, MA (T)
John Corrigan (DJG), 9 September 2009, Chicago
Sharon Corrigan (DJG), 16 July 2012, Madison, WI
Ernesto Cortes Jr. (Lynnell J. Burkett), 27 May 1994, San Antonio, TX (University of Texas at San Antonio)
Jim Covington (DJG), 12 May 2011, Springfield, IL
David Craven (DJG), 29 June 2013, 10 August 2013, Chicago
Bob Creamer (DJG), 30 June 2013, Evanston, IL
Dan Cronin (DJG), 10 November 2011, Wheaton, IL
Megan Crowhurst (DJG), 7 May 2013, Portland, OR
Lori Joyce Cullen (DJG), 7 November 2011, Springfield, IL
John Cullerton (DJG), 12 November 2010, Chicago
Jeffrey Cummings (DJG), 9 September 2009; 28 September 2009 and 14 June 2015 (E), Chicago
Barbara Flynn Currie (DJG), 9 September 2009, Chicago
Lacey Curry (DJG), 12 May 2010, Chicago (T)
Charles U. Daly (DJG), 5 March 2014, West Palm Beach, FL (T)
Jack Darin (DJG), 8 November 2011, Springfield, IL
Tom Dart (DJG), 1 July 2013, Chicago
Tom Daschle (Jim Gilmore), 10 June 2008, Washington, DC (*Frontline*)
Cindy (Huebner) Davidsmeyer (DJG), 13 May 2011, Springfield, IL
H. Ron Davidson (DJG), 2 (2) April 2014, Miami, FL (E)
Allison S. Davis (Robert Draper), 2009, Chicago
Allison S. Davis (DJG), 12 May 2010, Chicago
Andrew Davis (DJG), 26 July 2012, Chicago
Danny K. Davis (David Axelrod), 29 September 2003, Chicago (JKP)
Danny K. Davis (Terence Sims), 10 May 2010, Chicago (CAAMC, Columbia College)
Frank Marshall Davis (Chris Conybeare and Joy Chong), May 1987 (?), Honolulu (UH CLER)
Jennifer Davis (William Minter), 12 December 2004, Washington, D. (NoEasyVictories.org)
Lennard Davis (DJG), 13 and 14 December 2009 (E); 12 January 2010, 6 (2) December 2014 (E), Chicago
Rochelle Davis (DJG), 5 August 2013, Wilmette, IL
Michael Dawson (DJG), 5 November 2010, Chicago (E)
Michael Dees (DJG), 12 September 2012, Reno, NV (T)
Steve Delaney (DJG), 2 February 2010, Needham, MA
William Delgado (DJG), 13 May 2011, Springfield, IL
Tiffanie Cason De Liberty (DJG), 3 April 2014, Santa Rosa, CA (T)
Beth DeLisle (DJG), 5 May 2014, Boston (T)
Yvonne Delk (DJG), 5 March 2013, Norfolk, VA (E)
Michelle (Jacobs) DeLong (DJG), 27 February 2013, Scarsdale, NY (T)
Miguel del Valle (DJG), 27 May 2010, Chicago
Laura Demanski (DJG), 12 March 2004 (T), 29 March 2014 (E), Chicago
Steve Derks (DJG), 2 November 2010, Chicago
Mark Dery (DJG), 20 January 2010 (E); 30 January 2010, New York, NY
Diana Derycz-Kessler (DJG), 13 December 2011, Los Angeles

Alba De Souza (DJG), 26 June 2009, Cambridge, UK
Kurt DeWeese (DJG), 14 May 2011, Springfield, IL
Alice Dewey (DJG), 20 December 2011, Honolulu
Nathan Diament (DJG), 28 April 2011, Washington, DC
Kelly Dietrich (DJG), 25 June 2013, Chicago
Todd Dietterle (DJG), 26 June 2013, Chicago
Kirk Dillard (Mark DePue), 9 November 2009, Springfield, IL (ALPL)
Kirk Dillard (DJG), 18 November 2010, Springfield, IL
Marlene Dillard (DJG), 14 September 2009, Chicago
Greg Dingens (DJG), 1 November 2010, Chicago
Paolo DiRosa (DJG), 15 February 2013, Washington, DC (T)
Mary Dixon (DJG), 17 November 2010, Springfield, IL
Doug Dobmeyer (DJG), 12 March 2013, Chicago
Alan Dobry (DJG), 8 September 2009, Chicago
Lois Dobry (DJG), 8 September 2009, Chicago
Jim Dodge (DJG), 11 May 2011, Springfield, IL
Bernardine Dohrn (DJG), 28 June 2013, Chicago
Peter Dolotta (DJG), 16 September 2014, New York, NY (T)
Mark Donahue (DJG), 8 November 2011, Springfield, IL
Michael Dorf (DJG), 2 May 2011, Chicago
Michael Dorf (DJG), 15 February 2013, Ithaca, NY (T)
Kevin Downey (DJG), 19 October 2010, Washington, DC
John Drew (DJG), 14 December 2011 (T); 16 December 2011, Laguna Niguel, CA
Liz Drew (DJG), 25 July 2013, Washington, DC
Benjamin Dreyer (Robert Draper), 2009, New York, NY (E)
Timothy Driscoll (DJG), 13 March 2013, Mineola, NY (T)
Steven A. Drizin (DJG), 24 and 27 (2) October 2011, Chicago (E)
Tammerlin Drummond (DJG), 17 March 2014, Cambridge, MA (T)
Len Dubi (DJG), 18 March 2013, Homewood, IL
Jan Anne Dubin (DJG), 25 June 2013, Chicago
Walter Dudycz (DJG), 25 March 2014, Chicago (T)
Jim Duffett (DJG), 8 November 2011, Springfield, IL
Ralph Dunham (DJG), 28 April 2011, Springfield, VA
Anita Dunn (DJG), 22 July 2013, Washington, DC
Marv Dyson (DJG), 2 June 2015, Chicago (T)
Jane Dystel (Robert Draper), 2009, New York, NY (E)
John "Vince" Eagan (DJG), 17 May 2009, Covington, GA (T)
John Eason (DJG), 6 March 2014, College Station, TX (T)
Claire Eberle (DJG), 13 May 2011, Springfield, IL
Barack Echols (DJG), 2 April 2014 (E); 29 April 2014 (T), 6 November 2014 (E), Chicago
Romelle Rowe Ecung (DJG), 6 December 2014, Los Angeles (T)
Jim Edgar (Mark DePue), 28 May, 2 September, 9 September, 8 November, 18 November and 14 December 2010, Urbana, IL (ALPL)
Jim Edgar (DJG), 10 March 2014, Tempe, AZ (T)
Christopher Edley (DJG), 5 December 2011, Berkeley, CA
Erin Edmonds (DJG), 21 September 2014, Salt Lake City (T)
Nathaniel Edmonds (DJG), 20 March 2014 (E); 2 September 2014 (T), Washington, DC
Michael Edney (DJG), 13 March 2014, Washington, DC (T)
Joella Edwards (Jack McAdoo), 1, 4, and 10 November 2010, Orlando, FL (JMacRadio.com)
Joella Edwards (Earl Yates), 15 and 23 January 2011, Hilton Head, SC (Low Country View, YouTube)
Pal Eldredge (DJG), 21 December 2011, Honolulu
Robert Elia (DJG), 29 March, 2 April, and 17 October 2014, Warren, OH (T)
Mo Elleithee (DJG), 23 July 2013, Washington, DC
Jeffrey Ellman (DJG), 19 March 2013, Atlanta (T)
Lisa Ellman (DJG), 19 December 2014, Washington, DC
Leif Elsmo (DJG), 11 March 2013, Chicago
Susan Epstein (DJG), 30 March 2014, Santa Barbara, CA (T)
Jason Erkes (DJG), 4 May 2011, Chicago
James Esseks (DJG), 13 October 2010, New York, NY
Timothy C. Evans (DJG), 7 June 2011, Chicago
Barbara Eyman (DJG), 5 October 2010, Washington, DC (E)
Stanley Farier (DJG), 13 January 2010, South Holland, IL (T)

Ruth Fecych (DJG), 29 May 2015 (E), 1 June 2015 (T), New York, NY
Lynn Feekin (DJG), 7 May 2010, Eugene, OR
Brent Feigenbaum (DJG), 11 October 2010, New York, NY
Jeremy Feinberg (DJG), 26 December 2010 (T), 29 December 2010 (E), New York, NY
Leonard Feldman (DJG), 19 February 2013, Seattle (T)
Stephen Feldman (DJG), 17 April 2014, Raleigh, NC (T)
Dave Feller (DJG), 12 August 2013, Chicago
Edward Felsenthal (DJG), 5 September 2013, New York, NY (T)
Julie Fernandes (DJG), 21 March 2014, Washington, DC (T)
Henry Ferris (DJG), 28 May 2015, New York, NY (T)
Lew Finfer (DJG), 1 February 2010, Boston
Jim Finley (DJG), 6 February 2013, Henderson, NV (E)
Tom Finnegan (DJG), 17 May 2010, Blue Island, IL (T)
Sunny Fischer (DJG), 15 March 2013, Chicago
Lisa Fisher (DJG), 20 January 2010, Arlington, VA
Rob Fisher (DJG), 12 (2), and 13 (2) January 2010 (E); 20 January 2010; 29, and 31 July 2010, 2 (2), 3, 5 (2), and 24 August 2010, 6 September 2010 (E); 20 November 2010 (T); 27 April 2011; 29 April 2011, 12 September 2011, 12 February 2012, 17 May 2012, 17 (2), and 18 (3) February 2013, 27 (2), 29, and 29 March 2013, 11 and 21 May 2013 (E); 24 July 2013; 28 March 2015, 7 (2), 21, and 23 (2) April 2015, 24 May 2015, 16 August 2015 (E), Arlington VA and Washington, DC
William "Terry" Fisher (DJG), 10 April 2015, Cambridge, MA (E)
Peter G. Fitzgerald (DJG), 10 March 2014 (E); 14 March 2014 (T), McLean, VA
Tom FitzGibbon (DJG), 2 September 2014, Indianapolis, IN (T)
Jeremy Flynn (DJG), 30 March 2014, Springfield, IL (T)
Lula Ford (DJG), 27 March 2014, Chicago (T)
Fred Fortier (DJG), 16 August 2013, Oak Park, IL (T)
Andy Foster (DJG), 9 December 2011, Palo Alto, CA
Bunny Fourez (DJG), 17 November 2010, Springfield, IL
Homer D. Franklin (DJG), 6 February 2015, Pearland, TX (T)
Joel Freid (DJG), 6 April 2015 (E), 6 April 2015 (T), Oakland, CA
Susan Freiwald (DJG), 3 December 2011; 4 and 6 December 2011 (E), San Francisco
Charles Fried (DJG), 7 and 8 May 2015, Cambridge, MA (E)
Joe Frolik (DJG), 24 July 2015, Cleveland, OH (T)
Wilson Frost (Marie Scatena), 13 November 2014, Palm Desert, CA (T)
Gerald Frug (DJG), 18 April 2011, Cambridge, MA
Jackie Fuchs (DJG), 5 July 2010, Cambridge, UK
James Fugate (Robert Draper), 2009, Los Angeles
Bill Fuhry (DJG), 14 April 2014, Evanston, IL (T)
Kelli Furushima (DJG), 20 December 2011, Honolulu
George Galland (DJG), 19 May 2010, Chicago
Greg Galluzzo (Don Elmer), January 2009, Detroit, MI (Community Organizer Genealogy Project)
Greg Galluzzo (DJG), 31 March 2009, 2 November 2010, 16 May 2011, Chicago
Steve Ganis (DJG), 18 April 2011; 2 May 2011 (E), Boston
Robert T. Gannett (DJG), 3 November 2010, Chicago (E)
Jesus "Chuy" Garcia (Dale Rosen), 15 April 1989, Chicago (Blackside)
Jesus "Chuy" Garcia (Betty Brown-Chappell), 11 January 1990, Chicago (BBCP)
Jesus "Chuy" Garcia (DJG), 11 March 2013, Chicago
Romeo Garcia (DJG), 5 December 2011, Oakland, CA
Randon Gardley (DJG), 12 November 2010, Chicago
Edward G. Gardner (Davis Lacy Jr.), 1 June 1989, Chicago (Blackside)
Edward G. Gardner (Julieanna Richardson), 12 August 2002, Chicago (TheHistoryMakers.com)
Joseph Gardner (James A. DeVinney), 14 April 1989, Chicago (Blackside)
Molly Garhart (DJG), 17 March 2014, San Francisco (T)
Betty Garrett (DJG), 17 September 2009, Chicago
Selmo "Lucky" Garrett Jr. (DJG), 17 September 2009, Chicago
John Geanuracos (DJG), 4 December 2010, London, UK (E)
Michael Gecan (DJG), 1 October 2010 (E); 15 October 2010, Princeton, NJ

Julius Genachowski (DJG), 25 April 2011, Washington, DC
Harry Gendler (DJG), 14 September 2009, Chicago
Tom Geoghegan (DJG), 26 July 2012, Chicago
Christine Gervais (DJG), 12 May 2010, Monticello, IN (T)
Pete Giangreco (DJG), 12 March 2013, Evanston, IL
Jeanne M. Gills (DJG), 21 March 2014, Chicago (T)
Robert Ginsburg (DJG), 25 May 2010, Chicago
Douglas Glick (DJG), 21 September 2009; 22 September 2009 (E), Binghamton, NY
Virginia Dunham Goeldner (DJG), 20 November 2010, Maumelle, AR (T)
David Goldberg (DJG), 26 January 2010, New York, NY
Michael Golden (DJG), 26 June 2013, Chicago (E)
Jonathan Goldman (DJG), 25 June 2013, Chicago
Mary Gonzales (DJG), 7 May 2011, Chicago
Tasneem Goodman (DJG), 31 March 2014, Chicago (T)
Kent Goss (DJG), 1 (E) and 6 April 2009, Los Angeles
Josh Gottheimer (DJG), 18 September 2013, Ridgewood, NJ (T)
Arnie Graf (DJG), 20 October 2010, Silver Spring, MD
Patricia A. Graham (DJG), 23 January 2010 (E); 2 May 2010; 1 November 2010 (E), Cambridge, MA, and Denver, CO
Tom Grauman (DJG), 4 January 2012; 9 January 2012, 11 January 2013 (E), Friday Harbor, WA
Donita Powell Greene (DJG), 9 February 2015, Alexandria, VA (T)
Nadyne Griffin (DJG), 14 January 2010; 9 February 2015 (T), Chicago
Jacky Grimshaw (Betty Brown-Chappel), 23 October 1989, Chicago (BBCP)
Bill Grimshaw (DJG), 26 May 2010, Chicago
Jacky Grimshaw (DJG), 26 May 2010, Chicago
Erwin N. Griswold (Victoria L. Radd), 13 January 1992, Washington, DC (DCCHS.org)
Adam Gross (DJG), 12 November 2010, Chicago
Dave Gross (DJG), 16 Novemer 2010, Springfield, IL
Ed Grossman (DJG), 27 June 2013, Chicago
Andrew Gruber (DJG), 19 March 2013, Salt Lake City (T)
Carlton Guthrie (DJG), 19 March 2014, Chicago (T)
Michael Guzman (DJG), 20 October 2010, Washington, DC
Bill Haine (DJG), 17 November 2010, Springfield, IL
Marc Haine (DJG), 22 December 2011, Honolulu
Leslie Hairston (DJG), 14 May 2010, Chicago
Dodie Hajra (DJG), 22 February 2013, Orlando, FL (T)
Dan Hale (DJG), 19 October 2010, Falls Church, VA
Jay Halfon (DJG), 16 September 2011, New York, NY
Anne Hallett (DJG), 29 April 2009, Chicago
Charles Halpern (DJG), 5 December 2011; 6 December 2011 (E), Berkeley, CA
Debbie Halvorson (DJG), 5 May 2011, Homewood, IL
Charles V. Hamilton (DJG), 14 March 2013, Chicago (T)
Ian Haney-Lopez (DJG), 24 December 2011, Waialua, HI
Gary Hannig (DJG), 15 March 2014, Litchfield, IL (T)
Don Harmon (DJG), 9 May 2011, Springfield, IL
(Frank) Hill Harper (DJG), 14 December 2011, Studio City, CA
Monica Harris (DJG), 25 October 2016, Bigfork, MT (T)
Ray Harris (DJG), 20 May 2010, Chicago
Paul Harstad (DJG), 2 April 2014 (T); 22 April 2014, Boulder, CO
Brett Hart (DJG), 15 April 2014, Chicago (T)
Chester Hartman (DJG), 5 April 2012, Washington, DC (E)
Hermene Hartman (DJG), 30 March 2010, 12 April 2010, 4 November 2010 (E); 14 November 2011, 21 June 2013, 8 August 2013, Chicago
Carol Harwell (DJG), 28 April 2009, Chicago
Linda Hawker (DJG), 10 May 2011, Springfield, IL
Derek Hawkins (DJG), 28 October 2010, Newark, NJ (E)
Carl Hawkinson (DJG), 19 November 2010, Galesburg, IL
Lisa Hay (DJG), 19 March 2010 (E); 5 May 2010, Portland, OR
Tom Hazen (DJG), 14 November 2010, Evanston, IL
Joann Hebing (DJG), 30 December 2011, Mill City, OR
Mark Hebing (DJG), 30 December 2011; 31 December 2011, 1 January 2012, 30 April 2013 (E), Mill City, OR
Philip Hebing (DJG), 30 December 2011, Mill City, OR
Steven Heinen (DJG), 19 February 2013, Tulsa, OK (T)

Stephen Heintz (DJG), 16 September 2011, 5 September 2013, New York, NY
Beverly Helm-Renfro (DJG), 29 May 2010, Springfield, IL
Beverly Helm-Renfro (Mark DePue), 13 July 2010, Springfield, IL (ALPL)
Sondra A. Hemeryck (DJG), 1 November 2010; 1 November 2010 (E), Chicago
Rickey Hendon (Terry Martin), 13 January 2009, Springfield (Illinois Channel)
Rickey Hendon (Peter Slen), 12 June 2010, Chicago (C-Span Book TV)
Carter Hendren (Mark DePue), 28 April 2009, Springfield, IL (ALPL)
Carter Hendren (DJG), 8 November 2011, Springfield, IL
Sim Heninger (DJG), 22 October 2010; 12 and 13 January 2011, 14 (2) May 2011 (E), Durham, NC
Lisa (Shachtman) Hennessey (DJG), 29 January 2010, New York, NY
Mike Henry (DJG), 28 April 2011, Washington, DC
Paul Herrmannsfeldt (DJG), 16 and 21 May 2010, Santa Fe, NM (E)
Eileen Hershenov (DJG), 25 January 2010, Yonkers, NY
Marcie Hershman (DJG), 14 July 2015, Brookline, MA (T)
Bethann Hester (DJG), 25 June 2013, Chicago
Clyde Higa (DJG), 25 February 2013, Honolulu (E)
Susan M. Higgins (DJG), 24 October 2010, Charlotte, NC
Tracy Higgins (DJG), 14 October 2010, New York, NY
David Hill (DJG), 20 October 2010, Arlington, VA
Jay Hines-Shah (DJG), 3 March 2014 (E); 24 March 2014 (T), Chicago
Jeff Hoberman (DJG), 14 March 2014 (E); 20 March 2014 (T), San Paulo, Brazil
Glenn Hodas (DJG), 10 July 2012, Springfield, IL
Mike Hoffmann (DJG), 18 November 2010, Springfield, IL
Larry Hogue (DJG), 9 January 2010, San Diego, CA
Barbara Holt (DJG), 11 November 2011; 12 November 2011 (E), Chicago
James D. Holzhauer (DJG), 2 January 2010, Chicago (E)
John Hooker (DJG), 12 August 2013, Chicago
Elaine Horn (DJG), 27 March 2014, Washington, DC (T)
George M. Houser (Lisa Brock), 19 July 2004, Rockland County, NY (NoEasyVictories.org)
Desta Houston (DJG), 26 May 2010, Chicago
Jim Howard (DJG), 24 July 2013; 25 July 2013 (E), Washington, DC
Anne Howells (DJG), 22 April 2009, Seattle
Josh Hoyt (DJG), 9 November 2010, Chicago
Yi-Fun Hsueh (DJG), 23 February 2013, Bethesda, MD (T)
Craig Huffman (DJG), 25 July 2012, Chicago
Blair Hull (DJG), 7 May 2014 (T); 9 May 2014 (E), Chicago
Laura Russell Hunter (DJG), 24 July 2013, Washington, DC
Mattie Hunter (DJG), 9 November 2011, Springfield, IL
Chuck Hurley (DJG), 7 March 2014, Annapolis, MD (T)
Dennis Hutchinson (DJG), 4 February 2009, 9 (2) April 2014, Chicago (E)
Lee Hwang (DJG), 7 December 2011; 17 February 2012 (T), San Francisco
Dan Hynes (DJG), 7 May 2009, Chicago
James Iams (DJG), 22 December 2011 (E); 22 December 2011 (T), Honolulu
Scott Inoue (DJG), 30 October 2014, Stockton, CA (E)
Michael Isbell (DJG), 4 October 2010, New York, NY (E)
Adrienne Bitoy Jackson (DJG), 10 September 2009; 12 and 13 September 2009 (E), Chicago
Jesse Jackson Jr. (David Axelrod), 17 February 2004, Chicago (JKP and TPP)
Jesse Jackson Jr. (DJG), 1 May 2011, Chicago
Denny Jacobs (DJG), 16 November 2010, Springfield, IL
Sharon Jacobson (DJG), 5 November 2011, Houston, TX (T)
Sheila Miyoshi Jager (DJG), 6 (2), 7 (3), 8 (2), and 9 (2) August 2009, 8 (2), 9, 12, and 13 April 2010, 7 September 2011, 9 (2), 10, and 22 January 2012, 1 (3), 3, and 30 May 2012, 20 (4), 21 (2), 24, and 28 (2) June 2012, 28 July 2012, 9 and 29 August 2012, 12, 13, and 14 September 2012, 7 (2) November 2012, 3 and 4 December 2012, 29 (4) May 2013, 17, 20, and 21 August 2013, 23 March 2014, 30 and 31 August 2014, 12 (2) and 13 (2) September 2014, 10 (7) and 14 November 2014, 20 (2), 21 (2), 22 (2), and 23 January 2015, 7, 8 (2), 9, 11 (3), 12 (2), and 17 February 2015, Oberlin, OH, and Seoul, South Korea (E)
David James (DJG), 28 November 2011 (E); 7 March 2014 (T), Los Angeles
Ken Jania (DJG), 15 November 2010, Hammond, IN
Valerie Jarrett (Robert Draper), 2009, Washington, DC

Valerie Jarrett (Jim Gilmore), 20 August 2012, Washington, DC (*Frontline*)
Laura Jehl (DJG), 18 October 2010, Washington, DC
Alan Jenkins (DJG), 6 September 2013, New York, NY
Tim Jessup (DJG), 18 and 19 February 2015 (E), 21 February 2015 (T), 3 March 2015 (E), Jakarta, Indonesia
Dave Joens (DJG), 9 May 2011, Springfield, IL
Al Johnson (Betty Brown-Chappell), 14 December 1989, Chicago (BBCP)
Cheryl Johnson (DJG), 19 September 2009, Chicago
Don Johnson (DJG), 29 June 2013, Chicago
Hazel Johnson (DJG), 19 September 2009, Chicago
Jo Johnson (DJG), 13 May 2011, Springfield, IL
Thomas E. Johnson (DJG), 13 September 2009, Chicago
Dan Johnson-Weinberger (DJG), 10 May 2011, Springfield, IL
Kevin Jokisch (DJG), 11 (2) and 12 (3) February 2015, St. Louis, MO (E)
Brad Jonas (DJG), 6 July 2015, Chicago (T)
Alma Jones (DJG), 13 January 2010, Chicago
Emil Jones Jr. (Julieanna Richardson), 6 and 18 January 2000, Chicago (TheHistoryMakers.com)
Emil Jones Jr. (Bernard Sieracki), 23, February 2006, Springfield, IL (IllinoisChannel.org)
Emil Jones Jr. (DJG), 1 May 2009, Chicago
John Paul Jones (DJG), 26 May 2010, Chicago
Linzey Jones (DJG), 6 May 2011, Bridgeview, IL
Matt Jones (DJG), 18 November 2010, Springfield, IL
Nancy Jones (DJG), 18 May 2010, Chicago (T)
Ben Joravsky (Betty Brown-Chappell), 6 December 1989, Chicago (BBCP)
Gwendolyn M. Jordan (DJG), 2 June 2015, Washington, DC (T)
James V. Jordan (DJG), 19 September 2009, Chicago
Tom Joyce (DJG), 17 May 2010, Chicago
Asad Jumabhoy (DJG), 11 January 2011, London, UK
Scott Kaiser (DJG), 7 November 2011, Springfield, IL
Keith Kakugawa (DJG), 17 May 2009, Sacramento, CA (T)
Keith Kakugawa (Jack McAdoo), 28 and 29 October 2010, 4 and 10 November 2010, Sacramento (JMacRadio.com)
Tom Kaminski (DJG), 11 September 2009, Chicago
Nancy Kao (DJG), 25 April 2011, Washington, DC (E)
Sokoni Karanja (Larry Crowe), 7 January 2005, Chicago (TheHistoryMakers.com)
Sokoni Karanja (DJG), 6 May 2009, Chicago
Doris Karpiel (DJG), 14 March 2013, Carol Stream, IL (T)
Marilyn Katz (DJG), 7 May 2011, Chicago
Jan-Michele Lemon Kearney (DJG), 15 February 2013, Cincinnati, OH (T)
Joan F. Keleher (DJG), 24 February 2015 and 28 May 2015, Wilmette, IL (T)
Keith Kelleher (DJG), 7 May 2011; 8 and 9 June 2012 (E), Chicago
Alison Kelley (DJG), 27 January 2010, Brooklyn, NY
Jerry Kellman (Jim Gilmore), 24 July 2008, Chicago (*Frontline*)
Jerry Kellman (DJG), 17 (2) and 24 March 2009 (E); 3 April 2009; 6 April and 13 May 2009, 3 September 2009, 8 December 2009, 30 March 2010 (E); 15 September 2011 (T); 22 and 23 August 2014, 28 and 29 September 2014 (E), Winnetka, IL
John Kelly (DJG), 7 June 2011, Chicago
Mark Kende (DJG), 18 May 2010, Des Moines, Iowa (T)
Randall Kennedy (DJG), 23 November 2010, Cambridge, MA
Scott Kennedy (DJG), 25 July 2012, Chicago
Janis Kestenbaum (DJG), 26 April 2011, Washington, DC
Amy F. Kett (DJG), 28 April 2011, Washington, DC
Mona Khalidi (DJG), 6 September 2013, New York, NY
Joe Khan (DJG), 2 and 11 March 2014 (E); 24 March 2014 (T), Philadelphia
Julius G. Kiano (Harry Kreisler), 14 September 1989, Berkeley, CA
Lauren Kidwell (DJG), 23 July 2013, Washington, DC
Al Kindle (DJG), 28 April 2009, Chicago
David Kindler (DJG), 2 April 2009; 11 and 12 February 2015 (E), Oak Park, IL
Martha Kindred (DJG), 24 May 2010, Chicago
Jonathan Z. King (DJG), 26 February 2013, New York, NY (T)
Kraig King (DJG), 22 November 2014, Minneapolis (T)
Martin "Marty" King (DJG), 1 May 2011, Chicago
Tim King (DJG), 10 November 2011, Chicago
Ben Kiningham (DJG), 14 March 2003, Petersburg, IL (E)
Kathy Kish (DJG), 26 June 2013, Crete, IL (T)

Lauren Ezrol Klein (DJG), 1 March 2013, New York, NY (T)
Robert Klonowski (DJG), 4 April 2009, Chicago
Michael Klonsky (Mark Larson), July 2014, Chicago (AmericanStoriesContinuum.com)
Diana Mitsu Klos (DJG), 18 May 2010, Reston, VA (T)
Eddie L. Knox (DJG), 7 February 2015, Chicago (T)
Thomas Knutson (DJG), 14 January 2010, Harvey, IL
Jeffrey W. Kobrick (DJG), 21 and 24 May 2013 (E); 31 May 2013 (T), San Mateo, CA
Alex Kotlowitz (DJG), 15 June 2015, Oak Park, IL (E)
Madhuri Kommareddi (DJG), 6 September 2013, New York, NY
Mark Kozlowski (DJG), 26 January 2010; 26 February 2010 (E), New York, NY
Tom Krause (DJG), 20 October 2010, Falls Church, VA
Raja Krishnamoorthi (DJG), 25 June 2013, Chicago
Mike Kruglik (DJG), 1 April 2009, 5 May 2009; 12 May 2009 (E); 16 January 2010, 1 June 2010, 10 March 2015, Chicago and Phoenix, AZ
Mike Kulisheck (DJG), 22 April 2014, Boulder, CO
John Kupper (DJG), 30 June 2013; 10 July 2013 (E); 5 August 2013; 1 September 2014 (E), Wilmette, IL
Paula Miyashiro Kurashige (DJG), 23 December 2011 (E); 24 December 2011, Honolulu
Kitty Kurth (DJG), 11 November 2011, Chicago
Eric Kusunoki (DJG), 19 December 2011, Honolulu
Phil Lackman (DJG), 28 May 2010, Springfield, IL
Nicole Lamb-Hale (DJG), 11 September 2014, Washington, DC (T)
Leonard Lamkin (DJG), 8 July 2012 (E); 13 July 2012 (T), Evanston, IL
Bethany A. Lampland (DJG), 17 March 2014, New York, NY (T)
Ian Larkin (DJG), 21 June 2013, Chicago
Clinton Latimore Jr. (DJG), 12 April 2014, Chicago (T)
Mike Lawrence (Mark R. DePue), 2 and 3 July 2009, Carbondale, IL (ALPL)
Mike Lawrence (DJG), 29 May 2010, Springfield, IL
Cathy Lazere (DJG), 29 January 2010; 30 January 2010, 10 (3) and 11 February 2010, 7 and 8 December 2010, 18 (2) and 26 September 2012 (E), New York, NY
Howard Learner (DJG), 16 July 2013, Chicago (T)
Tracy Leary (DJG), 29 January 2010, Long Island City, NY
Fred Lebed (DJG), 3 May 2011; 3 May 2011 (E), Chicago
David Le Breton (DJG), 25 July 2012, Chicago
Christine M. Lee (DJG), 22 October 2010, Roanoke, VA
Daniel Lee (DJG), 15 January 2010, Chicago
Deborah Leff (DJG), 20 January 2010, Washington, DC
Sean Lev (DJG), 23 June 2011, Washington, DC
Lois Leveen (DJG), 19 March 2014, Portland, OR (T)
John Levi (DJG), 12 January 2010; 3 February 2010, 22 March 2010, 17 July 2013 (E), Chicago
Deneese Walia Levin (DJG), 20 March 2014, Chicago (T)
Beth Noymer Levine (DJG), 14 May 2010 (T); 7 November 2010, 5 and 8 December 2010 (E), Salt Lake City
Saul Levmore (DJG), 6 January 2016, Chicago (E)
Susan (Pierson) Lewers (DJG), 11 April 2014, Chicago (T)
Ted Liazos (DJG), 11 April 2014, Washington, DC (T)
Carol Platt Liebau (DJG), 14 March 2014, New Canaan, CT (T)
Mike Lieteau (DJG), 18 November 2010, Springfield, IL
Kimberly Lightford (Adele Hodge), 1 September 2000, Chicago (TheHistoryMakers.com)
Kimberly Lightford (DJG), 16 November 2010, Springfield, IL
Kevin Limbeck (DJG), 11 May 2010, Chicago
Daniel Lin (DJG), 3 March 2014, San Francisco (E)
Tom Lindenfeld (DJG), 5 February 2014; 7 April 2014 (E), Washington, DC
Tom Lininger (DJG), 25 (2) December 2011, Eugene, OR (E)
Terry Link (DJG), 17 November 2010, Springfield, IL
Kevin G. Little (DJG), 12 February 2013, Fresno, CA (T)
Yvonne Lloyd (DJG), 19 September 2009, Nashville, TN (T)
Tondra L. Loder-Jackson (DJG), 5 February 2013, Birmingham, AL (E)
Dario Longhi (DJG), 12 August 2009, Olympia, WA (T)
Richard C. Longworth (DJG), 22 February 2010, Chicago (E)

Ann F. Lopez (DJG), 11 February 2010, Boston (E)
Lourdes Lopez-Isa (DJG), 20 October 2010, Arlington, VA
John R. Lott Jr. (DJG), 8 March 2014, Burke, VA (T)
Ronald Loui (Jack McAdoo), 4 and 10 November 2010, Cleveland, OH (JMacRadio.com)
Debbie Lounsberry (DJG), 10 May 2011, Springfield, IL
Alvin Love (DJG), 2 April 2009, Chicago
Ruth Love (Felicia Ratliff Stewart), 6 January 1995, Chicago
Bernard Loyd (DJG), 6 May 2011, Chicago
Dave Luechtefeld (DJG), 11 May 2011, Springfield, IL
William Luking (DJG), 2 May 2014, Springfield, IL (T)
Alan Lum (DJG), 19 December 2011, Honolulu
Beatrice Lumpkin (DJG), 24 May 2010, Chicago
Lyndell Luster (DJG), 25 March 2014, Chicago (T)
Freddrenna Lyle (DJG), 15 March 2013, Chicago
Roberta Lynch (DJG), 9 September 2009; 23 February 2010, 10 August 2010 (E), Chicago
Scott Lynch-Giddings (DJG), 9 August 2013; 11 August 2013 (E), Chicago
Harvey Lyon (DJG), 26 June 2013, Beverly Shores, IN (T)
Kelly Jo MacArthur (DJG), 6 January 2012, Seattle
Kenneth Mack (DJG), 21 January 2010, Washington, DC
Kenneth Mack (Jim Gilmore), 13 June 2012, Cambridge, MA (*Frontline*)
Lisa Madigan (DJG), 1 June 2010, Chicago
Michael J. Madigan (Robert V. Remini), 10 August 2009, Chicago (UIC)
Mahboob "Beenu" Mahmood (DJG), 9 August 2010 (T); 24 and 26 September 2011 (E), Islamabad, Pakistan, and Singapore
Robert Mahnke (DJG), 12 March 2014, San Jose, CA (T)
Leo Mahon (Greg Sakowicz), 25 January 2008, Chicago (Catholic Community of Faith)
Leo Mahon (DJG), 22 May 2010, Chicago
Darwin Malloy (DJG), 9 December 2011, San Jose, CA
Ed Maloney (DJG), 26 June 2013, Chicago (T)
Steven Mange (DJG), 18 November 2010, Springfield, IL
Dave Manning (DJG), 16 November 2010, Springfield, IL
Samuel Yaw "Kofi" Manu (DJG), 24 March 2009, Cincinnati, OH (T)
Adelmo Marchiori (DJG), 13 March 2013, Edwardsville, IL (T)
Michael Marrs (DJG), 12 July 2012; 13 July 2012 (E), Chicago
Lawrence Marshall (DJG), 27 October 2011, Chicago (E)
Karla Martin (DJG), 25 April 2013, San Francisco
R. Eden Martin (DJG), 2 July 2013, Chicago
Iris Martinez (DJG), 12 May 2011, Springfield, IL
Luz Martinez (DJG), 2 and 4 February 2013, Quezon City, Philippines (E)
Peter Martinez (DJG), 10 May 2010, Chicago
Mark S. Martins (DJG), 22 February 2013, Alexandria, VA (T)
Connie Martinson (Robert Draper), 2009, Los Angeles
Paul Massari (DJG), 14 March 2014, Cambridge, MA (T)
Chris Mather (DJG), 21 June 2013, Chicago
Ian Mattoch (DJG), 8 December 2011 (E); 21 December 2011, Honolulu
Darin McAtee (DJG), 14 October 2010, New York, NY
Amy (Christian) McCormick (DJG), 12 February 2013 (E); 16 February 2013 (T), East Lansing, MI
Rod McCulloch (DJG), 2 July 2013, Chicago
Nancy McCullough (DJG), 6 January 2010, Los Angeles (T)
Michael McGann (DJG), 24 June 2013, Pittsfield, IL (T)
John McKnight (DJG), 5 August 1983, 20 February 1984; 5 April 2009 (T); 14 November 2010, Evanston, IL
Chris McLachlin (DJG), 23 December 2011, Honolulu
Joe McLean (DJG), 2 (2) October 2013 (E); 10 October 2013; 11 October 2013 (E), Washington, DC
William McNary (DJG), 17 November 2010, Springfield, IL
Alex McNear (DJG), 19 May 2010, 30 (2) July 2010 (E); 12 October 2010; 13 October 2010 (E); 17 September 2011; 18 November 2011 (E); 20 November 2011 (T); 20 and 21 November 2011, 27 January 2012, 8 April 2012, 1 May 2012, 19 and 20 June 2012 (E), 17 (T), 20 and 30 November 2014, 1 and 2 December 2014 (E), Sag Harbor and New York, NY
James Meeks (David Axelrod), 30 September 2003, Chicago (JKP)
Salil Mehra (DJG), 18 March 2014, Philadelphia (T)

David Melton (DJG), 26 March 2014, Chicago (T)
David Mendell (Jim Gilmore), 19 June 2008, Chicago (*Frontline*)
Peggy Mendelow (DJG), 20 March 2013, New York, NY (T)
Chris Meyer (DJG), 25 January 2010, Yonkers, NY
Rob Mier (Betty Brown-Chappell), 17 November 1989, Chicago (BBCP)
Margot Mifflin (DJG), 18, 19 (4), and 20 January 2010, 7 February 2010, 26 September 2010 (E); 16 September 2011; 18 September 2011, 5, 7 and 8 (2) May 2012, 29 June 2012, 28 (2) and 29 September 2012, 10, 11, and 12 January 2013, 2 April 2013, 22 (3) November 2014, 12 (2) and 13 (2) December 2016 (E), New York, NY
Abner Mikva (Stephen J. Pollak), 23 May 1996–19 May 1999, Washington, DC (DCCHS.org)
Abner Mikva (Harry Kreisler), 12 April 1999, Berkeley, CA (IIS, UCB)
Abner Mikva (DJG), 18 September 2009, Chicago
Abner Mikva (Mark DePue), 22 October 2014, Chicago (ALPL)
Amanda Fuchs Miller (DJG), 11 October 2013, Washington, DC
Arnie Miller (DJG), 2 February 2010; 2 February 2010 (E), Newton, MA
Cynthia K. Miller (DJG), 15 September 2009, Chicago
John Millner (DJG), 11 May 2011, Springfield, IL
Linda Mills (DJG), 28 December 2013 (E); 30 December 2013 (T), Chicago
Judson Miner (Betty Brown-Chappell), 22 February 1990, Chicago (BBCP)
Judson Miner (DJG), 5 May 2009; 18 April 2014 (E), Chicago
Chin See Ming (DJG), 5 May 2010, Portland, OR
Martha Minow (Liza Mundy), 3 May 2007, Cambridge, MA (T) (LMP).
Martha Minow (DJG), 12 January 2010 (E); 2 February 2010, Cambridge, MA
Newton Minow (Jim Gilmore), 19 June 2008, Chicago (*Frontline*)
Newton Minow (DJG), 12 January 2010, Chicago
Ned Mitchell (DJG), 14 March 2013, Sesser, IL (T)
Rob Mitchell (DJG), 11 July 2015, Concord, MA (T)
Kari Moe (Betty Brown-Chappell), 3 January 1990, Chicago (BBCP)
Robert Molaro (DJG), 10 May 2011, Springfield, IL
Jon Molot (DJG), 27 April 2011, Washington, DC
Mary Ellen Montes (DJG), 20 May 2010, "Arthurville, IL"
Alex Montgomery (DJG), 12 March 2004, Chicago (T)
James Montgomery (David Axelrod), 28 January 2004, Chicago (JKP and TPP)
Don Moore (DJG), 1 June 2010, Chicago
Eric Moore (DJG), 6 April 2009; 29 July 2009 (T); 30 March 2010 (E), Los Angeles
Eric Moore (Jim Gilmore), 29 June 2012, Los Angeles (*Frontline*)
Robert Moriarty (DJG), 19 May 2010, Chicago
John Morillo (DJG), 31 August 2010 (E); 22 October 2010; 28 October 2010 (E), Raleigh, NC
Eugene Morris (DJG), 13 August 2013, Chicago (T)
Jerry Morrison (DJG), 16 May 2011, Chicago
Bill Moyers (DJG), 22 January 2010, New York, NY (E)
Tom Moyes (DJG), 8 January 2010, Temecula, CA (T)
J. Herbert Muhammad (Sam Pollard), 4 June 1989, Chicago (Blackside)
Katrina (Emmons) Mulligan (DJG), 21 July 2013, Alexandria, VA
Phil Mullins (DJG), 10 September 2009, Chicago
George Munoz (DJG), 18 February 2013, Arlington, VA (E)
Glee Murray (DJG), 6 October 2014, Washington, DC (T)
Judith Myers (DJG), 14 March 2013, Danville, IL (T)
Verna Myers (DJG), 23 November 2010, Cambridge, MA
David Nahmias (DJG), 9 December 2010, Atlanta (T)
David Narefsky (DJG), 24 March 2014, Chicago (T)
Greg Naylor (DJG), 9 April 2014, Philadelphia (T)
Kim (Jones) Nelson (DJG), 15 March 2013, Minneapolis (T)
Marty Nesbitt (Robert Draper), 2009, Chicago
Felton Newell (DJG), 18 March 2014, Los Angeles (T)
Eric Newhall (DJG), 8 April 2009, Los Angeles
Sandy Newman (DJG), 22 January 2010, Washington, DC
John Nicolay (DJG), 8 November 2011, Springfield, IL

Bonnie Nitsche (DJG), 13 January 2010, Mokena, IL

Walter Nitsche (DJG), 13 January 2010, Mokena, IL

Lars Noah (DJG), 13 February 2013, Gainesville, FL (E)

Jane (Catler) Nober (DJG), 20 February 2013, Ft. Worth, TX (T)

Sean Noble (DJG), 1 July 2013, Chicago

Duane Noland (DJG), 18 March 2014, Blue Mound, IL (T)

Laura (Mullens) Nolen (DJG), 1 May 2014, Houston, TX (T)

Rolf Nordahl (DJG), 18 December 2011, Honolulu

Cynthia Norris (DJG), 1 April 2014, Silver Spring, MD (T)

Trent Norris (DJG), 2 January 2013, San Francisco

Courtney Nottage (DJG), 17 November 2010, Springfield, IL

Patricia Novick (DJG), 6 May 2011, Chicago

Juleigh Nowinski (DJG), 8 August 2013, Chicago

Barack Obama (Zeke Gonzalez), 27 October 1993, Chicago (*Becoming Barack: Evolution of a Leader*)

Barack Obama (Bill Thompson), 9 August 1995, Washington, DC (eyeonbooks.com)

Barack Obama (Connie Martinson), 11 August 1995, Van Nuys, CA (*Connie Martinson Talks Books,* United Artists Cable Television)

Barack Obama (Marc Strassman), 11 August 1995, Studio City, CA (Book Channel)

Barack Obama (Jack Van Der Slik), 3 December 1998, Springfield, IL (*Inside Illinois Government,* University of Illinois at Springfield)

Barack Obama (Edward McClelland), 4 February 2000, Chicago

Barack Obama, (Julieanna Richardson), 16 February 2001, Chicago (TheHistoryMakers.com)

Barack Obama (Jeff Berkowitz), 27 June 2002, 25 November 2002, 24 July 2003, and 18 July 2004, Chicago (*Public Affairs with Jeff Berkowitz*)

Barack Obama (Chinta Strausberg), 1 July 2002, Chicago (*The Strausberg Report*)

Barack Obama (Bruce Dumont), 17 August 2003, Chicago (*Beyond the Beltway,* MBC)

Barack Obama (David Axelrod), 2 October 2003, Chicago (JKP)

Barack Obama (Terry Martin), 6 November 2003, Springfield, IL (IllinoisChannel.org)

Barack Obama (David Axelrod), 7 February 2004, Chicago (JKP and TPP)

Barack Obama (Cathleen Falsani), 27 March 2004, Chicago (Beliefnet.com)

Barack Obama (Dave Davies), 10 August 2004, Philadelphia (Fresh Air, *NPR,* 12 August 2004)

Barack Obama (Jacob Weisberg), 29 June 2006, Washington, DC

Barack Obama (Anthony Mason), 17 August 2011, Washington, DC (CBSNews.com)

Barack Obama (David Blum), 30 July 2013, Chattanooga, TN (Amazon.com)

Michelle Obama (John Kupper), 6 February 2004, Chicago (JKP)

Michelle Obama (Jacob Weisberg), July 2006, Chicago (T)

Michelle Obama (Robert Draper), 2009, Washington, DC

Patrick O'Brien (DJG), 9 December 2011, Redwood Shores, CA

Maurice Obstfeld (DJG), 19 March 2015 and 4 April 2015, Washington, DC (E)

Margaret O'Dell (DJG), 28 June 2013, Chicago

Nia Odeoti-Hassan (DJG), 12 May 2011, Springfield, IL

Sandy O'Donnell (DJG), 27 April 2009 (2), Chicago (E)

Ursula Dudley Oglesby (DJG), 22 October 2010, High Point, NC

Chad Oldfather (DJG), 26 May 2010; 27 May 2010 (E), Chicago

Zeituni Onyango (DJG), 29 April 2010; 30 June 2011, 7 and 20 July 2011, 3 September 2011, 5 and 6 February 2012, 17 April 2012, 15, 18, 22, and 27 May 2012, 8 and 19 June 2012 (E), South Boston

Bruce Orenstein (DJG), 13 May 2010; 10 March 2015 (T), Chicago and Durham, NC

Peter Osnos (Robert Draper), 2009, New York, NY

Peter Osnos (DJG), 2 June 2015, New York, NY (E)

Andrea Dolan Owen (DJG), 12 May 2010 (E); 12 May 2010 (T), Coto de Caza, CA

John Owens (DJG), 30 April 2009; 7 December 2009 (E); 1 June 2010; 24 August 2010 (E), Chicago

Martha Pacold (DJG), 19 March 2014, Chicago (T)

Kaleshia Page (DJG), 10 November 2010, Chicago

Lisa Paget-Kahn (DJG), 26 February 2013, Chappaqua, NY (T)

Alice Palmer (Julieanna Richardson), 9 June 2000, Chicago (TheHistoryMakers.com)

Alice Palmer (Kate McAuliff), 29 March 2010, Chicago. (CAAMC, Columbia College)

Alice Palmer (DJG), 17 May 2011, Chicago
Lu Palmer (Davis Lacy Jr.), 14 April 1989, Chicago (Blackside)
Michael Parham (DJG), 6 January 2012, Seattle
Kathleen Parker (DJG), 24 July 2012, Northbrook, IL
Richard Parker (DJG), 4 March 2013, Little Compton, RI (T)
John Parry (DJG), 6 May 2010; 6 and 10 May 2010 (E), Portland, OR
Tyrone Partee (DJG), 4 November 2011, Hawthorne, CA (T)
Keith Patchel (DJG), 4 September 2013; 5 September 2013 (E), New York, NY
Carl Patten (DJG), 5 March 2014, Jacksonville, FL (T)
Ann Patty (DJG), 10 March 2014, Rhinebeck, NY (T)
Charles T. Payne (DJG), 7 June 2011, Chicago
Richard A. Pearson (DJG), 7 November 2011, Springfield, IL
Eric Pelander (DJG), 8 February 2013, Cleveland, OH (T)
Josh Pemstein (DJG), 7 March 2014, Boston (T)
Aurie Pennick (DJG), 3 May 2010 (E); 10 May 2010, Chicago
Deno Perdiou (DJG), 30 April 2014, Springfield, IL (T)
Bill Perkins (DJG), 19 May 2010, Chicago
Stephen Perkins (DJG), 2 May 2011, Chicago
Barry Perlstein (DJG), 12 December 2011, Santa Monica, CA
Tom Perrelli (DJG), 27 April 2011, Washington, DC
Howard Peters (Mark DePue), 25 November 2009, Springfield, IL (ALPL)
Howard Peters (DJG), 28 May 2010, Springfield, IL
Keith M. Peterson (DJG), 2 May 2010, Denver, CO
Tony Peterson (DJG), 27 October 2010, Nashville, TN
William "Bill" Peterson (DJG), 18 March 2014, Buffalo Grove, IL (T)
Ed Petka (DJG), 12 November 2011, Chicago
Alice Peurala (Elizabeth Balanoff), 30 September 1977, Chicago (Roosevelt University Labor Oral History Project)
Michael Pfleger (David Axelrod), 28 January 2004, Chicago (TPP)
Michael Pfleger (DJG), 9 November 2010, Chicago
Earl Martin Phalen (DJG), 18 April 2011, Boston
James "Pate" Philip (DJG), 12 November 2011, Chicago
James "Pate" Philip (Mark DePue), 11 July 2012, Springfield, IL (ALPA)
Terrie Pickerill (DJG), 14 April 2014, Chicago (T)
Greg Pierce (DJG), 17 April 2013, Chicago (E)
Matt Piers (Betty Brown-Chappell), 1 November 1989, Chicago (BBCP)
Matt Piers (DJG), 11 July 2012, Chicago
Bernard Pietrzak (DJG), 24 July 2012, Barrington, IL
Nat Piggee (DJG), 25 March 2014 (T); 27 March 2014 (E), Chicago
Richard Pildes (DJG), 26 January 2010, New York, NY
Alberta Pillows (DJG), 17 September 2009, Chicago (T)
Eugene Pincham (David Axelrod), 28 January 2004, Chicago, IL (TPP)
Steuart Pittman (DJG), 6 October 2011, Davidsonville, MD (T)
Gregory L. Poe (DJG), 5 March 2013 (T); 15 April 2013 (E), Washington, DC
Richard Poethig (DJG), 11 November 2010, Chicago
Jan Poledziewski (DJG), 4 September 2009 (E); 8 September 2009 (T); 9 (2) September 2009, 3 and 5 November 2009 (E), Munster, IN
Wendy Pollack (DJG), 2 May 2014, Chicago (T)
Jef Pollock (DJG), 3 September 2013, New York, NY
Rachael Pontikes (DJG), 10 March 2014, Chicago (T)
Julian Posada (DJG), 12 March 2013, Chicago
Jeremiah Posedel (DJG), 3 May 2011, Chicago
Glenn Poshard (DJG), 16 November 2010, Springfield, IL
Eric Posner (DJG), 12 May 2010, Chicago
Alison Powell (DJG), 14 January 2010, Country Club Hills, IL
Ernest Powell Jr. (DJG), 14 January 2010, Country Club Hills, IL
Shirley Powell (DJG), 14 January 2010, Country Club Hills, IL
Steve Preckwinkle (DJG), 7 November 2011, Springfield, IL
Toni Preckwinkle (DJG), 5 May 2009, Chicago
Alonzo C. Pruitt (DJG), 10 and 13 (2) March 2014, Richmond, VA (E)
Preston Pugh (DJG), 1 July 2013, Chicago
Stephen Pugh (DJG), 1 July 2013
Paula Purdue (DJG), 9 May 2011, Springfield, IL
Sue Purrington (DJG), 16 August 2013, Chicago (T)
Jonathan Putnam (DJG), 11 March 2014, Larchmont, NY (T)

Sheila Quinlan (DJG), 6 December 2011, San Francisco (T)
Dan Rabinovitz (DJG), 3 February 2010; 11 August 2010, 10 August 2014, 5 April 2015 (E), Cambridge, MA
Al Raby (DJG), 8 March 1984, Washington, DC
Christine Radogno (DJG), 12 May 2011, Springfield, IL
Mike Ragen (DJG), 6 March 2014 (E); 11 March 2014 (T), Springfield, IL
Edith Ramirez (DJG), 26 April 2011, Washington, DC
Jorge Ramirez (DJG), 15 February 2013, Lubbock, TX (T)
Gregory J. Ramos (DJG), 30 April 2010, Denver, CO
Michael J. Ramos (DJG), 30 April 2010, Denver, CO
Jane Ramsey (DJG), 25 July 2012, Chicago
Linda Randle (DJG), 28 April 2009, Chicago
Barbara Ransby (DJG), 8 December 2014, Chicago (E)
Susan Shadow Ransone (DJG), 18 March 2014, 7 April 2014 (T); 8 April 2014 (E), Tampa, FL
Radhika Rao (DJG), 25 April 2013, San Francisco
Miles Rapoport (DJG), 14 October 2010, New York, NY
Geoffrey Rapp (DJG), 9 September 2013, Toledo, OH (E)
Morris Ratner (DJG), 25 April 2013, San Francisco
Steve Rauschenberger (DJG), 16 May 2011, Elgin, IL
Chuck Rawlings (DJG), 31 August 2010, San Jose, CA (E)
Jim Rea (DJG), 15 March 2013, Christopher, IL (T)
Mark Real (DJG), 22 August 2013, Columbus, OH
Adolph Reed Jr. (DJG), 2 February 2009, Philadelphia (E)
Jim Reed (DJG), 7 November 2011, Springfield, IL
Thomas Reed (DJG), 23 July 2013, Washington, DC
Alan Reger (DJG), 13 November 2015, Chantilly, VA (T)
Mitchell Reich (DJG), 18 April 2011; 20 April 2011 (E), Cambridge, MA
Jessica Reighard (Robert Draper), 2009, Baltimore, MD
Boro Reljic (DJG), 16 November 2010, Springfield, IL
Ben Reyes (DJG), 9 and 13 July 2012, Chicago (E)
Tynia Richard (DJG), 30 January 2010, Brooklyn, NY
Maury Richards (DJG), 13 January 2010; 1 April 2010 (E), Chicago
Rick Ridder (DJG), 21 April 2014, Denver, CO
Vicky Rideout (DJG), 3 December 2011, San Francisco
Dale Righter (DJG), 13 May 2011, Springfield, IL
Diane Ring (DJG), 19 April 2011, Newton, MA
Michael Risch (DJG), 2 (2) March 2014 (E), Villanova, PA
Rebecca Rivera (DJG), 24 January 2010 (E); 1 February 2010, Cambridge, MA
Charlie Robb (DJG), 14 March 2013, Menlo Park, CA (T)
Janis Robinson (DJG), 10 November 2010, Chicago
Mary Beth Roche (Robert Draper), 2009, New York, NY
Jillayne Rock (DJG), 7 November 2011, Springfield, IL
Byron Rodriguez (DJG), 6 March 2014, San Francisco, CA (T)
Gwendolyn (LaRoche) Rogers (DJG), 3 November 2010, Chicago
Joel Rogers (DJG), 16 July 2012, Madison, WI
John W. Rogers Jr. (Julieanna Richardson), 13 June and 17 July 2002, Chicago (TheHistoryMakers.com)
John W. Rogers Jr. (DJG), 7 June 2011, Chicago
Ken Rolling (DJG), 7 May 2009, 10 November 2010; 16 November 2012 (E); 21 June 2013; 11 July 2013, 2 September 2014, 28 May 2015 (E), Chicago
Katie Romich (DJG), 18 January 2011 (E); 11 March 2011 (T), Rocky River, OH
Carol Ronen (DJG), 25 May 2010, Chicago
Don Rose (DJG), 15 September 2009, Chicago
Mark Rosen (DJG), 8 November 2010, Chicago (T)
David Rosenberg (DJG), 2 February 2010, Cambridge, MA
Neal Rosenstein (DJG), 29 January 2010, New York, NY
Andrew Roth (DJG), 5 December 2011; 3, 6, and 16 May 2012, 24, 25, and 26 September 2012 (E), San Francisco
Kenneth Rothwell (DJG), 10 April 2015, Newton, MA (T)
Pete Rouse (Jim Gilmore), 11 July 2008, Washington (*Frontline*)
Peter J. Rubin (DJG), 26 February 2013, Boston (E)
Jean Rudd (DJG), 25 March 2009 (E); 30 April

2009 and 28 July 2012, Chicago and Beverly Shores, IN
John G. Ruggie (DJG), 18 February 2012, Cambridge, MA (E)
Jesse Ruiz (DJG), 1 June 2010, Chicago
Bobby Rush (Terry Rockefeller), 20 October 1988, Chicago (Blackside)
Bobby Rush (Julieanna Richardson), 22 August 2000 and 18 January 2001, Chicago (TheHistoryMakers.com)
Bobby Rush (Julian Bond), 15 December 2005, Washington, DC (Institute for Public History, UVA)
Thomas D. Rush (DJG), 26 March 2014, Asheboro, NC (T)
Barry Rutizer (DJG), 28 January 2010, New York, NY
Mary Ryan (DJG), 2 February 2015, Hinsdale, IL (T)
Indira Saladi (DJG), 7 March 2014, Chicago (T)
Kathy Saltmarsh (DJG), 12 May 2011, Springfield, IL
Bettylu Saltzman (DJG), 12 January 2010, Chicago
Al Sampson (Julieanna Richardson), 19 August 2002, Chicago (TheHistoryMakers.com)
Al Sampson (DJG), 13 August 2013, Chicago (T)
Martin Sandoval (DJG), 10 May 2011, Springfield, IL
Sima Sarrafan (DJG), 5 January 2012, Redmond, WA
Greg Sater (DJG), 13 February 2013, Los Angeles (T)
Chris Sautter (DJG), 21 January 2010, Washington, DC
Eugene Sawyer (Betty Brown-Chappell), 18 December 1989, Chicago (BBCP)
Jan Schakowsky (David Axelrod), 17 February 2004, Chicago (JKP and TPP)
Jan Schakowsky (DJG), 30 June 2013, Evanston, IL
Andrew Schapiro (DJG), 12 November 2010, Chicago
Scott Scheper (DJG), 14 March 2013, San Diego (T)
Lisa (Hertzer) Schertler (DJG), 13 October 2010, Washington, DC (E)
Robbin Schiff (Robert Draper), 2009, New York, NY
Derick Schilling (DJG), 4 and 8 December 2014, New York, NY (E)
Jeanne Reynolds Schmidt (DJG), 5 February 2013, Augusta, GA (T)
John Schmidt (DJG), 27 April 2009; 27 April and 1 August 2009 (2) (E), Chicago
Paul Schmitz (DJG), 13 July 2012, Milwaukee, WI
Shannon Schmoyer (DJG), 26 February 2013, San Antonio, TX (T)
Barbara Schneider (DJG), 13 February 2013, Durham, ME (T)
Jeff Schoenberg (DJG), 11 May 2011, Springfield, IL
Bernard Schoenburg (Mark DePue), 9 March 2009, Springfield, IL (ALPL)
George Schopp (DJG), 16 January 2010, Chicago, IL
Patty Schuh (DJG), 12 May 2011, Springfield, IL
Ellen Schumer (DJG), 18 September 2009, Chicago
Michael Schwartz (DJG), 22 February 2013, Denver, CO (T)
Karen Schweickart (DJG), 24 March 2014, Chicago (T)
Janny Scott (DJG), 22 September 2011, New York, NY (E)
Jason Scott (DJG), 3 September 2013, New York, NY
Rob Scott (DJG), 31 July 2015, Springfield, IL (T)
Paul Scully (DJG), 5 May 2009; 6, 22, and 25 May 2009, 5 August 2009, 2, 3, 4, 5, and 9 March 2015 (E), Chicago
Scott Sederstrom (DJG), 6 (2) December 2010, Chicago (E)
Kristen Seeger (DJG), 20 March 2014, Chicago (T)
Jay Seidenstein (DJG), 20 December 2011 (E), 21 December 2011 (T), Honolulu
Gary Seidman (DJG), 3 January 2012, Seattle
Jeffrey Selbin (DJG), 5 December 2011, Berkeley, CA
Joseph Seliga (DJG), 1 July 2013, Chicago
Orin Sellstrom (DJG), 30 April 2013, San Francisco
Claire Serdiuk (DJG), 8 November 2010, Chicago
George Shadid (DJG), 29 May 2010, Peoria, IL
Tarak Shah (DJG), 6 February 2014, Arlington, VA
Carolyn Shapiro (DJG), 9 November 2010, Chicago (T)
David Shapiro (DJG), 9 March 2014, Cambridge, MA (E)
Robert Y. Shapiro (DJG), 14 March 2013, New York, NY (T)
John Sharick (Philip Bracy), 14 April 1981, Youngstown, OH (Youngstown State University)
Al Sharp (DJG), 28 June 2013, Chicago

Michael Sheehan (DJG), 6 February 2014, Washington, DC
Susan Sher (Robert Draper), 2009, Washington, DC
Adam Sherman (DJG), 27 March 2014 (T); 28 March 2014 and 20 November 2014 (E), Tulsa, OK
Scott Sherman (DJG), 19 February 2013, Houston, TX (T)
Nancy Shier (DJG), 12 May 2011, Springfield, I.
Helen Shiller (Jacob Lingan), 22 April 2010, Chicago (CAAMC, Columbia College)
Evie Shockley (DJG), 9 March 2014 (E); 17 March 2014 (T), New Brunswick, NJ
Dan Shomon (Jen Christensen), 22 July 2008, Chicago (CNN)
Dan Shomon (DJG), 2 and 13 November 2010; 12 December 2010 (E); 30 March 2011 (T); 2 and 10 May 2011, 12 July 2012, 21 June 2013; 19 (2) January 2014, 16 (2) August 2015, 17 August 2015 (E), 17 August 2015, 30 October 2015 (T), 3 December 2015 (E), Chicago and Springfield, IL
Sohale Siddiqi (Jim Gilmore), 30 June 2012, Seattle (*Frontline*)
Todd Sieben (DJG), 19 March 2014, Geneseo, IL (E)
Martin Siegel (DJG), 13 March 2013, Houston, TX (T)
Scott Siff (DJG), 20 October 2010, Washington, DC
Dina Silva (DJG), 1 March 2013, Hartford, CT (T)
Ira Silverstein (DJG), 17 November 2010, Springfield, IL
Fred Simari (DJG), 11 January 2010, Oak Park, IL
Adele Simmons (DJG), 4 May 2011, Chicago.
Monice Mitchell Simms (DJG), 30 June 2015, Los Angeles (T)
Linda Simon (DJG), 6 May 2014, Chicago (E)
Sheila Simon (DJG), 11 May 2011, Springfield, IL
Linda Singer (DJG), 18 October 2010, Washington, DC
Gerry Singsen (DJG), 13 November 2009, Watertown, MA (E)
Joe Sinsheimer (DJG), 30 October 2013, Saskatoon, SK, Canada (T)
Chris Sipes (DJG), 28 April 2011, Washington, DC
Sarah Siskind (DJG), 16 July 2012, Madison, WI
Chris Slaughter (DJG), 17 and 18 November 2011, Chicago (E)
Daniel Slifkin (DJG), 20 April 2011, New York, NY
David Smail (DJG), 19 February 2013, New York, NY (T)
Callie Smith (DJG), 14 July 2012, Chicago (E)
Kenny Smith (DJG), 24 October 2010, Charlotte, NC
Nathan Smith (DJG), 26 March 2014, Houston, TX (T)
Pam Smith (DJG), 21 November 2010, Chicago
Scott Smith (DJG), 5 May 2010, McMinnville, OR (T)
Tristé Lieteau Smith (DJG), 17 March 2014, Chicago (T)
Bill Snider (DJG), 23 October 2010, Greensboro, NC
Dan Sokol (DJG), 6 March 2014, Gainesville, FL (T)
Danny Solis (DJG), 26 May 2010, Chicago
Aaron Solomon (DJG), 3 March 2014, Denver, CO (E)
Donald M. Solomon (DJG), 5 April 2012, Boston (E)
Elysia Solomon (DJG), 9 and 17 March 2014, Louisville, KY (T)
Alan Solow (DJG), 21 June 2013, Chicago
Bronwen Solyom (DJG), 19 and 21 December 2011; 12 July 2015 (E), Honolulu
Garrett Solyom (DJG), 19 and 21 December 2011, Honolulu
Unmi Song (DJG), 1 July 2013, Chicago
Whitman Soule (DJG), 24 July 2012, Evanston, IL
Richard Speicher (Philip Bracy), 31 March 1981, Youngstown, OH (Youngstown State University)
David Spiegel (DJG), 23 June 2013, Chicago
Jennifer Amdur Spitz (DJG), 1 November 2010, Chicago
Bruce Spiva (DJG), 18 October 2010, Washington, DC
Howard Stanback (Betty Brown-Chappel), 31 October and 20 November 1989, Chicago (BBCP)
Howard Stanback (DJG), 5 November 2010 (T); 9 November 2010 (E), Hedgesville, WV
David Starrett (DJG), 2 February 2013, Springfield, IL (T)
Jeff Stauter (DJG), 14 March 2013, Dixon, IL (T)
Richard Steele (DJG), 1 November 2011, Chicago (E)
Peter Steffen (DJG), 11 March 2014 (E); 12 March 2004 (T), Chicago
Kathryn Stell (DJG), 19 April 2014, Chicago (T)
Bill Stenzel (DJG), 29 April 2009; 30 November 2011 (E), Oak Lawn, IL

Judy Stevens (DJG), 25 July 2013, Washington, DC
Ryan Stoll (DJG), 12 July 2012, Chicago
Adam Stolorow (DJG), 4 September 2013, New York, NY
Geoffrey Stone (DJG), 15 January 2010, Chicago
Dave Stover (DJG), 7 March 2014, Springfield, IL (T)
Ron Stradt (DJG), 28 July 2015, Springfield, IL (E)
Chinta Strausberg (DJG), 14 March 2013, Chicago
David Strauss (DJG), 27 March 2014, Chicago (T)
Paul Strauss (DJG), 15 January 2010, Chicago
Aletha Strong (Gibson) (DJG), 30 October 2010, New Orleans
Aleeza Strubel (DJG), 15 March 2014, Chicago (T)
Sarah (Eaton) Stuart (DJG), 26 February 2013, Canton, MA (T)
Eva Sturgies (DJG), 17 September 2009, Chicago
Stephen Subrin (DJG), 20 March 2015, Boston (E)
Larry Suffredin (DJG), 22 June 2013, Evanston, IL
Sung-Hee Suh (DJG), 21 March 2014, New York, NY (T)
Dave Sullivan (DJG), 18 November 2010, 9 November 2011, Springfield, IL
Greg Sullivan (DJG), 7 November 2011, Springfield, IL
Heather Sullivan (DJG), 9 September 2014, Kittery Point, ME (T)
John Sullivan (DJG), 12 May 2011, Springfield, IL
Julianne Alfred Sullivan (DJG), 24 March 2014, Portland, ME (T)
Kenneth Sulzer (DJG), 6 January 2010, Los Angeles
Ron Sunshine (DJG), 5 February 2013 (E); 6 February 2013 (T), Allentown, PA
Chuck Supple (DJG), 11 March 2014, Sacramento (T)
Pam Sutherland (DJG), 16 November 2010, Springfield, IL
Nathan Sutton (DJG), 20 March 2014, Washington, DC (T)
Dave Syverson (DJG), 16 November 2010, Springfield, IL
Madeline Talbott (DJG), 7 May 2011, Chicago
Nate Tamarin (DJG), 24 July 2013, Washington, DC
Larry Tavares (DJG), 20 December 2011, Honolulu
Kerry Taylor (DJG), 21 May and 13 June 2009, Charleston, SC (E)
Thom Thacker (DJG), 10 August 2010, Sleepy Hollow, NY (E)
Regina (Foran) Thibeau (DJG), 13 March 2013, Libertyville, IL (T)
Capril Bonner Thomas (Robert Draper), 2009, Los Angeles
Liisa M. Thomas (DJG), 19 March 2014, Chicago (T)
Rosa Thomas (DJG), 20 May 2010, Hobart, IN
Darrel Thompson (DJG), 15 December 2015, Washington, DC (T)
Kevin Thompson (DJG), 1 January 2012, Seattle
Laura Tilly (DJG), 9 September 2009, Chicago
Naranhkiri Tith (DJG), 17 and 23 January 2012, 23 October 2014, Washington, DC (E)
Anne S. Toker (DJG), 13 October 2010, New York, NY
Laurie Tom (DJG), 20 December 2011 (E); 22 December 2011, Honolulu
Tom Topolinski (DJG), 2 January 2012; 11 March 2012 and 18, 19 and 20 November 2014 (E), Everett, WA
Tom Topolinski (Jim Gilmore), 30 June 2012, Everett, WA (*Frontline*)
Gina Torielli (DJG), 15 February 2013 (T); 15, 18 (2), 19, and 22 February 2013, 4 March 2013 (E), Lansing, MI
Bob Torrey (DJG), 20 December 2011, Honolulu (T)
R. Kent Torrey (DJG), 16 February 2013, Carmel, CA (T)
Andrew Trask (DJG), 21 and 22 May 2014, London, UK (E)
Carole Travis (DJG), 28 April 2013; 29 April 2013 (E), Pleasant Hill, CA
Laurence H. Tribe (DJG), 18 April 2011, Cambridge, MA
Laurence H. Tribe (Jim Gilmore), 12 June 2012, Cambridge, MA (*Frontline*)
Donne Trotter (DJG), 11 November 2010, Chicago
David Troutt (DJG), 27 January 2010, Newark, NJ
Jennifer (Radding) Gardner Trulson (DJG), 13 February 2013, New York, NY (T)
Stacy Monahan Tucker (DJG), 3 (2) March 2014, Redwood City, CA (E)
Arthur Turner (Julieanna Richardson), 9 June 2000, Chicago (TheHistoryMakers.com)
Philip Turner (DJG), 15 (2) June 2015, New York, NY (E)
Scott Turow (Robert Draper), 2009, Evanston, IL
Scott Turow (DJG), 1 July 2013, Evanston, IL

Kevin Tyson (DJG) 4 November 2010 (E), 8 November 2010, Chicago
Howard Ullman (DJG), 30 April 2013, San Francisco
Roberto Mangabeira Unger (David Remnick), 1 and 6 March 2009, Cambridge, MA (E)
Roberto Mangabeira Unger (DJG), 2 and 3 February 2010, Cambridge, MA
Amiekoleh (Kim Kimbrew) Usafi (DJG), 6 January 2010, Los Angeles
Todd Vandermyde (DJG), 31 October 2011 (E); 7 November 2011, Springfield, IL
John (Doyle) Van Fossen (DJG), 19 and 20 February 2013, 26 September 2014, Naples, ID (E)
Donna Van Zanten (DJG), 14 September 2009, Chicago
Tony Van Zanten (DJG), 14 September 2009, Chicago
Andrew "Pete" Vayda (DJG), 6 March 2015 (E), 8 March 2015 (T), 8 March 2015 (E), New York, NY
Brenda Vinson (DJG), 13 October 2010, New York, NY (T)
Louis Viverito (DJG), 18 November 2010, Springfield, IL
Andrew Walder (DJG), 4 (2) March 2013, Palo Alto, CA (E)
Isabella Waller (DJG), 17 September 2009, Chicago
Larry M. Walsh (DJG), 21 May 2010, Joliet, IL
Terry Walsh (DJG), 7 August 2013, Evanston, IL
Tom Walsh (DJG), 17 November 2010, Springfield, IL
Pauline Wan (DJG), 4 December 2011, Saratoga, CA
Ty Wansley (DJG), 23 September 2014, Chicago (T)
Mark Warnick (DJG), 17 March 2014, Seattle (T)
Mark Warnsing (DJG), 13 May 2011, Springfield, IL
Kenneth Warren (DJG), 28 September 2011, Chicago (E)
Laura Washington (DJG), 12 January 2010, Chicago
Tom Wathen (DJG), 19 October 2010, Washington, DC.
Frank Watson (DJG), 9 November 2011, Springfield, IL
Frank Watson (Mark DePue), 10 October 2012, 8 and 13 May 2013, Springfield, IL (ALPA)
Kevin J. Watson (DJG), 1 June 2010, Chicago
Dawna Weatherly-Williams (DJG), 30 October 2014, Honolulu (T)
Shelley Marks Weathers (DJG), 13 February 2013, Houston, TX (T)
Lori-Christina Webb (DJG), 18 October 2010, Rockville, MD
Brad Weber (DJG), 17 May 2010, Los Angeles (E)
Rebecca Weber (DJG), 29 January 2010, New York, NY
John Webster (DJG), 30 November 2011, Chicago (T)
Wayne P. Weddington III (DJG), 25 March 2014, New York, NY (T)
Wayne Weightman (DJG), 13 February 2013 (E), 2 July 2015 (T), Phnom Penh, Cambodia, and Richmond, VA
Michael Weinberger (DJG), 13 October 2010, New York, NY
David Weisbaum (DJG), 11 April 2014, Springfield, IL (T)
Pat Welch (DJG), 11 May 2011, Springfield, IL
Randy Wells (DJG), 12 May 2011, Springfield, IL
G. Christopher Welton (DJG), 19 May 2014 (E); 26 May 2014, Rueil Malmaison, France
Ann West (DJG), 6 December 2012, Sydney, NSW, Australia
Robin West (DJG), 17 October 2010, Princeton, NJ
Tommy West (DJG), 16 January 2010, Chicago
Stephanie (Fulks) Westerman (DJG), 18 February 2013, Buffalo, NY (T)
Wayne Whalen (DJG), 12 July 2012, Chicago
Eric Whitaker (Robert Draper), 2009, Chicago
Gordon Whitman (DJG), 19 and 20 October 2010, Washington, DC
Sarah Leah Whitson (DJG), 18 February 2013, Brooklyn, NY (T)
Brad Wiegmann (DJG), 26 April 2011, Washington, DC
Don Wiener (DJG), 28 and 29 (2) May 2013 (E); 29 May 2013 (T); 30 (2) and 31 (2) May 2013, 2 and 11 (2) June 2013, 20 July 2013, 27 October 2013 (E), Madison, WI
David Wilhelm (DJG), 23 August 2013, Gahanna, OH
David Wilkins (DJG), 16 November 2010, Cambridge, MA (E)
Marni Willenson (DJG), 8 November 2010, Chicago
Grant Williams (DJG), 24 March 2014, St. Louis, MO (T)
Michael Williams (DJG), 12 February 2013 (T); 13 February 2013 (E), Geneva, Switzerland
Paul Williams (DJG), 14 January 2010, Chicago
Richard "Rick" Williams (DJG), 14 September 2009, Chicago

Vince Williams (DJG), 20 November 2010, Chicago
Alec Williamson (DJG), 12 May 2010 (T); 12 May 2010 (E), Central Point, OR
Audra Wilson (DJG), 2 November 2010, Chicago
Mary-Ann Wilson (DJG), 3 April 2009, Chicago
F. Scott Winslow (DJG), 15 January 2016 (2), Chicago (E)
Ed Wojcicki (DJG), 12 May 2011, Springfield, IL
Paula Wolff (DJG), 8 November 2010, Chicago
Robert Paul Wolff (DJG), 9 April 2012, Amherst, MA (E)
Carolyn Wortham (DJG), 14 March 2013, Chicago (T)
Jeremiah A. Wright Jr. (Julieanna Richardson), 11 January 2002, Chicago (TheHistory Makers.com)
Jeremiah A. Wright Jr. (John Kupper), 6 February 2004, Chicago (JKP)
Jeremiah A. Wright Jr. (Arlen Parsa), 24 March 2009, Chicago (CAAMC, Columbia College)
Jeremiah A. Wright Jr. (DJG), 3 November 2010; 4, 5 and 8 November 2010 (E), Chicago
Jeremiah A. Wright Jr. (Ed Klein), 11 November 2011, Chicago
Timothy W. Wright III (Jonathan Vogel), 25 November 2009, Chicago (CAAMC, Columbia College)
Timothy W. Wright III (DJG), 18 May 2010, Chicago
Addie L. Wyatt (Elizabeth Balanoff), 16 March 1977, Chicago (Roosevelt University Labor Oral History Project)
Addie L. Wyatt (Joan McGann Morris), 14 December 2002, Chicago (Working Women's History Project)
Alicia (Rubin) Yamin (DJG), 21 February 2013, Dar Es Salaam, Tanzania (T)
Annette Yee (DJG), 19 February 2013 (E); 22 February 2013 (T), Carmel, CA
Jonathan Yi (DJG), 11 March 2004, Atlanta (T)
Quentin Young (DJG), 11 November 2010, Chicago
Mary Yu (DJG), 6 January 2012, Seattle
Pake Zane (DJG), 17 December 2011 (E); 20 December 2011, Honolulu
Jonathan Zimmerman (DJG), 28 February 2013, Narberth, PA (T)
Joseph H. Zwicker (DJG), 31 January 2010, Boston (E)

INDEX